AF837447

Unsaturated Soil Mechanics in Engineering Practice

Unsaturated Soil Mechanics in Engineering Practice

D. G. Fredlund
H. Rahardjo
M. D. Fredlund

WILEY
JOHN WILEY & SONS, INC.

Cover Image: © iStockphoto.com/Qingyi
Cover Design: Holly Wittenberg

This book is printed on acid-free paper. ♾

Copyright © 2012 by John Wiley & Sons, Inc. All rights reserved

Published by John Wiley & Sons, Inc., Hoboken, New Jersey
Published simultaneously in Canada

No part of this publication may be reproduced, stored in a retrieval system, or transmitted in any form or by any means, electronic, mechanical, photocopying, recording, scanning, or otherwise, except as permitted under Section 107 or 108 of the 1976 United States Copyright Act, without either the prior written permission of the Publisher, or authorization through payment of the appropriate per-copy fee to the Copyright Clearance Center, 222 Rosewood Drive, Danvers, MA 01923, (978) 750-8400, fax (978) 646-8600, or on the web at www.copyright.com. Requests to the Publisher for permission should be addressed to the Permissions Department, John Wiley & Sons, Inc., 111 River Street, Hoboken, NJ 07030, (201) 748-6011, fax (201) 748-6008, or online at www.wiley.com/go/permissions.

Limit of Liability/Disclaimer of Warranty: While the publisher and author have used their best efforts in preparing this book, they make no representations or warranties with the respect to the accuracy or completeness of the contents of this book and specifically disclaim any implied warranties of merchantability or fitness for a particular purpose. No warranty may be created or extended by sales representatives or written sales materials. The advice and strategies contained herein may not be suitable for your situation. You should consult with a professional where appropriate. Neither the publisher nor the author shall be liable for damages arising herefrom.

For general information about our other products and services, please contact our Customer Care Department within the United States at (800) 762-2974, outside the United States at (317) 572-3993 or fax (317) 572-4002.

Wiley publishes in a variety of print and electronic formats and by print-on-demand. Some material included with standard print versions of this book may not be included in e-books or in print-on-demand. If this book refers to media such as a CD or DVD that is not included in the version you purchased, you may download this material at http://booksupport.wiley.com. For more information about Wiley products, visit www.wiley.com.

Library of Congress Cataloging-in-Publication Data:

Fredlund, D. G. (Delwyn G.), 1940-
Unsaturated soil mechanics in engineering practice / D.G. Fredlund, H. Rahardjo, Murray D. Fredlund.
p. cm.
Includes bibliographical references and index.
ISBN 978-1-118-13359-0 (hardback); ISBN 978-1-118-28049-2 (ebk); ISBN 978-1-118-28050-8 (ebk); ISBN 978-1-118-28051-5 (ebk); ISBN 978-1-118-28052-2 (ebk); ISBN 978-1-118-28053-9 (ebk); ISBN 978-1-118-28054-6 (ebk)
1. Soil mechanics. 2. Soil moisture. 3. Soils–Testing. 4. Soil mechanics–Mathematical models. I. Rahardjo, H. (Harianto) II. Fredlund, Murray D., 1968- III. Title.
TA710.5.F735 2012
624.1′5136–dc23

2012013543

10 9 8 7 6 5 4 3 2 1

To our wives and families for their patience, tolerance, and enduring appreciation of our dedication to the development of the science for unsaturated soil behavior in engineering practice.

CONTENTS

FOREWORD

In 1993, Professors Fredlund and Rahardjo published the first textbook solely concerned with the behavior of unsaturated soils: *Soil Mechanics for Unsaturated Soils*. That volume maintained the framework of classical soil mechanics, but extended it to incorporate the soil suction phenomenon as an independent variable that is amenable to measurement and calculation. It marked a major milestone in the evolution of unsaturated soil mechanics.

Professors Fredlund and Rahardjo have now collaborated with Murray Fredlund to publish their successor volume, *Unsaturated Soil Mechanics in Engineering Practice*. Murray Fredlund adds computational skills to the team and, in the view of the authors, these are essential to meet their objectives of presenting a volume that not only covers our present knowledge of unsaturated soil behavior, but also provides guidance on the manner in which practical problems involving unsaturated soil behavior are formulated and solved. Many flux-related problems in unsaturated soil behavior require the solution of nonlinear partial differential equations with associated boundary conditions and the volume adds guidance on these computational issues as applied to the formulation of water, air, and heat flow through unsaturated soils. Separate chapters concentrate on the shear strength of unsaturated soils and its application to earth pressure, bearing capacity, and stability problems, as well as the formulation of stress-deformation behavior and its application to heave- and stiffness-related problems.

A fundamental distinction between saturated and unsaturated soil behavior is the need to express the relationship in the latter between water content and soil suction, that is, the soil-water characteristic curve. Since 1993, there has been an explosion of studies into the measurement of soil suction and the development of soil-water characteristic curves. A particular effort has been made here to synthesize these developments in a manner that facilitates applications.

While most readers will concentrate on the technical contents of this book, I urge students of the subject to also reflect on the contents of Chapter 1 related to the emergence of unsaturated soil mechanics in a coherent form and the assessment of challenges to its implementation. The guiding spirit of this welcome volume is to give the reader confidence that all of these challenges can be addressed in a consistent and rational manner.

Understandably, given current research efforts in the field of unsaturated soil mechanics, not all researchers and practitioners will accept the total contents of this book in an uncritical manner. Science is the search for truth, predominantly by hypothetico-deductive methods, which drive its progression. However, engineering is the pursuit of functionality and it progresses by incremental improvements to enhance intended function. It is particularly in the latter context that the authors have made an important contribution to geotechnical engineering. I expect that *Unsaturated Soil Mechanics in Engineering Practice* will remain an essential reference for educators, researchers, and practitioners for a long time to come.

N.R. MORGENSTERN
Distinguished University Professor (Emeritus)
of Civil Engineering

and

Past President, International Society of Soil
Mechanics and Geotechnical Engineering (1989–1994)

University of Alberta
August, 2011

PREFACE

Soil mechanics is a relatively young applied science. Karl Terzaghi published his English version of *Theoretical Soil Mechanics* in 1943. The book provided a science-based context for analyzing the physical behavior of saturated soils.

Geotechnical engineering has changed in many ways since the 1940s. The procedures for performing subsurface investigations have undergone some changes, but the investigative procedures remain quite similar. Boreholes are still drilled with disturbed and undisturbed soil samples taken at intervals for later laboratory testing. However, the manner in which we obtain our geotechnical engineering solutions has changed dramatically. Terzaghi and his contemporaries assembled the context for soil mechanics at a time when the tools for solving mathematical problems were significantly different from the tools that are available today.

In the 1940s, the writers of soil mechanics textbooks attempted to take complex three-dimensional, real-world problems and reduce them to simplified, closed-form solutions. *Flownets* provided a graphical solution for the movement of water through an isotropic, homogeneous, two-dimensional porous continuum. *Methods of (vertical) slices* provided a solution for calculating the factor of safety of a two-dimensional slope. *Methods of (horizontal) layers* provided a solution for the calculation of settlement of a one-dimensional, compressible clay soil. The soil mechanics world contained a series of soil property constants (e.g., k, c', and ϕ'), and those soil properties that were not constants were converted to a linear form to be represented as constants (e.g., C_c and C_s).

It became clear in the 1960s and 1970s that unsaturated soil properties would need to be defined as nonlinear *unsaturated soil property functions* (USPFs). *Unsaturated soil mechanics* became a vibrant area of geotechnical research, and it was apparent that we were entering a new era that required a new paradigm for solving saturated-unsaturated soil mechanics problems. If unsaturated soil mechanics was to find its way into geotechnical engineering practice there needed to be reliable methodologies for obtaining the unsaturated soil property functions at reasonable cost and effort. Consequently, a wide variety of *estimation procedures* emerged from research in many countries. The estimation procedures relied heavily on the saturated soil properties and an understanding of the *soil-water characteristic curve* (SWCC), that is, the relationship between water content and soil suction.

The 1960s and 1970s were decades that witnessed rapid growth in our ability to solve complex mathematical formulations. The computer could be used to solve new mathematical formulations that described the physical behavior of saturated-unsaturated soil mechanics problems. Numerical methods of solution emerged for all areas of material behavior, areas that spanned well beyond classical soil mechanics. Soil mechanics problems were visualized as *boundary value problems* with the following conditions defined: (i) geometry and stratigraphy, (ii) initial conditions and boundary conditions, (iii) soil properties, and (iv) solution techniques. The physics of soil behavior was defined for a *referential elemental volume* (REV) of the saturated-unsaturated soil continuum and the mathematical formulation describing the physics of soil behavior took on the form of a *partial differential equation* (PDE). Generally the PDEs were found to be nonlinear because of the nonlinear unsaturated soil property functions required as part of the formulation. The type of equations that many of us disliked as undergraduate students became the heart of unsaturated soil problem solving. Fortunately, we were able to hide the PDE solver in advanced computer software tools.

Geotechnical engineers have benefited from research undertaken in two primary areas: (i) soil physics and agronomy and (ii) computer technology and mathematics. In particular, it was the rapid growth in computing capability (i.e., computer hardware and software) that made the solution of unsaturated soil problems possible. The stage was set for solving saturated-unsaturated soil mechanics problems within a boundary value context through use of numerical modeling techniques.

It is an understatement to say that the digital computer has revolutionized the way that soil mechanics is now implemented in engineering practice. It is safe to say that it would not be possible to model and solve saturated-unsaturated soil mechanics problems within a science framework without the power of the digital computer. Geotechnical engineering has moved into a *new paradigm*, a problem-solving environment involving SWCCs, USPFs, and PDEs. It is a world in which the challenge becomes the *convergence* and the *uniqueness* of the soil mechanics solution. It is a world in which computer software is no longer a luxury but a necessity for sound engineering practice.

During the course of writing this book, numerous example problems were analyzed using the SVOffice geotechnical software suite. The examples in this book are freely distributed as resources related to the learning process associated with unsaturated soil mechanics. Instructions for obtaining these examples may be found at www.soilvision.com/usmep. The examples include seepage (SVFLUX™), slope stability (SVSLOPE®), freeze/thaw (SVHEAT™), and stress/deformation (SVSOLID™) finite element numerical models for which the setup and solution information can be examined.

Eduardo Alonso and Antonio Gens (2011) put it well in the Preface to the Fifth International Conference on Unsaturated Soils, Barcelona, Spain, when they wrote, "The development of unsaturated soil mechanics in recent decades has been remarkable and it has resulted in momentous advances in fundamental knowledge, testing methods, computational procedures, prediction methodologies and geotechnical practice." As authors, we trust that the book *Unsaturated Soil Mechanics in Engineering Practice* will further advance the usage of the science of unsaturated soil behavior in engineering practice.

Unsaturated Soil Mechanics in Engineering Practice constitutes a substantial addition and reorganization of information from what was synthesized in the book *Soil Mechanics for Unsaturated Soils* by D. G. Fredlund and H. Rahardjo. *Unsaturated Soil Mechanics in Engineering Practice* more thoroughly covers our present knowledge of unsaturated soil behavior and better reflects the manner in which practical unsaturated soil engineering problems are solved. The fundamental physics of unsaturated soil behavior presented in *Soil Mechanics for Unsaturated Soils* has been retained in the present edition while greater emphasis has been placed on the importance of using the SWCC when solving engineering problems. Greater emphasis has also been placed on the quantification of thermal and moisture boundary conditions based on the use of weather data. In the end, the reader should find *Unsaturated Soil Mechanics in Engineering Practice* to be a practical book leading geotechnical engineers through to the implementation of unsaturated soil mechanics into engineering practice.

Delwyn G. Fredlund

ACKNOWLEDGMENTS

Many persons have contributed to the synthesis of information presented in *Unsaturated Soil Mechanics in Engineering Practice*. Graduate students and research fellows have contributed much as they have researched a wide range of topics associated with unsaturated soil mechanics. The graduate students have sifted through past research and attempted to assemble a coherent synthesis of the findings of many researchers worldwide. The theses and research papers of many persons have been used in the synthesis of information required for the development of the science for unsaturated soil mechanics.

A number of researchers in unsaturated soil mechanics have volunteered to review various chapters of the book. The authors are grateful for their comments, edits and suggestions. In particular, the authors wish to acknowledge the contributions from the following persons: J. Côté, G. Gitirana, S. G. Goh, S. Houston, E. C. Leong, G. Newman, M. Padilla, H. Pham, A. Satyanaga, J. Stianson, S. Vanapalli, H. Vu, M. Yuan, C. Zapata, Q. Zhai, J. Zhang, and L. M. Zhang. The ongoing mentorship of Dr. N. R. Morgenstern and Dr. R. L. Lytton is also acknowledged and greatly appreciated.

A number of manufacturers of unsaturated soil testing equipment and monitoring devices have generously made information available for this book. The contributions from Campbell Scientific Corporation, Logan, Utah; Decagon Devices, Pullman, Washington; GCTS, Geotechnical Consulting and Testing Services, Tempe, Arizona; GDS Instruments, London; SoilMoisture Equipment Corporation, Santa Barbara, California; UPC, Barcelona, Spain, and Wille Geotechnik, Germany, are acknowledged. The initiatives on the part of many soil testing companies have done much to promote the practice of unsaturated soil mechanics. Their initiatives to develop soil testing and monitoring equipment contributes significantly to the practice of unsaturated soil mechanics.

Others have contributed greatly to the preparation of the manuscript and the figures for this book. Their diligence and willingness to assist have been greatly appreciated. The authors would like to acknowledge the following persons: Esther McAleer, Felicitas Egunyu, Patti Sawchuk, Marilyn D'Souza, Maki Ito, and O. Zapata.

Behind each of the authors there has been a company or institution that has been supportive of the venture to write a new book that embraces more of the recent research findings and engineering protocols for engineering practice. Dr. Del Fredlund has been employed by Golder Associates after his retirement from the University of Saskatchewan, Saskatoon. Golder Associates have been supportive and promoted the development of the Golder Unsaturated Soils Group (GUSG) under the direction of Greg Misfeldt. The GUSG has functioned as a loosely knit Web-based organization of the Golder office with the intent of addressing unsaturated soil problems worldwide. The GUSG has provided a valuable platform for the implementation of practical engineering protocols for a wide range of unsaturated soil problems.

Dr. Harianto Rahardjo has been employed at Nanyang Technological University (NTU), Singapore, since 1990. NTU has shown strong support for unsaturated soils research over many years. The result has been the establishment of a world-class unsaturated soil research laboratory containing most of the latest equipment for testing unsaturated soils. The research program has been complimented with numerous field instrumentation studies on residual soils. There has also been an ongoing series of unsaturated soil studies that have resulted in numerous journal, conference, and research reports.

Dr. Murray Fredlund started the software computing company called SoilVision Systems in 1997. The vision of the company was to provide computer software capable of solving problems involving unsaturated soils as well as saturated soils. This meant that it was necessary to develop and use solvers capable of solving highly nonlinear partial differential equations. SoilVision has not only provided engineering practitioners with software capable of solving a wide range of unsaturated soils problems but has also embarked on numerous research studies in an attempt to determine the most satisfactory solutions to unsaturated soil problems.

The support of Golder Associates, Nanyang Technological University, and SoilVision Systems has provided a framework for developing high-level engineering protocols for unsaturated soil mechanics. The authors want to express their appreciation for the support received through their places of employment. A platform for a vision has been developed for the promotion of unsaturated soil mechanics in geotechnical practice, and the authors are grateful for the opportunities afforded to them.

Delwyn G. Fredlund
Harianto Rahardjo
Murray D. Fredlund

CHAPTER 1

Theory to Practice of Unsaturated Soil Mechanics

1.1 INTRODUCTION

Soil mechanics involves a combination of engineering mechanics, soil behavior, and the properties of soils. This description is broad and can encompass a wide range of soil types. These soils could either be saturated with water or have other fluids in the voids (e.g., air). The development of classical soil mechanics has led to an emphasis on particular types of soils. The common soil types are saturated sands, silts and clays, and dry sands. These materials have formed the primary emphasis in numerous soil mechanics textbooks. More and more, it is realized that attention must be given to a broader spectrum of soil materials.

There are numerous soil materials encountered in engineering practice whose behavior is not consistent with the principles and concepts of classical, saturated soil mechanics. The presence of more than one fluid phase, for example, results in material behavior that is challenging to engineering practice. Soils that are unsaturated (i.e., water and air in the voids) form the largest category of soils which do not adhere in behavior to classical saturated soil mechanics.

The general field of soil mechanics can be subdivided into the portion dealing with saturated soils and the portion dealing with unsaturated soils. The differentiation between saturated soils and unsaturated soils becomes necessary due to basic differences in the material nature and engineering response. An unsaturated soil has more than two phases, and the pore-water pressure is negative relative to pore-air pressure. Any soil near the ground surface, present in an environment where the water table is below the ground surface, will be subjected to negative pore-water pressures and possible reduction in degree of saturation.

The process of excavating, remolding, and compacting a soil requires that the material be unsaturated. It has been difficult to predict the behavior of compacted soils within the framework of classical soil mechanics.

Natural surficial deposits of soil are found to have relatively low water contents over a large portion of the earth. Highly plastic clays subjected to a changing environment have produced the category of materials known as swelling or expansive soils. The shrinkage of these soils may pose an equally severe situation. Loose silt soils often undergo collapse when subjected to wetting and possibly a change in the loading environment. The pore-water pressures in both of the above-mentioned cases are initially negative, and volume changes occur as a result of increases in the pore-water pressure. Residual soils have also been of particular concern since their engineering behavior appears to deviate from classical soil mechanics principles. Once again, the primary factor contributing to the unusual behavior of residual soils is negative pore-water pressures.

Unsaturated soil mechanics is herein presented in the context of having a limited number of physical areas of application, namely, water flow (and storage), air flow (storage and compressibility), heat flow (and storage), shear strength, and volume-mass change (including swelling and collapse). The unsaturated soil theories are applied to real-world problems and solutions are illustrated in the context of a "boundary value problem." The physical behavior of unsaturated soil is formulated as a partial differential equation(s) that must be solved using a numerical technique. The partial differential equations are generally slightly too highly nonlinear in character and as a result computer analyses play an important role in solving practical engineering problems.

1.1.1 Application of Unsaturated Soil Mechanics in Engineering Practice

The content of this book takes into consideration the history of classical soil mechanics and the significant impact that the computer has had on the practice of geotechnical engineering. It is fair to say that the computer has resulted in a paradigm shift in how geotechnical engineering problems in general and specifically unsaturated soils problems are analyzed. The significant role that the computer has played has also been taken into consideration in assembling the content for this book. The nature of unsaturated soil problems makes it essentially imperative to use numerical methods when solving geotechnical engineering problems.

Terzaghi (1943) contributed significantly toward our understanding of unsaturated soil behavior in two chapters of his textbook *Theoretical Soil Mechanics*. Chapter 14 on "Capillary Forces" and Chapter 15 on "Mechanics of Drainage" (with special attention to drainage by desiccation) illustrate the importance of unsaturated soils. These chapters emphasize the importance of the unsaturated portion of the soil profile and in particular provide insight into the fundamental nature and importance of the air-water interface [i.e., the contractile skin (Fredlund and Rahardjo, 1993a)]. Considerable discussion was directed toward soils with negative pore-water pressures. Figure 1.1 shows an earth dam illustrating the manner in which water flows above the phreatic line through the capillary zone (Terzaghi, 1943). The contributions of Karl Terzaghi toward unsaturated soil behavior were truly commendable and are still worthy of consideration.

Terzaghi (1943) stated that "the theories of soil mechanics provide us only with the working hypothesis, because our knowledge of the average physical soil properties of the subsoil and the orientation of the boundaries between the individual strata is always incomplete and often utterly inadequate." Terzaghi emphasized the importance of clearly stating all assumptions upon which the theories were based and pointed out that "almost every alleged contradiction between theory and practice can be traced back to some misconception regarding the conditions for the validity of the theory." Terzaghi's advice from the early days of soil mechanics is extremely relevant as the theories for unsaturated soil behavior are being brought to the "implementation" stage in geotechnical engineering. With such an early emphasis on unsaturated soil behavior, one might ask the question, "Why did unsaturated soil mechanics not emerge simultaneously with saturated soil mechanics?" Pondering this question leads to the realization

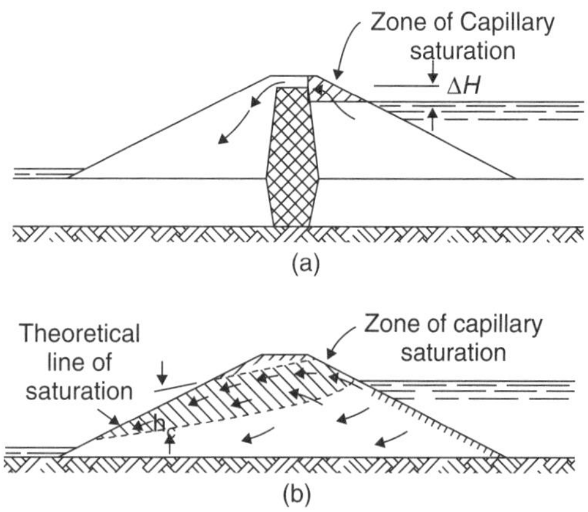

Figure 1.1 An earth dam shown by Terzaghi (1943) illustrating the flow of water above the phreatic line through the capillary zone. (a) Water siphoning over the core of the dam. (b) Water flow across the phreatic line.

that there were several theoretical and practical challenges associated with unsaturated soil behavior that needed further research before unsaturated soil mechanics could be implemented into engineering practice. In fact, unsaturated soil mechanics would need to wait several decades before it would take on the character of a science that could be used in routine geotechnical engineering practice.

Research within the agriculture-related disciplines strongly influenced the physical and hydraulic models that would later need to be brought into unsaturated soil mechanics (Baver, 1940). With time, numerous significant contributions have come from the agriculture-related disciplines (i.e., soil science, soil physics, and agronomy) into geotechnical engineering. It can be said that historically geotechnical engineers tended to test soils by applying total stresses to soils through the use of an oedometer or triaxial cell. On the other hand, agriculture-related counterparts tended to apply stresses to the water phase (i.e., tensions) through the use of pressure plate cells. Eventually, geotechnical engineers would realize that the information accumulated in agriculture-related disciplines was what was needed in geotechnical engineering with unsaturated soils. Careful consideration needed to be given to each of the test procedures and testing techniques when transferring the technology from agriculture into geotechnical engineering.

1.1.2 Scope of the Book

The scope of this book is limited to the field of unsaturated soil mechanics. An attempt is made to cover all aspects normally associated with soil mechanics. When the term "unsaturated soil mechanics" is used, the authors are referring to soils which have negative pore-water pressures.

The aspects of interest to geotechnical engineering fall into three main categories. These can be listed as problems related to (1) flow of fluids (i.e., air and water in liquid and vapor form) through porous media, (2) shear strength, and (3) volume-mass change behavior of unsaturated soils. An entire chapter is devoted to understanding the soil-water characteristic curve. A chapter is also devoted to heat flow through soils and another chapter is devoted to the establishment of boundary conditions. In particular, emphasis is placed on the quantification of the ground surface moisture flux boundary condition. Either moisture is falling to the ground, for example, in the form of rain or snow, or moisture is moving upward through evaporation and evapotranspiration. The quantification of both downward and upward moisture flow is pivotal to solving unsaturated soil mechanics problems. While this topic was largely absent from classical saturated soil mechanics, it has become an essential component related to solving unsaturated soil problems.

No attempt is made to duplicate or redevelop information already available in classical saturated soil mechanics books. This book should be used to assist the geotechnical engineer in understanding soil mechanics concepts unique to unsaturated soils. At the same time, these concepts have been developed and organized to appear as logical and relatively simple extensions of classical saturated soil mechanics concepts. Subjects such as clay mineralogy and physicochemical properties of soils are vitally important to understanding why soils behave in a certain manner. However, the readers are referred to other references for coverage of these subjects (Mitchell, 1993).

Most soil mechanics problems can be linked to a few key soil properties that are related to important processes. These properties relate to (1) the ease of flow through the multiphase material (e.g., liquid water, air, and heat), (2) the ability of the material to store (e.g., water storage, air storage and compression, and heat storage), (3) the shear strength characteristics, and (4) the volume-mass change soil properties (i.e., including the soil-water characteristic curve). The chapters in this book describe (1) theory related to each process and relevant soil properties, (2) measurement of each soil property, (3) estimation of each soil property (and soil property functions), and (4) application of the theory and soil properties to one or more soil mechanics problems.

The main objective of this book is to synthesize theories associated with the behavior of unsaturated soils and show how the theories can be applied in geotechnical engineering practice. The theoretical derivations are presented in considerable detail because unsaturated soil behavior is a relatively new area of study and many of the derivations are not readily available to engineers in a familiar context. The theory, measurement, and estimation of the soil-water characteristic curve are pivotal to the implementation of unsaturated soil mechanics. For this reason, they have been given special attention throughout the book. There is ongoing need for case histories, and it is anticipated that these will become more commonly reported in future decades. Hopefully, as the analyses illustrated in this book are put into engineering practice, case histories will emerge that verify the consistent theoretical context provided within this book.

1.1.3 Gradual Emergence of Unsaturated Soil Mechanics

Unsaturated soil mechanics did not emerge simultaneously with saturated soil mechanics. Rather, there were a number of important experimental findings and theoretical developments that led to the gradual emergence of a science for unsaturated soil mechanics. These developments took place over a period of several decades, and by the late 1970s it became clear that unsaturated soil mechanics would take the form of a natural extension of saturated soil mechanics. A few key findings that contributed significantly to the emergence of unsaturated soil mechanics are listed below.

Experimental laboratory studies in the late 1950s (Bishop et al., 1960) showed that it was possible to independently measure (or control) pore-water and pore-air pressures through the use of high-air-entry ceramic disks. Laboratory studies were reported over the next decade that revealed fundamental differences between the behavior of saturated and unsaturated soils. The studies also revealed that there were significant challenges that still needed to be addressed. The laboratory testing of unsaturated soils proved to be time consuming and demanding from a technique standpoint. The usual focus on soil property constants was diverted toward the study of nonlinear unsaturated soil property functions. Soil-water characteristic curves (SWCCs) were found to hold an important relationship to each of the unsaturated soil property functions (Croney and Coleman, 1954; Fredlund and Rahardjo, 1993a). The increased complexity of unsaturated soil behavior extended from the laboratory to theoretical formulations and solutions.

Originally, there had been a search for a single-valued effective stress equation for unsaturated soils, but by the late 1960s, there was increasing awareness that the use of two independent stress state variables would provide an approach more consistent with the principles of continuum mechanics (Fredlund and Morgenstern, 1977).

The 1970 decade was a period when constitutive relations for the classical areas of soil mechanics were proposed and studied with respect to uniqueness (Fredlund and Rahardjo, 1993a). Initially, constitutive behavior focused primarily on the study of seepage, shear strength, and volume change problems. Gradually it became apparent that the behavior of unsaturated soils could be viewed as a natural extension of saturated soil behavior (Fredlund and Morgenstern, 1976). Numerous studies attempted to combine volume change and shear strength in the form of elastoplastic models that were an extension from research on saturated soils (Alonso et al., 1990; Wheeler and Sivakumar, 1995; Blatz and Graham, 2003). The study of contaminant transport properties, thermal soil properties, and air flow properties for unsaturated soils also took on the form of nonlinear soil property functions (Newman, 1995; Lim et al., 1998; Pentland et al., 2001; Ba-Te et al., 2005).

The 1980 decade was a period when boundary value problems were solved using numerical, finite element and finite difference modeling methods. Computers were required and iterative, numerical solutions became the norm. The challenge was to find techniques that would ensure convergence of highly nonlinear partial differential equations on a routine basis (Thieu et al., 2001; Fredlund et al., 2002b). Saturated-unsaturated seepage modeling became the first of the unsaturated soil problems to come into common engineering practice. Concern for stewardship toward the environment further promoted interest in seepage and geoenvironmental, advection-dispersion modeling.

The 1990 decade and beyond have become a period when the emphasis has been on the implementation of unsaturated soil mechanics into routine geotechnical engineering practice. A series of international conferences have been dedicated to the exchange of information on the engineering behavior of unsaturated soils and it has become apparent that the time has come for increased application of unsaturated soil mechanics in engineering practice. Implementation can be defined as "a unique and important step that brings theories and analytical solutions into engineering practice" (Fredlund, 2000a). There are several stages in the development of a science that must be brought together in an efficient and appropriate manner in order for the implementation of a science to become reality. The primary stages are as follows (Fredlund, 2000a): (1) state variable stage, (2) constitutive stage, (3) formulation stage, (4) solution stage, (5) design stage, (6) verification and monitoring stage, and (7) implementation stage.

The "state variable" stage identifies each of the non–material variables that are required to understand material behavior on a scientific basis within the context of continuum mechanics. The "constitutive" stage identifies the basic equation of soil behavior for each physical process of interest. Constitutive equations provide a linkage between state variables and incorporate soil properties (or soil property functions), which must be measured or estimated. At the "formulation" stage a referential elemental volume, REV, is selected to which the conservative laws of physics must be satisfied. The "formulation" stage generally results in the derivation of a (partial) differential type equation which must then be solved. The "solution" stage is identified as an independent stage because it is sometimes possible to derive a partial differential equation that describes material behavior; however, it might not be possible to solve the derived equation. In other words, the solution of the formulated equation(s) of behavior is worthy of being referred to as an independent stage. The "design" stage takes the formulated and solved equations of behavior and determines how the equations can be used for design purposes. If it is possible to determine and implement cost-effective design procedures, then it is important to also go through a "verification" stage where all previous stages are tested for reliability in routine engineering practice.

The "implementation" stage suggests that conditions related to all previous stages have been met and engineering protocols can be clearly identified. Implementation suggests that a reliable engineering science is available for describing particular material behavior, and it is prudent and in the best interest of the public to utilize the engineering science in engineering practice. Research studies are necessary to develop practical, efficient, cost-effective, and appropriate technologies.

1.1.4 Challenges to Implementation

There are a number of major challenges that needed to be addressed before unsaturated soil mechanics could become a part of routine geotechnical engineering practice. Each challenge has an associated solution that has emerged through ongoing research studies. In some cases it has been necessary to adopt new approaches to solving problems involving unsaturated soils. Some of the key problems that needed to be solved for unsaturated soil mechanics to take on the form of a science are listed below. A brief explanation of each "challenge" area is given to show that these major problems have been essentially resolved or solved. Techniques and engineering procedures have been forthcoming from research in various parts of the world, thereby preparing the way for more widespread application of unsaturated soil mechanics.

Challenge No. 1: The development of a theoretically sound basis for describing the physical behavior of unsaturated soils starting with appropriate state variables.

Solution No. 1: The adoption of independent stress state variables based on multiphase continuum mechanics has formed the basis for describing the stress state independent of soil properties. The stress state variables can then be used to develop suitable constitutive models.

Challenge No. 2: Constitutive relations commonly accepted for saturated soil behavior needed to be extended to also describe unsaturated soil behavior.

Solution No. 2: Gradually it became apparent that all constitutive relations for saturated soil behavior could be extended to embrace unsaturated soil behavior and thereby form a smooth transition between saturated and unsaturated soil conditions. In each case, research studies needed to be undertaken to verify the uniqueness of the extended constitutive relations.

Challenge No. 3: Nonlinearity associated with the partial differential equations formulated for unsaturated soil behavior resulted in iterative procedures in order to arrive at a solution. The convergence of highly nonlinear partial differential equations proved to be an important challenge.

Solution No. 3: Computer solutions for numerical models have embraced automatic mesh generation, automatic mesh optimization, and automatic mesh refinement techniques (i.e., known as adaptive grid refinement, AGR). These techniques have proved to be of great assistance in obtaining convergence when solving nonlinear partial differential equations. The solution procedures were largely forthcoming from cooperative research in the mathematics and computer science disciplines.

Challenge No. 4: Greatly increased costs and time were required for the testing of unsaturated soils. As well, laboratory equipment for measuring unsaturated soil properties has proved to be technically demanding and quite complex to operate.

Solution No. 4: Indirect estimation procedures for the characterization of unsaturated soil property functions were developed. These procedures were related to the SWCC and saturated soil properties. Several estimation procedures have emerged for each of the unsaturated soil property functions. The computer has proven to be useful in calculating unsaturated soil property functions.

Challenge No. 5: Highly negative pore-water pressures (i.e., matric suctions greater than 100 kPa) have proven to be difficult to measure, particularly in the field.

Solution No. 5: New instrumentation such as the direct, high-suction tensiometer and the indirect thermal conductivity suction sensor has provided new measurement techniques for the laboratory and the field. These devices allow suctions to be measured over a considerable range of matric suctions. The null-type, axis translation technique remains a laboratory reference procedure for the measurement of matric suction.

Challenge No. 6: New technologies and engineering protocols such as those proposed for unsaturated soil mechanics are sometimes difficult to incorporate into engineering practice. The implementation of unsaturated soil mechanics into engineering practice has proven to be a challenge.

Solution No. 6: Educational materials, visualization systems, computer software, and research conferences have greatly assisted in effective technology transfer. It has been possible to demonstrate through teaching, software demonstrations, and research conferences that the concepts of unsaturated soil behavior can be implemented in engineering practice. Information on unsaturated soil mechanics is also being incorporated into the undergraduate and graduate curricula at universities. Protocols for engineering practice are being developed for all application areas of geotechnical engineering.

Changes are necessary in geotechnical engineering practice in order for unsaturated soil mechanics to be implemented in a routine manner. Each challenge in unsaturated soil mechanics has been met with a definitive and practical solution. A significant paradigm shift has been required with regard to the determination of unsaturated soil property functions (Houston, 2002). New approaches that have been developed have provided cost-effective procedures for the determination of unsaturated soil property functions for all classes of problems (Fredlund, 2002a).

1.1.5 Laboratory and Field Visualization of Degree of Saturation

Climatic conditions around the world range from very humid to dry. Climatic classification is based on the average annual net moisture flux at the ground surface (i.e., precipitation minus potential evaporation; Thornthwaite, 1948). The ground surface climate is an important factor that controls the depth to the groundwater table and therefore the thickness of the unsaturated soil zone (Fig. 1.2).

The zone between the ground surface and the water table is referred to as the unsaturated soil zone. This is somewhat of a misnomer since the capillary fringe immediately above the water table is essentially saturated. A more correct term for the entire zone above the water table is the "vadose zone" (Bouver, 1978). However, the entire zone subjected

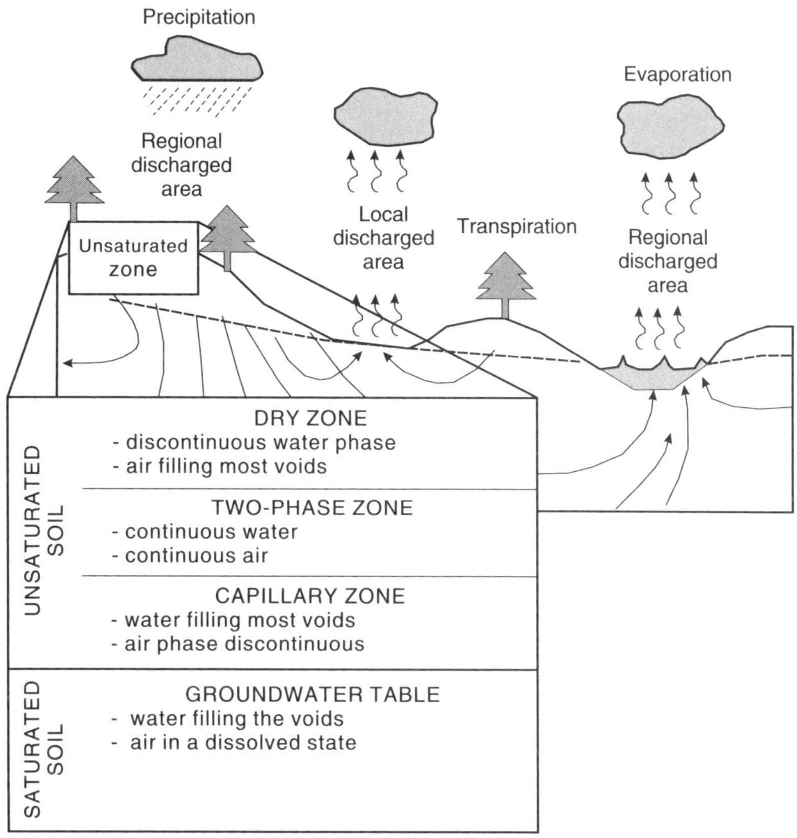

Figure 1.2 Subdivisions of unsaturated soil zone (vadose zone) on local and regional basis.

to negative pore-water pressures has become widely referred to as the unsaturated soil zone in geotechnical engineering. The unsaturated soil zone forms a transition between the water in the atmosphere and the groundwater (i.e., positive pore-water pressure zone).

The pore-water pressures in the unsaturated soil zone can range from zero at the water table to a maximum tension on the order of 1,000,000 kPa under dry soil conditions (Croney et al., 1958). The degree of saturation of the soil can range from 100% to zero. The changes in soil suction result in distinct zones of saturation. The zones of saturation have been defined in situ as well as in the laboratory (i.e., through the SWCC; Fig. 1.3). Table 1.1 compares the terminologies commonly used to describe saturation conditions in situ and in the laboratory. Soils in situ start at saturation at the water table and tend to become unsaturated toward the ground surface.

Soils near the ground surface are often referred to as problematic soils, but it is the handling of highly negative pore-water pressures that tends to present the most serious problem for geotechnical engineers. Common problematic soils are expansive soils, collapsible soils, and residual soils. Any of the above soils, as well as other soil types, can also be compacted, once again giving rise to a material with negative pore-water pressures.

Table 1.1 Terminology Commonly Used to Describe Degrees of Saturation in Field and Laboratory

Degree of Saturation, S, %	Field Description	Laboratory Description
~100	Saturated	Saturated
~90–100	Capillary zone	Boundary effect zone
~15–90	Two-phase zone	Transition zone
~15–0	Dry zone	Residual zone

1.2 MOISTURE AND THERMAL FLUX BOUNDARY CONDITIONS

Climate plays an important role in determining whether a soil is saturated or unsaturated. Water is removed from the soil either by evaporation from the ground surface or by evapotranspiration from a vegetative cover (Fig. 1.4). These processes produce an upward flux of water in the form of vapor from the soil. On the other hand, rainfall and other forms of precipitation recharge the soil through a downward liquid flux. The difference between these two flux conditions on a local scale largely dictates the pore-water pressure conditions in the soil profile.

A net upward moisture flux produces a gradual drying, cracking, and desiccation of the soil mass whereas a net downward flux tends to wet a soil mass. The depth of the water table is influenced, among other things, by the net surface flux. A hydrostatic line relative to the groundwater table represents an equilibrium condition representing zero moisture flux at the ground surface. During dry periods, the pore-water pressures become more negative than those represented

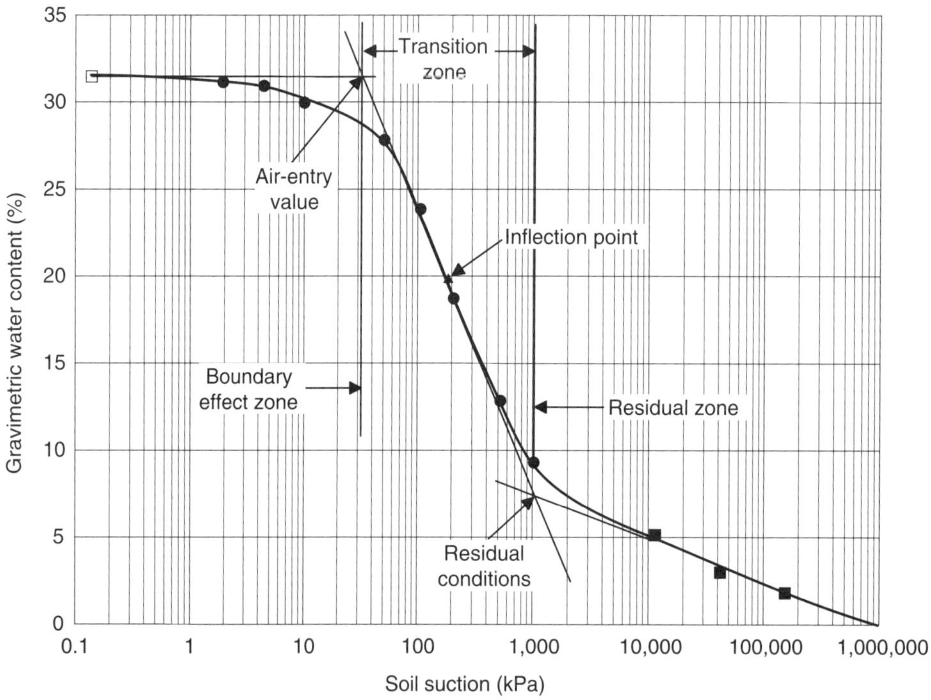

Figure 1.3 Zones of desaturation defined on SWCC.

by the hydrostatic line. The opposite condition occurs during wet periods.

Grasses, trees, and other plants growing on the ground surface dry the soil by applying a tension to the pore-water through evapotranspiration (Dorsey, 1940). Most plants are capable of applying 1000–2000 kPa (10–20 atm) of tension to the pore-water prior to reaching their wilting point (Taylor and Ashcroft, 1972). Evapotranspiration dries the soil resulting in desaturation and cracking and an overconsolidation of the soil mass.

The tension applied to the pore-water acts in all directions and can readily exceed the lateral confining pressure in the soil. When this occurs, a secondary mode of desaturation of the soil mass commences (i.e., cracking). Year after year, the deposit is subjected to varying and changing environmental conditions. The varying climate produces changes in the pore-water pressure distribution, which in turn results in shrinking and swelling of the soil deposit. The pore-water pressure distribution with depth can take on a variety of shapes as a result of environmental changes (Fig. 1.4).

Significant areas of the earth's surface are classified as arid and semiarid. The annual evaporation from the ground surface in these regions exceeds the annual precipitation. Figure 1.5 shows the climatic classification of the extremely arid, arid, and semiarid areas of the world. Meigs (1953) used the Thornthwaite moisture index (Thornthwaite, 1948; Thornthwaite and Mather, 1955) to map these zones while excluding the cold deserts. About 33% of the earth's surface is considered arid and semiarid (Dregne, 1976).

Arid and semiarid areas usually have a deep groundwater table. Soils located above the water table have negative pore-water pressures. The degree of saturation of the soils is reduced when evaporation and evapotranspiration exceed precipitation. Climatic changes highly influence the water content of the soil in the proximity of the ground surface. The pore-water pressures increase upon wetting, tending toward positive values. As a result, the volume and shear strength of the soil are changed. Many soils exhibit severe swelling or expansion when wetted while other soils exhibit significant collapse when wetted. Many soils are known for significant loss of shear strength upon wetting.

Changes in the negative pore-water pressures associated with heavy rainfalls are the cause of numerous slope failures. Reductions in the bearing capacity and resilient modulus of soils in roadways are also related to an increase in the pore-water pressures. These phenomena indicate the important role that negative pore-water pressures play in controlling the mechanical behavior of unsaturated soils.

1.2.1 Quantification of Moisture and Thermal Boundary Fluxes

Soil mechanics textbooks have been particularly silent on how thermal and moisture fluxes at ground surface are to be calculated. Problems involving the flow of water through soil have generally required that either a "hydraulic head" boundary condition be imposed or else a "zero flux" (i.e., an impervious boundary condition) be imposed when solving seepage problems. However, the ground surface is a boundary across which there is continuous moisture movement. Moisture is either going up to the atmosphere in the form of evaporation or evapotranspiration or coming down in the form of precipitation (i.e., rainfall, snowfall, or irrigation).

Weather stations have dotted the globe collecting large amounts of data on key variables related to moisture and thermal fluxes; however, little usage has historically been made of these data in geotechnical engineering. The design of soil cover systems (i.e., also called alternative covers) and other near-ground-surface engineered structures has provided an

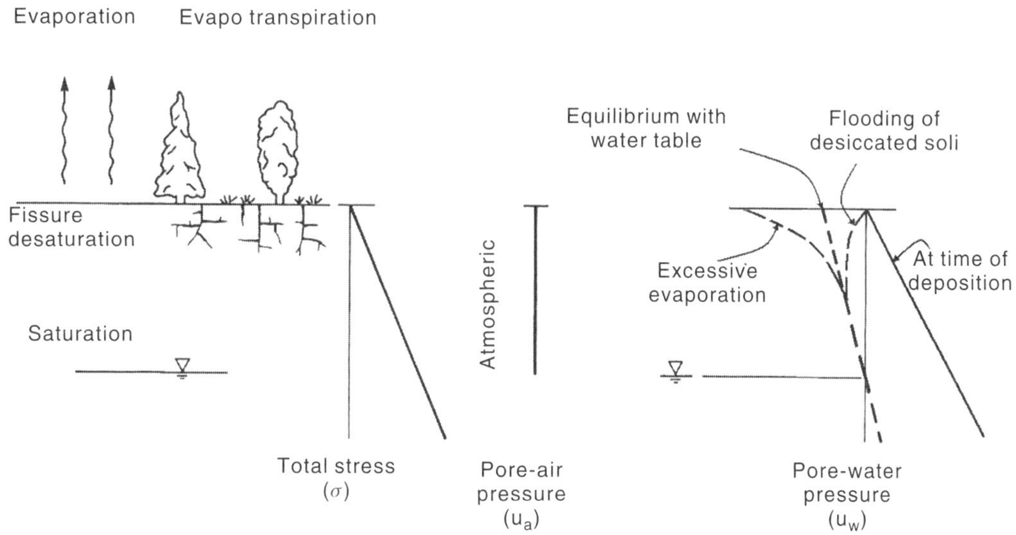

Figure 1.4 Total stress, pore-air pressure, and pore-water pressure distributions in unsaturated soil.

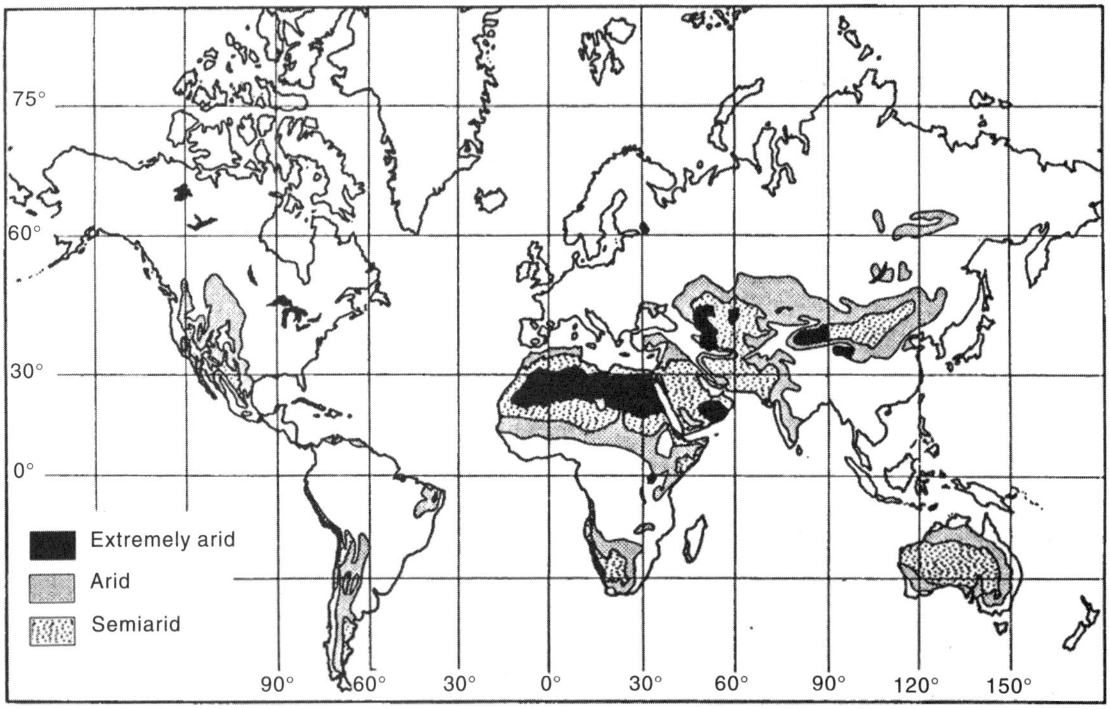

Figure 1.5 Extremely arid, arid, and semiarid areas of the world [from Meigs (1953) and Dregne (1976)].

impetus to utilize weather data for the calculation of moisture flux conditions at the ground surface.

There are a number of factors that affect the calculation of thermal and moisture fluxes at ground surface. However, temperature and precipitation data are the primary variables required for the calculation of flux boundary conditions. The runoff of water at ground surface is also a challenging variable to calculate. The net difference between the ground surface moisture fluxes is called "net infiltration" and is the amount that will flow into the unsaturated soil at ground surface. Often it is the infiltration past a particular depth (e.g., bottom of the "cover layer") that is of primary interest and its magnitude is referred to as "deep infiltration." Terms such as "percolation" and "recharge" are also used when describing the infiltration phenomenon. The procedures that can be used to calculate potential evaporation (PE), actual evaporation (AE), and other water balance variables will be described later in this book.

1.3 DETERMINATION OF UNSATURATED SOIL PROPERTIES

Advances in the development of laboratory testing equipment for unsaturated soils along with the ability to measure soil suction have prepared the way for the implementation of unsaturated soil mechanics. It is now possible to measure most unsaturated soil properties; however, the direct measurement of unsaturated soil properties can be time consuming and expensive. Consequently, a variety of "estimation techniques" for unsaturated soil property functions have been forthcoming from research studies in various countries (Fredlund, 2000a, Vanapalli et al., 1996a). It has become part of acceptable geotechnical engineering practice to utilize "estimation techniques" when applying unsaturated soil mechanics at the preliminary engineering design stage. Estimation techniques have particularly found widespread usage in the area of saturated-unsaturated seepage modeling where the permeability functions and the water storage functions are estimated from a SWCC and the saturated hydraulic conductivity (Thieu et al., 2000).

1.3.1 Estimation Procedures for Unsaturated Soil Properties

The estimation of unsaturated soil property functions provides a new philosophical framework (or paradigm) that has been effective in expediting the implementation of unsaturated soil mechanics. The challenge is to determine the estimation procedures for unsaturated soil property characterization that best describe the actual unsaturated soil properties. The estimation techniques are particularly attractive because direct unsaturated soil testing in the laboratory becomes too costly for many engineering projects.

Figure 1.6 shows that one of several general approaches can be used for the determination of unsaturated soil property functions. The direct measurement of unsaturated soil property functions appears to be possible only for special cases that are often of a research or extremely important

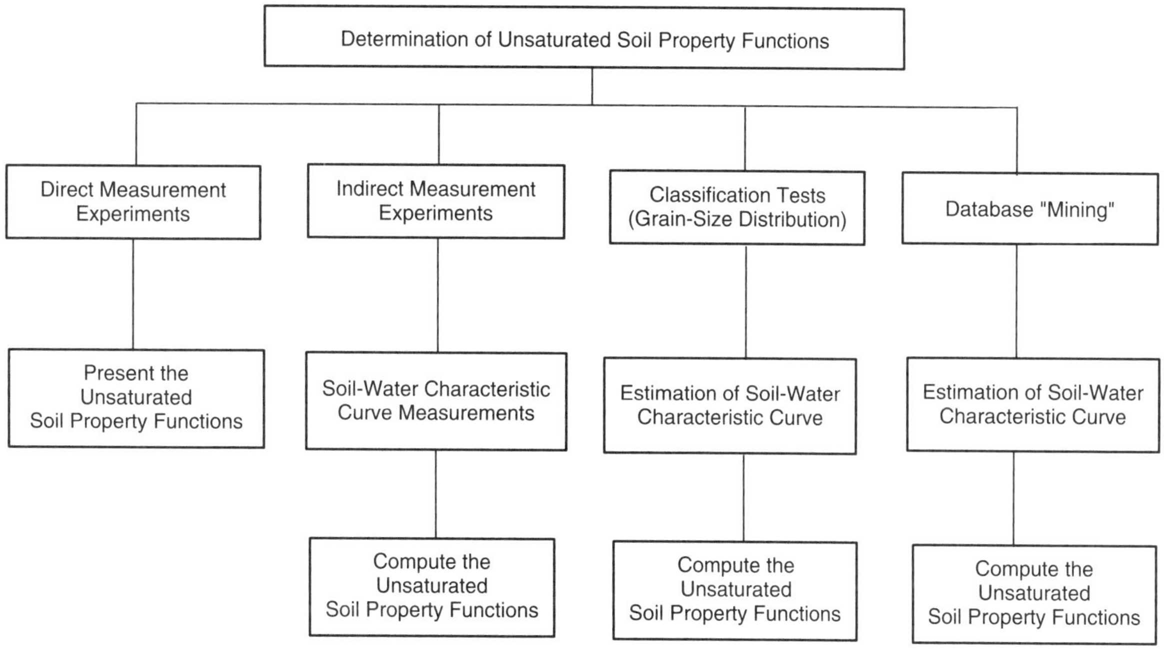

Figure 1.6 Methodologies for determination of unsaturated soil property functions.

nature. The direct application of unsaturated soil theories at the highest level involves unsaturated soil testing with soil suction measurements (or controls). These tests are relatively complex and time consuming to perform in comparison to saturated soil tests.

Other approaches for the implementation of unsaturated soils theories in practice have been suggested wherein the SWCC is used along with saturated soil parameters to estimate unsaturated soil properties (Fredlund, 1995b). These indirect measurements and estimation procedures open the way for a hierarchical approach to the application of unsaturated soil mechanics in geotechnical engineering practice. The assumption is made that saturated soil properties are known. The unsaturated soil property functions take the form of a smooth extension of the saturated soil properties.

Unsaturated soil property functions may be defined with reasonable accuracy for many soil mechanics problems by performing an indirect laboratory test (e.g., a pressure plate test to measure the SWCC). The data from the SWCC can be used to calculate the required unsaturated soil property functions. It is also possible to obtain an indication of the SWCC from classification soil properties, correlations, or "mining" databases containing the results of past laboratory soil tests.

Classification tests (e.g., grain-size distribution curve) can also be used for the estimation of a SWCC. The grain-size distribution curve is used to estimate the SWCC that is then used to compute the desired unsaturated soil property function (Fredlund et al., 1997a). There may be a reduction in the accuracy of the estimated unsaturated soil property function when using this procedure. The engineer must assess whether or not the approximated unsaturated soil property function is satisfactory for the analysis of the problem at hand. This estimation procedure is somewhat analogous to using the grain-size distribution curve to estimate the saturated hydraulic conductivity of a soil (Hazen, 1911).

1.3.2 Design Protocols for Unsaturated Soil Properties

Engineering design protocols can generally be placed within one of two primary categories: (1) preliminary design and (2) final design. Preliminary design protocols can make ample use of estimated unsaturated soil properties whereas final design protocols need to rely on measured values of the SWCC or direct measurement of unsaturated soil property functions. Further details on the estimation and measurement of unsaturated soil properties are provided throughout this book.

1.4 STAGES IN MOVING TOWARD IMPLEMENTATION

Developing a science basis for unsaturated soil mechanics can be viewed in terms of a series of stages leading toward implementation in engineering practice. The stages leading toward implementation are listed in Fig. 1.7 (Fredlund, 2000). Research studies over the past six decades have been directed at all stages leading toward an appropriate technology for implementation.

1.4.1 State Variable Stage

The state variable stage is the most basic and fundamental level at which a science for unsaturated soil behavior can be initiated. The most important state variables for an unsaturated

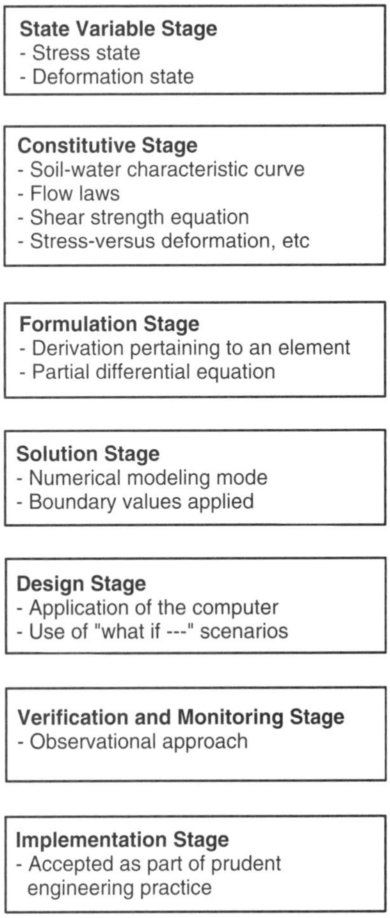

Figure 1.7 Stages leading toward implementation of unsaturated soil mechanics in engineering practice.

soil are the stress state variables: the net normal stress ($\sigma - u_a$, where σ = total stress and u_a = pore-air pressure) and the matric suction ($u_a - u_w$, where u_w = pore-water pressure). These stress state variables are later shown to take on the form of two stress tensors (i.e., 3×3 matrices). The stress state variables have become widely accepted and illustrate the need to separate the effects of total stress and pore-water pressure when pore-water pressures are negative. The stress state variables also satisfy the need for a smooth transition between the saturated and unsaturated states.

1.4.2 Constitutive Stage

The constitutive stage becomes the point at which empirical, semi empirical, and possibly theoretical relationships between state variables are proposed and verified. The verification of proposed constitutive relations must be conducted for a wide range of soils in order to ensure uniqueness and subsequent confidence on the part of practicing geotechnical engineers. Extensive research studies on constitutive relationships for unsaturated soils were made in the 1970s, but earlier and later developments have also contributed to our understanding of the constitutive behavior of unsaturated soils.

1.4.3 Formulation Stage

The formulation stage involves combining the constitutive behavior of a material with the conservation laws of physics applied to a representative elemental volume (i.e., a REV). The result is generally a partial differential equation that describes a designated physical process for an element of the continuum.

1.4.4 Solution Stage

The solution stage involves solving specific examples representative of a class of problems. The solving of formulated equations constitutes an independent stage since it is possible that fundamental equations can be derived; however, the solution to the equations may not be possible. The solution stage for unsaturated soils generally involves solving partial differential equation(s) that are placed into a numerical solution form (e.g., finite difference of finite element form) that can be solved using computer software. Other solvers may also be used as is the case for limit equilibrium solutions of slope stability problems.

1.4.5 Design Stage

There is a gradual increase in engineering confidence as research progresses from the formulation stage to the solution stage and on to the design stage. The design stage focuses on the primary unknowns that must be quantified from a practical engineering standpoint. The design stage generally involves the quantification of geometric and soil property variables that become part of an engineering design. The computer will generally be an important tool in the design of most earth structures. The computer has truly changed the manner in which geotechnical design is conducted. Nowadays, geotechnical engineering design generally takes the form of a series of parametric-type studies.

1.4.6 Verification and Monitoring Stage

There is need to "observe" the behavior of an engineered structure during construction and during operation of the structure in order to provide feedback to the designer. Only through field monitoring and feedback can confidence be firmly established in the design procedures. The "observational method" as defined by Peck (1969) extends beyond the verification of design and is considered to be part of the design process.

1.4.7 Implementation Stage

It is possible that the implementation stage may not be realized in engineering practice even when theoretical formulations and design procedures have been fully studied and verified. Implementation is the final stage in bringing an engineering science into routine engineering practice. There are other factors that also need to be addressed at the implementation stage, such as (i) the cost of undertaking any special site investigations, soil testing, and engineering analyses, (ii) the human resistance to change, and (iii) the political, regulatory, and litigation factors.

Slowness in the implementation of unsaturated soil mechanics appears to have been related to the cost of laboratory soil testing for the quantification of unsaturated soil properties. The original soil mechanics paradigm involving the direct measurement of soil properties becomes excessively costly when measuring unsaturated soil property functions. However, estimation procedures utilized for determining unsaturated soil property functions have significantly increased the use of unsaturated soil mechanics in engineering practice. Proposed estimation procedures allow for the assessment of unsaturated soil property functions in a cost-saving manner, as illustrated in Fig. 1.6 (Fredlund, 2000).

Unsaturated soil mechanics has developed through a series of stages and today engineers are faced with the challenge of implementing the science as part of geotechnical engineering practice. Implementation of a new science is always a challenge, but it is a challenge that can open the door to unique technological applications and financial rewards.

1.5 NEED FOR UNSATURATED SOIL MECHANICS

Success in the practice of soil mechanics can be traced largely to the ability of engineers to relate observed soil behavior to stress state conditions. This ability has led to the transferability of the soil mechanics science in a relatively consistent engineering manner. This has been the case for saturated soils, but unsaturated soil mechanics has lagged behind in its implementation. Difficulty has been experienced in extending classical soil mechanics to embrace unsaturated soils.

The question can be asked: "Why hasn't a practical science developed and flourished for unsaturated soils?" A cursory examination might suggest that there is no need for such a science. However, this is certainly not the case when the problems associated with expansive soils are considered. Jones and Holtz (1973) reported that, in the United States alone, "each year, shrinking and swelling soils inflict at least $2.3 billion in damages to houses, buildings, roads, and pipelines—more than twice the damage from floods, hurricanes, tornadoes, and earthquakes!" They also reported that 60% of the new houses built in the United States will experience minor damage during their useful lives and 10% will experience significant damage—some beyond repair.

In 1980, Krohn and Slosson estimated that $7 billion is spent each year in the United States as a result of damage to all types of structures built on expansive soils. Snethen (1986) stated: "While few people have ever heard of expansive soils and even fewer realize the magnitude of the damage they cause, more than one fifth of American families live on such soils and no state is immune from the problem they cause. Expansive soils have been called the 'hidden disaster': while they do not cause loss of life, economically these soils have become one of the United States costliest natural hazards." Expansive soil problems have occurred in many other countries of the world. In Canada, Hamilton (1977) stated: "Volume changing clay subsoils constitute the most costly natural hazard to buildings on shallow foundations in Canada and the United States. In the Prairie Provinces alone, a million or more Canadians live in communities built on subsoils of very high potential expansion."

It would appear that most countries in the world have had problems with expansive soils. Many countries have reported the nature of their problems at research conferences. Some countries reporting expansive soil problems are Australia, Argentina, Myanmar, China, Cuba, Ethiopia, Ghana, Great Britain, India, Iran, Israel, Kenya, Mexico, Morocco, South Africa, Spain, Turkey, and Venezuela. In general, the more arid the climate, the more severe is the shrinking and swelling soil problem.

There is also need for reliable engineering design associated with compacted soils, collapsing soils, and residual soils. Soils that collapse usually have an initially low density. The soils may or may not be subjected to additional load, but when they are given access to water, the soil structure collapse occurs. The water causes an increase in the pore-water pressures with the result that the soil volume decreases. The process is similar to that occurring in an expansive soil but the direction of volume change is opposite. Examples of soil collapse have been reported in the United States as well as other countries (Houston et al., 2001; Johnpeer, 1986). Leaky septic tanks and water lines are a common initiating factor in soil structure collapse.

Inhabited areas with steep slopes consisting of residual soils are sometimes the site of catastrophic landslides which claim many lives. A widely publicized case is the landslide at Po-Shan in Hong Kong which claimed 67 lives (Commission of Inquiry, 1972). Similar problems have been reported in South American countries and other parts of the world.

The soils involved are often residual genesis and have deep water tables. The surface soils have negative pore-water pressures which play a significant role in the stability of the slope. However, heavy, continuous rainfalls can result in increased pore-water pressures to a significant depth, resulting in the destabilizing of the slope. The pore-water pressures along the slip surface at the time of failure may be negative or positive.

There appear to be two main reasons why a practical science has been slow in being developed for unsaturated soils (Fredlund, 1979). First, there was lack of an appropriate science with a theoretical basis. There was a lack of appreciation of the engineering problems and an inability to place the solution within a theoretical continuum mechanics context. The stress conditions and mechanisms involved, as well as the soil properties that must be measured, did not appear to be fully understood. The boundary conditions were related to the environment and were difficult to quantify. Research work has largely remained empirical in nature with little coherence and synthesis. Engineering experience is often applied on a regional basis and there seems to have been poor communication among engineers regarding design procedures.

Second, there appears to have been a lack of a system for financial recovery for services rendered by the engineer. In the case of expansive soil problems, the possible liability to the engineer is often large relative to the financial remuneration. Other areas of geotechnical engineering practice are more profitable to consultants. The owner often reasons that the cost outweighs the risk. The hazard to life and injury is largely absent, and for this reason little attention has been given to expansive soil problems. Although the problem basically remains with the owner, it is the geotechnical engineer who has the greatest potential to provide an engineered solution.

Certainly, there is a need for an appropriate technology for unsaturated soil behavior. Such a technology must (1) be practical, (2) not be too costly to use, (3) have a sound theoretical basis, and (4) run parallel in concept to conventional saturated soil mechanics.

1.5.1 Application Areas for Unsaturated Soil Mechanics

Common to all unsaturated soil situations are the negative pressures in the pore-water. The class of unsaturated soil problems involving negative pore-water pressures that has received the most attention is that of swelling or expansive clays. An attempt is made in this book to broaden the scope of unsaturated soil problems by presenting the principles and concepts that can be applied to a wider range of unsaturated soil problems.

Several typical problems are described to illustrate relevant questions which might be asked of a geotechnical engineer. An attempt is made throughout this book to respond to general unsaturated soil mechanics questions from a theoretical standpoint.

1.5.2 Construction and Operation of a Dam

Let us consider the construction of a homogeneous rolled earth dam. The construction involves compacting soil in approximately 150-mm lifts from its base to the full height of the dam. The compacted soil would have an initial degree of saturation between 70 and 80%. Figure 1.8 shows a dam at approximately one-half of its design height, with a lift of soil having just been placed. The pore-air pressure in the layer of soil being compacted is approximately equal to the atmospheric pressure. The pore-water pressure is negative, often considerably lower than zero absolute pressure.

The soil at lower elevations in the fill is compressed by the placement of the overlying fill. Each layer of fill constitutes an increase in total stress to the embankment. Compression results in a change in the pore-air and pore-water pressures.

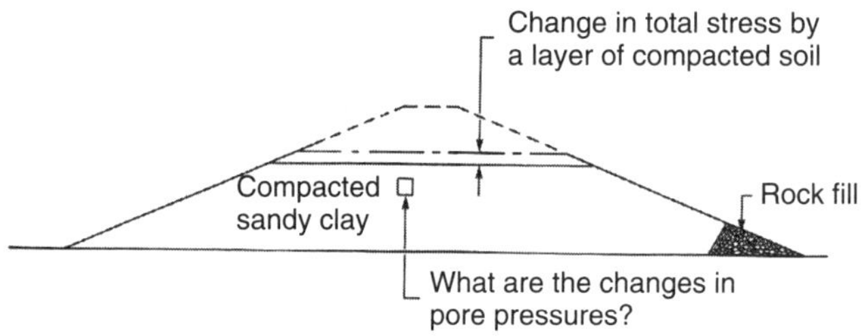

Figure 1.8 Changes in pore-water and pore-air pressure generated as a result of the placement of fill for a compacted earthfill dam.

The construction of the fill is generally sufficiently rapid that the soil undergoes volume change under undrained conditions. Contours of the pore-air and pore-water pressures can be drawn at any time during construction, as shown in Fig. 1.9. In reality, some dissipation of the pore pressures will occur as the fill is being placed. The pore-air pressure will dissipate to the atmosphere. The pore-air pressure may also be influenced by evaporation and infiltration at the surface of the dam. All pore pressure changes have the potential of producing volume changes since the stress state is being changed. There are many soil mechanics questions that can be asked, and there are many analyses that would be useful to the geotechnical engineer.

Once the construction of the dam is complete, the filling of the reservoir will change the pore pressures in a manner similar to that shown in Fig. 1.10. A transient process is taking place under new boundary conditions. Changes in the surrounding environment after steady-state conditions are established may give rise to further questions (Fig. 1.11).

The questions that might be asked of a geotechnical engineer involve analyses associated with saturated/unsaturated seepage, the change in volume of the soil mass, and the change in shear strength. The change in the shear strength could be related to a change in the factor of safety. The questions are similar to those asked when dealing with saturated soils. There is, however, a significant difference in the case of unsaturated soil problems since the moisture flux boundary conditions are produced by the climatic environment.

1.5.3 Natural Slopes Subjected to Environmental Changes

Natural slopes are subjected to a continuously changing environment (Fig. 1.12). An engineer may be asked to investigate the present stability of a slope and predict what would happen if the geometry of the slope is changed or if the environmental conditions should change. In this case, boreholes may be drilled and undisturbed samples obtained for laboratory tests. Most or all of the potential slip surfaces may lie above the groundwater table in this case. In other words, the potential slip surface may pass through unsaturated soils with negative pore-water pressures.

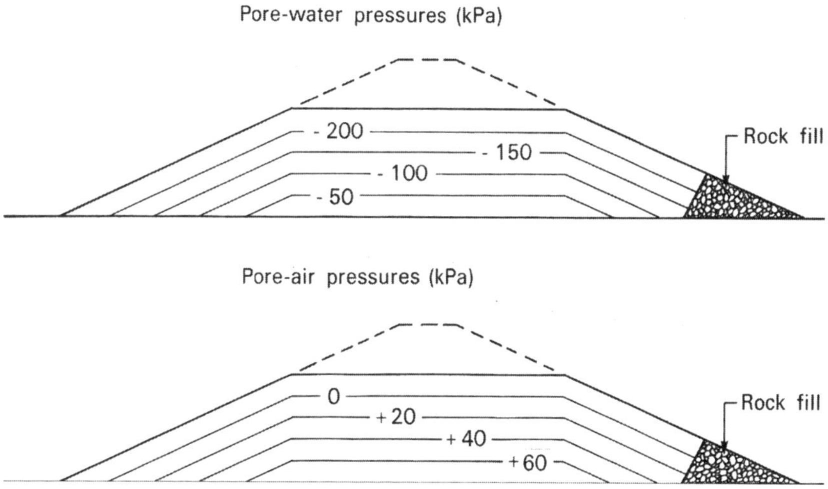

Figure 1.9 Hypothetical pore-water and pore-air pressure distributions after partial construction of an earthfill dam.

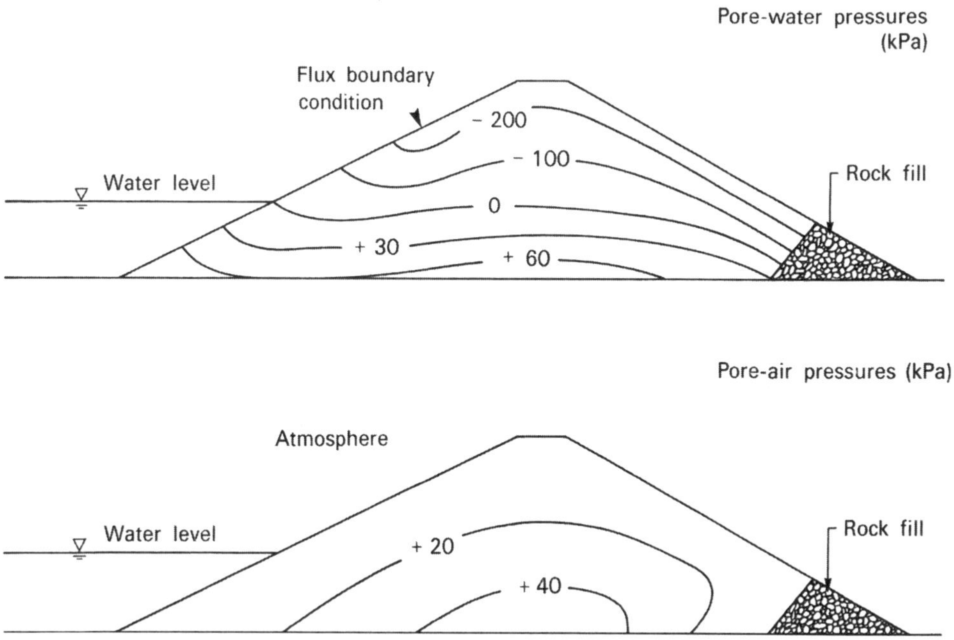

Figure 1.10 Hypothetical pore-water and pore-air pressure distributions after partial dissipation of pore-water and pore-air pressures.

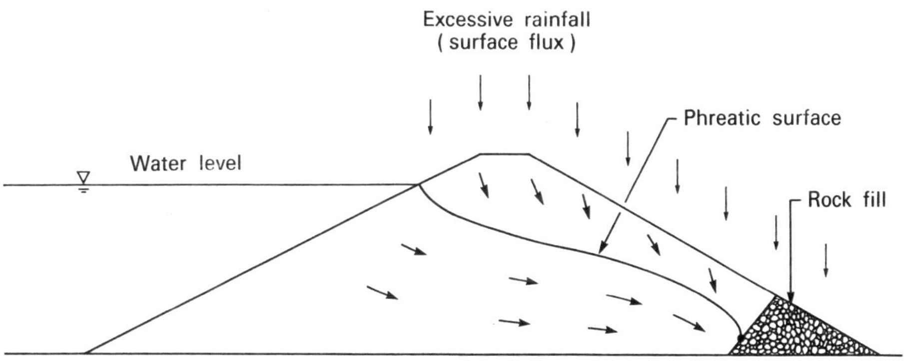

Figure 1.11 Effect of precipitation on long-term seepage flow through an earthfill dam.

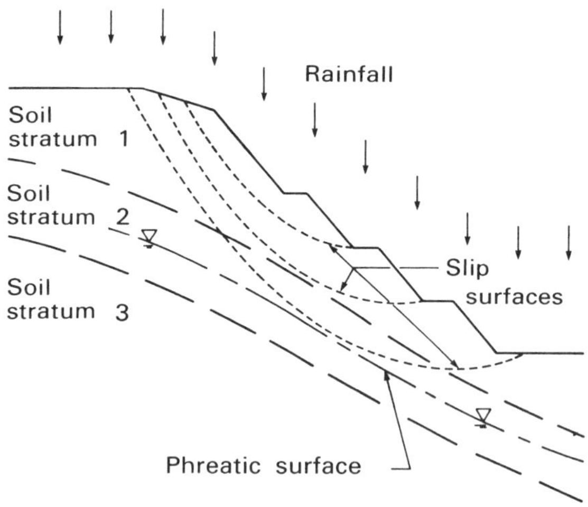

Figure 1.12 Effect of precipitation on man-made slope subjected to rainfall.

Surface sloughing commonly occurs on relatively flat slopes following prolonged periods of precipitation. These failures have received little attention from an analysis standpoint. One of the main difficulties appears to have been associated with the assessment of pore-water pressures in the zone above the groundwater table.

The slow, gradual, downslope creep of soil is another aspect which has not received much attention in the literature. It has been observed, however, that the movements occur in response to seasonal, environmental changes. Wetting and drying and freezing and thawing are known to be important factors. It would appear that an understanding of unsaturated soil behavior is imperative in formulating a mathematical assessment of these problems.

1.5.4 Mounding Below Waste Retention Ponds

Waste materials from mining and industrial operations are often stored as a liquid or slurry retained by low-level dykes (Fig. 1.13). Soil profiles with a deep water table have often been considered to be ideal locations for these waste ponds. The soils above the water table have negative pore-water pressures and may be unsaturated. The

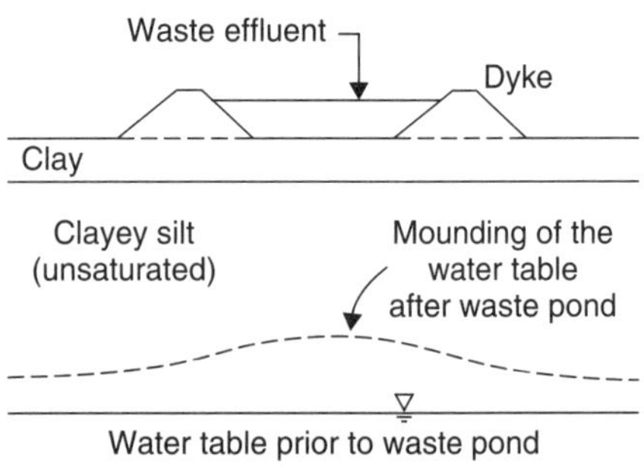

Figure 1.13 Mounding of water table below a waste pond due to seepage through unsaturated soil zone.

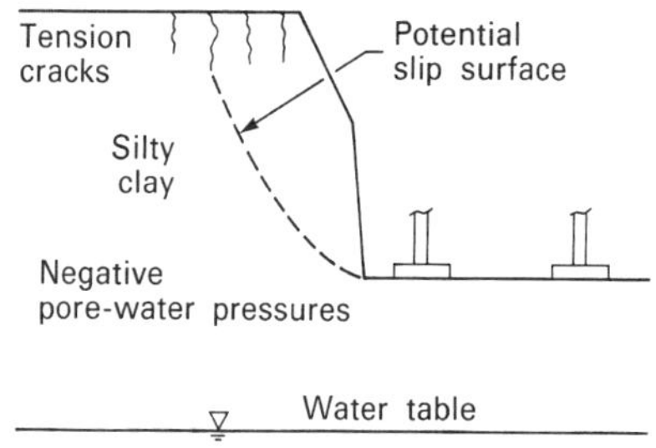

Figure 1.14 Potential instability of near-vertical excavation during construction of a foundation.

assumption is often made that as long as the pore-water pressure remains negative, there is little or no movement of fluids downward from the waste pond. However, it has been observed that there is often a mounding of the water table below the waste ponds even when the intermediate soil may remain unsaturated. Engineers now realize that significant volumes of water and contaminants can move through the unsaturated soil portion of the profile even though the soil has negative pore-water pressures.

1.5.5 Stability of Vertical or Near-Vertical Excavations

Vertical or near-vertical excavations are often used for the installation of a foundation or a pipeline (Fig. 1.14). It is well known that the backslope in a moist silt or clay soil will stand at a near-vertical slope for some time before failing. Failure of the backslope is a function of the soil type, the depth of the excavation, the depth of tension cracks, the amount of precipitation, as well as other factors. In the event that the contractor should leave the excavation open longer than planned or should a high precipitation period be encountered, the backslope may fail, causing damage and possible loss of life.

The excavations being referred to are in soils above the groundwater table where the pore-water pressures are negative. The excavation of soil for the trench also produces a temporary decrease in pore-water pressures and an increase in the shear strength of the soil. With time, there may be a gradual increase in the pore-water pressures in the backslope and a loss in shear strength. The increase in the pore-water pressure is the primary factor contributing to the instability of the excavation. Engineers often place the responsibility for ensuring backslope stability onto the contractor. Predictions associated with this problem require an in-depth understanding of unsaturated soil behavior and moisture flux conditions.

1.5.6 Bearing Capacity for Shallow Foundations

The foundations for light structures are generally shallow spread footings (Fig. 1.15). The bearing capacity of the underlying soils is often computed based on the unconfined compressive strength of the soil. Shallow footings can easily be constructed when the water table is below the elevation of

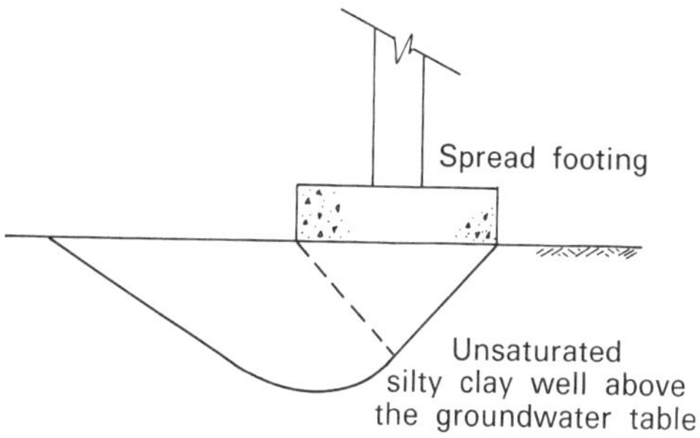

Figure 1.15 Analysis of bearing capacity for lightly loaded structure placed on unsaturated soil with negative pore-water pressures.

the footings. In many cases, the water table is at a considerable depth and the soil below the spread footings has a negative pore-water pressure. Undisturbed samples, held intact by negative pore-water pressures, are routinely tested in the laboratory to obtain a measurement of the shear strength of the soil. The assumption is often made that the pore-water pressure conditions in the field will remain relatively constant with time and therefore the unconfined compressive strength will also remain essentially unchanged. Based on this assumption and a relatively high design factor of safety, the bearing capacity of the soil is computed.

The above design procedure has involved soils with negative pore-water pressures. The engineer almost seems to have been oblivious to the fact that the design procedure assumes that there will be a long-term retention of negative pore-water pressure in the soil. The geotechnical engineer has taken a very different attitude toward the long-term retention of negative pore-water pressures when dealing with slope stability problems. In the case of a slope stability problem, the geotechnical engineer has generally assumed that long-term negative pore-water pressures cannot be relied upon to contribute to the shear strength of the soil. The two seemingly opposite attitudes or perceptions give rise to an important question, "How constant and stable are negative pore-water pressures with respect to time?" Or, a more probing question might be "Has the engineer's attitude toward negative pore-water pressures been strongly influenced by expediency?"

1.5.7 Ground Movements Involving Expansive Soils

There is no problem involving soils with negative pore-water pressures that has received more attention than the prediction of heave associated with the wetting of an expansive soil. Light structures such as a roadway or a small building are often subjected to severe distress as a result of changes in the surrounding environment subsequent to construction (Figs. 1.16 and 1.17). Changes in the climatic environment may occur as a result of the removal of trees and grass and the excessive watering of a lawn around a new structure. The zone of soil that undergoes volume change on an annual basis has been referred to as the "active zone." The higher the swelling properties of the soil, the greater will be the amount of heave to the structure.

It has been common practice to obtain undisturbed soil samples from the upper portion of the profile for one-dimensional oedometer testing in the laboratory. The laboratory results are used to provide quantitative estimates of potential heave. Numerous laboratory testing techniques and analytical procedures have been used in engineering practice. The prediction of heave in expansive soils is dealt with in considerable detail later in this book.

1.5.8 Design of Soil Cover Systems and Capillary Breaks

Soil cover systems have increasingly become a potential solution to mitigate environmental damage. Covers became

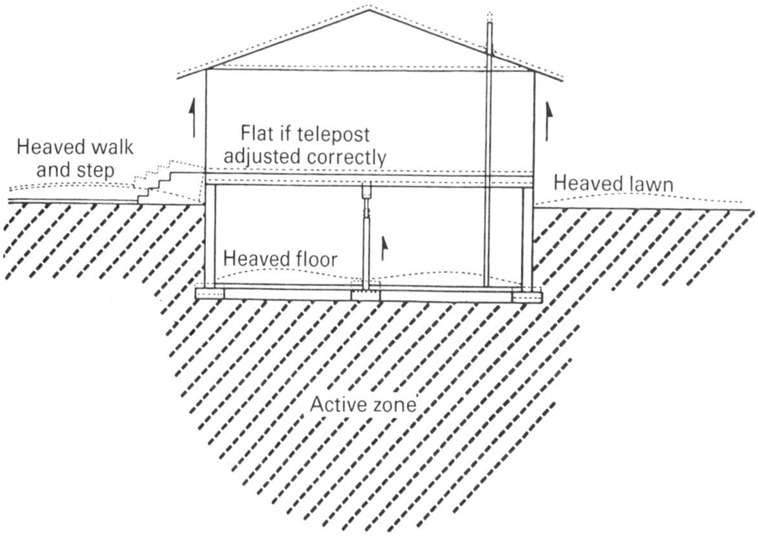

Figure 1.16 Common ground movements associated with a house with basement foundation constructed in expansive soils (Hamilton, 1977).

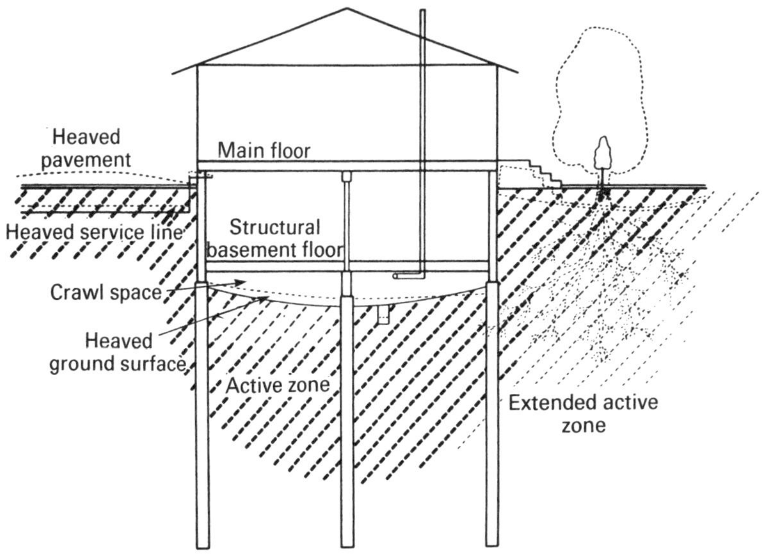

Figure 1.17 Potential remedial measures for a house founded upon piles placed through expansive soils (Hamilton, 1977).

a favored engineered solution around the 1980s, particularly for waste containment facilities and the remediation of contaminated sites. Mining operations have two streams of waste material that need to be properly handled to mitigate damage to the environment. These two streams are waste rock and mine tailings. Cover systems provide a potential solution for both streams of waste materials.

The design of cover systems is easy to understand. However, there are numerous issues and challenges associated with various aspects of design. There are issues related to the required input information and the solution of the moisture flow partial differential equation. There are a large number of assumptions that need to be made at various stages of the design process. These assumptions can significantly influence the final performance of the cover design. The quantification of unsaturated soil properties for each material involved (e.g., the permeability functions and the water storage functions) has proven to be a challenge for the geotechnical engineer (M.D. Fredlund Fredlund, 2000a, b; D.G. Fredlund, 2007b).

Shackelford (2005) presented a summary of the primary current and future environmental issues associated with cover systems. The issues mentioned were (1) long-term performance of waste containment systems, (2) alternative barriers or covers (i.e., alternative to clay covers) and barrier materials, (3) innovative barriers (covers) and barrier materials, (4) forms of waste materials, (5) significance of biological waste processes, (6) the role of numerical modeling in cover design, and (7) professional identity.

The design of a cover system depends on the ability to predict moisture fluxes in and out of the ground surface as well as moisture fluxes through the bottom of the cover system. The design analysis can be viewed as a "flux-driven" problem because of the importance of the surrounding weather conditions. Problems involving the predictions of hydraulic head are known to be easier to solve than problems involving the prediction of moisture fluxes (D.G. Fredlund, 2007b). The boundary conditions at ground surface need to be described in terms of a moisture flux when performing the cover design. The ground surface has moisture either coming down in the form of precipitation or going up in the form of evaporation and evapotranspiration. The quantification of the net moisture flux boundary conditions at the ground surface has also proven to be a challenging analysis.

A cover system can be viewed as a thin interface placed between the overlying atmosphere and the underlying soil strata, as shown in Fig. 1.18. The climate imposed on top of the cover can vary widely from arid to humid conditions. A particular cover system cannot function in a similar manner over a wide range of climatic conditions. Rather, it would be expected that a particular cover could only perform satisfactorily under certain climatic conditions.

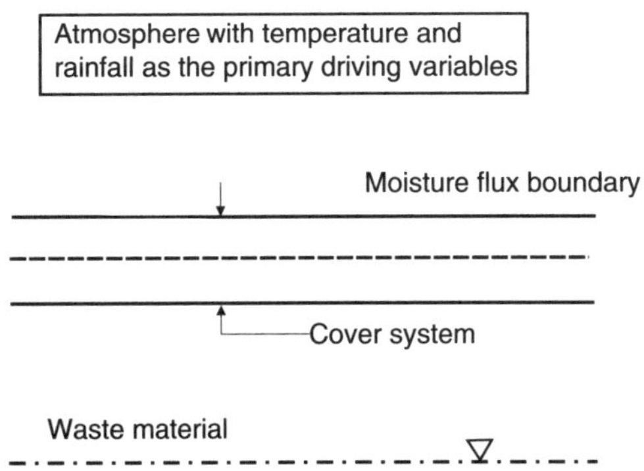

Figure 1.18 Soil cover system used as interface between waste material and atmosphere.

The purpose of the cover system can be viewed as functioning quite differently in differing situations. For example, the cover may operate as a "store-and-release" system in one case, but it might be desirable to have it function as a saturation oxygen barrier in another situation. The location of the water table in the underlying materials also has a strong influence on the functionality of the cover system. Elements of a soil cover system (i.e., atmosphere, cover, and underlying soils) are variable, highly nonlinear, hysteretic, and random in nature (Fredlund, 2006). The challenges related to solving such a problem are formidable but not impossible. At present, it is fair to say that several geotechnical engineers might produce somewhat different design solutions for a particular cover system.

1.5.9 Road and Railroad Structures

Some of the oldest infrastructural components of civilization are roads and railroads. The preparation of the subgrade and the placement of subbase and base materials have been largely determined over decades of trial and error. Consequently, the design of pavement and railroad structures is quite empirical in nature.

There are many aspects related to the performance of pavements and railroads which are not well understood because the underlying materials are most often unsaturated. Little research has been focused on understanding the physical behavior of unsaturated subgrade materials as well as subbase and base materials. Attempts to model pavement structures have provided new insight into the movement of water through the unsaturated granular materials (Barbour et al., 1992). Normally, granular materials are perceived of as being highly pervious; however, this is not true when the materials are unsaturated. These materials have the capability of forming capillary barriers and resisting the movement of water. Much more could be learned about the suitability of various granular materials from an engineering standpoint if the water movement properties were characterized and actual climate events were simulated using numerical saturated-unsaturated seepage models.

Matric suction in the subgrade of highways and railroads plays a major role in the performance of the structure (Khogali et al., 1991; Marjerison et al., 2001; Zapata et al., 2009). Measurements of matric suction in the subgrade have been shown to decrease in the spring of the year or during wet periods of the year. Sattler et al. (1989) proposed bearing capacity approaches to the design of railroads and highways based on the unsaturated shear strength of the subgrades.

1.5.10 Characteristics of Unsaturated Soil Examples

The above examples show that there are many situations that require an in-depth understanding of seepage, volume change, and shear strength characteristics of unsaturated soils. There is often a need to simultaneously consider all three aspects of unsaturated soil mechanics behavior when dealing with a particular engineering problem. Typically, a flux boundary condition produces an unsteady-state saturated/unsaturated flow situation which results in volume change and a change in the shear strength of the soil. The change in shear strength can be translated into a change in factor of safety. There may also be an interest in quantifying the change of other volume-mass soil properties (i.e., water content and degree of saturation).

1.6 PARTIAL DIFFERENTIAL EQUATIONS IN SOIL MECHANICS

Various classes of soil mechanics problems can be visualized as the solution of one or more partial differential equations (PDEs). The PDE is derived by applying appropriate constitutive relationships to a REV while adhering to the conservative laws of physics (i.e., conservation of mass and conservation of energy). The resulting PDE satisfies the physical conditions associated with the behavior of the soil for a particular class of geotechnical engineering problems.

The physics for a REV can then be applied to a finite-sized element of a continuum which is called a "finite element." Combining many finite elements together allows an entire continuum to be modeled. Boundaries or limits must be placed on the region that is being considered. This gives rise to what is called a "boundary value problem" such as is shown in Fig. 1.19. At the heart of boundary value formulation is a description of the physical soil behavior for the REV.

The laws of physics that pertain to the REV are extended throughout a region until the boundaries of the selected geometry are reached. A numerical solving technique (e.g., the finite element method) is used to solve a series of equations that embrace the physics for the entire region under consideration.

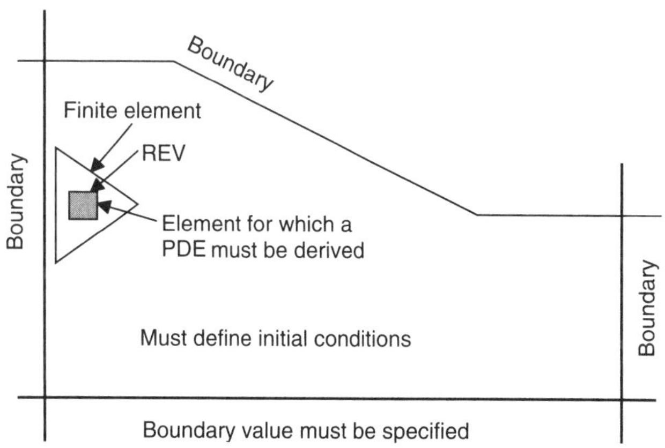

Figure 1.19 Use of REV to formulate PDE to solve a "boundary value problem" (BVP).

1.6.1 Components of Boundary Value Problem

The primary components of a BVP can be listed as follows:

1. Characterization of the ground surface geometry and stratigraphy that define the ground surface and the separation of soil strata (or other materials). The geometry and stratigraphy can take on a one-, two- or three-dimensional representation of the engineering problem. In some situations a one-dimensional representation is satisfactory; however, with advanced computing capabilities, two-, and three-dimensional representations become relatively easy, more realistic, and more representative of in situ conditions. The description of the geometry and stratigraphy for problems involving unsaturated soils is essentially the same as required for a saturated soil problem.
2. Assessment of the boundary and initial conditions for analyzing the problem at hand. For flow-type problems, specifying the head for water flow, pressure for air flow, or temperature for heat flow results in what is called a Dirichlet boundary condition. The specification of a flow rate across a boundary of the problem results in a Neumann type boundary condition. Other intermediate type boundary conditions are also possible. Similar type boundary conditions can be specified for stress-deformation types of analyses.

   The boundary condition that often needs to be assessed for many unsaturated soils problems is the ground surface moisture flux. This means the downward flux (e.g., rainfall), as well as "actual evaporation" and evapotranspiration needs to be quantified for one or more years. For soil cover design the length of weather records that may be used in simulations may exceed 10 years in length. Initial conditions also need to be determined because the problem being analyzed is nonlinear in character.

   The assessment of moisture flux conditions at the ground surface plays a primary role in performing reliable simulations of an unsaturated soil problem. It can be said that most unsaturated soil problems are moisture flux problems due to the proximity of unsaturated soil to the ground surface. Precipitation conditions need to be assessed based on measured weather station data. It might be necessary to determine precipitation records that capture (i) typical average year conditions, (ii) typical dry year conditions, and (iii) typical wet year conditions. As an example, the design of soil cover systems relies heavily upon actual measured weather conditions over numerous past years.

   A coupled moisture and thermal flow model can be used to compute actual evaporation at the ground surface (Wilson et al., 1994, 1997b). The solution takes the form of a "soil-atmosphere model." Evapotransportation involves the effect of vegetation and requires further approximations to represent the effect of "leaf area index," growing season, and other variables.
3. The soil properties for an unsaturated soil take on the character of unsaturated soil property functions (USPFs). Estimation procedures are widely accepted for the assessment of unsaturated soil property functions (Fredlund, 1996a; Fredlund et al., 1997a). However, not all estimation procedures are equally reliable and the geotechnical engineer needs to understand the limitations and assumptions associated with each USPF (M.D. Fredlund et al., 2003). Some of the factors that need to be taken into consideration when determining unsaturated soil property functions are discussed later in the book.

   The assessment of the SWCC forms the basis for the estimation of all other unsaturated soil property functions. Later discussions in the book deal with the factors that need to be taken into consideration in order to obtain reliable soil property estimations. There is a linkage between the SWCCs and the unsaturated soil property functions, and this linkage must always be maintained in order to obtain acceptable results from a numerical analysis.
4. The above information forms the basis for performing a numerical modeling simulation of a boundary value problem in geotechnical engineering (Fredlund, 1998a). The partial differential equation that describes the physics of the problem is nonlinear when unsaturated soils are involved, and consequently there can be challenges in achieving convergence to the correct solution. It is important to meet two criteria when solving nonlinear numerical problems: (i) the computer software must converge when solving the problem and (ii) the software must ensure convergence to the correct solution. In other words, convergence alone is not a sufficient criterion since there must be assurance that the solution has converged to the correct solution. Fortunately, there have been significant advances in solving highly nonlinear PDEs.

   Unsaturated soil problems are distinctive in that the soil properties always take on a nonlinear soil property form. The unsaturated soil properties are a function of the primary state variables for which a solution is being sought. Consequently, the PDE must be solved through use of an iterative technique. The PDEs being solved generally fall into the category of a nonlinear solution with complex boundary conditions.

   Not all computer software packages are equally capable of solving unsaturated soil problems. Researchers in the mathematics and computing science disciplines have made significant progress with respect to solving highly nonlinear PDEs. Software dedicated to solving unsaturated soil problems needs to take advantage of the latest findings in complimentary mathematics and computing disciplines.

1.6.2 Partial Differential Equation Solving

Partial differential equation solvers have been developed in mathematics and computing science disciplines and are increasingly being used to obtain solutions for geotechnical engineering problems. A single PDE solver can be used to solve several types of PDEs relevant to saturated-unsaturated soil mechanics problems. It is also possible for more than one physical phenomenon to be operative within a REV, resulting in the necessity to combine the solution of multiple PDEs in a "coupled" or "uncoupled" manner. For example, it is necessary for both an equilibrium and a continuity PDE to be simultaneously satisfied when solving consolidation (or swelling) problems. In other words, it is necessary for more than one PDE to be simultaneously solved in a coupled or uncoupled mode. Commonly available PDE solvers are generally capable of addressing these tasks.

Each PDE has a "primary" variable that can be computed, and when there is more than one PDE to be simultaneously solved, there will be more than one primary variable. Each PDE may have one or more soil property to be input. If the soil properties are a function of the primary variable being solved, the solution becomes nonlinear. This is the general situation when dealing with unsaturated soil problems and as a consequence it is necessary to iterate to get a converged solution. When the equations are highly nonlinear, it can be a challenge to obtain a converged solution.

The ability to solve a wide range of geotechnical engineering problems within a similar context gives rise to the possibility of developing a special problem solving computer platform called problem solving environment (PSE). Gallopoulos et al., (1994) described a PSE as " a computer system that provides all the computational facilities needed to solve a target class of problems. PSEs use the language of the target class of problems, so users can run them without specialized knowledge of the underlying computer hardware or software. PSEs create a framework that is all things to all people; they solve simple or complex problems, support rapid prototyping or detailed analysis, and can be used in introductory education or at the frontiers of science." Partial differential equation solvers form the basis for PSEs for solving saturated-unsaturated soil mechanics problems.

All classic soil mechanics problems can be viewed in terms of the solution of a PDE. Let us consider a few problems commonly encountered in unsaturated soil mechanics. The PDE for water flow through a saturated-unsaturated soil system in either two or three dimensions is probably the most common problem encountered. These solutions have been extensively applied in disciplines beyond geotechnical engineering such as agriculture, environmental engineering, and water resources. Hydraulic heads are the primary variables computed followed by solving for other variables of interest. The water coefficient of permeability (hydraulic conductivity) is dependent upon the negative pore-water pressure and this gives rise to a nonlinear equation with associated convergence challenges.

All flow processes (i.e., water, air, and heat) have similar PDEs that can be solved using a similar PDE solver. Heat flow problems can readily be solved, but there are added challenges associated with the freezing and thawing of water. Air flow problems add the challenge of dealing with a compressible fluid phase.

Analyses associated with slope stability, bearing capacity, and earth pressure calculations have historically been performed using plasticity and limit equilibrium methods (e.g., methods of slices). However, all these application areas are increasingly being viewed as "optimization" solutions imposed on the results of a stress analysis (Pham and Fredlund, 2003). Stresses computed from linear elastic analyses with approximate elastic parameters have been shown to be acceptable for subsequent usage in an optimization procedure (Stianson, 2008). Total stress fields are computed in an uncoupled manner by "switching on" gravity body forces. Techniques such as "dynamic programming" are then used to determine the shape and location of the rupture surface (Pham et al., 2001a, Pham and Fredlund 2003a; Stianson, 2008). Although the finite-element-based "optimization" solutions to slope stability problems have been published, long-standing limit equilibrium analyses are still important in geotechnical engineering practice. The limit equilibrium analyses that have been applied for saturated soil problems can be extended to unsaturated soil problems. The theory associated with both limit equilibrium methods of slices and optimization methods are presented in the book.

The prediction of volume changes associated with expansive clays and collapsible soils can also be performed using the solution of a stress-deformation analysis (Vu and Fredlund, 2000). The soil properties for this problem are nonlinear but can be converted to equivalent, incremental elastic parameters. These problems can also be solved using more elaborate elastoplastic models, but these solutions are considered to be outside the scope of this book. In the case of expansive or collapsible soils, the volume changes are commonly associated with changes in the negative pore-water pressures (or matric suctions). Consequently, it is necessary to combine a seepage analysis and a stress analysis in a coupled or uncoupled manner in order to solve transient volume change problems.

1.6.3 Convergence of Nonlinear Partial Differential Equations

The nonconvergence of nonlinear PDEs is probably the single greatest deterrent to the use of numerical methods in engineering practice. However, there are several techniques that have emerged that greatly assist in the solution of highly nonlinear PDEs (Mansell et al., 2002). The most successful technique appears to involve automatic, dynamic finite element mesh refinement as well as mesh optimization (Oden, 1989), which are referred to as adaptive grid refinement (ADR), methods.

The primary source of error when solving PDEs using numerical solutions is insufficient spatial resolution (Yeh, 2000). Traditional solutions have used fixed spatial grids that are generated prior to the onset of the numerical solution. It is highly unlikely that the grid generated without considerations of the types of errors associated with the numerical solution can produce a satisfactory solution for nonlinear PDEs.

There are two approaches that can be used to ensure that an accurate PDE solution is obtained. The first approach involves the mathematical alteration of the PDE and the second approach involves the incorporation of AGR algorithms (Mansell et al., 2002). Bern et al. (1999) stated: "Scientific and engineering computations has become so complex that traditional numerical computations on uniform meshes are generally not possible or too expensive." The use of finer and finer meshes becomes an increasingly impossible solution as three-dimensional and large geographical areas are modeled. This is particularly true when solving problems such as the infiltration of water into the unsaturated soil near ground surface.

The solution that has received the most attention for convergence involves the use of AGR based upon error estimates (Bern et al., 1999). Babuska (1989) stated: "The main purpose of any adaptive approach is to achieve, in a most effective way, an approximate solution which is in the range of the admissible accuracy (tolerance)." It would appear that AGR is often necessary when solving nonlinear PDEs common to unsaturated soil behavior (Fredlund et al., 2002).

1.6.4 Uncoupled Processes in Unsaturated Soil Mechanics

A few of the PDEs associated with basic processes common to unsaturated soil mechanics are shown in this section. The equations are formulated for the case of a two-dimensional slice through a soil continuum. The equations presented can be solved through use of PDE solvers (M.D. Fredlund et al., 2002; FlexPDE solutions, 1999). Equations for the three-dimensional case or the axisymmetric case can readily be formulated as well.

The PDE for the seepage of water through a saturated-unsaturated soil can be written as

$$k_{wx}\frac{\partial^2 h}{\partial x^2} + \frac{\partial k_{wx}}{\partial x}\frac{\partial h}{\partial x} + k_{wy}\frac{\partial^2 h}{\partial y^2} + \frac{\partial k_{wx}}{\partial y}\frac{\partial h}{\partial y} = -m_2^w \gamma_w \frac{\partial h}{\partial t} \tag{1.1}$$

where:

x, y = coordinate axes,
k_{wx} = hydraulic conductivity in the x-direction, $k_{wx} = f(u_a - u_w)$, m/s,
k_{wy} = hydraulic conductivity in y-direction, $k_{wy} = f(u_a - u_w)$, m/s,
$u_a - u_w$ = matric suction, kPa,
u_a = pore-air pressure, kPa,
u_w = pore-water pressure, kPa,
h = hydraulic head, m,
γ_w = unit weight of water, 9.81 kN/m^3,
t = time, s, and
m_2^w = coefficient of water volume change (i.e., water storage) with respect to matric suction, 1/kPa.

The coefficient of water volume change can be defined in terms of the volume of water content per total volume of soil:

$$m_2^w = \frac{d(V_w/V_0)}{d(u_a - u_w)} \tag{1.2}$$

where:

V_w = volume of water and
V_0 = total volume of soil element.

The PDEs for the flow of air through an unsaturated soil can be written

$$k_a\frac{\partial^2 u_a}{\partial x^2} + k_a\frac{\partial^2 u_a}{\partial y^2} + \frac{\partial k_a}{\partial x}\left(\frac{\partial u_a}{\partial x}\right) + \frac{\partial k_a}{\partial y}\left(\frac{\partial u_a}{\partial y}\right) = -\left(\frac{e}{1+e}S_a - u_a m_2^w\right)\frac{\omega_a g}{RT_K}\frac{\partial u_a}{\partial t} \tag{1.3}$$

where:

k_a = hydraulic conductivity, m/s,
e = void ratio,
ω_a = average molecular weight of air, 28.8 g/mol,
R = universal gas constant, 8.314 J/(mol·K)
T_K = temperature, K, and
S_a = degree of air saturation, unitless.

The PDEs for heat flow through an unsaturated soil can be written as

$$\lambda_x\frac{\partial^2 T}{\partial x^2} + \frac{\partial \lambda_x}{\partial x}\frac{\partial T}{\partial x} + \lambda_y\frac{\partial^2 T}{\partial y^2} + \frac{\partial \lambda_y}{\partial y}\frac{\partial T}{\partial y} = \left(\zeta + L\theta\frac{\partial \theta_u}{\partial T}\right)\frac{\partial T}{\partial t} \tag{1.4}$$

where:

λ_x, λ_y = thermal conductivity of soil in x- and y-directions, respectively, J/(m sK),
T = temperature, °C,
ζ = volumetric specific heat of soil, J/(m^3 K),
L = latent heat of fusion of water, 3.34×10^8, J/m^3,
θ = volumetric water content at initiation of freezing, and
$\partial\theta_u/\partial T$ = change in unfrozen water content of soil with temperature.

The PDEs for a stress-deformation analysis can be written as

$$\frac{\partial}{\partial x}\left[D_{11}\frac{\partial u}{\partial x}+D_{12}\frac{\partial v}{\partial y}\right]+\frac{\partial}{\partial y}\left[D_{44}\left(\frac{\partial u}{\partial y}+\frac{\partial v}{\partial x}\right)\right]=0 \quad (1.5)$$

$$\frac{\partial}{\partial x}\left[D_{44}\left(\frac{\partial u}{\partial y}+\frac{\partial v}{\partial x}\right)\right]+\frac{\partial}{\partial y}\left[D_{12}\frac{\partial u}{\partial x}+D_{11}\frac{\partial v}{\partial y}\right]+\gamma=0 \quad (1.6)$$

where:

u = displacement in x-direction,
v = displacement in y-direction,
$D_{11} = E(1-\mu)/[(1+\mu)(1-2\mu)]$,
$D_{12} = E\mu/[(1+\mu)(1-2\mu)]$,
$D_{44} = E/[2(1+\mu)]$,
E = Young modulus, kPa,
μ = Poisson ratio, and
γ = total unit weight or body force acting in y-direction (i.e., vertical), kN/m^3.

If the computed stress states are optimized with respect to failure conditions in the soil, the shape and location of the critical slip can be determined. This is, in essence, a slope stability analysis where the shape of the potential slip surface is not prescribed. A similar type of analysis is also possible for the calculation of bearing capacity and earth pressures.

Stress-versus-deformation analyses become more challenging when the objective of the analysis is to compute deformations. Predictions of deformation require accuracy for the nonlinear constitutive soil properties. The stress-deformation analysis needs to start with the calculation of the initial stress states and pore-water pressures.

It is possible to simultaneously solve more than one partial differential equation in a coupled or uncoupled manner. Some studies have shown that uncoupled solutions yield satisfactory results for most geotechnical engineering purposes (Vu 2003). It is also possible to use the AGR approach when solving two or more PDEs. In other words, there can be independent optimized grid (or mesh) being used for the solution of the seepage equation while a different mesh is simultaneously being used for the stress analysis. The two solvers (i.e., two meshes) are able to communicate with one another as the overall solution moves toward convergence.

The above equations are simply meant to illustrate how the classic areas of saturated-unsaturated soil mechanics can be solved in a similar manner through the solution of a limited number of PDEs.

1.6.5 Numerical Modeling of Saturated-Unsaturated Soils

A series of example problems involving unsaturated soils are presented to illustrate solutions that can be obtained through the use of numerical modeling. Problems can involve the solution of either a single PDE or more than one PDE in an uncoupled or coupled manner. The examples are typical of problems encountered in geotechnical engineering. Automatic, optimized mesh generation and grid refinement techniques are used in solving each of the unsaturated soil mechanics problems.

The saturated-unsaturated software programs from SoilVision Systems were used for the solution of the example problems in this book (M.D. Fredlund, 2005). The two- and three-dimensional seepage problems were solved using the software called SVFlux. The stress analysis and slope stability example problem was solved using SVSolid and SVSlope. The volume change problem involving expansive soils was solved using SVSolid and SVFlux. The unsaturated soil properties were estimated using the SoilVision database (M.D. Fredlund, 2003).

1.6.6 Example of Two-Dimensional Seepage Analysis

Figure 1.20a shows the final optimized mesh for the computation of steady-state saturated-unsaturated seepage through a two-dimensional earthfill dam. Concentrations of finite elements occur at locations where refinement of the mesh is required to obtain an accurate solution (e.g., locations of increased gradient). Figure 1.20b shows the computed hydraulic heads for steady-state seepage through the earthfill dam. A parametric study showed that variations in the permeability function for the shell of the dam had little effect on the phreatic line and the equipotential lines. Solving the entire saturated-unsaturated region removes the need to make any assumption regarding the location of the phreatic surface.

1.6.7 Finite Element Mesh for Three-Dimensional Tailings Pond

The generation of an appropriate finite element mesh for a three-dimensional problem is an extremely difficult task. Even when a three-dimensional mesh can be generated, there is no assurance that it will meet all the requirements for a correct solution to the nonlinear partial differential seepage equation. Consequently, it is extremely important that mathematically satisfied criteria be used in the automatically generated and refined meshes that are part of the three-dimensional solution.

Figure 1.21 illustrates the automatically generated mesh for steady-state seepage through the waste pond. A total of nine soil units were identified and permeability functions were input for all soils near the ground surface. The solution to this problem also involved the use of time-dependent flux boundary conditions that varied from one location to another. Further details related to data input and the solution of this problem can be found in Rykaart et al. (2001a). The primary intent of this example is to illustrate the optimized finite element mesh.

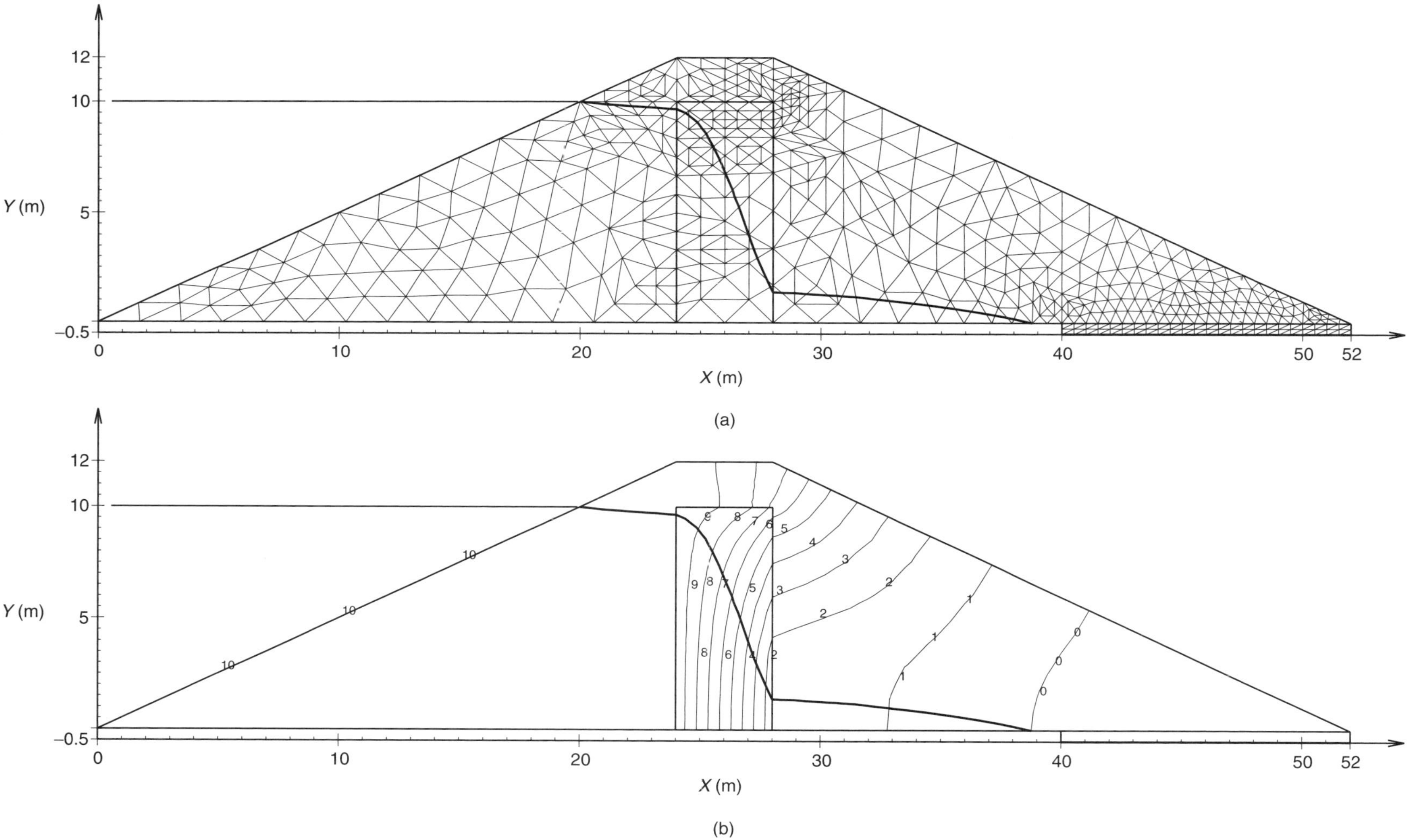

Figure 1.20 Two-dimensional analysis of seepage through an earthfill dam with a clay core. (a) Automatically generated finite element mesh. (b) Contours of computed hydraulic head or equipotential lines.

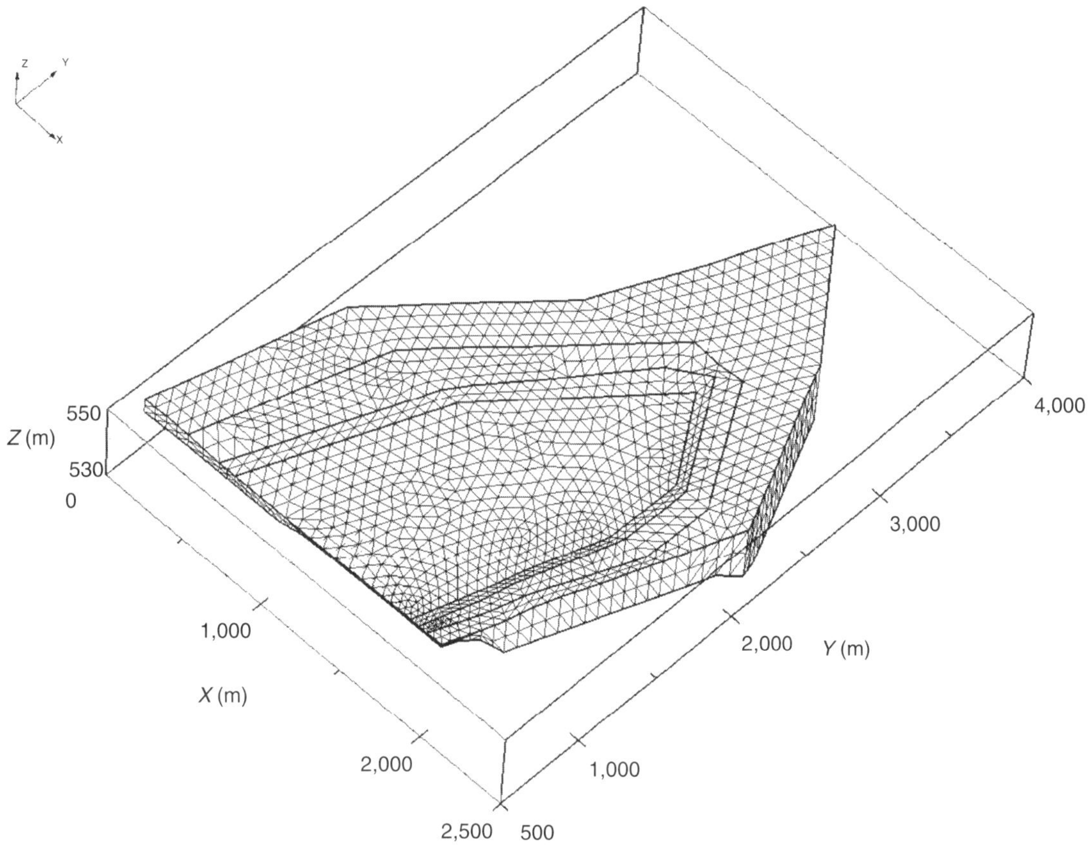

Figure 1.21 Finite element mesh for three-dimensional seepage analysis of a tailings pond.

1.6.8 Example of Stress and Shear Strength Applications

Slope stability analyses have conventionally been performed using one of several possible methods of slices where the normal stresses at the base of a slice are computed from static equilibrium of the complete slice. Each method of slices uses different assumptions and elements of statics to render the solution determinate. Consequently, each method computes a slightly different normal force at the base of each slice. More recently it has been shown that the normal force at the base of a slice can be computed from a stress analysis that involves "switching-on" the gravity body force (Yamagami and Ueta, 1988; Pham and Fredlund, 2003). The total unit weight of the soil is used in the analysis as the gravity body force. Positive and negative pore-water pressures are computed using an independent analysis.

The normal forces computed by "switching on" gravity are quite similar to those computed using the methods of slices, but the stress analysis approach provides some additional advantages (Pham and Fredlund, 2003). For example, the stress analysis needs to be performed one time and the nonlinearity associated with computing the normal forces at the base of a slice is removed. It is possible to use an optimization technique to determine the shape and location of the critical slip surface once the stress states have been computed. Pore-water pressure can be computed in an uncoupled manner using a saturated-unsaturated seepage analysis. Figure 1.22 shows the results of using a stress analysis on a slope to determine the location of the critical slip surface. The location of the critical slip surface is part of the "dynamic programming" technique.

Figure 1.23 shows the results of a comparative study between the "dynamic programming" approach and a conventional method-of-slices approach. Method-of-slices results and dynamic programming results have been shown to be quite similar (Pham et al., 2001). The computed factors of safety are slightly affected by the assumed Poisson's ratio for the soil. The dynamic programming technique has advantages from a theoretical, computational, and practical standpoint. The stress-deformation model utilized does not appear to significantly affect the computation of in situ stresses.

1.6.9 Example of Combined Stress, Seepage, and Deformation Analysis

A common and relevant problem in unsaturated soil mechanics involves the prediction of the rate and amount of swelling that an expansive soil might experience under various moisture flux boundary conditions. The rate and amount of collapse that a collapsible soil might experience are also

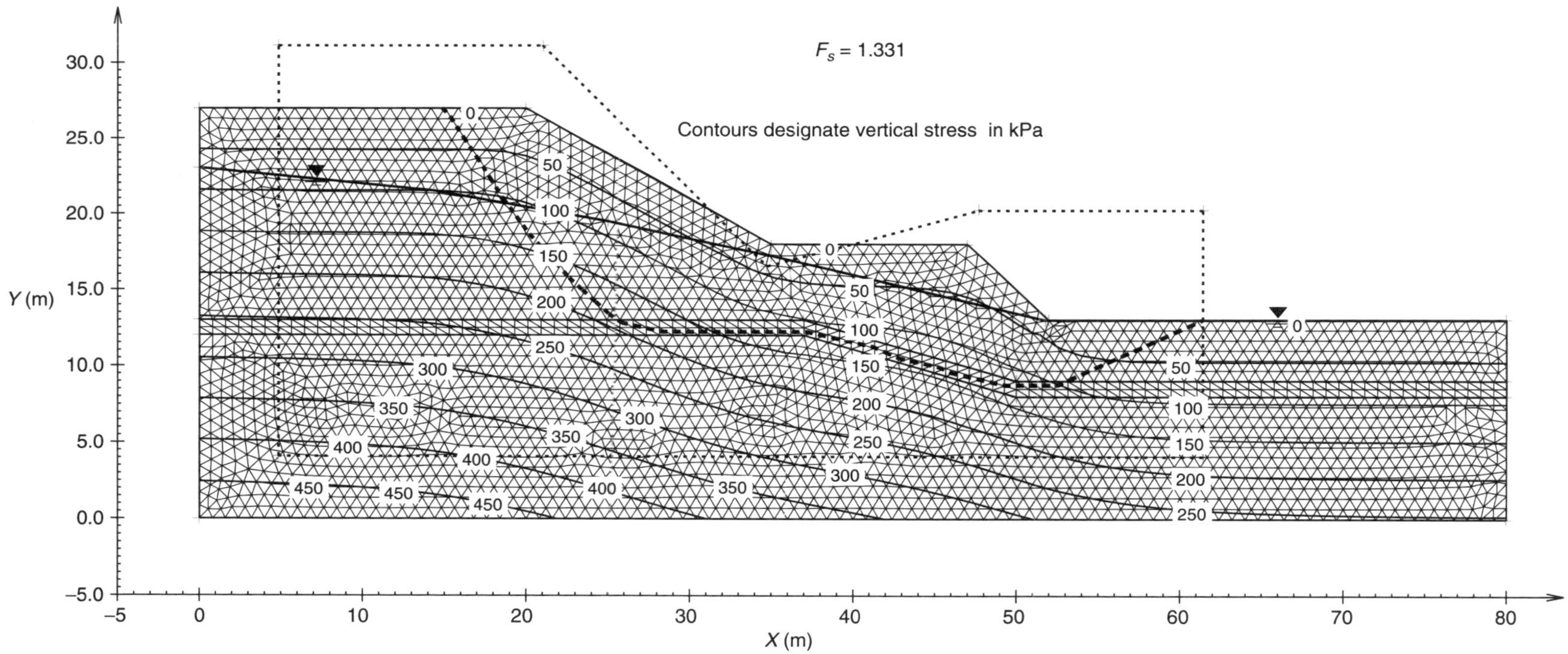

Figure 1.22 Dynamic programming slope stability analysis based on stress states computed by switching on gravity forces.

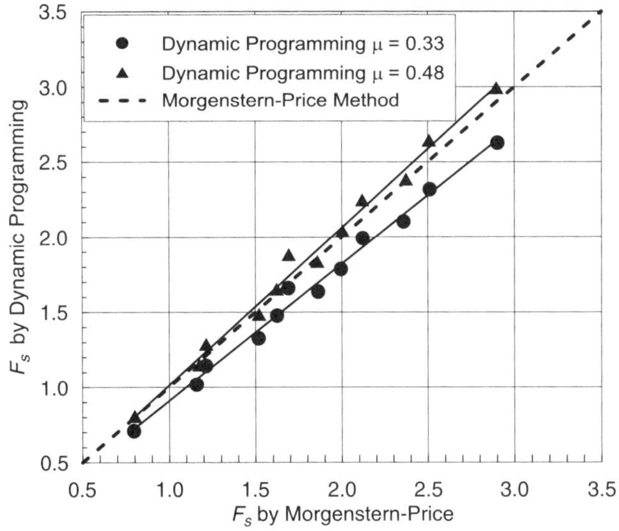

Figure 1.23 Comparison of factors of safety, F_s, computed using dynamic programming and limit equilibrium methods of analyses.

of considerable interest in geotechnical engineering. These problems bring together an unsaturated, transient seepage analysis and an unsaturated soil stress-deformation analysis. Saturated soils may also be a part of the overall analysis. The seepage and stress-deformation analysis can be brought together in either a coupled or an uncoupled manner (Vu et al., 2002). The nonlinearity associated with the unsaturated soil, stress-deformation analysis presents challenges in solving this problem. The initial stress state and pore-water pressures must be established prior to the start of the modeling process because of the nonlinearity in the soil properties. The seepage analysis boundary condition for an uncoupled analysis of a slab-on-ground problem are shown in Fig. 1.24.

The example problem involves a slab on ground subjected to the influences of a moisture flux boundary condition (Vu and Fredlund 2004, 2004). The expansive soil layer is 3 m deep and has a swelling index of 0.15. The initial matric suction is assumed to be 400 kPa at the base of the expansive soil layer. The swelling index C_s is converted to an equivalent elasticity function for the unsaturated soil. The elasticity function means that the incremental Young's modulus for the material is a function of the stress state. The analysis

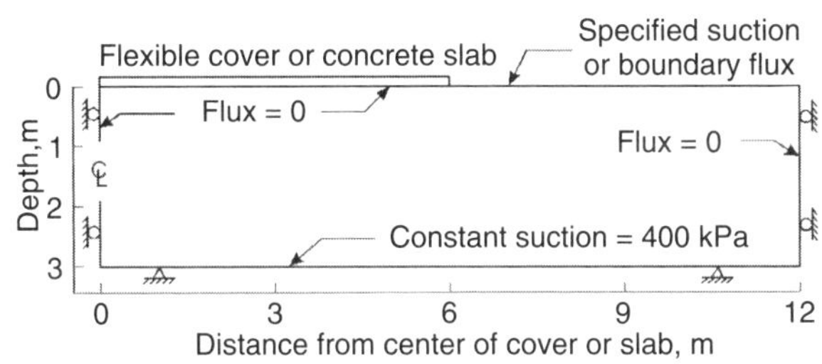

Figure 1.24 Seepage analysis boundary conditions for slab on ground placed on expansive soil (after Vu and Fredlund, 2004).

becomes nonlinear with respect to both net total stress and matric suction.

A permeability function must be specified for the swelling soil and a moisture flux is designated at ground surface to simulate either rainfall conditions or evaporation conditions from the soil around the flexible slab. A transient analysis was first run for the case of evaporation from the soil surrounding the impervious surface slab. Matric suction values along the ground surface are shown for elapsed times of 1, 3, and 5 days in Fig. 1.25. Matric suctions for the same elapsed times along a vertical section below the edge of the impervious slab are shown in Fig. 1.26. The above two figures provide an indication of the change in matric suction along vertical and horizontal directions from the edge of the slab. Numerous other graphs could be plotted to depict changes in matric suction with time at any location in the soil mass. Figure 1.27 shows the matric suction contours throughout the soil mass after 3 days of evaporation from the soil surface.

Changes in matric suction can be combined in a coupled or uncoupled manner with a stress-deformation analysis. Figure 1.28 shows the vertical displacements along the ground surface for 1, 3, and 5 days of evaporation from the ground surface. Figure 1.29 shows the vertical displacements along a vertical section below the edge of the impervious

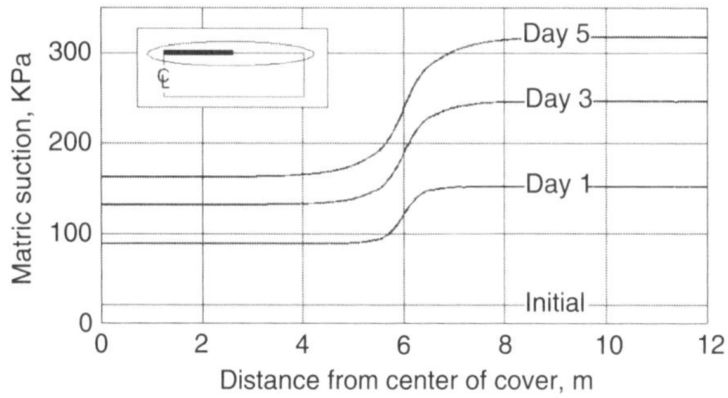

Figure 1.25 Matric suction values along ground surface after elapsed times of 1, 3, and 5 days (after Vu and Fredlund, 2004).

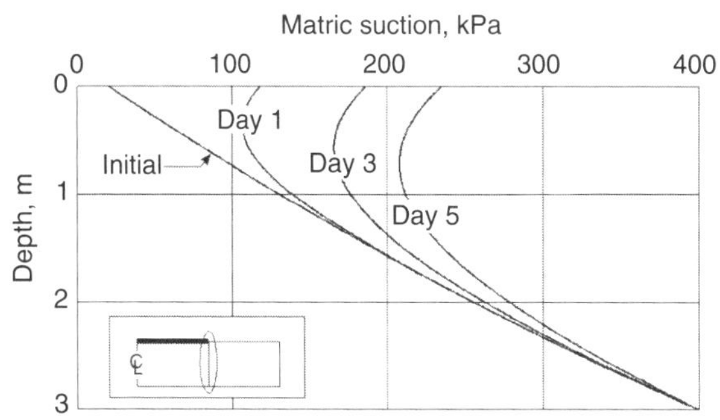

Figure 1.26 Matric suction values along vertical section below edge of foundation slab after 1, 3, and 5 days (after Vu and Fredlund, 2004).

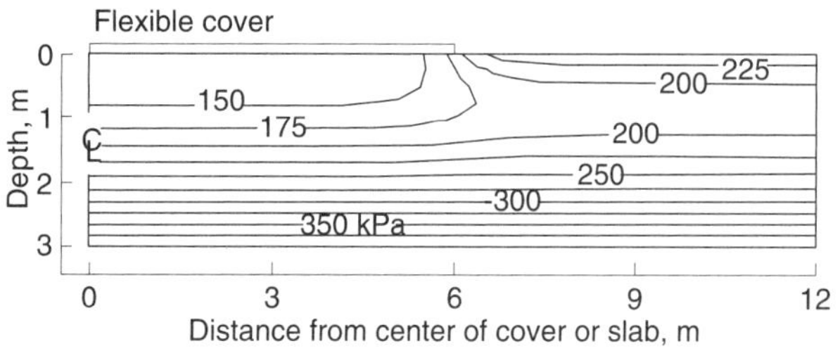

Figure 1.27 Contours of matric suctions throughout expansive soil mass after 3 days of moisture evaporation from ground surface (after Vu and Fredlund, 2004).

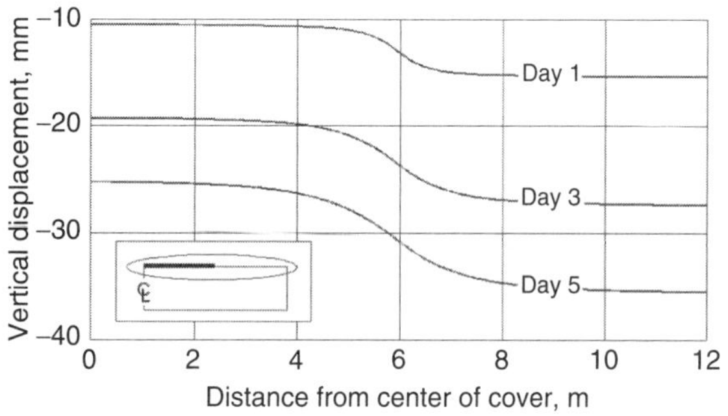

Figure 1.28 Vertical displacements along ground surface for 1, 3, and 5 days of evaporation to atmosphere (after Vu and Fredlund, 2004).

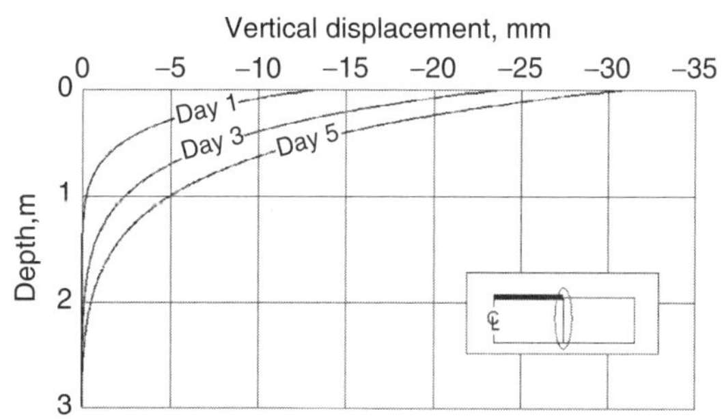

Figure 1.29 Vertical displacements along vertical section below edge of slab on ground after 1, 3, and 5 days of evaporation (after Vu and Fredlund, 2004).

slab. The soil is shown to shrink away from the concrete slab as evaporation takes place from the ground surface.

Similar plots to those presented above can be produced for the case of infiltration of water at ground surface. The boundary condition at the ground surface can be specified in terms of a moisture flux or a specified head (or pressure) condition. The microclimatic conditions at a specific site could also be analyzed to provide realistic boundary conditions for the site. There are a wide variety of unsaturated expansive soils and collapsible soil problems that can be studied as a result of combining a transient seepage analysis with a nonlinear stress-deformation analysis.

1.7 ENGINEERING PROTOCOLS FOR UNSATURATED SOILS

A natural and logical progression leading to the development of suitable engineering design protocols should involve (i) development of a fundamental science with suitable constitutive laws to describe the material behavior of concern, (ii) verification of suitable laboratory testing procedures to measure appropriate soil properties, and (iii) designation of design procedures (i.e., protocols) that should be followed in engineering practice (Fredlund and Rahardjo, 1993a). While this is the preferred approach, it is not the pattern for the development of protocols for unsaturated soil problems. More commonly, solutions have been proposed and then later the most appropriate engineering design procedures emerge. To a large extent, this has been the history associated with the design of "cover systems" that became an engineered solution for many situations in the 1980s. The emergence of similar engineering protocols can be observed for other application areas in unsaturated soil mechanics.

A variety of estimation techniques have become an acceptable part of engineering protocols when applying unsaturated soil mechanics. The question can be asked, "Are all estimation procedures for unsaturated soil property functions, equally suitable for all engineering problems?" Or, the question could be phrased, "Under what conditions should various estimating and soil testing procedures be used?" These are questions for which practicing engineers desire answers in order to utilize sound engineering protocols with confidence.

A lack of designated high-level engineering protocols makes practicing engineers more vulnerable to litigation. It is also difficult for engineers to insist on adopting high-level engineering practices in all situations. Engineering protocols can be used to provide guidance and recommendations for sound engineering practices for the implementation of unsaturated soil mechanics. There is need for general guidelines related to the estimation of the SWCC, and to a lesser extent to the USPF, since the SWCC is the basis for estimating unsaturated soil property functions.

1.7.1 Definition of Engineering Protocol

When considering the practice of unsaturated soil mechanics, the word "protocol" can be defined as "procedures and practices acceptable for prudent unsaturated soils engineering practice." The term protocol is commonly used in connection with computing procedures (e.g., telecommunications protocols or Internet protocols) or standards for global acceptance (e.g., the Kyoto Protocol). The definition has a sense of flexibility and is meant to set a temporary standard that could be changed with time. The term is meant to reflect a generally accepted, high standard of engineering practice.

Engineering protocols for unsaturated soil applications should provide general guidelines for the implementation of engineering procedures. These procedures should reflect the research published in journals and at research conferences. In other words, there needs to be a clear connection between research and engineering practice without turning engineering projects into research-type projects.

A complete engineering protocol needs to address all aspects of engineering design ranging from the geometric description of the problem to the assessment of suitable soil properties and boundary conditions. The protocol also needs to address the type of analysis that should be performed when solving a particular problem. Protocols in this book are mainly directed toward the determination of suitable unsaturated soil properties (i.e., SWCC and USPF).

1.7.2 Categorization of Engineering Design Protocols

Engineering design protocols can generally be placed within one of two primary categories: (i) preliminary design protocols and (ii) final design protocols, as shown in Fig. 1.30. It could be reasoned that there might be other intermediate protocol categories but there is benefit in keeping the categorization of design protocols as straightforward as possible.

1.7.3 Preliminary Design Protocols

Preliminary engineering design protocols should be as simple as possible while still taking advantage of the many procedures that have been proposed, tested, and verified in the research literature. It is suggested that "preliminary design protocols" utilize estimation procedures for the assessment of SWCCs. There are a variety of estimation procedures that have been proposed and some attributes of each procedure are later discussed.

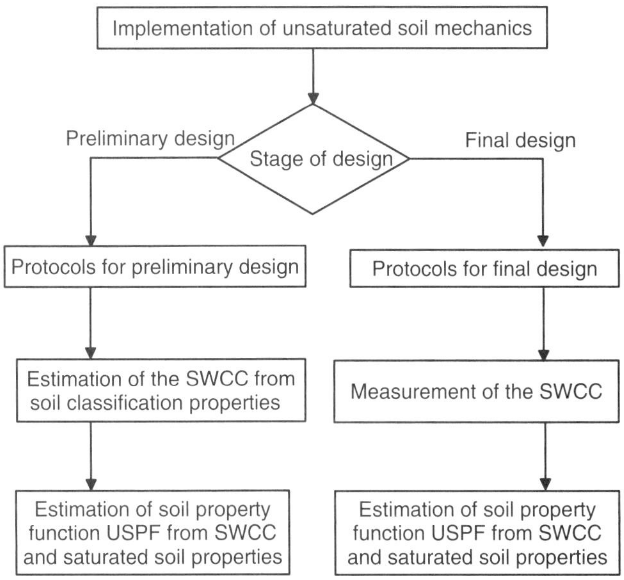

Figure 1.30 Classification of engineering protocols when dealing with unsaturated soil mechanics problems.

1.7.4 Final Design Protocols

It is suggested that "final design protocols" be based on the measurement of SWCCs. In other words, representative soil samples of the material involved in the field should be tested in the laboratory under the range of soil suctions that are anticipated to occur during the life of the structure. Only in situations of extremely low risk it may not be necessary to perform measurements of the SWCC.

It is the measurement or the estimation of the SWCC that becomes the primary determining factor separating a preliminary design protocol from a final design protocol. Estimation procedures for the determination of unsaturated soil property functions are later discussed but do not form the basis for whether a "preliminary design" or a "final design" is performed.

It is recommended that the most appropriate estimation techniques be used for the assessment of the SWCC of various materials when undertaking a preliminary design. The SWCC results should later be confirmed through laboratory measurements for the final design. It may also be necessary to confirm the performance of the design through in situ monitoring of water content and/or matric suction during the life of the structure.

1.7.5 Verification or Monitoring Category

There are engineering projects where verification through field monitoring should be undertaken to ensure that the engineered structure is performing in accordance with the design. Verification through field monitoring is left as an independent confirmation test on design rather than another design category.

Field monitoring generally requires the measurement of the same two variables that define the SWCC. In other words, it will be necessary to measure in situ water content and soil suction (i.e., matric suction). In measuring these two variables in the field it is possible to confirm that the in situ SWCCs are in reasonable agreement with the design curves.

1.7.6 Other Factors Affecting Engineering Design Protocol

The classification of soil type associated with a particular engineering problem must be known to the engineer. The saturated soil properties will also need to be measured or estimated. The detail and reliability sought in assessing soil properties depend upon a number of factors related to the engineering project. If the soil classification is known, as well as the saturated soil properties, it is always possible to obtain an estimation of the unsaturated soil properties.

All unsaturated soil property functions can be estimated from saturated soil properties and SWCCs. The SWCC can also be estimated from the soil classification properties, in particular, from the grain-size distribution curve. Consequently, there is no reason for an engineer to say that he or she cannot take the unsaturated soil portion of the problem into consideration in the analysis. A more relevant question is, "How accurate must the SWCC be known for the problem at hand?" To answer this question, the engineer must be aware of the factors that affect the assessment of the SWCC.

The risks associated with the engineering design need to be taken into consideration. If the risks associated with a hazard are low, an estimation of the SWCC and subsequent unsaturated soil properties are justifiable. If the risks are high, it is prudent to more accurately assess the SWCC (i.e., possibly the desorption and adsorption curves) and subsequent unsaturated soil properties.

Saturated-unsaturated soil problem analyses are generally viewed as the solution of a PDE within the context of a boundary value problem. As such, there will be a primary variable that is solved for in the PDE solution. In the case of a water flow seepage analysis, the primary variable is "hydraulic head" h. The hydraulic head variable generally has a reduced sensitivity to the unsaturated soil properties. Therefore, it is said that hydraulic heads can be computed with reasonable accuracy because heads are quite insensitive to the input unsaturated soil property functions. Hydraulic heads are used as input to a slope stability analysis and, as such, it is possible to perform meaningful saturated-unsaturated seepage analyses and use the results in a slope stability study. On the other hand, if the amount of water flow (i.e., water flux) in a seepage analysis is of interest to the engineer, then the saturated-unsaturated soil properties play a prominent role. Water infiltration or moisture flux is of primary importance in the design of "cover" systems and as such the soil properties need to be determined with greater accuracy.

The engineer must be aware that there are drying and wetting branches associated with the SWCC. The drying and wetting curves can be considerably different and this phenomenon is referred to as "hysteresis." It is necessary to determine a suitable drying and wetting curve for engineering problems where the variable of interest from the modeling exercise is significantly dependent upon the unsaturated soil property functions. It is important when performing numerical modeling to ensure that the appropriate bounding curve is used for solving the engineering problem. The desorption and adsorption curves can be assumed to have a similar shape (i.e., congruent) apart from a lateral shift of the inflection point on the drying and wetting curves.

1.7.7 Challenge for the Future

There are many challenges that still lie ahead before it can be said that engineers clearly understand how best to apply unsaturated soil mechanics in engineering practice. However, one challenge that is important to address is the need to establish protocols for various geotechnical engineering problems associated with unsaturated soil mechanics.

Reliance on past experience and empirical practices will be followed by engineers when there are no distinct standards of practice or protocol for addressing various unsaturated soil problems. The end result will be a lower level of engineering practice with difficulties being encountered that will likely be settled through the litigation process. It is the responsibility of those involved in geotechnical engineering practice to clearly define methodologies or protocols associated with prudent engineering practice for various unsaturated soil problems.

CHAPTER 2

Nature and Phase Properties of Unsaturated Soil

2.1 INTRODUCTION

The relationship of one phase of a mixture to another phase in terms of mass and volume is discussed under the topic of "volume-mass" relations. Generally each phase of an unsaturated soil is considered to remain chemically inert while the proportions of each phase may change as a result of a "process." It is possible, however, for air to move in and out of the liquid water phase. There are also situations where phase changes occur while processes are underway. Cases where there may be phase changes are briefly discussed in this chapter. The effect of temperature on each phase is also discussed in this chapter.

The classification properties (e.g., grain-size distribution and Atterberg limits) take on increased significance when dealing with unsaturated soils. The additional importance of classification properties to unsaturated soil behavior is discussed in this chapter.

The concept of "state variables" is briefly introduced; however, an entire chapter is later devoted to a more thorough presentation on this important topic. Numerous approaches could be taken when developing the discipline of unsaturated soil mechanics. The approaches could range from an empirical approach based primarily on experience to a particulate mechanics or quantum mechanics approach. The approach used throughout this book can be referred to as a macroscopic, phenomenological approach to unsaturated soil behavior. In other words, the science is developed around observable phenomena while adhering to the principles of continuum mechanics. The continuum mechanics approach starts with an understanding of the state variables for the material under consideration. The continuum mechanics approach has proven successful in *saturated* soil mechanics and it would appear reasonable to retain the same approach for *unsaturated* soil mechanics. An attempt should be made to ensure there is a smooth transition and consistency in rationale between saturated and unsaturated behavior.

2.1.1 What Is an Unsaturated Soil?

An unsaturated soil is commonly defined as having three phases: (1) solids, (2) water, and (3) air. It is, however, more correct to recognize the existence of a fourth phase, namely, the air-water interface or the contractile skin (Fredlund and Morgenstern, 1977). The justification and need for a fourth phase are discussed later in this chapter and the following chapter on state variables.

The presence of even the smallest amount of free air renders a soil unsaturated. Even a small amount of air, likely occurring as occluded air bubbles, renders the pore fluid compressible. It is possible for a soil to remain essentially saturated while the pore-water pressure becomes negative relative to the air phase.

The zone immediately above the water table is referred to as the capillary zone and is characterized as being essentially saturated while having negative pore-water pressures. Soil in the capillary zone is generally embraced within the context of an unsaturated soil. It is the negative pore-water pressure relative to the internal or external air pressure that qualifies the soil as an unsaturated soil.

Generally it is a larger amount of air (i.e., about 15% air by volume) that makes the air phase continuous throughout the soil. The principles and concepts required to describe the behavior of an unsaturated soil become necessary as the pore-air and pore-water pressures begin to become significantly different. Terzaghi (1943) clearly described the important role of the air-water interface in understanding unsaturated soil behavior (i.e., Chapters 14 and 15 of *Theoretical Soil Mechanics*).

2.1.2 Unsaturated Soil as Four-Phase Mixture

An unsaturated soil can be viewed as a four-phase system because of the unique role of the air-water interface or the contractile skin on soil behavior. The contractile skin acts like a thin membrane interwoven throughout the voids of the soil, forming a fixed partition between the air and water phases. Stress state changes in the contractile skin can cause an unsaturated soil to change water content, change volume, and change shear strength. It is the contractile skin that forms a barrier between air and water, determining the ratio of the volume of air to the volume of water in the voids (i.e., degree of saturation). It is particularly important to view an unsaturated soil as a four-phase mixture when considering the stress state of an unsaturated soil. Equilibrium stress equations can be written for each phase within the context of multiphase continuum mechanics. The equilibrium analysis reveals the stress state variables associated with maintaining equilibrium conditions in the unsaturated soil (see Chapter 3).

An unsaturated soil should be viewed as having two phases that flow under the influence of a stress gradient (i.e., air and water) and two phases that come to equilibrium under the influence of a stress gradient (i.e., structural arrangement of soil particles and the contractile skin forming a partition between the fluid phases) (Fredlund and Rahardjo, 1993a).

The contractile skin has physical properties differing from the contiguous air and water phases and interacts with the soil structure to influence soil behavior. The contractile skin can be considered as part of the water phase when considering changes in the volume-mass soil properties; however, the contractile skin must be considered as an independent phase when describing the stress state and phenomenological behavior of an unsaturated soil.

Terzaghi (1943) emphasized the important role played by surface tension property associated with the air-water interface (i.e., contractile skin). The important role played by the contractile skin can be demonstrated by simply observing the shrinkage of an initially saturated ball of clay from which water is allowed to evaporate. There may be essentially no changes in the total stress state as water evaporates; however, the contractile skin can result in a significant overall volume change of the soil mass. In other words, the contractile skin acts like an elastic membrane pulling the soil particles closer together as water is removed.

2.1.3 Distinctive Features of Contractile Skin

Numerous research studies on the nature of the contractile skin point toward its significant and independent role in unsaturated soil mechanics (Wang and Fredlund, 2003). Terzaghi (1943) suggested that the contractile skin might be on the order of 10^{-6} mm in thickness. More recent studies suggest that the thickness of the contractile skin is in the order of 1.5 to 2 water molecules in diameter (i.e., 5 Å) (Israelachvili, 1991; Townsend and Rice, 1991).

If the contractile skin is about 5 Å thick, then a lineal surface tension value of 75 mN/m translates into a unit stress in the order of 140,000 kPa (i.e., stress per unit area or surface tension divided by the thickness of the contractile skin). Kyklema (2000) showed that the distribution of water molecules across the contractile skin takes the form of a hyperbolic tangent function as shown in Fig. 2.1. Properties of the contractile skin are different from the properties

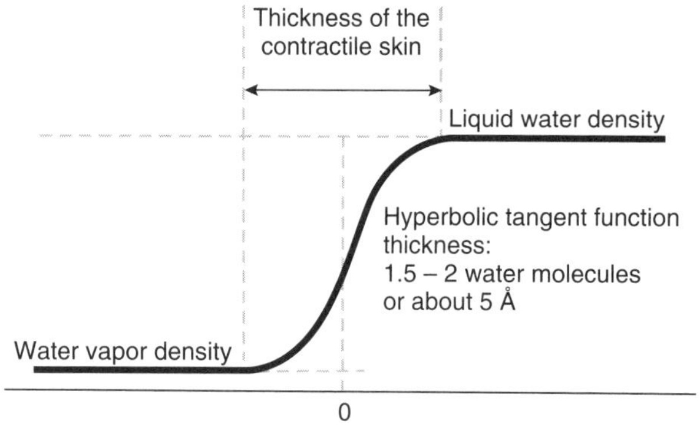

Figure 2.1 Density distribution across air-water interface (contractile skin) (after Kyklema, 2000).

of ordinary water and have a molecular structure similar to that of ice (Derjaguin and Churaev, 1981; Mitsuhiro and Kataoka, 1988).

The Young-Laplace and Kelvin equations describe fundamental behavioral aspects of the contractile skin, but both equations have limitations. The Young-Laplace equation is not able to explain why an air bubble can gradually dissolve in water without any apparent difference between the air pressure and the water pressure. The validity of the Kelvin equation becomes suspect as the radius of curvature reduces to the molecular scale (Adamson and Gast, 1997; Christenson, 1988).

Terzaghi (1943) recognized the limitations of the Kelvin equation and stated that if the radius of a gas bubble "approaches zero, the gas pressure—approaches infinity". However, within the range of molecular dimensions, "the equation loses its validity." Terzaghi (1943) recognized the limitations associated with the Kelvin equation but later researchers have attempted to incorporate the Kelvin equation into formulations for the compressibility of air-water mixtures to no avail (Schuurman, 1966). Details pertaining to the behavior of the contractile skin are still not fully understood, but the contractile skin is known to play an important role in unsaturated soil behavior. Terzaghi (1943) stated that surface tension "*is valid regardless of the physical causes.*" He went on to say, "The views concerning the molecular mechanism which produces the surface tension are still controversial. Yet the existence of the surface film was established during the last century beyond any doubt." The statements of Terzaghi in 1943 are still relevant today.

2.1.4 Terminology for Continuum Mechanics Variables of State

The authors are reticent to introduce many new variables and terms for the synthesis of a science for unsaturated soil mechanics; however, the terminology associated with saturated soil behavior has some limitations when considering unsaturated soil behavior. As a result, some universally acceptable terms from continuum mechanics and thermodynamics are introduced. Definitions for the following terms common to continuum mechanics and thermodynamics are expressed in a succinct form:

State variables: Non–material variables required for the characterization of a material system.
Stress state variables: Non–material variables required for the characterization of stress equilibrium conditions.
Deformation state variables: Non–material variables required for the characterization of deformation conditions or deviations from an initial state.
Constitutive relations: Single-valued equations expressing a mathematical relationship between state variables.

The International Dictionary of Physics and Electronics (Michels, 1961) defines state variables as "a limited set of dynamical variables of the system, such as pressure, temperature, volume, etc., which are sufficient to describe or specify the state of the system completely for the considerations at hand."

Fung (1965) describes the state of a system as that "information required for a complete characterization of the system for the purpose at hand." Typical state variables for an elastic body are given as those variables describing the strain field, the stress field, and its geometry. The state variables must be independent of the physical properties of the material system. Temperature is also a state variable. Temperature is of particular importance for the consideration of the behavior of unsaturated soils found near ground surface.

Constitutive relations, on the other hand, are single-valued expressions or equations that relate one state variable to one or more other state variables (Fung, 1977). As an example, a stress-versus-strain relationship is a constitutive relationship which describes the mechanical behavior of a material. The material properties involved may be Young's elastic modulus and Poisson's ratio. There are also many complex constitutive models that have been proposed for describing the behavior of both saturated and unsaturated soils. Common to all constitutive equations is the fact that selected state variables are linked to form a mathematical equation.

The ideal gas law equation relates gas pressure and temperature to density and is called a constitutive equation. The density term defines the relationship between volume and mass. The gas constant is the material property. Simple, idealized constitutive equations are well established for nonviscous fluids, Newtonian viscous fluids, and perfectly elastic solids (Fung, 1965).

Examples of constitutive relations for unsaturated soils are equations relating stress state variables to deformation state variables. Other examples of constitutive relations are equations for shear strength (e.g., Mohr-Coulomb equation) and pore pressure generation (i.e., pore-air and pore-water pressure parameter equations). Soil-water characteristic curve, SWCC, equations are also constitutive relationships.

Basic definitions from continuum mechanics clearly indicate that the physical properties of a system belong to the constitutive relationships of a system and are not to be part of the description of stress state.

2.1.5 Designation of Stress State Variables

State variables should be defined within the context of continuum mechanics as variables that are independent of soil properties. These variables are required for the characterization of a system or a mixture of phases (Fung, 1965). The theoretical basis for state variables lies within the fundamental conservative laws of physics. More specifically, the stress state variables arise from the conservation of energy.

The stress state variables associated with an unsaturated soil are related to equilibrium considerations (i.e., Newtonian summation of forces or conservation of momentum) of a multiphase system (Fredlund and Morgenstern, 1977). The stress state variables in the equilibrium equations can be extracted to form one or more tensors (i.e., 3×3 matrices). The stress state variables take on the form of tensors because of the three-dimensional Cartesian coordinate system generally used for the formulation of engineering problems (i.e., our three-dimensional world).

The description of the state variables for an unsaturated soil becomes the fundamental building block for developing an applied engineering science. It is the ability to describe soil behavior in terms of stress state variables that allows us to refer to unsaturated soil mechanics as a science. The universal acceptance of unsaturated soil mechanics as a science depends largely upon how well the stress state variables can be defined, justified, and measured. The universal acceptance of the description of the stress state for an unsaturated soil has unfortunately been slow. The end result has been a slow emergence of a consistent theoretical framework for unsaturated soil behavior.

2.1.6 Designation of Deformation State Variables

Deformation state variables must satisfy conditions prescribed by the conservation of mass. Deformation state variables allow the movement of various phases of a multiphase system to be described. Stated another way, deformation state variables describe the mapping of relative movements for each phase of a multiphase system.

Deformation state variables associated with the fluid phases describe relative changes in volume or amount of fluid in a soil element. It is necessary to independently define both shear and normal types of strain or deformation in the case of a solid particulate system constituting the soil structure. The term "deformation state variables" is used in this book to embrace both strain- and deformation-type variables.

A referential type of soil element has been most commonly used in saturated soil mechanics and is suitable for developing theories in unsaturated soil mechanics. A referential element is linked to a fixed number of soil particles. Deformation state variables should ensure that conservation of mass is maintained. Conservation of mass can also be written in terms of a volumetric requirement that must be maintained for a referential element.

2.1.7 Typical Profiles of Unsaturated Soils

Climatic conditions in an area constitute one of the main factors causing a soil deposit to be unsaturated. Unsaturated soils or soils with negative pore-water pressures can occur in essentially any geological deposit. An unsaturated soil can be a residual soil, a lacustrine deposit, a bedrock formation, or any other soil or rock type. However, there are certain geological soil categories that are commonly found to have negative pore-water pressures. A few examples will illustrate some of the features common to these deposits.

2.1.7.1 Tropical Residual Soil Profile

Tropical residual soils possess unique characteristics related to their composition and the environment under which they have been formed. Most distinctive is the microstructure which changes in a gradational manner with depth (Vargas, 1985; Brand, 1985). The in situ water content of residual soils is generally greater than its optimum water content for compaction. The density, plasticity index, and compressibility of residual soils are likely to be less than the corresponding values for temperate zone soils with comparable liquid limits. The strength and permeability of residual soils are likely to be greater than those of temperate zone soils with comparable liquid limits (Mitchell and Sitar, 1982).

Behavioral concepts in classical soil mechanics have mainly emerged in regions where the climate is temperate. Geotechnical engineers have often encountered difficulties in modeling residual soil behavior in a reasonable manner. Engineers are now recognizing that negative in situ pore-water pressures play an important role in understanding the behavior of residual soils. Much of the unusual behavior of residual soils is related to the suction changes imposed during laboratory testing (Fredlund and Rahardjo, 1985).

A typical deep, tropical weathering profile is shown in Fig. 2.2 (Little, 1969). Boundaries between layers and changes with depth are generally not clearly defined. Numerous systems of classification for residual soils have been proposed based mainly on the degree of weathering and engineering properties (Deere and Patton, 1971; Tuncer and Lohnes, 1977; Brand, 1982).

Zones of completely weathered or highly weathered rock that contain particulate soil but retain the original rock structure are termed saprolite. Once the deposit has essentially no resemblance to the parent rock, it is termed a lateritic or residual soil. Figure 2.3 shows the profile and soil properties for a porous saprolite soil from basalt in Brazil (Vargas, 1985). The region has a hot, humid summer and a mild, dry winter climate with an annual rainfall of less than 1500 mm. The structure is highly porous and in some cases may be unstable, resulting in collapse upon saturation. The soil

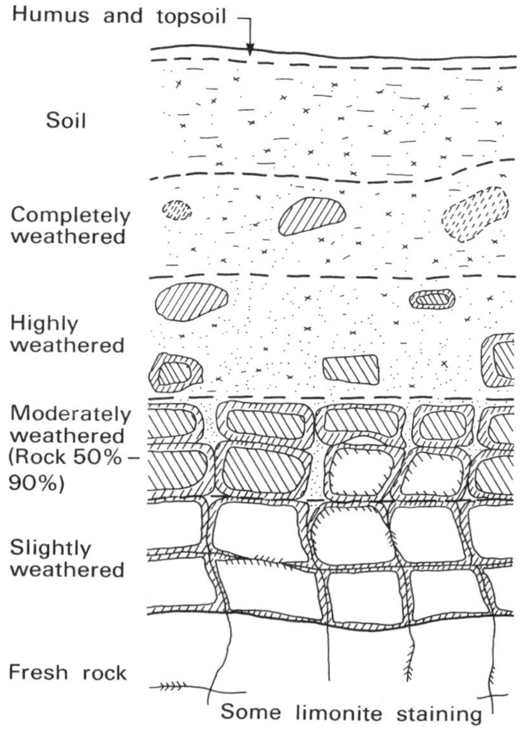

Figure 2.2 Schematic diagram showing typical tropical residual soil profile (after Little, 1969).

deposits are often unsaturated and the in situ pore-water pressures are negative.

2.1.7.2 Expansive Soil Profile

Expansive soil deposits commonly occur as lacustrine deposits or as bedrock shale deposits. In general, expansive soils have high plasticity (i.e., high liquid limit) and are relatively stiff or dense as a result of compression or desiccation. The above description is typical but not exclusive. The expansive nature of the soil is most obvious near ground surface where the profile is subjected to seasonal, environmental changes. The pore-water pressure may be initially negative and the deposit may be unsaturated. These soils often have montmorillonite clay minerals as part of the clay size particles. Monovalent cations absorbed to the clay mineral (e.g., sodium) can render the soil highly expansive. Expansive soil deposits and their related engineering problems have been reported and studied in many countries.

Extensive areas of western Canada are covered by preglacial, lacustrine clay sediments that are expansive in nature. Regina, Saskatchewan, is located in a semiarid climate where the annual precipitation is approximately 350 mm. A typical soil profile is shown in Fig. 2.4 (Fredlund, 1973a). The average liquid limit of the soil is 75% and the average plastic limit is 25%. The shrinkage limit is typically 15%. The lacustrine clay is classified as

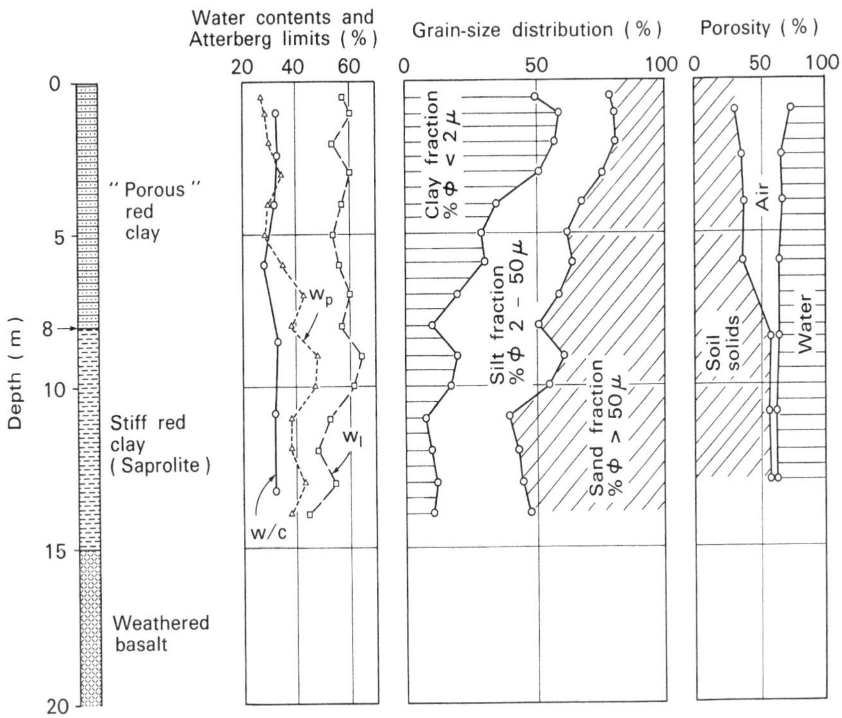

Figure 2.3 Porous saprolite soil from basalt near Londrina, Brazil (after Vargas, 1985).

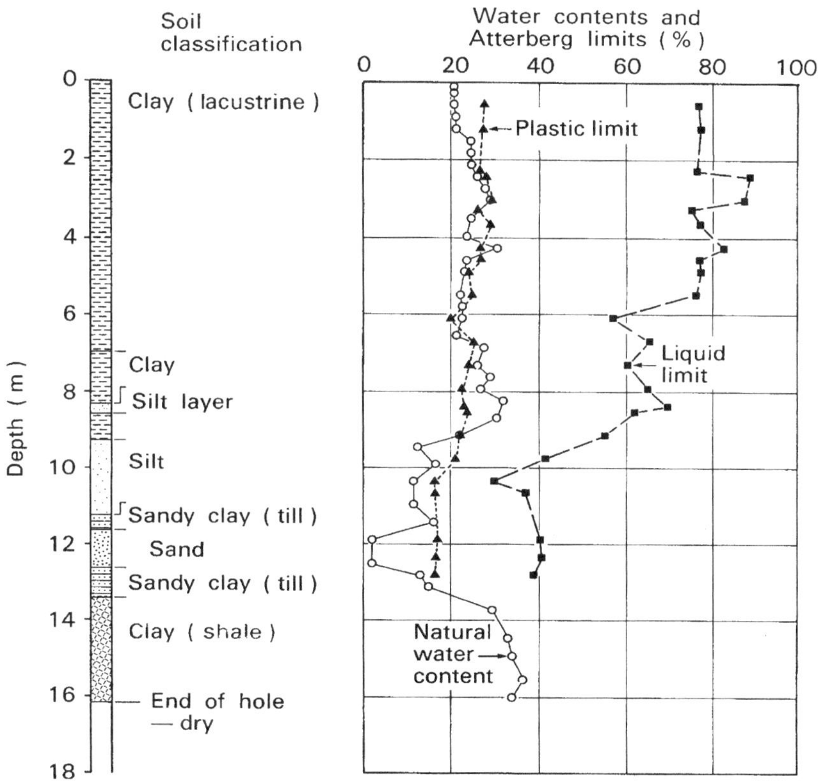

Figure 2.4 Profile of expansive soils from Regina, Canada (from Fredlund, 1973a).

a calcium montmorillonite. Buildings founded on shallow footings have been known to experience 50–150 mm of movement in the years subsequent to construction.

The above soil profile is typical of conditions which are found in many parts of the world. In each case, the natural water contents are relatively low and the pore-water pressures are highly negative. Throughout each season, and from season to season, the soil expands and contracts in response to changes in the net moisture flux at ground surface. It is the climatic and human-imposed surface flux conditions at the ground surface that strongly affect the potential expansion and contraction of the underlying soil.

2.2 SOIL CLASSIFICATION

There are two laboratory tests that are routinely used in the classification of soils: the grain-size distribution curve and the Atterberg limits. These tests are performed on reconstituted soil samples. However, when unsaturated soils are involved, the classification tests take on additional significance and meaning.

The grain-size distribution curve provides information on the distribution of the solid particles or the percentage of each particle size. The distribution of the solid particles bears a relationship to the distribution of pore sizes or void spaces. Information on the pore-size distribution can be used for the estimation of the water content–soil suction relationship for the soil (i.e., SWCC). Consequently, the grain-size distribution becomes of increased value for understanding unsaturated soil behavior.

There are three plasticity characteristics of a soil which are referred to as Atterberg limits: liquid limit, plastic limit, and shrinkage limit. Independent of the shrinkage limit is the "shrinkage curve," which is measured by allowing soil to dry while measurements are made of the volume and mass of the specimen. These measurements allow the calculation of void ratio (or specific volume) and water content as the soil dries.

Some key characteristics of unsaturated soil behavior can be visualized along the shrinkage curve. The shrinkage curve provides an indication of the air-entry value of the soil as well as residual moisture conditions. The shrinkage curve is generally measured on an initially slurry soil, but it is also possible to measure the shrinkage curve starting at unsaturated initial conditions.

Both of the commonly performed soil classification tests provide information on the unsaturated soil behavior of a soil and are further discussed in the following sections.

2.2.1 Grain-Size Distribution Curves

The grain-size distribution curve provides information on the distribution of the sizes of the solid particles in a soil and can be used to introduce concepts related to unsaturated soil behavior (M.D. Fredlund, 2000). It is the distribution of the voids (i.e., the inverse of the distribution of the solid particles) that forms the basis for numerous estimations techniques that have been proposed for estimating the SWCC for an unsaturated soil (M.D. Fredlund et al., 1997a).

Information from the grain-size distribution curve has been used for the estimation of the SWCC of a soil, as shown in Fig. 2.5. The grain-size distribution curve provides an indication of the air-entry value and the rate of desaturation of the soil. These estimation procedures are referred to as pedo-transfer functions. Details pertaining to several pedo-transfer estimation techniques for the SWCC are presented later in Chapter 5.

2.2.2 Equation for Grain-Size Distribution Curve

A mathematical equation to represent the entire range of measured grain-size particles provides a basis for further analyses. A grain-size distribution equation provides a flexible means of searching databases for similar soils. The grain-size equation provides a continuous mathematical function that can be used for estimating the SWCC. Most grain-size distribution curves have a unimodal distribution of particle sizes while other soils have a bimodal distribution of particle sizes. Both types of distributions can be described by a mathematical equation.

Gardner (1956) used a two-parameter, lognormal distribution to fit grain-size distribution data. The two-parameter fit of the grain-size distribution was performed using a geometric mean parameter, x_g, and a geometric standard deviation, σ_g. The method of fitting lognormal equations to the grain-size distribution has some limitations in engineering practice. Hagen et al., (1987) presented a computerized, iterative procedure that required two sieve sizes to determine the parameters for a standard, two-parameter lognormal distribution. The lognormal distribution, however, often fails to provide a close fit of the actual grain-size distribution near the extremities of the curve (Gardner, 1956; Hagen et al., 1987). Wagner and Ding (1994) later attempted to improve upon the log normal equation through use of three and four lognormal parameters.

Campbell (1985) presented a classification diagram based on the assumption that the particle-size distribution was approximately lognormal. The particle-size distribution was approximated using a Gaussian distribution function. Any combination of sand, silt, and clay could be represented by a geometric (or log) mean particle diameter and a geometric standard deviation. A modified U.S. Department of Agriculture (USDA) textural classification chart was suggested by Shirizi and Boersma (1984).

The use of a lognormal-type equation has a limitation in that the grain-size distribution is assumed to be symmetric. However, grain-size distribution curves are often nonsymmetric and can be better fit using other forms of equations. A method is required to more accurately represent soils that are bimodal or gap graded.

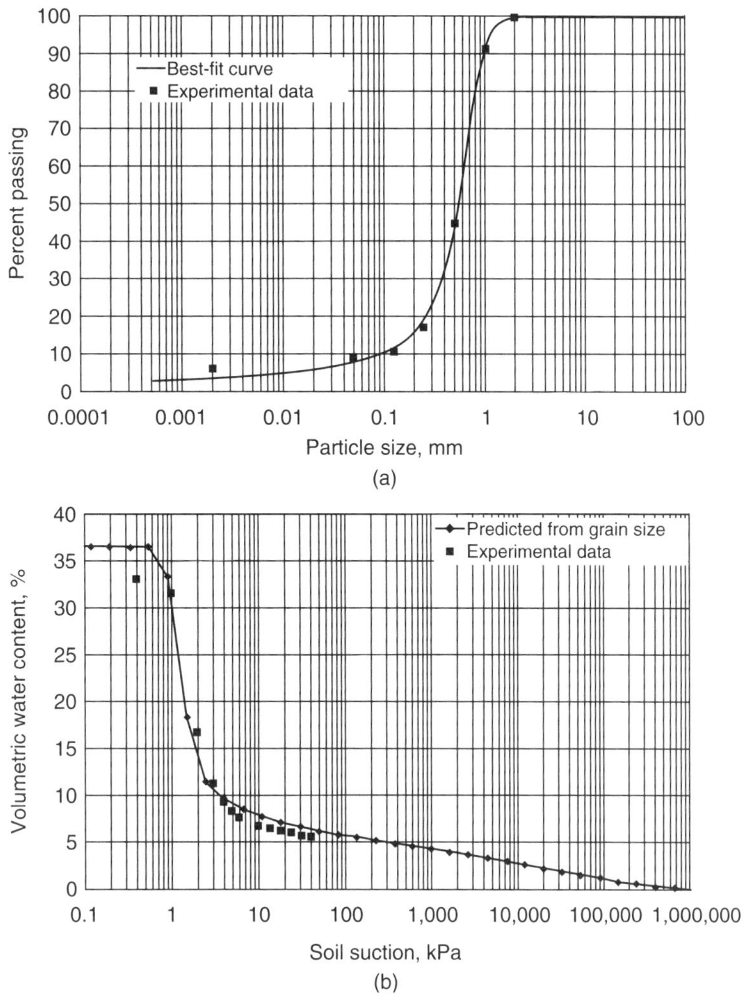

Figure 2.5 Comparison between experimental and predicted SWCCs for sand (from M.D. Fredlund, 2000). (a) Grain-size distribution curve for sand. (b) Measured and predicted SWCC for sand.

There are three general categories of grain-size distributions (Holtz and Kovacs, 1981): *well-graded* soils, *uniform* soils, and *gap-graded* soils. Well-graded and uniform soils are unimodal in character while gap-graded soils are bimodal.

2.2.2.1 *Unimodal Equation for Grain-Size Distribution*

Grain-size distribution curves have shapes that appear to be similar to SWCCs when plotted using a semilog scale. The similarity in shape is anticipated since the SWCC provides a representation of the void distribution in a soil while the grain-size distribution curve represents the distribution of the solid particles. The solid particles plus the voids add up to the total soil volume. It would be anticipated that the distribution of the soil particles (i.e., represented by the grain-size distribution) would bear an inverse relationship to the distribution of voids (i.e., represented by the SWCC) (M.D. Fredlund, 2000).

There are several empirical equations that have been proposed to represent the SWCC. Fredlund and Xing (1994) proposed a four-parameter equation that provided a reasonable fit over the entire soil suction range. The similarity in shape between the SWCC and the grain-size distribution curve suggests that the SWCC equation (Fredlund and Xing, 1994) could be used to describe the grain-size distribution curve.

The Fredlund and Xing (1994) equation allows for independent control of the shape of the lower end of the grain-size distribution curve (i.e., the fine particle size range). The reversed grain-size distribution scale requires that the original Fredlund and Xing (1994) SWCC equation be modified as

$$P_p(d) = \frac{1}{\left\{\ln\left[\exp(1) + \left(\dfrac{a_{gr}}{d}\right)^{n_{gr}}\right]\right\}^{m_{gr}}} \times \left\{1 - \left[\frac{\ln\left(1 + \dfrac{d_r}{d}\right)}{\ln\left(1 + \dfrac{d_r}{d_m}\right)}\right]^7\right\} \tag{2.1}$$

where:

$P_p(d)$ = percentage by mass of particles passing a particular particle size,
a_{gr} = parameter designating the inflection point on the grain-size distribution curve and is related to the initial breaking point near the coarse particle sizes,
n_{gr} = parameter related to the steepest slope on the grain-size distribution curve (i.e., uniformity of the particle size distribution),
m_{gr} = parameter related to the shape of the grain-size curve as it approaches the fine-grained region,
d_r = parameter related to the particle size in the fine-grained region and is referred to as the residual particle size,
d = diameter of any particle size under consideration, and
d_m = diameter of the minimum allowable size particle (e.g., 0.0001 mm).

Equation 2.1 is unimodal in character and can be used to fit a wide variety of soils. A quasi-Newton fitting algorithm can be used to best fit three fitting parameters for a soil (M.D. Fredlund, 2000). The best-fit particle size distribution function can be plotted along with the grain-size distribution data as shown in Fig. 2.6a. The unimodal sigmoidal equation provides an excellent fit of the grain-size data.

The particle-size distribution equation provides information on the amount and distribution of particle sizes. Differentiation of the particle size distribution curve produces a particle-size probability density function (PDF) (M.D. Fredlund, 2000). The differentiated form of the unimodal grain-size distribution equation is called the arithmetic PDF and is shown in Eq. 2.2 and Fig. 2.6b:

$$\frac{\mathrm{d}P_p}{\mathrm{d}d} = \frac{1}{\left\{\ln\left[\exp(1) + \left(\frac{a_{gr}}{d}\right)^{n_{gr}}\right]\right\}^{m_{gr}}}\left\{1 - \left[\frac{\ln\left(1+\frac{d_r}{d}\right)}{\ln\left(1+\frac{d_r}{d_m}\right)}\right]^7\right\} \times m_{gr}\left(\frac{a_{gr}}{d}\right)^{n_{gr}} \times \frac{n_{gr}}{d\left\{\left[\exp(1)+\left(\frac{a_{gr}}{d}\right)^{n_{gr}}\right]\ln\left[\exp(1)+\left(\frac{a_{gr}}{d}\right)^{n_{gr}}\right]\right\}} + \frac{7}{\left\{\ln\left[\exp(1) + \left(\frac{a_{gr}}{d}\right)^{n_{gr}}\right]\right\}^{m_{gr}}}\frac{\ln\left(1+\frac{d_r}{d}\right)^6}{\ln\left(1+\frac{d_r}{d_m}\right)^7} \times \frac{d_r}{\left[d^2\left(1+\frac{d_r}{d}\right)\right]} \tag{2.2}$$

where:

$\frac{\mathrm{d}P_p}{\mathrm{d}d}$ = arithmetic slope of the grain-size distribution curve in units equal to the inverse of particle size.

The arithmetic particle-size distribution exaggerates the clay portion of the particle distribution curve and does not appear to provide meaningful information for engineering purposes. It is more meaningful to represent the PDF on a logarithmic scale rather than as an arithmetic PDF. Differentiating the grain-size distribution equation on a logarithm scale produces a PDF that is more meaningful, as shown in the equation,

$$p_1(d) = \frac{\mathrm{d}P_p}{\mathrm{d}\log(d)} = \frac{\mathrm{d}p_p}{\mathrm{d}d}\ln(10)d \tag{2.3}$$

where:

$p_1(d)$ = logarithmic PDF

The peak value in Eq. 2.3 represents the most frequent particle size. The probability of the logarithmic PDF can be calculated according to

$$\text{probability } (d_1 < d < d_2) = \int_{x=\log(d_1)}^{x=\log(d_2)} p_1(x)dx \tag{2.4}$$

Figure 2.6c shows the logarithmic PDF for clayey silt. The unimodal equation fit for silt sand is shown in Fig. 2.7 along with the logarithmic PDF.

The unimodal equation (i.e., Eq. 2.1) shows characteristics that are similar to the Fredlund and Xing (1994) equation for the SWCC (Fig. 2.8). The a_{gr} parameter is related to the initial break in the grain-size distribution equation (in the coarse particle size region) and represents the inflection point on the curve. Its effect on the grain-size distribution curve can be seen in Fig. 2.8, where a_{gr} is varied from 0.1 to 10 while the other equation parameters are held constant. The a_{gr} parameter is related to the largest particle sizes in the soil sample.

Figure 2.9 shows how the parameter n_{gr} influences the slope of the grain-size distribution. The point of maximum slope along the grain-size distribution curve (i.e., through the inflection point) provides an indication of the dominant particle size in the soil (i.e., on a logarithm scale). Figure 2.9 shows the effect of varying n_{gr} from 1 to 4.

The parameter m_{gr} influences the break onto the finer particle size region of the soil sample. The effect of varying the m_{gr} parameter from 0.3 to 0.9 can be seen in Fig. 2.10. The parameter d_r affects the shape along the finer particle size portion of the curve, but its influence on the curve is quite minimal, as shown in Fig. 2.11. The d_r variable can be manually adjusted to improve the fit of the curve to the data when using a best-fit regression analysis. It has been observed that a value of 0.001 mm for d_r can provide a reasonable fit for most soils.

2.2.2.2 *Bimodal Equation for Grain-Size Distribution*

The unimodal grain-size distribution equation (i.e., Eq. 2.1), does not provide a satisfactory fit when soils are gap-graded as shown in Fig. 2.12. In this case it is necessary to use a bimodal equation when performing the best-fit regression

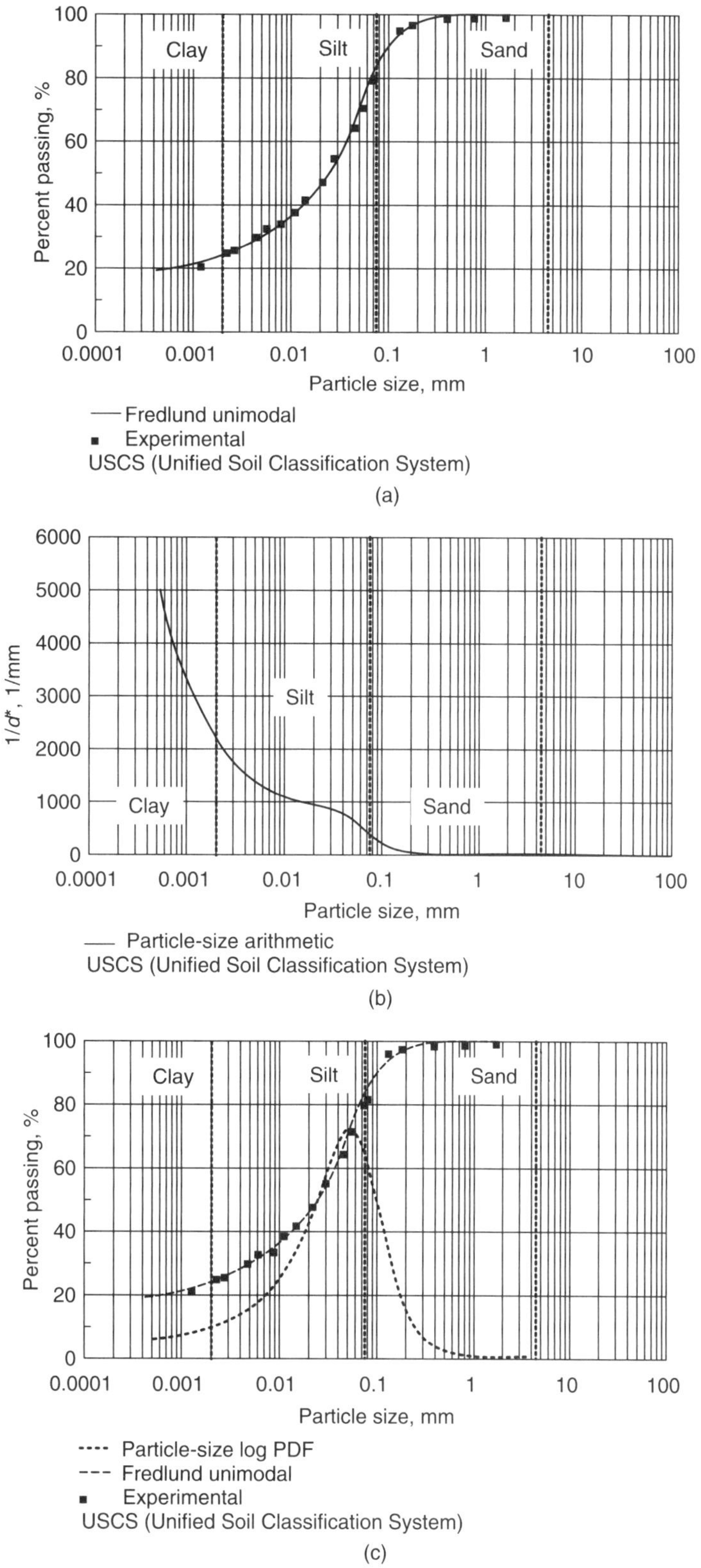

Figure 2.6 Grain-size distribution data with unimodal fit for clayey silt (from M.D. Fredlund et al., 2000a). (a) Best fit of grain-size distribution data for clayey silt. (b) Arithmetic probability density function. (c) Logarithmic probability density function.

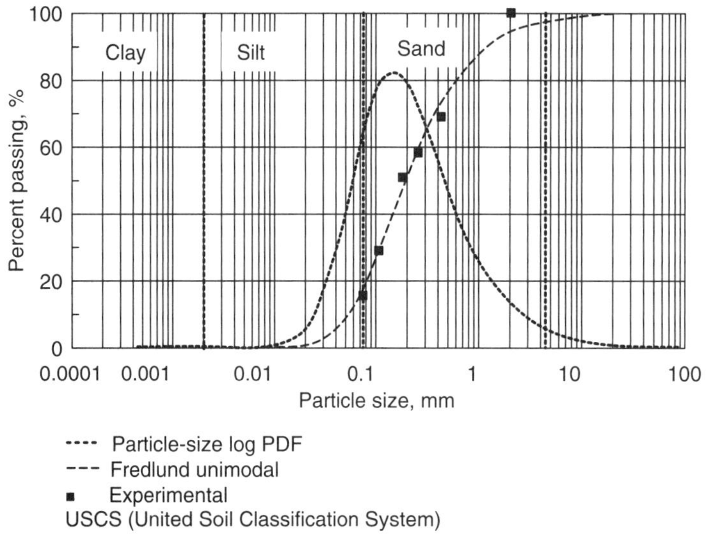

Figure 2.7 Grain-size distribution data fit with unimodal equation for silty sand along with logarithmic PDF (from M.D. Fredlund et al., 2000).

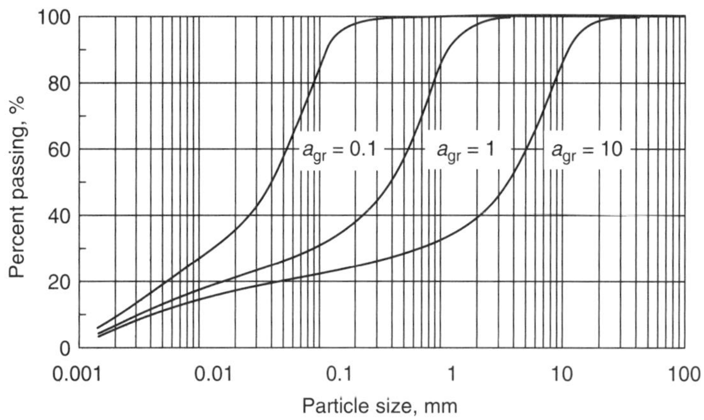

Figure 2.8 Effect of varying a_{gr} parameter on grain-size distribution curve (from M.D. Fredlund et al., 2000).

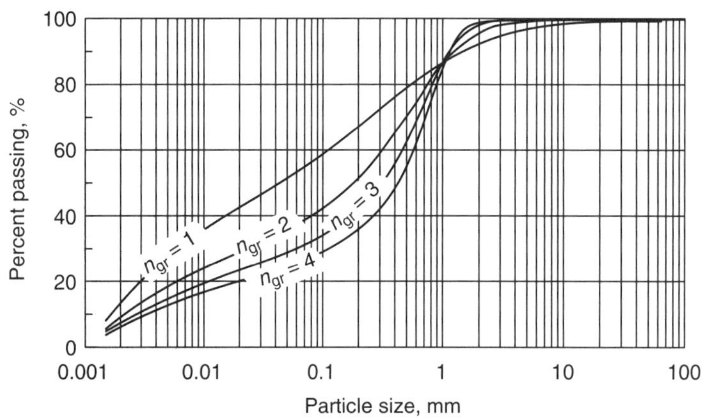

Figure 2.9 Effect of varying n_{gr} parameter while holding other parameters constant; $a_{gr} = 1.0$, $m_{gr} = 0.5$, $d_{r,gr} = 1000$, and $d_m = 0.001$ (from M.D. Fredlund et al., 2000).

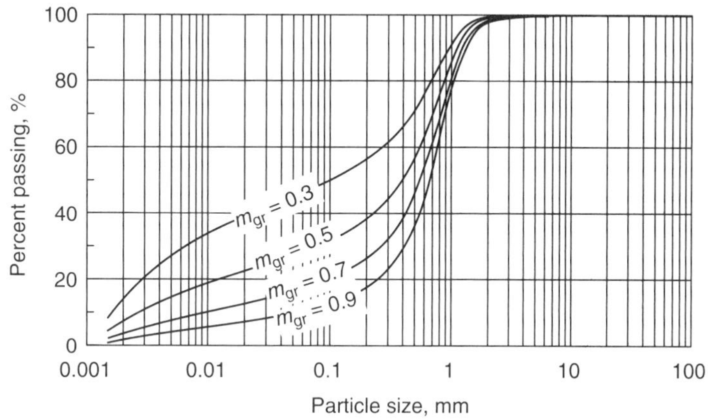

Figure 2.10 Effect of varying m_{gr} parameter while holding other parameters constant; $a_{gr} = 1.0$, $n_{gr} = 4.0$, $d_{r,gr} = 1000$, and $d_m = 0.001$ (from M.D. Fredlund et al., 2000).

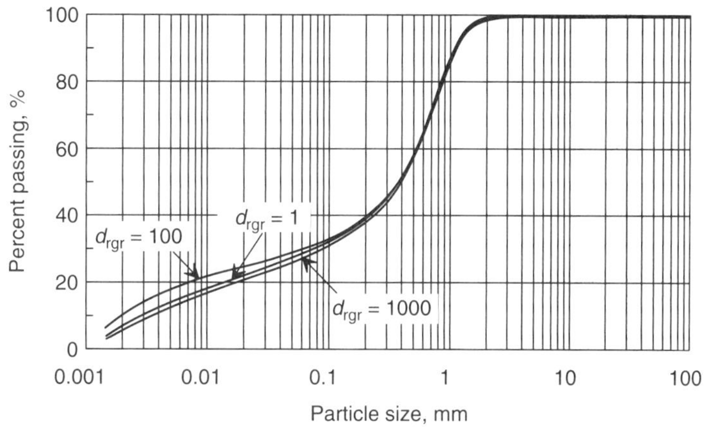

Figure 2.11 Effect of varying $d_{r,gr}$ parameter on grain-size distribution curve while holding other parameters constant; $a_{gr} = 1.0$, $n_{gr} = 4.0$, $m_{gr} = 0.5$, and $d_m = 0.001$ (from M.D. Fredlund et al., 2000).

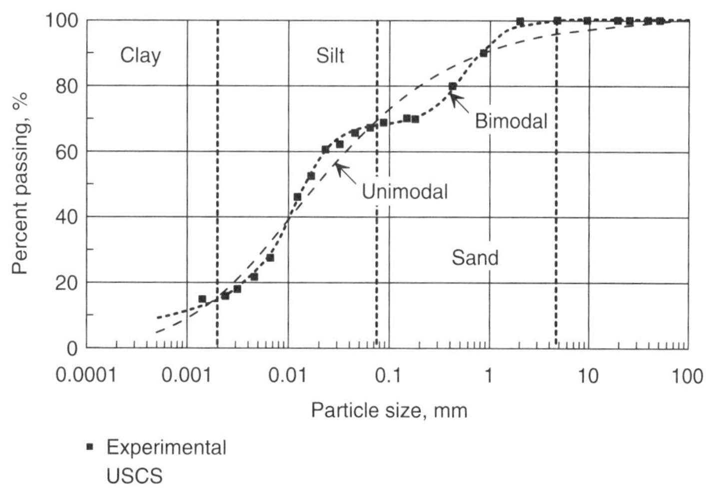

Figure 2.12 Fitting of gap-graded soil with unimodal and bimodal grain-size distribution equation (from M.D. Fredlund et al., 2000).

analysis. The characteristic shape of a bimodal or gap-graded soil is the double hump observed from experimental data. The hump indicates that the particle-size distributions are concentrated around two separate particle sizes. A gap-graded soil can be viewed as a combination of two or more soils (Durner, 1994). The gap-graded soil can be best fit by "stacking" two (or more) unimodal equations:

$$P_p(d) = \left[\sum_{i=1}^{k} w_i \left\{ \frac{1}{\left\{ \ln \left[\exp(1) + \left(\frac{a_{gr}}{d} \right)^{n_{gr}} \right] \right\}^{m_{gr}}} \right\} \right] \times \left\{ 1 - \left[\frac{\ln \left(1 + \frac{d_r}{d} \right)}{\ln \left(1 + \frac{d_r}{d_m} \right)} \right]^7 \right\} \tag{2.5}$$

where:

$P_p(d)$ = percentage by mass of particles passing a particular size,

k = number of "subsystems" for the total particle-size distribution, and

w_i = weighting factors for each of the subcurves adhering to the condition; $0 < w_i < 1$ and $\sum w_i = 1$.

For a bimodal curve, $k = 2$ and the number of parameters to be determined is $4k + (k - 1)$ (i.e., 9). A unimodal equation is used as the basis for the prediction of the bimodal equation. The final bimodal curve equation in its extended form can be written as

$$P_p(d) = \left\{ w \left[\frac{1}{\{\ln[\exp(1) + (a_{bi}/d)^{n_b}]\}^{m_b}} \right] + (1 - w) \times \left[\frac{1}{\{\ln[\exp(1) + (j_{bi}/d)^{k_{bi}}]\}^{l_{bi}}} \right] \right\} \times \left[1 - \left(\frac{\ln(1 + d_r/d)}{\ln(1 + d_r/d_m)} \right)^7 \right] \tag{2.6}$$

where:

a_{bi} = parameter related to the first breaking point along the coarse-grained portion of the grain-size distribution curve,

n_{bi} = parameter related to the first steep slope along the coarse-grained part of the grain-size distribution curve,

m_{bi} = parameter related to the first portion of the grain-size distribution curve,

j_{bi} = parameter related to the second breaking point of the curve,

k_{bi} = parameter related to the second steep slope along the curve,

l_{bi} = parameter related to the second shape along the curve, and

d_r = parameter related to the particle size in the fines region, referred to as the residual particle size.

A total of nine parameters must be computed when fitting the bimodal equation to experimental data (i.e., stacking unimodal curves). Seven parameters can be determined using a nonlinear least-squares fitting algorithm while two parameters can be fixed, namely, d_r and d_m. The superposition method provides a robust method of fitting bimodal data sets.

The bimodal grain-size distribution curve equation has been found to closely fit essentially any data set. If the data set tends to be unimodal in character, it is better to best fit the data using the unimodal grain-size equation. Figure 2.13 shows a bimodal fit for clayey silt sand along with its logarithmic PDF.

2.2.2.3 *Application of Grain-Size Distribution Function*

A mathematical equation for grain-size distribution data is of considerable value in geotechnical engineering. First, the unimodal and bimodal grain-size equations provide a continuous mathematical function for the grain-size distribution curve. Second, grain-size distributions can be identified by equations that are best fit to the experimental data. The grain-size information can be stored in a database and used for identification purposes. Third, grain-size distribution equations provide a consistent method for determining physical indices such as percent clay, percent sand, percent silt, and particle diameter variables such as d_{10}, d_{20}, d_{30}, d_{50}, and d_{60}.

Grain-size distribution curves have also been used for estimating SWCCs (Gupta and Larson, 1979a; Arya and Paris, 1981; Haverkamp and Parlange, 1986; Ranjitkar and Sunder, 1989; M.D. Fredlund et al., 2000a). An accurate representation of the grain-size distribution curve forms a reliable basis for the estimation of the SWCC (i.e., pedo-transfer estimation functions).

The percentages of each range of particle sizes can be calculated from the grain-size distribution equation. The grain-size distribution equations are in the form of percent passing a particular particle size, $P_p(d)$, where d is particle diameter (mm). The percent clay, percent silt, and percent sand can be read directly from the curve by designating appropriate particle diameters. The input diameters used depend upon the criteria associated with the various soil classification systems. For example, the USDA classification designates clay-sized particles as being less than 0.002 mm. Silt-sized particles are in the range between 0.002 and 0.05 mm and sand-sized particles are between 0.05 and 2.0 mm. The Unified Soil Classification System (USCS) uses a boundary of 0.005 mm for the start of clay size particles. Silt-sized particles are between 0.005 and 0.075 mm, and sand-sized particles are between 0.075 and 4.75 mm. The divisions can be determined for any classification method by substituting the appropriate particle sizes into the grain-size equation as shown in Fig. 2.14.

Diameter variables can also be read from the grain-size distribution curve equation in an inverse manner. Particle-size diameters answer to the question "What particle diameter has 10 percent of the total mass smaller than this size?" A "half-length" algorithm can be used to read diameters from the grain-size distribution equation as shown in Fig. 2.15.

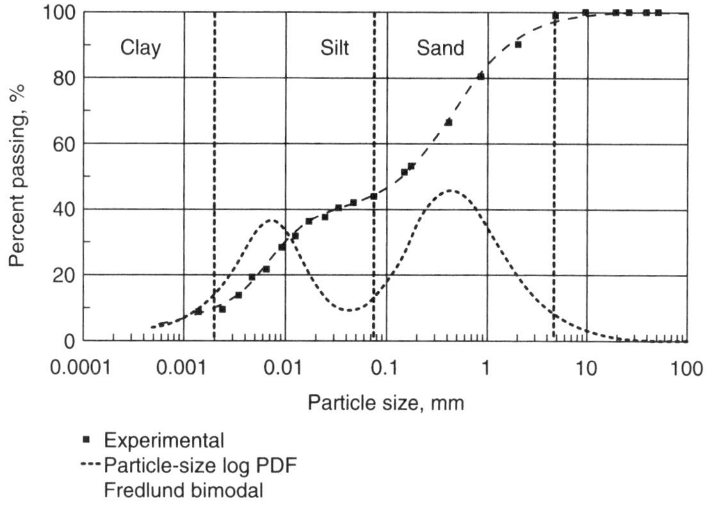

Figure 2.13 Fitting gap-graded clayey silt sand with bimodal equation along with its logarithmic PDF (from M.D. Fredlund et al., 2000).

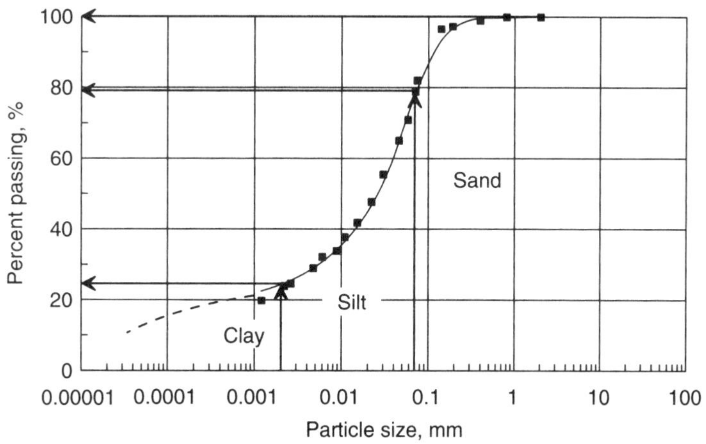

Figure 2.14 Determination of soil-size fractions (i.e., % clay, % silt, and % sand) according to USCS (from M.D. Fredlund et al., 2000).

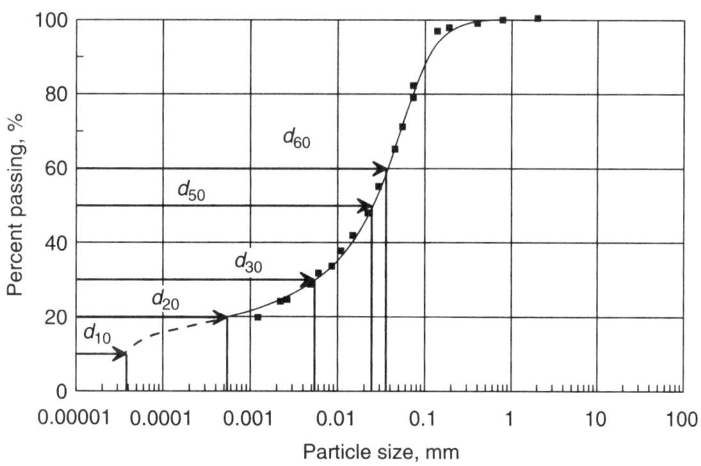

Figure 2.15 Determination of percent passing for selected particle sizes, *d* (from M.D. Fredlund et al., 2000).

2.2.3 Atterberg Limits

The shrinkage curve associated with an initially slurry soil provides information on the consistency of the soil in terms of gravimetric water content. The shrinkage curve starts near the liquid state, passes through the plastic state, and into the semisolid state and the solid state as the soil is allowed to dry. The soil passes through the various states in response to an increase in soil suction caused by the drying of the soil. It is useful to understand the shrinkage curve in response to changes in the stress state (i.e., soil suction or negative pore-water pressures) (Fig. 2.16).

A clay soil gradually increases in shear strength as it is dried from initial water contents above the liquid limit. The shear strength becomes approximately 1.7 kPa at the liquid limit of the soil (Aitkinson, 2007). Wood (1990) suggested that the average effective stress in the soil was about 8 kPa at the liquid limit. The matric suction continues to increase as the water content decreases as a result of evaporation.

Most soils start to desaturate as the water content approaches the plastic limit of the soil. The shear strength at the plastic limit is estimated to be around 170 kPa (Atkinson, 2007). In other words, there is approximately a two-order-of-magnitude increase in undrained shear strength between the liquid limit and the plastic limit of a clayey soil (Wood and Wroth, 1978; Budhu, 2010). As the water content of a clayey soil decreases below the plastic limit, the volume of the soil decreases at a reduced rate while the shear strength increases substantially. Eventually there is no further volume change as the soil reaches its limit of shrinkage. Further drying reduces the water content of the soil to zero under essentially constant-volume conditions. At the point of zero water content the soil suction approaches about 1,000,000 kPa.

There are also zones along the shrinkage curve that bear a relationship to zones along the SWCC for an initially slurry soil. The SWCC allows engineers to view soil behavior from the standpoint of stress state changes.

2.2.4 Shrinkage Curve Equation

The "shrinkage curve" for a soil is distinct from the "shrinkage limit" of a soil. The *shrinkage limit* is the water content corresponding to all voids being filled with water when the soil has been dried to zero water content (and minimum void ratio) without any external loading applied. In other words, the shrinkage limit represents the maximum amount of water that could be injected into a completely dried soil without allowing any volume change. The shrinkage limit has received limited application in routine geotechnical engineering practice while the *shrinkage curve* is now proving to be of considerable value in constructing the volume-mass constitutive relationships for unsaturated soils. Historically, the volume of a drying soil was measured using the mercury immersion technique. This procedure is now discouraged for health safety reasons. A series of caliper readings of

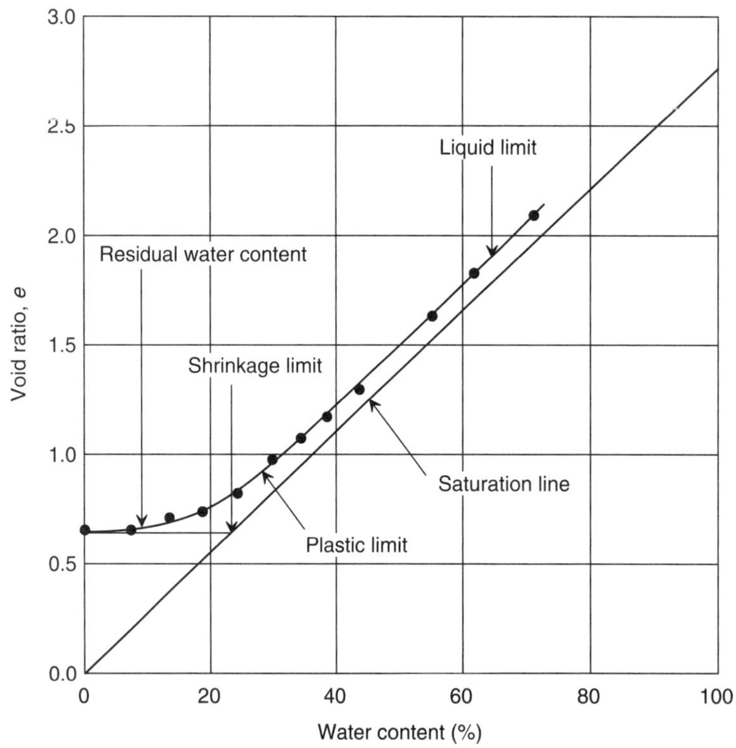

Figure 2.16 Relationship of Atterberg limits to shrinkage curve for initially slurry highly plastic clay (data points generated).

diameter and soil specimen thickness has been shown to provide satisfactory accuracy for the volume of small drying specimens (Ho, 1988; D.G. Fredlund et al., 2011). The shrinkage curve provides an important visual link between changes in the consistency of a soil and the stress state in the soil.

The shrinkage curve can be measured for a soil that is either initially in a slurry state or in some other unsaturated soil state (e.g., compacted soil). The water content and void ratio of the soil are measured as the soil is gradually dried to zero water content. A mathematical representation of the shrinkage curve provides a relationship between volume and mass on the constitutive surfaces in response to an increase in soil suction. The shrinkage curve is part of constitutive behavior for an unsaturated soil and can be used in conjunction with a SWCC to determine the relationship between volume change (i.e., void ratio or specific volume), and soil suction.

Shrinkage curves can be represented as either the specific volume versus water content or void ratio versus water content (Haines, 1923). A typical shrinkage curve is shown in Fig. 2.17. Particle-size distribution and stress history are the primary factors controlling shrinkage behavior. There are three main initial states from which soil shrinkage can be measured: (i) undisturbed samples from the field, (ii) compacted soil specimens, and (iii) slurry specimens prepared near the liquid limit.

Drying of a saturated soil (e.g., starting at point A) follows the saturation line until air begins to enter the largest soil voids at point B (Fig. 2.17). Point B is an indication of the air-entry value (AEV) of an initially slurry soil. As the soil continues to dry, it reaches a minimum void ratio beyond which there is no further volume change. When the water content of the soil reaches zero, the soil suction has increased to 1,000,000 kPa (i.e., point D).

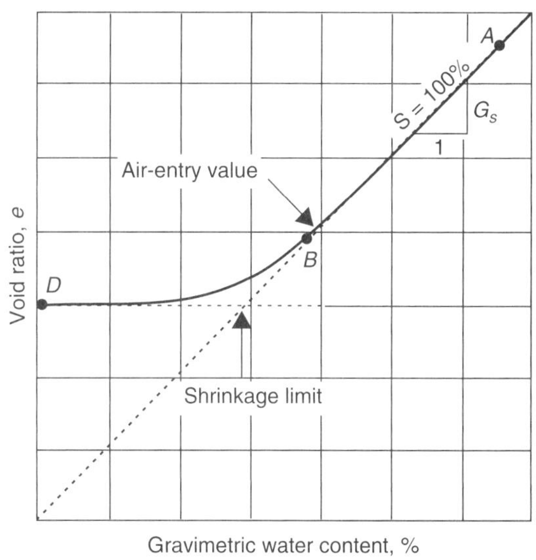

Figure 2.17 Drying phenomenon showing relationship between air-entry value and shrinkage limit of a soil (after Marinho, 1994).

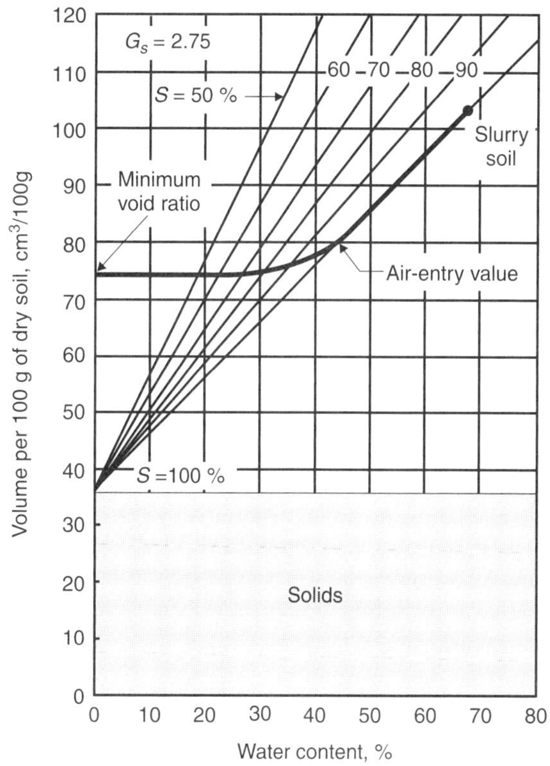

Figure 2.18 Volume-mass relationships for drying of initially slurry soil.

The water content at the intersection between the minimum void ratio and the saturation line is defined as the "shrinkage limit" (Fig. 2.18). A somewhat different shrinkage pattern is observed when soils are dried from degrees of saturation less than 100% (Fig. 2.19). Point A in Fig. 2.19 is a hypothetical starting point for drying a soil that is initially unsaturated. If the soil maintains a constant ratio of air to water in the voids as the soil dries, a constant degree of saturation line will be followed that is less than 100%. The line will pass through point B. If the volume of air remains constant during the drying process (i.e., passing through point C), the degree of saturation decreases and move toward zero degree of saturation (point D).

2.2.4.1 Mathematical Representation of Shrinkage Curve

Figure 2.20 shows a collection of shrinkage curves for soils of differing plasticity and different initial states, namely, (i) undisturbed, (ii) compacted near the plastic limit, or (iii) slurry near the liquid limit. A hyperbolic mathematical form captures the shape of the shrinkage curves. The following equation was proposed by M.D. Fredlund et al. (2002a) for the shrinkage curve:

$$e(w) = a_{\text{sh}} \left[\frac{w^{c_{\text{sh}}}}{b_{\text{sh}}^{c_{\text{sh}}}} + 1 \right]^{1/c_{\text{sh}}} \tag{2.7}$$

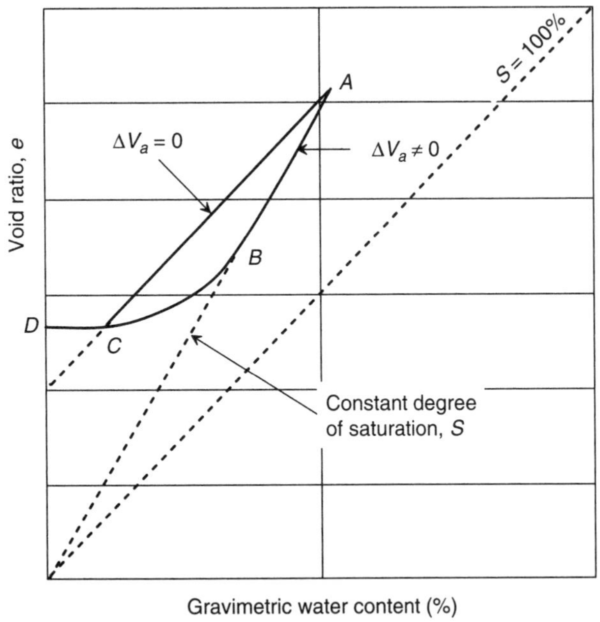

Figure 2.19 Volume-mass shrinkage curve for initially unsaturated soil specimen (from M.D. Fredlund et al., 2002a).

where:

a_{sh} = minimum void ratio, e_{min},
b_{sh} = slope of the line of tangency,
c_{sh} = curvature of the shrinkage curve,
a_{sh}/b_{sh} = G_s/S = constant for a specific soil, and
w = gravimetric water content.

A least-squares fitting algorithm can be used to fit the shrinkage curve equation to various data sets. A typical fit of the data for an initially saturated soil is shown in Fig. 2.21. The proposed shrinkage curve equation can describe a wide variety of shrinkage curves. The effects of soil structure and initial degree of saturation can be accommodated through soil parameters which have physical meaning. The shrinkage curve equation is meant for the simulation of the drying of a soil.

The a_{sh} parameter represents the minimum void ratio corresponding to the shrinkage limit of the soil. The relationship between void ratio and gravimetric water content is linear if the degree of saturation, S, remains constant during shrinkage. The slope of the relationship between void ratio and water content is related to the initial volume-mass properties. The slope of the line tangent to the initial portion of the shrinkage curve is equal to a_{sh}/b_{sh}. The effect of varying the a_{sh} parameter can be seen in Fig. 2.22:

$$\frac{a_{sh}}{b_{sh}} = \frac{G_s}{S} \tag{2.8}$$

The influence of changing the b_{sh} parameter is shown in Fig. 2.23. The b_{sh} parameter is equal to the slope of the degree of saturation line on the shrinkage plot. Equation 2.8 shows that once the minimum void ratio of a soil is known, the b_{sh} parameter can be calculated for a specific degree of saturation.

The c_{sh} equation parameter controls the rate of curvature of the shrinkage curve as the soil begins to desaturate, as shown in Fig. 2.24. The parameter, c_{sh} varies somewhat depending on the soil type and the initial state of the soil.

2.2.4.2 *Estimation of Shrinkage Curve*

The volume change of a soil specimen is commonly not measured when performing a laboratory test for the SWCC. The basic experimental data from a SWCC test generally consists of gravimetric water content versus soil suction. The shrinkage curve can be used in conjunction with the SWCC equation to compute the void ratio–soil suction relationship.

The ratio between the a_{sh} and b_{sh} shrinkage curve equation parameters is fixed by the initial degree of saturation S and the specific gravity G_s. There are two additional parameters to determine: a_{sh} and c_{sh}. Values for the curvature of the shrinkage curve, c_{sh}, can be estimated based on regression analyses

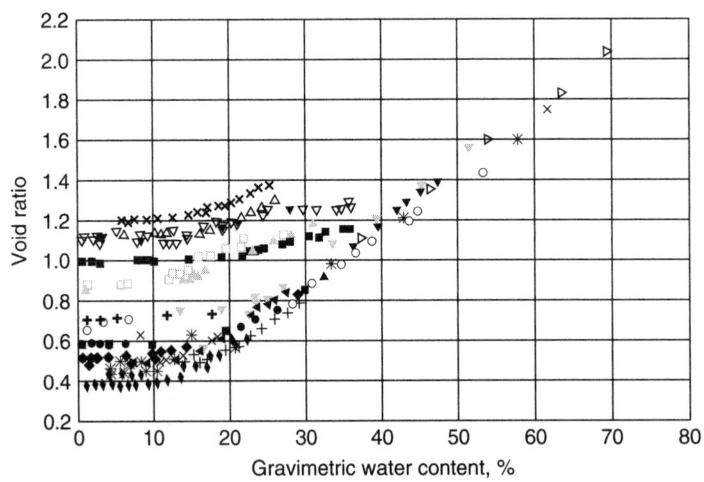

Figure 2.20 Typical soil shrinkage data for variety of soils with different initial volume-mass properties (from M.D. Fredlund, 2002a).

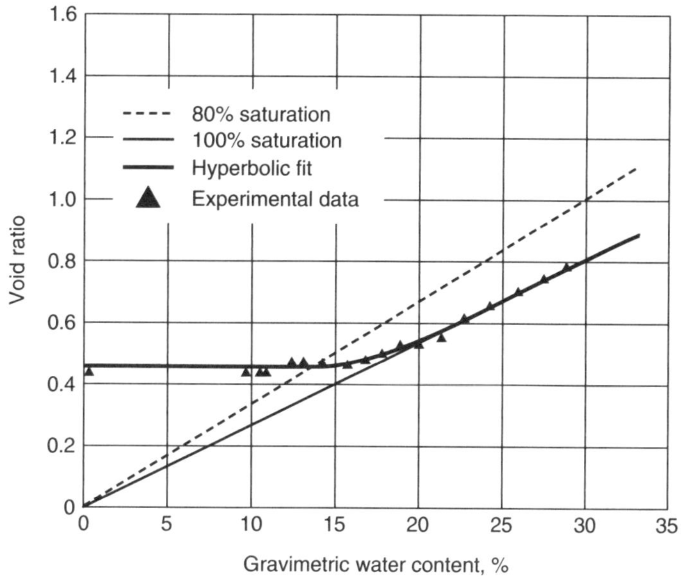

Figure 2.21 Best fit of shrinkage curve data for London clay; fitting parameters: $a_{sh} = 0.47$, $b_{sh} = 0.176$, $c_{sh} = 10.56$ (data from Croney and Coleman, 1953).

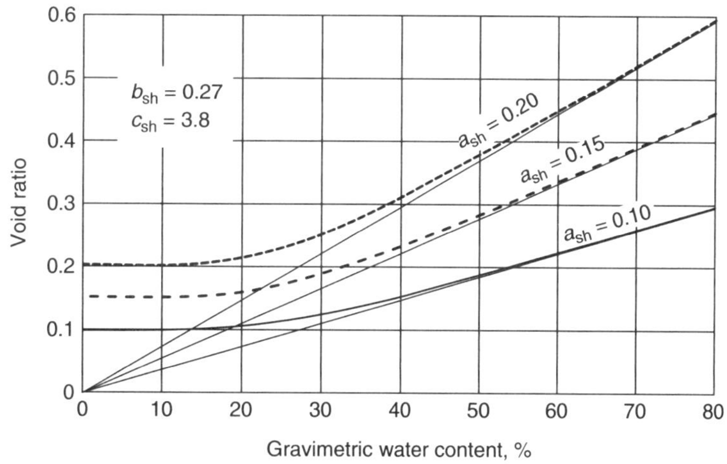

Figure 2.22 Effect of varying a_{sh} parameter on shrinkage curve equation (from M.D. Fredlund et al., 2002a).

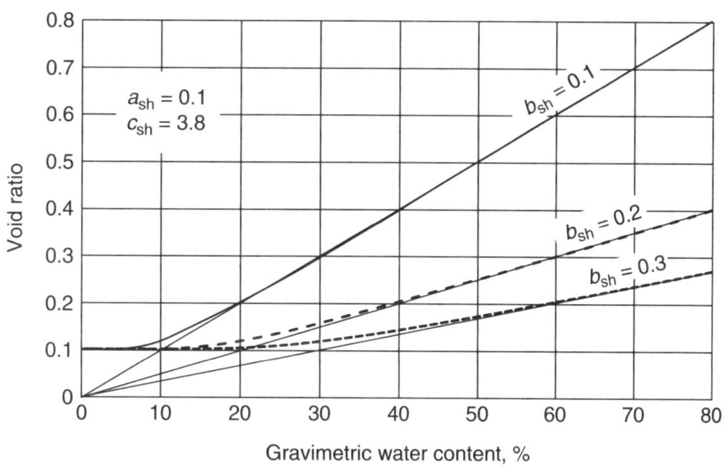

Figure 2.23 Effect of varying b_{sh} parameter on shrinkage curve equation (from M.D. Fredlund et al., 2002a).

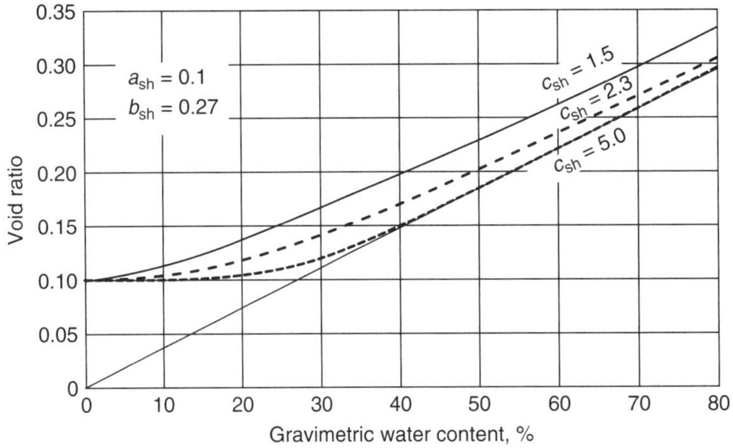

Figure 2.24 Effect of varying c_{sh} parameter on shrinkage curve equation (from M.D. Fredlund et al., 2002a).

of published test results (M.D. Fredlund et al., 2002a). The average value for the c_{sh} parameter for *undisturbed* soil specimens was 9.57 with a standard deviation of ±9.3. The average value for the c_{sh} parameter for initially slurry soil specimens was 25.3 with a standard deviation of ±25.4. The average value for the c_{sh} parameter for *compacted* soil specimens was 8.47 with a standard deviation of ±3.6.

Approximate values for the minimum void ratio or a_{sh} equation parameter have been presented by Holtz and Kovacs (1981) and are shown in Fig. 2.25. The data allow the shrinkage limit to be estimated and converted to an equivalent void ratio when the plastic limit of the soil is known. Data in Fig. 2.25 provide a means of estimating the a_{sh} shrinkage curve parameter.

The estimation procedure described provides an approximate shape for the shrinkage curve. However, it is preferable to measure the entire shrinkage curve in the laboratory. The largest variation in the shrinkage curve occurs as a result of estimating the minimum void ratio, a_{sh}, of the soil. It is relatively easy to measure the minimum void ratio of a soil experimentally. The c_{sh} parameter provides the general curvature for the shrinkage curve. The variation in the shrinkage curve produced by varying the c_{sh} parameter is quite minimal and as such it is possible to obtain an estimate of the shrinkage curve.

2.2.4.3 Calculation of Void Ratio versus Soil Suction

The SWCC describes the relationship between the water content of a soil and soil suction. The volume change occurring when drying a clayey soil can be substantial and is relevant to the interpretation of SWCC data. Figure 2.26 shows SWCC data for highly plastic Black clay (Dagg et al., 1967). It is possible to determine the void ratio–soil suction relationship when the volume changes are known during the drying process (i.e., from the shrinkage curve). The volume changes can be estimated through the use of the shrinkage curve. The shrinkage curve for the Black clay is shown in Fig. 2.27. Figure 2.28 shows the void ratio–soil suction constitutive relationship computed from the SWCC and the

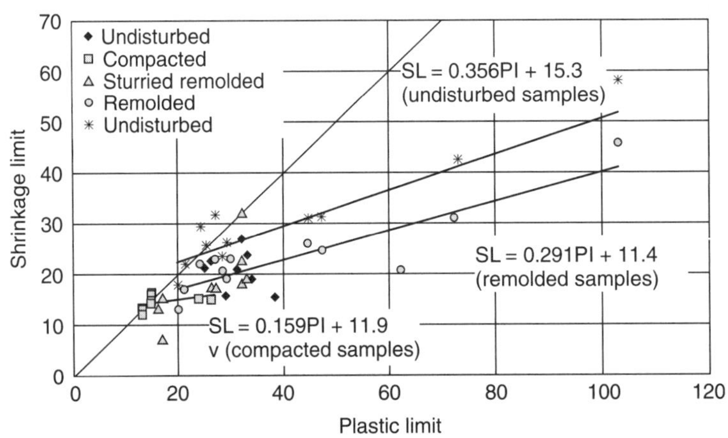

Figure 2.25 Relationship between shrinkage limit and plastic limit for variety of soils (after Holtz and Kovacs, 1981).

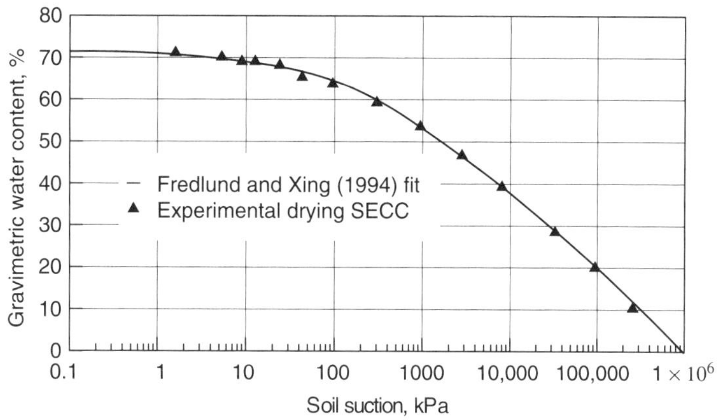

Figure 2.26 Experimental SWCC data for Black clay (data from Dagg et al., 1967).

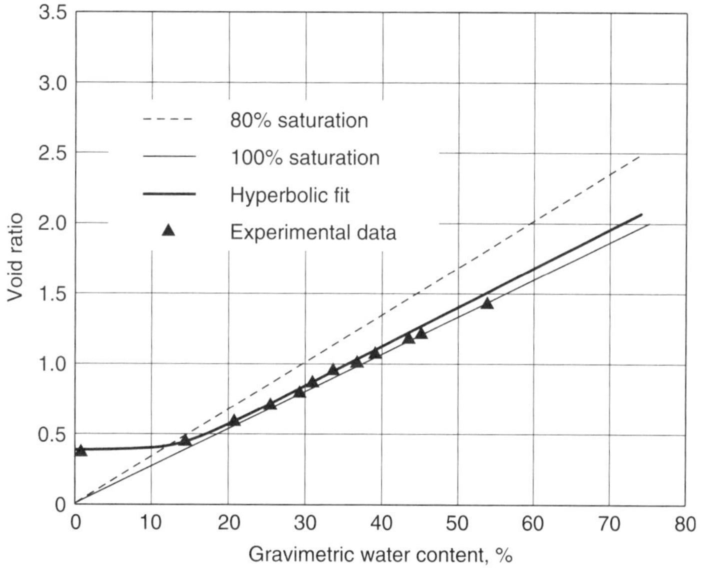

Figure 2.27 Experimental shrinkage curve data for Black clay; $a_{sh} = 0.386$, $b_{sh} = 0.14$, $c_{sh} = 5.04$ (data from Dagg et al., 1967).

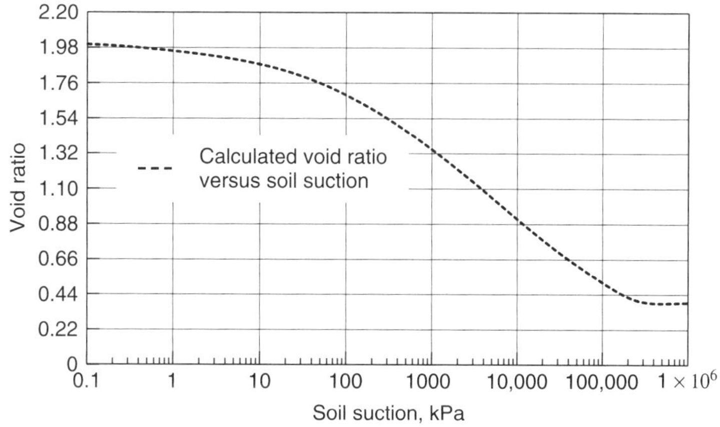

Figure 2.28 Void ratio versus soil suction computed from SWCC and shrinkage curve for Black clay (data from Dagg et al., 1967).

shrinkage curve. The void ratio–soil suction represents the limiting boundary condition associated with the void ratio constitutive surface.

The degree of saturation versus soil suction can also be plotted through use of the basic volume-mass relationship (i.e., $S = wG_s/e$). The degree-of-saturation plot versus soil suction provides the best measure of the air-entry value for the soil. Consequently, the degree-of-saturation curve versus soil suction becomes an important relationship for the estimation of other unsaturated soil property functions.

2.3 PHASE PROPERTIES

A portion of a mixture qualifies as an independent phase if it has the following characteristics: (1) differing properties from the contiguous materials and (2) definite bounding surfaces. These two conditions must be met in order for a portion of a multiphase system to qualify as an independent phase. It is easy to understand how a saturated soil consists of two phases (i.e., soil solids and water). It is also quite understandable that the air becomes another independent phase when the soil becomes unsaturated. Each of these phases (i.e., soil solids, water, and air) clearly meets the requirements for designation as a phase.

It is also possible for a phase to change "state," as is the case when water freezes or vaporizes. Ice then becomes an independent phase from that of the water when analyzing engineering problems. Water can also exist in vapor form and, as such, water can exist in all three states in an unsaturated soil.

Some special properties and interactions associated with the interfaces between two phases are also discussed in this chapter. The phase diagram (Fig. 2.29) shows the mass and volume quantities associated with each phase of an unsaturated soil. The thickness of the contractile skin is in the order of only a few molecular layers. Therefore, the physical subdivision of the contractile skin is unnecessary when establishing volume-mass relations for an unsaturated soil. The contractile skin is considered as part of the water phase without producing significant error.

An unsaturated soil is a mixture of several phases. It is important to establish the number of phases comprising the soil since it has an influence on how the stress state of the mixture is defined. It is important to first understand the properties of each phase of an unsaturated soil. On the basis of the definition of a phase, it is proposed that an unsaturated soil actually consists of four phases rather than the commonly referred to three phases. It is postulated that the air-water interface (i.e., contractile skin) can legitimately be referred to as a fourth phase in an unsaturated soil.

An unsaturated soil is a multiphase system comprising (1) soil solids (particles), (2) water, (3) air, and (4) contractile skin. The contractile skin plays an important role from the standpoint of stress state considerations. However, when considering volume-mass relations for an unsaturated soil, it is not necessary to separate the water in the contractile skin from the remaining water mass (and volume). In other words, an unsaturated soil can be considered to be a three-phase system with respect to volume-mass properties but needs to be considered as a four-phase system with respect to stress state conditions (Fig. 2.30).

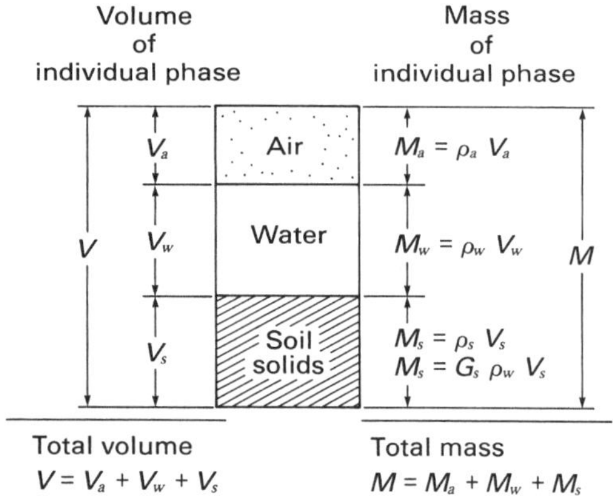

Figure 2.29 Phase diagram for unsaturated soil.

2.3.1 Solid Phase

There are several basic material properties that can be defined for each phase. A few typical values are also presented for each of the properties. Some of the main material properties are (i) density (and specific volume) and (ii) thermal properties.

2.3.1.1 Density of Solids

Density and specific volume are used to define the volume-mass relationship property of each phase. Density, ρ is defined as the ratio of mass to volume. Each phase of a soil mixture has a specific density. The density of a solid can also be referenced to the density of water to give

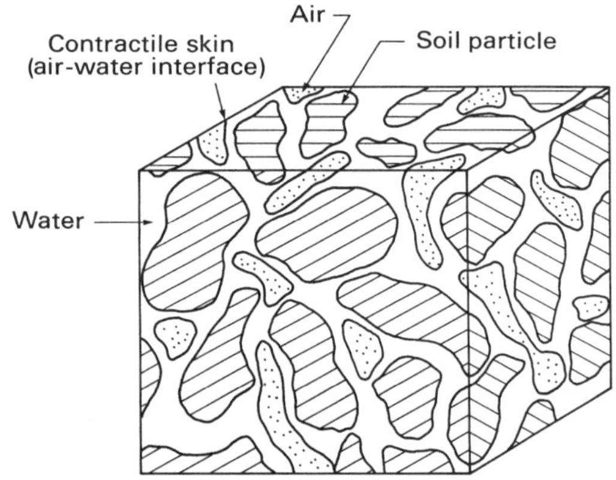

Figure 2.30 Element of unsaturated soil with continuous air phase.

Table 2.1 Specific Gravity of Several Minerals

Mineral	Specific Gravity, G_s
Quartz	2.65
K feldspars	2.54–2.57
Na-Ca feldspars	2.62–2.76
Calcite	2.72
Dolomite	2.85
Muscovite	2.7–3.1
Biotite	2.8–3.2
Chlorite	2.6–2.9
Pyrophyllite	2.84
Serpentine	2.2–2.7
Kaolinite	2.61[a]; 2.64±0.02
Halloysite ($2H_2O$)	2.55
Illite	2.84[a]; 2.60–2.86
Montmorillonite	2.74[a]; 2.75–2.78
Attapulgite	2.30

Source: From Lambe and Whitman, 1979.
[a]Calculated from crystal structure.

the specific gravity of the solid phase. Specific volume v_0 is generally defined as the inverse of density; therefore, specific volume is the ratio of volume to mass.

The density of the soil particles, ρ_s, is defined as

$$\rho_s = \frac{M_s}{V_s} \tag{2.9}$$

where:

M_s = mass of solids and
V_s = volume of solids.

The specific gravity of the soil particles is defined as the ratio of the density of the soil particles to the density of water at a temperature of 4°C under standard pressure conditions (i.e., 101.3 kPa). In the International System of Units (SI), the specific gravity is referred to as the relative density of the soil particles:

$$G_s = \frac{\rho_s}{\rho_w} \tag{2.10}$$

The density of water at 4°C and 101.3 kPa is 1000 kg/m^3. Some typical values of specific gravity G_s for soil solids are presented in Table 2.1.

2.3.1.2 Thermal Properties of Solids

There are two thermal properties that are of primary interest in geotechnical engineering: specific heat capacity and thermal conductivity. There are also phase change properties that need to be taken into consideration when there is a phase change as temperature is changed. Rocks and minerals exist only in the solid phase for most geotechnical engineering problems.

Specific heat capacity C_s is defined as the quantity of heat required to raise the temperature of a unit mass of material by one degree Kelvin. The SI units for specific heat capacity are kilojoules per kilogram per degree Kelvin (kJ/kg K). An average value for the specific heat capacity for rocks and minerals is 0.85 kJ/kg K. Specific heat capacity values for a series of rocks and minerals are presented in Table 2.2 (Tarnawski and Wagner, 1991).

Thermal conductivity λ_s is defined as the ability of the material to transmit heat through conduction. The SI units for thermal conductivity are Watts per meter per degree Kelvin (W/m/ K). Typical values for the thermal conductivity of various rocks and minerals are presented in Table 2.2. The thermal conductivity can be seen to vary from about 1 to 7 for most materials with hematite showing a value of about 11.3 W/m K.

2.3.2 Water Phase

The water phase plays an important role in the behavior of soils. There are several physical properties that are of particular interest when dealing with soils. Water may exist

Table 2.2 Specific Gravity and Thermal Properties of Solid Minerals

Minerals and Rocks	Specific Heat Capacity, C_s (kJ/kg K)	Thermal Conductivity, λ_s (W/m K)	Specific Gravity, G_s
Quartz	0.698	7.69	2.65
Calcite	0.793	3.57	2.71
Dolomite	0.930	5.50	2.86
Orthoclase	0.610	2.31	2.58
Muscovite	—	2.32	2.85
Biotite	—	1.17	2.98
Chlorite	—	5.14	2.64
Talc	0.870	6.10	2.82
Hematite	0.610	11.28	5.14
Granite	0.880–1.382	1.65–2.83	2.60–2.65
Gneiss	0.766–0.871	2.58–2.94	2.70–2.73
Marble	0.750	2.79–2.89	2.60–2.69
Limestone	0.825–0.950	1.70–2.68	2.41–2.67
Sandstone	0.762–1.072	2.18–5.10	2.35–2.97
Slate	0.779	1.89–2.59	2.70–2.76
Organics	1.923	0.25	1.30

Source: From Tarnawski and Wagner, 1991.
Note: The thermal properties of solids, water, and air can be combined to provide typical thermal properties for various soil types and mixtures.

in one of three states (i.e., liquid, solid, and vapor), and its properties are a function of temperature and pressure.

2.3.2.1 *Density of Water*

The density of water, ρ_w, is defined as

$$\rho_w = \frac{M_w}{V_w} \tag{2.11}$$

Water is essentially a homogeneous substance the world over, except for variations produced by salts and isotopes of hydrogen and oxygen (Dorsey, 1940). Distilled water under the pressure of its saturated vapor is called pure, saturated water. The density of pure, saturated water can be measured experimentally. Figure 2.31 shows the density of pure water under various applied pressures and temperatures.

The variation in the density of water due to temperature differences is more significant than its variation due to applied pressure for geotechnical engineering problems. McCutcheon et al. (1993) proposed an empirical equation for the density of water with respect to changes in temperature. The closed-form equation proposed for the density of pure water as a function of temperature is as follows:

$$\rho_w = 1000\left[1 - \frac{(T + 288.9414)(T - 3.9863)^2}{508929.2(T + 68.12963)}\right] \tag{2.12}$$

where:

ρ_w = density of water, kg/m^3, and
T = temperature, °C.

The density of water under isothermal conditions is commonly taken as 1000 kg/m^3 for most geotechnical engineering problems. Table 2.3 shows more precise densities of water over a range of temperatures.

Table 2.3 Density of Water at Various Temperatures

Temperature (°C)	Density (kg/m^3)
+100	958.4
+80	971.8
+60	983.2
+40	992.2
+30	995.65
+25	997.05
+20	998.21
+15	999.10
+10	999.70
+4	999.97
0	999.84
−10	998.12
−20	993.55
−30	983.85

Source: From Lide, 1990.

2.3.2.2 *Salt-Water Mixture*

The density of salty water increases from that of pure water. Either the mass or the volume of water needs to change in order for the density of a mixture of pure water and salt to change. The density of pure water is about 1000 kg/m^3 and common salt (NaCl) has a considerably higher density of 2160 kg/m^3.

Let us suppose that a 1000-mL flask is filled to the 1000-mL mark with distilled water at 23°C. Now let us add 30 g NaCl. The weight of the flask plus its contents will obviously increase by 30 g, but it is interesting to observe what will happen to the volume of water plus the salts. It might come as a surprise to discover that there is a reduction in overall volume of about 0.66%. The density of the salt-water

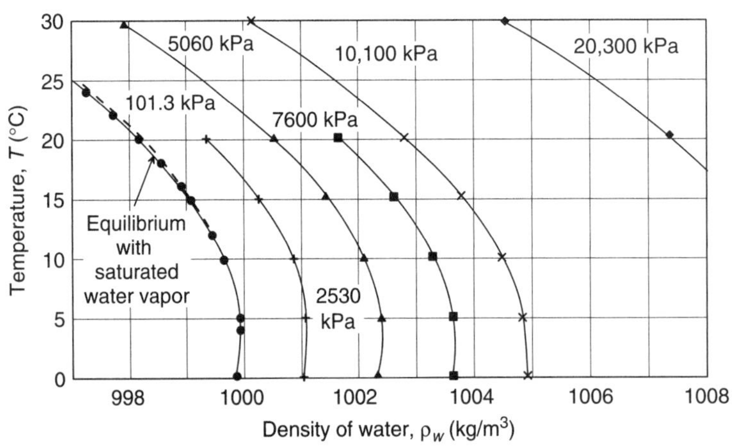

Figure 2.31 Density of pure water for various applied pressures and temperatures (from Dorsey, 1940).

mixture is increased for two reasons: (i) the density of salt is greater than the density of water and (ii) the addition of salt causes a decrease in overall volume. This phenomenon is possible because of the dipolar nature of water and the slight changes in molecular arrangement and bonding that occur when salts are added to water.

McCutcheon et al. (1993) proposed an empirical equation for the density of water when salts are added in varying concentrations. The density of water when temperature and salt concentration, s, are varied can be written as

$$\rho_{ws} = \rho_w + As + Bs^{1.5} + Cs^2 \tag{2.13}$$

where:

ρ_{ws} = density of water containing salts, kg/m^3,
s = salt concentration, g/kg,
$A = 0.824493 - 0.0040899T + 7.6438 \times 10^{-5}T^2 - 8.2467 \times 10^{-7}T^3 + 5.3675 \times 10^{-9}T^4$
$B = -0.005724 + 1.0227 \times 10^{-4}T - 1.6546 \times 10^{-6}T^2$,
$C = 0.00048314$, and
T = temperature, °C.

Figure 2.32 shows a graph of water density versus temperature for salt concentrations varying from zero to 100 g/kg. The results show that (i) the relative effect of a temperature change from 40 to 10° Celsius is comparable to a salt concentration change from zero to 10 g/kg and (ii) water density changes are relatively small until the salt concentration exceeds 10 g/kg.

2.3.2.3 *Thermal Properties of Water*

Water can exist in three phases in a soil; namely, liquid phase, solid phase and vapor phase. Each phase has its own thermal properties of *specific heat capacity* and *thermal conductivity*. Typical values of specific heat capacity for water are 4.19 kJ/kg K for the liquid phase, 2.11 kJ/kg K for the solid phase (i.e., ice), and 1.97 kJ/kg K for the vapor phase. The specific heat of water, C_w, is a function of temperature and typical values are presented in Table 2.4. The thermal conductivity of water, λ_w, also depends on temperature and typical values are presented in Table 2.4.

The change from the liquid phase to the solid phase occurs at 0°C in the absence of soil particles. The phase change from the liquid to the vapor phase occurs at 100°C at a pressure state of 1 atm. The latent heat of melting is 334 kJ/kg when there is a phase change from a solid to a liquid. The latent heat of evaporation is 2270 kJ/kg when there is a phase change from liquid to the vapor phase. Water also has a thermal expansion coefficient of 0.042 when the temperature range is between 4 and 100°C.

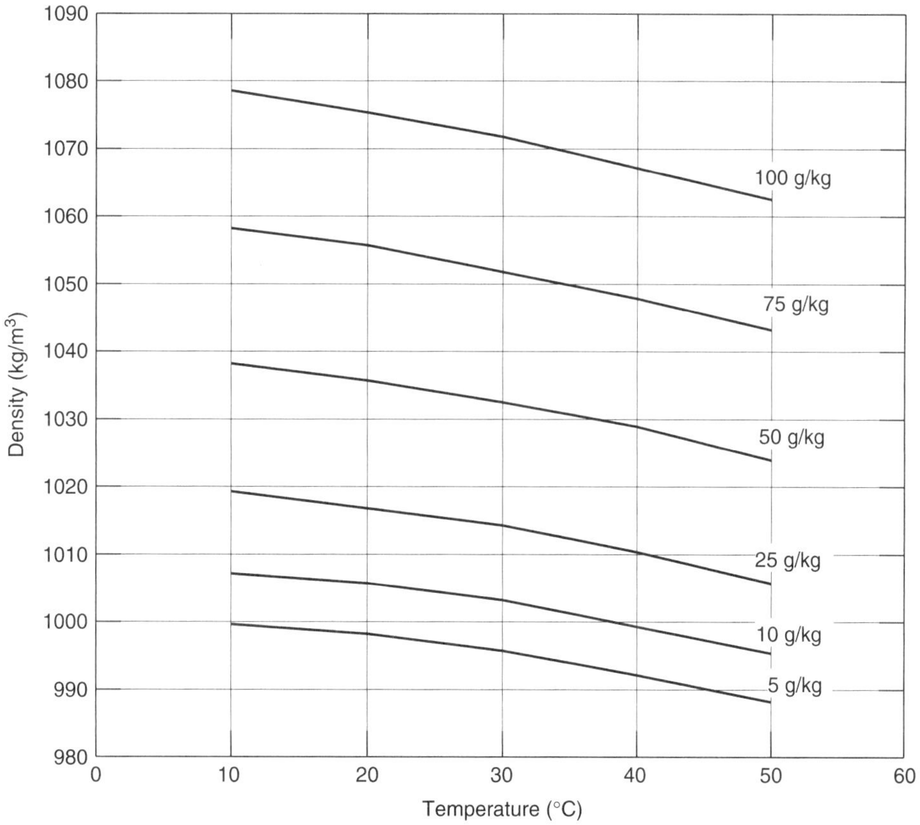

Figure 2.32 Density versus temperature for water with various concentrations of salt.

Table 2.4 Thermal Properties of Water at Various Temperatures

Temperature (°C)	Specific Heat Capacity, C_w(kJ/kg K)[a]	Thermal Conductivity, λ_w(W/m K)[b]
5	4.204	
10	4.193	0.582
15	4.186	
20	4.183	0.560
25	4.181	0.608
30	4.179	0.615
35	4.178	
40	4.179	0.629
50	4.182	0.640
60	4.185	0.651
70	4.191	0.659
80	4.198	0.667
90	4.208	0.673
100	4.219	0.677

[a]The Engineering ToolBox, www.EngineeringToolBox.com Water–Thermal Properties.
[b]Data from Sengers and Watson, 1986.

2.3.2.4 *Viscosity of Water*

All fluids resist a change of form or the action of shearing, and this property is called viscosity. The absolute (dynamic) viscosity, ν, of a fluid is defined as the resistance of the fluid to shearing force applied by sliding one plate over another with the fluid placed in between. Absolute viscosity depends on pressure and temperature. However, the influence of pressure is negligible for the range of pressures commonly encountered in civil engineering applications.

The viscosity of water under atmospheric pressure conditions (i.e., 101.3 kPa) and at various temperatures is given in Table 2.5. The viscosity of liquids is shown to decrease with an increase in temperature.

2.3.2.5 *Cavitation of Water*

The cavitation point of a liquid is generally assumed to be equal to the absolute vapor pressure of the liquid under consideration. If the vapor pressure of water at a particular temperature is 4.0 kPa and the absolute air pressure is 101.3 kPa, then cavitation should occur at $-101.3 + 4.0$ or -97.3 kPa on the gauge pressure scale. This is the vacuum pressure under which the water should start to vaporize (Pearsall, 1972).

The above example suggests that water will cavitate and commence boiling once the tension in the water approaches the absolute vapor pressure. However, water in a soil is known to be able to sustain pore-water pressures (i.e., tensions) that are much more negative (Young, 1989). It has also been possible to produce conditions in the laboratory where water can be shown to sustain tensions of more than 10 atm (Ridley and Burland, 1993; Guan and Fredlund, 1997a). The water used in the laboratory must first be conditioned through a pressurization process where the cavitation nuclei in the water have been collapsed.

Table 2.5 Viscosity of Water at 101.3 kPa

Temperature (°C)	Absolute (Dynamic) Viscosity, ν ($\times 10^{-3}$ N · s/m^2)
5	1.519
10	1.310
20	1.009
30	0.800
40	0.654
50	0.548

Source: Modified from Tuma, 1976.

Water can sustain high tensile stresses and this phenomenon would at first appear to be in conflict with our fundamental understanding of the behavior of water. The topic of water cavitation warrants closer examination. It is important to understand the conditions under which water can sustain a high tensile stress. The topic of cavitation is of particular interest in geotechnical engineering since it has been possible to produce suction sensors that can provide a direct measurement of relatively high soil suctions (i.e., 500 – 1500 kPa). Direct, high-suction-measuring devices take advantage of the high tensile strength associated with "conditioned" or pressurized water (Ridley and Burland, 1993; Guan and Fredlund, 1997b).

Cavitation is the creation of a new cavity or the expansion of a preexisting cavity in a liquid (Young, 1989). The cavity may commence in the form of a bubble suspended in the liquid or it may be trapped in tiny cracks along the boundary between a liquid and a solid. The bubbles may contain gases or water vapor. If the bubble contains water vapor, then reducing the pressure at a constant temperature results in an "explosive" vaporization called cavitation. If the bubble expands as a result of an increase in temperature the process is known as boiling.

Young (1989) described four ways by which bubble growth can be induced: (1) *gaseous cavitation* can occur as a result of pressure reduction or an increase in temperature, (2) *vaporous cavitation* can occur when a vapor-filled bubble is subjected to pressure reduction, (3) *degassing* can occur when gas comes out of a bubble through the process of diffusion, and (4) *boiling* can occur when there is sufficient temperature rise.

Boiling and cavitation may be initiated either from a macroscopic bubble or a microscopic bubble. If the gas or vapor bubbles are collapsed, the fluid (water) is said to no longer have any cavitation nuclei. It is estimated that water can withstand tensions of 500,000 – 1,000,000 kPa in the absence of cavitation nuclei (Harvey et al., 1947). Plesset (1969) suggested that the tensile strength of water without cavitation nuclei could be in excess of 1,000,000 kPa. The primary process for collapsing the cavitation nuclei in water has been pressurization.

The cavitation nuclei in water can be collapsed through a pressurization process; however, it must be realized that the water is not in a stable state. The water has crossed the boundary from a stable liquid state into the vapor state without changing phase. In other words, the water is in a metastable state (Apfel, 1970). It is simply a matter of time and surrounding conditions until nucleation will produce vapor cavities within the boundaries of the liquid (Trevena, 1987). Even though it has not been possible to produce a suction sensor containing water in a stable state (over a long time period), there has still been significant success in producing direct, high-suction sensors for laboratory usage through use of the pressurization procedure. Further information is provided on direct, high-suction sensors in Chapter 4.

2.3.3 Air Phase

The air phase has physical properties that vary significantly with temperature and pressure.

2.3.3.1 Density of Air

The density of air, ρ_a, can be expressed as

$$\rho_a = \frac{M_a}{V_a} \tag{2.14}$$

The specific volume of air, v_{ao}, is given as

$$v_{ao} = \frac{V_a}{M_a} \tag{2.15}$$

Air behaves as a mixture of several gases (Table 2.6) and also varying amounts of water vapor.

The mixture of gases is called dry air when no water vapor is present and moist air when water vapor is present. It is important to note that water vapor behaves as a gas and not as small liquid drops of water. Dry and moist air can be considered to behave as an ideal gas under pressures and temperatures commonly encountered in geotechnical engineering. The ideal gas law can be written as

$$\bar{u}_a V_a = \frac{M_a}{\omega_a} R T_K \tag{2.16}$$

where:

$\bar{u}_a$ = absolute air pressure where the overbar indicates absolute pressure (i.e., $\bar{u}_a = u_a + \bar{u}_{atm}$) kN/m^2 or kPa,
u_a = gauge air pressure, kN/m^2 or kPa,
$\bar{u}_{atm}$ = atmospheric pressure (i.e., 101.3 kPa or 1 atm),
V_a = volume of air, m^3,
M_a = mass of air, kg,
ω_a = molecular mass of air, kg/kmol,
R = universal (molar) gas constant [i.e., 8.31432 J/(mol K)],
T_K = absolute temperature (i.e., $T_K = T + 273.16$), K, and
T = temperature, °C.

Table 2.6 Composition of Dry Air

	Percentage by Volume	Density (kg/m^3)	Molecular Mass (Basis of Natural Scale, $O = 16$) (kg/kmol)
Nitrogen (N_2)	78.08	1.25055	28.016
Oxygen (O_2)	20.95	1.42904	32.000
Other gases	0.97		
Air	100.0	1.2929	28.966

[a]Under standard conditions (i.e., 101.3 kPa and 0°C) with no water vapor.

The right-hand side of Eq. 2.16 [i.e., $(M_a/\omega_a)RT_K$] is a constant for a gas in a closed system with constant mass and temperature. Under closed-system conditions, Eq. 2.16 can be written as Boyle's law:

$$\bar{u}_{a1} V_{a1} = \bar{u}_{a2} V_{a2} \tag{2.17}$$

where:

$\bar{u}_{a1}$ = absolute air pressure at condition 1,
V_{a_1} = volume of air (m^3) at condition 1,
$\bar{u}_{a2}$ = absolute air pressure at condition 2, and
V_{a_2} = volume of air (m^3) at condition 2.

Rearranging the ideal gas equation (i.e., Eq. 2.17) gives

$$\frac{M_a}{V_a} = \frac{\omega_a}{RT_K} \bar{u}_a \tag{2.18}$$

Substituting Eq. 2.14 into Eq. 2.18 gives an equation for the density of air:

$$\rho_a = \frac{\omega_a}{RT_K} \bar{u}_a \tag{2.19}$$

The molecular mass of air, ω_a, depends on the composition of the mixture of dry air and water vapor. Dry air has a molecular mass of 28.966 kg/kmol (Table 2.6), and the molecular mass of the water vapor (H_2O) is 18.016 kg/kmol. The composition of air, namely, nitrogen (N_2) and oxygen (O_2), is essentially constant in the atmosphere. The carbon dioxide (CO_2) content in air may vary depending on environmental conditions such as the rate of consumption of fossil fuels.

The constituent of air that can vary most is water vapor. The volumetric percentage of water vapor in air can range from as little as 0.000002% to as high as 4–5% (Harrison, 1965). The molecular mass of air is affected by the change in each constituent. Consequently, the molecular mass affects the density of air.

2.3.3.2 *Properties of Water Vapor in Air*

The concentration of water vapor in the air is commonly expressed in terms of relative humidity:

$$\text{RH} = h_r = \frac{u_v(100)}{u_{vo}} \tag{2.20}$$

where:

- RH = relative humidity, h_r = %,
- u_v = partial pressure of water vapor in the air, kPa, and
- u_{vo} = saturation pressure of water vapor at the same temperature, kPa.

Table 2.7 shows values of air density for different absolute air pressures and temperatures. The numbers in the top portion of Table 2.7 are computed for air with a relative humidity of 50% and 0.04% carbon dioxide by volume. For air having a relative humidity other than 50%, a correction should be applied as shown in the bottom portion of Table 2.7. The correction for relative humidity is small, but it should be noted that the density of air decreases as the relative humidity increases. This indicates that moist air is lighter than dry air.

There are other variables used for defining the amount of water vapor in an air mixture. The variables take the form of a mass ratio or a mass-volume ratio. Some variables bear a relationship to the partial pressure of water vapor.

Humidity ratio, mixing ratio, or *moisture content,* MR_i, is defined as the ratio of the mass of water vapor, M_w, to the mass of dry air, M_d, at a given pressure:

$$\text{MR}_i = \frac{M_w}{M_d} \tag{2.21}$$

The moisture content of air is a common measure used on psychrometric charts. The moisture content of air can also be expressed in terms of the partial pressure of water and atmospheric air pressure:

$$\text{MR}_i = \delta\left[\frac{u_v}{\bar{u}_a - u_v}\right] \tag{2.22}$$

where:

- δ = 0.62197 or the ratio of specific gas constants (i.e., water vapor to dry air),
- u_v = partial pressure of water vapor, and
- $\bar{u}_a$ = absolute atmospheric pressure of moist air.

Table 2.7 Density of Air at Different Absolute Pressures, Temperatures, and Relative Humidity

Density of Air, ρ_a(kg/m^3)

Absolute Air Pressure, $\bar{u}_a$(kPa)	10°C	20°C	30°C
80	0.982	0.946	0.910
85	1.043	1.005	0.968
90	1.105	1.065	1.025
95	1.167	1.124	1.083
100	1.228	1.184	1.140
101	1.240	1.196	1.152
105	1.290	1.243	1.198

Density Adjustments for Humidity (kg/m^3)

Relative Humidity, RH (%)	10°C	20°C
20	+0.003	+0.006
30	+0.002	+0.004
40	+0.001	+0.002
50	0	0
60	−0.001	−0.002
70	−0.002	−0.004
80	−0.003	−0.006

Source: From Kaye and Laby, 1973.

Specific humidity, SH, is defined as the ratio of the mass of water vapor, M_w, to the mass of moist air (i.e., $M_a + M_w$):

$$\text{SH} = \frac{M_w}{M_a + M_w} \tag{2.23}$$

Specific humidity, SH, is related to the moisture content of air and can be written as

$$\text{SH} = \frac{\text{MR}_i}{1 + \text{MR}_i} \tag{2.24}$$

Substituting Eq. 2.21 into Eq. 2.24 gives

$$\text{SH} = \frac{\delta}{\left(\bar{u}_a - u_v\right)/u_v + \delta} \tag{2.25}$$

Absolute humidity, AH, is defined as the mass of water vapor, M_w, in a particular volume of air, V_a:

$$\text{AH} = \frac{M_w}{V_a} \tag{2.26}$$

The most common units for absolute humidity are grams per cubic meter with values ranging from 0 to 30 g of water per cubic meter of air. Absolute humidity changes as air pressure changes.

2.3.3.3 Calculation of Saturated Vapor Pressure

Saturated vapor pressure becomes the reference for water vapor gradients. Saturated vapor pressures are a function of temperature and can be calculated using one of several empirical equations such as those presented by Tetens (1930) and Lowe (1977). The Tetens (1930) equation for saturation vapor pressure as a function of temperature is as follows:

$$u_{vo}^{\text{air}} = 0.6108 \exp\left[\frac{17.27T}{T + 237.7}\right] \tag{2.27}$$

where:

u_{vo}^{air} = saturated vapor pressure, kPa, and
T = air temperature, °C.

The Lowe (1977) equation for saturated vapor pressure as a function of temperature is as follows:

$$u_{vo}^{\text{air}} = a_0 + a_1 T + a_2 T^2 + a_3 T^3 + a_4 T^4 + a_5 T^5 \tag{2.28}$$

where:

u_{vo}^{air} = saturated vapor pressure, kPa,
T = air temperature, °C
a_0 = 0.6183580754,
a_1 = 0.0411427320,
a_2 = 0.0017217473,
a_3 = 0.0000174108,
a_4 = 0.0000003985, and
a_5 = 0.0000000022.

The saturation vapor pressures calculated by the above equations are shown in Fig. 2.33. Relative humidity changes nonlinearly with temperature. Changes in relative humidity may occur because the saturated water vapor pressure increases with temperature. Temperature generally increases toward noon each day and the relative humidity goes down.

2.3.3.4 Calculation of Relative Humidity from Dew Point

Relative humidity of an air-water mixture can be calculated if the actual air temperature T and the dew-point temperature t_d are known. A number of equations have been proposed to perform the conversion between dew point and relative humidity.

There is a simple approximation that allows the calculation of relative humidity h_r based on the dew-point temperature t_d and the dry bulb temperature T in degrees Celcius. The equation is as follows:

$$h_r = 100 - 5\left(T - t_d\right) \tag{2.29}$$

The equation is reasonably accurate as long as the relative humidity is above 50%. The equation states that the relative humidity changes by 5% for every 1°C difference in temperature between the dew point and the dry bulb temperature, starting at 100% relative humidity.

The August-Roche-Magnus equation was developed for the calculation of dew point when the specific relative humidity and dry bulb temperature are known. This equation produces results that are essentially the same as those computed using the Tetens (1930) equation.

The August-Roche-Magnus equation can be rearranged and solved for relative humidity when the dew-point temperature is known. The August-Roche-Magnus equation is considered valid under the following conditions:

$$0^\circ\text{C} < T < 60^\circ\text{C}$$
$$1\% < h_r < 100\%$$
$$0^\circ\text{C} < t_d < 50^\circ\text{C}$$

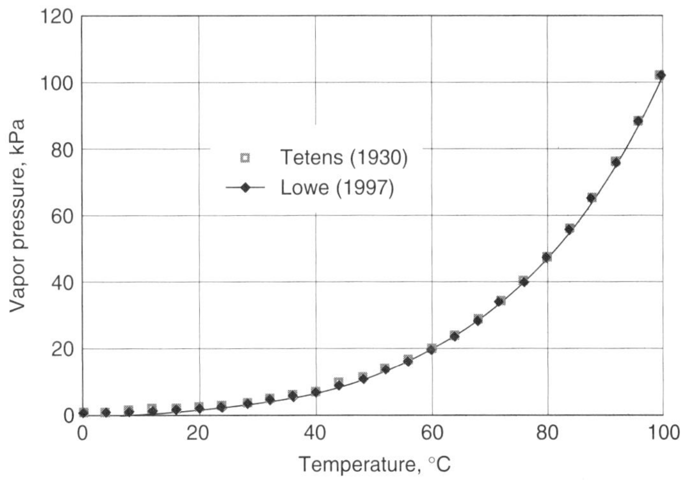

Figure 2.33 Curve of saturation vapor pressure versus temperature.

The August-Roche-Magnus equation for the calculation of dew point can be written as

$$t_d = \frac{b\left[aT/(b+T) + \ln\left({}^{h_r}/_{100}\right)\right]}{a - \left[aT/(b+T) + \ln\left({}^{h_r}/_{100}\right)\right]} \tag{2.30}$$

where:

a = 17.271 and
b = 237.7°C.

Rearranging Eq. 2.30 and solving for relative humidity h_r give

$$h_r = 100\exp\left[\frac{ab\left(t_d - T\right)}{(b+T)\left(t_d + b\right)}\right] \tag{2.31}$$

where:

h_r = relative humidity, %,
T = dry bulb temperature, °C,
t_d = dew-point temperature, °C, and
a, b = constants with values equal to those shown above.

The Bosen (1958) equation was also developed for the approximation of dew-point temperature when relative humidity and dry bulb temperature are known. The Bosen (1958) equation for the calculation of dew point is as follows:

$$t_d = (a + 0.9T)\,h_r^{0.125} - a + 0.1T \tag{2.32}$$

where:

a = 112 when temperature is measured, °C.

The Bosen (1958) equation can be rearranged such that the relative humidity is computed when the dew-point temperature and the dry bulb temperature are known:

$$h_r = 100\left[\frac{t_d - 0.1T + 112}{112 + 0.9T}\right]^8 \tag{2.33}$$

where:

T = dry bulb temperature, °C,
t_d = dew-point temperature, °C, and
h_r = relative humidity, %.

Table 2.8 contains calculated relative humidity values for typical values for dry bulb temperature and dew-point temperature. The results show that the proposed simplified equation provides reasonable estimates of the relative humidity as long as the computed values are higher than 50%. The August-Roche-Magnus equation and the Bosen equation result in similar values for the computed relative humidity throughout the entire range of temperatures. Either the August-Roche-Magnus equation or the Bosen equation can be used to compute relative humidity from dew-point temperature measurements.

2.3.3.5 *Thermal Properties of Air*

The specific heat of air, C_a, is a function of temperature and typical values are presented in Table 2.9. The thermal conductivity of air, λ_a, is also a function of temperature and typical values for λ_a are presented in Table 2.9.

Table 2.8 Comparisons of Computed Relative Humidity Values

Dry Bulb Temperature (°C)	Dew-Point Temperature (°C)	Relative Humidity, h_r		
		Simplified Equation	August-Roche-Magnus Equation	Bosen Equation
25.0	24.5	97.5	97.06	97.06
25.0	24.0	95.0	94.20	94.20
25.0	22.0	85.0	83.48	83.49
25.0	20.0	75.0	73.84	73.85
25.0	18.0	65.0	65.19	65.21
25.0	16.0	55.0	57.44	57.46
25.0	12.0	35.0	44.33	44.34
25.0	10.0	25.0	38.82	38.83
25.0	6.0	5.0	29.57	29.57
20.0	19.0	95.0	93.98	94.01
20.0	16.0	80.0	77.79	77.88
20.0	12.0	60.0	60.03	60.16
20.0	10.0	50.0	52.56	52.71
20.0	6.0	30.0	40.04	40.19

Table 2.9 Thermal Properties of Air at Various Temperatures

Temperature, T (°C)	Specific Heat Capacity, C_a (kJ/kg K)	Thermal Conductivity, λ_a (W/m K)
−50	1.005	0.0204
0	1.005	0.0243
20	1.005	0.0257
40	1.005	0.0271
60	1.009	0.0285
80	1.009	0.0299
100	1.009	0.0314

Source: www.vaisala.com.

2.3.3.6 *Viscosity of Air*

The viscosity of air at atmospheric pressure (i.e., 101.3 kPa) and different temperatures is given in Table 2.10.

Figure 2.34 shows a comparison of the absolute viscosities of air, water, and other materials at different temperatures. The viscosity of air increases as the temperature increases.

2.3.4 Air-Water Interface or Contractile Skin

The most distinctive property of the contractile skin is its ability to exert a tensile pull. It behaves as if it is an elastic membrane under tension interwoven throughout the soil structure.

Table 2.10 Viscosity of Air at 101.3 kPa

Temperature (°C)	Absolute (dynamic) Viscosity, $\nu(\times 10^{-5}\ \mathrm{N \cdot s/m^2})$
−10	1.667
0	1.705
10	1.761
20	1.785
30	1.864
40	1.909

Source: Modified from Tuma, 1976.

Most properties of the contractile skin appear to be different from those of the contiguous water phase (Davies and Rideal, 1963). For example, the density of water in the contractile skin is reduced, its heat conductance is increased, and its birefringence data (i.e., a measure of the water structure) are similar to those of ice. The transition from liquid water to the contractile skin appears to be distinct and jumpwise according to Derjaguin (1965). It is interesting to note that insects such as the "water spider" are able to walk on top of the contractile skin, while insects such as the "backswimmer" are able to walk on the bottom of the contractile skin (Milne and Milne, 1978). The water strider would sink into the water were it not for the contractile skin, whereas the backswimmer would pop out of the water if the contractile skin did not exist.

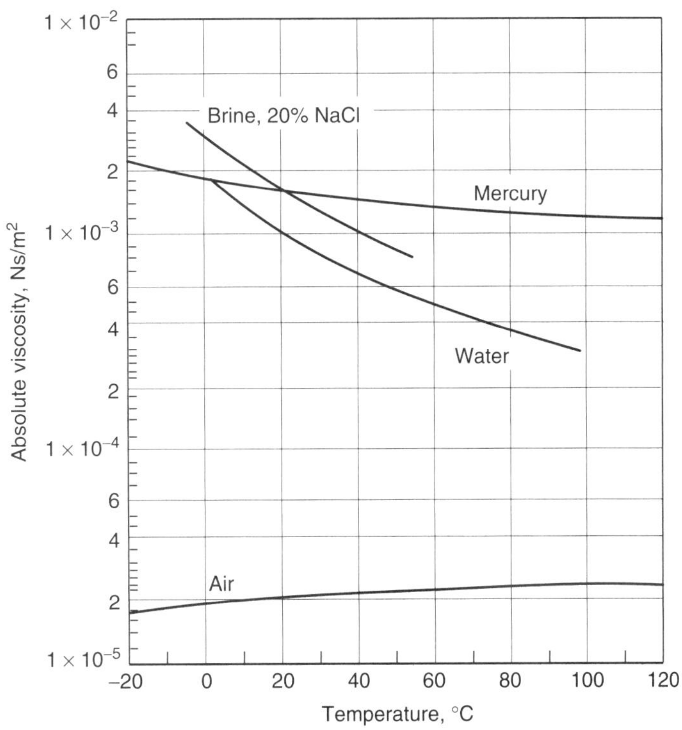

Figure 2.34 Viscosity of fluids at different temperatures (from Streeter and Wylie, 1975).

Recognizing the uniqueness of the fourth phase (i.e., contractile skin) assists in understanding the stress state variables for an unsaturated soil (Fredlund and Morgenstern, 1977). From a physical behavioral standpoint, an unsaturated soil can be visualized as a mixture with two phases that come to equilibrium under applied stress gradients (i.e., soil particles and contractile skin) and two phases that flow under applied stress gradients (i.e., air and water). From the standpoint of the volume-mass relations for an unsaturated soil, it is possible to consider the soil as a three-phase system since the volume of the contractile skin is small and its mass can be considered as part of the mass of the water.

Geotechnical engineers are familiar with a shrinkage type of experiment where a small soil specimen (i.e., initially saturated) is allowed to dry by exposure to the atmosphere. The total stresses on the specimen remain unchanged near zero while the specimen undergoes a decrease in volume. The pore-water pressure goes increasingly negative during the experiment. It is the contractile skin (or air-water interface) that acts like a thin rubber membrane that pulls the particles together, causing volume change.

2.3.5 Interaction of Air and Water

Air and water behave as both an immiscible mixture and as a miscible mixture. The *immiscible* mixture is a combination of free air and pure water without any interaction. The immiscible mixture is characterized by a separation of liquid and gas produced by the contractile skin. A *miscible* air-water mixture can have two forms. First, air dissolves in water and can occupy approximately 2% of the water by volume (Dorsey, 1940). Second, water vapor can be present in the air. Various types of air-water mixtures are discussed in the following sections. Consideration is also given to the possible "states" for water.

2.3.5.1 Solid, Liquid, and Vapor States of Water

Water can be found in one of three "states": the solid state as ice, the liquid state as water, or the gaseous state as water vapor (Fig. 2.35). The state of water depends on the pressure and temperature environment. Three lines are drawn on the water state diagram (Fig. 2.35): the vaporization curve *AB*, the fusion curve *AC*, and the sublimation curve *AD*. The vaporization curve *AB* is also called the vapor pressure curve of water. It gives combination values of temperature and pressure for which the liquid and vapor states of water can coexist in equilibrium. The fusion curve *AC* separates the solid and liquid states of water, and the sublimation curve *AD* separates the solid and the vapor states of water. The solid state can coexist in equilibrium with the liquid state along the fusion curve and with the vapor state along the sublimation curve.

The vaporization, fusion, and sublimation curves meet at point *A*. This point is called the triple point of water where the solid, liquid, and vapor states of water can coexist in equilibrium. The triple point of water is achieved at a temperature of 0°C and a (gauge) pressure of 0.61 kPa.

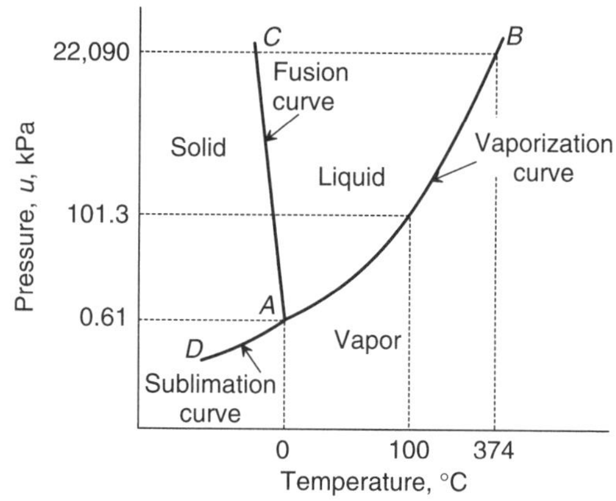

Figure 2.35 State diagram for water (not to scale; from van Haveren and Brown, 1972).

2.3.6 Water Vapor

The vaporization curve *AB* in Fig. 2.35 represents an equilibrium condition between the liquid and vapor states of water. In this equilibrium state, evaporation and condensation processes occur simultaneously at the same rate. The rate of condensation depends on the pressure in the water vapor which reaches its saturation value on the vaporization line. On the other hand, the evaporation rate depends only on temperature. Therefore, a unique relationship exists between the saturation water vapor pressure and temperature, which is referred to as the vaporization curve. Saturation water vapor pressures u_{v0} are presented in Table 2.11.

Water vapor is mixed with air in the atmosphere. The presence of the air has no effect on the behavior of the water vapor. This phenomenon is expressed by Dalton's law of partial pressures. Dalton's law states that the pressure of a mixture of gases is equal to the sum of the partial pressures at which each individual component of the gas would exert if it alone filled the entire volume. In other words, the behavior

Table 2.11 Saturation Pressures of Water Vapor at Various Temperatures

Temperature (°C)	Saturation Vapor Pressure (kPa)
0	0.6107
10	1.2276
20	2.3384
30	4.2451
40	7.3812
60	19.933
80	47.375
100	101.325
120	198.49

Source: From Kaye and Laby, 1973.

of a gas component of a mixture of gases is independent of the other gases. Therefore, the partial pressure of water vapor in the atmosphere which is in equilibrium with water is the saturation pressure given in Table 2.11. Similarly the presence of air above water does not change the equilibrium state of water (Fig. 2.35).

In nature, the water vapor in air is usually not in equilibrium with adjacent bodies of water. This means that the partial pressure of the water vapor in air, u_v, is usually not the same as the saturation pressure of the water vapor, u_{vo}, at the corresponding temperature. The water vapor in air at a given temperature is therefore said to be undersaturated, saturated, or supersaturated when the partial pressure of water vapor, u_v, is less than, equal to, or greater than the saturation water vapor pressure, respectively. The saturated condition represents equilibrium conditions between the water vapor and the water where evaporation and condensation take place at the same rate. On the other hand, the undersaturated and supersaturated states of water vapor are not at equilibrium conditions.

The supersaturated state indicates an excess of water vapor which will eventually condense. In this case, the rate of condensation exceeds the evaporation rate until the partial pressure of the water vapor, u_v, has been reduced to the saturation vapor pressure, u_{vo}. In the undersaturated state, there is a lack of water vapor relative to the equilibrium condition. Therefore, the rate of evaporation exceeds the rate of condensation until the partial pressure of the water vapor, u_v, has reached the saturation water vapor pressure, u_{vo}.

The partial pressure of the water vapor in air defines the degree to which the air is saturated with water vapor at a particular temperature. The degree of saturation with respect to water vapor is referred to as the relative humidity, RH.

2.3.7 Air Dissolving in Water

Water molecules form a lattice structure with openings referred to as "cages" that can be occupied by a gas (Rodebush and Buswell, 1958), as illustrated in Fig. 2.36. Air dissolves into the water and fills the "cages" which have a volume of approximately 2% by total volume.

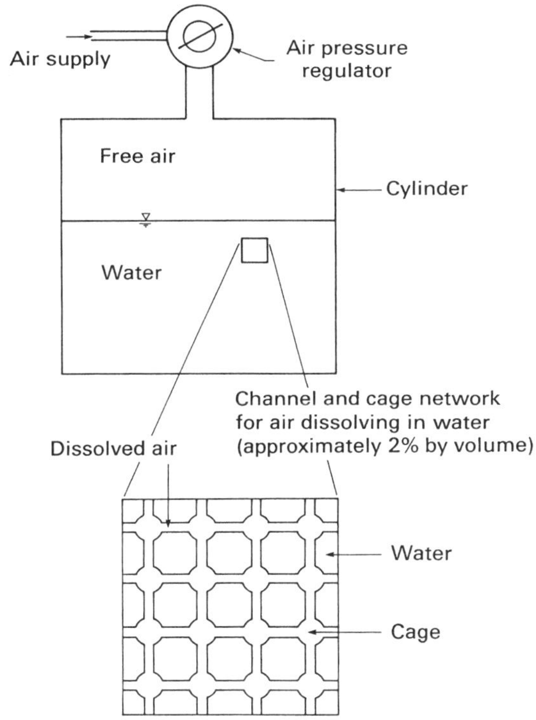

Figure 2.36 Visualization aid for understanding how air dissolves in water.

The water lattice is relatively rigid and stable (Dorsey, 1940), and the density of water changes little as a consequence of the presence of the dissolved air. An analogy using a cylinder with a piston and a porous stone is useful in visualizing the behavior of an air-water mixture. Consider a cylinder with a porous stone at its base and a frictionless piston at the top (Fig. 2.37). The porous stone has pores equaling 2% of its total volume. The porous stone is used to simulate the behavior of water. In this model, the cylinder contains free air above the porous stone.

Let us suppose there is an initial pressure applied equally to the free air and to the air in the porous stone in the cylinder. If

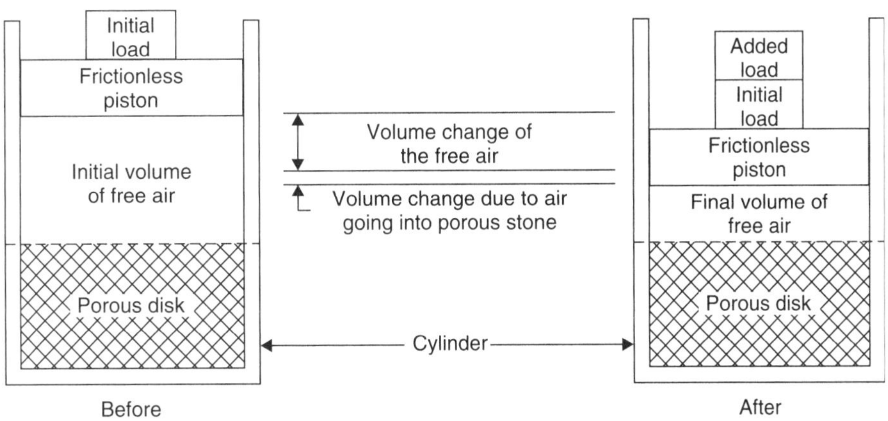

Figure 2.37 Piston and porous stone analogy for air dissolving in water.

the load on the piston is increased, the free air above the porous stone will instantaneously be compressed in accordance with Boyle's law. Then some air will move into the porous stone in accordance with Henry's law. Henry's law states that the mass of gas dissolved in a fixed quantity of liquid at a constant temperature is directly proportional to the absolute pressure of the gas above the solution (Sisler et al., 1953). The process of air flow into the porous stone will continue over time until the free air pressure is equal to the pressure of the air in the porous stone. Each time the piston load is increased, there is an immediate volume change in the free air followed by a slow process where air flows into the porous stone. Eventually, all the free air will have moved into the porous stone, and any additional applied load will be carried by the porous stone (i.e., water).

The above analogy cannot totally simulate the processes that occur in an unsaturated soil. In the presence of a solid, such as soil particles, the air and water pressures can have different magnitudes. The air and water pressures in a soil can also change at differing rates during the loading process. In the analogy, the free air and the water (i.e., porous stone) have the same pressure. Differences between the air and water pressures are later discussed and shown to be of significance to the compressibility formulation of air-water mixtures.

The mass of air going into or coming out of water is time dependent. This time-dependent process can either be ignored or taken into consideration, depending upon the engineering problem under consideration. The amount of air that can be dissolved in water is referred to as solubility, and the rate of solution is referred to as diffusivity. The volume of dissolved air in water is essentially independent of air or water pressures. This can be demonstrated using the ideal gas law and Henry's law. The ideal gas law can be rearranged and applied to a gas dissolving in water at a particular temperature and pressure:

$$V_d = \frac{M_d}{\bar{u}_a} \frac{RT}{\omega_a} \tag{2.34}$$

where:

V_d = volume of dissolved air in water,
M_d = mass of dissolved air in water, and
$\bar{u}_a$ = absolute pressure of the dissolved air.

The absolute pressure of dissolved air is equal to the absolute pressure of the free air under equilibrium conditions. Referring to the piston and porous stone analogy, an increase in the piston load will instantaneously increase the pressure in the free air, and therefore more free air will commence entering the porous stone (i.e., water). After some time, an equilibrium condition will be reached where the pressures in the free air and the dissolved air are equal. If the piston load is then increased, the process will be repeated.

The mass of dissolved air under equilibrium conditions is dependent upon the corresponding absolute air pressure as stated by Henry's law. If the temperature remains constant throughout the process, the ratio between the mass of air and the absolute pressure in the dissolved air is constant:

$$\frac{M_{d1}}{\bar{u}_{a1}} = \frac{M_{d2}}{\bar{u}_{a2}} = \text{const} \tag{2.35}$$

where:

$M_{d_1}, \bar{u}_{a1}$ = mass and absolute pressure of the dissolved air, respectively, at condition 1, and
$M_{d_2}, \bar{u}_{a2}$ = mass and absolute pressure of the dissolved air, respectively, at condition 2.

The volume of dissolved air in water, V_d, is computed from the gas law. At a constant temperature, the volume of dissolved air in water is a constant for different pressures. The ratio between the mass of each gas that can be dissolved in a liquid and the mass of the liquid is called the coefficient of solubility, H. Table 2.12 presents the coefficients

Table 2.12 Solubility of Gases in Water (Under a Pressure of 101.3 kPa)

	Coefficient of Solubility, H^a		Volumetric Coefficient of Solubility, h^b	
Temperature (°C)	Oxygen	Nitrogen, Argon, etc.	Air[c]	Air[d]
0	14.56×10^{-6}	23.87×10^{-6}	38.43×10^{-6}	0.02918
4	13.06×10^{-6}	21.59×10^{-6}	34.65×10^{-6}	0.02632
10	11.25×10^{-6}	18.82×10^{-6}	30.07×10^{-6}	0.02284
20	9.11×10^{-6}	15.51×10^{-6}	24.62×10^{-6}	0.01868
30	7.55×10^{-6}	13.10×10^{-6}	20.65×10^{-6}	0.01564

Source: From Dorsey, 1940.
[a] At standard atmospheric pressure.
[b] $h = (\rho_w/\rho_a) H$.
[c] Units of grams of air per gram of water.
[d] Units of volume of air per volume of water.

of solubility of oxygen, nitrogen, and air in water for various temperature conditions. All coefficients of solubility are referenced to an absolute standard pressure of 101.3 kPa.

The ratio of the volume of dissolved gas, V_d, in a liquid to the volume of the liquid is called the volumetric coefficient of solubility, h, which varies slightly with temperature. Values for the volumetric coefficient of solubility for air in water under various temperatures are given in Table 2.12. The volumetric coefficient of solubility of a gas in water is approximately 2%. In other words, about 2% of water can be considered to be voids that can be occupied by a gas.

2.3.8 Diffusion of Air through Water

The rate at which air can pass through water is described by Fick's law of diffusion. The rate at which mass is transferred across a unit area is equal to the product of the coefficient of diffusion, D, and the concentration gradient. When considering the diffusion of air through water, the concentration difference is equal to the difference in density between the free air external to the water and the dissolved air in the water.

Under constant-temperature conditions, the density of air is a function of the (absolute) air pressure (Eq. 2.19). An increase in pressure in the free air will develop a pressure difference between the free and dissolved air. The difference in air pressures becomes the driving potential for the free air to diffuse into (or dissolve in) water.

The gases composing air diffuse as individual components into water. The coefficients of diffusion, D, for each component of air through water are tabulated in Table 2.13. The combined gases comprising air dissolve in water at a rate of approximately 2.0×10^{-9} m^2/s (U.S. Research Council, 1933).

Barden and Sides (1967) measured the coefficient of diffusion for air through the water phase of both saturated and compacted clays and the results are presented in Table 2.14. The study concluded that the coefficient of diffusion appears to decrease with decreasing water content of the soil. The coefficient of diffusion for air through the water in a soil appears to differ by several orders of magnitude from the coefficient of diffusion for air through free water.

Table 2.13 Coefficients of Diffusion for Certain Gases in Water

Gas	Temperature (°C)	Coefficient of Diffusion, D (m^2/s)
CO_2	20	1.7×10^{-9}
N_2	22	2.0×10^{-9}
H_2	21	5.2×10^{-9}
O_2	25	2.92×10^{-9}

Source: From Kohn, 1965.

Table 2.14 Coefficient of Diffusion for Air through Different Materials

Material	Water Content, w (%)	Coefficient of Diffusion, D (m^2/s)
Free water	—	2.2×10^{-9}
Natural rubber	—	1.1×10^{-10}
Kaolin consolidated at 483 kPa	47	3.0×10^{-10}
Kaolin consolidated at 34.5 kPa	75	6.2×10^{-10}
Derwent clay (Illite) consolidated at 34.5 kPa	53	4.7×10^{-10}
Jackson clay and 4% bentonite consolidated at 34.5 kPa	39	$<1.0 \times 10^{-11}$
Compacted Westwater clay	16	1.0×10^{-11}
Saturated coarse stone	21	2.5×10^{-5}
Saturated ceramic	49	1.6×10^{-10}

Source: From Barden and Sides, 1967.

2.3.9 Surface Tension

The air-water interface (i.e., contractile skin) possesses a property called surface tension. The phenomenon of surface tension results from intermolecular forces acting on molecules in the contractile skin. These forces are different from those that act on molecules in the interior of the water (Fig. 2.38a).

A molecule in the interior of the water experiences equal forces in all directions, which means there is no unbalanced force. A water molecule within the contractile skin experiences an unbalanced force toward the interior of the water. A tensile pull is generated along the contractile skin in order for the contractile skin to be in equilibrium. The property of the contractile skin that allows it to exert a tensile pull is called its surface tension, T_s. Surface tension is measured as the tensile force per unit length of the contractile skin (i.e., units of mN/m). Surface tension is tangential to the contractile skin surface. Its magnitude decreases slightly as temperature increases. Table 2.15 gives surface tension values for the contractile skin at different temperatures.

The surface tension causes the contractile skin to behave like an elastic membrane. The behavior of the contractile skin is similar to an inflated balloon which has a greater pressure inside the balloon than outside. If a flexible two-dimensional membrane is subjected to different pressures

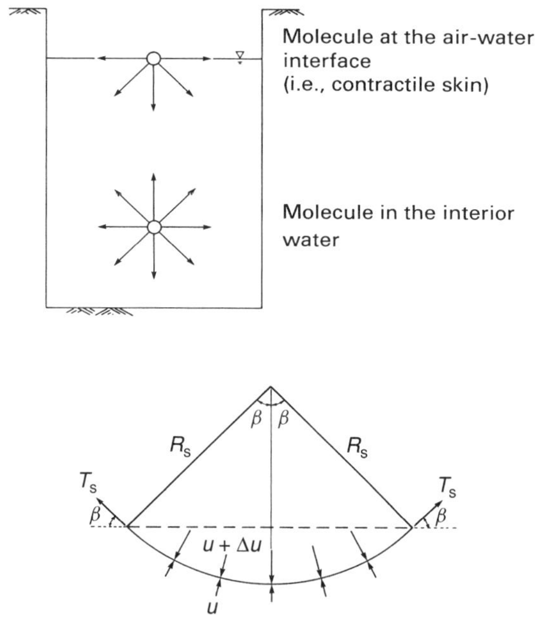

Figure 2.38 Surface tension phenomenon at air-water interface. (a) Intermolecular forces acting on contractile skin. (b) Surface tension forces associated with curved two-dimensional surface.

on each side, the membrane must assume a concave curvature toward the larger pressure and exert a tension in the membrane in order to be in equilibrium. The pressure difference across the curved surface can be related to the surface tension and the radius of curvature of the membrane (Fig. 2.38b).

The pressures acting on the membrane are u and $u + \Delta u$. The membrane has a radius of curvature R_s and a surface tension T_s. The horizontal forces along the membrane balance each other. Force equilibrium in the vertical direction requires that

$$2T_s \sin\beta = 2\Delta u R_s \sin\beta \tag{2.36}$$

Table 2.15 Surface Tension of Contractile Skin (i.e., Air-Water interface) at Various Temperatures

Temperature (°C)	Surface Tension, T_s (mN/m)
0	75.7
10	74.2
20	72.75
30	71.2
40	69.6
60	66.2
80	62.6
100	58.8

Source: From Kaye and Laby, 1973.

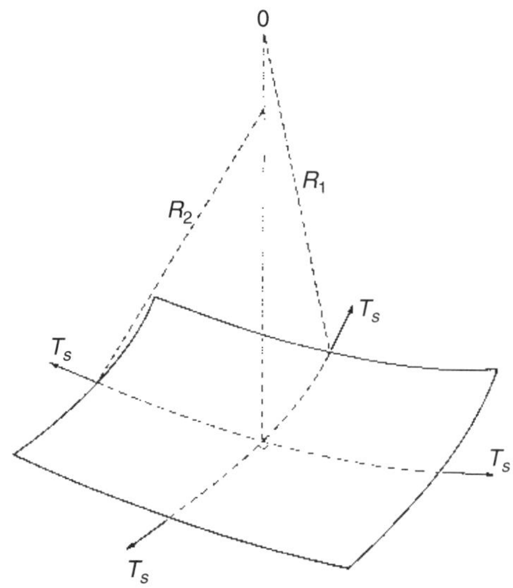

Figure 2.39 Surface tension on three-dimensional warped membrane.

where:

$2R_s \sin\beta$ = length of the membrane projected onto the horizontal plane.

Rearranging Eq. 2.36 gives

$$\Delta u = \frac{T_s}{R_s} \tag{2.37}$$

Equation 2.37 gives the pressure difference across a two-dimensional curved surface with a radius R_s and a surface tension T_s. For a warped or saddle-shaped surface (i.e., three-dimensional membrane), Eq. 2.37 can be extended using the Laplace equation:

$$\Delta u = T_s\left(\frac{1}{R_1} + \frac{1}{R_2}\right) \tag{2.38}$$

where:

R_1, R_2 = radii of curvature of a warped membrane in two orthogonal principal planes.

If the radius of curvature is the same in all directions (i.e., R_1 and R_2 are equal to R_s), Eq. 2.38 becomes

$$\Delta u = \frac{2T_s}{R_s} \tag{2.39}$$

The contractile skin in an unsaturated soil is subjected to an air pressure u_a which is greater than the water pressure u_w. The pressure difference $u_a - u_w$ is referred to as matric suction (or the capillary pressure) in a soil. The pressure difference causes the contractile skin to bend to a curvature shown by Eq. 2.40:

$$u_a - u_w = \frac{2T_s}{R_s} \tag{2.40}$$

Equation 2.40 is referred to as Kelvin's capillary model equation. The radius of curvature of the contractile skin decreases as the matric suction of a soil increases. The curved contractile skin is often called a meniscus. When the difference between the pore-air and pore-water pressure goes to zero, the radius of curvature R_s goes to infinity. Therefore, a flat air-water interface exists when the matric suction goes to zero. Interestingly, even under these conditions the surface tension property of water remains at a constant value.

2.3.10 Capillary Phenomenon

The capillary phenomenon is associated with the matric suction component of total suction. The height of water rise in a capillary tube and the radius of curvature of the air-water interface have direct implications to the water content–matric suction relationship in soils (i.e., the soil-water characteristic curve). The capillary rise is different for the wetting and drying processes in a soil due to variations in capillary pore size.

2.3.10.1 Capillary Height

Consider a small glass tube inserted into water under atmospheric conditions (Fig. 2.40). Water rises up in the tube as a result of the surface tension of the contractile skin and the tendency of water to wet the surface of the glass tube (i.e., hygroscopic properties). Capillary behavior can be analyzed by considering the surface tension T_s acting around the circumference of the meniscus. The surface tension T_s acts at an angle α_1 from the vertical wall of the capillary tube. The angle is known as the contact angle, and its magnitude depends on the adhesion between the molecules in the contractile skin and the material comprising the tube (i.e., glass).

Let us consider the vertical force equilibrium of the capillary water in the tube shown in Fig. 2.40. The vertical component of surface tension (i.e., $2\pi r T_s \cos\alpha_1$) is responsible for holding the weight of the water column, which has a height h_c (i.e., $\pi r^2 h_c \rho_w g$):

$$2\pi r T_s \cos\alpha_1 = \pi r^2 h_c \rho_w g \qquad (2.41)$$

where:

r = radius of the capillary tube,
T_s = surface tension of water,
α_1 = contact angle,
h_c = capillary height, and
g = gravitational acceleration.

Equation 2.41 can be rearranged to give the maximum height of water in the capillary tube, h_c:

$$h_c = \frac{2T_s}{\rho_w g R_s} \qquad (2.42)$$

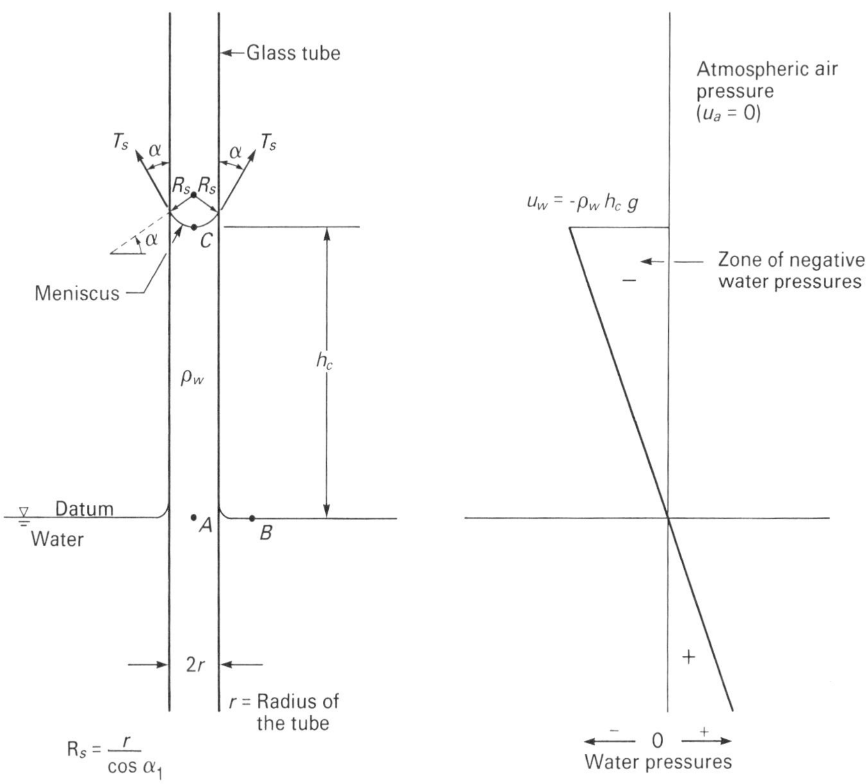

Figure 2.40 Physical model of capillarity.

where:

R_s = radius of curvature of the meniscus (i.e., $r/\cos\alpha_1$).

The contact angle between the contractile skin for pure water and clean glass is zero (i.e., $\alpha_1 = 0$). When the angle α is zero, the radius of curvature R_s is equal to the radius of the tube, r (Fig. 2.40). Therefore, the capillary height of pure water in a clean glass tube is

$$h_c = \frac{2T_s}{\rho_w g r} \tag{2.43}$$

The radius of the tube is analogous to the pore radius in soils. Equation 2.43 shows that the capillary height increases as the pore radius gets smaller. The capillary height can be plotted against the pore radius using Eq. 2.43 where the contact angle is assumed to be zero (Fig. 2.41).

2.3.10.2 *Capillary Pressure*

Points A, B, and C in the capillary system shown in Fig. 2.40 are in hydrostatic equilibrium. The water pressures at points A and B are atmospheric (i.e., the pore-water pressure at point A is equal to the pore-water pressure at point B, which is equal to 0.0). The elevation of points A and B on the water surface is considered as the datum for the system (i.e., zero elevation). As a result, the hydraulic heads (i.e., elevation head plus pressure head) at points A and B are equal to zero.

Point C is located at a height h_c from the datum (i.e., elevation head is equal to h_c). The hydrostatic equilibrium among points C, B, and A requires that the hydraulic heads at all three points be equal. In other words, the hydraulic head at point C is also equal to zero. This means that the pressure head at point C is equal to the negative value of the elevation head at point C. The water pressure at point C can be calculated as

$$u_w = -\rho_w g h_c \tag{2.44}$$

where:

u_w = water pressure.

The water pressures above point A in the capillary tube are negative, as shown in Fig. 2.40. The water in the capillary tube is said to be under tension. On the other hand, the water pressures below point A (i.e., water table) are positive due to hydrostatic pressure conditions. At point C, the air pressure is atmospheric (i.e., $u_a = 0$) and the water pressure is negative (i.e., $u_w = -\rho_w g h_c$). The matric suction $(u_a - u_w)$ at point C can then be expressed as follows:

$$u_a - u_w = \rho_w g h_c \tag{2.45}$$

Substituting Eq. 2.42 into Eq. 2.45 allows matric suction to be written in terms of surface tension:

$$u_a - u_w = \frac{2T_s}{R_s} \tag{2.46}$$

Equation 2.46 applies to the pressure difference across a contractile skin. The radius of curvature R_s can be considered analogous to the pore radius r in a soil by assuming a zero contact angle (i.e., $\alpha_1 = 0$). As the pore radius gets smaller, the matric suction in the soil gets larger, as shown in Fig. 2.41.

Surface tension has the ability to support a column of water, h_c, in a capillary tube. The surface tension associated with the contractile skin places a reaction force on the wall of the capillary tube (Fig. 2.42). The vertical component of this reaction force produces compressive stresses on the wall of the tube. In other words, the weight of the water column is transferred to the tube through the contractile skin. The contractile skin results in an increased compression on the soil structure in the capillary zone. As a result, the presence of matric suction in an unsaturated soil produces a volume decrease and an increase in the shear strength of the soil.

2.3.10.3 *Height of Capillary Rise and Radius Effects*

The effect of the height of capillary rise and the radius of curvature on capillarity is illustrated in Fig. 2.43 (Taylor, 1948). A clean capillary tube of radius r allows pure water to rise to a maximum capillary height h_c (Fig. 2.43a). However, the water rise in a capillary tube may be restricted by the limited length of the tube (Fig. 2.43b). A decrease in the capillary height results in an increase in the radius of curvature R_s, as indicated by Eq. 2.42 [i.e., $h_c = 2T_s/(\rho_w g R_s)$]. The increase in R_s for a tube with a constant radius causes an increase in the contact angle since R_s is equal to $r/\cos\alpha_1$.

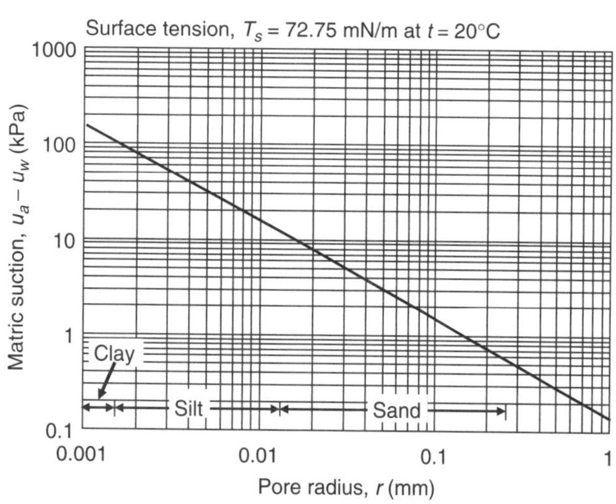

Figure 2.41 Relationship of matric suction to pore size for various soils.

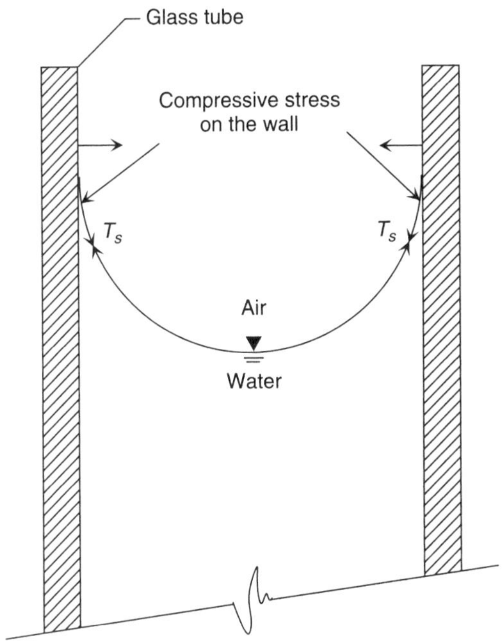

Figure 2.42 Forces acting on capillary tube.

The radius or opening of the tube is a significant factor in the development of capillary rise, as illustrated in Figs. 2.43c and 2.43d. In both cases, the tube has a bulb with a radius r_1 which is larger than the radius of the tube, r. The presence of the bulb at the midheight of the capillary height h_c prevents water from rising above the base of the bulb (Fig. 2.43c). In other words, nonuniform openings along the capillary tube can prevent the full development of capillary height. On the other hand, the capillary height h_c can be fully developed if the bulb is filled by submerging it below the water surface and then raising it above the surface (Fig. 2.43d).

The development of capillary rise in a soil is also affected by the pore size distribution in the soil, as shown in Fig. 2.43e. The water surface in the soil can rise to the capillary height h_c through continuous soil pores with radii that are smaller than or equal to r. Capillary heights greater than h_c may also develop if the height of the soil column is extended. The higher capillary rise corresponds to the pore radii that are smaller than r. However, the water surface cannot rise within the large openings at the center of the soil column (Fig. 2.43e).

The above capillary tube analogy also applies to soil conditions in situ. Nonuniform pore size distribution in a soil can result in hysteresis in the soil-water characteristic curve. The soil-water contents during the wetting and drying processes are different at particular matric suction values, as illustrated by the examples shown in Figs. 2.43c and 2.43d, respectively. The contact angle at an advancing interface during the wetting process is also different from that at a receding interface during the drying process (Bear, 1979). The above factors as well as the presence of entrapped air in the soil are considered to be the main causes for hysteresis in the drying and wetting soil-water characteristic curves.

The capillary model is conceptually simple but has some limitations when used to explain the mechanical behavior of unsaturated soils. For example, an apparent anomaly arises when the capillary model is incorporated into the formulation of pore fluid compressibility, as will be later explained. The pore radius in the capillary equation (i.e., Eq. 2.46) cannot be measured and as a result the capillary model is quite impractical in engineering practice. There are also other factors, such as the adsorptive forces between clay particles, that contribute to soils being able to sustain negative pore-water pressures greater than 1 atm.

2.3.10.4 Capillary Model and Air-Entry Value of a Soil

The air-entry value of a soil consisting of similar-sized spheres can be estimated using the capillary model (Gvirtzman and Roberts, 1991). The point of air entry into an arrangement of spheres occurs when the largest pore allows air to displace the water within the pore. The air-entry value is a function of the size of the largest opening between a set of particles.

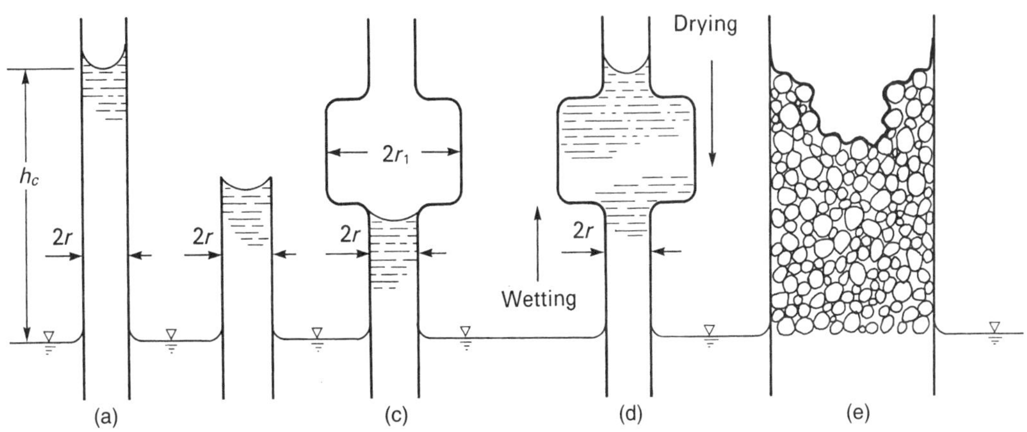

Figure 2.43 Height, radius, and shape effects on capillarity (after Taylor, 1948).

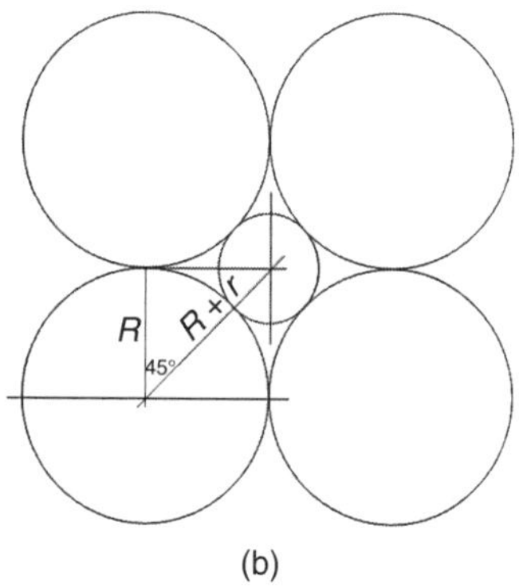

Figure 2.44 Arrangement of particles used to calculate open space between particles. (a) Cubic packing arrangement (after Graton and Fraser, 1935). (b) Geometry of a simple cubic layer.

Let us simply consider the case where a series of spheres are the same size. The largest pore in a closely packed system of uniform spheres occurs with a cubic arrangement of spheres, as shown in Fig. 2.44. Figure 2.44b shows the geometry of a simple cubic layer. The capillary theory can be used to calculate the matric suction corresponding to the air-entry value. A right-angled triangle can be formed by drawing a line from the contact point of two of the spheres to the center of the inner capillary circle, as shown in Fig. 2.44b. A geometric relationship is formed between the radius of the capillary circle, r, and the radius of the spherical particle, R:

$$r = \frac{R}{\cos 45} - R \tag{2.47}$$

where:

r = radius of the pore opening, and
R = radius of the spherical particles.

The pressure difference across a three-dimensional surface can be described using the capillary equation (i.e., Eq. 2.46), assuming the contact angle, α_1, is zero. The matric suction at the air-entry point can be calculated by substituting Eq. 2.47 into the capillary equation:

$$u_a - u_w = \frac{2T_s}{R/(\cos 45) - R} \tag{2.48}$$

Equation 2.48 provides an approximation of the relationship between the air-entry value of a soil and the radius of the largest soil particles in a soil. Equation 2.48 can be abbreviated as a ratio of the soil particle size if the surface tension of water is taken to be 0.07275 N/m:

$$u_a - u_w = \frac{0.3513}{R} \tag{2.49}$$

The radius of the particles is in millimeters and matric suction is in kilopascals for a constant of proportionality of 0.3513. The above simplified analysis involving uniform spheres forms the basis of Fig. 2.41. The diameter of the particles and the pore sizes are shown on the horizontal axis.

2.4 VOLUME-MASS VARIABLES

There are a series of volume relations, mass relations, and volume-mass relations that form useful properties in geotechnical engineering practice. The volume relations are presented first, followed by the mass relations and then the volume-mass relations.

2.4.1 Porosity

Porosity n in percent is defined as the ratio of the volume of voids, V_v, to the total volume V (Fig. 2.45):

$$n = \frac{V_v\,(100)}{V} \tag{2.50}$$

The term porosity can also be applied as the volumetric portion of each phase referenced to the total volume. Therefore, water and air porosities can be defined as follows:

$$n_w = \frac{V_w(100)}{V} \tag{2.51}$$

$$n_a = \frac{V_a(100)}{V} \tag{2.52}$$

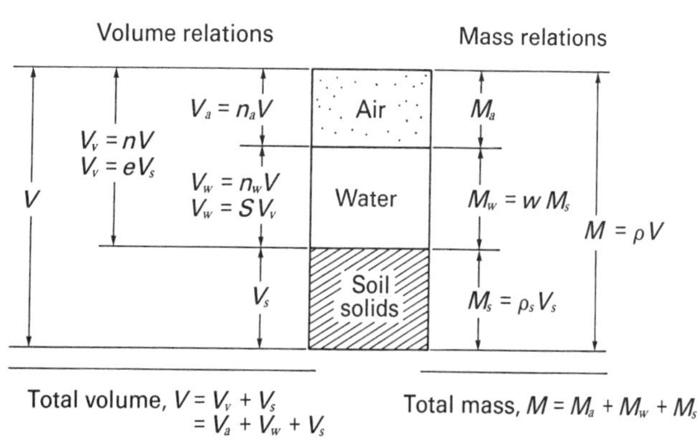

Figure 2.45 Relationship between volume and mass of unsaturated soil.

where:

n_w = water porosity, %, and
n_a = air porosity, %.

The volume associated with the contractile skin can be assumed to be negligible or part of the water phase.

The water porosity n_w, expressed in decimal form, is synonymous with the volumetric water content θ used in unsaturated soil mechanics The volumetric water content notation is used throughout this book, primarily because of its association with the SWCC. Much of the early research related to the SWCC has been conducted within the soil physics and agriculture disciplines where the amount of water in the soil has been defined in terms of volumetric water content. Consequently, it has become necessary for geotechnical engineers to become familiar with both the mass and volumetric definitions of water content. Typical values of porosity for some soils are given in Table 2.16.

2.4.2 Void Ratio

Void ratio e is defined as the ratio of the volume of voids, V_v, to the volume of soil solids, V_s:

$$e = \frac{V_v}{V_s} \tag{2.53}$$

The relationship between overall soil porosity and void ratio can be obtained by equating the volume of voids, V_v in Eq. 2.50 and Eq. 2.53:

$$n = \frac{e}{1+e} \tag{2.54}$$

Typical values for void ratio are shown in Table 2.16.

2.4.3 Degree of Saturation

The percentage of the void space which contains water is expressed as the degree of saturation S (%):

$$S = \frac{V_w(100)}{V_v} \tag{2.55}$$

Table 2.16 Typical Values of Porosity, Void Ratio, and Dry Density

Soil Type	Void Ratio, e		Porosity, n (%)		Density, ρ (kg/m^3)	
	maximum	minimum	maximum	minimum	maximum	minimum
Granular Materials: Uniform Materials						
(a) Equal spheres (theoretical values)	0.92	0.35	47.6	26.0	—	—
(b) Standard Ottawa sand	0.80	0.50	44.0	33.0	1762	1474
(c) Clean, uniform sand (fine or medium)	1.0	0.40	50.0	29.0	1890	1330
(d) Uniform, inorganic silt	1.1	0.40	52.0	29.0	1890	1281
Granular Materials: Well-Graded Materials						
(a) Silt sand	0.90	0.30	47.0	23.0	2034	1394
(b) Clean, fine to coarse sand	0.95	0.20	49.0	17.0	2210	1362
(c) Micaceous sand	1.20	0.40	55.0	29.0	1922	1217
(d) Silt sand and gravel	0.85	0.14	46.0	12.0	2239	1426
Mixed Soils						
(a) Sandy or silty clay	1.8	0.25	64.0	20.0	2162	961
(b) Skip-graded silt clay with stones or rock fragments	1.0	0.20	50.0	17.0	2243	1346
(c) Well-graded gravel, sand, silt, and clay mixture	0.70	0.13	41.0	11.0	2371	1602
Clay Soils						
(a) Clay (30–50% clay sizes)	2.4	0.50	71.0	33.0	1794	801
(b) Colloidal clay (−0.002 mm ≥ 50%)	12.0	0.60	92.0	37.0	1698	308
Organic Soils						
(a) Organic silt	3.0	0.55	75.0	35.0	1762	641
(b) Organic clay (30–50% clay sizes)	4.4	0.70	81.0	41.0	1602	481

Source: Modified from Hough, 1969.
Note: Tabulation is based on $G_s = 2.65$ for granular soils, $G_s = 2.70$ for clays, and $G_s = 2.60$ for organic soils

The degree of saturation S can be used to subdivide soils into three main groups which can be described as follows. Dry soils (i.e., $S = 0\%$) consist of soil particles and air where essentially no water is present. Saturated soils (i.e., $S = 100\%$) consist of a soil where the voids are filled with water. Unsaturated soils (i.e., $0\% < S < 100\%$) consist of a soil where both air and water occupy the void space of the soil.

An unsaturated soil can be further subdivided depending upon the continuity of the air phase and the water phase. For example, at high degrees of saturation (i.e., $\sim 85\% < S < 100\%$), the water phase may contain occluded air bubbles and as a consequence the air phase is discontinuous. Under these conditions the pore-air and pore-water combine to form a compressible fluid phase. On the other hand, it is possible at low degrees of saturation (i.e., $0\% < S <\sim 15\%$) that the water phase becomes discontinuous and liquid flow essentially ceases through the soil.

2.4.4 Volumetric Water Content

Volumetric water content θ is defined as the ratio of the volume of water, V_w, to the total volume of the soil, V:

$$\theta = \frac{V_w}{V} \tag{2.56}$$

The volumetric water content can also be expressed in terms of porosity, degree of saturation, and void ratio. This allows volumetric water content to be written as

$$\theta = \frac{SV_v}{V} \tag{2.57}$$

The above equation can be rewritten since V_v/V is equal to the porosity of the soil.

$$\theta = Sn \tag{2.58}$$

The volumetric water content equation can also be written as

$$\theta = \frac{Se}{1+e} \tag{2.59}$$

The relationship between the gravimetric water content w and the volumetric water content θ can be written using the basic volume-mass relationship (i.e., $Se = wG_s$):

$$\theta = \frac{wG_s}{1+e} \tag{2.60}$$

Recognizing that specific gravity G_s can be written as ρ_s/ρ_w (i.e., ρ_s = density of soil solids and ρ_w = density of water) and using the definition for the dry density of a soil, volumetric water content can be written in terms of dry density and the density of water:

$$\theta = w\frac{\rho_d}{\rho_w} \tag{2.61}$$

There is one further arrangement of the volume-mass properties that can be used to define the volumetric water content of a soil:

$$\theta = \frac{SwG_s}{S + wG_s} \tag{2.62}$$

Geotechnical engineers are most familiar with measuring gravimetric water content while agriculture-related disciplines have commonly used volumetric water content. Volumetric water content becomes the preferred term to use when considering many unsaturated soil problems, and it is for this reason that several relationships for converting between volumetric water content and other volume-mass properties have been presented.

Volumetric water content references the volume of water to the total volume of the soil. However, there are situations where the total volume of the soil may change during the process under consideration. In this case, care must be taken to ensure that there is consistency between the manner in which the soil properties are measured and the manner in which the soil properties are applied in an analysis.

When performing a test for the water content–soil suction relationship (i.e., SWCC), the amount of water in the soil is usually referenced to the original volume of the soil specimen. However, if the soil specimen changes volume as soil suction is increased, there would be a discrepancy between the original total volume and the instantaneous total volume. This issue is further discussed when analyzing test results associated with the measurement of the SWCC.

2.4.5 Gravimetric Water Content

Gravimetric water content w is defined as the ratio of the mass of water, M_w, to the mass of soil solids, M_s (Fig. 2.45). Gravimetric water content is generally presented as a percentage [i.e., w (%)]:

$$w = \frac{M_w(100)}{M_s} \tag{2.63}$$

Gravimetric water content has been most commonly used in geotechnical engineering. It is important to know whether reference is being made to gravimetric water content or volumetric water content when dealing with unsaturated soils. The SWCC data are often plotted using both volumetric water content and gravimetric water content. However, many geotechnical engineering formulations, particularly those associated with fluid flow, use a volumetric water content representation when computing unsaturated soil property functions or performing further process analyses. The degree of saturation designation can be the preferred representation of water in the soil when there is volume change as soil suction is increased.

2.4.6 Volume-Mass Relations

Density and unit weight variables provide information on the relationship between volume and mass designations of a soil. Saturated and completely dry conditions are special limiting cases of a wide range of possible soil conditions.

2.4.6.1 Soil Density

Two commonly used soil density definitions are total density and dry density. The total density of a soil, ρ, is the ratio of the total mass M to the total volume of the soil, V:

$$\rho = \frac{M}{V} \tag{2.64}$$

The total density is also called the bulk or moist density. The dry density of a soil, ρ_d, is defined as the ratio of the mass of the soil solids, M_s, to the total volume of the soil, V:

$$\rho_d = \frac{M_s}{V} \tag{2.65}$$

Typical minimum and maximum dry densities for various soils are presented in Table 2.16. Other soil density definitions are the saturated density and the buoyant density. The saturated density of a soil is equal to the total density of the soil for the case where the voids are filled with water (i.e., $V_a = 0$ and $S = 100\%$).

The total and dry densities of the overall soil mass can also be expressed in terms of other volume-mass properties of the soil (e.g., S, e, w, and G_s). Assuming that the mass of the air phase, M_a, is negligible, the total mass of the soil is the sum of the mass of the water, M_w, and the mass of the soil solids, M_s. The total volume of the soil, V, is given by the volume of the soil solids, V_s, and the volume of the voids, V_v. The density of a soil, ρ, can be rewritten in the following forms:

$$\rho = \frac{M_s + M_w}{V_s + V_v} \tag{2.66}$$

$$\rho = \frac{G_s \rho_w V_s + w G_s \rho_w V_s}{V_s + e V_s} \tag{2.67}$$

$$\rho = \frac{G_s(1 + w)}{1 + e} \rho_w \tag{2.68}$$

Substituting the *basic volume-mass relationship* (i.e., $Se = wG_s$) into Eq. 2.68 gives the following equation for the total density:

$$\rho = \frac{G_s + Se}{1 + e} \rho_w \tag{2.69}$$

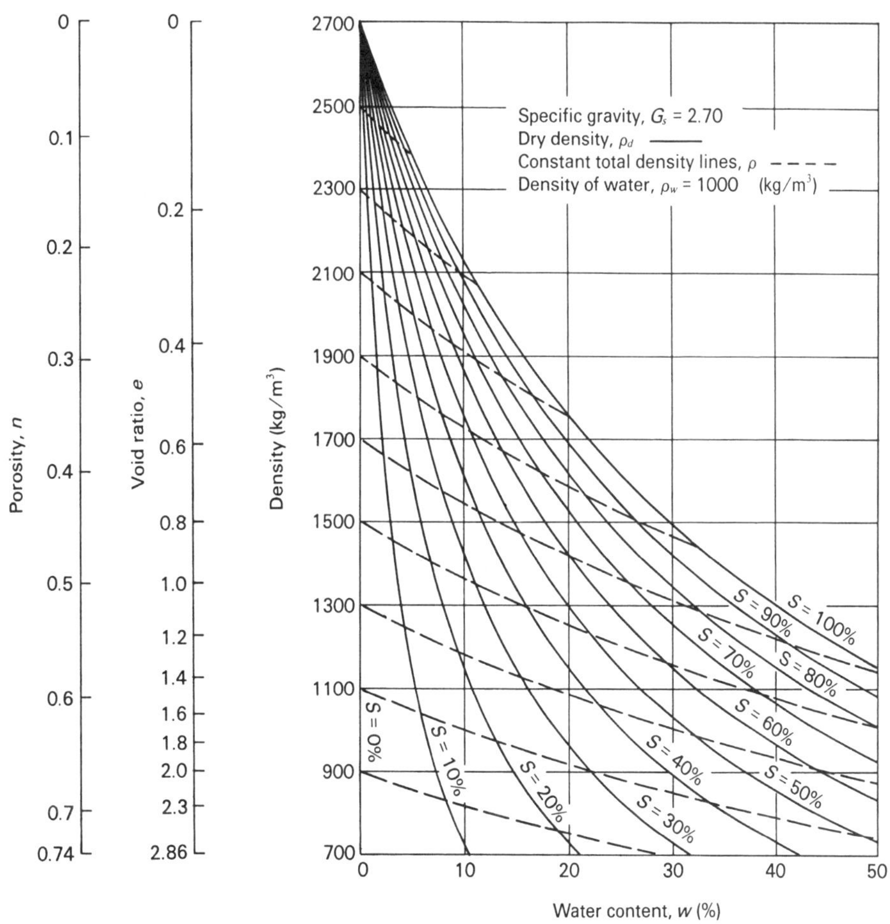

Figure 2.46 Relationship between volume-mass properties of unsaturated soil.

The dry density of a soil, ρ_d, is obtained by eliminating the mass of the water, M_w, from Eq. 2.66:

$$\rho_d = \frac{G_s}{1+e}\rho_w \tag{2.70}$$

The relationship between total density ρ and dry density ρ_d for different water contents is presented graphically in Fig. 2.46. If any two of the volume-mass properties of a soil (e.g., e, w, or S) are known, the total density of the soil, ρ, can be computed in accordance with Fig. 2.46 or Eq. 2.69. The dry density of the soil, ρ_d, is computed using Eq. 2.70 provided the void ratio e or the porosity n of the soil are known.

The dry density curve corresponding to a degree of saturation of 100% is called the "zero air voids" curve. The dry density curves for various degrees of saturation are commonly presented in connection with soil compaction data.

2.4.6.2 *Unit Weight*

The unit weight of a soil mass, γ, can be computed by multiplying the total density of a soil by gravitational acceleration g (i.e., ρg). The unit weight of a soil is often referred to as the body force involved when gravity acts on the soil. The above equations for density can be written in the following unit-weight forms:

$$\gamma = \frac{G_s(1+w)}{1+e}\gamma_w \tag{2.71}$$

$$\gamma = \frac{G_s + Se}{1+e}\gamma_w \tag{2.72}$$

When solving soil mechanics problems such as the calculation of the factor of safety for a slope stability analysis, the unit weight is used to calculate the weight of a slice of soil. Likewise, the unit weight of a soil is used when "switching on" the gravity force (or body force) of a soil. The unit weight also allows for the calculation of the total stress state in a soil continuum.

2.4.6.3 *Unit Weight as Function of Volumetric Water Content*

The unit weight of a soil changes as the amount of water in a soil changes. The amount of water in a soil can be defined in terms of gravimetric water content w, volumetric water content θ, or degree of saturation S. Unsaturated soil problems involving the calculation of the total stress state often involve switching on the gravity forces. This means that the unit weight of the soil is instantaneously applied and the corresponding total stress states are computed. It is possible to write the unit weight as a function of the amount of water for an unsaturated soil rather than designating a single value for the unit weight of the unsaturated soil. The water content may fluctuate because of imposed moisture fluxes at the ground surface or for other reasons.

Changes in total unit weight can be accommodated through incorporation of the SWCC, which relates volumetric water content θ to soil suction ψ:

$$\gamma = \frac{G_s + \theta(\psi)(1+e)}{1+e}\gamma_w \tag{2.73}$$

$$\gamma = \frac{G_s + w(\psi)\left(\rho_d/\rho_w\right)}{1+e}\gamma_w \tag{2.74}$$

If a mathematical equation is known for the SWCC (i.e., in terms of either volumetric water content or gravimetric water content), the unit weight of the soil can be written as a function of the stress state, namely, soil suction. Consequently, variations in the total unit weight due to changes in water content in the unsaturated soil zone can be taken into account. Equation 2.73 could also take volume changes into account during a process (i.e., depending on how the SWCC is defined), whereas Eq. 2.74 assumes that no volume change occurs during a process.

2.4.6.4 *"Basic" Volume-Mass Relationship*

The volume and mass for each phase can be related to one another using basic relations from the phase diagram. The mass of water in a soil, M_w, is the product of the volume and density of water:

$$M_w = \rho_w V_w \tag{2.75}$$

The volume of water, V_w, can also be computed from the volume relations shown in Fig. 2.45 (i.e., left-hand side):

$$V_w = SeV_s \tag{2.76}$$

The relationship given in Eq. 2.76 is shown in Fig. 2.45 (i.e., left-hand side). Equation 2.75 can then be rewritten as

$$M_w = \rho_w SeV_s \tag{2.77}$$

The mass of the water, M_w, can also be related to the mass of the soil solids, M_s:

$$M_w = wM_s \tag{2.78}$$

The mass of the soil solids, M_s, is also defined in the phase diagram in Fig. 2.45:

$$M_s = G_s\rho_w V_s \tag{2.79}$$

Substituting Eq. 2.79 into Eq. 2.78 yields

$$M_w = wG_s\rho_w V_s \tag{2.80}$$

Equating Eqs. 2.80 and 2.77 results in a *basic volume-mass relationship* for soil:

$$Se = wG_s \tag{2.81}$$

The basic volume-mass relationship is an equation that is commonly used to calculate phase relationship properties in an unsaturated soil.

2.4.6.5 Changes in Volume-Mass Properties

The basic volume-mass relationship (Eq. 2.81) provides a relationship among commonly used variables in geotechnical engineering (e.g., S, e, and w). A change in any one of the three volume-mass properties may produce changes in the other two properties. Changes in two of the volume-mass quantities must be determined through an independent analysis or independent measurements in order to compute the change in the third variable. If changes in the void ratio e and the water content w are known, the change in degree of saturation S can be computed. Similarly, if changes in S and e or S and w are known, then the change in w or e, respectively, can be computed.

The relationship between changes in the volume-mass properties can be derived from the basic volume-mass relationship expressed in Eq. 2.81. Consider a soil that undergoes a process such that changes occur in the volume-mass properties of the soil. The volume-mass properties of the soil have the following relationship prior to the initiation of a process:

$$S_i e_i = w_i G_s \tag{2.82}$$

where:

S_i = initial degree of saturation,
e_i = initial void ratio, and
w_i = initial water content.

The final volume-mass properties at the end of a process are also related by the basic volume-mass relationship:

$$S_f e_f = w_f G_s \tag{2.83}$$

where:

S_f = final degree of saturation,
e_f = final void ratio, and
w_f = final water content.

The relationships between initial and final conditions for each variable can be written as follows:

$$S_f = S_i + \Delta S \tag{2.84}$$

$$e_f = e_i + \Delta e \tag{2.85}$$

$$w_f = w_i + \Delta w \tag{2.86}$$

where:

ΔS = change in the degree of saturation,
Δe = change in the void ratio, and
Δw = change in the water content.

The final properties can be written as follows using the initial properties and changes in each property:

$$S_i \,\Delta e = \Delta S e_i + \Delta S\,\Delta e = \Delta w\, G_s \tag{2.87}$$

Equation 2.87 can be rearranged such that the degree of saturation ΔS can be written in terms of change in void ratio, Δe, and the change in gravimetric water content, Δw:

$$\Delta S = \frac{\Delta w\, G_s - S_i\,\Delta e}{e_f} \tag{2.88}$$

Similarly, the change in the void ratio, Δe, is obtained by substituting Eq. 2.85 into Eq. 2.87 and solving for Δe:

$$\Delta e = \frac{\Delta w\, G_s - \Delta S\, e_i}{S_f} \tag{2.89}$$

The change in water content, Δw, can similarly be written as

$$\Delta w = \frac{S_f \Delta e - \Delta S\, e_i}{G_s} \tag{2.90}$$

Each of the above three equations reveals that if changes in any two volume-mass properties are known as a result of a process that has occurred, then it is possible to compute all other volume-mass properties. These equations also reveal that it is necessary to have two independent volume-mass constitutive relationships when solving problems involving unsaturated soils.

2.4.7 Volume-Mass Relations When Pore Fluid Is Not Water

There are situations where the pore fluid may be heavier or lighter than "pure" water. Let us consider the situation where the pore fluid is miscible with water. A couple of examples where the pore fluid might differ in density from that of pure water are (i) the application of liquors to heap leach and (ii) brine solutions seeping into a freshwater aquifer. The heap leach process commonly involves the application of fluids referred to as liquor that is added to tailings. The application of the liquor produces a reaction which releases minerals from the tailings that can subsequently be recovered in the leachate. The specific gravity (and density) of the pore fluid is different than that of pure water and also changes with time as a result of minerals released from the solid phase that go into solution.

Another example involves the refining of potash which results in the accumulation of piles of salt tailings as well as brine ponds where the concentration of salts may be considerably higher than that of seawater. Laboratory testing of these materials or analyses on field situations often requires the calculation of volume-mass properties corresponding to various times in the life of the facility.

Calculations used in conjunction with a heap leach simulation are used as an example to illustrate changes in the volume-mass equations when the pore fluid has a density other than that of pure water.

2.4.7.1 Volume-Mass Relationships (Pore Fluid Other Than Water)

Figure 2.47 shows the volume and mass designations for a multiphase material that is comprised of a solid, a fluid, and a gas. The fluid could be water, an acid (e.g., liquor), or some other liquid whose density may vary from that of pure water. The gas is assumed to be air and conditions are assumed to remain isothermal. The mass variables can be written as the respective volumes multiplied by the density of the corresponding phase. The subscript f is used to refer to a fluid phase where the fluid has a density that differs from that of water (i.e., an alternate fluid).

The term "specific gravity" is used for the components of the multiphase system. The attached subscript designates the material to which reference is being made. The specific gravity of the solids is defined as

$$G_s = \frac{\rho_s}{\rho_w} \tag{2.91}$$

where:

ρ_s = density of solids or particles and
ρ_w = density of water at 4°C.

The specific gravity of a liquid other than water is; G_f. given as

$$G_f = \frac{\rho_f}{\rho_w} \tag{2.92}$$

where:

ρ_f = density of the fluid other than water.

The definition for the degree of saturation is not influenced by the fluid in the pores:

$$S = \frac{V_f}{V_v} \tag{2.93}$$

where:

V_f = volume of fluid and
V_v = volume of voids.

The calculation of void ratio is likewise unaffected by the nature of the pore fluid:

$$e = \frac{V_v}{V_s} \tag{2.94}$$

where:

V_s = volume of solids.

Gravimetric water content is defined as the ratio of the mass of water to the mass of solids when the pore fluid is water:

$$w = \frac{M_w}{M_s} \tag{2.95}$$

where:

M_w = mass of water and
M_s = mass of solids.

The gravimetric fluid content when the fluid is other than water can be defined as the ratio of the mass of alternate fluid to the mass of solids:

$$w_f = \frac{M_f}{M_s} \tag{2.96}$$

where:

M_f = mass of the alternate fluid.

2.4.7.2 Basic Volume-Mass Relationship When Pore Fluid Is Not Water

A basic volume-mass relationship can be derived by comparing volume and mass representations for the amount of fluid in a material. The mass of fluid, M_f, in a material can be written as the volume of fluid multiplied by the density of the fluid:

$$M_f = Se\rho_f V_s \tag{2.97}$$

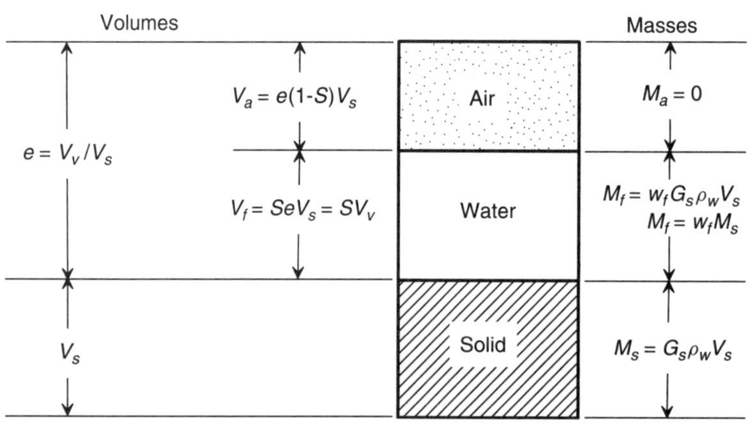

Figure 2.47 Three-phase soil model relating mass and volume portions.

The mass of fluid in a material, M_f, can also be written as

$$M_f = w_f G_s \rho_w V_s \tag{2.98}$$

Equating Eqs. 2.97 and 2.98 gives rise to a basic volume-mass relationship for a material where the pore fluid has a density other than that of pure water,

$$w_f G_s \rho_w = Se\rho_f \tag{2.99}$$

or

$$Se = \frac{w_f G_s \rho_w}{\rho_f} \tag{2.100}$$

The density of the pore fluid appears in the basic volume-mass relationship. If the pore fluid is pure water, the density of the pore fluid is 1.0 and can be omitted from the equation. If the pore fluid has a density different from that of water, the density of water and the alternate pore fluid must be retained in all calculations.

Bulk density is defined as the ratio of total mass of the materials involved to the total volume and can be written in one of several forms. The mass of each component of the multiphase system can be written in terms of the respective volumes multiplied by the density of each phase:

$$\rho = \frac{V_s \rho_s + V_f \rho_f}{V_s(1+e)} \tag{2.101}$$

The volume of the pore fluid must be converted to the pore fluid in the mass, giving rise to the following equations for density of the overall mass:

$$\rho = \frac{G_s + SeG_f}{1+e}\rho_w \tag{2.102}$$

$$\rho = \frac{\rho_s/\rho_w + Se\rho_f + \rho_w}{1+e}\rho_w \tag{2.103}$$

The unit-weight equations for a three-phase system are equal to the density equations multiplied by acceleration due to gravity, g.

2.5 SOIL COMPACTION

Compaction is the mechanical process used to increase the density of a soil through the elimination of air (i.e., densification). Soil compaction is widely used in geotechnical engineering practice for the construction of roads, dams, landfills, airfields, foundations, hydraulic barriers, and ground improvement. A laboratory compaction curve determined using a standard test procedure is used as a guide for control of the construction process in the field. The objective of laboratory testing is to determine the optimum water content that will produce the maximum density. Also of interest is an understanding of the changes in the soil properties as the compaction water content varies from optimum conditions.

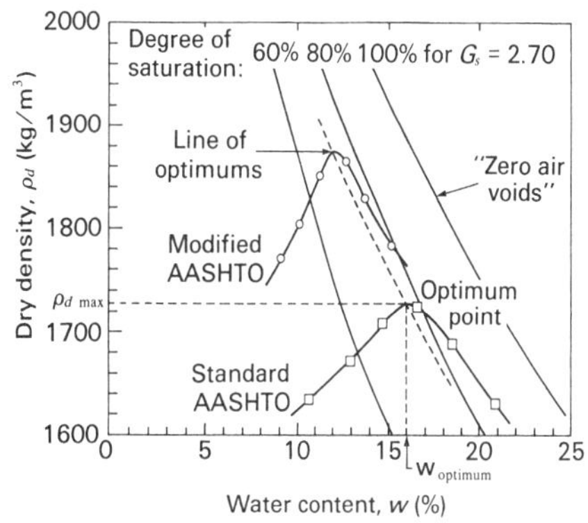

Figure 2.48 Standard and modified American Association of State Highway and Transportation Organization (AASHTO) compaction curves.

The objective of the compaction process in the field is generally to increase density, decrease soil compressibility, increase soil shearing strength, and reduce water permeability. Typical compaction curves are shown in Fig. 2.48 illustrating the relationship between compaction effort and the density of the soil. The line corresponding to a degree of saturation of 100% is called the "zero air voids" curve. The compaction curves show that different states of density are attained when the soil is compacted with the same apparent energy input but different water contents. The water content at the peak of the compaction curve is called optimum water content and represents the water content at which dry density is a maximum for a given compaction energy.

Since the pioneering work of Proctor in 1933, many researchers have attempted to qualitatively explain the fundamental mechanisms involved in the densification process. The desire has been to understand the shape of the compaction curve, particularly on the dry side of optimum water content. The compaction curve has been explained in terms of a capillarity and lubrication theory (Proctor, 1933), a viscous water theory (Hogentogler, 1936), a pore pressure theory (Hilf, 1956), a physicochemical interaction theory (Lambe, 1960b), and an effective stress theory (Olson, 1963). More recently, Barden and Sides (1970) undertook an experimental study of the relationship between the engineering performance of compacted, unsaturated clay and microscopic observations of clay structure. Lee and Suedkamp (1972) also conducted research on the shape of the compaction curve for different soils.

Various theories have been proposed to assist in understanding the compaction process in soils. In general, these theories have utilized little information related to the unsaturated soil mechanics principles. Modeling of the soil compaction curve within the context of unsaturated soil mechanics has been undertaken by Kurucuk et al. (2007). The research provides insight into the role of various unsaturated soil variables that appear to affect the shape of the compaction curve. It appears that the fundamental theories of unsaturated soil mechanics (Fredlund and Rahardjo, 1993a) can assist in understanding the compaction process and explain the inverted parabolic shape of the compaction curve.

The prediction of the soil compaction curve has been studied for static loading conditions (Kurucuk et al., 2007, 2008). Theoretically based estimations of the compaction curve have been compared to data found in the literature and found to compare favorably. Only the theoretical aspects associated with compaction are discussed in this book. The application of compaction in geotechnical engineering practice can be found in many other textbooks.

The main unsaturated soil mechanics theories utilized in formulating a theoretical framework for the compaction process are the undrained pore pressure generation theory described by Hilf (1948) and the volume change theory for unsaturated soils described by Fredlund and Morgenstern (1976). The theory of pore pressure generation under undrained conditions is further discussed in Chapter 15. The theory related to volume change is discussed in Chapters 13 and 14.

2.5.1 Pore Pressure Development During Static Compaction

The compaction process is assumed to involve the undrained loading of the soil with respect to the air and water phases. Hilf (1948) developed a relationship between changes in pore-air pressure and the applied total stress. Other assumptions made by Hilf (1948) were that loading conditions could be simulated as one-dimensional (K_0) compression, and Boyle's law and Henry's law were applicable. The derived equation became known as the Hilf (1948) equation and can be written as

$$\Delta u_a = \left[\frac{1}{1 + \dfrac{(1 - S_o + hS_o)\, n_o}{[(\bar{u}_{ao} + \Delta u_a)\, m_v]}} \right] \Delta \sigma_y \qquad (2.104)$$

where:

Δu_a = change in pore-air pressure,
S_o = initial degree of saturation,
h = coefficient of solubility (i.e., 0.02 by volume),
n_o = initial porosity,
$\bar{u}_{ao}$ = initial absolute air pressure, and
$\Delta \sigma_y$ = change in applied vertical stress.

The coefficient of volume change, m_v, was used to represent the volume change property of the soil. Overall volume change was assumed to be equal to the amount of air dissolving in the water and the compression of free air. The liquid and solid phases were considered to be incompressible. It was also assumed that the change in pore-air pressure was equal to the change in pore-water pressure, and therefore, changes in matric suction were insignificant. Experimental results show that during compaction suction changes can be large.

Following compaction the soil is unloaded. Unloading the soil forces matric suction from the wetting SWCC toward the drying SWCC (Tarantino and De Col, 2008; Tarantino and Tombolato, 2005). Hilf's (1948) assumption of suction remaining constant during compaction appears to be reasonable based on later laboratory tests performed by Kurucuk et al. (2007).

2.5.2 Computation of Volume Change and Dry Density

The volume change relationship for K_0 loading of an unsaturated soil can be written in terms of two independent stress variables as proposed by Fredlund and Morgenstern (1976):

$$\varepsilon_v = \frac{\Delta V_v}{V_o} = m_1^s \Delta(\sigma_y - u_a) + m_2^s \Delta(u_a - u_w) \qquad (2.105)$$

where:

ε_v = volumetric strain,
ΔV_v = overall volume change of soil element,
V_o = initial total volume of soil element,
m_1^s = compressibility of soil structure with respect to net applied stress $(\sigma_y - u_a)$,
m_2^s = compressibility of soil structure with respect to matric suction $(u_a - u_w)$,
$\Delta(\sigma_y - u_a)$ = change in net applied stress, and
$\Delta(u_a - u_w)$ = change in matric suction.

Deformation during undrained loading of the soil mass is assumed to be due primarily to compression of the pore-air phase and air going into solution. The volume change constitutive relationship along with Hilf's (1948) equation provides the basic physical relationship for formulating a compaction model for unsaturated soils. The challenge is to determine the appropriate soil parameters for the prediction of the compaction curve.

2.5.3 Compaction Model Assumptions

Initially the assumption was made that the compressibility coefficients of the soil remained constant during the compaction process. However, this assumption did not produce the required parabolic shape for the compaction curve, particularly on the dry side of optimum water content (Kurucuk et al., 2007). The results showed that volume change during compaction was primarily controlled by the compressibility of the soil with respect to the total stress state variable for the soil structure (i.e., m_1^s). There appeared to be some influence due to changes in matric suction, but its influence was relatively small. Superior results were obtained when the volumetric compressibility of the soil was assumed to be linear on a semilog plot (i.e., slope of the normal compression line).

2.5.4 Comparison of Predicted and Measured Compaction Curves

The performance of the compaction model has been demonstrated by comparing predicted compaction curves with the experimental results (Montanez, 2002; Kenai et al., 2006). Figures 2.49 and 2.50 show compaction curves for sand-bentonite mixtures with 5 and 15% bentonite content (by weight). The specific gravity of the mixture was 2.66. Montanez (2002) used the standard Proctor test procedure with an energy input of 637 kJ/m^3). The compaction model predictions are also shown in Figs. 2.49 and 2.50.

The curves shown as dashed lines represent static compaction curves predicted using the undrained model loading up to an external quasi-static pressure of 637 kPa. The curves shown as solid lines are for equal-energy input that was calculated by integrating the applied stress with respect to volumetric strain. The actual energy input into the soil was computed based on values applicable at optimum water content. The results show a reasonable shape for the predicted compaction curves.

Figure 2.51 shows the compaction curve for a sandy clay soil with a liquid limit of 39% and a plasticity index of 15% (Kenai et al., 2006). Experimental results correspond to a static pressure of 2100 kPa and a dynamic energy input of 3000 kJ/m^3. Both predicted compaction curves were produced using quasi-static compaction pressures of 2100 and 4000 kPa, respectively.

Compaction curves computed using undrained K_o loading conditions appear to be similar to laboratory measured results. Unsaturated soil mechanics theories appear to produce the familiar inverted parabolic shape for the compaction curve. A number of assumptions are necessary when predicting the compaction curve. The controlling soil property governing the compaction process appears to be the coefficient of compressibility with respect to net applied

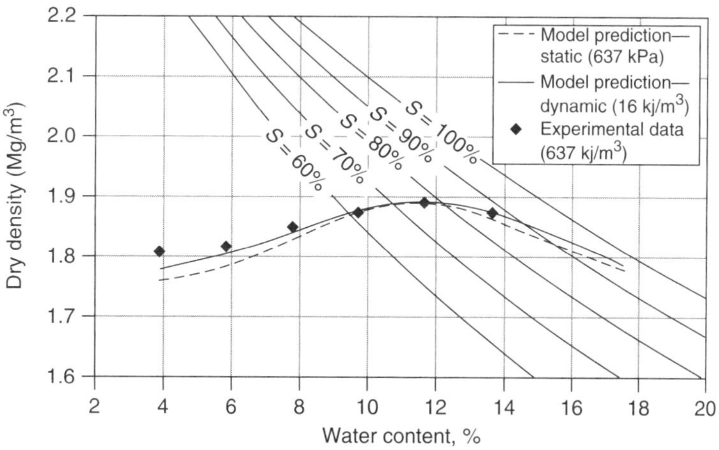

Figure 2.49 Comparison of the predicted and experimental compaction curves for well-graded sand with 5% bentonite (data from Montanez, 2002).

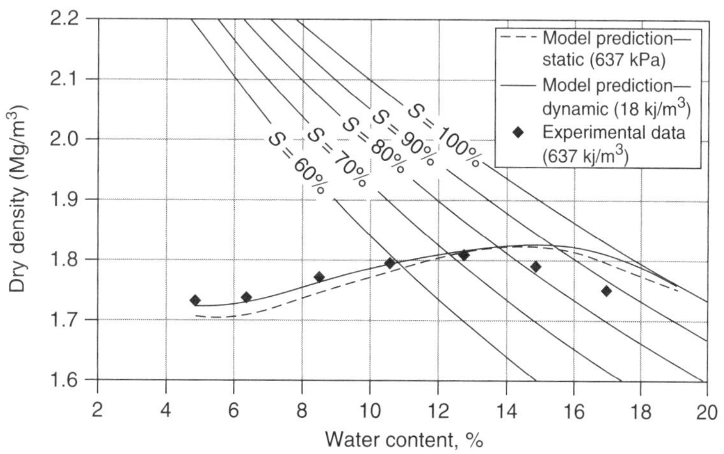

Figure 2.50 Comparison of the predicted and experimental compaction curves for well-graded sand with 15% bentonite (data from Montanez, 2002).

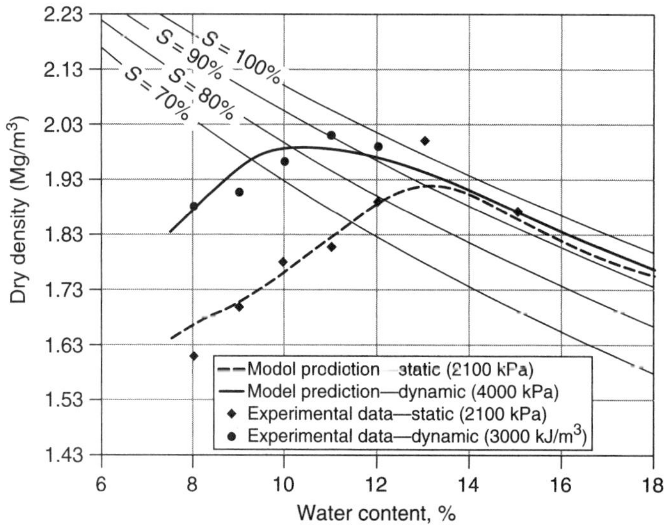

Figure 2.51 Comparison of the predicted and experimental compaction curves for clayey sand soil (data from Kenai et al., 2006).

total stress, m_1^s. Further research is necessary in order to better understand all factors influencing compaction curves.

2.6 VOLUME-MASS RELATIONS WHEN MASS IS LOST FROM SYSTEM

A special class of geotechnical engineering problems exists where mass is lost from the system while one or more processes are taking place. One such situation involves heap leach processes. Another example involves the processes associated with the management of municipal solid waste (MSW). While these two examples may appear to be quite different, there are a number of features that are similar to both processes. Both problems have features that are not commonly addressed in classical soil mechanics.

Some material properties involved in problems involving the loss of mass are different than those commonly encountered in geotechnical engineering. However, it is the change in solids content with time that makes these problems unique from most other modeling simulations common to geotechnical engineering practice. Certain material properties change with time when the solids content changes with time. There are independent processes associated with the loss of material that must be taken into consideration when formulating the mathematical equations to be solved.

2.6.1 Similarities between Heap Leach and Municipal Solid Waste Problems

Let us consider some similarities between heap leach problems and municipal solid waste problems:

1. There is a loss of mass from the system with time in both situations. This is a new feature that must be taken into consideration.
2. There is a combining (or coupling) of more than one process. In geotechnical engineering, the term "coupled" is reserved for situations where more than one type of partial differential equation is solved simultaneously.
3. The processes involved may be thermally driven, particularly in the case of MSW.
4. The principles and theories of unsaturated soil mechanics must be incorporated into the formulation of the partial differential equations that describe the physical processes. In other words, the materials involved in both situations are commonly subjected to varying degrees of saturation as the processes proceed.
5. The partial differential equations being solved are nonlinear in form, meaning that special "numerical solvers" will likely be needed to ensure that the solutions converge to the "correct" answer.
6. Consideration of air flow and heat flow may be relevant to both situations with air flow playing a key role in the MSW problem.
7. Modeling may be of the transient form, which means that starting conditions (or initial conditions) must be determined or assumed.
8. There are significant geoenvironmental issues that need to be addressed for both the MSW and heap leach problems.

2.6.2 Differences between Heap Leach and Municipal Solid Waste Problems

The differences between heap leach problems and municipal solid waste problems are as follows:

1. The generation of gas (predominantly methane) is a dominant and significant process associated with processes involved with MSW.

2. Heterogeneity of MSW can be extreme, particularly when there has not been prior sorting of the waste materials.
3. The void ratios are commonly in the range of 2–4 for MSW but are much lower for heap leach materials.
4. The specific gravity of MSW materials can vary widely. The amount of organic matter can vary widely from one site to another and the specific gravity of organics can be quite low.
5. The natural water contents can be extremely high for MSW. Special laboratory testing procedures are required in order to obtain meaningful measures of the water content in organic-rich MSW.

These are a few of the similarities and differences associated with engineering heap leach and MSW materials.

2.6.3 Continuity When Mass Is Lost

The modeling of heap leach problems and municipal solid waste problems is somewhat different from routine seepage problem simulations in that the mass of solids can be removed during the modeling period. From a continuum mechanics standpoint, there are two primary issues to ensure: (i) that the conservation of mass and energy is satisfied at all times and (ii) that the physical model accounts for the removal of solid mass from the system with time.

The conservation of mass is usually satisfied in soil mechanics through use of a volumetric requirement referenced to a referential elemental volume, REV. The REV is attached to a fixed mass of soil solids. The mass of solids remains constant during the processes under consideration.

Let us consider the case of an unsaturated material where there can be a loss of solids from the system. The conservation of mass states that the sum of the components constituting the REV must be equal to the total mass at all times. The conservation of mass can be stated as follows:

$$M_t = M_s + M_w + M_a \tag{2.106}$$

where:

M_s = mass of the solids in the REV,
M_w = mass of water,
M_a = mass of air, and
M_t = total mass of the overall REV.

When processes or changes occur with time, the sum of the changes in any component of the system must also equal the change in the overall mass of the REV. Mathematically, this requirement can be stated as

$$\frac{\Delta M_t}{M_t} = \frac{\Delta M_s}{M_t} + \frac{\Delta M_w}{M_t} + \frac{\Delta M_a}{M_t} \tag{2.107}$$

The mass of solids lost from the system is zero for most conventional soil mechanics problems, and so Eq. 2.97 reduces to the form

$$\frac{\Delta M_t}{M_t} = \frac{\Delta M_w + \Delta M_a}{M_t} \tag{2.108}$$

A "volume-based" formulation is most commonly used in soil mechanics rather than a "mass-based" formulation of the derived partial differential equations that describe each process. Consequently, it is possible to write the conservation of mass in terms of a "volumetric requirement" that must be satisfied. The volumetric requirement can be written as

$$V_t = V_s + V_w + V_a \tag{2.109}$$

During a process, the sum of the volume changes associated with any component of the system must equal the change in the overall volume of the REV. Mathematically this principle can be written as

$$\frac{\Delta V_t}{V_t} = \frac{\Delta V_s}{V_t} + \frac{\Delta V_w}{V_t} + \frac{\Delta V_a}{V_t} \tag{2.110}$$

Usually the change in volume of soil solids is zero for soil mechanics problems, and as a result Eq. 2.110 reverts to the following volumetric requirement that must be satisfied:

$$\frac{\Delta V_t}{V_t} = \frac{\Delta V_w}{V_t} + \frac{\Delta V_a}{V_t} \tag{2.111}$$

For heap leach problems and MSW problems, volume (and mass) will be lost from the system with time. Therefore, it is necessary to retain the change in volume of solids term in the volumetric requirement. It may also be more appropriate to directly satisfy the conservation of mass (i.e., Eq. 2.106).

2.6.4 Incorporation of Constitutive Relations into Physical Processes

Constitutive relations, flow laws, compressibility laws, and dissolution laws can now be used to determine how each of the components in Eq. 2.106 can be represented.

2.6.4.1 Water Flow Processes

Changes in the amount of water in a soil (i.e., an REV) at any time during a process can be determined by solving the saturated-unsaturated seepage partial differential equation, PDE. The seepage equation is solved by setting the time derivative of the volumetric water content constitutive equation equal to the spatial variation of water velocity.

The water-phase PDE is used to compute the water leaving and entering the system. Solving the PDE provides information on changes in hydraulic head, which can then be used to calculate the changes in water volume. The water-phase volumetric constitutive relationship takes the form:

$$\frac{\Delta V_w}{V_t} = m_w^1 \Delta(\sigma - u_a) + m_w^2 \Delta(u_a - u_w) \tag{2.112}$$

where:

σ = total stress,
u_a = pore-air pressure,
u_w = pore-water pressure,
m_w^1 = coefficient of water volume change associated with the net normal stress state variable, and
m_w^2 = coefficient of water volume change associated with the matric suction.

The total stress state may also be more complex than what has been suggested. The assumption can generally be made that changes in the total stress do not significantly change the volume of water in the soil for heap leach and MSW problems. Under this assumption, Eq. 2.112 reverts to the equation for the SWCC written in terms of volumetric water content:

$$\frac{\Delta V_w}{V_t} = m_w^2 \Delta(u_a - u_w) = \Delta\theta \tag{2.113}$$

2.6.4.2 Air Flow Processes

The air flow PDE can also be derived in a similar manner to the water flow PDE; however, air is also compressible and involves additional terms. The air phase constitutive relationship is used in deriving the air phase PDE.

2.6.4.3 Overall Volume Change

There must be a constitutive relationship that relates overall volume change to the stress state variables on the left-hand side of Eq. 2.111. The constitutive relationship needs to change when dealing with heap leach or municipal solid waste problems. In other words, there will be an additional volume change associated with the loss of mass (or solids) from the system. The regular constitutive relationship for an unsaturated soil takes the form

$$\frac{\Delta V_t}{V_t} = m_t^1 \Delta(\sigma - u_a) + m_t^2 \Delta(u_a - u_w) \tag{2.114}$$

where:

m_t^1 = coefficient of overall volume change associated with the net normal stress state variable and
m_t^2 = coefficient of overall volume change associated with the matric suction.

Overall volume change can be computed using a stress state–strain analysis (e.g., incremental elasticity model).

There is an additional term that needs to be added to the overall volume change constitutive relationship when dealing with problems where mass is lost from the system. Consequently, the overall constitutive relationship will take the form

$$\frac{\Delta V_t}{V_t} = m_t^1 \Delta(\sigma - u_a) + m_t^2 \Delta(u_a - u_w) + m_d \Delta M_s \tag{2.115}$$

The m_d term in Eq. 2.115 is a coefficient that designates the volume change associated with a particular amount of mass loss. The term ΔM_s designates a loss of solid mass. In other words, a change in the mass of solids will produce a certain amount of volume change. If the term m_d is zero, it means that there is no volume change even when there is a change in the mass of solids. It is necessary to consider the term m_d as an additional soil parameter that must be estimated, measured, or determined in some way.

The loss of solid mass appears in two places in the formulation: on the left-hand side of the volumetric requirement equation and in the last term on the right of the equal sign in Eq. 2.115.

2.6.5 Model to Predict Removal of Solid Mass

The prediction of solids lost from the system involves an independent physical process. There needs to be another partial differential equation (or some other form of equation) that is solved independent of the other partial differential equations. The partial differential equation must describe the chemical reactions taking place and in the process predict the solid mass lost from the system with time. It is the change in the solids volume which in turn produces a change in the overall volume in the REV:

$$dV_s = \frac{dM_s}{\rho_w G_s} \tag{2.116}$$

where:

dM_s = change in mass of solids that is predicted using an independent physical and chemical model and
dV_s = corresponding change in volume of solids.

There is not a one-to-one correspondence between the change in volume of solids and the change in overall volume. There must be a constitutive relationship to relate these two volume changes:

$$\frac{dV_{ts}}{V_t} = \frac{m_{ts} dV_s}{V_t} \tag{2.117}$$

where:

dV_{ts} = total or overall volume change produced by a change in volume of solids lost from the system and
m_{ts} = soil property linking a change in volume of solids to an overall volume change.

If $m_{ts} = 1.0$, the overall volume change bears a one-to-one correspondence to the volume of solids lost. If

$m_{ts} = 0.0$, it means there will be no overall volume change as the volume of solids is lost from the system. The real situation will likely be somewhere in between these two extremes and it may also be a nonlinear function.

Equation 2.117 can also be written in a more appropriate form but should have the same meaning as shown above:

$$\frac{\Delta V_t}{V_t} = m_t^1 \Delta(\sigma - u_a) + m_t^2 \Delta(u_a - u_w) + m_{ts} \Delta V_s \tag{2.118}$$

Both the heap leach and municipal solid waste problems have an important environmental component. It is anticipated that there will be increased public pressure to be responsible stewards of the environment. Geotechnical engineering will likely play a key role in performing advanced numerical model simulations of complex systems.

CHAPTER 3

State Variables for Unsaturated Soils

3.1 INTRODUCTION

The ability to describe the "state" of a material sets the stage for the development of a science to describe the physical behavior of the material. The term "state variables" was defined in previous chapters as the non–material variables required to describe the physical behavior of the material under consideration, in this case, an unsaturated soil. The identification of satisfactory state variables forms the basic building block for the emergence of a scientific continuum mechanics framework for unsaturated soil behavior.

The state variables associated with stress conditions play the most important role in formulating a science for material behavior. The discovery of the effective stress variable for saturated soils set the stage for the development of a science for saturated soil mechanics (Terzaghi, 1936). The identification of the stress state variables for an unsaturated soil likewise forms the necessary building blocks for the development of a science for unsaturated soil mechanics. This chapter is devoted to presenting the basis and the importance of state variables for an unsaturated soil.

There are state variables other than those related to the stress state that are required when formulating a science for unsaturated soils. For example, it is also important to have acceptable designations for deformation state variables. It should also be mentioned that temperature and time are also state variables. The primary state variables of importance when developing a science for unsaturated soil mechanics are those variables that can be used to describe volume change, distortion, and shear strength. The state variables control the behavior of the soil structure as well as other phases of the multiphase system.

The fluid phases also have independent state variables that need to be used when describing flow. For example, hydraulic head (i.e., consisting of the components of pressure head and elevation head) is the state variable that is used to describe the flow of water through a saturated or unsaturated soil. The state variable for describing air flow can be either a concentration or a pressure. In the case of a gas such as air there is a fixed relationship between air concentration and air pressure. When dealing with contaminant transport problems it is the concentration of each of the chemical components that constitute the state variables for describing chemical movement.

Water can also move in the form of a vapor. For example, water leaves the ground surface as vapor and the state variable used to describe vapor flow is vapor pressure. In other words, vapor pressure is the state variable for vapor movement.

It is a temperature gradient that causes heat flow. Therefore, temperature is the state variable that can be used when describing heat flow.

State variables for describing flow are readily apparent when considering a variety of fluids. On the other hand, the state variables that describe the mechanical behavior of multiphase materials are key identifiers for the formation of a science. It is the stress state variables for the solid phase of a multiphase system such as an unsaturated soil that have been the most difficult to determine and become universally agreed upon.

The mechanical behavior of a soil (i.e., the volume change, distortion, and shear strength) needs to be described in terms of fundamental state variables. The state of stress in a soil consists of a combination of stress state variables. By definition, these variables must be independent of the physical properties of the soil (Fung, 1965). The stress state variables need to be put into the form of a stress tensor because of the three Cartesian coordinate system generally used when solving engineering problems. Also, it is possible for there to be more than one stress state tensor for a multiphase system. It is possible for there to be two or more independent tensors containing stress state variables for a multiphase system. The number of phases involved in a multiphase system provides an indication of the number of independent stress tensors that might be required for the description of the stress state.

3.1.1 Basis for State Variables

The state of a system can be defined within the context of thermodynamic considerations. The mechanical behavior of saturated soils has historically been described mainly within an isothermal environment. Temperature is known to be a state variable, but the assumption is usually made that isothermal conditions exist. When considering unsaturated soil behavior, however, it becomes increasingly important to be aware that temperature is a state variable that plays a prominent role in some physical processes (e.g., evaporation from ground surface). Unsaturated soils are usually found near to the ground surface in an environment subjected to significant changes in thermal conditions on a daily basis.

The basis for the state variables for a material can be understood by considering the basic conservative laws of physics, namely, the conservation of energy and the conservation of mass. If the system under consideration is assumed to be in an isothermal environment, then the state variables embedded within the conservation of energy are those related to the stress state. Thermodynamic considerations are therefore reduced to the conservation of momentum. A multiphase system can be assumed to be in static equilibrium or subjected to negligible velocities with the exception of earthquake or seismic loadings. Consequently, the system reduces to the consideration of Newton's second law of statics (i.e., summation of forces and moments).

The stress state variables associated with a multiphase material can be obtained by considering Newtonian equilibrium for each phase of a multiphase system. Free-body diagrams with designated spatial variation can be drawn for each phase of a multiphase system. The superposition of coincident equilibrium stress fields for a multiphase system such as saturated or unsaturated soil systems is presented in this chapter.

Newtonian equilibrium considerations become a theoretical justification for the choice of acceptable stress state variables. There are specific requirements that must be satisfied when performing an equilibrium type of analysis. For example, the summation of forces across a wavy plane cannot be considered as an equilibrium analysis since a wavy plane does not constitute a legitimate free-body diagram. Rather, the starting free-body diagram must be multidimensional and have unbiased, smooth surfaces to be acceptable within the context of continuum mechanics (Fung, 1965).

The state variables associated with states of deformation must satisfy the conservation of mass. As such, systems of mapping the movement of each phase of a multiphase system must satisfy the conservation-of-mass requirements for the system as a whole. There are different types of elements that can be used when defining the deformation state variables (e.g., spatial element or referential element). An REV will be used in this book when describing both the stress and deformation state variables for saturated and unsaturated soil systems.

Continuum mechanics has provided a valuable context in which to study material behavior for engineering purposes. Continuum mechanics principles provide an important relationship between the conservative laws of physics and the development of a science for the behavior of a multiphase system. The conservative laws reveal the basic "building blocks" required for the development of a science. These building blocks are called state variables.

3.1.2 Background on Stress State for Saturated Soils

The effective stress variable $\sigma - u_w$ commonly used in saturated soil mechanics is a stress state variable to which saturated soil behavior can be related. The effective stress variable is applicable to sands, silts, or clays and it is independent of the soil properties. The volume change process and the shear strength characteristics of a saturated soil are both controlled by effective stress variables. The effective stress state variable can be independently applied in each of the three Cartesian coordinate directions. In so doing, effective stress takes on the form of a stress tensor (i.e., a 3×3 matrix).

Soil mechanics as a science has been successfully applied to many geotechnical problems involving saturated soils. The success of the stress state variables is largely due to the ability of engineers to uniquely relate observed soil behavior to stress conditions in the soil. Terzaghi (1936) described the stress state variables controlling the behavior of saturated soils as follows: "The stresses in any point of a section through a mass of soil can be computed from the total principal stresses, σ_1, σ_2, σ_3, which act at this point. If the voids of the soil are filled with water under a stress, u_w, the total principal stresses consist of two parts. One part, u_w, acts in the water and in the solid in every direction with equal intensity. It is called the neutral (or the pore-water) pressure. The balance, $\sigma_1' = \sigma_1 - u_w$, $\sigma_2' = \sigma_2 - u_w$, and $\sigma_3' = \sigma_3 - u_w$ represents an excess over the neutral stress, u_w, and it has its seat exclusively in the solid phase of the soil. All the measurable effects of a change in stress, such as compression, distortion, and a change in shearing resistance, are exclusively due to changes in the effective stress σ_1', σ_2', and σ_3'."

Effective stress is a stress state variable for a saturated soil. Effective stress has also been expressed in the form of an equation:

$$\sigma' = \sigma - u_w \tag{3.1}$$

where:

σ' = effective normal stress,
σ = total normal stress, and
u_w = pore-water pressure.

Effective stress is the term used to describe the general stress state that can subsequently be used to describe the physical behavior of saturated soils. Experimental evidence has shown that the effective stress variable (i.e., $\sigma - u_w$) is sufficient to describe the mechanical behavior of a saturated soil. A more complete understanding of the stress state for a saturated soil is achieved by writing the effective stress variables for each of the three orthogonal directions to form a tensor. Shear stress components arise from consideration of moment equilibrium and should also be added to the stress tensor, as shown in the equation

$$\begin{bmatrix} \sigma_x - u_w & \tau_{yx} & \tau_{zx} \\ \tau_{xy} & \sigma_y - u_w & \tau_{zy} \\ \tau_{xz} & \tau_{yz} & \sigma_z - u_w \end{bmatrix} \tag{3.2}$$

The effective stress concept provides a fundamental basis for studying saturated soil mechanics. The effective stress concept states that all mechanical behavior in a saturated soil is governed by effective stresses (and shear stresses) in each of the three Cartesian coordinate directions. Changes in volume and shear strength are controlled by changes in effective stress. An effective stress change (i.e., a change in pore-water pressure or a change in total stresses) will alter the equilibrium state of a saturated soil. Consequently, the effective stress variables qualify as stress state variables.

3.1.3 Background of Stress State for Unsaturated Soils

In 1941, Biot proposed a general theory of consolidation for an unsaturated soil with occluded air bubbles. The constitutive equations relating stress and strain were formulated in terms of two independent stress state variables, namely, effective stress $(\sigma - u_w)$ and pore-water pressure, u_w. It was recognized that there needed to be a separation between the effects of total stress changes and pore-water pressure changes when attempting to describe unsaturated soil constitutive behavior.

Coleman (1962) suggested the use of three "reduced" stress variables, namely, $\sigma_1 - u_a$, $\sigma_3 - u_a$, and $u_w - u_a$, to represent the axial, confining, and pore-water pressures, respectively, in the interpretation of triaxial test results. Constitutive relations for volume change in unsaturated soils were then formulated in terms of the above stress variables.

In 1963, Bishop and Blight reevaluated their previously proposed effective stress equation for unsaturated soils and noted that a variation in matric suction $(u_a - u_w)$ did not result in the same change in soil behavior as did a change in the net normal stress $(\sigma - u_a)$. Laboratory test results were presented using three-dimensional graphical plots with matric suction and net normal stress forming independent orthogonal axes. In other words, net normal stress and matric suction were presented as independent stress variables.

Matyas and Radhakrishna (1968) introduced the concept of "state parameters" in describing the volume change behavior of unsaturated soils. Volume change was presented as a three-dimensional graphical surface with respect to two state parameters, namely, $\sigma - u_a$ and $u_a - u_w$. Changes in degree of saturation were likewise plotted versus the matric suction and net normal stress indicating that two independent constitutive relations were required for the complete characterization of volume-mass behavior. Barden et al. (1969a) also suggested that volume change in unsaturated soils be analyzed in terms of separate components of net applied stress $(\sigma - u_a)$ and matric suction $(u_a - u_w)$.

In the 1970s (Fredlund, 1973; Fredlund and Morgenstern, 1977), a theoretical equilibrium analysis was formulated for an unsaturated soil element using concepts consistent with multiphase continuum mechanics. An unsaturated soil had generally been viewed as a three-phase system; however, it was shown that the contractile skin (i.e., the air-water interface) should be introduced as a fourth and independent phase when studying the equilibrium conditions for each phase. The equilibrium analysis on an unsaturated soil element provided justification for the use of independent stress state variables for an unsaturated soil. The soil particles were assumed to be incompressible and the soil was treated as being chemically inert. These assumptions have been historically applied in saturated soil mechanics.

The analysis concluded that any two of three possible stress state variables can be used to describe the stress state of an unsaturated soil. The three possible combinations which can be justified as stress state variables for an unsaturated soil are (1) $\sigma - u_a$ and $u_a - u_w$, (2) $\sigma - u_w$ and $u_a - u_w$, and (3) $\sigma - u_a$ and $\sigma - u_w$. Out of the three possible combinations of stress state variables that can be justified, it is the $\sigma - u_a$ and $u_a - u_w$ combination that received the widest acceptance in formulating unsaturated soil mechanics problems.

The stress state variables for an unsaturated soil take on the form of two independent stress tensors when considering a three-dimensional Cartesian coordinate system. The proposed stress state variables for unsaturated soils were experimentally tested by Fredlund (1973a) and subsequently used to formulate constitutive equations to describe shear strength behavior and volume change behavior.

Stress tensors that contain stress state variables form the basis for developing a science for both saturated and unsaturated soils. It is possible to write first, second, and third stress invariants for each stress tensor. While it is not imperative that the stress invariants be used in developing constitutive models, the stress invariants should be given consideration because all three Cartesian coordinates are independently taken into consideration.

In summary, it is the two independent stress tensors containing stress state variables [e.g., net normal stress $(\sigma - u_a)$, matric suction $(u_a - u_w)$, and shear stresses] that form a fundamental basis for the development of a science for unsaturated soil mechanics. Constitutive relationships connecting various state variables can then be used in conjunction with soil properties (and soil property functions) to solve practical engineering problems. All proposed constitutive relationships must be tested for uniqueness in the laboratory on a variety of soil types. The laboratory equipment must be able to independently control each stress component of the stress state variables.

3.1.3.1 What about Effective Stress Equations?

There have been numerous equations proposed that relate one or more of the stress variables to other stress variables through the inclusion of a soil property. It is important to differentiate between the role of these equations and the description of the stress state at a point in a continuum. There are theoretical and formulation limitations associated with the use of effective stress equations. Little attention is given to these equations in this book; however, a brief summary is given of effective stress equations that have been proposed.

The oldest and most-often-referred-to single-valued effective stress relationship is that proposed by Bishop (1959). The equation is commonly referred to as Bishop's effective stress equation for unsaturated soils and has the form

$$\sigma' = (\sigma - u_a) + \chi (u_a - u_w) \qquad (3.3)$$

where:

σ' = effective stress and
χ = a soil parameter related to degree of saturation and ranging from 0 to 1.

Bishop's equation relates net normal stress to matric suction through the incorporation of a single-valued soil property, χ. Bishop's equation should not be referred to as a fundamental description of stress state for an unsaturated soil. The equation contains a soil property and should be referred to as a constitutive equation. Within the context of continuum mechanics it is not proper to elevate the Bishop equation to the status of a stress state variable for an unsaturated soil.

Morgenstern (1979) explained the limitations of Bishop's effective stress equation as follows: Bishop's effective stress equation "proved to have little impact on practice. The parameter, χ, when determined for volume change behavior was found to differ when determined for shear strength. While originally thought to be a function of degree of saturation and hence bounded by 0 and 1, experiments were conducted in which χ was found to go beyond these bounds. The effective stress is a stress variable and hence related to equilibrium considerations alone."

Morgenstern (1979) went on to explain: Bishop's effective stress equation "contains the parameter, χ, that bears on constitutive behavior. This parameter is found by assuming that the behavior of a soil can be expressed uniquely in terms of a single effective stress variable and by matching unsaturated soil behavior with saturated soil behavior in order to calculate χ. Normally, we link equilibrium considerations to deformations through constitutive behavior and do not introduce constitutive behavior into the stress state."

Other equation forms similar to the Bishop (1959) equation have been proposed by several researchers for the development of unsaturated soil behavior models (e.g., Jommi, 2000).

$$\sigma^*_{ij} = \sigma_{ij} - (Su_w + (1 - S)u_a)\delta_{ij} \qquad (3.4)$$

where:

σ_{ij} = total stress tensor,
δ_{ij} = Kroneker delta or substitution tensor,
σ^*_{ij} = Bishop's average soil skeleton stress, and
S = degree of saturation.

The degree of saturation has been substituted for the χ soil parameter in Eq. 3.4. The equation is once again empirical and constitutive in character. Consequently, the equation must face the rigor of tests for "uniqueness" in order to determine whether it proves satisfactory for geotechnical engineering practice. It is the philosophical and theoretical justifications that present the greatest hurdles with regard to acceptance of Eqs. 3.3 and 3.4.

It can be reasoned that the need for an effective stress equation for unsaturated soils has been replaced through the use of independent stress tensors containing stress state variables. Reexamination of proposed effective stress equations has led many researchers to conclude that independent stress state variables (e.g., $\sigma - u_a$ and $u_a - u_w$) provide the increased flexibility needed for describing unsaturated soil behavior. However, there continues to be ongoing attempts to form a simple, so-called effective stress relationship between total stresses and soil suction. The simplified stress component relationship is then used to develop shear strength and volume change constitutive relations. This approach places serious constraints on subsequent formulations and violates the basic assumption inherent to classical continuum mechanics.

3.1.4 Designation of Deformation State Variables

Deformation state variables are necessary for describing relative volume changes and distortions of the various phases comprising an unsaturated soil. Deformation state variables may take on a variety of forms but must always satisfy continuity, compatibility, and conservation of mass requirements for a multiphase system (i.e., Fredlund, 1973a).

The mapping of deformation state changes has historically started with the definition of selected volume-mass soil properties such as void ratio e, gravimetric water content w, and degree of saturation S. These variables are related to one another through the following basic volume-mass relationship:

$$Se = wG_s \qquad (3.5)$$

where:

G_s = specific gravity of the soil solids.

The basic volume-mass relationship shows that it is necessary to have at least two independent constitutive relations in order to predict phase deformation state changes for an unsaturated soil. Changes in void ratio (i.e., distortion and volume changes) are related to directional changes in the soil structure (i.e., arrangement of the soil solids), forming an REV. Volume changes and distortions can be written in the form of a strain tensor in a manner consistent with continuum mechanics (or theory of elasticity or plasticity):

$$\begin{bmatrix} \varepsilon_{xx} & \gamma_{yx} & \gamma_{zx} \\ \gamma_{xy} & \varepsilon_{yy} & \gamma_{zy} \\ \gamma_{xz} & \gamma_{yz} & \varepsilon_{zz} \end{bmatrix} \tag{3.6}$$

where:

$\varepsilon_{xx}, \varepsilon_{yy}, \varepsilon_{zz}$ = longitudinal strain in the x-, y-, and z-directions, respectively, and
$\gamma_{yx}, \gamma_{zx}, \gamma_{xy}, \gamma_{zy}, \gamma_{xz}, \gamma_{yz}$ = shear strains on the x-, y-, and z-planes.

The trace of the strain tensor (i.e., $\varepsilon_{xx} + \varepsilon_{yy} + \varepsilon_{zz}$) yields volumetric strain ε_v:

$$\varepsilon_v = \varepsilon_{\mathrm{xx}} + \varepsilon_{\mathrm{yy}} + \varepsilon_{\mathrm{zz}} \tag{3.7}$$

The amount of air and water in the same REV can be described on a volumetric basis, θ_a (volumetric air content) and θ (volumetric water content). The overall volume change ε_v of a REV must equal changes in the volumes representing air and water:

$$\varepsilon_v = d\theta + d\theta_a \tag{3.8}$$

Equation 3.8 can be referred to as a volumetric requirement for a referential element. Volume changes associated with the air and water phases refer to changes in the volume of the referential element that is represented by air and the volume that is represented by water. In other words, when reference is made to a change in water volume, $d\theta$, in the REV, reference is being made to a change in the portion of the REV that is comprised of water. The same logic is true for the air phase. Even though the air phase is highly compressible, a change in the volumetric air content, $d\theta_a$, simply refers to a change in the portion of the REV that is comprised of air. A change in the volume of air in the REV, for example, could be the result of overall volume changes in the REV or the result of flow of either water or air from the element.

The overall volume change of a REV can be due to strains in the x-, y-, and z-directions; however, the terms $d\theta$ and $d\theta_a$ should not be considered as strains. It is for this reason that the term *deformation state variables* is used rather than *strain state variables* when referring to the water and air phases.

3.1.5 Constitutive Relations Stage

Once the stress state variables and the deformation state variables are defined for a multiphase system (e.g., an unsaturated soil), the stage is set for proposing constitutive relationships that can be used to represent changes in independent state variables. Constitutive relations require the incorporation of soil properties. The types of geotechnical engineering problems that are addressed will dictate the number of required constitutive relations. The constitutive relations stage of development of a science brings in the need to test the "uniqueness" of any proposed constitutive relation on a variety of soil types. In other words, there is no need to test for uniqueness at the state variable stage; rather, tests for uniqueness are a characteristic associated with the constitutive stage of the development of a science.

Constitutive relations are largely empirical in nature. Constitutive relations can also be thought of as the stage in the development of a science where the geotechnical engineer becomes involved in establishing workable empirical laws for solving practical engineering problems.

A variety of empirical constitutive models can be proposed for the relationship between the stress and deformation state variables. In each case assumptions need to be made with respect to anticipated soil behavior and limitations in the type of soil behavior being represented. Volume-mass models can range from those consistent with historical soil mechanics properties (e.g., coefficient of compressibility and coefficient of volume change) to incremental elasticity models (i.e., Young's modulus and Poisson's ratio) (Fredlund and Rahardjo, 1993a). There can also be elastoplastic constitutive models that use other soil properties (e.g., λ and κ) to represent initial compression and rebound-reloading compression, respectively, on a semilog scale (Alonso et al, 1990; Wheeler and Sivakumar 1995, 2003; Blatz and Graham, 2003). A complete critical state model must also embrace shear strength behavior.

Shear strength models may take the form of an extension of a Mohr-Coulomb representation (Fredlund et al., 1978) or the representation of shear strength at critical state conditions (Wheeler and Sivakumar, 1995; Sheng et al., 2009).

3.2 BASIS FOR STRESS STATE VARIABLES

Let us further examine the justification for the proposed stress state variables. There does not appear to be any theory which can be used to prove that a particular set of stress state variables must be used as the basis for formulating a science for unsaturated soils. The selection of the most suitable stress state variables should be judged based on the following criteria: (1) variables that can be experimentally tested in the laboratory, (2) variables that can be theoretically justified using equilibrium considerations, (3) variables where the component stresses can be measured in engineering practice, and (4) variables that meet the requirements set out by the definition of state variables in continuum mechanics.

It is possible that the above conditions might be met by more than one set of stress state variables. The final selection of state variables may come back to personal preference and in the end constitute a philosophical preference. In any case, it is not prudent to incorporate soil properties into the selection of stress state variables. Developments within continuum mechanics have, in general, adhered to this condition and the framework of continuum mechanics has proven to be an all-encompassing and unifying theory embracing the behavior for many types of materials.

3.2.1 Experimental Evidence for Stress State Variables

Numerous laboratory experiments have been performed which assist in understanding the stress state description of unsaturated soils. The laboratory tests are not a test for "uniqueness" but rather are a test based on deviations from equilibrium conditions. Stress state variables can only be "tested" for equilibrium considerations.

A key experiment on unsaturated soils was performed by Bishop and Donald (1961) (Fig. 3.1). In this experiment, a specimen of loose Braehead silt was subjected to drained triaxial compression where there was independent control of σ_3 (i.e., the all-round confining pressure), u_a, and u_w. The pressure components were changed during the experiment with the restriction that changes in net confining pressure [i.e., $\Delta\left(\sigma_3 - u_a\right)$] remained at zero and changes in matric suction [i.e., $\Delta\left(u_a - u_w\right)$] remained at zero while the individual components of stress were changed. Deviatoric stresses and volumetric strain were monitored and found to vary monotonically with axial strain, as shown in Fig. 3.1. The results can be interpreted as confirming that each of the stress state variables (i.e., $\sigma_3 - u_a$, $u_a - u_w$, and $\sigma_3 - u_w$), was not altered and therefore the unsaturated soil behavior remained monotonic. It is interesting to note that the experimental results can also be used to lend support to the Bishop (1959) type of effective stress equation in this case because the material soil property linking the state variables remained constant.

The experimental results support the validity of using independent stress state variables but do not allow exclusive interpretation of the data in favor of a particular set of stress state variables. It is also possible to suggest that any one of three possible combinations of the stress state variables is experimentally justifiable.

In 1956, Hilf proposed the axis translation testing technique along with test results on several soils. The axis translation technique made it possible to measure matric suctions greater than 1 atm. Unsaturated soil specimens were placed onto a high-air-entry disk and the negative pore-water pressure was measured (i.e., provided the matric suction was less than 1 atm). Results in Fig. 3.2 show that when the air pressure above the soil specimen was changed by a particular amount, the measured pore-water pressure reaction was always equal to the air pressure change. This

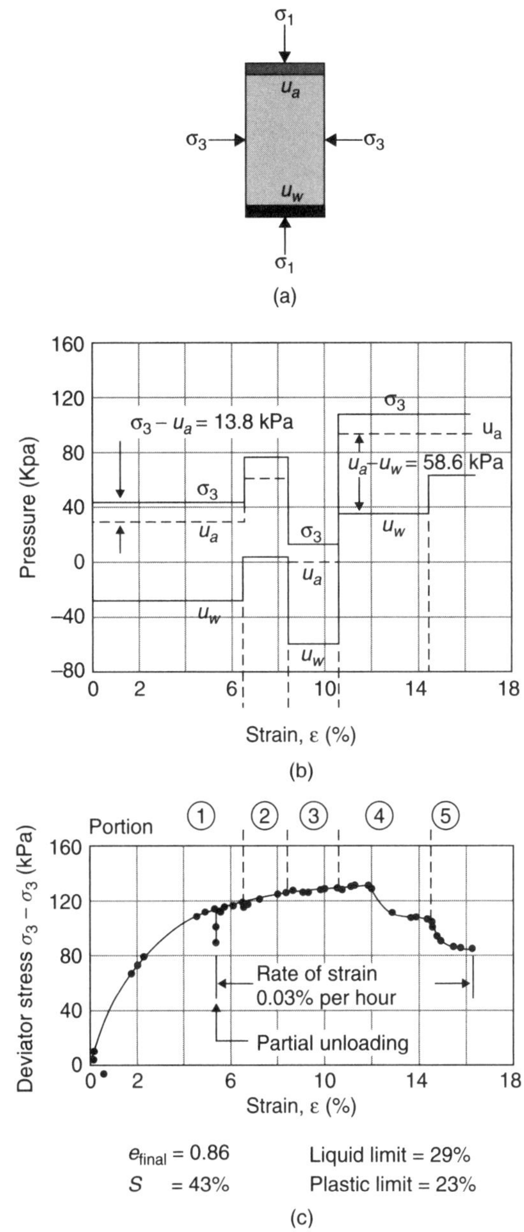

Figure 3.1 (a) Drained triaxial test on unsaturated loose silt with σ_3, u_a, and u_w, varied while keeping $\sigma_3 - u_a$ and $u_a - u_w$ constant. (b) Pressure changes versus strain. (c) Deviator stress versus strain. (Data from Bishop and Donald, 1961.)

meant that the measured matric suction remained at a constant value.

Examination of the axis translation procedure reveals that the test constituted a special "null" type of test under which the stress state remains constant. It should be noted that in this test the total pressure is equal to the applied air pressure [i.e., $\Delta\left(\sigma - u_a\right) = 0$]. Therefore, the stress variables $\sigma - u_a$ and $u_a - u_w$, remain constant throughout the axis translation test. As a consequence, the axis translation type

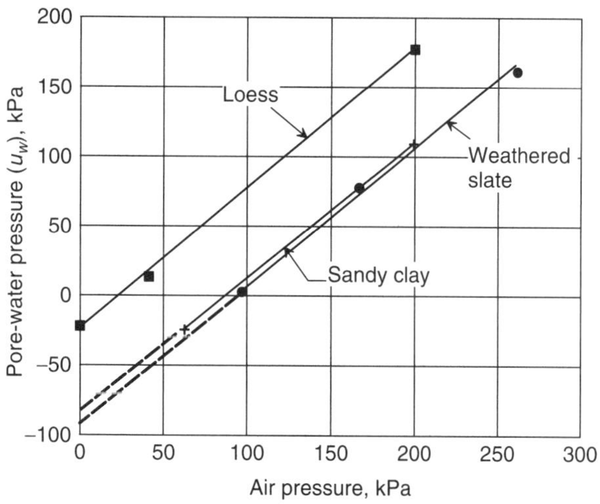

Figure 3.2 Determination of matric suction using axis translation technique (after Hilf, 1956).

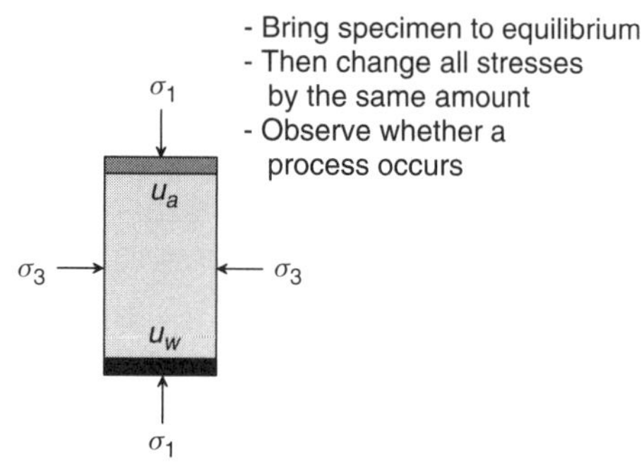

Figure 3.3 Null-type triaxial test for verification of stress state variables.

of test can be successfully used to measure the stress state of an unsaturated soil.

In 1977, Fredlund and Morgenstern reported a series of null-type tests on compacted and natural clays. The null tests subjected an unsaturated soil specimen to equal-magnitude changes in total pressure, pore-air pressure, and pore-water pressure while the overall volume and water volume of the specimens were monitored (Fig. 3.3). The stress components were varied by equal magnitudes but the stress state variables [i.e., $(\sigma - u_a)$, $(u_a - u_w)(\sigma - u_w)$] were kept constant:

$$\Delta\sigma_x = \Delta\sigma_y = \Delta\sigma_z = \Delta u_w = \Delta u_a \tag{3.9}$$

The stress components were either increased or decreased by an equal amount while overall volume changes and water volume changes were monitored. If the proposed stress state variables are valid, there should not be any change in the overall volume of the soil specimen, and the degree of saturation of the soil specimen should remain constant throughout the null test. In other words, positive results from the null test should show zero overall volume change and zero water volume change. The null tests were interpreted in view of the following criteria justifying independent stress state variables.

A suitable set of independent stress state variables are those variables that produce no distortion or volume change of an element when the individual components of the stress state variables are modified but the stress state variables themselves are maintained constant (Fredlund, 1973a). Thus the proposed stress state variable should maintain equilibrium conditions. The equilibrium conditions should apply to the overall soil structure and the contractile skin: the two solid phases of an unsaturated soil. Equilibrium relative to the contractile skin is maintained if there is no change in the degree of saturation when the null-type test is performed. In other words, it is the contractile skin which constitutes an equilibrium boundary between the air and water phases.

A total of 19 null tests were performed on compacted kaolin (Fredlund, 1973a). The soil specimens were compacted according to the standard AASHTO procedure. Two types of equipment were used in performing the null tests. The first apparatus was a modified one-dimensional enclosed oedometer. The second apparatus used isotropic loading in a modified triaxial cell. The axis translation technique was used in both cases.

The measured volume changes of the overall soil specimen and the water inflow or outflow were measured during each test. The results from one test are presented in Fig. 3.4. The results show essentially no volume change in the overall specimen and little water flow during the null tests. The stress state variables are therefore "tested" in the sense that equilibrium conditions were not changed for the unsaturated soil specimen while the stress components were changed by equal amounts. Consequently, the suggested stress state variables qualify for describing the mechanical behavior of unsaturated soils.

The measured overall volume changes and water volume changes were approximately 2% of the volume changes that would have occurred if only one stress component had been changed. These tests demonstrate that it is the differences between the stress components that form stress state variables that can be used to describe the behavior of an unsaturated soil. Stated another way, it is the difference between the stress components that forms stress state variables that maintain equilibrium conditions in the unsaturated soil specimen.

It should also be noted that it appears to be impossible to experimentally verify that there is an exclusive combination of stress variables for the description of the stress state. Other shear strength and volume change laboratory studies on unsaturated soils have also supported the use of independent stress state variables (Escario, 1980; Gulhati and Satija, 1981).

3.2.2 Theoretical Evidence of Stress State Variables

A multiphase system can be subdivided into a series of component phases where the sum of the parts must constitute the combined or overall system. A total stress field applies to

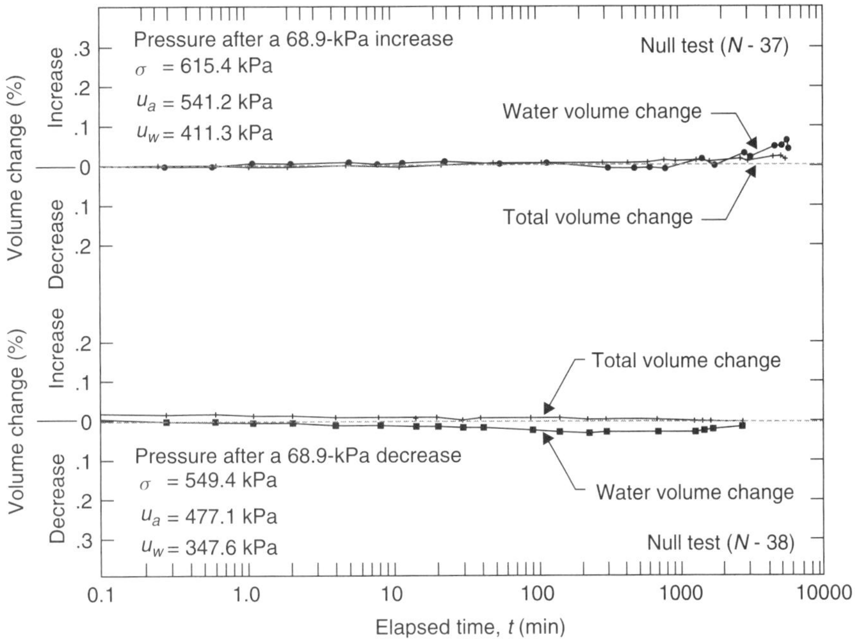

Figure 3.4 Results of null tests N-37 and N-38 on compacted kaolin (from Fredlund, 1973a).

the combination of all the component stress fields. In 1977, Fredlund and Morgenstern used the concept of multiphase continuum mechanics to write the equilibrium equations that act on the soil particle phase (or soil structure) of an unsaturated soil. The principle of superposition of coincident equilibrium stress fields was used as described in continuum mechanics (Truesdell and Toupin, 1960; Green and Naghdi, 1965; Truesdell, 1966).

The assumption was made that an independent continuous stress field was associated with each phase of the multiphase system. The number of independent force equilibrium equations that can be written for a multiphase system is equal to the number of Cartesian coordinate directions multiplied by the number of phases constituting the continuum. Equilibrium equations can also be written for various combinations of phases, but the number of independent equations is limited by the number of phases.

The stress variables appearing in the equilibrium equations for the soil structure can be taken as the stress state variables. The stress state variables must be expressed in terms of the measurable stresses, such as total stress σ, pore-water pressure u_w, and pore-air pressure u_a in the case of an unsaturated soil. It was observed that the surface tractions appearing in the force equilibrium equations for the soil structure were the same as those for the equilibrium equations for the contractile skin (i.e., air-water interface). It was also observed that three possible sets of stress state variables could be combined to form two stress tensors where each set of tensors depended on the reference phase used when deriving the equilibrium equations.

The force equilibrium equations for the multiphase system support the concept of independent stress state variables. However, the derivation cannot be viewed as exclusive proof of a particular set of stress state variables. It would appear that the selection of the most suitable combination of stress state variables must be based on practical preference considerations rather than theoretical considerations. There has been a general consensus that $\sigma - u_a$ and $u_a - u_w$ form the most easy-to-use combination of stress state variables for geotechnical engineering practice. This combination of stress state variables allows for the independent consideration of the effects of external total stresses and the effect of internal pore-water pressure u_w. Each of these stress variables are referenced to the pore-air pressure u_a, which in most real-world situations is equal to atmospheric pressure.

3.3 STRESS STATE VARIABLES FOR UNSATURATED SOILS

The mechanical behavior of soils is controlled by the same stress variables that control the equilibrium of the soil structure. Therefore, the stress variables required to describe the equilibrium of the soil structure can be taken as the stress state variables for the soil. The following sections illustrate the role of equilibrium considerations in providing a clear understanding of stress state variables in the development of an engineering science for multiphase systems.

3.3.1 Equilibrium Analysis for One-Phase Solid

There are two types of forces that can act on an element of soil. These are body forces and surface forces (i.e., surface tractions). Body forces act through the centroid of a soil element and are expressed as a force per unit volume. Gravitational and interaction forces between phases are examples of body forces. Surface forces such as external loads act only on the boundary surfaces of a soil element. The average value of a surface force per unit area tends to a limiting value as the surface area approaches zero. This limiting value is called the stress vector or the surface traction on a particular face of the element. The component of the stress vector perpendicular to a face of the element is defined as a normal stress, σ. The stress components parallel to a face of an element are referred to as shear stresses, τ.

There are an infinite number of planes that can be passed through a *point* in a soil mass. The stress state at a point can be analyzed by considering all the stresses acting on the planes that form a cubical element of infinitesimal dimensions. Body forces act through the centroid of the soil element and also need to be considered.

A cubical element that is completely enclosed by imaginary, unbiased boundaries can be used as a free-body diagram for a stress equilibrium analysis (Fung, 1977). The cubical element is referred to as a referential elemental volume. Figure 3.5 shows a cubical soil element with infinitesimal dimensions of *dx*, *dy*, and *dz* in the Cartesian coordinate system. The normal and shear stresses on each surface of the element are shown. The body forces are not shown.

Normal and shear stresses act on every plane in the x-, y-, and z-directions. The normal stress, σ, has one subscript to denote the plane on which it acts. Soils are most commonly subjected to compressive normal stresses. In soil mechanics, a positive normal stress is used to indicate compression in the soil. All the normal stresses shown on the element are positive or compressive. Opposite directions would indicate negative normal stresses or tensions. The shear stress τ has two subscripts. The first subscript denotes the plane on which the shear stress acts, and the second subscript refers to the direction of the shear stress. As an example, the shear stress τ_{yz} acts on the y-plane and in the z-direction. All of the shear stresses shown have positive signs. Opposite directions would indicate negative shear stresses. Equating the summation of moments about the x-, y-, and z-axes to zero results in the following shear stress relationships:

$$\tau_{xy} = \tau_{yx} \tag{3.10}$$

$$\tau_{xz} = \tau_{zx} \tag{3.11}$$

$$\tau_{yz} = \tau_{zy} \tag{3.12}$$

The stress components can vary from plane to plane across an element. The spatial variation of a stress component can be expressed as its derivative with respect to space. The stress variations in the x-, y-, and z-directions are expressed as stress fields.

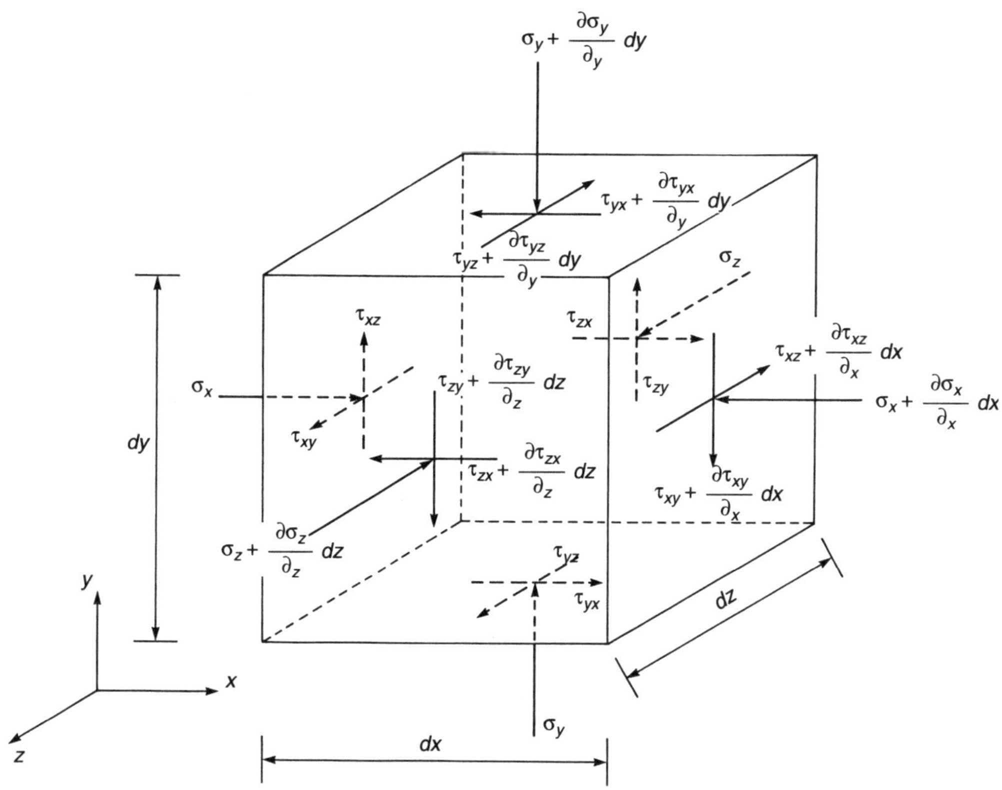

Figure 3.5 Normal and shear surface tractions on cubical soil element of infinitesimal dimensions, *dx*, *dy*, and *dz*.

The equilibrium analysis of a one-phase solid can be written based upon the conservation of linear momentum. The conservation of linear momentum in the y-direction can be written as

$$\left(\frac{\partial \tau_{xy}}{\partial x}+\frac{\partial \sigma_y}{\partial y}+\frac{\partial \tau_{zy}}{\partial z}+\rho g\right) dx\,dy\,dz=\left(\rho\frac{Dv_y}{Dt}\right) \tag{3.13}$$

where:

τ_{xy} = shear stress on the x-plane in the y-direction,
σ_y = total normal stress on the y-plane,
τ_{zy} = shear stress on the z-plane in the y-direction,
ρ = total density of the soil,
g = gravitational acceleration,
dx, dy, dz = dimension of the element in the x-, y-, and z-directions, respectively,
$\frac{Dv_y}{Dt}=\frac{\partial v_y}{\partial t}+\frac{\partial v_y}{\partial y}\frac{\partial y}{\partial t}$ = acceleration in the y-direction, and
v_y = velocity in the y-direction.

Let us assume that the REV does not undergo acceleration. Then the right-hand side of the above equation becomes zero and is referred to as the Newtonian equation of equilibrium:

$$\left(\frac{\partial \tau_{xy}}{\partial x}+\frac{\partial \sigma_y}{\partial y}+\frac{\partial \tau_{zy}}{\partial z}+\rho g\right) dx\,dy\,dz=0 \tag{3.14}$$

Similarly, equilibrium equations can be written for the x-direction,

$$\left(\frac{\partial \sigma_x}{\partial x}+\frac{\partial \tau_{yx}}{\partial y}+\frac{\partial \tau_{zx}}{\partial z}\right) dx\,dy\,dz=0 \tag{3.15}$$

and the z-direction,

$$\left(\frac{\partial \tau_{xz}}{\partial x}+\frac{\partial \tau_{yz}}{\partial y}+\frac{\partial \sigma_z}{\partial z}\right) dx\,dy\,dz=0 \tag{3.16}$$

Once the equilibrium equations have been written for each of the Cartesian coordinate directions, it is possible to extract the surface tractions and place them in the form of a tensor (i.e., a 3×3 matrix). The tensor represents the stress state at a point for a one-phase solid:

$$\begin{bmatrix} \sigma_x & \tau_{yx} & \tau_{zx} \\ \tau_{xy} & \sigma_y & \tau_{zy} \\ \tau_{xz} & \tau_{yz} & \sigma_z \end{bmatrix} \tag{3.17}$$

The stresses in the uppermost row in the matrix (i.e., σ_x, τ_{yx}, τ_{zx}) were extracted from the x-direction equilibrium equation. The middle row of stresses was from the y-direction equilibrium equation and the lowermost row of stresses was from the z-direction equilibrium equation. The 3×3 matrix is related to the three Cartesian coordinate directions commonly used to formulate engineering problems and is called a stress tensor.

The stress tensor shown in Eq. 3.17 can be used as the building block for a science for any one-phase material (e.g., steel, concrete). It can also be viewed as the total stress or overall stress tensor for a multiphase material.

3.3.2 Equilibrium Equations for Multiphase System

An equilibrium analysis can also be applied to a cubical element of a multiphase system such as a saturated soil or an unsaturated soil. Newton's second law is applied to the soil element by summing forces in each direction (i.e., x-, y-, and z-directions). A saturated soil can be considered as a two-phase system and equilibrium equations can be written for each of the phases. Equilibrium equations can also be written for each phase of the four phases of an unsaturated soil element (i.e., air, water, contractile skin, and soil particles). Each phase is assumed to behave as an independent, linear, continuous, and coincident stress field in each of the three Cartesian coordinate directions.

Equilibrium equations can also be overlain or summed using the principle of superposition. However, this may not give rise to equilibrium equations with stresses that can be measured. For example, the interparticle stresses between the soil solids cannot be directly measured. The force equilibrium equations for the air phase, water phase, and contractile skin together with the total equilibrium equation for the soil element can be used in formulating the equilibrium equation for the soil structure (Fredlund, 1973b). It is necessary to combine the independent phases in such a way that measurable stresses appear in the equilibrium equation for the soil solids.

The total number of independent equilibrium equations will be equal to 3 times the number of phases in a multiphase system. The stress variables that control the equilibrium of the soil structure also control the equilibrium of the contractile skin (Fredlund, 1973b).

3.3.3 Independent Phase Equilibrium

The soil particles and the contractile skin are assumed to behave as solids in an unsaturated soil. In other words, it is assumed that these phases come to equilibrium under applied stress gradients. The arrangement of soil particles is referred to as the soil structure. The water phase and the air phase qualify as fluids since they flow under an applied stress gradient. Each phase is assumed to have an independent, linear, continuous, and coincident stress field in each direction. The principle of superposition can be applied to the equilibrium equations for each of the phases because the stress fields are linear. The sum of the equilibrium equations for each phase is equal to the total or overall equilibrium of the soil element (Westergaard, 1952; Green and Naghdi, 1965; Truesdell, 1966; Faizullaev, 1969).

The equilibrium equations for the water phase, air phase, and contractile skin can be written independently. The three individual phase equilibrium equations together with the total equilibrium equation form the basis for formulating the equilibrium equation for the soil structure.

The areas or volumes associated with each phase can be written relative to the dimensions of the soil element (i.e., dx, dy, and dz). This is done using a porosity designation for the volume and area associated with individual phases. The porosity n of a phase is the ratio of the volume (or surface area) of each phase relative to the total volume. The volume porosity is assumed to be equal to the area ratio of each phase relative to the total cross-sectional area. This concept is referred to as the "theorem of the equality of volume and surface porosities in a homogeneous porous medium" (Faizullaev, 1969). Therefore, the same porosity term can be applied to the body force as to the surface tractions.

3.3.4 Water Phase Equilibrium

The water phase is used to illustrate how the equilibrium equation can be written for an individual phase (Fig. 3.6). Let us consider the case of a saturated soil (i.e., a two-phase system). It is necessary to include an interaction force which acts between the soil particles and the water when the water phase is separated from the remainder of the unsaturated soil element. The interaction force is a body force and can be given the designation F_{sy}^{w}. The water phase equilibrium in the y-direction also includes the gravity force due to the volume of water. Summing forces in the y-direction for the water phase gives

$$\left(n_w \frac{\partial u_w}{\partial y} + n_w \rho_w g + F_{sy}^{w}\right) dx\,dy\,dz = 0 \qquad (3.18)$$

where:

u_w = pore-water pressure,
ρ_w = density of water,
n_w = porosity with respect to the water phase, and
F_{sy}^{w} = interaction force (i.e., body force) between the water phase and the soil particles in the y-direction.

The interaction force between the water and the soil solids, F_{sy}^{w}, has the form of a "seepage force" associated with water flow through a saturated soil. Similar equilibrium equations can be written in the x- and z-directions for the water phase.

Since the sum of the parts (i.e., each of the phases) is equal to the overall element, it is possible to subtract the water phase equilibrium equation from the total or overall equilibrium equation. The resulting equation is the equilibrium equation for the soil solids. The resulting equation contains the surface tractions known as the effective stress variable.

3.3.5 Air Phase Equilibrium

Let us consider the case where a soil is completely dry. Once again the soil is a two-phase system consisting of soil solids and air. Equilibrium equations can be written for the air phase in a manner similar to that shown for the water phase (Fredlund and Rahardjo, 1993a).

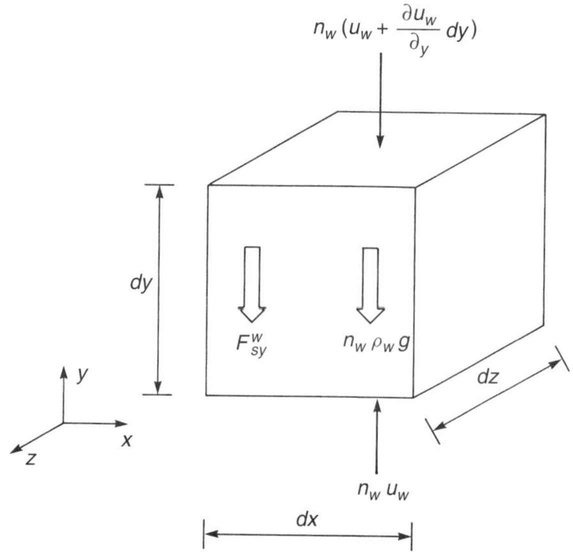

Figure 3.6 Surface tractions and body forces associated with water phase force equilibrium in y-direction for saturated soil.

Since the sum of the parts is equal to the overall element, it is possible to subtract the air phase equilibrium equation from the total or overall equilibrium equation. The resulting equation is the equilibrium equation for the soil solids when the pore fluid is air. The resulting equation contains the surface tractions equal to the difference between total stress and pore-air pressure.

3.3.6 Contractile Skin Equilibrium

The contractile skin is only a few molecular layers in thickness. However, its presence affects the equilibrium conditions in an unsaturated soil. The contractile skin influences the soil structure through its ability to exert surface tension, T_s. The contractile skin also allows for equilibrium conditions to be maintained between the air and water phases. That means that constant-degree-of-saturation conditions can be established in an unsaturated soil. It also means that it is possible for air and water to flow through the soil in an independent manner. The independent flow of air and water takes place while the contractile skin maintains a constant degree of saturation in the soil.

The magnitude of the unit surface tension is constant at a particular temperature. The interdependency of the contractile skin with the other phases causes its equilibrium equation to be somewhat complex (Fredlund and Rahardjo, 1993a). Details of the analysis are not presented herein (Fredlund and Morgenstern, 1977).

3.3.7 Equilibrium for the Soil Structure (Arrangement of Soil Particles)

The total equilibrium of an unsaturated soil element is equivalent to the resultant of the equilibrium equations for the individual phases. The equilibrium of the soil structure in the y-direction can be written as the difference between the total equilibrium equation and the sum of the water, air, and contractile skin equilibrium equations. The following equation is obtained when using the air phase as a reference phase during the derivation of the equilibrium equation for the soil structure in the y-direction:

$$\left\{\frac{\partial \tau_{xy}}{\partial x} + \frac{\partial \sigma_y}{\partial y} - n_a \frac{\partial u_a}{\partial y} - n_w \frac{\partial u_w}{\partial y} + n_c f^* \frac{\partial (u_a - u_w)}{\partial y} + \frac{\partial \tau_{zy}}{\partial z} + \rho g - n_a \rho_a g - n_w \rho_w g - n_c \rho_c g - F_{sy}^w - F_{sy}^a + n_c \left(u_a - u_w\right) \frac{\partial f^*}{\partial y}\right\} dx\, dy\, dz = 0 \tag{3.19}$$

where:

n_a = porosity with respect to the air phase,
n_c = porosity with respect to the contractile skin,
f^* = an interaction function between the contractile skin and the soil structure (Fredlund and Rahardjo, 1993a), and
F_{sy}^a = an interaction body force between the solids and the air phase in the y-direction.

Similar equilibrium equations can be written for the x- and z-directions, respectively:

$$\left\{\frac{\partial \sigma_x}{\partial x} + \frac{\partial \tau_{yx}}{\partial y} - n_a \frac{\partial u_a}{\partial y} - n_w \frac{\partial u_w}{\partial y} + n_c f^* \frac{\partial (u_a - u_w)}{\partial y} + \frac{\partial \tau_{zx}}{\partial z} - F_{sx}^w - F_{sx}^a + n_c (u_a - u_w) \frac{\partial f^*}{\partial y}\right\} dx\, dy\, dz = 0 \tag{3.20}$$

$$\left\{\frac{\partial \tau_{xz}}{\partial x} + \frac{\partial \tau_{yz}}{\partial y} - n_a \frac{\partial u_a}{\partial y} - n_w \frac{\partial u_w}{\partial y} + n_c f^* \frac{\partial (u_a - u_w)}{\partial y} + \frac{\partial \sigma_z}{\partial z} - F_{sz}^w - F_{sz}^a + n_c (u_a - u_w) \frac{\partial f^*}{\partial y}\right\} dx\, dy\, dz = 0 \tag{3.21}$$

The stress variables controlling the equilibrium of the soil structure are the stress state variables that control the mechanical behavior of soils. Fredlund and Morgenstern (1977) also showed that the stress variables controlling the equilibrium of the soil structure were the same as those controlling the equilibrium of the contractile skin.

There are three independent sets of normal stresses (i.e., surface tractions) that can be extracted from the equilibrium equations for the soil structure to form the stress state variables. The three stress state variables are $\sigma - u_a$, $u_a - u_w$, and u_a. The stress variable u_a can be eliminated if the soil particles are assumed to be incompressible. Therefore, the stress state variables for the soil structure and the contractile skin in an unsaturated soil are $\sigma - u_a$ and $u_a - u_w$.

The stress state variables act in three Cartesian coordinate directions and the variables can be collected to form two independent stress tensors. The two independent stress tensors can be written as the stress state at a point in an unsaturated soil:

$$\begin{bmatrix} \sigma_x - u_a & \tau_{yx} & \tau_{zx} \\ \tau_{xy} & \sigma_y - u_a & \tau_{zy} \\ \tau_{xz} & \tau_{yz} & \sigma_z - u_a \end{bmatrix} \tag{3.22}$$

$$\begin{bmatrix} u_a - u_w & 0 & 0 \\ 0 & u_a - u_w & 0 \\ 0 & 0 & u_a - u_w \end{bmatrix} \tag{3.23}$$

The pore-air pressure appears in both stress tensors; however, it is the difference between stress components that allows the two tensors to qualify as independent stress state variables. The stress variables in Eqs. 3.22 and 3.23 can be placed on the surface of a cube to give the stress state at a point as shown in Fig. 3.7.

As an unsaturated soil approaches saturation, the degree of saturation S approaches 100%. The pore-water pressure u_w approaches the pore-air pressure u_a and the matric suction term $u_a - u_w$ goes toward zero. The second stress tensors for the unsaturated soil disappear because the matric suction $u_a - u_w$ becomes zero. It should be noted that it is not necessary for the pore-water pressure to go to zero in order for the soil to behave as a saturated soil. Rather, it is necessary for the pore-water pressure to increase until it becomes equal to the pore-air pressure for the soil to behave as a saturated soil (i.e., a two-phase system).

Once pore-water pressure is equal to pore-air pressure, only the first stress tensor is left to represent the stress state for a saturated soil. The pore-air pressure term in the first stress tensor becomes equal to the pore-water pressure u_w as the soil becomes saturated. The stress state at a point in a saturated soil is illustrated in Fig. 3.8. The stress tensor for a saturated soil is consistent with that presented by Terzaghi (1936) as the effective stress variable $\sigma - u_w$.

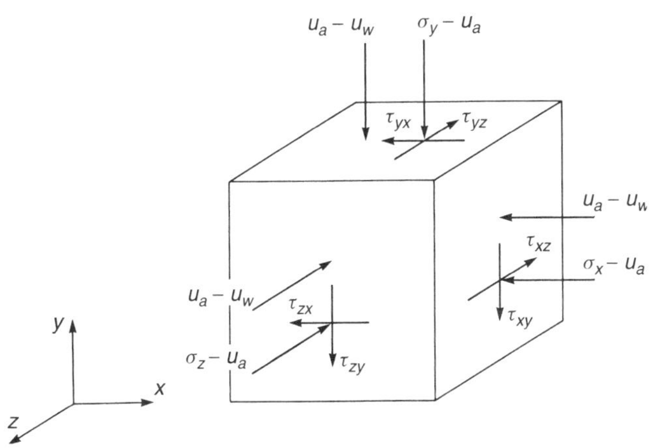

Figure 3.7 Stress state at point based on surface tractions extracted from soil structure force equilibrium equations for unsaturated soil (i.e., x-, y-, and z-directions).

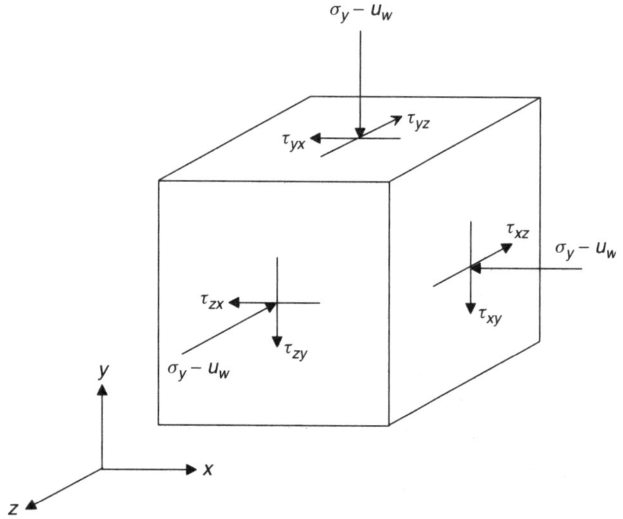

Figure 3.8 Stress state at a point based on the surface tractions extracted from the soil structure force equilibrium equations for a saturated soil (i.e., *x*-, *y*-, and *z*-directions).

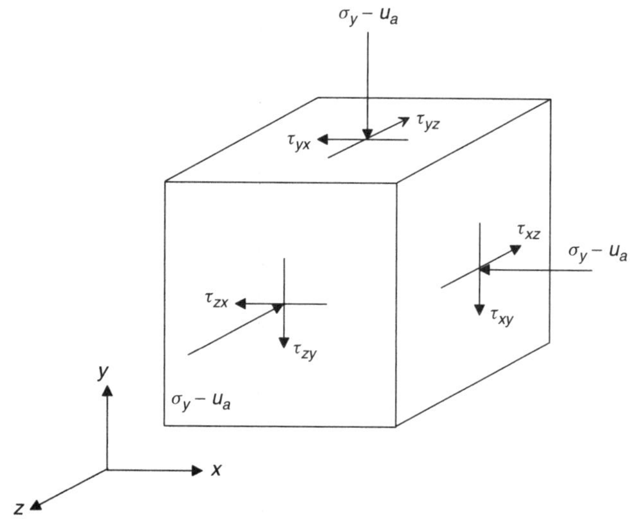

Figure 3.9 Stress state at point based on surface tractions extracted from soil structure force equilibrium equations for dry soil (i.e., *x*-, *y*-, and *z*-directions).

3.3.8 Other Combinations of Stress State Variables

The soil structure equilibrium equations can have three different forms depending upon the reference phase used during the derivation of the equilibrium equations. Equation 3.17 is the result of using the air phase as the reference phase during the derivation. Other forms of the derivation can use the water phase or else the total stress field as the reference. Each form of the equilibrium equations contains a combination of two stress state variables. In other words, any two of three possible stress variables (i.e., $\sigma - u_a$, $\sigma - u_w$, and $u_a - u_w$) can be used to describe the stress state for the soil structure and contractile skin in an unsaturated soil.

3.3.9 Saturated Soils as Special Case of Unsaturated Soils

A saturated soil can be viewed as a two-phase system. The equilibrium equations for the soil structure (i.e., the arrangement of soil solids) of a saturated soil can be derived in the manner described above. Equilibrium equations can be written for the water phase and the overall total stress field. The equilibrium equation for the soil solids will be the difference between the total or overall stress field and the water phase stress field. The extraction of the surface tractions from the resulting equilibrium equations gives rise to the following stress tensor containing effective stress variables which can be placed on a cube to represent the stress state at a point (Fig. 3.8):

$$\begin{bmatrix} \sigma_x - u_w & \tau_{yx} & \tau_{zx} \\ \tau_{xy} & \sigma_y - u_w & \tau_{zy} \\ \tau_{xz} & \tau_{yz} & \sigma_z - u_w \end{bmatrix} \qquad (3.24)$$

It is interesting to note that no area correction appears in the stress tensor when the analysis is performed in a manner consistent with multiphase continuum mechanics principles.

3.3.10 Case of Dry Soil (Solids and Air)

As a soil becomes extremely dry, a change in soil suction may no longer produce significant changes in mechanical properties. However, soil suction remains as a stress state variable even if it is not required to describe the behavior of the dry soil. The first stress tensor with $\sigma - u_a$ will have the most significant influence on the shear strength and volume change of the dry soil. The stress tensor for a dry soil can be written as

$$\begin{bmatrix} \sigma_x - u_a & \tau_{yx} & \tau_{zx} \\ \tau_{xy} & \sigma_y - u_a & \tau_{zy} \\ \tau_{xz} & \tau_{yz} & \sigma_z - u_a \end{bmatrix} \qquad (3.25)$$

The stress state at a point in a dry soil is illustrated in Fig. 3.9. There are some physical processes where it is necessary to consider soil suction as a stress state variable when there is a change from the completely dry state (e.g., volume increase or swelling).

3.3.11 Smooth Transition from Saturated to Unsaturated Conditions

There is a smooth transition between the stress state for a saturated soil and the stress state required for an unsaturated soil. As a soil approaches saturation, the pore-water pressure u_w approaches the pore-air pressure u_a and the matric suction term $u_a - u_w$ goes toward zero. In other words, the second stress tensor (i.e., the soil suction tensor) disappears as pore-water pressure becomes equal to pore-air pressure. This is true regardless of the pore-air pressure. When matric suction goes to zero, the pore-air pressure term in the first stress tensor becomes equal to the pore-water pressure u_w and soil behavior can be described using the stress state variables for a saturated soil (i.e., $\sigma - u_w$).

There is a hierarchy of limiting stress conditions related to the magnitude of the individual stress components for an unsaturated soil. The total stress must always be the largest value and the pore-water pressure must be the smallest value. The pore-air pressure must be intermediate or equal to the pore-water pressure or the total stress:

$$\sigma \geq u_a \geq u_w \tag{3.26}$$

The above hierarchy must be maintained in order to ensure stable equilibrium conditions. Limiting stress state conditions occur when one of the stress state variables becomes zero. For example, if the pore-air pressure u_a is momentarily increased in excess of the total stress σ, an "explosion" of the soil may occur. In other words, once the $\sigma - u_a$ variable goes to zero, a limiting stress state condition is reached. This limiting stress condition occurs when testing unsaturated soils in a pressure plate apparatus.

Another limiting stress state condition occurs when matric suction $u_a - u_w$ vanishes. If the pore-water pressure is increased and approaches the pore-air pressure, the degree of saturation of the soil approaches 100% and matric suction goes toward zero. The behavior of the soil can then be described in terms of one stress state variable (i.e., $\sigma - u_w$). A smooth transition from the unsaturated case to the saturated case takes place under the limiting stress state condition where the pore-air pressure becomes equal to the pore-water pressure.

A limiting condition also occurs in saturated soils when the $\sigma - u_w$ stress state variable (i.e., effective stress) reaches zero. Any attempt to force effective stress to become negative results in the soil structure of the saturated soil becoming unstable. The soil is said to "quick." A further increase in the pore-water pressure results in a "boil" being formed.

3.3.12 Consideration of Equilibrium across Wavy Plane

There have been numerous attempts to prove the form of an appropriate effective stress equation for both saturated and unsaturated soil by passing a wavy plane through a soil mass (Fig. 3.10). The wavy plane verification analysis is unacceptable for a number of reasons even though it is found in many soil mechanics textbooks. Consideration of a biased plane or a wavy plane is unacceptable within the context of a Newtonian equilibrium-type analysis (Fig. 3.11).

Force equilibrium across a plane is comparable to Newton's third law, which states that "action must be equal to reaction." A wavy plane allows one set of stresses to be set equal to another variable; however, this does not qualify

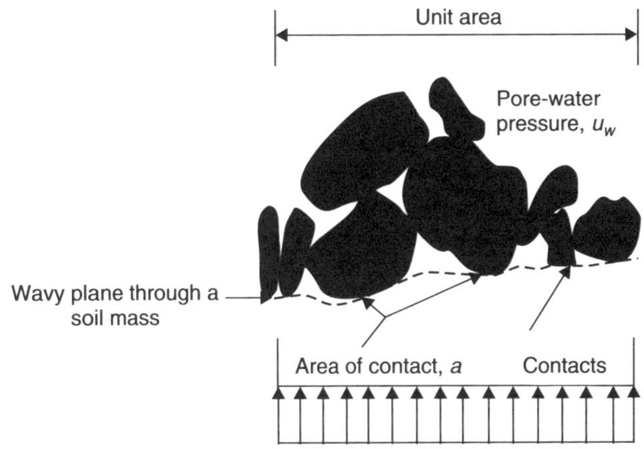

Figure 3.10 Consideration of force equilibrium across wavy plane.

as a proof or statical equilibrium. Rather, it is necessary to introduce spatial variation into the stress fields in order to write legitimate force equilibrium equations.

Continuum mechanics and the mechanics of statics require that equilibrium considerations must commence with a legitimate free-body diagram. The starting point must make use of an REV that has unbiased, planar sides in three orthogonal directions. In a multiphase system, it is also necessary for the volume phase percentages of the element to be equal to the area phase percentages. Only in this way can the fundamental definition of a continuum be maintained.

The definition of a continuum dictates that the density function for the mixture under consideration must remain constant as the size of the REV is reduced. Then the results of an equilibrium analysis cannot be biased by the analyst.

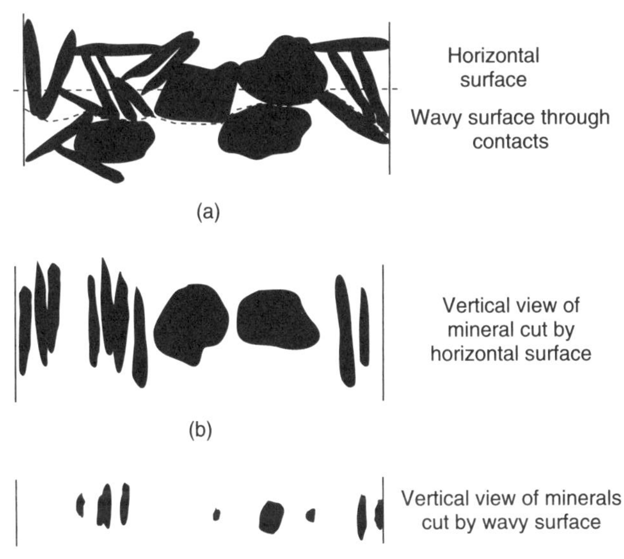

Figure 3.11 Shortcomings associated with passing plane through multiphase medium.

The use of a wavy plane through a soil mass can bias the final outcome toward a preconceived notion.

The stress state variables for a multiphase system must rely on accepted principles of statics and continuum mechanics where spatial variation is brought into the analysis through the use of stress field representations applied to the phases of an REV. When stress field equations are written for the overall element and the various phases of a multiphase system, the area of contact and other soil properties do not appear in the designation of stress state variables.

3.4 REPRESENTATION OF STRESS STATES

Stress state variables become part of the mathematical equations used in the engineering design process. The stress state conditions are required for three dimensions of the Cartesian coordinate system. In some cases it might be possible to reduce engineering formulations to a two-dimensional or one-dimensional formulation.

It is necessary to be able to take the components of the stress state variables and determine how each stress component can be computed under field conditions (i.e., in situ stress states) on a practical engineering project. The question being addressed can be stated as follows, "How can the magnitude and distribution of total stresses and pore fluid stresses be determined in the field?"

3.4.1 In Situ Designation of Stress Components

The magnitude and distribution of the stress components in the field need to be computed as part of a geotechnical engineering analysis. The distribution of the components of stress allows the computation of in situ profiles for net normal stress $\sigma - u_a$ and matric suction $(u_a - u_w)$. Procedures are necessary for the assessment of each of the stress components. The stress components can then be used to compute the stress state variable distributions with space.

Totals stresses in three orthogonal directions need to be computed as well as the pore-water and pore-air pressure values with respect to space (i.e., depth and laterally). Soil behavior can then be visualized in terms of the magnitudes of the stress state variables and potential changes in the stress state variables. The components of stress revert to effective stress states $(\sigma - u_w)$ as the soil becomes saturated. The effect of in situ stress history is often assessed on the basis of laboratory oedometer measurements. Possible final (or future) stress state profiles are often assumed on the basis of engineering experience or computed based on other theoretical considerations.

The total vertical stress under geostatic conditions is a function of the density or the total unit weight of the soil. The magnitude and distribution of the total normal stress are also affected by the application of external loads such as buildings or the removal of soil through excavation.

Let us consider geostatic conditions where the ground surface is horizontal and the vertical and horizontal planes do not have shear stresses (Lambe and Whitman, 1979). The net normal stresses in the vertical and horizontal directions are related to the density of soil. The net vertical stress $\sigma_v - u_a$ is called the overburden pressure, which can be computed as follows (Fig. 3.12):

$$\sigma_v - u_a = \int_{y_2}^{y_1} \rho\,(y)\, g\, dy - u_a \tag{3.27}$$

where:

$\sigma_v - u_a$ = vertical net normal stress,
u_a = pore-air pressure,
dy = incremental distance in the vertical direction,
$\rho\,(y)$ = density of the soil as a function of depth,
y_1 = ground surface elevation,
y_2 = elevation under consideration, and
g = gravitational acceleration.

The vertical net normal stress distribution with respect to depth is a straight line as long as the soil density is a

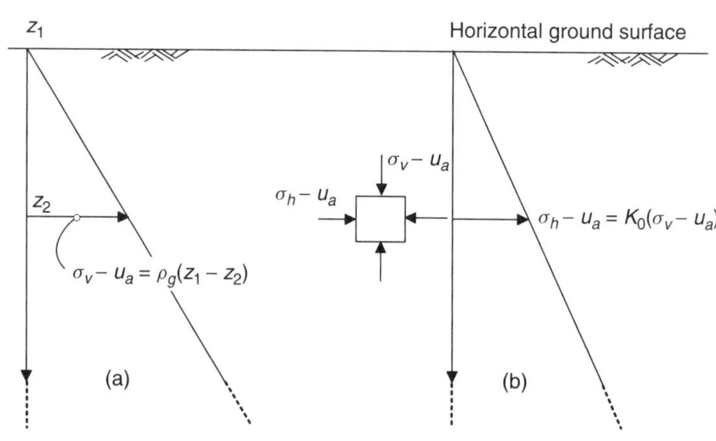

Figure 3.12 In situ net normal stress profile under geostatic conditions: (a) vertical net normal stress; (b) horizontal net normal stress.

constant value with depth. The pore-air pressure is generally assumed to be in equilibrium with atmospheric pressure (i.e., zero gauge pressure). Figure 3.12a shows a typical profile of the vertical net normal stress for geostatic conditions. Total stresses can be computed by integration of Eq. 3.27 across each layer when there are soil strata with distinctly different densities. The total vertical stress in a level soil mass is computed in the same manner for both saturated and unsaturated soils (Fig. 3.12). The total vertical stress σ_v can be computed as follows if only one soil is involved:

$$\sigma_v = \int_0^H \rho g \, dy \tag{3.28}$$

where:

y = vertical distance from ground surface and
H = depth of soil under consideration.

The total vertical stress at any depth can be written as follows if the soil mass is homogeneous:

$$\sigma_v = \rho g H \tag{3.29}$$

The pore-air pressure is generally at equilibrium with atmospheric pressure. The pore-water pressure above the groundwater table can be either estimated or measured. The negative pore-water pressure profile is largely controlled by the location of the water table in the immediate region above the water table. The negative pore-water pressures are largely controlled by the net moisture influx at the ground surface in the region near the ground surface. In other words, the local climatic conditions significantly influence the near-ground-surface pore-water pressures. The negative pore-water pressure profile tends toward hydrostatic conditions if the annual net moisture flux at ground surface approaches zero.

The matric suction profile is closely related to the near-ground-surface climatic environment. The approximation or determination of the initial soil suction profile is required when analyzing a geotechnical engineering problem. The in situ profile of pore-water pressures (and thus soil suction) can vary significantly throughout each year. The effect of climatic conditions in a region should be analyzed by taking moisture flux boundary conditions into consideration throughout a year.

3.4.2 Coefficient of Lateral Earth Pressure

It is difficult to theoretically quantify the coefficient of earth pressure at rest due to complexities arising from the stress history to which the soil mass has been subjected. Consideration of elastic equilibrium within a soil mass can provide some insight into the coefficient of earth pressure at rest. Elastic equilibrium can also provide some indication of the depth of potential cracking in a soil mass. Further details on elastic equilibrium can be found in Chapter 13.

The coefficient of lateral earth pressure K_0 can be defined as the ratio of horizontal net normal stress to vertical net normal stress (Fig. 3.12). This is a slight variation from saturated soil mechanics since the horizontal and vertical stresses are now referenced to the pore-air pressure (which is generally at atmospheric pressure):

$$K_0 = \frac{\sigma_h - u_a}{\sigma_v - u_a} \tag{3.30}$$

where:

$\sigma_h - u_a$ = horizontal net normal stress.

The geostatic stress condition where there is no horizontal strain is referred to as the coefficient of lateral earth pressure at rest, K_0 (Terzaghi, 1925). The coefficient of lateral earth pressure at rest depends on several factors such as the type of soil, its stress history, and its density. Saturated soils commonly have K_0 values ranging from as low as 0.4 to values in excess of 1.0. Unsaturated soils are often over consolidated and can have coefficients of earth pressure at rest greater than 1.0 (Brooker and Ireland, 1965). On the other hand, the coefficients of earth pressure at rest can go to zero for the case where the soil becomes desiccated and cracked. A profile of the horizontal net normal stresses under at-rest conditions is shown in Fig. 3.12b.

The application of the theory of lateral earth pressures to cracked, unsaturated soils is challenging to understand. The horizontal total stress in the crack through a soil mass is zero. However, the soil immediately adjacent to the crack is being held together by soil suction. Consequently, one horizontal stress condition can be observed in the crack of the soil mass while another completely different stress condition is observed in the adjacent intact soil.

The theory of elasticity can be used to compute changes in total stress states. Elastic theory can be applied in a similar manner for both saturated and unsaturated soils. The total stress state within a continuum (i.e., including the horizontal stress state) can be solved as a "boundary value" problem when the ground surface is not horizontal and where several soil types may be present. In this case, the ground surface geometry and the soil strata can be modeled using a finite element numerical stress analysis. Elasticity parameters, including Young's modulus and Poisson's ratio, need to be input for each soil layer. The computed stress state is quite insensitive to the input soil parameters and as a result it is possible to calculate reasonable estimates of the total stress state. Figure 3.13 shows contours of vertical stresses and horizontal stresses computed by "switching on" the gravity forces in a numerical modeling analysis. The switching-on type of analysis can readily be performed even for complex geometric and soil conditions. Further details pertaining to the stress-versus-strain relationship for an unsaturated soil are presented in Chapter 13.

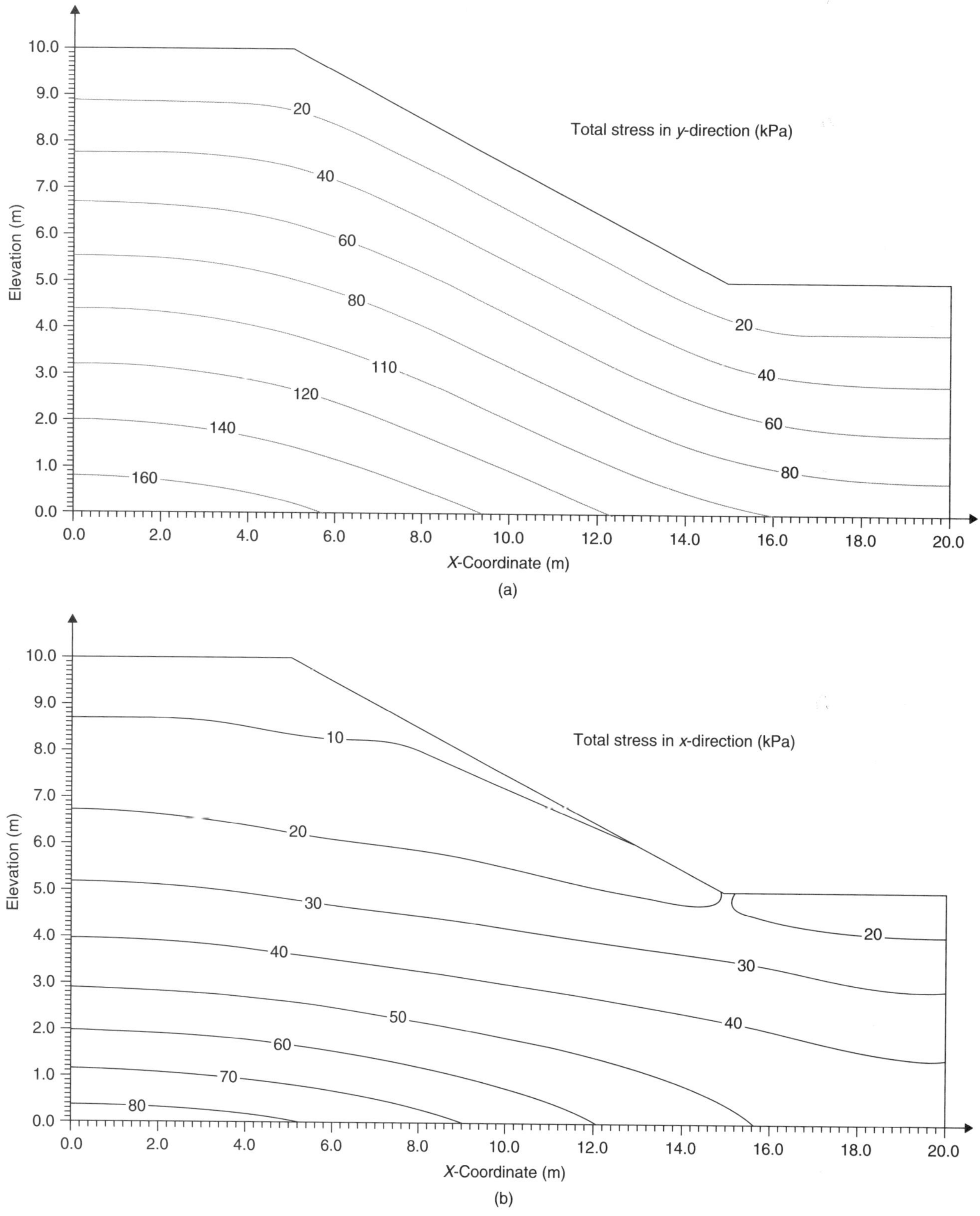

Figure 3.13 Total (a) vertical and (b) horizontal stresses in soil mass with sloping ground surface with stresses computed by switching on gravity forces.

The stress-versus-strain equation in the vertical direction for a homogeneous, isotropic, linear elastic, unsaturated soil can be written as

$$\varepsilon_v = \frac{\sigma_v - u_a}{E} - \frac{2\mu}{E}\left(\sigma_h - u_a\right) + \frac{u_a - u_w}{H} \tag{3.31}$$

where:

ε_v = normal strain in the vertical direction,
σ_v = total normal stress in the vertical direction,
σ_h = total normal stress in the horizontal direction,
μ = Poisson's ratio,
E = elastic modulus with respect to a change in $\sigma - u_a$,
H = elastic modulus with respect to a change in $u_a - u_w$,
u_a = pore-air pressure, and
u_w = pore-water pressure.

The stress-versus-strain equation in the horizontal direction is written as

$$\varepsilon_h = \frac{\sigma_h - u_a}{E} - \frac{\mu}{E}\left(\sigma_v + \sigma_h - 2u_a\right) + \frac{u_a - u_w}{H} \tag{3.32}$$

The above equation applies to both horizontal directions. For the at-rest or K_0 condition in an intact, homogeneous, unsaturated soil mass, the strain in the horizontal directions can be set to zero (i.e., $\varepsilon_h = 0$). The net horizontal stress can then be written in terms of the vertical stress:

$$\sigma_h - u_a = \frac{\mu}{1-\mu}\left(\sigma_v - u_a\right) - (1-\mu)\frac{E}{H}\left(u_a - u_w\right) \tag{3.33}$$

The above equation can be normalized to the net vertical stress, and the equation takes the form for the coefficient of earth pressure at rest, K_0:

$$K_0 = \frac{\mu}{1-\mu} - \frac{E}{(1-\mu)\,H}\,\frac{u_a - u_w}{\sigma_v - u_a} \tag{3.34}$$

The above equation reverts to the form common to a saturated soil when matric suction goes to zero. When matric suction is present in the soil, the horizontal stress is reduced as the soil tends toward cracking. The reduction in horizontal stress is also a function of the depth under consideration. At shallow depths, the second part of Eq. 3.34 shows that a small matric suction may cause the net horizontal stress to go to zero, or tend to go negative. If the soil cannot sustain tensile strain, cracking of the soil will occur commencing at ground surface.

Typical K_0 values for the first loading of clay ranges between approximately 0.3 and 0.7, depending upon Poisson's ratio. Let us consider the slow drying of a lacustrine clay deposit. For illustrative purposes, the following properties are assumed; $\mu = 0.35$, $E/H = 0.17$, and $\rho = 1886$ kg/m^3. Figure 3.14 illustrates the relationship between the coefficient of earth pressure at rest and matric suction for various overburden pressures. When the soil is saturated with zero pore-water pressure, the at-rest coefficient of earth pressure is 0.538. The at-rest coefficient then decreases as the matric suction of the soil increases. This behavior applies for all depths, but the rate of reduction in the at-rest coefficient is greater at shallow depths. An at-rest coefficient K_0 of zero indicates a tendency for cracking.

The above example is a simplification which does not take into consideration the effects of prior processes such as wetting and drying, freezing and thawing, and loading and

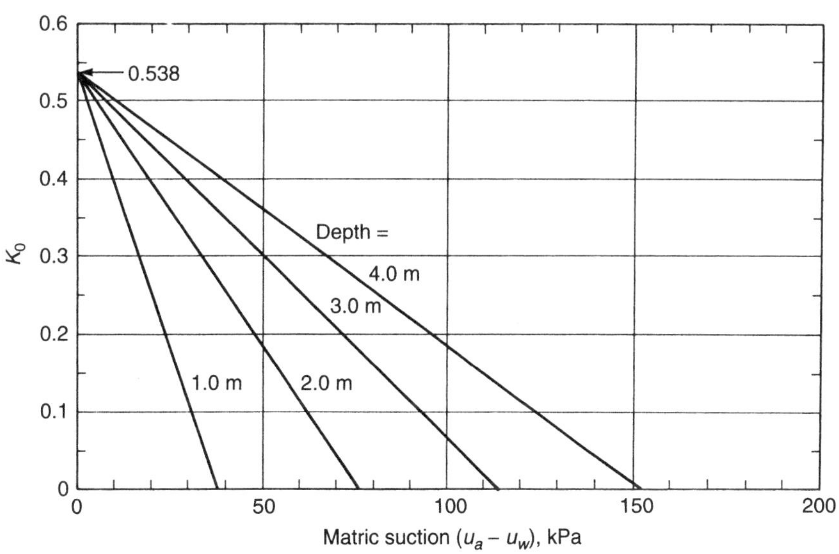

Figure 3.14 Relationship between coefficient of earth pressure at rest, K_0, and matric suction.

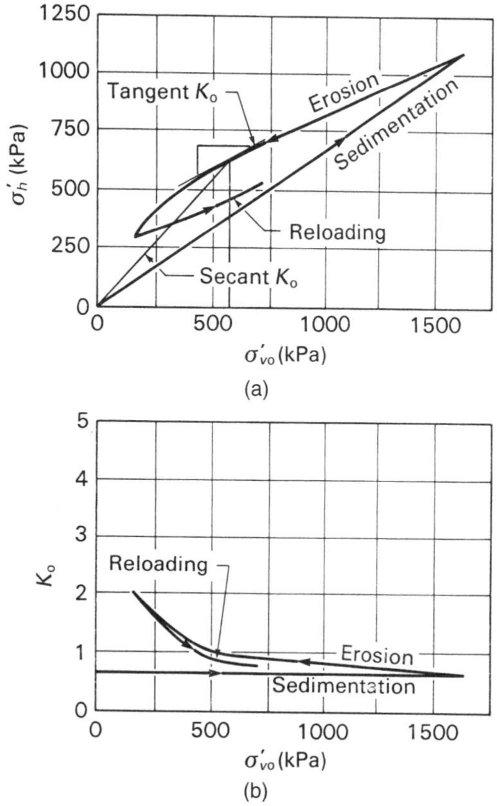

Figure 3.15 Relationship between horizontal stresses and vertical overburden stresses during sedimentation, erosion, and reloading of saturated soil. (a) Effective horizontal stress for various effective vertical stresses. (b) Coefficient of earth pressure at rest, K_0, with changing overburden stresses.

unloading cycles. Figure 3.15 illustrates typical horizontal and vertical effective stress paths where a saturated soil is subjected to a history of sedimentation followed by erosion and subsequent reloading. The stress paths can become even more complex for unsaturated soils subjected to cycles of drying and wetting and freezing and thawing. The coefficients of earth pressure can go from as low as zero to as high as the coefficient of passive earth pressure.

If all of the elastic parameters were known and the above analyses were applied to a soil that had undergone a complex stress history, the coefficient of earth pressure should be a tangent value of K_0 as opposed to a secant value of K_0 (Fig. 3.15). The engineer is generally interested in the secant value. The above analysis is most relevant immediately following sedimentation.

It must be recognized that the soil behaves in a nonlinear manner and the incremental elastic parameters are difficult to assess accurately. The assessment of the horizontal stress state of an unsaturated soil is a complex topic. It is primarily for this reason that empirical expressions have been proposed for at rest coefficient of earth pressure. Bishop (1957) presented the results of a laboratory study into the

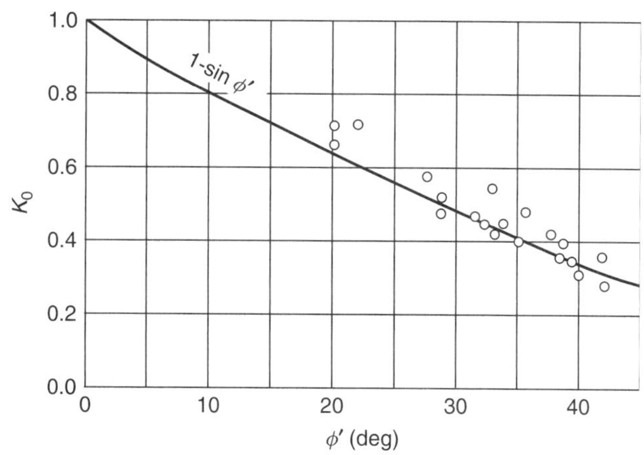

Figure 3.16 Relationship between effective angle of internal friction and coefficient of earth pressure at rest, K_0 (after Bishop, 1957).

coefficient of earth pressure at rest. The results supported Jaky's (1944) earth pressure expression (Fig. 3.16):

$$K_0 = 1 - \sin\phi' \tag{3.35}$$

where:

ϕ' = effective angle of internal friction.

The equation applies for initial or first loading of the soil. Other researchers have also lent support for this equation (Simons, 1958; Brooker and Ireland, 1965). Test results obtained by Bishop (1957, 1961b) and Simons (1958) are shown in Table 3.1.

The compaction of granular soils against an unyielding wall can produce horizontal pressures greater than the vertical pressure (i.e., K_0 greater than 1.0). Expansive soils can also exert lateral pressures greater than the vertical pressure. It is even possible to reach the passive pressure state, at which time the soil may fail in shear. This condition is illustrated in the oedometer specimen shown in Fig. 3.17. The K_0 value for soils has also been shown to be a function of the over consolidation ratio of the soil. An increase in the over consolidation ratio produces an increase in K_0.

3.5 EQUATIONS FOR MOHR CIRCLE

The state of stress at a point in the soil is three dimensional but often a practical engineering problem can be represented in a two-dimensional form. In two dimensions, there always exists a set of two mutually orthogonal principal planes with real-valued principal stresses. The principal planes are the planes on which there are no shear stresses. The direction of the principal planes depends on the general stress state at a point. The largest principal stress is called the major principal stress and is given the symbol σ_1. The smallest

Figure 3.17 Passive-type failure in compacted clay till resulting from swelling in confining oedometer ring.

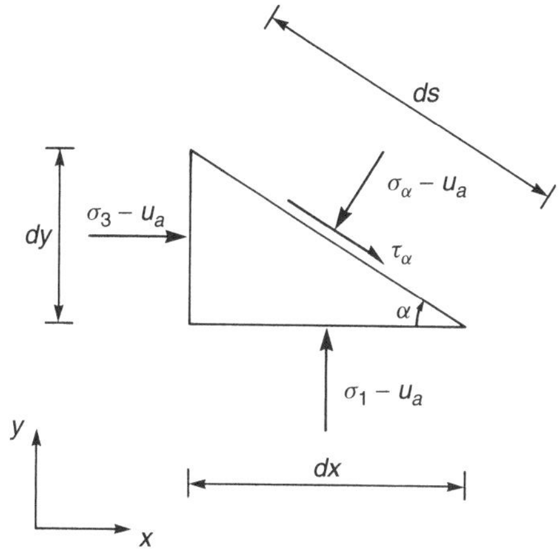

Figure 3.18 Net normal and shear stresses on inclined plane at point in soil mass below horizontal ground surface.

principal stress is called the minor principal stress and is given the symbol σ_3. The horizontal and vertical planes constitute the principal planes in the case of a horizontal ground surface. The vertical net normal stress is often near the net major principal stress $(\sigma_1 - u_a)$ and the horizontal net normal stress is near to the net minor principal stress $(\sigma_3 - u_a)$.

If the magnitude and the direction of the stresses acting on any two mutually orthogonal planes (e.g., the principal planes) are known, the stress condition on any inclined plane can be determined. In other words, the net normal stress and shear stress on any inclined plane can be computed from the known net principal stresses. The matric suction $u_a - u_w$ on every inclined plane through a point is constant since it is an isotropic tensor. Therefore, only the net normal stress and shear stress on an inclined plane need to be considered.

Let us consider an unsaturated soil under at rest conditions beneath a horizontal ground surface. The net normal stress and shear stress on a plane with an inclination angle α from the horizontal plane are illustrated in Fig. 3.18. The inclined plane has an infinitesimal length ds and results in a triangular free-body element with horizontal and vertical planes. The horizontal plane has an infinitesimal length of dx. Its length can be written in terms of the sloping length ds and the angle α.

$$dx = ds \cos\alpha \tag{3.36}$$

The vertical plane has an infinitesimal length of dy:

$$dy = ds \sin\alpha \tag{3.37}$$

All the planes have a unit thickness in the perpendicular direction. The equilibrium of the triangular element requires that the summation of forces in the horizontal and vertical directions be equal to zero. Summing forces horizontally gives

$$-\left(\sigma_\alpha - u_a\right) ds \sin\alpha + \tau_\alpha ds \cos\alpha + \left(\sigma_3 - u_a\right) dy = 0 \tag{3.38}$$

Summing forces vertically gives

$$-\left(\sigma_\alpha - u_a\right) ds \cos\alpha - \tau_\alpha ds \sin\alpha + \left(\sigma_1 - u_a\right) dx = 0 \tag{3.39}$$

Table 3.1 Coefficients of Earth Pressure At Rest

Type of Soil	Liquid Limit, w_L	Plastic Limit, w_p	Plasticity Index, PI	Activity	K_0
Loose, saturated sand	—	—	—	—	0.46
Dense, saturated sand	—	—	—	—	0.36
Compacted residual clay	—	—	9.3	0.44	0.42
Compacted residual clay	—	—	31	1.55	0.66
Undisturbed, organic, silty clay	74.0	28.6	45.4	1.2	0.57
Remolded kaolin	61	38	23	0.32	0.66
Undisturbed marine clay	37	21	16	0.21	0.48
Quick clay	34	24	10	0.18	0.52

Source: From test results of Bishop (1957, 1961b) and Simons (1958).

Substituting dx and dy (i.e., Eqs. 3.36 and 3.37) into Eqs. 3.39 and 3.38, respectively, and multiplying Eq. 3.38 by $\sin\alpha$ and Eq. 3.39 by $\cos\alpha$ gives

$$-(\sigma_\alpha - u_a)\, ds \sin^2\alpha + \tau_\alpha\, ds \sin\alpha\cos\alpha + (\sigma_3 - u_a)\, ds \sin^2\alpha = 0 \tag{3.40}$$

$$-(\sigma_\alpha - u_a)\, ds \cos^2\alpha - \tau_\alpha\, ds \sin\alpha\cos\alpha + (\sigma_1 - u_a)\, ds \cos^2\alpha = 0 \tag{3.41}$$

Summing Eqs. 3.40 and 3.41 gives

$$-(\sigma_\alpha - u_a)\, ds(\sin^2\alpha + \cos^2\alpha) + (\sigma_3 - u_a)\, ds \sin^2\alpha + (\sigma_1 - u_a)\, ds \cos^2\alpha = 0 \tag{3.42}$$

Using trigonometric relations to solve for $\sigma_a - u_a$ gives

$$\sigma_\alpha - u_a = (\sigma_1 - u_a)\left(\frac{1+\cos 2\alpha}{2}\right) + (\sigma_3 - u_a)\left(\frac{1-\cos 2\alpha}{2}\right) \tag{3.43}$$

Rearranging Eq. 3.43 gives

$$\sigma_\alpha - u_a = \left(\frac{\sigma_1+\sigma_3}{2} - u_a\right) + \left(\frac{\sigma_1-\sigma_3}{2}\right)\cos 2\alpha \tag{3.44}$$

The shear stress τ_α is obtained by substituting dx and dy (i.e., Eqs. 3.36 and 3.37) into Eqs. 3.39 and 3.38, respectively, and multiplying Eq. 3.38 by $\cos\alpha$ and Eq. 3.39 by $\sin\alpha$:

$$-\left(\sigma_\alpha - u_a\right) ds \sin\alpha\cos\alpha + \tau_\alpha\, ds\cos^2\alpha + \left(\sigma_3 - u_a\right) ds \times \sin\alpha\cos\alpha = 0 \tag{3.45}$$

$$-\left(\sigma_\alpha - u_a\right) ds \sin\alpha\cos\alpha - \tau_\alpha\, ds\sin^2\alpha + \left(\sigma_1 - u_a\right) ds \times \sin\alpha\cos\alpha = 0 \tag{3.46}$$

Subtracting Eq. 3.46 from Eq. 3.45 gives

$$\tau_a\, ds\left(\sin^2\alpha + \cos^2\alpha\right) + \left(\sigma_3 - u_a\right) ds \sin\alpha\cos\alpha - \left(\sigma_1 - u_a\right) ds \sin\alpha\cos\alpha = 0 \tag{3.47}$$

Using trigonometric relations, it is possible to solve for τ_α:

$$\tau_\alpha = \left(\frac{\sigma_1-\sigma_3}{2}\right)\sin 2\alpha \tag{3.48}$$

Equations 3.44 and 3.48 give the net normal stress and the shear stress on an inclined plane through a point. The term $\sigma_1 - \sigma_3$ is called the deviator stress and is an indication of the shear stress in the soil. The largest shear stress for a given stress state $(\sigma_1 - \sigma_3)/2$ occurs on a plane with an inclination angle α such that $\sin 2\alpha$ is equal to unity.

The net normal stress and shear stress at a point can also be determined using a graphical method. If Eqs. 3.44 and 3.48 are squared and added, the result is the equation of a circle:

$$\left[(\sigma_\alpha - u_a) - \left(\frac{\sigma_1+\sigma_3}{2} - u_a\right)\right]^2 + \tau_\alpha^2 = \left(\frac{\sigma_1-\sigma_3}{2}\right)^2 \tag{3.49}$$

The circle is known as the Mohr diagram and represents the stress state at a point. In saturated soils, the Mohr diagram is often plotted in terms of the principal effective normal stress on the abscissa and the shear stress on the ordinate. For unsaturated soils, an extended form of the Mohr diagram can be used as shown in Fig. 3.19.

The extended Mohr diagram uses a third orthogonal axis to represent matric suction. The circle described in Eq. 3.49 is drawn on a plane with the net normal stress $\sigma - u_a$ as the abscissa and the shear stress τ as the ordinate. The center of the circle has an abscissa defined by $(\sigma_1 + \sigma_3)/2 - u_a$ and a radius defined by $(\sigma_1 - \sigma_3)/2$.

The matric suction must also be included as part of the description of the stress state. The matric suction determines the position of the Mohr diagram along the third axis. Matric suction goes to zero as the soil becomes saturated, and the Mohr diagram reverts to a single $(\sigma - u_w)$-versus-τ plane.

3.5.1 Construction of Mohr Circles for Unsaturated Soils

The construction of the Mohr diagram on the $(\sigma - u_a)$-versus-τ plane is shown in Fig. 3.20. A compressive net normal stress is plotted as a positive net normal stress in accordance with the sign convention for the Mohr diagram. A shear stress that produces a counterclockwise moment about a point within the element is plotted as a positive shear stress. This shear stress sign convention is different from the convention normally used in continuum mechanics (Desai and Christian, 1977). This convention should only be used for plotting the Mohr diagram. The major and minor net principal stresses, $\sigma_1 - u_a$ and $\sigma_3 - u_a$ are plotted on the abscissa while the center of the Mohr circle is located at $(\sigma_1 + \sigma_3)/2 - u_a$. The radius of the circle is $(\sigma_1 - \sigma_3)/2$. The Mohr circle represents the net normal stress and shear stress on any plane through a point in an unsaturated soil.

The net normal stress and shear stress on any plane can be determined if the pole point or the origin of planes is known. Any plane drawn through the pole point will intersect the Mohr diagram and give the net normal stress and the shear stress acting on that plane. On the other hand, if the net normal stress and shear stress on a plane are known and plotted as a stress point on the Mohr circle, the direction of the plane under consideration is given by the orientation of a line joining the stress point and the pole point.

The pole point for the condition shown in Fig. 3.20 is determined from the known net normal stress and shear stress on a particular plane. Consider, for example, the case where the major principal stress acts on a horizontal plane. The stress

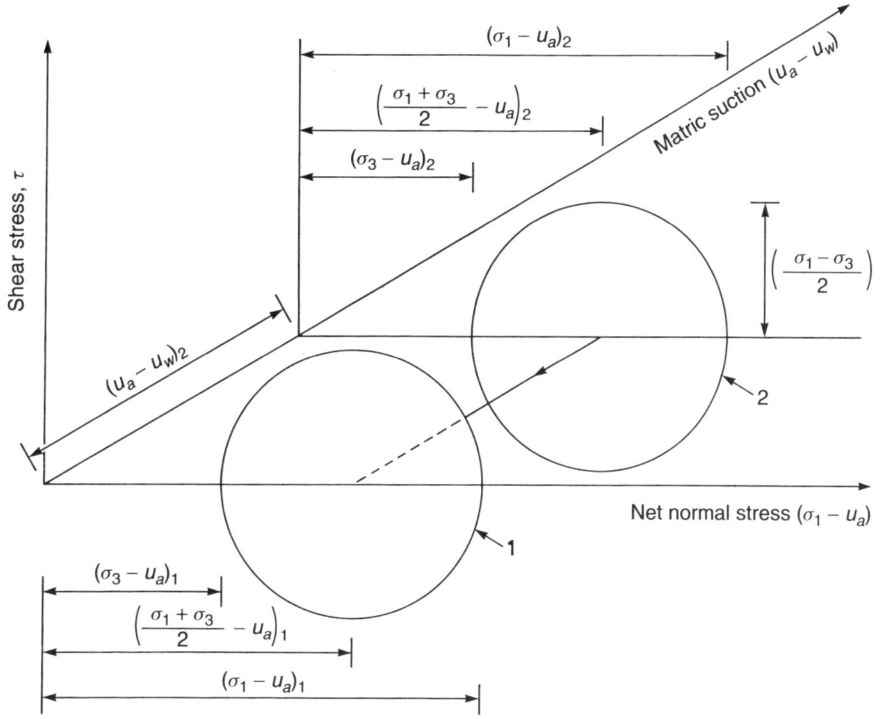

Figure 3.19 Use of extended Mohr diagram to represent stress states in unsaturated soil.

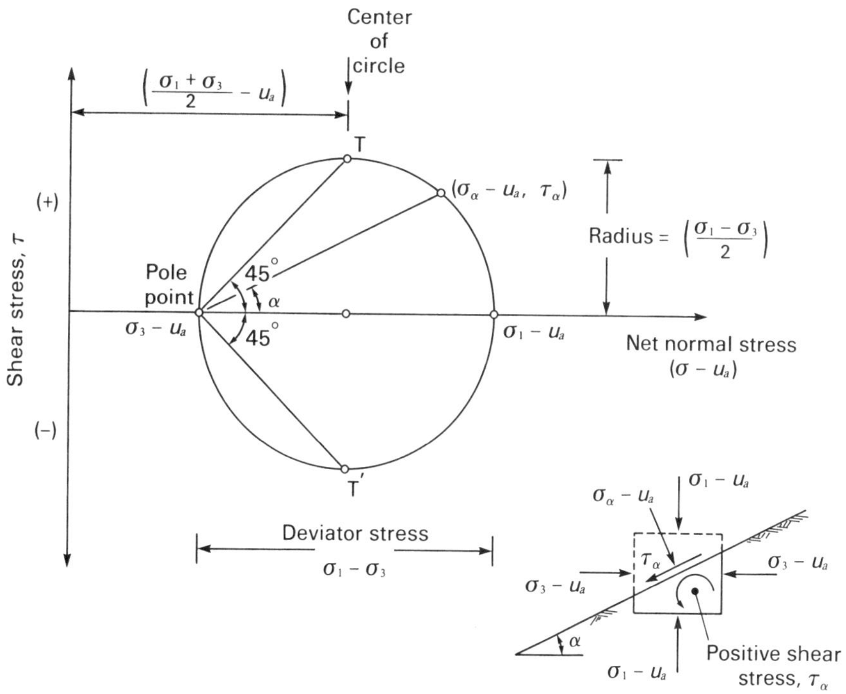

Figure 3.20 Construction of Mohr circle using net normal stresses.

condition on the horizontal plane is represented by the stress point $\sigma_1 - u_a$ on the Mohr circle. If a horizontal line is drawn through the stress point $\sigma_1 - u_a$, the line will intersect the Mohr circle at the stress point $\sigma_3 - u_a$. This is the pole point. The net normal stress and shear stress on the inclined plane shown in Fig. 3.20 can then be determined using the same pole point. A line at an orientation α can be drawn through the pole point to intersect the Mohr circle at the stress point $(\sigma_\alpha - u_a)$, τ_α. The horizontal coordinate of the intersection point is the net normal stress $\sigma_a - u_a$ acting on the inclined plane (Fig. 3.20). The shear stress τ_α on the plane is positive and is given by the ordinate of the intersection point.

The plane with the maximum shear stress $+(\sigma_1 - \sigma_3)/2$ goes through the top point of the Mohr circle (i.e., stress point T in Fig. 3.20). The maximum negative shear stress $-(\sigma_1 - \sigma_3)/2$ occurs at the bottom point T' on the Mohr circle. The planes with the maximum positive and negative shear stresses are oriented at an angle of 45° from the principal planes or from the horizontal and vertical planes in this case.

The principal planes are not always the vertical and horizontal planes. A more general case is shown in Fig. 3.21 where shear stresses may be present on the vertical and horizontal planes. The principal stresses and principal planes can be found graphically using the known stresses on the vertical and horizontal planes. The vertical net normal stress $\sigma_y - u_a$ is a compressive stress, and the horizontal net normal stress $\sigma_x - u_a$ is negative because it is in tension. The matric suction $u_a - u_w$ acts on every plane with equal magnitude. The shear stresses τ_{xy} and τ_{yx} are always equal in magnitude and opposite in sign.

The extended Mohr circle for the stress state shown in Fig. 3.21 is presented in Fig. 3.22. The Mohr circle is drawn on the τ and $(\sigma - u_a)$ plane. Its position along the $u_a - u_w$ axis is determined by the magnitude of the matric suction. The first step in plotting the Mohr diagram is to plot the stress points which represent the stresses corresponding to the vertical and horizontal planes (i.e., $(\sigma_x - u_a)$, τ_{xy} and $(\sigma_y - u_a)$, τ_{yx}, respectively). A line joining the two stress points intersects the $\sigma - u_a$ axis at a point $(\sigma_x + \sigma_y)/2 - u_a$. The intersection point is the center for the Mohr circle. The Mohr circle can then be drawn with the two stress points forming the diameter of the circle. The intersection points between the Mohr circle and the $\sigma - u_a$ axis (i.e., where the shear stress is zero) are the net major and net minor principal stresses (i.e., $\sigma_1 - u_a$ and $\sigma_3 - u_a$) (Fig. 3.22). The net minor principal stress is negative, which indicates that the net minor principal stress is in tension.

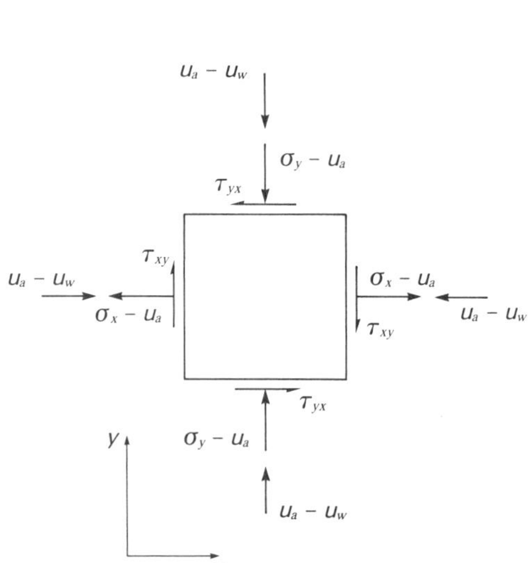

Figure 3.21 Stress state at point in unsaturated soil.

The second step is to locate the pole point by drawing a horizontal plane through the stress point $(\sigma_y - u_a)$, τ_{yx}. The intersection of the horizontal line and the Mohr circle is the pole point. The pole point can also be obtained by drawing a vertical line from the stress point corresponding to the vertical plane (i.e., $(\sigma_x - u_a)$, τ_{xy}). A line joining the pole point and the net major or net minor principal stress point gives the orientation of the major or minor principal plane (Fig. 3.22). The major and minor principal planes are at an angle of α and β with respect to the horizontal, respectively.

The top and the bottom stress points on the Mohr circle correspond to the planes on which the maximum and minimum shear stresses occur. The maximum and minimum shear stress planes are oriented at an angle of 45° from the principal planes.

3.5.2 Stress Invariants

There are three principal stresses on three mutually orthogonal principal planes in a three-dimensional analysis. The three principal stresses are named according to their magnitudes, namely, the net major stress, the net intermediate stress, and the net minor principal stress. The symbols used for the net major, net intermediate, and net minor principal stresses are $\sigma_1 - u_a$, $\sigma_2 - u_a$, and $\sigma_3 - u_a$, respectively. A corresponding Mohr circle is shown in Fig. 3.23. The matric suction acts equally on all three principal planes.

The principal stresses at a point can be visualized as the characterization of the physical state of stress. These principal stresses are independent of the selected coordinate system. The independent properties of principal stresses are expressed in terms of constants called stress invariants. There are three stress invariants that can be derived from each of the two independent stress tensors for an unsaturated soil. The first stress invariants of the first and second stress tensors, respectively, are

$$I_{11} = \sigma_1 + \sigma_2 + \sigma_3 - 3u_a \tag{3.50}$$

and

$$I_{12} = 3(u_a - u_w) \tag{3.51}$$

where:

I_{11} = first stress invariant of the first tensor and
I_{12} = first stress invariant of the second tensor.

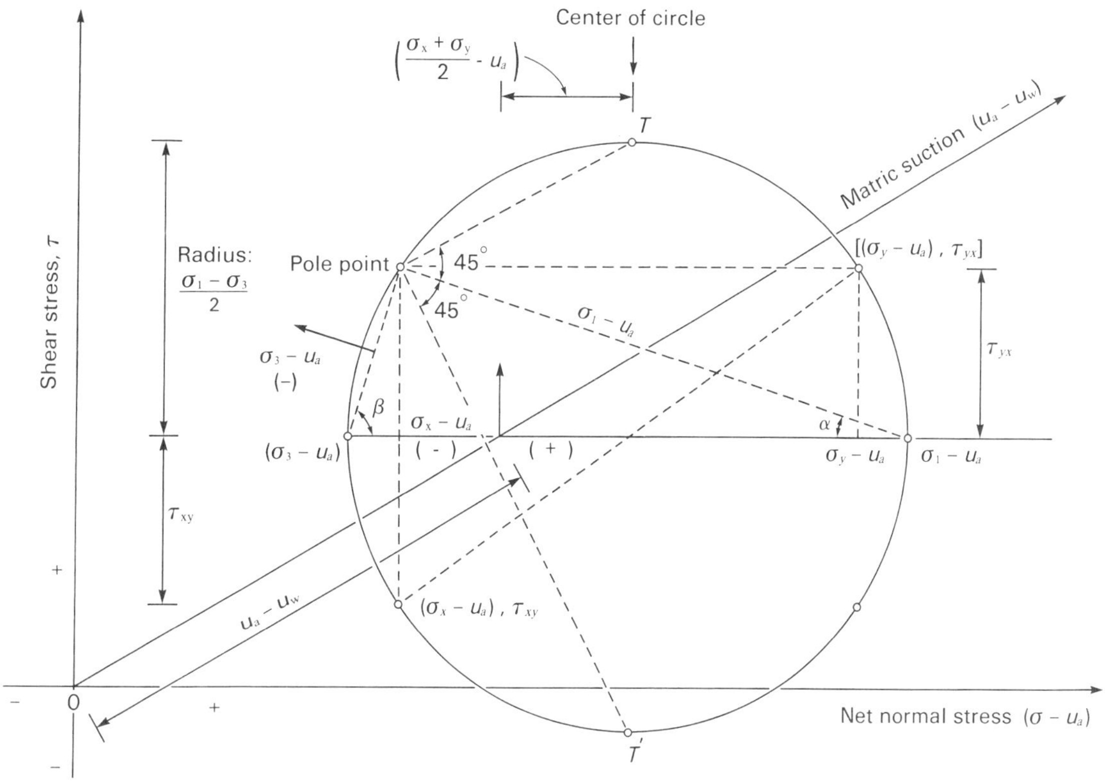

Figure 3.22 Extended Mohr diagram showing general stress state for unsaturated soil.

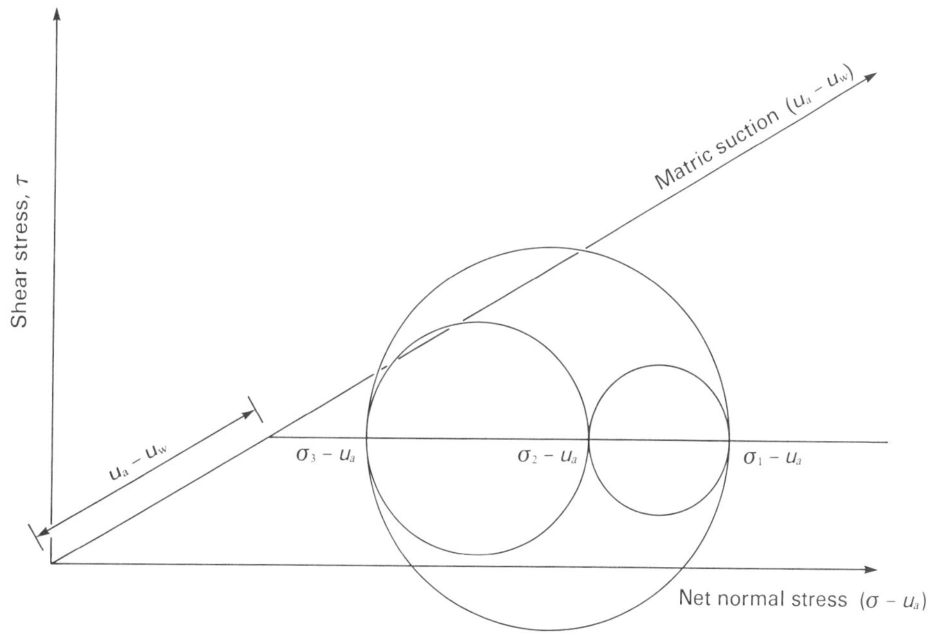

Figure 3.23 Mohr diagram for representing three-dimensional stress state of unsaturated soil.

The second stress invariants of the first and second stress tensors, respectively, are

$$I_{21} = (\sigma_1 - u_a)(\sigma_2 - u_a) + (\sigma_2 - u_a)(\sigma_3 - u_a) + (\sigma_3 - u_a)(\sigma_1 - u_a) \quad (3.52)$$

and

$$I_{22} = 3(u_a - u_w)^2 \quad (3.53)$$

where:

I_{21} = second stress invariant of the first tensor and
I_{22} = second stress invariant of the second tensor.

The third stress invariants of the first and second stress tensors, respectively, are

$$I_{31} = (\sigma_1 - u_a)(\sigma_2 - u_a)(\sigma_3 - u_a) \quad (3.54)$$

$$I_{32} = (u_a - u_w)^3 \quad (3.55)$$

where:

I_{31} = third stress invariant of the first tensor and
I_{32} = third stress invariant of the second tensor.

The stress invariants of the second tensor, I_{12}, I_{22}, and I_{32}, are related as follows:

$$I_{32} = \left(\tfrac{1}{3}I_{12}\right)^3 = \tfrac{1}{9}I_{12}I_{22} \quad (3.56)$$

Therefore, only one stress invariant is required to represent the second tensor. In other words, a total of four stress invariants are required to characterize the stress state of an unsaturated soil as opposed to three stress invariants for a saturated soil.

3.5.2.1 *Stress Points (Lambe, 1967)*

Geotechnical engineering analyses often require an understanding of the development or change in the stress state resulting from following various loading patterns. These changes can be visualized by drawing a series of Mohr circles which follow the loading process. However, the pattern of the stress state change may become confusing when the loading pattern is complex. There are benefits associated with using an approximate single-stress-point representation from a Mohr circle to indicate the stress path followed during a process (Lambe, 1967). A selected stress point at the top of the Mohr circle (i.e., point of maximum deviator stress) can be used to represent the stress path being followed. The stress point suggested by Lambe (1967) is representative of stress changes in a two-dimensional plane strain problem.

Figure 3.24 shows a Mohr circle for a two-dimensional case where the vertical and horizontal planes are principal planes. The stress point selected to represent the Mohr circle has coordinates of (p, q, r), where

$$p = \left(\frac{\sigma_v + \sigma_h}{2} - u_a\right) \quad \text{or} \quad p = \left(\frac{\sigma_1 + \sigma_3}{2} - u_a\right) \quad (3.57)$$

$$q = \left(\frac{\sigma_v - \sigma_h}{2}\right) \quad \text{or} \quad q = \left(\frac{\sigma_1 - \sigma_3}{2}\right) \quad (3.58)$$

$$r = u_a - u_w \quad (3.59)$$

where:

$(\sigma_v - u_a)$ = vertical net normal stress and
$(\sigma_h - u_a)$ = horizontal net normal stress.

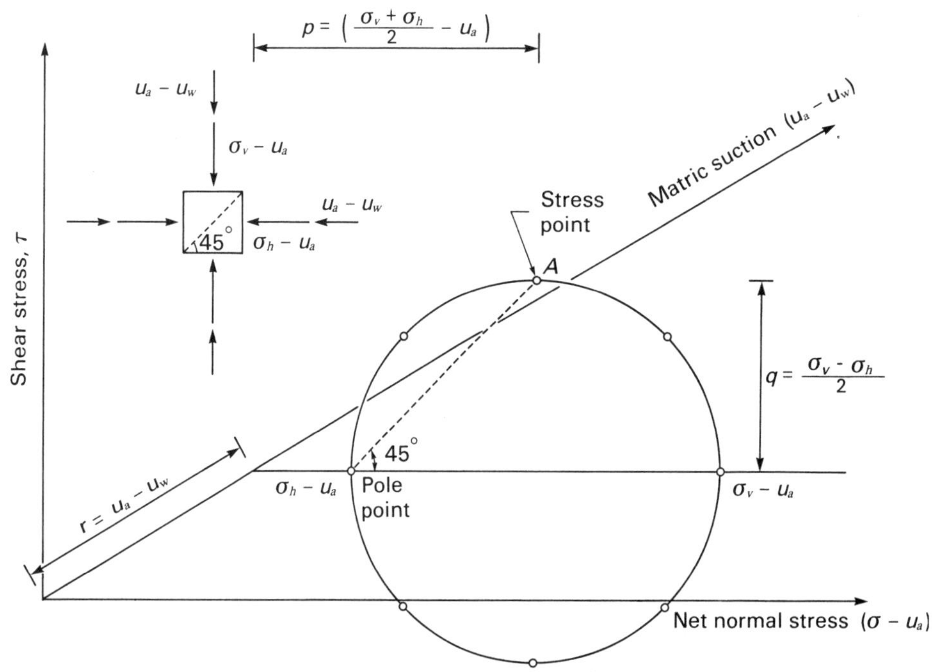

Figure 3.24 Designation of stress point for unsaturated soil using extended Mohr circle.

The q-coordinate is one half the deviator stress $(\sigma_v - \sigma_h)$. The selected stress point represents the state of stress on a plane with an orientation of 45° from the principal planes, as shown in Fig. 3.24.

The vertical net normal stress for the condition shown in Fig. 3.24 is greater than the horizontal net normal stress (i.e., $\sigma_v - u_a$ is greater than $\sigma_h - u_a$). This results in a positive q-coordinate. A negative q-coordinate would indicate the condition where $\sigma_h - u_a$ is greater than $\sigma_v - u_a$. For the hydrostatic or isotropic stress state (i.e., $\sigma_h - u_a = \sigma_v - u_a$), the q-coordinate is equal to zero. A zero q-coordinate means the absence of shear stresses.

There are three independent stress points and as a result it is somewhat difficult to graphically present the stress paths being followed in practical engineering problems.

3.5.2.2 Stress Points (Elastoplastic system; Wheeler and Sivakumar, 1995)

The laboratory testing of saturated and unsaturated soils is most commonly performed in a triaxial test. The mean net total stress can be defined for the triaxial test specimen as the vertical total stress plus two times the horizontal stresses. In other words, the stress state is defined for the special three-dimensional coordinate system defined by the triaxial test.

The triaxial test representation of the stress state can then be used as the reference condition for describing the physical properties of the soil. Wheeler (1996) suggested the use of the net mean total stress, the maximum principal stress difference, and the matric suction to describe the stress path followed when testing and analyzing an unsaturated soil problem:

$$p = \left(\frac{\sigma_v + 2\sigma_h}{3} - u_a\right) \quad \text{or} \quad p = \left(\frac{\sigma_1 + \sigma_2 + \sigma_3}{3} - u_a\right) \tag{3.60}$$

$$q = \sigma_v - \sigma_h \quad \text{or} \quad q = \sigma_1 - \sigma_3 \tag{3.61}$$

$$r = u_a - u_w \tag{3.62}$$

where:

σ_2 = intermediate principal stress.

The shear stress is defined in terms of the difference between the maximum and minimum principal stresses, $\sigma_1 - \sigma_3$.

Wheeler (1996) selected specific volume as the deformation state variable for overall volume change or soil structure deformations. The state variables proposed were used to advance an elastoplastic framework for unsaturated soil behavior. Once again, it is a challenge to graphically illustrate multidimensional stress and deformation changes due to the number of variables involved.

3.6 ROLE OF OSMOTIC SUCTION

Total suction ψ is made up of two primary components, namely, matric suction $u_a - u_w$ and osmotic suction π, and can be mathematically represented by

$$\psi = (u_a - u_w) + \pi \tag{3.63}$$

Matric suction is known to vary with time mainly as a result of conditions imposed by environmental changes. Any change in suction affects the overall equilibrium of the soil mass. Changes in soil suction may be caused by a change in either one or both components of soil suction.

The role of osmotic suction has been associated more with unsaturated soils than with saturated soils. However, osmotic suction is related to the salt content in the pore-water. It should be remembered that salts are present in the pore-water of both saturated and unsaturated soils. The role of osmotic suction is therefore equally applicable to both unsaturated and saturated soils. Osmotic suction changes can have an effect on the mechanical behavior of a soil. There can be a change in the overall volume and shear strength of the soil if the salt content changes.

Most engineering problems involving unsaturated soils are commonly the result of environmental changes. The accumulation of water below a house may result in a reduction in matric suction and subsequent heaving of the structure. Similarly, the stability of an unsaturated soil slope may be endangered by excessive rainfall that reduces the suction in the soil. These weather-imposed conditions primarily affect the matric suction component. Osmotic suction changes are generally less significant and consequently receive less attention. However, there are situations where the salt content of a soil is changed and, as a result, there can be a change in the physical properties of the soil.

Figure 3.25 shows the relative importance of changes in osmotic suction as compared to changes in matric suction when water content is varied. The total and matric suction curves are almost congruent one to another, particularly in the higher water content range. In other words, a change in total suction is essentially equivalent to a change in the matric suction [i.e., $\Delta\psi \approx \Delta(u_a - u_w)$]. Matric suction changes are almost the same as total suction changes, and vice versa, for many geotechnical problems involving unsaturated soils as long as additional salts are not added to the soil.

There is another reason why it is generally not necessary to take osmotic suction into account. The reason is related to the laboratory test procedures adopted when solving geotechnical engineering problems. Changes in osmotic suction that occur in the field are essentially simulated during the laboratory testing for pertinent soil properties. For example, let us consider the swelling process of a soil as a result of rainfall infiltration into the soil. The rainfall, which is essentially distilled water, dilutes the pore-water and changes the osmotic suction. In the laboratory, the soil

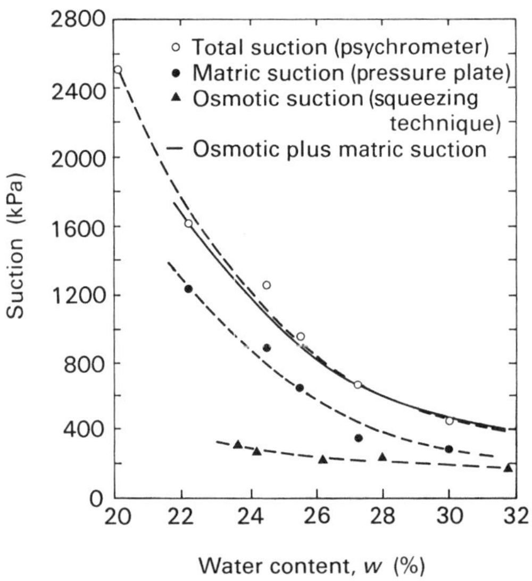

Figure 3.25 Total, matric, and osmotic suction measurements on compacted Regina clay (from Krahn and Fredlund, 1972).

specimen is generally immersed in distilled water at the start of the test (e.g., volume change test in an oedometer). The matric suction is reduced to zero by immersing the soil specimen. The osmotic suction in the sample may also be changed in the process. It is not necessary to know the change in osmotic suction provided changes occurring in the field are simulated in the laboratory test.

In the case where the salt content of the soil is altered by chemical contamination or deliberate chemical change, the effect of the osmotic suction change on the soil behavior may be significant. In this case, it is necessary to consider osmotic suction as part of the stress state or as an independent stress state variable. The significance of the osmotic suction component can apply equally for saturated and unsaturated soils.

The role played by osmotic suction in influencing the mechanical behavior of a soil may or may not bear a one-to-one correspondence with the role played by matric suction. The osmotic suction is more closely related to the diffuse double layer around the clay particles whereas matric suction is mainly associated with the air-water interface (i.e., contractile skin). It is possible and likely more reasonable to consider osmotic suction π as an independent, isotropic stress state variable.

In the case where matric and osmotic suctions have the same quantitative influence on the behavior of a soil, the soil suction stress tensor can be written in the form

$$\begin{bmatrix} (u_a - u_w) + \pi & 0 & 0 \\ 0 & (u_a - u_w) + \pi & 0 \\ 0 & 0 & (u_a - u_w) + \pi \end{bmatrix} \tag{3.64}$$

Research has been published that suggests it may be possible to algebraically combine matric suction and osmotic suction when analyzing some geotechnical problems (Bailey, 1965; Chattopadhyay, 1972).

3.6.1 Independence of Components of Soil Suction

Total suction ψ has two primary components of suction, namely, matric suction $u_a - u_w$ and osmotic suction π. There may be a tendency to think that the components of suction can always be added together and then used in a combined form. However, each component of soil suction may act in independent manners in the soil.

For example, it is unacceptable to use total suction to model water flow in an unsaturated soil. It is possible for the matric suction to result in water flow in a particular direction while the chemical concentration gradient may result in flow in another direction. The independent flow mechanisms can only be taken into consideration if the matric suction and osmotic suction are handled in an independent manner.

Equation 3.63 simply provides a mathematical relationship between the components of suction and can only be used for that purpose. Through use of thermodynamic considerations, Edlefsen and Anderson (1943) showed that Lord Kelvin's equation applied to both capillary phenomena as well as osmotic phenomenon. However, this does not mean that the components of suction can be added and then used to describe all physical processes.

Total suction is measured through the assessment of vapor pressure. Evaporation from a soil is controlled through a vapor pressure gradient. The calculation of actual evaporation or potential evaporation from a soil surface requires a vapor pressure gradient in the soil or external to the soil.

Matric suction is written as the difference between the pore-air pressure u_a and the pore-water pressure u_w. Matric suction does not cause liquid or vapor flow. Rather, it is the independent components of air and water that can be used to calculate the flow of the independent phases. A gradient in the air density or an air concentration gradient can cause the flow of air through a soil. Water pressure combines with elevation head to produce hydraulic head, which causes the flow of water through a soil. It is possible to have water flowing in one direction through a soil while having air flow in another direction. In other words, total suction or total potential should not be used to model the process of water flow.

Osmotic suction is related to the salt concentration in the free pore-water of a soil. Differences in salt concentration at different points in a soil can cause the movement of salts due to a concentration gradient. However, there may or may not be any movement of water.

Equation 3.63 is of significant value when solving unsaturated soil mechanics problems; however, it is important that the relationship be used in a proper manner. Formulations that use total suction to model the flow of liquid water through a soil would constitute a questionable usage of Eq. 3.63. The physics associated with a particular physical mechanism must be adhered to in order to ensure the correct usage of soil suction.

3.6.2 Description of Osmosis

Osmosis is the term used to describe the phenomenon by which a solvent passes from a solution of lower solute concentration through a semipermeable membrane into a solution of higher solute concentration. A membrane is described as semipermeable if it allows the passage of solvent but not solute. If the flow of water is restricted, a pressure imbalance equal to the osmotic pressure difference between the two solutions needs to be present. Osmotic pressure can be calculated from thermodynamic principles (Robinson and Stokes, 1968). Osmotic pressure can also be approximated by the van't Hoff equation (Metten, 1966):

$$\pi = RT_K C \tag{3.65}$$

where:

π = osmotic pressure, kPa,
C = sum of the molar concentrations in solution, mol/L,
R = universal gas constant = 8.314 J/(mol·K), and
T_K = absolute temperature, K.

3.6.3 Mechanisms Associated with Osmotic Suction

Conceptual models have been used to explain the influence of pore fluid chemistry on the mechanical behavior of clays in a quantitative way (Barbour, 1987). Macroscopic descriptions of physical processes such as the flow of the pore fluid, volume change, and change in shear strength must all commence with the selection of suitable state variables. Studies on two particular processes related to the physicochemical processes in clay soils have assisted in understanding the state variables that should be used to describe mechanical processes. The processes can be referred to as (i) osmotic consolidation and (ii) osmotically induced consolidation (Barbour and Fredlund, 1989a).

Osmotic consolidation occurs as a result of a change in the electrostatic repulsive stress minus the attractive stress ($R - A$) between clay particles. Osmotically induced consolidation occurs as a result of fluid flow from the clay in response to osmotic gradients. The volume of a clay specimen will change as a result of the combined processes of osmotic flow and osmotic compressibility when clay is exposed to a concentrated salt solution. Changes occur in the interparticle repulsive stresses as salt moves into the clay. These changes lead to the suppression of the double layer and changes in void ratio. Changes in the osmotic suction π in the pore fluid are referred to as osmotic consolidation.

There is another volume change process that occurs as a result of fluid flow responding to osmotic gradients. An osmotic gradient between the electrolyte concentration inside the clay and the external electrolyte solution results in water flow in or out of the clay soil. An outward flow causes negative pore fluid pressures to develop within the specimen, which then leads to increases in the intergranular stresses between particles and possible volume changes in the soil. Changes in pore-water pressure are the result of osmotic pressures and are called osmotically induced consolidation.

The osmotic flow of water through a soil can be described using a flow law:

$$q_\pi = k_\pi \frac{\Delta \pi}{\Delta x} = o_e k_h \frac{\Delta \pi}{\Delta x} \tag{3.66}$$

where:

q_π = water flux, m/s,
k_π = coefficient of osmotic permeability, m/s,
k_h = coefficient of (hydraulic) permeability, m/s,
π = osmotic pressure head ($\pi_o/\rho_f g$), m,
π_o = osmotic pressure, kPa,
ρ_f = pore fluid density, kg/m^3,
g = gravitational acceleration, m/s^2,
o_e = osmotic efficiency, and
x = distance, m.

The coefficient of osmotic permeability, k_π, is similar to the coefficient of hydraulic permeability if the soil behaves as a perfect semipermeable membrane. In this case, pure water will flow in response to osmotic gradients. However, if the membrane is "leaky," the osmotic permeability k_π will be the hydraulic permeability multiplied by an osmotic efficiency. The moving water will carry some dissolved salts in this situation. Osmotic efficiency is a measure of the degree to which the clay behaves as a perfect semipermeable membrane. The above description clearly defines the state variables associated with flow even though the efficiency of the water flow system is not perfect.

Force equilibrium considerations can be applied to a saturated clay-water system within the context of multiphase continuum mechanics. The procedure is similar to that used for determining the state variables associated with a saturated or unsaturated soil. Barbour and Fredlund (1989b) analyzed a clay-water system from a continuum mechanics equilibrium standpoint. The soil was considered as a three-phase system consisting of soil particles, pore fluid, and the diffuse double hull surrounding each clay particle. The resulting stress state tensor was as follows:

$$\begin{bmatrix} (\sigma_x - u_f) - (R - A) & \tau_{yx} & \tau_{zx} \\ \tau_{xy} & (\sigma_y - u_f) - (R - A) & \tau_{zy} \\ \tau_{xz} & \tau_{yz} & (\sigma_z - u_f) - (R - A) \end{bmatrix} \tag{3.67}$$

where:

$\sigma_x, \sigma_y, \sigma_x$ = total normal stresses,
$\tau_{xy}, \tau_{yx}, \tau_{yz}, \tau_{zy}, \tau_{zx}, \tau_{xz}$ = shearing stresses,
u_f = pore fluid pressure, and
$R - A$ = net interparticle repulsive stress minus attractive stress.

The stress state variable for the normal stress in Eq. 3.67 can be subdivided into two components. The first component is the conventional effective stress for a saturated soil ($\sigma - u_f$). The second component is the net interparticle repulsive stress minus attractive stress ($R - A$). The two stress components can also be visualized as a single stress state variable ($(\sigma - u_f) - (R - A)$) provided there is a means of measuring the $R - A$ stress variable. This expression is similar to that proposed equation for "true" effective stress by Balasubramonian (1972) and Chattopadhyay (1972):

$$\sigma^* = \sigma - u_f - (R - A) \tag{3.68}$$

where:

σ^* = true effective stress.

Two requirements need to be met to justify the usage of a combination of stresses as a single stress state variable. The first requirement is that a particular change in one component of the stress state variable be equally as effective in changing physical soil properties as changing the other component of the stress state variable. Fredlund and Morgenstern (1977) demonstrated that a null-type test could be used to confirm the correct components for the stress state variables for unsaturated soils using this technique. Balasubramonian (1972), Morgenstern and Balasubramonian (1980), and Chattopadhyay (1972) experimentally demonstrated that calculated values of $R - A$ produced equal changes in effective stress when constant-volume conditions were maintained.

The second requirement for the use of a combined stress state variable is that all components of the proposed stress state variable be measurable. When using the true effective stress σ^*, it is not possible to directly measure $R - A$. The computation of $R - A$ is possible only under ideal conditions. This results in a severe restriction to the use of $R - A$ as a component of stress state. However, it is still possible to use an independent approximation of $R - A$ as a state variable provided it is used in a manner independent of the other components shown in Eq. 3.68.

The osmotic suction π concept relates the change in repulsive stress between clay particles to the change in the osmotic suction between the interparticle fluid and the bulk pore fluid. Mitchell (1962) suggested that the double layer interactions and the subsequent repulsions between particles are a reflection of osmotic pressure.

Osmotic pressure can be visualized as having two components: one component associated with the free salt in the bulk solution and the other component caused by the additional ions required to satisfy the double diffuse layer. The osmotic suction π of the bulk solution can be evaluated using samples of the pore fluid; however, the net interparticle repulsive stress minus attractive stress ($R - A$) is difficult to measure.

CHAPTER 4

Measurement and Estimation of State Variables

4.1 INTRODUCTION

The state variable of greatest relevance to unsaturated soil mechanics is soil suction. Soil suction is the general term which may be used when referring to matric suction, osmotic suction, or total suction. There are also other state variables associated with unsaturated soils that may need to be measured or estimated, namely, the amount of water in the soil or its natural water content.

The variables of soil suction and water content measurements combine to form the SWCC, a function that is used for the assessment of unsaturated soil properties. Unsaturated soil mechanics engineering designs can be verified in the field through in situ measurement of soil suction and water content. In situ measurements can verify that the actual response of an unsaturated soil in the field is consistent with the SWCC used during design. There are some situations where it is sufficient to simply measure soil suction in situ or water content in situ. These situations are discussed later.

There are other state variables that might need to be measured in the field for verification purposes. There are variables that may be related to ground movements in the horizontal and/or vertical directions. These measurements are related to the deformation state variables. The instrumentation required for the measurement of field movements is quite similar for both saturated and unsaturated soils. Limited attention will be given to the measurement of ground movements in this chapter.

4.2 MEASUREMENT OF SOIL SUCTION

It is paramount to be able to measure soil suction in the laboratory and in the field. The application of unsaturated soil mechanics in engineering practice is closely related to the ability to measure soil suction. Soil suction (i.e., matric suction and total suction) has proven to be a difficult variable to measure. The cavitation of water in soil suction measuring systems has made it difficult to measure soil suctions that exceed 1 atm of tension. While much progress has been made with regard to the measurement of soil suction, there are still significant challenges to overcome. The geotechnical engineer should understand the physics associated with soil suction and the range of suction values that can be measured with various suction devices.

4.2.1 Theory Related to Measurement of Soil Suction Components

The theoretical concept of soil suction was developed in soil physics in the early 1900s (Buckingham, 1907; Gardner and Widtsoe, 1921; Richards, 1928; Schofield, 1935; Edlefsen and Anderson, 1943; Childs and Collis-George, 1948; Bolt and Miller, 1958; Corey and Kemper, 1961; Corey et al., 1967). The soil suction theory was mainly developed in relation to soil-water-plant systems. The importance of soil suction in explaining the mechanical behavior of unsaturated soils encountered in civil engineering was introduced at the Road Research Laboratory in England (Croney and Coleman, 1948; Croney et al., 1950). In 1965, the review panel for the soil mechanics symposium "Moisture Equilibria and Moisture Changes in Soils" (Aitchison, 1965) provided quantitative definitions of soil suction and its components from a thermodynamic standpoint. These definitions have become accepted concepts in geotechnical engineering (Krahn and Fredlund, 1972; Wray, 1984; Fredlund and Rahardjo, 1988).

The components of soil suction have a common context when considering the free-energy state of soil-water (Edlefsen and Anderson, 1943). The free energy of the soil-water can be measured in terms of the partial vapor pressure adjacent to the soil-water (Richards, 1965). The thermodynamic relationship between soil suction (or the free energy of the soil-water) and the partial pressure of the pore-water vapor can be written as follows:

$$\psi = -\frac{RT_K}{v_{w0}\omega_v}\ln\left(\frac{\bar{u}_v}{\bar{u}_{v0}}\right) \tag{4.1}$$

where:

ψ = soil suction or total suction, kPa,
R = universal (molar) gas constant [i.e., 8.31432 J/(mol K)],

T_K = absolute temperature [i.e., $T_K = 273.15 + T$ (K)],
T = temperature, °C,
v_{w0} = specific volume of water or the inverse of the density of water (i.e., $1/\rho_w$), m^3/kg,
ρ_w = density of water (i.e., 998 kg/m^3 at temperature $T = 20$ °C),
ω_v = molecular mass of water vapor (i.e., 18.016 kg/kmol),
$\bar{u}_v$ = partial pressure of pore-water vapor, kPa, and
$\bar{u}_{v0}$ = saturation pressure of water vapor over a flat surface of pure water at the same temperature, kPa.

The reference state for quantifying the components of suction is the vapor pressure above a flat surface of pure water (i.e., water with no salts or impurities). The relative vapor pressure in the air immediately adjacent to the water, $\bar{u}_v/\bar{u}_{v0}$, is called relative humidity (h_r or RH, %). If we select a reference temperature of 20 °C, the constants in Eq. 4.1 give a value of 135,022 kPa. Equation 4.1 can now be written as a fixed relationship between total suction in kilopascals and relative vapor pressure:

$$\psi = -135{,}022 \ln\left(\frac{\bar{u}_v}{\bar{u}_{v0}}\right) \tag{4.2}$$

Figure 4.1 shows a plot of Eq. 4.1 for three different temperatures. Soil suction ψ approaches zero when the relative humidity (i.e., $\bar{u}_v/\bar{u}_{v0}$) is equal to 100%. A relative humidity value less than 100% in a soil indicates the presence of suction in the soil. Figure 4.1 also shows that suction can become extremely large. For example, a relative humidity of 94.24% at a temperature of 20 °C corresponds to a soil suction of 8000 kPa. The range of soil suctions of most common interest in geotechnical engineering correspond to high-relative-humidity values.

4.2.2 Components of Soil Suction

Soil suction can be quantified in terms of the relative humidity immediately adjacent to the water surface and is called "total suction." There are two primary components to soil suction, namely, matric* and osmotic suction. There also appears to be a component of soil suction related to the adsorptive forces associated with water and a solid surface (Yong and Warkentin, 1966). However, it is difficult to isolate and quantify the adsorptive component of suction. It has been possible to formulate solutions for most geotechnical engineering problems through use of matric suction and osmotic suction, recognizing the sum of the two components as total suction.

Total, matric, and osmotic suctions can be defined as follows (Aitchison, 1964):

Matric or capillary component of free energy: Matric suction is the equivalent suction derived from the measurement of the partial pressure of the water vapor in equilibrium with the soil-water relative to the partial pressure of the water vapor in equilibrium with a solution identical in composition with the soil-water.

Osmotic (or solute) component of free energy: Osmotic suction is the equivalent suction derived from the measurement of the partial pressure of the water vapor in equilibrium with a solution identical in composition

*The spelling of the term "matric" is in accordance with the recommendation of the Committee on Terminology of the Society of Soil Science of America. The definition is from their *Glossary of Soil Science Terminology* (1963, 1970, and 1979).

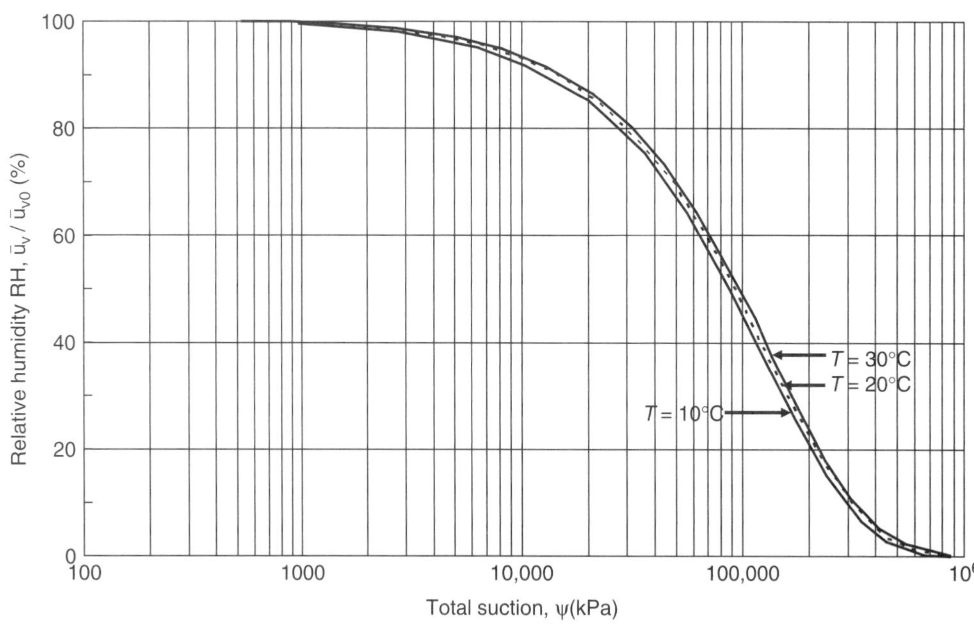

Figure 4.1 Thermodynamic equilibrium between relative humidity and total suction.

with the soil-water relative to the partial pressure of water vapor in equilibrium with free pure water.

Total suction or free energy of soil-water: Total suction is the equivalent suction derived from the measurement of the partial pressure of the water vapor in equilibrium with the soil-water relative to the partial pressure of water vapor in equilibrium with free pure water.

The above definitions state that total suction corresponds to the free energy of the soil-water, while the matric and osmotic suctions are the components of the free energy. The mathematical relationship between the components of soil suction can be written as follows:

$$\psi = (u_a - u_w) + \pi \tag{4.3}$$

where:

$u_a - u_w$ = matric suction, kPa,
u_a = pore-air pressure, kPa,
u_w = pore-water pressure, kPa, and
π = osmotic suction, kPa.

Figure 4.2 illustrates the concept of total suction and its components relative to the free energy of the soil-water. Matric suction is commonly associated with the capillary phenomenon arising from the surface tension of water. The capillary phenomenon can be illustrated by the rise of water in a capillary tube (Figure 4.2).

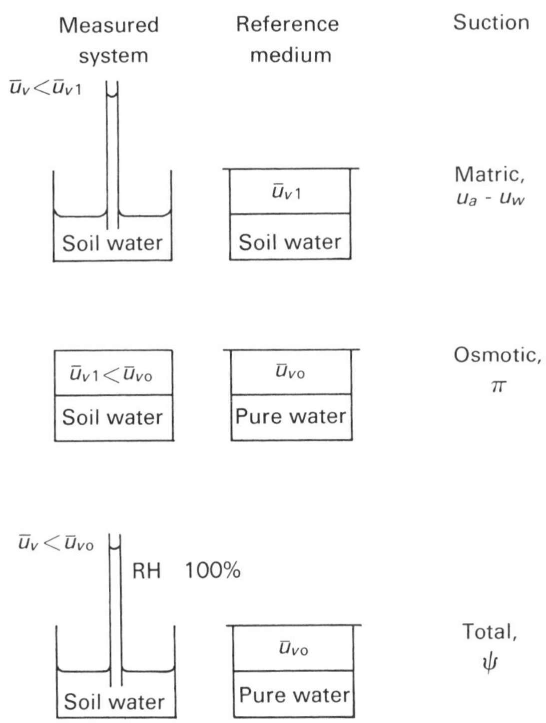

Figure 4.2 Total suction and components of matric suction and osmotic suction.

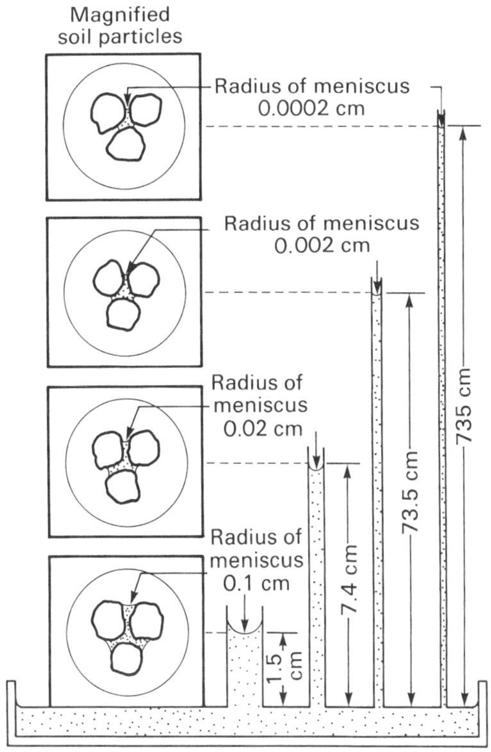

Figure 4.3 Capillary tubes showing air-water interfaces at varying heights for different radii of curvature (after Janssen and Dempsey, 1980).

The pores in a soil are analogous to capillary tubes with small radii. Soil-water rises above the water table because of the capillaries created by the soil (Fig. 4.3). The capillary water has a negative pressure with respect to the air pressure. The air pressure in the field is generally atmospheric (i.e., $u_a = 0$). The pore-water pressures can be highly negative at low degrees of saturation (Olson and Langfelder, 1965). The adsorptive forces between soil particles and water are believed to play an important role in sustaining highly negative pore-water pressures.

Let us consider a capillary tube filled with water squeezed from a soil. The surface of the water in the capillary tube is curved and is called a meniscus. The same soil water will have a flat surface when placed in a large container. The partial pressure of the water vapor above the *curved* surface of soil water, $\bar{u}_v$, is less than the partial pressure of water vapor above a *flat* surface of the same soil water, $\bar{u}_{v1}$ (i.e., $\bar{u}_v < \bar{u}_{v1}$). In other words, the relative humidity in a soil will decrease due to the presence of the curved water surface produced by the capillary phenomenon. The water vapor pressure or the relative humidity decreases as the radius of curvature of the water surface decreases. The radius of curvature is inversely proportional to the difference between the air and water pressures across the surface (i.e., $u_a - u_w$) and is called matric suction. Therefore, one component of total suction is matric suction, and it contributes to a reduction in the relative humidity.

The pore-water in a soil generally contains dissolved salts. The water vapor pressure over a flat surface of solvent (i.e., salt solution), $\bar{u}_{v1}$, is less than the water vapor pressure over a flat surface of pure water, $\bar{u}_{v0}$. In other words, the relative humidity decreases with increasing dissolved salts in the pore-water of the soil. The decrease in relative humidity due to the presence of dissolved salts in the pore-water is referred to as the osmotic suction π.

4.2.3 Typical Suction Values and Measuring Devices

Table 4.1 shows typical matric, osmotic, and total suction values for two soils often used as subgrade material for roads built in the province of Saskatchewan, Canada (Krahn and Fredlund, 1972). Regina clay is highly plastic, inorganic clay with a liquid limit (LL) of 78% and a plastic limit (PL) of 31%. Glacial till has a liquid limit of 34% and a plastic limit of 17%. Suction values are given in Table 4.1 for soils compacted to standard AASHTO conditions at water contents near optimum water content and 2% dry of optimum water content.

Figure 4.4 shows experimental data illustrating that matric suction plus osmotic suction are essentially equal to the total suction in a soil. The glacial till data presented in Fig. 4.4 are for specimens compacted to modified AASHTO conditions at various initial water contents. The components of soil suction (i.e., matric suction and osmotic suction) and total suction were independently measured.

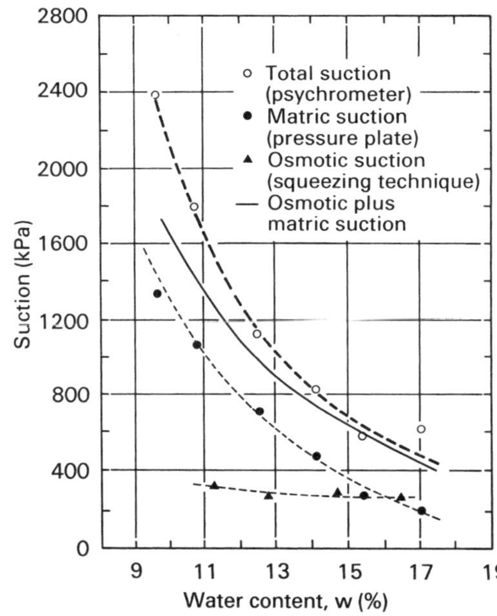

Figure 4.4 Total, matric, and osmotic suctions for glacial till (from Krahn and Fredlund, 1972).

Several devices commonly used for measuring total, matric, and osmotic suctions are listed in Table 4.2 along with the range over which measurements can be made.

4.2.4 Measurement of Matric Suction

Matric suction is the difference between the pore-air and pore-water pressures ($u_a - u_w$) and can be measured either in a direct or indirect manner. Direct methods measure the negative pore-water pressure. The pore-air pressure component of matric suction is generally atmospheric in the field. Indirect methods measure a variable other than the negative pore-water pressure (e.g., thermal conductivity) through use of a specially designed ceramic material. A calibration of the measuring device is required to determine the matric suction in the soil when using an indirect method.

High-air-entry ceramic disks are used as the primary separator between air and water pressures when *direct* measurements are made of negative pore-water pressure. Properties of high-air-entry ceramic disks are presented to assist in understanding the operation of direct measurement devices. Conventional tensiometers measure matric suctions up to almost 1 atm. More recently, direct high-suction tensiometers have been developed based on a special treatment procedure where the water in the measuring device is subjected to high pressures prior to usage in the measuring device.

Several types of porous material sensors have been used for making *indirect* measurements of matric suction. A feature common to all indirect methods of measuring matric suction is the need to calibrate the sensors. Calibration is required because a property other than the pore-water pressure is being measured. The thermal or electrical properties of a specially designed porous material are a function of its water content. The water content of the ceramic, in turn, is a function of matric suction or the tension in the water phase.

A measurement of the electrical or thermal properties of the porous sensor provides an indication of the matric suction in the sensor and the surrounding soil through use of a calibration

Table 4.1 Typical Suction Values for Compacted Soils

Soil Type	Water Content ω, %	Matric Suction $(u_a - u_w)$, kPa	Osmotic Suction, π, kPa	Total Suction, ψ, kPa
Regina clay, $\gamma_{max} = 13.81$ kN/m^3	30.6 (optimum)	273	187	460
	28.6	354	202	556
Glacial till, $\gamma_{max} = 19.24$ kN/m^3	15.6 (optimum)	310	290	600
	13.6	556	293	849

Table 4.2 Devices for Measuring Soil Suction and Its Components

Name of Device	Suction Component Measured	Range, kPa	Comments
Psychrometers (Peltier type)	Total	100[a] to ~8000	Constant-temperature environment required
Filter paper	Total	Entire range	May measure matric suction when in good contact with moist soil
Tensiometers	Negative pore-water pressures or matric suction when pore-air pressure is atmospheric	0–90	Difficulties with cavitation and air diffusion through ceramic cup
Null-type pressure plate (axis translation)	Matric	0–1500	Range of measurement is a function of the air-entry value of ceramic disk
Thermal conductivity sensors	Matric	10 to ~1500	Indirect measurement using variable-pore-size ceramic sensor
Pore fluid squeezer	Osmotic	Entire range	Used in conjunction with psychrometer or electrical conductivity measurement

[a]Controlled temperature environment to ±0.001°C.

curve. The electrical output of the indirect sensor must first be calibrated against a series of applied or known suction values. The sensors are then ready for usage in measuring matric suction in the soil. The indirect methods of measuring matric suction based on the thermal properties of a standard porous ceramic are described later in this chapter.

4.2.4.1 High-Air-Entry Ceramic Disks

High-air-entry ceramic disks have small pores of relatively uniform size. The disk acts as a membrane or separator between air and water (Fig. 4.5). The disk is generally made of a ceramic* material such as sintered kaolin. Once the high-air-entry disk is saturated with water, air cannot pass through the disk since the contractile skin resists the flow of free air.

The ability of the ceramic disk to withstand the flow of free air is related to the surface tension T_s of the contractile skin. The contractile skin acts as a thin membrane joining the small pores of radius R_s on the surface of the ceramic disk. The difference between the air pressure above the contractile skin and the water pressure below the contractile skin is equal to matric suction. The maximum matric suction that can be maintained across the surface of the disk is called its air-entry value $(u_a - u_w)_d$. The air-entry value of the disk can be illustrated using Kelvin's capillary equation:

$$(u_a - u_w)_d = \frac{2T_s}{R_s} \tag{4.4}$$

*Manufactured by Soilmoisture Equipment Corporation Santa Barbara, CA.

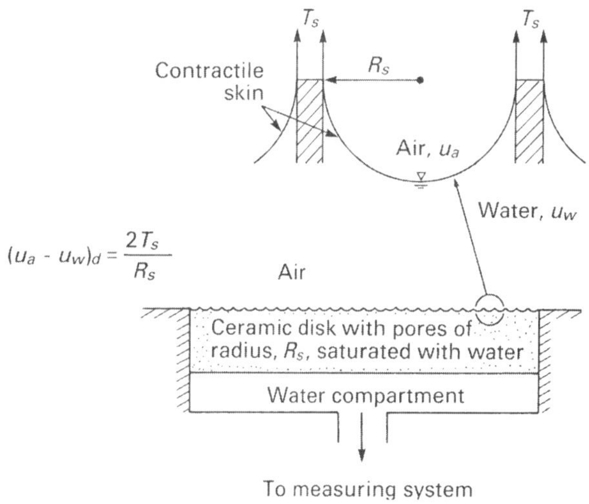

Figure 4.5 Operating principle of high-air-entry disk described in terms of capillary model.

where:

$(u_a - u_w)_d$ = air-entry value of the high-air-entry disk, kPa,

T_s = surface tension of the contractile skin or the air-water interface (e.g., T_s = 73.75 mN/m at 20 °C), and

R_s = radius of curvature of the contractile skin or the radius of the maximum pore size, m.

Surface tension T_s changes slightly with temperature (see Chapter 2). The air-entry value of a disk is largely controlled by the radius of curvature R_s of the largest pore in the disk. The pore sizes in the ceramic are controlled by the manufacturing procedure and the sintering technique used to produce the ceramic disks. The air-entry value of a ceramic disk increases as the largest pore size in a disk is decreased.

The ability of the high-air-entry disk to withstand a difference between air and water pressures makes the disk a suitable separator for the direct measurement of negative pore-water pressures in an unsaturated soil. The disk is used as an interface between the unsaturated soil and the pore-water pressure measuring system. The water in the ceramic disk provides continuity between the pore-water in the soil and the water in the measuring system. At the same time, free air is not allowed to pass through the high-air-entry disk into the measuring system.

The separation of air and water pressures by a high-air-entry disk can be achieved only as long as the matric suction in the soil does not exceed the air-entry value of the disk. Once the air-entry value of the disk is exceeded, free air will freely pass through the disk and enter the measuring system, as shown in Fig. 4.6. The release of air in the measuring system forces water through the ceramic disk back to the soil specimen. The end result is an erroneous measurement of the pore-water pressure in the soil.

The properties of several high-air-entry disks manufactured by Soilmoisture Equipment Corporation are tabulated in Table 4.3. The disks are identified by specified air-entry values expressed in bars (i.e., 1 bar = 100 kPa). The water coefficient of permeability of a ceramic disk can be measured by mounting the disk in a triaxial apparatus and placing water above the disk. It is important that there is a seal between the sides of the ceramic disk and the triaxial cell pedestal. A gradient can be applied across the high-air-entry disk by applying an air pressure to the water above the ceramic disk. The volume of water flowing through the disk is measured using a water volume change indicator.

The air-entry value and the coefficient of permeability of high-air-entry disks should be measured prior to usage in testing unsaturated soils. Figure 4.6 presents the air passage characteristics of high-air-entry disks from Soilmoisture Equipment Corporation. The measured air-entry values appear to be higher than the nominal values specified by the manufacturer. The results of measurements of the air-entry values and the coefficients of permeability for various high air-entry disks are summarized in Table 4.4.

4.2.5 Direct Measurements (Low Suctions up to 100 kPa)

There are several suitable devices that can be used for the direct measurement of negative pore-water pressures. For example, there are several low-range tensiometers that are commercially available. Tensiometers utilize a high-air-entry ceramic cup as the interface between the measuring system and the negative pore-water pressure in the soil. The axis translation technique can be used in the laboratory to extend the range over which matric suctions can be measured. The axis translation technique is discussed later in this chapter.

4.2.5.1 Tensiometers

Tensiometers measure the negative pore-water pressure in a soil. The tensiometer consists of a high-air-entry, porous ceramic cup connected to a pressure measuring device through a small bore tube. The tube is usually made from

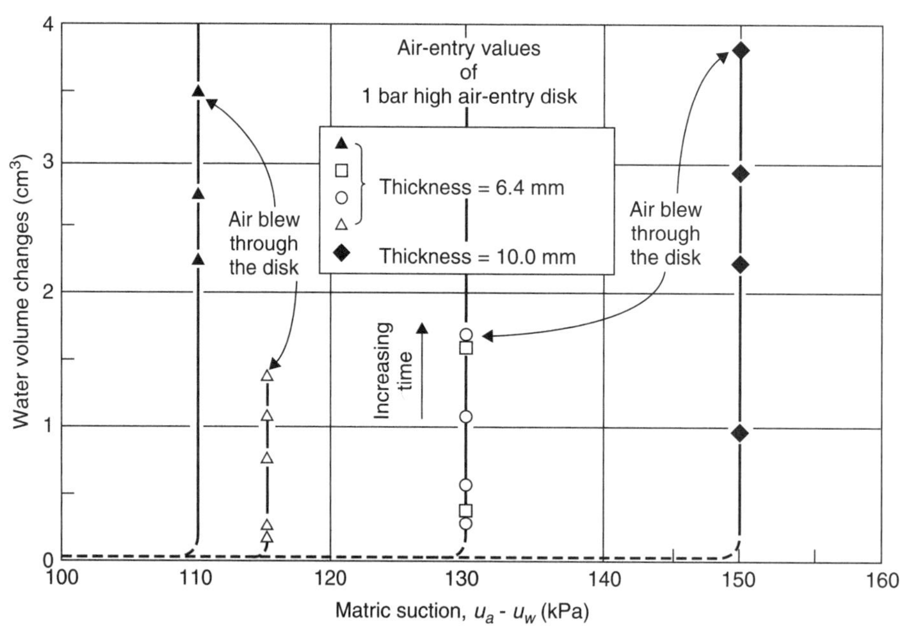

Figure 4.6 Air passage characteristics of 1-bar, high-air-entry disks (from Rahardjo, 1990).

Table 4.3 Properties of High-Air-Entry Disks Manufactured by Soilmoisture Equipment Corporation (Manufacturer's Results)

Type of Disks	Approximate Pore Diameter ($\times 10^{-3}$ mm)	Coefficient of Permeability with Respect to Water, k_d, m/s	Air-Entry Value, $(u_a - u_w)_d$, kPa
½ Bar, high flow	6.0	3.11×10^{-7}	48–62
1 Bar	2.1	3.46×10^{-9}	138–207
1 Bar, high flow	2.5	8.60×10^{-8}	131–193
2 Bars	1.2	1.73×10^{-9}	241–310
3 Bars	0.8	1.73×10^{-9}	317–483
5 Bars	0.5	1.21×10^{-9}	>550
15 Bars	0.16	2.59×10^{-11}	>1520

Note: Soilmoisture Equipment Corporation, Santa Barbara, CA.

plastic due to its low heat conduction and noncorrosive nature. The tube and the cup are filled with deaired water. The cup can be inserted into a predrilled hole. It is important that there be good contact between the high-air-entry ceramic and the soil. It is also important that the soil particles not be too large; otherwise there may be inadequate contact between the ceramic and the water in the soil.

The water in the tensiometer will have the same negative pressure as the pore-water in the soil once equilibrium is achieved between the soil and the measuring system. The pore-water pressure that can be measured in a tensiometer is limited to approximately negative 90 kPa due to the possibility of cavitation of the water in the tensiometer. The measured negative pore-water pressure is numerically equal to the matric suction when the pore-air pressure is atmospheric (i.e., u_a = zero gauge pressure). When the pore-air pressure is greater than atmospheric pressure (i.e., during axis translation), the tensiometer reading can be added to the ambient pore-air pressure reading to give the matric suction of the soil. The measured matric suction must not exceed the air-entry value of the ceramic cup on the tensiometer. The osmotic component of soil suction is not measured with tensiometers since soluble salts are free to move through the porous cup.

There are several types of tensiometers commercially available from Soilmoisture Equipment Corporation and other companies. The tensiometer tube in Fig. 4.7 has a diameter of approximately 20 mm and various lengths up to 1.5 m. The tensiometer cups can be installed in the field to a depth of 1.5 m below the ground surface. However, the negative water pressure recorded at the ground surface must be corrected for the elevation head difference between the tensiometer tip in the soil and the gauge at ground surface. The elevation correction results in a more negative water pressure being measured at ground surface than that being experienced by the soil. A length of 1.5 m corresponds to a pressure correction of 15.2 kPa.

The tensiometer must be properly serviced prior to usage in order to obtain reliable results. Details regarding the preparation, installation, and usage of a tensiometer are presented by Cassel and Klute (1986). The ceramic cup should be checked for signs of cracks and air bubbles should be removed from the tensiometer prior to installation. The response time of the tensiometer should be checked by placing the empty tensiometer upright in a pail of water and allowing the cup to soak in water. Later, the ceramic tip of the tensiometer can

Table 4.4 Permeability and Air-Entry Value Measurements on High-Air-Entry Disks from Soilmoisture Equipment Corporation

Type of Disks	Diameter of Disk, mm	Thickness of Disk, mm	Air-Entry Value of Disk, $(u_a - u_w)_d$, m/s	Coefficient of Permeability of Disk, k_d, m/s
1 Bar, high flow	19.0	6.4	115	5.12×10^{-8}
	19.0	6.4	130	3.92×10^{-8}
	19.0	6.4	110	3.98×10^{-8}
	19.0	6.4	130	5.09×10^{-8}
	19.0	6.4	150	5.60×10^{-8}
	101.6	10.0	>200	4.20×10^{-8}
5 Bars	56.8	6.2		1.30×10^{-9}
15 Bars	56.8	3.1		8.41×10^{-9}
	57.0	3.1		6.82×10^{-10}

Source: From Fredlund, 1973a; Rahardjo, 1990.

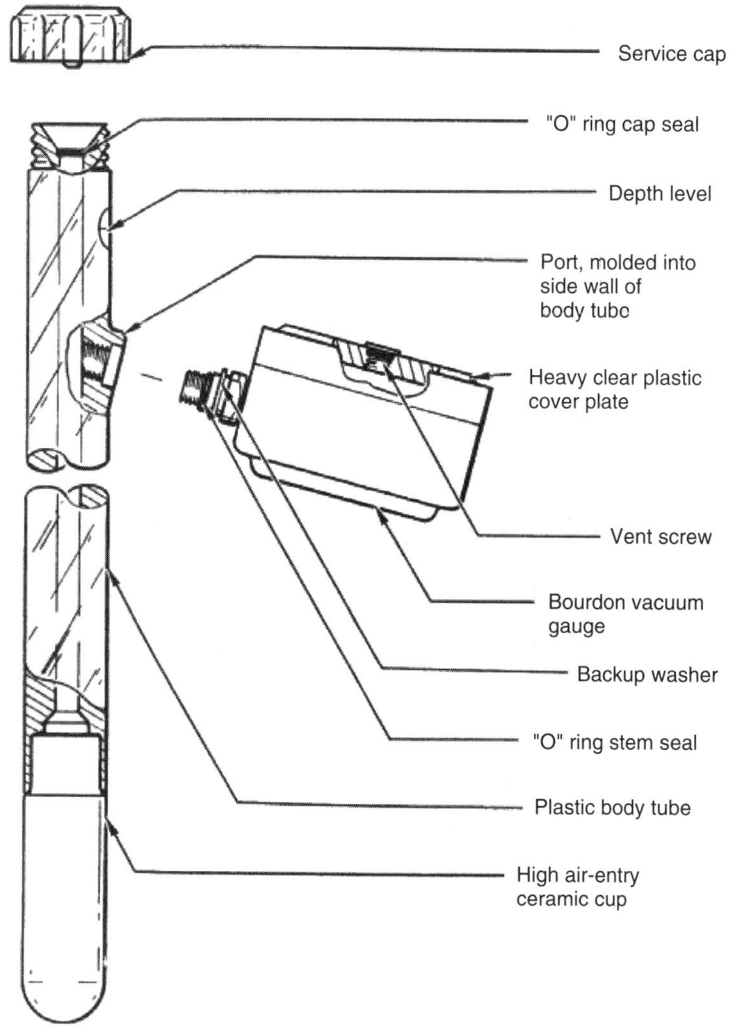

Figure 4.7 Conventional tensiometer manufactured by Soilmoisture Equipment Corporation.

be exposed to the air and evaporation can be allowed for the buildup of tension in the water in the tensiometer.

The response time of a tensiometer can be checked by allowing the development of a negative water pressure of approximately 80 kPa. The negative pressure is developed by allowing evaporation from the ceramic cup. The ceramic cup is then immersed in water. The negative water pressure in the tensiometer should increase toward atmospheric pressure within 5 min after immersion of the ceramic tip in water. A sluggish response may indicate a plugged porous cup, the presence of entrapped air in the system, or a faulty gauge that requires rezeroing.

The ceramic cup must be kept immersed in water prior to its installation in order to avoid desaturation due to evaporation from the cup. The prepared tensiometer can then be installed in a predrilled hole in the field or in a soil specimen (e.g., a compacted soil specimen) in the laboratory. It is important to ensure good contact between the ceramic cup and the soil in order to establish continuity between the pore-water in the soil and the water in the tensiometer tube.

4.2.5.2 Jet-Fill Tensiometers

A jet-fill-type tensiometer is shown in Fig. 4.8 and is an improvement on the regular tensiometer. A water reservoir is provided at the top of the tensiometer tube for the purpose of removing air bubbles. The jet-fill mechanism is similar to the action of a vacuum pump. The accumulated air bubbles are removed by pressing the button at the top to activate the jet-fill action. The jet-fill action causes water to be injected from the water reservoir to the tube of the tensiometer, and air bubbles move upward to the reservoir. The jet-fill action also leaves the tensiometer reading at zero, and therefore, time must be allowed for the tensiometer to return to a value indicative of the suction in the soil.

4.2.5.3 Small-Tip Tensiometer

A small-tip tensiometer with flexible coaxial tubing is shown in Fig. 4.9. The tensiometer is prepared for installation using a similar procedure to that described for the regular tensiometer tube. A vacuum pump can be initially used to

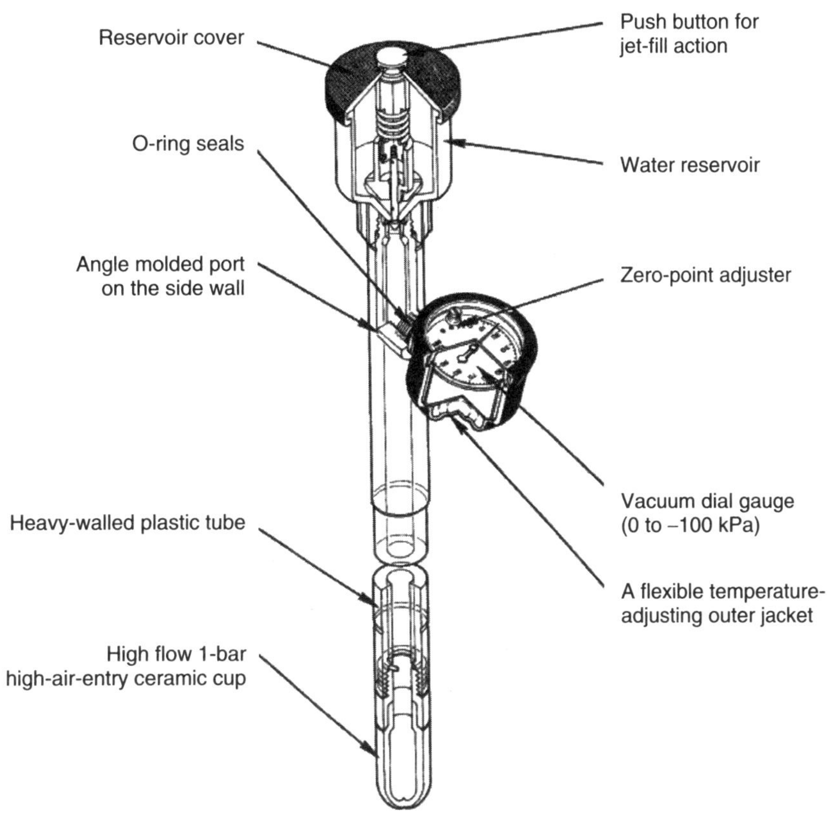

Figure 4.8 Jet-fill tensiometer manufactured by Soilmoisture Equipment Corporation.

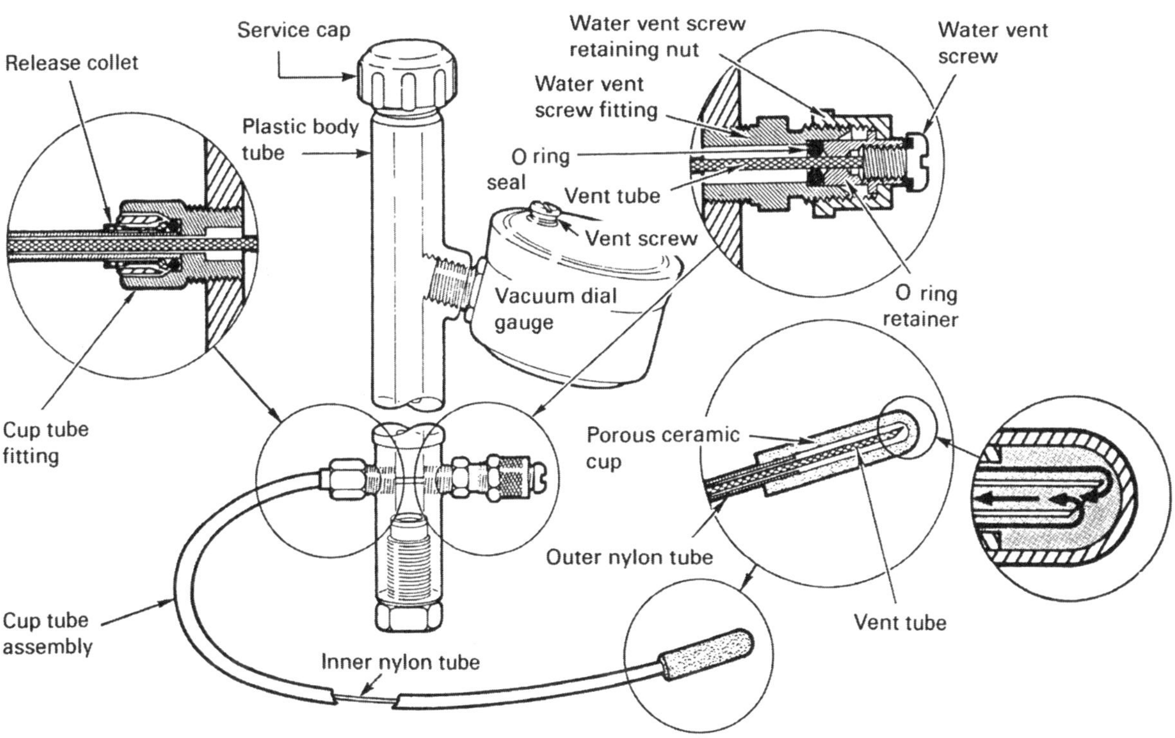

Figure 4.9 Small-tip tensiometer with flexible coaxial tubing (from Soilmoisture Equipment Corporation).

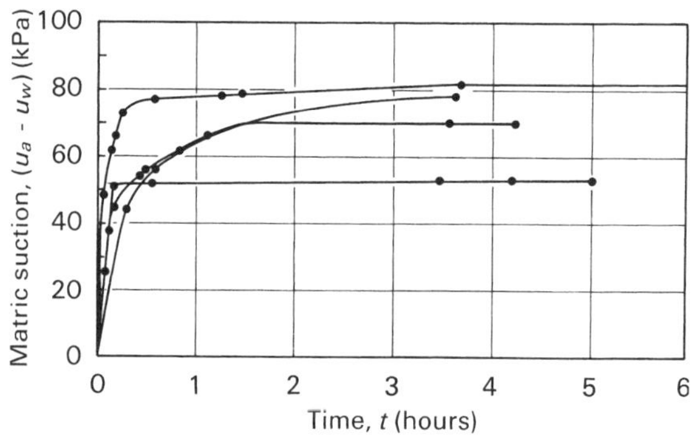

Figure 4.10 Typical time responses of small-tip tensiometer placed in decomposed volcanic soil (after Sweeney, 1982).

remove air bubbles from the top of the tensiometer tube. Subsequent removal of air bubbles can be performed by flushing water through the coaxial tube.

Flushing is conducted by circulating water from the top of the tube and opening the water vent screw to the atmosphere. Water containing air bubbles will be forced into the inner nylon tube and released to the atmosphere through the opened water vent. This procedure may be required on a daily basis when matric suctions are greater than about 50 kPa. It is sometimes difficult to obtain an accurate suction measurement because water moves from the tensiometer into the soil each time the tensiometer is serviced. The movement of water into the soil reduces the soil suction.

The small ceramic tip with the flexible tube allows the installation of the tensiometer into a relatively small soil specimen during laboratory experiments. As an example, the small-tip tensiometer has been installed in a consolidation specimen to measure changes in matric suction during the collapse of compacted silt (Tadepalli, 1990).

Figure 4.10 illustrates typical time responses of a small-tip tensiometer inserted into decomposed volcanic soil. The response is plotted in terms of matric suctions by referencing measurements to atmospheric air pressures. Sweeney (1982) found that this type of tensiometer could only maintain matric suction equilibrium for about one or two days before the suction readings began to drop. The time that equilibrium can be maintained is a function of the matric suction value being measured.

4.2.5.4 Quick Draw Tensiometers

Figure 4.11 shows a Quick Draw tensiometer equipped with a coring tool and a carrying case. The Quick Draw tensiometer has proven to be a particularly useful portable tensiometer to rapidly measure negative pore-water pressures. The water in the tensiometer is subjected to tension for only a short time while each measurement is being made. Therefore, air diffusion through the ceramic cup over time is minimized.

The Quick Draw tensiometer can repeatedly measure pore-water tensions approaching 1 atm when it has been properly serviced. The probe is maintained in water-saturated cotton in a carrying case when the Quick Draw tensiometer is not in use. The rapid response of the Quick Draw tensiometer in a sand soil is illustrated in Fig. 4.12.

4.2.5.5 Tensiometer Performance for Field Measurements

Tensiometers have been used to measure negative pore-water pressures for numerous geotechnical engineering applications. One example is a cut slope consisting of 5–6 m of colluvium overlying deep weathered granite in Hong Kong. Figure 4.13 illustrates the construction of the shaft, along with

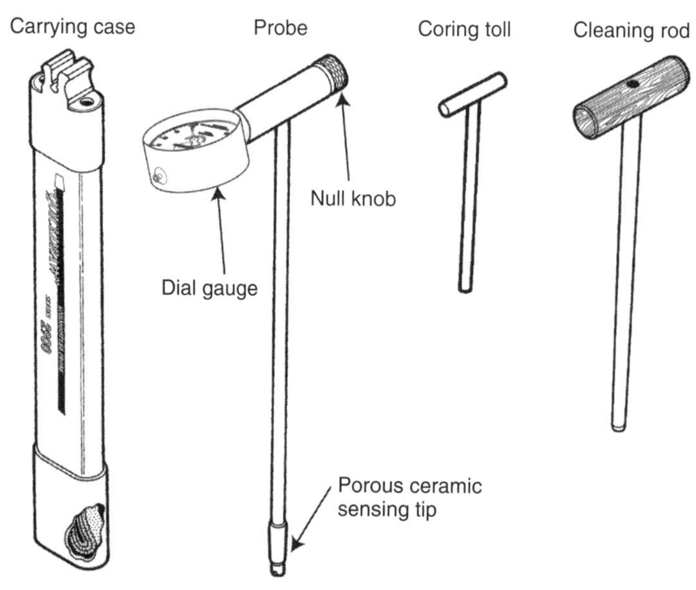

Figure 4.11 Quick Draw tensiometer with coring tool and carrying case manufactured by Soilmoisture Equipment Corporation.

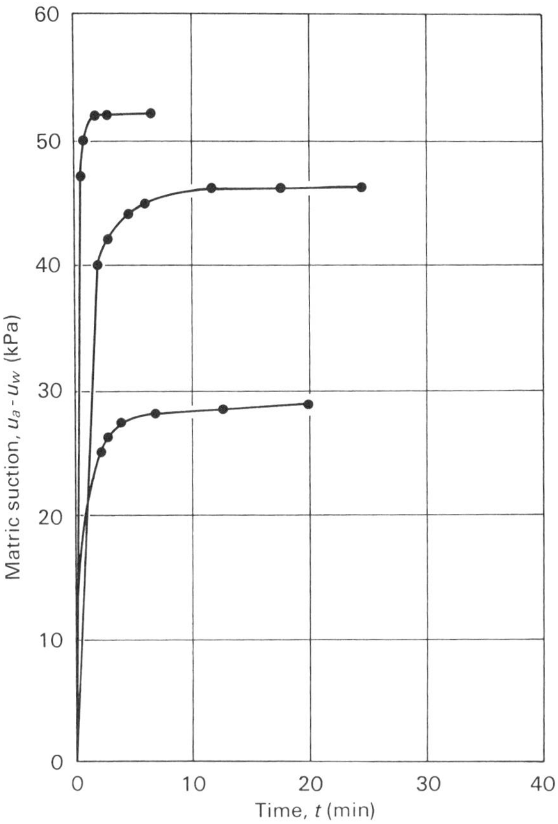

Figure 4.12 Typical responses of Quick Draw tensiometer in sandy soil (after Sweeney, 1982).

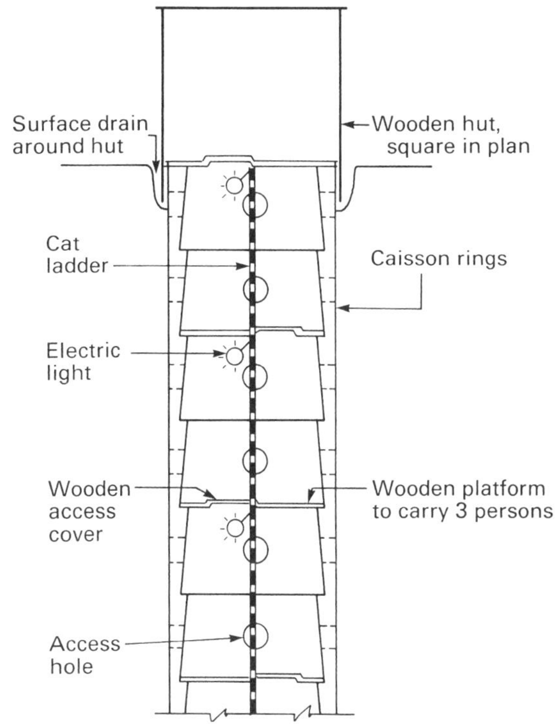

Figure 4.13 Observation shaft used to install tensiometers at various depths in the field (after Sweeney, 1982).

four circular openings along each caisson ring. The openings functioned as access holes for installing tensiometers into the soil at various locations along the shaft. The shaft was equipped with wood platforms at various elevations, a ladder, and lighting facilities, as shown in Figure 4.13.

Measurements of negative pore-water pressures along the shafts are presented in Fig. 4.14. The results are plotted in terms of matric suction and hydraulic head profiles throughout the depth of the shaft. The results indicate the fluctuation in matric suction values with respect to the time of the year. The largest variations occur near the ground surface (Fig. 4.14). The hydraulic heads along the depth of the shaft are plotted by adding the negative pore-water pressure head to the elevation head. The hydraulic head plots indicate a net downward water flow toward the water table through the unsaturated zone.

4.2.6 Direct Measurements (High Suctions Exceeding 100 kPa)

The direct measurement of high suction values suffers from the fact that water tends to cavitate whenever its pressure drops below about −100 kPa. However, the tensile strength of water has been measured by many physicists and shown to be greater than several atmospheres (Knapp et al., 1970; Trevena, 1987).

In recent years, direct high-suction tensiometers have been developed and used primarily at a research level. The water in the high-suction tensiometer must initially be subjected to a number of high-pressure cycles that are sustained for a period of time. Studies have shown that water in a small chamber can sustain tensions greater than 1 atm for a period of time.

It should be noted that the term *tensile strength of water* is somewhat of a misnomer since the tensile strength of water has never been directly measured. The measured tensile strength of water is significantly influenced by a number of conditions including the surface condition of the walls of the container and the manner in which the tension is applied to the water. The term tensile strength of water is used in the sense that it represents the overall ability of the water-container system to withstand tension applied to the water.

High positive pressures need to first be used to dissolve potential cavitation nuclei and increase the tensile strength of the water in the suction probe. A pressure plate cell can be used to test the accuracy and response time of suction probes.

4.2.6.1 Historical and Theoretical Background

The theoretical tensile strength of water has been estimated to be on the order of several hundred atmospheres (Vincent 1941; Temperley, 1947; Harvey et al., 1947; Fisher, 1948). The tensile strength of a liquid is the tensile stress at which the liquid ruptures or cavitates. Cavitation starts when vapor

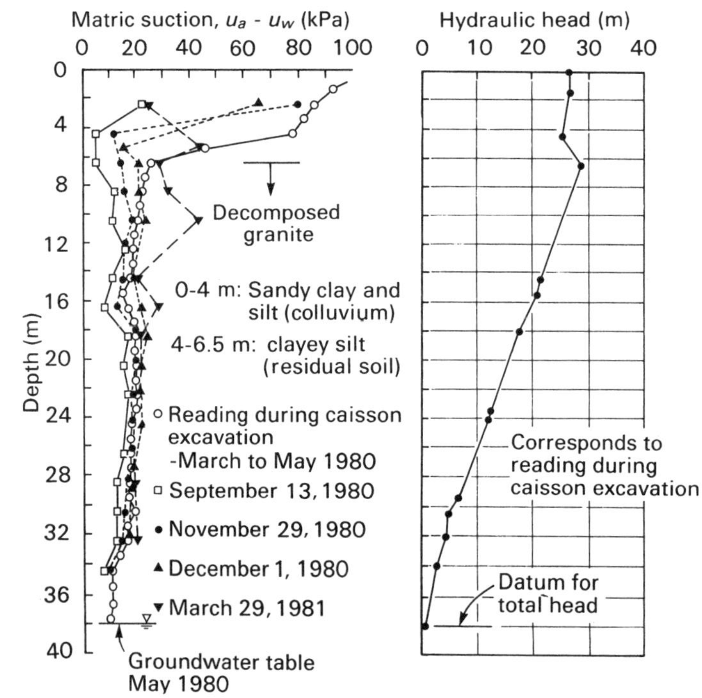

Figure 4.14 Measured matric suction profiles along shaft through colluviums and residual soil in Hong Kong (after Sweeney, 1982).

bubbles begin to form in water. Vapor bubbles are triggered at gaseous or other hydrophobic surfaces called potential cavitation nuclei.

It is normally assumed that water will began to boil (or cavitate) when the absolute pressure above the water surface approaches the vapor pressure. This is true; however, if the radii of the cavitation nuclei in the water are sufficiently small, the water will be able to sustain a high tensile stress without undergoing cavitation. For example, if the radius is assumed to be the size of a water molecule (i.e., about 3×10^{-7} mm), at the inception of the formation of a gas bubble, the tensile strength of water should be about 480,000 kPa.

The tensile strength of water measured by physicists is widely scattered with most values greater than 500 kPa. Liquid water under high tension is in a metastable state (Temperley and Chambers, 1946; Temperley, 1947; Chapman et al., 1975; Green et al., 1990). Rupture of the water can occur due to the presence of potential cavitation nuclei in the liquid and on the container walls. Tiny crevices on the surfaces of container walls and dust particles are a common source of cavitation nuclei. High positive pressures can suppress cavitation nuclei in the water and on container walls. The end result is a significant increase in the tensile strength of the water (Knapp, 1958; Winterton, 1977). Research studies have focused mainly on the maximum tensile strength of water as opposed to how long a tensile stress can be sustained by the water.

Chapman et al., (1975) and Winterton (1977) measured high tensions in water which were sustained for a few minutes. Henderson and Speedy (1980) reported a tension of 10,000 kPa that was sustained for over a week. Guan (1996) studied the behavior of a small volume of water subjected to high tensions using a high-air-entry ceramic disk and a high-range-pressure transducer. Based on past studies it appears that there is a cavitation tension that is somewhat related to each prepressurization procedure and each particular suction probe. A tension can be sustained for several hours or days if the applied tension does not exceed the cavitation tension.

The methods used by physicists for measuring high tensions in water are not directly adaptable to the measurement of soil suction. Gilbert (1960) attempted to measure high tensions in sucrose solutions in a triaxial cell. Ridley (1993) prepressurized a small volume of water in a tensiometer-type device (Fig. 4.15) and measured matric suctions up to 1500 kPa. Ridley and Burland (1993) found reasonable agreement between matric suction values in London clay measured using the Ridley-tensiometer-type probe and the filter paper method. It was suggested that the microstructure of the soil, particularly the size of the voids, may affect the measurement of matric suction when using the high-suction probe. The device was later modified to measure matric suction in osmotically controlled oedometer tests (Dineen and Burland, 1995).

A direct high-suction tensiometer was developed at the University of Saskatchewan for the direct measurement of high soil suctions. Studies by Guan (1996) indicated that the direct high-suction tensiometer could sustain tensile stresses of 1250 kPa. The suction probe has been tested for measuring suctions in various types of soils.

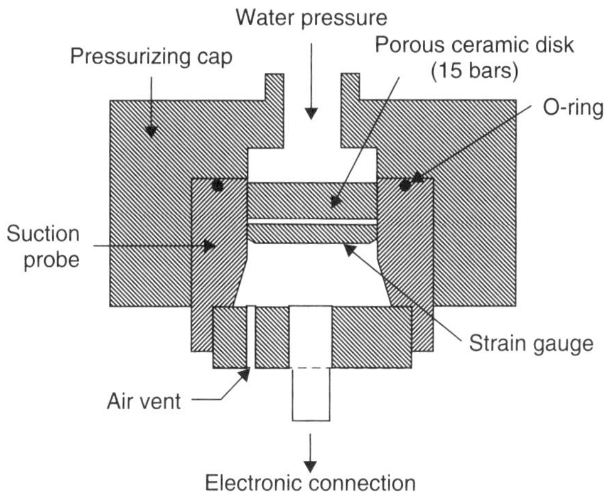

Figure 4.15 Schematic of Imperial College, London, high-suction range tensiometer (after Ridley, 1993).

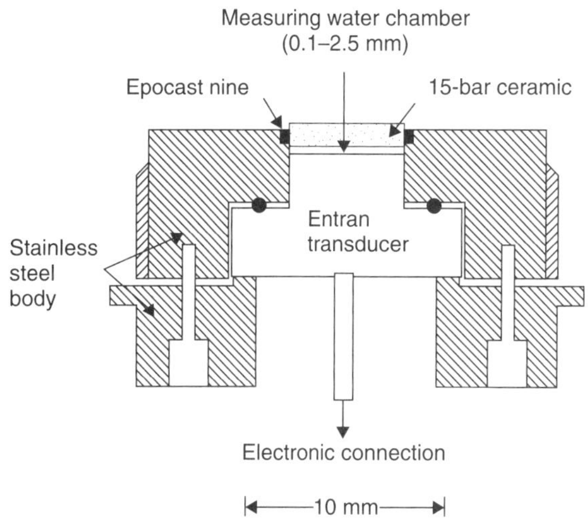

Figure 4.16 Schematic of University of Saskatchewan, Canada, high-suction-range tensiometer (from Guan, 1996).

4.2.6.2 Design of Direct High-Suction Tensiometers

The Guan and Fredlund (1997b) direct high-suction tensiometer was developed independent of the research undertaken at Imperial College by Ridley (1993). Although the two research studies were independent, the outcome of the studies was quite similar. Both studies produced similar direct high-suction tensiometers and both used Entran pressure transducers.

The Guan and Fredlund (1997b) direct high-suction tensiometer consisted of a high-range-pressure transducer (i.e., 15,000 kPa) and a stainless steel shroud which was precisely machined to embrace the transducer (Fig. 4.16). The sensing area of the transducer had a smooth, circular surface with a diameter of 7.0 mm. A 15-bar high-air-entry ceramic disk was fitted into the shroud. Shrouds of different sizes were manufactured and tested. The transducer and shroud were assembled while being held under water. The separation between the pressure transducer and the ceramic disk ranged from 0.1 to 0.5 mm.

A mini direct high-suction probe was also developed at Nanyang Technological University in Singapore by Meilani et al., (2002). The suction probe used a 1500-kPa miniature PDCR-81 pressure transducer manufactured by Druck (Leicester, England). The miniature pressure transducer was able to operate under water. The suction probe was placed in a steel tube as shown in Fig. 4.17. The miniature suction probe consisted of a 0.09-mm-thick silicon diaphragm mounted within 1.2 mm of a 5-bar high-air-entry disk. The high-air-entry disk was ground to a thickness of only 1.0 mm in order to achieve a rapid response when measuring negative pore-water pressures. The probe had negligible weight (i.e., 3 g) and could be mounted on the side of a triaxial soil specimen. The ceramic disk was glued in place using Araldite 2021 epoxy.

Three miniature direct high-suction sensors were mounted along the side of a triaxial specimen as shown in Fig. 4.18.

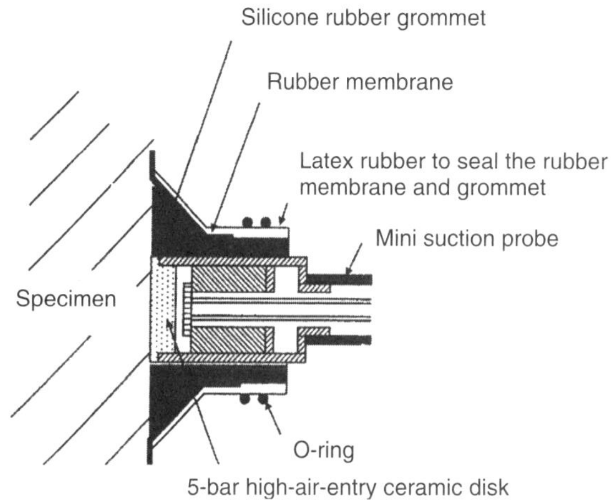

Figure 4.17 Schematic of Nanyang Technological University, Singapore, mini high-suction-range tensiometer (from Meilani et al., 2002).

Matric suction measurements were made as the triaxial specimen was slowly sheared to failure.

Several laboratory research studies have led to conclusions on a number of issues that need to be given careful consideration when designing a direct high-suction sensor. Marinho and Chandler (1994) stated that the following factors could affect the successful performance of a direct high-suction sensor:

1. The water and all surfaces internal to the high-suction sensor must be pure and clean (Henderson and Speedy, 1980).
2. The surfaces in contact with the internal water system must be smooth in order to reduce or avoid the number and size of small crevices that could initiate the cavitation process.

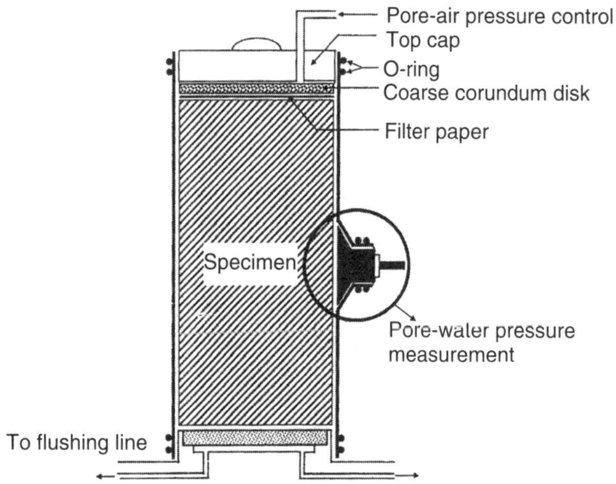

Figure 4.18 Mounting Nanyang Technological University mini high-suction-range tensiometer on side of triaxial specimen (from Meilani et al., 2002).

3. A vacuum should be applied to the water in the system to remove entrapped air (Jones et al., 1981).
4. Prepressurization of the water to a high pressure is required to dissolve all free air (Harvey, 1944). The prepressurization has become a key part of ensuring the successful functioning of direct high-suction tensiometers.

4.2.6.3 Conditioning of Water for a High-Suction Tensiometer

A manual pressurization system can be used to prepressurize the water in the high-suction tensiometer. The high-suction tensiometer can be mounted into a pressurizing device. A cylinder and piston arrangement can be used to produce pressures as high as 15,000 kPa. A vacuum inlet should also be part of the design in order that a negative pressure of about 85 kPa can be applied.

Experiments by Guan and Fredlund (1997b) indicated that six cycles of pressurization produced the maximum sustainable tension in the high-suction tensiometer. Each cycle included the application of a positive pressure of 12,000 kPa for 1 h followed by a tensile stress of 85 kPa for 1 h. The sustainable tension in the high-suction tensiometer appeared to also be partly related to the characteristics of the ceramic disk. The size of the water chamber did not appear to have an appreciable effect on the sustainable tension in the high-suction tensiometer.

Laboratory studies have shown that it is possible to use ordinary water in a high-suction tensiometer. The pressurization cycling technique appeared to be quite important. A small water compartment between the pressure transducer and the high-air-entry ceramic disk appears to be preferable; however, it has been shown that relatively high suctions can also be achieved when the water compartment is enlarged. A 15-bar ceramic stone has generally been used when constructing the high-suction probes except for the mini suction probes developed by Meilani et al. (2002) where a thin 5-bar ceramic disk was used with success.

There has been discussion in the literature regarding whether a 15-bar ceramic disk may serve as a low-efficiency osmotic membrane. The concern was that the high-suction sensor might actually measure total suctions. Ridley and Burland (1993) showed results that indicated that the high-suction probes only measure the matric suction in a soil.

4.2.6.4 Comparison between Applied and Measured Suctions

An experimental study was undertaken by Guan and Fredlund (1997b) to test the accuracy and response time of a high-suction tensiometer. A modified pressure plate apparatus was used to apply a specified suction to a soil specimen (Fig. 4.19). The soil specimen was placed on a 5-bar ceramic plate at the bottom of the pressure plate apparatus. Water drained from the bottom of the soil specimen until the matric suction in the soil was equal to the applied air pressure.

A high-suction tensiometer was then placed on top of the soil specimen of Regina clay. The high-suction tensiometer should read the matric suction induced by the applied air pressure once equilibrium is achieved with the soil specimen. When the applied air pressure is released to atmospheric pressure conditions, the soil should have a negative pore-water pressure numerically equal to the air pressure previously applied to the soil once equilibrium conditions are achieved.

The air pressure initially applied to the soil was 200 kPa. Water flowed out from the soil through the 5-bar ceramic

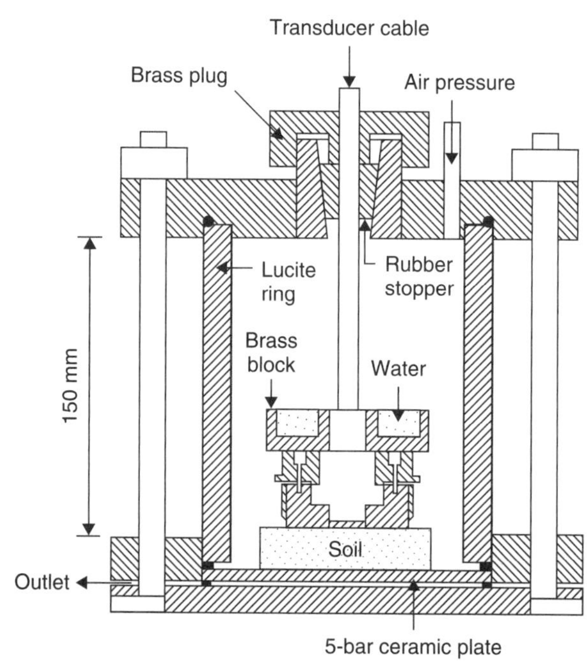

Figure 4.19 Modified pressure plate apparatus used to apply known matric suctions to a soil (from Guan and Fredlund, 1997).

plate. The water below the 5-bar ceramic plate was removed using an air pump when outflow from the soil had ceased. The air pressure in the pressure chamber was quickly released to zero and the reduction in the water pressure in the high-suction tensiometer was recorded.

When the water pressure recorded on the high-suction tensiometer ceased to change significantly, the applied air pressure was increased to 300 kPa and the above procedure was repeated. The same test procedure was repeated using pressures of 400 and 500 kPa. The above procedure was also used for tests on a reconstituted silt specimen.

Figure 4.20 shows the changes in measured water pressures with time for the tests on Regina clay using the high-suction tensiometer. When an air pressure of 200 kPa was applied to the pressure chamber, the suction probe recorded an almost instantaneous increase in water pressure from 0 to 195 kPa. The water pressure gradually reduced to about 25 kPa after 42 h.

Reducing the air pressure from 200 to 0 kPa caused the water pressure to immediately fall from 25 kPa to a reading of –158 kPa. The change in air pressure (i.e., –200 kPa) was close to the change in water pressure [i.e., $-(25 + 158) = -183$ kPa]. The water pressure measured by the suction probe gradually increased because water slowly flowed into the soil.

When the water pressure reached –151 kPa, an increase in air pressure to 300 kPa produced an immediate jump of water pressure to about 145 kPa, an increase of 296 kPa. Further tests also showed that a change in air pressure in the pressure chamber immediately produced an approximately equal change in water pressure in the high-suction tensiometer. The difference between the change in the air pressure inside the pressure chamber and the change in water pressure in the suction probe was generally less than 5%. The Regina clay remained saturated throughout the test.

4.2.6.5 Measurement of Preapplied Matric Suction

Figure 4.21 shows the laboratory setup that was used to measure the matric suction on reconstituted specimens of

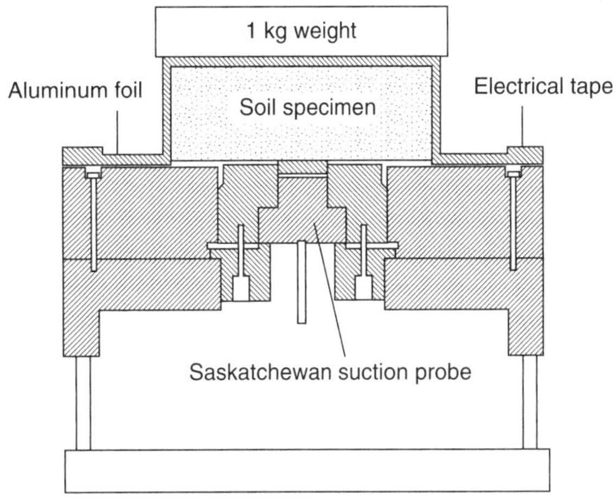

Figure 4.21 Laboratory setup for measuring soil suction with direct, high-suction range tensiometer (from Guan and Fredlund, 1997a).

Regina clay and reconstituted silt. Matric suction values were applied to Regina clay and silt using a pressure plate apparatus. The soil specimens were then placed onto the high-suction tensiometer. The soil specimens were immediately covered with tin foil to prevent the loss of water from the soil.

The results of matric suction measurements on Regina clay placed in direct contact with the ceramic disk of the high-suction tensiometer are shown in Fig. 4.22. The preapplied matric suctions varied from 150 to 550 kPa. The recorded curves are in reasonable agreement with the preapplied matric suctions up to 450 kPa. The recorded negative pressures for preapplied matric suctions of 500 and 550 kPa developed more slowly than those of the lower suction specimens. It was noted that a crack appeared in the soil specimen when matric suctions exceeded 450 kPa.

Figure 4.23 shows the results of measurements of matric suction on the reconstituted silt specimens. There was a

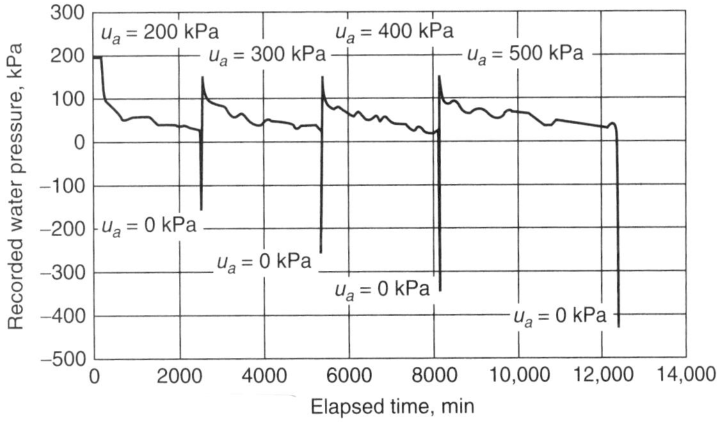

Figure 4.20 Verification of high-suction measurements on Regina clay using University of Saskatchewan direct high-suction range tensiometer (from Guan and Fredlund, 1997).

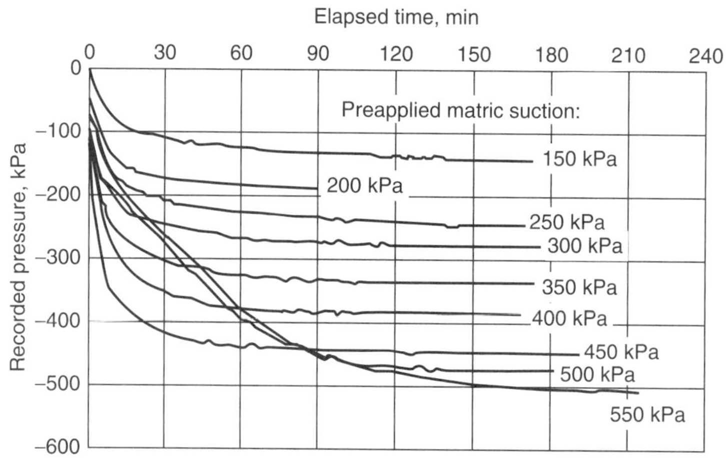

Figure 4.22 Comparison of applied and measured matric suctions on Regina clay using University of Saskatchewan direct high-suction range tensiometer (from Guan and Fredlund, 1997).

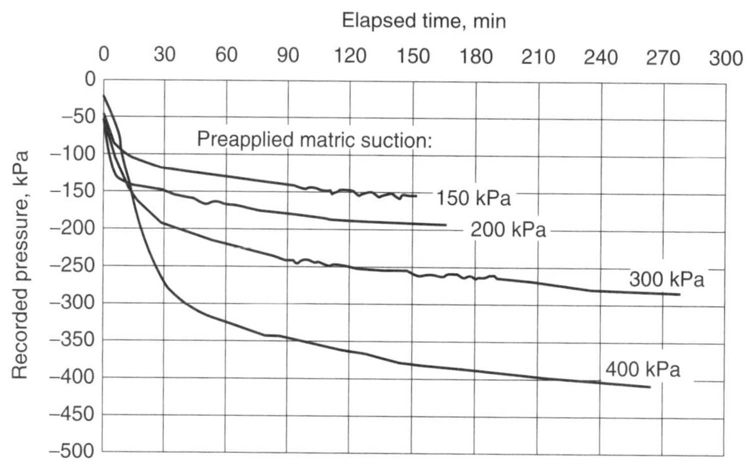

Figure 4.23 Comparison of applied and measured matric suctions on reconstituted silt using University of Saskatchewan direct, high-suction tensiometer (Guan and Fredlund, 1997).

slight drift in the recorded suctions when the preapplied matric suction exceeded 300 kPa. A matric suction of 300 kPa corresponded to a degree of saturation of about 38% for the silt specimen.

Differences between applied matric suctions and the measured negative pressures were less than 5%. The accuracy of matric suction measurements appear to have been limited in part by the accuracy of the transducer in the high-suction tensiometer. The response time of the high-suction probe to a change in the applied matric suction was on the order of seconds. The fast response of the high-suction tensiometer is attributable to the rigidity of the chamber of the suction probe and the high degree of saturation of the soil.

The mini probe developed by Meilani et al. (2002) was successfully used to measure suction up to 400 kPa along a triaxial specimen. Suctions on the order of 400 kPa were maintained for about 15 h while suctions of 200 kPa were maintained without cavitation for about 155 h (i.e., ~6 days). The time period over which a particular matric suction can be maintained is believed to be related to factors affecting the diffusion of dissolved air through the water in the high-air-entry disk.

Figure 4.24 shows the results from one experiment where the soil specimen was subjected to a net confining pressure of 25 kPa and a matric suction of 100 kPa. The matric suction of 100 kPa was achieved by applying an air pressure of 290 kPa to the top of the soil specimen. The water pressure at the bottom of the specimen was then lowered from 290 to 190 kPa. The difference between the air and water pressures then produced a matric suction of 100 kPa which remained in the soil specimen for 130 h until the shearing of the specimen was complete. The use of the mini high-suction probe meant that the matric suction in the soil specimen could be directly measured without use of the axis translation technique.

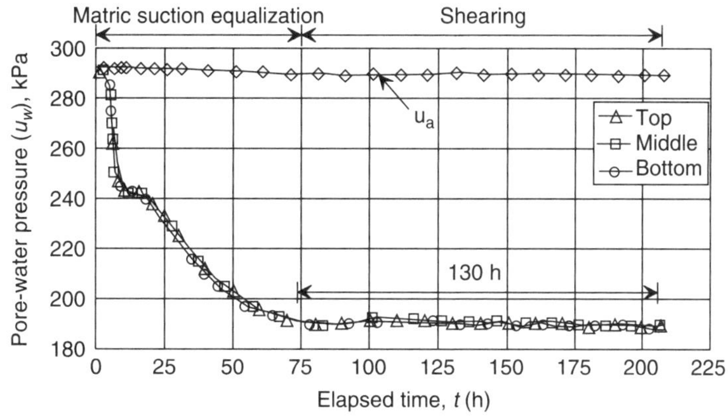

Figure 4.24 Use of Nanyang Technological University mini high-range tensiometers at three locations along triaxial specimen during shear at net confining pressure of 25 kPa (from Meilani et al., 2002).

4.2.6.6 Moisture Transfer When Using High-Suction Tensiometer

Moisture can transfer between the high-suction tensiometer and the soil in both the liquid and vapor phases. Moisture from the suction probe can evaporate into the soil in the vicinity of the ceramic disk. Liquid water transfer occurs through direct contact between the soil and the ceramic disk of the high-suction tensiometer. Tension appears to be developed in the high-suction tensiometer as a result of combined vapor and liquid flow between the soil and the ceramic disk. If moisture is continuously lost from the system through evaporation during the measurement of suction, there is a continuous increase in observed pore-water tension. The amount of moisture lost through evaporation may be minuscule; however, this loss can have a significant effect on the suction measurement. The response of the high-suction tensiometer can be divided into the following three phases (Fig. 4.25):

Phase I: This is the initial phase of tension development. Significant tension is rapidly developed during a relatively short period of time (e.g., 5–10 m). In phase I, liquid water transfer is more dominant than the evaporation process (curve 1). Curve 2 (Fig. 4.25) occurs when tension in the high-suction tensiometer has already developed due to evaporation that took place before the suction probe was put in contact with the soil. However, the process of suction equalization due to liquid water transfer will quickly become dominant when there is good contact between the soil and the high-suction tensiometer. The tension developed due to evaporation decreases and equilibrium is reached with the suction in the soil.

Phase II: Following phase I there is a process during which suction equalization due to liquid water transfer is dominant but slow. At the end of this phase, equilibrium between the high-suction tensiometer and the suction in the soil is reached. The liquid water transfer from the soil to the high-suction tensiometer can offset the loss of moisture from the suction probe due to evaporation.

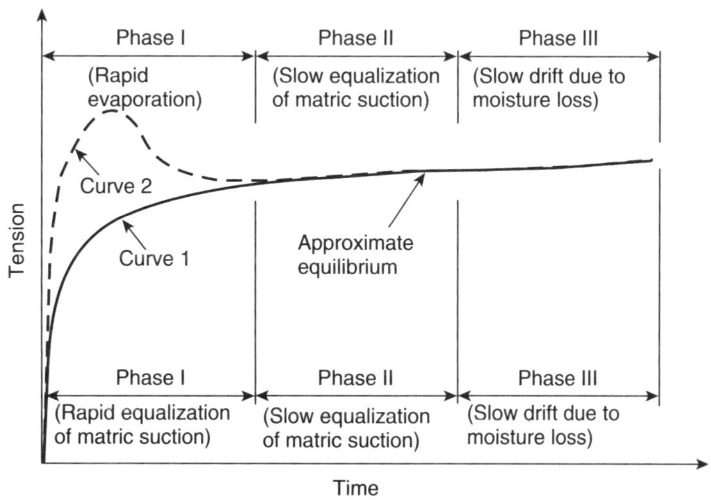

Figure 4.25 Typical suction response curves when using direct, high-suction-range tensiometer.

Phase III: After the suction response curve has come to equilibrium (i.e., curve is essentially horizontal), there is a slow increase in the observed suction measurement for both saturated and unsaturated soils. The slow drift in suction appears to be mainly attributable to moisture loss due to evaporation from both the soil and the high-suction tensiometer. Measurements showed that the water content generally decreased by up to 0.1% during the measurement of matric suction.

Test results to-date show that it is possible to sustain tensile stresses in high-suction tensiometers for a considerable time. It is important that the tension in the water phase not exceed the air-entry value of the ceramic disk. The slow drift in measured negative pressures should be further studied. The high-suction tensiometers are extremely sensitive to minute losses of water during the measurement process. The measurement of matric suction should be made in a constant-temperature, high-humidity environment. The soil specimens should also be at the same temperature as the high-suction tensiometer at the start of the test (Fredlund et al., 1997b).

High-suction tensiometers appear to have potential for routinely measuring soil suction in the laboratory. Direct measurements can be made with suctions in excess of 1000 kPa. It is not likely that the high-suction probes will find extensive application for field usage in its present form because of the need to occasionally recharge the high-suction tensiometer.

4.2.7 Axis Translation Technique for the Laboratory

Laboratory measurements of negative pore-water pressure can be made using the axis translation technique. The measurement is performed on either undisturbed or compacted specimens. This technique was originally proposed by Hilf (1956) and is illustrated in Fig. 4.26. Hilf placed an unsaturated soil specimen in a closed pressure chamber. A pore-water pressure measuring probe consisting of a needle with a saturated high-air-entry ceramic tip was inserted into the soil. The probe was connected to a null-type pressure measuring system through a tube filled with deaired water and a mercury plug in the middle of the line.

The water in the tube tended to go into tension and the Bourdon gauge began registering a negative pressure as soon as the probe was inserted into the specimen. The tendency of the water in the measuring system to become negative was countered by increasing the air pressure in the chamber. Eventually, an equilibrium condition was achieved with the mercury plug (i.e., the null indicator) remaining stationary. The difference between the air pressure in the chamber and the measured water pressure at equilibrium was taken to be the matric suction of the soil $(u_a - u_w)$.

The matric suction value was numerically equal to the negative pore-water pressure when the air pressure was atmospheric (i.e., $u_a = 0$). The axis translation technique simply translates the origin of reference for the pore-water pressure from standard atmospheric condition to the final air pressure in the chamber (i.e., axis translation). The water pressure in the measuring system does not become highly negative and as a result the problem of cavitation is prevented.

Hilf (1956) demonstrated that the measured pore-water pressure increased by an amount equal to the increase in the ambient chamber air pressure. In other words, the matric suction in the soil $(u_a - u_w)$ remained constant when measured at various ambient air pressures. The condition of no water flow during the measurement of matric suction justifies the use of the axis translation technique.

The axis translation technique has also been used by other researchers (Olson and Langfelder, 1965; Fredlund, 1989c). The procedure used is as follows. A soil specimen is placed on top of a saturated high-air-entry disk inside an air pressure chamber. The air-entry value of the disk must be higher than the matric suction in the soil. A 1-kg mass can be placed on top of the specimen to ensure good contact between the soil specimen and the high-air-entry disk. The placement of the specimen onto the ceramic disk and the assemblage of the cell are performed as rapidly as possible (i.e., within approximately 30 s).

The water pressure in the compartment below the high-air-entry disk is maintained as close as possible to zero pressure by increasing the air pressure in the chamber. The pressure transducer connected to the water compartment is used as a null indicator. It is important for the apparatus to have a system to flush air bubbles from below the high-air-entry disk in order to keep the compartment above the transducer saturated with water.

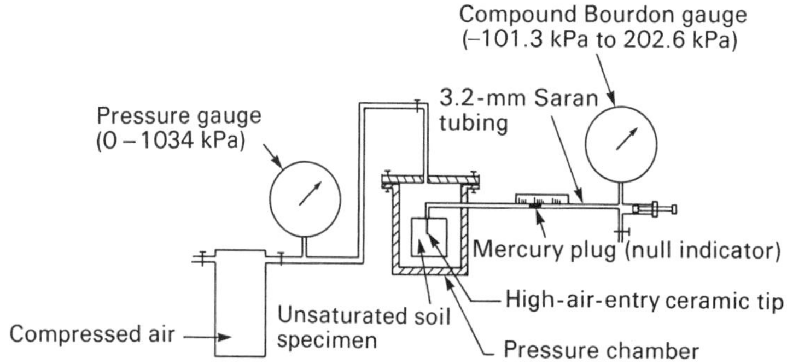

Figure 4.26 Null-type, axis translation device used to measure negative pore-water pressures (after Hilf, 1956)

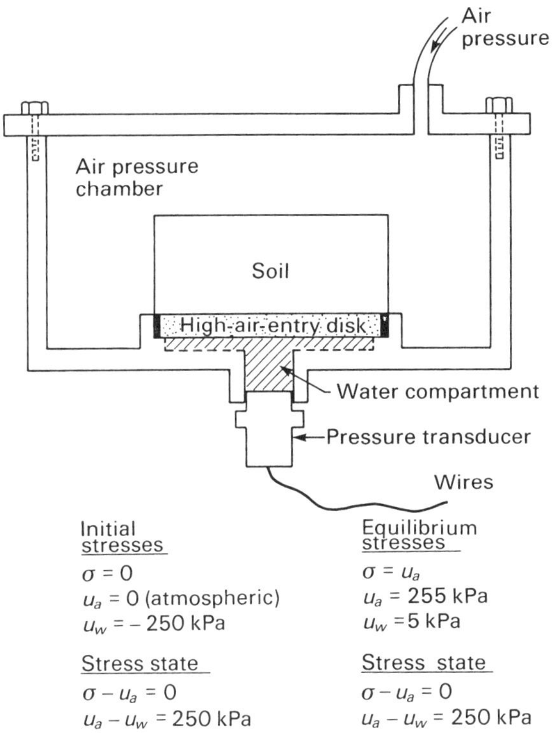

Figure 4.27 Pressure changes associated with measurement of matric suction when using null-type pressure plate apparatus (from Fredlund, 1989c).

Pressure values shown in Fig. 4.27 illustrate how highly negative pore-water pressures can be measured when using this apparatus. Let us suppose that a soil specimen has an initial pore-water pressure of −250 kPa when placed onto the saturated, high-air-entry disk. The specimen will immediately tend to draw water up through the ceramic disk, causing the pressure transducer to commence registering a negative value. The cover must be quickly placed on top of the device, and the air pressure in the chamber can be increased until there is no further tendency for the upward movement of water through the high-air-entry disk. Let us assume that at equilibrium conditions the chamber air pressure is reading 255 kPa, while the water compartment is registering 5 kPa. Therefore, the matric suction of the soil is 250 kPa (i.e., 255 − 5 = 250 kPa).

Typical response time curves when using the axis translation technique for the measurement of matric suction are shown in Figs. 4.28 and 4.29 (Widger, 1976; Filson, 1980). The response curves generally exhibit an S shape and a relatively fast equilibration time. Pressure response versus time is a function of the permeability characteristic of the high-air-entry disk and the soil. The results shown in Figs. 4.28 and 4.29 were obtained when making matric suction measurements on highly plastic clay (i.e., Regina clay).

Figures 4.30 and 4.31 present water content versus matric suction relationships for compacted specimens of Regina clay and glacial till, respectively. The matric suctions were measured using the axis translation technique. Reasonably good agreement and reproducibility in the measurement of matric suction has been obtained by many researchers indicating the reliability of the axis translation technique.

Similar water content versus matric suction relationships were obtained for several compacted soil types by Olson and Langfelder (1965), as shown in Fig. 4.32. Again, there is a distinct relationship between decreasing water contents and increasing matric suctions for as-compacted soils.

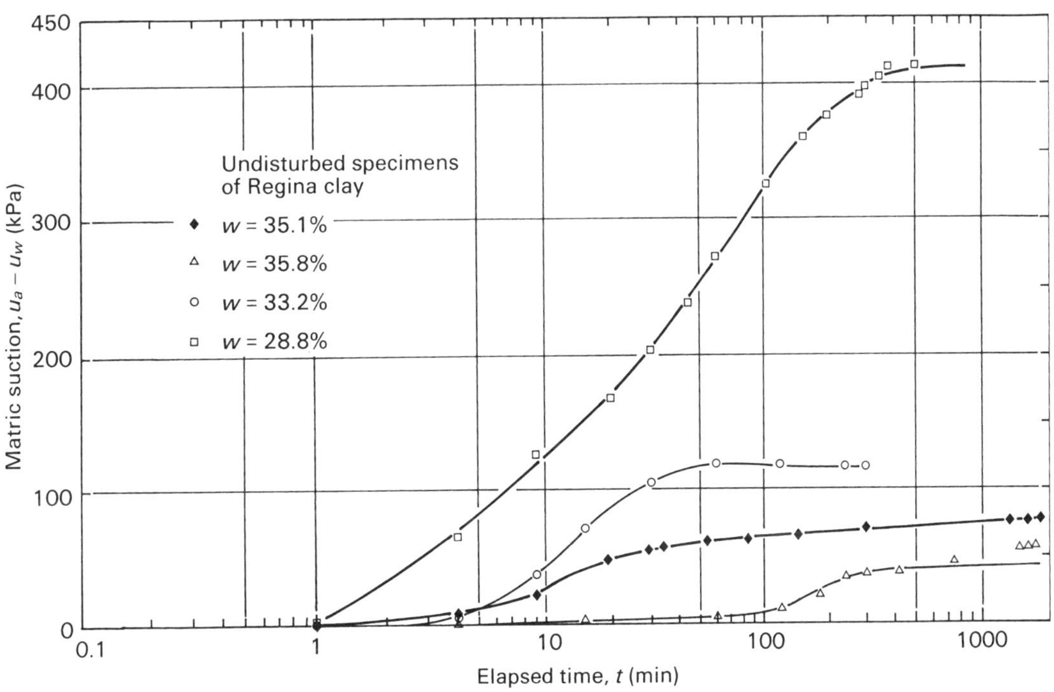

Figure 4.28 Matric suction response versus time for axis translation measurements on Regina clay (from Widger, 1976).

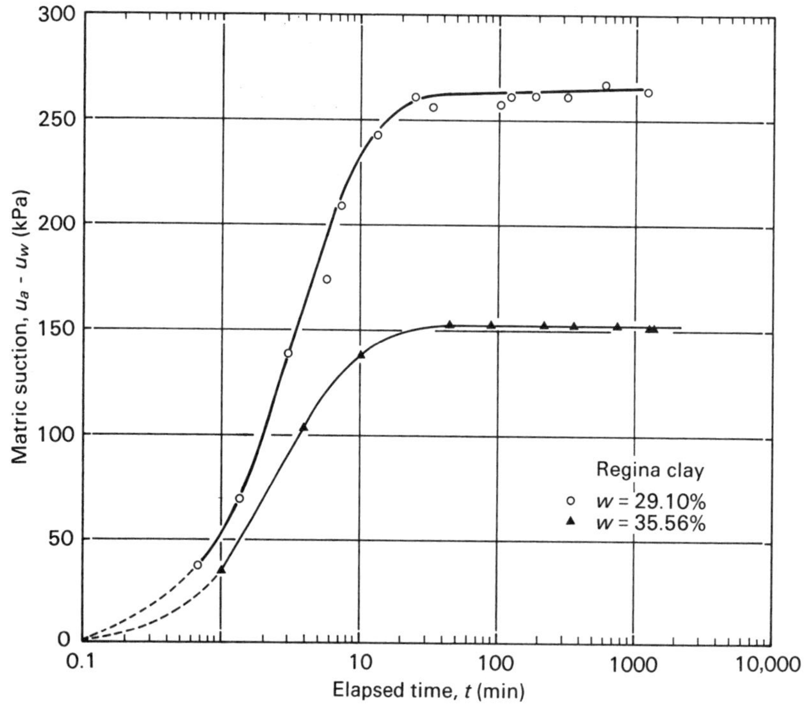

Figure 4.29 Matric suction response versus time for axis translation measurements on Regina clay (after Filson, 1980).

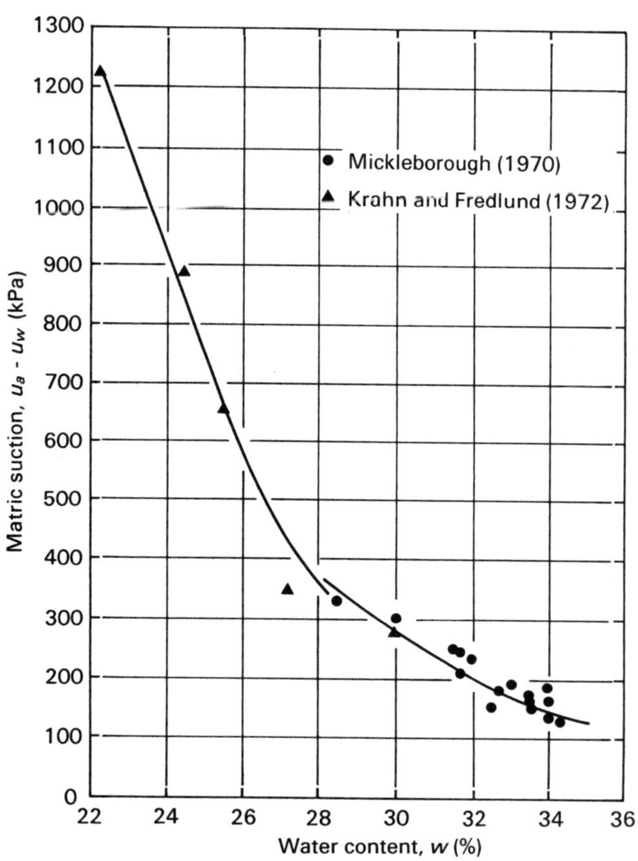

Figure 4.30 Water content versus matric suction for as-compacted specimens of Regina clay.

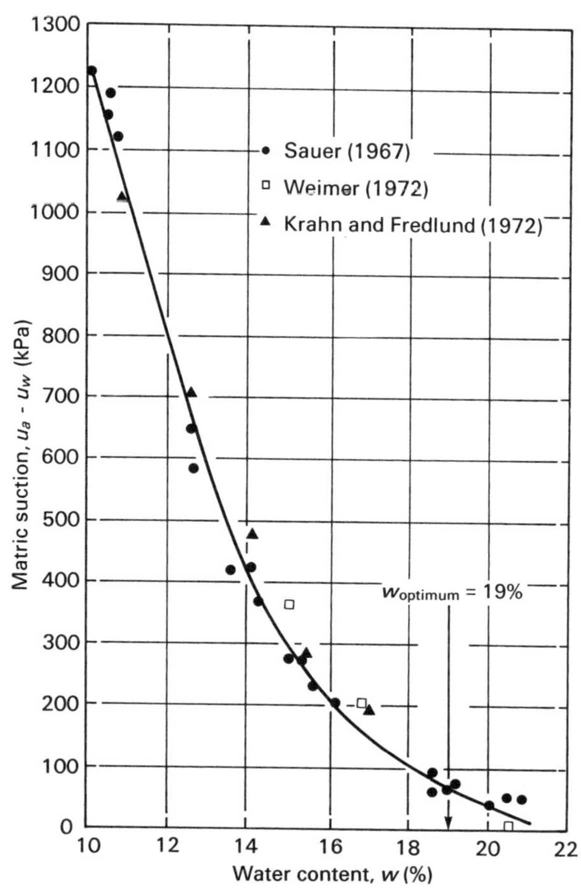

Figure 4.31 Water content versus matric suction for as-compacted specimens of glacial till.

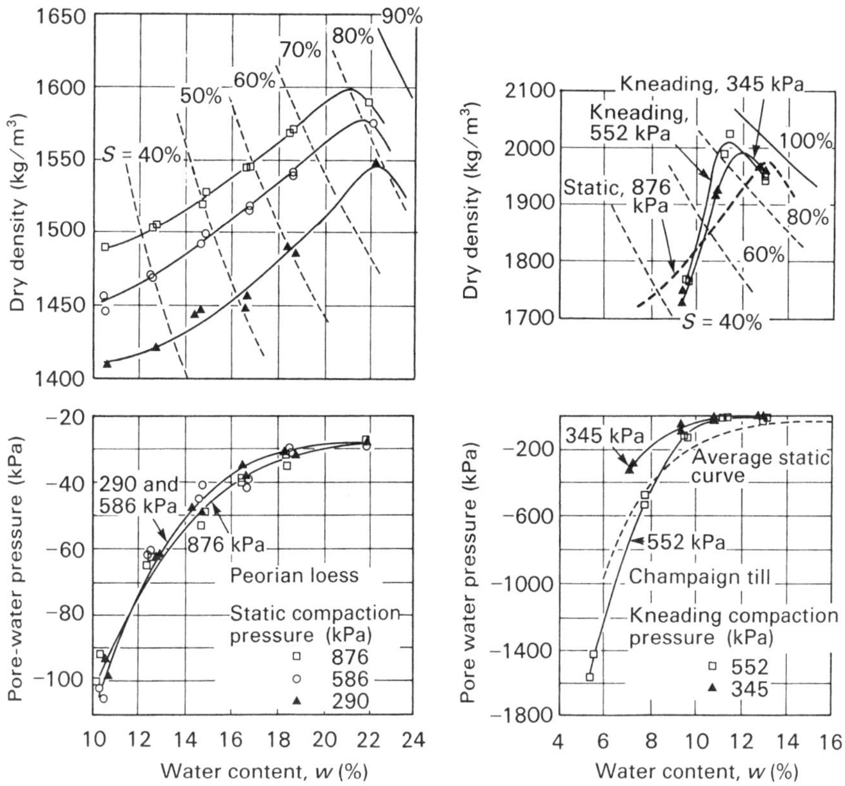

Figure 4.32 Negative pore-water pressure measurements on as-compacted soil specimens using axis translation technique (after Olson and Langfelder, 1965).

The null-type axis translation technique can be used to measure negative pore-water pressures in the laboratory with reasonable accuracy. High-air-entry disks with a maximum air-entry value up to 1500 kPa are commercially available. Theoretical studies on the axis translation technique suggested that the technique is best suited to soils with a continuous air phase (Bocking and Fredlund, 1980). The presence of occluded air bubbles in the soil specimen can result in an over estimation of the measured matric suction. In addition, air diffusing through the high-air-entry disk can cause an underestimation of the measured matric suction.

4.2.8 Indirect Measurement of Soil Suction

Indirect measurements of matric suction can be made using a standard porous block as a measuring sensor. A range of porous materials have been examined for their water retention characteristics in order to select the most appropriate material for making the sensor. These materials have included nylon, fiberglass, gypsum plaster, clay ceramics, sintered glass, and porous metals.

The porous block sensor must be brought into equilibrium with the matric suction of the soil. At equilibrium, the matric suctions in the porous block and the soil are equal. The matric suction is inferred from the water content of the porous block through a calibration process. The water content of the porous block can be determined by measuring the electrical or thermal properties of the porous block. These properties are a function of the water content of the porous block and matric suction can be established through a calibration process. In the calibration process, the porous block is subjected to various applied matric suction values, and its electrical or thermal properties are measured. The measured electrical or thermal property of the porous block under each equilibrium condition with the soil can be used to determine the matric suction of the soil through use of the calibration curve.

Indirect measurements of the water content of the sensor based on electrical properties have been found to be unsatisfactory due to the presence of dissolved salts in the pore-water (Richards, 1974). On the other hand, indirect measurements based of the thermal properties of a porous block sensor show little effect from the dissolved salts in the pore-water. As a result, the thermal-conductivity-type sensor appears to be the most promising device for the indirect measurements of matric suction (Richards, 1974).

4.2.8.1 Electrical Resistance Sensors

Electrical resistance sensors constitute the simplest type of indirect sensor for the measurement of soil suction. Unfortunately, these relatively inexpensive sensors do not have the accuracy or the long-term stability in the calibration curve for applications in geotechnical engineering. Gypsum blocks with an internal electrical resistor are the oldest type of electrical resistance sensor. The gypsum blocks are based on the principle that changes in water content produce changes in electrical resistance. Unfortunately, electrical resistance is even more sensitive to the presence of salts. With time salts can move into the gypsum and result in a substantial change in the calibration of the sensor. Gypsum block sensors have generally been found to be unsuitable for geotechnical engineering applications.

A more recent variation of electrical-resistance-type suction sensor is the Watermark soil suction sensor. The Watermark sensors consist of a granular matrix with embedded electrodes for the measurement of electrical resistance. The granular matrix is next to a gypsum wafer and forms a partial buffer from salts in the soil. The entire sensor is placed within a stainless steel enclosure. The measurement range of the Watermark sensor is 0–200 kPa. The Watermark sensor provides an indication of the wetness of the soil but does not provide a reliable indication of the matric suction in the soil. The Watermark sensor was patented in 1985 and is manufactured by Irrometer Company. The Watermark suction sensors are relatively inexpensive and can be connected to a data acquisition system where they are used to trigger irrigation systems.

4.2.9 Thermal Conductivity Suction Sensors

The thermal properties of a soil have been found to be indicative of the water content of a soil. Water is a better thermal conductor than air. The thermal conductivity of a soil increases with increasing water content. This is particularly true where the change in water content is associated with a change in the degree of saturation of the soil.

Shaw and Baver (1939) developed a device consisting of a temperature sensor and heater which could be installed directly into the soil to measure thermal conductivity. It was found that the presence of salts did not significantly affect the thermal conductivity measurements. However, different soils required different calibrations in order to relate thermal conductivity measurements to the water contents of the soil. Johnston (1942) suggested that the thermal conductivity sensor could be enclosed in a porous medium that would have a standard calibration curve. The porous medium could then be brought into equilibrium with the (negative) water pressure in the soil. Johnston (1942) used plaster of Paris as the porous medium to encase a heating element.

In 1955, Richards patented an electrothermal element for measuring moisture in porous media. The element consisted of a resistance thermometer which was wrapped with a small heating coil. The electrothermal element was then mounted in a porous cup and sealed with ceramic cement. Richards proposed the use of a sandy silt material for the porous block. It was suggested that the porous cup should have an air-entry value less than 10 kPa.

Bloodworth and Page (1957) studied three materials for use as a porous cup for the thermal conductivity sensors. Plaster of Paris, fired clay, or ceramic and castone (i.e., a commercially available dental stone powder) were used in the study. The castone was found to be the best material for the porous cup.

Phene et al. (1971) developed a thermal conductivity sensor using a germanium *p-n* diode as a temperature sensor. The sensor was wrapped with 40-gauge Teflon-coated copper wire that served as the heating coil. The sensing unit was embedded in a porous block. The optimum dimensions of the porous block were calculated based on a theoretical analysis. The block must be large enough to contain the heat pulse without interference from the surrounding soil. The distribution of the pore sizes in the sensor was important.

Gypsum, ceramics, and mixtures of ceramics and castone were examined as potential porous block materials by Phene et al. (1971). Ceramic blocks appeared to exhibit a linear response and provided a stable solid matrix.

In the mid-1970s, Moisture Control System (MCS) of Findlay, Ohio, manufactured the MCS 6000 thermal conductivity sensor. The sensor was built using the design and construction procedures used by Phene et al. (1971). The manufactured sensors were subjected to a two-point calibration. The suggested calibration curves were assumed to be linear from zero suction to a suction of 300 kPa. Above 300 kPa, the calibration curves were empirically extrapolated. In the region above 300 kPa, the calibration curves became highly nonlinear and less accurate.

The MCS 6000 sensors were used for making some matric suction measurements in the laboratory and in the field (Picornell et al., 1983; Lee and Fredlund, 1984). The sensors appeared to be quite suitable for field usage, being insensitive to temperature and salinity changes. Relatively accurate measurements of matric suction were obtained in the range from 0 to 300 kPa. Curtis and Johnston (1987) used the MCS 6000 sensors in a groundwater recharge study. The sensors were found to be quite responsive and sensitive. However, Moisture Control System discontinued production in early 1980, and the MCS 6000 sensor is no longer commercially available.

In 1981, Agwatronics in Merced, California, commenced production of the AGWA thermal conductivity sensors. The design of the sensor was changed from previous designs but was based on the research by Phene et al., (1971). There were several difficulties associated with the AGWA sensor that resulted in a new design called the AGWA-II sensor in 1984.

A detailed calibration study on the AGWA-II sensors was undertaken at the University of Saskatchewan, Canada (Wong et al., 1989; Fredlund and Wong, 1989). Several difficulties were reported with the use of the AGWA-II sensors. These include the deterioration of the electronics and the porous block with time. The AGWA-II sensors have been used for laboratory and field measurements of matric suctions on several research studies (van der Raadt et al., 1987; Sattler and Fredlund, 1989; Rahardjo et al., 1989).

In 2000, the GCTS company (Tempe, Arizona) undertook the production and marketing of the heat dissipation suction sensors that had been researched and developed at the University of Saskatchewan, Canada. The production of the heat dissipation sensors continues and is referred to as the FTC-100 thermal conductivity suction sensor.

Around the year 2000, Campbell Scientific commenced the production of another heat dissipation sensor that is referred to as the CS-229 sensor. It operates on the same basic principles as the sensor produced by Geotechnical Consulting Testing Systems (GCTS); however, a number of features are different. The diameter of the ceramic stone is 15 mm in diameter, and the characteristics of the ceramic stone are quite different from those of the FTC-100 sensor. The heat pulse is delivered to the interior of the ceramic stone through a needle that is inserted into the ceramic stone. Initially, these thermal conductivity suction sensors were reported to be independent of the ambient temperature. The calibration curve for the CS-229 sensor was later found to be quite sensitive to ambient temperature, and as a result, a study was initiated to better understand how a temperature correction could be applied to the basic calibration curves for the sensors. The results of the temperature affect study on the CS-229 sensors were reported by Flint et al. (2002).

The primary application of heat dissipation suction sensors has been for the monitoring of soil suctions in situ, mainly in soil cover systems. Both the CS-229 and FTC-100 heat dissipation suction sensors have been installed in soil cover systems and have been used to monitor soil suctions over extended time periods. The FTC-100 suction sensors have been installed and used for monitoring soil suctions in subgrade soil under thin asphalt pavements in the province of Saskatchewan, Canada. This study has been ongoing for more than 8 years and the suction sensors are still functioning satisfactorily.

4.2.9.1 Theory of Operation for Thermal Conductivity Suction Sensors

A thermal conductivity sensor consists of a porous ceramic block containing a temperature sensing element and a miniature heater (Fig. 4.33). The thermal conductivity of the porous block varies in accordance with the water content of the block. The water content of the porous block is dependent upon the matric suctions applied to the block by the surrounding soil (Fig. 4.34). The thermal conductivity of the porous block must be calibrated against applied matric suctions in the laboratory prior to their usage.

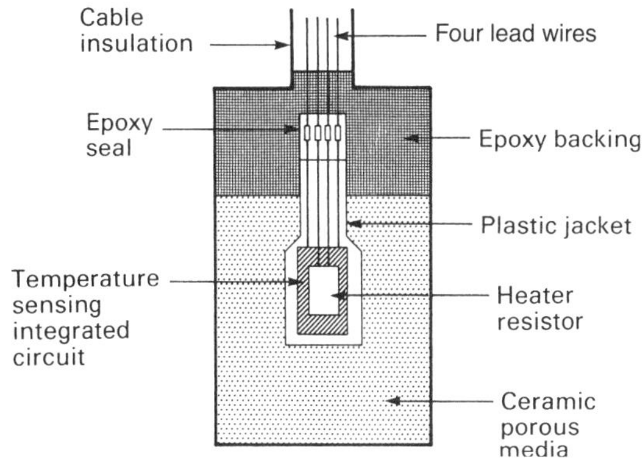

Figure 4.33 Cross section of AGWA-II thermal conductivity suction sensor (after Phene et al., 1971).

A calibrated sensor can then be used to measure matric suction by placing the sensor in the soil and allowing it to come to equilibrium with the state of stress in the pore-water (i.e., the matric suction of the soil). Thermal conductivity measurements at equilibrium are related to the matric suction in the soil through use of a calibration curve.

Thermal conductivity measurements are performed by measuring heat dissipation within the porous block. A controlled amount of heat is generated by the heater at the center of the block. A portion of the generated heat will be dissipated throughout the block. The amount of heat dissipation is controlled by the presence of water within the porous block. A *change* in the thermal conductivity of the porous block is directly related to a change in water content of the block. In other words, more heat will be dissipated as the water content of the block increases.

The heat that is not dissipated results in a temperature rise at the center of the block. The temperature rise is measured

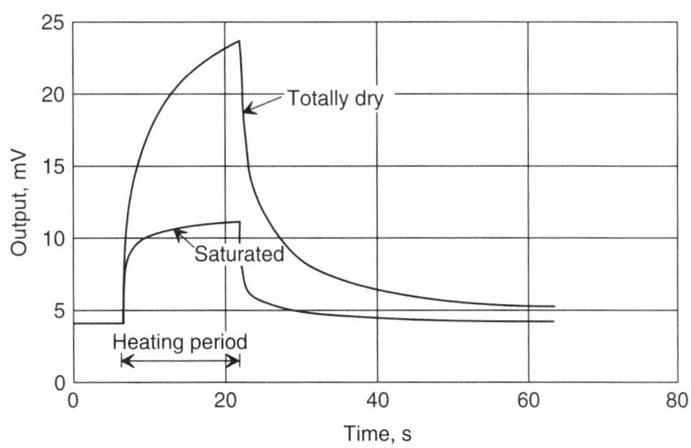

Figure 4.34 Temperature rise in mV with time of heating as function of ratio of air to water in ceramic.

by the temperature sensing element after a specified time interval, and its magnitude is inversely proportional to the water content of the porous block. The measured temperature rise can be expressed in terms of a change in voltage output.

4.2.9.2 Calibration of Thermal Conductivity Suction Sensors

Thermal conductivity sensors are usually subjected to limited calibration prior to shipment from the factory. One calibration reading can be taken with the sensors placed in water (i.e., zero matric suction). A second calibration reading can be taken when the sensors are dry. Other calibration points at specified applied suctions may also be obtained. This calibration procedure may be adequate for some applications. However, it is suggested that a more rigorous calibration procedure should be used for geotechnical engineering applications (Fredlund and Wong, 1989).

A more thorough calibration of thermal conductivity sensors can be performed by applying a range of matric suction values to the sensors. Readings of the change in voltage output, after the application of a heat impulse, provide a measure of the thermal conductivity (or the water content) of the porous block under the applied matric suction. Matric suctions can be applied to the thermal conductivity sensors using either a modified pressure plate apparatus (Fig. 4.35; Wong et al., 1989; Fredlund and Wong, 1989) or a specially designed calibration device (Fig. 4.36; GCTS, 2008). The maximum suction that can be applied through use of the axis-translation technique depends on the air-entry value of the ceramic disk in the calibration device. The air-entry value of the calibration device is generally 500 kPa.

The use of the small independent calibration chamber for each thermal conductivity suction sensor appears to be preferable for calibration purposes. Figure 4.37 shows a series of cells being used for the calibration of the thermal conductivity suction sensors.

There have been a number of equations that have been used to best fit calibration data for thermal conductivity suction sensors. It is relatively easy to obtain two

Figure 4.35 Calbration of thermal conductivity suction sensors in modified pressure plate apparatus.

Figure 4.36 Calibration cells and FTC-100 thermal conductivity suction sensors (Manufactured by GCTS, Tempe, AZ).

"anchor-type" points on the calibration curve for thermal conductivity suction sensors. The first calibration anchor point represents the reading from the sensor when the ceramic stone is completely dry (or dry to atmospheric conditions). In other words, the measured thermal conductivity will represent the thermal conductivity of the ceramic stone with air filling the voids. It is also important to define what is meant by a dry ceramic stone. In other words, it is possible for the "dry ceramic stone" to be initially air dried, initially oven dried, or initially placed above salts that create an extremely high total suction environment. The relative humidity should be recorded that is associated with the dry condition. This allows the computation of a total suction value based on Lord Kelvin's equation.

The second calibration anchor point represents the reading on the thermal conductivity suction sensor when the ceramic stone is saturated with water. In this case, the

Figure 4.37 Calibration of series of thermal conductivity suction sensors (Manufactured by GCTS, Tempe, AZ).

measured thermal conductivity will represent the thermal conductivity of the ceramic stone with water filling the voids. This is an easy reading to obtain, but once again, it is important to define how the ceramic stone has been saturated with water. It is possible for the "wet or saturated ceramic stone" to be immersed in water for a period of time or placed in a back-pressured environment to ensure that all air has been dissolved in water. Previous research studies (Feng, 1999; Feng et al., 2002; Feng and Fredlund, 2003) have shown that simply immersing the ceramic stones in water for a period of time does not fully saturate the ceramic stone. Rather, occluded air bubbles become entrapped in some voids of the ceramic stone. It was found that back-pressuring the ceramics resulted in higher degrees of saturation and a change in the thermal conductivity measurement.

The complete calibration of a thermal conductivity suction sensor generally involves placing the sensors onto a pressure-plate-type device where three to seven additional suctions are applied to the sensor while monitoring the output from the sensors. The applied suction values usually range from 10 to 500 kPa along the desorption calibration curve. The complete set of data points can then be best fit using an empirical calibration equation.

Figure 4.38 shows a typical set of calibration curve data for the GCTS thermal conductivity (TC) suction sensors. Figure 4.39 shows a typical set of calibration curve data for the CS-229 TC suction sensors.

Hu et al. (2007) undertook the calibration of 16 GCTS thermal conductivity suction sensors. The measured temperature changes under the suction applied to each of the 16 TC suction sensors are shown in Fig. 4.40.

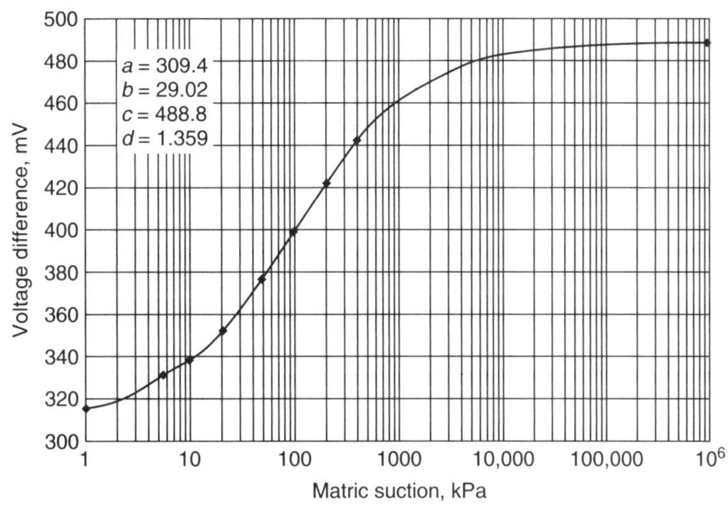

Figure 4.38 Typical calibration curve for FTC-100 thermal conductivity suction sensor (manufactured by GCTS, Tempe, AZ).

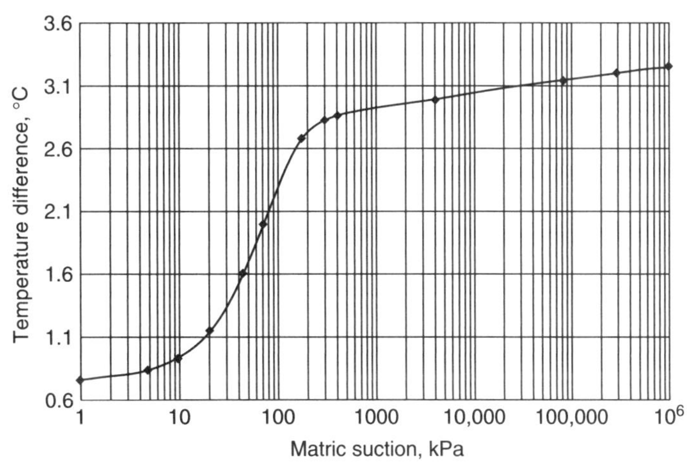

Figure 4.39 Typical calibration curve for CS-229 thermal conductivity suction sensor (manufactured by Campbell Scientific 2006).

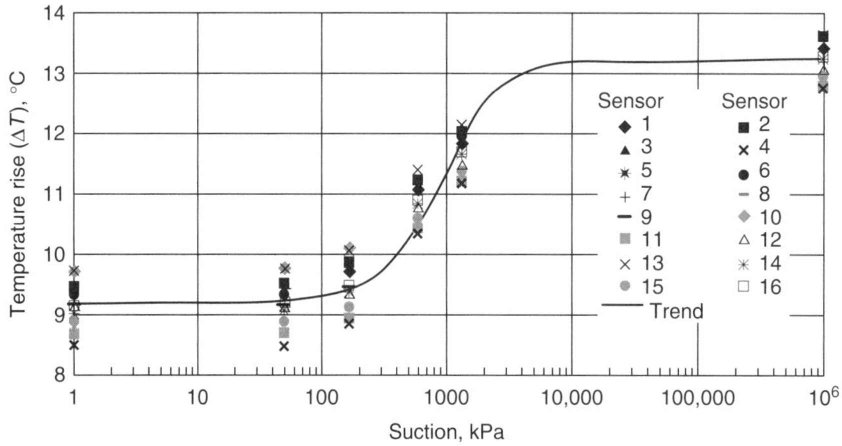

Figure 4.40 Measured temperature changes for 16 FTC-100 thermal conductivity suction sensors (after Hu et al., 2007).

4.2.9.3 Calibration Equations for Thermal Conductivity Suction Sensors

The electrical response from a thermal conductivity suction sensor can be plotted versus the logarithm of soil suction. The resulting curve is S shaped with anchor-type points at minimum suction (i.e., assumed zero suction) and a maximum suction (i.e., near 10^6 kPa). The calibration curve has a sigmoidal shape.

In 1999, Feng and Fredlund proposed the following form for the calibration equation for thermal conductivity suction sensors:

$$\psi = \frac{b(\Delta T - a)}{c - \Delta T} \tag{4.5}$$

where:

ψ = applied suction, kPa,
ΔT = temperature rise, °C, and
a, b, c, d = four parameters to define the shape of the calibration curve.

Of the four fitting parameters, a and c are equivalent to the temperature rise ΔT for the thermal conductivity sensors under saturated and dry conditions, respectively. The parameter b defines a particular horizontal position on the calibration curve and the parameter d defines the slope of the curve at point b. However, point b was not the most appropriate point around which to perform the best-fit regression analysis. Rather, it is better to use the inflection point on the semilogarithm calibration curve for the definition of the b parameter.

In 2007, Hu et al. proposed a modification to Eq. 4.5 that would give more significant physical meaning to each of the fitting parameters. The new proposed calibration equation was written as

$$\psi = b_1 \left[\frac{(\Delta T - a_1)}{c_1 - \Delta T} \right]^{d_1 \frac{c_1 - a_1}{4 b_1}} \tag{4.6}$$

The a_1 and c_1 fitting parameters in Eq. 4.6 have the same meaning as the a and c parameters in Eq. 4.5. The fitting parameter b_1 is the suction when ΔT is at the inflection point of the calibration curve. In other words, it is midway between the "dry" reading and the "saturated" sensor reading. Stated another way, b_1 is equal to the soil suction at the inflection point (i.e., suction when $\Delta T = (a_1 + c_1)/2$). Then, d_1 is the slope of the calibration curve at point b_1 (i.e., $d\psi/dT|_{\Delta T = ((a_1 + c_1)/2)}$) (see Fig. 4.41).

The Hu et al. (2007) revision to the calibration equation was used to analyze the calibration data of 16 FTC-100 thermal conductivity suction sensors. The calibration measurements are summarized in Table 4.5. The Shuai et al. (2002) temperature correction was applied to the measured data prior to determining the best-fit equation parameters.

The average values for the four calibration parameters on the 16 sensors are as follows: $a_1 = 9.19$°C, $b_1 = 261.1$ kPa, $c_1 = 13.3$°C, and $d_1 = 185.0$ kPa/°C. The a_1 parameter is the thermal conductivity sensor reading when the ceramic is saturated with water, and c_1 is the thermal conductivity sensor reading when the ceramic is dry. The b_1 variable is the soil suction corresponding to a temperature change halfway between dry and water saturated conditions (i.e., the inflection point along the calibration curve on a logarithm scale). The d_1 variable is the slope corresponding to point b_1.

The following points are noteworthy from the calibration of the 16 thermal conductivity suction sensors. First, the a_1 and c_1 parameters provide the most consistent measurements on the calibration curve. The standard deviation is similar for both variables with values of approximately 0.3–0.4°C. A typical value for the a_1 parameter was 9.2°C and a typical value for the c_1 parameter was 13.2°C. These values are primarily a function of the constant current applied during calibration. The b_1 parameter provides an indication of the range where the sensor has the greatest sensitivity to suction changes. The 16 suction sensors calibrated show the b_1 parameter to be around 260 kPa. The b_1 and d_1 parameters

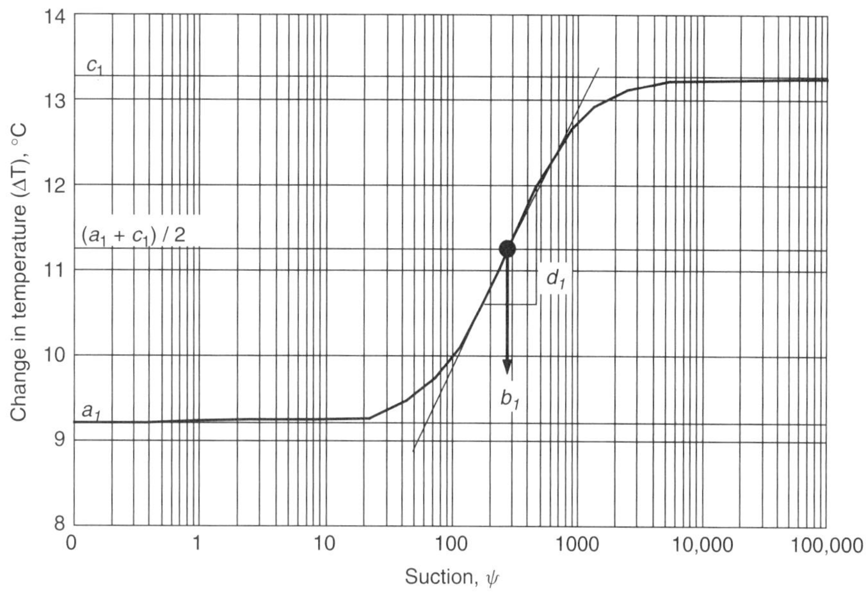

Figure 4.41 Variables associated with calibration equation for FTC-100 thermal conductivity suction sensors (after Hu et al., 2007).

Table 4.5 Calibration Parameters with Shuai et al., (2002) Temperature Correction Applied to Laboratory-Measured Data

Sensor	a_1 (°C)	b_1 (kPa)	c_1 (°C)	$d\psi/d\Delta T$ (kPa/°C)	Power Term
FTC1	9.546	265.197	13.301	187.915	0.665
FTC2	9.466	249.972	13.502	193.114	0.780
FTC3	9.466	251.427	13.683	146.760	0.615
FTC4	8.502	271.992	12.874	200.325	0.805
FTC5	9.111	247.115	13.512	162.070	0.722
FTC6	9.399	284.507	13.720	214.919	0.816
FTC7	9.391	265.434	13.479	199.341	0.768
FTC8	9.284	247.859	13.473	154.341	0.652
FTC9	9.187	272.298	13.184	208.399	0.765
FTC10	9.725	282.024	13.346	237.945	0.764
FTC11	8.708	259.302	12.788	169.213	0.666
FTC12	9.131	263.353	13.158	149.147	0.570
FTC13	9.758	259.571	13.593	207.638	0.767
FTC14	8.995	261.762	13.216	210.067	0.847
FTC15	8.833	263.482	12.928	173.399	0.674
FTC16	8.509	231.937	12.599	145.282	0.640
Maximum value	9.758	284.507	13.720	237.945	0.847
Minimum value	8.502	231.937	12.599	145.282	0.570
Average value	9.188	261.077	13.272	184.992	0.720
Standard deviation	0.394	13.468	0.333	28.374	0.080
Coefficient of variation	4.283	5.159	2.508	15.338	11.169

are related to a common point along the calibration curve. A typical value for the d_1 parameter was about 190 kPa/°C.

Figure 4.42 plots the response of the dry sensor to the response of the wet sensor. The results show that whenever the dry sensor response is a little higher than average, the wet sensor response likewise seems to increase. Figure 4.43 compares the difference between the wet and dry response of the FTC sensors to the dry sensor reading. The difference between the wet and dry response remains quite similar for all sensors, with an average value of 4.02°C. The temperature change is a function of the applied constant current. Figure 4.44 compares the slope of the calibration curve at the inflection point with the suction reading at the inflection point.

The results from the calibrations study suggest that increased quality control in constructing the thermal conductivity sensors should lead to multiple sensors having a similar response.

Thermal Conductivity Suction Measurements in the Laboratory. Results of laboratory measurements of soil suction using thermal conductivity suction sensors on highly plastic clays are shown in Figs. 4.45, 4.46, and 4.47. The soils were sampled in the field using Shelby tubes near Sceptre and Regina, Saskatchewan.

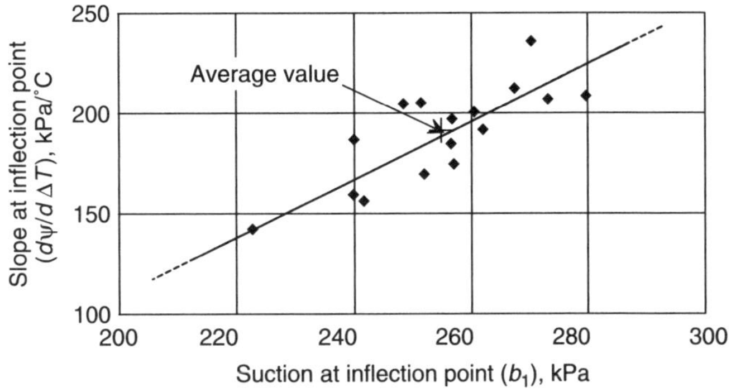

Figure 4.44 Slope of calibration curve at inflection point versus inflection point suction for FTC-100 thermal conductivity suction sensors (after Hu et al., 2007)

Several laboratory measurements were conducted using two sensors inserted into each soil specimen. One sensor was initially air dried, and the other was initially water saturated. The sensors were inserted into predrilled holes at either end of the Shelby tube sample. The specimen with the inserted thermal conductivity sensors was wrapped in aluminum foil to prevent moisture loss as the sensors came

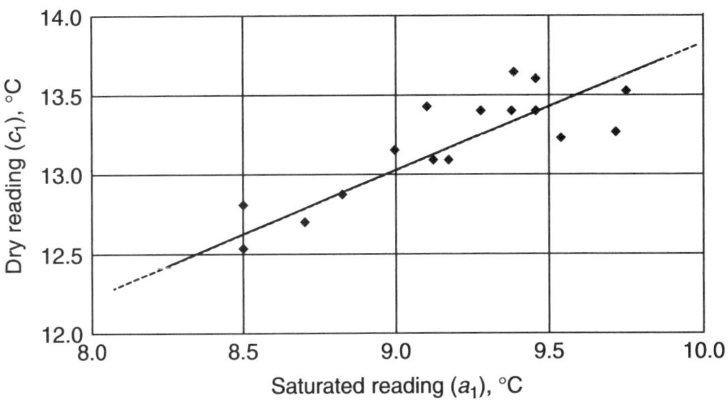

Figure 4.42 Relationship between wet sensor readings and dry sensor readings for 16 FTC-100 thermal conductivity suction sensors (after Hu et al., 2007).

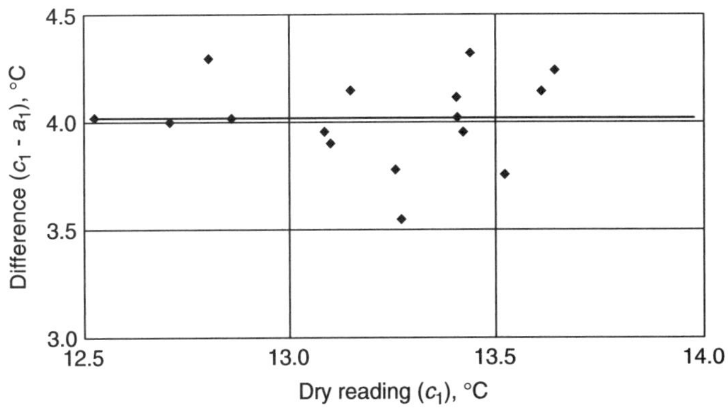

Figure 4.43 Comparison of difference between wet and dry sensor readings with dry sensor readings for FTC-100 suction sensors (after Hu et al., 2007).

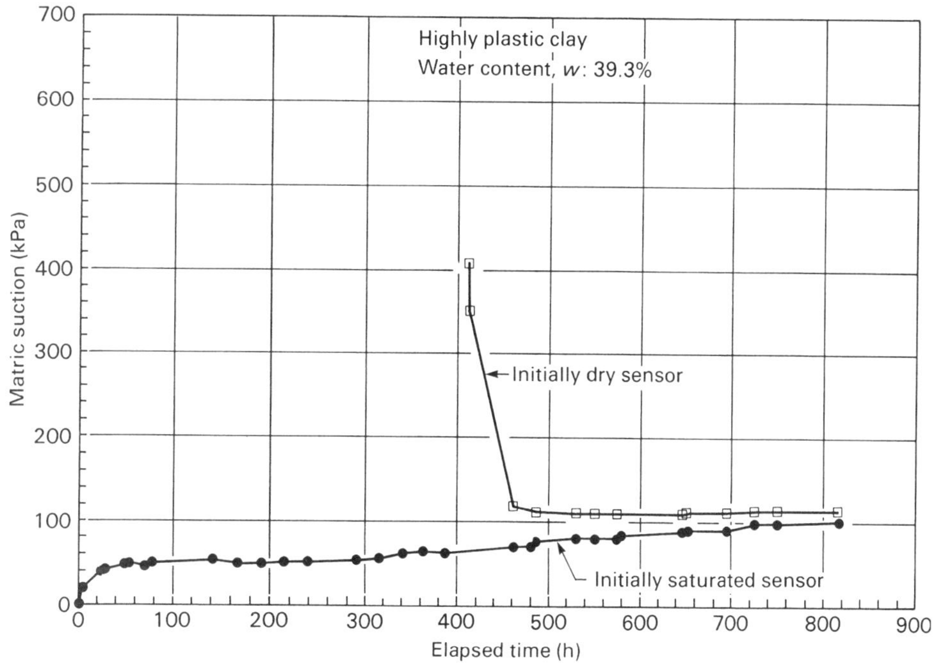

Figure 4.45 Laboratory measurement of matric suction on high-plasticity clay from Sceptre, Canada, using thermal conductivity suction sensors ($w = 39.3\%$).

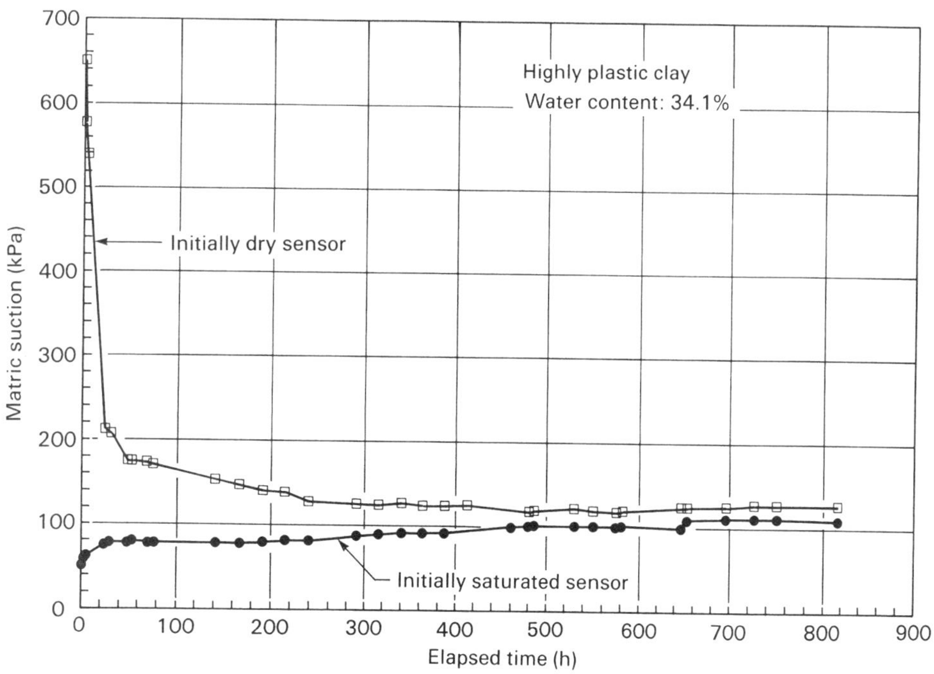

Figure 4.46 Laboratory measurement of matric suctions on high-plasticity clay from Sceptre, Canada, using thermal conductivity suction sensors ($w = 34.1\%$).

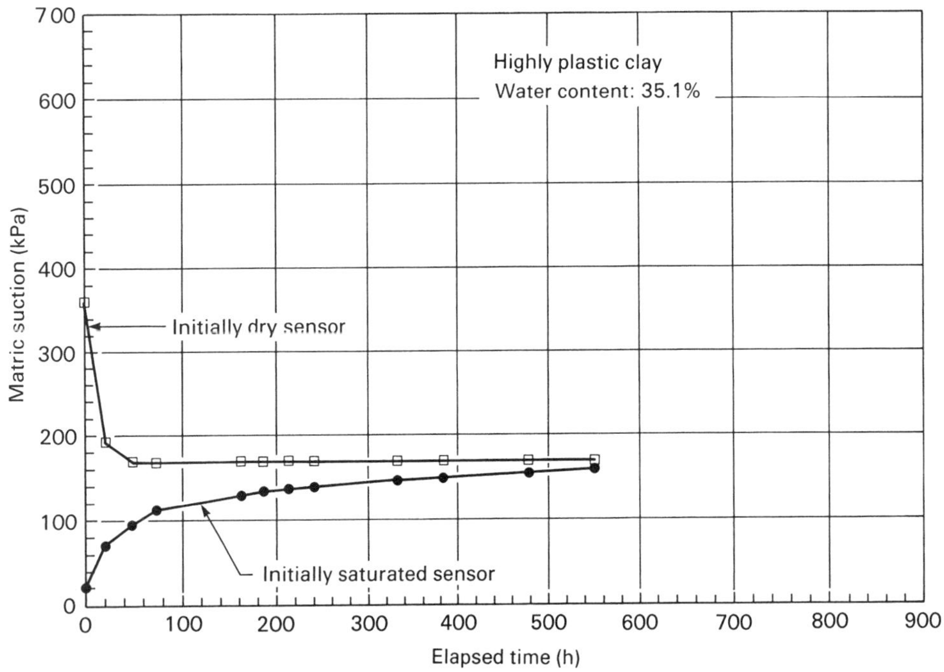

Figure 4.47 Laboratory measurement of matric suction on high-plasticity clay from Darke Hall, Regina, Canada, using thermal conductivity suction sensors ($w = 35.1\%$).

to equilibrium with the water in the soil. The responses of both sensors were monitored at various elapsed times after their installation. The results indicate that the time required for the initially dry sensor to come to equilibrium with the soil specimen was less than the equilibrium time required for the initially saturated sensor to come to equilibrium.

On the basis of numerous laboratory experiments, it would appear that the thermal conductivity suction sensors that were initially dry yielded a matric suction value which was close to the correct value. In general, the initially dry sensor should yield a value which is slightly high. On the other hand, the initially wet sensor should yield a value which is lower than the actual soil suction. Table 4.6 gives the interpretation of the results presented in Figs. 4.45, 4.46, and 4.47.

It is suggested that if only one sensor is installed in an undisturbed sample, the sensor should be initially dry. It may take 4–7 days before suction equilibrium is achieved. If the sensors are left in situ for a long period of time, there should be no need to take readings any more often than about once every 3 h. Thermal conductivity suction sensors are not fast-acting sensors and as such are not meant to register rapid changes in pore-water pressure.

Table 4.6 Interpretation of Laboratory Matric Suction Measurements

Soil Type	Water Content, %	Initially Dry Sensor, kPa	Initially Wet Sensor, kPa	Best Estimate, kPa
Sceptre clay	39.3	120	100	114
	34.1	136	108	126
Regina clay	35.1	160	150	157

Laboratory measurements of matric suction have been used to establish the negative pore-water pressures in undisturbed samples of Winnipeg clay taken from various depths within a railway embankment (Sattler et al., 1990). The samples were brought to the laboratory for matric suction measurements using thermal conductivity suction sensors. The measured matric suctions were corrected for the removal of the overburden stress and plotted as a negative pore-water pressure profile (Fig. 4.48). The results indicated that the negative pore-water pressures approached zero at the average water table. In general, the pore-water pressures tended to be more negative than the hydrostatic line through the water table.

4.2.9.4 Thermal Conductivity Suction Measurements in the Field

Field measurements of matric suction under a controlled environment have been conducted in the subgrade soils of a Department of Highways indoor test track at Regina, Saskatchewan (Loi et al., 1992). The temperature and the relative humidity within the test track facility were controlled. Twenty-two thermal conductivity suction sensors were installed in the subgrade of the test track. The subgrade consisted of highly plastic clay and glacial till. The sensors were initially air dried and installed into predrilled holes at various depths in the subgrade. The sensor outputs were recorded twice daily.

Typical matric suction measurements made on compacted Regina clay and glacial till soils at the test track are shown in Fig. 4.49. Consistent readings of matric suction ranging from 50 to 400 kPa were monitored over a period of more

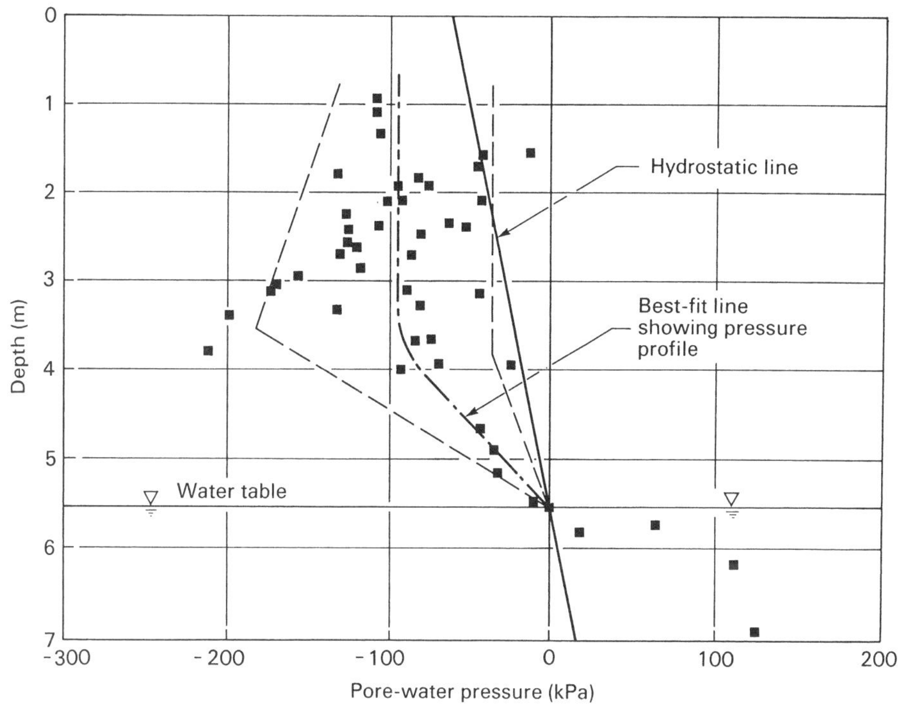

Figure 4.48 Negative pore-water pressure measured using AGWA-II thermal conductivity sensors.

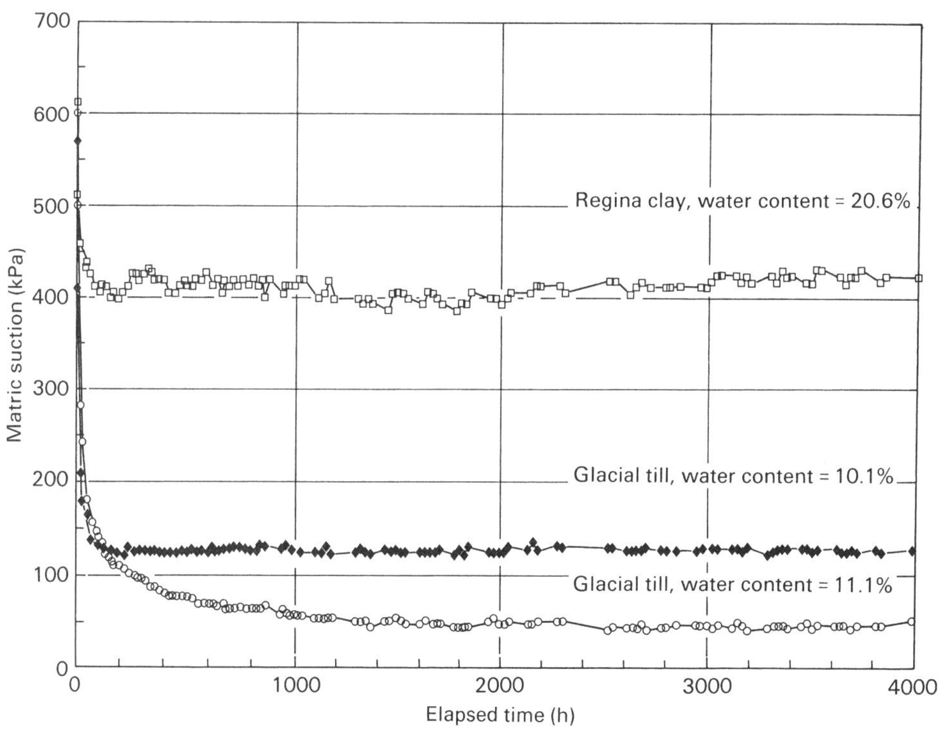

Figure 4.49 Measurement of soil suctions using AGWA-II thermal conductivity sensors under controlled environment (Test Track facility of the Department of Highways, Regina, Canada).

than 5 months prior to flooding the test track. The results demonstrated that the thermal conductivity suction sensors can provide stable measurements of matric suction over relatively long time periods.

Matric suction variations in the field can be related to environmental changes. Several AGWA-II sensors have been installed at various depths in the subgrade of roads and railroads. Matric suctions in the soil were monitored at various times of the year. The results clearly indicate seasonal variations of matric suctions in the field with the greatest variation occurring near ground surface.

4.2.9.5 Influence of Ambient Temperature

The thermal conductivity suction sensor measurement is dependent upon the thermal conductivity of the individual components of the sensor, namely, the ceramic, the water, the air, and the internal electrical and epoxy components. Several research studies have been undertaken to assess the effect of ambient temperature on the calibration curve.

The thermal conductivity λ of the ceramic portion is about 1.7 W/m °C. Published thermal conductivities have shown the thermal conductivity of the ceramic to be essentially independent of the ambient temperature. The thermal conductivity of air is about 0.025 W/m °C. Published values show that the thermal conductivity of air has a slight dependence on temperature.

The thermal conductivity of water is known to change with temperature. Table 4.7 provides a list of thermal conductivities for water, λ over the range from the freezing point to the boiling point of water.

Shuai et al. (2002) graphically presented results of the thermal conductivity of water versus temperature (Fig. 4.50). The results show some nonlinearity in the relationship between the thermal conductivity of water and temperature, particularly in the low temperatures near freezing and the high temperatures near the boiling point. The thermal conductivity of water versus temperature can be best fit with the following linear equation:

$$\lambda = 0.0014T + 0.5743 \tag{4.7}$$

where:

λ = thermal conductivity of water, W/m °C, and
T = temperature, °C.

Campbell et al., (1994) proposed the following parabolic equation for the thermal conductivity of water:

$$\lambda = 0.024 + (0.0000773)\,T - (0.000000026)\,T^2 \tag{4.8}$$

Figure 4.51 provides a comparison of the thermal conductivities of the ceramic, air, and water over a range of temperatures. The water phase appears to show the greatest dependency on ambient temperature. A number of studies have been undertaken to assess the influence of temperature on the measurements of soil suction when using thermal conductivity suction sensors. The effect of ambient temperature on thermal conductivity suction sensor readings has been observed since the 1970s (Phene et al., 1971; Wong et al., 1988). Concern over the effect of ambient temperature changes has been expressed when using both the GCTS thermal conductivity suction sensor and the Campbell Scientific CS229 thermal conductivity suction sensors.

In 2002, Shuai et al. conducted a series of tests in an attempt to quantify the effect of ambient temperature on the measurement of soil suction when using the GCTS type of thermal conductivity sensors. (These tests were performed on suction sensors manufactured at the University of Saskatchewan but are of the same design as used by GCTS.) Figure 4.52 shows the effect of temperature change on the observed readings from the GCTS-type suction sensors. A thermometer was placed next to the GCTS thermal conductivity suction sensor embedded in a soil sample that was wrapped in tin foil to maintain a constant suction value. The results showed that when the ambient temperature increased, the measured suction (i.e., obtained from the 23°C calibration curve) decreased. When the temperature decreased, the opposite behavior was observed.

It was suggested that the inter relationship between temperature and the thermal conductivity suction measurement might be due to the change in thermal conductivity of the water in the soil. The dependency of the thermal conductivity of water on temperature led Shuai et al., (2002) to propose a correction equation that could be applied to the observed measurement, taking ambient temperature into consideration. The proposed temperature correction took the form of Eq. 4.5. It should be noted that the temperature variation was only about 3°C and the sensor reading change

Table 4.7 Thermal conductivity of Water at Various Temperatures

Temperature, °C	Thermal Conductivity, λ, W/m °C
0.01	0.5610
10	0.5800
20	0.5984
30	0.6155
40	0.6306
50	0.6436
60	0.6543
70	0.6631
80	0.6700
90	0.6753
100	0.6791

Source: Cooper and Dooley, 2008.

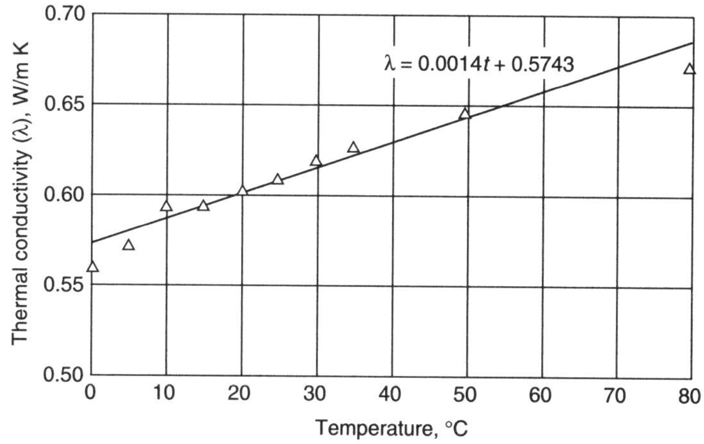

Figure 4.50 Thermal conductivity of water versus temperature with linear best-fit equation (after Shuai et al., 2002).

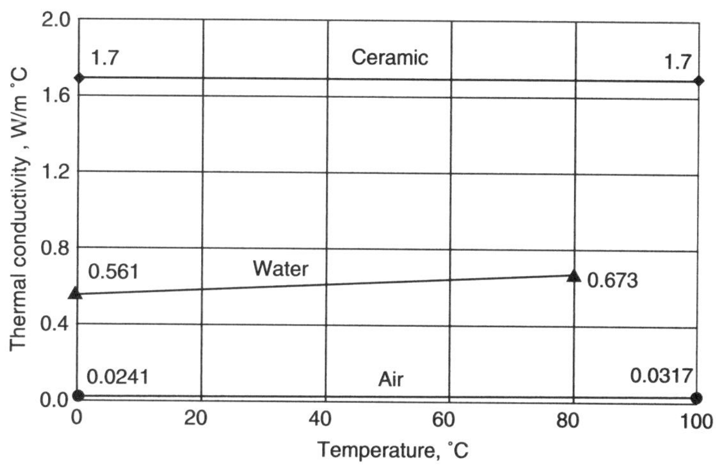

Figure 4.51 Comparison of thermal conductivity of ceramic, air, and water (after Shuai et al., 2002).

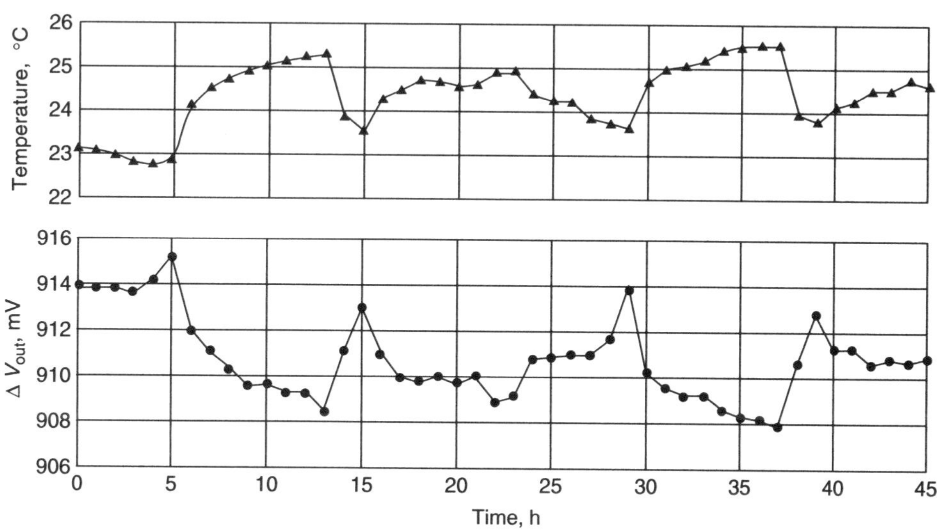

Figure 4.52 Observed reaction of thermal conductivity suction sensors to ambient temperature changes (after Shuai et al., 2002).

was about 7 mV. The thermal conductivity suction sensors were calibrated in the laboratory at a temperature of 23°C:

$$\Delta T_{23^\circ\mathrm{C}} = \frac{0.0014T + 0.5743}{0.6065} \Delta T_T \tag{4.9}$$

where:

$\Delta T_{23^\circ\mathrm{C}}$ = corrected rise in temperature corresponding to 23°C and

ΔT_T = rise in temperature measured by the TC sensor at the ambient temperature T.

The proposed correction equation was then applied to the data presented in Fig. 4.52. The temperature-corrected soil suction measurements are shown in Fig. 4.53. The results show that the variation in the sensor reading was reduced to about 2 mV when the ambient temperature correction was applied.

Shuai et al., (2002) undertook a series of tests where the ambient temperature was varied from 8 to 23°C in a series of steps. The output voltage was measured and the temperature correction was applied to the recorded readings. The constant suctions applied in each case were 10, 50, and 200 kPa. The results obtained when the applied suction was 50 kPa are shown along with the measured temperatures in Fig. 4.54. The results indicate that the suctions remained essentially constant when the proposed temperature correction was applied.

Nichol et al. (2003) observed that ambient temperatures influenced the thermal conductivity suction sensor readings and proposed a more rigorous model to take into account the effect of ambient temperatures. His studies involved the use of the GCTS thermal conductivity suction sensors and the following temperature change correction equation was proposed:

$$\Delta T_{23^\circ\mathrm{C}} = \Delta T \left(s_1 + s_2 T + s_3 T^2\right) \tag{4.10}$$

where:

ΔT = change in temperature measured by the TC suction sensor,

$\Delta T_{23^\circ\mathrm{C}}$ = corrected temperature change that can be applied to the 23°C calibration curve,

T = ambient temperature recorded by the TC suction sensor, and

s_1, s_2, and s_3 = constants that depend on the range of suction being measured.

The temperature correction proposed by Nichol et al. (2003) is dependent not only on the ambient temperature but also on the magnitude of the soil suction. It is apparent that more variables are taken into consideration in the development of the Nichol et al. (2003) correction. The correction constants vary with the level of soil suction and suggested values are presented in Table 4.8.

Table 4.8 Temperature Correction Constants for Nichol et al. (2003) Equation

Suction Range kPa	s_1, °C	s_2, °C^{-1}	s_3, °C^{-2}
0–10	0.960	1.9×10^{-3}	-0.8×10^{-6}
10–30	0.962	1.8×10^{-3}	-0.7×10^{-6}
30–75	0.968	1.5×10^{-3}	-0.6×10^{-6}
75–300	0.973	1.3×10^{-3}	-0.5×10^{-6}
300–10^6	0.986	7.0×10^{-4}	-0.3×10^{-6}

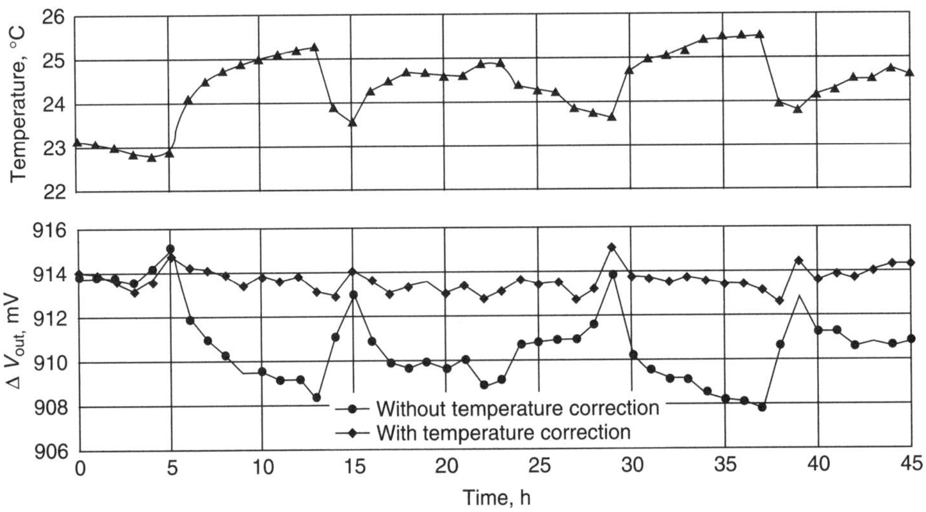

Figure 4.53 Soil suction readings "with" and "without" applied temperature corrections (after Shuai et al., 2002).

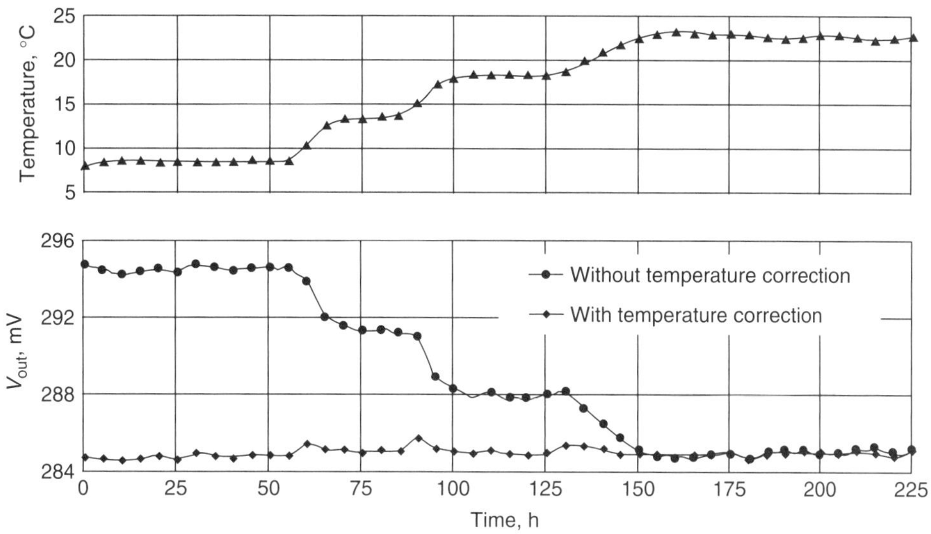

Figure 4.54 Effect of applying temperature correction as ambient temperature is increased (after Shuai et al., 2002).

The measured GCTS thermal conductivity suction sensor reading is then multiplied by the correction factor to obtain the temperature rise that would have been measured at 23°C. The correction factor is largest near zero suction when the ceramic has the highest water content. This is different from the Shuai et al. (2002) correction, which multiplies the measured suction data by a constant value for all soil suctions.

Figure 4.55 shows a plot of the Nichol et al. (2003) temperature corrections for the range from zero degrees to 40°C. The plot shows that the largest corrections are applied when the soil suctions are the lowest (i.e., 0–10 kPa). The correction factor in the low-suction range is calculated to be about 0.96 when the temperature is close to zero °C. When the temperature increases to 40°C the temperature correction becomes 1.024. The calibration temperature was 23°C. Once again the temperature corrections are quite modest.

The Nichol et al. (2003) correction factors appear to be more reasonable from a theoretical standpoint; however, Hu et al. (2007) undertook a study in which he applied the Shuai et al. (2002) and the Nichol et al. (2003) correction techniques and found that the results were quite similar. The results of tests on 14 GCTS thermal conductivity suction sensors can be observed in Fig. 4.56. A study was also undertaken by Tan et al. (2004b) where the Shuai et al. (2002) temperature correction and the Nichol et al. (2003) temperature correction were applied to 7 years of field measurement data. The temperature corrections turned out to be quite similar using both suggested procedures.

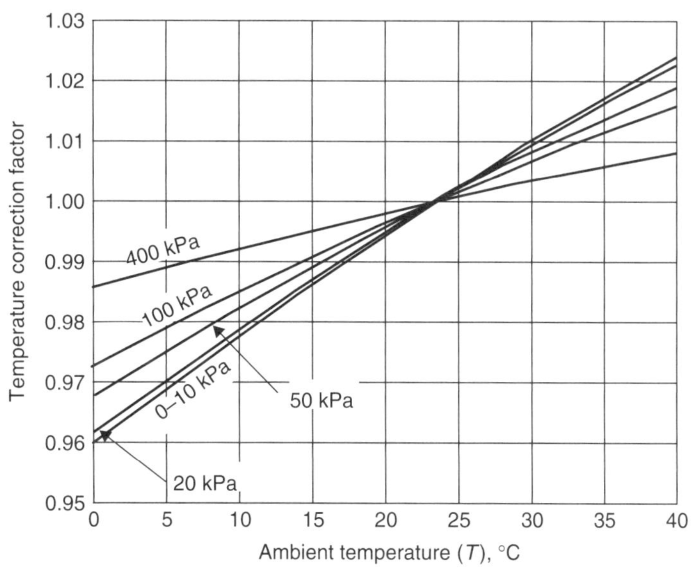

Figure 4.55 Ambient temperature correction suggested by Nichol et al. (2003).

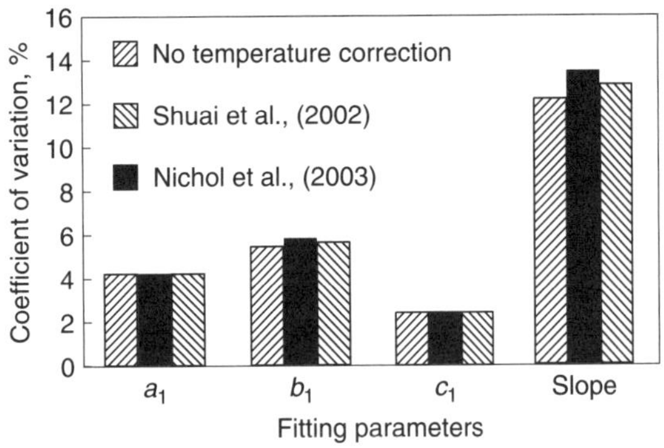

Figure 4.56 Comparison of coefficients of variation in fitting parameters when using different temperature correction equations (after Hu et al., 2007).

4.2.9.6 Influence of Temperature on the CS-229 Suction Sensors

Campbell Scientific manufactures a thermal conductivity soil suction sensor referred to as the CS-229 Water Matric Potential Sensor (Fig. 4.57). The thermal conductivity suction sensor measures soil water potential from 10 to 1000 kPa. The sensor consists of a heating element and thermocouple placed in epoxy in a hypodermic needle which is encased in a porous ceramic matrix.

The effect of the ambient temperature on the suction measurement has also been realized for the CS 229 thermal conductivity sensors. Reece (1996) suggested the use of a normalization procedure to reduce the influence of varying ambient temperature. Flint et al., (2002) suggested using the following normalization equation to give a dimensionless temperature rise, ΔT^*:

$$\Delta T^* = \frac{\Delta T_d - \Delta T}{\Delta T_d - \Delta T_w} \tag{4.11}$$

where:

ΔT = temperature rise of the source in the ceramic in a specified elapsed time,
ΔT_d = temperature rise for a dry ceramic, and
ΔT_w = temperature rise for a fully saturated ceramic.

Equation 4.11 normalizes the sensor output reading between completely dry and fully saturated ceramic conditions. The normalized temperature change would be equal to zero for a dry sensor and unity for a water-saturated ceramic.

It was reasoned that the most significant effect of ambient temperature would be realized when the ceramic was dry. In other words, the largest temperature corrections would occur when the ceramics contain considerable air. (Note that this is opposite to the observations made with the GCTS thermal conductivity sensors.) It was reasoned that the gas phase thermal conductivity changed because of changes in the latent heat of distillation across the pores of the ceramic.

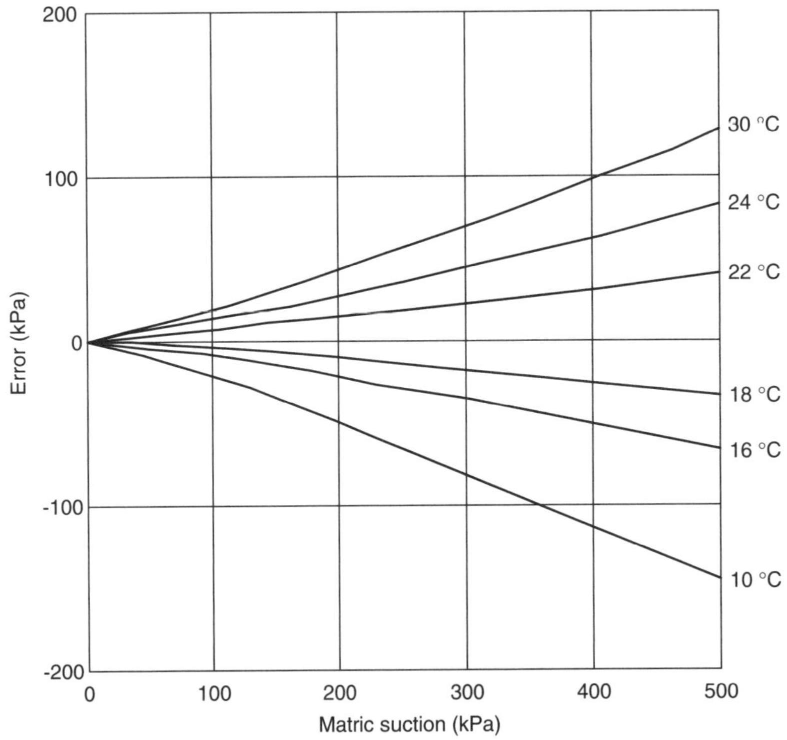

Figure 4.57 Possible measurement error for range of ambient temperatures and soil suctions up to 500 kPa (after Campbell Scientific Instruction Manual, 2009).

Campbell et al. (1994) reported that a correction needed to be applied to the thermal conductivity of dry air in order to accommodate the latent heat transport term when moisture vapor was present in the air. A procedure was proposed for the calculation of the soil suction corrected for temperature and the thermal conductivity of water vapor in the air phase. Figure 4.58 shows the possible variation in thermal conductivity reading when the ambient temperature is different than 20°C. The results show, for example, that if the soil suction were 400 kPa and the temperature varied by plus or minus 10°C, the error in the soil suction measurement could be plus or minus 100 kPa (i.e., values between 300 and 500 kPa).

The research results reported by Flint et al., (2002) raise a number of questions related to the calibration and usage of heat dissipation sensors. The research suggests that the most significant temperature effects occur as the temperatures and soil suctions increase. The results assume that no temperature correction is required when the thermal conductivity suctions sensors are essentially filled with water. When the measured soil suctions are about 250 kPa, it is estimated that the measured suction could be as high as 310 kPa if the ambient temperature were 30°C, and as low as 180 kPa if the ambient temperature were 10°C.

4.2.9.7 *Effect of Freezing Temperature on Thermal Conductivity Suction Measurements*

Shuai et al., (2002) conducted other tests to assess the effect of temperature on suction measurements when using thermal conductivity suction sensors. Figure 4.59 shows the influence of changing the ambient temperature from about 23°C to the freezing point. The Shuai et al., (2002) temperature corrections were applied to the measurements. The results showed that soil suction remained essentially constant except when the temperature was reduced to freezing conditions. The soil suction readings dropped to zero when the soil became frozen. Later the soil suctions returned to their original values when the soil thawed.

There was no apparent damage to the sensors upon freezing. It was reasoned that the soil suction readings dropped to zero because the imposed heat pulse was absorbed by latent heat.

4.2.9.8 *Effect of Hysteresis in the TC Suction Sensors*

One of the limitations associated with any indirect measurement system is the issue of hysteresis. The water content versus matric suction curves for any porous material are not the same during wetting and drying. The hysteresis associated with the water retention curves for the ceramic results in hysteresis in the suction sensor response upon wetting and drying. Attempts have been made to take the capillary hysteresis of the porous ceramic tip into consideration when measuring matric suction (Feng, 1999).

Desorption and adsorption curves have been measured for the ceramic used for the GCTS thermal conductivity suction sensors. A typical set of drying and wetting water retention curves for the porous ceramic used for the sensing tip is shown in Fig. 4.60. The results show that there is

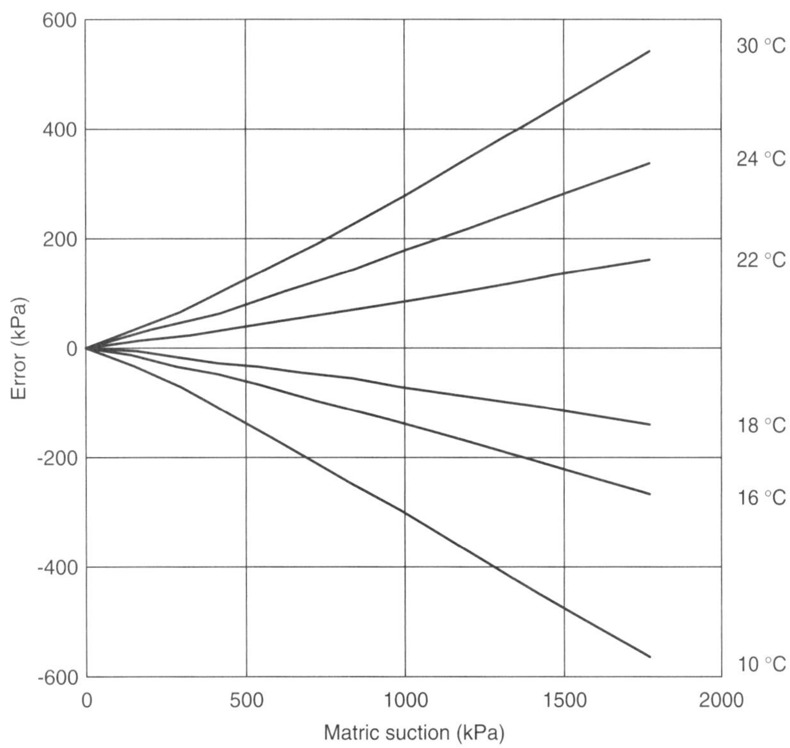

Figure 4.58 Possible measurement error for range of ambient temperatures and soil suctions up to 1750 kPa (after Campbell Scientific Instruction Manual, 2009).

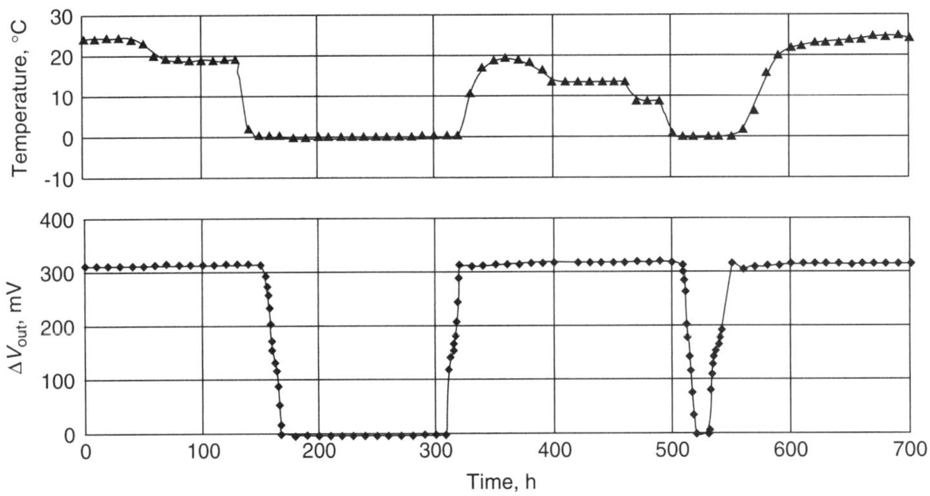

Figure 4.59 Effect of imposing freezing temperatures to thermal conductivity suction sensors (after Shuai et al., 2002).

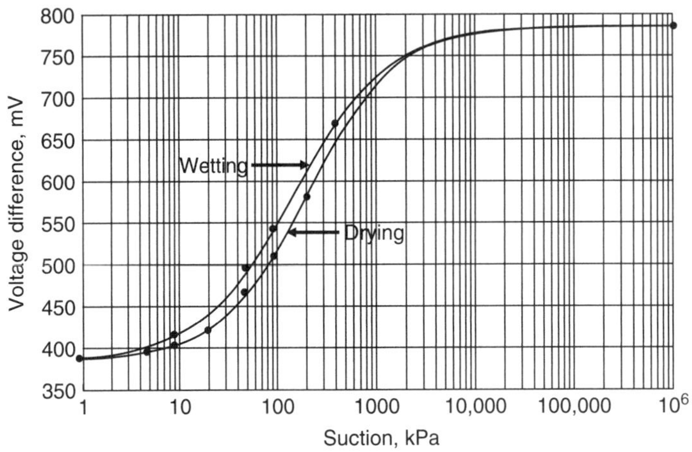

Figure 4.60 Typical hysteresis measured on FTC-100 thermal conductivity soil suction sensors (from Feng et al., 2002).

a difference between the drying and wetting curves; however, the end points of the calibration curves are similar. The maximum difference between the two curves occurs when the thermal conductivity suction sensors are approximately 50% saturated.

Feng (1999) studied the difference between the drying and the wetting calibration curves of the GCTS-type thermal conductivity suction sensors. Several suction sensors were calibrated to measure the difference between the drying and wetting curves. The difference between the two calibration curves was mathematically defined as follows:

$$\Delta\psi^* = \frac{\psi_d - \psi_w}{\psi_w} \times 100 \tag{4.12}$$

where:

$\Delta\psi^*$ = percent change in suction between the drying and wetting calibration curves,
ψ_d = suction on the main drying calibration curve, and
ψ_w = suction on the main wetting calibration curve.

The measured difference between the drying and wetting calibration curves, $\Delta\psi^*$, was about 40% of the drying curve value when suction values were greater than about 30 kPa. As a result, the wetting calibration curve can be approximated using Eq. 4.13 based on the test results of Feng (1999):

$$\psi_w = 0.70\psi_d \tag{4.13}$$

Equation 4.13 states that a drying curve suction of 50 kPa corresponds to an average wetting calibration curve suction of 35 kPa. Likewise, a drying curve suction of 500 kPa would correspond to a wetting curve suction of 350, and so on. Laboratory tests by Feng (1999) showed that the hysteresis curves were stable and reproducible. In other words, the drying and wetting calibration curves did not appear to change with time and were similar for all ceramic stones tested.

A hysteresis model was proposed by Feng (1999) to mitigate the influence of hysteresis associated with the ceramic block. The model was designed to trace scanning curves both during the wetting process and during the drying process. Calculated and measured scanning curves associated with changing from the drying curve to the wetting curve are shown in Fig. 4.61. The model can be used to subsequently modify a dataset of measurements obtained during field monitoring.

The monitoring readings are collected and soil suction values are first computed through use of the drying calibration curve. Once the data are compiled in a database, it is possible to further process the data by examining whether the past data points have been undergoing a drying or wetting process. This information is used to determine which calibration curve is best to use for the final calculation of soil suction at any particular time. In this way it is possible to increase the accuracy of the measured soil suctions. Figure 4.62 shows the difference between using the drying calibration curve or the wetting calibration curve when interpreting field monitoring suction data. The improved approach utilizes the drying calibration curve when the soil is undergoing drying and the wetting calibration curve when the soil is undergoing a wetting process.

Figure 4.63 shows a comparison between matric suctions measured using a tensiometer and a GCTS thermal conductivity suction sensor where hysteresis was taken into consideration. Both sensors were installed in the Indian Head till and the soil was subjected to a series of drying and wetting processes. During the drying process, the drying calibration curve was used to calculate soil suction, and during the wetting process, the wetting calibration curve

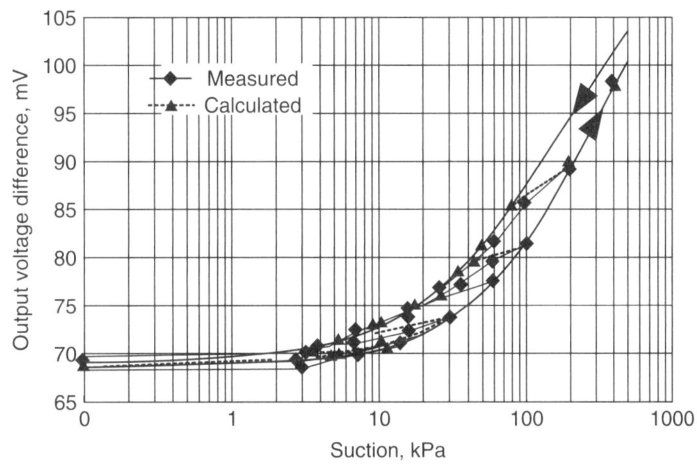

Figure 4.61 Comparison of applied suctions and predicted suctions when using hysteresis model proposed by Feng et al. (2002).

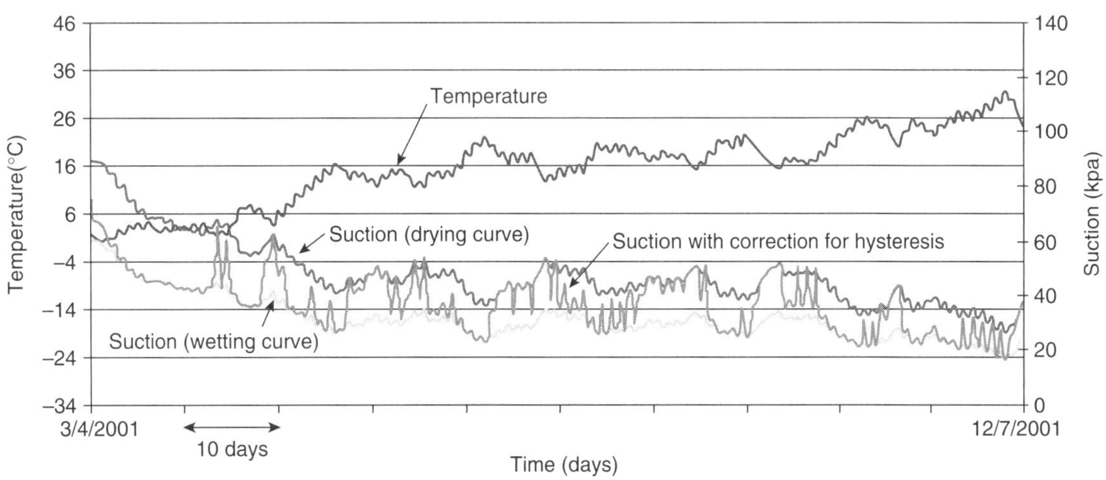

Figure 4.62 Effect of using both drying and wetting calibration curves to interpret thermal conductivity soil suction measurements in the field (from Tan et al., 2007).

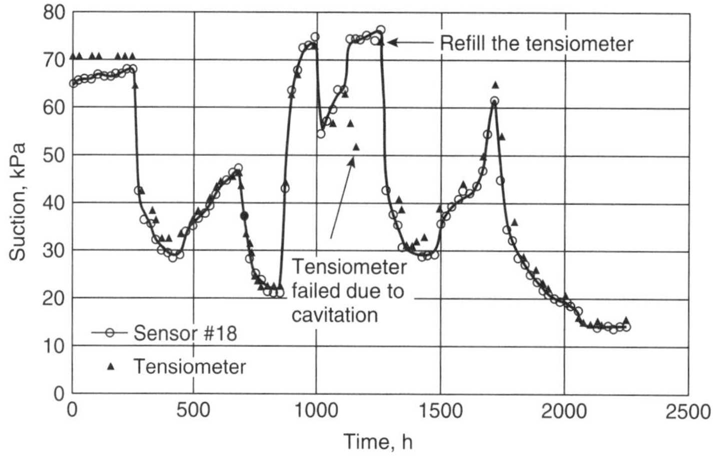

Figure 4.63 Drying and wetting soil suction readings taking hysteresis in calibration curves into account when using thermal conductivity suction sensors (from D.G. Fredlund et al., 2000b)

was used. The results show that the soil suction measurement made using the thermal conductivity suction sensor agrees well with those measured by the tensiometer. Some differences occurred because of the water cavitation in the tensiometer.

4.2.9.9 Effect of Chemicals on the Thermal Conductivity Suction Sensors

It is possible that in situ monitoring might be required under adverse acidity and alkalinity conditions. The effect of adverse chemical conditions was studied by Shuai et al. (2002). Following saturation, the thermal conductivity suction sensors were calibrated as increased suction values were applied in a calibration cell. The thermal conductivity sensors were calibrated from zero suction to 400 kPa.

Three thermal conductivity suction sensors were submerged in a hydrochloric acid (HCl) solution with a pH value of 4.25 after calibration. Another three thermal conductivity suction sensors were placed in a solution of sodium hydroxide (NaOH) with a pH value of 10.9. The sensors remained submerged in acid and base solutions for 3 weeks after which time the sensors were recalibrated.

Typical calibration and recalibration curves for one of the thermal conductivity suction sensors are shown in Fig. 4.64. The results showed that the original calibration and recalibration curves were essentially the same after submergence in the acid solution for 3 weeks. Similar results were observed for the thermal conductivity suction sensors immersed in the NaOH solution as shown in Fig. 4.65. Figure 4.66 shows the results of regression analyses performed on all six thermal conductivity suction sensors. While the tests continued for only 3 weeks, the results showed negligible differences between the initial calibrations and the recalibration values for all sensors.

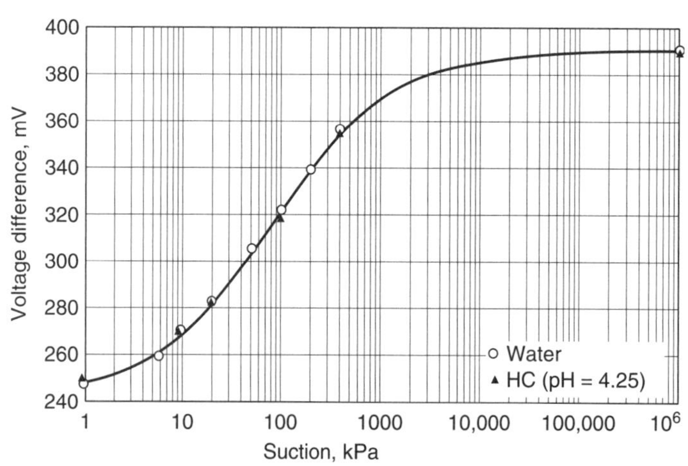

Figure 4.64 Initial and recalibration curves for University of Saskatchewan thermal conductivity suction sensors subjected to HCl (pH 4.25) for 3 weeks (after Shuai et al., 2002).

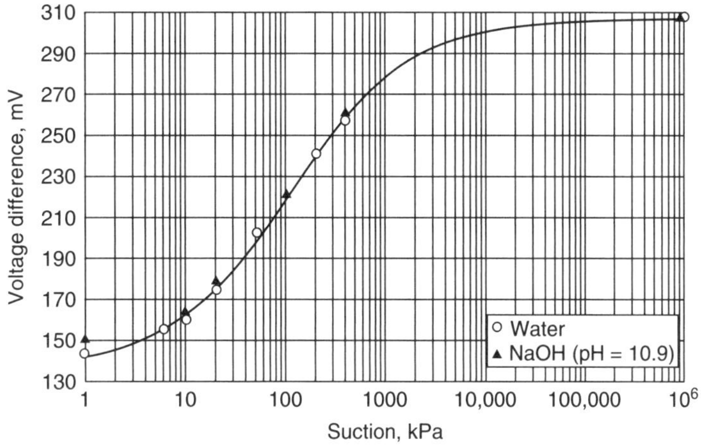

Figure 4.65 Initial and recalibration curves for University of Saskatchewan thermal conductivity suction sensors submerged in NaOH (pH 10.9) for 3 weeks (after Shuai et al., 2002).

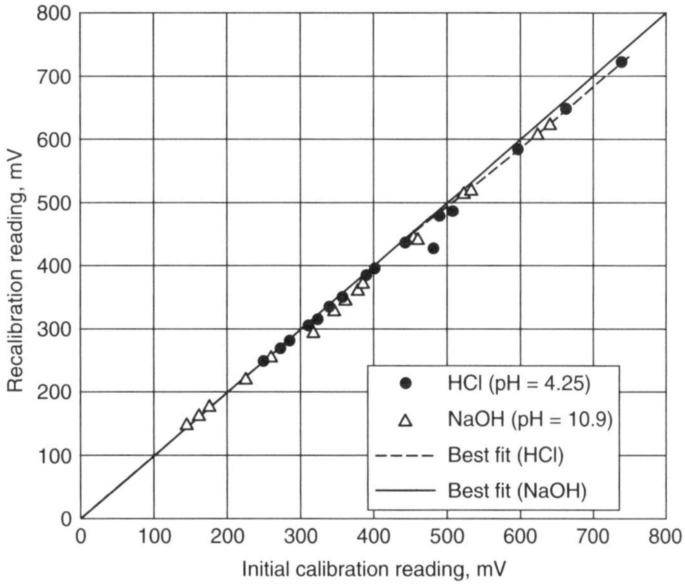

Figure 4.66 Regression analysis of "before" and "after" calibrations on University of Saskatchewan thermal conductivity suction sensors subjected to extreme chemical environments (after Shuai et al., 2002).

4.3 MEASUREMENT OF TOTAL SUCTION

Environmental changes (i.e., weather at the ground surface) are the primary cause of a change in the pore-water pressures in a soil mass. Matric suctions in a soil mass change as a result of moisture infiltration and evaporation at the ground surface. Osmotic suction in the soil does not appear to be highly sensitive to modest changes in the water content of the soil. As a result, a change in the total suction is quite representative of a change in the matric suction. Total suction measurements can sometimes be used in geotechnical engineering to indicate changes in matric suction. This is mainly true in the high-suction range when temperature fluctuations are small.

The free energy of the soil water (i.e., total suction) can be determined by measuring the vapor pressure adjacent to the soil-water or the relative humidity in the soil. The direct measurement of relative humidity in a soil is performed using a device called a psychrometer. The relative humidity in a soil can be indirectly measured using filter paper as a measuring sensor. The water content of the filter paper is calibrated against total suction. One or more filter papers are then equilibrated with the vapor pressure in the soil (i.e., total suction).

4.3.1 Psychrometers

Thermocouple psychrometers can be used to measure the total suction in a soil by measuring the relative humidity in the air phase of the soil pores or the environment immediately around the soil surface. The relative humidity can be expressed in terms of relative vapor pressure ($\bar{u}_v/\bar{u}_{v0}$), which is related to total suction.

There are two basic types of thermocouple psychrometers: (i) the wet loop type (Richards and Ogata, 1958) and (ii) the Peltier type (Spanner, 1951). Both types of psychrometers operate on the basis of temperature difference measurements between a nonevaporating surface (i.e., dry bulb) and an evaporating surface (i.e., wet bulb). The difference in the temperatures between these surfaces is related to relative humidity. The wet-loop and the Peltier-type psychrometers differ in the manner by which the evaporating junction is wetted in order to induce evaporation.

4.3.1.1 Peltier-Type Psychrometers

A Peltier current causes a thermocouple junction to cool below the dew point with the result that a minute quantity of water is condensed onto the thermocouple junction. The Peltier-type psychrometer has been quite commonly used in geotechnical engineering. The Seebeck and the Peltier effects constitute the main principles behind the operation of the Peltier psychrometer.

Seebeck (1821) discovered that an electromotive force (emf) was generated in a closed circuit of two dissimilar metals when the two junctions of the circuit are at different temperatures. This phenomenon is referred to as the Seebeck effect. The Seebeck effect allows the use of two dissimilar wires (i.e., a thermocouple) to measure temperature.

Peltier (1834) discovered that when a current is passed through a circuit of two dissimilar metals, one of the junctions becomes warmer, while the other junction becomes cooler. This phenomenon allows the use of thermocouples for the measurement of relative humidity.

The Peltier effect can be used to cool a thermocouple junction to reach the dew-point temperature corresponding to the surrounding atmosphere. As a result, water vapor condenses on the junction. After terminating the passage of the applied current, the condensed water tends to evaporate to the surrounding atmosphere, causing a further reduction in the temperature at the junction. The temperature reduction is a function of the evaporation rate, which is in turn affected by the water vapor pressure in the atmosphere. If the ambient temperature and the temperature reduction due to evaporation are measured using the Seebeck effects, the relative humidity of the atmosphere can be computed. There is a maximum degree of junction cooling that can be achieved using the Peltier current (Spanner, 1951).

The Peltier psychrometer is often called a Spanner psychrometer and details of the design of the sensor are shown in Fig. 4.67. The thermocouple consists of 0.025-mm-diameter wires of constantan (i.e., copper-nickel) and chromel (i.e., chromium-nickel). The wires are welded together to form an evaporating or a measuring junction. The other ends of the wires are connected to 26 American Wire Gauge (AWG) copper lead wires to form a reference junction. The highly conductive copper wires have a large diameter (i.e., large thermal mass), in order to serve as heat sinks that can maintain a constant temperature at the reference junction. The heat sink absorbs the heat generated near the reference junction when the measuring junction is being cooled.

The maximum degree of cooling generated by the chromel-constantan thermocouple is about 0.6°C below the ambient temperature (Brown and Bartos, 1982). This maximum cooling represents the lowest relative humidity (i.e., 94%) or the upper limit of the total suction (i.e., 8000 kPa), which can be measured using the thermocouple psychrometer. The lowest suction which can be measured using a thermocouple psychrometer is somewhat less than 100 kPa provided the temperature environment is controlled to within ± 0.001°C (Krahn and Fredlund, 1972). The lower limit corresponds to a relative humidity approaching 100%. A slight lowering of the temperature as the 100% relative humidity is approached will produce condensation on the thermocouple.

A protective housing is usually provided around the thermocouple wires. The protective cover may take the form of a ceramic cup, a stainless steel screen, or a solid tube (stainless steel or teflon), with a screen end window. The selection of the type of the protective cover depends on the application of the psychrometer. The time required for water vapor equilibration is significantly affected by the type of the protective cover. The ceramic cup appears to have the longest equilibration time and may not be practical in some situations.

Measurements of total suction are conducted by suspending a psychrometer near a soil specimen in a closed environment. The relative humidity is measured after equilibrium is attained between the air near the thermocouple

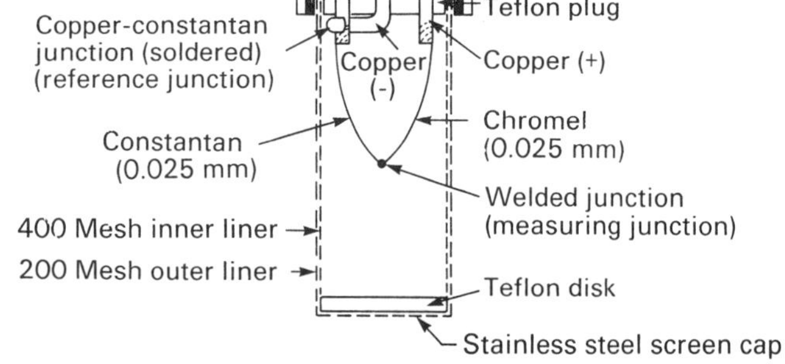

Figure 4.67 Screen-caged single-junction Peltier thermocouple psychrometer (after Brown and Collins, 1980).

and the pore-air in the soil specimen. Isothermal equilibrium must be achieved between the soil and the air before the psychrometer reading is taken. A temperature environment controlled to within ±0.001°C is required in order to measure total suctions to an accuracy of ±10 kPa (Krahn and Fredlund, 1972). Thermal equilibrium within the psychrometer is assured by obtaining a zero reading on the microvoltmeter.

4.3.1.2 *Calibration of Peltier-Type Psychrometers*

The calibration of a psychrometer consists of determining the relationship between microvolt outputs from the thermocouple for a salt solution with a known total suction value. The psychrometer can be mounted within a sealed chamber.

Filter papers can be saturated with a solution of NaCl or KCl and placed at the base of the chamber. The osmotic suctions for NaCl and KCl solutions at different molalities and temperatures are summarized in Tables 4.9 and 4.10.

The water vapor pressure or the relative humidity in the calibration chamber under isothermal equilibrium conditions is related to the osmotic suction of the salt solution. The psychrometer can be calibrated at various suction values by using different molalities (or osmotic suctions) for the salt solution. Isothermal conditions are maintained by placing the chamber in a constant-temperature bath, as illustrated in Fig. 4.68.

4.3.1.3 *Psychrometer Performance*

In situ measurements of total suction using psychrometers are generally not recommended because the thermal environment is not sufficiently constant in the field. However, laboratory measurements can be conducted under controlled temperature environments using undisturbed soil specimens from the field.

A soil specimen is placed into a stainless steel or Lucite chamber together with the thermocouple psychrometer, as illustrated in Fig. 4.69. The entire assembly is then placed

Table 4.9 Osmotic Suctions for NaCl Solutions

Molality	0°C	7.5°C	15°C	25°C	35°C
0	0.0	0.0	0.0	0.0	0.0
0.20	836	860	884	915	946
0.50	2070	2136	2200	2281	2362
0.70	2901	2998	3091	3210	3328
1.00	4169	4318	4459	4640	4815
1.50	6359	6606	6837	7134	7411
1.70	7260	7550	7820	8170	8490
1.80	7730	8035	8330	8700	9040
1.90	8190	8530	8840	9240	9600
2.00	8670	9025	9360	9780	10,160

Source: From Lange, 1967.

Table 4.10 Osmotic Suctions for KCl Solutions

Molality	10°C	15°C	20°C	25°C	30°C
0	0.0	0.0	0.0	0.0	0.0
0.10	436	444	452	459	467
0.20	859	874	890	905	920
0.30	1277	1300	1324	1347	1370
0.40	1693	1724	1757	1788	1819
0.50	2108	2148	2190	2230	2268
0.60	2523	2572	2623	2672	2719
0.70	2938	2996	3057	3116	3171
0.80	3353	3421	3492	3561	3625
0.90	3769	3846	3928	4007	4080
1.00	4185	4272	4366	4455	4538

Source: From Campbell and Gardner, 1971.

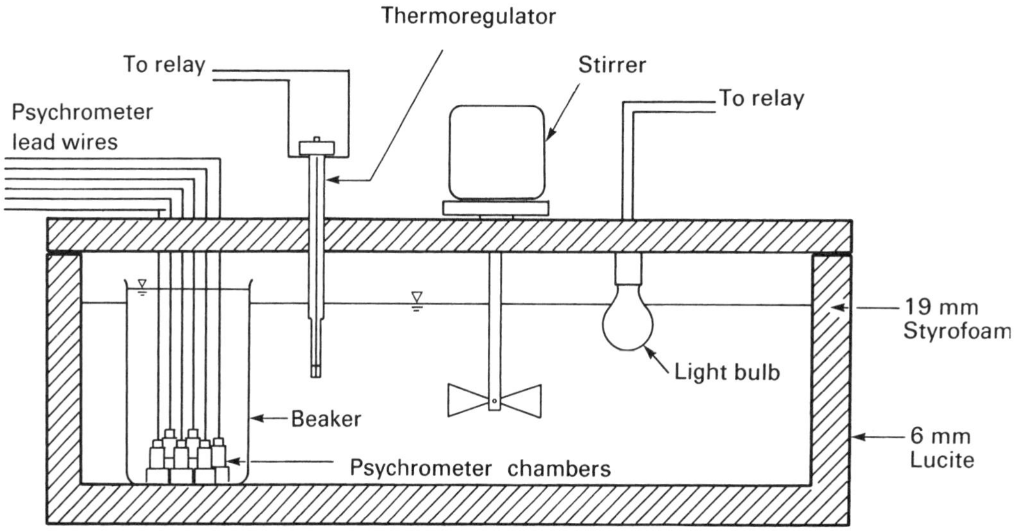

Figure 4.68 Schematic diagram of double-buffered constant-temperature bath (from Krahn and Fredlund, 1972).

in a constant-temperature bath, as shown in Fig. 4.68. The temperature of the bath should be maintained at a constant temperature with variations on the order of ±0.001°C (Krahn and Fredlund, 1972). As a result, the thermoregulator must be able to respond to a fluctuation in temperature of ± 0.001°C. The soil temperature is expected to be maintained within the same degree of accuracy of the water bath (or greater) due to the buffering effect of the glass beaker.

Figures 4.70 and 4.71 present measured data between total suction and initial water content for compacted glacial till and compacted Regina clay, respectively. The total suction measurements were conducted using thermocouple psychrometers. It should be noted that these relationships do not constitute a SWCC for the soils since the results were obtained from various soil specimens compacted at different water contents. A SWCC describes the water content versus suction relationship for a single soil specimen where the applied soil suction is changed on one sample. Figures 4.70 and 4.71 indicate a rather unique relationship between soil suction and initial water content for a particular compacted soil, regardless of its dry densities. The in situ total suction

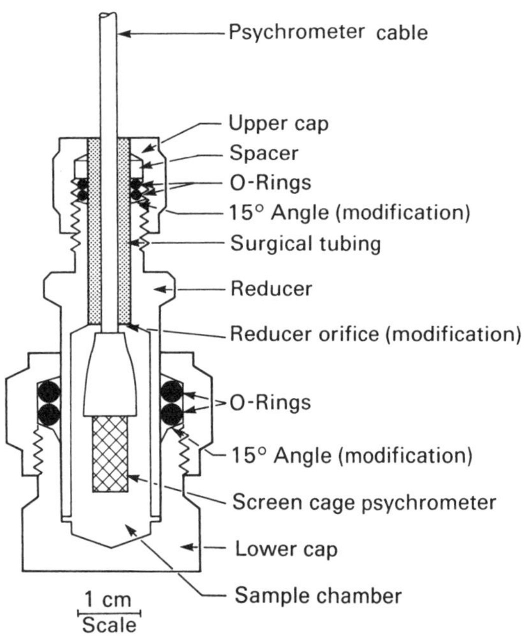

Figure 4.69 Stainless steel sample chamber with sealed psychrometer in place (after Brown and Colllins, 1980).

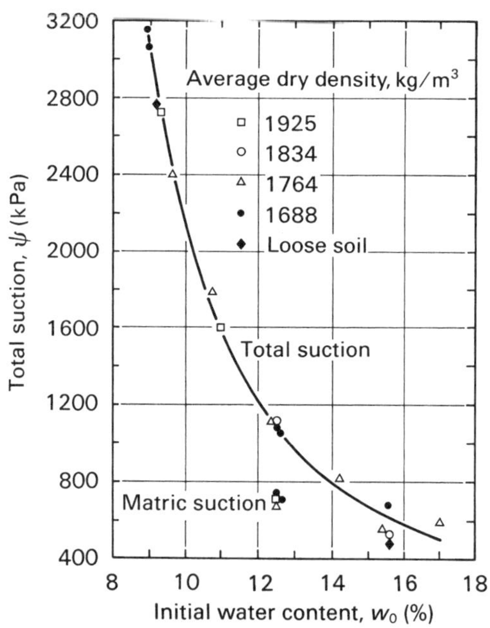

Figure 4.70 Total suction versus as-compacted water content for glacial till specimens (from Krahn and Fredlund, 1972)

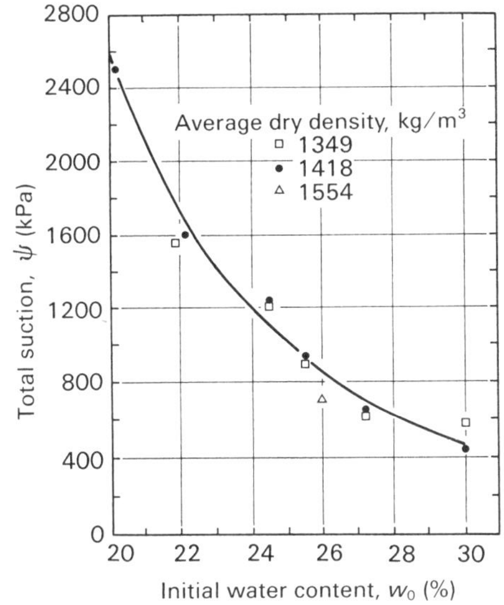

Figure 4.71 Total suction versus as-compacted water content for Regina clay (from Krahn and Fredlund, 1972)

of the same soil compacted in the field can be inferred from the relationships illustrated in Figs. 4.70 and 4.71.

Comparisons between suction measurements using thermocouple psychrometers and suction measurements using the filter paper technique are shown in Fig. 4.72. The results indicate reasonably close agreement between suction measurements made using psychrometers or filter paper.

4.3.1.4 Wet-Loop-Type Psychrometers

The wet-loop psychrometer was described by Richards and Ogata (1958). It differs from the Peltier-type psychrometer in the manner by which the evaporating junction is wetted. The evaporating junction in the wet-loop psychrometer is mechanically wetted by placing a drop of water

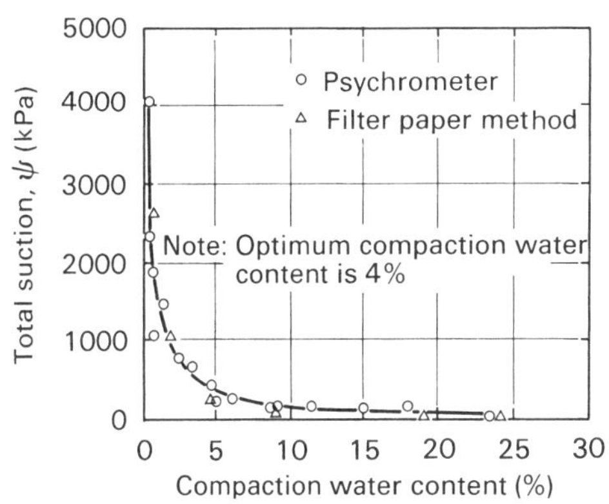

Figure 4.72 Comparison of psychrometer and filter paper measurements of total suction on compacted silty sand (after Daniel et al., 1981).

into a small silver ring. The water then evaporates, cooling the thermocouple junction. This procedure allows for a more stable reading to be taken over a longer period of time.

The Peltier psychrometer has an upper range for measuring total suction of about 8000 kPa. This limit is governed by an ability to condense a drop of water onto the tip of a thermocouple. The wet-loop psychrometer does not rely on the condensation of a water droplet and therefore extends the range over which total suction can be measured. The upper limit for measuring total suction with the wet-loop psychrometer is 300,000 kPa.

4.3.2 Chilled-Mirror Psychrometer

The chilled-mirror psychrometer uses a dew-point measurement to determine the total suction in a soil. The measurement is performed under isothermal conditions in a sealed container. Gee et al., (1992) introduced the chilled-mirror technique that can be used for the measurement of total suction in the range from 3000 to 300,000 kPa. This allowed the measurement of total suction in the mid- to high-suction range. The device was first introduced as a water activity meter and later commercialized by Decagon Co.

The chilled-mirror Water PotentiaMeter, WP4, from Decagon has been studied by several geotechnical researchers (Leong et al., 2003b; Thakur et al., 2006; Agus and Schanz, 2005; Campbell et al., 2007; Cardoso et al., 2007). A testing procedure has been set forth in the American Society Testing and Materials (2008) standard D 6836–02. Some new features have been added to the chilled-mirror Water PotentiaMeter to minimize the effect of ambient temperature fluctuations when making a total-suction measurement. The new device with the internal temperature control is called the WP4-T.

Figure 4.73 shows a schematic diagram of a dew-point Water PotentiaMeter, WP4 (Leong et al., 2003b). The device consists of a sealed chamber with a fan, a mirror, a photoelectric cell, and an infrared thermometer. A soil specimen is placed in a stainless steel or plastic container that has a diameter of 40 mm. The specimen is then moved into the chamber on a tray. The specimen thickness may vary between 1 and 5 mm. The chamber is closed and the soil specimen is thermodynamically equilibrated with the chamber environment. A fan accelerates the equilibration process.

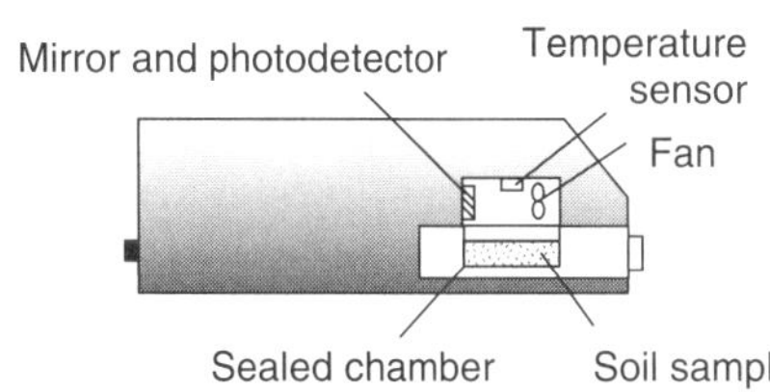

Figure 4.73 Schematic diagram of chilled-mirror WP4 Water PotentiaMeter (after Leong et al., 2003).

A cooling system is used to reduce the temperature on the surface of mirror to the dew-point temperature. A photoelectric cell detects the first sign of condensation on the mirror, a sign that signals the dew point. The dew-point temperature is measured using a thermocouple. An infrared thermometer is used to measure the temperature of the chamber, which is assumed to be the same as the temperature of the soil specimen.

The vapor pressure above the soil specimen in the chamber and the saturated vapor pressure at the same temperature are computed using the dew-point and specimen temperatures, respectively. Kelvin's equation is then used to calculate the total suction of the soil specimen.

The WP4-T Water PotentiaMeter has proven to be a valuable laboratory tool for measuring total suction in the mid- to high-total-suction range. The WP4-T apparatus can be calibrated using a standard solution of 0.5 M KCl, which yields a total suction value of 2190 ± 100 kPa at 25°C. The device is reliable and easy to use for total-suction measurements as high as 300,000 kPa. It is particularly useful for studying unsaturated soil behavior in the residual soil suction range.

4.3.3 Filter Paper Method

The filter paper method can be applied over a wide range of soil suctions, namely, from a few kilopascals to several hundred thousand kilopascals (Fawcett and Collis-George, 1967; McQueen and Miller, 1968a). However, the measurements must be performed with great care. Only the *noncontact* filter paper procedure (i.e., the filter paper is not in contact with the soil) can be assured of measuring total suction. The *contact* filter paper procedure (i.e., the filter paper is in contact with the soil) may measure either the total or the matric suction depending on the degree of contact between the soil and the filter paper.

The filter paper method for measuring soil suction was developed in the soil science discipline and has mainly been used in agriculture and agronomy (Gardner, 1937; Fawcett and Collis-George, 1967; McQueen and Miller, 1968a; Al-Khafaf and Hanks, 1974). Attempts have been made to use the filter paper method in geotechnical engineering (Ho, 1979; Tang, 1978; McKeen, 1981; Ching and Fredlund, 1984a; Gallen, 1985; McKeen, 1985; Chandler and Gutierrez, 1986). The filter paper method has been used, for example, in the study of airport pavement subgrades and the swelling potential of profiles in expansive soils (McKeen, 1985). The filter paper method has gained limited acceptance in geotechnical engineering, but the technique is deserving of further research and application.

4.3.3.1 Principle of Measurement (Filter Paper Method)

The filter paper method has been used to measure either total suction or matric suction of a soil. The filter paper is used as a sensor. The filter paper method is classified as an "indirect method" of measuring soil suction since it is necessary to calibrate the water content of the filter paper against soil suction. The filter paper method is based on the assumption that a filter paper will come to soil suction equilibrium with the soil suction in the water phase.

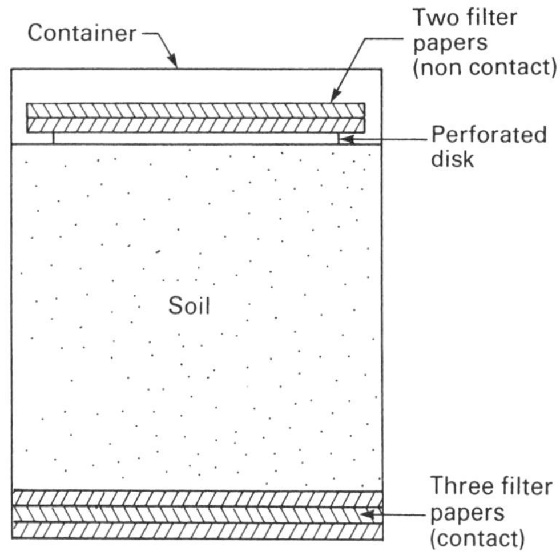

Figure 4.74 Use of contact and noncontact filter paper technique for measuring matric and total suction, respectively (after Al-Khafaf and Hanks, 1974).

Equilibrium can be reached by either liquid or vapor moisture exchange between the soil and the filter paper. When a dry filter paper is placed in direct contact with a soil specimen, it is assumed that moisture will flow from the soil to the paper until equilibrium is achieved (Fig. 4.74). When filter paper is suspended above a soil specimen (i.e., no contact with the soil), vapor transfer can occur between the filter paper and the soil until vapor pressure equilibrium is achieved. The water content of the filter paper is measured once vapor pressure equilibrium is established.

The water content of the filter paper corresponds to a suction value provided by the filter paper calibration curve. McQueen and Miller (1968a) have made extensive measurements of the calibration curve for filter paper, and some results are shown in Fig. 4.75. Several methods have been

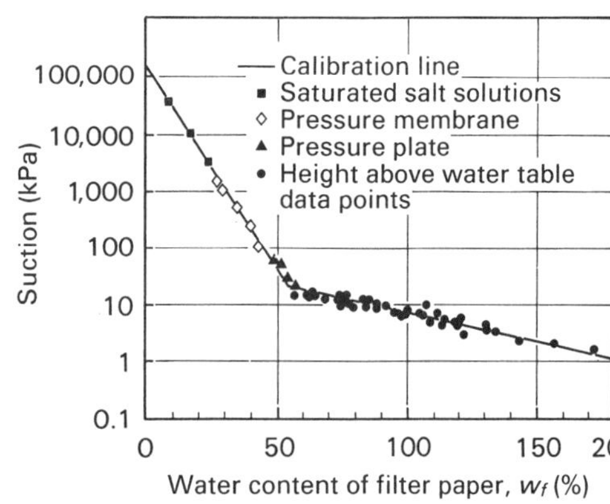

Figure 4.75 Suggested calibration curve for filter paper (after McQueen and Miller, 1968a).

used to produce a controlled soil suction environment, as shown in Fig. 4.75.

The equilibrium water content of the filter paper should correspond to the matric suction in the soil if the filter paper has been placed in contact with the soil-water. On the other hand, the equilibrium water content of the filter paper should correspond to the total suction of the soil if the paper has had no contact with the soil-water. Originally it was assumed that the same calibration curve could be used for determining matric suction and total suctions. More recent information on the filter paper calibration curves is presented later.

There appears to have been greater success in measuring total suction than matric suction. The filter paper method can be used to measure soil suction over a wide range of values. The measurements are generally performed in the laboratory by equilibrating a filter paper with an undisturbed or disturbed soil sample obtained in the field. Measurements should be made in a controlled temperature environment.

4.3.3.2 Measurement and Calibration Techniques for the Filter Paper Method

The measurement and calibration technique for the filter paper method was written into a tentative ASTM (1997b) standard (D5298–94, Committee D18 on Soil and Rock). The filter paper sensor must be of the ash-free, quantitative type II as specified by ASTM standard E833. Whatman No. 42 and Schleicher and Schuell No. 589 white ribbon are two commonly used brands of filter paper. A typical filter paper has a disk diameter of 55 mm. Filter papers from the same brand are considered to be "identical" in the sense that all filter paper disks are assumed to have the same calibration curve.

The equipment associated with the filter paper method consists of large and small metal containers, an insulated box, a balance, and a drying oven. The large container must be able to contain a soil specimen of approximately 200 g. The container should be treated to prevent rusting. The large container must have an air-tight lid. The container is used to equilibrate the soil specimen and the filter paper for a period of several days. The small container with a volume of approximately 60 cm^3 is used to contain the filter paper during oven drying for its water content measurement. The small container should be as light as possible because of the small mass of the filter paper.

An insulated box can be used to store the large containers with the soil specimens and filter papers during the equilibration period. The box should be maintained at a constant temperature within $\pm 1^{\circ}C$. An accurate balance with a minimum capacity of 20 g and a readability of 0.0001 g should be used when weighing the filter paper for determination of water content.

The filter papers should be pretreated prior to usage to prevent fungal and bacterial growth during the equilibration period (Fawcett and Collis-George, 1967; McQueen and Miller, 1968a). Solutions of 3% pentachlorophenol, C_6Cl_5OH, or 0.005% HgCl have been used to pretreat the filter papers. There have also been other studies that have suggested that the results from pretreated and untreated filter papers have given the same results (Hamblin, 1984; Chandler and Gutierrez, 1986).

The most common practice is to have the filter paper initially dry and then allow water to be absorbed from the soil specimen during equilibration. Filter paper calibration curves have historically been established using initially dry filter papers. It has been suggested that hysteresis in the filter papers may affect the calibration curve (Lykov, 1961). More recent studies on hysteresis and other factors are presented in the next section.

The filter paper should be oven dried for several hours prior to usage. The dry filter paper should then be cooled and stored in a desiccant container. A soil specimen on which suction is to be measured can be placed in a container. The soil specimen should almost fill the container in order to reduce the time to equilibration. The no-contact procedure can be achieved by placing two dry filter papers on a perforated disk seated on top of the soil specimen, as shown in Fig. 4.74. The contact procedure should have three stacked filter papers placed in direct contact with the soil specimen, as illustrated in Fig. 4.74. The contact filter papers can be placed below the soil specimen or the soil specimen can be cut in half and the filter papers placed such that there is contact both above and below the filter papers. The center filter paper is generally used for the measurement of soil suction while the outer filter papers are primarily used to protect the center paper from soil contamination.

The sample container can be sealed with plastic electrical tape once the filter paper and the soil specimen are placed in the large container. The sealed container is then stored in an insulated box for the equilibration period. The ambient temperature should remain within $\pm 1^{\circ}C$ during the equilibration period (Al-Khafaf and Hanks, 1974).

The filter papers can be removed from the large container at the end of the equilibration period. A pair of tweezers can be used when handling the filter papers. Each filter paper is placed into a small metal container for the measurement of water content. The water contents of each of the filter papers used for the no-contact procedure can be independently measured. The water content of the center filter paper is of primary importance for the contact measurement.

Extreme care must be exercised when measuring the small mass associated with the filter papers. The filter paper should be transferred from the large container to the small container within a short period of time (e.g., 3–5 *s*). A short period for transferring the filter paper is necessary to minimize water loss or gain between the filter paper and the surrounding atmosphere. The small container containing the filter paper must be closed and weighed immediately in order to determine the mass of the filter paper and the adsorbed water.

The container along with the filter paper is then placed in an oven at a temperature of $110 \pm 5^{\circ}C$. The lid of the container should be removed to allow water to escape from the filter paper. After drying, the container and the dry filter paper are weighed with the lid in place. The difference between the dry and wet mass of the filter paper is used to compute the equilibrium water content of the filter paper.

The equilibrium soil suction is obtained from the calibration curve using the measured equilibrium water content of the filter paper. The equilibrium suction in the filter paper is assumed to be equal to the suction in the soil specimen.

It is also possible to measure the calibration curve for a specific filter paper by measuring the water content of filter papers that have been equilibrated over salt solutions having a known osmotic suction. In principle, the filter paper calibration is similar to the calibration of psychrometers. The filter paper should be suspended above at least 50 cm^3 of a salt solution. The procedure for ensuring equilibration and the measurement of water content is the same as that used for the measurements of soil suction. Various filter paper water contents can then be plotted against the respective osmotic suction value to produce a calibration curve.

The calibration curve for filter papers exhibits a bilinear shape, as shown in Fig. 4.75. The lower part of the curve represents the high range of filter paper water contents where the water is believed to be held by the influence of capillary forces. On the other hand, the upper part of the calibration curve represents lower water contents where the water is believed to be held as an adsorbed water film within the filter paper (Miller and McQueen, 1978).

The filter paper technique is highly user dependent, and great care must be taken when measuring the water content of the filter paper. Each dry filter paper has a mass of about 0.52 g, and at a water content of 30%, the mass of water in the filter paper is about 0.16 g.

The above filter paper procedure is primarily based on ASTM standard E833. Recent research on the filter paper method (Leong et al., 2002a; Rahardjo and Leong, 2006) has provided further insight into variations and discrepancies that have been reported when using the filter paper method.

4.3.3.3 Further Details on the Calibration Technique and Factors Affecting Calibration of Filter Paper

Leong et al. (2002a) and Rahardjo and Leong (2006) undertook studies on the factors influencing the calibration of filter paper for the measurement of soil suction. The calibration curve needs to be accurately measured and needs to be reproducible. Gardner (1937) suggested that improved control in manufacturing the filter paper might result in greater accuracy and repeatability when using the filter paper methodology. However, relatively consistent calibration curves have been observed when using different batches of a particular brand and filter paper number.

Figure 4.76a shows the average calibration curves measured on Whatman No. 42 filter paper by a variety of researchers. All of the calibration curves are similar, suggesting that the quality of the Whatman filter paper

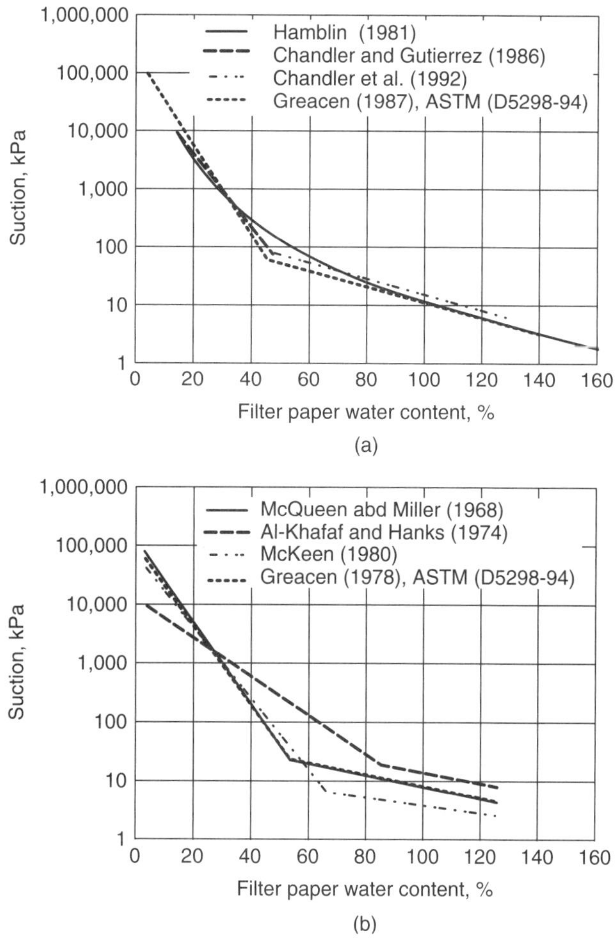

Figure 4.76 Proposed calibration curves for various filter papers (from Rahardjo and Leong, 2006): (a) calibration of Whatman No. 42 filter paper; (b) calibration of Schleicher and Schuell No. 589 filter paper.

is consistent in terms of its adsorptive characteristics. Figure 4.76b shows the average calibration curves measured on Schleicher and Schuell No. 589 in several research studies. There appears to be slightly more variation in the calibration curves measured on the Schleicher and Schuell No. 589 filter paper.

The calibration curves are bilinear in character on a plot of gravimetric water content versus the logarithm of soil suction. The transition between the wet and dry portions of the calibration curve occurs around a water content of about 47% for the Whatman filter paper. The transition between the wet and dry portions of the calibration curve on the Schleicher and Schuell No. 589 filter paper appears to range between water contents of 54 and 85%. The change in slope on the filter paper calibration curve appears to occur at the residual suction point on the water retention curve.

Two equations can be written for the bilinear calibration curve for filter paper. One equation can be written for the high-water-content range (i.e., low suction range), and

Table 4.11 Calibration Curves for Whatman No. 42 and Schleicher and Shuell No. 589 Filter Papers

(a) Whatman No. 42 Filter Paper		
References	Calibration curves	
Hamblin (1981)	$\log \psi = 8.022 - 3.683 \log w_f$	
Chandler and Gutierrez (1986)	$\log \psi = 4.84 - 0.0622 \log w_f$	$w_f < 47$
Chandler et al. (1992)	$\log \psi = 6.05 - 2.48 \log w_f$	$w_f \geq 47$
Greacen et al. (1987)	$\log \psi = 5.327 - 0.0779 \log w_f$	$w_f < 45.3$
ASTM (1997b)	$\log \psi = 2.413 - 0.0135 \log w_f$	$w_f \geq 45.3$
(b) Schleicher and Schuell No. 589 Filter Paper		
	$\log \psi = 5.238 - 0.0723 \log w_f$	$w_f < 54$
McQueen and Miller (1968a)	$\log \psi = 1.8966 - 0.01025 \log w_f$	$w_f \geq 54$
	$\log \psi = 4.136 - 0.0337 \log w_f$	$w_f < 85$
Al-Khafaf and Hanks (1974)	$\log \psi = 2.0021 - 0.009 \log w_f$	$w_f \geq 85$
	$\log \psi = 4.9 - 0.0624 \log w_f$	$w_f < 66$
McKeen (1980)	$\log \psi = 1.25 - 0.0069 \log w_f$	$w_f \geq 66$
Greacen et al. (1987)	$\log \psi = 5.056 - 0.0688 \log w_f$	$w_f < 54$
ASTM (1997b)	$\log \psi = 1.882 - 0.0102 \log w_f$	$w_f \geq 54$

Note: ψ = suction (kPa), w_f = filter paper water content (%).

another equation can be written for the low-water-content range (i.e., high suction range). Various proposed calibration equations by different researchers yield quite similar values for soil suction provided consideration is given to the breaking point on the calibration curve. Table 4.11 shows a summary of calibration equations that have been proposed for Whatman No. 42 and Schleicher and Shuell No. 589 filter papers. In each case [except for the Hamblin (1981) case], two equations are written for the filter paper calibration curves.

There are several factors that appear to have an influence on the calibration curves for filter paper: the quality of filter paper, the method used to control the applied soil suction, hysteresis, and equilibration time.

Quality of the Filter Paper. Differences in filter paper quality between batches of filter paper could possibly affect the measured calibration curves. Hamblin (1981) found that the calibration curves for batches of Whatman No. 42 filter paper produced two years apart were almost identical. Similar results were found by Fawcett and Collis-George (1967), Chandler and Gutierrez (1986), and Swarbrick (1995). The quality of the filter paper was studied by comparing the calibration data using pressure plate and pressure membrane apparatuses for soil suctions less than 1000 kPa. The vapor pressure generated by salt solutions at high suctions was used to establish suctions greater than 1000 kPa. Whatman No. 42 and Schleicher and Schuell No. 589 filter papers are shown in Figs. 4.77a and 4.77b, respectively. Measured data are confined to a relatively narrow band even though the results were obtained by different researchers, at different times, with different batches of filter paper. The Whatman No. 42 data appear to be slightly more consistent than the Schleicher and Schuell No. 589 data. It can be concluded that variations in the source of the filter paper do not appear to be a serious problem.

Bacterial Growth on the Filter Paper. Gardner (1937) pretreated Schleicher and Schuell No. 589 filter papers with 0.2% $HgCl_2$ to prevent contamination by soil organisms prior to installation. McQueen and Miller (1968a) and Al-Khafaf and Hanks (1974) used a 3% solution of pentachlorophenol in ethanol to prevent the growth of bacteria and other organisms. Fawcett and Collis-George (1967) pretreated Whatman No. 42 filter papers with 0.005% $HgCl_2$ and oven dried the filter paper at 105°C, while Hamblin (1981) did it with 0.005% $HgCl_2$. Hamblin (1981) and Chandler and Gutierrez (1986) reported that there was no need to pretreat the filter paper before usage. There does not appear to be any reports of serious concern over bacteria or algae growth on filter papers when using filter paper to measure soil suction.

Control of Applied Suction during Calibration. Several methods have been used by various researchers to create a controlled suction environment for filter paper calibration with suggested methods listed in Table 4.12. Ridley (1995) suggested that the calibration of filter paper be performed under conditions close to expected experimental conditions.

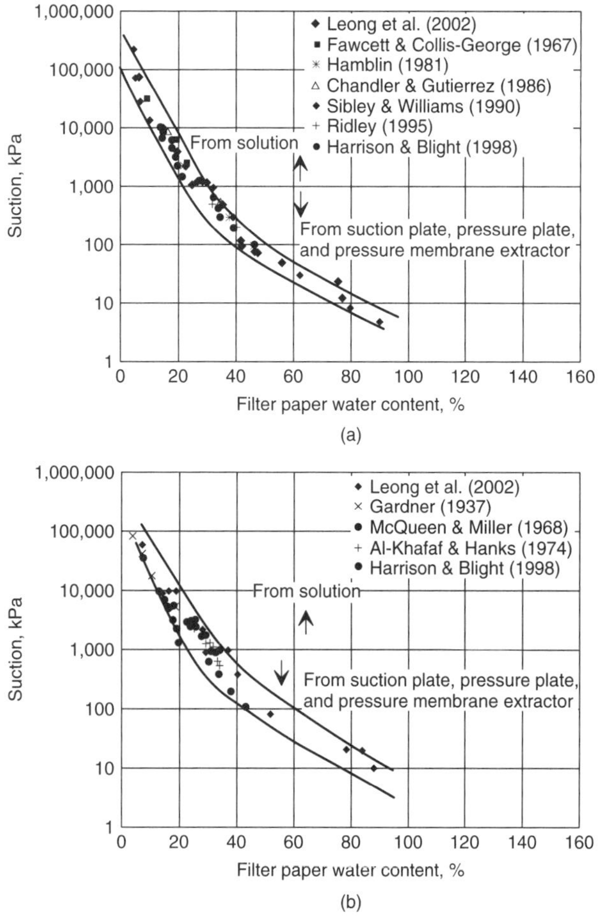

Figure 4.77 Calibration data obtained from different batches of filter paper (from Rahardjo and Leong, 2006): (a) data from Whatman No. 42 filter paper; (b) data from Schleicher and Schuell No. 589 filter paper.

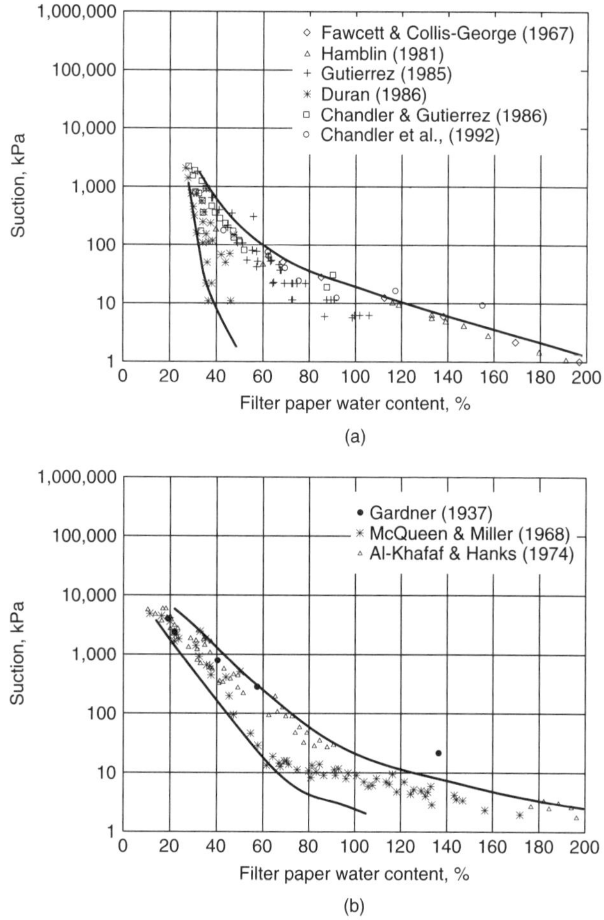

Figure 4.78 Calibration data obtained when equilibrating filter paper with soils of known soil suction (from Rahardjo and Leong, 2006): (a) data from Whatman No. 42 filter paper; (b) data from Schleicher and Schuell No. 589 filter paper.

Soil samples of "known" soil suction have also been used for the calibration of filter paper. Soil samples conditioned at known heights above the groundwater table have been used for suctions less than 20 kPa (Gardner, 1937; McQueen and Miller, 1968a; Al-Khafaf and Hanks, 1974). Saturated soil samples have also been consolidated in an oedometer or triaxial apparatus and then rapidly unloaded. The matric suction generated in a soil sample is assumed to be equal to the applied effective stress. Matric suctions up to 2000 kPa can be generated in this manner provided the soil sample remains saturated (Gutierrez 1985; Chandler and Gutierrez, 1986; Duran, 1986; Chandler et al., 1992).

Matric suctions can also be generated in soil samples using a pressure membrane extractor, a pressure plate apparatus, or a suction plate apparatus (McQueen and Miller, 1968a; Al-Khafaf and Hanks, 1974; Hamblin, 1981; Chandler et al., 1992). Calibration data obtained using soil samples with applied matric suctions are plotted in Fig. 4.78a for Whatman No. 42 filter paper and Fig. 4.78b for Schleicher and Schuell No. 589 filter paper. The data show considerable scatter, suggesting that the use of soil samples with applied matric suction may not be the most reliable procedure to use for the calibration of filter paper.

Hysteresis. The drying and wetting of porous materials are known to show hysteretic behavior. It is possible that filter paper may also show hysteresis during drying and wetting. Experimental evidence of filter paper hysteresis was suggested by Williams and Sedgley (1965) and Fawcett and Collis-George (1967). Al-Khafaf and Hanks (1974) suggested that filter paper should always be calibrated in the same manner to avoid issues related to hysteresis. Sibley and Williams (1990) and Swarbrick (1995) suggested that soil suction measurements with filter paper should be consistent with whether a wetting or drying procedure was used during calibration.

Ridley (1995) showed matric suction calibration data on Whatman No. 42 filter paper where hysteresis was noticeable. Harrison and Blight (1998) showed calibration data for Whatman No. 42 and Schleicher and Schuell No. 589 filter papers under drying and wetting conditions and hysteresis

Table 4.12 Suction Control Procedure for Filter Paper Calibration

Suction Generation Method	Suction Range	References
Suction plate, pressure membrane extractor, pressure plate	0–1500 kPa	Fawcett and Collis-George (1967) McQueen and Miller (1968a) Al-Khafaf and Hanks (1974) McKeen (1980) Hamblin (1981) Greacen et al. (1987) Sibley and Williams (1990) Sibley et al. (1990) Houston et al. (1994) Harrison and Blight (1998)
Salt solutions	0–340,000 kPa	Gardner (1937) Fawcett and Collis-George (1967) McQueen and Miller (1968a) Al-Khafaf and Hanks (1974) McKeen (1980) Hamblin (1981) Chandler and Gutierrez (1986) El-Ehwany and Houston (1990) Sibley and Williams (1990) Sibley et al. (1990) Houston et al. (1994) Harrison and Blight (1998)
Field soil samples	0–2000 kPa	McQueen and Miller (1968a) Al-Khafaf and Hanks (1974)
Oedometer and triaxial samples	100–2000 kPa	Gutierrez (1985) Duran (1986) Chandler and Gutierrez (1986) Chandler et al. (1992)
Centrifuge	100–3100 kPa	Gardner (1937)

was observed. Houston et al. (1994) reported no measurable differences between the wetting and drying calibration curves for matric suctions between 8 and 2500 kPa. Leong et al. (2002a) conducted drying and wetting calibrations of the Whatman No. 42 and Schleicher and Schuell No. 589 filter papers. Wetting and drying calibration data from Leong et al., (2002a) along with other data from the research literature are plotted in Figs. 4.79a and 4.79b for Whatman No. 42 and Schleicher and Schuell No. 589 filter paper, respectively. The data were obtained by either suspending the filter paper above salt solutions (i.e., no contact) or placing the filter paper on a suction plate, a pressure plate, or a pressure membrane apparatus (i.e., contact).

No-contact hysteresis (i.e., total suction) of Whatman No. 42 and Schleicher and Schuell No. 589 filter papers on drying and wetting was investigated using three methods: (i) distilled water in a vacuum desiccator (i.e., suction of 0 kPa), (ii) salt solution in a vacuum desiccator (i.e., total suction equal to 9700 kPa), and (iii) air environment with a relative humidity of 60% (i.e., total suction equal to 70,000 kPa). The changes in the water content of the filter paper, w_f, with time under the three conditions are shown in Figs. 4.80a and 4.80b for Whatman No. 42 and Schleicher and Shuell No. 589, respectively. Hysteresis can be observed in the filter paper responses. Both types of filter paper show 12% differences in water content for distilled water equilibration; 1% differences in water content for salt solution equilibration; and 0.5% differences in water content for the air environment equilibration. The equilibration times when using the drying mode were longer than the equilibration times required when using the wetting mode. The equilibration time was about 28 days for drying and 14 days for wetting when using the distilled water method. The results showed that inadequate equilibration times can produce the appearance of significant hysteresis between the wetting and drying equilibrium conditions for the filter paper.

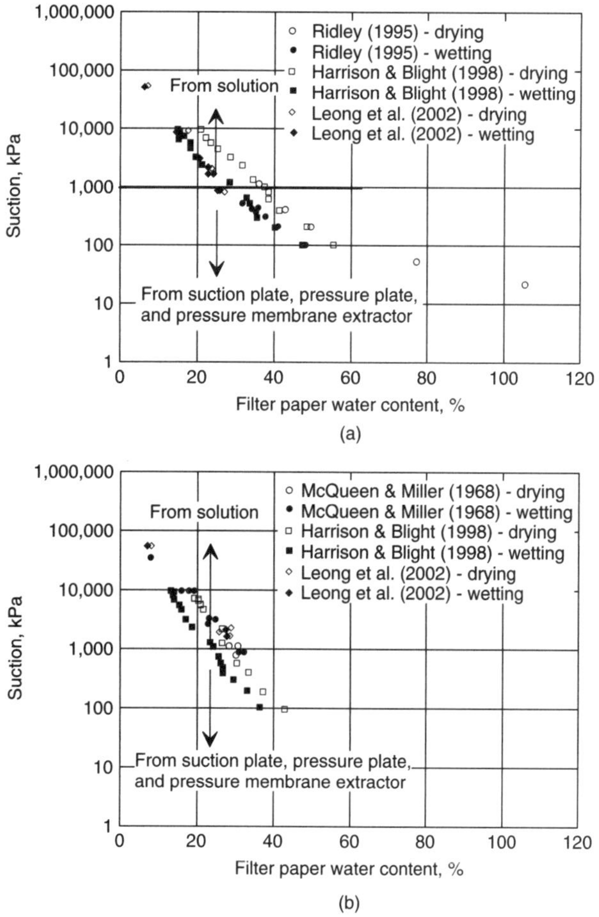

Figure 4.79 Drying and wetting processes in coming to equilibrium (from Rahardjo and Leong, 2006): (a) data for Whatman No. 42 filter paper; (b) data for Schleicher and Schuell No. 589 filter paper.

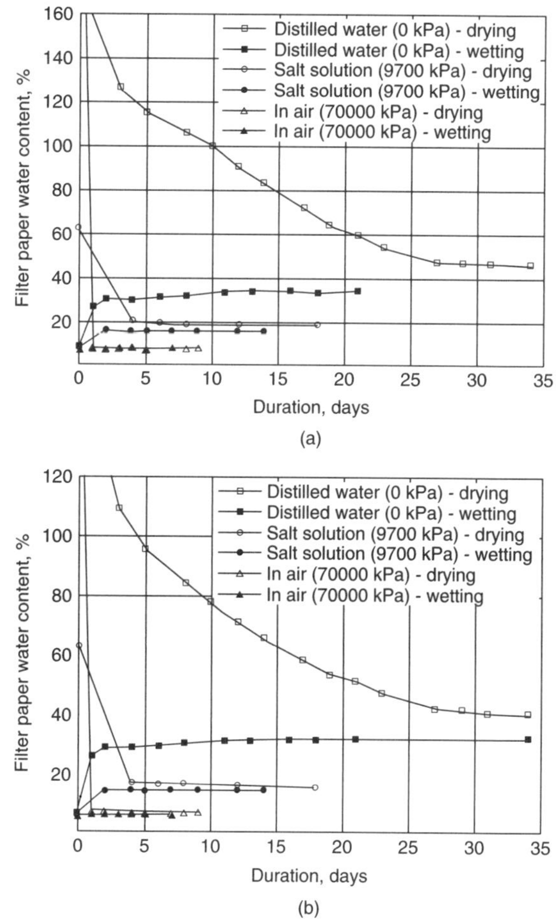

Figure 4.80 Time required for water content equilibration in filter paper (from Rahardjo and Leong, 2006): (a) equilibrium for Whatman No. 42 filter paper; (b) equilibrium for Schleicher and Schuell No. 589 filter paper.

Figures 4.81a and 4.81b present the equilibrium drying and wetting water contents, w_f, for Whatman No. 42 and Schleicher and Schuell No. 589 filter papers, respectively, that were held over varying concentration salt solutions. The equilibrium water content was taken as the water content of the filter paper when there was no measurable change in the weight of the filter paper with time. The hysteresis observed was small (i.e., 1–5% in terms of the water content of the filter paper), with the largest differences observed for suctions less than 100 kPa. Hysteresis appeared to be minimal when sufficient equilibration time was allowed.

Equalization Time. Table 4.13 presents various equilibration times that have been used when making suction measurements with filter paper. ASTM D5298–94 recommends a minimum equilibration time of seven days. Two points are important when examining the time required for the equilibration of the filter paper with a suction source. Time is required for equilibration between the environment and suction source. The water vapor in the air space of a closed container takes time to reach equilibrium. The equilibrium time is particularly important when calibrating filter papers using salt solutions. Time is also required for equilibration between the filter paper and the suction source. The water vapor in the air space must first reach equilibrium and then the filter paper will come to equilibrium with the water vapor in the air space.

Matric suction is applied instantly when filter paper is calibrated using a pressure plate apparatus. The equilibration time is the time required for the filter paper to achieve equilibrium with the applied matric suction. The filter paper equilibration time depends on suction source, contact condition, and suction level, as illustrated by the differences in equilibration time reported in Table 4.13.

The equilibration of water vapor with the suction source appears to be particularly important when an enclosed environment with a soil specimen is used. Water must be lost or gained from the soil specimen to the environment in order to establish an equilibrium condition. A small enclosed environment enables equilibrium conditions to be achieved more rapidly (McQueen and Miller 1968a).

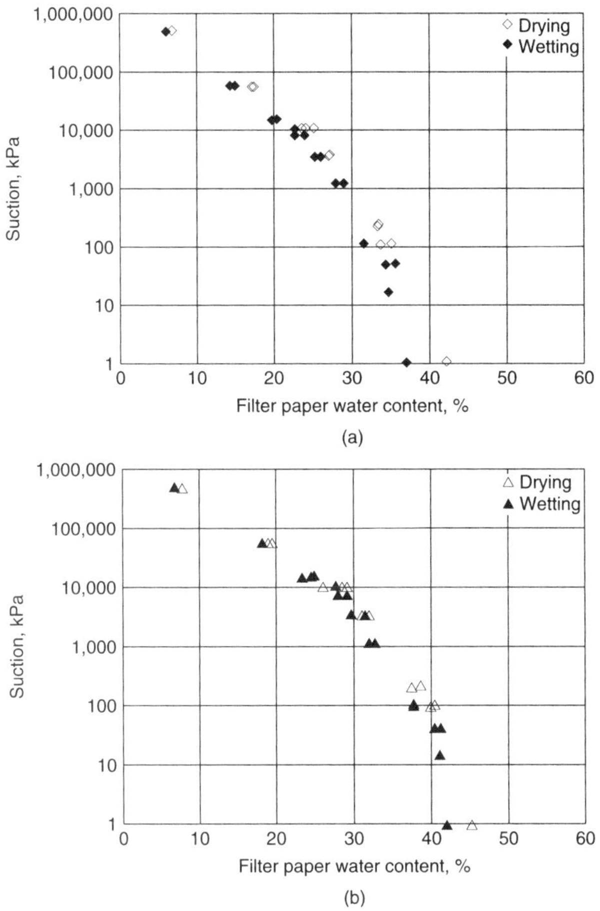

Figure 4.81 Wetting and drying responses of filter paper over different concentrations of salt solutions in desiccators at 24°C (from Rahardjo and Leong, 2006): (a) calibration for Whatman No. 42 filter paper; (b) calibration for Schleicher and Schuell No. 589 filter paper.

Filter papers absorb water from the soil specimen when the wetting method is used to measure soil suction. The soil specimen becomes slightly drier and exhibits a higher suction. Figures 4.82a and 4.82b show the equilibration times for Whatman No. 42 and Schleicher and Schuell No. 589 filter papers, respectively. The wetting method was used in conjunction with (i) a pressure plate apparatus, (ii) equilibration over salt solutions, and (iii) equilibration over distilled water. Equilibration times when using a 500-mL container were between 2 and 5 days. The longest equilibration time was 14 days which was required for filter paper placed over distilled water (Fig. 4.82).

4.3.3.4 *Total Suction and Matric Suction Calibration Curves*

Leong et al. (2002a) and Rahardjo and Leong (2006) developed total and matric suction calibration curves for Whatman No. 42 and Schleicher and Schuell No. 589 filter paper. Total suction data were obtained by calibrating initially dry

Table 4.13 Equilibration Times When Using Filter Paper Method

References	Equilibrium Time	Filter Paper Method
Fawcett and Collis-George (1967)	6–7 days	Contact
McQueen and Miller (1968a)	7 days	Contact
Al-Khafaf and Hanks (1974)	2 days	Contact and uncertain contact
Hamblin (1981)	Minutes–36 days	Contact
Chandler and Gutierrez (1986)	5 days	Contact
Duran (1986)	7 days	No contact
Greacen et al., (1987)	7 days	Contact
Sibley and Williams (1990)	3 days	Contact
	10 days	Noncontact
Lee and Wray (1992)	14 days	Contact and no contact
Houston et al., (1994)	7 days	Contact and no contact
Harisson and Blight (1998)	7–10 days	Wetting and no contact
	21 days	Drying and no contact
	10 days	Wetting and no contact
	25–30 days	Drying and no contact

filter papers over salt solutions of various concentrations in vacuum desiccators. Matric suction data for initially dry filter papers were obtained using the pressure plate apparatus. The calibration data are shown in Fig. 4.83. The results indicate that the filter paper responses are different for total and matric suctions calibration curves. This behavior was also observed by Houston et al., (1994) and Harrison and Blight (1998).

The suction data indicate that for suctions less than 1000 kPa the water content of the filter paper has a lower and different response to total suction than to matric suction. Above 1000 kPa, the total and matric suction calibration curves appear to converge. Houston et al. (1994) found that filter paper buried in a soil had essentially the same water content as filter papers suspended above the soil for suction values of about 98,100 kPa. Fredlund (1992) and Al-Khafaf

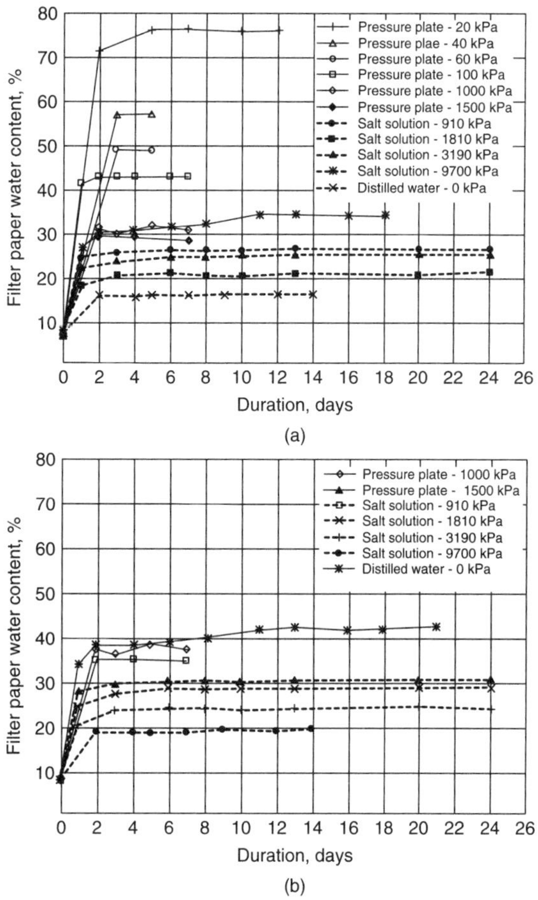

Figure 4.82 Equilibration of filter papers placed in pressure plate apparatus over salt solution or distilled water (from Rahardjo and Leong, 2006): (a) response of Whatman No. 42 filter paper; (b) response of Schleicher and Schuell No. 589 filter paper.

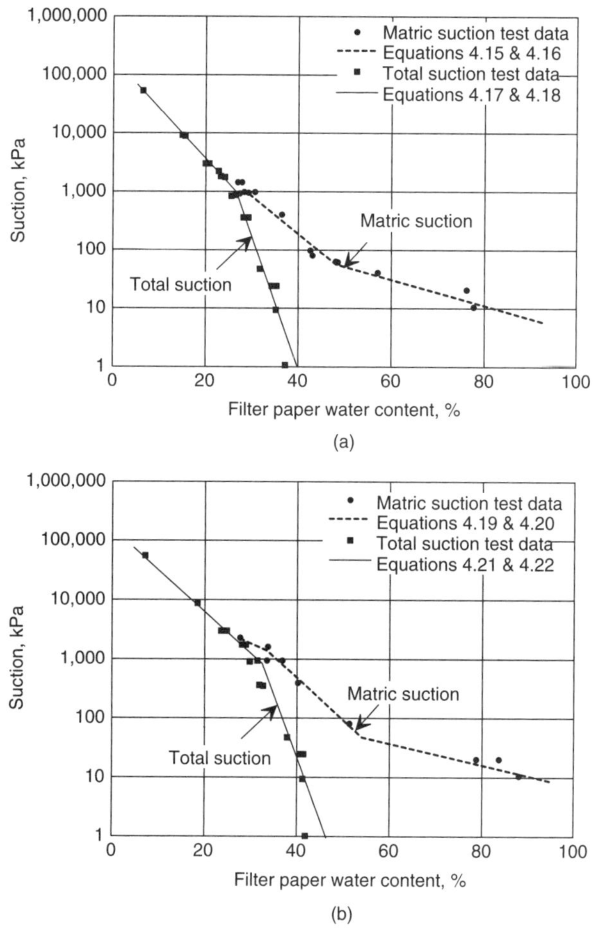

Figure 4.83 Proposed total and matric suction calibration curves (from Rahardjo and Leong, 2006): (a) calibration of Whatman No. 42 filter paper; (b) calibration of Schleicher and Schuell No. 589 filter paper.

and Hanks (1974) also reported that matric suction curves were essentially the same as total suction curves in the high-suction range.

Field measurements of soil suction using filter paper by van der Raadt et al. (1987) showed that filter paper placed in contact and no contact were similar for suctions greater than 1000 kPa. It was suggested that at suctions greater than 1000 kPa most moisture movement occurred through vapor transfer rather than capillary transfer. It was suggested that the contact filter paper method can be used to reliably measure matric suction as long as the suctions were less than 1000 kPa. It was suggested that the no-contact method would tend to measure total suctions when the suction values were greater than 1000 kPa.

Calibration Equations. A number of equations have been suggested for Whatman No. 42 and Schleicher and Schuell No. 589 filter paper calibration curves over the years (Table 4.10). Hamblin (1981) suggested that the calibration equation obtained by other researchers can be used for Whatman No. 42 filter paper unless particular accuracy demands recalibration. The filter paper method is particularly attractive for the measurement of soil suction when the calibration of the filter paper can be avoided. The calibration results presented in Fig. 4.76 indicate that the performance of the Whatman No. 42 filter paper may be slightly more consistent than the Schleicher and Shuell No. 589 filter paper.

The overall calibration curve is usually represented by two equations that represent different sensitivity of the filter paper response in the higher and lower suction range. All the equations presented in Table 4.10, except that of Hamblin (1981), are of the form

$$\log \psi = a - bw_f \qquad (4.14)$$

where:

$\log \psi$ = logarithm of suction (kPa) to base 10,
a, b = fitting constants for the calibration curve, and
w_f = filter paper water content, %.

The experimental calibration curve evidence indicates that separate equations should be used for the contact filter paper and for the no-contact filter paper. Total and matric suction calibration data can be best fit in the region above and below the changes in slope on the calibration curves.

The matric suction calibration curves for Whatman No. 42 filter paper can be written as follows:

$$\log \psi = 2.909 - 0.0229w_f \quad \text{when } w_f \geq 47\% \tag{4.15}$$

$$\log \psi = 4.945 - 0.0673w_f \quad \text{when } w_f < 47\% \tag{4.16}$$

The total suction calibration equations can be written as

$$\log \psi = 8.778 - 0.222w_f \quad \text{when } w_f \geq 26\% \tag{4.17}$$

$$\log \psi = 5.31 - 0.0879w_f \quad \text{when } w_f < 26\% \tag{4.18}$$

The matric suction calibration curves for Schleicher and Schuell No. 589 filter paper can be written as follows:

$$\log \psi = 2.659 - 0.018w_f \quad \text{when } w_f \geq 54\% \tag{4.19}$$

$$\log \psi = 5.438 - 0.069w_f \quad \text{when } w_f < 54\% \tag{4.20}$$

The total suction calibration equations can be written as

$$\log \psi = 8.778 - 0.191w_f \quad \text{when } w_f \geq 32\% \tag{4.21}$$

$$\log \psi = 5.26 - 0.0705w_f \quad \text{when } w_f < 32\% \tag{4.22}$$

Two sets of calibration equations are required for each of the filter papers because of changes in sensitivity of the filter paper in the different suction ranges and whether matric suction or total suction is measured.

The filter paper calibration curves can also be viewed as a water retention curve for filter paper. Consequently, it is possible to use either the van Genuchten (1980) or the Fredlund and Xing (1994) equations proposed for a SWCC as a single equation to fit either the total suction or matric suction calibration data. Applying the van Genuchten (1980) equation to the Whatman No. 42 filter paper calibration data yields the following equation for matric suction:

$$\psi = 0.051\left[\left(\frac{248}{w_f}\right)^{9.615} - 1\right]^{0.473} \tag{4.23}$$

The van Genuchten (1980) equation for total suction can be written as

$$\psi = 56{,}180\left[\left(\frac{37}{w_f}\right)^{0.44} - 1\right]^{2.361} \tag{4.24}$$

The van Genuchten (1980) equation for the Schleicher and Schuell No. 589 filter paper for matric suction can be written as

$$\psi = 0.048\left[\left(\frac{251}{w_f}\right)^{5.621} - 1\right]^{0.896} \tag{4.25}$$

The van Genuchten (1980) equation for total suction can be written as

$$\psi = 64{,}940\left[\left(\frac{44}{w_f}\right)^{0.464} - 1\right]^{2.516} \tag{4.26}$$

Applying the Fredlund and Xing (1994) equation to the Whatman No. 42 filter paper calibration data yields the following equation for matric suction:

$$\psi = 0.23\left[e^{(268/w_f)^{0.629}} - e\right]^{2.101} \tag{4.27}$$

The Fredlund and Xing (1994) equation for total suction can be written as

$$\psi = 18,500\left[e^{(37/w_f)^{0.242}} - e\right]^{2.248} \tag{4.28}$$

The Fredlund and Xing (1994) equation for the Schleicher and Schuell No. 589 filter paper for matric suction can be written as

$$\psi = 0.844\left[e^{(282/w_f)^{1.071}} - e\right]^{0.779} \tag{4.29}$$

The Fredlund and Xing (1994) equation for total suction can be written as

$$\psi = 20,000\left[e^{(42/w_f)^{0.246}} - e\right]^{2.058} \tag{4.30}$$

Equations 4.23–4.30 require only one equation for either matric or total suction. The van Genuchten (1980) and the Fredlund and Xing (1994) provide a reasonable fit of the calibration data and are convenient to use. However, the equations used for an independent fit for the low- and high-suction range are slightly more accurate (i.e., Eqs. 4.15–4.22).

4.3.3.5 Use of the Filter Paper Method in Engineering Practice

A question in the mind of the practicing geotechnical engineers is, "What is the accuracy of suction measurement made when using the filter paper method?" This question is still being addressed, but comparisons between measurements with psychrometers and filter paper provide an indication of the accuracy that can be anticipated when using the filter paper method (Fig. 4.84). The results from the no-contact filter paper agree fairly closely with the psychrometer results, indicating that total suction was measured.

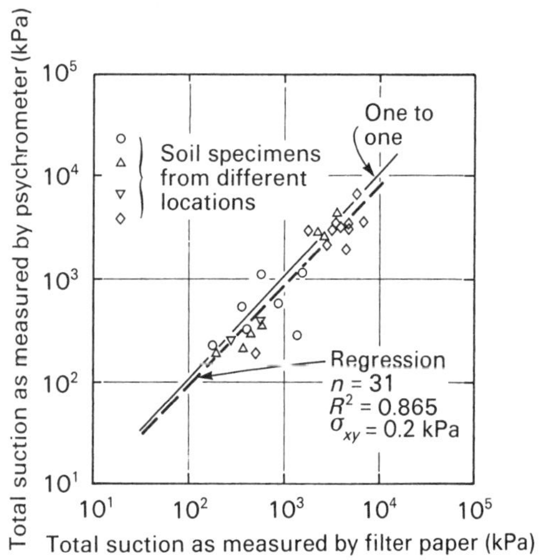

Figure 4.84 Comparisons of total-suction measurements obtained using filter paper or psychrometers (after McKeen, 1981).

It appears to be difficult to ensure adequate contact between the soil specimen and the filter paper for the measurement of matric suction. Consequently, total suction is generally measured when using the filter paper method. Figure 4.84 demonstrates reasonably close agreement between total-suction measurements obtained when using the filter paper method and thermocouple psychrometers.

4.4 MEASUREMENT OF OSMOTIC SUCTION

Osmotic suction is related to the amount of salts dissolved in the free pore-water. Several procedures can be used to measure the osmotic suction of a soil. For example, it is possible to mix distilled water with a soil until the soil is in a near-fluid condition and then drain off some effluent and measure its electrical conductivity. This procedure is referred to as the saturation extraction method. The electronic conductivity measurement then needs to be extrapolated back to the natural water content in order to obtain the osmotic suction for the soil. Although the test procedure is simple, there is presently no reliable procedure for ensuring an accurate measurement of in situ osmotic suction when using the saturation extraction method (Krahn and Fredlund, 1972).

A psychrometer can also be placed over the fluid extract to measure the vapor pressure; however, this procedure likewise does not produce reliable osmotic suction values. Studies have shown that the use of a pore fluid squeezer provides the most reasonable values for osmotic suction (Krahn and Fredlund, 1972).

4.4.1 Squeezing Technique

The *osmotic suction* of a soil can be indirectly computed based on the electrical conductivity of a small amount of pore-water extracted from the soil. The intent of the squeezing technique is to obtain a sample of the free pore-water which contains dissolved salts.

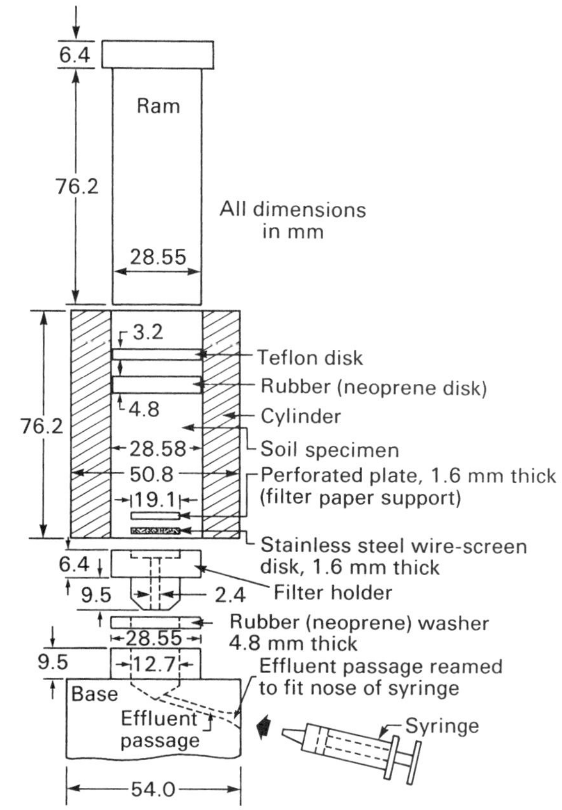

Figure 4.85 Design of pore fluid squeezer (after Manheim, 1966).

Pure water has a low electrical conductivity in comparison to pore-water with dissolved salts. The electrical conductivity of the pore-water from the soil can be used to indicate the total concentration of dissolved salts, which is related to osmotic suction.

The pore-water in the soil can be extracted using a pore fluid squeezer which consists of a heavy-walled cylinder and piston squeezer (Fig. 4.85). The electrical conductivity of the pore-water is then measured. A calibration curve (Fig. 4.86) relates the electrical conductivity to the osmotic pressure of the soil. The calibration relationship shown in Fig. 4.86 applies for most salts commonly encountered in soils. The following equation can be written to fit the line showing the relationship between osmotic suction π in kilopascals, and the measured electrical conductivity EC in millimhos per centimeter:

$$\pi = 0.024\,\mathrm{EC}^{1.08} \tag{4.31}$$

Results obtained when using the squeezing technique are known to be somewhat affected by the magnitude of the applied extraction pressure. Krahn and Fredlund (1972) used an extraction pressure of 34,500 kPa when collecting fluid for osmotic suction measurements on the glacial till and Regina clay.

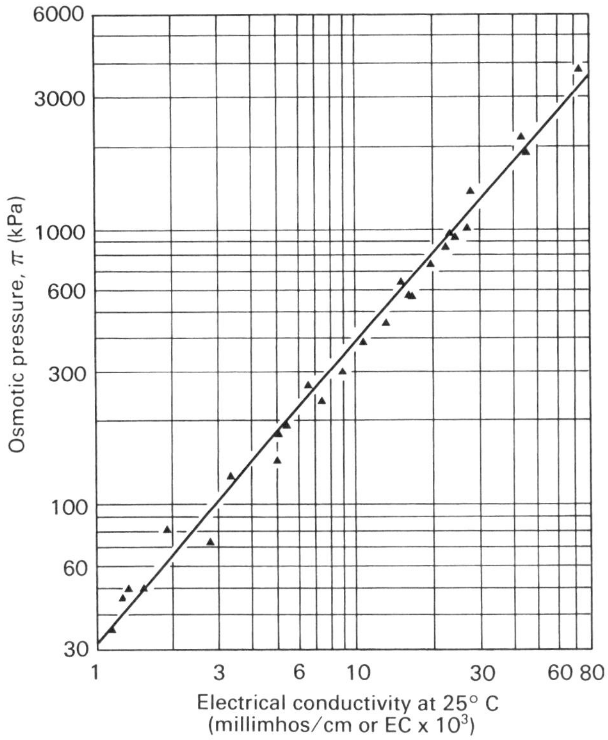

Figure 4.86 Osmotic suction versus electrical conductivity for pore-water containing mixtures of dissolved salts (from USDA Agricultural Handbook No. 60, 1950).

Figures 4.87 and 4.88 present the results of osmotic suction measurements on glacial till and Regina clay, respectively. The measurements were conducted using the squeezing technique. The measured osmotic suctions are shown to agree closely with the difference between independently measured total suction and matric suction values. The discrepancies shown at low water contents for the glacial till (Fig. 4.87) appear to be primarily related to inaccurate measurements of matric suction (Krahn and Fredlund, 1972).

Figures 4.87 and 4.88 indicate that the squeezing technique can be used for the measurement of osmotic suction. The results also support the validity of the relationship between matric and osmotic components of soil suction. It

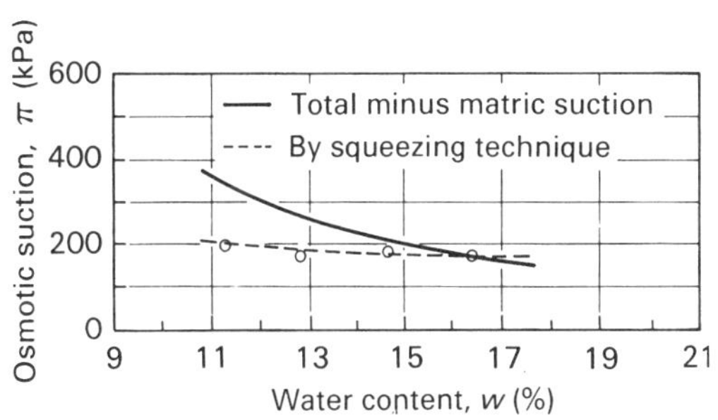

Figure 4.87 Osmotic suction versus as-compacted water content for glacial till (from Krahn and Fredlund, 1972).

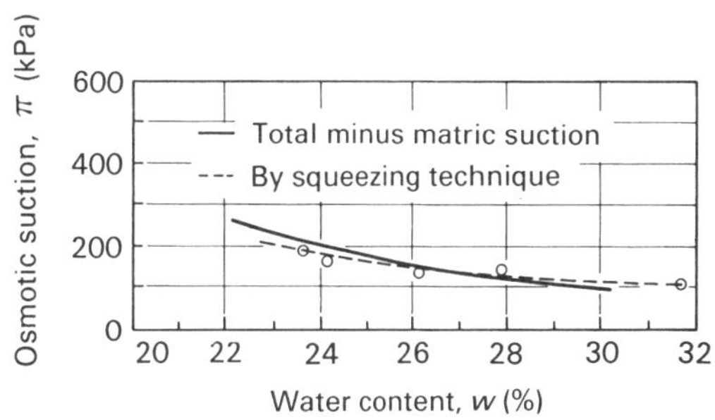

Figure 4.88 Osmotic suction versus as-compacted water content for Regina clay (from Krahn and Fredlund, 1972).

appears that osmotic suction is relatively constant over a range of water contents. It would appear to be reasonable to use osmotic suction as a relatively fixed value that can be subtracted from total suction in order to provide an indication of matric suction.

Water is being removed from a soil as it dries. However, the salts are left behind in the soil. Consequently, there can be a significant increase in the salt content of the pore-water near the evaporation front. As a result, the osmotic component of suction increases. The increase in salts on or near the evaporation surface of a soil is known to "shut off" or greatly decrease the evaporation rate of water from the soil surface. The decrease in evaporation rate is closely related to a significant increase in the osmotic component of soil suction. The increase in osmotic suction increases total suction of the soil.

4.5 MEASUREMENT OF IN SITU WATER CONTENT

The void ratio and the water content of a saturated soil bear a fixed relationship by the specific gravity of the soil. However, water content and void ratio become independent variables for unsaturated soils. Consequently, it is often necessary to measure or monitor water content when dealing with unsaturated soils.

Both soil suction and the water content may need to be measured in the field when monitoring performance of soil cover systems. However, some projects may only require that the water content be monitored while on other projects it might only be necessary to measure soil suction. For example, it might be preferable and sufficient to measure soil suction when dealing with clay soil and it might be preferable and sufficient to measure water content when dealing with sand soils. On other occasions it might be preferable to measure the changes in both soil suction and water content. Each situation needs to be studied and an engineered instrumentation plan designed for each situation.

4.5.1 Measurement of Water Content

Gravimetric water content (i.e., ratio of mass of water to mass of solids) is the easiest variable to measure in the laboratory to represent the amount of water in a soil. However, most field measurement of water content provides an indication of the volume of water relative to the total volume of soil. Gravimetric water content is commonly used in soil mechanics; however, volumetric water content becomes a common representation of the amount of water in a soil when dealing with unsaturated soil problems. It is important to be able to convert between gravimetric and volumetric water content.

There are three primary ways in which water content is measured in geotechnical engineering. First, the measurement of gravimetric water content is a simple and direct means; however, it is a destructive measurement in the sense that a soil sample must be taken into the laboratory for testing. Other methods for the measurement of water content are indirect methods and are commonly referred to as surrogate methods. Second, there are two radiological techniques that can be used for the measurement of water content: the neutron scattering technique and the gamma absorption technique. Third, water content can be derived from the dielectric properties of a soil using time-domain reflectometry. The radiological and dielectric procedures determine volumetric water content and do not require an independent measure of soil density. Details pertaining to technologies available for the measurement of water content can be found in the training course manual of the International Atomic Energy Agency (IAEA, 2008).

4.5.2 Radiological-Based Methods

One of the radiological methods utilizes the neutron scattering methodology to measure volumetric water content. The technique is based on the interaction of high-energy neutrons and the nuclei of the hydrogen atoms in the soil. The other radiological methodology utilizes the attenuation of gamma rays that are passed through the soil. Both methodologies have been packaged as portable devices that can be taken to a field site. Safety and health precautions are necessary when using any radiological method because of the potential for receiving doses of radiation. The only time that the radioactive probe should leave its shield is when it is being lowered into the access tube for measurements.

4.5.2.1 Neutron Scattering Methodology

The hydrogen nuclei have about the same mass as neutrons in the soil. The nuclei are effective in slowing down neutrons upon collision with other nuclei in the soil. The slowed neutrons in the vicinity of the probe are returned to the sensor in the form of a "count" and can be empirically related to the volumetric water content of the soil.

Figure 4.89 shows a field setup with the probe and shield resting on top of an access tube. The probe is lowered into the access tube and readings are taken at various depths. The radius of influence in wet soil may be about 150 mm whereas in dry soil it is increased to about 500 mm. This technique

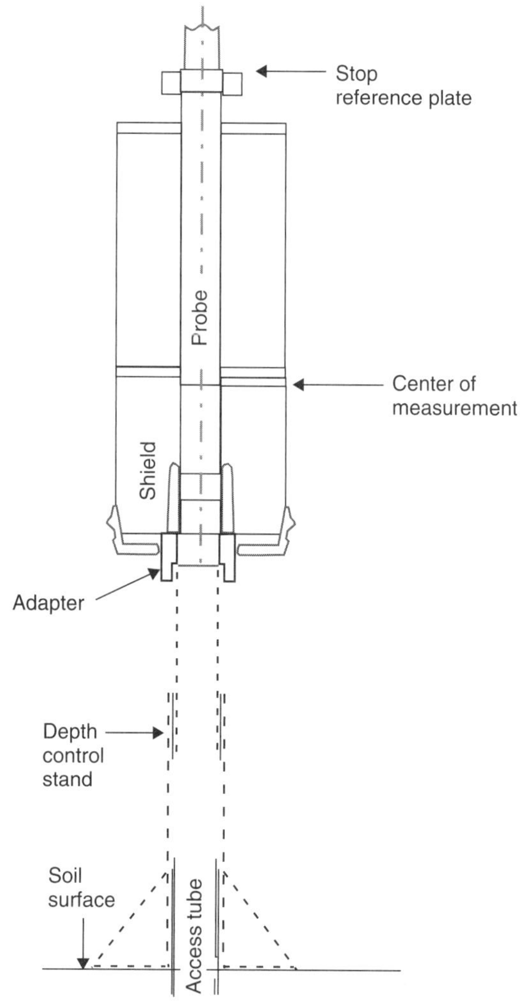

Figure 4.89 Schematic of nuclear neutron moisture meter placed on depth control stand.

does not have the ability to measure the water content in thin layers within the soil. The access tubes can be made of a variety of noncorrosive materials ranging from stainless steel to plastic, although polyvinylchloride should be avoided. The diameter of the access tubes are usually about 50 mm and the thickness of the tubes is about 1.25 mm. For geotechnical engineering purposes the neutron probe should be calibrated for the soils encountered at a particular site.

Nuclear moisture meters are manufactured by Campbell Pacific Nuclear (model 503DR) and Troxler Electronics Laboratories (model 4300). Nuclear moisture meters appear to provide the most reliable measurements of water content based on comparative studies involving a variety of surrogate methods (IAEA, 2008). There appears to have been some reluctance toward the use of nuclear moisture meters in geotechnical engineering. In part, this may be related to safety issues associated with the usage of a radioactive probe and also the inability to automate the recording of periodic readings.

4.5.2.2 Gamma Ray Attenuation Methodology

In contrast to the neutron method, the gamma ray adsorption technique scans a thin layer of soil. Dual-probe gamma devices have been quite commonly used for research studies in the soil laboratory. The water content of the soil is determined from the difference between the total density and the dry density attenuation values. The attenuation of the gamma rays is related to mass and therefore it is necessary to know the dry density of the soil in order to monitor changes in water content.

4.5.3 Dielectric-Based Methods

The dielectric constant (or permittivity) of a material is defined as the ratio of the amount of electrical energy stored when a voltage is applied through a material in comparison to the amount of electrical energy stored in a vacuum. A dielectric material is a substance that is a poor conductor of electricity but an efficient supporter of electrostatic fields.

The dielectric constant of water is approximately one order of magnitude less than that of dry soil. Table 4.14 presents the dielectric constants for the components found in soils. The contrast in dielectric constant among water, air, and soil forms the basis for the measurement of volumetric water content in a soil.

The measurement of the dielectric constant for a soil allows for a comparison between the value for a dry soil and a wet soil. Consequently, the volumetric water content of a soil can be determined. The technique is rapid and safe to use and has increasingly been used for field measurements of water content. Two methodologies have emerged which make use of soil dielectric measurements for determining the water content of a soil. The two methods are referred to as (i) time-domain reflectrometry measurements and (ii) frequency-domain measurements.

4.5.3.1 Time-Domain Reflectometry Methodology

The time-domain reflectometry (TDR) method determines the dielectric constant of the soil by measuring the travel time of an electronic pulse. The electrical pulse is in the microwave (gigahertz) frequency range. The pulse is passed along a waveguide consisting of two parallel rods that are embedded in the soil. The pulse is reflected at the end of the waveguide and its propagation velocity is measured. Theoretically, the propagation velocity is inversely proportional to the square root of the dielectric constant of the material in which the rods are embedded.

Table 4.14 Typical Values for Dielectric Constants of Soil Components

Material	Dielectric Constant (unitless)	Temperature, °C
Air	1.0	
Water	88	0.0
Water	80.1	20.0
Water	55.3	100
Dry silica sand	2.2	
Quartz	4.3	
Soil	2.4–2.9	

The parallel rods are commonly separated by 50 mm and vary in length from 100 to 500 mm. The rods can be made of any metallic material. A sampling volume a few centimeters in radius is formed around each of the metal rods. It is preferable to place the rods in a horizontal manner when attempting to measure the water content at a particular depth.

The volumetric water content versus dielectric constant relationship is quite similar for many soils; however, it is not unique for all soils and may require calibration for particular soils (Hu et al., 2010). A widely used relationship between the dielectric constant and volumetric water content, θ, was that published by Topp et al. (1980) and can be written as follows:

$$\theta = -0.053 + 0.029K - 0.00055K^2 + 0.0000043K^3 \tag{4.32}$$

where:

K = dielectric constant for the soil-water system.

Equation 4.32 has been found to be applicable for many soils. It is relatively independent of soil texture and gravel content (Drungil et al., 1989). The dielectric constant can also be written in terms of the volumetric water content:

$$\sqrt{K} = 3.03 + 9.3\theta + 146\theta^2 - 76.7\theta^3 \tag{4.33}$$

Various types of time-domain reflectometry sensors have been developed that measure the dielectric properties of a soil. Conventional time-domain reflectometry systems have been manufactured by (i) Tektronix Inc. (TDR Cable Tester), (ii) Campbell Scientific Inc. (TDR100), (iii) Dynanax Inc. (Vadose TDR), (iv) Soil Moisture Corporation (Trase TDR), and (v) Delta-T (Thetaprobe). The TRIME FM3 moisture meter is manufactured in Germany and is classified as a quasi-time-domain reflectometry system.

The conventional time-domain reflectometry devices have a pulse generator that generates a square-wave pulse. The pulse travels along the cable to the header connected to the rods that are inserted into the soil. The pulse reaches the end of the TDR rods and then returns. An oscilloscope electronic system captures the pulse as shown in Fig. 4.90. The water content of the soil influences the travel time of the pulse along the rods surrounded by soil. It is the permittivity of the soil that affects the travel time and the permittivity of the soil is related to the volumetric water content of the soil.

Originally it was suggested that a single calibration curve might exist for all soils. However, it is now suggested that

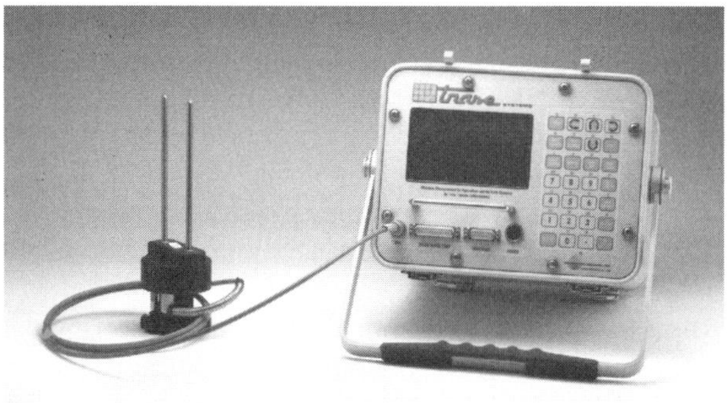

(a) Readout device for TRASE

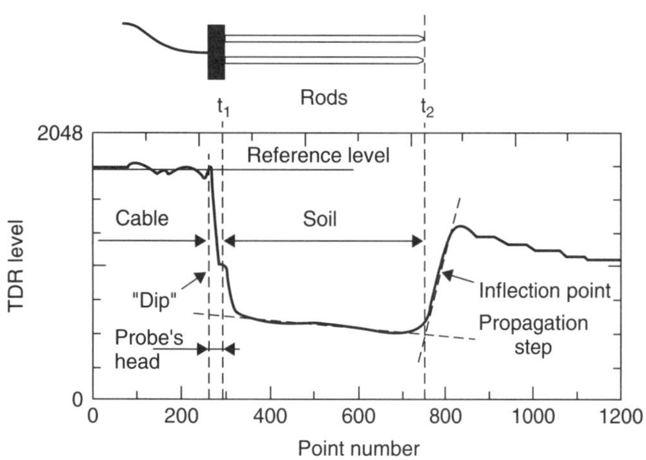

(b) Typical signal from TRASE reading

Figure 4.90 Characteristics of TRASE (manufactured by Soilmoisture Corporation) signal using 30-cm-long rods (after Laurent, 1998).

the TDR method be calibrated for the specific soil where measurements are being made. The oscilloscope electronic system is quite costly and there have been attempts to develop other technologies for the measurement of water content. The result has been the development of a series of capacitance sensor systems that still depend on the permittivity of the soil.

4.5.3.2 Frequency-Domain Methodology

The frequency-domain or capacitance-type sensors measure the dielectric constant using a single microwave (gigahertz) frequency. A single open-ended coaxial cable is used that has a single reflectometer at the end of the probe. The amplitude and phase are measured at a particular frequency. The device is generally calibrated in air and in blocks of material and liquids with known dielectric constant properties.

A relatively small volume of soil is involved in the dielectric constant measurement when using the frequency-domain method. Consequently, the contact between the soil and the probe becomes a critical factor. It has been suggested that this method performs well in the laboratory but may be subject to spatial variability in the field (Dirksen, 1999).

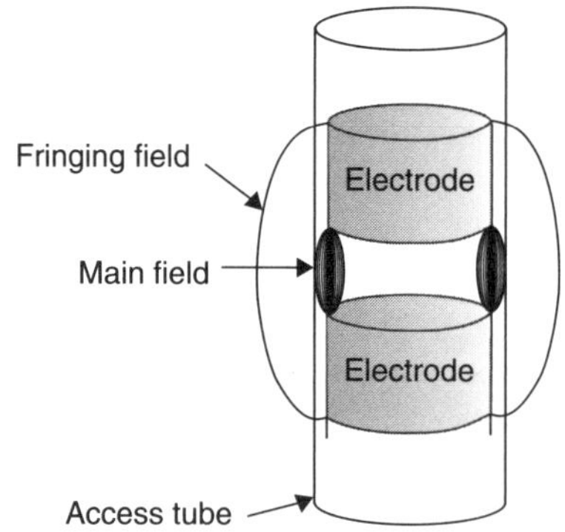

Figure 4.91 Schematic of capacitance probe lowered into access tube.

Several manufacturers produce capacitance-type sensors that are used in a plastic access tube. For example, there is a capacitor consisting of two hollow cylindrical metal electrodes that are arranged coaxially but separated by several millimeters by an insulating plastic (Fig. 4.91). An electronic oscillator is used to produce a sinusoidal waveform. The capacitor forms part of the oscillating circuit and the electrodes are arranged close to the inside of the access tube. The intent is that the field of the capacitor will interact with the soil outside the tube. The soil bulk permittivity is measured and correlated with the water content of the soil.

There are a number of sensors that come within the capacitance or frequency-domain category. The Sentek Diviner 2000 and the Sentek Enviro-SCAN manufactured by Sentek Sensor Technology are of the frequency-domain type. Another capacitance-type sensor is the Delta-T PR1/6 sensor. The capacitance-type sensors are practical and easy to use for in situ monitoring of changes in water content in soil cover systems and other near-ground-surface applications.

4.5.3.3 Comparative Study on the Measurement of Water Content

A comparative study involving highly plastic clay (i.e., Regina clay with a liquid limit of 74%) selected three types of time-domain sensors and undertook to measure the calibration curves. It was found that the measured volumetric water contents differed considerably from the commonly used calibration equation (Hu et al., 2010). The three types of dielectric probes used in the study are known by their brand names: CS616, ML2x, and SM200 (Fig. 4.92). The CS616 device is known as a water content reflectometry (WCR) sensor (Hansson and Lundin, 2006). The ML2x (or Thetaprobe) and SM200 devices are known as amplitude-domain reflectometry sensors (Munoz-Carpena, 2004).

Clay soils have high electrical conductivity because of the electrically charged clay colloids and the electrolytes in the soil solution. The electrolytes affect the measurement of the dielectric constant. The square root of the bulk dielectric constant and the volumetric water content tested on Regina

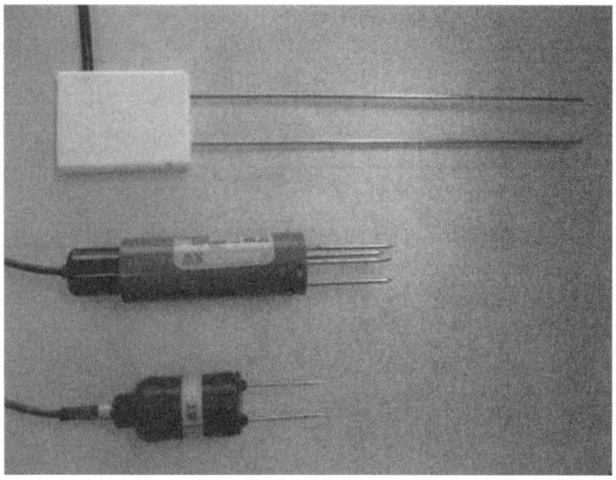

Figure 4.92 CS616 (top), ML2x (middle), and SM200 (bottom) probes for measuring volumetric water content.

clay is shown in Fig. 4.93 for the CS616 device. Similar calibration results are shown for the ML2x and SM200 devices in Figs. 4.94 and 4.95. All three dielectric constant devices showed a standard error between the measured and predicted volumetric water contents of slightly more than 3%. A second-order polynomial equation was found to fit the relationship between the dielectric constant and volumetric water content.

Another comparative study was reported by the IAEA (2008) where several surrogate water content measuring systems were installed in two irrigated test plots (Fig. 4.96). Dry and wet irrigation conditions are shown along with variation of readings with depth. The most consistent results were obtained when using the nuclear moisture meter.

Geotechnical engineers become involved with the field monitoring of water content on a wide range of materials. As a result, it is important to take into consideration the material in which the water contents are being monitored. Calibration checks should be undertaken to confirm the reliability of the company-published calibration curves.

4.6 ESTIMATION OF SOIL SUCTION

Estimation techniques play an important role in approximating soil variables that might be needed as part of an engineering analysis involving unsaturated soils. It is not always possible to measure every variable of relevance; however, in many situations it is sufficient to estimate certain variables or possibly estimate a likely range of values that might be encountered. Estimation procedures are particularly useful when assessing the components of soil suction.

4.6.1 Estimation of Matric Suction

There are a number of observations that can provide an engineer with an indication of the matric suction in the field. The matric suction in the field will be less than the residual suction for the soil. Sandy soils may typically have an air-entry value between 1 and 10 kPa. There is often about one order of magnitude between the air-entry value and the residual matric suction. Therefore, it would be anticipated that the residual suction for a sand soil would often be between about 10 and 100 kPa. The classification of the soil can provide

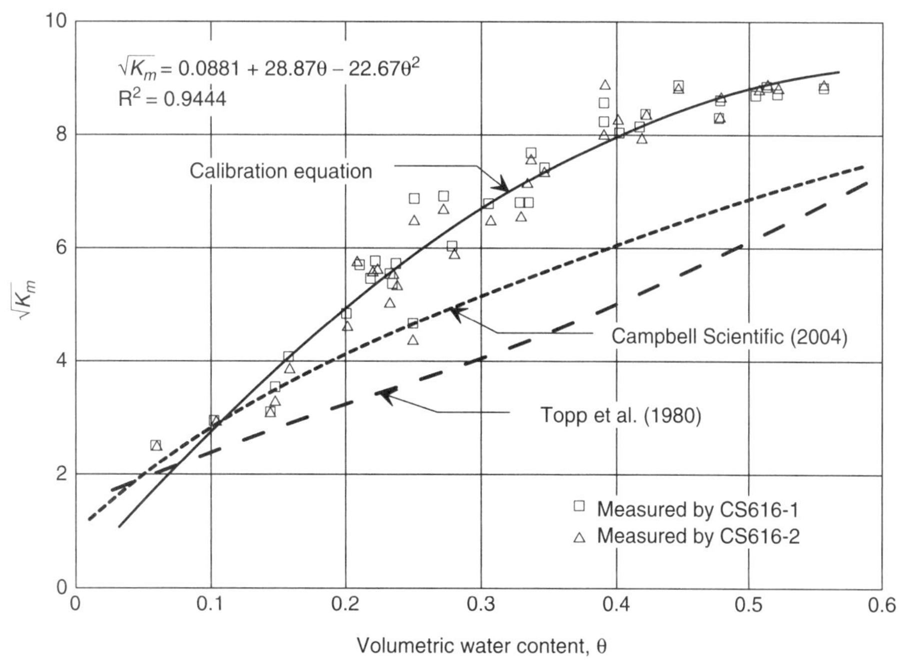

Figure 4.93 Dielectric constant as function of volumetric water content when using CS616 probe on Regina clay (after Hu et al., 2010).

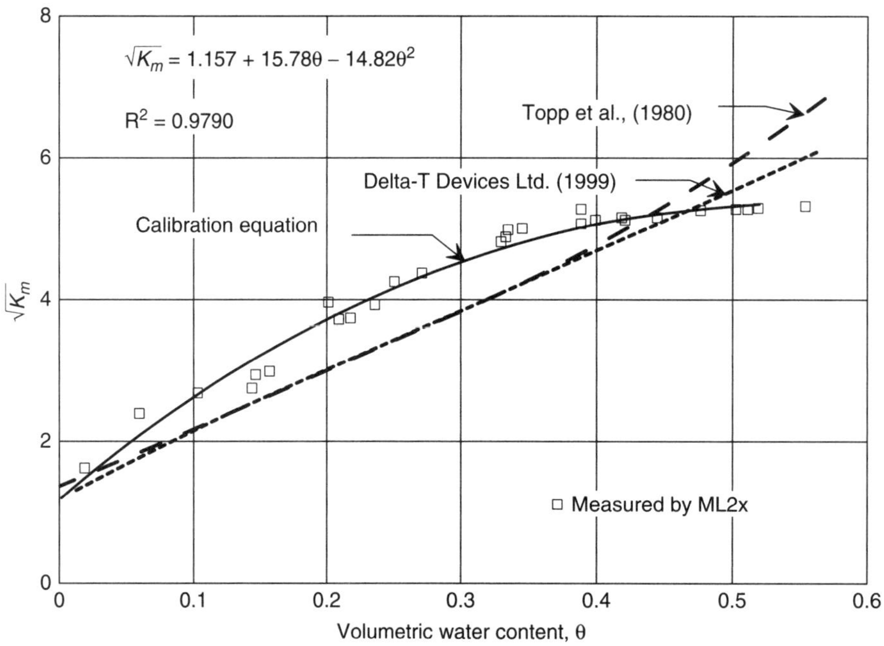

Figure 4.94 Dielectric constant as function of volumetric water content when using ML2x probe on Regina clay (after Hu et al., 2010).

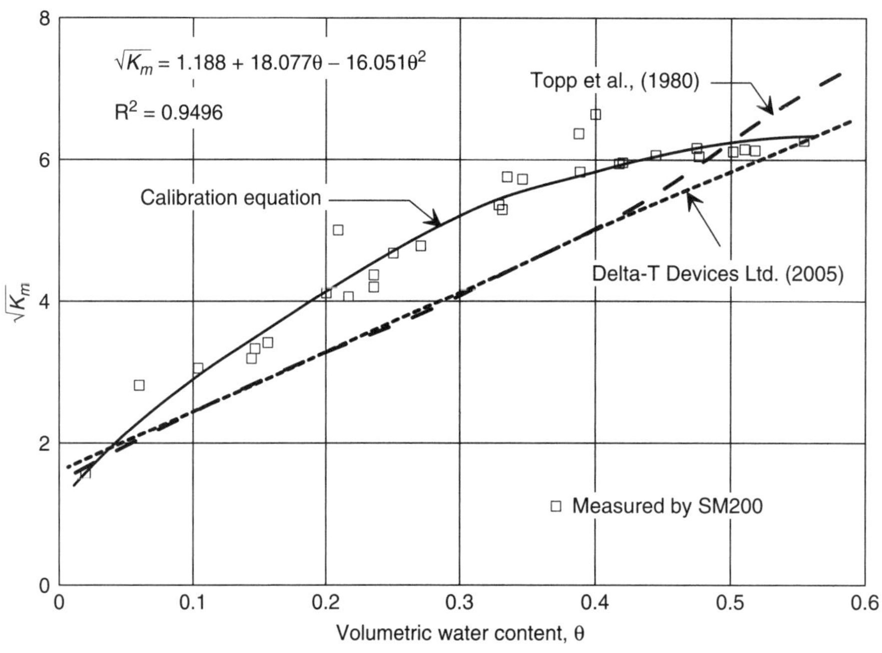

Figure 4.95 Dielectric constant as function of volumetric water content when using SM200 probe on Regina clay (after Hu et al., 2010).

an indication of the air-entry value and the residual suction conditions. The air-entry value and the residual suction provide an indication of the most likely suction range to be encountered in the field.

The soil classification properties provide the first indication of the likely SWCC for a material. The shape of the SWCC likewise provides an indication of the likely range of matric suctions to be encountered in the soil profile.

4.6.1.1 Reference to Hydrostatic Conditions

The depth of the water table can provide valuable information on the likely matric suctions that might be encountered in the field. The matric suction at the water table must be zero. In the region immediately above the water table, the negative pore-water pressure will tend to be hydrostatic and this provides an indication of the likely matric suction values. The matric suction profile will tend to be strongly

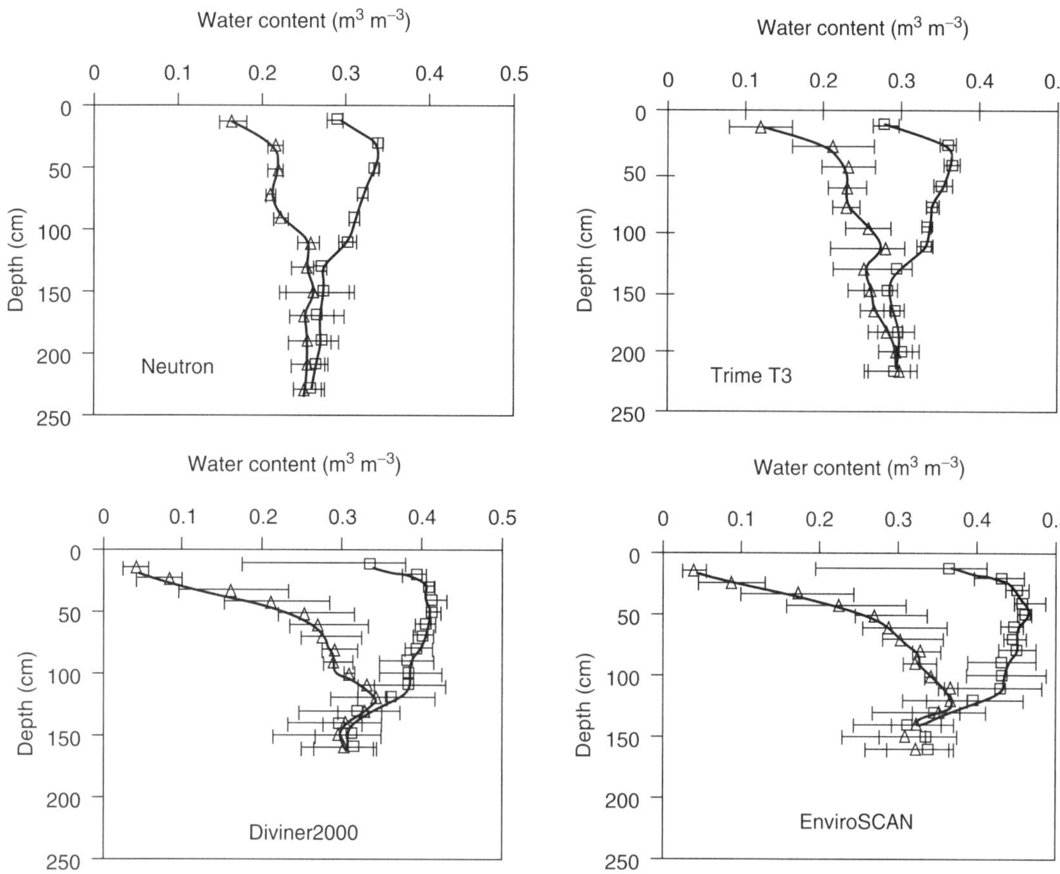

Figure 4.96 Soil-water contents measured using several different measurement sensors (from Training Course No. 30, IAEA, 2008).

influenced by hydrostatic pressure conditions if the water table is within a few meters of the ground surface. If it is assumed that the net infiltration at the ground surface is zero (i.e., precipitation is approximately equal to actual evaporation over each year), then the matric suction profile will be controlled by the hydrostatic pore-water pressure line.

The depth to the water table and the hydrostatic pore-water pressure line provide an indication of the matric suction values to likely be encountered. Figure 4.97 illustrates the hydrostatic line corresponding to a water table at various depths below ground surface. The hydrostatic line is an equilibrium pressure line corresponding to either no net infiltration at ground surface or nearly equal inflow and outflow of moisture at the ground surface.

Excessive infiltration of water at the ground surface will tend to reduce the matric suction at ground surface. During a rainstorm water may temporarily accumulate on the ground surface, reducing the matric suction to zero. The infiltration of water at the ground surface will also affect the underlying matric suction profile. In other words, the matric suction profile can be thought of as being strongly influenced by the water table and the ground surface boundary conditions. Conversely, excessive evaporation of water from the ground surface can result in a drying of the soil at the ground surface with the result that the near-ground-surface matric suction increases beyond what would be anticipated based on hydrostatic conditions.

4.6.2 Use of SWCC to Estimate Soil Suction

The SWCCs are used extensively for the estimation of unsaturated soil property functions (Fredlund, 1995b; 2000a), and have been pivotal to the implementation of unsaturated soil mechanics into geotechnical engineering practice. Estimation procedures for unsaturated soil property functions have been proposed for virtually every physical process involving unsaturated soils (Fredlund et al., 1997b; Fredlund, 2006b).

Engineers have desired to use SWCCs as a means to estimate in situ soil suctions, but their usage for this purpose has been discouraged (Fredlund et al., 2001b). SWCCs can be used for the estimation of in situ suction but the user needs to understand (i) the level of accuracy associated with using the SWCC and (ii) the likely range of soil suction values associated with a measured natural water content of the soil.

The use of SWCCs to estimate in situ soil suction must take into consideration hysteresis associated with the drying (desorption) and wetting (adsorption) curves. Generally the drying curve is measured in the laboratory. The asymptotic nature of most empirical equations used to represent the SWCC makes it possible to only estimate soil

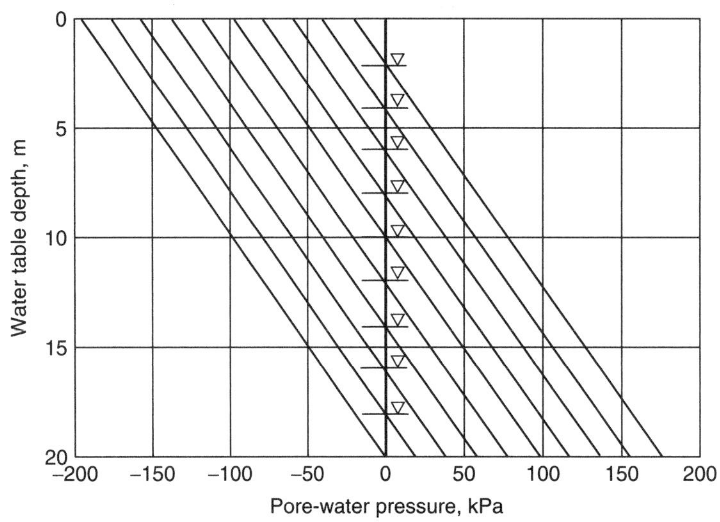

Figure 4.97 Equilibrium hydrostatic pore-water pressures for various depths of water table.

suction between the air-entry value and residual conditions (Fredlund et al., 2007). The retrieval of water content samples from the field does not provide an indication of whether the in situ soil is on the drying curve, the wetting curve, or somewhere in between these two limiting conditions. Soil suction varies on a logarithm scale with water content, which also makes it more difficult to obtain an accurate estimation of in situ soil suction.

The assumption must be made that the geotechnical engineer has either measured or estimated the drying SWCC for a particular soil deposit. The wetting SWCC branch can either be measured in the laboratory or estimated based on suggested empirical guidelines (Pham, 2002). The SWCC can be plotted as (gravimetric) water content versus logarithm of soil suction (Fig. 4.98). There are two points along the SWCC that are pivotal to describing the SWCC: the "air-entry value" of the soil and "residual conditions."

The SWCC is not a single-valued, unique relationship but rather is hysteretic, as shown in Fig. 4.99. Therefore, it is not possible to determine a single soil suction value based on a single water content measurement. Several observations can be made from the drying and wetting SWCCs. First, the hysteresis loops appear to be generally quite reproducible during repeated drying and wetting cycles. Second, some air becomes entrapped in the soil after it is wetted from a high-suction stress state. Third, the slope of the desorption curve is approximately parallel to that of the adsorption curve when the soil has been dried to a point beyond residual suction. Each of the bounding SWCCs has an inflection point of maximum slope on a logarithmic scale. The inflection point can be used as a point to define the lateral shift between the drying and wetting SWCCs.

4.6.2.1 Accommodation of SWCC Hysteresis Effect

There are a number of approaches that can be used for the estimation of in situ suction (Fredlund et al. 2011). It is possible to:

1. Ignore the effect of hysteresis and solely use desorption SWCCs for the estimation of soil suction (i.e., the upper bounding branch of the SWCCs in Fig. 4.99). This would give an estimate of the maximum soil suction value corresponding to the measured water content.
2. Measure the desorption SWCC and approximate the adsorption SWCC by estimating the lateral shift of the hysteresis loop at the inflection point on the drying curve. This would allow for an estimation of the maximum and minimum soil suctions corresponding to the measured water content.
3. Measure the desorption branch and the adsorption branch of the SWCCs. Often this type of laboratory test is considered too costly. This approach would tend to provide a more accurate estimate of maximum and minimum soil suctions.
4. Determine a "median" SWCC midway between the drying and wetting SWCCs (i.e., midway on a logarithmic scale). The water content corresponding to the inflection point on the drying curve can be used as the reference point for the lateral shift of the SWCC. The median SWCC can then be used to estimate the soil suction corresponding to the median curve.
5. Use a more rigorous mathematical equation that describes desorption, adsorption, and scanning SWCCs (Pham et al., 2003). Such an approach would provide little additional value since it is not possible to take the wetting/drying history of the soil into consideration.

4.6.2.2 Rewriting SWCC Equations with Soil Suction as the Dependent Variable

Most of the empirical equations that have been proposed to best fit water content versus soil suction data can be rearranged such that soil suction can be computed if the water content of the soil is known. Let us first consider the drying or desorption SWCC and then consider how best to

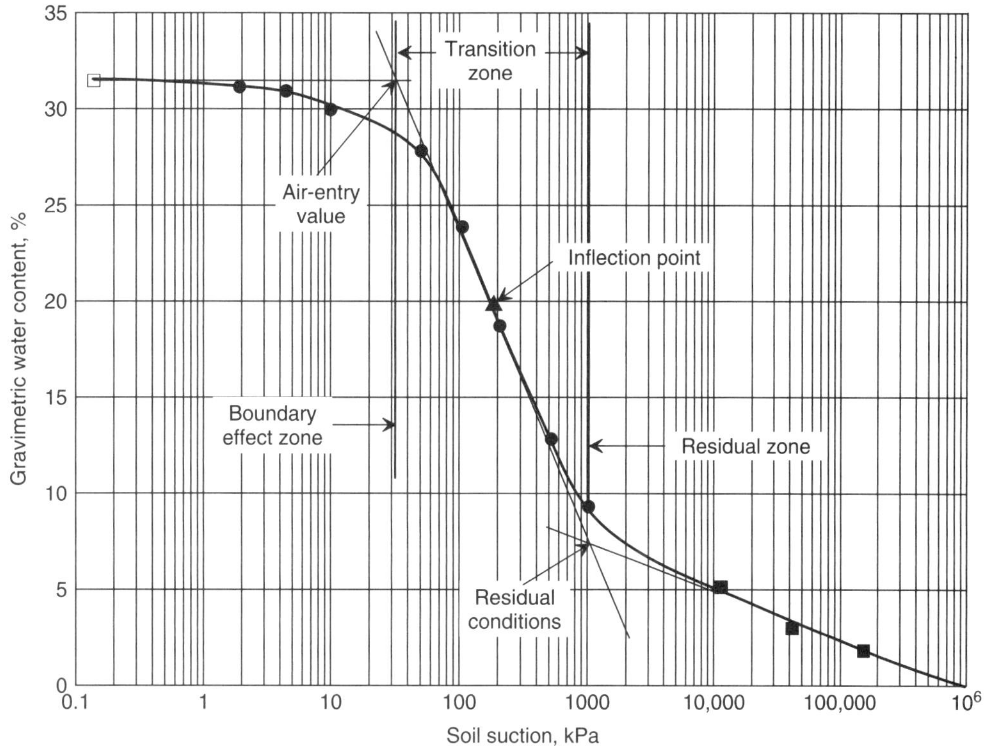

Figure 4.98 Desorption branch with definition of variables for SWCC.

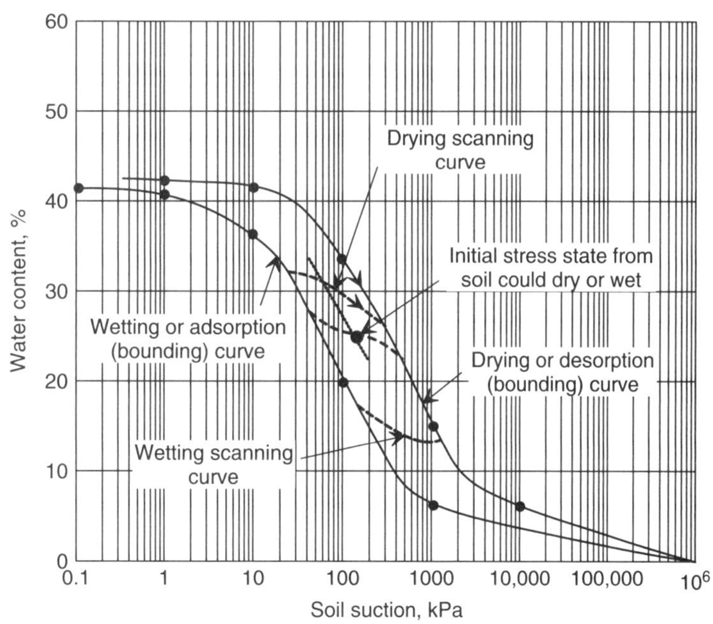

Figure 4.99 Relationship of initial stress state to hysteresis loops of SWCC (from Fredlund, 2003).

accommodate hysteresis. The SWCC equations are written in terms of gravimetric water content; however, each of the rearranged equations would have the same form if written in terms of volumetric water content.

The parameters for the SWCC must be known when using a SWCC to compute soil suction. Sigmoidal shapes for the SWCC can be defined using three fitting parameters, generally identified as *a, n,* and *m*. The *a*-type parameter in most SWCCs is a soil suction value that is somewhat greater than the actual air-entry value. The *n*-type parameter is related to the rate of desaturation of the soil as suction exceeds the air-entry value, and in some cases, a further *m*-type parameter is used to give greater flexibility in fitting SWCC data. In addition, the saturated gravimetric water content w_s must be known.

Gardner (1958a). The Gardner (1958a) equation was originally proposed to describe the coefficient of permeability function for an unsaturated soil. However, the mathematical form proposed for the permeability equation has also been applied to the water content versus soil suction relationship. The Gardner equation can be written as

$$w(\psi) = \frac{w_s}{1 + a\psi^n} \tag{4.34}$$

where:

ψ = any soil suction value, kPa,
$w(\psi)$ = water content at any soil suction, %,
w_s = saturated water content, %, and
a, n = fitting parameters.

Equation 4.34 can be rearranged such that soil suction ψ is dependent upon the measured water content w:

$$\psi = \left[\frac{1}{a}\left(\frac{w_s}{w} - 1\right)\right]^{1/n} \tag{4.35}$$

The soil suction can be calculated corresponding to any other water content, *w,* if the fitting parameters for the Gardner equation are known.

Brooks and Corey (1964). Brooks and Corey (1964) divided the SWCC into two zones: one zone where the soil suctions are less than the air-entry value and the other zone where soil suctions are greater than the air-entry value. This gives rise to equations of the following form:

$$\begin{cases} w(\psi) = w_s & \psi < \psi_{aev} \\ w(\psi) = w_s\left(\dfrac{\psi}{a}\right)^{-n} & \psi \geq \psi_{aev} \end{cases} \tag{4.36}$$

where:

ψ_{aev} = air-entry suction, kPa.

It is not possible to use the proposed SWCC equation for estimating suctions prior to the air-entry value. The Brooks and Corey equation can be rearranged to compute soil suction corresponding to the measured water content:

$$\psi = a\left(\frac{w_s}{w}\right)^{1/n} \tag{4.37}$$

Two fitting parameters, a and n, and the saturated water content w_s are required along with the measured water content for the calculation of soil suction.

Van Genuchten (1980). Van Genuchten (1980) proposed a three-parameter SWCC equation which has the following form:

$$w(\psi) = \frac{w_s}{\left[1 + (a\psi)^n\right]^m} \tag{4.38}$$

The van Genuchten (1980) equation can be rearranged to solve for soil suction in terms of water content:

$$\psi = \frac{1}{a}\left[\left(\frac{w_s}{w}\right)^{1/m} - 1\right]^{1/n} \tag{4.39}$$

The soil suction of a soil can be calculated if the three fitting parameters a, m, and n for the van Genuchten equation are known along with the saturated water content w_s. The usage of this equation is again limited to the range between the air-entry value and the residual suction due to the asymptotic nature of the equation.

Van Genuchten (1980)–Mualem (1976). In 1976, Mualem had suggested that the n and m parameters in the SWCC equation bear a fixed relationship with $m = (n - 1)/n$. This suggestion reduces the three-parameter equation of van Genuchten (1980) to a two-parameter SWCC equation:

$$w(\psi) = \frac{w_s}{\left[1 + (a\psi)^n\right]^{1-1/n}} \tag{4.40}$$

The van Genuchten (1980)–Mualem (1976) equation can be rearranged to solve for soil suction in terms of water content:

$$\psi = \frac{1}{a}\left[\left(\frac{w_s}{w}\right)^{n/(n-1)} - 1\right]^{1/n} \tag{4.41}$$

The van Genuchten (1980)–Mualem (1976) equation can be applied between the air-entry value and residual conditions.

Van Genuchten (1980)–Burdine (1953). In 1953, Burdine suggested that the n and m parameters for a SWCC equation could bear a fixed relationship with $m = (n - 2)/n$. This resulted in a two-parameter equation:

$$w(\psi) = \frac{w_s}{\left[1 + (a\psi)^n\right]^{1-2/n}} \tag{4.42}$$

The van Genuchten (1980)–Burdine (1953) equation can be rearranged to solve for soil suction in terms of water content:

$$\psi = \frac{1}{a}\left[\left(\frac{w_s}{w}\right)^{n/(n-2)} - 1\right]^{1/n} \tag{4.43}$$

The van Genuchten (1980)–Burdine (1953) equation can be applied between the air-entry value and residual conditions.

Fredlund and Xing (1994)—Without the Correction Factor. The Fredlund and Xing (1994) equation for the SWCC has a correction factor $C(\psi)$ that extends the range of suctions beyond residual suction to completely dry conditions:

$$w(\psi) = C(\psi)\frac{w_s}{\{\ln[e + (\psi/a)^n]\}^m} \tag{4.44}$$

where:

a, n, m = fitting soil parameters for the SWCC,
e = base of the natural logarithm,
$C(\psi)$ = correction factor, $1 - \dfrac{\ln\left(1+\psi/\psi_r\right)}{\ln\left[1+\left(1000000/\psi_r\right)\right]}$, and
ψ_r = soil suction at residual conditions, kPa.

It is not easy to rearrange the Fredlund and Xing (1994) equation and solve for soil suction when the "correction factor" is kept within the equation. Therefore, the correction factor for suctions greater than residual suction is set to 1.0. The Fredlund and Xing (1994) equation for calculating soil suction in terms of water content can then be written as follows:

$$\psi = a\left(e^{(w_s/w)^{1/m}} - e\right)^{1/n} \tag{4.45}$$

A numerical technique must be used to compute soil suction from a water content measurement if the entire Fredlund and Xing (1994) equation along with the correction factor is used.

Fredlund and Pham (2006). Fredlund and Pham (2006) divided the SWCC into three zones: (i) a low-suction portion from a small suction value (e.g., 1 kPa) to the air-entry suction ψ_{aev}, (ii) an intermediate portion from the air-entry suction ψ_{aev} to the residual suction ψ_r, and (iii) a high-suction portion from the residual suction ψ_r to 10^6 kPa, as shown in Fig. 4.100. The water contents corresponding to the air-entry value and residual suction must be known as well as the saturated water content under low-suction conditions (e.g., 1 kPa) in order to use these equations over the entire suction range.

Each suction range has a SWCC equation of a similar form. The three equations cover the entire water content range from completely saturated conditions to completely dry conditions with each portion having a meaningful slope representing water content versus suction. It is necessary to know the water contents at the start and end points for each line segment. The slope of the straight-line portions of the three zones are defined as S_1, S_2, and S_3, when going from the low-suction range of the SWCC curve to the high-suction range.

The SWCC equation for the low-suction range, intermediate-suction range, and high-suction range can be written as follows:

$$\begin{aligned} w_1(\psi) &= w_u - S_1 \log(\psi) & 1 \le \psi < \psi_{aev} \\ w_2(\psi) &= w_{aev} - S_2 \log\left(\frac{\psi}{\psi_{aev}}\right) & \psi_{aev} \le \psi < \psi_r \\ w_3(\psi) &= S_3 \log\left(\frac{10^6}{\psi}\right) & \psi_r \le \psi < 10^6 \text{ kPa} \end{aligned} \tag{4.46}$$

where:

w_u = water content corresponding to a suction of 1 kPa and
w_{aev} = water content corresponding to the air-entry value, %.

Each of the above equations can be rearranged to solve for suction in terms of water content. The equations for the low

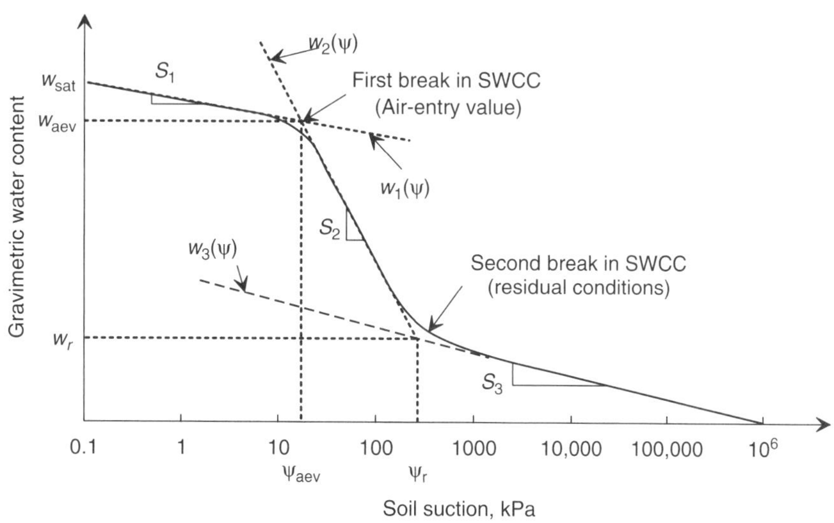

Figure 4.100 Three primary zones of desaturation for drying SWCC (Fredlund and Pham, 2006).

suction zone, the intermediate zone, and the high-suction zone of the SWCC are, respectively,

$$\begin{aligned} &\psi = (10)^{(w_u - w)/S_1} && 1 \le \psi < \psi_{aev} \\ &\psi = \psi_{aev}(10)^{(w_{aev} - w)/S_2} && \psi_{aev} \le \psi < \psi_r \\ &\psi = (10)^{(6 - w/S_3)} && \psi_r \le \psi < 10^6 \text{ kPa} \end{aligned} \quad (4.47)$$

The three equations proposed by Fredlund and Pham (2006) cover the entire range of suctions from a small value to a value of 10^6 kPa.

4.6.2.3 Percent Error in Soil Suction Depending on the Hysteresis Shift in the SWCC

The desorption (or drying) SWCC is generally the first curve determined for a soil (Fredlund et al., 2001b, 2007). Then the adsorption (or wetting) curve is usually estimated by assuming an appropriate lateral shift for the bounding SWCC. The drying and wetting curves are assumed to be congruent or parallel to one another.

The *adsorption* and median SWCC is always shifted to the left on a plot of the type shown in Fig. 4.99. The *a* fitting parameter in the SWCC equations generally controls the lateral shift of the boundary SWCCs. The Fredlund and Xing (1994) SWCC equation can be used to illustrate the lateral translation of the SWCCs. Figure 4.101 shows the effect of changing the *a* fitting parameter. The *n* and *m* fitting parameters were kept constant.

The "percent shift" of the SWCC boundary curves, ξ, can be defined on a logarithmic scale. Therefore, a 100% shift of the inflection point corresponds to one log scale of change. This means that the *a* fitting parameter will change by one order of magnitude. Similarly, a 50% shift means that the *a* fitting parameter is shifted ½ log cycle to the left (i.e., to a lower value). The percent shift of the SWCC boundary curves, ξ, can be mathematically written as follows:

$$\xi = 100\left(\log \psi_{ad} - \log \psi_{aw}\right) \quad (4.48)$$

where:

ψ_{ad} = suction at the point of inflection, kPa,
ad = subscript for the *a* fitting parameter on the drying curve,
aw = subscript for the *a* fitting parameter on the wetting curve, and
ψ_{aw} = suction at the point of inflection, a_w, on the wetting curve, kPa.

The same equation can be used when moving from the drying SWCC to the median SWCC:

$$\xi_m = 100\left(\log \psi_{ad} - \log \psi_{am}\right) \quad (4.49)$$

where:

ψ_{ad} = suction at the point of inflection on the drying curve, kPa,
ψ_{am} = suction at the point of inflection, a_m, on the "median" curve, kPa, and
ξ_m = percent shift between the point of inflection on the drying curve and the point of inflection on the median SWCC.

Since the drying SWCC and the wetting SWCC (and the median SWCC) are assumed to be congruent, the lateral shift defined by Eq. 4.48 (and Eq. 4.49), ξ, applies not only

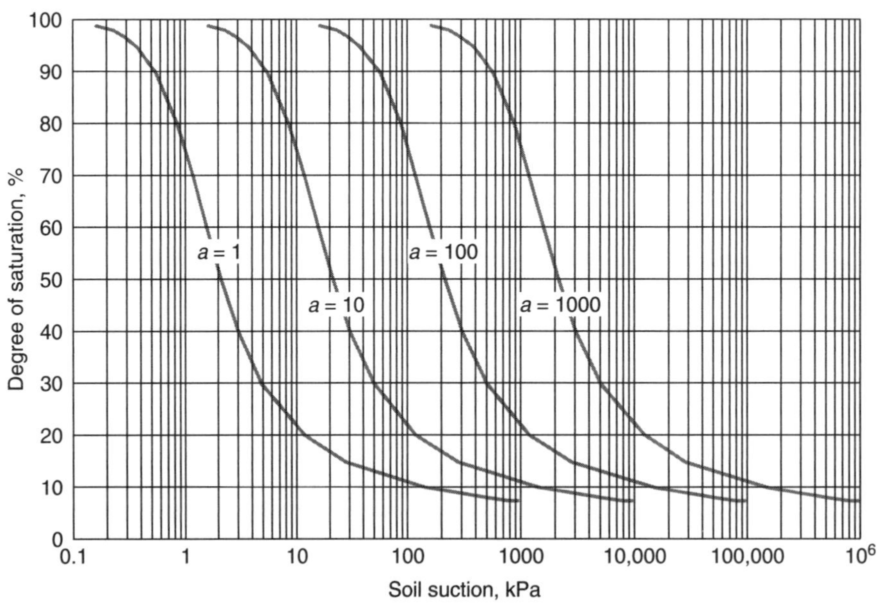

Figure 4.101 Effect of changing inflection point fitting parameter *a* on SWCC.

at the inflection points but at all points along the SWCCs. Therefore, Eq. 4.48 can be written as follows:

$$\xi = 100\left(\log \psi_d - \log \psi_w\right) \tag{4.50}$$

where:

ψ_d = suction at any point along the drying SWCC, kPa, and

ψ_w = suction at any corresponding water content on the wetting SWCC, kPa.

The equations representing the lateral shift of the SWCC can be rearranged such that the suction on a congruent SWCC can be computed from the drying curve. Then Eq. 4.50 can be written as follows:

$$\psi_w = 10^{\left(\log \psi_d - \frac{\xi}{100}\right)} \tag{4.51}$$

Likewise, the suction on the median curve can be written as

$$\psi_m = 10^{\left(\log \psi_d - \frac{\xi_m}{100}\right)} \tag{4.52}$$

Let us define the percent change in suction between any two congruent SWCCs, ζ, as the suction difference between any two curves referenced to the suction on the drying curve,

$$\zeta = 100(\psi_d - \psi_w)/\psi_\text{d} \tag{4.53}$$

where:

ψ_d = suction on the desorption (or drying) curve, kPa, and

ψ_w = suction on the adsorption (or wetting) curve, kPa.

The same equation applies for a shift to the median curve at the same water content. Equation 4.51 (or Eq. 4.52) can be substituted into Eq. 4.53 to provide a relationship between the percent change in suction between two congruent SWCCs, ζ, and the percent shift in the curves, ξ:

$$\zeta = 100\left[\psi_d - 10^{\left(\log \psi_d - \frac{\xi}{100}\right)}\right]\psi_\text{d} \tag{4.54}$$

Equation 4.54 shows that there is a fixed relationship between the percent change in suction, ζ, and the percent lateral shift between the drying and the wetting (or median) curves, ξ, as shown in Fig. 4.102. Figure 4.102 shows that for a lateral shift of 25% the percent error in suction would be 43.8% for a particular water content. Similarly, for a lateral shift of 50%, the percent change in suction would be 68.4%, and for a lateral shift of 100%, the percent change in suction would be 90%.

Figure 4.102 can also be used to compare the suction value computed from the median SWCC and the drying curve. In this case, a percent shift between the drying and wetting curves of 25% would correspond to a percent shift of 12.5% in going from the drying curve to the median SWCC. Consequently, the median SWCC significantly reduces the percent error in the estimated soil suction. Table 4.15 provides example calculations for the case where the lateral shift is 50% between the drying and wetting curves. The results show that the soil suctions on each drying curve produce a 68.4% change (reduction) in the suction on the wetting curve.

Figure 4.103 illustrates another way to visualize the relationship between suctions on the drying curve and any other congruent curve. The log-log plot provides a quick estimation of the relationship between suctions on the drying curve and suctions on any other SWCC if the percent shift between the two curves is known. The plot applies for any two congruent SWCCs regardless of the a and n fitting parameters. Table 4.16 provides a summary of the percent reduction in moving from the drying curve to the wetting (or median) curve for each percent lateral shift between the SWCCs.

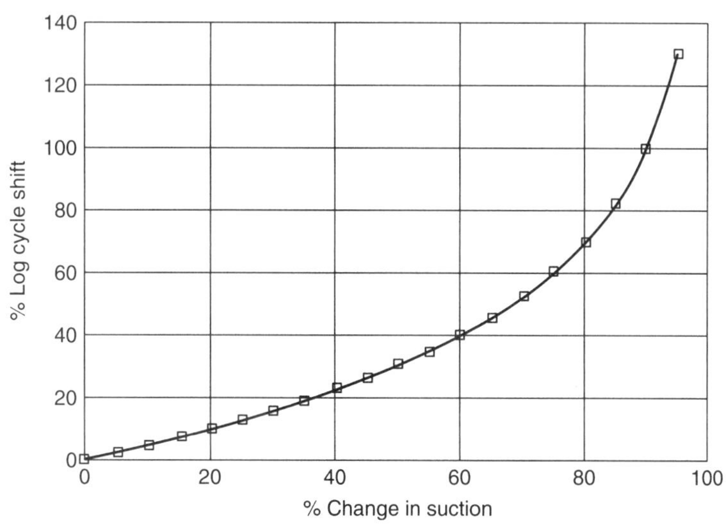

Figure 4.102 Percent change in calculated suction as function of lateral shift between drying and wetting SWCCs.

Table 4.15 Calculation of Soil Suction on Drying and Wetting (or Median) SWCC Drying Curve

Water Content, %	Dimensionless Water Content	Suction on Drying Curve, kPa	Suction on Wetting Curve, kPa	Change in Suction, %
35.50	0.986	11.4	3.61	68.38
33.50	0.931	35.4	11.2	68.38
31.50	0.875	55.8	17.7	68.38
29.50	0.819	76.6	24.2	68.38
27.50	0.764	99.0	31.3	68.38
25.50	0.708	124.	39.3	68.38
23.50	0.653	154.	48.7	68.38
21.50	0.597	190.	60.1	68.38
19.50	0.542	236.	74.5	68.38
17.50	0.486	296.	93.8	68.38
15.50	0.431	383.	120.	68.38
13.50	0.375	515.	163.	68.38
11.50	0.319	741.	234.	68.38
9.50	0.264	1199.	379.	68.38
7.50	0.208	2417.	764.	68.38
5.50	0.153	7834.	2477.	68.38
3.50	0.097	95059.	30060.	68.38
2.50	0.069	1476477.	466900.	68.38

Note: Inflection Point $a_d = 100$ kPa; n parameter $= 1.5$; m parameters $= 1.0$; 50% lateral shift; and wetting curve inflection point $a_w = 31.623$ kPa.

Table 4.16 Percent Change in Soil Suction for Lateral Shifts between Drying and Wetting (or Median) SWCCs

Lateral Shift, %	Change (Reduction) in Suction, %
0.0	0.00
10.0	20.57
20.0	30.90
25.0	43.77
30.0	49.88
40.0	60.19
50.0	68.38
60.0	74.88
70.0	80.05
75.0	82.22
80.0	84.15
90.0	87.41
100.0	90.00
120.0	93.69
150.0	96.84

4.6.2.4 Approximate Shift and Estimation Procedure When Using the Drying and Wetting SWCCs

It is possible to estimate the maximum suction, the minimum suction, and the likely median suction for soil suction if the drying and wetting SWCCs are known. Pham (2002) analyzed the lateral shift between the drying and wetting SWCCs for

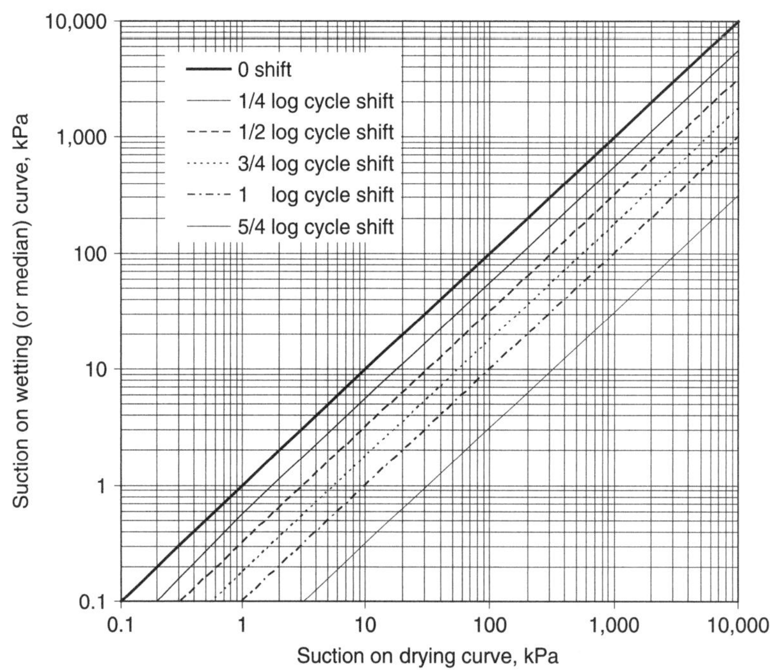

Figure 4.103 Relationship between suction on drying curve and suction on any other congruent SWCC.

Table 4.17 Suggested Lateral Shifts of Inflection Point between Drying and Wetting Curves for Various Soils

Soil Type	Range of Typical Shifts (% of log cycle)	Average Shift (% of log cycle)
Sand	15–35%	25%
Silt and loam	35–60%	50%
Clay	—	Up to 100%

published data from various researchers. Experimental results showed that the largest shift between the drying and wetting curves occurred for clay soils and the smallest shift occurred for uniform sand soils. Table 4.17 summarizes typical (average) lateral shifts at the inflection point of the SWCC for various soils (Pham et al., 2002, 2003a).

Let us assume that the soil under consideration is sand with the a parameter on the drying curve equal to 10.0 kPa. Let us also select a lateral shift of 25% of a log cycle to move to the wetting curve. The a parameter for the wetting SWCC can be calculated using Eq. 4.55:

$$0.25 = \log(10) - \log(\psi_{aw}) \quad (4.55)$$

Therefore, the suction corresponding to the a parameter at the inflection point on the wetting curve is

$$\psi_{aw} = 10^{0.75} = 5.62 \quad (4.56)$$

The suction at the inflection points went from 10 kPa on the drying curve to 5.62 kPa on the wetting curve. The suction at the inflection point on the median SWCC can be calculated by using a lateral shift of 12.5% (i.e., 0.125). The n fitting parameter remains the same for all SWCCs because of congruency.

Three soil suction values can be computed for any measured water content. The following format can be used to represent soil suction values corresponding to the measured water content:

$$\text{Wetting curve suction} < \text{median curve suction} < \text{drying curve suction} \quad (4.57)$$

As an example, a water content corresponding to the inflection points would have soil suctions written as 5.62 < 7.5 < 10.0 and read as

Most likely or median suction = 7.5 kPa,
Maximum estimated suction = 10.0 kPa,
Minimum estimated suction = 5.62 kPa, and
Estimated range of suctions = 5.62 – 10 kPa.

The above format provides a representation of the range and central tendencies that can be anticipated. The geotechnical engineer should view all three suction values and ask, "Does an understanding of the suction range and central tendencies assist me in making engineering judgements?" In some case, the answer may be "yes" while in other cases it may be "no."

4.6.2.5 Example Problem to Illustrate Typical Values and Ranges of Values from the Proposed Procedure to Obtain Soil Suction

Let us assume the soil parameters shown in Table 4.18 for sand soil, silt, and clay soil. The Fredlund and Xing (1994) equation is used to illustrate typical differences between various soil classifications.

Drying, wetting, and median SWCCs are computed for three different soil types. The soil parameters are typical values for a sand, silt, and clay but should not be taken as fixed values for each soil classification. Figure 4.104 illustrates the relationship among the three SWCCs (i.e., drying curve, median curve, and wetting curve) for a soil with a lateral shift of 25%. The n parameter was set to 4, which represents rather steep SWCCs typical of uniform sand.

Figures 4.105 and 4.106 illustrate the relationship among the three SWCCs for soils with a lateral shift of 50 and 100%, respectively. The n parameters were reduced to 2.0 and 1.5, respectively. All three sets of SWCCs are shown in Fig. 4.107 to show the wide range of SWCCs that can occur for various soils.

4.6.2.6 Suggestions and Recommendations for Using the SWCC to Estimate In Situ Soil Suction in Geotechnical Engineering Practice

The following guidelines are suggested for usage of the SWCCs for the estimation of in situ suction: (1) The

Table 4.18 Soil Properties Associated with Drying and Wetting SWCCs for Three Soils

Soil Type	Saturated Water Content	a_d Drying SWCC	n_d Drying SWCC	% Shift	a_w Wetting SWCC	a_m Median SWCC
Sand	30.0	10.0	4.0	25	5.623	7.500
Silt	40.0	200	2.0	50	63.25	112.5
Clay	60.0	3000	1.5	100	300.0	948.7

Note: SWCC variable $m = 1.0$ and correction factor $C(\psi) = 1.0$

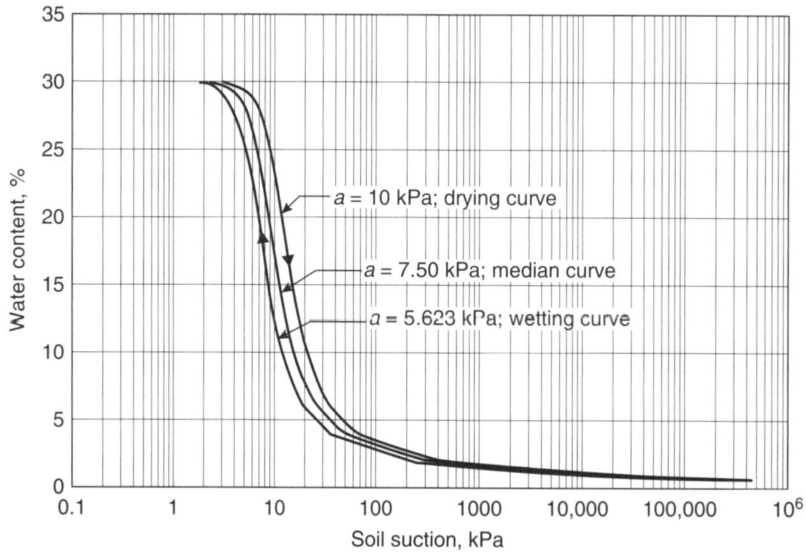

Figure 4.104 Comparison of drying curve, wetting curve, and median curve for soil with lateral shift of 25% between drying and wetting curves (e.g., sand) and saturated gravimetric water content of 30%.

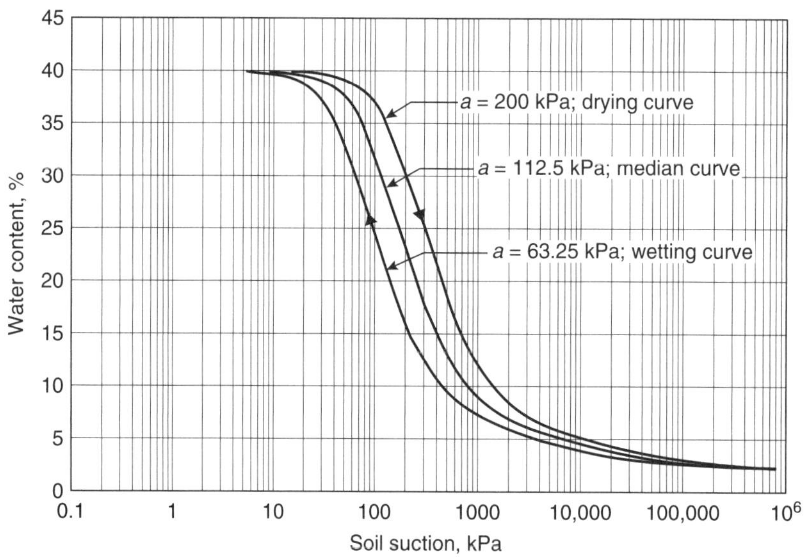

Figure 4.105 Comparison of drying curve, wetting curve, and median curve for soil with lateral shift of 50% between drying and wetting curves (e.g., silt) and saturated gravimetric water content of 40%.

desorption curve provides an estimate of the maximum value for in situ soil suction. (2) The estimated adsorption curve provides an estimate of the minimum value for in situ soil suction. (3) The estimated median curve provides an estimate closest to the most likely (or middle) in situ soil suction.

The percent error in the estimated soil suction can be referenced to the drying curve. Because of the logarithmic nature of the SWCCs, it is possible for the errors in the estimated soil suction to be quite large. The potential errors are much smaller for sand soils than for soils with high clay content.

The above analysis indicates that it is difficult to obtain an accurate indication of in situ soil suction through use of SWCCs and the measured natural water content of the soil. The proposed procedure merely allows the engineer to obtain a crude approximation of in situ suction conditions. The analysis provides a warning to geotechnical engineers who desire to indiscriminately use the SWCC to estimate in situ suction. The analysis also provides a guide to engineers who desire to assess the likely range of in situ suctions that might correspond to a single water content measurement. Users of the SWCCs should also be aware that in situ suctions in any soil can range between zero and 10^6 kPa.

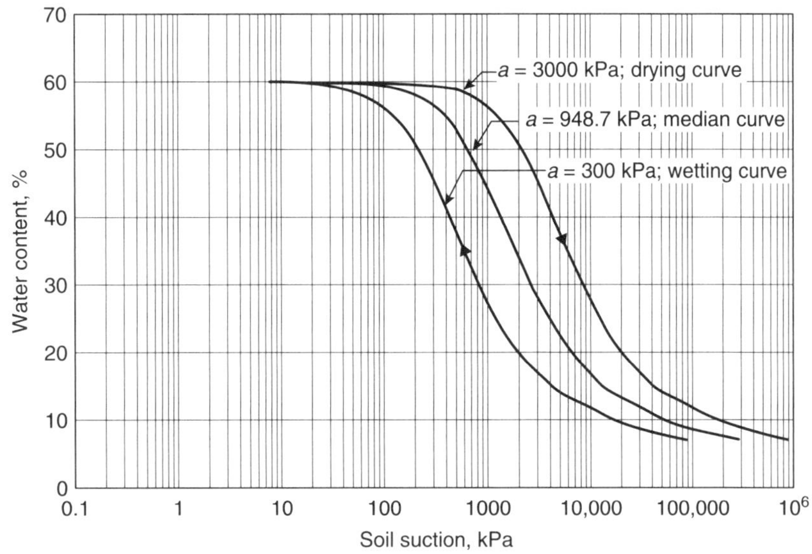

Figure 4.106 Comparison of drying curve, wetting curve, and median curve for soil with lateral shift of 100% between drying and wetting curves (e.g., clay) and saturated gravimetric water content of 60%.

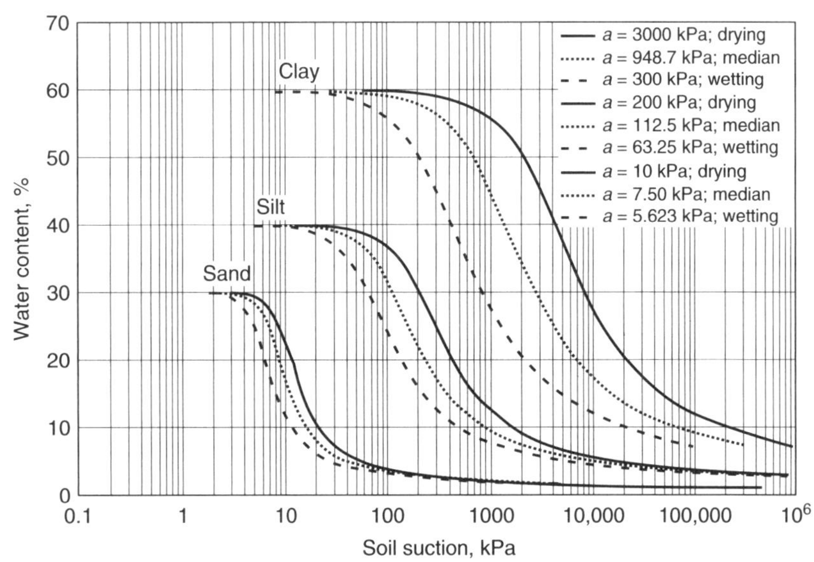

Figure 4.107 Comparison of three sets of drying curves, wetting curves, and median curves for wide range of possible soils.

4.6.3 Estimation of Total Suction

Total suction is equal to the sum of the matric suction and the osmotic suction. The osmotic suction is related to the salt content of the pore-water. The salts generally remain in the soil and as a result there is always an osmotic component of suction in the soil. Even when the soil is fully saturated with positive pore-water pressures, there is still an osmotic suction in the soil. Therefore, the minimum total suction in a soil is equal to the osmotic suction of the saturated soil.

The largest total suction in a soil is controlled by the minimum possible vapor pressure in the air next to the soil. The minimum vapor pressure corresponds to zero relative humidity and Lord Kelvin's equation indicates that the total suction goes to approximately 10^6 kPa. Lord Kelvin's equation can also provide an estimation of the total suction in any soil at equilibrium with a particular relative humidity environment. For example, if a soil was left in a relative humidity environment of 30% for a period of time, Lord Kelvin's equation can be used to compute the equilibrium total suction, which would be about 150,000 kPa. In other words, the vapor pressure in the air surrounding a soil becomes an equilibrium boundary condition for the soil.

4.6.4 Estimation of Osmotic Suction

The salt content of a soil from a particular geological stratum is usually found to be relatively constant, and as a result, the osmotic suction component is relatively constant over a considerable range of water contents. For example, the osmotic suction of glacial till deposits in western Canada is commonly on the order of about 200 kPa and the osmotic suction for Regina clay ranges from 150 to 200 kPa. Osmotic suction measurements are not a routine part of geotechnical engineering practice; rather, typical values are often obtained from special studies where the salt content has been measured.

The salt content of soils near ground surface can be altered as a result of excessive evaporation or excessive infiltration. The conditions associated with excessive evaporation of water at ground surface is a common occurrence. As water evaporates from the ground surface, the salts are left behind as a white deposit. The increased salts at ground surface have the effect of increasing the total suction and reducing the vapor pressure gradient between the soil and the atmosphere. As a result, moisture loss from the ground surface can be greatly reduced or "shut off." White salt deposits on the ground surface are an indication of an increased osmotic suction (and total suction) and a subsequent reduction in evaporation of water from the ground surface.

Special engineering studies might be encountered where the salt content in the soil is altered over time. Under these conditions it is osmotic suction that is the key variable being altered and attention needs to be given to the engineering significance of changing the salt content.

The concentration of salt in water can be designated in several ways. Table 4.19 provides some typical values for salt solutions found in bodies of water.

Osmotic suction can be estimated from the concentration of salts found in the water. The same theory applies whether the salt water is squeezed from the soil or exists in a large body of water. Osmotic suction can be designated in terms of the molarity of a salt solution through use of the van't Hoff equation (Campbell, 1985):

Table 4.19 Typical Salt Concentrations in Bodies of Water

Water Source	Concentration, s (g/g), %	Concentration, s_s (g/L)	Electrical Conductivity (mmhos/cm)
Distilled water	0.0	0.0	5.5×10^{-8}
Freshwater	0.02	0.2	0.005 – 0.05
Ocean water	3.5	36	48
Semisaline water	8.0	88	
Potash brine	20	250	
Dead-sea water	22	270	

$$\pi = \nu C R T_K \tag{4.58}$$

where:

π = osmotic suction, kPa,
R = universal (molar) gas constant [i.e., 8.31432 J/(mol K)],
T_K = absolute temperature, K (i.e., $T_K = 273.16 + T$, where T = temperature, °C),
C = molarity of the salt solution, and
ν = number of osmotic active particles per molecule (e.g., 2 for NaCl).

The amount of salt in a solution can vary over a wide range, and therefore, it is useful to plot both the molarity and the osmotic suction on a logarithm scale. Figure 4.108 shows

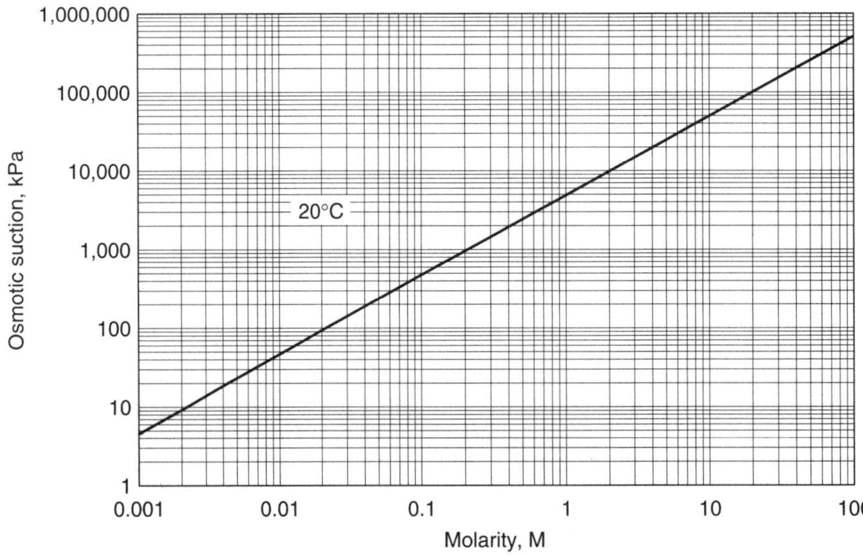

Figure 4.108 Osmotic suction versus molarity of NaCl solution.

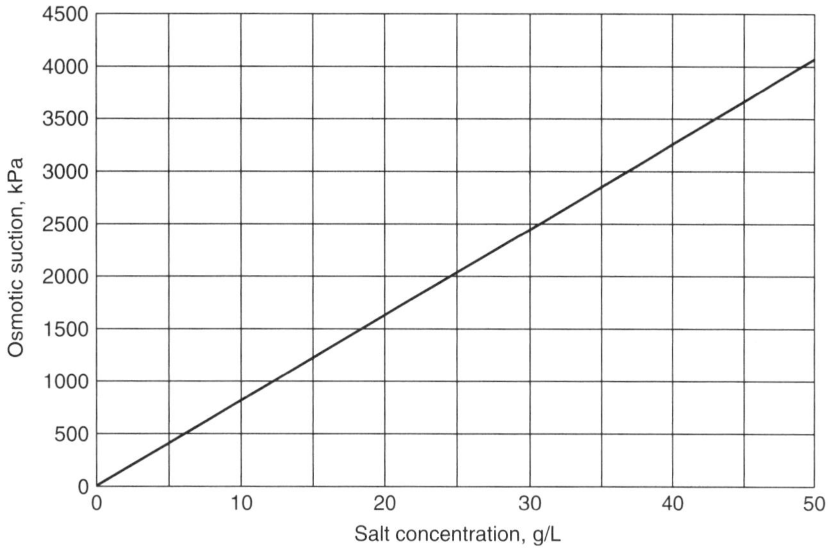

Figure 4.109 Osmotic suction versus salt concentration in grams per liter of NaCl.

the linear increase in osmotic suction with the molarity of a salt solution of NaCl ranging from 0.001 M to 100 M. The effect of temperature is small and only the computed results corresponding to 20°C are shown. The osmotic suction is about 4870 kPa at 1 M.

The molecular weight of a salt corresponds to the weight of Avagadro's number of atoms (i.e., 6.022×10^{23}). Let us consider the molecular weight of NaCl. Sodium, Na, has an atomic number of 11 and an atomic weight of 22.990 g. Chlorine, Cl, has an atomic number of 17 and an atomic weight of 35.453 g. Added together, the atomic weight of NaCl molecules is 58.443 g. One mole is equal to Avagadro's number of atoms. Therefore, the molecular weight is the weight of 1 mol of molecules. Or stated another way, the atomis weight of 1 mol of NaCl is 58.443 g.

Figure 4.109 shows osmotic suction plotted versus the salt concentration of NaCl in grams per liter. A salt solution of 10 g/L gives an osmotic suction of about 810 kPa. Comparing Figs. 4.108 and 4.109 shows that 58.4 g/L of NaCl is equal to a 1 M solution of NaCl.

CHAPTER 5

Soil-Water Characteristic Curves for Unsaturated Soils

5.1 INTRODUCTION

The SWCC provides a conceptual understanding between the mass (and/or volume) of water in a soil and the energy state of the water phase. The SWCC has proven to be an interpretive model that uses the elementary capillary model to provide an understanding of the distribution of water in the voids. The effects of soil texture, gradation and, void ratio have also become part of the interpretation of measured laboratory SWCC data.

The SWCCs have an important role in the determination of unsaturated soil property functions. The procedures that have been proposed for unsaturated soil properties are approximate but are generally satisfactory for analyzing unsaturated soil mechanics problems. General agreement has emerged within the geotechnical engineering profession with regard to procedures for the measurement and interpretation of SWCC data measured in the laboratory.

5.1.1 Background of Unsaturated Soil Mechanics

Early soil mechanics studies (e.g., 1930s and 1940s) focused mainly on the behavior of saturated soils with positive pore-water pressures. Common geotechnical engineering examples involved the application of soil mechanics to deep foundations and structures where positive pore-water pressures were involved. Over time, it became common practice to simply ignore negative pore-water pressures and assume that consideration of positive pore-water pressures would ensure a conservative and safe design. Many regions of development were located in humid regions where the groundwater table was high. Successes associated with the application of saturated soil mechanics further promoted the sole consideration of saturated soils. The ability to describe the stress state of saturated soils in terms of effective stress and the ability to model time-dependent behavior (e.g., the consolidation process) provided technical tools that were successful and financially rewarding.

Approximately one-third of the earth's surface is situated in arid and semiarid regions; however, this did not appear to provide sufficient impetus for the development of unsaturated soil mechanics. Rather, it was a shift toward a global economy and an emphasis on resource development that began to bring about consideration of a science-based understanding of unsaturated soil behavior. Many countries of the world also began to press to be better stewards of the environment. The design of containment facilities, cover designs for landfills and mine wastes, the reclamation of mining areas, and the remediation of spills were but a few examples of unsaturated soil problems that needed to be better understood. There was also a revisitation of compacted soil behavior as well as more detailed consideration of the behavior of residual soils, expansive soils, and collapsible soils. These soils had become known as "problematic soils," and it was clear that there was need for a theoretical framework that embraced all unsaturated soil behavior.

The progress toward establishing a scientific basis for unsaturated soil behavior took time for two main reasons. First, there was a lack of an appropriate science with a theoretical basis and, second, there was the lack of an appropriate technology to render engineering practice financially viable (Fredlund, 1979). It was the theoretically based, multiphase continuum mechanics analysis of an unsaturated soil system (Fredlund and Morgenstern, 1977) that provided the verification of stress state variables for an unsaturated soil. The analysis revealed that two independent stress state variables should be used when describing unsaturated soil behavior. Formulations that described each physical process associated with unsaturated soils were published. Fredlund and Morgenstern (1976) formulated volume change behavior in terms of independent stress state variables. Shear strength behavior was also formulated in terms of stress state variables (Fredlund et al., 1978). Papagiannakis et al. (1984) and Lam et al., (1987) published the formulations for steady-state and transient seepage analysis for unsaturated-saturated soil systems. Later, Wilson et al. (1994) published the solution of ground surface, climate-driven, moisture flux boundary condition problems.

Unsaturated soil mechanics emerged in the 1960s and 1970s but there needed to be an improved understanding of the linkages between theory, laboratory measurements, and methods for solving practical engineering problems. A more in-depth understanding of the SWCC provided a constitutive framework for combining the theory of unsaturated soil behavior and unsaturated soil properties. The SWCC provided the necessary linkages for the emergence of a science for unsaturated soil mechanics along with economical procedures for the implementation of unsaturated soil mechanics.

5.1.2 Early Developments on the SWCC in Soil Physics

Much of the theory and behavioral understanding of the SWCC has arisen out of research in agricultural-related disciplines (e.g., soil science, soil physics, and agronomy). It is important to acknowledge the important contributions made in agriculture-related disciplines to unsaturated soil mechanics. At the same time, it was important that there be a reinterpretation of scientific findings to better fit geotechnical engineering practice. Barbour (1998) provided a review of the early understanding of the SWCC within soil physics. Numerous researchers made significant contributions to our understanding of the SWCC. Schumacher and Wollny, considered as the fathers of soil physics in Europe, made significant contributions in the 1800s to the understanding of soil moisture and its movement in soils. In North America, King (1899) and Slichter (1899) were forerunners in soil physics while working for the U.S. Geological Survey.

Slichter (1899) developed a model for pore spaces in a soil through use of a series of spheres packed together in various arrangements. The geometry of the pore spaces was then calculated. Briggs (1897) developed a centrifuge method for determining the "moisture equivalent" of soils while working for the Bureau of Soils of the U.S. Department of Agriculture. An attempt was made to classify soils based on the measured moisture equivalent. Briggs (1897) stated that "the storage capability of soil is due to surface tension within films of water around numerous capillaries." Water in a soil was classified on the basis of the mechanism by which the water was held in the pores. Hygroscopic water was assumed to be held on soil particles by attractive forces. Capillary water was assumed to be held by surface tension forces. Gravitational water was assumed to be water that could freely drain from the soil because it was not held by the soil particles.

Haines (1927) further developed the concept of the air-water interface introduced by Briggs (1897). The air-water film or contractile skin was viewed as being wrapped around the soil particles, thereby applying increased stress between the particles. Consideration of the air-water interface led to the use of a capillary model to describe soil behavior. The capillary model was compared to a series of glass tubes of varying diameters. Water would rise to differing heights in the capillary tubes depending upon their respective diameters. The capillary model was further expanded to include individual tubes with varying diameter, thereby providing an explanation of hysteresis associated with wetting and drying processes.

The capillary model provided a mathematical relationship between the radius of curvature of the air-water meniscus and the pressure difference between the air and water phases. The difference between the air pressure u_a and the water pressure u_w was shown to be balanced by the surface tension T_s acting at the wetting angle α along the solid surface. Terzaghi (1943) and Taylor (1948) used the capillary model to explain the differences in physical behavior between saturated and unsaturated soils. The capillary model assisted in explaining some aspects of unsaturated soil behavior but was later found to create serious limitations in the general application to unsaturated soil mechanics (Fredlund and Rahardjo, 1993a). The capillary model was a micromechanics-type model that could not be directly incorporated into a macrocontinuum mechanics framework for solving practical geotechnical engineering problems.

Buckingham (1907) is credited with introducing the concepts of "capillary potential" and "capillary conductivity." These concepts were based on the ability of an unsaturated soil to transmit and store water. Buckingham emphasized the "energy" or "potential" of water in the soil. Water could be stored and it was also assumed that it could flow in response to a capillary potential. Buckingham (1907) held the view that "the energy of the water in soil (was) representative of its ability to hold water." Buckingham (1907) held a phenomenological view of total potential suggesting that soil-water could have various components. One component of energy was that due to mechanical stresses across the air-water interface and this led to the idea of capillary suction or matric suction, which was defined as the difference between the pore-air, u_a, and pore-water, u_w, pressures $(u_a - u_w)$. Energy due to the chemistry of the soil-water system was considered as an adsorption component or the osmotic suction, π. The sum of the matric suction component and the osmotic suction component was referred to as the total energy or the total potential, ψ, of the pore-water. Aitchison (1964) made use of the concepts of Buckingham (1907) when introducing soil suction into geotechnical engineering.

The concept of water having potential energy or water potential was useful, but there appears to have been some misuse of the physics associated with various processes (Barbour, 1998). For example, it was suggested that components of water potential were additive in a linear manner. Some of the reported components of water potential were (i) gravitational potential, (ii) osmotic potential, (iii) vapor potential, (iv) matric potential, (v) hydrostatic pressure potential, and (vi) overburden pressure potential, as well as other potentials. In soil mechanics, the flow of water through a saturated soil has been strictly related to hydraulic head, which was a combination of a pressure head and an elevation (or gravity) head. These two components were linearly combined and operative in the vertical direction. On the other hand, air flow, which is part of matric suction, was considered to experience flow in response to its own pressure gradient and was independent of water flow.

The chemical potentials could also produce independent processes.

The contributions of soil physics have been of significant value in the development of unsaturated soil mechanics. It has also been necessary to ensure that the successful contextual framework for saturated soil mechanics be retained when embracing unsaturated soil behavior. In general, a parallelism between saturated soil mechanics and unsaturated soil mechanics has been maintained. Unsaturated soil mechanics is largely taught as an extension of the concepts fundamental to saturated soil mechanics. The SWCC has played an important role in extending saturated soil mechanics to embrace unsaturated soil behavior.

5.1.3 Early Equipment for Measuring the SWCC

Buckingham (1907) measured SWCCs for a variety of soils. Long cylinders were filled with soil and water was allowed to come to equilibrium with a water reservoir at the base of the column. The gravitational potential energy within each column of soil was calculated based on the distance above the free-water surface in the reservoir and was referred to as the "capillary potential." Figure 5.1 shows two SWCCs plotted as continuous functional relationships between capillary potential and water content.

Richards (1928) placed thin soil specimens on a high-air-entry ceramic disk that was connected to an underlying water reservoir. A vacuum tank was used to place the water in the underlying water reservoir under tension. The soil specimens were subjected to a negative water pressure head until equilibrium was established between the soil and the pressure in the water reservoir. The water content of the soil was measured once equilibrium conditions were established. The equipment became known as the *Richards pressure plate* apparatus (Richards, 1928).

Haines (1930) used a Buchner funnel with a high-air-entry disk sealed in the base of the funnel to measure the SWCC. A tensile stress was applied to the underlying water reservoir and the soil was allowed to come to equilibrium with the applied negative water pressure. The maximum tensile stress that could be applied to the soil-water was somewhat less than 1 atm.

Valle-Rodas (1944) performed open-tube and capillarimeter tests on uniform sands. Various sand particle size ranges were separated using sieves. Each material with a limited range of grain sizes was placed in the open tube and the capillarimeter. The bottom end of the sand in the open tube was immersed in water. Experimental results showed a gradual change in the water content in the sand with distance above the water reservoir (Fig. 5.2). Further tests were performed that illustrated hysteretic behavoir in the water content versus matric suction relationship during wetting and drying. Other open-tube and capillarimeter tests were also performed on cohesionless soils ranging from silt sizes to gravel sizes by Lane and Washburn (1946).

Richards (1928) also allowed soil samples to establish equilibrium water content conditions with the water vapor above a saturated salt solution (i.e., a fixed relative humidity environment). The vapor pressures established above the salt solutions could be computed using the Lord Kelvin energy equation.

5.1.4 Early Conceptual Models of Flow in Unsaturated Soils

In 1931 Richards derived a water flow equation for unsaturated soils. The flow equation was based on capillary conduction and water storage was assumed to be a function of capillary potential. Capillary potential was perceived to be equivalent to the negative pore-water pressure in the soil.

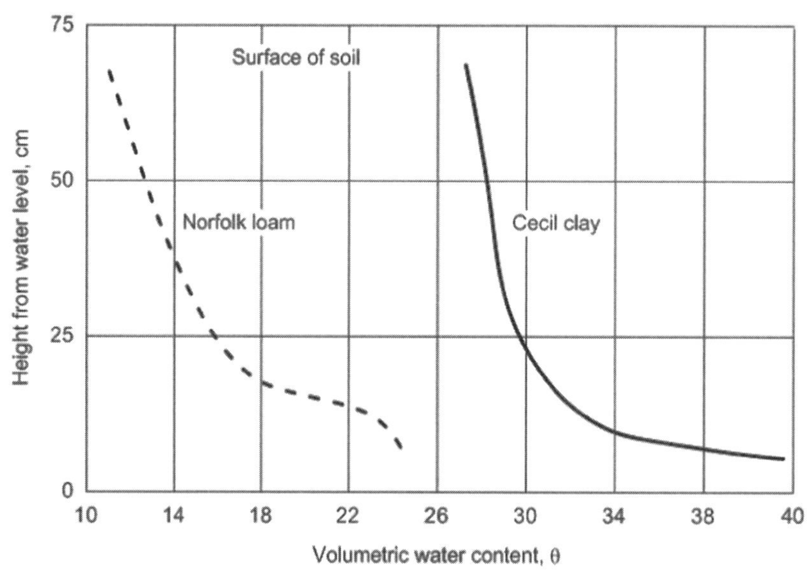

Figure 5.1 Early measurements of water content versus matric suction on two soils (after Buckingham, 1907).

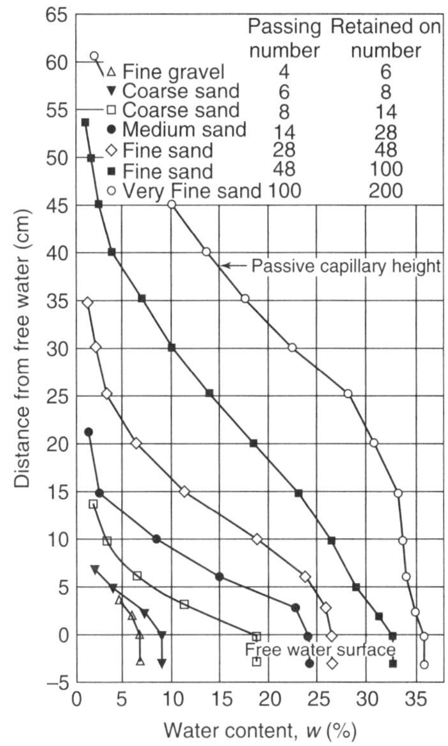

Figure 5.2 Distribution of capillary water in sands of varying gradation (after Valle-Rodas, 1944).

The capillary model visualized water flow as occurring through a series of capillary tubes. The air-entry value of the soil corresponded to the difference between the air and water pressure at which the largest diameter capillary drained. Childs (1940) and Childs and Collis-George (1948, 1950) expanded the capillary model to represent a random distribution of pore sizes to represent the SWCC. Childs (1940) suggested that there was a relationship between the SWCC and the coefficient of permeability for an unsaturated soil. The SWCC was assumed to contain valuable information on the soil structure and the pore size geometry and distribution. Childs (1940) noted, "Use of the soil moisture characteristics in this way depends on the fact that these characteristics may be interpreted as showing the pore size distribution within the soil. Thus they play a part analogous and complementary to mechanics analyses: they give the same sort of information about the pores as that given by mechanical analysis about the particles."

Combining the capillary model and the SWCC led to the definition of an effective pore-size diameter. The water was viewed as being held within the smaller diameter pores at specific soil suction values. The slope of the SWCC represented the water-filled pores of the soil. An illustration of pore-size distribution interpreted from the SWCC for a sand sample and an aggregated clay sample is shown in Fig. 5.3. The void spaces can be seen to be similar between the sand and the aggregated clay soil.

Early research in soil physics provided a conceptual model showing the relationship between the amount of water in the soil and the energy within the pore-water (Barbour, 1998). The relationship formed a continuous function from saturated soil conditions to completely dry soil conditions. Principles of thermodynamic equilibrium between the pore-water and a reservoir of water at a known energy state were used to measure the stress state in the water phase. The energy concept associated with the water phase provided a useful model for describing unsaturated soil behavior. The capillary tube model was easy to visualize, but it was a microtype model that proved to have serious limitations and apparent anomalies when used to interpret certain unsaturated soil phenomena (Fredlund and Rahardjo, 1993a).

5.1.5 The SWCC in Early Soil Mechanics

Some of the earliest soil mechanics articles by Karl Terzaghi and other soil mechanics leaders contained discussions on unsaturated soil behavior in engineering practice. Terzaghi (1942) stated, "If the groundwater were subject to no forces other than gravity there would be a sharp boundary between dry and saturated soil. This boundary would be located at the level at which water rises in observation wells. Below this surface, which is called the watertable or the phreatic

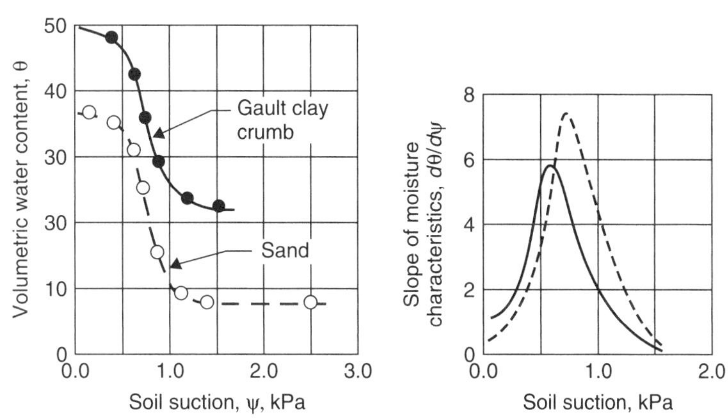

Figure 5.3 SWCCs and pore-size distribution curves for two soils (after Child, 1940).

surface, the water is acted upon solely by the force of gravity and behaves according to the laws of groundwater flow."

Soil is capable of permanently retaining considerable quantities of water within its voids above the phreatic surface. The permanent presence of water within this zone supports the existence of forces that counteract the mechanical effect of the force of gravity. These forces are referred to as capillary forces.

Terzaghi (1942) used the complexities of the SWCC to explain the relationship between capillarity and negative pore-water pressure or soil suction. Terzaghi's writings made it clear that water in the pores of a soil above the phreatic line could participate in groundwater flow. Water was shown to rise through capillary action to heights greater than 9 m.

The First International Conference on Soil Mechanics and Foundation Engineering (1936) contained research papers that discussed a wide range of unsaturated soil mechanics problems. Vreedenburgh and Stevens (1936) studied unsaturated soil seepage problems and explained that the coefficient of permeability was a function of the degree of saturation of the soil. Van Mourik Broekman and Buisman (1936) noted that negative pore-water pressures played an important role in the stability of slopes. The early developments in soil mechanics appear to have included consideration of soils in the unsaturated zone. Unfortunately, many of the early findings seemed to become lost during subsequent decades. Terzaghi (1943) devoted significant portions of two chapters of his book to unsaturated soil behavior. Later textbooks on soil mechanics gave less attention to the behavior of unsaturated soils.

Terzaghi was well aware of the research work undertaken in soil physics and often quoted from their research findings. Terzaghi recognized the nature of the SWCC and its value in understanding the permeability function for an unsaturated soil. His perception was that the uppermost part of the moistened zone is in a state of partial capillary saturation. The rate of flow of the water appeared to be determined by capillary potential which was a function of the degree of saturation of the sand (Baver, 1940). However, the methods of computation which are based on the concept of the capillary potential had not yet reached a stage such that they could be applied to the solution of engineering problems. Simplifications were made for solving seepage problems. For example, the saturated coefficient of permeability was used as long as the soils were within the capillary fringe zone. In other cases, the phreatic surface was assumed to form the uppermost boundary of flow, as was the case for flownet solutions proposed by Casagrande (1937).

Lambe (1958b) attempted to find a single soil property for flow in the negative pore-water pressure zone. "Capillary head" was used to describe "draining" and "wetting" conditions. Geotechnical engineers attempted to further understand and apply water flow theories to unsaturated soils during the 1950s and 1960s. A common misconception appears to have prevailed which assumed that water could only flow in saturated soils within the "capillary fringe" and in the positive pore-water pressure range (Barbour, 1998).

The 1960 Conference on Pore Pressure and Suction in Soils witnessed a difference of opinion with respect to formulating mathematical equations for solving unsaturated soil problems within the engineering profession. British engineers attempted to explain unsaturated soil behavior through use of a single-stress-state variable. Australian engineers emphasized an approach that advocated the use of a continuous mathematical function to represent the SWCC. The Australian approach found its basis in earlier research in soil physics (Aitchison, 1961). The Australians felt that the historical research in soil physics had much to contribute toward establishing an acceptable fundamental, contextual framework for solving unsaturated soil problems in geotechnical engineering.

Gardner (1961) combined the concepts of water potential and its relationship to the coefficient of permeability for an unsaturated soil. This gave rise to the application of a continuous function for describing seepage through unsaturated soils. The result was a nonlinear analysis that would need to wait for the development of the computer in order to solve highly nonlinear engineering problems. The finite element numerical methodology would later utilize continuous permeability functions to solve steady-state seepage formulations (Freeze, 1971; Papagiannakis and Fredlund, 1984). Water storage functions were added shortly thereafter for solving transient seepage analyses (Lam et al., 1987). The computational power of the computer provided a powerful tool for the solution of geotechnical engineering problems that contained nonlinear soil property functions.

In 1977, Fredlund and Morgenstern began to establish a common theoretical stress state basis for describing and formulating geotechnical engineering problems involving unsaturated soils. A macroscopic phenomenological approach based on principles of multiphase continuum mechanics was advocated and used to provide a theoretical basis for defining the stress state variables for an unsaturated soil. The stress state variables formed the basis for the formulation of volume-mass, shear strength, and water flow constitutive relations. In the 1980s and 1990s, the SWCC would become the basis for the estimation of nonlinear unsaturated soil property functions for all types of geotechnical engineering problems.

5.1.6 Need for Unsaturated Soil Property Functions

The SWCC has had an important role to play in the implementation of unsaturated soil mechanics (D.G. Fredlund, 2002a). The SWCC was initially viewed as a means of estimating in situ soil suctions by measuring the natural water content and using the SWCC as a fixed relationship between suction and water content. Laboratory data showed that the drying and wetting SWCCs were significantly different (i.e., hysteretic), giving rise to a wide range of possible soil suction values. The SWCC later proved to be of significant value for the estimation of unsaturated soil property functions (M.D. Fredlund, 2000).

One of the obstacles hindering the implementation of unsaturated soil mechanics was the excessive cost and demanding laboratory testing techniques associated with the experimental measurement of unsaturated soil properties. Nonlinear soil properties meant that it was necessary to make a series of soil property measurements on a particular soil specimen. In other words, the unsaturated soil properties needed to be measured under various controlled stress state conditions.

High-air-entry ceramic disks have played an important role in testing unsaturated soils for more than 50 years and still remain the basic means of separating pore-air and pore-water pressures (i.e., axis translation technique). High-air-entry disks have two primary limitations: an extremely low coefficient of permeability and a gradual diffusion of air through the water phase. The low coefficient of permeability places a limit on the rate of pore pressure response as well as the impeded release of water from the soil. The high-air-entry disk places a restriction on the largest coefficient of permeability that can be measured in the laboratory. The air-entry value of the ceramic disk also limits the maximum matric suction that can be applied.

The measurement of unsaturated soil property functions becomes unacceptably costly for routine geotechnical engineering purposes. Consequently, the need arises for practical solutions for the determination of unsaturated soil property functions. SWCCs have emerged as a practical and sufficiently accurate tool for the estimation of unsaturated soil property functions for most geotechnical engineering problems. The SWCC has become viewed as the "key" to the implementation of unsaturated soil mechanics in engineering practice (D.G. Fredlund, 2002a). For this reason, an entire chapter is devoted to better understanding the theory related to the SWCC as well as procedures that can be used for its measurement and estimation. Each subsequent chapter will show how the SWCC can be used for the estimation of a variety of unsaturated soil property functions.

5.1.7 Terminology and Definitions

There have been a number of terms that have been used to describe the relationship between the amount of water in the soil and soil suction. The variation in terminology is somewhat related to the variety of disciplines that have made use of the water content–soil suction relationship. An attempt has been made to establish a level of consistency with regard to the terminology used within the disciplines of geotechnical engineering. Some of the terms used when referring to the relationship between the amount of water in the soil and soil suction are as follows: (i) SWCC, (ii) suction–water content relationship, (iii) retention curves, (iv) moisture retention curves, (v) soil moisture retention curves, and (vi) water retention curves (WRCs). These terms can be independently applied to the drying and wetting water content–soil suction curves.

The term soil-water characteristic curve has been selected as the preferred terminology for this book on geotechnical engineering with unsaturated soils. The preference is to use "soil-water" because it is the water content of the soil that is measured. The preference is to use the word "characteristic" simply because it appears to have been the most common term historically used in engineering. In addition, the term implies that the curve describes some characteristic of the soil. The word "retention" is more closely related to retaining water for plant growth in agriculture or the retention of water in materials other than soil (e.g., mine tailings).

SWCCs are pivotal to the solution of unsaturated soil problems. SWCCs constitute one of the bounding relationships of the volume-mass constitutive properties for an unsaturated soil (Pham, 2005). Figure 5.4 shows the volume-mass constitutive surfaces for a clay soil. The degree of saturation–soil suction relationship provides a definitive value for the air-entry value for a clay soil that may undergo considerable volume change prior to desaturation. If the volume change of the soil is small, it is possible to use the gravimetric water content versus soil suction relationship to determine the air-entry value and residual water content for a particular soil. The volume-mass constitutive surfaces for a sand soil are illustrated in Fig. 5.5 (Pham, 2005). Gravimetric water content does not require a volume measurement in the laboratory whereas degree of saturation and volumetric water content require that the volume of the soil specimen be known. It is, however, the volumetric water content that is most directly required to represent the water stored in an unsaturated soil and therefore consideration of the overall volume of the soil is important in the determination of the SWCC.

Unsaturated soil properties are a function of all volume-mass properties. The gravimetric water content versus soil suction relationship is the easiest relationship to measure, and for many applications (e.g., sands), it is an adequate representation of the SWCC. Various measures of the amount of water in the soil can be used in the estimation of the unsaturated soil properties if the soil exhibits little volume change as soil suction is changed. On the other hand, if the soil exhibits substantial volume change when soil suction is changed, then greater care must be exercised when establishing key variables associated with unsaturated soil behavior.

Water storage is always measured as the slope of the volumetric water content versus soil suction relationship. Most permeability functions assume that it is the change in degree of saturation that initiates the calculation of a significant decrease in the coefficient of permeability. Figure 5.6 shows a typical SWCC with its three distinct zones of desaturation. The reliability of estimated unsaturated soil property functions is quite closely related to the accuracy with which the SWCC is defined.

The key transition points on the SWCC are the air-entry value and the residual value for suction and water content. These transition points are defined on the degree of

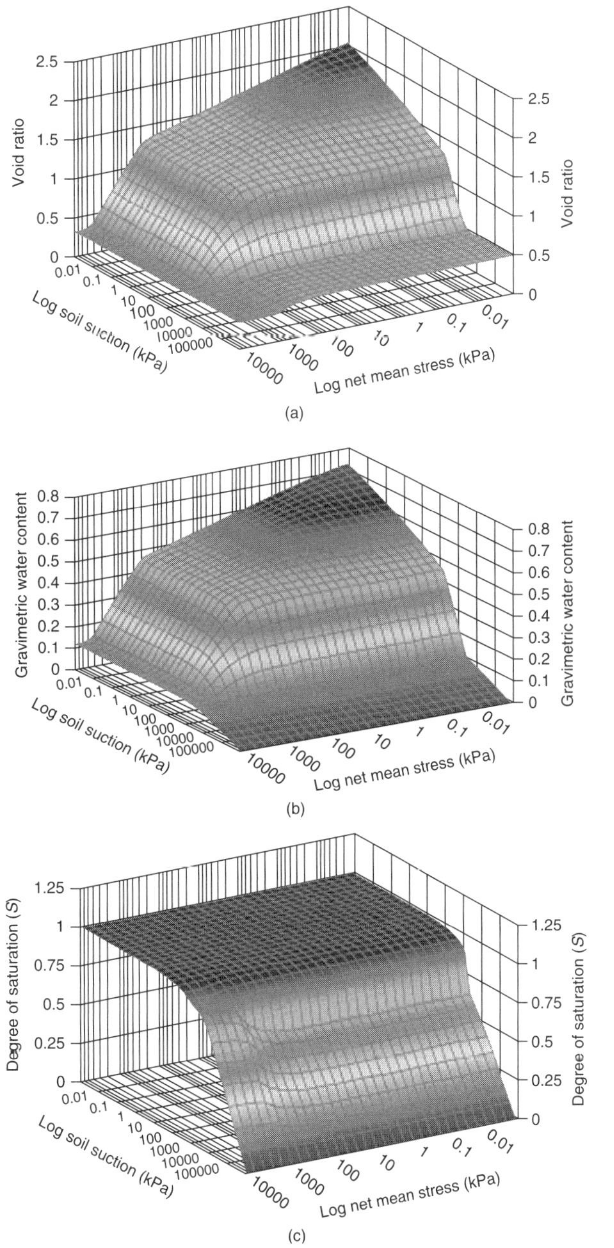

Figure 5.4 Volume-mass constitutive surfaces for highly plastic clay preconsolidated to 200 kPa (from Pham, 2005a): (a) void ratio constitutive surface, (b) gravimetric water content constitutive surface, and (c) degree-of-saturation constitutive surface.

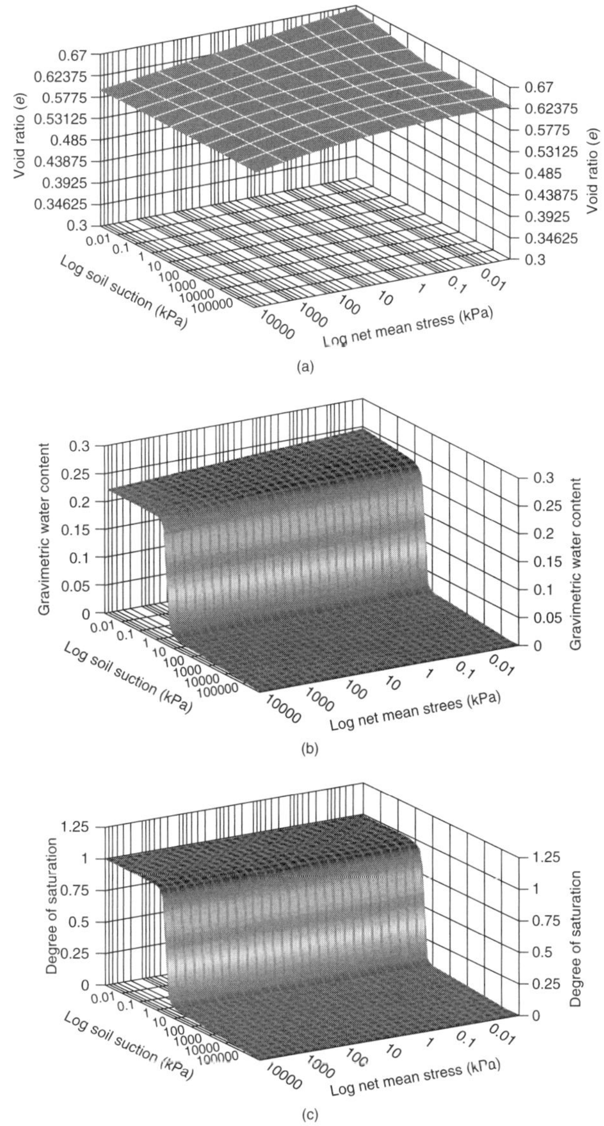

Figure 5.5 Volume-mass constitutive surfaces for sand soil (from Pham, 2005a): (a) void ratio constitutive surface, (b) gravimetric water content constitutive surface, and (c) degree-of-saturation constitutive surface.

saturation versus soil suction plot and subdivide the SWCC into the "boundary effect" zone, the "transition" zone, and the "residual" zone. The same three zones of desaturation can be defined for the drying (or desorption) branch and the wetting (or adsorption) branch.

5.2 VOLUME-MASS CONSTITUTIVE RELATIONS

The SWCC constitutes the primary soil information required for the analysis of seepage, shear strength, volume change, air flow, and heat flow problems involving unsaturated soils (Fredlund, 1997a; M.D. Fredlund, 2000). The permeability function and the water storage function for an unsaturated soil are related to the SWCC. In each case, it is important that the proper form of the SWCC be used to estimate the unsaturated soil properties. The form of the SWCC that should be used depends upon whether or not the soil undergoes volume change as soil suction is changed.

The unsaturated soil property functions are estimated using the SWCC and the saturated soil properties. The SWCC becomes the single most valuable piece of soil information for geotechnical engineering practice involving unsaturated soils. It is necessary that the SWCC be properly estimated (or measured) and interpreted. The SWCC can be

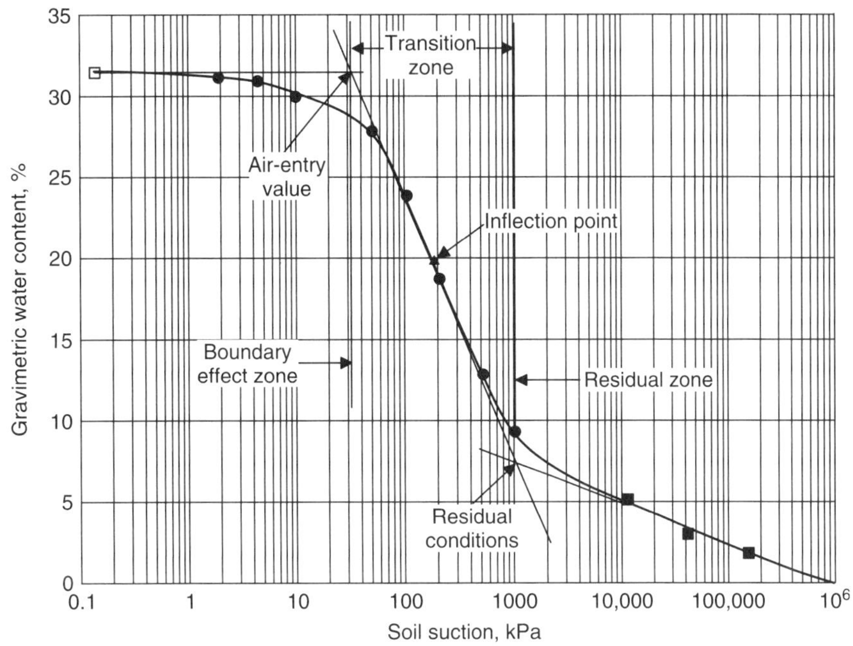

Figure 5.6 Typical desorption SWCC showing distinct zones of desaturation.

viewed as a part of the water content constitutive surface for the overall volume-mass constitutive relationships, as shown in Figs. 5.4 and 5.5.

The SWCC is broadly defined as the relationship between the amount of water in a soil and soil suction. This simplistic definition means that the term SWCC refers to a class of relationships and immediately gives rise to a number of questions:

1. How should the amount of water in the soil be quantified?
2. Are different ways of quantifying the amount of water used for different estimation procedures?
3. What is the influence of deformation (or overall volume change) on the interpretation of the data for geotechnical engineering applications?
4. What influence does "initialization of stresses" at the start of the SWCC test have on the interpretation of test data? For example, the test results from an initial slurry soil specimen is quite different from a compacted or undisturbed soil specimen.
5. What component(s) of soil suction are plotted on the SWCC?
6. How is the hysteresis of the SWCC taken into account for various geotechnical engineering applications?
7. What is the primary SWCC information that is used when solving geotechnical engineering problems?

5.2.1 Designation of Amount of Water in Soil

The SWCC defines the amount of water in a soil versus soil suction. The amount of water in the soil can be defined using more than one variable. Variables used to define the amount of water in the soil are (i) gravimetric water content w, (ii) volumetric water content θ, (iii) degree of saturation S, and (iv) volume of water, V_w, referenced to the original volume of the specimen, V_0 (i.e., V_w/V_0).

Gravimetric water content w is the most common term used in geotechnical engineering and is defined as follows:

$$w = \frac{M_w}{M_s} \tag{5.1}$$

where:

w = gravimetric water content,
M_w = mass of water, and
M_s = mass of soil solids.

Volumetric water content θ is commonly used in agriculture-related disciplines and relates the amount of water in the soil to the total volume of the soil:

$$\theta = \frac{V_w}{V_v + V_s} \tag{5.2}$$

where:

θ = volumetric water content,
V_w = volume of water,
V_v = volume of voids, and
V_s = volume of solids.

It is difficult to measure the total volume of the soil specimen when testing an unsaturated soil. It is easier to reference the amount of water in the soil to the original volume of the soil specimen at the start of the test. Using the original volume of the specimen as a reference gives rise to another definition for volumetric water content. The reference volume for defining volumetric water content becomes important in situations where the soil specimen undergoes significant volume change as soil suction is changed.

The degree of saturation S references the volume of water to the instantaneous volume of voids and therefore requires a measurement of the instantaneous total volume of the soil specimen:

$$S = \frac{V_w}{V_v} \tag{5.3}$$

where:

S = degree of saturation.

Gravimetric water content is referenced to a mass of soil solids while volumetric water content and degree of saturation are referenced to volumes. Volumetric water content and degree of saturation are referenced to instantaneous volumes; however, these variables have often been referenced back to the original volume with the assumption that overall soil specimen volume changes are negligible.

All three designations of the amount of water in a soil are the same if the volume of the soil specimen remains constant as soil suction increases. If the soil specimen is essentially nondeformable (i.e., $V_v = V_{v0}$, where V_{v0} is the original volume of voids), the measured amount of water in the soil is referenced to a constant value. Only the gravimetric water content is referenced to a constant value if the soil structure is deformable.

Gravimetric water content has been the preferred designation for the amount of water in a soil since the inception of soil mechanics in the 1930s. Agriculture-related disciplines have mainly used volumetric water content. The amount of water in the soil is generally referenced to the original soil specimen volume and overall volume changes during a test were not a major concern in the interpretation and application of the test results. Each of the terms used to represent the amount of water in a soil has advantages and disadvantages when used in soil mechanics (see Table 5.1).

There are other designations of water content that have also been used in unsaturated soil mechanics when describing unsaturated soil property functions (D.G. Fredlund, 2002a). Dimensionless gravimetric water content Θ_{dg} is defined as any gravimetric water content divided by the original saturated water content:

$$\Theta_{dg} = \frac{w}{w_s} \tag{5.4}$$

where:

w = any gravimetric water content and
w_s = gravimetric water content at saturation

The saturated water content corresponds to a somewhat arbitrary value where the soil is allowed to imbibe water and

Table 5.1 Advantages and Disadvantages of Various Designations for Amount of Water in Soil

Designation	Advantages	Disadvantages
Gravimetric water content, w	• Consistent with usage in classic soil mechanics • Most common means of measurement • Does not require a volume measurement • Reference value is a “mass of soil” which remains constant	• Does not allow differentiation between change in volume and change in degree of saturation • Does not yield the correct air-entry value when the soil changes volume upon drying
Volumetric water content, θ	• Is the basic form that emerges in the derivation of transient seepage and fluid storage in unsaturated soils • Commonly used in databases of results obtained in soil science and agronomy	• Requires a volume measurement • Rigorous definition requires a volume measurement at each soil suction • Is the designation least familiar and least used historically in geotechnical engineering
Degree of saturation, S	• Most clearly defines the air-entry value • Appears to be the variable most closely controlling unsaturated soil property functions	• Requires a volume measurement • Although volume measurements are required, the degree of saturation variable does not quantify overall volume change

approach saturated conditions under near-zero confinement. Dimensionless gravimetric water content can be written in terms of the individual measured mass values as follows:

$$\Theta_{dg} = \frac{w}{w_s} = \frac{M_w/M_s}{M_{w0}/M_s} \tag{5.5}$$

where:

M_{w0} = mass of water in the saturated soil at the start of the test,
M_s = mass of the soil solids, and
M_w = mass of water at any point under consideration.

Dividing the top and bottom of Eq. 5.5 by the unit weight of water, γ_w, allows dimensionless water content to be written in terms of volumes:

$$\Theta_{dg} = \frac{M_w/\gamma_w}{M_{w0}/\gamma_w} = \frac{V_w}{V_{w0}} \tag{5.6}$$

where:

V_{w0} = volume of water in the saturated soil at the start of the test.

Dimensionless volumetric water content Θ_{dv} can be written in equation form by dividing any water content by the volumetric water content corresponding to initial (saturated) conditions:

$$\Theta_{dv} = \frac{\theta}{\theta_0} = \frac{V_w/(V_v + V_s)}{V_{w0}/(V_v + V_s)} \tag{5.7}$$

Dimensionless gravimetric and volumetric water contents can be used in either decimal or percent form. Dimensionless gravimetric and volumetric water contents are equivalent in magnitude. Dimensionless water contents are also equal to the degree of saturation S of the soil provided (i) there is essentially no deformation of the soil during the laboratory test and (ii) the volumetric water contents are redefined at each stage of applied suctions.

Normalized gravimetric water content Θ_{ng} has the amount of water in the soil normalized between residual water content and saturated water content:

$$\Theta_{ng} = \frac{w - w_r}{w_s - w_r} \tag{5.8}$$

where:

w_r = residual gravimetric water content.

Normalized volumetric water content form Θ_{nv} has the amount of water in the soil normalized between the residual volumetric water content and the initial saturated volumetric water content:

$$\Theta_{nv} = \frac{\theta - \theta_r}{\theta_s - \theta_r} \tag{5.9}$$

where:

θ_r = residual volumetric water content.

Normalized water contents isolate physical soil behavior between saturated conditions and residual water content conditions. Figure 5.7 shows a typical data set for a sandy soil plotted in terms of dimensionless water content and normalized water content. Residual conditions generally reflect a significant change in the behavior of the soil. For example, the permeability function appears to describe the hydraulic head flow of water through an unsaturated soil as long as the water content is greater than the residual water content (Brooks and Corey, 1964). Vapor flow begins to dominate moisture flow at soil suctions higher than residual suction conditions. Both the dimensionless and normalized water content forms of the SWCC serve a useful purpose in the interpretation of SWCC data and the computation of unsaturated soil property functions.

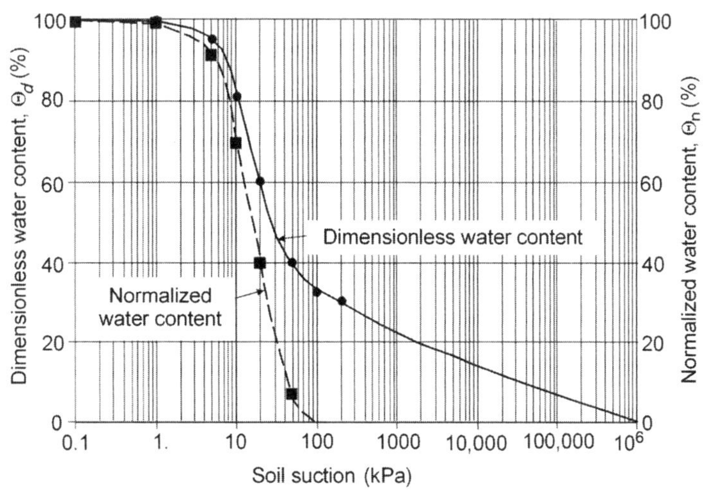

Figure 5.7 Data from a sandy soil plotted in terms of dimensionless water content and normalized water content.

5.2.2 Deformable and Nondeformable Soils

The terms "deformable" and "nondeformable" are being used mainly with respect to the material response to changes in soil suction. Sands are essentially nondeformable materials because of their low compressibility. Sands are essentially nondeformable even when prepared as slurry. Clays can be either deformable or essentially nondeformable depending upon the initial water content, stress history, and nature of the clay minerals. An initially slurry clay will be highly compressible while compacted, low-activity clay may have relatively low compressibility. An undisturbed clay soil can have low to high deformability.

The influence of the applied net normal stress should also be taken into consideration. It is possible for soils to undergo volume decrease (i.e., collapse) upon wetting and there is also the possibility for soils to increase in volume (i.e., swelling) upon wetting at the start of the SWCC test. Volume changes during the initial stages of the SWCC test setup also need to be taken into consideration in the interpretation of the final test results. There is justification for using slight variations in the SWCC test procedure depending upon the material being tested.

The initial state of the soil specimen should be recorded whenever the SWCC is measured. The initial state of the soil has an influence on data reduction and data interpretation. SWCC datasets should be classified into one of the following categories: slurry, compacted, or undisturbed (i.e., natural structured soils or "real" soils).

Figure 5.8 suggests a categorization system for SWCC datasets. A particular soil initially prepared in different ways is expected to result in SWCCs with distinctly differing shapes. The initial state should be recorded and in this way the dataset will be of increased value when subsequently used as part of a knowledge-based system (Fredlund et al., 1997b). Compaction and other complex stress histories make the estimation of a representative SWCC difficult. Estimations of SWCCs from grain-size distribution data is most feasible for silt and sand soils as well as soils prepared in an initial slurry condition (Fredlund et al., 1997a).

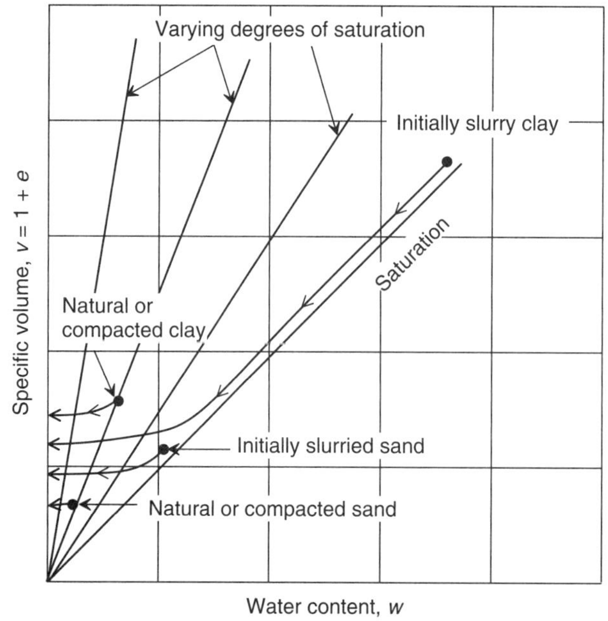

Figure 5.9 Shrinkage curves corresponding to various initial conditions of soil specimen.

Shrinkage curves provide information on the amount of deformation or volume change that is likely to occur as a soil dries. Figure 5.9 illustrates the shrinkage characteristics for various soils. A primary differentiation can be made between soils where there is essentially no deformation as soil suction increases and soils where large deformations take place as soil suction increases. Figure 5.10 illustrates differences that can be observed in SWCCs for a single soil with differing stress histories and methods of specimen preparation.

Theoretical formulations for the flow of water through an unsaturated soil are generally written in terms of volumetric water content because a referential element is used during the derivation of the partial differential flow equation. The water storage input data should be in terms of volumetric water content referenced to the instantaneous overall volume. The numerical modeler should be aware of the need

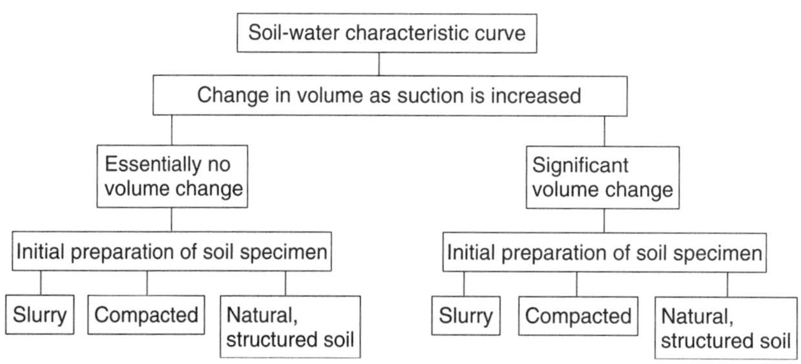

Figure 5.8 Classification of SWCCs based on amount of volume change that occurs as soil suction is increased.

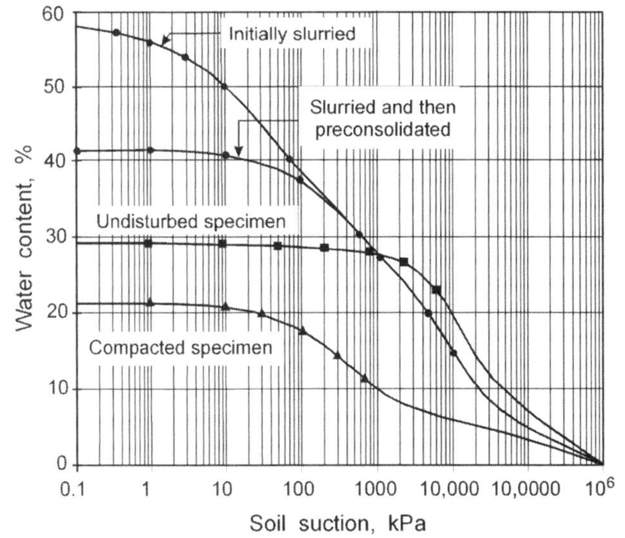

Figure 5.10 Effect of stress history and method of specimen preparation on measured SWCC.

for consistency between measurements of the SWCC in the laboratory and the theoretical requirements of derived partial differential formulation. The results of a numerical modeling simulation are often presented in terms of soil suction, water contents, or degree of saturation. This is particularly true in connection with soil cover design. Often a unique SWCC relationship is used to move between the stress state and the amount of water in the soil. The modeler must be aware of the simplications and assumptions associated with presenting model simulation results in this manner.

5.2.3 Designation of Stress State

The term "soil suction" has been used to designate matric suction, osmotic suction, and total suction. Soil suction can range from zero to 10^6 kPa. The low-suction range needs to be expanded and the high-suction range needs to be compressed, and consequently, a logarithmic scale is suitable for plotting laboratory results. It has become acceptable practice to plot matric suction in the lower soil suction range of the SWCC (e.g., up to approximately 1500 kPa). Above 1500 kPa it has become acceptable practice to plot total suction when graphing SWCC data.

It is the test procedures used when measuring soil suction that indicate whether matric or total suction is being measured. A change has been made from matric suction to total suction along the abscissa. There is an apparent inconsistency in the soil suction variables that are plotted, but the practice has worked quite well for most geotechnical engineering applications. Most geotechnical engineering processes are associated with either matric suction in the lower suction range (e.g., permeability and shear strength) or total suction in the higher suction range (e.g., actual evaporation).

Laboratory desorption data for highly plastic clay (i.e., Regina clay with LL = 75% and PL = 25%) are used to illustrate the influence of high volume changes as soil suction is increased. The clay soil was initially placed in a slurry state and then consolidated to various pressures, as shown in Fig. 5.11. The components of soil suction when the clay is consolidated to 6.2 and 400 kPa is shown in Fig. 5.12. The difference between using gravimetric water content and degree of saturation for the interpretation of the test results on a highly compressible clay soil is shown in Fig. 5.13. The results show that while there is a yield point related to the preconsolidation pressure of the soil (e.g., about 400 kPa), the air-entry value is approximately 1500 kPa for the highly plastic soil prepared as a slurry. The Regina clay soil shows a consistent air-entry value of about 1500 kPa as long as the preconsolidation pressure is lower than the air-entry value.

The desaturation of a soil is commonly divided into three main zones: boundary effect zone, transition zone, and

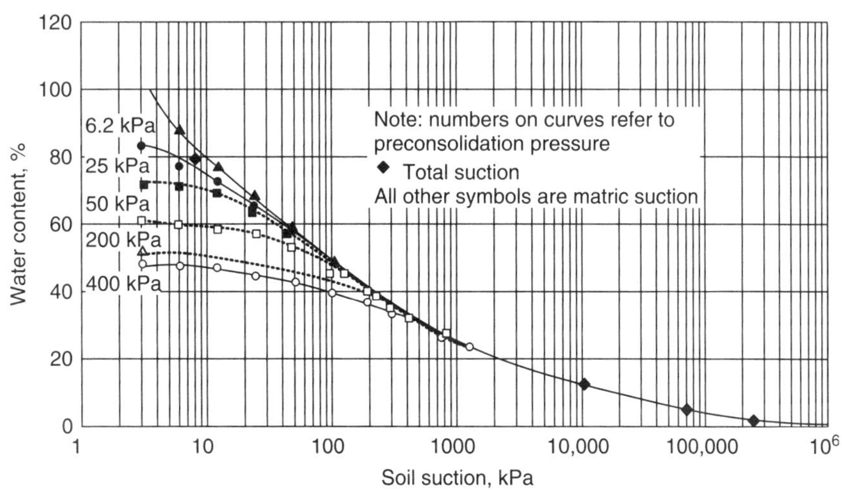

Figure 5.11 Soil suction versus gravimetric water content for initially slurry Regina clay (from Fredlund, 1964).

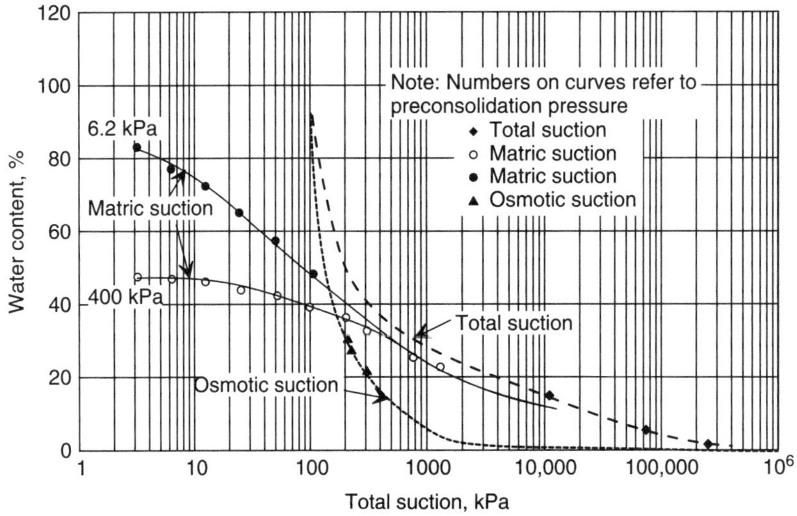

Figure 5.12 Components of soil suction and total suction for Regina clay (from Fredlund, 2002a).

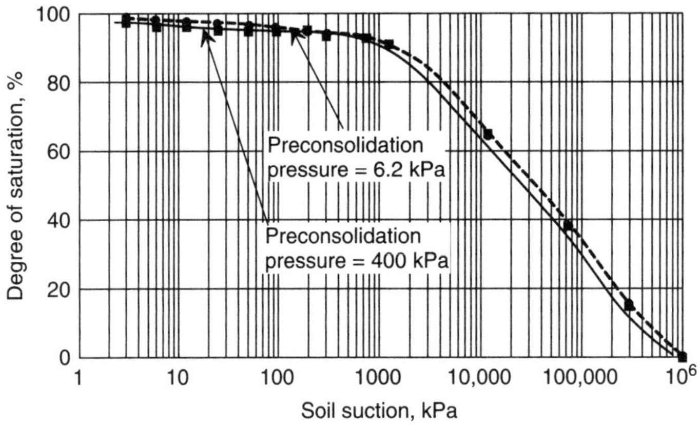

Figure 5.13 Degree of saturation versus soil suction for a highly plastic clay prepared as a slurry (from Fredlund, 1964).

residual zone. These zones are delineated by two distinct points along the SWCC: the AEV of the soil and residual conditions. Unsaturated soil behavior can likewise be subdivided into three zones. The "boundary effect zone" up to the air-entry value can be simulated using saturated soil mechanics principles. The "transition zone" between the air-entry value and residual conditions requires the usage of the theories of unsaturated soil mechanics. The unsaturated soil properties in the transition zone become nonlinear, and as a result, subsequent mathematical formulations are nonlinear. There are also differences in soil behavior that occur once the "residual zone" is entered.

Unsaturated soil theories have mainly been proposed and tested for uniqueness for seepage, shear strength, and volume change in each of the degree of saturation zones. The permeability function for a soil can be used to illustrate differences in behavior as a soil desaturates. At soil suctions below the air-entry value, water flows through the soil in response to the soil property called the saturated coefficient of permeability. Between the air-entry value and residual conditions, the coefficient of permeability decreases in essentially a linear manner on a plot of the logarithm of soil suction versus the logarithm of coefficient of permeability. There is little information and few measurements regarding liquid water flow at soil suctions beyond residual conditions. However, the general consensus appears to be that liquid flow essentially ceases and vapor flow begins to dominate moisture flow once residual suction is exceeded.

Van Genuchten (1980) suggested that residual conditions correspond to soil suctions beyond 1500 kPa. A residual suction of 1500 kPa does not reflect a distinct break in every SWCC but it does separate laboratory measurements of suction made using pressure plate apparatuses from those made using vapor pressure equalization. An arbitrary construction procedure can be used for defining the residual point on the SWCC (Fredlund, 1997a; Vanapalli et al., 1998). The construction procedure identifies the distinct residual suction point common to most SWCCs.

Nonlinear unsaturated soil property functions are commonly used for seepage, shear strength, and volume change problems when soil suction is less than 1500 kPa. The SWCC for a soil is defined over the entire range of soil suction; however, estimated unsaturated soil properties may not be extended to suctions higher than residual suction.

The entire SWCC is used in the estimation of actual evaporation from a soil surface (Wilson et al., 1994). Evaporation from the soil surface occurs as a result of a vapor pressure gradient between the air and the soil. Vapor pressure, in turn, is affected by both matric suction and osmotic suction (i.e., total suction). Coupled heat and moisture flow analyses make use of the entire SWCC when predicting liquid and vapor flow. The osmotic effect of salts at the soil surface is particularly important in the high-soil-suction range.

Certain phenomena appear to be dominant in the low-suction range but disappear in the high-suction range. Other phenomena become dominant in the high-suction range. The matric suction range has been applied to many classic geotechnical engineering analyses and the total suction has likewise been applied to other geotechnical engineering analyses.

A single SWCC with its mixed components of matric and total suction has been successfully used in unsaturated soil mechanics. It is anticipated that this will continue to be the way in which the SWCC is plotted and used in geotechnical engineering. There may be some exceptions that will arise where it is important to understand the independent roles of the components of soil suction.

5.2.4 Upper Limit for Soil Suction

The Gibbs free-energy state equation for water vapor has formed the basis for the idea that the maximum soil suction is approximately 10^6 kPa (Gibbs, 1873; Edlefsen and Anderson, 1943). The free-energy state equation is a thermodynamic relationship between soil suction and the partial pressure of water vapor (i.e., relative humidity). The Gibbs free-energy equation can be viewed as the water retention curve for air and can be written as follows:

$$\psi = -\frac{RT_K}{v_{w0}\omega_v}\ln\left(\frac{u_v}{u_{v0}}\right) \qquad (5.10)$$

where:

ψ = total suction, kPa,
R = universal gas constant [8.31432 J/(mol K)],
T_K = absolute temperature, K ($T_K = 273.16 + T$, where T = temperature, °C),
v_{w0} = specific volume of water, m^3/kg,
ω_v = molecular mass of water vapor (18.016 kg/kmol),
u_v = partial pressure of water, kPa, and
u_{v0} = saturation pressure of water vapor over a flat surface of pure water.

Values of relative humidity (i.e., $RH = u_v/u_{v0}$) can be substituted into Eq. 5.10 and presented graphically (Fig. 5.14). The semilogarithm graph shows an asymptotic behavior as relative humidity approaches 100% and also as relative humidity approaches zero.

Figure 5.14 shows that soil suction changes from zero to 2700 kPa as relative humidity changes from 100 to 98%. Soil suction changes rapidly in the high-relative-humidity range, the range of significant interest in geotechnical engineering.

The free-energy curve in the low-relative-humidity range (i.e., below 50% relative humidity) is shown in an expanded form in Fig. 5.15. The extrapolation of a straight-line portion on the semilog plot of relative humidity values below 50% gives a total suction of approximately 400,000 kPa at zero relative humidity. However, the low-relative-humidity portion of the curve bends toward 10^6 kPa and beyond. A soil suction value of 10^6 is not a singular total-suction value since the curve theoretically extends to infinity. Figure 5.16 shows the data from Fig. 5.15 plotted using a log–log format. The results show that a total suction (or free-energy value) of 10^6 kPa corresponds to a relative humidity value of 0.06%.

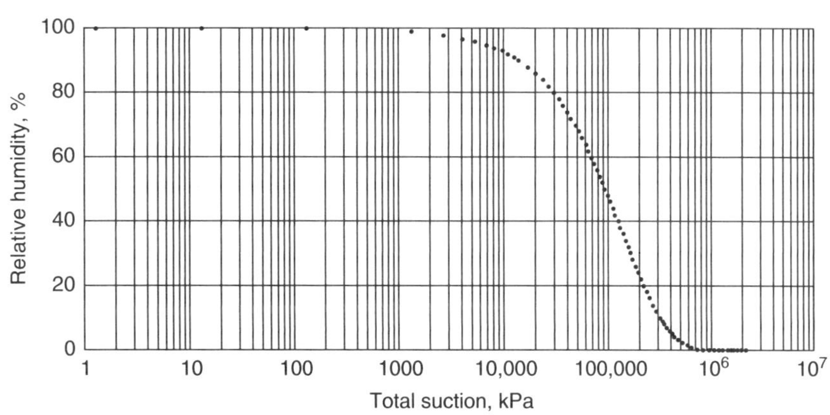

Figure 5.14 Soil suction as a function of relative humidity in the ambient or internal pore-air.

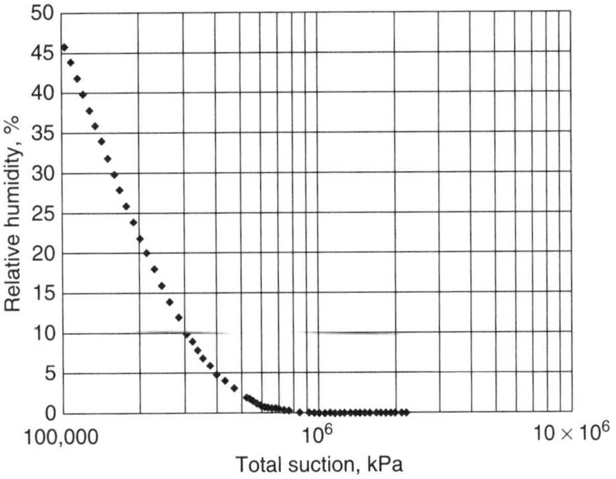

Figure 5.15 Theoretical soil suctions corresponding to relative humidities in extremely high total-suction range.

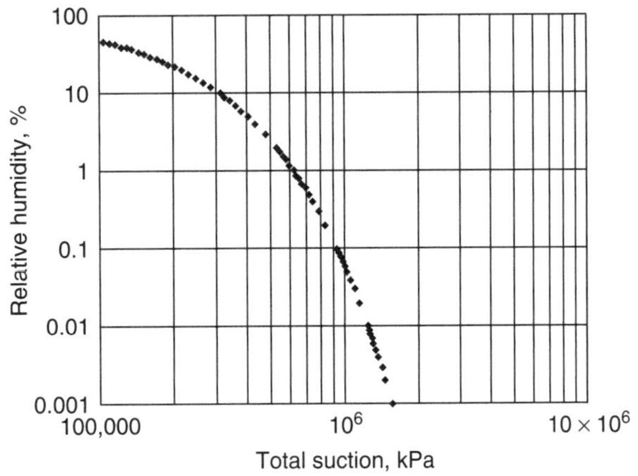

Figure 5.16 Theoretical soil suctions versus relative humidity in extremely high total-suction range plotted on a log–log scale.

The maximum soil suction value (i.e., total suction) is controlled by free-energy consideration of the water vapor in air immediately adjacent to the air-water interface. This means that high total-suction values are related to the free energy associated with water in the soil. A total-suction value of 10^6 kPa is an easy value to remember, and this value is often assumed as the soil suction value representing zero relative humidity.

The maximum total suction is related to minimum relative humidity conditions. When a soil is dried in an oven, it is assumed that all water is driven from the soil. A water content of zero is assumed equal to a relative humidity close to zero percent. This is not necessarily true because the environment inside the drying oven may not be zero relative humidity. However, the differences are generally considered insignificant.

The Gibbs free-energy equation is of particular interest when solving problems related to "actual evaporation" from the ground surface. The Gibbs equation can be viewed as the water retention curve for air above the soil and as such defines the evaporative potential at the ground surface. Evaporation from the soil surface is "shut off" when the relative humidity (i.e., water vapor pressure) in the soil at ground surface becomes equal to the relative humidity in the air above the ground surface. In this way, the Gibbs free-energy equation defines the ground surface boundary condition for vapor flow.

5.2.5 Volume-Mass Constitutive Relationships

The relationship between the volume-mass properties of a soil and stress state provides a conceptual framework for visualizing the SWCC. Two volume-mass constitutive relations are required in order to relate all volume-mass soil properties to the stress state (Fredlund and Morgenstern, 1976). The most common volume-mass properties used in geotechnical engineering are void ratio e, gravimetric water content w, and degree of saturation S.

The volume-mass constitutive relationships make use of saturated soil conditions as a reference for the development of unsaturated soil models. Overall volume change has historically been defined in terms of void ratio change, de, which is then related to effective stress in a saturated soil. Critical state models have used changes in specific volume, $d(1+e)$, as the deformation state variable. Figure 5.17 shows the reference compression curve relationship for a saturated soil under K_0 loading (plotted to the base-10 logarithm scale) and isotropic loading conditions (plotted to a natural logarithm scale).

Loading along the virgin compression line, as well as the unloading and reloading lines, is approximated as straight lines on the semilogarithm plot. Isotropic loading conditions act as a reference for elastoplastic models and provide a separation from the application of deviator stresses (or shear stresses). The K_0 loading conditions are also used because the equipment is commonly available in soil mechanics laboratories. The equation representing the reference deformation-stress state line for a saturated soil under K_0 loading conditions can be written as follows:

$$e = e_o - C_c \log \frac{p - u_w}{p_o - u_w} \tag{5.11}$$

where:

e_o = initial or reference void ratio at $p_o - u_w$,
u_w = pore-water pressure,
p_o = initial or reference total stress (i.e., vertical stress for K_0 loading),
p = total stress state under consideration, and
C_c = compressive index (i.e., slope of the virgin compression branch).

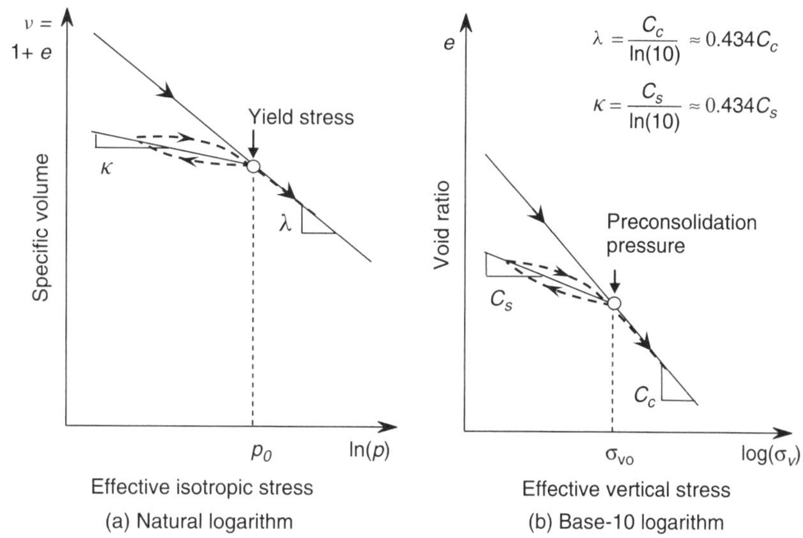

Figure 5.17 Reference compression curves for saturated soil (from Fredlund, 2006): (a) natural logarithm scale; (b) base-10 logarithm scale.

One-dimensional oedometer results have conventionally been plotted as void ratio versus logarithm of vertical stress (to the logarithm base 10). Within the elastoplastic framework, the specific volume $(1 + e)$ is generally plotted versus the logarithm of mean applied stress (i.e., mean of three principal stresses plotted using a natural logarithm base). The mathematical relationship between the soil properties on the two logarithm scales can be written as follows:

$$\lambda = \ln(10)C_c \sim 0.434C_c \tag{5.12}$$

$$\kappa = \ln(10)C_c \sim 0.434C_s \tag{5.13}$$

where:

λ = slope of the virgin compression line on a plot of specific volume and the natural logarithm of effective stress and

κ = slope of the rebound or reloading compression line on a plot of specific volume and the natural logarithm of effective stress.

The volume change versus effective stress equations for a saturated soil can be converted to an incremental elasticity form with respect to stress state. The elasticity modulus of the soil changes with stress state and the solution of practical problems becomes nonlinear. Stress reversals and complex loading paths can also be simulated through use of more rigorous elastoplastic models.

The constitutive relations for an unsaturated soil can be visualized as an extension of the saturated soil model. The extended models for an unsaturated soil include the effect of changes in matric suction which results in further nonlinearities in the soil properties. Fredlund and Morgenstern (1976) used three-dimensional surfaces to represent the void ratio change de and water content change dw constitutive relations for unsaturated soils. Figure 5.18 illustrates the limiting (or bounding) relationships associated with a typical clayey silt soil. General differential equations can be written for small stress changes from a stress point on the void ratio constitutive surface (Eq. 5.14) and water content constitutive surface (Eq. 5.15):

$$de = \frac{\partial e}{\partial(\sigma_m - u_a)} d(\sigma_m - u_a) + \frac{\partial e}{\partial(u_a - u_w)} d(u_a - u_w) \tag{5.14}$$

$$dw = \frac{\partial w}{\partial(\sigma_m - u_a)} d(\sigma_m - u_a) + \frac{\partial w}{\partial(u_a - u_w)} d(u_a - u_w) \tag{5.15}$$

where:

σ_m = mean total stress, $\sigma_m = \frac{1}{3}(\sigma_1 + \sigma_2 + \sigma_3)$, and

$\sigma_m - u_a$ = mean net stress which is referenced to pore-air pressure, u_a.

Equations 5.14 and 5.15 have one part that designates the stress point under consideration (i.e., $\sigma_m - u_a$ and $u_a - u_w$) and another part that provides a general form for associated soil properties [e.g., $\partial e/\partial(\sigma_m - u_a)$, $\partial w/\partial(\sigma_m - u_a)$, $\partial e/\partial(u_a - u_w)$, and $\partial w/\partial(u_a - u_w)$]. The soil properties at a point on the constitutive surface can be approximated as a linear compressibility modulus provided the stress increments are relatively small. The compressibility modulus is also a function of the stress state. There is need for

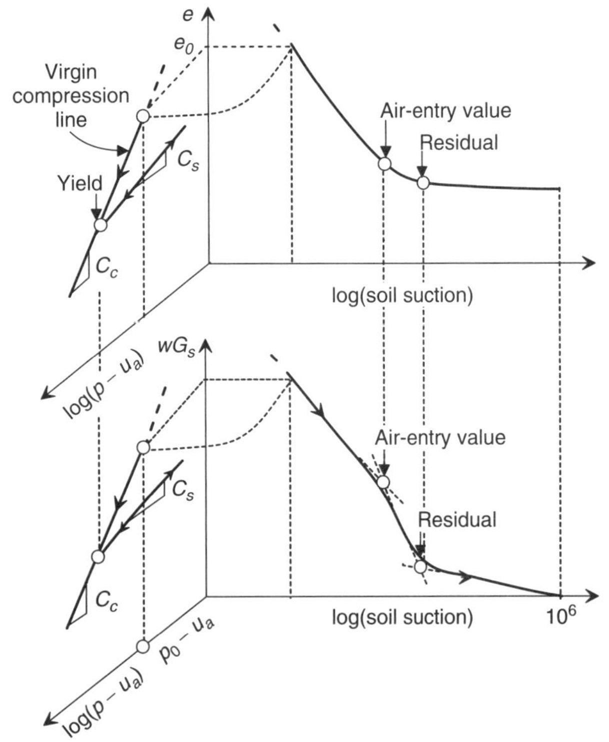

Figure 5.18 Limiting or bounding relationships for typical clayey silt soil (from Fredlund, 2006).

equations that can represent the entire constitutive surfaces for unsaturated soils. Once such equations are known, it is possible to differentiate the equation to obtain the soil properties at any stress state.

The volume-mass constitutive surfaces have distinctly different characteristics for clay soil and sand soil. Unsaturated soils have distinct features such as the air-entry value and residual conditions (i.e., residual suction and residual water content), which are identifiable along the soil suction stress state. Figures 5.4 and 5.5 previously showed typical volume-mass constitutive surfaces (i.e., void ratio, gravimetric water content, and degree of saturation) generated from measured data on Beaver Creek sand and Regina clay (Pham, 2005). Each of the constitutive surfaces are uniquely influenced by the yield stress (or preconsolidation pressure), the AEV, and the residual conditions of the soil. Each term is a function of stress state.

The SWCC constitutes the water content–soil suction relationship portion of the overall volume-mass constitutive surfaces. The interpretation of the SWCC data on a clay soil is quite different from the interpretation of SWCC data on a sand soil. Either the water content or the degree-of-saturation plot can be used to identify the air-entry value for sand; however, it is the degree-of-saturation plot that must be used to identify the correct air-entry value for clay soil. Further information on the volume-mass constitutive surfaces is later provided during the discussion of volume change.

5.3 EQUATIONS FOR SWCC

A large number of closed-form, empirical equations have been proposed to best fit laboratory data for SWCCs. A list of some of the equations appearing in the research literature is shown in Table 5.2. The equations can be classified as two-parameter SWCC equations and three-parameter SWCC equations. Each of these equations can be best fit to laboratory data using a least-squares regression analysis (Fredlund and Xing, 1994).

Each SWCC equation is written in terms of gravimetric water content with the realization that the equations can also be written in terms of volumetric water content or degree of saturation. Some equations are written in terms of dimensionless water content Θ_d, and others are written in terms of normalized water content Θ_n. The saturated gravimetric water content is designated as w_s, and the gravimetric residual water content is designated as w_r.

Each of the SWCC equations shown in Table 5.2 has one variable that bears a relationship to the air-entry value of the soil and a second variable that is related to the rate at which the soil desaturates. A third variable is used for some equations and it allows the low-suction range near the air-entry value to have a shape that is independent of the shape of the SWCC in the high-suction range near residual conditions. The use of three parameters for the SWCC provides greater flexibility for the best-fitting analysis. It should be noted that the three-parameter equations may not have parameters that are fully independent of one another.

Each of the SWCC equations can be best fit to either the drying (desorption) branch or the wetting (adsorption) branch. There are two difficulties common to the empirical equations shown in Table 5.2. The first problem occurs in the low-suction range where the SWCC equations become asymptotic to a horizontal line. In other words, a differentiation of the equation gives a water storage value, written in terms of volumetric water content, m_2^w that approaches zero. This is not realistic and results in numerical instability when modeling unsaturated soil behavior in the low-suction range. The last empirical equation shown in Table 5.2 (i.e., Pham and Fredlund, 2006) does not provide a continuous function for the SWCC, but it does provide a more reasonable representation of the SWCC in the low-suction range.

The second problem with the empirical SWCC equations occurs at high-soil-suction values beyond residual conditions where the results become asymptotic to a horizontal line as soil suction goes to infinity. The Fredlund and Xing (1994) equation overcomes this second problem by applying a correction factor that directs the SWCC equation to a soil suction of 10^6 kPa at zero water content. The Fredlund and Xing (1994) equation is written as follows:

$$w(\psi) = C(\psi) = \frac{w_s}{\left\{\ln\left[e + (\psi/a_f)^{n_f}\right]\right\}^{m_f}} \tag{5.16}$$

Table 5.2 Empirical Equations Used to Best Fit SWCC Data

Reference	Equation	Description
Gardner (1958b)	$\Theta_d = \dfrac{1}{1 + a_g \psi^{n_g}}$ where $\Theta_d = \dfrac{w(\psi)}{w_s}$	a_g = fitting parameter which is a function of air-entry value of the soil n_g = fitting parameter which is a function of rate of water extraction from soil once air-entry value of soil has been exceeded
Brooks and Corey (1964)	$w(\psi) = w_s$ or $\Theta_n = 1$ for $\psi \le \psi_{\text{aev}}$ $\Theta_n = \left[\dfrac{\psi}{\psi_{\text{aev}}}\right]^{-\lambda_{bc}}$ $\psi > \psi_{\text{aev}}$ where $\Theta_n = \dfrac{w(\psi) - w_r}{w_s - w_r}$	ψ_{aev}= air-entry value of soil λ_{bc} = pore size distribution index w_r = residual water content located through trial-and-error process that yields straight line on semi log plot of degree of saturation versus suction
Brutsaert (1967)	$\Theta_n = \dfrac{1}{1 + [\psi/a_b]^{n_b}}$ where $\Theta_n = \dfrac{w(\psi) - w_r}{w_s - w_r}$	a_b = fitting parameter which is a function of air-entry value of soil n_b = fitting parameter which is a function of rate of water extraction from soil once air-entry value has been exceeded
Laliberte (1969)	$\Theta_n = \dfrac{1}{2}\text{erfc}\left[a_l - \dfrac{b_l}{c_l + (\psi/\psi_{\text{aev}})}\right]$ where $\Theta_n = \dfrac{w(\psi) - w_r}{w_s - w_r}$	a_l, b_l, c_l = parameters assumed to be unique functions of pore-size distribution index λ
Campbell (1974)	$w = w_s, \psi < \psi_{\text{aev}}$ $w = w_s\left[\dfrac{\psi}{\psi_{\text{aev}}}\right]^{-1/b_c}, \psi \ge \psi_{\text{aev}}$	ψ_{aev} = air-entry value of soil b_c = fitting parameter
van Genuchten (1980)	$\Theta_n = \dfrac{1}{[1 + (a_{vg}\psi)^{n_{vg}}]^{m_{vg}}}$ where $\Theta_n = \dfrac{w(\psi) - w_r}{w_s - w_r}$	a_{vg}, a_{vm}, a_{vb} = fitting parameters primarily related to inverse of air-entry value (units equal to 1/kPa) n_{vg}, n_{vm}, n_{vb} = fitting parameters primarily related to rate of water extraction from soil once air-entry value has been exceeded m_{vg}, m_{vm}, m_{vb} = fitting parameters that are primarily related to residual water content conditions
van Genuchten (1980)–Mualem (1976)	$\Theta_n = \dfrac{1}{[1 + (a_{vm}\psi)^{n_{vm}}]^{m_{vm}}}$ where $m_{vm} = 1 - \dfrac{1}{n_{vm}}$	

(*continued overleaf*)

Table 5.2 *(Continued)*

Reference	Equation	Description
van Genuchten (1980)–Burdine (1953)	$\Theta_n = \frac{1}{[1+(a_{vb}\psi)^{n_{vb}}]^{m_{vb}}}$ where $m_{vb} = 1 - \frac{2}{n_{vb}}$	
McKee and Bumb (1984) (Boltzmann distribution)	$\Theta_n = \exp\left[\frac{a_{m1}-\psi}{n_{m1}}\right]$ where $\Theta_n = \frac{w(\psi)-w_r}{w_s-w_r}$	a_{m1} = curve-fitting parameter n_{m1} = curve-fitting parameter
McKee and Bumb (1987) (Fermi distribution)	$\Theta_n = \frac{1}{1+\exp[(\psi - a_{m2})/n_{m2}]}$ where $\Theta_n = \frac{w(\psi)-w_r}{w_s-w_r}$	a_{m2} = curve-fitting parameter n_{m2} = curve-fitting parameter
Fredlund and Xing (1994)	$w(\psi) = C(\psi)\frac{w_s}{\{\ln[e+(\psi/a_f)^{n_f}]\}^{m_f}}$ where $C(\psi) = 1 - \frac{\ln(1+\psi/\psi_r)}{\ln[1+(10^6/\psi_r)]}$ $\Theta_d = \frac{w(\psi)}{w_s}$	a_f = fitting parameter which is primarily a function of air-entry value of soil n_f = fitting parameter which is primarily a function of rate of water extraction from soil once air-entry value has been exceeded m_f = fitting parameter which is primarily a function of residual water content $C(\psi)$ = correction factor which is primarily a function of suction corresponding to residual water content.
Pereira and Fredlund (2000)	$w(\psi) = w_r + \frac{w_s - w_r}{[1+[\psi/a_p]^{n_p}]^{m_p}}$	a_p = fitting parameter which is primarily a function of air-entry value of soil n_p = fitting parameter which is primarily a function of rate of water extraction from soil, once air-entry value has been exceeded m_p = fitting parameter which is primarily a function of residual water content
Pham and Fredlund (2005)	$\begin{cases} w_1(\psi) = w_u - S_1\log(\psi) & 1 \le \psi < \psi_{aev} \\ w_2(\psi) = w_{aev} - S_2\log\left(\frac{\psi}{\psi_{aev}}\right) & \psi_{aev} \le \psi < \psi_r \\ w_3(\psi) = S_3\log\left(\frac{10^6}{\psi}\right) & \psi_r \le \psi < 10^6 \text{ kPa} \end{cases}$	S_1, S_2, S_3 = slope of straight line portions of SWCC within each of three zones w_u = water content at 1 kPa w_{aev} = water content at air-entry value w_1, w_2, w_3 = water content in line segments 1, 2, and 3, respectively.

where:

$w(\psi)$ = water content at any soil suction, ψ,
$C(\psi)$ = correction factor directing the SWCC to 10^6 kPa at zero water content,
e = irrational constant equal to 2.71828, used when taking the natural logarithm,
a_f = fitting parameter indicating the inflection point that bears a relationship to and is greater than the air-entry value,
n_f = fitting parameter related to the rate of desaturation, and
m_f = fitting parameter related to the curvature near residual conditions.

Laboratory studies have shown that there is a relationship between the SWCC for a particular soil and the unsaturated soil properties (Fredlund and Rahardjo, 1993a). For example, the permeability function for an unsaturated soil bears a relationship to the SWCC. It has become acceptable practice in geotechnical engineering (and other disciplines) to empirically estimate the permeability function for an unsaturated soil by using the saturated coefficient of permeability and the SWCC (Marshall, 1958; Mualem, 1986; van Genuchten, 1980; Fredlund et al., 1994a). Similar procedures have been proposed for estimating the shear strength properties of an unsaturated soil (Vanapalli et al., 1996a) as well as other unsaturated soil property functions.

Figure 5.19 shows a typical SWCC for a silt soil along with identification of some of its main characteristics. The air-entry value of the soil is the matric suction where air starts to displace water in the largest pores in the soil. The residual water content is the water content where a larger suction change is required to remove additional water from the soil. In other words, there is a change in the rate at which water can be extracted from the soil (on a logarithm scale). These definitions are quite vague and empirical construction procedures have been proposed for the quantification of air-entry value and residual conditions.

The empirical procedure for defining residual water content and residual soil suction is shown in Fig. 5.19 (Fredlund, 1997; Vanapalli et al., 1998). A tangent line is drawn through the inflection point of the SWCC. The SWCC relationship in the high-suction range is approximated using another line that extends through zero water content and 10^6 kPa. The residual water content θ_r is approximated as the ordinate of the point at which the two lines intersect (Fig. 5.19).

The total suction corresponding to zero water content appears to be essentially the same for all porous materials. A value slightly below 10^6 kPa has been experimentally supported for a variety of soils (Croney and Coleman, 1961). The 10^6 kPa value is also supported by thermodynamic considerations (Richards, 1965).

The main curve shown in Fig. 5.19 is a desorption curve. The adsorption curve differs from the desorption curve as a result of hysteresis related to wetting and drying. The end point of the adsorption curve may differ from the starting point of the desorption curve because of air entrapment in the soil. The drying and wetting SWCCs have similar forms (i.e., are essentially congruent).

Typical SWCCs (i.e., desorption curves) for soils ranging from sands to clays are shown in Fig. 5.20. The saturated water content and the air-entry value (or bubbling pressure), $(u_a - u_w)_b$, generally increase with the plasticity of the soil. Other factors such as stress history and secondary soil structure also affect the shape of the SWCCs.

5.3.1 Comments on Some Empirical SWCC Equations

Gardner (1958b), proposed an empirical equation for the permeability function. The equation emulates the SWCC and has subsequently been used to best fit the SWCC:

$$\Theta_d = \frac{1}{1 + a_g \psi^{n_g}} \tag{5.17}$$

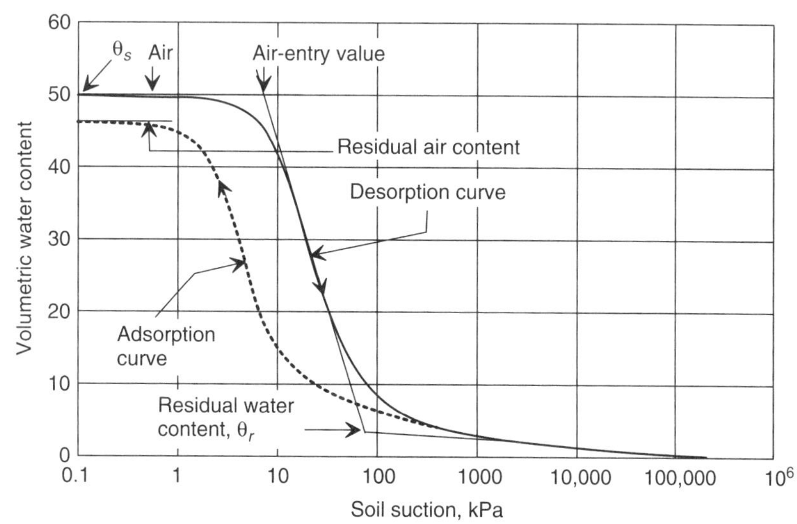

Figure 5.19 Typical SWCC for silt soil.

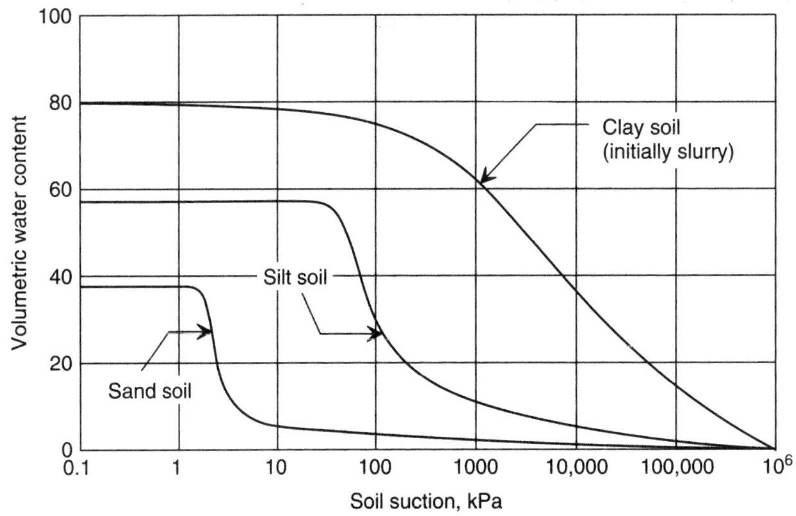

Figure 5.20 Comparative desorption SWCCs for sand, silt, and clay soils.

where:

Θ_d = dimensionless volumetric water content, equal to θ/θ_s where θ = any volumetric water content and θ_s = saturated volumetric water content.
a_g = curve-fitting parameter related to the air-entry value of the soil and
n_g = curve-fitting parameter related to the slope at the inflection point on the SWCC.

Among the earliest equations proposed for the SWCC is the power law proposed by Brooks and Corey (1964). The proposed equation represents desaturation of the soil when the soil suction is greater than the air-entry value:

$$\Theta_n = \left[\frac{\psi}{\psi_{\text{aev}}}\right]^{-\lambda_{bc}} \tag{5.18}$$

where:

Θ_n = normalized volumetric water content [e.g., $\Theta_n = (\theta - \theta_r)/(\theta_s - \theta_r)$, where θ_s and θ_r are the saturated and residual volumetric water contents, respectively] and the fit can also be made on the degree of saturation or the gravimetric water content SWCC if there is negligible volume change when measuring the SWCC,
ψ = soil suction,
ψ_{aev} = air-entry value (or bubbling pressure), and
λ = pore-size distribution index.

The degree of saturation S can be used in place of the normalized water content. Equation 5.18 describes the SWCC in the region between the air-entry value and residual conditions (Campbell, 1974; Clapp and Hornberger, 1978; Gardner et al., 1970a, 1970b; Rogowski, 1971; Williams et al., 1983; McCuen et al., 1981). The equation was proposed for describing moisture movement in agricultural soils.

McKee and Bumb (1984) suggested an exponential function (i.e., Boltzmann distribution) for the relationship between the normalized water content and soil suction:

$$\Theta_n = \exp\left[\frac{a_{m1} - \psi}{n_{m1}}\right] \tag{5.19}$$

where:

a_{m1} and n_{m1} = fitting parameters.

Equation 5.19 is valid for suction values greater than the air-entry value of the soil but is not valid near maximum desaturation or under fully saturated conditions. McKee and Bumb (1987) and Bumb (1987) suggested the following equation to overcome some of the mathematical concerns of Eq. 5.19:

$$\Theta_n = \frac{1}{1 + \exp[(\psi - a_{m2})/n_{m2}]} \tag{5.20}$$

where:

a_{m2}, n_{m2} = fitting parameters.

The McKee and Bumb (1987) equation provides a better approximation of the SWCC in the low-suction range but it is not suitable in the high-suction range where the curve drops exponentially to zero.

The Brooks and Corey (1964) equation as well as other proposed equations suggest that there is a sharp discontinuity in the soil suction versus water content near the air-entry value for the soil. Although some coarse-grained sands may have a rapid change in water content at low soil suctions, most medium- and fine-textured soils show a gradual curvature in the air-entry region of the SWCC.

Van Genuchten (1980) proposed the use of a continuous function between soil suction and normalized water content:

$$\Theta_n = \frac{1}{\left[1 + (a_{vg}\psi)^{n_{vg}}\right]^{m_{vg}}} \tag{5.21}$$

where:

a_{vg}, n_{vg}, m_{vg} = three soil fitting parameters and
Θ_n = normalized water content (i.e., normalized between saturated water content, θ_s and residual water content, θ_r).

The van Genuchten (1980) equation provides additional flexibility to the SWCC. In an attempt to obtain a closed-form expression for hydraulic conductivity, van Genuchten (1980) fixed the relationship between the m_{vg} and n_{vg} parameters through use of the equations shown in Table 5.2. However, the linkage between the m_{vg} and n_{vg} parameters reduced the flexibility of Eq. 5.21. More accurate best fits to measured laboratory data can be obtained by leaving m_{vg} and n_{vg} parameters without a fixed relationship.

5.3.2 Theory of Pore-Size Distribution and the SWCC

Equations that have been proposed for the SWCC are empirical in nature. Each equation was originally applied to a particular group of soils. There are many different forms of SWCC equations that could be tested to assess their fit of experimental data. The pore-size distribution or slope of the SWCC appears to have the form of a normal distribution curve. Consequently, the following equation might be used to approximate the SWCC:

$$\Theta_n = a_5 e^{-(b_5\psi)^m} \tag{5.22}$$

where:

a_5, b_5, m = curve-fitting parameters.

Equation 5.22 is not suitable as a general form for the SWCC, but it might apply for some soils over a limited range of suction values. The pore-size distribution curve for soils provides a theoretical basis for the shape of the SWCC. Soils can be regarded as a set of interconnected pores that are randomly distributed. The pores can be characterized by a pore radius r and described by a function $f(r)$, where $f(r)\,dr$ is the relative volume of pores between radius r to $r + dr$. In other words, $f(r)$ is the density of pore volume corresponding to radius r. Since $f(r)\,dr$ represents the pores between radius r to $r + dr$ that are filled with water, the volumetric water content can be expressed as follows:

$$\theta(R) = \int_{R_{\min}}^{R_{\max}} F(r)\,dr \tag{5.23}$$

where:

$\theta(R)$ = volumetric water content when all the pores with radius less than or equal to R are filled with water,
$R_{\min}$ = minimum pore radius in the soil, and
$R_{\max}$ = the maximum pore radius, and for the saturated case, $\theta(R_{\max}) = \theta_s$.

The capillary law indicates that there should be an inverse relationship between matric suction and the radius of curvature of the air-water interface (i.e., the meniscus). The air-water interface should bear an inverse relationship to the pore sizes that are desaturated at a particular suction as indicated in the equation

$$r = \frac{C}{\psi} \tag{5.24}$$

where:

$C = 2T_s \cos\alpha_1$ = constant,
T_s = surface tension of water, and
α_1 = angle of contact between water and soil.

Two particular suction conditions can be defined as follows:

$$\psi_{\max} = \frac{C}{R_{\min}} \tag{5.25}$$

$$\psi_{\text{aev}} = \frac{C}{R_{\max}} \tag{5.26}$$

where:

$\psi_{\max}$ = suction value corresponding to the minimum pore radius and
ψ_{aev} = air-entry suction value of the soil.

Using the capillary law, Eq. 5.23 can be expressed in terms of soil suction:

$$\theta(\psi) = \int_{\psi_{\max}}^{\psi} f\left(\frac{C}{h}\right) d\left(\frac{C}{h}\right) = \int_{\psi}^{\psi_{\max}} f\left(\frac{C}{h}\right) \frac{C}{h^2}\,dh \tag{5.27}$$

where:

h = a dummy variable of integration representing soil suction.

Equation 5.27 is a general equation form that can be used to describe the relationship between volumetric water content and soil suction. The SWCC can be uniquely determined using Eq. 5.27 if the pore-size distribution $f(r)$ of a soil is known. Several special cases can be defined as follows:

1. Case of a constant-pore-size function. The pore sizes can be assumed to be uniformly distributed, that is,

$f(r) = A$, where A is a constant. Equation 5.27 can then be written as

$$\theta(\psi) = \int_{\psi}^{\psi_{\max}} \frac{AC}{h^2} dh = AC\left(\frac{1}{\psi} - \frac{1}{\psi_{\max}}\right) = \frac{B}{\psi} - D \tag{5.28}$$

where:

$B = AC =$ constant and
$D = AC/\psi_{\max} =$ constant.

2. Case where pore-size function varies inversely as r^2. For the case where $f(r) = A/r^2$, the relationship between volumetric water content and suction can be written as

$$\theta(\psi) = \int_{\psi}^{\psi_{\max}} \frac{Ah^2}{C^2}\frac{C}{h^2} dh = B - D\psi \tag{5.29}$$

where:

$B = A\psi_{\max}/C =$ constant and
$D = A/C =$ constant.

Equation 5.29 represents a linear variation in the distribution of pore sizes. In other words, there is a linear relationship between volumetric water content and soil suction.

3. Case where pore-size function varies inversely as $r^{(m+1)}$. For the case where $f(r) = A/r^{(m+1)}$, with m as an integer, the relationship between volumetric water content and soil suction can be written as

$$\theta(\psi) = \int_{\psi}^{\psi_{\max}} \frac{Ah^{m+1}}{C^{m+1}}\frac{C}{h^2} dh = B - D\psi^m \tag{5.30}$$

where:

$B = A(\psi_{\max})^m/(mC^m) =$ constant and
$D = A/(mC^m) =$ constant.

The power law relationship (i.e., Eq. 5.19) proposed by Brooks and Corey (1964) is a special case of Eq. 5.30. In other words, the Brooks and Corey (1964) power law relationship is valid when the pore-size distribution is close to the distribution function $f(r) = A/r^{m+1}$.

The volumetric water content θ over the entire soil suction range from 0 to 10^6 kPa can be referenced to zero water content; otherwise, the normalized water content becomes negative at water contents less than the residual value θ_r. In this case, water content is written in a dimensionless form (i.e., $\Theta_d = \theta/\theta_s$). Equation 5.27 suggests that the following integration form can be used as a general form to approximate the SWCC:

$$\theta(\psi) = \theta_s \int_{\psi}^{\infty} f(h)\, dh \tag{5.31}$$

where:

$f(h) =$ pore-size distribution as a function of soil suction.

Equation 5.31 will generally produce a nonsymmetrical S-shaped curve. Several special cases follow.

Let us consider the case of a normal distribution curve where the $f(h)$ function is assumed to take the form of a normal distribution:

$$f(h) = \frac{1}{\sqrt{2\pi}\sigma} e^{-(h-\mu)^2/2\sigma^2} \tag{5.32}$$

where:

$\mu =$ mean value of the distribution of $f(h)$,
$e =$ irrational constant equal to 2.71828, and
$\sigma =$ standard deviation of the distribution of $f(h)$.

The SWCC defined by Eq. 5.31 can be expressed as follows:

$$\theta(\psi) = \theta_s \int_{\psi}^{\infty} f(h)\, dh = \frac{\theta_s}{2}\frac{2}{\sqrt{\pi}} \int_{(\psi-\mu)/\sqrt{2}\sigma}^{\infty} e^{-y^2} dy = \frac{\theta_s}{2} \operatorname{erfc}\left(\frac{\psi - \mu}{\sqrt{2}\sigma}\right) \tag{5.33}$$

where:

$$\operatorname{erfc}(x) = \frac{2}{\sqrt{\pi}} \int_{x}^{\infty} e^{-y^2}/dy = 1 - \operatorname{erf}(x) = 1 - \frac{2}{\sqrt{\pi}} \int_{0}^{\infty} e^{-y^2}/dy \text{ and}$$

$\operatorname{erfc}(x) =$ complement of the error function, $\operatorname{erf}(x)$.

Equation 5.33 takes the form of a symmetrical S-shaped curve. The SWCC of the soil will be close to a symmetrical S-shaped curve if the pore-size distribution of a soil can be approximated as a normal distribution. Equation 5.33 can then be used as a model to describe the SWCC relationship.

The two fitting parameters (i.e., the mean value μ and the standard deviation σ) in Eq. 5.33 are related to the air-entry value of the soil and the slope at the inflection point on the SWCC. If the slope at the inflection point is s and the air-entry value is ψ_{aev}, then the standard deviation σ can be written as

$$\sigma = \frac{\theta_s}{\sqrt{2\pi}s} \tag{5.34}$$

and the mean value μ can be calculated as

$$\mu = \psi_{\text{aev}} + \frac{\theta_s}{2s} \tag{5.35}$$

Let us consider the case where a gamma-type distribution is used for the function $f(r)$. In this case, $f(h)$ takes the form

$$f(h) = \frac{h^{\alpha-1}e^{-h/\beta}}{\beta^{\alpha}\Gamma(\alpha)} \quad \alpha, \beta > 0 \quad 0 \le h < \infty$$
$$f(h) = 0 \text{ elsewhere} \tag{5.36}$$

where

$$\Gamma(\alpha) = \int_0^{\infty} h^{\alpha-1}e^{-h}\,dh \tag{5.37}$$

The SWCC defined by Eq. 5.31 has a smaller air-entry value, a steeper slope near saturation, and a gentler slope near the residual water content. In the special case when α is an integer the SWCC defined by Eq. 5.31 becomes

$$\theta(\psi) = \theta_s \int_{\psi}^{\infty} \frac{h^{\alpha-1}e^{-h/\beta}}{\beta^{\alpha}\Gamma(\alpha)}\,dh \tag{5.38}$$

$$\theta(\psi) = \frac{\theta_s}{\Gamma(\alpha)} \int_{\psi/\beta}^{\infty} h^{\alpha-1}e^{-h}\,dh \tag{5.39}$$

$$\theta(\psi) = \theta_s \sum_{i=0}^{\alpha-1} \frac{\psi^i e^{-\psi/\beta}}{i!\beta^i} \tag{5.40}$$

When $\alpha = 1$, the gamma distribution becomes an exponential type distribution:

$$f(h) = \frac{1}{\beta}e^{-h/\beta} \quad \beta > 0 \quad 0 \le h < \infty$$
$$f(h) = 0 \text{ elsewhere} \tag{5.41}$$

The SWCC defined by Eq. 5.31 can then be written as

$$\theta(\psi) = \theta_s e^{-\psi/\beta} \tag{5.42}$$

Equation 5.42 has the same form as Eq. 5.19, which was the form used by McKee and Bumb (1984) to describe the SWCC. Therefore, Eq. 5.19 gives good results when the pore-size distribution of the soil reflects a gamma distribution.

Let us consider the case of a beta-type distribution for the function $f(r)$:

$$f(h) = \frac{h^{\alpha-1}(1-h)^{\beta-1}}{B(\alpha,\beta)} \quad \alpha, \beta > 0 \quad 0 \le h \le 1$$
$$f(h) = 0 \text{ elsewhere} \tag{5.43}$$

where

$$B(\alpha,\beta) = \int_0^1 h^{\alpha-1}(1-h)^{\beta-1}dh = \frac{\Gamma(\alpha)\Gamma(\beta)}{\Gamma(\alpha+\beta)} \tag{5.44}$$

The SWCC given by Eq. 5.31 has greater flexibility when using the beta-type distribution function. For $\alpha = \beta$, Eq. 5.31 generates a symmetrical S-shaped curve. For $\alpha > \beta$, the curve is nonsymmetrical and has a higher air-entry value, a gentler slope near saturation, and a steeper slope near the residual water content. For $\alpha < \beta$, the curve has a smaller air-entry value, a steeper slope near saturation, and a gentler slope near the residual water content. When α and β are integers, the SWCC defined by Eqs. 5.31 and 5.44 is related to the binomial probability function as follows (Mendenhall et al., 1981):

$$\theta(\psi) = \theta_s \int_{\psi}^{1} \frac{h^{\alpha-1}(1-h)^{\beta-1}}{B(\alpha,\beta)}\,dh$$
$$= \theta_s - \theta_s \sum_{i=\alpha}^{\alpha+\beta-1} \binom{\alpha+\beta+1}{i} \psi^i (1-\psi)^{\alpha+\beta-1-i} \tag{5.45}$$

Defining the pore radius r over the interval (0, 1) does not restrict its usage. The beta density function can be applied to any interval by translation and a change in the scale.

5.3.3 Proposed Fredlund-Xing (1994) Equation

The pore-size distribution function $f(h)$ can also be written in the form

$$f(\psi) = \frac{mnp(p\psi)^{n-1}}{[1+(p\psi)^n]^{m+1}} \tag{5.46}$$

Figure 5.21 shows a sample probability distribution for Eq. 5.46 along with its integration. It can be seen that the integration drops to zero over a narrow suction range. Therefore, Eq. 5.21 is not suitable in the high-soil-suction region. Experimental data show that beyond residual water content the plot should decrease linearly (on a logarithm scale), to a value of about 10^6 kPa. To describe the SWCC more accurately, the following distribution can be used:

$$f(\psi) = \frac{mn(\psi/a)^{n-1}}{a[e+(\psi/a)^n]\{\log[e+(\psi/a)^n]\}^{m+1}} \tag{5.47}$$

Equation 5.47 and its integration form are shown in Fig. 5.22 for the same set of parameters (i.e., $a = 1/p, n, m$). This distribution function drops more slowly than shown by Eq. 5.46 as soil suction increases. Therefore, Eq. 5.47 produces a nonsymmetrical curve that more closely emulates experimental data.

Integrating Eq. 5.47 using Eq. 5.31 gives the Fredlund and Xing (1994) relationship between volumetric water content and soil suction:

$$\theta = \theta_s \left[\frac{1}{\ln[e+(\psi/a_f)^{nf}]}\right]^{m_f} \tag{5.48}$$

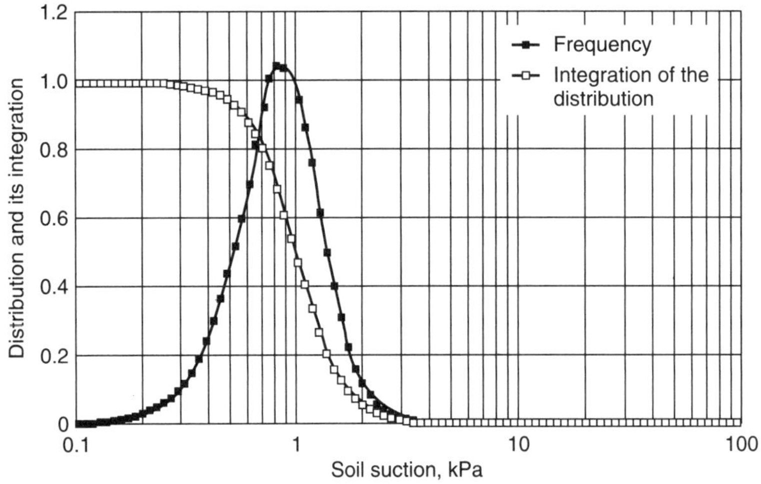

Figure 5.21 Pore-size distribution function along with corresponding integration function for soil with little water storage in high-suction range.

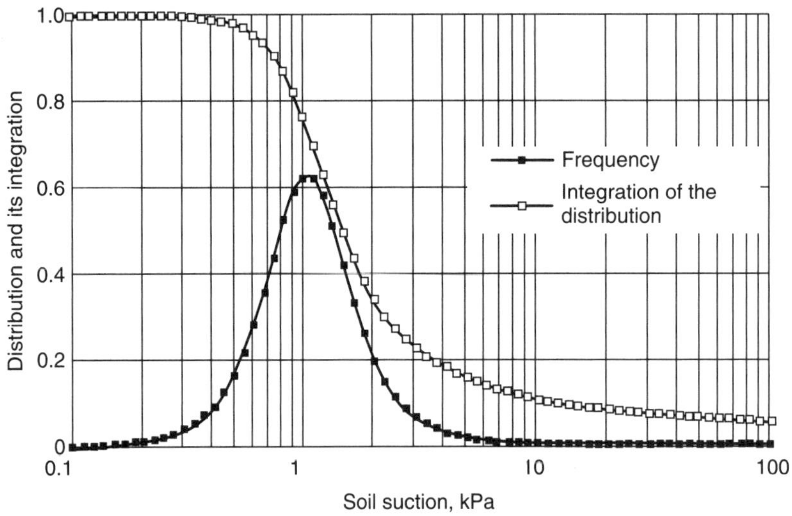

Figure 5.22 Pore-size distribution function along with integration function for soil where water storage does not drop to zero in high-suction range.

where:

a_f, n_f, m_f = three fitting parameters,
θ = volumetric water content corresponding to a selected soil suction,
θ_s = saturated volumetric water content, and
e = irrational constant equal to 2.71828.

Figures 5.23–5.25 show the effect of varying each of the three parameters a_f, n_f, and m_f on the shape of the SWCC. Figure 5.23 shows that when n_f and m_f are fixed, the a_f parameter (with units of kPa), is related to the air-entry value. In general, the value for the parameter a_f is higher than the air-entry value. However, for small values of m_f, the air-entry value is close to the parameter a_f.

Figure 5.24 shows that when a_f and n_f are fixed, the m_f parameter provides independent curvatures in the low- and high-suction ranges.

Figure 5.25 indicates that parameter n_f controls the slope of the SWCC. The slope or differential of Eq. 5.48 attains its maximum value approximately at the value of a_f. Therefore, the point $(a, \theta(a_f))$ can be used as the inflection point. A graphical estimation for the three parameters can be obtained from the SWCC. First, locate the inflection point (ψ_i, θ_i) on the SWCC plot and draw a tangent line through this point (Fig. 5.26). Let s denote the slope of the tangent line. The three parameters a_f, n_f, and m_f can then be determined as follows:

$$a_f = \psi_i \tag{5.49}$$

$$m_f = 3.67 \ln\left(\frac{\theta_s}{\theta_i}\right) \tag{5.50}$$

$$n_f = \frac{1.31^{m+1}}{m_f \theta_s} 3.72 s \psi_i \tag{5.51}$$

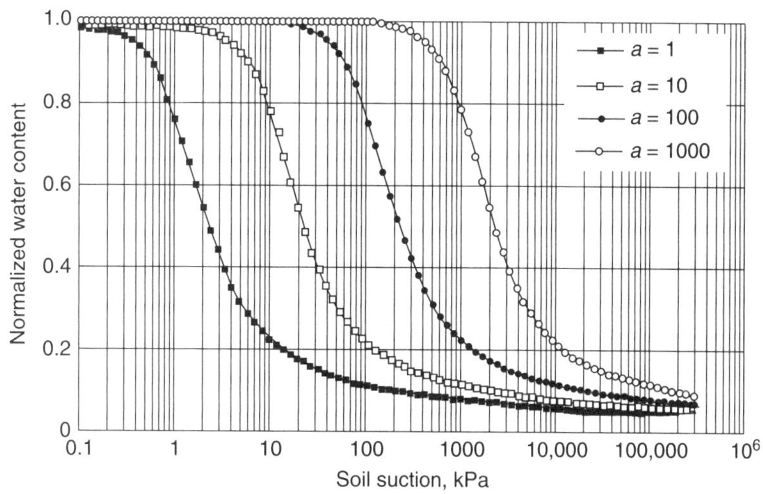

Figure 5.23 Effect of a_f soil parameter on SWCC when $n_f = 2$ and $m_f = 1$ (from Fredlund and Xing, 1994).

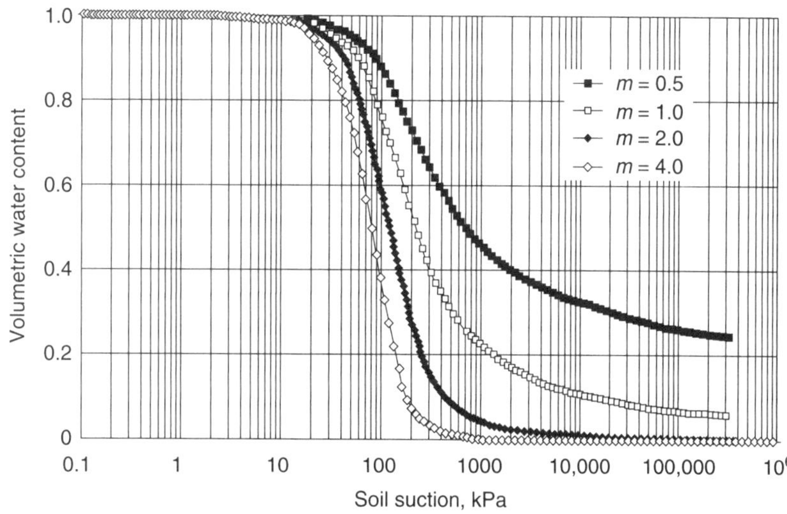

Figure 5.24 Effect of m_f soil parameter on SWCC when $a_f = 100$ and $n_f = 2$ (from Fredlund and Xing, 1994).

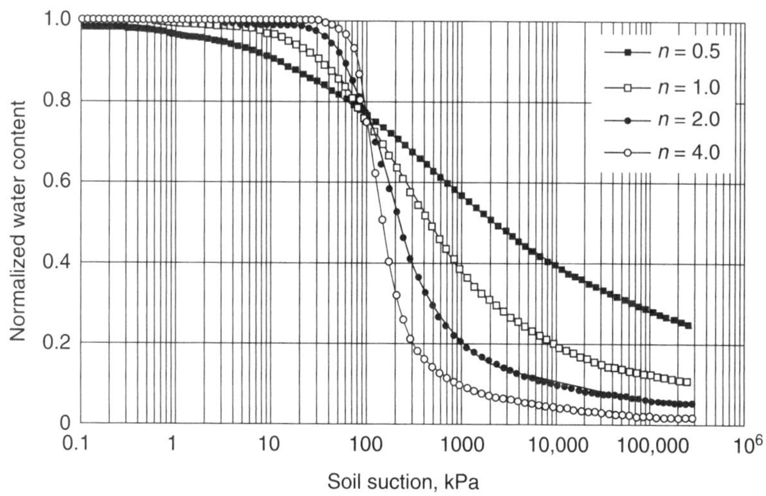

Figure 5.25 Effect of n_f soil parameter on SWCC when $a_f = 100$ and $m_f = 1$ (from Fredlund and Xing, 1994).

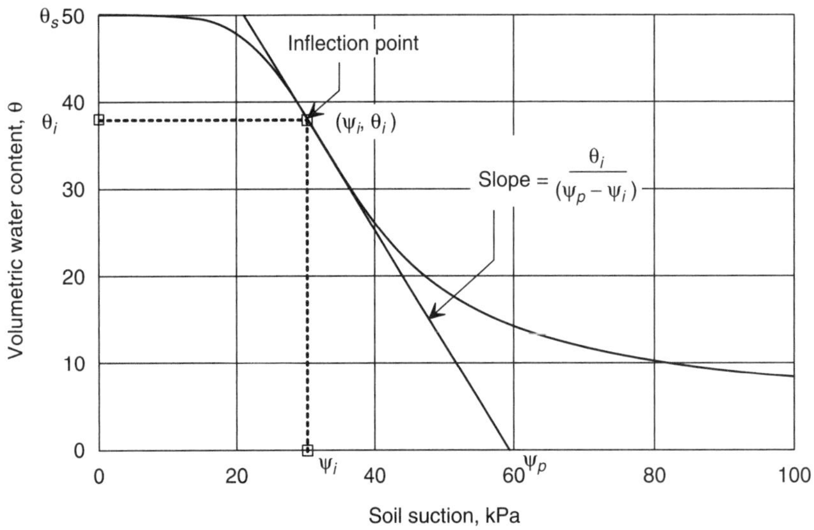

Figure 5.26 Graphical solution for estimation of three fitting parameter (a_f, n_f, and m_f) (from Fredlund and Xing, 1994).

The slope s of the tangent line can be calculated as

$$s = \frac{\theta_i}{\psi_p - \psi_i} \tag{5.52}$$

where:

ψ_p = suction at intercept of the tangent line with the zero water content line (Fig. 5.26) and
ψ_i = suction at inflection point on the SWCC.

Small values of m_f result in a moderate slope in the high-suction range, and large values of n_f produce a sharp curvature near the air-entry value (see Fig. 5.25). An example of a best-fit curve to the experimental data for a silt soil from Brooks and Corey (1964) is shown in Fig. 5.27.

In Eq. 5.48, volumetric water content θ becomes equal to the saturated volumetric water content θ_s when soil suction is zero. On the other hand, when volumetric water content θ becomes zero, soil suction goes to infinity. It is also possible to use the degree of saturation for curve fitting, since the degree of saturation varies from 0 to 100%. Gravimetric water content w can be similarly normalized for curve fitting. Three plots are shown for the same silt soil in Fig. 5.28 using different variables to represent the water content of the soil (i.e., degree of saturation, volumetric water content, and gravimetric water content). It should be noted that the above comparisons can be made because the silt soil is a relatively low volume change soil.

Experimental data has shown that the suction of a soil reaches a maximum value of approximately 10^6 kPa at zero

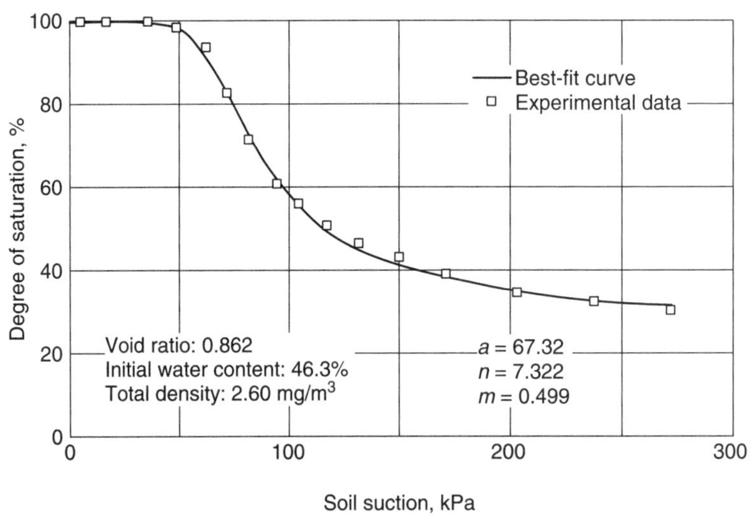

Figure 5.27 Best-fit SWCC fitted to experimental data on a silt soil (data from Brooks and Corey, 1964).

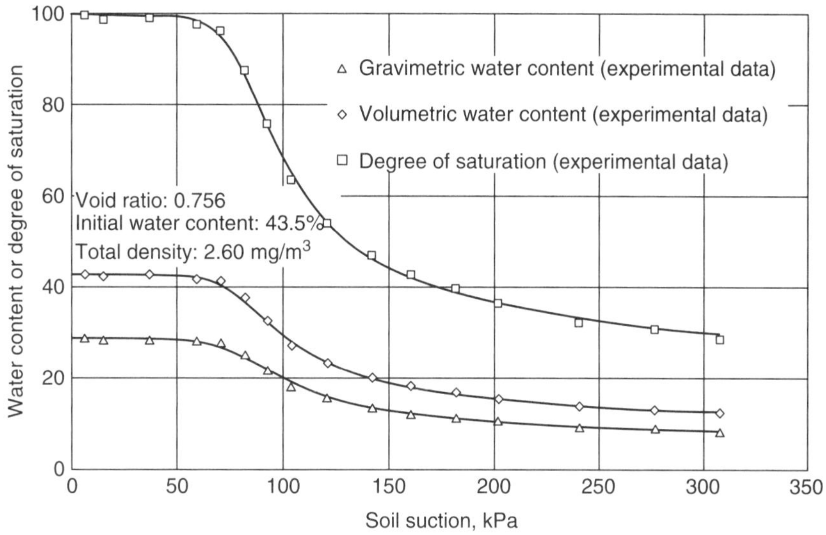

Figure 5.28 Best-fit curves fitted to experimental data of silt soil using three representations of water content (i.e., degree of saturation, volumetric water content, and gravimetric water content) (data from Brooks and Corey, 1964).

water content. An upper limit of 10^6 kPa can be built into Eq. 5.48; however, it should be noted that it is also possible to use a different maximum value for soil suction. Equation 5.48 can have a correction factor $C(\psi)$ applied that can control the shape of the SWCC in the high-soil-suction range:

$$\theta(\psi, a_f, n_f, m_f) = C(\psi)\frac{\theta_s}{\{\ln[e + (\psi/a_f)^{n_f}]\}^{m_f}} \tag{5.53}$$

The correction factor $C(\psi)$ can be written as follows:

$$C(\psi) = 1 - \frac{\ln(1 + \psi/\psi_r)}{\ln[1 + (10^6/\psi_r)]} \tag{5.54}$$

where:

ψ_r = soil suction corresponding to the residual water content θ_r.

Equation 5.54 shows that the correction factor $C(\psi)$ is equal to zero when soil suction is 10^6 kPa. A sample SWCC plot using Eq. 5.54 is shown in Fig. 5.29. The curve in the low-suction range is not significantly affected since the correction function is approximately equal to 1.0 in this region.

Figure 5.30 shows a best-fit curve to the experimental data obtained for glacial till without using the correction function $C(\psi)$ (i.e., Eq. 5.48). A best-fit curve to the same experimental data when using the correction function

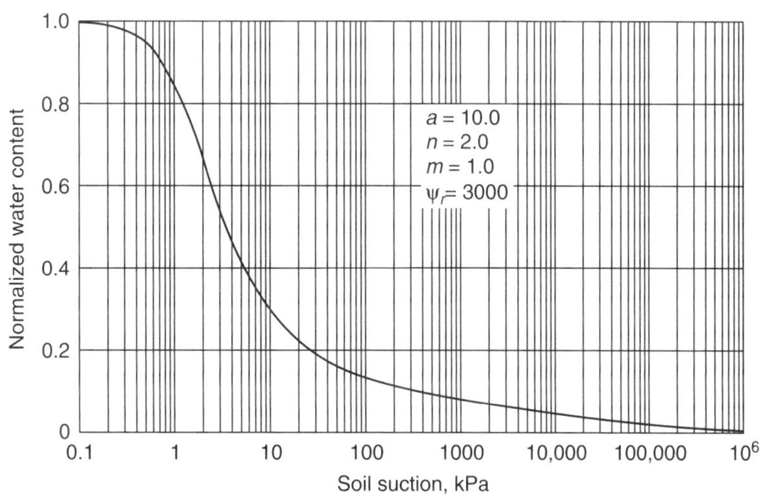

Figure 5.29 Fredlund and Xing (1994) SWCC equation with the correction factor $C(\psi)$ applied for 10^6 kPa at zero water content.

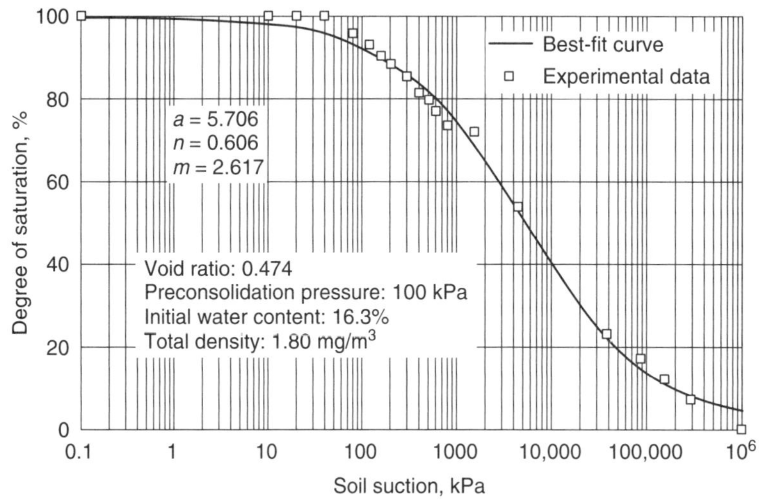

Figure 5.30 Best-fit SWCC for glacial till (Data from Vanapalli, 1994).

(i.e., Eq. 5.53) is shown in Fig. 5.31. It can be seen that the application of the correction function improves the fit of the SWCC equation to the measured data. The SWCC curve is forced by the correction function $C(\psi)$ to go to zero at a suction of 10^6 kPa.

A graphical estimation of the four soil fitting parameters a_f, n_f, m_f, and ψ_r for a SWCC with the correction function (i.e., Eq. 5.53) can be obtained as follows. The SWCC information should be plotted on a semilog form. First, determine the suction corresponding to the residual water content ψ_r by locating a point where the curve changes slope in the high-suction range (Fig. 5.32). Numerical results show that, in most cases, an approximation of residual suction, ψ_r, is satisfactory in Eq. 5.53. The magnitude of the residual suction will generally be in the range 1500–3000 kPa for most silt and clay soils. Sand soil can have residual suction values that are in the range of 10–100 kPa. Figure 5.32 uses a residual suction value ψ_r equal to 3000 kPa for illustration purposes. Next, locate the inflection point (ψ_i, θ_i) on the semilog plot and draw a tangent line through this point (Fig. 5.32). Let s denote the slope of the tangent line on the semilog plot. Then, the fitting parameters a_f, n_f, and m_f can be determined as follows:

$$a_f = \psi_i \tag{5.55}$$

$$m_f = 3.67 / \ln\left(\frac{\theta_s C(\psi_i)}{\theta_i}\right) \tag{5.56}$$

$$n_f = \frac{1.31^{m_f+1}}{m_f C(\psi_i)} 3.72 s^* \tag{5.57}$$

where

$$s^* = \frac{s}{\theta_s} - \frac{\psi_i}{1.31^m(\psi_i + \psi_r)\ln[1 + (10^6/\psi_r)]} \tag{5.58}$$

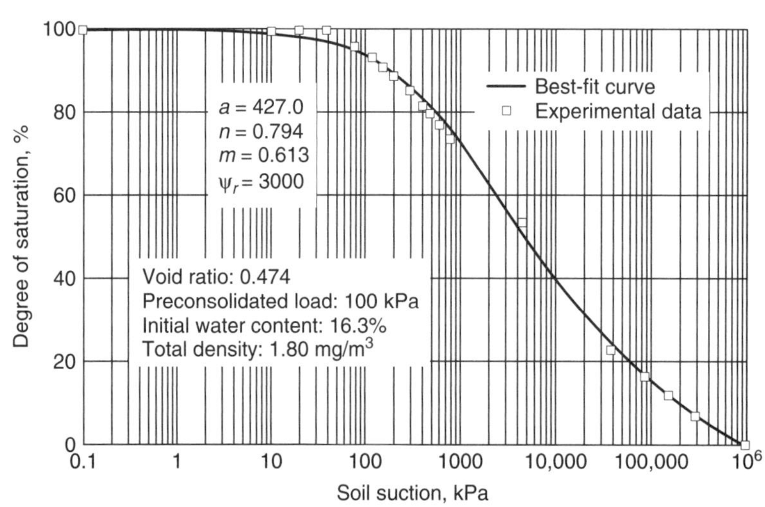

Figure 5.31 Best-fit SWCC for glacial till with correction factor $C(\psi)$ applied (Data from Vanapalli, 1994).

The slope s of the tangent line can be calculated as follows:

$$s = \frac{\theta_i}{\log(\psi_p/\psi_i)} \tag{5.59}$$

where:

ψ_p = suction at intercept of the tangent line on the semilog plot and the zero-water-content line (Fig. 5.32) and
ψ_i = suction at inflection point on the SWCC.

The graphical procedure provides an approximation of each of the fitting parameters. To obtain a closer fit to experimental data, the three parameters (a_f, n_f, and m_f) in Eq. 5.54 can be determined using a least-squares regression method. When performing the best-fit regression analysis, reasonable initial values should be selected for each of the three parameters. The following objective function (i.e., sum of the squared deviations of the measured data from the calculated data) can then be minimized with respect to the three parameters a_f, n_f, and m_f:

$$O(a_f, m_f, n_f) = \sum_{i=1}^{M} [\theta_i - \theta(\psi_i, a_f, m_f, n_f)]^2 \tag{5.60}$$

where:

$O(a_f, n_f, m_f)$ = objective function,
M = total number of measurements, and
θ_i, ψ_i = measured values.

The best-fit regression analysis is a nonlinear minimization process. A curve-fitting utility can be used based on Eqs. 5.53 and 5.60 along with a quasi-Newton method. Best-fit curves for tailings sand, silt, and clay are shown in Figs. 5.33–5.35. The regression analyses show that Eq. 5.53

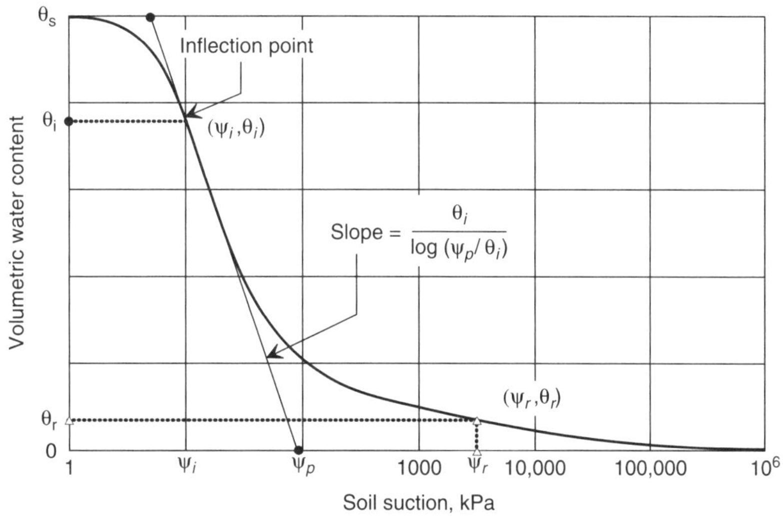

Figure 5.32 Graphical estimation of four SWCC soil fitting parameters (a_f, n_f, m_f, and ψ_r).

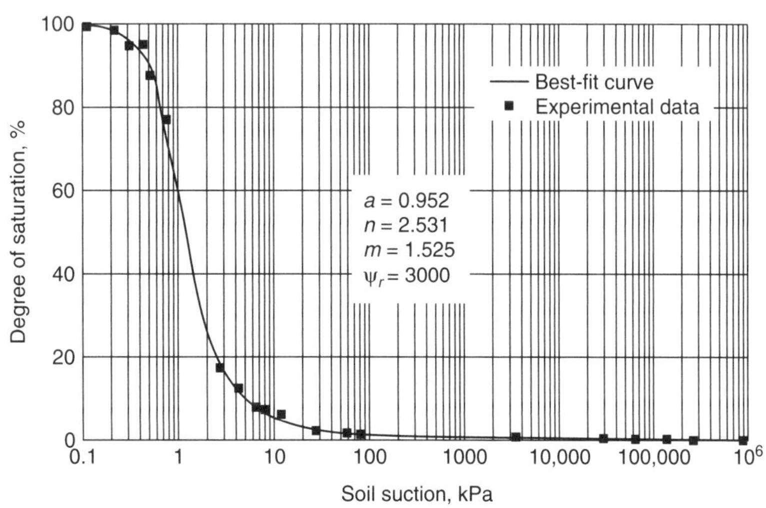

Figure 5.33 Best-fit Fredlund and Xing (1994) equation applied to experimental data on sand.

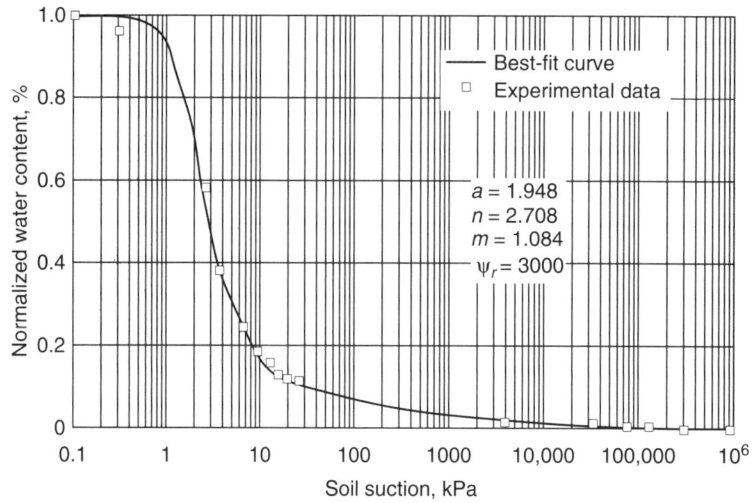

Figure 5.34 Best-fit Fredlund and Xing (1994) equation applied to experimental data on silt.

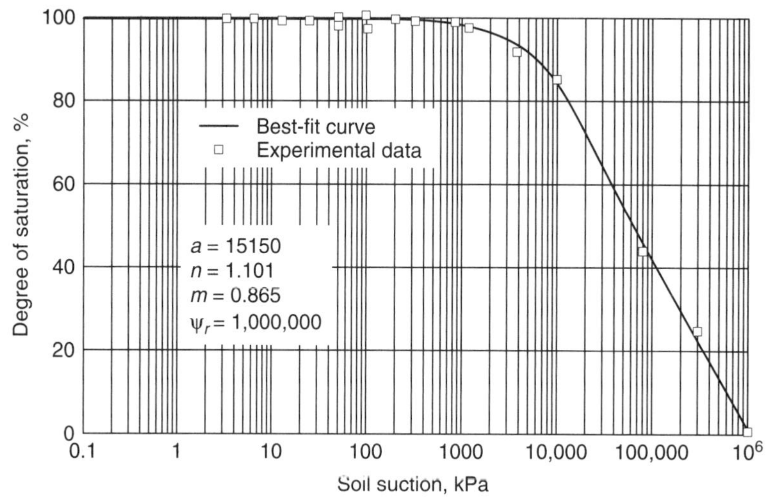

Figure 5.35 Best-fit Fredlund and Xing (1994) equation applied to experimental data on initially slurry Regina clay (Data from Fredlund, 1964).

can be used to fit the experimental data reasonably well over the entire suction range from 0.1 to 10^6 kPa.

If the SWCC needs to only be fit between saturated conditions and residual conditions, another form of Eq. 5.48 can be used to best fit to the data, as shown in the following equation:

$$\theta = \theta_r + \frac{\theta_s - \theta_r}{\left\{\ln[e + (\psi/a_f)^{n_f}]\right\}^{m_f}} \tag{5.61}$$

where:

θ_r = residual volumetric water content.

The volumetric water contents θ_r and θ_s can be treated as two additional parameters. The five parameters a_f, n_f, m_f, θ_s, and θ_r in Eq. 5.61 can be systematically identified through a best-fit analysis on experimental data. Equation 5.61 is for the Fredlund and Xing (1994) fit of the data; however, the form is similar to that often shown for the van Genuchten (1980) equation and other equations that only fit the experimental data between saturated and residual water content conditions.

5.4 REGRESSION ANALYSIS ON SWCC EQUATIONS

The desorption (or drying) branch of the SWCCs is most commonly measured in the laboratory. SWCCs are presented in terms of one of three variables to describe the amount of water in the soil: gravimetric water content, volumetric water content, or degree of saturation (D.G. Fredlund, 2002a; 2006). Various regression analysis studies have been performed to assess the closeness of fit of the proposed SWCC equations to experimental data sets.

5.4.1 Equations for the SWCC

Several of the commonly used SWCC equations are sigmoidal in character and as such do not provide a reasonable description of water storage in the low-suction range (i.e., below the air-entry value). Many of the SWCC equations have been derived for usage in disciplines other than geotechnical engineering. The requirements for numerical modeling in geotechnical engineering problems need to be realized when selecting the SWCC equation.

SWCC equations can be categorized according to the portion of the data that is adequately described by the proposed equation. Figure 5.36 shows a typical SWCC along with the manner in which a sigmoidal function can be best fit to laboratory data. A sigmoidal curve is a continuous function that can fit the SWCC from near the air-entry value to residual water content. A sigmoidal function does not perform well in the region below the air-entry value and above residual conditions. Figure 5.37 provides a categorization of some SWCC equations that have been proposed (Sillers, 1997; Sillers et al., 2001). Some equations have two fitting soil parameters, while others have three and four fitting soil parameters. Some equations are continuous while others are discontinuous. The larger the number of fitting parameters, the more likely the SWCC equation will closely fit the laboratory data. However, even with the use of three fitting parameters, the extremities of the SWCC may still not be adequately defined. It is also useful for the SWCC functions to have parameters that have physical meaning with respect to the storage of water in a porous medium.

The van Genuchten (1980) equation is typical of proposed sigmoidal equations. It has been widely used in agriculture-related disciplines. The continuous function is defined over the entire suction range; however, experimental data are only adequately represented between the air-entry values to residual suction. Equations have been proposed by Mualem (1976a) and Burdine (1953) that place an empirical relationship between the m and n fitting parameters.

The Fredlund and Xing (1994) equation is sigmoidal in character up to residual suction conditions after which the equation is extended to a suction value of 10^6 kPa at zero suction. It is also possible to have soil suction values other than 10^6 kPa corresponding to zero water content.

Fredlund and Pham (2006) proposed a SWCC equation that more clearly captures the independent effects of isotropic net normal stress and soil suction for the water content and void ratio (and degree of saturation) constitutive surfaces. The volume-mass constitutive model becomes

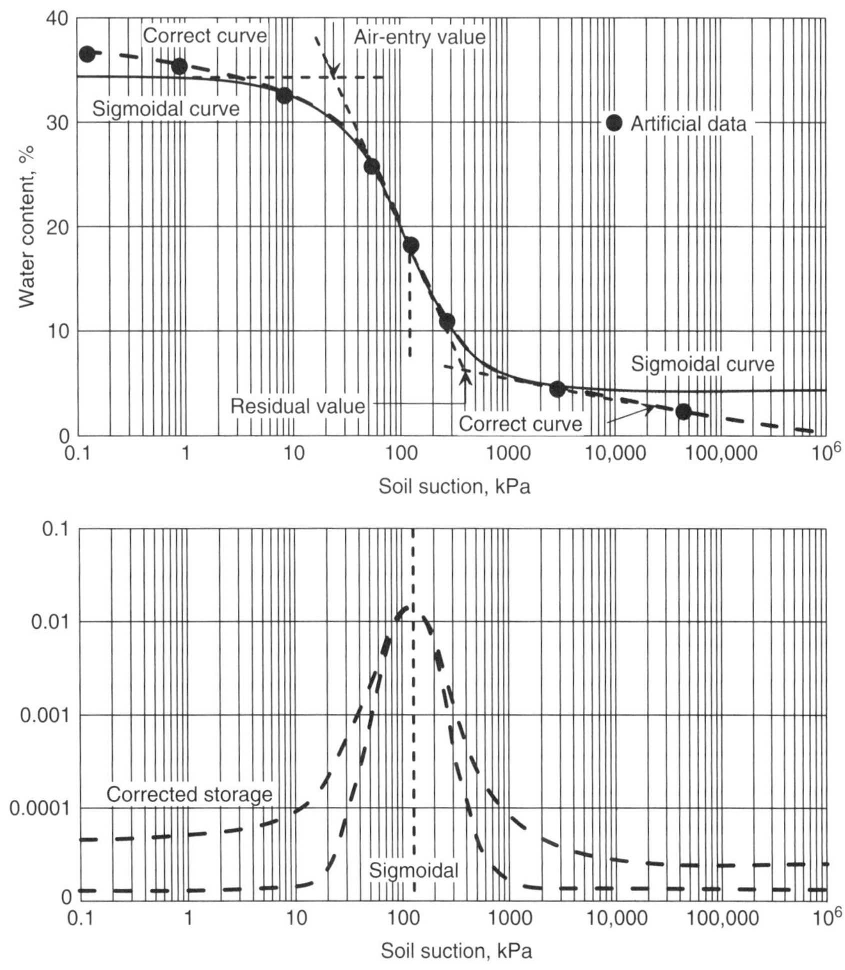

Figure 5.36 Limitations associated with sigmoidal SWCC equations.

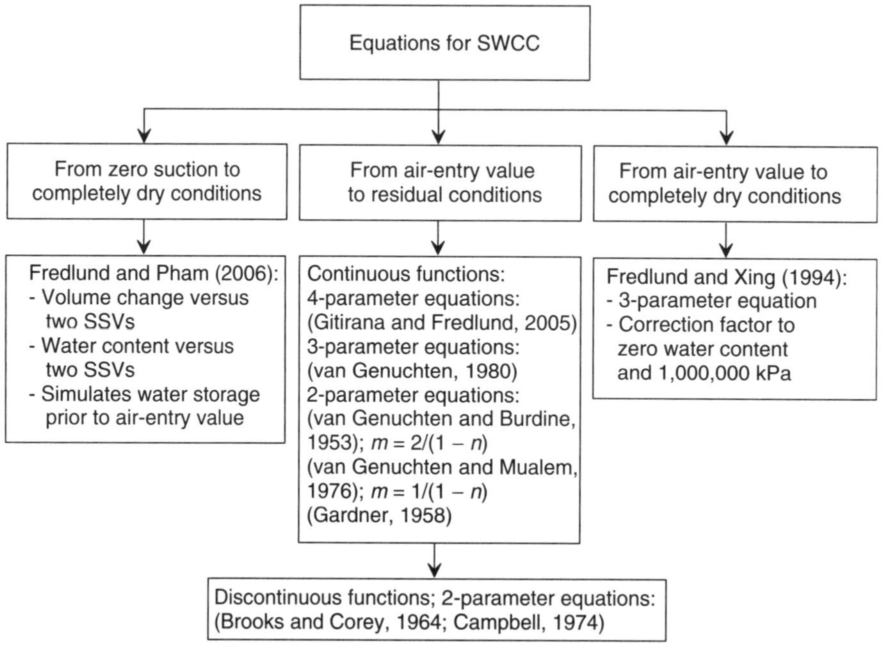

Figure 5.37 Categorization of SWCC equations based on portion of SWCC that is well defined for engineering applications.

stress path dependent and is able to more clearly describe the water storage character of the soil for a wider range of loading and soil suction conditions. The SWCC model commences with an initially slurry material.

5.4.2 Comparative Studies of SWCCs

A number of statistical studies have been undertaken where a variety of SWCC equations have been best fit to data sets of laboratory test results. Tinjum et al. (1997) undertook a study that applied several empirical SWCC equations to measured data sets. A conclusion from the study was that both two-parameter and three-parameter SWCC equations could be best fit to laboratory data; however, the three-parameter equations provided a better fit due to the greater flexibility of the equation. Tinjum et al. (1997) also suggested that it was appropriate to use estimated SWCCs for many geotechnical and geoenvironmental engineering applications.

Leong and Rahardjo (1997) undertook a review of several empirical SWCC equations by applying each equation to a database of laboratory measurements. It was noted that the SWCC equations with the highest number of soil fitting parameters provided the closest fit to the data sets. The Fredlund and Xing (1994) SWCC equation was found to "perform marginally better" than the van Genuchten (1980) equation. The Fredlund and Xing (1994) equation is found to perform much better than other SWCC equations when the data set contained measurements where some of the soil suction measurements were greater than residual suction.

Figure 5.38 shows the results of a study in which a total of nine SWCC equations were best fit to a database containing 231 soils (Sillers, 1997). The soil data sets contained eight USDA soil classification system groups. The Akaike (1974) information criterion statistical indicator was used to assess the "fit" of each of the SWCC equations. The average Akaike information was obtained from fitting each SWCC equation to the SWCC data sets (Sillers, 1997).

The results showed that the Fredlund and Xing (1994) equation performed the best with an Akaike number of −1007. Note that the more negative the Akaike number, the better the equation fit. The next best performing SWCC

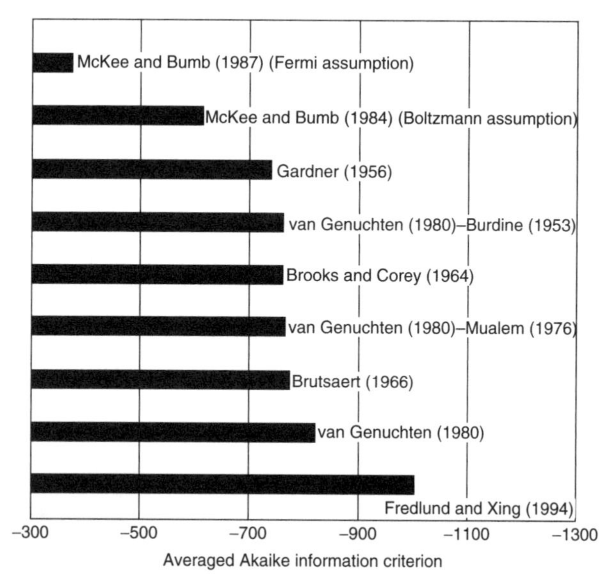

Figure 5.38 Averaged Akaike information criterion for various empirical SWCC equations.

equation was the van Genuchten (1980) equation where the Fredlund and Xing (1994) "correction function" was applied beyond residual suction, giving an Akaike number of −824. The Akaike number was poorer when the Fredlund and Xing (1994) correction function was not added to the van Genuchten equation.

The Sillers (1997) study also showed that there was no advantage associated with fixing the relationship between the m_{vg} and n_{vg} variables in the van Genuchten (1980) equation [i.e., using the fixed relationships suggested by Burdine (1953) and Mualem (1976a)]. The best-fit analysis showed that the same quality of fit was obtained by leaving the m_{vg} variable at 1.0 in the van Genuchten (1980) equation. The study also showed that the Fredlund and Xing (1994) "correction factor" could be applied to other SWCC equations and in each case there was an improvement in the quality of fit.

An analysis of the *SoilVision Systems* database of results showed that neither the Mualem (1976a) nor the Burdine (1953) relationship between the m and n variables was close to values computed using the van Genuchten (1980) equation with independent m and n values. The results showed that the n parameter generally varies between a small number and 10 with most of the n values being less than 3. The best-fit m parameters appear to extend beyond values of 50, but most of the values were less than 20. The data presented in Fig. 5.39 suggest that it is best to use m and n as independent variables.

5.5 HYSTERESIS, INITIALIZATION, AND INTERPRETATION OF SWCC

Typical features of the drying and wetting portions of the SWCCs are shown in Fig. 5.40. The hysteresis loop indicates that there is no single or unique SWCC. Consequently, special precautions must be taken when applying the SWCC in geotechnical engineering. The wetting and drying branches form extreme bounds of the water content versus soil suction relationship. There are an infinite number of intermediate drying and wetting scanning curves that bridge between the wetting and drying curves, as shown in Fig. 5.41. The drying and wetting scanning curves become asymptotic to the bounding curves.

An engineer cannot determine whether the soil is on a drying path or a wetting path when the soil is sampled in the field. The difference between the drying and wetting branches of the SWCC may be as much as an order of magnitude in terms of soil suction. This means that water content could correspond to a soil suction ranging between, for example, 10 and 100 kPa. Slight changes in the nature of the soil can also produce a lateral shift in the SWCCs. Consequently, it is difficult to use the SWCC for the estimation of in situ soil suctions, other than to obtain a wide range of approximate values. On the other hand, the SWCCs have proven to be useful for the estimation of unsaturated soil property functions in geotechnical engineering practice.

5.5.1 Hysteresis in SWCC

The drying and wetting SWCCs are significantly different, and in many cases it becomes necessary to differentiate the soil properties associated with the drying curve from those associated with the wetting curve. This means that the geotechnical engineer must make a decision regarding the process that is to be simulated (i.e., the drying or wetting process) and then use the appropriate estimated unsaturated soil property function (Tami et al., 2004). More elaborate soil models that include scanning curves and hysteresis have been developed (Mualem, 1974, 1976a, 1976b; Pham et al., 2003a, b) but it might be more practical at present to use simpler unsaturated soil property models in geotechnical engineering practice. It might also be appropriate in some cases to use an average SWCC (i.e., between the drying and wetting SWCCs) when estimating unsaturated soil property functions.

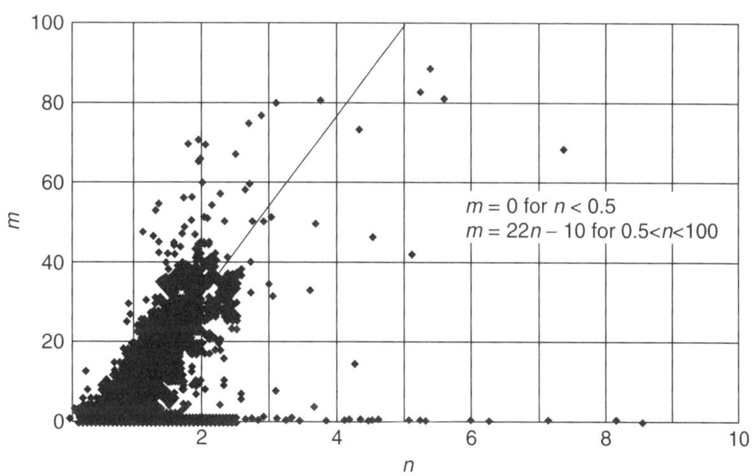

Figure 5.39 Best-fit regression n and m parameters for range of soils best fit with van Genuchten (1980) SWCC equation.

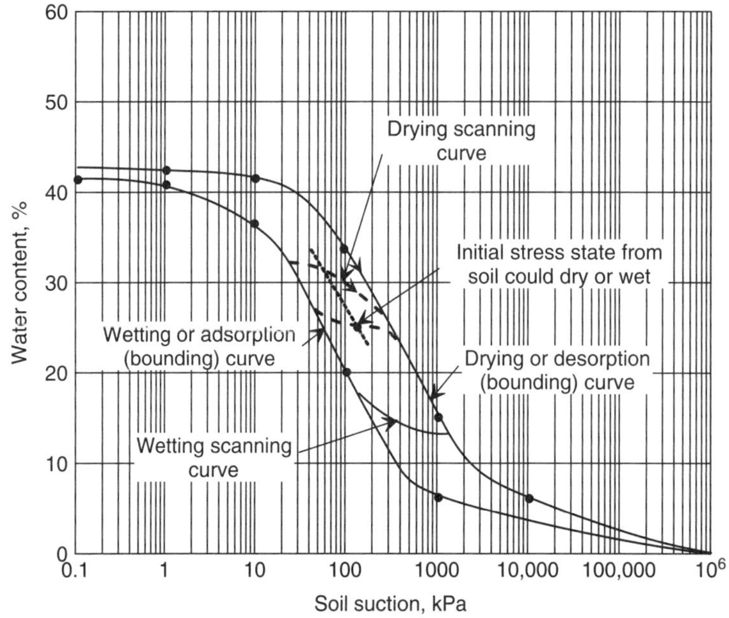

Figure 5.40 Wetting and drying scanning curves and designation of initial stress state.

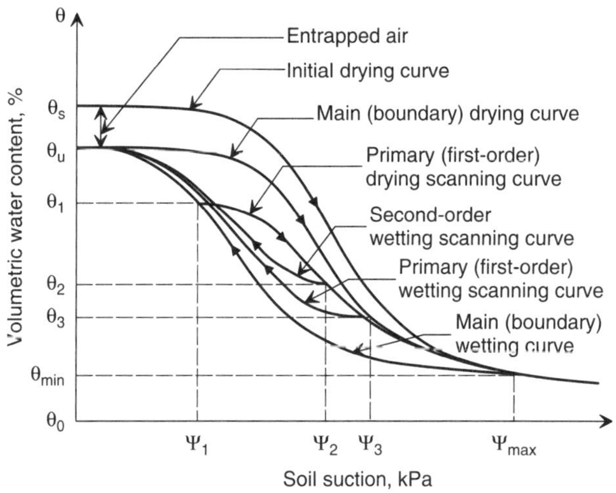

Figure 5.41 Bounding and scanning curves used to define drying and wetting of unsaturated soil (after Pham et al., 2003a).

Desorption curves are the easiest SWCC to measure in the laboratory and, consequently, are most commonly measured. Desorption and adsorption curves have, however, been measured on numerous soil types. Pham et al., (2002) measured drying and wetting curves on sand (Fig. 5.42) and silt (Fig. 5.43). Three specimens were tested in each case and the results were found to be reproducible and consistent. The air-entry value for the sand specimen was 2 kPa and the residual suction was 13 kPa. The air-entry value for the silt was 10 kPa and the residual suction was 120 kPa.

Pham et al., (2002, 2003a, b) analyzed the drying and wetting curves for 34 data sets on a variety of soils reported in the literature. The difference between the hysteresis loops at the inflection points was used as the primary indicator of the magnitude of hysteresis.

The bounding drying curve tends to be approximately congruent (i.e., parallel on a semilog plot) to the bounding wetting curves. Table 5.3 shows that the distance between the main drying and wetting curves varied between 0.15 and 0.35 of a log cycle for sands. The bounding curve spacing for more well-graded silt soils varied between 0.35 and 0.60 of a log cycle. It was suggested that the approximate spacing between the drying and wetting SWCCs could be assumed to be 25% of a log cycle for sands and 50% of a log cycle for well-graded silts.

Hysteretic behavior observed in the laboratory while performing wetting and drying SWCCs has also been observed

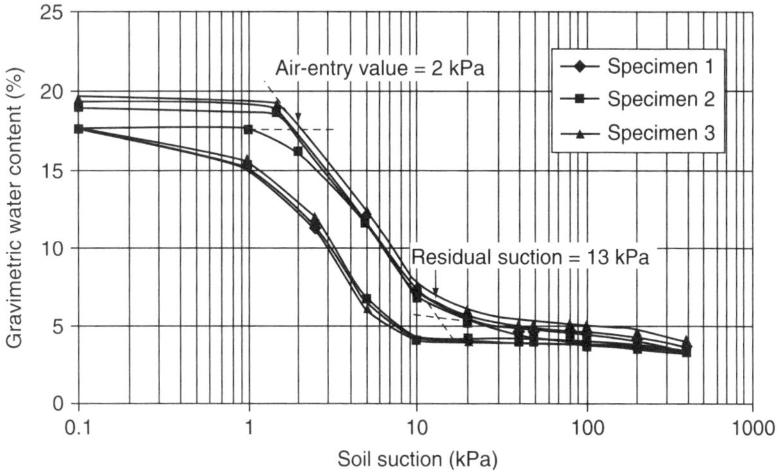

Figure 5.42 Measured drying and wetting SWCCs on Beaver Creek sand (from Pham, 2002).

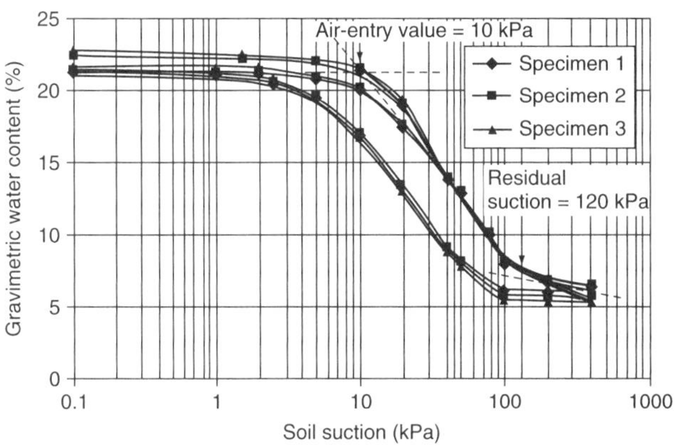

Figure 5.43 Measured drying and wetting SWCCs on processed silt (from Pham, 2002).

Table 5.3 Approximate Shift between Drying and Wetting SWCCs

Soil Type	Range of Shift at Inflection Point as Ratio of Log Cycle	Approximate Average Shift as Percentage of Log Cycle
Sand	0.15–0.35	25%
Silt	0.35–0.60	50%
Clay	0.60 to >1.0	Highly variable

in large-scale laboratory and field tests. Tami et al., (2004b, c) constructed and instrumented a 2-m long model of a slope consisting of two cohesionless soils. The experiment was undertaken to study the performance of capillary barriers on slopes. A flux boundary condition was imposed while water contents were measured using TDR devices and matric

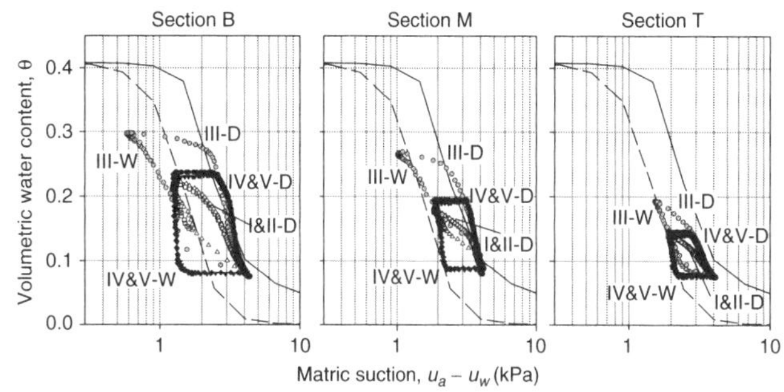

Figure 5.44 SWCCs defined from independent water content and matric suction measurements during wetting and drying model simulations (after Tami et al., 2004b, 2004c).

suctions were measured using tensiometers. The SWCC for each of the soils was independently measured in the laboratory. Figure 5.44 presents some of the model test results and clearly indicates that the SWCC relationships independently measured in the laboratory provided a reasonable representation of the water content versus matric suction bounding curves followed under in situ drying and wetting conditions (Tami et al., 2004b).

5.5.2 Initialization of Stress State in SWCC Test

The suction in a soil specimen is released to zero at the start of a SWCC test. Soil specimens can be placed within a confining ring and given access to water. The water can be imbibed from the bottom of the specimen or given excess water from the top of the specimen. It is preferable to wet the specimen from the bottom. In this way air is not as likely to be entrapped in the soil specimen. The soil specimen is usually allowed one or more days to take on as much water as possible. The soil specimen should be covered with metallic foil during the equalization process to prevent loss of moisture by evaporation.

The soil specimen is usually 20–30 mm in thickness. The prepared soil specimen might also be placed into a pressure plate apparatus and allowed to imbibe water for one or more days. The initial suction in the soil is assumed to be zero once the soil has fully imbibed water. The stress state of the soil has now been altered from in situ conditions and the stress path followed during setup of the soil specimen is illustrated in Fig. 5.45. The specimen may undergo a change in volume if the soil has considerable clay content.

Once the specimen has fully imbibed water, the test data for the SWCC can be obtained by applying a series of applied suction values. Each applied soil suction should be maintained until the specimen has reached an equilibrium state (i.e., no further release of water). The drying (or desorption) bounding curve of the water content versus soil suction relationship is most commonly measured. Figure 5.46 illustrates the change in state experienced by a sand soil as it is given access to water prior to testing.

Understanding the changes in stress state that occur during the setup of the soil specimen assists in the interpretation of the test data. It is clear that the laboratory SWCC data do not represent a soil being dried from its in situ stress state. Rather, the laboratory-measured SWCC data corresponds to a particular test procedure that does not currently represent in situ conditions and may not represent later conditions. However, the SWCC has proven to be of significant value as a guide for the estimation of unsaturated soil property functions.

The test procedure used to measure the SWCC is based on a long history of measurements in agriculture-related disciplines. It could be reasoned that there might be merit in modifying the test procedure to more accurately represent in situ total stress conditions when performing a laboratory SWCC test. The possibility of applying total stresses has been taken into consideration in the design of recent laboratory testing apparatuses.

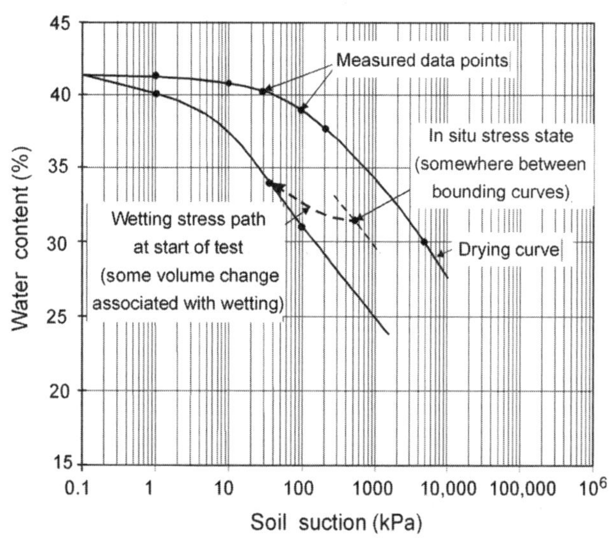

Figure 5.45 Stress path followed at start of pressure plate test on clay soil.

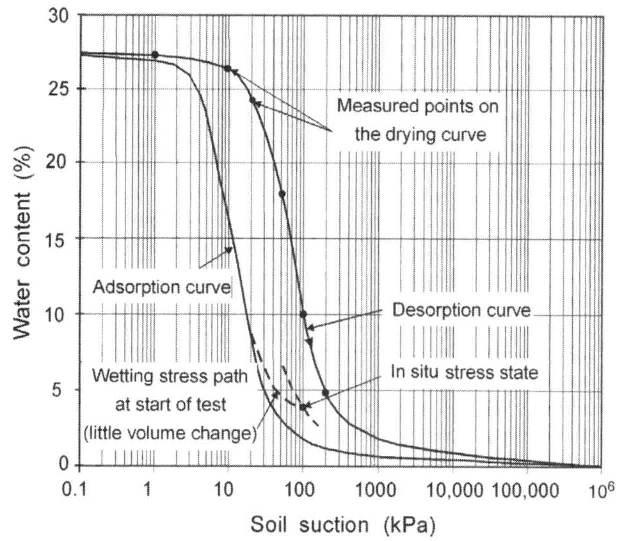

Figure 5.46 Stress path followed during initial stages of pressure plate test on sand.

Geotechnical engineers have been successful in using SWCC data obtained using test procedures developed in agriculture-related disciplines. It has also been possible to successfully use existing data sets for the estimation of unsaturated soil properties (Fredlund, 1998b). Most of the existing data represent the drying or desorption curve subsequent to saturating a soil specimen. Records should also be kept related to details of the soil specimen preparation and test procedure (Fig. 5.47). It is important for the geotechnical engineer to know whether the soil has been initially remolded near the LL, initially compacted near PL-conditions, or tested from an undisturbed or disturbed sample state.

5.5.3 Desaturation Stages along SWCC

The SWCC laboratory test procedure commences with the saturation of the soil specimen. The soil then begins to

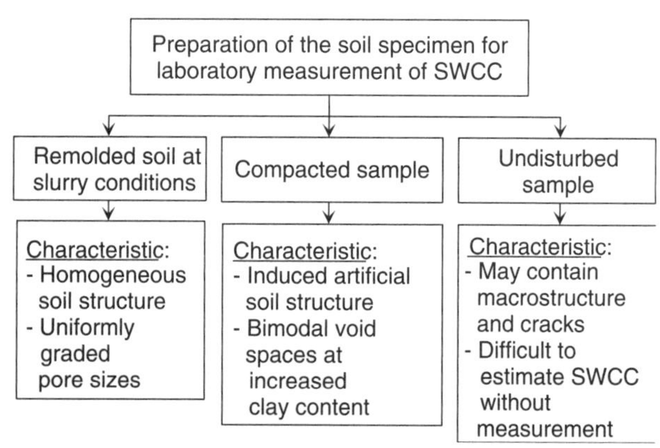

Figure 5.47 Characteristics associated with initial conditions of soil specimen for SWCC test.

desaturate as it is subjected to ever-increasing applied soil suctions. Figure 5.48 shows the desaturation stages along the desorption branch of the SWCC (White et al., 1970). There are three identifiable stages of desaturation: the boundary effect stage, the transition stage (i.e., with primary and secondary transition stages), and the residual stage of desaturation. The air-entry value separates the boundary effect stage and the transition stage. The residual value separates the transition stage and the residual stage. These stages can be clearly defined on a plot of degree of saturation versus logarithm of soil suction. The entire picture of unsaturated soil behavior is most clearly understood when the SWCC data are plotted from a low suction value (e.g., 0.1 kPa) to the maximum value of 10^6 kPa.

A line joining the air-entry value and the residual value (i.e., on a degree of saturation versus logarithm of soil suction plot) defines the primary information required from SWCC. Similar features can also be observed on the wetting SWCC branch. However, the first break on the wetting curve (starting from a high-suction value to a lower suction value) is called the water-entry value and the next break on the wetting curve defines the point where only occluded bubbles remain in the soil. The occluded bubble point on the wetting curve is referred to as the "air closure" soil suction value.

5.5.4 Interpretation of SWCC Data for Low-Volume-Change Soil

The interpretation of pressure plate data is often based on the assumption that the overall volume change of the soil specimen is negligible during the laboratory test. In agriculture-related disciplines, the water content versus soil suction relationship is of greatest importance and the volume change of the soil is of little consequence. In geotechnical engineering applications, the amount of volume change in the soil can be of significance when interpretating laboratory data.

Most sand soils (and even most silt soils) undergo only a small amount of volume change over a wide range of applied soil suctions. Figure 5.49 shows the SWCCs for a sand soil and a graded silt soil. The distinctive features that must be identified along the desorption branch are the air-entry value and the residual conditions (i.e., residual suction and residual water content). It is useful to plot all data sets to a similar abscissa and ordinate scale. It is suggested that the abscissa be a logarithm scale in kilopascals that commonly ranges from 0.1 to 10^6 kPa. For some coarse-grained soils it might also be necessary to extend the soil suction scale lower than 0.1 kPa. The variable and units used to designate the amount of water in the soil should be clearly designated. Each SWCC should be viewed as a desorption "fingerprint" for the soil. When the volume change of the soil is small, any of the three measures for the amount of water in the soil (i.e., gravimetric water content, volumetric water content, and degree of saturation) will yield a similar interpretation of key breaking points on the SWCC.

The sand soil shown in Figure 5.49 has an air-entry value of 1.3 kPa and a residual suction of 5 kPa with a residual gravimetric water content of 7%. The graded silt has an air-entry value of 3 kPa and a residual suction of approximately 100 kPa with a residual gravimetric water content of 11%. The slope of a line joining the air-entry value and the residual point (i.e., slope at the inflection point) provides a measure of the rate at which water is removed from the soil as suction is increased past the air-entry value. The slope of the SWCC indicates the rate of change of water storage in the soil with respect to soil suction. The interpretation of the SWCC for a low-volume-change soil is based on the assumption that changes in void ratio approach zero as suction is increased. Consequently, dw or $d\theta$ or dS portray similar information regarding unsaturated soil behavior.

5.5.5 Interpretation of SWCC Data for High-Volume-Change Soil

The interpretation of SWCC data for a soil that undergoes a high volume change in response to changes in soil suction

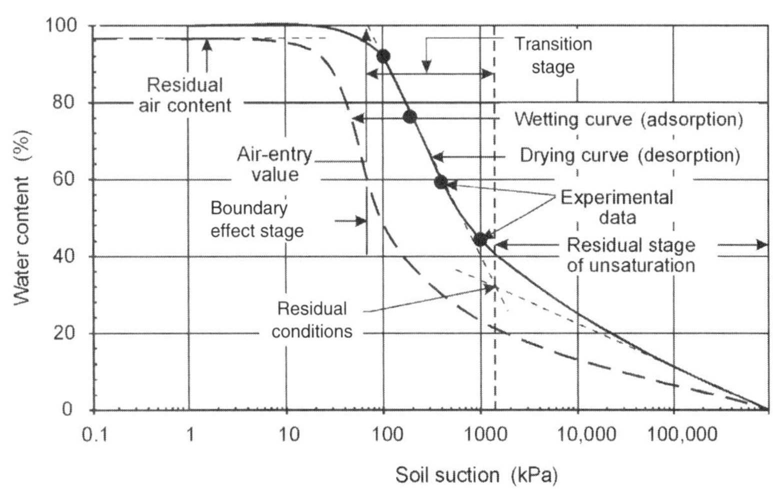

Figure 5.48 Definition of variable associated with typical SWCC.

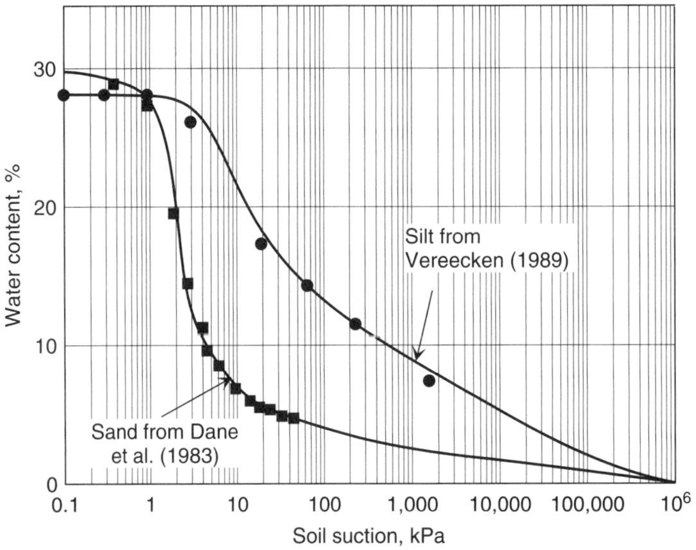

Figure 5.49 Desorption SWCC data for sand (data from Dane and Hruska, 1983) and silt (data from Vereecken et al., 1989).

is considerably different from that of a low-volume-change soil. A highly plastic clay (i.e., Regina clay) was prepared in an initial slurry state (i.e., slurry water content of 88%) and then preconsolidated to various pressures. The test results on the preconsolidated clay are used to illustrate the interpretation of SWCC data on a high-volume-change soil. There are changes in the amount of water in the soil (i.e., dw or $d\theta$ or dS) and also changes in overall volume (i.e., de or dn or dv where n is porosity and v is specific volume).

The liquid limit of Regina clay is 75%, the plastic limit is 25%, and the soil has 50% clay size particles. Figure 5.50 shows the gravimetric water content versus soil suction plot for Regina clay soil specimens that were preconsolidated to 6.2 and 400 kPa. The initial saturated water content of the soil specimen preconsolidated to 6.2 kPa was 70% and the water content of the soil specimen preconsolidated to 400 kPa was 56%. The preconsolidated specimens show a gradual curvature onto the virgin compression branch for the soil as soil suction is increased. The curves lack a distinct break in curvature common to specimens loaded in one-dimensional compression. Fredlund (1964) observed that the gradual curvature of the SWCC onto the virgin compression branch is probably due to the three-dimensional volume change during suction change and is certainly not related to the air-entry value of the soil. Figure 5.51 shows a shrinkage curve for the same soil and Fig. 5.52 shows a plot of degree of saturation versus water content for specimens of Regina clay. Both plots show that the soil begins to desaturate near the plastic limit of the soil (i.e., 25%). The soil suction at the plastic limit is close to the limiting suction that can be applied in the pressure plate apparatus (i.e., about 1500 kPa).

The above information can be cross-plotted to give two additional graphs of the volume-mass soil properties versus soil suction. A plot of void ratio e versus soil suction yields

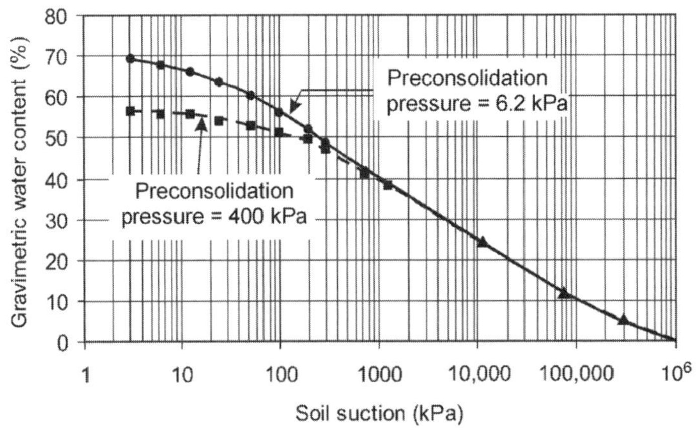

Figure 5.50 Gravimetric water content versus soil suction for Regina clay prepared at two preconsolidation pressure values (from Fredlund, 1964).

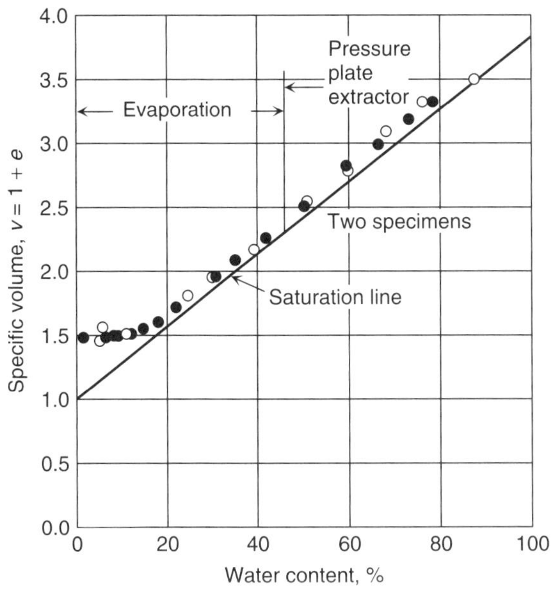

Figure 5.51 Specific bulk volume versus gravimetric water content for Regina clay (from Fredlund, 1964).

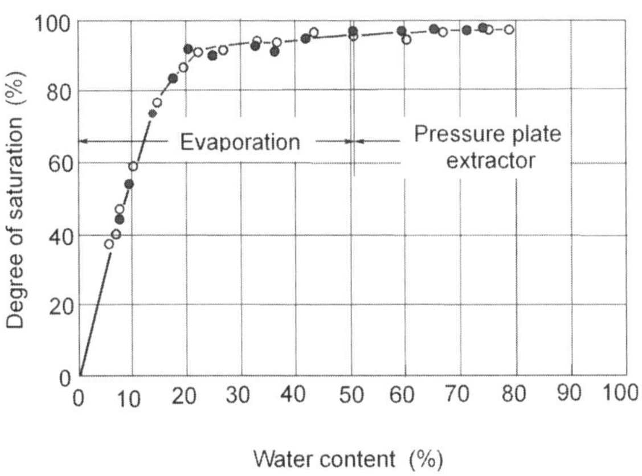

Figure 5.52 Degree of saturation versus gravimetric water content for Regina clay (from Fredlund, 1964).

a graph that is similar in shape to the gravimetric water content SWCC with the exception that the high-suction portion remains horizontal at a void ratio corresponding to the shrinkage limit (i.e., $e = 0.45$). The large volume changes that occur as soil suction is increased tend to mask the actual air-entry value of the soil. The shrinkage curve of the soil can be measured and used to assist in the interpretation of the SWCC test results. It is possible to plot degree of saturation versus the logarithm of soil suction and determine the correct air-entry value of Regina clay.

Figure 5.13 (i.e., previously shown in Chapter 5) is a plot of the degree of saturation versus soil suction for Regina clay soil specimens. The results show that all soil specimens produce essentially one unique line with a distinct break for the air-entry value around 1500 kPa. In other words, the correct air-entry value must be determined from the degree of saturation versus logarithm of soil suction plot. The air-entry value of clay soils can be extremely high, particularly when the soil is in an initial slurry state.

Residual conditions may not be clearly defined on any of the forms of the SWCC plot. Residual conditions would appear to be near to (or below) the shrinkage limit of the soil. The measured shrinkage limit for Regina clay was 13% and the corresponding degree of saturation would appear to be around 50%. In this case, residual suction would appear to be on the order of 30,000 kPa. The air-entry values and the residual values for the highly plastic clay illustrate the significantly different SWCC behavior that can be observed for sand soils and clay soils. The degree of saturation versus soil suction plot provides the most valuable information for the estimation of the air-entry value of a soil. The degree of saturation versus soil suction plot is generally the most meaningful relationship to use for the estimation of unsaturated soil property functions.

It is also possible to plot volumetric water content versus soil suction for Regina clay samples tested. The differentiation (or slope) of the volumetric water content versus soil suction plot provides the water storage function required as input for a transient unsaturated soil seepage analysis. Volumetric water content can be calculated in one of two ways: with reference to either the original total volume or the instantaneous total volume. The volumetric water content is defined as follows when calculations are referenced to the original total volume:

$$\theta_1 = \frac{V_w}{V_{t0}} \tag{5.62}$$

where:

θ_1 = referential volumetric water content,
V_w = volume of water, and
V_{t0} = total volume of the soil specimen at the start of the test.

Volumetric water content (i.e., referenced to the original total volume) versus soil suction results in similar limitations to those encountered when using gravimetric water content. On the other hand, a plot of instantaneous volumetric water content versus soil suction exhibits the character of the degree of saturation versus soil suction plot. It is important to specify the manner in which volumetric water content is computed when high-volume-change soils are being analyzed.

The air-entry value and the residual conditions provide the most important information for generating unsaturated soil property functions. Consequently, a plot of degree of saturation (or instantaneous volumetric water content) is of greatest significance to the geotechnical engineer. This is disconcerting since it is difficult to directly measure the overall volume of the soil specimen when using conventional pressure plate equipment.

There are some pressure plate apparatuses available that attempt to measure overall volume change as well as water volume change. The equipment is more elaborate and more costly to operate but may provide measurements of the degree of saturation versus soil suction. A normal stress can be applied to the soil specimen while vertical deformation is measured. Volume changes can be measured provided the soil specimen does not separate from the sides of the K_0 ring during increases in soil suction. The alternative is to measure the shrinkage curve for the soil in conjunction with the SWCC test. The degree of saturation versus soil suction relationship can then be computed and used in the interpretation of the test results.

5.5.6 Linkage between SWCC and Unsaturated Soil Properties

The SWCC can be used to undertake a subsequent set of estimations related to unsaturated soil property functions (USPFs). Examples of subsequent estimations of functions are the hydraulic conductivity function, the water storage function, the shear strength function, and the volume-mass change functions. The accuracy of the computed USPFs are dependent upon the accuracy with which the SWCC is characterized.

There is an interrelationship between the SWCC and the estimated USPFs (Fredlund et al., 1995b; 1997b). A consistent linkage to the SWCC must be respected and maintained throughout all estimations for modeling purposes. In other words, it is not possible to use the SWCC to estimate a USPF and then subsequently change any aspect of the USPF in order to alleviate convergence problems associated with solving a highly nonlinear partial differential equation. The numerical modeler must be aware that a change in a USPF means that there is a change in the soil being analyzed. Consequently, any change in one of the USPFs produces an inconsistency with respect to other USPFs used in the analysis.

USPFs can produce highly nonlinear water storage and permeability functions, particularly for sand soils. The highly nonlinear functions may subsequently create convergence difficulties when solving partial differential equations such as the unsaturated soil seepage equation. Convergence difficulties may tempt the modeler to adjust one or more of the unsaturated soil parameters in order to obtain a converged solution. This is "ill-advised practice" and should not be advocated. Rather, it is important to use partial differential equation solvers that can ensure convergence under highly nonlinear conditions.

Misfeldt et al., (2006) reported on the "preliminary design" of a cover system where it was shown that a change in one of the soil parameters associated with the hydraulic conductivity function significantly affected the outcome of the cover design. Changing any one soil parameter changes the soil that is being analyzed. It should also be noted that while estimation techniques can be used to obtain SWCCs for preliminary design purposes, it is important that individual unsaturated soil property parameters not be arbitrarily changed when performing a numerical modeling simulation.

5.6 PHAM AND FREDLUND (2011) EQUATION FOR ENTIRE SWCC

Early empirical equations proposed for the best fit of water content–soil suction data (i.e., prior to about the year 2000) failed to adequately define the portion of the SWCC below the air-entry value. The shortcoming in the empirical SWCCs was pronounced when the soil exhibited significant volume change as soil suction was increased (e.g., initially slurry soils). An asymptotic equation was not appropriate for the low-soil-suction range because water storage tended toward zero and created numerical instability problems during the modeling of water flow problems.

Pham and Fredlund (2011) proposed two SWCC equations that assist in modeling the low-suction range as well as the entire soil suction range for a soil (Fredlund and Pham, 2006). The two proposed equations for the SWCC are referred to as (i) the meaningful parameter SWCC equation, and (ii) the simplified SWCC equation. The meaningful parameter SWCC equation has curve-fitting parameters that have physical significance and meaning. The simplified SWCC equation is particularly useful when performing analytical solutions for unsaturated soil engineering applications.

The SWCC (i.e., in terms of gravimetric water content) typically is S shaped (i.e., for unimodel behavior). Soil suction changes have a similar influence on the soil to changes in net mean total stress when suctions are less than the air-entry value. The first portion of the gravimetric SWCC should therefore have a slope consistent with the volume change properties of the soil. The magnitude of the initial slope depends on the stress history of the soil (Fig. 5.53).

Let us consider the drying or wetting of a soil under zero net mean total stress. When soil suction is less than the air-entry value, the volume change due to a soil suction change is equal to the application of an isotropic net mean total stress. The slope of the initial portion of the SWCC at soil suctions less than the air-entry value can be calculated as in Eq. 5.63. The differentiation applies for the virgin compression branch (or the recompression branch) as long as the applied suctions are less than the air-entry value:

$$\frac{dw}{d\psi} = \frac{d(e/G_s)}{d\psi} = \begin{cases} \dfrac{C_c}{G_s} & \text{for } \psi > p_c \\[2ex] \dfrac{C_s}{G_s} & \text{for } \psi \le p_c \end{cases} \tag{5.63}$$

where:

w = gravimetric water content,
e = void ratio,
G_s = specific gravity,
C_s = unloading-reloading index,
C_c = virgin compression index,
p_c = preconsolidation pressure, and
ψ = soil suction.

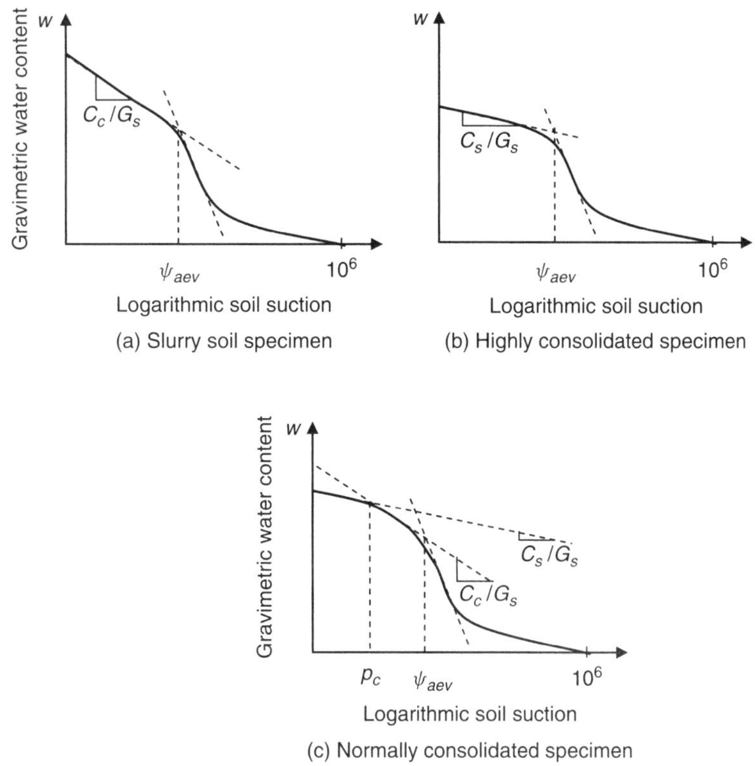

Figure 5.53 Typical shapes for SWCCs (i.e., gravimetric water content) of a soil subjected to various stress histories (after Pham, 2005).

Several typical forms for the gravimetric water content versus soil suction curve are shown in Fig. 5.53. A clay soil, for example, can be subjected to different stress histories with the result that the SWCC changes; however, the general form of the equation remains the same. A slurry soil will be highly compressible in the low-effective-stress range. Preconsolidated soils may have been subjected to a stress state that is higher than the air-entry value of the soil. Most originally proposed empirical equations for the SWCC are sigmoidal in shape. Consequently, these equations cannot adequately represent the SWCC in the region below the air-entry value since the equations are asymptotic to a horizontal line. The sigmoidal equations are more suitable for best fitting the degree of saturation form of the SWCC. It is prudent to have a curve-fitting equation that can be applied to soils that exhibit significant volume change (Mbonimpa et al., 2006).

5.6.1 SWCC Equation with Meaningful Parameters for Entire Suction Range

It is useful to have SWCC equations written in terms of variables that have physical significance (e.g., air-entry value; residual soil suction; residual water content; maximum slope of the SWCC). Such an equation is particularly useful when conducting sensitivity studies where the fundamental properties of the soil can be varied. The basic properties of a SWCC are shown in Fig. 5.54. Most SWCCs can be represented by three straight lines with slopes of S_1, S_2, and

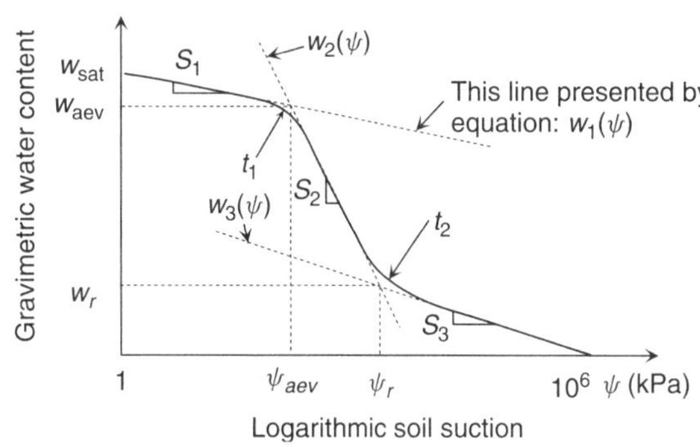

Figure 5.54 Typical SWCC for a soil that changes volume along with a designation of the variables required to define the curve (after Pham, 2005).

S_3. The water content at a soil suction of 10^6 kPa is assumed to be equal to zero (Fredlund and Xing, 1994). Parameters t_1 and t_2 can be used to control the transitions between various portions of the SWCC. Three equations correspond to the three sloping lines that can be written as three mathematical functions [i.e., $w_1(\psi)$, $w_2(\psi)$, and $w_3(\psi)$]:

$$w_1(\psi) = w_s - S_1 \log\left(\frac{\psi}{1}\right) = S_3 \log(10^6) + (S_1 - S_2) \times \log(\psi_{aev}) + (S_2 - S_3)\log(\psi_r) - S_1 \log(\psi) \tag{5.64}$$

where:

w_s = gravimetric water content at soil suction of 1 kPa and
S_1, S_2, S_3 = slopes of the three parts of the SWCC.

An arbitrary starting reference point for the $w_1(\psi)$ equation is the water content corresponding to 1 kPa. Equation 5.64 can be modified so that reference suctions other than 1 kPa are preferred:

$$w_2(\psi) = w_{aev} - S_2 \log\left(\frac{\psi}{\psi_{aev}}\right) = S_3 \log(10^6) + (S_2 - S_3)\log(\psi_r) - S_2 \log(\psi) \quad (5.65)$$

where:

w_{aev} = gravimetric water content at the air-entry value,
ψ_{aev} = soil suction at the air-entry value of the soil,
S_2 = slope of the SWCC for the portion between the air-entry value and the residual soil suction, and
S_3 = slope of the portion beyond the residual soil suction,

and

$$w_3(\psi) = S_3 \log\left(\frac{10^6}{\psi}\right) \quad (5.66)$$

A mathematical technique can be used to connect the above three sloping line equations into a single SWCC equation. A function f can be used to connect any two straight lines. The function varies between 0 and 1 with an inflection point at the intersection of the two straight lines. This technique has been implicitly used in some SWCC equations. A general form of the combined functions for the SWCC-fitting equation is as follow:

$$w(\psi) = [w_1(\psi) f_1(\psi, \psi_{aev}) + w_2(\psi) f_2(\psi, \psi_{aev}) + f_3(\psi, \psi_{aev})] f_4(\psi, \psi_r) + w_3(\psi) f_5(\psi, \psi_r) + f_6(\psi, \psi_r) \quad (5.67)$$

where:

ψ = soil suction and
f_1, f_2, f_3, f_4 = function that must satisfy the following conditions:
f_1 is increased to 1 when ψ is decreased from ψ_{aev} to 0,
f_1 is decreased to 0 when ψ is increased from ψ_{aev} to $+\infty$,
$f_2 = 1 - f_1$,
f_4 is increased to 1 when ψ is decreased from ψ_r to 0,
f_4 is decreased to 0 when ψ is increased from ψ_r to $+\infty$,
$f_5 = 1 - f_4$,
f_3 is added to control the transition at the air-entry value, and
f_6 is added to control the transition at the residual soil suction.

The following equation can be used to represent the functions f_1 and f_4:

$$f(s, s_1) = \frac{s_1^{n'}}{s^{n'} + s_1^{n'}} \quad (5.68)$$

where:

s, s_1, and n' = three arbitrary variables.

Plots of Eq. 5.68 with $s_1 = 10$ and varying n' values (i.e., $n' = 1$, 4, and 10) are shown in Fig. 5.55. Function f_3 controls the transition at the air-entry value of the SWCC. Function f_3 can be obtained by differentiating function f_1 on a logarithmic scale:

$$f_3(s) = \frac{d[f_1(s)]}{\log(s)}\alpha \quad (5.69)$$

where:

α = a scaling factor.

Function f_6 can be obtained in a similar manner. Substituting Eqs. 5.64, 5.65, 5.66, 5.68, and 5.69 into Eq. 5.67 yields the following SWCC equation containing the slopes of three parts of the SWCC:

$$w(\psi) = \begin{bmatrix} A(\psi) \times (S_2 - S_1)\left[\log\left(\frac{\psi}{\psi_{aev}}\right) - \frac{\ln(10)}{2t_1} \right. \\ \left. [1 - A(\psi)]\right] + (S_3 - S_2)\left[\log\left(\frac{\psi}{\psi_r}\right) \right. \\ \left. - \frac{\ln(10)}{2t_2}(1 - B(\psi))\right] \end{bmatrix} \times B(\psi) + S_3 \log\left(\frac{10^6}{\psi}\right) \quad (5.70)$$

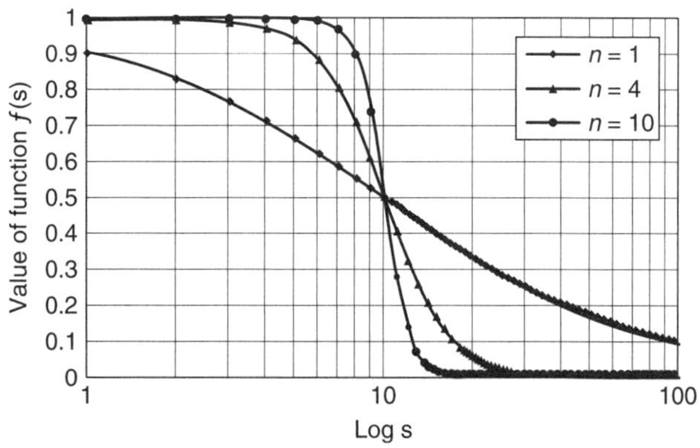

Figure 5.55 Plots of meaningful parameter SWCC equation used to represent functions f_1 and f_4 (after Pham, 2005).

where

$$A(\psi) = \frac{\psi_{\text{aev}}^{t_1}}{\psi_{\text{aev}}^{t_1} + \psi^{t_1}} \qquad B(\psi) = \frac{\psi_r^{t_2}}{\psi_r^{t_2} + \psi^{t_2}}$$

and

t_1, t_2 are curve-fitting parameters that control the transitions of the curve.

The slope of the curve at and beyond residual soil suction is of particular interest in geotechnical engineering when considering problems involving the calculation of actual evaporation. The variable S_3 can be replaced by a more meaningful variable, w_s (i.e., gravimetric water content at a soil suction of 1 kPa). The relationship between S_3 and w_s can be written as follows:

$$S_3 = \frac{w_{\text{sat}} + (S_2 - S_1)\log(\psi_{\text{aev}}) - S_2 \log(\psi_r)}{\log\left(\dfrac{10^6}{\psi_r}\right)} \tag{5.71}$$

Equation 5.70 can now be reduced to the form shown in Eq. 5.72. Figure 5.56 shows typical plots of Eq. 5.72 for various soil properties (i.e., curve-fitting parameters). The transition parameters used for plotting Fig. 5.56 are $t_1 = 5$ and $t_2 = 5$:

$$w(\psi) = w_s(M_1 + M_3) + S_1[-\log(\psi_{\text{aev}})(M_1 + M_3) - M_2] + S_2\left(-M_1 \log\left(\frac{10^6}{\psi_{\text{aev}}}\right) - \log\left(\frac{\psi_r}{\psi_{\text{aev}}}\right) M_3 + M_2\right) \tag{5.72}$$

where

$$M_1(\psi) = \frac{\left[\log\left(\dfrac{\psi}{\psi_r}\right) - \dfrac{\ln(10)}{2t_2}(1 - \dfrac{\psi_r^{t_2}}{\psi_r^{t_2} + \psi^{t_2}}\right] \times \dfrac{\psi_r^{t_2}}{\psi_r^{t_2} + \psi^{t_2}}}{\log\left(\dfrac{10^6}{\psi_r}\right)}$$

$$M_2(\psi) = \frac{\left[\log\left(\dfrac{\psi}{\psi_{\text{aev}}}\right) - \dfrac{\ln(10)}{2t_1}(1 - \dfrac{\psi_{\text{ae}}^{t_1}}{\psi_{\text{aev}}^{t_1} + \psi^{t_1}}\right]}{\dfrac{\psi_{\text{aev}}^{t_1}\psi_r^{t_2}}{(\psi_{\text{aev}}^{t_1} + \psi^{t_1})(\psi_r^{t_2} + \psi^{t_2})}}$$

$$M_3(\psi) = \frac{\log(10^6) - \log(\psi)}{\log(10^6) - \log(\psi_r)}$$

The curve-fitting parameter t_1 controls the radius of curvature at the air-entry value and curve-fitting parameter t_2 controls the radius of curvature at residual suction. These two curve-fitting parameters have little influence on the shape of the remainder of the SWCC. The transition parameters t_1 and t_2 can be chosen as constants prior to undertaking a curve-fitting process (e.g., $t_1 = 5$ and $t_2 = 5$). Higher values for t_1 and t_2 result in more distinct breaks along the SWCC. When best fitting the degree-of-saturation SWCC, the curve-fitting parameter S_1 can be set to zero. The curve-fitting parameter w_s (i.e., saturated water content) should be replaced by a constant value (i.e., 100% for the degree of saturation).

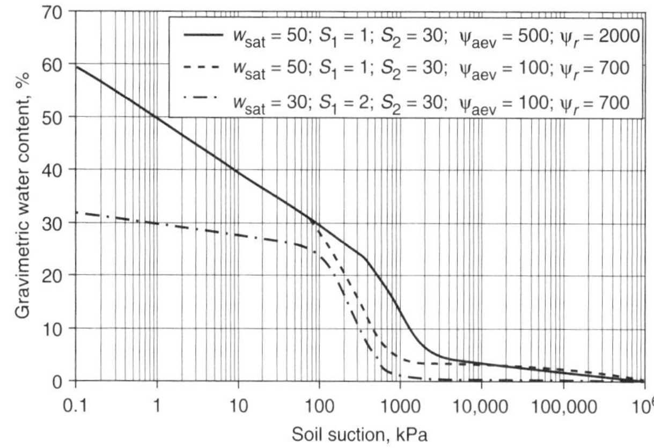

Figure 5.56 Sample plots of the meaningful parameter SWCC equation using various curve-fitting parameters (after Pham, 2005).

5.6.2 Pham and Fredlund (2009) Simplified Equation for Entire SWCC

The simplified SWCC equation for high-volume-change soils also has curve-fitting parameters that are independent of one another; however, the parameters may not have physical meaning. The equation has two parts: (i) a part of the simplified SWCC equation that is for soil suction values less than residual soil suction (i.e., ψ_r) and (ii) a part that makes use of the Fredlund and Xing (1994) correction factor. The equation has the following form:

$$w(\psi) = \left([w_s - S_1 \log(\psi) - w_r]\frac{a}{\psi^b + a} + w_r\right) \times \left(1 - \frac{\ln[1 + \psi/\psi_r]}{\ln[1 + 10^6/\psi_r]}\right) \tag{5.73}$$

where:

w_s = gravimetric water content at 1 kPa soil suction,
S_1 = slope of the portion of the curve in the low soil suction,
w_r = residual water content,
a, b = curve-fitting parameters,
ψ^b = soil suction at the b fitting parameter, and
ψ_r = residual soil suction [i.e., can be approximated as $(2.7a)^{1/b}$]

The water content of a soil at a soil suction of 1 kPa is used as a reference starting point for the SWCC. The slope of the portion of the SWCC for soil suctions less than the air-entry value is defined back to 1 kPa of suction, assuming that the air-entry value is greater than 1 kPa. Therefore, w_s and S_1 can be considered as two known values.

The bending point on an SWCC at residual conditions occurs at a suction of approximately $(2.7a)^{1/b}$. The user can either choose to set the residual soil suction value ψ_r as $(2.7a)^{1/b}$ or select the residual suction value as an independent curve-fitting parameter. If ψ_r is replaced by $(2.7a)^{1/b}$, the curve has only three additional curve-fitting parameters: w_r, a, and G_s). Plots of Eq. 5.73 with various soil parame ters are shown in Fig. 5.57.

5.6.3 Pham and Fredlund (2011) Equation Fit of Soil Data

The performance of the two Pham and Fredlund (2008) equations can be illustrated using published SWCC results on several soils. The selected SWCC results are for (i) Regina clay (Fredlund, 1964), (ii) Jossigny silt (Fleureau et al., 1995), (iii) kaolin (Fleureau et al., 2004), (iv) Beaver Creek sand, (v) processed silt, (vi) Saskatchewan silty sand, and (vii) Indian Head till (Pham, 2005). All soils used in the testing program were prepared as slurry. The classification properties of some of the soils tested at the University of Saskatchewan, Saskatoon, are presented in Table 5.4. The grain-size distribution curves for these soils are presented in Fig. 5.58.

The seven soils were best fit using the meaningful parameter SWCC equation (Eq. 5.72). The parameter w_s was estimated for each soil and a curve-fitting procedure was used to find the following soil properties: (i) slope of the SWCC at soil suctions less than the air-entry value, S_1; (ii) slope of the main portion of the SWCC (i.e., desaturation zone), S_2; (iii) air-entry value of the soil, ψ_{aev}; and (iv) residual soil suction ψ_r. Best-fitting curves for Regina clay (Fredlund, 1964), Jossigny silt (Fleureau et al., 1995), and kaolin (Fleureau et al., 2004) are shown in Fig. 5.59. Best-fitting curves for the four soils tested at the University of Saskatchewan, Saskatoon, are shown in Fig. 5.60. Curve-fitting parameters associated with the seven soils are summarized in Table 5.5.

The best-fit results for the seven soils using the simplified SWCC equation (i.e., Eq. 5.73) were also computed. The slopes of the SWCCs at soil suctions less than the air-entry values for the seven soils were estimated prior to

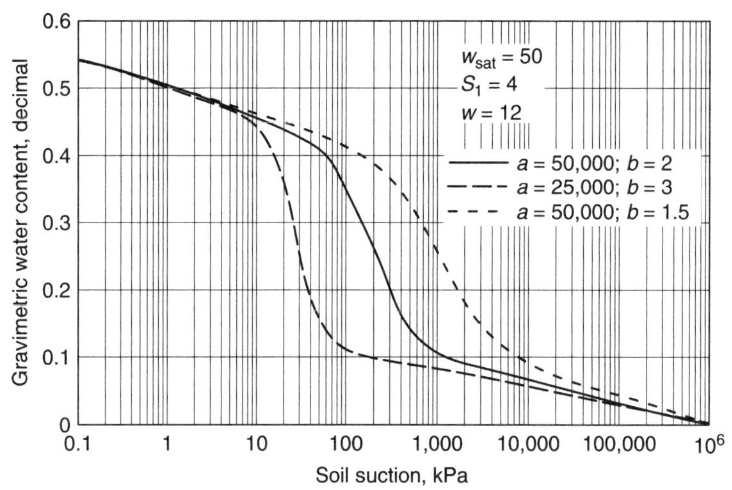

Figure 5.57 Plot of SWCC equation when using different curve-fitting parameters (after Pham, 2005).

Table 5.4 Soil Properties for Four Soils Tested at University of Saskatchewan

Soil	Liquid Limit (LL)	Plastic Limit (PL)	Specific Gravity (G_s)
Beaver Creek sand	N/A	N/A	2.65 (Bruch, 1993)
Saskatchewan silty sand	N/A	N/A	2.65
Processed silt	26.8 (Wilson, 1990)	25.4 (Wilson, 1990)	2.70 (Wilson, 1990)
Indian Head till	36.1	16.4	2.73

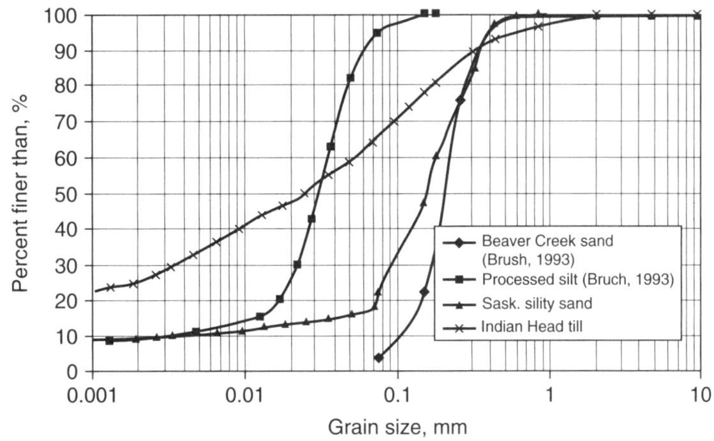

Figure 5.58 Grain-size distributions for four soils for which SWCCs were measured (after Pham, 2005).

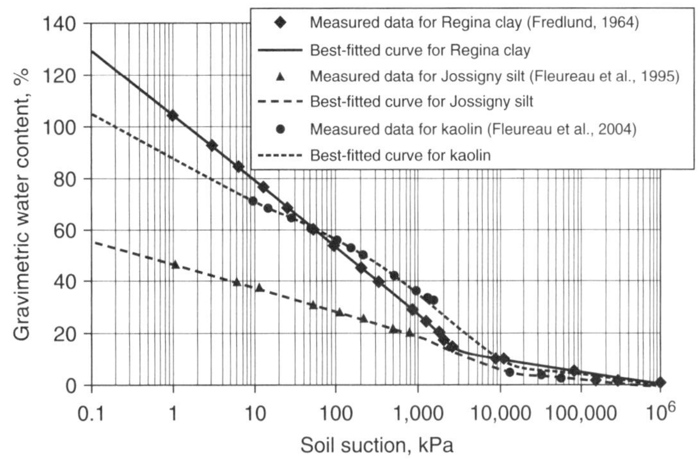

Figure 5.59 Best-fitting curves using meaningful parameter SWCC equation along with published data sets for three soils (Data fits by Pham, 2005).

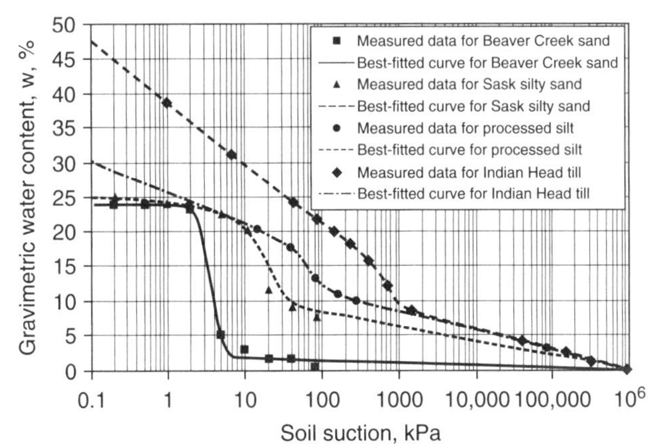

Figure 5.60 Best-fitting curves using meaningful parameter SWCC equation along with measured data sets for four soils (after Pham, 2005).

undertaking the best-fitting procedure. The fitting procedure was used to find the parameters a, b, S_1, w_s, and w_r. Best-fit SWCC parameters for three soils are shown in Fig. 5.61. Best-fit SWCC parameters corresponding for the soils tested at the University of Saskatchewan, Saskatoon, are shown in Fig. 5.62. A summary of the curve-fitting parameters corresponding to the simplified SWCC equation for seven soils is shown in Table 5.6.

5.6.4 Comments on Pham and Fredlund (2011) SWCC Equations

The Pham and Fredlund (2011) SWCC equations (i.e., Eqs. 5.72 and 5.73) make use of the curve-fitting parameters w_s and S_1 corresponding to saturated initial slurry conditions. The variable w_s represents the gravimetric water content at a soil suction of approximately 1 kPa (provided the soil

Table 5.5 Curve-Fitting Parameters for Seven Soils Using Meaningful Parameter SWCC Equation

ID	Soil	Curve-Fitting Parameters				
		w_s	S_1	S_2	ψ_{aev}, kPa	ψ_r, kPa
1	Regina clay[b]	1.042	0.256	0.677	2310	2780
2	Jossigny silt	0.463	0.089	0.169	1010	4820
3	Kaolin	0.881	0.168	0.303	883	12,500
4	Beaver Creek sand	0.239	0.003	0.652	2.46	5.33
5	Saskatchewan silty sand	0.241	0.010	0.200	8.01	40.2
6	Processed silt	0.255	0.044	0.140	33.7	118
7	Indian Head till[c]	0.387	0.088	0.498	801	958

[a]Parameters t_1 and t_2 are chosen as constant values prior to the fitting procedure with $t_1 = 4$ and $t_2 = 8$.
[b]Several empirical data points were added along the desaturation part for the Regina Clay to assist in curve fitting.
[c]Data points for the SWCC at suctions higher than 1500 kPa were measured by Sillers (1997).

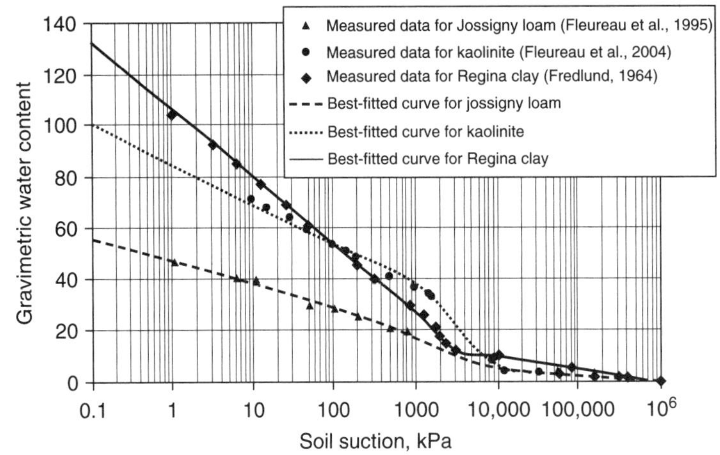

Figure 5.61 Best-fit SWCCs using simplified SWCC along with published data sets for three soils (after Pham, 2005).

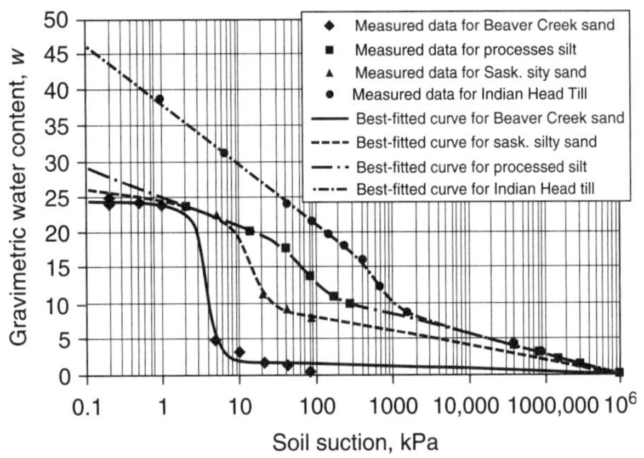

Figure 5.62 Best-fit SWCCs using simplified SWCC equation along with measured data sets for four soils (after Pham, 2005).

has an air-entry value greater than 1 kPa). For soils with an air-entry value less than 1 kPa, w_s represents the water content at a soil suction of 1 kPa on an extension of the line through a suction value less than the air-entry value. The curve-fitting parameter S_1 represents the slope of the gravimetric SWCC at suction values less than the air-entry value.

Figures 5.59–5.62 show that the two SWCC equations proposed by Pham and Fredlund (2009) result in close fits of measured SWCC data for a wide range of soil types (i.e., from sand to clay). Both equations make use of the curve-fitting parameters w_s and S_1 (i.e., representing the gravimetric water content at 1 kPa soil suction and the slope of the SWCC at soil suctions less than the air-entry value).

The meaningful parameter SWCC equation requires five curve-fitting parameters (i.e., in addition to the two transitional parameters for curvature at the air-entry and residual points along the SWCC). The five curve-fitting parameters

Table 5.6 Curve-Fitting Parameters for Seven Soils Using Simplified SWCC Equation

ID	Soil	Curve-Fitting Parameters					
		w_s	S_1	a	b	w_r	ψ_r, kPa
1	Regina clay[b]	1.057	0.261	1.07×10^{33}	9.838	0.131	2520
2	Jossigny silt	0.467	0.086	7.13×10^{4}	1.404	0.064	5810
3	Kaolin	0.830	0.145	1.92×10^{9}	2.527	0.049	6990
4	Beaver Creek sand	0.242	0.0046	3.70×10^{3}	6.186	0.020	4.43
5	Saskatchewan silty sand	0.244	0.015	5.30×10^{4}	4.052	0.098	18.7
6	Processed silt	0.250	0.036	3.01×10^{5}	2.824	0.114	124.
7	Indian Head till[c]	0.382	0.084	1.47×10^{8}	2.627	0.078	1880

[a]Residual soil suctions were calculated using the curve-fitting parameters a and b.
[b]Several empirical data points were added along the desaturation part of the Regina clay for ease in curve fitting.
[c]Data points for the SWCC at suctions higher than 1500 kPa were measured by Sillers (1997).

are independent and have physical soil property meaning (e.g., water content at 1 kPa soil suction, slope of the curve at soil suctions less than the air-entry value, the air-entry value, residual soil suction, maximum slope of the SWCC). The curve-fitting parameters for the SWCC can also be estimated from basic soil classification properties. The proposed SWCC equations closely represent gravimetric water content–soil suction conditions over the entire range of soil suction values.

The simplified SWCC equation also requires five curve-fitting parameters, but only four parameters are truly independent (i.e., since residual soil suction can be approximated using the curve-fitting parameters a and b). Of the four curve-fitting parameters, only two parameters have physical meaning (i.e., the gravimetric water content at 1 kPa soil suction and the slope of the SWCC at soil suctions less than the air-entry value). The advantage of this equation is its mathematical simplicity while still retaining an accurate representation of the entire SWCC. The simplified SWCC equation retains a defined slope at soil suctions less than the air-entry value. The equation is simple to use and can be readily implemented in analytical solutions for geotechnical engineering problems.

5.7 GITIRANA AND FREDLUND (2004) SWCC

Gitirana and Fredlund (2004) proposed an equation with independent physical parameters for the SWCC. The equation is a combination of two rotated and translated hyperboles. The equation was intended for use with the degree-of-saturation SWCCs.

Three types of equations were proposed to fit various shapes of the SWCC: (1) a unimodal equation with one bending point, (2) a unimodal equation with two bending points, and (3) a bimodal equation. The equations are based on the general hyperbolic equation in the coordinate system, log ψ versus degree of saturation S. The equation parameters are defined as the coordinates where the hyperbola asymptotes meet. A consistent geometric relationship exists between the shape of the SWCC and the equation parameters.

Wetting SWCCs that achieve a maximum degree of saturation that is less than 100% can be represented by multiplying the proposed equations by the maximum achievable degree of saturation.

5.7.1 Unimodal Equation with One Bending Point

A rotated and translated hyperbola was used to represent the first type of SWCC curve: the two straight lines defined by the coordinates (0, 1) (ψ_{aev}, 1) and (10^6, 0) for the hyperbola asymptotes. The degree of saturation, S, equation is written as follows:

$$S = \frac{\tan\theta_g(1+r^2)\ln(\psi/\psi_{aev})}{(1-r^2\tan^2\theta_g)} - \frac{1+\tan^2\theta_g}{1-r^2\tan^2\theta_g}\sqrt{r^2\ln^2\left(\frac{\psi}{\psi_{aev}}\right)+\frac{a^2(1-r^2\tan^2\theta_g)}{(1+\tan^2\theta_g)}} \tag{5.74}$$

where:

ψ = any soil suction value,
ψ_{aev} = air-entry value for the soil,
θ_g = $-\lambda/2$, the hyperbola rotation angle,
r = $\tan(\lambda/2)$, the aperture angle tangent, and
λ = $\arctan[1/\ln(10^6/\psi_{aev})]$, the desaturation slope.

The first derivative of Eq. 5.74 with respect to suction, ψ, is required to define the water storage function for a transient seepage analysis and can be written as follows:

$$\frac{dS}{d\psi} = \frac{1}{\psi}\left[\frac{\tan\theta_g(1+r^2)}{1-r^2\tan^2\theta_g} - \frac{r^2\ln(\psi/\psi_{aev})(1+\tan^2\theta_g)/(1-r^2\tan^2\theta_g)}{\sqrt{r^2\ln^2(\psi/\psi_{aev})+a^2(1-r^2\tan^2\theta_g/(1+\tan^2\theta_g)}}\right] \tag{5.75}$$

5.7.2 Unimodal Equation with Two Bending Points

Two rotated and translated hyperbola are needed to define an entire unimodal SWCC with two bending points. The three straight lines defined by the coordinates (0, 1) (ψ_{aev}, 1) (ψ_{res}, S_r) and (10^6, 0) are the asymptotes of the hyperbola. These two hyperbola are merged through a third equation to produce a continuous function with a smooth transition. The proposed equation is as follows:

$$S = \frac{S_1 - S_2}{1 + (\psi/\sqrt{\psi_{aev}\psi_{res}})^d} + S_2 \quad (5.76)$$

where

$$S_i = \frac{\tan\theta_{gi}(1 + r_i^2)\ln(\psi/\psi_i^a)}{1 - r_i^2\tan^2\theta_{gi}} + (-1)^i \frac{1 + \tan^2\theta_{gi}}{1 - r_i^2\tan\theta_{gi}} \times \sqrt{r_i^2 \ln\left(\frac{\psi}{\psi_i^a}\right) + \frac{a^2(1 - r_i^2\tan^2\theta_{gi})}{(1 + \tan^2\theta_{gi})}} + S_i^a$$

and

$i = 1, 2$;
$\theta_{gi} = -(\lambda_{i-1} + \lambda_i)/2$, hyperbola rotation angles,
$r_i = \tan[(\lambda_{i-1} - \lambda_i)/2]$, aperture angle tangents;
$\lambda_0 = 0$ and $\lambda_i = \arctan\{(S_i^a - S_{i+1}^a)/[\ln(\psi_{i+1}^a/\psi_i^a)]\}$ are the desaturation slopes;
$S_1^a = 1$, $S_2^a = S_r$, $S_3^a = 0$, where S_r is the degree of saturation at residual conditions,
$\psi_1^a = \psi_{aev}$, $\psi_2^a = \psi_{res}$, $\psi_3^a = 10^6$,
$d = 2\exp[1/ln(\psi_{res}/\psi_{aev})]$, a weight factor for S_1 and S_2 that produces a continuous and smooth curve, and,
ψ_{res} = residual suction

The first derivative of Eq. 5.76 with respect to ψ is

$$\frac{dS}{d\psi} = \frac{dS_1/d\psi - dS_2/d\psi}{1 + (\psi/\sqrt{\psi_{aev}\psi_{res}})^d} - \frac{S_1 - S_2}{[1 + (\psi/\sqrt{\psi_{aev}\psi_{res}})^d]^2}\left(\frac{\psi}{\sqrt{\psi_{aev}\psi_{res}}}\right)^d \frac{d}{\psi} + \frac{dS_2}{d\psi} \quad (5.77)$$

where

$$\frac{dS_i}{d\psi} = \frac{1}{\psi}\left[\frac{\tan\theta_{gi}(1 + r_i^2)}{1 - r_i^2\tan^2\theta_{gi}} + (-1)^i \times \frac{r_i^2\ln(\psi/\psi_i^a)(1 + \tan^2\theta_{gi})/(1 - r_i^2\tan^2\theta_{gi})}{\sqrt{r_i^2\ln^2(\psi/\psi_i^a) + a^2(1 - r_i^2\tan^2\theta_{gi})/(1 + \tan^2\theta_{gi})}}\right]$$

$i = 1, 2$

5.7.3 Bimodal SWCC Equation

Four hyperbolae are needed to model a bimodal SWCC delineated by the five asymptotes that are defined by coordinates (0, 1) (ψ_{aev1}, 1)(ψ_{r_1}, S_{r_1})(ψ_{aev2}, S_{aev})(ψ_{r_2}, S_{r_2}) and 10^6, 0):

$$S = \frac{S_1 - S_2}{1 + (\psi/\sqrt{\psi_{aev}\psi_{res1}})^{d_1}} + \frac{S_2 - S_3}{1 + (\psi/\sqrt{\psi_{res1}\psi_{aev2}})^{d_2}} + \frac{S_3 - S_4}{1 + (\psi/\sqrt{\psi_{aev2}\psi_{res2}})^{d_3}} + S_4 \quad (5.78)$$

where:

S_i, θ_{gi}, r_i, and λ_i are as defined in Eq. 5.76,
$i = 1, 2, 3, 4$,
$S_1^a = 1$, $S_2^a = S_{r_1}$, $S_3^a = S_{aev}$, $S_4^a = S_{r_2}$, $S_5^a = 0$,
S_{aev2} = degree of saturation at the air-entry value for the second part of the SWCC.
$\psi_1^a = \psi_{aev1}$,
$\psi_2^a = \psi_{res1}$,
$\psi_3^a = \psi_{aev2}$,
$\psi_4^a = \psi_{res2}$,
$\psi_5^a = 10^6$, and
$d_j = 2\exp[1/\ln(\psi_{j+1}^a/\psi_j^a)]$ are weight factors, $j = 1$, 2, 3.

The derivative of Eq. 5.78 with respect to ψ can be obtained in a manner similar to that shown above for Eq. 5.76.

5.7.4 Parametric Analysis of Gitirana and Fredlund (2004) SWCC Equations

A parametric analysis shows the influence of the fitting properties on the proposed equations. Figures 5.63–5.65 illustrate the independent influence of each of the equation parameters. The SWCC curve parameters are shown to be mathematically independent.

Figure 5.63 illustrates that larger values of a provide a smoother curve. The air-entry value might appear to be reduced as a increases but that is not the case. The apparent reduction is simply a smoothing effect evenly distributed to suction values lower and higher than both ψ_{aev} and ψ_{res}. The curve limits may start deviating from S = 100% and S = 0% when values of a are greater than 0.2 (Fig. 5.63). It is suggested that a values should lie within the range from 0 to 0.15.

5.7.5 Fitting Gitirana and Fredlund (2004) SWCC Equation to Experimental Data

Figure 5.66 shows the fitting of two data sets using the Gitirana and Fredlund (2004) SWCC equation. The laboratory test data was from Patience Lake silt (Bruch 1993) and Beaver Creek sand (Sillers, 1997). The best-fit parameters are shown in Fig. 5.66.

Figure 5.67 illustrates how the Gitirana and Fredlund (2004) SWCC equation can also be used to fit bimodal SWCCs. Experimental data for two bimodal soils are used to show the fitting capability of the proposed bimodal equation.

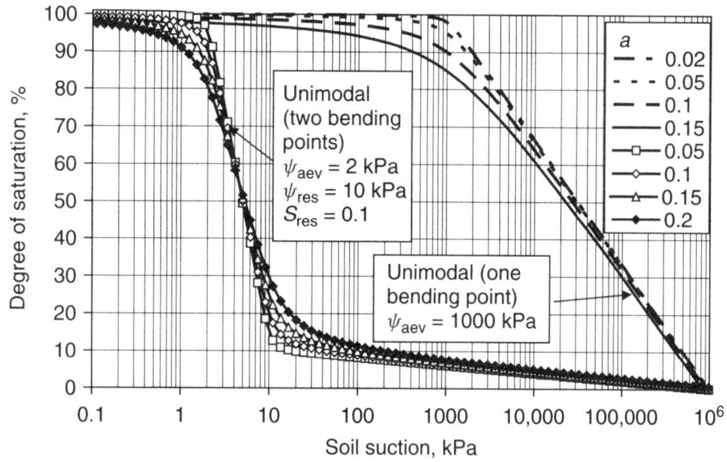

Figure 5.63 Effect of changing a variable on unimodal curves (after Gitirana and Fredlund, 2004).

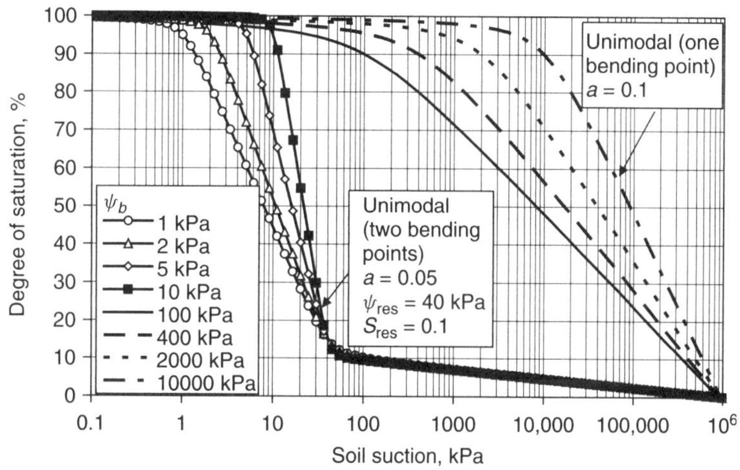

Figure 5.64 Effect of changing ψ_b variable on unimodal SWCC curves (after Gitirana and Fredlund, 2004).

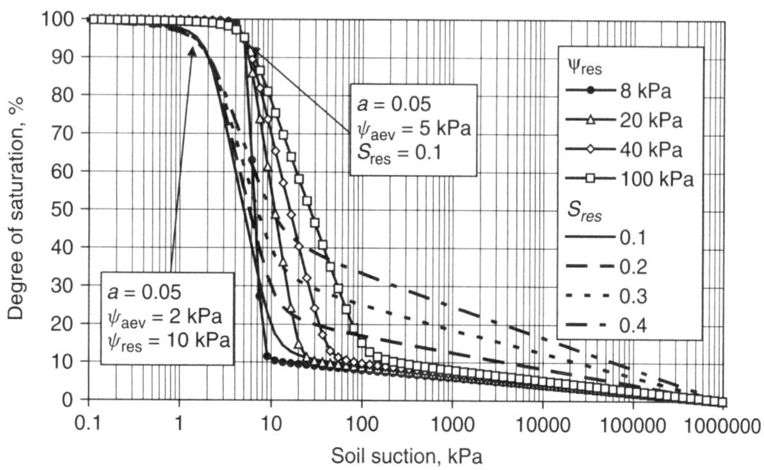

Figure 5.65 Effect of changing ψ_{res} and S_{res} variables on unimodal SWCC curve with two bending points (after Gitirana and Fredlund, 2004).

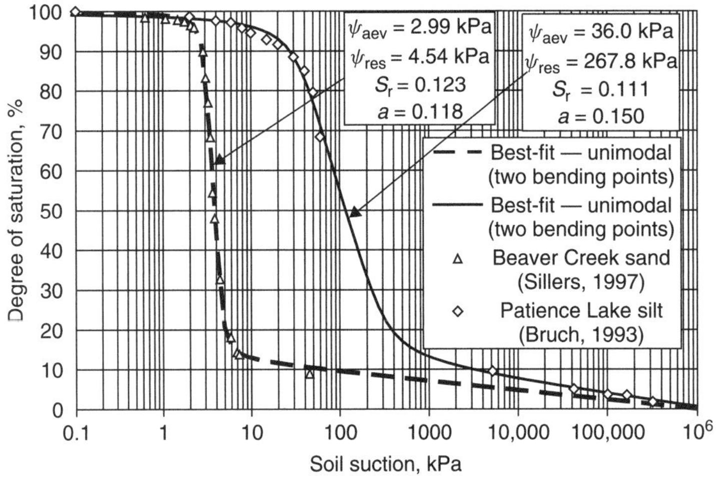

Figure 5.66 Best-fit SWCC curves to experimental data for Patience Lake silt and Beaver Creek sand (after Gitirana and Fredlund, 2004).

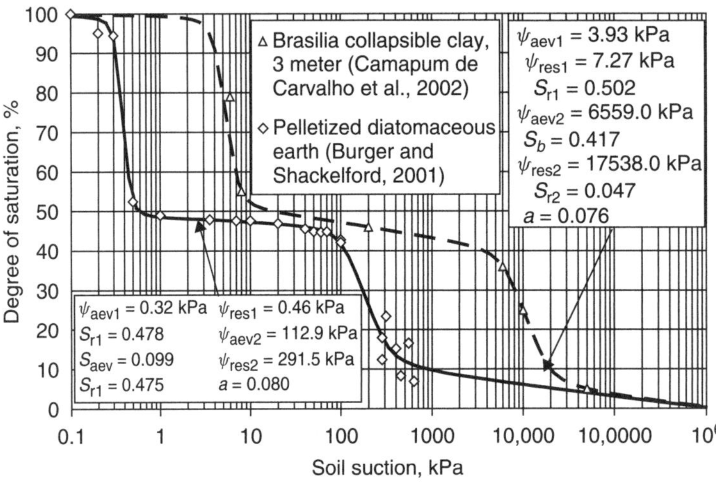

Figure 5.67 Best-fit SWCC curves for experimental data on pelletized diatomaceous earth and Brasilia collapsible clay (after Gitirana and Fredlund, 2004).

Figure 5.67 shows the best-fit curve, the fitting parameters, and the experimental data for a pelletized diatomaceous earth (Burger and Shackelford, 2001) and for residual, highly collapsible clay from Brasilia (Camapum de Carvalho et al., 2002). Close fits are observed for both materials.

5.8 MEASUREMENT OF SWCC USING PRESSURE PLATE DEVICES

Several types of test equipment and test procedures have been used for measuring the SWCC. Much of the original laboratory testing equipment for measuring the SWCC was developed in the agricultural discipline. The early measurement of the SWCC generally involved the use of large pressure chambers that tested several soil specimens at one time (Fredlund and Rahardjo, 1993a). A number of smaller pressure-plate-type cells have more recently been developed in geotechnical engineering that test individual soil specimens. Laboratory equipment for measuring the SWCC can broadly be divided into equipment that provides an applied matric suction and equipment that provides a controlled total suction.

Matric suctions are applied to a soil specimen through use of a high-air-entry disk (or a cellulose membrane). The maximum value of most high-air-entry disks is 1500 kPa. The axis translation technique is used to develop a differential air and water pressure without producing cavitations in the water phase. Figure 5.68 shows a single-specimen pressure plate device developed at the University of Saskatchewan, Saskatoon, for applied suctions up to 500 kPa. An air pressure can be applied to the specimen chamber through use of an air pressure regulator. Water drains against atmospheric pressure conditions and as a result a matric suction is applied to the soil specimen (i.e., $u_a - u_w$, where u_a = pore-air pressure and u_w = pore-water pressure).

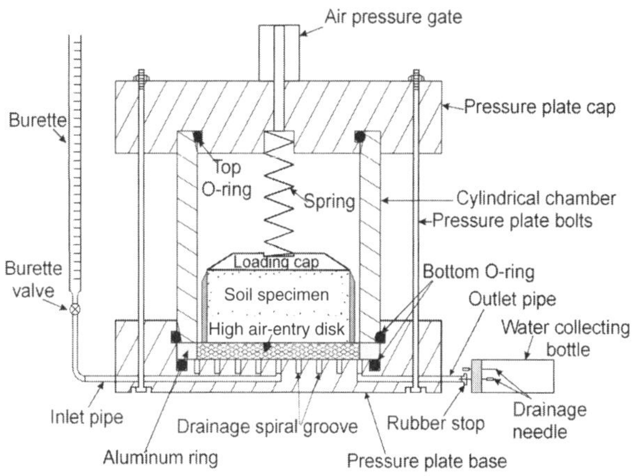

Figure 5.68 Single specimen pressure plate cell developed at University of Saskatchewan, Saskatoon, Canada.

The use of pressure regulators is satisfactory for matric suctions greater than about 5 kPa. At matric suctions less than about 5 kPa, an inverted column of water can be used to apply small negative pressures to the water below the base of the high-air-entry disk. The air pressure in the chamber is maintained at atmospheric conditions when using an inverted column to apply small suction values.

Matric suctions can be applied as high as about 1500 kPa using pressure plate equipment. Generally 1500 kPa has formed the dividing line between matric suction measurements and total suction measurements.

A controlled relative humidity environment is used to establish a fixed total suction. A small soil specimen is brought into equilibrium with the surrounding vapor pressure. Figure 5.69 shows the cross section of a desiccator with a salt solution placed in the bottom to create a constant-relative-humidity environment. The relative humidity environment is converted to total suction through use of the Lord Kelvin equation (Fredlund and Rahardjo, 1993a).

Tens of thousands of SWCC data sets have been measured in various countries of the world (Fredlund,

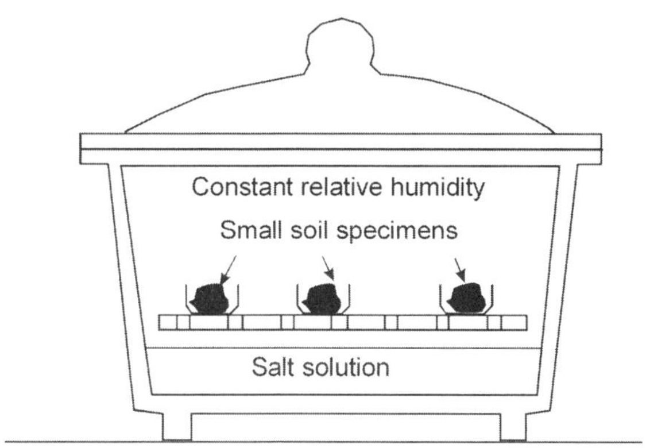

Figure 5.69 Vacuum desiccator for equilibrating small soil specimens in constant-relative-humidity environment created by controlled salt solution.

1997). Most of the laboratory measurements have been made within agriculture-related disciplines. Over the years it became quite acceptable to perform SWCC measurements in agricultural disciplines without applying total stresses to the soil specimen. The data was primarily used to assess the availability of water for plant growth near ground surface. Most laboratory testing equipment has been developed without consideration of applying total stresses and without the ability to measure volume change as matric suction is applied. The design of some pressure plate apparatuses is somewhat restrictive when obtaining data for geotechnical engineering applications (Ng and Pang, 1999, 2000a). Geotechnical engineers have attempted to utilize soil testing equipment used in other disciplines, but at the same time several new devices have been developed that are better suited for geotechnical engineering applications.

It has become common practice to measure matric suctions below 1500 kPa and total suctions above about 1500 kPa. Data sets for a particular soil are generally plotted with matric suction and total suction on the same graph (i.e., along the abscissa on a logarithm scale). It has also become common practice to simply measure the desorption branch of SWCCs. The desorption branch is the easiest and most rapid branch of the SWCC to measure. Soil specimens are commonly placed into a pressure plate apparatus, initially saturated, and allowed to come to equilibrium prior to the start of the test. Therefore, the initial matric suction in the soil specimen is released to zero at the start of the test.

The SWCC has proven to be the most important unsaturated soil property function to measure in unsaturated soil mechanics. The SWCC provides the key relationship needed to implement unsaturated soil mechanics in geotechnical engineering practice. SWCCs have been measured in agriculture-related disciplines since the 1930s. A number of devices have been developed for applying a wide range of soil suction values. Typical pressure plate apparatuses are Tempe cells (100 kPa) (Reginato and van Bavel, 1962) (Figs. 5.70 and 5.71), volumetric pressure plates (200 kPa) (Figs. 5.72 and 5.73), and large pressure plate apparatuses (500 and 1500 kPa) (Figs. 5.74, 5.75, and 5.76); (Fredlund and Rahardjo, 1993a). These apparatuses use the measurement of either a water mass or a water volume to allow the back calculation of equilibrium water contents corresponding to applied matric suctions.

ASTM designation D6836–02 (2008) provides a detailed description for the determination of the SWCCs using several testing procedures: (i) hanging column, (ii) pressure plate extractor (with volumetric measurements and gravimetric water content measurements of water content), (iii) chilled-mirror hygrometer, and (iv) centrifuge. For suction values greater than 1500 kPa, small soil specimens are allowed to come to equilibrium in a fixed relative humidity environment (i.e., vacuum desiccators). Various molar salt solutions are used to create the fixed relative humidity environments.

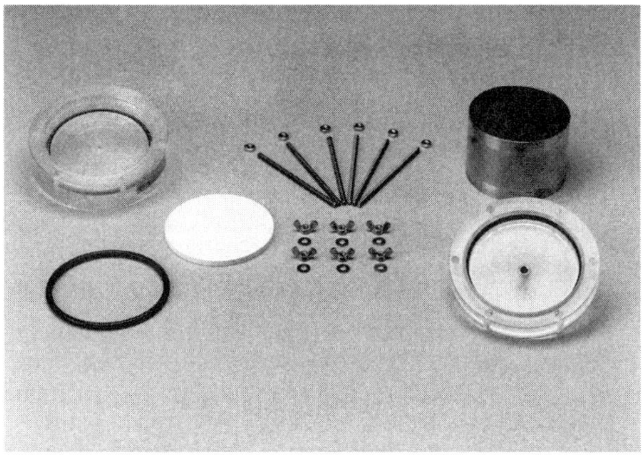

(a)

(b)

Figure 5.70 Assemblage of a Tempe pressure cell: (a) disassembled components of a Tempe cell; (b) assembled Tempe cell. (Courtesy of SoilMoisture Equipment Corporation, Santa Barbara, CA).

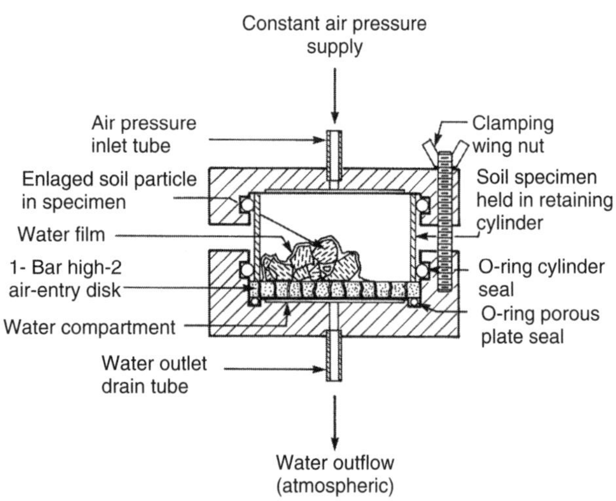

Figure 5.71 Cross-section of a Tempe pressure plate cell manufactured by SoilMoisture Equipment Corporation.

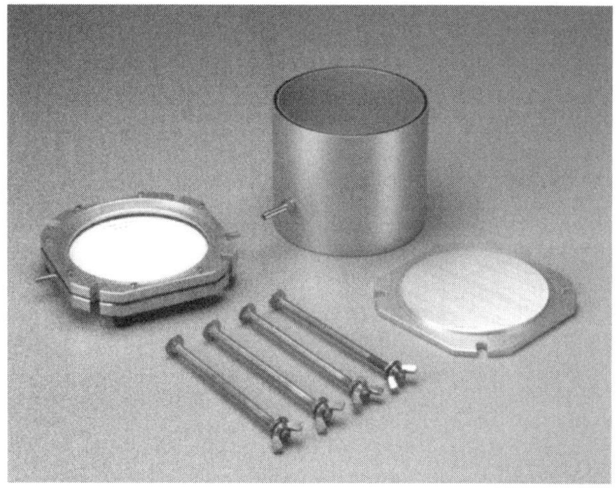

(a)

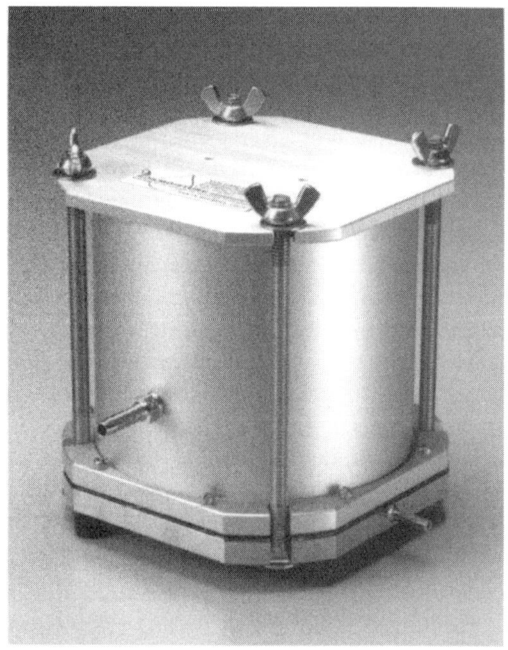

(b)

Figure 5.72 Assemblage of volumetric pressure plate extractor: (a) disassembled components of volumetric pressure plate extractor; (b) assembled volumetric pressure plate extractor. (Courtesy of SoilMoisture Equipment Corporation, Santa Barbara, CA).

There is no single, unique SWCC; rather, there is an infinite number of scanning curves contained within a drying (desorption) boundary curve and an adsorption (wetting) boundary curve (Fig. 5.77). It is the primary drying and wetting curves that are of greatest relevance in unsaturated soil mechanics. The drying SWCC is easier to measure in the laboratory and has become the primary curve used in estimating unsaturated soil property functions.

5.8.1 Tempe Cell Apparatus and Test Procedure

The term "Tempe cell" refers to a particular pressure plate cell that was designed and built by Reginato and van Bavel (1962)

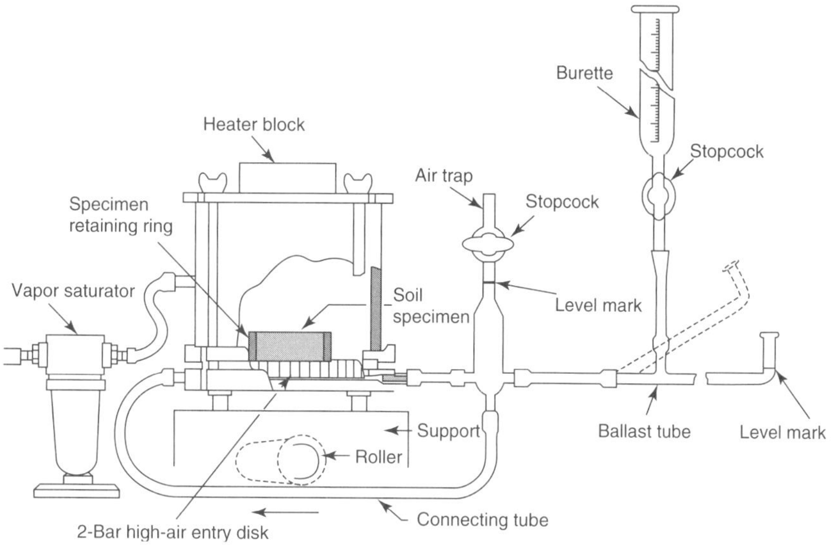

Figure 5.73 Setup of volumetric pressure plate extractor along with its hysteresis attachment (from SoilMoisture Equipment Corporation, 1985).

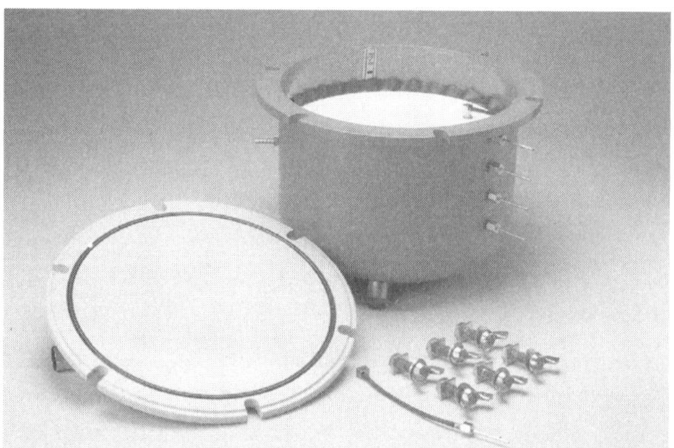

Figure 5.74 A 5-bar ceramic plate extractor (i.e., maximum applied matric suction is 500 kPa). (Courtesy of SoilMoisture Equipment Corporation, Santa Barbara, CA.)

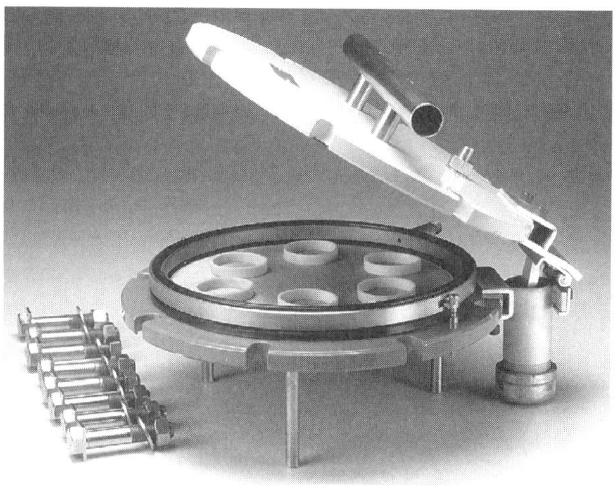

Figure 5.76 Pressure membrance extractor (from SoilMoisture Equipment Corporation, Santa Barbara, CA).

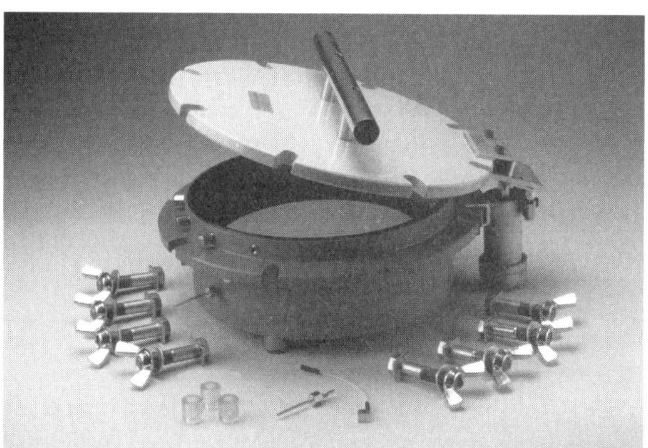

Figure 5.75 A 15-bar ceramic plate extractor (i.e., maximum applied matric suction is 1500 kPa). (Courtesy of SoilMoisture Equipment Corporation, Santa Barbara, CA.)

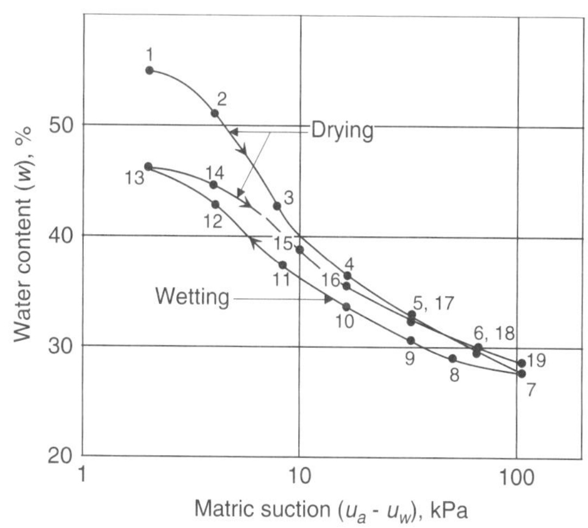

Figure 5.77 SWCCs for drying and wetting processes for Aiken clayey silt (after Richards and Fireman, 1943).

at the Water Conservation Laboratory in Tempe, Arizona. The Tempe cell has been subsequently manufactured by SoilMoisture Equipment Corporation of Santa Barbara, California.

A Tempe cell assemblage is shown in Fig. 5.70 and a cross-sectional view is given in Fig. 5.71. A soil specimen is placed onto the high-air-entry disk inside the retaining cylinder of the Tempe pressure cell. An outlet tube located at the base plate underneath the high-air-entry disk allows the drainage of water from the soil specimen. Air pressure is supplied through the inlet tube on the top plate.

A test is started by saturating the high-air-entry disk. The soil specimen is also allowed to fully imbibe water at the start of the test. Excess water is removed from the cell after saturation of the soil specimen. The top plate is then mounted and tightened into place. The lowest air pressure is applied to the Tempe cell. The applied air pressure is equal to the desired lowest matric suction value.

Water starts draining from the specimen through the high-air-entry disk once the air pressure is applied. At equilibrium, the matric suction in the soil is equal to the applied air pressure. The time required to reach equilibrium depends upon the thickness and permeability of the specimen and the permeability of the high-air-entry disk. The apparatus continues to be used for sand soils that have an air-entry value (and residual water content) less than 100 kPa. The air-entry value of the ceramic disk is slightly greater than 100 kPa. Changes in water content are measured by weighing the specimen and the cell after equilibrium is reached with the applied air pressure.

The procedure is repeated at higher applied air pressures (i.e., high matric suctions). Once the highest matric suction has been applied, the air pressure in the cell is released and the soil specimen is removed. The water content corresponding to the highest matric suction is measured by oven drying the soil specimen. The final water content, together with the previous changes in weight, is used to back calculate the water contents corresponding to other applied suction values. The matric suction values are plotted versus the corresponding water contents to give the SWCC.

5.8.2 Volumetric Pressure Plate Extractor Apparatus and Test Procedure

Figure 5.72 shows an assemblage of a volumetric pressure plate extractor manufactured by SoilMoisture Equipment Corporation. The maximum matric suction that can be attained with the volumetric pressure plate extractor is 200 kPa. This apparatus can also be used to study hysteresis effects associated with the drying and wetting of the soil. A hysteresis attachment is required, as shown in Fig. 5.73.

The hysteresis attachments allows for the measurement of the volume of water that flows in or out of the soil specimen. The hysteresis attachments consist of a heater block, vapor saturator, air trap, ballast tube, and burette. The heater block is attached to the top plate to prevent water from condensing on the inside walls of the extractor. The heater block maintains the walls of the extractor at a slightly higher temperature than the soil temperature. Water condensation on the walls would introduce an error in the determination of equilibrium water contents, particularly for long-term tests. The vapor saturator is used to completely saturate the air that flows into the volumetric pressure plate extractor. The saturated air surrounding the soil specimen prevents the soil from drying by evaporation.

The air trap is provided to collect air that may diffuse through the high-air-entry disk. A "level mark" is provided on the stem of the air trap as a reference point in measuring the volume of water. A ballast tube is provided as a horizontal storage for water flowing in or out of the soil specimen under atmospheric conditions. A "level mark" on the ballast tube also serves as a reference point. A burette is used to store or supply water. The change in the volume of water in the burette during the equalization process is equal to the water volume change in the soil specimen.

Drying and wetting processes can be performed on the same soil specimen when using the volumetric pressure plate extractor. During the drying process, matric suction is increased, and pore-water drains from the specimen into the ballast tube. During the wetting process, matric suction is decreased, and water in the ballast tube is slowly absorbed back into the soil specimen. The drying branch of the SWCC is first measured, followed by measurements for the wetting branch. Further drying and wetting SWCCs can be measured subsequent to the first cycle, if desired.

5.8.3 Test Procedure for Volumetric Pressure Plate Extractor

The test procedure commences with the placement of the soil specimen into the retaining ring. The ring and soil specimen are then placed on top of the water-saturated high-air-entry disk. The specimen is first saturated, and the hysteresis attachments are connected to the extractor. The hysteresis attachments are filled with water to the level mark. Air bubbles should be flushed from below the ceramic disk at the start of the test by running the roller over the connecting tube. This action will push water from the air trap through the grooves beneath the disk and cause air bubbles to move into the air trap. The trapped air bubbles are then released by adjusting the water level in the air trap to the level mark. This is accomplished by opening the stopcock at the top of the air trap and applying a small vacuum to the outlet stem of the air trap. The vertical position of the ballast tube is made level with the top surface of the ceramic disk or with the center of the specimen by placing the extractor on a support.

5.8.4 Drying Portion of SWCC

A starting point on the first drying curve is established by applying a low matric suction to the soil specimen (i.e., raising the air pressure in the extractor to a low pressure). Water starts to drain from the soil specimen through the high-air-entry disk to the ballast tube. Water should drain to the burette when the ballast tube is full of water. This is accomplished by applying a small vacuum to the top of the burette and carefully opening the stopcock on the burette. The outflow of water from the specimen stops when equilibrium is reached.

Diffused air is removed from the grooves underneath the ceramic disk using the pumping procedure described above. Air may diffuse through the pore-water and the water in the high-air-entry disk during the test. The air is called "diffused air" and it comes out of the solution in the grooves beneath the ceramic disk. The removal of diffused air should be performed each time equilibrium is reached with a particular applied suction. The accuracy of the measurement of water flow from the soil specimen is increased through use of this technique. Also, the accumulation of diffused air below the ceramic disk may prevent the uptake of water by the soil during the wetting process.

The above testing procedure can be repeated at each increasing matric suction (i.e., increasing air pressure in the extractor) until the drying curve is complete. The water volume readings from the burette will provide the information necessary for the calculation of the equilibrium water contents for each applied pressure.

5.8.5 Wetting Portion of SWCC

The volumetric pressure plate test can be continued through a wetting process upon completion of the drying portion of the test. The soil matric suction is reduced by decreasing the air pressure in the extractor. A decrease in the air pressure causes water to flow from the ballast tube back into the soil specimen. The water volume required for the backflow may be in excess of the water volume in the ballast tube. In this case, water should be added to the ballast tube by opening the burette stopcock. Equilibrium is reached when the water flow from the ballast tube into the specimen has stopped. Following equilibrium, the water levels in the air trap and the ballast tube are again adjusted to the level mark, and the burette reading is taken for computation of water volume change. The procedure is repeated at decreasing matric suctions until the desired range of the wetting curve is obtained. Subsequent cycles of drying and wetting can also be performed if desired.

The final water content of the specimen corresponding to the last matric suction is measured at the end of the test. The final water content and the water volume changes between two successive applied pressures are used to calculate the equilibrium water contents. A typical plot of matric suction versus water content for the drying and wetting processes is shown in Fig. 5.77. The plot illustrates the effect of hysteresis between the drying and wetting curves.

The volumetric pressure plate extractor has also been used to measure the coefficient of permeability in unsaturated soils using the outflow method described by Klute (1965a). The coefficient of permeability of the high-air-entry disk must be measured to ensure that there is no flow impedance resulting from the low permeability of the ceramic disk. It is quite common that impedance from the high-air-entry disk will not allow the coefficient of permeability of the soil to be measured.

5.8.6 Pressure Plate Extractors for Drying Tests

Pressure plate extractors such as the type manufactured by SoilMoisture Equipment Corporation can be used to measure the SWCC (see Figs. 5.74–5.76). The extractors are commonly used to apply various matric suctions to one or more soil specimens, and the test is called a pressure plate test [ASTM D2325]. The pressure plate extractor consists of a high-air-entry ceramic disk contained in an air pressure chamber. The high-air-entry disk is saturated and is always in contact with water in a compartment below the disk. The compartment is maintained at zero water pressure.

Tempe cells and volumetric pressure plate extractors are two types of extractors which can be used on single soil specimens in the low range of matric suction. Pressure plate extractors for the high-matric-suction range (i.e., up to 1500 kPa) are presented in Figs. 5.74, 5.75, and 5.76. The pressure membrane extractor utilizes cellulose membranes instead of a high-air-entry ceramic disk. In addition, a compression diaphragm is provided on the lid of the pressure membrane extractor. A pressure regulator maintains a slightly higher pressure behind the diaphragm than inside the extractor in order to keep the soil specimens in contact with the cellulose membrane. The increased contact pressure is particularly important for the high range of matric suctions where the soil specimens may shrink and an air gap may tend to form at the contact with the high-air-entry disk.

Soil specimens are placed on top of the high-air-entry disk, and the airtight chamber is pressurized to the desired matric suction. The disk does not allow the passage of air as long as the applied matric suction does not exceed the air-entry value of the disk. The air-entry value of the disk is related to the diameter of the pores in the ceramic disk. The air-entry value of the disk and the strength of the pressure chamber control the maximum air pressure which can be applied to a soil specimen.

The application of matric suction to a soil causes the pore-water to drain to the water compartment through the disk. At equilibrium, the soil will have a reduced water content corresponding to the increased matric suction. Since more than one soil specimen is generally tested, it is necessary to dismantle the chamber and measure the weight of each specimen after equilibrium at the applied pressure. This procedure is commonly used with 5- and 15-bar ceramic plate extractors when several specimens are tested at the same time. The plot of equilibrium water content versus the logarithm of the corresponding soil suction constitutes the SWCC.

5.8.7 Typical Results from Pressure Plate Extractor Tests

The SWCCs obtained from pressure plate tests provide information regarding the water phase constitutive surface for an unsaturated soil. An unsaturated soil in the field is often subjected to significant and frequent changes in matric suction which result in changes in the unsaturated soil properties. Changes in soil suction are often more significant than changes in total stress. The soil undergoes processes of drying and wetting because of the imposed climate. On the other hand, the applied total stress on the soil is seldom altered. It is important to know the SWCC of an unsaturated soil in order to predict soil behavior associated with drying or wetting.

Croney and Coleman (1954) show SWCCs which compare the behavior observed for incompressible and compressible soils. Figure 5.78 compares the SWCCs of soft and hard chalks, which are considered relatively incompressible. The drying curves of the incompressible soils exhibit essentially constant water contents at low matric suctions and rapidly decreasing water contents at higher suctions. The point where

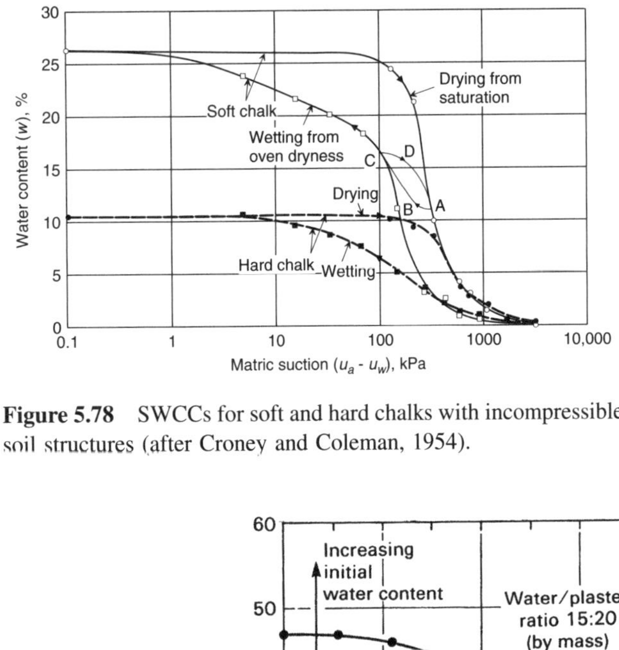

Figure 5.78 SWCCs for soft and hard chalks with incompressible soil structures (after Croney and Coleman, 1954).

the water content starts to decrease significantly indicates the air-entry value of the soil. The data show that the hard chalk has a higher air-entry value than the soft chalk. The high pressure applied during the formation of the hard chalk bed results in smaller average pore sizes than in the soft chalk.

Another noticeable characteristic is that the drying curves for both hard and soft chalks become identical at high matric suctions (Fig. 5.78). This indicates that under high-suction conditions both soils have similar pore size distributions. There is marked hysteresis between the drying and wetting curves for both soils.

The effect of initial water content on the drying curves of incompressible mixtures is demonstrated in Fig. 5.79. An increase in the initial water content of the soil results in a decrease in the air-entry value. The decrease can be attributed to the larger pore sizes in the high-initial-water-content mixtures. The soils with the high initial water contents drain quickly at relatively low matric suctions. As a result, the water content in the soil with the large pores is less than the water content in the soil with small pores at matric suctions beyond the air-entry value. The soils with low initial water content (i.e., small pore sizes) require larger matric suction values to commence desaturation. There is a slower rate of water drainage from the pores at higher soil suctions.

The initial dry density of incompressible soils has a similar effect on the SWCC, as was illustrated by the initial water contents. As the dry density of a soil increases, the pore sizes become small and the air-entry value becomes higher, as illustrated in Fig. 5.80. The high-density specimens desaturate at a slower rate than the low-density specimens. As a result, the high-density specimens have higher water contents than the low-density specimens at matric suctions beyond their air-entry values. Hysteresis associated

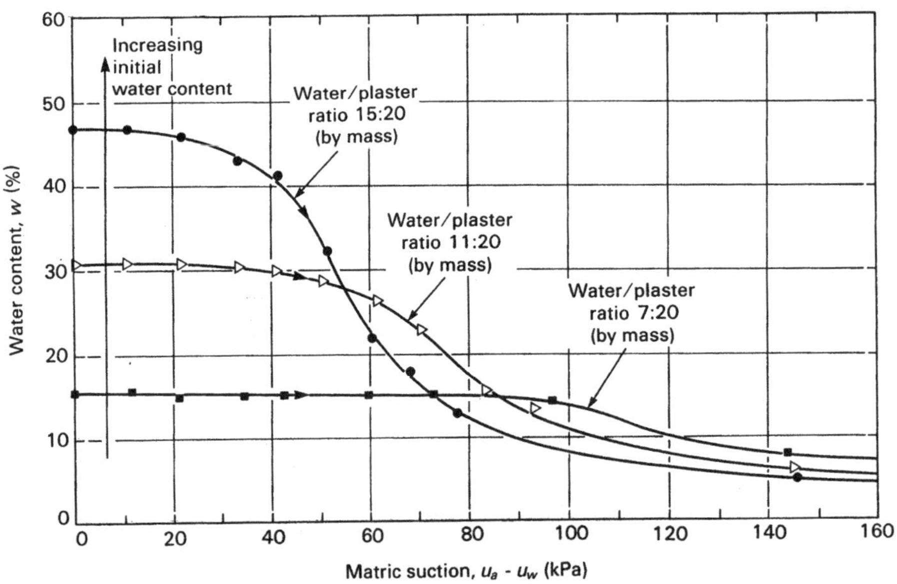

Figure 5.79 Effects of initial water content on the drying curves of incompressible mixtures (after Croney and Coleman, 1954).

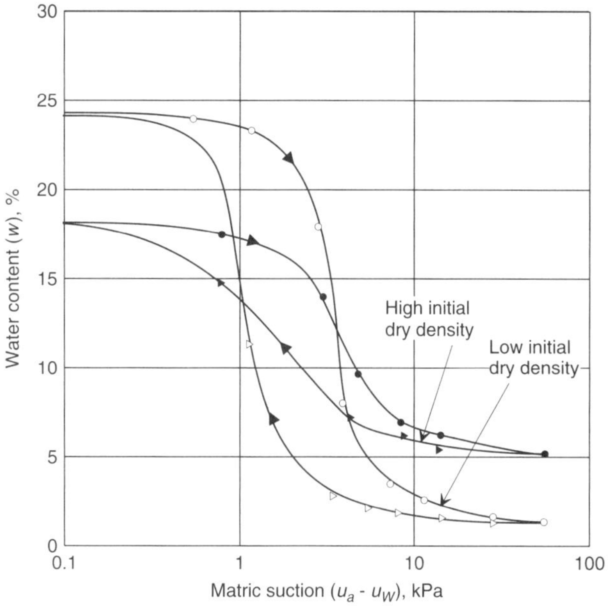

Figure 5.80 Effect of initial dry density on SWCCs of compacted silty sand (after Croney and Coleman, 1954).

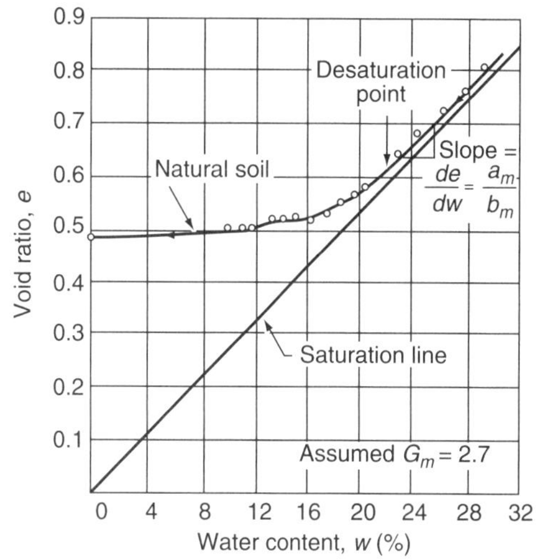

Figure 5.82 Shrinkage curve for London clay (after Croney and Coleman, 1954).

with the high-density specimens is less than the hysteresis exhibited by the low-density specimens.

Croney and Coleman (1954) used the SWCC for London clay (Fig. 5.81) to illustrate the behavior of a compressible soil upon wetting and drying. The gradual decrease in water content upon drying does not provide a clear indication of the air-entry value. In this case, the shrinkage curve of the soil (Fig. 5.82) must be used together with the SWCC in order to determine the correct air-entry value of the soil. The shrinkage curve clearly indicates the compressible nature of the soil. The total and water volume changes caused by an increase in matric suction are essentially equal until the water content reaches 22%. The shrinkage curve above a water content of 22% is parallel and close to the saturation line indicating essentially a saturation condition. The soil starts to desaturate when the water content goes below 22%, causing the shrinkage curve to deviate from the saturation line. The void ratio of the soil reaches a limiting value (i.e., $e = 0.48$), corresponding to a water content of 0%. A water content of 22% in the natural soil corresponds to a matric suction of approximately 1000 kPa according to the SWCC (Fig. 5.81).

Some irreversible structural changes causing an irreversible volume change occur primarily during the first drying process, as indicated by curve *A* in Fig. 5.81. Subsequent wetting and drying cycles follow curves *B* and *C* (Fig. 5.81), respectively.

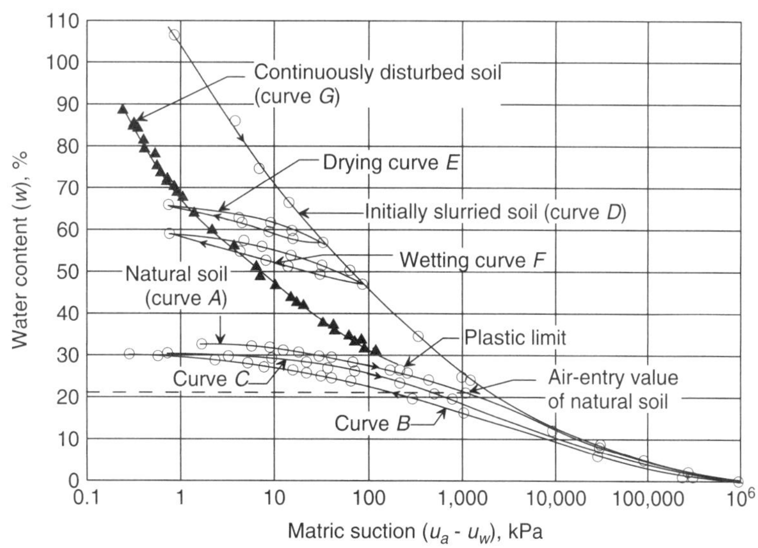

Figure 5.81 SWCCs for London clay (after Croney and Coleman, 1954).

Curves B and C have lower water contents than curve A, with the difference indicating irreversible volume change.

Curve D in Fig. 5.81 was obtained from initially slurry specimens where the soil structure was slightly disturbed. Curve A for the natural soil joined curve D at a matric suction of 6300 kPa, indicating the maximum suction to which the clay may have been subjected during its geological history. The deviations of the natural soil curves A, B, and C from the initially slurry soil curve D represent the influence of past drying and wetting cycles.

Another curve plotted in Fig. 5.81 is curve G, which relates the water content to the matric suction of a continuously disturbed soil specimen. Curve G is different from the other measured SWCCs mainly because the soil structure was continuously being changed throughout the test. An SWCC should be measured on a single specimen or several specimens with "identical" initial soil structures. Curve G appears to be unique for London clay. State points along the drying curve D or curve A will move to corresponding points on curve G when disturbed at constant water content. Similarly, state points along any wetting curve will move to curve G when disturbed at constant water content. Disturbance can take the form of remolding or thoroughly mixing the specimens. Similar relationships to curve G have also been found for other soils. There appears to be a unique relationship between water content and matric suction for a disturbed or remolded soil, regardless of the stress path (Croney and Coleman, 1954).

5.9 SINGLE-SPECIMEN PRESSURE PLATE DEVICES FOR GEOTECHNICAL ENGINEERING

Most pressure plate apparatuses were originally developed for testing soils for agricultural applications. Pressure plate equipment and laboratory test procedures have been transferred into geotechnical engineering as unsaturated soil mechanics emerged. However, the testing of unsaturated soils in geotechnical engineering has a number of additional requirements that should be addressed. Some of the desirable features beneficial to geotechnical engineering applications can be listed as follows:

1. It is desirable to test individual soil specimens.
2. It is desirable to test undisturbed soil samples.
3. It is desirable to be able to apply in situ total stresses in the laboratory when testing the soil samples.
4. It is desirable to be able to measure changes in overall volume as well as changes in the amount of water in the soil specimen.
5. It is desirable to be able to measure both the drying and wetting SWCCs.
6. Increased accuracy is required with respect to performing and interpreting the SWCC tests.
7. The apparatus must be easy to assemble and safe to operate under high pressures.
8. Attention must be paid to the measurement of air diffused through the high-air-entry disks.

Some of the features desirable for engineering applications have been met in equipment developed by geotechnical engineers in recent years. There has been an attempt to continuously monitor water volume change and overall volume change in order that all volume-mass soil properties can be measured (e.g., w, S, and e). There are several recent apparatuses that have been developed that better meet the needs for measuring the SWCC in geotechnical engineering. Some of the desired specifications for an apparatus to measure the volume-mass properties along matric suction and applied total stress paths are as follows. The suction range of operation should be up to at least 500 kPa and possibly up to 1500 kPa. The air pressure source to operate the device will also need to be at least 500 or 1500 kPa.

It is desirable to independently apply total stresses to the soil. The application of isotropic stresses may be preferable; however, it is considerably more economical to develop equipment for K_0 loading. It is also desirable that both the water volume change and overall volume change of the specimen be measured in order that all volume-mass soil properties can be determined (e.g., w, S, and e).

It is important that provision is made to independently measure the volume of air which might diffuse through the water in the high-air-entry disk. The measured water flowing from or to the soil specimen must be corrected for diffused air flow.

It appears preferable to test individual soil specimens for geotechnical engineering purposes. The initial state of the soil should be recorded (i.e., initially remolded at a high water content, compacted or undisturbed). It is preferable if the apparatus can accommodate both drying and wetting procedures.

Several pressure plate cells have been developed over the past few years that meet many of the above-mentioned requirements. One cell was developed at the Technical University of Catalonia (UPC) in Barcelona, Spain (Gens et al., 1995; Romero et al., 1995). Wille Geotechnik (Germany) has also developed a pressure plate cell with many of the above-mentioned features (Lins and Schanz, 2004). As well, GCTS United States has developed a similar pressure plate cell (Pham et al., 2004).

5.9.1 Barcelona Cell

The Barcelona K_0 pressure plate was developed in 1999 for testing sand-bentonite mixtures at UPC (Hoffmann et al., 2005). The suction-controlled oedometer can also accommodate the application of vertical load to the soil specimen. Figure 5.83 shows the design drawings of the cell. The base

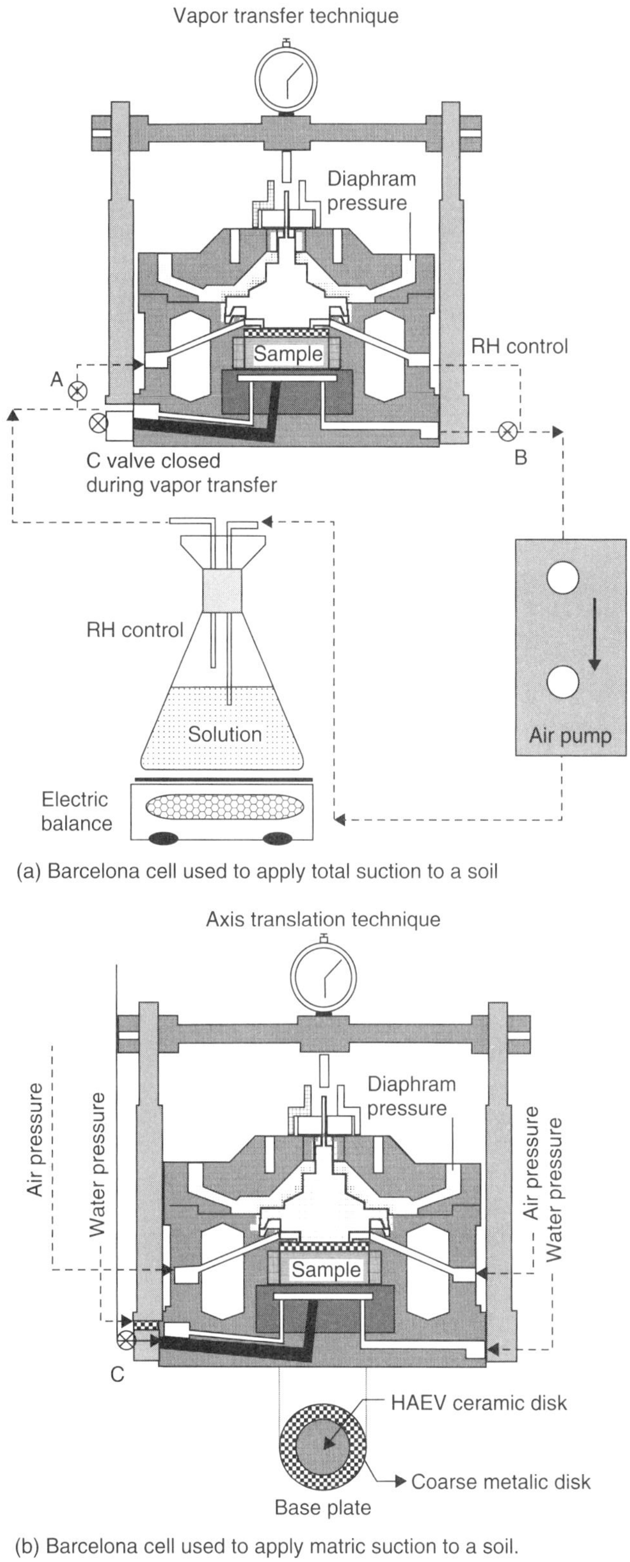

Figure 5.83 Barcelona K_0 pressure plate cell. (Courtesy of CIMNE, Barcelona, Spain.)

plate of the cell has a high-air-entry ceramic disk and uses the axis translation technique to apply various matric suction values.

Total suctions can also be controlled using the vapor circulation from above a saturated salt solution (Fig. 5.83). A coarse disk is substituted for the high-air-entry disk in the base plate. Air at a constant relative humidity can be circulated through the coarse disk. In this way, it has been possible to apply extremely high suctions to a soil specimen.

5.9.2 Wille Geotechnik Cell

A single-soil-specimen pressure plate apparatus was designed and built by Wille Geotechnik in Gottingen, Germany. The soil specimen is contained inside a metal ring for K_0 loading conditions. Both water content changes and volume changes can be measured provided the specimen does not shrink away from the sides of the metal ring. Figure 5.84 shows the cell placed in the hydraulic loading frame. Figure 5.85 shows drying and wetting curve water contents measured on sand using the Wille cell (Lins and Schanz, 2004). Volume changes were also recorded and void ratio was calculated; however, the sand showed small changes in void ratio, as shown in Fig. 5.86. Figure 5.87 shows the computed degrees of saturation as applied suctions were changed. The apparatus is capable of measuring water content and void ratio changes with applied matric suction and total stresses. These results show that it is possible for a variety of stress paths to be followed and in this way the entire constitutive surfaces for water content and void ratio can be measured. The Wille cell can be placed within a loading frame for the application of vertical stresses.

5.9.3 Fredlund Pressure Plate Cell (D.G. Fredlund, 2001)

The Fredlund cell shown in Fig. 5.88 was manufactured by GCTS in Tempe, Arizona. The pressure plate cell accommodates the measurement of water content and overall volume change on a single soil specimen under various K_0 total stress loading conditions (Pham et al., 2002). Desorption SWCC can readily be measured. More time is required and greater experimental care is needed when measuring the adsorption SWCC curve. Soil suctions as high as 1500 kPa can be applied. A small air pressure booster system can also be purchased for the application of high air pressures.

The Fredlund SWCC device comes with interchangeable high-air-entry disks. The device consists of a pressure cell assembly and a pressure panel as shown in Fig. 5.89. The pressure plate device can accommodate various sizes of specimens ranging between 62 and 75 mm in diameter. The height of the specimen can be between 12 and 25 mm.

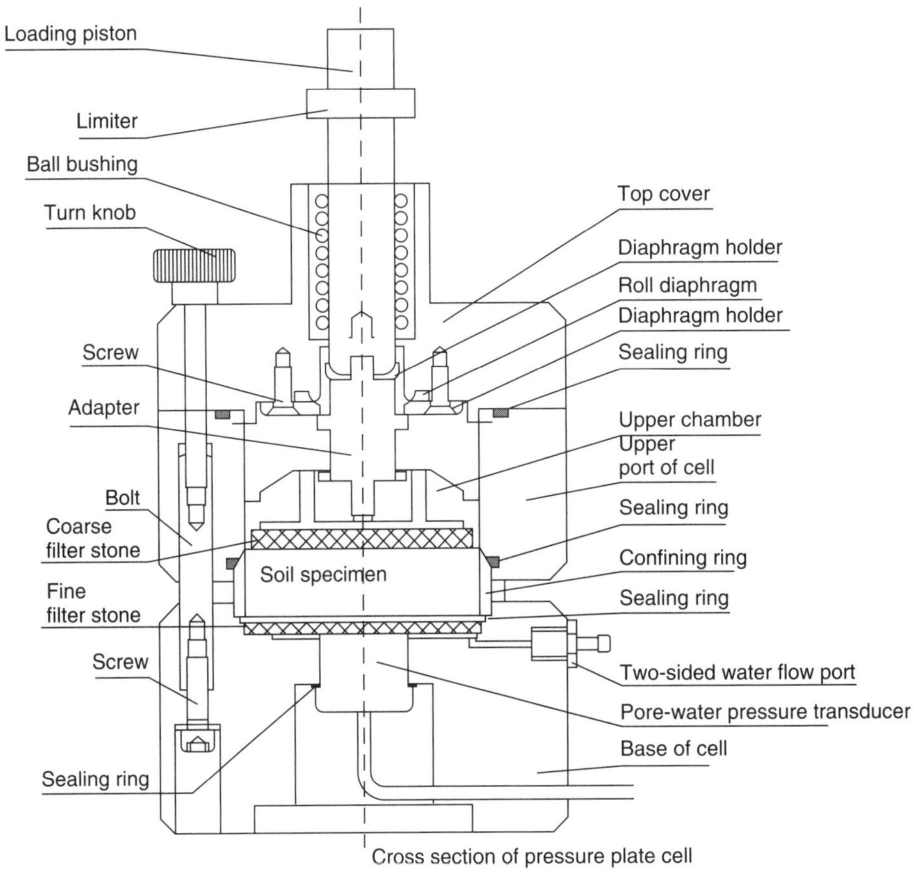

Figure 5.84 Wille pressure plate cell. (Courtesy of Wille Geotechnik, Germany.)

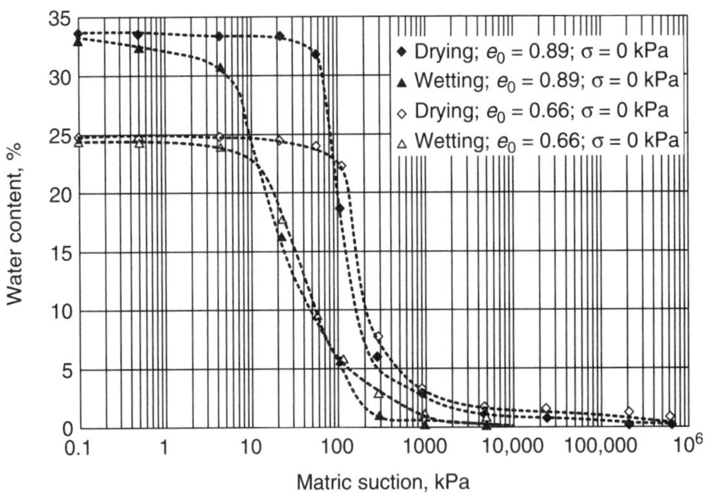

Figure 5.85 Drying and wetting water content SWCC measured on sand using the Wille pressure plate cell (after Lins and Schanz, 2004).

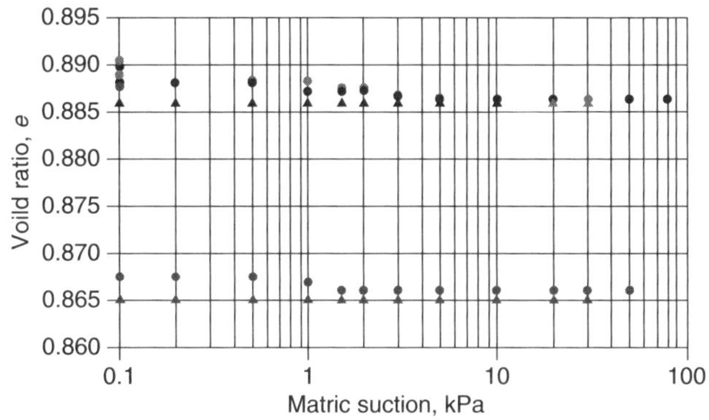

Figure 5.86 Drying and wetting void ratio curves measured on a sand soil using the Wille pressure plate cell with volume change measurements (after Lins and Schanz, 2004).

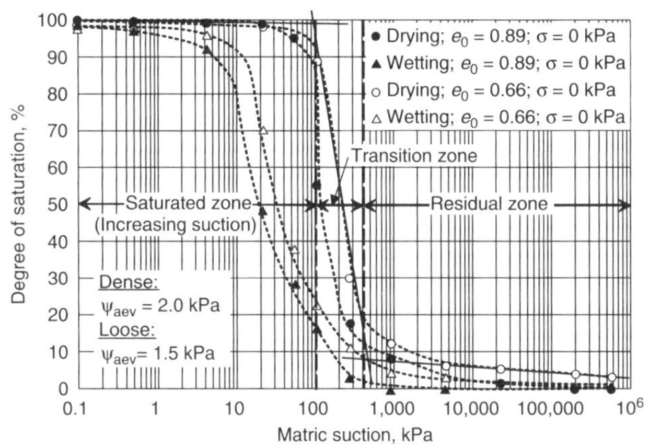

Figure 5.87 Drying and wetting degree-of-saturation data measured on sand using the Wille pressure plate cell (after Lins and Schanz, 2004).

Vertical loads can be applied to the soil specimen either through use of dead weights placed on the loading plate or through use of a pneumatic pressure loading system (Fig. 5.89). A pressure "compensator" is used to balance the uplift force generated by internal air pressure acting on the loading shaft. The loading shaft slides up and down through the compensator on a set of O-rings with minimal friction.

The apparatus has the ability to measure changes in the volume of water in the soil specimen without dismantling the apparatus during testing. This is accomplished by measuring water released (or absorbed) from the specimen using two volume indicator tubes on the pressure panel. The base-plate compartment is connected to two volume indicator tubes on the pressure panel via plastic tubing. The volume indicator tubes are graduated to read the amount of water released or absorbed during the test.

The diffusion of air into the compartment below the high-air-entry disk (during a test) is registered as a water volume change. The diffused air bubbles can be removed from the system through use of a flushing technique. The amount of diffused air is measured by recording the volume change burettes before and after flushing the base-plate compartment. An alternating air pulse is applied to the top of one of the burettes through use of a squeeze bulb. The pulsing motion flushes the air from the compartment below the high-air-entry disk.

The Fredlund pressure plate cell has an ability to measure the overall volume change of the soil specimen during the test. The movement of the load plate indicates the movement of the top surface of the soil specimen. The movement of the load plate can be measured using a dial gauge or a linear vertical displacement transducer (LVDT). The volume change of the soil specimen can be computed based on vertical movement of the load plate provided the specimen does not contract diametrically. The pressure plate cell is equipped with either a small heater or a heating jacket that can maintain the chamber temperature slightly above ambient temperature in the laboratory. The heater helps prevent condensation of water vapor inside the chamber.

A "hanging-column" accessory is also available for the application of low suctions (Fig. 5.90).

Soil specimens can be saturated either in the pressure plate cell or prior to testing by partial submergence in water. The saturated soil specimen is placed on a saturated ceramic stone. A vertical load (i.e., token or higher loads) can be applied to the soil specimen.

Water moves out from the soil specimen and drains through the ceramic disk until equilibrium is established with the applied suction. The water level in the volume indicator tubes shows the amount of water released and movement ceases when equilibrium is attained. Following equilibrium, the water levels in the volume indicator tubes are recorded, and generally diffused air is flushed from the compartment below the soil specimen. The volume indicator tubes should be read before and after flushing. The tops of the water volume change tubes should be covered with metal foil in order to minimize the evaporation of water. The water content of the soil specimen is computed based on back calculation of the volume indicator tube readings. The overall volume change of the soil specimen is also measured using the initial and final vertical movement measurements.

The above procedure can be repeated by applying pressure decrements to obtain the "wetting" curve. At the end of the test, the soil specimen is removed from the cell and its water content is determined by oven drying the specimen. The volume of the soil specimen can be measured using calipers.

Four different ceramic stones rated as 1, 3, 5, and 15 bars, are capable of withstanding air pressures of 100, 300, 500, and 1500 kPa, respectively. The appropriate ceramic stone should be selected based on the type of soil being tested and the suction range of greatest interest. It may be necessary to change the high-air-entry disks during the course of one test.

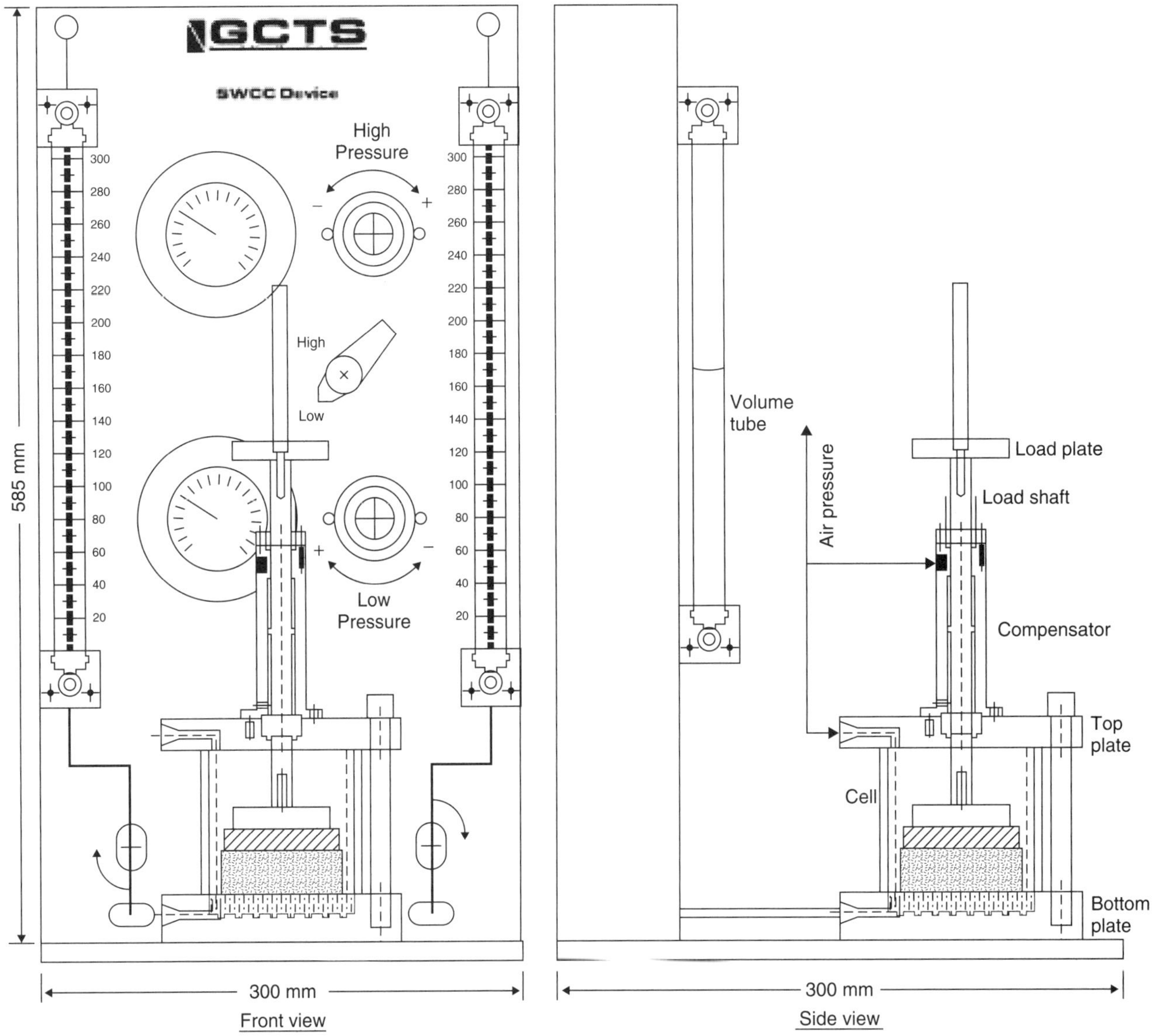

Figure 5.88 Front and side views of the GCTS pressure plate cell and the control panel. (Courtesy of GCTS, Tempe, AZ.)

5.9.4 Golder Basic Pressure Plate Cell (Fredlund, and Stone, 2011)

A pressure plate device was designed by members of the Golder Unsaturated Soils Group, Golder Associates, Saskatoon, Canada. The primary design specifications are as follows:

1. The pressure plate device should accommodate a single soil specimen.
2. The pressure plate cell should have a minimum of parts.
3. The device must have a high design factor of safety against breakage under high air pressures.
4. The pressure plate design should accommodate various ceramic disks with air-entry values up to 500 and 1500 kPa.
5. The device only needs to be designed to measure the drying curve.
6. The effect of air diffusion should be minimized.
7. The device must be sufficiently light such that the device and soil specimen can be weighed to ± 0.01 g on a balance.
8. The air supply at the top of the device should have a quick connector that maintains the internal air pressure in the cell when disconnected.

Figure 5.91 shows the SWCC cell designed by the Golder Unsaturated Soils Group. The device meets the desired specifications for safety and simplicity. The pressure plate cell has been manufactured by GCTS. Silt, sand, and clay soils have been tested to verify the functionality of the new cell.

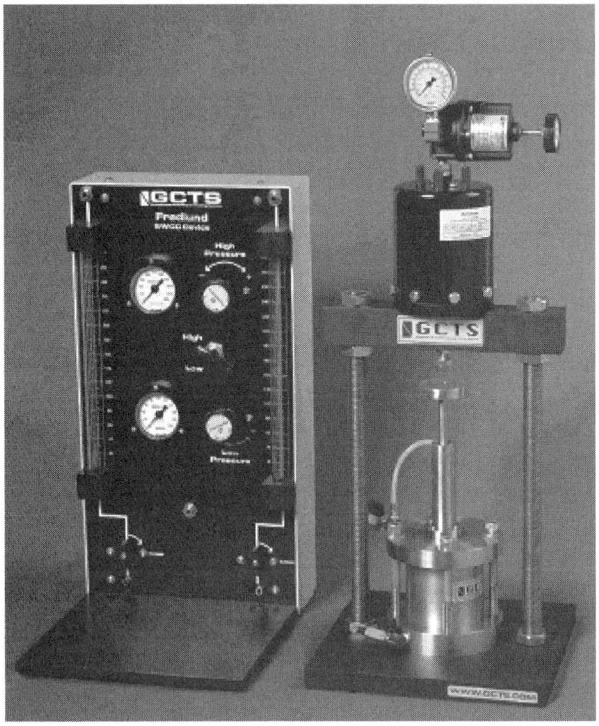

Figure 5.89 GCTS pressure plate cell with pneumatic loading frame. (Courtesy of GCTS, Tempe, AZ.)

Figure 5.91 Golder (Fredlund and Stone, 2011) basic pressure plate cell connected to air pressure supply. (Courtesy of GCTS, Tempe, AZ.)

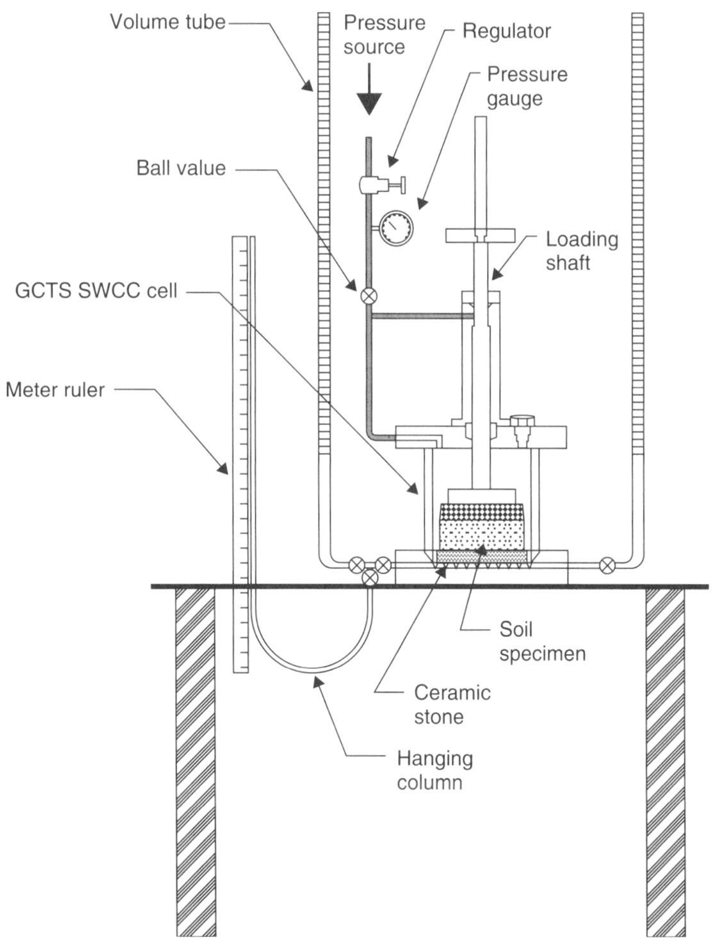

Figure 5.90 Hanging-column assembly for GCTS pressure plate cell. (Courtesy of GCTS, Tempe, AZ.)

Ten small holes (i.e., 0.08 mm in diameter) were drilled in the base of the apparatus to provide drainage for water from the high-air-entry ceramic disk. The holes also needed to be sufficiently small to remain filled with water. Some diffusion of air through the ceramic disk did not affect the test results.

5.9.5 Modified Triaxial Tests

A modified triaxial cell is a rather elaborate apparatus that can be used to measure the SWCC. The triaxial equipment needs to have a high-air-entry disk sealed onto the base pedestal and a low-air-entry material used to control the air pressure at the top of the soil specimen.

There are a number of advantages associated with the use of the modified triaxial equipment. First, the triaxial apparatuses provide greater flexibility in terms of the stress path that can be followed. It is possible to apply confinement to the soil specimen while data are collected for the SWCC. Second, it is possible to measure both overall volume change as well as water volume change and thereby obtain more accurate measurements of the volume-mass properties. It is also possible to use volume change measuring devices (e.g., GDS pressure-volume controllers) and thereby independently measure changes in water volume and air volume (Bankole et al., 1996). When air volume change is measured, it is not necessary to measure overall volume changes in the soil specimen.

A series of triaxial tests were performed by Thu et al., (2007b) where the net confining pressure was altered. The study involved tests on several statically compacted kaolin specimens that were prepared in the same manner. The soil specimens were compacted to the maximum dry density and optimum water content in 10 layers, each layer being 10 mm in thickness using the standard Proctor method. The height and diameter of the specimens were approximately 100 and 50 mm, respectively. Mini high-suction probes (Meilani et al., 2002) were installed along the height of the specimens to observe pore-water pressures during the SWCC tests.

Liquid limit, grain size analysis, specific gravity, and hydraulic conductivity tests were conducted to determine the index properties of the compacted kaolin. The kaolin was of high plasticity (i.e., MH according to the USCS). Compaction tests on the kaolin showed a maximum dry density of 1.35 Mg / m^3 and an optimum water content of 22%.

A modified triaxial apparatus (Fredlund and Rahardjo, 1993a) was used and a series of net confining stresses were applied to the soil specimens. The axis translation technique (Hilf, 1956) was adopted for the application of matric suction to the soil specimens. The soil specimens were initially saturated. The air pressure line was replaced by a water pressure line connected to a digital pressure and volume controller (GDS pressure volume device) to inject water from the top of the specimen. The specimen was saturated while applying a cell pressure σ_3 and a back pressure u_w. Full saturation was achieved with B parameters greater than 0.97 (Head, 1986).

The specimens were then consolidated under selected isotropic net confining stresses. Each specimen required about 1 h for the consolidation stage to be complete. After isotropic consolidation was complete, the first matric suction was applied to the soil specimen through control of the pore-air and pore-water pressures.

The GDS pressure-volume controller was connected to the pore-water pressure at the base plate. The pore-water pressure was controlled through a 5-bar high-air-entry ceramic disk sealed onto the base pedestal of the cell. The soil specimen was placed in contact with the high-air-entry disk. The pore-water pressure in the soil specimen was held constant while water volume change was measured. The net confining stress was maintained at a constant value. During the drying stage, matric suction was increased by decreasing the pore-water pressure at the base plate while maintaining the pore-air pressure. For the wetting stage, the matric suction was decreased by increasing the pore-water pressure at the base plate.

The amount of water that drained from the soil specimen and the total volume change of the specimen during drying and wetting were recorded. Equalization of matric suction was deemed to be complete when there was negligible water flow either into or out of the soil specimen. The SWCCs were conducted under several net confining stresses (i.e., $\sigma - u_a = 10, 50, 100, 150, 200, 250, 300$ kPa).

Figure 5.92 shows the SWCC in terms of degree of saturation with respect to matric suction under a net confining stress of 100 kPa. Figure 5.92 also indicates the procedure for determining the air-entry value. Figure 5.92 shows how the air-entry value of the soil increases with increasing net confining stress. The increase in air-entry value appears to be related to the reduced void ratio associated with an increased net confining stress.

5.9.6 Column Tests

The SWCCs are often required for coarse, cohesionless soils and it is possible to obtain satisfactory results through use of a simple column test (Fig. 5.93). The distance above the water level at the base of the column can be converted into an equivalent matric suction value by assuming hydrostatic conditions for the water within the column. Columns approximately 1 m in height have proven satisfactory when the air-entry value of the soil is less than about 7 kPa and residual conditions are around 10 kPa. It is possible to run the column test in a wetting mode or a drying mode and thereby measure the wetting and drying SWCCs, respectively.

Wetting SWCCs are most commonly measured using column tests. The wetting SWCC is measured by allowing water to be drawn up into the soil in the column from water supplied at the base of the column. Water content measurements can be made at various heights in the column once equilibrium has been achieved. Figure 5.93c shows the wetting SWCC for three coarse sands tested as a column test. A Tempe cell test was used to obtain the drying curve (Yang et al., 2004b).

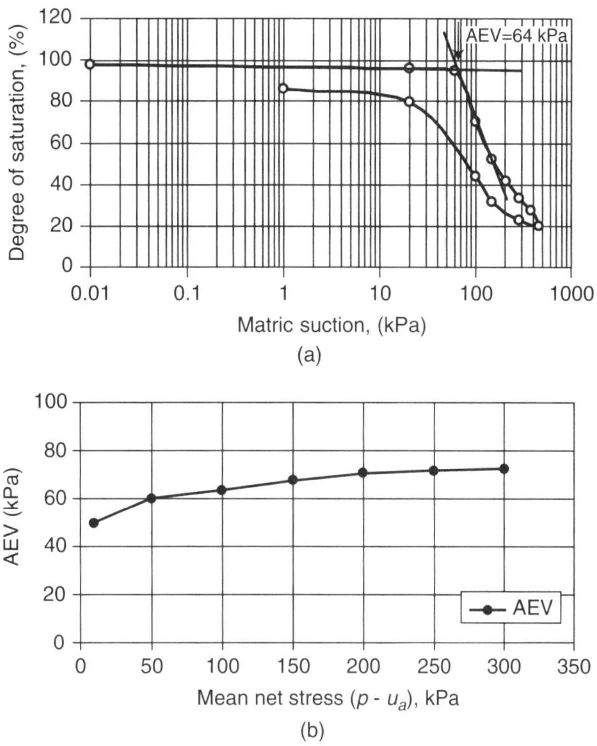

Figure 5.92 SWCCs for compacted kaolin specimens tested in a modified triaxial apparatus: (a) degree of saturation for specimen consolidated at 100 kPa, (b) change in AEV for various consolidation pressures (after Thu et al., 2007).

Drying SWCCs can be measured in a column test by saturating the soil from the top of the column and then allowing the water to drain down to an equilibrium hydrostatic state. Equilibrium is generally attained within a few days. Samples of the soil can be retrieved for water content measurements. The distance from the water level at the bottom of the column is used to calculate respective matric suction values.

5.10 VACUUM DESICCATORS FOR HIGH SUCTIONS

The vapor pressure equilibrium technique can be used to measure the water content versus total suction relationship in the high-suction range. Desiccators can be used to establish a partial vapor pressure in the soil (Edlefsen and Anderson, 1943; Richards, 1965). The thermodynamic relationship between total suction (or the free energy of the soil-water) and the partial pressure of the pore-water vapor can be written as follows:

$$\psi = \frac{-RT_K \rho_w}{\omega_v} \ln\left(\frac{u_v}{u_{v0}}\right) \tag{5.79}$$

where:

ψ = soil suction (or total suction), kPa,
R = universal (molar) gas constant [i.e., 8.31432 J/(mol K)],
T_K = absolute temperature (i.e., $T_K = 273.16 + T, K$),
T = temperature, °C,
ρ_w = density of water, kg/m^3,
ω_v = molecular mass of water vapor (i.e., 18.016 kg/kmol),
u_v = partial pressure of pore-water vapor, kPa, and
u_{v0} = saturation pressure of water vapor over a flat surface of pure water at the same temperature, kPa.

The term u_v/u_{v0} is called relative humidity RH. At a particular temperature the variables in front of the natural logarithm of relative humidity (i.e., free-energy gas constant) become a constant. The free-energy constant is −135,053 and it is possible to write total suction as a function of relative humidity (or relative vapor pressure h_r):

$$\psi = -135,053 \ln\left(\frac{u_v}{u_{v0}}\right) \tag{5.80}$$

5.10.1 Use of Salts to Create Constant-Relative-Humidity Environments

Constant-relative-humidity environments can be created in the laboratory through use of either saturated or unsaturated salt solutions. Any salt solution at a particular concentration and a constant temperature results in a fixed vapor pressure environment under equilibrium conditions. Sulfuric acid solutions can also be used.

Saturated salt solutions provide a convenient, inexpensive, and accurate controlled relative humidity environment.

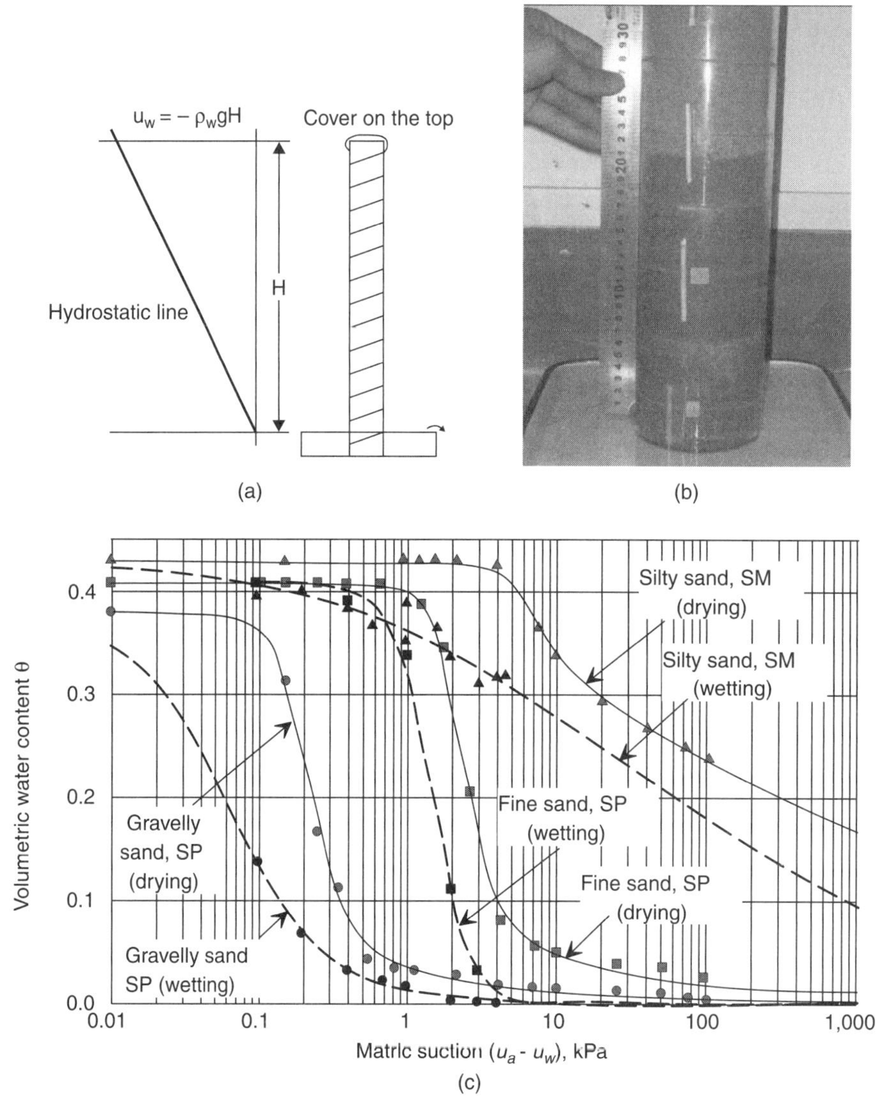

Figure 5.93 Column test performed on coarse-grained materials to measure the soil-water characteristic curve. (a) Equilibrium hydrostatic matric suctions in the column test, (b) Column test showing capillary rise for the wetting test. (c) SWCCs for coarse sands measured using simple soil column for adsorption curve and pressure plate apparatus for desorption curve (after Yang et al., 2004a).

Saturated salt solutions are able to liberate and adsorb considerable quantities of water with negligible changes to the equilibrium relative humidity. Consequently, the use of saturated salt solutions is preferable to the use of *unsaturated* salt solutions.

Salt solutions that are commonly used to create a range of relative humidity environments can be found in ASTM (2003) Designation E 104–02. Table 5.7 provides a list of suggested saturated salt solutions along with the general range of relative humidity created.

The generated relative humidity values created are also dependent on the temperature of the salt solutions.

5.10.2 Laboratory Test Procedure

A convenient sized, leak-proof container for creating a constant relative environment is the common laboratory desiccator. Salt solutions can be placed in the bottom of a desiccator of the type shown in Fig. 5.94. The salts are mixed with water until there are some free salts that appear in the bottom of the desiccator in an undissolved state. The saturated salt solution should be prepared at room temperature (or at a temperature slightly higher than room temperature) in order to ensure that the salt solution is saturated. The salt solutions suggested in the ASTM (2003) Standard E 104–02 can be kept and used to create fixed

Figure 5.94 Vacuum desiccator with soil specimens coming to equilibrium in controlled relative humidity environment.

Table 5.7 Approximate Equilibrium Relative Humidity for Selected Saturated Salt Solutions

Salt Solution	Average Relative Humidity, %
Cesium fluoride	4
Lithium bromide	7
Lithium chloride	12
Potassium acetate	23
Magnesium chloride	33
Potassium carbonate	43
Sodium bromide	59
Potassium iodide	70
Sodium chloride	75
Potassium chloride	85
Potassium sulfate	98

relative humidity environments for more than one year. Some other salt solutions can only be used for a few months.

The principal factors affecting the time required to establish vapor pressure equilibrium are (1) the ratio of free surface area of the solution to the chamber volume, (2) the amount of air circulation, (3) the absorbing properties of the sample, and (4) the agitation of the salt solution.

Soil specimen sizes can be small or quite large. The larger the specimen size, the longer the time that is required to attain equilibrium conditions. Specimen sizes ranging from 1 to 5 g are sufficient for purposes of measuring the water content corresponding to high-total-suction values on the SWCC.

Specimens with initial water contents above the final equilibrium water content will define the drying curve branch of the SWCC. Specimens with an initial water content that is below the final equilibrium water content will define the wetting curve branch of the SWCC. However, the difference between the drying and wetting portions of the SWCC at high total suctions is generally quite small.

The total-suction value corresponding to any temperature for a specified salt can be computed from the Gibbs free-energy equation. Young (1967) published equilibrium relative humidity data for a series of saturated salt solutions.

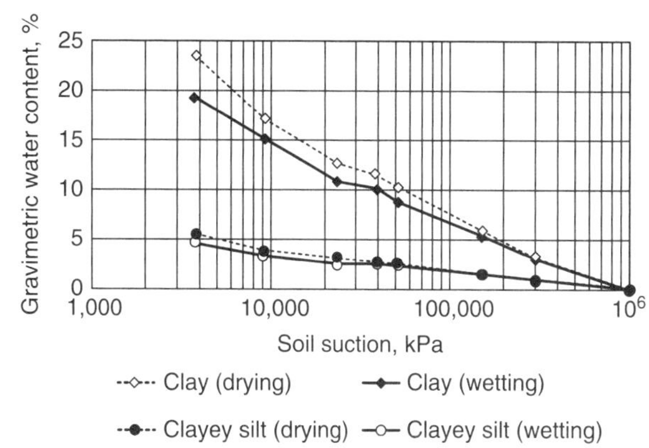

Figure 5.95 Drying and wetting SWCCs for clay and clayey silt placed in vacuum desiccators (Ebrahimi–Birang et al., 2007).

5.10.3 Typical Laboratory Results

Small soil specimens consisting of a few grams of soil can be suspended above a salt solution in the desiccator. It is possible to periodically weigh the soil specimens to determine whether moisture equilibrium has been attained. Equilibrium conditions are achieved more rapidly under low-relative-humidity conditions than under high-relative-humidity conditions. Low-relative-humidity equilibrium conditions can be attained in about one or two weeks. However, when the equilibrium relative humidity is high (e.g., 97% RH), several months may be required for equilibrium to be established. The time for equalization is also dependent upon the size of the soil specimen and constant temperature control becomes more important at high-relative-humidity values.

Figure 5.95 shows the wetting and drying curves for a clay soil and a clayey silt soil. Soil specimens were initially prepared in both wet and dry conditions. Laboratory results showed that there was some hysteresis in the SWCC relationship. The classification properties of the soils tested are shown in Table 5.8.

5.11 USE OF CHILLED-MIRROR OR DEW-POINT METHOD

Gee et al., (1992) introduced a method to measure the total suction in the medium- to high-suction range through measurement of water activity. The device was first introduced

Table 5.8 Properties of Clayey Silt and Clay Test Using Desiccator and Chilled-Mirror Device

Soil Properties	Clayey Silt	Clay
Liquid limit, %	20.3	81
Plastic limit, %	15.1	23.1
Plasticity index, %	5.2	57.9
Specific gravity	2.71	2.7
Sand, %	35	5
Silt, %	45	21
Clay, %	20	74

as a water activity meter by Gee et al., (1992) and later commercialized to measure total suction by Decagon Devices, Pullman, Washington. An overall view of the WP4 apparatus is shown in Figs. 5.96 and 5.97.

The chilled-mirror Water PotentiaMeter, WP4, has been studied by several geotechnical researchers (Leong et al., 2003b; Thakur et al., 2006; Agus and Schanz, 2005; Campbell et al., 2007; Cardoso et al., 2007). A testing procedure has been set forth in ASTM D 6836–02 (2008). Recently, the Water PotentiaMeter has been upgraded to provide reduced temperature fluctuations in the ambient environment when suction measurements are made. The Water PotentiaMeter with internal temperature control is called WP4-T.

Figure 5.96 WP4 chilled-mirror Water PotentiaMeter. (Courtesy of Decagon Devices, Pullman, WA.)

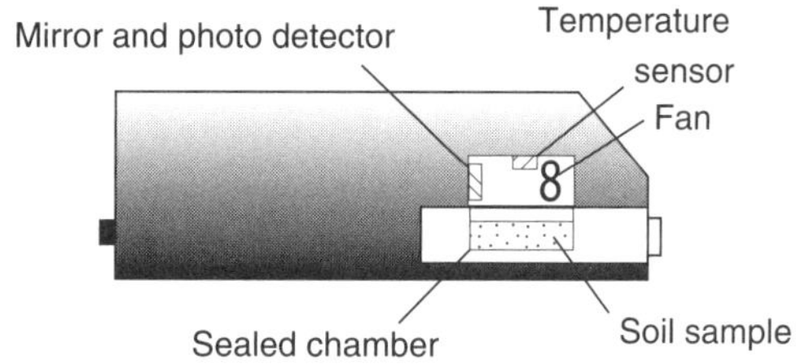

Figure 5.97 Schematic of WP4 chilled-mirror Water PotentiaMeter (after Decagon Devices Inc., 2009).

The device consists of a sealed chamber with a fan, a mirror, a photoelectric cell, and an infrared thermometer. The soil specimen is placed in a stainless steel or plastic container with a diameter of 40 mm. The specimen is then placed on a tray and moved into the temperature-controlled chamber. The specimen thickness may vary between 1 and 5 mm. The chamber is closed and the soil specimen is thermodynamically equilibrated with the chamber environment. The fan accelerates the equilibration process. A Peltier cooling system is used to reduce the temperature on the surface of a mirror to the dew-point temperature. A photoelectric cell detects the first sign of condensation on the mirror. The temperature at which moisture appears on the mirror corresponds to the dew point. The dew-point temperature is measured using a thermocouple. An infrared thermometer is used to measure the temperature of the chamber which is assumed to be the same as the temperature of the soil specimen at equilibrium.

The vapor pressure above the soil specimen in the chamber and the saturated vapor pressure at the same temperature are computed using the dew-point and specimen temperatures, respectively. Kelvin's equation (Eq. 5.79) is then used to calculate the total suction of the soil specimen. The calculations are performed internal to the device and the total suction is displayed along with the specimen temperature.

5.11.1 Calibration of Chilled-Mirror Apparatus

The WP4-T apparatus should be calibrated prior to usage by using a standard solution of 0.5 M KCl. This solution should yield a total suction value of 2.19 ± 0.1 MPa at 25°C. The standard solution of 0.5 M KCl is usually provided by the manufacturer. The calibration of the device can be improved by measuring the osmotic suction of known saturated salt solutions. Theoretical values of total suction can be calculated using Eq. 5.79. Table 5.9 shows the list of some solutions suggested by ASTM standards (ASTM E 104–02, 2003).

Figure 5.98 shows the typical theoretical suction values versus the measured values using the WP4-T device. The slope of the line should be 45°. If the slope of the line is not 45°, the suction values obtained using WP4-T might be overestimating or underestimating the correct total-suction values. If the line deviates slightly from 45°, it is possible to correct the measured values using a fitting equation.

Table 5.9 Saturated Salt Solutions and Corresponding Relative Humidity

Salt Solution	Relative Humidity, %	Suction, kPa
Lithium chloride	11.3 ± 0.3	297,321
Magnesium chloride	32.8 ± 0.2	152,517
Potassium iodide	68.9 ± 0.3	48,835
Sodium chloride	75.3 ± 0.2	38,738
Potassium chloride	84.3 ± 0.3	23,187
Potassium nitrate	93.6 ± 0.6	8,635
Potassium sulfate	97.3 ± 0.5	3,544

Source: ASTM E 104–85 (1998) and E 104–02 (2003).

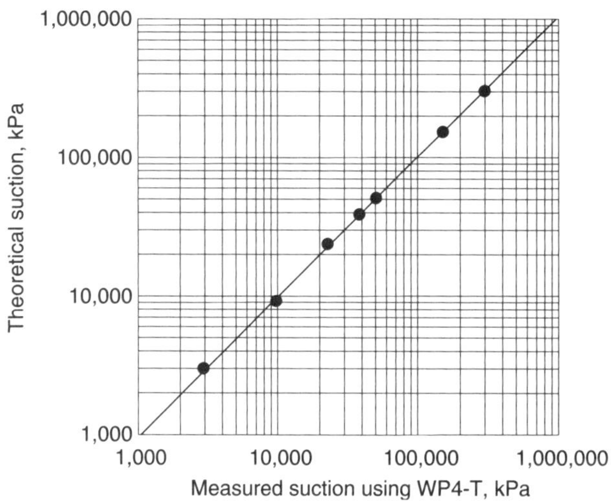

Figure 5.98 Comparison of theoretical calculated suction and measured suction (after Ebrahimi–Birang et al., 2007).

5.11.2 Suggested Procedure for Measuring SWCC using WP4

The following procedure can be used to measure the wetting SWCC:

1. Measure the mass of the clean stainless steel WP4 soil specimen cup.
2. Mix an appropriate amount of soil and water to obtain the desired water content.
3. Seal the sample using a disposable plastic lid that comes with the apparatus.
4. Perform Steps 1–3 for various water contents.
5. Let the prepared soil specimens equilibrate for at least 16 h.
6. Measure the water potential of each soil specimen and then immediately measure the mass of the wet soil and the specimen cup.
7. Place each soil specimen with their lids off in a drying oven. Leave the soil specimens in the oven for at least 16 h at 105°C. When the soil specimens are removed from the oven, immediately place them into desiccators to prevent the collection of moisture from the surrounding air.
8. As soon as the soil specimens are cool, measure the mass of the dry soil and the soil specimen cup. The water content of each soil specimen can then be calculated.

Similar steps to those listed above can be followed for the measurement of the drying branch of the SWCC. However, in this case, it is important that the initial water content of the soil be much higher than the final equilibrium state.

The drying and wetting SWCCs are often assumed to be the same. The results obtained for the drying branch of the SWCC are generally quite similar to those from the wetting curve except when the soil is highly plastic. Highly plastic soils appear to exhibit more hysteresis in the high-suction range than sand and silt soils (Fredlund, 1964; Ebrahimi-Birang et al., 2007).

5.12 ESTIMATION OF SWCC

Unsaturated soil property functions have been developed as extensions of saturated soil properties. In most situations it is acceptable to use estimations of unsaturated soil property functions (e.g., permeability function) when solving unsaturated soil problems in geotechnical engineering (M.D. Fredlund et al., 2003). Unsaturated soil property functions play an important role in the solution of saturated-unsaturated soil systems. Numerical modeling solutions may not be exact but can provide a reasonable indication of likely trends. The numerical modeling results provide the geotechnical engineer with valuable information that assists in decision making.

Let us consider the case of a homogeneous dam where the engineer desires to know the hydraulic heads in the saturated and unsaturated soil zones. The selected permeability function has a reduced effect on the calculated hydraulic heads. Consequently, much can be learned regarding the behavior of an engineered structure through the use of approximate unsaturated soil property functions.

There is one variable that has a reduced dependence on the soil properties when solving partial differential "field" equations within a continuum mechanics framework. In the case of seepage problems, the primary variable that can be computed is the independent variable called hydraulic head h. One of the dependent variables is water flux q_w, and its prediction requires an increased accuracy in the determination of the permeability function. Consequently, the accuracy required in the measurement or estimation of the unsaturated soil property functions depends on the variable of primary importance for the problem at hand. The same rationale is true for a stress analysis problem such as the loading of a foundation footing. The stresses below the footing are almost independent of the soil properties while the displacements are highly dependent upon the soil properties.

This pattern can be observed when solving all partial differential "field" equations.

The variable that is of primary interest when solving a geotechnical engineering problem has a significant influence on how accurately the unsaturated soil property functions need to be determined. A crude approximation of the SWCC (and subsequently the unsaturated soil property function) is all that may be required for the computation of a particular variable (e.g., the independent variable). On the other hand, a more accurate assessment of the soil properties may be required for the prediction of some other variables (e.g., some dependent variables). Possibly no single factor more strongly influences the engineering solution than an understanding of whether it is the dependent or independent variable that is of primary importance for solving the problem at hand. An understanding of the primary variable that is of greatest interest for the problem at hand may also influence whether hysteresis effects need to be taken into consideration in the assessment of saturated soil property functions.

The above rationale applies for both saturated and unsaturated soil systems. There are also other factors such as the level of risk that may influence the accuracy with which the unsaturated soil property functions need to be determined. For example, when the level of risk is high, the unsaturated soil properties need to be assessed with greater accuracy.

Unsaturated soil properties are an extension of the saturated soil properties. In many cases it is the saturated soil properties that have the most significant influence on the computed engineering solution.

5.12.1 Use of Grain-Size Distribution for Estimation of SWCC

The grain-size distribution curve provides information on the distribution of solids in a soil (i.e., percent passing certain particle sizes). The SWCC is related to the distribution of voids in the soils and the amount of water in the voids. Various studies have been undertaken that show that the grain-size distribution curve can be used to estimate the SWCC. These estimation techniques are attractive, but the user must be aware of the assumptions and limitations associated with each of the estimation procedures. The volume of voids in a soil is estimated and then used to estimate a likely desorption curve for the soil. The effects of stress history, soil fabric, confinement, and hysteresis are difficult to address when using estimation procedures associated with the grain-size distribution curve.

Several models have been proposed for the estimation of the SWCC from the grain-size distribution curve. The starting point for most methods involves the representation of the grain-size distribution curve as shown by M.D. Fredlund (2000) (Fig. 5.99). The methodology proposed by M.D. Fredlund (2000) hypothesized that the grain-size distribution curve can be viewed as having incremental particle sizes ranging from the smallest particles to the largest particles.

Each of the particle sizes can be assembled to build a SWCC. Small increments (on a logarithmic scale) of uniform-sized particles can be transposed to form an SWCC. Once the entire grain-size distribution curve has been analyzed as a series of incremental particle sizes, the individual SWCCs can be combined using a superposition technique to give the overall SWCC for the soil.

The SWCC for each uniform particle size range is assumed to be relatively unique when building the overall SWCC using the M.D. Fredlund (2000) methodology. Typical SWCCs for various mixtures of sand, silt, and clay have been studied. Representative SWCCs for various soils can be fitted using the Fredlund and Xing (1994) equation. There are approximate curve-fitting parameters representative of various effective grain-size diameters.

The shape of an estimated SWCC is predominantly controlled by the grain-size distribution and secondarily influenced by soil density (i.e., initial porosity). The M.D. Fredlund (2000) unimodal and bimodal fitting of grain-size distribution curves can be used as the starting point for the estimation of the SWCC.

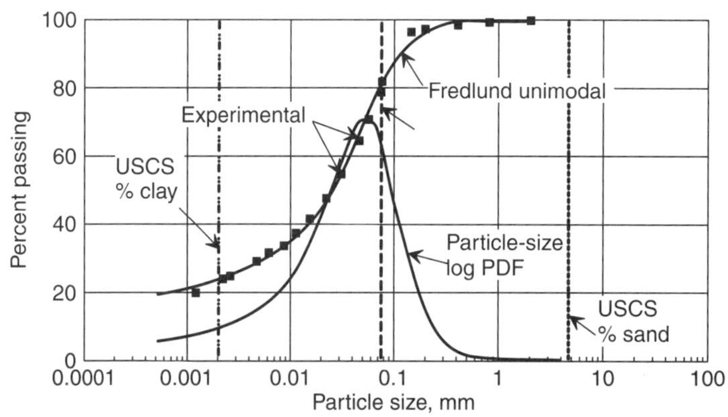

Figure 5.99 Fit of grain-size distribution curve for uniform silt (data from Ho, 1988) with M.D. Fredlund (2000) grain-size equation.

5.12.2 Review of SWCC Estimation Procedures from Grain-Size Distribution Curves

A number of methods have been proposed for the estimation of SWCCs from grain-size distribution data. There are three broad categories of estimation techniques based on: (i) statistical estimates of water contents at various soil suctions (Gupta and Larson, 1979a), (ii) estimation of soil parameters for an algebraic function for the SWCC (Rawls and Brakensiek, 1985; Vereecken et al., 1989; Scheinost et al., 1997), and (iii) physico-empirical models where a grain-size distribution curve is used in the prediction of the SWCC (Arya and Paris, 1981; Tyler and Wheatcraft, 1989; M.D. Fredlund et al., 1997a).

Arya and Paris (1981) used a physico-empirical type model for the estimation of the SWCC from grain-size distribution curves. The procedure has become quite widely used in agricultural disciplines. M.D. Fredlund et al. (1997a) used the capillary model equation along with the M.D. Fredlund et al. (1997a) equation for the grain-size distribution curve to develop a new physico-empirical equation for calculating the SWCC. Parametric studies were performed using several SWCC data sets. A combination of the physico-empirical model and the parametric study information formed the basis for the M.D. Fredlund et al. (1997a) estimation technique for the SWCC.

5.12.3 Definition of Terms Associated with Estimation Techniques

A few definitions are useful for describing the conceptual models for the estimation of the SWCC:

1. A *soil property function* is a relationship between a physical soil property and either soil suction, the complete stress state, or the volume-mass properties of a soil.
2. A *pedo-transfer function*, PTF (Bouma, 1989), is a function that has as its arguments basic soil data such as the grain-size distribution and dry density (or porosity) and yields a soil property function.
3. An estimated SWCC is either:
   i. a *monotonic, single-valued function* that yields water content (i.e., volumetric or gravimetric) for a given scale of soil suction or
   ii. *a relationship* consisting of two functions including a drying and wetting branch (Tietje, 1993). In other words, the function has hysteresis and the drying and wetting branches form limiting conditions.

5.12.4 Representation of Grain-Size Distribution

The presented PTF uses the M.D. Fredlund et al., (1997a) unimodel (Eq. 2.1) and bimodal (Eq. 2.6) equations to represent the grain-size distribution (M.D. Fredlund et al., 2000a). The unimodal and bimodal Fredlund (1997) grain-size distribution equations can be found in Chapter 2. Most PTFs use grain-size distribution information in some form as the basis for the estimation of the SWCC.

5.12.5 Description of Pedo-Transfer Functions

Two methodologies have been used to estimate the SWCC from grain-size distribution curves: (i) the functional parameter regression methodology and (ii) the physico-empirical methodology.

5.12.6 Functional Parameter Regression Method

The functional parameter regression method assumes that parameters of the SWCC equation can be correlated to basic physical properties. An example is the correlation between the air-entry parameter of an SWCC equation and soil properties such as percent sand or porosity. Rawls and Brakensiek (1985) presented regression equations for estimating the parameters for the Brooks and Corey (1964) equations. The regression equations estimated the air-entry pressure a_{aev} (i.e., bubbling pressure), the pore size index λ_{be}, and the residual water content w_r for the Brooks and Corey (1964) equation.

The Vereecken et al. (1989) method involved fitting a data set of 40 Belgian soils with the van Genuchten (1980) equation. A sensitivity analysis was performed on the optimized parameters of the SWCC to assess the relative importance and uniqueness of the parameters. A principal factorial analysis was used to examine the relationship between the estimated SWCC and the basic measured soil properties. It was concluded that the SWCC could be estimated to a reasonable level of accuracy using soil properties such as grain-size distribution, dry density, and carbon content. The study focused mainly on the agricultural discipline where organic soils were involved and the emphasis is on water availability for plant growth.

5.12.7 Description of Modified Kovacs Model (Aubertin et al., 2003)

The modified Kovacs (MK) model is a functional parameter-type model developed for the estimation of SWCCs (or WRCs) by Aubertin et al., (2003). The model has been found to perform well for tailing materials as well as granular and cohesive soils.

The MK model assumes that the amount of water held in a soil is attributable to two primary mechanisms: (i) capillary saturation S_c and (ii) adhesive saturation S_a. Capillary saturation is believed to dominate water adsorption in the low-suction range while adhesive saturation dominates in the high-saturation range. The two components of saturation are combined using the following equation:

$$S = \frac{\theta}{n} = 1 - \langle 1 - S_a \rangle (1 - S_c) \tag{5.81}$$

where:

S = any degree of saturation,
θ = any volumetric water content,
n = initial porosity of the soil,
S_c = saturation associated with the capillary component,

S_a = saturation associated with the adhesive component, and
$\langle\cdot\rangle$ = Macaulay brackets defined as $\langle x\rangle = 0.5(x + |x|)$, used to describe a ramp function.

The capillary component of saturation is empirically related to soil suction using the following equation:

$$S_c = 1 - \left[\left(\frac{h_{c0}}{\psi}\right)^2 + 1\right]^m \exp\left[-m(h_{co}\psi)^2\right] \quad (5.82)$$

where:

h_{c0} = equivalent capillary height which is related to an equivalent pore diameter and the solid surface area,
ψ = soil suction represented as a head or length, and
m = pore-size coefficient, unitless.

The adhesive component of saturation is empirically related to soil suction through the following equation:

$$S_a = a_c\left(1 - \frac{\ln(1 + \psi/\psi_r)}{\ln(1 + \psi_0/\psi_r)}\right)\frac{(h_{c0}/\psi_n)^{2/3}}{e^{1/3}(\psi/\psi_n)^{1/6}} \quad (5.83)$$

where:

a_c = adhesion coefficient, unitless,
e = void ratio,
ψ_n = normalization parameter introduced to maintain consistency in the units, ψ_n = 1 cm when ψ is in cm, and
ψ_0 = suction head equal to 10^7 cm of water corresponding to a dry soil condition.

Four parameters, h_{co}, ψ_r, m, and a_c, are required when solving the MK model. The parameters are defined as follows when the soil is granular in nature:

$$h_{c0}\ (\text{cm}) = \frac{0.75}{[1.17\log(C_u) + 1]eD_{10}} \quad (5.84)$$

$$\psi_r\ (\text{cm}) = 0.86h_{c0}^{1.2} \quad (5.85)$$

$$m = \frac{1}{C_u} \quad (5.86)$$

$$a_c = 0.01 \quad (5.87)$$

where:

D_{10} = diameter corresponding to 10% passing on the grain-size curve and
C_u = uniformity coefficient equal to D_{60}/D_{10}.

The four parameters are defined as follows when the soil is plastic and essentially incompressible:

$$h_{c0}\ (\text{cm}) = \frac{0.15\rho_s}{e}w_L^{1.45} \quad (5.88)$$

$$\psi_r\ (\text{cm}) = 0.86h_{c0}^{1.2} \quad (5.89)$$

$$m = 0.00003 \quad (5.90)$$

$$a_c = 0.0007 \quad (5.91)$$

where:

ρ_s = density of the soil solids, kg/m^3,
w_L = liquid limit, %.

5.12.8 Description of Ayra and Paris (1981) Method

Arya and Paris (1981) presented the first physio-empirical method to estimate the SWCC. The model made use of basic information from the grain-size distribution curve. Volumetric water contents were calculated based on an estimation of the pore sizes in the soil. The pore radii were converted to equivalent soil suctions through use of the capillary theory. The estimation method used empirical factors to account for uncertainties in the procedure. The pore radius estimation was based on the assumption of spherical particles and cylindrical pores. Arya and Paris (1981) assumed the pore-size distribution and the grain-size distribution of soils to be approximately congruent. In other words, larger particles produce larger interparticle voids than smaller particles and vice versa.

Various models have been proposed to estimate the random packing nature of spherical particles in an attempt to improve on the estimation of the pore-size distribution in a heterogeneous system (Iwata et al., 1998). The Arya and Paris (1981) model was later modified by Haverkamp and Parlange (1986), who applied the concept of shape similarity between the SWCC and the cumulative grain-size distribution for sand soils. Gupta and Ewing (1992) applied the Arya-Paris model to (i) the grain-size distributions in order to model intra-aggregate pores and (ii) the aggregate-size distributions to model inter-aggregate pores. Nimmo (1997) presented a method of accounting for the influence of fabric and soil structure through the use of aggregrate-size distribution.

5.12.9 Description of M.D. Fredlund (2000) Model

The physio-empirical method forms the basis for the development of the M.D. Fredlund (2000) model. It was hypothesized that the grain-size distribution provides a physical description that could be used as the basis for the SWCC estimation technique. The grain-size distribution is limited, however, in that it does not provide an indication of the in situ density (or porosity) of a soil or the fabric of the soil. In addition, the packing arrangement of various grain sizes constitutes another important factor. No attempt was made to represent complex soil fabrics in the estimation of the SWCC.

The M.D. Fredlund (2000) method first divides the grain-size distribution into small particle groupings of relatively uniform particle sizes. It is hypothesized that for each uniform group of particles there exists a somewhat unique desorption SWCC and that this curve represents conditions indicative of an initially slurry soil.

5.12.10 M.D. Fredlund (2000) Model to Predict SWCC

The M.D. Fredlund (2000a) unimodal and bimodal equations (see Chapter 2) can be used to best fit grain-size distribution data. The M.D. Fredlund (2000) grain-size equations provide a continuous fit of the entire grain-size distribution curve, including the coarse and fine extremes (M.D. Fredlund et al., 2000a). The mathematical fit of the grain-size distribution data provides the basis for developing an algorithm to predict the SWCC. The predictive model makes use of a combination of the capillary model and an understanding of changes in SWCCs associated with particle sizes. Volume-mass properties and grain-size distribution data form the basic information required for the estimation of the SWCC.

The methodology behind the M.D. Fredlund (2000) method can be expressed in terms of a series of theorems:

Theorem 1—A soil composed entirely of uniform, homogeneous particle sizes has a unique drying (or desorption) SWCC.

Theorem 2—The capillary model can satisfactorily estimate the air-entry value of each collection of uniform, homogeneous particle sizes.

Theorem 3—The SWCC for soils composed of more than one particle size can be represented as the summation of the SWCCs for each of the individual particle sizes.

The above theorems form the basic information for the method based on the capillary model and the packing of each uniform particle-size range. The Fredlund and Xing (1994) equation was used to model SWCCs typical of each range of soil particle sizes. The a_f fitting parameter (i.e., a_f is the inflection point on the SWCC) in the Fredlund and Xing (1994) equation can be loosely related to the air-entry value of the soil. Figure 5.100 shows the relationship between the air-entry value of many soils and the a_f fitting parameter. A data set was used to "train" the proposed pedotransfer function. The a_f parameter is typically higher than the actual (or construction-based) air-entry value. The a_f parameter is the primary variable that represents the lateral position of the SWCC.

The variation of the Fredlund and Xing (1994) equation is shown in Fig. 5.101 for a range of n_f and m_f parameters when the a_f parameter is held constant at 100 kPa. The representative SWCC can be shifted laterally for other values of a_f. An equivalent air-entry value (i.e., soil suction) can be calculated based on the capillary model when the pore radius between the soil particles is known. The soil suction corresponding to the equivalent air-entry value for a soil with uniform particle sizes can be written as follows:

$$\psi = 2T_s \frac{\cos \alpha_1}{\rho_w g r} \tag{5.92}$$

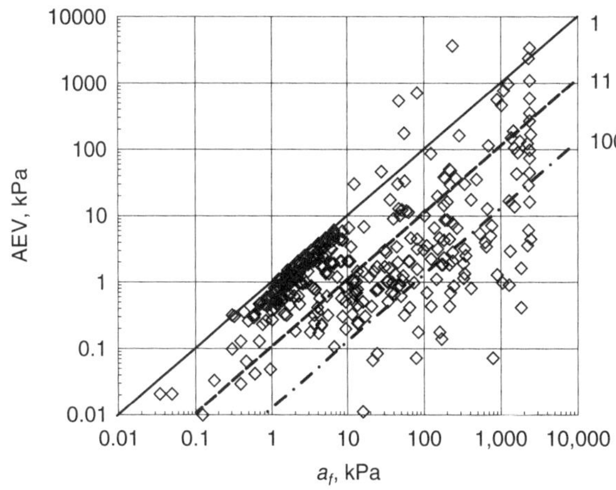

Figure 5.100 Relationship between AEV obtained using Fredlund (1997b) construction and parameter a_f from Fredlund and Xing (1994) equation for 311 soils.

where:

T_s = surface tension of water,
α_1 = contact angle,
ρ_w = density of water,
g = acceleration due to gravity,
r = pore radius, and
ψ = soil suction.

An approximate shape for the SWCC can be computed for each particle size. Estimating the shape for coarse sand or fine silt can be accomplished with reasonable certainty. Measured SWCCs for glass beads of various sizes were used as a reference (Nimmo, 1997). The SWCC for a fine-grained material was estimated from the results of soils with increasing clay content. The glass beads and the clay soil provide limiting values for groups of soils consisting of uniformly sized particles (Fig. 5.102).

Typical SWCCs for uniform particle-size materials ranging from coarse sand to clay-sized particles were computed by incrementally altering the parameters of the Fredlund and Xing (1994) equation. The n_f and m_f soil parameters for the Fredlund and Xing (1994) equation are also required for each uniform collection of particle sizes. A data set involving soils from Rawls et al., (1992), Sillers (1997), and the Cecil soil survey (Bruce et al., 1983) was used to determine approximate trends in the n_f and m_f parameters.

The grain-size distribution curve was divided into small uniform soil divisions. A packing porosity n_p was estimated, starting with the smallest particle divisions (Harr, 1977) and computing an SWCCs, as illustrated in Fig. 5.103. The divisional SWCCs were then summed starting from the smallest particle sizes and continuing until the volume of the pore space was equal to that of the combination of all particle sizes. The end result is an estimation of a predicted SWCC representative of the desorption curve for a slurry soil.

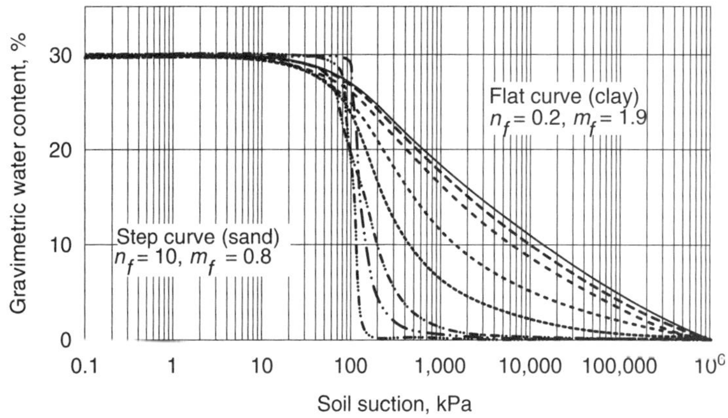

Figure 5.101 Variation of parameters n_f and m_f according to particle-size diameter while holding a_f constant at 100 kPa (from Fredlund, 1997b).

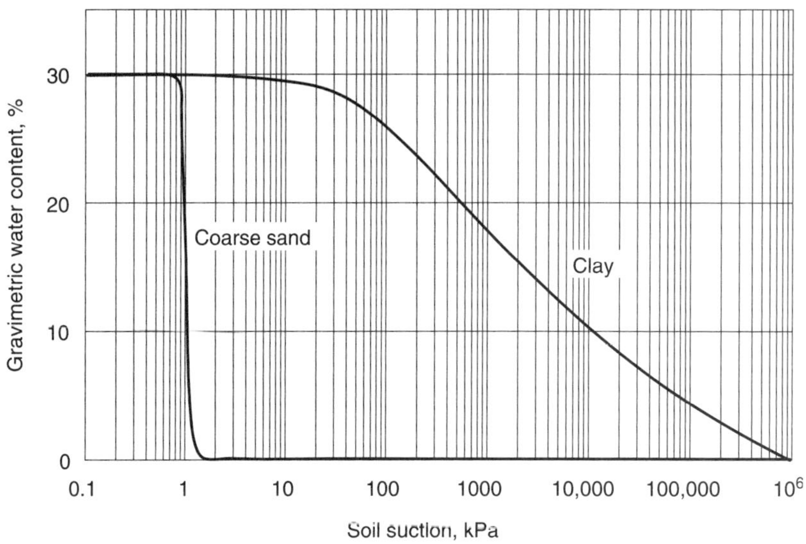

Figure 5.102 Assumed limits for SWCCs for uniform coarse sand ($a_f = 1, n_f = 20, m_f = 2, h_r = 3000$) and a clay ($a_f = 100, n_f = 1, m_f = 0.5, h_r = 3000$) (from Fredlund, 1999b).

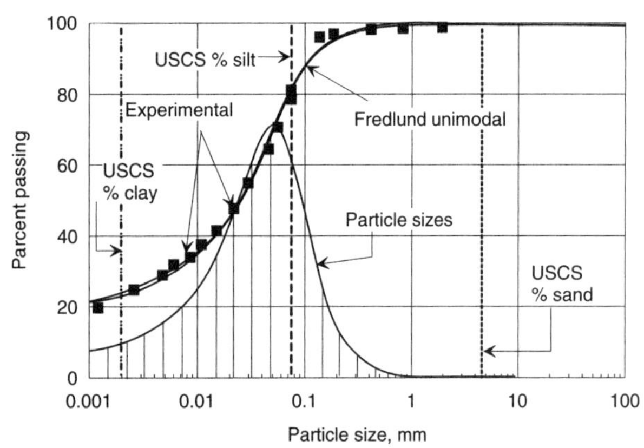

Figure 5.103 Small divisions of particle sizes used to build overall SWCC.

5.12.11 Designation of Porosity and Packing Arrangement for M.D. Fredlund (2000) Model

The grain-size distribution curve can be divided into n fractions of uniform sized particles. Each fraction has a packing arrangement and porosity. The summation of the individual fraction porosity values may be greater than the overall porosity of the combined soil fractions. In the assemblage of soil particles, the voids created between larger particles will be filled with smaller particles. Consequently, the influence of the larger particles on the SWCC is reduced (Yazdani et al., 2000). The porosity of the individual fractions can be summed until the overall porosity of the soil is reached. After this point the remaining particle fractions are ignored.

The "packing" porosity n_p for each uniform particle size needs to be approximated. It can be assumed that the n_p

variable is the same for each successive particle fraction, but it would appear that n_p should be a function of particle diameter.

5.12.12 Data Sets Analyzed in Testing M.D. Fredlund (2000) Model

A sample data set of 188 soils was selected to test the proposed model. The soils were from Sillers (1997), Rawls and Barkensiek, et al., (1985), and Williams (1992). All soils selected had a measured grain-size distribution and a measured SWCC.

The reliability of the M.D. Fredlund (2000) pedo-transfer function was evaluated using cross-validation (Hjorth, 1994). The reliability of a pedo-transfer function was assessed by (1) drawing a random sub sample from the data set, (2) developing a pedo-transfer function for the sub sample, and (3) testing the accuracy of the pedo-transfer function against the data left after subsampling. Data sets were first selected based on the availability of grain-size distribution curves and volume-mass properties (i.e., void ratio, dry density, and specific gravity). The database was split into two parts with one part used to train the proposed M.D. Fredlund (2000) pedo-transfer function and the other part used to test the pedo-transfer function.

5.12.13 Estimation of SWCC Using M.D. Fredlund (2000) model

Comparisons of experimental and predicted results are shown in Figs. 5.104–5.106. Ten optimal estimations of the SWCC were performed using the M.D. Fredlund (2000) method and the results can be seen in Fig. 5.107. The estimation of a SWCC from grain-size distribution was attempted for all soil types. There was greater difficulty in estimating the SWCC for clay, till, and well-graded soils than for silt and sands,

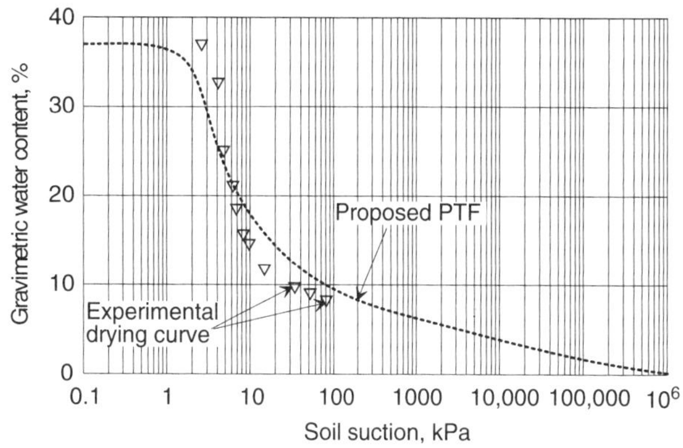

Figure 5.105 Comparison of experimental and predicted SWCC for sandy silt (Data after Schuh et al., 1991).

although all estimated SWCCs appear to be quite reasonable. Results tend to be somewhat sensitive to the assumed packing porosity n_p. More research is required to better understand the influence of the packing porosity.

The M.D. Fredlund (2000) pedo-transfer function provided a reasonable estimation of the SWCC for a wide range of textural classes. There are several groups of soils for which it is particularly difficult to estimate the SWCC. These general categories of soils include (i) soils having a large amount of clay-sized particles, (ii) soils that contain large amounts of coarse-size particles mixed with fines, (iii) soils that exhibit bimodal behavior such as sand-bentonite mixtures, and (iv) man-made soils such as mine tailings and waste rock.

The assumed minimum particle size has an influence on the estimation of the SWCC. If the minimum particle size is set too small, there will be an overabundance of clay-size particles that will influence the estimation of the SWCC. If

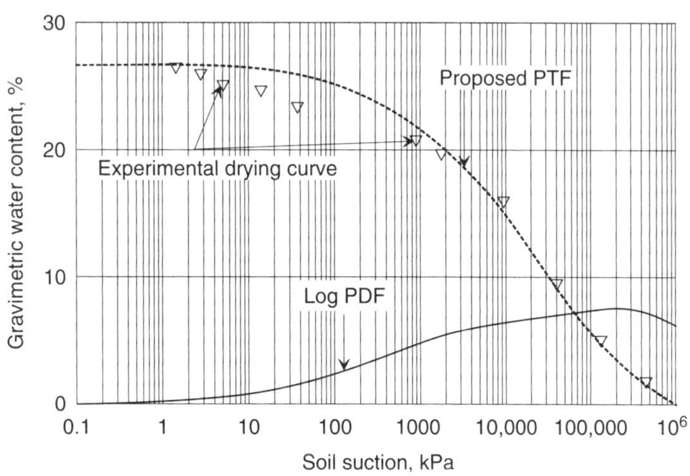

Figure 5.104 Comparison of experimental and predicted SWCC and logarithmic probability density curve (log PDF) for clay soil (Data from Russam, 1958).

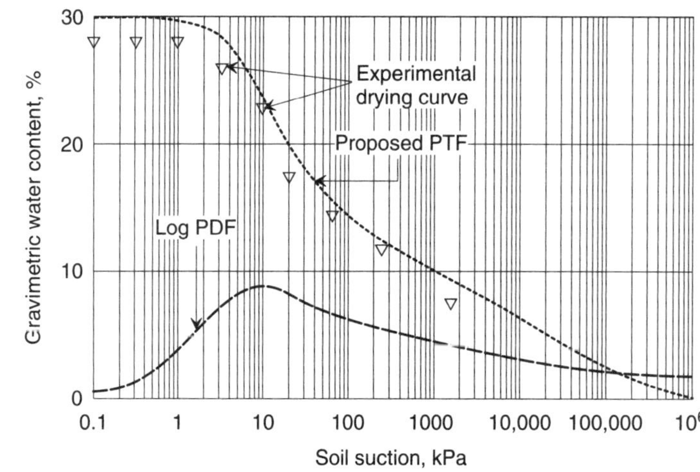

Figure 5.106 Comparison of experimental and predicted SWCC and logarithmic probability density curves for silt (after Vereecken et al., 1989).

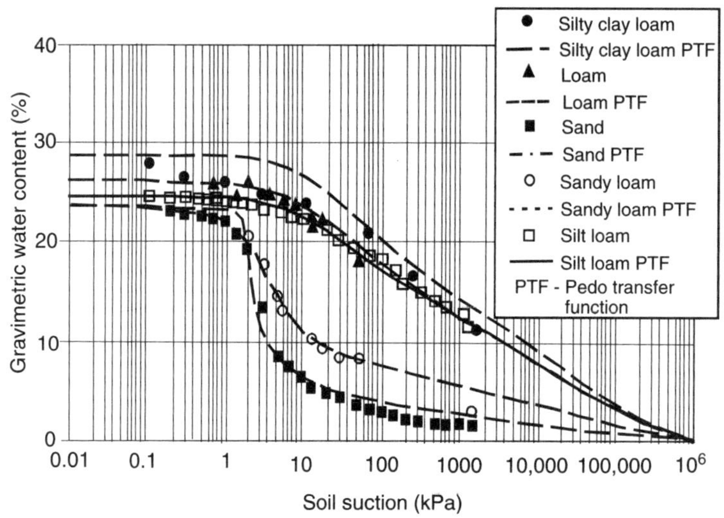

Figure 5.107 SWCCs computed from grain-size distribution curves for variety of soil types (from M.D. Fredlund, 2000).

the minimum particle size is too large, there is an absence of smaller particles which resulted in the soil drying prematurely.

The initial packing porosity n_p is the only volume-mass variable used in the estimation of the SWCC. The packing porosity is important to the estimation of the SWCC and two techniques have been proposed for its estimation: (i) a statistical method and (ii) a neural net method.

The statistical method involved examining the frequency distribution of the packing porosities for various textural categories. A mean and variance value was computed for the packing porosity for different soil types.

The use of a neural net appeared to improve the estimation of the SWCC. The neural net has an algorithm that is trained to respond to various input conditions. The effect of the packing porosity is shown in Fig. 5.108 for sand and in Fig. 5.109 for silt. These examples show that the packing porosity influences the entire SWCC. Fig. 5.109 shows that the estimated SWCC does not reach 100% saturation with packing porosity $n_p = 0.36$. This condition occurs when the packing porosity falls too far below the actual porosity of a soil.

5.12.14 Comparison of Various Pedo-Transfer Function Methodologies

The M.D. Fredlund (2000) estimation technique can be compared to other estimation models, namely, Arya and Paris (1981), Scheinost et al., (1997), Rawls and Brakensiek (1985), Vereecken et al., (1989), and Tyler and Wheatcraft (1989).

Arya and Paris (1981). The Arya and Paris (1981) pedo-transfer function was originally developed from a modest database and then extrapolated to larger databases. A value of 1.38 is commonly accepted as a reasonable estimate for the α variable. Later investigations by Arya et al. (1982)

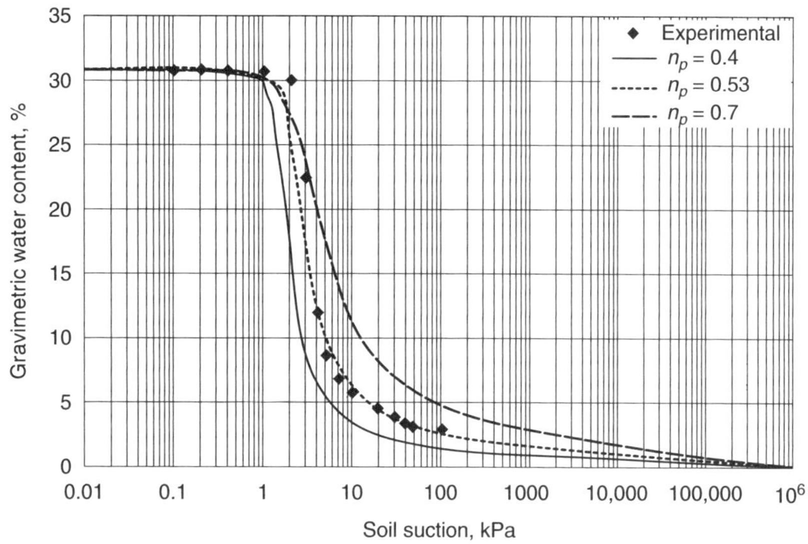

Figure 5.108 Effect of varying packing factor n_p for sand (data from Mualem, 1984).

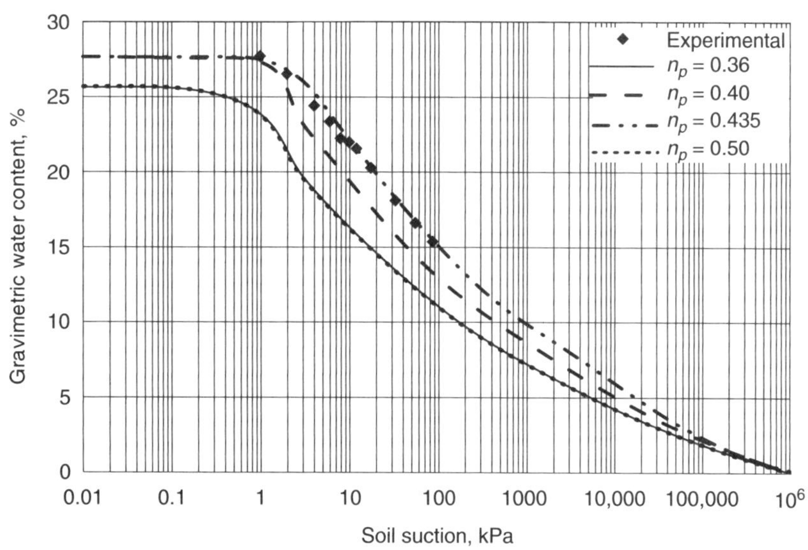

Figure 5.109 Effect of varying packing factor n_p for silt (data from Schuh et al., 1991).

showed that the α value varied among textural classes and ranged from 1.1 for fine-textured soils to 2.5 for coarse-textured soils. The α value was estimated in accordance with the values shown in Table 5.10. The Arya and Paris (1981) method requires a well-defined grain-size distribution curve. Several estimations performed using the Arya and Paris (1981) pedo-transfer function are shown in Fig. 5.110.

Scheinost et al., (1997). The Scheinost et al., (1997) pedo-transfer function uses a linear regression analysis to estimate the parameters for a van Genuchten (1980) equation. The Scheinost pedo-transfer function was developed to account for extreme variations in the soil parameters, with textures varying from gravels to clays, organic contents over a wide range, and bulk densities from 0.80 to 1.85 mg/m^3. The parameters for the pedo-transfer function were obtained using a soil data set from Germany. The Scheinost et al. (1997) pedo-transfer function was able to estimate the desaturation rate of most soils with reasonable accuracy.

Rawls and Brakensiek (1985). The Rawls and Brakensiek (1985) pedo-transfer function is based on a regression analysis that estimates parameters for the Brooks and Corey (1964) equation. While the estimation of the air-entry value for most soils seems reasonable, the desaturation rate appears to generally be overestimated.

Vereecken et al., (1989). The Vereecken et al. (1989) pedo-transfer function uses a statistical regression analysis to estimate the parameters for the van Genuchten (1980) equation. The Vereecken et al. (1989) pedo-transfer function has been applied to a wide range of soils and has the ability to account for high-organic-matter contents. In

Table 5.10 Values of Alpha Proposed by Arya et al. (1981)

USDA Texture	α
Sand	1.285
Sandy loam	1.459
Loam	1.375
Silt loam	1.15
Clay	1.16

[a]Dimensionless.

general, the method performs well for the estimation of desaturation rates.

Tyler and Wheatcraft (1989). Tyler and Wheatcraft (1989) use a fractal dimension to estimate the Arya and Paris (1981) α input parameter. The fractal dimension is calculated through the use of a linear regression analysis over particles associated with the grain-size fractions. The method does not appear to improve on the performance of the Arya and Paris (1981) estimation.

5.12.15 Summary Comments on the Pedo-Transfer Estimation Models

A statistical analysis study appears to show that the Fredlund (1997) PTF performed reasonably well. A comparison of the frequency distribution for six PTFs is shown in Fig. 5.111. The bell-shaped frequency response indicates a positive predictive method. The estimation of an SWCC from a grain-size distribution has been found to give reasonable approximations for sands and silts. The SWCC for clays, tills, and well-graded fine soils are more difficult to predict although the estimation algorithm still appears quite reasonable. The predicted SWCCs tend to be sensitive to the packing porosity n_p.

The M.D. Fredlund (2000) pedo-transfer function was compared to five other pedo-transfer functions. Results of the comparisons between the measured and estimated air-entry values indicated significant improvement when using the M.D. Fredlund (2000) pedo-transfer function over other methods. The M.D. Fredlund (2000) pedo-transfer function and the Arya and Paris (1981) methods showed the highest level of confidence in estimating the air-entry value of a soil. The M.D. Fredlund (2000) pedo-transfer function showed reasonable accuracy in estimating the maximum slope of the SWCC.

5.12.16 Other Procedures for Estimating SWCC

The grain-size distribution curves for a soil can be used in other ways to select an approximate SWCC. It is possible to use the classification of a soil when searching a database of test results for an appropriate SWCC (M.D. Fredlund et al., 1996). Figure 5.112 illustrates approaches that can be used to obtain an SWCC. The approaches range from the direct measurement of the SWCC to its estimation from classification soil properties. Figure 5.113 further categorizes some of the estimation procedures that can be used to obtain an estimated SWCC. The estimated SWCC can subsequently be used for the determination of unsaturated soil property functions.

The classification soil properties and previously measured SWCCs can be incorporated into a knowledge-based database to assist users in estimating a reasonable SWCC. Correlations between classification soil properties (e.g., grain-size distribution and Atterberg limits) can provide approximate parameters for SWCC equations. The estimated SWCCs are satisfactory for many engineering applications (Aubertin et al., 2003; Zapata et al., 2000).

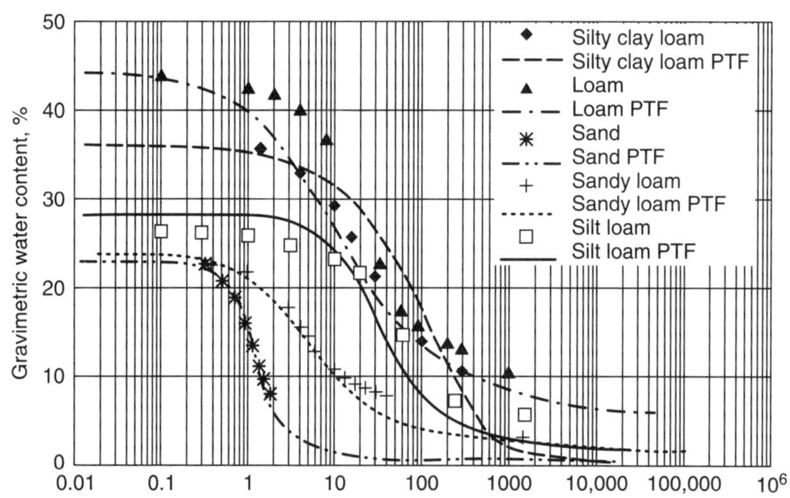

Figure 5.110 Best estimation of SWCC for five soil textures using Arya and Paris (1981) pedo-transfer function: silty clay, loam, sand, sandy silt, silt.

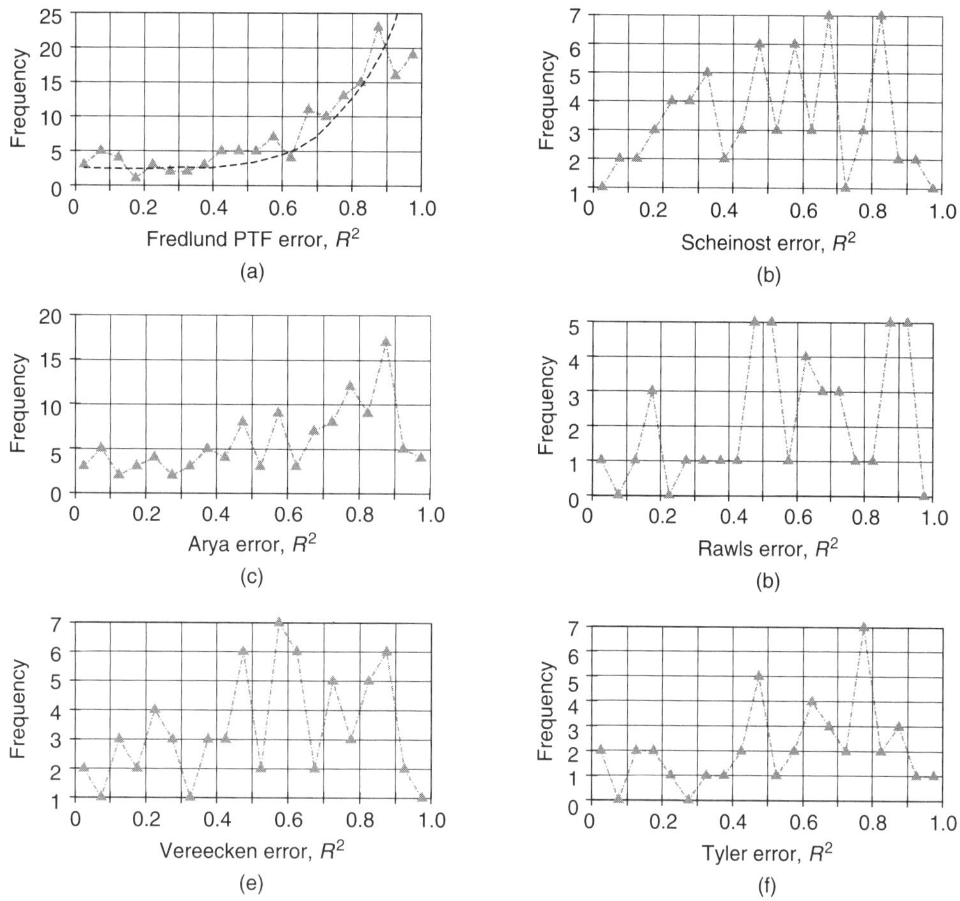

Figure 5.111 Comparison of frequency distribution of R^2 for values between 0.0 and 1.0 for six PTFs.

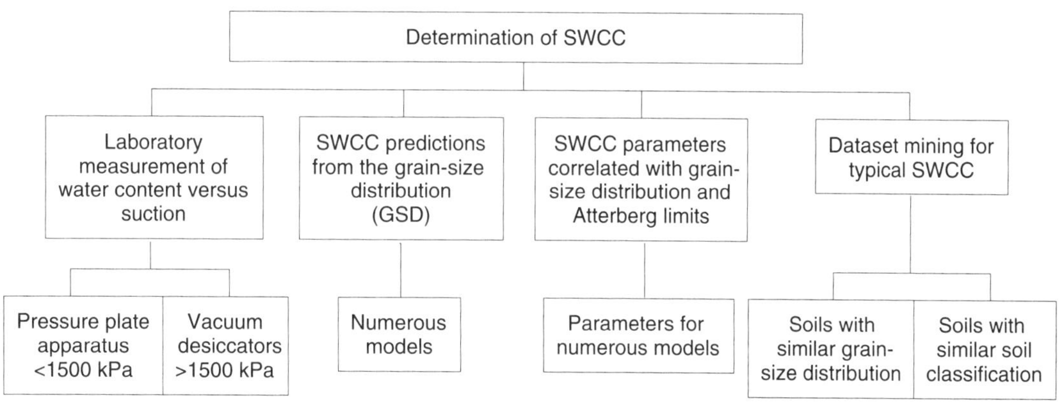

Figure 5.112 Methodologies for determination of SWCC.

5.13 TWO-POINT METHOD OF ESTIMATING SWCC

Let us consider a soil that does not undergo volume change as soil suction is increased. In this case, the SWCC has two primary defining points: (1) the water content and soil suction at the air-entry value for the soil and (2) the water content and soil suction at residual conditions. There are also two additional points that define the extreme limits on the SWCC: completely saturated conditions under zero suction (i.e., saturated water content) and completely dry conditions (i.e., zero water content and a soil suction of 10^6). The SWCC can either be measured or estimated using one of several available techniques. The final intent is to arrive at an appropriate representation of the air-entry value and the residual conditions for each soil involved in the engineering problem at hand.

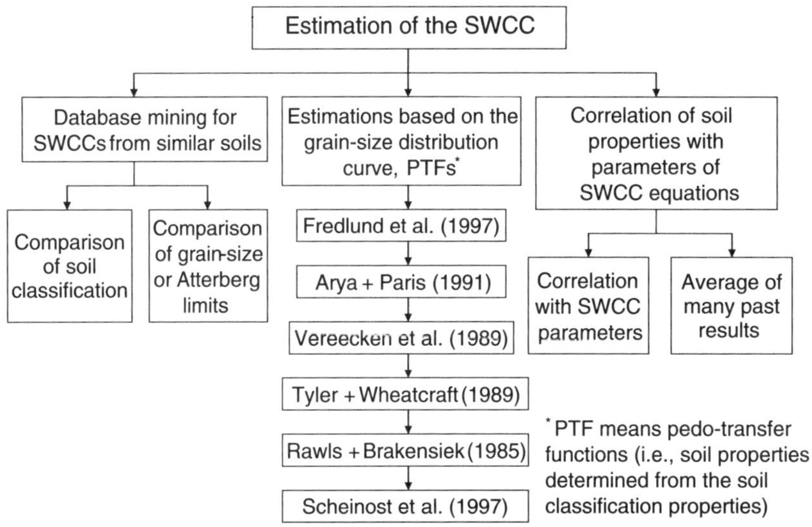

Figure 5.113 Procedures for estimation of the SWCC.

Let us suppose that a number of grain-size distribution curves may have been measured on soil samples taken in the field. It is possible to utilize one of several proposed pedo-transfer functions for the estimation of the SWCC (M.D. Fredlund et al., 2002a). An analysis of available information allows the engineer to arrive at a "best guess" for the air-entry value, the residual conditions, as well as the saturated initial porosity for the soil.

A computer spreadsheet can readily be used to generate a series of data points along an entire SWCC by simply entering the soil suction and water content for two points on the SWCC. The artificially generated data set can then be used for the estimation of unsaturated soil property functions.

Figure 5.114 shows the definition of variables used in the calculation of the SWCC. The degree of saturation versus soil suction graph provides the most definitive way to identify the air-entry value ψ_{aev} and the residual conditions ψ_r for a soil. However, if the soil undergoes negligible volume change as soil suction is increased, then the air-entry value and residual suction occur at the same suction values on the gravimetric and volumetric plots versus soil suction.

The SWCC consists of three straight lines with slopes of S_1, S_2, and S_3 on a logarithmic scale (Pham, 2005). The slope S_1 refers to the slope between the saturated conditions at a low suction (e.g., 0.1 or 1.0 kPa) and the air-entry value computed on the gravimetric water content scale. The slope S_2 refers to the slope of a line between the air-entry value and residual conditions computed on the gravimetric water content scale. The slope S_3 refers to the slope of a line between residual conditions and completely dry conditions corresponding to soil suction of 10^6 kPa. The

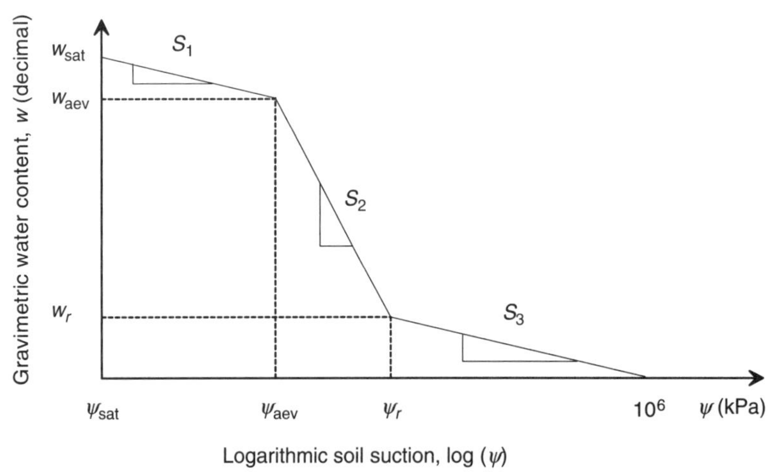

Figure 5.114 Definition of variables used to compute three main branches of a SWCC presented in terms of gravimetric water content (after Pham, 2005).

equation of the three straight lines of the SWCC can be written as

$$w = \begin{cases} w_s + S_1 \log \dfrac{\psi}{\psi_s} & \text{when } \psi < \psi_{\text{aev}} \\ w_{\text{aev}} + S_2 \log \dfrac{\psi}{\psi_{\text{aev}}} & \text{when } \psi_{\text{aev}} < \psi < \psi_r \\ w_r + S_3 \log \dfrac{\psi}{\psi_r} & \text{when } \psi_r < \psi \end{cases} \tag{5.93}$$

where:

w_s = gravimetric saturated water content (decimal),
ψ_s = low (arbitrary) suction corresponding to saturated conditions, kPa,
w_{aev} = gravimetric water content at the air-entry value,
ψ_{aev} = suction at air-entry value, kPa,
w_r = gravimetric residual water content, and
ψ_r = residual suction.

The slope variables in the above equations can be defined as follows:

$$S_1 = \frac{w_{\text{aev}} - w_s}{\log(\psi_{\text{aev}}) - \log(\psi_s)} \tag{5.94}$$

$$S_2 = \frac{w_r - w_{\text{aev}}}{\log(\psi_r) - \log(\psi_{\text{aev}})} \tag{5.95}$$

$$S_3 = \frac{-w_r}{\log(10^6) - \log(\psi_r)} \tag{5.96}$$

Typical input data for a spreadsheet are as follows: specific gravity $G_s = 2.65$; saturated gravimetric water content, $w_s = 37\%$; saturated suction, $\psi_s = 0.1$ kPa; gravimetric water content at the air-entry value, $w_{\text{aev}} = 36.5$ %; air-entry value for the soil, $\psi_{\text{aev}} = 5.0$ kPa; gravimetric water content at residual conditions, $w_r = 10\%$, and residual suction $\psi_r = 100$ kPa. Other values shown on the spreadsheet are calculated from the basic volume-mass information. The suction increment ratio controls the spacing of the data points along the SWCC (e.g., a value of 2 results in a doubling of suction between each data point).

The spreadsheet can generate three representations of the SWCC (Fig. 5.115). Figure 5.116 shows gravimetric water content versus soil suction; Fig. 5.117 shows volumetric water content versus soil suction; and Fig. 5.118 shows the degree of saturation versus soil suction for the same soil. Values can be cut and paste from the spreadsheet. It is possible to best fit the data set with any of the commonly used mathematical equations for the SWCC. Unsaturated soil property functions can also be computed using the soil suction versus water content data along with the saturated soil properties.

Parameter	Value
Specific gravity, G_s	2.65
Void ratio, e	0.981
Dry density, γ_d (kg/m^3)	1338
Water contents in decimal from	
Saturated water content, w_s	0.37
Saturated suction, ψ_{sat} (kPa)	0.1
Air-entry water content, w_a	0.365
Air-entry suction, ψ_{aev} (kPa)	5.0
Residual water content, w_r	0.1
Residual suction, ψ_r (kPa)	100
Slope S_1	−0.003
Slope S_2	−0.204
Slope S_3	−0.025
Suction increment ratio	2

Soil suction (kPa)	w	θ_w	S
0.1	0.370	0.495	1.000
0.2	0.369	0.494	0.998
0.4	0.368	0.493	0.995
0.8	0.367	0.492	0.993
1.6	0.366	0.490	0.990
3.2	0.366	0.489	0.988
6.4	0.343	0.459	0.927
12.8	0.282	0.377	0.762
25.6	0.221	0.295	0.596
51.2	0.159	0.213	0.430
102.4	0.100	0.133	0.270
204.8	0.092	0.123	0.249
409.6	0.085	0.113	0.229
819.2	0.077	0.103	0.209
1,638.4	0.070	0.093	0.188
3,276.8	0.062	0.083	0.168
6,553.6	0.055	0.073	0.148
13,107.2	0.047	0.063	0.127
26,214.4	0.040	0.053	0.107
52,428.8	0.032	0.043	0.087
104,857.6	0.024	0.033	0.066
209,715.2	0.017	0.023	0.046
419,430.4	0.009	0.013	0.025
838,860.8	0.002	0.003	0.005
1,000,000.0	0.000	0.000	0.000

Figure 5.115 Spreadsheet showing computed SWCC data set based on estimations of air-entry value and residual conditions.

5.14 CORRELATION OF FITTING PARAMETERS TO SOIL PROPERTIES

The fitting parameters for any of the proposed SWCC equations can be correlated with the classification properties of a soil. Each SWCC equation has either two or three fitting parameters and each parameter can bear a correlation with one or more of the soil classification properties. Unfortunately, independent correlations are required for each of the proposed empirical SWCC equations.

Several researchers have undertaken correlations between fitting parameters for the SWCC equation and the classification properties of the soil (Ghosh, 1980; Williams et al., 1983; Ahuja et al., 1985; Rawls et al., 1992; Cresswell and

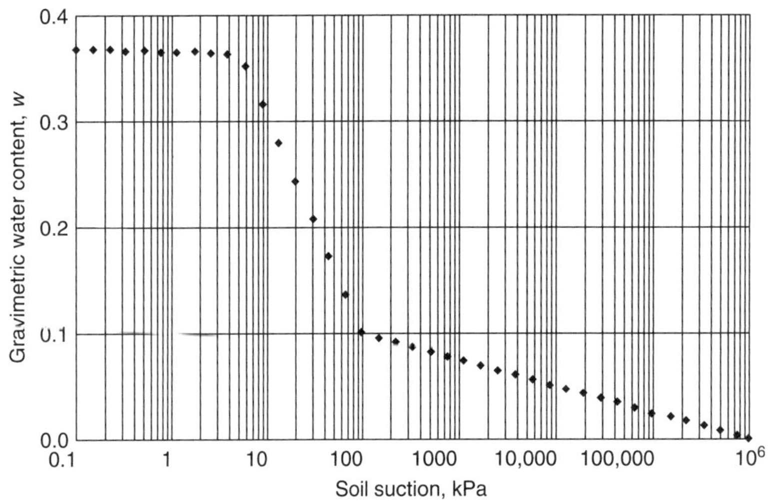

Figure 5.116 Computed gravimetric water content versus soil suction for soil with air-entry value of 5 kPa and residual suction of 100 kPa.

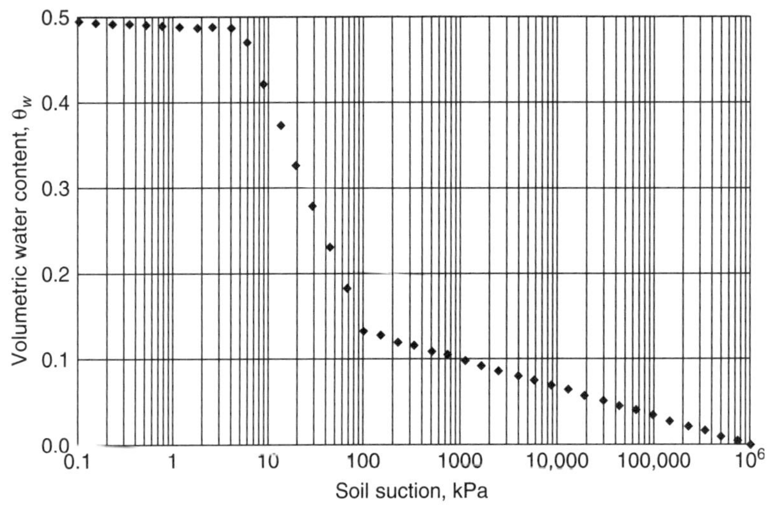

Figure 5.117 Computed volumetric water content versus soil suction for soil with air-entry value of 5 kPa and residual suction of 100 kPa.

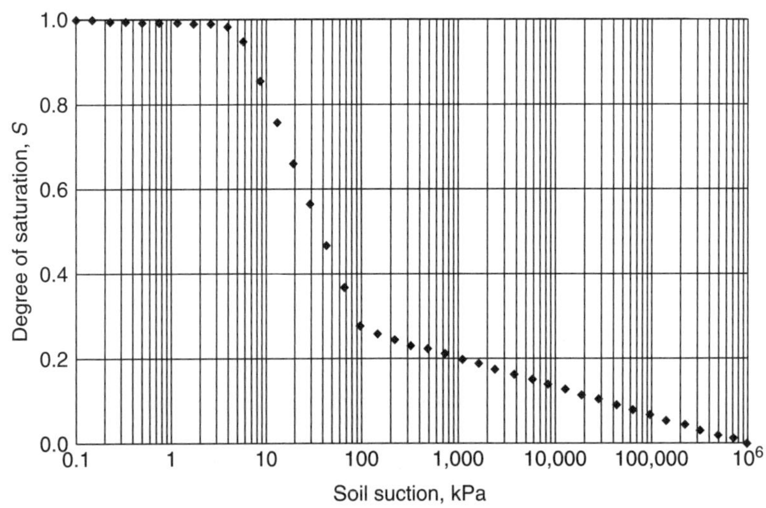

Figure 5.118 Computed degree of saturation versus soil suction for soil with air-entry value of 5 kPa and residual suction of 100 kPa.

Paydar, 1996; Tomasella and Hodnett, 1998; Zapata, 1999). The correlation of SWCC fitting parameters published by Zapata (1999) is for fitting parameters used in the Fredlund and Xing (1994) equation. The correlations were based on a statistical analysis of soil classification properties and the SWCC fitting parameters.

5.14.1 Zapata (1999) Correlation Model for SWCC

Zapata (1999) used a database of approximately 190 soils collected from research papers as well as information on a number of soils found in the knowledge-based database developed by SoilVision Systems. The soils collected were divided into two categories: soils exhibiting plasticity (i.e., positive plasticity index, PI) and soils with a PI equal to zero. The database used in the analysis consisted of approximately 70 soils with PI greater than zero and 120 soils with PI equal to zero. The data collected for the soils with PI greater than zero consisted of the percentage passing the No. 200 sieve and the Atterberg limits, in particular, the plasticity index. The grain-size diameter D_{60} was used to represent nonplastic soils. Each soil used in the statistical correlation had a measured and well-defined SWCC.

The product of the percentage passing the No. 200 sieve (used as a decimal value) was multiplied by the plasticity index to form a weighted PI value (i.e., w_pPI) for soils with a plasticity index greater than zero. The weighted PI was used as the primary variable for correlation. The D_{60} variable was used as the primary variable for correlation purposes when the soil was nonplastic. The assumption was made in the analysis that all soils remained at a constant volume as soil suction was increased. The Zapata (1999) correlations of SWCC were presented in terms of the degree of saturation of the soil.

The correlation study yielded a family of SWCCs for both plastic and nonplastic soils. The results of the Zapata (1999) study have been used as part of the Enhanced Integrated Climatic Model (EICM, version 2.6) in the pavement design guide (Houston et al., 2006). The family of curves has also been used as an approximate guide to a potential desorption curve for soils.

5.14.2 SWCC for Granular Nonplastic Materials (Zapata, 1999)

The grain-size distribution curve for each soil was analyzed when the soil was classified as nonplastic. Each SWCC was best fit with the Fredlund and Xing (1994) equation where four fitting parameters were generated for each soil: a_f, n_f, m_f, and h_r. The four parameters were correlated with the particle diameters corresponding to 60% passing. The correlations for nonplastic soils were represented by the following equations:

$$a_f = 0.8627(D_{60})^{-0.751} \tag{5.97}$$

where:

a_f = fitting parameter related to the inflection point on the SWCC and

D_{60} = grain-size diameter corresponding to 60% passing by mass.

$$m_f = 0.1772[\ln(D_{60})] + 0.7734 \tag{5.98}$$

where:

m_f = fitting parameter influencing the curvature of the SWCC at low and high suctions.

$$n_f = 7.5 \tag{5.99}$$

where:

n_f = fitting parameter equal to the slope at the inflection point of the SWCC.

$$h_r = a_f\left(\frac{1}{D_{60} + 9.7e^{-4}}\right) \tag{5.100}$$

where:

e = constant equal to 2.71828, base of the natural logarithm.

The above suggested equations can be used to generate a family of curves based on D_{60} values as shown in Fig. 5.119. The grain-size D_{60} is limited to the range between 0.1 and 1.0 mm.

5.14.3 SWCC for Plastic Materials (Zapata, 1999)

Soils exhibiting plasticity were analyzed as a separate group of soils. The fitting parameters (i.e., a_f, m_f, n_f, and h_r) were found to correlate well with the percent passing the No. 200 sieve (i.e., w_p) multiplied by the plasticity index of the soil (i.e., w_pPI). The following equations were generated for the fitting parameters of plastic soils:

$$a_f = 0.00364(w\mathrm{PI})^{3.35} + 4(w\mathrm{PI}) + 11 \tag{5.101}$$

$$m_f = 0.0514(w\mathrm{PI})^{0.465} + 0.5 \tag{5.102}$$

$$n_f = m_f(-2.313(w\mathrm{PI})^{0.14} + 5) \tag{5.103}$$

$$h_r = a_f(32.44e^{0.0186(w\mathrm{PI})}) \tag{5.104}$$

where:

e = constant equal to 2.71828, the base of the natural logarithm.

The above equations can be used to generate a family of SWCCs for soils with w_pPI values ranging from 0.1 to 50, as shown in Fig. 5.120.

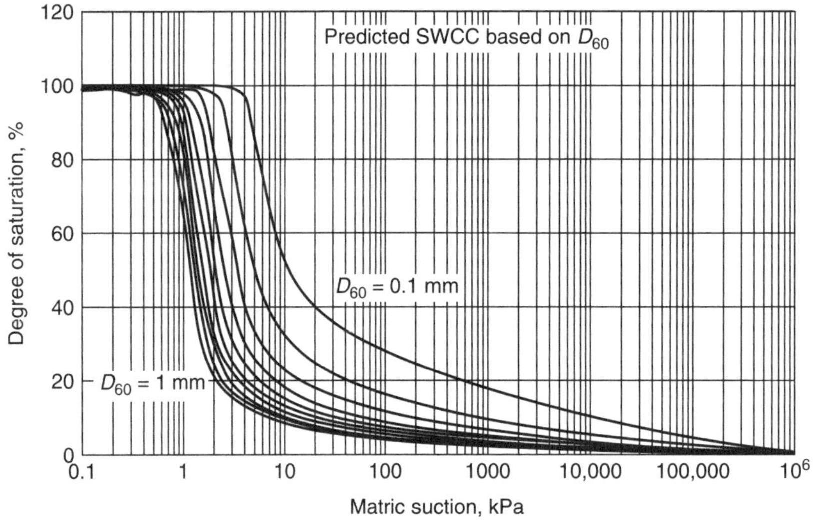

Figure 5.119 Family of SWCCs for nonplastic soils based on Zapata (1999) model when using D_{60} as representative particle size.

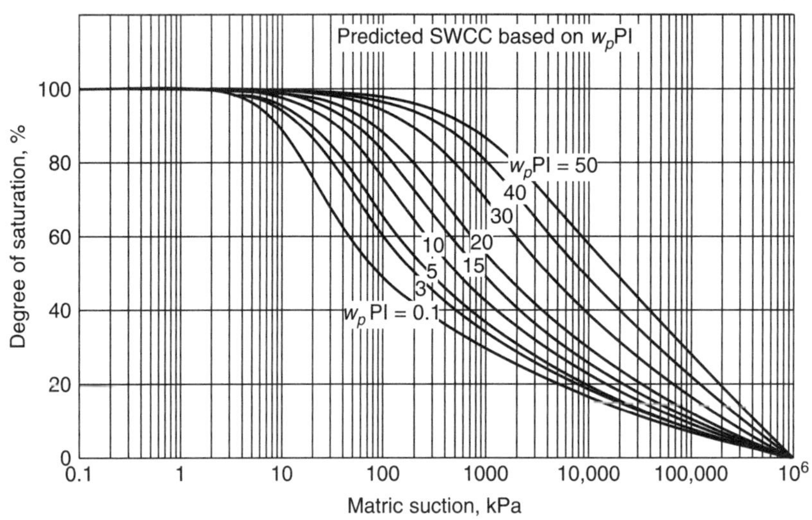

Figure 5.120 Family of SWCCs for plastic soils based on the Zapata (1999) model when using w_pPI as the correlation parameter.

5.14.4 Estimated Desorption SWCC for all Soils (Zapata, 1999)

The families of SWCCs can be combined and presented as shown in Fig. 5.121. All SWCCs converge at a soil suction value of 10^6 kPa. It should be noted that the family of curves shown in Fig. 5.121 are the average of many test results.

These curves provide an indication of anticipated results but should not be viewed as a replacement for the measurement of the SWCC in the laboratory. It is reasonable to use the family of curves based on the Zapata (1999) correlation study along with the pedo-transfer from grain-size distribution curves at the preliminary design stage in order to obtain a crude approximation of the SWCC for a particular soil. It is advocated that the SWCC be measured in the laboratory as part of the final design stage of engineering.

5.14.5 Torres (2011) Model for Granular Materials

Torres (2011) analyzed a database of about 4500 granular soils and found that the D_{10} particle size produced improved correlation coefficients to those obtained when using the D_{60} particle size. The correlation coefficients were obtained for the Fredlund and Xing (1994) SWCC equation. The correlation for the a_f fitting parameter was represented by the following equation:

$$a_f = -967.2D_{10}^2 + 218.4D_{10} - 2.70 \qquad (5.105)$$

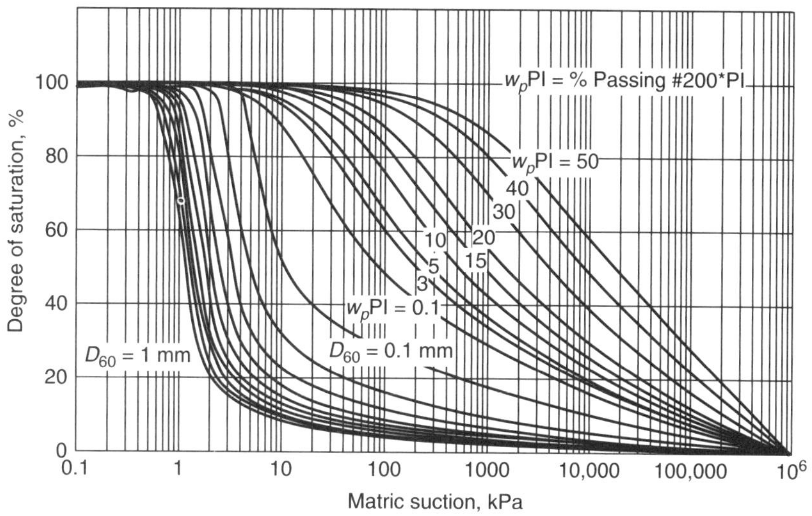

Figure 5.121 Combined family of SWCCs for plastic and nonplastic soils (after Zapata, 1999).

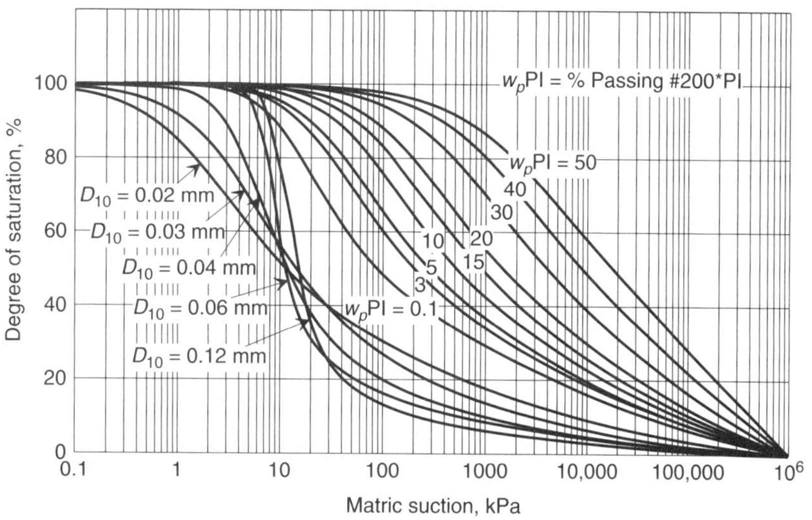

Figure 5.122 Combined family of SWCCs when using Torres model when D_{10} is the representative particle size for nonplastic soils (after Torres, 2011).

It was also suggested that if the D_{10} particle diameter was less than 0.020, the a_f fitting parameter should be set to 1.28.

The slope of the SWCC n_f was found to be related to the a_f fitting parameter (i.e., the inflection point):

$$\log n_f = -0.0075a_f^3 + 0.1133a_f^2 - 0.3577a_f + 0.3061 \quad (5.106)$$

The m_f fitting parameter was also found to be related to the a_f fitting parameter:

$$m_f = 0.0058a_f^3 - 0.0933a_f^2 + 0.4069a_f + 0.3481 \quad (5.107)$$

The soil suction at residual conditions, ψ_r, was found to be relatively constant for all granular soils:

$$\psi_r = 100.0 \quad (5.108)$$

The above suggested fitting parameters for granular materials can be combined with the previously proposed equation for plastic soils to produce the family of SWCCs shown in Fig. 5.122.

5.15 APPLICATION OF SWCC

It is not practical to undertake rigorous and costly tests on unsaturated soils for routine geotechnical engineering practice. As a result, a variety of practical, indirect methodologies and test procedures have emerged for quantifying unsaturated soil property functions. These indirect procedures have proven to be adequate for most routine geotechnical engineering problems (M.D. Fredlund, 2000a).

Indirect procedures for the estimation of unsaturated soil property functions have primarily made use of the

SWCC, along with the saturated soil properties (Fredlund and Rahardjo, 1993a; Vanapalli et al., 1996a; Barbour, 1998; D.G. Fredlund, 2000a; Fredlund et al., 2003). The SWCC is relatively easy to measure and has become the key unsaturated soil information for obtaining unsaturated soil property functions. Various empirical procedures have been proposed and tested for the estimation of essentially all unsaturated soil property functions (e.g., permeability, shear strength, volume change, and others). In each case, the estimation procedure involves the use of the saturated soil properties in conjunction with the SWCC.

Numerous estimation procedures have been proposed for the water permeability function (Fredlund et al., 1994b). The procedures differ primarily in the basic assumptions involved in the development of the proposed model and the mathematical manner (e.g., method of integration), by which the SWCC is used in conjunction with the saturated soil properties. As an example, several permeability functions have been proposed, each one using the SWCC in a somewhat different manner.

Constitutive relations for unsaturated soils can be written using the SWCC equation. The SWCC is also dependent upon the total stress state; however, in most cases the constitutive equations for an unsaturated soil are sufficiently accurate when written simply as a function of soil suction.

The development of unsaturated soil property functions has made it possible to perform numerical modeling studies in application areas such as seepage, shear strength, and volume change (M.D. Fredlund, 2000). Even though the unsaturated soil property functions are not measured, it is still possible to estimate reasonable soil properties and then observe anticipated soil-structure interactions. The numerical modeler can ask a series of plausible, "What if . . . .?" questions and thereby observe a range of possible responses. An understanding of the statistical variability associated with the SWCC likewise provides an indication of the variability that might be anticipated in the unsaturated soil property functions.

5.15.1 Unsaturated Soil Properties When Soil Suction Exceeds Residual Suction

There is limited information on the form of the classic constitutive relations for an unsaturated soil as residual suction is exceeded. There is little experimental evidence related to the coefficient of permeability of a soil as residual suction is exceeded. However, the coefficient of permeability is known to become extremely low and hydraulic flow would appear to essentially cease near residual conditions, giving way to vapor flow. In other words, there appears to be a change from liquid flow to vapor flow as residual suction is exceeded.

The effective angle of internal friction can be applied to a soil up to the air-entry value. The friction angle then decreases until it becomes essentially zero at soil suctions exceeding residual soil suction. Recent research on a number of soil types would indicate that the angle of friction tends toward zero at high soil suctions (Nishimura and Fredlund, 1999). Therefore, the angle of internal friction with respect to soil suction tends to zero at residual soil suction.

The shrinkage curve for a soil indicates that the change in void ratio tends toward zero at high soil suctions. It would appear that the residual soil suction of a soil is closely related to the shrinkage limit of a soil. The volume change function (i.e., change in void ratio or specific volume) is equal to the compressibility of the saturated soil up to the air-entry value. The compressibility of the soil then decreases throughout the transition zone becoming zero as residual soil suction conditions are exceeded.

5.15.2 Important Role of SWCC

The unsaturated soil property functions can be related to the SWCC because variations in the unsaturated soil properties are primarily a function of the amount of water in the soil. For some unsaturated soil properties it would appear that the unsaturated soil property functions are primarily related to the amount of water on an unbiased plane passing through the soil. The coefficient of permeability of an unsaturated soil is a function of the volume of water in the soil while the shear strength and volume change behavior is primarily a function of the area of water on an unbiased cross section. The area of water on an unbiased plane should be the same as the volume of water in a representative elemental volume (Fung, 1965). Figure 5.123 illustrates the comparison between the volumetric and area representations of the amount of water in a soil.

5.15.3 Application of Hysteretic SWCC in Geotechnical Engineering

Geotechnical engineers have attempted to use the SWCC to obtain an estimate of the in situ soil suction. In taking this

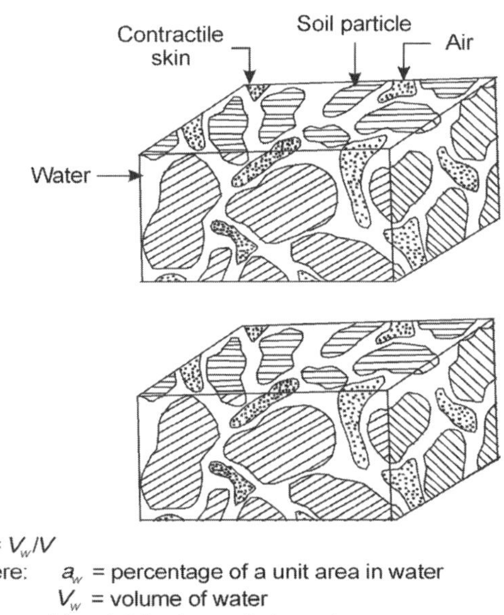

$a_w = V_w/V$

where: a_w = percentage of a unit area in water

V_w = volume of water

V = total volume of element

Figure 5.123 Continuum mechanics equivalence of area and volume representations of water in a soil.

approach, it is assumed that water content measurements on disturbed soil samples can be used to provide an indication of in situ soil suction. All such attempts have proven to be difficult for a number of reasons. The drying and wetting portions of the hysteresis loops may be shifted laterally as much as one (or more) orders of magnitude. It is difficult to use this procedure to estimate the in situ soil suction to within an order of magnitude on fine-grained soils.

The engineer cannot determine with any confidence whether the in situ soil is on the drying or wetting loop (or one of the scanning curves) because there is no indication of stress history. It is difficult to use SWCCs for the estimation of in situ soil suction, but the same is not true when the laboratory results are used for the estimation of unsaturated soil property functions.

The quantification of unsaturated soil property functions, more than any other single factor, holds potential for the implementation of unsaturated soil mechanics in geotechnical engineering practice.

The SWCC can be plotted as degree of saturation versus the logarithm of soil suction and used to determine two distinctive points along the curve: the air-entry value and residual conditions. It is possible to obtain the necessary information for the estimation of unsaturated soil property functions. A similar interpretation procedure can be applied to the wetting (or adsorptive) branch of the SWCC.

The geotechnical engineer must decide whether the process being simulated in the field is a wetting or drying process. The appropriate wetting or drying curve needs to be determined for the soil and then the appropriate unsaturated soil property function can be estimated. It is possible to theoretically simulate both wetting and drying processes through the use of more complex models that incorporate the effects of hysteresis; however, it currently appears to be better to separate the drying and wetting processes when simulating practical situations.

5.15.4 Estimation of Unsaturated Soil Property Functions

The accuracy of the USPF used in the solution of partial differential equations for unsaturated soil systems depends upon (i) the accuracy of the SWCC (i.e., SWCC equation) and (ii) the accuracy of the empirical estimation procedure for the USPFs (Fredlund et al., 1994b; 1996b; Vanapalli and Fredlund, 2000b). In other words, it is important to achieve the closest possible representation of the physical behavior of an unsaturated soil.

Let us consider the example of modeling a transient saturated-unsaturated seepage problem (Thieu et al., 2001). It is necessary to have the best possible representation of (i) the water storage function and (ii) the hydraulic conductivity function. A typical SWCC can be used to generate both the hydraulic conductivity function and the water storage function. It is particularly important to estimate the most accurate permeability function to represent the actual hydraulic conductivity of the unsaturated soil.

5.15.5 Selection of Most Appropriate Estimation Procedures

Numerous estimation procedures are available for solving unsaturated soil problems. The large number of estimation procedures can lead to confusion among practicing engineers. It is important to adequately estimate the SWCC for preliminary design purposes and the SWCC needs to be properly measured for final design purposes.

It cannot be assumed that all estimation procedures are of equal accuracy when estimating the SWCC (Fredlund, 2002a). It is incumbent upon a prudent engineer to utilize the estimation techniques that have been proven to be most accurate. When the best estimation techniques have been utilized for obtaining the SWCC, then the engineer needs to correspondingly utilize the most reliable estimation procedures for computing the necessary unsaturated soil property functions.

5.16 GUIDELINES AND RECOMMENDATIONS FOR ENGINEERING PRACTICE

A number of general guidelines and recommendations are suggested when attempting to establish appropriate engineering protocols. The following suggestions are made with regard to the assessment of unsaturated soil property functions:

1. The SWCC can be estimated for preliminary design purposes but must be measured for final design purposes. It is recommended that laboratory measurements be made on undisturbed soil samples whenever possible.
2. The estimation of unsaturated soil property functions is related to the drying (or desorption) branch of the SWCC. It is the responsibility of the geotechnical and geo-environmental engineer to determine whether the process being simulated in the field corresponds to the drying or wetting process. If the process corresponds to the wetting (or adsorption) process, then the SWCC should be shifted in accordance with the average suggested values.
3. It is important to use the SWCC estimation procedures that have been shown to be the most accurate when using the grain-size distribution curve.
4. It is important to utilize a SWCC equation that covers the entire range of soil suctions that are of interest for a particular engineering problem.
5. It is important to use the estimation techniques that have been shown to provide the most accurate estimations for unsaturated soil properties.
6. There is a strong linkage between the SWCC and the unsaturated soil properties and this relationship must be maintained throughout the numerical modeling process. It is not considered acceptable engineering practice to adjust some of the variables associated with unsaturated

soil property functions in order to obtain convergence of the solution.

7. When a soil undergoes significant volume change during the change in soil suction, it necessary to use the shrinkage curve (or measure volume change in the SWCC test) and use the appropriate water content versus suction relationship for the estimation of the unsaturated soil property functions. For example, volumetric water content versus soil suction should be used when computing the water storage function for a soil. On the other hand, the degree of saturation versus soil suction relationship must be used for the interpretation of the correct air-entry value for the soil and the computation of the unsaturated soil property functions.
8. The practicing geotechnical engineer needs to recognize that the reliability of theoretical or statistical estimation techniques becomes progressively less reliable as the clay content of a soil increases.

CHAPTER 6

Ground Surface Moisture Flux Boundary Conditions

6.1 INTRODUCTION

The quantification of moisture flux boundary conditions is an important topic which was not historically addressed as part of classic soil mechanics. There are two main types of boundary conditions that can be applied when solving moisture movement problems. It is possible to specify a hydraulic head or a flow rate along the boundaries of the geometry. Hydraulic heads specified at the boundary of a problem are called Dirichlet boundary conditions. Flow rates specified across the boundary of a problem are referred to as Neumann boundary conditions. Textbooks on saturated soil mechanics explain how to apply a hydraulic head along the geometric boundary or a zero-flux condition which designates an impervious boundary. However, these are not sufficient boundary conditions when dealing with near-ground-surface problems involving unsaturated soils.

The ground surface forms the uppermost boundary for most soil mechanics problems. This boundary is periodically subjected to precipitation in the form of rain or snow. At other times moisture moves upward from the ground surface through evaporation and transpiration. The soil immediately below the ground surface is generally unsaturated. The ground surface forms a continuously changing moisture flux boundary. There are other components such as "runoff" that also need to be evaluated in conjunction with precipitation in order to determine the "net moisture flux" at the ground surface.

The determination of the net moisture flux at the ground surface is complex and involves numerous assumptions as a series of calculations are performed. There are several other disciplines such as agriculture and surface hydrology that have had an interest in calculating the net flux at the ground surface. Extensive research has been undertaken in other disciplines and the results are of great interest to applications in soil mechanics. It is important for geotechnical engineers to reevaluate theories from other disciplines to ensure that the assumptions used as part of mathematical formulations are realistic for geotechnical engineering problems.

Weather forecasting has been an area of intense research over the past decades. Weather forecasting has as its primary objective the ability to predict the temperature, wind, storms, and precipitation of various areas of the world. Weather forecasting has proven to be a challenging science due largely to the randomness of the processes involved and the complexities of compressible mixture flow.

There are many near-ground-surface engineered structures (e.g., soil covers) that require the input of likely weather conditions as part of the design process. These engineered structures rely on the statistical evaluation of past weather conditions recorded at nearby weather stations. An evaluation of the past 10 or more years of weather information is necessary in order to establish a weather "finger print" for a particular location under consideration.

The evaluation of moisture movement upward from the ground surface can first be viewed in terms of the potential amount of evaporation that could occur. Potential evaporation, PE, from a body of water or a saturated ground surface is easier to compute than the calculation of actual evaporation, AE, from a soil surface. The state of stress in the soil at ground surface can significantly affect the actual rate of evaporation. It is important to understand the difference between PE calculations and AE calculations. It is the determination of AE that is of primary interest to the geotechnical engineer when attempting to determine the net moisture flux at the ground surface. Net moisture flux becomes a Neumann-type boundary condition that can be applied during the numerical simulation of many unsaturated soil mechanics problems.

The calculation of net moisture flux (i.e., actual infiltration) at the ground surface is required as part of the assessment of the water balance at a site. The area under consideration might be a mining site or some other resource development operation. Soil cover systems have proven to be invaluable in limiting and controlling the rate of water infiltration into underlying soils during the past several decades. The design of soil cover systems depends on the calculation of the components of moisture movement that give rise to net infiltration at ground surface.

Cover systems have increasingly become a viable solution to mitigate environmental damage. Cover systems became a common solution around the 1980s for the remediation of contaminated sites and waste containment facilities. Mining operations have two main streams of waste material that need to be properly handled to reduce distress to the environment. These two streams of materials can be referred to as waste rock and mine tailings. Cover systems provide a potential solution for both streams of waste materials.

The design of cover systems is conceptually easy to understand but there are numerous issues and challenges associated with various aspects of the engineering analysis. There are issues related to the required input information and the solution of the mathematical equations (e.g., the moisture flow partial differential equation). There are also a large number of assumptions that must be made at various stages of the design process. These assumptions can significantly influence the outcome of the cover system design. The quantification of unsaturated soil properties for each material involved (e.g., the permeability functions and the water storage functions) has also proven to be a challenge for geotechnical engineers (Fredlund, 2007b).

The design of cover systems depends upon the ability to predict moisture fluxes in and out of the ground surface as well as moisture movement through the unsaturated soils comprising the cover system. The analysis can be viewed as a "flux-driven" problem. Problems involving the predictions of hydraulic head are generally easier to solve than problems involving the prediction of moisture fluxes. The boundary conditions at the ground surface must be described in terms of a moisture flux when undertaking a soil cover design. The quantification of the ground surface moisture flux boundary conditions is coupled with the solution of moisture movement in the underlying unsaturated soil.

A cover system can be viewed as a thin interface placed between the atmosphere and the underlying soil strata, as shown in Fig. 6.1. The climate imposed on top of the cover can vary widely from arid to humid conditions. A particular soil cover system cannot be expected to function in a similar manner when subjected to a wide range of climatic conditions. Rather, it would be anticipated that a particular type of soil cover can only perform satisfactorily when subjected to a limited range of climatic conditions. Consequently, soil cover systems need to be designed for each site by taking into consideration the available soil types for the soil cover and the imposed weather conditions.

Cover systems can function quite differently under different climatic situations. For example, a cover system may operate in a "store-and-release" mode in one situation but might function as a saturation oxygen barrier in another case. The location of the water table in the underlying materials can also have an influence on the functionality of the soil cover system. All elements of the cover system (i.e., atmosphere, soil cover, and underlying soils) are highly variable and material responses are generally highly nonlinear and hysteretic (Shackelford, 2005; Fredlund, 2006).

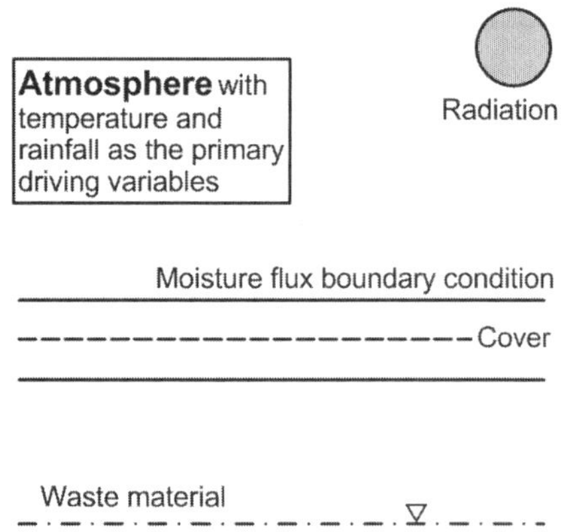

Figure 6.1 Covers used as an interface between material to be protected and climatic environment.

The challenges associated with the analytical design of a soil cover system are formidable but not impossible to accommodate. It might be difficult to obtain the same decision from two geotechnical engineers regarding whether a cover system will function satisfactorily in a particular climate. The design of soil cover systems involves the use of SWCCs. The SWCCs are hysteretic in character and as such engineering protocols need to be established for the calculation of water movement in and out of the unsaturated soils near the ground surface.

6.2 CLIMATIC CLASSIFICATION FOR A SITE

The climatic context should first be evaluated for any site for which water balance or water infiltration is to be quantified. The average climatic conditions as well as likely variations in the local climate should be understood prior to embarking on the numerical modeling of moisture movement.

Store-and-release soil covers placed at the ground surface are known to have the potential to function satisfactorily in arid or semiarid environments. However, there are situations where soil cover systems have been designed and used where the climate is not arid or semiarid. The geotechnical engineer should have an appreciation of the significance of variations in the climate from one year to the next.

Climatic classifications are based primarily upon an approximate water balance calculation referenced to the ground surface. Calculations of climate classification provide the geotechnical engineer with guidance regarding the suitability of a particular type of cover system. The climatic classification provides information on whether a particular type of cover system has the potential to operate in a satisfactory manner. It is possible to proceed with the cover design once it has been decided that a particular cover system might function in a satisfactory manner.

The climate in a particular area is commonly evaluated using the Thornthwaite climate classification system (Thornthwaite, 1931, 1948). Thornthwaite published the first climate classification system in 1931. The classification system was revised in 1948 and it is the latter climate classification system that is generally used in geotechnical engineering. The climate near a site under consideration can be classified based on average temperature and precipitation measurements. Temperature along with global setting (i.e., latitude) is used to estimate PE. Potential evaporation is primarily driven by average temperatures (or more accurately by net radiation). Figure 6.2 shows temperature contours for the world for 2 months of the year. Figure 6.2a shows the mean global temperatures for January and Fig. 6.2b shows the mean global temperatures for July. Similar plots can be made for each month of the year. Figure 6.3 shows average precipitation contours for the world for the 12 months of the year. It is also possible to plot the monthly average precipitation values.

Thornthwaite (1948) developed a classification system for climate based on the calculation of a moisture index I_m that takes into consideration the total annual precipitation and the annual potential evapotranspiration. In 1948 it was assumed that the water surplus from one season might be carried forward to another season and therefore a weighting factor

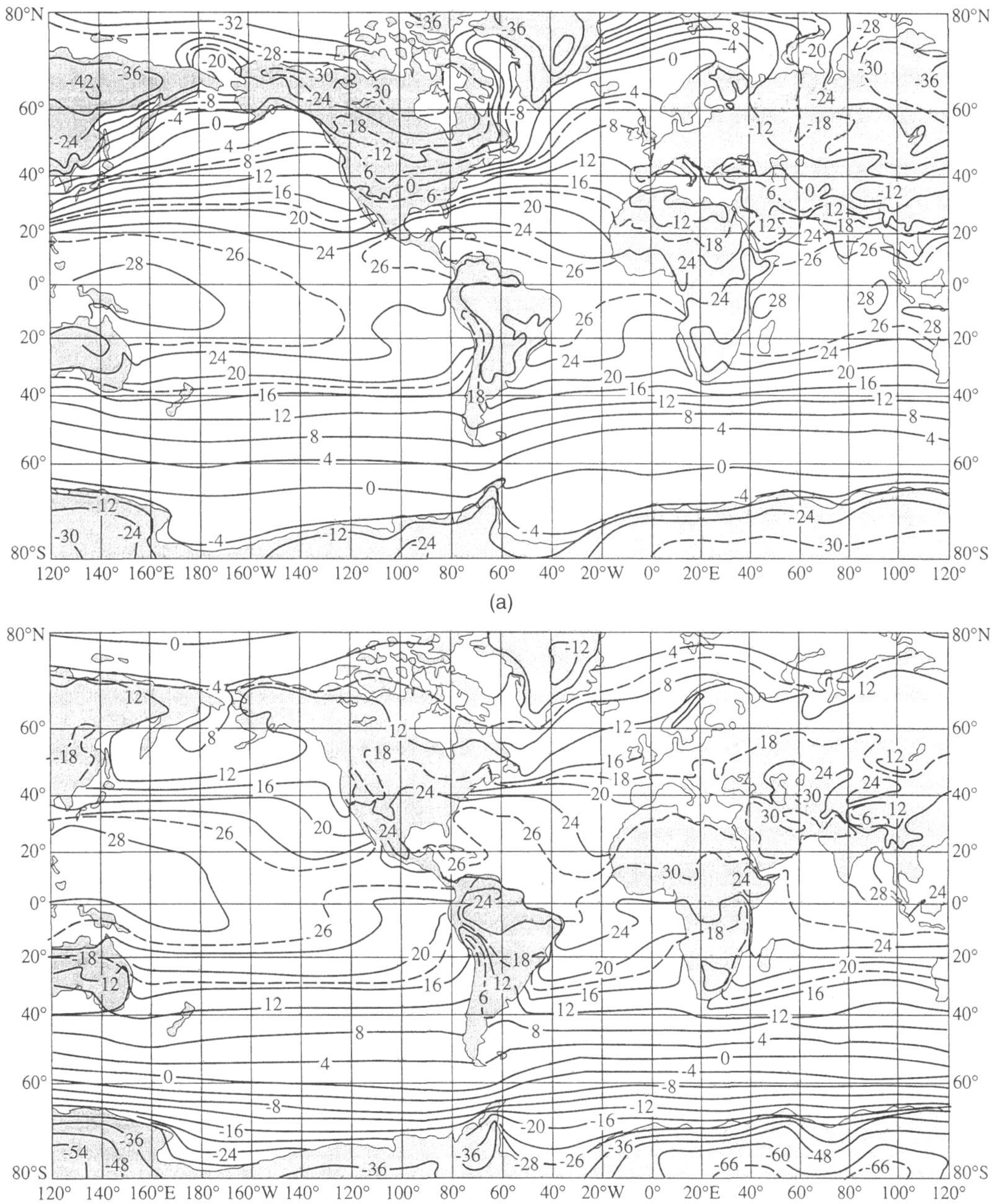

Figure 6.2 Temperature is the primary factor controlling the evaporative flux from the earth's surface. (a) Mean temperatures for January. (b) Mean temperatures for July. From *The Physics of Climate*, by J. P. Peixoto and by A.H. Oort © 1992, used with permission of American Institute of Physics.

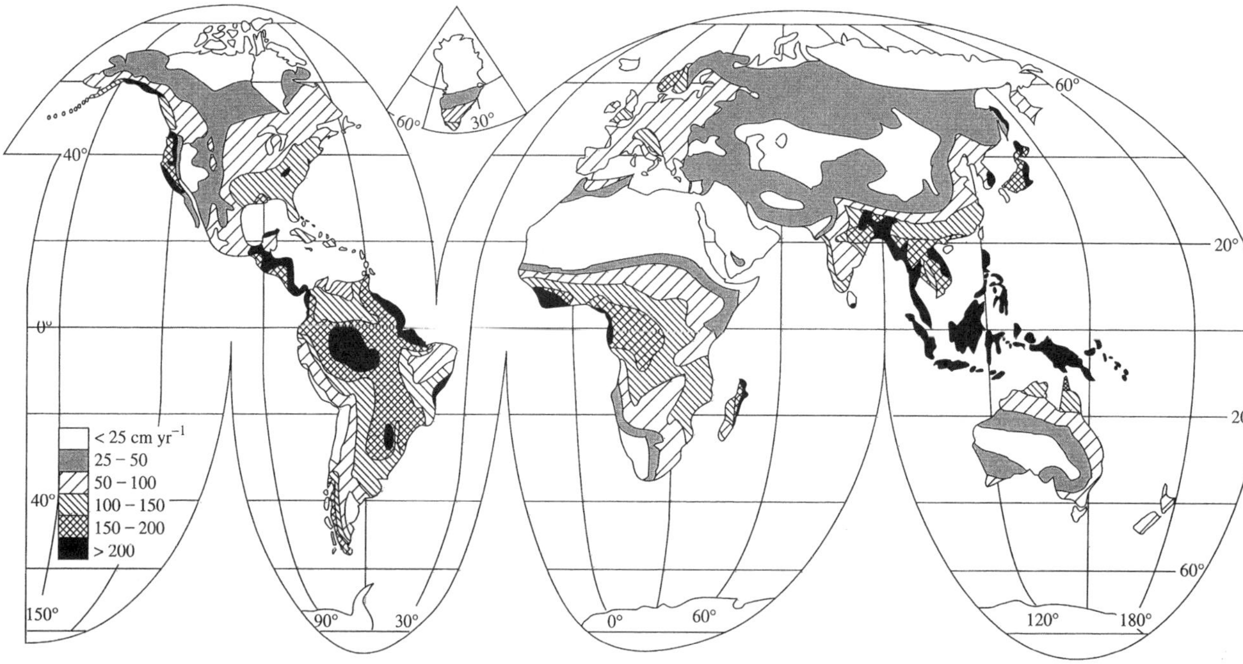

Figure 6.3 Rainfall and snowfall are contributors to precipitation falling on the earth's surface. Reprinted with permission of Prentice-Hall from *Climatology* by John E. Oliver and John J. Hidore, © 1984 by Bell & Howell Co.

of 0.6 was applied to the water deficit calculation. However, in 1955, the 0.6 factor was dropped from the calculations and moisture index is now defined as follows (Thornthwaite and Mather, 1955; Thornthwaite and Hare, 1955):

$$I_m = 100\left(\frac{P}{\mathrm{PE}} - 1\right) \tag{6.1}$$

where:

I_m = 1955 Thornthwaite moisture index,
P = total annual precipitation, and
PE = total annual potential evapotranspiration calculated as the summation of the Thornthwaite (1948) monthly potential evaporations.

Thornthwaite (1948) incorporated the variables of length of daylight hours, mean monthly temperature, and an empirical constant into the prediction of potential evaporation. The simplicity of the Thornthwaite equation has led to its usage for the classifications of global climate as well as numerous other agricultural and engineering applications. The potential daily evaporation can be written as follows using the Thornthwaite (1948) equation:

$$\mathrm{PE}_d = 0.5333\left(\frac{L}{12}\right)\left(\frac{N}{30}\right)\left(\frac{10T_a}{I}\right)^{a_t} \tag{6.2}$$

where:

PE_d = potential evaporation, mm/day,
L = length of daylight, h,
N = number of days in the month,
T_a = mean monthly air temperature, °C,
I = sum for 12 months of the function $(T_a/5)^{1.514}$; (i.e., $\sum_{\text{month}=1}^{12} (T_a/5)^{1.514}$), based on correlations to "pan evaporation" measurements, and
a_t = complex function of the variable I [i.e., $a_t = (6.75 \times 10^{-7})\ I^3 - (7.71 \times 10^{-5})\ I^2 + (1.79 \times 10^{-2})\ I + 0.492$].

For a 30-day month and assuming that there are 12 h of the day for evaporation, the potential evaporation equation can be written as

$$\mathrm{PE}_m = 16.0\left(\frac{10T_a}{I}\right)^{a_t} \tag{6.3}$$

where:

PE_m = potential evaporation, mm/month.

The potential monthly evapotranspiration can be calculated for each of the 12 months. For 1 year, the annual potential Thornthwaite evaporation can be written as follows:

$$\mathrm{PE} = \sum_{i=1}^{12} \mathrm{PE}_m \tag{6.4}$$

The total annual evapotranspiration is used in conjunction with the total annual precipitation to classify the climate through use of Eq. 6.1. The climate classification criteria

Table 6.1 Criteria for Climate Classification[a]

I_m	Category of Climate
> 100	Perhumid
20–100	Humid
0–20	Moist humid
−33–0	Dry subhumid
−67 to −33	Semiarid
−100 to −67	Arid

[a]Moisture index I_m as defined in 1955. (i.e., Eq. 6.1).

are shown in Table 6.1. The Thornthwaite classification is an empirical equation that was developed from limited climatic data collected in the United States. However, it forms an adequate basis for establishing the climatic type and variations in climate.

Climatic classification and the likely variation in climate should be established prior to considering a particular cover system solution for the site under consideration. It is important to determine whether a particular engineered system will be suitable for the site under consideration. It is the ratio of the total annual precipitation P to the total annual PE, which provides a guide to the suitability of a particular type of cover system. Figure 6.4 shows contours of average potential evaporation for the world.

6.2.1 Partitioning of Each Year

Normally a year is viewed as starting on January 1 and ending on December 31. However, the evaluation of net moisture flux at the ground surface can be quite different at various times of the year, and as a result there is a more meaningful way to partition each year. In the northern areas of Canada, for example, it is reasonable to select the autumn freeze-up date as the beginning of a new year (Stianson et al., 2010). This is the time of the year when the "near-ground-surface" water content conditions are essentially "locked in" for several months. Other regions of the world might have other times that should be used for starting the analysis of moisture flux at the ground surface.

A common freeze-up date is November 1 in northern Saskatchewan, Canada (no. 1 in Fig. 6.5). The winter period is referred to as the inactive season and has been assumed to extend to April 18 of the following year (nos. 2 and 3 in Fig. 6.5). Most of the precipitation is in the form of snow during the inactive season. The snow accumulated during the inactive season can be converted to an equivalent amount of rain at the weather station.

The period from April 19 to October 31 is referred to as the active season (no. 2 to no. 1 in Fig. 6.5). The terms inactive and active are designated in Fig. 6.5 and are used in a relative magnitude sense since there can be some moisture movement activity all year round at the ground surface. The snow and rain from the time of freeze-up accumulates over winter and then infiltrates the ground surface in the spring of the year (no. 3 to no. 4 in Fig. 6.5). There is also considerable runoff which can occur during the springtime and throughout the summer months.

6.2.2 Analysis of 28-Year Weather Data Set

The analysis of a 28-year weather data set from a site in eastern Canada is shown in Table 6.2. The site was at a

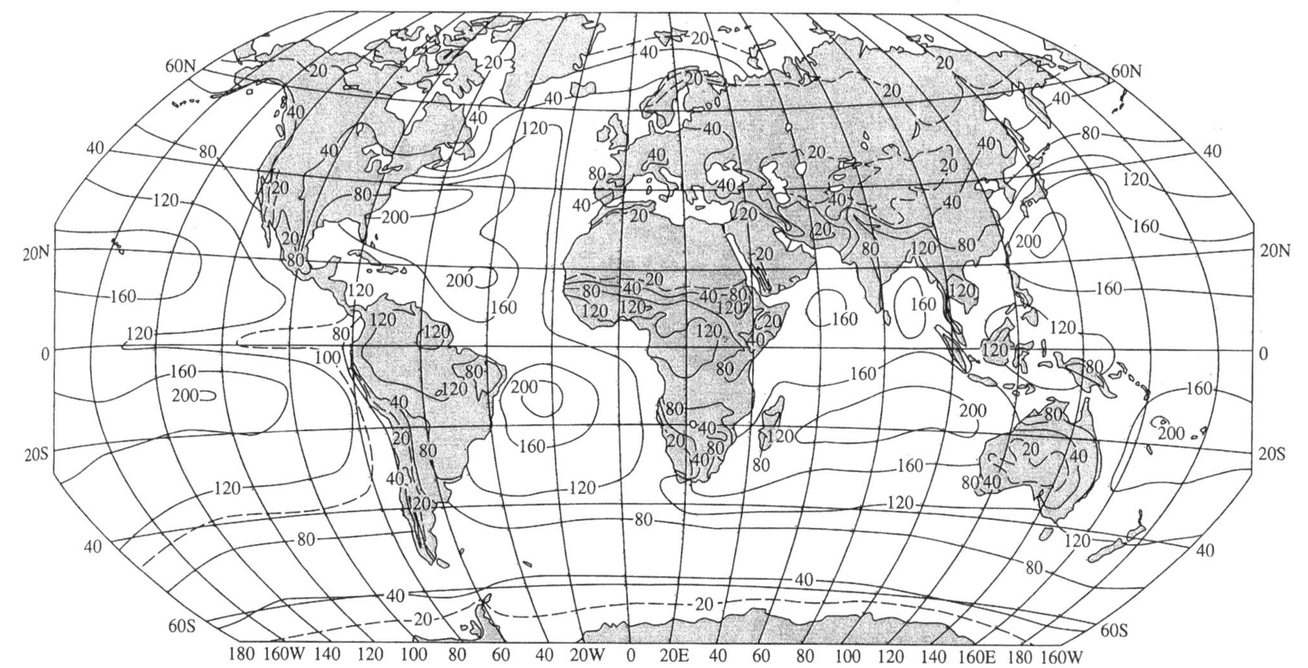

Figure 6.4 Average potential evaporation contours for the world. From *The Physics of Climate*, by J. P. Peixoto and by A.H. Oort © 1992, used with permission of American Institute of Physics.

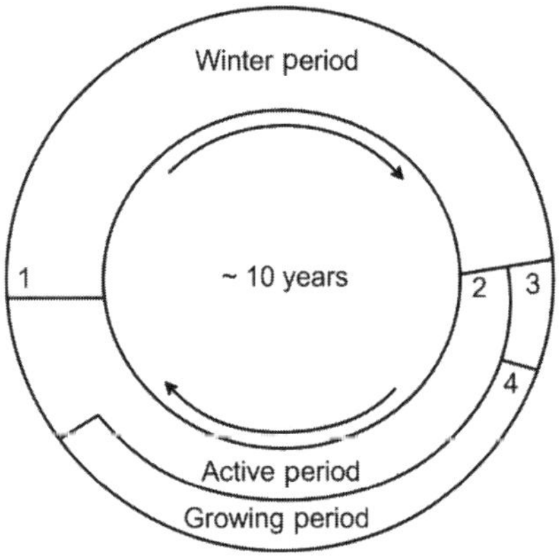

Figure 6.5 Active and Inactive portions of each year with respect to modeling near ground surface moisture movement.

latitude of approximately 47° (Stianson et al., 2010). The precipitation was separately computed for the inactive and active portions of each year. Likewise potential evaporation was computed for the inactive and active portions of each year. The results of these calculations are presented in Table 6.2 along with the calculation of the Thornthwaite moisture index for each year (i.e., from November 1 of one year to October 31 of the next year).

The results show that the average precipitation at this site during the inactive portion of the year was 321 mm. The average precipitation during the active portion of each year was 514 mm, giving rise to a total annual precipitation of 835 mm. The precipitation results are graphically presented in Fig. 6.6.

The average potential evaporation during the inactive portion of each year was 110 mm. The average potential evaporation during the active portion of each year was 574 mm, giving rise to a total annual potential evaporation of 684 mm. The potential evaporation results are graphically presented in Fig. 6.7.

Based on average weather conditions, the climate at the site can be classified as humid with an average moisture index of +24.8 (Fig. 6.8). However, for 5 of the 28 years the climate classifies as "moist humid," and for 4 years the climate classifies as "dry subhumid." The results show that there are considerable differences in the ground surface water balance conditions from one year to another. The standard deviation in the moisture index at this site was computed to be 19.8. The mean plus one standard deviation gives a moisture index of 44.6 while the mean minus one standard deviation gives a moisture index of 5.0.

The tabulation of annual climatic conditions in this manner assists the engineer in selecting meaningful scenarios for numerical simulation of near-ground-surface conditions. Typical dry years and typical wet years, as well as a so-called average year, can readily be selected based on the climatic classification.

6.3 BOUNDARY VALUE FRAMEWORK FOR NEAR-GROUND-SURFACE DESIGN

The design and performance of near-ground-surface engineered systems can be considered within the context of a "boundary value" problem. There are many factors that need to be taken into consideration during the design process. An attempt is made to illustrate the complexities associated with the design of near-ground-surface engineered structures. Care must be exercised in the assessment of all variables related to the quantification of ground surface moisture flux conditions.

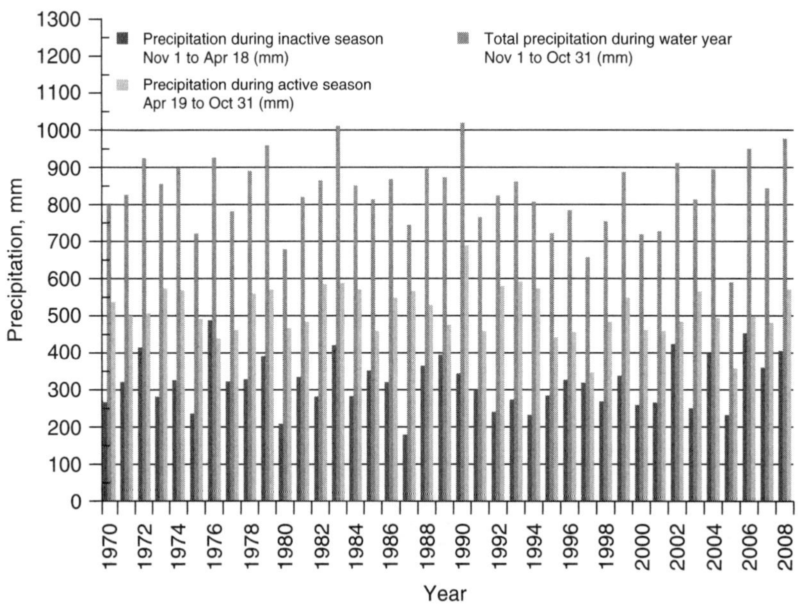

Figure 6.6 Precipitation during inactive and active portions of each year.

Table 6.2 Annual Precipitation and Potential Evaporation Conditions for Inactive and Active Portions of Each Year

Year[a]	Precipitation—Inactive Season, Nov. 1–Apr. 18 (mm)	Precipitation—Active Season, Apr. 19–Oct. 31 (mm)	Total Precipitation—Nov. 1–Oct. 31 (mm)	Potential Evaporation—Inactive Season, Nov. 1–Apr 18 (mm)	Potential Evaporation—Active Season, Apr. 19–Oct. 31 (mm)	Total Potential Evaporation—Nov. 1–Oct. 31 (mm)	Moisture Index	Climate Classification
1969 1970	265.2	536.5	801.7	78.9	575.5	654.4	22.5	Humid
1970 1971	323.2	503.1	826.3	99.9	639.5	739.4	11.8	Moist humid
1971 1972	416.2	508.8	925.0	79.1	569.3	648.4	42.7	Humid
1972 1973	280.9	576.3	857.2	114.4	580.0	694.4	23.4	Humid
1973 1974	328.6	570.2	898.8	83.9	550.4	634.3	41.7	Humid
1974 1975	233.5	489.8	723.3	87.7	605.9	693.6	4.3	Moist humid
1975 1976	490.4	437.2	927.6	108.1	646.1	754.1	23.0	Humid
1976 1977	322.4	459.7	782.1	93.6	527.4	621.0	25.9	Humid
1977 1978	330.5	560.0	890.5	92.0	476.0	568.0	56.8	Humid
1978 1979	393.1	568.5	961.6	132.0	578.6	710.6	35.3	Humid
1979 1980	212.5	467.2	679.7	109.0	631.7	740.7	−8.2	Dry subhumid
1980 1981	331.9	485.3	817.2	132.4	604.1	736.4	11.0	Moist humid
1981 1982	281.8	584.6	866.4	130.5	529.6	660.1	31.3	Humid
1982 1983	422.0	590.4	1012.4	99.0	571.8	670.8	50.9	Humid
1983 1984	283.0	572.8	855.8	120.7	519.0	639.8	33.8	Humid
1984 1985	349.8	463.0	812.8	100.3	567.6	667.9	21.7	Humid
1985 1986	323.2	548.3	871.5	110.1	597.5	707.7	23.2	Humid
1986 1987	177.3	566.3	743.6	150.6	542.9	693.5	7.2	Moist humid
1987 1988	365.1	528.7	893.8	103.5	657.8	761.3	17.4	Moist humid
1988 1989	396.0	478.5	874.5	116.8	543.6	660.3	32.4	Humid
1989 1990	343.2	691.1	1034.3	110.3	498.4	608.7	69.9	Humid
1990 1991	302.2	460.9	763.1	143.6	681.7	825.3	−7.5	Dry subhumid
1991 1992	243.0	581.1	824.1	112.8	564.6	677.4	21.7	Humid
1992 1993	271.4	589.3	860.7	100.9	468.1	569.0	51.3	Humid
1993 1994	233.6	573.3	806.9	105.2	529.4	634.6	27.2	Humid
1994 1995	283.4	440.0	723.4	128.8	693.2	822.0	−12.0	Dry subhumid
1995 1996	326.9	458.1	785.0					
1996 1997	318.3	343.2	661.5					
1997 1998	271.1	484.5	755.6	123.4	552.3	675.8	11.8	Moist humid

(*continued overleaf*)

Table 6.2 (*Continued*)

Year[a]	Precipitation—Inactive Season, Nov. 1–Apr. 18 (mm)	Precipitation—Active Season, Apr. 19–Oct. 31 (mm)	Total Precipitation—Nov. 1–Oct. 31 (mm)	Potential Evaporation—Inactive Season, Nov. 1–Apr 18 (mm)	Potential Evaporation—Active Season, Apr. 19–Oct. 31 (mm)	Total Potential Evaporation—Nov. 1–Oct. 31 (mm)	Moisture Index	Climate Classification
1998 1999	339.1	547.1	886.2					
1999 2000	257.6	462.0	719.6					
2000 2001	266.2	459.6	725.8					
2001 2002	426.5	485.0	911.5					
2002 2003	250.3	564.4	814.7					
2003 2004	399.8	495.0	894.3					
2004 2005	236.6	356.4	593.0			817.7	−27.5	Dry subhumid
2005 2006	454.7	496.7	951.4					
2006 2007	361.4	483.8	845.2					
2007 2008	406.6	570.1	976.7					
Minimum	177.3	343.2	593.0	78.9	468.1	568.0	−12.0	Dry subhumid
Maximum	490.4	691.1	1034.3	150.6	693.2	825.3	69.9	Humid
Mean	321.0	513.8	834.8	109.9	574.1	684.1	24.8	Humid

[a]Each year starts November 1 and ends October 31.

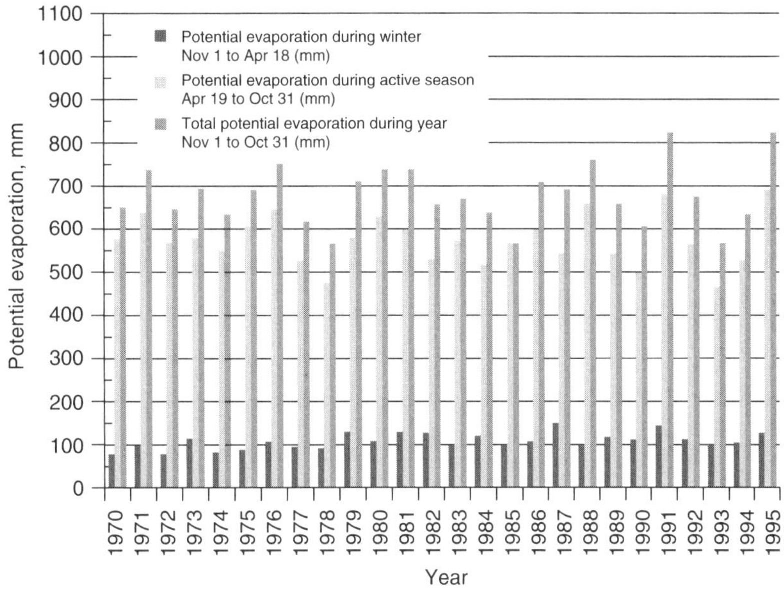

Figure 6.7 Annual potential evaporation for inactive and active portions of each year.

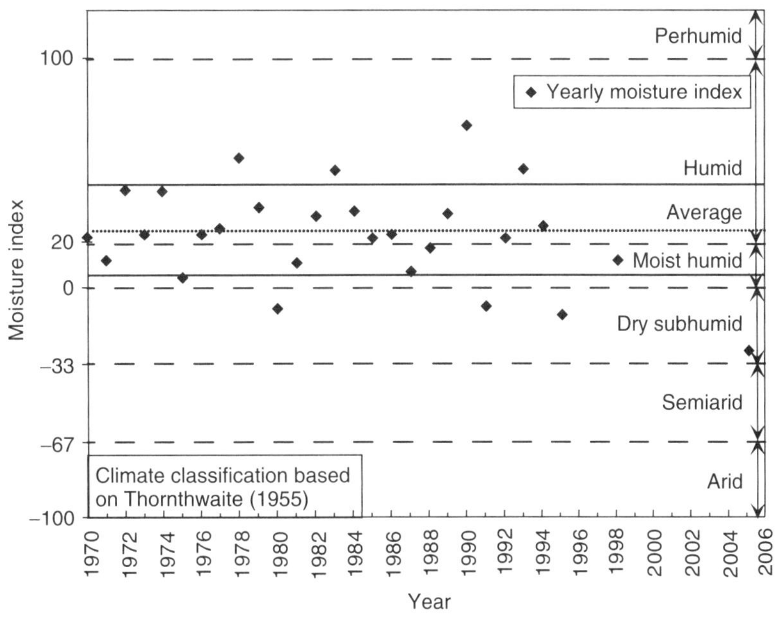

Figure 6.8 Thornthwaite moisture index for each of the 28 years at the site under consideration.

The computer provides a means whereby moisture infiltration can be studied as the solution of a partial differential seepage equation where ground surface moisture flux conditions are specified. The information required for analyzing problems within a boundary value context is quite standard. It is necessary to know the ground surface geometry and have a fairly well-informed appreciation of the underlying stratigraphic sequences.

Approximate conditions must be designated for all boundaries of the problem. For unsaturated soils problems it is the characterization of the ground surface (net) moisture flux boundary condition that is of primary importance. The assessment of the ground surface moisture flux boundary conditions largely relies on past measured weather conditions at the site. Starting, or "initial," conditions must also be specified for the analysis.

The characterization of the (unsaturated) soil properties forms another important part of the required input information. These are the primary factors that must be known. And it is within the context of a boundary value analysis that the geotechnical engineer can investigate a series of questions that might be of interest to the future performance of the engineered structure.

6.3.1 Geometry and Stratigraphy

Let us consider a cover system that forms the surface geometry of a problem to be analyzed. The stratigraphy associated with the waste material pile has been built into the pile through the methods used for disposal. Cover systems are often analyzed using a one-dimensional type of analysis. That would seem reasonable when the cover over waste material areas is relatively flat. Waste rock piles often have steep side slopes making it more difficult to construct and analyze a cover system on these surfaces. Both waste rock piles and tailings deposits can take on a three-dimensional shape (e.g., Fig. 6.9); however, design considerations are often performed using a one-dimensional analysis to simulate a relatively flat portion at the top of the pile. A two-dimensional analysis may sometimes be used for the simulation of the side slopes.

The original one-dimensional modeling of a cover system was solved using the SoilCover computer code (MEND, 1993) based on the soil-atmosphere formulation proposed by Wilson (1990). Two-dimensional analysis of covers on a sloping surface was performed by Bussière and Aubertin (2003). A three-dimensional solar energy–based approach has also been developed by Weeks and Wilson (2005). The Weeks and Wilson model focuses on obtaining values for net radiation that take into consideration the orientation of the ground surface (i.e., the dip and strike) as well as the location of the sun relative to the ground surface location. The sophistication of a three-dimensional analysis of a cover system may not be necessary for routine engineering designs. However, studies by Weeks (2006) have shown that the computations of net evaporative flux from the soil surface can vary significantly depending upon the angle between the sun's rays and the orientation of the surface of the ground. Consequently, the performance of a soil cover can vary significantly from one part of the pile to another part.

The geotechnical engineer needs to be aware that the ground surface may not be level and that there is a potential for runoff and ponding. The unevenness of the ground surface can be the result of differential settlement of the waste rock (or the tailings) with the result that the ground surface becomes susceptible to ponding. In other words, the performance of the cover from one location to another can differ considerably from a simple one-dimensional characterization. There may not need to be much change in the orientation of the ground surface to bring about significant changes to the water balance over the surface of the waste pile.

Figure 6.9 Illustration of varying evaporation conditions on the three-dimensional cover at Equity Silver (Weeks and Wilson, 2005).

Cover systems, particularly when viewed in a one-dimensional setting, have the appearance of being relatively simple problems to numerically simulate. This notion is somewhat deceptive in that, while the geometry is simple, the characterization of the soil properties is highly nonlinear and can lead to difficulties in obtaining convergence to a correct solution (Shackelford, 2005). "Checks" should be made on water balance calculations against published benchmarked example solutions (Gitirana et al., 2005).

6.3.2 Physics of Saturated-Unsaturated Water Flow for REV

An REV can be selected from within the soil mass. The physics of saturated-unsaturated water flow can be derived while satisfying the conservation of mass requirement. The substitution of the constitutive behavior for water flow and water storage, while satisfying conservation of mass, results in a PDE for saturated-unsaturated liquid flow through soil. The saturated-unsaturated water flow partial differential equation for transient flow can be written as follows for the two-dimensional case:

$$k_{wx}\frac{\partial^2 h_w}{\partial x^2} + \frac{\partial k_{wx}}{\partial x}\frac{\partial h_w}{\partial x} + k_{wy}\frac{\partial^2 h_w}{\partial y^2} + \frac{\partial k_{wy}}{\partial y}\frac{\partial h_w}{\partial y} = m_2^w \rho_w g \frac{\partial h_w}{\partial t} \qquad (6.5)$$

where:

h_w = hydraulic head comprised of the pressure head and elevation head,
k_{wx} = water coefficients of permeability in the x-direction which is a function of soil suction,
$\partial h_w/\partial x$ = hydraulic head gradient in the x-direction,
k_{wy} = water coefficients of permeability in the y-direction which is a function of soil suction,
$\partial h_w/\partial y$ = hydraulic head gradient in the y-direction,
ρ_w = density of water,
g = acceleration due to gravity,
t = time, and
m_2^w = water storage coefficient which is a function of soil suction.

Further details on the derivation of saturated-unsaturated water flow through soils are presented in Chapters 7 and 8.

The primary variable that must be computed during the solution of the PDE is the hydraulic head. In order to solve the saturated-unsaturated water flow equation (Eq. 6.5), it is necessary to have information on two soil properties, namely, the coefficient of permeability k_w (or hydraulic conductivity) of the soil and the water storage coefficient m_2^w for the soil. Both the coefficient of permeability and water storage coefficient soil properties are nonlinear functions of soil suction (Fredlund et al., 1994b).

The pore-water pressure term of soil suction is one component of hydraulic head and as a result the partial differential equation being solved is nonlinear. The nonlinearity requires that the soil properties first be estimated (i.e., hydraulic conductivity and water storage are usually set to saturated soil parameters), while the hydraulic heads are computed. Then the soil properties must be adjusted in consideration of the newly computed hydraulic heads. A new set of hydraulic heads are then computed. This process is repeated until the nonlinear partial differential equation for seepage has converged (i.e., the hydraulic heads used in estimating the soil properties match the computed hydraulic heads within a specified level of accuracy).

Convergence means that reasonably accurate soil properties (i.e., coefficients of permeability and water storage) were used when computing the hydraulic heads. The iterative process may need to be repeated many times when the soil properties are highly nonlinear. It is also possible that the nonlinear partial differential equation may not converge. It is also possible for the partial differential equation to appear to have converged but the convergence may be to the wrong values of hydraulic head. The solving of highly nonlinear partial differential equations has become an area of extensive research in mathematics and computing science and much can be learned from research in these disciplines.

The permeability and water storage functions are actually more complex than shown in Fig. 6.10 since both functions exhibit hysteresis. In other words, one set of relationships correspond to conditions when the soil is drying and another set of conditions apply when the soil is wetting, as shown in Fig. 6.11 (Pham et al., 2003b). While hysteresis is known to exist in all soils, its effect is often not taken into account in the design of cover systems. This is just one of many approximations that are made in the design of covers systems.

The material(s) used for a cover system may change considerably with time because of environmental influences. Cracking resulting from settlement and volume change is quite certain to occur. Furthermore, the growth of vegetation creates a network of root holes, fissures, and cracks. Freezing and thawing cycles tend to produce a nugget-type structure particularly in fine-grained soils. The end result is a change in the near-ground-surface soil properties with time.

There may also be microbial contamination and other bio-intrusions that affect the soil structure. Changes in the soil structure tend to significantly change the SWCC for the materials involved. Figure 6.12 illustrates the type of changes that might occur in a measured SWCC on a soil that contains clay-size particles. It is possible that the SWCC will take on a bimodal character and the saturated hydraulic conductivity may increase by one or more orders of magnitude as a result of freeze-thaw cycles or drying-wetting cycles. Consequently, numerical modeling simulations based on the properties of originally intact materials can be considerably different from the properties of soils that develop near the ground surface over time.

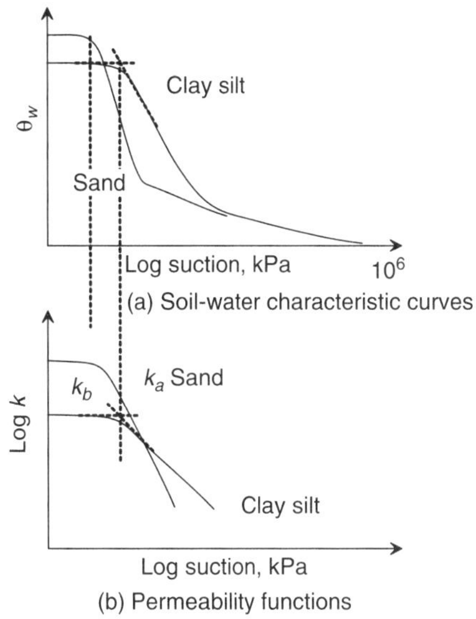

Figure 6.10 Typical SWCCs and permeability functions for two soil types.

There are challenges associated with solving the nonlinear partial differential equations defining the flow of water through the unsaturated soil. There are also challenges associated with determining the net moisture flux that should be applied at the ground surface. Much of this chapter is devoted to describing how the ground surface moisture flux can be quantified for engineering purposes.

6.3.3 Water Balance at Ground Surface

Methodologies for the computation of net moisture flux conditions at the ground surface were not part of historical soil mechanics. The calculation of ground surface moisture flux based on climatic data has become an important part of unsaturated soil mechanics. The calculation of net moisture flux involves a number of assumptions and approximations. Some inherent difficulties are mentioned in the following sections along with suggested engineering protocols for their solution. Other factors such as freezing and thawing are often not adequately taken into account and are not discussed in detail in this chapter.

The ground surface forms a flux boundary which interacts with atmospheric weather. Water is either entering the

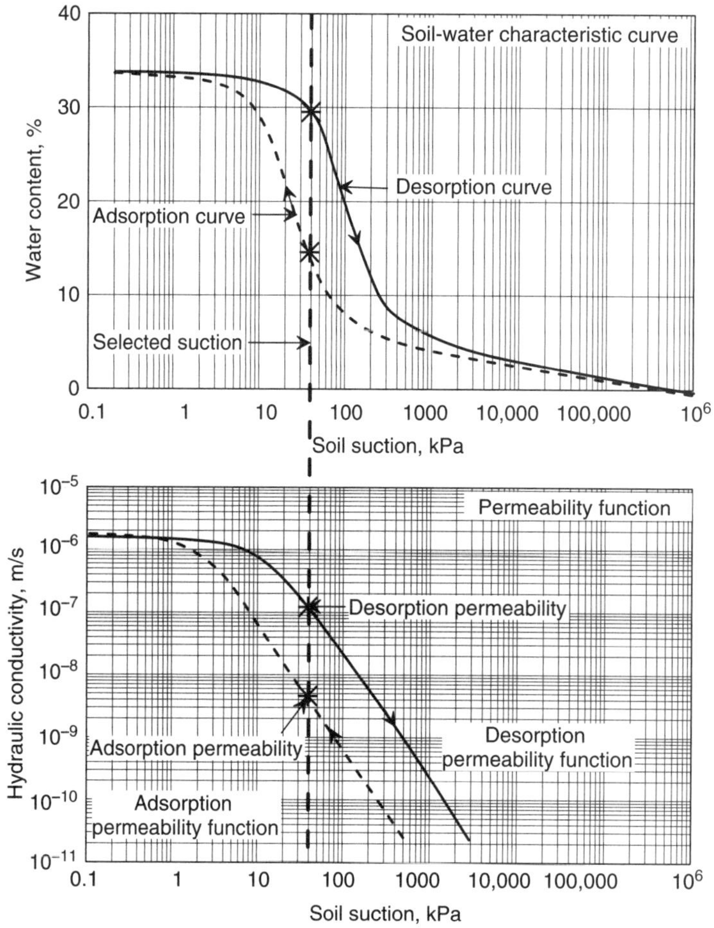

Figure 6.11 Effect of hysteresis upon drying and wetting of a soil.

ground surface boundary as precipitation or it is leaving the ground surface through actual evaporation AE or transpiration T. Water may also be shed from the ground surface through runoff R. The components of moisture flux at the ground surface and the net infiltration component, I, can be written in the following form:

$$\text{Net infiltration } (I) = \text{precipitation } (P) - \text{actual evaporation (AE)} - \text{transpiration } (T) - \text{runoff } (R) \quad (6.6)$$

In abbreviated form, the net infiltration at ground surface can be written as

$$I = P - \text{AE} - T - R \quad (6.7)$$

There has not been a long history of calculating ground surface moisture flux conditions in geotechnical engineering. There are complexities associated with the analysis and numerous assumptions must be made as part of the computational procedures.

The calculations associated with determining AE have proven to be the most challenging. PE occurs when there is an ample supply of water at the ground surface. The AE can be thought of as evaporation from a ground surface where the soil is resisting evaporation because of soil suction or the affinity of the soil for water. The AE requires that the effect of suction in the soil be taken into consideration as part of the computational procedure. It is important that the physics associated with the determination of PE be fully understood prior to attempting to calculate AE.

Each of the components of net infiltration must be assessed as part of the determination of moisture entering the soil at ground surface. The components of net moisture flux are discussed along with a brief description of the calculations and main assumptions required when performing the calculations.

The assessment of most factors influencing net infiltration is made using approximate soil properties and average daily moisture fluxes. The average effects may not provide a complete and accurate picture of long-term performance of a cover system. Extreme events or high-intensity storms can have a significant influence and may cause serious damage to cover systems.

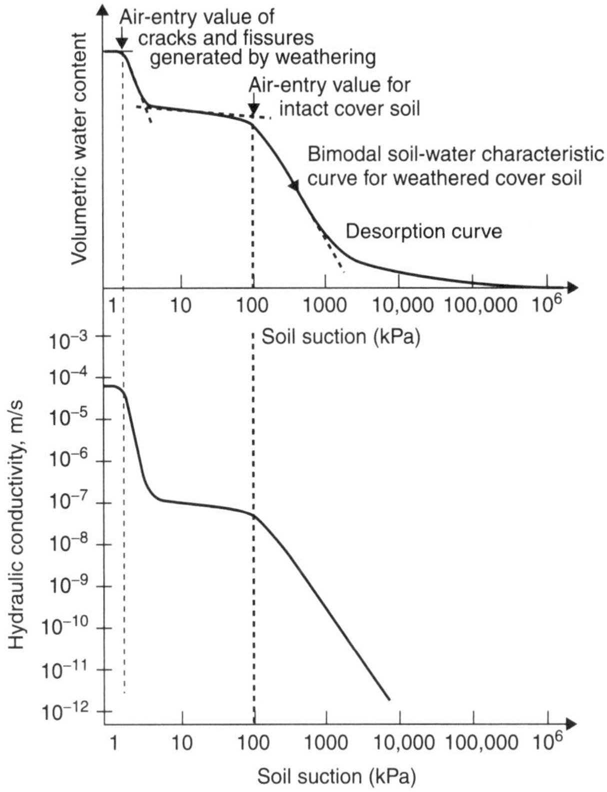

Figure 6.12 Effect of cracking that may occur as a result of weathering of near-surface soils.

6.3.4 Atmospheric Flux Conditions

An atmospheric moisture flux balance and a thermal flux balance exist at the ground surface. Both moisture and temperature flux conditions are involved in the calculation of AE. Moisture can fall to the ground surface in the form of rain or snow and is referred to as precipitation. The precipitation either infiltrates the soil or runs off. Precipitation information can be obtained from weather station records and is usually provided on a daily basis. Moisture flow in saturated and unsaturated soils can be described using Darcy's law.

Moisture also rises to the sky through the process of evaporation. The evaporative flux of primary interest for unsaturated soil modeling is AE. The ground surface moisture flux balance equation is shown in Eq. 6.7.

The thermal flux balance equation can be written as

$$R_n = R_h + R_l + R_g \tag{6.8}$$

where:

R_n = net radiation, $J/m^2/day$, but can also be converted to units of mm/day (i.e., $Q_n = 1000R_n/L_v$, where Q_n = net radiation, mm/day; L_v = volumetric latent heart of vaporization, J/m^3, and 1000 is the conversion between mm and m),

R_h = sensible heat transferring from ground surface to air, $J/m^2/day$, but can be converted to mm/day (i.e., $Q_h = 1000R_h/L_v$, where Q_h = sensible heat, mm/day),

R_l = latent heat associated with the water phase change including evaporation or freezing, $J/m^2/day$, but can be converted to mm/day (i.e., $Q_l = 1000R_l/L_v$, where Q_l = latent heat, mm/day), and

R_g = ground heat flux, $J/m^2/day$, but can be converted to mm/day (i.e., $Q_g = 1000R_g/L_v$, where Q_g = ground heat flux, mm/day).

Net radiation Q_n can either be obtained from weather station records or it can be approximated using equations such as those suggested by Penman (1948) and others (e.g., Irmak et al., 2003). The latent heat component, Q_l, can be estimated using AE or the formation of ice near the ground surface during freezing. The sensible heat component, Q_h, reflected from the ground surface to the air was described by Penman (1948) and has been incorporated into geotechnical engineering practice (Wilson, 1990):

$$Q_h = C_f \eta f(u) \left(T_{soil} - T_a\right) \tag{6.9}$$

where:

Q_h = sensible heat, mm/day,
C_f = conversion factor (i.e., 1 kPa = 0.0075 mHg),
η = psychometric constant, 0.06733 kPa/°C at 20°C,
T_{soil} = temperature of the soil surface, °C,
T_a = temperature of the air above the soil, °C,
$f(u)$ = function depending on wind speed, mm/day/kPa, $f(u) = 0.35(1 + 0.146\ W_w)$, and
W_w = wind speed, km/h.

The AE is difficult to measure directly but can be calculated using thermodynamic considerations. Equations 6.7 and 6.8 are fundamental to describing the coupling of moisture and heat flow processes. The AE depends on the water content and temperature of the soil at ground surface. The rate of evaporation of water also depends on the air temperature. The air temperature above the ground surface and the soil temperature at the ground surface are generally not the same but are interrelated by the net radiation Q_n, latent heat Q_l, and sensible heat Q_h. The availability of surface water is controlled by total precipitation, actual evaporation, and runoff. These variables play an important role in partitioning convective heat flux into sensible heat and latent heat (Wetzel and Boone, 1995).

6.3.4.1 Some Procedures for Combining and Coupling Heat and Moisture Boundary Conditions

The coupling process between moisture flow and heat flow is quite complex. There are several procedures that can be used to satisfy the conditions of coupling between heat and moisture flow. In addition, a number of procedures have been proposed for the calculation of the AE moisture flux; however, there have been few studies to assess the quantitative relationship between each of the suggested solution procedures. The reliability of all proposed analytical procedures has not been fully studied to assess situations under which various analyses give reliable computations of AE.

Calculation procedures for AE can be divided into two broad categories, namely, "moisture-flux-only" solutions and "coupled moisture-and-heat-flow" solutions. Moisture-flux-only solutions strictly apply when the ground surface soil temperature is equal to the air temperature or the ground surface temperature is calculated by some empirical means. The coupled solution means that the moisture and heat flow partial differential equations are solved simultaneously. In the moisture-flux-only solution the ground surface soil temperature is either assumed or calculated in an empirical manner. The ground surface temperature becomes the boundary condition for heat flow throughout the soil domain.

Three independent calculation procedures for AE have been used in geotechnical engineering applications: (i) the "limiting function" equation proposed by Wilson et al. (1997a), (ii) the Wilson-Penman equation, (Wilson et al. 1994), and (iii) the "experimental-based" equation proposed by Wilson et al. (1997a). Each of the three methodologies can be solved in a coupled or uncoupled manner. This gives rise to six possible procedures for the calculation of AE, as illustrated in Fig. 6.13.

The solution procedures involve varying amounts of input detail and calculation rigor. The advantages and disadvantages of each of the calculation procedures have not been fully studied to assess their applicability in geotechnical engineering practice. The components associated with heat and moisture flux boundary can be independently studied. Then procedures can be considered for the formulation of coupled and uncoupled solutions.

6.3.5 Precipitation (Rainfall and Snowfall)

Precipitation can take the form of rainfall and snowfall. The magnitude of precipitation must be measured at or near the site under consideration. Temperature, relative humidity, wind speed, and rainfall are the basic variables usually measured on an elementary weather station. Daily precipitation measurements may have been recorded over a period of many years.

The precipitation data for each year can be considered as an independent record and used as such for analysis purposes. A cumulative annual precipitation record can be plotted for each year (Fig. 6.14). The cumulative annual precipitation provides information on the distribution of precipitation within each year, as shown in Fig. 6.14. Two differing scenarios are shown: one scenario that shows an even distribution of precipitation throughout the year and

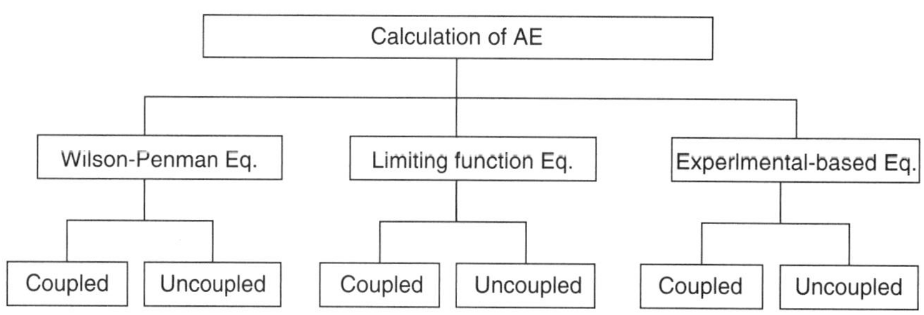

Figure 6.13 Possible coupled and uncoupled solutions to compute AE.

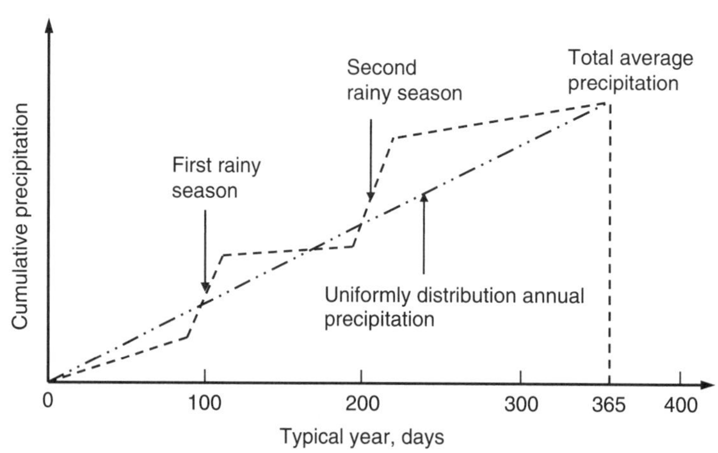

Figure 6.14 Effect of variation in precipitation distribution throughout year.

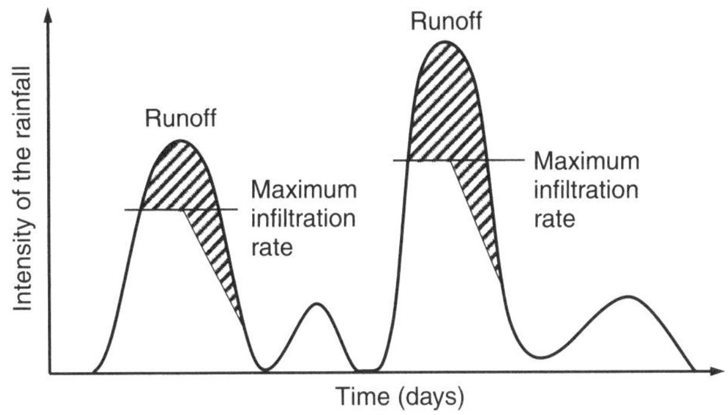

Figure 6.15 Partitioning of rainfall based on ability of soil to accept water.

another scenario that shows two distinct rainy seasons. Even though the total precipitation during a year might be the same for the two cases, the response of a cover system (or the near-ground-surface soils) to each situation would be considerably different.

An unsaturated soil can only accept water at a particular rate and the rate is mainly dependent upon the hydraulic conductivity and water storage capabilities near the ground surface. It is possible for the intensity of rainfall during a storm to exceed the ability of the soil to accept water. Figure 6.15 shows a hypothetical plot of the intensity of rainfall versus time along with the capacity of the soil to accept water. Moisture begins to run off or pond on the ground surface when the intensity of precipitation exceeds the infiltration capacity.

The conventional collection of precipitation data may not allow for the character of a particular storm to be quantified. Rainfall is usually recorded on a daily basis. The rainfall may be spread over an entire day with no consideration as to whether the storm was 10 min long or 10 h long, as shown in Fig. 6.16. If the rainfall for each day is spread over the entire day, most of the precipitation will appear to infiltrate the soil. This is an inherent weakness in the conventional manner in which weather station rainfall data are collected.

The primary driver for net ground surface moisture flux is precipitation and snowfall. This information is obtained from weather stations. Figure 6.17 shows a typical plot of daily rainfall records and the cumulative precipitation over the period of one year.

6.3.6 Calculation of Relative Humidity

The primary driving mechanism of evaporation of water from the ground surface is the water vapor pressure gradient. Air consists of a mixture of gases and water vapor. The partial pressure of water vapor in air forms the basis for water vapor movement. The water vapor pressure gradient between two points causes moisture to move through the air phase of an unsaturated soil. Water vapor pressure gradients at the ground surface result in the evaporation of moisture from the surface of the soil. In other words, evaporation is fundamentally driven by a water vapor pressure gradient.

Most of the mathematical formulations describing water vapor movement in unsaturated soil mechanics are written in terms of changes in relative humidity RH or h_r. A number of other variables are also used as a measure of the amount of water vapor in air.

Relative humidity h_r is defined as the ratio of the partial pressure of water vapor in an air mixture, u_v^{air}, to the saturated vapor pressure of water vapor, u_{v0}^{air}, at the same temperature. Relative humidity h_r can be written either in decimal form or as a percentage:

$$h_r = \frac{u_v^{air}}{u_{v0}^{air}} \tag{6.10}$$

Relative humidity is a fundamental variable associated with water vapor flow, namely, the gradient of water vapor partial pressure. Relative humidity is one of the variables generally collected at weather stations.

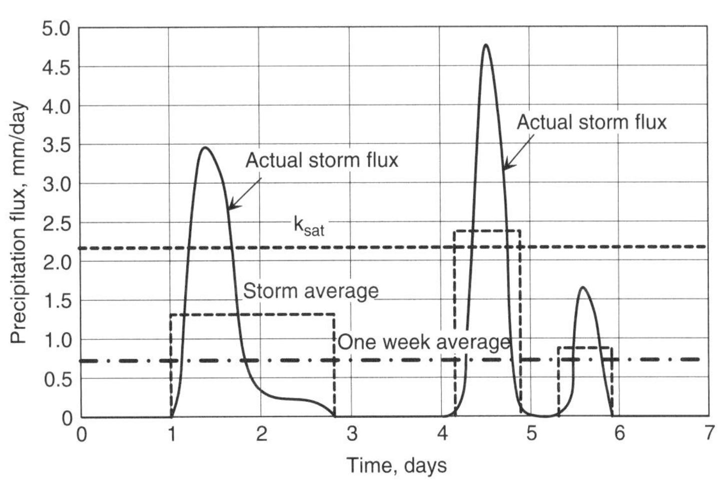

Figure 6.16 Distribution of typical rainfall intensity diagrams.

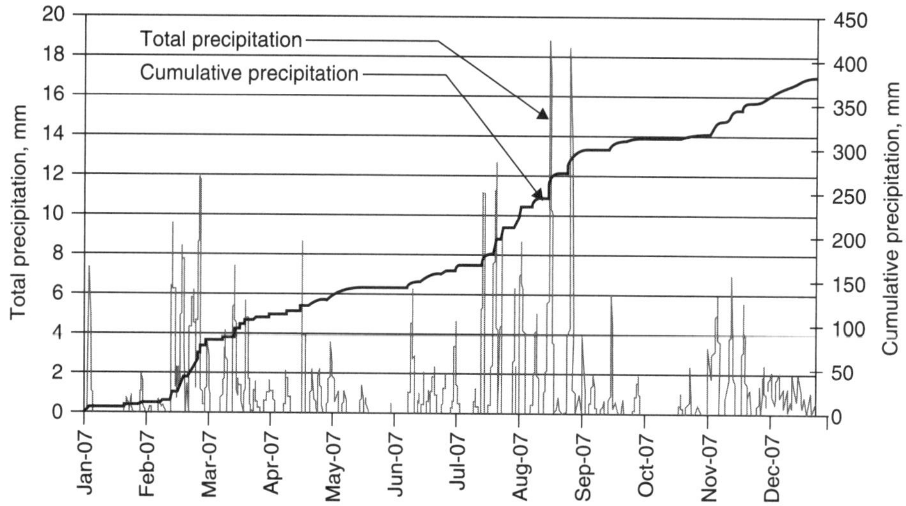

Figure 6.17 Typical weather station record showing daily precipitation and cumulative rainfall at particular site in Canada.

6.3.6.1 Measurement of Relative Humidity

A device used to measure relative humidity is called a *hygrometer.* There are three major types of hygrometers: mechanical, electrical, and cold-spot or dew-point hygrometers. A simple mechanical type of hygrometer has a sensing element that is usually an organic material that expands and contracts with changes in the water vapor in the surrounding air (Fig. 6.18). The material most commonly used is human hair.

Electrical hygrometers use the change in electrical resistance of a substance that can be calibrated against relative humidity. The third type of hygrometer is called a cold-spot or dew-point apparatus. The dew-point temperature is measured and used to compute relative humidity. Dew point is the temperature at which water vapor comes to the point of saturation (i.e., 100% relative humidity). The usual procedure involves the chilling of a polished surface until a film of moisture just starts to appear. The temperature of the polished surface is then measured.

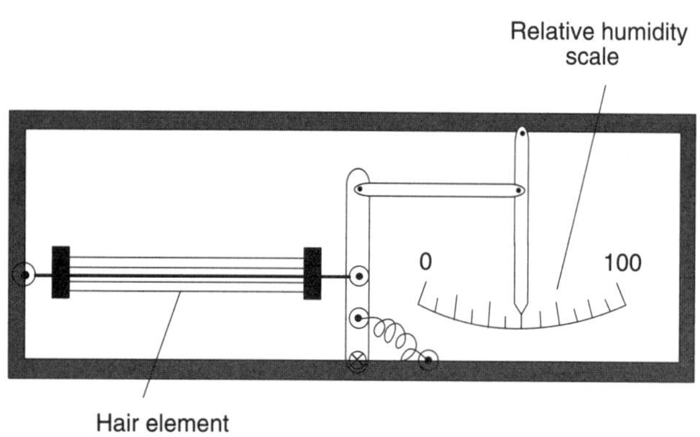

Figure 6.18 Hygrometer using hair as sensing element for measuring humidity. (After D. M. Considine and S. D. Ross, *Process Instruments and Controls Handbook*, 2nd ed., McGraw-Hill, New York, 1974.)

Dew-point hygrometers are commonly used in weather stations and as a result the raw data on humidity from a weather station may be expressed in the form of a dew-point reading. Consequently, it is necessary to convert dew-point readings into relative humidity.

6.3.6.2 Vapor Pressure under Saturated Conditions

The saturated vapor pressure is the reference for water vapor gradients. Saturated vapor pressures are a unique function of temperature and can be calculated using one of several empirical equations such as those presented by Tetens (1930) and Lowe (1977).

Weather conditions above the ground surface measure the relative humidity and temperature corresponding to a particular point above the ground surface. The air in the soil at ground surface has a relative humidity that may differ from that of the air above the ground surface. As a result there is a water vapor pressure gradient that causes the movement of water vapor flow between the soil surface and the air above the ground surface.

6.3.6.3 Relationship of Relative Humidity to Temperature

Relative humidity changes nonlinearly with temperature. Changes in relative humidity occur because the partial pressure of water vapor decreases with temperature. Air temperatures generally increase toward noonday, and as a result, relative humidity goes down. The nonlinear relationship between temperature and humidity makes it necessary to give consideration to changes in relative humidity throughout each day when modeling evaporation from soil surfaces.

Figure 6.19 shows a typical relationship between temperature and humidity throughout one day. These diurnal

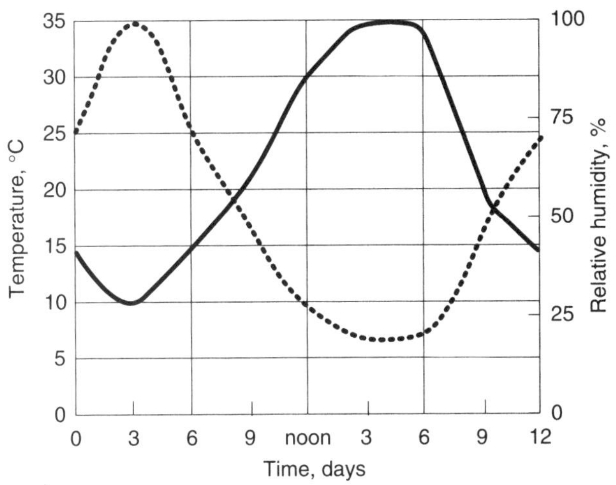

Figure 6.19 Illustration of the typical inverse relationship between temperature and relative humidity throughout a day.

changes in humidity influence the actual evaporation from a soil surface. It has been suggested that the changes in relative humidity throughout each day should be taken into consideration when modeling actual evaporation (Wilson, 1990). Diurnal variations can be based on either assumed daily variations or actual measured changes in relative humidity and temperature throughout each day.

6.3.6.4 Calculation of Relative Humidity from Dew-Point Measurements

The relative humidity of an air-water mixture can be calculated if the air temperature T and the dew-point temperature T_d are known. There are a number of equations that have been proposed for the calculation of relative humidity based on the dew-point temperature. There is a simple approximation that allows the relative humidity h_r to be calculated based on the dew-point temperature t_d and the dry bulb temperature T. The equation is as follows:

$$h_r = 100 - 5\left(T - t_d\right) \tag{6.11}$$

where:

T = air temperature, °C, and
t_d = dew-point temperature, °C.

The equation is quite accurate as long as the relative humidity is above 50%. The equation states that the relative humidity changes by 5% for every 1°C difference in temperature between the dew point and the dry bulb temperature starting at 100% relative humidity.

6.3.6.5 August-Roche-Magnus Calculation of Relative Humidity

The August-Roche-Magnus equation was developed for the calculation of the dew-point temperature when a specific relative humidity and dry bulb temperature were known. The results from this equation are comparable to those obtained when using the equation proposed by Tetens (1930).

The August-Roche-Magnus equation can be rearranged and solved for relative humidity when the dew-point temperature is known. The August-Roche-Magnus equation is considered valid under the following conditions:

$$\begin{aligned} 0^\circ\mathrm{C} &< T < 60^\circ\mathrm{C} \\ 1\% &< h_r < 100\% \\ 0^\circ\mathrm{C} &< t_d < 50^\circ\mathrm{C} \end{aligned} \tag{6.12}$$

The August-Roche-Magnus equation for the calculation of dew point is as follows:

$$t_d = \frac{b\left[aT/\left(b+T\right) + \ln\left(h_r/100\right)\right]}{a - \left[aT/\left(b+T\right) + \ln\left(h_r/100\right)\right]} \tag{6.13}$$

where:

h_r = relative humidity, %,
T = dry bulb temperature, °C,
t_d = dew point temperature, °C
a = 17.271, and
b = 237.7°C.

Rearranging Eq. 6.13 and solving for relative humidity h_r gives

$$h_r = 100\ \exp\left[\frac{ab\left(t_d - T\right)}{\left(b+T\right)\left(t_d+b\right)}\right] \tag{6.14}$$

6.3.6.6 Bosen (1958) Calculation of Relative Humidity

The Bosen (1958) equation was developed for the approximation of the dew-point temperature when the relative humidity and dry bulb temperature were known. The Bosen (1958) equation for the calculation of dew point is as follows:

$$t_d = \left(a + 0.9T\right) h_r^{0.125} - a + 0.1T \tag{6.15}$$

where:

a = constant equal to 112 when temperature is in °C.

The Bosen (1958) equation can be rearranged such that the relative humidity h_r can be computed when the dew-point temperature and the dry bulb temperature are known:

$$h_r = 100\left[\frac{t_d - 0.1T + 112}{112 + 0.9T}\right]^8 \tag{6.16}$$

Table 6.3 Comparison of Computed Relative Humidity

Dry Bulb Temperature, °C	Dew-Point Temperature, °C	RH Simplified Equation, %	RH August-Roche-Magnus Equation, %	RH Bosen Equation, %
25.0	24.5	97.5	97.06	97.06
25.0	24.0	95.0	94.20	94.20
25.0	22.0	85.0	83.48	83.49
25.0	20.0	75.0	73.84	73.85
25.0	18.0	65.0	65.19	65.21
25.0	16.0	55.0	57.44	57.46
25.0	12.0	35.0	44.33	44.34
25.0	10.0	25.0	38.82	38.83
25.0	6.0	5.0	29.57	29.57
20.0	19.0	95.0	93.98	94.01
20.0	16.0	80.0	77.79	77.88
20.0	12.0	60.0	60.03	60.16
20.0	10.0	50.0	52.56	52.71
20.0	6.0	30.0	40.04	40.19

6.3.6.7 Comparison of Equations for Calculating Relative Humidity

Table 6.3 contains calculated relative humidity values when typical values are selected for the dry bulb temperature and the dew-point temperature. The results show that the proposed simplified equation provides reasonable estimates of the relative humidity as long as the computed values are above about 50%. The August-Roche-Magnus and the Bosen (1958) equations give similar computed humidity values over the entire range of relative humidity.

6.3.7 Runoff

Runoff R is defined as the rainwater that cannot gain entrance into the soil during the period of rainfall. The amount of moisture leaving the ground surface by actual evaporation can also be taken into consideration. Rainfall is generally assumed to fall vertically, and therefore the slope of the ground surface should be taken into consideration. The water balance equation at ground surface takes on the general form shown below, and its meaning is illustrated in Fig. 6.20:

$$R = P \cos \alpha_2 - I - \mathrm{AE} \tag{6.17}$$

where:

I = net moisture infiltration, mm/day,
P = precipitation, mm/day,
α_2 = ground surface slope, deg,
AE = actual evaporation, mm/day, and
R = runoff, mm/day.

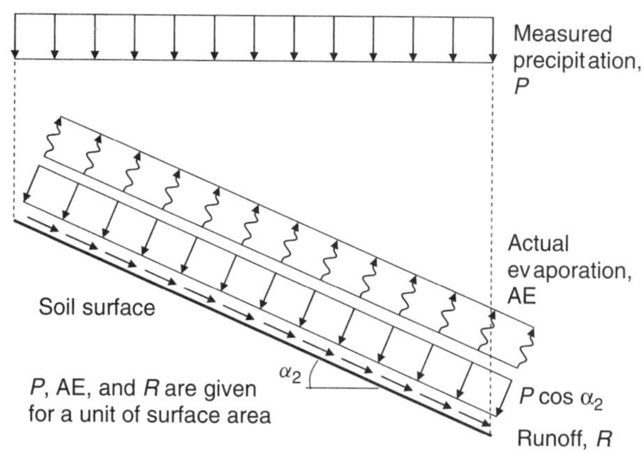

Figure 6.20 Ground surface moisture flux associated with runoff.

Equation 6.17 applies only during periods of rainfall and does not represent a rigorous analysis of the runoff process. Runoff has no functional relationship with precipitation other than as part of the water balance at ground surface. Rather, runoff appears as the missing term in the water mass balance equation (i.e., the term that is not generally measured). Runoff as presented in Eq. 6.17 is relevant only during precipitation periods where there is excess water to generate runoff. The assumption is sometimes made that there is no evaporation from the ground surface during precipitation events (i.e., $R = P \cos \alpha_2 - I$).

Actual evaporation is a function of weather parameters and the soil suction at the soil-atmosphere boundary. Actual evaporation is discussed later and can be computed based on one of several proposed procedures (Wilson et al., 1997a).

Runoff must be computed in an interactive manner with actual evaporation.

Runoff water is removed from the ground surface if a drainage system exists. For runoff to occur, net moisture flux NF should not produce pore-water pressures at ground surface. The following set of equations can be used to represent runoff conditions (Gitirana et al., 2005):

$$\mathrm{NF} = \begin{cases} P\cos\alpha_2 - \mathrm{AE} & \text{if } P\cos\alpha_2 - \mathrm{AE} > 0 \text{ and } u_{ws} < 0 \\ \mathrm{EF}\left(0 - u_{ws}\right) & \text{if } P\cos\alpha_2 - \mathrm{AE} > 0 \text{ and } u_{ws} \geqslant 0 \\ P\cos\alpha_2 - \mathrm{AE} & \text{if } P\cos\alpha_2 - \mathrm{AE} \le 0 \end{cases} \tag{6.18}$$

where:

u_{ws} = pore-water pressure at the ground surface and
EF = a large number relative to the saturated coefficient of permeability (e.g., 1000 times the saturated coefficient of permeability). The EF variable acts as a "review" boundary condition that triggers the change between a moisture flux boundary condition and a zero pore-water pressure boundary condition.

Runoff R may take place when the value of $P\cos\alpha_2 -$ AE is larger than the saturated hydraulic conductivity. The amount of runoff corresponds to the difference between the water available ($P\cos\alpha_2 -$ AE) and the amount of infiltration.

The area flux boundary condition [i.e., NF = EF(0 − u_{ws})] becomes mathematically equivalent to a hydraulic head boundary condition at the ground surface (i.e., $u_{ws} = 0$) if the multiplier EF tends to infinity (Gitirana et al., 2005). This approach tends to avoid numerical oscillation problems associated with switching boundary conditions that produce an instantaneous change in the value at a node. An instantaneous change in boundary conditions does not represent real-world conditions. Instantaneous changes in node values tend to cause extreme mesh refinements (Gitirana et al., 2005;). Equation 6.18 imposes a realistic and stable boundary condition.

The above-mentioned approach to the simulation of a rainfall event allows both the infiltration and the runoff components to be accurately simulated. Figure 6.21 illustrates an accurate simulation of infiltration and runoff at the ground surface boundary (Gitirana et al., 2006).

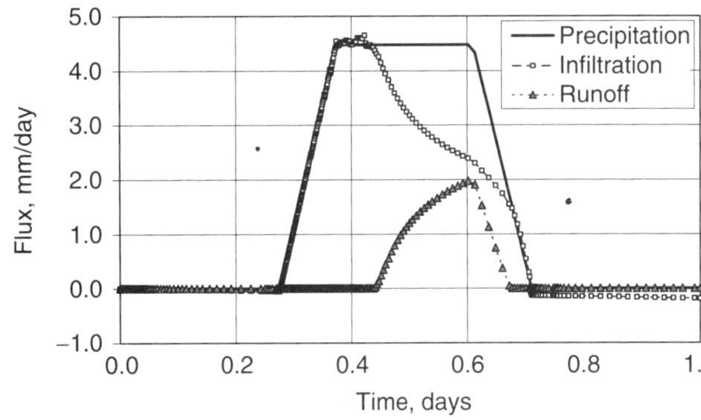

Figure 6.21 Illustration of the ability to simulate infiltration and runoff conditions.

6.3.8 Potential Evaporative Flux

The word "evaporation" generally refers to moisture movement from a water surface or a soil surface, while the word "transpiration" refers to upward moisture movement through vegetation on the ground surface. The PE is the amount of water that can evaporate if ample water is available at the ground surface. The PE takes into consideration the thermal energy available for evaporation and the ability of the lower atmosphere to transport water vapor away from the ground surface. PE is lower when the sky is cloudy due to a reduction in net radiation reaching the soil surface. PE is higher in the summer and at locations nearer to the equator. PE is also higher on windy days because the evaporated moisture is quickly removed from the soil surface, allowing further evaporation to take place.

The PE can either be measured using an evaporation pan or calculated based on climatic factors. Weather stations commonly record precipitation events, net radiation, temperature, wind velocity, relative humidity, and possibly some other variables. The above-mentioned weather information can be used to calculate PE.

The PE also depends on a number of surface factors (e.g., free water surface, pan of water, soil type, and vegetation). Average annual potential evaporation can be compared to the average annual precipitation P, and the ratio of precipitation to potential evaporation, P/PE, is called the aridity index.

Evaporation from the ground surface can be quantified through consideration of the physical processes involved. Numerous studies have been conducted since the 1920s in an attempt to predict potential evaporation from the ground surface. It is, however, the AE and transpiration that are of primary interest in geotechnical engineering. The factors controlling PE need to be understood before attempting to calculate AE. PE is the amount of water removed by the atmosphere through evaporation if water is freely available at the ground surface.

There are three terms commonly used when discussing upward moisture movement from a soil surface or vegetated surface: (i) PE, (ii) AE, and (iii) PT. The terms AE and PT are required for quantification of the water balance at the ground surface. Figure 6.22 illustrates the components of evaporation from the ground surface. The AE is less than the PE mainly because of the affinity that soil has for water (i.e., soil suction in the soil at ground surface).

In general, about 80% of the energy required for evaporation comes from the sun in the form of "net radiation," while wind (in the form of a "mixing term") and the vapor deficit of the air form a second important component contributing

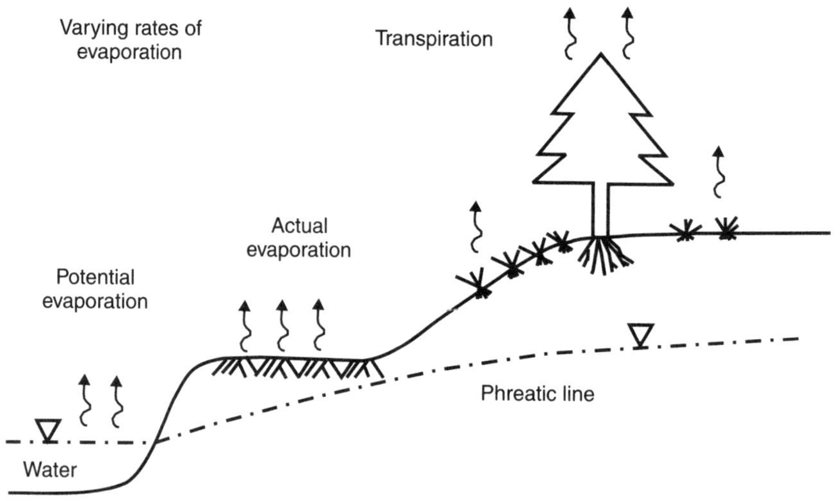

Figure 6.22 Illustration of thc various types of evaporative fluxes.

to evaporation. *Pan evaporation* measurements (i.e., an open water surface) provide a measure of PE.

Researchers have attempted to develop mathematical relationships that embrace the primary variables controlling the rate of evaporation from a free water surface (i.e., PE). Each proposed PE equation uses specific weather-recorded data. The calculation of PE can be presented in units of millimeters per day, and the range of values is generally between zero and 10 mm/day. These units of measurement are particularly useful in geotechnical engineering.

Rowher (1931) incorporated the variables of wind speed, vapor pressure at the evaporating surface, and vapor pressure above an evaporating surface in an attempt to predict PE. The vapor pressure gradient between the water surface and the air above the water was seen as the primary driving mechanism for evaporation. Vapor pressure at the water surface was assumed to correspond to saturated vapor pressure conditions which were uniquely related to temperature. The vapor pressure above the evaporating surface was dependent upon the relative humidity and temperature of the air.

Sutton (1943) incorporated the variables of wind speed, vapor pressure at the evaporating surface, vapor pressure above the evaporating surface, evaporating area, and a constant related to temperature into the prediction of PE.

Following is a summary of some equations that are used in geotechnical engineering for the estimation of PE.

6.3.8.1 *Thornthwaite (1948) Equation*

The Thornthwaite (1948) equation is commonly used to assess climatic conditions of aridity and humidity. Thornthwaite incorporated the variables of length of daylight hours, mean monthly temperature, and an empirical constant into the calculation of PE. The Thornthwaite equation has been extensively used for the classifications of global climate as well as numerous other agricultural and engineering applications. Daily PE can be written as follows:

$$\text{PE} = 0.533\left(\frac{L_d}{12}\right)\left(\frac{N}{30}\right)\left(\frac{10T_a}{I}\right)^{a_t} \tag{6.19}$$

where:

PE = potential evaporation, mm/day,
L_d = length of daylight, h,
N = number of days in the month,
T_a = mean monthly air temperature, °C,
I = summation for 12 months of the function $(T_a/5)^{1.514}$; (i.e., $I = \sum_{\text{month}=1}^{12}\left[T_a/5\right]^{1.514}$,
a_t = complex function of the variable I (i.e., $a_t = (6.75 \times 10^{-7})\ I^3 - (7.71 \times 10^{-5})\ I^2 + (1.79 \times 10^{-2})\ I + 0.492$), based on correlations to pan evaporation measurements.

6.3.8.2 *Penman (1948) Equation*

Penman (1948) incorporated a number of variables commonly collected at weather stations into the prediction of PE. The Penman equation combines a Dalton-type formulation with the heat budget equation, and as a result, the temperature at the evaporating surface is no longer required. The Penman equation uses the following routine weather data as input, namely, relative humidity, air temperature, wind speed, and net radiation. The Penman equation for PE can be written as follows:

$$\text{PE} = \frac{\Gamma Q_n + \eta E_a}{\Gamma + \eta} \tag{6.20}$$

where:

PE = potential evaporation, mm/day,
Γ = slope of saturation vapor pressure versus temperature curve, kPa/°C,

Q_n = net radiation at the water (or saturated ground) surface, mm/day,
η = psychrometric constant, kPa/°C,
$E_a = 2.625(1 + 0.146W_w)(u_{v0}^{air} - u_v^{air})$, mm/day,
W_w = wind speed, km/h,
u_v^{air} = vapor pressure in the air above the water (or saturated ground), surface, kPa, and
u_{v0}^{air} = saturated vapor pressure at the mean air temperature, kPa.

The Penman (1948) equation shows that the vapor pressure gradient between the water surface and the air above the water becomes the primary driving mechanism for evaporation. There are two terms in the numerator of Eq. 6.20. The first term involves net radiation and characterizes the power of the sun to evaporate water. Net radiation quantifies the net effect of short- and long-wave radiation from the sun, surface reflectance (i.e., albedo), and surface temperature. The second term involves "mixing" of the air above the water or the drying power associated with air movement.

The Penman (1948) equation assumes there is no heat exchange with the ground, no advective energy within the water and no exchange of heat energy. The Penman equation is applicable to free-water surfaces. The Penman equation has been shown to produce comparable results to pan evaporation and evaporation from lake surfaces. The air temperature above the water surface is used in the calculation of PE.

6.3.8.3 Blaney and Criddle (1950) Equation

Blaney and Criddle (1950) used the variables of mean daily air temperature and the fraction of the day in sunlight to compute potential evaporation:

$$PE = p(0.46\,T_{mean} + 8) \tag{6.21}$$

where:

PE = potential evaporation, mm/day,
p = percentage of daylight hours each day, and
T_{mean} = mean daily temperature, °C, calculated as $T_{mean} = (T_{max} + T_{min})/2$.

Jensen and Haise (1963) used air temperature and incident solar radiation to predict potential evaporation from a water surface.

6.3.8.4 Monteith (1965) Equation

Monteith (1965) made modifications to the Penman (1948) equation and the subsequent equation became known as the Penman-Monteith equation (Monteith, 1965). The Penman-Monteith equation contained the main variables from the Penman (1948) equation but added variables such as air volumetric heat capacity, vapor pressure deficit, fraction of the day in sunlight as well as canopy and aerodynamic resistance factors. The Penman-Monteith equation (Penman, 1965) better represents evapotranspiration from a vegetated surface by incorporationg canopy conductance.

The Penman-Monteith equation appears to be an enhancement of the original Penman equation and has become widely used in agriculture-related disciplines. However, the Penman (1948) equation appears to still be widely used for geotechnical engineering applications. The Penman-Monteith equation is

$$PE = \frac{\Gamma\left(R_n - G\right) + \rho_a C_a \left(u_{v0}^{air} - u_v^{air}\right) g_a}{\left[\Gamma + \eta\left(1 + g_s/g_a\right)\right] L\rho_w}\left(86.4 \times 10^6\right) \tag{6.22}$$

where:

L = latent heat of vaporization or the energy required to evaporate a unit mass of water (i.e., ~2.42×10^6 J/kg),
Γ = slope of the saturation vapor pressure versus air temperature, kPa °C^{-1},
R_n = net radiation (J/m^2/day), from the external source of energy,
G = soil heat flux, J/m^2/day,
C_a = specific heat capacity of air, MJ kg^{-1} C^{-1},
ρ_a = dry air density, kg m^{-3},
ρ_w = water density, kg m^{-3},
u_v^{air} = vapor pressure in the air above the saturated ground surface, kPa,
u_{v0}^{air} = saturated vapor pressure at the mean air temperature, kPa,
g_a = atmospheric conductance, m s^{-1},
g_s = surface conductance, m s^{-1},
η = psychrometric constant ($\eta \approx 0.066$ kPa°C^{-1}), and
86.4×10^6 = conversion of m/s to mm/day.

The soil heat flux can be assumed to be close to zero in some cases.

6.3.8.5 Priestley-Taylor (1972) Equation

The Priestley-Taylor (1972) equation is quite similar to the Penman equation but contains some different variables:

$$PE = \alpha_3 \frac{\Gamma}{\Gamma + \eta}\left(Q_n - G\right) \tag{6.23}$$

where:

α_3 = a constant taken as 1.26 for open and well-watered surfaces,
Γ = slope of saturation vapor pressure versus temperature curve, kPa°C^{-1},
Q_n = net radiation, mm/day,
η = psychrometric constant, kPa°C^{-1}, and
G = soil heat flux, mm/day.

Hargreaves (1975) incorporated the variables—extraterrestrial radiation, temperature, and the difference between the mean monthly maximum and minimum temperatures—into an equation to compute potential evaporation.

6.3.9 Use of Penman (1948) Equation for Calculating Potential Evaporation

The vapor pressure in the air above the water and the saturated vapor pressure at the water surface are the dominant variables driving evaporation. The saturated vapor pressure is a function of temperature while the actual vapor pressure in the air is related to relative humidity. The two variables on the bottom of the Penman (1948) equation are also related to the physical properties of water vapor.

It is generally assumed that it is not sufficient to use average daily values for variables such as temperature and relative humidity when solving the Penman (1948) equation. Rather, minimum and maximum values for each day are used to provide an improved calculation of the changes in vapor pressure in the air throughout each day. Figure 6.23 shows the minimum, maximum, and average air temperature readings at a particular site in Canada for the year 2007. Figure 6.24 shows the minimum, maximum, and average relative humidity values for the same site in the same year. Minimum, maximum, and average values can be used to approximate the nonlinear variation in vapor pressure throughout each 24-h day.

Figure 6.25 shows a plot of wind speed and Fig. 6.26 shows a plot of the net radiation for each day at a particular site. Net radiation values are not always measured at weather stations, and therefore it is sometimes necessary to estimate net radiation values based on the latitude of the site under consideration as well as other variables.

The PE calculations provide the engineer with an indication of the maximum evaporation that might be anticipated from a water-saturated surface. As the soil at the ground surface dries, it holds onto the water while the sun and wind are pulling water skyward. The "struggle" between the climate and the soil gives rise to the AE from the ground surface.

The contribution of each component of PE can be computed by writing the Penman (1948) equation in the following form. The net radiation component on potential evaporation, PE_{netrad}, is written as

$$PE_{netrad} = \frac{\Gamma Q_n}{\Gamma + \eta} \tag{6.24}$$

and the mixing component, PE_{mixing}, is written as

$$PE_{mixing} = \frac{\eta E_a}{\Gamma + \eta} \tag{6.25}$$

Equations 6.24 and 6.25 allow the separation of the evaporative moisture flux due to net radiation and that due to mixing when using the Penman (1948) equation. Calculations are presented for a typical climate in northern Canada. The analysis of the weather data for the years 2005 and 2006 are shown in Figs. 6.27 and 6.28. The total evaporative flux can vary significantly from one day to the next; however, the figures show the portion of total evaporation for each day that is due to each of the components of potential evaporation. The climate at the site under consideration can be classified as perhumid to moist humid during the fall and spring period and as semiarid to arid during the summer period from April to October.

The calculations from all years showed that the mixing component of potential evaporation dominated during the

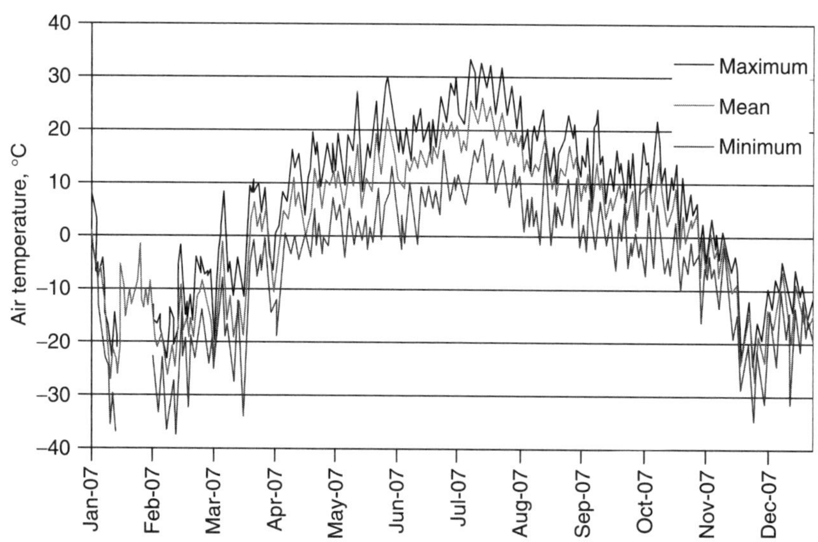

Figure 6.23 Minimum, maximum, and average air temperature readings for a particular site in the year 2007.

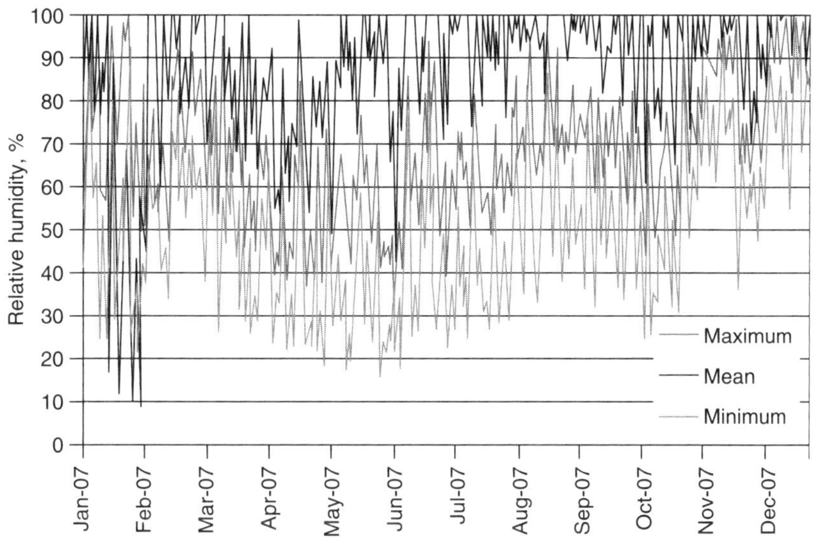

Figure 6.24 Minimum, maximum, and average relative humidity readings for a particular site in the year 2007.

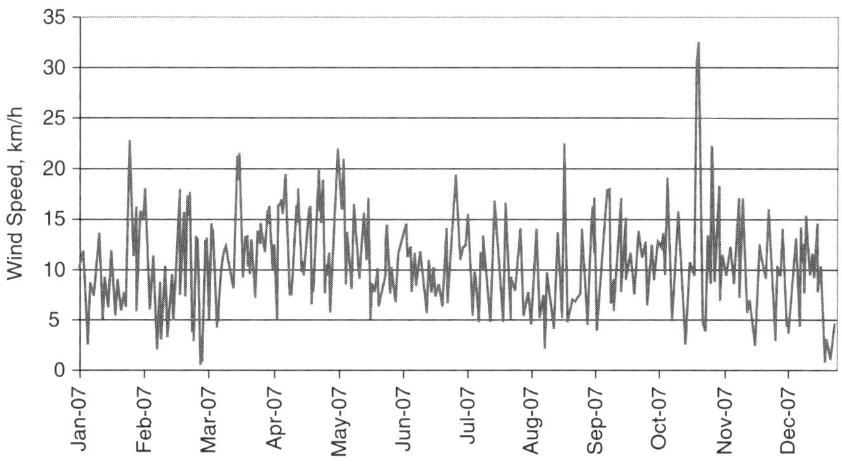

Figure 6.25 Average wind speed readings for a particular site in the year 2007.

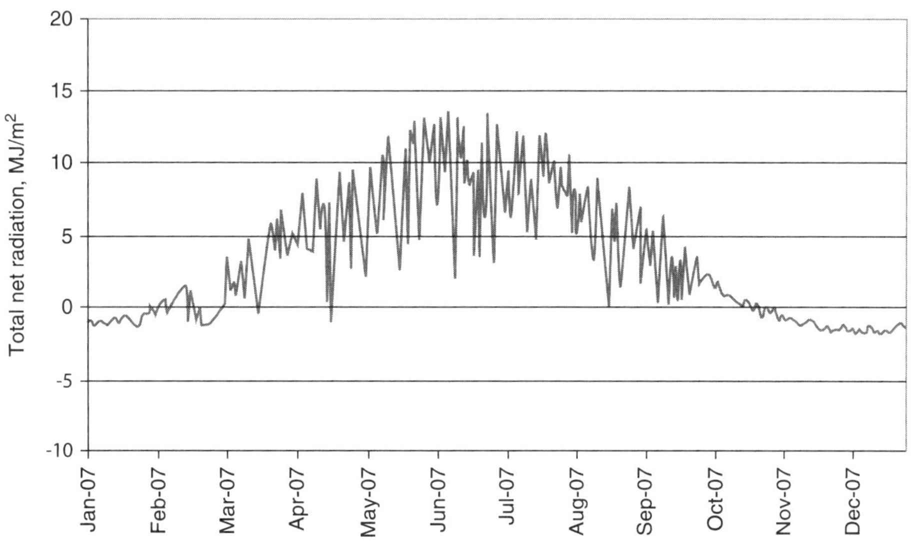

Figure 6.26 Total net radiation readings for the same site in the year 2007.

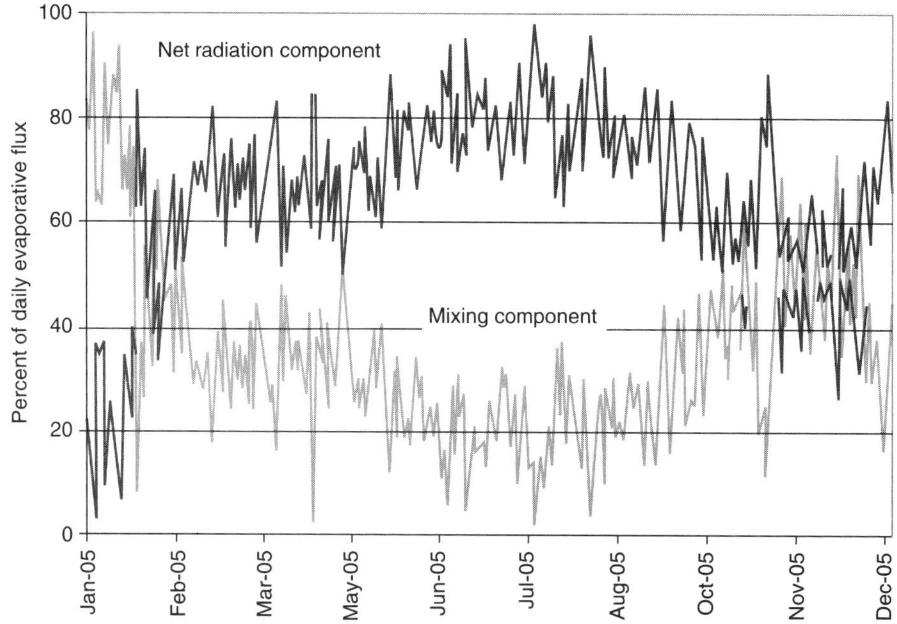

Figure 6.27 Breakdown between net radiation and mixing components of PE for the year 2005 at a site in northern Canada.

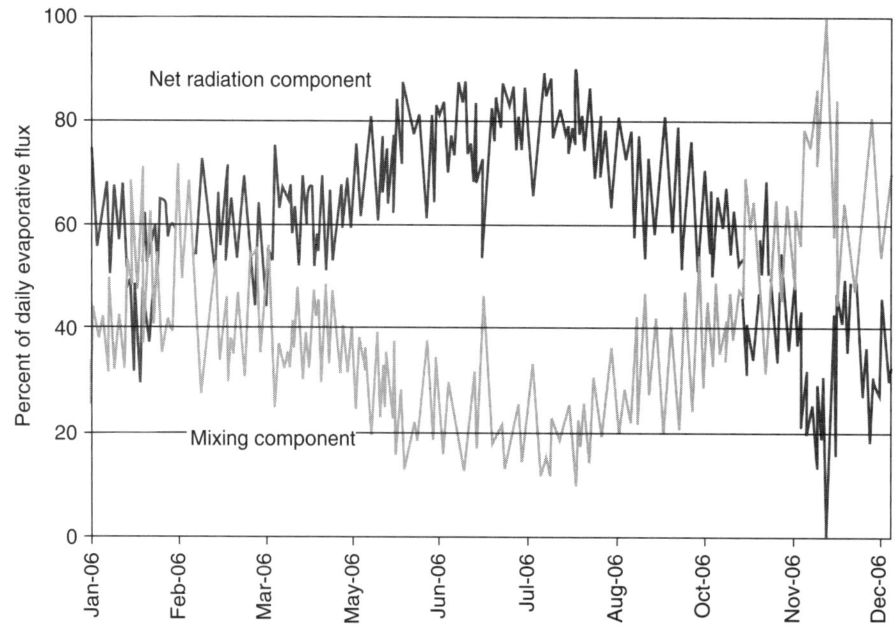

Figure 6.28 Breakdown between net radiation and mixing components of PE for the year 2006 at a site in northern Canada.

period from January to February and from November to December. The mixing component comprised 70–100% of total evaporation during these periods of the year. However, it must also be noted that the total daily evaporation was relatively small during these months.

During the months from April to October, the net radiation component dominates potential evaporation. Net radiation accounted for 65–85% of total potential evaporation. During the months from February to April and October to November the net radiation and mixing components of evaporation are similar with each component fluctuating between 40 and 60%.

When the total amount of potential evaporation is taken into consideration, it is the net radiation that is the primary contributor to total moisture flux. For the 13-year period analyzed in Canada, net radiation contributed to most of the total evaporative flux.

6.3.10 Calculation of Saturation Vapor Pressure

Several empirical equations have been proposed relating temperature to saturation vapor pressure. Only two equations are presented below: the Tetens (1930) equation and the Lowe (1977) equation. The slope of the saturation vapor pressure versus temperature curve is also computed since it is required as the Γ variable when solving the Penman (1948) equation.

The Tetens (1930) equation for saturation vapor pressure as a function of temperature is as follows:

$$u_{v0}^{air} = 0.6108 \exp\left[\frac{17.27T_a}{T_a + 237.3}\right] \quad (6.26)$$

where:

u_{v0}^{air} = saturation vapor pressure, kPa, and
T_a = air temperature, °C.

The Lowe (1977) equation for calculating saturation vapor pressure as a function of temperature is written as follows:

$$u_{v0}^{air} = a_0 + a_1T_a + a_2T_a^2 + a_3T_a^3 + a_4T_a^4 + a_5T_a^5 \quad (6.27)$$

where:

u_{v0}^{air} = saturation vapor pressure, kPa,
T_a = air temperature, °C
a_0 = 0.6183580754,
a_1 = 0.0411427320,
a_2 = 0.0017217473,
a_3 = 0.0000174108,
a_4 = 0.0000003985, and
a_5 = 0.0000000022.

The saturation vapor pressures calculated using the above equations are shown in Table 6.4 and graphically presented in Fig. 6.29.

Table 6.4 Values of Saturated Vapor Pressure and Slope of Vapor Pressure versus Temperature

	Tetens (1930)		Lowe (1977)	
Temperature °C	Saturation Vapor Pressure, kPa	Slope Vapor Pressure Γ, kPa/°C	Saturation Vapor Pressure, kPa	Slope Vapor Pressure Γ, kPa/°C
0	0.6108	0.04445	0.6184	0.04114
4	0.8133	0.05724	0.8117	0.05586
8	1.0728	0.07306	1.0683	0.07289
12	1.4026	0.09248	1.3989	0.09297
16	1.8183	0.11614	1.8171	0.11686
20	2.3383	0.14474	2.3400	0.14542
24	2.9839	0.17909	2.9879	0.17956
28	3.7799	0.22008	3.7852	0.22026
32	4.7548	0.26868	4.7602	0.26859
36	5.9410	0.32595	5.9456	0.32565
40	7.3756	0.39307	7.3786	0.39263
44	9.1005	0.47130	9.1015	0.47079
48	11.1628	0.56201	11.1616	0.56145
52	13.6153	0.66666	13.6116	0.66600
56	16.5170	0.78683	16.5100	0.78589
60	19.9331	0.92418	19.9212	0.92265
64	23.9360	1.08050	23.9159	1.07788
68	28.6053	1.25766	28.5711	1.25322
72	34.0282	1.45764	33.9708	1.45042
76	40.3001	1.68251	40.2059	1.67125
80	47.5249	1.93443	47.3748	1.91760
84	55.8152	2.21566	55.5833	2.19138
88	65.2931	2.52855	64.9451	2.49459
92	76.0899	2.87551	75.5820	2.82930
96	88.3470	3.25906	87.6243	3.19764
100	102.2157	3.68177	101.2109	3.60182

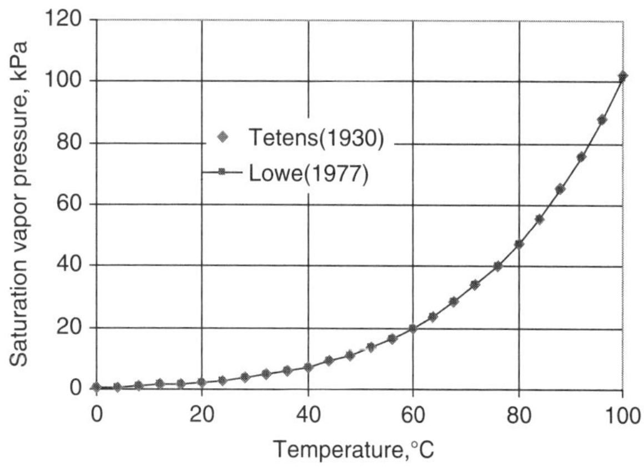

Figure 6.29 Curve of saturation vapor pressure versus temperature.

6.3.11 Calculation of Slope of Saturation Vapor Pressure versus Temperature Curve, Γ

The Tetens (1930) equation for calculating the slope of the saturation vapor pressure versus temperature curve is as follows:

$$\Gamma = \frac{4098 u_{v0}^{air}}{(T_a + 237.3)^2} \tag{6.28}$$

where:

Γ = slope of the saturation vapor pressure versus temperature, kPa/°C,
T_a = air temperature, °C, and
u_{v0}^{air} = saturation vapor pressure, kPa.

The Tetens (1930) equation (i.e., Eq. 6.28) can also be used with the saturation vapor pressures calculated using the Lowe (1977) equation (i.e., Eq. 6.27) to calculate the slope of the saturation vapor pressure versus temperature curve. The results from the Tetens (1930) equation and the Lowe (1977) are essentially the same as shown in Fig. 6.30.

6.3.12 Calculation of Psychrometric Constant, η

The psychrometric constant η is not truly a constant since it varies slightly with elevation but is generally taken as a constant in the equation relating the standard wet and dry bulb hygrometer equation. The psychrometric constant can be calculated as follows:

$$\eta = \frac{C_a \bar{u}_a}{\varepsilon L_v} \tag{6.29}$$

where:

η = psychrometric constant, kPa/°C,
C_a = specific heat of moist air equal to 1.013 kJ / (kg °C),
$\bar{u}_a$ = absolute atmospheric pressure equal to $(u_{atm} + u_a)$, where u_{atm} is standard pressure (sea level) at 101.3 kPa,

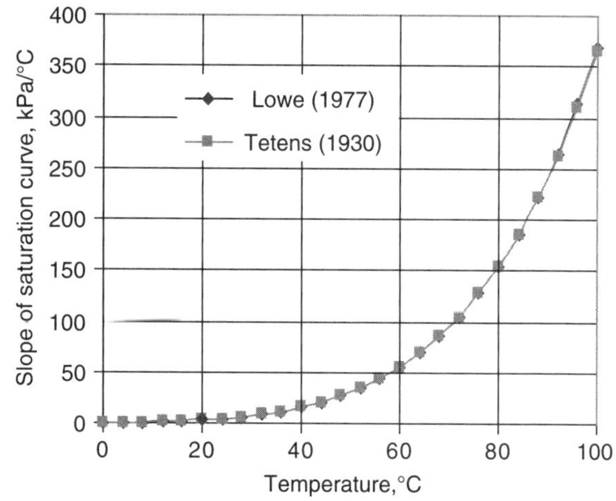

Figure 6.30 Slope of saturation vapor pressure versus temperature relationship.

ε = ratio of molecular weight of water vapor to dry air equal to 0.622, and
L_v = latent heat of vaporization, MJ/kg.

At standard atmospheric pressure conditions (i.e., 101.3 kPa) and a temperature of 20°C, the psychrometric constant is equal to 0.06733 kPa/°C.

6.3.13 Calculation of Latent Heat of Vaporization, L_v

Latent heat of vaporization, L_v, varies with temperature in accordance with the following equation (Harrison, 1963):

$$L_v = 2.501 - 0.002361 T_a \tag{6.30}$$

where:

L_v = latent heat of vaporization, MJ/kg, and
T_a = air temperature, °C.

The value of latent heat varies only slightly over normal temperature ranges. A value of 2.45 MJ/kg, corresponding to 20°C, is often used.

6.3.14 Calculation of Atmospheric Pressure, $\bar{u}_a$

Absolute atmospheric pressure varies with elevation and temperature in accordance with the following equation (Burman et al., 1987):

$$\bar{u}_a = u_{atm} \left[\frac{T_{K0} - \alpha_1 (z - z_0)}{T_{K0}} \right]^{g/(R\alpha_1)} \tag{6.31}$$

where:

$\bar{u}_a$ = absolute atmospheric pressure at elevation, z, kPa,
u_{atm} = atmospheric pressure at sea level, taken as 101.3 kPa,

z = elevation, m,
z_0 = reference elevation (usually at sea level), m,
g = gravitational acceleration, 9.807 m/s^2,
R = specific gas constant, 287 J/(kg K),
α_1 = constant lapse rate for moist air, 0.0065 K/m,
T_{K0} = reference temperature, K, at reference elevation (i.e., $T_{K0} = 273.16 + T_a$), and
T_a = mean air temperature for the time period, °C.

Assuming that $u_{\text{atm}} = 101.3$ kPa at $z_0 = 0.0$ and $T_{K0} = 293$ K (i.e., corresponding to a temperature of 20°C), Eq. 6.31 can be written as follows:

$$\bar{u}_a = 101.3 \left[\frac{293 - 0.0065z}{293} \right]^{5.26} \tag{6.32}$$

At an elevation of 400 m the calculated absolute atmospheric pressure is 96.65 kPa. The refinement shown in Eq. 6.31 is not always warranted when calculating potential evaporation.

6.3.15 Determination of Net Radiation

Net radiation is of primary importance when computing evaporation from water, soil, and plant surfaces. Net radiation is the driving force for many natural processes (e.g., evaporation, evapotranspiration, air fluxes, and photosynthesis). Net radiation is the difference between the incoming solar radiation and outgoing radiation from the sun and can be viewed as a measure of the energy available at the ground surface (Irmak et al., 2003). The temperature of the earth surface increases when the surface of the earth absorbs more energy than it loses, and vice versa.

Figure 6.31 illustrates the energy balance of the earth-atmosphere system. The components of radiation associated with the incoming short-wave solar radiation as well as the components associated with the terrestrial long-wave radiation are shown. Net radiation R_a is equal to the difference between the incoming short-wave radiation R_{ns} and the outgoing net long-wave radiation R_{nl}.

Net radiation is equal to the solar radiation absorbed plus the terrestrial radiation absorbed minus the terrestrial radiation emitted. For example, if the solar radiation absorbed by the earth surface was 600 W/m^2, the terrestrial radiation absorbed was 80 W/m^2, and the terrestrial net radiation emitted by the earth's surface was 300 W/m^2, then the net radiation would be 380 W/m^2 (i.e., 600–300 + 80 = 380). The daily cycle of net radiation is the primary factor determining the normal daily temperature cycle. When net radiation is positive, the temperature usually increases, and when net radiation is negative, the temperature decreases, as shown in Fig. 6.32.

Figure 6.32 shows that at early morning (e.g., 2:30 a.m.) the terrestrial radiation emitted by the surface exceeds the terrestrial radiation emitted to the surface by the atmosphere, and as a result, the earth's surface temperature

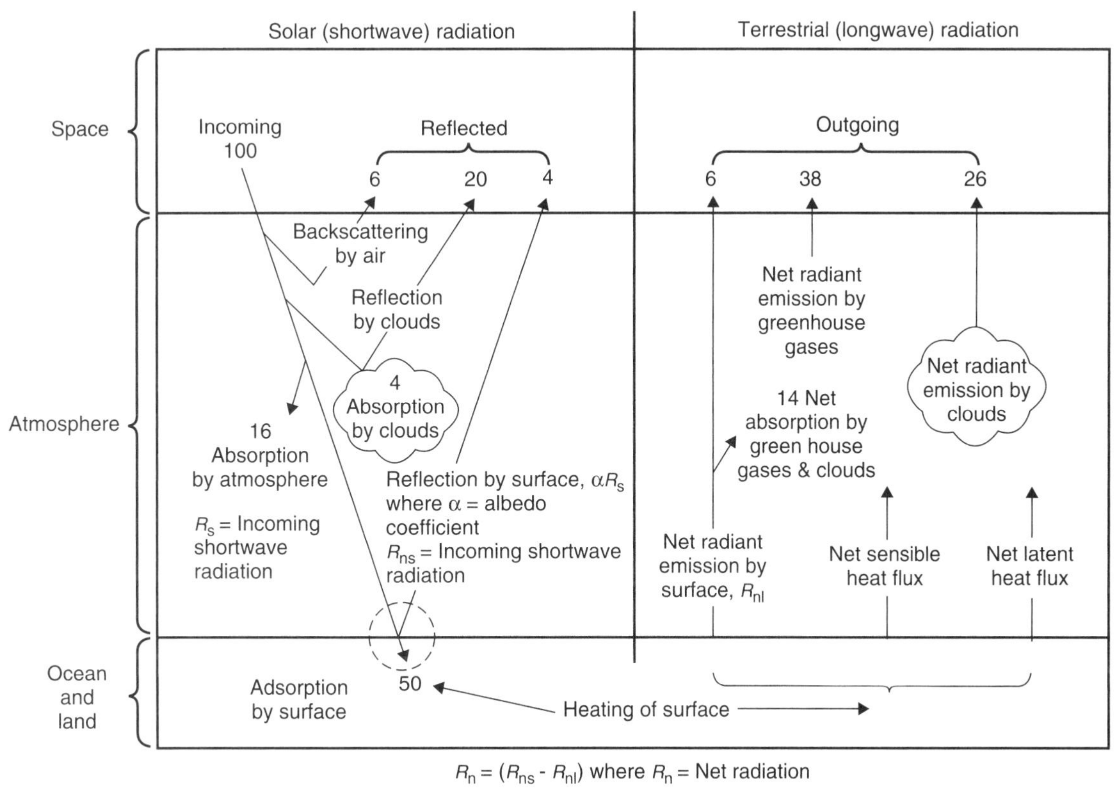

Figure 6.31 Energy balance of earth-atmosphere system.

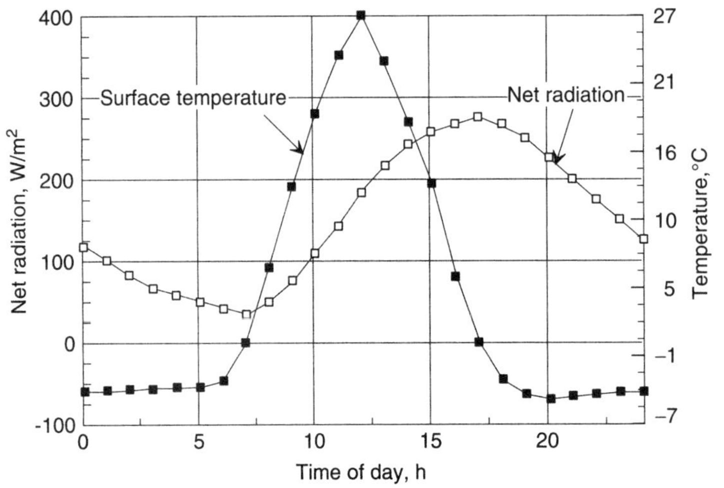

Figure 6.32 Example of relationship between variation in net radiation and ground surface temperature.

decreases. Around sunrise (e.g., 6:00 a.m.), solar radiation starts to reach the ground surface and the net radiation begins to increase. However, the net radiation is still negative and so the temperature continues to decrease. At some point after sunrise (e.g., 7:00 a.m.), enough solar radiation reaches the ground that the net radiation becomes zero. This is the time when the minimum temperature occurs. As the sun moves higher in the sky (e.g., 10:00 a.m.), increasing solar radiation reaches the ground surface, the net radiation becomes positive, and the temperature increases.

Solar radiation peaks at around noon when the sun is at its highest point in the sky. This is also the time when net radiation is a maximum value. After noon the sun starts to move lower in the sky and the net radiation then decreases. However, as long as the net radiation remains positive, temperature at the ground surface will continue to increase. Net radiation will again be zero as the sun gets lower in the sky (e.g., 5:00 p.m.). At this point the temperature stops increasing and the maximum temperature occurs. Once again, net radiation goes negative and temperature starts to decrease. This illustrates how net radiation changes throughout a typical day.

The integration of net radiation throughout a day gives the average daily net radiation in terms of joules per square meter per day. The average daily net radiation has a significant role in determining the pattern of ground surface temperatures. Latent energy transfer also affects the ground surface temperature. Energy is required for the evaporation of water and the melting of snow. Energy that is absorbed during the phase change of water is not available to increase the ground (or water) surface temperature.

The daily net radiation that reaches ground surface constitutes primary information that is required for calculating potential and actual evaporation. It is preferable to measure net radiation in the field at a weather station by measuring the components of radiation. There are a number of instruments that can be used to quantify net radiation.

One of the procedures that can be used to obtain net radiation involves the measurement of the Bowen ratio B along with the total radiation reaching ground surface. The Bowen ratio is defined as the ratio of the sensible heat Q_h and the latent heat Q_e:

$$B = \frac{Q_h}{Q_e} \tag{6.33}$$

When the Bowen ratio B is less than 1.0, it means that a greater proportion of the available energy at the ground surface is passing to the atmosphere as latent heat than as sensible heat. When latent heating approaches zero, as is the case in arid regions, the Bowen ratio is a poor choice of instrument for determining net radiation. For arid regions it is better to compute the *evaporative fraction* E_vF, which can be written as

$$E_vF = \frac{Q_e}{Q_e + Q_h} \tag{6.34}$$

6.3.16 Estimation of Net Radiation

Limited attention seems to have been historically given to making routine measurements of net radiation when measuring weather-related variables. Lack of measurements of net radiation is mainly due to financial reasons. There are also theoretical reasons why other approaches have been used as an indirect means of obtaining values for net radiation. Net radiation can be estimated based on a number of meteorological parameters such as total solar radiation R_s, air temperature, and other variables, including relative humidity and extraterrestrial radiation R_a (Reddy, 1971; Wright, 1982; Dong et al., 1992). Several procedures have been proposed for the determination of net radiation and a few methodologies are mentioned below.

6.3.16.1 Use of Penman (1948) Equation

The Penman (1948) equation can be used to calculate net radiation Q_n for a soil-atmosphere heat budget. Net radiation can be written in terms of a flux designated as millimeters per day:

$$Q_n = (1 - r) R_c - \sigma_s T_a^4 \left(0.56 - 0.092 \left(u_v^{\text{air}}\right)^{0.5}\right) \left(0.10 + \frac{0.90n}{N}\right) \tag{6.35}$$

where:

Q_n = net radiation budget, mm/day,
r = reflection coefficient (unitless),
T_a = air temperature, °C,
R_c = short-wave radiation equal to $R_a(0.18 + 0.55$ n/N), mm/day,
R_a = solar radiation, MJ/m²/day,
σ_s = Stefan-Boltzmann constant, W/m²/K⁴,
u_v^{air} = vapor pressure in the air above ground surface, kPa, and
n/N = sunshine ratio (i.e., actual/possible hours of bright sunshine).

Net radiation values calculated using the Penman (1948) equation have been found to be greater than measured net radiation values.

6.3.16.2 FAO56-Penman-Monteith Method

The International Commission for Irrigation and Drainage and the Food and Agriculture Organization of the United Nations have recommended that the Penman-Monteith (FAO56-PM) method be used for the calculation of PE. Allen et al. (1998) outlined a procedure for the determination of net radiation R_a required for the FAO-56 method. Net radiation was calculated as follows:

$$R_n = R_{ns} - R_{nl} \tag{6.36}$$

where:

R_n = net radiation, J/m²/d,
R_{ns} = incoming net short-wave radiation, J/m²/d, and
R_{nl} = outgoing net long-wave radiation, J/m²/d.

The incoming net short-wave radiation R_{ns} is a result of the difference between incoming and reflected solar radiation and can be written as follows:

$$R_{ns} = R_s - \alpha_4 R_s \tag{6.37}$$

where:

α_4 = albedo or canopy reflection coefficient (see Table 6.5 for typical values of α_4) and
R_s = total incoming solar radiation, J/m²/d.

Table 6.5 Albedo Coefficients for Water and Various Earth Materials

Surface	Conditions	Albedo, α_4
Clouds	Low overcast, 100 m thick	0.40
	Low overcast, 200 m thick	0.50
	Low overcast, 500 m thick	0.70
Liquid water	Smooth, solar angle 60°	0.05
	Smooth, solar angle 30°	0.10
	Smooth, solar angle 10°	0.35
	Smooth, solar angle 5°	0.60
	Wavy, solar angle 60°	0.10
Solid water	Fresh snow, low density	0.85
	Fresh snow, high density	0.65
	Old snow, clean	0.55
	Old snow, dirty	0.45
	Glacier ice, clean	0.35
	Glacier ice, dirty	0.25
Sand	Dry, light; high sun	0.35
	Dry, light; low sun	0.60
	Gray, wet	0.10
	Gray, dry	0.20
	White, wet	0.25
	White, dry	0.35
Soil	Organic, dark	0.10
	Clay	0.20
	Sandy, light	0.30
Grass	Typical fields	0.20
	Dead, wet	0.20
	Dead, dry	0.30
Tundra		0.15
Crops	Cereals	0.25
	Cotton, potato, tomato	0.20
	Sugar cane	0.15
Trees	Rain forest	0.15
	Eucalyptus	0.20
	Red pine forest	0.10
	Mixed hardwoods in leaf	0.18

Source: Data from Lee (1980).

The albedo coefficient is a measure of how strongly a material reflects a light source such as the sun. Albedo (or whiteness) is defined as a dimensionless ratio of the total reflected radiation to incident radiation. The average albedo for the earth's surface is about 0.30.

The outgoing net long-wave radiation R_{nl} is proportional to the absolute temperature of the surface raised to the fourth power and can be expressed as follows:

$$R_{nl} = \sigma_s T_{aK}^4 \left(0.56 - 0.092 \left(u_v^{\text{air}}\right)^{0.5}\right) \left(0.10 + \frac{0.90n}{N}\right) \tag{6.38}$$

where:

σ_s = Stefan-Boltzmann constant (4.903×10^{-9} MJK^{-4} m^{-2} d^{-1}),
T_{aK} = average absolute daily temperature [i.e. $(T_{\max,K} + T_{\min,K})/2$]
$T_{\max,K}$ = daily maximum absolute temperature [K = °C + 273.16),
$T_{\min,K}$ = daily minimum absolute temperature (K = °C + 273.16), and
u_v^{air} = actual vapor pressure, kPa.

Daily values for the clear sky radiation, R_{s0}, are a function of elevation, $z(m)$, and the extraterrestrial radiation, R_a (Doorenbos and Pruitt, 1977):

$$R_{s0} = (0.75 + 0.00002)\, R_a \tag{6.39}$$

where:

R_a = extraterrestrial radiation, MJ/m^2/d.

Extraterrestrial radiation can be calculated on a daily basis as a function of the day of the year, the solar constant, the solar inclination, and the latitude in accordance with the following empirical equation:

$$R_a = \frac{1440}{\pi} G_{sc} d_r \left[\omega_s \sin\phi \sin\delta + \cos\phi \cos\delta \sin\omega_s\right] \tag{6.40}$$

where:

G_{sc} = solar constant (i.e., 0.0820 MJ/m^2/min),
ϕ = latitude [rad, where 1 rad = $\pi/180$ × (decimal degrees)],
d_r = inverse relative distance from the earth to sun,
ω_s = sunset hour angle, rad, and
δ = solar declination, rad.

The *inverse relative distance* can be computed as

$$d_r = 1 + 0.033 \cos\left(\frac{2\pi}{365} D_j\right) \tag{6.41}$$

The solar inclination δ can be computed as

$$\delta = 0.409 \sin\left(\frac{2\pi}{365} D_j - 1.39\right) \tag{6.42}$$

where:

D_j = day of the year.

The sunset hour angle can be computed as follows:

$$\omega_s = \arccos\left[-\tan(\phi)\tan(\delta)\right] \tag{6.43}$$

Equations 6.36–6.43 provide information for the solution of the Penman-Monteith (FAO56-PM) method. The above procedure for computing net radiation is quite demanding and other simpler equations have been proposed for the estimation of net radiation. Two of the simpler empirical equations that can be used for calculating net radiation are shown in the following sections.

6.3.16.3 Equation 1 from Irmak et al. (2003) for Approximating Net Radiation

Irmak et al. (2003) proposed an empirical equation for the estimation of net radiation that is based on a minimal amount of climate data. The data required are site latitude, maximum air temperature, minimum air temperature, and mean relative humidity. It was proposed that net radiation could be estimated using the following equation in cases where solar radiation is measured at the site:

$$R_n = -0.054T_{\max} + 0.101T_{\min} + 0.462R_{S,\text{measured}} - 49.243d_r + 50.831 \tag{6.44}$$

where:

R_n = net radiation, MJ/m^2/day,
$T_{\max}$ = maximum daily air temperature, °C,
$T_{\min}$ = minimum daily air temperature, °C,
$R_{S,\text{measured}}$ = measured solar radiation, MJ/m^2/day, and
d_r = inverse relative distance from the earth to the sun, ranges from 0.967 to 1.033.

6.3.16.4 Equation 2 from Irmak et al. (2003) for Approximating Net Radiation

Net radiation R_n can also be estimated using the following equation when measured data are not available:

$$R_n = -0.09T_{\max} + 0.203T_{\min} - 0.101RH_{\text{mean}} + 0.687R_{S,\text{predicted}} + 3.97 \tag{6.45}$$

where:

$R_{S,\text{predicted}}$ = predicted solar radiation, MJ/m^2/day.

The solar radiation can be predicted as follows:

$$R_{S,\text{predicted}} = (K_t)\, R_a \left(T_{\max} - T_{\min}\right)^{0.5} \tag{6.46}$$

where:

K_t = empirical coefficient equal to 0.162 for interior regions and 0.190 for coastal regions and
R_a = extraterestial radiation, MJ/m^2/day.

Samani (2000) used average monthly data for each year to obtain an indication of the empirical coefficient K_t. Data were taken from 65 weather stations located between 7° and 50° North latitude in the United States. The following relationship between temperature difference during each day $(T_{max} - T_{min})$ and the empirical coefficient K_t was found to be related by the equation

$$K_t = 0.00185\left(T_{max} - T_{min}\right)^2 - 0.0433\left(T_{max} - T_{min}\right) + 0.4023 \tag{6.47}$$

6.3.17 Relative Humidity of Air above Water Surface

The vapor pressure in the air above a saturated soil surface is required in the calculation of the net radiation term and the mixing term. The relative humidity in the air above a saturated soil surface provides a measure of the vapor pressure in the air:

$$h_r = \frac{u_v^{air}}{u_{v0}^{air}} \tag{6.48}$$

or

$$u_v^{air} = h_r u_{v0}^{air} \tag{6.49}$$

where:

h_r = relative humidity of the air above the saturated ground (or water) surface,
u_{v0}^{air} = saturated vapor pressure, kPa, and
u_v^{air} = vapor pressure in the air above ground surface, kPa.

The above equations use the saturated vapor pressure computed from the measured air temperature while the actual vapor pressure in the air is computed from the measured relative humidity of the air above the ground surface.

Other assumptions can be made when solving the Penman equation for potential evaporation. For example, it is sometimes assumed that the vapor pressure is the same in the air and at the saturated soil surface (e.g., Priestley-Taylor, Eq. 6.23). When this assumption is made, the calculated potential evaporation is typically reduced in drier climate regimes.

The Penman (1948) equation has been widely used in geotechnical engineering, and as a result it is important to understand the role played by each of the variables in the equation. It has been noted that the Penman equation generally overestimates the evaporation rate for conditions of high wind and low humidity.

6.3.18 Comparison of Potential Evaporation Calculations

The Thornthwaite (1948) equation and the Penman (1948) equation rely on data collected at a weather station; however, the Penman formulation is considered to be a more rigorous formulation for PE. The Thornthwaite equation depends on average monthly temperatures whereas the Penman equation depends on daily temperatures and other weather-related factors.

Thornthwaite (1948) incorporated the variables of length of daylight hours, mean monthly temperature, and an empirical constant into the prediction of PE. Penman (1948) utilized daily records of relative humidity, air temperature, wind speed, and net radiation for the calculation of PE. The vapor pressure gradient between the water surface and the air above the water becomes the primary driving mechanism for evaporation in the Penman equation. Two components drive evaporation: the *net radiation* component that characterizes the power of the sun to evaporate water and the *mixing* component associated with the power of wind to remove vapor from the ground surface.

The Penman equation would intuitively appear to have a distinct advantage over the Thornthwaite equation in predicting PE. A comparison of the two calculation methods provides an indication of the reliability of the calculation methods.

Figures 6.33 and 6.34 present daily PE calculations for two years (i.e., 2006 and 2007) at a site in northern Canada (latitude of approximately 50°). The daily PE calculations vary considerably from day to day when using the Penman (1948) equation. This is as anticipated since temperature, relative humidity, and wind speed can vary quite widely on a daily basis.

The Thornthwaite (1948) equation simply uses the average monthly temperature when calculating PE. Daily PE is a single value for each month of the year when using the Thornthwaite equation. The daily PE calculations can be accumulated over the entire year in order to determine whether the monthly averaging of temperature is a reasonable assumption to make when calculating PE.

The Thornthwaite (1948) equation is not defined when the average monthly temperature is less than 0°C. The Thornthwaite calculations show that the winter period (i.e., an inactive period), is between November 1 and April 1 of each year at this site. The Penman (1948) equation does not restrict the calculation of PE to the same period. As much as 25% of PE may occur outside the period between November 1 and April 1 in each year.

Figures 6.33 and 6.34 also show that the PE calculations from the Thornthwaite equation lag behind the Penman calculations for the period from spring to midsummer. The reverse is true for the period from midsummer to the fall of each year.

Figures 6.35 and 6.36 show the accumulated PE calculations for two years (i.e., 2006 and 2007) at the site in northern Canada. The cumulative precipitation for each year is also shown. Cumulative precipitation values illustrate the relationship between PE and precipitation. The cumulative PE at the end of the year is quite similar for the two methods of calculation. However, the difference between the results

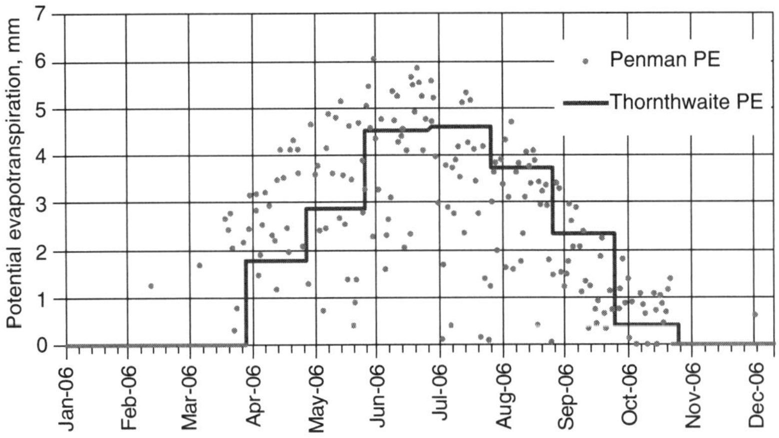

Figure 6.33 PE calculation for 2006 using Penman and Thornthwaite equations.

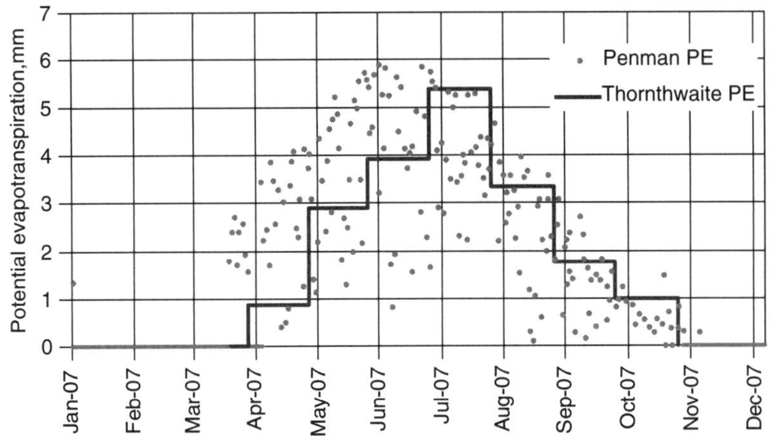

Figure 6.34 PE calculation for 2007 using Penman and Thornthwaite equations.

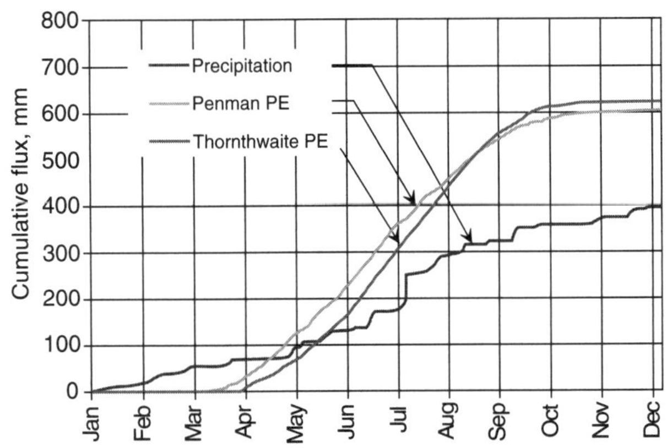

Figure 6.35 Cumulative PE calculation for 2006 using Penman and Thornthwaite equations.

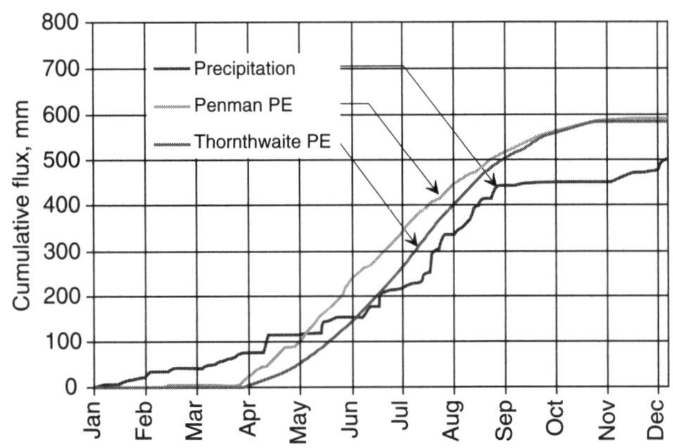

Figure 6.36 Cumulative PE calculation for 2007 using Penman and Thornthwaite equations.

obtained from the two equations may differ from one year to the next, as shown in Table 6.6.

The weather data at this site in northern Canada were analyzed for a period of 14 years in order to see if there was a consistent trend between the Penman equation calculations and the Thornthwaite equation calculations of PE. Table 6.6 presents the cumulative PE results. The Penman equation showed an average PE over the 14-year period of 652 mm. The Thornthwaite equation showed an average PE of 577 mm. This means that on average the PE calculated

Table 6.6 Comparison of Calculated PE over a 14-Year Period at Site in Northern Canada

	Penman (1948)		Thornthwaite (1948)	
Year	Cumulative PE (mm)	Percent Variation	Cumulative PE (mm)	Percent Variation
1994	707.5	8.54	597.4	3.60
1995	635.0	−2.59	556.4	−3.51
1996	644.9	−1.06	545.9	−5.32
1997	666.6	2.27	565.2	−1.98
1998	739.6	13.46	638.8	10.79
1999	624.2	−4.25	591.2	2.54
2000	677.0	3.86	551.8	−4.30
2001	709.1	8.78	593.1	2.87
2002	666.3	2.21	522.2	−9.44
2003	688.3	5.59	583.6	1.22
2004	583.0	−10.57	536.3	−6.98
2005	591.5	−9.26	582.6	1.04
2006	602.7	−7.53	622.3	7.93
2007	590.2	−9.45	585.5	1.55
Average	651.9		576.6	
Standard deviation	49.7		32.5	

using the Penman equation was about 12% higher than the results obtained using the Thornthwaite (1948) equation. There was only one year in the 14-year record when the Thornthwaite equation gave PE values that were higher than those calculated by the Penman (1948) equation.

Table 6.6 also shows the standard deviation of the calculated cumulative PE. The standard deviation was 50 mm for the Penman method and 33 mm for the Thornthwaite method. It would appear that the Penman method better captures the effect of the individual components of weather and would be the preferred calculation procedure for use in engineering practice.

The engineer requires the AE to quantify water balance at ground surface (i.e., in addition to potential evaporation). Actual evaporation can be calculated by modifying the Penman (1948) equation and taking the underlying soil conditions into consideration. Wilson (1990) made modifications to the Penman equation and developed the Wilson-Penman method (Wilson, 1990) for calculating actual evaporation. Two other procedures were also proposed for the calculation of actual evaporation: the *limiting function* procedure (Wilson et al., 1997a) and the *experimental-based* procedure (Wilson et al., 1997a).

6.3.19 Actual Evaporative Flux

The AE from a soil surface can be considerably less than PE. The quantification of AE at a particular site allows the geotechnical engineer to better establish water balance conditions at the ground surface.

The concepts associated with AE can be described as follows. The net radiation effect of the sun and the mixing effect of wind form the basis for PE from a wet soil surface. At the same time as the sun and the wind are removing water from the soil, the soil begins to dry and hold onto the water more strongly. In other words, the soil suction increases as the soil dries. The net result is a struggle between the climate and the soil and the consequence is a reduced evaporation from the ground surface (i.e., AE). If the vapor pressure in the air above ground surface and the vapor pressure in the soil at ground surface are the same, then evaporation from the ground surface will cease due to lack of a vapor pressure gradient. It is the "total suction" in the soil at the ground surface (i.e., total suction = matric suction + osmotic suction) that creates the equilibrium vapor pressure (or relative humidity) in the soil at the ground surface.

The AE from a soil surface might be considerably less than the PE. The AE equations are primarily the outcome of research by Wilson (1990), who used evaporation from thin soil layers and sand column drying tests to verify the fundamental physical relationships required to extend the Penman (1948) equation to the calculation of the AE from drying soil surfaces.

Several methodologies have been proposed for calculating AE. Two of the proposed methods have a thermodynamic basis while one has an experimental basis. The equations for AE along with different temperatures above the soil surface and at the soil surface give rise to at least six possible methodologies for the calculation of AE from the soil surface.

The proposed methodologies differ mainly in the assumptions related to the determination of air and soil temperatures. Differences between the soil surface temperature and the air temperature give rise to various "coupled" and "uncoupled" moisture and heat flow formulations. The temperature of the soil and the air generally differ; however, the effect of the temperature difference does not appear to have been fully studied.

Figure 6.37 shows the temperature profiles measured near the surface of a soil layer that was being dried by evaporation. Figure 6.37a shows that the air temperature coming into the chamber above the soil had an average temperature of about 23°C. Immediately above the soil the average temperature was about 21°C. However, directly on top of the soil layer the average measured temperature dropped to about 17.5°C. At 50 mm below the soil surface the average soil temperature was about 19°C. The results show the cooling effect associated with evaporation and illustrate why the heat and moisture partial differential equations should be coupled to represent the physical processes associated with evaporation.

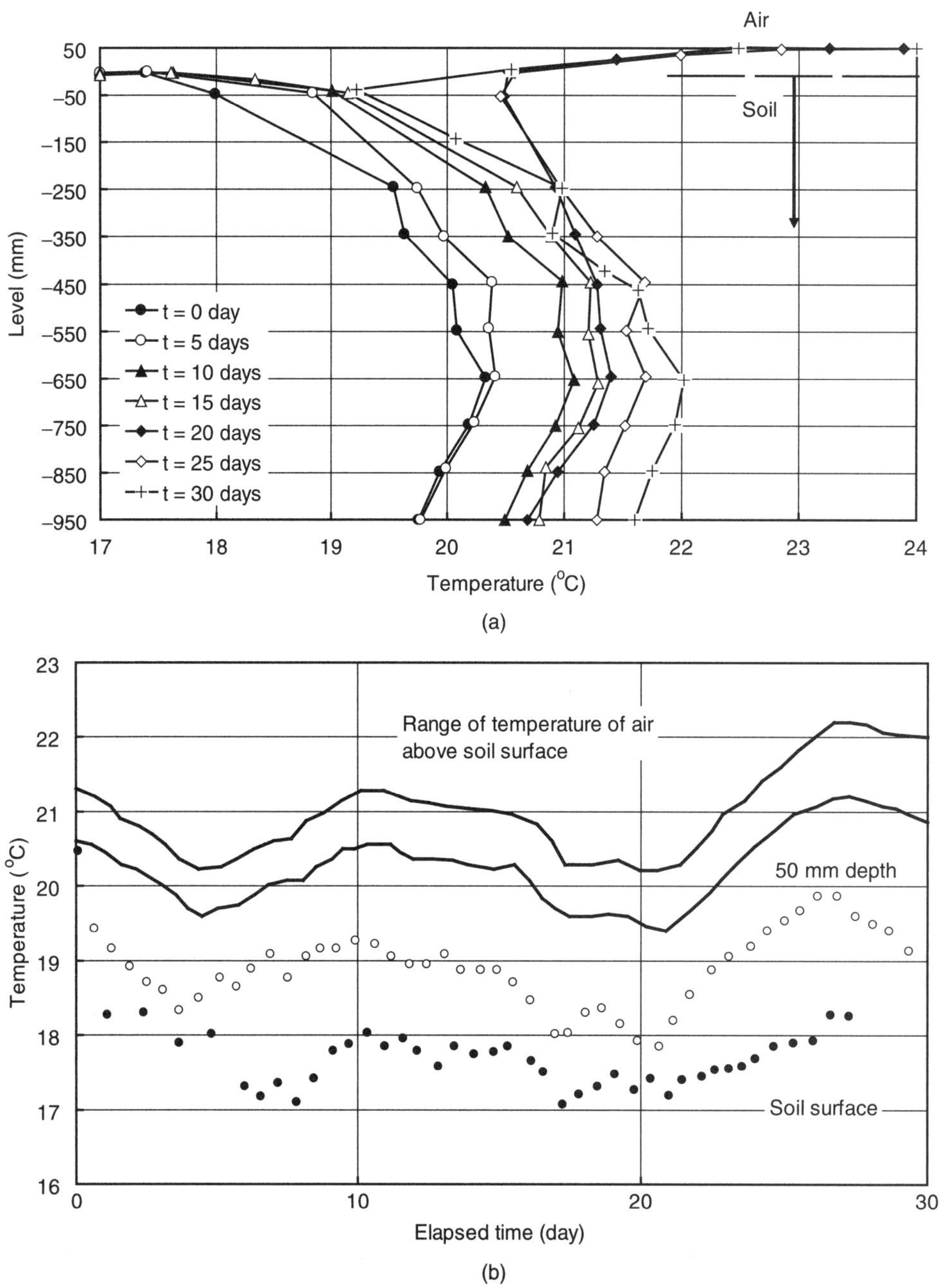

Figure 6.37 (a) Temperature versus depth throughout soil layer. (b) Temperature changes resulting from evaporation of water from soil surface (Ta et al., 2010).

The term coupled, associated with the calculation of AE, means that the moisture flow partial differential equation is solved simultaneously with the heat flow partial differential equation. If an assumption is made regarding the relationship between the soil temperature and the air temperature, then the analysis for AE reduces to the solution of the moisture partial differential equation and the solution is referred to as an uncoupled analysis. The analysis is also referred to as an uncoupled analysis if a closed-form, empirical relationship is used to designate the relationship between the soil and air temperature.

Three methods are also considered for relating the AE to the PE: (i) the Wilson-Penman method, (Wilson, 1990), for calculating AE (ii) the limiting function method (Wilson et al., 1997a) and (iii) the experimental-based method (Wilson et al., 1997a). The three relationships between AE and PE along with the possibility of using a coupled or uncoupled solution give rise to six procedures for the assessment of AE. These calculation procedures are outlined in the following sections.

6.3.19.1 Wilson-Penman Equation for Computing AE (Coupled Solution)

Wilson et al., (1994) proposed that the Penman (1948) equation for PE be modified and then used to calculate AE. The modified equation, which became known as the Wilson-Penman equation, (Wilson et al., 1994) took into consideration the difference in temperature and relative humidity (and therefore vapor pressure) between the soil surface and the overlying air. The difference in vapor pressure conditions between the air and the water at ground surface formed the basis for the development of the *soil-atmospheric model* implemented into the SoilCover (MEND, 1993) and SVFlux computer code (M.D. Fredlund, 2001).

The following assumptions are made when combining moisture flow and thermal flow:

- Moisture flow and vapor flow beneath the soil surface are governed by hydraulic head gradients, vapor pressure gradients, and temperature gradients.
- Heat transfer beneath the soil surface (i.e., ground thermal flux in the equation) is mainly governed by thermal conduction. Heat transfer by thermal convection can be included, but it can generally be neglected for most applications.
- Soil freezing/thawing processes can be considered when the soil temperature goes below the freezing point. For frozen soils there must be a reduction in the hydraulic conductivity because of the existence of ice in the pores of the soil.
- The latent heat due to phase change, including evaporation and freezing/thawing, can be considered in the heat transfer equation beneath the soil surface.
- The soil surface temperature can be different from the air temperature and is determined in accordance with the heat energy balance equation (Eq. 6.8).
- Actual evaporation is calculated using the Wilson-Penman equation, (Wilson et al., 1994).

Partial Differential Equation Governing Moisture Flow. Liquid water and vapor flow through an unsaturated soil is presented in detail in Chapter 7. Equation 6.50 is a simplified form for liquid and vapor flow and can be used to solve for AE (Jame, 1977; Wilson, 1990; Gitirana, 2005; M.D. Fredlund et al., 2011):

$$\frac{\partial}{\partial y}\left[k_{y11}\frac{\partial\left(u_w\right)}{\partial y}+k_{wy}+k_{y12}\frac{\partial T}{\partial y}\right]+S_{sk}=\frac{\partial\theta_u}{\partial t}+\frac{\rho_i}{\rho_w}\frac{\partial\theta_i}{\partial t} \tag{6.50}$$

where

$$k_{y11}=\frac{k_{wy}+k_{vh}}{\gamma_w} \tag{6.51}$$

$$k_{y12}=k_{vT} \tag{6.52}$$

and

k_{wy} = hydraulic conductivity, m/s,
k_{vh} = water vapor conductivity by vapor diffusion within the air phase, m/s,
k_{vT} = water vapor conductivity due to temperature gradient, m/s,
u_w = pore-water pressure, a component of hydraulic head, kPa,
T = soil temperature, °C,
t = time, s,
y = vertical direction or elevation, m,
γ_w = unit weight of water, kN/m^3,
θ_u = volumetric unfrozen water content in soil, m^3/m^3,
θ_i = volumetric ice content in soil, m^3/m^3,
ρ_w = density of water, kg/m^3,
ρ_i = density of ice, kg/m^3, and
S_{sk} = water sink or source (i.e., water added or water taken away), (m^3/m^3)/t.

Partial Differential Equation Governing Heat Flow. The heat flow partial differential equation is derived in detail in Chapter 10. The following simplified heat flow equations can be used to solve for AE. The heat flow equation includes the soil freezing/thawing process and is based on the derivations by Jame (1977), Wilson (1990), Gitirana (2005), and M.D. Fredlund et al. (2011):

$$\frac{\partial}{\partial y}\left[k_{y21}\frac{\partial\left(u_w\right)}{\partial y}+k_{y22}\frac{\partial T}{\partial y}\right]=\left(C+L_f m_2^i\right)\frac{\partial T}{\partial t} \tag{6.53}$$

where

$$k_{y21}=\frac{L_f k_y^w+Lk_{vh}}{\gamma_w} \tag{6.54}$$

$$k_{y22} = \lambda_y + Lk_{vT} \quad (6.55)$$

$$C = C_s\theta_s + C_w\theta_u + C_v\theta_v + C_i\theta_i + C_a\theta_a \quad (6.56)$$

$$m_2^i = \frac{\partial \theta_u}{\partial T} \quad (6.57)$$

and

λ_y = thermal conductivity, J/s m °C,
C = volumetric heat capacity of soil mixture, J/m^3,
C_s = volumetric heat capacity of solid phase of soil, J/m^3,
C_w = volumetric heat capacity of liquid water of soil, J/m^3,
C_v = volumetric heat capacity of vapor phase of soil, J/m^3,
C_a = volumetric heat capacity of dry air phase of soil, J/m^3,
θ_s = volumetric content solid phase in soil, m^3/m^3,
θ_v = volumetric vapor content in soil, m^3/m^3,
θ_a = volumetric air content in soil, m^3/m^3,
L = volumetric latent heat of water vaporization or condensation, J/m^3, $L = 2.5 \times 10^9$ J/m^3 if $T > T_{ef}$, otherwise $L = 0$,
L_f = volumetric latent heat of water freezing or thawing, J/m^3,
= $L_f = 3.34 \times 10^8$ J/m^3 if $T_{ef} \geqslant T \geqslant T_{ep}$, otherwise $L_f = 0$,
T_{ef} = temperature at soil freezing point, °C,
T_{ep} = temperature at the end of soil phase change during freezing, °C, and
m_2^i = slope of the SFCC (i.e., relationship between unfrozen water content and soil temperature).

The term for pore-water vapor conductivity, k_{vh}, associated with vapor diffusion within the air phase can be expressed in terms of the universal gas law as follows:

$$k_{vh} = \frac{\beta\theta_a g D_v \omega_v u_{sv0}}{\rho_w R\,(273.15 + T)} \quad (6.58)$$

where:

u_{sv0} = saturation vapor pressure at the soil surface, kPa,
ρ_{sv0} = saturation vapor density in soil, kg/m^3,
ω_v = molecular weight of vapor, 0.018016 kg/mol,
β = soil tortuosity,
θ_a = volumetric air content in soil, m^3/m^3,
R = universal gas constant, 8.3144 J/mol/K,
D_v = molecular diffusivity of vapor in soil pore air, m^2/s, and
g = gravitational acceleration, m/s^2.

The vapor diffusion term D_v can be written as a function of temperature:

$$D_v = 2.29 \times 10^{-5}\left(1 + \frac{T}{273.15}\right)^{1.75} \quad (6.59)$$

The soil tortuosity term β is dimensionless and can be expressed in terms of the volume of air in the soil as follows:

$$\beta = (\theta_a)^{2/3} \quad (6.60)$$

The term for water vapor conductivity due to the temperature gradient can also be expressed as follows:

$$k_{vT} = \frac{k_{vh}}{\gamma_w}\left[\frac{\rho_w}{\rho_{sv0}}\left(\frac{\partial u_{sv0}}{\partial T} - \frac{u_{sv0}}{273.15 + T}\right) + \frac{u_a - u_w}{273.15 + T}\right] \quad (6.61)$$

where:

u_a = pore-air pressure of soil, kPa, with pore-air in the soil equal to zero if it is connected to the atmosphere.

The temperature dependence of saturated vapor pressure is omitted in some earlier formulations. Including the term $\partial u_{sv0}/\partial T$ in Eq. 6.61 takes the temperature dependence on saturated vapor pressure into consideration.

Initial Water Content Conditions. An analysis of moisture movement must start by having an initial set of designated values. The initial values may be designated on the basis of (i) the location of the water table and assumed hydrostatic conditions, (ii) equilibrium with a SWCC for the soil, (iii) computed steady-state water contents, (iv) measured water contents in situ, or (v) some other selected procedure.

Initial Temperature Conditions. The soil temperature in the soil domain needs to be initialized to a specific value or an expression, T_0. For example, the average air temperature T_a might be used for the initial soil temperature T:

$$T = T_0 \quad (6.62)$$

Thermal Boundary Condition at Soil Surface. The thermal boundary condition at the ground surface can be specified as a constant temperature or a temperature expression (i.e., Dirchlet condition):

$$T|_{\text{surface}} = T_{\text{soil}} \quad (6.63)$$

Temperatures within the soil profile are required for the solution of the vapor flow component in the moisture flow equation. The temperature profile in the soil can be computed by applying an estimated ground surface temperature to the soil. The ground surface temperature can differ from the air temperature.

The thermal boundary condition at the soil surface can be specified as a heat flux expression (i.e., Neumann-type boundary condition). The thermal flux into the soil ground at the soil surface can be rewritten in accordance with the thermal flux balance (Eq. 6.8):

$$R_g|_{\text{surface}} = R_n - R_h - R_l \quad (6.64)$$

where:

R_g = thermal flux into soil ground, J/(m^2/day),
R_n = net radiation, J/m^2/day
R_h = sensible heat flux, J/m^2/day
R_l = latent heat flux, J/m^2/day, where $R_l = (L_v)(\text{AE})/1000$ (1000 = unit conversion factor, where 1 mm = 1/1000 m),
L_v = volumetric latent heat of vaporization, J/m^3, and
AE = actual evaporation, mm/day.

Boundary Condition for Moisture Flow. The atmospheric moisture flux at the ground surface, $q_y|_{\text{surface}}$, must satisfy the water balance equation (Eq. 6.7):

$$q_y|_{\text{surface}} = P - \text{AE} - R \tag{6.65}$$

where:

q_y = moisture flow rate at soil surface, mm/day,
P = precipitation flux (i.e., rainfall plus snowmelt water), mm/day,
R = water runoff, mm/day, and
AE = actual evaporation, mm/day.

The net infiltration flux at the ground surface, q_y, is defined in terms of $k\ \partial h/\partial y$.

Actual Evaporation. The Wilson-Penman equation (Wilson et al., 1994), provides one procedure for the calculation of AE:

$$\text{AE} = \frac{\Gamma Q_n + \eta E_a}{\Gamma + \eta/h_s} \tag{6.66}$$

where:

E_a = flux associated with "mixing"; $f(u)C_f u_v^{air}(1/h_r - 1/h_s)$, mm/day,
h_r = relative humidity in the air above the ground (i.e., $h_r = u_v^{\text{air}}/u_{v0}^{\text{air}}$),
h_s = relative humidity at the soil surface (i.e., $h_s = u_v^{\text{soil}}/u_{v0}^{\text{soil}}$),
u_v^{air} = water vapor pressure in the air above ground surface, kPa,
u_{v0}^{air} = saturated vapor pressure at the mean air temperature, kPa,
u_{v0}^{soil} = saturated vapor pressure in the soil at ground surface, kPa,
Γ = slope of saturation vapor pressure versus temperature curve, kPa/°C,
Q_n = net radiation at the water surface, mm/day, and
η = psychrometric constant, kPa/°C, $\eta = 0.06733$ kPa/°C.

The saturated vapor pressure for the air, u_{v0}^{air}, in Eq. 6.66 can be calculated based on air mean temperature, in accordance with Eq. 6.27. The water vapor pressure in the air, u_v^{air}, is defined as follows:

$$u_v^{\text{air}} = u_{v0}^{\text{air}} h_r \tag{6.67}$$

where:

h_r = relative humidity of the air above the soil surface.

The daily air temperature T_a and daily relative humidity h_r of the air above the ground surface are usually measured at a weather station. Minimum and maximum daily temperature and relative humidity values are generally recorded on a daily basis. Sometimes the hourly values of temperature and relative humidity are also recorded. The amount of data becomes quite excessive when hourly values are recorded. When minimum and maximum values are given for temperature and relative humidity, then an assumption can be made regarding the distribution of these variables throughout each day.

Figures 6.38 and 6.39 show two possible assumptions that can be made regarding the daily patterns of air temperature and relative humidity based on minimum and maximum daily readings. Figure 6.38 illustrates an asymmetric variation of air temperature and relative humidity throughout each day. Figure 6.39 illustrates a symmetrical variation of air temperature and relative humidity throughout each day. In the asymmetric daily patterns shown in Fig. 6.38, the relative humidity is assumed to have a maximum value at night and a minimum value at about noon. The daily distributions of temperature and relative humidity are known to have some effect on the calculation of AE.

The symmetric patterns shown in Fig. 6.39 have a maximum value of relative humidity at about midnight and a minimum value of relative humidity occurring near noon. The daily changing pattern of air temperature is in general opposite to the pattern for relative humidity.

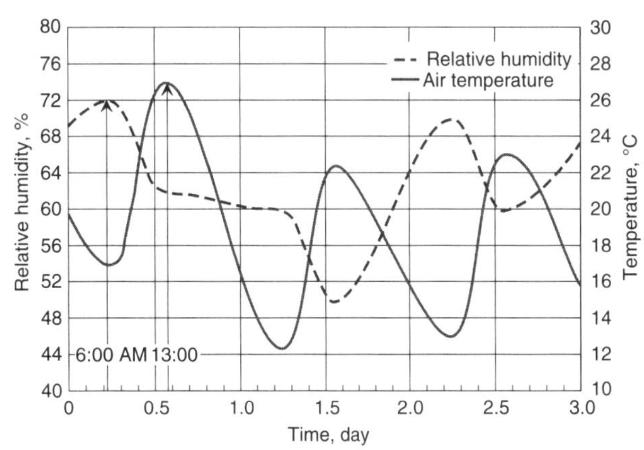

Figure 6.38 Asymmetric distributions of daily changing pattern of air temperature and relative humidity based on maximum and minimum values.

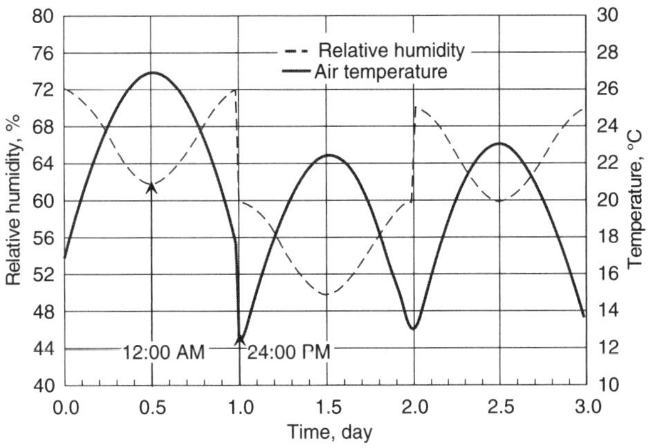

Figure 6.39 Symmetric distributions of daily changing of air temperature and relative humidity based on maximum and minimum values.

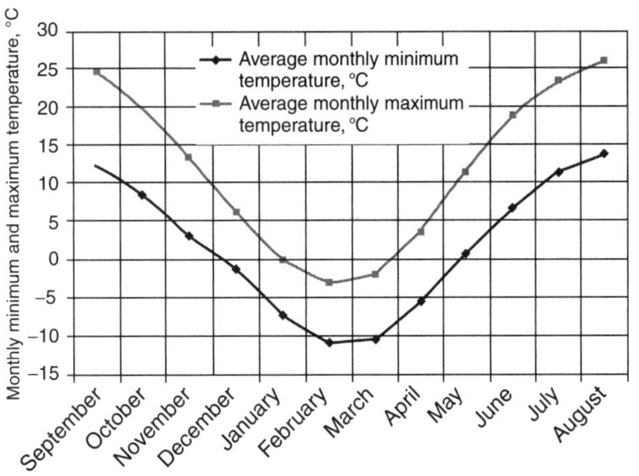

Figure 6.40 Monthly minimum and maximum air temperatures.

In some cases, the minimum and maximum daily air temperature can be assumed when only the mean values of air temperature are available. Figure 6.40 is an example illustrating the change of minimum and maximum air temperature with the month of a year. The average difference in Fig. 6.40 between minimum and maximum temperature is about 10.2 °C. The data are typical of the central northern part of Canada.

Figure 6.41 illustrates the effect of using three different daily patterns of air temperature and relative humidity on the calculation of PE. The data shown in Table 6.7 were used in the analysis. In Fig. 6.41, "asymmetric air temperature and relative humidity" means the air temperature and relative humidity are interpolated using the daily patterns shown in Fig. 6.38. Symmetric air temperature and relative humidity refer to the changing pattern shown in Fig. 6.39, and average air temperature and relative humidity refer to the use of average values of air temperature and relative humidity throughout the day, as given in Table 6.7. The calculated PE

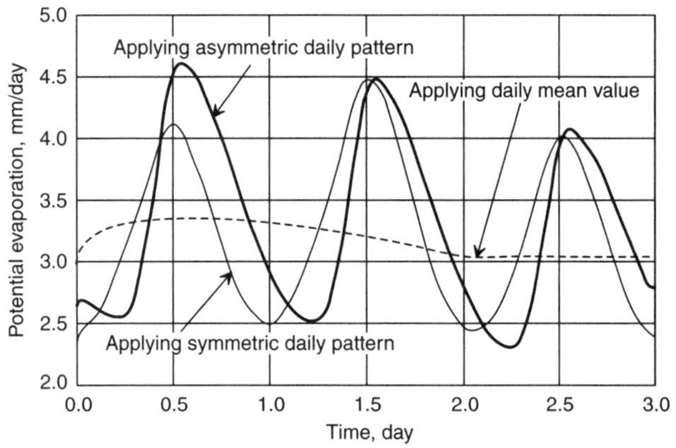

Figure 6.41 Comparison of PE calculated using different changing patterns of daily air temperature and relative humidity.

Table 6.7 Air Temperature and Relative Humidity Used in Calculation of PE

Time (day)	Avg. Temp.	Min. Temp.	Max. Temp.	Avg. RH	Min. RH	Max. RH
0	22	17	27	67	62	72
1	17.4	12.4	22.4	55	50	60
3	18.1	13.1	23.1	65	60	70

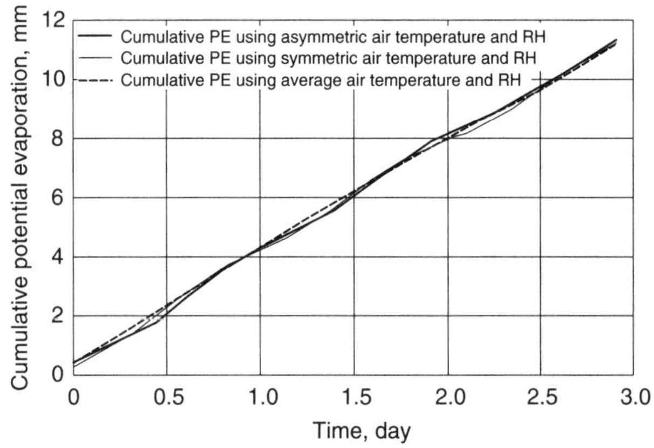

Figure 6.42 Comparison of cumulative PE calculated using different changing patterns of daily air temperature and relative humidity.

values presented in Fig. 6.42 show slight differences when using different daily patterns of temperature and relative humidity; however, the cumulative PEs are quite similar for all three patterns.

The relative humidity at the soil surface, h_s, and the water vapor pressure, u_v^{soil}, at the soil surface can be calculated using the relative humidity equation proposed by Edlefsen and Anderson (1943) provided the total suction in the soil is known:

$$h_s = \frac{u_v^{\text{soil}}}{u_{v0}^{\text{soil}}} = \exp\left(\frac{-\psi g \omega_v}{\gamma_w R\left(273.15 + T_{\text{soil}}\right)}\right) \tag{6.68}$$

where:

ψ = total suction, kPa (i.e., total suction is equal to matric suction plus osmotic suction, $\psi = (u_a - u_w) + \pi$),
ω_v = molecular weight of water vapor, 0.018016 kg/mol,
T_{soil} = soil surface temperature, °C, and
π = osmotic suction, kPa.

Osmotic suction in a soil is related to the salt content in the soil. For typical field water content conditions, osmotic suction may range from less than 100 to 1000 kPa or more. The osmotic component of suction increases as a soil dries because the salt concentration increases. The relative humidity at the soil surface approaches 100% when total suction in the soil is less than about 3000 kPa (see Fig. 6.43). Under these conditions the AE is approximately equal to the PE (i.e., AE = PE).

As total suction in the soil increases, the rate of evaporation from the soil surface decreases. Consequently, the ratio of AE to PE (i.e., AE/PE), bears a similar shape to that of relative humidity (see Fig. 6.44). It is noteworthy that the line describing the ratio of AE/PE falls slightly below the relative humidity line. The equation for the relative humidity line can be used to describe the ratio AE/PE; however, it is necessary to apply an adjustment factor to the suction values, as shown in the following equation:

$$h_s = \exp\left(\frac{-10^{\delta}\psi g \omega_v}{\gamma_w R\left(273.15 + T_{\text{soil}}\right)}\right) \tag{6.69}$$

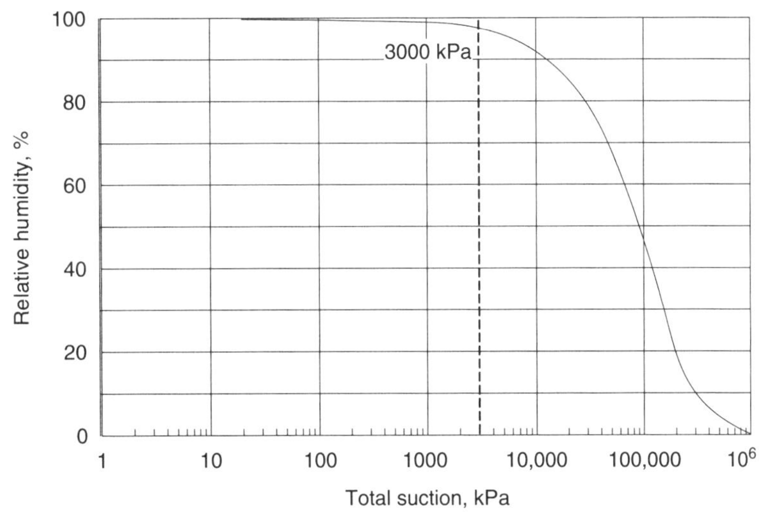

Figure 6.43 Relationship between relative humidity and total suction.

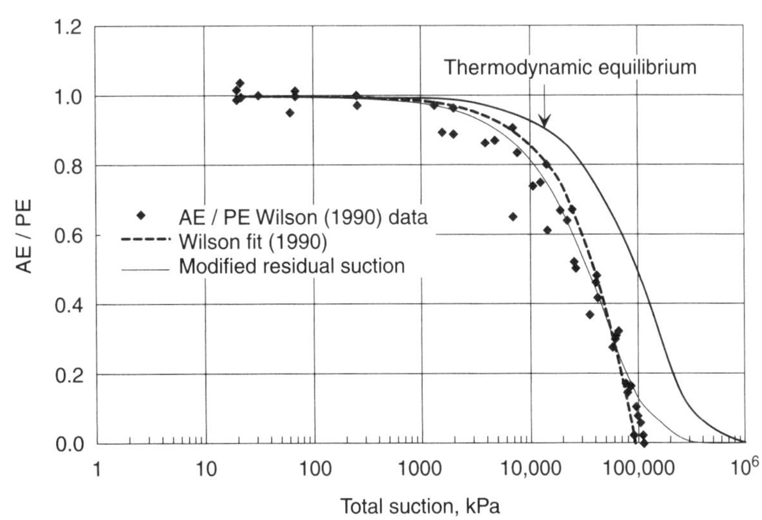

Figure 6.44 Comparison of predicted values of AE/PE using different suggested equations.

where:

δ = a dimensionless correction factor of suction, varying from 0 to 2.

An adjustment factor of 1.8 is based on the analysis of data later presented on evaporation from sand (i.e., Figs. 6.45 and 6.46). It is anticipated that the adjustment factor will vary for different soils with the most negative values applicable for coarse-grained soils. The adjustment of the negative pore-water pressures must be made for both coupled and uncoupled solutions.

The vapor pressure in the soil, u_v^{soil}, used in the diffusive flow equation (i.e., Eq. 6.50) and the heat flow equation (i.e., Eq. 6.51) can be calculated using the following equation:

$$u_v^{\text{soil}} = u_{v0}^{\text{soil}} h_r^{\text{soil}} = u_{v0}^{\text{soil}} \exp\left(\frac{-10^{\delta}\psi g \omega_v}{\gamma_w R\left(273.15 + T_{\text{soil}}\right)}\right) \tag{6.70}$$

The saturated vapor pressure, u_{v0}^{soil}, is the function of soil surface temperature and can be calculated using Eq. 6.27. Net radiation at the water surface, $R_{n,}$ can be measured or approximated using the Penman (1948) equation (i.e., Eq. 6.66), or some other procedure.

The AE must be solved by satisfying the governing moisture and heat flux balance equations 6.50 and 6.53. The initial thermal boundary condition equation 6.62, the thermal boundary condition equation 6.63 or 6.64, together with the heat flow equation 6.50, the moisture flow equation 6.53, and the AE equation 6.66, must also be satisfied.

The calculation procedure can be outlined as follows:

1. Initialize the water content profile.
2. Initialize the soil temperature profile using Eq. 6.62.
3. Apply the moisture flux boundary conditions according to Eq. 6.65.
4. Apply thermal boundary conditions at the ground surface using Eq. 6.63 or 6.64.
5. Ensure the soil model domain is initialized.
6. Calculate AE using Eq. 6.66 based on the soil temperature at the ground surface.
7. Calculate ground surface temperature using the thermal flux equation 6.64 in terms of the known *AE*.
8. Solve the moisture flow 6.50 and the thermal flow 6.53 based on initial conditions and boundary conditions.
9. Calculate net percolation I or infiltration at the soil surface. The net infiltration flux at the ground surface is defined as $k\ \partial h/\partial y$.
10. Calculate runoff R using Eq. 6.7 with the known value for precipitation P, AE, and net infiltration I at the ground surface.
11. Check the solution to see if the equations have converged with respect to hydraulic head h and soil temperature T_{soil}. The difference between iteration must be less than the designated tolerances.
12. If the solution has not yet converged, repeat steps 6–11 until convergence has been achieved.

The above calculation procedure indicates that the surface temperature and the AE equation 6.66 are used as boundary conditions and solved simultaneously with the calculation of moisture flow (i.e., liquid and vapor) and heat flow partial differential equations.

6.3.19.2 Wilson-Penman Equation for Computing AE (Uncoupled Solution)

The uncoupled solution of the modified Wilson-Penman equation (Wilson et al., 1994) can be achieved using the moisture flow partial differential equation along with the following assumptions:

- Moisture flow and vapor flow occurs through the unsaturated soil near the ground surface.
- The soil temperatures in the entire domain remain constant. Therefore, the ground thermal flux is neglected (i.e., $Q_g = 0$).
- The soil temperature at the ground surface can be different from the air temperature. The heat exchanged between the air and soil surface is constant but obeys the convection law (Eq. 6.9). The ground surface temperature is not imposed as a boundary condition for heat distribution through the underlying soil. The solution can be referred to as an isothermal model with atmospheric coupling.
- Actual evaporation is calculated using the Wilson-Penman equation 6.66 (Wilson et al., 1994).

Partial Differential Equation Governing Moisture Flow. Moisture flow in one dimension is described using the same equation as used for the coupled solution (i.e., Eq. 6.50). It is necessary to use another equation to determine the soil temperature since it is different from the air temperature. Wilson (1990) suggested that the ground surface temperature might be computed using the following empirical heat balance equation. Substituting Eq. 6.9 into Eq. 6.8 results in the following thermal balance equation:

$$C_f \eta f(u)\left(T_{\text{soil}} - T_a\right) = \frac{1000\left(R_n - R_g\right)}{L_v} - \text{AE} \tag{6.71}$$

where:

T_{soil} = soil temperature at soil surface, °C,
T_a = air temperature, °C,
C_f = conversion factor (i.e., 1 kPa = 0.00750 mHg),
η = psychrometric constant, 0.06733 kPa/°C,
$f(u)$ = function dependent upon wind speed, $f(u) = 0.35(1.0 + 0.146\ W_w)$,

W_w = wind speed, km/h,
R_n = net radiation, J/m^2/day,
R_g = ground surface thermal flux, J/m^2/day,
L_v = volumetric latent heat of vaporization, J/m^3, and
AE = actual evaporation, mm/day.

Equation 6.71 can be rearranged and solved for the temperature at the surface of the soil, T_{soil} assuming ground surface flux is zero.

$$T_{soil} = T_a + \frac{1000\left(R_n\right) - L_v\text{AE}}{C_f \eta f\left(u\right) L_v} \qquad (6.72)$$

Equation 6.72 does not take the ground thermal flux into consideration (i.e., $Q_g = 0$) and can be considered as an empirical estimation for the calculation of the soil temperature. It is not fully known whether there are limitations associated with the use of Eq. 6.72 in geotechnical engineering practice.

Initial Soil-Water Content Profile. The initial water content profile needs to be set to an initial set of values. The water content profile may correspond to the initial pore-water pressures dictated by the SWCC or the initial water content may be measured (or estimated) based on the location of the water table.

Initial Soil Surface Temperature. The soil temperatures throughout the soil profile can be initialized to the air temperature in accordance with Eq. 6.62.

Boundary Condition for Moisture Flow. The ground surface boundary condition for moisture flow can be defined using Eq. 6.65.

Actual Evaporation. The Wilson-Penman equation (Wilson et al., 1994) can be used for calculating AE (i.e., Eq. 6.66). The uncoupled solution must be solved satisfying the governing moisture PDE (i.e., Eq. 6.50) and soil temperature equation 6.72 using the initial moisture and thermal conditions. The moisture flow boundary condition equation 6.65 must be satisfied along with the AE equation 6.66.

It is necessary to know the soil temperature at the ground surface when calculating AE (Eq. 6.66). When using Eq. 6.72 to calculate the soil temperature, the AE must also be known. In other words, Eqs. 6.72 and 6.66 must be solved simultaneously to get AE and soil temperature.

Calculation Procedure.

1. Initialize water content profile.
2. The soil surface temperatures are made equal to the air temperature.
3. Apply the moisture flux boundary condition in accordance with Eq. 6.65.
4. Ensure the model domain is initialized.
5. Calculate AE using Eq. 6.66 based on a soil surface temperature T_{soil} given by Eq. 6.72.
6. Solve the moisture flow PDE (i.e., Eq. 6.50) and soil temperature equation 6.72 simultaneously based on the initial conditions and moisture flux boundary condition designated by Eq. 6.65.
7. Calculate net percolation I or infiltration at the soil surface. The net infiltration flux at the ground surface is defined as $k\ \partial h/\partial y$.
8. Calculate runoff R using Eq. 6.7 using the previously calculated AE and infiltration I.
9. Check the solution to see if the convergence has been achieved with respect to hydraulic head h. The difference between subsequent iteration must be less than the designated tolerance.
10. If the solution has not yet converged, repeat steps 5–10 until convergence has been achieved.

Atmospheric (quasi) coupling has been solved using a closed-form solution for soil surface temperature as one of the governing equations (i.e., Eq. 6.72).

6.3.19.3 Limiting Function for AE (Coupled Solution)

A limiting function relationship was proposed by Wilson et al. (1997a) that provided a relationship between AE and PE. The vapor pressures associated with the relative humidity at ground surface and the relative humidity in the air above ground surface are scaled based on the assumption that the air and soil temperatures were the same.

The limiting function solution is based on the following assumptions:

- The soil temperature at the ground surface is equal to the air temperature.
- The moisture flow and heat transfer beneath the ground surface are the same as in the coupled rigorous solution.
- Soil freezing/thawing processes are the same as in the coupled rigorous solution.
- The latent heat due to phase change, including evaporation and freezing/thawing, must be considered in the heat transfer beneath the ground surface.
- AE is calculated using the limiting function equation proposed by Wilson et al. (1997a).

Partial Differential Equations Governing Moisture and Heat Flow. The moisture mass transfer and heat transfer partial differential equations are described by Eqs. 6.50 and 6.53.

Initial Water Content Conditions. Initial water contents throughout the soil profile must be initialized as previously described.

Initial Temperature Conditions. The soil temperatures throughout the soil profile can be initialized to the air temperature (Eq. 6.62).

Thermal Boundary Condition at Soil Surface. The thermal boundary condition at the soil surface can be specified as a constant temperature or a temperature expression (i.e., Dirchlet boundary condition) or computed from a thermal flux (Neumann boundary condition), in accordance with Eqs. 6.63 and 6.64, respectively.

Boundary Condition for Moisture Flow. The atmospheric moisture flux at the ground surface needs to satisfy Eq. 6.65.

Actual Evaporation. Actual evaporation is computed using the limiting function (Wilson et al., 1997a), written as follows:

$$AE = PE \left[\frac{u_v^{soil} - u_v^{air}}{u_{v0}^{soil} - u_v^{air}} \right] \tag{6.73}$$

where:

PE = potential evaporation, mm/day,
u_v^{soil} = vapor pressure in the soil at the ground surface temperature, kPa,
u_{v0}^{soil} = saturated vapor pressure in the soil at the ground surface temperature, kPa, and
u_v^{air} = vapor pressure in the air above the soil surface, kPa.

If the air and soil temperatures are assumed to be approximately equal, Eq. 6.73 can be written as follows:

$$AE = PE \frac{\exp\left(\dfrac{-10^{\delta} \psi g \omega_v}{\gamma_w R \left(273.15 + T_s\right)} \right) - u_v^{air}/u_{v0}^{air}}{1 - u_v^{air}/u_{v0}^{air}} \tag{6.74}$$

where:

ψ = total suction (i.e., matric suction plus osmotic suction), kPa, and
δ = dimensionless factor for the adjustment of suction.

Potential Evaporation. The PE can be determined in a number of different ways, including using the following:

1. Measured data (e.g., pan evaporation)
2. The Penman (1948) equation 6.20
3. The Thornthwaite (1948) equation 6.19
4. The Priestley-Taylor (1972) equation 6.23
5. The Monteith (1965) equation 6.22

Calculation Procedure.

1. Initialize the water content profile.
2. Initialize the soil temperature profile.
3. Apply the moisture flux boundary conditions according to Eq. 6.65.
4. Apply thermal boundary conditions at the ground surface using Eq. 6.63 or 6.64.
5. Ensure the model domain is initialized.
6. Determine PE using measured data or calculate from one of the above-mentioned equations.
7. Calculate the AE using Eq. 6.73 or 6.74 together with Eq. 6.70.
8. Solve the moisture flow equation 6.50 and the thermal flow equation 6.53 based on initial conditions and boundary conditions.
9. Calculate net infiltration I at the soil surface. The net infiltration flux at the ground surface is defined as $k\ \partial h/\partial y$.
10. Calculate runoff R using Eq. 6.7 using the known value of precipitation P AE, and net infiltration I at ground surface.
11. Check the solution to see if the convergence has been reached for hydraulic head h and temperature T. The difference between iterations must be less than the designated tolerances.
12. If the solution has not yet converged, repeat steps 6–11 until convergence is achieved.

6.3.19.4 Limiting Function for AE (Uncoupled Solution)

The Limiting Function can be used in an uncoupled manner when solving for AE. In this case, the following assumptions are made:

- Liquid flow and vapor flow through the soil are governed by hydraulic head gradients and vapor pressure gradients.
- The soil temperature in the entire domain is assumed to be the same and equal to the air temperature above the soil surface (i.e., ground thermal flux is neglected, $Q_g = 0$).
- The ground surface temperature is assumed to be equal to the air temperature.
- The AE is calculated using the limiting function proposed by Wilson et al. (1997a).

Partial Differential Equation Governing Moisture Flow. Moisture flow in one dimension beneath the soil surface can be described using Eq. 6.50.

Initial Soil-Water Content. The initial water content profile can be assigned using one of the procedures mentioned above.

Boundary Condition for Moisture Flow. The ground surface boundary condition for moisture flow can be defined using Eq. 6.65.

Actual Evaporation. The limiting function proposed by Wilson et al., (1997a) can be used to calculate AE (i.e., Eq. 6.73 or 6.74).

Calculation Procedure.

1. Initialize the water content profile.
2. Initialize soil temperatures with the air temperature.
3. Apply the seepage boundary condition (i.e., Eq. 6.65), which might involve precipitation or evaporation.
4. Determine PE through either measurement or one of the empirical methods previously discussed [e.g., Penman (1948) equation].
5. Calculate AE using Eq. 6.73 or 6.74.
6. Solve the moisture flow PDE (Eq. 6.50) and calculate the soil temperature (Eq. 6.72) based on the initial conditions and the specified boundary conditions.
7. Calculate net percolation I or infiltration at the ground surface. The net infiltration is equal to $k\ \partial h/\partial y$,

8. The runoff R can be calculated using Eq. 6.7. The variables AE and I should be known when this calculation is performed.
9. Check the solution to see if the convergence has been reached for hydraulic head h. The difference in hydraulic head between the last two iterations must be less than the designated tolerance.
10. If the solution has not converged, repeat steps 4–9 while checking for convergence on hydraulic head.

6.3.19.5 Experimental-Based Relationship between AE and PE (Coupled Solution)

Wilson (1997) also presented experimental results that showed a unique relationship for all soils between total suction at the soil surface and the ratio of actual evaporation to potential evaporation, AE/PE. In 1997, Wilson et al., (1997a) presented an equation that provided a reasonable fit through the experimental data. Wilson (1990) showed that it is primarily the soil suction at the ground surface of any soil that mainly controls the rate of evaporation. Consequently, the soil type at ground surface is not a controlling factor when assessing the actual rate of evaporation.

Figure 6.44 shows a plot of the experimental data along with the proposed mathematical relationships for calculating AE. The laboratory tests involved the evaporation of water from thin layers of sand, silt, and clay soils. The soils were initially deposited as slurry (i.e., high water contents). The rates of evaporation from the soils were compared to the rate of evaporation from a pan of water at the same temperature. The comparison of the evaporation rates from the soil and the water allowed the computation of the relative evaporation or the ratio of AE to PE (i.e., AE / PE). The rate of evaporation from the soils became increasingly slower with time than evaporation from a pan of water.

The water content versus soil suction relationship was independently measured for each of the soils. This allowed the relative rates of evaporation to be compared to the suction in the soil. The results in Fig. 6.44 show that the actual evaporation data for sand, silt, and clay essentially form a single unique curve. This means that the rate of water evaporation from any soil is the same when soil suction is used as the basis of comparison. In other words, if the soil suction is known at the ground surface, then the rate of evaporation from the ground surface can be estimated from the unique relationship shown in Fig. 6.44.

The following assumptions are made when performing a coupled analysis along with the *experimentally based AE/PE* equation:

- The temperature of the soil at ground surface is equal to the air temperature.
- The moisture flow and heat transfer beneath the ground surface are the same as in the previously mentioned coupled solutions.
- Soil freezing/thawing processes are the same as in the previously mentioned coupled solutions.
- The latent heat due to phase change including evaporation and freezing/thawing is considered for heat transfer beneath the soil surface.
- The AE is based on an empirical mathematical equation that fits the data shown in Fig. 6.44.

Partial Differential Equation Governing Moisture and Heat Flow. The moisture mass and heat transfer governing equations are defined by Eqs. 6.50 and 6.53.

Initial Water Content Conditions. The starting water content profile must be initialized using one of the previously mentioned procedures.

Initial Temperature Conditions. The temperatures throughout the soil profile can be initialized to the air temperature in accordance with Eq. 6.62.

Boundary Condition for Moisture Flow. The ground surface boundary condition for moisture flow can be defined using Eq. 6.65.

Thermal Boundary Condition at Soil Surface. The thermal boundary condition at the soil surface can be specified as a constant temperature or a temperature expression (Dirchlet boundary condition, Eq. 6.63) or computed from a thermal flux (Neumann boundary condition, Eq. 6.64).

Actual Evaporation. The rate of evaporation from the ground surface can be estimated using the Wilson et al., (1997a) empirical experimental-based relationship. The ratio AE/PE is approximated using an equation similar in form to the thermodynamic equilibrium relationship between relative humidity and total suction (Edlefsen and Anderson, 1943). The AE can be written as a function of PE using the following experimental-based equation:

$$\frac{\mathrm{AE}}{\mathrm{PE}} = \exp\left(\frac{-\psi g \omega_v}{\zeta\left(1-h_a\right)\gamma_w R\left(T_{\mathrm{soil}}+273.15\right)}\right) \qquad (6.75)$$

where:

ζ = a dimensionless empirical parameter equal to about 0.7 based on experimental data (Wilson, 1990).

Potential Evaporation. The PE can be either measured or calculated using one of several empirical methods [e.g., Penman (1948) equation 6.20, Thornthwaite (1948) equation 6.19, Priestley-Taylor (1972) equation 6.23, or Monteith (1965) equation 6.22].

Calculation Procedure.

1. Initialize the water content profile.
2. Initialize the soil temperature profile.
3. Apply the moisture flux boundary condition according to Eq. 6.65.
4. Apply the thermal boundary condition at the ground surface using Eq. 6.63 or 6.64.

5. Ensure the model domain is initialized.
6. Determine PE using pan evaporation data that are measured or calculated using one of the above-mentioned methods.
7. Calculate the AE using Eq. 6.75.
8. Solve the moisture flow equation 6.50 and the thermal flow equation 6.53, based on initial conditions and designated boundary conditions.
9. Calculate net infiltration I at ground surface. The net infiltration flux at the ground surface is defined as $k\ \partial h/\partial y$.
10. Calculate runoff R using Eq. 6.7 using the known value of precipitation P, AE, and net infiltration NP at the ground surface.
11. Check the solution to see if the convergence has been reached for both seepage head and soil temperature, T_{soil}. The difference in the calculated variables between the last two iterations must be less than the designated convergence tolerances.
12. If the solution has not yet converged, repeat steps 6–11 iteratively until the convergence is achieved.

6.3.19.6 Experimental-Based Relationship between AE and PE (Uncoupled Solution)

The experimental-based solution is the same as the previously presented solution except that the thermal regime partial differential equation for conductive and convective flow is not solved. The following assumptions can be made when performing the uncoupled solution that utilizes the experimental-based relationship (i.e., Eq. 6.75):

- Moisture flow (i.e., liquid and vapor) in the soil is driven by hydraulic head and vapor pressure gradients.
- The soil temperature throughout the model domain is constant and is assumed to be equal to the air temperature above the soil surface. In other words, the ground thermal flux can be neglected (i.e., $R_g = 0$).
- The soil surface temperature is assumed to be equal to the air temperature.
- The actual evaporation can be calculated using the empirical experimental-based equation proposed by Wilson et al., (1997a).

Partial Differential Equation Governing Moisture and Heat Flow. The moisture flow governing equation is Eq. 6.50 and the soil temperature is computed using Eq. 6.72.

Initial Water Content Conditions. The initial water content profile must be initialized.

Initial Temperature Conditions. The soil temperatures throughout the soil profile can be initialized to the air temperature.

Boundary Condition for Moisture Flow. The ground surface boundary condition for moisture flow can be defined using Eq. 6.65.

Empirical Experimental-Based Equation for AE. The AE is obtained through use of the empirical experimental-based equation 6.75.

Potential Evaporation. The PE can be either measured or calculated using one of several empirical methods [e.g., Penman (1948), Eq. 6.20; Thornthwaite (1948), Eq. 6.19; Priestley-Taylor (1972), Eq. 6.23; or Monteith (1965), Eq. 6.22].

Calculation Procedure.

1. Initialize the water content profile.
2. Apply the moisture flux boundary condition according to Eq. 6.65.
3. Determine PE using either measured pan evaporation or one of the above-mentioned PE equations.
4. Calculate AE using Eq. 6.75.
5. Solve the moisture flow equation 6.50 and the soil temperature equation 6.72 based on the initial conditions and the boundary conditions.
6. Calculate net percolation I at the ground surface. The net infiltration flux at the ground surface is defined as $k\ \partial h/\partial y$.
7. Calculate runoff R using Eq. 6.7. At this time, AE and I are known.
8. Check the solution to determine whether convergence has been achieved with respect to hydraulic head. The difference between the last two iterations must be less than the designated convergence tolerance. If the solution has not converged, repeat steps 3–8 iteratively until convergence is achieved.

6.3.20 Example Calculations of AE

Measurements of actual evaporation from a soil column were presented by Wilson (1990). The results serve as benchmark data for comparison with numerical model calculations. Figure 6.45 compares the measured evaporation rates with computed results obtained with an uncoupled analysis using SVFlux (SoilVision, 2010). The uncoupled analyses were performed using three procedures: (i) Wilson-Penman model (Wilson et al., 1994) (ii) limiting function model (Wilson et al., 1997a), and (iii) experimental-based model (Wilson et al., 1997a). Also shown are the PE and AE measurements by Wilson (1990). Figure 6.45 shows that the measured evaporation rates are similar to the computed evaporation rates obtained when using the three uncoupled models. The results from the three uncoupled models are also quite similar to the measured results.

Figure 6.46 shows the actual evaporation calculations performed using a coupled mode with SVFlux and SVHeat (SoilVision, 2010). The coupled analyses were performed using: (i) Wilson-Penman model (Wilson et al., 1994) (ii) limiting function model (Wilson et al., 1997a), and (iii) experimental-based model (Wilson et al., 1997a). Once again, the three models show similar amounts of computed

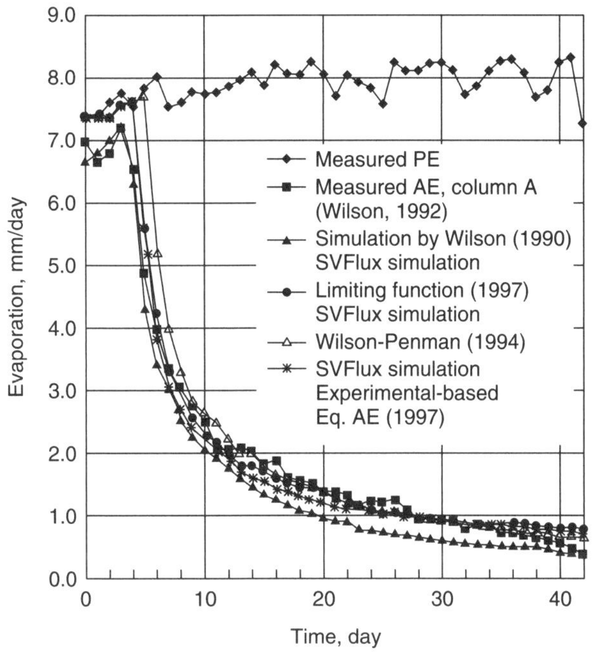

Figure 6.45 Comparison of the measured evaporation rates to predicted evaporation rates using uncoupled moisture movement models.

evaporation. The coupled and uncoupled results were also similar.

The similarity between the coupled and uncoupled model results may be due to the fact that the ambient air temperature did not undergo significant change during the experiment. The temperatures were recorded to have changed from

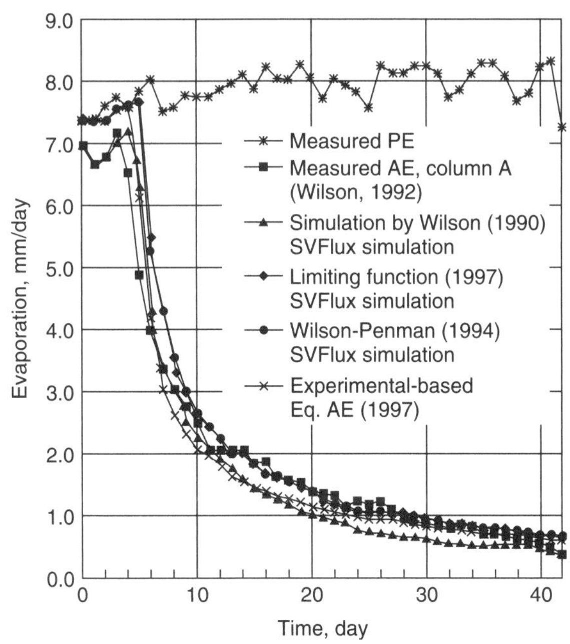

Figure 6.46 Comparison of measured evaporation rates to predicted evaporation rates using coupled moisture movement models.

37.4 to 38.1 °C during the experiment. Consequently, thermal transfer in the soil column was not a dominant term in the coupled modeling of SVFlux and SVHeat.

Actual evaporation calculations based on the Wilson-Penman model (Wilson et al., 1994), and the limiting function (Wilson et al., 1997a) model most closely match experimental data. The experimental-based (Wilson et al., 1997a) model results slightly underpredict the measured AE.

There was a slight overprediction of AE in the first 5 days. The overprediction appears to come from the overprediction of relative humidity at the soil surface. There should not be a significant decrease in relative humidity until the soil suction at the soil surface is larger than about 3000 kPa. However, the sand soil will go past the residual water content well below a suction of 3000 kPa. Under these conditions there appears to be a surface resistance to evaporation that is more dominant (Alvenas and Jansson, 1997; Lee and Pielke, 1992; Kondo and Saigusa, 1992). An adjustment factor was introduced to compensate for the vapor pressure calculations at ground surface and to improve numerical stability. The adjustment factor appears to range from 0 to 2. It is anticipated that the adjustment factor will vary for different soils with the largest values applicable for coarse-grained soils.

Figure 6.47 presents the measured air temperature and the temperature at the soil surface that is observed during the experiment (Wilson, 1990). The soil surface temperatures calculated during each of the numerical simulations are also presented. The predicted temperature at the soil surface shows reasonable agreement with the measured values.

There is some difference between the predicted and measurement values of temperature in the first 5 days when using the uncoupled (or quasi-coupled) Wilson-Penman

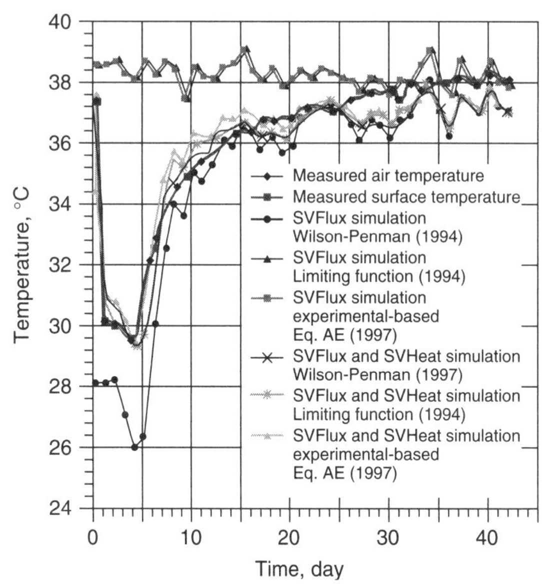

Figure 6.47 Comparison of predicted soil surface temperatures and measured soil temperatures.

model (Wilson et al., 1994). The uncoupled limiting function and experimental-based models assume that the soil surface temperature is equal to the air temperature.

A large amount of heat (i.e., latent heat of vaporization, which is about 2.5×10^9 J/m^3), is absorbed during evaporation from the soil surface. The result is a significant drop in temperature at the soil surface in the first few days. The temperature at the soil surface subsequently increases and becomes close to the air temperature when evaporation is reduced.

It has been demonstrated that there are a number of different possible ways in which AE from a soil surface can be calculated. A total of six calculation procedures have been described. The solution procedures were first separated into coupled and uncoupled solutions. The uncoupled procedures require considerably less computational effort. All six calculation procedures appear to give quite similar results. It is difficult to say whether uncoupled solutions are satisfactory in all situations for computing AE from a soil surface. There is need for further comparative studies to be undertaken on the computational procedures.

6.3.21 Transpiration Flux

Transpiration T is the term used to describe evaporation from vegetated surfaces. The term *transpiration* is usually separated from the terms *actual evaporation* and *potential evaporation* in geotechnical engineering. Actual evaporation plus transpiration from plants gives rise to the term *evapotranspiration*, as illustrated in Fig. 6.48. There are advantages in separating the components of evaporation in this manner because of the independent calculations required for each component.

Evaporation accounts for the movement of water from the surface of the earth to the air from sources such as the soil and vegetated surfaces. Transpiration accounts for the movement of water within plants and the subsequent loss of water as vapor through stomata in its leaves (Fig. 6.49). Transpiration is part of the overall water cycle and therefore part of the ground surface water balance boundary condition. Trees, shrubs, and grass contribute to transpiration and are called transpirators.

Plants can remove significant amounts of water through transpiration. Computer simulation of a vegetated ground surface is usually handled in two steps: first, the assumption is made that there is no vegetation on the ground surface and, second, further simulations are performed while considering the presence of vegetation.

Numerous assumptions must be made when considering the simulation of vegetation on evaporative flux. Plants can be visualized as small pumps (i.e., sinks) that remove water from a zone of soil below ground surface (Tratch et al., 1995) (Fig. 6.49). The effect of vegetation has proven to be quite complex. Transpiration is a function of the soil zone from which water is extracted as well as the leaf area characteristics of the plants. The growing season for the vegetation must be assumed and there must be sufficient nutrients available in the soil to sustain plant growth. The long-term nutrient sustainability for plant growth can be a significant factor when evaluating transpiration.

The simulation of plant-covered ground surfaces requires a number of assumptions related to the vegetation. The simulation of transpiration has been a topic of considerable interest in agriculture-related disciplines. Much of the agricultural research has direct interest and application to problems encountered in geotechnical engineering. Tratch et al. (1995) gave consideration to the research in agriculture and proposed protocols for geotechnical engineering practice.

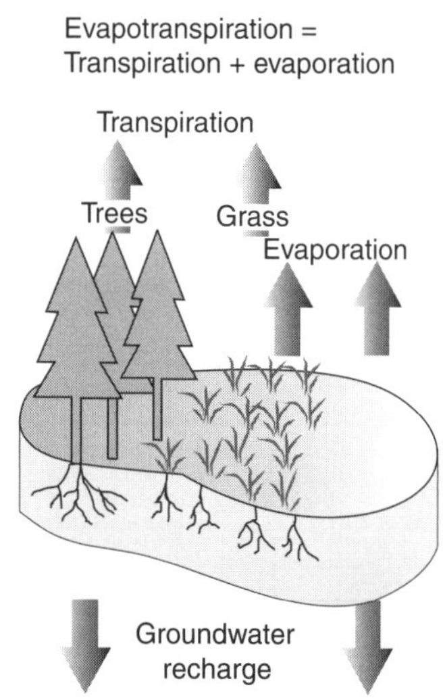

Figure 6.48 Illustration of evaporative components contributing to evapotranspiration.

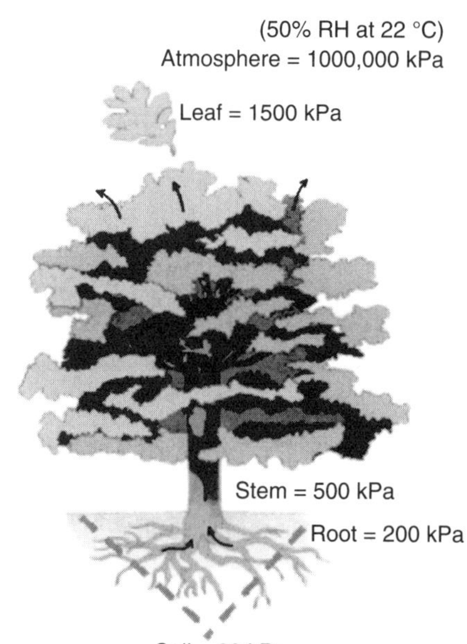

Figure 6.49 Movement of moisture through vegetation.

The transpiration process can be applied as a sink term below a specified evaporation boundary. There are a number of factors that control the amount of water that can be transpired by the vegetation, including bare soil evaporation, leaf area index (LAI), plant limiting function (PLF) as it relates to soil suction, and the root zone profile (Saxton, 1982). These are the key factors that need to be taken into consideration when simulating transpiration for geotechnical engineering applications.

The lack of available water for plant growth and high evaporative demands may cause plants to biologically react by closing stoma, reducing transpiration, and reducing metabolic reactions (Saxton, 1982). The plant will reach its wilting point if subjected to continuous stress. The wilting point results in the dropping of leaves and tissue death.

Tratch et al. (1995) suggested a four-step methodology for the prediction of moisture uptake via transpiration from plants:

1. Determine PE, which is a measure of the maximum evapotranspiration rate possible under the specified atmospheric conditions.
2. Determine the potential transpiration (PT) rate based on PE and the characteristics of the plant population at the site. This is a measure of the maximum transpiration rate possible under the specified atmospheric conditions.
3. Distribute the PT, which is a surface flux, into a potential root uptake profile within the active root zone.
4. Modify the potential root uptake based upon the availability of moisture within the actual root uptake profile.

The PE can be applied as a boundary flux that is dependent on the atmospheric conditions. Potential evaporation is the cumulative sum of bare soil evaporation and plant transpiration (Granger, 1989).

Determination of the PT constitutes the second step in evaluating the transpiration flux. Potential transpiration is the PE modified by the LAI.

6.3.21.1 Leaf Area Index

The LAI represents the effect of the vegetation cover on the energy available to extract water from the ground surface. A plant cover with a large LAI has a larger potential to extract water. There is an adjustment in the evaporation rate as the plants absorb more energy.

The LAI, by definition, is the surface area of the leaves divided by the surface area covered by the soil. The LAI can be specified as a function of time (e.g., days of each year) (Fig. 6.50). There is a growing season and there is an inactive season when no growth takes place. For example, the growing season may be assumed to start on April 1 and end on August 15, in which case the length of the growing season is 223 days. The LAI will control the role that vegetation plays on overall evaporation from the ground surface.

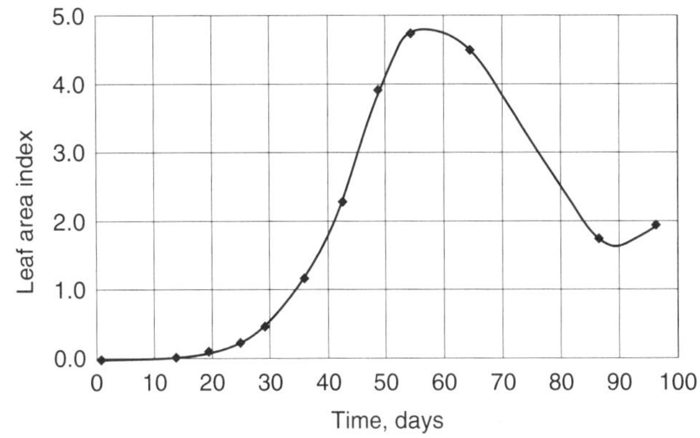

Figure 6.50 LAI as function of days of year.

Ritchie (1972) observed that the transpiration component of evaporation was dependent upon the LAI. The following controls were suggested for the calculation of potential transpiration PT:

$$PT = 0 \quad \text{when LAI} < 0.1 \tag{6.76}$$

$$PT = PE\left(-0.21 + 0.70\ \text{LAI}^{0.5}\right) \quad 0.1 <= \text{LAI} < 2.7 \tag{6.77}$$

$$PT = PE \quad \text{when } 2.7 <= \text{LAI} \tag{6.78}$$

where:

PT = potential transpiration rate per unit time, mm/day,
LAI = leaf area index (unitless), and
PE = potential evapotranspiration rate per unit time, mm/day.

Tratch et al. (1995) suggested the use of one of three plant quality covers: excellent cover, good cover, and poor cover. Figure 6.51 shows these curves for a one-year period. The mass flux due to transpiration can first be defined as a ground surface flux. Then the ground surface moisture flux is distributed through the soil profile. Two shape functions were suggested by Tratch et al. (1995) for describing root distribution: a triangular shape and a rectangular shape.

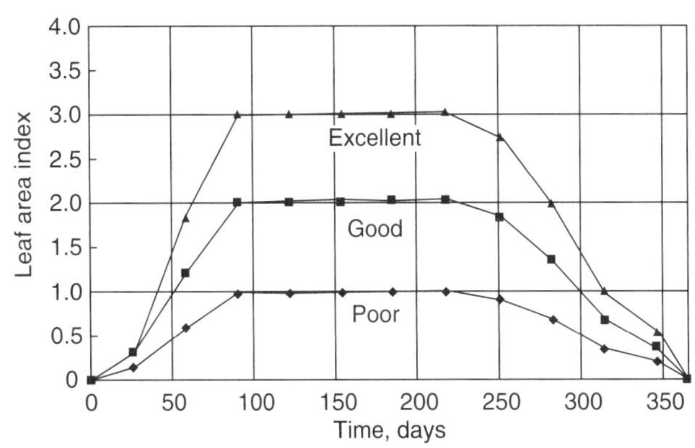

Figure 6.51 Quality of plant growth in terms of LAI.

6.3.21.2 Root Distribution

The zone over which plant transpiration is assumed to extract water is dependent upon the depth of the roots and the distribution of the roots. Two functions for root zone distribution appear to be sufficient for geotechnical engineering applications. The root shape zones are classified as a triangular root zone or a rectangular root zone. Both functions extract the same volume of water from the soil, but the water is extracted differently with respect to depth.

The triangular root distribution shape function describes a decreasing root uptake of water with depth, as illustrated in Fig. 6.52a. The rectangular root distribution shape function describes a constant root uptake of water with depth, as illustrated in Fig. 6.52b. The potential root uptake at a point under consideration can be defined as follows:

$$\mathrm{PRU} = \frac{\mathrm{RSF\ PT}}{R_T}\left(1 - \frac{R_n}{R_T}\right) \tag{6.79}$$

where:

PRU = potential root uptake rate per unit time, m^3/day,
RSF = root distribution shape factor (i.e., triangular or rectangular),
PT = potential transpiration rate per unit time, m/day,
R_T = total thickness of the root zone in length units m, and
R_n = depth to the given point in length units, m.

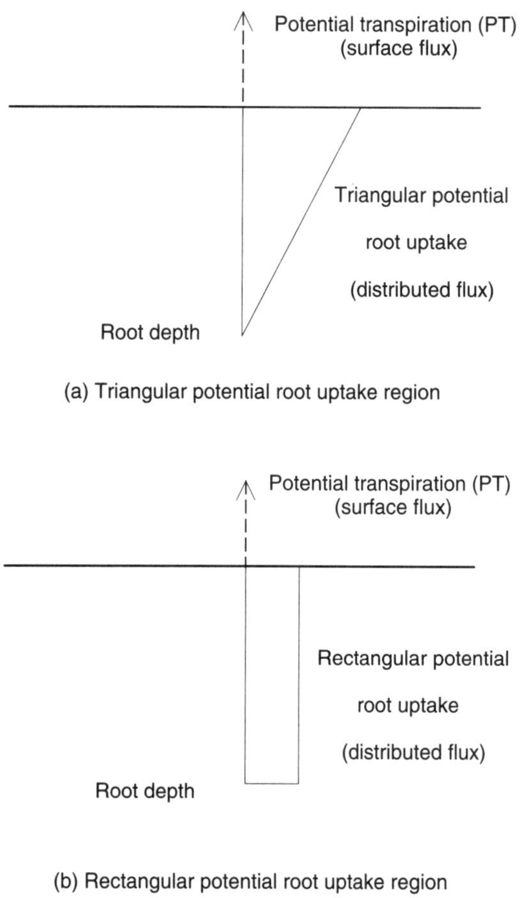

Figure 6.52 Common shapes for root uptake areas.

The PRU variable must be integrated over the root zone when a transpiration boundary is specified. The PRU is applied as a potential transpiration sink.

6.3.21.3 Plant Limiting Factor, PLF

A lack of available water reduces the ability of a plant to transpire water. The PLF is a function of the suction in the soil in the root zone. When soil suction is low, the PLF is 1.0, and when the wilting point of the plant is reached, the PLF is reduced to zero. Figure 6.53 shows a typical plot of PLF versus soil suction.

The root transpiration sink S_{root} is calculated by modifying the potential transpiration sink by a reducing term based on the suction in the soil, as shown in the following equation:

$$S_{\mathrm{root}} = (\mathrm{PRU})(\mathrm{PLF}) \tag{6.80}$$

The actual transpiration sink is maintained at the potential transpiration values when soil suction is low. Once a specified limiting value is exceeded, the actual transpiration flux decreases exponentially (i.e., straight line on a semilog plot) until the wilting point is reached. Transpiration will be zero at soil suctions above the wilting point. Tratch et al. (1995) suggested defaults for the limiting value and the wilting point to be 100 and 1500 kPa, respectively. The transpiration sink term is part of the moisture flow partial differential equation (i.e., Eq. 6.53). A negative value indicates that water is being pulled from the system.

6.3.22 Net Evaporative Flux at Ground Surface

Each of the components that influence the moisture flux at the ground surface has been described. Each of the moisture flux components can be plotted for the year and then the net infiltration curve can be computed (Fig. 6.54). The net infiltration component becomes the ground surface moisture flux boundary condition. It is possible to proceed with the calculation of infiltration of water into the soil once the net moisture flux at the ground surface boundary is known.

The above-mentioned calculations for net moisture flux are not independent of modeling soil infiltration. The AE is dependent upon knowing the total suction at ground surface. Actual evaporation is computed as part of the infiltration model. There is a “coupling effect” between the infiltration model and the calculation of AE. Stated another way, the calculations combine the climatic ground surface moisture flux conditions with the solution of the nonlinear partial differential equation of unsaturated soil seepage. The combination of the unsaturated soil moisture flow model and the climatic boundary conditions is called the soil-atmospheric model.

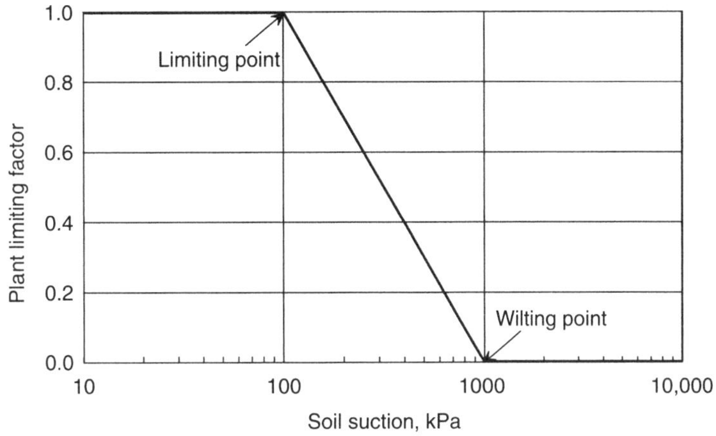

Figure 6.53 PLF versus soil suction.

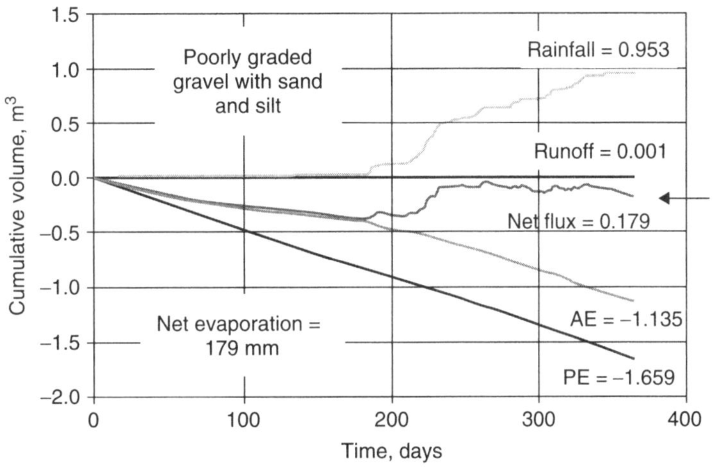

Figure 6.54 Net infiltration computed after ground surface moisture flux has been applied to soil-atmospheric model.

The soil-atmospheric model must be solved on an elapsed time scale that might be on the order of a few minutes. Each day is modeled as a series of time steps and the time scale is continued for the entire year. However, one year may not be sufficient for design purposes in some cases (e.g., the cover system). Rather, it may be necessary to perform these calculations for as much as 10 years or more. Needless to say, the design of soil cover systems is computationally demanding even without consideration of two- and three-dimensional geometries. As well, the high nonlinearity of the partial differential moisture flow equation makes convergence of the solution a challenge.

The net effect of the components discussed above is called the "net infiltration" at the soil surface. Infiltration at the bottom of a soil cover system is commonly referred to as "deep infiltration"; however, other terms are also used. The magnitude of deep infiltration provides an indication of the amount of water that is likely to pass below the cover system into the underlying materials. Needless to say, there are many assumptions and extensive calculations that go into the estimation of moisture infiltration.

6.4 CHALLENGES OF NUMERICAL MODELING GROUND SURFACE MOISTURE FLUX CONDITIONS

Shackelford (2005) presented a summary of some of the current and future environmental issues of special interest to cover systems. The issues mentioned were (1) long-term performance of waste containment systems, (2) alternative barriers or covers (i.e., alternative to clay covers) and barrier materials, (3) innovative barriers (covers) and barrier materials, (4) forms of waste materials, (5) significance of biological waste processes, (6) the role of numerical modeling in cover design, and (7) professional identity. Of the mentioned issues, it is the role of numerical modeling as it relates to cover systems that is of primary interest in the application of unsaturated soil mechanics.

Shackelford (2005) made several comments that are noteworthy with regard to the difficulties associated with the modeling of cover systems. He stated:

> The ability of models to accurately predict field performance of engineered systems has been and will continue to be an issue facing geotechnical engineers. This issue is particularly important when dealing with environmental problems because models are often used as a tool to predict the future impacts and resulting risk from engineering activities related to environmental protection.
>
> The uncertainty in the accuracy of the model predictions can be considerable, particularly in cases where the timeframe for the prediction is long, such as the disposal of radioactive wastes with design lives ranging from hundreds to thousands of years.
>
> – uncertainty is associated not only with the comprehensiveness and accuracy of the physical, chemical, and biological processes upon which the model is based, but also with the accuracy of the input data and the lack of knowledge concerning the changes in the material properties and processes that may occur with time. As a result, predictions made with existing models generally cannot be considered reliable until and unless the predictions are verified by comparison with field data.
>
> – the problems in trying to predict the field performance of engineered systems, particularly when the processes involved are highly nonlinear and the properties of the systems are time dependent. These examples also illustrate why evaluation and verification of numerical models will continue to be an important issue facing geotechnical engineers, particularly within the realm of environmental issues.

Numerical modeling associated with cover design has presented many challenges to geotechnical engineers. There appears to have been significant problems related to convergence when solving the partial differential moisture flow equation. There are also questions related to the assumptions made at various stages of the analysis. The numerical modeling analysis is handled as a transient seepage analysis that may extend over a time period of several years (e.g., typically 10 years) using time steps on the order of a few minutes. The weather conditions can vary from near-average conditions to extreme conditions and the soil properties can vary from being almost constant to taking on the form of highly nonlinear mathematical functions. Needless to say, the numerical modeling challenges are significant and appear to exceed those of most other unsaturated soil mechanics applications.

6.4.1 Basic Functionality of Cover Systems

The first designed cover systems were predominantly made of compacted clays. The intent was to construct a relatively impervious cover over waste materials. However, the primary problem encountered with clay covers was the formation of cracks associated with drying and desiccation. Cracks had the potential to make the clay covers quite permeable to water. As a result, a new generation of cover systems called "alternatives covers" became more common. The name alternative covers was used because they were an alternative to clay covers. These cover systems were also referred to as store-and-release covers and ET (evapotranspirative) covers.

The engineering design of a Store-and-Release cover system involved the application of unsaturated soil mechanics principles (Fig. 6.55). The covers could change in degree of saturation with time and function in a manner that accommodated environmental fluctuations. Reductions in the degree of saturation reduced the hydraulic conductivity (or coefficient of permeability) of the cover system as long as the surface soil did not crack due to desiccation drying. A reduction in the degree of saturation increased the water storage capacity of the cover soil. The intent was to have a soil cover that buffered extreme climate forcing factors by storing water during wet periods and releasing it back to the atmosphere during dry periods.

Alternative covers can be constructed using a variety of soil types and often make use of sand and silt soils. The covers are designed on the basis of water storage and water release (i.e., a water balance design). There must be sufficient capability for the water that infiltrates the cover on an annual basis to essentially be removed during the same year. Store-and-release soil covers are most suitable in arid and semiarid climates. The cover design must also take into consideration the distribution of precipitation throughout the year as well as the distribution of thermal energy to drive evaporation. Stated

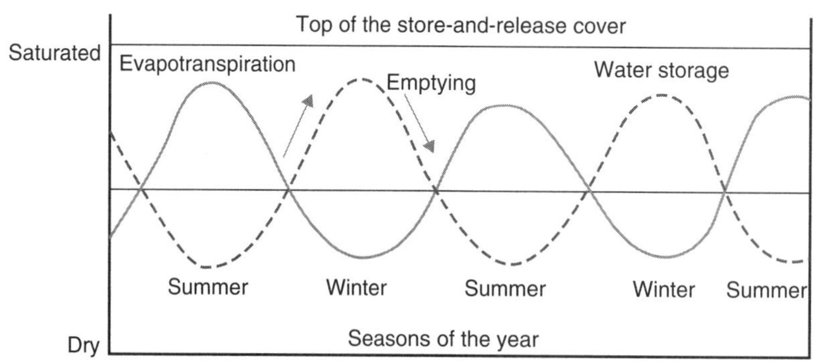

Figure 6.55 Concept of store and release used in the design of alternative cover systems.

another way, the cover material must be able to provide sufficient water storage capacity and thermal energy to evaporate water. These conditions must be met throughout each year. It is necessary to test the functionality of the cover by subjecting a proposed design to several years of past climatic conditions. The cover may be subjected to 10 or more years of past climatic record data. The numerical simulations may be reduced to time steps and as a result the analysis becomes computationally intensive. The concept of "storing water" and "releasing water" throughout the year is relatively simple; however, the analysis is dependent on the assessment of many variables as well as the estimation of several nonlinear unsaturated soil property functions.

6.4.2 Classification of Alternative Cover Systems

Alternative covers can be classified into two main categories: (i) monolithic covers (MCs) and (ii) capillary barrier covers (CBCs). Monolithic covers are also referred to as store-and-release covers, soil-plant covers, ET covers, or phytocovers or vegetated covers.

The MCs usually consist of a single, relatively thick layer of fine-textured soil with a relatively high water storage capacity. Alternative cover design requires there to be sufficient evaporative flux potential for the removal of water from the storage layer.

The CBCs usually take the form of two main layers. There will often be an additional organic layer placed on top of the basic cover layers. The primary cover material usually consists of a fine-textured soil overlying a relatively coarse soil layer. The coarse layer is meant to function as a capillary break layer in order for the cover to function properly. The design criteria that must be met are shown in Fig. 6.56. The first design criterion states that the matric suction at the interface between the fine and coarse soil layers must always remain higher than the intersection of the permeability functions for the two materials. Failure to meet this criterion results in unwanted infiltration. The second design criterion states that the matric suction at the interface between the fine and coarse soil layers must be kept greater than the residual suction of the coarse layer. These two criteria are illustrated on the permeability functions shown for the fine and coarse soil layers (Fig. 6.56).

Alternative covers have rapidly gained popularity due to lower susceptibility to cracking and ability to accommodate a wider range of materials. The concept behind the design of alternative covers is simple and easy to grasp since the focus is on the storage of water. However, the design verification analysis is quite complex and involves a day-by-day numerical simulation of typical climatic data. There is a wide range of possible soil cover materials that might possibly be used for soil layers. Once the alternative cover is constructed it usually has low postclosure maintenance requirements. There should, however, be postclosure studies to verify that the cover system is indeed functioning as it was designed to function.

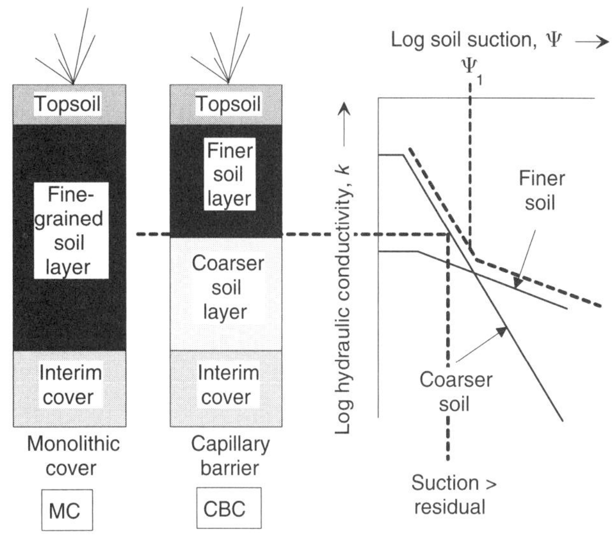

Figure 6.56 Concept of a "capillary break" used in the design of alternative cover systems.

SWCCs become the primary soil information required for alternative cover designs. There are two breaks on the SWCC that largely define the behavior of the soils upon desaturation: (i) air-entry value of the soil and (ii) residual suction of the soil. The SWCC for the soil defines the water storage capabilities of the cover system.

The infiltration and transmission abilities of the cover system are largely dependent upon the permeability function for the soil (i.e., hydraulic conductivity versus soil suction). The permeability function for the cover material (and the underlying material) is also a function of the saturated hydraulic permeability and the SWCC of the soil.

The design of a cover system appears to be a relatively straightforward application of unsaturated soil mechanics principles. The inherent simplicity and straightforward concepts associated with the theory and design of cover systems can lead to unwarranted confidence in the design procedure. The cover system must serve the purpose of being an interceptor between the soil and the atmosphere. The cover systems are called upon to perform satisfactorily under an extremely wide range of diverse moisture and thermal flux boundary conditions.

Another class of cover systems is known as "innovative covers." Innovative covers are classified mainly in terms of their state of development. In other words, innovative covers are still deemed to be at a fundamental stage of research and development. Examples of innovative covers are clay membrane barriers and polymer-clay nanocomposites with superior mechanical, thermal, and electrical properties.

6.4.3 Geotechnical Aspects That Give Rise to Modeling Challenges

There are several aspects of the cover design procedure that make the procedure an engineering challenge. The infiltration modeling involves the solution of a highly nonlinear

PDE that describes the seepage and water storage processes. The soil properties that produce nonlinearity in the moisture flow PDEs are the permeability function and the water storage function. Both of these soil properties are related to the SWCC.

If the SWCC has a relatively high air-entry value and if the rate of desaturation of the soil is relatively slow, then there are usually few difficulties associated with convergence of the moisture flow PDE. On the other hand, if the air-entry value is low (e.g., less than 10 kPa), or extremely low (e.g., less than 1.0 kPa), then convergence difficulties can be encountered with some software packages. The rate of desaturation also affects the nonlinearity of the PDE solution. There can be considerable difficulties encountered in obtaining a converged solution if the rate of desaturation defined by the SWCC is steep.

There is also the possibility that the solution may have converged but the computer code has made compromising assumptions to assist in the convergence. It is possible that these assumptions will result in convergence to the wrong answer. However, there are special computational techniques that should be used when dealing with highly nonlinear PDEs. The technique is referred to as adaptive grid refinement (AGR) and can be applied to the finite element method.

The criterion to ensure convergence is similar to the criteria imposed when solving finite difference numerical solutions using longhand calculations. The time-and-space relationship for the numerical solution must meet certain criteria in order to ensure convergence. The convergence criteria for a finite difference calculation are defined in terms of a β value which must be less than 0.5. The convergence criteria can be achieved by changing either the spatial distance between nodes or the time step. When solving a highly nonlinear, multidimensional finite element solution, it is necessary to make changes to the time steps and/or the size of the elements in order to ensure convergence to the correct solution. These changes in the finite element mesh must be performed in a dynamic manner that may not even be observable to the modeller.

6.4.4 Initial Conditions to Start Analysis

It is necessary to have a procedure for starting a nonlinear analysis. In other words, it is necessary to define the initial conditions for commencing the analysis. Some possible procedures for commencing a nonlinear analysis are suggested.

One of the procedures that can be used to establish initial conditions is to create a situation that simulates average conditions over the past several years. The intent in this case is to create a fictitious long-term steady-state situation. There are a number of ways a near steady state can be achieved. One procedure involves averaging the precipitation and weather conditions over a period of several years. For example, average conditions can be calculated over a 10-year period if weather records are available.

Figure 6.57 shows the average precipitation conditions calculated from a 10-year record of weather conditions. The total precipitation for the year is also equal to the average precipitation over the 10-year period. The average conditions can be input into a preliminary infiltration model along with an arbitrary set of initial conditions for other variables. Calculations can be performed over a number of years until the suction profile at the end of back-to-back years remains the same.

It often takes between 2 and 4 years of simulation for the suction profile to converge in simple cases and upward of 10 or more years of simulation in complex cases. It might even be necessary to "assist" the solution toward convergence onto the steady-state suction profile. Under steady-state conditions the suction profile with respect to depth can be used as the initial conditions for further modeling of actual weather conditions.

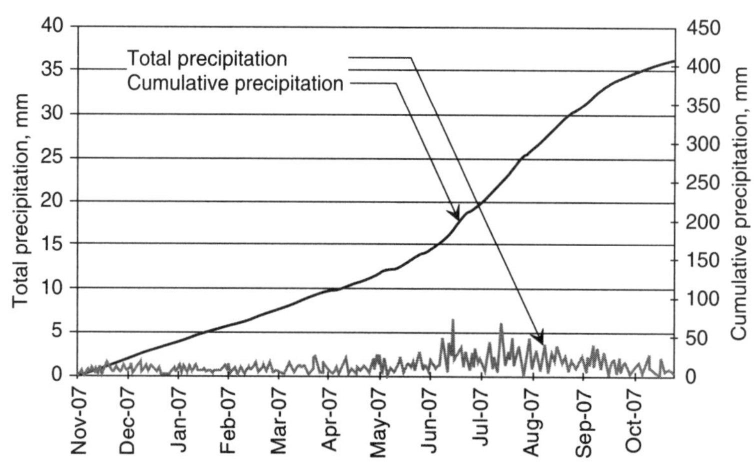

Figure 6.57 Average daily precipitation over 10 years used to generate an initial steady-state condition for the start of modeling.

6.4.5 Convergence Difficulties

The convergence of the highly nonlinear partial differential moisture flow equation is the most pressing challenge facing soil cover modelers (Shackelford, 2005). However, significant advances have been made in resolving problems of nonconvergence. There are two conditions that need to be satisfied in order to ensure that the correct modeling solution is attained. First, the solution of the PDE must converge for every time step of the lengthy design modeling period (e.g., 10 years). Second, the modeler needs assurance that the solution has converged to the correct solution. There are mathematical criteria that can be used to ensure that both of the mentioned criteria are met while solving the PDEs.

It is possible to plot tornado diagrams of the profiles of minimum and maximum values of soil suction or saturation levels on an annual basis. The tornado diagrams will be quite consistent from one year to the next once steady-state conditions have been reached.

The most successful solution to date has involved the use of the AGR technique (Oden, 1989; Yeh, 2000). When using this technique, the finite element mesh is continually refined, as necessary, in order to meet the conditions that promote convergence to the correct solution.

A simple finite difference formulation can be used to illustrate the adaptive time-and-space technique for solving nonlinear PDEs. Figure 6.58 illustrates the finite difference solution of the one-dimensional consolidation equation. When solving the finite difference equation, it is possible to define a variable β to ensure convergence to the correct solution (i.e., similar to the coefficient of consolidation times the time step and divided by the square of the spatial distance between nodes). In order for the finite difference solution to be correct, the defined variable β needs to be kept less than a value of 0.5. If β becomes greater than 0.5, errors can be incurred, as described in Fig. 6.58.

The coefficient of consolidation of a soil can change as the coefficient of permeability of the soil changes. Changes can be substantial and the β variable can readily go outside

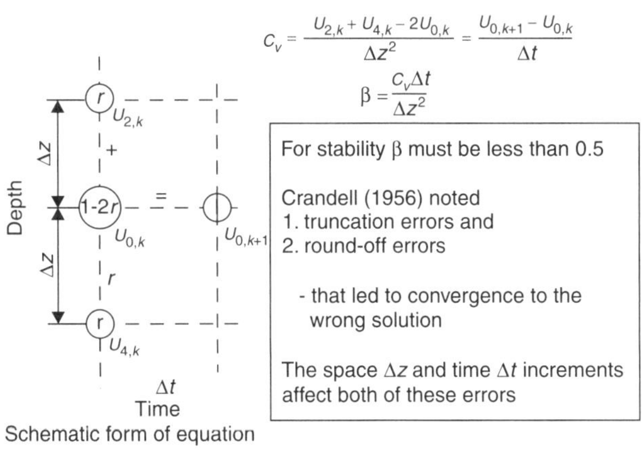

Figure 6.58 Controlling convergence through use of adjustment of the time-and-space steps.

acceptable limits and cause the solution to be in error. The problem should be solved by changing either the time interval or the spatial interval in order to bring the β variable into an acceptable range. It is easy to see how this can be accomplished when using a finite difference solution. When performing a finite element solution, the spatial changes need to be accomplished by changing the size of the finite elements. Changing element sizes while the solution is underway is a challenge, but this technique has been incorporated into the solvers of all finite element solutions used by SoilVision Systems (Fredlund and Thode, 2010). The AGR technique appears to generally meet the conditions necessary for convergence to the correct solution when solving highly nonlinear PDEs.

6.4.6 Verification of Cover Designs

Field monitoring is an essential part of quality assurance (QA) for cover systems. Measurements need to be taken to ensure that the cover system is performing as anticipated from the design. Field monitoring can provide verification that the cover performance is satisfactory.

A typical field monitoring system for a cover system may involve (i) the installation of a lysimeter, (ii) the installation of a weather station, (iii) soil suction measurements within the soil cover, and (iv) water content measurements within the soil cover (Fig. 6.59). Equipment is available commercially for each of the mentioned measurements. All types of instrumentation do not necessarily need to be installed in all cover systems. Rather, there is often a hierarchy on the instrumentation monitoring items. For example, a lysimeter might be considered to be the most important item of instrumentation.

Design details for the lysimeter are important in order to obtain accurate measurements of infiltration. The lysimeter may need to be on the order of 10–20 m in length, of proper depth, and filled with appropriate material in order to yield accurate results (Benson, 2002).

Weather stations are reasonably priced and can be placed at the site of a proposed cover system. While the cost of the weather station is modest, its maintenance and servicing may entail considerable expenditure. It must also be remembered that visual field inspections and reporting need to be an ongoing part of the monitoring process.

6.4.7 Uniqueness of Cover Design Procedure

There are many assumptions that need to be made as part of the design procedure for a cover system. The soil conditions can change with time due to the effects of weathering and freeze-thaw conditions with the result that the soil properties become far from the initially measured or assumed values. The changes can prove to differ by orders of magnitude from initial compacted or placement conditions. This does not make a realistic design impossible but simply shows that much greater care and detail must be given to the assessment of the unsaturated soil properties.

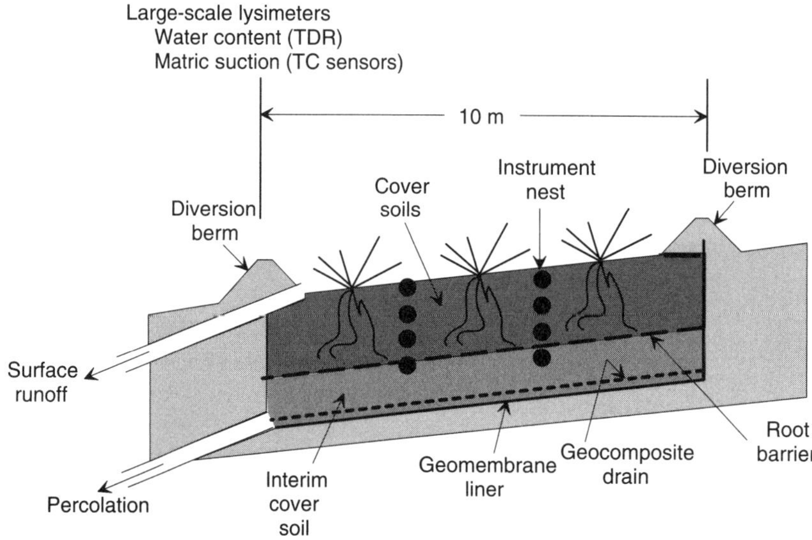

Figure 6.59 Types of measurements required for verification monitoring of cover systems.

The climatic quantification that provides the final "net moisture flux" at ground surface has many inherent broad assumptions. The tendency to focus the analysis on average conditions may cause a deviation from the real performance of the cover system. In other words, extreme weather conditions may need to be more carefully evaluated during cover design. For example, extreme conditions where there is significant water runoff can result in serious erosion problems.

The effect of cracks forming in a clay-rich soil can radically change the soil response to infiltration and exfiltration. Unsaturated soil properties are highly nonlinear and may even change to be bilinear in character. These extreme conditions need to be given more attention and may even turn out to be a controlling factor once the clay content reaches a particular percentage.

6.4.8 Long-Term Integrity

The assessment of long-term integrity of soil cover systems may increasingly become an activity of concern within geotechnical engineering. Soil cover designs have grown rapidly and the evaluation of their performance is essential for ensuring their long-term integrity and acceptability.

There appears to have been a strong dependence upon average or typical conditions for design purposes; however, it is the influence of extreme conditions that may have the most dramatic effect. This is true for both soil properties and weather conditions. Soils crack and the properties of the overall soil mass can change dramatically. Extreme weather conditions can be destructive, causing failures to cover systems.

CHAPTER 7

Theory of Water Flow through Unsaturated Soils

7.1 INTRODUCTION

The rate of flow of water through a porous medium is regulated by the hydraulic conductivity or coefficient of permeability of the soil. The coefficient of permeability is the primary soil property required when analyzing steady-state and transient (or unsteady-state) flow of an incompressible fluid through a porous medium. The water storage properties must also be defined when analyzing transient flow problems.

The coefficient of permeability is generally assumed to be a constant when analyzing flow through saturated soil. However, the coefficient of permeability for an unsaturated soil can vary widely depending on the stress state (or degree of saturation) of the soil. The coefficient of permeability of an unsaturated soil takes on the form of a mathematical function. Any change in the stress state of a soil can affect the coefficient of permeability, but it is changes in soil suction beyond the air-entry value that have the greatest effect on the coefficient of permeability.

The coefficient of permeability generally changes by several orders of magnitude as a soil desaturates. Water can only flow through that portion of a porous medium that consists of water. As the amount of water in a soil decreases (i.e., a decrease in the degree of saturation), the coefficient of permeability decreases because there is less cross-sectional area through which water can flow. However, there is not a one-to-one relationship between the amount of water in the soil and the coefficient of permeability. The coefficient of permeability decreases at a much faster rate than the degree of saturation. A reduction in the amount of water in the soil also increases the tortuosity of the flow path. As a consequence, an arithmetic decrease in the volume of water in the soil generally results in a logarithmic decrease in the coefficient of permeability. A dry soil has a much lower coefficient of permeability than a wet soil.

Mualem (1976b) performed a laboratory experiment that assists in understanding the coefficient of permeability of unsaturated materials. The experiment involved flowing water through glass beads subjected to varying negative pore-water pressures (Fig. 7.1). The glass beads were initially saturated and then subjected to a series of applied negative water pressures (or matric suctions). The coefficient of permeability of the glass beads was measured starting at saturated conditions. The results showed that the coefficient of permeability of the glass beads started to decrease when the matric suction was 3 kPa. A further increase in suction caused the coefficient of permeability to drop by several orders of magnitude.

It was also noted that the permeability function was different when the beads were in a drying mode than when the beads were in a wetting mode. In other words, the permeability function was hysteretic in the sense that it was dependent upon whether the soil was undergoing a drying mode or a wetting mode. The water content of the glass beads was also measured during the drying and wetting processes, as shown in Fig. 7.2. The hysteresis in the water content versus matric suction relationship was related to the hysteresis in the water permeability function.

The results from the Mualem (1976b) experiment showed that a decrease in the coefficient of permeability commenced at the air-entry value of the soil. The relationship between the permeability function and the SWCC would subsequently become the basis for a variety of estimation procedures proposed for the estimation of the permeability function. The SWCC along with the saturated coefficient of permeability became the basis for estimating the water permeability function.

7.2 THEORY OF FLOW OF WATER

Several concepts have been used to explain the flow of water through an unsaturated soil. For example, it has been suggested that a water content gradient, a matric suction gradient, or a hydraulic head gradient could form the driving potential for water flow through a porous media. Geotechnical engineers have historically described water flow through a saturated soil in terms of hydraulic head gradient. It is important to also use hydraulic head gradient for water flow through

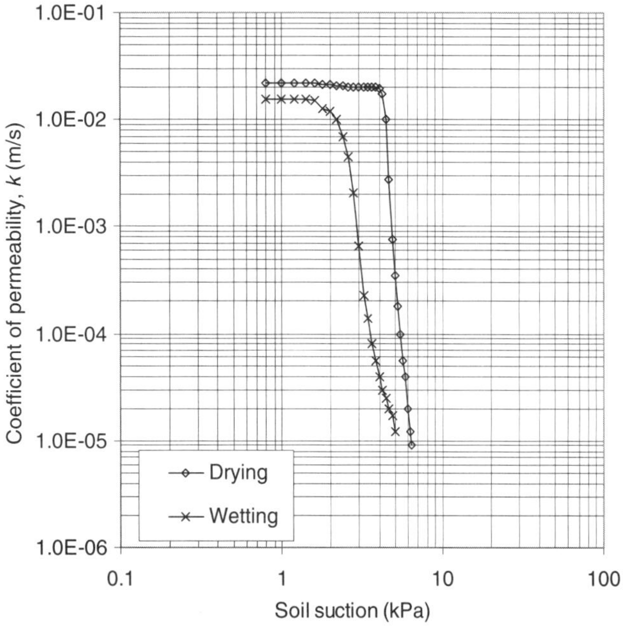

Figure 7.1 Shape of drying and wetting permeability functions for glass beads (after Mualem, 1976b).

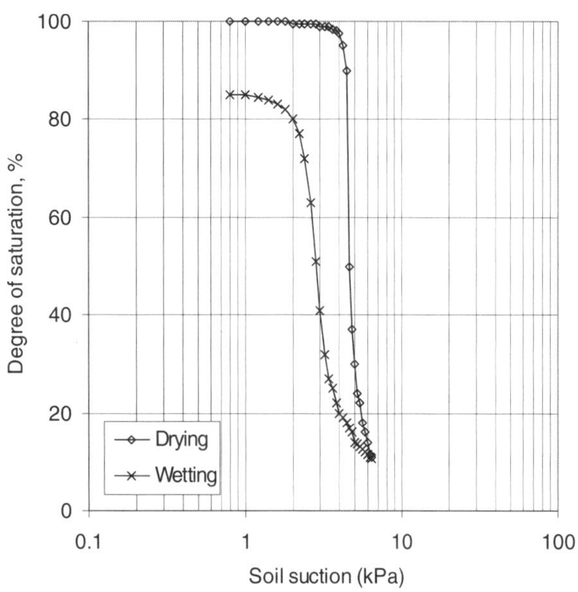

Figure 7.2 SWCCs for glass beads showing drying and wetting hysteresis (after Mualem, 1976b).

an unsaturated soil provided hydraulic head is shown to be an acceptable and fundamental description of flow.

Let us first consider some alternatives to using a hydraulic head gradient to define water flow through an unsaturated soil. A gradient of (volumetric) water content has sometimes been used to describe the flow of water through unsaturated soils. It is assumed that water flows from a point of high water content to a point of lower water content. This form of flow law, however, does not provide a fundamental basis for describing water flow since water can also flow from a region of low water content to a region of high water content when there are variations in the soils involved, hysteretic effects, or stress history variations. Consequently, the water content gradient should not be used as a fundamental driving potential for the flow of water (Fredlund, 1981b).

A matric suction gradient has sometimes been used as the driving potential for water flow in an unsaturated soil. However, the flow of water does not fundamentally and exclusively depend upon a gradient in matric suction (i.e., $u_a - u_w$, where u_a is pore-air pressure and u_w is pore-water pressure). Figure 7.3 demonstrates three hypothetical cases where the air and water pressure gradients are controlled across an unsaturated soil element at a constant elevation. In all cases, the air and water pressures on the left-hand side are greater than the pressures on the right-hand side.

The matric suction on the left-hand side may be smaller than on the right-hand side (case 1), equal to the right-hand side (case 2), or larger than on the right-hand side (case 3). However, air and water will flow from left to right in response to the (pressure) head gradient in the individual phases, regardless of the matric suction gradient. Even in case 2, where the matric suction gradient is zero, air and water will still flow as two independent phases. It can be concluded that water cannot always be assumed to flow

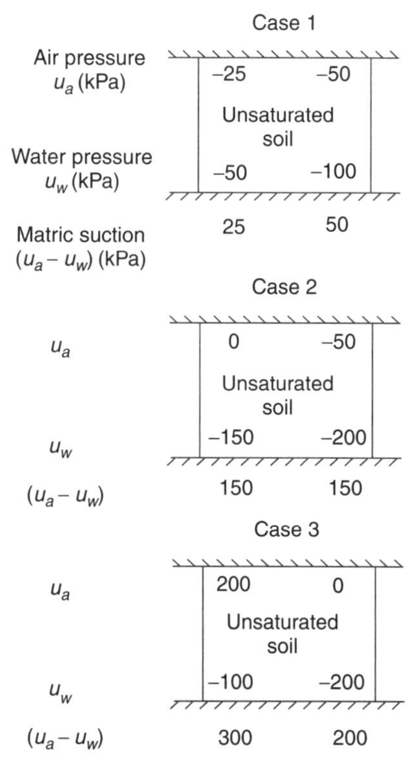

Figure 7.3 Air pressures, water pressures, and matric suction gradients across unsaturated soil specimen.

from a point of low suction to a point of higher suction, not even when the elevation head is constant.

Water flow can most appropriately be defined in terms of a hydraulic head gradient (i.e., a pressure head plus elevation head gradient in the water). In other words, the matric suction gradient is not the fundamental driving potential for the flow of water in an unsaturated soil. In the special case where the air pressure gradient is zero, the matric suction gradient becomes numerically equal to the pressure gradient in the water. Air pressure in a soil continuum is generally a constant, and this is probably the reason for the suggestion that water flows in response to a matric suction gradient. The elevation head component has then been omitted if the soil suction gradient is used to describe water flow.

The flow of water through any soil is governed by the pressure head gradient in the water phase in addition to a gradient due to elevation head differences. The pressure and elevation gradients are linearly combined to give the hydraulic head gradient that constitutes the fundamental driving potential for water flow. The hydraulic head gradient can be applied to each fluid phase of a multi-fluid-phase system. Hydraulic head forms the fundamental driving potential for flow in saturated and unsaturated soils.

7.2.1 Driving Potential for Water Phase

The driving potential for the flow of water can be defined in terms of the energy or capacity to do work. The energy at a point is computed relative to an elevation datum. The datum can be chosen arbitrarily because it is the gradient in the energy head between two points that is of importance in describing water flow.

A point in the water phase has three primary components of energy, namely, gravitational, pressure, and velocity. Figure 7.4 shows point A in the water phase which is located at an elevation y above an arbitrary datum. Let us consider the energy state of point A. Point A has a gravitational energy E_g which can be written as

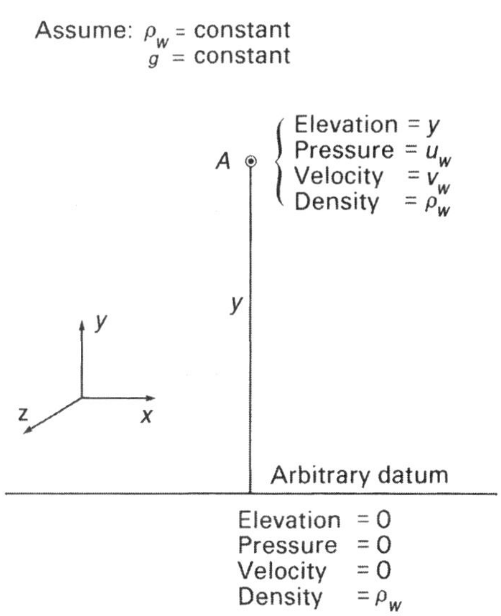

Figure 7.4 Energy at point A in y-direction relative to arbitrary datum.

$$E_g = M_w g y \tag{7.1}$$

where:

E_g = gravitational energy,
M_w = mass of water at point A,
g = gravitational acceleration, and
y = elevation of point A above the datum.

The component of energy due to the pressure at point A is written as follows (Freeze and Cherry, 1979):

$$E_p = M_w \int_0^{u_w} \frac{V_w}{M_w} du_w \tag{7.2}$$

where:

E_p = pressure energy,
u_w = pore-water pressure at point A, and
V_w = volume of water at point A.

The pressure energy component can also be written:

$$E_p = M_w \int_0^{u_w} \frac{du_w}{\rho_w} \tag{7.3}$$

where:

ρ_w = density of water.

The above equation takes the following form when water density ρ_w is constant:

$$E_p = \frac{M_w u_w}{\rho_w} \tag{7.4}$$

The flow rate of water at point A gives rise to a velocity energy E_v:

$$E_v = \frac{M_w v_w^2}{2} \tag{7.5}$$

where:

E_v = velocity energy and
v_w = flow rate of water at point A (i.e., in the y-direction).

The total potential energy at point A is the summation of the gravitational, pressure, and velocity components:

$$E = M_w g y + \frac{M_w u_w}{\rho_w} + \frac{M_w v_w^2}{2} \tag{7.6}$$

where:

E = total energy.

The total energy at point A can be expressed as energy per unit weight, which is called a potential or a hydraulic head. The hydraulic head h_w at point A is obtained by dividing the total energy by the unit weight of water at point A (i.e., $M_w g$):

$$h_w = y + \frac{u_w}{\rho_w g} + \frac{v_w^2}{2g} \tag{7.7}$$

where:

h_w = total energy head.

The total energy head consists of three components, namely, the gravitational head y, the pressure head $(u_w/\rho_w g)$, and the velocity head $(v_w^2/2g)$. The velocity head in a soil is negligible in comparison with the gravitational and the pressure heads. The hydraulic head at any point in the soil mass is generally written as the addition of the gravitational head and the water pressure head, as shown in the following equation:

$$h_w = y + \frac{u_w}{\rho_w g} \tag{7.8}$$

The components of hydraulic head have the dimension of length. Hydraulic head is a measurable quantity and the difference in hydraulic head between two points provides a gradient value that describes the available energy to cause water flow in saturated and unsaturated soils. Devices such as piezometers and tensiometers can be used to measure the in situ pore-water pressure at a point (Fig. 7.5). The distance between the elevation of the point under consideration and the datum indicates the elevation head (i.e., y_A and y_B).

A piezometer can be used to measure the pore-water pressure at a point when the pore-water pressure is positive (e.g., point B in Fig. 7.5). A tensiometer can be used to measure the pore-water pressure when the pressure is negative (e.g., point A in Fig. 7.5).

The water level in the measuring device will rise or drop, depending upon the pore-water pressure at the point under consideration. For example, the water level in the piezometer can rise above the elevation of point B a distance equal to the positive pore-water pressure head at point B. The water level in the tensiometer can alternatively drop below the elevation of point A a distance equal to the negative pore-water pressure head at point A. The distance between the water level in the measuring device and the datum is the sum of the gravitational and pressure heads (i.e., the hydraulic head).

In Fig. 7.5, point A has a higher total head than point B [i.e., $h_w(A) > h_w(B)$]. Water will flow from point A to point B due to the total head gradient between these two points. The driving potential causing flow in the water phase has the same form for both saturated soils (i.e., point B) and unsaturated soils (i.e., point A) (Freeze and Cherry, 1979). Water will flow from a point of high total head to a point of low total head, regardless of whether the pore-water pressures are positive or negative.

Osmotic suction has sometimes been included as a component in the total head equation for water flow. However, it is better to visualize the osmotic suction gradient as an independent driving potential for the osmotic diffusion process (Corey and Kemper, 1961). Osmotic diffusion is a process where ionic or molecular constituents move as a result of their kinetic activity. An osmotic gradient across a semipermeable membrane causes the movement of water through the membrane. On the other hand, the bulk flow of solutions (i.e., pure water and dissolved salts) in the absence of a semipermeable membrane is governed by the hydraulic head gradient. Therefore, it is better to analyze bulk flow of water separately from the osmotic diffusion process since the two mechanisms are independent (Corey, 1977).

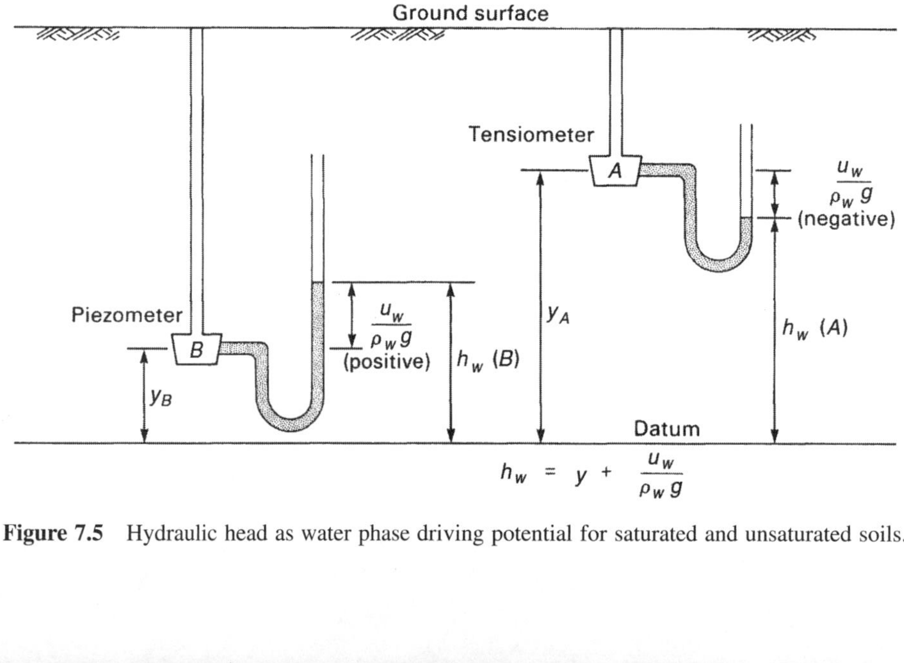

Figure 7.5 Hydraulic head as water phase driving potential for saturated and unsaturated soils.

7.3 DARCY'S LAW FOR UNSATURATED SOILS

The flow of water in a saturated soil can be described using Darcy's law. Darcy (1856) postulated that the rate of water flow through a soil mass was proportional to the hydraulic head gradient and could be described using the following equation:

$$v_w = -k_w \frac{\partial h_w}{\partial y} \tag{7.9}$$

where:

v_w = flow rate of water,
k_w = coefficient of permeability with respect to the water phase, and
$\partial h_w/\partial y$ = hydraulic head gradient in the y-direction designated as i_{wy}.

The coefficient of proportionality between the flow rate of water and the hydraulic head gradient is called the coefficient of permeability, k_w. The coefficient of permeability is relatively constant for a specific saturated soil. Darcy's law can also be written for the x- and z-directions. The negative sign in the flow equation indicates that water flows in the direction of a decreasing hydraulic head.

Darcy's law also applies for the flow of water through an unsaturated soil (Buckingham, 1907; Richards, 1931; Childs and Collis-George, 1950). However, the coefficient of permeability in an unsaturated soil cannot generally be assumed to be constant. Rather, the coefficient of permeability is a variable which is predominantly a function of the water content or the matric suction of the unsaturated soil.

Water can be visualized as flowing only through the pore spaces that are filled with water. Air-filled pores are non-conductive channels to the flow of water. Therefore, the air-filled pores in an unsaturated soil can be considered as behaving similarly to the solid phase, and the soil can be treated as a saturated soil having reduced water content (Childs, 1969). The validity of Darcy's law can be verified for an unsaturated soil using an experiment similar to that used for the verification of the flow law for saturated soils. However, the volume of water (or water content) must remain constant while the hydraulic head gradient is varied.

Experiments to verify Darcy's law for unsaturated soils have been performed, and key experimental results are presented in Fig. 7.6 (Childs and Collis-George, 1950). A column of unsaturated soil held at a uniform water content and a constant-water-pressure head was subjected to various gradients of gravitational head. The results indicate that at a specific water content the coefficient of permeability k_w is constant for various hydraulic head gradients applied to the unsaturated soil. In other words, the rate of water flow through an unsaturated soil is linearly proportional to the hydraulic head gradient, with the coefficient of permeability being a constant. This behavior is similar to that observed for saturated soils.

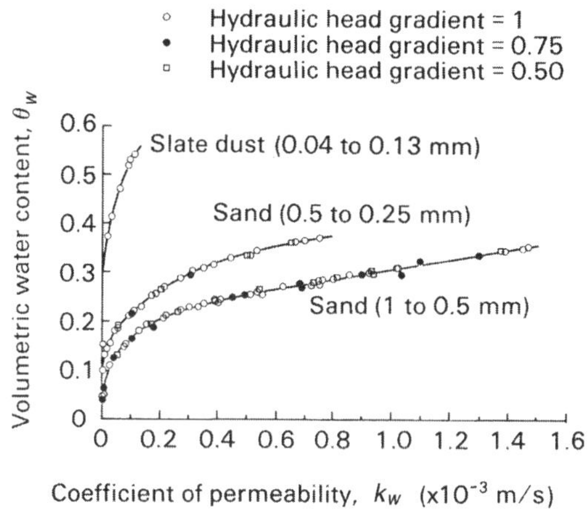

Figure 7.6 Experimental verification of Darcy's law for water flow through an unsaturated soil (after Childs and Collis-George, 1950).

The results of the Childs and Collis-George (1950) experiment confirm that Darcy's law can also be applied to unsaturated soils. However, it needs to be noted that the magnitude of the coefficient of permeability will differ at various volumetric water contents θ in an unsaturated soil.

7.3.1 Coefficient of Permeability with Respect to Water Phase

The coefficient of permeability with respect to the water phase, k_w, is a measure of the ease with which water can flow in the space available for water flow in the unsaturated soil. The coefficient of permeability depends upon the properties of the fluid and the properties of the porous medium. Different types of fluid (e.g., water and oil) or different types of soil (e.g., sand and clay) produce different values for the coefficient of permeability k_w.

7.3.2 Fluid and Porous Medium Components

The coefficients of permeability with respect to water, k_w, can be expressed in terms of the intrinsic permeability K, which can mathematically be described as follows:

$$k_w = \frac{\rho_w g}{\nu} K = \frac{\gamma}{\nu} K \tag{7.10}$$

where:

K = intrinsic permeability of the soil, m^2,
g = gravitational acceleration, m/s^2,
ν = dynamic viscosity, cP,
ρ = density, kg/m^3, and
γ = unit weight of the medium, N/m^3.

Fluids properties can be "scaled" to the properties of water at standard conditions (Parker et al., 1987). The intrinsic permeability is approximately 10^{-12} m^2 for a soil that has an assumed hydraulic conductivity of 10^{-5} m/s. The density of water is assumed to be 1000 kg/m^3 and the dynamic viscosity v is selected for a temperature of 20°C. (Note that 1 cP is equal to 10^{-3} Pa·s or 1 mPa·s.)

Intrinsic permeability is often used to convert flow properties from one discipline to another discipline (e.g., between petroleum, agriculture disciplines, and geotechnical engineering). The concept of intrinsic permeability also has an important role to play in unsaturated soil mechanics because the soil has two fluid phases (i.e., water and air). For example, measurements of saturated water hydraulic conductivity can be used to calculate a value for air hydraulic conductivity for a dry soil at the same void ratio. Saturated water hydraulic conductivity measurements can also be used along with a SWCC to estimate an air permeability function for an unsaturated soil. The properties of fluids in a porous medium can provide an estimate of the relative magnitude differences between the ease of flow of the two fluids (e.g., air and water).

The intrinsic permeability of a soil, K, represents the characteristics of the porous medium and is independent of the fluid properties. The porous medium, in turn, is a function of the volume-mass properties of the soil. In geotechnical engineering, the coefficient of permeability k_w is used to embraces the overall effect of variables related to the porous medium and the pore fluid. The coefficient of permeability k_w is the variable form most commonly used in geotechnical engineering and is used throughout this book.

7.3.3 Relationship between Permeability and Volume-Mass Properties

The coefficient of permeability k_w is a function of any two of three possible volume-mass properties (Lloret and Alonso, 1980; Fredlund, 1981b):

$$k_w = k_w(S, e) \tag{7.11}$$

or

$$k_w = k_w(e, w) \tag{7.12}$$

or

$$k_w = k_w(w, S) \tag{7.13}$$

where:

S = degree of saturation,
e = void ratio, and
w = gravimetric water content.

The coefficient of permeability in a *saturated* soil is a function of the void ratio (Lambe and Whitman, 1979). However, the coefficient of permeability is generally assumed to be a constant when analyzing most steady-state and transient flow problems in saturated soils.

The coefficient of permeability in an *unsaturated* soil is significantly affected by combined changes in void ratio and degree of saturation (or water content) of the soil. Water flows through the pore space filled with water; therefore, the percentage of the voids filled with water is an important factor. As a soil begins to desaturate, air first replaces some of the water in the large pores. The onset of desaturation causes water to flow through the smaller pores. The pathway through the smaller pores leads to increased tortuosity and consequently much slower flow. An increase in matric suction in the soil leads to a further decrease in the pore volume occupied by water. The air-water interface is drawn closer and closer to the soil particles (Fig. 7.7). As a result, the coefficient of permeability with respect to the water phase decreases rapidly as the space available for water flow is reduced.

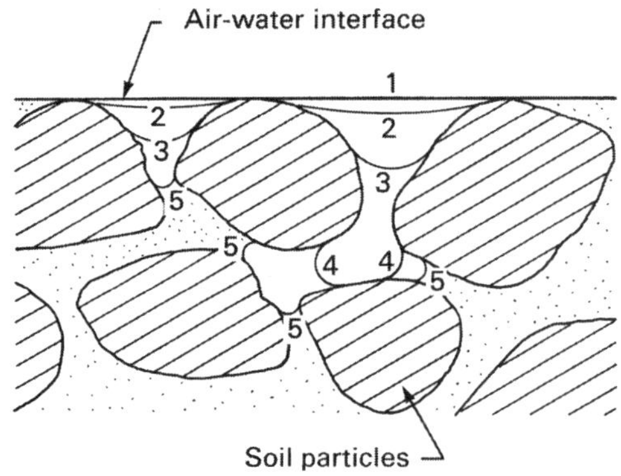

Figure 7.7 Development of unsaturated soil by retraction of air-water interface under increasing matric suctions (i.e., stages 1–5) (after Childs, 1969).

7.3.4 Effect of Variations in Degree of Saturation on Coefficient of Permeability

The coefficient of permeability of an unsaturated soil can vary considerably during a transient process as a result of changes in the volume-mass properties. The change in void ratio in an unsaturated soil may be small in soils of low compressibility and its effect on the coefficient of permeability may be secondary. However, changes in degree of saturation generally produce large changes in the coefficient of permeability. As a result, the coefficient of permeability is often described as a singular function of the degree of saturation, S. The degree of saturation is related to matric suction and the relationship is referred to as the SWCC (Fig. 7.8a).

Numerous semiempirical equations for the coefficient of permeability have been derived using the matric suction versus degree of saturation curve. The soil pore-size distribution forms the basis for predicting the coefficient of permeability. The pore-size distribution concept originated in the soil sciences and has been introduced into geotechnical engineering. The pore-size distribution has also been used in other disciplines to provide an estimate of the rate at which the coefficient of permeability decreases with the degree of saturation of a soil.

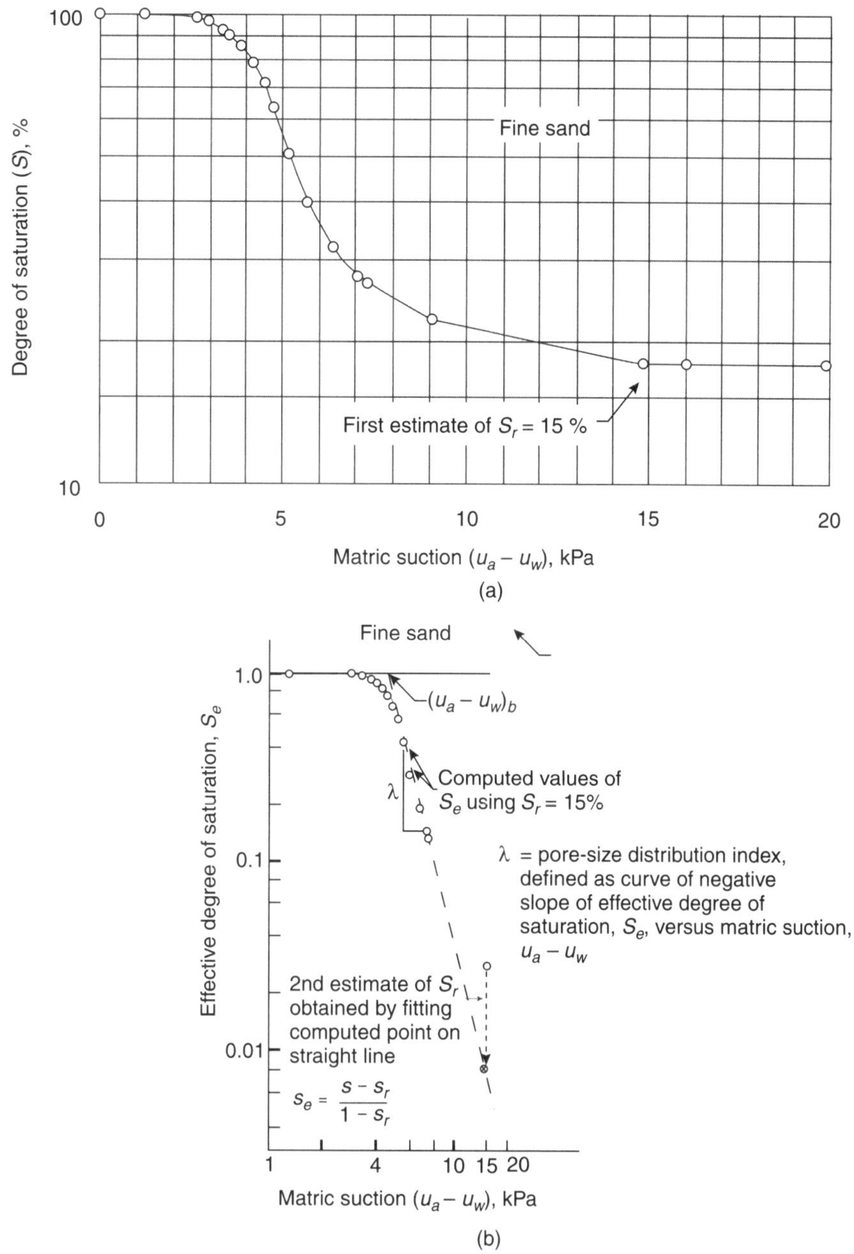

Figure 7.8 Determination of air-entry value $(u_a - u_w)_b$, residual degree of saturation S_r, and pore-size distribution index λ : (a) matric suction versus degree of saturation curve; (b) effective degree of saturation versus matric suction curve (after Brooks and Corey, 1964).

Further details on how the coefficient-of-permeability function can be estimated from the matric suction versus degree of saturation curve (or the SWCC) are presented in the following sections as well as under estimation procedures for the permeability function.

7.3.5 Relationship between Coefficient of Permeability and Degree of Saturation

Burdine (1953) and Brooks and Corey (1964) suggested that the permeability function could be obtained using the saturated coefficient of permeability and the matric suction versus degree of saturation relationship. The matric suction versus degree of saturation curve exhibits hysteresis. Only the drainage or drying curve is used to illustrate the concepts and theory behind the estimation of the permeability function. The soil structure is assumed to be incompressible.

There are three soil parameters that can be identified from the matric suction versus degree of saturation curve: the air-entry value of the soil, $(u_a - u_w)_b$, the residual degree of saturation, S_r, and the pore-size distribution index, λ. Corey (1954) suggested visualizing these parameters on a plot of effective degree of saturation S_e versus matric suction (Fig. 7.8b):

$$S_e = \frac{S - S_r}{1 - S_r} \qquad (7.14)$$

where:

S_e = effective degree of saturation,
S_r = residual degree of saturation (decimal form), and
S = any degree of saturation (decimal form).

The residual degree of saturation S_r was defined as the degree of saturation at which an increase in matric suction did not produce a significant change in the degree of saturation. While the intent of the definition of residual degree of saturation is clear, the numerical determination of this value is left somewhat vague.

The effective degree of saturation S_e can be computed by first estimating a value for the residual degree of saturation S_r (Fig. 7.8a). The effective degree of saturation was then plotted against the matric suction, as illustrated in Fig. 7.8b. A horizontal and a sloping line can be drawn through the data points. However, points at high-matric-suction values may not lie on the straight line used for the first estimate of the residual degree of saturation (Fig. 7.8b). It was reasoned that the point corresponding to high matric suctions should be forced to lie on the straight line by estimating a new value for the residual degree of saturation S_r. A second estimate of the residual degree of saturation was then used to recalculate values for effective degree of saturation S_e. A new plot of the matric suction versus effective degree of saturation curve can then be obtained. The above procedure is repeated until all of the points on the sloping line constitute a straight line. This usually occurs after the second estimate of the residual degree of saturation. The use of the above theory is based on the assumption that coefficient of permeability asymptotically approaches zero as the residual degree of saturation of the soil is exceeded.

The air-entry value of the soil, $(u_a - u_w)_b$, is the matric suction value that must be exceeded before air recedes into the soil pores. The air-entry value is also referred to as the "displacement pressure" in petroleum engineering or the "bubbling pressure" in ceramics engineering (Corey, 1977). The term "air-entry value" has been most commonly used in the agricultural sciences and in geotechnical engineering. The air-entry value is a measure of the maximum pore size in a soil. The intersection point between the straight sloping line and the complete saturation ordinate (i.e., S_e = 1.0) in Fig. 7.8b defines the air-entry value of the soil. The sloping line passing through points where the matric suction is greater than the air-entry value can be described by the following equation:

$$S_e = \left\{ \frac{(u_a - u_w)_b}{(u_a - u_w)} \right\}^{\lambda} \quad \text{for } (u_a - u_w) > (u_a - u_w)_b \qquad (7.15)$$

where

λ = pore-size distribution index defined as the negative slope of the effective degree of saturation S_e versus matric suction $(u_a - u_w)$.

Soils with a wide range of pore sizes have a small value for the pore-size distribution index λ. Soils with a uniform pore-size distribution have a larger value for λ. Figure 7.9 presents typical λ values for various soils which have been obtained from matric suction versus degree of saturation curves.

The coefficient of permeability with respect to the water phase, k_w, can be predicted from the matric suction versus degree of saturation curves as follows (Brooks and Corey, 1964):

$$k_w = k_s \quad \text{for} \quad u_a - u_w \le (u_a - u_w)_b \qquad (7.16)$$

$$k_w = k_s S_e^{\delta} \quad \text{for} \quad u_a - u_w > (u_a - u_w)_b \qquad (7.17)$$

where:

k_s = coefficient of permeability with respect to the water phase for the saturated soil (i.e., S = 100%) and
δ = an empirical constant for the permeability function.

The empirical constant δ can be related to the pore-size distribution index λ:

$$\delta = \frac{2 + 3\lambda}{\lambda} \qquad (7.18)$$

Table 7.1 presents several δ values and their corresponding pore-size distribution indices λ measured for various soil types.

7.3.6 Relationship between Water Coefficient of Permeability and Matric Suction

A closed-form equation for the permeability function can be obtained by substituting Eq. 7.15 for the effective degree of saturation into Eq. 7.17. The resulting equation defines the coefficient of permeability of the soil at suction values

Table 7.1 Suggested Values of Constant δ and Pore-Size Distribution Index λ for Various Soils

Soils	δ	λ	Source
Uniform sand	3.0	∞	Irmay (1954)
Soil and porous rocks	4.0	2.0	Corey (1954)
Natural sand deposits	3.5	4.0	Averjanov (1950)

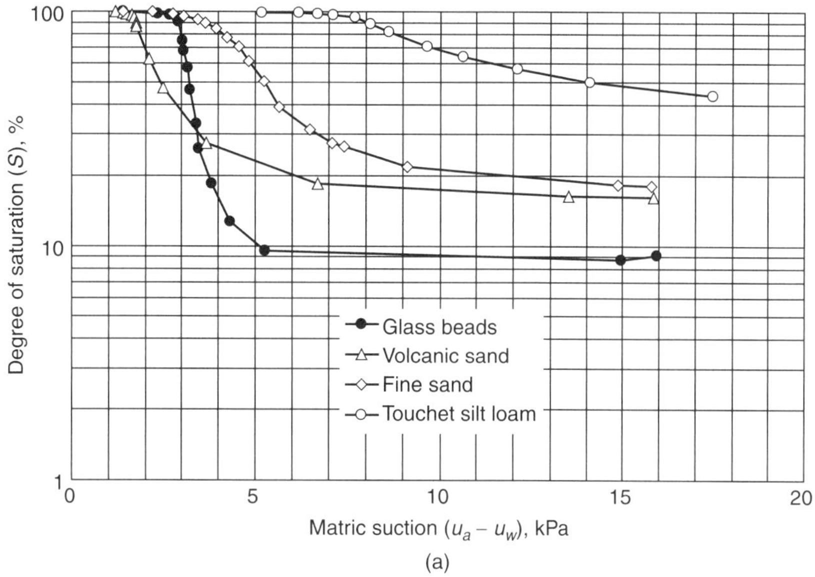

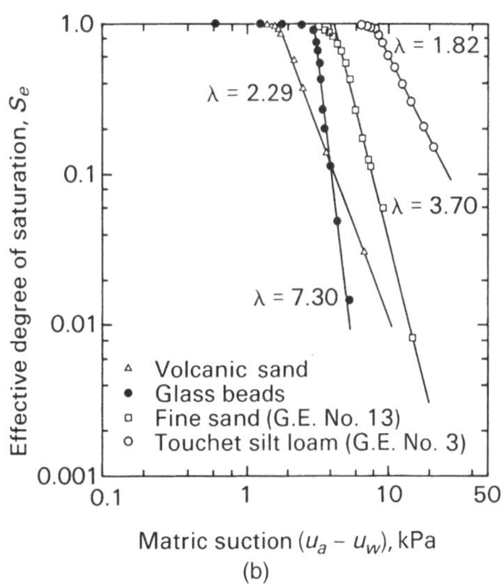

Figure 7.9 Typical matric suction versus degree of saturation curves for soils with varying pore-size distribution indices λ: (a) matric suction versus degree of saturation; (b) effective degree of saturation versus matric suction (after Brooks and Corey, 1964).

greater than the air-entry value of the soil (Brooks and Corey, 1964):

$$k_w = k_s \left\{ \frac{(u_a - u_w)_b}{(u_a - u_w)} \right\}^{2+3\lambda} \quad \text{for } u_a - u_w > (u_a - u_w)_b \tag{7.19}$$

Several other relationships between the coefficient of permeability and matric suction have also been proposed and are shown in Table 7.2 (Gardner, 1958a; Arbhabhirama and Kridakorn, 1968).

The water coefficients of permeability k_w corresponding to various degrees of saturation can be expressed as the relative water phase coefficient of permeability k_{rw} in percent form:

$$k_{rw} = \frac{k_w \times 100}{k_s} \tag{7.20}$$

where:

k_s = saturated soil coefficient of water permeability.

Experimental data for sandstone expressed in terms of the relative permeability are shown in Fig. 7.10. A hydrocarbon liquid was used in the experiments instead of water in order to produce a more stable soil structure and consistent fluid properties. The results are essentially the same as for water flow since the relative permeability is not a function of the fluid properties.

The shape of the relative permeability plot shows an exponential reduction in the relative coefficient of permeability with respect to the degree of saturation of the soil. This type of plot allows a visualization of the change in the coefficient of permeability over about two orders of magnitude.

Table 7.2 Relationships between Water Coefficient of Permeability and Matric Suction

Equations	Number	Source	Symbols
$k_w = k_s$ for $u_a - u_w \leq (u_a - u_w)_b$ $k_w = k_s \left\{ \frac{(u_a - u_w)_b}{u_a - u_w} \right\}^{\eta}$ for $u_a - u_w > (u_a - u_w)_b$		Brooks and Corey (1964)	η = empirical constant $= 2 + 3\lambda$
$k_w = \frac{k_s}{1 + a\left\{(u_a - u_w)/(\rho_w g)\right\}^n}$		Gardner (1958a)	a, n = constant
$k_w = \frac{k_s}{\left\{(u_a - u_w)/(u_a - u_w)_b\right\}^{n'} + 1}$		Arbhabhirama and Kridakorn (1968)	n' = constant

In reality, the coefficient of permeability can change several more orders of magnitude as the degree of saturation of the soil is reduced.

An empirical mathematical form for a relationship between coefficient of permeability and matric suction was proposed by Gardner (1958a), and typical shapes for the function are shown in Fig. 7.11. The equation provides a continuous permeability function extending from near-saturated conditions under low matric suctions to high-suction conditions. The permeability function is defined by two constants or fitting parameters: a and n. The constant n defines the slope of the permeability function while the a constant bears a relationship to the breaking point on the permeability function (i.e., the air-entry value of the soil). Four typical functions with differing values for a and n are illustrated in Fig. 7.11.

The Gardner permeability equation does not meet the requirements for the estimation of a permeability function since a and n are determined through use of a best-fit regression analysis of permeability measurements. In other words, coefficient-of-permeability data must first be measured for various matric suction values. It is logical to infer a relationship between Gardner's a parameter and the air-entry value of the soil. However, it does not necessarily follow that the n parameter bears a unique relationship to the slope of the SWCC.

The term "permeability function" is used whenever the coefficient of permeability is written as a function of any volume-mass variable or stress state variable. In geotechnical engineering, permeability function most commonly refers to a mathematical relationship between coefficient of permeability and soil suction. The permeability function term closely fits the manner in which saturated-unsaturated seepage problems are solved using a numerical modeling procedure.

Numerous estimation procedures have been developed for the calculation of a permeability function. Essentially all estimation procedures for obtaining the permeability function make use of the SWCC (i.e., the degree of saturation versus soil suction relationship or the volumetric water content versus soil suction relationship). A variety of procedures are later discussed under the topic of "estimation procedures" for obtaining the permeability function.

7.3.7 Hysteresis of Permeability Functions

The degree of saturation or volumetric water content shows significant hysteresis when plotted versus soil suction (Fig. 7.12a). Let us assume that there is no significant volume change as soil suction is increased. The coefficient of permeability is directly related to the volumetric water content (or degree of saturation) and will also show significant hysteresis when plotted versus soil suction. Figures 7.12a and 7.12b demonstrate a similar hysteresis form for both the volumetric water content θ and the coefficient of permeability k_w when plotted versus soil suction. However, when the coefficient of permeability is cross-plotted against volumetric water content, there is essentially no hysteresis, as demonstrated in Fig. 7.13.

The coefficient of permeability k_w is generally assumed to be uniquely related to the degree of saturation S or the volumetric water content θ. This assumption is reasonable since the volume of water flow is a direct function of the volume of water in the soil. The relationships between the degree of saturation (or volumetric water content) and the coefficient of permeability appear to exhibit little hysteresis (Nielsen and Biggar, 1961; Topp and Miller, 1966; Corey, 1977; Hillel, 1982). Nielsen et al., (1972) stated: "The function $k_w(\theta)$ is well-behaved, inasmuch as for coarse-textured soils, it is approximately the same for both wetting and drying." However, this is not the case for the relationship between the water coefficient of permeability k_w and matric suction $(u_a - u_w)$. Since there is hysteresis in the relationship between the volume of water in a soil and the stress state (i.e., namely, $u_a - u_w$), there will also be hysteresis in

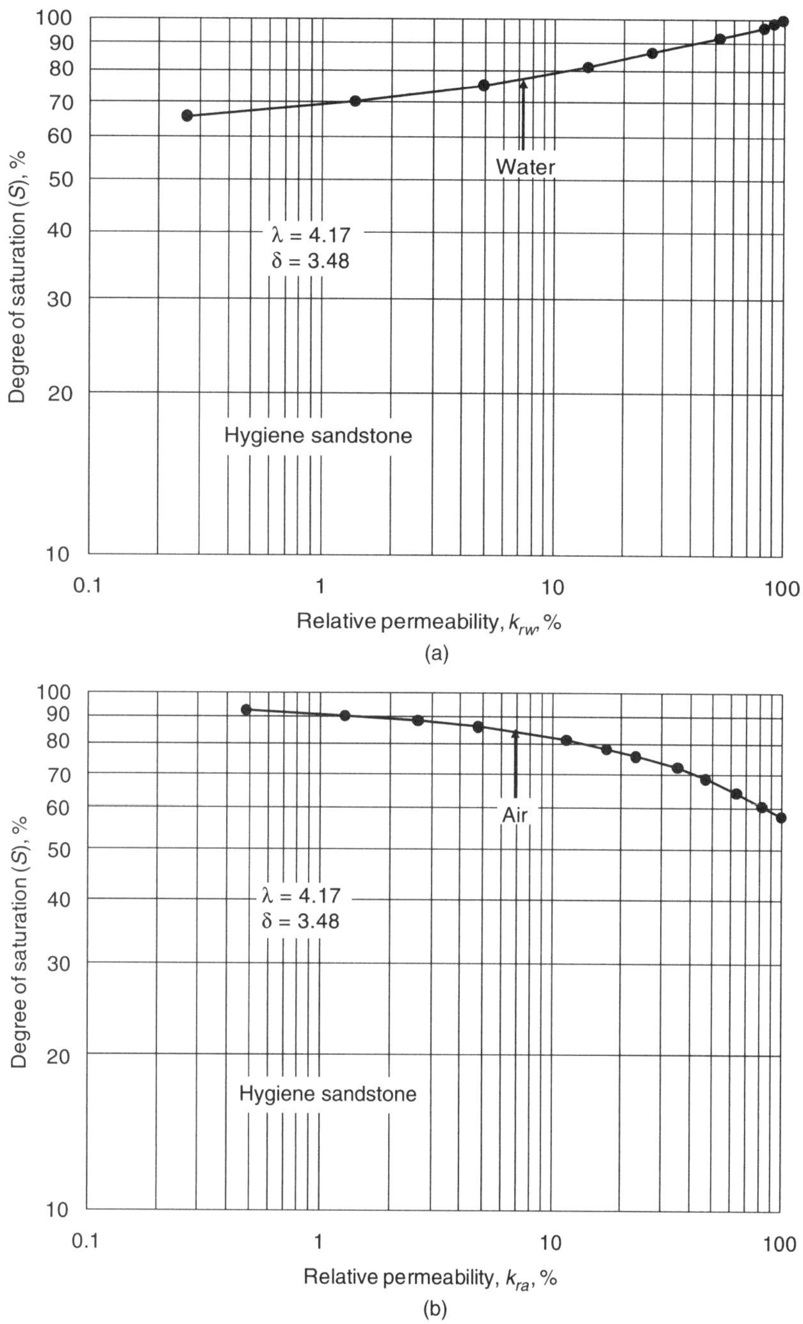

Figure 7.10 (a) Relative water and (b) relative air permeabilities as function of degree of saturation during desorption (after Brooks and Corey, 1964).

the relationship between the coefficient of permeability and matric suction.

The question might well be asked, "Why not simply define the water content versus water coefficient of permeability (which has essentially no hysteresis) and use this relationship when solving practical engineering problems?" Examination of the solution of practical problems shows there is nothing is to be gained through use of the water content versus water coefficient of permeability relationship because the partial differential seepage equation also requires a relationship between water content and soil suction during the solution. The water content versus soil suction then encounters nonlinearity and hysteresis. Consequently, it is just as easy to use the permeability function (i.e., soil suction versus coefficient of permeability) when solving the partial differential seepage equation for hydraulic heads in an unsaturated soil.

Thc cngineer must decide how the effects of hysteresis will be accommodated during the solution of a practical problem. Drying and wetting boundary curves produce hysteresis

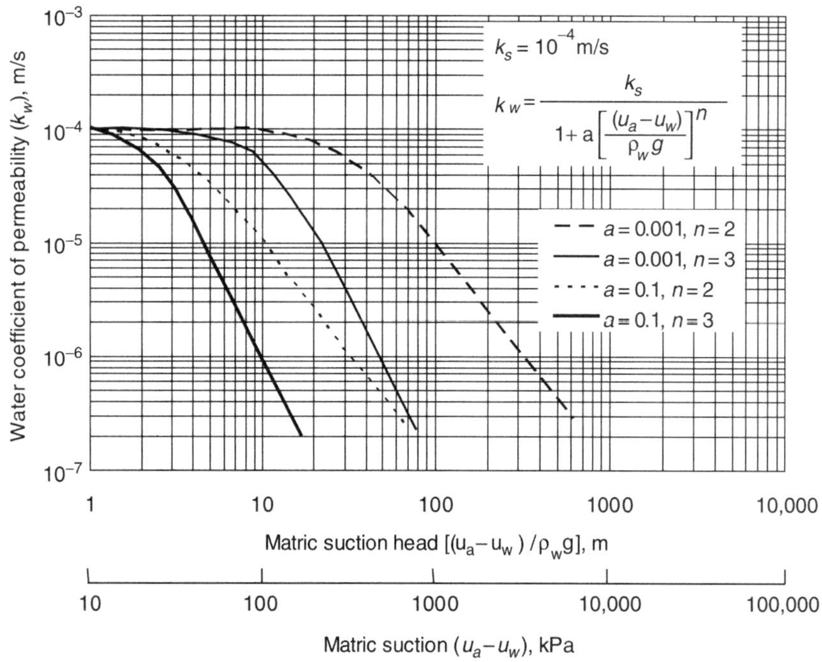

Figure 7.11 Gardner's (1958a) equation for water coefficient of permeability as function of matric suction.

in the permeability function. There are also an infinite number of scanning curves passing from the wetting to the drying curve and vice versa. Coefficient-of-permeability models have been proposed for unsaturated soils that include scanning paths between the wetting and drying curves (Watson and Sardana, 1987); however, geotechnical engineers generally decide whether it is a drying or wetting process that is being modeled and then select the appropriate permeability function.

The permeability function for an unsaturated soil is not routinely measured in the laboratory. Rather, the saturated coefficient of permeability and the SWCC are measured (or estimated) and combined through use of an estimation procedure that yields a permeability function. The drying (or desorption) branch of the SWCC is most often measured in the laboratory, and consequently, the permeability function is first computed for the drying curve. The permeability function for the wetting curve is then estimated based on measured or estimated hysteresis loops associated with the SWCCs (Pham et al., 2003b).

Anisotropic soil conditions add another variation to the permeability function, as shown in Fig. 7.14. The primary change in the permeability function is related to the difference between the saturated major and minor coefficients of permeability (Freeze and Cherry, 1979). The air-entry value observed on the SWCC corresponds to the point where both the major and minor coefficients of permeability start to decrease. Consequently, the mathematical form for the permeability function is similar for both the drying and wetting branches. The water storage function is also obtained from the SWCC and will also have hysteretic behavior.

7.3.8 Water Vapor Flow

The movement of water in the form of water vapor is particularly important when considering the loss of water from a soil through evaporation. Diffusion of water vapor is driven by a gradient in water vapor pressure. Vapor flow can also be driven as air flow in the form of vapor advection. Vapor diffusion follows Fick's law and is further discussed in the chapter on air flow. Vapor advection is governed by the flow of air and is a function of vapor content in the air. Water vapor behaves as a gas and in fact can be considered as one component of air.

Understanding the mechanisms related to water vapor flow is important for the development of a soil-atmospheric model as well as other applications. As a soil desaturates there is a point at which water vapor flow becomes more significant than liquid water flow. The challenge is to know when the predominant component of moisture flow changes from liquid to vapor flow and how each component of flow can be quantified. The water mass flux by liquid flow is traditionally described using Darcy's law. The mass flux by water vapor and advection within bulk air can be described using a modified form of Fick's law (Philip and de Vries, 1957; Dakshanamurthy and Fredlund, 1981) as follows:

$$q_y^v = -D^v \frac{\partial C_v}{\partial y} - \frac{\rho_v}{\rho_a} D^a \frac{\partial C_a}{\partial y} = -D^{v*} \frac{\partial u_v^{\mathrm{air}}}{\partial y} - \frac{\rho_v}{\rho_a} D^{a*} \frac{\partial \bar{u}_a}{\partial y} \quad (7.21)$$

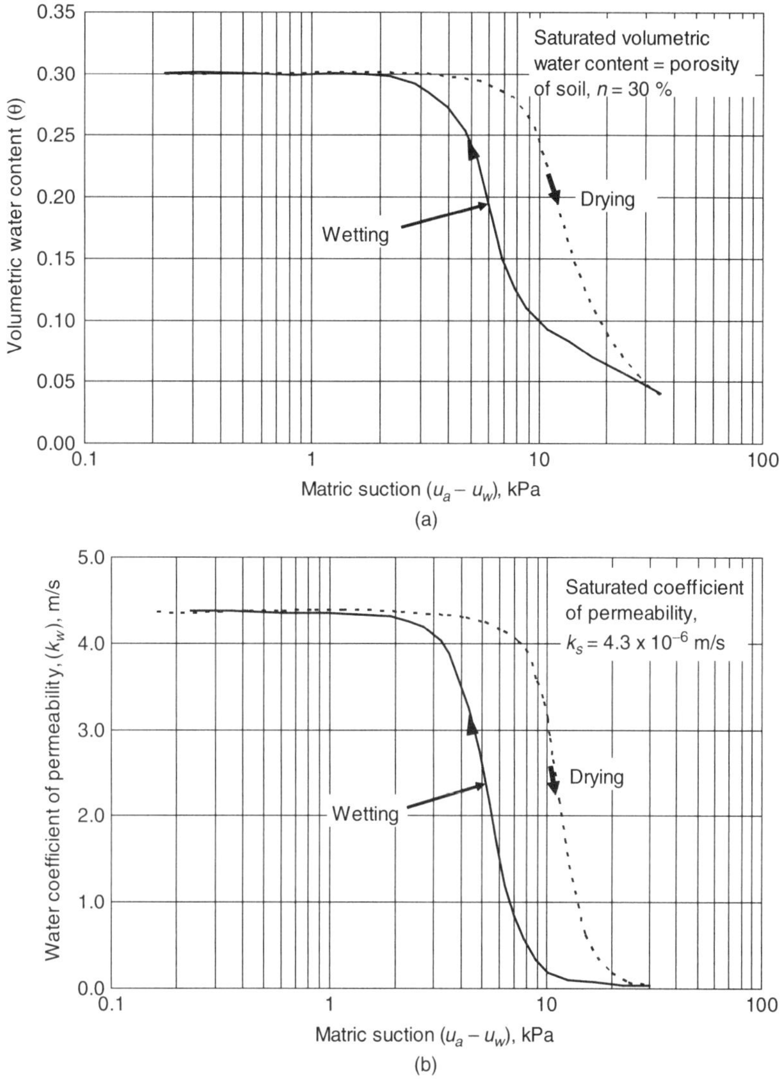

Figure 7.12 Similar forms of hysteresis for volumetric water content and water coefficient of permeability when plotted as function of matric suction $(u_a - u_w)$ for sand: (a) volumetric water content versus matric suction; (b) water coefficient of permeability versus matric suction.

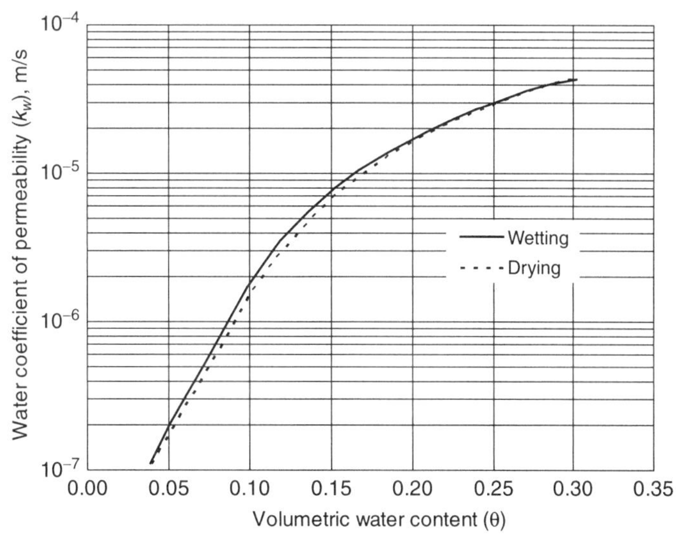

Figure 7.13 Elimination of hysteresis in drying and wetting relationship of water coefficient of permeability to volumetric water content.

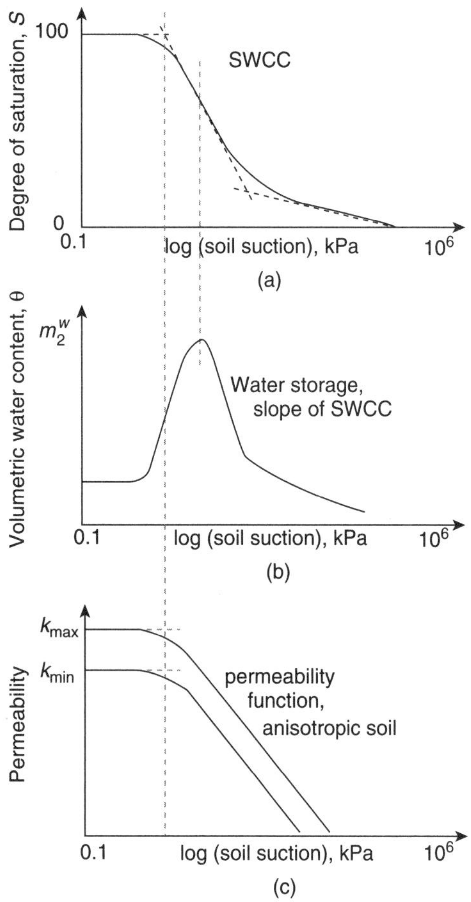

Figure 7.14 Typical permeability and water storage functions for anisotropic unsaturated soil: (a) degree of saturation versus soil suction; (b) water storage versus soil suction; (c) coefficient of permeability versus soil suction .

where:

q_i^v = flux rate of mass of water vapor within the air phase in the i-direction per unit of total area, kg/m^2/s,

D^v = molecular diffusivity of water vapor in air [i.e., $D^v = 0.229 \times 10^{-4}(1 + T_K/273.15)^{1.75}$, m^2/s (Kimball et al., 1976)],

D^a = coefficient of diffusion of air, $D^a \approx D^v$ (Wilson et al., 1997a),

T_K = temperature, K,

C_v = concentration of water vapor in terms of the mass of vapor per unit volume of soil, $C_v = \rho_v(1 - S)n$,

C_a = concentration of air in terms of the mass of air per unit volume of soil, $C_a = \rho_a(1 - S)n$,

S = degree of saturation,

n = porosity,

ρ_v = density of the water vapor, kg/m^3,

$\rho_v = W_v u_v^{air}/(RT)$, kg/m^3,

ρ_a = density of the bulk air phase, kg/m^3,

$\rho_a = W_a \bar{u}_a/(RT)$, kg/m^3,

W_v = molecular weight of water vapor, 18.016 kg/kmol,

W_a = molecular weight of pore-air, kg/kmol,

u_v^{air} = partial pressure of water vapor, kPa,

$\bar{u}_a$ = absolute pressure in the bulk air phase (i.e., $u_a + u_{atm}$), kPa,

u_a = pore-air pressure, kPa,

u_{atm} = atmospheric pressure, 101.325 kPa,

R = universal gas constant, 8.314 J/(mol·K),

D^{v*} = diffusion coefficient of vapor through soil (kg m)/(kN s), equal to $(1 - S)nD^v W_v/RT$ (kg·m)/kN s), and

$D^{a*} = (1 - S)nD^a W_a/RT$ (kg·m)/(kN s).

Equation 7.21 defines the diffusion coefficient of vapor through soil, D^{v^*}, based on Fick's law. However, an important factor has been omitted, namely, the tortuosity variable for the diffusion of water vapor through the soil pores (Lai et al., 1976). The diffusion coefficient of water vapor through soil, D^{v*}, can be written in the following form when tortuosity is taken into account:

$$D^{v*} = \alpha\beta D^v W_v/RT \tag{7.22}$$

where:

α = tortuosity factor of the soil; $\alpha = \beta^{2/3}$ (Lai et al., 1976), where β is the cross-sectional area of soil available for vapor flow per total area [i.e., $\beta = (1 - S)n$].

Neglecting the gradient in atmospheric pressure and assuming the air phase is continuous and in direct contact with the atmosphere, gradients in the free air will be equal to gradients in the partial pressure of water vapor, u_v^{air}. Therefore, Eq. 7.21 can be rewritten as follows:

$$q_y^v = -D^{v^*}\frac{\partial u_v^{air}}{\partial y} - \frac{u_v^{air}}{\bar{u}_a}D^{v^*}\frac{\partial u_v^{air}}{\partial y} = -\frac{\bar{u}_a + u_v^{air}}{\bar{u}_a}D^{v^*}\frac{\partial u_v^{air}}{\partial y} \tag{7.23}$$

Equation 7.23 describes vapor flow based on partial vapor pressure gradients. It is not possible to make a comparison between the liquid flow equation and the vapor flow equation (i.e., Eq. 7.23) because different gradients exist in each equation. Consequently, a direct comparison between the values of k_w and D^{v^*} is not meaningful.

Based on the thermodynamic theory of soil moisture (Edlefsen and Anderson, 1943), the vapor pressure in the air, u_v^{air}, can be expressed as a function of the total potential of the liquid pore-water and temperature. The amount of vapor in the pore-air phase depends on total suction and soil temperature. Changes in total suction or temperature result in mass transfer between the gaseous and liquid phases.

Phase transfers are usually assumed to be instantaneous in soil systems. Assuming local thermodynamic equilibrium, neglecting the effects of the osmotic suction, and assuming that the air pressure is equal to the atmospheric pressure, the following relationship can be written (i.e., Lord Kelvin's equation):

$$u_v^{\text{air}} = u_{v0}^{\text{air}} e^{\frac{u_w g W_v}{\gamma_w R T_K}} \tag{7.24}$$

where:

u_{v0}^{air} = saturation vapor pressure, kPa, of the soil-water at temperature T_K and
g = acceleration of gravity, 9.81 m/s^2.

Values of saturation soil vapor pressure have been experimentally obtained by Kaye and Laby (1973) for various temperatures. Other parameters in Eq. 7.24 were previously defined. A relationship between the vapor pressure gradients u_v^{air} and the gradients of the other two variables, u_w and T_k, are determined by differentiating Eq. 7.24 with respect to space, y:

$$\frac{\partial u_v^{\text{air}}}{\partial y} = \frac{g W_v u_v^{\text{air}}}{\gamma_w R T_K}\left(\frac{\partial u_w}{\partial y} - \frac{u_w}{T_K}\frac{\partial T_K}{\partial y}\right) \tag{7.25}$$

The total moisture flow (liquid and vapor), q_y^m [kg/(m^{2}s)], is obtained by summing Eq. 7.25 and Darcy's liquid flow equation and the water vapor equation 7.23. Equation 7.25 can then be used to express vapor flow in terms of pore-water pressure and temperature gradients:

$$q_y^m = -\frac{1}{g}\left(k_w + \gamma_w D^m\right)\frac{\partial u_w}{\partial y} - \rho_w k_w \frac{\partial Y}{\partial y} + \rho_w D^m \frac{u_w}{T_K}\frac{\partial T_K}{\partial y} - \frac{\bar{u}_a + u_v^{\text{air}}}{\bar{u}_a} D^{v*} \frac{\partial u_v^{\text{air}}}{\partial y} \tag{7.26}$$

where

$$D^m = \frac{\bar{u}_a + u_v^{\text{air}}}{\bar{u}_a}\,\frac{g}{\gamma_w}\,\frac{W_v u_v^{\text{air}}}{R T_K}\,\frac{D^{v*}}{\rho_w} \quad \text{(m/s)/(kN/m}^3\text{)}$$

Equation 7.26 governs one-dimensional flow of liquid water and water vapor flow based on a single gradient. The comparison of the terms k_w and k_v (where $k_v = \gamma_w D^m$) gives a comparison between the amount of liquid and vapor water flow. The terms k_w and k_v can be compared if isothermal conditions exist, and the gravitational component is neglected since both terms are then based on the gradient of pore-water pressure.

Temperature gradients are required for the solution of Eq. 7.26 for nonisothermal conditions. The nonisothermal solution would require that the heat flow partial differential equation also be solved (Gitirana and Fredlund, 2003b).

Equation 7.22 shows that the diffusion coefficient of water vapor through soil, D^{v*}, is a function of degree of saturation S and porosity n, which in turn are functions of soil suction. The diffusion coefficient D^{v*} is equal to zero when the soil is saturated and increases as air starts to occupy part of the soil pore space. On the other hand, the hydraulic conductivity k_w is at its highest value when the soil is saturated and starts declining as air begins entering the soil pore spaces. As the soil dries, the water coefficient of permeability k_w eventually becomes lower than the vapor diffusion coefficient k_v. At this point, vapor flow will become equal to liquid water flow.

7.3.9 Lower Limit for Water Coefficient of Permeability

The liquid water coefficient of permeability k_w decreases with increasing soil suction. There is, however, a "shutoff" of liquid water flow at some value of suction. The challenge is to determine the smallest possible water coefficient of permeability. Most of the proposed procedures for the calculation of the water permeability function result in an indefinite decrease in the water coefficient of permeability on a logarithmic scale. Liquid water flow may occur at relatively high soil suctions, but there should be some point where there is a transfer from liquid water flow to predominantly vapor water flow.

Extremely small values of the liquid water coefficient of permeability k_w can create serious difficulties when undertaking numerical modeling of water seepage. Ebrahimi-Birang et al.(2004) presented two approaches for the determination of a minimum value for the water coefficient of permeability k_w. The first approach used residual conditions to calculate minimum values for the liquid coefficient of permeability k_w. The second approach was based on water vapor flow theory.

7.3.9.1 Liquid Water Permeability at Residual Conditions

Ebrahimi-Birang et al.(2004) selected 45 soils from the literature and compiled the measured SWCCs, saturated coefficients of permeability, and volume-mass properties. The soils were classified as sand, silty sand, sandy silt, and silt. Several predictive permeability models were used to calculate the water permeability coefficient at residual conditions for each soil. The unsaturated coefficient of permeability corresponding to residual water content conditions was determined. The determination of the water coefficient of permeability corresponding to residual conditions is illustrated in Fig. 7.15.

Water coefficients of permeability corresponding to residual conditions were determined using the procedure proposed by Fredlund et al. (1994b), as shown in Fig. 7.16. Similar distributions of the water coefficient of permeability were also found for other proposed procedures for predicting the water permeability functions. The results show a considerable variation in the water permeability coefficient near residual conditions. In general, the water coefficients of permeability corresponding to residual suction appeared to be too large.

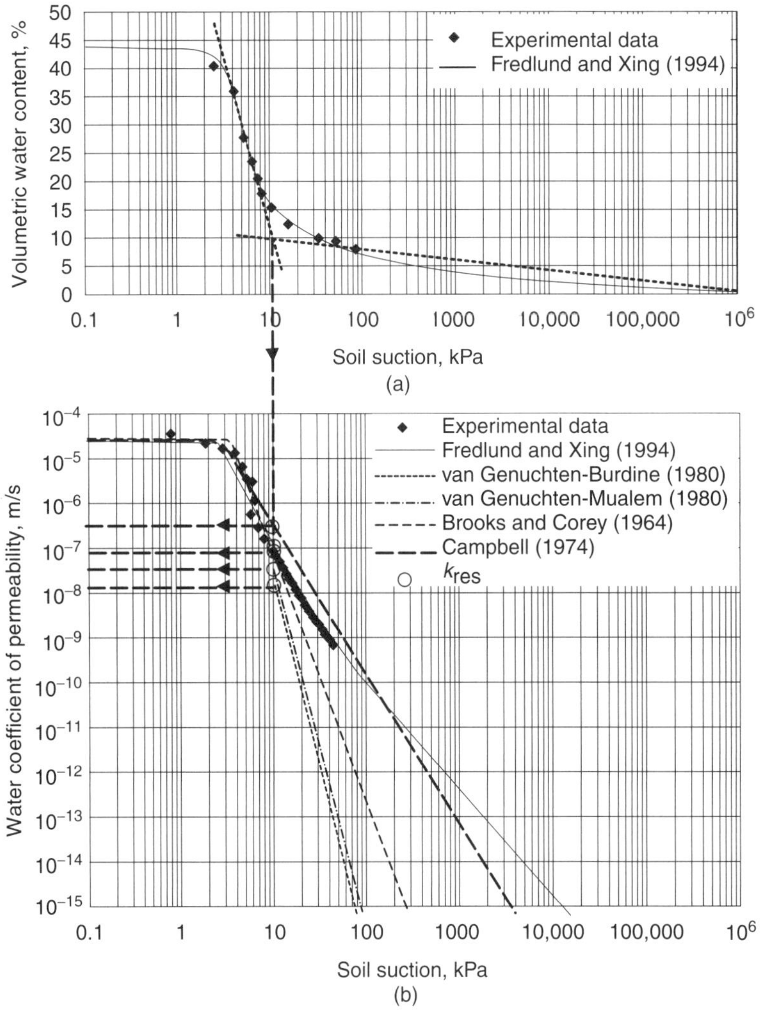

Figure 7.15 Calculation of water permeability functions using various estimation procedures: (a) Fredlund and Xing (1994) fit to the SWCC; (b) various water permeability functions for silty sand and their relationship to residual coefficient of water permeability (after Ebrahimi-Birang et al., 2004).

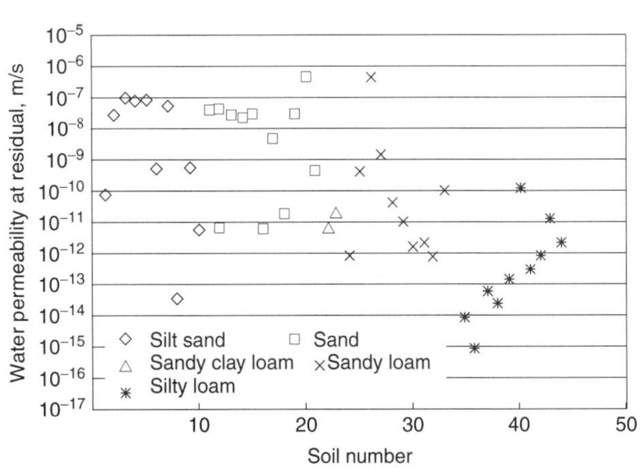

Figure 7.16 Coefficients of water permeability at residual conditions calculated using Fredlund et al. (1994b) estimation of permeability function (after Ebrahimi-Birang et al., 2004).

7.3.9.2 Water Vapor Permeability Functions

Six procedures were used to calculate tortuosity factors for water vapor flow through soils and the results are shown in Table 7.3 and in Fig. 7.17. The vapor permeability coefficient is negligible in the low-suction range where liquid water flow is dominant. The tortuosity model proposed by Lai et al. (1976) was selected and results are shown for the sand, silty sand and, silt soils in Fig. 7.18.

All tortuosity functions produce similar vapor coefficients of permeability in the high-suction range. The equation by Lai et al. (1976) produces the highest values for the vapor permeability coefficient with results presented in Table 7.4.

Figure 7.19 shows the variations of water, vapor, and overall permeability coefficients for a silty sand soil. The coefficient of vapor permeability k_v has its lowest value as the soil approaches saturation and begins to increase as air

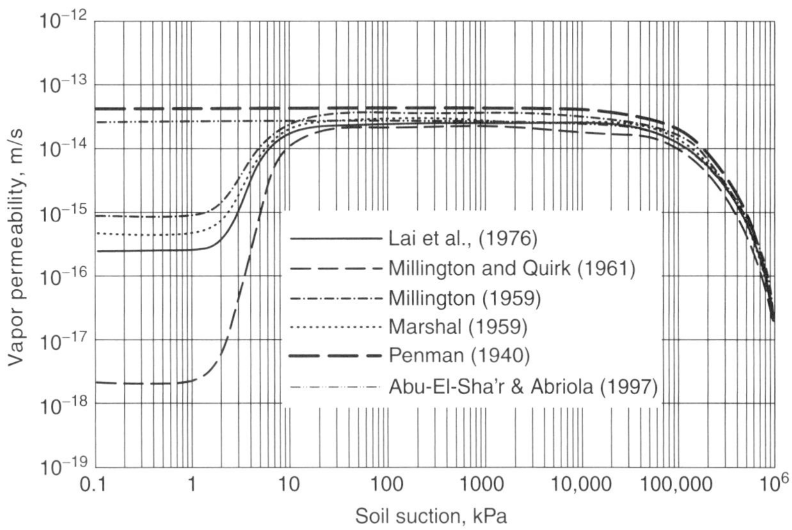

Figure 7.17 Coefficient of vapor permeability with respect to soil suction when assuming various tortuosity functions (after Ebrahimi-Birang et al., 2004).

Table 7.3 Tortuosity Coefficients from the Literature

Reference	$\alpha\ \beta$
Lai et al. (1976)	$\beta^{5/3}$
Millington and Quirk (1961)	$\beta^{10/3}/n^2$
Millington (1959)	$\beta^{\frac{4}{3}}$
Marshal (1959)	$\beta^{\frac{3}{2}}$
Penman (1940)	0.66 N
Abu-El-Sha'r and Abriola (1997)	0.435 N

Note: $\alpha\beta$ = computed using Eq. 7.22, n = porosity.

Table 7.4 Maximum Values of Vapor Coefficients of Permeability

Soil Texture	$(k_v)_{max}$, m/s
Silty sand	2.42×10^{-14}
Silt	1.76×10^{-14}
Sand	1.63×10^{-14}

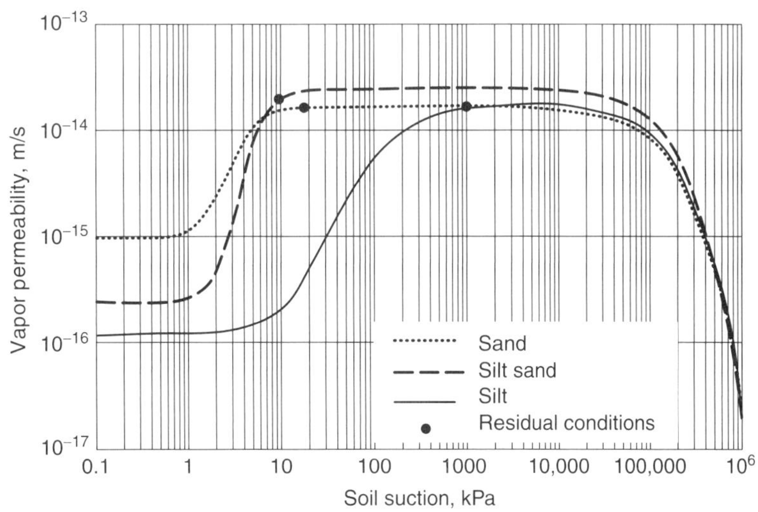

Figure 7.18 Vapor permeability measurements for three soils (after Ebrahimi-Birang et al., 2004).

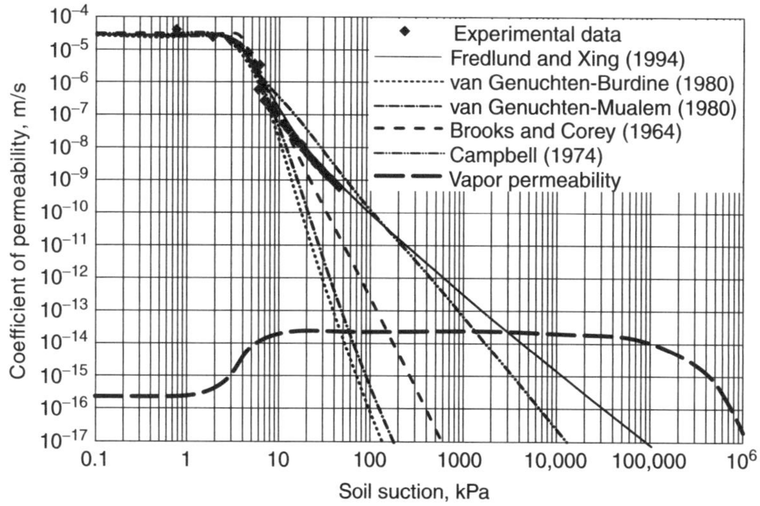

Figure 7.19 Liquid water, vapor, and overall permeability coefficients with respect to soil suction for silty sand (after Ebrahimi-Birang et al., 2004).

occupies a portion of the soil pores. The liquid water permeability starts to decline as air enters the soil pore space. When the liquid water coefficient of permeability k_w reaches a value near the residual water content conditions, the vapor coefficient of permeability k_v becomes a maximum value. However, the k_w at residual conditions still appears to be dominant over the k_v term.

The intersection point of liquid water and water vapor coefficients of permeability is the point after which the vapor permeability coefficient begins to be dominant. The intersection point provides an estimation of the minimum value for the liquid water coefficient of permeability. Ebrahimi-Birang et al.(2004) suggested that a reasonable lower limit for the water coefficient of permeability is 1×10^{-14} m/s.

7.4 PARTIAL DIFFERENTIAL EQUATIONS FOR STEADY-STATE WATER FLOW

Geotechnical engineers are often interested in knowing the direction and quantity of flow through porous media as well as the pore-water pressures (or hydraulic heads). Seepage problems can be categorized as steady-state or unsteady-state flow analyses. The hydraulic head and the coefficient of permeability at any point in the soil mass remain constant with respect to time for steady-state analyses. For an unsteady-state analysis, the hydraulic head (and possibly the coefficient of permeability) changes with respect to time. Changes in hydraulic head are usually in response to a change in the boundary conditions with respect to time. The formulation of steady-state flow analyses are considered first, followed by unsteady-state flow.

The quantity of flow of an incompressible fluid such as water is expressed in terms of a flux q. Flux is equal to a flow rate v multiplied by a cross-sectional area A. The governing partial differential seepage equations are derived while satisfying the conservation of mass. The conservation of mass for steady-state seepage of an incompressible fluid dictates that the flux into an element must equal the flux out of an element. In other words, the net flux must be zero at any point in the soil mass.

The slow movement of water through soil is commonly referred to as seepage or percolation. Seepage analyses may form an important part of studies related to slope stability, groundwater contamination control, earth dam design, and many other geotechnical engineering applications. Seepage analyses involve the computation of the rate and direction of water flow and the pore-water pressure distributions within the flow regime.

Seepage analyses have historically focused on the flow of water in the saturated zone. However, water from the atmosphere (precipitation) passes through the unsaturated soil zone on its way to becoming part of the groundwater system. Seepage through dams used to be analyzed by considering only the zone below an empirically computed line of seepage or phreatic line (Casagrande, 1937). With our present understanding of the physics of flow through the unsaturated soil zone, along with the available computational ability of the computer, it is possible to model the continuous flow of water between the saturated and unsaturated zones, as shown in Fig. 7.20a (Freeze, 1971; Papagiannakis and Fredlund, 1984). Another example is shown in Fig. 7.20b, which illustrates the effect of infiltration and evaporation on pore-water pressures above the phreatic surface within a slope. Generally the infiltration and evaporation processes are considered as transient processes which respond to a moisture flux boundary condition. Steady-state water flux through the unsaturated zone of the slope requires a constant water flux across the ground surface boundary.

Water flow through unsaturated soils is governed by the same flow law as through saturated soils (i.e., Darcy's law).

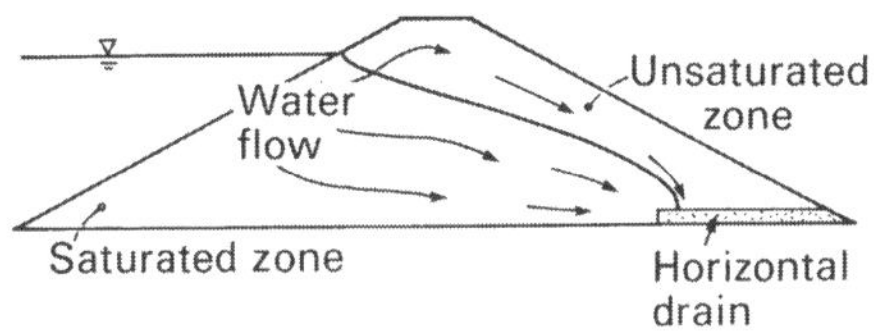

(a) Water flow above the phreatic line in an earthfill dam.

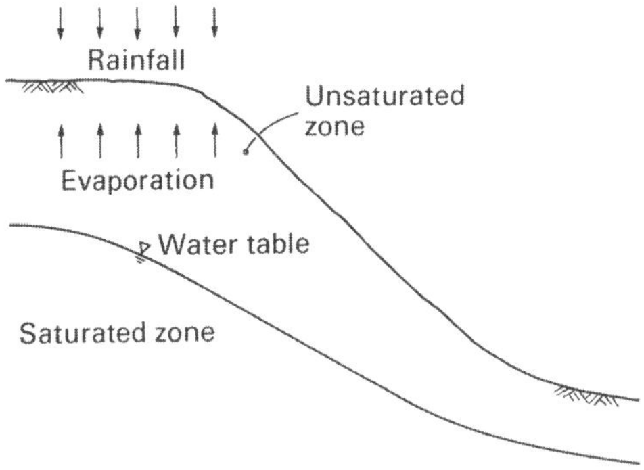

(b) Downward moisture movement (infiltration) and upward moisture movement (evaporation) through the unsaturated zone.

Figure 7.20 Examples involving flow through unsaturated soils.

The main difference between saturated and unsaturated soil problems is the difference in the water coefficient of permeability which is assumed to be constant for saturated soils while it is necessary that it be considered as a function of suction, water content, or some other variable for unsaturated soils. The pore-water pressure generally has a positive gauge value in a saturated soil and a negative gauge value in an unsaturated soil. The formulation of differential equations of flow can be derived in a similar manner for both the saturated and unsaturated soils even though there are significant differences between the two cases. In other words, there is a smooth transition when going from the unsaturated to the saturated case (Fredlund, 1981b).

7.4.1 Variation of Coefficient of Permeability with Space for Unsaturated Soil

The coefficient of permeability is a constant with respect to time at each point in a soil continuum for steady-state seepage analyses. However, the coefficient of permeability usually varies from one point to another in an unsaturated soil. A spatial variation in permeability in a saturated soil can be attributed to a heterogeneous distribution of the *soil solids*. The variation in the volume distribution of the *pore-fluid* (i.e., pore-water) gives rise to a heterogeneous system in an unsaturated soil. Consequently, there is a spatial variation in the coefficient of permeability in an unsaturated soil system. Although the soil solid distribution may be homogeneous, the pore-fluid volume distribution can be heterogeneous due to spatial variations in matric suction. A point of high matric suction (or low water content) will have a lower water coefficient of permeability than a point having low matric suction.

Several functional relationships between the water coefficient of permeability and matric suction [i.e., $k_w(u_a - u_w)$] have been described. The coefficients of permeability at different points in a soil mass are obtained from the permeability function where the magnitude of the coefficient of permeability depends on matric suction.

The coefficient of permeability at a point may also vary with respect to direction, and this condition is referred to as anisotropy. The largest coefficient of permeability is called the major coefficient of permeability. The smallest coefficient of permeability is in a direction perpendicular to the largest permeability and is called the minor coefficient of permeability.

7.4.2 Heterogeneous, Isotropic Steady-State Seepage

Permeability conditions in unsaturated soils can be classified into three groups, as illustrated in Fig. 7.21. The

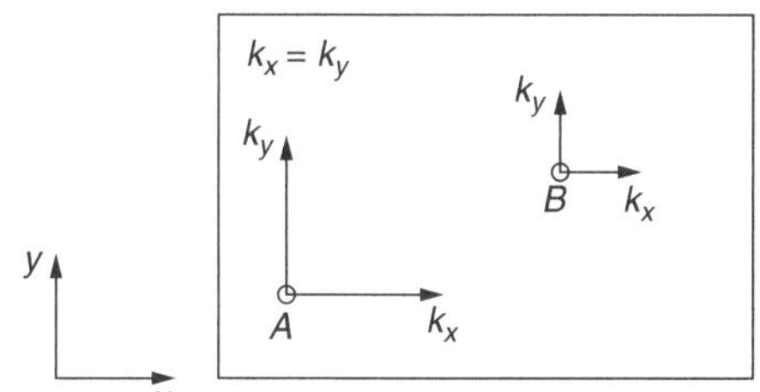

(a)

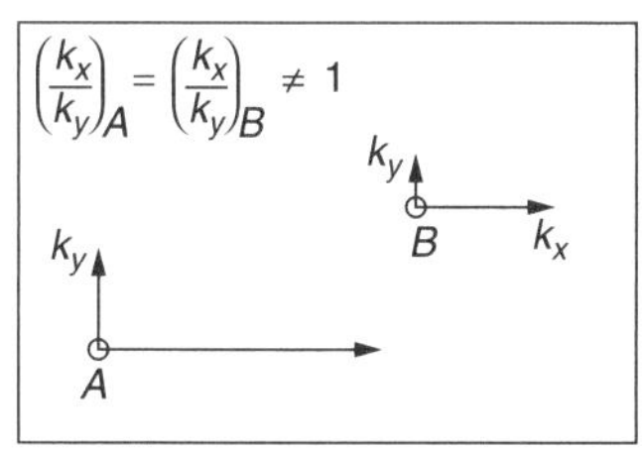

(b)

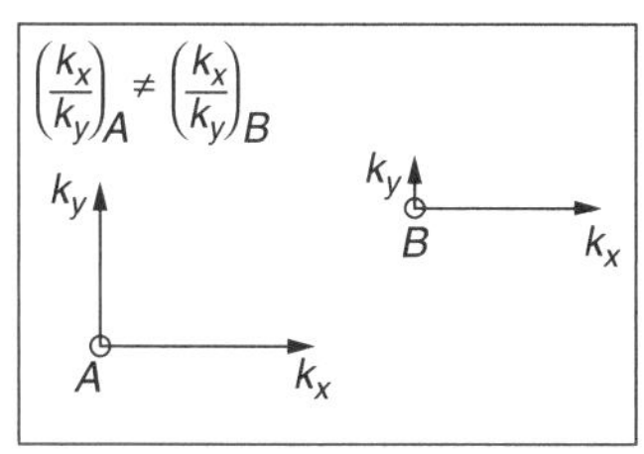

(c)

Figure 7.21 Principal coefficient-of-permeability variations in unsaturated soil: (a) heterogeneous, isotropic conditions; (b) heterogeneous, anisotropic conditions; (c) continuous variation of permeability with space.

classification is based on the pattern of permeability variation. A soil is called heterogeneous, isotropic if the coefficient of permeability in the x-direction, k_x, is equal to the coefficient of permeability in the y-direction at any point within the soil mass (i.e., $k_x = k_y$ at A and $k_x = k_y$ at B) (Fig. 7.21a). However, the magnitude of the coefficient of permeability in either direction can vary from point A to point B since its magnitude depends on matric suction. The variation in the coefficient of permeability with respect to matric suction is often assumed to follow a single-valued functional relationship even though it is known that the soil is hysteretic with respect to the wetting and drying processes.

7.4.3 Heterogeneous, Anisotropic Steady-State Seepage

Figure 7.21b illustrates the heterogeneous, anisotropic case. The ratio of the coefficient of permeability in the x-direction, k_x, to the coefficient of permeability in the y-direction, k_y, is a constant at any point (i.e., k_x/k_y at $A = k_x/k_y$ at $B =$ a constant not equal to unity). The magnitude of the coefficients of permeability k_x and k_y can also vary with matric suction from one location to another, but their ratio is assumed to remain constant. Anisotropic conditions can also be oriented in any two perpendicular directions.

The third case of coefficients of permeability occurs when there is a continuous variation in the coefficient of permeability (Fig. 7.21c). The permeability ratio k_x/k_y may not be a constant from one location to another (i.e., k_x/k_y at $A \neq k_x/k_y$ at B), and different directions may have different permeability functions.

The following steady-state seepage formulations deal with heterogeneous, isotropic and heterogeneous, anisotropic cases. The case where there is a continuous variation in permeability with space is not considered. All steady-state seepage analyses assume that the pore-air pressure has reached a constant equilibrium value. When the equilibrium pore-air pressure is atmospheric, the water coefficient-of-permeability function with respect to matric suction, $k_w(u_a - u_w)$, has the same absolute value as the permeability function with respect to pore-water pressure, $k_w(-u_w)$.

7.4.4 Ability of Unsaturated Soil to Store and Release Water

The simulation of transient flow processes requires a characterization of a water storage property that changes with the degree of saturation (or volumetric water content) of the soil. The water storage property is part of the partial differential equation describing a transient process.

The water storage soil property associated with water flow through an unsaturated soil is given the symbol m_2^w and is equal to the arithmetic slope of the (volumetric water content) SWCC. The differentiation of the mathematical equation for the SWCC serves as a measure of the water storage soil property. Figure 7.14 showed the form of the water storage function for a typical soil. There is strong nonlinearity to the water storage function at the point of inflection on the SWCC. The nonlinearity of the water storage soil property can give rise to numerical instability and errors in computing water balances if not properly taken into account during the solution of the seepage partial differential equation. There is also an independent water storage curve for the drying and wetting processes since there is hysteresis in the SWCC.

7.4.5 One-Dimensional Flow in Unsaturated Soil

There are numerous situations where the water flow is predominantly in one direction. Let us consider a covered ground surface with the water table located at a specified depth, as shown in Fig. 7.22. The surface cover prevents the vertical flow of water from the ground surface.

The pore-water pressures are negative with respect to the water table under static equilibrium conditions. The negative pore-water pressure head has a linear distribution with depth (i.e., line 1). Its magnitude is equal to the gravitational head (i.e., elevation head) measured relative to the water table. In other words, the hydraulic head (i.e., the gravitational head plus the pore-water pressure head) is zero throughout the soil profile. This means that the change in head with respect to depth, and likewise the hydraulic gradient, is equal to zero. Therefore, there can be no flow of water in the vertical direction (i.e., $q_{wy} = 0$).

The soil surface would be exposed to the environment if the covering were removed from the ground surface. Environmental changes could produce flow in a vertical direction and subsequently alter the negative pore-water pressure head profile. Steady-state evaporation would cause the pore-water pressures to become more negative, as illustrated by line 2 in Fig. 7.22. Hydraulic head changes to a negative value since the gravitational head remains constant. The hydraulic head has a nonlinear distribution from a zero value at the water table to a more negative value at ground surface. An assumption is made that the water table remains at a constant elevation for this example. The nonlinearity of the hydraulic head profile is caused by the spatial variation in the coefficient of permeability. Water flows in the direction of the decreasing hydraulic head, which means that water flows from the water table upward to the ground surface. The upward constant flux of water is designated as being positive for steady-state evaporation.

Steady-state infiltration causes a downward water flow. The negative pore-water pressure increases from the static equilibrium condition. This condition is indicated by line 3 in Fig. 7.22. The hydraulic head profile starts with a positive value at ground surface and decreases to zero at the water table. Therefore, water flows downward with a constant negative flux for steady-state infiltration.

The above one-dimensional illustration of flow involves the application of moisture flux boundary conditions. A steady rate of evaporation or infiltration is used as the boundary

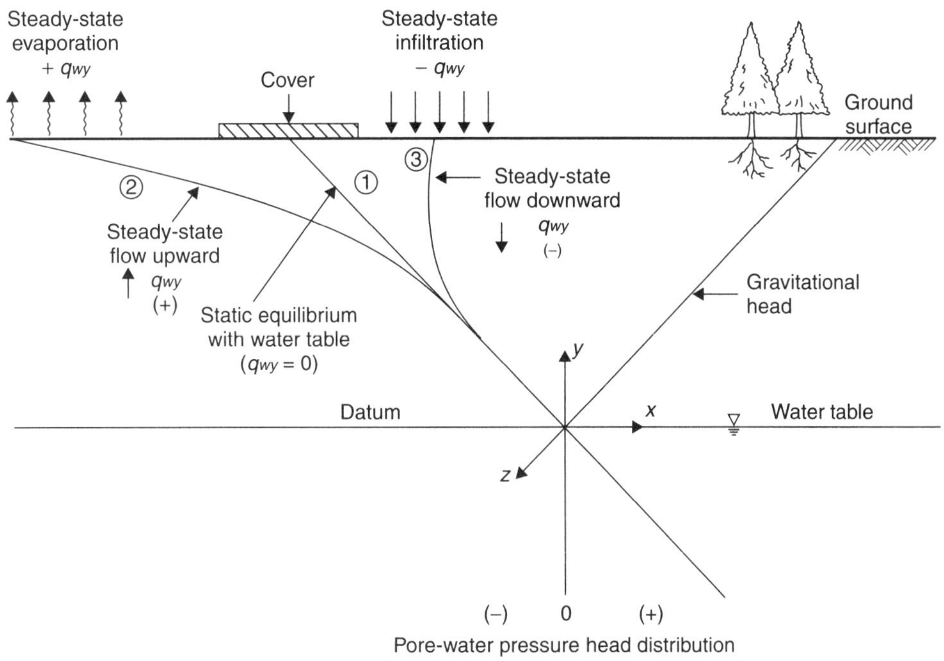

Figure 7.22 Hydrostatic equilibrium and steady-state flow conditions in zone of negative pore-water pressures.

condition at ground surface. The water table acts as the lower boundary condition and is given a fixed zero pore-water pressure head.

In the laboratory measurement for the coefficient of permeability, hydraulic heads are controlled as boundary conditions at the top and bottom of the soil specimen. Let us consider an unsaturated soil element with one-dimensional water flow in the y-direction (Fig. 7.23). The element has infinitesimal dx, dy, and dz dimensions. The flow rate v_{wy} is assumed to be positive when water flows upward in the y-direction. Continuity requires that the volume of water flowing in and out of the element be equal for steady-state conditions:

$$\left(v_{wy} + \frac{dv_{wy}}{dy}dy\right) dx\,dz - v_{wy}\,dx\,dz = 0 \tag{7.27}$$

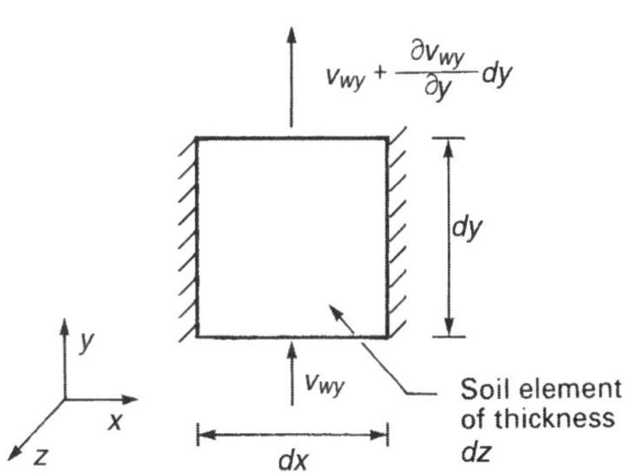

Figure 7.23 One-dimensional water flow through an unsaturated soil element.

where:

v_{wy} = water flow rate across a unit area of the soil in the y-direction and

dx, dy, dz = dimensions in the x-, y-, and z-directions, respectively.

The net flow through the element can be written as follows:

$$\frac{dv_{wy}}{dy}dx\,dy\,dz = 0 \tag{7.28}$$

Substituting Darcy's law into the above equation yields

$$\frac{d\left\{-k_{wy}\left(u_a - u_w\right)dh_w/dy\right\}}{dy}dx\,dy\,dz = 0 \tag{7.29}$$

where:

$k_{wy}(u_a - u_w)$ = water coefficient of permeability, which varies in the y-direction since it is a function of matric suction,

dh_w/dy = hydraulic head gradient in the y-direction, and

h_w = hydraulic head (i.e., gravitational head plus pore-water pressure head).

Equation 7.29 can be used to solve for the hydraulic head distribution in the y-direction. Since matric suction varies from one location to another, the coefficient of permeability also varies. However, for the remainder of the formulation, the coefficient-of-permeability $k_{wy}(u_a - u_w)$ term will be written as k_{wy} for simplicity. Rewriting Eq. 7.29 and considering the nonzero dimensions for dx, dy, and dz gives the following nonlinear differential equation:

$$k_{wy}\frac{d^2h_w}{dy^2} + \frac{dk_{wy}}{dy}\frac{dh_w}{dy} = 0 \qquad (7.30)$$

where:

dk_{wy}/dy = change in water coefficient of permeability in the y-direction due to changes in matric suction.

The nonlinearity in the differential equation for flow is caused by its second term, which accounts for the variation in permeability with respect to space. When the soil becomes *saturated*, the water coefficient of permeability k_{wy} can be taken as being equal to a single-valued, saturated coefficient of permeability k_s. If the saturated soil is heterogeneous (e.g., layered soil), the coefficient of permeability will again vary with respect to location. In a saturated soil, the heterogeneous distribution of the *soil solids* is the primary factor producing a varying coefficient of permeability. As a result, the flow equation can be written as follows:

$$k_s\frac{d^2h_w}{dy^2} + \frac{dk_s}{dy}\frac{dh_w}{dy} = 0 \qquad (7.31)$$

where:

k_s = saturated coefficient of permeability.

Comparing Eqs. 7.30 and 7.31 reveals a similar form. In other words, the nonlinearity in the unsaturated soil flow equation produces the same form of equation as that required for a heterogeneous, saturated soil. In an *unsaturated* soil, the variation in the coefficient of permeability is caused by the heterogeneous distribution of the pore-fluid volume occurring as a result of different matric suction values.

If a saturated soil is *homogeneous*, the coefficient of permeability is constant for the entire soil mass. Substituting a nonzero, constant coefficient of permeability into Eq. 7.31 produces a linear differential equation:

$$\frac{d^2h_w}{dy^2} = 0 \qquad (7.32)$$

Equations similar to Eq. 7.31 can also be derived for one-dimensional flow in the x- and z-directions, but in these cases there is no elevation component contributing to the gradient. In other words, it can be stated that flow in the x- and z- (horizontal) directions occurs as a result of pressure head gradients while flow in the y-direction occurs due to hydraulic head gradients.

7.4.6 Two-Dimensional Flow through Unsaturated Soil

Seepage through an earth dam constitutes a common example of two-dimensional flow. Let us assume that water flows through the cross-sectional plane of the dam while flow in the perpendicular plane is negligible. Historically, it was conventional practice to neglect the flow of water in the unsaturated zone of the dam. The seepage analysis presented herein assumes that water flows through both the saturated and unsaturated zones in response to a hydraulic head driving potential.

The following two-dimensional formulation is an expanded form of the previous one-dimensional flow equation. The formulation is called an uncoupled solution since only the continuity of the water phase is satisfied. For a more rigorous formulation of two-dimensional water flow, continuity would need to be coupled with the force equilibrium equations (see Chapter 16).

The following derivation is for the general case of a *heterogeneous*, *anisotropic*, unsaturated soil (Fig. 7.21b). The coefficients of permeability in the x-direction, k_{wx}, and the y-direction, k_{wy}, are assumed to be related to matric suction by the same form for the permeability function, $k_w(u_a - u_w)$. The ratio of the coefficients of permeability in the x- and y-directions (k_{wx}/k_{wy}) is assumed to be constant at any point within the soil mass.

A soil element with infinitesimal dimensions of dx, dy, and dz is considered, but flow is assumed to be two dimensional (Fig. 7.24). The flow rate v_{wx} is positive when water flows in the positive x-direction. The flow rate v_{wy} is positive for flow in the positive y-direction. Continuity for two-dimensional, steady-state flow can be expressed as follows:

$$\left(v_{wx} + \frac{\partial v_{wx}}{\partial x}dx - v_{wx}\right)dy\,dz + \left(v_{wy} + \frac{\partial v_{wy}}{\partial y}dy - v_{wy}\right)dx\,dz = 0 \qquad (7.33)$$

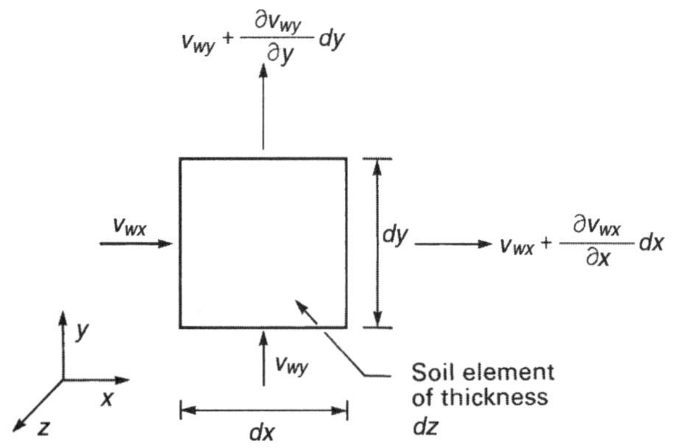

Figure 7.24 Two-dimensional water flow through an unsaturated soil element.

where:

v_{wx} = water flow rate across a unit area of the soil in the x-direction.

Therefore, the net flux in the x- and y-directions is

$$\left(\frac{\partial v_{wx}}{\partial x} + \frac{\partial v_{wy}}{\partial y}\right) dx\, dy\, dz = 0 \quad (7.34)$$

Substituting Darcy's laws into the net flux equation results in the following nonlinear partial differential equation:

$$\frac{\partial}{\partial x}\left\{k_{wx}(u_a - u_w)\frac{\partial h_w}{\partial x}\right\} + \frac{\partial}{\partial y}\left\{k_{wy}(u_a - u_w)\frac{\partial h_w}{\partial y}\right\} = 0 \quad (7.35)$$

where:

$k_{wx}(u_a - u_w)$ = water coefficients of permeability as a function of matric suction, the permeability is able to vary with location in the x- direction, and

$\partial h_w/\partial x$ = hydraulic head gradient in the x-direction.

For the remainder of the formulations, the coefficients of permeability $k_{wx}\left(u_a - u_w\right)$ and $k_{wy}\left(u_a - u_w\right)$ are written as k_{wx} and k_{wy}, respectively, for simplicity. Equation 7.35 describes the hydraulic head distribution in the x-y plane for steady-state water flow. The nonlinearity of the partial differential seepage equation becomes clearer when it is written in an expanded form:

$$k_{wx}\frac{\partial^2 h_w}{\partial x^2} + k_{wy}\frac{\partial^2 h_w}{\partial y^2} + \frac{\partial k_{wx}}{\partial x}\frac{\partial h_w}{\partial x} + \frac{\partial k_{wy}}{\partial y}\frac{\partial h_w}{\partial y} = 0 \quad (7.36)$$

where:

$\partial k_{wx}/\partial x$ = change in water coefficient of permeability in the x-direction.

Equation 7.36 contains two unknowns, namely, hydraulic head and the coefficient of permeability. These two variables are related by another equation referred to as the permeability function. The two equations can be solved through use of an iterative process where the coefficient of permeability is first assumed and the hydraulic heads are solved. The process is repeated with continuously improved assumptions regarding the coefficients of permeability (through use of the permeability function). The process is repeated until convergence is reached with respect to the assumed and calculated coefficient of permeability and hydraulic heads.

For the *heterogeneous*, *isotropic* case, the coefficients of permeability in the x- and y-directions are equal (i.e., $k_{wx} = k_{wy} = k_w$). Therefore, Eq. 7.36 can be written as follows:

$$k_w\left(\frac{\partial^2 h_w}{\partial x^2} + \frac{\partial^2 h_w}{\partial y^2}\right) + \frac{\partial k_w}{\partial x}\frac{\partial h_w}{\partial x} + \frac{\partial k_w}{\partial y}\frac{\partial h_w}{\partial y} = 0 \quad (7.37)$$

where:

k_w = water coefficient of permeability in the x- and y-directions.

Table 7.5 summarizes the relevant equations for two-dimensional steady-state flow through unsaturated soils.

Seepage through a dam involves flow through the unsaturated and saturated zones. For the *saturated* portion, the water coefficient of permeability becomes equal to the saturated coefficient of permeability k_s. The saturated coefficients of permeability in the x- and y-directions, k_{sx} and k_{sy}, respectively, may not be equal due to anisotropy. The saturated coefficients of permeability may vary with respect to location and give rise to heterogeneity (i.e., different soil zones).

7.4.7 Three-Dimensional Flow through Unsaturated Soil

The real world is defined in terms of three spatial dimensions. It is increasingly becoming appropriate to simulate real-world flow systems using a three-dimensional flow analysis. Three-dimensional flow can be formulated by expanding the two-dimensional flow equation to include the third dimension. The three-dimensional equation is once again derived based on continuity, and the equation is referred to as an uncoupled equation of saturated-unsaturated flow.

Let us consider an unsaturated soil having heterogeneous, anisotropic conditions (Fig. 7.21b). The coefficient of permeability at a point varies in the x-, y-, and z-directions. However, the permeability variations in the three directions will be assumed to be governed by the same permeability function. Figure 7.25 shows a cubical soil element with water flow in the x-, y-, and z-directions. The soil element has infinitesimal dimensions of dx, dy, and dz. The flow rates v_{wx}, v_{wy}, and v_{wz} are assumed to be positive when water flows in the positive x-, y-, and z-directions. Continuity for three-dimensional, steady-state flow can be satisfied as follows:

$$\left(v_{wx} + \frac{\partial v_{wx}}{\partial x}dx - v_{wx}\right)dy\, dz + \left(v_{wy} + \frac{\partial v_{wy}}{\partial y}dy - v_{wy}\right) \times dx\, dz + \left(v_{wz} + \frac{\partial v_{wz}}{\partial z}dz - v_{wz}\right)dx\, dy = 0 \quad (7.38)$$

Table 7.5 Two-Dimensional Steady-State Equations for Unsaturated Soils

Heterogeneous, anisotropic:

$$k_{wx}\frac{\partial^2 h_w}{\partial x^2} + k_{wy}\frac{\partial^2 h_w}{\partial y^2} + \frac{\partial k_{wx}}{\partial x}\frac{\partial h_w}{\partial x} + \frac{\partial k_{wy}}{\partial y}\frac{\partial h_w}{\partial y} = 0$$

Heterogeneous, isotropic:

$$k_w\left(\frac{\partial^2 h_w}{\partial x^2} + \frac{\partial^2 h_w}{\partial y^2}\right) + \frac{\partial k_w}{\partial x}\frac{\partial h_w}{\partial x} + \frac{\partial k_w}{\partial y}\frac{\partial h_w}{\partial y} = 0$$

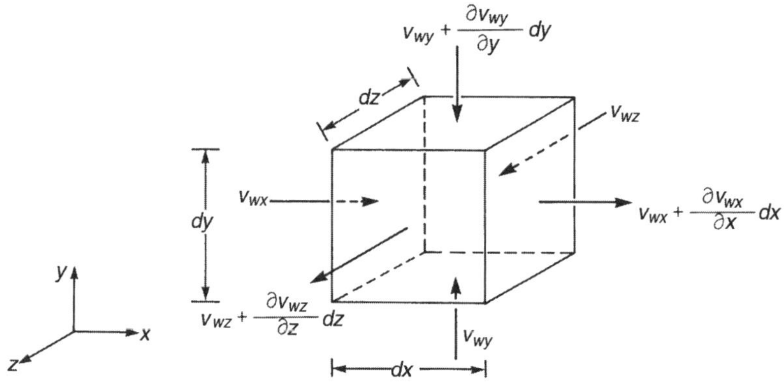

Figure 7.25 Three-dimensional water flow through an unsaturated soil element.

where:

v_{wz} = water flow rate across a unit area of the soil in the z-direction.

The above steady-state equation reduces to the following form:

$$\left(\frac{\partial v_{wx}}{\partial x} + \frac{\partial v_{wy}}{\partial y} + \frac{\partial v_{wz}}{\partial z} \right) dx\, dy\, dz = 0 \qquad (7.39)$$

Substituting Darcy's law into Eq. 7.39 yields the following nonlinear partial differential equation:

$$\frac{\partial}{\partial x} \left\{ k_{wx} \left(u_a - u_w \right) \frac{\partial h_w}{\partial x} \right\} + \frac{\partial}{\partial y} \left\{ k_{wy} \left(u_a - u_w \right) \frac{\partial h_w}{\partial y} \right\} + \frac{\partial}{\partial z} \left\{ k_{wz} \left(u_a - u_w \right) \frac{\partial h_w}{\partial z} \right\} = 0 \qquad (7.40)$$

where

$k_{wz}(u_a - u_w)$ = water coefficient of permeability as a function of matric suction and

$\partial h_w / \partial z$ = hydraulic head gradient in the z-direction.

For the remainder of the formulations, the permeability function terms $k_{wx}\left(u_a - u_w\right)$, $k_{wy}\left(u_a - u_w\right)$, and $k_{wz}\left(u_a - u_w\right)$ are written as k_{wx}, k_{wy}, and k_{wz}, respectively, for simplicity. The hydraulic head distribution in a soil mass during three-dimensional steady-state flow is described by Eq. 7.40, which can be rewritten as follows:

$$k_{wx} \frac{\partial^2 h_w}{\partial x^2} + k_{wy} \frac{\partial^2 h_w}{\partial y^2} + k_{wz} \frac{\partial^2 h_w}{\partial z^2} + \frac{\partial k_{wx}}{\partial x} \frac{\partial h_w}{\partial x} + \frac{\partial k_{wy}}{\partial y} \frac{\partial h_w}{\partial y} + \frac{\partial k_{wz}}{\partial z} \frac{\partial h_w}{\partial z} = 0 \qquad (7.41)$$

where:

$\partial k_{wz} / \partial z$ = change in water coefficient of permeability in the z-direction.

The fourth, fifth, and sixth terms in Eq. 7.41 account for the spatial variation in the coefficient of permeability. In the case of two-dimensional flow, the hydraulic head gradient in the third direction is assumed to be negligible (e.g., $\partial h_w / \partial z = 0$), and Eq. 7.41 reverts to Eq. 7.36.

For the heterogeneous, isotropic case, the coefficients of permeability in the x-, y-, and z-directions are equal and Eq. 7.41 takes the following form:

$$k_w \left(\frac{\partial^2 h_w}{\partial x^2} + \frac{\partial^2 h_w}{\partial y^2} + \frac{\partial^2 h_w}{\partial z^2} \right) + \frac{\partial k_w}{\partial x} \frac{\partial h_w}{\partial x} + \frac{\partial k_w}{\partial y} \frac{\partial h_w}{\partial y} + \frac{\partial k_w}{\partial z} \frac{\partial h_w}{\partial z} = 0 \qquad (7.42)$$

where:

k_w = water coefficient of permeability in the x-, y-, and z-directions.

Table 7.6 summarizes the three-dimensional steady-state equations for unsaturated soils. The three-dimensional steady-state flow equations can be solved using finite difference and finite element numerical procedures.

Table 7.6 Three-Dimensional Steady-State Equations for Unsaturated Soils

Heterogeneous, anisotropic:

$$k_{wx} \frac{\partial^2 h_w}{\partial x^2} + k_{wy} \frac{\partial^2 h_w}{\partial y^2} + k_{wz} \frac{\partial^2 h_w}{\partial z^2} + \frac{\partial k_{wx}}{\partial x} \frac{\partial h_w}{\partial x} + \frac{\partial k_{wy}}{\partial y} \frac{\partial h_w}{\partial y} + \frac{\partial k_{wz}}{\partial z} \frac{\partial h_w}{\partial z} = 0$$

Heterogeneous, isotropic:

$$k_w \left(\frac{\partial^2 h_w}{\partial x^2} + \frac{\partial^2 h_w}{\partial y^2} + \frac{\partial^2 h_w}{\partial z^2} \right) + \frac{\partial k_w}{\partial x} \frac{\partial h_w}{\partial x} + \frac{\partial k_w}{\partial y} \frac{\partial h_w}{\partial y} + \frac{\partial k_w}{\partial z} \frac{\partial h_w}{\partial z} = 0$$

7.5 PARTIAL DIFFERENTIAL EQUATIONS FOR TRANSIENT SEEPAGE

The steady-state water flow partial differential equations can be expanded to include transient or unsteady-state formulations. It is necessary to take changes in water storage into account when considering a transient analysis. A transient analysis allows the geotechnical engineer to give consideration to a wider range of possible questions that can be asked. The computational power of the computer also allows the engineer to perform parametric-type analyses where many of the questions involve possible changes that might occur over time.

Many situations are encountered in engineering practice where the ground surface is a moisture flux boundary. In other words, the climatic conditions at a site give rise to a varying moisture flux at the ground surface. The conversion of thermal (and other variable) boundary conditions into an actual evaporative flux becomes of significant interest in solving geotechnical engineering problems. Moisture flux boundary conditions open the way for the consideration of a variety of new design possibilities (e.g., design of soil cover systems). This book has devoted one chapter to the assessment of the moisture flux conditions at the ground surface. Unsteady-state or transient formulations are required when solving problems where there is a net moisture flux imposed at the ground surface as a boundary condition changing with time.

7.5.1 Uncoupled Two-Dimensional, Unsteady-State Formulations

Water flow through an earth dam during the filling of its reservoir is an example of two-dimensional, unsteady-state or transient flow. Eventually, the flow of water through the dam will reach a steady-state condition. Subsequent fluctuations of water level in the reservoir will again initiate further unsteady-state water flow conditions. Infiltration and evaporation cause a continuously changing flow regime. Transient analyses of water flow are strongly influenced by conditions in the unsaturated zone (Freeze, 1971).

The following uncoupled formulation satisfies the continuity equation for the water phase. Analyses that take the interaction between the fluid flows (e.g., water and air) and the soil structure equilibrium into consideration are part of combined analyses and are given consideration in Chapter 16. The derivation of the partial differential equations for two-dimensional, unsteady-state water flow in two directions (i.e., the x- and y-directions) is first considered in this chapter. Fluid flow in the third direction (i.e., the z-direction) is assumed to be negligible.

Consideration is first given to the handling of anisotropic permeability properties before deriving the equations for transient water flow through saturated-unsaturated soil systems. Water flow properties are generally defined with respect to major and minor coefficients of permeability. However, practical engineering problems are generally solved with respect to the x-, y-, and z-Cartesian coordinate system. As a result it is necessary to perform a transformation of the major and minor coefficients of permeability into the Cartesian coordinate system.

7.5.2 Unsteady-State Seepage in Anisotropic Soil

The term anisotropy is used to refer to soil conditions where the coefficient of permeability varies with respect to direction. The coefficients of permeability in the x- and y-directions are assumed to be different at any point in the soil mass. The conditions associated with anisotropy will be first discussed followed by the derivation of the transient saturated-unsaturated soil seepage. The coefficient of permeability also varies with respect to space (i.e., heterogeneity) due to variations in matric suction throughout the soil mass.

Unsteady-state water flow through an anisotropic soil is analyzed by considering the continuity of the water phase. The pore-air pressure is assumed to remain constant with time (i.e., $\partial u_a/\partial t = 0$). According to Freeze and Cherry (1979), "The single-phase approach to unsaturated flow leads to techniques of analysis that are accurate enough for almost all practical purposes, but there are some unsaturated flow problems where the multiphase flow of air and water must be considered." The water phase partial differential equation can be obtained in a similar manner to that used for the formulation of water flow through an isotropic soil.

7.5.2.1 Anisotropic Permeability

Let us consider the general case of a variation in the coefficient of permeability with respect to space (heterogeneity) and direction (anisotropy) in an unsaturated soil, as illustrated in Fig. 7.26. At a particular point, the largest or major coefficient of permeability, k_{w1}, occurs in the direction s_1, which is inclined at an angle α to the x-axis (i.e., horizontal). The smallest coefficient of permeability is in a perpendicular direction to the largest permeability (i.e., in the direction s_2)

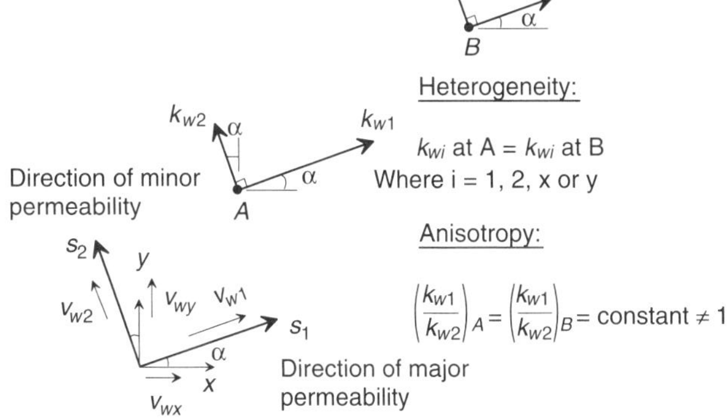

Figure 7.26 Consideration of the orientation of the principal coefficients of permeability in heterogeneous and anisotropic unsaturated soil.

and is called the minor coefficient of permeability k_{w2}. The ratio of the major to the minor coefficients of permeability is a constant greater than unity at any point within the soil mass. The magnitudes of the major and minor coefficients of permeability, k_{w1} and k_{w2}, can vary with matric suction from one location to another (i.e., heterogeneity), but their ratio is assumed to remain constant at every point.

An unsteady-state seepage formulation and subsequent solution are generally performed with respect to the x and y Cartesian coordinate directions. Therefore, it is necessary to write the coefficients of permeability for the x- and y-directions in terms of the major and minor coefficients of permeability. The relationship between the two permeability systems can be derived by first writing the water flow rates in the major and minor permeability directions (i.e., directions s_1 and s_2, respectively):

$$v_{w1} = -k_{w1}\frac{\partial h_w}{\partial s_1} \tag{7.43}$$

$$v_{w2} = -k_{w2}\frac{\partial h_w}{\partial s_2} \tag{7.44}$$

where:

v_{w1} = water flow rate across a unit area of the soil element in the s_1-direction,
v_{w2} = water flow rate across a unit area of the soil element in the s_2-direction,
k_{w1} = major coefficient of permeability with respect to water as a function of matric suction which varies in the s_1-direction [i.e., $k_{w1}(u_a - u_w)$],
k_{w2} = minor coefficient of permeability with respect to water as a function of matric suction which varies in the s_2-direction [i.e., $k_{w2}(u_a - u_w)$],
h_w = hydraulic head (i.e., gravitational plus pore-water pressure head, or $y + u_w/\rho_w g$),
y = elevation,
s_1 = direction of major coefficient of permeability, k_{w1},
s_2 = direction of minor coefficient of permeability, k_{w2},
$\partial h_w/\partial s_1$ = hydraulic head gradient in the s_1-direction, and
$\partial h_w/\partial s_2$ = hydraulic head gradient in the s_2-direction.

The chain rule can be used to express the hydraulic head gradients in the s_1- and s_2-directions (i.e., $\partial h_w/\partial s_1$ and $\partial h_w/\partial s_2$, respectively) in terms of the gradients in the x- and y-directions (i.e., $\partial h_w/\partial x$ and $\partial h_w/\partial y$, respectively):

$$\frac{\partial h_w}{\partial s_1} = \frac{\partial h_w}{\partial x}\frac{\partial x}{\partial s_1} + \frac{\partial h_w}{\partial y}\frac{\partial y}{\partial s_1} \tag{7.45}$$

$$\frac{\partial h_w}{\partial s_2} = \frac{\partial h_w}{\partial x}\frac{\partial x}{\partial s_2} + \frac{\partial h_w}{\partial y}\frac{\partial y}{\partial s_2} \tag{7.46}$$

where:

$\partial h_w/\partial x$ = hydraulic head gradient in the x-direction and
$\partial h_w/\partial y$ = hydraulic head gradient in the y-direction.

From trigonometric relations, the following relationships can be obtained (see Fig. 7.26):

$$\frac{dx}{ds_1} = \cos\alpha \tag{7.47}$$

$$\frac{dy}{ds_1} = \sin\alpha \tag{7.48}$$

$$\frac{dx}{ds_2} = -\sin\alpha \tag{7.49}$$

$$\frac{dy}{ds_2} = \cos\alpha \tag{7.50}$$

Substituting Eqs. 7.45–7.50 into Eqs. 7.43–7.44 gives

$$v_{w1} = -k_{w1}\left(\cos\alpha\frac{\partial h_w}{\partial x} + \sin\alpha\frac{\partial h_w}{\partial y}\right) \tag{7.51}$$

$$v_{w2} = -k_{w2}\left(-\sin\alpha\frac{\partial h_w}{\partial x} + \cos\alpha\frac{\partial h_w}{\partial y}\right) \tag{7.52}$$

The water flow rates in the x- and y-directions can be written by projecting the flow rates in the major and minor directions to the x- and y-directions:

$$v_{wx} = v_{w1}\cos\alpha - v_{w2}\sin\alpha \tag{7.53}$$

$$v_{wy} = v_{w1}\sin\alpha + v_{w2}\cos\alpha \tag{7.54}$$

where:

v_{wx} = water flow rate across a unit area of the soil element in the x-direction,
v_{wy} = water flow rate across a unit area of the soil element in the y-direction, and
α = angle between the direction of the major coefficient of permeability and the x-direction.

Substituting Eqs. 7.51 and 7.52 for v_{w1} and v_{w2}, respectively, into Eqs. 7.53 and 7.54 results in the following relations:

$$v_{wx} = -k_{w1}\cos^2\alpha\frac{\partial h_w}{\partial x} - k_{w1}\sin\alpha\cos\alpha\frac{\partial h_w}{\partial y} - k_{w2}\sin^2\alpha\frac{\partial h_w}{\partial x} + k_{w2}\sin\alpha\cos\alpha\frac{\partial h_w}{\partial y} \tag{7.55}$$

$$v_{wx} = -k_{w1}\sin\alpha\cos\alpha\frac{\partial h_w}{\partial x} - k_{w1}\sin^2\alpha\frac{\partial h_w}{\partial y} + k_{w2}\sin\alpha\cos\alpha\frac{\partial h_w}{\partial x} - k_{w2}\cos^2\alpha\frac{\partial h_w}{\partial y} \tag{7.56}$$

Rearranging Eqs. 7.55 and 7.56 yields the expressions for the rate of water flow in the x- and y-directions:

$$v_{wx} = -\left(k_{wxx}\frac{\partial h_w}{\partial x} + k_{wxy}\frac{\partial h_w}{\partial y}\right) \quad (7.57)$$

$$v_{wy} = -\left(k_{wyx}\frac{\partial h_w}{\partial x} + k_{wyy}\frac{\partial h_w}{\partial y}\right) \quad (7.58)$$

where:

$$k_{wxx} = k_{w1}\cos^2\alpha + k_{w2}\sin^2\alpha$$
$$k_{wxy} = \left(k_{w1} - k_{w2}\right)\sin\alpha\cos\alpha$$
$$k_{wyx} = \left(k_{w1} - k_{w2}\right)\sin\alpha\cos\alpha$$
$$k_{wyy} = k_{w1}\sin^2\alpha + k_{w2}\cos^2\alpha$$

Equations 7.57 and 7.58 provide the flow rates in the x- and y-directions in terms of the major and minor coefficients of permeability. These flow rate expressions can then be used in the formulation for unsteady-state seepage analyses.

7.5.3 Water Phase Partial Differential Equation for Transient Flow

The water phase partial differential equation for transient two-dimensional seepage can be derived by considering the continuity for the water phase. The assumption is made that there are no changes in total stress with time in this analysis. The net flux of water through an element of unsaturated soil (Fig. 7.24) can be computed from the volume rates of water entering and leaving the element over a period of time:

$$\frac{\partial\left(V_w/V_0\right)}{\partial t} = \frac{\partial v_{wx}}{\partial x} + \frac{\partial v_{wy}}{\partial y} \quad (7.59)$$

where:

V_w = volume of water in the element,
V_0 = initial overall volume of the element (i.e., dx, dy, dz),
dx, dy, dz = dimensions in the x-, y-, and z-directions, respectively, and
$\partial(V_w/V_0)/\partial t$ = rate of change in the volume of water in the soil element with respect to the initial volume of the element.

The difference between a steady-state and unsteady-state water flow formulation lies in the representation of changes in the amount of water in the soil, V_w, with respect to time. The amount of water in the soil for any applied suction is represented by the SWCC. The SWCC must be written in terms of the volumetric water content when calculating water storage. Any mathematical representation of the (volumetric) SWCC can be differentiated to yield the change in volumetric water content, θ, of the soil with respect to soil suction [i.e., $d\theta/d\left(u_a - u_w\right)$]. The differentiation must be performed on an arithmetic scale and gives rise to the water storage modulus m_2^w.

The volume-mass constitutive equations required for derivation of the transient water flow equation is presented in Chapter 13 on volume change (i.e., stress-deformation). The net flux of water can be computed from the time derivative of the water phase constitutive equation. The time derivatives of the total stress and the pore-air pressure are equal to zero since both pressures are assumed to remain constant with time.

The continuity condition can be satisfied by equating the divergence of the water flow rates ($\partial v_x/\partial x + \partial v_y/\partial y$) to the time derivative of the constitutive equation for the water phase:

$$\frac{\partial}{\partial x}\left(k_{wxx}\frac{\partial h_w}{\partial x} + k_{wxy}\frac{\partial h_w}{\partial y}\right) + \frac{\partial}{\partial y}\left(k_{wyx}\frac{\partial h_w}{\partial x} + k_{wyy}\frac{\partial h_w}{\partial y}\right) = m_2^w \rho_w g\frac{\partial h_w}{\partial t} \quad (7.60)$$

Equation 7.60 is the governing partial differential equation for unsteady-state water seepage in an anisotropic soil. The pore-air pressure is assumed to remain constant with time. In many cases, the directions for the *major and minor* coefficients of permeability coincide with the x- and y-directions, respectively. In this case, the α angle is equal to zero, and the governing equation can be simplified by setting the α angle to zero:

$$\frac{\partial}{\partial x}\left(k_{w1}\frac{\partial h_w}{\partial x}\right) + \frac{\partial}{\partial y}\left(k_{w2}\frac{\partial h_w}{\partial y}\right) = m_2^w \rho_w g\frac{\partial h_w}{\partial t} \quad (7.61)$$

The k_{w1} and k_{w2} terms in Eq. 7.60 are the major and minor coefficients of permeability in the x- and y-directions, respectively. These permeability coefficients are a function of matric suction that can vary with location in the x- and y-directions [i.e., $k_{w1}\left(u_a - u_w\right)$ and $k_{w2}\left(u_a - u_w\right)$]. For *isotropic* soil conditions, the k_{w1} and k_{w2} terms are equal [i.e., $k_{w1} = k_{w2} = k_w\left(u_a - u_w\right)$], and the transient seepage equation can be further simplified as follows:

$$\frac{\partial}{\partial x}\left(k_w\frac{\partial h_w}{\partial x}\right) + \frac{\partial}{\partial y}\left(k_w\frac{\partial h_w}{\partial y}\right) = m_2^w \rho_w g\frac{\partial h_w}{\partial t} \quad (7.62)$$

Rearranging the above equation gives

$$k_w\frac{\partial^2 h_w}{\partial x^2} + \frac{\partial k_w}{\partial x}\frac{\partial h_w}{\partial x} + k_w\frac{\partial^2 h_w}{\partial y^2} + \frac{\partial k_w}{\partial y}\frac{\partial h_w}{\partial y} = m_2^w \rho_w g\frac{\partial h_w}{\partial t} \quad (7.63)$$

Substituting $y + u_w/\rho_w g$ for the hydraulic head h_w and rearranging the equation result in an equation written in terms of pore-water pressure:

$$c_v^w\frac{\partial^2 u_w}{\partial x^2} + \frac{c_v^w}{k_w}\frac{\partial k_w}{\partial x}\frac{\partial u_w}{\partial x} + c_v^w\frac{\partial^2 u_w}{\partial y^2} + \frac{c_v^w}{k_w}\frac{\partial k_w}{\partial y}\frac{\partial u_w}{\partial y} + c_g\frac{\partial k_w}{\partial y} = \frac{\partial u_w}{\partial t} \quad (7.64)$$

where:

$c_v^w = k_w / \left(\rho_w g m_2^w\right)$, coefficient of consolidation for the unsaturated soil, and

$c_g = 1/m_2^w$, the gravity term.

The general governing equation for an anisotropic soil can be simplified to an anisotropic case which coincides with the Cartesian coordinates by setting the α angle to zero. Isotropic soil conditions are obtained by setting $k_{wx} = k_{wy}$.

The general governing equation can therefore be used to solve transient or unsteady-state water seepage problems through saturated-unsaturated flow systems. The water coefficient of permeability becomes equal to the saturated coefficient of permeability k_s for the *saturated* portion. The saturated coefficient of permeability may vary with respect to direction (i.e., anisotropy) or with respect to location (i.e., heterogeneity). In this case, both anisotropy and heterogeneity with respect to permeability can be taken into account. The coefficient of water volume change, m_2^w, approaches the value of the coefficient of volume change, m_v, as the soil becomes saturated.

7.6 DIRECT MEASUREMENT OF WATER FLOW PROPERTIES

The application of flow laws to engineering problems requires the determination of the hydraulic properties of a soil. The water phase coefficient of permeability for an unsaturated soil can be determined either using direct laboratory measurements or through indirect estimation techniques based on the SWCC. Direct measurements of the water coefficient of permeability can be performed either in the laboratory or in the field. The direct measurements are referred to as permeability tests and the apparatus used to perform the test in the laboratory is called a permeameter.

Indirect estimation techniques make use of the volume-mass properties, the SWCC, and the saturated coefficient of permeability. The coefficient of permeability is computed for a range of soil suctions and the results are referred to as a permeability function for the soil.

There is no theoretical procedure whereby the coefficient of permeability for a soil can be computed with the same confidence that is obtained from a direct measurement (Hillel, 1982). The hydraulic head gradient and the flow rate are determined from pore-water pressure head and quantity of flow measurements when directly measuring the coefficient of permeability. Measurements of the coefficient of permeability for an unsaturated soil must be measured at a variety of soil suction values.

There are a number of other procedures that have been proposed for measuring the coefficient of permeability of an unsaturated soil. Sometimes the pore-water pressure or the water content is measured while other variables are inferred from the SWCC. Measurements can be performed in situ, but the methods are not as advanced and standardized as laboratory methods. Laboratory tests are more economical to perform.

7.6.1 Laboratory Test Methods

Various laboratory methods can be used for measuring the coefficient of permeability with respect to the water phase, k_w, in unsaturated soils (Klute, 1972). All methods assume the validity of Darcy's law, which states that the coefficient of permeability is the ratio of the flow rate to the hydraulic head gradient. The flow rate and the hydraulic head gradient are the variables usually measured during a permeability test. The flow rate and the hydraulic head gradient can either be kept constant with time (i.e., time independent) or varied with time during the test. Correspondingly, the various testing procedures can be categorized into two primary groups: steady-state methods where the quantity of flow is time independent and unsteady-state methods where the quantity of flow is time dependent.

The steady-state method is described first, followed by a description of the unsteady-state, instantaneous profile method. Some recent applications of the instantaneous profile method for measuring the unsaturated permeability function are also presented. The method is referred to as the "wetting front advancing method" and was developed by Li et al.(2009). In this method, a large-scale soil column is used to simulate the flow of water into an unsaturated soil column. The rate of water rise in the column along with other data is monitored and used to calculate the permeability function. Results when using this test procedure are later reported for several coarse-grained soils. The measurement of the saturated coefficient of permeability has been described in numerous soil mechanics books and is not repeated herein.

7.6.2 Steady-State Methods to Measure Permeability

The steady-state method for the measurement of the water coefficient of permeability is performed by maintaining a constant hydraulic head gradient across an unsaturated soil specimen. The matric suction and water content of the soil are also maintained constant. The constant hydraulic head gradient produces a steady-state water flow across the specimen. Steady-state conditions are achieved when the flow rate entering the soil is equal to the flow rate leaving the soil. The coefficient of permeability k_w corresponding to the applied matric suction is computed. The experiment can be repeated for different magnitudes of matric suction (or water content). The steady-state method can be used for both compacted and undisturbed specimens.

7.6.2.1 Apparatus for Steady-State Method

An assemblage of the equipment used for the steady-state method is shown in Fig. 7.27 (Klute, 1965a). A cylindrical soil specimen is placed in a permeameter between two high-air-entry ceramic plates. The pressure applied at the

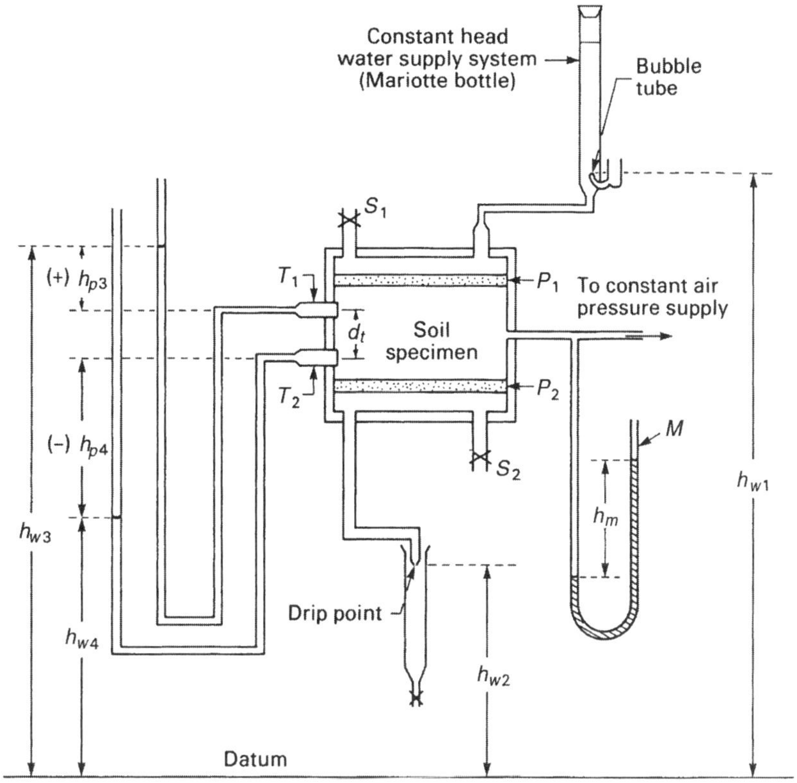

Figure 7.27 Apparatus for measurement of unsaturated soil coefficient of permeability using steady-state method (after Klute, 1965a)

top of the specimen is denoted as P_1, and P_2 is the pressure applied at the bottom of the specimen. Two tensiometers, T_1 and T_2, are installed along the height of the specimen for the measurement of pore-water pressures at two points in the soil. An air supply maintains a constant pore-air pressure to the soil specimen. The air pressure is measured using the manometer, M.

A water supply is applied to the top of the upper porous plate, P_1, to develop a constant hydraulic head gradient across the soil. The water supply provides a constant hydraulic head, h_{w1}, by means of a Mariotte bottle. It is also possible to use a simple overflow system. Water flows through the ceramic plate P_1, the soil specimen, and the ceramic plate P_2. The outflow of water is maintained at a constant hydraulic head h_{w2}, below the porous plate. Plate P_2 controls the outflow elevation. Valves S_1 and S_2, are used to flush air bubbles that may accumulate in the water compartment adjacent to the porous plate.

The permeability test is started using a low matric suction value. The matric suction is increased in steps as the permeability is measured. In other words, the test is commenced at a condition near saturation and proceeds through a drying process in accordance with the following procedure. The matric suction $(u_a - u_w)$ is set to a specified value by controlling the pore-air pressure, as indicated by the height in the manometer fluid, h_m. The pore-water pressure is measured using the tensiometers installed along the specimen. The pore-water pressure head is determined from the elevation of the fluid interface in the attached manometer relative to the tensiometer elevation. The pressure head is positive when the fluid interface is above the elevation of the tensiometer and is negative when the interface is below the tensiometer elevation. The tensiometer readings in Fig. 7.27 indicate pressure heads of $(+)h_{p3}$ and $(-)h_{p4}$ from tensiometers T_1 and T_2, respectively.

A constant hydraulic head gradient is then applied to the specimen as depicted by hydraulic heads h_{w1} and h_{w2}. Steady-state flow conditions are achieved when the inflow and outflow rates are equal and the tensiometer readings are constant with time. The volume of water, Q, flowing across the cross-sectional area of the soil, A, in a period of time t is measured and used to compute the flow rate. The hydraulic heads h_{w3} and h_{w4} correspond to tensiometers T_1 and T_2, which are placed a distance d_t apart. By knowing the distance d_t, it is possible to compute the hydraulic head gradient in the soil. The measured hydraulic head gradient is used in calculating the coefficient of permeability owing to uncertainties associated with the hydraulic head changes across the high-air-entry ceramic disks. The contact planes between the specimen and the porous plates also produce uncertainties in predicting heads throughout the specimen (Klute, 1972).

7.6.2.2 *Computations of Unsaturated Coefficient of Permeability Using Steady-State Method*

The coefficient of permeability k_w is computed based on Darcy's law:

$$k_w = \left(\frac{Q}{At}\right)\left(\frac{d_t}{h_{w3} - h_{w4}}\right) \quad (7.65)$$

The pore-air pressure is assumed to be uniform throughout the specimen:

$$u_a = \rho_m g h_m \quad (7.66)$$

where:

ρ_m = density of the manometer fluid,
g = gravitational acceleration, and
h_m = height of the manometer fluid.

The applied hydraulic head gradient causes the pore-water pressures to differ at tensiometers T_1 and T_2. The average pore-water pressure is computed as follows:

$$(u_w)_{\text{ave}} = \left(\frac{h_{p3} + h_{p4}}{2}\right)\rho_w g \quad (7.67)$$

where:

ρ_w = density of water and
$(u_w)_{\text{ave}}$ = average of pore-water pressure.

The calculated coefficient of permeability k_w corresponds to the average matric suction in the soil:

$$(u_a - u_w)_{\text{ave}} = \rho_m g h_m - \left(\frac{h_{p3} + h_{p4}}{2}\right)\rho_w g \quad (7.68)$$

where:

$(u_a - u_w)_{\text{ave}}$ = average matric suction.

The water content of the soil specimen can be measured directly using either destructive or nondestructive techniques. Using a destructive technique, water contents are measured after each stage in the permeability test. The soil specimen can therefore be used for only one stage of the permeability test. Several "identical" specimens must be prepared in order to obtain the coefficient of permeability at different matric suctions or water contents.

Using a nondestructive technique, water contents at different matric suctions are measured using a single soil specimen. This can be done using a gamma attenuation technique or by weighing the soil specimen (together with the apparatus) in order to obtain the change in water content after each stage of the test. The initial and final water contents are computed from the initial and final masses of the soil specimen. The mass of soil solids is obtained by oven drying the specimen at the end of the test.

An indirect technique has also been used to infer the water content at particular matric suctions by referring to an independently measured SWCC. The SWCC and the coefficient of permeability can be obtained simultaneously using the apparatus (Ingersoll, 1981). The change in water content during each increment of matric suction needs to be measured.

Having measured the coefficient of permeability k_w corresponding to a particular matric suction (or water content), the permeability test is then repeated for higher values of matric suction. The matric suction of the soil specimen is increased by increasing the applied air pressure or by decreasing the applied water pressures. The average pore-water pressure can be decreased by reducing the values of h_{w1} and h_{w2}. At no time should the matric suction exceed the air-entry value of the ceramic plates or the matric suction limit that can be read on the tensiometers (i.e., approximately 90 kPa). The use of higher air-entry ceramic plates allows testing at higher values of suction. High-air-entry ceramic plates with the highest possible coefficient of permeability should be selected in the design of the apparatus to prevent impeded flow (Klute, 1972).

7.6.2.3 *Water Coefficients of Permeability or Hydraulic Conductivity Function*

A series of coefficient-of-permeability values is obtained by the end of the test. A typical set of data for the coefficient of permeability k_w versus matric suction is shown in Fig. 7.28. The measured water coefficients of permeability correspond to the drying curve. It is noted that there is hysteresis in the relationship of the water coefficient of permeability and matric suction. Hysteresis in the permeability function can be observed when the permeability test is performed for both the drying and the wetting processes. However, the steady-state method is normally performed only when going from a saturated to an unsaturated condition.

Another apparatus which uses a steady-state flow system is shown in Fig. 7.29 (Klute, 1965b). A cylindrical soil specimen is placed between two high-air-entry ceramic plates located within an air pressure chamber. The soil specimen is subjected to an all-around controlled air pressure u_a. The pore-water pressure u_w is measured by means of tensiometers placed along the length of the specimen. A constant-hydraulic-head gradient is applied across the specimen, as described in Fig. 7.29. The permeability test is performed using the procedure previously described. The matric suction in the specimen is increased by increasing the all-around air pressure.

The specimen diameter is typically on the order of 25–100 mm and its length ranges from 10 to 500 mm. The longer the length of the specimen, the longer is the time required to reach steady-state conditions. Klute (1965b) suggested a specimen height on the order of 10–50 mm. The specimen should not be too short because space is required between the two tensiometers inserted in the soil specimen. A long specimen may also require more than

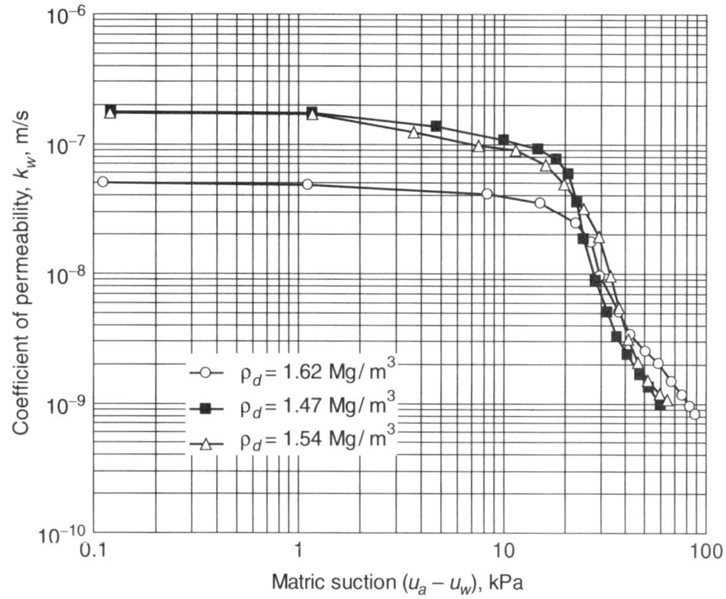

Figure 7.28 Water permeability functions measured using steady-state method (modified from Ingersoll, 1981).

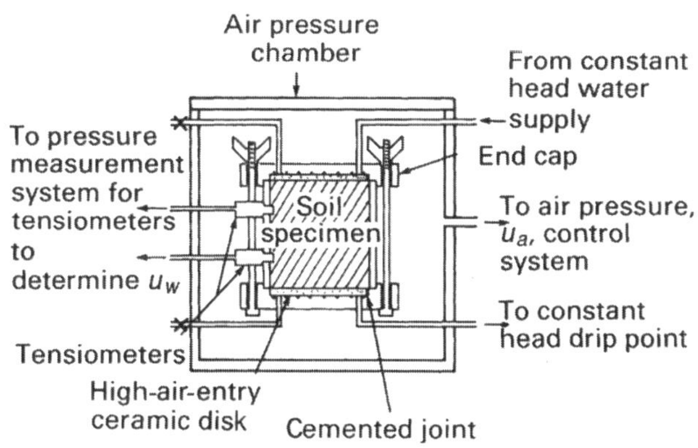

Figure 7.29 Enclosed and modified Tempe cell used for measurement of water coefficient of permeability using steady-state method (after Klute, 1965b).

two tensiometers for accurate measurements of pore-water pressure. The steady-state method requires quite a long time for testing; however, the results are generally considered to be more accurate than those from the unsteady-state method (Hillel, 1982).

7.6.2.4 Limitations and Difficulties with Steady-State Method

There are several difficulties associated with the steady-state method (Klute, 1965b, 1972; Olson and Daniel, 1981). The main difficulty arises from the low coefficient of permeability of unsaturated soils, particularly at high matric suctions. As a result, the water flow rates during the test are extremely low, and a long time is required to complete a series of permeability measurements. The low flow rates necessitate an extremely accurate mass of water measurements. The possibility of water loss within the apparatus must be minimized. Air diffusion through the water and water loss from the apparatus (e.g., plastic container) can cause errors in the measurements of the volume of water. Air bubbles below the porous disks must be periodically removed by flushing. Diffused air inside the tensiometers should also be flushed from time to time. A diffused air volume indicator can be used to measure the amount of diffused air in order to apply the necessary corrections to the water volume change measurements.

Another testing difficulty is related to the contact between the soil specimen and the permeameter. As matric suction increases, the soil specimen may shrink and separate from the permeameter wall and from the high-air-entry ceramic plates. The air gap between the soil and the wall of the permeameter does not allow the seepage of water since air is nonconductive to the flow of water. Therefore, the separation of soil from the permeameter wall generally does not create a serious problem in testing unsaturated soils (Daniel, 1983). However, a good contact between the soil specimen and the porous plates is required in order to ensure the continuity of water flow. The use of loaded porous plates has been suggested to overcome the problem of separation of the soil from the porous plates (Elrick and Bowman, 1964; Richards and Wilson, 1936). It is also important to maintain good contact between the soil and the tensiometers for accurate measurements of pore-water pressure.

The use of a high air-entry ceramic disk results in two limitations when attempting to measure the permeability function. These limitations are shown in Fig. 7.30. First, the soil suction will be limited by the air-entry value of the ceramic disk.

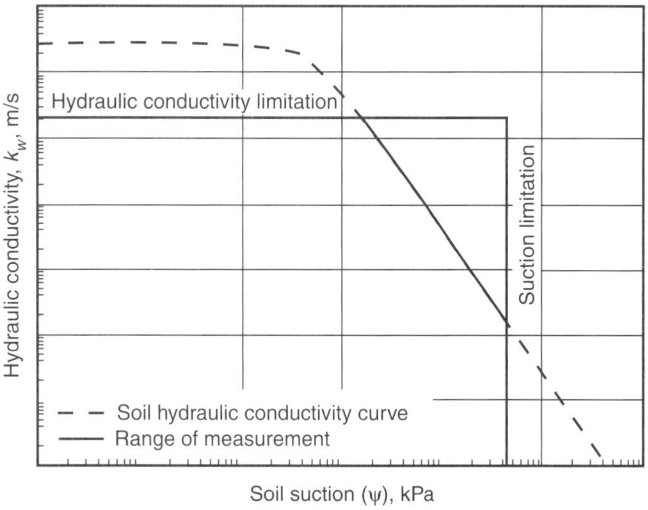

Figure 7.30 Design considerations when using direct measurement of unsaturated coefficient of permeability.

Second, the measured permeability of the soil must be smaller than the saturated permeability of the high-air-entry ceramic disk. Consequently, the steady-state method is limited to a narrow measurement zone, as illustrated in Fig. 7.30. The above limitations also reveal that the permeability measurements made with low applied matric suctions are likely to be inaccurate. Therefore, it is more acceptable practice to independently measure the saturated coefficient of permeability of any soil for which the permeability function is required.

7.6.2.5 Simplified Steady-State Laboratory Measurement of Permeability

There is a rather simple procedure that can be used to measure the unsaturated coefficient of permeability of a soil. Let us suppose that a permeameter is filled with dry sand and then a constant water flux is applied at the top of the permeameter. Let us also start with water fluxes that are well below the saturated coefficient of permeability of the soil. If a tensiometer were placed near the top and bottom of the per meameter, it would be found that both tensiometers would read essentially the same value under steady-state seepage conditions. The two tensiometers would read the same value because the infiltration is occurring under a gradient of 1.0.

The applied flux becomes the unsaturated coefficient of permeability for the soil. The tensiometer reading is the matric suction in the soil corresponding to the applied flux. The above experiment can be repeated for different applied water fluxes, and each time the applied water flux becomes the unsaturated coefficient of permeability of the soil. This is essentially a steady-state permeability test where there is no need for a high-air-entry disk.

As long as the matric suction in the soil is quite small, there is no need for a high-air-entry disk. The theory related to this simple permeability test can be understood by considering the infiltration of water at ground surface, as shown

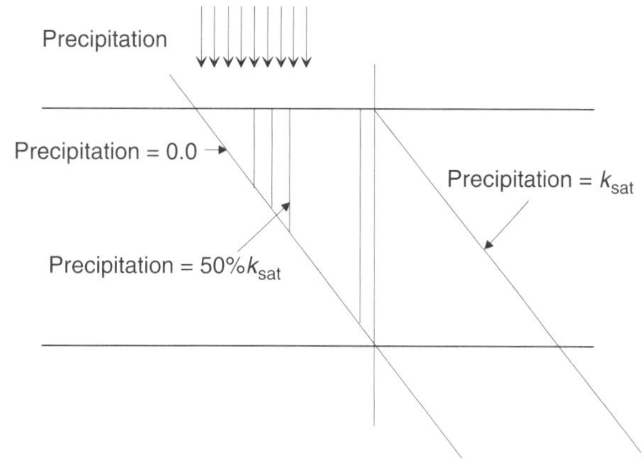

Figure 7.31 Field illustration of a steady-state test procedure that can be used in the laboratory.

in Fig. 7.31. As long as the applied rainfall is well below the saturated coefficient of permeability, infiltration will take place under a gradient close to 1.0.

7.6.3 Instantaneous Profile Method

The instantaneous profile method has been described by Richards and Weeks (1953), Hamilton et al., (1981), Meerdink et al., (1996), Fujimaki and Inoue (2003), Cui et al., (2008), and others. The instantaneous profile method is an unsteady-state method that can be used either in the laboratory or in situ. The method uses a cylindrical specimen of soil that is subjected to a continuous water flow from one end of the specimen. The test method has several variations. The methodologies differ mainly in the flow process used and in the measurement of hydraulic head gradient and flow rate. The flow process can be a wetting process where water flows into the specimen or a drying process where water flows out of the specimen under gravity.

The hydraulic head gradient and flow rate at various points along the specimen can be obtained using one of several procedures (Klute, 1972). For one of the procedures, the water content and pore-water pressure head distributions can be measured independently. The water content distribution can be used to compute flow rates. The pore-water pressure head gradient can be calculated from the measured pore-water pressure head distribution. The gravitational head gradient is obtained from the elevation difference.

Using another procedure, the water content distribution is measured while the pore-water pressure head is inferred from the SWCC. There is still another possible procedure where the pore-water pressure head distribution is measured, and the water content is inferred from the SWCC. Tensiometers and psychrometers have been used to measure the pore-water pressure distribution along the permeameter. Of the above procedures, the first one appears to be most satisfactory. All variations of the instantaneous profile procedure are based on the same theoretical principles. The hydraulic head gradient and the flow rate are determined concurrently and instantaneously at various elapsed times, after the flow of water has commenced.

7.6.3.1 Laboratory Instantaneous Profile Method

The apparatus and procedure for the instantaneous profile method proposed by Hamilton et al. (1981) is shown in Fig. 7.32. A water flux is controlled at one end of the soil specimen while the other end is vented to the atmosphere. Water flows in the horizontal direction as a result of a gradient in the pore-water pressure head. Gravitational head gradient effects are thereby negligible. Hamilton et al. (1981) elected to measure the pore-water pressure head distribution during unsteady-state water flow and obtained water contents from the soil-water characteristic curve. The hydraulic head gradient and the flow rate vary with time during the test. The pore-water pressures, and therefore the hydraulic head gradients, are measured at several points along the soil specimen.

The change in water content is related to a change in negative pore-water pressure (or matric suction) through use of the SWCC. Flow rates are then computed from the change in volumetric water content. The ratio between the flow rate and the hydraulic head gradient gives the coefficient of permeability. Measurements at different locations along the specimen at different times during the unsteady-flow process produce a series of coefficients of permeability measurements. Each measured coefficient of permeability corresponds to a particular matric suction (or water content). The method does not require an assumption of uniform hydraulic properties in the soil, as is the case in the steady-state method.

The following procedure is used to illustrate a permeability test performed using the wetting process. A compacted or undisturbed soil specimen is inserted into a cylindrical permeameter (Fig. 7.32). Both ends of the permeameter are covered by end plates with O-ring seals. Water is supplied to the left-end plate using a hypodermic needle and is distributed across the soil surface through the use of several sheets of filter paper. Air in the specimen is vented to the atmosphere

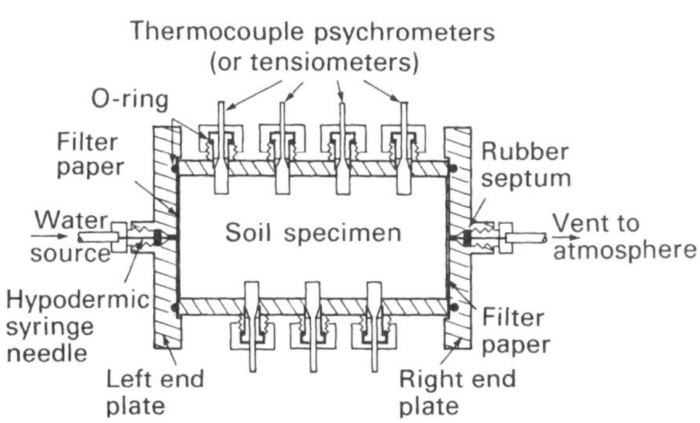

Figure 7.32 Apparatus for measuring the coefficient of water permeability using the instantaneous profile method (after Hamilton et al., 1981).

using a hypodermic needle at the right-end plate. Filter paper is also placed across the right-end surface of the permeameter. Several ports are provided along the permeameter wall for the installation of tensiometers or psychrometers.

Tensiometers are used for a relatively moist soil having matric suctions less than approximately 90 kPa. Thermocouple psychrometers can be used for measuring suctions ranging from approximately 100 to 8000 kPa. The tensiometers or psychrometers are inserted through the ports in the permeameter and extended into small holes drilled into the soil specimen. The entire apparatus should be placed in a temperature-controlled chamber with a high relative humidity when psychrometers are used.

The test commences with an unsaturated specimen and proceeds toward a saturated condition. The initial suction is first measured under equilibrium conditions. The equilibrium condition is then altered by slowly injecting water into the specimen using the hypodermic needle. The water inflow rate should be selected such that the pore-water pressure is always negative across the length of the specimen. Flow rates in the range of 0.2–5 cm^3/day are commonly satisfactory (Daniel, 1983). Soil suctions are measured at various time intervals (e.g., every 24 h). Psychrometers can replace the tensiometers when soil suction increases above 90 kPa.

The test is terminated when the pore-water pressure at the entrance of the permeameter (i.e., the left end) becomes positive. Positive pore-water pressures may produce flow along the inside wall of the permeameter or around the tensiometers. Upon completion of the test, the soil specimen is divided into several sections for final water content measurements. The final volumetric water contents along the specimen are plotted against the corresponding suctions to produce a SWCC. The entire test may take as long as 2–3 weeks. Results from independent measurements of the water content versus soil suction relationship (i.e., the wetting curve) can also be used in the determination of the SWCC.

7.6.3.2 Computations for Instantaneous Profile Method

The coefficient-of-permeability calculations are performed by plotting the pore-water pressure head versus the volumetric water content from various points along the specimen, as illustrated in Figs. 7.33a and 7.33b, respectively. The pore-water pressures are obtained from either the tensiometer or psychrometer readings. The total suction profile obtained from psychrometers can be taken as the negative pore-water pressure profile when the osmotic suction gradient is negligible and the air pressure is atmospheric. The pore-water pressure can be divided by the unit weight of water (i.e., $\rho_w g$) to give the pressure head. The hydraulic head gradient (i.e., pore-water pressure head gradient) at a point in the specimen for a specific time is equal to the slope of the hydraulic head profile:

$$i_w = \frac{dh_w}{dx} \tag{7.69}$$

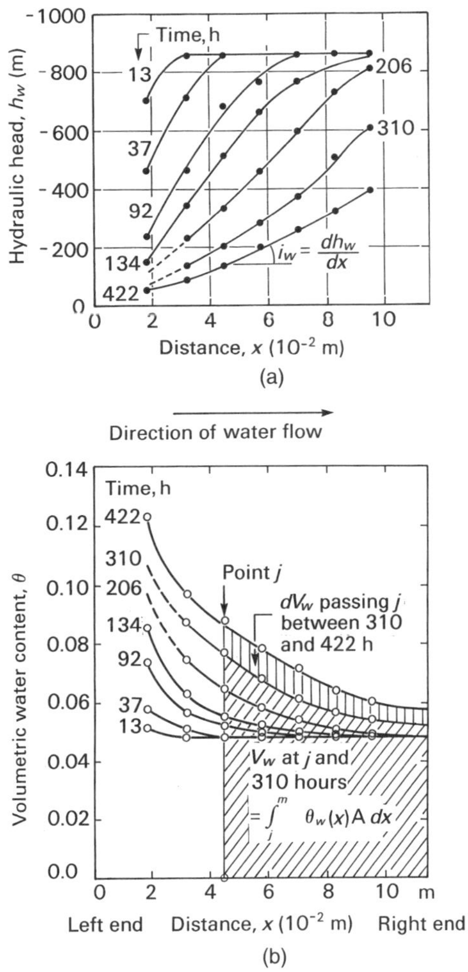

Figure 7.33 Hydraulic head and water content profiles generated during unsteady-state permeameter test: (a) profile of hydraulic head along permeameter; (b) Profile of volumetric water content along permeameter (after Hamilton et al., 1981).

where:

i_w = hydraulic head gradient at a point for a specific time and

dh_w/dx = slope of the hydraulic head profile at the point under consideration.

The volumetric water content profile is obtained from the measured pore-water pressures and the SWCC. The absolute value of the negative pore-water pressure recorded by the tensiometer is equal to the matric suction. The flow rate v_w at a point is equal to the volume of water that flows across the cross-sectional area of the specimen, A, during a time interval dt. Water flows from the left end of the permeameter to the right end, and there is no water flow out of the right end of the specimen. Therefore, the total volume of water passing through a point in the soil specimen during a period of time is equal to the water volume change that occurs between the point under consideration and the right end of the specimen during the specified time period.

The total volume of water present between point j and the right end of the specimen (i.e., point m) at a specific time is obtained by integrating the volumetric water content profile over the specified time interval:

$$V_w = \int_j^m \theta(x)\, A\, dx \tag{7.70}$$

where:

V_w = total volume of water in the soil between point j and the right end of the specimen designated as point m,

$\theta(x)$ = volumetric water content profile as a function of distance x for a specific time, and

A = cross-sectional area of the specimen.

The difference in volumes of water, dV_w, computed between two consecutive times (i.e., an interval dt) is the quantity of water flowing past point j during the time interval under consideration (Fig. 7.33b). The flow rate at the point is then computed as follows:

$$v_w = \frac{dV_w}{A\, dt} \tag{7.71}$$

The flow rate corresponds to an average value for the hydraulic head gradients obtained at two consecutive times. The coefficient of permeability k_w is computed by dividing the flow rate v_w by the average hydraulic head gradient i_{ave}:

$$k_w = \frac{v_w}{i_{ave}} \tag{7.72}$$

Computations for the coefficient of permeability can be repeated for different points and different times. As a result, many coefficients of permeability can be computed corresponding to various water contents or suction values obtained from one test.

Figure 7.34 shows typical laboratory results of coefficient of permeability k_w for a clay as a function of suction. The permeability test was performed using the instantaneous profile method, and the data are presented in Fig. 7.33. Test procedural problems that may be encountered using the unsteady-state method are similar to those described previously for the steady-state method.

7.6.3.3 Comparison between Drying Permeability Function Obtained from Instantaneous Profile Method and Indirect Statistical Methods

Experimental studies assist in assessing the reliability of various methods that can be used for the determination of the permeability function for an unsaturated soil. Following are the results of a comparison between one of the proposed methods for the estimation of the permeability functions and results obtained using an instantaneous profile method analysis on a model slope. The study was reported by Krisdani

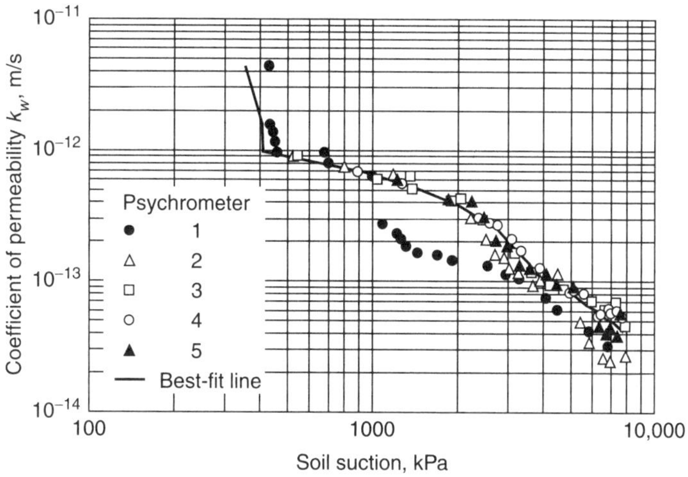

Figure 7.34 Unsaturated coefficients of permeability obtained in laboratory using instantaneous profile method (after Hamilton et al., 1981).

et al., (2009). Independent measurements were made of the SWCCs followed by an estimation of associated permeability functions. An independent model study was performed using the same soils, and the instantaneous profile method was used to analyze the measurements obtained during water infiltration and evaporations from the model slope.

Krisdani et al. (2009) first measured the saturated coefficient of permeability and the SWCC on three types of soil: a residual soil, a mixture of the residual soil and fine sand, and gravelly sand. The residual soil (RS) was derived from the weathering of Bukit Timah granite which covers about one-third of Singapore. The soil mixture contained 60% residual soil and 40% fine sand by dry mass, MRF 60–40. The gravelly sand (GS) was a commercially available mixture of 50% coarse sand and 50% fine gravel. The grain-size distributions of the soils used in the study are presented in Fig. 7.35 and Table 7.7. The drying SWCCs of thc soils were obtained using Tempe cell and pressure plate apparatuses. The wetting SWCCs were obtained using the capillary rise open-tube method (Yang et al., 2004a). The experimental data were fitted using the Fredlund and Xing (1994) equation with the correction factor $C(\psi)$ set to 1. The results of drying (D) and wetting (W) SWCC tests together with the fitted curves are shown in Fig. 7.36, while the fitting parameters

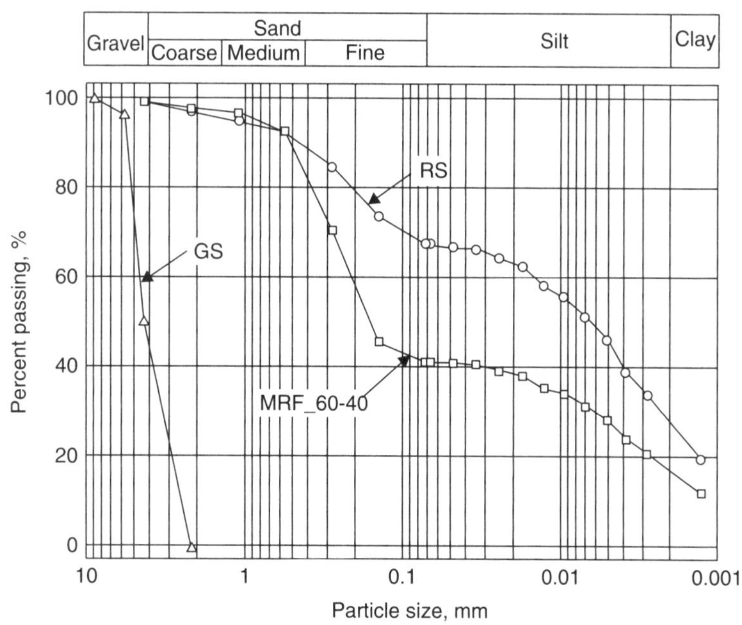

Figure 7.35 Grain-size distribution of soils used in unsaturated flow study at Nanyang Technological University (NTU), Singapore (after Krisdani, et al., 2009).

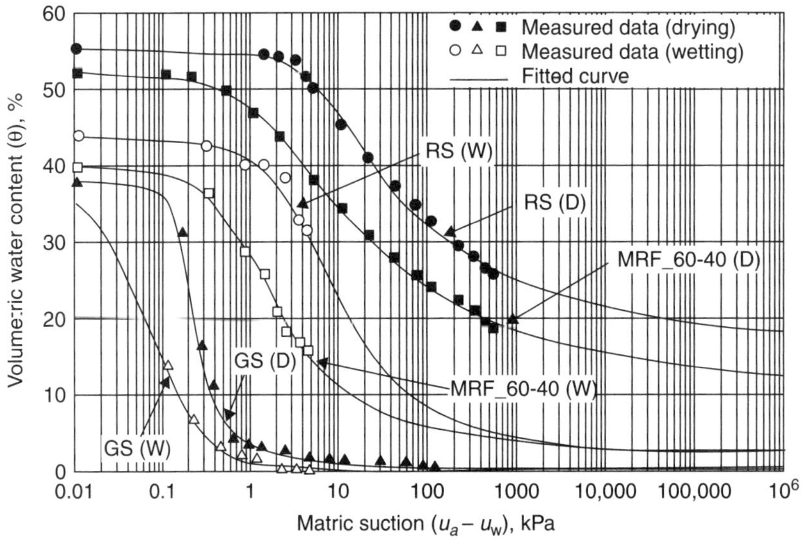

Figure 7.36 Independently Measured SWCCs for Soils in NTU Water Flow study Krisdani et al. (2009).

Table 7.7 Basic Properties of Soils in NTU Study

No.	Description	Soil RS	Soil MRF 60–40	Soil GS
1	Unified Soil Classification System (ASTM, 2006)	Sandy silt (ML)	Silty sand (SM)	Gravelly sand (SP)
2	Specific gravity, G_s	2.65	2.65	2.62
3	Grain-size analysis			
	Gravel content (>4.75 mm) (%)	0.6	0.1	49.9
	Sand content (0.075–4.75 mm) (%)	32.1	59.1	50.1
	Fine content (<0.075 mm) (%)	67.3	40.8	0.0
4	Atterberg limits			
	Liquid limit, LL (%)	47.0	34	—
	Plastic limit, PL (%)	32.0	22	—
	Plasticity index, PI (%)	15.0	12	—
5	Saturated coefficient of permeability, k_s (m/s)[a]	8.6×10^{-8}	2.7×10^{-7}	7.6×10^{-2}

Source: After Krisdani et al., (2009)
[a]Obtained from falling-head tests.

of the SWCCs are listed in Table 7.8. The basic properties and SWCC of the gravelly sand were measured by Krisdani et al. (2009).

A large-scale model slope was set up and filled with the above-mentioned soils. The model was used to study the infiltration and evaporation of water from the slope. The infiltration box was fabricated using stainless steel frames and acrylic sheets as the sidewalls. The size of the infiltration box was 2.45 m in length, 2.0 m in height, and 0.4 m in width. A number of holes were drilled along the sidewalls for the installation of measuring devices. Some additional holes were provided for collection of runoff. Drainage was also provided at the front panel of the infiltration box. A PVC (polyvinyl chloride) plate was installed as the base of the slope model to form an impervious boundary condition during the experiments.

The measuring devices consisted of tensiometers for the measurement of negative pore-water pressures and time domain reflectometry (TDR) for the measurement of volumetric water content. In addition, weighing balances

Table 7.8 Parameters for SWCCs

No.	Description	Symbol	Soil RS	Soil MRF 60–40	Soil GS
1	Saturated volumetric water content	θ_s	0.555	0.525	0.382
2	Air-entry value	ψ_a (kPa)	2.748	0.468	0.110
3	Residual volumetric water content	θ_r	0.300	0.258	0.021
4	Residual matric suction	ψ_r (kPa)	163.598	40.209	0.401
5	Water-entry value	ψ_w (kPa)	42.949	5.568	0.285
6	Best-fitting parameters (drying curve)	a_f (kPa)	4.079	0.870	0.176
		n_f	2.231	1.446	4.440
		m_f	0.230	0.314	1.130
7	Best-fitting parameters (wetting curve)	a_f (kPa)	6.277	0.484	0.071
		n_f	1.204	1.935	1.232
		m_f	1.808	0.615	2.710

and a digital flowmeter were installed. The amount of applied precipitation was monitored. The measuring devices were connected to data acquisition systems for continuous monitoring during the experiments. The arrangement of tensiometers and TDR probes in the slope models is shown in Fig. 7.37.

A rainfall simulator was used to apply a rainfall event to the slope model. The rainfall was produced using two nozzles installed at both ends of the infiltration box. A high air pressure (i.e., 300 kPa), was used to distribute water as a mist along the slope model. Three infrared light bulbs, or three fans, or a combination of both was used to simulate different rates of evaporation. The light bulbs were 175 W and the air flow capacity of each fan was 84 L/s. More details on the experimental setup used in this study are given in Tami et al., (2004c), and Krisdani et al. (2005) (Table 7.9).

The instantaneous profile method (Watson, 1966; Hillel et al., 1972; Olson and Daniel, 1981; Fredlund and Rahardjo, 1993a; Meerdink et al., 1996) was used to compute the permeability functions for the residual soil and the soil mixture

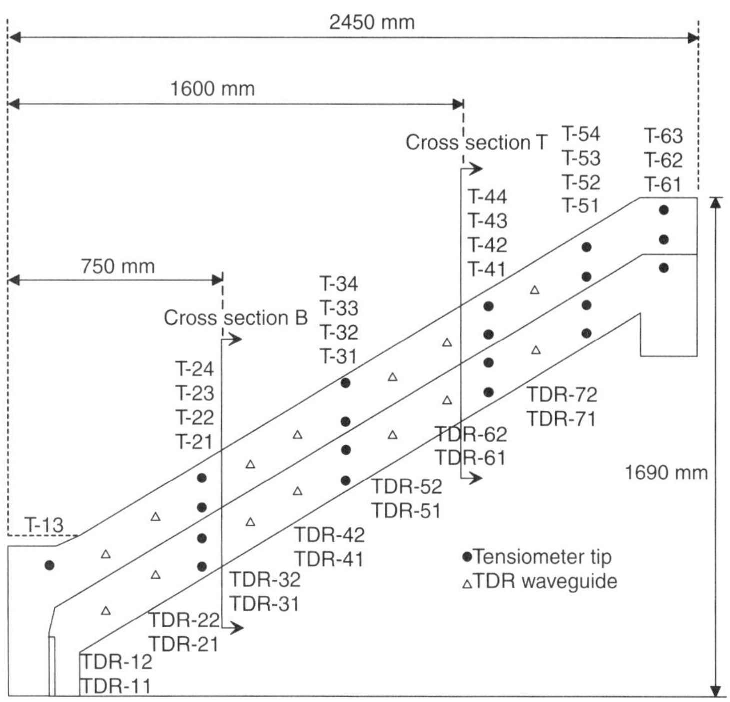

Figure 7.37 Location of measuring devices and designated cross sections along slope model (after Krisdani et al., 2009).

Table 7.9 Moisture Flux Conditions Applied to Slope Models

		Rainfall		Evaporation		
Name of Model	Name of Test	Intensity (mm/h)	Duration (mm/h)	Mode of Evaporation	Measured Potential Evaporation (mm/h)	Duration (h)
SM1	SM1 T1	8.95	3	Ambient condition	0.02	475
	SM1 T2	9.01	3	Lamp	0.20	120
	SM1 T3	8.88	3	Fan	0.12	120
	SM1 T4	9.04	3	Lamp and fan	0.22	80
SM1	SM2 T1	9.68	3	Lamp and fan	0.43	100
	SM2 T2	10.09	3	Ambient condition	0.02	120
	SM2 T3	9.88	3	Lamp	0.19	120
	SM2 T4	9.85	6	Lamp and fan	0.52	120

used in the slope models. The instantaneous profile method was used to obtain the permeability function during drying processes. Water flowed upward from the bottom of the upper layers during the drying process. Pore-water pressure data from tensiometers T-23, T-24, T-33, T-34, T-43, and T-44 (see Fig. 7.38) were used for the calculation of the drying permeability functions. Volumetric water contents were computed based on the SWCC and measured pore-water pressures (Fig. 7.39). Table 7.8 summarizes the unsaturated soil parameters used for the calculation of unsaturated coefficient of permeability by the instantaneous profile method. The unsaturated coefficients of permeability k_w of RS and MRF 60–40 were plotted against corresponding matric suctions and are shown in Figs. 7.40 and 7.41.

The unsaturated coefficient-of-permeability functions were also computed using a statistical estimation method described by Childs and Collis-George (1950). The calculation of the permeability function requires the saturated coefficient of permeability and SWCC. The saturated coefficient of permeability of the RS was 9×10^{-7} m/s. The MRF 60–40 soil had a saturated coefficient of permeability of 5×10^{-6} m/s. These values appear to be about one order of magnitude higher than the saturated coefficient of permeability obtained from the independent falling-head tests (Table 7.7).

The instantaneous profile method and the statistical estimation method were used to obtain permeability functions for the RS and the MRF 60–40, comprising the slope models. The estimated permeability functions appeared to provide a reasonable representation of the actual permeability

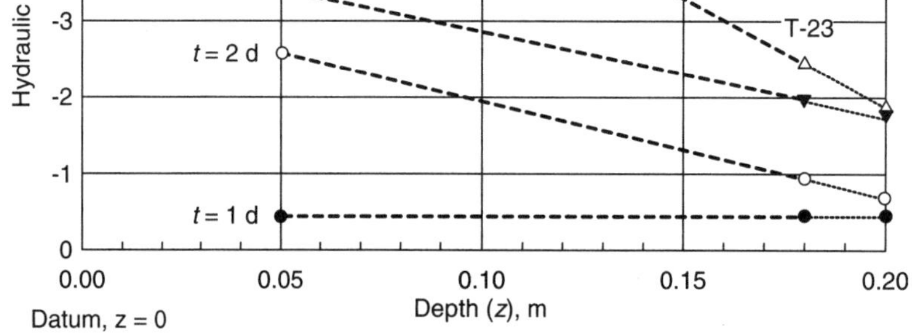

Figure 7.38 Hydraulic head profiles for Test SM1-T1 measured by tensiometers T-23 and T-24 during drying process (after Krisdani et al., 2009).

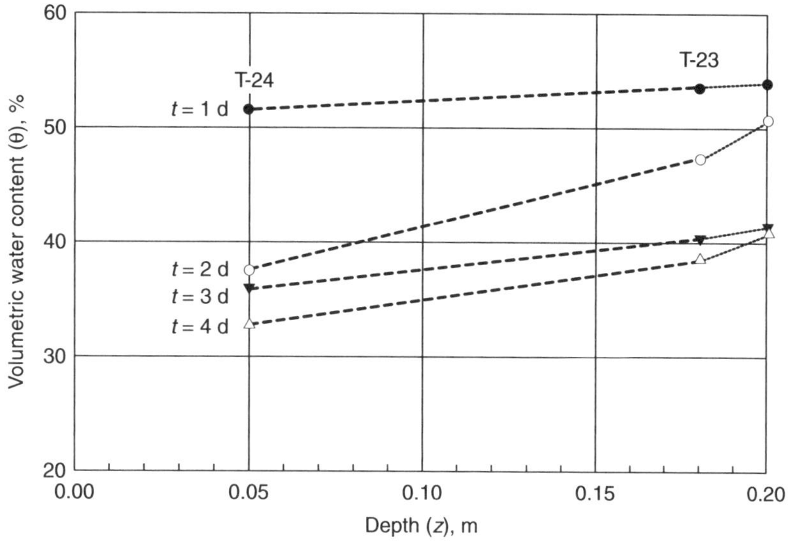

Figure 7.39 Volumetric water content profiles of Test SM1-T1 based on SWCC and pore-water pressure data measured on tensiometers T-23 and T-24 during drying process (after Krisdani et al., 2009).

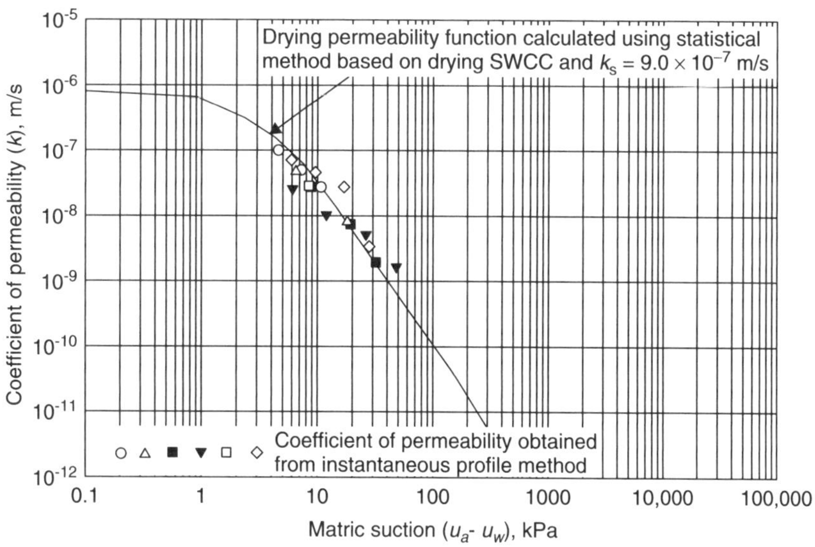

Figure 7.40 Permeability function for RS computed using instantaneous profile and statistical method (after Krisdani et al., 2009).

function for the unsaturated soil. Obtaining the correct saturated coefficient of permeability appeared to be the greatest challenge.

7.6.3.4 In Situ Instantaneous Profile Method

Nonhomogeneity and anisotropy of soils in the field make in situ measurements for the coefficient of permeability superior to laboratory measurements. Fissures, fractures, tension cracks, and root holes commonly encountered in unsaturated soils cannot be properly represented in small-scale laboratory specimens. Furthermore, laboratory specimens are subjected to sampling disturbance. On the other hand, in situ permeability tests have not been developed to the same extent as laboratory tests. Also, laboratory tests cost less than field tests. These are the main reasons why most testing has been performed in the laboratory.

The instantaneous profile method is generally considered to be the best method for permeability testing in the field (Klute, 1972). The procedure used in the field is similar in concept to the instantaneous profile method described for the laboratory.

The advantage of the in situ instantaneous profile method is that relatively simple equipment is required. The method is not applicable when there is significant water flow in the horizontal direction. Horizontal flow may occur as a result of either a low-permeability layer or a highly permeable

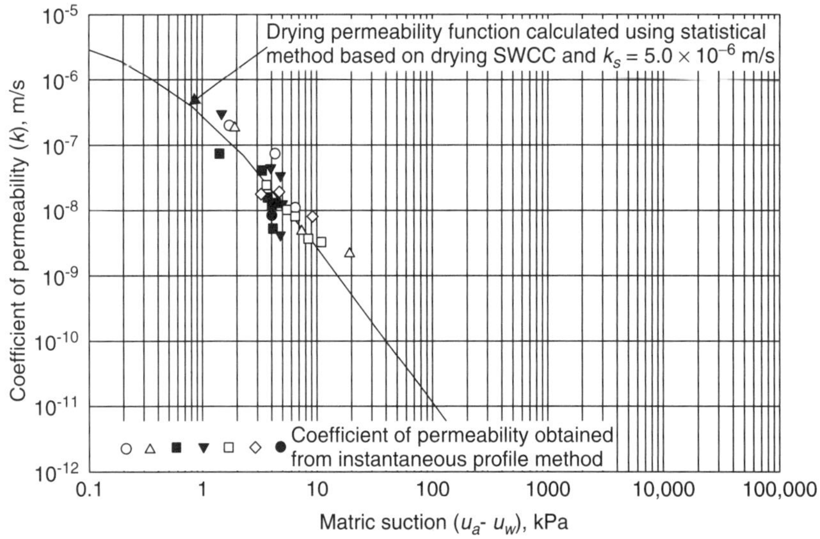

Figure 7.41 Permeability function for the MRF 60–40 mixture computed using instantaneous profile and statistical method (after Krisdani et al. 2009).

layer. The method is not practical when the matric suctions exceed approximately 50 kPa because of the extremely slow drainage process.

The instantaneous profile method also has a number of disadvantages:

1. The method is time consuming and may take several months to conduct.
2. A proper flow rate for wetting the soil is difficult to choose. The flow rate limitation particularly applies when performing the test in the laboratory. If the flow rate is too high, the gradually changing suction versus water content change profiles cannot be discerned. Daniel (1983) suggested flow rates in the range of 0.2–5 cm^3/day. Flow rates of this magnitude are difficult to control.
3. The accuracy of the test is related to the space between water content versus suction monitoring points. Theoretically, the closer the water content versus suction monitors, the more accurate is the calculated unsaturated permeability function. However, the use of too many sensors may cause soil disturbance and affect the water infiltration process.

7.6.4 Measurement of Wetting Permeability Function Using Wetting-Front Column Technique (Li et al., 2009)

The "wetting-front" advancing column test has been performed as either an infiltration test or a capillary rise test. Only the capillary rise procedure and analysis are described in this section. Figure 7.42 shows the soil column device developed by Li et al. (2009) with water content and soil suction measuring devices mounted along the walls of the column. A water flow control and measurement system was also part of the apparatus. The acrylic column was 120 mm in diameter and 1000 mm high.

Four Theta probes were installed for water content measurement as well as four tensiometers for suction measurement. The sensors were installed along the column at heights of 100, 300, 600, and 800 mm above the pedestal. Miniature tip tensiometers with a high-air-entry ceramic cup were connected to a pressure measuring device through a small plastic tube. The ceramic cups were saturated and installed in the soil. The working suction range of the tensiometers was limited to between 0 and 90 kPa. Theta probe devices (i.e., ML2x model) from Delta-T Devices (Miller and Gaskin, 1997) were used to monitor water content. The voltage signal from the sensor was transformed to the volumetric water content through use of a calibration curve.

The soil is initially placed into the column in a dry state. During the wetting process, the dry soil in the column turned to a darker color as water rose in the column. The change in color can be used to provide information on the rate at which the wetting front advanced. This information, along with either the water content or soil suction measurements, was used to calculate the permeability function for the soil. A series of unsaturated capillary rise tests were conducted using the soil column apparatus. The grain-size distribution curves for five soils identified as GW-GM with sand, SM with gravel, SC with gravel, sandy ML, and CL with sand are shown in Fig. 7.43. The SWCCs for each of the soils were independently measured and the results are shown in Fig. 7.44.

7.6.4.1 Method of Data Interpretation from Wetting-Front Column Tests

Figure 7.45a shows the capillary rise of water from the bottom of the column. The top of the soil column is at section *A*. The water content versus suction measurement sensors were installed at section *B*. The observed wetting front rises with

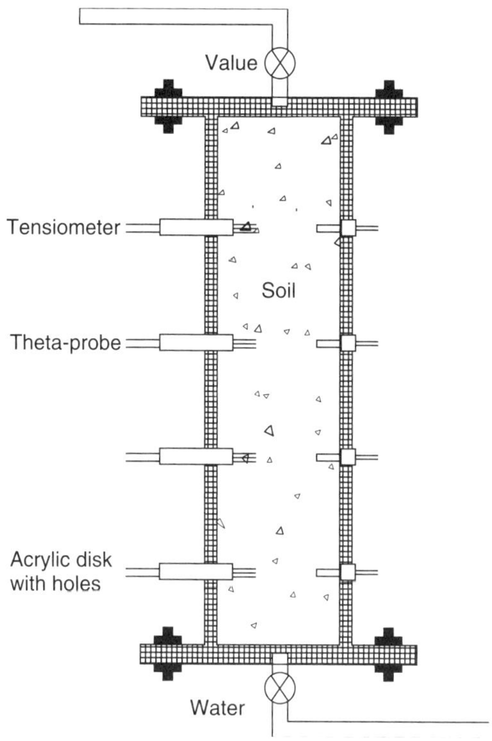

Figure 7.42 Column apparatus used for wetting-front advancing column tests (after Li et al., 2009).

time and is recorded as shown in Fig. 7.45b during the capillary rise process. Figure 7.45a shows that at time t_1 the wetting front reaches section F_1 and at time t_2 the wetting front reaches section F_2. The flux passing through section B between two time intervals can be calculated as

$$Q_B = \Delta W_{B-A} + Q_A \tag{7.73}$$

where:

Q_B = flux passing through section B from t_1 to t_2,
Q_A = flux passing through section A from t_1 to t_2, which is assumed to be zero during the capillary rise process, and
ΔW_{B-A} = water storage change between section B and section A from t_1 to t_2.

The water storage ΔW_{B-A} can be calculated as

$$\Delta W_{B-A} = \int_{h_B}^{h_A} \theta\left(h, t_2\right) A\, dh - \int_{h_B}^{h_A} \theta\left(h, t_1\right) A\, dh \tag{7.74}$$

where:

$\theta(h, t)$ = volumetric water content distribution function,
A = cross-sectional area of the soil column,
t = time,
h = section height from the bottom of the column,
h_B = height at section B, and
h_A = height at section A.

Figure 7.45a shows that in a short time period Δt (i.e., $t_2 - t_1$) the wetting front advances a distance of Δh (i.e., height difference between F_1 and F_2). If the water content contour is assumed to advance smoothly during the time interval, then the water content distribution function can be written as follows

$$\theta(h, t + \Delta t) = \theta(h - \Delta h, t) \tag{7.75}$$

and the suction distribution function can be written as

$$\psi(h, t + \Delta t) = \psi(h - \Delta h, t) \tag{7.76}$$

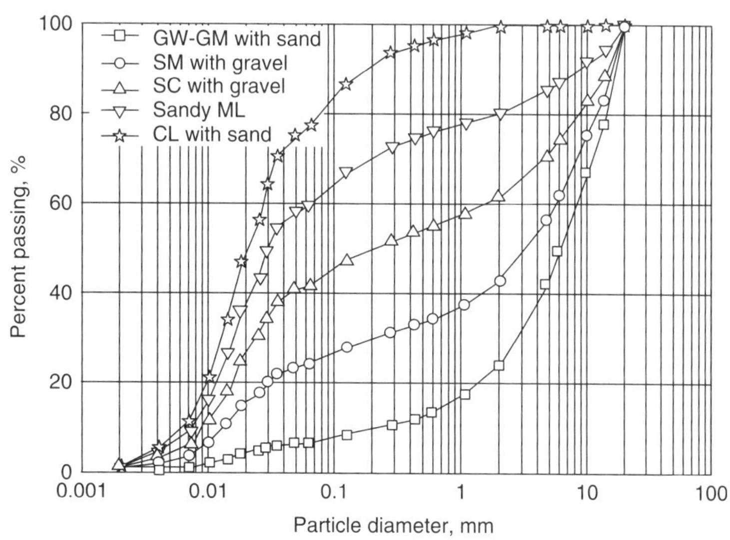

Figure 7.43 Grain-size distribution curves for colluvial soils used in wetting-front advancing column tests (after Li et al., 2009).

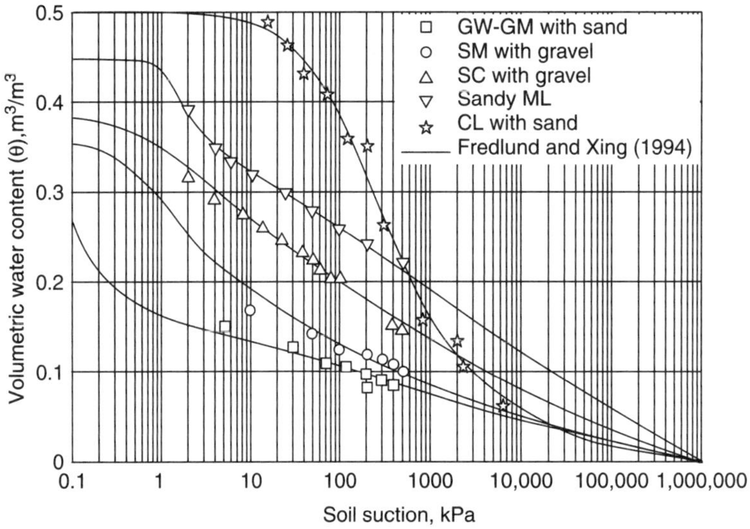

Figure 7.44 SWCCs for colluvial soils (after Li et al., 2009).

The total flow volume of water passing through section B in Eq. 7.73 can then be written as

$$Q_B = \int_{h_B}^{h_A} \theta(h - \Delta h, t_1) A\, dh - \int_{h_B}^{h_A} \theta(h, t_1) A\, dh = \int_{h_B-\Delta h}^{h_B} \theta(h, t_1) A\, dh - \int_{h_A-\Delta h}^{h_A} \theta(h, t_1) A\, dh \tag{7.77}$$

The first part of Eq. 7.77 can be calculated as

$$\int_{h_B-\Delta h}^{h_B} \theta(h, t_1) A\, dh \approx 0.5\left[\theta(h_B, t_1) + \theta(h_B - \Delta h, t_1)\right] \times A\Delta h \approx 0.5\left[\theta(h_B, t_1) + \theta(h_B, t_2)\right] A\Delta h \tag{7.78}$$

If the initial water content is low (and therefore the coefficient of permeability is extremely small), the water content change in the initially air-dried soil zone can be ignored during the limited wetting period. Consequently, the water contents in this zone remain at the initial water content θ_i during the entire wetting process. The second part of Eq. 7.77 can be calculated as follows:

$$\int_{h_A-\Delta h}^{h_A} \theta(h, t_1) A\, dh \approx 0.5[\theta(h_A, t_1) + \theta(h_A - \Delta h, t_1)] A\Delta h \approx \theta_i A\Delta h \tag{7.79}$$

By substituting Eqs. 7.78 and 7.79 into Eq. 7.77, Q_B can be calculated

$$Q_B \approx 0.5\left[\theta(h_B, t_1) + \theta(h_B, t_2) - 2\theta_i\right] A\Delta h \tag{7.80}$$

The flow quantity can be written using Darcy's law,

$$Q = k_w i A dt \tag{7.81}$$

where:

k_w = instantaneous coefficient of permeability at a specific section along the soil column and
i = hydraulic gradient at that section.

The hydraulic gradient can be calculated as

$$i = \frac{-dh + d\psi/\gamma_w}{dh} = \frac{d\psi}{\gamma_w dh} - 1 \tag{7.82}$$

The hydraulic gradient at time t_2 at section B can be obtained by substituting Eq. 7.76 into Eq. 7.82,

$$i = \frac{-\psi(h_B, t_2) + \psi(h_B + \Delta h, t_2)}{\gamma_w \Delta h} - 1 = \frac{\psi(h_B, t_1) - \psi(h_B, t_2)}{\gamma_w \Delta h} - 1 \tag{7.83}$$

The average coefficient of permeability k_w from t_1 to t_2 at section B can be calculated by substituting Eqs. 7.80 and 7.83 into Eq. 7.81:

$$k_w = \frac{Q_B}{iC(t_2 - t_1)} = \frac{\left[\theta(h_B, t_1) + \theta(h_B, t_2) - 2\theta_i\right]\gamma_w \Delta h^2}{2\left[\psi(h_B, t_1) - \psi(h_B, t_2) - \gamma_w \Delta h\right](t_2 - t_1)} \tag{7.84}$$

The wetting-front advancing velocity (see Fig. 7.45b) is defined as

$$v = \frac{\Delta h}{\Delta t} \tag{7.85}$$

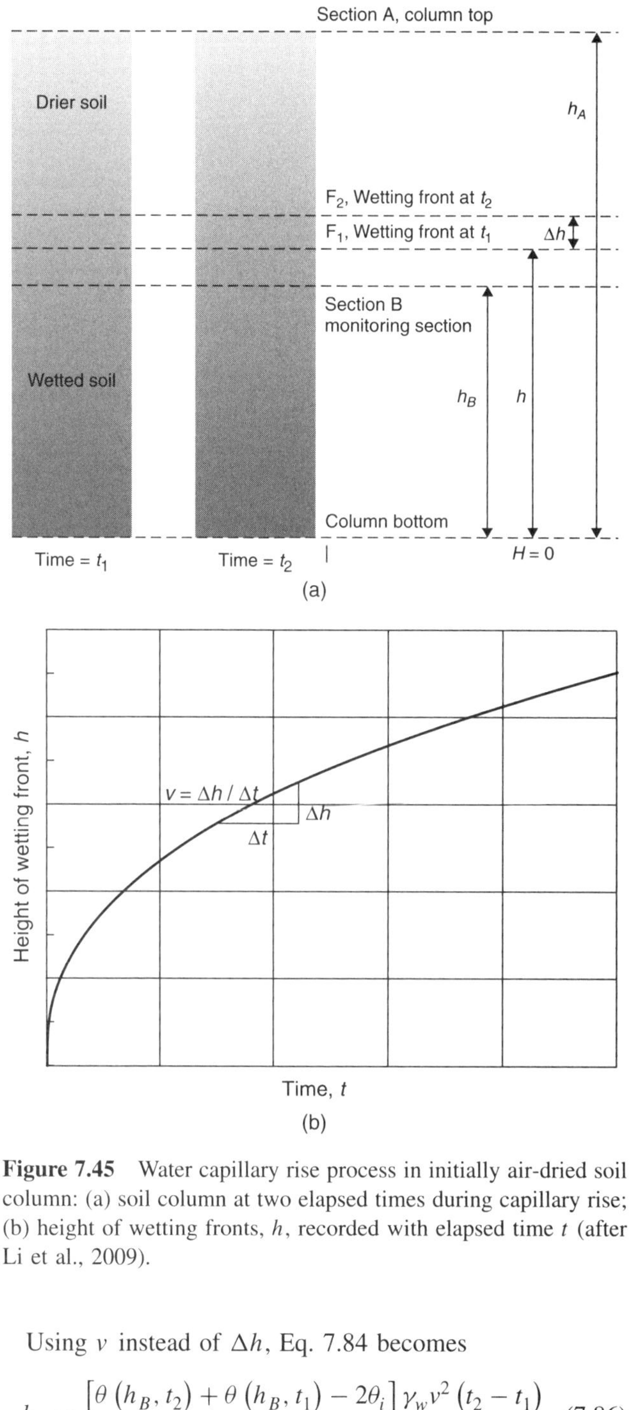

Figure 7.45 Water capillary rise process in initially air-dried soil column: (a) soil column at two elapsed times during capillary rise; (b) height of wetting fronts, h, recorded with elapsed time t (after Li et al., 2009).

Using v instead of Δh, Eq. 7.84 becomes

$$k_w = \frac{\left[\theta\left(h_B, t_2\right) + \theta\left(h_B, t_1\right) - 2\theta_i\right]\gamma_w v^2\left(t_2 - t_1\right)}{2\left[\psi\left(h_B, t_1\right) - \psi\left(h_B, t_2\right) - \gamma_w v\left(t_2 - t_1\right)\right]} \quad (7.86)$$

The calculated coefficient of permeability k_w corresponds to a mean suction value (i.e., $0.5[\psi(h_B, t_1) + \psi(h_B, t_2)]$). The monitored data at one section along with the wetting-front advancing velocity is required to calculate a series of unsaturated coefficients of permeability. It is not necessary to have a series of measurement sections (as is the case in the instantaneous profile method); however, each section allows for the calculation of the unsaturated coefficient of permeability.

The method depends on being able to discern the movement of the wetting front. Therefore, it is important to start with the soil in a dry state and the soils need to show a distinct change in color when wetted. It should be noted that the wetting front in clays may not be discernible.

7.6.4.2 *Presentation of Results from Capillary Rise Column Tests*

The capillary rise processes in the column tests for five soils tested are presented in Fig. 7.46a. The wetting front in the GW-GM with sand rose to a height of 113 mm above the water table and remained constant. The wetting-front advancing velocity at a height of 113 mm became extremely slow. The results showed that the unsaturated coefficient of permeability above that section was extremely low. The GW-GM with sand contained 35% sand, 58% gravel, and 7% fines. The GW-GM with sand contained large particles that formed a structure with numerous macropores. The macropores in the soil were filled with air and few capillary flow channels were available above a height of 113 mm for water flow. The corresponding soil suction was 1.1 kPa at steady-state conditions.

The wetting-front advancing process for the other four soils (i.e., SC with gravel, SM with gravel, sandy ML, and CL with sand) showed similar behavior when capillary water rose to the top of the soil column (i.e., 700–1000 mm). It took 2–4 weeks for water to rise in these columns. The fines content in these four soils was larger than 24% and numerous micropores were formed that filled with water under low soil suctions (i.e., smaller than 100 kPa in the capillary rise process). There was an abundance of continuous-flow channels for the capillary rise process.

The wetting-front advancing velocity (see Fig. 7.46b) generally increased as the fines content increased. In particular, the advancing velocity in the GW-GM with sand was much slower (i.e., about $\frac{1}{5}$–$\frac{1}{10}$ the rate) in the four finer soils. The wetting-front advancing velocity shows a loglinear relationship with time as described by the following equation:

$$\log v = a + \lambda \log t \quad (7.87)$$

where:

$\lambda = -0.63$ and
$a = 0.18$ for the four finer soils (when using units of mm/s with reference to Fig. 7.46b).

Based on Eq. 7.87, the rising height of the wetting front, h, is

$$h = 10^a t^{\lambda+1} / (\lambda + 1) \quad (7.88)$$

According to Eq. 7.82, if the column length is doubled, the time needed for the wetting front to reach the column top will be $2^{1/(\lambda+1)}$ times longer (i.e., 6.5 times as long when $\lambda = -0.63$). Consequently, the column length influences the time required for the experiment. Figure 7.46a shows that

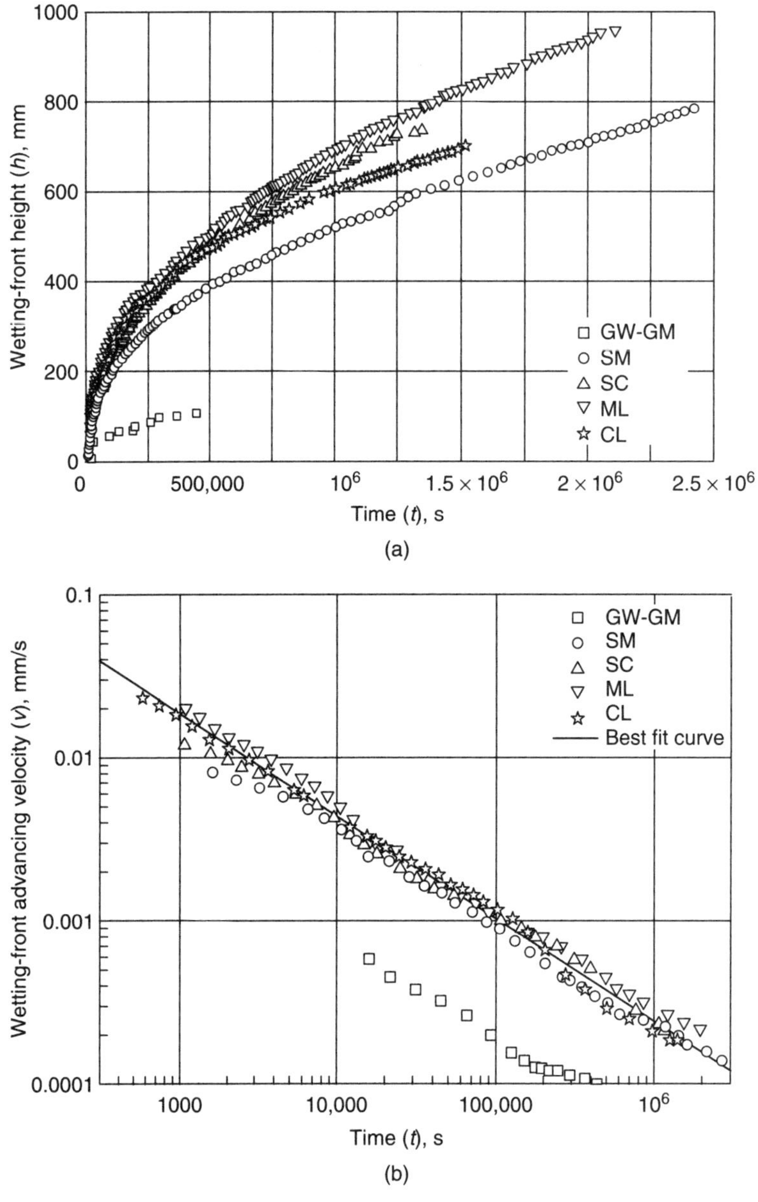

Figure 7.46 Capillary rise process for several soils: (a) wetting-front height recorded over time; (b) wetting-front velocity on a log-log scale (after Li et al., 2009).

the time for an experiment in a soil column 400 mm high is about 3–6 days. The time for the experiment for a soil column 600 mm high is about 7–14 days while the time for an experiment for a soil column 800 mm high will increase to 15–30 days.

The water content changes at heights of 100, 300, 600, and 800 mm were monitored during the experiments. The water content changes for two of the soils are shown in Fig. 7.47. The water contents in the soil columns show three distinct stages: the initial stage, the wetting stage, and the wetted stage.

In the initial stage, before the wetting front reached a particular section, the water content in the section remained equal to the initial water content corresponding to placement conditions. Since the soil was initially oven dried and had a low water content (and an extremely low coefficient of permeability), the hydraulic gradient in the soil was approximately 1.0 (i.e., the gradient was controlled by elevation change).

During the wetting stage, after the wetting front reached a particular section, the suctions in the wetted soil decreased rapidly and large hydraulic gradients were present. Therefore, water was able to rapidly enter the initially dried soil and the water content quickly increased.

During the wetted stage, after the soil at a particular section was wetted, the suction at that section would change to a similar value as found at a lower adjacent section.

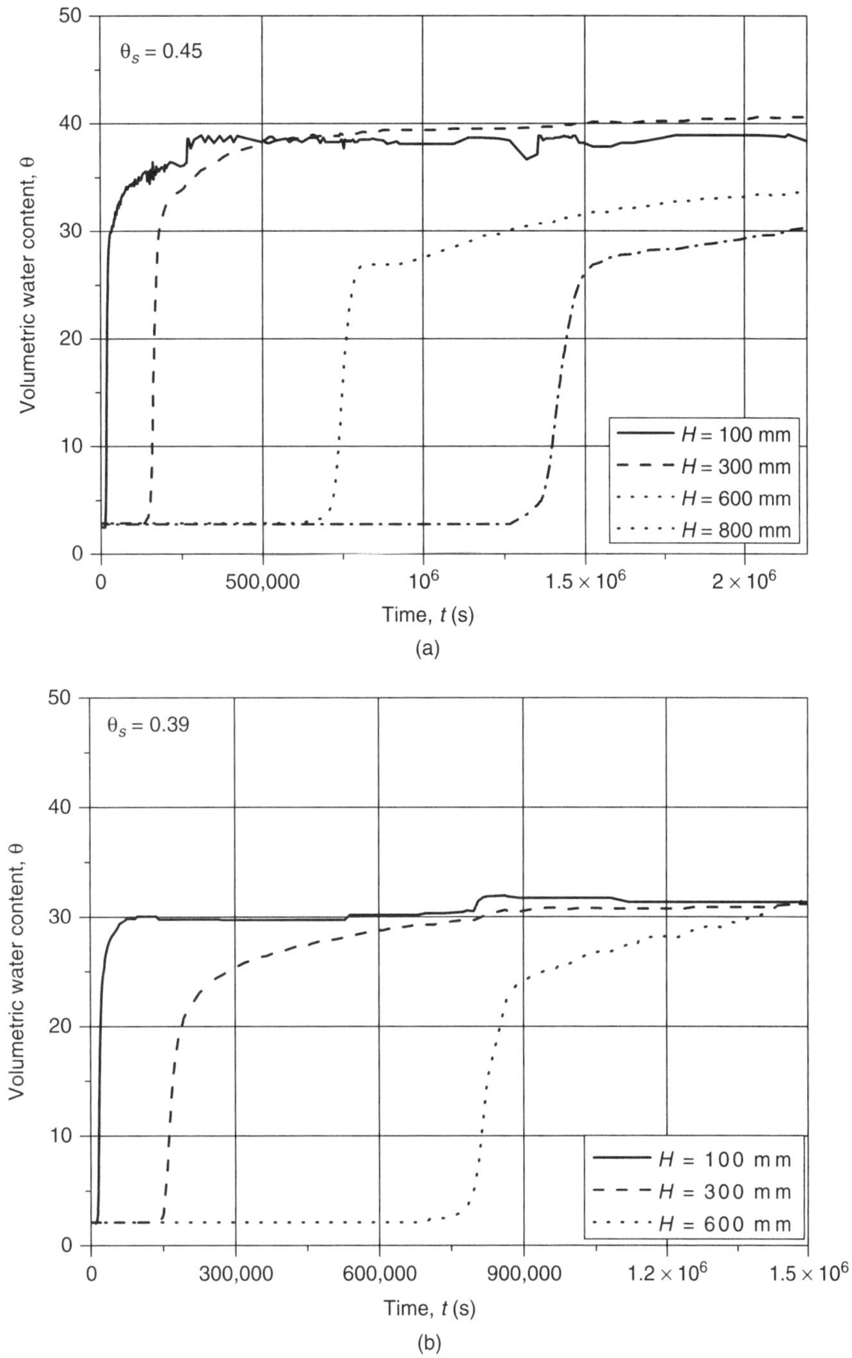

Figure 7.47 Water content measurements during capillary rise process: (a) monitored water content profiles at various sections of ML soil; (b) monitored water content profiles at various sections of SC soil (Li et al., 2009).

The hydraulic gradient would decrease to a low value. Therefore, the water content in the wetted soil could only increase slowly as it approached steady-state water contents. The water contents in the wetted zone corresponded to the degrees of saturation of about 80% for CL with sand and sandy ML, about 75% for SM with gravel, and about 40% for SC with gravel.

Tensiometers were inserted into the side of the soil column after the wetting front passed through the respective sections. The monitored suctions were relatively stable and decreased slowly with time. One of the monitored suction profiles is shown in Fig. 7.48, where suctions decreased with time and approached a steady-state condition. At the 100-mm-high section, matric suction decreased to a constant

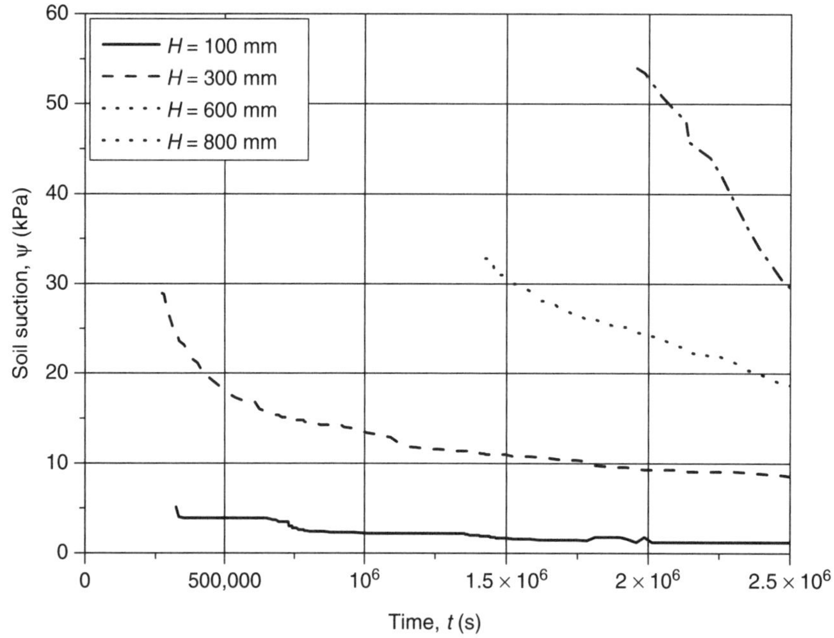

Figure 7.48 Measured matric suction profiles during capillary rise for ML soil (after Li et al., 2009).

value of 1.2 kPa, which was close to the steady-state value of 1.0 kPa at that section.

7.6.4.3 Calculation of Unsaturated Soil Permeability Functions

The water contents and soil suctions during the infiltration versus capillary rise process can be used to calculate the permeability function using Eq. 7.80. If measured suction values are inserted directly into Eq. 7.80, the permeability functions can be calculated. Figures 7.49 to 7.51 show the calculated permeability functions for the CL, ML, and SC soils (i.e., using open symbols). Soil suctions larger than 80 kPa could not be monitored due to the limited range of tensiometers. Therefore, the directly calculated permeability function values are limited to a narrow suction range.

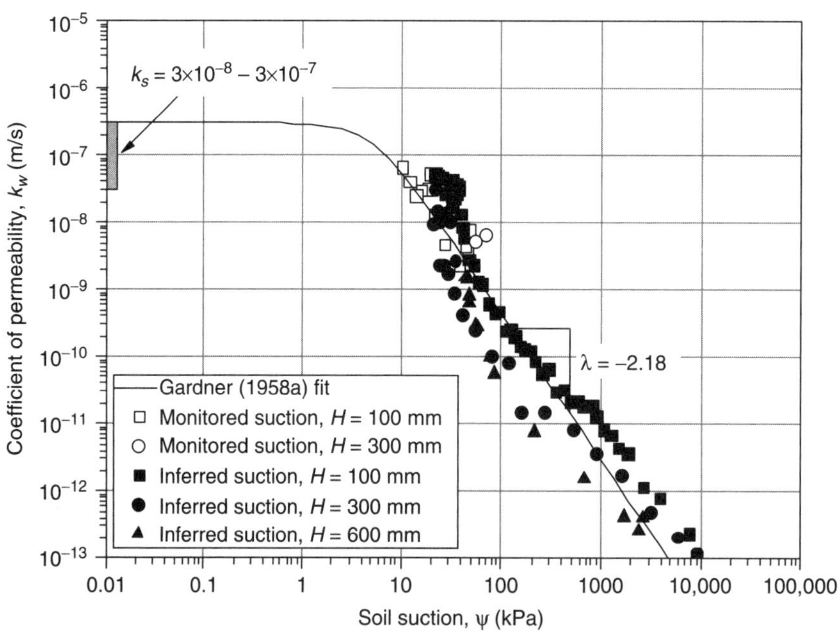

Figure 7.49 Calculated unsaturated permeability functions for CL soil (after Li et al., 2009).

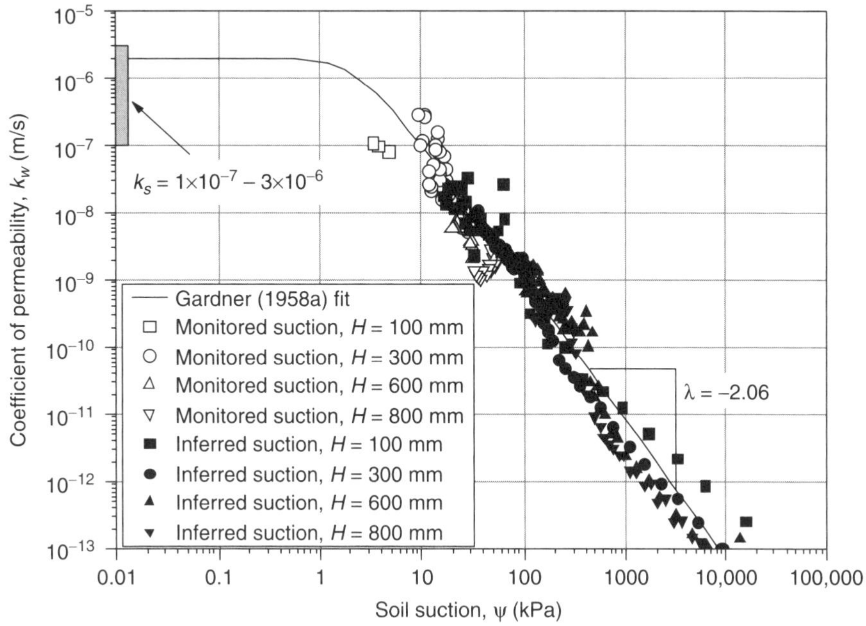

Figure 7.50 Calculated unsaturated permeability functions for ML soil (after Li et al., 2009).

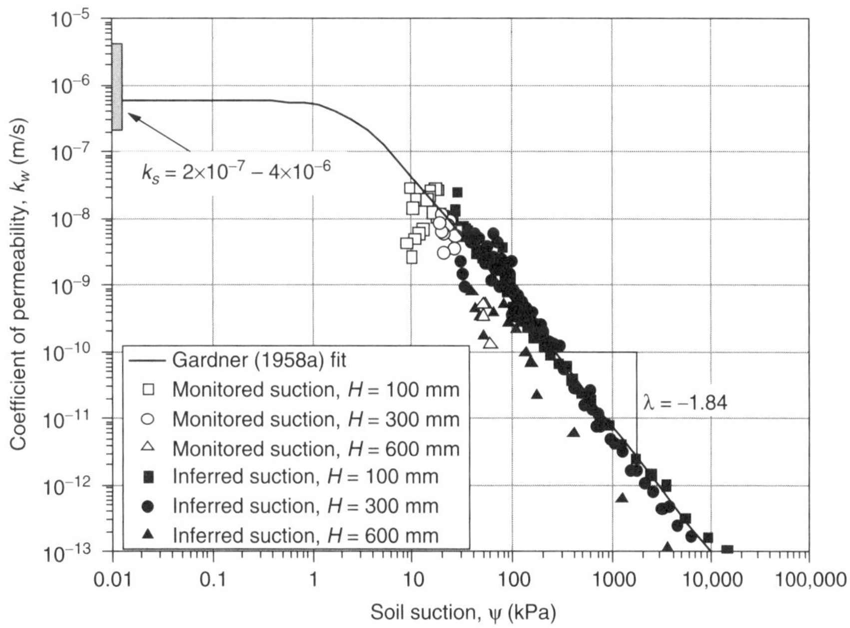

Figure 7.51 Calculated unsaturated permeability functions for SC with gravel (after Li et al., 2009).

The SWCCs can also be used in Eq. 7.80 to extend the suction range. The soil suctions can be estimated from the SWCCs (see Fig. 7.44) through use of the monitored water contents. The permeability functions can then be calculated over a much wider suction range, as shown in Figs. 7.49 to 7.51 (i.e., see the solid symbols).

There is a distinct overlap between the permeability function computed from the suction measurements and that from inferred suctions taken from the SWCC. The permeability functions from the measurements at section heights of 100, 300, and 600 mm tend to form a loglinear band. The upper values tend to be about 5–10 times higher than that of the lower values at the same suction. The Gardner (1958a) equation can be used to best fit the calculated data on the CL sand using a saturated coefficient of permeability of 3.0×10^{-7} m/s. The laboratory-measured saturated coefficients of permeability for the CL sand ranged from 3×10^{-8} to 3.0×10^{-7} m/s. The slope in the loglinear part was 2.18. As the soil suction increased from 10 to 10,000 kPa, the permeability decreased from 9.0×10^{-8} to 1.0×10^{-13} m/s.

The permeability functions for sandy ML and SC with gravel are shown in Figs. 7.50 and 7.51, respectively. The permeability functions calculated from the monitored suctions agree well with the permeability function obtained when suctions are inferred from the SWCC. The permeability functions interpreted based on the measurements at all four monitoring sections appear to be quite similar. The calculated permeability functions is located along a loglinear permeability band, where the upper value was about 3–8 times greater than the lower values at the same suction. The Gardner (1958a) equation also best fit the calculated data when using a saturated coefficient of permeability of 2×10^{-6} m/s for the ML soil and a saturated coefficient of permeability of 6×10^{-7} m/s for the SC soil.

The results of the wetting-front advancing column tests appear to provide an acceptable procedure for the measurement of the unsaturated permeability function for a range of coarse-grained materials. The permeability values interpreted based on measurements at different sections along the column appear to follow a consistent trend. It appears that the interpreted permeability function is independent of the location at which the suction or water content measurements are made. The calculated permeability function based on measured suctions appears to agree well with the permeability values calculated from the inferred SWCC. The permeability functions show a loglinear relationship in the suction range between 10 and 10,000 kPa. The slope of the loglinear portion of the permeability function appears to increase with an increase in the proportion of fines content in the soil. The permeability functions obtained are located in a relatively narrow loglinear permeability band with the upper permeability values about 5–10 times greater than the lower permeability values at the same suction. A column length of 400–600 mm appears to be appropriate when using the wetting-front advancing column test. Each wetting-front column test can be performed in 1 or 2 weeks.

7.6.5 From Measurement to Estimation of PermeabilityFunction

The theory of water flow through unsaturated soils has been verified through experimental testing. The hydraulic conductivity or coefficient of permeability can be presented as a function of soil suction. While it is possible to experimentally measure the permeability function for a soil, the time and costs are generally considered to be too prohibitive in geotechnical engineering practice. As a result, considerable attention has been given to the use of "estimation" techniques for obtaining the permeability function.

A significant portion of the next chapter is devoted to methodologies for the estimation of the permeability function and its usage for solving practical geotechnical engineering problems.

CHAPTER 8

Solving Saturated/Unsaturated Water Flow Problems

8.1 INTRODUCTION

Direct experimental measurements of the water coefficient-of-permeability function for an unsaturated soil are difficult and time consuming. Attempts have been made to calculate the coefficient of permeability through use of theoretical considerations of the soil pore-size distribution. Some of these procedures were mentioned earlier as part of the theory of flow through an unsaturated soil. A limited number of proposed estimation procedures have experienced the greatest success for application in geotechnical engineering practice. These procedures are either directly or indirectly related to the features of the SWCC, and are commonly referred to as indirect methods for the estimation of the water permeability function.

The estimation procedures have found increasing acceptance as it has become evident that (i) the cost of directly measuring the permeability function is prohibitive on most engineering projects and (ii) the estimation techniques are providing adequate information for engineering design purposes. The success of applying unsaturated soil mechanics in engineering practice has been closely related to the use of the SWCC for estimating the water coefficient of permeability. There is a clear linkage between the commencement of desaturation of a soil (i.e., the air-entry value) and the decrease in its coefficient of permeability. The relationship between the SWCC and the water coefficient of permeability is illustrated in Fig. 8.1 for two soil types, namely, sand and clayey silt.

The water storage characteristics of the soil are also required when modeling transient seepage problems. The slope of the SWCC on the arithmetic soil suction scale gives the water storage modulus m_2^w at designated soil suctions.

The soil suction at the air-entry value of a soil can range from a low value for sand (e.g., <10 kPa) to a high value for intact clay soils (e.g., >1000 kPa). The wide range of potential soil suction values can be more clearly observed on a plot of the logarithm of soil suction versus the logarithm of coefficient of permeability, as shown in Fig. 8.2. A soil has the highest coefficient of permeability when it is saturated, and it starts to reduce when soil suction exceeds the air-entry value of the soil (i.e., the degree of saturation reduces below 100%). There is an exception to this rule for the condition where the soil cracks as a result of increased soil suction. In this case, the cracks can produce a saturated coefficient of permeability that is greater than the saturated intact coefficient of permeability.

Most soils show a gradual decrease in the coefficient of permeability as soil suction is increased once soil suction exceeds the air-entry value. The log-log representation of the water permeability function generally produces a straight-line relationship once the air-entry soil suction is exceeded. All soils appear to become essentially impervious to liquid water flow once soil suction exceeds residual suction conditions.

8.2 ESTIMATION OF PERMEABILITY FUNCTION

The estimation technique for describing the water permeability functions can be subdivided into four categories of models: (i) empirical models, (ii) statistical models, (iii) correlation models, and (iv) regression models. These categories provide a practical classification for estimated permeability functions.

"Empirical" models recognize that there is a relationship between the character of the SWCC and the permeability function and proceed to utilize this relationship in an empirical manner. The Brooks and Corey (1964) estimation technique determines a pore-size distribution index and uses the computed index for the estimation of the water permeability function.

"Statistical" models start with a physical model of the assemblage of pore channels through which water can flow. The most common physical model used is that of Childs and Collis-George (1950). A number of permeability functions have been derived based on the suggested physical assemblage of flow channels suggested by Childs and Collis-George. Most permeability functions derived in this manner

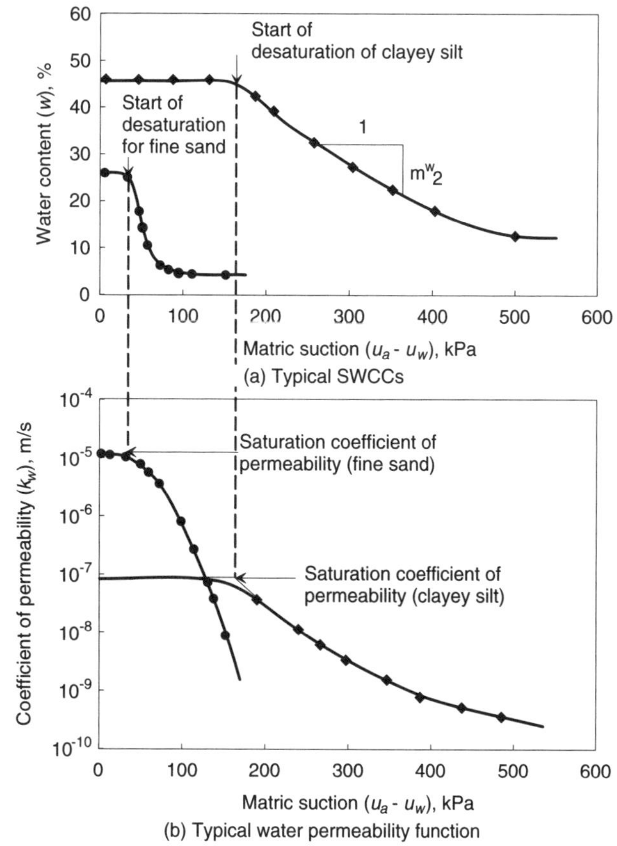

Figure 8.1 Typical relationship between SWCCs and estimated permeability functions for sand and clayey silt.

utilize an integration procedure along the SWCC, starting from saturated soil conditions.

"Correlation" models recognize that there is an apparent correlation between aspects of the SWCC and the permeability function. The apparent correlation then uses an additional soil parameter to relate the SWCC to the water permeability function for a particular soil type. The correlation procedure was used by Leong and Rahardjo (1997a) to generate a water phase permeability function from the SWCC. The additional soil parameter for the permeability function was estimated from a correlation with the classification properties of soils.

In "regressions" models, a series of coefficient-of-permeability data points may be available as a result of either laboratory permeability testing or an estimations technique that calculates coefficients of permeability. For example, it is possible to use the Fredlund et al., (1994) statistical model to generate a series of data points corresponding to a range of soil suction values. However, there is no closed-form representation of the data points because of the integration form for the permeability model. Consequently, after the data points have been generated using the Fredlund et al. (1994) permeability model, it is useful to best fit the data points using a regression or best-fit analysis.

As an example, the Gardner (1958a) empirical equation with two soil parameters can be used to best fit to the data points and thereby provide a mathematical equation to represent the permeability function. It is also possible to use other closed-form permeability functions to best fit the generated data (i.e., water coefficient of permeability versus soil suction). The Fredlund and Xing (1994) equation for the SWCC can also be used along with a regression analysis to best fit the water coefficient of permeability versus soil suction data points. In this way, a closed-form equation can be generated for the permeability function.

8.2.1 Designation of Amount of Water in Soil

There is a range of water contents over which liquid water can flow through the soil. Liquid water flow becomes smaller as the amount of water in the soil decreases. It is somewhat difficult to ascertain a definite point at which liquid water ceases to flow. Two primary assumptions have been made regarding the range over which liquid water flows through a porous medium.

Some of the earlier permeability models were based on the assumption that liquid water flow occurred over the range between full saturation of the soil and residual water content (Brooks and Corey, 1964). An "effective degree of saturation" variable was proposed that changed from 1.0 to 0.0 as the degree of saturation of the soil varied between fully saturated conditions and residual water content conditions. In other words, the designation of the amount of water in the soil was normalized between the water content at saturation and residual water content. Other models have been proposed that allow the water content to vary from the water content at full saturation and zero water content. In other words, the water content is rendered dimensionless by dividing any water content by the saturation water content. In this case, the dimensionless water content can once again range from 1.0 to 0.0. Ebrahimi-Birang et al. (2004) presented research results that suggested that liquid water flow appeared to cease at some point beyond the residual water content.

When the entire water content range is referenced to the saturated water content, the amount of water in a soil can be written in a dimensionless form that is equivalent to the degree of saturation provided the volume change of the soil structure is negligible as soil suction changes:

$$\Theta_d = \frac{\theta}{\theta_s} = \frac{V_w/V_t}{V_v/V_t} = \frac{V_w}{V_t} = S \tag{8.1}$$

where:

Θ_d = dimensionless water content,
S = degree of saturation,
V_w = volume of water in the soil specimen,
V_v = volume of void in the soil specimen, and
V_t = total volume of the soil specimen.

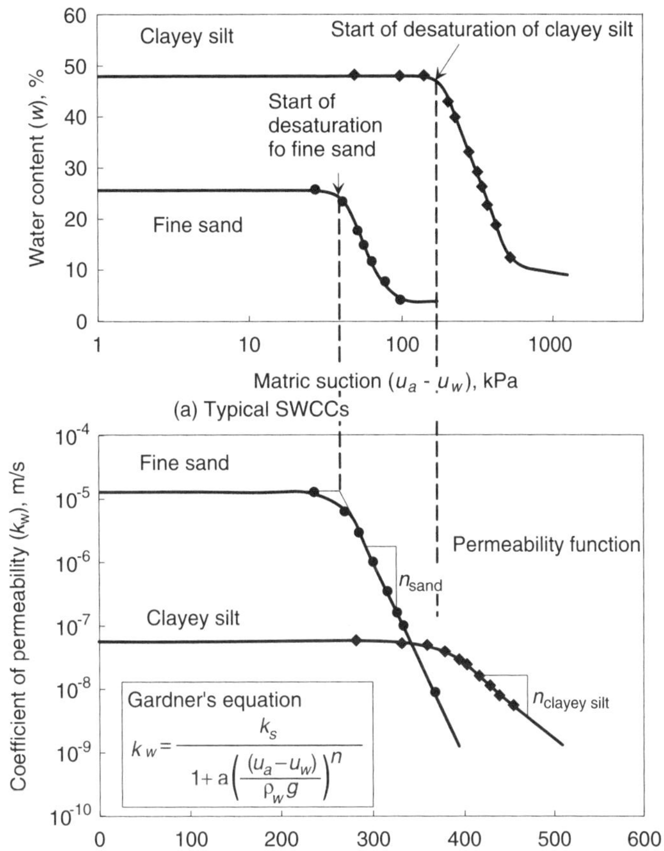

Figure 8.2 Relationship between SWCCs and permeability function for sand and clayey silt plotted as log–log function.

Similarly, gravimetric water content w can be referenced to zero water content to give a dimensionless water content Θ_d that can be written as follows:

$$\Theta_d = \frac{w}{w_s} = \frac{m_w/m_p}{m_{ws}/m_p} = \frac{m_w}{m_{ws}} = \frac{V_w\rho_w}{V_v\rho_w} = \frac{V_w}{V_v} = S \quad (8.2)$$

where:

w_s = saturated gravimetric water content,
m_w = mass of water,
m_p = mass of soil particles,
m_{ws} = mass of water when the soil is saturated, and
ρ_w = density of water.

Equations 8.1 and 8.2 indicate that the dimensionless (gravimetric or volumetric) water contents are identical to the degree of saturation when water contents range between dry soil conditions and the saturated water content, provided the overall soil structure volume change is negligible when soil suction is changed.

The effective degree of saturation defined by Brooks and Corey (1964) normalized the amount of water in a soil between saturated soil conditions and the residual degree of saturation:

$$S_e = \frac{S - S_r}{1.0 - S_r} \quad (8.3)$$

where:

S_e = effective degree of saturation,
1.0 = saturation of the soil, and
S_r = residual degree of saturation.

Volumetric water content θ can also be written in a normalized form:

$$\Theta_n = \frac{\theta - \theta_r}{\theta_s - \theta_r} \quad (8.4)$$

where:

Θ_n = normalized volumetric water content,
θ_s = saturated volumetric water content, and
θ_r = residual volumetric water content.

The normalized water content Θ_n and the effective degree of saturation S_e are the same variable when volume changes are negligible as soil suction is increased. The above-mentioned water content variables (i.e., volumetric water content, gravimetric water content, and degree of saturation) are used somewhat interchangeably along with the dimensionless and normalized forms, and it is important to note the conditions under which errors can be introduced when using the improper designation of the amount of water in a soil for the estimation of the unsaturated permeability function. On the other hand, it should be noted that volumetric water content θ must be used when differentiating the SWCC (i.e., expressed as volumetric water content versus soil suction) to obtain the water storage modulus.

8.2.2 Definitions Related to Designation of Water Coefficient of Permeability

The coefficient of permeability depends on the amount of water in the soil, which in turn depends on soil suction ψ. Soil suction may be in terms of either matric suction (i.e., $u_a - u_w$, where u_a is pore-air pressure and u_w is pore-water pressure) or total suction (i.e., matric plus osmotic suctions). The term "permeability function" for unsaturated soils is generally used to represent the relationship between the coefficient of permeability and soil suction. The relative coefficient of permeability, $k_r(\psi)$, is equal to the coefficient of permeability at a selected soil suction, $k_w(\psi)$, referenced to the saturated coefficient of permeability, k_s, as follows:

$$k_r(\psi) = \frac{k_w(\psi)}{k_s} \tag{8.5}$$

The relative coefficient of permeability can also be written as a function of volumetric water content, $k_r(\theta)$. However, for geotechnical engineering applications the permeability function is generally used as a scalar quantity which is a function of soil suction, $k_r(\psi)$.

8.2.3 Empirical Equations for Water Permeability Function

All permeability functions appear to provide essentially the same approximations of the coefficient of permeability from saturated conditions through the air-entry value of the soil and into the start of the transition zone. All equations produce a similar overall functional form that responds to the air-entry value and the rate of desaturation of the soil. All empirical and statistical estimation procedures for the prediction of the water permeability function involve the use of the SWCC.

Empirical permeability functions describe the variation in the water coefficient of permeability with soil suction, $k_w(\psi)$ [or volumetric water content, $k_w(\theta)$]. The parameters for the equations are generally determined from the SWCC through use of a curve-fitting procedure. The Brooks and Corey (1964) equation is one example of an empirical model. The equation is sometimes also referred to as a macroscopic model because of consideration given to the pore-size distribution within the soil. The Brooks and Corey (1964) permeability function consists of two parts: the permeability below the air-entry value of the soil and the permeability above the air-entry value. The permeability below the air-entry value is equal to the saturated coefficient of permeability:

$$k_w = k_s \quad \text{for} \quad u_a - u_w \leq (u_a - u_w)_b \tag{8.6}$$

Above the air-entry value of the soil, the Brooks and Corey (1964) function can be written as

$$k_w = k_s \left\{ \frac{(u_a - u_w)_b}{(u_a - u_w)} \right\}^{2+3\lambda} \quad \text{for} \quad u_a - u_w > (u_a - u_w)_b \tag{8.7}$$

where:

λ = pore-size distribution index defined as the negative slope of the effective degree of saturation.

The pore-size distribution index is obtained from the SWCC. Figure 8.3a shows a typical SWCC while Fig. 8.3b shows the same plot as degree of saturation versus soil suction. Figure 8.3c shows the same information plotted as the effective degree of saturation (logarithm scale) versus soil suction (logarithm scale). The degree of saturation has been normalized between full-saturation conditions and a point representing residual saturation. The slope of the normalized log-log plot using the effective degree of saturation yields the pore-size distribution index λ. The pore-size distribution index along with an understanding of the air-entry value and the saturated coefficient of permeability of the soil provides the necessary information for the estimation of the water permeability function.

The Brooks and Corey (1964) empirical permeability functions allow the unsaturated coefficient of permeability of a soil to be computed without measuring the unsaturated coefficient of permeability. All estimation procedures for the permeability function have some similar features. First, all estimated permeability functions start at fully saturated soil conditions. Second, all estimated permeability functions are also intimately related to the air-entry value of the soil. As a result, it is to be anticipated that all estimation procedures provide similar estimations of the unsaturated coefficient of permeability in the low soil suction range. It is to be expected that different permeability estimation procedures will begin to show divergence in the estimated permeability functions as soil suctions increase well beyond the air-entry value.

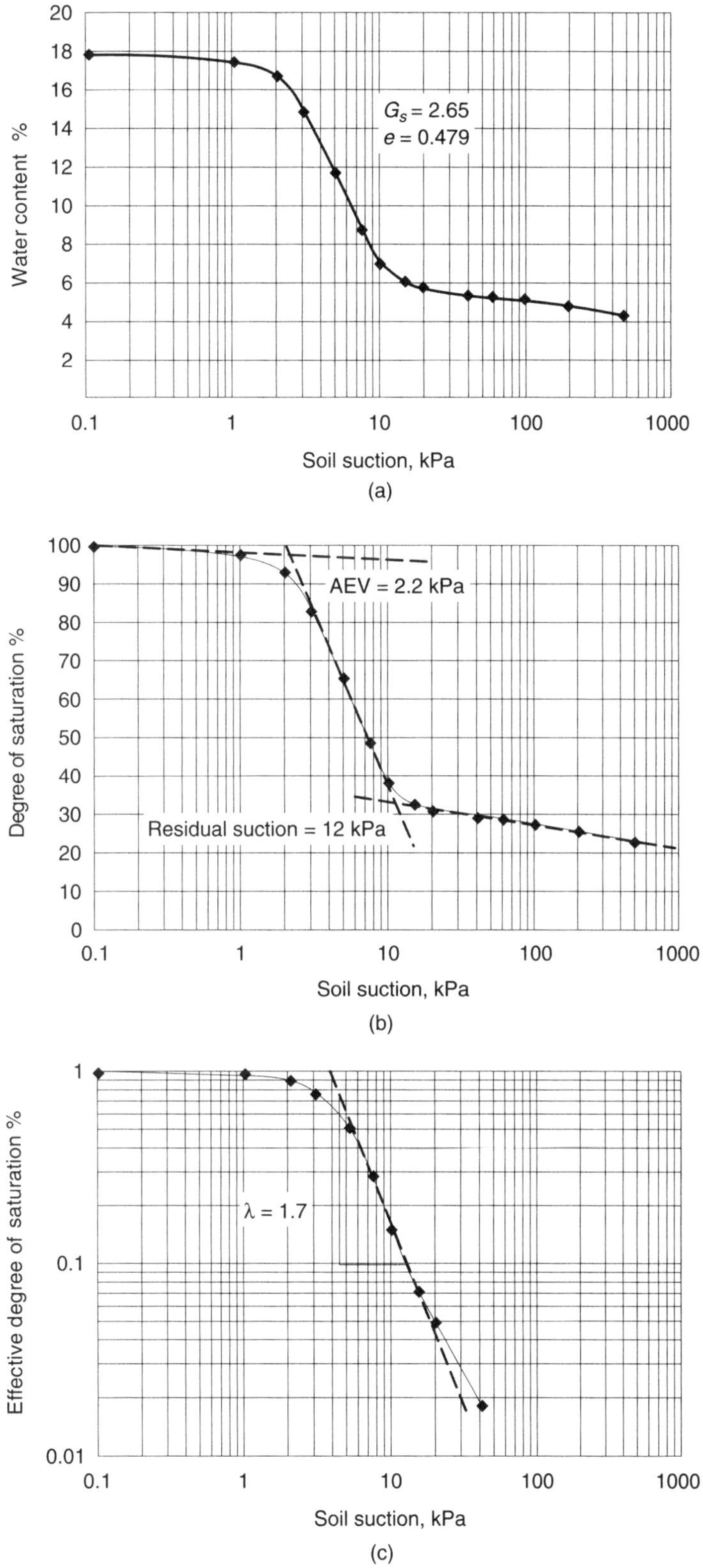

Figure 8.3 Determination of pore-size distribution index for sand soil: (a) gravimetric water content versus soil suction; (b) degree of saturation versus soil suction; (c) effective degree of saturation versus soil suction.

8.2.4 Statistical Models for Estimation of Permeability Function

There are groups of proposed estimation models that are based on statistical assumptions regarding the pore distributions. The statistical models are developed based on the interpretation of the SWCC. These models are usually categorized as either macroscopic approaches or microscopic approaches. The macroscopic models provide a closed-form analytical equation for the unsaturated permeability function and are related to either the normalized water content Θ_n or the dimensionless water content Θ_d.

8.2.4.1 Childs and Collis-George Model with Modifications by Marshall and Others

Childs and Collis-George (1950) proposed a model for estimating the coefficient of permeability based on a random variation in pore sizes. The assumption is made that the overall volume change of the soil is negligible as soil suction is increased. The model was later improved by Marshall (1958) and further modified by Kunze et al. (1968). A permeability function $k_w(\theta)$ was derived based on Poiseuille's equation. The following permeability function is similar in form to the function presented by Kunze et al. (1968). The function has been slightly modified to accommodate SI units and a matric suction designation rather than a pore-water pressure head. Variables used in the equation are illustrated in Fig. 8.4. The matric suction corresponding to the midpoint of each interval is used to calculate the coefficient of permeability:

$$k_w(\theta_i) = \frac{k_s}{k_{sc}} \frac{T_s^2 \rho_w g}{2\upsilon_w} \frac{\theta_s^p}{N^2} \sum_{j=i}^{m} \left[(2j + 1 - 2i)(u_a - u_w)_j^{-2}\right]$$

$$i = 1, 2, \ldots, m \qquad (8.8)$$

where:

$k_w(\theta_i)$ = calculated coefficient of permeability for a specified volumetric water content θ_i corresponding to the ith interval,
i = interval number that increases with decreasing water content (e.g., $i = 1$ identifies the first interval that closely corresponds to the saturated water content θ_s and $i = m$ identifies the last interval corresponding to the lowest water content θ_L on the experimental SWCC),
j = counter from i to m,
k_{sc} = calculated saturated coefficient of permeability,
T_s = surface tension of water,
ρ_w = density of water,
g = gravitational acceleration,
υ_w = absolute viscosity of water,
p = constant that accounts for the interaction of pores of various sizes,
m = total number of intervals between the saturated volumetric water content θ_s and the lowest water content θ_L on the experimental SWCC
N = total number of intervals computed between the saturated volumetric water content θ_s and zero water content (i.e., $\theta = 0$) (note that $N = m\left[\theta_s/(\theta_s - \theta_L)\right]$, $m \le N$, and $m = N$ when $\theta_L = 0$), and
ψ_j = soil suction (kPa) corresponding to the midpoint of the jth interval (Fig. 8.5).

The calculation of the water coefficient of permeability, k_w, at a specific volumetric water content, $(\theta)_i$, involves the summation of the matric suction values that correspond to the volumetric water contents at and below a specific volumetric water content, $(\theta)_i$. Several procedures have been

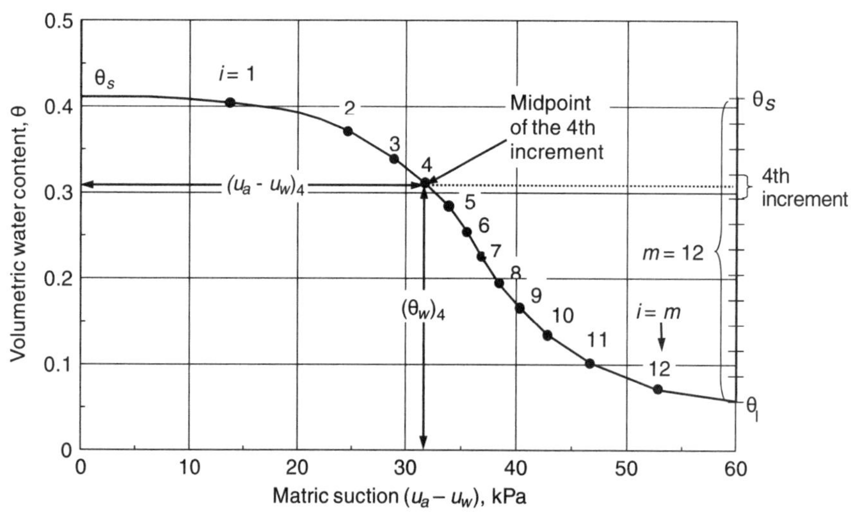

Figure 8.4 Calculation of water coefficient of permeability function from the SWCC (after Marshall, 1958; Kunze et al., 1968).

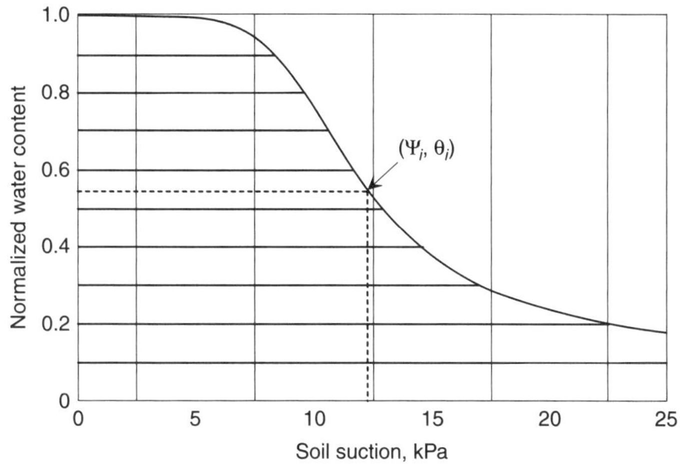

Figure 8.5 Typical SWCC for calculating water permeability function (after Marshall, 1958; Kunze et al., 1968).

proposed for the calculation of the permeability function $k_w(\theta)$ using Eq. 8.8. The main difference between the various procedures lies in the interpretation of the pore interaction term (i.e., θ_s^p/N in Eq. 8.8) (Green and Corey, 1971a).

The matching factor $\left(k_s/k_{sc}\right)$ based on the saturated coefficient of permeability is used to provide a smooth transition between the measured saturated coefficient of permeability and the remainder of the permeability function. The above computational procedure for obtaining the permeability function has been found to be most successful for sandy soils having a relatively narrow pore-size distribution (Nielsen et al., 1972). A comparison between the permeability function $k_w(\theta)$ computed from Eq. 8.8 and experimental data is shown in Fig. 8.6 for a fine sand. The SWCC for the sand and the comparison of its permeability function are shown in Figs. 8.6a and 8.6b, respectively. If the information shown in Figs. 8.6a and 8.6b is cross plotted, a more conventional form for the permeability function is obtained (i.e., logarithm of permeability versus logarithm of soil suction), as shown in Fig. 8.7.

The coefficient of permeability k_w at a specific volumetric water content θ is computed directly from Eq. 8.8. The shape of the permeability function is determined by the terms inside the summation sign portion of the equation which are obtained from the SWCC. However, the magnitude of the permeability function needs to be adjusted to coincide with the measured saturated coefficient of permeability k_s when soil suction is zero. The permeability function can be computed directly from the SWCC if the saturated coefficient of permeability is measured because all of the terms in front of the summation sign in Eq. 8.8 can be considered as an adjustment factor. The permeability function $k_w(\theta)$ can therefore be written as follows:

$$k_w(\theta)_i = \frac{k_s}{k_{sc}} A_d \sum_{j=i}^{m} \left[(2j+1-2i)\left(u_a-u_w\right)_j^{-2}\right]$$

$$i = 1, 2, \ldots, m \qquad (8.9)$$

where:

A_d = adjustment factor [i.e., $\left(T_s^2 \rho_w g/2v_w\right)\left(\theta_s^p/N^2\right)$ (m·s^{-1}·kPa^2)].

The terms in the "adjustment factor" in Eq. 8.8 result from attempting to calculate the saturated coefficient of permeability from Poiseuille's theory. In geotechnical engineering practice the saturated coefficient of permeability of the soil is usually measured in the laboratory. Then the above integration procedure is used to simply calculate the shape of the unsaturated permeability function. The unsaturated coefficient of permeability is therefore equal to the saturated coefficient of permeability multiplied by the summation terms shown in Eq. 8.9.

The data for a SWCC are all that is required for the computation of the shape of the permeability function [i.e., $k_w(\theta)$]. The following example is used to illustrate the technique by which the coefficient of permeability $k_w(\theta)$ can be computed as a function of volumetric water content.

Let us consider the SWCC shown in Fig. 8.8. The computation of $k_w(\theta)$ is illustrated for the drying curve. The drying curve is first divided into m equal intervals of volumetric water content, as shown in Fig. 8.9. In this case, the drying curve has maximum and minimum volumetric water contents (percentage form) of 38.8 and 10.9%, respectively. The drying curve is divided into 20 intervals with 20 midpoints. The first volumetric water content corresponds to saturated conditions. Each volumetric water content midpoint, $\left(\theta_w\right)_i$, corresponds to a particular matric suction, $\left(u_a-u_w\right)_i$. The midpoints are numbered starting from point 1 (i.e., $i=1$) to point 20 (i.e., $i=m$). The permeability function $k_w(\theta)$ is computed in accordance with the following equation:

$$k_w(\theta)_i = k_s \sum_{j=i}^{m} \left\{(2j+1-2i)\left(u_a-u_w\right)_j^{-2}\right\} \qquad (8.10)$$

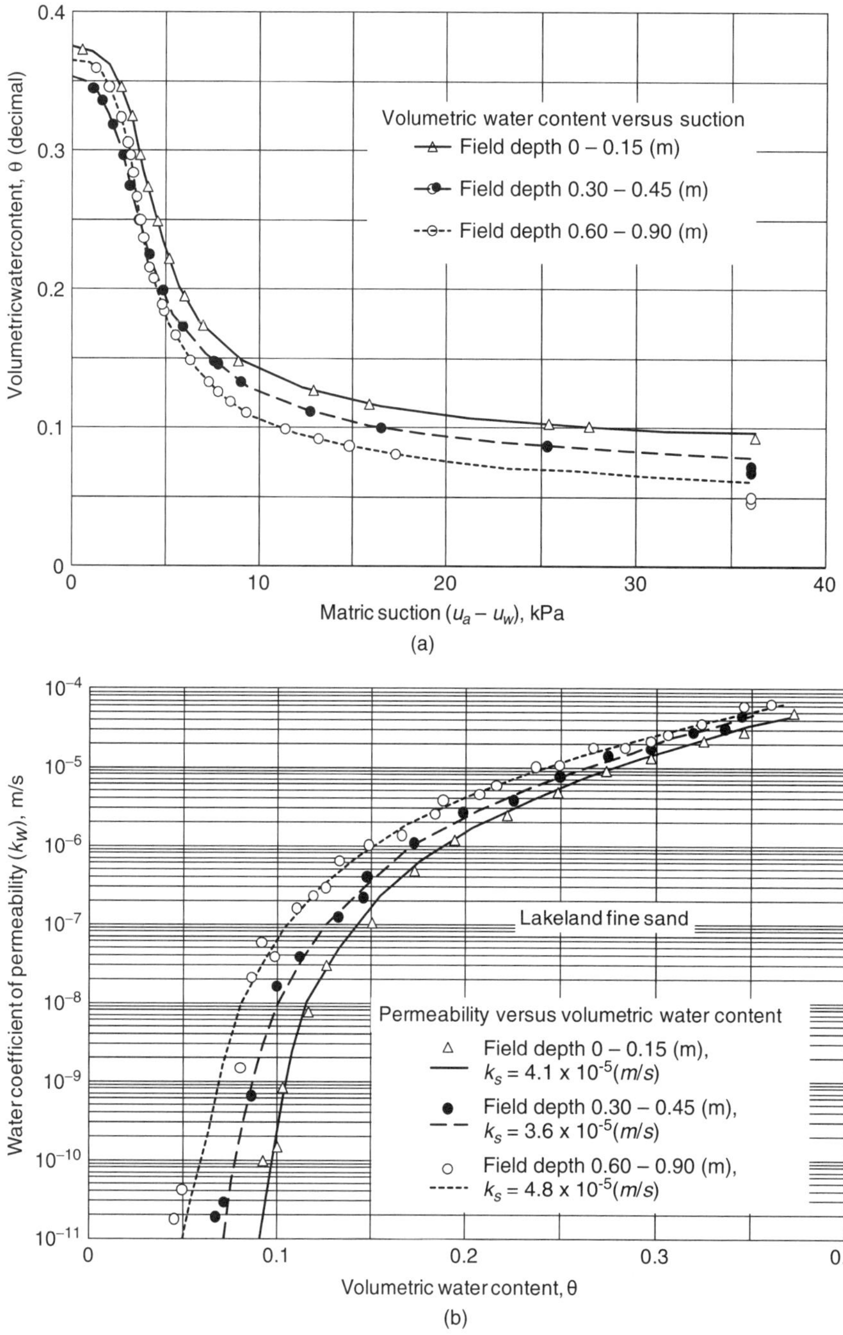

Figure 8.6 Comparisons between calculated and measured coefficients of permeability for Lakeland fine sand: (a) SWCC; (b) water coefficient for permeability as function of volumetric water content (after Elzeftawy and Cartwright, 1981).

The term $\sum_{j=i}^{m}\left\{(2j+1-2i)\left(u_a-u_w\right)_j^{-2}\right\}$ in Eq. 8.10 describes the shape of the permeability function. The coefficient-of-permeability values k_w are adjusted in accordance with the saturated coefficient of permeability k_s. The A_d term does not need to be computed and is assumed to be unity when the saturated coefficient of permeability is independently measured.

The saturated coefficient of permeability k_s in the example shown has a value of 5.83×10^{-8} (m/s) (Gonzalez and Adams, 1980). The permeability function $k_w(\theta)_i$ is calculated by substituting matric suction values from the midpoints along the drying curve into Eq. 8.10. The saturated coefficient of permeability k_s was measured in a laboratory test using the steady-state method (Gonzalez and Adams, 1980). Similar computations of the permeability

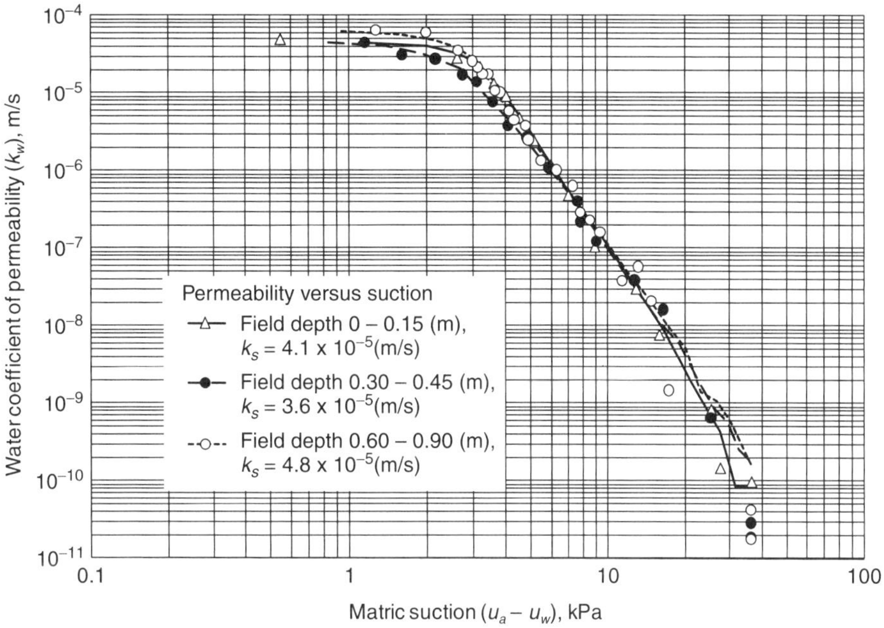

Figure 8.7 Plot of measured permeability data on Lakeland fine sand to show water coefficient of permeability as function of soil suction.

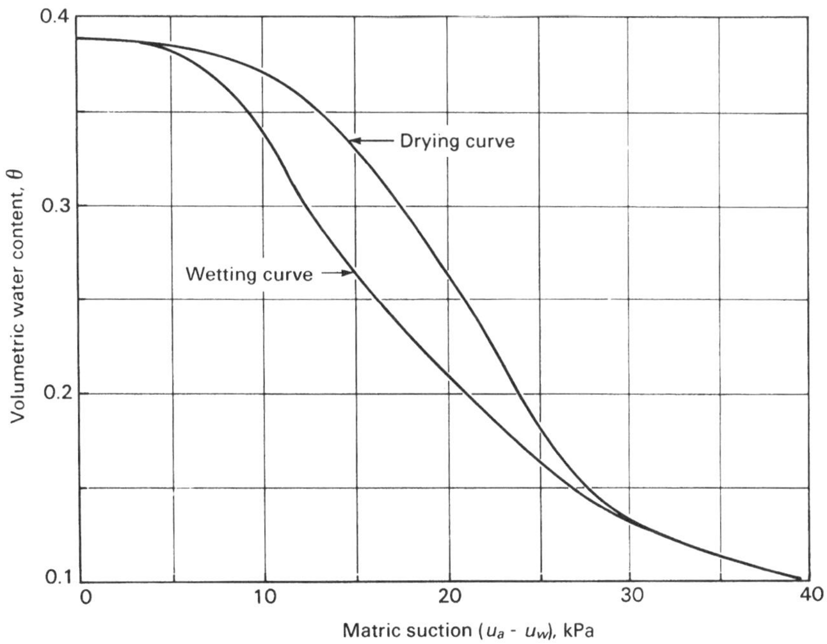

Figure 8.8 SWCC of fine tailings sand (after Gonzalez and Adams, 1980).

function $k_w(\theta)_i$ can also be performed for the wetting curve. The $k_w(\theta)_i$ calculations should proceed from low to high matric suction values for both the wetting and drying curves.

The water coefficients of permeability k_w are computed as a function of volumetric water content θ. It is also possible to plot the data shown in Fig. 8.10 in the more conventional form used for the permeability function (i.e., logarithm of permeability versus logarithm of soil suction). Figure 8.11 shows the measured data from Gonzalez and Adams (1980) plotted along with the computed values for the coefficient of permeability.

8.2.4.2 Burdine Model (1953)

Several statistical models have been proposed which make use of the Childs and Collis-George (1950) visualization of water flow through an unsaturated soil. Examples of

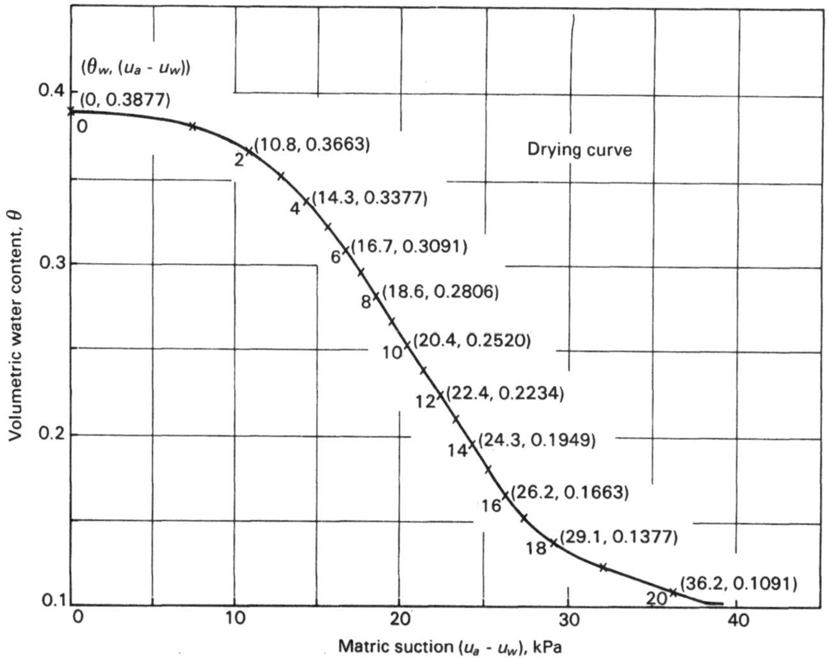

Figure 8.9 Subdivision of SWCC for calculation of water coefficient of permeability from SWCC (after Marshall, 1958; Kunze et al., 1968).

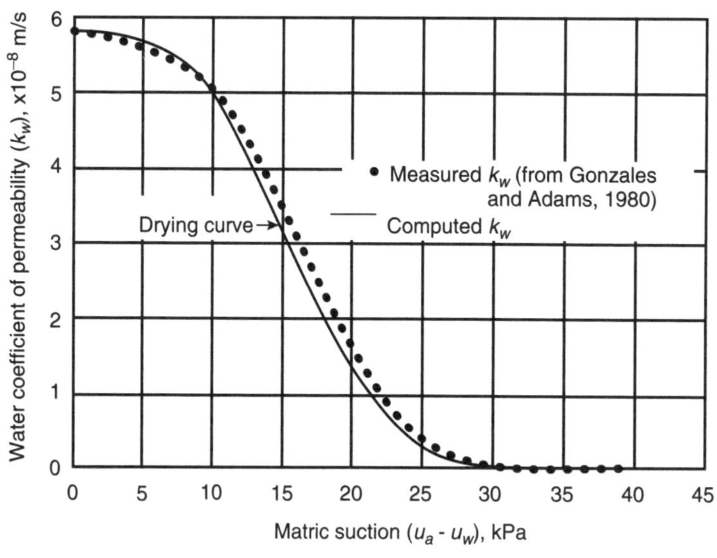

Figure 8.10 Comparisons between the computed and measured coefficients of permeability (after Gonzalez and Adams, 1980).

permeability models based on the physical model suggested by Childs and Collis-George are the Burdine (1953) and Mualem (1976a) permeability models. The saturated coefficient of permeability and the SWCCs are used to solve the integral form of the statistical models and thereby compute a water permeability function.

Statistical models can be used to calculate the permeability function when the saturated coefficient of permeability k_s and the SWCC are known. This approach assumes that the permeability function and the SWCC are both closely related to the pore-size distribution of the soil. Figure 8.12 shows a typical SWCC for a silt soil along with its permeability function. Burdine (1953) proposed the following equation for the relative coefficient of permeability based on pore-size distribution:

$$k_w(\theta) = k_s \Theta_n^q \frac{\int_{\theta_r}^{\theta} d\theta / [\psi^2(\theta)]}{\int_{\theta_r}^{\theta_s} d\theta / [\psi^2(\theta)]} \tag{8.11}$$

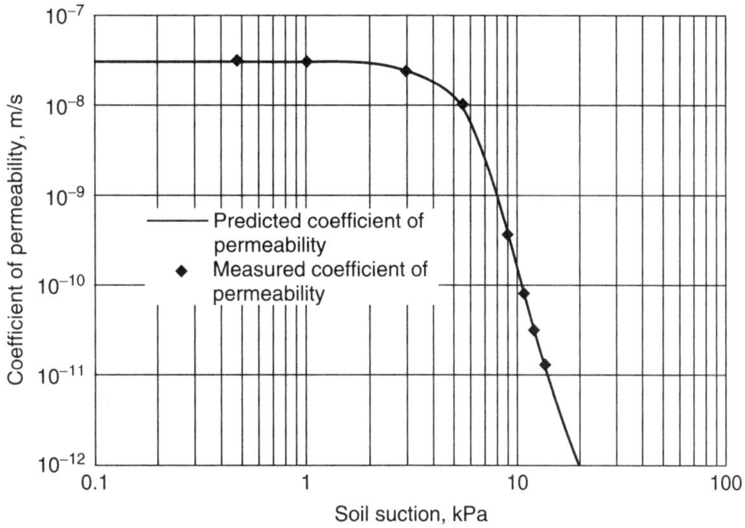

Figure 8.11 Computed and measured coefficients of permeability plotted as conventional water permeability function (modified from Gonzalez and Adams, 1980).

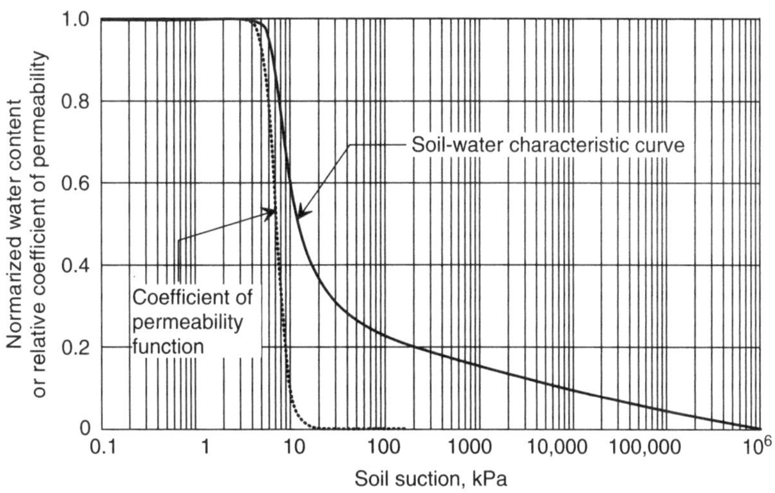

Figure 8.12 Typical SWCC and permeability function for a silt soil.

where:

q = a constant assumed to be equal to 2, and

Θ_n = normalized water content (i.e., equal to $[(\theta - \theta_r)/(\theta_s - \theta_r)]$ with θ_s equal to saturated volumetric water content and θ_r equal to residual volumetric water content).

The square of the normalized water content is meant to account for the effect of tortuosity of the water flow path. This model appears to be significantly more accurate than the same equation used without the correction factor, Θ_n^q.

In 1980, van Genuchten proposed a three-parameter equation for the SWCC. It was then noted that a closed-form solution could be obtained from the Burdine integration procedure provided the three parameters in the van Genuchten SWCC equation could be reduced to a two-parameter equation. A closed-form permeability equation was obtained from the Burdine integration procedure when n_{vb} and m_{vb} SWCC fitting parameters were related as follows: $m_{vb} = 1 - 2/n_{vb}$. It was then possible to perform the integration of the Burdine (1953) equation and arrive at the following closed-form equation for the permeability function:

$$k_r(\psi) = \frac{k_w(\psi)}{k_s} = \frac{1 - (a_{vb}\psi)^{n_{vb}-2}\left[1 + (a_{vb}\psi)^{n_{vb}}\right]^{-m_{vb}}}{\left[1 + (a_{vb}\psi)^{n_{vb}}\right]^{2n_{vb}}} \tag{8.12}$$

where:

m_{vb} = fitting parameter set equal to $1 - 2/n_{vb}$,

n_{vb} = fitting parameter from the van Genuchten (1980) equation best fit to the SWCC,

k_w = unsaturated coefficient of permeability,
k_s = saturated permeability coefficient,
ψ = soil suction, and
a_{vb} = fitting parameter related to the inverse of the air-entry value.

Equation 8.12 is the result of substituting the van Genuchten (1980) equation into the statistical integration model proposed by Burdine (1953). The resulting closed-form expression for the permeability function is commonly referred to as the van Genuchten-Burdine permeability equation. A closed-form equation for the coefficient of permeability has the advantage of simplicity over an integration form for the permeability function.

The accuracy of the closed-form van Genuchten-Burdine equation is in part dependent on the reasonableness of the proposed relationship between the m_{vb} and n_{vb} variables. It should be noted that Burdine (1953) did not write an equation for the SWCC. Rather, van Genuchten (1980) made an assumption related to the three parameters for the SWCC equation, thereby reducing his equation to a two-parameter equation that could yield a closed-form permeability function consistent with the Burdine (1953) integral form of the equation.

8.2.4.3 Mualem (1976a) Model

Mualem (1976a) analyzed a conceptual model of a porous medium similar to that of the Childs and Collis-George (1950) model and derived the following equation for predicting the coefficient of permeability:

$$k_r(\theta) = \Theta_n^q \left(\frac{\int_{\theta_r}^{\theta} d\theta / [\psi(\theta)]}{\int_{\theta_r}^{\theta_s} d\theta / [\psi(\theta)]} \right)^2 \tag{8.13}$$

where:

q = a constant set to 0.5,
Θ_n = normalized water content for the SWCC, and
$k_r(\theta)$ = relative coefficient of permeability [i.e., $k_r(\theta) = k(\theta)/k_s$, where $k(\theta)$ is any coefficient of permeability as a function of volumetric water content and k_s is the saturated coefficient of permeability of the soil].

The value of q depends on the soil-fluid properties and may vary considerably for different soils. Based on the permeability data for 45 soils, Mualem (1976a) suggested an optimal value of 0.5 for the q variable.

Soil suction must be expressed as a function of volumetric water content when using a statistical model. The integrations are performed along the volumetric water content axis and the permeability is expressed as a function of volumetric water content. The van Genuchten (1980) equation for the SWCC can be used in combination with the Mualem (1976a) integration equation form for the permeability function.

Van Genuchten suggested that the three fitting parameters for the SWCC equation could be reduced to two fitting parameters. The reduction to two fitting parameters was accomplished by assuming a fixed relationship between the n_{vm} and m_{vm} parameters; namely, $m_{vm} = 1 - 1/n_{vm}$ with $0 < m_{vm} < 1$. It was now possible to produce a closed-form permeability function for an unsaturated soil based on soil parameters associated with the SWCC. By substituting the van Genuchten (1980) equation into the statistical integration model proposed by Mualem (1976a), a closed-form expression was derived for the permeability function:

$$k_r(\psi) = \frac{\left\{1 - (a_{vm}\psi)^{n_{vm}-1}\left[1 + (a_{vm}\psi)^{n_{vm}}\right]^{-m_{vm}}\right\}^2}{\left[1 + (a_{vm}\psi)^{n_{vm}}\right]^{0.5}} \tag{8.14}$$

where:

m_{vm} = fitting parameter set equal to $1 - 1/n_{vm}$.

Equation 8.14 is commonly referred to as the van Genuchten-Mualem permeability equation (van Genuchten, 1980). The closed-form equation for the coefficient-of-permeability function has the advantage of simplicity over an integration form for the permeability function when it comes to solving saturated-unsaturated seepage problems. The accuracy of the closed-form van Genuchten-Mualem permeability equation is in part dependent upon the reasonableness of the assumption relating the m_{vm} and n_{vm} variables.

There does not appear to have been an attempt by van Genuchten (1980) to use laboratory test results to justify the empirical relationship between the n_{vm} and m_{vm} soil parameters. The relationship between n_{vm} and m_{vm} is a mathematical restriction and the m_{vm} parameter was limited to varying between zero and 1. The n_{vm} variable also needed to be a positive value. The mathematical restriction was necessary in order to ensure that the "power" term was positive and less than 1 (e.g., $0 < m_{vm} < 1$).

It is possible to test the reasonableness of the relationship between m and n by performing best-fit regression analyses on SWCC data on unsaturated soils. A large number of test results were retrieved from the SoilVision database and independent van Genuchten (1980) m_{vg} and n_{vg} parameters were back calculated using a best-fit regression analysis. The results indicate that the empirical relationships suggested by Burdine (1953) and Mualem (1976a) are not representative of most experimental data. The Fredlund, et al. (1994b) permeability function retains independence among the a_f, n_f, and m_f fitting parameters when using the SWCC in the calculation of the permeability function. It is suggested that retaining independence of the SWCC fitting variables leads to superior estimations of the permeability functions.

8.2.4.4 Fredlund et al. (1994b) Model

Fredlund et al. (1994b) used the Fredlund and Xing (1994) SWCC equation along with the Childs and Collis-George (1950) physical model to compute a water permeability function. The procedure involved numerical integration along the SWCC. Independent permeability functions can be computed for the drying and wetting SWCC. It is assumed that the volume change of the soil structure is negligible when soil suction is changed. Equation 8.8 can be expressed in the following integration form:

$$k(\theta) = k_s \int_{\theta_L}^{\theta} \frac{\theta - x}{\psi^2(x)} dx \bigg/ \int_{\theta_L}^{\theta_s} \frac{\theta_s - x}{\psi^2(x)} dx \tag{8.15}$$

where:

ψ = soil suction as a function of volumetric water content θ,
x = dummy variable of integration representing water content,
θ = volumetric water content,
θ_s = volumetric water content at saturation, and
θ_L = lower limit for the volumetric water content.

Equation 8.15 can also be expressed as a relative coefficient of permeability, $k_r(\theta_i)$, in summation form:

$$k_r(\theta_i) = \sum_{j=i}^{m} \frac{2(j-i)+1}{\psi_j^2} \bigg/ \sum_{j=i}^{m} \frac{2j-1}{\psi_j^2} \tag{8.16}$$

The integration form of Eq. 8.16 can be written as follows:

$$k_r(\theta) = \int_{\theta_L}^{\theta} \frac{\theta - x}{\psi^2(x)} dx \bigg/ \int_{\theta_L}^{\theta_s} \frac{\theta_s - x}{\psi^2(x)} dx \tag{8.17}$$

The residual volumetric water content θ_r is the water content below which large increases in suction are required to remove additional water. The residual water content can generally be observed as a break in the SWCC; however, the break in curvature is not always visible for high-plasticity clay soils. It is commonly assumed that the coefficient of permeability of a soil is essentially zero when its water content is below the residual water content. Kunze et al. (1968) investigated the effect of using only a portion of the SWCC for the prediction of coefficient of permeability. It was concluded that the accuracy of the prediction is significantly improved when the SWCC extended at least to residual conditions. Therefore, the x integration should be carried out at least over the interval from θ_s to θ_r:

$$k_r(\theta) = \int_{\theta_r}^{\theta} \frac{\theta - x}{\psi^2(x)} dx \bigg/ \int_{\theta_r}^{\theta_s} \frac{\theta_s - x}{\psi^2(x)} dx \tag{8.18}$$

Most proposed models for the relative coefficient of permeability k_r (e.g., Eqs. 8.11, 8.13, and 8.18) require the designation of residual water content and the SWCCs are rarely used over the entire range of water contents. Brooks and Corey (1964), Mualem (1976a), and van Genuchten (1980) suggest different methods for the extrapolation of the measured soil-water characteristic data and the determination of the residual water content. The conventional procedure for the prediction of the coefficient of permeability using Eq. 8.18 consists of two steps. First, the residual water content of the soil under consideration was estimated from experimental data. Then the measured soil-water characteristic data are fitted using a mathematical equation for the interval between saturated water content θ_s and residual water content θ_r. Subsequent integrations were performed using the best-fit curve between these two water content conditions. It is not clearly known under what water content condition the coefficient of permeability goes to zero. Therefore, it is better to estimate the coefficient-of-permeability function without being restricted to an empirical estimation of zero water flow.

The total suction corresponding to zero water content (i.e., oven-dried water content) is essentially the same for all types of soils. A value approaching 10^6 kPa has been supported by thermodynamic considerations (Richards, 1965) and experimental results (Croney and Coleman, 1961). A general equation describing the SWCC over the entire suction range (i.e., $0 - 10^6$ kPa) was proposed by Fredlund and Xing (1994):

$$\theta = C(\psi) \frac{\theta_s}{\left\{\ln\left[e + (\psi/a_f)^{n_f}\right]\right\}^{m_f}} \tag{8.19}$$

where:

e = the natural number 2.71828,
a_f = inflection point, related to the air-entry value of the soil,
n_f = parameter that controls the slope at the inflection point in the SWCC,
m_f = parameter that is related to the residual water content,
$C(\psi)$ = correction function defined as

$$C(\psi) = 1 - \frac{\ln(1 + \psi/\psi_r)}{\ln(1 + 1{,}000{,}000/\psi_r)}$$

ψ_r = constant related to the suction at residual water content, θ_r.

Equation 8.19 fits experimental data over the entire suction range from 0 to 10^6 kPa (Fredlund and Xing, 1994). It is convenient to perform the integration along the soil suction axis when calculating the coefficient of permeability using Eqs. 8.18 and 8.19. Equation 8.18 can also be transformed into the following form:

$$k_r(\psi) = \int_{\psi}^{\psi_r} \frac{\theta(y) - \theta(\psi)}{y^2} \theta'(y)\, dy \Bigg/ \int_{\psi_{\text{aev}}}^{\psi_r} \frac{\theta(y) - \theta_s}{y^2} \theta'(y)\, dy \tag{8.20}$$

where:

ψ_{aev} = air-entry value of the soil under consideration (i.e., the suction where air starts to enter the largest pores in the soil),
ψ_r = suction corresponding to the residual water content, θ_r,
y = dummy variable of integration representing suction, and
θ' = derivative of Eq. 8.19.

Equation 8.19 fits the experimental data over the entire suction range and the integrations in Eq. 8.20 can be performed from ψ_{aev} to 10^6 for all types of soils. This simplifies the prediction procedure for the coefficient of permeability since the residual value (θ_r or ψ_r) does not have to be accurately determined for each soil.

It is necessary to independently determine the air-entry value ψ_{aev} of the soil. The air-entry value can be determined using the empirical construction procedure shown in Chapter 5. Consequently, the soil is assumed to remain at the saturated coefficient of permeability at suction values below the air-entry value of the soil. The coefficient of permeability of the soil then decreases at soil suctions higher than the air-entry value in accordance with the proposed integration procedure. A series of discrete data points are calculated corresponding to the relationship between the coefficient of permeability and soil suction. It is also possible to best fit the permeability data points with another empirical equation, if so desired.

To avoid numerical difficulties associated with performing integration over the soil suction range from ψ_{aev} to 10^6 kPa on an arithmetic scale, it is more convenient to perform the integration on a logarithmic scale. Therefore, the following variation of Eq. 8.20 is preferred:

$$k_r(\psi) = \int_{\ln(\psi)}^{b} \frac{\theta(e^y) - \theta(\psi)}{e^y} \theta'(e^y)\, dy \Bigg/ \int_{\ln(\psi_{\text{aev}})}^{b} \frac{\theta(e^y) - \theta_s}{e^y} \theta'(e^y)\, dy \tag{8.21}$$

where:

b = upper limit of integration [i.e., $\ln(10^6)$],
y = dummy variable of integration representing the logarithm of suction,
θ' = derivative of the SWCC equation, and
e^y = natural number raised to the dummy variable power.

8.2.5 Numerical Integration for Relative Coefficient of Permeability

The permeability function is expressed in a form where integration can be performed on a logarithm scale (Eq. 8.21). The two integrals in Eq. 8.21 can be evaluated using the following numerical integration method. Let a and b denote the lower and upper limits of the integration. Then $a = \ln(\psi_{\text{aev}})$ and $b = \ln(10^6)$. Let us divide the range $[a, b]$ into N subintervals of the same size and let Δy denote the length of each subinterval. Then,

$$a = y_1 < y_2 < \cdots s < y_N < y_{N+1} = b \quad \Delta y = \frac{b-a}{N} \tag{8.22}$$

The denominator of Eq. 8.21 can be evaluated as follows:

$$\int_{\ln(\psi_{\text{aev}})}^{b} \frac{\theta(e^y) - \theta_s}{e^y} \theta'(e^y)\, dy \approx \Delta y \sum_{i=1}^{N} \frac{\theta(e^{\bar{y}_i}) - \theta_s}{e^{\bar{y}_i}} \theta'(e^{\bar{y}_i}) \tag{8.23}$$

where $\bar{y}_i$ is the midpoint of the ith interval, $[y_i, y_{i+1}]$, and θ' is the derivative of Eq. 8.19 given as follows:

$$\theta'(\psi) = C'(\psi) \frac{\theta_s}{\{\ln[e + (\psi/a_f)^{n_f}]\}^{m_f}} - C(\psi) \frac{\theta_s}{\{\ln[e + (\psi/a_f)^{n_f}]\}^{m_f+1}} \times \frac{m_f n_f (\psi/a_f)^{n_f-1}}{a[e + (\psi/a_f)^{n_f}]} \tag{8.24}$$

where

$$C'(\psi) = \frac{-1}{(\psi_r + \psi)\ln(1 + 10^6/\psi_r)} \tag{8.25}$$

For any suction value ψ in the range between the air-entry value, ψ_{aev}, and 10^6, the value of $\ln(\psi)$ is between a and b. Let us assume that $\ln(\psi)$ is in the jth interval, $\lfloor y_j, y_{j+1} \rfloor$. Then, the numerator of Eq. 8.21 can be evaluated as follows:

$$\int_{\ln(\psi)}^{b} \frac{\theta(e^y) - \theta(\psi)}{e^y} \theta'(e^y)\, dy \approx \Delta y \sum_{i=j}^{N} \frac{\theta(e^{\bar{y}_i}) - \theta(\psi)}{e^{\bar{y}_i}} \theta'(e^{\bar{y}_i}) \tag{8.26}$$

Therefore, the relative permeability $k_r(\psi)$ at suction ψ is

$$k_r(\psi) \approx \sum_{i=j}^{N} \frac{\theta(e^{\bar{y}_i}) - \theta(\psi)}{e^{\bar{y}_i}} \theta'(e^{\bar{y}_i}) \Bigg/ \sum_{i=1}^{N} \frac{\theta(e^{\bar{y}_i}) - \theta_s}{e^{\bar{y}_i}} \theta'(e^{\bar{y}_i}) \qquad (8.27)$$

The lower limit of the integration given by Eq. 8.21 corresponds to the saturated water content θ_s. Although the air-entry value is used as the lower limit of the integration, any other value between 0 and ψ_{aev} can be used. In other words, the air-entry value does not have to be known precisely. However, the value must be positive to perform the integration on a logarithm scale.

8.2.6 Numerical Results and Comparisons with Experimental Data

The numerical integration of Eq. 8.21 can be accomplished as follows. First, it is necessary to determine the four fitting parameters for the SWCC (i.e., a_f, n_f, m_f, and ψ_r), as shown in Eq. 8.19. These parameters can be manually determined or computed using a nonlinear least-squares routine. Neither of the residual values θ_r and ψ_r are explicitly required as part of the curve-fitting procedure. The permeability function can be computed once the SWCC fitting parameters are known. Equation 8.21 is used in conjunction with the saturated coefficient of permeability.

Following are some comparisons between measured and predicted coefficient of permeability curves for several soils. The soil properties and values of the fitting parameters for Eq. 8.19 for each soil are listed in Table 8.1.

Figure 8.13 shows a curve that is best-fit to the experimental data for Touchet silt (GE3) from Brooks and Corey (1964). The predicted coefficient-of-permeability function based on the best-fit curve in Fig. 8.14 is compared with the measured permeability data. Results show that the predicted permeability function closely matches the measured permeability data.

Table 8.1 Soil Properties and Fitting parameters for Fredlund and Xing (1994) Equation on Several Soils

Soil Type	θ_s	$k_s \times 10^{-6}$ (m/s)	a_f	n_f	m_f	ψ_r
Touchet silt (GE3)	0.430	—	8.34	9.90	0.44	30.0
Columbia sandy silt	0.458	—	6.01	11.86	0.36	30.0
Yolo light clay	0.375	0.123	2.70	2.05	0.36	100.0
Guelph silt drying	0.520	3.917	5.61	2.24	0.40	300.0
wetting	0.430	—	3.12	4.86	0.23	100.0
Superstition sand	—	18.3	2.77	11.20	0.45	300.0

Permeability function calculations for Columbia sandy silt (Brooks and Corey, 1964) are presented in Figs. 8.15 and 8.16. The calculated relative coefficient-of-permeability function closely matches the experimentally measured values.

A best-fit SWCC for the experimental data for Superstition sand (Richards, 1952) is shown in Fig. 8.17. The relative coefficients of permeability computed from the measured SWCC are shown in Fig. 8.18. The calculated relative coefficient-of-permeability values deviate slightly from the measured coefficients of permeability as soil suction increases. Residual conditions are reached at soil suctions around 10 kPa.

The differences between the calculated and measured coefficients of permeability may, in part, be due to inadequacies in the fit of the experimental SWCC data, as shown in Fig. 8.17. To test this hypothesis, a second and more accurate fit was performed on the experimental permeability data. The improved best-fit curve to the experimental soil-water characteristic for the Superstition sand is shown in Fig. 8.19. The predicted permeability function based on the improved best-fit curve provides a more accurate fit of the experimental permeability data, as shown in Fig. 8.20. The results indicate that it is important to obtain an accurate fit of the experimental soil-water characteristic data in order to accurately compute the coefficient-of-permeability function.

The SWCC and permeability function results for Yolo light clay (Moore, 1939) are presented in Figs. 8.21 and 8.22, respectively. The predicted permeability function is close to the measured data in the low-soil-suction range. The unusual shape of the measured permeability function data suggests that there may have been errors in the measured permeability data. The clay material may have experienced some volume change as soil suction was changed. If this is the case, then the degree-of-saturation SWCC would have provided a better indication of the actual air-entry value for Yolo light clay. The estimation of the coefficient-of-permeability function for clayey soils is generally less accurate than permeability function estimations for sandy soils.

SWCC and permeability data on Guelph silt (Elrick and Bowman, 1964) are shown in Figs. 8.23 and 8.24. This example illustrates the effect of hysteresis on the SWCCs and the permeability function. The results indicate that, although there is considerable hysteresis observed in the volumetric water content SWCC, hysteresis is less pronounced in the permeability function $k(\theta)$.

8.2.7 Comparison of Data Fit and Theory

The examples of experimental soil-water characteristic data and measured permeability data presented in the previous

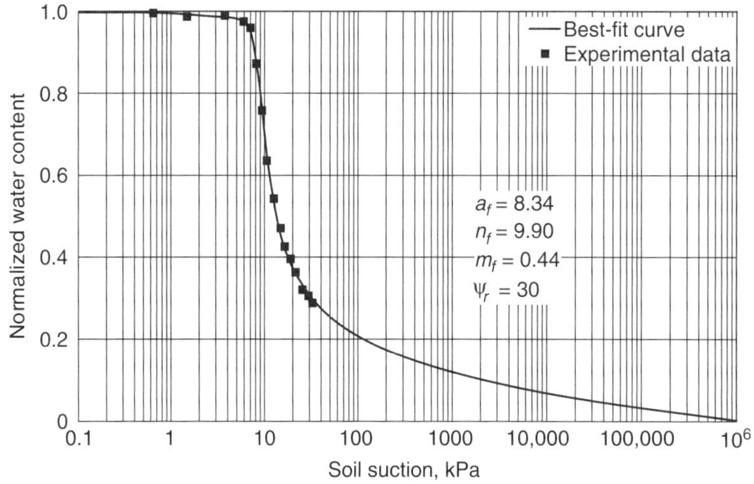

Figure 8.13 Fredlund and Xing (1994) SWCC fit to experimental data for GE3. (Data from Brooks and Corey, 1964.)

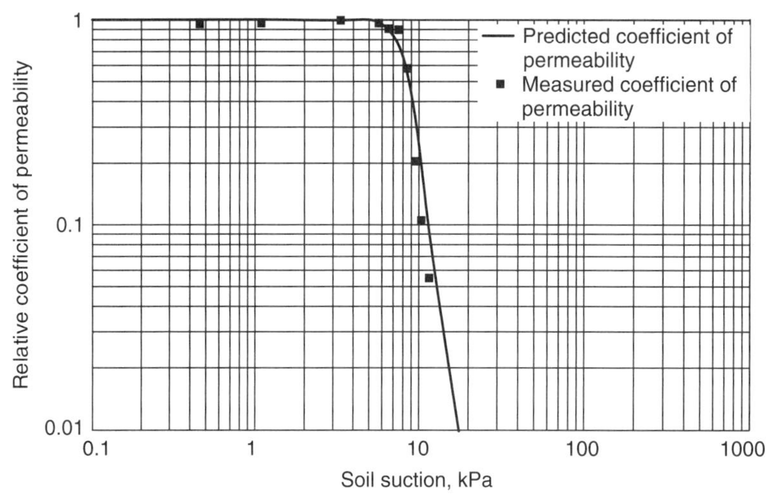

Figure 8.14 Comparison between predicted and measured relative permeability for GES. (Data from Brooks and Corey, 1964.)

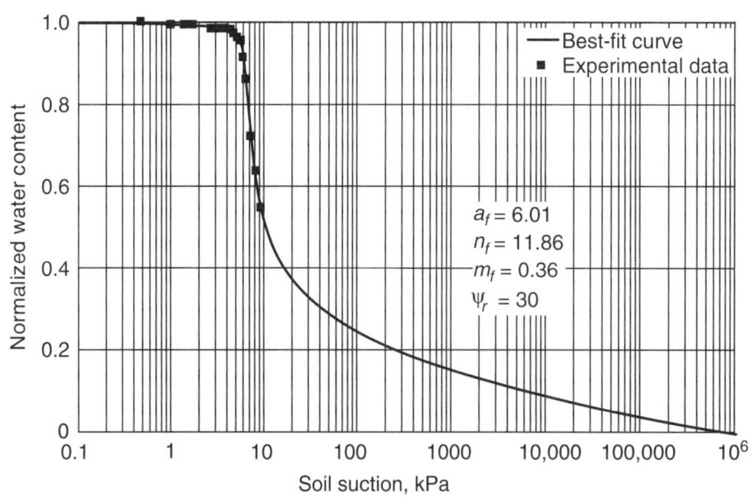

Figure 8.15 Fredlund and Xing (1994) SWCC fit experimental data for Columbia sandy silt. (Data from Brooks and Corey, 1964.)

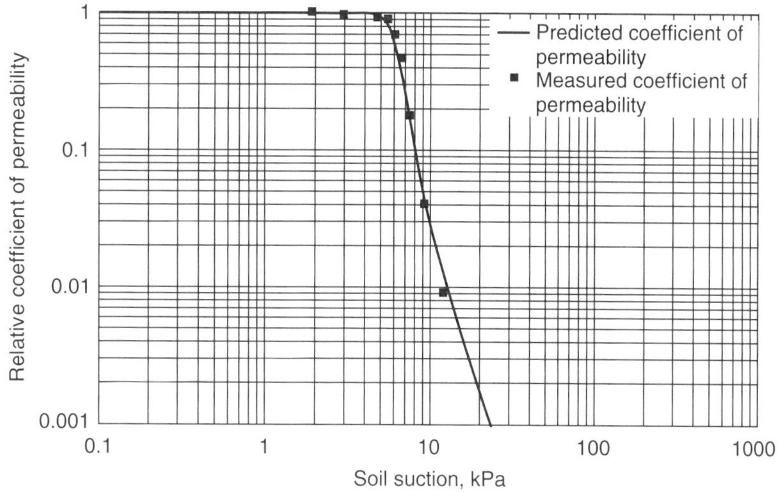

Figure 8.16 Comparison of predicted and measured relative permeability function for Columbia sandy silt. (Data from Brooks and Corey, 1964.)

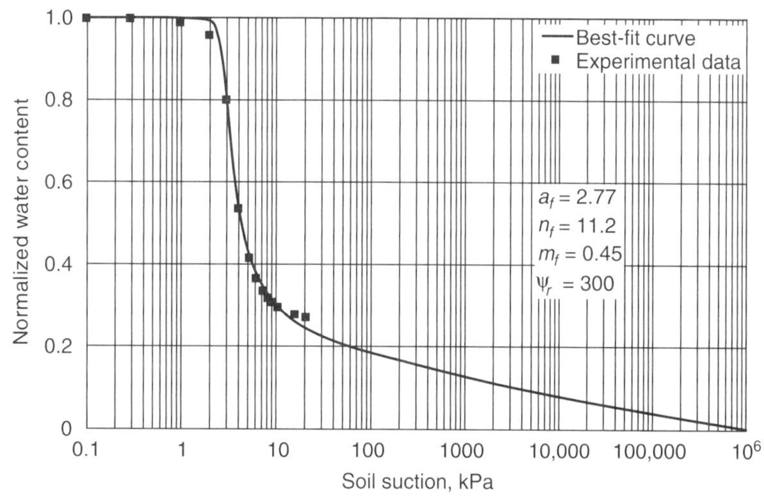

Figure 8.17 Fredlund and Xing (1994) SWCC fit to experimental data for Superstition sand. (Data from Richards, 1952.)

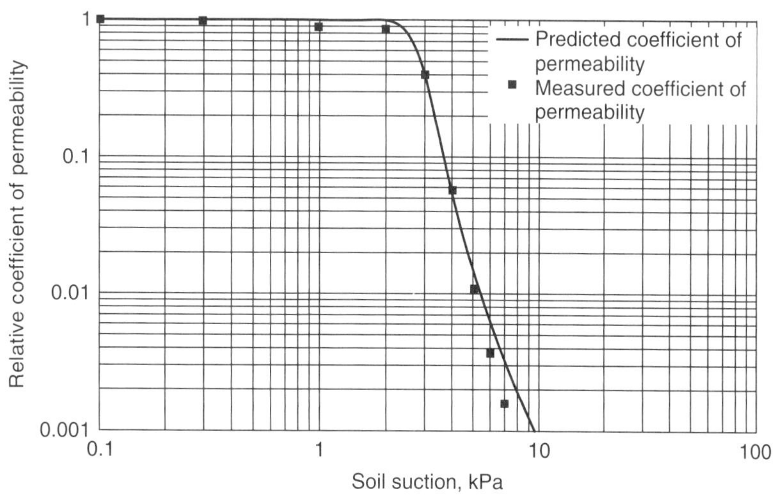

Figure 8.18 Comparison of predicted and measured relative permeability for Superstition sand. (Data from Richards, 1952.)

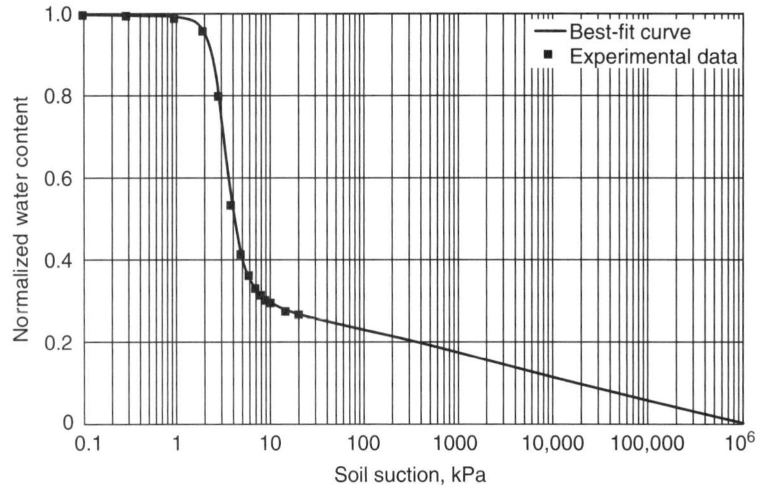

Figure 8.19 Improved fit of Fredlund and Xing (1994) SWCC data for Superstition sand. (Data from Richards, 1952.)

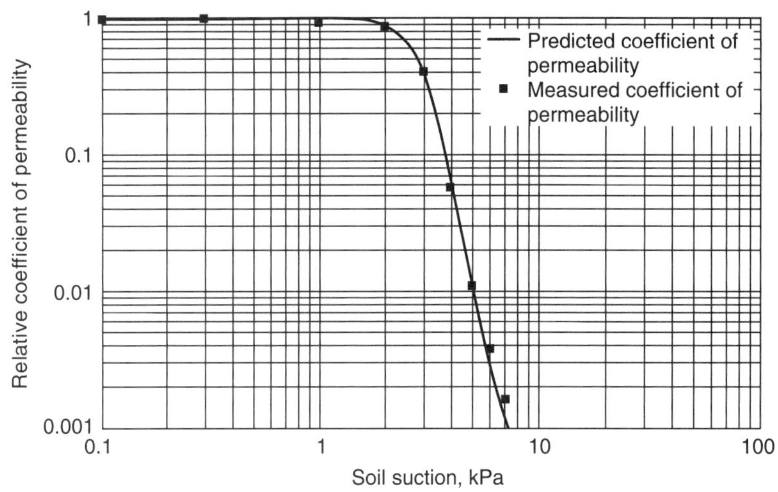

Figure 8.20 Comparison of predicted and measured relative permeability function using improved best-fit analysis of SWCC experimental data on Superstition sand. (Data from Richards, 1952.)

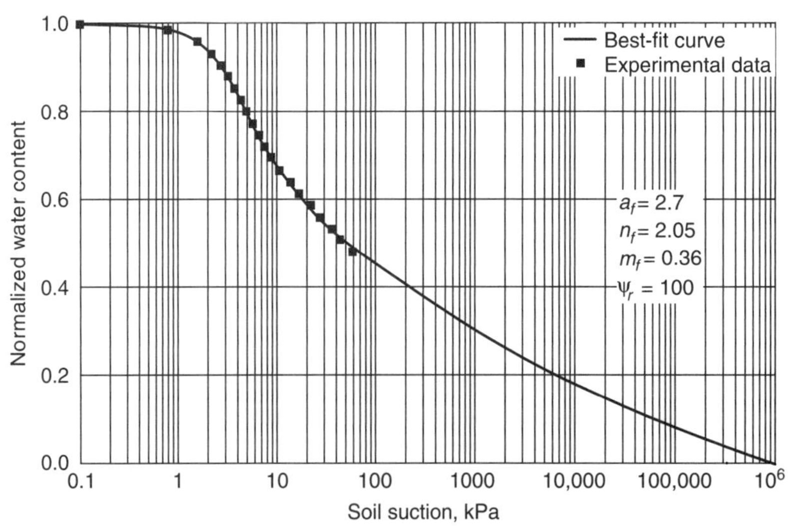

Figure 8.21 Fredlund and Xing (1994) SWCC fit experimental data for Yolo light clay. (Data from Moore, 1939.)

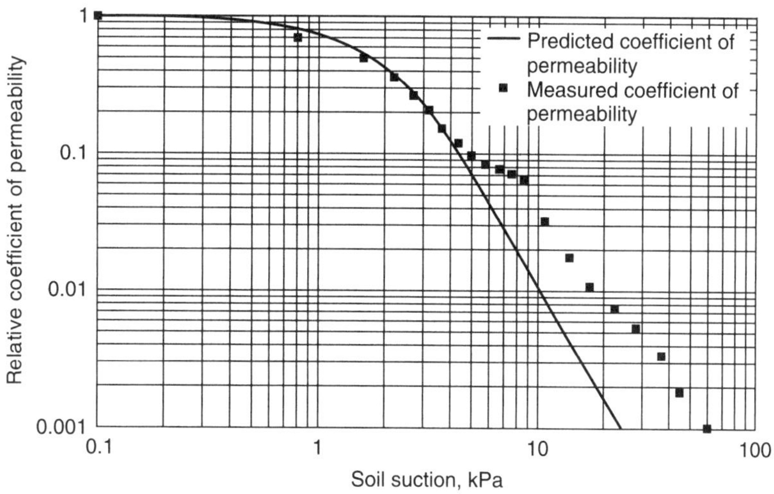

Figure 8.22 Comparison of predicted and measured relative permeability function for Yolo light clay. (Data from Moore, 1939.)

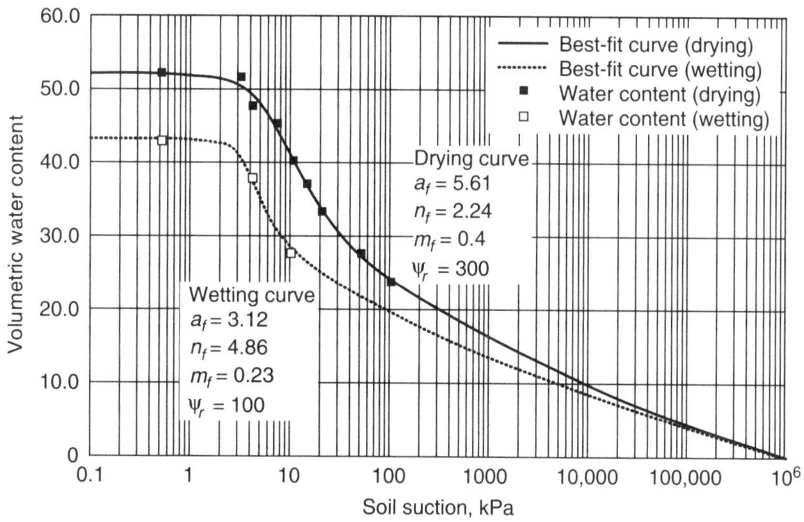

Figure 8.23 Fredlund and Xing (1994) SWCC fit of experimental data for Guelph silt. (Data from Elrick and Bowman, 1964.)

section are primarily in the low-soil-suction range. Equation 8.19 for the SWCC can also be used to fit experimental data in the high-soil-suction range. It is no longer necessary to estimate the residual volumetric water content θ_r for each soil when predicting the coefficient-of-permeability function in the high-soil-suction range. Reasonable predictions of the coefficient of permeability function appear to be possible for a wide range of SWCCs.

It would appear that volume changes associated with changes in soil suction are the primary reason for deviations between measured permeability functions and estimated permeability functions. It appears to be better to give consideration to the degree-of-saturation SWCC when performing integration along the SWCC when there are volume changes associated with soil suction changes. It is recommended that the SWCC also be measured in the high-soil-suction range (i.e., at soil suction higher than the residual suction).

The accuracy of the estimated coefficient-of-permeability function depends not only on the closeness of the best-fit curve to the experimental soil-water characteristic data but also on the precision of the model adopted. Mualem (1986) concluded that there was no single physical model that fits every soil. Proposed permeability estimation models have been found to be most satisfactory for sandy soils, whereas agreement with experimental data may not be satisfactory for fine-grained soils. The flexibility of Eq. 8.21 can be increased by multiplying the equation by a correction factor Θ^q which takes tortuosity into consideration (Mualem, 1986) and increases the flexibility of the

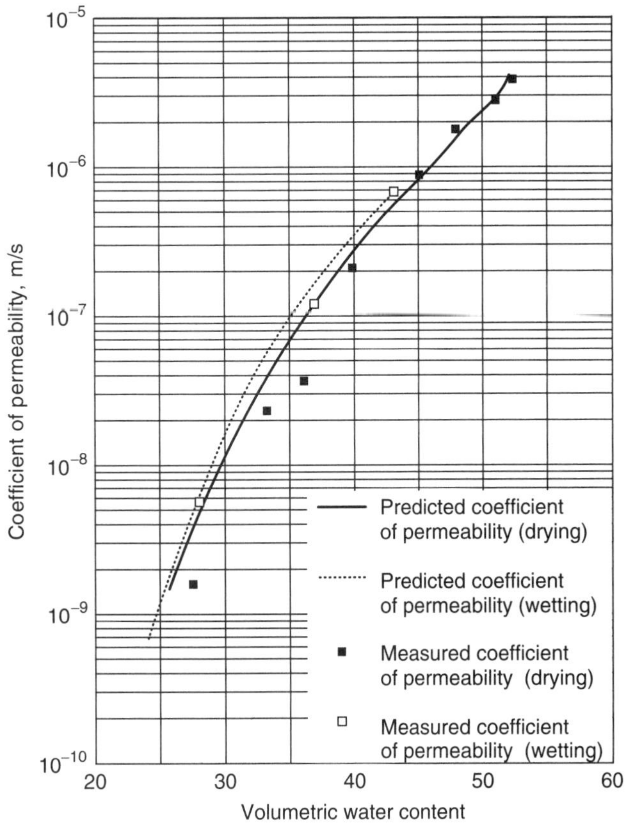

Figure 8.24 Comparisons of predicted and measured coefficients of permeability for Guelph silt. (Data from Elrick and Bowman, 1964.)

permeability estimation equation. Therefore, a more accurate estimation permeability equation can be written using the following modification to Eq. 8.21:

$$k_r(\psi) = \Theta^q(\psi) \int_{\ln(\psi)}^{b} \frac{\theta(e^y) - \theta(\psi)}{e^y} \theta'(e^y)\, dy \Bigg/ \int_{\ln(\psi_{aev})}^{b} \frac{\theta(e^y) - \theta_s}{e^y} \theta'(e^y)\, dy \qquad (8.28)$$

Kunze et al. (1968) suggested that the value of the power q could be set equal to 1.

8.2.8 Correlation of Soil Parameters for Permeability Function

A number of possible correlations have been undertaken which assist in the estimation of an unsaturated soil permeability function. Correlations have been undertaken between the fitting parameters for a particular permeability function or SWCC and soil classification properties. Unfortunately, the correlations related to a particular permeability equation can only be used with the equation used when performing the correlation study. The correlations cannot be transferred to other permeability functions.

Probably the most general correlation that has been undertaken for permeability functions is related to the functional form suggested by Leong and Rahardjo (1997a). Details of parameters for various soil types are presented below along with statistical variation of the single required fitting parameter.

8.2.8.1 Leong and Rahardjo (1997a) Permeability Function

Leong and Rahardjo (1997a) suggested using a dimensionless equation for the SWCC that was then raised to a power q to obtain the permeability function. The permeability function was written in the following form:

$$k_r(\psi) = k_s \left[\Theta_d(\psi)\right]^q \qquad (8.29)$$

where:

$\Theta_d(\psi)$ = dimensionless water content form θ/θ_s for SWCC, and
q = correlation-based soil fitting parameter.

Equation 8.29 is attractive for estimating the permeability function due to its simplicity. The equation is easy to use and reflects the relationship between the SWCC and the permeability function. It is necessary to know the value to use for the q soil fitting parameter when using Eq. 8.29. Leong and Rahardjo (1997a) used the data from six soils to assess the magnitude of the q soil fitting parameter. The results of the fitting exercise are shown in Table 8.2.

A later study undertook the analysis of approximately 300 permeability data sets (Fredlund et al., 2001a). Each data set consisted of (i) experimental results where the SWCC was measured, (ii) experimental results where the coefficient of permeability under various applied soil suctions was measured, and (iii) the saturated coefficient of permeability was measured. All experimental results were extracted from the SoilVision knowledge-based system. The soils were divided into a number of categories dependent upon

Table 8.2 Summary of Typical q Parameters from Study by Leong and Rahardjo (1997a)

Soil Type	a_f(kPa)	n_f	m_f	q
Beit Netofa clay	389	0.69	1.176	52.12
Rehovot sand	2.25	4.32	1.235	6.04
Touchet silt	7.64	7.05	0.506	4.55
Columbia sandy silt	5.81	10.59	0.381	5.79
Superstition sand	2.66	6.86	0.525	6.21
Yolo light clay	2.93	2.11	0.379	9.57

Table 8.3 Summary of Soil Types Analyzed for Permeability Function

USDA Textural Classification	Number of Samples
Clay	21
Clay silt	18
Silt	12
Silty sand	29
Sand	49
Sandy clay silt	17
Sandy silt	30
Silt	74
Silty clay	34
Silty clay	18

USDA textural classification. The textural categories along with the number of data sets in each category are shown in Table 8.3.

Soils in each textural category were first analyzed and then all soils were combined to obtain overall average values for q. Each SWCC data set was first best fit to determine the a_f, n_f, and m_f parameters associated with the Fredlund and Xing (1994) equation. Then the SWCC and the saturated coefficient of permeability were used to best fit the permeability data in accordance with the equation proposed by Leong and Rahardjo (1997a). The best-fit analysis of the permeability data allowed the calculation of the q soil parameter.

A typical best fit of the Leong and Rahardjo (1997) procedure for a sand soil is shown in Fig. 8.25. The best-fit q parameter for the sand was 2.54. A similar best fit for silt is shown in Fig. 8.26 and the q parameter was 3.62. Both plots illustrate that the predicted permeability function quite closely fits the experimental data when an appropriate q parameter is selected for the permeability function.

Figure 8.27 shows the frequency distribution associated with the q parameter when all soil textural categories are combined (Fredlund et al., 2001a). The statistical analysis of the q parameter is presented in Table 8.4. The results show that the overall mean value for q is 3.29 and the standard deviation of q is 1.40. These values yield an overall coefficient of variation of 43%, which is quite high. The overall median value is 2.96 and the mode is 5.61.

Frequency distributions can be drawn for each of the soil textural categories. The frequency distribution of q for sand soil is shown in Fig. 8.28 and the frequency distribution of q for silt is shown in Fig. 8.29. Table 8.5 summarizes the statistical properties for the computed q parameters for each of the soil textural categories.

The standard deviation was 0.49 for the sand soil and 1.50 for the clay soil. The standard deviation was 0.84 for the

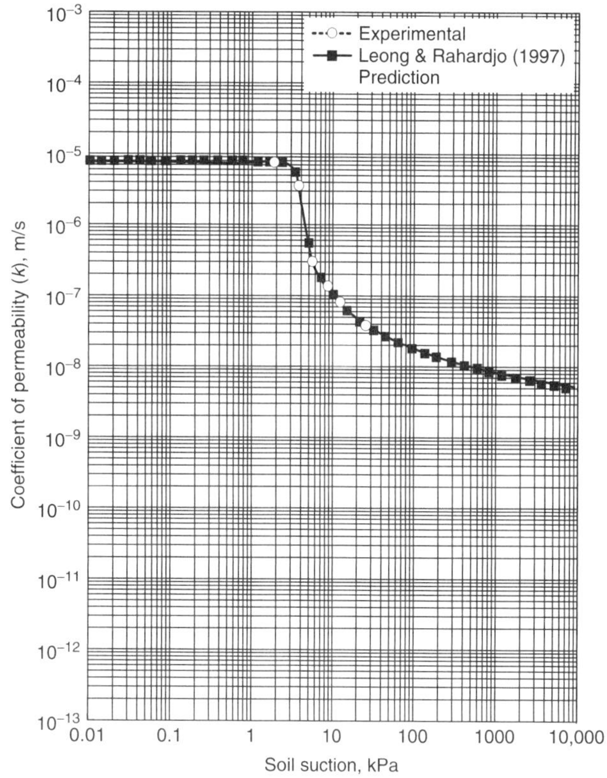

Figure 8.25 Fit of proposed Leong and Rahardjo (1997a) permeability function for experimental data on sand.

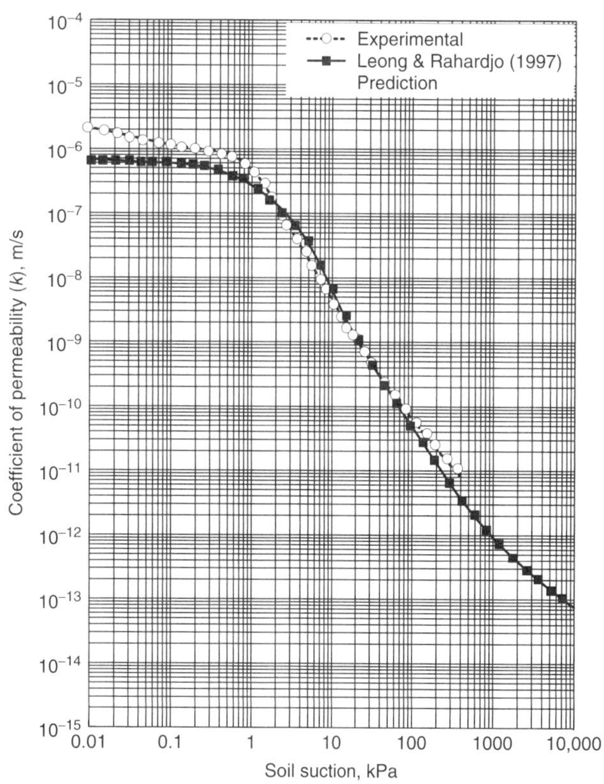

Figure 8.26 Fit of proposed Leong and Rahardjo (1997a) permeability function for experimental data on silt.

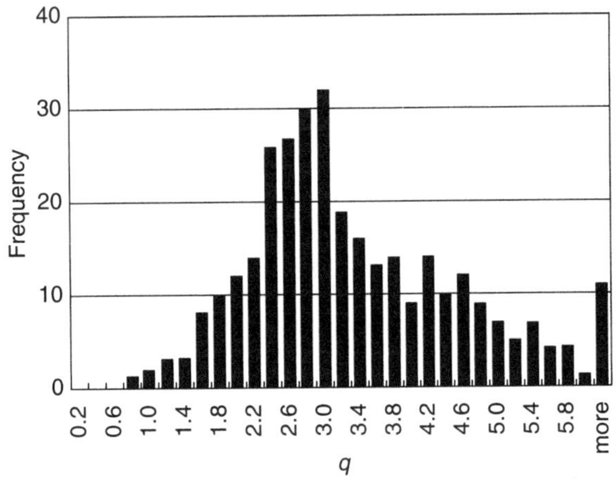

Figure 8.27 Frequency distribution for q permeability parameter for all soils analyzed (from Fredlund et al., 2001a).

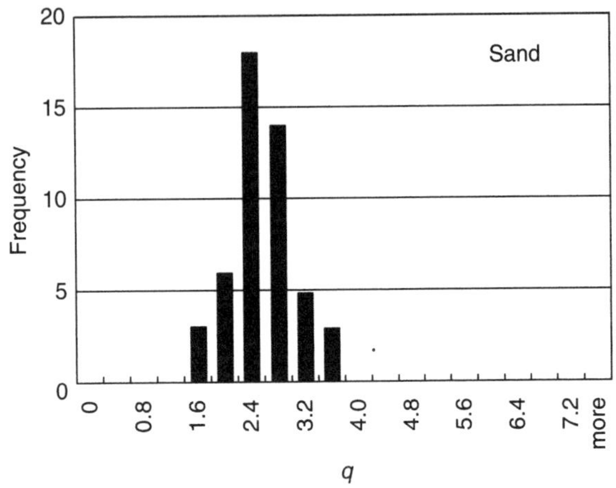

Figure 8.28 Frequency distribution for q permeability parameter for sand soils analyzed (from Fredlund et al., 2001a).

sandy silt and 1.81 for the clayey silt. The results indicate that there is less scatter in the fitting parameter q as the soil becomes closer to a sand material. The fitting parameter q tends toward 1.0 for sand soils.

The sand showed a mean q value of 2.37 while the clay soils showed a mean q value of 4.34. The sandy silt showed a mean q value of 2.86 and the clay silt showed a mean q value of 3.58. The results indicate a trend toward a larger q value for soils with higher plasticity and a wider distribution of particle sizes. The mean fitting parameters for all soil types ranged from 2.4 to 5.6. The statistical results provide a general indication of the range and scatter that can be anticipated when using the Leong and Rahardjo (1997a) equation (i.e., Eq. 8.29) for estimation of the permeability function for an unsaturated soil.

The analysis of an even larger database of measured SWCC and permeability function measurements would be useful to better understand the mean and variation

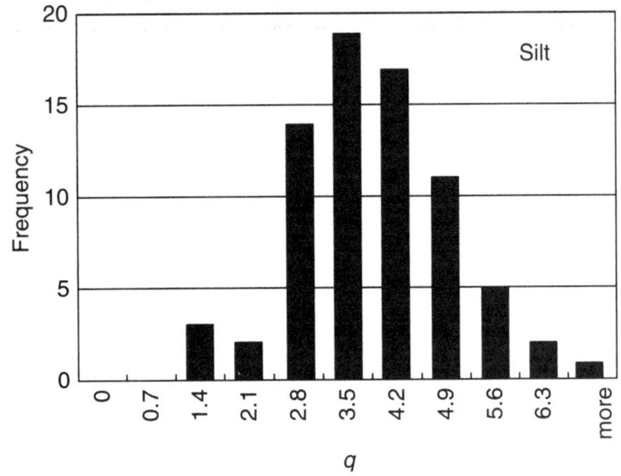

Figure 8.29 Frequency distribution for q permeability parameter for silt soils analyzed (from Fredlund et al., 2001a).

Table 8.4 Statistical Analysis of q Soil Parameter for All Soils

Statistical Variable	q Soil Parameter
Mean	3.29
Standard error	0.08
Median	2.96
Mode	5.61
Standard deviation	1.40
Sample variance	1.96
Kurtosis	14.80
Skewness	2.35
Range	14.39
Minimum	0.64
Maximum	15.03
Number of soils	323

associated with the q parameter. However, it must also be recognized that there will always be considerable scatter in the q parameter. The estimation procedure proposed by Leong and Rahardjo (1997a) is of value, but it does not appear to provide a singular value for the q parameter.

8.2.9 Regression Techniques for Best-Fit Permeability Function

Gardner (1958a) published two empirical forms for a permeability equation. The first equation was presented earlier and takes the following form:

$$k_w\left(u_a-u_w\right)=\frac{k_s}{1+a_g\left\{\left(u_a-u_w\right)/\left(\rho_w g\right)\right\}^{n_g}} \tag{8.30}$$

Table 8.5 Statistics of Permeability Fitting Parameter q for Various Soil Textures

Statistics	Clay	Clay loam	Loam	Sand	Sandy clay loam	Silty clay	Silty clay loam	Silt loam	Sandy loam	Loam sand
Mean	4.34	3.58	3.78	2.37	2.80	5.59	3.22	3.52	2.86	2.67
Median	4.71	2.62	3.56	2.36	2.62	4.77	3.18	3.46	2.85	2.59
Mode	3.00	2.80	3.25	2.20	2.75	4.60	4.05	3.15	2.75	4.63
Standard deviation	1.50	1.81	1.16	0.49	1.00	1.31	1.36	1.09	0.84	0.68
Sample variance	2.25	3.29	1.34	0.24	0.99	1.73	1.86	1.19	0.71	0.46
Range	5.42	5.92	4.14	2.25	4.71	6.33	4.97	5.99	4.20	4.01
Minimum	1.52	1.84	1.63	1.25	0.64	1.11	1.28	0.83	1.02	1.41
Maximum	6.94	7.76	5.76	3.49	5.35	7.44	6.25	6.82	5.22	5.42
Number of sets	21	18	12	49	17	34	18	74	30	29

where:

$k_w(u_a - u_w)$ = coefficient of permeability as a function of soil suction,
k_s = saturated coefficient of permeability,
a_g = fitting parameter related to the inverse of the air-entry value,
n_g = fitting parameter related to the rate of desaturation of the soil,
$u_a - u_w$ = matric suction,
ρ_w = density of water, and
g = acceleration due to gravity.

The second Gardner (1958a) permeability equation has the following form:

$$k_w(u_a - u_w) = k_s \exp\left[-c_g(u_a - u_w)\right] \tag{8.31}$$

where:

c_g = soil parameter related to the exponential decrease in permeability with respect to soil suction.

Inherent in the form of the latter equation is the assumption that the air-entry value for the soil is near zero. Equation 8.31 should only be used when it is known that the air-entry value of the soil under consideration is approaching zero. Neither of the above permeability functions should be used unless it has been possible to determine the a_g and n_g fitting parameters through use of an independent analysis or through direct measurements of the coefficient of permeability at various applied soil suction values.

8.2.10 Estimation of Minimum Coefficient of Permeability

Limited research has been undertaken on the form that the permeability function should take once residual water content conditions are reached and exceeded. A recent study (Ebrahimi-Birang et al., 2004) has suggested that there should be a lower limit for the water coefficient of permeability and that the magnitude of potential vapor diffusion be taken into consideration.

Figure 8.30 shows the results of the method proposed by Ebrahimi-Birang et al. (2004) applied to silty sand. The lower limit that was suggested for liquid water flow was 1×10^{-14} m/s. It was suggested that the same lower limit for the coefficient of permeability might be applicable for all soils. The lower limit for the water coefficient of permeability is of importance with respect to numerical modeling of water flow. The lack of a lower limit for the water coefficient of permeability can give rise to numerical convergence problems during seepage modeling.

8.2.11 Water Storage Modulus for Transient Modeling

The water storage property of a soil, m_2^w, is defined as the relationship of the slope of the (volumetric) water content and soil suction. The water storage variable is required whenever a transient seepage analysis is performed. The water storage modulus can be obtained through the (arithmetic) differentiation of any of the equations designated for the SWCC (Fig. 8.31).

8.3 APPLICATION TO SATURATED-UNSATURATED WATER FLOW PROBLEMS

Steady- and unsteady-state water seepage through a saturated-unsaturated soil system can be analyzed by solving the governing partial differential equation of seepage. The solution is obtained through use of the numerical modeling methods such as finite difference and finite element methods. The approach is similar for steady-state seepage and unsteady-state seepage with the exception that unsteady-state seepage is time dependent, requiring discretization of time (i.e., elapsed time is handled in an incremental elapsed time manner for unsteady-state water seepage).

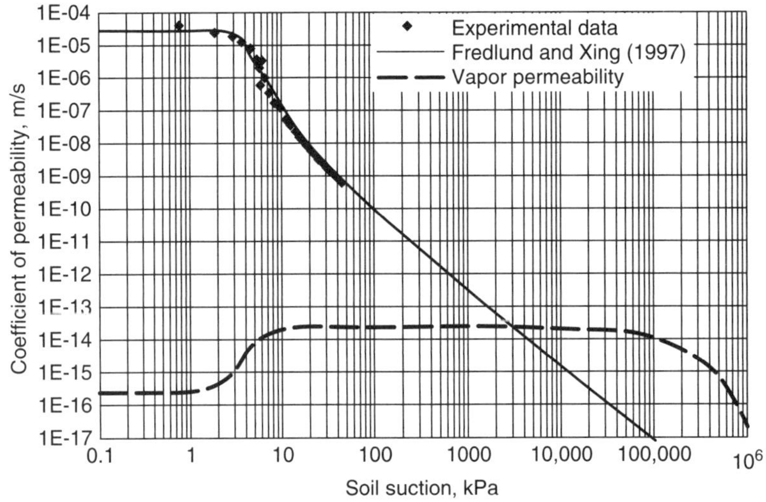

Figure 8.30 Variations of liquid water and vapor permeability function for a silty sand soil.

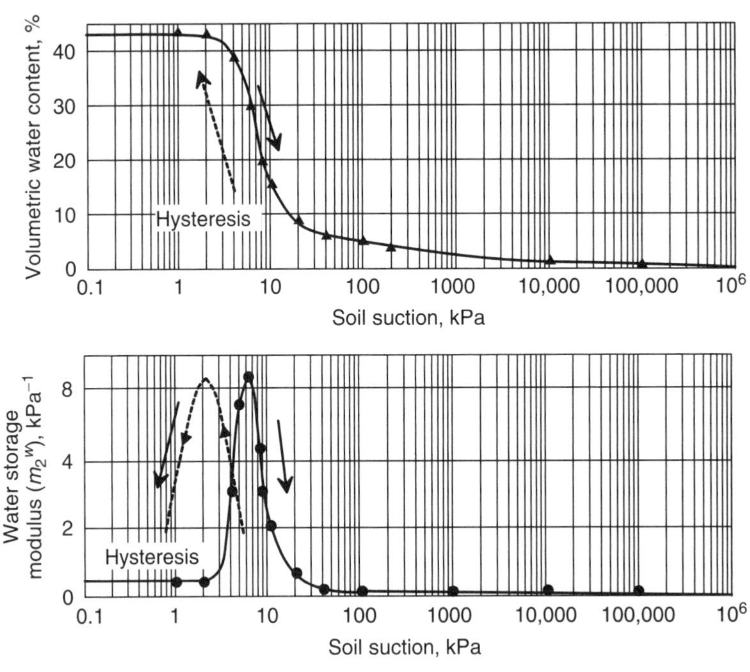

Figure 8.31 Relationship of water storage functions to drying and wetting SWCCs.

The solution of saturated-unsaturated seepage problems is presented through a series of problems that are progressively more complex. The first seepage problems are of a steady-state nature and these are followed by several unsteady-state (or transient) analysis problems. One-dimensional steady-state problems are first solved and these are followed by two-dimensional examples. The one-dimensional seepage examples illustrate the application of hydraulic head and moisture flux boundary conditions. The one-dimensional problems are solved using a finite difference technique. Two-dimensional steady-state example problems are then solved using the finite element method. These example problems are followed by a series of two-dimensional, unsteady-state examples that are solved using the finite element methodology.

8.3.1 Solution of One-Dimensional Flow Problems

The differential equation for one-dimensional steady-state flow through a *homogeneous*, *saturated* soil can be solved by integrating the differential equation two times. The equation for one-dimensional steady-state flow through an *unsaturated* soil requires a more complex solution than that for a saturated soil. A numerical solution must be used rather than a closed-form solution. The finite difference method

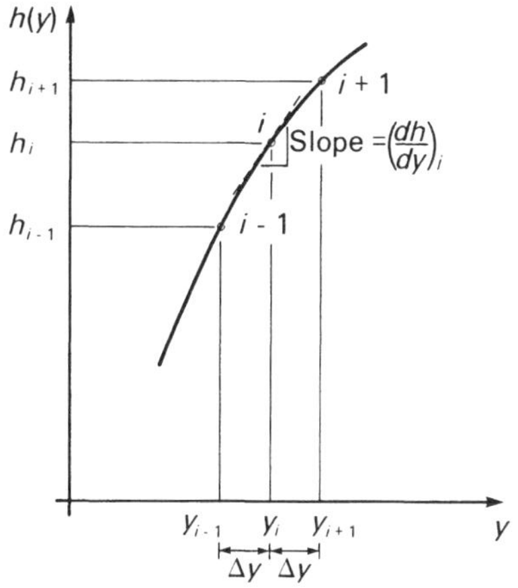

Figure 8.32 Hydraulic head function $h(y)$ written in finite difference form.

can be used to illustrate the solution to the steady-state flow equation for an unsaturated soil.

The seepage differential equation can be written in a finite difference form. Consider the situation where a function $h(y)$ varies in space, as shown in Fig. 8.32. Values of the function at points along the curve describing hydraulic head can be computed using a Taylor series to write forward-difference and backward-difference equations. The forward difference-equation can be written as

$$h_{i+1} = h_i + \Delta y \left(\frac{dh}{dy}\right)_i + \frac{\Delta y^2}{2!}\left(\frac{d^2h}{dy^2}\right)_i + \frac{\Delta y^3}{3!}\left(\frac{d^3h}{dy^3}\right)_i + \cdots \tag{8.32}$$

The backward-difference equation can be written as

$$h_{i-1} = h_i - \Delta y \left(\frac{dh}{dy}\right)_i + \frac{\Delta y^2}{2!}\left(\frac{d^2h}{dy^2}\right)_i - \frac{\Delta y^3}{3!}\left(\frac{d^3h}{dy^3}\right)_i + \cdots \tag{8.33}$$

where:

$i-1, i, i+1$ = three consecutive points spaced at increments Δy.

Subtracting the forward- and backward-difference equations and neglecting the higher order derivatives result in the first derivative of the function at point i:

$$\left(\frac{dh}{dy}\right)_i = \frac{h_{i+1} - h_{i-1}}{2\,\Delta y} \tag{8.34}$$

Summing the forward- and backward-difference equations and again neglecting the higher order derivatives give the second derivative of the function at point i:

$$\left(\frac{d^2h}{dy^2}\right)_i = \frac{h_{i+1} + h_{i-1} - 2h_i}{\Delta y^2} \tag{8.35}$$

Equations 8.34 and 8.35 are called the central-difference approximations for the first and second derivatives of the function $h(y)$ at point i. These approximations can be used to solve the differential seepage equation. Similar approximations can be derived for a function $h(x)$ in the x-direction when there is flow in the x-direction. The use of an iterative finite difference technique in solving seepage problems is illustrated in the following sections.

8.3.2 Head Boundary Conditions Applied

Steady-state evaporation under a fixed head boundary condition from a column of unsaturated soil is illustrated in Fig. 8.33. A tensiometer is installed near the ground surface to measure the negative pore-water pressure. One-dimensional, steady-state flow is assumed when the tensiometer reading remains constant with respect to time. The pore-water pressure at the base of the column (i.e., the water table) is equal to zero.

The hydraulic head distribution along the length of the column is given by Eq. 7.30 [Chapter 7; i.e., $k_{wy}\left(d^2h_w/dy^2\right) + \left(dk_{wy}/dy\right)\left(dh_w/dy\right) = 0$]. This equation can be solved using the finite difference approximations shown in Eqs. 8.34 and 8.35. The column length is first subdivided into n equally spaced nodal points at a distance Δy apart (Fig. 8.33). A central-difference approximation is then applied to the hydraulic head and coefficient-of-permeability derivatives. The finite difference form can be written as follows for point i:

$$k_{wy(i)}\left\{\frac{h_{w(i+1)} + h_{w(i-1)} - 2h_{w(i)}}{(\Delta y^2)}\right\} + \left\{\frac{k_{wy(i+1)} - k_{wy(i-1)}}{2\,\Delta y}\right\}\left\{\frac{h_{w(i+1)} - h_{w(i-1)}}{2\,\Delta y}\right\} = 0 \tag{8.36}$$

where:

$k_{wy(i)}, k_{wy(i-1)}, k_{wy(i+1)}$ = water coefficients of permeability in the y-direction at points i, $i-1$, and $i+1$, respectively, and

$h_{w(i)}, h_{w(i-1)}, h_{w(i+1)}$ = hydraulic heads at points i, $i-1$, and $i+1$, respectively.

Equation 8.36 can be rearranged after assuming equal Δy increments:

$$-\left\{8k_{wy(i)}\right\}\cdot h_{w(i)} + \left\{4k_{wy(i)} + k_{wy(i+1)} - k_{wy(i-1)}\right\}\cdot h_{w(i+1)} + \left\{4k_{wy(i)} + k_{wy(i-1)} - k_{wy(i+1)}\right\}\cdot h_{w(i-1)} = 0 \tag{8.37}$$

The hydraulic heads at the external points (i.e., points 1 and n) become the boundary conditions. The hydraulic head at point 1 is zero. The elevation of point n relative to the datum, h_{gn}, gives the gravitational head at point n. The tensiometer reading near the ground surface indicates

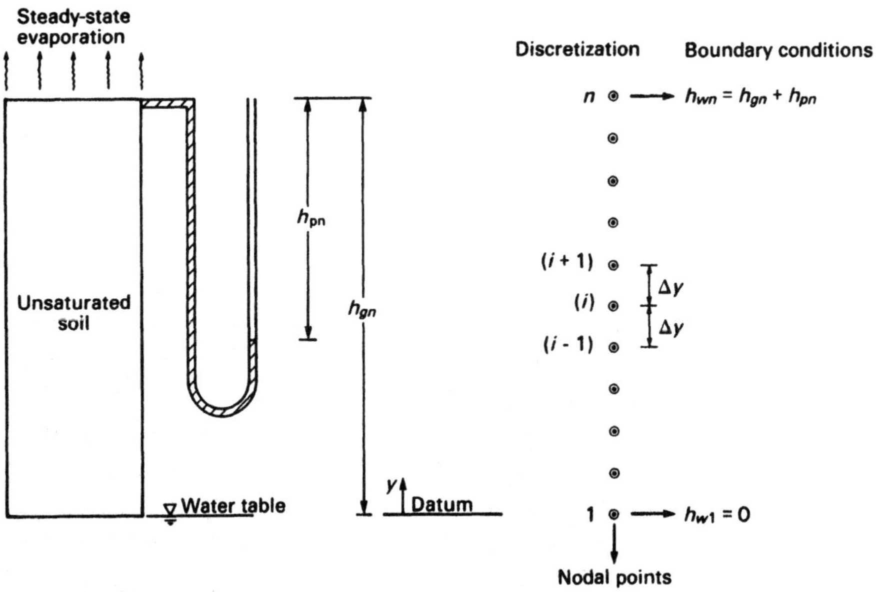

Figure 8.33 One-dimensional, steady-state water flow through an unsaturated soil with a constant-head boundary condition.

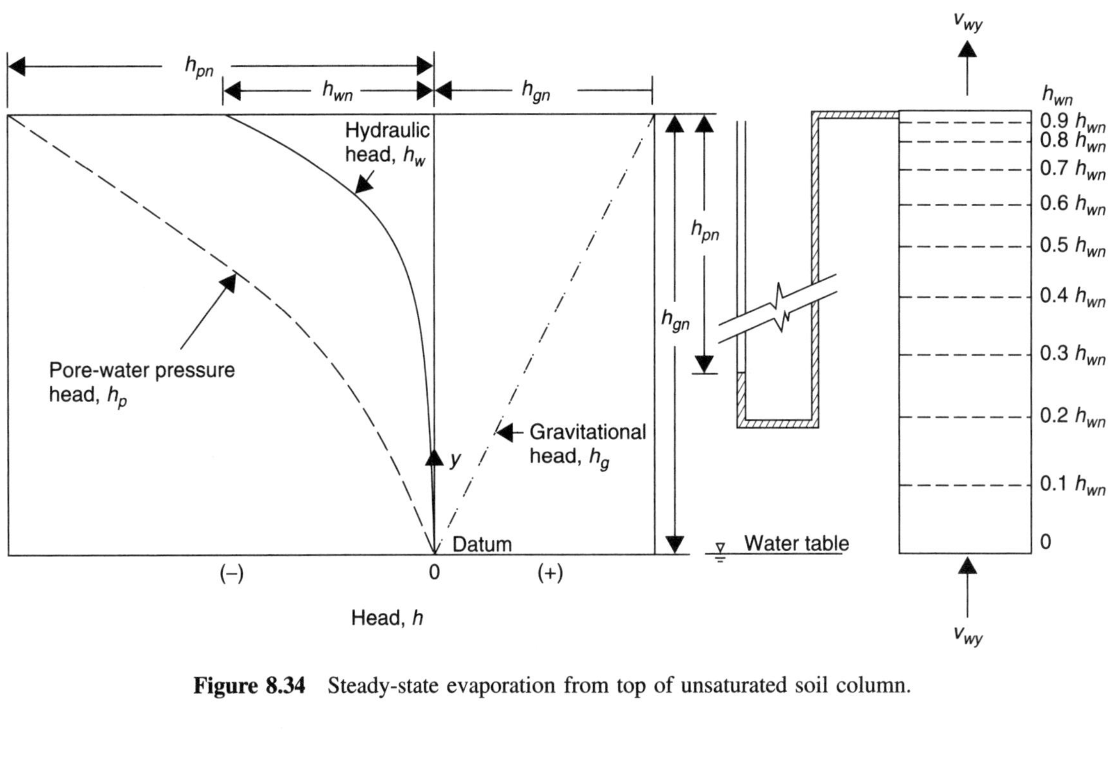

Figure 8.34 Steady-state evaporation from top of unsaturated soil column.

the negative pore-water pressure head at point n (i.e., h_{pn}). The hydraulic head boundary condition at the top and the base of the soil column can be expressed mathematically as follows:

$$h_{w(1)} = 0.0 \quad \text{at } y = 0.0 \text{ (base)}$$

$$h_{w(n)} = h_{gn} + h_{pn} \quad \text{at } y = h_{gn} \text{ (top)}$$

The finite difference seepage equation can be written for the $n - 2$ internal points (i.e., points $2, 3, \ldots, n - 1$). As a result, there are $n - 2$ equations that must be solved simultaneously for $n - 2$ hydraulic heads at intermediate points. The illustrated finite difference scheme is called an implicit form. The equation is also nonlinear since the coefficients of permeability k_{wy} are a function of matric suction, which in turn is a function of pore-water pressure, which is a component of hydraulic head h_w. The nonlinear equations require several iterations to produce convergence. The water coefficient of permeability at each node is revised to a new constant value following each solution for the hydraulic heads.

For the first iteration, the k_{wy} values at all points can be set equal to the saturated coefficient of permeability k_s. The $n - 2$ linear equations can then be solved simultaneously using a procedure such as the Gaussian elimination technique. The computed hydraulic heads are used to calculate new values for the water coefficient of permeability. The coefficient-of-permeability values at each point must be in agreement with the coefficient of permeability versus soil suction function. The revised coefficient-of-permeability values k_{wy} are then used for the second iteration. New hydraulic heads are then computed for all depths. The iterative procedure is repeated until there is no longer a significant change in the computed hydraulic heads and the computed coefficients of permeability.

Figure 8.34 illustrates typical distributions for the pore-water pressure and the hydraulic head along the unsaturated soil column. Flow is occurring under steady-state evaporation conditions. The nonlinearity of the flow equation results in a nonlinear distribution of the hydraulic head and the pore-water pressure head. The equipotential lines are not equally spaced along the column. This is different from the uniformly spaced equipotential lines for a homogeneous, saturated soil column. The difference is the result of the varying coefficient-of-permeability values throughout the unsaturated soil column. The above analysis can also be applied to the steady-state downward flow of water through an unsaturated soil. Once again, the hydraulic head boundary conditions at two points along the soil column must be known.

8.3.3 Flux Boundary Conditions Applied

Infiltration into an unsaturated soil column is another example which can be used to illustrate the solution of the nonlinear differential seepage equation (Fig. 8.35). Steady-state infiltration can be established as a result of sprinkling irrigation. Let us assume a constant downward water flux q_{wy} which is less than the saturated coefficient of permeability of the soil. Steady-state flow can be described using Eq. 7.30. The hydraulic head distribution can be determined by solving the finite difference form of the steady-state flow equation (i.e., Eq. 8.37). The hydraulic head boundary condition at the ground surface is a computed value. However, the water flux q_{wy} is known and is constant throughout the soil column for steady-state conditions.

The soil column can be discretized into n nodal points with an equal spacing Δy (Fig. 8.35). The water flux at point i can be expressed in terms of the hydraulic heads at points $i + 1$ and $i - 1$ using Darcy's law:

$$q_{wy} = -k_{wy(i)} \frac{h_{w(i+1)} - h_{w(i-1)}}{2\ \Delta y} A \tag{8.38}$$

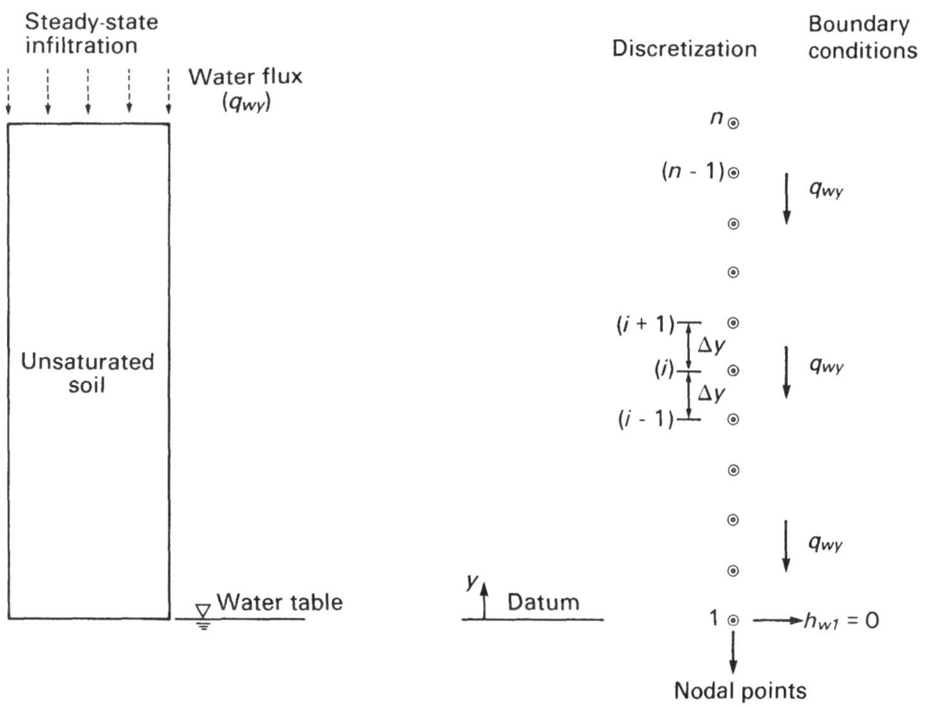

Figure 8.35 One-dimensional steady-state water flow through an unsaturated soil with a flux boundary condition.

where:

q_{wy} = water flux through the soil column during the steady-state flow, where flux is assumed to be positive in an upward direction and negative in a downward direction, and
A = cross-sectional area of the soil column.

Equation 8.38 can be rearranged as follows:

$$h_{w(i+1)} = h_{w(i-1)} - \frac{2\,\Delta y}{A k_{wy(i)}} q_{wy} \tag{8.39}$$

Substituting $h_{w(i+1)}$ into the flow equation for point i yields the following form:

$$-\left\{8k_{wy(i)}\right\} h_{w(i)} + \left\{4k_{wy(i)} + k_{wy(i+1)} - k_{wy(i-1)}\right\} \left\{h_{w(i-1)} - \frac{2\,\Delta y}{A k_{wy(i)}} q_{wy}\right\} + \left\{4k_{wy(i)} + k_{wy(i-1)} - k_{wy(i+1)}\right\} \cdot h_{w(i-1)} = 0 \tag{8.40}$$

The hydraulic head can now be solved for point i:

$$h_{w(i)} = h_{w(i-1)} - \left\{\frac{4k_{wy(i)} + k_{wy(i+1)} - k_{wy(i-1)}}{8k^2_{wy(i)}}\right\} \frac{2\,\Delta y}{A} q_{wy} \tag{8.41}$$

The finite difference hydraulic head equation (i.e., Eq. 8.41) is in an explicit form. The hydraulic heads can be solved directly starting from a known boundary condition such as point 1 (Fig. 8.35), which has a zero hydraulic head. The base of the soil column is a suitable point to commence solving for the remaining hydraulic heads. Hydraulic heads can subsequently be solved point by point up to the ground surface. Equation 8.41 is nonlinear since the coefficient of permeability k_{wy} varies with the pore-water pressure component of hydraulic head h_w. The equation must be solved iteratively with the coefficient of permeability set to constant values that are consistent with the computed pore-water pressures.

The coefficient of permeability at each node, k_{wy}, is assumed to be equal to the saturated coefficient of permeability, k_s, for the first iteration. The computed hydraulic heads, and subsequently the negative pore-water pressures, are used to revise the coefficients of permeability for the second iteration. The iterative procedure can be repeated until there is convergence with respect to the hydraulic heads and the coefficients of permeability. When computing the hydraulic head at the ground surface, the $k_{wy(n+1)}$ value can be assumed to be equal to the $k_{wy(n)}$ value.

Typical distributions of pore-water pressure and hydraulic head during steady-state infiltration are illustrated in Fig. 8.36. The nonlinear distribution of the pore-water pressure and hydraulic head is produced by the nonlinearity of the differential seepage equation. The equipotential lines are not uniformly distributed along the soil column. The above analysis is also applicable to steady-state, upward flow (e.g., evaporation from ground surface), where the flux q_{wy} is known.

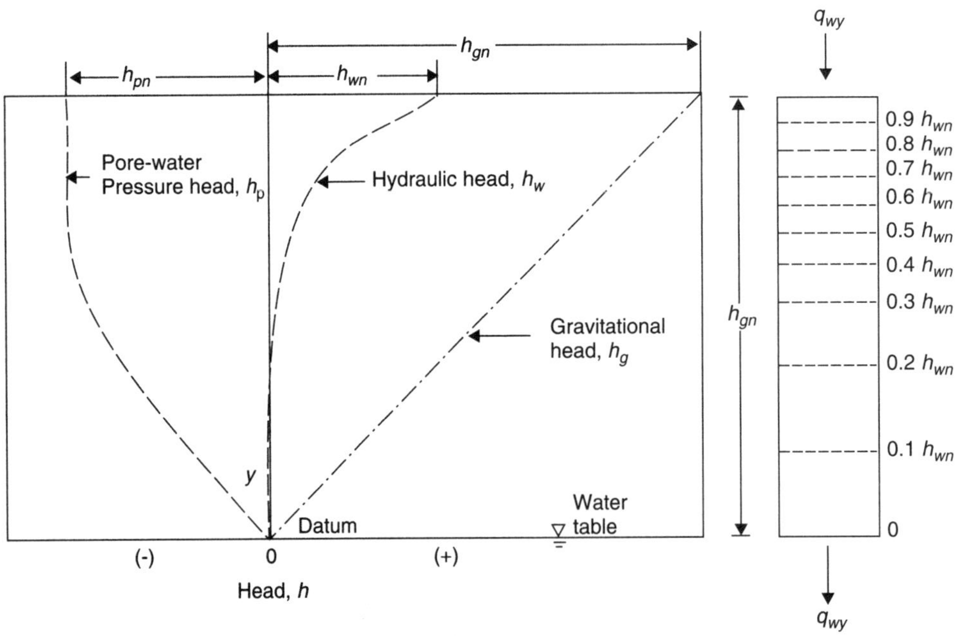

Figure 8.36 Steady-state infiltration through an unsaturated soil with a designated head at the ground surface.

In the case of a *heterogeneous*, *saturated* soil, the coefficients of permeability can be replaced by k_s in Eq. 8.41:

$$h_{w(i)} = h_{w(i-1)} - \left\{ \frac{4k_{s(i)} + k_{s(i+1)} - k_{s(i-1)}}{8k_{s(i)}^2} \right\} \frac{2\,\Delta y}{A} q_{wy} \tag{8.42}$$

Equation 8.42 becomes linear when the soil is homogeneous:

$$h_{w(i)} = h_{w(i-1)} - \frac{\Delta y}{Ak_s} q_{wy} \tag{8.43}$$

Equation 8.43 specializes to a linear distribution of the hydraulic head for a homogeneous, saturated soil column subjected to one-dimensional steady-state flow.

8.3.4 Solution of Two-Dimensional Water Flow Problems

Two-dimensional steady-state flow through a homogeneous, isotropic saturated soil must have the x- and y-direction components added together. The result is a partial differential equation referred to as the Laplacian equation. The solution of this equation describes the head at all points in a soil mass. The solution can be obtained using closed-form analytical methods, analog methods, or numerical methods. Historically, a graphical method referred to as drawing a *flownet* has been used to solve the Laplacian equation (Casagrande, 1937).

The flownet solution results in two families of curves referred to as flow lines and equipotential lines. The flownet solution has been used extensively to analyze problems involving seepage through saturated soils and is explained in most soil mechanics textbooks. Boundary conditions for the soil domain must be known prior to the construction of the flownet. Either a head or zero flux is prescribed along the boundary. A boundary condition exception is the case of a free surface. A network of flow lines and equipotential lines is sketched by trial and error in order to satisfy the boundary conditions and the requirement of right-angled, equal-dimensional elements.

A hydraulic head boundary condition or an impermeable boundary condition can readily be imposed for most saturated soil problems. For example, steady-state seepage beneath a sheet pile wall has the boundary conditions shown in Fig. 8.37a. However, the boundary conditions can be more difficult to assign when dealing with unsaturated soils.

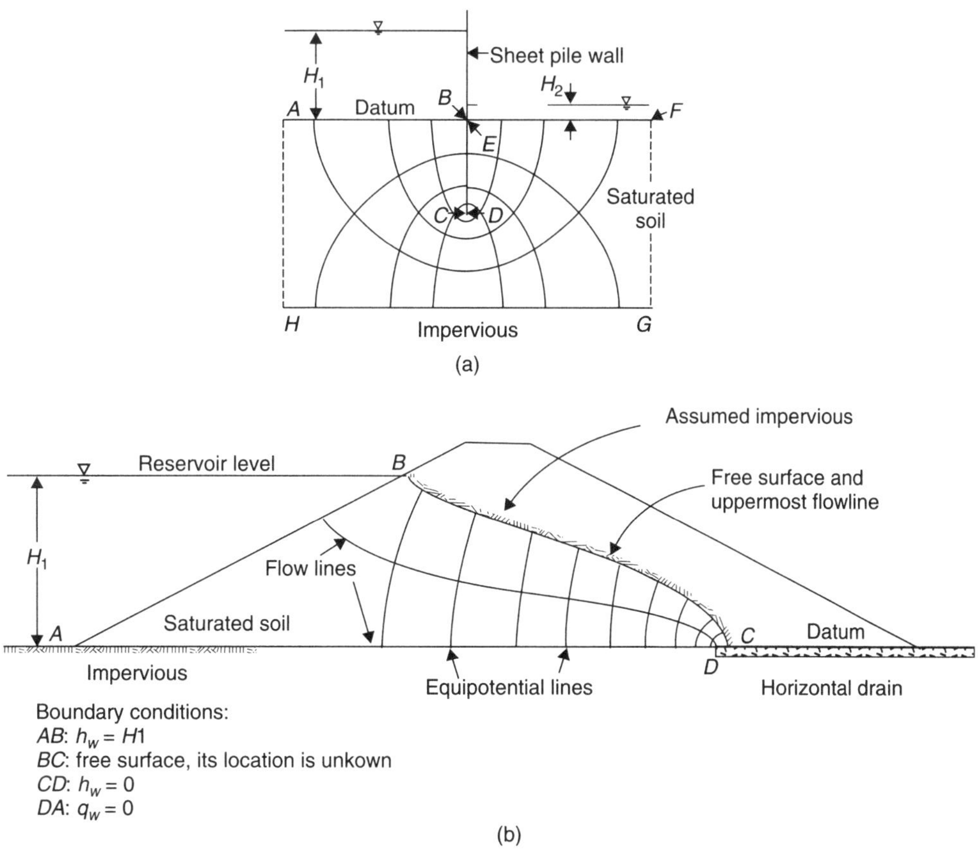

Figure 8.37 Flownet constructions to solve the Laplacian partial differential equation: (a) steady-state seepage throughout homogeneous, isotropic saturated soil, (b) steady-state seepage throughout homogeneous, isotropic earth dam.

Let us consider steady-state seepage through an earth dam (Fig. 8.37b). In early soil mechanics, the assumption was generally made that there was negligible water flow through the unsaturated soil zone due to its low coefficient of permeability. The phreatic line was assumed to behave as an impervious, uppermost boundary when constructing a flownet. This uppermost boundary (i.e., line BC in (Fig. 8.37b) is considered to be not only a phreatic line but also an uppermost flow line. The uppermost boundary was referred to as a free surface under these special conditions (Freeze and Cherry, 1979). However, the position of the free surface was unknown and its location needed to be approximated prior to constructing the flownet.

The position of the free surface was usually determined using an empirical procedure (Casagrande, 1937). The assumption that the free surface is a phreatic line requires that the pore-water pressures be zero along this line. Equipotential lines must intersect the free surface (which is also an uppermost flow line) at right angles. In other words, it was assumed that there was no flow across the free surface. The *flownet* could then be drawn.

The flownet technique was mainly suitable for analyzing steady-state seepage through isotropic, homogeneous, saturated soils. The flownet technique becomes complex and difficult to use when analyzing anisotropic, heterogeneous soil systems. There is an inherent problem associated with applying the flownet technique to saturated-unsaturated flow. Freeze (1971) stated that "the boundary conditions that are satisfied on the free surface specify that the pressure head must be atmospheric and the surface must be a streamline. Whereas the first of these conditions is true, the second is not." The incorrect assumption regarding the uppermost boundary condition can be avoided by recognizing that there can be water flow between the saturated and unsaturated zones (Freeze, 1971; Papagiannakis and Fredlund, 1984).

Steady-state flow in the saturated and unsaturated zones can be solved simultaneously using the general partial differential equation governing water flow through saturated-unsaturated soils. Both zones are treated as a single domain. The water coefficient of permeability in the saturated zone is equal to k_s. The water coefficient of permeability k_w varies with respect to the matric suction in the unsaturated zone. The flownet technique is not applicable to saturated-unsaturated flow modeling since the governing flow equation is not of the Laplacian form. The general flow equation can be solved using a numerical technique such as the finite difference method or the finite element method. Figure 8.38 shows several typical solutions by Freeze (1971) involving saturated-unsaturated flow modeling.

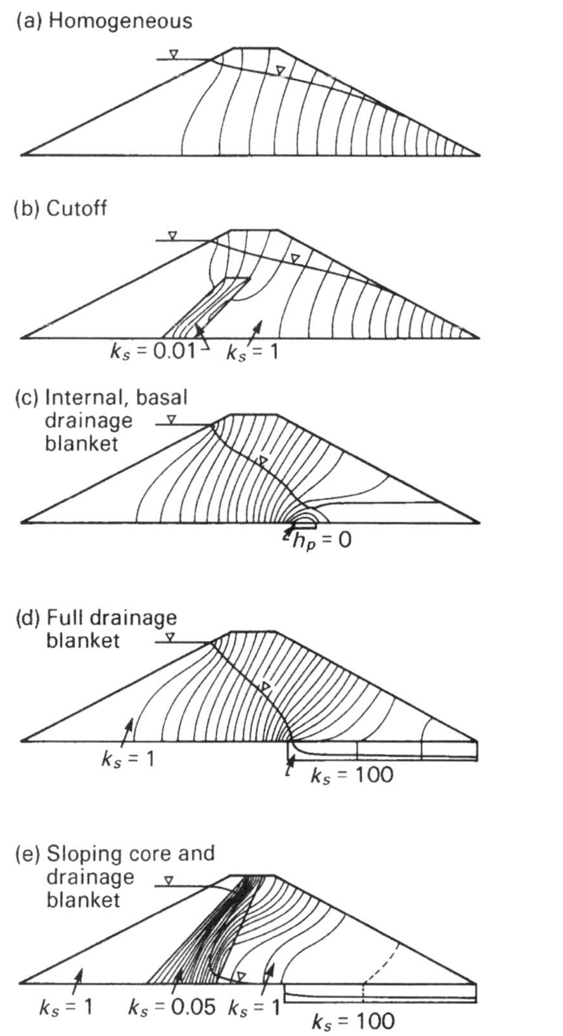

Figure 8.38 Typical solutions for saturated-unsaturated flow modeling of various dam sections (after Freeze, 1971).

8.3.5 Steady-State Seepage Analysis Using Finite Element Method

The application of the finite element method requires the discretization of the soil mass into elements. Triangular and quadrilateral shapes of elements are commonly used for two-dimensional problems. Figure 8.39 shows the cross section of a dam that has been subdivided into triangular elements. The lines separating the elements intersect at nodal points. The hydraulic head at each nodal point is obtained by solving the governing flow equation for the applied boundary conditions.

The finite element formulation for steady-state seepage in two dimensions has been derived using the Galerkin principle of weighted residuals (Papagiannakis and Fredlund, 1984):

$$\int_A \begin{bmatrix} \dfrac{\partial}{\partial x}\{L\} \\ \dfrac{\partial}{\partial y}\{L\} \end{bmatrix}^{\mathrm{T}} \begin{bmatrix} k_{wx} & 0 \\ 0 & k_{wy} \end{bmatrix} \times \begin{bmatrix} \dfrac{\partial}{\partial x}\{L\} \\ \dfrac{\partial}{\partial y}\{L\} \end{bmatrix} dA \{h_{wn}\} \int_S \{L\}^{\mathrm{T}} \bar{v}_w \, dS_p = 0 \qquad (8.44)$$

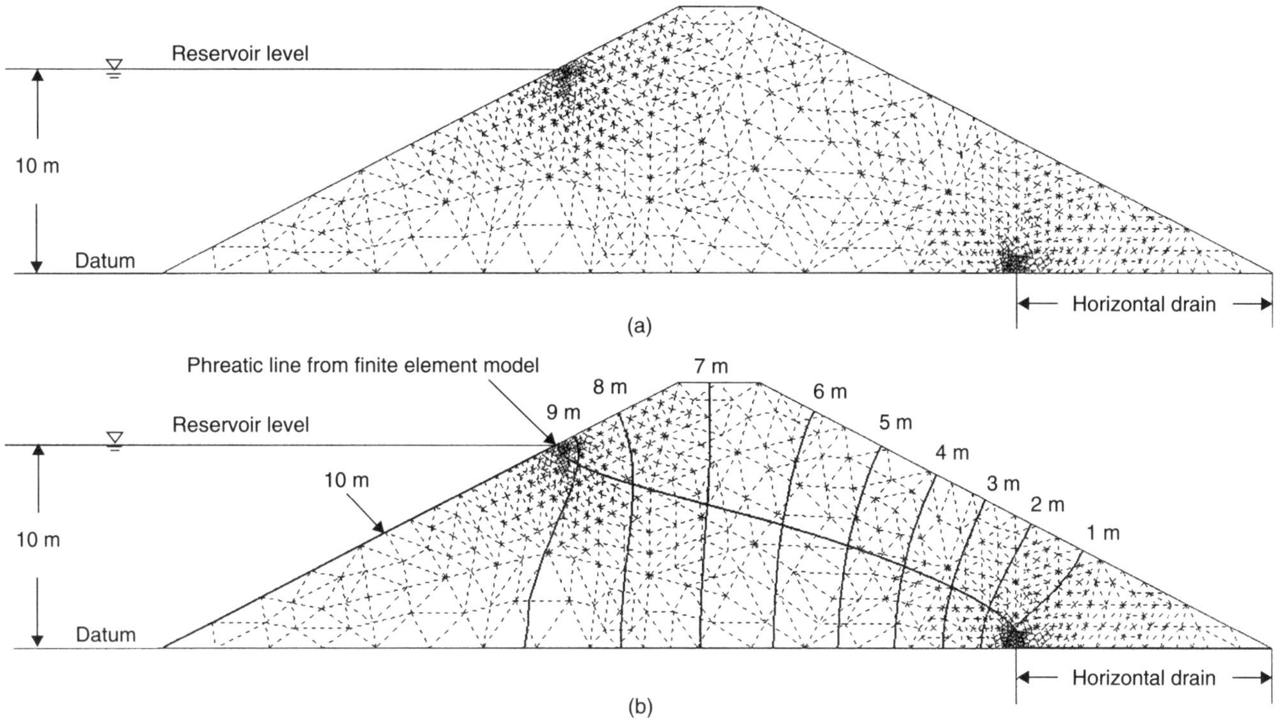

Figure 8.39 Discretization of cross section of earthfill dam for finite element analysis. (a) Optimized finite element mesh. (b) Finite element mesh with superimposed hydraulic head solution.

where:

$\{L\}$ = matrix of the element area coordinates (i.e., $\{L_1 L_2 L_3\}$) and
L_1, L_2, L_3 = area coordinates of points in the element.

The area coordinates of points in the element are related to the Cartesian coordinates of nodal points as follows (Fig. 8.40):

$$L_1 = 1/2A\left\{\left(x_2y_3 - x_3y_2\right) + \left(y_2 - y_3\right)x + \left(x_3 - x_2\right)y\right\}$$
$$L_2 = 1/2A\left\{\left(x_3y_1 - x_1y_3\right) + \left(y_3 - y_1\right)x + \left(x_1 - x_3\right)y\right\}$$
$$L_3 = 1/2A\left\{\left(x_1y_2 - x_2y_1\right) + \left(y_1 - y_2\right)x + \left(x_2 - x_1\right)y\right\}$$

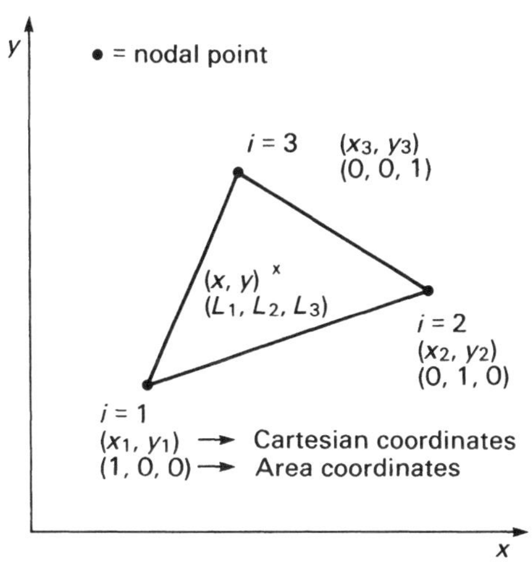

Figure 8.40 Area coordinates in relation to the Cartesian coordinates for a triangular element.

where:

x_i, y_i $(i = 1, 2, 3)$ = Cartesian coordinates of the three nodal points of an element,
x, y = Cartesian coordinates of a point within the element,
A = area of the element,
$\begin{bmatrix} k_{wx} & 0 \\ 0 & k_{wy} \end{bmatrix}$ = matrix of the water coefficients of permeability (i.e., $[k_w]$),
$\{h_{wn}\}$ = matrix of hydraulic heads at the nodal points, that is, $\begin{Bmatrix} h_{w1} \\ h_{w2} \\ h_{w3} \end{Bmatrix}$,
$\bar{v}_w$ = external water flow rate in a direction perpendicular to the boundary of the element, and
S_p = perimeter of the element.

A simplified form of Eq. 8.44 can be written as follows:

$$\int_A [B]^T [k_w] [B]\, dA \{h_{wn}\} - \int_S [L]^T \bar{v}_w\, dS_p = 0 \quad (8.45)$$

where, $[B]$, the matrix of the derivatives of the area coordinates can be written as

$$\frac{1}{2A}\left\{\begin{array}{ccc} y_2 - y_3 & y_3 - y_1 & y_1 - y_2 \\ x_3 - x_2 & x_1 - x_3 & x_2 - x_1 \end{array}\right\} \tag{8.46}$$

Either a hydraulic head or a flow rate must be specified at boundary nodal points. Specified hydraulic heads at the boundary nodes are called Dirichlet boundary conditions. A specified flow rate across the boundary is referred to as a Neumann boundary condition. The second term in Eq. 8.45 accounts for the specified flow rate measured in a direction normal to the boundary. For example, a specified flow rate v_w in the vertical direction must be converted to a normal flow rate $\bar{v}_w$, as illustrated in Fig. 8.41. The normal flow rate is converted to a nodal flow Q_w (Segerlind, 1984). Figure 8.41 shows the computation of the nodal flows Q_{wi} and Q_{wj} at the boundary nodes i and j, respectively. A positive nodal flow signifies that there is infiltration at the node or that the node acts as a "source." A negative nodal flow indicates evaporation or evapotranspiration at the node and that the node acts as a "sink." When the flow rate across a boundary is zero (e.g., impervious boundary), the second term in Eq. 8.45 disappears.

The finite element equation 8.45 can be written for each element and assembled to form a set of global flow equations. The global equations are solved while satisfying compatibility at each node (Desai and Abel, 1972). Nodal compatibility requires that a particular node shared by the surrounding elements must have the same hydraulic head in all of the elements (Zienkiewicz, 1971).

Equation 8.45 is nonlinear because the coefficients of permeability are a function of matric suction, which is related to the hydraulic head at each nodal point, $\{h_{wn}\}$. The hydraulic heads are unknown variables in Eq. 8.45 which are solved by using an iterative procedure. The coefficient of permeability within an element is set to a value depending upon the average matric suction at the nodal points. In this way, the global flow equations are linearized and can be solved simultaneously using a Gaussian elimination technique. The computed hydraulic head at each nodal point is again averaged to determine a new coefficient of permeability from the permeability function $k_w(u_a - u_w)$. The above steps are repeated until the hydraulic heads and the coefficients of permeability no longer change by a significant amount.

The hydraulic head gradients in the x- and y-directions can be computed for an element by taking the derivative of the element hydraulic heads with respect to x and y, respectively:

$$\left\{\begin{array}{c} i_x \\ i_y \end{array}\right\} = [B]\{h_{wn}\} \tag{8.47}$$

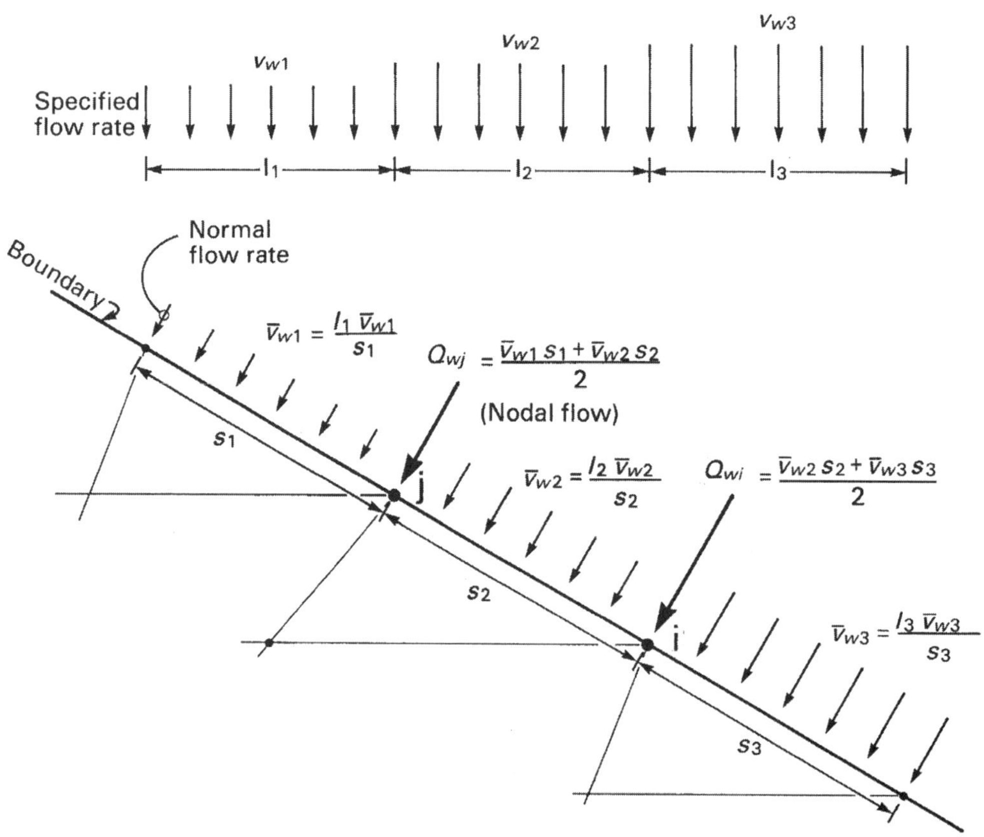

Figure 8.41 Vertical rainfall converted to applied moisture flow rate across sloping boundary expressed as nodal flows.

where:

i_x, i_y = hydraulic head gradient within an element in the x- and y-directions, respectively.

The element flow rates v_w can be calculated from the hydraulic head gradients and the coefficients of permeability in accordance with Darcy's law:

$$\left\{ \begin{array}{c} v_{wx} \\ v_{wy} \end{array} \right\} = [k_w][B][h_{wn}] \tag{8.48}$$

where:

v_{wx}, v_{wy} = water flow rates within an element in the x- and y-directions, respectively.

The hydraulic head gradient and the flow rate at nodal points are computed by averaging the corresponding quantities from all elements surrounding the node. The weighted average is computed in proportion to the element areas.

8.3.6 Examples of Two-Dimensional, Steady-State Water Flow Problems

The following examples are presented to demonstrate the application of the finite element method to steady-state seepage through saturated-unsaturated soils. Papagiannakis and Fredlund (1984) solved several examples of seepage through a dam using a saturated-unsaturated finite element seepage analysis. The cross section and discretization of the problem are illustrated in Fig. 8.39. The discretization of the soil continuum is designed in an optimum manner. In other words, there are areas where it is advantageous to have smaller elements and there are other areas where larger elements are adequate. The nonlinear soil properties also influence the design of the finite element mesh. An automatic finite element mesh design algorithm has been used for all finite element example problems shown in this book (M.D. Fredlund, 2005). Consequently, each example problem has its own finite element mesh.

Water 10 m high is applied to the upstream of the dam. The permeability function used in the analysis is shown as function A in Fig. 8.42. The saturated coefficient of permeability k_s is 1.0×10^{-7} m/s. The pore-air pressure is assumed to be atmospheric. Therefore, the matric suction values in Fig. 8.42 are numerically equal to the pore-water pressures. The base of the dam is chosen as the datum.

The first example is an isotropic earth dam with a horizontal drainage layer near the downstream toe, as shown in Fig. 8.43. The 10-m-high water on the upstream of the dam gives a 10-m hydraulic head at each node along the upstream face. A zero hydraulic head is specified at nodes along the horizontal drain. Zero nodal flow is specified at nodes along the remaining boundaries. The results of the finite element analysis are presented in Figs. 8.43a and 8.43b.

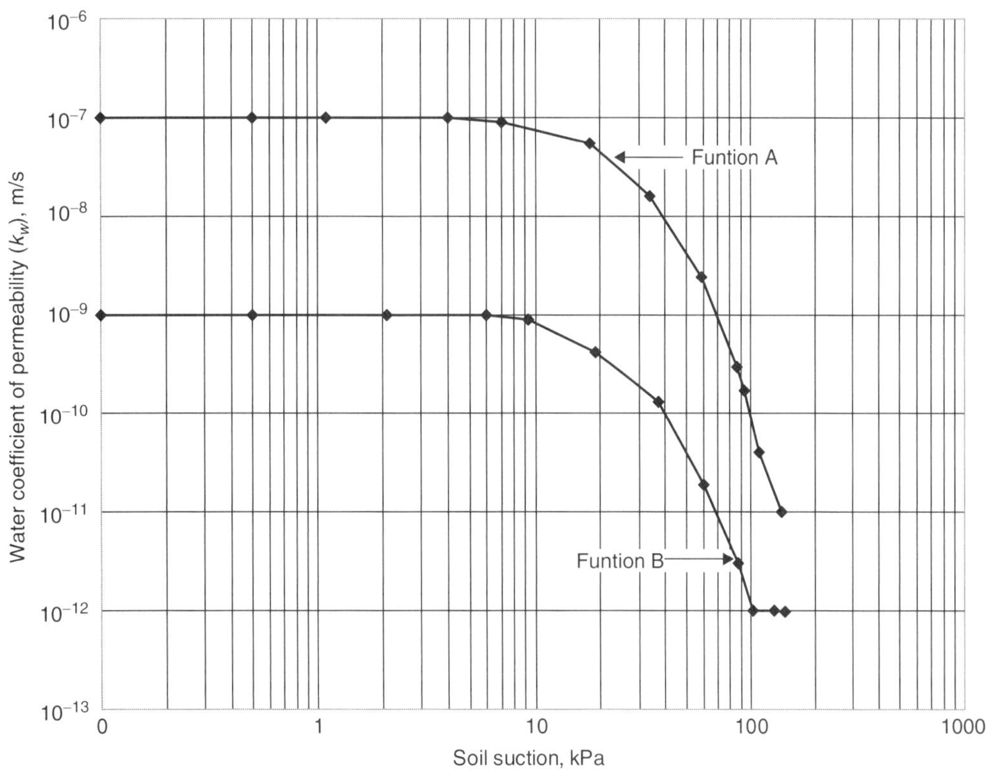

Figure 8.42 Specified permeability functions for analyzing steady-state seepage through earthfill dam.

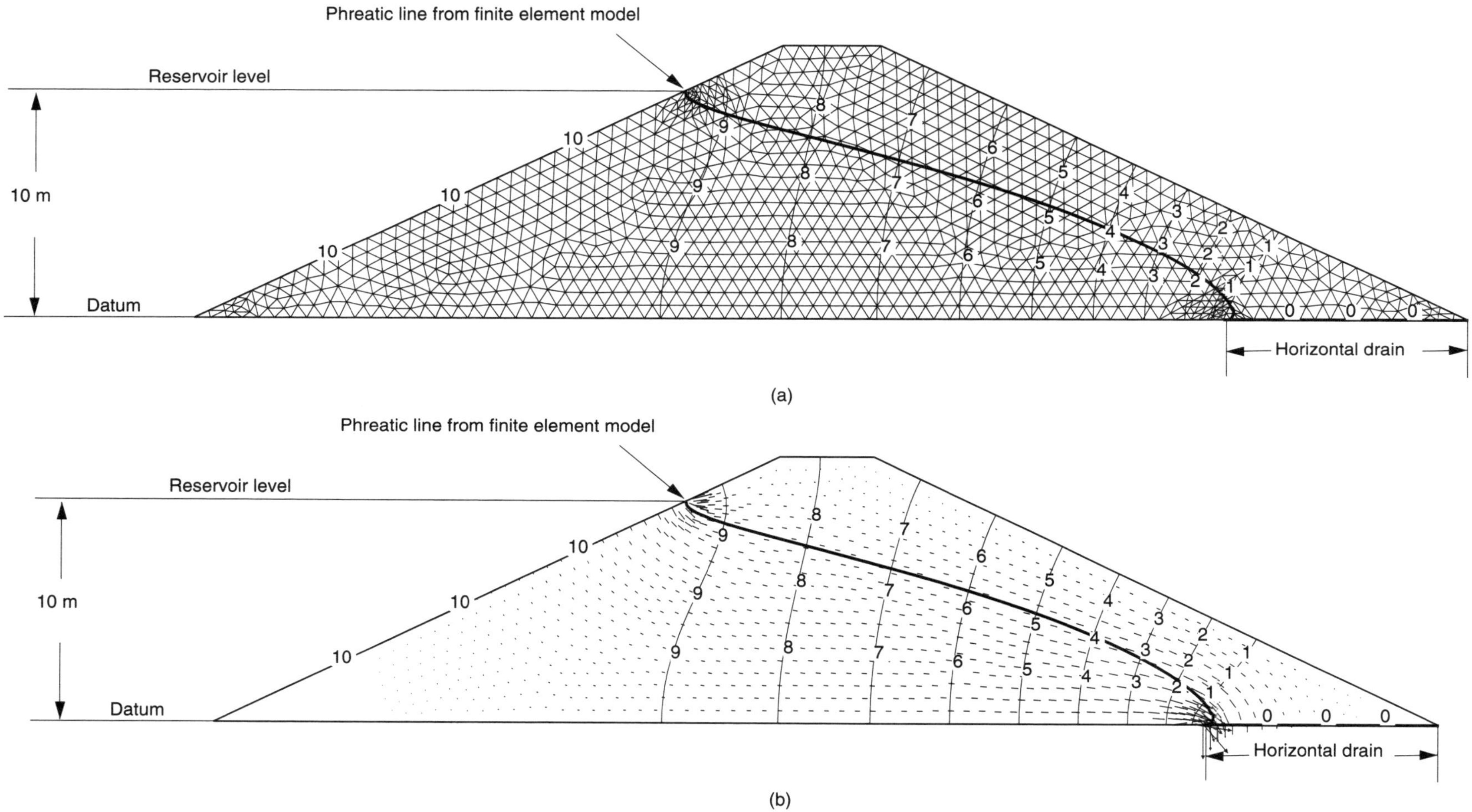

Figure 8.43 Saturated-unsaturated, steady-state seepage through isotropic earthfill dam with horizontal drain: (a) equipotential lines, phreatic line, and finite element mesh in background; (b) equipotential lines, phreatic line, and flow vectors.

The phreatic line resulting from the saturated-unsaturated seepage analysis is in close agreement with the empirical free surface from the conventional flownet construction (Casagrande, 1937). This observation supports the assumption that the empirical free surface is approximately equal to a phreatic line. However, water can flow across the phreatic line, as indicated by the nodal flow rate vectors. Water flow across the phreatic line into the unsaturated zone indicates that the phreatic line is not the uppermost flow line as assumed in the flownet solution technique.

The difference between the phreatic line (from the finite element analysis) and the free surface (from the flownet technique) decreases as the permeability function for the unsaturated zone becomes steeper. A steep permeability function indicates a rapid reduction in the water coefficient of permeability for a small increase in matric suction. In this case, the quantity of water flow into the unsaturated zone is considerably reduced. This condition approaches the assumption associated with the conventional flownet technique.

Equipotential lines extend from the saturated zone through the unsaturated zone, as shown in Fig. 8.43. Changes in hydraulic head between equipotential lines demonstrate that water flows in both the saturated and unsaturated zones. The amount of water flowing in the unsaturated zone depends upon the rate at which the coefficient of permeability changes with respect to matric suction.

The pore-water pressure heads are computed at all nodes throughout the dam. Pore-water pressure heads are computed by subtracting the elevation head from the hydraulic head. Contour lines of equal-pressure heads or isobars are of little value for the interpretation of the computer results for this problem (apart from the phreatic line or zero-pressure line).

The flow of water in the saturated and unsaturated zones is approximately parallel to the phreatic line as observed from the flow rate vectors in the central section of the dam. This is not the situation for sections close to the upstream face and the toe of the dam. Near the upstream face of the dam, water flows across the phreatic line from the saturated to the unsaturated zone and continues to flow in the unsaturated zone. The water in the saturated and unsaturated zones then flows essentially parallel to the phreatic line in the central section of the dam. The water in the saturated zone then flows across the phreatic line into the unsaturated zone at the toe of the dam.

Figure 8.44 shows steady-state seepage through the above dam cross section when the soil is assumed to be anisotropic. The water coefficient of permeability in the horizontal direction is assumed to be six times larger than in the vertical direction (i.e., $k_{wx} = 6k_{wy}$). The anisotropy ratio is assumed to be constant throughout the dam. One permeability function (i.e., function A in Fig. 8.42) is used for the x- and y-directions. The phreatic line is elongated in the direction of the major coefficient of permeability for the anisotropic case (Fig. 8.44). The saturated zone could exit on the downstream face of the dam for higher anisotropy ratios for the horizontal-to-vertical coefficients of permeability.

The third example shows an isotropic earth dam having a core with a lower coefficient of permeability and a horizontal drain, as illustrated in Fig. 8.45. The soil forming the shell of the dam has a saturated coefficient of permeability k_s of 1.0×10^{-7} m/s, and the permeability function is in accordance with function A in Fig. 8.42. The core has a saturated coefficient of permeability k_s of 1.0×10^{-9} m/s and a permeability function in accordance with function B in Fig. 8.42. The boundary conditions used in the analysis are the same as those applied to the previous problems.

The results show that most of the hydraulic head change occurs in the region around the core as depicted by the concentrated distribution of equipotential lines in the core zone. As the difference in the coefficients of permeability between the soil and the core increases, greater hydraulic head changes take place in the core. The nodal flow rate vectors also indicate that a significant amount of water flows upward into the unsaturated zone and over the top of the relatively impermeable core (i.e., the siphon effect), as shown in Fig. 8.45b.

The fourth example demonstrates the effect of a moisture flux boundary condition (i.e., infiltration) placed on the isotropic earth dam (Fig. 8.43). The seepage analysis results are presented in Fig. 8.46. Steady-state infiltration is simulated by applying a positive nodal flow Q_w of 1.0×10^{-8} m^2/s to each of the nodes along the upper boundary of the dam. The results can be compared to the case of zero flux across the upper boundary by comparing Figs. 8.43 and 8.46. Infiltration results in a rise in the phreatic line. Consequently, the pore-water pressures in the unsaturated zone increase relative to the case where there is zero infiltration at the ground surface.

The fifth example demonstrates the development of a seepage face on the downstream of the dam. In this case, there is no horizontal drain and zero nodal flows are specified along the entire lower boundary. There is close agreement between the phreatic line obtained from the finite element analysis and the free surface obtained using the flownet technique (Fig. 8.47). The phreatic line extends to the downstream face of the dam. The phreatic line exits on the downstream face and the portion below the exit point is called the seepage face. The seepage face has a zero pore-water pressure (i.e., atmospheric pressure), boundary condition. In other words, the hydraulic head is equal to the gravitational or elevation head along the seepage face.

The location of the exit point is not known prior to performing the analysis. Therefore, the location of the exit point must be assumed in order to commence the analysis. The exit point is revised following each solution of the finite element solver. The seepage face boundary condition is reevaluated and the solution is repeated.

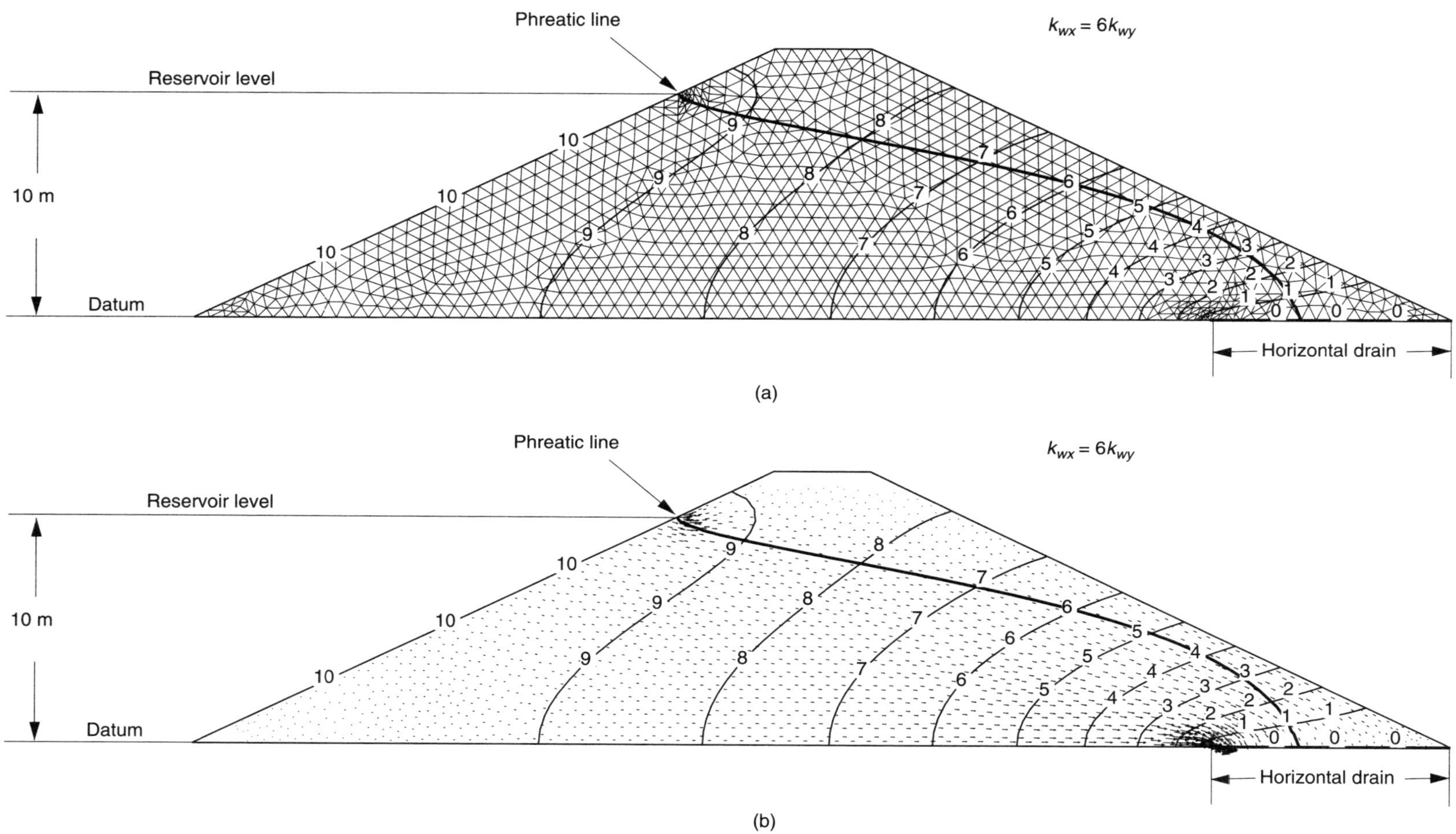

Figure 8.44 Saturated-unsaturated, steady-state seepage through an anisotropic earthfill dam with a horizontal drain: (a) equipotential lines, phreatic line, and finite element mesh in the background; (b) equipotential lines, phreatic line, and flow vectors.

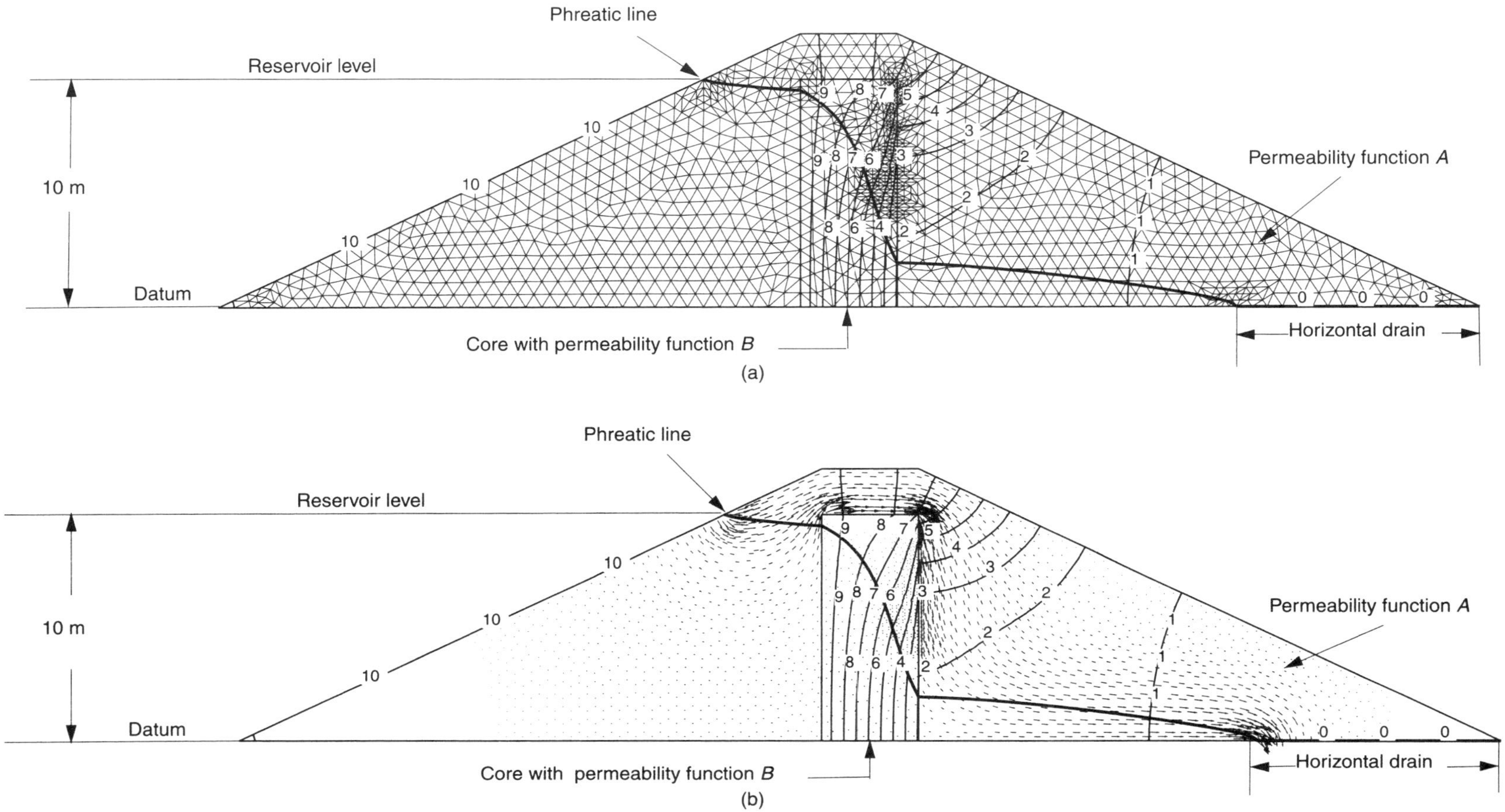

Figure 8.45 Saturated-unsaturated, steady-state seepage through earthfill dam with core and horizontal drain: (a) equipotential lines, phreatic line and finite element mesh in background; (b) equipotential lines, phreatic line, and flow vectors.

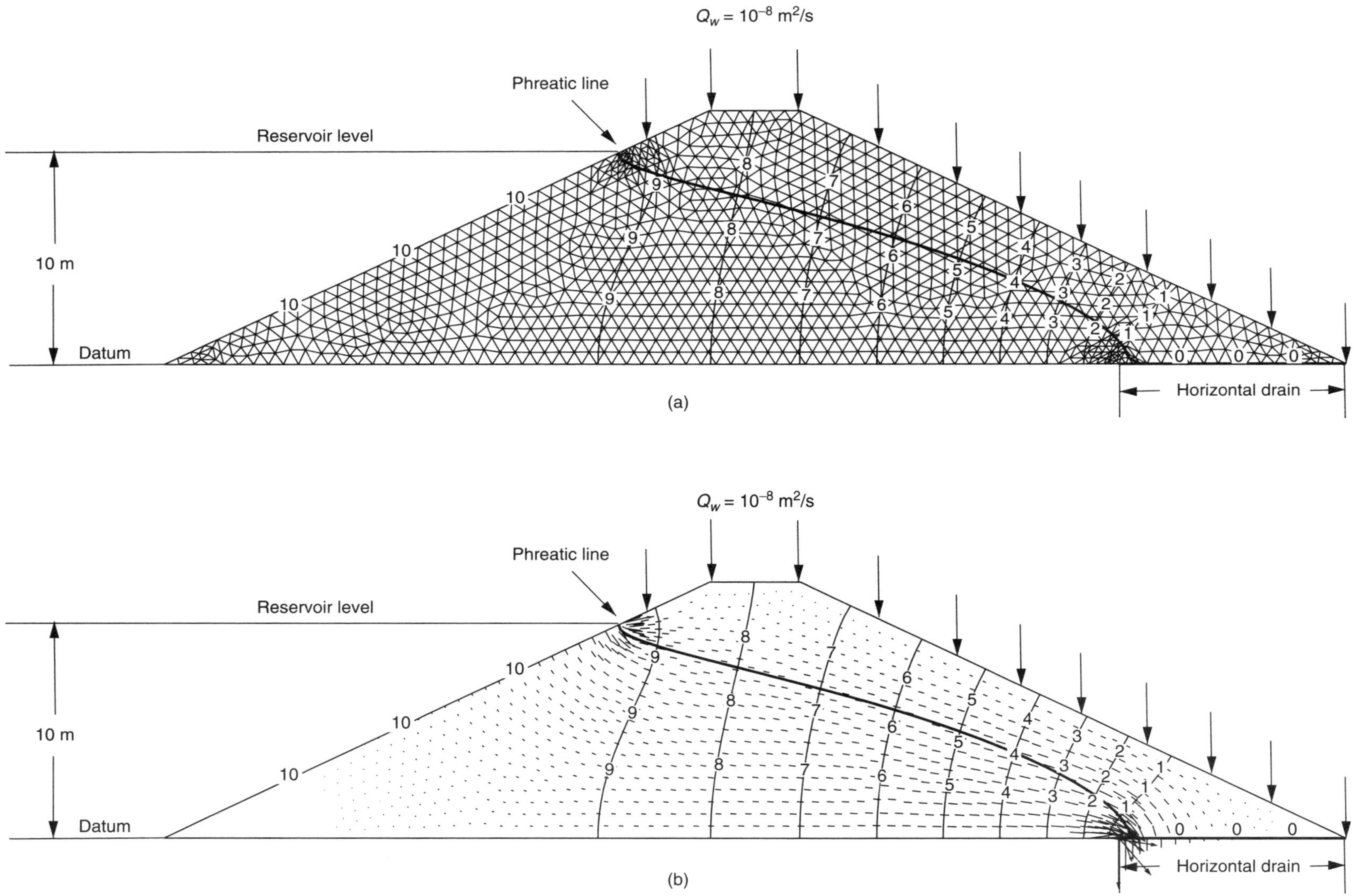

Figure 8.46 Saturated-unsaturated steady- state seepage through isotropic earth dam with horizontal drain under steady-state rainfall at ground surface: (a) equipotential lines, phreatic line, and finite element mesh in background; (b) equipotential lines, phreatic line, and flow vectors.

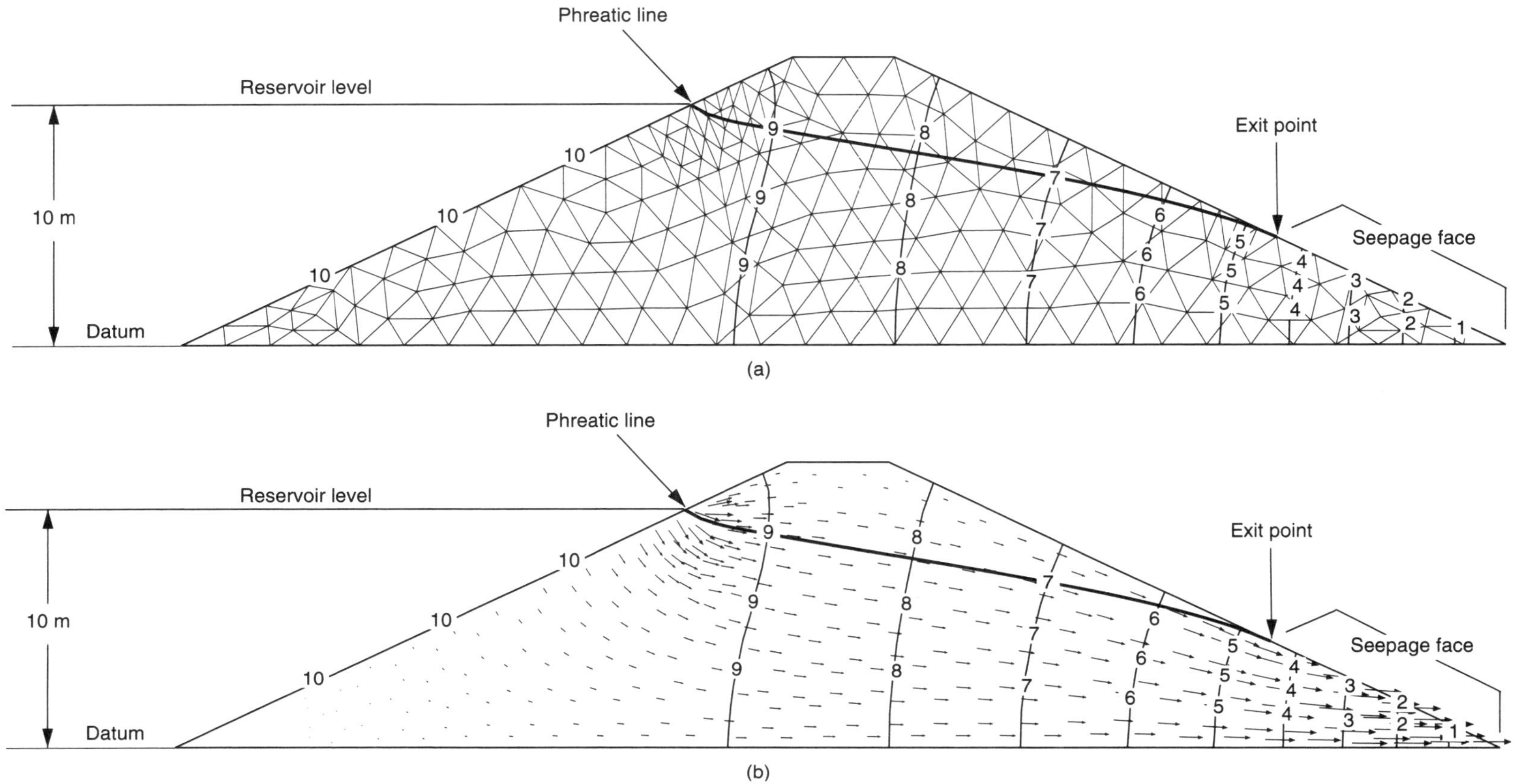

Figure 8.47 Saturated-unsaturated steady-state seepage through isotropic earthfill dam with impervious lower boundary: (a) equipotential lines, phreatic line, and finite element mesh in background; (b) equipotential lines, phreatic line, and flow vectors.

The nodal flows above the assumed exit point are set at zero while iterating toward a solution. The hydraulic heads at or below the assumed exit point are specified as being equal to the gravitational heads. After convergence, the pore-water pressure head at the node directly above the assumed exit point is examined. A negative pore-water pressure head at this point indicates that the assumed exit point is correct. Otherwise, the seepage face boundary needs to be revised by assuming a higher exit point for the phreatic line. The above procedure is repeated until the correct exit point is obtained. The convergence of the solution to the correct exit point on the downstream face adds one more degree of nonlinearity that needs to be satisfied when using a numerical solution for saturated-unsaturated seepage problems.

The above examples deal with seepage through earth dams. However, steady-state analyses can be applied to many other problems involving saturated-unsaturated flow.

8.3.7 Example of Infinite Slope

A slope of infinite length is illustrated in Fig. 8.48. Let us consider the case where steady-state water flow is established within the slope and the phreatic line is parallel to the ground surface. Water flows through both the saturated and unsaturated zones and is parallel to the phreatic line. The direction of the water flow indicates that there is no flow perpendicular to the phreatic line. In other words, the hydraulic head gradient is equal to zero in a direction perpendicular to the phreatic line. In this case, the lines drawn normal to the phreatic line are equipotential lines.

Isobars are parallel to the phreatic line and this condition is similar to that of the central section of a homogenous dam shown earlier. The coefficient of permeability is essentially independent of the pore-water pressure in the saturated zone. Therefore, the saturated zone can be subdivided into several flow channels of equal size. An equal amount of water (i.e., water flux q_w) flows through each channel. Lines separating the flow channels are referred to as flow lines.

The water coefficient of permeability depends on the negative pore-water pressure or the matric suction in the unsaturated zone. The pore-water pressure decreases from zero at the phreatic line to some negative value at ground surface. Similarly, the permeability decreases from the phreatic line to ground surface. Increasingly larger flow channels are therefore required in order to maintain the same quantity of water flow, q_w, as ground surface is approached.

The water flow in each channel is one dimensional, in a direction parallel to the phreatic line. The coefficient of permeability varies in the direction perpendicular to flow. This condition can be compared to the previous case of water flow through a vertical column. In the case of the vertical column, the coefficient of permeability varied in the flow direction and the equipotential lines were not equally distributed throughout the soil column.

The above examples illustrate that equipotential lines and flow lines intersect at right angles for unsaturated flow

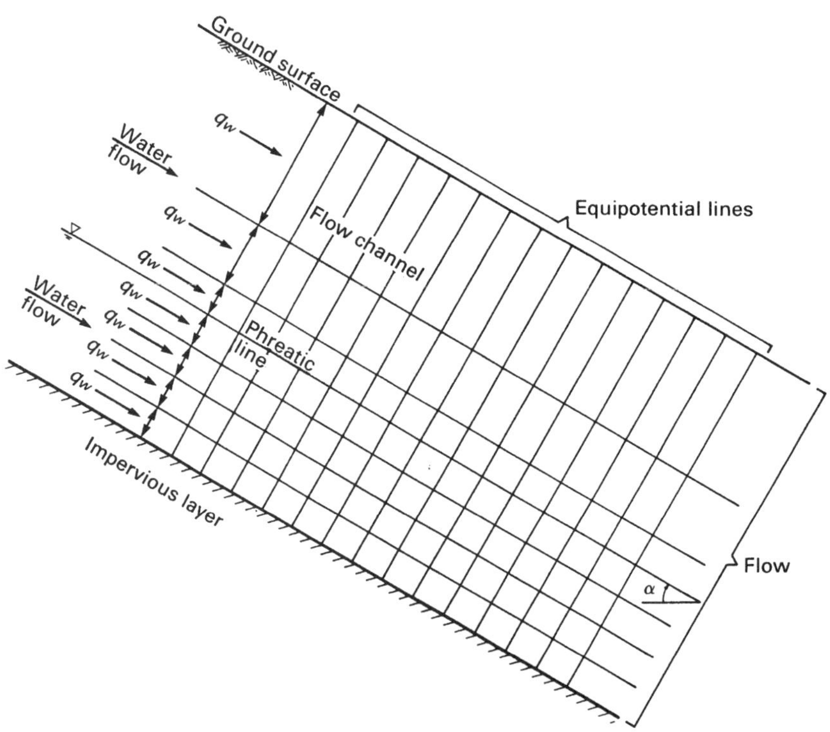

Figure 8.48 Saturated-unsaturated, steady-state seepage through infinite slope with impervious lower boundary.

problems as long as the soil is isotropic. Heterogeneity with respect to the coefficient of permeability results in varying distances between either flow lines or equipotential lines; however, the respective lines must cross at 90°.

The pore-water pressure distribution in the unsaturated zone can be analyzed by considering a horizontal datum through an arbitrary point (e.g., point A in Fig. 8.49) on the phreatic line. The pore-water pressure distribution in a direction perpendicular to the phreatic line (i.e., in the a-direction) is first examined. The results are then used to analyze the pore-water pressure distribution in the y-direction (i.e., vertically). The gravitational head distribution in the a-direction is zero at point A (i.e., datum) and increases linearly to a gravitational head of $H\cos^2\alpha$ at ground surface. The pore-water pressure head at a point in the a-direction must be negative and equal in magnitude to its gravitational head because the hydraulic heads are zero in the a-direction. Therefore, the pore-water pressure head distribution in the a-direction must start at zero at the datum (i.e., point A) and decrease linearly to $-H\cos^2\alpha$ at ground surface. A pore-water pressure head of $-H\cos^2\alpha$ applies to any point along the ground surface since every line parallel to the phreatic line is also an isobar.

The pore-water pressure head distribution in a vertical direction also commences with a zero value at point A and decreases linearly to a head of $-H\cos^2\alpha$ at ground surface. However, the pore-water pressure head is distributed along a length $H\cos\alpha$ in the a-direction, while the head is distributed along a length H in the vertical direction. The negative pore-water pressure head at a point on a vertical plane can therefore be expressed as follows:

$$h_{pi} = -y\cos^2\alpha \tag{8.49}$$

where:

h_{pi} = negative pore-water pressure head on a vertical plane (i.e., the y-direction) for an infinite slope,
y = vertical distance from the point under consideration to the datum (i.e., point A), and
α = inclination angle of the slope and the phreatic line.

When the ground surface and the phreatic line are horizontal (i.e., $\alpha = 0$ or $\cos\alpha = 1$), the negative pore-water pressure head at a point along a vertical plane, h_{ps}, is equal to $-y$. This is the condition of static equilibrium above and below a horizontal water table. The ratio between the pore-water pressure heads on a vertical plane through an infinite slope (i.e., $h_{pi} = -y\cos^2\alpha$) and the pore-water pressure heads associated with a horizontal ground surface (i.e., $h_{ps} = -y$) is plotted in Fig. 8.50. The ratio of pressure heads indicates the reduction in the pore-water pressures on a vertical plane as the slope α becomes steeper.

The gravitational head at a point along a vertical plane is equal to its elevation from the datum, y (Fig. 8.49). The hydraulic head is computed as the sum of the gravitational and pore-water pressure heads:

$$h_{wi} = \left(1 - \cos^2\alpha\right)y \tag{8.50}$$

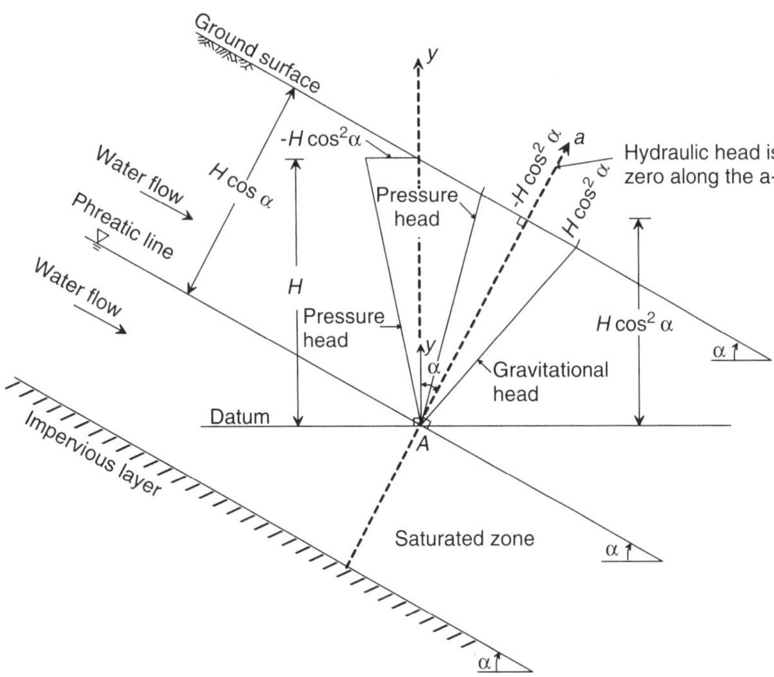

Figure 8.49 Pore-water pressure distribution above phreatic line in unsaturated zone during steady-state seepage through infinite slope.

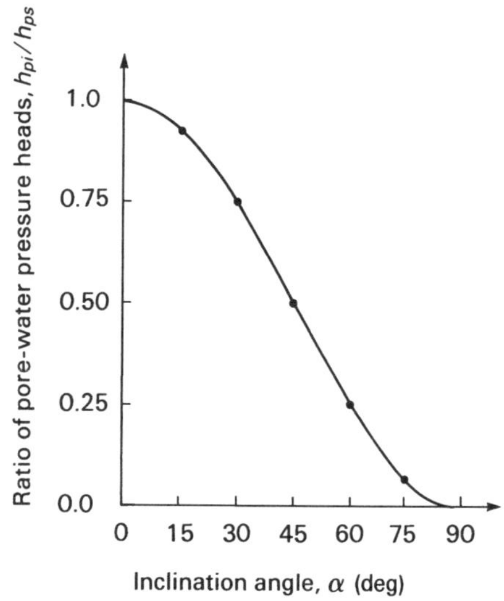

Figure 8.50 Effect of slope inclination on the pore-water pressure distribution along a vertical plane.

The pore-water pressure head equation indicates that there is a decrease in the hydraulic head as the datum is approached. In other words, there is a vertical downward component of water flow. The above analysis also applies to the pore-water pressure conditions below the phreatic line. Using the same horizontal line through point A, positive pore-water pressure heads along a vertical plane can be computed in accordance with Eq. 8.49. The hydraulic head (Eq. 8.50) is zero at the phreatic line and decreases linearly with depth along a vertical plane.

8.3.8 Examples of Two-Dimensional, Unsteady-State Water Flow Problems

Example problems are presented to illustrate unsteady-state seepage solutions when using the finite element method. The finite element formulation for two-dimensional, saturated-unsaturated seepage analyses is first presented followed by the solution of several example problems. Unsteady-state problems require that the permeability function and the water storage function (i.e., k_w and m_2^w) be known for each soil strata that might become desaturated at any elapsed time. Both the permeability function and the water storage function take the form of a nonlinear mathematical function for unsaturated soils.

8.3.9 Unsteady-State Seepage Analysis Using Finite Element Method

The finite element formulation for unsteady-state seepage in two dimensions can be derived using the Galerkin principle of weighted residuals (Lam et al., 1987). The Galerkin solution to the governing seepage equation is given by the following integrals over the area and the boundary surface of a triangular element:

$$\int_A [B]^{\mathrm{T}} [k_w] [B]\, dA \{h_{wn}\} + \int_A [L]^{\mathrm{T}} \lambda [L] dA \frac{\partial \{h_{wn}\}}{\partial t} - \int_S [L]^{\mathrm{T}} \bar{v}_w\, dS = 0 \tag{8.51}$$

where:

$[B]$ = matrix of the derivatives of the area coordinates, as shown for the steady-state formulation,
$[L]$ = matrix of the element area coordinates (i.e., $\{L_1 L_2 L_3\}$), and
$\lambda = \rho_w g m_2^w$.

Boundary conditions must also be defined for an unsteady-state finite element analysis. Either the hydraulic head or the flow rate must be specified at the boundary nodal points. Specified hydraulic heads at the boundary nodes are called Dirichlet boundary conditions. A specified flow rate across the boundary is referred to as a Neuman boundary condition. The third term in Eq. 8.51 (i.e., the terms just before the equal sign) accounts for the specified flow rates across the boundary. The specified flow rates at the boundary must be projected to a direction normal to the boundary. As an example, a specified flow rate v_w in the vertical direction must be converted to a normal flow rate $\bar{v}_w$.

The normal flow rate must in turn be converted to a nodal flow Q_w (Segerlind, 1984). A positive nodal flow signifies that there is infiltration at the node or that the node acts as a source. A negative nodal flow indicates evaporation or evapotranspiration at the node and shows that the node acts as a sink. When the flow rate across a boundary is zero (e.g., impervious boundary), the third term in Eq. 8.51 disappears.

The numerical integration of Eq. 8.51 results in the following expression for the saturated-unsaturated seepage equation:

$$[D]\{h_{wn}\} + [E]\{h_{wn}\} = [F] \tag{8.52}$$

where:

$[D]$ = stiffness matrix, that is, $[B]^{\mathrm{T}} [k_w] [B] A$,
$[E]$ = capacitance matrix, that is, $\dfrac{\lambda A}{12} \begin{bmatrix} 2 & 1 & 1 \\ 1 & 2 & 1 \\ 1 & 1 & 2 \end{bmatrix}$,
$\{h_{wn}\}$ = matrix of the time derivatives of the hydraulic heads at the nodal points (i.e., $\partial \{h_{wn}\} / \partial t$), and
$[F]$ = flux vector reflecting the boundary conditions (i.e., $\int_S [L]^{\mathrm{T}} \bar{v}_w\, dS$).

The time derivative in Eq. 8.52 can be approximated using a finite difference form. The relationship between the nodal heads of an element at two successive time steps Δt can be expressed using the central-difference approximation

$$\left([D] + \frac{2[E]}{\Delta t}\right)\{h_{wn}\}_{t+\Delta t} = \left(\frac{2[E]}{\Delta t} - [D]\right)\{h_{wn}\}_t + 2[F] \tag{8.53}$$

or the backward-difference approximation

$$\left([D] + \frac{[E]}{\Delta t}\right)\{h_{wn}\}_{t+\Delta t} = \frac{[E]}{\Delta t}\{h_{wn}\}_t + [F] \qquad (8.54)$$

The above time derivative approximations are considered to be unconditionally stable. The central-difference approximation generally gives a more accurate solution than that obtained from the backward-difference approximation. However, the backward-difference approximation is found to be more effective in reducing numerical oscillations commonly encountered in highly nonlinear systems of flow equations (Neuman and Witherspoon, 1971; Neuman, 1973).

The finite element seepage equation can be written for each element and assembled to form a set of global flow equations. The finite element formulation is solved while satisfying nodal compatibility (Desai and Abel, 1972). Nodal compatibility requires that a particular node shared by the surrounding elements has the same hydraulic head in all shared elements (Zienkiewicz, 1971; Desai, 1975b).

The global flow equations for the whole system are solved for the hydraulic heads at the nodal points, $\{h_{wn}\}$. However, Eq. 8.52 is nonlinear because the coefficients of permeability are a function of matric suction, which is related to the hydraulic head at the nodal points.

The hydraulic heads are unknown variables in Eq. 8.52. The equation must be solved using an iterative procedure that involves a series of successive approximations. For the first approximation, the coefficients of permeability are estimated in order to calculate the first set of hydraulic heads at the nodal points. The computed hydraulic heads are used to calculate the average matric suction within an element. In the subsequent approximations, the coefficient of permeability is adjusted to a value corresponding to the computed average matric suction in the element. The adjusted permeability value is then used to calculate a new set of nodal point hydraulic heads. The above procedure is repeated until both the hydraulic head and the permeability differences for each element are smaller than a specified tolerance between two successive iterations.

The above iterative procedure allows the global flow equations to be solved using a Gaussian elimination technique. The convergence on hydraulic head and coefficient of permeability is dependent on the degree of nonlinearity of the permeability function and the spatial discretization of the continuum. A steep permeability function generally requires an increased number of iterations to achieve convergence. A finer discretization in both element size and time step assists in obtaining more rapid convergence with a smaller tolerance. The solution often converges to within a tolerance of less than 1% in a few iterations provided the permeability and water storage functions are not too highly nonlinear. The use of PDE solvers with optimized mesh design and automatic mesh refinement is of great value in solving highly nonlinear PDEs. The design of the finite element mesh becomes dynamic, changing as necessary to meet the requirements for convergence to a reliable solution.

The unsteady-state seepage equation is assumed to be solved for one time step once the converged nodal hydraulic heads of the system have been obtained. Having reached convergence at a particular time step, other secondary quantities, such as pore-water pressures, hydraulic head gradients, and water flow rates, can then be calculated using the converged nodal hydraulic heads. The equation for nodal pore-water pressures is

$$\{u_{wn}\} = (\{h_{wn}\} - \{y_n\})\,\rho_w g \qquad (8.55)$$

where:

$\{u_{wn}\}$ = matrix of pore-water pressures at the nodal points $\left(\text{i.e., } \begin{Bmatrix} u_{w1} \\ u_{w2} \\ u_{w3} \end{Bmatrix}\right)$,

$\{y_n\}$ = matrix of elevation heads at the nodal points $\left(\text{i.e., } \begin{Bmatrix} y_1 \\ y_2 \\ y_3 \end{Bmatrix}\right)$.

The hydraulic head gradients in the x- and y-directions can be computed for an element by taking the derivative of the element hydraulic heads with respect to x and y, respectively:

$$\begin{Bmatrix} i_x \\ i_y \end{Bmatrix} = [B]\{h_{wn}\} \qquad (8.56)$$

where:

i_x, i_y = hydraulic head gradient within an element in the x- and y-directions, respectively.

The element flow rates v_w can be calculated from the hydraulic head gradients and the coefficients of permeability in accordance with Darcy's law:

$$\begin{Bmatrix} v_{wx} \\ v_{wy} \end{Bmatrix} = [k_w][B]\{h_{wn}\} \qquad (8.57)$$

where:

v_{wx}, v_{wy} = water flow rates within an element in the x- and y-directions, respectively.

The hydraulic head gradient and the flow rate at nodal points are computed by averaging the corresponding quantities from all elements surrounding the node. The weighted average is computed in proportion to the respective elemental areas.

8.3.10 Example of Water Flow through Earth Dam

The example problem involving water flow through an earthfill dam is first discussed. The soil is assumed to be isotropic with respect to its coefficient of permeability and the permeability function along with the SWCC and the water storage modulus function are shown in Fig. 8.51. The saturated coefficient of permeability k_s is 1.0×10^{-7} m/s. The pore-air pressure is assumed to remain at atmospheric pressure. The base of the dam is selected as the datum for elevation head.

The reservoir is initially assumed to be empty so the water level is at 0 m (i.e., the datum). The water level in the reservoir is then instantaneously raised to a level of 10 m above datum. The water level remains constant at 10 m as transient or unsteady-state seepage takes place in the dam. The rise of the phreatic line from the initial steady-state condition to an elapsed time of 1500 days is computed.

The development of equipotential lines, the phreatic surface, and water flow vectors through the dam are illustrated for four elapsed times during the transient process. Figure 8.52 shows the development of the phreatic lines, the flow vectors, and the equipotential lines after an elapsed time of 25 days. The dynamically generated finite element mesh corresponding to the converged solution is also shown. Similar plots of the phreatic lines, the flow vectors, and the equipotential lines after an elapsed time of 60 days are shown in Fig. 8.53. Likewise, the development of seepage toward steady-state conditions is shown for elapsed times of 120 and 1500 days in Figs. 8.54 and 8.55, respectively.

The increase in the reservoir level results in an increase in pore-water pressures with time. This is demonstrated by the advancement of equipotential lines from upstream to downstream of the dam with elapsed time. It should also be noted that the equipotential lines extend from the saturated through the unsaturated zones. In other words, water flows in both the saturated and the unsaturated zones as a result of the hydraulic head differences between the equipotential lines. The flow of water in both zones can be observed directly from the flow rate vectors that exist in both the saturated and unsaturated zones. The amount of water flowing in the unsaturated zone depends on the rate of change in the coefficient of permeability with respect to matric suction changes.

8.3.11 Example of Groundwater Seepage below Lagoon

The second example problem illustrates unsteady-state groundwater seepage below a lagoon. The lagoon is placed on top of a 1-m-thick soil liner. The geometry of the problem is symmetrical, and the liner and the surrounding soil are assumed to be isotropic with respect to coefficient of permeability. The problem is analyzed by considering half of the geometry. The SWCC is assumed to be the same for the soil liner and the underlying soil and is shown in Fig. 8.56a. The permeability functions for the soil liner and the surrounding soil are shown in Fig. 8.56b. The saturated coefficients of permeability are equal to 5.0×10^{-6} and 1.0×10^{-5} m/s for the liner and the surrounding soil, respectively. A water storage modulus m_2^w of for both soils is shown in Fig. 8.56c.

The discretized cross section of the soil liner and its surrounding soil are depicted in Fig. 8.57. Initially the groundwater table was located 5 m below the ground surface. The lagoon was assumed to be instantaneously filled with water to a depth of 1 m. No-flow boundary conditions were assumed along the ground surface outside the lagoon, the bottom boundary, and the vertical axis of symmetry. On the right-hand boundary, a hydrostatic condition was assumed to exist below the groundwater table. A no-flow condition was assumed to exist above the groundwater table to the ground surface. The exit point of the groundwater table was assumed to be fixed with elapsed time.

The transient modeling process commenced as the lagoon was filled with water to the 1-m height which gave rise to a 1-m pore-water pressure head. The water commenced seeping from the lagoon, causing the groundwater to gradually mound to its final steady-state condition. Figure 8.57 illustrates the transient positions of the water table at an elapsed time of 24 h. Figure 8.57a shows the pressure head contours, including the phreatic line, along with the automatically generated finite element mesh required for solving the problem at 24 h of elapsed time. It can be seen that there is a slight rise in the groundwater table 5 m below the ground surface. At the same time there is a phreatic line that has developed at the 0.5-m depth below ground surface as water moves downward from the lagoon. Figure 8.57b shows the equipotential heads along with the flow vectors corresponding to an elapsed time of 24 h. The water level in the lagoon remains constant at a 1-m height throughout the transient process.

The seepage flow pattern and the development of pore-water pressures in the soil below the lagoon after an elapsed time of 40 h are shown in Figs. 8.58a and 8.58b. At the beginning of the transient process, water moved downward from the lagoon, while the position of the groundwater table showed little effect. With time, the wetting front moves deeper into the soil mass while the groundwater begins to mound upward. After 52 h, the downward wetting front from the lagoon joins the rising groundwater table, as shown in Fig. 8.59. Steady-state conditions are approached after an elapsed time of about 200 h, as shown in Fig. 8.60. The length of time required to approach steady-state seepage conditions is largely controlled by the saturated coefficients of permeability for the soils.

8.3.12 Example of Seepage within Layered Hill Slope

The third example problem illustrates unsteady-state seepage within a layered hill slope under constant-infiltration conditions. Rulon and Freeze (1985) studied this problem using a sandbox model of the layered hill slope. Details of the geometry of the slope and the applied surface moisture flux are

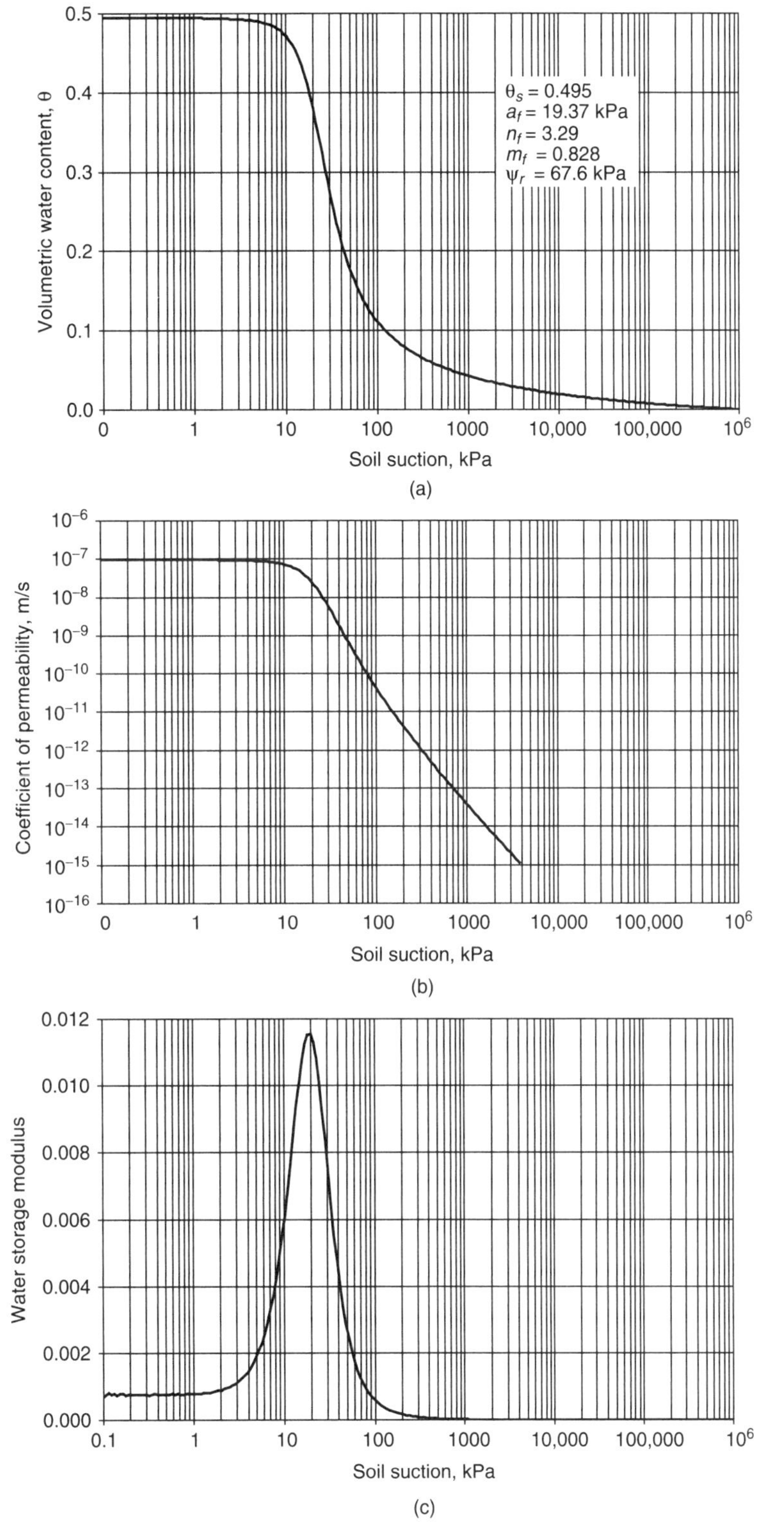

Figure 8.51 Permeability and water storage function for analyzing unsteady-state seepage through an isotropic earthfill dam. (a) SWCC. (b) Permeability function. (c) Water storage function.

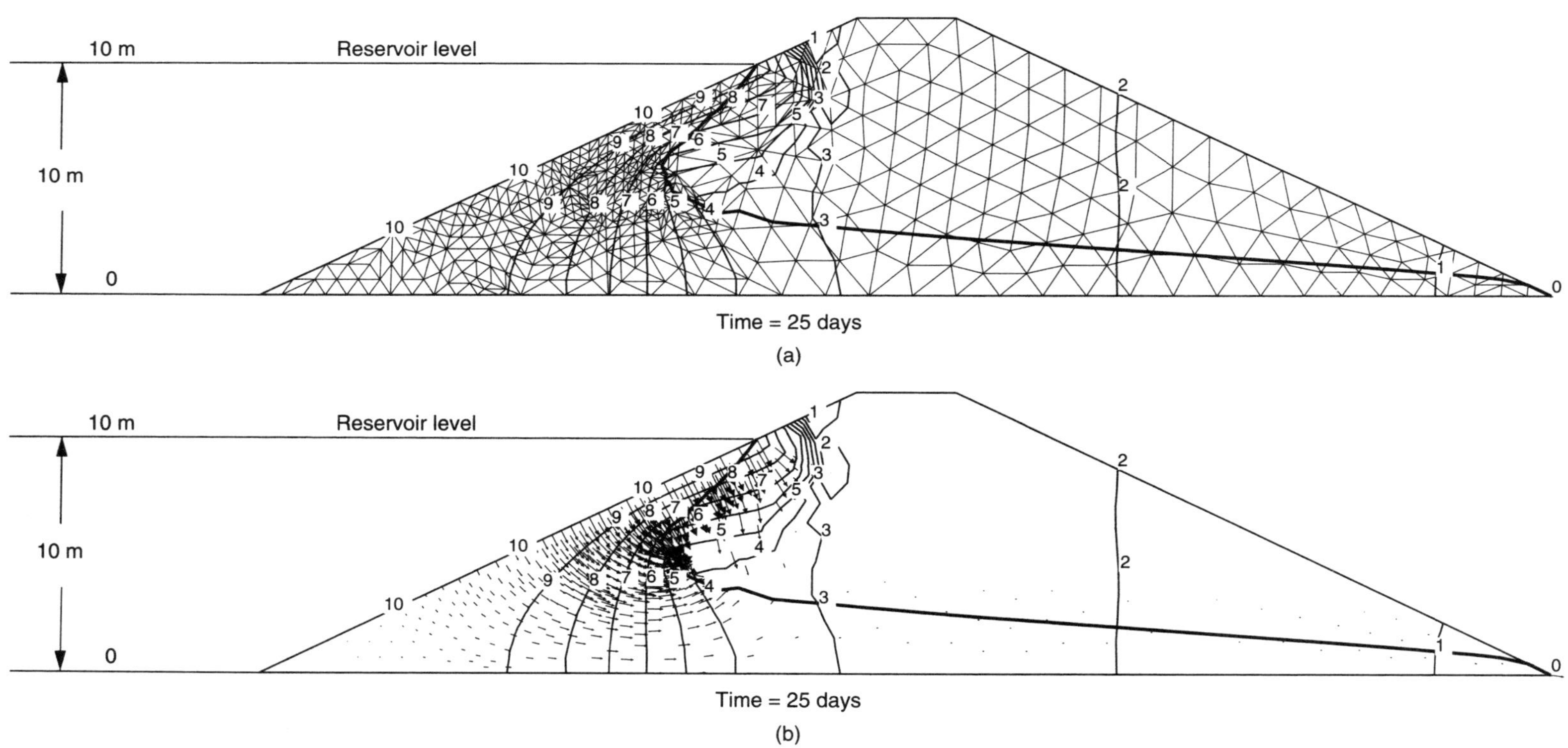

Figure 8.52 Hydraulic heads after an elapsed time of 25 days. (a) Finite element mesh and hydraulic heads. (b) Velocity vectors and hydraulic heads.

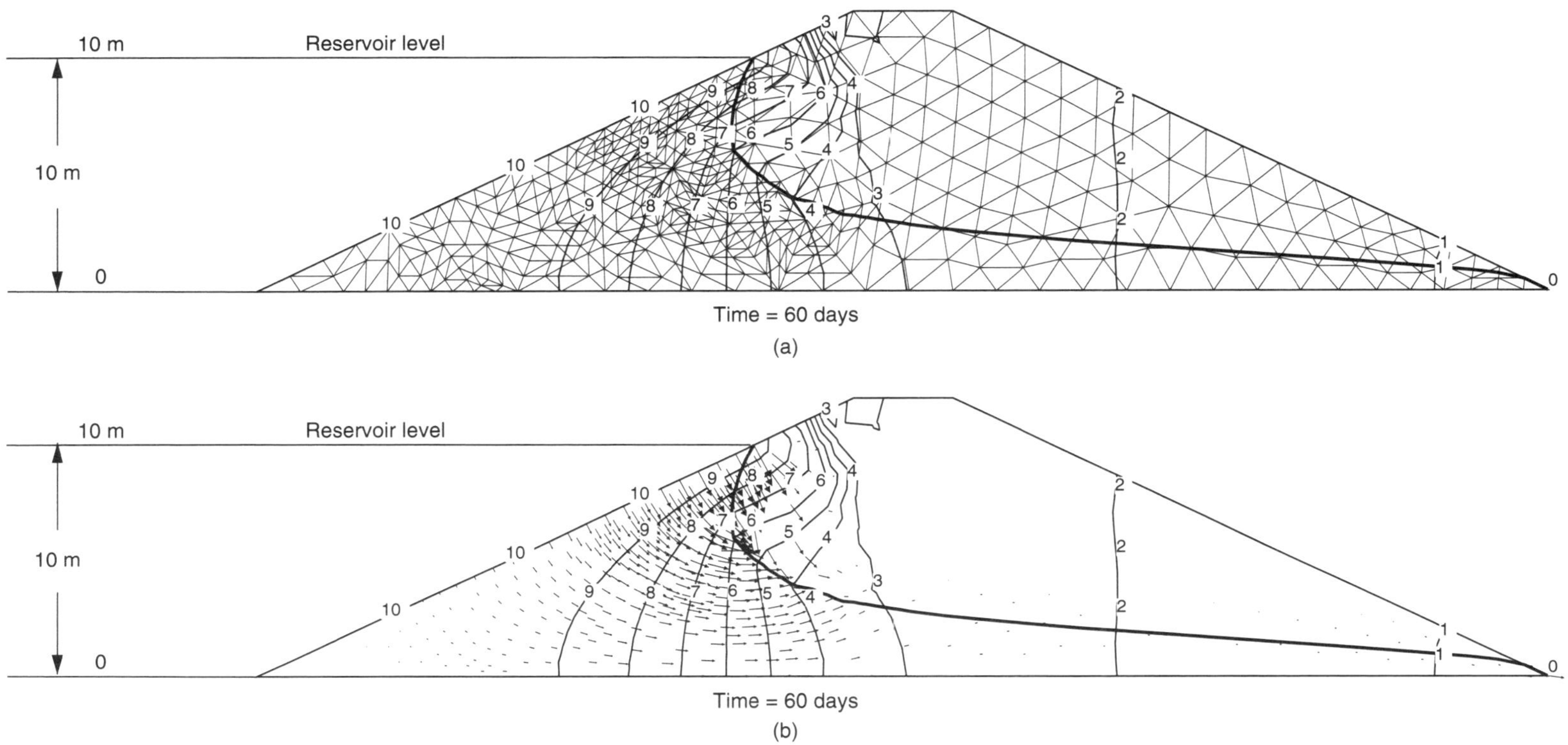

Figure 8.53 Hydraulic heads after an elapsed time of 60 days. (a) Finite element mesh and hydraulic heads. (b) Velocity vectors and hydraulic heads.

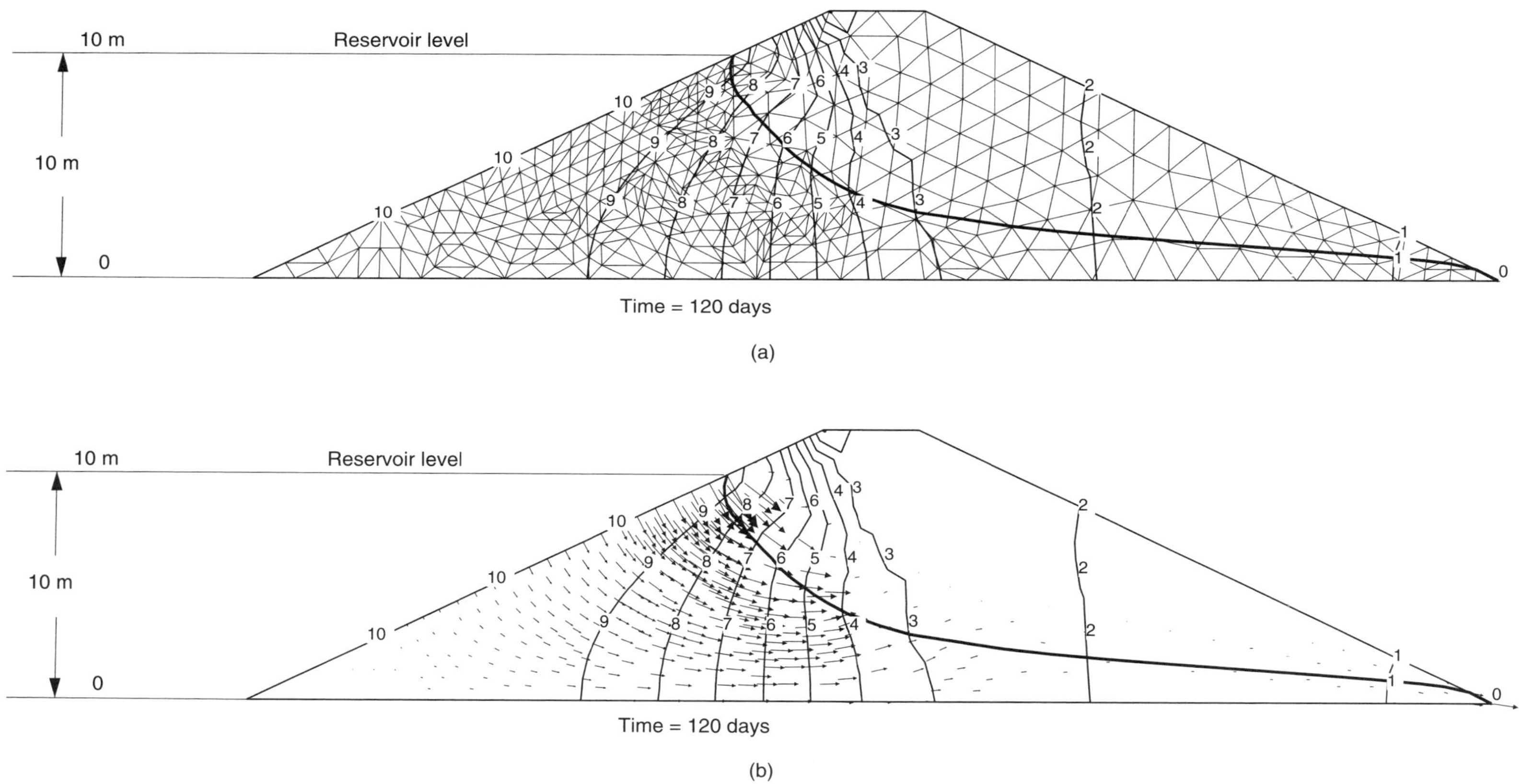

Figure 8.54 Hydraulic heads after an elapsed time of 120 days. (a) Finite element mesh and hydraulic heads. (b) Velocity vectors and hydraulic heads.

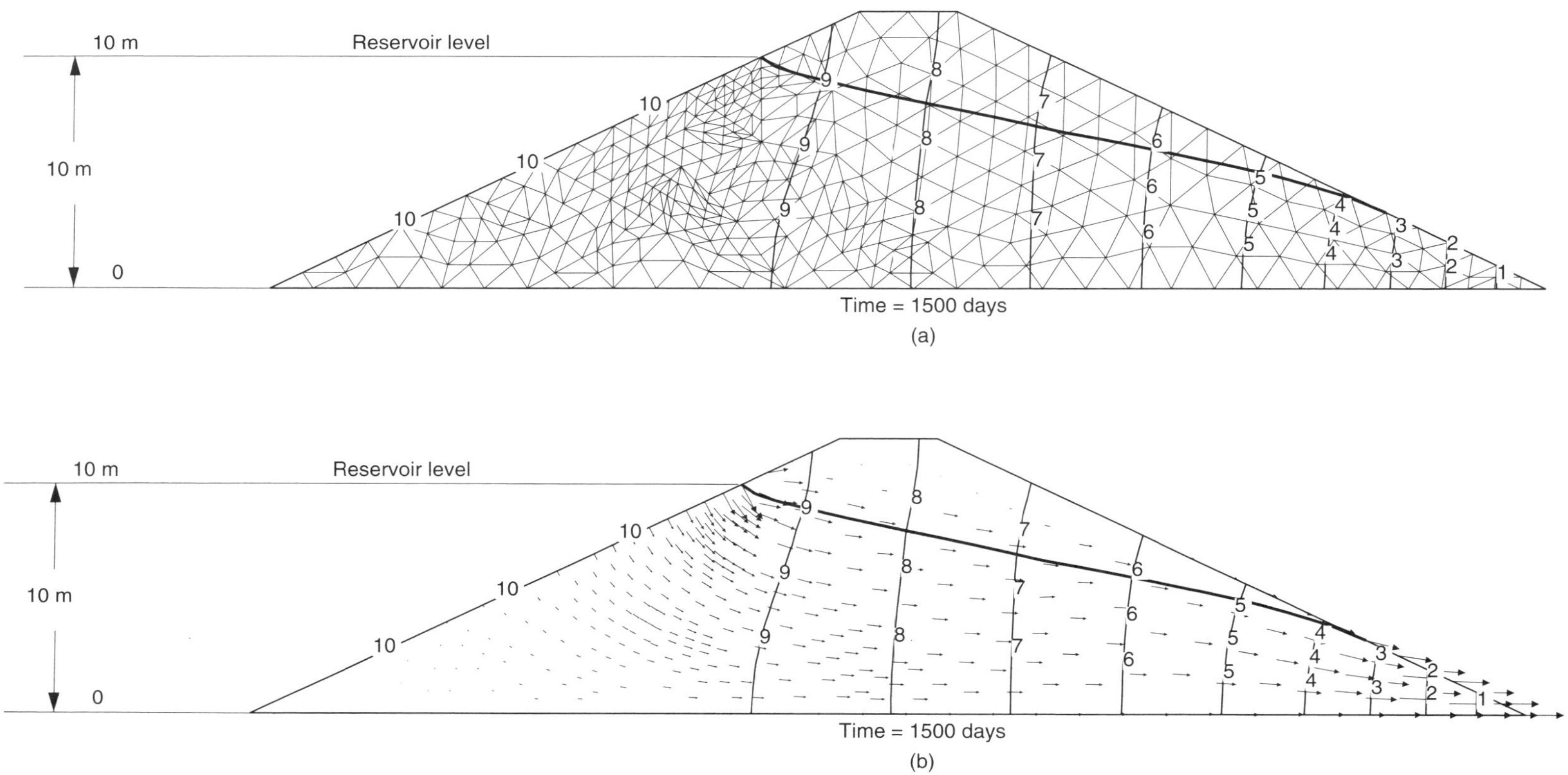

Figure 8.55 Hydraulic heads after an elapsed time of 1500 days. (a) Finite element mesh and hydraulic heads. (b) Velocity vectors and hydraulic heads.

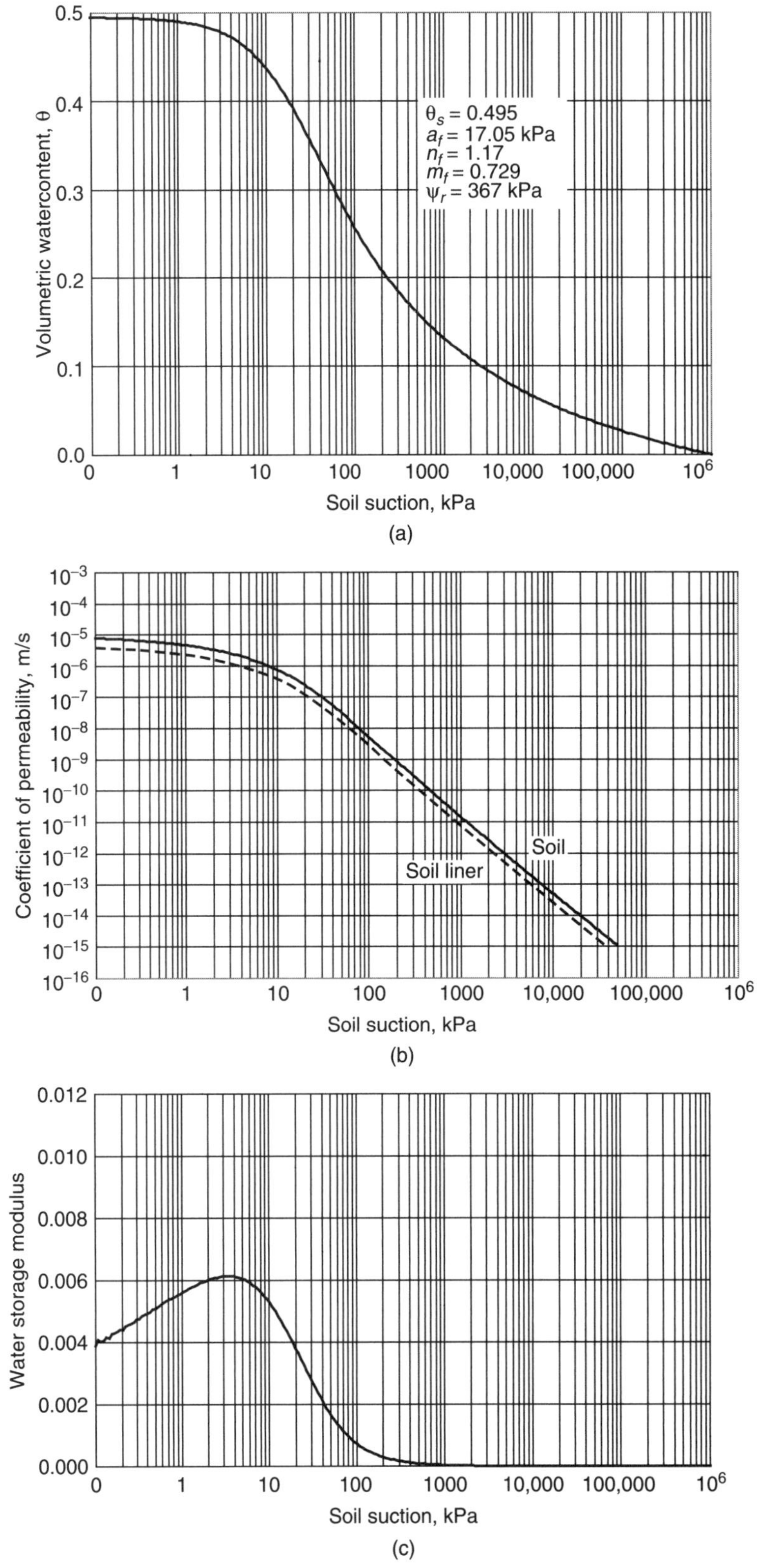

Figure 8.56 Permeability and water storage function for analyzing unsteady-state seepage below a lagoon. (a) SWCC. (b) Permeability function. (c) Water storage function.

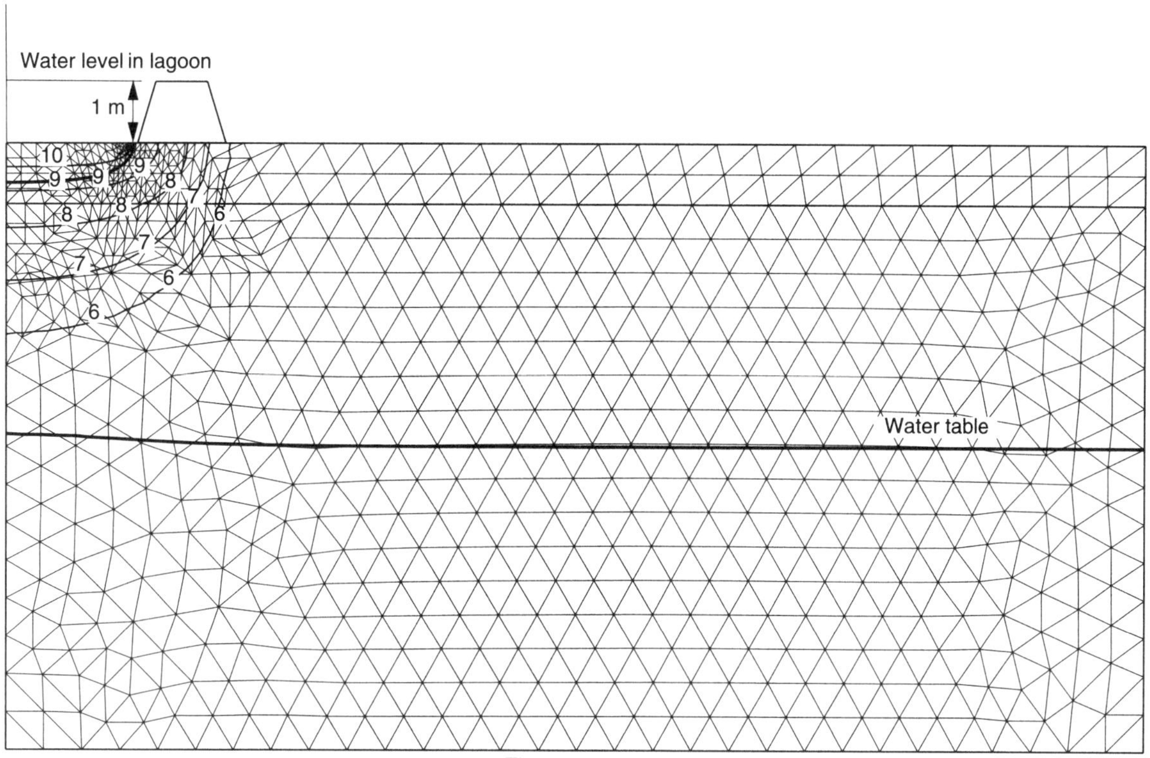

Time = 24 h

(a)

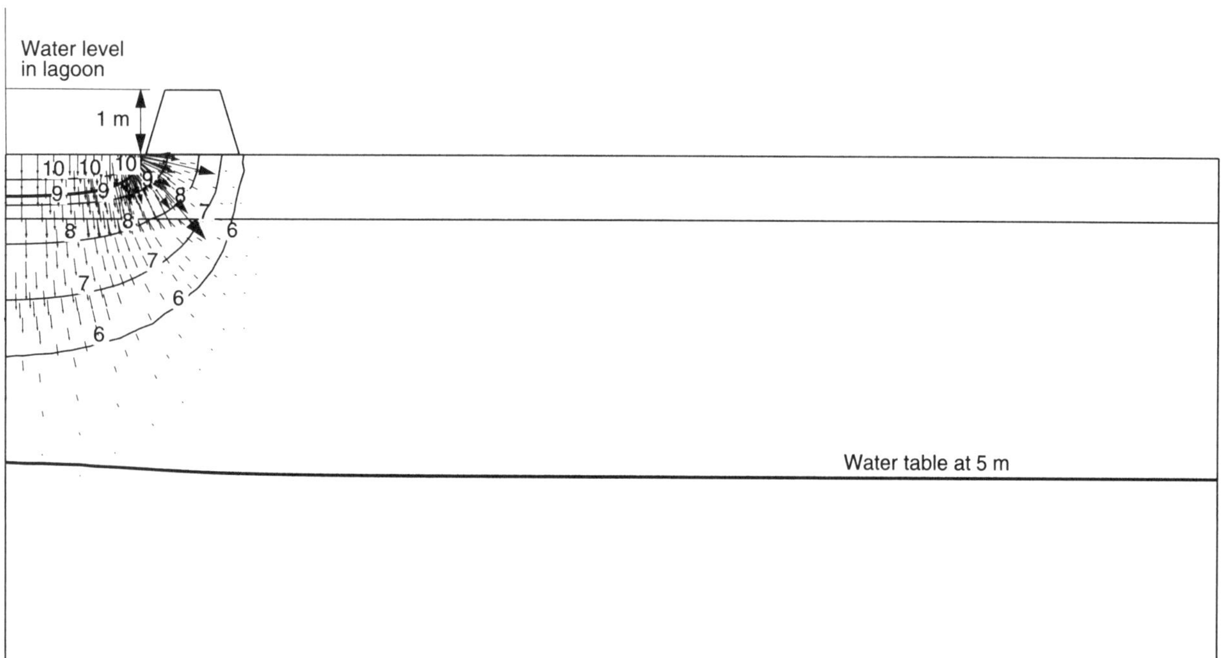

Time = 24 h

(b)

Figure 8.57 Hydraulic heads after an elapsed time of 24 hours. (a) Finite element mesh and hydraulic heads. (b) Velocity vectors and hydraulic heads.

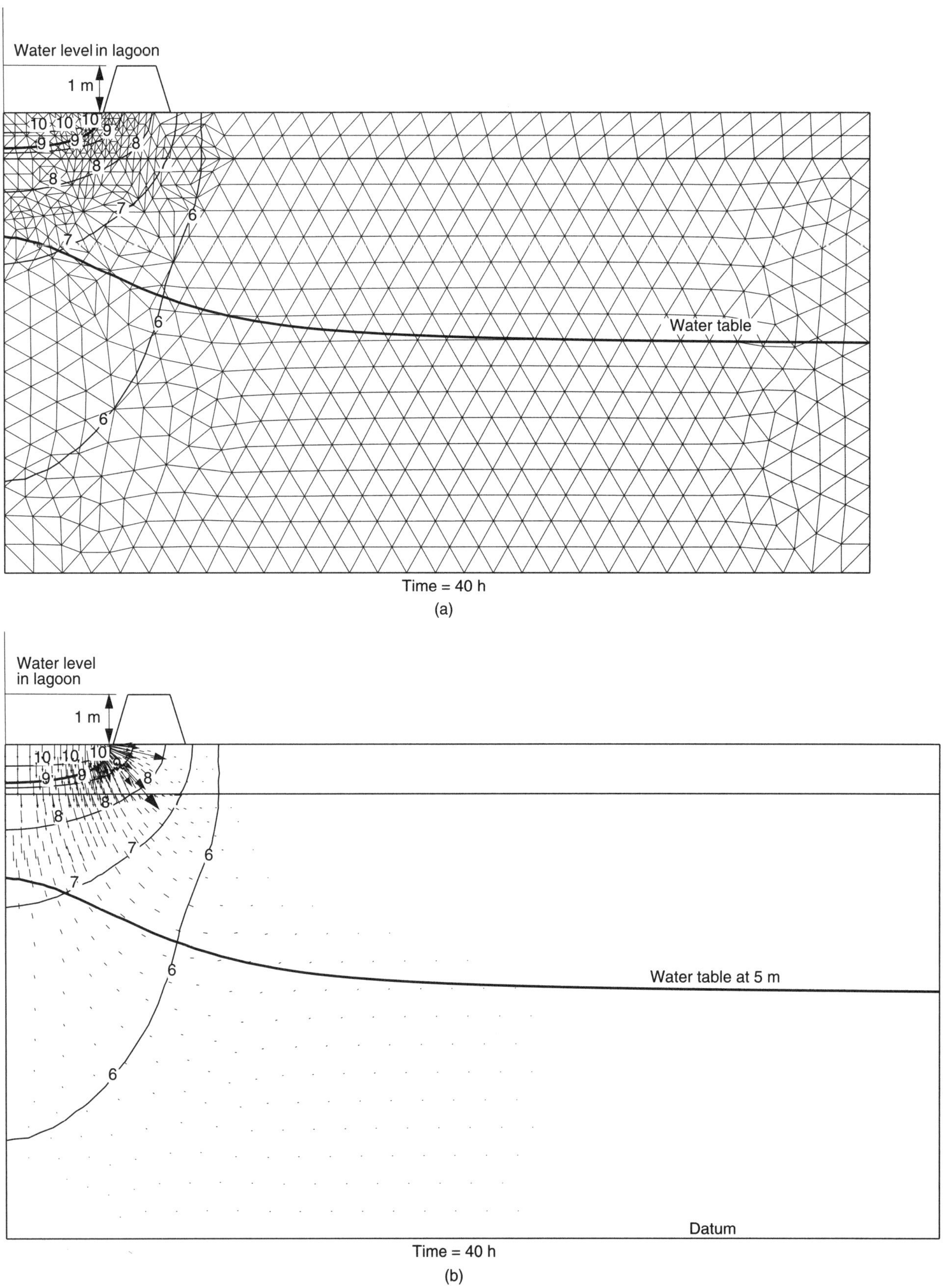

Figure 8.58 Hydraulic heads after an elapsed time of 40 hours. (a) Finite element mesh and hydraulic heads. (b) Velocity vectors and hydraulic heads.

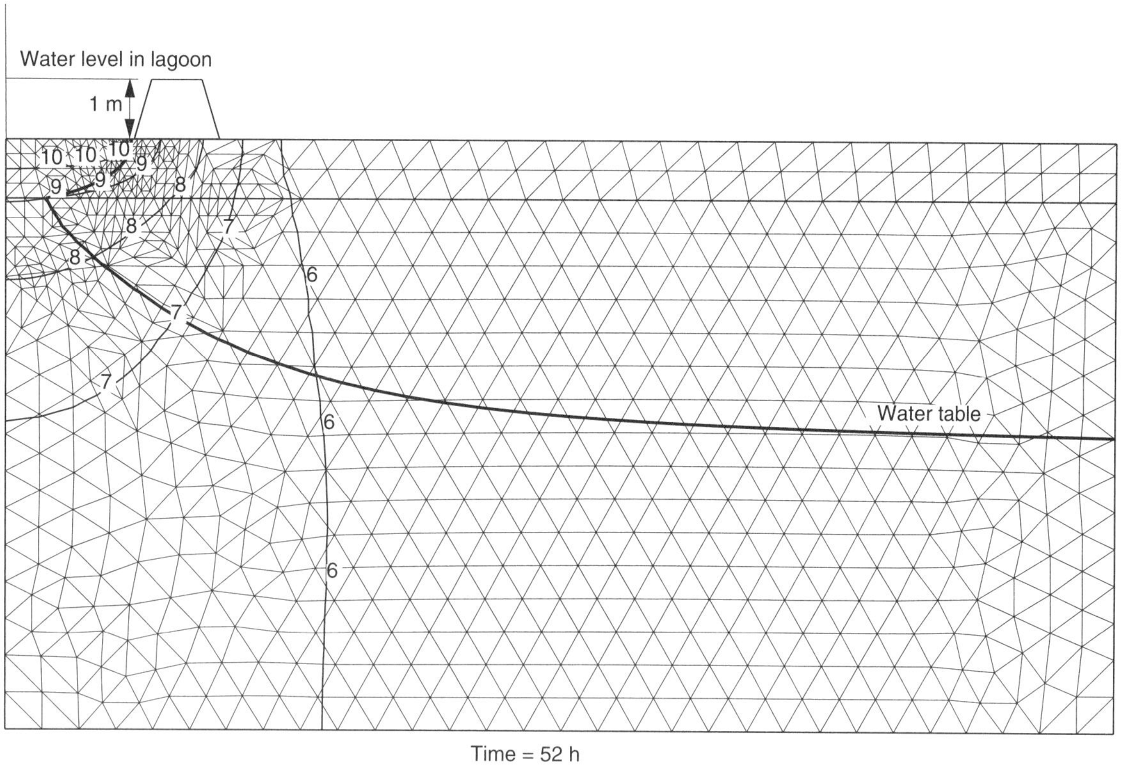

(a)

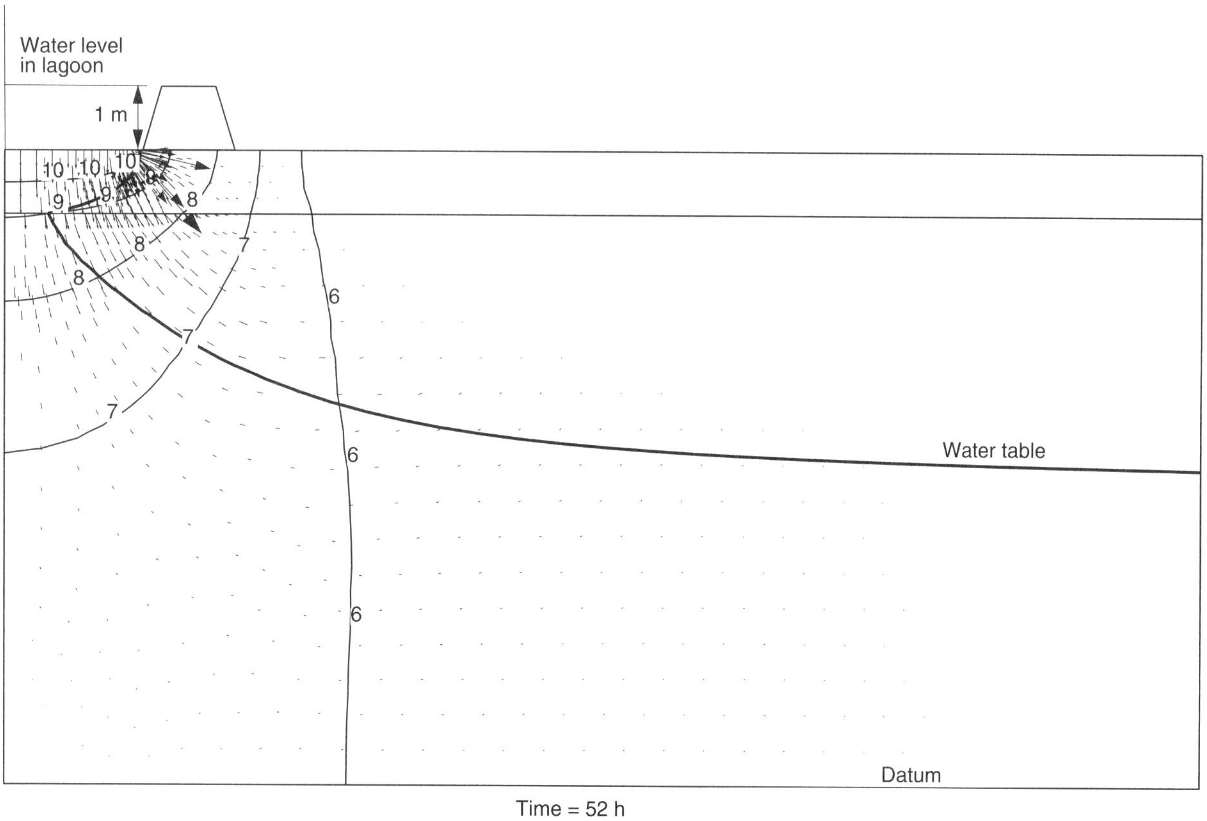

(b)

Figure 8.59 Hydraulic heads after an elapsed time of 52 hours. (a) Finite element mesh and hydraulic heads. (b) Velocity vectors and hydraulic heads.

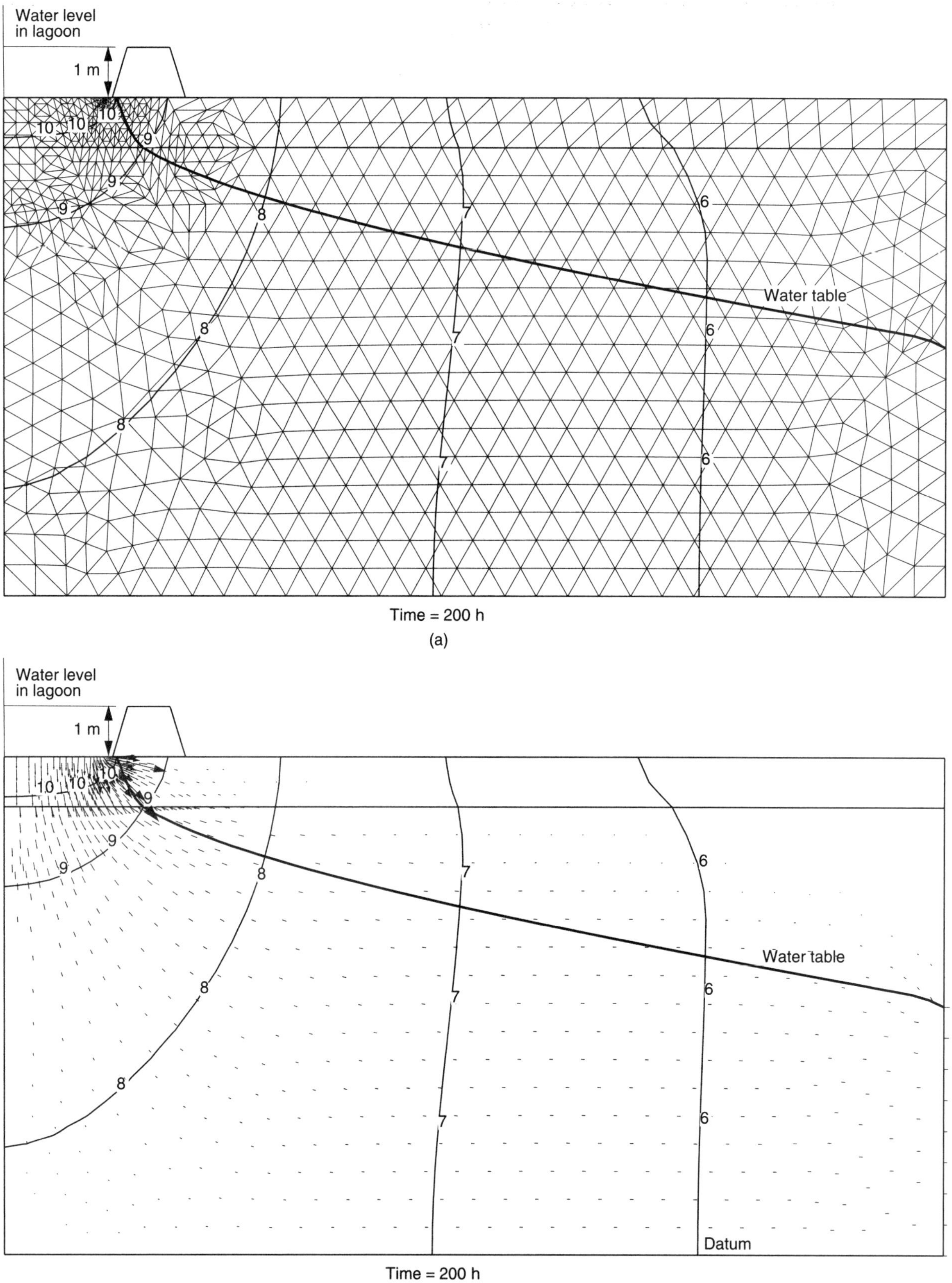

Figure 8.60 Hydraulic heads after an elapsed time of 200 hours. (a) Finite element mesh and hydraulic heads. (b) Velocity vectors and hydraulic heads.

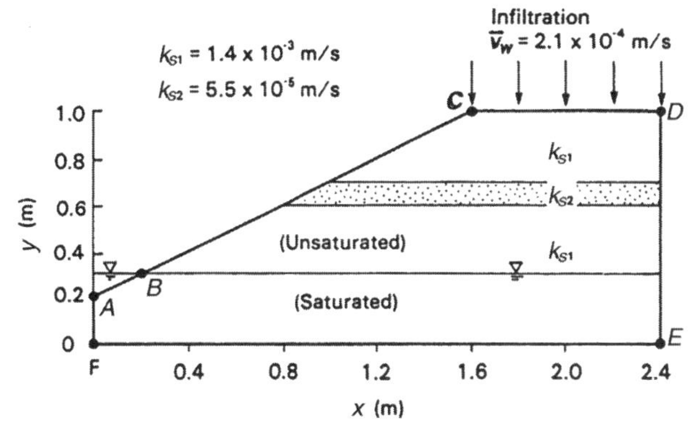

Figure 8.61 Geometry and boundary conditions for an example involving layered soils in a hillslope.

shown in Fig. 8.61. The physical model was instrumented with pore-water pressure measuring devices. The slope consisted of medium sand with a horizontal layer of fine-grained sand. The fine sand had a lower coefficient of permeability than the medium-grained sand. The intent was to study whether the fine sand layer would impede water flow and create a seepage face on the slope. The steady-state results from the experimental measurements are shown in Fig. 8.62. The steady-state equipotential lines and the phreatic water surface are presented. The experimental study forms a valuable benchmark problem against which numerical models can be tested.

The soil properties of the materials involved are shown in Fig. 8.63. The two sand materials are assumed to have the same SWCCs (Fig. 8.63a); however, the saturated coefficients of permeability for the two sand materials were different (Fig. 8.63b). The saturated coefficient of permeability for the medium sand was 1.4×10^{-3} m/s and the fine sands had a saturated coefficient of permeability of 5.5×10^{-5} m/s (Rulon and Freeze, 1985). The water storage function for both sands was obtained by differentiating the volumetric SWCC and the result is shown in Fig. 8.63c.

The constant rate of infiltration of 2.1×10^{-4} m/s was applied at the top of the hill slope in the experimental model

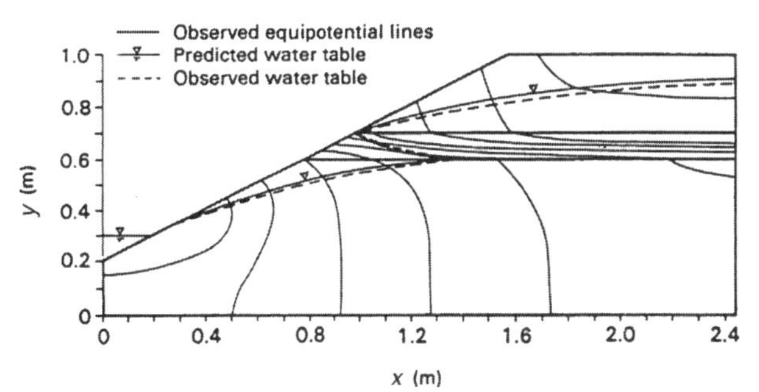

Figure 8.62 Comparison of observed water table and pore-water pressures in experimental tank with that predicted using Neumann model under steady-state conditions (from Rulon and Freeze, 1985).

and in the numerical simulation. The results of pore-water pressure measurements can be compared with the computed results from a numerical model. Close agreement has been observed between pore-water pressure measurements and the results of an unsteady-state finite element analysis.

An initial steady-state condition was assumed with the water table located at a height of 0.3 m from the toe of the slope. At an elapsed time taken as zero, a constant infiltration rate of 2.1×10^{-4} m/s was applied to the top portion of the hill slope. The development of the groundwater table from its initial steady-state condition (i.e., at time equal to zero) to an elapsed time of 100 s is shown in Fig. 8.64. The water level at the toe of the slope and the infiltration rate at the top of the slope were assumed to remain constant throughout the transient process. No flow boundary conditions were assumed along the remainder of the slope boundaries.

The seepage flow pattern and the development of the equipotential lines within the slope continued to change with elapsed time. The finite element mesh was also automatically refined as necessary, internal to the analysis. The equipotential lines and the phreatic line corresponding to an elapsed time of 200 s are shown in Fig. 8.65. The results for an elapsed time of 220 s are shown in shown in Fig. 8.66. The results show a gradual rise in the lower phreatic water surface and a gradual change in the equipotential lines. At the start of the infiltration process, water infiltrates vertically toward the impeding layer. As the infiltration of water continues, water moves through the impeding layer, causing the groundwater table to rise. After 240 s a perched water table develops on the impeding layer and moves toward the face of the slope (see Fig. 8.67). After 260 s a wedge-shaped unsaturated zone is formed, and the perched water table moves toward the edge of the slope, as shown in Fig. 8.68.

Now, two seepage faces began to develop. One seepage face is near the toe of the slope and the other seepage face develops at the top of the impeding layer. In other words, the presence of the impeding layer results in a complex configuration for the groundwater table and the position of the equipotential lines. A steady-state condition is established after about 280 s. The steady-state results are shown in Fig. 8.69. There is close agreement between the results of the physical model measurements (Rulon and Freeze, 1985) and the numerical model results. The positions of the developed water table, the seepage faces, and pore-water pressures are similar.

The finite element numerical analyses presented for the three example problems have illustrated the way in which unsteady-state modeling can be performed for saturated-unsaturated flow problems. The application of a saturated flow model to each of these problems would be virtually impossible to undertake. However, the use of a combined saturated-unsaturated flow is quite straightforward. The flow systems that developed throughout the unsaturated soil system can be complex, depending upon the contrast in the coefficients of permeability and the water storage modulus for the different soils.

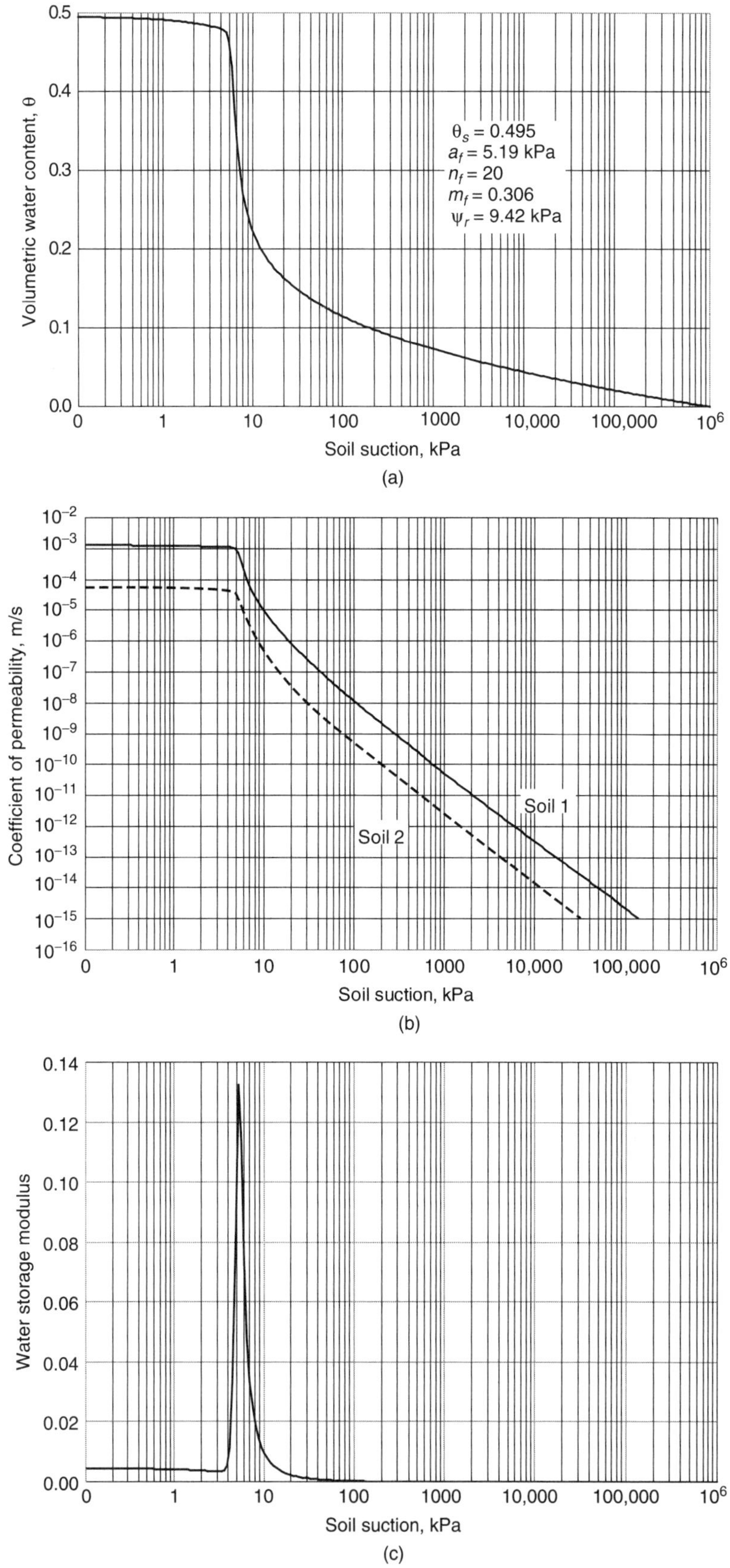

Figure 8.63 Permeability and water storage functions for analyzing unsteady-state seepage below a lagoon. (a) SWCC. (b) Permeability functions. (b) Water storage function.

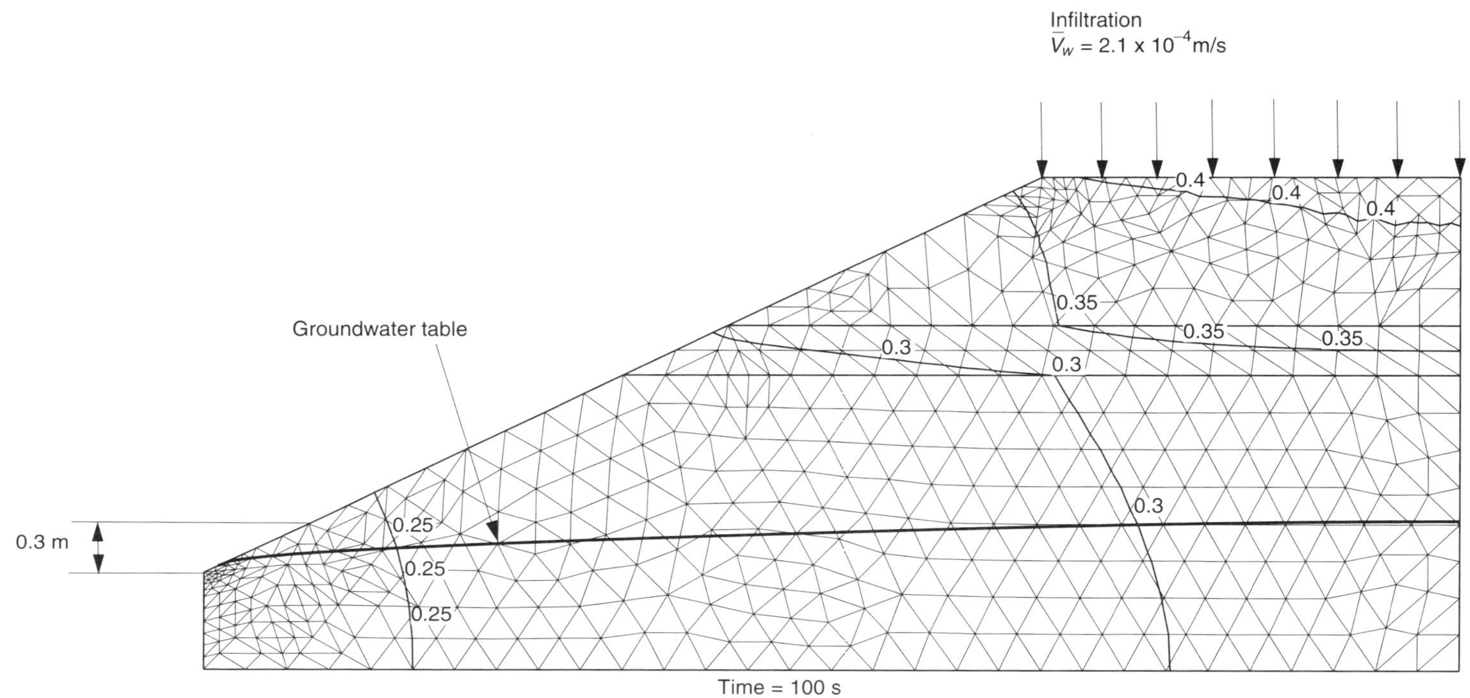

Figure 8.64 Finite element mesh and hydraulic heads after an elapsed time of 100 s.

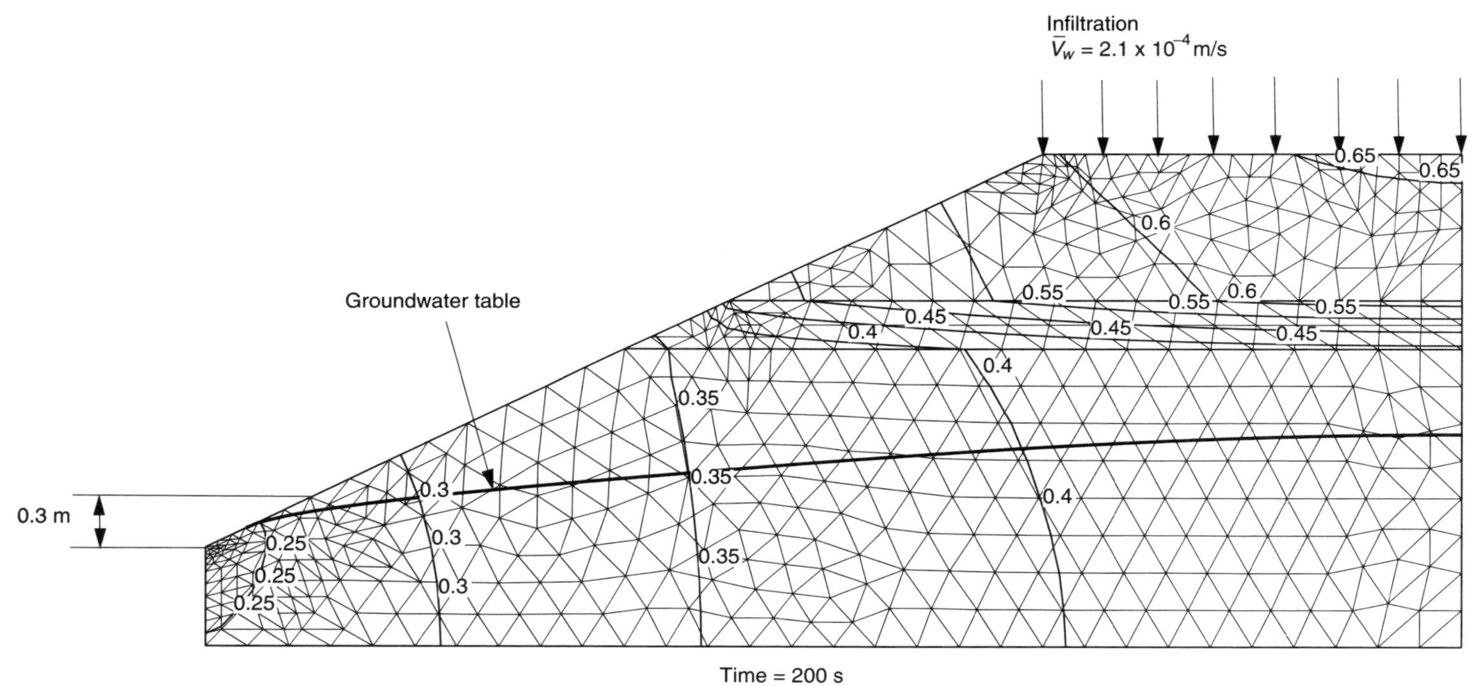

Figure 8.65 Finite element mesh and hydraulic heads after an elapsed time of 200 s.

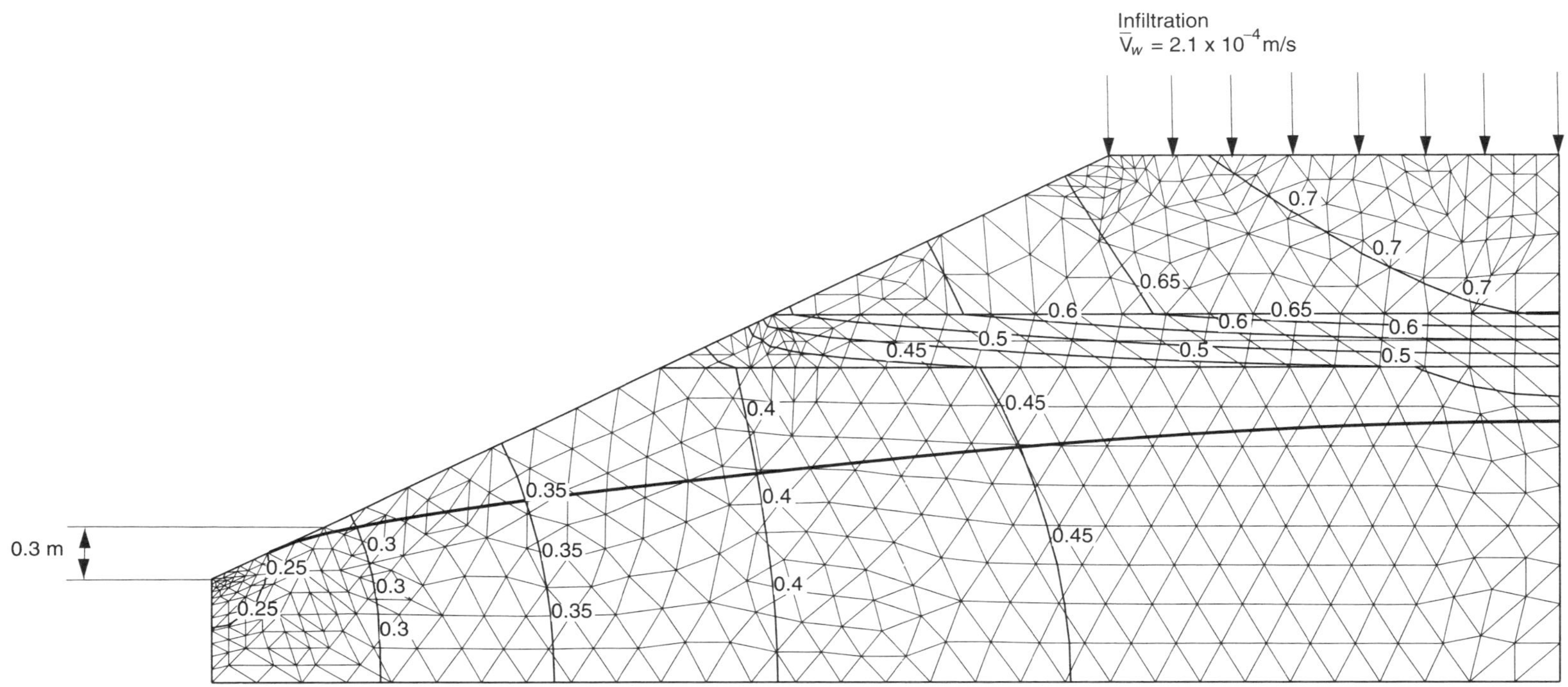

Figure 8.66 Finite element mesh and hydraulic heads after an elapsed time of 220 s.

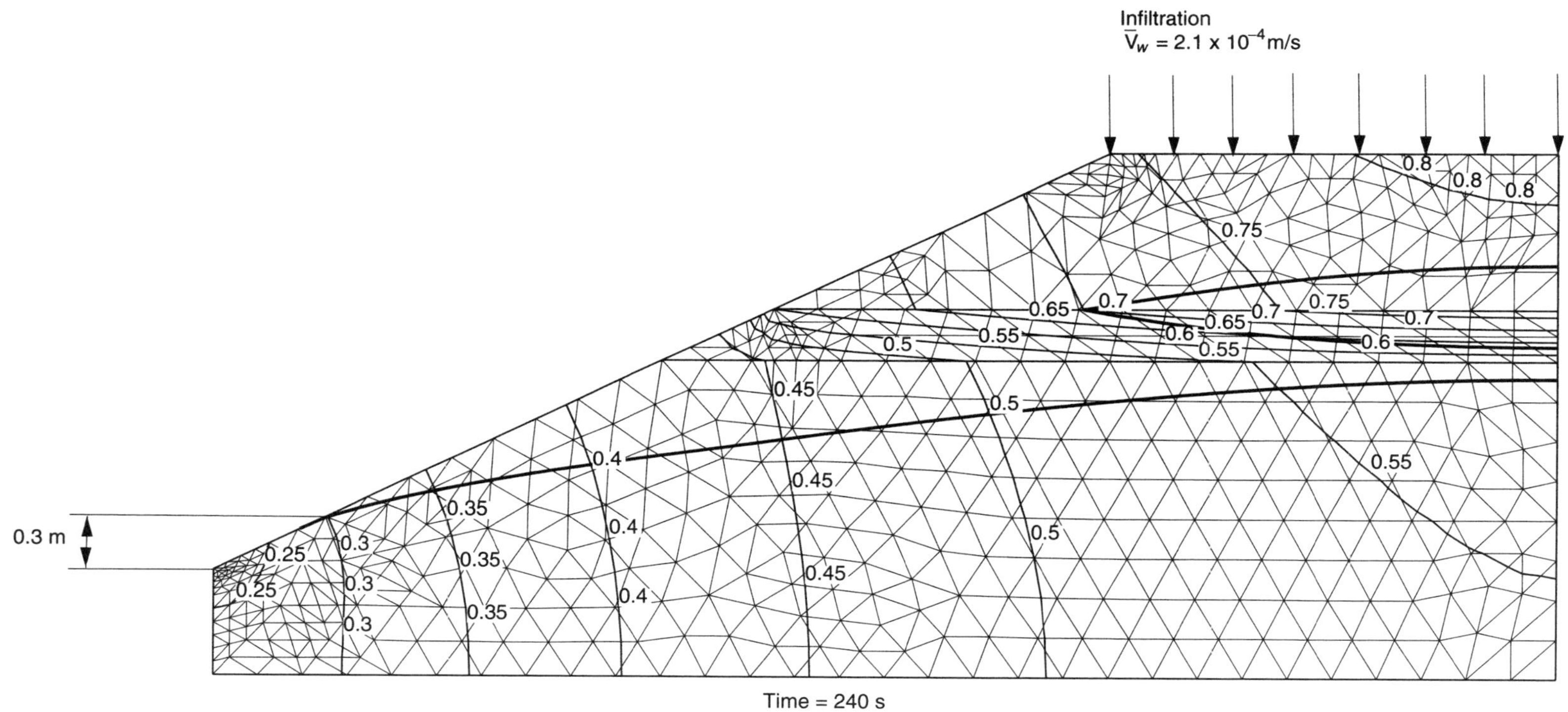

Figure 8.67 Finite element mesh and hydraulic heads after an elapsed time of 240 s.

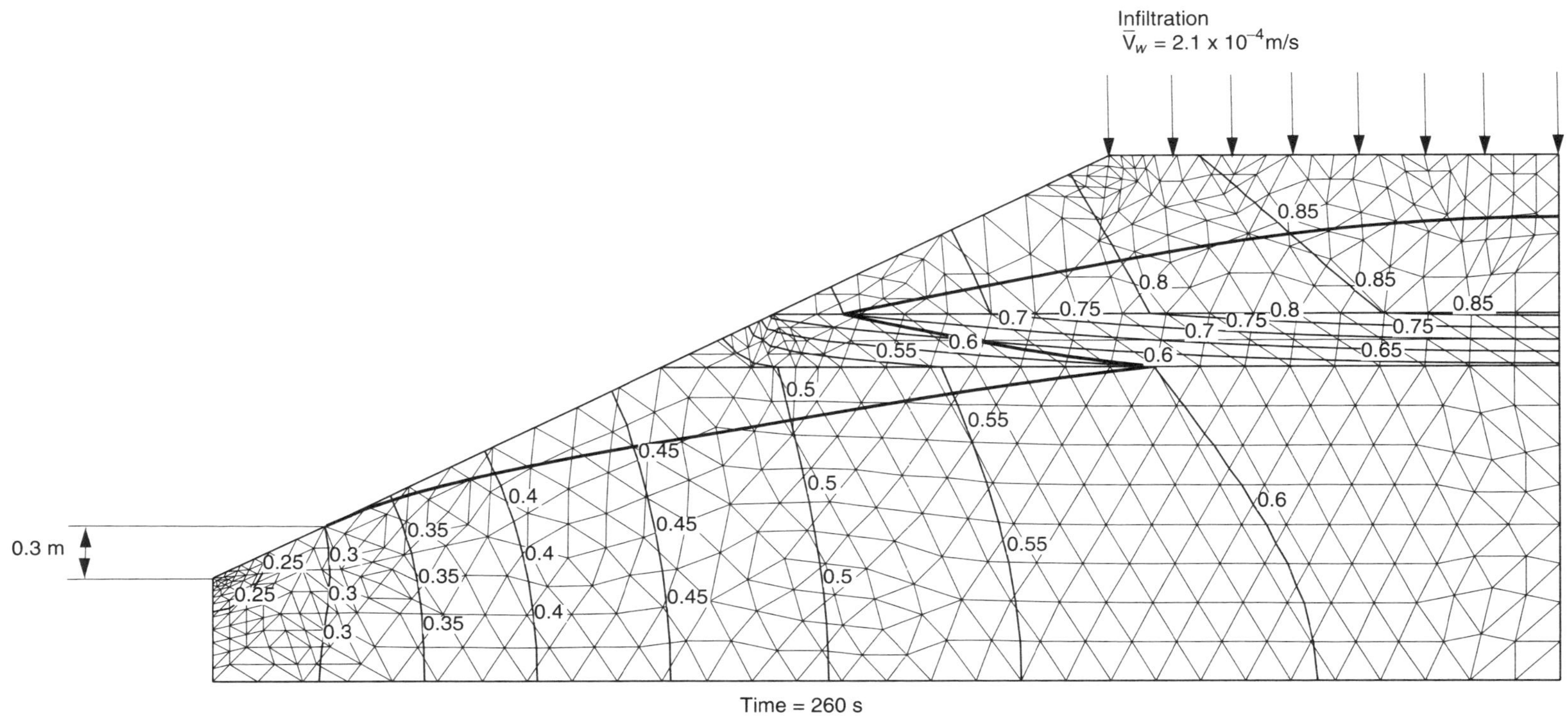

Figure 8.68 Finite element mesh and hydraulic heads after an elapsed time of 260 s.

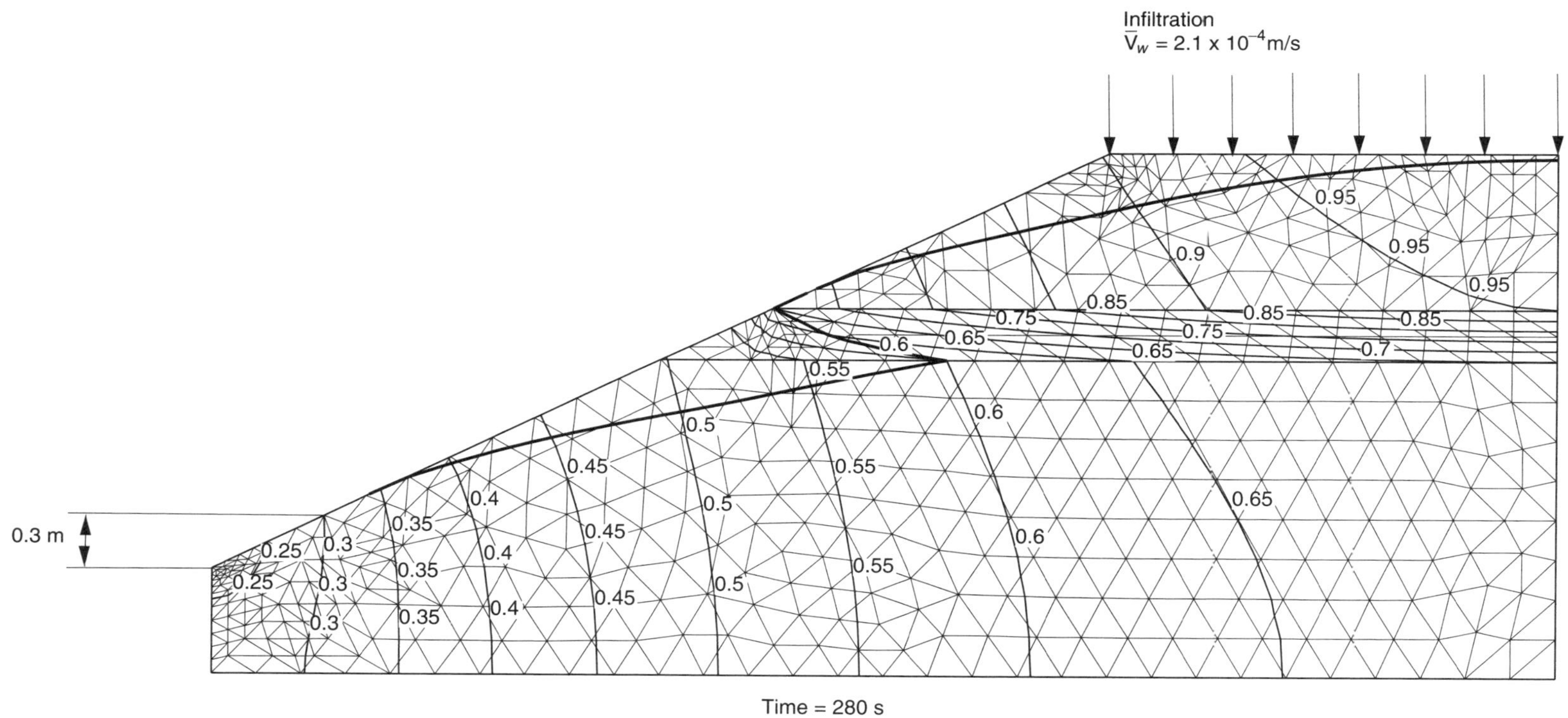

Figure 8.69 Finite element mesh and hydraulic heads after an elapsed time of 280 s.

There are many natural slope conditions where the pore-water pressures in the unsaturated soil portion of the slope are strongly influenced by the infiltration of precipitation. Modeling the unsaturated soil portion becomes important in understanding changes in matric suction in a slope and consequential changes in the computed factor of safety.

8.3.13 Ground Surface Moisture Flux Conditions

The solution for nonisothermal flows can be obtained by solving the partial differential equations for the air phase, the liquid water and water vapor phases, and the heat flow equation simultaneously. The pore-air pressure changes can sometimes be assumed to be negligible during a transient process (Rahardjo, 1990; Wilson, 1990). As a result, only water and heat flow differential equations need to be solved simultaneously.

Solving the partial differential equations for nonisothermal moisture flow at the ground surface is closely related to modeling evaporation from the ground surface. Further information on coupled heat and moisture flow was presented in Chapter 6 where ground surface boundary conditions where addressed.

8.4 CONDITIONS UNDER WHICH MATRIC SUCTION CAN BE MAINTAINED

The effect of negative pore-water pressure on the stability of slopes is sometimes ignored in geotechnical engineering practice. While this might be a reasonable assumption in some situations, it does not provide the geotechnical engineer with an in-depth understanding of potential failure mechanisms. This is particularly true when the water table is deep and a reduction in matric suction near ground surface becomes the trigger mechanism for slope instability.

It is often rationalized that matric suction can be ignored in slope stability studies because there is a perception that the infiltration of rainfall will produce a wetting front that gradually moves downward and causes the disappearance of matric suction throughout the soil profile. Consequently, it is prematurely concluded that matric suction cannot be relied upon to improve the stability of a slope in the long term.

The infiltration of water into a soil slope depends not only on the rainfall intensity and duration but also on the saturated coefficient of permeability of the near-ground-surface soil, the permeability function, and the water storage capacity of the soil. An unsaturated soil can have a greatly reduced coefficient of permeability from that of a saturated soil and also possesses an increased capacity for water storage.

Rainfall must be sustained over a considerable time period and the rainfall intensity must approach the saturated coefficient of permeability of the soil at the ground surface in order to eliminate matric suction from the soil profile. In situ suction measurements of negative pore-water pressures have shown that in some cases matric suctions do not disappear following intense rainstorms while in other cases the negative pore-water pressures are eliminated in a relatively short time following a rainfall. Sweeney (1982) presented in situ suction measurements from a slope instrumented with tensiometers in Hong Kong. The slope was inclined at 60° with an average height of 30 m. From the slope surface to a depth of 5–10 m the soil consisted of a completely weathered rhyolite. The saturated coefficient of permeability for in situ completely decomposed rhyolite was in the range of $10^{-5} - 10^{-7}$ m/s (GEO, 1993). The stratum immediately below the completely decomposed rhyolite was a completely to highly weathered rhyolite layer that was 5–10 m in thickness. The underlying soil was a slightly weathered rhyolite followed by bedrock at 20–30 m.

Suction measurements were made throughout the year of 1980 and the following observations were made from the results. First, matric suction showed a gradual reduction near ground surface during the rainy season. However, the pore-water pressures remained negative even during the long rainy season. Second, matric suctions in the 5–17 m depth remained constant throughout the year. The groundwater table at a depth of about 30 m rose and fell by 9 m throughout the year. The results illustrate the significant water storage capacity of the soil and its role in maintaining matric suctions throughout most of the soil profile. The matric suctions decreased at shallow depths into the slope but remained almost constant throughout the intermediate depths.

A second example where in situ matric suction measurements were made involved a slope consisting of colluvium material in Hong Kong (Anderson, 1983). The saturated coefficient of permeability of the colluvium material ranged from 10^{-4} to 10^{-7} m/s (GEO, 1993). The measured matric suctions in the colluviums were relatively low prior to a rainstorm and decreased to small values after heavy rainstorms. This example showed that it is also possible for the matric suctions to essentially disappear under certain conditions.

The above two examples show that there are certain conditions under which matric suctions in the soil might disappear during prolonged rainfall while there are other conditions under which matric suctions are maintained throughout the soil profile. There are rainfall and soil conditions under which soil suction can be maintained in a soil profile. Situations where soil suction may be lost or maintained during rainfall can be studied using saturated-unsaturated soil infiltration modeling. The physical processes associated with infiltration have been the focus of numerous studies (Zhang et al., 2004).

The wetting-front or wetting-band concept was introduced by Lumb (1962) in relation to the investigation of slope failures in Hong Kong. Figure 8.70 shows the variation of degree of saturation with depth during rainfall. The soil is assumed to become saturated near ground surface during heavy rainfall and essentially saturated to a depth h under ongoing infiltration. The wetting front was assumed to be defined by a sharp separation between the initial conditions

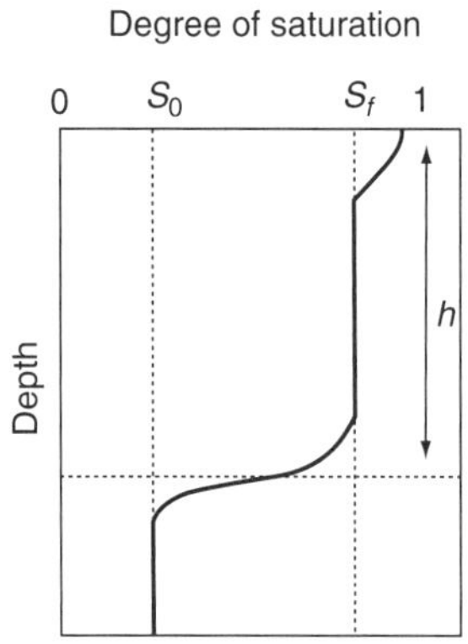

Figure 8.70 Variation of degree of saturation with depth during rainfall infiltration (after Lumb, 1962).

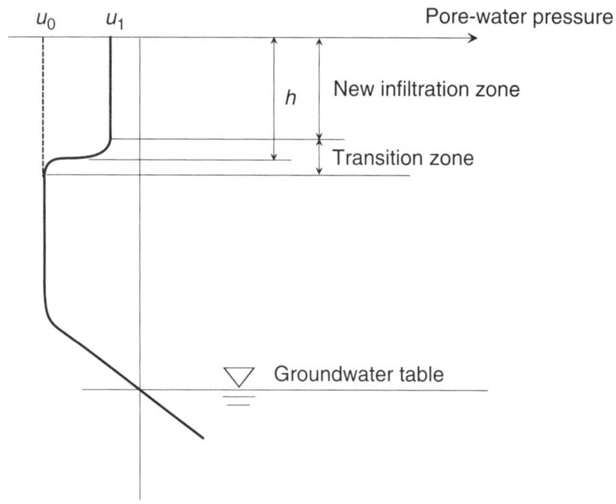

Figure 8.71 Transient infiltration into an unsaturated soil, (modified from Sun et al., 1998).

and the wetted zone. Under prolonged and heavy rainfall the depth of the wetting front, h, was defined as

$$h = \frac{k_s t}{n\left(S_f - S_0\right)} \tag{8.58}$$

where:

k_s = saturated coefficient of permeability,
S_f = final degree of saturation,
S_0 = initial degree of saturation,
n = porosity of the soil, and
t = elapsed time.

Sun et al. (1998) proposed a generalized wetting-band equation based on the Lumb (1962) wetting-band approach and presented the results of a series of one-dimensional finite element analyses. Figure 8.71 shows a typical variation of negative pore-water pressure with depth in an unsaturated soil. For a given ground surface moisture flux q_0 less than the saturated coefficient of permeability k_s, the matric suction is $(u_a - u_w)_0$ under steady-state conditions. If the infiltration rate is increased to q_1, a new infiltration zone is formed that gradually progresses downward. The depth of the wetting front, h, corresponding to a particular elapsed time was calculated using the following equation:

$$h = \frac{\left(k_1 - k_0\right)t}{\theta_1 - \theta_0} \tag{8.59}$$

where:

$k_0 = q_0$, initial moisture flux,
$k_1 = q_1$, subsequent moisture flux,
θ_0 = initial volumetric water content, and
θ_1 = subsequent volumetric water contents.

Equations 8.58 and 8.59 are similar, but the latter equation allows for a flux change from one steady-state condition to a new steady-state condition. The wetting-front concept provides a simplified methodology for considering changes in the degree of saturation (or matric suction) under a change in rainfall conditions. However, there will not always be a distinct change between the infiltration zone and the zone

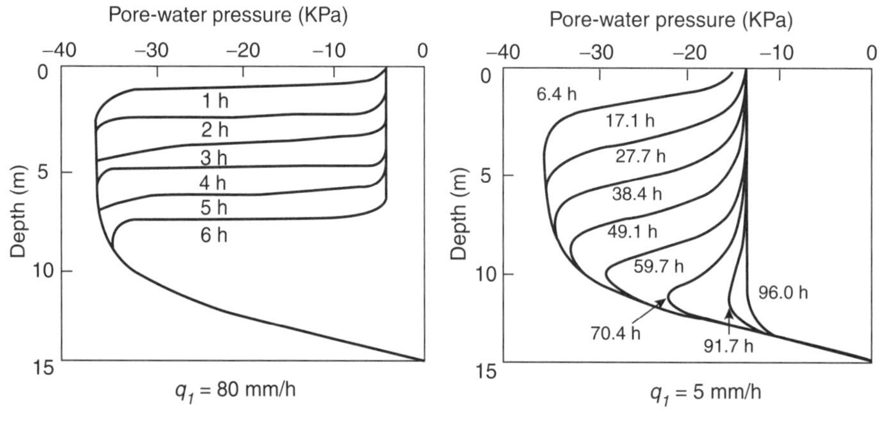

Figure 8.72 Transient infiltration at 80 mm/hr and 5 mm/hr based on a one-dimensional analysis (after Sun et al., 1998).

where the negative pore-water pressures are maintained. The numerical study conducted by Sun et al. (1998) showed that the transition zone is less distinct when the new infiltration rate is quite small (Fig. 8.72).

Matric suction in an unsaturated soil may not be destroyed even under long periods of rainfall infiltration. Kasim et al., (1998a) studied the relationship between steady-state rainfall, the permeability function, and the SWCC on equilibrium matric suction conditions. The study was carried out for both horizontal ground surfaces and sloping ground surfaces. Results of the analysis showed that steady-state rainfall does not necessarily eliminate matric suctions in the soil. The long-term matric suction in the soil is largely maintained when the steady-state rainfall is one or more orders of magnitude less than the saturated coefficient of permeability. It appears that matric suction will not disappear unless the steady-state rainfall flux approaches the saturated coefficient of permeability of the soil near the ground surface.

There are two soil properties that need to be taken into consideration when analyzing the infiltration of water into an unsaturated soil. These two properties are the permeability function and the water storage function. Both of these properties bear a relationship to the SWCC as shown in Fig. 8.73. Steady-state seepage conditions can be analyzed by giving consideration to the permeability function. However, both the permeability function and the water storage function need to be taken into consideration when analyzing transient seepage.

8.4.1 Infiltration during Rainfall

The infiltration and the seepage of water through an unsaturated soil can be analyzed through use of saturated-unsaturated seepage modeling. The results of numerical studies, however, can be difficult to interpret because several variables are involved. The primary soil parameters are the coefficient of permeability function, water storage function, rainfall intensity, and rainfall duration. The unsaturated soil properties are related to the SWCC. The fundamental processes associated with infiltration under both steady-state and unsteady-state conditions need to be considered separately in order to interpret results from a numerical analysis.

Kisch (1959) derived an equation for one-dimensional steady-state seepage into a soil subjected to a constant water flux q at the ground surface. The magnitude of the water flux, q, was calculated for a vertical soil column of unit cross-sectional area:

$$q = k_w \left(1 + \frac{du_w}{\rho_w g\, dy}\right) \tag{8.60}$$

The moisture flux q is constant under steady-state conditions and Eq. 8.60 can be written as

$$\frac{d\left(u_w/\rho_w g\right)}{dy} = \left(\frac{q}{k_w} - 1\right) \tag{8.61}$$

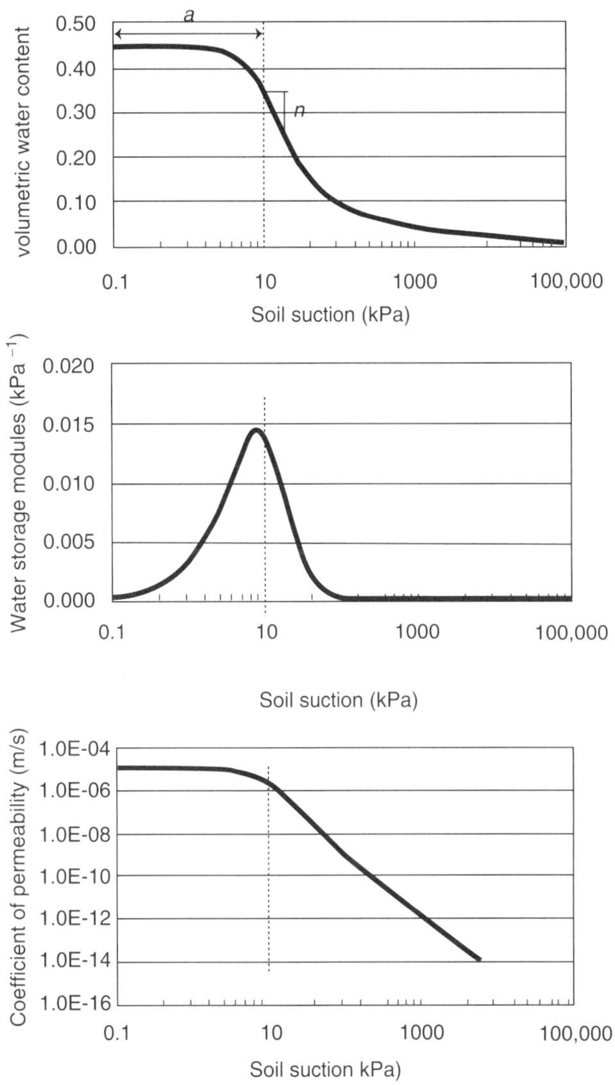

Figure 8.73 Relationship between the SWCC, water storage function and permeability function for an unsaturated soil.

If the magnitude of the ground surface flux, q, is in the unsaturated coefficient of permeability range for a soil, k_w, the pore-water pressure head gradient $d\left(u_w/\rho_w g\right)/dy$ becomes zero [i.e., pressure head $u_w/\left(\rho_w g\right)$ tends toward a constant].

Figure 8.74 shows steady-state pressure head profiles in a clay liner and an underlying sand layer published by Kisch (1959). The surface water flux q was equal to the saturated coefficient of permeability of the clay liner, $k_{s,\text{clay}}$. The saturated coefficient of permeability of the sand, $k_{s,\text{sand}}$, was 140 times $k_{s,\text{clay}}$. The thickness of the clay liner was 0.5 m. The water table was at depths of 2.5 and 4.5 m for cases I and II, respectively.

The gradient of pressure head is zero at the upper portion of the sand layer under both steady-state cases. The pore-water pressures at the top of the sand remain at a constant value of −5.5 kPa when the ground surface flux q

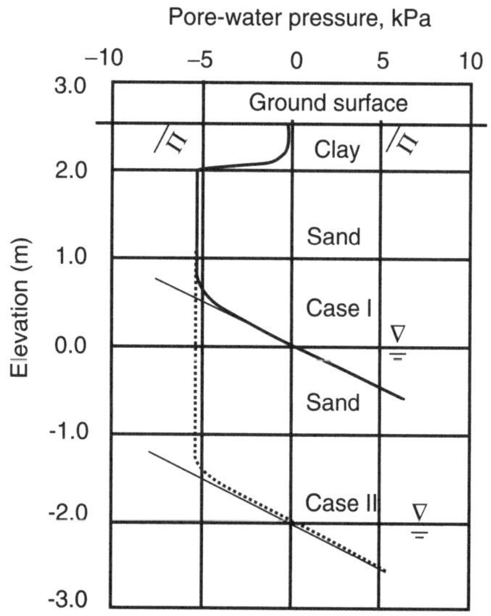

Figure 8.74 Variation of pore-water pressure with depth in the clay liner and the underlying sand layer for a surface water flux, $q = k_{s,\text{clay}} = 0.0068\ k_{s,\text{sand}}$. (modified from Kisch, 1959).

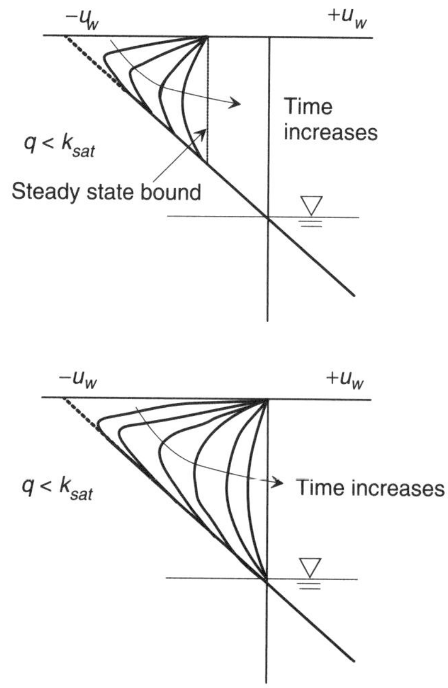

Figure 8.76 Infiltration into unsaturated soil under transient seepage conditions with two different ground surface moisture fluxes.

(equal to $k_{s,\text{clay}}$), is equal to the coefficient of permeability of the unsaturated sand. Figure 8.75 shows typical pore-water pressure distributions under steady-state conditions. Hydrostatic conditions correspond to the situation where there is no moisture flux at the ground surface. Under steady-state infiltration conditions the gradient of pore-water pressure is −1, as shown in Fig. 8.75. The water flux in and out of the soil becomes equal. If the water flux is the same as the coefficient of permeability of the unsaturated soil at a particular value of matric suction, the pore-water pressure gradient is zero as shown in Fig. 8.75.

Figure 8.76 shows typical pore-water pressure distributions under transient conditions. Infiltration under transient conditions can be considered as a transitional state between the initial and final steady states. The time to reach steady-state conditions is a function of the ground surface flux, the coefficient of permeability of the soil, and the water storage of the soil. The matric suction in the unsaturated soil can decrease but will not disappear when the flux is less than the saturated coefficient of permeability. Only when the ground surface moisture flux is equal to or greater than the saturated coefficient of permeability can the matric suctions be eliminated with depth. The conceptual framework set forth by Kisch (1959) sets steady-state bounds for the infiltration of water into unsaturated soils.

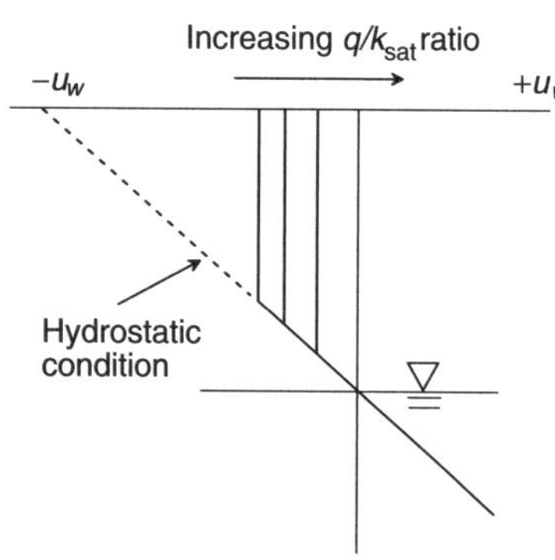

Figure 8.75 Infiltration into an unsaturated soil under steady state conditions with various ground surface moisture fluxes expressed as a ratio of the saturated coefficient of permeability at ground surface.

8.4.2 Numerical Modeling of Steady-State Conditions

Let us consider a slope of 20 m in height inclined at 30°. The slope is assumed to consist of a homogenous, isotropic soil. A series of finite element saturated-unsaturated seepage analyses were conducted in order to study the influence of various unsaturated soil properties on the infiltration of water into the slope. The geometry and boundary conditions are shown in Fig. 8.77 (Zhang et al., 2004). Constant hydraulic heads were applied along the left and right boundaries below the groundwater table. A zero flux boundary was applied along the left and right boundaries above the groundwater table. The groundwater table was fixed as a straight line between the boundaries in order to consider the same cross section for each set of soil and boundary conditions. The lowermost material in the slope was assumed to be impermeable.

Precipitation was modeled as a moisture flux boundary q applied along the ground surface. Section *X-X* located in the middle of the slope was used to illustrate the pore-water pressure profile in response to applied precipitation.

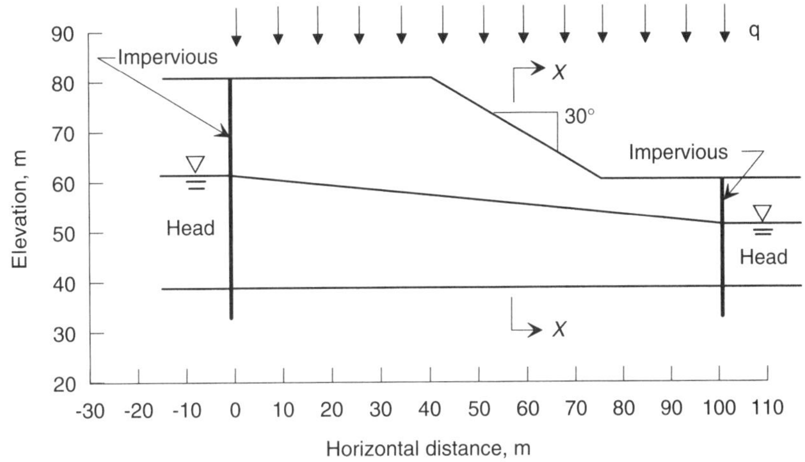

Figure 8.77 Geometry and boundary conditions for analyzing moisture infiltration to a slope.

Various a_f values for the Fredlund and Xing (1994) SWCC were used (e.g., 1, 5, 10, 20, 50, 100, and 200 kPa) and fixed values of n_f, m_f and saturated coefficient of permeability k_s values (i.e., $n_f = 2$, $m_f = 1$, and $k_{sat} = 10^{-5}$ m/s) were used in the parametric study. The SWCCs and the corresponding coefficient-of-permeability functions for the soils are shown in Fig. 8.78. The effect of varying n_f and k_s was also studied.

The results shown in Figs. 8.79a to 8.79f were obtained for soils having the same saturated coefficient of permeability of 10^{-5} m/s, with $n_f = 2$ and $m_f = 1$. The pore-water pressure profiles correspond to soils with a_f equal to 5, 10, 20, 50, 100, and 200 kPa, respectively (Fredlund and Xing, 1994). Different ratios of ground surface flux to saturated coefficient of permeability (i.e., q/k_s equal to 0.001, 0.01, 0.1, 0.2, 0.5) were applied to the ground surface of the slope.

The results illustrate that matric suction at the surface of the slope approaches zero as the rainfall flux approaches the saturated coefficient of permeability of the soil. There is essentially a vertical matric suction profile [i.e. $d\left(u_w/\rho_w g\right)/dy$ is zero] established under steady-state conditions as the ground surface flux is increased toward the saturated coefficient of permeability. The depth to which there is a constant-matric-suction profile decreases with decreasing ratio of the steady-state rainfall flux to the saturated coefficient of permeability q/k_s.

Matric suction values along a vertical portion of the pressure profiles decrease with decreasing a_f values for SWCC. The depth of the constant-matric-suction portion decreases with an increasing a_f value for the soil. The long-term matric suction remains close to the hydrostatic profile when the steady-state rainfall flux is two or more orders of magnitude less than the saturated coefficient of permeability and the a_f value is greater than 100 kPa (Fig. 8.79f). The reduction in matric suction is more significant when the a_f parameter is small.

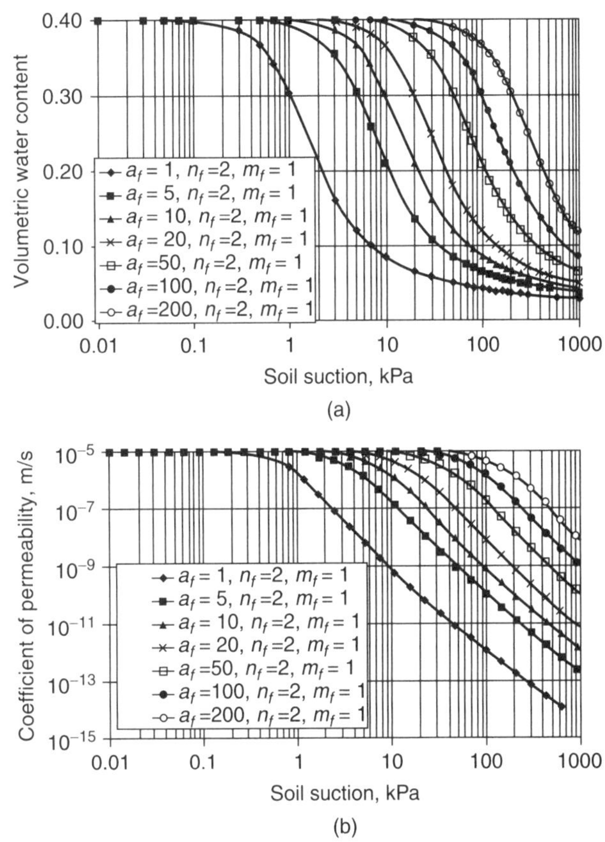

Figure 8.78 SWCCs and permeability functions for parametric study with varying a_f values: (a) SWCCs with various a_f values; (b) estimated permeability functions based on SWCCs (after Zhang et al., 2004).

Figure 8.80 presents the pore-water pressure profiles for a soil with the same SWCC (i.e., $a_f = 10, n_f = 2, m_f = 1$) but different saturated coefficients of permeability (i.e., 10^{-3}, 10^{-5}, and 10^{-7} m/s, respectively). These results confirm the observations of Kasim et al., (1998a) which

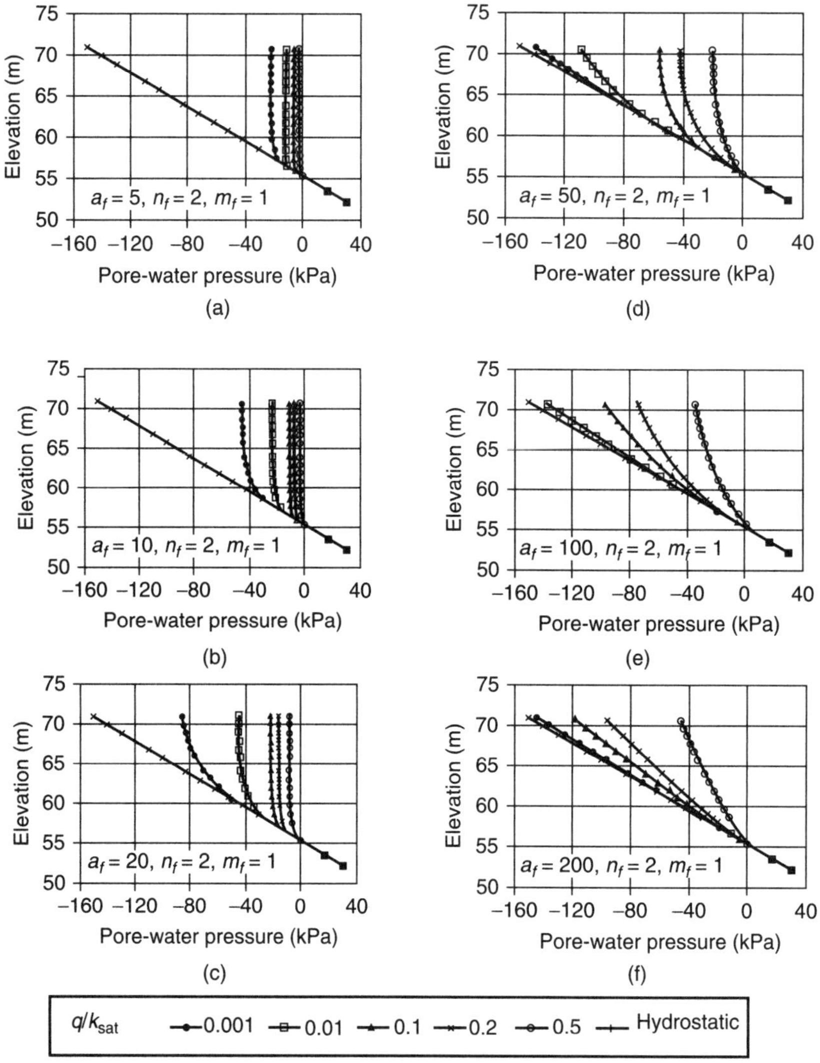

Figure 8.79 Pore-water pressure profiles in slopes with various a values subjected to rainfall fluxes under steady-state conditions: (a) $a_f = 5$ kPa, (b) $a_f = 10$ kPa, (c) $a_f = 20$ kPa, (d) $a_f = 50$ kPa, (e) Parameter $a_f = 100$ kPa, and (f) $a_f = 200$ kPa (after Zhang et al., 2004).

show that it is not the absolute value of the ground surface moisture flux, but rather the ratio of q/k_s, that influences the pore-water pressure profiles under steady-state conditions. Therefore, the soil response to a moisture flux is better understood when the rainfall flux is expressed as a percentage of the saturated coefficient of permeability.

Figure 8.81 illustrates that the long-term matric suction profiles in slopes with different n_f values are comparable. However, the values of matric suction along the vertical portion of the pore-water pressure profiles at the same q/k_s do not decrease or increase monotonically with the n_f values.

Kasim et al. (1998a) illustrated that the long-term matric suction at ground surface could be estimated using the coefficient-of-permeability function when the ground surface was horizontal. The long-term matric suction distribution under hydrostatic conditions can be represented by a straight line with a hydraulic gradient of zero. When a steady-state rainfall flux q_1 is applied at the ground surface, a hydraulic gradient of unity (or close to unity) will be established for a portion of the unsaturated soil profile. The long-term matric suction $(u_a - u_w)_1$ associated with a constant moisture flux q_1 can be determined directly from the coefficient-of-permeability function of the soil, as illustrated in Fig. 8.82.

Kasim (1997) showed that the difference between the matric suction values from numerical modeling and the estimated values from the permeability functions were related to the depth of the groundwater table. The matric suction profile for an unsaturated soil is cut off from the constant-matric-suction profile for an infinite thickness of unsaturated soil. The estimated values for matric suction increase with an increasing thickness of unsaturated soil (Fig. 8.83).

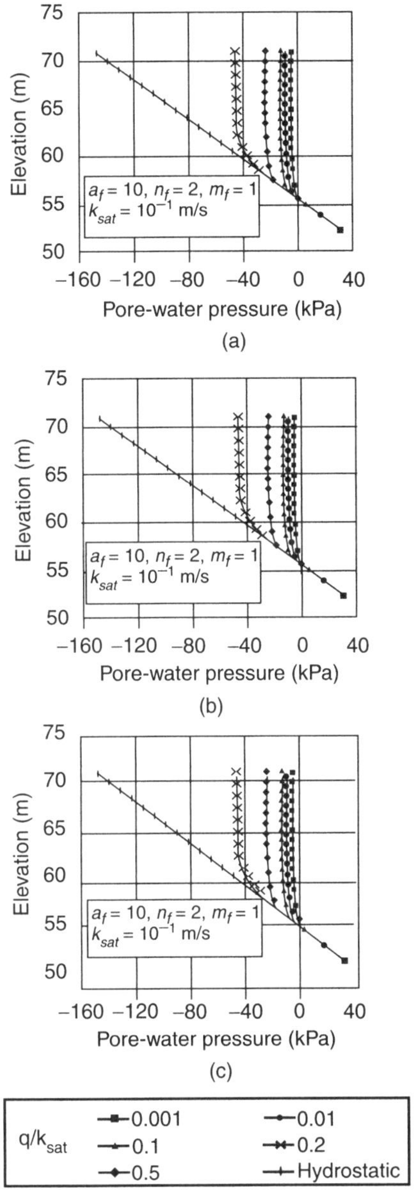

Figure 8.80 Comparison of pore-water pressure profiles for soils with varying saturated coefficients of permeability k_{sat}: (a) $k_{sat} = 10^{-3}$ m/s, (b) $k_{sat} = 10^{-5}$ m/s, and (c) $k_{sat} = 10^{-7}$ m/s (after Zhang et al., 2004).

Figure 8.84 presents the SWCCs and the corresponding permeability functions for soils with varying n_f values. The theoretical long-term matric suction along the vertical portion of the pore-water pressure profiles for soils with varying n_f values does not decrease or increase monotonically by a significant amount as n_f varies with a constant q/k_{sat} ratio. These results assist in explaining the pore-water pressure profiles shown in Fig. 8.81.

8.4.3 Numerical Modeling of Transient Conditions

The effects of air-entry value, water storage capacity, saturated coefficient of permeability, and groundwater boundary conditions on the pore-water pressure profiles can be illustrated by starting from a hydrostatic condition. It should be noted that varying initial soil moisture conditions can significantly influence rainfall infiltration and pore-water pressure profiles in a slope (Freeze, 1969; Tsaparas et al., 2002).

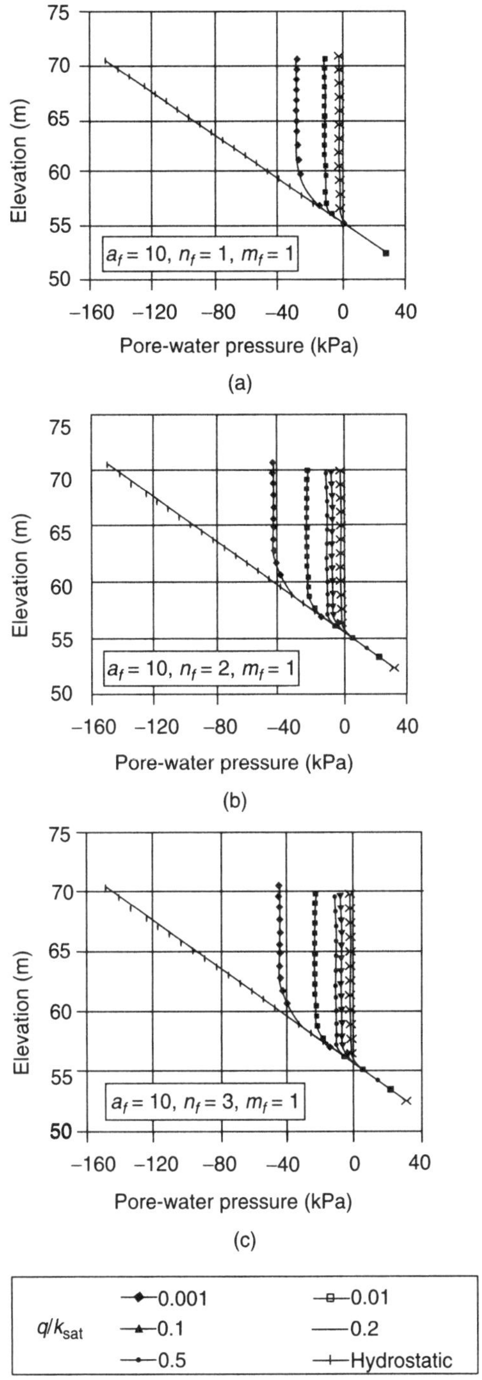

Figure 8.81 Comparison of pore-water pressure profiles for soils with various n values: (a) $n_f = 1$, (b) $n_f = 2$, and (c) $n_f = 3$ (after Zhang et al., 2004).

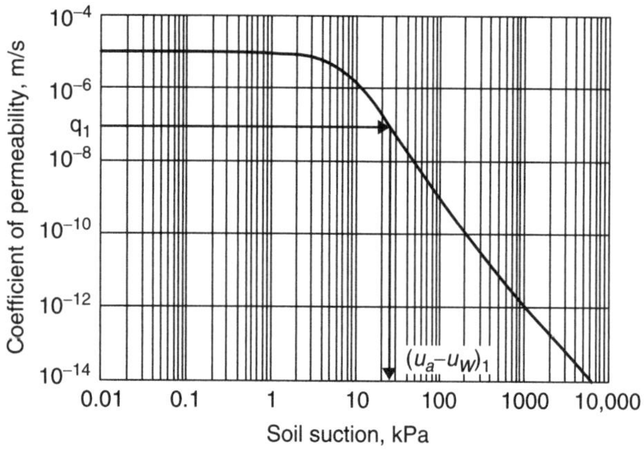

Figure 8.82 Determination of matric suction values at horizontal ground surface for soil subjected to steady-state rainfall flux q_1 (after Kasim, et al., 1998).

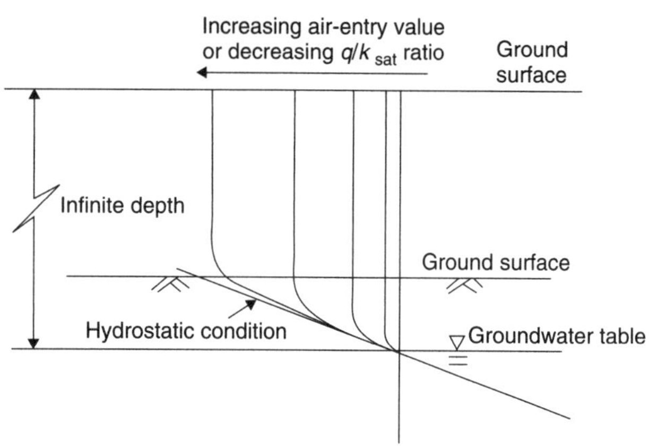

Figure 8.83 Cutoff matric suction profiles in unsaturated soil with horizontal ground surface (after Kasim et al., 1998).

8.4.4 Effect of Air-Entry Value on Wetting Front during Transient Modeling

Figures 8.85a–8.85f show the pore-water pressure profiles under transient condition with various a_f values for the Fredlund and Xing (1994) SWCC. The rainfall flux is set equal to the saturated coefficient of permeability, which means that the ground surface is subjected to the maximum moisture flux it can accept. The results illustrate that different infiltration patterns than those calculated for steady-state seepage are generated for the pore-water pressure profile. When the a_f value is equal to 1 kPa, the wetting front is sharp and distinct. The matric suction near the ground surface decreases with time, but the rate of downward movement of the wetting front is small. The depth of the wetting front is only about 1 m below the ground surface after 50 days of rainfall with a flux equal to the saturated coefficient of permeability.

When $a_f = 5$ kPa, the transition zone between the infiltration zone and the unaffected zone is still quite sharp and

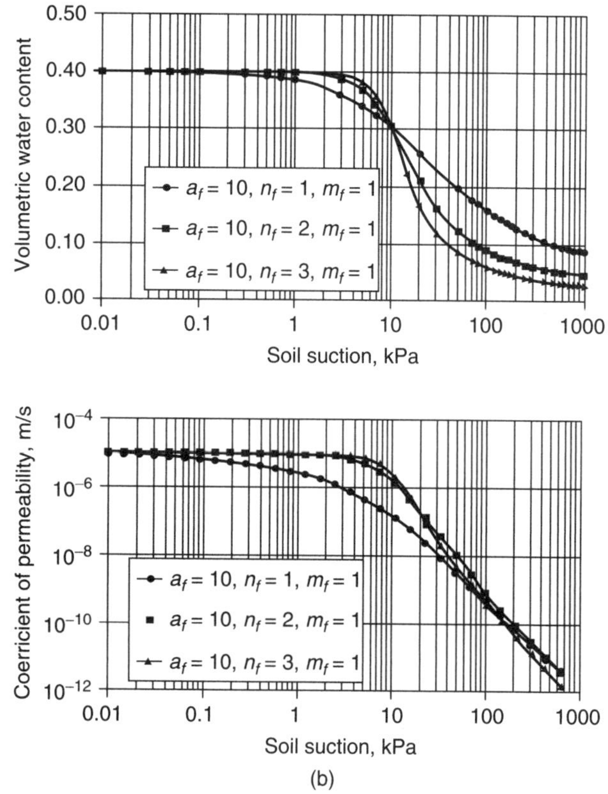

Figure 8.84 SWCCs and permeability functions with varying n values of soils used in parametric study (after Zhang et al., 2004).

distinct. The matric suction decreases at shallow depths into the slope but remains essentially constant at greater depths. It takes about 1 day for the wetting front to move to a depth of 1 m. After 3 days the infiltration depth of the wetting front is about 4 m. The negative pore-water pressure essentially disappears after 6 days of rainfall.

The pore-water pressure profiles in the soil when $a_f =$ 10 kPa resembles the profiles of the soil with $a_f =$ 5 kPa. However, the negative pore-water pressures almost disappear after 4 days of rainfall infiltration. The transition between the infiltration zone and the underlying zone becomes less distinct for $a_f > 10$ kPa. The remaining matric suctions in the soil decrease at a more rapid rate when the a_f value is increased.

The time required to eliminate matric suction varies significantly with the air-entry value of the soil. Figure 8.86 shows the matric suction profiles after 24 h for soils with differing air-entry values. After 24 h of rainfall, soils with a low air-entry value (i.e., $a_f < 5$ kPa) have an average infiltration depth in the order of a few millimeters. However, soils with a greater air-entry value (e.g., $a_f > 10$ kPa) have an infiltration depth of wetting that can extend a few meters. When the a_f value is even greater (i.e., $a_f > 100$ kPa), the soil can be fully saturated after 24 h of rainfall when a flux rate equal to the saturated coefficient of permeability is applied.

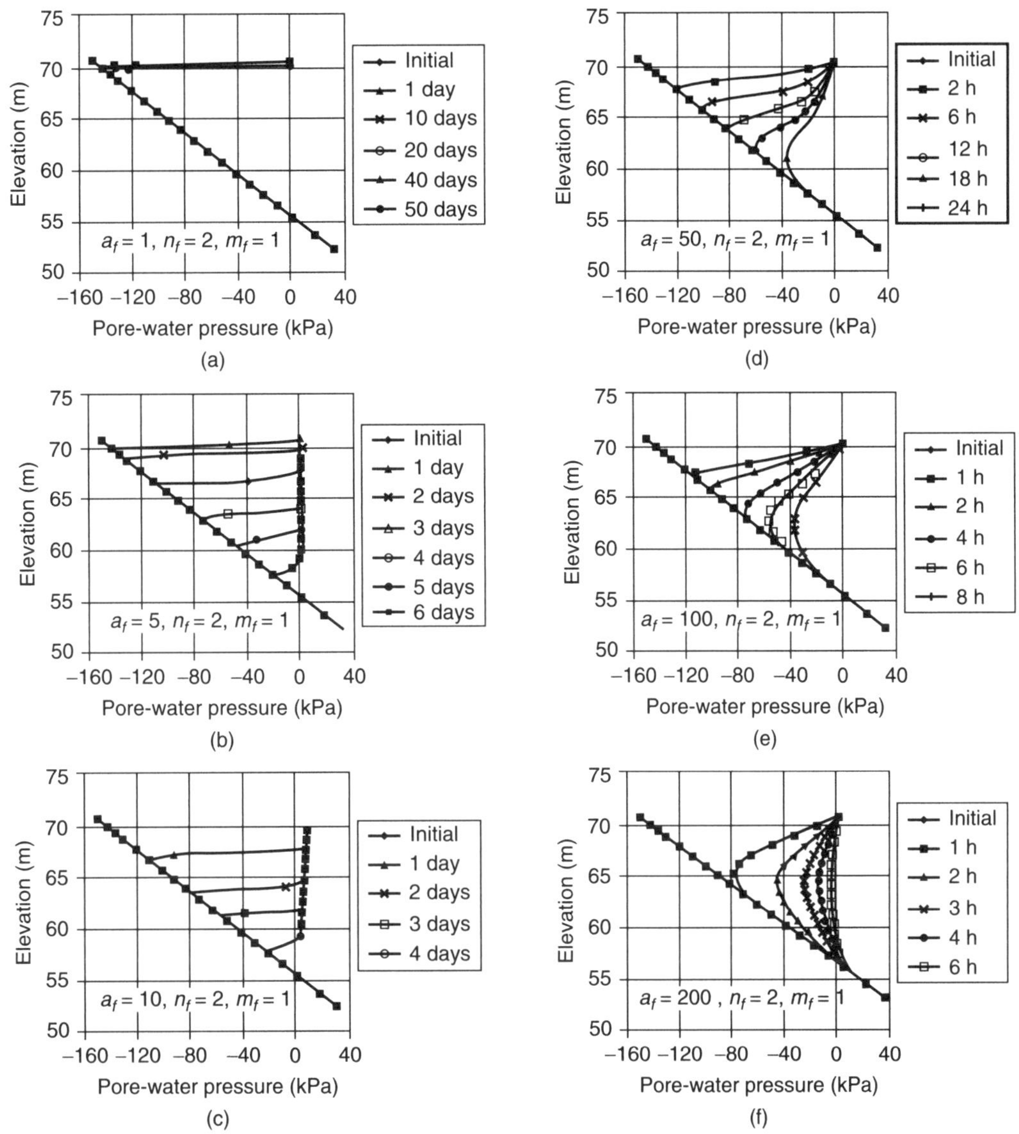

Figure 8.85 Pore-water pressure profiles for soils with various air-entry values (i.e., a parameters) and $k_{sat} = 10^{-5}$ m/s subjected to flux $q = 10^{-5}$ m/s under transient seepage conditions: (a) $a_f = 1$ kPa, (b) $a_f = 5$ kPa, (c) $a_f = 10$ kPa, (d) $a_f = 50$ kPa, (e) $a_f = 100$ kPa, and (f) $a_f = 200$ kPa (after Zhang et al., 2004).

The gradient of pore-water pressure in the transition zone becomes smaller as a_f increases (Fig 8.85). For example, the gradient of the pore-water pressures in the transition zone approaches infinity for soils with $a_f < 10$ kPa. Conversely, the wetting fronts for soils with $a_f > 100$ kPa no longer appear as horizontal lines. According to Eq. 8.60, the gradient of the pore-water pressure depends on the ratio of moisture flux and the coefficient of permeability of the unsaturated soil.

Figure 8.87 shows two examples of pore-water pressure profiles when the rainfall flux is 10^{-6} m/s, which is 10% of the saturated coefficient of permeability of the soils. The most significant difference is seen in the shape of the wetting front. For soils with $a_f = 10$ kPa, the wetting front is approximately horizontal and the infiltration rate is much greater than the unsaturated coefficient of permeability, according to Eq. 8.61. However, for soils with $a_f = 100$, the initial coefficient of permeability in the soil is comparable to the flux rate, and the pore-water pressure gradient approaches zero.

8.4.5 Effect of Varying Saturated Coefficient of Permeability

The pore-water pressure profiles for soils with the same SWCC (i.e., $a_f = 100$, $n_f = 2$, $m_f = 1$) but different saturated coefficients of permeability (i.e., 10^{-7}, 10^{-5}, and 10^{-3} m/s, respectively) are shown in Figures 8.88a–8.88c.

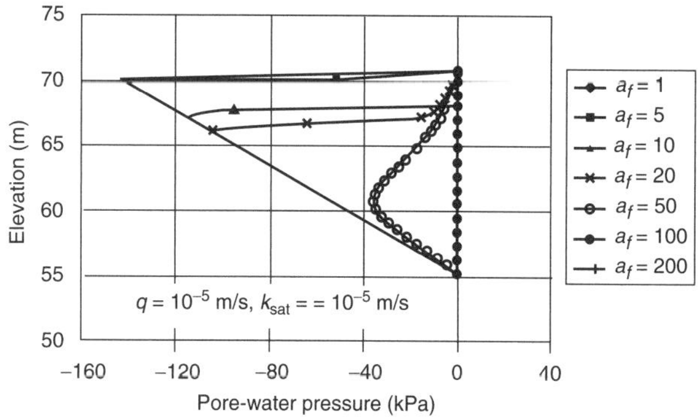

Figure 8.86 Comparison of pore-water pressure profiles for soils with various air-entry values after 24 h of rainfall (after Zhang et al., 2004).

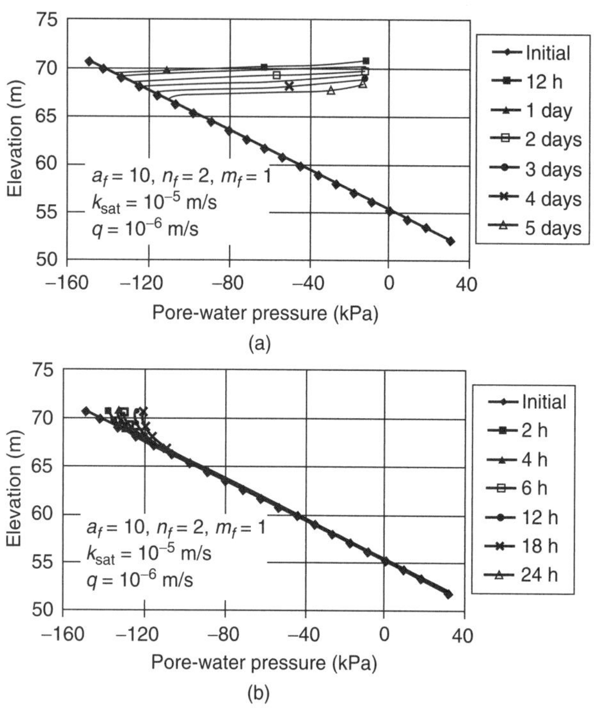

Figure 8.87 Examples of pore-water pressure profiles in soil with two air-entry values subjected to surface flux, q = 10^{-6} m/s: (a) a_f = 10 kPa; (b) a_f = 100 kPa (after Zhang et al., 2004).

The rainfall fluxes are assumed to be equal to the saturated coefficient of permeability for the three cases. The pore-water pressure profiles are similar, but the rates of downward movement of the wetting fronts are distinct, indicating that the rainfall infiltration can be related to the absolute intensity of the rainfall and the soil properties.

8.4.6 Influence of Water Storage on Pore-Water Pressure Profile

Unsaturated soil property functions corresponding to two air-entry-values are shown in Figure 8.89. The infiltration of water into soils with varying air-entry values is shown in

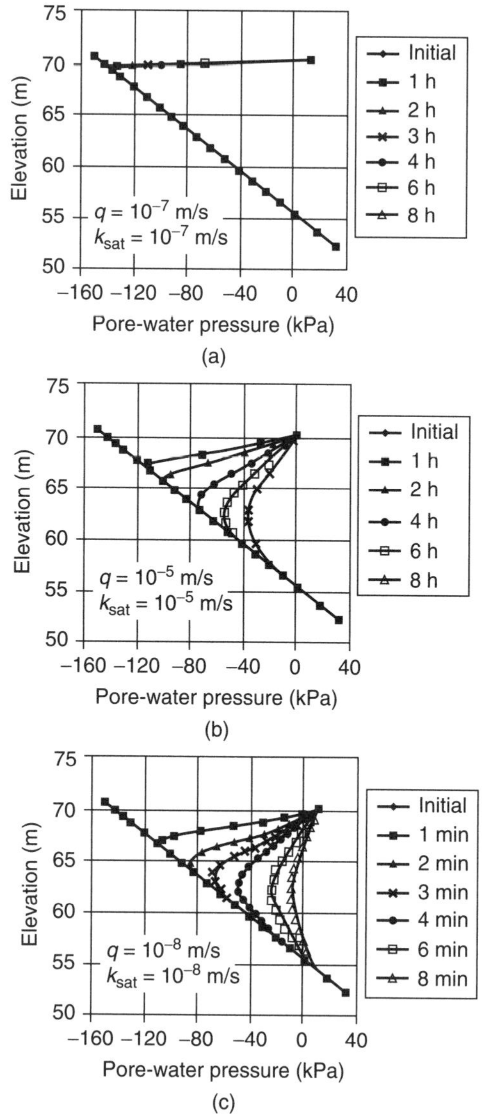

Figure 8.88 Pore-water pressure profiles in slope subjected to the same ratio of $q/k_{sat} = 1$ for soils with $a_f = 100$, $n_f = 2$, $m_f = 1$: (a) $q = 10^{-7}$ m/s and $k_{sat} = 10^{-7}$ m/s; (b) $q = 10^{-5}$ m/s and $k_{sat} = 10^{-5}$ m/s, (c) $q = 10^{-3}$ m/s and $k_{sat} = 10^{-3}$ m/s (after Zhang et al., 2004).

Figure 8.90. The primary difference between the two SWCCs is the air-entry value. The desaturation rate (i.e., $n_f = 2$) is the same for both soils, but the maximum water storage modulus for the soil with $a_f = 1$ is 0.1445 while the soil with $a_f = 10$ kPa is only 0.0145. These values show that the desaturation rate n_f represents the slope at the inflection point of the SWCC and is expressed on a logarithm scale of matric suction while the water storage modulus m_2^w is expressed as the arithmetic slope of the SWCC. The change of matric suction, $\Delta(u_a - u_w)$, is much smaller when $a_f = 1$ than when $a_f = 10$ because the desaturation part is in the low-matric-suction portion for $a_f = 1$ and as a result it has a larger water storage modulus.

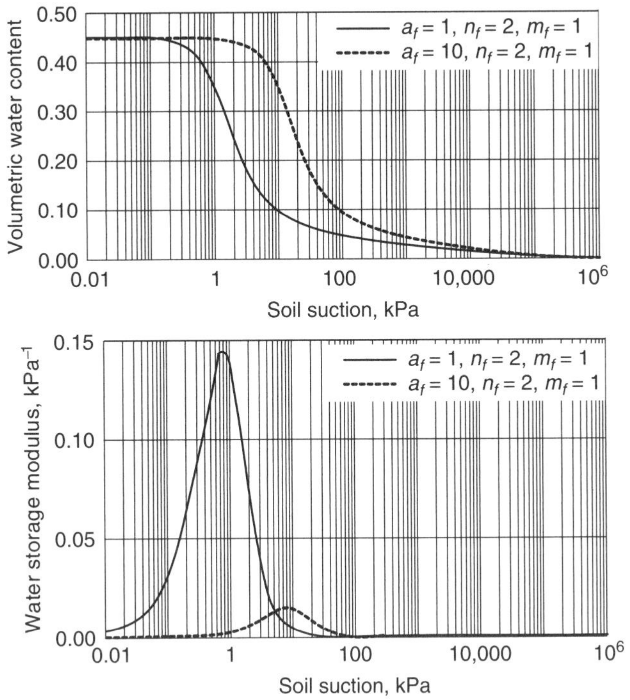

Figure 8.89 Comparison of water storage functions for soils with $a_f = 1$ and $a_f = 10$.

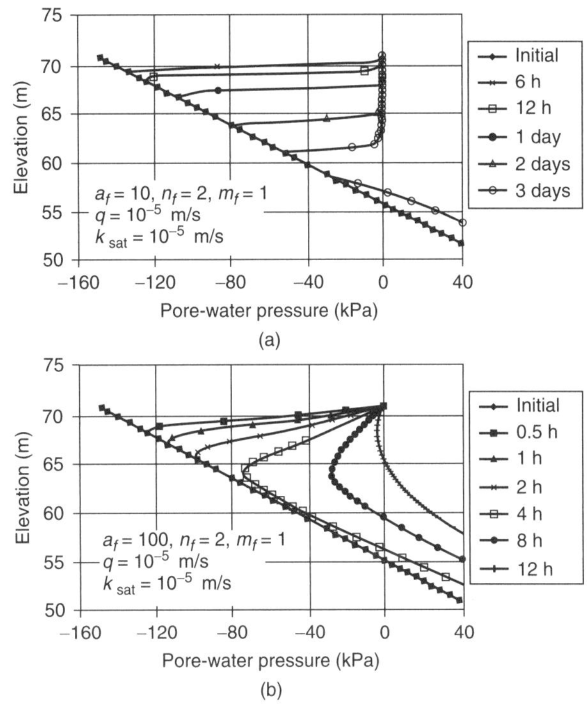

Figure 8.90 Pore-water pressure profiles in a slope (with a water table) subjected to rainfall flux $q = 10^{-5}$ m/s: (a) $a_f = 10$ kPa; (b) $a_f = 100$ kPa (after Zhang et al., 2004).

Differences in the water storage functions help explain the rate of movement of the wetting front for soils with different air-entry values (see Figs 8.85a–8.85f). For soils with the same coefficient of permeability and the same desaturation rate subjected to the same magnitude of rainfall flux, the soil with a lower air-entry value has a greater water storage capacity than the soil with a larger air-entry value. Consequently, the movement of the wetting front is much slower for the soil with the low air-entry value. On the other hand, if the soils have the same SWCC (i.e., the same water storage function), the soil with a smaller saturated coefficient of permeability allows less infiltration than the soil with a larger coefficient of permeability even though both soils have the same water holding capacity. These results illustrate why the rate of downward movement of the wetting front for a soil with $k_s = 10^{-7}$ m/s is much slower than for a soil with $k_s = 10^{-3}$ m/s (See Figs. 8.88a–8.88c).

Freeze (1969) studied the mechanism of groundwater recharge and discharge using a one-dimensional finite difference model involving transient flow through a saturated-unsaturated soil system. Three soils, Del Monte sand, Rehovot sand, and Grenville silt, were used in the study. The permeability function, water storage capacity, and air-entry value were shown to be controlling parameters for the pore-water pressure response of the soil. The wetting front was shown to propagate more slowly through the pore-water pressure profile in the Rehovot sand, which had a larger water storage capacity, than through the Del Monte sand and the Grenville silt.

8.4.7 Effect of Varying Groundwater Boundary Conditions

The groundwater tables in soil slopes can vary significantly depending on the imposed boundary conditions. Figures 8.90 and 8.91 show the pore-water pressure profiles when the groundwater table is allowed to rise during infiltration. The selected rainfall fluxes were 100 and 10% of k_s, respectively. Comparing Fig. 8.90 with Fig. 8.85 for similar pore-water pressure profiles, it can be observed that the rates of downward movement of the wetting front are comparable. The rising of the groundwater table is more obvious for the soils with larger air-entry values because the soil has less water storage capacity. Similar observations can be seen when comparing Fig. 8.91 and Fig. 8.87.

The pore-water pressure profiles presented above were for section *X-X* at the midpoint of the slope (see Fig. 8.75). However, the pore-water pressure distribution across the entire slope appeared to a give a similar message with regard to the permanence of matric suction.

8.4.8 Geotechnical Engineering Implications

Numerous research studies have been undertaken that show the changes in pore-water pressure profiles in slopes subjected to various surface moisture flux conditions. Much has been learned about the response of the soil in the unsaturated soil zone to rainfall conditions, but this seems to be where most of the studies have ended. Consequently,

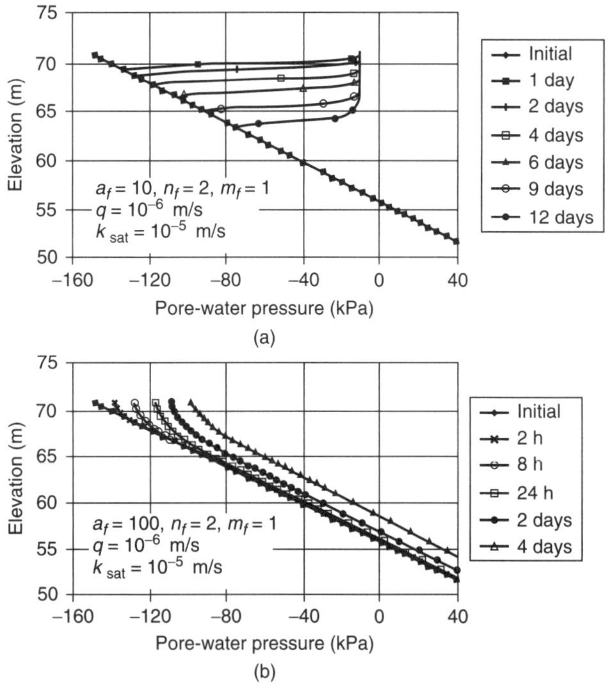

Figure 8.91 Pore-water pressure profiles in a slope (with a water table) subjected to rainfall flux $q = 10^{-6}$ m/s: (a) $a_f = 10$ kPa; (b) $a_f = 100$ kPa (after Zhang et al., 2004).

the application of numerical modeling of rainfall infiltration has not significantly influenced engineering design. The use of matric suction in slope design has not yet received general acceptance in routine engineering practice. There appears to be a perception that there are insufficient field measurements as well as modeling studies to demonstrate the long-term reliance upon matric suction in soil slopes during and after rainstorms. The results of the above parametric numerical study contribute to a better understanding of infiltration processes. The studies assist geotechnical engineers in formulating more appropriate engineering design protocols and practices where the permanence of matric suction is part of the functionality of slopes.

Numerical studies allow for a wide variety of factors to be considered during the design of a soil slope. Once the SWCCs and the saturated coefficients of permeability for the materials involved are known, the times required for the dissipation of negative pore-water pressure under various rainstorm conditions can be modeled. If the matric suction requires more time to dissipate than the possible duration of heavy rainfalls, then the matric suction profiles should be taken into consideration in the design of the soil slope. On the other hand, if the dissipation or loss of matric suctions occurs in a relatively short time, the soil slope needs to be designed accordingly.

The long-term matric suction changes in a slope are controlled by factors related to the ground surface moisture flux as well as the hydraulic properties of the soils near the ground surface. The ratio of the ground surface moisture flux to the saturated coefficient of permeability of the soil near the ground surface appears to be the primary variable that should be considered when assessing the permanence of matric suction.

It is important to use design features that tend to reduce the infiltration flux at ground surface. A common practice in Hong Kong is to provide a layer of soil cement and lime plaster cover called "chunam" on the soil slopes. Any near-ground-surface layer of reduced permeability assists in maintaining the matric suction in the underlying soils. Lim et al. (1996) carried out a field instrumentation program to monitor negative pore-water pressures in residual soil slopes in Singapore. The slopes were protected by various types of surface covers. Changes in matric suction due to changes in ground surface moisture flux were found to be most significant below a bare slope and least significant under a canvas-covered slope. Other relatively impermeable surface covers can also be adopted depending on the saturated coefficients of permeability of the material at the surface of a slope.

It is also effective to maintain negative pore pressures in a slope through the use of a soil of reduced saturated permeability (see Fig. 8.91). After the 1976 Sau Mau Ping failure in Hong Kong, the investigation panel (Government of Hong Kong, 1976) recommended that the minimum stabilization of a suspect slope should consist of removing the loose surface soil by excavating the slope to a vertical depth of not less than 3 m for slopes that were not steeper than 1 on 1.5 and recompacting the soil to 95% of standard compaction density. The average saturated coefficient of permeability for the Sau Mau Ping earthfill material decreased by three times when the relative compaction of the fill was increased from 82 to 95%.

In some cases it might not be practical to prevent the infiltration of rainfall into a soil slope through use of a relatively impermeable surface cover. It still may not mean that the matric suctions will go toward zero with time. Steady-state simulations might reveal a permanence of the negative pore-water pressure conditions. On the other hand, it may be important to assume zero matric suction in the slope for design purposes when the results of numerical modeling simulations considering realistic site and weather conditions show that negative pore-water pressures cannot be maintained. The risks associated with a possible slope failure also need to be taken into consideration.

The initial pore-water pressure conditions in the slope and the boundary conditions of the slope are important factors to be considered in the quantification of possible changes in pore-water pressures. Numerical studies provide an opportunity for greater ingenuity and design considerations for soil slopes. Field measurements and monitoring are needed to provide greater confidence with respect to the use of computed changes in matric suction.

Matric suction can be eliminated after a rainfall in some situations while in other situations matric suction may not disappear over a reasonable time span. This phenomenon has been observed based on in situ measurements of negative pore-water pressures in soil slopes as well as from numerical studies. The fundamental behavior of infiltration under both long-term steady-state rainfall and unsteady-state rainfall conditions needs to be understood in order to identify the most important factors controlling rainfall infiltration. Engineers need to formulate appropriate questions to consider when analyzing the long-term stability of a slope. Measures can often be taken to maintain matric suctions in slopes.

CHAPTER 9

Air Flow through Unsaturated Soils

9.1 INTRODUCTION

Air behaves as a compressible fluid whose density (or the concentration of air molecules per unit volume) changes significantly with a change in air pressure or temperature. The effect of changes in air density due to pressure changes is taken into consideration in the PDEs that describe air flow. The effect of changes in air density due to temperature changes is considered as a mechanism that couples with heat flow.

Air is a gas comprised mainly of oxygen and nitrogen along with small percentages of other gases and water vapor. Each component of air behaves as an independent gas, but for most geotechnical engineering problems it is possible to consider the combination of gases as a single gas, namely, air. It is possible to write the constitutive law of air flow in the form of Fick's law or Darcy's law. It is useful to be able to use either Fick's law or Darcy's law in geotechnical engineering and be able to convert between the intrinsic permeability of a soil and the measured hydraulic conductivity (or coefficient of permeability). The Klinkenberg effect associated with the absolute pressure at which air flow measurements are made is also described in this chapter.

A basic understanding of the physical concepts of air flow is presented along with example problems involving air movement. The need to accommodate air phase compressibility makes the formulation of the air flow PDEs appear to be more complex than the PDEs for water flow. On the other hand, the low density of air makes it possible for the gravity term (i.e., unit weight of air) to be omitted.

The geotechnical engineer does not frequently encounter air flow problems; however, when air flow problems are encountered, it is important to understand the differences between the physics of water flow and the physics of air flow. One of the applications of air flow involves forcing air through the soil for the purpose of removing volatile soil contaminants. If a positive air pressure is used to force air through the soil, the process is referred to as *air sparging*. If a negative air pressure (i.e., a vacuum) is used to produce air flow through the soil, the process is referred to as *soil vapor extraction*. Increased air pressure (i.e., air pressure translation) is sometimes used during the construction of tunnels under water and is referred to as an *air cushion*. There are certain conditions that must be met in order for the air pressure translation process to be successful. For example, the air-entry value of the material around the tunnels should be greater than the required air pressure translation.

Ba-Te (2005) studied the effect of pore-air pressure reduction on the stability of slopes. It was found that even a relatively small reduction in pore-air pressure can result in a significant increase in the factor of safety of a slope. The results were confirmed as part of an analytical study as well as laboratory experiments utilizing a tilting box of sand soil.

The air in an unsaturated soil consists of two forms: "free air" and "dissolved air." Water has the appearance of being a permeable material with a diffusion coefficient that has the form of a coefficient of permeability of water with respect to air flow. The flow of air in and out of solution obeys the laws of a compressible fluid flowing through a porous medium. The volume-mass relations for free air and dissolved air were described in Chapter 2. The axis translation technique in the laboratory requires that consideration be given to the effect of air diffusion through water.

There are certain problems where air pressure changes and temperature changes are interrelated and occur simultaneously. Interrelated air flow and heat flow are common to the processes that take place in solid waste containment (i.e., landfills). The same is true for waste materials from mining operations (e.g., waste rock). Chemical reactions often generate heat which, in turn, changes air pressure and causes the flow of air by convection. Air flow and heat flow often occur simultaneously, and it becomes necessary to consider the respective processes in either an uncoupled or a coupled manner. However, it is useful to first understand air flow as an independent physical process prior to combining the process with heat flow.

9.2 THEORY OF FREE AIR FLOW

Air flow can occur through the mechanisms of convection, conduction, and radiation. This chapter considers air flow due to convection. Air flow by conduction is assumed to be negligible because air acts as an insulator. Air flow resulting from radiation was discussed in Chapter 6 for the assessment of actual evaporative flux for climatic boundary conditions. Air flow by diffusion (e.g., air dissolving in water) and air flow by advection (e.g., air swept along by water flow) are briefly discussed.

Air in an unsaturated soil can exist either as a continuous phase or in the form of occluded air bubbles (Fig. 9.1). The air phase generally becomes continuous as the degree of saturation becomes less than about 85% (Corey, 1957). The flow of free air through an unsaturated soil commences as the air phase becomes continuous. When the degree of saturation is above about 90%, the air phase generally exists as occluded air bubbles and air flow is reduced to the diffusion of air through the pore-water (Matyas, 1967) as well as dissolved air flow by advection.

The flow of air through a soil can be caused by factors such as variations in barometric pressure, rainwater infiltration that compresses the air in the pores of the soil, temperature changes, and externally applied loads. There are also some situations where human impact results in changes in air pressure and subsequent convective air flow.

9.2.1 Driving Potential for Air

The flow of free air is governed by a concentration, density, or pressure gradient when the air phase is continuous. The elevation gradient is negligible. The pressure gradient is most commonly considered as the only driving potential for the air phase. Both Fick's law and Darcy's law can be used to describe the flow of air through a porous medium. Fick's law may appear to be more fundamentally correct, but it is also useful to be able to apply a Darcy-type formulation with its associated air coefficient of permeability.

9.3 FICK'S LAW AND DARCY'S LAW FOR AIR FLOW

Fick's (1855) law is often used to describe the diffusion of gases through liquids. A modified form of Fick's law can be applied to the air flow process. Fick's first law states that

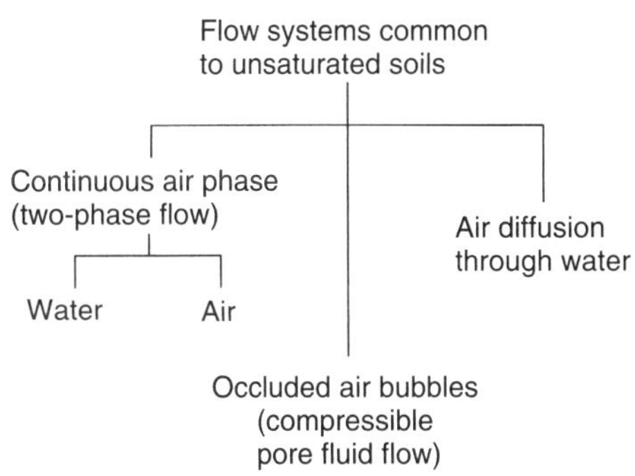

Figure 9.1 Subdivision of air flow systems in unsaturated soil.

the rate of mass transfer of a diffusing substance across a unit area is proportional to the concentration gradient of the diffusing substance. It is reasonable to apply Fick's law for the flow of free air through an unsaturated soil, the flow of water vapor in response to a vapor pressure gradient, and the diffusion of air through fluids such as water.

Air has a low density and as a result the driving potential for flow can be taken as the air pressure gradient. The mass of air flow, m_a (as opposed to the volume of air flow), can be written using Fick's (1855) constitutive flow relationship:

$$m_{ax} = -D_{ax}\frac{du_a}{dx} \tag{9.1}$$

$$m_{ay} = -D_{ay}\frac{du_a}{dy} \tag{9.2}$$

$$m_{az} = -D_{az}\frac{du_a}{dz} \tag{9.3}$$

where:

m_{ax}, m_{ay}, m_{az} = mass flow rate in the x-, y-, and z-directions, respectively, and
D_{ax}, D_{ay}, D_{az} = air diffusivity in the x-, y- and z-directions, respectively.

The air flow law can also be written as a velocity of air flow, v_a, similar to Darcy's (1856) law, thereby taking on the following form (Blight, 1971):

$$v_{ax} = -k_{ax}\frac{du_a}{dx} \tag{9.4}$$

$$v_{ay} = -k_{ay}\frac{du_a}{dy} \tag{9.5}$$

$$v_{az} = -k_{az}\frac{du_a}{dz} \tag{9.6}$$

The air coefficient of permeability k_a (and air diffusivity) is a mathematical function in the sense that the transmission of air varies with the amount of air in the soil. The amount of air can be described using the SWCC. Figure 9.2a shows a SWCC for sand and Fig. 9.2b illustrates the form of the air permeability function (Ba-Te et al., 2005). The air permeability function takes on an inverse form to that of the water permeability function.

The magnitude of the air coefficient of permeability tends toward the diffusion of air through water when soil suction becomes less than the air-entry value of the soil. The air coefficient of permeability increases by several orders of magnitude as suction is increased once the suction in the soil exceeds the air-entry value. The low viscosity of air means that air can flow through a soil with much greater ease than water flow through a soil of equal porosity.

The void space of a porous medium (i.e., porosity of the soil) can be used as the reference. There is one extreme situation where the void space is filled with water and another

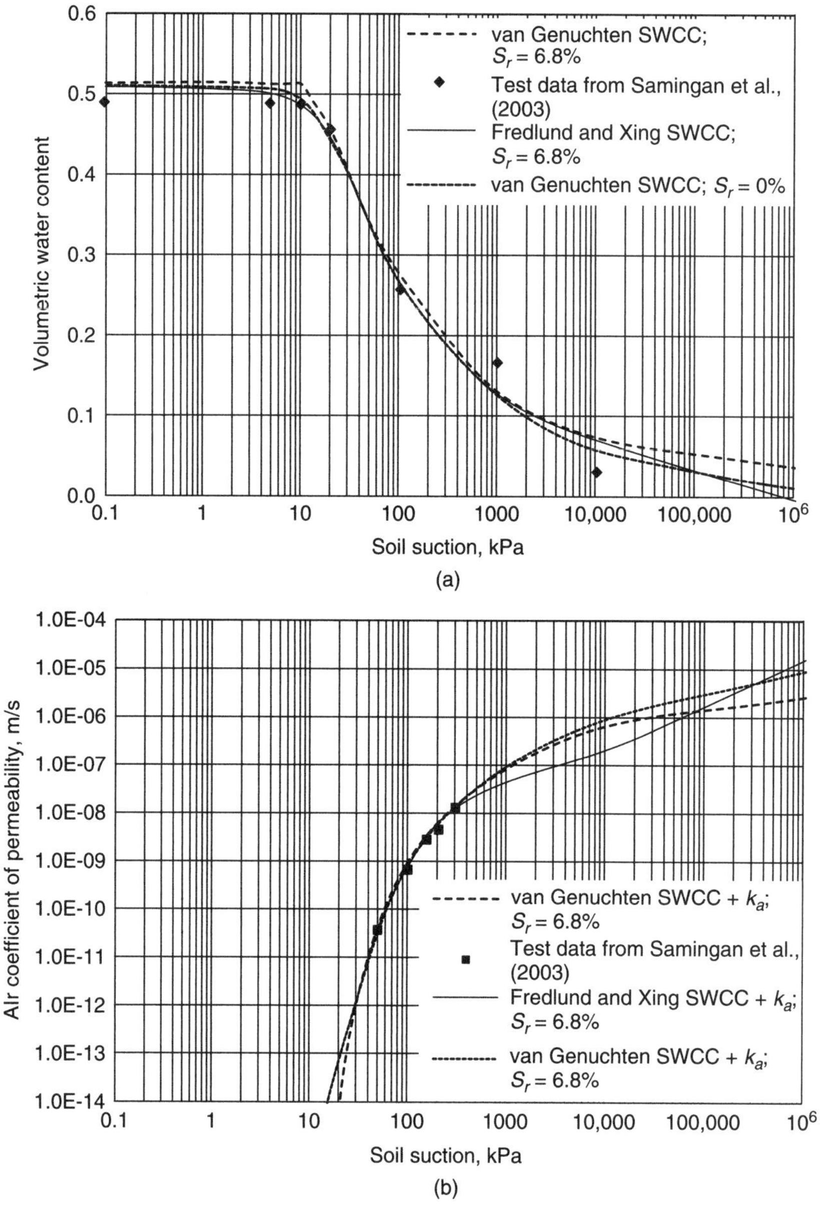

Figure 9.2 SWCC and air permeability function for sand tested by Samingan et al. (2003): (a) SWCC data and best-fit SWCC for sand; (b) measured and calculated air permeability function for sand (after Ba-Te et al., 2005).

extreme situation where the void space is filled with air. It is possible to visualize the soil as having a coefficient of permeability with respect to water and a coefficient of permeability with respect to air.

The mass rate of air flow and the concentration gradient in the air are computed with respect to a unit area and a unit volume of the soil:

$$J_a = -D_a \frac{\partial C}{\partial y} \tag{9.7}$$

where:

J_a = mass rate of air flowing across a unit area of the soil,
D_a = transmission constant for air flow through a soil,
C = concentration of the air expressed in terms of the mass of air per unit volume of soil, and
$\partial C/\partial y$ = concentration gradient in the y-direction.

The negative sign in the air flow equation indicates that air flows in the direction of a decreasing concentration gradient. Air flow equations can similarly be written for each of the three Cartesian coordinate directions.

The concentration of air with respect to a unit volume of the soil can be written as

$$C = \frac{M_a}{V_a/(1 - S)n} \tag{9.8}$$

where:

M_a = mass of air in the soil,
V_a = volume of air in the soil,
S = degree of saturation, and
n = porosity of the soil.

Substituting the density-of-air term ρ_a for M_a/V_a in the above equation gives

$$C = \rho_a(1 - S)n \tag{9.9}$$

Air density is related to the absolute air pressure in accordance with the universal gas law [i.e., $\rho_a = (\omega_a \overline{u}_a)/RT_K$, where, ω_a = molecular mass of air, R = universal (molar) gas constant, and T_K = absolute temperature; see Eq. 2.19]. Therefore, the concentration gradient can also be expressed with respect to a pressure gradient in the air. The gauge air pressure can be used in reformulating Eq. 9.7:

$$J_a = -D_a \frac{\partial C}{\partial u_a} \frac{\partial u_a}{\partial y} \tag{9.10}$$

where:

u_a = pore-air pressure and
$\partial u_a/\partial y$ = pore-air pressure gradient in the y-direction (or similarly in the x- and z-directions).

A modified form of Fick's law is obtained by defining a coefficient of transmission for air flow through soils, D_a^* :

$$D_a^* = D_a \frac{\partial C}{\partial u_a} \tag{9.11}$$

or

$$D_a^* = D_a \frac{\partial[\rho_a(1 - S)n]}{\partial u_a} \tag{9.12}$$

The coefficient of transmission D_a^* is a function of the volume-mass properties of the soil (i.e., S and n) and air density. Substituting D_a^* into Eq. 9.10 results in an equation for the mass of air flow:

$$J_a = -D_a^* \frac{\partial u_a}{\partial y} \tag{9.13}$$

A modified form of Fick's law has been used in geotechnical engineering to describe air flow through soils (Blight, 1971). The coefficient of transmission D_a^* can be related to the air coefficient of permeability k_a, which can be measured in the laboratory.

Steady-state air flow can be established through an unsaturated soil specimen that has a fixed degree of saturation (or constant matric suction). The soil specimen is treated as an element of soil having an air coefficient of permeability corresponding to the selected degree of saturation. The air coefficient of permeability is assumed to be constant throughout the soil specimen. Steady-state air flow is produced by applying an air pressure gradient across the soil specimen. The amount of air flowing through the soil specimen is measured at the exit point. The flow rate must be measured at a designated constant-pressure condition (i.e., usually at 101.3 kPa absolute or zero gauge pressure) since air is compressible (Matyas, 1967).

The mass rate of the air flow is measured at a constant air density ρ_{ma}. Equation 9.13 can be rewritten as a volume rate of air flow for the case of constant air pressure:

$$\rho_{ma} \frac{\partial V_a}{\partial t} = -D_a^* \frac{\partial u_a}{\partial y} \tag{9.14}$$

or

$$v_a = -D_a^* \frac{1}{\rho_{ma}} \frac{\partial u_a}{\partial y} \tag{9.15}$$

where:

ρ_{ma} = constant air density corresponding to the pressure used in the measurement of the mass rate (i.e., at the exit point of flow) and
$\partial V_a/\partial t$ = volume rate of the air flow across a unit area of the soil at the exit point of flow, designated as the flow rate v_a.

The pore-air pressure u_a can also be expressed in terms of a pore-air pressure head h_a using a constant air density ρ_{ma}:

$$v_a = -D_a^* g \frac{\partial h_a}{\partial y} \tag{9.16}$$

where:

h_a = pore-air pressure head (i.e., $u_a/\rho_{ma}g$) and
$\partial h_a/\partial y$ = pore-air pressure head gradient in the y-direction, designated as i_{ay}.

Equation 9.16 has the same form as Darcy's equation for the air phase and can be written as follows:

$$v_a = -k_a \frac{\partial h_a}{\partial y} \tag{9.17}$$

Equation 9.17 was earlier written for each of the Cartesian coordinate directions (i.e., Eqs. 9.4, 9.5, and 9.6). The relationship between the air coefficient of transmission D_a^* and the air coefficient of permeability k_a is as follows:

$$k_a = D_a^* g \tag{9.18}$$

The hydraulic head gradient in Eq. 9.17 consists of the pore-air pressure head gradient as the driving potential and has been used to compute the air coefficient of permeability k_a (Barden, 1965; Matyas, 1967; Langfelder et al., 1968; Barden et al., 1969b; Barden and Pavlakis, 1971).

Air permeability measurements can be performed at various applied matric suction values (or different degrees of saturation) to establish an air permeability functional relationship $k_a(u_a - u_w)$ or $k_a(S)$. The air permeability function also applies to the air coefficient of transmission D_a^* since the two coefficients are related by the gravitational constant, g.

Experimental verifications have been performed for Fick's law and Darcy's law and some results are presented in Fig. 9.3 (Blight, 1971). A series of permeability tests were performed by establishing steady-state air flows through dry soils. The soils were assumed to have a rigid structure with no measurable volume change during the tests. The air flow measurements were referenced to the air-filled pore space (Blight, 1971). The mass rate of air flow must be multiplied by the air porosity, n_a, as shown in Fig. 9.3a, when using the bulk soil as the reference state.

The applicability of Fick's law to air flow is demonstrated in Fig. 9.3b. The mass flow rate J_a is almost linearly proportional to the pore-air pressure gradient $(\partial u_a/\partial y)$, with D_a^* being the coefficient of proportionality corresponding to a small change in the pore-air pressure gradient. It should be noted that the air pressure gradients used in the above experiment were relatively high. The magnitudes of D_a^* and k_a vary with the volume-mass properties of an unsaturated soil.

9.3.1 Coefficient of Permeability with Respect to Air Flow

Several relationships have been proposed between the air coefficient of permeability and the volume-mass properties of a soil. The coefficient of transmission D_a^* can either be computed in accordance with Eq. 9.18 or measured directly in experiments. The coefficient of permeability for the air phase, k_a, is a function of the fluid (i.e., air) and soil volume-mass properties. The fluid properties are generally considered to be constant during the flow process. Therefore, the air coefficient of permeability can be expressed as a function of the volume-mass properties of the soil. The volumetric percentage of air in the pores largely controls the flow rate of air through a soil. The air coefficient of permeability increases as soil suction increases (or the degree of saturation decreases), giving rise to the terminology of an air permeability function.

9.3.2 Relationship between Air Coefficient of Permeability and Degree of Saturation

The air coefficient of permeability of an unsaturated soil is strongly influenced by the degree of saturation of the soil. The air coefficient of permeability approaches its maximum value when the water degree of saturation is low. The air coefficient of permeability decreases as the water degree of saturation increases until the suction reaches the air-entry value of the soil. The air-entry value is the point where air starts to enter the largest pores of the soil. Air flow takes the form of air diffusion through the soil-water when soil suction is below the air-entry value of the soil. At this point, the air coefficient of permeability is extremely small.

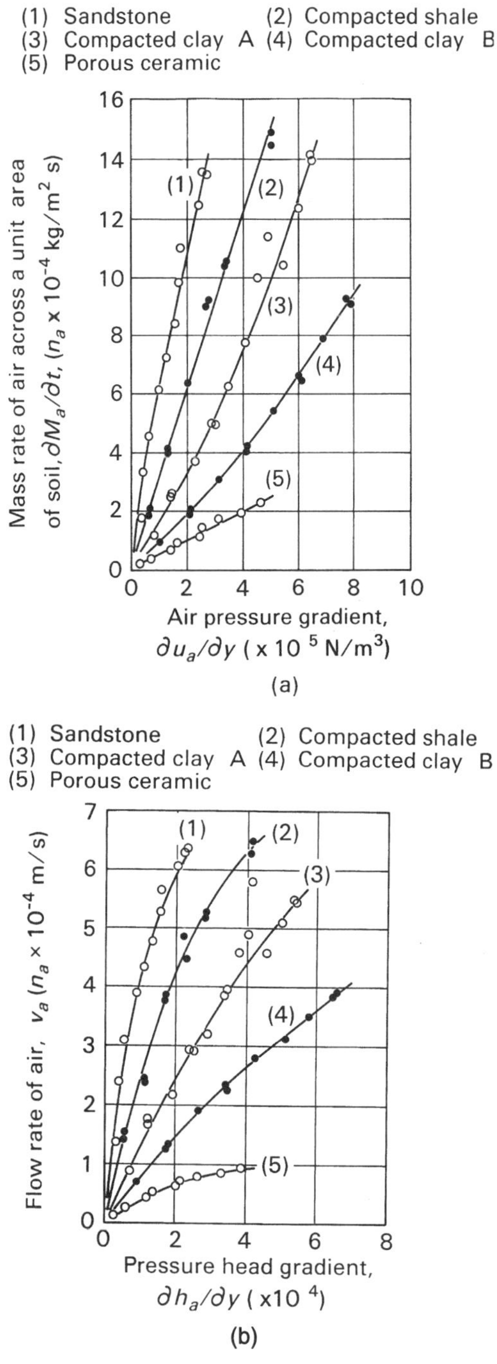

Figure 9.3 Verifications of Fick's and Darcy's laws for flow of air through dry porous medium: (a) mass of air flow versus air pressure gradient (Fick's law); (b) flow rate of air versus pressure gradient (Darcy's law) (after Blight, 1971).

The prediction of the air coefficient of permeability based on the pore-size distribution and the SWCC has also been proposed for the air phase. The relative air coefficient permeability function k_{ra} is essentially the inverse of the relative water coefficient-of-permeability function k_{rw}, as shown in Fig. 9.4. The data are for Hygiene sandstone, and the respective air

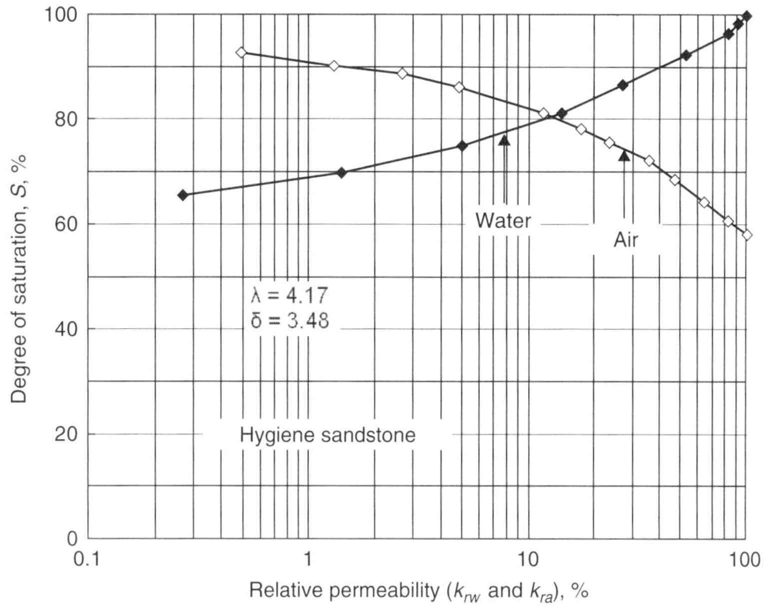

Figure 9.4 Relative permeability of water and air as function of degree of saturation during drainage (modified from Brooks and Corey, 1964).

and water coefficients of permeability are plotted relative to completely dry and wet conditions. In other words, the relative air coefficient of permeability is presented as a ratio of the air permeability corresponding to the completely dry condition. Similarly, the relative water coefficient of permeability is plotted relative to the water permeability corresponding to the water-saturated conditions.

The following equations were used by Brooks and Corey (1964) to describe the $k_a(S_e)$ function:

$$k_a = 0.0 \qquad \text{for } u_a - u_w \leq (u_a - u_w)_b \tag{9.19}$$

$$k_a = k_d(1 - S_e)^2(1 - S_e^{(2+\lambda)/\lambda}) \qquad \text{for } u_a - u_w > (u_a - u_w)_b \tag{9.20}$$

where:

k_d = coefficient of permeability with respect to the air phase for a soil at a degree of saturation of zero.

The air coefficient of permeability k_a corresponding to various degrees of saturation can be computed using Eq. 9.20. The relative coefficient of permeability of air permeability, k_{ra} (%), can be written as follows:

$$k_{ra} = \frac{k_a \times 100}{k_d} \tag{9.21}$$

A comparison between the computed relative air permeability for several soils with measured experimental data is shown in Fig. 9.5. The data from each of the soils show that the air coefficient of permeability increases with respect to the applied soil suction until a maximum value is obtained. The data also show that at under low applied soil suctions the soil becomes saturated and the air coefficient of permeability tends to a small value (i.e., diffusivity of air through the water in the soil).

9.3.3 Relationship between Air Coefficient of Permeability and Matric Suction

Equation 9.20 can also be expressed in terms of matric suction $(u_a - u_w)$. The functional form for the air coefficient of permeability takes on an inverse form of the equation suggested by Brooks and Corey (1964) for the water permeability function:

$$k_a = 0.0 \quad \text{for } u_a - u_w \leq (u_a - u_w)_b \tag{9.22}$$

$$k_a = k_d\left[1 - \left(\frac{(u_a - u_w)_b}{u_a - u_w}\right)^{\lambda}\right]^2\left[1 - \left(\frac{(u_a - u_w)_b}{u_a - u_w}\right)^{2+\lambda}\right] \quad \text{for } u_a - u_w > (u_a - u_w)_b \tag{9.23}$$

Figure 9.2 illustrates the agreement between measured data and the theoretical air coefficient-of-permeability function described using Eq. 9.23.

Several studies have been undertaken to measure the air permeability of compacted soils. The coefficient of permeability with respect to air, k_a, decreases as the degree of saturation of the soil increases (Ladd, 1960; Olson, 1963; Langfelder et al., 1968; Barden and Pavlakis, 1971). Figure 9.6 presents air and water coefficients of permeability for a soil compacted at different water contents or matric suction values by Barden and Pavlakis (1971). The air and water coefficients of permeability k_a and k_w were measured on the same soil specimen

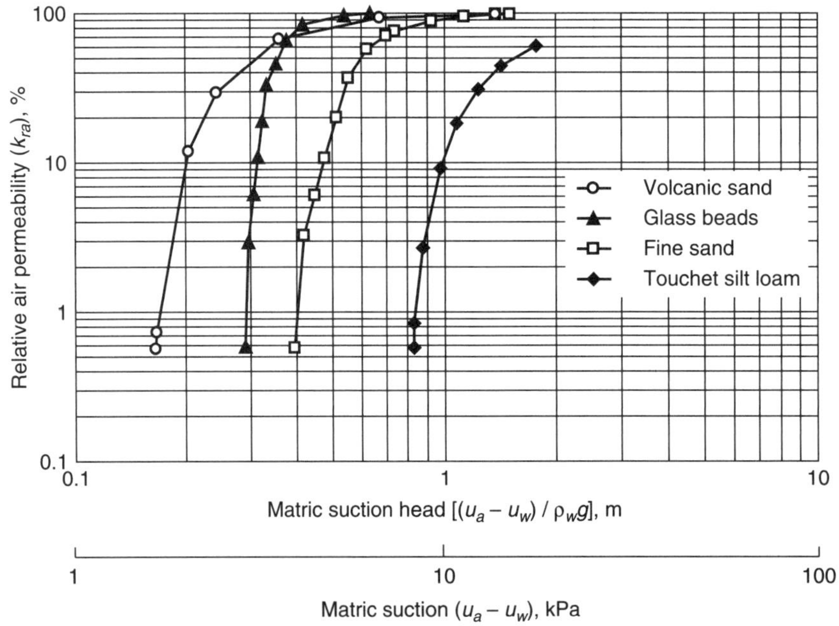

Figure 9.5 Measured data and calculated relative permeability for air as function of matric suction (modified from Brooks and Corey, 1964).

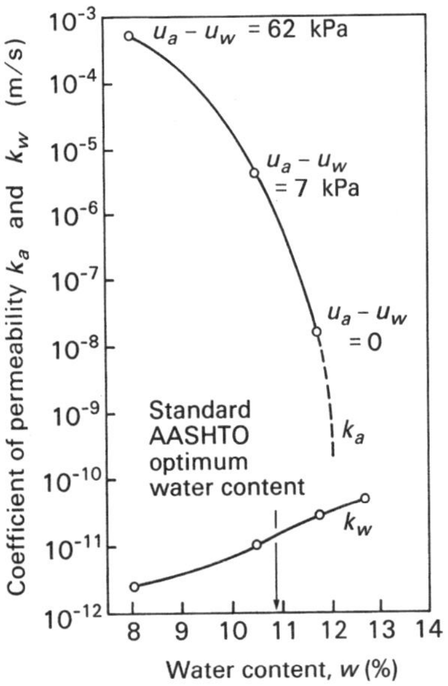

Figure 9.6 Air coefficient of permeability k_a and water coefficients of permeability k_w as function of gravimetric water content for Westwater soil (after Barden and Pavlakis, 1971).

under steady-state flow conditions subjected to small pressure gradients. The air coefficient of permeability k_a decreased rapidly as optimum water content was approached. At this point, the air phase becomes occluded, and the flow of air takes place as the diffusion of air through water. The occluded stage for soils with high clay content usually occurs at water contents slightly above optimum water content (Matyas, 1967; Barden and Pavlakis, 1971).

Although the air coefficient of permeability decreases and the water coefficient of permeability increases with increasing water content, the air permeability values remain significantly larger than the water permeability values for all water contents (Fig. 9.6). The difference in the viscosity of air and water is one of the main reasons for the air coefficient of permeability being greater than the water coefficient of permeability.

The coefficient of permeability is inversely proportional to the absolute (i.e., dynamic) viscosity of the fluid, υ. The absolute viscosity of water, υ_w, is approximately 56 times the absolute viscosity of air, υ_a, at an absolute pressure of 101.3 kPa and a temperature of 20°C. Assuming that the volume-mass properties of a soil do not differ for completely saturated and completely dry conditions, the saturated water coefficient of permeability would be expected to be about 56 times smaller than the air coefficient of permeability at dry condition (Koorevar et al., 1983). It should be noted that other factors also have an influence on the measured air and water coefficients of permeability.

Another factor affecting the measured air coefficient of permeability is the method of compaction. A dynamically compacted soil usually has a higher air coefficient of permeability than a statically compacted soil at the same density.

The air coefficient of transmission D_a^* can be obtained by dividing the air coefficient of permeability k_a by the gravitational acceleration g. If the gravitational acceleration is assumed to be constant, D_a^* functions are similar to the above air permeability k_a functions.

9.3.4 Intrinsic Permeability

Intrinsic permeability K is a fluid flow property related to the nature of a porous medium. Intrinsic permeability is computed by taking the properties of the pore fluid into consideration. Bear (1972) called intrinsic permeability the "medium's permeability" and described it as "an average (or macroscopic) medium property that measures the ability of the porous medium to transmit fluid through it."

Two primary fluid properties that influence the hydraulic conductivity k are commonly measured in the laboratory (Fredlund and Rahardjo, 1993a). The fluid properties are (1) density of the fluid, ρ (kg m^{-3}), and (2) dynamic viscosity of the fluid, υ (Pa s or kg m^{-1} s^{-1}). The mathematical relationship between intrinsic permeability K(m^2) and hydraulic conductivity k (m s^{-1}) can be written as follows:

$$k = \frac{\rho g}{\upsilon} K = \frac{\gamma}{\upsilon} K \tag{9.24}$$

where:

g = gravitational acceleration, m s^{-2},
ρ = density, kg/m^3, and
γ = unit weight of the medium, N m^{-3}.

The pore fluid properties can be "scaled" to the properties of water at standard conditions (Parker et al., 1987). Using the properties of water at 20°C a hydraulic conductivity of 10^{-5} m/s measured using water will give an intrinsic permeability of approximately 10^{-12} m^2. The density of water was taken to be 1000 kg/m^3 and the dynamic viscosity υ was taken to be 1 cP (where 1 cP = 10^{-3} Pa s = 1 mPa s).

Intrinsic permeability can be used to convert flow properties used in one discipline (e.g., geotechnical engineering) to flow properties used in another discipline (e.g., petroleum and soil science disciplines). The concept of intrinsic permeability also has an important role to play in unsaturated soil mechanics. For example, the measurement of saturated water coefficient of permeability can be used to calculate a value for air coefficient of permeability for a dry soil at the same void ratio. Or vice versa, air flow measurements in the laboratory or field on a dry soil can be used to compute saturated coefficient of permeability. Saturated soil-water coefficient-of-permeability measurements can also be used in conjunction with a SWCC to estimate the air permeability function for an unsaturated soil. The properties of the pore fluid can provide an estimate of the relative ease with which air and water can flow through a soil.

The viscosity of water and air are dependent upon temperature as mentioned in Chapter 2 (Tuma, 1976). Dynamic viscosity is presented in *centipoises*, where 1 cP = 10^{-3} kg m^{-1} s^{-1}. At 20°C, the viscosity of water is 1.009 cP whereas the viscosity of air is only 0.0179 centipoise; a viscosity ratio of 56.37. The density of water changes with temperature and typical values were shown in Chapter 2 (McCutcheon et al., 1993).

Let us consider air flow and water flow through a horizontal tube, porous medium with a cross-sectional area A and length L under an applied entrance pressure p_1 and an exit pressure p_2 where $p_1 > p_2$. The test is performed under isothermal conditions at 20°C. The air flow rate can be written in accordance with Darcy's law:

$$Q_a = \frac{K}{\upsilon_a} \frac{p_1 - p_2}{L} A \tag{9.25}$$

where:

Q_a = volumetric flow rate, m^{3}s^{-1},
υ_a = dynamic viscosity of air, and
K = intrinsic permeability of the porous medium.

Similarly, the water flow rate (m^3 s^{-1}) can be written using Darcy's law:

$$Q_w = \frac{K}{\upsilon_w} \frac{p_1 - p_2}{L} A \tag{9.26}$$

where:

Q_w = volumetric flow rate and
υ_w = dynamic viscosity of water.

It is possible to compare the flow volumes of air and water since the intrinsic permeability is a constant. The air volume flow rate would be 56.4 times more than the water volume flow rate using dynamic viscosities corresponding to 20°C:

$$\frac{Q_a}{Q_w} = \frac{\upsilon_w}{\upsilon_a} = \frac{1.009}{0.0179} = 56.4 \tag{9.27}$$

The ratio of air coefficient of permeability k_a to water coefficient of permeability k_w must take into consideration the relative densities of air and water. The ratio of the air coefficient of permeability to water coefficient of permeability corresponding to 20°C and standard pressure (i.e., 101.3 kPa) is

$$\frac{k_a}{k_w} = \frac{\rho_a}{\upsilon_a} \frac{\upsilon_w}{\rho_w} = \frac{1.196}{0.0179} \frac{1.009}{998.2} = \frac{1}{14.8} \tag{9.28}$$

Equation 9.28 shows that the air coefficient of permeability is 14.8 times less than the water coefficient of permeability. The information from Eqs. 9.27 and 9.28 may at first seem to be contradictory. Perhaps that is the reason why it is possible to encounter seemingly contradictory statements regarding the ease of air and water flow through soils in the literature.

The ratios of mass flow rates of air and water, dM_a/dt and dM_w/dt, respectively, can be written as follows:

$$\frac{dM_a/dt}{dM_w/dt} = \frac{\rho_a Q_a}{\rho_w Q_w} = \frac{\rho_a \upsilon_w}{\rho_w \upsilon_a} = \frac{1}{14.8} \tag{9.29}$$

A comparison of the air mass flow rate and the water mass flow rate shows that the mass flow rate of water will be 14.8 times more than the mass flow rate of air. However, more air (by volume) will flow through a porous medium than water (by volume) under the same pressure gradient. Equation 9.29 shows that the mass flow rate of water will be higher than the mass flow rate of air under the same pressure gradients.

The volume flow ratio of air to water under a soil temperature of 10°C can be written as follows:

$$\frac{Q_a}{Q_w} = \frac{\upsilon_w}{\upsilon_a} = \frac{1.310}{0.0176} = 74.3 \tag{9.30}$$

The mass flow ratio of air to water under the same soil temperature can be written as follows:

$$\frac{k_a}{k_w} = \frac{dM_a/dt}{dM_w/dt} = \frac{1}{9.18} \tag{9.31}$$

The volume flow rate of air is 74.3 times more than the volume flow rate of water. However, the mass flow rate of water will be 9.18 times more than the mass flow rate of air.

9.3.5 Klinkenberg Effect

The intrinsic permeability measured by passing air through a dry porous medium, K_a, should be similar to the intrinsic permeability measured by passing water through the same saturated porous medium, K_w. However, air passage does not behave quite the same as water flow because the air velocity at the solid boundaries of a pore is not zero as it is for the case when water is used. The intrinsic permeability measured using air, K_a, will appear to be larger than the intrinsic permeability measured when using water, as shown in Fig. 9.7. This difference in flow behavior is known as "gas slippage" or the "Klinkenberger effect" (Klinkenberger, 1941). Klinkenberger demonstrated that the permeability measured using gases was approximately a linear function of the reciprocal of the applied gas pressure.

Klinkenberger (1941) showed that the effect of gas slippage can be evaluated by measuring the air intrinsic permeability K_a for a range of pressures and extrapolating the results to an infinite mean pressure. The extrapolated value will be similar to the intrinsic permeability measured when using water,

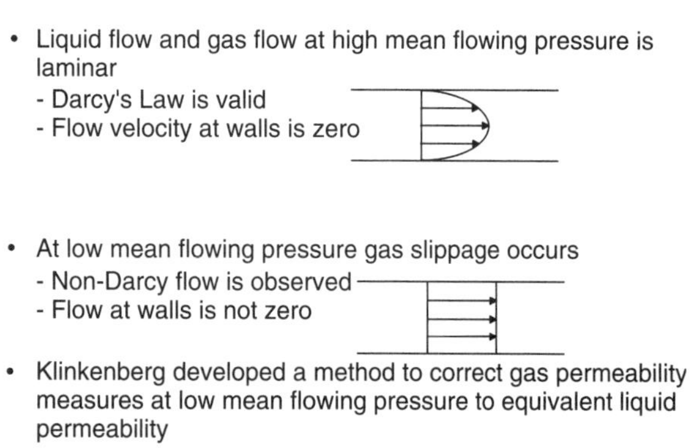

Figure 9.7 Difference in intrinsic permeability observed when using water and a gas such as nitrogen.

K_w. It has been noted that the intrinsic permeability measured when using water is still smaller than the intrinsic permeability measured when using air even after the air permeability results have been corrected for gas slippage.

The Klinkenberger effect becomes important when the measured intrinsic permeability is lower than 10^{-18} m^2 and the applied differential pressures are low during the measurement of permeability. When the rate of air flow is high and the applied pressures are high, the two intrinsic measurements come closer to one another. The correction of air intrinsic permeability for the Klinkenberger effect becomes relevant as the permeability measurements become extremely low. In general, the application of a correction for gas slippage is not necessary because soils have a sufficiently high coefficient of permeability.

The air coefficient of permeability under dry conditions, K_a, is often considered to be a constant for simplicity. However, the intrinsic permeability varies with the absolute pore-air pressure, and this is also part of the Klinkenberg effect (Klinkenberg, 1941). The Klinkenberg effect can be approximated using the following relationship:

$$K_a = K_\infty\left(1 + \frac{b}{u_a}\right) \tag{9.32}$$

where:

K_a = intrinsic permeability measured using air,
K_∞ = air phase intrinsic permeability when using a high air pressure, and
b = parameter representing the porous medium.

Heid et al. (1950) proposed an empirical relationship between K_∞ and b based on experimental data:

$$b = 0.147K_\infty^{-0.39} \tag{9.33}$$

where b is in kilopascals and K_∞ is in meters squared.

Equation 9.33 has been adopted by the American Petroleum Institute for correcting for the Klinkenberg effect (Baehr and Hult, 1991). The maximum percent error in estimating the air permeability by ignoring the Klinkenberg effect was studied by Baehr and Hult (1991). The Klinkenberg effect was shown to produce an error greater than 10% when measuring the air permeability of a porous media where the intrinsic permeability was 10^{-13} m^2 (or 10^{-7} mm^2) or less.

9.4 DIFFUSION OF AIR THROUGH WATER

The diffusion process occurs in response to a concentration gradient. Ionic or molecular movement will take place from regions of higher concentration to regions of lower concentration. Air and water in the voids of a soil are the conducting media for diffusion processes. On the other hand, the soil structure determines the path length and cross-sectional area available for diffusion. The transport of gases (e.g., O_2 and CO_2), water vapor, and chemicals provides examples of diffusion processes in soils.

There are two examples of air diffusion common to unsaturated soils. The first example of diffusion involves the flow of air through the pore-water in a saturated or unsaturated soil (Matyas, 1967; Barden and Sides, 1967). Another example of air diffusion involves the passage of air through water in a high-air-entry ceramic disk of a pressure plate apparatus. This type of diffusion involves gases dissolving into the water on one side of the ceramic disk and subsequently coming out of the water on the other side of the ceramic disk.

The second type of diffusion involves the movement of constituents through the water phase due to a chemical concentration gradient or an osmotic suction gradient. Chemical diffusion comes under the topic of contaminant transport phenomena and is not discussed in this book.

9.4.1 Air Diffusion through Water

Fick's law can be used to describe the diffusion process. The concentration gradient provides the driving potential for the diffusion process and can be expressed with respect to the soil voids. In other words, the mass rate of diffusion and the concentration gradient are expressed with respect to a unit area and a unit volume of the soil voids. The formulation of Fick's law for diffusion in the y-direction is as follows:

$$\frac{\partial M}{\partial t} = -D\frac{\partial C}{\partial y} \tag{9.34}$$

where:

$\partial M/\partial t$ = mass rate of the air diffusing across a unit area of the soil voids,
D = coefficient of diffusion,
C = concentration of the diffusing air expressed in terms of mass per unit volume of the soil voids, and
$\partial C/\partial y$ = concentration gradient in the y-direction (or similarly in the x- or z-direction).

The diffusion equation can appear in several forms, similar to the forms presented for the flow of air through a porous medium. The concentration gradient for gases or water vapor (i.e., $\partial C/\partial y$) can be expressed in terms of partial pressures. Consider a constituent diffusing through the pore-water in a soil. Fick's law for diffusion can be rewritten with respect to the partial pressure of the diffusing constituent:

$$\frac{\partial M}{\partial t} = -D\frac{\partial C}{\partial \bar{u}_i}\frac{\partial \bar{u}_i}{\partial y} \tag{9.35}$$

where:

$\bar{u}_i$ = partial pressure of the diffusing constituent,
$\partial C/\partial \bar{u}_i$ = change in concentration with respect to a change in partial pressure, and
$\partial \bar{u}_i/\partial y$ = partial pressure gradient in the y-direction (or similarly in the x- or z-direction).

The mass rate of the constituent diffusing across a unit area of the soil voids (i.e.,$\partial M/\partial t$) can also be determined by measuring the volume of the diffusing constituent under constant-pressure conditions. The ideal gas law is applied to the diffusing constituent in order to obtain the mass flow rate:

$$\frac{\partial M}{\partial t} = \bar{u}_{fi}\frac{\omega_i}{RT_K}\frac{\partial V_{fi}}{\partial t} \tag{9.36}$$

where:

$\bar{u}_{fi}$ = absolute constant pressure used in the volume measurement of the diffusing constituent,
ω_i = molecular mass of the diffusing constituent,
R = universal (molar) gas constant,
T_K = absolute temperature,
$\partial V_{fi}/\partial t$ = flow rate of the diffusing constituent across a unit area of the soil voids, and
V_{fi} = volume of the diffusing constituent across a unit area of the soil voids.

The change in concentration of the diffusing constituent relative to a change in partial pressure (i.e.,$\partial C/\partial \bar{u}_i$) is obtained by considering the change in density of the dissolved constituent in the pore-water. The density of the dissolved constituent in the pore-water is the ratio of the mass of dissolved constituent to the volume of water:

$$\frac{\partial C}{\partial \bar{u}_i} = \frac{\partial(M_{di}/V_w)}{\partial \bar{u}_i} \tag{9.37}$$

where:

M_{di} = mass of the dissolved constituent in the pore-water and
V_w = volume of water.

Applying the ideal gas law yields the following equation:

$$\frac{\partial C}{\partial \bar{u}_i} = \frac{\partial[(V_{di}/V_w)\bar{u}_i(\omega_i/RT)]}{\partial \bar{u}_i} \tag{9.38}$$

where:

V_{di} = volume of the dissolved constituent in the pore-water.

The ratio of the volume of dissolved constituent to the volume of water (i.e.,V_{di}/V_w) is referred to as the volumetric coefficient of solubility, h. Under isothermal conditions, h is essentially a constant:

$$\frac{\partial C}{\partial \bar{u}_i} = h\frac{\omega_i}{RT_K}. \tag{9.39}$$

where:

h = volumetric coefficient of solubility for the constituent in water.

Substituting Eq. 9.36 into Eq. 9.39 results in the following diffusion equation (van Amerongen, 1946):

$$v_{fi} = -\frac{Dh}{\bar{u}_{fi}}\frac{\partial \bar{u}_i}{\partial y} \tag{9.40}$$

where:

v_{fi} = flow rate of the diffusing constituent across a unit area of the soil voids (i.e.,$\partial V_{fi}/\partial t$).

The above equation can be applied to air or gas diffusion through water in a soil or through other materials such as a rubber membrane (Poulos, 1964). The partial pressure term can be expressed in terms of the partial pressure head h_{fi} (i.e.,$h_{fi} = \bar{u}_i/\rho_{fi}g$) with respect to the density of the constituent, ρ_{fi}. The constituent density corresponds to the absolute constant pressure $\bar{u}_{fi}$ used in the measurement of the diffusing constituent volume. The absolute constant pressure $\bar{u}_{fi}$ is usually selected to correspond to atmospheric conditions (i.e., 101.3 kPa), and ρ_{fi} is the constituent density at the corresponding pressure:

$$v_{fi} = -Dh\frac{\rho_{fi}g}{\bar{u}_{fi}}\frac{\partial h_{fi}}{\partial y} \tag{9.41}$$

where:

ρ_{fi} = constituent density at the constant pressure $\bar{u}_{fi}$ used in the volume measurement of the diffusing constituent and

h_{fi} = partial pressure head $(\bar{u}_{fi}/\rho_{fi}g)$.

The above equation has a form similar to Darcy's law. The equation can be considered as a modified form of Darcy's law for air flow through an unsaturated soil with occluded air bubbles where the air flow is of the diffusion form:

$$v_{fi} = -k_{fi}\frac{\partial h_{fi}}{\partial y} \tag{9.42}$$

where:

k_{fi} = diffusion coefficient of permeability for air through an unsaturated soil with occluded air bubbles.

The diffusion coefficient of permeability can then be written as follows:

$$k_{fi} = Dh\frac{\rho_{fi}g}{\bar{u}_{fi}} \tag{9.43}$$

Substituting the ideal gas law into Eq. 9.43 results in another form for the diffusion coefficient of permeability:

$$k_{fi} = D\frac{h\omega_i g}{RT_K} \tag{9.44}$$

The above equation indicates that under isothermal conditions the coefficient of permeability (i.e., diffusion type) is directly proportional to the coefficient of diffusion since the term $h\omega_i g/RT_K$ is a constant.

The coefficients of diffusion for several gases through water and for air through different materials were presented in Chapter 2. The diffusion coefficients D for air through pure water were computed in accordance with Eq. 9.42 (Barden and Sides, 1967). The diffusion values for soils appear to be significantly smaller than the diffusion values for free water. The difference has been attributed to factors such as the tortuosity within the soil and the higher viscosity of the adsorbed water close to the soil particle clay surface. Measured diffusion coefficients have been observed to decrease as the water content of the soil decreases.

9.5 OTHER COMPONENTS OF AIR FLOW

Air flow may take place by two other mechanisms: (i) air flow within the liquid pore-water by diffusion and (ii) dissolved pore-air being carried by water flow. In the latter case the air movement is referred to as advective flow. If air flows through the water in the soil, it is driven by gradients in the dissolved pore-air concentration and is referred to as flow by diffusion.

The mass flux of free air can be described in terms of Fick's law and Darcy's law. The flow of dissolved air is also driven by gradients in dissolved air concentration. Dissolved air flow can also be described using Fick's law:

$$v_y^{ad} = -\frac{D^{ad}}{\rho_a}\frac{\partial C^{ad}}{\partial y} \tag{9.45}$$

where:

v_y^{ad} = dissolved air flow rate in the y-direction across a unit area of the soil due to pore-air concentration gradients, m/s,

D^{ad} = molecular diffusivity of dissolved air through water, m^2/s (also referred to as coefficient of transmission of air),

C^{ad} = concentration of dissolved air in terms of the mass of vapor per unit volume of soil, $C^{ad} = \rho_a Snh$,

h = Henry's volumetric coefficient of solubility, V_{ad}/V_w,
V_{ad} = volume of air dissolved in pore-water, and
ρ_a = density of air, kg/m^3.

Equation 9.45 can be rewritten as follows:

$$v_y^{ad} = -\frac{D^{ad}}{\rho_a}\frac{\partial C^{ad}}{\partial \bar{u}_a}\frac{\partial \bar{u}_a}{\partial y} = -\frac{D^{ad*}}{\rho_a}\frac{\partial \bar{u}_a}{\partial y} \tag{9.46}$$

where:

D^{ad*} = coefficient of transmission, (kg m)/(kN s), and
$\bar{u}_a$ = absolute air pressure, kPa.

The soil properties D^{ad} and D^{ad*} have been defined and can be measured or estimated. The diffusion of dissolved air through liquid pore-water decreases as the degree of saturation of a soil decreases. Dissolved air movement becomes insignificant when compared with the flow of free air through the soil. The dissolved air component v^{ad} can be incorporated into the prediction of the coefficient of transmission D^{ad*} through use of a tortuosity coefficient.

The flow of dissolved pore-air carried by water flow (i.e., advection) can be described using Darcy's law for water flow and taking the amount of dissolved air into account:

$$v_y^{aa} = -hk_w\frac{\partial h}{\partial y} \tag{9.47}$$

where:

v_y^{aa} = dissolved pore-air flow rate in the y-direction across a unit area of the soil due to bulk pore-liquid water flow, m/s.

The total flow of pore-air, v_y^a, is the summation of the three flow mechanisms described by Eqs. 9.42, 9.45, 9.46, and 9.47:

$$v_y^a = v_y^{af} + v_y^{ad} + v_y^{aa} = -\frac{k_a}{\gamma_a}\frac{\partial u_a}{\partial y} - \frac{k^{ad}}{\gamma_a}\frac{\partial u_a}{\partial y} - hk_w\frac{\partial h}{\partial y} \tag{9.48}$$

where:

k_a = air coefficient of permeability, $\gamma_a D^{a*}/\rho_a$, m/s,
k^{ad} = diffusion coefficient of air through water, $\gamma_a D^{a*}/\bar{u}_a$, m/s, and
γ_a = unit weight of air, kN/m^3.

Equation 9.48 provides a smooth transition between unsaturated and saturated conditions. The soil becomes saturated as soil suction decreases and k_a decreases toward zero. However, the flow of air does not completely cease for saturated conditions. Pore-air conductivity by diffusion within the pore-water and the flow of dissolved air carried by bulk water flow increase as the pores become filled with water.

9.6 PARTIAL DIFFERENTIAL EQUATIONS FOR AIR FLOW THROUGH UNSATURATED SOILS

Air is compressible and, as such, changes in density must be taken into consideration when describing the flow of air through a referential element. The mass flow of air through a referential element takes the form of a partial differential equation for compressible fluid flow.

A series of partial differential equations are presented that describe the components of air flow under various conditions. Steady-state air flow is first presented followed by transient or unsteady-state flow. The partial differential equations associated with one-, two-, and three-dimensional situations are presented.

9.6.1 One-Dimensional Steady-State Air Flow

The bulk flow of air (i.e., free air) can occur through an unsaturated soil when the air phase is continuous. Air is compressible and the derivation of the partial differential equations for air flow must take into consideration changes in density as a result of pressure changes.

The air coefficient of transmission D_a^* or the air coefficient of permeability k_a (i.e., $D_a^* g$) is a function of the volume-mass properties (or stress state) of the soil. The relationship between the air coefficient of permeability k_a and matric suction [i.e., $k_a(u_a - u_w)$] or degree of saturation [i.e., $k_a(S_e)$] was previously described. The value of k_a or D_a^* may vary with location, depending upon the distribution of the pore-air volume in the soil. The air coefficient of permeability in an unsaturated soil can vary with direction and location. The air coefficient of permeability at a point can be assumed to be constant with respect to time during steady-state air flow provided soil suction remains constant.

Steady-state formulations for one-dimensional air flow are first presented using Fick's law. Heterogeneous, isotropic, and anisotropic conditions are taken into consideration in the derivation of the flow equations. Steady-state air flow is analyzed assuming that the pore-water pressures have reached equilibrium and are not influenced by the pore-water pressures. One-dimensional air flow equations can be solved using numerical methods such as the finite difference method or the finite element method.

Consider an *unsaturated soil* (i.e., heterogeneous), element with air flow in the y-direction (Fig. 9.8). The air flow has a mass rate J_{ay} under steady-state conditions. The mass rate is assumed to be positive for upward air flow. The principle of continuity requires that the mass of air flowing into the soil element must be equal to the mass of air flowing out of the element:

$$\left(J_{ay} + \frac{dJ_{ay}}{dy}dy\right)dx\,dz - J_{ay}\,dx\,dz = 0 \tag{9.49}$$

where:

J_{ay} = mass rate of air flow across a unit area of the soil in the y-direction.

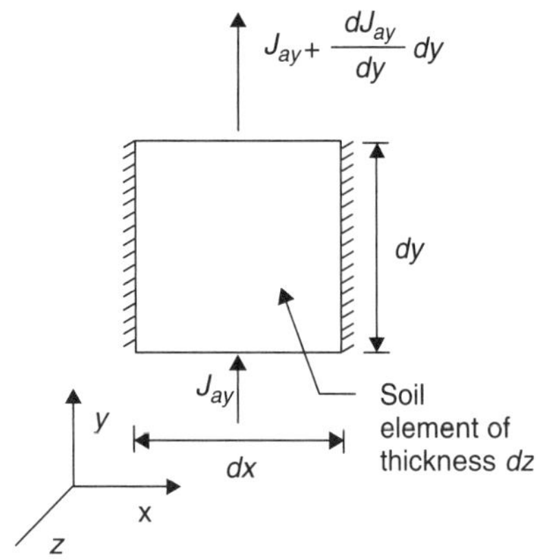

Figure 9.8 One-dimensional steady-state air flow through unsaturated soil element.

The net mass rate of air flow can be written as

$$\left(\frac{dJ_{ay}}{dy}\right) dx\,dy\,dz = 0 \tag{9.50}$$

Substituting Fick's law for the mass rate of flow into the above equation yields a nonlinear differential equation:

$$\frac{d\{-D^*_{ay}(u_a - u_w)du_a/dy\}}{dy} = 0 \tag{9.51}$$

where:

$D^*_{ay}(u_a - u_w)$ = air coefficient of transmission as a function of matric suction,
du_a/dy = pore-air pressure gradient in the y-direction,
u_a = pore-air pressure, and
u_w = pore-water pressure.

The coefficient of transmission $D^*_{ay}(u_a - u_w)$ will be written as D^*_{ay} for simplicity, and it must be remembered that the soil property is a function of soil suction. The spatial variation of the variable, D^*_{ay}, results in nonlinearity in the mass flow equation:

$$D^*_{ay}\frac{\partial^2 u_a}{dy^2} + \frac{dD^*_{ay}}{dy}\frac{du_a}{dy} = 0 \tag{9.52}$$

where:

dD^*_{ay}/dy = change in the air coefficient of transmission in the y-direction.

Equation 9.52 describes the pore-air pressure distribution in the soil mass in the y-direction. The second term in the equation accounts for the spatial variation in the coefficient of transmission. The coefficient of transmission is obtained by dividing the air coefficient of permeability k_{ay} by gravitational acceleration (i.e., $D^*_{ay} = k_{ay}/g$). In other words, the coefficients D^*_{ay} and k_{ay} have similar functional relationships with respect to soil suction.

The measurement of the air coefficient of permeability in a triaxial cell is an application of one-dimensional steady-state air flow. The air coefficient of permeability is assumed to be constant throughout the soil specimen. The air flow equation is reduced to a linear differential equation by neglecting changes in the air coefficient of permeability with respect to location:

$$\frac{\partial^2 u_a}{dy^2} = 0 \tag{9.53}$$

The pore-air pressure distribution in the y-direction is obtained by integrating Eq. 9.53 two times:

$$u_a = C_1 y + C_2 \tag{9.54}$$

where:

C_1, C_2 = constants of integration related to the boundary conditions
y = distance in the y-direction.

Figure 9.34 illustrates the pore-air pressure distribution within a soil specimen during an air permeability test. The air pressures at both ends of the specimen (i.e., $u_a = u_{ab}$ at $y = 0.0$ and $u_a = u_{at} = 0.0$ at $y = h_s$) are the boundary conditions. Substituting the boundary conditions into the flow equation results in a linear equation for the pore-air pressure along the soil specimen [i.e., $u_a = (1 - y/h_s)u_{ab}$].

9.6.2 Two-Dimensional Steady-State Air Flow

Two-dimensional steady-state air flow is first formulated for the case where the air coefficient of transmission is a function of matric suction. The air coefficients of transmission in the x- and y-directions, D^*_{ax} and D^*_{ay}, are related to matric suction using the same transmission function, $D^*_a(u_a - u_w)$. The D^*_{ax}/D^*_{ay} ratio is assumed to be constant at any point within the soil mass. An element of soil subjected to two-dimensional air flow is shown in Fig. 9.9. Satisfying continuity for steady-state flow yields the following equation:

$$\left(J_{ax} + \frac{\partial J_{ax}}{\partial x}dx - J_{ax}\right) dy\,dz + \left(J_{ay} + \frac{\partial J_{ay}}{\partial y}dy - J_{ay}\right) dx\,dz = 0 \tag{9.55}$$

where:

J_{ax} = mass rate of air flowing across a unit area of the soil in the x-direction.

Rearranging Eq. 9.55 results in the following equation:

$$\left(\frac{\partial J_{ax}}{\partial x} + \frac{\partial J_{ay}}{\partial y}\right) dx\,dy\,dz = 0 \tag{9.56}$$

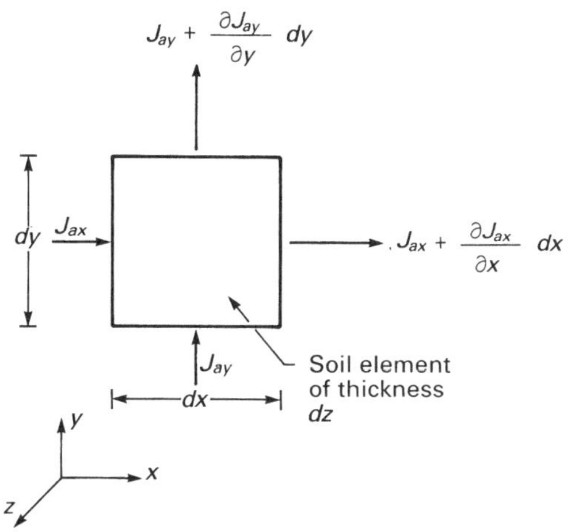

Figure 9.9 Two-dimensional steady-state air flow through an unsaturated soil element.

Substituting Fick's law for the mass rates J_{ax} and J_{ay} gives the following nonlinear partial differential equation:

$$\frac{\partial}{\partial x}\left\{D_{ax}^*(u_a - u_w)\frac{\partial u_a}{\partial x}\right\} + \frac{\partial}{\partial y}\left\{D_{ay}^*(u_a - u_w)\frac{\partial u_a}{\partial y}\right\} = 0 \quad (9.57)$$

where:

$D_{ax}^*(u_a - u_w)$ = air coefficient of transmission as a function of matric suction and

$\partial u_a/\partial x$ = pore-air pressure gradient in the x-direction.

Let us write $D_{ax}^*(u_a - u_w)$ and $D_{ay}^*(u_a - u_w)$ simply as D_{ax}^* and D_{ay}^*, respectively. The coefficient of transmission D_{ax}^* is related to the air coefficient of permeability k_{ax} by the gravitational acceleration (i.e., $D_{ax}^* = k_{ax}/g$). Expanding Eq. 9.57 results in the following flow equation:

$$D_{ax}^*\frac{\partial^2 u_a}{\partial x^2} + D_{ay}^*\frac{\partial^2 u_a}{\partial y^2} + \frac{\partial D_{ax}^*}{\partial x}\frac{\partial u_a}{\partial x} + \frac{\partial D_{ay}^*}{\partial y}\frac{\partial u_a}{\partial y} = 0 \quad (9.58)$$

where:

$\partial D_{ax}^*/\partial x$ = change in air coefficient of transmission in the x-direction.

Spatial variations in the coefficients of transmission are accounted for by the third and fourth terms in Eq. 9.58. These two terms produce the nonlinearity in the flow equation. Equation 9.58 describes the pore-air pressure distribution in the x-y plane of the soil mass during two-dimensional steady-state air flow.

For the *heterogeneous*, *isotropic* case, the coefficients of transmission in the x- and y-directions are equal (i.e., $D_{ax}^* = D_{ay}^* = D_a^*$), and the air flow equation becomes

$$D_a^*\left(\frac{\partial^2 u_a}{\partial x^2} + \frac{\partial^2 u_a}{\partial y^2}\right) + \frac{\partial D_a^*}{\partial x}\frac{\partial u_a}{\partial x} + \frac{\partial D_a^*}{\partial y}\frac{\partial u_a}{\partial y} = 0 \quad (9.59)$$

where:

D_a^* = air coefficient of transmission in the x- and y-directions.

The partial differential equations for steady-state air flow bear a similarity in form to those previously presented for water flow.

9.6.3 Three-Dimensional Steady-State Air Flow

The three-dimensional steady-state partial differential equation for air flow requires that a third dimension be added to the previously presented two-dimensional air flow equation:

$$D_a^*\left(\frac{\partial^2 u_a}{\partial x^2} + \frac{\partial^2 u_a}{\partial y^2} + \frac{\partial^2 u_a}{\partial z^2}\right) + \frac{\partial D_a^*}{\partial x}\frac{\partial u_a}{\partial x} + \frac{\partial D_a^*}{\partial y}\frac{\partial u_a}{\partial y} + \frac{\partial D_a^*}{\partial z}\frac{\partial u_a}{\partial z} = 0 \quad (9.60)$$

The three-dimensional geometry must be input to the computer and a numerical technique must be used to solve three-dimensional problems.

9.6.4 One-Dimensional Transient Air Flow

The derivation for the differential equation for transient air flow is derived using Darcy's flow law for air. The flow of air in and out of a referential element must take density changes into consideration. Both the density of air and the velocity take the form of a field representation across the soil element. The velocity of air flow multiplied by the density of air is equal to the mass of air flow:

$$\left(\rho_a + \frac{\partial \rho_a}{\partial y}\right)\left(v_{ay} + \frac{\partial v_{ay}}{\partial y}dy\right) dx\,dz - \rho_a v_{ay}\, dx\,dz = \frac{\partial M_a}{\partial t} \quad (9.61)$$

where:

ρ_a = density of air,
v_{ay} = velocity of air flow in the y-direction, and
M_a = mass of air in the element.

The net air flux in the y-direction is obtained by multiplying out the terms in the above equation:

$$\left(\rho_a\frac{\partial v_{ay}}{\partial y} + v_{ay}\frac{\partial \rho_a}{\partial y}\right) dx\,dy\,dz = \frac{\partial M_a}{\partial t} \quad (9.62)$$

Equation 9.62 can be written as follows since a unit volume is being considered:

$$\rho_a\frac{\partial v_a}{\partial y} + v_{ay}\frac{\partial \rho_a}{\partial y} = \frac{\partial M_a}{\partial t} \quad (9.63)$$

The density of air can be written using the ideal gas law:

$$\rho_a = \frac{\bar{u}_a \omega_a}{RT_K} \tag{9.64}$$

where:

$\bar{u}_a$ = absolute air pressure,
ω_a = molecular weight of air, 28.8 g/mol,
R = universal gas constant, 8.314 g/(s^2 cm^{2}mol K), and
T_K = absolute temperature, K.

The mass of air for a unit volume of soil can be written in terms of the volume of air multiplied by the density of air:

$$M_a = \frac{e}{1+e} S_a \frac{\bar{u}_a \omega_a}{RT_K} \tag{9.65}$$

where:

e = void ratio and
S_a = amount of voids filled with air, equal to $1 - S$, where S is the degree of saturation with respect to water.

Darcy's law can be written for air flow in the y-direction:

$$v_{ey} = \frac{k_{ay}}{\rho_{ma} g} \frac{\partial u_a}{\partial y} \tag{9.66}$$

where:

v_{ey} = volume rate of air flow across a unit area at the exit point under constant-pressure conditions (i.e., usually 101.3 kPa absolute pressure) and
ρ_{ma} = constant air density.

For a particular mass of air, the following condition exists:

$$\rho_a v_{ay} = \rho_{ea} v_{ey} \tag{9.67}$$

where:

ρ_{ev} = density of air at the exit point under constant pressure conditions.

Darcy's equation can be written as

$$v_{ay} = \frac{k_{ay}}{\rho_a g} \frac{\partial u_a}{\partial y} \tag{9.68}$$

Substituting Darcy's law into the net air flux equation (i.e., Eq. 9.63) gives the following flow equation:

$$\rho_a \frac{\partial v_{ay}}{\partial y} + v_{ay} \frac{\partial \rho_a}{\partial y} = \frac{\partial M_a}{\partial t} \tag{9.69}$$

9.6.5 Two- and Three-Dimensional Transient Air Flow

The above differential equation can be expanded to form a partial differential equation for transient air flow in two and three dimensions. The two-dimensional transient air flow equation can be written as follows:

$$\rho_a \frac{\partial v_{ax}}{\partial x} + v_{ax} \frac{\partial \rho_a}{\partial x} + \rho_a \frac{\partial v_{ay}}{\partial y} + v_{ay} \frac{\partial \rho_a}{\partial y} = \frac{\partial M_a}{\partial t} \tag{9.70}$$

The three-dimensional transient air flow equation can be written as follows:

$$\rho_a \frac{\partial v_{ax}}{\partial x} + v_{ax} \frac{\partial \rho_a}{\partial x} + \rho_a \frac{\partial v_{ay}}{\partial y} + v_{ay} \frac{\partial \rho_a}{\partial y} + \rho_a \frac{\partial v_{az}}{\partial z} + v_{az} \frac{\partial \rho_a}{\partial z} = \frac{\partial M_a}{\partial t} \tag{9.71}$$

The air flow M_a is the total mass of air flowing in all directions.

9.6.6 Transient Air Flow Considering Other Components of Air Flow

Fredlund and Gitirana (2005) presented partial differential equations that satisfy the conservation of air mass when three components of air flow are considered (i.e., convection, advection, and diffusion). The flow law equations and an air volume change constitutive equation must be combined to satisfy the continuity for air mass.

The partial differential equation for air mass flow is formed using three main partial derivatives for the flow in each direction. The presence of the partial derivatives is a result of the assumption that the mass rate of air flow across a representative elemental volume is continuously distributed in space. Therefore, the spatial distribution of air flow rate can be described by the partial derivative of flow for each direction.

Considering a constant referential volume V_0, the following expanded form of the air flow equation is obtained under conditions that the soil element does not change volume:

$$\begin{aligned} &\frac{\partial}{\partial x}\left[\frac{k_a}{g}\frac{\partial u_a}{\partial x} + \frac{k^{ad}}{g}\frac{\partial u_a}{\partial x} + \rho_a h k_w \frac{\partial}{\partial x}\left(\frac{u_w}{\gamma_w}\right)\right] \\ &+ \frac{\partial}{\partial y}\left[\frac{k_a}{g}\frac{\partial u_a}{\partial y} + \frac{k^{ad}}{g}\frac{\partial u_a}{\partial y} + \rho_a h k_w \frac{\partial}{\partial y}\left(\frac{u_w}{\gamma_w} + y\right)\right] \\ &+ \frac{\partial}{\partial z}\left[\frac{k_a}{g}\frac{\partial u_a}{\partial z} + \frac{k^{ad}}{g}\frac{\partial u_a}{\partial z} + \rho_a h k_w \frac{\partial}{\partial z}\left(\frac{u_w}{\gamma_w}\right)\right] \\ &= (1 - S + hS)n \frac{W_a}{RT_K}\frac{\partial u_a}{\partial t} \end{aligned} \tag{9.72}$$

Equation 9.72 shows that the flow and storage of air within a saturated-unsaturated soil is a function of the pore-air and pore-water pressures provided the soil element remains at constant volume. Therefore, the pore-water partial differential equation also needs to be solved. The pore-air conductivity function, the pore-air conductivity by diffusion within the pore-liquid water, and the coefficient-of-water-permeability function are related to the SWCC. The soil property functions vary with soil suction, and therefore, the partial differential equation is mathematically nonlinear. The above equation provides a smooth transition between unsaturated and saturated conditions when the soil properties are written as functions of state variables.

The air flow properties have a smooth transition from completely dry, unsaturated soil conditions to saturated soil conditions. A dry soil has high pore-air permeability. The pore-air conductivity decreases and eventually may become negligible as soil suction decreases and the soil moves toward saturation. However, air flow continues to take place as air diffusion within the pore-water phase.

9.6.7 Boundary Conditions for Air Flow

Several types of boundary conditions can be applied at the boundaries of an air flow problem. The boundary conditions for air flow bear a similarity to the boundary conditions previously discussed for water seepage problems. The *Neumann* boundary condition corresponds to the total air flowing normal to the boundary surface. The Neumann boundary condition does not differentiate between the type of air flow (i.e., whether it is free air flow or dissolved air flow).

The other type of primary boundary condition that can be applied is called a predetermined pore-air pressure value (i.e., *Dirichlet* boundary condition).

9.7 DIRECT MEASUREMENT OF AIR COEFFICIENT OF PERMEABILITY

There are direct and indirect methods available for obtaining the coefficient of permeability with respect to air, k_a. Indirect methods for calculating the air coefficient-of-permeability function make use of the SWCC. The indirect methods of computing the air permeability function are discussed later in this chapter.

The air permeability of a soil can vary over a wide range. Fine-grained soils such as silts and clays may experience low air flow rates and the system for measuring air flow may need to be quite different from a system required for a coarse-grained soil that may have a high flow rate. Direct steady-state methods are preferable to transient testing procedures due to the compressibility of air.

9.7.1 Steady-State Air Permeability Measurement in Fine-Grained Soils

The testing procedure used for the direct measurement of the water coefficient of permeability k_w can also be used in the measurement of the air coefficient of permeability k_a. In principle, the air and water coefficients of permeability can be measured at different water contents by changing the matric suction of the soil. A change in either the pore-air pressure u_a or the pore-water pressure u_w causes a change in matric suction accompanied by a change in the volume of water in the soil. The total volume or the void ratio of the soil may or may not change significantly, depending upon the compressibility characteristics of the soil. Changes in the soil volume-mass properties (i.e., S, e, or w) alter the coefficients of permeability for the soil.

An alternative permeability testing procedure involves changing the net confining pressure $(\sigma - u_a)$ and the matric suction $(u_a - u_w)$ in order to change the volume-mass properties. This procedure has been used to obtain direct measurements of the air and water coefficients of permeability. An increase in the all-around stress σ on an unsaturated soil will result in a decrease in total volume, even when the water phase is undrained. The result is an increase in the degree of saturation and a decrease in matric suction. These changes affect the coefficients of permeability with respect to water, k_w, and air, k_a.

The net normal stress of an unsaturated soil can be altered using a one-dimensional odometer [Massachusetts Institute of Technology (MIT), 1963; Barden et al., 1969b] or a triaxial permeameter cell (Matyas, 1967; Barden et al., 1969b; Barden and Pavlakis, 1971; Blight, 1971). Permeability measurements performed using an odometer sometimes encounter difficulties in obtaining a reliable seal between the soil specimen and the odometer ring when the matric suction is increased (Barden and Pavlakis, 1971).

9.7.2 Triaxial Permeameter Cell for Measurement of Low Air Coefficients of Permeability

The triaxial permeameter cell shown in Fig. 9.10 was developed by Matyas (1967) to measure the air coefficient of permeability. The soil specimen was placed between two dry coarse porous stones. A soil specimen was subjected to an isotropic confining pressure σ (i.e., cell pressure). A constant air pressure was applied to the base of the specimen. The outflow of air from the top of the specimen was

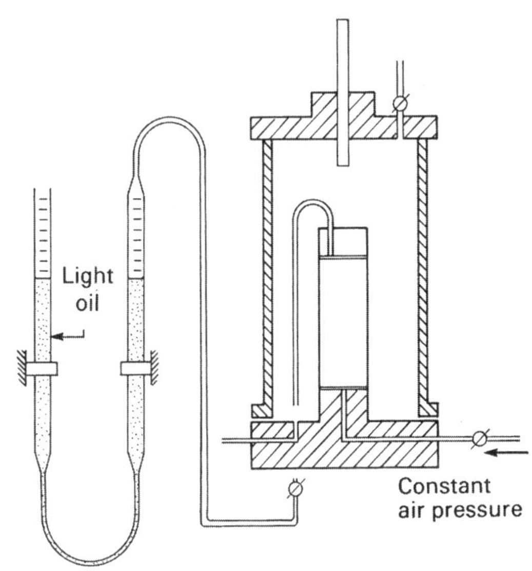

Figure 9.10 Triaxial permeameter for measuring air coefficient of permeability (after Matyas, 1967).

collected in a graduated burette that had an air-oil interface. The air flow was measured under atmospheric pressure conditions by adjusting the legs of the U-tube and maintaining the air-oil interfaces at the center elevation of the specimen.

The air coefficient of permeability for Sasumua clay was measured at various confining pressures using the triaxial permeameter. A steady-state air flow was maintained during measurements of the coefficient of permeability. Two specimens were compacted at an optimum water content of 60.5%. The Sasumua clay had high clay content and was compacted at standard AASHTO compaction. Values for the air coefficient of permeability were obtained by applying Darcy's law to the test results. The results are plotted in Fig. 9.11 as a function of the confining pressure σ. Decreasing air permeability values were observed as the confining pressure was increased or the matric suction was decreased.

Air and water permeability tests have been measured on Backwater boulder clay at increasing values of net confining pressure $(\sigma - u_a)$ from 35 to 1000 kPa and at decreasing matric suction values from 94 to 14 kPa. Several sets of air and water permeability measurements are shown in Fig. 9.12. The results showed decreasing air coefficients of permeability and increasing water coefficients of permeability as the soil specimen was compressed. The matric suction in the soil decreased as the confining pressure was increased.

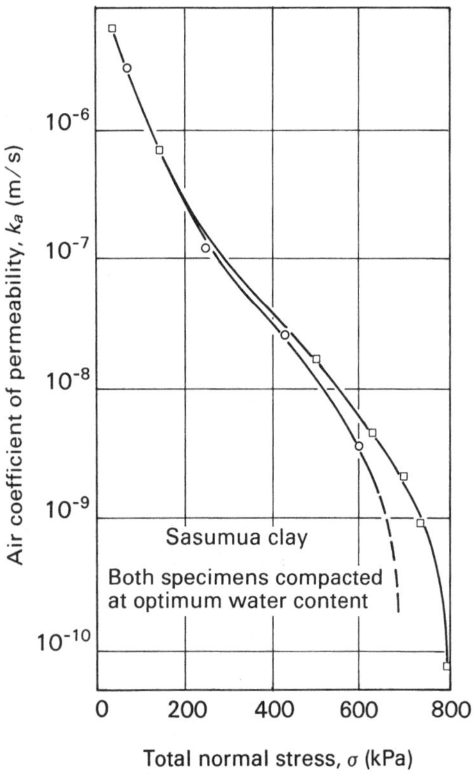

Figure 9.11 Air coefficients of permeability for two specimens of Sasumua clay measured using a triaxial permeameter (after Matyas, 1987).

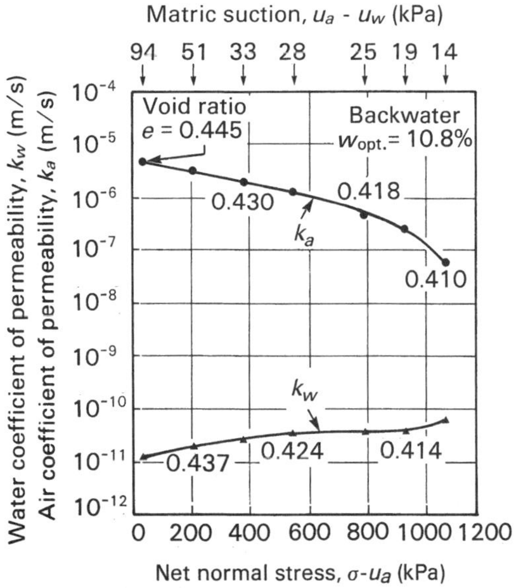

Figure 9.12 Water and air coefficients of permeability measured on Backwater boulder clay using triaxial permeameter cell (from Barden and Pavlakis, 1971).

9.7.3 Triaxial Permeameter Cell for Measurement of High Air Coefficients of Permeability

Ba-Te (2005) used a triaxial cell to measure the air permeability of soils under the application of both positive and negative air pressures. The laboratory tests were performed on fine and uniform crushed silica sand (i.e., Leighton Buzzard sand), with grain sizes ranging between 0.09 and 0.15 mm. A flowmeter with a maximum air flow rate of 2000 (standard) cm^3/min was used. The test procedure was similar to that used for the measurement of water coefficient of permeability.

The air permeability apparatus consists of three main parts: the pressure source, the triaxial cell, and the flowmeter (Fig. 9.13). Pressure transducers were also used to measure the air pressure at the top and bottom of the soil specimen. The platens placed at the top and bottom of the soil specimen were specially designed to ensure that there was no impedance to air flow through the soil. Details of the platen design are shown in Fig. 9.14. The holes in the stainless steel plate were 0.075 mm in diameter. The air permeability of the top and bottom disks was at least two orders of magnitude greater than that of the soil. The mass flowmeter (i.e., GFM air flow meter) had a range from 0 to 2000 (standard) cm^3/min with an accuracy of $\pm$ 1.5% of full scale.

The soil specimens were 70 mm in diameter and 60 mm in height and were confined using a flexible rubber membrane. The cell pressure used to confine the soil specimen was 140 kPa.

Test results on dry Leighton Buzzard sand ranged from 2.0×10^{-6} to 9.0×10^{-6} m/s, as shown in Fig. 9.15. The air coefficient of permeability was found to increase as the

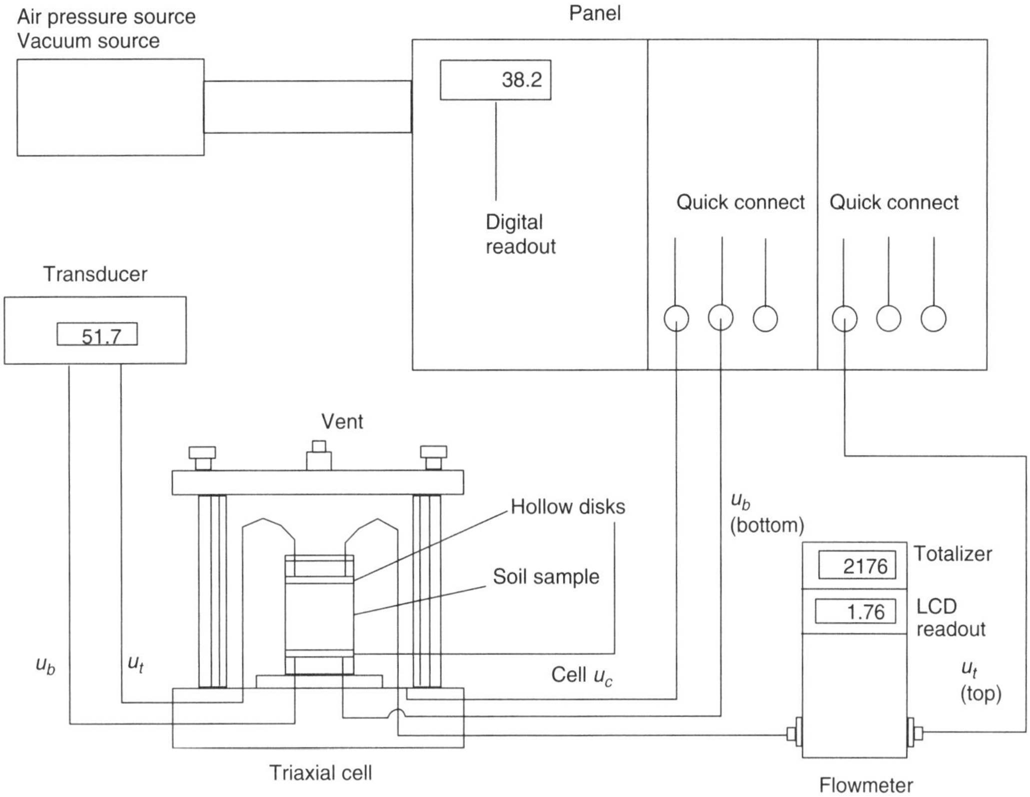

Figure 9.13 Permeability apparatus used to measure air coefficients of permeability (after Ba-Te, 2005).

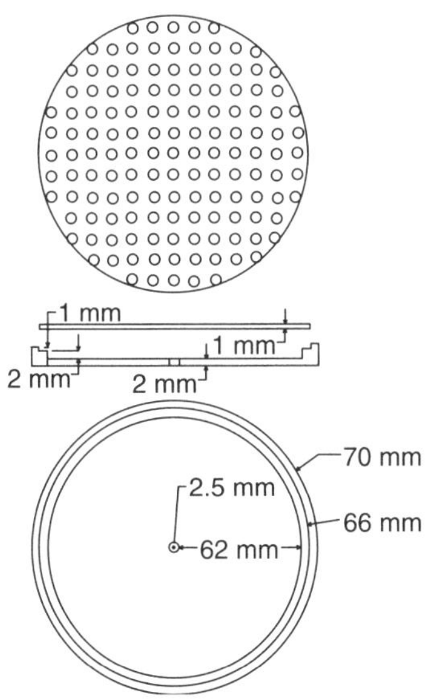

Figure 9.14 High-flow disk placed at top and bottom of air permeability soil specimen (after Ba-Te, 2005).

absolute air pressure increased. The water coefficient of permeability was also measured and found to range between 1.3×10^{-4} and 1.4×10^{-4} m/s. Intrinsic permeability values computed from the measured air and water coefficients of permeability were found to be comparable.

9.8 DIRECT MEASUREMENT OF AIR DIFFUSION THROUGH WATER

Long periods of time are often required for the laboratory testing of unsaturated soils. These long time periods allow air to diffuse through water. Although a high-air-entry disk is placed between the soil specimen and the measuring system to resist the passage of free air, air can diffuse through the water within the ceramic disk (Bishop and Donald, 1961). The diffused air accumulates beneath the disk and introduces an error in either the pore-water pressure measurement associated with undrained tests or the water volume change measurement associated with drained tests. Air diffusion is a common and important problem which must be given serious consideration when testing unsaturated soils (Bishop, 1969).

9.8.1 Mechanism of Air Diffusion through High-Air-Entry Disks

Undrained shear strength tests with pore-water pressure measurements can be performed by maintaining the water phase as a closed system. However, air can diffuse through the high-air-entry disk and come out of solution in the "compartment" below the disk, as illustrated in Fig. 9.16a. Diffused air displaces the water in the compartment (Fig. 9.16b), and water is forced upward through the disk back into the soil specimen.

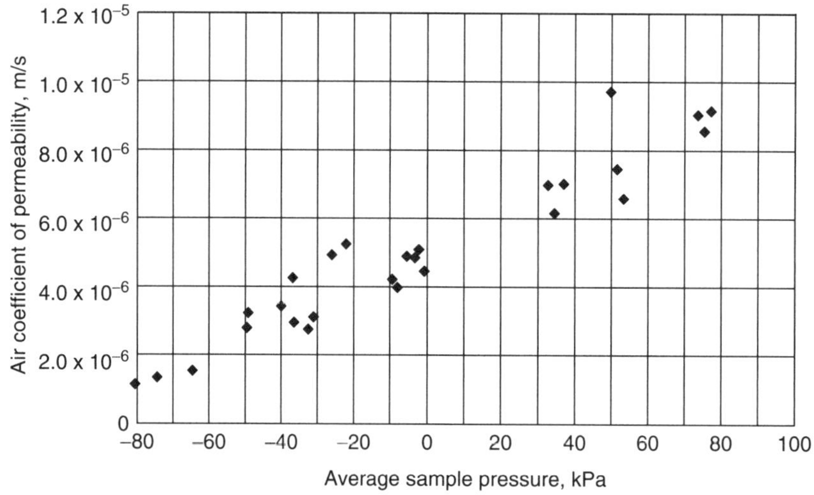

Figure 9.15 Air coefficients of permeability measured for various average applied air pressures (after Ba-Te, 2005).

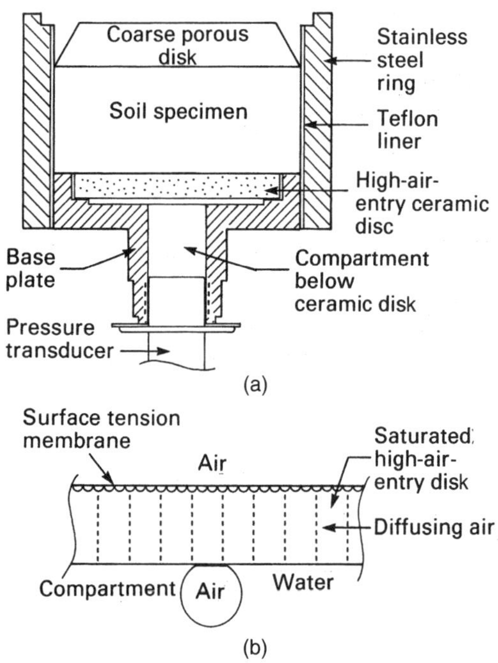

Figure 9.16 Replacement of water in compartment below high-air-entry disk by diffused air (from Fredlund, 1975c): (a) assembly showing compartment below high-air-entry disk; (b) air diffusion through saturated high-air-entry disk.

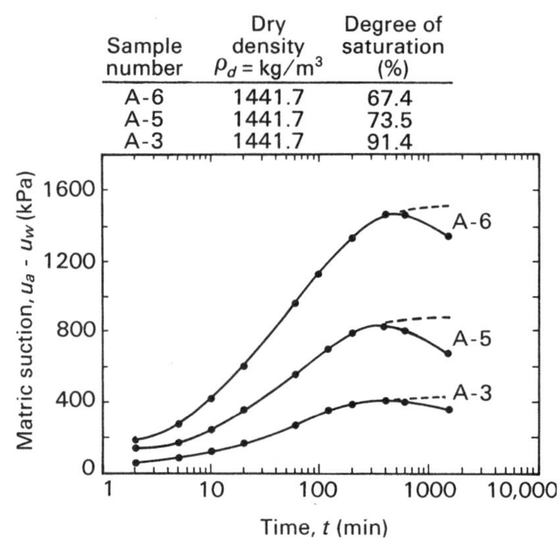

Figure 9.17 Development of matric suction in statically compacted Regina clay using axis translation technique (from Pufahl, 1970).

The water pressure measurement gradually changes from its original equilibrium value to the applied air pressure. The matric suction of the soil specimen does not tend to a constant value but gradually decreases with time. The air diffusion problem is common to null-type measurements of matric suction, as shown by the results in Fig. 9.17. The measured matric suction increases until the flow of diffused air forces water into the soil specimen with the result that the matric suction in the soil is decreased. Air diffusion imposes a limitation on the closed system measurement of pore-water pressure.

Drained shear strength tests and volume change tests on unsaturated soils are performed by controlling the pore-water pressure u_w and the pore-air pressure u_a. Determining the change in water volume involves measuring the flow of water to and from the soil specimen. Diffused air also moves through the water in the high-air-entry disk. Therefore, a diffused air volume indicator should be used to independently measure the diffused air volume. A correction needs to be applied to the measured overall fluid volume change (i.e., water and dissolved air). The diffused air beneath the disk can also prevent or retard the uptake of water if the soil specimen should tend to expand or dilate during the test.

The volume of air diffusing through a high-air-entry disk over a period of several days can, in some cases, exceed the total volume of water in the soil specimen (Fredlund, 1973a). A drained test extending more than one day should take into account the volume of diffused air if it is necessary to know the change in water content or degree of saturation of the soil (Fredlund, 1975c).

The time required for air diffusion depends on the thickness of the high-air-entry disk and the matric suction being applied or measured. Attempts have been made to theoretically calculate the volume of diffused air; however, there are many factors affecting the diffusion rate, making a theoretical prediction difficult. Consequently, it is better to measure the volume of diffused air in laboratory experiments.

9.8.2 Diffused Air Volume Indicators

Several devices have been proposed to measure the volume of diffused air. The intent in each case is to provide an independent measurement of the volume of diffused air. One example of a diffused air volume measurement is the air trap and roller used in conjunction with the volumetric pressure plate apparatus (see Chapter 5). The measured inflow or outflow of water is adjusted by releasing the trapped diffused air. This adjustment is applied as a diffused air correction to the overall water volume change measured.

Two other devices used for the measurement of the diffused air volume are the "bubble pump" (Bishop and Donald, 1961) and the "diffused air volume indicator (DAVI)," consisting of an inverted burette (Fredlund, 1975c).

Figure 9.18 shows a perspex bubble pump connected to the compartment below a high-air-entry disk in a triaxial cell. A U-tube containing mercury is used to circulate water through the base plate in order to flush and collect diffused air in a calibrated bubble trap. The collection of diffused air was accomplished by rocking the U-tube to establish a differential pressure of about 1.0 kPa across the base plate. The rocking action was continued for approximately 30 s. Flexible connecting tubes and perspex valves with soft rubber washers enabled the mercury in the U-tube to act as a piston in a double-acting pump during the rocking process. The volume of diffused air is measured in the 5-cm^3 bubble trap.

The diffused air volume indicator proposed by Fredlund (1975c) differs from the bubble pump in that it is relatively simple to construct and can function under an applied backpressure. A variable pressure gradient of at least 7 kPa was used to establish a gradient across the base plate. Figure 9.19 shows the layout of the diffused air volume indicator which can be connected to the base plate of a triaxial cell.

The diffused air volume indicator consists of a 10-cm^3 graduated burette and a Lucite exit tube placed inside an 80-mm-diameter Lucite cylinder. A vent on the top of the burette can be opened when filling the burette with water. The vent remains closed during operation of the

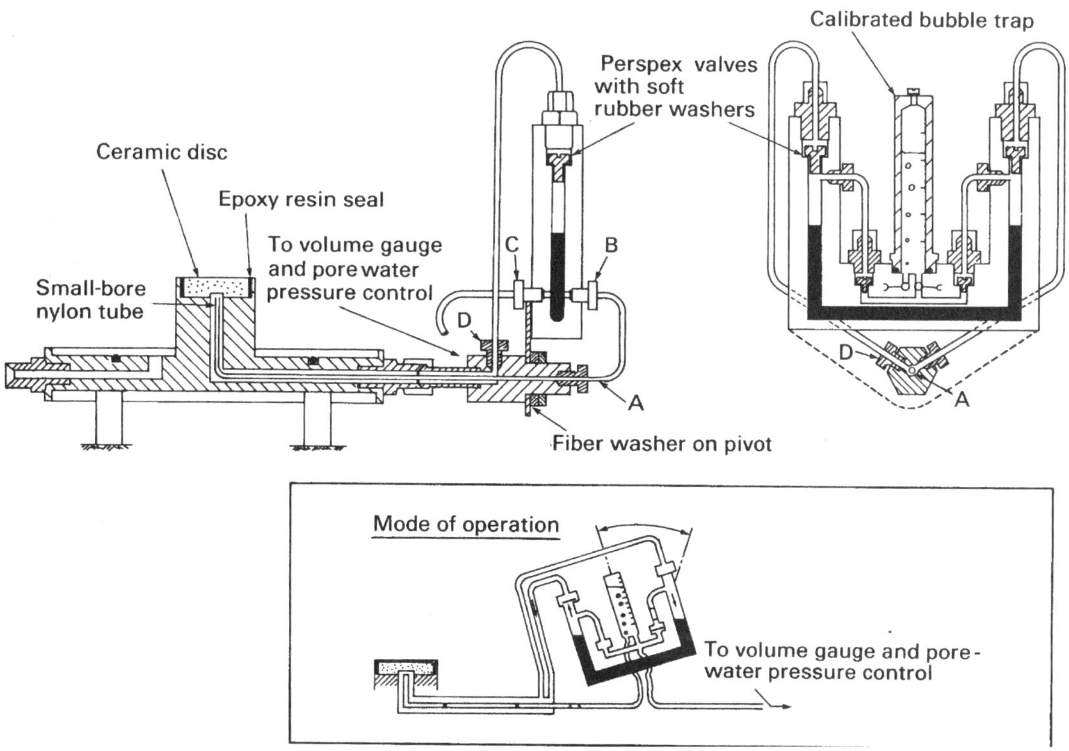

Figure 9.18 Bubble pump and air trap constructed at Imperial College, United Kingdom (from Bishop and Donald, 1961).

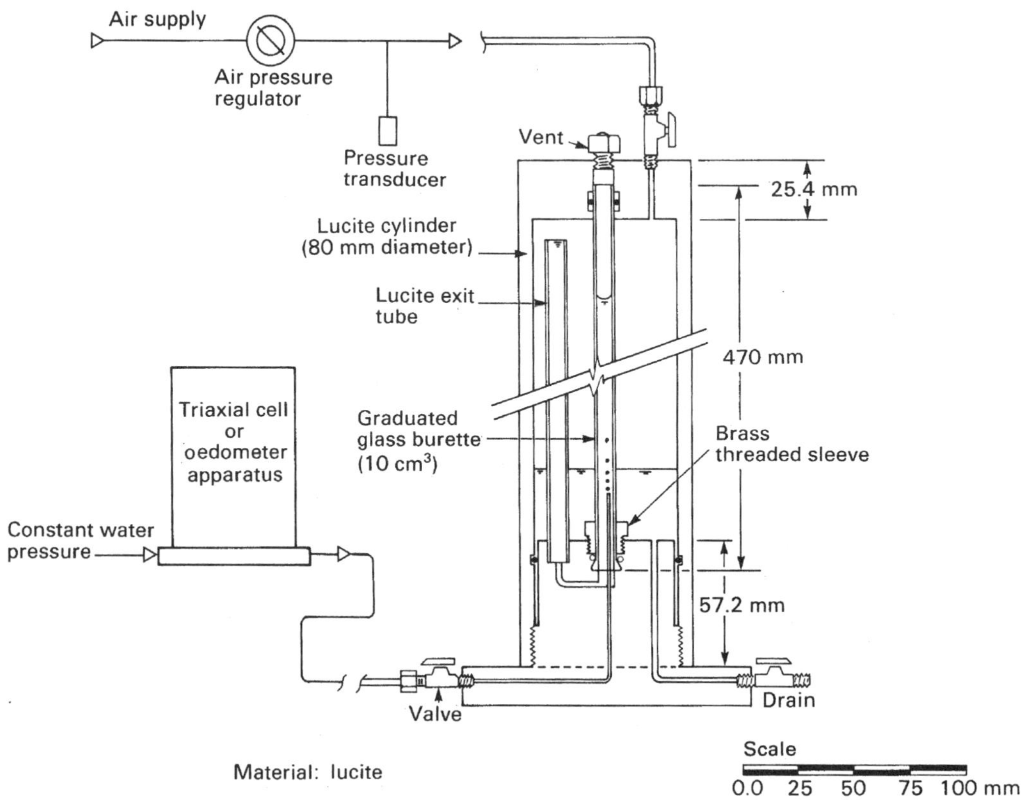

Figure 9.19 Diffused air volume indicator (from Fredlund, 1975c).

indicator. Two O-rings are placed around the top of the burette to form a seal with the Lucite cylinder. The base of the burette is sealed against the Lucite cylinder using a threaded brass sleeve tightened against an O-ring. The exit tube is connected to the burette to form a U-tube. A constant air pressure u_{ab} is supplied to the Lucite cylinder to establish a pressure gradient relative to the controlled water pressure below the ceramic disk.

The burette is first filled with water to any desired level. A few drops of the commercial cleaner Fantastik should be added to the water in the burette to reduce surface tension and promote the upward movement of air bubbles in the burette.

The valve connecting the base plate and the diffused air volume indicator is only opened when making a diffused air volume measurement. The water pressure below the disk is controlled at a constant pressure u_c. A backpressure u_{ab} (i.e., $u_{ab} < u_c$) can be applied to the air in the Lucite cylinder in order to establish a suitable pressure gradient across the base plate. The pressure gradient between the water pressure in the base plate and the air pressure in the diffused air volume indicator is the driving force for flushing water containing air bubbles from the base plate into the burette. The pressure gradient should not exceed 70 kPa to avoid causing bubbles to form in the water due to the spontaneous release of dissolved air from the water below the base plate.

9.8.3 Procedure for Measuring Diffused Air Volume

The initial water level reading in the burette, h_{a1}, can be taken when the air pressure in the Lucite cylinder is at atmospheric pressure or at the applied backpressure (i.e., at the air pressure u_{ab}) (Fig. 9.20). The connecting valve is then opened momentarily to apply a pressure gradient across the base plate. The pressure gradient flushes the diffused air into the burette. The flushed diffused air displaces the water in the burette while water flows out of the exit tube. After

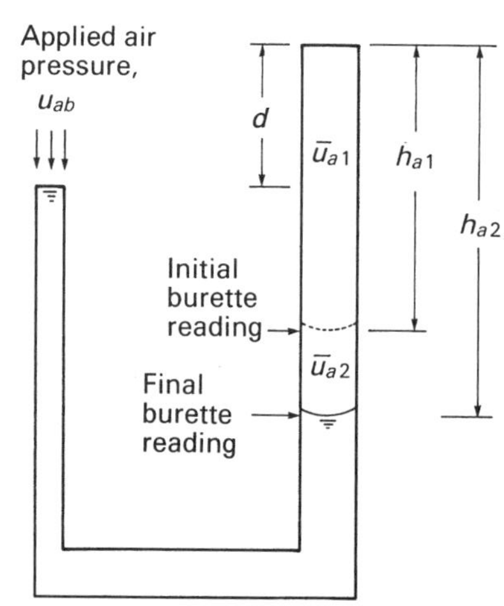

Figure 9.20 Computation of volume of diffused air under isothermal conditions.

removing all of the diffused air from the base plate, a new water level reading, h_{a2}, is recorded.

9.8.4 Computation of Diffused Air Volume

The diffused air volume in the base plate, V_{fa}, is registered on the water volume change indicator as though it were water leaving the soil specimen. The measured water volume changes from the soil specimen must therefore be corrected for the volume of diffused air (i.e., diffused air volume must be subtracted from the total water volume).

The diffused air pressure below the high-air-entry disk is assumed to be at a pressure equal to the water pressure in the base-plate compartment, u_c. The diffused air mass M_{fa} can be expressed as

$$M_{fa} = \frac{\omega_a}{RT_K}\bar{u}_{fa}V_{fa} \tag{9.73}$$

where:

M_{fa} = mass of diffused air,
ω_a = molecular mass of diffused air,
R = universal (molar) gas constant,
T_K = absolute temperature (i.e., $T_K = 273.16 + T$, where T = temperature, °C),
$\bar{u}_{fa}$ = absolute diffused air pressure in the base plate (i.e., $\bar{u}_{fa} = \bar{u}_{atm} + u_c$),
$\bar{u}_{atm}$ = atmospheric pressure (i.e., 101.3 kPa),
u_c = water pressure in the base plate, and
V_{fc} = volume of diffused air in the base plate.

The diffused air volume in the base plate, V_{fa}, is computed by applying the ideal gas law to the air volume measured in the diffused air volume indicator. Let us consider the initial and final conditions in the diffused air volume indicator shown in Fig. 9.20. The initial mass of air in the burette, M_{a1}, can be computed from its absolute pressure $\bar{u}_{a1}$ and its volume V_{a1}:

$$M_{a1} = \frac{\omega_a}{RT_K}\bar{u}_{a1}V_{a1} \tag{9.74}$$

where:

M_{a1} = initial mass of air in the burette,
$\bar{u}_{a1}$ = absolute initial air pressure in the burette [i.e., $\bar{u}_{atm} + u_{ab} + (h_{a1} - d)\rho_w g$],
u_{ab} = gauge applied air pressure in the Lucite cylinder,
h_{a1} = initial reading of the water level in the burette,
d = height difference between the burette and the exit tube,
V_{a1} = initial volume of air in the burette (i.e., $h_{a1}A$), and
A = cross-sectional area of the burette.

Similarly, the final mass of air in the burette can be expressed as

$$M_{a2} = \frac{\omega_a}{RT_K}\bar{u}_{a2}V_{a2} \tag{9.75}$$

where:

M_{a2} = final mass of air in the burette,
$\bar{u}_{a2}$ = absolute final air pressure in the burette [i.e., $\bar{u}_{atm} + u_{ab} + (h_{a2} - d)\rho_w g$],
h_{a2} = final reading of the water level in the burette, and
V_{a2} = final volume of air in the burette (i.e., $h_{a2}A$).

When the initial and final burette readings are at atmospheric pressure, $u_{ab} = 0$ in Eqs. 9.74 and 9.75. The difference between the initial and the final mass of air in the burette (i.e., $M_{a2} - M_{a1}$) is the amount of diffused air removed from the base plate, M_{fa}. The gas law is used to compute the diffused air volume V_{fa} under isothermal conditions:

$$\bar{u}_{a2}V_{a2} - \bar{u}_{a1}V_{a1} = \bar{u}_{fa}V_{fa} \tag{9.76}$$

The diffused air volume is

$$V_{fa} = \frac{\bar{u}_{a2}V_{a2} - \bar{u}_{a1}V_{a1}}{\bar{u}_{fa}} \tag{9.77}$$

It is also possible to compute the amount of diffused air removed from the base plate, M_{fa}, by reading the initial and final water levels in the burette at the same applied air pressure (Fredlund, 1975c). For example, if the initial burette reading was taken at atmospheric pressure, the air in the Lucite cylinder should be depressurized to atmospheric pressure conditions before the final reading is recorded. The diffused air volume is computed as

$$V_{fa} = \frac{\bar{u}_{a1}(V_{a2} - V_{a1})}{\bar{u}_{fa}} \tag{9.78}$$

The base plate is generally flushed once or twice a day. Corrections to intermediate water volume change readings can be made using a linear interpolation of diffused air flow with respect to time. No temperature correction is necessary provided the temperature of the diffused air volume indicator and the base plate of the triaxial cell are the same.

The accuracy of the diffused air volume indicator has been tested by diffusing air through a saturated high-air-entry ceramic disk. An air pressure is applied to the top of a saturated ceramic disk. The saturated disk is connected to the water volume change indicator. A change in the water volume reading with time indicates the accumulation of diffused air below the high-air-entry disk. The ceramic disk remains saturated since the air-entry value of the disk is not exceeded.

The volume of diffused air can be measured either periodically using the diffused air volume indicator or continuously using the water volume change indicator. The two results should be essentially the same if the diffused air volume indicator performs satisfactorily. Figure 9.21 compares the volume of diffused air measured using the water volume change indicator and the diffused air volume indicator.

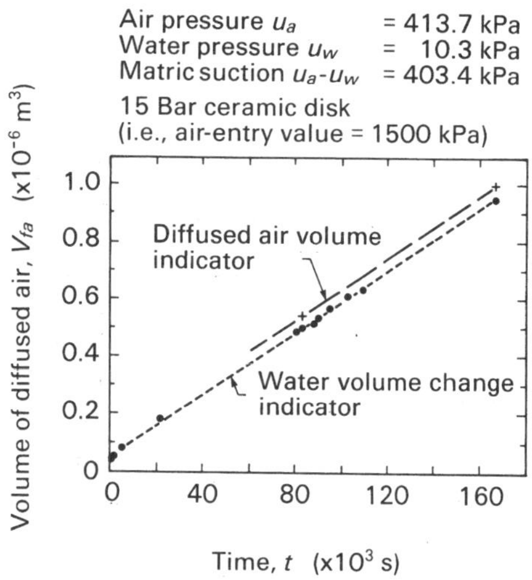

Figure 9.21 Measurement of steady-state diffusion of air through saturated ceramic disk using water volume change indicator and diffused air volume indicator (from Fredlund, 1973a).

The air diffusion test was performed on a saturated, 15-bar ceramic disk (i.e., 1500 kPa air-entry value). Both the water volume change indicator and the diffused air volume indicators recorded essentially the same volumes of diffused air over a period of 2 days. The straight lines indicate steady-state diffusion of air through water in the high-air-entry disk.

Figure 9.22 shows an automated diffused air volume indicator developed by GCTS. The device can be set to automatically flush the compartment below the high-air-entry disk of a triaxial cell at prescribed time intervals. The appropriate valves are closed and diffused air is flushed from below the high-air-entry disks at predetermined times. Measured results of air diffusion through 5- and 15-bar ceramic disks are shown in Figs. 9.23 and 9.24.

The diffusion of air through the high-air-entry disks appears to accelerate as the air-entry value of the disk is approached.

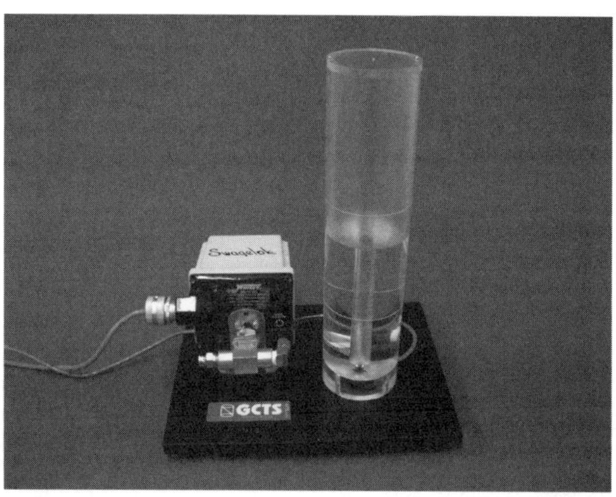

Figure 9.22 Automated diffused air volume indicator developed by GCTS. (Photo courtesy of GCTS, Tempe, AZ.)

The 5-bar disks showed that there was about 0.15 cm^3 of diffused air that collected in the compartment below the high-air-entry disk each day as the applied air pressure approached 500 kPa. The 15-bar disks showed that there was more than 2.0 cm^3 of diffused air per day as the applied air pressure approached 1500 kPa. Laboratory tests on unsaturated soils often run for many days and even weeks, and as a result it is important that diffused air volumes be measured and used to correct water volume changes in the soil specimen.

9.9 INDIRECT ESTIMATION OF AIR FLOW PROPERTIES

The SWCC can be used for the indirect estimation of the air permeability function. The air permeability function provides a relationship between the air coefficient of permeability and either soil suction or the air content of the soil. The minimum air permeability occurs as the soil approaches saturation. The assumption is made that air can only diffuse through the water phase at a rate defined by the diffusivity of air through water and the tortuosity of the water paths around the soil particles. The soil will have the maximum coefficient of air permeability when the soil is completely dry and soil suction is approaching a value of 10^6 kPa. The concept of intrinsic permeability can be used to relate the saturated hydraulic conductivity of a soil to its maximum coefficient of air permeability.

9.9.1 Soil-Air Characteristic Curve

The concept of a soil-air characteristic curve (SACC), can be used to describe the relationship between the volume of air in the soil (i.e., volumetric air content) and soil suction. The volumetric air content can also be referred to as the air degree of saturation. The SACC can be used to compute the air permeability function (Ba-Te et al., 2005). The amount of air in the voids can be written as the air degree of saturation $S_a(\psi)$, which is related to the water degree of saturation:

$$S_a(\psi) = 1.0 - S(\psi) \tag{9.79}$$

where:

ψ = soil suction and
S = degree of water saturation.

The degree of saturation is expressed in decimal form in Eq. 9.79. The amount of air in the soil can also be written as a percentage of the total volume of the soil sample and the volumetric water content:

$$\theta_a(\psi) = \frac{e}{1+e} - \theta(\psi) \tag{9.80}$$

where:

θ_a = volumetric air content,
θ = volumetric water content, and
e = void ratio.

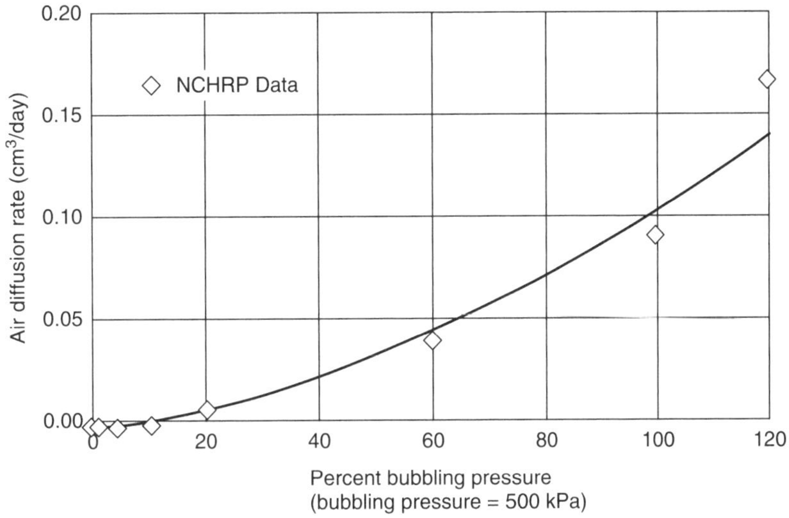

Figure 9.23 Air diffusion rates through 5-bar ceramic disks (after Padilla et al., 2006a).

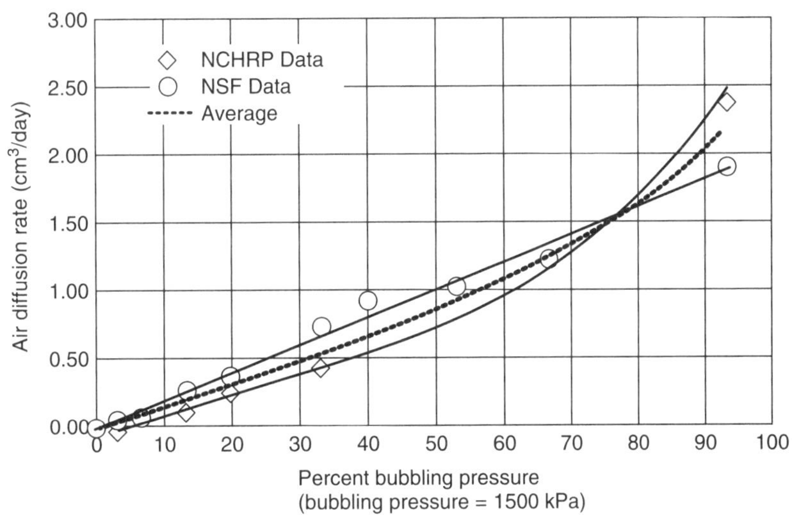

Figure 9.24 Air diffusion through 15-bar ceramic disks (after Padilla et al., 2006a).

Equations 9.79 and 9.80 show two different forms for writing the SACC equation. There is more than one way to define the amount of air in a soil just as there was more than one way to define the amount of water in a soil. Some soil-water characteristic curve (SWCC) empirical equations have been defined over the entire water content range while others have been defined between saturated conditions and residual conditions. The use of various SWCC equations is investigated for the estimation of the air permeability function.

Numerous empirical equations have been used to describe the SWCC (Brooks and Corey, 1964; van Genuchten 1980; Fredlund and Xing, 1994). The Fredlund and Xing (1994) equation for the SWCC is written as follows:

$$\theta(\psi) = \theta_s \left(1 - \frac{\ln(1 + \psi/\psi_f)}{\ln(1 + 10^6/\psi_r)}\right) \frac{1}{\{\ln[e + (\psi/a_f)^{n_f}]\}^{m_f}} \tag{9.81}$$

where:

a_f, n_f, m_f = fitting parameters for the SWCC,
e = irrational constant equal to 2.71828 for taking the natural logarithm,
ψ = any soil suction value, and
ψ_r = soil suction at residual condition.

Residual suction can vary over a wide range depending upon the soil type. Residual suction can exceed 3000 kPa for clay soils and be less than 100 kPa for coarse-grained soils. The Fredlund and Xing (1994) equation uses three fitting parameters, namely, a_f, n_f, and m_f. Equation 9.81 can be combined with Eq. 9.80 to yield an expression for the SACC. The SACC can be developed for both the drying and wetting curves.

The van Genuchten (1980) equation defines volumetric water content between saturation and residual water content conditions (i.e., normalized volumetric water content):

$$\Theta_n = \frac{1}{(1 + (a_v \psi)^{n_v})^{mv}} \quad (9.82)$$

where:

a_v, n_v, m_v = soil fitting parameters for the SWCC,
Θ_n = normalized volumetric water content defined as $\Theta_n = \frac{\theta - \theta_r}{\theta_s - \theta_r}$,
θ = any selected volumetric water content,
θ_s = saturation volumetric water content, and
θ_r = residual volumetric water content.

The van Genuchten (1980) model describes the SWCC over the range between saturation and residual suction conditions. The equation uses three fitting parameters; namely, a_v, n_v and m_v. Other SWCC equations could also be used in conjunction with Eqs. 9.79 and 9.80 to write equations for the SWCCs. However, only the Fredlund and Xing (1994) equation and the van Genuchten (1980) equation are used to illustrate the estimation procedure.

9.9.2 Relationships between Air Coefficient of Permeability and Degree of Air Saturation

A number of empirical relationships have been proposed for the coefficient of air permeability and the amount of air in a soil. Published air permeability relationships fall into two main categories, namely, those that assume power functions for the air content and those based on the SWCC (Stylianou and DeVantier, 1995). Power relations include equations proposed by Irmay (1954), Wyllie (1962), Falta et al. (1989), and Delage et al. (1998). Relations based on the SWCC include equations proposed by Brooks and Corey (1966), Parker et al., (1987), and Cary et al., (1989).

Brooks and Corey (1964) proposed the following empirical relationship relating the coefficient of air permeability k_a to the effective degree of water saturation S_e. The equation applies for soil suctions greater than the air-entry value of the soil:

$$k_a = k_d(1 - S_e)^2(1 - S_e^{(2+\lambda)/\lambda}) \quad (9.83)$$

where:

S_e = effective degree of saturation with respect to the water phase, defined as $S_e = (1.0 - S)/(1.0 - S_r)$, where degree of saturation is in decimal form,
S_r = residual degree of saturation,
k_d = coefficient of air permeability when the soil is dry (i.e., zero water content), and
λ = fitting coefficient.

Van Genuchten (1980) proposed a coefficient-of-air-permeability equation based on an effective degree-of-saturation equation that embraces the Mualem (1976) unsaturated permeability model. The end result is a closed-form expression for relative air permeability. The relative coefficient of air permeability is assumed to be a function of degree of air saturation. The van Genuchten-Mualem (1980) model for the coefficient of air permeability can be expressed as follows:

$$k_a = k_d(1 - S_e)^{1/2}(1 - S_e^{1/q})^{2q} \quad (9.84)$$

where:

k_a = air coefficient of permeability,
k_d = air coefficient of permeability of the dry soil,
q = fitting parameter related to the pore-size distribution; van Genuchten (1980) suggested limiting q values to the range between 0 and 1.0.

The van Genuchten (1980) air permeability equation can be written by substituting Eq. 9.82 into Eq. 9.84. Ba-Te et al., (2005) extended the van Genuchten (1980) model (i.e., Eq. 9.84) for use with the Fredlund and Xing (1994) equation which accepts the entire degree-of-saturation range (i.e., zero to 100%):

$$k_a = k_d(1 - S)^{1/2}(1 - S^{1/q})^{2q} \quad (9.85)$$

9.9.3 Relationship between Air Coefficient of Permeability and Soil Suction

The Fredlund and Xing (1994) air permeability model and the van Genuchten (1980) model are used to illustrate the application of the SWCC to the computation of the coefficient-of-air permeability function. The Fredlund and Xing (1994) SWCC equation can be used in conjunction with Eq. 9.85 to compute the air permeability function. The van Genuchten (1980) SWCC equation can be modified and the degree of saturation can be written as follows:

$$S = \frac{1+e}{e}\left[\frac{1}{[1 + (a_v\psi)^{n_v}]^{m_v}}\left(\frac{e}{1+e} - \theta_r\right) + \theta_r\right] \quad (9.86)$$

where:

θ_r = residual volumetric water content and
e = void ratio.

The Fredlund and Xing (1994) SWCC equation can be written as

$$S = \frac{1+e}{e}\theta_s\left(1 - \frac{\ln(1 + \psi/\psi_r)}{\ln(1 + 10^6/\psi_r)}\right)\frac{1}{\{\ln[e + (\psi/a_f)^{n_f}]\}^{m_f}} \quad (9.87)$$

where:

e = void ratio,
θ_s = saturated volumetric water content,

e = irrational constant equal to 2.71828,
ψ = any soil suction value, and
ψ_r = soil suction at residual conditions.

The air permeability equation is obtained by substituting Eq. 9.87 [i.e., Fredlund and Xing (1994) SWCC equation] into Eq. 9.57.

Other empirical relationships between the coefficient of air permeability and the degree of saturation have also been suggested. Different air permeability functions can also be obtained by using different equations to describe the SWCC. The reliability of various proposed air permeability functions can best be validated through comparisons with experimental measurements of air permeability. However, data from air permeability measurements are scarce.

9.9.4 Comparison of Estimated and Measured Air Permeability Functions

The results obtained from indirect computation of the air permeability (i.e., through use of the SWCC) can be compared with published experimental data (Ba-Te et al., 2005). Laboratory data from SWCC tests can be best fit with the Fredlund and Xing (1994) SWCC equation or the van Genuchten (1980) SWCC equation. The relationship between the air content of the soil and soil suction (i.e., SACC) can then be calculated. The SACC results are then substituted into the proposed air permeability functions to give the relationship between the coefficient of air permeability and soil suction. A comparison can then be made between the experimental data and the estimates made using the indirect, empirical air permeability functions.

Two data sets on air permeability measurements were analyzed by Ba-Te (2005). The results of the analysis can be used to compare the measured and estimated air permeability functions. One set of results is from a series of air permeability tests undertaken by Samingan et al., (2003) on Singapore residual soils. The second set of air permeability measurements was undertaken by Moldrup et al., (2003) on a number of soils from Japan. Laboratory measurements on the soils were compared to estimations of the air permeability functions based on measured SWCCs. The results were interpreted using the residual degree of saturation as well as a zero degree of saturation.

9.9.4.1 Samingan et al. (2003) Results

Samingan et al., (2003) conducted a series of air permeability tests and made SWCC measurements on four Singapore residual soil samples. The soil consisted of 43% sand, 45% silt, and 12% clay. The initial void ratio and saturated volumetric water content of the specimen were 1.05 and 0.51, respectively. The air permeability test results on soil specimen UP-2 showed hysteresis; however, only the drying curve results are presented.

The Fredlund and Xing (1994) equation and the van Genuchten (1980) equation were used to fit the experimental results using a least-squares method. The residual volumetric water content was estimated at 3.5% and the residual suction was estimated at 10,000 kPa for the Samingan soil specimen. The SWCC for the Samingan sample is shown in Fig. 9.2a.

The best-fit soil parameters for both soils are listed in Table 9.1. The fitting parameters for the SWCC data are shown for the case where the observed residual conditions were used as well as the case where the minimum degree of saturation was zero. Dry air permeability values k_d are also presented along with an estimated value for the q parameter.

The volumetric air content for the Samingan et al., (2003) soil as well as the Moldrup et al., (2003) soil is shown in Fig. 9.25. At low soil suctions the graphs show zero air content and the air-entry value indicates the point at which free air commences flowing through the soil sample.

Substituting Eq. 9.87 [i.e., Fredlund and Xing (1994) SWCC equation] into Eq. 9.85 yields the functional relationships for the air coefficient of permeability and soil suction. Other functional relationships between the air coefficient of permeability and soil suction can be calculated using the

Table 9.1 Parameters for Fitting Test Data from Samingan et al. (2003)

θ_r (S_{res}) (decimal)	Model	a	n	m	ψ_r (kPa)	k_d (m/s)	q
0.035 (0.068)	Fredlund and Xing + $k_a(S)$	30.68	0.80	1.54	10,000	1.79×10^{-5}	2.923
	Fredlund and Xing + $k_a(S_e)$					6.92×10^{-6}	2.850
	van Genuchten $+k_a(S)$	0.07	16.62	0.02	—	3.38×10^{-4}	3.528
	van Genuchten $+k_a(S_e)$					1.01×10^{-4}	3.430
0 (0)	Fredlund and Xing $+k_a(S)$	30.68	0.80	1.54	10,000	1.70×10^{-5}	2.914
	Fredlund and Xing $+k_a(S_e)$					1.73×10^{-5}	2.917
	van Genuchten $+k_a(S)$	0.06	2.53	0.13	—	1.71×10^{-4}	3.320
	van Genuchten $+k_a(S_e)$					1.68×10^{-4}	3.317

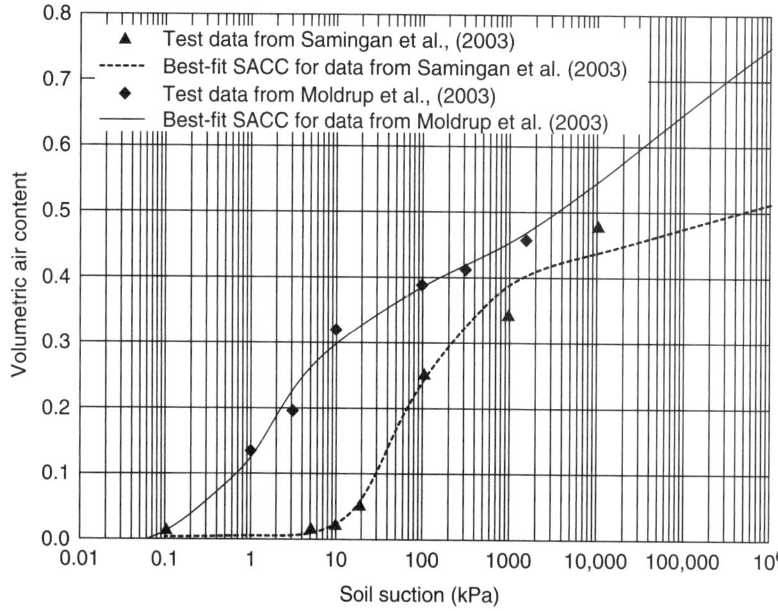

Figure 9.25 SACCs best fit with Fredlund and Xing (1994) SWCC equation for test data from Samingan et al., (2003) and Moldrup et al., (2003).

above equations. In total, four air permeability functions were computed. The variable q can be determined using a least-squares technique for each of the four equations. The coefficient of air permeability under dry conditions was not independently measured for the soil tested by Samingan et al., (2003). Therefore, the dry soil coefficient of air permeability was considered as an independent variable and obtained using a least-squares method. The air permeability functions were then computed.

The estimated air permeability results were plotted together with the experimental data for the Samingan soil in Fig. 9.26. The parameters for the air permeability functions are shown in Table 9.1. Reasonable agreement between the measured and estimated air coefficient of permeability was found in the range over which the tests were performed (i.e., 50–300 kPa).

9.9.4.2 Moldrup et al. (2003) Results

Moldrup et al. (2003) presented data on SWCCs and gas diffusion coefficients as a function of water content for 22 soils from Japan. The measurements were originally conducted to evaluate factors that influence plant performance.

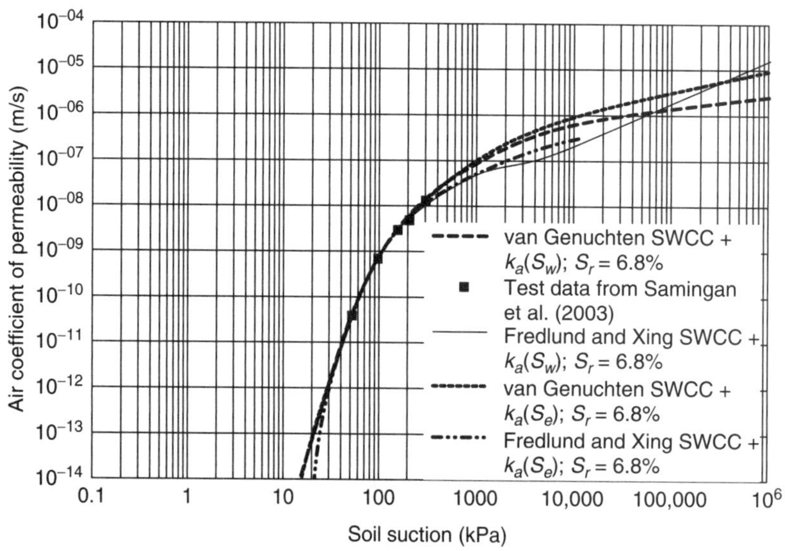

Figure 9.26 Calculated and measured air coefficient of permeability functions for soil tested by Samingan et al., (2003).

Besides the SWCCs, gas diffusivity and void ratio were tabulated (i.e., $e = 3.0$). Results from the test labeled "Miura 4 Andisol" were analyzed by Ba-Te et al., (2005).

The Fredlund and Xing (1994) equation and the van Genuchten (1980) equation were fit to the experimental results using a least-squares method. The residual volumetric water content was estimated at 29.2 % and the residual suction was estimated at 1554 kPa for the Moldrup sample. The SWCC for the Moldrup sample is shown in Fig. 9.27.

The best-fit soil parameters for both soils are listed in Table 9.2. The fitting parameters for the SWCC data are shown for the case where the observed residual conditions were used as well as the case where the minimum degree of saturation was set to zero. The coefficients of air permeability for the dry soil, k_d, are also presented along with estimated values for the q parameter.

The permeability data from Moldrup et al. (2003) were presented in terms of the relative air coefficient of permeability. An arbitrary value of 2×10^{-9} m/s was assigned for the dry coefficient of air permeability k_d.

The generated air permeability equations are plotted together with the experimental data for the Moldrup soil in Fig. 9.28. The parameters for the air permeability functions are shown in Table 9.2. Reasonable agreement was found over the range of soil suction applied to the soil (i.e., 50–300 kPa). It appears that leaving the dry permeability k_d as a fixed value in Eq. 9.85 yields a closer fit to thc data than Eq. 9.84, as shown in Fig. 9.28.

9.9.4.3 Comparison of Samingan et al., (2003) and Moldrup et al., (2003) Analysis

The coefficient air permeability equations tend to asymptotically approach a zero value when soil suction was low

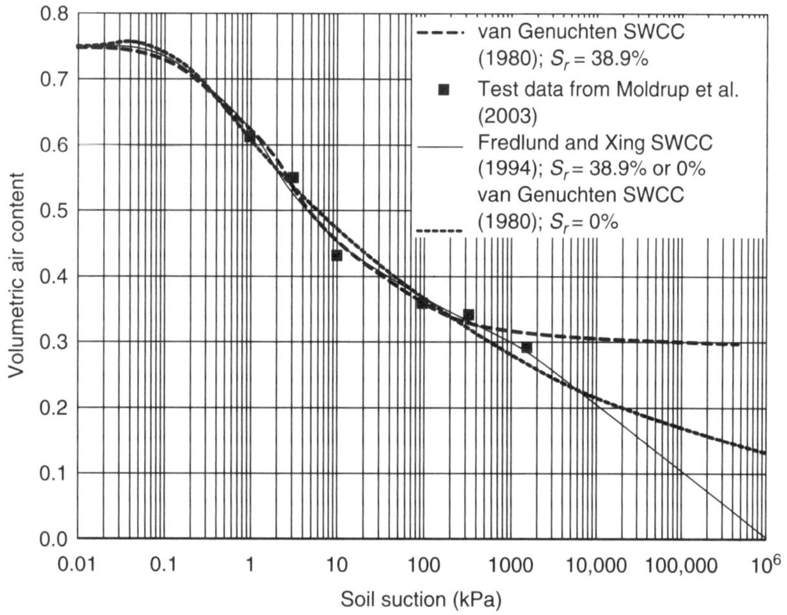

Figure 9.27 SWCCs best fit to test data from Moldrup et al. (2003).

Table 9.2 Parameters for Fitting Test Data from Moldrup et al., (2003)

θ_r (S_{res}) (decimal)	Model	a	n	m	ψ_r (kPa)	k_d (m/s)	q
0.292 (0.068)	Fredlund and Xing + $k_a(S)$	0.47	0.37	1.26	1554	2×10^{-9}	1.126
	Fredlund and Xing + $k_a(S_e)$					2×10^{-9}	1.606
	van Genuchten $+k_a(S)$	1.19	0.96	0.43	—	2×10^{-9}	1.126
	van Genuchten $+k_a(S_e)$					2×10^{-9}	1.604
0 (0)	Fredlund and Xing $+k_a(S)$	0.47	0.37	1.26	1544	2×10^{-9}	1.126
	Fredlund and Xing $+k_a(S_e)$					2×10^{-9}	1.126
	van Genuchten $+k_a(S)$	6.70	3.05	0.04	—	2×10^{-9}	1.132
	van Genuchten $+k_a(S_e)$					2×10^{-9}	1.132

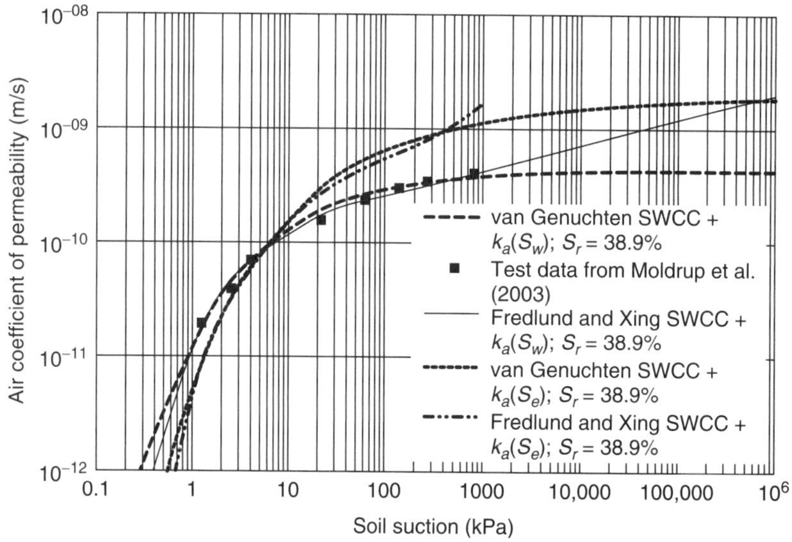

Figure 9.28 Calculated and measured air coefficient-of-permeability functions for soil tested by Moldrup et al., (2003).

(e.g., < 50 kPa). A zero coefficient of air permeability should not occur because there is still the diffusion of air through water. There should be a limiting value of coefficient of air permeability where the coefficient of air permeability tends toward the diffusion of air through water.

At high soil suction values (e.g., over ψ_r), the volumetric water content should extend to be less than the residual value ($\theta < \theta_r$). However, when using the van Genuchten (1980) SWCC equation, the volumetric water content is asymptotic to θ_r. The volumetric water content must therefore be held larger than residual water content.

Four air permeability equations were best fit to the data from Samingan et al. (2003) in order to determine the coefficients of air permeability k_a that were closest to the dry soil permeability k_d in the high-suction range when using van Genuchten's SWCC equation (Ba-Te et al., 2005). Different values for residual volumetric water content were also selected. Figure 9.29 shows that when the residual water content θ_r was 0.035, the combination of Eqs. 9.82 and 9.84 produced results that were close to the dry soil permeability. Similar trends were found regarding the best fit for the air permeability equations applied to the Moldrup et al. (2003) data, as shown in Fig. 9.30. Setting the residual water content to zero brings the predicted air permeability k_a closer to the dry air permeability k_d for the combination of Eqs. 9.85 and 9.86. It appears that

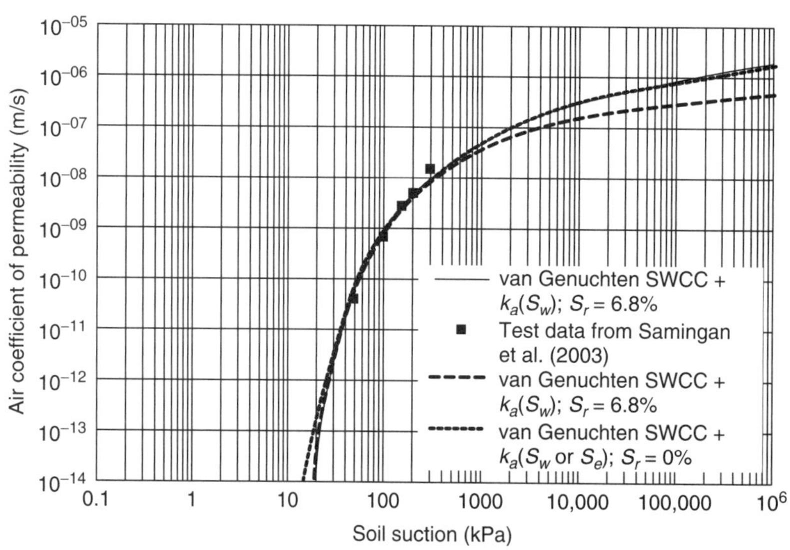

Figure 9.29 Calculated and measured air coefficient-of-permeability functions for soil tested by Samingan et al. (2003) using van Genuchten (1980) SWCC model.

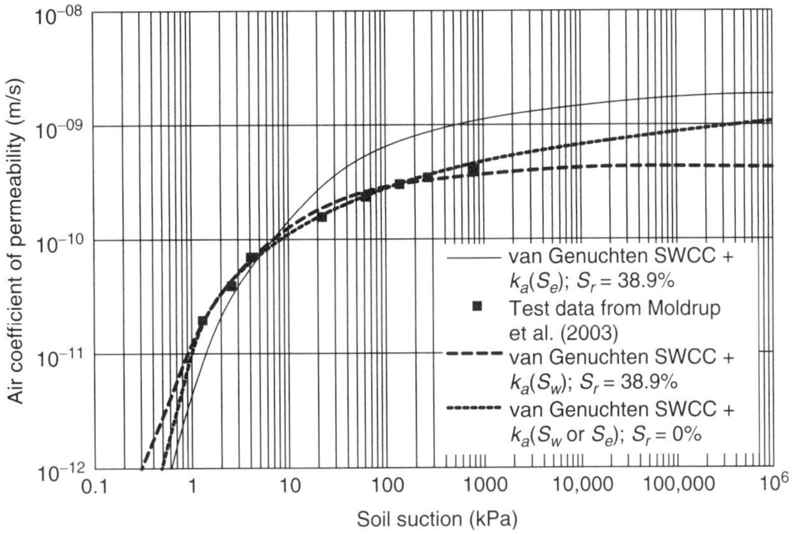

Figure 9.30 Calculated and measured air coefficient-of-permeability functions for soil tested by Moldrup et al., (2003) using van Genuchten (1980) SWCC model.

when using the van Genuchten's (1980) SWCC equation, Eq. 9.85 provides a better fit of the data. Setting the residual volumetric water content to zero appears to improve its performance in the high-suction range.

The air coefficient of permeability appears to reach the dry air permeability k_d when soil suction reaches 10^6 kPa. It should be noted that the usage of Eqs. 9.84 and 9.87 can result in the calculation of negative numbers. Setting the residual water content at zero avoids this difficulty. Figures 9.31 and 9.32 also show that setting the residual water content to zero provides an air permeability function that is close to the combination of Eqs. 9.85 and 9.87. In summary, the Fredlund and Xing (1994) SWCC equation appears to provide the closest fit of the air permeability data over the entire soil suction range.

9.9.5 Air Storage Function

The volumetric air content of a soil is, in essence, the inverse of the SWCC defined using volumetric water content. Volumetric air content versus soil suction (i.e., the SACC) defines the amount of air in a soil at any particular soil suction. The slope of the SACC defines the rate of change of the volume of air in the soil with respect to a change in soil suction and can be described using the symbol m_2^a:

$$m_2^a = \frac{d\theta_a}{d\psi} \tag{9.88}$$

where:

θ_a = volumetric air content in the soil and
ψ = soil suction.

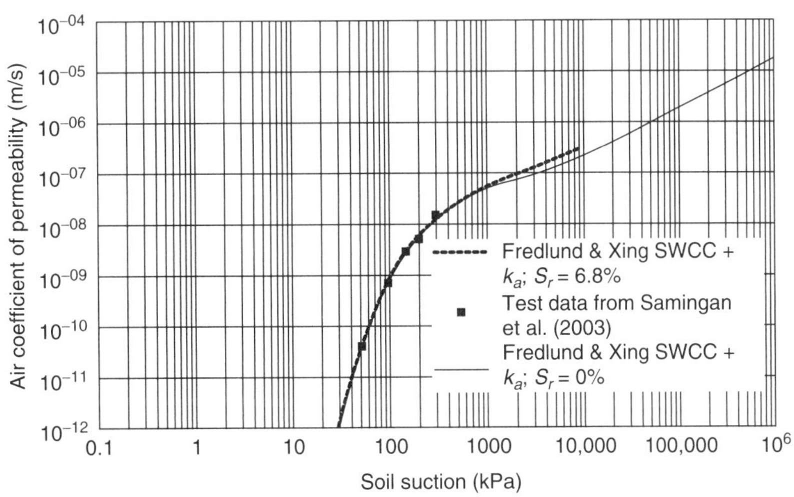

Figure 9.31 Calculated and measured air coefficient-of-permeability functions for soil tested by Samingan et al., (2003) using Fredlund and Xing (1994) SWCC model.

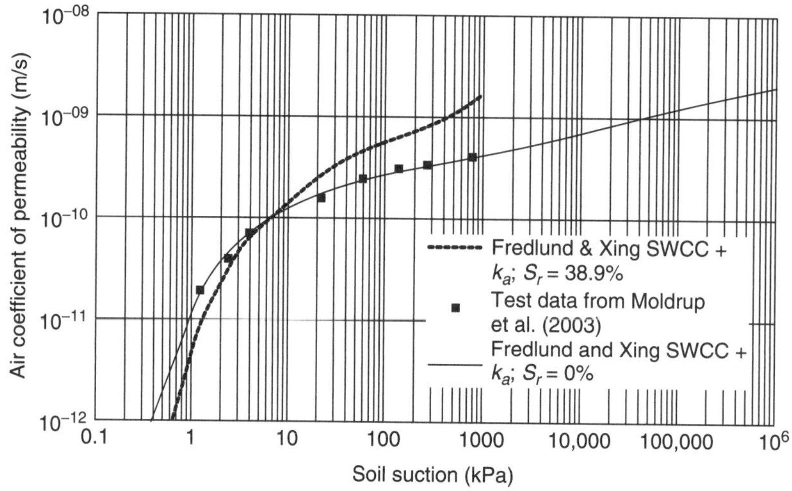

Figure 9.32 Calculated and measured air coefficient-of-permeability functions for soil tested by Moldrup et al., (2003) using Fredlund and Xing (1994) SWCC model.

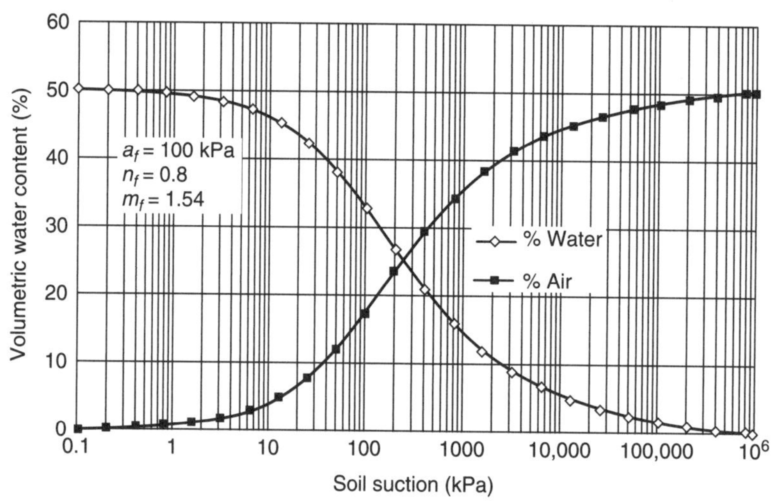

Figure 9.33 Inverse relationship between SWCC and SACC.

The symbol m_2^a is not a compressibility of the air phase; rather, it is a symbol that indicates the change in air storage with respect to the soil suction. The compressibility of the air phase is defined in terms of the general gas law, as shown in Chapter 15.

Figure 9.33 shows a plot of both a SWCC and a SACC illustrating the reciprocal nature of these two curves. Both curves can be generated using the Fredlund and Xing (1994) SWCC equation. The initial porosity was assumed to be 50% and the void ratio was assumed to be 0.5. The SWCC parameters are shown on Fig. 9.33. The assumption is made that there is no change in void ratio as suction is changed. The reference for the coefficient of air permeability becomes the completely dry soil state. The air-entry value (on the desorption curve) becomes the point at which air can only diffuse through the water phase of the soil.

9.10 APPLICATIONS TO SATURATED-UNSATURATED AIR FLOW PROBLEMS

The application section illustrates a few of the concepts associated with air flow. Relatively simple air flow problems are used to illustrate the flow of air through an unsaturated soil. The conditions defining each problem are described along with the soil properties needed to solve the problem. All problems are solved within the context of a boundary value problem.

9.10.1 Solution to Air Flow Boundary Value Problem

A boundary value problem requires that the following information be available to solve the problem at hand. First, the ground surface geometry and the stratigraphic soil units must be defined. A decision must be made regarding whether the problem at hand should be solved in a one-, two-, or three-dimensional context.

Second, the geotechnical engineer must be satisfied that an appropriate partial differential equation can be written for a representative elemental volume of the continuum. The partial differential equation must adequately describe the physical processes that control air flow through a porous media.

Third, it must be possible to measure or estimate the air phase soil properties and other soil properties relevant to the problem being solved.

Fourth, boundary conditions must be assessed for all boundaries. Initial conditions also need to be established when modeling time-dependent flow problems. Transient air flow problems can be highly nonlinear and the computed heads can be dependent upon the selected initial values. The modeler should also assess whether a steady-state or transient analysis is required. It is generally advisable to first perform a steady-state analysis.

Fifth, the solutions should be performed and the results should be studied from the standpoint of being reasonable. There are many potential sources of error when undertaking numerical modeling studies and it is important that "checks" be performed to ensure that the computer output is correct.

9.10.2 Types of Boundary Conditions for Air Flow Problems

The first air flow problem illustrates the diffusion of air through a saturated ceramic disk. The analysis is one dimensional and the initial air pressures are known. It is not necessary to use numerical modeling software to solve the first problem.

Other example problems are more complex and require numerical modeling software in order to obtain a solution. Air fluxes near ground surface may be part of the input information that needs to be provided by the modeler. Often it is the air pressure isochrones that are of greatest interest.

9.10.3 Example 1: One-Dimensional Air Flow Examples

The diffusion of air through a saturated ceramic high-air-entry disk is an example of steady-state air diffusion through water. Another example is the diffusion of air through a saturated soil specimen. In each case, the diffused air pressure is dissipated across a region of water.

The measurement of the coefficient of diffusion can be used as an example of steady-state air diffusion through water. The coefficient is assumed to be a constant. The partial differential equation describing air diffusion takes the same form as that for air flow through an unsaturated soil (i.e., $d^2u_a/dy^2 = 0$). The pore-air pressure distribution through the soil specimen can be assumed to be linear.

An example showing the pore-air and pore-water pressure distributions across a soil specimen during the measurement of the coefficient of diffusivity is shown in Fig. 9.34. The air

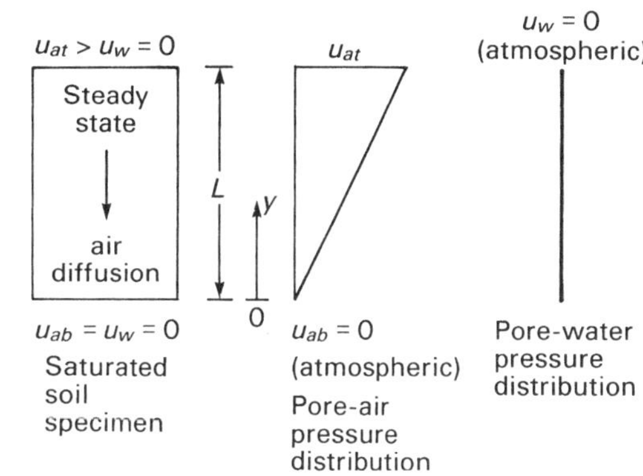

Figure 9.34 Pore-air pressure distribution during measurement of coefficient of diffusion D.

pressures at each end of the specimen (i.e., $u_a = u_{ab} = 0.0$ at $y = 0.0$ and $u_a = u_{at}$ at $y = L$) are the boundary conditions. Substituting the boundary conditions into the differential equation yields a linear equation for the diffusing pore-air pressure distribution in the y-direction [i.e., $u_a = (y/L)\, u_{atm}$].

9.10.4 Two-Dimensional Air Flow Well (or Trench) Example

A two-dimensional cross section is passed through a 2-m-wide trench that was dug to a depth of 20 m. The lower 6 meters of the trench was sealed off and subjected to a pressure of 20 kPa (gauge). Steady-state air flow conditions were established from the bottom of the trench to the surrounding atmosphere.

Relevant coordinates of the geometry of the problem are shown in Fig. 9.35. The soil is fine sand with an air coefficient of permeability of 2.18×10^{-4} m/s. The right and left sides of the model were assumed to be impervious. The ground surface and the top portion of the trench were exposed to atmospheric pressure (i.e., 101.0 kPa absolute). The objective was to calculate the dissipation of pore-air pressures from the bottom of the trench to the surrounding atmosphere.

Figure 9.36 shows the generated finite element mesh along with the contour of pore-air pressure under steady-state air flow using the SVAir software. The highest pore-air pressure gradients are near the packers that were placed at a distance of 6 m above the base of the trench. The highest pore-air pressure gradient locations are also the locations where the finite element mesh needs to be refined (i.e., smaller elements) in order to ensure a correct numerical solution. Also shown are the air pressure contours throughout the sand.

9.10.5 Two-Dimensional Air Flow Well with Clay Layer

The above example involving the trench in sand was reanalyzed with one change made to the geometry of the problem, as shown in Fig. 9.37. In this case, a 2-m-thick layer of fine silt was shown to be located 2 m below the bottom of

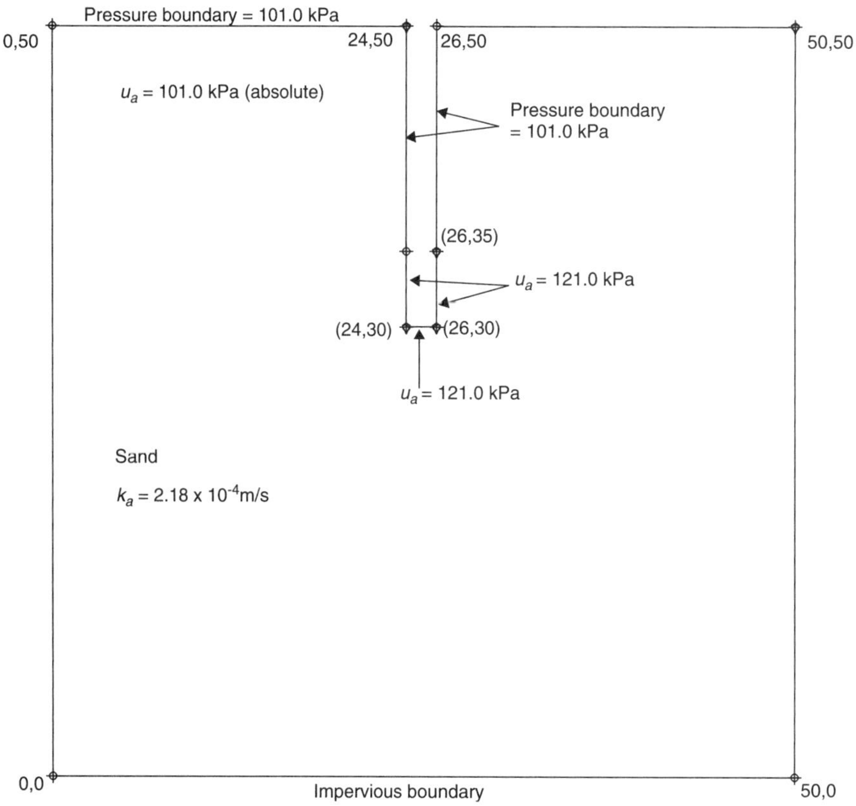

Figure 9.35 Geometry of trench with air pressure applied near bottom of trench.

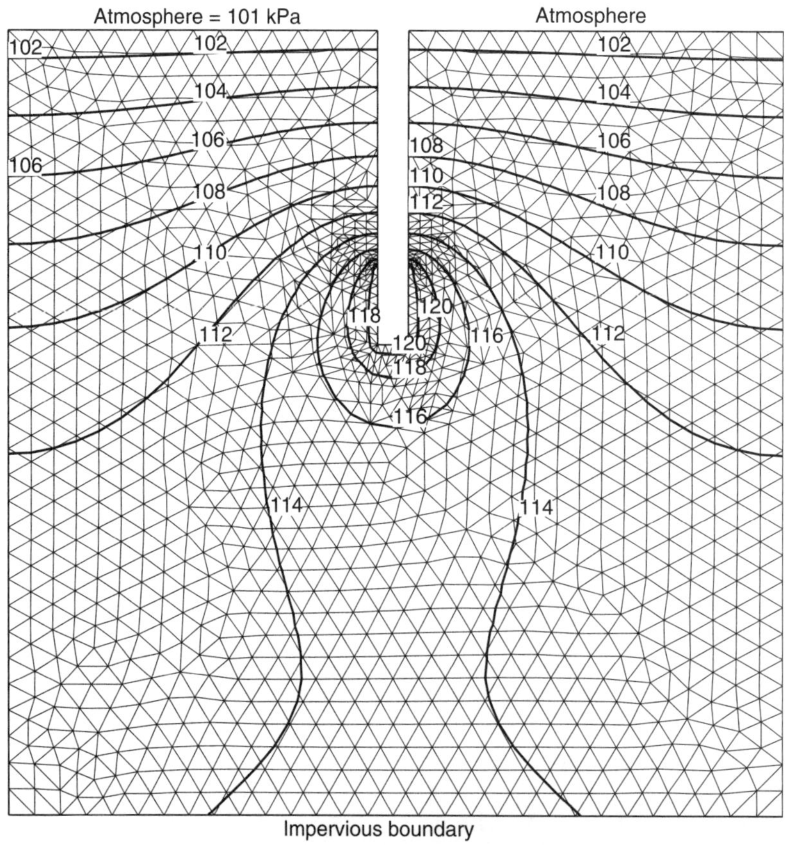

Figure 9.36 Numerical model results showing air pressure contours in soil around trench.

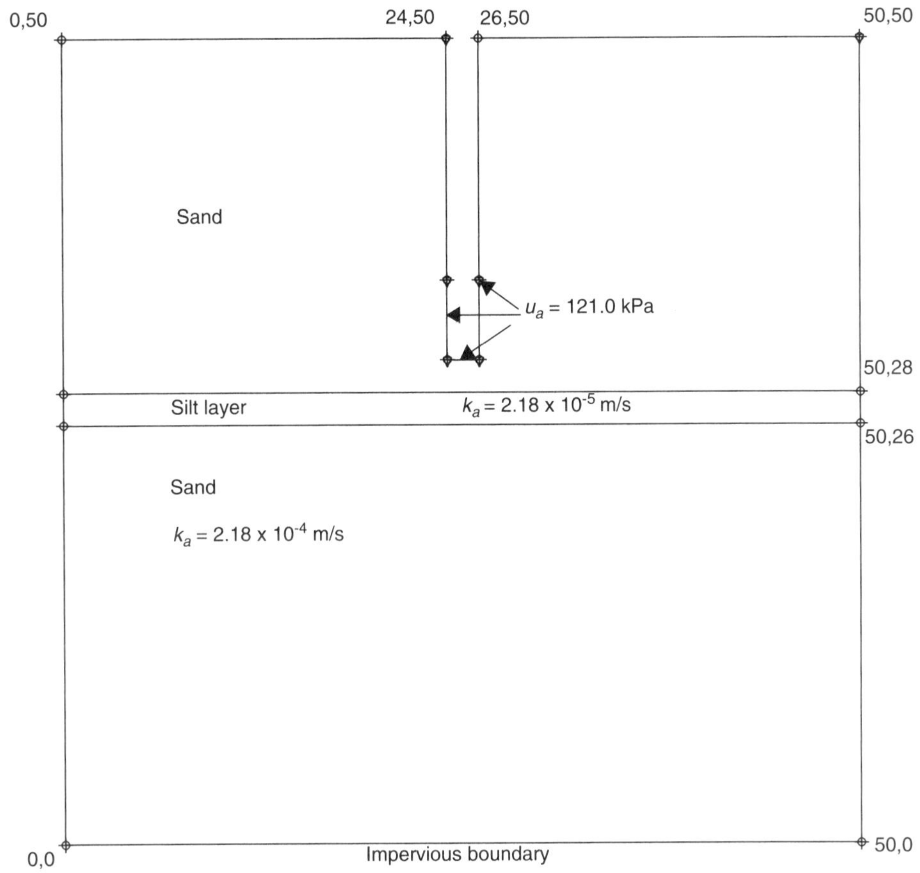

Figure 9.37 Geometry of trench subjected to air pressure when less permeable layer exists below bottom of trench.

the trench. The silt layer has a coefficient of air permeability of 2.18×10^{-5} m/s and the sand has a coefficient of air permeability of 2.18×10^{-4} m/s. The bottom portion of the trench was again subjected to a positive air pressure of 20 kPa above atmospheric pressure.

The trench was again approximated as a two-dimensional plane passing through the soil mass. The model is meant to study the influence of a less pervious soil layer located below the bottom of the trench. The contours of pore-air pressure were calculated throughout the soil region. The dense clay layer beneath the borehole is meant to inhibit the flow of air. The pore-air pressure contours around the trench computed using SVAir are shown in Fig. 9.38. The air flux results show that the clay layer has significantly impeded the flow of air.

The above example problems have been analyzed with the assumption that a positive pressure has been applied to the zone near the bottom of the trench. It is also possible to solve the same problem with a vacuum applied to the bottom of the trench (i.e., vacuum extraction).

9.10.6 Baehr and Hult (1991) Two-Dimensional Steady-State Air Flow

Baehr and Hult (1991) analyzed an air flow problem involving a single material layer exposed to the atmosphere. A well was drilled to a depth of 2.88 m and a horizontal water table was encountered at a depth of 3.37 m. The well depth was 2.88 m and the well screen height was 0.6 m. The well radius was 0.115 m, the model radius was 20 m, and the model depth was 3.37 m.

The absolute air pressure at the ground surface was 101.5 kPa (i.e., atmospheric pressure). The side and base of the well screen had an applied air flux (or standard condition velocity) of 0.1379 m/s. The remainder of the well had zero air flux designated as the boundary condition. The intrinsic permeability of the soil was 1.4×10^{-10} m^2. The vertical anisotropy ratio was 0.3142. An initial air pressure of 101.5 kPa was applied to the soil through the well screen. The soil was exposed to the atmosphere.

Baehr and Hult (1991) presented two analytical solutions for steady-state axisymmetric air flow into a single well that was partially screened in the unsaturated soil zone. Input data were collected from a series of pneumatic tests. One model assumed that there was a stratum with a relatively low coefficient of air permeability that created a separation of the soil from the atmosphere. In another scenario, it was assumed that there was no separating layer between the soil and the atmosphere. The second analytical model with no separating layer was solved using a closed-form solution and presented by Baehr and Hult (1991).

Figures 9.39–9.42 show the normalized air pressure versus radial distance from the pumping well at various

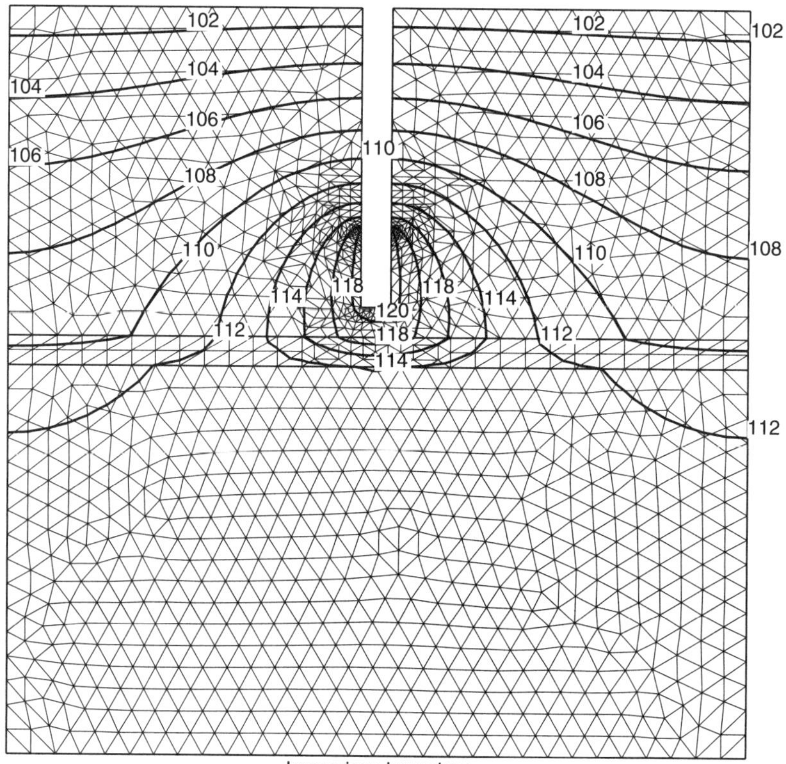

Figure 9.38 Numerical model results showing air pressure contours around trench when less permeable layer is present below trench.

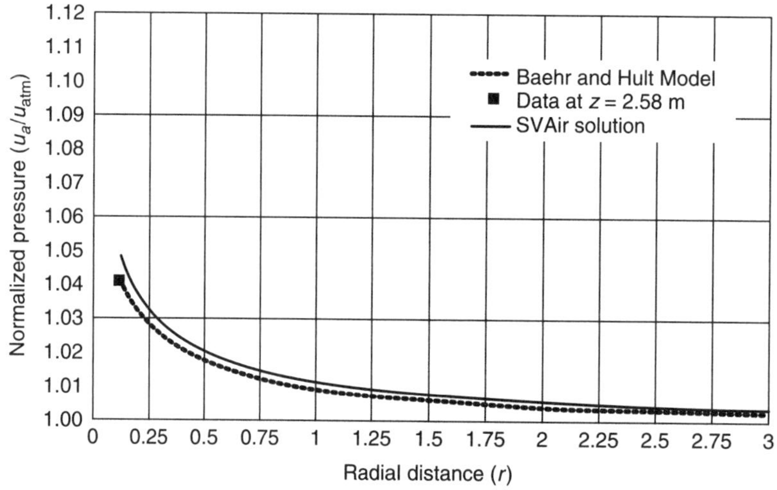

Figure 9.39 Normalized pressure as function of cylindrical coordinates (r, z) in upper unsaturated soil zone for depth $z = 2.58$ m.

depths. The square symbols represent measured values obtained from field tests. The dashed lines indicate the analytical solution. The computer program SVAir (M.D. Fredlund, 2011) was used to solve the same problem and the results are presented as solid lines on each of the graphs. There is good agreement with both the analytical solution and the measured field results.

9.10.7 Coupled Air Flow with Other Physical Processes

There are numerous time-dependent analyses (i.e., transient flow) and coupled solutions that could be used to illustrate more complex problems that could be solved using numerical procedures. The volume or mass of air flowing through

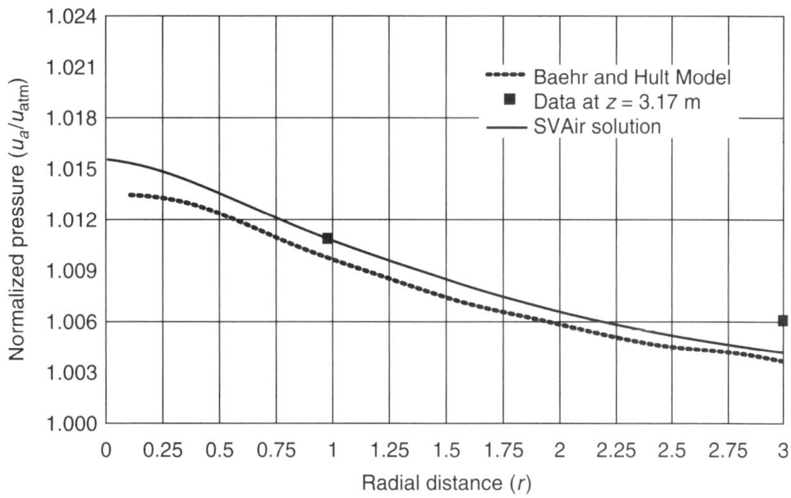

Figure 9.40 Normalized pressure as function of cylindrical coordinates (r, z) in upper unsaturated soil zone for depth z = 3.17 m.

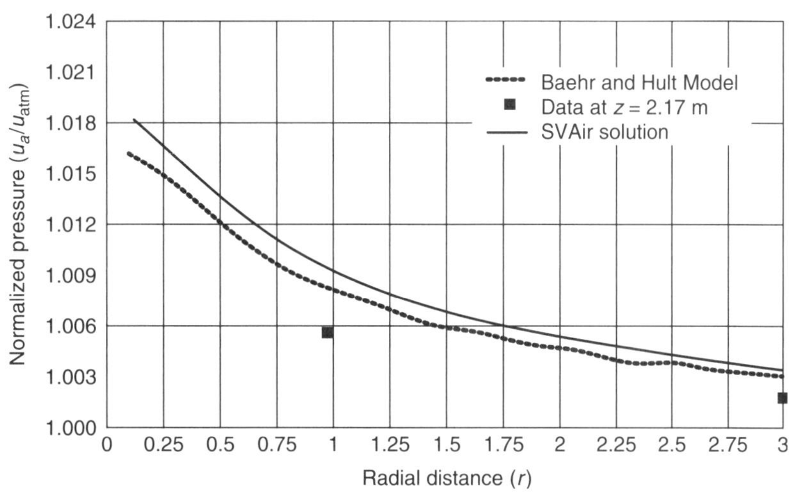

Figure 9.41 Normalized pressure as function of cylindrical coordinates (r, z) in upper unsaturated soil zone for depth z = 2.17 m.

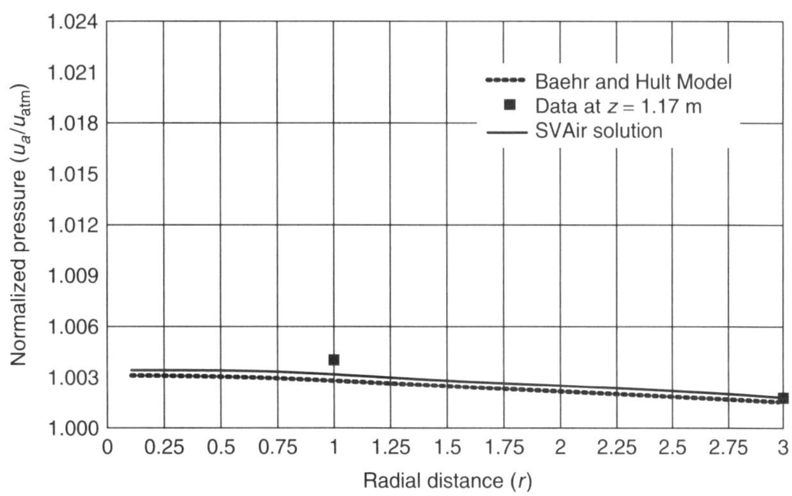

Figure 9.42 Normalized pressure as function of cylindrical coordinates (r, z) in upper unsaturated soil zone for depth z = 1.17 m.

any plane passed through a soil (or at the surface of the soil) can be calculated in addition to air pressures.

Air flow numerical analyses can be performed in terms of either gauge pressures or absolute pressures when steady-state conditions are being analyzed. Numerical analyses should be performed in terms of absolute pressures when transient or unsteady-state conditions are being analyzed.

Convection air flow problems can also be coupled with either heat flow or water flow. It is possible for there to also be coupling with stress–deformation soil behavior as is required in the study of long-term storage of radioactive materials. These types of analyses are considered to be beyond the scope of this book.

CHAPTER 10

Heat Flow Analysis for Unsaturated Soils

10.1 INTRODUCTION

Heat transfer in soils occurs by three primary mechanisms: conduction, convection, and radiation. Heat transfer by conduction occurs as thermal energy is transferred between molecules as a result of direct contact. Heat transfer by convection occurs when kinetic energy in the pore fluid of a soil is transported from one location to another location. Heat transfer by convection is usually considered to be much smaller than heat transfer by conduction (Milly, 1984a) and can be neglected in many situations (Jame and Norum, 1980; Wilson, 1990).

Heat transfer associated with radiation occurs as electromagnetic energy is emitted from one object and transmitted through space to another object. Radiation from the sun is the driving mechanism for phase change in the form of vaporization and condensation at the ground surface. The phase change may also be in the form of freezing and thawing of the water in soils. Radiation has an important role in unsaturated soils, but it is generally considered as small in comparison to heat flow by conduction.

Unsaturated soil mechanics problems can involve heat transfer through each of the three mentioned mechanisms. It is possible for the water phase in soils to exist in the liquid phase and solid phase as well as being in the vapor phase and condensing back to the liquid phase. This chapter will focus primarily on unsaturated soil mechanics problems where there is heat flow by conduction. Chapter 6 dealt with the calculation of moisture flux in vapor form at ground surface. As such, Chapter 6 describes the role of net radiation in the calculation of evaporation and transpiration.

Many environmental problems involve the complex interaction of several physical processes. For example, it is common for heat fluxes, air fluxes, and vapor fluxes to occur simultaneously in an unsaturated soil. This means that there are several physical processes that are coupled processes. It is difficult to model interrelated soil problems without fully understanding the physical processes associated with each independent process. It is often possible that one (or possibly two) of the physical processes are dominant for a particular geotechnical engineering problem and therefore a rigorous solution of all processes is unwarranted.

The constitutive relations associated with heat flow are first presented for the situation where heat flow is above the freezing phase change isotherm (e.g., 0°C). This is followed with a description of the constitutive relations related to freezing and thawing. Partial differential equations are then derived for heat flow problems of increasing complexity.

There is a section on the measurement of thermal properties followed by a section on estimation procedures for the determination of thermal conductivity properties. The estimation of saturated and unsaturated thermal properties contains a summary of properties and equations that have been published by various researchers. Tables and plots of previously measured thermal properties are useful for estimating specific thermal properties for each phase of the soil. A number of example problems are solved using the solution of the partial differential equation for heat flow.

10.1.1 Heat Flow Processes Involving Unsaturated Soils

Conductive heat flow is often viewed as being somewhat analogous to the flow of water through soil with the obvious exception that there is no gravity component in the total energy state that drives heat flow. There are some similarities between heat flow by conduction and water flow through soils, but it is important to also understand the fundamental differences.

Heat flow by conduction is similar to water flow in that both processes involve two primary soil properties. One soil property defines the rate at which flow occurs and the other property defines the capacity for storage. In the case of heat flow, the rate at which flow occurs is defined as the thermal conductivity λ of the material (e.g., unsaturated soil). The capacity to store thermal energy is defined in terms of the volumetric heat capacity ζ of the material. Objects get warmer as heat is stored in much the same manner that a soil gets wetter as water is stored. The ability of a soil to store heat (e.g., the rate at which heat can be taken on or given off) is a function of its temperature. A soil is less likely to take on more heat as it gets warmer.

There are also a number of key differences between the physics associated with liquid water flow and heat flow. For example, heat flow analyses quite commonly encompass water in both the liquid and solid state. The analysis may also involve a phase change as is the case when the soil water freezes or thaws. There may also be a phase change as water evaporates at ground surface. The phase changes involve a material property referred to as the latent heat of fusion or the latent heat of evaporation. Heat flow in unsaturated, unfrozen soils is presented as well as situations where freezing and thawing may take place.

The assessment of soil properties for water flow problems (i.e., saturated or unsaturated seepage) is quite different from the procedure used for heat flow analyses. Unsaturated soil water flow problems generally require that the saturated coefficient of permeability be measured and that the permeability function be estimated from the SWCC. The water storage function is also computed from the SWCC. However, for heat flow analyses the thermal conductivity and heat storage soil parameters can generally be indirectly calculated based on an understanding of the compositions of the soil (i.e., amount of air and water) and the types of minerals present in the soil. Heat flow soil properties can be measured in the laboratory, but the validity of such measurements depends quite strongly on the skill of the person performing the test and the quality of the equipment. Consequently, estimation procedures are satisfactory for many heat flow engineering problems.

It is noteworthy that while the hydraulic conductivity of a soil may vary over several orders of magnitude, thermal conductivity typically varies by about three times for all types of soil solids (i.e., about one-third of one order of magnitude). The estimation of heat flow parameters constitutes an acceptable approach for many geotechnical engineering problems. Sensitivity and parametric studies play an important complimentary role when analyzing heat flow problems.

The SWCC curve also has a role to play in conductive heat flow analyses since it is the relative components of air, water, and solids that largely dictate thermal properties. However, the proportion of air and water in a soil often remains essentially constant during a conductive heat flow analysis. Therefore, the thermal properties can often be assumed to be constant for a particular analysis.

Consideration is given to the case where the thermal properties remain constant during the heat flow analysis as well as situations where the thermal properties are nonlinear functions of the SWCC. Writing the thermal properties in terms of the SWCC is particularly relevant when considering the solution of coupled analyses where more than one physical process is involved (e.g., simultaneous flow of water, air, and heat).

10.2 THEORY OF HEAT FLOW

The constitutive relations associated with heat flow are described for each of the thermal soil properties. Typical properties are given along with the units of measurement. The freezing and thawing of soils involve the latent heat of fusion, and modeling these conditions is more complex than modeling conductive heat flow through unfrozen soils. The theory section discusses heat flow when the temperatures are above freezing as well as thermal modeling when temperatures are reduced below the phase change isotherm (i.e., 0°C).

10.2.1 Heat Flow Constitutive Relations

The state variable that creates the driving potential for conductive heat flow is temperature T. Temperature is a state variable. Instruments designed to measure temperature are referred to as thermometers. Some of the devices for measuring temperature are mercury thermometers, liquid-in-glass thermometers, bimetallic thermometers, bourdon thermometers, electrical resistance thermometers, and thermocouples.

Heat flow by conduction is the result of a difference in temperature between two points or what is termed a thermal gradient. Heat flow q_h can be described using Fourier's law, which has a form similar to Fick's law. The soil property controlling conductive heat flow is thermal conductivity λ

$$q_{hx} = -\lambda \frac{dT}{dx} \tag{10.1}$$

$$q_{hy} = -\lambda \frac{dT}{dy} \tag{10.2}$$

$$q_{hz} = -\lambda \frac{dT}{dz} \tag{10.3}$$

where:

T = temperature, °C, and
λ = thermal conductivity of the soil (W/m/K or J/m/°C since 1 W = 1 J/s)

Standard SI units of measurement are used to represent each of the thermal properties for soils and other materials. Thermal conductivity is generally assumed to be an isotropic soil property although particle shape effects can be taken into account.

10.2.2 Thermal Conductivity of Soils

Thermal conductivity governs the rate of flow of heat through a material (e.g., soil with solids, water, and air). Conduction is responsible for the flow of heat through solid materials such as soil particles. Heat also flows by conduction through water and air. It is also possible that the mechanisms of convection and radiation may be involved. The emphasis of this chapter is on heat transfer by conduction.

Table 10.1 shows thermal conductivity values for water over a range of temperatures between freezing and 80°C. Table 10.2 shows typical thermal conductivity values for a

Table 10.1 Thermal Conductivity versus Temperature for Water

Temperature, °C	Thermal Conductivity, λ_w (W/m/K)
0	0.5610
4	0.5694
5	0.5736
10	0.5862
15	0.5945
20	0.6029
25	0.6113
30	0.6196
35	0.6280
40	0.6322
45	0.6406
50	0.6448
60	0.6543
70	0.6631
80	0.6700

Source: Data compiled from Lide (1992).

Table 10.2 Thermal Properties of Common Materials and Soils

Substance	Thermal Conductivity, λ (W/m/K)
Air (20°C)	0.025
Water (20°C)	0.58
Ice (0°C)	2.2
Quartz	8.8
Clay minerals	2.9
Soil organic matter	0.25
Light soil with roots	0.11
Wet sand ($\theta = 40\%$)	1.8

Source: Data from de Vries (1963) and Rosenberg et al. (1983).

range of common materials. Average values for the thermal conductivity of various rock and soil types are given in Table 10.3.

The thermal conductivity λ of a soil consisting of soil particles, water, and air can be expressed as follows (de Vries, 1963):

$$\lambda = \frac{f_p\theta_p\lambda_p + f_w\theta\ \lambda_w + f_a\theta_a\lambda_a}{f_p\theta_p + f_w\theta + f_a\theta_a} \tag{10.4}$$

where:

f_p, f_w, f_a = weighting factors for the solid particles, water, and air phases, respectively,

$\theta_p, \theta, \theta_a$ = percentage of total volume that is solid particles, water, and air, respectively, and

$\lambda_p, \lambda_w, \lambda_a$ = thermal conductivities of the solid particles, water, and air phases, respectively.

De Vries (1963) showed from experiments that the thermal conductivity of quartz sand could vary from 0.25 to 1.9 W/m/K when the temperature was 20°C and increased to 2.5 when the temperature increased to 60°C. The volumetric water content of water of the sand varied from zero to 10%. Riha et al., (1980) measured thermal conductivity valucs ranging from 0.1 to 1.0 W/m/K for silt loam where the volumetric water content varied between zero and 55%, respectively.

A typical thermal conductivity value for water is 0.57 W/m/K while air has a thermal conductivity of 0.025 W/m/K (de Vries, 1963; Jame, 1977). The amount of water in the soil has a significant effect on the thermal conductivity of a soil. The difference in thermal conductivity between air and water (i.e., 0.025–0.57 W/m/K) gives rise to the change in overall thermal conductivity. The latent heat of water vapor also has an effect on thermal conductivity when there is a temperature gradient.

The thermal conductivity of air, λ_a, consists of two components (Jame, 1977):

$$\lambda_a = \lambda_{\text{dry air}} + \lambda_{\text{water vapor}} \tag{10.5}$$

Table 10.3 Average Values of Thermal Conductivity of Solid Particles

Material	ρ_s (g/cm^3)	λ_p (W/m/K)
Rocks		
Anorthosite	2.73	1.8
Basalt	2.90	1.7
Diabase	2.98	2.3
Dolostone	2.90	3.8
Gabbro	2.92	2.2
Gneiss	2.75	2.6
Granite	2.75	2.5
Limestone	2.75	2.5
Marble	2.80	3.2
Quartzite	2.65	5.0
Sandstone	2.80	3.0
Schist	2.65	1.5
Shale	2.65	2.0
Syenite	2.80	2.0
Trap rock	2.90	2.0
Coal	1.35	0.26
Peat	1.50	0.25
Silt and clay	2.75	2.90

Source: Summarized by Côté and Konrad (2005) from various published sources.

where:

$\lambda_{\text{dry air}}$ = 0.025 W/m/K and
$\lambda_{\text{water vapor}}$ = varies linearly between 0 and 0.0736 W/m/K for volumetric water contents, θ , ranging between 0 and 12.1%.

The weighting factors f are calculated using the assumption that the soil particles are ellipsoidal in shape. The weighting factor for a continuous medium (i.e., air or water) is equal to 1.0. Water can be selected as the continuous medium with a weighting factor f_w equal to 1.0. The weighting factors for the solid particles and air can then be calculated in accordance with the following relationship:

$$f = \frac{1}{3}\sum_{i=1}^{3}\left[1 + \left(\frac{\lambda}{\lambda_w} - 1\right) g_i\right]^{-1} \tag{10.6}$$

where:

$f = f_a$ or f_p,
$\lambda = \lambda_a$ or λ_p, and
g = depolarization factors for the ellipsoid (i.e., g_1, g_2, and g_3, where $g_1 + g_2 + g_3 = 1$); the values of g_1, g_2, and g_3 are independent of particle size but dependent on the ratio of the length of the ellipsoid axes.

Wilson (1990) used equal depolarization factors of 1/3 in computing the weighting factors for a sandy soil. The sand particles were assumed to have spherical shapes. The depolarization factors g_1 and g_2 for the air phase were assumed to decrease linearly from 0.333 to 0.105 for volumetric water contents θ ranging from 23.6 to 12.1%, respectively, and from 0.105 to 0.015 for volumetric water contents ranging from 12.1 and 0%, respectively (Jame, 1977). Figure 10.1 illustrates the thermal conductivity λ variation for Beaver Creek sand with respect to gravimetric water contents.

The thermal conductivity of an unsaturated soil is a function of the relative amounts of air, water, and solids in the soil and therefore can be considered to be a function of the SWCC (Aldrich, 1956). Figure 10.2 shows the shape of the thermal conductivity–suction relationship with the SWCC.

A specific thermal conductivity value can be computed for a soil when the degree of saturation is a constant value. However, if the partial differential equations for moisture flow and heat flow equations are solved as "combined" processes, the thermal conductivity will change as the amount of water in the soil changes.

The thermal conductivity of water bears a relationship to temperature, as shown in Fig. 10.3. The thermal conductivity of ice is different from that of water. It is as if an additional phase is slowly added to the soil when part of the pore-water becomes frozen. Not all water in the pores of the soil freezes when the temperature is lowered to 0°C. The amount of water that freezes is a function of the temperature below 0°C. The unfrozen water content in the soil can be calculated from the SWCC and the Clapeyron equation (Newman, 1995). The latent heat of fusion L must be taken into consideration when the water in a soil either freezes or thaws.

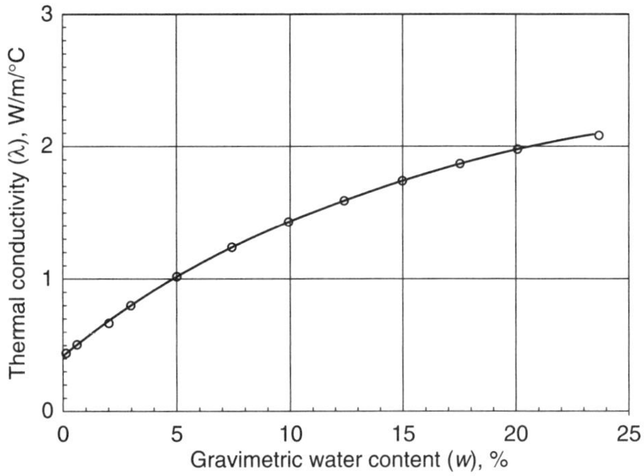

Figure 10.1 Thermal conductivity versus water content for the Beaver Creek sand (after Wilson, 1990).

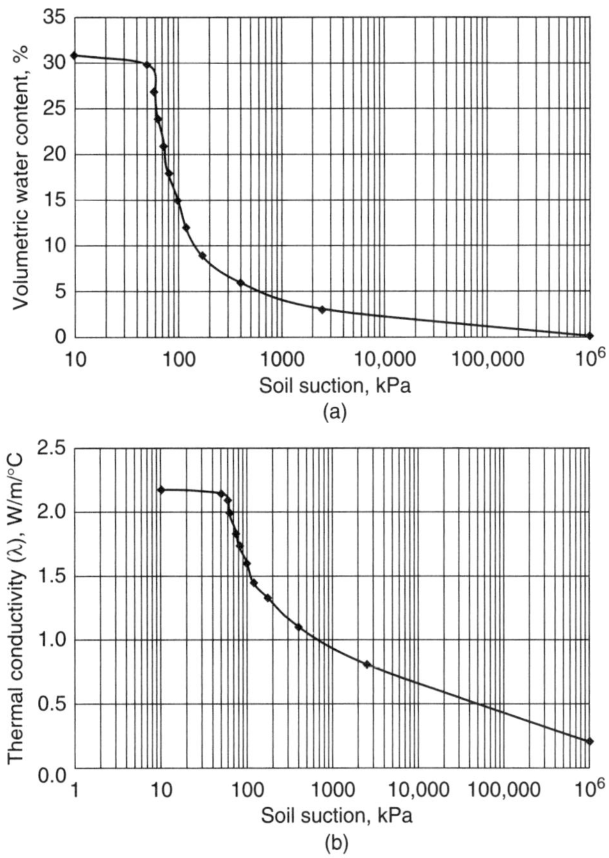

Figure 10.2 Volumetric water content and thermal conductivity versus soil suction for silica flour (from Jame, 1972): (a) volumetric water content versus soil suction; (b) thermal conductivity versus soil suction.

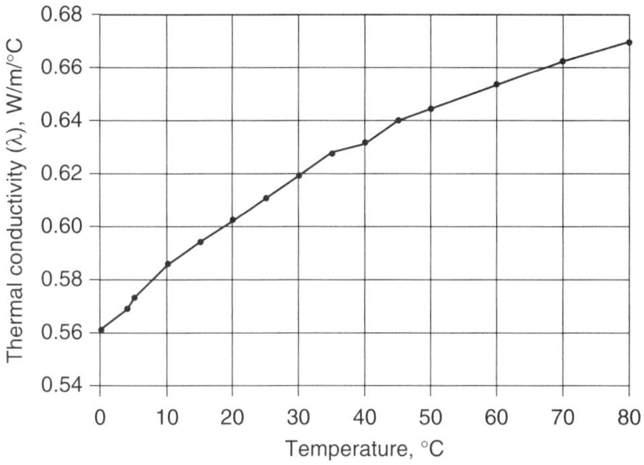

Figure 10.3 Thermal conductivity of water versus temperature.

10.2.3 Heat Capacity and Specific Heat for Unsaturated Soils

Temperature change in a soil occurs in response to the adsorption or release of heat. The soil property describing the ability of a material to adsorb or release heat is called volumetric heat capacity. The heat capacity of a soil is dependent upon the type of minerals present, the porosity of the soil, and the relative portion of water and air in the voids (Baver, 1956). Volumetric heat capacity is the term commonly used when describing the heat storage capability of a soil-water-air mixture. Specific heat is the term often used when referring to each of the constituents of a multiphase system.

It is possible to calculate the volumetric heat capacity of a multiphase system if the specific heat of each constituent and the amount of each constituent is known. Considerable attention has been given to the accurate measurement of the specific heat of each constituent of a multiphase system. It is also useful to know the specific heat of water as a function of the various states in which it can exist (i.e., liquid, solid, and vapor).

The heat capacity of a soil can be designated either with reference to a unit volume or with respect to a unit mass. Specific heat is commonly designated with respect to a unit mass of material. Volumetric heat is referenced to a unit volume. The volumetric heat capacity ζ of an unfrozen soil consisting of solids, water, and air can be calculated using the relationship given by de Vries (1963):

$$\zeta = \zeta_p\theta_p + \zeta_w\theta + \zeta_a\theta_a \qquad (10.7)$$

where:

ζ_p = heat capacity of the soil solids [i.e., a typical value is 2.235×10^6 J/m^3/K for fine sands (de Vries, 1963)],
θ_p = volumetric solid content (i.e., V_s/V),
V_s = volume of soils solids in the soil,
V = total volume of soil,
ζ_w = heat capacity for the water phase (i.e., 4.2×10^6 J/m^3/K for water at 20°C),
θ = volumetric water content (i.e., V_w/V),
V_w = volume of water in the soil,
ζ_a = heat capacity for the air phase (i.e., 1.2×10^3 J/m^3/K), and
θ_a = volumetric air content (i.e., V_a/V).

Typical values for specific heat of water at various temperatures are shown in Table 10.4.

It can be observed in Table 10.5 that volumetric heat is equal to the heat capacity C_w multiplied by the density ρ_w of the material. For example, the volumetric heat capacity of water can be written as follows:

$$\zeta_w = \rho_w C_w \qquad (10.8)$$

10.2.4 Heat of Vaporization

The volumetric latent heat of vaporization or condensation, L, is 2.5×10^9 J/m^3 (Andersland and Anderson, 1978). The latent heat of vaporization in terms of mass is 2.42×10^6 J/kg. The latent heat of vaporization can be applied during a vaporization process which may occur when the temperature is greater than the freezing temperature. Condensation is the reverse of vaporization and the latent heat term for condensation is the same as for vaporization.

10.2.5 Diffusivity Constant

Thermal diffusivity is another term that is often used when considering the analysis of heat flow. Thermal diffusivity

Table 10.4 Specific Heat Capacity C_w versus Temperature for Water (Liquid Phase)

Temperature, °C	Specific Heat Capacity C_w (J/g/K)
0	4.2161
4	4.2077
5	4.2035
10	4.1910
15	4.1868
20	4.1826
25	4.1784
30	4.1784
35	4.1784
40	4.1784
45	4.1784
50	4.1826
60	4.1843
70	4.1895
80	4.1963

Source: Data compiled from Lide (1992).

Table 10.5 Specific Heat and Volumetric Heat Capacity for Common Materials and Soils

Substance	Density (k/m^3)	Specific Heat, Capacity, C (J/kg/K)	Volumetric Heat Capacity, ζ (J/m^3/K)
Air (20°C)	1.2	1000	1200
Water (20°C)	1000	4200	4.2×10^6
Ice (0°C)	9200	2100	1.9×10^6
Quartz	2660	800	2×10^6
Mineral clay	2650	800	2×10^6
Soil organic matter	1300	2500	2.7×10^6
Light soil with roots	1400	1300	500,000
Wet sand ($\theta = 40\%$)	1600	1700	2.7×10^6

Source: Data from de Vries (1963) and Rosenberg et al. (1983).

is the ratio of the thermal conductivity of a material to the product of specific heat and material density. The thermal diffusivity term allows the mathematical equations of heat flow to be written in a more convenient form for obtaining closed-form solutions. When undertaking numerical modeling solutions in geotechnical engineering, it is usually better to keep the heat flow and heat storage soil properties as separate entities.

Diffusivity α is defined as the ratio of the thermal conductivity to volumetric heat capacity:

$$\alpha = \frac{\lambda}{\zeta} \tag{10.9}$$

where:

λ = thermal conductivity of the soil, W/m/K, and
ζ = volumetric heat capacity, J/m^3/K.

Combining thermal conductivity and heat storage into a single parameter is comparable to combining hydraulic conductivity and the coefficient of volume change into the coefficient of consolidation when solving consolidation problems. Using a single constant value for the material properties has distinct advantages when performing long-hand calculations; however, when solving numerical models in geotechnical engineering, it is generally advantageous to maintain independence between the heat transfer and heat storage properties, particularly when dealing with unsaturated soil conditions.

10.3 THEORY OF FREEZING AND THAWING SOILS

The Arctic and Antarctic regions have soils that remain frozen for the entire year. Moving away from the poles, the soils may be frozen for part of the year and then remain unfrozen for the remainder of the year. The portion of the soil profile that is frozen for part of the year and unfrozen for the remainder of the year is referred to as the "active zone." The prediction of the depth of the active zone and the time period over which the soil will remain frozen is relevant to geotechnical engineering designs.

The thermal energy of a soil-water system becomes discontinuous when the temperature passes through the zero isotherm. The discontinuity is the result of the latent heat of water, as shown in Fig. 10.4. The thermal energy change is essentially linear with respect to temperature when the temperature is either above or below the freezing point. Heat is either adsorbed or given off at the freezing point as a result of the phase change in the water. The heat quantity associated with the phase change in water is called the latent heat of fusion.

The freezing front in a soil is, in reality, more of a freezing zone, as illustrated in Fig. 10.5. The freezing zone has a particular thickness and the water within this zone goes from being ice in the completely frozen portion to being water at the completely unfrozen portion. Jumikis (1966), Dirksen and Miller (1966), and Hoekstra (1966) viewed a freezing soil as having three zones: a frozen zone, a freezing zone, and an unfrozen soil. The water coefficient of permeability gradually changes across the freezing zone. The unfrozen water content varies across the freezing zone and, likewise, the coefficient of water permeability changes across the freezing zone.

The freezing zone can also be viewed as a zone with a water coefficient of permeability versus unfrozen water content function. High soil suction values are also generated in the freezing zone. The soil suctions are related to the Clapeyron equation (Newman, 1995). Consequently, the permeability in the freezing zone can also be viewed as a relationship between soil suction and water coefficient of permeability. While the general mechanisms associated with the freezing of soils is understood, there has not been general agreement on how to predict the water coefficient of permeability in the freezing zone (Newman, 1995). The water permeability function forms an important part of mechanistic models that have been

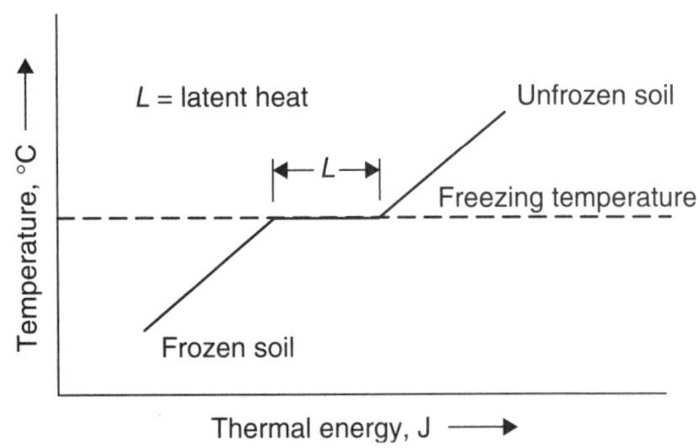

Figure 10.4 Thermal energy of soil-water system illustrating role of latent heat of water.

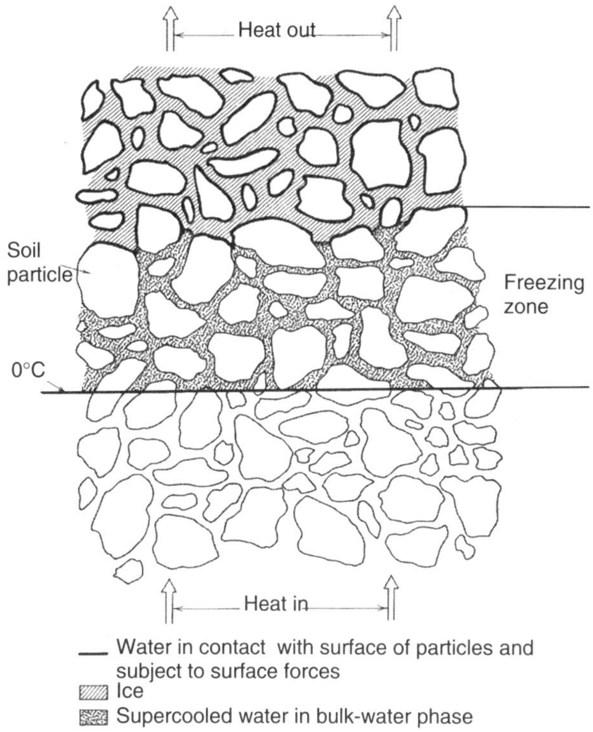

Figure 10.5 Idealization of partially frozen zone in soil (i.e., transition from water to ice in the voids).

proposed for describing the freezing behavior of fine-grained soils (Harlan, 1973; Guymon and Luthin, 1974; Jame, 1977; Konrad and Morgenstern, 1990; Nixon, 1992).

10.3.1 Latent Heat of Fusion

The volumetric latent heat of fusion L is the amount of energy associated with phase change during the freezing or thawing of a soil. The volumetric latent heat of water-ice is 3.34×10^8 J/m^3 (Andersland and Anderson, 1978). The latent heat consumption must be taken into consideration when water is converted to ice and vice versa.

The latent heat of fusion applies only to the water phase but can be written in terms of the total soil mass (Andersland and Anderson, 1978):

$$L = \frac{3.34 w \rho_d}{\rho_w} \times 10^8 \tag{10.10}$$

where:

L = latent heat written in terms of the total soil sample, J/m^3,
w = gravimetric water content,
ρ_d = dry density, kg/m^3, and
ρ_w = density of water, kg/m^3.

The freezing of water in a soil does not occur instantaneously. Rather, some of the water is converted to ice as heat is extracted from the soil system. Ice starts to form at 0°C. The temperature may be from −5 to −10°C before all the water in the soil is frozen.

The latent heat of fusion is often viewed as a discontinuity that occurs as water freezes or thaws. However, in reality, water in a soil gradually freezes as the temperature is lowered. The reverse process occurs as ice thaws.

10.3.2 Freezing of Soils

The freezing of soils is usually driven by thermal conditions in the atmosphere above the ground surface. A thermal gradient develops between the cold atmosphere and the underlying warmer soil (Fig. 10.5). A transient process is initiated that involves conductive, convective, and radiation heat transfer. However, the conductive heat transfer is generally dominant.

As the temperature of the atmosphere drops below the freezing point, water in the pores of the soil begins to freeze and releases latent energy into the system. The amount of liquid water is reduced in the freezing zone and negative pore-water pressures are generated in the freezing zone. The thermal and suction gradients result in water flow toward the freezing zone. The affinity of the ice-water interface for more water can result in the buildup of ice lenses in the soil.

The frost front advances downward into the soil as long as the frozen zone can remove the latent energy released by the phase change. The suction gradients are high, calculated in accordance with the Clapeyron equation. At the same time, there is a reduction in the water permeability across the freezing zone. The reduced water coefficient of permeability limits the amount of water that can be delivered to the freezing front. Harlan (1973) suggested that the frozen water reduces the water coefficient of permeability in a similar manner to the way that desaturation of a soil reduces the water coefficient of permeability for an unsaturated soil. Partial freezing of the soil gives rise to the soil suction versus water coefficient of permeability relationship for the freezing soil.

10.3.3 Unfrozen Water Content

Boyoucous (1920) showed that part of the water in clay remains unfrozen at temperature values as low as −78°C. Other researchers have attempted to explain the unfrozen water content in terms of freezing point depression and the freezing temperature as a function of the unfrozen water content (Williams, 1964, 1966; Nersesova and Tsytovich, 1966; Miller, 1966). One of the most common forms for representing the freezing mechanism has been the relationship of unfrozen water content to temperature below the freezing point of pure water (Yong, 1965; Nersesova and Tsytovich, 1966). Figure 10.6 shows temperature below freezing versus water content for silt and clay soils compacted at various water contents (Yong, 1965). Similar plots for a variety of soils are shown in Fig. 10.7 (Nersesova and Tsytovich, 1966). Jame (1972) showed that the relationship between freezing point depression and water content was

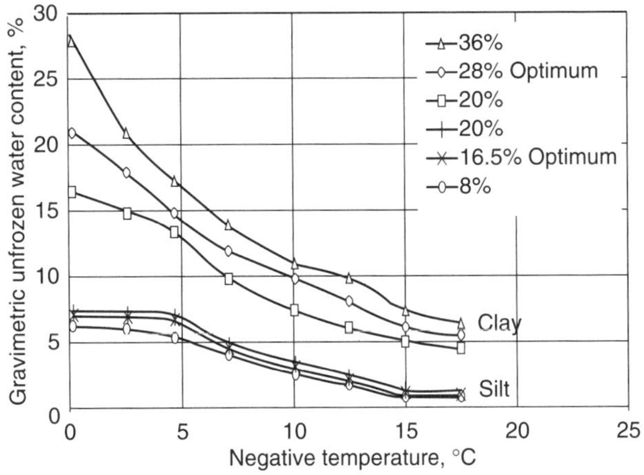

Figure 10.6 Unfrozen water content versus temperature below 0°C for silt and clay with various initial water contents (after Yong, 1965).

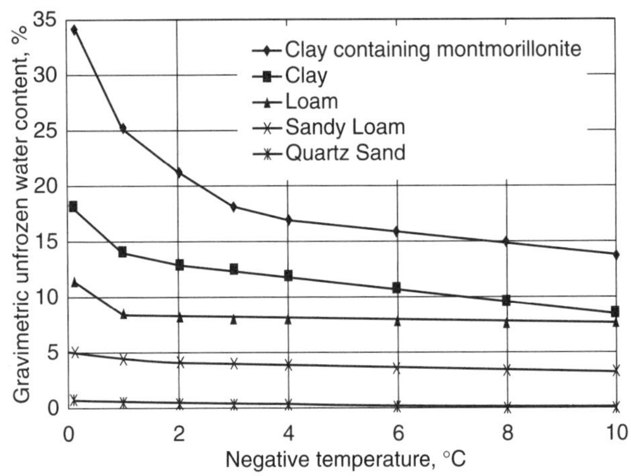

Figure 10.7 Unfrozen water content versus temperature below 0°C for several nonsaline soils (after Nersesova and Tsytovich, 1965).

similar in character to the relationship of temperature below freezing and water content. Consequently, there is a relationship between water content, soil suction, and temperature which is related to the SWCC for the soil.

10.3.4 Clapeyron Equation

The Clapeyron equation relates a change in pressure between two phases of a substance (e.g., liquid-solid) to the change in temperature of the system. The Clapeyron equation has been of limited value for the solution of practical engineering problems but has been used in the estimation of the amount of unfrozen water in a freezing soil as a function of the temperature below freezing (i.e., 0°C).

Various forms of the Clapeyron equation have been presented in the research literature. Jame (1972) plotted soil suction versus temperature below freezing for a coarse-grained soil, as shown in Fig. 10.8. The Clapeyron equation has been used in the development of theories for ice segregation and frost heave models (Harlan, 1973; Taylor and Luthin, 1978; Nixon, 1991, 1992).

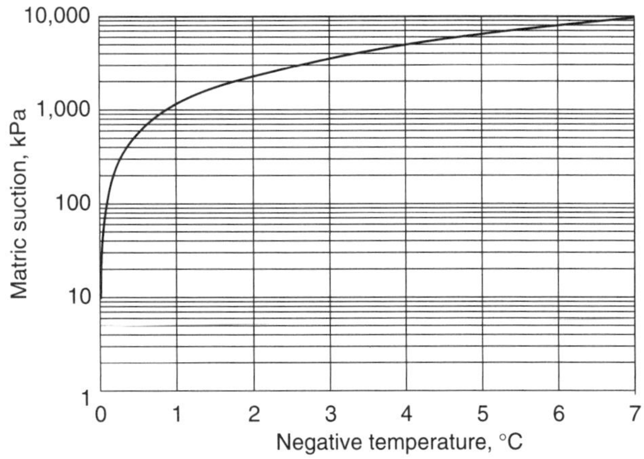

Figure 10.8 Relationship between soil suction and temperature below 0°C for coarse-grained soil (after Jame, 1972).

10.3.5 Soil-Freezing Characteristic Curve

Williams (1964) presented data that showed a relationship between SWCCS measured at room temperature and a soil-freezing curve for the same material (i.e., temperature below freezing and unfrozen water content). The Clapeyron equation provides the freezing temperature for the unfrozen water content. Koopmans and Miller (1966) noted a similarity between the soil drying and wetting curves and the freezing and thawing curves, as shown in Fig. 10.9. It was noted that the freezing/thawing curve exhibited hysteresis just as the wetting/drying curve exhibits hysteresis.

The Clapeyron equation allows a relationship to be developed between soil suction and the temperature below freezing when the SWCC is known for a particular soil. The SFCC is therefore inherently linked to the SWCC.

The temperature range over which water becomes frozen is defined by the SFCC. The temperature at which water in the soil starts to freeze is designated by the variable T_{ef} (i.e., generally 0°C), and the temperature at which freezing is complete is designated as T_{ep}. The latent heat of freezing is applied over the temperature range between T_{ef} and T_{ep}.

Figure 10.10 shows a typical SFCC for silica flour tested by Jame (1972). The slope of the SFCC is defined by the variable m_2^i. The slope of the SFCC is defined by the change in unfrozen volumetric water content with respect to a change in temperature below freezing.

10.3.6 Coefficient of Permeability for Freezing Zone

The soil-freezing curve can be used in conjunction with the saturated coefficient of permeability for a soil to estimate the water coefficient of permeability versus soil suction relationship for a freezing soil (Newman, 1995). Figure 10.11

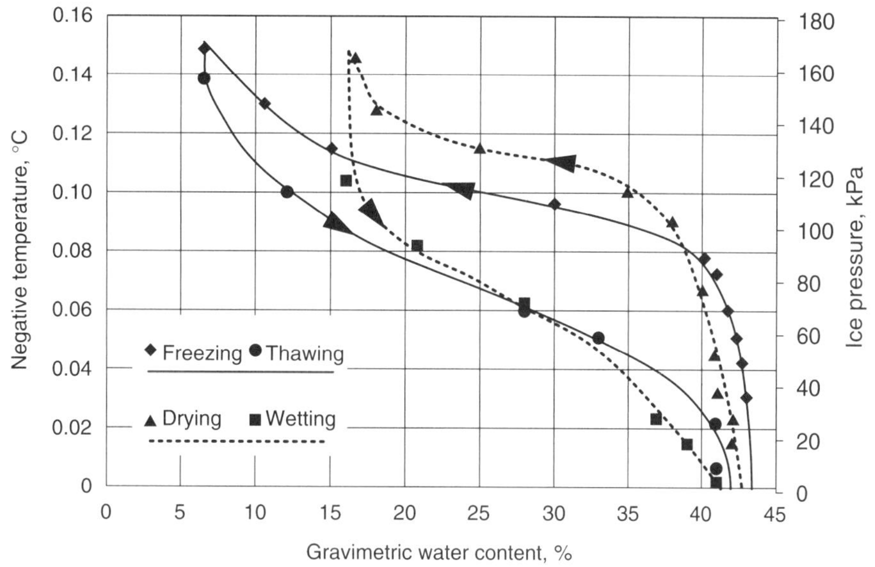

Figure 10.9 Experimental data showing similarity between SFCCs and SWCCs, including hysteresis effects (after Koopmans and Miller, 1966).

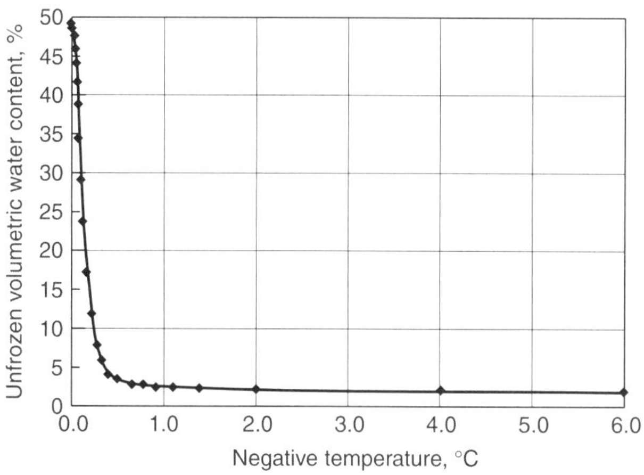

Figure 10.10 SFCC for silica flour (after Jame,1972).

shows the results for the silica flour tested by Jame (1977). The saturated coefficients of permeability ranged from 2.5×10^{-6} to 3.0×10^{-7} m/s when the porosity ranged from 0.52 to 0.48, respectively. The SWCCs for the silica flour are shown in Fig. 10.11a and the relative coefficients of permeability versus soil suctions computed using the Fredlund and Xing (1994) equation are shown in Fig. 10.11b.

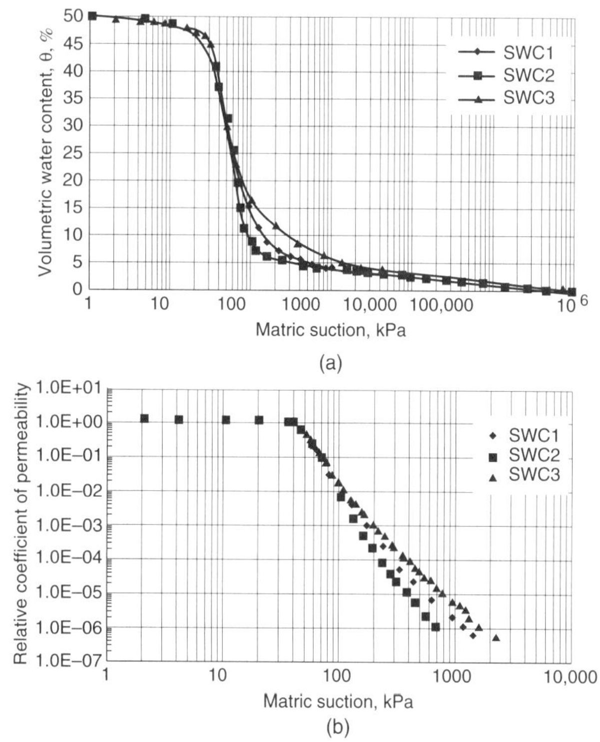

Figure 10.11 Relative coefficient-of-permeability functions computed for three specimens of silica flour using Fredlund and Xing (1994) SWCC equation: (a) SWCCs for silica flour (Jame, 1972); (b) relative coefficient-of-permeability curves for silica flour.

10.4 FORMULATION OF PARTIAL DIFFERENTIAL EQUATIONS FOR CONDUCTIVE HEAT FLOW

Conductive heat flow differential equations can be derived satisfying the conservation of thermal energy. The formulations are similar in form to those previously derived for water flow and air flow. The analysis starts with the selection of a REV. The conservation of thermal energy under steady-state conditions requires that the heat flow into an element must be equal to the outward heat flow.

Partial differential equations can be written for one-, two- and three-dimensional geometries. The partial differential equations can be solved for situations where the boundary conditions are constant as well as situations where the boundary conditions change with time. These two situations are referred to as steady-state conditions and transient (or unsteady-state) conditions, respectively. The divergence of heat flow ($\partial/\partial x + \partial/\partial y + \partial/\partial z$) is equated to the change in heat storage in the elemental volume with respect to time.

The PDEs for heat flow are solved in a manner similar to the procedures used for water flow and air flow in unsaturated soils. Heat flow formulations have the appearance of being relatively simple to solve. Heat flow problems can also be complex when the water undergoes a phase change. The heat flow formulations are presented in a hierarchical manner from the simplest situation to more complex situations. Each situation is considered in terms of one, two, and three dimensions and also in terms of steady-state conditions and transient conditions.

The first set of formulations is for the analysis of heat flow when the temperature conditions in the soil range from just above freezing conditions to near the boiling point of water. The formulation of the partial differential equation does not need to take any phase changes into consideration. The second set of formulations is for the case where the soil may experience freezing and thawing conditions, and as a result, the latent heat of fusion needs to be taken into consideration. The third set of formulations is for the case where water may move into the vapor phase and as a result the latent heat of vaporization needs to be taken into consideration. Finally, some special cases are also presented.

10.4.1 Steady-State Heat Flow Partial Differential Equation for Unfrozen Soils

One-Dimensional Heat Flow: Consider the one-dimensional flow of heat in the y-direction through a REV of unsaturated soil (Fig. 10.12). The element has infinitesimal dx, dy, and dz dimensions. The dy and dz sides of the element are assumed to be perfectly insulated, and therefore, heat flow occurs only in the y-direction. One-dimensional heat flow could also occur when the soil on the dy and dz sides is identical to the soil in the column. Boundary conditions on the soil adjacent to the element being analyzed must also be the same as for the element being analyzed.

The heat flow rate in the y-direction is designated as q_{hy}. The conservation of thermal energy requires that the amount of heat flowing in and out of the element must be equal for steady-state conditions:

$$\left(q_{hy} + \frac{dq_{hy}}{dy}dy\right) - q_{hy} = 0 \qquad (10.11)$$

where:

q_{hy} = heat flow rate across a unit area of the soil in the y-direction and
dx, dy, dz = dimensions in the x-, y-, and z-directions, respectively.

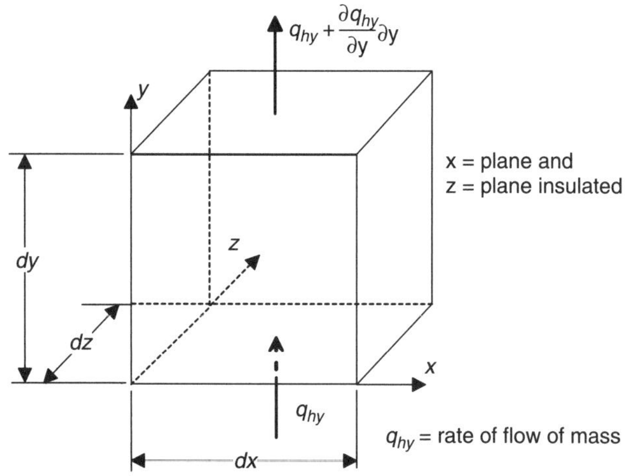

Figure 10.12 Representative elemental volume showing heat fluxes for one-dimensional heat flow, other sides insulated.

The net heat flow can be written as follows:

$$\frac{dq_{hy}}{dy} = 0 \qquad (10.12)$$

Let us assume that the thermal conductivity λ of the soil might vary from one location to another in a soil column due to a variation in the degree of saturation of the soil. Substituting Fourier's law for heat flow into the net heat flow equation yields

$$\frac{d\{-\lambda dT/dy\}}{dy} = 0 \qquad (10.13)$$

where:

λ = thermal conductivity of the soil, W/m/K,
dT/dy = temperature gradient in the y-direction, and
T = temperature, °C.

The above equation can be solved for temperature T provided there are independent, known values for thermal conductivity. For example, the variations in the degree of saturation in the soil might he known as a result of an independent unsaturated water seepage analysis. The variation in thermal conductivity of the soil from one location to another would need to be taken into consideration when solving Eq. 10.13. The above equation can be rewritten in the following nonlinear differential form:

$$\lambda\frac{d^2T}{dy^2} + \frac{d\lambda}{dy}\frac{dT}{dy} = 0 \qquad (10.14)$$

The above equation is nonlinear since the thermal conductivity is, in general, variable throughout the soil column. If the unsaturated soil varied in soil suction in the y-direction, then it would be necessary to take into account the variation of thermal conductivity with matric suction, water content, or degree of saturation. It is also possible for the thermal analysis to be combined with a saturated-unsaturated soil seepage analysis. The results of the seepage analysis could be used to calculate the thermal conductivity throughout the soil mass.

Only one soil thermal property is required when solving a steady-state heat flow analysis, namely, the thermal conductivity of the soil. In some cases the thermal conductivity can be represented as a constant value while in other cases the thermal conductivity may need to be considered as a function of the stress state or one of the volume-mass soil properties.

The nonlinearity in the heat flow equation arises in the second term of Eq. 10.14, which accounts for the variation in thermal conductivity with respect to space. If the thermal conductivity is a constant value and the soil is homogeneous, then Eq. 10.14 becomes linear, as shown in the equation

$$\lambda \frac{d^2T}{dy^2} = 0 \tag{10.15}$$

Two-Dimensional Heat Flow: The partial differential equation for steady-state heat flow through a saturated-unsaturated soil can readily be extended to a two-dimensional continuum. One face of the soil continuum (i.e., the third dimension) would have an insulated surface or else it is possible that the soil adjacent to the soil being analyzed is subjected to the same boundary conditions and the same soil properties as the two-dimensional plane of soil. The partial differential equation for two-dimensional steady-state heat flow can be written as follows:

$$\lambda \frac{\partial^2 T}{\partial x^2} + \frac{\partial \lambda}{\partial x}\frac{\partial T}{\partial x} + \lambda \frac{\partial^2 T}{\partial y^2} + \frac{\partial \lambda}{\partial y}\frac{\partial T}{\partial y} = 0 \tag{10.16}$$

There might be a variation in the thermal conductivity of the soil in two-dimensional space. In this case, the solution of the PDE for heat flow would rely upon an independent analysis for the determination of appropriate thermal conductivity values. Thermal conductivity is most significantly affected by the proportion of air and water in the soil voids. The SWCC can be used to relate thermal conductivity to the degree of saturation of the soil (or soil suction).

Three-Dimensional Heat Flow: The three-dimensional, steady-state heat flow equation can be written as follows:

$$\lambda \frac{\partial^2 T}{\partial x^2} + \frac{\partial \lambda}{\partial x}\frac{\partial T}{\partial x} + \lambda \frac{\partial^2 T}{\partial y^2} + \frac{\partial \lambda}{\partial y}\frac{\partial T}{\partial y} + \lambda \frac{\partial^2 T}{\partial y^2} + \frac{\partial \lambda}{\partial y}\frac{\partial T}{\partial y} = 0 \tag{10.17}$$

There may be situations where the soil is not isotropic, and this situation can lead to different thermal conductivity values in different principal directions. It is also possible that the principal directions for the thermal conductivity properties may not coincide with the Cartesian coordinate directions.

10.4.2 Transient Heat Flow Partial Differential Equation for Unfrozen Soils

A transient analysis requires that the net heat flux through an element be equal to the change in heat storage with respect to time. The volumetric heat capacity of the soil, ζ, must be specified when performing a transient heat flow analysis.

One-Dimensional Heat Flow: The conservation of thermal energy requires that the net heat flux through an elemental volume be equal to the volumetric heat capacity of the soil multiplied by the change in temperature with respect to time. The heat flux equation satisfying transient conditions can be written as follows:

$$\frac{\partial q_{hy}}{\partial y} = \zeta \frac{\partial T}{\partial t} \tag{10.18}$$

where:

ζ = volumetric specific heat, J/m^3/°C.

The volumetric heat capacity ζ (i.e., in units of J/m^3/°C) can be written as the mass specific heat (i.e., in units of J/g/°C) multiplied by soil density ρ (i.e., in units of g/cm^3). The above equation applies as long as the temperature of the soil is above freezing.

A transient analysis occurs when one or more of the boundary conditions are changed. Consequently, there are changes in temperature within the continuum with respect to time. Fourier's heat flow equation can be substituted for the quantity of heat flow to give the following differential equation:

$$\lambda \frac{\partial^2 T}{\partial y^2} + \frac{\partial \lambda}{\partial y}\frac{\partial T}{\partial y} = \zeta \frac{\partial T}{\partial t} \tag{10.19}$$

The second term on the left-hand side of the equation takes into account variations in thermal conductivity in the y-direction. For example, variations in the degree of saturation of a soil would influence the thermal conductivity of the soil and thereby affect the results of a heat flow analysis.

If there is no change in thermal conductivity with respect to the y-dimension, then Eq. 10.19 takes on the standard diffusion form

$$\lambda \frac{\partial^2 T}{\partial y^2} = \zeta \frac{\partial T}{\partial t} \tag{10.20}$$

The above equation can be applied to heat flow in either an unfrozen soil or a frozen soil provided the respective soil properties are inserted for thermal conductivity and volumetric heat capacity. However, the temperatures must not pass through the zero isotherm during the analysis.

Two-Dimensional Heat Flow: A two-dimensional transient analysis assumes that there is no heat loss in the third dimension. The continuum is subjected to a change in boundary conditions (or an internal thermal source or sink), and the temperatures throughout the remainder of the continuum are calculated:

$$\lambda \frac{\partial^2 T}{\partial x^2} + \frac{\partial \lambda}{\partial x}\frac{\partial T}{\partial x} + \lambda \frac{\partial^2 T}{\partial y^2} + \frac{\partial \lambda}{\partial y}\frac{\partial T}{\partial y} = \zeta \frac{\partial T}{\partial t} \tag{10.21}$$

Three-Dimensional Heat Flow: A three-dimensional transient thermal analysis involves an extension of the previous partial differential equation into the third dimension. Each dimension that is added must have the heat flux in that direction taken into consideration. However, the right-hand side of the equation involving the heat storage term remains the same:

$$\lambda\frac{\partial^2 T}{\partial x^2}+\frac{\partial\lambda}{\partial x}\frac{\partial T}{\partial x}+\lambda\frac{\partial^2 T}{\partial y^2}+\frac{\partial\lambda}{\partial y}\frac{\partial T}{\partial y}+\lambda\frac{\partial^2 T}{\partial y^2}+\frac{\partial\lambda}{\partial y}\frac{\partial T}{\partial y}=\zeta\frac{\partial T}{\partial t} \quad (10.22)$$

The above partial differential equation satisfies transient and steady-state conditions as long as the soil temperature is in the unfrozen soil range. The thermal soil properties have also been assumed to be isotropic.

10.4.3 Heat Flow Partial Differential Equations Including Freezing and Thawing of Soils

The phase change from liquid water to ice produces what appears to be a discontinuity in the heat flow partial differential equations. The phase change can be modeled as a process that occurs as the water in the soil gradually freezes over a relatively small temperature range.

The heat storage portion of the formulation is influenced by the freezing and thawing of a soil. There is no need to write partial differential equations for steady-state conditions involving freezing and thawing because it is only the transient heat flow process that is affected by the latent heat of fusion.

As temperature is lowered to zero degrees (Celcius), the latent heat of fusion L must be satisfied before the soil-water temperature can decrease further. The reverse process occurs as ice in the soil reverts to liquid water. The latent heat of fusion behaves as a heat sink during the freezing process and as a heat source during the thawing process.

The heat and mass transfer formulations proposed by Harlan (1973) have been extensively used for modeling heat flow when passing through the freezing isotherm. Nixon (1975), Taylor and Luthin (1978), and Fuch et al., (1978) have shown that the convective heat term proposed by Harlan (1973) is negligible and can be omitted from the formulation.

The volumetric heat capacity term ζ needs to be modified to accommodate the latent heat of fusion, which can now be written as follows:

$$\overline{\zeta}=\zeta-L\rho_i\frac{\partial\theta_i}{\partial T} \quad (10.23)$$

where:

$\overline{\zeta}$ = modified volumetric heat capacity as the soil freezes or thaws, J/m^3/°C,
ζ = volumetric heat capacity of the unfrozen soil, J/m^3/°C,
L = latent heat of fusion, J/kg,
ρ_i = density of ice, kg/m^3, and
θ_i = volumetric fraction of ice, m^3/ m^3.

The modified volumetric heat capacity term applies only as the soil is going through either the freezing process or the thawing process.

One-Dimensional Heat Flow: The transient differential equation that includes the freezing process in a soil must use the equation with the modified volumetric heat capacity of the soil. The modified volumetric heat capacity equation (i.e., Eq. 10.23) can be substituted into the transient heat flow equations presented in the previous section to yield the following equation:

$$\lambda\frac{\partial^2 T}{\partial y^2}+\frac{\partial\lambda}{\partial y}\frac{\partial T}{\partial y}=\left(\zeta-L_f\rho_i\frac{\partial\theta_i}{\partial T}\right)\frac{\partial T}{\partial t} \quad (10.24)$$

The terms on the right-hand side of Eq. 10.24 can be expanded to separate the heat flow components that are due to latent heat from the remainder of the heat flow equation:

$$\lambda\frac{\partial^2 T}{\partial y^2}+\frac{\partial\lambda}{\partial y}\frac{\partial T}{\partial y}=\zeta\frac{\partial T}{\partial t}-L\rho_i\frac{\partial\theta_i}{\partial T}\frac{\partial T}{\partial t} \quad (10.25)$$

The term, $\partial\theta_i/\partial T$, represents the slope of the SFCC, m_2^i. Therefore, Eq. 10.25 can incorporate the soil-freezing function (i.e., the slope of the ice content versus temperature below 0°C):

$$\lambda\frac{\partial^2 T}{\partial y^2}+\frac{\partial\lambda}{\partial y}\frac{\partial T}{\partial y}=\zeta\frac{\partial T}{\partial t}-L\rho_i m_2^i\frac{\partial T}{\partial t} \quad (10.26)$$

The latent heat term applies when the temperature lies within the zone of temperatures over which the water in the soil freezes (or thaws). The zone of freezing temperatures ranges from 0°C to a lower temperature where all the water in the soil becomes frozen.

The thermal conductivity λ and the volumetric heat capacity ζ of the soil must correspond to the unfrozen soil when the temperature is above 0°C and the frozen soil properties must be used when the temperature corresponds to conditions where the pore-water is frozen.

Two-Dimensional Heat Flow: The two-dimensional transient heat flow equation for freezing conditions can be written as follows:

$$\lambda\frac{\partial^2 T}{\partial x^2}+\frac{\partial\lambda}{\partial x}\frac{\partial T}{\partial x}+\lambda\frac{\partial^2 T}{\partial y^2}+\frac{\partial\lambda}{\partial y}\frac{\partial T}{\partial y}=\zeta\frac{\partial T}{\partial t}-L_f\rho_i m_2^i\frac{\partial T}{\partial t} \quad (10.27)$$

The soil properties must represent either the frozen or unfrozen conditions in the soil.

Three-Dimensional Heat Flow: The three-dimensional transient heat flow equation for freezing conditions can be written as follows:

$$\lambda\frac{\partial^2 T}{\partial x^2}+\frac{\partial\lambda}{\partial x}\frac{\partial T}{\partial x}+\lambda\frac{\partial^2 T}{\partial y^2}+\frac{\partial\lambda}{\partial y}\frac{\partial T}{\partial y}+\lambda\frac{\partial^2 T}{\partial z^2}+\frac{\partial\lambda}{\partial z}\frac{\partial T}{\partial z}=\zeta\frac{\partial T}{\partial t}-L_f\rho_i m_2^i\frac{\partial T}{\partial t} \quad (10.28)$$

The soil properties must represent either the frozen or the unfrozen conditions in the soil.

10.4.4 Heat Flow Equations Including Vapor Flow

There can also be vapor flow associated with a thermal gradient; however, vapor flow is closely associated with flow in the pore-air and pore-water phases. The components related to vapor flow are mentioned below, but further details related to the theory and solution to practical engineering problems are mentioned under the topic of coupling liquid and vapor flow through unsaturated soils.

Fredlund and Gitirana (2005) presented the vapor flow in terms of two components: one due to vapor diffusion within the air phase and a second due to advection within the free pore-air. The velocity of vapor flow can be written as follows:

$$v_y^v = v_y^{vd} + v_y^{va} = -\frac{k^{vd}}{\gamma_w}\frac{\partial u_w}{\partial y} + \frac{k^{vd}}{\gamma_w}\frac{u_w}{T + 273.15}\frac{\partial T}{\partial y} - \frac{k^{va}}{\gamma_a}\frac{\partial u_a}{\partial y} \quad (10.29)$$

where:

v_y^v = total velocity of vapor flow, m/s,
v_y^{vd} = velocity of vapor flow by diffusion through the air phase, m/s,
v_y^{va} = velocity of vapor flow by advection in the air phase, m/s,
k^{vd} = pore-water vapor conductivity by vapor diffusion within the air phase; which can be written as $\gamma_w \dfrac{\omega_v u_v^{air}}{\rho_w R(T + 273.15)} \dfrac{D^{v*}}{\rho_w}$, m/s
k^{va} = pore-water vapor conductivity by advection within the free pore-air, which can be written as $\gamma_a \dfrac{\rho_v}{\rho_a} \dfrac{D^{a*}}{\rho_w}$, m/s
γ_w = unit weight of water, kN/m^3,
ρ_v = density of the vapor, kg/m^3,
ρ_w = density of water, kg/m^3,
ρ_a = density of air, kg/m^3,
u_v^{air} = vapor pressure, kPa,
u_a = pore-air pressure, kPa,
u_w = pore-water pressure, kPa,
T = temperature, °C,
ω_v = molecular weight of water vapor, 18.016 kg/kmol,
R = universal gas constant, 8.314 J/(mol.K),
D^{v*} = molecular diffusivity of vapor through soil, m^2/s, and
D^{a*} = $(1-S)nD^v\omega_v/RT$, kg·m/kN·s.

The above equation shows that the flow and storage of heat within a saturated-unsaturated soil can be a function of three primary variables: u_w, u_a, and T. Volume changes associated with the relative phases of the soil are not explicitly presented as primary variables; however, volume-mass changes affect thermal conductivity and volumetric heat capacity. Several unsaturated soil property functions can be identified in the above equation: the thermal conductivity function, the vapor conductivity function (diffusion), and the vapor advection function. These soil property functions can vary with the stress state (e.g., soil suction), and therefore, the partial differential equations are nonlinear.

10.4.5 Temperature and Heat Flux Boundary Conditions

There are two common thermal boundary conditions associated with the solution of heat flow problems: Dirichlet-type boundary conditions (i.e., the primary variable specified is temperature T) and Neumann-type boundary conditions (i.e., the derivative of the primary variable or the heat flux q_h). The Neumann heat flux can be specified as zero when the boundary is perfectly insulated. The boundary condition can also be considered as a zero heat flux boundary under conditions of symmetry for the material being modeled (e.g., a two-dimensional section out of a laterally extending continuum).

Most geotechnical engineering problems involve exposure to the ground surface where weather conditions are rapidly and continuously changing. The nighttime and morning periods are generally cool relative to the midday temperatures. Also, the temperatures from one day to the next can change significantly, as can the conditions from one month to the next. In other words, the temperature boundary condition at ground surface is in an unsteady or transient state.

Some geotechnical engineering problems can be modeled using average daily temperature values. However, other engineering problems may require a more rigorous representation of the thermal boundary conditions (e.g., a mathematical function approximating changes in temperature throughout the day). Weather stations commonly collect data and report the information on a daily basis (Fig. 10.13). Quite often the average daily temperature is recorded along with values for the maximum and minimum temperatures for the day. Temperature data can also be collected on an hourly basis but the amount of data becomes excessive.

Temperature as a Dirichlet-type boundary condition can readily be applied at the ground surface. The temperature may be assumed to vary according to a particular pattern during the day. For example, temperature may take the form of a sine function with the peak temperature near noontime. It is also possible to use measured hourly temperatures as input for numerical modeling purposes.

Temperature at a weather station is generally recorded at 1 or 2 m above ground surface. It is the ground surface temperature, however, that forms the more accurate boundary condition for modeling geotechnical engineering problems. The soil temperature at the ground surface may not be the same as the air temperature above the ground surface. The difference in temperature between the air and the soil can be taken into consideration when performing coupled heat and moisture flow analyses for the calculation of actual evaporation at ground surface. Wilson (1990) suggested a

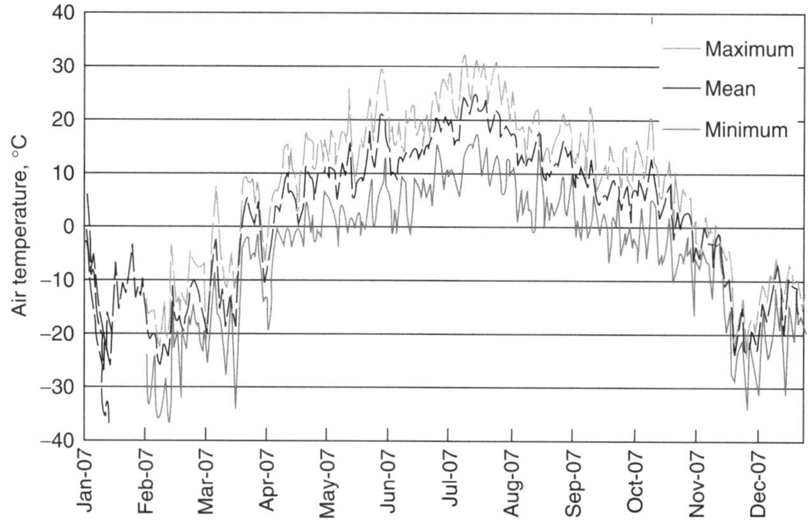

Figure 10.13 Minimum, maximum, and average air temperature readings at a site in northern Canada in 2007.

closed-form equation that can be used for the estimation of the soil temperature when the air temperature above the soil is measured:

$$T_s = T_a + \frac{1}{C_f \eta f(u)}(Q_n - \text{AE}) \tag{10.30}$$

where:

T_s = soil temperature, °C,
T_a = air temperature, °C,
C_f = conversion factor (i.e., 1 kPa = 0.0075 mHg),
η = psychrometric constant, 0.06733 kPa/°C,
$f(u)$ = wind speed function [i.e., $f(u) = 0.35(1. + 0.146W_w)$, where W_w is the wind speed in km/hr],
Q_n = net radiation, mm/day, and
AE = actual evaporation, mm/day.

Further information on the above equation is presented in Chapter 6, where consideration is given to coupled heat and moisture flow analysis for the assessment of actual evaporation.

Thermal boundary conditions can also be applied internal to the overall geometry. As an example, consider the case of a pipeline installed underground. The pipeline may operate at a temperature different from the temperature of the surrounding soil. Therefore, heat may be extracted from the surrounding soil to the pipeline or vice versa. If heat is extracted from the pipeline to the soil, the pipeline operates as a source. If the reverse is true, the pipeline operates as a sink. If the pipeline passes through a permafrost area and is operated at a temperature above water-freezing conditions, there will be a gradual thawing of the surrounding soil.

It is also possible to have Neumann-type boundary conditions that are based on a computed heat flux from the system. A mathematical function describing a boundary condition that is intermediate between the Dirichlet- and Neumann-type conditions can also be used when solving problems.

10.5 DIRECT MEASUREMENT OF THERMAL PROPERTIES

The following section presents information on the measurement of the thermal properties of soils. Quite often the thermal properties are computed using estimation procedures, and as a result the thermal properties may not be measured for geotechnical engineering purposes. However, it is important to understand the principles involved with the measurement of thermal properties for saturated and unsaturated soils.

10.5.1 Measurement of Thermal Conductivity

Thermal conductivity is the most important soil property associated with the analysis of conductive heat flow. It is possible to estimate the thermal conductivity of a soil mixture, but sometimes it is preferable to undertake measurements in the laboratory. The principles behind the laboratory measurement of thermal conductivity are similar to those associated with the measurement of hydraulic conductivity. However, in this case, a thermal gradient is applied across a soil specimen rather than a hydraulic head gradient. Also, the sides of the soil specimen must be thermally insulated rather than being impervious, as in the case of permeability measurements.

10.5.1.1 Laboratory Thermal Conductivity Apparatus

Numerous researchers have measured the thermal conductivity of dry soils as well as soils with varying water contents (Smith, 1942; Kersten, 1949; van Rooyen and Winterkorn, 1959; Johansen, 1975). The apparatus recently developed

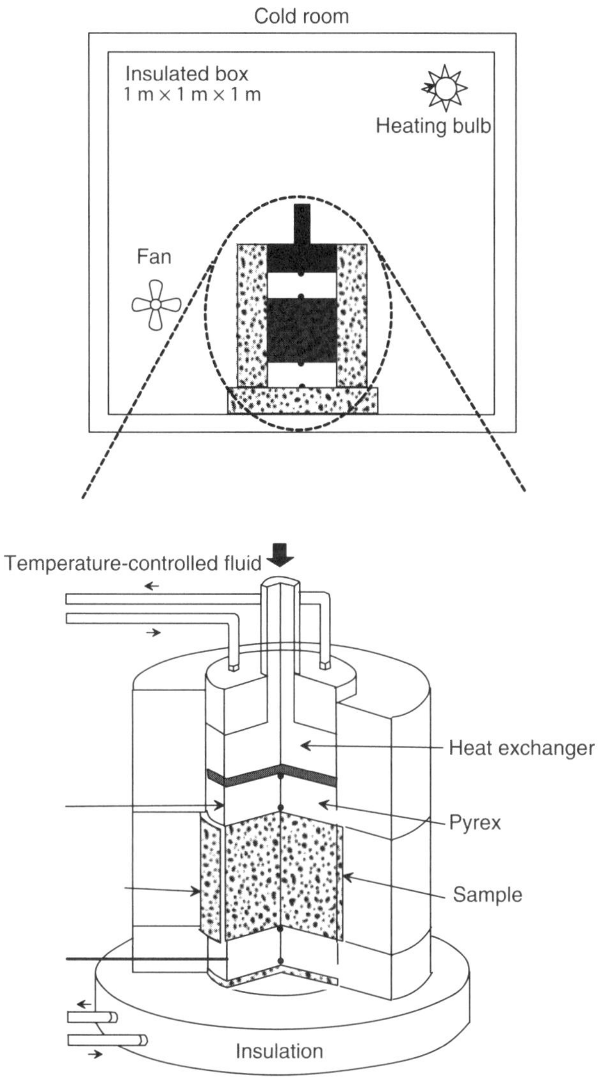

Figure 10.14 Experimental setup used to measure the thermal conductivity of soils (after Côté and Konrad, 2005).

by Côté and Konrad (2005) is described to illustrate the procedure that can be used to measure the thermal conductivity of unfrozen as well as frozen soils.

Figure 10.14 shows the experimental apparatus used by Côté and Konrad (2005) to measure the thermal conductivity of a range of materials. The equipment consists of a thermal conductivity cell surrounded with an insulated, temperature-controlled box. The entire apparatus was placed in a room where the temperature was maintained at 4°C. The cylinder had an inside diameter of 101.6 mm and a height of 75 mm. Soil specimens were placed between Pyrex disks that were 30 mm in thickness. Each Pyrex disk had thermistors embedded just below the top and bottom surfaces of the disk. The thermal conductivity of the Pyrex disks was 1.015 W/m/°C at −20°C and 1.090 W/m/°C at +20°C. The instrumented Pyrex disk acted as a heat flux meter. The temperature boundary conditions at the top and bottom of the three-layer system were maintained at constant values using independent heat exchangers. The heat exchangers created a constant vertical heat flux through the Pyrex disks and the soil specimen.

The soil specimen and the Pyrex disks were surrounded by a 50-mm-thick polystyrene jacket to reduce radial heat losses. The temperatures at the top and bottom of the Pyrex disks were recorded periodically until steady-state heat flux conditions were established, as shown in Fig. 10.15 (Côté and Konrad, 2005). The thermal conductivity of the soil specimen, λ, was determined from the temperature measurements at the surfaces of the Pyrex disks along with the following equation:

$$\lambda = \frac{1}{2}\left[\left(\lambda_{\text{uf}}\frac{\Delta T_{\text{uf}}}{\Delta h_{\text{uf}}} + \lambda_{\text{lf}}\frac{\Delta T_{\text{lf}}}{\Delta h_{\text{lf}}}\right)\frac{\Delta h}{\Delta T}\right] \tag{10.31}$$

where:

Δh = distance between two temperature measurements, m,
ΔT = temperature difference, °C,
λ = thermal conductivity, W/m/°C,
uf = upper heat flux meter, and
lf = lower heat flux meter.

The Côté and Konrad (2005) apparatus was used for measuring the thermal conductivity of both unfrozen and frozen soil specimens. The temperature difference between the top and the bottom of the soil specimen was about 8°C, resulting in a thermal gradient of 0.2–0.6°C/cm. Figure 10.16 shows typical thermal conductivity values measured on a variety of coarse materials. The specimens were tested at various initial water contents. Figure 10.17 shows thermal conductivity measurements on the same materials in a frozen state.

10.5.1.2 Use of Thermal Needle Probe

The use of a thermal needle probe to measure thermal conductivity originated with Stalhane and Pyk (1931) and was further developed by van der Held and van der Drunen (1949). ASTM Designation D5334–08 (2008b) provides a standard method for the measurement of thermal conductivity when using the thermal needle probe. The proposed transient heating method can be used when the temperature range is between 0 and 100°C. The method fails once there is a phase change involved.

Figure 10.18 shows the layout of the thermal needle probe along with the electrical components used to apply a constant current. The temperature change is measured with respect to elapsed time. The needle probe must have a large length-to-diameter ratio which is assumed to simulate an infinitely long but thin heating source. The thermal needle consists of a stainless steel hypodermic needle which contains a heater element and a thermocouple.

A known constant current is applied inside the needle and the rise in temperature with time is monitored. The needle

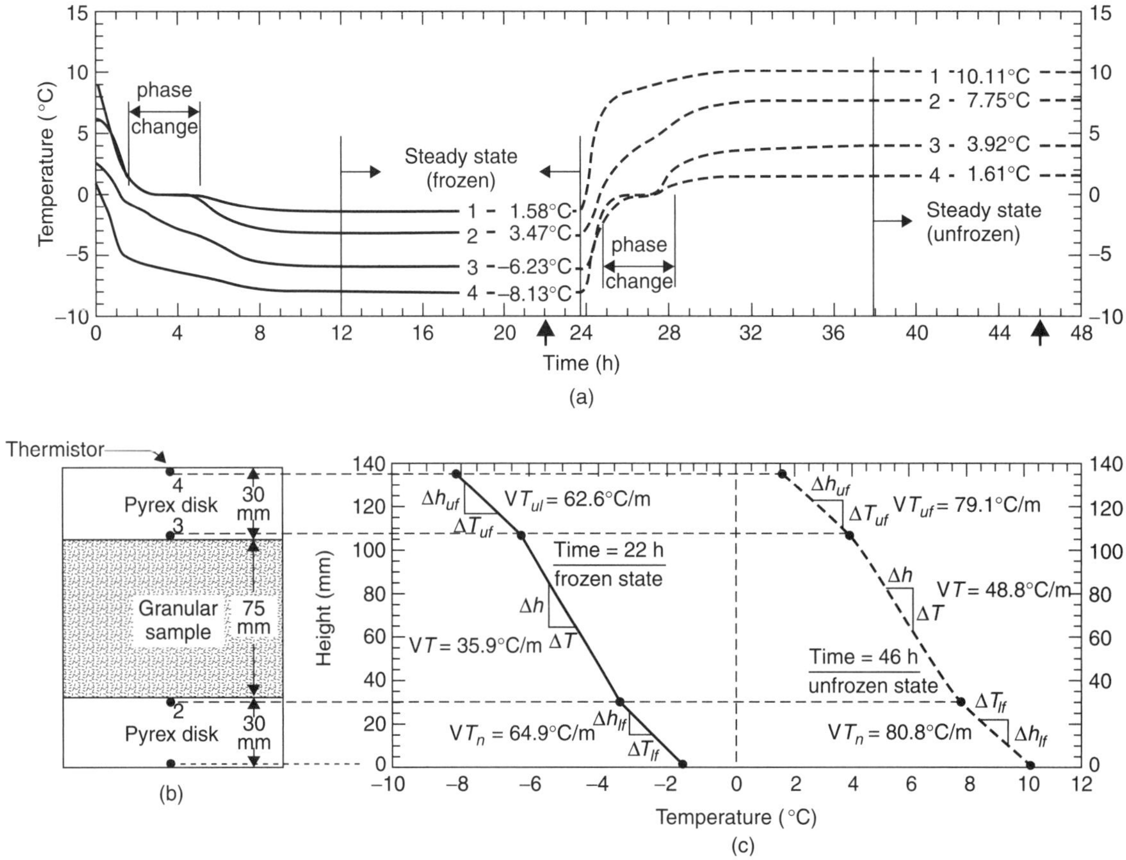

Figure 10.15 Typical data obtained during measurement of thermal conductivity of granular soil (Côté and Konrad, 2005): (a) evolution of temperature with time measured using thermisters; (b) specimen layout; (c) temperature distribution at frozen steady states and unfrozen steady states.

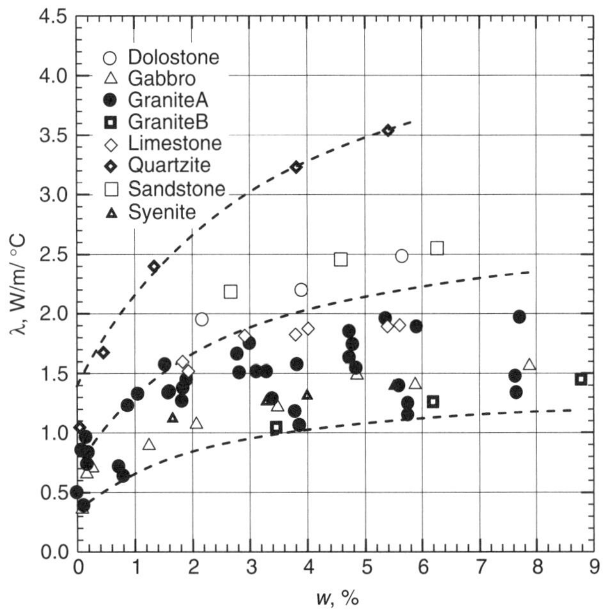

Figure 10.16 Thermal conductivity of unfrozen base course materials as function of water content (data compiled by Côté and Konrad, 2005).

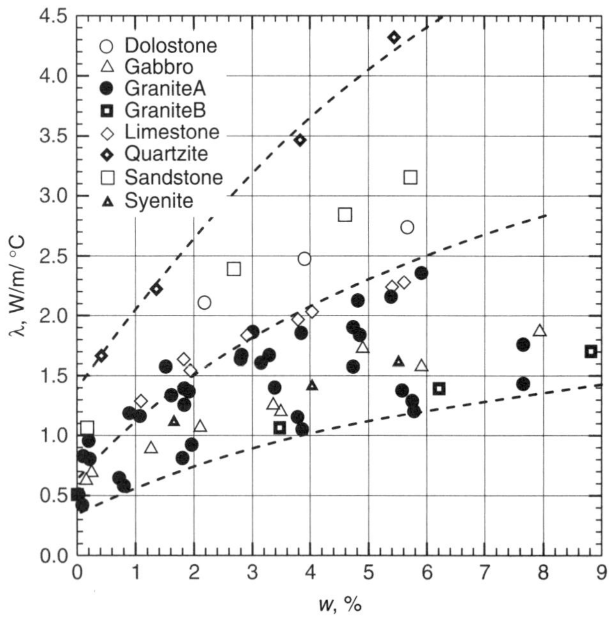

Figure 10.17 Thermal conductivity of frozen base course materials as function of water content (data compiled by Côté and Konrad, 2005).

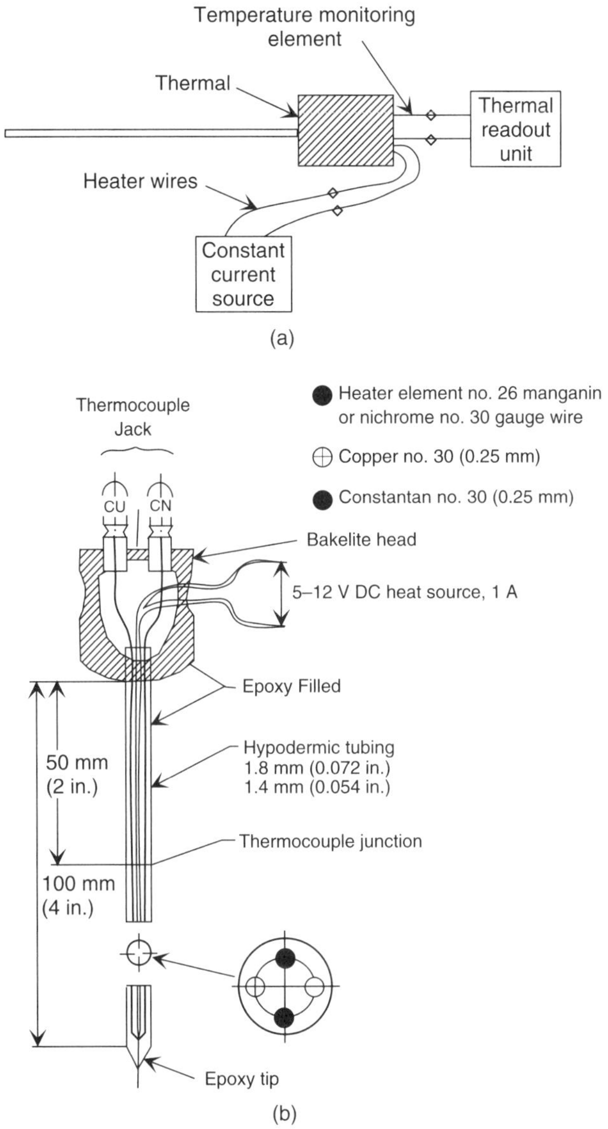

Figure 10.18 Details of needle probe used to measure thermal conductivity of soils: (a) thermal probe setup; (b) components of needle probe (ASTM D5334-08, 2008b).

probe can be calibrated against a material with a known thermal conductivity such as water (i.e., thermal conductivity of 0.607 W/m/°K at 25°C) or glycerine (i.e., thermal conductivity of 0.292 W/m/°K). The calibration constant for the needle, C_n, can be calculated as follows:

$$C_n = \frac{\lambda_{\text{material}}}{\lambda_{\text{measured}}} \tag{10.32}$$

where:

$\lambda_{\text{material}}$ = thermal conductivity of the known material and
$\lambda_{\text{measured}}$ = thermal conductivity measured with the needle probe on the calibration material.

There is no need for a calibration constant if the heat input to the needle is well controlled. The calibration constant remains a constant value and the thermal needle probe can be used to measure the thermal conductivity of other materials, λ, through use of the following equation from ASTM (2008b) D5334–08:

$$\lambda = \frac{C_n Q}{4\pi S} \tag{10.33}$$

where:

Q = heat input, W/m,
C_n = calibration factor, and
S_t = slope of the temperature change versus natural logarithm of elapsed time.

Figure 10.19 shows typical results of temperature change versus the natural logarithm of elapsed time, the slope of which is the variable S_t. It is also possible to use the measurement of the drop in temperature as the needle probe is allowed to cool when the application of the constant current is terminated.

10.5.2 Measurement of Volumetric Heat Capacity and Specific Heat of Soils

Heat capacity controls the amount of temperature change that will occur as heat is absorbed or released from a material. It is possible to measure the specific heat of the constituents of a soil as well as the volumetric heat capacity of the combined soil-water system. It is possible to estimate the volumetric heat capacity of a soil-water system if the specific heat of each of the constituents of a soil is known. Specific heat is usually defined with respect to mass while volumetric heat capacity is defined with respect to volume. It is quite common to refer to the specific heat of each of the constituents of a mixture while referring to the volumetric heat capacity of a mixture.

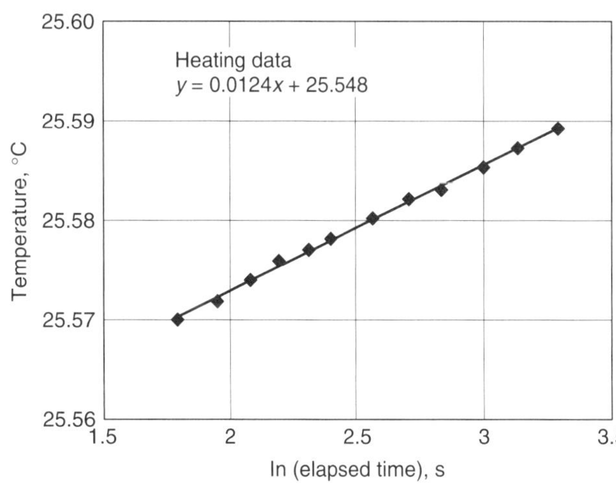

Figure 10.19 Data obtained during heating cycle with thermal needle probe.

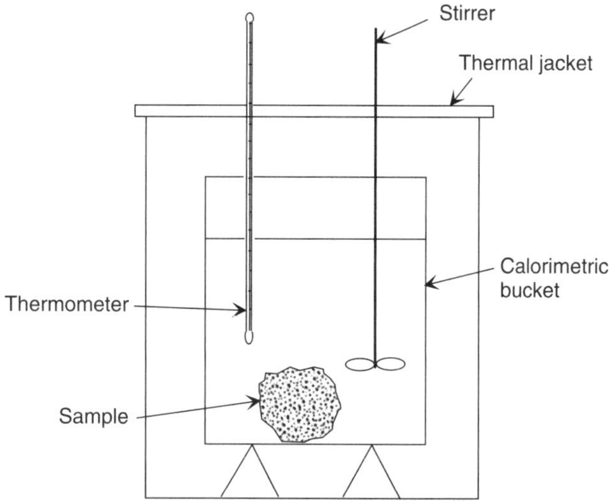

Figure 10.20 General layout of laboratory calorimeter.

Specific heat and volumetric heat capacity can be measured using a commercially available calorimeter. Figure 10.20 shows the general layout of a calorimeter. The important features of the calorimeter are the insulated walls, top, and bottom along with a thermometer capable of measuring temperature to within 0.01°C.

The heat capacity of the calorimeter should be measured prior to its usage for measuring the heat capacity of a material. This is done by adding a known amount of water at a slightly higher temperature than the temperature of the water already in the calorimeter. The heat capacity of the calorimeter, C_{cal}, is calculated using the following equation:

$$C_{\mathrm{cal}} = M_{wc}C_w - M_{wa}C_w \left[\frac{\Delta T_w}{\Delta T_{cf}}\right] \tag{10.34}$$

where:

M_{wc} = mass of water initially in the calorimeter, g,
M_{wa} = mass of water added to the calorimeter, g,
C_w = specific heat of water at the mean temperature when determining the calorimeter calibration (i.e., 4184 J kg^{-1} K^{-1} at 15°C),
ΔT_w = temperature change of the water in degrees Celcius between its temperature before and after being added to the calorimeter,
ΔT_{cf} = temperature change for the water already in the calorimeter, and
C_{cal} = mean specific heat or the calorimeter constant in J kg^{-1} K^{-1}.

The specific heat capacity of the calorimeter is measured once the mixture has come to a new equilibrium temperature. It is then possible to place a mass of soil into the calorimeter for the measurement and calculation of the specific heat of the soil. The specific heat of the soil, C_p, is computed using the following equilibrium equation:

$$C_p = C_w\left(\frac{M_{wa}}{M_s}\right)\left[\frac{\Delta T_w}{\Delta T_{cf}}\right] - \frac{M_{wc}C_w + C_{cal}}{M_s} \tag{10.35}$$

where:

M_s = mass of soil used in the experiment.

It may be necessary to correct the change in temperature, ΔT_{cf}, for thermal losses and to compensate for energy added by the stirrer. A detailed test procedure can be found in Taylor and Jackson (1986).

10.5.3 Measurement of Unfrozen Water Content

A calorimeter can also be used to measure the unfrozen water content of frozen soils. The test procedure is similar to that used for the measurement of the specific heat capacity of a soil. The test procedure requires that a soil specimen frozen to a particular temperature below 0°C be placed inside the calorimeter. The calorimeter is generally filled with water. The calorimeter consists of an inner and outer jacket and an internal stirrer, as shown in Fig. 10.20.

The temperature is recorded when a new equilibrium temperature is achieved after the partly frozen soil has been placed in the calorimeter. Radiation losses may need to be taken into account when measuring the unfrozen water content. An equation satisfying thermal equilibrium can be used to calculate unfrozen water content. It is necessary to know the specific heat of each component of the soil mixture in order to compute the unfrozen water content.

Thermal equilibrium can be written for each component of the soil mixture. A soil specimen is prepared at a temperature slightly below 0°C for testing purposes. A portion of the water in the soil is frozen and a portion is unfrozen (i.e., a partly frozen soil). The objective of the experiment is to determine the percentage of the water that was unfrozen at the initial temperature at which the soil specimen was prepared.

Several soil specimens can be prepared at slightly different temperatures below 0°C and the unfrozen water content corresponding to each initial temperature is computed based on the calorimeter measurements. The results from several measurements give rise to an unfrozen water content function.

The heat associated with each component of the soil mixture can be quantified. The initial quantity of heat associated with the soil solids, H_{is}, can be written as the mass of the solids times the specific heat of the solids and the absolute temperature in Kelvin:

$$H_{is} = M_s C_p T_{Ki} \tag{10.36}$$

where:

M_s = mass of the soil solids, kg,
C_p = specific heat of the soil solids, J kg^{-1} K^{-1}, and
T_{Ki} = initial sample temperature, K.

The final quantity of heat, H_{fs}, associated with the soil solids, H_s, can be written in a similar manner:

$$H_{fs} = M_s C_p T_{Kf} \quad (10.37)$$

where:

T_{Kf} = final sample temperature, K (i.e., equilibrium temperature).

The change in stored heat, ΔH_s, in the soil solids between the initial and final temperatures can be written as

$$\Delta H_s = M_s C_p (T_{Kf} - T_{Ki}) \quad (10.38)$$

The quantity of heat associated with the frozen water (ice), H_i, has a stored heat term for ice between the initial temperature and zero degrees as well as a latent-heat-of-fusion term. The two components of change in heat for ice can be written as

$$\Delta H_i = (M_w - M_{uw}) C_i (273.15 - T_{Ki}) + (M_w - M_{uw}) L \quad (10.39)$$

where:

M_w = mass of water, kg,
M_{uw} = mass of unfrozen water, kg,
C_i = specific heat of ice, J kg^{-1} K^{-1}, and
L = latent heat of fusion, J/kg.

The change in the quantity of heat associated with water in the soil specimen, ΔH_w, can be written as

$$\Delta H_w = (M_w - M_{uw}) C_w \Delta T_{Kf} + M_{uw} C_w (T_{Kf} - T_{Ki}) \quad (10.40)$$

where:

C_w = specific heat of water, J kg^{-1} K^{-1}, and
ΔT_{Kf} = $T_{Kf} - 273.15$, difference between the final temperature and 273.15 K.

The quantity of heat associated with the calorimeter, ΔH_c, can be written as

$$\Delta H_c = (M_1 - M_2) C_w \, \Delta T_1 \quad (10.41)$$

where:

M_1 = mass of water in the calorimeter, kg,
M_2 = mass of calorimeter, kg, and
ΔT_1 = temperature change of the calorimeter and contents during the test.

The above-mentioned changes in heat can be combined into one equation to give

$$\Delta H_s + \Delta H_i + \Delta H_w = \Delta H_c \quad (10.42)$$

The thermal equilibrium equation can be written by substituting the individual heat components into Eq. 10.42 to give

$$M_s C_p (T_{Kf} - T_{Ki}) + (M_w - M_{uw}) C_i \, \Delta T_{Ki} + (M_w - M_{uw}) L + (M_w - M_{uw}) C_w \, \Delta T_{Kf} + M_{uw} C_w (T_{Kf} - T_{Ki}) = (M_1 - M_2) C_w \, \Delta T_1 \quad (10.43)$$

Rearranging Eq. 10.43 and solving for the mass of unfrozen water gives

$$M_{uw} = \left[\frac{M_w (C_i \Delta T_{Ki} + L + C_w \Delta T_{Kf}) + M_s C_p (T_{Kf} - T_{Ki}) - (M_1 - M_2) C_w \Delta T_1}{C_i \Delta T_{Ki} + L + C_w \Delta T_{Kf} - C_w (T_{Kf} - T_{Ki})} \right] \quad (10.44)$$

The unfrozen water content can be calculated as follows:

$$w_{uw} = \frac{M_{uw}}{M_s} \quad (10.45)$$

Soil samples tested for unfrozen water content must be initially frozen to various temperatures (e.g., −2°C, −4°C, −8°C, etc.) in order to obtain a function describing the relationship between the temperature below 0°C and the amount of unfrozen water in the soil.

10.6 ESTIMATION PROCEDURES FOR THERMAL PROPERTIES

Estimation procedures are of particular interest when dealing with heat flow problems since it is quite common to use estimated thermal properties when performing thermal analyses in geotechnical engineering. Some of the common procedures used for the estimation of unsaturated soil thermal properties are described in the following sections.

10.6.1 Thermal Conductivity of Soils

Estimation procedures for the determination of the thermal conductivity of soil mixtures have been proposed by several researchers. The predictive models take into consideration soil properties such as water content, degree of saturation, and dry density (or porosity). Three estimation procedures are presented for thermal conductivity: (i) the Kersten (1949) model, (ii) the Johansen (1975) model, and (iii) the Côté and Konrad (2005) model. The Johansen (1975) model requires that the thermal conductivity of the saturated material be part of the input to the predictive model.

10.6.1.1 Kersten Estimation Model

Kersten (1949) undertook an extensive study involving the measurement of thermal conductivity on more than 1000 soils. The soils included gravels, crushed rocks, sands, silts, clays, and peat. The results from a portion of the measured values were used to develop an empirical relationship between thermal conductivity and other volume-mass soil properties such as water content and dry density. The empirical relationships were developed for both unfrozen and frozen soil conditions. Soil fabric was found to have an important influence on thermal conductivity giving rise to different equations for coarse- and fine-grained materials.

The empirical thermal conductivity equations proposed by Kersten (1949) for sandy soils were converted to SI units by Andersland and Anderson (1978). The proposed empirical equation for unfrozen sandy soils is as follows:

$$\lambda_u = 0.1442\left[0.7\ \log(w) - 0.4\right] \times 10^{0.6243\rho_d} \tag{10.46}$$

where:

λ_u = thermal conductivity of the unfrozen soil, W/m/K,
w = gravimetric water content, %, and
ρ_d = dry density, g/cm^3.

The above equation shows that the water content and the dry density of the soil are the dominant factors influencing thermal conductivity.

The proposed empirical equation for frozen sandy soils is

$$\lambda_f = 0.001096 \times 10^{0.8116\rho_d} + 0.00461w \times 10^{0.9115\rho_d} \tag{10.47}$$

where:

λ_f = thermal conductivity of the frozen soil, W/m/K.

Kersten (1949) proposed the following empirical equation for unfrozen fine-grained soils:

$$\lambda_u = 0.1442\left[0.9\ \log(w) - 0.2\right] \times 10^{0.6243\rho_d} \tag{10.48}$$

Another empirical equation was proposed for frozen fine-grained soils:

$$\lambda_f = 0.001442 \times 10^{1.373\rho_d} + 0.01226w \times 10^{0.4994\rho_d} \tag{10.49}$$

Côté and Konrad (2005) applied the Kersten (1949) model to a variety of coarse-grained materials and found the model generally predicted thermal conductivities that were higher than the measured values. The closest predicted thermal conductivity values were for crushed sandstone and quartzite. It should be noted that the Kersten (1949) model does not include the influence of the thermal conductivity of the solid particles which can have a significant influence on the overall thermal conductivity.

10.6.1.2 Johansen Model

Johansen (1975) used the concept of a normalized (or relative) thermal conductivity and reanalyzed the database generated by Kersten (1949). The mineralogy of the soil was taken into consideration as well as the degree of saturation. A normalized thermal conductivity equation was developed for unfrozen and frozen soils. The normalized thermal conductivity equation for unfrozen sandy soils is as follows:

$$\lambda_{ru} = 0.7\ \log(S_u) + 1 \tag{10.50}$$

where:

λ_{ru} = normalized thermal conductivity of the unfrozen soil, W/m/K, and
S_u = degree of saturation of the unfrozen soil.

The normalized thermal conductivity equation for frozen sandy soils was written as follows:

$$\lambda_{rf} = S_u \tag{10.51}$$

where:

λ_{rf} = normalized thermal conductivity of the frozen soil, W/m/K, and
S_u = degree of saturation in decimal form.

The normalized thermal conductivity equation for unfrozen fine-grained soils was written as follows:

$$\lambda_{ru} = \log(S_u) + 1 \tag{10.52}$$

The normalized thermal conductivity equation for frozen fine-grained soils was written as follows:

$$\lambda_{rf} = S_u \tag{10.53}$$

The normalized thermal conductivity, λ_r, equation with respect to saturated and fully dried conditions is as follows:

$$\lambda_r = \frac{\lambda - \lambda_{dry}}{\lambda_s - \lambda_{dry}} \tag{10.54}$$

where:

λ_r = normalized thermal conductivity of the soil at a particular water content, W/m/°C,
λ_s = thermal conductivity of the saturated soil, W/m/°C, and
λ_{dry} = thermal conductivity of the dry soil, W/m/°C.

Johansen (1975) normalized thermal conductivity with respect to the amount of water in the soil (i.e., normalized water content or degree of saturation, S). This infers that the following limiting condition must be observed:

$$S = 0; \quad \lambda = \lambda_{dry} \quad \text{or} \quad \lambda_r = 0 \tag{10.55}$$

$$S = 1.0; \quad \lambda = \lambda_s \quad \text{or} \quad \lambda_r = 1 \tag{10.56}$$

It should be noted that the equations proposed by Johansen (1975) do not satisfy the equation for the unfrozen soil when the degree of saturation goes to zero.

Equation 10.54 can be rewritten as follows for an unfrozen soil:

$$\lambda_u = (\lambda_{s(u)} - \lambda_{\text{dry}})\lambda_{ru} + \lambda_{\text{dry}} \qquad (10.57)$$

where:

$\lambda_{s(u)}$ = thermal conductivity of the saturated, unfrozen soil, W/m/°C.

Equation 10.54 can also be rewritten as follows for a frozen soil:

$$\lambda_f = (\lambda_{s(f)} - \lambda_{\text{dry}})\lambda_{rf} + \lambda_{\text{dry}} \qquad (10.58)$$

where:

$\lambda_{s(f)}$ = thermal conductivity of the saturated, frozen soil, W/m/°C.

Johansen (1975) used the concept of a geometric mean for the soil particles and obtained the following thermal conductivity equations for saturated, unfrozen soils, $\lambda_{s(u)}$:

$$\lambda_{s(u)} = \lambda_p^{1-n}\,\lambda_w^n \quad \text{when } T \geq 0°\text{C} \qquad (10.59)$$

The equivalent equation proposed for the thermal conductivity equations for saturated, frozen soils, $\lambda_{s(f)}$, was as follows:

$$\lambda_{s(f)} = \lambda_p^{1-n}\,\lambda_i^n \quad \text{when } T < 0°\text{C} \qquad (10.60)$$

where:

λ_p = thermal conductivity of the soil solids, W/m/°C,
λ_w = thermal conductivity of water, W/m/°C,
λ_i = thermal conductivity of ice, W/m/°C, and
n = porosity of the soil in decimal form.

It was noted that the amount of quartz mineral in a soil has a significant effect on the thermal conductivity of the soil solids. Johansen (1975) proposed the following equations to compensate for the quartz content.

$$\lambda_p = 2.0^{1-q} \times 7.7^q \quad \text{when } q > 0.2 \text{ (or 20\%)} \qquad (10.61)$$

$$\lambda_p = 3.0^{1-q} \times 7.7^q \quad \text{when } q \leq 0.2 \text{ (or 20\%)} \qquad (10.62)$$

where:

q = amount of quartz expressed as a volume ratio referenced to the solids portion.

The Johansen (1975) model takes the amount of quartz mineral into consideration in the estimation of the overall thermal conductivity. However, it does not take a wider variety of possible mineral solids into consideration.

The thermal conductivity for natural soils that were rounded and subrounded particles was estimated using a derivation from the de Vries (1963) equation:

$$\lambda_{\text{dry}} = \frac{0.135\rho_d + 64.7}{2700 - 0.947\rho_d} \qquad (10.63)$$

It was noted that crushed rocks (angular and subangular) yielded a different equation for the thermal conductivity for dry materials:

$$\lambda_{\text{dry}} = 0.39n^{-2.2} \qquad (10.64)$$

There are also other predictive models that have been proposed for the estimation of thermal conductivity (Farouki, 1981); however, the Johansen (1975) model has been shown to perform quite well.

10.6.1.3 Côté and Konrad (2005) Model

Côté and Konrad (2005) undertook an examination of the factors influencing the prediction of thermal conductivity of coarse soils and have suggested a series of refinements to the Johansen (1975) model. The first refinement took into consideration the change in porosity that occurs when water in a saturated soil freezes. The water in the voids expands by 9% upon freezing, causing a change in the porosity of a saturated soil:

$$n_f = \frac{1.09n_u}{1 + 0.09n_u} \qquad (10.65)$$

where:

n_u = porosity of the saturated unfrozen soil and
n_f = porosity of the saturated frozen soil.

No volume change occurs when a soil is dry. If the assumption is made that volume change occurs linearly with the amount of water in the soil, the above equation can then be written in terms of the degree of saturation:

$$S_f = \frac{1.09S_u}{1 + 0.09S_u} \qquad (10.66)$$

where:

S_f = degree of saturation of the frozen soil and
S_u = degree of saturation of the unfrozen soil.

Côté and Konrad (2005) also suggested that the geometric mean method of Sass et al., (1971) and Woodside and Messmer (1961) be used to determine the thermal conductivity of soil solids:

$$\lambda_p = \prod_{j=1}^{z} \lambda_{mj}^{x_j} \qquad \sum_{j=1}^{z} x_j = 1 \qquad (10.67)$$

where:

$\prod$ = product of the thermal conductivity of a particular mineral, k_m
x = volumetric proportion of a particular mineral,
j = refers to the jth mineral, and
z = total number of minerals.

Table 10.6 provides thermal conductivity values for most common minerals. The thermal conductivity values were compiled by Côté and Konrad (2005) from the database of Horai (1971).

Further measurements of thermal conductivity on dry minerals, λ_{dry}, have been made by Côté and Konrad (2005). The thermal conductivity of dry soils heavily relies on structural effects (e.g., cemented particles, angular particles, and rounded particles) and the thermal conductivity ratio. For example, a soil with a porosity of 40% could have a thermal conductivity for natural quartz sand of 0.3 W/mK in the dry state. For the same porosity, dry crushed quartzite has a value of 0.65 W/mK and for naturally porous quartzite rock the thermal conductivity could be 1.15 W/mK. The materials have the same origin and porosity but different particle arrangements. The lower the ratio of fluid thermal conductivity to particle thermal conductivity, λ_f/λ_p, the higher will be the effect of structural characteristics.

Côté and Konrad (2002) proposed the following equation for two-phase porous materials (e.g., solid particles and a fluid such as water in the pores) with any fluid filling the pores:

$$\lambda_{2P} = \frac{(\kappa_{2P}\lambda_p - \lambda_f)(1-n) + \lambda_f}{1 + (\kappa_{2P} - 1)(1-n)} \tag{10.68}$$

where:

κ_{2P} = an empirical factor that depends on structural effects (two-phase thermal conductivity),
λ_p = thermal conductivity of the solid particles,
λ_f = thermal conductivity of the pore fluid such as water, and
n = porosity.

The factor κ_{2P}, was empirically determined for a number of soil types including porous rock and concrete. Figure 10.21 shows the variation of κ_{2P} as a function of the ratio, λ_f/λ_p and particle types. The HSU and HSL lines refer to the Hashin-Shtrikman (1962) theoretical bounds for thermal conductivity of two-phase porous materials. The relationship between κ_{2P} and the ratio, λ_f/λ_p can be expressed in the form

$$\kappa_{2P} = 0.29\left(\frac{15\lambda_f}{\lambda_p}\right)^{\beta} \tag{10.69}$$

where:

β = factor related to particle type and depending on the trigger value of $\frac{1}{15}$ of the ratio of λ_f/λ_p, (Table 10.7)

Table 10.6 Thermal Conductivity Values λ_p for Some Rock-Forming Minerals

Mineral	λ_p (W/m/°C)
Amphibole	3.46
Calcite	3.59
Chlorite	5.15
Dolomite	5.51
Feldspar	2.25
Mica	2.03
Olivine	4.57
Plagioclase	1.84
Plagioclase (labradorite)	1.53
Pyroxene	4.52
Quartz	7.69

Source: Côté and Konrad (2005).

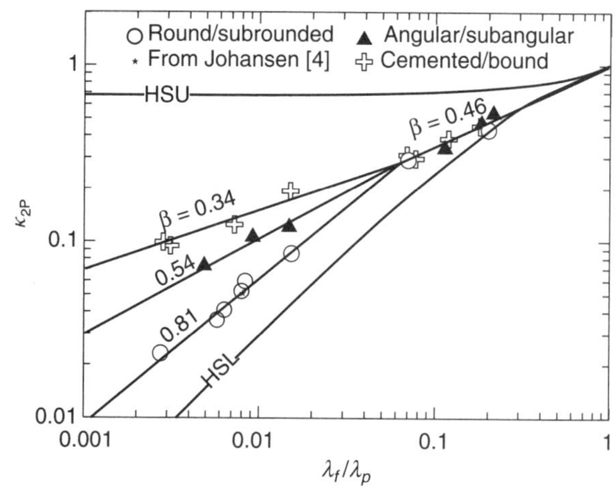

Figure 10.21 Values for the structural effects factor, κ_{2P}, for various two phase porous materials.

Table 10.7 Empirical β Factors for Materials with Various Particle Types

Particle Type	$\lambda_f/\lambda_p < \frac{1}{15}$, Dry	$\lambda_f/\lambda_p < \frac{1}{15}$, Saturated
Natural (rounded)	0.81	0.46
Crushed (angular)	0.54	0.46
Cemented (bound)	0.34	0.46

Source: Côté and Konrad (2005).

The following equation by Côté and Konrad (2005) can be used to estimate the thermal conductivity of dry soil:

$$\lambda_{\text{dry}} = \frac{(\kappa_{2P}\lambda_p - \lambda_f)(1-n) + \lambda_f}{1 + (\kappa_{2P} - 1)(1-n)} \qquad (10.70)$$

where:

κ_{2P} = from Eq. 10.69 with $\beta = 0.81$ for natural soils and 0.54 for crushed rock.

When the pore fluid in a soil is water, the ratio λ_f/λ_p is generally larger than $\frac{1}{15}$ and there are no structural effects. Consequently, the β factor can be set equal to 0.46 for all particle types. The results are essentially the same as those obtained by Johansen (1975) using the geometric mean procedure.

The relative or generalized thermal conductivity model by Côté and Konrad (2005) can be written as follows:

$$\lambda_{r(u,f)} = \frac{\kappa_{(u,f)} S_{(u,f)}}{1 + (\kappa_{(u,f)} - 1) S_{(u,f)}} \qquad (10.71)$$

where:

$S_{(u,f)}$ = degree of saturation, unfrozen or frozen,
$\lambda_{r(u,f)}$ = thermal conductivity of either unfrozen or frozen soils, u designating unfrozen and f designating frozen soils, and
$\kappa_{(u,f)}$ = an empirical factor that depends on the type of soil, (Table 10.8)

A comparison of the predicted and measured values of thermal conductivity for an unfrozen gravelly material when using the Côté and Konrad (2005) model are shown in Fig. 10.22. A comparison of the predicted and measured thermal conductivities for a frozen gravelly material is shown in Fig. 10.23.

There has been a gradual evolution of the equations used for the estimation of thermal conductivity of both frozen and unfrozen soils. The estimated saturated and unsaturated thermal properties are adequate for solving most geotechnical engineering problems.

Table 10.8 Empirical Factors $\kappa_{(u,f)}$ for Unfrozen and Frozen Soils

Material	$\kappa_{(u)}$ Unfrozen	$\kappa_{(u)}$ Frozen
Gravel and coarse sand	4.6	1.7
Medium and fine sand	3.55	0.95
Silts and clays	1.9	0.85
Peat	0.3	0.25

Source: Côté and Konrad (2005).

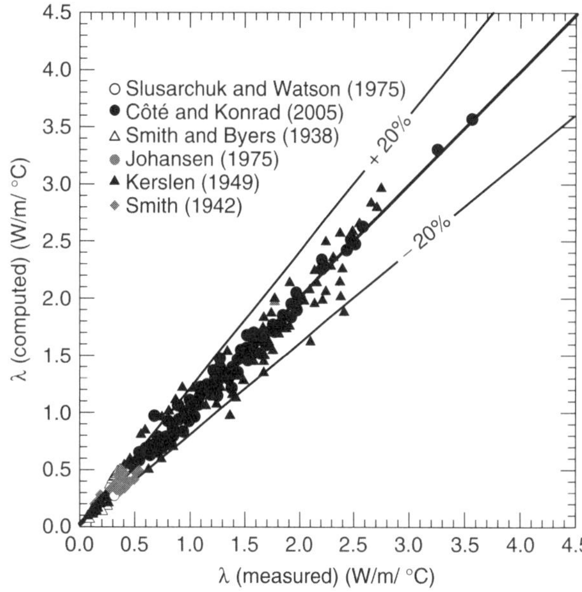

Figure 10.22 Comparison of measured and predicted thermal conductivity of unfrozen gravelly materials using Cote and Konrad (2005) model.

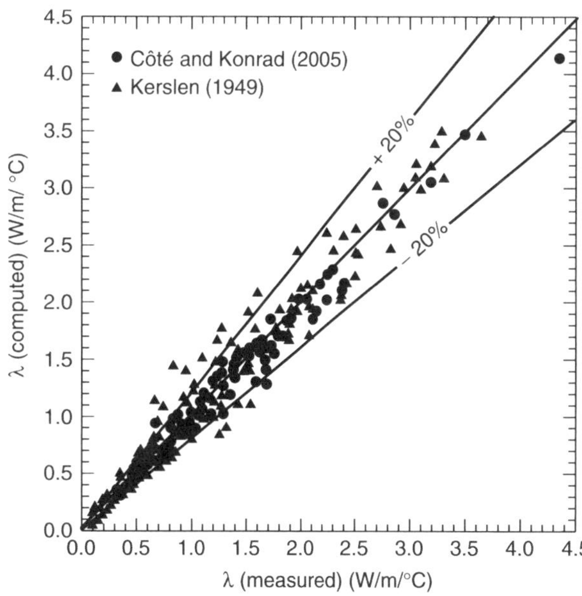

Figure 10.23 Comparison of measured and predicted thermal conductivity of frozen gravelly materials using Côté and Konrad (2005) model.

10.6.1.4 Degree of Saturation versus Matric Suction

The equations proposed for the estimation of the thermal conductivity of a soil show the importance of the degree of saturation. The degree of saturation of a soil is also related to soil suction, and therefore the thermal conductivity functions can also be written in terms of soil suction (see Fig. 10.2). The relationship between soil suction and thermal conductivity is particularly useful when solving heat flow problems where the degree of saturation of the soil varies throughout the geometry or with elapsed time.

The relationship between thermal conductivity and soil suction becomes obvious when coupling a thermal analysis with a moisture flow analysis. The SWCC can be used for the calculation of all unsaturated soil heat flow and moisture flow properties.

10.6.2 Volumetric Specific Heat of Soils

The volumetric heat capacity ζ of a soil consisting of solids, water, and air can be calculated by proportioning the volumetric heat capacities of each phase according to the volume of each phase, as shown in Eq. 10.7. Typical values for the volumetric heat capacity of water, soil solids, and air have been presented earlier in this chapter. The volumetric heat capacity of air can be assumed to be negligible in comparison to that of water or soil solids. Therefore the volumetric heat capacity of a soil mixture can be written as

$$\zeta = \zeta_p \theta_p + \zeta_w \theta \qquad (10.72)$$

where:

ζ_p = volumetric heat capacity of the soil solids, J/m^3/°C,
θ_p = volumetric solid content (i.e., V_s/V),
V_s = volume of solid particles in a total soil element, m^3,
ζ_w = volumetric heat capacity for the water phase, J/m^3/°C, and
θ = volumetric water content (i.e., V_w/V), and
V_w = volume of water in the soil element.

A typical value for the volumetric heat capacity of soil solids is 2.2×10^6 J/m^3/°C and a typical volumetric heat capacity for water is 4.2×10^6 J/m^3/°C. Therefore, the volumetric heat capacity of a soil mixture can be approximated as follows:

$$\zeta = \left(2.2 \times 10^6\right)\theta_p + \left(4.2 \times 10^6\right)\theta \qquad (10.73)$$

where:

ζ = volumetric heat capacity for the overall soil mass, J/m^3/°C.

The volumetric phase representations can also be written in terms of other volume-mass soil properties:

$$\zeta = \left(2.2 \times 10^6\right) G_s + \left(4.2 \times 10^6\right) w\rho_d \qquad (10.74)$$

where:

w = gravimetric water content,
ρ_d = dry density of the soil, and
G_s = specific gravity of the soil solids.

10.6.3 Relationship of Thermal Properties to SWCC

The SWCC provides a relationship between the amount of water in a soil and soil suction. The amount of water in the soil can be designated in terms of gravimetric water content, volumetric water content, or degree of saturation.

The heat capacity of a soil is usually proportioned according to the proportion of the overall volume that is water and the amount that is comprised of soil solids. Under saturated soil conditions, the volumetric water content represents the porosity of the soil which partitions the soil mass into solids and water. At any other soil suction the volumetric water content designates the portion of the total volume that is occupied by water.

Volumetric heat capacity of a soil will vary with soil suction. The SWCC can be used to calculate the heat capacity as a function of soil suction. The relationship between soil suction and heat capacity is hysteretic in character. The need to use the relationship between heat capacity and soil suction becomes more apparent when solving problems where the amount of water in the soil is changing with time (e.g., coupled heat and moisture flow).

The thermal conductivity of a soil is also a function of the proportion of the phase making up the soil mixture. The amount of air in the soil also has a significant influence on the thermal conductivity. The plot of degree of saturation versus soil suction can be used to write the thermal conductivity of a soil mixture in terms of soil suction. The thermal conductivity relationship is hysteretic, as is the SWCC. There are distinct advantages in relating all soil properties to the SWCC when solving problems that couple more than one physical process.

The SWCC is also of value in estimating the amount of frozen water in the freezing zone of a soil.

10.7 APPLICATIONS TO THERMAL PROBLEMS

The application section illustrates some of the concepts associated with modeling heat flow. A limited number of relatively simple heat flow problems are used for this purpose. The conditions associated with each problem are described along with the soil properties needed to solve the problem.

10.7.1 Solution to Boundary Value Heat Flow Problem

The solution of a heat flow boundary value problem requires the following information. First, it is necessary that the ground surface geometry and the stratigraphic soil units be defined. It must also be decided whether the problem at hand should be solved as a one-dimensional, two-dimensional, or three-dimensional geometry.

Second, the geotechnical engineer must be satisfied that a partial differential equation can be written for a representative elemental volume. The partial differential equation must adequately describe the physical processes that control soil behavior.

Third, it must be possible to measure or estimate the thermal soil properties and other soil properties relevant to the problem being solved.

Fourth, boundary conditions must be applied to the boundaries of the geometry. Initial conditions also need to be established when modeling situations where the partial differential equation is time dependent. Many geothermal analyses are not highly nonlinear, and as a result it is quite straightforward to assess initial or starting conditions when analyzing a transient heat flow problem. It is generally advisable to first perform a steady-state analysis, even when a transient analysis is required. Heat flow problems often involve performing numerous transient-type analyses to capture the sensitivity associated with changing a number of variables related to the problem at hand.

Fifth, the numerical solutions should be performed and each set of results studied from the standpoint of reasonableness. There are many potential sources of error when undertaking numerical modeling studies, and it is important that the computer output be properly interpreted. Most studies involve the use of a series of analyses (i.e., a parametric or sensitivity analysis) in order to ascertain the range of results that might be anticipated in situ.

10.8 ONE-DIMENSIONAL HEAT FLOW IN UNFROZEN AND FROZEN SOILS

The first heat flow problem illustrates the solution of a transient one-dimensional geometry where the soil is unfrozen. The initial temperature is assumed to be constant with depth, and it is assumed that a sinusoidal temperature variation is imposed at the ground surface over a period of one year. The temperature profile with depth is plotted every 2 weeks of the year.

10.8.1 Description of One-Dimensional Heat Flow

Let us consider the case where the soil profile is unfrozen and the initial temperature is constant with depth at 15°C. The geometry consists of a column of soil 5 m high. The ground surface is level and the geometry is one dimensional. The upper meter of soil is somewhat different from the underlying soil. Therefore, different thermal properties are input for the zone above and below the 1-m depth.

It is necessary to input both the thermal conductivity and the heat capacity properties for the soil since the problem under consideration is transient in nature. The thermal conductivity in the upper meter of soil is 1.1 W/m/K. The remaining soil profile has a thermal conductivity of 1.9 W/m/K. The volumetric heat capacity of the soil in the upper meter is 2×10^6 J/m^3/K and the remaining soil profile has a volumetric heat capacity of 2.7×10^6 J/m^3/K.

The temperature imposed at the ground surface varies from a low of +5°C to a high of +35°C, in accordance

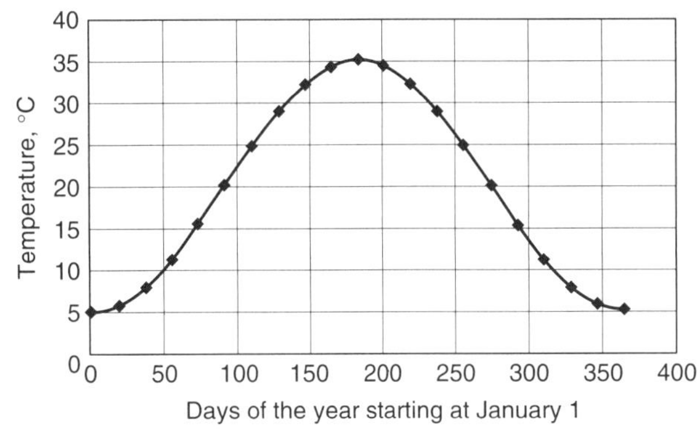

Figure 10.24 Sinusoidal variation of temperature throughout one year.

with the graph in Fig. 10.24. The temperature throughout 365 days can be written in equation form as follows:

$$T = 20 + 15 \sin\left(-\frac{\pi}{2} + \frac{2\pi t}{365}\right) \qquad (10.75)$$

where:

T = temperature, °C, and
t = time in days from January 1.

The highest temperature occurs in July and the coldest temperature occurs in January.

10.8.2 Solutions of Partial Differential Heat Flow Equation

The heat flow problem requires the solution of the partial differential equation 10.21. The finite element mesh generated for the solution of this problem consists of a series of nodes that are almost equally distributed with depth. Figure 10.25 shows the temperature isotherms with time (Côté, 2009). The isotherms have a bell shape with the highest variations at ground surface and virtually no change in temperature at a depth of 2.5 m.

10.9 TWO-DIMENSIONAL HEAT FLOW EXAMPLE INVOLVING CHILLED PIPELINE

Coutts and Konrad (1994) presented an example conductive heat flow problem that is common to the pipeline industry. The example considers the transfer of oil or gas through a pipeline from a northern cold region to a warmer southern region. It is assumed that the fluid in the pipeline is chilled to avoid thawing of the permafrost terrain. A problem may be encountered when the pipeline passes through discontinuous permafrost or when the terrain is not frozen.

10.9.1 Description of Soil Freezing around Pipeline

Consider the case of a pipeline where the inside fluid has a temperature of −2°C. The pipeline is embedded in a soil with

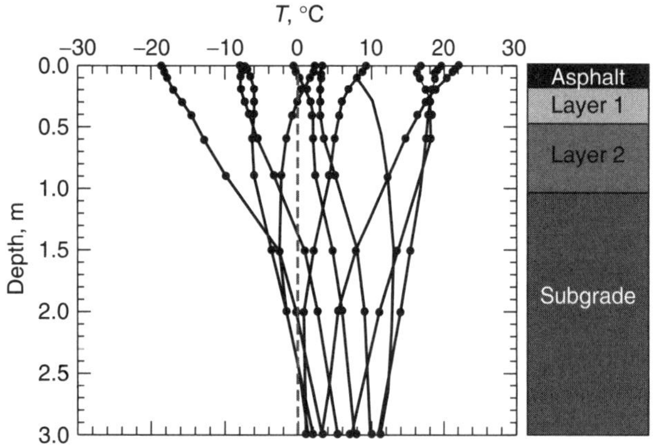

Figure 10.25 Temperature isotherms below highway surface for each month of the year (from Cote, 2009; Colloquium address)

a temperature of 3°C. Figure 10.26 shows the unfrozen water content versus temperature graph, referred to as the SFCC. The slope of the SFCC is given the symbol m_2^i. The SFCC forms a continuous function with respect to temperature. Small temperature change steps must be used for the soil-freezing curve in order to ensure convergence to the correct solution.

Figure 10.27 shows a plot of thermal conductivity versus temperature for the soil surrounding the pipeline. The volumetric water content of the soil is 37.7%. The volumetric

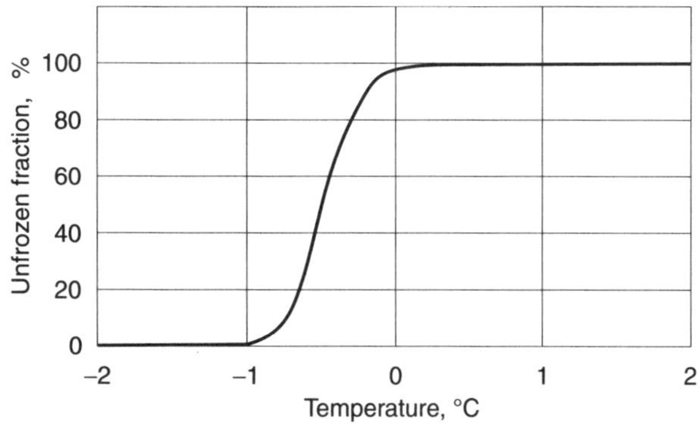

Figure 10.26 SFCC for the soil.

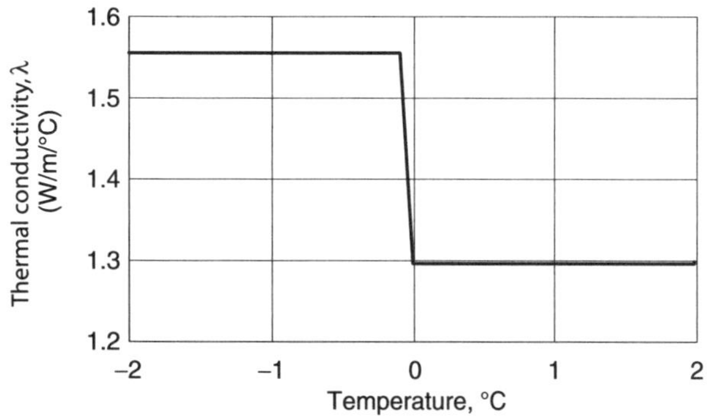

Figure 10.27 Thermal conductivity of unfrozen and frozen soil.

heat capacity of the unfrozen soil is estimated at 2.5 × 10^6 J/m^3/°K. The volumetric heat capacity of the frozen soil is estimated at 2.070 × 10^6 J/m^3/°K. The thermal conductivity of the unfrozen soil is estimated to be 1.3 W/m/K and the thermal conductivity of the frozen soil is 1.55 W/m/K.

The diameter of the pipeline is 0.3 m and the top of the pipeline is 0.3 m below ground surface. The temperature of the ground surface is assumed to remain constant at 3.0°C. Figure 10.28 shows the geometry of the problem along with boundary and initial conditions. Also shown is the final automatically generated mesh associated at an elapsed time of 7 days. Figure 10.29 shows the isotherms around the pipeline corresponding to an elapsed time of 7 days. The freezing front is on average about 0.1 m outside the pipeline.

Figure 10.30 shows the isotherms around the pipeline after 730 days. The freezing front has now extended about 0.6 m below the invert of the pipe and about 0.1 m above the pipe. The soil and the pipeline appear to have reached a steady-state condition after an elapsed time of 730 days. The slope of the SFCC, m_2^i, no longer has any influence on the solution after 730 days.

The above heat flow example problem has involved the use of constant specific heat soil properties. The thermal conductivity is also assumed to have constant values in the unfrozen and frozen states. There is a change in thermal conductivity as the soil changes phases from liquid to ice, and this change is taken into account in the analysis.

The thermal conductivity and volumetric heat capacity properties do not need to be changed during the heat flow process unless there is a change in the volume-mass properties of the soil.

10.10 TWO-DIMENSIONAL HEAT FLOW EXAMPLE WITH SURFACE TEMPERATURES ABOVE AND BELOW FREEZING

The next example problem has a two-dimensional geometry and involves consideration of the latent heat of fusion associated with the freezing and thawing of soil. The results were published by Harlan and Nixon (1978) and have been used as a benchmark solution.

10.10.1 Description of Two-Dimensional Heat Flow Problem

Harlan and Nixon (1978) performed a steady-state analysis involving the heat transfer between two adjacent areas with a ground surface temperature of +4°C on the left side and −5°C on the right side, as shown in Fig. 10.31. The thermal conductivity of the soil was estimated to be 1.0 W/m/K.

The computer simulation was set up as two adjacent semi-infinite regions. One region was subjected to a surface temperature of +4°C while the other surface had a temperature of −5°C. The soil temperatures were computed for

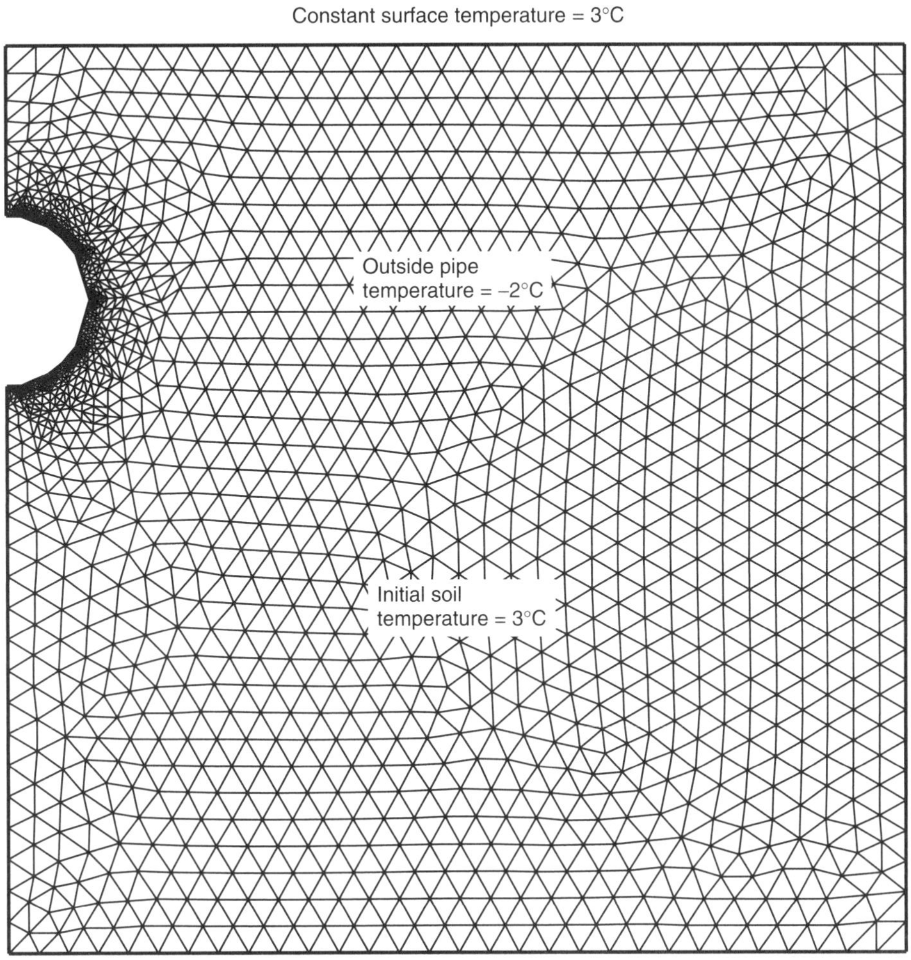

Figure 10.28 Geometry and boundary conditions for chilled pipe placed in unfrozen soil along with generated finite element mesh.

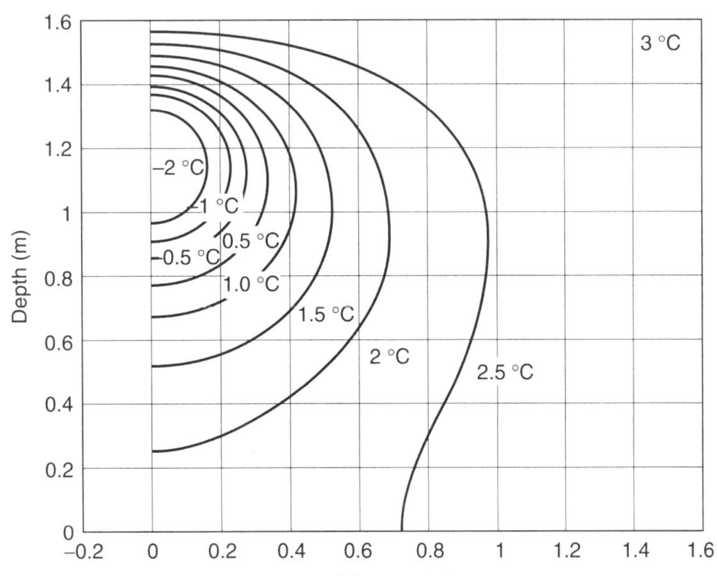

Figure 10.29 Temperature contours, °C, for heat transfer around pipe after 7 days.

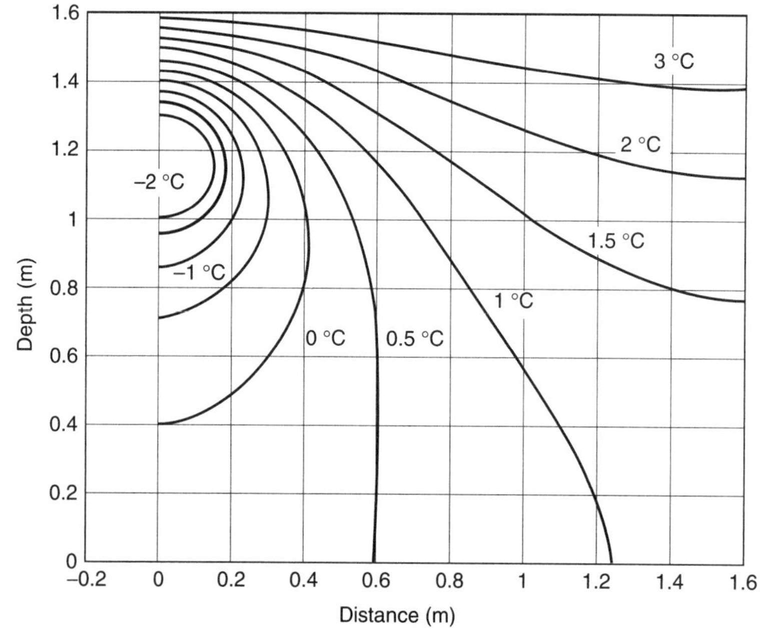

Figure 10.30 Temperature contours, °C, around pipe after 730 days.

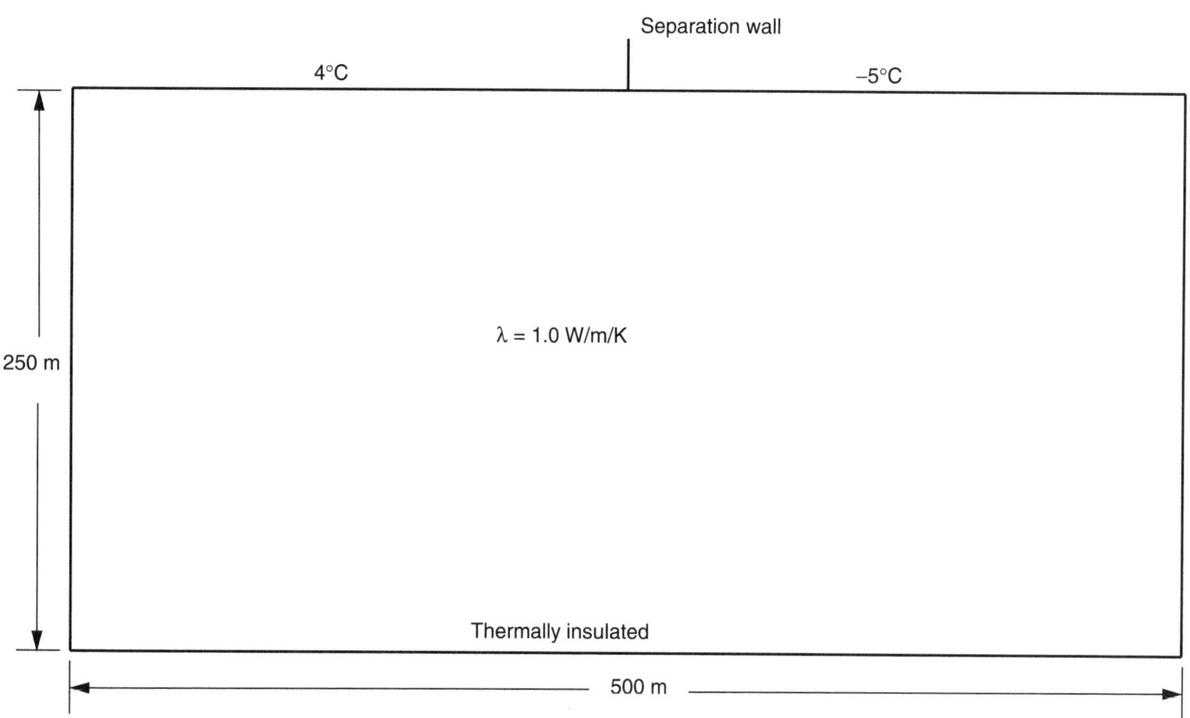

Figure 10.31 Partition separating above and below freezing conditions at ground surface, Harlan and Nixon (1978) heat flow problem.

steady-state conditions under the applied boundary conditions. The computer-generated finite element mesh is shown in Fig. 10.32.

The analytical solution and the finite element simulation are in close agreement (Fig. 10.33). Both solutions showed the same location for the freezing front as well as the temperature contours in both the frozen and thawed soil.

10.10.2 Placement of Warm Building on Soil

Let us consider a variation of the above example problem where the effect of a building maintained at a temperature

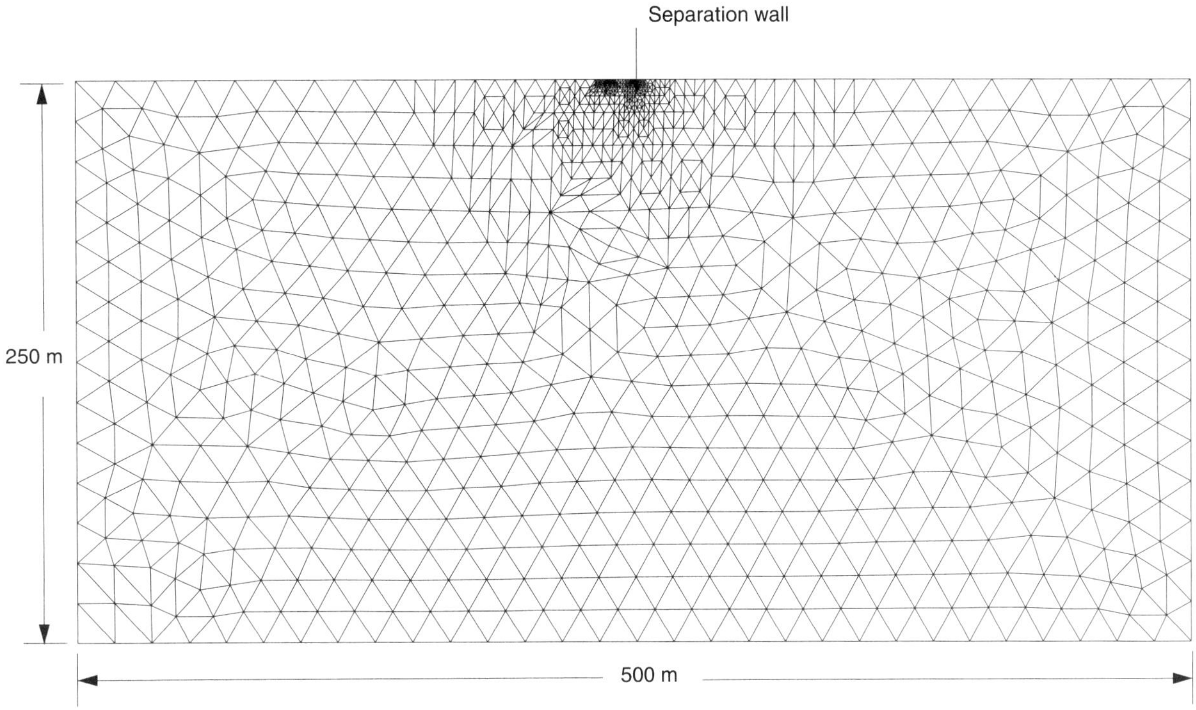

Figure 10.32 Computer-generated finite element mesh generated for solution of Harlan and Nixon (1978) heat flow problem.

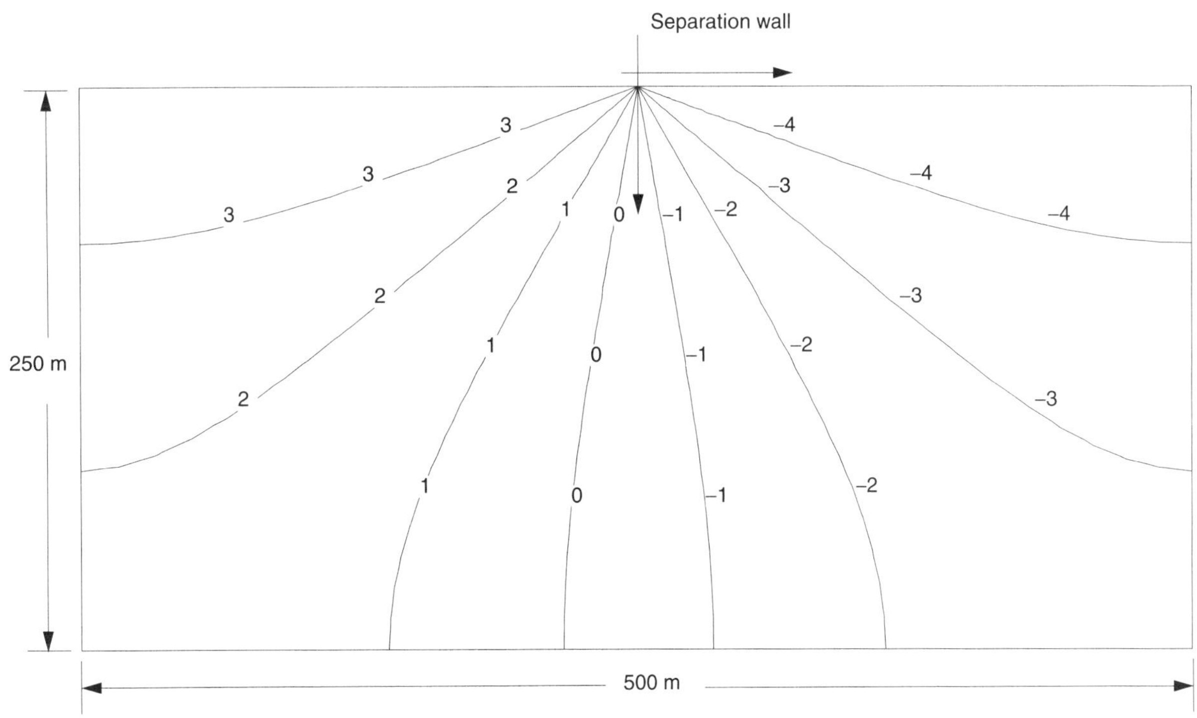

Figure 10.33 Temperature contours computed using SVHeat software.

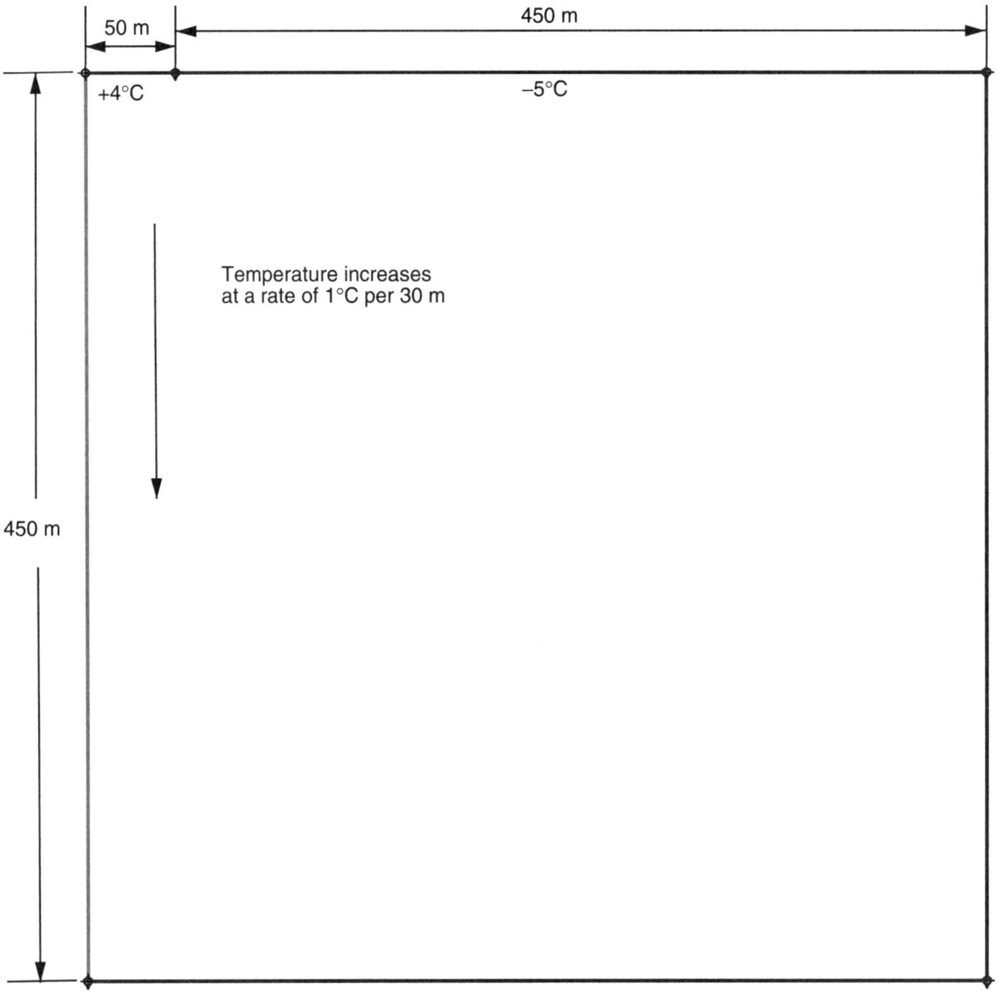

Figure 10.34 Geometry and boundary conditions for warm building placed over frozen soil problem.

above freezing is placed on frozen soil. Once again the building is modeled as a heated strip at a temperature of +4°C. Only steady-state conditions are considered in the analysis. The ground surface temperature on either side of the building is maintained at −5°C, as shown in Fig. 10.34. However, it is assumed that the temperature below ground surface increases at a rate of 1°C per 30 m. The thermal conductivity of the soil is 1.0 W/m/K. Only half of the geometry is modeled due to symmetry.

The finite element mesh should be refined in the vicinity of the edge of the building to ensure accuracy of the solution (Fig. 10.35). The refinement of the computer-generated mesh can be controlled by specifying the desired tolerance for the final solution. Figure 10.36 shows contours of the computed constant temperatures below and around the heated building.

10.11 ALDRICH (1956) EXAMPLE OF VERTICAL COLUMN

Aldrich (1956) and Li and Koike (2001) published a closed-form analytical heat flow for the calculation of frost depth. Only conductive heat flow was modeled. The geometry consisted of a one-dimensional vertical column 5 m long, as shown in Fig. 10.37. The soil is assumed to be homogeneous.

The freezing temperature applied at ground surface was −5°C and later the thawing temperature was set at +5°C. The intermittent temperatures were applied in accordance with the graph shown in Fig. 10.38. The graph shows that the freezing temperatures were applied for 120 days. It was assumed that 10 days were required to change to freezing conditions and vice versa. The bottom of the soil column has a unit gradient imposed and its initial temperature was 5°C.

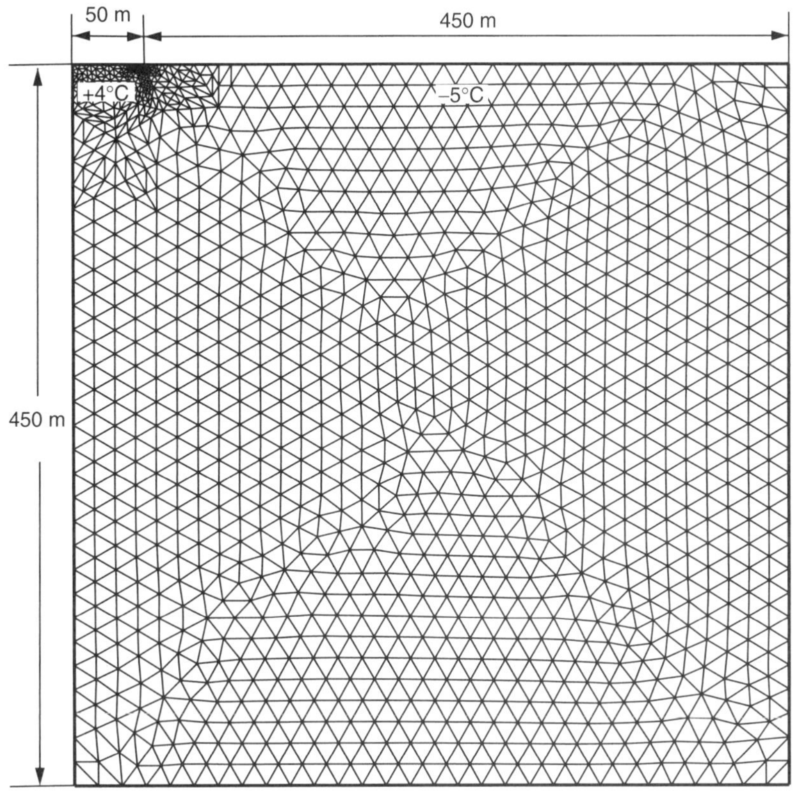

Figure 10.35 Computer-generated finite element mesh for warm building placed over frozen soil problem.

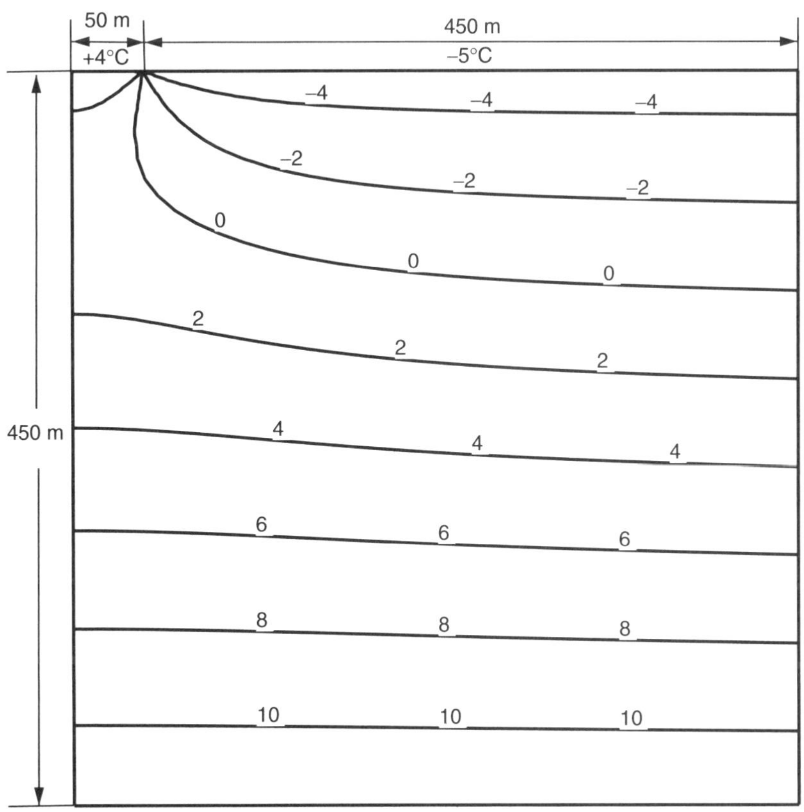

Figure 10.36 Contours of constant temperature below and around heated building.

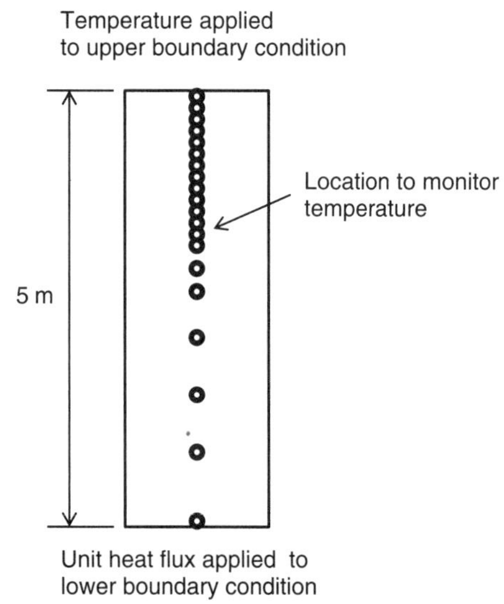

Figure 10.37 One-dimensional column with nodes used in solving Aldrich (1956) problem.

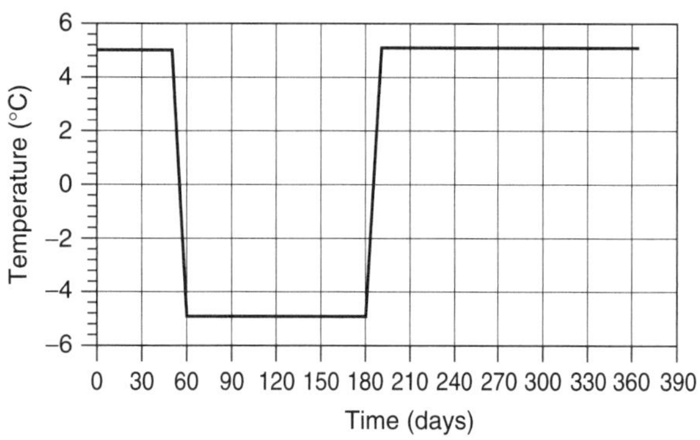

Figure 10.38 Temperature conditions applied on upper boundary.

The material was assumed to be silica flour with material properties equal to those defined by Jame (1977). The thermal conductivity was calculated based on the Johansen method (Johansen, 1975) and a value of 2.873 W/m/K was obtained. The volumetric heat capacity was calculated based on the specific heat of the solid component, water, ice, soil dry density, and respective volumetric fractions (Jame, 1977; Newman, 1995). The thermal conductivity, volumetric heat capacity, and dry density of the soil are required as input when solving this problem. The volumetric heat capacity of the solid component was 837 J/kg/K. The latent heat of fusion was taken as 3.34×10^8 J/m^3. The volumetric water content was 47% and the dry density was 1330 kg/m^3.

The SFCC was estimated using the Fredlund and Xing (1994) SWCC equation. It was assumed that the phase change from water to ice commenced at −0.1°C and that the phase change was complete when the temperature reached −0.607°C. The Fredlund and Xing (1994) equation was fitted to laboratory data of unfrozen water content as a function of soil suction.

The computer results are presented in terms of temperature contours and are presented in Fig. 10.39. The frost depth (i.e., zero degree isotherm) is shown in Fig. 10.40. The closed-form analytical solution could only be applied for the freezing period; however, the numerical solution was continued throughout the subsequent thawing period.

10.11.1 Solving Complex Coupled Heat Flow Problems

There is a wide variety of example problems involving heat flow that could be solved for illustration purposes; however, only the above few examples have been presented. It is also possible to couple heat flow with the flow of air and/or water; however, these topics are outside the scope of this book. Other phase flows require the coupling (or simultaneous solution) of more than one partial differential equation.

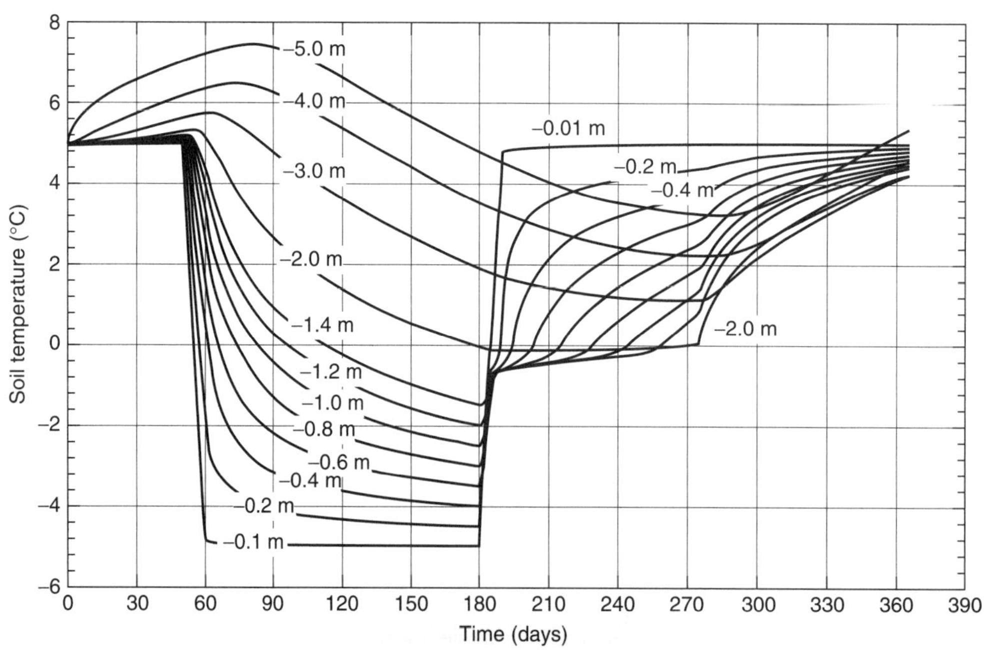

Figure 10.39 Soil temperature contours that develop with elapsed time.

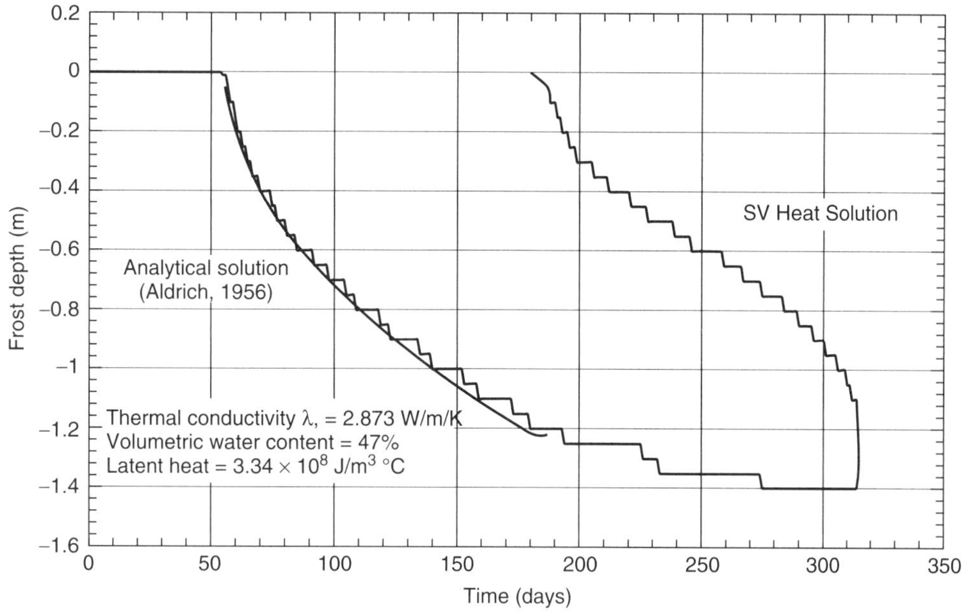

Figure 10.40 Depth of freezing during soil freezing and subsequent thawing.

CHAPTER 11

Shear Strength of Unsaturated Soils

11.1 INTRODUCTION

The safety of many engineered structures is dependent upon the strength of the underlying soil. Bearing capacity, lateral earth pressures, and slope stability are examples of common geotechnical applications that depend on the shear strength of the soil. The soils used for the construction of engineered structures are often unsaturated soils. It is important to quantify the shear strength of the unsaturated soil and to also be able to quantify changes in shear strength that might occur as a result of water infiltration into the soil.

The general principles associated with the shear strength of a saturated soil also apply to unsaturated soils. For example, the strength of a soil changes when the stress state of a soil is changed. Also, concepts of dilation and changes in strength with strain are similar for both saturated and unsaturated soils. Shear strength can be related to the stress state of the soil. In the case of an unsaturated soil the stress state can be described in terms of the net normal stresses $(\sigma - u_a)$ and the matric suction $(u_a - u_w)$.

The term *matric suction* is generally used when describing the shear strength of an unsaturated soil; however, the term *soil suction* could also be applied since the shear strength concepts are of importance over the entire range of suction values. The influence of soil suction on shear strength in the high-suction range may be significantly reduced; however, general concepts pertaining to the influence of soil suction are important to understand.

The shear strength of an unsaturated soil is presented both in an analytical and a graphical manner. The presentations of laboratory test procedures assist in visualizing the changes that occur when going from unsaturated to saturated conditions and vice versa. The possibility of nonlinearity and hysteresis in the description of shear strength is also taken into consideration.

Each soil sampled from in situ conditions has an inherent stress history. The stress history has an influence on the soil fabric or soil structure. The geotechnical engineer must discern when soils can be assumed to be the "same soil" and when they must be considered as "different" soils. Soils with the same classification properties and similar stress histories can generally be considered to be the same soil and their shear strength properties can be quantified by testing several soil specimens.

Several soil specimens that qualify as the same soil (i.e., according to their classification properties, volume-mass properties, and stress history) are required for the measurement of shear strength parameters. The specimens should be prepared at the same initial water content and with the same compaction effort if the strength parameters for a compacted soil are being measured. The compacted soil specimens can then be allowed to equalize under a range of applied stress conditions and then tested for shear strength. Conversely, soils compacted at different water contents result in different densities and are generally considered as different soils (Fig. 11.1). Soils compacted to similar densities at water contents above and below optimum water content conditions can have quite different soil fabrics. It is anticipated that the compacted soils from below and above optimum water content would likely have different "peak" and possibly "critical state" shear strength parameters. Soils compacted above and below optimum water content conditions should be viewed as different soils.

Two apparatuses commonly used to measure shear strength are triaxial equipment and direct shear equipment. The theory associated with various types of triaxial tests and direct shear tests on unsaturated soil is presented along with the methodology for interpreting the test results. The laboratory test should reasonably represent the in situ and loading conditions that are likely to occur in the field. Various stress paths can be simulated when using triaxial test equipment.

11.2 THEORY OF SHEAR STRENGTH

Theories of shear strength for an unsaturated soil have been proposed as extensions to the concepts and mathematical equations that have been used for shear strength theories for saturated soils. The shear strength section commences with

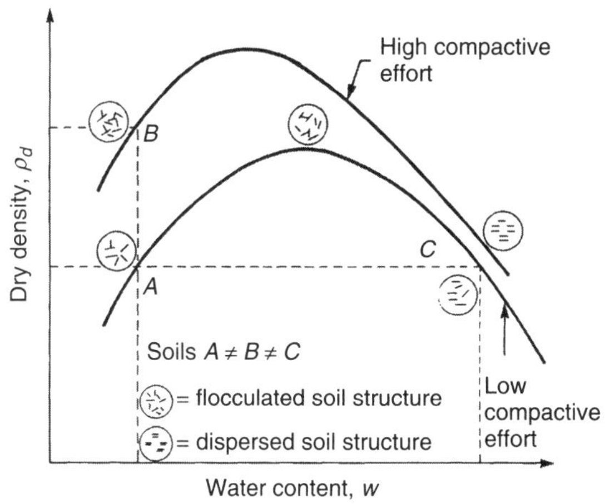

Figure 11.1 Particle structure or fabric of clay specimens compacted at various dry densities and water contents (after Lambe, 1958).

a brief review of the Mohr-Coulomb failure criteria (Mohr, 1914) for a saturated soil. Other failure criteria such as the Tresca (1868) criterion and the von Mises (1913) criterion could also be extended to consider unsaturated soil behavior; however, these extensions have not been included in this book. Rather, the presentation will be limited to extensions to the Mohr-Coulomb failure conditions for unsaturated soils. The extensions to the Mohr-Coulomb criterion are referred to as the "extended Mohr-Coulomb" failure criterion for unsaturated soil.

11.2.1 Shear Strength of Saturated Soil

The shear strength of a *saturated* soil can be described using the Mohr-Coulomb failure criterion and the effective stress variable (Terzaghi, 1936):

$$\tau_{ff} = c' + \left(\sigma_f - u_w\right)_f \tan \phi' \tag{11.1}$$

where:

τ_{ff} = shear stress on the failure plane at failure,
c' = effective cohesion, which is the shear strength intercept when effective normal stress is equal to zero,
$(\sigma_f - u_w)_f$ = effective normal stress on the failure plane at failure,
σ_{ff} = total normal stress on the failure plane at failure,
u_{wf} = pore-water pressure at failure, and
ϕ' = effective angle of internal friction.

Equation 11.1 defines a linear relationship between shear strength and effective stresses as illustrated in Fig. 11.2. The line tangent to the Mohr circles is commonly referred to as a failure envelope since it represents possible combinations of shear stress and effective normal stress on the failure plane at failure. The shear and normal stresses are given the subscript f. The f subscript within the parentheses refers to the failure plane, and the f subscript outside of the parentheses indicates the failure stress condition. One subscript f is given to the pore-water pressure when referring to failure conditions. The pore-water pressure is isotropic and therefore acts equally on all planes. The shear stress described by the failure envelope indicates the shear strength of the soil for each effective normal stress.

The failure envelope for a saturated soil is obtained by plotting a line tangent to a series of Mohr circles representing the stress state at failure. The slope of the line defines the effective angle of internal friction, ϕ', and its intercept on the ordinate is called the effective cohesion, c'. The direction of the failure plane in the soil is obtained by joining the pole point to the point of tangency between the Mohr circle and the failure envelope. The tangent point on the Mohr circle at failure represents the stress state on the failure plane at failure.

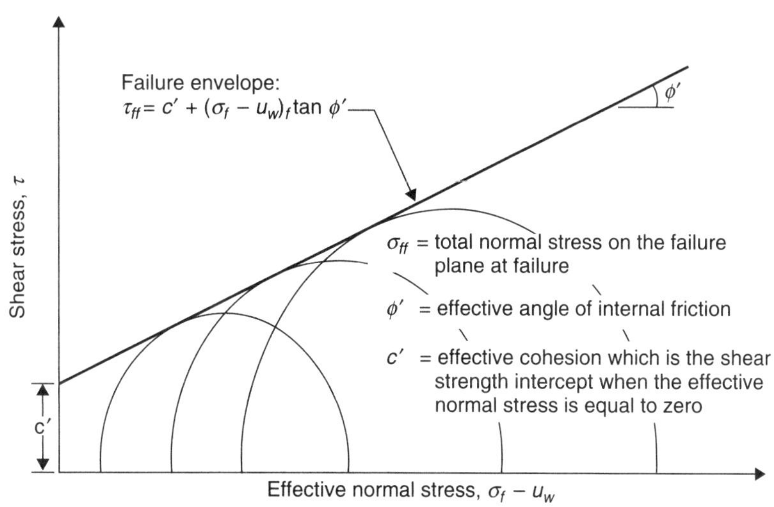

Figure 11.2 Mohr-Coulomb failure envelope for saturated soil.

The use of effective stresses with the Mohr-Coulomb failure criterion has proven to be satisfactory when dealing with saturated soils in engineering practice. Two independent stress state variables are required to define the stress state and consequently the shear strength of an unsaturated soil (Fredlund et al., 1977).

Numerous shear strength tests and other related studies on unsaturated soils have been conducted during the past 30 years. This section presents a brief review of studies related to the shear strength of unsaturated soils. The shear strength testing of an unsaturated soil can be viewed as occurring in two stages. The testing procedures are similar to those used when testing saturated soils. The first stage takes place prior to shearing when the soils can either be consolidated to a specific set of stresses or left unconsolidated. The second stage pertains to the control of drainage as the soil specimen is sheared. The pore-air and pore-water phases can be independently maintained in an undrained or drained mode during shear.

The pore fluid is allowed to completely drain from the specimen under drained conditions. The desire is to have essentially no excess pore pressures in the soil during the shear process. In other words, the pore pressure is externally *controlled* at a constant value during shear. No drainage of pore fluid is allowed under undrained conditions. Changes in pore pressures during shear may or may not be measured. It is important, however, to measure or control the pore-air and pore-water pressures when it is necessary to know the net normal stress and the matric suction at failure. The stress state variables at failure must be known in order to assess the shear strength of the soil in a fundamental manner.

Many shear strength tests on unsaturated soils have been performed without either controlling or measuring the pore-air and pore-water pressures during shear. In some cases, the matric suction of the soil has been measured at the beginning of the test. These results serve only as an indicator of the present soil shear strength since the actual stress state at failure is unknown.

A high-air-entry disk with an appropriate air-entry value should be used when measuring or controlling pore-water pressures in an unsaturated soil. The absence of a high-air-entry disk limits the ability to measure the difference between the pore-air and pore-water pressure. The interpretation of the results from shear strength tests on unsaturated soils becomes ambiguous when the stress state variables at failure are not known. On the other hand, shear strength tests where the pore-air and pore-water pressures have been controlled or measured have provided the greatest insight into unsaturated soil behavior.

11.2.2 Concept of Strain

The concept of "strain" is used in presenting triaxial test results in the form of stress versus strain curves. Normal or longitudinal strain is defined as the ratio of the change in length to the original length. When a soil specimen is subjected to an axial normal stress, the normal strain in the axial direction is defined as follows (Fig. 11.3):

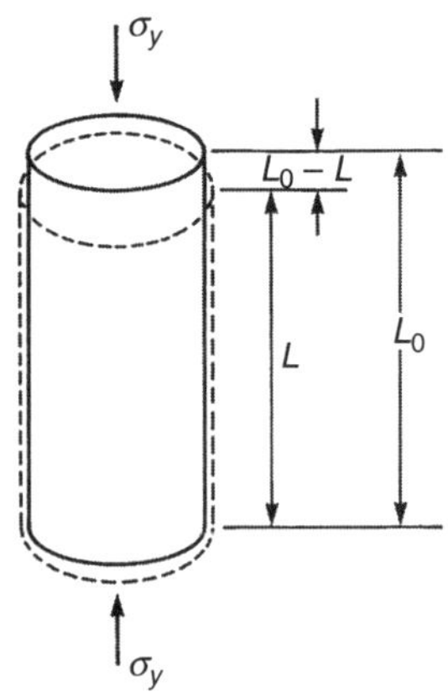

Figure 11.3 Strain concept applied to triaxial test specimen.

$$\varepsilon_y = \frac{L_0 - L}{L_0} \times 100 \tag{11.2}$$

where:

ε_y = axial normal strain in the y-direction expressed as a percentage,
L_0 = original length of the soil specimen, and
L = any subsequent or final length of the soil specimen.

The measurement of lateral displacement on a triaxial soil specimen allows the normal or longitudinal strains in the x- and z-directions to also be computed, that is, ε_x and ε_z, respectively. It is also possible to perform simple shear types of tests where the rotation of the specimen allows the direct measurement of shear strain.

11.2.3 Failure Criteria and Failure Envelope

The shear strength envelope is a measure of the ability of the soil to withstand applied shear stresses. Consequently, the shear strength envelope is a description of failure conditions. The soil will fail when the applied shear stresses equal or exceed the shear strength of the soil. The above-mentioned definition of shear strength or failure conditions may seem straightforward, but selecting the point of failure can prove to be a challenge.

Figure 11.4 shows typical behavioral plots for an initially loose and an initially dense soil. The initially dense soil shows a stress versus strain plot with a peak shear stress that then reduces to a critical state value. The other plots show the corresponding volume changes (i.e., void ratio changes) that occur as deviator stress is applied to the specimen. The question can be asked, "When did the dense soil fail?" Did failure occur at "peak" deviator stress conditions or under strains corresponding to the critical state? The geotechnical community has wrestled over the definitions of what

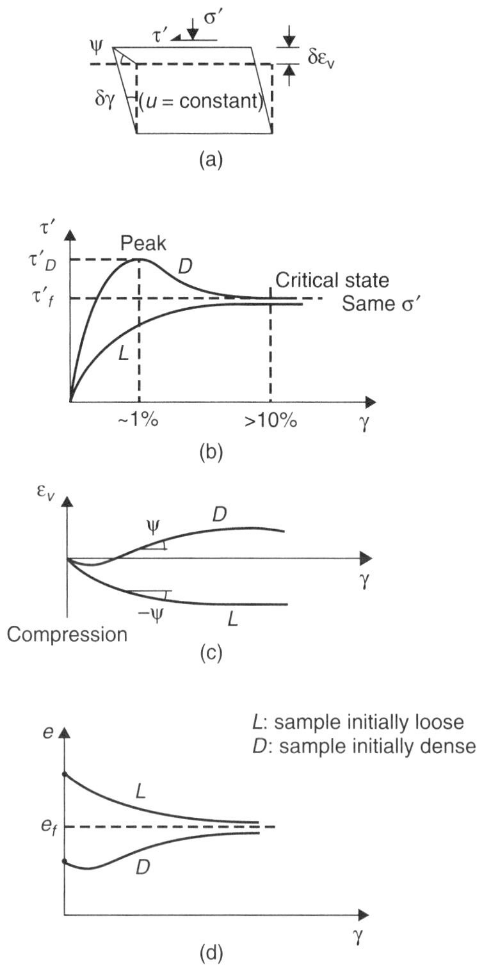

Figure 11.4 Typical behavior of loose and dense soils under drained shear conditions: (a) soil specimen subjected to simple shear; (b) peak and critical shear strength; (c) volume change upon strain; (d) void ratio change during strain.

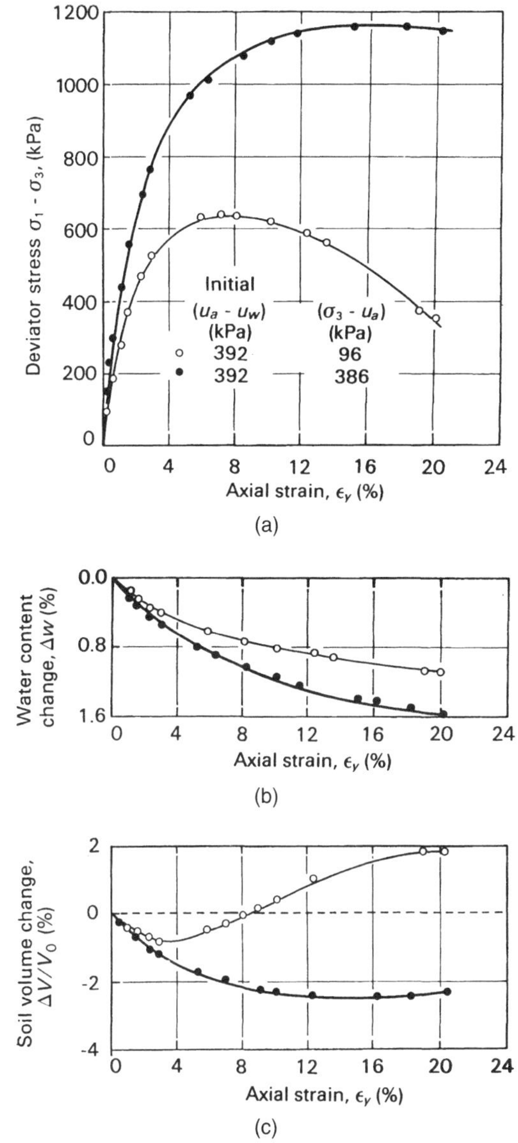

Figure 11.5 Consolidated drained triaxial test results on Dhanauri clay: (a) stress versus strain curves; (b) water content change versus strain curves; (c) soil volume change versus strain curves (after Satija, 1978).

constitutes the definition of failure. The authors of this book attempt to embrace the concept of critical state shear strength as well as the concept of peak strengths in describing unsaturated soil shear strength. There are several criteria that have been proposed for defining soil failure in unsaturated soils just as there are for saturated soils.

A triaxial shear strength test is performed by loading a soil specimen with increasing applied loads until a condition of failure is reached. There are several ways to perform the triaxial test, and likewise, there are several ways to define the state of failure. Consider a consolidated drained triaxial compression test where the pore pressures in the soil specimen are allowed to drain (i.e., remain constant). The soil specimen is subjected to a constant matric suction and is surrounded by a constant net confining pressure (i.e., the net minor normal stress) $(\sigma_3 - u_a)$. The specimen is failed by increasing the net axial pressure (i.e., the net major normal stress) $(\sigma_1 - u_a)$. The difference between the major and minor normal stresses is commonly referred to as the deviator stress $(\sigma_1 - \sigma_3)$ and is a measure of the shear stress developed in the soil specimen. As the soil is compressed, the deviator stress increases gradually until a maximum value is obtained. The applied deviator stress can be plotted versus the axial strain ε_y and the plot is referred to as a "stress versus strain" curve. Figure 11.5 shows two stress versus strain curves for Dhanauri clay performed as consolidated drained triaxial tests at two different net confining pressures. The maximum deviator stress, $(\sigma_1 - \sigma_3)_{max}$, is an indicator of the shear strength of the soil and can be used as a failure criterion. The net principal stresses corresponding to failure conditions are called the net major and net minor normal stresses at failure [i.e., $(\sigma_1 - u_a)_f$ and $(\sigma_3 - u_a)_f$, respectively].

The principal stress ratio is defined as $(\sigma_1 - \sigma_3)_f / (\sigma_3 - u_w)_f$ (Bishop et al. 1960) and can also be used as a failure criterion when pore-water pressures are measured during the test. A plot of the principal stress ratio versus the axial strain for an undrained triaxial test on compacted shale is shown along with the corresponding stress versus strain curve in Fig. 11.6a. The maximum deviator stress, $(\sigma_1 - \sigma_3)_{max}$, and the maximum principal stress ratio, $(\sigma_1 - \sigma_3)_{max} / (\sigma_3 - u_w)_f$, may not occur at the same axial strain in the undrained test, as illustrated in Fig. 11.6a. The maximum principal stress ratio is a function of the pore-water pressure measured during an undrained test (Fig. 11.6b). On the other hand, the maximum deviator stress is not a direct function of the pore pressures. For the results presented in Fig. 11.6a, the authors selected the maximum principal stress ratio as the failure criterion since it occurred prior to the maximum deviator stress. It is not presently clear whether pore-air pressure or pore-water pressure should be used in calculating the principal stress ratio. It is also possible that other definitions of principal stress ratios could be used to define the state of failure under undrained conditions.

The above-mentioned failure criteria describe some maximum stress combinations that the soil can resist. Researchers have also suggested that the amount of strain in the soil specimen should play a role in assessing the state of failure. Consideration of amount of strain at failure is somewhat related to whether peak or critical state stresses should be used when selecting failure conditions for engineering design purposes.

A plot of stress versus strain does not always exhibit a distinct maximum value even at quite large strains, as shown in Fig. 11.7. In this case, an arbitrary strain (e.g., 12%) may be selected to represent the failure criterion. The shape of triaxial test specimens can become quite distorted at high strain values, and this can jeopardize the quality of the test data. A limiting displacement definition for failure is sometimes used when interpreting direct shear test results.

The above-mentioned failure criteria have been proposed for the analysis of shear strength data for unsaturated soils, often with limited corroborating evidence. In general, the different failure criteria produce quite similar shear strength parameters. Probably the largest difference in shear strength parameters occurs as a result of the difference between peak conditions and critical state conditions.

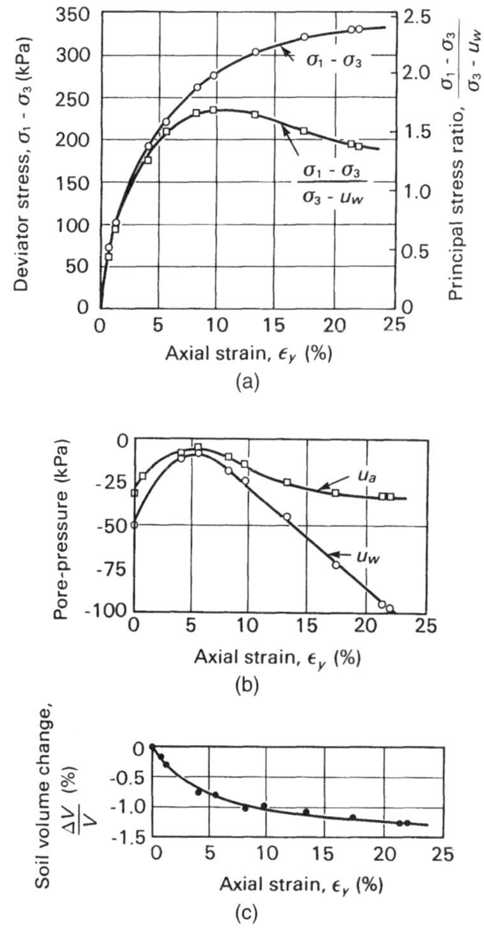

Figure 11.6 Undrained triaxial tests on compacted shale: (a) stress versus strain curves; (b) pore pressure versus strain curves; (c) soil volume change versus strain curves (after Bishop et al., 1960).

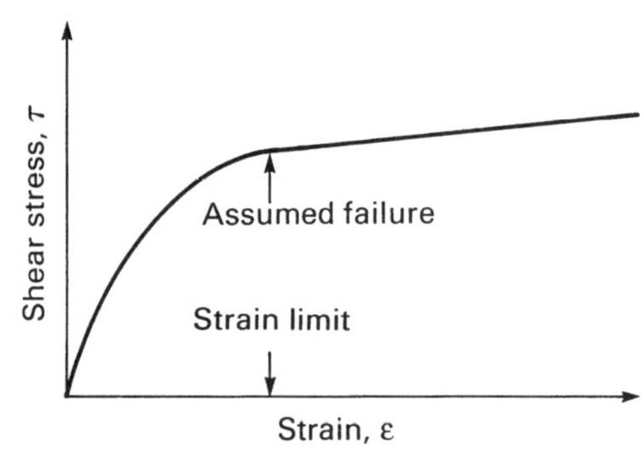

Figure 11.7 Defining failure criterion in terms of limiting strain value.

11.2.4 Shear Strength Equations for Unsaturated Soils

The shear strength constitutive relationship provides a mathematical equation relating the normal and shear components of the stress tensors. Any one of several shear strength failure criteria could be used to extend the saturated soil shear strength representation to unsaturated soil conditions. The Mohr-Coulomb failure criterion was extended to embrace unsaturated soils by Fredlund et al., (1978). The extended shear strength equation for an unsaturated soil can be written as follows:

$$\tau' = c' + (\sigma_n - u_a) \tan \phi' + (u_a - u_w) f_1 \qquad (11.3)$$

where:

τ = shear strength,
c' = effective cohesion intercept,
σ_n = total normal stress on the failure plane at failure,
ϕ' = effective angle of internal friction, and
f_1 = soil property function defining the relationship between shear strength and soil suction, the derivative of which [i.e., $df_1/d(u_a - u_w)$] gives the instantaneous rate of change in shear strength with respect to soil suction (Fredlund, 2005).

Figure 11.8 shows Eq. 11.3 plotted as a three-dimensional constitutive surface with matric suction perpendicular to the conventional two-dimensional Mohr-Coulomb plot. The soil properties c' and ϕ' are presented as saturated soil constants, but the soil property f_1 is shown as a function of soil suction (i.e., or the amount of water filling the voids of the soil). The form of Eq. 11.3 allows the shear strength envelope with respect to matric suction to be either linear or curved, depending upon the experimental test results. Shear strength envelopes that are measured over a wide range of matric suction values tend to show a curved shape.

Under low-suction conditions (i.e., less than the air-entry value of the soil), the derivative of f_1 tends to a value equal to the tangent of the effective angle of internal friction of the saturated soil (i.e., $\tan \phi'$). The shear strength envelope starts to become curved as the air-entry value for the soil is exceeded. There generally appears to be a gradual curvature in the shear strength envelope up to residual suction conditions. At high suction conditions (i.e., greater than residual soil suction), the derivative of f_1 has been shown to tend toward zero for several soils with varying silt and clay contents (Nishimura and Fredlund, 2001). Sandy soils have shown that the slope may even become negative at suctions greater than the residual value (Donald, 1956; Gan and Fredlund, 1996).

A linear form of the general shear strength equation (i.e., Eq. 11.3) was published by Fredlund et al. (1978). Some

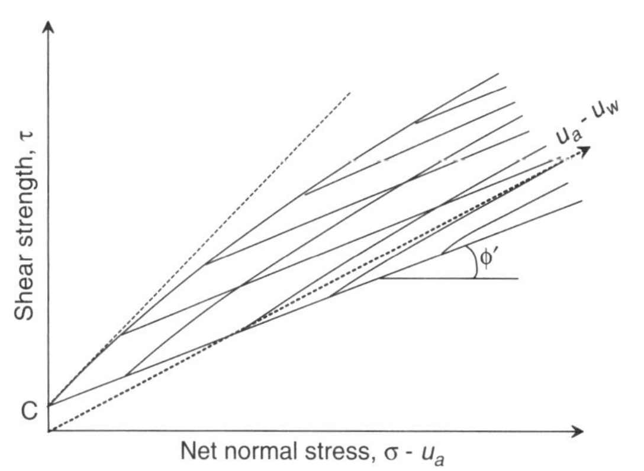

Figure 11.8 Extended Mohr-Coulomb failure surface depicted in terms of independent stress state variables (from Fredlund et al., 1978).

of the earlier unsaturated soil shear strength data sets (e.g., Bishop et al., 1960) show a close fit to the linear shear strength equation (Fredlund and Rahardjo, 1993a). The linear form (i.e., $f_1 = \tan \phi^b$) is convenient to use for illustrating the effect of matric suction on shear strength.

11.2.5 Shear Strength Equations at Critical State

Several critical state shear strength equations have also been proposed for unsaturated soils (Alonso et al., 1990; Wheeler and Sivakumar, 1995; Toll, 1990; Maatouk et al., 1995). The proposed equations attempt to describe the shear strength of an unsaturated soil under critical state conditions in terms of q-p-r space. In general, the results show an increase in shear strength as matric suction is increased, but the character of the results varies among researchers. The stress state variables used to present critical state shear strength are as follows:

$$q = \sigma_1 - \sigma_3; \quad \text{deviator stress} \tag{11.4}$$

$$p = \frac{\sigma_1 + \sigma_2 + \sigma_3}{3}; \quad \text{mean total stress} \tag{11.5}$$

$$r = u_a - u_w; \quad \text{matric suction} \tag{11.6}$$

where:

σ_1 = major principal stress,
σ_2 = intermediate principal stress,
σ_3 = minor principal stress,
u_a = pore-air pressure, and
u_w = pore-water pressure.

A general equation form for incorporating matric suction into the critical state shear strength equation can be written as follows:

$$q = Mf_2(p - u_a, u_a - u_w) \tag{11.7}$$

where:

M = material characteristic independent of stresses and
f_2 = an independent function of $p - u_a$ and $u_a - u_w$.

The critical state shear strength model proposed by Alonso et al. (1990) has the following form:

$$q = M(p - u_a) + \kappa(u_a - u_w) \tag{11.8}$$

where:

κ = soil parameter.

Joomi (2000) combined the net mean stress and matric suction using the degree of saturation and suggested a similar equation form:

$$q = M[(p - u_a) + S(u_a - u_w)] \tag{11.9}$$

where:

S = degree of saturation of the soil at failure.

Wheeler and Sivakumar (1995) suggested a critical shear strength equation where each of the shear strength properties is a function of matric suction:

$$q = M_s(p - u_a) + \mu_s \tag{11.10}$$

where:

M_s = material characteristic that is a function of suction, and

μ_s = material characteristic that is a function of suction.

A summary of critical state shear strength models can be found in Leong et al. (2003a).

11.2.6 Approximate Linear Shear Strength Equation

A linear form of the shear strength for an unsaturated soil can be formulated in terms of independent stress state variables (Fredlund et al., 1978). Any two of three possible stress state variables can be used to write an appropriate shear strength equation. The stress state variables $\sigma - u_a$ and $u_a - u_w$ have been found to be the most practical combination of stress state variables for solving practical engineering problems. The linear form of the shear strength equation can be written as follows when using $\sigma - u_a$ and $u_a - u_w$ as the stress state variables:

$$\tau_{ff} = c' + (\sigma_f - u_a)_f \tan\phi' + (u_a - u_w)_f \tan\phi^b \tag{11.11}$$

where:

c' = intercept of the "extended" Mohr-Coulomb failure envelope on the shear stress axis where the net normal stress and the matric suction at failure are equal to zero, also referred to as "effective cohesion,

$(\sigma_f - u_a)_f$ = net normal stress state on the failure plane at failure,

u_{af} = pore-air pressure on the failure plane at failure,

ϕ' = angle of internal friction associated with the net normal stress state variable, $(\sigma_f - u_a)_f$,

$(u_a - u_w)_f$ = matric suction on the failure plane at failure, and

ϕ^b = angle indicating the rate of increase in shear strength with respect to a change in matric suction, $(u_a - u_w)_f$.

A comparison of Eqs. 11.1 and 11.11 reveals that the shear strength equation for an unsaturated soil takes the form of an extension of the shear strength equation for a saturated soil. Two stress state variables are used to describe the shear strength of an unsaturated soil while only one stress state variable [i.e., effective normal stress $(\sigma_f - u_w)_f$] is required for a saturated soil.

The proposed shear strength equation for an unsaturated soil exhibits a smooth transition to the shear strength equation for a saturated soil. The pore-water pressure u_w approaches the pore-air pressure u_a as the soil approaches saturation and the matric suction $u_a - u_w$ goes to zero. As the matric suction component vanishes, Eq. 11.11 reverts to the equation for a saturated soil.

11.2.7 Extended Mohr-Coulomb Failure Envelope

The failure envelope for a saturated soil is obtained by plotting a series of Mohr circles corresponding to failure conditions on a two-dimensional plot as shown in Fig. 11.2. The line tangent to the Mohr circles is called the failure envelope and is defined by Eq. 11.1. In the case of an unsaturated soil, Mohr circles corresponding to failure conditions can be plotted in a three-dimensional manner, as illustrated in Fig. 11.9. The three-dimensional plot has the shear stress τ as the ordinate and the two stress state variables $\sigma - u_a$ and $u_a - u_w$ as abscissas. The frontal plane represents saturated soil conditions where matric suction is zero. The $\sigma - u_a$ axis reverts to the $\sigma - u_w$ axis on the frontal plane since the pore-air pressure becomes equal to the pore-water pressure at saturation.

The Mohr circles for an unsaturated soil are plotted with respect to the net normal stress axis $\sigma - u_a$ in the same manner as the Mohr circles are plotted for saturated soils with respect to effective stress axis $\sigma - u_w$. The location of the Mohr circle plot in the third dimension is a function of the matric suction (Fig. 11.9). The surface tangent to the Mohr circles at failure is referred to as the extended Mohr-Coulomb failure envelope for unsaturated soils. The extended Mohr-Coulomb failure envelope defines the shear strength of an unsaturated soil. The intersection line between the extended Mohr-Coulomb failure envelope and the frontal plane is the failure envelope for saturated conditions.

The inclination of the failure plane is defined by joining the tangent point on the Mohr circle to the pole point. The tangent point on the Mohr circle at failure represents the stress state on the failure plane at failure.

The extended Mohr-Coulomb failure envelope may be a planar surface or it may be somewhat curved. The theory is first presented with the assumption that the failure envelope is planar and can be described by Eq. 11.11. A curved failure envelope can also be described by Eq. 11.11 for limited changes of the stress state variables. Techniques for handling the nonlinearity of the failure envelope are discussed later.

Figure 11.9 shows a planar failure envelope that intersects the shear stress axis, giving a cohesion intercept c'. The envelope has slope angles of ϕ' and ϕ^b with respect to $\sigma - u_a$ and $u_a - u_w$ axes, respectively. Both angles are

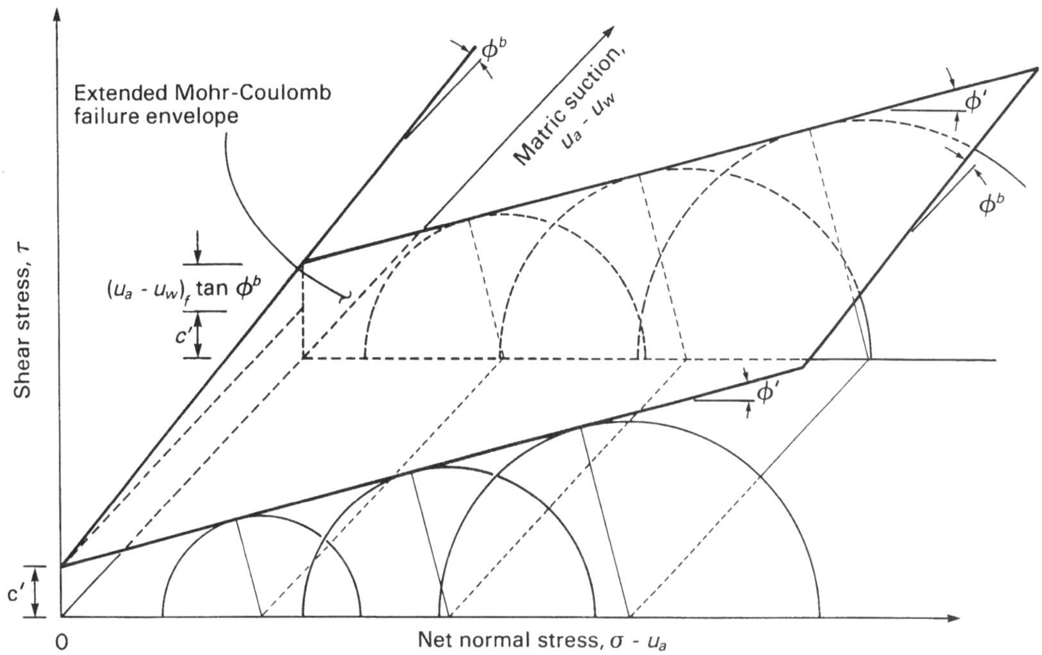

Figure 11.9 Extended Mohr-Coulomb failure envelope for unsaturated soils.

assumed to be constants. The cohesion intercept c' and the slope angles ϕ' and ϕ^b are the strength parameters used to relate shear strength to the stress state variables. The shear strength parameters represent many factors which have influenced the results of the shear strength test. Some of these factors are density, void ratio, degree of saturation, mineral composition, stress history, and strain rate. In other words, these factors have been combined and are expressed mathematically in terms of the shear strength parameters.

The mechanical behavior of an unsaturated soil is affected differently by changes in net normal stress than by changes in matric suction (Jennings and Burland, 1962). The increase in shear strength due to an increase in net normal stress is characterized by the friction angle ϕ'. The increase in shear strength caused by an increase in matric suction is characterized by the angle ϕ^b when assuming linear failure conditions. The value of ϕ^b appears to be consistently equal to or less than ϕ', as indicated in Table 11.1 The soils represented in Table 11.1 are from a variety of geographic locations.

The failure envelope intersects the shear stress versus matric suction plane along a line of intercepts, as illustrated in Fig. 11.10. The line of intercepts represents the increase in strength as matric suction increases. The shear strength increase with respect to an increase in matric suction is defined by the angle ϕ^b. A linear form for the line of intercepts can be written as follows:

$$c = c' + \left(u_a - u_w\right)_f \tan \phi^b \qquad (11.12)$$

where:

c = intercept of the extended Mohr-Coulomb failure envelope with the shear stress axis at a specific matric suction $\left(u_a - u_w\right)_f$. At zero net normal stress, the intercept can be referred to as the "total cohesion intercept."

The extended Mohr-Coulomb failure envelope can be presented as a horizontal projection onto the shear strength τ versus $\sigma - u_a$ plane. The horizontal projection can be shown for various matric suction values $\left(u_a - u_w\right)_f$. The horizontal projection of the failure envelope onto the τ versus $\sigma - u_a$ plane results in the series of contours shown in Fig. 11.11. The lines have cohesion intercepts that depend upon the magnitude of the corresponding matric suctions. The cohesion intercept reverts to the effective cohesion c' when matric suction goes to zero. All lines of equal suction have the same slope angle ϕ' as long as the failure plane is planar. The equation for each of the contour lines can be written as

$$\tau_{ff} = c + \left(\sigma_f - u_a\right)_f \tan \phi' \qquad (11.13)$$

where:

c = total cohesion intercept.

Substituting Eq. 11.12 into Eq. 11.13 yields the equation for the extended Mohr-Coulomb failure envelope. Equation 11.11 is the same as Eq. 11.13 and Fig. 11.11 is a two-dimensional representation of the extended Mohr-Coulomb failure envelope. The failure envelope projections illustrate the increase in shear strength as matric suction is increased at a specific net normal stress. Equation 11.13 is a convenient form of the shear strength equation to use when performing simple analytical studies involving unsaturated soils.

Table 11.1 Experimental Values Measured for ϕ^b

Soil Type	c' (kPa)	ϕ' (deg)	ϕ^b (deg)	Test Procedure	Reference
Compacted shale; $w = 18.6\%$	15.8	24.8	18.1	Constant water content triaxial	Bishop et al. (1960)
Boulder clay; $w = 11.6\%$	9.6	27.3	21.7	Constant water content triaxial	Bishop et al. (1960)
Dhanauri clay; $w = 22.2\%$, $\rho_d = 1580$ kg/m^3	37.3	28.5	16.2	Consolidated drained triaxial	Satija (1978)
Dhanauri clay; $w = 22.2\%$, $\rho_d = 1478$ kg/m^3	20.3	29.0	12.6	Constant drained triaxial	Satija (1978)
Dhanauri clay; $w = 22.2\%$, $\rho_d = 1580$ kg/m^3	15.5	28.5	22.6	Consolidated water content triaxial	Satija (1978)
Dhanauri clay; $w = 22.2\%$, $\rho_d = 1478$ kg/m^3	11.3	29.0	16.5	Constant water content triaxial	Satija (1978)
Madrid grey clay; $w = 29\%$	23.7	22.5[a]	16.1	Consolidated drained direct shear	Escario (1980)
Undisturbed decomposed granite; Hong Kong	28.9	33.4	15.3	Consolidated drained multistage triaxial	Ho and Fredlund (1982a)
Undisturbed decomposed rhyolite; Hong Kong	7.4	35.3	13.8	Consolidated drained multistage triaxial	Ho and Fredlund (1982a)
Tappen-Notch Hill silt; $w = 21.5\%$, $\rho_d = 1590$ kg/m^3	0.0	35.0	16.0	Consolidated drained multistage triaxial	Krahn et al. (1989)
Compacted glacial till; $w = 12.2\%$, $\rho_d = 1810$ kg/m^3	10.0	25.3	7–25.5	Consolidated drained multistage direct shear	Gan et al. (1988)

[a] Average value.

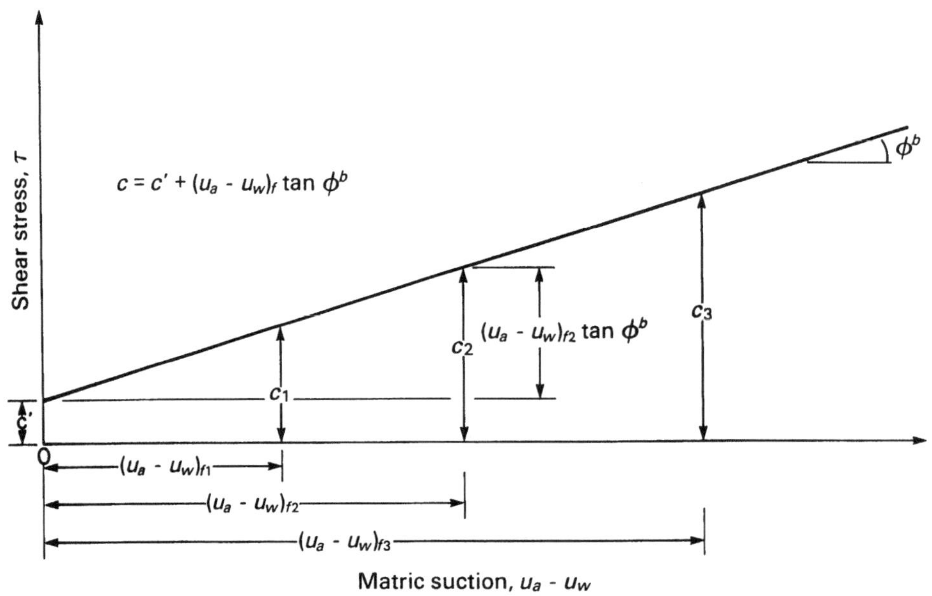

Figure 11.10 Line of intercepts along failure plane on τ versus $u_a - u_w$ plane.

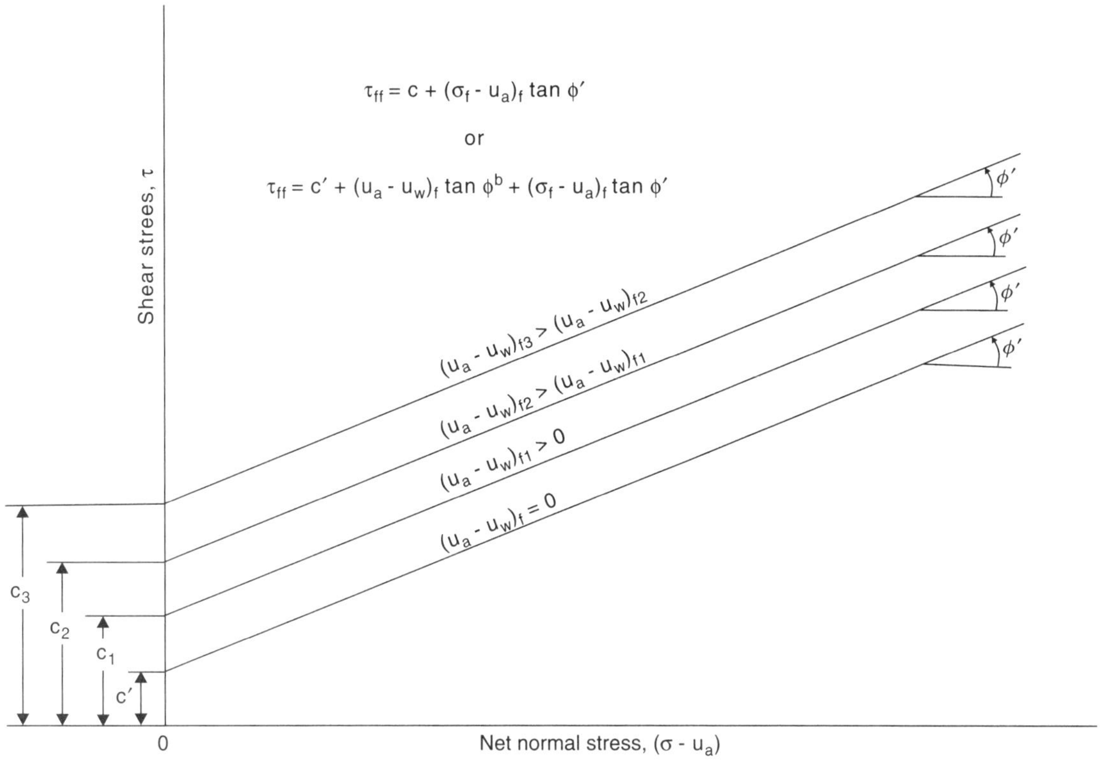

Figure 11.11 Horizontal projection of failure envelope onto τ versus $\sigma - u_a$ plane viewed parallel to matric suction $u_a - u_w$ axis as contour lines of failure envelope on $\sigma - u_a$ plane.

The inclusion of matric suction in the definition of the cohesion intercept does not necessarily suggest that matric suction is a cohesion component of shear strength. Rather, the matric suction component [i.e., $(u_a - u_w)\tan\phi^b$] is lumped with effective cohesion c' for the purpose of translating the three-dimensional failure envelope onto a two-dimensional representative plot. The suction component of shear strength has also been called the apparent or total cohesion (Taylor, 1948).

A smooth transition from an unsaturated condition to saturated conditions can be demonstrated using the extended Mohr-Coulomb failure envelope shown in Fig. 11.9. As the soil becomes saturated, matric suction goes to zero and the pore-water pressure approaches the pore-air pressure. The three-dimensional failure envelope is consequently reduced to the two-dimensional envelope of τ versus $\sigma - u_w$. The smooth transition between unsaturated and saturated soil conditions can also be observed in Fig. 11.11. The failure envelope projection gradually lowers onto the failure envelope for the saturated soil as matric suction decreases. The cohesion intercept c then becomes equal to the effective cohesion c'.

The extended Mohr-Coulomb failure envelope can also be projected horizontally onto the τ versus–$u_a - u_w$ plane (Fig. 11.12). The horizontal projection is made for various net normal stresses at failure, $(\sigma_f - u_a)_f$. The resulting contour lines have an ordinate intercept of $c' + (\sigma_f - u_a)_f$ $\tan\phi'$ and a slope angle of ϕ^b. The horizontal projection shows that there is an increase in shear strength as the net normal stress is increased with matric suction held constant.

11.2.8 Use of $\sigma - u_w$ and $u_a - u_w$ to Define Shear Strength

The shear strength equation has thus far been expressed using $\sigma - u_a$ and $u_a - u_w$ as the stress state variables. However, the shear strength equation for an unsaturated soil can also be expressed in terms of other combinations of stress state variables, such as $\sigma - u_w$ and $u_a - u_w$:

$$\tau_{ff} = c' + (\sigma_f - u_w)_f \tan\phi' + (u_a - u_w)_f \tan\phi'' \quad (11.14)$$

where:

$(\sigma_f - u_w)_f$ = net normal stress state with respect to pore-water pressure on the failure plane at failure and

ϕ'' = friction angle associated with the matric suction stress state variable $(u_a - u_w)_f$ when using the $\sigma - u_w$ and $u_a - u_w$ stress state variables in formulating the shear strength equation.

It should be noted that the stress variables and the soil properties in Eq. 11.14 are changed in such a way that the

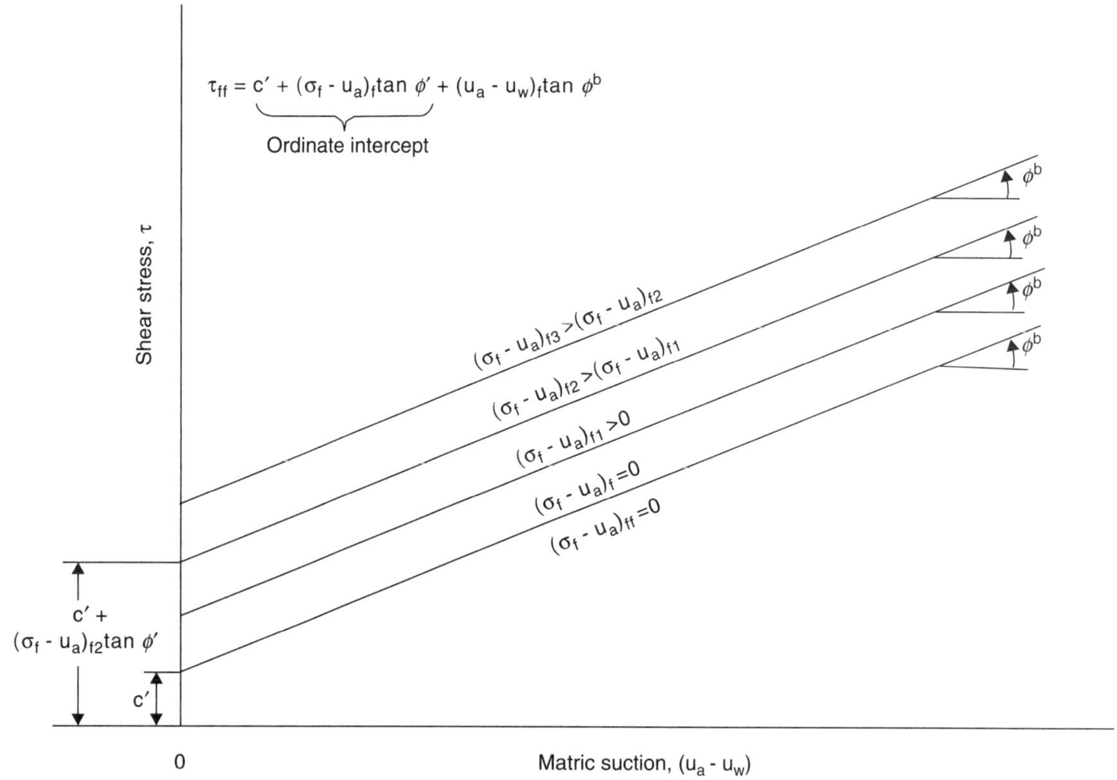

Figure 11.12 Horizontal projection of failure envelope onto τ versus $u_a - u_w$ plane viewed parallel to $\sigma - u_a$ axis as contour lines of failure envelope on τ versus $u_a - u_w$ plane.

computed shear strength for the unsaturated soil remains consistent with the shear strength previously calculated. As suction goes to zero, the third terms in Eqs. 11.14 and 11.11 disappear, and the pore-water pressure approaches the pore-air pressure. As a result, both equations revert to the shear strength equation for a saturated soil. The second term in both equations should have the same friction angle parameter ϕ' [i.e., $(\sigma_f - u_a)_f \tan \phi'$ and $(\sigma_f - u_w)_f \tan \phi'$].

Equations 11.14 and 11.11 give the same shear strength for a soil at a specific stress state; therefore, the two equations can be equated:

$$- u_{af} \tan \phi' + (u_a - u_w)_f \tan \phi^b = -u_{wf} \tan \phi' + (u_a - u_w)_f \tan \phi'' \quad (11.15)$$

Rearranging Eq. 11.15 gives the relationship between the respective angles of friction:

$$\tan \phi'' = \tan \phi^b - \tan \phi' \quad (11.16)$$

Equation 11.16 shows that the friction angle ϕ'' will generally be negative since the magnitude of ϕ^b is less than or equal to ϕ'. Figure 11.13 shows the extended Mohr-Coulomb failure envelope when failure conditions are plotted with respect to $\sigma - u_w$ and $u_a - u_w$ (i.e., Eq. 11.14) and with respect to $\sigma - u_a$ and $u_a - u_w$ (i.e., Eq. 11.11).

11.2.9 Mohr-Coulomb and Stress Point Envelopes

The extended Mohr-Coulomb envelope has been defined as a surface tangent to the Mohr circles at failure. Each Mohr circle is constructed using the net minor and net major principal stresses at failure (i.e., $\sigma_{3f} - u_{af}$ and $\sigma_{1f} - u_{af}$), as shown in Fig. 11.14a. The difference between the net minor and net major principal stresses at failure is called the maximum deviator stress.

The top point of a Mohr circle with coordinates (p_f, q_f, r_f) can be used to represent the stress conditions at failure. A stress point surface (i.e., stress point envelope) can be drawn through the stress points at failure (Fig. 11.14b). The stress point envelope constitutes another representation of the stress state of the soil at failure conditions. However, the stress point envelope and the extended Mohr-Coulomb failure envelope are different surfaces. Nevertheless, the stress point envelope is mathematically related to stress state at failure.

The stress point envelope can be mathematically defined using the following equation:

$$q_f = d' + p_f \tan \psi' + r_f \tan \psi^b \quad (11.17)$$

where:

q_f = half of the deviator stress at failure [i.e., $(\sigma_1 - \sigma_3)_f /2$],

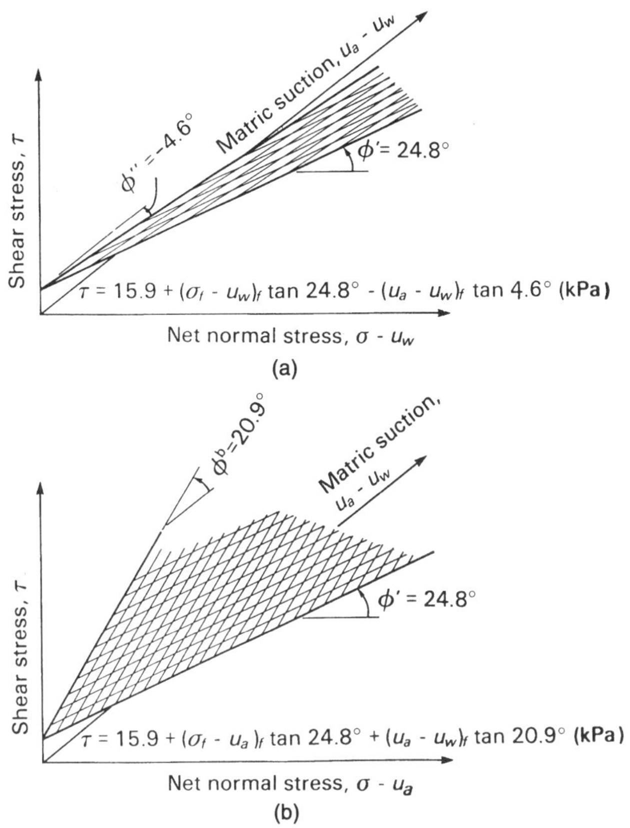

Figure 11.13 Extended Mohr-Coulomb failure envelope plotted with respect to two possible combinations of stress state variables: (a) failure envelope defined in terms of $\sigma - u_w$ and $u_a - u_w$ stress state variables; (b) failure envelope defined in terms of $\sigma - u_a$ and $u_a - u_w$ stress state variables.

σ_{1f} = major principal stress at failure,
σ_{3f} = minor principal stress at failure,
d' = intercept of the stress point envelope on the q axis when p_f and r_f are equal to zero,
p_f = $\left[(\sigma_1 + \sigma_3)/2 - u_a\right]_f$; mean net normal stress at failure,
ψ' = slope angle of the stress point envelope with respect to the stress variable p_f,
r_f = matric suction at failure [i.e., $(u_a - u_w)_f$], and
ψ^b = slope angle of the stress point envelope with respect to the stress variable r_f.

Figure 11.14b presents a planar stress point envelope corresponding to the planar extended Mohr-Coulomb failure envelope shown in Fig. 11.14a. Equation 11.17 defines the stress point envelope. The frontal planes shown in Fig. 11.14 represent the saturated condition where matric suction is zero. As a result, the $(\sigma_1 + \sigma_3)/2 - u_a$ axis reverts to the $(\sigma_1 + \sigma_3)/2 - u_w$ axis on the frontal plane. The intersection line between the stress point envelope and the frontal plane is a line commonly referred to as the K_f line in saturated soil mechanics (Lambe and Whitman, 1979). The K_f line passes through the top points of the Mohr circles for saturated soils at failure. The K_f line has a slope angle ψ' with respect to the p axis and an ordinate intercept d' on the q axis. Any line parallel to the K_f line on the planar stress point envelope will have a slope angle ψ' with respect to the p axis. The stress point envelope reverts to the K_f line as the soil becomes saturated or when matric suction r_f becomes equal to zero.

The intersection line between the stress point envelope and the q-versus-r plane has a slope angle ψ^b with respect to the r axis (Fig. 11.14b). The intersection line indicates that there is an increase in strength as the matric suction at failure, r_f, increases. The equation for the intersection line can be written as follows:

$$d = d' + r_f \tan \psi^b \tag{11.18}$$

where:

d = ordinate intercept of the stress point envelope on the q axis at a r_f and p_f value equal to zero.

The ordinate intercept of the stress point envelope on the q-versus-r plane is equal to d' when r_f is zero. The ordinate intercept is equal to d (i.e., Eq. 11.18) when r_f is not zero. The above variables, d', ψ', and ψ^b, are the required parameters for Eq. 11.18. The stress point envelope can also be represented by contour lines when the surface is projected onto the q-versus-p plane. The equation for the contour lines is obtained by substituting Eq. 11.18 into Eq. 11.17:

$$q_f = d + p_f \tan \psi' \tag{11.19}$$

The stress point envelope can be related to the extended Mohr-Coulomb failure envelope by obtaining the relationships between the parameters used to define both envelopes (i.e., c, ϕ', ϕ^b and d, ψ', ψ^b). Figure 11.15 presents Mohr circles on the τ versus $\sigma - u_a$) plane for a specific matric suction. The extended Mohr-Coulomb failure envelope is drawn tangent to the Mohr circles (e.g., at point A), whereas the stress point envelope passes through the top points of the Mohr circles (e.g., through points B). The extended Mohr-Coulomb failure envelope and the stress point envelope have slope angles of ϕ' and ψ', respectively, with respect to the $\sigma - u_a$ axis. The distance between the tangent point A and the top point B (i.e., $\overline{AB}$) can be computed from triangle ADC as being equal to $q_f \sin \phi'$. The radius q_f decreases and eventually goes to zero as the Mohr circle moves to the left. As a result, the distance between the tangent and the top points (i.e., $q_f \sin \phi'$) also decreases and eventually goes to zero. This means that the extended Mohr-Coulomb failure envelope and the stress point envelope converge to a common point on the $\sigma - u_a$ axis (i.e., point T).

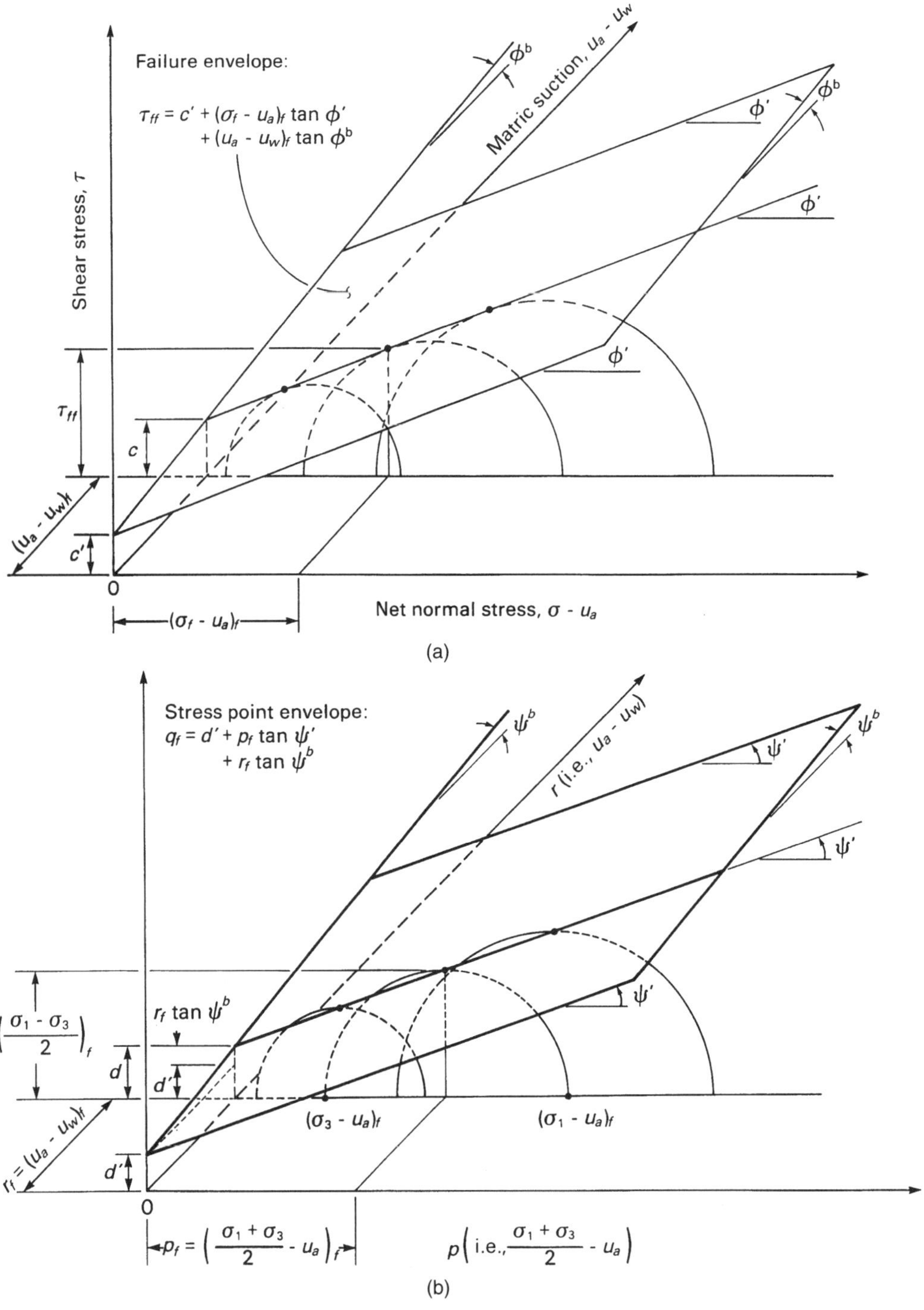

Figure 11.14 Comparisons of failure envelope and corresponding stress point envelope: (a) extended Mohr-Coulomb failure envelope; (b) extended stress point envelope.

The relationship between the slope angles ϕ' and ψ' is obtained by equating the lengths $\overline{TC}$ computed from triangles *TBC* and *TAC* as follows:

$$\frac{q_f}{\tan \psi'} = \frac{q_f}{\sin \phi'} \tag{11.20}$$

The q_f variable can be canceled to give

$$\tan \psi' = \sin \phi' \tag{11.21}$$

The relationship between the cohesion intercept c and the ordinate intercept d can be computed by considering the

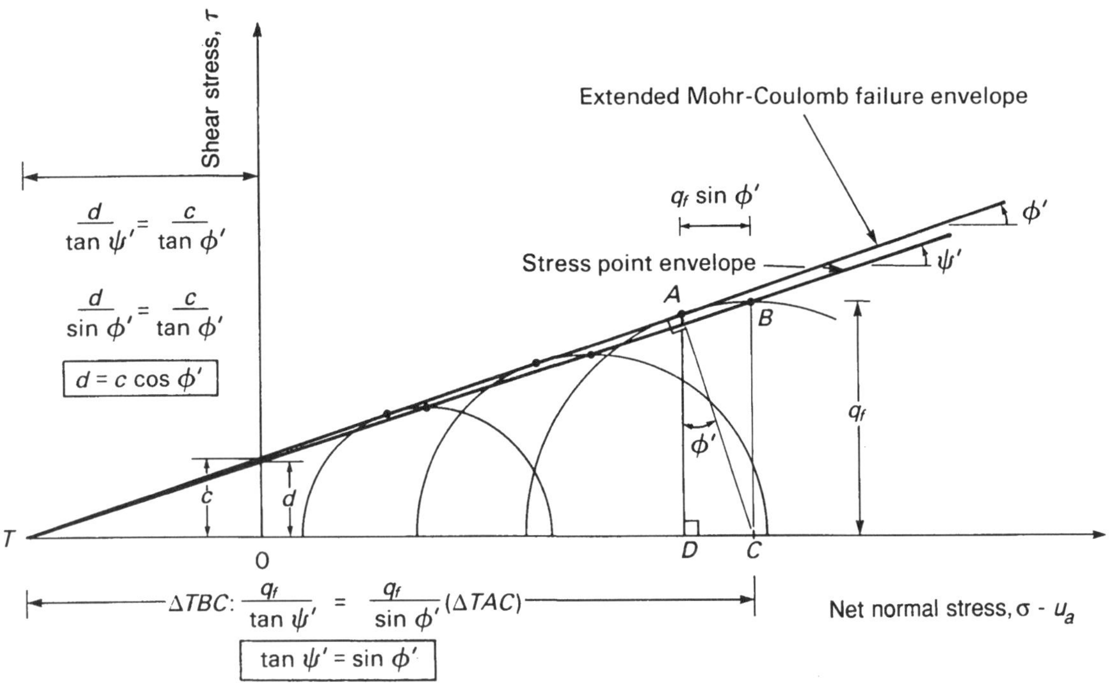

Figure 11.15 Relationships between c, d, ϕ', and ψ''.

distance between points T and O (i.e., $\overline{TO}$):

$$\frac{d}{\tan \psi'} = \frac{c}{\tan \phi'} \tag{11.22}$$

Substituting Eq. 11.21 into Eq. 11.22 and rearranging Eq. 11.22 yields

$$d = c \cos \phi' \tag{11.23}$$

When matric suction at failure is equal to zero (i.e., the saturated condition), Eq. 11.23 becomes

$$d' = c' \cos \phi' \tag{11.24}$$

Figure 11.16 shows the intersection lines of the extended Mohr-Coulomb failure envelope and the stress point envelope on the shear strength versus matric suction plane. The intersection lines associated with the extended Mohr-Coulomb failure envelope and the stress point envelope are defined by Eqs. 11.12 and 11.18, respectively. The ratio between the d and c values is always constant and equal to $\cos \phi'$ (i.e., Eq. 11.23) at various matric suctions. As a result, the difference between the d and c values is not constant for different matric suctions. In other words, the intersection lines are not parallel, or put another way, $\phi^b \neq \psi^b$. Substituting Eq. 11.12 and Eq. 11.18 into Eq. 11.23 gives the following relationship:

$$d' + r_f \tan \psi^b = c' \cos \phi' + \left(u_a - u_w\right)_f \tan \phi^b \cos \phi'. \tag{11.25}$$

The above equation can be rearranged by substituting Eq. 11.24 for d' and substituting $\left(u_a - u_w\right)_f$ for r_f in order to obtain the relationship between, ψ^b, ϕ^b, and ϕ':

$$\tan \psi^b = \tan \phi^b \cos \phi' \tag{11.26}$$

The above relationships can be used to define the stress point envelope corresponding to an extended Mohr-Coulomb failure envelope or vice versa. The extended Mohr-Coulomb failure envelope can be established by testing a soil under saturated and unsaturated conditions. The Mohr-Coulomb failure envelope for the saturated condition gives the angle of internal friction ϕ' and the effective cohesion c'.

The cohesion intercept c can be obtained from a single Mohr circle if a planar failure envelope is assumed at an angle ϕ' and the matric suction is known. Figure 11.17 illustrates the construction of a Mohr circle at failure with its corresponding p_f and q_f values. A failure envelope with a slope angle ϕ' is drawn tangent to the Mohr circle at point A. The envelope intersects the shear strength axis at point B and the $\sigma - u_a$ axis at point T. The cohesion intercept c is computed from triangle TBO (Fig. 11.17):

$$c = \left[\frac{q_f}{\sin \phi'} - p_f\right] \tan \phi' \tag{11.27}$$

Rearranging Eq. 11.27 gives

$$c = \frac{q_f}{\cos \phi'} - p_f \tan \phi' \tag{11.28}$$

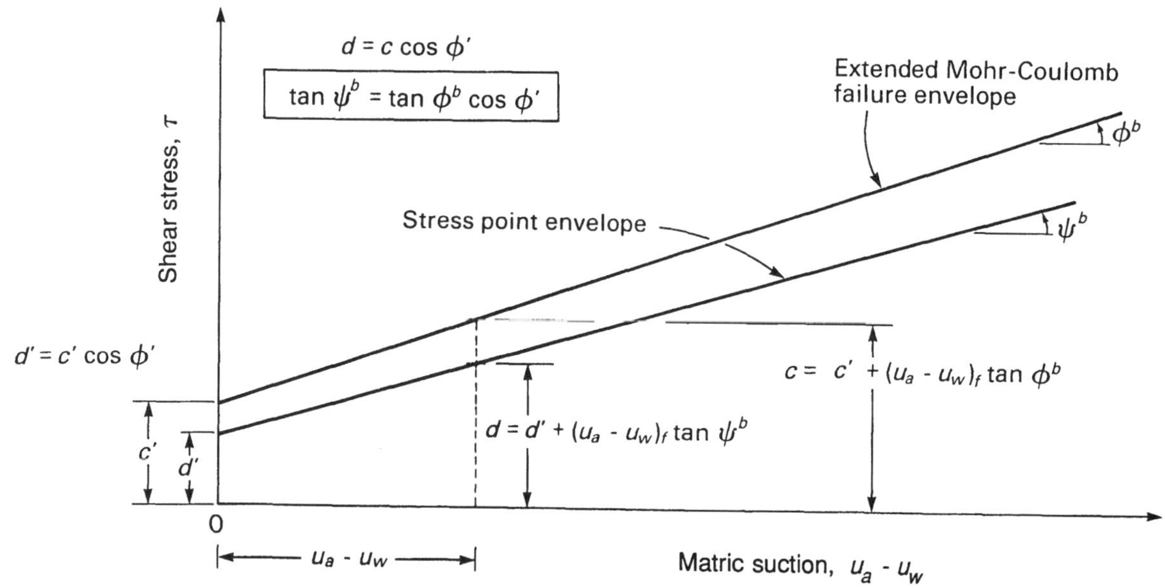

Figure 11.16 Relationship between ϕ', ϕ^b, and ψ^b angles.

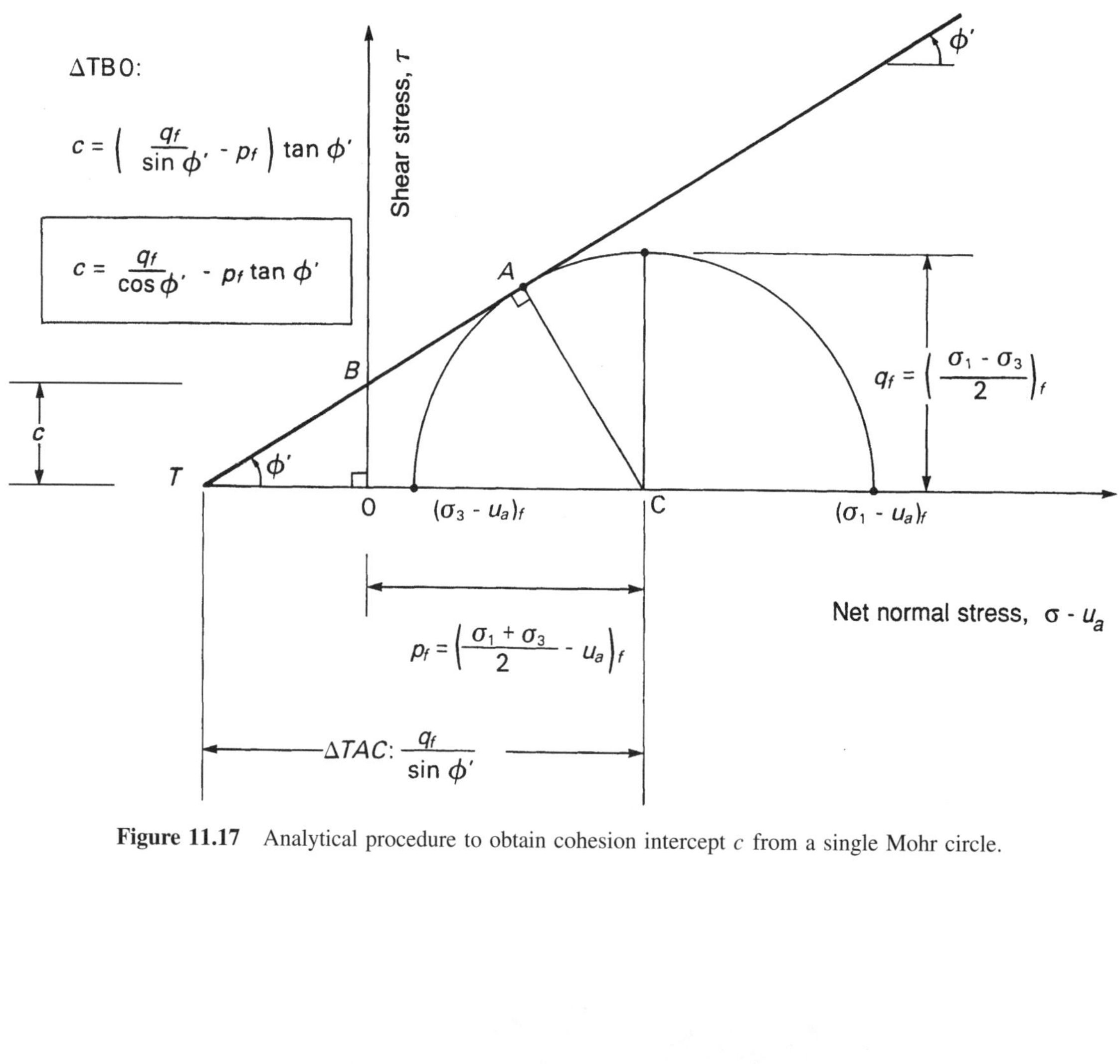

Figure 11.17 Analytical procedure to obtain cohesion intercept c from a single Mohr circle.

The cohesion intercepts c at various matric suctions can be computed using Eq. 11.28 and plotted on the shear strength versus matric suction plane (Fig. 11.16) in order to obtain the angle ϕ^b. Knowing the strength parameters c', ϕ', and ϕ^b, the parameters for the stress point envelope (i.e., d', ϕ', and ψ^b) can also be computed.

11.2.10 Nonlinearity of Failure Envelope

A linear shear strength theory has been presented for an unsaturated soil using an extended Mohr-Coulomb failure envelope. As a wider variety of soil types have been tested over a wider range of soil suctions, it has become increasingly apparent that the shear strength versus matric suction relationship should not be limited to a linear relationship.

Shear strength test results on compacted glacial till by Gan (1986) showed that the shear strength envelope was curved. Figure 11.18 illustrates the nonlinear matric suction failure envelope measured by Gan (1986). The ϕ^b angle appears to be equal to ϕ' at low matric suctions and decreases to a lower value at high matric suctions. Consequently, the ϕ^b angle appears to be a function of matric suction. Similar curved suction versus shear strength envelopes were earlier presented by Escario and Sáez (1986).

An examination of the effect of increasing matric suction from an initially saturated condition suggests that it is reasonable for the shear strength versus matric suction relationship to be nonlinear. Let us consider a shear strength test performed using the axis translation technique over a wide range of matric suctions. The tests can start from an initially saturated condition with the soil consolidated under a designated confining pressure σ. The initial matric suction is maintained at zero (i.e., pore-water pressure in the soil is equal to externally applied pore-air pressure).

The total confining pressure can be referenced to the external air pressure $\sigma - u_a$. Assuming that the ϕ' and c' parameters are known, the soil shear strength under the initial conditions is equal to $c' + (\sigma - u_a) \tan \phi'$. Let us now increase the applied air pressure (i.e., producing a positive matric suction) while keeping the net normal stress $\sigma - u_a$ constant.

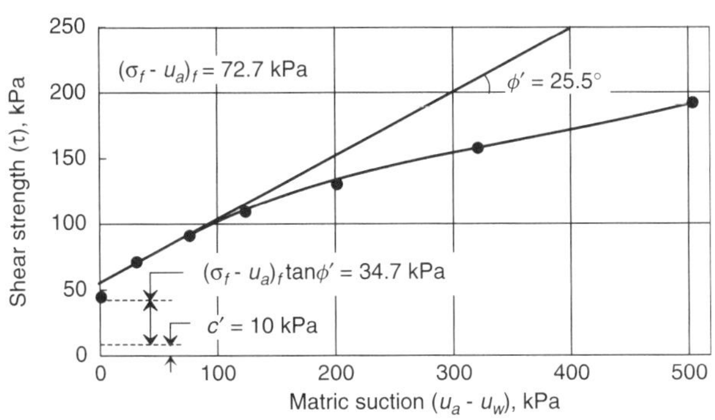

Figure 11.18 Direct shear test results exhibiting nonlinear behavior for failure envelope projected onto τ versus $u_a - u_w$ plane (after Gan, 1986).

At low matric suctions, the soil specimen remains saturated. Under these conditions the effect of pore-water pressure and total normal stresses on the shear strength are characterized by the friction angle ϕ'. Both the pore-water pressure and the total normal stresses are referenced to the same external air pressure, as expressed by the stress state variables $u_a - u_w$ and $\sigma - u_a$. The shear strength behaves as a saturated soil with a friction angle ϕ'. The shear stress versus matric suction envelope has a slope angle ϕ^b equal to ϕ'. This condition is maintained as long as the soil is saturated.

Water is drawn out of the pore spaces once the air-entry value of the soil specimen, $(u_a - u_w)_b$, is exceeded. As the matric suction in the soil is increased, the soil begins to desaturate. The pore-water pressure is referenced to the pore-air pressure (i.e., $u_a - u_w$), which is now both external and internal to the soil since the soil becomes unsaturated. The total normal stress is similarly referenced to the pore-air pressure through the $\sigma - u_a$ stress state variable.

The pore-water occupies only a portion of the pore spaces in the soil as indicated by a degree of saturation which is less than 100%. A further increase in matric suction is not as effective in increasing shear strength as is an increase in net normal stress. As a result, it is necessary for the ϕ^b angle to reduce to a value lower than ϕ' when matric suction is increased beyond the air-entry value of the soil, $(u_a - u_w)_b$. The effects of net normal stress and pore-water pressure on the shear strength of an unsaturated soil take on independent roles in terms of the stress state variables $\sigma - u_a$ and $u_a - u_w$.

The air-entry value of a soil largely depends on the grain size distribution of the soil. Coarse-grained soils such as sands desaturate at lower matric suctions than fine-grained clayey soils. Test results have shown that there is a relationship between the air-entry value of the soil and the pore sizes in the soil. Sands often have an air-entry value less than 10 kPa whereas clayey soils can have an air-entry value well beyond 100 kPa. The air-entry value of a soil may also depend to some extent on the net confining pressure applied to the soil. The air-entry value provides an indication of the point where the shear strength versus matric suction starts to exhibit nonlinear shear strength behavior.

The study of shear strength results on soils tested over a wide range of suction values has clearly revealed that the suction versus shear strength envelope bears a relationship to the SWCC for a soil. The ϕ^b angle with respect to matric suction is more clearly understood by considering the amount of water in the pores of the soil. Curvature of the shear strength versus matric suction envelope is related to the air-entry value of the soil and residual conditions.

Several empirical procedures have been proposed whereby the SWCC can be used to estimate the shape of the shear strength function for an unsaturated soil. The proposed procedures are presented later under the section on the estimation of the unsaturated soil shear strength envelope (i.e., the beginning of Chapter 12).

11.2.11 Relationship between tan ϕ^b and χ

The shear strength equation proposed by Bishop et al. (1960) and that proposed by Fredlund et al. (1978) appears to be similar in form. The χ parameter emerged in the shear strength equation from an extension of the effective stress equation. Equating the linear form of the Fredlund et al. (1978) shear strength equation and the Bishop et al. (1960) equation yields the following relationship:

$$\left(u_a - u_w\right)_f \tan \phi^b = \chi \left(u_a - u_w\right)_f \tan \phi' \qquad (11.29)$$

Equation 11.29 can be solved for the parameter χ:

$$\chi = \frac{\tan \phi^b}{\tan \phi'} \qquad (11.30)$$

A graphical comparison between the ϕ^b representation of strength and the χ representation of strength is shown in Fig. 11.19. The increase in shear strength due to matric suction is represented as an upward translation from the saturated failure envelope when using the Fredlund et al., (1978) designation. The magnitude of the upward translation is equal to $\left\{\left(u_a - u_w\right) \tan \phi^b\right\}_f$ (i.e., point A in Fig. 11.19). In this case, the failure envelope for the unsaturated soil is viewed as an upward translation of the failure envelope for a saturated soil.

The Bishop et al. (1960) equation with the χ parameter suggests that the same failure envelope applies for saturated and unsaturated conditions. Matric suction is assumed to produce an increase in the net normal stress. The increase in net normal stress is a fraction of the matric suction at failure [i.e., $\chi \left(u_a - u_w\right)_f$]. The shear strength at point A using the ϕ^b method is equivalent to the shear strength at point A' when using the χ parameter method.

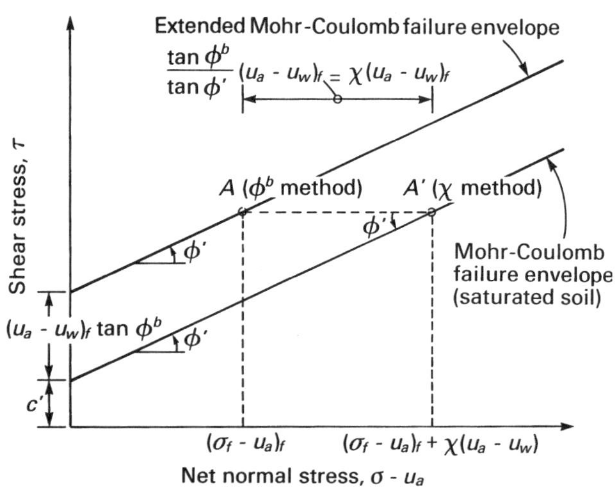

Figure 11.19 Comparison between ϕ^b and χ methods of designating shear strength.

One χ value is theoretically obtained for a particular soil when the failure envelope is planar. A planar failure envelope uses one value of ϕ' and one value of ϕ^b. If the failure envelope is nonlinear with respect to ϕ^b, there will be two or more values for χ. A χ value equal to 1.0 corresponds to the condition where $\phi^b = \phi'$. A χ value less than 1.0 corresponds to the condition when $\phi^b < \phi'$. There will be various χ values corresponding to different matric suction values for envelopes which are curved with respect to matric suction.

Attempts have been made to correlate the χ parameter with the degree of saturation of the soil (Bishop et al., 1960). Curvature in the shear strength envelope results in nonlinearity of the χ parameter. The ϕ^b and χ relationship applies to initially "identical" soils. These may be soils compacted at the same water content to the same dry density. The use of $\sigma - u_a$ and $u_a - u_w$ as independent stress state variables provides greater flexibility in terms of characterizing nonlinear shear strength behavior corresponding to a variety of stress paths.

11.3 MEASUREMENT OF SHEAR STRENGTH

The shear strength of unsaturated soils has been studied in the laboratory using both triaxial and direct shear testing equipment. It is necessary to make modifications to conventional shear strength testing equipment in order to independently control pressures in the air and water phases. Modifications to shear strength testing apparatuses to accommodate unsaturated soils were commenced in the 1950s. Following is a brief summary of some of the test results obtained.

11.3.1 Background of Shear Strength Measurements

A series of direct shear tests on unsaturated fine sands and coarse silts were conducted by Donald (1956). The tests were performed on a modified direct shear box as shown in Fig. 11.20a. The pore-air and pore-water pressures were controlled during shear. The top of the direct shear box was open to the atmosphere, thereby controlling the pore-air pressure u_a (i.e., zero gauge pressure). The pore-water pressure u_w was controlled at a negative value by applying a constant negative water head below the lower porous stone. The specimen was in contact with water below the base of the shear box through use of a colloidon membrane. The water in the base of the shear box was then connected to a constant-head-overflow tube to produce the desired negative water pressure (Fig. 11.20b). The negative pore-water pressures applied to the soils were generally less than 50 kPa.

The soil specimens of uniform initial density were consolidated under a total stress of approximately 48 kPa. The desired negative pore-water pressure was applied for several hours while the specimens came to equilibrium. The

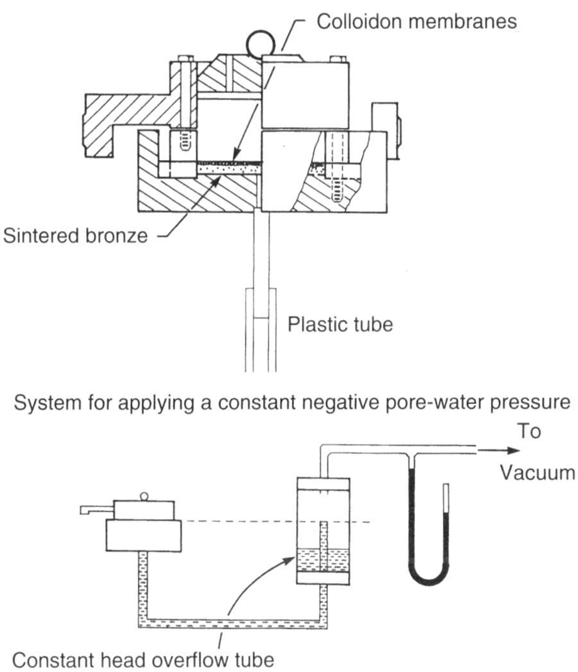

Figure 11.20 Modified direct shear equipment for testing soils under low-matric-suction conditions: (a) modified direct shear box with colloidon membrane; (b) system for applying constant negative pore-water pressure (after Donald, 1956).

specimens were then sheared at a rate of 0.071 mm/s. The results from four types of sand are presented in Fig. 11.21. The shear strength at zero matric suction is the strength due to the applied total stress. The shear strength increases to a peak value as matric suction is increased. Following the peak value for shear strength there was a decrease to a fairly constant shear strength value. The shear strength of the sands appears to increase at the same rate as for an increase in total stress as long as the specimens remain saturated. The increase in strength decreases once the sands start to desaturate. The strength of the sand is decreased once matric suction increases beyond some limiting value.

The water content versus matric suction desorption curves (i.e., SWCCs) were also measured for each of the sand materials. A comparison of the SWCCs and the shear strength envelopes showed that there was a correlation between the air-entry value for each soil and the point at which the shear strength envelope became nonlinear. The residual conditions on the SWCC were also shown to correspond quite closely with the point where the shear strength became essentially horizontal after passing the peak strength. These relatively simple direct shear strength tests provided considerable insight into the behavior of unsaturated soils.

An extensive research program on unsaturated soils was conducted at Imperial College, London, in the 1950s and early 1960s. Bishop et al. (1960) proposed testing techniques and presented the results of five types of shear strength tests on unsaturated soils classified as (1) consolidated drained, (2) consolidated undrained, (3) constant water content, (4) undrained, and (5) unconfined compression tests.

The tests were performed using a modified triaxial cell. The pore-air and pore-water pressures were either measured or controlled during the strength tests. Bishop (1961b) provided details on how the pore pressures could be controlled and measured using a modified triaxial apparatus. Tests confirmed that pore-water pressures could be measured directly through a saturated porous ceramic disk sealed onto the base pedestal below a soil specimen. Pore-water pressure measurements were made by balancing the pressure in the measuring system through use of a null-type, no-flow indicator. This direct measurement, however, was limited to a negative gauge pressure above 90 kPa. Bishop and Eldin (1950) successfully measured pore-water pressures down to negative 90 kPa in a saturated soil specimen while performing a consolidated undrained test with a carefully deaired measuring system. It was suggested that the axis translation technique was required when the pore-water pressures became lower than 1 atm negative.

The axis translation technique translates highly negative pore-water pressure to a pressure that can be measured without cavitation occurring in the water of the measuring system. A ceramic disk with an air-entry value greater than the matric suction being measured was required in order to prevent the passage of pore-air into the measuring system. A single layer of coarse glass fiber cloth with a low attraction for water was placed on top of the soil specimen for pore-air pressure measurement or control.

Constant-water-content (or CW) triaxial tests were performed on compacted shale and compacted boulder clay by Bishop et al. (1960). The shale had a clay fraction of 22% and was compacted at a water content of 18.6%. A series of triaxial tests on the saturation specimens of the compacted shale gave an angle of internal friction ϕ' of 24.8° and an effective cohesion c' of 15.8 kPa. The boulder clay had a clay fraction of 18% and was compacted at a water content of 11.6%. The saturated boulder clay showed an effective angle of internal friction ϕ' of 27.3° and an effective cohesion c' of 7.6 kPa. The tests on the compacted boulder clay were performed at a strain rate of 3.5×10^{-5}%/s, and 15 % strain was considered to represent failure. The laboratory results were presented in terms of stress points designated as $\{(\sigma_1 + \sigma_3)/2 - u_w\}_f$ and $\{(\sigma_1 + \sigma_3)/2 - u_a\}_f$ at failure (where σ_1 is the major principal stress and σ_3 is the minor principal stress). Figure 11.22 shows a typical plot of constant-water-content test results on compacted shale. The condition when the $\{(\sigma_1 - \sigma_3)/(\sigma_3 - u_w)\}_f$ ratio reached a maximum value was considered to be the failure condition.

Pore-air diffusion through the rubber membrane around the triaxial specimen was prevented by surrounding the membrane on the specimen with mercury rather than water. Triaxial test results of a consolidated drained test on unsaturated loose silt were used to verify the significance and application of $\sigma - u_a$ and $u_a - u_w$ as stress variables.

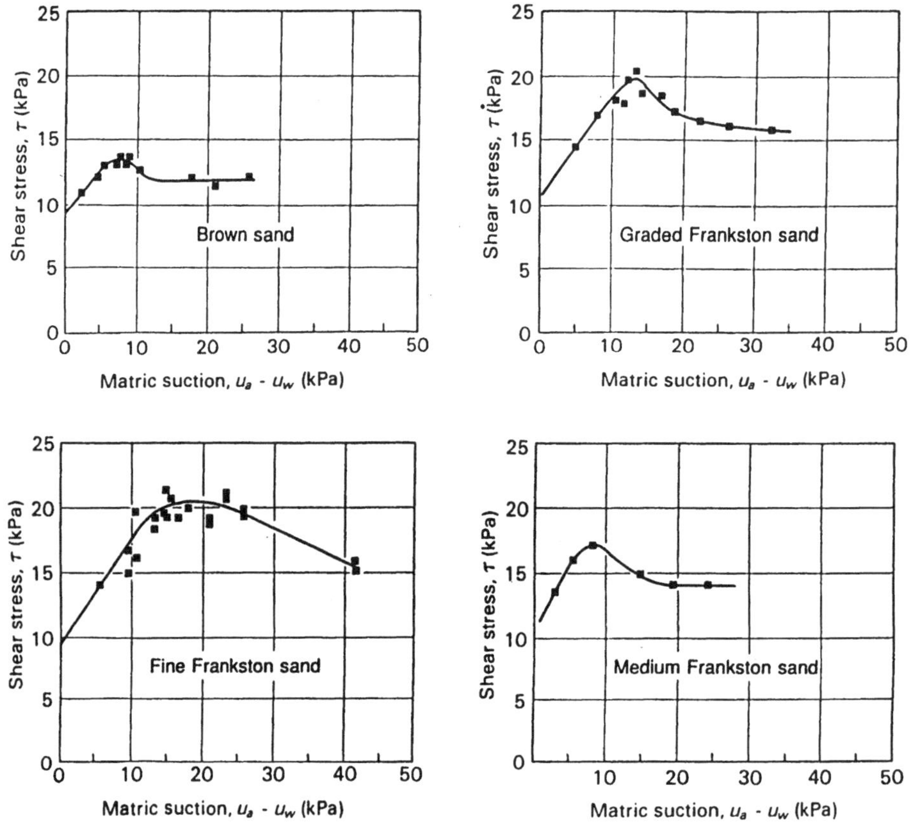

Figure 11.21 Result of direct shear tests on sands under low applied matric suctions (after Donald, 1956).

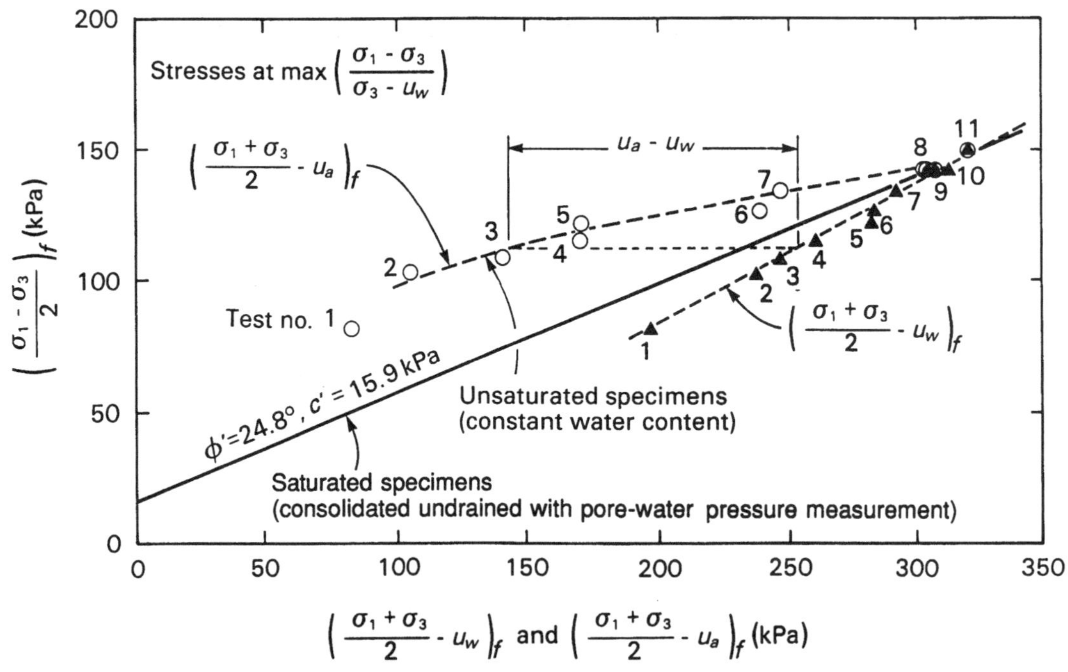

Figure 11.22 Results of constant-water-content triaxial tests on shale (clay fraction 22%) compacted at water content of 18.6% (after Bishop et al., 1960).

Laboratory testing techniques and details pertaining to various types of triaxial tests were explained and summarized by Bishop and Henkel (1962).

The acceptability of the axis translation technique when testing unsaturated soils was verified by Bishop and Blight (1963). Compression tests were conducted on a compacted Selset clay and Talybont clay. The experiments confirmed the applicability of the axis translation technique for laboratory tests on unsaturated soils. The ability of the pore-water to withstand absolute tensions greater than 1 atm (i.e., 101.3 kPa) was also confirmed.

The development of pore-air and pore-water pressures during undrained tests was also studied by Bishop and Blight (1963). Typical results of constant-water-content tests were presented and discussed. Donald (1963) presented further results of undrained tests on compacted Talybont clays with pore-air and pore-water pressure measurements.

Blight (1967) reported the results of several consolidated drained tests performed on unsaturated silt specimens. All specimens were compacted and brought to equilibrium at three matric suction values in a triaxial cell. In addition to measuring the shear strength of the soil, the overall volume change and water volume change were independently monitored. The results showed an increase in shear strength with increasing matric suction and also with increasing net confining pressure.

The shear strength of two unsaturated, compacted soils from India, namely, Delhi silt and Dhanauri clay, were tested by Satija and Gulhati (1979). Consolidated drained tests were performed with the pore pressures being maintained in a modified triaxial cell. Constant-water-content tests with pore-air pressure control and pore-water pressure measurement were also performed.

Fredlund et al., (1978) proposed the use of $\sigma - u_a$ and $u_a - u_w$ as independent stress state variables for the description of the shear strength of unsaturated soils. The shear strength of an unsaturated soil was considered to consist of an effective cohesion intercept c' and independent contributions from the net normal stress $\sigma - u_a$ and matric suction $u_a - u_w$. The effective angle of internal friction, ϕ', was associated with the shear strength contribution from the net normal stress state variable. Another angle, namely ϕ^b, was introduced to describe the increase in shear strength as matric suction was increased. The failure envelope was then viewed as a three-dimensional surface with $\sigma - u_a$ and $u_a - u_w$ as variables on the abscissas. The failure surface was visualized as an extension of the conventional Mohr-Coulomb failure envelope (Fredlund, 1979). Data from the research literature were used to examine the fit of the theory to laboratory test results.

Satija (1978) conducted an experimental study on the shear strength behavior of unsaturated Dhanauri clay. Constant-water-content and consolidated drained tests were conducted on compacted specimens for various values of $\sigma - u_a$ and $u_a - u_w$ stresses. The triaxial test apparatus was similar to that used in the Imperial College research program. Pore pressures were either controlled or measured throughout the shear test. The appropriate strain rate was found to decrease with decreasing degrees of saturation (Satija and Gulhati, 1979). The results were presented as a three-dimensional surface where half of the deviator stress at failure, $\{(\sigma_1 - \sigma_3)/2\}_f$, was plotted versus the net minor principal stress at failure, $(\sigma_3 - u_a)_f$, and the matric suction at failure, $(u_a - u_w)_f$ (Gulhati and Satija, 1981).

A series of consolidated drained direct shear tests on unsaturated Madrid grey clay were reported by Escario in 1980. The tests were performed under controlled matric suction conditions using the axis translation technique. A modified shear box device enclosed in a pressure chamber was used to apply a controlled air pressure to the soil specimen. The specimen was placed on a high-air-entry disk in contact with water at atmospheric pressure. The soil specimens were statically compacted and brought to the desired matric suction under an applied vertical normal stress. Typical results obtained from direct shear tests are presented in Fig. 11.23. The failure envelopes exhibit almost a parallel upward translation, indicating a continuous increase in shear strength as soil matric suction was increased. The results of triaxial tests by Escario (1980) on Madrid clay are shown in Fig. 11.24. The triaxial cell was modified to allow the use of the axis translation technique. The results show

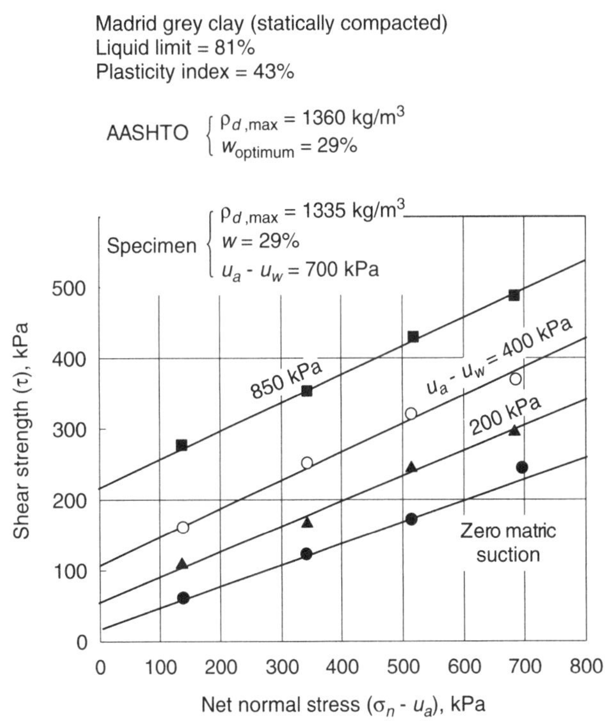

Figure 11.23 Direct shear tests showing increase in shear strength for Madrid clay as result of increasing matric suction (after Escario, 1980).

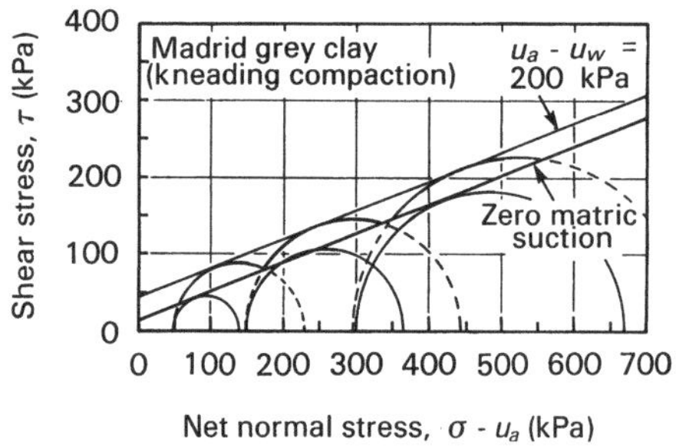

Figure 11.24 Triaxial tests showing increase in shear strength due to increase in matric suction for Madrid clay (modified from Escario, 1980).

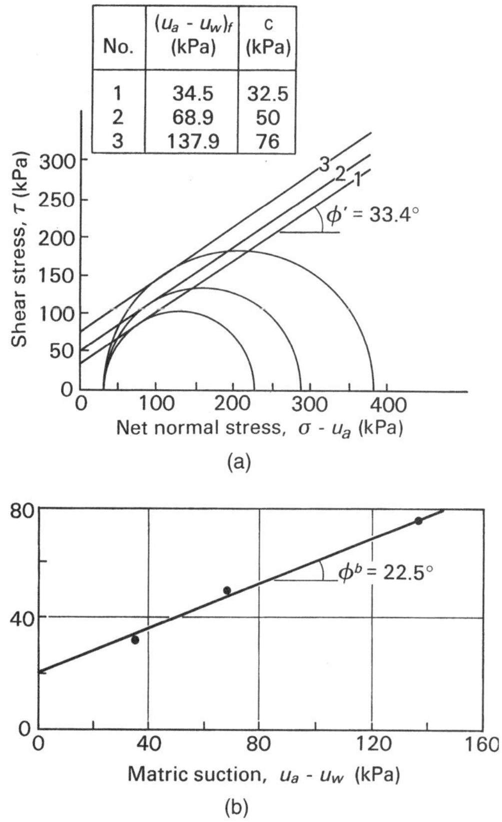

Figure 11.25 Two-dimensional presentation of the failure envelope for decomposed granite specimen No. 22: (a) failure envelope projected onto τ versus $\sigma - u_a$ plane; (b) Intersection line between failure envelope and τ versus $u_a - u_w$ plane (from Ho and Fredlund, 1982b).

an increase in shear strength resulting from the increase in matric suction.

A series of multistage triaxial tests was performed by Ho and Fredlund (1982b) on unsaturated soils. Undisturbed specimens of two residual soils from Hong Kong were used in the testing program. The soils were decomposed rhyolite and decomposed granite. The program consisted of consolidated drained tests with the pore-air pressure controlled from the top of the specimen through a coarse porous disk. The pore-water pressure was controlled from the bottom of the specimen using a high-air-entry disk sealed around the perimeter onto the base pedestal. The desired matric suction in the specimen was obtained by controlling the pore-air and pore-water pressures using the axis translation technique.

The triaxial test results showed that a single soil specimen could be used to provide an indication of the increase in strength as a result of increasing matric suction. The assumption was made that the effective angle of internal friction ϕ' remained a constant value. Typical two-dimensional projections of the failure envelope onto the shear stress τ versus $\sigma - u_a$ plane are presented in Fig. 11.25a. The intersections between the failure envelope and the ordinate are plotted in Fig. 11.25b. The shear strength at failure increased for a constant net confining pressure, as shown by the ϕ^b angle in Fig. 11.25b.

Gan (1986) conducted a multistage direct shear testing program on an unsaturated glacial till. A modified direct shear box allowed the control of the pore-air and pore-water pressures during the test. The shear box was enclosed in an air pressure chamber in order to control the pore-air pressure. The pore-water pressure was controlled through the base of the specimen using a high-air-entry disk. Consolidated drained direct shear tests were performed with matric suction being controlled during shear (i.e., axis translation technique). Applied matric suctions ranged from 0 to 500 kPa while the net normal stress was maintained at approximately 72 kPa. Typical test results are presented in Fig. 11.18. The shear stress was plotted with respect to matric suction (i.e., τ versus $u_a - u_w$ plane) for constant net normal stress at failure, $(\sigma_f - u_a)_f$. The results showed that there was nonlinearity in the failure envelope on the shear stress versus matric suction plane. The ϕ^b angle commences at a value equal to ϕ' (i.e., 25.5° when measured under saturated conditions) at low matric suctions. The ϕ^b angle decreased to a final value of 7° at high matric suctions.

There have been many more recent studies on a wide range of soil types. In general, the results corroborate the test results from earlier studies. At the same time the shear strength studies have led to a more indepth understanding of shear strength behavior of unsaturated soils. In particular, a number of studies have focused on the critical state shear strength behavior of unsaturated soils.

11.3.2 Incomplete Stress Variable Measurements

Numerous shear strength tests on unsaturated soils have been conducted without the measurement of pore-air and pore-water pressures at failure. Unfortunately, these results are of limited value in terms of obtaining a better understanding of the shear strength of an unsaturated soil. The extended Mohr-Coulomb shear strength envelope for unsaturated soils requires that three shear strength parameters be defined, namely, c', ϕ', and a linear or nonlinear angle with respect to matric suction, ϕ^b. These parameters can be measured in the laboratory using triaxial and direct shear apparatuses that have been modified for testing unsaturated soils.

The c' and ϕ' parameters can be measured using saturated soil specimens. The equipment and techniques used for testing saturated soils are well documented by Bishop and Henkel (1962) and Head (1986). The tests on unsaturated soil specimens are primarily performed to obtain the ϕ^b soil parameter. General procedures for testing unsaturated soils using a modified triaxial and direct shear apparatus are described in this book. Computations and data reduction associated with analyzing the test results are similar to those required for saturated soils.

11.4 SPECIAL EQUIPMENT DESIGN CONSIDERATIONS

Conventional triaxial and direct shear equipment requires that a number of modifications be made to the equipment prior to attempting to test unsaturated soils. Several factors related to the nature of an unsaturated soil must be considered when modifying the equipment. The presence of air and water in the pores of the soil causes the testing procedures and techniques to be considerably more complex than when testing saturated soils. The primary modifications are associated with the independent measurement and/or control of the pore-air and pore-water pressures. The pore-water pressure is usually negative (gauge) and can result in cavitation of the water in the measuring system when the water pressures are too low.

Triaxial testing methods are defined on the basis of the drainage conditions imposed following the application of the confining pressure σ_3 and the drainage conditions imposed during the application of the deviator stress, $\sigma_1 - \sigma_3$. For undrained testing conditions, the pore fluid (i.e., pore-air or pore-water) is not allowed to drain. Excess pore-air and pore-water pressures developed during undrained loading conditions must be independently measured.

For drained testing conditions, the pore-air and pore-water pressures are allowed to drain to prescribed pressures imposed on the top or bottom of the soil specimen. Volume changes associated with the pore fluid flow during a drained test should be measured.

11.4.1 Axis Translation Technique

The axis translation technique is commonly used when testing unsaturated soils in the laboratory. The axis translation technique has proven to be particularly useful for controlling matric suction values in excess of 100 kPa. The pore-water pressure u_w in an unsaturated soil can be measured or controlled through use of a ceramic disk with fine pores (i.e., high-air-entry disks). The functionality of the ceramic disk can be observed while applying various differential air and water pressures across the disk.

The ceramic disk is hydrophilic and as such has a high affinity for water. Water can move into and through the ceramic disk. However, free air cannot flow through the disk unless a limiting air-entry value is exceeded. The limiting value of air pressure that can be applied to the ceramic disk is called the "air-entry value." Even prior to exceeding the air-entry value of the disk, it is possible for air to dissolve in the water in the ceramic disk and then slowly diffuse through the water in the disk. The ceramic disk serves as a separator between the air and water phases, but unfortunately it is not a "perfect" separator.

Figure 11.26 illustrates the direct measurement of pore-water pressure relative to atmospheric air pressure conditions (i.e., $u_a = 0$). The soil specimen is placed on top of a saturated porous disk and a U-tube is connected to the water compartment below the disk. The U-tube can be moved up

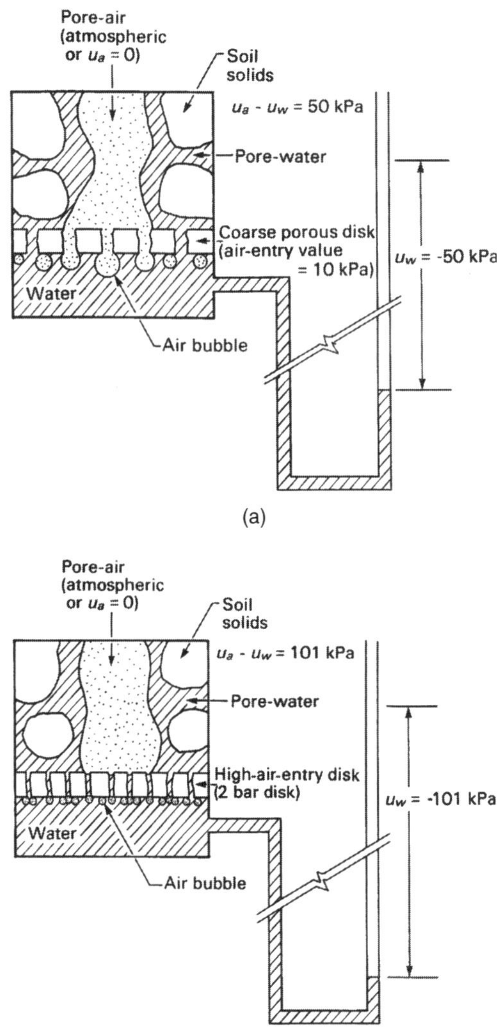

Figure 11.26 Direct measurement of pore-water pressure in an unsaturated soil specimen (not to scale): (a) air movement through the porous disk when its air-entry value is exceeded; (b) air diffusion through the high-air-entry disk and water cavitation in the measurement of matric suction.

and down until there is no tendency for flow in or out of the specimen. The negative pressure head at equilibrium will indicate the matric suction in the soil specimen (Fig. 11.26a). However, this direct measurement of suction has limitations.

The porous disk in contact with the soil must have an air-entry value greater than the matric suction in the soil. The air-entry value of a porous disk is the matric suction at which air commences passing through the largest voids in the saturated porous disk. Air enters the water compartment once the pressure differential between the air and water phases exceeds the air-entry value of the ceramic disk. Consequently, the water in the measuring system can no longer indicate the pore-water pressure in the soil. The measuring system then becomes filled with air.

Another limitation associated with attempting to directly measure the pore-water pressure arises from the fact that water cavitates as a gauge pressure of −1 atm is approached (i.e., −101 kPa gauge). As an example, consider an unsaturated soil specimen with a matric suction of 101 kPa. The pore-water pressure in the specimen is being measured relative to atmospheric air pressure (i.e., $u_a = 0$), as shown in Fig. 11.26b. The lowest gauge pore-water pressure that can be measured should be about −101 kPa. A saturated ceramic disk with an air-entry value of 202 kPa would meet the requirement for not allowing air flow through the ceramic disk. Nevertheless, the water in the compartment below the ceramic disk would start to cavitate as the gauge water pressure approaches −101 kPa. Occluded air bubbles would then start to accumulate below the disk in the water compartment. The air bubbles would cause an error in the measurement of pore-water pressure. The direct measurement of pore-water pressure is limited to a value of −101 kPa. The limiting value for the direct measurement of pore-water pressure is true regardless of the air-entry value of the ceramic disk.

The limitations of direct control or measurement of pore-water pressure in an unsaturated soil can be overcome by translating the reference axis for the measurement of air pressure. The procedure is referred to as the axis translation technique. The axis translation technique is particularly useful for testing unsaturated soils at high matric suctions. Basically, both the pore-air and pore-water pressures are translated into the positive-pressure range when applying the axis translation technique.

The translation of pore-air pressure can be considered as an artificial increase in the atmospheric pressure under which the test is performed (Bishop et al., 1960). Consequently, the negative-gauge pore-water pressure is raised by an equal amount to a positive-gauge pressure. The matric suction of the soil specimen remains constant regardless of the magnitude of the pore-air pressure. A fine, porous disk with an air-entry value greater than the matric suction of the soil must be used to prevent air from entering the water compartment.

The application of an axis translation of 202 kPa to an unsaturated soil specimen is illustrated in Fig. 11.27. Let

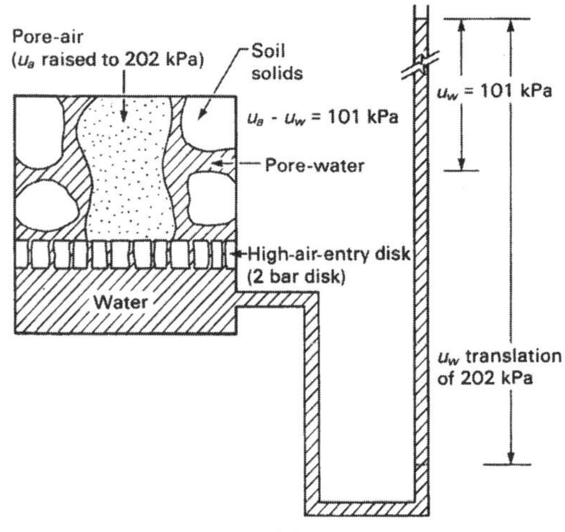

(a)

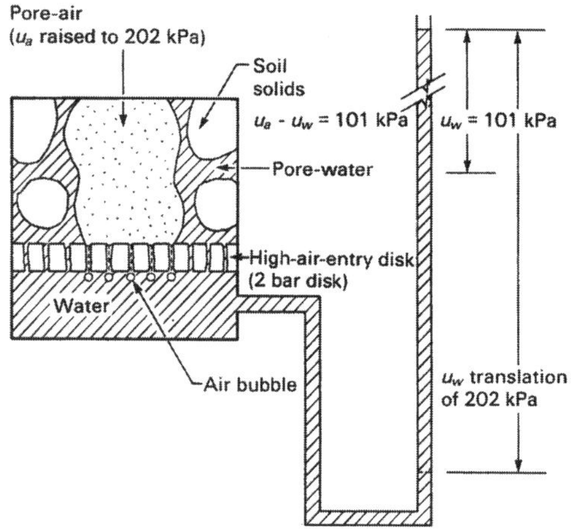

(b)

Figure 11.27 Measurement of pore-water pressure in an unsaturated soil specimen using axis translation technique (not to scale): (a) axis translation of 101 kPa; (b) air diffusion through high-air-entry disk.

us assume that the matric suction in the soil specimen is known to be 101 kPa. The pore-water pressure is measured below a saturated porous disk which has an air-entry value of 202 kPa. An air pressure of 202 kPa is applied directly to the specimen in order to translate the pore-air (and pore-water) pressures. In this example, the pore-air pressure u_a is raised to 202 kPa, which in turn increases the pore-water pressure by an equal amount. As a result, the pore-water pressure u_w is now at a positive value of 101 kPa, and there is no longer a problem associated with cavitation of the water in the measuring system. The pore-water pressure can now be measured since continuity is maintained between the pore-water in the soil and the water in the measuring system. The matric suction of the soil (Fig. 11.27a) remains constant at 101 kPa during the axis translation procedure. However, the pore-air pressure which is used as a reference pressure has been translated from 0 to 202 kPa.

The use of the axis translation technique over a long period of testing does not guarantee that the water pressure measuring system (i.e., compartment) will remain free of air bubbles. Unsaturated soil testing often requires extended periods of time due to low coefficients of permeability of the soil and the ceramic disk. Air may diffuse through the water in the high-air-entry disk and appear as air bubbles beneath the disk as the test progresses.

If the measuring system is being used in a closed-system mode, there will be serious problems with measuring the correct pore-water pressure due to diffused air. If the measuring system is operated in an open-system mode, a diffused air volume indicator, DAVI, can be used to periodically flush and measure the volume of diffused air. The diffused air volume can be used to correct the water volume change measurements under drained testing conditions. It is not possible to correct for pressure changes resulting from air diffusion under undrained testing conditions.

The use of the axis translation technique in the shear strength measurement of unsaturated soils was advocated by Bishop and Blight (1963). The results of an "unconfined triaxial" test on compacted Selset clay using the axis translation technique are presented in Fig. 11.28. The unconfined triaxial test was a special case of a constant-water-content test where the net confining pressure $\sigma_3 - u_a$ was maintained at zero pressure. The test was performed by applying an all-around air pressure σ_3 equal to the pore-air pressure u_a. However, the all-around air pressure was varied using a series of steps, as shown in Fig. 11.28a. The air pressure increase in each step was applied gradually in order for the pore-air pressure to equalize throughout the soil specimen. The pore-water pressure was measured throughout the test and the measurements indicate that the pore-water pressure also increased in a stepwise manner. The results indicate monotonic curves of deviator stress and matric suction versus axial strain regardless of the stepwise axis translation during the test (Fig. 11.28b). The shearing of the unsaturated soil specimen did not appear to be affected by the translation of the reference pressure.

The acceptability of using the axis translation technique to measure properties of an unsaturated soil was also demonstrated through two unconfined triaxial tests conducted on compacted Talybont clay (Bishop and Blight, 1963). The test results are shown in Fig 11.29. The first test was performed using an axis translation technique such that the gauge pore-air pressure was elevated to 483 kPa and the pore-water pressure was measured. The measured pore-water pressure was approximately 69 kPa, which indicated a

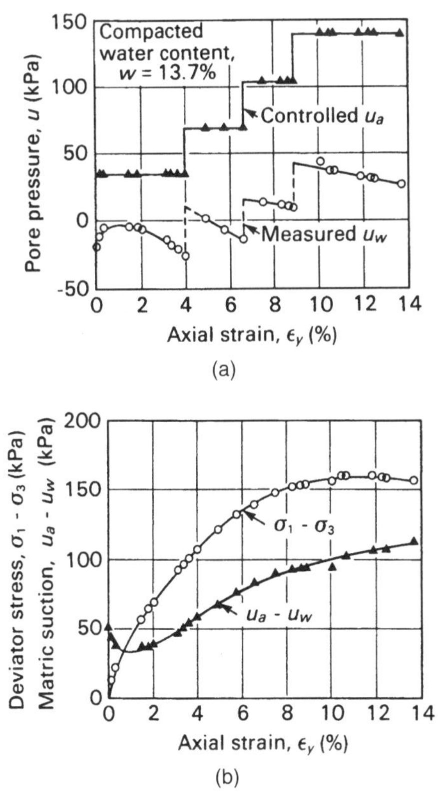

Figure 11.28 Unconfined triaxial test on compacted Selset clay specimen subjected to varying pore-air pressures: (a) pore pressures versus axial strain curve; (b) curves of deviator stress and matric suction versus strain (after Bishop and Blight, 1963).

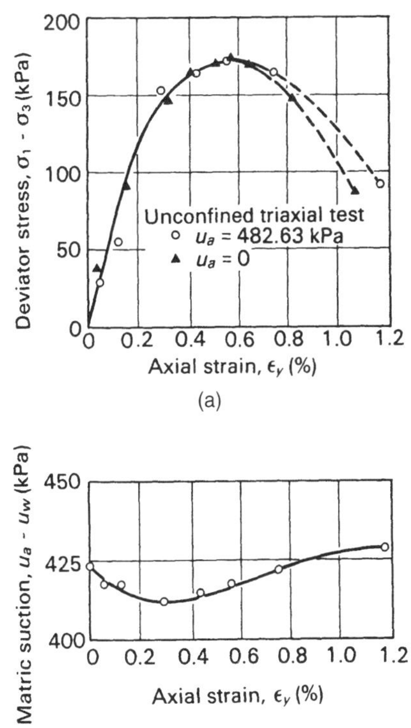

Figure 11.29 Unconfined triaxial tests on compacted Talybont clay with and without using axis translation technique: (a) deviator stress versus axial strain curve; (b) matric suction versus axial strain during test with axis translation (after Bishop and Blight, 1963).

matric suction of approximately 414 kPa during the test (Fig. 11.29b). The second test was performed under atmospheric air pressure conditions (i.e., $u_a = 0$). The stress versus strain curves from both compressions tests showed essentially the same results up to the maximum deviator stress and only differed slightly subsequent to the peak strengths. In other words, the measured shear strength remained essentially unaffected by the axis translation procedure. The results also illustrate that the pore-water in the soil was able to withstand high tensions without rupturing. Since the soil matric suction in both tests was about 414 kPa, the pore-water pressure in the second test specimen would be −414 kPa gauge pressure or −313 kPa in terms of absolute pressure.

11.4.2 Pore-Water Pressure Control and/or Measurement

Pore-water pressures in an unsaturated soil specimen can be controlled at a prescribed value when the water phase is maintained under drained conditions during a shear strength test (e.g., consolidated drained test). Pore-water pressures can be measured when the water phase is maintained in an undrained mode throughout the test (e.g., constant water content or in a consolidated undrained test). The key element that provides the separation between the pore-air and pore-water pressures is the high-air-entry ceramic disk.

The high-air-entry disk acts as a semipermeable membrane that separates the air and water phases. The separation of the air and water phases can be achieved only when the air-entry value of the disk is greater than the matric suction of the soil. The air-entry value of the disk depends on the maximum size of pores in the ceramic disk. Figure 11.30 illustrates how air commences to flow through the ceramic disks once a limiting air-water differential pressure is applied across a ceramic disk. The theoretical maximum pore size for any air-water pressure differential can be estimated using the capillary model.

$$\left(u_a - u_w\right)_d = \frac{2T_s}{R_s} \tag{11.31}$$

where:

$\left(u_a - u_w\right)_d$ = air-entry value of the high-air-entry disk,
T_s = surface tension of the contractile skin or the air-water interface (e.g., $T_s = 72.75$ mN/m at 20°C), and
R_s = radius of curvature of the contractile skin or the pore radius.

The air-entry value in Eq. 11.31 refers to the maximum matric suction to which the high-air-entry disk can be subjected before free air passes through the disk. The maximum matric suction that can be sustained is associated with the radius of the largest pore in the ceramic disk. The pore sizes in the ceramic are controlled by the preparation and sintering process used to manufacture the ceramic disk. The smaller the pore sizes in a disk, the larger will be the air-entry value.

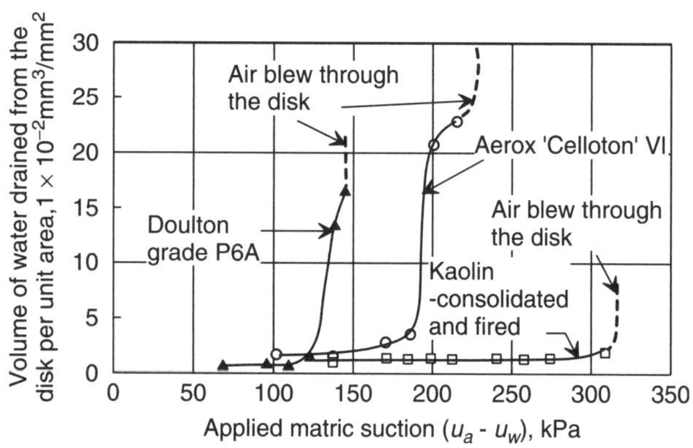

Figure 11.30 Air passage characteristics of three high-air-entry disks (after Bishop and Henkel, 1962).

The properties of high-air-entry disks manufactured by Soilmoisture Equipment Corporation in Santa Barbara, California, are tabulated in Table 11.2. The disks are commonly identified by the air-entry value of the ceramic disk and are usually expressed in the unit of bars (i.e., 1 bar = 100 kPa). The water coefficient of permeability of a disk can be measured by mounting the disk in a triaxial apparatus and placing water above the disk. The disk must be sealed on the sides (i.e., around the perimeter) in order to prevent the passage of water around its circumference. An applied air pressure then produces a pressure gradient across the high-air-entry disk. The volume of water flowing through the disk can be measured using a water volume change indicator.

Table 11.2 High-Air-Entry Disks Manufactured by Soilmoisture Equipment Corporation

Type	Approximate Pore Diameter ($\times 10^{-3}$ mm)	Coefficient of Permeability, k_d (m/s)	Range of Air-Entry Values, $(u_a - u_w)_d$ (kPa)
½ bar (high flow)	6.0	3.11×10^{-7}	48–62
1 bar	2.1	3.46×10^{-9}	138–207
1 bar (high flow)	2.5	8.6×10^{-8}	131–193
2 bars	1.2	1.73×10^{-9}	241–310
3 bars	0.8	1.73×10^{-9}	317–483
5 bars	0.5	1.21×10^{-9}	>550
15 bars	0.16	2.59×10^{-11}	>1520

The selection of a high-air-entry disk for testing an unsaturated soil should be primarily based upon the maximum possible matric suction that can occur during the laboratory test. The initial matric suction in an unsaturated soil specimen should be taken into consideration when selecting a ceramic disk for constant-water-content or undrained tests. It must also be understood that the matric suction may increase as the specimen is loaded to failure conditions.

The water coefficient of permeability and the thickness of a high-air-entry disk are factors that affect the time required for the pore-water pressure to equalize across the ceramic disk. This information is of importance when assessing the time required for consolidation prior to shearing the specimen as well as assessing the time required for shearing the specimen. It is desirable to have the highest possible coefficient of permeability (i.e., the largest possible pore size) for the high-air-entry disk. This will ensure the most rapid equilibration of the pore-water pressure in an undrained test. It will also minimize impeded drainage of the pore-water in a drained test.

The use of a thin air-entry disk reduces the time required for pore-water pressure equalization throughout the soil specimen. However, a thin ceramic disk means there is a shorter pathway for air to diffuse and reach the underlying water compartment. Air bubbles collect below the high-air-entry disk as diffused air enters the water compartment. The use of a thin ceramic disk results in a greater accumulation of diffused air in the water compartment.

A thin ceramic disk also has a tendency to more easily crack. Care must be taken to ensure that the total stress and the pore-air pressure are applied to the soil specimen before pressurizing the water compartment beneath the ceramic disk. The removal of all pressures from the top of the disk while a water pressure is applied to the bottom side of the disk produces an upward bending moment that can readily crack the ceramic disk. Experience has shown that an excess uplift pressure greater than about 70 kPa on a 3.2-mm-thick ceramic disk (63.5 mm in diameter) can produce cracks in the ceramic disk (Fredlund, 1973a). The cracks may not be visible to the naked eye but are evident when checking the air-entry value of the disk or testing for the permeability of the disk. A 6.4-mm-thick disk can withstand higher uplift pressures, but all ceramic disks should be checked for cracks whenever subjected to an excess pressure greater than about 70 kPa.

11.4.2.1 Installation of High-Air-Entry Disk in Triaxial Cell

Figure 11.31 shows the primary modifications that must be made to a conventional triaxial cell when testing unsaturated soil. The primary modification involves sealing a high-air-entry disk onto the base pedestal of the triaxial cell. There are several ways in which the high-air-entry ceramic disk can be mounted, but in each case it is paramount that air not be allowed to bypass the ceramic disk and find its way into the water measuring system (i.e., the water compartment).

The installation of a high-air-entry disk onto the base pedestal of a triaxial cell is illustrated in Fig. 11.32. The most important feature of this modification is related to the epoxy seal that is placed between the high-air-entry disk and the base pedestal (see Fig. 11.32b). A tight seal between the ceramic disk and the base pedestal ensures that air will not leak into the water compartment. A series of grooves, generally in a spiral pattern, should be machined onto the pedestal. The grooves form a part of the water compartment and serve as water channels for flushing air bubbles that may become trapped below the ceramic disk or that may have

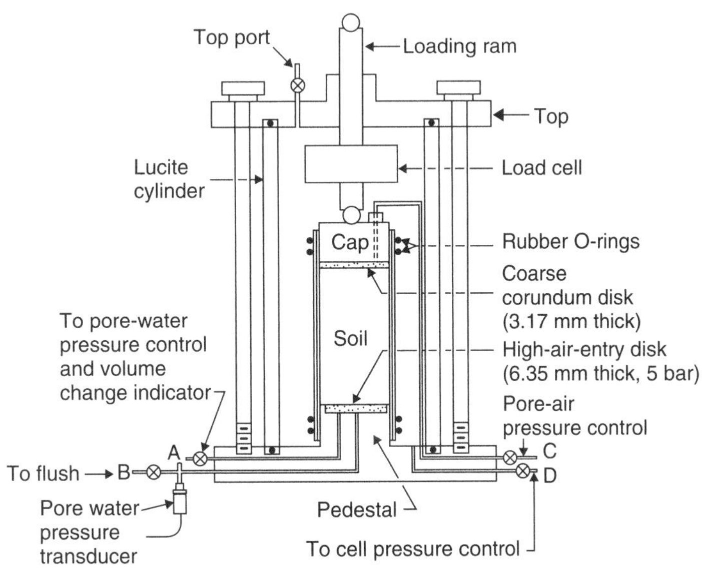

Figure 11.31 Modifications to convert conventional triaxial cell for testing unsaturated soils.

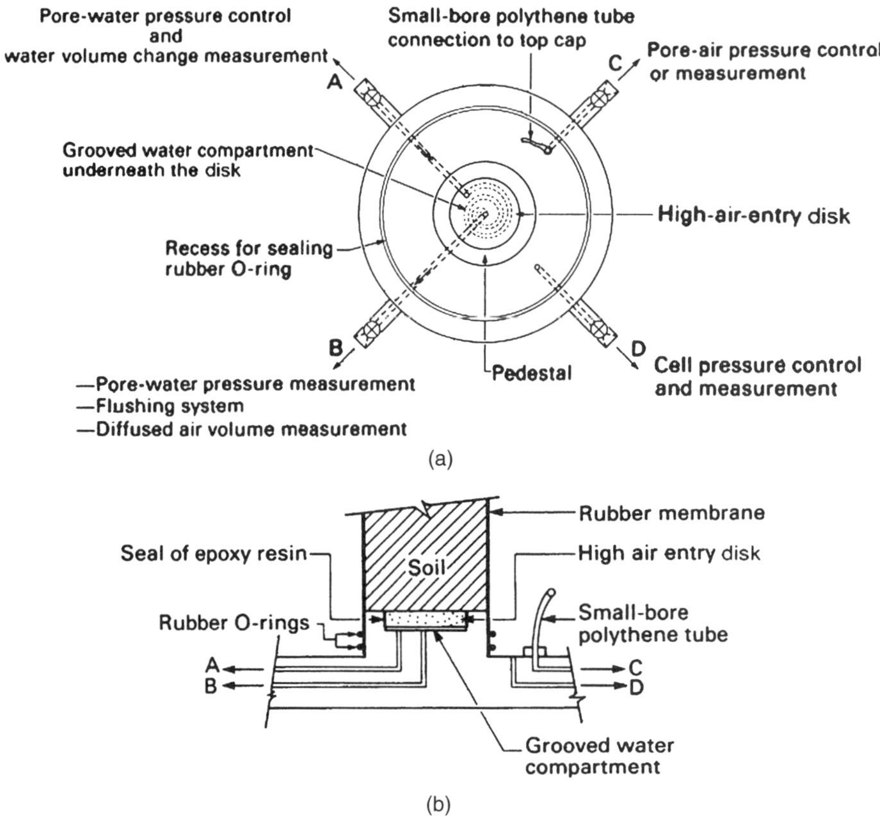

Figure 11.32 Triaxial base plate for unsaturated soil testing: (a) plan view of base plate with its outlet ports; (b) cross section of base plate with high-air-entry disk.

accumulated as a result of air diffusion through the ceramic disk. A conventional triaxial base plate may also require a modification in order to accommodate the application of air pressure to the top of the soil specimen. Figure 11.32a shows valve C being used for this purpose.

Valve A in the base plate (Fig. 11.32) can be used to control the pore-water pressure and to measure the water volume change during a *drained* shear test. Valve A can be connected to a water volume change indicator to measure water flow in or out of the soil specimen (Bishop and Donald, 1961; Bishop and Henkel, 1962; Head, 1986). Valve B on the base plate can be used to measure the pore-water pressures during an *undrained* shear test. It can also be used to measure the applied pore-water pressure during a drained test. The pore-water pressure can be measured using either a null-type indicator (Bishop and Henkel, 1962) or a pressure transducer. Pressure transducers are now most commonly used for measuring pore-water pressure (Head, 1986).

A diffused air volume indicator can be connected to valve B. It is used to measure the volume of diffused air accumulating below the high-air-entry disk. Valve C is connected to the pore-air pressure control or measuring system. The cell pressure is controlled through valve D.

11.4.3 Saturation Procedure for High-Air-Entry Disks

It is preferable to not boil the ceramic disk and the base plate in an attempt to produce complete saturation of the high-air-entry disk. Boiling may produce fine cracks between the epoxy and the high-air-entry disk as a result of differential thermal properties of the materials involved. The high-air-entry disk can best be saturated by first passing water through the disk and then using pressurization to force air into solution. Once the disk has been saturated, it appears to remain saturated as long as it is kept in contact with water.

Ceramic disks are relatively easy to saturate because of their hydrophilic nature. The following procedure is suggested to ensure saturation of a high-air-entry disk mounted in the triaxial or direct shear apparatus (Fredlund, 1973a). The ceramic disk should first be properly mounted onto the testing apparatus. Distilled, deaired water should be added to the chamber to a height of about 25 mm above the ceramic disk. The triaxial cell or the direct shear box chamber should then be assembled. The water in the cell can be subjected to an air pressure of approximately 600 kPa. Water is allowed to flow through the porous disk for approximately 1 h.

Air bubbles that collect below the ceramic disk can be periodically flushed. The valves connecting the water compartment to the measuring system should then be closed.

The water above and below the ceramic disk equalizes to the pressure applied in the cell. It is suggested that the air pressure be applied for approximately 1 h, during which time the air in the porous disk dissolves in water. The valves connected to the water compartment can then be opened for approximately 10 min to allow the water in the disk to flow into the compartment. The air bubbles are then flushed from below the high-air-entry disk. The above procedure can be repeated several times after which the high-air-entry disk should be saturated. The disk should remain covered with water until a soil specimen is ready to be mounted onto the disk. The suggested procedure for saturating the ceramic disk is similar to that described by Bishop and Henkel (1962).

The coefficient of permeability of the ceramic disk can be measured during the saturation process by recording the quantity of water passing through the high-air-entry disk. The coefficient of permeability is of interest in evaluating the compliance of the pore pressure measuring system and ensuring that the ceramic disk has no cracks. The measured coefficient of permeability should dramatically increase once the ceramic disk has even a small crack.

11.4.4 Pore-Air Pressure Control or Measurement

Pore-air pressure is controlled at a specified value when performing a drained shear test (e.g., consolidated drained or constant-water-content test). Pore-air pressure is measured when performing an undrained shear test (e.g., consolidated undrained test where the air and water phases are not allowed to drain).

Pore-air pressure can be controlled or measured through use of a porous element which provides continuity between the air voids in the soil and the air pressure system. The porous element must have a low attraction for water or an extremely low air-entry value in order to prevent water from entering the pore-air pressure system. The porous element can be a fiberglass cloth disk (Bishop, 1961b) or a coarse porous disk (Ho and Fredlund, 1982a).

The arrangement for applying a controlled air pressure to the soil specimen in a triaxial apparatus is shown in Fig. 11.33. A 3.2-mm-thick coarse corundum disk is placed between the soil specimen and the loading cap. The disk is connected to the pore-air pressure control through a hole drilled in the loading cap. A small-bore polythene tube then connects the loading cap to the base plate of the triaxial cell. Pore-air pressure can be controlled at a desired pressure using a pressure regulator from an air supply. In the case of a modified direct shear apparatus, the air pressure line is connected to the top of the chamber enclosing the shear box.

The measurement of pore-air pressures can be achieved using a small pressure transducer mounted on the loading cap. The volume of the measuring system should be kept to a minimum when measuring pore-air pressures in an undrained mode. Pore-air pressure is also difficult to measure in a closed system because of the ability of air to diffuse through rubber membranes, water, polythene tubing, and other materials. This is particularly true when considering the long test periods required when testing unsaturated soils.

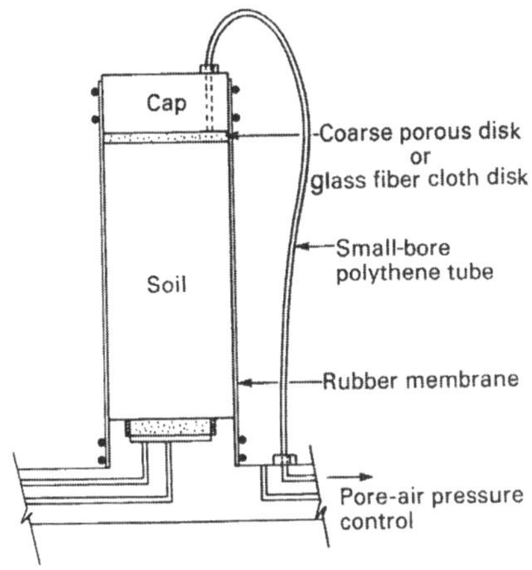

Figure 11.33 System for controlling pore-air pressure.

11.4.5 Water Volume Change Measurement

The past few decades have witnessed a number of new apparatuses for the measurement of the volume of water flowing in or out of soil specimens. The volume of water flowing in or out of unsaturated soils can be quite small. It is important that the volume of water movement in or out of the soil specimen be accurately measured. The quantity of water flowing in or out of the soil specimen can be quite small while at the same time possible leakages and air diffusion can produce significant errors. Consequently, measuring water volume changes has proven to be a challenge when testing unsaturated soils.

Conventional twin-burette volume change indicators should be slightly modified prior to usage for testing unsaturated soil specimens (Fig. 11.34). A small-bore burette (e.g., 10 mL volume) should be used as the central tube in order to achieve a volume measurement accuracy of 0.01 mL. Leakage needs to be essentially eliminated for tests that take long periods of time. There is a preferred procedure for making water volume change measurements when using twin burettes (Fredlund, 1973a). The burette opposite to the direction that the values point gives the most reliable volume change reading. The other burette also measures volume changes associated with water leakage and air diffusion.

The diffusion of pore-air through membranes around the soil specimen can be greatly reduced by using a composite membrane consisting of one or two sheets of slotted aluminum foil (Dunn, 1965). Silicone grease can be placed between the two aluminum sheets. Rubber membranes can be placed next to the soil specimen and on the outside of the aluminum sheets. Further details on the control of water

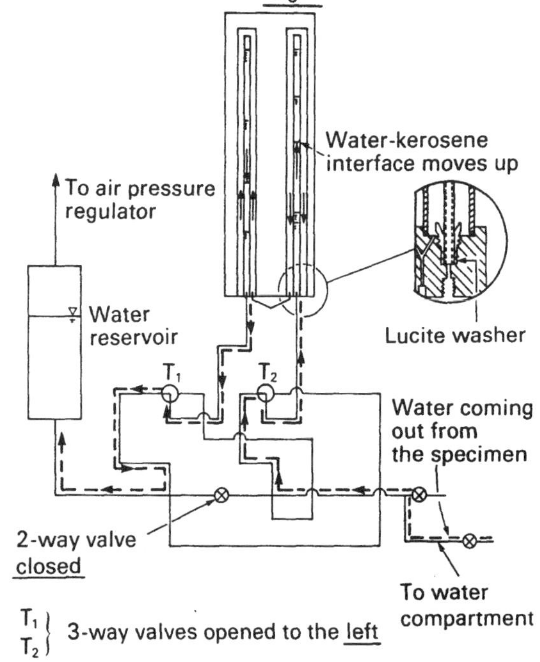

Figure 11.34 Direction of water movement for flow out from specimen with three-way valves opened to left.

and diffused air leakage in the triaxial test are given by Poulos (1964). Figure 11.35 shows the overall layout of the apparatus components associated with water volume change and the measurement of diffused air. Water volume changes must be corrected for the volume of air that diffuses through

Figure 11.36 Pressure-volume controller manufactured by GDS Instruments. (Courtesy of GDS Instruments, London.)

the high-air-entry disk and appear as water volume change on the water volume change indicator.

Several soil-testing equipment manufacturers have developed controlled-pressure piston- and cylinder-type apparatuses for measuring the flow of water in or out of soil specimens. The separation between air and water is still maintained through a high-air-entry disk at the base of the soil specimen. Figure 11.36 shows one such device (i.e., pressure/volume controller) manufactured by GDS Instruments, London. The controller can be used either in a pressure control mode or a volume change mode. Measurements of water volume change can be made when the controller is operated in a pressure control mode. Details of operation of the controller are illustrated in Fig. 11.37.

Air can also diffuse through high-air-entry disks, and it is necessary to have a device that can measure diffused air. Air diffusion should be taken into consideration whenever the

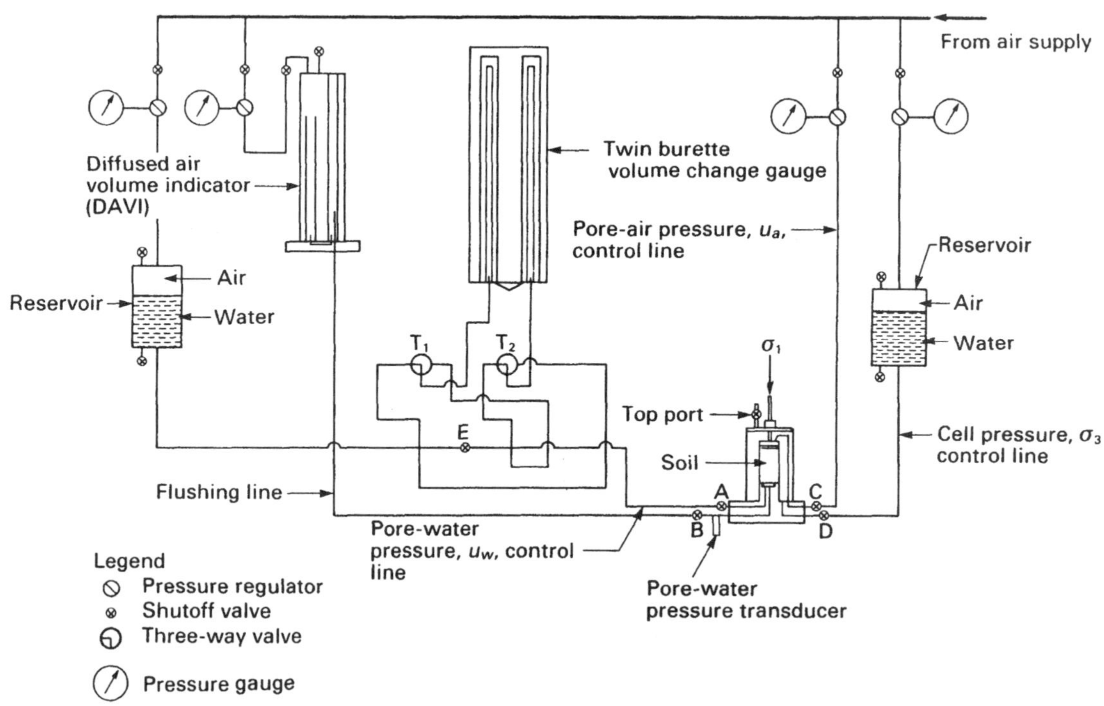

Figure 11.35 Schematic of control board and plumbing layout for modified triaxial apparatus.

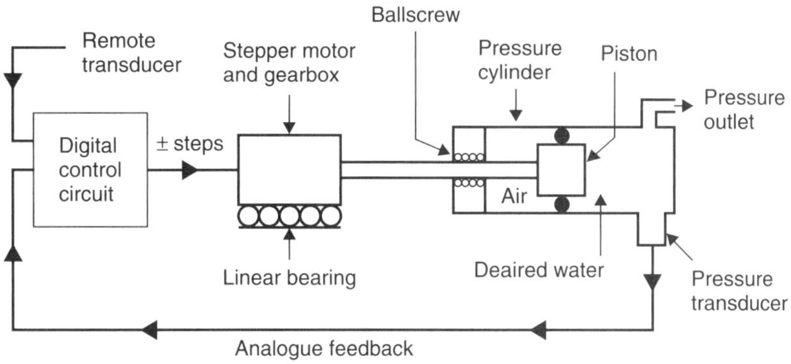

Figure 11.37 Details of operation of pressure-volume controllers. (Courtesy of GDS Instruments, London.)

duration of the laboratory test exceeds 1 or 2 days. Several other manufacturers also produce pressure-volume control devices similar to those manufactured by GDS Instruments.

11.4.6 Air Volume Change Measurement

The overall volume change of an unsaturated soil specimen is equal to the sum of the water and air volume changes. The soil particles and water can be assumed to be incompressible. The measurements of any two of the above-mentioned volume changes (i.e., overall volume change of the specimen, water volume change, or air volume changes) are necessary for describing the volume change behavior of an unsaturated soil. The overall volume change and water volume changes are generally measured while the air volume change is computed as the difference between the measured volume changes. Air volume change is somewhat difficult to measure due to its high compressibility and sensitivity to temperature change. Adams et al. (1996) showed, however, that it was possible to use a GDS apparatus to also measure air volume changes.

Figure 11.38 illustrates the use of two burettes to measure air volume change under atmospheric air pressure conditions (Bishop and Henkel, 1962; Matyas, 1967). Air moves out of a soil specimen through a coarse porous disk and is collected in a graduated burette. The burettes can be adjusted to maintain the air-water interface at the midheight of the specimen. The pore-air pressure is therefore maintained at atmospheric pressure conditions. The changing elevation of the air-water interface in the graduated burette indicates the air volume change in the soil specimen. Water is prevented from entering the burette because the gauge pore-water pressure is negative. In addition, the coarse porous disk has a low air-entry value. Water losses due to evaporation from the open burette can be prevented by covering the water surface with a layer of light oil (Head, 1986) or replacing the water entirely with a light oil (Matyas, 1967).

The above method for measuring air volume change is somewhat cumbersome. The apparatus shown is limited to making measurements under atmospheric conditions. However, the apparatus can be modified such that it will operate

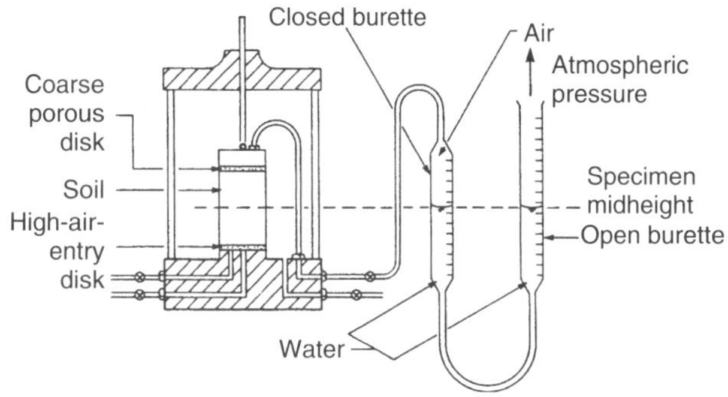

Figure 11.38 Air volume change measurements from triaxial cell under atmospheric pressure conditions.

under an applied backpressure. The apparatus has been primarily used to measure air coefficients of permeability.

11.4.7 Overall Volume Change Measurement

Overall volume change of a saturated soil specimen is equal to the measured water volume change. For an unsaturated soil, the water volume change constitutes only a part of the overall volume change of a soil specimen. Overall volume change in an unsaturated soil could be due to a change in the amount of either air or water in the soil specimen. Consequently, it is necessary to make a second volume change measurement independent of the water volume change measurement.

It is generally preferable to measure the overall volume change and also be able to separate horizontal strain components from the vertical strain component. The air phase is highly compressible, making its volume change somewhat difficult to measure. It would appear to be possible to measure total volume change by measuring the flow of water in or out from the triaxial cell fluid. The triaxial cell fluid can be connected to a volume change indicator that measures overall volume change of the soil specimen. Most such attempts to measure overall volume change by measuring fluid flow from the cell have met with limited success. These measurements are susceptible to temperature change and errors introduced by leakage and diffusion.

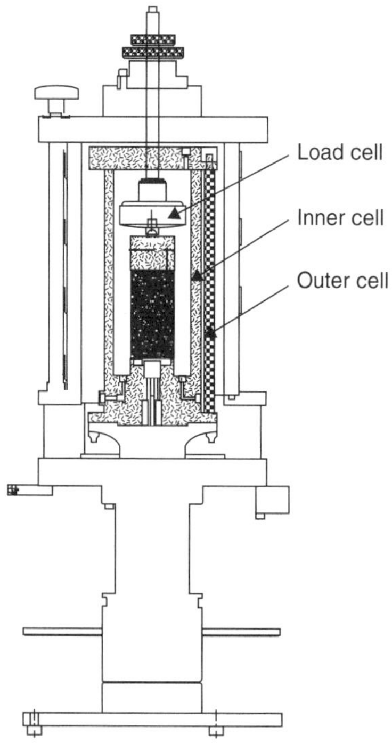

Figure 11.39 Double-walled triaxial cell to increase accuracy in measuring total volume change of soil specimen.

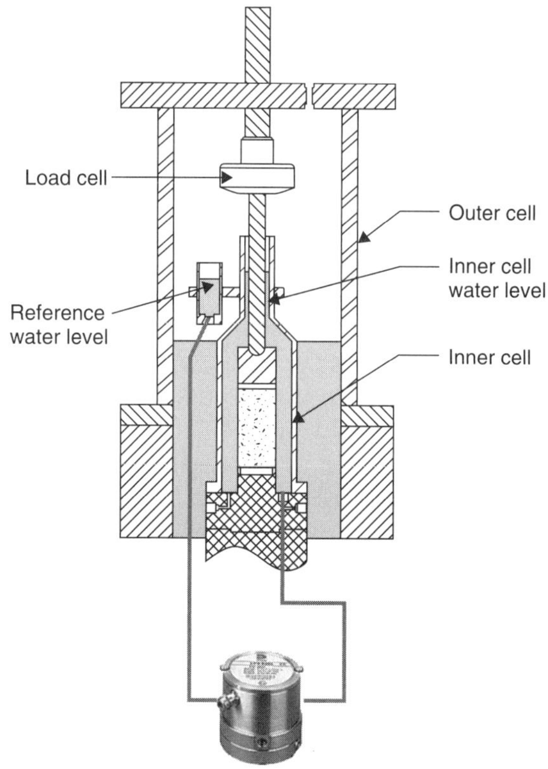

Figure 11.40 Double-walled triaxial cell that uses differential pressure measurements to determine total volume change of soil specimen. (Courtesy of GDS Instruments, London.)

Successful cell fluid measurements have been made using mercury as the cell fluid (Bishop et al., 1960). These measurements were also made using a double-walled triaxial cell in a temperature-controlled room. However, mercury is hazardous to health and its use should be avoided. Fluids other than mercury have also been used in conjunction with double-walled triaxial cells, as shown in Fig. 11.39 (Wheeler and Sivakumar, 1992).

There have been ongoing attempts to further improve measurements of total volume change of unsaturated soil specimens. One such system was developed at Hong Kong University of Science and Technology and later manufactured by GDS Instruments (GDS Instruments, 2011). Figure 11.40 shows how a double-walled cell can be used in conjunction with an accurate differential pressure transducer to measure soil specimen volume changes. The differential pressure transducer measures changes in water level between a reference water level and fluctuations in a reduced-diameter cylinder surrounding the vertical piston. The system shown in Fig. 11.40 is presently marketed by GDS Instruments, London.

The overall volume change of a triaxial soil specimen has also been attempted by independently measuring the vertical deflection and radial deformation of the specimen during the test. The vertical deflection of the soil specimen can be measured using a conventional dial gauge or an LVDT (i.e., linear variable differential transformer). LVDTs have accuracy comparable to that of dial gauges. Some LVDTs can be submerged in oil or water. Various applications of LVDTs in triaxial testing are described by Head (1986).

Noncontacting transducers have been used during triaxial testing programs (Cole, 1978; Khan and Hoag, 1979; Drumright, 1987). The device consists of a sensor and an aluminum target (Fig. 11.41). The sensor is a displacement transducer clamped to a post and connected to the electronic measuring system through a port in the base plate. The aluminum target can be attached to the rubber membrane (e.g., using silicon grease), near the midheight of the specimen. Three sensors and targets can be placed around the circumference of the specimen at 120° intervals (Fig. 11.41).

The noncontacting transducers operate on an eddy current loss principle. An eddy current is induced in the aluminum target by a coil in the sensor. The magnitude of the induced eddy current is a function of the distance between the sensor and the aluminum target. The distance changes between the aluminum target and the sensor as the specimen deforms laterally. The distance change causes a change in the magnitude of the generated eddy current. Impedance in the coil changes, resulting in a change in the DC voltage output.

Another system that has proven to work well involves the use of a combined radial and axial set of LVDTs, as shown in Fig. 11.42. The measuring system is clamped onto the soil specimen near midheight. The average axial strain is

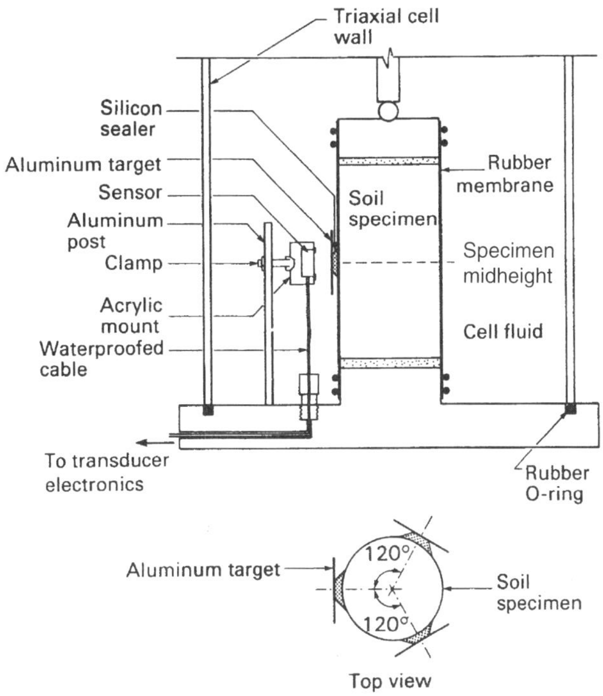

Figure 11.41 Installation of noncontacting radial deformation transducer (after Drumright, 1987).

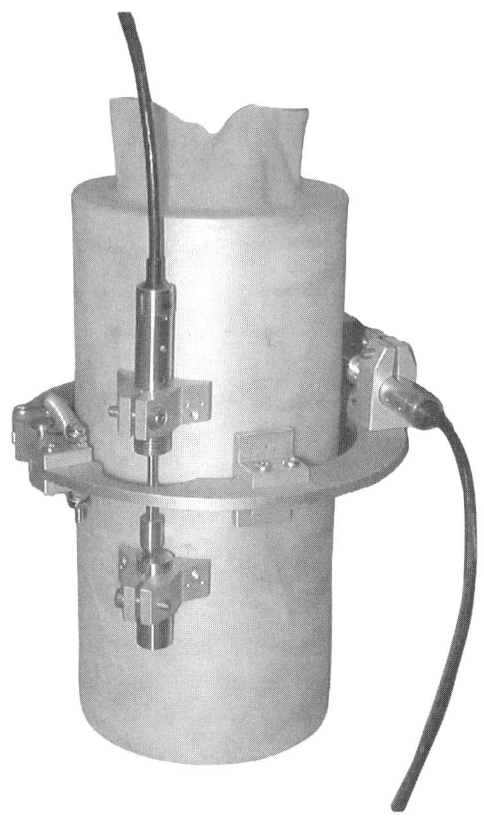

Figure 11.42 Local axial and radial measurement of strain. (Courtesy of GDS Instruments, London.)

measured along with the local axial strain near the middle of the specimen.

11.5 TRIAXIAL TEST PROCEDURES FOR UNSATURATED SOILS

The testing of unsaturated soils requires great care on the part of the technician in order to obtain reliable results. There are a number of technician-related factors that influence the outcome of the test. The duration of unsaturated soil triaxial tests is generally considerably longer than the duration of comparable saturated soil tests. The difference in duration is mainly related to the reduced coefficient of permeability of unsaturated soils.

There is a comparable unsaturated soil triaxial test for each type of commonly performed saturated soil triaxial test. However, the test procedures for unsaturated soils are generally more demanding. Since there are two fluid phases involved (i.e., water and air), it is possible for one fluid phase to be tested in a drained mode while the other fluid phase is tested in an undrained mode. This is the situation when performing a constant-water-content test.

The following sections give consideration to each of the types of triaxial tests that can be performed on an unsaturated soil. In each case, the computation of the stress states associated with each test procedure is illustrated.

11.5.1 Specimen Preparation

Unsaturated soil specimens obtained from either undisturbed or compacted samples can be used for shear strength testing. The unsaturated soil may initially have high matric suction, but the test might be performed at lower matric suction. For example, a multistage test on an unsaturated soil often has the matric suction reduced for the first stage while subsequent stages are conducted at higher applied matric suctions (Ho and Fredlund, 1982b; Gan et al., 1988).

It is often necessary to relax the high initial matric suction in the specimen prior to performing the test. One way to reduce the initial matric suction is to impose an applied air and water pressure such that the corresponding matric suction approaches zero. The water from the compartment below the high-air-entry disk must flow upward into the specimen toward a new equilibrium condition. The equilibration process, however, may require a long time due to the low permeability of the high-air-entry disk. The relaxation of the initial matric suction can be expedited by wetting the specimen from the sides and top of the soil specimen. The relaxation of the initial matric suction is not required for some modes of testing such as undrained and unconfined compression.

11.5.1.1 Matric Suction Relaxation for Triaxial Test

A procedure used to relax the initial matric suction for multistage triaxial testing was described by Ho and

Fredlund (1982b). The specimen was first trimmed to the desired diameter and height and then mounted on the presaturated high-air-entry disk. Appropriate measurements of the volume-mass properties of the specimen were taken during the setup of the specimen. A coarse porous disk and loading cap were placed on top of the specimen. The specimen was then enclosed in two rubber membranes. The specimen had a composite membrane consisting of two slotted aluminum foil sheets between rubber membranes. The purpose of the aluminum foil is to greatly minimize air diffusion from the specimen. O-rings were placed over the membranes on the bottom pedestal.

Spacers (i.e., consisting of pieces of 3.2-mm plastic tubing) were inserted between the membranes and the loading cap to allow air within the specimen to escape while water was added to the sides of the specimen. The cylinder of the triaxial cell was installed and the cell was partially filled with water (i.e., about one-half way up the specimen). Water was allowed access to the soil specimen either by being added manually or by being applied through the air pressure line connected to the loading cap.

The specimen was left for several hours to allow the distribution of water throughout the specimen. The relaxation process was continued until air could no longer be seen escaping from around the top of the specimen. The matric suction in the soil was assumed to be essentially zero at the end of the wetting process. The desired matric suction value was then applied to the specimen. Time was allowed for equalization to the new applied stress conditions.

11.5.1.2 Matric Suction Relaxation for Direct Shear Test

A similar relaxation procedure has been used to prepare soil specimens for direct shear testing (Gan, 1986; Escario and Sáez, 1986). The two halves of the direct shear box are lightly sealed together using vacuum grease. The vacuum grease ensures that water applied to the top of the soil specimen will flow toward the high-air-entry disk. It is important to *not* smear vacuum grease onto the surface of the high-air-entry disk.

The soil specimen is mounted into the shear box. Then the coarse porous stone and loading cap are installed. The initial matric suction in the soil specimen can be relaxed by adding water to the top of the soil specimen.

11.5.2 Triaxial Test Procedures

Triaxial testing equipment allows flexibility in that a variety of test procedures can be used. The triaxial cell must be modified in accordance with using the special design considerations previously explained. Several procedural checks should be conducted prior to performing each test. The high-air-entry disk should be saturated. Attempts should be made to thoroughly flush water through the compartment below the high-air-entry disk and all the connecting lines to ensure the expulsion of air bubbles. The volume change measuring devices, including the diffused air volume indicator, should be initialized (Fredlund, 1973a).

The soil specimen should then be prepared for testing. The initial confining air and water pressures to be applied to the soil specimen can be set on the pressure regulators. The confining pressure along with the air and water pressures can be applied to the soil specimen. An initial water pressure of 30 kPa or greater is desirable in order to provide sufficient differential pressure for flushing air from the base plate. The diffused air volume that accumulates while performing the test can be measured on the diffused air volume indicator.

11.5.3 Consolidated Drained Test

The initial stress state for the soil specimen is established by applying a prescribed confining pressure σ_3, pore-air pressure u_a, and pore-water pressure u_w. The confining, pore-air, and pore-water pressures are applied by opening the appropriate valves on the triaxial control panel in the mentioned respective order. Other valves on the base pedestal remain closed during the test except during the flushing of diffused air from below the high-air-entry disk.

The vertical deflection and the radial deformation are periodically monitored to measure the overall volume change of the specimen. The volume of water flowing in or out of the specimen is recorded on the water volume change indicator. The water volume change indicator valves remain open during a consolidated drained test, except when flushing diffused air from the base plate. Air volume change is generally not measured. Consolidation is assumed to have reached an equilibrium condition when there is no longer a tendency for the overall volume to change or the flow of water from the specimen.

Upon attaining an equilibrium condition under the applied pressures (i.e., σ_3, u_a, and u_w), the specimen is *sheared* by compression at an appropriate strain rate. The magnitude of the axial load applied to the specimen can be recorded using a load cell. The axial load measurements are converted to a deviator stress $\sigma_1 - \sigma_3$. The shearing process is conducted under drained (or open-valve) conditions for the applied pore-air and pore-water pressures. The overall and water volume changes are monitored throughout the shear process. The shearing process is terminated when the selected failure criterion (e.g., maximum deviator stress) has been achieved.

Diffused air is generally flushed from the base plate once a day during both consolidation and shearing. The frequency of the diffused air measurement depends on the applied air pressure and the air-entry value of the ceramic disk. The diffused air volume can be measured less frequently when the applied air pressure is low. The water volume change correction associated with diffused air becomes necessary whenever tests extend over a period of several days.

The diffused air in the base plate can be flushed into the DAVI by applying a pressure differential of 7 kPa or higher between the base plate and the diffused air volume indicator. Each apparatus needs to be tested to assess the differential pressure at which diffused air can readily be removed from the base compartment. If the applied differential pressure is too high, the water from below the compartment may tend to cavitate and form small air bubbles. It is also desirable to not significantly alter the water pressure in the base plate. It may be necessary to elevate the air backpressure in the diffused air volume indicator during the flushing process.

The valves on the water volume change indicator are closed during the flushing process so that the water volume change indicator is bypassed. Diffused air can be flushed from the base plate using a series of water surges through the compartment below the high-air-entry disk. Diffused air moves into the diffused air volume indicator and displaces the water in the burette. A few seconds may be required between each surge. The water pressure in the base plate only deviates momentarily from its set value when using this procedure. The valves to the water volume change indicator are opened while the valves associated with the flushing process are closed after the diffused air volume has been measured.

The procedure involving consolidation and shearing is repeated during each stage of a multistage test. The consolidation for each stage should preferably be performed with the deviator stress reduced to zero. The deviator stress can be brought to zero by release of the axial load. The shearing process for each stage of a multistage test should be stopped when the maximum deviator stress is imminent, except for the last stage, where the specimen can be sheared to a larger strain.

11.5.4 Constant-Water-Content Test

The initial consolidation process is carried out in the same manner for the constant-water-content test as for the consolidated drain test. When equilibrium conditions have been achieved under the applied pressures (i.e., σ_3, u_a, and u_w), the soil specimen is sheared under drained conditions for the pore-air phase and undrained conditions for the pore-water phase. The pore-air pressure is maintained at the value applied during consolidation. The pore-air pressure valve remains open during consolidation and shear. On the other hand, the pore-water pressure valve remains closed during shear in order to produce undrained pore-water conditions. The pore-water pressure is measured using a pressure transducer mounted near the base plate of the triaxial cell.

Diffused air volume should also be measured during shear under undrained water phase conditions. The water pressure in the base plate should be recorded prior to the flushing process and reset to the same value after flushing. The water in the pore-water pressure control line should first be subjected to the same pressure as recorded in the base plate. The air backpressure in the diffused air volume indicator should be adjusted to a pressure slightly lower than the water pressure recorded in the base plate. When the valve on the water line to the base plate is opened, there will be a quick equalization of the water pressure in the compartment below the high-air-entry disk.

Diffused air is removed from the base plate by momentarily opening the water pressure valve, which produces a pressure difference across the base plate. The valves to the diffused air volume indicator are closed at the end of the diffused air volume measurement. The undrained pore-water pressure is then returned to the value existing prior to the flushing process. Disturbance to the undrained condition of the soil specimen should be minimal if the flushing time is short.

11.5.5 Consolidated Undrained Test with Pore Pressure Measurements

The soil specimen is consolidated using the procedure described for the consolidated drained test. After equilibrium conditions have been established under the applied pressures (i.e., σ_3, u_a, and u_w), the soil specimen is sheared under undrained conditions with respect to the air and water phases. Undrained conditions during shear are achieved by closing valves on the pore-air line and the pore-water line.

The pore-water pressure developed during shear can be measured on the pressure transducer mounted on the base plate. A pore-air pressure transducer should be mounted on the loading cap, if possible, when measuring pore-air pressure changes. It may be somewhat difficult to maintain undrained air conditions due to air diffusion.

The diffused air volume can be measured during the constant-water-content test; however, there is no way of correcting the measured pore-water pressures for air diffusion. Difficulties associated with air diffusion may be the main reason why few consolidated undrained tests with pore-air and pore-water pressure measurements have been performed. Thicker high-air-entry disks help to reduce the effect of air diffusion.

11.5.6 Undrained Test

The procedure for performing an undrained test on an unsaturated soil specimen is similar to the procedure used when performing an undrained test on a saturated soil specimen. The unsaturated soil specimen is tested at its initial water content or matric suction. In other words, the initial matric suction in the specimen is not relaxed or changed prior to performing the shear strength test.

There is no consolidation process allowed since the confining pressure σ_3 is applied under undrained conditions for both the pore-air and pore-water phases. The specimen is axially compressed under undrained conditions with respect to both the air and water phases. The undrained test is usually run at a strain rate between 0.017 and 0.03% per second, and no attempt is made to measure the pore-air and pore-water pressures. Conventional triaxial equipment can be used to perform the undrained test on unsaturated soils.

The porous disks are usually replaced by metal or plastic disks on the top and bottom of the soil specimen. The specimen is enclosed in a rubber membrane during the test. The undrained test results on unsaturated soils can be interpreted within the context of unsaturated soil shear strength theory.

11.5.7 Unconfined Compression Test

The unconfined compression test procedure is similar to the undrained test procedure with the exception that no confining pressure is applied to the soil specimen (i.e., $\sigma_3 = 0$). The test is commonly performed in a simple loading frame by applying an axial load to the soil specimen.

11.6 INTERPRETATION OF TRIAXIAL TEST RESULTS

Various test procedures can be used for triaxial shear strength testing of unsaturated soils. It is important to interpret the meaning of the test results within the context of unsaturated soil shear strength theory. There are basic principles of interpretation that can be used for each of the triaxial test procedures. The triaxial test is usually performed on a cylindrical soil specimen that is enclosed in a rubber membrane and placed in a triaxial cell. The cell is filled with water and pressurized in order to apply a constant all-around pressure or confining pressure. The soil specimen can be subjected to an axial stress through a loading ram in contact with the top of the soil specimen.

The application of the confining pressure is the first stage of a triaxial test. Either the soil specimen can be allowed to drain (i.e., consolidate) during the application of the confining pressure or drainage can be prevented. The term *consolidation* is used to describe the process whereby excess pore pressures due to the applied stress are allowed to dissipate, resulting in volume change. The pore fluids are allowed to drain following the application of the confining pressure. On the other hand, the consolidation process will not occur if the pore fluids are maintained in an undrained mode. The terms "consolidated" and "unconsolidated" are used as the first designation for categorizing triaxial tests.

The application of the axial stress is considered as the second stage or the shearing stage in the triaxial test. Shear stresses are generally applied to the soil specimen through the application of a vertical compressive load. The total confining pressure generally remains constant during shear. The axial stress is continuously increased until a failure condition is reached or a limiting strain condition is achieved. The axial stress generally becomes the total major principal stress σ_1 in the axial direction while the isotropic confining pressure becomes the total minor principal stress σ_3 in the lateral direction. The total intermediate principal stress σ_2 is equal to the total minor principal stress σ_3 (i.e., $\sigma_2 = \sigma_3$).

Figure 11.43 illustrates the stress conditions associated with a consolidated drained triaxial test. The pore fluid drainage conditions during the shearing process are used as the second designation in categorizing triaxial tests. The test is referred to as a drained test when the pore fluid is allowed to flow in or out of the soil specimen during shear. On the other hand, a test is called an undrained test if the flow of pore fluid is prevented. The pore-air and pore-water phases can also have differing drainage conditions during shear.

Triaxial test procedures for unsaturated soils are designated based upon the drainage conditions adhered to during the first and second stages of the triaxial test. The

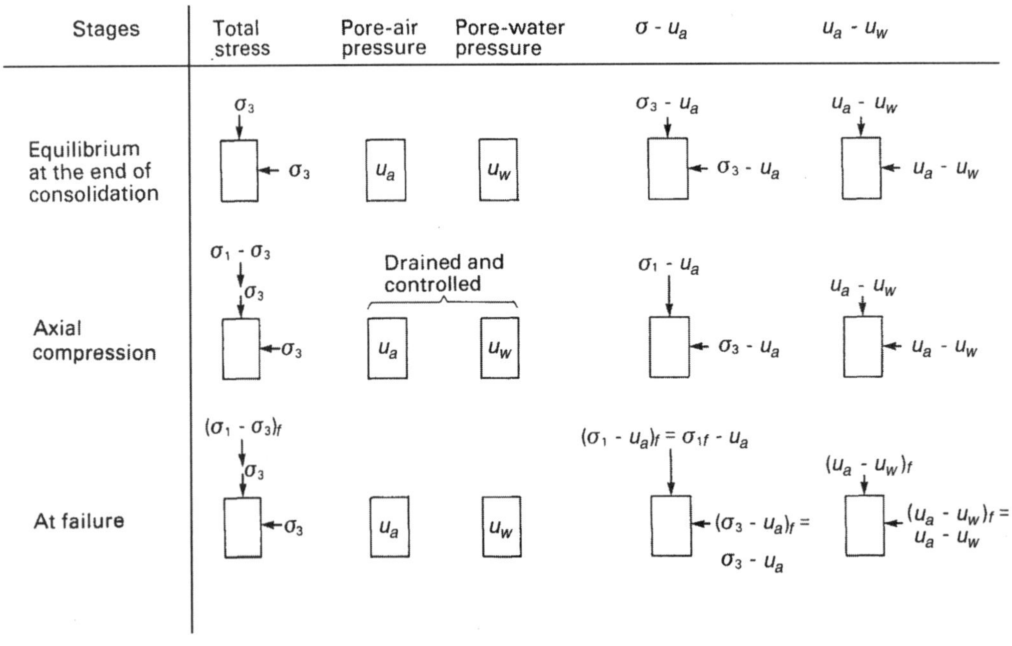

Figure 11.43 Stress conditions at various stages of consolidated drained triaxial compression test.

triaxial test methods are usually given a two-word designation or abbreviated to a two-letter symbol. The designations are (1) consolidated drained (CD) test, (2) constant-water-content or (CW) test, (3) consolidated undrained (CU) test with pore pressure measurements, (4) undrained (UU) test, and (5) unconfined compression (UC) test.

In the case of CD and CU tests, the first letter refers to the drainage condition prior to shear while the second letter refers to drainage condition during shear. The constant-water-content test is a special case where only the pore-air is maintained in a drained mode while the pore-water phase is kept in an undrained mode during shear (i.e., constant water content). The pore-air and pore-water are not allowed to drain throughout the test for the undrained triaxial test. The unconfined compression test is a special loading condition for an undrained triaxial test. These five testing procedures are explained in the following sections. A summary of triaxial testing conditions along with the measurements performed is given in Table 11.3. The air, water, or total volume changes may or may not be measured during the triaxial tests.

The shear strength data obtained from triaxial tests can be analyzed using the stress states at failure or the total stresses at failure when the pore pressures are not known. This concept is similar to the effective stress approach and the total stress approach used in saturated soil mechanics. The pore pressures are controlled at a desired value during shear in a drained test. Any excess pore pressures caused by the applied loads are dissipated by allowing the pore fluids to flow in or out of the soil specimen. The pore pressures at failure are known values. The stress state variables at failure can be used to analyze the shear strength data.

The excess pore pressures generated in an undrained test can build up because pore fluid flow is prevented during shear. The pore pressures at failure are known if the changing pore pressures during shear are measured. In this case, the stress state variables can be computed. However, the stress state variables are unknown if pore pressure measurements are not made during undrained shear. In this case, the shear strength can only be interpreted in terms of total stresses at failure.

The total stress approach should be applied in the field only in situations where it is reasonable to assume that the strength measured in the laboratory has relevance to the drainage conditions being simulated in the field. In other words, the applied total stresses that cause failure in the soil specimen are assumed to be the same as the applied total stresses that will cause failure in the field. The above simulation assumes that the stress state variables control the shear strength of the soil. It is difficult, however, to closely simulate field loading conditions with an undrained test in the laboratory.

11.6.1 Consolidated Drained Test

The CD test refers to a test condition where the soil specimen is first consolidated and then sheared under drained conditions for both the pore-air and pore-water phases, as illustrated in Fig. 11.43. The soil specimen is consolidated to a stress state corresponding to the range of total stresses likely to be encountered in the field or in design. The soil is generally consolidated under an isotropic confining pressure of σ_3, while the pore-air, u_a, and pore-water, u_w, pressures are controlled at selected values. The pore-air and pore-water pressures can be controlled at positive values to overcome the possibility of cavitation in the pore-water pressure measuring system (i.e., the axis translation technique). The soil specimen has a net confining pressure of $\sigma_3 - u_a$ and a matric suction of $u_a - u_w$ at the end of the consolidation process.

The soil specimen is compressed in the axial direction by applying a deviator stress (i.e., $\sigma_1 - \sigma_3$) during the shearing process. The drainage valves for both pore-air and pore-water remain open (i.e., under drained conditions). The pore-air and pore-water pressures are controlled at constant values (i.e., equal to the pressures at the end of consolidation). The deviator stress is applied slowly in order to prevent the generation of excess pore-air or pore-water pressure in the soil. The net confining pressure $\sigma_3 - u_a$, and the matric suction, $u_a - u_w$, remain constant throughout the test until failure conditions are reached, as indicated in Fig. 11.43 [i.e., $(\sigma_3 - u_a)_f = \sigma_3 - u_a$ and $(u_a - u_w)_f = u_a - u_w$]. Only the deviator stress $\sigma_1 - \sigma_3$, keeps increasing during shear until

Table 11.3 Various Triaxial Tests for Unsaturated Soils

Test Methods	Consolidation Prior to Shearing Process	Drainage		Shearing Process		
		Pore-Air	Pore-Water	Pore-Air Pressure, u_a	Pore-Water Pressure, u_w	Soil Volume Change, ΔV
Consolidated drained	Yes	Yes	Yes	C	C	M
Constant water content	Yes	Yes	No	C	M	M
Consolidated undrained	Yes	No	No	M	M	
Undrained	No	No	No			
Unconfined compression	No	No	No			

[a]*Note:* M = measurement, C = controlled.

the net major principal stress reaches a value of $(\sigma_1 - u_a)_f$ at failure.

Blight (1967) reported the results of several consolidated drained tests performed on unsaturated silt specimens. All specimens were compacted at a water content of 16.5% using the standard AASHTO compaction effort. The specimens were then brought to equilibrium at three matric suction values in a triaxial cell. Two specimens with constant matric suction values were tested at two different net confining pressures $\sigma_3 - u_a$ (i.e., 13.8 and 27.6 kPa). The deviator stress versus strain curves are shown in Fig. 11.44a. The results indicate an increase in shear strength with increasing matric suction and increasing net confining pressure.

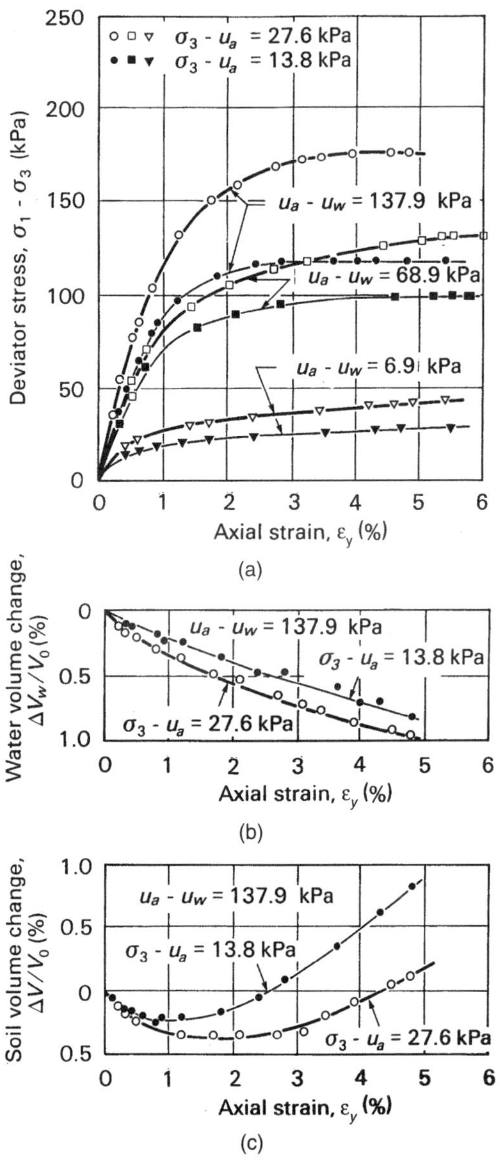

Figure 11.44 Consolidated drained tests on unsaturated silt: (a) typical deviator stress versus strain curves; (b) water volume change versus strain relations; (c) specimen volume change versus strain relations (after Blight, 1967).

The water volume change (i.e., $\Delta V_w/V_0$) and overall specimen volume change (i.e., $\Delta V/V_0$) during compression are presented in Figs. 11.44b and 11.44c, respectively, for the specimens sheared under two different net normal stresses.

Typical stress paths followed during consolidated drained tests subjected to constant matric suction are illustrated in Fig. 11.45. The tests are performed on several specimens at various net confining pressures. For example, stress point A represents the stress state at the end of consolidation when the soil specimen had a net confining pressure $\sigma_3 - u_a$, and a matric suction $u_a - u_w$. As the soil is compressed during shear, the stress point moves from point A to point B along the stress path $\overline{AB}$. Stress point B represents the stress state at the condition of failure. When moving from stress point A to stress point B, the Mohr circle diameter or the deviator stress increases until the failure condition is reached at stress point B. However, the net confining pressure and the matric suction remain constant throughout the stress path $\overline{AB}$. A line drawn tangent to the Mohr circles at failure (i.e., through stress points C, C_1, and C_2) represents the failure envelope corresponding to the matric suction used for the tests. The failure envelope has a slope angle of ϕ' with respect to the $\sigma - u_a$ axis. The friction angle appears to be essentially equal to the effective angle of internal friction obtained from shear strength tests on saturated soil specimens. The friction angle ϕ' for compacted soils commonly lies in the range of 25°–35°, as shown in Table 11.1. The effect of compactive effort on the saturated shear strength parameters ϕ' and c' for a clayey sand is illustrated in Fig. 11.46. The results indicate that densification of a soil using different compaction efforts results in a soil with differing saturated shear strength parameters; that is, the soils should be considered as different soils.

Figure 11.47 presents the stress paths followed during consolidated drained tests under a constant net confining pressure and various matric suctions. The Mohr circle at failure increases in diameter as the matric suction increases at failure. The Mohr circle at failure is tangent to the failure envelope corresponding to the matric suction used in the test (e.g., at stress points C_1, C_2, and C_3). However, stress points C_1, C_2, and C_3 do not occur at the same net normal stress. Therefore, a line joining the stress points C_1, C_2, and C_3 will not give the angle ϕ^b. Rather, it is suggested that the failure envelope be extended to intersect the shear strength versus $u_a - u_w$ plane to give cohesion intercepts. A line joining the cohesion intercepts at various matric suctions gives the angle ϕ^b.

11.6.2 Constant-Water-Content Test

The specimen is first consolidated and then sheared for CW triaxial tests. The pore-air phase is allowed to drain while the pore-water phase is maintained in an undrained mode. The consolidation procedure is similar to that of the consolidated drained test. The axis translation technique can be used to impose matric suctions greater than 1 atm. The soil specimen

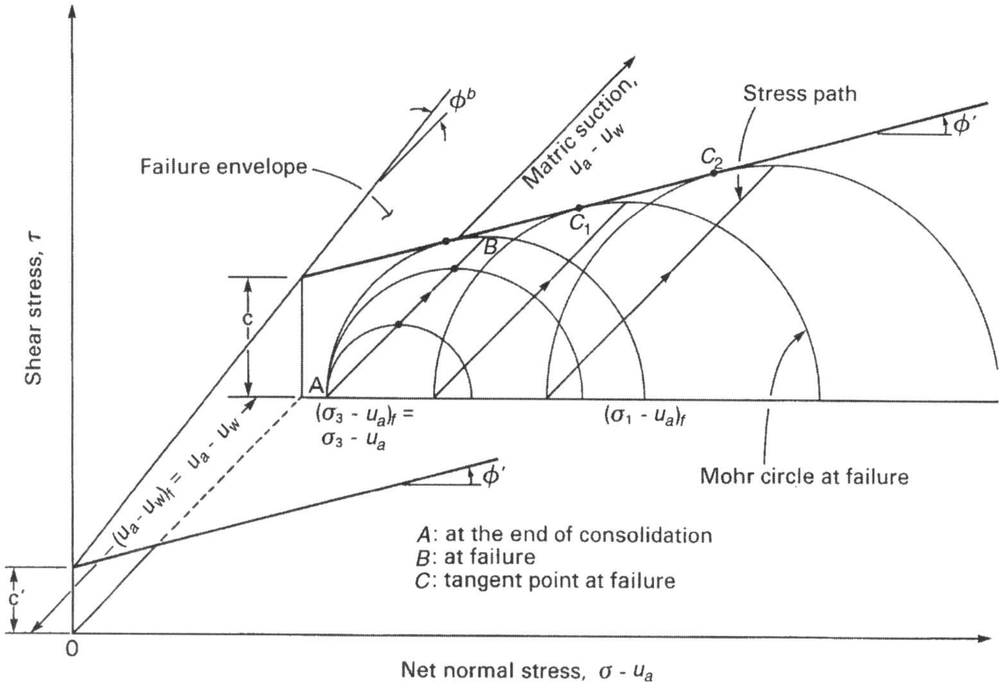

Figure 11.45 Stress paths followed during consolidated drained test at various net confining pressures under same applied matric suction.

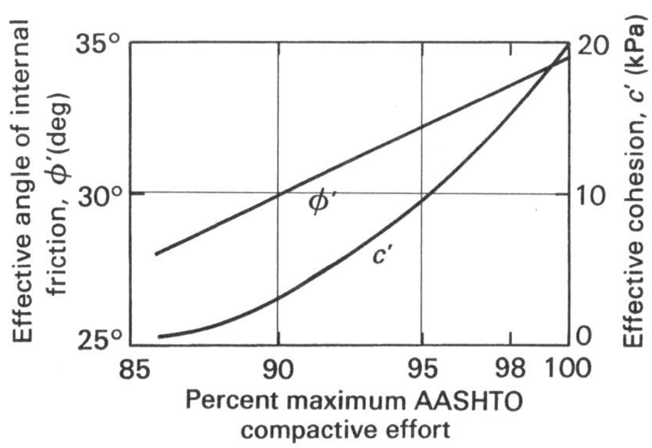

Figure 11.46 Effect of compactive effort on effective shear strength parameters ϕ' and c' for clayey sand (after Moretto et al., 1963).

has a net confining pressure of $\sigma_3 - u_a$ and a matric suction of $u_a - u_w$ when equilibrium is reached at the end of consolidation. The specimen is sheared by increasing the deviator stress $\sigma_1 - \sigma_3$, until failure is reached. During shear, the drainage valve for the pore-air remains open (i.e., under drained conditions), while the drainage valve for the pore-water is closed (i.e., under undrained conditions). The pore-air pressure u_a is maintained at the pressure applied during consolidation. The pore-water pressure u_w changes during shear under undrained loading conditions. The excess pore pressure is related to deviator stress using the D pore pressure parameter. The net confining pressure $\sigma_3 - u_a$ remains constant throughout the test while the matric suction $u_a - u_w$ changes, as illustrated in Fig. 11.48 [i.e., $(\sigma_3 - u_a)_f = \sigma_3 - u_a$ and $(u_a - u_w)_f = u_a - u_w - \Delta u_{wf}$]. The net major principal stress reaches a value of $(\sigma_1 - u_a)_f$ at failure.

Typical stress versus strain curves for constant-water-content tests are shown in Fig. 11.49a. The stress versus strain curves obtained from the constant-water-content tests are similar in shape to those obtained from the consolidated drained triaxial test. The maximum deviator stress increases with an increase in the net confining pressure $\sigma_3 - u_a$ for soil specimens prepared at the same initial matric suction. The matric suction and the soil volume changes during shear are shown in Figs. 11.49b and 11.49c, respectively. Matric suction decreases during shear since the pore-air pressure is maintained at a constant value while the pore-water pressure increases. The degree of saturation of the soil specimen increased as the pore voids compressed in response to pore-air being squeezed out of the soil. The water content remained constant. Figure 11.49c indicates that the soil specimens undergo compression until the maximum deviator stress is reached. The soil specimen with the lower net confining pressure dilates after reaching the maximum deviator stress. Dilation is accompanied by a slight increase in matric suction.

A hypothetical stress path that may be followed by a soil specimen during a constant-water-content test is shown in Fig. 11.50. Stress point A represents the stress state at the end of consolidation where the soil specimen has a net confining pressure $\sigma_3 - u_a$ and a matric suction $u_a - u_w$. The

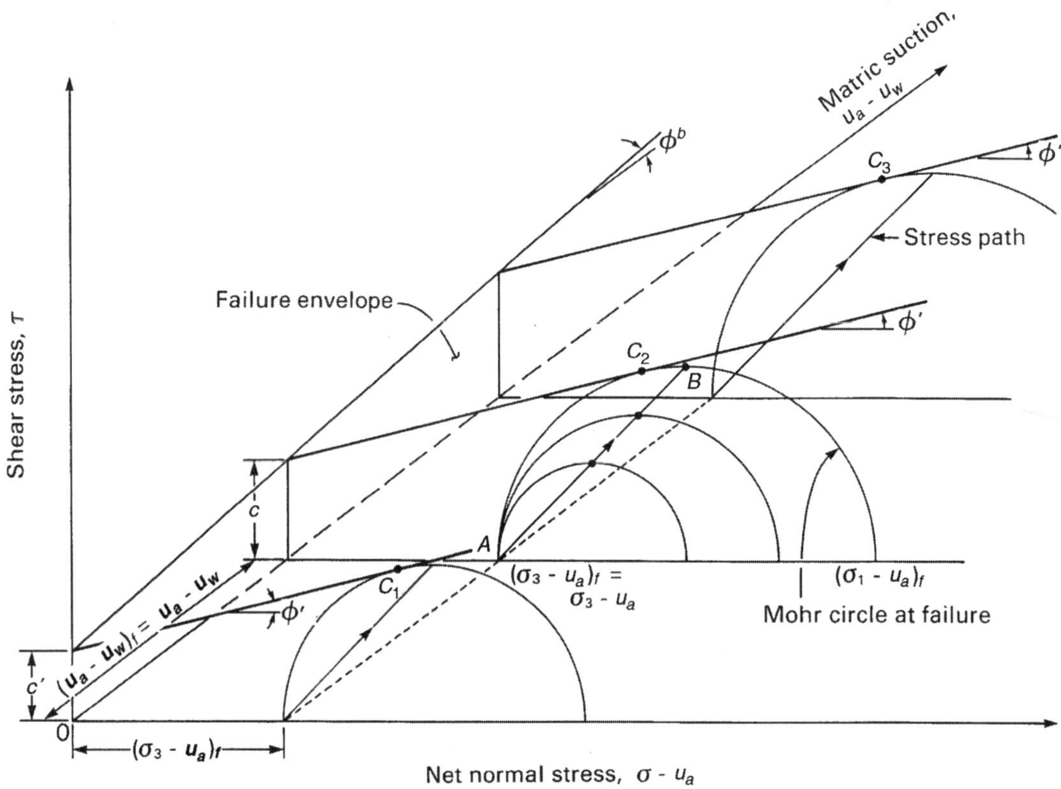

Figure 11.47 Stress paths followed during consolidated drained tests at varying applied matric suctions with constant net confining pressure.

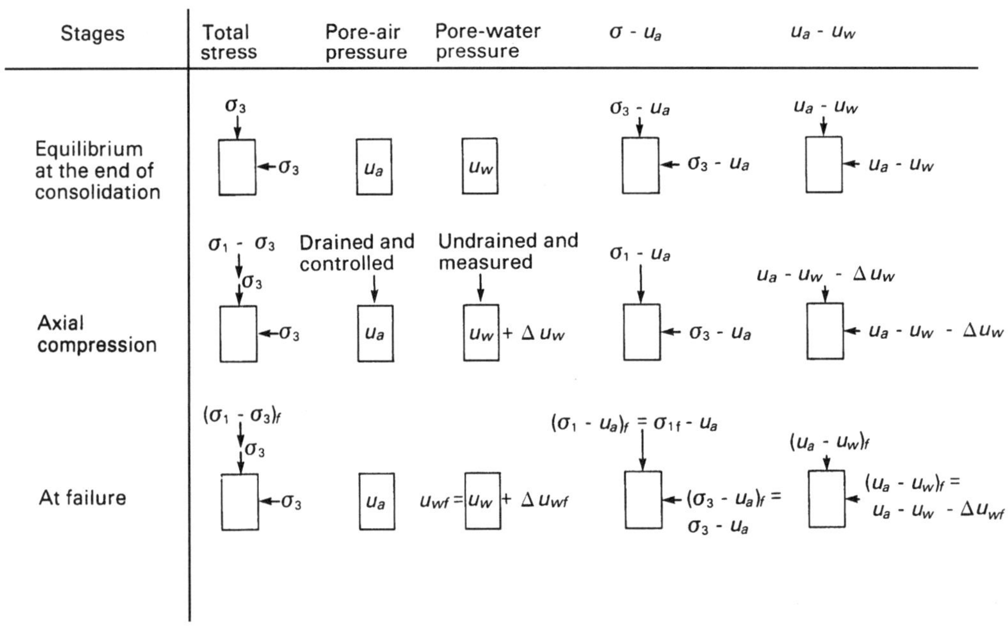

Figure 11.48 Stress conditions at various stages of constant-water-content triaxial compression test.

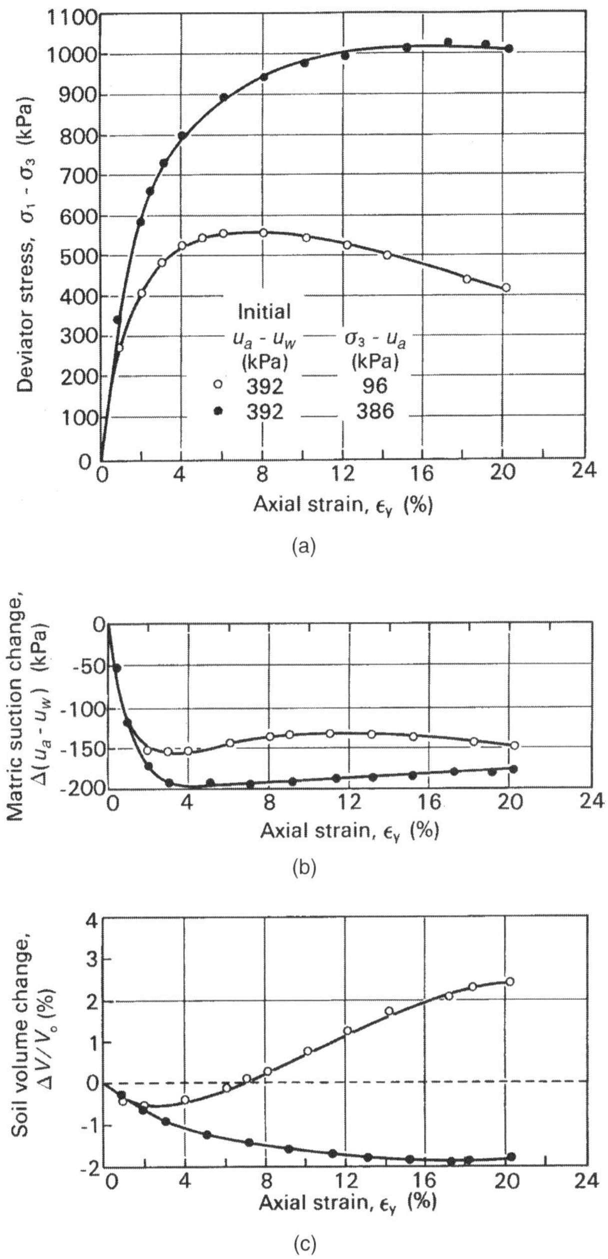

Figure 11.49 Constant-water-content triaxial tests on Dhanauri clay: (a) stress versus strain curve; (b) matric suction change versus strain; (c) soil specimen volume change versus strain (after Satija, 1978).

stress point is assumed to move from point A to point B along stress path $\overline{AB}$ as the soil is compressed during shear. Stress point B represents the stress state at failure.

The net confining pressure remains constant at $\sigma_3 - u_a$ along stress path $\overline{AB}$ since the pore-air pressure is maintained at the pressure used during consolidation. The pore-water pressure is assumed to increase continuously during shear. The result is a reduction in matric suction [i.e., $(u_a - u_w)_f < u_a - u_w$]. The failure envelope with an angle ϕ' can be drawn tangent to the Mohr circle at failure (e.g., at stress point C). The failure envelope intersects the shear strength versus matric suction plane at a cohesion intercept c. The cohesion intercepts obtained at various matric suctions can be joined to give the ϕ^b angle.

11.6.3 Consolidated Undrained Test with Pore Pressure Measurements

CU tests are performed with the soil specimen first consolidated and then sheared under undrained conditions. The pore-air and pore-water phases are undrained during the shearing process, as shown in Fig. 11.51. The consolidation process brings the soil specimen to the desired initial stress state (i.e., $\sigma_3 - u_a$ and $u_a - u_w$). The axis translation technique is used to establish matric suctions greater than 1 atm. Once equilibrium conditions are achieved, the soil specimen is sheared by increasing the axial load $\sigma_1 - \sigma_3$ until failure is reached.

The drainage valves for the pore-air phase and the pore-water phase are closed (i.e., undrained conditions) during shear. Excess pore-air and pore-water pressures are developed during undrained loading. The excess pore pressures are related to the deviator stress through use of the D pore pressure parameter (see Chapter 15). The pore-air and pore-water pressures should be measured during the shear process. The net confining pressure $\sigma_3 - u_a$ and matric suction $u_a - u_w$, are altered throughout the test due to changing pore-air and pore-water pressures. The magnitudes of the net major and minor principal stresses as well as the matric suction are a function of the pore pressures at failure.

A typical stress path for a consolidated undrained test is illustrated in Fig. 11.52. The stress state at the end of consolidation is represented by stress point A where the net confining pressure is $\sigma_3 - u_a$ and the matric suction is $u_a - u_w$. Shear causes the stress state to move from point A to point B along stress path $\overline{AB}$. The stress state at failure is represented by stress point B, corresponding to a different net confining pressure and matric suction from those associated with stress point A. In the example shown, the pore-air pressure is assumed to increase continuously during shear. This causes the net confining pressure to decrease [i.e., $(\sigma_3 - u_a)_f < \sigma_3 - u_a$]. Matric suction is also assumed to decrease continuously [i.e., $(u_a - u_w)_f < u_a - u_w$]. The failure envelope is tangent to the Mohr circle at failure (e.g., at stress point C) and inclined at an angle ϕ' with respect to the $\sigma - u_a$ axis. The failure envelope intersects the shear strength versus $u_a - u_w$ plane at a cohesion intercept c. The intersection line joining the cohesion intercepts produced by tests at different matric suctions gives the angle ϕ^b.

It should be noted that it may be difficult to maintain a fully undrained condition for the pore-air phase since air can diffuse through the pore-water, the rubber membrane, and other parts of the triaxial apparatus.

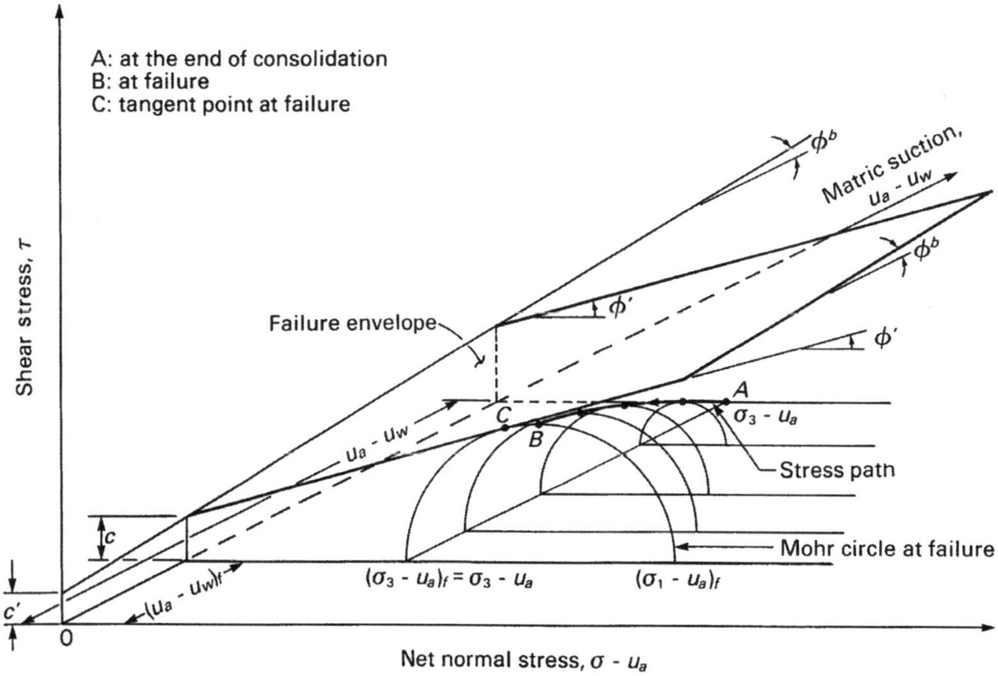

Figure 11.50 Stress path followed during constant-water-content triaxial test.

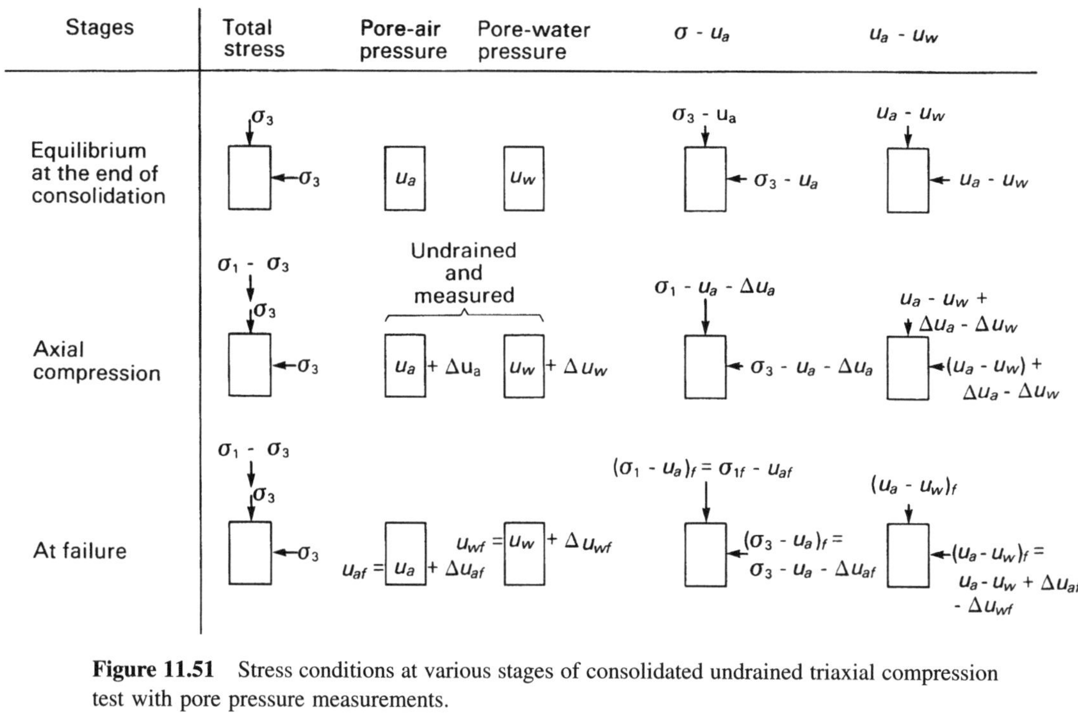

Figure 11.51 Stress conditions at various stages of consolidated undrained triaxial compression test with pore pressure measurements.

11.6.4 Undrained Test

The pore-air and pore-water phases are not allowed to drain in the undrained test. The lack of drainage applies when the confining pressure is applied and when the deviator stress is applied to the soil specimen (Fig. 11.53). The excess pore-air and pore-water pressures developed during the application of the confining pressure can be related to the isotropic confining pressure through use of the B pore pressure parameters provided the pore pressures are measured. The soil specimen may change volume due to compression of pore-air during the application of confining pressure. The soil has a net confining pressure $\sigma_3 - u_a$ and a matric suction $u_a - u_w$ after

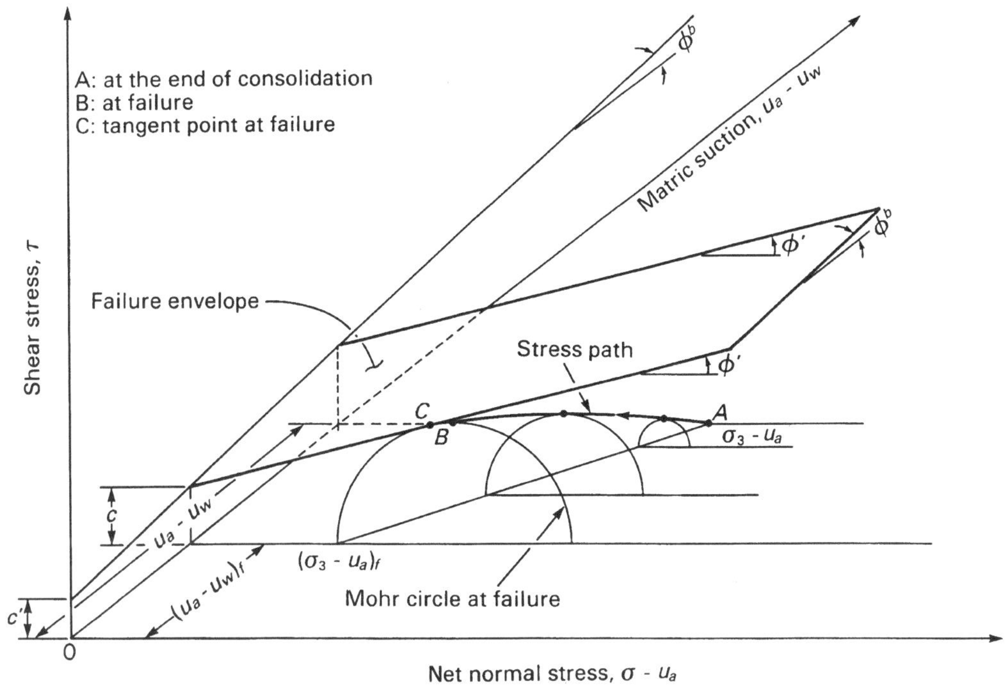

Figure 11.52 Typical stress path followed during consolidated undrained test.

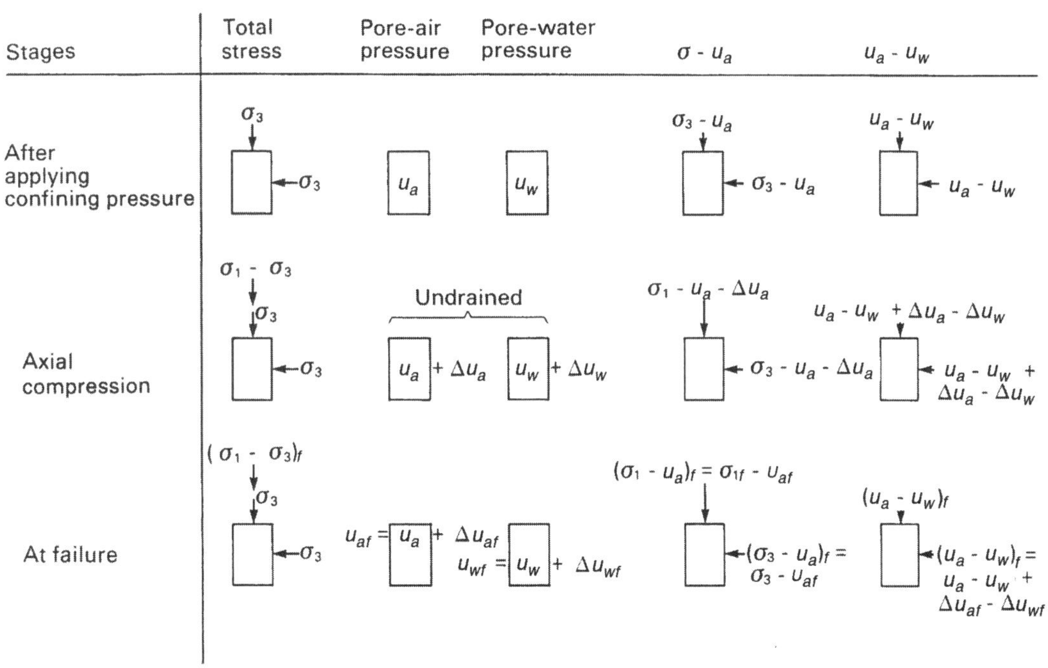

Figure 11.53 Stress conditions at various stages of undrained triaxial compression test.

the application of the confining pressure; however, the stress state will not be known unless the pore fluid pressures are measured.

The soil specimen is sheared by applying an axial stress $\sigma_1 - \sigma_3$ until failure is reached. Undrained loading during shear causes a further change in the excess pore-air and pore-water pressures. The excess pore pressures can be related to the deviator stress by the D pore pressure parameters for triaxial loading conditions provided the pore fluid pressures are measured. Generally, the pore pressures are not measured during an undrained test. Undrained test results are commonly used in conjunction with a total stress representation of a problem. The shear strength is therefore related to total stresses without an understanding of the pore pressures at failure.

The net confining pressure $\sigma_3 - u_a$ and matric suction $u_a - u_w$ vary throughout the shear process. However, the stress state variables during shear and at failure are unknown when the pore fluid pressures are not measured. Figure 11.53 illustrates how the stress state variables might change during an undrained triaxial test (i.e., assuming the pore fluid pressure had been measured). Only the total confining pressure σ_3 and the deviator stress $\sigma_1 - \sigma_3$ are values generally measured during an undrained test. Typical stress-strain curves for the undrained test are shown in Fig. 11.6a. These are the results of a special test in the sense that the pore-air and pore-water pressures were measured during shear (Fig. 11.6b). The pore-water pressure measurements were limited to −101.3 kPa (Bishop et al., 1960). The plot of overall volume change in Fig. 11.6c indicates that the soil specimen compressed during shear.

Hypothetical stress paths that could represent changes during the undrained test are illustrated in Fig. 11.54. Consider four identical specimens that are initially confined at four different confining pressures. The hypothetical results are represented by stress points A, A_1, A_2, and A_3, where σ_3 at $A < \sigma_3$ at $A_1 < \sigma_3$ at $A_2 < \sigma_3$ at A_3. The application of the total confining pressure under undrained conditions results in the compression of the pore fluids and the development of excess pore-air and pore-water pressures. The pore pressure increases in an unsaturated soil are less than the total stress increment applied. This is in keeping with a B pore pressure parameter which must be less than 1.0. Higher confining pressures result in higher net confining pressures (i.e., $\sigma_3 - u_a$ at $A < \sigma_3 - u_a$ at $A_1 < \sigma_3 - u_a$ at $A_2 < \sigma_3 - u_a$ at A_3) and lower matric suctions (i.e., $u_a - u_w$ at $A > u_a - u_w$ at $A_1 > u_a - u_w$ at $A_2 > u_a - u_w$ at A_3). In other words, the four identical soil specimens are brought to four different initial stress states which are represented by stress points A, A_1, A_2, and A_3 (Fig. 11.54). As the soil is sheared under undrained conditions, the pore fluids are further compressed and the pore pressures may further increase. The stress point moves from point A to point B along the stress path $\overline{AB}$. The net confining pressure and matric suction of the soil specimen decrease when going from stress point A to stress point B. Stress point B represents the stress state of the soil at failure. Similar stress paths are followed by the soil specimens at stress points A_1, A_2, and A_3 (i.e., $\overline{A_1B_1}$, $\overline{A_2B_2}$, and $\overline{A_3B_3}$).

Figure 11.54 shows an increase in the diameter of the Mohr circle at failure as the initial total confining pressure increases [i.e., $(\sigma_1 - \sigma_3)_f$ at $B < (\sigma_1 - \sigma_3)_f$ at $B_1 < (\sigma_1 - \sigma_3)_f$ at $B_2 < (\sigma_1 - \sigma_3)_f$ at B_3]. In other words, the shear strength of the soil increases with increasing initial total confining pressure even though the initial matric suction decreases. This occurs because the rate of shear strength increase caused by an increase in confining pressure is greater than the reduction in shear strength caused by a decrease in matric suction. This phenomenon can also be visualized as occurring because the ϕ^b angle is lower than the effective angle of internal friction, ϕ'.

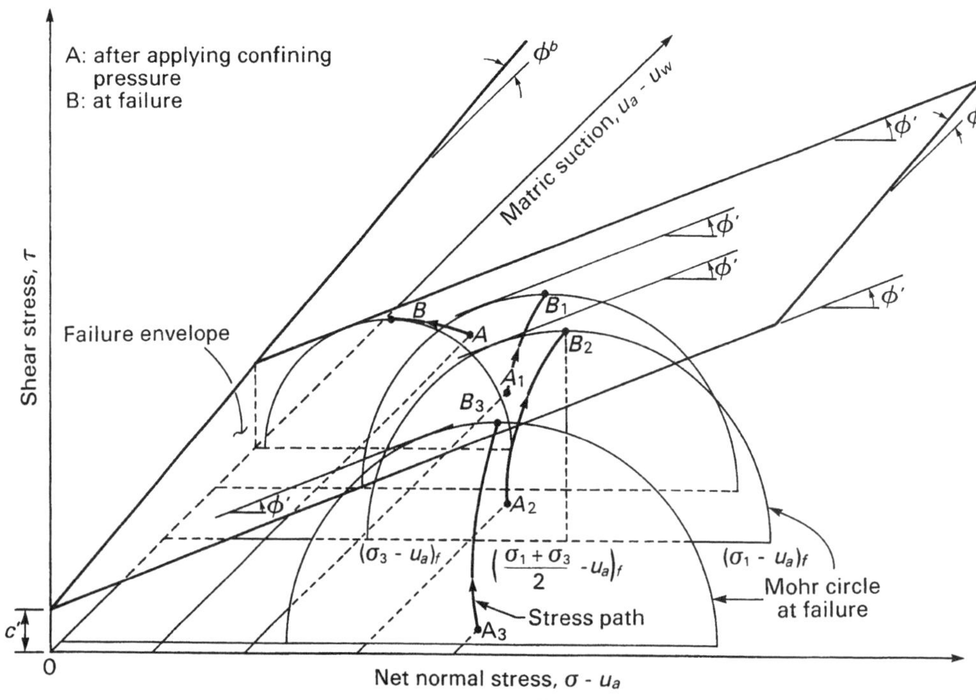

Figure 11.54 Stress paths followed during undrained triaxial test.

The increase in shear strength due to the increase in the initial total confining pressure can also be demonstrated using a shear stress versus total normal stress plot (i.e., τ versus σ plot), as shown in Fig. 11.55. The Mohr circles at failure are drawn using the total confining pressure at failure (i.e., $\sigma_{3f} = \sigma_3$) and the total major principal stress at failure, σ_{1f}. The diameter of the Mohr circle, $(\sigma_1 - \sigma_3)_f$, remains the same when the circle is plotted with respect to either the net normal stress $\sigma - u_a$ or the total normal stress σ. However, the position of the Mohr circles on the horizontal axis differs by a magnitude equal to the pore-air pressure at failure, u_{af}. This behavior can be seen by comparing the Mohr circle associated with stress point B_2 in Figs. 11.54 and 11.55.

A typical curved envelope can be drawn tangent to the Mohr circles for unsaturated soil specimens at failure (Fig. 11.55). The envelope defines a curved relationship between shear strength and total normal stress for unsaturated soils tested under undrained conditions. The curve indicates an increase in shear strength as the applied total stress increases. Care should be taken when interpreting the meaning of the curvature of the total shear strength envelope. Matric suction decreases and the degree of saturation increases as the confining pressure increases. Eventually, a point is reached where the soil approaches saturation. In the saturated condition, an increase in the confining pressure will be equally balanced by a pore-water pressure increase since the B_w pore pressure parameter will equal 1.0. In other words, the effective confining stress $\sigma_3 - u_w$ remains constant regardless of the total confining pressure σ_3.

The shear strength of the saturated soil should remain constant for different total confining pressures. The undrained tests on saturated soil specimens at various confining pressures result in a series of Mohr circles at failure with equal diameters. The failure envelope is horizontal and has a slope of zero (i.e., $\phi_u = 0$; see Fig. 11.55). The ordinate intercept of the horizontal envelope on the shear strength axis is half the deviator stress at failure and is referred to as the undrained shear strength c_u. The application of the undrained shear strength to the analysis of saturated soil is referred to as the $\phi_u = 0$ concept.

The shear strength versus total normal stress relationship for an unsaturated soil does not produce a horizontal line. Rather, the undrained shear strength is a function of the applied total normal stress, as shown in Fig. 11.55.

11.6.5 Unconfined Compression Test

The UC test is a special case of the undrained test. No confining pressure is applied to the soil specimen throughout the test (Fig. 11.56). The test can be performed by applying a deviator stress to a soil specimen in a simple loading frame. The unsaturated soil specimen has a negative pore-water pressure at the start of the test and the pore-air pressure can be assumed to be atmospheric. The soil matric suction $u_a - u_w$ is therefore numerically equal to the pore-water pressure.

The soil specimen is sheared by applying an axial load until failure is reached. The deviator stress $\sigma_1 - \sigma_3$ is equal to the major principal stress σ_1 since the confining pressure σ_3 is equal to zero. The compressive load is applied quickly in order to maintain undrained conditions for the pore-air and pore-water phases. The pore-air and pore-water pressures are not measured during compression. The excess pore pressures developed during an unconfined compression

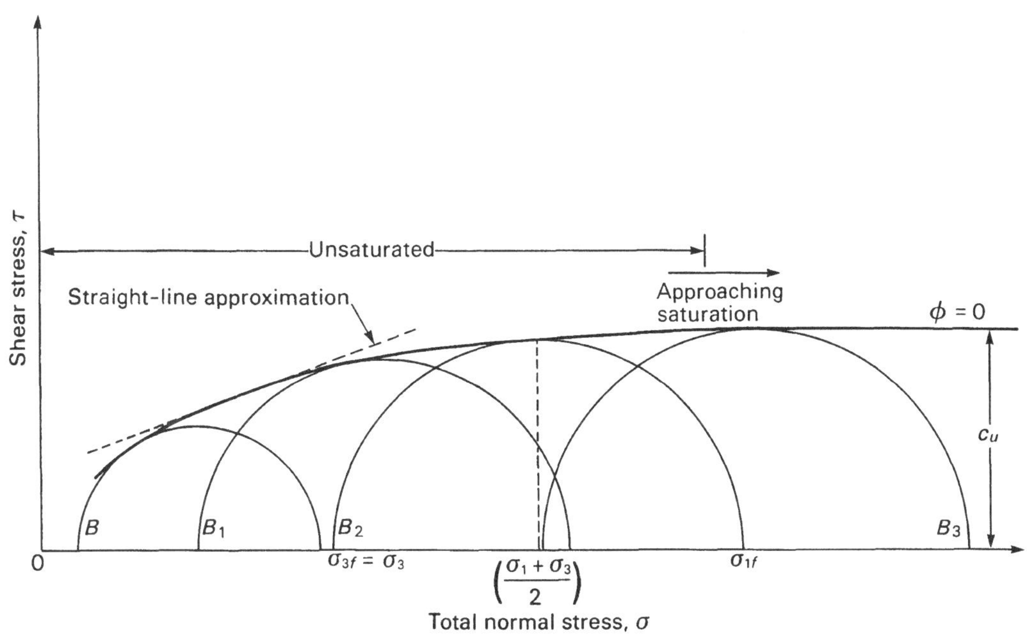

Figure 11.55 Shear stress versus total normal stress at failure for undrained triaxial tests.

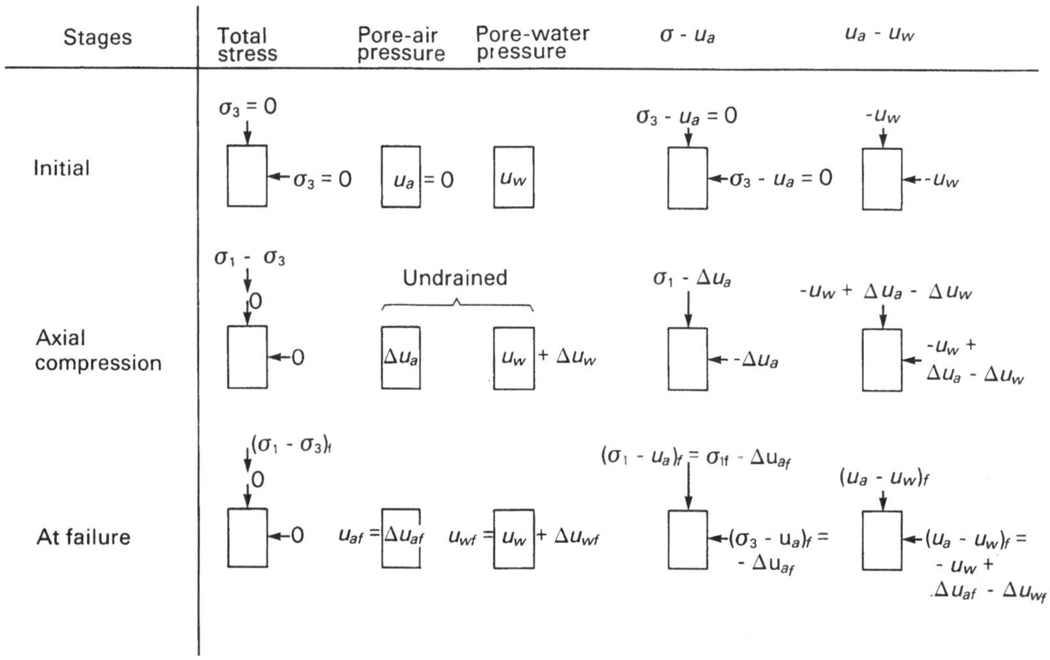

Figure 11.56 Stress conditions at various stages of unconfined compression test.

test can be theoretically related to the major principal stress through use of the D pore pressure parameter. Figure 11.56 illustrates typical changes in the stress state variables that would occur during unconfined compression if the pore pressures were measured.

Figure 11.57 illustrates two possible stress paths that could be followed by an unsaturated soil specimen during the unconfined compression. The initial stress state is represented by stress point A where the soil has a zero net confining pressure (i.e., $\sigma_3 - u_a = 0$) and a matric suction $u_a - u_w$. The matric suction can increase, decrease, or remain constant during undrained compression depending on the A parameter of the soil. Matric suction will most commonly decrease during undrained compression and the stress state in the soil will move forward from point A to point B along the stress path $\overline{AB}$ (Fig. 11.57). The stress state of the soil at failure is represented by point B. The pore-air pressure is assumed to increase slightly during compression. This causes the net confining pressure to decrease along stress path $\overline{AB}$ to a negative value [i.e., $(\sigma_3 - u_a)_f = -\Delta u_{af}$]. The matric suction at failure will be less than the initial matric suction [i.e., $(u_a - u_w)_f$ at $B < (u_a - u_w)$ at A].

In the case of constant matric suction being maintained during compression, the stress state of the soil could move from point A to point B_1 along the stress path $\overline{AB_1}$. The stress path therefore lies in a plane of constant matric suction. Stress point B_1 represents the stress state of the soil at failure. The pore-air and pore-water pressures are assumed to remain constant during compression in order for the matric suction to be constant. As a result, the net confining pressure remains equal to zero until failure is reached [i.e., $\sigma_3 - u_a = (\sigma_3 - u_a)_f = 0$].

For the case of increasing matric suction during compression (e.g., the soil dilates), the stress state in the soil will move backward from point A to a point (or plane) somewhere behind point A. This stress path is not shown in Fig. 11.57.

The stress paths illustrated in Fig. 11.57 are associated with unsaturated soil specimens. The soil may be unsaturated in the field or become unsaturated during sampling due to the release of the overburden pressure. Unconfined compression test results on the unsaturated soil apply to the condition where the total confining pressure σ_3 is zero (Fig. 11.58a). The deviator stress at failure, $(\sigma_1 - \sigma_3)_f$, is referred to as the unconfined compressive strength q_u. The unconfined compressive strength is commonly taken as being equal to twice the undrained shear strength c_u. However, the undrained shear strength for the unsaturated soil increases as the confining pressure increases. As a result, the compressive strength of the unsaturated soil may not satisfactorily approximate the undrained shear strength c_u at a confining pressure greater than zero.

Soil samples may remain saturated in the field in some cases even though the pore-water pressure is negative. The pore-water pressure may be negative because the soil is from some distance above the groundwater table, or it may be negative due to the release of overburden pressure, or both. The shear strength equation for saturated soils can be used even though the pore-water pressures are negative. The undrained shear strength c_u for the soil remains constant even when a variety of confining pressures are applied (Fig. 11.58b). For a saturated soil, the unconfined compressive strength q_u is twice the undrained shear strength c_u regardless of the magnitude of the confining pressure.

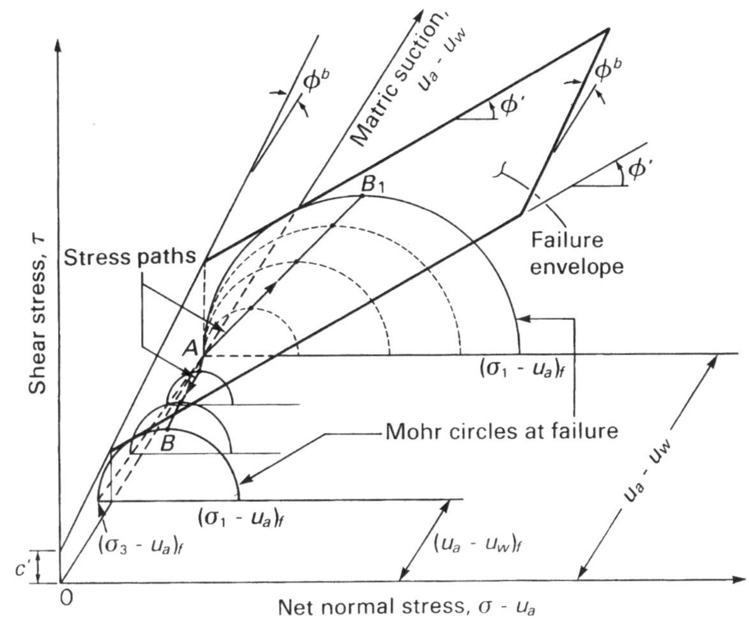

Figure 11.57 Possible stress paths followed during unconfined compression test.

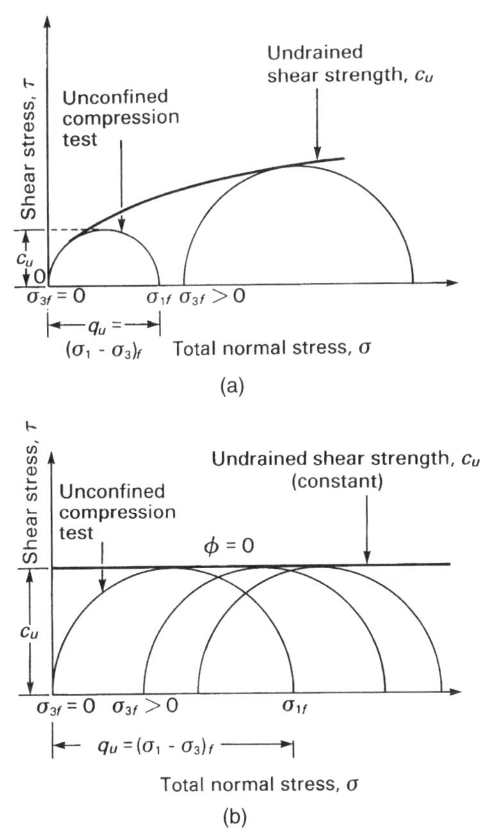

Figure 11.58 Use of unconfined compressive strength q_u to approximate undrained shear strength c_u for unsaturated soil and saturated soil: (a) relationship between q_u and c_u for unsaturated soil; (b) relationship between q_u and c_u for saturated soil.

It is quite likely that unconfined compression tests on unsaturated soils may greatly underestimate the available shear strength in the field. Unconfined compression tests are routinely performed for bearing capacity, slope stability, and other engineering studies. The engineer must understand that an approximate simulation of the field confining pressure is strongly recommended when performing undrained laboratory strength tests. The undrained tests should not be performed as a confined compression test when the soils are unsaturated.

11.7 DIRECT SHEAR TESTS

Direct shear tests have one distinct advantage over the triaxial test, in that the length of the drainage path associated with pore-water pressure equalization is considerably shorter in the direct shear test than in the triaxial test. Consequently, direct shear strength testing of an unsaturated soil can considerably expedite shear strength testing. The basics of the test procedure and the interpretation of the test results are explained in the following sections.

11.7.1 Direct Shear Test Procedure

The consolidated drained direct shear test on an unsaturated soil specimen can be conducted using the modified direct shear apparatus shown in the cross-sectional view in Fig. 11.59. The direct shear test apparatus consists of a split box with a top and bottom portion. The soil specimen is sheared by moving the lower portion of the shear box

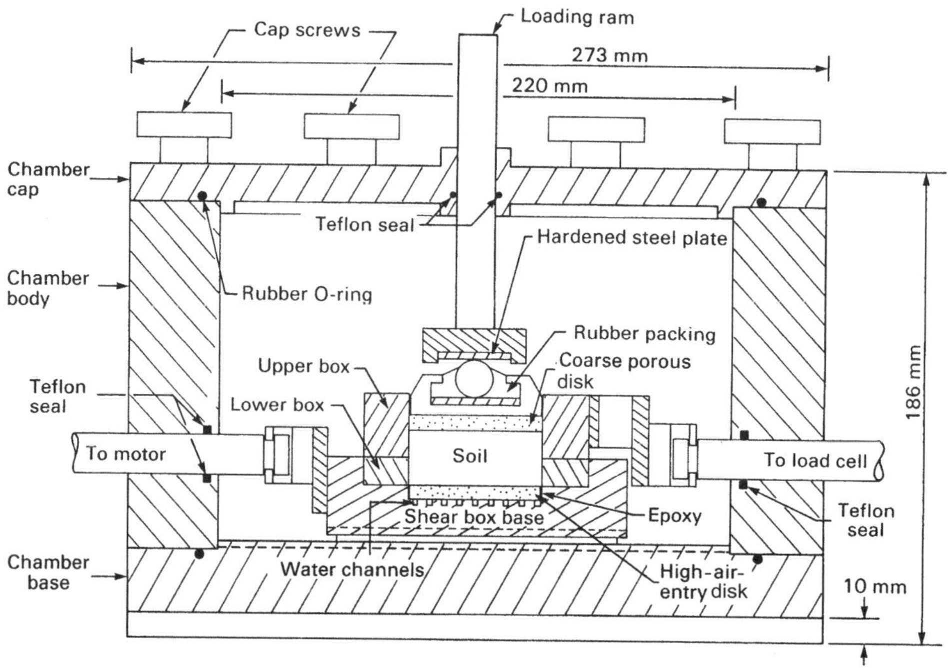

Figure 11.59 Modified direct shear apparatus for testing unsaturated soils (from Gan and Fredlund, 1988).

relative to the upper portion of the box. The test procedure is similar to that used for a conventional direct shear apparatus. A motor that provides a constant horizontal shear displacement rate is connected to the shear box base. The shear box is seated on a pair of rollers that can move along grooved tracks on the chamber base. The top box is connected to a load cell which measures the shear load resistance. Vacuum grease is placed between the two halves of the shear box prior to mounting the soil specimen in the shear box.

The plumbing layout for the control board of the modified direct shear apparatus is similar to that used for a triaxial test; however, it is difficult to measure water volume changes because of the small water volumes involved. The saturation of the high-air-entry disk and the flushing of entrapped air from the base plate and its connecting lines should be performed prior to commencing the test. The initial air and water pressures to be applied to the soil specimen can be set on the pressure regulators while the soil specimen is being assembled.

The tests are generally performed as consolidated drained direct shear tests. The soil specimen is usually given access to water after being placed into the direct shear box. The chamber cap is applied following saturation of the soil. The predetermined vertical normal load, air pressure, and water pressure are then applied to the specimen in this sequence. The vertical normal load is applied through the loading ram at the same time as the air and water pressures are applied. It is important to ensure there are no leaks in the system. For example, the leakage of air from the chamber surrounding the soil specimen will cause a continuous water vapor loss from the specimen. The water pressure applied at the base plate is measured using a pressure transducer. Measurements of vertical deflection and water movement from the specimen can be taken at various time increments. Consolidation takes place under the applied vertical normal stress, the air pressure, and the water pressure. Equilibrium is assumed to be complete when there is no further tendency for overall volume change and water volume change.

The soil specimen is sheared at an appropriate horizontal shear displacement rate after equilibration has been reached. The horizontal shear load resistance is measured using a load cell. Readings are also taken of vertical deflection, horizontal shear displacement, and water volume change during shear. Shearing can be terminated either when the horizontal shear stress resistance has reached its peak value or when the horizontal shear displacement has reached a designated limiting value. In the case of a multistage test, the shearing process for each stage should be stopped when the peak horizontal shear stress appears to be imminent.

11.7.2 Interpretation of Direct Shear Test Results

A soil specimen is placed in the direct shear box and consolidated under a vertical normal stress σ. The pore-air and pore-water pressures must be controlled at selected pressures during consolidation. The axis translation technique can be used to impose matric suctions greater than 1 atm. The direct shear test can be conducted in an air-pressurized chamber in order to elevate the pore-air pressure to a magnitude above atmospheric pressure. The pore-water pressure can be controlled below the soil specimen using a high-air-entry disk. The soil specimen has a net vertical normal stress of $\sigma_n - u_a$ and a matric suction of $u_a - u_w$ at the end of the consolidation process (Fig. 11.60).

Shearing of the soil specimen is achieved by horizontally displacing the top half of the direct shear box relative to the bottom half. The soil specimen is sheared along a horizontal plane between the top and bottom halves of the direct shear box. The horizontal load required to shear the specimen divided by the nominal area of the specimen gives the shear stress on the shear plane. The pore-air and pore-water pressures are controlled at constant values during the shear process. The shear stress is increased until the soil specimen has failed. The failure plane has a shear stress designated as τ_{ff}, corresponding to a net vertical normal stress of $(\sigma_f - u_a)_f$ (i.e., equal to $\sigma - u_a$) and a matric suction of $(u_a - u_w)_f$(i.e., equal to $u_a - u_w$) (Fig. 11.60). Figure 11.61 shows a typical plot of shear stress versus horizontal displacement for a direct shear test.

The failure envelope can be obtained from the results of direct shear test results without constructing Mohr circles. The shear stress at failure, τ_{ff}, is plotted as the ordinate, and $(\sigma_f - u_a)_f$ and $(u_a - u_w)_f$ are plotted as the abscissas to give a point on the failure envelope (Fig. 11.62). A line joining points of equal magnitude of $(\sigma_f - u_a)_f$ determines the ϕ^bangle (e.g., a line joining points A, B, and C in Fig. 11.62). Similarly, a line can be drawn through the points of equal $(u_a - u_w)_f$ to give the angle of internal friction ϕ' (e.g., a line drawn through points A_1, A_2, and A in Fig. 11.62).

The direct shear test is particularly useful for testing unsaturated soils because of the reduced length of the drainage path. The low coefficient of permeability for unsaturated soils results in "times to failure" that are reduced by direct shear testing. Other problems associated with testing unsaturated soils in a direct shear apparatus are similar to those common to testing saturated soils (e.g., stress concentrations, definition of the failure plane, and rotation of principal stresses).

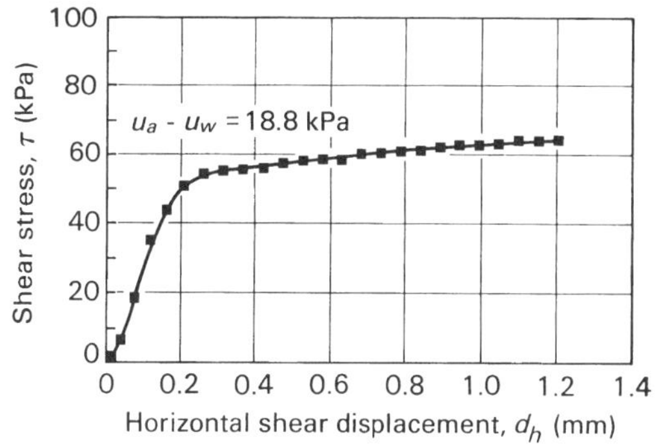

Figure 11.61 Typical shear stress versus shear displacement plot from direct shear test (from Gan, 1986).

11.8 TYPICAL LABORATORY TEST RESULTS

The presentation of shear strength results on unsaturated soils focuses mainly on the shear stress versus matric suction relationship (i.e., τ versus $u_a - u_w$ plane). The failure envelope on the shear strength versus matric suction plane is used to obtain the ϕ^b shear strength parameter or the nonlinear relationship between shear strength and matric suction

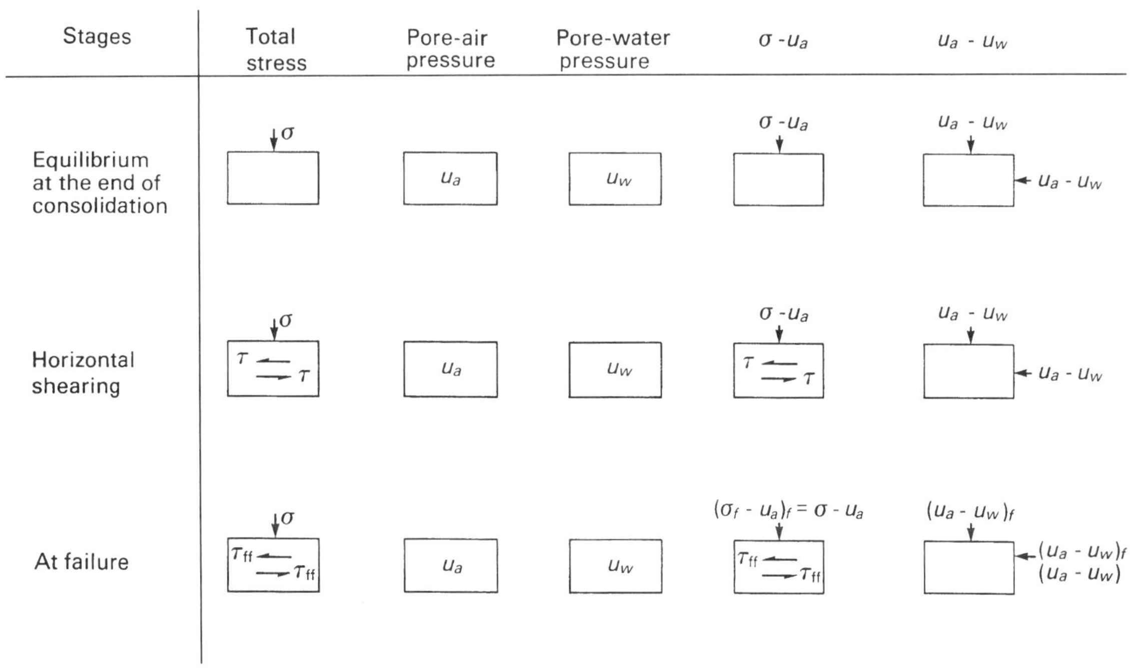

Figure 11.60 Stress conditions at various stages of consolidated drained direct shear test.

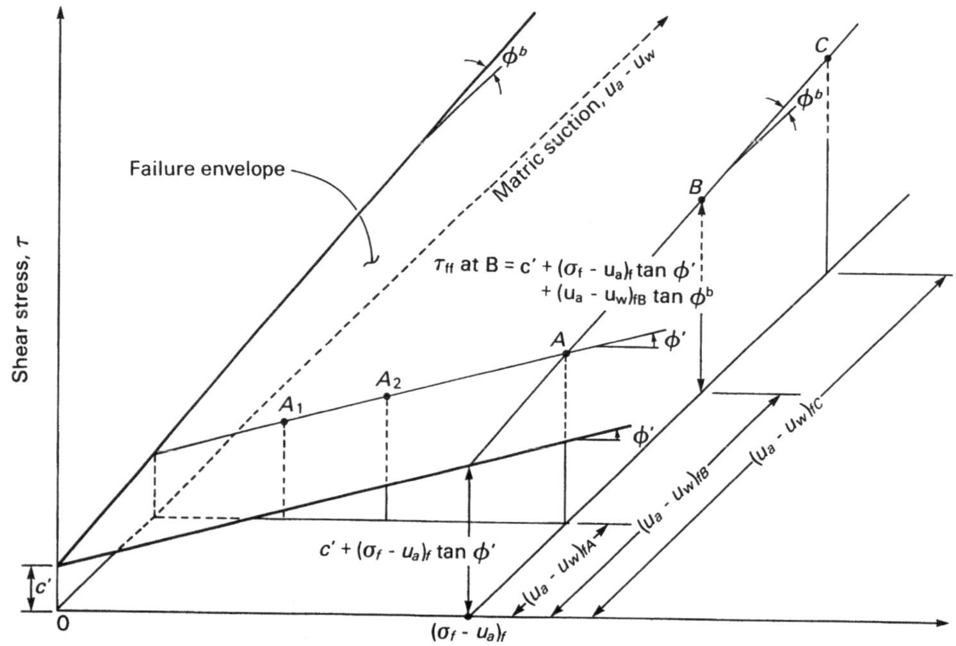

Figure 11.62 Extended Mohr-Coulomb failure envelope based on direct shear test results.

(i.e., unsaturated soil shear strength function). The nature of the shear strength versus net normal stress failure envelope for saturated soils is explained in many soil mechanics books and is not given further consideration here.

Typical laboratory test results obtained from undisturbed and compacted, unsaturated soil specimens are presented in this section. Only results from initially similar undisturbed or compacted soil specimens having the same dry density and water content can be analyzed to obtain the ϕ^b shear strength parameter or function.

Triaxial test data are presented in the following sections followed by direct shear test data. The triaxial test data are categorized as (1) consolidated drained test results and (2) constant-water-content test results. Both cases are used to illustrate unsaturated soil behavior. Data are presented to illustrate the linear and nonlinear relationship between shear strength and matric suction. Limited data on undrained and unconfined compression tests are also presented.

11.8.1 Consolidated Drained Triaxial Tests

A series of multistage consolidated drained triaxial tests on undisturbed specimens were performed by Ho and Fredlund (1982b). The specimens were from two residual soil deposits in Hong Kong, namely, decomposed granite and decomposed rhyolite. The decomposed granite specimens are mainly silty sand with an average specific gravity G_s of 2.65. The decomposed rhyolite specimens are sandy silt with an average specific gravity of 2.66. The mineral compositions of these two soils are similar. Both soils are brittle and highly variable. Undisturbed specimens were sampled tube samples from boreholes and block samples from open cuts.

Seventeen undisturbed specimens 63.5 mm in diameter and approximately 140 mm in height were tested. The tests were performed using the consolidated drained triaxial testing procedure. The deviator stress was removed from the soil specimen once a maximum value for deviator stress was obtained for a particular stage (i.e., multistage testing procedure). Then a new stress state was applied for the next stage of testing. The strain rate used in the testing program ranged from 1.7×10^{-5} to 6.7×10^{-5} %/s. A 5-bar high-air-entry disk (i.e., 505 kPa) was used for all tests. The effective angles of internal friction, ϕ', were 33.4° and 35.3° for the decomposed granite and rhyolite, respectively.

Figure 11.63 presents typical test results from a decomposed granite specimen using a cyclic multistage testing procedure. The test was performed by maintaining a constant net confining pressure $\sigma_3 - u_a$ and varying the matric suction $u_a - u_w$ from one stage to the next. The test results were interpreted based on the assumption that the effective angle of internal friction remained as a constant value. Similar multistage test results from two rhyolite specimens are shown in Fig. 11.64.

The average ϕ^b angles from all of the test results were found to be 15.3° for the decomposed granite and 13.8° for the rhyolite. It was observed that the soil structure in the soil specimens appeared to become disturbed to some degree during the second and third stages of testing. As a result, the measured peak deviator stress for the last stage (i.e., stage No. 3) may actually be smaller than that which would be obtained from a test on a single specimen under the same stress conditions. Part of the reduction in strength may also be due to nonlinearity in the shear strength versus matric suction relationship.

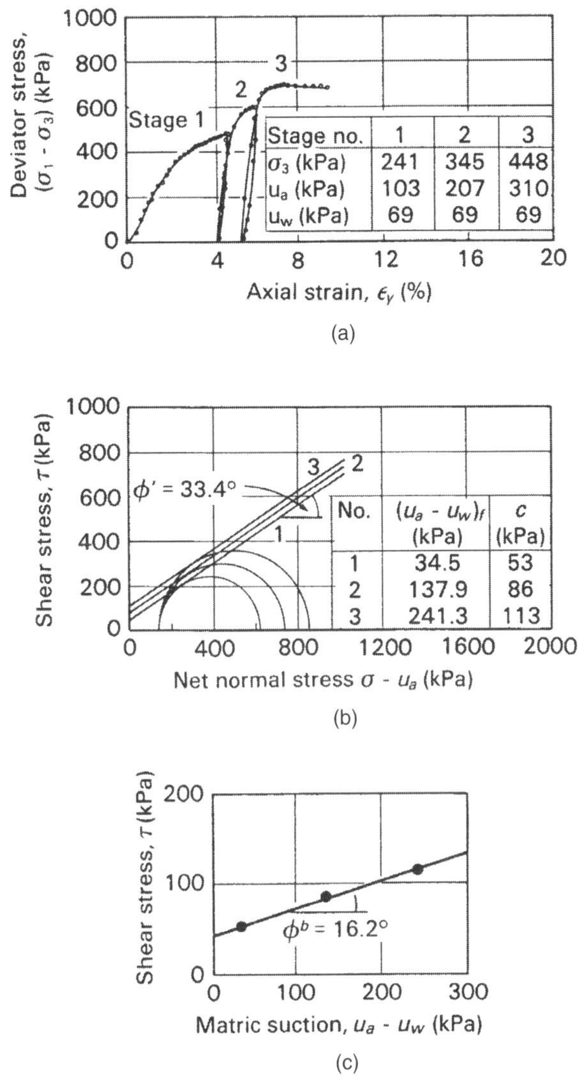

Figure 11.63 Stress versus strain curves and presentations of failure envelope for decomposed granite specimen No. 10: (a) deviator stress versus strain curve; (b) failure envelope projected onto τ versus $\sigma - u_a$ plane; (c) line defining slope of shear strength with respect to matric suction ϕ^b (from Ho and Fredlund, 1982a).

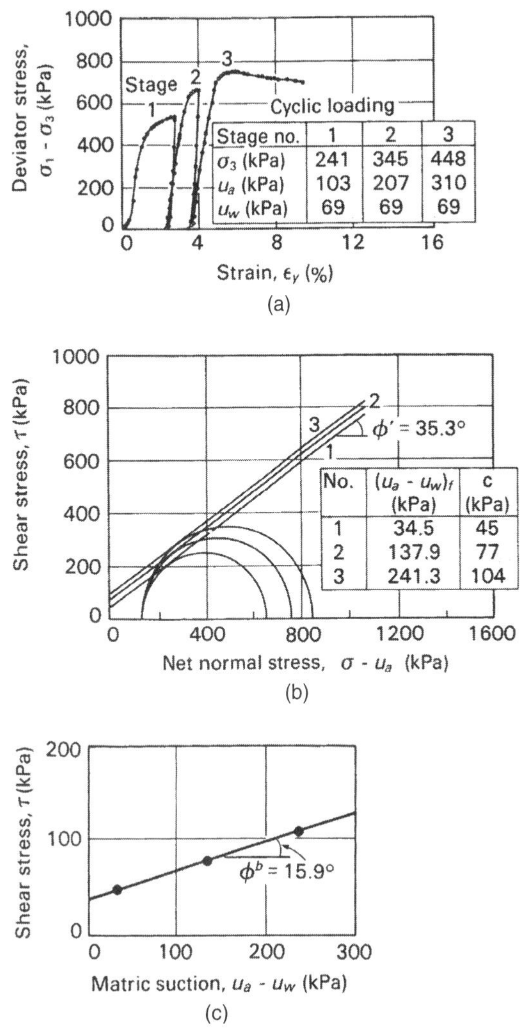

Figure 11.64 Stress versus strain curves and presentations of failure envelope for decomposed rhyolite specimen No. 11C: (a) deviator stress versus strain curve; (b) failure envelope projected onto τ versus $\sigma - u_a$ plane; (c) line defining slope of shear strength with respect to matric suction (from Ho and Fredlund, 1982a).

Two multistage triaxial tests on compacted specimens were reported by Krahn et al., (1989). The soil was sampled from a railway embankment at Notch Hill, British Columbia, and consisted of 10% clay, 85% silt, and 5% fine sand. The optimum water content was 21.5%, and the maximum dry density was 1590 kg/m^3 when the soil was compacted in accordance with the standard AASHTO procedure. Specimens with a diameter of 38 mm and a height of 75 mm were trimmed for triaxial testing from the compacted soil. Consolidated undrained triaxial tests were performed on four compacted saturated specimens with pore-water pressure measurements. The test results on the saturated specimens showed an angle of internal friction ϕ' of 35° and an effective cohesion c' equal to 0.0.

The multistage triaxial tests on the unsaturated compacted specimens were conducted using the consolidated drained test procedure. The tests were conducted with a constant net confining pressure while varying the applied matric suction (Fig. 11.65). The test results showed the ϕ^b angle to be 16° when a constant effective angle of internal friction was assumed during the interpretation of the data.

11.8.2 Constant-Water-Content Triaxial Tests

Constant-water-content triaxial tests were performed on a compacted shale and a compacted boulder clay by Bishop

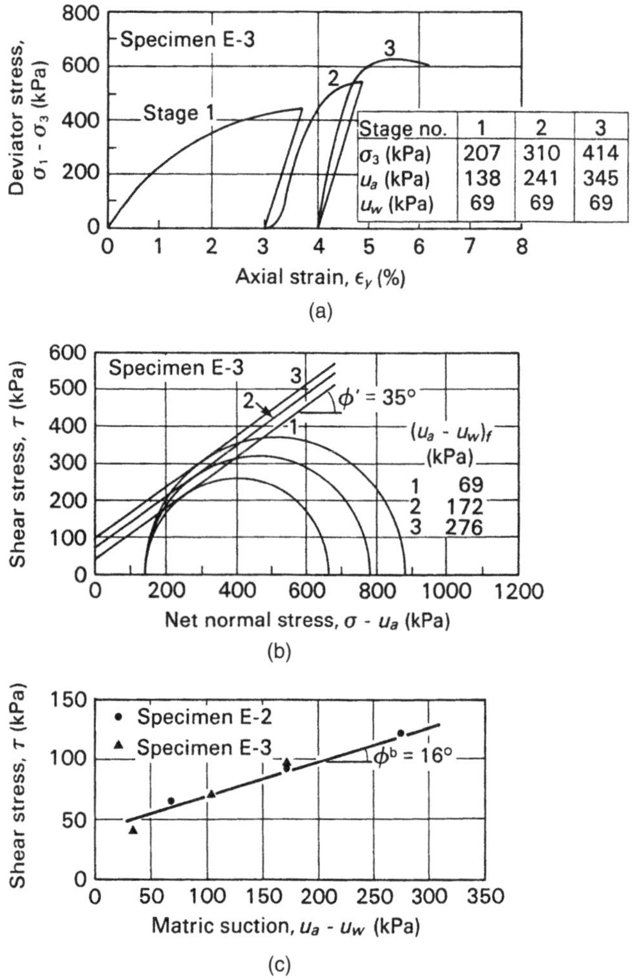

Figure 11.65 Stress versus strain curves and presentations of failure envelope for Tappen-Notch Hill Silt specimen No. E-3: (a) deviator stress versus strain curve; (b) failure envelope projected onto τ versus $\sigma - u_a$ plane; (c) line defining slope of shear strength with respect to matric suction (from Krahn et al., 1989).

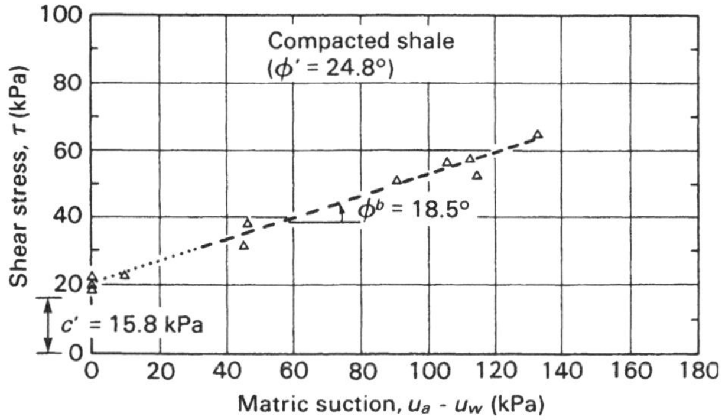

Figure 11.66 Intersection line between failure envelope and the τ versus $u_a - u_w$ plane for compacted shale (data from Bishop et al., 1960).

et al. (1960). The shale had a clay fraction of 22% and was compacted at a water content of 18.6%. A series of triaxial tests on saturation specimens of compacted shale gave an effective angle of internal friction ϕ' of 24.8° and an effective cohesion c' of 15.8 kPa. The laboratory results were reanalyzed by Fredlund et al. (1978), and the results are presented in Fig. 11.66. The ϕ^b angle was 18.5° for the compacted shale when assuming a constant effective angle of internal friction during the interpretation of the data.

The boulder clay had a clay fraction of 18% and was compacted at a water content of 11.6%. The saturated boulder clay showed an effective angle of internal friction ϕ' of 27.3° and an effective cohesion c' of 7.6 kPa. The tests on the compacted boulder clay were performed at a strain rate of 3.5×10^{-5} %/s, and 15% strain was considered to represent failure conditions. The laboratory results were reanalyzed by Fredlund et al. (1978) and the results are presented in Figs. 11.67. The average ϕ^b angle was 22.0° for the boulder clay when assuming a constant effective angle of internal friction during the interpretation of the data.

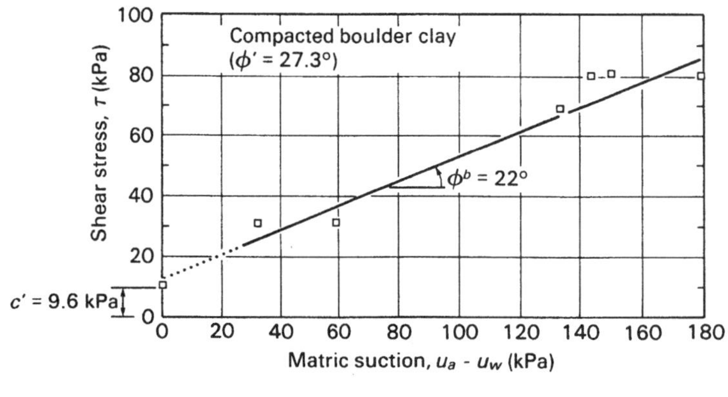

Figure 11.67 Intersection line between failure envelope and τ versus $u_a - u_w$ plane for compacted boulder clay (data from Bishop et al., 1960).

11.8.3 Nonlinear Shear Strength versus Matric Suction

The nonlinearity of the shear strength versus matric suction relationship became increasingly apparent as unsaturated soils were tested over a wider suction range (Fredlund et al., 1987). The reanalyses of the triaxial test results on the compacted Dhanauri clay using a planar and a curved failure envelope were compared. The CD and CW triaxial tests on Dhanauri clay compacted at two densities were presented by Gulhati and Satija (1981). The Dhanauri clay consisted of 5% sand, 70% silt, and 25% clay. The liquid limit of the soil was 48.5% and the plastic limit was 25%. The saturated effective shear strength parameters (i.e., c' and ϕ') for specimens compacted at two different densities were measured using consolidated undrained triaxial tests (Table 11.4). The consolidated drained and constant-water-content tests on the unsaturated compacted specimens were performed at strain rates of 1.3×10^{-4} and $6.7 \times 10^{-4}\%/s$, respectively.

The test results on the unsaturated soil specimens were reanalyzed by Ho and Fredlund (1982a) using the assumption of a planar failure envelope and the results are summarized in Table 11.4. It appears that the linear interpretation of the failure envelope results in different c' and ϕ^b parameters for the same soil tested using different test procedures (i.e., CD and CW tests). In other words, the results give the impression that different test procedures may yield different shear strength parameters. It appears that the assumption of a planar failure envelope when analyzing the data suggests there may be nonuniqueness in the shear strength parameters. In addition, the c' values obtained from the analysis of the unsaturated soil tests do not agree with values obtained from triaxial tests on saturated soil specimens (Table 11.4).

The reanalysis of the shear strength data on Dhanauri clay showed that the failure envelope with respect to the matric suction was nonlinear (Fredlund et al., 1987). Figures 11.68 and 11.69 present the results for compacted Dhanauri clay at low and high densities, respectively. The results are plotted on the shear strength versus matric suction plane corresponding to zero net normal stress at failure [i.e., $(\sigma_f - u_a)_f = 0$]. The shear strength parameters c' and ϕ' obtained from the consolidated undrained tests on the saturated specimens (see Table 11.4) were used in the analysis. The curved failure envelopes have a cohesion intercept of c' and a slope angle ϕ^b equal to ϕ' starting at zero matric suction. The ϕ^b angle begins to decrease significantly at matric suction values greater than 50 kPa for the low-density specimens. The decrease in ϕ^b begins at matric suction values of 75–100 kPa for the high-density specimens. The ϕ^b angle reaches a relatively constant value of 11° when the matric suction exceeds 150 kPa for the low-density specimens. The ϕ^b angles for the high-density specimens appear to reach a relatively constant value of 9° when the matric suction exceeds 300 kPa.

There is good agreement between the failure envelopes for the consolidated drained and constant-water-content test results when assuming a curved failure envelope with respect to matric suction. In other words, the assumption of a curved failure envelope leads to the conclusion that there is a unique failure envelope when the soil is tested using different stress paths. The uniqueness of the curved failure envelope is demonstrated at both densities. It is also apparent that specimens prepared at different densities should be considered as different soils.

Thu et al., (2006) studied the characteristics of the shear strength, volume change and pore-water pressure of compacted silt (i.e., coarse kaolin) during shearing under constant-water-content conditions. A series of CW triaxial tests were carried out on statically compacted silt specimens.

Table 11.4 Triaxial Tests on Compacted Dhanauri Clay

Initial Volume-Mass Properties	CU Tests on Saturated Specimens		Analysis of Test Results on Unsaturated Specimens (Ho and Fredlund, 1982a)		
	c' (kPa)	ϕ' (Deg)	Type of Test	c' (kPa)	ϕ^b (Deg)
Low density $\rho_d = 1478$ kg/m^3 $w = 22.2$ %	7.8	29.0	CD	20.3	12.6
			CW	11.3	16.5
High density $\rho_d = 1580$ kg/m^3 $w = 22.2$ %	7.8	28.5	CD	37.3	16.2
			CW	15.5	22.6

Note: ρ_d = dry density, w = water content, CU = consolidated undrained, CD = consolidated drained, CW = constant water content.
Source: Data from Satija (1978).

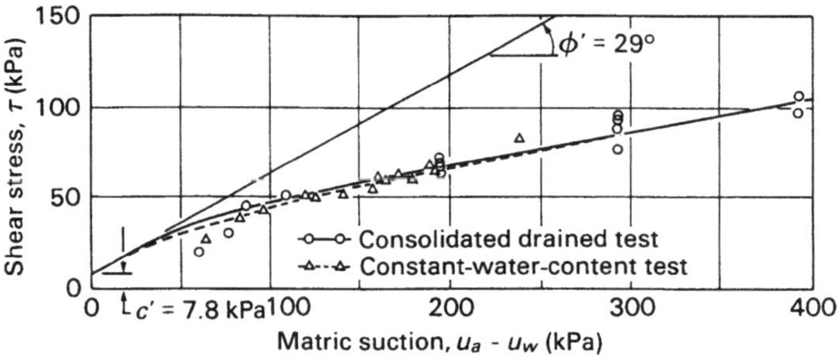

Figure 11.68 Nonlinearity in matric suction failure envelope for Dhanauri clay compacted to low density (data from Satija (1978).

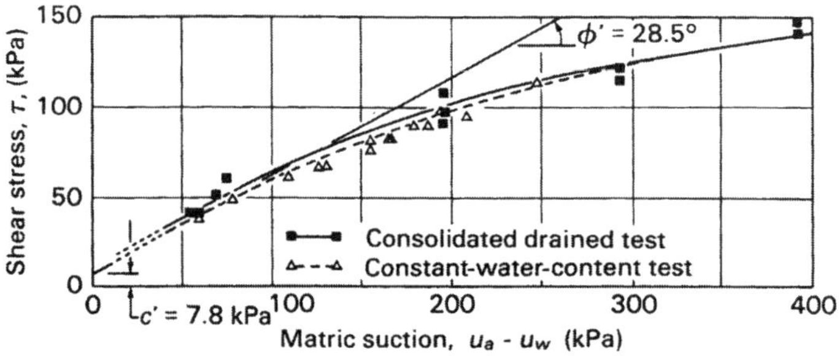

Figure 11.69 Nonlinearity in matric suction failure envelope for Dhanauri clay compacted to high density (data from Satija (1978).

The properties of the coarse kaolin are presented in Table 11.5. A modified triaxial apparatus similar to the modified triaxial apparatus described by Fredlund and Rahardjo (1993a) was used for performing the CW tests. The specimens were statically compacted in 10 layers 10 mm thick. Identical specimens were obtained by compacting kaolin at a maximum dry density of 1.35 Mg/m^3 and an optimum water content of 22%. The height and diameter of the specimens were 100 and 50 mm, respectively. Three mini–suction probes (Meilani et al., 2002) were installed along the length of the triaxial specimen. Details of the mini–suction probes developed at Nanyang Technological University, Singapore, were presented in Chapter 4.

The soil specimens were initially saturated in order to obtain consistent initial water content conditions. Each soil specimen required about 4 days for the saturation process. Each specimen required another 3–5 days for matric suction equalization to a new set of stress conditions. The specimens were sheared by applying an axial load at a constant strain rate once equilibrium conditions were achieved under the applied net confining stress $\sigma_3 - u_a$ and matric suction.

Specimens were tested in the modified triaxial apparatus under net confining stresses of 25, 50, 100, 150, 200, 250, 300, and 350 kPa with different initial matric suctions (i.e., matric suctions of 0, 100, 150, 200, and 300 kPa). A two-number naming convention was used to designate each CW triaxial test. The first number was the net confining pressure in kilopascals while the second number was the initial matric suction in kilopascals.

Table 11.5 Index Properties of Compacted Coarse Kaolin Specimens

Soil Properties	Value
Specific gravity, G_s	2.65
Liquid limit, LL (%)	51.0
Plastic limit, PL (%)	35.6
Plastic index, PI (%)	15.4
Silt (%)	85
Clay (%)	15
Unified Soil Classification System (USCS)	MH (silt of high plasticity)
Maximum dry unit weight, $\rho_{d,\max}$ (Mg/m^3)	1.35
Optimum water content, w_{opt} (%)	22.0
Saturated coefficient of permeability for water phase, k_s (m/s)	6.4×10^{-8}

Typical CW test results under different net confining stresses but the same matric suction values of 100 and 200 kPa are shown in Figs. 11.70 and 11.71, respectively. The experimental results show that at high initial matric suctions the soil behaves in a manner typical of a highly overconsolidated soil. The soil specimens exhibit

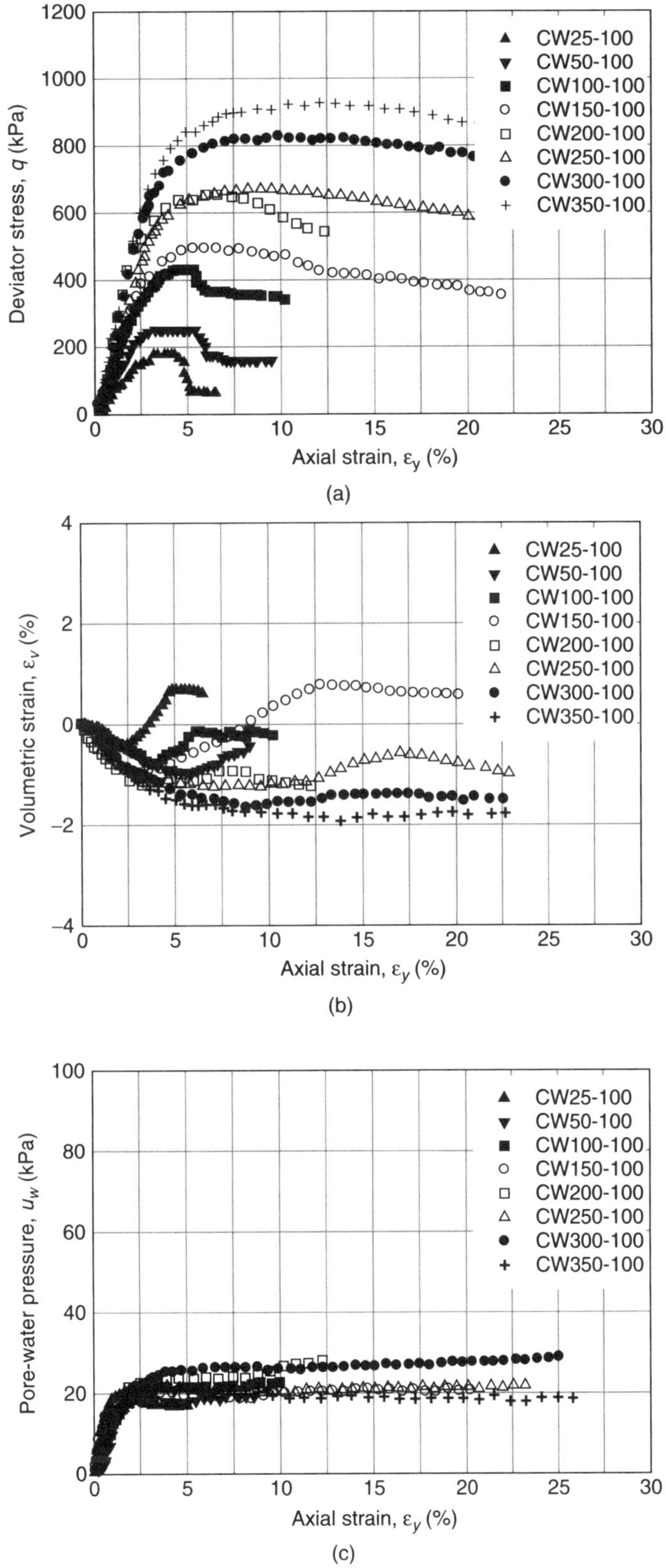

Figure 11.70 Results of constant-water-content triaxial tests under different net confining stresses but with the same initial matric suction of 100 kPa: (a) deviator stress versus strain; (b) volume change versus strain; (c) change in pore-water pressure versus strain.

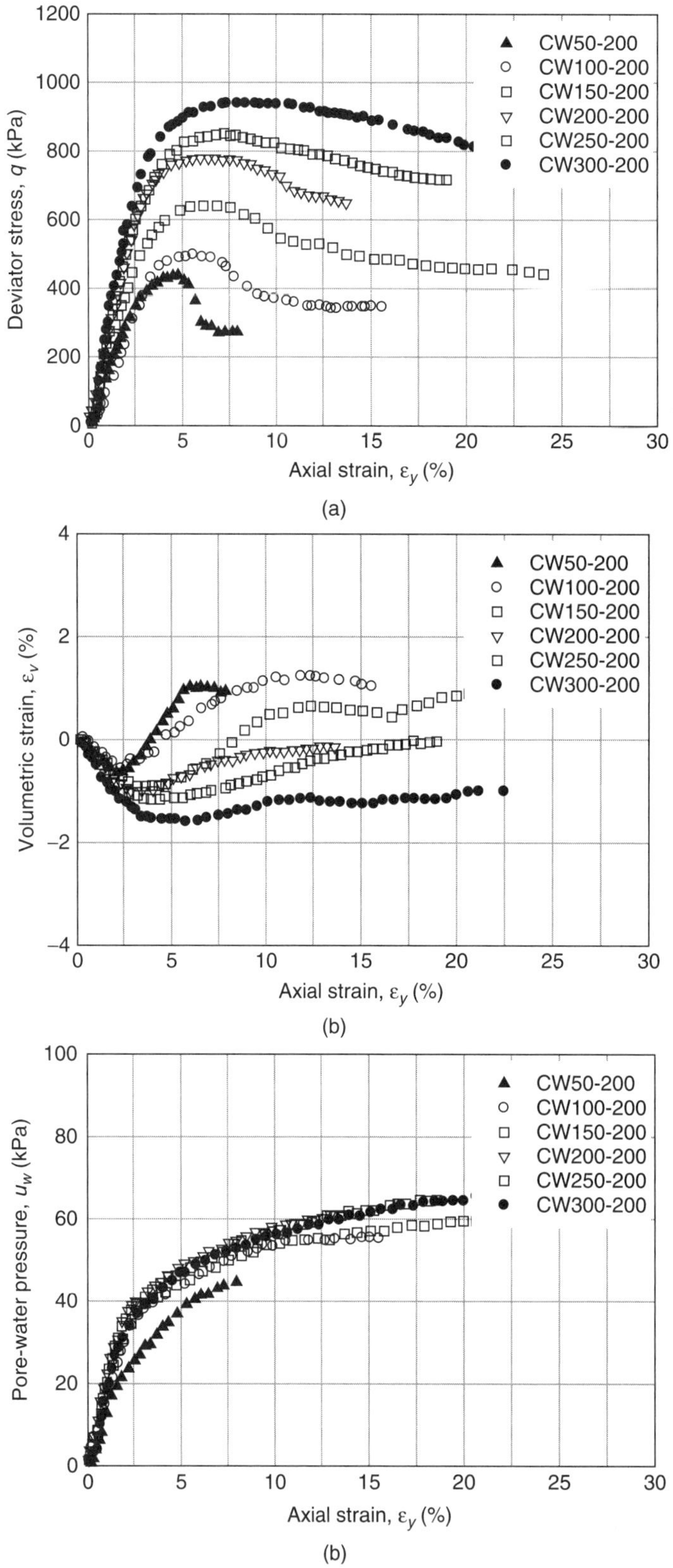

Figure 11.71 Results of constant-water-content triaxial tests under different net confining stresses but with the same initial matric suction of 200 kPa: (a) deviator stress versus strain; (b) volume change versus strain; (c) change in pore-water pressure versus strain.

considerable dilation and postpeak strain softening. The volumetric strain remained essentially constant when the deviator stress was constant.

Figure 11.72 shows the projection of shear stresses at failure, τ_f, onto the τ versus $u_a - u_w$ plane for the CW tests. The effective angle of internal friction, ϕ', of the compacted kaolin was found to be 32°. The ϕ^b angle was found to be the same as the effective angle of internal friction, ϕ' (i.e., 32°), at low matric suction values. The ϕ^b angle decreased to as low as 12° at high matric suctions. The matric suction at the inflection point of the relationship between ϕ^b and matric suction increased with an increase in the net confining stress, as shown in Fig. 11.73. In other words, the net confining stress had an effect on the nonlinearity of the ϕ^b angle for the CW tests on compacted kaolin specimens. The shear strength versus matric suction relationship started to deviate from the effective angle of internal friction (i.e., 32°) as the air-entry value of the soil was exceeded (Fig. 11.72).

11.8.4 Undrained and Unconfined Compression Tests

Six series of undrained and unconfined compression tests (i.e., UC) on unsaturated compacted specimens were performed by Chantawarangul (1983). The soil was clayey sand consisting of 52% sand, 18% silt, and 30% clay. The soil had a liquid limit of 30%, a plastic limit of 19%, and a shrinkage limit of 16%. The soil was compacted using a miniature Harvard apparatus to give high- and low-density specimens at various water contents. The water contents were generally on the dry side of optimum. The high- and low-density specimens correspond to dry density values of approximately 1800 and 1700 kg/m^3, respectively.

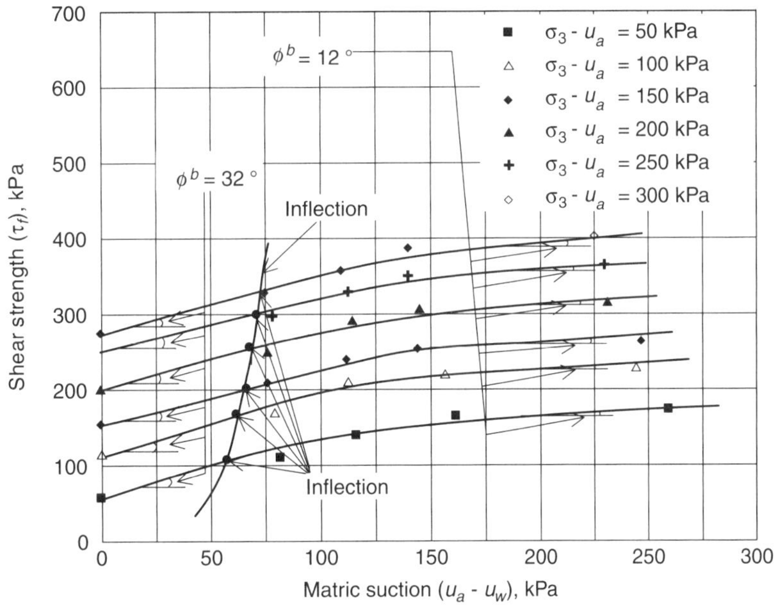

Figure 11.72 Intersection line between failure envelope and the τ_f versus matric suction plane.

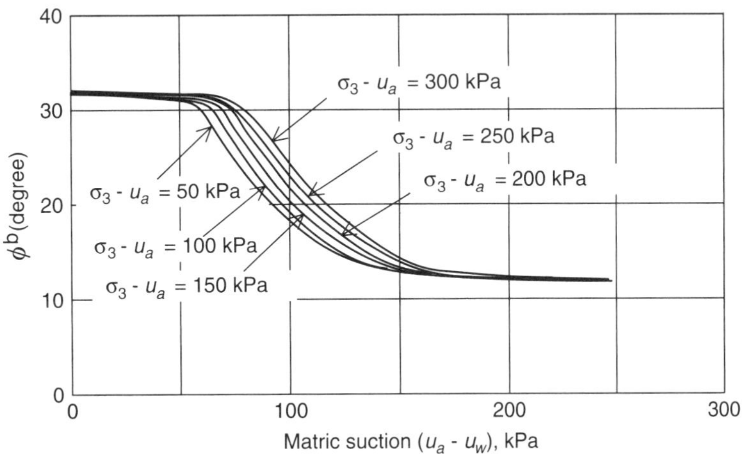

Figure 11.73 Nonlinearity in relationship between ϕ^b and matric suction for constant-water-content triaxial tests.

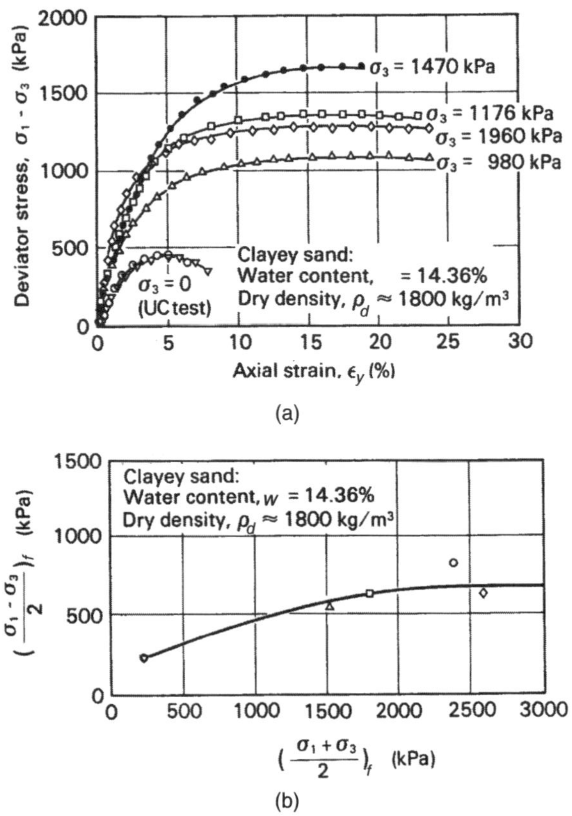

Figure 11.74 Undrained triaxial and unconfined compression tests on clayey sand compacted to high density: (a) deviator stress versus strain for various confining pressures; (b) total stress point envelope (after Chantawarangul, 1983).

The specimens were sheared under undrained conditions at a constant strain rate of 0.0017%/s. Test results are presented in Figs. 11.74 and 11.75 for the high- and low-density specimens, respectively. The results show a curved total stress point envelope which becomes a horizontal envelope at high confining pressures. The total stress point envelopes for specimens at various water contents are plotted in Fig. 11.76a for the high-density specimens and in Fig. 11.76b for the low-density specimens. The envelopes also show a decrease in shear strength as the water content of the specimens increased.

11.8.5 Direct Shear Test Results

Multistage direct shear tests have been performed on saturated and unsaturated specimens of compacted glacial till by Gan et al., (1988). The glacial till was sampled from the Indian Head area of Saskatchewan, and material passing the No. 10 sieve was used to form specimens for testing. The soil consisted of 28% sand, 42% silt, and 30% clay. The liquid and plastic limits of the soil are 35.5 and 16.8%, respectively. The soil specimens were prepared by compaction in accordance with the AASHTO standard. The

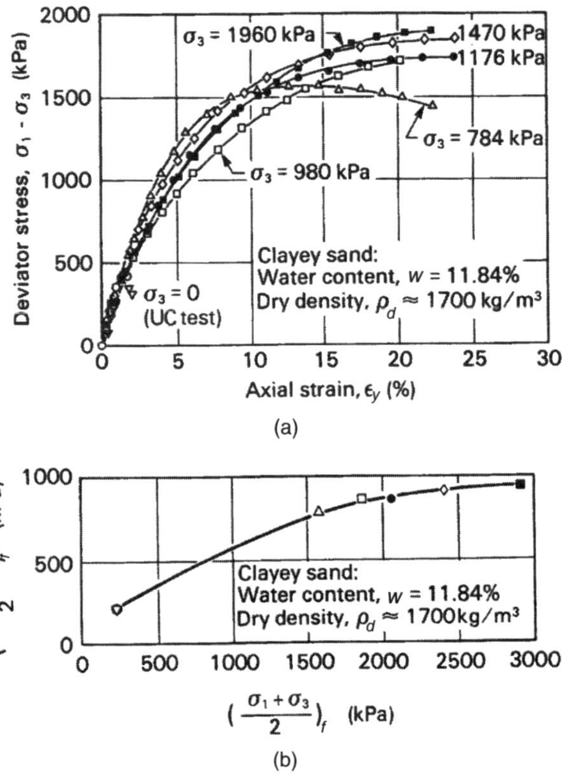

Figure 11.75 Undrained triaxial and unconfined compression tests on clayey sand compacted to low density: (a) deviator stress versus strain for various confining pressures; (b) total stress point envelope (after Chantawarangul, 1983).

maximum dry density and the optimum water content are 1815 kg/m^3 and 16%, respectively.

The shear strength parameters c' and ϕ' were obtained from several single-stage and multistage direct shear tests on compacted specimens that were saturated. The Indian Head glacial till was found to have an effective cohesion intercept of 10 kPa and an effective angle of internal friction of 25.5°. Multistage direct shear test results on saturated specimens showed that a total shear displacement of 1.2 mm was sufficient to mobilize the peak shear strength (Gan, 1986). The specimens were 50 × 50 mm and a total displacement of 1.2 mm was selected as the failure displacement for the multistage direct shear tests.

Five multistage consolidated drained direct shear tests were performed on five compacted specimens. The initial volume-mass properties of the five specimens are tabulated in Table 11.6. The tests were performed using the axis translation technique on a modified direct shear apparatus (Gan and Fredlund, 1988). A displacement rate of 1.7 × 10^{-4} mm/s was selected. Each specimen had six or seven stages of shearing. The tests were performed by maintaining a constant net normal stress $\sigma_n - u_a$ of 72 kPa while varying

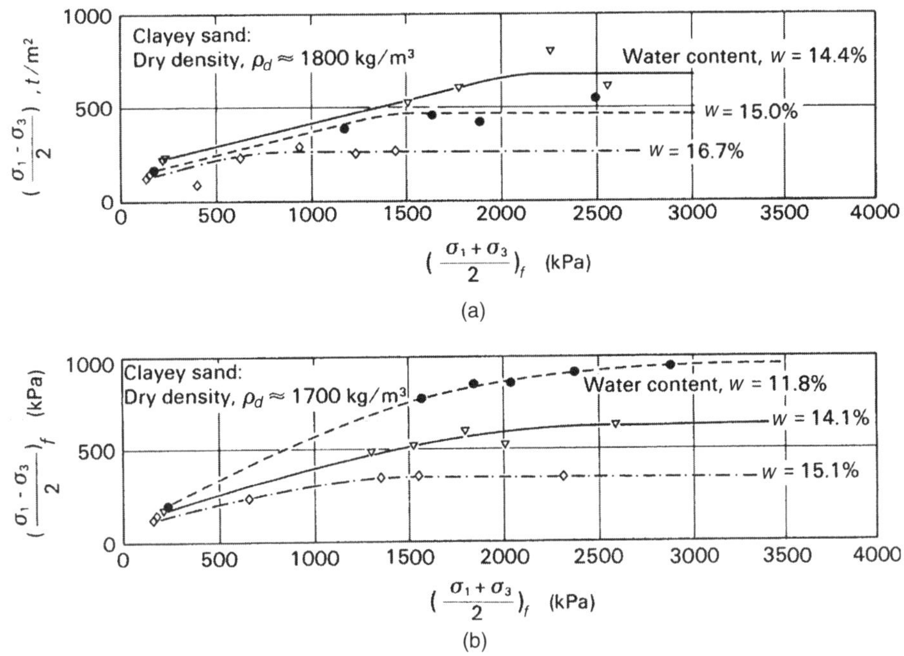

Figure 11.76 Total stress point envelopes obtained from undrained triaxial and unconfined compression tests on clayey sand: (a) total stress point envelopes for high-density specimens; (b) total stress point envelopes for low-density specimens (after Chantawarangul, 1983).

Table 11.6 Volume-Mass Properties for Five Speciemens Tested Using Multistage Direct Shear Tests (from Gan et al., 1987)

Specimen No.	GT-16-N1	GT-16-N2	GT-16-N3	GT-16-N4	GT-16-N5
Intitial Properties:					
Void ratio, e_0	0.77	0.53	0.69	0.51	0.54
Degree of saturation, S_0 (%)	42	59	48	65	61
Water content, w_0 (%)	11.8	11.5	12.3	12.2	12.1

the matric suction $u_a - u_w$ between stages. The matric suction ranged from 0 to 500 kPa. Failure envelopes of shear strength versus matric suction were plotted based on near-failure conditions at the end of each stage of testing. The results showed that the shear strength versus matric suction was nonlinear when the soil was tested over a wide matric suction range.

Matric suction equalization was generally attained in about 1 day for these particular tests. Typical results from the multistage direct shear tests on unsaturated specimens are illustrated in Figs. 11.77 and 11.78 for two test specimens. The vertical deflection versus horizontal displacement curves generally showed that the soil dilated during shear, except during the initial stages at low matric suctions. The curves showed an increase in the specimen height with increasing horizontal displacement as matric suction increased.

A plot of shear stress versus matric suction for sample GT-16-N5 is shown in Fig. 11.79. The plotted shear strength values correspond to a shear displacement of 1.2 mm. The line joining the data points forms the shear stress versus matric suction failure envelope. The envelope corresponds to an average net normal stress of 72 kPa at failure (Fig. 11.79a).

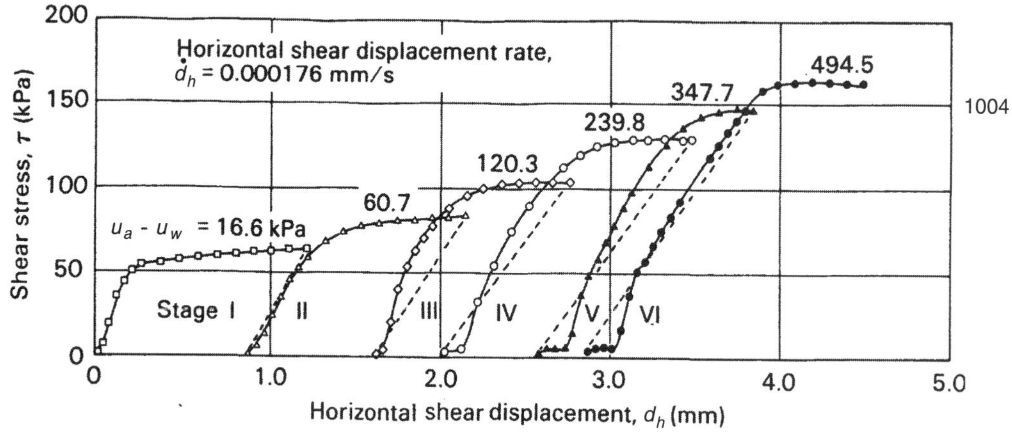

Figure 11.77 Multistage direct shear test results on compacted glacial till specimen No. GT16-N4; shear stress versus horizontal displacement (from Gan et al., 1988).

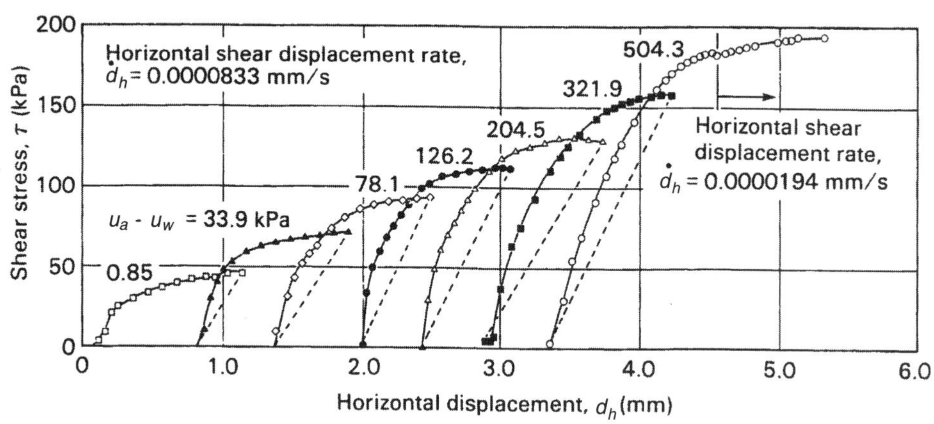

Figure 11.78 Multistage direct shear test results on compacted glacial till specimen No.GT-16-N5; shear stress versus horizontal displacement (from Gan et al., 1988).

The test results on the Indian Head glacial till exhibit significant nonlinearity in the failure envelope with respect to matric suction. The slope of the shear strength envelope (i.e., nonlinear ϕ^b) is plotted in Fig. 11.79b for various matric suction values.

Figure 11.80 presents a summary of the results obtained from five unsaturated specimens tested using the multistage direct shear test procedure. The results fall within a narrow band that forms a curved failure envelope. The ϕ^b angle starts at a slope equal to ϕ' (i.e., 25.5°) at matric suctions close to zero and decreases significantly at matric suctions in the range of 50–100 kPa. The ϕ^b angles reach a fairly constant value ranging from 5° to 10° when the matric suction exceeds 250 kPa. The scatter in the failure envelopes appears to be primarily due to slight variations in the initial void ratios of the soil specimens.

Nonlinearity of the failure envelope was also observed by Escario and Sáez (1986) on direct shear tests performed on several compacted soils. The properties, initial conditions, consolidation times, and displacement rates associated with two of the soils tested are tabulated in Table 11.7. The tests were performed using a modified direct shear apparatus. The consolidated drained testing procedure was used along with the axis translation technique.

Figures 11.81 and 11.82 present the results on compacted red clay of Guadalix de la Sierra and Madrid clayey sand specimens, respectively. The test results exhibited nonlinearity in the failure envelope with respect to matric suction. The failure envelopes with respect to the net normal stress appear to be essentially linear.

11.9 SELECTION OF STRAIN RATE

The strength testing of unsaturated soils is commonly performed at a constant rate of strain. An appropriate strain rate should be selected prior to commencing the test. The selected strain rate must ensure the equalization of induced

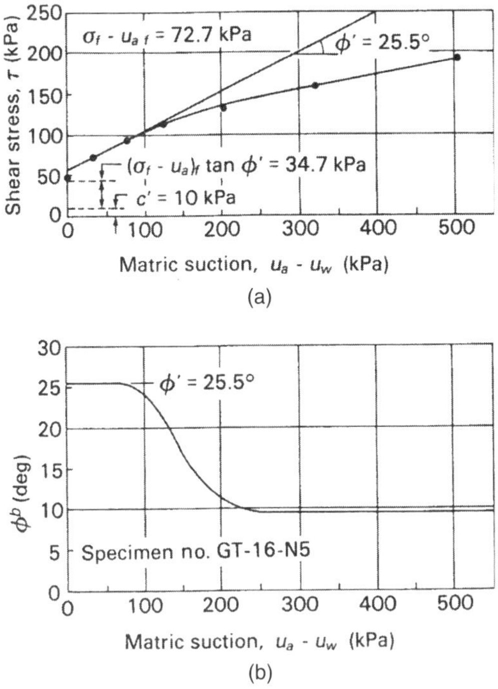

Figure 11.79 Failure envelope obtained from direct shear tests on glacial till specimen No. GT-16-N5: (a) failure envelope on τ versus $u_a - u_w$ plane; (b) slope of failure envelope (i.e., ϕ^b) for various levels of applied suction (from Gan and Fredlund, 1988).

pore pressures throughout the specimen during undrained shear. The selected strain rate during drained shear must ensure dissipation of induced pore pressures. The estimation of the strain rate for testing soils can be made partly on the basis of experimental evidence and partly on the basis of pore pressure dissipation theory.

11.9.1 Background on Strain Rates for Triaxial Testing

Prior to the 1960s, unsaturated soils were tested in shear using similar procedures and equipment to those used for saturated soils. Coarse ceramic disks were placed next to the soil specimens and the strain rates were relatively high. In the early 1960s, the test procedures and equipment used for testing unsaturated soils were modified. A high-air-entry ceramic disk was placed at the bottom of the soil specimen in order to measure or control the pore-water pressure independently from the pore-air pressure. Several empirical procedures were proposed for estimating the strain rate for testing unsaturated soils (Bishop et al., 1960; Lumb, 1966; Ruddock, 1966).

Laboratory testing equipment was also modified to accommodate the independent control of the pore-air and pore-water pressures in unsaturated soils. The pore-air pressure was controlled or measured through the use of a coarse ceramic disk generally placed on the top of the soil specimen. The control or measurement of the pore-water pressure was made possible through the use of a high-air-entry ceramic disk. The disk was sealed onto the base pedestal. The modified laboratory testing apparatuses allowed the use of the axis translation technique (Hilf, 1956). The axis translation technique continues to be used for the shear strength testing of unsaturated soils.

Shear strength tests on unsaturated soils have generally been performed at relatively slow rates in order to ensure equalization or dissipation of induced pore pressures. Trial-and-error procedures have been used by several researchers to assess the appropriate strain rate. Donald (1961) recommended that the effect of strain rate on the maximum deviator stress be used as a criterion in assessing an

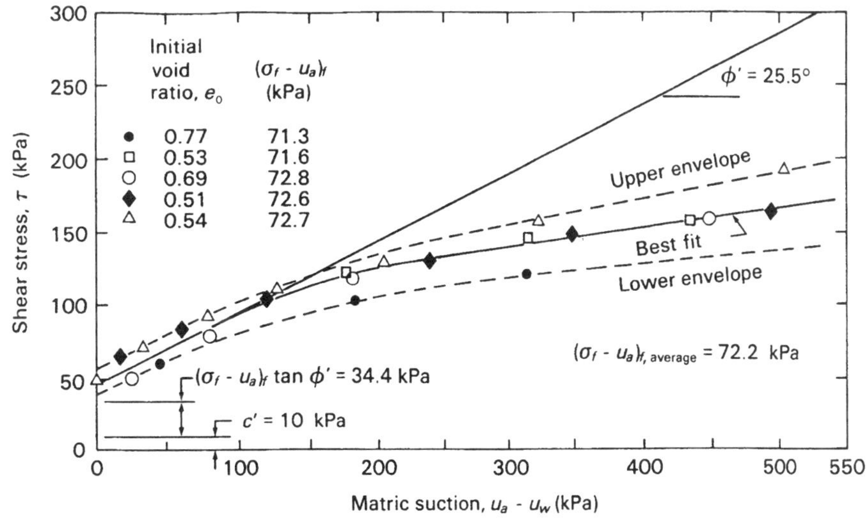

Figure 11.80 Failure envelopes on τ versus $u_a - u_w$ plane obtained from several direct shear tests on compacted glacial till specimens (from Gan et al., 1988).

Table 11.7 Direct Shear Tests on Unsaturated Compacted Soils

Properties	Red Clay of Guadalix de la Sierra	Madrid Clayey Sand
Liquid limit	33	32
Plasticity index	13.6	15
Sieve analysis: % passing		
10	—	100
16	100	94
40	97	48
200	86.5	17
AASHTO standard compaction		
ρ_d maximum (kg/m^3)	1800	1910
w optimum (%)	17	11.5
Initial conditions		
ρ_{d0} (kg/m^3)	1800	1910
w_0 (%)	13.6	9.2
$(u_a - u_w)_0$ (kPa)	2.8	0.7
Consolidation time under applied total stress and matric suction (days)	4	4
Displacement rate, d_h (mm/s)	2.8×10^{-5}	2.8×10^{-5}
Time to failure, t_f (days)	2–3	1–2

Source: From Escario and Sáez (1986).

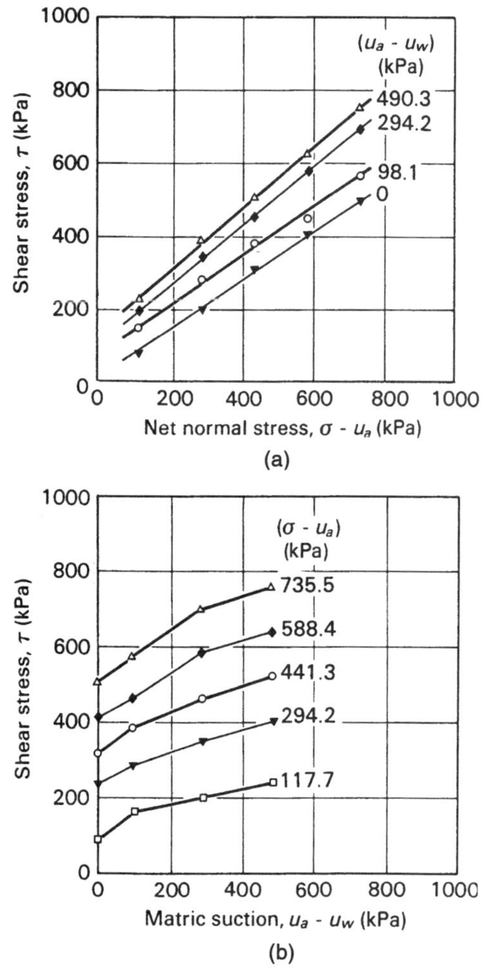

Figure 11.81 Direct shear tests on compacted red clay of Guadalix de la Sierra: (a) horizontal projection of failure envelope onto τ versus $\sigma - u_a$ plane; (b) horizontal projection of failure envelope onto τ versus $u_a - u_w$ plane (after Escario and Saez, 1986).

appropriate strain rate. Gibson and Henkel (1954) and Bishop and Henkel (1962) presented test data showing the variation in strength (i.e., the deviator stress at failure) with strain rate. The shear strength tended to a constant value below a particular strain rate, indicating a suitable strain rate for failing the soil specimen. No measurements were made to ensure that this strain rate limit was slow enough for complete pore pressure equalization or dissipation.

Satija and Gulhati (1979) concluded from measured test results that deviator stress is not highly sensitive to the effect of varying strain rates (Figs. 11.83a and 11.84a). It was suggested that changes in matric suction be used for CW tests (Fig. 11.83b), and changes in water content be used for CD tests (Fig. 11.84b), as a reasonable indicator of an appropriate strain rate. Tests were performed on specimens of compacted Dhanauri clay using strain rates varying from 5.3×10^{-5} to 3.3×10^{-3} (%/s). It was concluded that a strain rate of 6.7×10^{-4} (%/s) was adequate for CW tests on Dhanauri clay. The strain rate for CD tests was chosen to be one-fifth of the strain rate for CW tests (i.e., $1.3 \times 10^{-4}\%/s$).

The effect of strain rate on the equalization of pore-water pressures for undrained tests was studied by Bishop et al. (1960). Two compacted clay shale specimens 101.6 mm in diameter and 203.2 mm in height were tested at two different strain rates (i.e., 6.9×10^{-4} and 4.6×10^{-5} $\%/s$). The pore-water pressure was measured at the base of the specimen and at the midheight of the specimen. The results are presented in Figs. 11.85 and 11.86 for strain rates of 6.9×10^{-4} and 4.6×10^{-5} $\%/s$ respectively. The higher strain rate resulted in significantly different pore-water pressures across the specimen (Fig. 11.85b). The base measurement of pore-water pressure was higher than the midheight measurement. As a result, the principal stress ratios computed from the two pore-water pressure measurements were also different (Fig. 11.85a). This difference affects the assessment of the shear strength properties. Closer agreement between the pore-water measurements at the base and at the midheight point was obtained when using a slower strain rate (Fig. 11.86b).

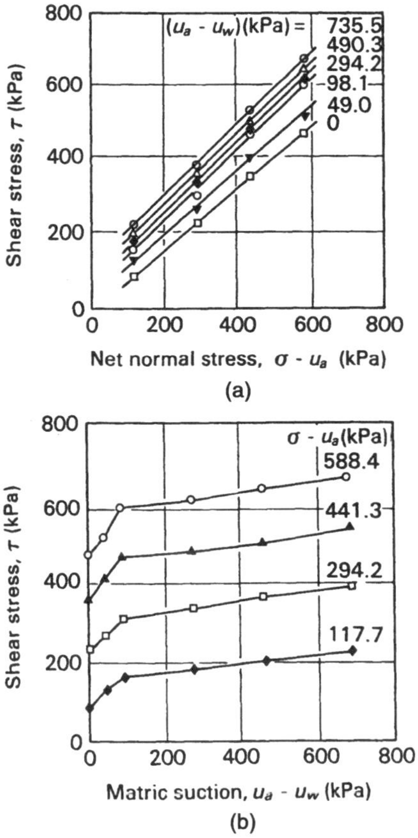

Figure 11.82 Direct shear tests on compacted Madrid clayey sand: (a) horizontal projection of failure envelope onto shear strength versus $\sigma - u_a$ plane; (b) horizontal projection of failure envelope onto shear strength versus $u_a - u_w$ plane (after Escario and Saez, 1986).

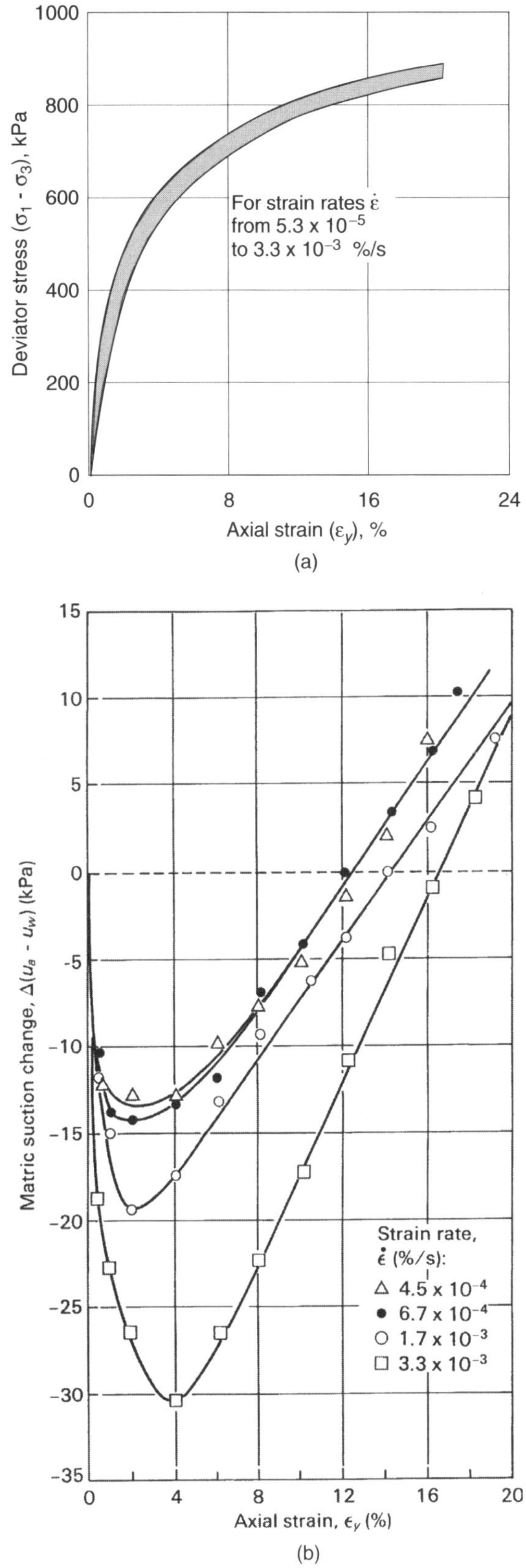

Figure 11.83 Effect of strain rate for constant-water-content tests on Dhanauri clay: (a) effect of strain rate on deviator stress; (b) effect of strain rate on matric suction change (after Satija, 1978).

11.9.2 Strain Rates for Triaxial Tests

Strain rate can be defined as the rate at which a soil specimen is axially compressed:

$$\dot{\varepsilon} = \frac{\varepsilon_f}{t_f} \tag{11.32}$$

where:

$\dot{\varepsilon}$ = strain rate for shearing a specimen in the triaxial test,
ε_f = strain of the soil specimen at failure, and
t_f = time required to fail the soil specimen, or "time to failure."

The strain at failure, ε_f, depends on the soil type and the stress history of the soil. Table 11.8 presents typical values of strain at failure, ε_f, obtained from numerous triaxial testing programs on unsaturated soils. The information in Table 11.8 can be used as a guide when determing a suitable strain rate $\dot{\varepsilon}$ for shear strength testing.

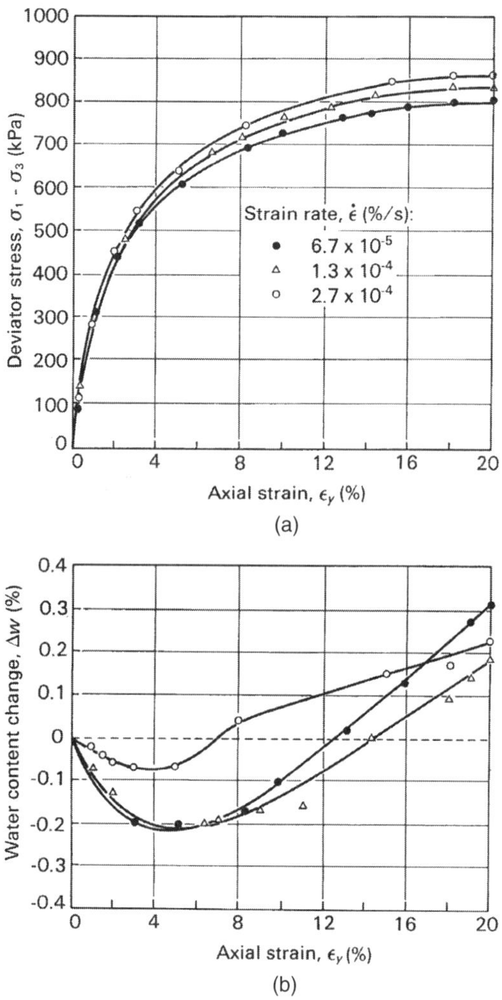

Figure 11.84 Effect of strain rate on consolidated drained tests on Dhanauri clay: (a) effect of strain rate on deviator stress; (b) effect of strain rate on water content change (after Satija, 1978).

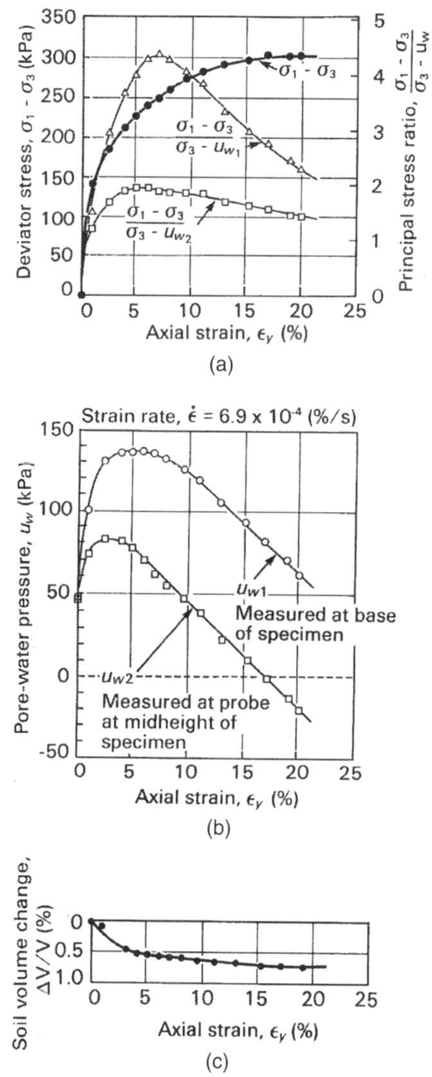

Figure 11.85 Undrained triaxial test on compacted clay shale specimen tested at strain rate of $6.9 \times 10^{-4}\%/s$: (a) shear stress versus strain curve; (b) pore-water pressure versus strain at two points of measurement; (c) total volume change versus strain (after Bishop et al., 1960).

Gibson and Henkel (1954) used the theory of consolidation to formulate a theoretical method for approximating the time required to fail a specimen under drained conditions. The theory is applicable for both triaxial and direct shear tests. The strain rate used in the triaxial testing of saturated soils has commonly been estimated using the Gibson and Henkel (1954) procedure (Bishop and Henkel, 1962). For unsaturated soils, it is necessary to consider the possibility of impeded flow through the high-air-entry disk at the base of the specimen as well as the physical properties of the soil (Ho and Fredlund, 1982c). The high-air-entry disk has a low coefficient of permeability k_d. The disk not only prevents the passage of air but also impedes the flow of water in and out of the specimen during shear. Another factor affecting the time to failure t_f is the low coefficient of permeability of the unsaturated soil.

Analytical considerations of the impeded flow problem for an unsaturated soil specimen can be solved in a manner similar to the problem of impeded flow caused by plugged porous stones placed next to saturated soil specimens (Bishop and Gibson, 1963). The solution gives the time required to fail the specimen when using a drained test. A somewhat smaller value for time to failure t_f should be used for an undrained test. The high-air-entry disk may control pore pressure equalization in the soil. It is suggested that the time-to-failure value computed for a drained test should be considered as a conservative estimate for the time to failure for an undrained test.

The ability of an unsaturated soil specimen and the measuring system to dissipate the excess pore pressures developed during drained shear is the main consideration in computing the time to failure. The excess pore pressure dissipation is essentially a one-dimensional consolidation process. Generally, the coefficient of permeability with respect to the air phase, k_a, is much larger than the coefficient of permeability with respect to the water phase,

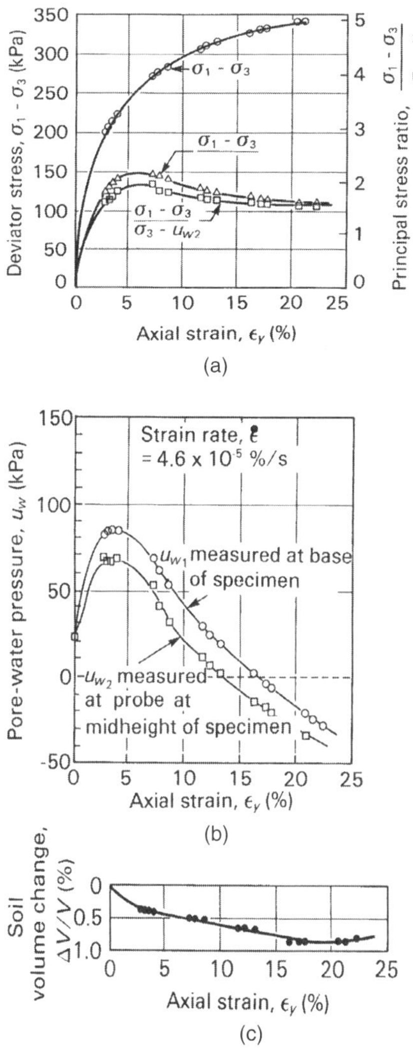

Figure 11.86 Undrained test on compacted clay shale specimen tested at strain rate of $4.6 \times 10^{-5}\%/s$: (a) shear stress versus strain curve; (b) pore-water pressure versus strain at two points of measurement; (c) total volume change versus strain (after Bishop et al., 1960).

k_w. When the air phase is occluded, the air coefficient of permeability reduces to the diffusivity of air through water. It is assumed that the dissipation of the excess pore-water pressure requires more time than the dissipation of the excess pore-air pressure. The time required to fail an unsaturated soil specimen can be computed by considering the dissipation of excess pore-water pressure.

The solution to the problem of excess pore-water pressure dissipation with impeded flow is mathematically equivalent to the heat conduction of a slab initially at a uniform temperature being cooled at its surface in accordance with Newton's law (Carslaw and Jaeger, 1959). The solution can be written as a Fourier series (Gray, 1945). Figure 11.87 shows the dimensions and property variables of an unsaturated soil specimen and high-air-entry disk required in the solution to the impeded dissipation of excess pore-water pressures. The high-air-entry disk is assumed to

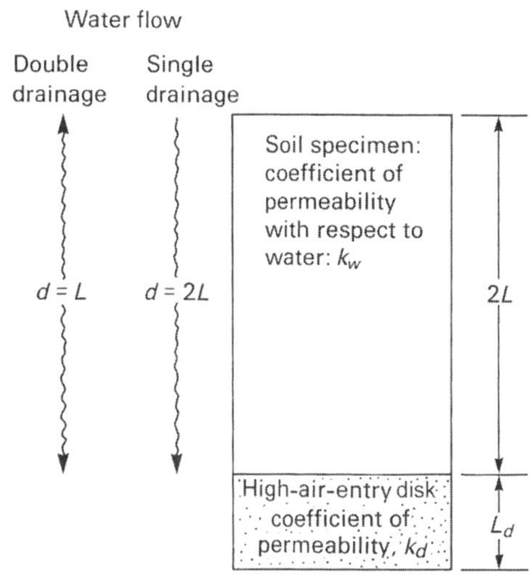

Figure 11.87 Properties and variables required to model pore pressure equalization for unsaturated soil specimen placed on high-air-entry disk.

be incompressible, and its properties appear in the form of an impedance factor λ:

$$\lambda = \frac{k_d d}{k_w L_d} \tag{11.33}$$

where:

λ = impedance factor,
k_d = water coefficient of permeability of the high-air-entry disk,
k_w = coefficient of permeability of the unsaturated soil with respect to the water phase,
d = length of drainage path in the soil, and
L_d = thickness of the high-air-entry disk.

Let us consider the case of single drainage where pore-water drains only to the bottom (or top) of the soil specimen. The length of the drainage path, d, is equal to the height of the soil specimen, $2L$ (i.e., $d = 2L$; see Fig. 11.87). If the pore-water is allowed to drain from the top and bottom of the specimen (i.e., double drainage), the length of the drainage path becomes half of the specimen height (i.e., $d = L$). Water is usually permitted to drain through the high-air-entry disk at the bottom of the specimen when testing unsaturated soils.

Equation 11.33 is plotted in Fig. 11.88 in terms of dimensionless ratios k_w/k_d and L_d/d for single drainage dissipation (i.e., $d = 2L$). The dimensionless form allows the assessment of the impedance factor λ for a wide range of soils and high-air-entry disks. The impedance factor λ can be obtained directly from Fig. 11.88 when the properties of the specimen and the high-air-entry disk are known. The coefficient of permeability with respect to the water phase, k_w, may vary depending upon the matric suction of the specimen. The coefficient of permeability of the soil can be expressed as a function of matric suction, $k_w(u_a - u_w)$. It would appear that the rate of strain should be decreased as matric suction is increased.

Table 11.8 Strain Rate and Strain at Failure for Triaxial on Unsaturated Soils

Soil Type	Triaxial Test	Strain Rate, $\dot{\epsilon}$ (% /s)	Approximate Strain at Failure, ϵ_f (%)	References
Boulder clay; $w = 11.6\%$ and % clay = 18%	CW	3.5×10^{-5}	15	Bishop et al. (1960)
Braehead Silt	CW	4.7×10^{-5}	11	Bishop and Donald (1961)
	CD	8.3×10^{-6}	12	
Talybont boulder clay; $w =$ 9.75% and % clay = 6%	Undrained with pore pressure measurements	4.7×10^{-7}	$\sigma_3 = 83$ kPa : 8.5 $\sigma_3 = 207$ kPa : 11	Donald (1963)
Dhanauri clay; $w = 22.2\%$ and % clay = 25 %	CW	6.7×10^{-4}	20	Satija and Gulhati (1979)
	CD	1.3×10^{-4}	20	
Undisturbed decomposed granite and rhyolite	CD	1.7×10^{-5}	Stage I: 3–5	Ho and Fredlund (1982a)
	Multistage	6.7×10^{-5}	Stage II: 1–3 Stage III: 1–3	
Clayer sand; $w = 14$–17% and % clay = 30%	Undrained and unconfined	1.7×10^{-3}	15–20	Chantawarangul (1983)

The time required to fail the soil specimen, t_f, can be expressed in terms of a desired degree of dissipation of excess pore-water pressure (Bishop and Gibson, 1963):

$$t_f = \frac{L^2}{\eta c_v^{\,w}\left(1 - \bar{U}_f\right)} \tag{11.34}$$

where:

$L =$ half of the actual length of the soil specimen, which is the same for single or double drainage,
$\eta = 0.75 / (1 + 3/\lambda)$ for single drainage,
$\eta = 3/(1 + 3/\lambda)$ for double drainage,
$c_v^w = k_w/\left(\rho_w g m_2^w\right)$, the average coefficient of consolidation with respect to the water phase,
$\rho_w =$ density of water,
$g =$ gravitational acceleration,
$m_2^w =$ slope of the plot of water volume change, $\left(V_w/V\right)$ versus suction, and
$\bar{U}_f =$ average degree of dissipation of the excess pore-water pressure at failure.

The effect of impeded flow on the time to failure t_f is taken into account through the impedance factor λ in the η parameter in Eq. 11.34. The properties of an unsaturated soil specimen are taken into account through the L and c_v^w terms. Approximate properties for the unsaturated soil (i.e., k_w and m_2^w) can be used to estimate the average coefficient of consolidation c_v^w. Any desired value for the average degree of dissipation $\bar{U}_f$ can be used, but a value of 95% is recommended (Gibson and Henkel, 1954; Bishop and Henkel, 1962). An average degree of dissipation of 95% means that 95% of the excess pore-water pressure developed during shear will be dissipated at the time of failure.

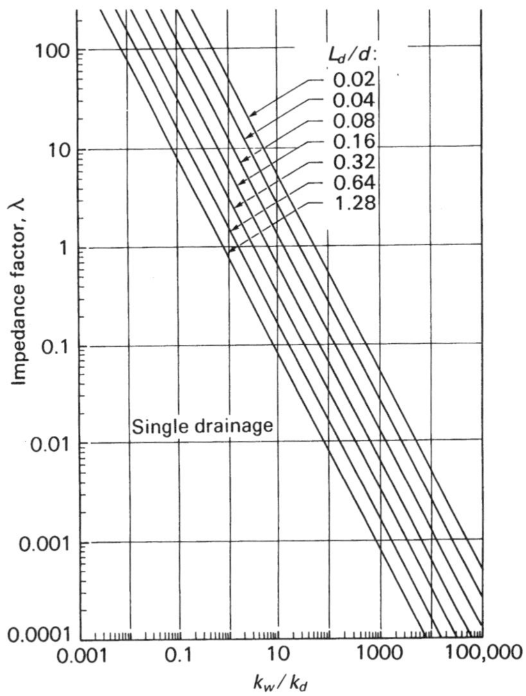

Figure 11.88 Impedance factors λ for various ratios k_w/k_d and L_d/d for single drainage.

The time to failure t_f (i.e., Eq. 11.34) can be plotted with respect to the impedance factor λ for various values of the coefficient of consolidation c_v^w, as shown in Fig. 11.89. For single drainage testing the length of the drainage path is equal to the height of the soil specimen (i.e., 140 mm). The graph indicates that more time is required to reach failure as the coefficient of consolidation decreases (i.e., a decrease

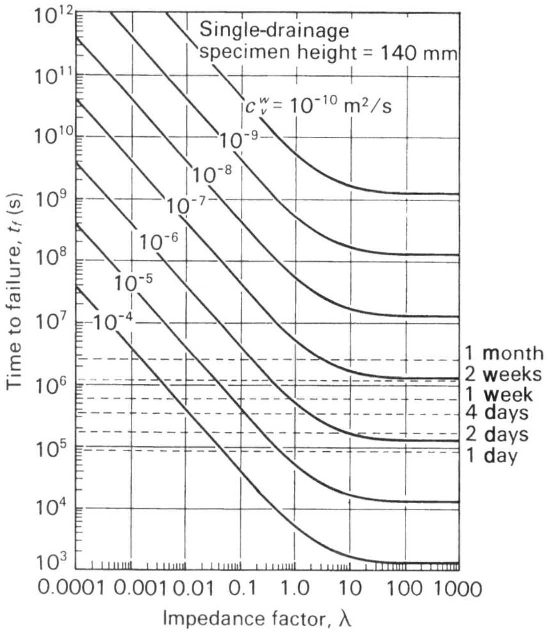

Figure 11.89 Time to failure t_f for drained triaxial test specimen 140 mm in height.

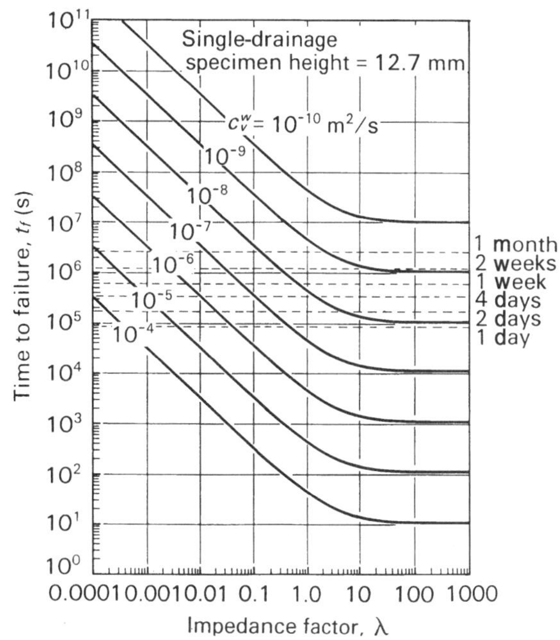

Figure 11.90 Time to failure t_f for drained direct shear tests with specimen height of 12.7 mm.

in the impedance factor λ). The increase in time to failure due to a decrease in the coefficient of consolidation c_v^w is uniform at various λ values. However, the increase in time to failure due to a decrease in the impedance factor becomes more significant for impedance factors less than 10. When the impedance factor is greater than 10, the time to failure is almost constant for a particular coefficient of consolidation.

An impedance factor of 10 corresponds to a specific permeability ratio k_w/k_d (Fig. 11.88). As an example, an impedance factor of 10 corresponds to a k_w/k_d ratio of 0.7 when the L_d/d ratio is 0.16. Impedance values less than 10 correspond to k_w/k_d ratios greater than 0.7 (Fig. 11.88). A comparison between Figs. 11.89 and 11.90 indicates that the time to failure t_f is significantly affected by impeded flow when the impedance factor is less than 10. In other words, the impeded flow problem exists even when the high-air-entry disk has a coefficient of permeability equal to the coefficient of permeability of the soil specimen. In many cases, the coefficient of permeability of the disk is lower than that of the soil. On the other hand, the effect of impeded flow on the time to failure t_f remains essentially unchanged when the impedance factor is greater than 10. This means that little improvement can be made in shortening the time to failure t_f when the impedance factor has reached a value greater than 10.

It is evident from the above discussion that impeded flow due to the presence of the high-air-entry disk is often the governing factor influencing the time to failure. One way to reduce the time to failure is by using a thin high-air-entry disk. High-air-entry disks generally range in thickness from 3.2 to 9.5 mm. For a specific k_w/k_d ratio and a specific specimen height, a decrease in the thickness of the high-air-entry disk causes an increase in the impedance factor (Fig. 11.88). An increase in the impedance factor reduces the time to failure t_f (Fig. 11.89). However, thicker disks are superior due to their ability to reduce the diffusion rate of air through the ceramic disk.

The strain rate $\dot{\varepsilon}_f$ for triaxial testing of unsaturated soils can be computed in accordance with Eq. 11.34. The computed strain rate is only an estimate because of the assumptions involved in the theory and the difficulties in accurately assessing relevant soil properties. Nevertheless, the theory does provide a somewhat rational approach to obtaining the strain rate for unsaturated soil testing. Typical strain rates that have been used for triaxial tests for unsaturated soil testing are tabulated in Table 11.7. The strain rate for testing should be relatively slow, even when the soil appears to be quite pervious, since the strain rate is largely controlled by the high-air-entry disk. Equation 11.34 can also be used to obtain a conservative estimate of strain rate for undrained shear strength testing.

It appears that water can more readily flow out from a soil specimen through the high-air-entry disk than into the specimen through the high-air-entry disk. If a soil tends to dilate during shear, it may be difficult to ensure the upward flow of water through the low-permeability disk into the soil specimen. This problem can be somewhat alleviated by using the axis translation technique and periodically flushing diffused air from below the high-air-entry disk.

11.9.3 Displacement Rate for Direct Shear Tests

The rate of horizontal shear displacement in a direct shear test is analogous to the strain rate in a triaxial test. The horizontal shear displacement rate can be defined as the relative rate at which the top and bottom halves of the direct shear box are displaced:

$$\dot{d}_h = \frac{d_h}{t_f} \tag{11.35}$$

where:

$\dot{d}$ = horizontal shear displacement rate for a direct shear test, and

d_h = horizontal displacement of the soil specimen at failure.

The horizontal displacement at failure, d_h, depends on the soil type and the soil stress history. An estimate of the horizontal displacement at failure is required in order to compute the horizontal shear displacement rate $\dot{d}_h$.

The theoretical estimation for the time to failure t_f used in triaxial testing (i.e., Eq. 11.34) is also applicable to drained direct shear tests (Eq. 11.35). The excess pore pressures developed during a drained direct shear test are also dissipated one dimensionally in the vertical direction. Pore-water is commonly drained out of the bottom of the specimen through the high-air-entry disk. The impedance factor for direct shear tests can be computed using Eq. 11.33 (Fig. 11.88). The time to failure t_f for direct shear tests can be computed using Eq. 11.34 (Fig. 11.90). Figure 11.90 is a plot of the time to failure for various impedance factors and coefficients of consolidation. The plot is generated for a specimen height of 12.7 mm.

The horizontal shear displacement rate $\dot{d}_h$ can be computed from Eq. 11.35 using estimated values of d_h and t_f. Typical horizontal shear displacement rates that have been used in direct shear testing on unsaturated soils are listed in Table 11.9.

A comparison of Figs. 11.89 and 11.90 reveals that the time required to fail the soil specimen is substantially reduced by using a thinner soil specimen. This can be observed from Eq. 11.34 where the time to failure is a function of the square of half the specimen height h. The direct shear test has the advantage that a thin soil specimen can significantly reduce testing times for shear strength.

11.9.4 Multistage Testing

Multistage testing can be performed using triaxial or direct shear equipment. Benefit is derived from being able to alter the stress path during the testing of a single soil specimen. Multistage testing is used to maximize the amount of shear strength information that can be obtained from one specimen. It also assists in eliminating the effect of soil variability on the test results. Multistage triaxial testing of saturated soils usually consists of using several effective confining pressures $\sigma_3 - u_w$ when testing one soil specimen. A Mohr-Coulomb failure envelope can be drawn tangent to Mohr circles corresponding to failure at the various effective confining pressures.

Multistage triaxial testing has been used in conjunction with a consolidated drained type of test on unsaturated soils. Multistage triaxial testing is also applicable when using the constant-water-content or consolidated undrained test

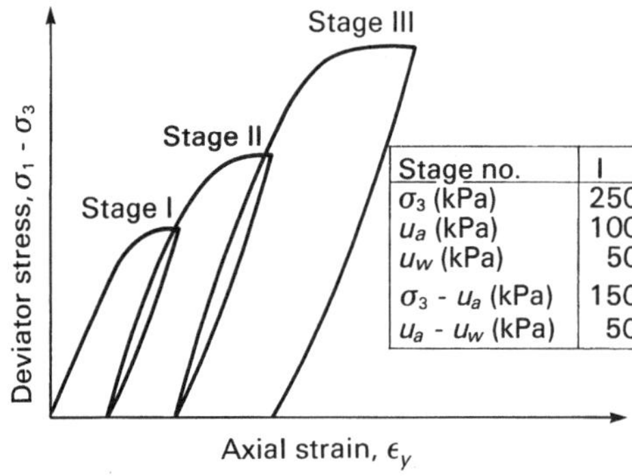

Stage no.	I	II	III
σ_3 (kPa)	250	350	500
u_a (kPa)	100	200	350
u_w (kPa)	50	50	50
$\sigma_3 - u_a$ (kPa)	150	150	150
$u_a - u_w$ (kPa)	50	150	300

Figure 11.91 Idealized stress-strain curve for multistage triaxial testing.

Table 11.9 Horizontal Displacement Rate and Horizontal Displacements at Failure for Several Direct Shear Tests on Unsaturated Soils

	Direct Shear Test	Displacement Rate, $\dot{d}_h$(mm/s)	Displacement at Failure, d_h(mm)[a]	References
Madrid grey clay	CD	1.4×10^{-4}	3.5–5	Escario (1980)
Madrid grey clay	CD	2.8×10^{-5}	6.0–7.2	Escario and Sáez (1986)
Red clay of Guadalix de la Sierra	CD	2.8×10^{-5}	4.8–7.2	Escario and Sáez (1986)
Madrid clayey sand	CD	2.8×10^{-5}	2.4–4.8	Escario and Sáez (1986)
Glacial till	CD (multistage)	1.7×10^{-4}	1.2	Gan (1986)

[a]Square specimen of 50 × 50 mm.

procedure. For multistage consolidated drained triaxial tests, the net confining pressure $\sigma_3 - u_a$ is usually maintained constant while the matric suction $u_a - u_w$ is varied from one stage to the next. Figure 11.91 illustrates a multistage triaxial test with three stages. Each stage corresponds to a different matric suction value. A multistage consolidated drained triaxial test can also be performed by maintaining the matric suction constant and varying the net confining pressure.

Each stage of the multistage test should be terminated when failure is imminent. Near-failure conditions can generally be determined by observing a plot of deviator stress versus strain. The soil specimen should not be subjected to excessive deformation, particularly during the early stages of loading. Difficulties related to excessive deformation in multistage triaxial testing have been reported by several researchers (Kenney and Watson, 1961; Wong, 1978; Ho and Fredlund, 1982b). When the specimen undergoes excessive deformation during the early stage of loading, the specimen will tend to undergo soil structure breakdown, and the strength may be reduced for subsequent stages of testing. The shear strength measured at successive stages may tend toward a reduced ultimate strength value. Excessive strain accumulation can be a problem in multistage testing, particularly in cases where the soil structure is sensitive to disturbance.

CHAPTER 12

Shear Strength Applications in Plastic and Limit Equilibrium

12.1 INTRODUCTION

The analysis of lateral earth pressure problems (i.e., active and passive earth pressures), bearing capacity problems (i.e., plasticity theory), and slope stability problems (i.e., limit equilibrium analyses) all require the shear strength properties of the soil (i.e., failure criterion). The equations of static equilibrium must also be satisfied when solving the above-mentioned geotechnical engineering problems.

The shear strength envelope becomes nonlinear when soils are unsaturated and soil suction varies over a wide range. The nonlinearity of the shear strength envelope is related to the changes in degree of saturation of the soil as soil suction changes. The previous chapter provided information on the measurement of unsaturated soil shear strength parameters. There are numerous situations, however, where it is sufficient and satisfactory to estimate the unsaturated shear strength parameters. This is particularly true at the preliminary design stage of a project. Numerous empirical equations have been proposed for the estimation of the shear strength of unsaturated soils. In most cases the estimated shear strength equations are based on the saturated shear strength parameters and the SWCC of the soil.

The first portion of this chapter is devoted to the "estimation" of the shear strength envelope for an unsaturated soil. The estimation procedures are of significant importance to geotechnical engineering practice. Mention is also made of another set of shear strength equations that fall under the category of fitting equations. In this case, it is assumed that shear strength data with respect to soil suction is available and one of the proposed fitting equations is then best fit through the data points. Figure 12.1 illustrates the difference between estimation procedures and fitting techniques.

12.2 ESTIMATION OF SHEAR STRENGTH FUNCTIONS FOR UNSATURATED SOILS

Unsaturated soil properties take the form of nonlinear mathematical functions with respect to soil suction. This is true for the shear strength properties of unsaturated soils as well as virtually every physical property of unsaturated soils. It is costly and time consuming to measure the unsaturated soil property functions in the laboratory because of the number of data points that are required. There has consequently been an emphasis on the estimation of unsaturated shear strength functions.

Many engineering projects may not be able to justify the measurement of the unsaturated shear strength relationship for an unsaturated soil. However, an estimation of the unsaturated soil shear strength envelope may provide the engineer with a significantly improved ability to analyze practical problems where a portion (or all) of the soil profile involves unsaturated soils. The contribution of soil suction to shear strength generally results in a significant increase in bearing capacity and an increase in the factor of safety of slopes. At the same time, the active earth pressure for an unsaturated soil is decreased by soil suction while the passive pressure in increased as soil suction is increased.

Most estimation shear strength equations proposed for an unsaturated soil have been related to the SWCC or the classification properties of the soil. Consideration has mainly been given to the use of shear strength estimation procedures that use two independent stress state variables and the SWCC. Empirical estimation equations associated with critical state constitutive models are not discussed in this book. Much of the published shear strength data are associated with the peak shear strength of the soil.

The method of preparation of soil specimens should be noted along with shear strength laboratory results. It is important to know the preparation procedures for soil specimens tested in the laboratory. Soil specimens can be prepared as (i) slurry soil specimens consolidated to specified applied pressures and then desaturated by increased soil suction, (ii) compacted specimens prepared by premixing at a specified water content and density and subsequently varying the soil suction that is applied, and (iii) undisturbed samples from the field that are then conditioned with various applied soil suctions.

A number of general observations can be made based on various shear strength studies that have been undertaken

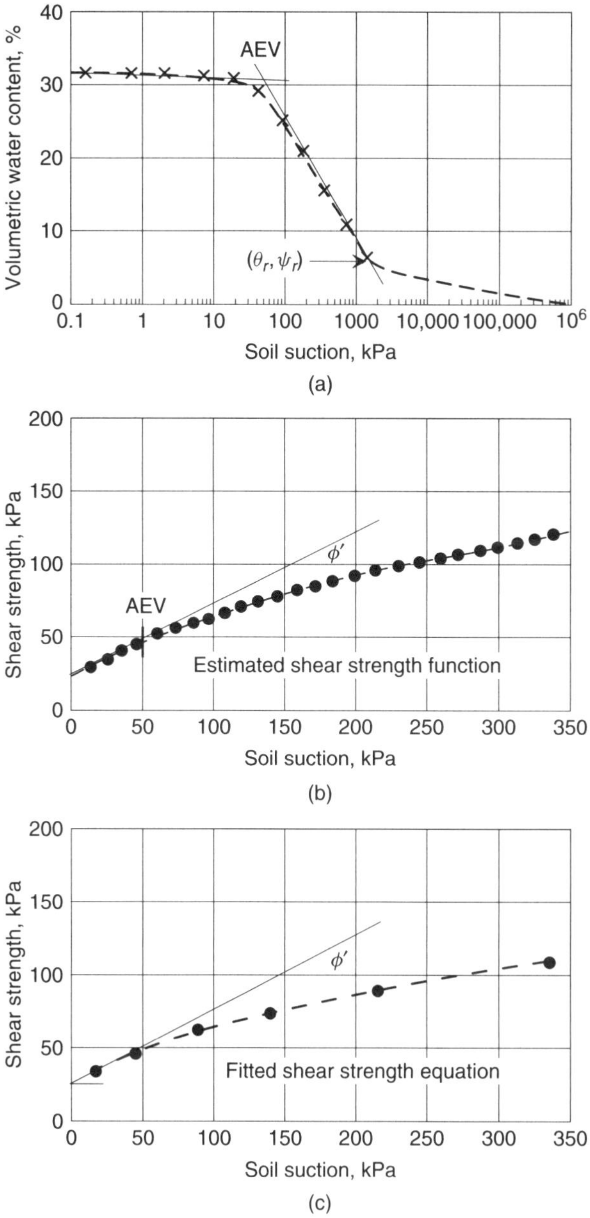

Figure 12.1 Fitting experimental shear strength data and estimating a soil suction shear strength function: (a) fitting and (b) estimation procedures to obtain shear strength function.

(Escario and Saez, 1986; Fredlund and Rahardjo, 1993a; Vanapalli et al., 1996; Cunningham et al., 2003):

1. Higher matric suctions result in higher shear strengths under the same confining pressures.
2. Higher confining pressures result in higher shear strengths under the same matric suction.
3. The relationship between shear strength and soil suction is nonlinear when the applied soil suctions are increased beyond the air-entry value of the soil. The shear strength increases most rapidly in the low-matric-suction range and then gradually flattens (or even decreases) at high suctions (i.e., suctions approaching residual conditions).

The shear strength of an unsaturated soil has been shown to bear a relationship to changes in the degree of saturation (or water content) of a soil. Therefore, it is not surprising that proposed shear strength functions for an unsaturated soil have been related to the SWCC. The shear strength function for an unsaturated soil appears to adhere to the following limiting conditions. The shear strength of an unsaturated soil increases in response to the effective angle of internal friction, ϕ', for matric suctions up to the air-entry value of the soil. Once the air-entry value is exceeded, the increase in shear strength responds to soil suction at a continuously decreasing rate throughout the transition region. There appears to be no significant increase (or decrease) in shear strength for most soils once residual suction conditions are exceeded.

12.2.1 Observed Relationship between SWCC and Shear Strength

Donald (1956) performed a series of direct shear tests on various gradations of Frankston sand. This is probably the earliest set of shear strength tests performed where the associated SWCC desorption curves were also measured. Figure 12.2b shows the change in strength that occurred for the Frankston sand soils when soil suction was increased past the air-entry value through the transition zone and beyond residual suction conditions. Each shear strength plot shows that the shear strength rises quickly as the negative pore-water pressure is increased. However, in each case there is a point at which shear strength levels off and even decreases as matric suction is further increased.

Changes in the shear strength with matric suction bear a relationship to the corresponding SWCCs for each of the sand soils. Figure 12.2a shows the SWCCs corresponding to the Frankston sand specimens. In each case, there appears to be a relationship between the changes on the shear strength envelope and key points on the SWCC, namely, the air-entry value and residual suction conditions.

The first deviation from the effective angle of internal friction, ϕ', occurs just prior to the air-entry value. There is a leveling off of the shear strength envelope just after residual suction conditions are exceeded. The peak values of shear strength and the decrease in shear strength occur in the transition zone of the SWCC. The test results on the sand specimens show a relationship between the SWCC and shear strength. These are some of the earliest results that form the basis for the development of estimation procedures for the unsaturated shear strength envelope. The saturated shear strength parameters and the SWCC provide the basic information required for formulating a shear strength function. Consequently, there is continuity between the shear strength equations for saturated and unsaturated soils.

Numerous shear strength testing programs have been performed over the years, and each time, when shear strength

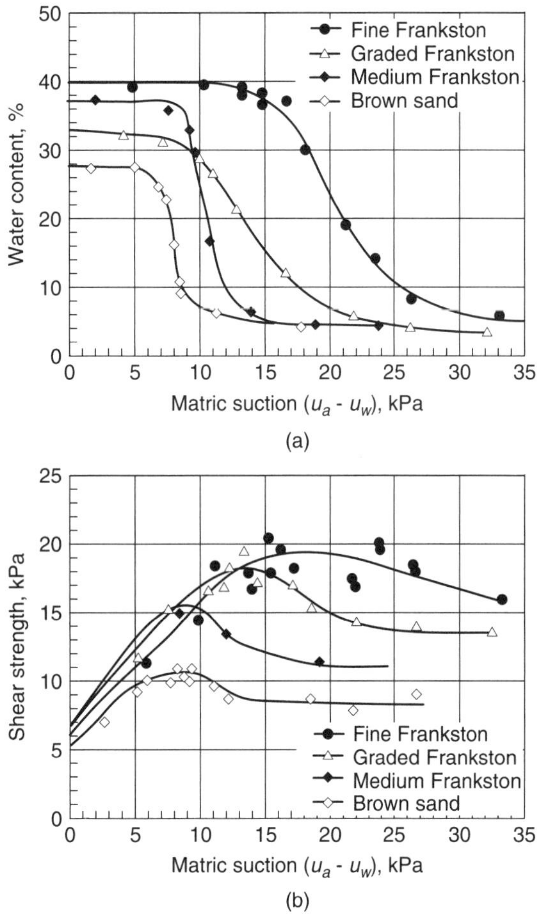

Figure 12.2 Result of series of direct shear tests on sands subjected to range of matric suction (Donald, 1956, with modifications by Vanapalli, 2009).

was measured over a wide range of soil suction values, it has been clear that there was a relationship between the SWCC and the shape of the unsaturated shear strength envelope. Figure 12.3 shows the results of a series of direct shear tests performed on glacial till. The SWCC shows that the air-entry value of the soil is approximately equal to 60 kPa. The shear strength envelopes show that the air-entry value is near the point where shear strength starts to deviate from the effective angle of internal friction, ϕ'. It is difficult to observe a distinct break in the SWCC that represents the start of residual conditions; however, residual suction appears to be beyond 10,000 kPa for the glacial till soil. Residual suction is well beyond the range of the shear strength tests. The shear strength measurements extended to a matric suction of 500 kPa and the envelope continued to increase at a reduced rate as soil suction increased.

The above-mentioned results along with other laboratory test results have shown that there is a clear relationship between the unsaturated shear strength of a soil and the corresponding SWCC. Figure 12.4 summarizes the general anticipated shear strength responses that might be anticipated from a variety of soils. The shear strength of all soil types appears to respond as a saturated soil as long as the matric suction is less than the air-entry value of the soil. There is curvature in the shear strength envelope once the air-entry value is exceeded. The shear strength envelope for most soils with silt- and clay-size particles bends toward an almost horizontal line near residual suction conditions. In other words, the shear strength envelope shows an increase in strength up to the residual suction for the soil. Sandy soils generally show a leveling off in strength even prior to residual suction being reached and tend to fall off in strength at higher soil suctions.

12.2.1.1 Shear Strength beyond Residual Suction

The early shear strength tests by Donald (1956) showed that it was possible for the shear strength of sand to decrease once residual suction conditions were exceeded. It is also known that clay soils generally continue to increase in strength as soil suction is increased beyond residual suction conditions. Some estimation models use the entire SWCC when estimating the unsaturated shear strength function. Other estimation models normalize water content between saturated conditions and residual water content conditions. There is little information to confirm the accuracy of any of the estimation models in the high-soil-suction range.

Nishimura and Fredlund (2002) performed unconfined compression tests on two soils that were conditioned to total suction values well beyond the respective residual suctions. The two soils were a nonplastic silt soil and kaolin. The statically compacted soil specimens were 100 mm in height with a diameter of 50 mm. The specimens were placed into a chamber that controlled the temperature and relative humidity and allowed equilibration under the applied total suction environment. Total suctions were controlled using saturated salt solutions.

Some of the soil specimens were prepared dry of residual total suction and then wetted to equilibrium at various relative humidity values (i.e., 80, 70, 60, 50 and 40). Other soil specimens were prepared wet of residual total suctions and allowed to dry to the equilibrium state. The specimens remained in the relative humidity environment for about 1 month. Unconfined compression tests were then conducted at an axial strain rate of 0.5 mm/min.

The SWCCs for each of the soils were measured using a pressure plate apparatus, salt solution desiccators, and a relative humidity chamber. The SWCCs for the silt soil and kaolin are shown in Fig. 12.5. The air-entry value and the residual conditions were selected from the SWCC. The estimated air-entry value for the silt soil and kaolin were 10 and 100 kPa, respectively. The residual state for the silt soil occurred at a water content of 2.5% and the residual suction was 2000 kPa. The residual suction for the kaolin appears to be around 5000 kPa and the corresponding water content was around 18%.

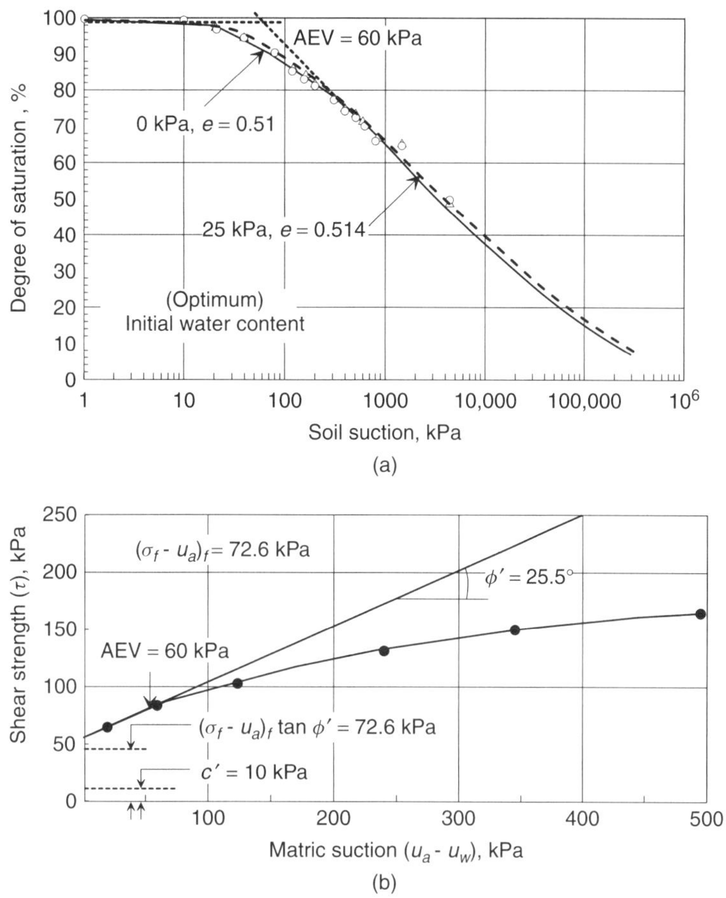

Figure 12.3 Curvature to shear strength envelope with respect to matric suction: (a) SWCC for compacted glacial till (after Vanapalli et al., 1996a); (b) multistage direct shear strength tests on compacted glacial till (from Gan et al., 1988).

Figures 12.6 and 12.7 show the relationship between unconfined compression strength and total suction for the silt soil and kaolin soil, respectively. There was a slight increase in the shear strength at soil suctions beyond residual conditions for both the silt and kaolin. The friction angle, $\phi^{b}_{(drying)}$, beyond residual conditions for the silt soil and kaolin were 0.02° and 0.3°, respectively. In other words, the failure envelope with respect to soil suction is essentially horizontal in the total suction range beyond residual suction.

When soil suctions were decreased from 124,000 to 94,000 kPa (i.e., a wetting process), there was a reduction in shear strength for both the silt soil and kaolin. A decrease in shear strength was measured with decreasing soil suction, and the failure envelope during wetting is shown in Figs. 12.6 and 12.7. The friction angle, $\phi^{b}_{(wetting)}$, with respect to the wetting portion were 0.01° and 0.05° for the silt soil and kaolin, respectively. Once again, the failure envelope angles were essentially 0° in the region beyond residual suction. There were slight differences in strength measured for the drying and wetting portions of the shear strength envelope; however, in both cases the overall angles of shearing resistance were close to 0°.

Further shear strength test data from the suction range beyond residual suction would be useful; however, it is usually the changes in shear strength in the suction range below residual suction that are of greatest interest in engineering practice. In general, it would appear to be reasonable to assume that the ϕ^b angle is 0° beyond residual suction conditions. The primary exception would be sand soils where the ϕ^b angle might decrease with increasing suction (i.e., become a negative value).

12.2.2 Categorization of Shear Strength Equations for Unsaturated Soils

Numerous equations have been proposed for describing the shear strength of unsaturated soils. Figure 12.8 subdivides shear strength equations for unsaturated soils into those that can be used to "best fit" a data set and those that can be used to estimate the unsaturated soil shear strength function based on the SWCC for a soil as well as other soil properties. The theory and measurement of unsaturated shear strength properties were presented in Chapter 11. This chapter primarily focuses on the estimation of the shear strength of an

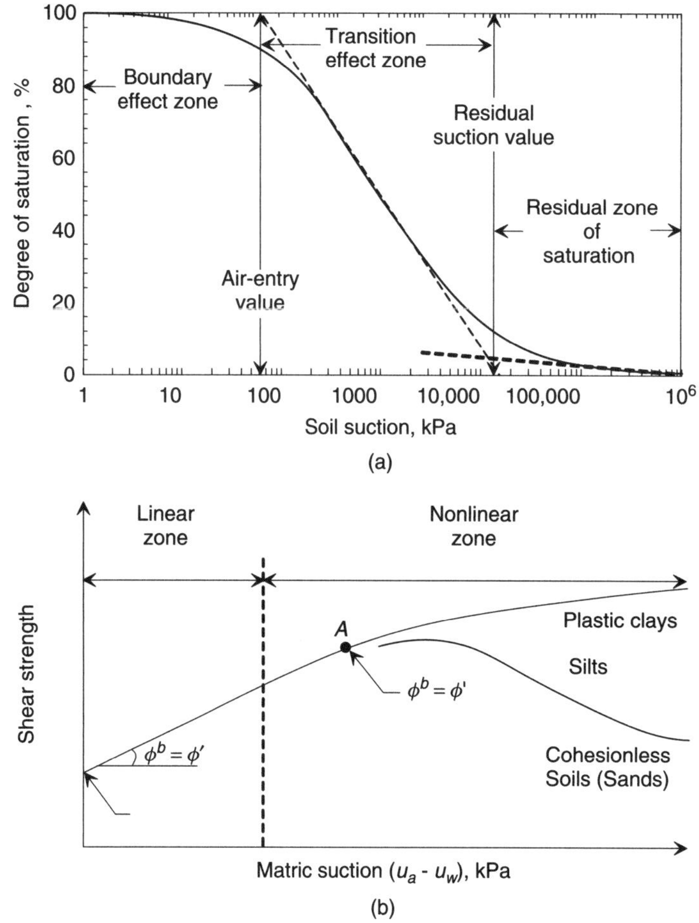

Figure 12.4 Relationship of SWCC to shear strength envelope for various soil types: (a) various zones along SWCC; (b) variation in shear strength envelopes for different soils in different zones of saturation.

unsaturated soil and then on the application of shear strength to practical engineering problems.

Shear strength equations have also been suggested for describing critical state shear strength with respect to soil suction. The critical shear strength equations are not presented herein.

The SWCC plays an important role in the estimation of the unsaturated shear strength of a soil. The air-entry value of the soil, ψ_{aev}, and the residual suction, ψ_r, are pivotal points that can be identified on the SWCC. These pivotal points subsequently influence the form of the proposed shear strength equations. Some estimation shear strength equations cease to apply at residual suction conditions while other equations apply for the entire soil suction range (Fig. 12.8). Most shear strength equations fall into either the best-fit category or the estimation category. The D.G. Fredlund et al., (1996) equation was originally a best-fit equation where the κ parameter was determined through use of a regression analysis. Later, Garven and Vanapalli (2006) correlated the κ parameter to the plasticity of the soil with the result that the shear strength equation now qualifies as an estimation equation.

12.2.3 Some Fitting Equations for Shear Strength versus Soil Suction

A brief summary is first given of some best-fit shear strength equations before a more detailed presentation is provided on equations that have been proposed for the estimation of the unsaturated soil shear strength. The Lee et al., (2005) best-fit unsaturated soil shear strength equation is discussed at the end of this section.

12.2.3.1 Fredlund et al., (1978) Linear Shear Strength Equation

Fredlund et al., (1978) proposed a shear strength equation for unsaturated soils based on two independent stress state variables. The friction angle associated with net normal stress, ϕ', was assumed to be independent of soil suction and the friction angle, ϕ^b, was also assumed to be independent of net normal stress. The shear strength equation was linear in form and was the basis for subsequent nonlinear equations that were to follow. The linear form of the shear strength equation proposed by Fredlund et al. (1978)

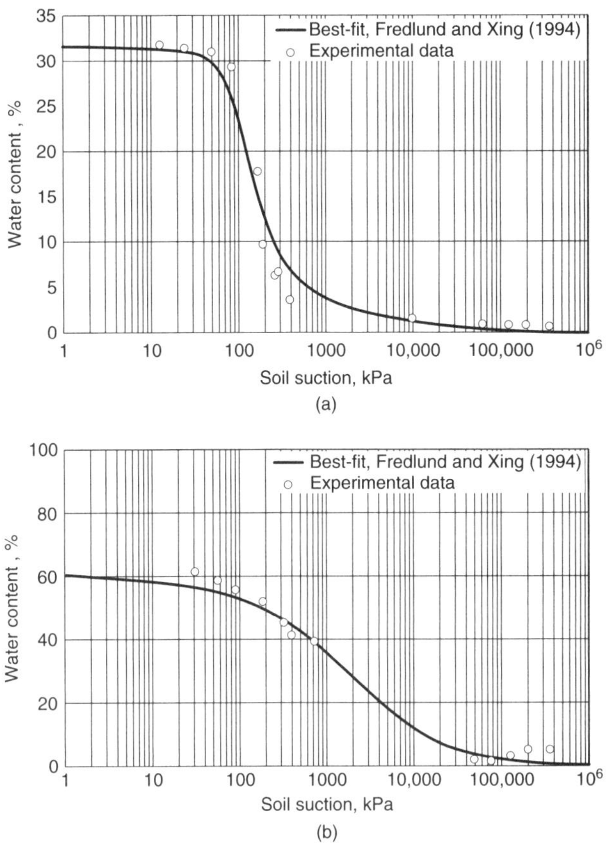

Figure 12.5 SWCCs for (a) silt soil and (b) kaolin (after Nishimura and Fredlund, 2002).

took the form of an extension of the Mohr-Coulomb failure criterion:

$$\tau = c' + (\sigma - u_a)\tan\phi' + (u_a - u_w)\tan\phi^b \qquad (12.1)$$

where:

c' = effective cohesion intercept,
u_a = pore-air pressure,
u_w = pore-water pressure,
$u_a - u_w$ = matric suction, also shown as ψ for soil suction,
$\sigma - u_a$ = net normal stress on the failure plane,
ϕ' = effective angle of internal friction for the saturated soil,
ϕ^b = contribution to the shear strength due to soil suction (i.e., matric suction or total suction).

Equation 12.1 shows that there are independent contributions to shear strength from soil suction and net normal stress. Early data sets indicated that the ϕ^b angle might be relatively constant over a limited range of matric suctions that might be encountered in situ. Later data sets revealed that there was significant nonlinearity in the ϕ^bangle when larger soil suction ranges were taken into consideration. This gave rise to nonlinear equations for the unsaturated soil shear strength criterion.

12.2.3.2 D.G. Fredlund et al. (1996) Nonlinear Shear Strength Equation

The D.G. Fredlund et al., (1996) equation incorporated the SWCC written in terms of dimensionless water content Θ_d into the unsaturated soil shear strength equation. The SWCC was raised to a power κ which was used as a fitting parameter. The resulting shear strength equation was nonlinear in form due to the nonlinearity of the SWCC:

$$\tau = c' + (\sigma - u_a)\tan\phi' + (u_a - u_w)[\Theta_d^{\kappa}]\tan\phi' \qquad (12.2)$$

where:

Θ_d = dimensionless water content defined as θ/θ_s, where θ is any volumetric water content and θ_s is the saturated volumetric water content, and
κ = fitting parameter for the SWCC.

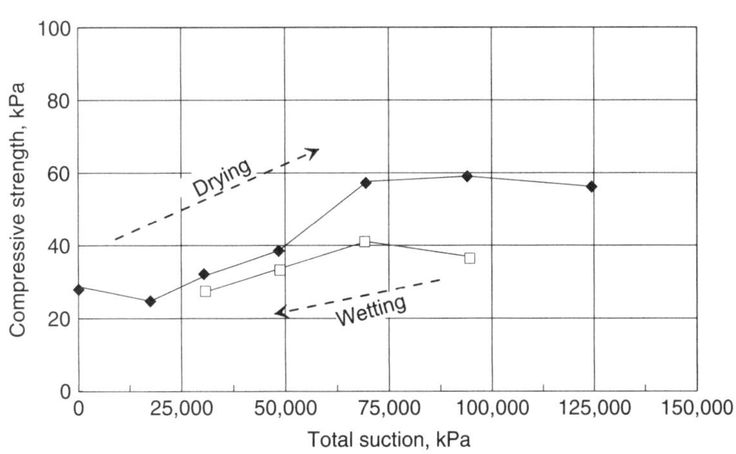

Figure 12.6 Relationship between compressive strength and total suction for silt soil (Nishimura and Fredlund, 2002).

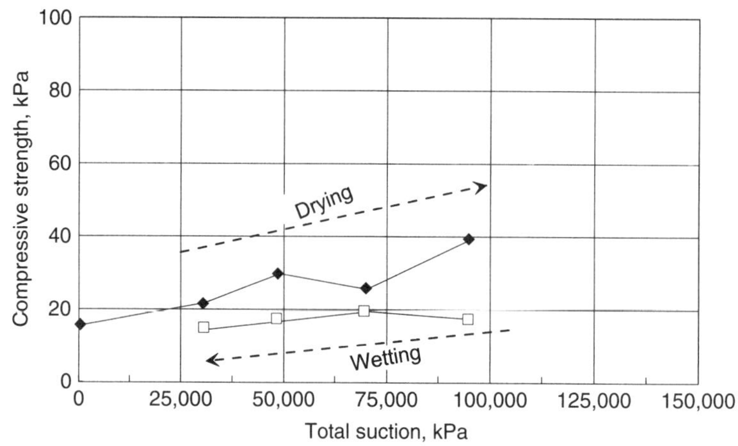

Figure 12.7 Relationship between compressive strength and total suction for kaolin (after Nishimura and Fredlund, 2002).

In 2006 Garven and Vanapalli published a correlation between the κ parameter and the plasticity index I_p of a soil. As a result, the D.G. Fredlund et al., (1996) equation could then be used as an estimation equation for the shear strength of an unsaturated soil.

12.2.4 Proposed Estimation Equations for Unsaturated Soil Shear Strength

There are numerous shear strength equations that have been proposed for the estimation of shear strength for an unsaturated soil. Most of the equations are empirical and phenomenological in nature. The equations are based on the assumption that the shear strength of an unsaturated soil can be viewed as an extension of saturated shear strength properties such as $c'$and ϕ'. The shear strength parameters c' and $\phi'$of a soil are considered to be fundamental soil properties but are also dependent upon initial conditions such as dry density, water content, and soil structure (Fredlund, 1989a).

Figure 12.8 shows two categories of estimation equations for unsaturated soil shear strength. Some equations either directly or indirectly make use of the properties of the SWCC and others utilize additional information such as soil classification information or the shear strength at residual suction conditions. It is clear that the air-entry value of the soil and the residual suction (or the shear strength at residual suction) form anchor points for the estimation of unsaturated soil shear strength.

There are also some shear strength estimation equations that are continuous from zero suction to a high-suction value whereas in other cases there is one equation from zero suction to the air-entry value and another equation from the air-entry value to a high-suction value. Most shear strength equations are limited to maximum soil suctions less than residual suction.

12.2.4.1 Vanapalli et al. (1996b) Estimation Shear Strength Equation

Vanapalli et al., (1996b) suggested a shear strength equation that involved a normalization of the SWCC between saturated conditions and residual volumetric water content conditions. The proposed relationship took the following form:

$$\tau = c' + (\sigma - u_a)\tan\phi' + (u_a - u_w)\left[\left(\frac{\theta - \theta_r}{\theta_s - \theta_r}\right)\right]\tan\phi' \quad (12.3)$$

where:

θ_r = volumetric water content at residual suction and

$\dfrac{(\theta - \theta_r)}{(\theta_s - \theta_r)}$ = normalized water content Θ_n (i.e., between saturated volumetric water content and residual volumetric water content).

The normalized water content Θ_n serves as a reduction term for the soil suction term. The $\tan\phi^b$ term from the linear Fredlund et al., (1978) equation can be written in a nonlinear form using normalized water content Θ_n:

$$\tan\phi^b = \left(\frac{\theta - \theta_r}{\theta_s - \theta_r}\right)\tan\phi' = \Theta_n \tan\phi' \quad (12.4)$$

The shear strength associated with soil suction, τ_s, can be written as

$$\tau_s = (u_a - u_w)\left(\frac{\theta - \theta_r}{\theta_s - \theta_r}\right)\tan\phi' \quad (12.5)$$

12.2.4.2 D.G. Fredlund et al. (1996) Estimation Shear Strength Equation

Unsaturated soil shear strength equations are related to the shape of the SWCC. The nonlinearity of the SWCC results in a nonlinear shear strength relationship (D.G. Fredlund et al., 1996). The volumetric water content of the soil was written in a dimensionless form by dividing by the volumetric water content at saturation (i.e., $\Theta_d = \theta/\theta_s$). The definition of unsaturated soil shear strength extends beyond residual suction conditions. The unsaturated soil shear strength property takes the form of a nonlinear function:

$$\tan\phi^b = \left(\frac{\theta}{\theta_s}\right)^{\kappa}\tan\phi' = \Theta_d^{\kappa}\tan\phi' = (S)^{\kappa}\tan\phi' \quad (12.6)$$

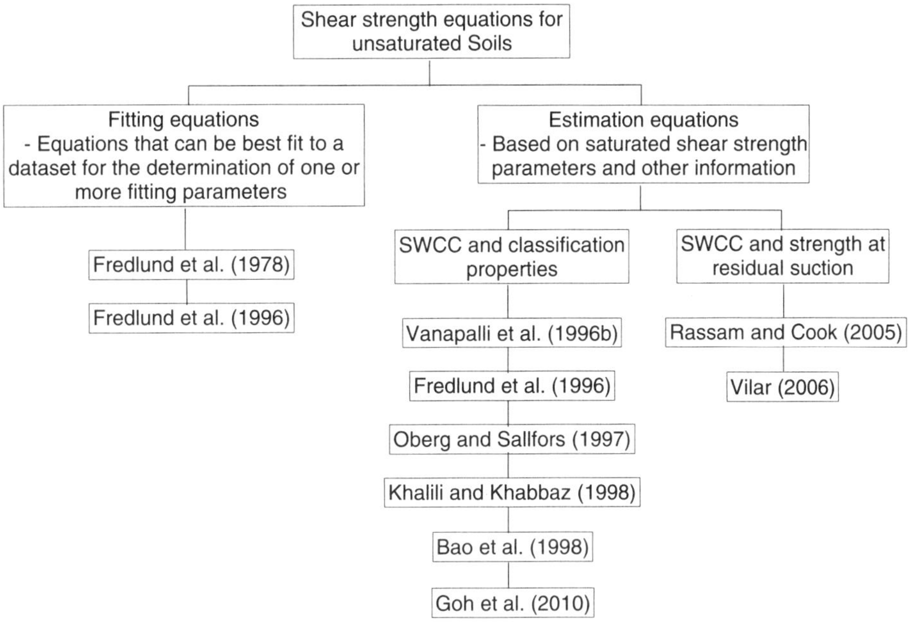

Figure 12.8 Categorization of shear strength equations for unsaturated soils.

The shear strength associated with soil suction, τ_s, can be written as

$$\tau_s = (u_a - u_w)\left(\frac{\theta}{\theta_s}\right)^{\kappa} \tan\phi' \qquad (12.7)$$

Garven and Vanapalli (2006) provided an empirical relationship between the fitting parameter κ and the plasticity index of the soil, PI. The proposed empirical equation linking the κ factor to the plasticity index has been revised a couple times but Figure 12.9 shows the latest experimental results and the form for the κ factor equation.

$$\kappa = -0.0016(\text{PI})^2 + 0.0975(\text{PI}) + 1 \qquad (12.8)$$

Equation 12.8 was originally developed through comparisons of measured and predicted values of shear strength for determining the fitting parameter κ This process has been continued as further experimental shear strength data have become available.

12.2.4.3 Oberg and Sallfors (1997) Estimation Shear Strength Equation

Oberg and Sallfors (1997) proposed an unsaturated soil shear strength equation that used the instantaneous degree of saturation S of the soil to describe shear strength. The contribution of soil suction to shear strength was assumed to

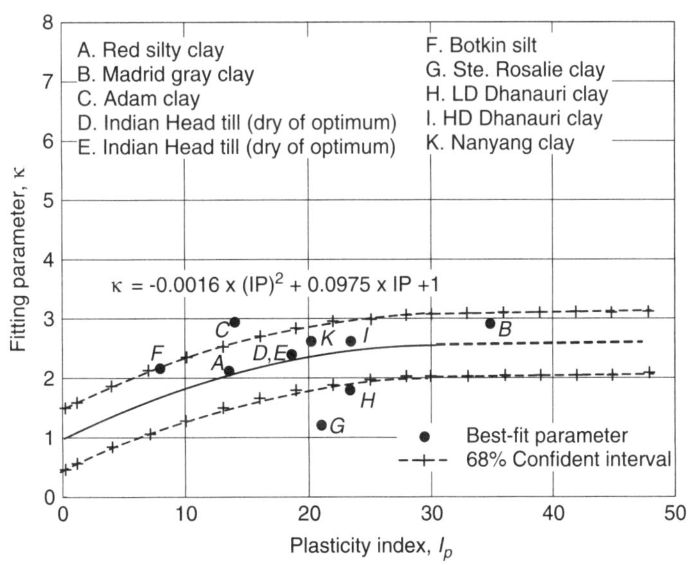

Figure 12.9 Relationship between parameter κ and plasticity index for natural and statically compacted soils (after Garven and Vanapalli, 2006).

be equal to ($S \tan \phi'$) and the shear strength equation was written as

$$\tau = c' + (\sigma - u_a) \tan \phi' + (u_a - u_w)[S] \tan \phi' \quad (12.9)$$

12.2.4.4 *Khalili and Khabbaz (1998) Estimation Shear Strength Equation*

Khalili and Khabbaz (1998) assumed that the soil behaved as a saturated soil as long as the matric suction was less than the air-entry value of the soil:

$$\tau = c' + (\sigma - u_a) \tan \phi' + (u_a - u_w) \tan \phi' \quad (12.10)$$

Once the air-entry value was exceeded, the suction component of shear strength was reduced by multiplying soil suction by the variable λ', and the shear strength equation was written as

$$\tau = c' + (\sigma - u_a) \tan \phi' + (u_a - u_w)[\lambda'] \tan \phi' \quad (12.11)$$

The parameter λ' was defined as

$$\lambda' = \left[\frac{u_a - u_w}{(u_a - u_w)_b}\right]^{-0.55} \quad (12.12)$$

Figure 12.10 shows a plot of λ' versus soil suction for two air-entry values. The λ' value is 1.0 for shear strengths up to the air-entry value and then decreases without any influence from residual suction. The rate of decrease of λ' is simply influenced by the air-entry value of the soil. It is assumed that soil suction always provides a positive increase in strength up to 10^6 kPa. In each case, the λ' parameter decreases in response to an exponent power of –0.55 for all soil types.

12.2.4.5 *Bao et al. (1998) Estimation Shear Strength Equation*

Bao et al., (1998) assumed that the soil behaved as a saturated soil as long as the matric suction was less than the air-entry value of the soil:

$$\tau = c' + (\sigma - u_a) \tan \phi' + (u_a - u_w) \tan \phi' \quad (12.13)$$

Once the air-entry value was exceeded, the suction component of shear strength was reduced by multiplying soil suction by the variable ζ, as shown in the equation

$$\tau = c' + (\sigma - u_a) \tan \phi' + (u_a - u_w)[\zeta] \tan \phi' \quad (12.14)$$

The parameter ζ was defined as

$$\zeta = \frac{\log(u_a - u_w)_r - \log(u_a - u_w)}{\log(u_a - u_w)_r - \log(u_a - u_w)_b} \quad (12.15)$$

Figure 12.11 shows a plot of ζ versus soil suction for two air-entry values. The ζ value is 1.0 at any air-entry value and decreases to a value of zero at residual suction conditions. In other words, the influence of soil suction on the shear strength of an unsaturated soil was normalized between the air-entry value and residual suction. It was assumed that the shear strength of the soil remained constant beyond residual suction. This behavior is quite consistent with shear strength behavior observed for fine-grained soils.

12.2.4.6 *Goh et al. (2010) Estimation Shear Strength Equation*

Goh et al. (2010) proposed an equation to estimate the shear strength under drying and wetting conditions. The Goh et al. (2010) equation is a modification of the D.G. Fredlund et al. (1996) equation and the Lee et al. (2005) equation. Volumetric water content was written in a dimensionless form

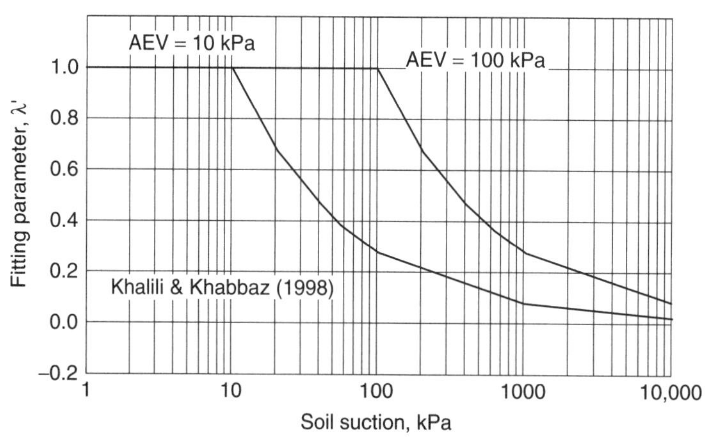

Figure 12.10 The λ' parameter associated with Khalili and Khabbaz (1998) estimation equation corresponding to air-entry values of 10 and 100 kPa.

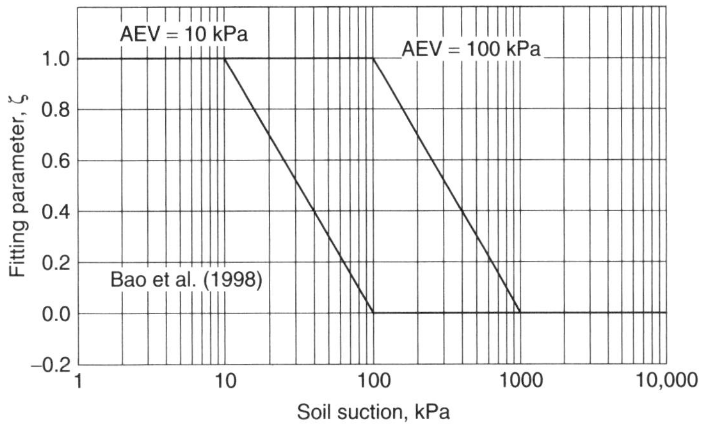

Figure 12.11 The ζ parameter associated with Bao et al. (1998) estimation equation corresponding to air-entry values of 10 and 100 kPa and residual suction values of 100 and 1000 kPa, respectively.

by dividing by the saturated volumetric water content (i.e., $\Theta_d = \theta/\theta_s$). The SWCC was used to control the secant slope between soil suction and unsaturated shear strength. The assumption was made that the ϕ^b angle is equal to the effective angle of internal friction for matric suctions lower than the air-entry value of the soil. Matric suction values below the air-entry value contribute to the effective stresses of a saturated soil.

The empirical equation proposed by Goh et al. (2010) was developed based on three data sets where shear strengths corresponding to the drying curve (i.e., drying shear strength) were measured and three data sets where shear strengths corresponding to the drying and wetting were measured along with the SWCCs. The data sets used by Goh et al. (2010) are shown in Table 12.1.

Table 12.1 Summary of Published Soil Data Sets Used in Development of Goh et al. (2010) Equation

Authors	Materials	USCS Classification
Escario and Juca (1989)	Madrid clay sand	CL
Han (1996)	Bukit Timah granitic residual soil	CL
Lee et al. (2005)	Korea weathered granitic residual soil	SM
Miao et al. (2002)	Nanyang expansive soil	CH
Rahardjo et al. (2004)	Jurong sedimentary formation residual soil	CL
Thu et al. (2006)	Compacted kaolin	MH

The fitting parameter κ used in the Fredlund et al. (1996) shear strength equation was designated as a constant value. Values for the κ parameter were obtained by Garven and Vanapalli (2006) through use of a curve-fitting correlation with the plasticity index of the soil. Later, Goh et al. (2010) suggested that the fitting parameter κ used in the Fredlund et al. (1996) equation [and the Lee et al. (2005) equation], varied nonlinearly with matric suction. This conclusion was arrived at after performing back-calculations on a series of published shear strength and SWCC data. Shear strength data were separated into two categories: namely, soils tested after drying towards an equilibrium state and soils tested after wetting towards an equilibrium state. The observed κ values for several soils dried to an equilibrium state are shown in shown in Fig. 12.12. The new fitting parameter was designated as κ' and had a value of zero when matric suction was equal to or lower than the air-entry value of the soil. The κ' value was found to increase in accordance with the logarithm difference between a designated matric suction value and the air-entry value.

A parametric study was undertaken by Goh et al. (2010) to incorporate the κ' parameter into the proposed shear strength equation. Some parameters were used together with the logarithm term to represent the κ' parameter. The parameters cover a relatively wide range of soil properties as well as a wide range of soil suction values. The regression analysis showed that two parameters: namely, y and b were needed to determine the parameters for estimating the shear strength of an unsaturated soil. The y and b variables were found to be related to other soil properties through use of a correlation with experimental data. The estimation of the unsaturated soil shear strength was divided into two parts: namely, one equation was applied up to the air-entry value of the soil [i.e., $(u_a - u_w)_b$], and another equation was applied for suctions beyond the air-entry value. The equations are

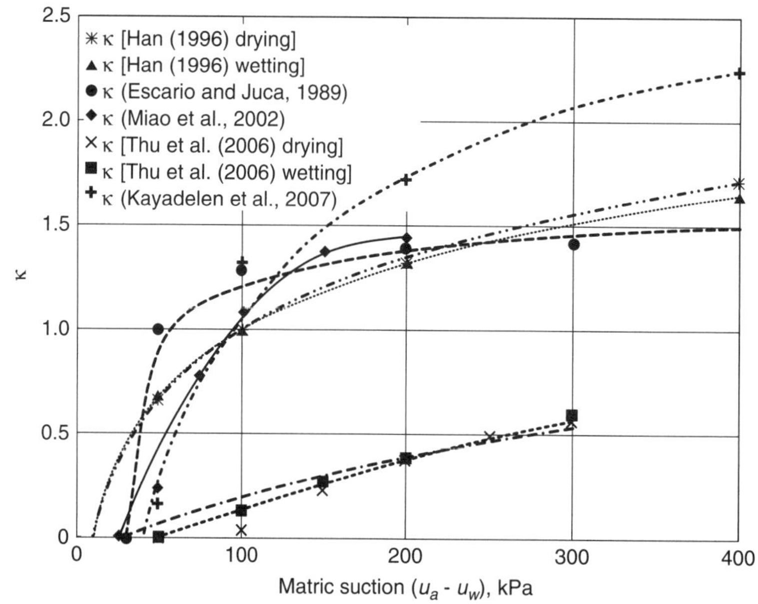

Figure 12.12 Trend of the back-calculated κ with respect to matric suction for various soils. (after Goh, 2012).

written as follows

$$\tau = c' + (\sigma - u_a)\tan\phi' + (u_a - u_w)\tan\phi^b \qquad (12.16)$$

where:

$\phi^b = \phi'$ when $(u_a - u_w) < (u_a - u_w)_b$.

The following shear strength equation is used once the air-entry value of the soil is exceeded

$$\tau = c' + \left[(\sigma - u_a)(u_a - u_w)\right]\tan\phi' + \left[(u_a - u_w) - (u_a - u_w)_b\right] b\Phi_n^{\kappa'}\tan\phi' \qquad (12.17)$$

when $(u_a - u_w) \geq (u_a - u_w)_b$

The κ' parameter is zero for suction values lower than the air-entry value and increases nonlinearly as soil suction increases in accordance with the following equation.

$$\kappa' = \left[\log(u_a - u_w) - \log(u_a - u_w)_b\right]^y \qquad (12.18)$$

where:

y and b = controlling parameters.

It should be noted that the κ' parameter used by Goh et al. (2010) and the κ parameter used by Fredlund et al. (1996) are defined differently. The κ' parameter in the Fredlund et al. (1996) equation is a best-fit variable obtained from the drying shear strength experimental data along with the drying SWCC results. The κ' parameter used by Goh et al. (2010) is used together with the variable b for estimating the drying and wetting shear strengths along with drying and wetting SWCCs.

The subscripts d and w were used to designate whether the controlling parameters were related to the drying and wetting SWCC, respectively (i.e., y_d, b_d, and y_w, b_w). The regression analysis showed that the parameter y_d (i.e., y_d is on the drying curve) was related to plasticity index I_p of the soil, following a natural logarithm (ln) scale. The parameter b_d (i.e., b for the drying curve) was related to I_p and the fitting parameter n_d [i.e., n fitting for the drying curve of the Fredlund and Xing (1994) SWCC equation]. The y_d parameter is graphically shown in Fig. 12.13 and can be written as follows:

$$y_d = 0.502\ln(PI + 2.7) - 0.387 \qquad (12.19)$$

The b_d parameter is graphically shown in Fig. 12.14 and can be written as follows:

$$b_d = -0.245\{\ln[n_d(\mathrm{PI} + 4.4)]\}^2 + 2.114\{\ln[n_d(\mathrm{PI} + 4.4)]\} - 3.522 \qquad (12.20)$$

The shear strength equation corresponding to the wetting SWCC was also predicted using the y_w and b_w (i.e.,both y_w and b_w are on the wetting curve). The wetting curve parameters were found to be related to the parameters for the drying SWCC (i.e., y_d and b_d) as well as the soil properties. The parameters y_w and b_w were found to adhere to the following relationships:

$$y_w = 3.55y_d - 3.00 \qquad (12.21)$$

$$b_w = 0.542b_d\left(\frac{n_d}{n_w}\right) + 0.389 \qquad (12.22)$$

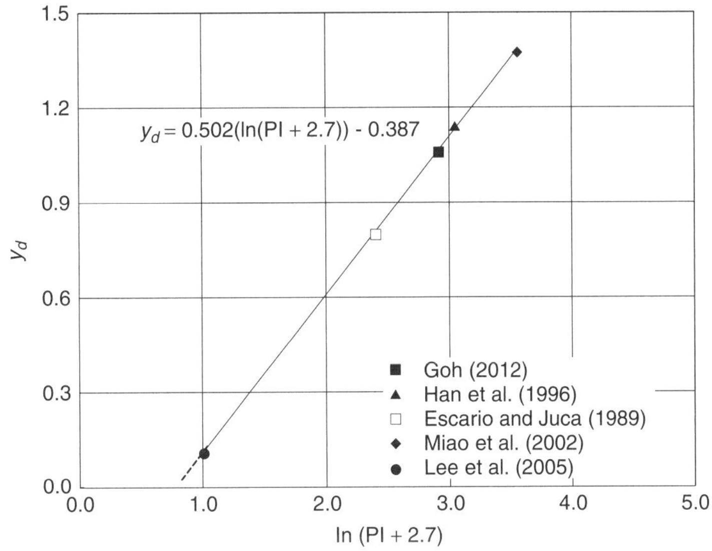

Figure 12.13 Relationship between y_d and PI of soils (after Goh et al., 2010).

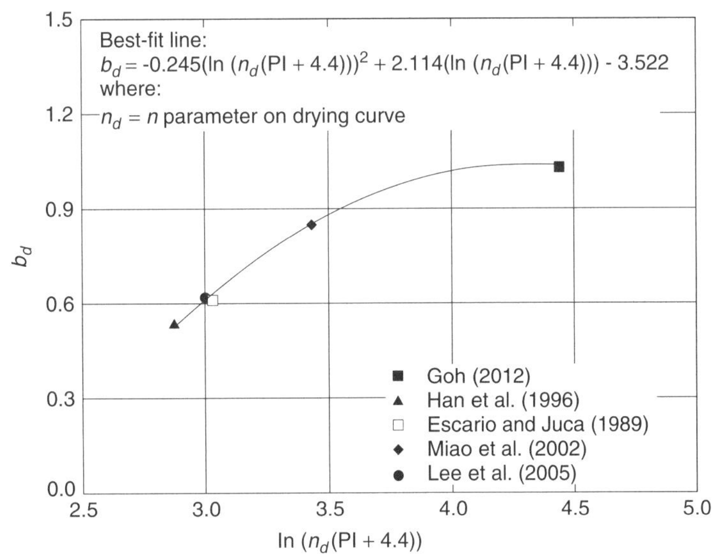

Figure 12.14 Relationship between b_d and PI and n_d of soils (after Goh et al., 2010).

The relationships between the shear strength parameters for a soil that has come to equilibrium along the wetting SWCC (i.e., y_w and b_w) and a soil that has come to equilibrium along the drying SWCC (i.e., y_d and b_d) are shown in Figs. 12.15 and 12.16, respectively.

12.2.4.7 Rassam and Cook (2002) Estimation Shear Strength Equation

Rassam and Cook (2002) proposed an unsaturated soil shear strength equation with four terms. The first three terms assume that the soil with negative pore-water pressures behaves essentially as a saturated soil and the fourth term applies a correction factor to account for the fact that increases in soil suction are not as effective in increasing shear strength as is net normal stress. The proposed shear strength equation is written as follows:

$$\tau = c' + (\sigma - u_a)\tan\phi' + \psi\tan\phi' - \phi(\psi - \psi_{\text{aev}})^{\beta} \tag{12.23}$$

The correction factor portion of the equation incorporates two new variables, ϕ' and β, defined by the following equations:

$$\phi = \frac{\psi_r\tan\phi' - \tau_{Sr}}{(\psi_r - \psi_{\text{aev}})^{\beta}} \tag{12.24}$$

$$\beta = \frac{\tan\phi'(\psi_r - \psi_{\text{aev}})}{\psi_r\tan\phi' - \tau_{Sr}} \tag{12.25}$$

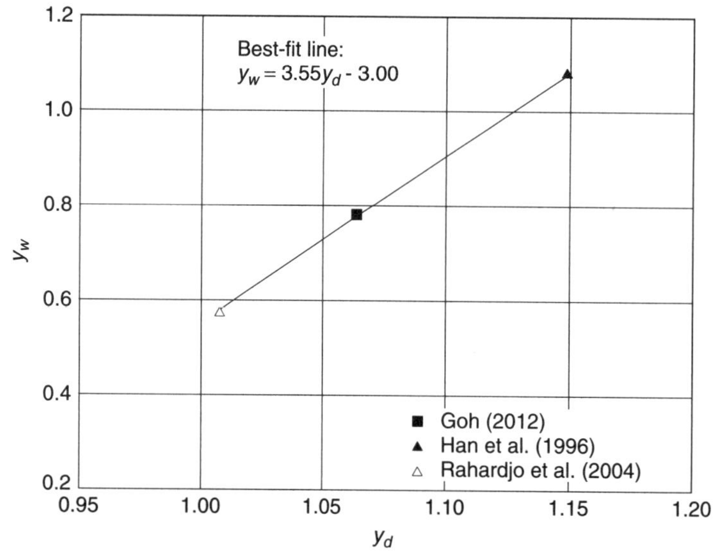

Figure 12.15 Linear relationship between y_w and y_d (after Goh, 2012).

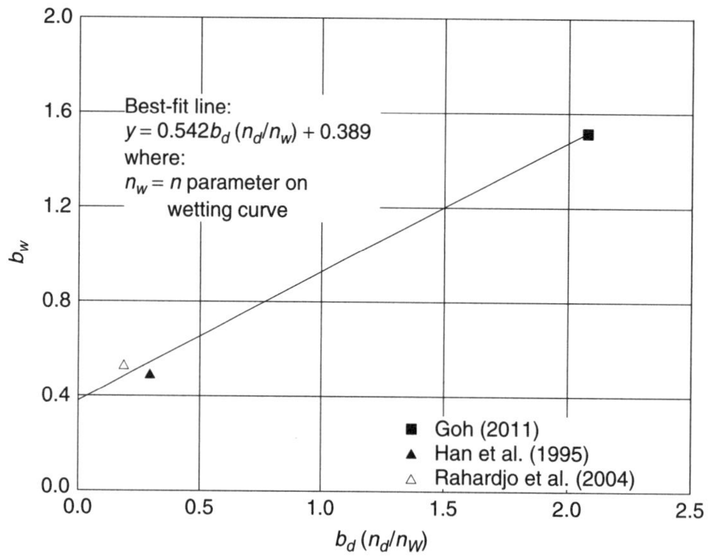

Figure 12.16 Relationship between b_w and b_d (after Goh, 2012).

where:

τ_{Sr} = matric suction contribution to shear strength at residual suction,
ψ_r = residual soil suction, and
ψ_{aev} = suction at the air-entry value of the soil.

The ϕ and β terms are part of the correction factor that reduces the shear strength as a result of desaturation of the soil. A new term, τ_{Sr}, is introduced into the shear strength equation. The τ_{Sr} term refers to the contribution of matric suction to the shear strength at residual suction conditions, ψ_r. Figure 12.17 shows the meaning of the τ_{Sr} term for the case where the residual suction is 200 kPa. In this case the shear strength at residual suction is 96 kPa.

12.2.4.8 Vilar (2006) Estimation Shear Strength Equation

Vilar (2006) proposed the use of a hyperbolic equation to describe the shear strength versus soil suction relationship for an unsaturated soil. Details of the Vilar (2006) model are presented as a fitting model, but it is possible that with further verification studies the model might be used for the estimation of unsaturated shear strength.

The proposed model is based on the general shape of the SWCC. It was assumed that the soil behaved as a saturated

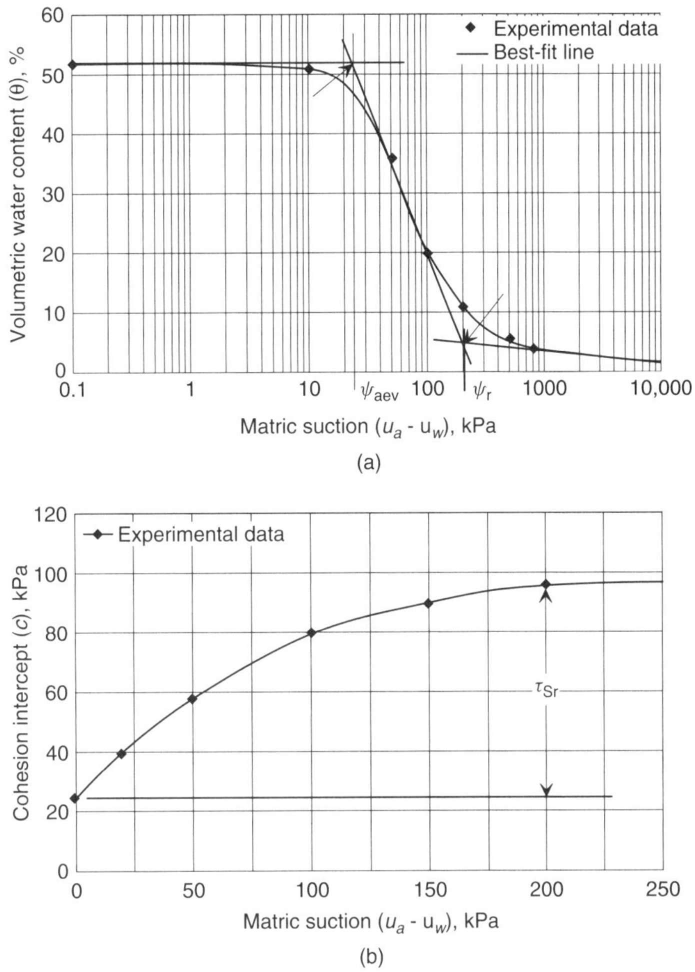

Figure 12.17 The τ_{Sr} term in Rassam and Cook (2002) shear strength equation: (a) SWCC defining air-entry value and residual suction; (b) shear strength versus soil suction showing shear strength at residual suction.

soil from the point of zero suction up to the air-entry value (Fig. 12.18). The total cohesion of the soil (i.e., effective cohesion plus cohesion due to soil suction) was written as a hyperbolic function of soil suction:

$$c = c' + \frac{u_a - u_w}{a + b(u_a - u_w)} \quad (12.26)$$

where:

a, b = fitting parameters.

The slope of the shear strength envelope was assumed to become equal to the tangent of the effective angle of internal friction, ϕ', as matric suction approaches zero:

$$\frac{dc}{d(u_a - u_w)|_{\psi\to 0}} = \frac{1}{a} = \tan\phi' \quad (12.27)$$

It was assumed that there was no significant change in the shear strength of the soil once the soil reaches suction values equal to or greater than the residual suction (Fig. 12.18). In other words, the shear strength remains constant at an ultimate value:

$$\lim_{\psi\to\infty} c = c_{\text{ultimate}} = c' + \frac{1}{b} \quad (12.28)$$

Consequently, b becomes equal to $1/(c_{\text{ultimate}} - c')$. It was also suggested that a soil sample be allowed to come to equilibrium in air and then the undrained strength of the soil be measured and used to represent the c_{ultimate} value. It was suggested that the laboratory temperature and relative humidity would be suitable for conditioning a soil sample to a high total suction. Another procedure involved conditioning a soil sample under a known total suction near 1500 kPa. In this way, an improved value for the b soil parameter could be computed using the following equation:

$$b = \frac{1}{c_{\text{measured}} - c'} - \frac{a}{\psi_{\text{measured}}} \quad (12.29)$$

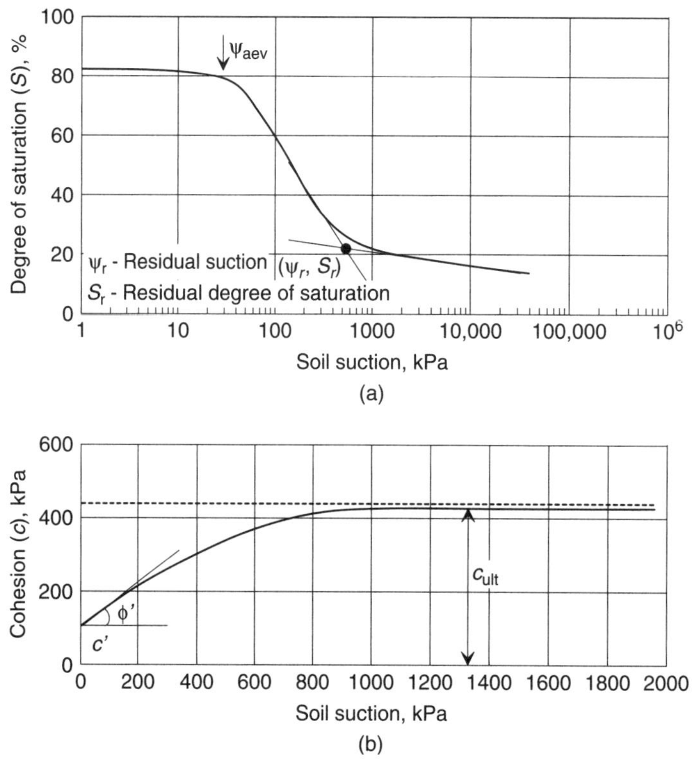

Figure 12.18 Assumed relationship between SWCC and shear strength envelope for unsaturated soil: (a) SWCC and key features; (b) hyperbolic function and assumed conditions to derive a and b (after Vilar, 2006).

where:

$c_{measured}$ = shear strength at the measured high soil suction and

$\psi_{measured}$ = high measured suction representing equilibrium conditions.

It should be noted that no soil parameters from the SWCC are directly used in the development of the Vilar (2006) model. However, the general character of the SWCC was used as justification for the form of the unsaturated soil shear strength equation.

One shear strength test on a soil specimen that is at or above the residual suction state is all that is required to use the Vilar (2006) model as long as the saturated effective shear strength parameters are known. Figure 12.18 illustrates the definition of the c_{ult} term that is used in estimating the unsaturated soil shear strength envelope. Comparisons between the Vilar (2006) model estimations and laboratory test results have been shown to be reasonable (Fig. 12.19).

12.2.4.9 Lee et al. (2005) Unsaturated Soil Shear Strength Equation

Lee et al., (2005) suggested using one linear shear strength equation for soil suctions up to the air-entry value. The shear strength equation prior to the air-entry value has the shear strength parameters of a saturated soil. Another nonlinear equation was proposed for the soil suctions from the air-entry value up to the high-soil-suction range. The Lee et al. (2005) equation uses the SWCC and two fitting parameters, namely, κ'' and λ'' to provide a best fit for shear strength beyond the air-entry value of the soil:

$$\tau = c' + [(\sigma - u_a) + (u_a - u_w)_b] \tan \phi' + [(u_a - u_w) - (u_a - u_w)_b]\Theta_d^{\kappa''}[1 + \lambda''(\sigma - u_a)] \tan \phi' \quad (12.30)$$

where:

κ'' and λ'' = fitting parameters,

$(u_a - u_w)_b$ = air-entry value of the soil, and

Θ_d = dimensionless volumetric water content.

12.2.5 Comparison of Estimated and Measured Unsaturated Soil Shear Strengths

Each of the proposed equations for estimating unsaturated soil shear strength has been verified using a limited number of data sets. It is always desirable to verify any proposed shear strength equations with many sets of laboratory results tested over a wide range of soil suctions. Independent laboratory data sets are presently somewhat limited. Therefore, it is difficult to undertake detailed verification studies on all equations that have been proposed for the estimation of unsaturated shear strength.

Table 12.2 summarizes equations that utilize the SWCC for the estimation of the shear strength relationship.

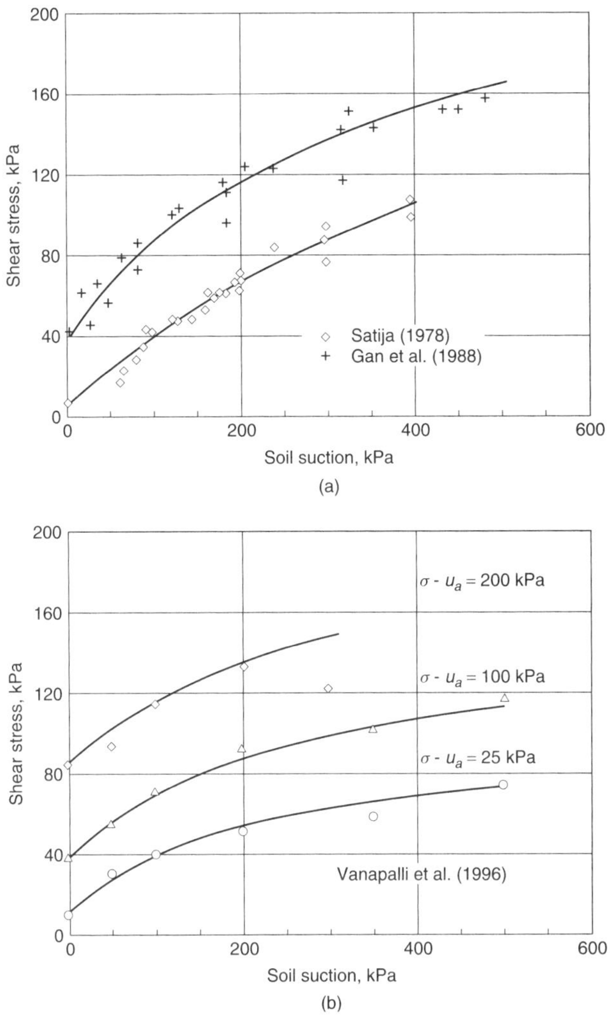

Figure 12.19 Comparison of experimental data with Vilar (2006) estimation of shear strength envelopes: (a) shear strength data from Satija (1978) and Gan et al. (1988); (b) shear strength data from Vanapalli (1994).

Comparisons are made between predicted shear strengths and measured shear strengths when using some of the proposed estimation equations. The soil data sets used for comparative purposes are independent of the data used in the development of the empirical shear strength equations.

Table 12.2 Estimation Equation Results Compared with Measured Test Results

Equation	Suction Soil Parameters
Oberg and Sallfors (1997)	$S \tan\phi'$
D.G. Fredlund et al. (1996)	$\left(\frac{\theta}{\theta_s}\right)^{\kappa} \tan\phi'$
Vanapalli et al. (1996b)	$\left(\frac{\theta - \theta_r}{\theta_s - \theta_r}\right) \tan\phi'$

Comparisons can be made between measured shear strength data and the estimation of unsaturated soil shear strengths (Sheng et al., 2009). The estimation equations used in the comparison are those proposed by (i) Oberg and Sallfors (1997), (ii) D.G. Fredlund et al., (1996), (iii) Vanapalli et al., (1996b), and (iv) Goh et al., (2010). Comparisons between four shear strength data sets are presented: (i) reconsituted silty clay (Cunningham, et al., 2003), (ii) compacted kaolin assuming no volume change (Thu et al., 2007a), (iii) compacted kaolin measuring volume change (Thu et al., 2007a), and (iv) Madrid grey clay (Escario and Juca, 1989).

12.2.5.1 Data on Reconstituted Silty Clay (Cunningham et al., 2003)

Cunningham et al., (2003) presented triaxial compression tests performed using various confining pressures on the reconstituted silty clay comprised of a mixture of 20% pure Speswhite kaolin, 10% London clay, and 70% silica silt. The slurry soil was isotropically preconsolidated to 130 kPa. The SWCC in terms of degree of saturation versus soil suction is presented in Fig. 12.20. The air-entry value of the soil determined using an empirical construction procedure was approximately 450 kPa. However, the degree of saturation starts to reduce at about 250 kPa. The a_f fitting parameter is 800 kPa, and it designates the inflection point associated with fitting the Fredlund and Xing (1994) equation. Figures 12.21 and 12.22 show the SWCC for the reconstituted silty clay plotted in terms of gravimetric water content and volumetric water content, respectively.

The original shear strength data published by Cunningham et al. (2003) for the reconstituted silty clay were presented in terms of the maximum deviator stress applied to the soil. Figure 12.23 shows the shear strength data presented in terms of the apparent cohesion intercept versus the applied matric suction for net confining pressures ranging from 50 to 400 kPa. The matric suction versus shear strength is quite nonlinear, tending to level off before residual suction conditions are reached. The data can be compared with the shear strengths estimated when using various proposed estimation equations.

The soil parameters from the SWCC required for each of the estimation procedures are tabulated in Table 12.3. It is noted that while all procedures rely upon the SWCC, the estimation equations depend upon different variables associated with the SWCC. The parameters for the saturated soil are $\phi' = 32°$, $c' = 0$ kPa, as given in Cunningham et al. (2003).

The results from triaxial compression tests on the reconstituted silty clay tested with a net confining pressure of 100 kPa are used for comparison with the estimated shear strength results. Figure 12.24 shows the estimations of shear strength from the procedures listed in Table 12.3 along with the shear strengths measured when the confining pressure was 100 kPa. The measured shear strength increases with matric suction in a nonlinear manner as anticipated.

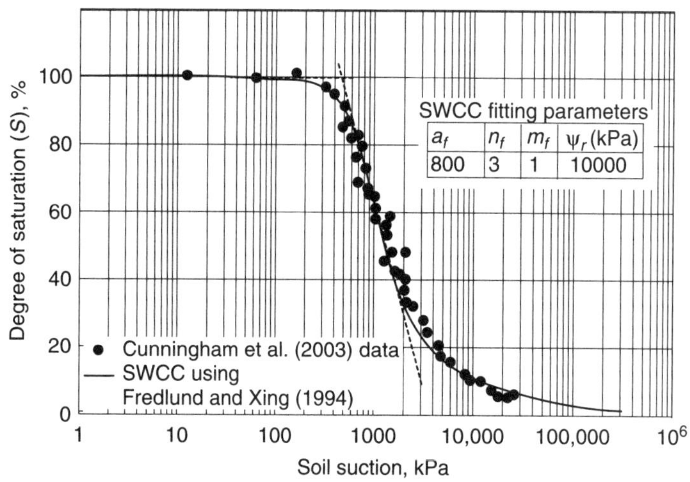

Figure 12.20 SWCC of reconstituted silty clay in terms of degree of saturation versus soil suction (data from Cunningham et al., 2003).

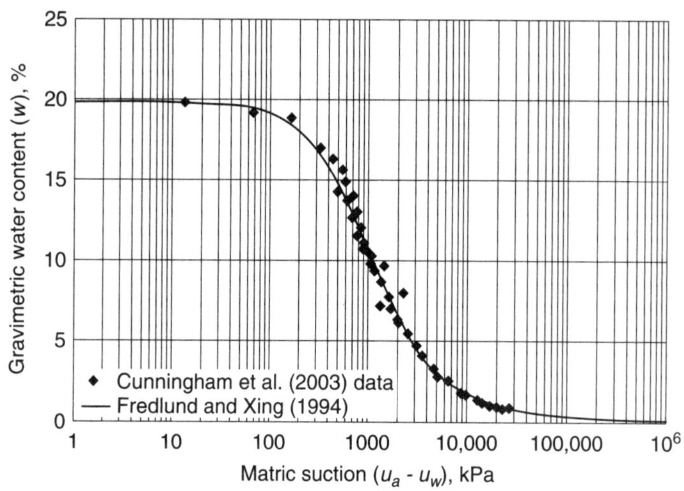

Figure 12.21 SWCC of reconstituted silty clay in terms of gravimetric water content versus soil suction (data from Cunningham et al., 2003).

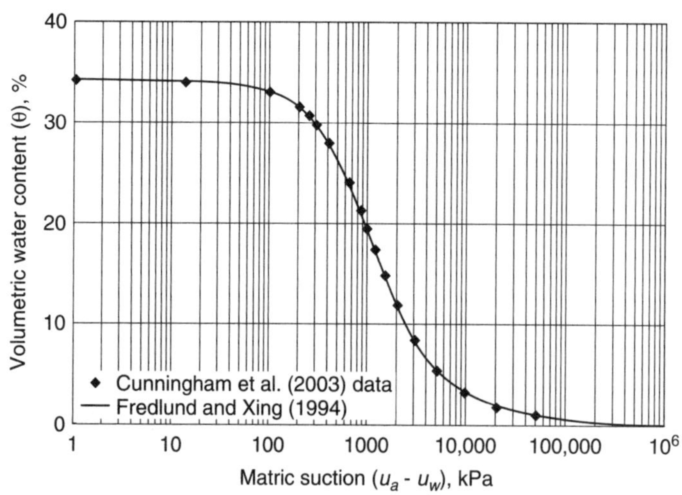

Figure 12.22 SWCC of reconstituted silty clay in terms of volumetric water content versus soil suction (data from Cunningham et al., 2003).

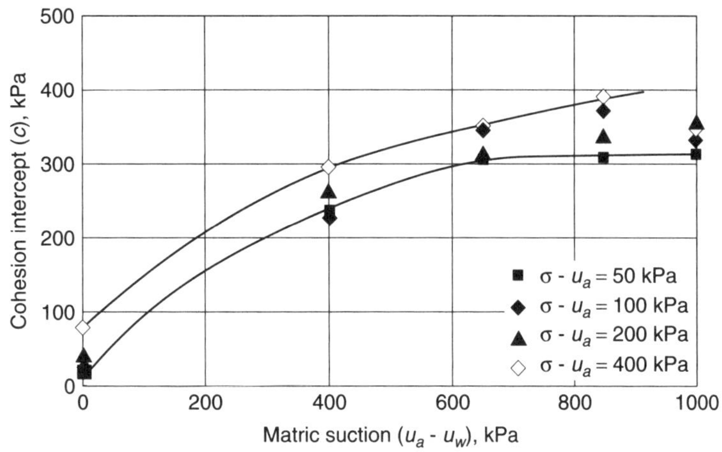

Figure 12.23 Measured shear strength data in terms of apparent cohesion intercept versus soil suction for various net confining pressures (after Cunningham et al., 2003). (Open diamonds have confining pressure of 400 kPa.)

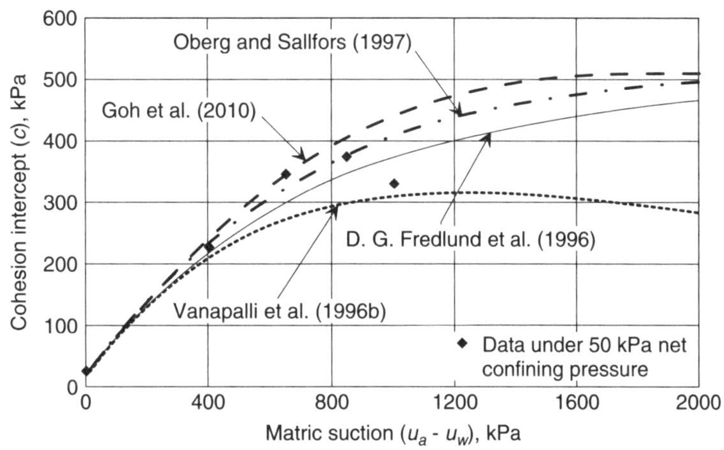

Figure 12.24 Comparison of estimations of shear strength from various proposed equations and measured data for reconstituted silty clay for net confining pressure of 100 kPa.

Table 12.3 Estimation Soil Parameters for Reconstituted Silty Clay

Equations	Soil Parameters
Oberg and Sallfors (1997)	SWCC, S
D.G. Fredlund et al. (1996)	$\kappa = 1.16$, $\theta_s = 0.6032$, SWCC
Vanapalli et al. (1996b)	$\theta_s = 0.6032$, $\theta_r = 0.055$, SWCC
Goh et al. (2010)	$n_d = 1.44$, $\psi_{\text{aev}} = 250$ kPa, PI, SWCC

The estimated shear strengths are quite consistent with the measurements in Cunningham et al., (2003), where it was observed that desaturation commences at a suction of about 250 kPa and becomes more significant at a suction of 400 kPa. The air-entry value and the saturation suction are assumed to be independent of the confining pressure.

The predictions of shear strength for the unsaturated silty clay tend to separate and spread out over quite a wide range once the inflection point on the SWCC is reached (see Fig. 12.24). Oberg and Sallfors (1997) and Goh et al., (2010) gave the highest strength estimations. Vanapalli et al., (1996b) gave the lowest shear strength estimations. The D.G. Fredlund et al., (1996) estimation of shear strength was near the middle of the results. Other test results obtained when using higher net confining pressures were quite consistent with the estimation of unsaturated soil shear strength.

12.2.5.2 Data on Compacted Kaolin Assuming No Volume Change (Thu et al., 2007a)

Thu et al. (2007a) measured the shear strength of compacted kaolin under different soil suctions using a modified triaxial

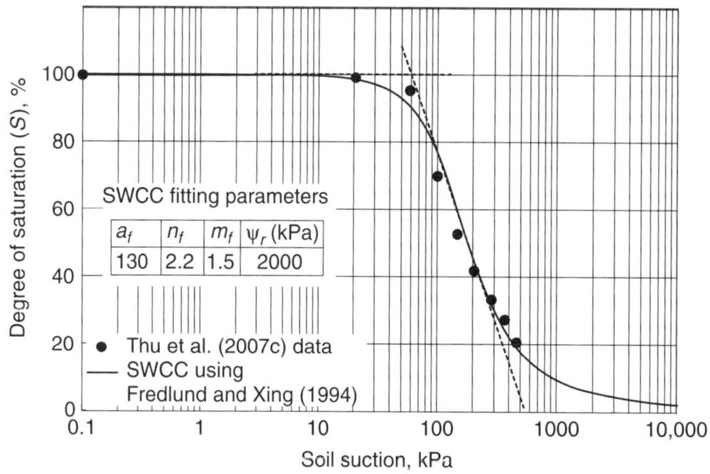

Figure 12.25 SWCC data for compacted kaolin in terms of degree of saturation versus soil suction (after Thu et al., 2007a).

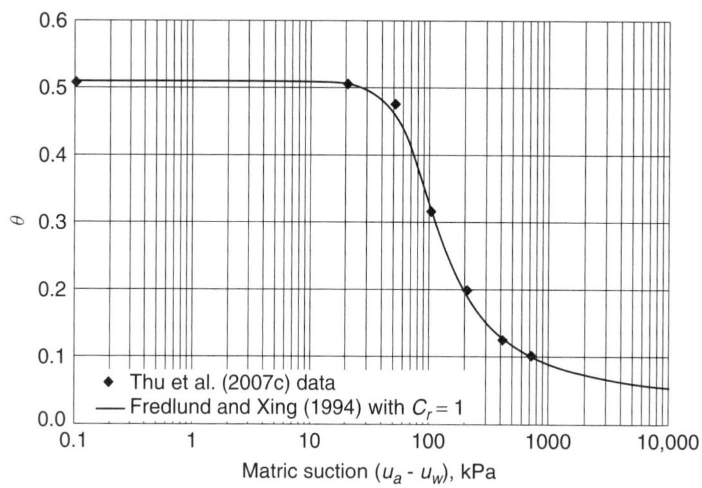

Figure 12.26 SWCC for compacted kaolin in terms of volumetric water content versus soil suction (after Thu et al., 2007a).

apparatus. The soil was a mixture of kaolin clay (15%) and silt (85%) compacted to the maximum dry density of 1350 kg/m^3 at an optimum water content of 22%. All specimens were statically compacted in 10 layers of equal thickness of 10 mm. The SWCC of the soil was obtained by desaturating the soil under zero net vertical pressure. The degree of saturation versus soil suction SWCC is shown in Fig. 12.25 along with the best-fit SWCC parameters for the Fredlund and Xing (1994) equation. The plot of volumetric water content versus soil suction is shown in Fig. 12.26.

The shear strength parameters for the saturated compacted kaolin were $\phi' = 31.1^\circ$ and $c' = 0$ kPa. Soil parameters from the SWCC that are required for the estimation of shear strength with respect to soil suction are listed in Table 12.4.

The measured shear strength data along with estimations of the shear strength envelope for the compacted kaolin are shown in Fig. 12.27. The measured shear strength increases with matric suction in a nonlinear manner. The equations by Goh et al., (2010) and Fredlund et al., (1996) provide estimations that are quite close to the measured shear strengths. The estimation procedures based on the SWCC appear to reflect

Table 12.4 Parameters for Shear Strength Equations for Compacted Kaolin Clay

Equations	Soil Parameters
Oberg and Sallfors (1997)	SWCC, S
Fredlund et al. (1996)	$\kappa = 0.531$, $\theta_s = 0.51$, SWCC
Vanapalli et al. (1996b)	$\theta_s = 0.51$, $\theta_r = 0.140$, SWCC
Goh et al. (2010)	$n_d = 2.908$, $\psi_{aev} = 47$ kPa, PI, SWCC

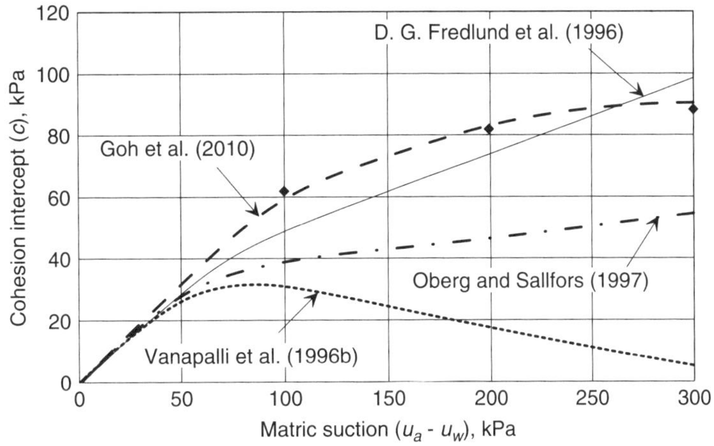

Figure 12.27 Comparison of estimations of shear strength from various proposed equations and measured data for compacted kaolin (after Thu et al., 2007a).

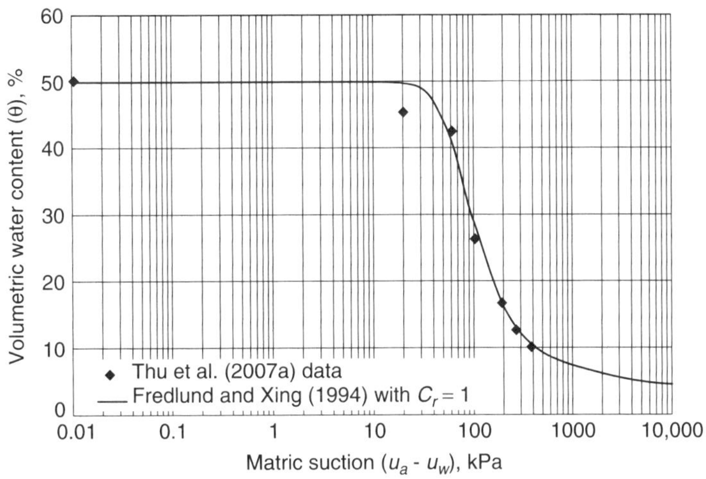

Figure 12.28 SWCC for compacted kaolin clay in terms of volumetric water content versus soil suction when net confining pressure is 10 kPa (after Thu et al., 2007a).

quite well the flattening of the strength envelope at higher suctions.

12.2.5.3 Data on Compacted Kaolin When Measuring Volume Change (Thu et al., 2007a)

Thu et al. (2007a) provided another set of shear strength results on compacted kaolin using a modified triaxial apparatus. The soil properties were the same as the above-mentioned values. The SWCC of the soil was obtained by desaturating the soil under a net confining pressure of 10 kPa. The volumetric water content versus soil suction SWCC is shown in Fig. 12.28.

The shear strength parameters for the saturated compacted kaolin were $\phi' = 31.1^\circ$ and $c' = 0$ kPa. Soil parameters from the SWCC required for the estimation of shear strength with respect to soil suction are listed in Table 12.5.

The measured shear strength data on the compacted kaolin along with estimations of the shear strength envelope are shown in Fig. 12.29. The measured shear strength increases with soil suction in a nonlinear manner. The equations by Goh et al., (2010) and D.G. Fredlund et al., (1996) provide relatively close estimations of the measured shear strengths. The estimation procedures based on the SWCC appear to reflect quite well the flattening of the strength envelope at high suctions.

Table 12.5 Soil Parameters for Shear Strength Equations for Compacted Kaolin Clay

Equations	Soil Parameters
Oberg and Sallfors (1997)	SWCC, S
Fredlund et al. (1996)	$\kappa = 0.446$, $\theta_s = 0.50$, SWCC
Vanapalli et al. (1996b)	$\theta_s = 0.50$, $\theta_r = 0.120$, SWCC
Goh et al. (2010)	$n_d = 3.271$, $\psi_{\text{aev}} = 50$ kPa, PI, SWCC

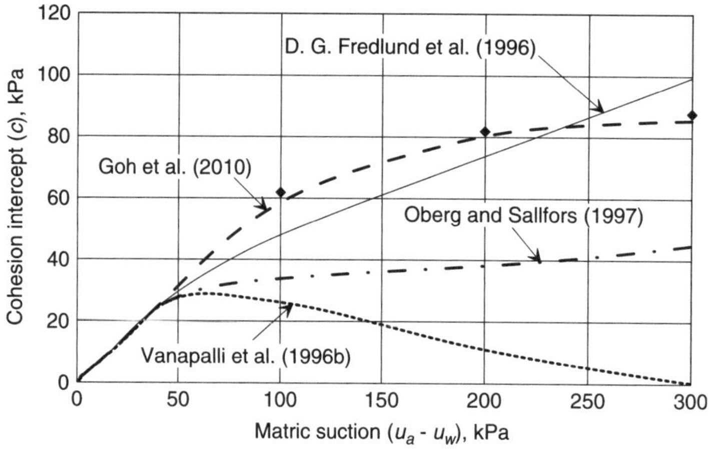

Figure 12.29 Comparison of estimations of shear strength from various proposed equations and measured data for compacted kaolin with volume change measurements (after Thu et al., 2007a).

12.2.5.4 Data on Madrid Grey Clay (Escario and Juca, 1989)

Escario and Juca (1989) used a modified direct shear box to measure the shear strength of three types of clay over an extended suction range up to 15,000 kPa. All specimens were statically compacted at the same water content and density conditions. The specimens were then allowed to come to equilibrium under predetermined total normal stresses as well as applied air and water pressures. The specimens were then sheared to failure under constant suction in the modified direct shear apparatus. One set of data is shown that compares the measured results with the estimation procedures for Madrid grey clay when the net normal stress was 300 kPa. The SWCC of the Madrid grey clay was obtained from drying tests and is shown in Fig. 12.30 along with the best-fit Fredlund and Xing (1994) parameters. Figure 12.31 shows a plot of the volumetric water content versus soil suction for Madrid grey clay.

The shear strength parameters for saturated Madrid grey clay are as follows: $\phi' = 25.3°$, $c' = 30$ kPa (Escario and Juca, 1989). The additional parameters used for comparing various strength equations are listed in the Table 12.6.

The predictions of shear strength under net normal stresses of 300 kPa are shown in Fig. 12.32. The measured shear strength appears to reach a peak around suction values of 10,000 kPa and then decreases slightly. The D. G. Fredlund et al. (1996) provides a reasonably close estimation of the unsaturated soil shear strength. The prediction using the Vanapalli et al., (1996) equation proves to be quite sensitive to the estimation of residual suction. The results in Fig. 12.32 correspond to an estimated residual suction of 30,000 kPa.

There are other experimental data on unsaturated soil shear strengths in the literature; however, the results

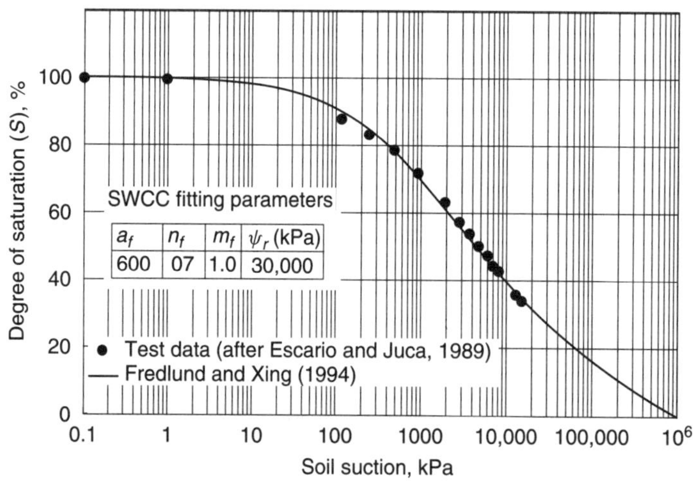

Figure 12.30 SWCC for Madrid grey clay in terms of degree of saturation versus soil suction (after Escario and Juca, 1989).

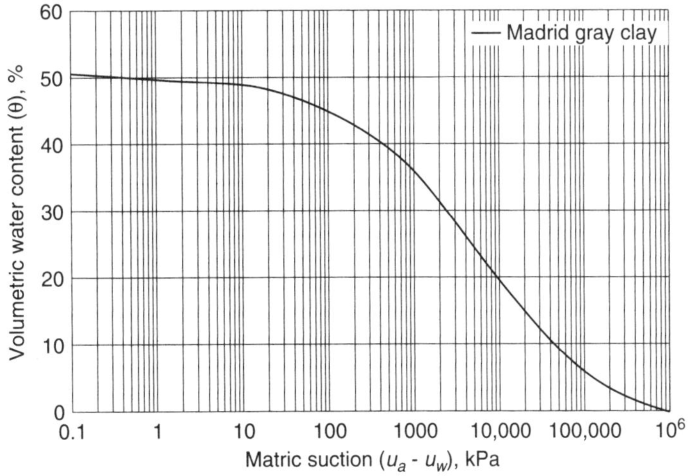

Figure 12.31 SWCC for Madrid grey clay in terms of volumetric water content versus soil suction (after Escario and Juca, 1989).

Table 12.6 Soil Parameters for Estimating the Shear Strength Equations for Madrid Grey Clay

Equations	Additional Parameters
Oberg and Sallfors (1997)	SWCC, S
D.G. Fredlund et al. (1996)	$\kappa = 2.69$, $\theta_s = 0.507$, SWCC
Vanapalli et al. (1996b)	$\theta_s = 0.507, \theta_r = 0.108$, SWCC
Goh et al. (2010)	$n_d = 0.55$, $\psi_{\text{aev}} = 110$ kPa, PI, SWCC

presented illustrate the level of confidence that can be anticipated when using estimation procedures for the shear strength of unsaturated soils. It appears that the proposed estimation equations will serve as an adequate tool for assessing the unsaturated soil shear strength in engineering practice. However, it should also be remembered that laboratory equipment is available for the measurement of unsaturated soil shear strength properties. The geotechnical engineer should insist on performing laboratory shear strength tests in cases where the expenditure is warranted.

In summary, the shear strength equation by Oberg and Sallfors (1997) appears to capture the general pattern of shear strength variation with suction but may over estimate the shear strength of unsaturated soils. The equations by Goh et al. (2010) and D.G. Fredlund et al. (1996) appear to give reasonable estimations of shear strength, particularly in the intermediate- to high-suction range. These equations appear to estimate the peak shear strength in a reasonable manner for the intermediate soil suction range. The predictions of shear strength when using the Vanapalli et al. (1996b) equation are somewhat sensitive to the selection of residual suction. It can be difficult to accurately determine residual suction on the logarithmic scale of soil suction.

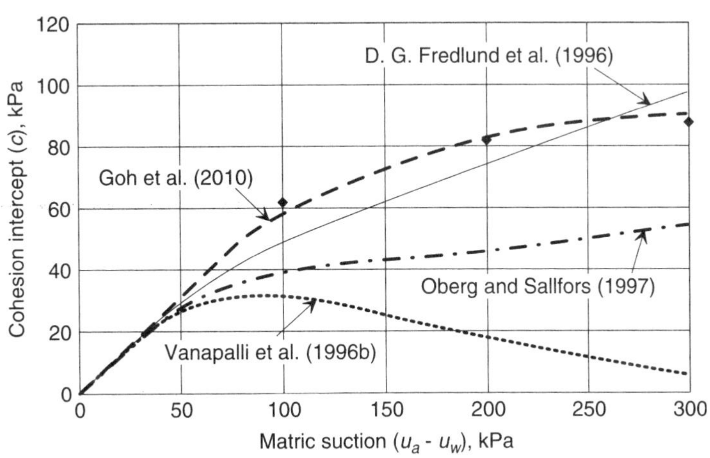

Figure 12.32 Comparison of estimations of shear strength from various proposed equations and direct shear strength measurements on Madrid grey clay (after Escario and Juca, 1989).

12.2.6 Procedures for Calculation of Unsaturated Soil Shear Strength Functions

The computations associated with the estimation of the shear strength for an unsaturated soil are not complex. The procedures described for the estimation of unsaturated soil shear strength involve the substitution of parameters from the SWCC into the shear strength equation.

The following steps are suggested for the estimation of the unsaturated soil shear strength envelope. The saturated soil shear strength parameters must be known (i.e., c' and ϕ'):

1. Information must be available on the SWCC. This information might be in the form of experimental data or best-fit parameters for an equation such as the Fredlund and Xing (1994) equation that can best fit the experimental data.
2. Select a series of matric suctions (e.g., 12 values) that reflect the range over which the unsaturated soil shear strength envelope is to be defined.
3. When using the D.G. Fredlund et al., (1996) shear strength estimation equation, it is necessary to select a κ value that is dependent upon the plasticity of the soil under consideration. When using the Vanapalli et al., (1996b) equation, it is necessary to determine the residual suction of the soil and then calculate the normalized water content versus soil suction. There are correlations with controlling parameters related to the SWCC that must be selected when using the Goh et al., (2010) shear strength equation.
4. Substitute the selected matric suction values into the respective estimation shear strength equations and calculate a series of unsaturated soil shear strengths.

The calculated shear strength values can be used in computer codes in one of several ways. For example, the unsaturated shear strength envelope could be defined using the same equations as were used for the calculation of shear strength (i.e., D.G. Fredlund et al., 1996; Vanapalli et al., 1996b; Oberg and Sallfors, 1997; Goh et al., 2010) or the shear strength information could be input as a table of values. In the latter case, an interpolation technique would need to be used to get the desired shear strength. Some computer codes can accept unsaturated soil shear strength equations.

12.2.7 Linearization of Unsaturated Soil Shear Strength

There are situations where it may be sufficient to describe the shear strength of an unsaturated soil in terms of a constant ϕ^b value. The nonlinear shear strength envelope would be approximated as linear segments along the shear strength envelope. Following is an example to illustrate the estimation of the unsaturated shear strength envelope for an unsaturated soil. In this case the SWCCs were generated from grain-size distribution curves and are shown in Fig. 12.33.

The dry density of the soil samples was estimated to be 1550 kg/m^3. These dry densities correspond to an initial void ratio of 0.774 or an initial saturated porosity of 43.6%. The specific gravity of the soil, G_s, was assumed to be 2.75. Each SWCC starts with a suction approaching zero and ends with a suction of 10^6 kPa. The air-entry value for the soils appears to be around 20 kPa.

The D.G. Fredlund et al., (1996) procedure is used to calculate the unsaturated soil shear strength envelope. The effective cohesion for the soils used in this study was estimated to be 5.0 kPa. The effective angle of internal friction was assumed to be 27°. The plasticity index was 22 and this value corresponds to a κ factor of 2.4, according to the correlation by Garven and Vanapalli (2006b).

Figures 12.34 and 12.35 show the computed unsaturated soil shear strength envelopes for two soils. The shear strength envelopes are clearly nonlinear in shape. It is now possible to select a slope on the shear strength envelope that best represents the suction range of interest in the field. A line can

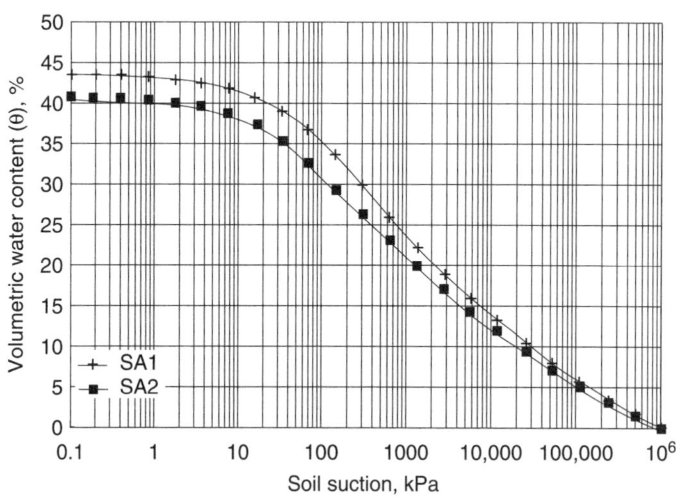

Figure 12.33 SWCCs estimated from grain-size distribution curves.

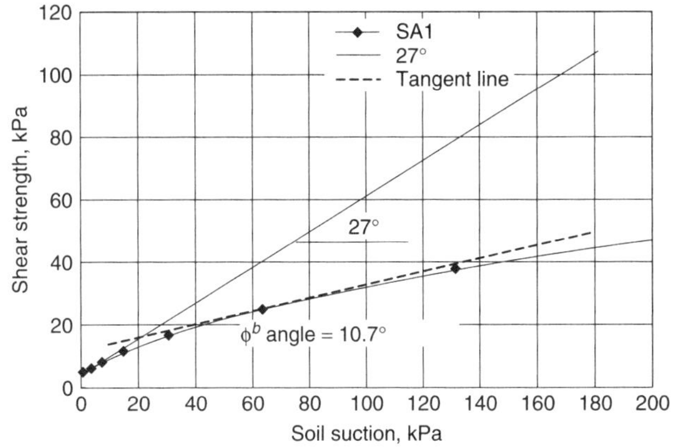

Figure 12.34 Shear strength estimation based on saturated shear strength parameters and estimated SWCC for sample no. SA1.

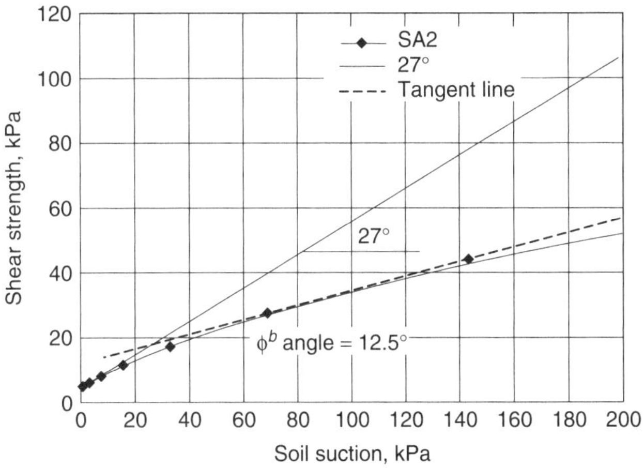

Figure 12.35 Shear strength estimation based on saturated shear strength parameters and SWCC for sample no. SA2.

be drawn tangent to the shear strength envelope representing average in situ suction conditions. Typical straight lines shown on each shear stength envelope correspond to ϕ^b values of 10.7° and 12.5°. It should be noted that the use of these ϕ^b values in an analysis is somewhat conservative since a portion of the shear strength is being neglected.

12.2.8 Application of Shear Strength of Unsaturated Soils in Engineering Practice

Measuring the shear strength parameters in the laboratory is time consuming and costly. As a result, there may not be many engineering projects where the client is prepared to pay for the costs associated with measuring the unsaturated shear strength properties. On the other hand, research studies to date have shown that any one of several proposed estimation procedures may provide reasonable estimations of the unsaturated soil shear strength properties.

Many engineering analyses involve soils from above and below the groundwater table. Consequently, a part of the soil will be unsaturated with negative pore-water pressures while another part of the soil profile may be below the water table. The geotechnical engineer may choose to ignore the benefits associated with the negative pore-water pressures above the water table in some cases. However, it has become increasingly apparent that it is possible to improve on most analyses by taking negative pore-water pressures into consideration in the analysis.

Soil mechanics problems associated with plastic and limit equilibrium analyses can be studied while making a variety of assumptions relative to the handling of the unsaturated soil portion of the profile. Common plastic and limit equilibrium analyses are related to (i) bearing capacity, (ii) lateral earth pressures, and (iii) slope stability. The analyses of these problems are discussed later in this chapter. It is possible to ignore the benefits related to the unsaturated soil portion; however, there is generally value in asking the question, "How does the inclusion of unsaturated soil properties and negative pore-water pressures influence the analysis?" If the engineer desires to study the "trigger" mechanism for potential slope stability failures, it becomes imperative to ascertain potential changes in matric suction in the slope. The computed changes in matric suction become input information to a simulated slope stability analysis.

A range of assumptions can be made when analyzing the unsaturated soil portion of problems involving the assessment of shear strength. For example, it might be possible to conduct a bearing capacity analysis assuming that all matric suctions go to zero. This analysis could be followed by subsequent analyses where the estimated unsaturated soil properties are input along with more realistic assumptions related to long-term pore-water pressures. If this design protocol is followed, the geotechnical engineer becomes aware of the significant effect that modest matric suctions have on the computed results. It is common to experience a twofold or more increase in bearing capacity when even modest matric suction is present. A realization of the significant influence of matric suction can also lead to the consideration of design details that tend to retain soil suction with time.

Lateral earth pressures associated with the "active" mode can be greatly reduced with modest matric suctions in the soil. On the contrary, the "passive" earth pressure resistance can be greatly increased with matric suction. A more complex mode involves the analysis of soil-structure interaction that arises as a soil dries and wets due to moisture movement. The changes in earth pressures on a wall are directly influenced by changes in matric suction over time.

The stability of slopes involving unsaturated soils is probably the engineering application area that is of greatest relevance in geotechnical engineering practice. If the water table is quite deep, the entire potential slip surface may pass through unsaturated soil. However, it is rainfall infiltration that is most likely to produce a catastrophic landslide with dire consequences. Even when the water table is quite high in the slope, it is possible to study the effect that changes in negative and positive pore-water pressures will have on the computed factor of safety. It is useful from an engineering perspective to have a quantitative assessment of the influence that the unsaturated soil portion has on computed factors of safety.

The unsaturated soil shear strength properties of a soil can readily be estimated. It is important to quantify the effect of matric suction (and changes in matric suction) on the analysis of the problem at hand. The influence of matric suction at shallow depths is often much more significant than anticipated. The engineer should carefully consider whether there are decisions that can be made to better ensure the maintenance of negative pore-water pressures over the life of the structure.

12.3 APPLICATION TO PRACTICAL SHEAR STRENGTH PROBLEMS IN GEOTECHNICAL ENGINEERING

Deformation or ground movement problems encountered in soil mechanics can be divided into one of two categories dependent upon the stress level involved. When deviator stress levels are relatively low, the problems are considered to be in the elastic range and problems can generally be analyzed using theories of elasticity. When the deviator stress levels are relatively high, the problems are considered to be in the plastic range and problems are analyzed using the theories of plasticity or limit equilibrium. The elastoplastic categories can be visualized on an idealized representation of a stress-strain curve (Fig. 12.36).

There is a class of problems in geotechnical engineering where the analysis is based on the assumption that the soil behaves in a perfectly plastic manner. When the state of plastic equilibrium is limited to a specific, thin zone, the problems are referred to as limit equilibrium analyses. The thin plastic zone is called a slip surface zone or a slip plane. The assumptions of plasticity and elasticity are a simplification of actual soil behavior that form a useful categorization of the types of analyses that can be performed in soil mechanics. Theories of elasticity and plasticity provide categories with theoretical limits within which the behavior of a soil mass can be studied.

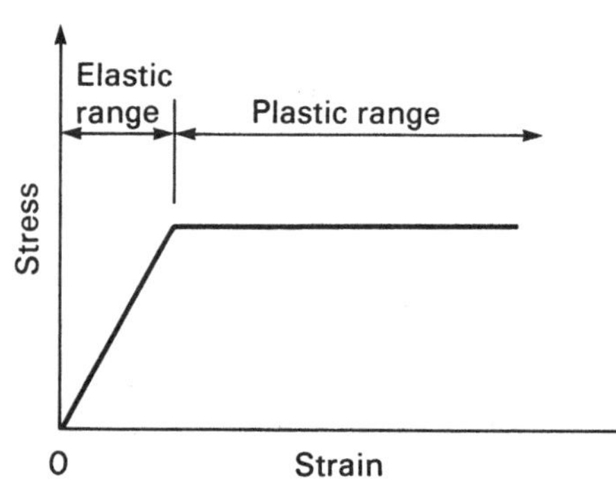

Figure 12.36 Idealized elastic-plastic behavior giving rise to two categories of deformation analysis.

There are three main types of soil mechanics analyses where plastic equilibrium forms the basis of analysis: (1) lateral earth pressure analyses, (2) bearing capacity analyses, and (3) slope stability analyses. Plastic equilibrium satisfies equations of equilibrium and a condition of failure (i.e., the failure criterion). Typical soil mechanics analyses are derived in this chapter based on plastic equilibrium. In each case, the soil is assumed to have negative pore-water pressures (or matric suctions).

Plasticity analyses can be subdivided into situations where the pore-water pressure effects are simulated in some manner during the testing of the soil (i.e., total stress approach) and situations where pore pressure designations become a part of the analysis (i.e., stress state variable approach). Most of the consideration in this chapter is given to the latter situation where actual or predicted pore pressures are designated.

12.3.1 Prediction of Earth Pressures

Some of the earliest work in soil mechanics dealt with earth pressures on retaining walls. However, there is little information on the earth pressures exerted on engineering structures backfilled and founded upon unsaturated soils. Engineers are well aware of pressures that can be exerted by expansive soils, but little attention has been given to developing a general earth pressure theory for these soils. It is not adequate to request that cohesive backfill not be used behind earth-retaining structures. Experience indicates that problems associated with the performance of earth structures can often involve compacted, clayey soils. Ireland (1964) stated that 68% of unsatisfactory retaining wall performance involved cases where clay was used as a backfill or where the retaining wall was founded on clay.

Some of the problems encountered with earth-retaining structures result from the tendency of clayey expansive soils to undergo substantial changes in volume as a result of changing environmental conditions. There are situations where unsaturated soils are used as backfill or where structural members are cast in place against the soil. It is apparent that only limited consideration has been given to the behavior of the retaining structure under these circumstances.

A limited but theoretical analysis of earth pressures is presented for soils with negative pore-water pressures. The active and passive earth pressures in a soil mass are discussed along with several example problems that illustrate the concepts involved. The meaning of the at-rest earth pressure condition was discussed in Chapter 3 in the section on the stress state in an earth mass. The at-rest pressure state is more appropriately related to elastic stress theories (i.e., with zero lateral strain).

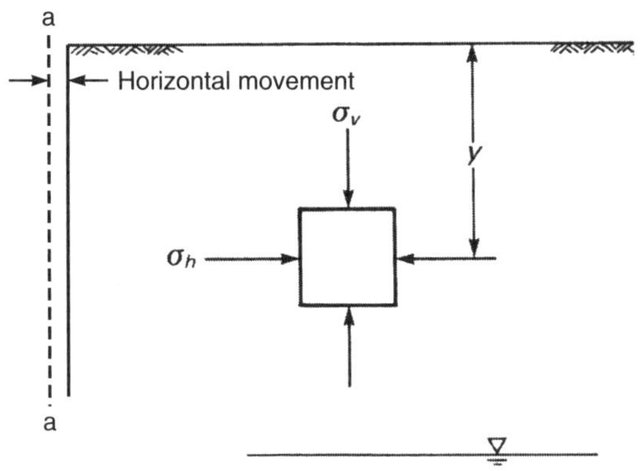

Figure 12.37 Stresses on an element in soil mass behind frictionless wall.

12.3.2 Extended Rankine Theory of Earth Pressures

The active and passive earth pressures for an unsaturated soil can be determined by assuming that the soil is in a state of plastic equilibrium. Let us first review the stresses in a soil mass when the surfaces of failure are planar. The major and minor principal planes at all points are assumed to have similar directions. The solution is known as the Rankine earth pressure theory. It is necessary to extend the conventional earth pressure concepts when dealing with unsaturated soils. For this reason, the earth pressure theories presented herein are referred to as the extended Rankine theory of earth pressures.

Figure 12.37 shows a vertical, frictionless plane passed through a soil mass of infinite depth. An element of unsaturated soil at any depth is subjected to a vertical stress σ_v and a horizontal stress σ_h. These planes are assumed to be principal planes and the vertical and horizontal stresses are the principal stresses. The ground surface is horizontal and the vertical stress is written in terms of the density of the soil.

The state of stress for an element of unsaturated soil can be shown on an extended Mohr diagram. The equation corresponding to the limiting or failure condition (i.e., shear strength equation) can be written as follows assuming a linear shear strength equation:

$$\tau = c' + (\sigma_n - u_a)\tan\phi' + (u_a - u_w)\tan\phi^b \qquad (12.31)$$

The above linear shear strength equation can also be written in a form similar to that used for saturated soils:

$$\tau = c + (\sigma_n - u_a)\tan\phi' \qquad (12.32)$$

Consequently, the shear strength equation for an unsaturated soil can be viewed as a two-parameter shear strength equation. The total cohesion c has two components and can be written as

$$c = c' + (u_a - u_w)\tan\phi^b \qquad (12.33)$$

Equation 12.32 for the shear strength of an unsaturated soil has the advantage that derivations relevant to saturated soils can readily be modified to accommodate situations where the soil is unsaturated. It is simply necessary to realize that the cohesion of a soil has two components (i.e., effective cohesion and cohesion due to matric suction).

The initial, total vertical stress in a soil mass is equal to the overburden pressure, $\rho g y$. The total horizontal pressure is equal to the coefficient of earth pressure at rest multiplied by the overburden pressure. The assumption is made that the soil has an initial matric suction equal to $(u_a - u_w)_0$. The pore-water pressure in the soil is influenced by environmental conditions but is assumed to remain constant during the analysis. Changes in total stress are assumed to not influence pore pressures.

12.3.3 Active Earth Pressure

Let us suppose that the wall a-a in Fig. 12.37 is allowed to move away from the soil mass. The horizontal stress is reduced until a limiting value corresponding to the plastic equilibrium state is attained. Failure is induced by reducing the horizontal stress. The horizontal stress is the minor principal stress while the vertical stress is the major principal stress. The horizontal stress at any point corresponding to the active state can be computed from the vertical stress and the failure criterion for the soil.

Figure 12.38 illustrates how the active and passive pressures in a soil change as the matric suction changes. The active pressure is shown to decrease as matric suction increases. In other words, as the pore-water pressure in the soil becomes more negative, the soil becomes stronger. Consequently, less force would be carried by the retaining wall.

Let us consider a vertical plane with the soil having constant matric suction as shown in Fig. 12.37. An element from a depth y has an overburden stress σ_v. The element of soil is shown in Fig. 12.39 along with the definition of pertinent variables. If the wall moves away from the soil, the active earth pressure is developed. The active pressure is designated as $\sigma_h - u_a$. The horizontal pressure can be written in terms of the vertical pressure $\sigma_v - u_a$ by considering the geometrics of the Mohr circle:

$$\sin\phi' = \frac{[(\sigma_v - u_a) - (\sigma_h - u_a)]/2}{[(\sigma_h - u_a) + (\sigma_v - u_a)] + c\cot\phi'} \qquad (12.34)$$

where:

c = total cohesion [i.e., $c = c' + (u_a - u_w)\tan\phi^b$].

Rearranging the above equation and solving for the net horizontal stress $\sigma_h - u_a$ gives

$$\sigma_h - u_a = (\sigma_v - u_a)\frac{1 - \sin\phi'}{1 + \sin\phi'} - 2c\frac{\cos\phi'}{1 + \sin\phi'}. \qquad (12.35)$$

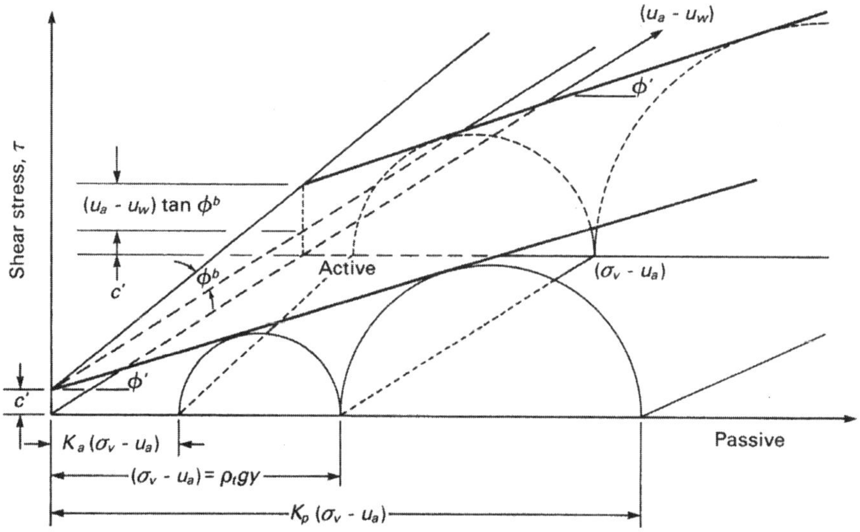

Figure 12.38 Active and passive earth pressure states for soil with matric suction.

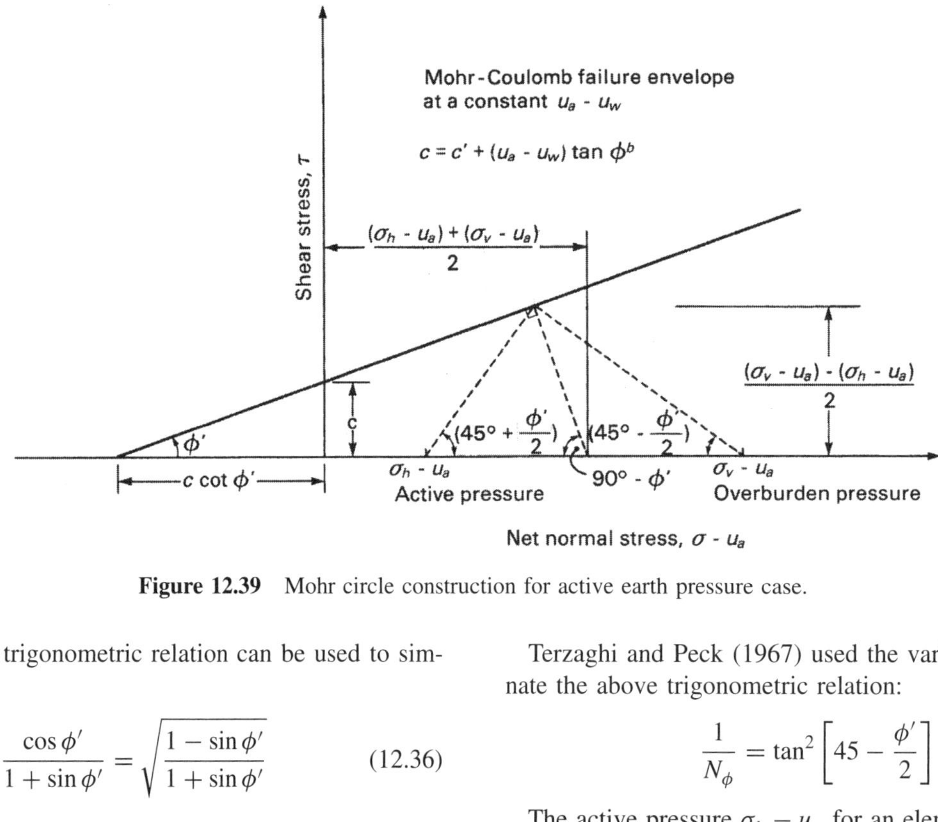

Figure 12.39 Mohr circle construction for active earth pressure case.

The following trigonometric relation can be used to simplify Eq. 12.35:

$$\frac{\cos\phi'}{1+\sin\phi'} = \sqrt{\frac{1-\sin\phi'}{1+\sin\phi'}} \tag{12.36}$$

Equation 12.36 can now be written as:

$$\sigma_h - u_a = (\sigma_v - u_a)\frac{1-\sin\phi'}{1+\sin\phi'} - 2c\sqrt{\frac{1-\sin\phi'}{1+\sin\phi'}}. \tag{12.37}$$

The trigonometric function appearing in Eq. 12.37 can be written in terms of the angle of the slip planes from a vertical plane:

$$\frac{1-\sin\phi'}{1+\sin\phi'} = \tan^2\left[45 - \frac{\phi'}{2}\right] \tag{12.38}$$

Terzaghi and Peck (1967) used the variable N_ϕ to designate the above trigonometric relation:

$$\frac{1}{N_\phi} = \tan^2\left[45 - \frac{\phi'}{2}\right] \tag{12.39}$$

The active pressure $\sigma_h - u_a$ for an element of soil at any depth can be written as

$$\sigma_h - u_a = (\sigma_v - u_a)\frac{1}{N_\phi} - 2c\frac{1}{\sqrt{N_\phi}} \tag{12.40}$$

The above equation can also be rewritten by taking into account the individual components of cohesion:

$$\sigma_h - u_a = (\sigma_v - u_a)\frac{1}{N_\phi} - 2c'\frac{1}{\sqrt{N_\phi}} - 2(u_a - u_w)\tan\phi^b\frac{1}{\sqrt{N_\phi}} \tag{12.41}$$

12.3.4 Coefficient of Active Earth Pressure

Let us define the coefficient of active earth pressure as the ratio of the net horizontal pressure to the net vertical pressure:

$$K_a = \frac{\sigma_h - u_a}{\sigma_v - u_a} \tag{12.42}$$

Normalizing the net horizontal stress by the vertical stress allows the coefficient of active earth pressure to be written as follows:

$$K_a = \frac{1}{N_\phi} - \frac{2c'}{\sigma_v - u_a}\frac{1}{\sqrt{N_\phi}} - 2\frac{u_a - u_w}{\sigma_v - u_a}\tan\phi^b\frac{1}{\sqrt{N_\phi}} \tag{12.43}$$

12.3.5 Active Earth Pressure Distribution (Constant Suction with Depth)

The horizontal pressure corresponding to the active state can be computed for various depths and plotted as shown in Fig. 12.40a. Conjugate planes are formed in the soil mass at angles of $45 + \phi'/2$ to the horizontal, as shown in Fig. 12.40b for the active case. The earth pressures for the saturated soil are designated using effective cohesion c' and effective angle of internal friction ϕ'. Let us suppose that matric suctions were a constant value with depth. Then the total cohesion is also a constant with respect to depth, and the active pressure distribution is translated to the left, parallel to the saturated soil case. Figure 12.41 shows the breakdown of the active pressure into its three stress distribution components.

12.3.6 Tension Zone Depth

The tension zone depth y against the wall can be computed by setting the total horizontal pressure to zero and assuming an atmospheric air pressure (i.e., $u_a = 0$) in Eq. 12.40 or 12.41:

$$y_t = \frac{2c'}{\rho g}\sqrt{N_\phi} + 2\frac{(u_a - u_w)\tan\phi^b}{\rho g}\sqrt{N_\phi} \tag{12.44}$$

The tension zone depth y_t is equal to the depth of the vertical cracking, y_c, when the tensile strength of the soil is assumed to be negligible. The tension zone depth increases as the matric suction of the soil increases. This depth corresponds to the zone which would pull away from the wall as the wall moved horizontally away from the soil. Figure 12.42 illustrates how matric suction causes a soil to pull away from the wall.

12.3.7 Active Earth Pressure Distribution (Linear Decrease in Suction to Water Table)

Let us assume that the matric suction in the soil decreases linearly with depth to a value of zero at the water table (Fig. 12.43). The negative pore-water pressure at ground surface for hydrostatic conditions can be written as a function of the distance from the groundwater table:

$$(u_a - u_w)_h = \rho_w g D \tag{12.45}$$

where:

$(u_a - u_w)_h$ = matric suction at ground surface and
D = depth from ground surface to the water table.

A simple relationship can be used to define the variation in matric suction with depth for this profile. For depth $y \le D$, the matric suction can be written as

$$(u_a - u_w)_y = (u_a - u_w)_h\left(1 - \frac{y}{D}\right) \tag{12.46}$$

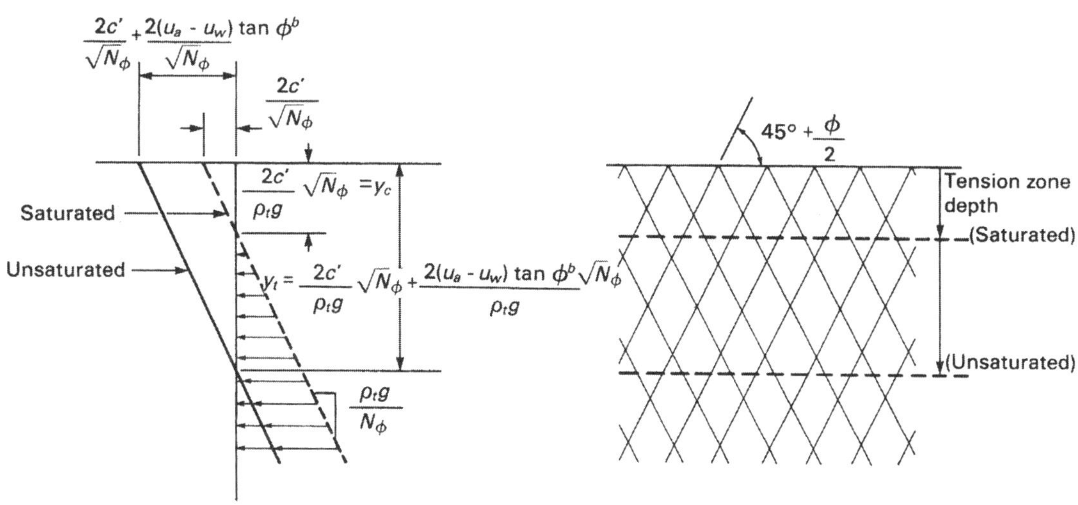

Figure 12.40 Rankine active earth pressure distribution for saturated soil and soil with constant matric suction with depth: (a) active pressure distribution corresponding to active pressure case; (b) conjugate planes corresponding to active pressure case.

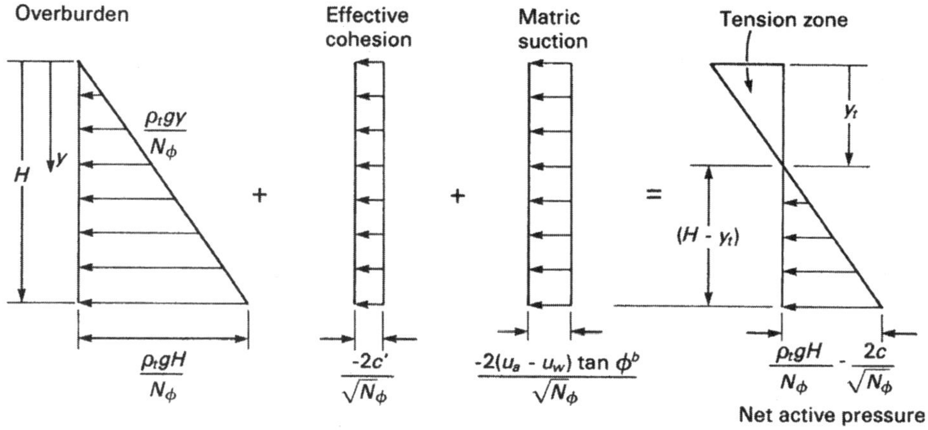

Figure 12.41 Components of active earth pressure distribution when matric suction is constant with depth.

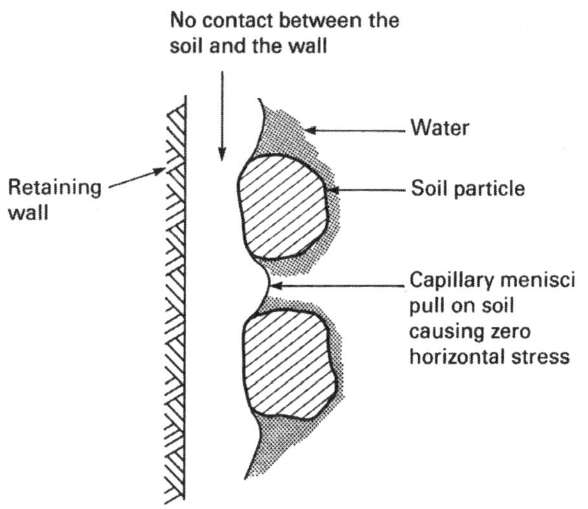

Figure 12.42 Matric suction causes soil to pull away from retaining wall.

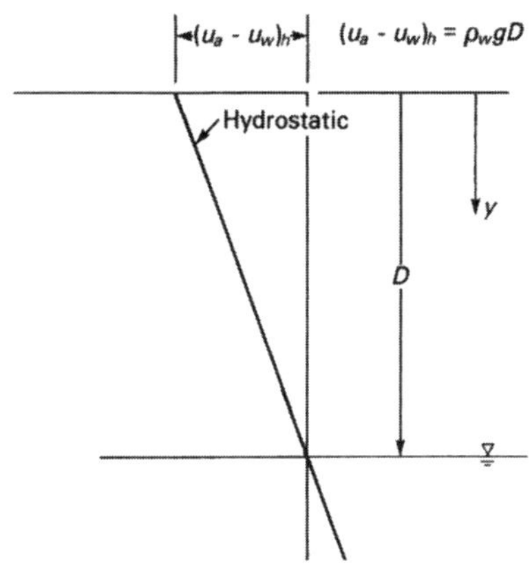

Figure 12.43 Designation of equilibrium matric suction profile.

The active pressure p_a at any depth above the water table is

$$p_a = (\sigma_v - u_a)\frac{1}{N_\phi} - \frac{2c'}{\sqrt{N_\phi}} - \frac{2(u_a - u_w)_h \tan\phi^b}{\sqrt{N_\phi}}\left(1 - \frac{y}{D}\right) \quad (12.47)$$

The active pressure distribution diagram along with the plot of each of the components is shown in Fig. 12.44.

The tension zone depth y_t can be computed by setting the total horizontal stress to zero and assuming an atmospheric air pressure (i.e., $u_a = 0$) in Eq. 12.41:

$$y_t = \frac{2c'\sqrt{N_\phi} + 2(u_a - u_w)_h \tan\phi^b \sqrt{N_\phi}}{2\sqrt{N_\phi}} \rho g + \frac{2\sqrt{N_\phi}}{D}(u_a - u_w)_h \tan\phi^b \quad (12.48)$$

where $y_t < D$.

12.3.8 Active Earth Pressure Distribution with Tension Cracks

The soil behind a retaining wall often has tension cracks. It should be noted that the depth of tension cracks in the soil must be considered as being analytically independent from the tension zone depth against the retaining wall. The depth of cracking is given the variable y_c, and the soil above this depth can be considered as a surcharge load applied to the underlying soil (Fig. 12.45). Equation 12.46 applies to the case under consideration as long as the water table is below the bottom of the tension cracks.

The surcharge load q_s must be applied below a depth y_c and is equal to the overburden pressure (i.e., $q_s = \rho g y_c$). An appropriate total density ρ must be used for the upper soil with the tension cracks. The active pressure above the water table can be derived in a manner similar to the previous cases. The active pressure at any depth can be written as

$$p_a = \frac{\sigma_v - u_a}{N_\phi} - \frac{2c'}{\sqrt{N_\phi}} - 2\frac{(u_a - u_w)_h \tan\phi^b}{\sqrt{N_\phi}}\left[1 - \frac{y}{D - y_c}\right] + \frac{q_s}{N_\phi} \quad (12.49)$$

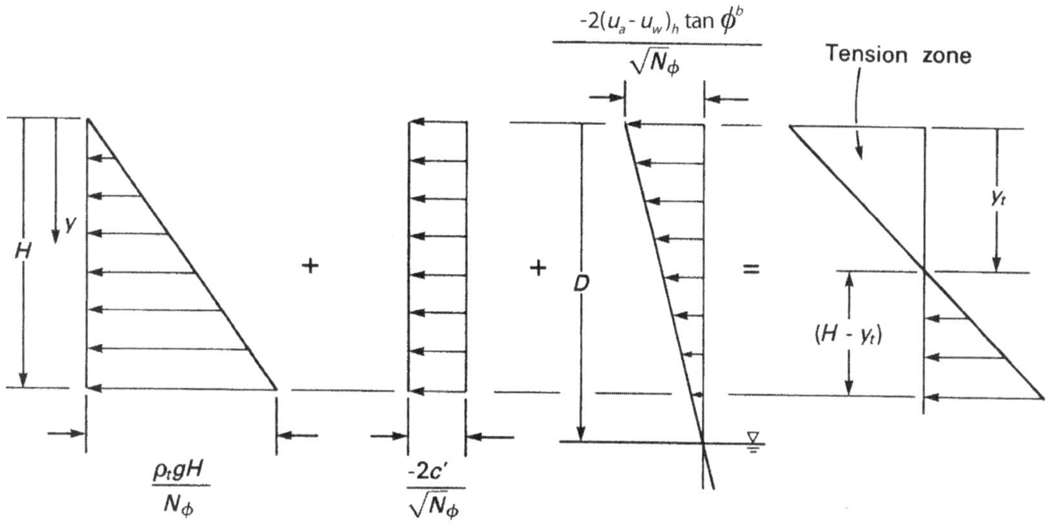

Figure 12.44 Components of active earth pressure distribution when matric suction decrease linearly with depth.

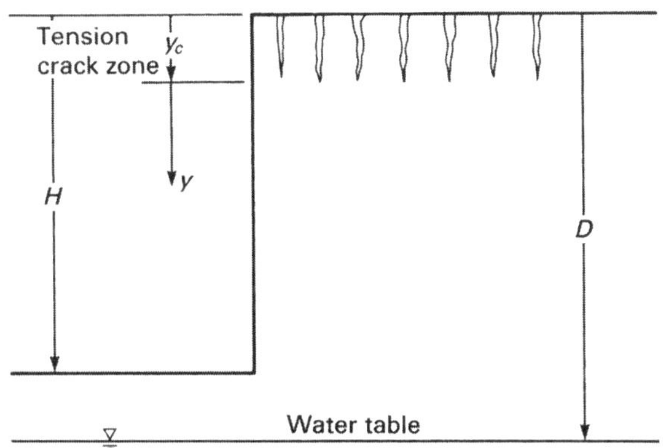

Figure 12.45 Variables that must be designated when soil behind retaining wall has tension cracks.

Equation 12.49 is an extension of Eq. 12.47, and shows the role of a surcharge load in affecting the active earth pressure. The components which make up the active earth pressure diagram for all depths are shown in Fig. 12.46. Other types of surcharge load can be applied in a similar manner.

The tension zone depth y_t can be computed as the depth where the horizontal stress is zero:

$$y_t = \frac{2c'\sqrt{N_\phi} + 2(u_a - u_w)_h \tan \phi^b \sqrt{N_\phi} - q_s}{\rho g + 2[\sqrt{N_\phi}/(D - y_c)](u_a - u_w)_h \tan \phi^b} \quad (12.50)$$

The above equation applies as long as the tension zone depth y_t is less than the distance, $D - y_c$. Other assumptions can be made regarding the distribution of the pore-water pressures with respect to depth. The same analytical procedure can be used to compute the active earth pressure state at any depth.

12.3.9 Passive Earth Pressure

If the wall in Fig. 12.37 is moved toward the soil (i.e., the soil is compressed), the horizontal pressure is increased to be greater than the vertical pressure. The total horizontal stress is the major principal stress when failure occurs in the soil mass. The total vertical stress is the minor principal stress. The horizontal stress corresponding to the passive state can be computed from the vertical stress and the failure criterion for the soil. Figure 12.38 illustrates the passive pressure in a soil mass as a function of matric suction. The passive pressure is shown to increase as matric suction increases.

Let us consider a vertical plane corresponding to a specific matric suction plane, as shown in Fig. 12.38. An element from a depth y has an overburden pressure σ_v. An element of soil is shown in Fig. 12.47, along with the definition of pertinent variables. The passive pressure can be defined as $\sigma_h - u_a$ for the case where the wall moves into the soil mass. The horizontal pressure can be derived in terms of the vertical pressure $\sigma_v - u_a$ in a manner similar to that used for the active pressure derivation:

$$\sin \phi' = \frac{[(\sigma_h - u_a) - (\sigma_v - u_a)]/2}{[(\sigma_h - u_a) + (\sigma_v - u_a)]/2} + c \cot \phi' \quad (12.51)$$

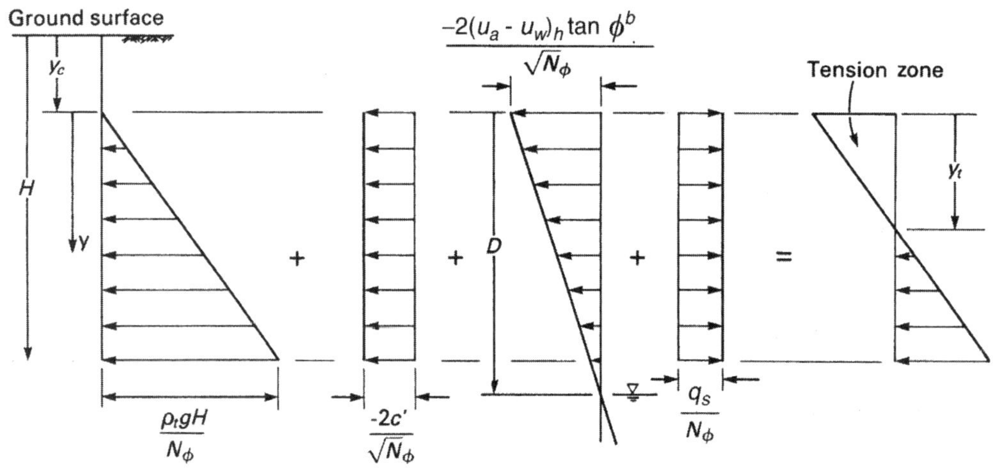

Figure 12.46 Components of active earth pressure distribution when soil has tension cracks and matric suction decreases linearly with depth.

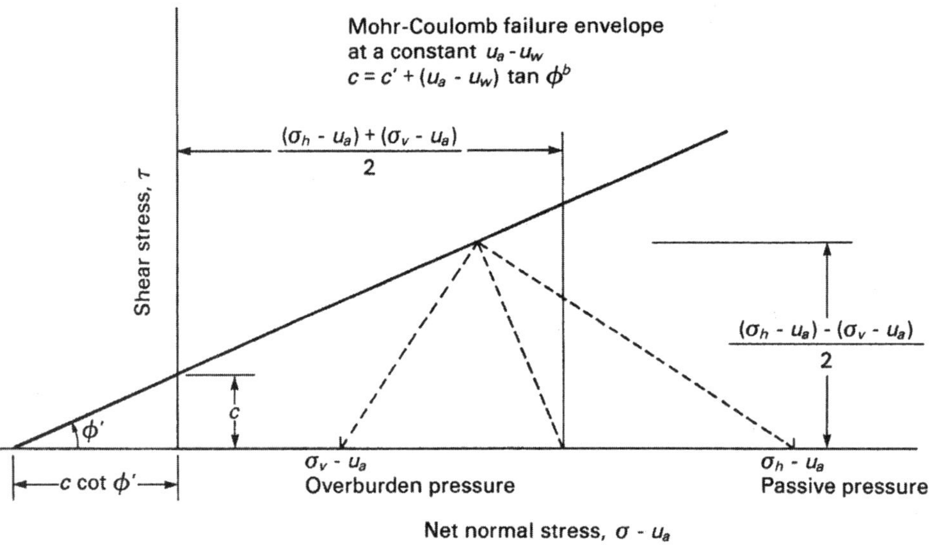

Figure 12.47 Mohr circle construction for passive earth pressure case.

where:

c = total cohesion [i.e., $c = c' + (u_a - u_w)\tan\phi^b$].

Rearranging Eq. 12.51, the horizontal stress can be written as

$$\sigma_h - u_a = (\sigma_v - u_a)\frac{1+\sin\phi'}{1-\sin\phi'} - 2c\frac{\cos\phi'}{1-\sin\phi'} \qquad (12.52)$$

Similar trigonometric functions to those used in the active pressure analysis can be used to rewrite Eq. 12.52:

$$\sigma_h - u_a = (\sigma_v - u_a)N_\phi + 2c'\sqrt{N_\phi} + 2(u_a - u_w)\tan\phi^b\sqrt{N_\phi} \qquad (12.53)$$

where

$N_\phi = 1 - \sin\phi'/1 + \sin\phi'$ or $N_\phi = \tan^2(45 + \phi'/2)$.

12.3.10 Coefficient of Passive Earth Pressure

The coefficient of passive earth pressure can be written as the ratio of the net horizontal pressure to the net vertical pressure. Dividing Eq. 12.53 by the net vertical pressure gives the passive earth pressure coefficient:

$$K_p = N_\phi + \frac{2c'\sqrt{N_\phi}}{\sigma_v - u_a} + \frac{2(u_a - u_w)\tan\phi^b\sqrt{N_\phi}}{\sigma_v - u_a} \qquad (12.54)$$

Equations 12.43 and 12.54 show that both active and passive earth pressure coefficients vary with overburden pressure.

12.3.11 Passive Earth Pressure Distribution (Constant Suction with Depth)

The horizontal pressure corresponding to the passive state can be computed for various depths and plotted as shown in Fig. 12.48a. For the passive case, conjugate planes are formed at $45 - \phi'/2$ to the horizontal, as shown in Fig. 12.48b. The saturated soil case is designated using the effective cohesion term c'. The passive pressure is translated to the right as the suction is increased since the total cohesion is assumed to be constant with depth. Figure 12.49 shows a breakdown of the passive pressure distribution into its three pressure distribution components.

The entire soil mass is in a state of compression for passive pressure conditions. The total horizontal pressure at ground surface is a function of total cohesion:

$$p_p = 2c'\sqrt{N_\phi} + 2(u_a - u_w)\tan\phi^b\sqrt{N_\phi} \tag{12.55}$$

12.3.12 Passive Earth Pressure Distribution (Linear Decrease in Suction to Water Table)

It is possible to select various distributions for matric suction with respect to depth. Let us assume that the matric suction linearly decreases with depth to a value of zero at the water table (Fig. 12.43). The matric suction at any depth can be

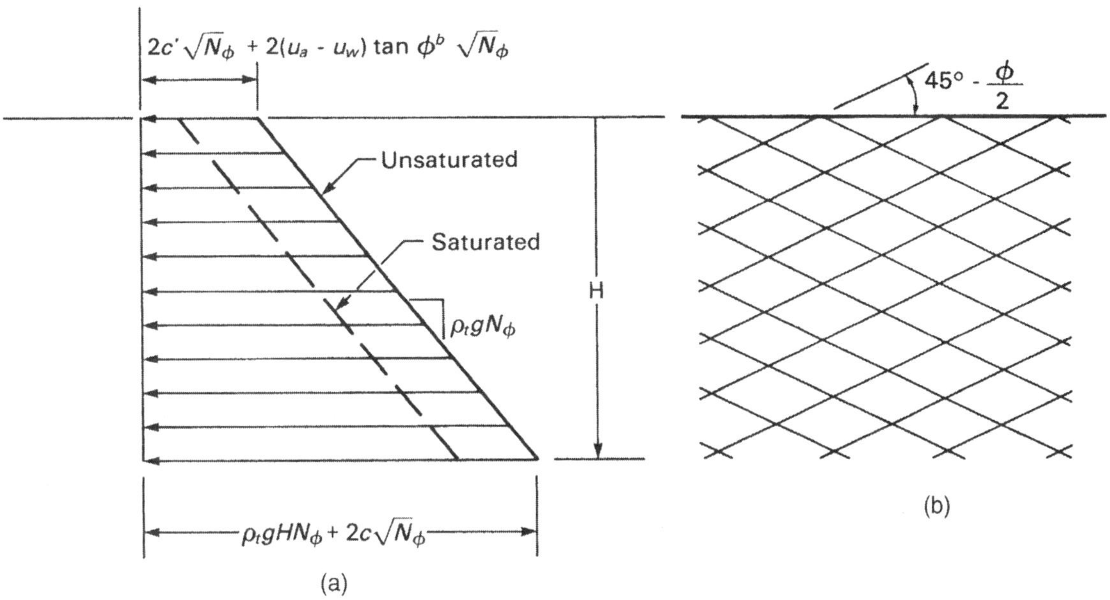

Figure 12.48 Rankine passive earth pressure distribution for saturated soil and soil with a constant matric suction: (a) passive pressure distribution corresponding to passive pressure case; (b) conjugate planes corresponding to passive pressure case.

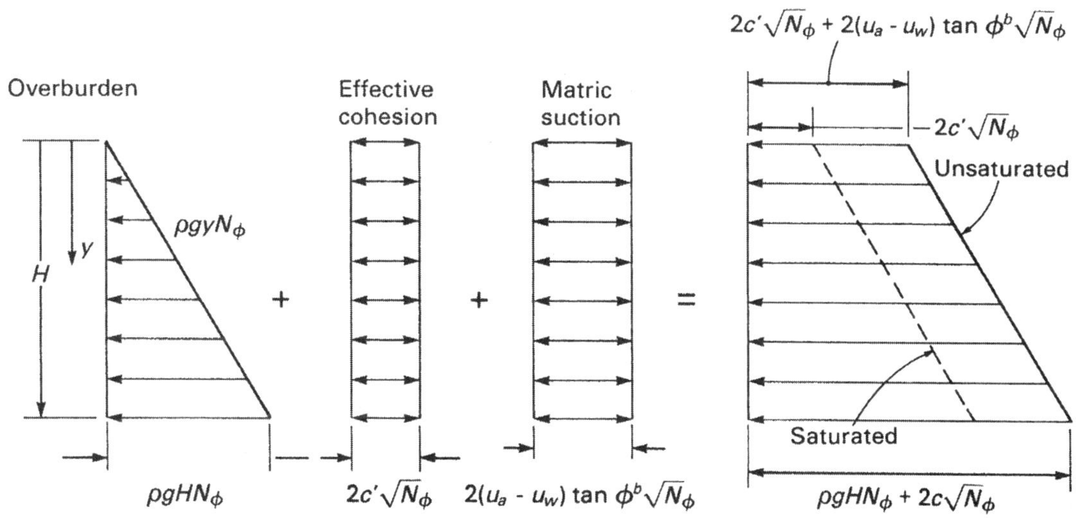

Figure 12.49 Components of passive earth pressure distribution when matric suction is constant with respect to depth.

written as a linear equation (Eq. 12.49). For this matric suction distribution, the passive pressure p_p can be written as

$$p_p = (\sigma_v - u_a)N_\phi + 2c'\sqrt{N_\phi} + 2(u_a - u_w)_h \tan\phi^b \left(1 - \frac{y}{D}\right)\sqrt{N_\phi} \quad (12.56)$$

The passive pressure distribution along with a plot of the components is shown in Fig. 12.50. The existence of tension cracks in the soil is not of relevance for the passive pressure case. The assumption is made that the cracks would close and the soil mass would become intact as the passive pressure is applied.

12.3.13 Deformations Associated with Active and Passive States

Studies by Terzaghi (1954) showed the relationship between the movement of a wall and the earth pressure coefficients for dense and loose sands (Fig. 12.51). Most unsaturated soils behave similar to a dense soil in that relatively low displacements are required to develop the active and passive states. The results indicate that displacements as low as $0.001H$ (where H is the height of the wall) are required to develop the active state in a dense soil. More displacement is required to develop the passive state.

Lambe and Whitman (1968) presented triaxial test results on a dense sand and summarized the percent strains required for the active and passive states. It was concluded that strains on the order of 0.5% can produce the active state. Under compressive loading, strains of 0.5% develop about one-half of the passive resistance. As much as 2% was required to reach the full passive resistance for dense sand. Figure 12.52 illustrates the difference between the active and passive stress paths. The larger stress changes associated with the development of the passive state is one of the reasons for a larger strain at failure.

Strains at failure in unsaturated soils might be quite different from those mentioned above because of unusual K_0

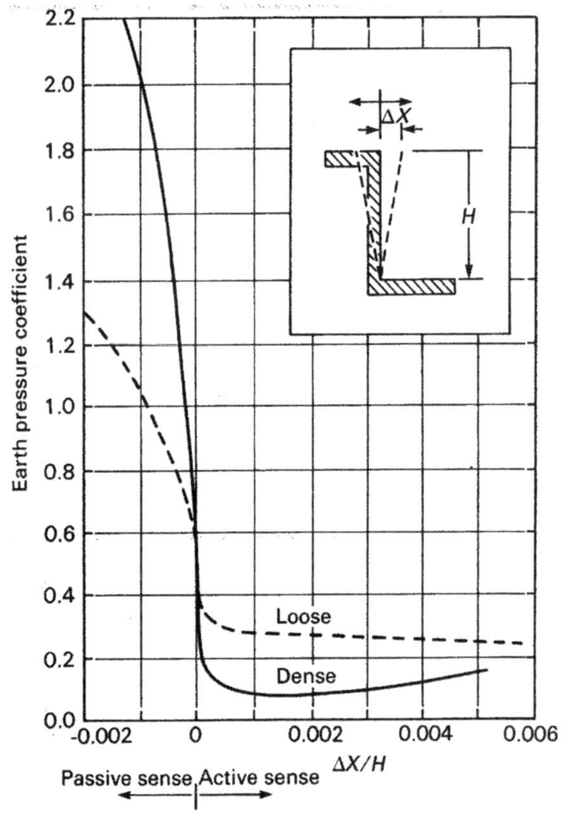

Figure 12.51 Relationship between movement of wall and earth pressures developed for varying soil densities (from Terzaghi, 1954).

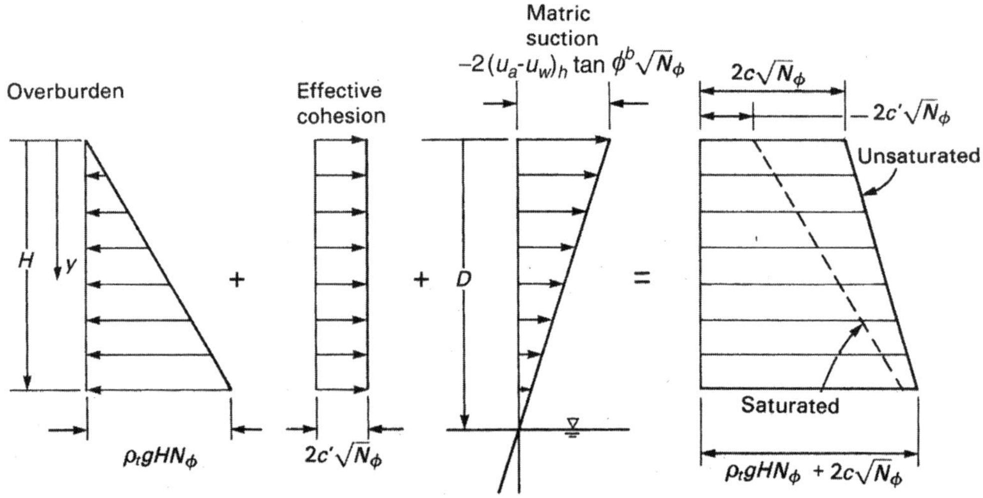

Figure 12.50 Components of passive earth pressure distribution when matric suction decreases linearly with respect to depth.

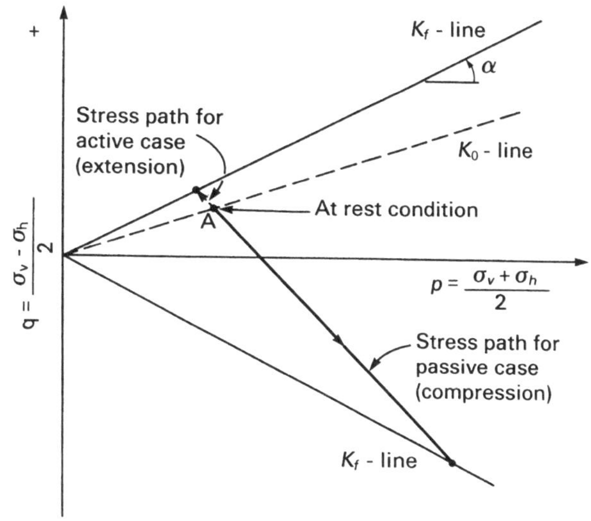

Figure 12.52 Stress paths for Rankine active and passive earth pressure conditions.

conditions. Little research has been done on this subject for unsaturated soils.

12.3.14 Total Lateral Earth Force

Retaining walls serve to retain the backfill placed behind the wall (Fig. 12.53). The earth pressure tends toward the at-rest pressure state if the wall is fixed. However, movement of the wall away from the soil gives rise to the development of the active earth force against the wall. Movement of the wall into the soil results in the development of the passive earth force against the wall. If the wall is vertical and smooth, the lateral force is equal to the sum of the active (or passive) pressures at all depths. There may be an active earth force developed against one side of the wall and a passive earth force developed against the other side, as shown in Fig. 12.53.

Analyses such as those proposed by Coulomb can be used when friction is considered between the soil and the wall. The shape of the slip surface is changed and is dependent upon the vertical movement of the wall as well as the horizontal movement when friction is considered along the wall. Only the case of an unsaturated soil placed against a smooth, vertical wall is considered in this chapter. The analysis of frictional walls in unsaturated soils is essentially the same as those for a saturated soil. The difference in analyzing a wall adjacent to saturated and unsaturated soils lies mainly in the fact that cohesion has a component related to matric suction in an unsaturated soil.

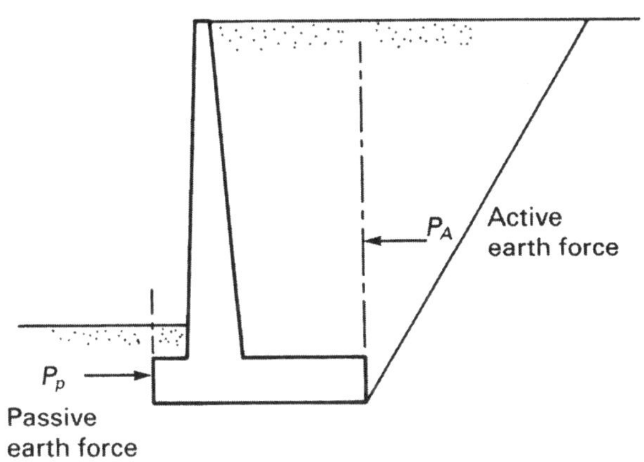

Figure 12.53 Approximations of earth pressures acting on a retaining wall.

12.3.15 Active Earth Force

The active earth force P_A against a smooth wall is equal to the active earth pressure integrated from the bottom of the tension zone depth y_t to the bottom of the wall H. The integration of the active pressure equation over the entire depth gives the total active force. The upper portion of the net pressure diagram is in a state of tension. It can be assumed that the soil in this zone cannot adhere to the retaining wall. Figure 12.54 illustrates the components of the pressure diagram and the limits of integration on the active pressure diagram:

$$P_A = \int_{yt}^{H} p_a \, dy \tag{12.57}$$

The depth of the tension cracks, y_t, must be less than H in order for there to be a positive total force on the wall. Let us consider the case where matric suction is a constant value with respect to depth:

$$P_A = \int_{y_t}^{H} \frac{\rho g y dy}{N_\phi} - \int_{y_t}^{H} \frac{2c' \, dy}{\sqrt{N_\phi}} - \int_{y_t}^{H} \frac{2(u_a - u_w) \tan \phi^b \, dy}{\sqrt{N_\phi}} \tag{12.58}$$

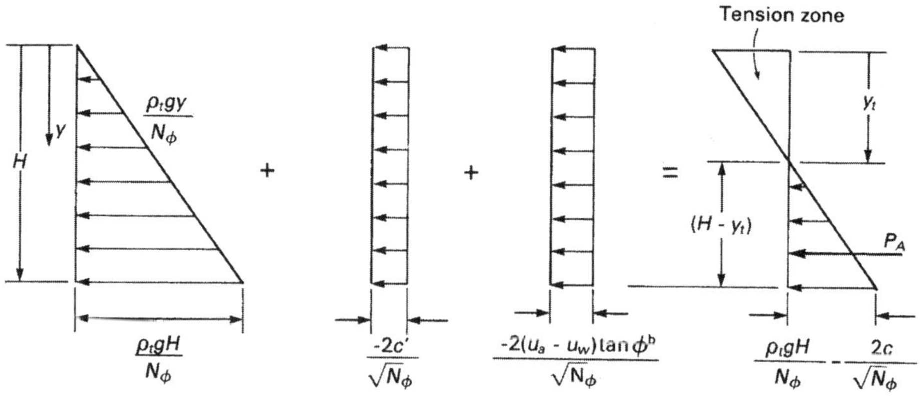

Figure 12.54 Components of active earth pressure showing limits of integration.

Integrating the above equation gives an equation which takes on an extended form of the active force equation for saturated soils:

$$P_A = \frac{\rho g}{N_\phi}\frac{(H^2 - y_t^2)}{2} - \frac{2c'}{\sqrt{N_\phi}}(H - y_t) - \frac{2(u_a - u_w)\tan\phi^b}{\sqrt{N_\phi}}(H - y_t) \qquad (12.59)$$

The active force decreases as matric suction increases. It is possible to reach a condition where the active force is zero. This situation is referred to as an unsupported excavation.

Equation 12.59 assumes that the soil density is a constant with respect to depth. Figure 12.55 illustrates a typical active earth force against a retaining wall as a function of the matric suction of the soil. Two heights of wall are considered (i.e., 8 and 12 m). The density of the soil is assumed to be 1800 kg/m^3. The shear strength parameters of the soil are an effective cohesion of 5 kPa, an effective angle of internal friction of 22°, and a ϕ^b angle of 14°. As the matric suction is increased from 0 to 100 kPa on the 8-m-high wall, the total active pressure is reduced from 206 to 35 kN/m, respectively. In other words, matric suction in the soil has a large influence on the active force against a retaining wall.

The matric suction in a soil may not remain at a constant value with time, and therefore, the active force against the wall may vary. The highest force will occur when the matric suction goes to zero. Even higher earth pressures are exerted on the wall if water is allowed to collect in back of the wall. The results illustrate the importance of maintaining matric suction in the soil. At the same time, it is readily recognized that it may be difficult to ensure the permanency of matric suction in the soil.

Equations for the active earth force can also be written for other situations. Let us consider the case where matric suction decreases linearly with depth. The water table is assumed to be 1 m below the base of the wall. The negative

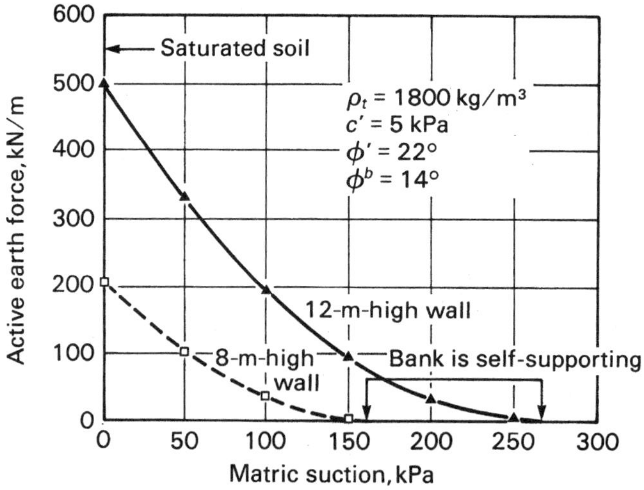

Figure 12.55 Active earth force when matric suction is constant with respect to depth and other variables are as designated.

pore-water pressures are assumed to decrease in a hydrostatic manner:

$$P_A = \frac{\rho g(H^2 - y_t^2)}{2N_\phi} - \frac{2c'(H - y_t)}{\sqrt{N_\phi}} - \left\{\frac{2(u_a - u_w)_h \tan\phi^b}{\sqrt{N_\phi}}\right\}\left(H - \frac{H^2}{2D} - y_t - \frac{y_t^2}{2D}\right) \tag{12.60}$$

Typical results using the above equation are shown in Fig. 12.56. The soil properties and the heights of the wall are the same as for the previous example. The results show that when matric suction decreases linearly with depth (which is often the case), the active earth force does not decrease as rapidly with changing matric suction as when the suctions are constant with depth. Equation 12.60 represents the compressive components of the active force diagram.

A surcharge term must be included in Eq. 12.60 when the soil has tension cracks to a depth y_c. Figure 12.57 shows the variation in the active earth force for the above example with tension cracks 2 m deep. The results show that when other variables remain constant, tension cracks increase the active earth force. In all of the above cases, the active earth force corresponding to zero matric suction refers to a saturated soil with zero pore-water pressures.

The point of application of the resultant force can be computed by considering the line of action of the force associated with each component of the active earth pressure diagram. The point of application becomes lower on the wall as the matric suction of the soil increases. The active earth force is zero when the tension zone depth is equal to the height of the wall. That is, the soil should stand without any support if the height of the vertical excavation is less than the tension zone depth.

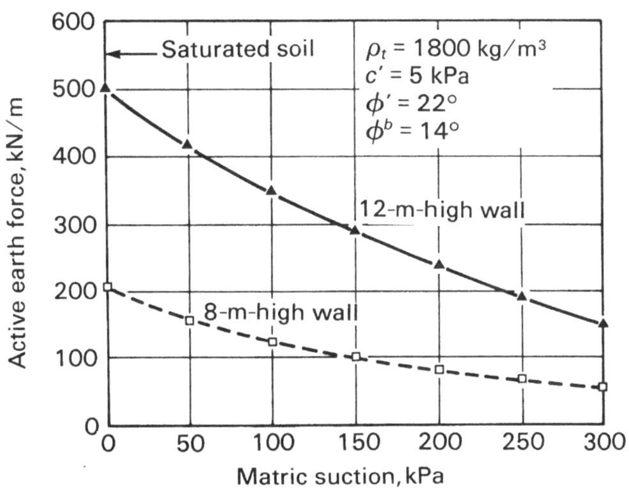

Figure 12.56 Active earth force when matric suction decreases linearly to a water table that is 1 m below base of wall.

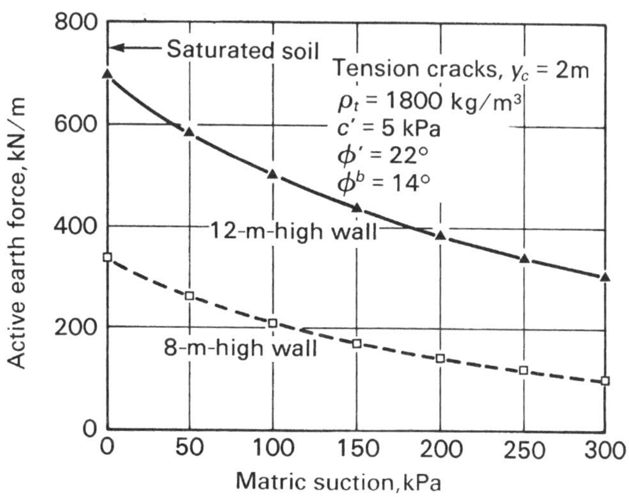

Figure 12.57 Active earth force when soil has tension cracks and linear decrease in matric suction to water table that is 1 m below base of wall.

12.3.16 Passive Earth Force

The passive earth force P_p against a smooth wall is equal to the passive pressure integrated over the depth under consideration. Equation 12.53 can be integrated over its entire depth when the matric suction is constant with respect to depth:

$$P_p = \frac{\rho g N_\phi H^2}{2} + 2c'\sqrt{N_\phi}H + 2(u_a - u_w)\tan\phi^b\sqrt{N_\phi}H \tag{12.61}$$

If the matric suction decreases linearly with respect to depth in a hydrostatic manner, Eq. 12.61 can be integrated with respect to depth to get the total passive force:

$$P_p = \frac{\rho g N_\phi H^2}{2} + 2c'\sqrt{N_\phi}H + 2(u_a - u_w)_h \tan\phi^b\sqrt{N_\phi}\left(H - \frac{H^2}{2D}\right) \tag{12.62}$$

Tension cracks are assumed to close for the passive earth force case. Equations 12.61 and 12.62 can be solved using typical soil properties to illustrate the effect of matric suction on the passive earth force.

The soil properties and wall dimensions used to illustrate the variation in active earth force with matric suction are again used to compute the passive earth force. Figure 12.58 shows the passive earth force for the case of constant matric suction with respect to depth. The passive earth force is shown to increase significantly as the matric suction of the soil increases. The point of application of the resultant force can be computed using the line of action of the force associated with each component of the passive earth pressure diagram.

Figure 12.59 shows the passive earth force developed against a smooth wall when the suction decreases linearly with depth. The water table is 1 m below the bottom of the

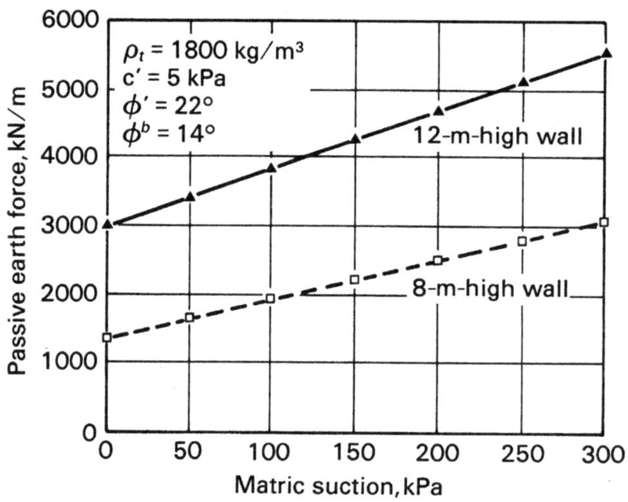

Figure 12.58 Passive earth force when matric suction is constant with respect to depth and other variables are as designated.

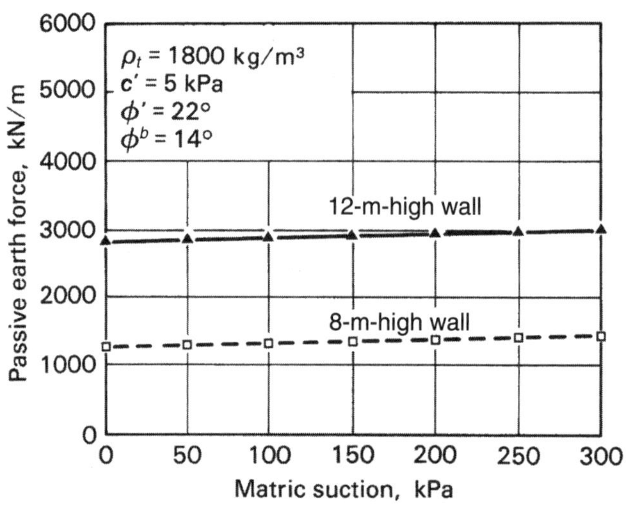

Figure 12.59 Passive earth force when matric suction decreases linearly to a water table that is 1 m below base of wall.

wall and the pore-water pressure distribution is assumed to be hydrostatic. It can be seen that the passive earth force in this case is less sensitive to suction changes than when the suction is a constant value with respect to depth.

12.3.17 Effect of Changes in Suction on Active and Passive Earth Pressure

Changes in the environment may result in water entering the retaining wall backfill. The water results in a change in the active or passive pressure. A decrease in matric suction results in an increase in the pressure against the wall for the active earth pressure case. The magnitude of the pressure developed depends on whether the wall moves in response to the pressure change. If the wall cannot move or does not move by a sufficient amount, the pressure developed against the wall can become even greater than the computed active pressure corresponding to the saturated soil case.

Let us consider compacted clay placed behind a retaining wall as shown in Fig. 12.60. Changes in stresses at a specific depth can be visualized using an extended Mohr-Coulomb diagram (Fig. 12.61). Circle A shows the net vertical and net horizontal pressures at a depth y for the active pressure state. Matric suction decreases as the water content in the soil increases. The horizontal pressure will follow a stress path over to the stress circle A_1 under unrestricted movement of the wall. The magnitude of the pressure against the wall is increased to the active pressure state for a saturated soil.

Let us suppose, on the other hand, that the retaining structure is fixed and the initial stress state is represented by the at-rest condition shown by circle C. As the soil becomes wet and the matric suction decreases, the horizontal pressure change will follow the stress path to circle C_1. The horizontal pressure becomes greater than the vertical pressure at this point.

Let us also consider the possibility where the initial at-rest coefficient of earth pressure is 1.0. The stress changes in the soil could now follow a path to circle D_1 where the final

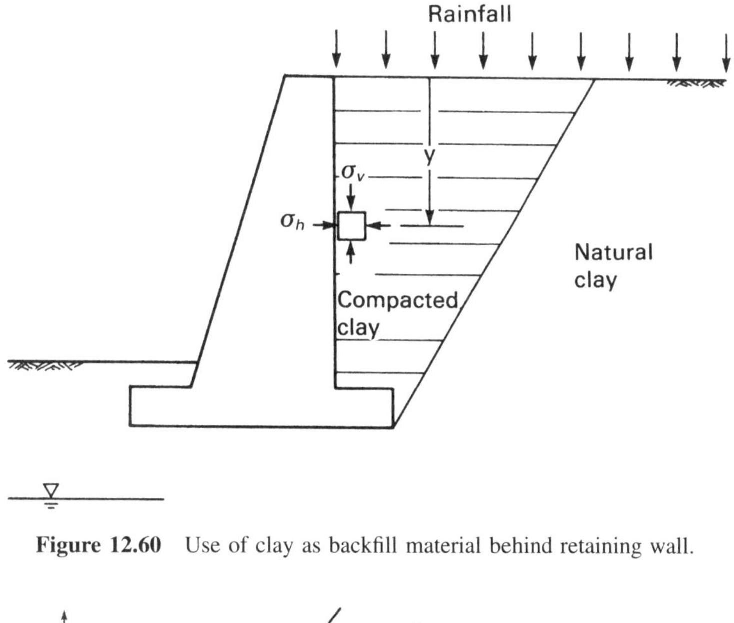

Figure 12.60 Use of clay as backfill material behind retaining wall.

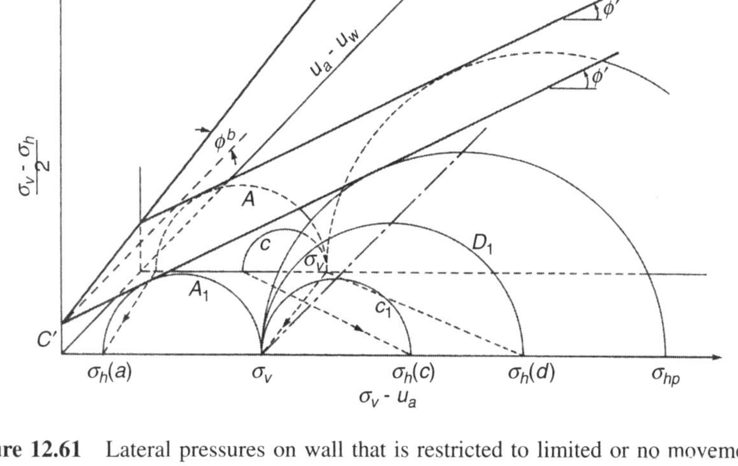

Figure 12.61 Lateral pressures on wall that is restricted to limited or no movement.

horizontal pressure is even higher than that shown for the previous case (i.e., C_1).

The exact stress path followed by the soil during wetting is dependent upon (1) the initial at-rest pressure state, (2) the deformation moduli of the soil, and (3) the rigidity of the retaining structure. In other words, the problem is one involving soil-structure interaction. It would be necessary to use a numerical technique such as the finite element method to more closely model changes in stress state.

There are undoubtedly conditions in practice where backfill with a material that has high matric suction is placed against a rigid wall or where a heavy structural member is cast against clay with an at-rest coefficient greater than 1.0. The horizontal stresses can become as large as the passive resistance of the soil in these circumstances. Generally, this situation would correspond to the passive resistance of the saturated soil. When this happens, either the structural member may fail or the soil may fail in shear. A structural member placed against an expansive soil must be designed to not move or resist the passive earth pressure.

12.3.18 Relationship between Swelling Pressures and Earth Pressures

The question can be asked, "What is the relationship between the above earth pressures and the swelling pressure of a soil?" Suppose the swelling pressure of the soil was measured in an oedometer using the "constant-volume" procedure. The measured swelling pressure corresponds to conditions of no volume change (i.e., vertically or horizontally). The swelling pressure will be a function of the initial stress state and the change in matric suction. However, it is difficult to relate the swelling pressure to the active or passive earth pressure states. The stress paths followed in the two situations are different.

The vertical stress will be the overburden pressure in the case of an earth-retaining structure. Its magnitude is constant while both the vertical and horizontal movement is restricted in the laboratory oedometer test (i.e., using the constant-volume testing technique). To more closely simulate the in situ retaining structure, it would be necessary to subject the laboratory specimen to a vertical pressure corresponding to a particular depth and immerse the specimen in water. It would then be necessary to measure the lateral swelling pressure of the soil in order to obtain an indication of the horizontal pressure against a retaining structure. Clearly, it is difficult to apply conventional swelling pressure measurements to retaining wall design. Also, several oedometer tests would need to be run to represent the respective depths.

The stress paths shown in Fig. 12.61 assist in understanding why walls with cohesive backfills may undergo large movements during the life of the structure. As the soil dries, a crack may open between the soil and the wall. Dust and debris collect in the opening, partially filling the crack. As the matric suction decreases during wetter seasons, the soil swells and pushes the wall until the resistance of the wall is in equilibrium with the soil mass. This cycle may be repeated many times over the years, gradually moving the wall.

12.4 BEARING CAPACITY

The bearing capacity of unsaturated soils, like other plasticity problems, can be viewed as an extension of the approach used in saturated soil mechanics. The unsaturated soil can be visualized as having a cohesion consisting of two components. One component is the effective cohesion, and the other component is cohesion associated with matric suction. With this concept in mind, the conventional bearing capacity theory can be applied to unsaturated soils.

Prandtl (1921) analyzed the case of a strip footing with a smooth base located at ground surface (Fig. 12.62). The load Q was increased until the strip footing penetrated the soil. The unit pressure q_f at this point was called the ultimate bearing capacity. The loading of the strip footing gives rise to an active pressure zone immediately below the footing and a passive zone where the soil pushes laterally and upward. The intermediate portion of the slip surface is defined by a logarithmic spiral. The soil is assumed to be weightless. The rigorous and demanding nature of the solution to this idealized problem has led to the consideration of other approximate solutions.

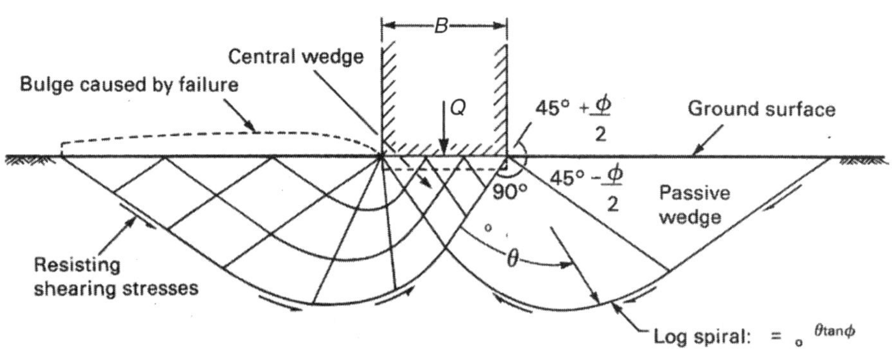

Figure 12.62 Idealized bearing capacity for failure geometry below footing using Prandtl log spiral (from Prandtl, 1921).

12.4.1 Terzaghi Bearing Capacity Theory

The Terzaghi bearing capacity formulation assumes that the base of the footing is rough and the slip surface is bounded by a straight line and a logarithmic spiral (Fig. 12.63a). The load Q at failure is calculated by considering the forces on the sliding mass of soil (Terzaghi, 1943).

The soil above the footing base was treated as a surcharge pressure for foundations embedded at some depth (Fig. 12.63b). The bearing capacity equation consists of three components: The first component, q_γ, takes into account the weight of the soil and the passive earth pressure block. This portion of bearing capacity can be written as follows:

$$q_\gamma = \tfrac{1}{2}\rho g B N_\gamma \qquad (12.63)$$

where:

N_γ = proportionally bearing capacity factor.

The second component is related to the cohesion of the soil:

$$q_c = cN_c \qquad (12.64)$$

where:

N_c = cohesion bearing capacity factor and
c = total cohesion of the soil.

The cohesion value used in the analysis is dependent on how the shear strength of the soil is defined. The third component takes into account the surcharge effect of the soil above the base of the footing:

$$q_q = \rho g D_f N_q \qquad (12.65)$$

where:

N_q = surcharge bearing capacity factor.

The ultimate bearing capacity of a soil can be expressed as the sum of the above-mentioned components:

$$q_f = \tfrac{1}{2}\rho g B N_\gamma + cN_c + \rho g D_f N_q \qquad (12.66)$$

where:

N_γ, N_c, N_q = dimensionless coefficients related to angle of internal friction.

The bearing capacity factors computed by Terzaghi (1943) are shown in Fig. 12.64. The bearing capacity equation 12.66 was derived for strip footings but was further refined to accommodate various shapes of the footing:

$$q_f = \lambda_\gamma \rho g \frac{B}{2} N_\gamma + \lambda_c c N_c + \rho g D_f N_q \qquad (12.67)$$

where:

$\lambda_\gamma, \lambda_c$ = shape factors.

The shape factors for circular footings are $\lambda_\gamma = 0.6$ and $\lambda_c = 1.3$ (Terzaghi, 1943). The shape factors for rectangular footings are $\lambda_\gamma = 1 - 0.2(B/L)$ and $\lambda_c = 1 + 0.2(B/L)$ (Skempton, 1951), where B is the width of the footing and L the length of the footing.

The ultimate bearing capacity can be reduced to an allowable bearing capacity through use of an appropriate factor of safety. Shallow foundations are of significant interest in

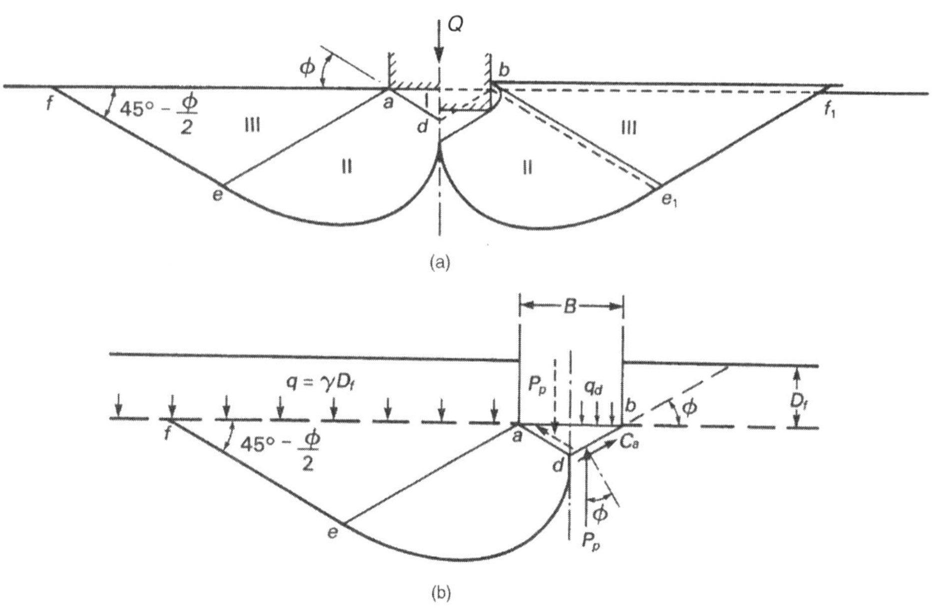

Figure 12.63 Boundaries of zone of plastic equilibrium after failure of soil beneath continuous footing: (a) rough-based footing; (b) rough-based footing with surcharge (from Terzaghi, 1943).

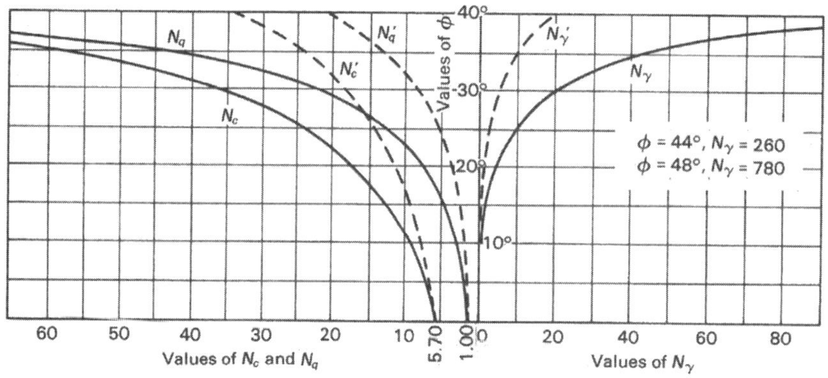

Figure 12.64 Relationship between angle of internal friction, ϕ, and Terzaghi's bearing capacity factors (from Terzaghi, 1943).

unsaturated soil mechanics. The above discussion illustrates that the same plasticity theory applies for both saturated and unsaturated soils. The most important information required when using the bearing capacity theory for unsaturated soils pertains to the selection of appropriate shear strength parameters and a reasonable design matric suction value.

12.4.2 Selection of Soil Parameters and Design Suction

The shear strength parameters used in engineering practice for bearing capacity design for saturated soils have ranged from effective shear strength parameters to the undrained shear strength, c_u. In other words, the design can be based on either an effective stress analysis or a total stress analysis. Likewise, it is possible to use either a stress state variable approach or a total stress approach when dealing with unsaturated soils.

12.4.3 Stress State Variable Approach

The shear strength parameters for a soil with matric suction are (1) effective angle of internal friction ϕ', (2) effective cohesion c', and (3) angle of shear strength change with respect to matric suction, ϕ^b. It is also possible to use a nonlinear shear strength function with respect to matric suction; however, a linear shear strength parameter will be used to present the primary concepts of bearing capacity in unsaturated soils. The linear shear strength parameters are based on the assumption that the failure surface of the soil is planar.

Two of the shear strength parameters can be combined with the result that the shear strength equation has a form similar to that used for saturated soils. The required shear strength parameters for a conventional bearing capacity design are the total cohesion c [i.e., $c = c' + (u_a - u_w) \tan \phi^b$] and the effective angle of internal friction ϕ'. It is necessary to estimate the ϕ^b angle for the soil and have an approximate design matric suction value. It is difficult to assess an appropriate design value for matric suction. It is reasonable to use a measured value of matric suction only if there were some assurance that the matric suction in the soil can be maintained. Generally, a factor of safety would be applied to measured values of matric suction. Another option is to consider the likely long-term hydrostatic condition of the groundwater table (Fig. 12.65).

Footings are commonly placed well above the groundwater table. It may be reasonable to assume that negative pore-water pressures will be maintained immediately below a footing if adequate surface and subsurface drainage is provided around the structure. It must be realized that there may be fluctuations in the groundwater table as a result of building the structure. The groundwater table may also be lowered, but

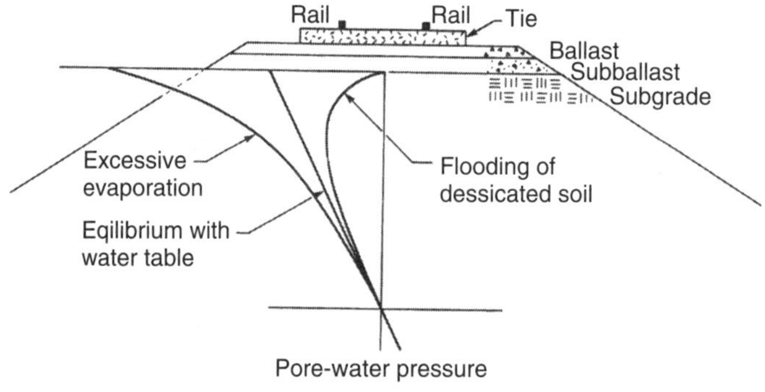

Figure 12.65 Pore-water pressure profiles that might exist below a spread footing.

more commonly, the water table will rise due to excessive watering of the vegetation surrounding the building.

There are situations where the groundwater table is far below the ground surface and the hydrostatic profile is not reasonable to apply for design purposes. Measurements of in situ suction below the footings of existing structures in the vicinity may provide some indication for a reasonable design matric suction value. Average matric suctions are generally found to be a function of the microclimate in the vicinity of the structure (Richards, 1967). The long-term suctions immediately below shallow footings are often 100 kPa or greater. These suctions can significantly contribute to the shear strength of the soil. Decisions regarding the design matric suction value depend on local experience and the microclimate in the region.

Figure 12.66 illustrates the effect of various matric suction values on bearing capacity. The following soil parameters were selected for the analysis: (1) an effective angle of internal friction ϕ' of 20°, (2) an effective cohesion c' of 5 kPa, and (3) a friction angle with respect to matric suction, ϕ^b, of 15°. The density of the soil is 1830 kg/m^3. The design is for strip footings with widths of 0.5 and 1.0 m. The footings are assumed to be placed at a depth of 0.5 m. The bearing capacity factors from Fig. 12.64 are as follows: $N_\gamma = 5.0$, $N_c = 17.5$, and $N_q = 8.0$.

The computations show that for a footing width of 0.5 m and no matric suction in the soil, the bearing capacity is 182 kPa. About 48% of the bearing capacity arises from the effective cohesion of the soil. When the matric suction is increased to 100 kPa (i.e., total cohesion equal to 32 kPa), the bearing capacity increases to 655 kPa. Now, 85% of the bearing capacity is due to the total cohesion component. The results show a similar trend for the wider footing.

The main observation is that the matric suction dramatically increases the bearing capacity of the soil. It is useful to construct a plot similar to that shown in Fig. 12.66 when attempting to select a suitable allowable bearing capacity in unsaturated soils.

12.4.4 Total Stress Approach

The total stress approach is similar to the procedure commonly used in geotechnical engineering practice. The geotechnical engineer may not normally view his or her design as one involving the behavior of an unsaturated soil. Let us assume that the site under consideration involves a clayey soil with a groundwater table well below the proposed depth for the footings. Typical engineering practice can be described as follows. A field investigation is conducted in which samples are obtained at a variety of depths. The samples are brought to the laboratory, extruded, and tested for unconfined compressive strength. The data are interpreted and a design compressive strength is selected.

The compressive strength is divided by 2 to give an undrained shear strength for the soil of c_u. The angle of friction is taken as zero and the bearing capacity equation is solved. It is easy for the engineer to lose sight of the fact that the soil had matric suction (or negative pore-water pressure) that was holding the specimen together. The shear strength of the soil specimen tested in the laboratory was a function of the negative pore-water pressure in the soil (i.e., matric suction). There was also a change in pore-water pressure resulting from unloading the soil during sampling. Changes in pore-water pressure during sampling may be small relative to the in situ negative pore-water pressure, but the suction changes result in an increase in the strength of the soil.

The shear strength measured in the laboratory reflects the matric suction in the soil. The extended Mohr-Coulomb failure envelope can be used to visualize the relationship between the unconfined compression test results and the shear strength defined in terms of matric suction. Figure 12.67 illustrates a possible stress path followed during an unconfined compression test (i.e., stress path $\overline{AB}$), where the matric suction is assumed to remain constant during the test. The unconfined compression test results can be translated onto the strength envelope to represent the stress state at failure. The undrained shear strength c_u can be mathematically written in terms of the extended Mohr-Coulomb fail-

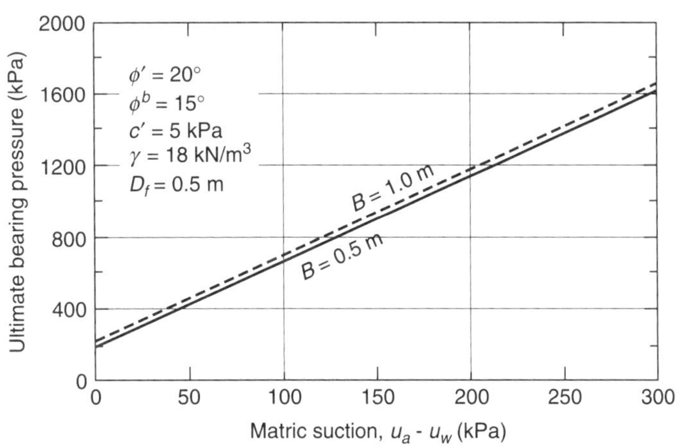

Figure 12.66 Bearing capacity of strip footing for various values of matric suction.

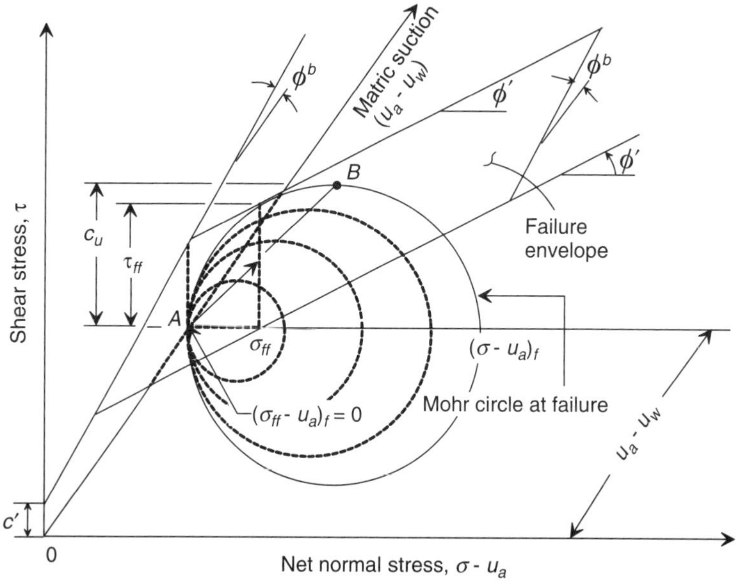

Figure 12.67 Typical stress path followed during unconfined compression test on unsaturated soil with soil suction.

ure envelope:

$$c_u \approx c' + (\sigma_f - u_a)_f \tan\phi' + (u_a - u_w)_f \tan\phi^b \quad (12.68)$$

where:

c_u = undrained shear strength.

The in situ matric suction can later increase or decrease in response to changes in the climatic conditions such as evaporation, precipitation, and lawn watering. The undrained shear strength of the soil will also change and its change can be expressed as follows:

$$\Delta c_u = \Delta(u_a - u_w)\tan\phi^b \quad (12.69)$$

where:

Δc_u = change in undrained shear strength due to matric suction change and

$\Delta(u_a - u_w)$ = change in matric suction due to drying and wetting.

The bearing capacity of clayey soil is often computed using the undrained shear strength c_u in accordance with the total stress approach (i.e., the $\phi = 0$ approach). Applying the $\phi = 0$ condition to Eq. 12.68 gives the ultimate bearing capacity of the clay in terms of its undrained shear strength (i.e., $q_f = c_u N_c$).

Let us consider a clay with an initially measured undrained shear strength of c_{u0} and an initial ultimate bearing capacity of q_{f0} (i.e., $N_c c_{u0}$). A change in matric suction in the field, $\Delta(u_a - u_w)$ (i.e., an increase or decrease), will result in a change in the undrained shear strength, Δc_u, as expressed in Eq. 12.69. There will also be a change in the ultimate bearing capacity, Δq_f (i.e., $N_c\ \Delta c_u$), and the final bearing capacity can be written as the sum of the initial capacity and the change in capacity (i.e., $q_f = q_{f0} + \Delta q_f$).

Figure 12.68 illustrates the possible variation in the ultimate bearing capacity of clay due to matric suction changes. The clay had an initial measured undrained shear strength c_{u0} of 50 kPa and a ϕ^b angle of 15°. The initial computed bearing capacity of the clay, q_{f0}, was 285 kPa for a strip footing or 342 kPa for a square footing. An average change in matric suction is assumed in order to compute the changes in the ultimate bearing capacity of the clay. The ϕ^b angle is assumed to remain constant. Figure 12.68 shows that an increase in soil matric suction increases bearing capacity, while a decrease in matric suction reduces bearing capacity.

The percent change in the ultimate bearing capacity can be related to the change in matric suction as follows:

$$\frac{\Delta q_f}{q_{f0}} = \frac{\Delta(u_a - u_w)}{c_{u0}}\tan\phi^b \quad (12.70)$$

where:

$\Delta q_f / q_{f0}$ = percent change in the ultimate bearing capacity and

$\Delta(u_a - u_w)/c_{u0}$ = percent change in the matric suction with respect to the initial undrained shear strength.

Results of the above equation are plotted in Fig. 12.69 for various ϕ^b values. The relationship is applicable to all shapes of footings since the analysis does not depend on N_c. For a ϕ^b value of 15°, the ultimate bearing capacity will increase or decrease by 27% when the matric suction

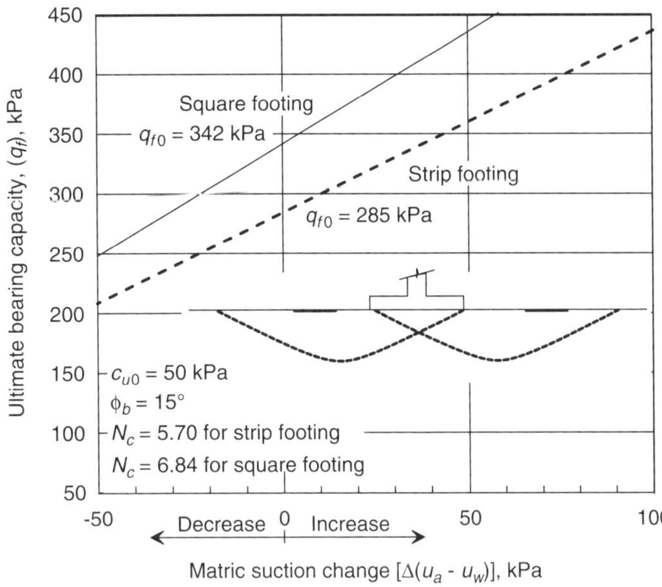

Figure 12.68 Variations in ultimate bearing capacity as a result of matric suction changes (from Rahardjo and Fredlund, 1992).

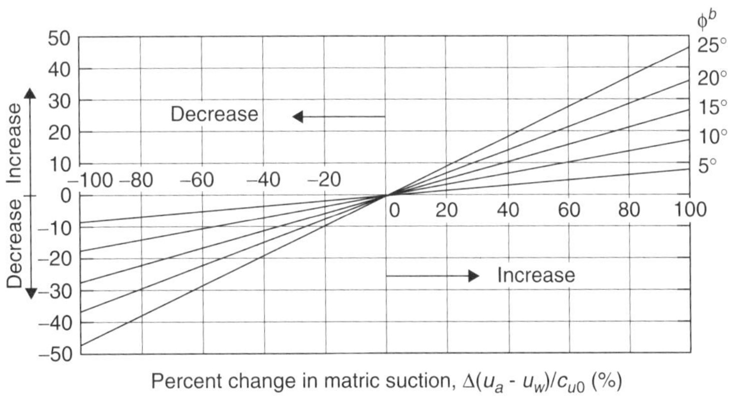

Figure 12.69 Variation in ultimate bearing capacity with variation in matric suction (from Rahardjo and Fredlund, 1992).

changes as much as the initial undrained shear strength [i.e., $\Delta(u_a - u_w) = 100\%$ of c_{u0}]. The percent change in the ultimate bearing capacity will be higher for higher ϕ^b values.

12.4.5 Bearing Capacity of Layered Systems

The concepts of bearing capacity are also relevant for same man-made earth structures. These applications involve the design of highways, railways, and airport systems. Highways and airport runways commonly consist of an asphalt layer, a base, and a subbase layer overlying the subgrade soil. The railway system consists of the track structure (i.e., rails and ties) resting on ballast and subballast layers. In both cases, it is the bearing capacity of the subgrade which is of primary importance.

The environmental conditions change throughout the year for the highway, railway, or airport, resulting in a change in the matric suction in the subsoil. "Road bans" may be placed on highways in the spring of the year because of the low matric suction in the subgrade (i.e., the low bearing capacity of the road).

Highway, railway, and airport designs have not commonly been viewed as bearing capacity designs. There appears to have been a lack of understanding of how to assess the shear strength of a compacted subgrade soil with matric suction. More recently there has been renewed interest in a bearing capacity approach in transportation applications.

The wheel loads applied to a highway surface are transmitted through the base and subbase onto the underlying subgrade. The theory of elasticity can be used to compute the stresses transmitted to the top of the subgrade. The subgrade may undergo a bearing capacity failure if the transmitted stresses are high. Broms (1965) suggested the use of a modified bearing capacity approach for the design of highway structures.

Train loads are transmitted through the rail-tie system and the ballast to the subgrade. The rail-tie system is a complex loading system. The stresses applied to the subgrade are of primary interest. Hanna and Meyerhof (1980) analyzed a layered system similar to a railway structure and suggested simplifications that could be applied to a conventional bearing capacity analysis in order to accommodate the rail system.

The shear strength of the subgrade of a highway, railway, or airport structure can be expressed in terms of the proposed equation for the shear strength of an unsaturated soil. An important aspect when describing unsaturated soil shear strength is the assessment of the design matric suction. For design purposes, it is necessary to have an indication of how low the matric suction may go during any year. Even matric suction values below 100 kPa can produce a substantial increase in the bearing capacity of a soil. Studies by van der Raadt (1988), Sattler et al., (1989), and Majerison et al., (2001) have provided a better understanding of seasonal variations in matric suction in highway and railway subgrades.

12.5 SLOPE STABILITY

Slope stability analyses have become a common analytical tool for assessing the factor of safety of natural and man-made slopes. Any one of several methods can be used to analyze the stability of a slope. There are a series of two-dimensional methods of slices that are commonly used in engineering practice. These methods are based upon the principles of statics (i.e., static equilibriums of forces and/or moments) without giving consideration to the amount of movement in the soil mass.

Several basic assumptions and principles are used in formulating the limit equilibrium methods of analysis. These methods of analysis can readily be extended to embrace the behavior of unsaturated soil. There are also more recent methods of analysis that involve the calculation of the stress state throughout the overall soil mass through the process of "switching on" the gravity forces. The calculation of gravity forces methods are discussed following the presentation of the method of slices for unsaturated soils.

Effective shear strength parameters (i.e., c' and ϕ') are generally used when performing slope stability analyses on saturated soils. The shear strength contribution from the negative pore-water pressures above the groundwater table are often ignored by setting negative pore-water pressures to zero. It may be a reasonable assumption to ignore negative pore-water pressures for situations where the major portion of the slip surface is below the groundwater table. However, for situations where the groundwater table is deep or where the concern is over the possibility of a shallow failure surface, negative pore-water pressures should not be ignored.

A clearer understanding of the role of negative pore-water pressures (or matric suctions) in increasing the shear strength of the soil has been forthcoming in recent years. Reasonable assumptions can be made with regard to plausible distributions of negative pore-water pressures that might exist over a time period, and each situation can be analyzed as part of a parametric slope stability study. It is quite appropriate to perform slope stability analyses that include the shear strength contribution from the negative pore-water pressures. Comparisons can be made of the calculated factors of safety corresponding to situations where negative pore-water pressures are included and ignored. Slope stability analyses that include negative pore-water pressures can be formulated as an extension of conventional limit equilibrium analyses.

Several aspects of a slope stability study remain the same for soils with positive pore-water pressures (e.g., saturated soils) and soils with negative pore-water pressures (e.g., unsaturated soils). For example, the nature of the site investigation, the identification of the soil strata, and the measurement of the total unit weight remain the same for saturated and unsaturated soils. On the other hand, extensions to conventional testing procedures are required for the characterization of the shear strength properties of the soil. Estimation procedures can be used to approximate the shear strength envelope for unsaturated soils. The analytical tools used to incorporate negative pore-water pressures and calculate the factor of safety need to be extended to embrace unsaturated soil behavior.

12.5.1 Free-Body Diagram for Slope Stability Analysis

The computation of the factor of safety of a slope must start with specifying geometric details related to the free-body diagram. The need for a free-body diagram should go without saying; however, the free-body diagram is the place where a slope stability analysis encounters its first difficulty. The boundaries of the sliding mass are not known at the start of the method-of-slices analysis. The shape and location of the most critical slip surface are unknowns.

The ground surface and the stratigraphic divisions between soil layers may be known for the two-dimensional cross section under consideration. However, the shape and location of the most critical slip surface that can be passed through the soil mass are not known, and as a result the limits of the free-body diagram are not known.

An assumption must be made regarding the shape of the slip surface. The slip surface may be planar, circular, or composite in shape. The selection of the shape of the potential slip surface is mainly dependent on whether the slope consisted of a purely frictional or cohesive material. The effect of geological considerations also leads to consideration of slip surfaces that were noncircular in shape (i.e., composite).

The second feature of the free-body diagram that needs to be determined is the location of the critical slip surface. Early slope stability studies sought to find the critical slip

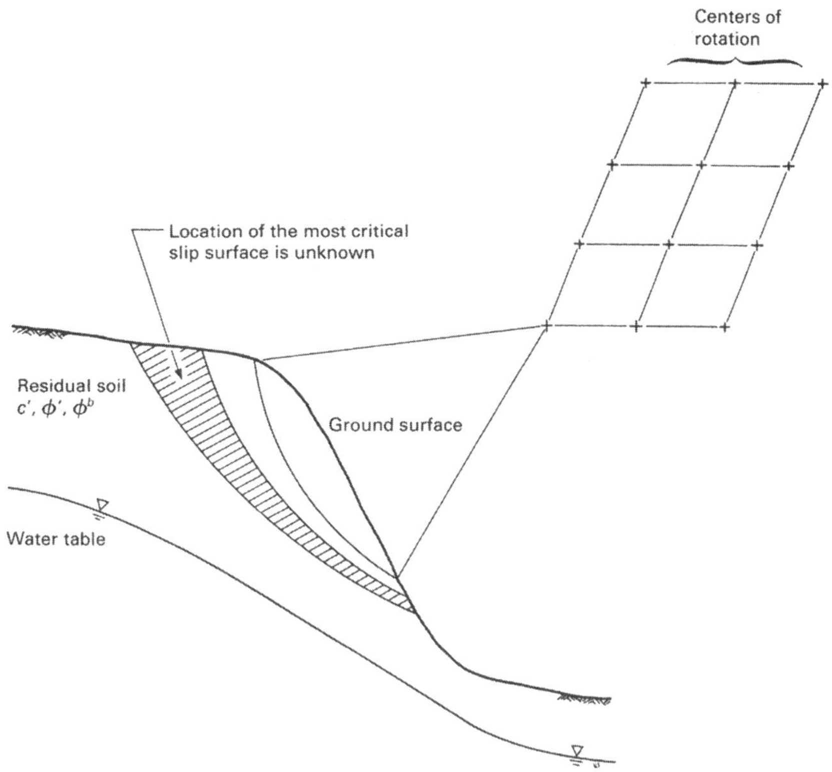

Figure 12.70 Typical steep natural slope with deep groundwater table.

surface through use of a trial-and-error search technique. A large number of potential slip surfaces were analyzed through the soil mass, and the slip surface which yielded the minimum computed factor of safety was selected as the critical slip surface. There were a variety of methods of slices that made use of a similar procedure to calculate the minimum factor of safety of a slope.

The analytical formulations proved to be indeterminate and as a result there were numerous methods of slices proposed over time. The formulation showed that there were more unknowns than there were equations to solve the problem. Consequently, a variety of assumptions were made to render the slope stability analysis determinate. The name of researcher(s) who made each of the assumptions was usually attached to the method for calculating the factor of safety. The summation of forces in orthogonal directions, as well as the moment equilibrium, were given consideration in each analysis; however, it is possible to consider all the methods of slices within a general limit equilibrium framework (Fredlund and Krahn, 1977). The primary difference between the various methods of slices proposed can be traced to differences in the manner in which the normal force at the base of a slice was calculated. There was also a difference in the manner in which overall static equilibrium was satisfied.

Computational advancements have led to the realization that other approaches can also be used to perform a slope stability analysis. For example, it was realized that it was possible to use numerical modeling and the "switch-on" of gravity forces to determine the state of stress in a soil mass. It was also possible to utilize optimization techniques developed in other areas of engineering for the determination of the minimum factor of safety of a slope. It also became possible for the shape of the slip surface and the critical location of the slip surface to become part of the overall solution for the minimum factor of safety.

The optimization procedure that has been most commonly used to locate the critical slip surface is referred to as the "dynamic programming" technique. There are distinct advantages associated with the use of optimization techniques for the calculation of the factor of safety of a slope; however, engineers have gained extensive experience in the method-of-slices procedures. It is important that studies be undertaken to verify the similarities and differences that may exist between the two approaches to the analysis of a slope.

Figure 12.71 subdivides methods of slope stability analysis into two broad categories: methods of slices and optimization methods. Methods of slices have been extensively used in engineering practice, but the optimization technique lends itself well to computational procedures. At the core of the optimization technique is the solution of two or more partial differential equations for the computation of the stress states in a soil mass.

The trend in solving all saturated-unsaturated soil problems is toward the solution of one or more partial differential

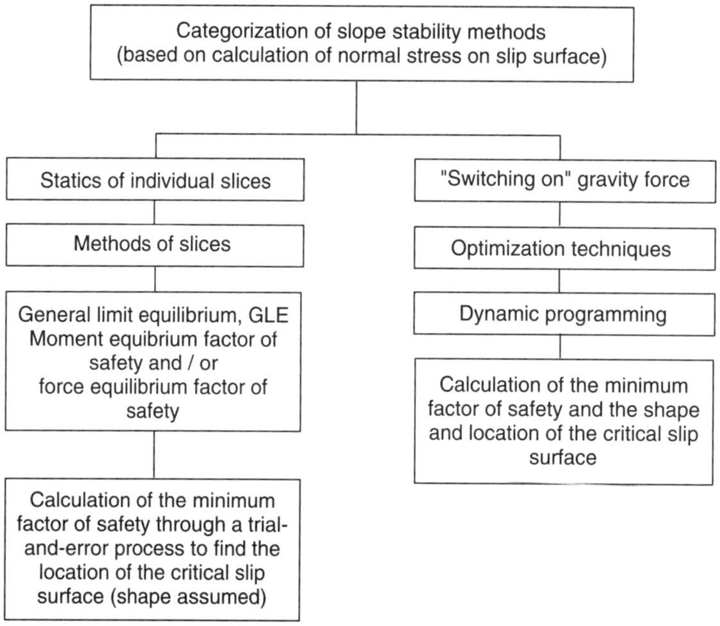

Figure 12.71 Categorization of slope stability methods based on methodology for calculating normal stress at various points along slip surface.

equations through numerical modeling. It is anticipated that with time optimization techniques will increase in popularity for the computation of the factor of safety of a slope. Pore-water pressures are also commonly assessed using numerical modeling and the results can readily be combined with the slope stability analysis.

12.5.2 Determination of Normal Stress on Slip Surface for Methods of Slices

There are a number of differences between each of the methods of slope stability analysis; however, probably of most significant difference is the manner in which the normal force is computed at the base of a slice. The various methods of slices give rise to differences in the calculated normal force at the base of a slice because of the assumptions that are made with regard to interslice forces. Each method of slices considers vertical force equilibrium on a slice and the primary variable influencing the calculation of the normal force is gravity force (i.e., unit weight of the soil), immediately above the base of the slip surface.

The optimization technique switches on the gravity force over the entire soil mass and uses a typical stress-strain constitutive model for the computation of the stress state at all points in the soil mass. This process can be performed one time and the analysis for the factor of safety becomes determinate from the standpoint of calculating the normal force at the base of a slice.

The commonly used methods of slices can be extended to embrace the analysis involving unsaturated soils. Optimization techniques can also be extended to consider unsaturated soils. The use of the optimization procedure is explained after the methods-of-slices procedure.

12.5.3 General Procedure for Methods of Slices

A study of the stability of a slope with negative pore-water pressures involves the following steps: (1) a survey of the elevation of the ground surface on a selected section perpendicular to the slope, (2) advancement of several boreholes to identify the stratigraphy and obtain undisturbed soil samples, (3) laboratory testing of undisturbed soil specimens to obtain suitable shear strength parameters for each stratigraphic unit (i.e., c', ϕ', and ϕ^b), and (4) measurement or estimation of negative pore-water pressures above the groundwater table. These steps provide the input data for performing a stability analysis. However, the location and shape of the most critical slip surface is unknown (Fig. 12.70). A particular combination of actuating and resisting forces along a slip surface of unknown shape and location will produce the lowest factor of safety for the slope. The location of the slip surface is known in the case of a slope that has already failed.

The methods of slices are presented within the context of the general limit equilibrium (GLE) method which considers force and moment equilibrium. The GLE equations can then be specialized to each of the common methods of slices. Common to all the methods of slices is an assumption regarding the shape of the slip surface.

12.5.4 Trial-and-Error Search for Critical Slip Surface When Using Methods of Slices

The shape of the unknown slip surface is generally assumed. The location of the critical slip surface is determined using a trial-and-error procedure. Let us first assume that the shape of the slip surface is circular. A grid of centers can be selected and the radius is varied at each center of rotation. The end result provides coverage of all possible circular geometries. When the slip surface takes on a composite shape (i.e., part circular and part linear), it is still possible to use a grid of centers and varying radii to search for the critical slip surface (Fredlund, 1981a). In addition, a general shape can also be assumed using a series of straight lines to define the slip surface.

Various automatic search routines have been programmed to reduce the number of computations. Some routines start with an assumed center and seek to find the critical center by moving in a zigzag manner (Wright, 1974). Others use an initially coarse grid of centers which rapidly converge to a critical center (Fredlund, 1981a).

There is probably no analysis conducted by geotechnical engineers which has received more attention than the limit equilibrium methods of slices for the computation of the factor of safety (Fredlund, 1980a). The limit equilibrium method of slices has proved to be a useful and reliable tool for assessing the stability of a slope. Its "track record" is impressive for cases where the shear strength properties of the soil and the pore-water pressure conditions have been adequately assessed (Sevaldson, 1956; Kjaernsli and Simons, 1962; Skempton and Hutchinson, 1969; Chowdhury, 1980). The limit equilibrium methods of slices require a limited amount of computer input but can quickly perform extensive trial-and-error searches for the critical slip surface.

12.5.5 General Limit Equilibrium Method

The GLE method provides a general theory wherein other methods can be viewed as special cases. These elements of statics along with the failure criterion are insufficient to make the slope stability problem determinate (Morgenstern and Price, 1965; Spencer, 1967). Either additional elements of physics or an assumption regarding the direction or magnitude of some of the forces is required to render the problem determinate. The GLE method utilizes an assumption regarding the direction of the interslice forces. This approach has been widely adopted in limit equilibrium methods (Fredlund and Krahn, 1977). The various limit equilibrium slope stability methods that assume a direction for the interslice forces have been shown to be special cases of the GLE method (Fredlund et al., 1981).

Calculations for the stability of a slope are performed by dividing the soil mass above the assumed slip surface into vertical slices. The forces acting on a slice within the sliding soil mass are shown in Fig. 12.72 for a composite slip surface. The forces are designated for a unit width (i.e., direction perpendicular to movement) of the slope. The variables are defined as follows:

W = total weight of the slice of width b and height h,
N = total normal force on the base of the slice,
S_m = shear force mobilized on the base of each slice,
E = horizontal interslice normal forces (the L and R subscripts designate the left and right sides of the slice, respectively),
X = vertical interslice shear forces (the L and R subscripts designate the left and right sides of the slice, respectively),
R = radius for a circular slip surface or the moment arm associated with the mobilized shear force S_m for any shape of slip surface,
f = perpendicular offset of the normal force from the center of rotation or from the center of moments,
x = horizontal distance from the centerline of each slice to the center of rotation or to the center of moments,
h = vertical distance from the center of the base of each slice to the uppermost line in the geometry (i.e., generally ground surface),
a = perpendicular distance from the resultant external water force to the center of rotation or the center of moments (the L and R subscripts designate the left and right sides of the slope, respectively),

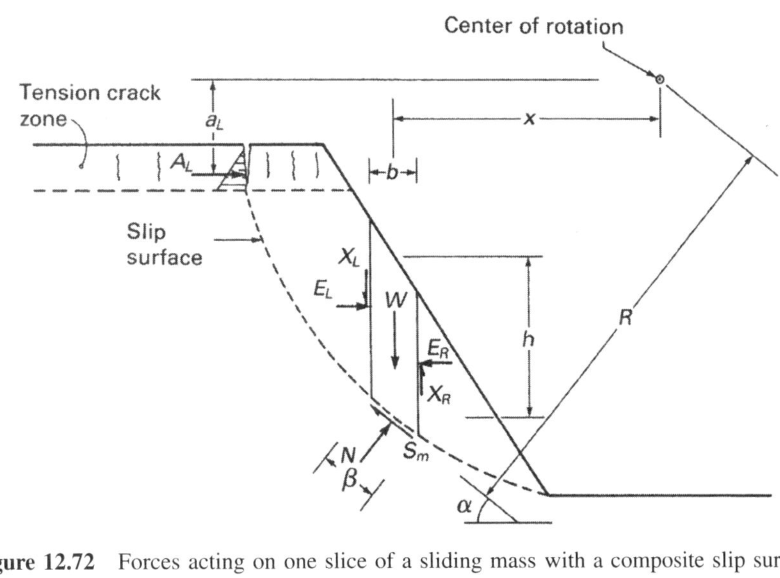

Figure 12.72 Forces acting on one slice of a sliding mass with a composite slip surface.

A = resultant external water forces (the L and R subscripts designate the left and right sides of the slope, respectively),

α = angle between the tangent to the center of the base of each slice and the horizontal, (when the angle slopes in the same direction as the overall slope of the geometry, α is positive, and vice versa), and

β = sloping distance across the base of a slice.

The example shown in Fig. 12.72 is typical of a steep slope with a deep groundwater table. The crest of the slope is highly desiccated and the tension cracks are assumed to be filled with water. The tension crack zone is assumed to have no shear strength and the presence of water in this zone produces an external water force A_L. The assumed slip surface in the tension crack zone is a vertical line. The depth of the tension crack is generally estimated (Spencer, 1968, 1973). The weight of the soil in the tension crack zone acts as a surcharge on the crest of the slope. The external water force A_L is computed as the hydrostatic force on a vertical plane. An external water force can also be present at the toe of the slope as a result of partial submergence. This water force is designated as A_R.

12.5.6 Shear Force Mobilized Equation

The mobilized shear force at the base of a slice can be written using the shear strength equation for an unsaturated soil:

$$S_m = \frac{\beta}{F_s}\left\{c' + (\sigma_n - u_a)\tan\phi' + (u_a - u_w)\tan\phi^b\right\} \tag{12.71}$$

where:

σ_n = total stress normal to the base of a slice, and

F_s = factor of safety, defined as the factor by which the shear strength parameters must be reduced in order to bring the soil mass into a state of limiting equilibrium along the assumed slip surface.

The factors of safety for the cohesive parameter (i.e., c') and the frictional parameters (i.e., $\tan\phi'$ and $\tan\phi^b$) are assumed to be equal for all soils involved and for all slices. The components of the mobilized shear force at the base of a slice are illustrated in Fig. 12.73. The contributions from the total stress and the negative pore-water pressures are separated by using the friction angles, ϕ' and ϕ^b.

It is possible to consider the matric suction term as part of the cohesion of the soil since matric suction can then be visualized as increasing the cohesion of the soil. As a result, the conventional factor-of-safety equations take on a form similar to that of a saturated soil. The mobilized shear force at the base of a slice, S_m, will then have the following form:

$$S_m = \frac{\beta}{F_s}\left\{c + (\sigma_n - u_a)\tan\phi'\right\} \tag{12.72}$$

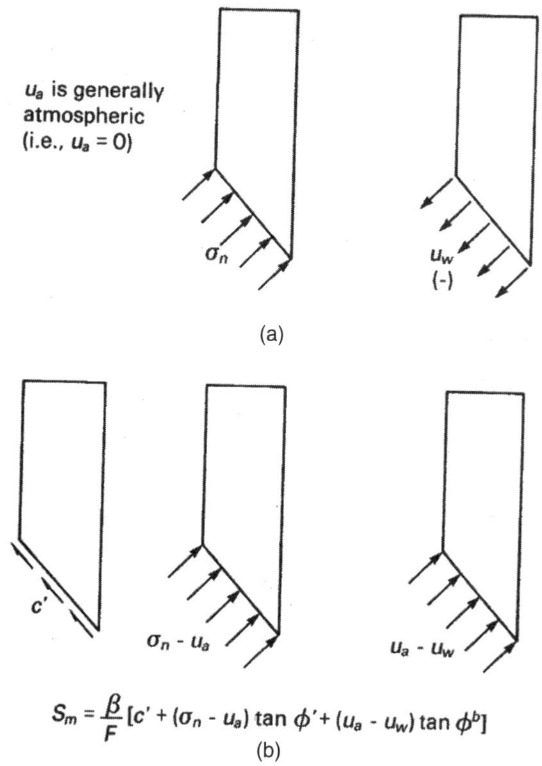

Figure 12.73 (a) Acting total and pore-water stresses and (b) shearing resistance at the base of a slice.

where:

c = total cohesion of the soil, which has two components [i.e., $c' + (u_a - u_w)\tan\phi^b$].

Combining the cohesion components has the advantage that the shear strength equation retains the conventional form used for saturated soils. Consequently, it is possible to use a computer program written for analyzing saturated soils when considering the analysis of an unsaturated soil slope. The soil in the negative pore-water pressure region must be subdivided into several discrete layers with each layer having a constant cohesion value. The pore-air and pore-water pressures must be set to zero in this case. This approach has the disadvantage that cohesion is not considered as a continuous function and total cohesion must be manually computed for each layer.

The slope stability derivations directly calculate the shear strength contribution from negative pore-water pressures. The mobilized shear force defined using Eq. 12.71 is used throughout the derivation. The effect of partial submergence at the toe of the slope, the effect of seismic loading, and external line loads are not shown in the derivations.

12.5.7 Normal Force Equation

The normal force at the base of a slice, N, is derived by summing forces in the vertical direction:

$$W - (X_R - X_L) - S_m \sin\alpha - N\cos\alpha = 0 \tag{12.73}$$

Substituting Eq. 12.71 into Eq. 12.73 and replacing the $\sigma_n\beta$ term with N gives

$$W-(X_R-X_L)-\left\{\frac{c'\beta}{F_s}+\frac{N\tan\phi'\beta}{F_s}-\frac{u_a\tan\phi'\beta}{F_s}+\frac{(u_a-u_w)\tan\phi^b\beta}{F_s}\right\}\sin\alpha-N\cos\alpha=0 \quad (12.74)$$

Rearranging Eq. 12.74 gives rise to the normal force equation:

$$N=\frac{W-(X_R-X_L)-\dfrac{c'\beta\sin\alpha}{F_s}+u_a\dfrac{\beta\sin\alpha}{F_s}(\tan\phi'-\tan\phi^b)+u_w\dfrac{\beta\sin\alpha}{F_s}\tan\phi^b}{m_\alpha} \quad (12.75)$$

where

$$m_\alpha=\cos\alpha+F_s/\sin\alpha\tan\phi' \quad (12.76)$$

The factor of safety F_s in Eq. 12.75 is equal to the moment equilibrium factor of safety F_m when solving for moment equilibrium and is equal to the force equilibrium factor of safety F_f when solving for force equilibrium. In most cases, the pore-air pressure u_a is atmospheric, and as a result, Eq. 12.75 reduces to the form

$$N=\frac{W-(X_R-X_L)-\dfrac{c'\beta\sin\alpha}{F_s}+u_w\dfrac{\beta\sin\alpha}{F_s}\tan\phi^b}{m_\alpha} \quad (12.77)$$

If the base of the slice is located in saturated soil (i.e., soil with positive pore-water pressures), the $\tan\phi^b$ term in Eq. 12.77 becomes equal to $\tan\phi'$. The normal force equation then reverts to the conventional normal force equation used in saturated slope stability analysis. The angle ϕ^b can be used whenever the pore-water pressure is negative and the angle ϕ' can be used whenever the pore-water pressure is positive. The ϕ^b angle can also be considered to be equal to ϕ' at low-matric-suction values up to the air-entry value of the soil while a lower ϕ^b angle is used at high matric suctions (Fredlund et al., 1987).

The vertical interslice shear forces X_L and X_R in the normal force equation can be computed using an interslice force function.

12.5.8 Factor of Safety with Respect to Moment Equilibrium

Two independent factor-of-safety equations can be derived: one with respect to moment equilibrium and the other with respect to horizontal force equilibrium. Moment equilibrium can be satisfied with respect to an arbitrary point above the central portion of the slip surface. The center of rotation is an obvious center for moment equilibrium for a circular slip surface. The center of moments would appear to be immaterial when both force and moment equilibrium is satisfied (e.g., in the GLE method). The computed factor of safety can vary with the point selected for the summation of moments when only moment equilibrium is satisfied.

Let us consider moment equilibrium for a composite slip surface (Fig. 12.72) with respect to the center of moments:

$$A_La_L+\sum Wx-\sum Nf-\sum S_mR=0 \quad (12.78)$$

Substituting Eq. 12.71 for S_m into Eq. 12.78 and replacing the $(\sigma_n\beta)$ term with N yields

$$A_La_L+\sum Wx-\sum Nf=\frac{1}{F_m}\sum\{c'\beta R+[N\tan\phi'-u_a\tan\phi'\beta+(u_a-u_w)\tan\phi^b\beta]R\} \quad (12.79)$$

where:

F_m = factor of safety with respect to moment equilibrium.

Rearranging Eq. 12.79 yields

$$F_m=\frac{\sum\left\{c'\beta R+\left[N-u_w\beta\dfrac{\tan\phi^b}{\tan\phi'}-u_a\beta\left(1-\dfrac{\tan\phi^b}{\tan\phi'}\right)\right]R\tan\phi'\right\}}{A_La_L+\sum Wx-\sum Nf} \quad (12.80)$$

In the case where the pore-air pressure is atmospheric (i.e., $u_a=0$), Eq. 12.80 has the form

$$F_m=\frac{\sum\left[c'\beta R+\left(N-u_w\beta\dfrac{\tan\phi^b}{\tan\phi'}\right)R\tan\phi'\right]}{A_La_L+\sum Wx-\sum Nf} \quad (12.81)$$

The ϕ^b value can be set equal to the ϕ' value when the pore-water pressure is positive. Equation 12.81 can also be simplified for a circular slip surface as follows:

$$F_m=\frac{\sum[c'\beta+(N-u_w\beta-u_a\beta)\tan\phi']R}{A_La_L+\sum Wx} \quad (12.82)$$

The radius R is constant for all slices, and the normal force N acts through the center of rotation (i.e., $f=0$) when a circular slip surface is considered.

12.5.9 Factor of Safety with Respect to Force Equilibrium

The factor of safety with respect to force equilibrium is derived from the summation of forces in the horizontal direction for all slices:

$$-A_L+\sum S_m\cos\alpha-\sum N\sin\alpha=0 \quad (12.83)$$

The horizontal interslice normal forces E_L and E_R cancel when summed over the entire sliding mass. Substituting Eq. 12.71 for the mobilized shear force S_m into Eq. 12.83 and replacing the $\sigma_n\beta$ term with N gives

$$\frac{1}{F_f}\sum[c'\beta\cos\alpha + (N\tan\phi' - u_a\tan\phi'\beta + (u_a - u_w)\tan\phi^b\beta)] = A_L + \sum N\sin\alpha \quad (12.84)$$

where:

F_f = factor of safety with respect to force equilibrium.

Rearranging Eq. 12.84 yields

$$F_f = \frac{\sum\left\{c'\beta\cos\alpha + \left[N - u_w\beta\frac{\tan\phi^b}{\tan\phi'} - u_a\beta\left(1 - \frac{\tan\phi^b}{\tan\phi'}\right)\right]\tan\phi'\cos\alpha\right\}}{A_L + \sum N\sin\alpha} \quad (12.85)$$

In the case where the pore-air pressure is atmospheric (i.e., $u_a = 0$), Eq. 12.85 reverts to the following form:

$$F_f = \frac{\sum\left[c'\beta\cos\alpha + \left(N - u_w\beta\frac{\tan\phi^b}{\tan\phi'}\right)\tan\phi'\cos\alpha\right]}{A_L + \sum N\sin\alpha} \quad (12.86)$$

The ϕ^b value is equal to the ϕ' value when the pore-water pressure is positive. Equation 12.86 is the same for both circular and composite slip surfaces.

12.5.10 Interslice Force Function

The interslice normal forces E_L and E_R are computed from the summation of horizontal forces on each slice:

$$E_R - E_L = N\cos\alpha\tan\alpha - S_m\cos\alpha \quad (12.87)$$

Substituting Eq. 12.73 for the $N\cos\alpha$ term in Eq. 12.87 gives the following equation:

$$E_R - E_L = [W - (X_R - X_L) - S_m\sin\alpha]\tan\alpha - S_m\cos\alpha \quad (12.88)$$

Rearranging Eq. 12.88 gives

$$E_R = E_L + [W - (X_R - X_L)]\tan\alpha - \frac{S_m}{\cos\alpha} \quad (12.89)$$

The interslice normal forces are calculated by integrating across the slope. The left interslice normal force on the first slice is equal to any external water force which may exist, A_L, or it is set to zero when there is no water present in the tension crack zone.

The assumption is made that the interslice shear force X can be related to the interslice normal force E by a mathematical function (Morgenstern and Price, 1965):

$$X = \lambda f(x)E \quad (12.90)$$

where:

$f(x)$ = a functional relationship which describes the manner in which the magnitude of X/E varies across the slip surface and
λ = a scaling constant which represents the percentage of the function $f(x)$ used for solving the factor-of-safety equations.

There are a number of reasonable mathematical functions that can be assumed to describe the direction of the interslice forces. However, an unrealistic assumption of the interslice force function can result in convergence difficulties when solving nonlinear factor-of-safety equations (Ching and Fredlund, 1983). Morgenstern and Price (1967) suggested that the interslice force function should be related to the shear and normal stresses on vertical slices through the soil mass. Maksimovic (1979) used the finite element method and a nonlinear characterization of the soil to compute stresses in a soil mass. These stresses were then used in the limit equilibrium slope stability analysis.

A generalized interslice force function $f(x)$ was proposed by Fan et al. (1986). The function was based on two-dimensional finite element analyses of a linear elastic continuum. The normal stresses in the x-direction and the shear stresses in the y-direction were integrated along vertical planes within a sliding mass in order to obtain normal and shear forces, respectively. The ratio of the shear force to the normal force along each vertical section provided the direction of the resultant interslice forces. Figure 12.74 illustrates a typical interslice force function for one slip surface through a relatively steep slope.

The analysis of many slopes showed that the interslice force function could be approximated by an extended form of an error function equation. Inflection points were close to the crest and toe of the slope. The slope of the resultant interslice forces was steepest at the midpoint and tended toward zero at some distances behind the crest and beyond the toe of the slope. The mathematical form for the empirical interslice force function can be written as follows:

$$f(x) = e^{-(C^n\omega^n)/2} \quad (12.91)$$

where:

e = base of the natural logarithm, 2.71828,
C = variable to define the inflection points of the function,
n = variable to specify the flatness or sharpness of curvature of the function, and
ω = dimensionless x-position relative to the midpoint of the slope.

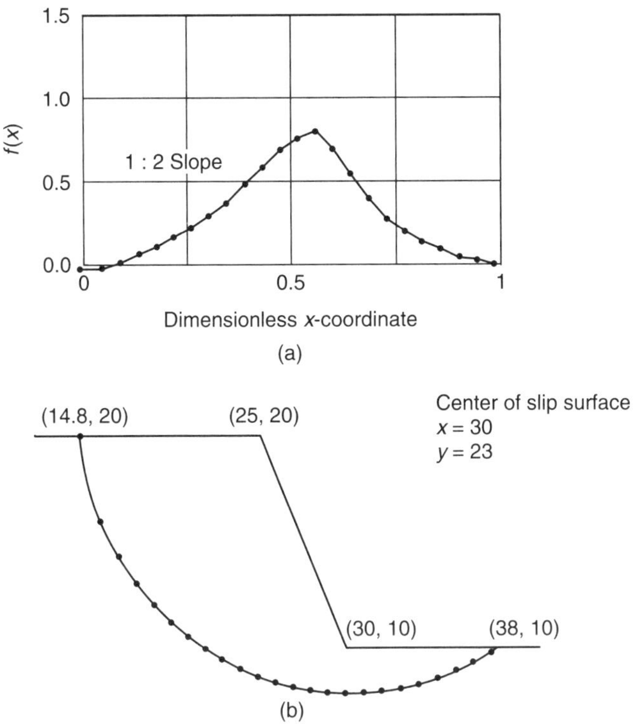

Figure 12.74 (a) Interslice force function for (b) a deep-seated slip surface computed by switching on gravity forces.

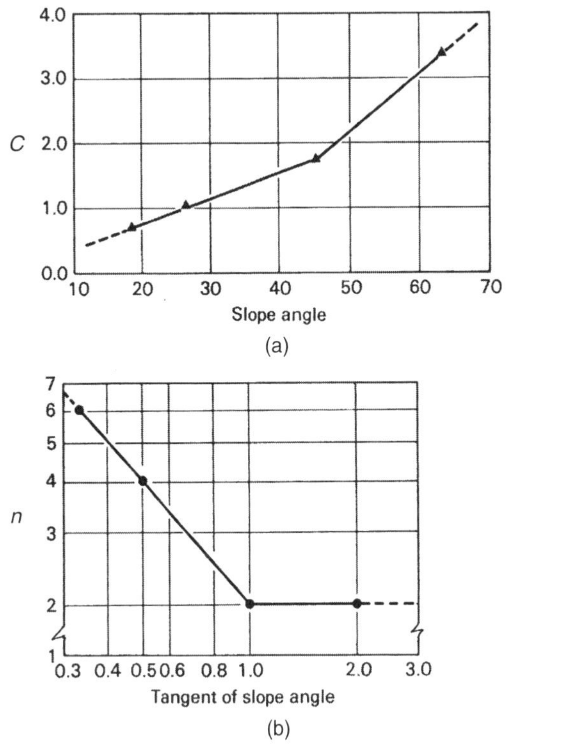

Figure 12.76 Values of C and n coefficients for interslice force function: (a) C versus slope angle; (b) n versus tangent of slope angle.

Figure 12.75 illustrates the definition of the dimensionless distance ω. Elastic, finite element functions have been computed for slip surfaces which are circular. However, the shape of the function is quite similar for composite slip surfaces. The magnitude of the C parameter varies with the slope angle, as does the n variable (Fig. 12.76).

12.5.11 Solving Factor-of-Safety Equations

The factor-of-safety equations with respect to moment and force equilibriums are nonlinear. The factors of safety F_m or F_f appear on both sides of the equations, with the factor of safety being included through the normal force equation (i.e., Eq. 12.77). The nonlinear factor-of-safety equations can be solved using an iterative technique. The factors of safety with respect to moment and force equilibriums can be calculated when the normal force N on each slice is known. The computation of the normal force requires a magnitude for the interslice shear forces X_L and X_R and an estimate of the factor of safety F_s.

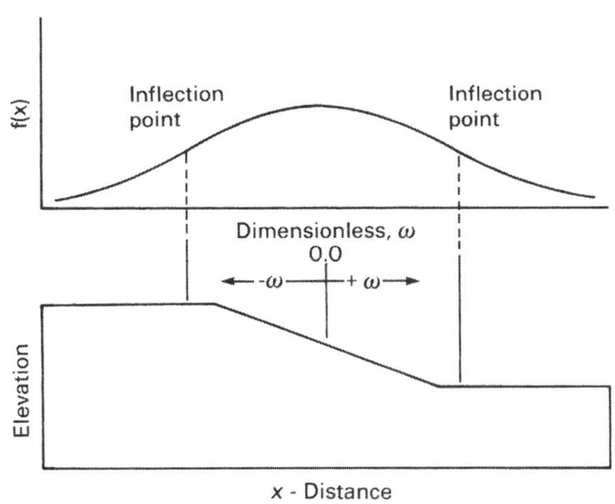

Figure 12.75 Definition of dimensionless distance ω for interslice force function.

For the first iteration, the factor of safety F_s in the normal force equation can be set to 1.0 or estimated from the ordinary method (Fredlund, 1987). The interslice shear forces can be set to 0.0 for the first iteration when computing the normal force. The computed normal force is then used to calculate the factors of safety with respect to moment and force equilibriums (i.e., F_m and F_f).

The next step is to compute the interslice normal forces E_L and E_R in accordance with Eq. 12.89. There are two sets of interslice force calculations: one associated with moment equilibrium and the other associated with force equilibrium. The interslice shear forces in Eq. 12.75 can be set to zero for the first iteration. The computed interslice normal forces E_L and E_R can then be used in the calculation of the interslice shear forces X_L and X_R for each slice according to Eq. 12.91. The selected interslice force function along with a specified λ value is used during the iterative procedure until convergence is achieved.

The computed moment equilibrium factor of safety F_m and the corresponding interslice forces are used to recalculate new values for the normal force N and the moment of equilibrium factor of safety F_m. The updated values for the normal force N and the moment equilibrium factor of safety F_m are then used to revise the interslice normal forces and interslice shear forces (i.e., E_L, E_R, X_L, and X_R) associated with moment equilibrium. The computed force equilibrium factor of safety F_f and the corresponding interslice forces from the first iteration are used to revise the magnitudes of the variables N, F_f, E_L, E_R, X_L, and X_R associated with force equilibrium.

Calculations are stopped when the difference in the factor of safety between two successive iterations is less than the desired tolerance. When the solution has converged, moment and force equilibrium factors of safety corresponding to the selected interslice force function $f(x)$ and the selected λ value are obtained. The analysis can proceed using the same interslice force function $f(x)$ but varying the λ value. Several factor of safety values F_m and F_f associated with different λ values can be obtained and plotted as shown in Fig. 12.77. The moment equilibrium factor of safety F_m does not vary significantly with respect to the λ values for circular slip surfaces. The force equilibrium factor of safety F_f varies with λ values for a circular slip surface. Curves joining the F_m and F_f data intersect at a point where total equilibrium (i.e., moment and force equilibrium) is satisfied.

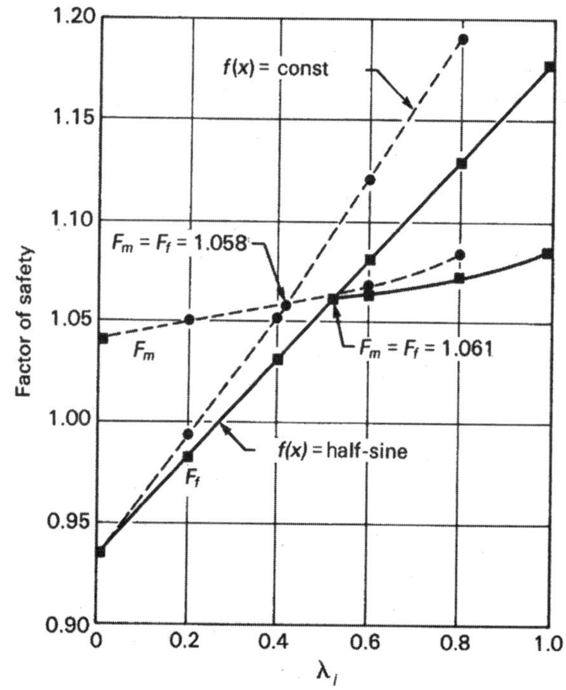

Figure 12.77 Variation of moment and force equilibrium factors of safety with respect to λ.

It is most practical and appropriate to perform an independent analysis for the assessment of the pore-water and pore-air pressures. The computed pore pressures can then be imported into the slope stability analysis as a grid of designated values, as shown in Fig. 12.78. Pore-water pressures can be either positive or negative and an interpolation technique can be used to obtain the pore-water pressure at any designated point (e.g., the base of a slice).

Piezometric lines can also be used to designate the pore-water pressures in a slope (Fig. 12.79). However, care must be exercised when using a piezometric line to designate negative pore-water pressures. The vertical distance from the piezometric line down to a point below the line is equal to the positive pore-water pressure head (i.e., $u_w = h_w \rho_w g$). On the other hand, the vertical distance from the piezometric line up to a point above the line can be considered as the negative pore-water pressure head [i.e., $u_w(-) = h_w \rho_w g$]. The use of a piezometric line can lead to pore-water pressures which are in considerable error when slopes are steep and the gradient along the water table is high.

12.5.12 Input of Pore-Water and Pore-Air Pressures

Several procedures have been used to designate the pore-water pressures when performing a slope stability analysis where saturated soils are involved. Some of the suggested procedures are not appropriate when using unsaturated soils with negative pore-water pressures. For example, the pore pressure coefficients r_u become a negative value that tends to negative infinity as the ground surface is approached.

12.5.13 Relationship between GLE and Other Methods of Slices

The GLE method of slices can be specialized to correspond to various limit equilibrium methods. The various methods of slices can be categorized in terms of the conditions of static equilibrium satisfied and the assumption used with respect to the interslice forces. Table 12.7 summarizes the conditions of static equilibrium satisfied by the various

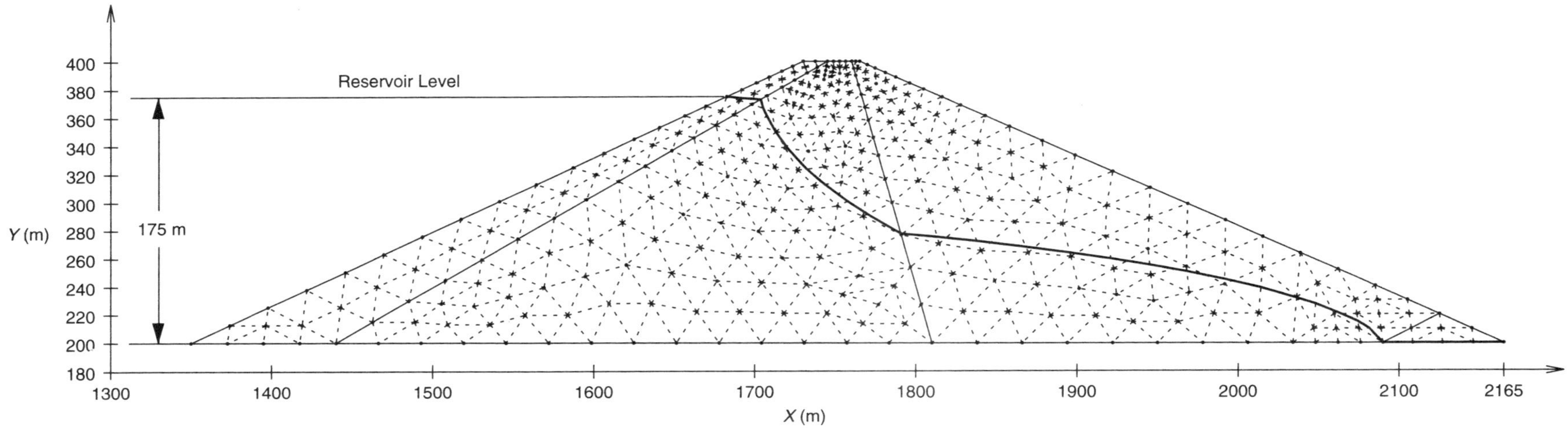

Figure 12.78 Grid of pore-water pressures designated at nodal points of finite element mesh.

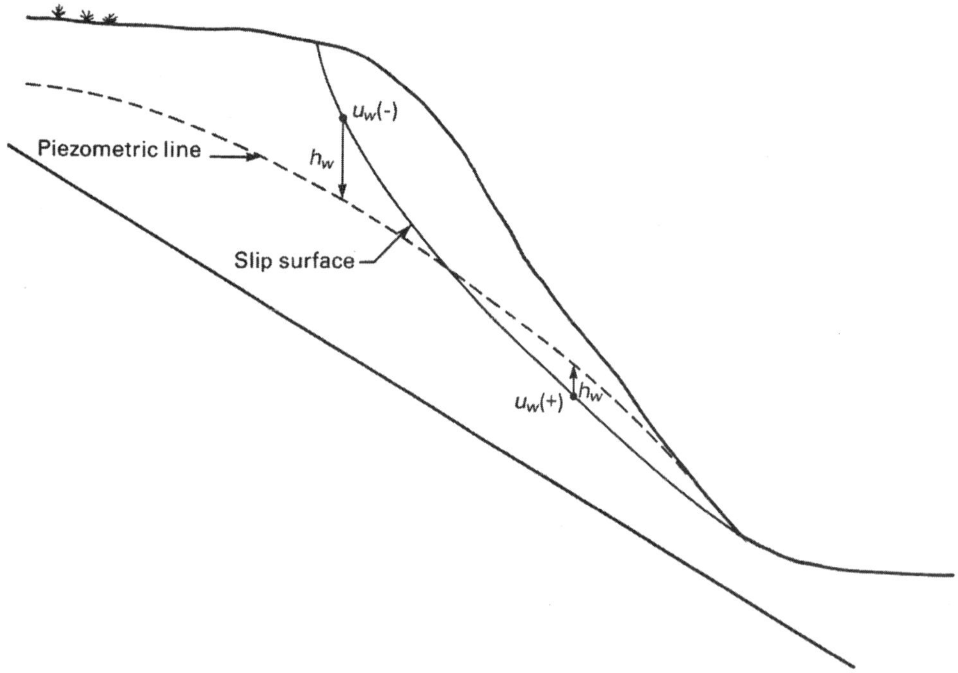

Figure 12.79 Use of piezometric line to designate pore-water pressures.

Table 12.7 Elements of Static Equilibrium Satisfied by Various Limit Equilibrium Methods

Method	Force Equilibrium		Moment Equilibrium
	First Direction[a] (e.g., Vertical)	Second Direction[a] (e.g., Horizontal)	
Ordinary or Fellenius	Yes	No	Yes
Bishop's simplified	Yes	No	Yes
Janbu's simplified	Yes	Yes	No
Janbu's generalized	Yes	Yes	[b]—
Spencer	Yes	Yes	Yes
Morgenstern-Price	Yes	Yes	Yes
Corps of Engineers	Yes	Yes	No
Lowe-Karafiath	Yes	Yes	No

[a]Any of two orthogonal directions can be selected for the summation of forces.
[b]Moment equilibrium is used to calculate interslice shear forces.

methods of slices. The statics used in each of the methods of slices for computing the factor of safety are summarized in Table 12.8. The ordinary and Bishop's simplified methods use moment equilibrium, while Janbu's simplified, Janbu's generalized, the Lowe and Kariafiath, and the Corps of Engineers methods use force equilibrium in computing the factor of safety. The Spencer and Morgenstern-Price methods satisfy both moment and force equilibrium when computing the factor of safety. The Spencer and Morgenstern-Price methods are similar in principle to the GLE method which satisfies force and moment equilibrium when calculating the factor of safety. The interslice force assumptions used in simulating the various methods are given in Table 12.8.

12.6 OPTIMIZATION PROCEDURES TO SOLVE FOR FACTOR OF SAFETY

The assumptions related to the interslice force function in limit equilibrium methods are not required when a finite element stress-strain analysis is used to obtain the normal and shear stresses acting at the base of slices (Fredlund and Scoular, 1999). A stress analysis provides normal and shear stresses through the use of the finite element numerical method where gravity forces are switched on. The equation for calculating the factor of safety becomes linear. Assumptions regarding the shape of the critical slip surface are not necessary when an appropriate optimization technique is introduced into the analysis.

Table 12.8 Comparison of Commonly Used Methods of Slices

Method	Factors of Safety: Moment Equilibrium	Factors of Safety: Force Equilibrium	Interslice Force Assumption
Ordinary	x		$X = 0,\ E = 0$
Bishop's Simplified	x		$X = 0,\ E \geq 0$
Janbu's Simplified		x	$X = 0,\ E \geq 0$
Janbu's Generalized		x	$X_R = E_R \tan \alpha_t - (E_R - E_L)t_R/b$[a]
Spencer	x	x	$X/E = \tan \theta$[b]
Morgenstern-Price	x	x	$X/E = \lambda f(x)$
Lowe-Karafiath		x	X/E =Average slope of ground and slip surface
Corps of Engineers		x	X/E =Average ground surface slope

[a] α_t = angle between the line of thrust across a slice and the horizontal.
t_R = vertical distance from the base of the slice to the line of thrust on the right side of the slice.
[b] θ = angle of the resultant interslice force from the horizontal.

There have been a number of optimization techniques that result in a variety of approaches for the determination of the shape and the location of the critical slip surface (Celestino and Duncan, 1981; Nguyen, 1985; Chen and Shao, 1988; Greco, 1988). Each approach has its own advantages and shortcomings.

The "dynamic programming" method combines a finite element stress analysis with an optimization technique to provide a slope stability solution. Dynamic programming overcomes some of the difficulties associated with limit equilibrium methods of slices. The dynamic programming approach requires the designation of stress-strain soil properties such as Poisson's ratio μ and modulus of elasticity E for the soil mass.

The dynamic programming method for a slope stability analysis is more complex than the method of slices, but the method lends itself well to computer simulation. In 1980, Baker introduced an optimization procedure that utilized the algorithm of the dynamic programming method to determine the critical slip surface. The factor of safety was calculated using the Spencer (1967) method of slices. Yamagami and Ueta (1988) enhanced Baker's (1980) approach by combining the dynamic programming method with a finite element stress analysis for the calculation of the factor of safety. The critical slip surface was assumed to be a chain of linear segments connecting two *state* points located in two successive *stages* (Fig. 12.80). The resisting and the actuating forces used to calculate an *auxiliary function* were determined from stresses interpolated from Gaussian points within the domain of the problem. Zou et al., (1995) proposed an improved dynamic programming technique that used essentially the same method as introduced by Yamagami and Ueta (1988).

The analytical procedure involves the use of a partial differential solver to compute the stress states throughout a soil continuum. The shape and location of the critical slip surface corresponding to a minimum factor of safety is determined through use of the dynamic programming optimization technique. A number of example problems were studied by Pham and Fredlund (2003b), and the results were compared to conventional method-of-slices solutions.

12.6.1 Background to Optimization Theory for Slope Stability

Bellman (1957) introduced a mathematical method called the dynamic programming method. One of the objectives of the dynamic programming method was to maximize or minimize a function. The dynamic programming method has been widely used in various fields other than geotechnical engineering. Baker (1980) applied the optimization technique to slope stability analysis. The definition for the factor of safety when using optimization methods can be written as follows.

For an arbitrary slip surface AB, as shown in Fig. 12.80, the equation for the factor of safety F_s can be written as

$$F_s = \frac{\int_A^B \tau_f \, dL}{\int_A^B \tau \, dL} \tag{12.92}$$

where:

τ = mobilized shear stress along the slip surface,
τ_f = shear strength of the soil, and
dL = an increment of length along the slip surface.

The critical slip surface is approximated by an assemblage of linear segments. Each linear segment connects two

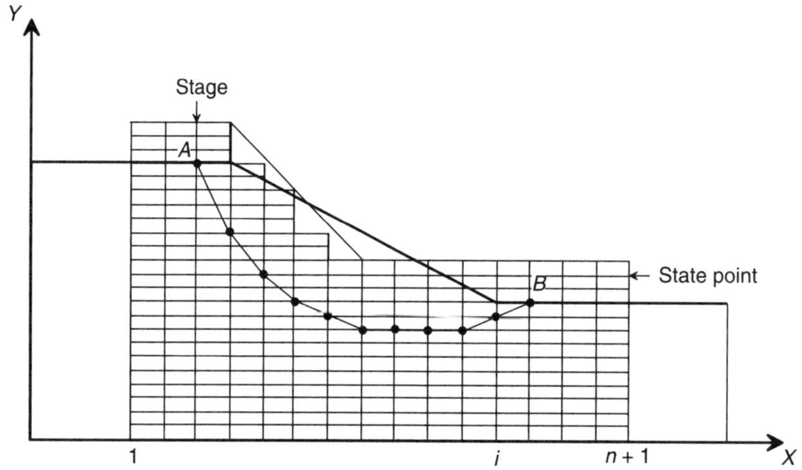

Figure 12.80 Slip surface *AB* selected through use of an optimization technique such as dynamic programming.

state points located in two successive stages. The *stage-state* system forms a grid consisting of rectangular elements called the *search grid.* The rectangular elements formed by the search grid are called *grid elements.* The overall factor of safety for the slip surface *AB* is defined as follows for the discretized soil mass:

$$F_s = \frac{\sum_{i=1}^{n} \tau_{f_i} \Delta L_i}{\sum_{i=1}^{n} \tau_i \Delta L_i} \tag{12.93}$$

where:

n = number of discrete segments,
τ_i = mobilized shear stress,
τ_{f_i} = shear strength, and
ΔL_i = length of ith segment.

12.6.1.1 Theory of Dynamic Programming Method

A minimization technique is necessary to obtain the factor of safety F_s (Eq. 12.93). Baker (1980) showed that the minimum of F_s in Eq. 12.93 can be found by using an *auxiliary function G*. The auxiliary function is also known as the *return function* and can be defined as follows (Fig. 12.81):

$$G = \sum_{i=1}^{n} (R_i - F_s S_i) \tag{12.94}$$

where:

S_i = actuating forces acting on the ith segment of the slip surface,
R_i = resisting forces acting on the ith segment of the slip surface, and
n = total number of discrete segments comprising the slip surface.

The minimum value of the auxiliary function, $G_{\min}$, is defined as

$$G_{\min} = \min \sum_{i=1}^{n} (R_i - F_s S_i) \tag{12.95}$$

The shear strength for a saturated-unsaturated soil can be calculated and applied along the ith segment (Fredlund and Rahardjo, 1993a):

$$\tau_{f_i} = c' + (\sigma_n - u_a) \tan \phi' + (u_a - u_w) \tan \phi^b \tag{12.96}$$

where:

c', ϕ', ϕ^b = shear strength parameters of a saturated-unsaturated soil,
$\sigma_n - u_a$ = net normal stress acting on the ith segment, and
$u_a - u_w$ = matric suction.

The normal and shear stresses acting on the ith segment can be computed from a stress analysis as follows:

$$\sigma_n = \sigma_x \sin^2 \theta + \sigma_y \cos^2 \theta - \tau_{xy} \sin 2\theta \tag{12.97}$$

$$\tau_n = \tau_{xy} (\sin^2 \theta - \cos^2 \theta) - \frac{\sigma_y - \sigma_x}{2} \sin 2\theta \tag{12.98}$$

where:

σ_n = normal stresses acting on the ith segment,
τ_n = shear stresses acting on the ith segment,
θ = inclined angle of the ith segment with the horizontal direction, and
$\sigma_x, \sigma_y, \tau_{xy}$ = normal and shear stresses acting in the x- and y-coordinate directions.

The total stresses within the soil mass can be determined from a finite element stress analysis using any particular soil behavioral model. It can be assumed that stresses are constant within a small grid element if the density of the search grid is sufficiently fine. The constant stresses are signified by stresses at the center points of the grid element. The resisting and actuating forces acting on the ith segment of a slip surface can be calculated as follows (Fig. 12.81):

$$R_i = \sum_{ij=1}^{ne} R_{ij} = \sum_{ij=1}^{ne} \tau_{f_{ij}} l_{ij}$$
$$= \sum_{ij=1}^{ne} [c'_{ij} + (\sigma_n^{ij} - u_a)\tan\phi'_{ij} + (u_a - u_w)\tan\phi^b_{ij}] l_{ij} \quad (12.99)$$

$$S_i = \sum_{ij=1}^{ne} S_{ij} = \sum_{ij=1}^{ne} \tau_{ij} l_{ij} \quad (12.100)$$

where:

ij = a grid element traveled by the ith segment,
$\tau_{f_{ij}}, \tau_{ij}$ = shear strength and shear stress acting at the center point of ij, respectively,
$c'_{ij}, \phi'_{ij}, \phi^b_{ij}$ = strength parameters of the saturated-unsaturated soil within ij,
ne = number of ij, and
l_{ij} = length of the ith segment limited by the boundary of ij.

An *optimal function* $H_i(j)$ obtained at state point $\{j\}$ located in stage $[i]$ is introduced. The optimal function $H_i(j)$ is defined as the minimum of the return function G calculated from a state point for the initial stage to state point $\{j\}$ located in stage $[i]$. According to *the principle of optimality* (Bellman, 1957), the optimal function $H_{i+1}(k)$ obtained at state point $\{k\}$ located in stage $[i+1]$ is defined as

$$H_{i+1}(k) = H_i(j) + G_i(\overline{j,k}) \quad (12.101)$$

where:

$G_i(\overline{j,k})$ = return function calculated from state point $\{j\}$ of stage $[i]$ to state point $\{k\}$ of stage $[i+1]$.

At the initial stage, the value of the optimal function $H_1(j)$ is equal to zero:

$$H_1(j) = 0 \qquad j = 1, \ldots, \mathrm{NP}_1 \quad (12.102)$$

where:

NP_1 = number of state points in the initial stage.

At the final stage (i.e., $i = n+1$), the optimal function $H_{n+1}(k)$ must be equal to the minimum value of the return function G_m, that is;

$$H_{n+1}(j) = G_m = \min \sum_{i=1}^{n} (R_i - F_s S_i) \qquad j = 1, \ldots, \mathrm{NP}_{n+1} \quad (12.103)$$

where:

NP_{n+1} = number of state points located in the final stage.

The *optimal point* in the final stage is defined as the state point at which the calculated optimal function is a minimum. From the optimal state point $\{k\}$ found in the final stage, the optimal state point $\{j\}$ located in the previous stage is also determined. The *optimal path* defined by connecting optimal

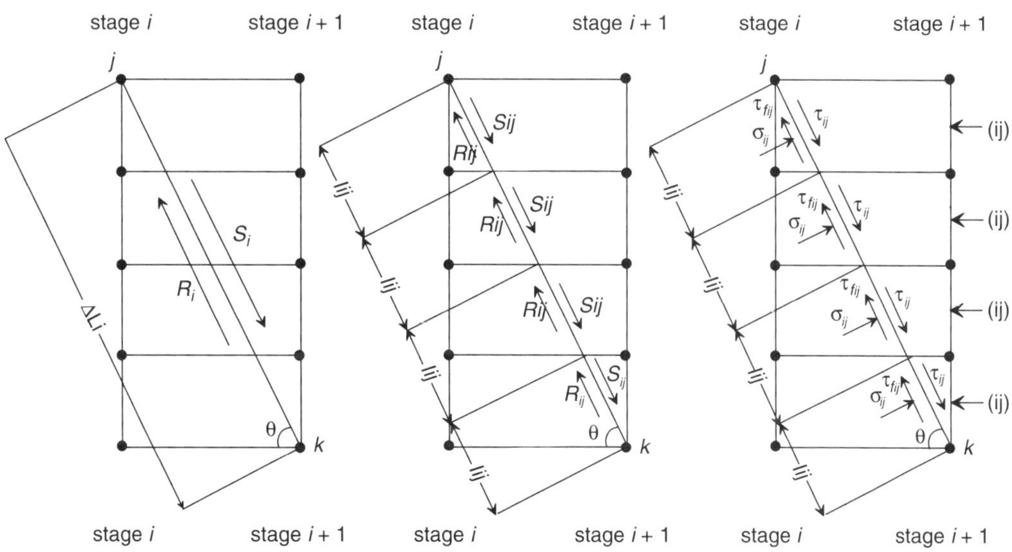

Figure 12.81 Actuating and resisting forces acting on ith segment between segments.

state points located in every stage is eventually found by tracing back from the final stage to the initial stage. This optimal path defines the *critical slip surface.*

The value of the overall factor of safety F_s in Eq. 12.94 has not been defined in advance, and therefore an initial value must be assumed. The trial value of F_s is updated using the value of F_s evaluated after each trial of the search. The optimization process will stop when a predefined convergence is reached.

12.6.1.2 Finite Element Analysis for Stress State

A general partial differential equation solver can be used to solve the partial differential equations for the stress states throughout the soil mass. (The computed strain values are of no interest to this analysis.) For a two-dimensional, plane strain analysis (i.e., $\varepsilon_z = 0$), a soil element subjected to its body forces has partial differential equations representing the stress balance in the x- and y-directions:

$$\frac{\partial \sigma_x}{\partial x} + \frac{\partial \tau_{yx}}{\partial y} + F_x = 0 \tag{12.104}$$

$$\frac{\partial \tau_{xy}}{\partial x} + \frac{\partial \sigma_y}{\partial y} + F_y = 0 \tag{12.105}$$

where:

σ_x, σ_y = normal stress in the Cartesian x- and y-coordinate directions,
τ_{yx}, τ_{xy} = shear stress in x-y planes, and
F_x, F_y = body forces in the x- and y-coordinate directions.

Partial differential equations 12.104 and 12.105 can be solved in conjunction with appropriate specified boundary conditions. The stresses are computed and stored at the Gaussian points over the domain of the problem. The stresses can then be interpolated from the Gaussian nodes to any location on the arbitrary grid that is superimposed for the slope stability analysis. The stresses at the center point of each grid element are determined using the interpolation shape functions. The stress interpolation process is undertaken prior to the performance of the dynamic programming search.

12.6.1.3 Dynamic Programming Solution Procedure

The analytical scheme of the dynamic programming method for performing a slope stability analysis is illustrated in Fig. 12.82 (Pham and Fredlund 2003). A search boundary is defined such that it extends outside the geometry of the problem at the top and bottom of the slope. There needs to be one point near midheight where the search boundary dips below the soil surface of the geometry. This point will force all possible slip surfaces to lie within the soil mass. The stresses at each grid intersection point are computed from the nodal stresses.

The initial factor of safety is assumed to be 1.0 and a search is launched through state points located in the initial stage. Trial segments of the slip surface are formed by connecting all state points of the initial stage to all state points located in the second stage. Values of the optimal function are obtained at all state points of the second stage using Eqs. 12.101 and 12.102 and the assumed factor of safety F_s. The number of optimal functions to be calculated at one state point of the second stage is equal to the number of state points located in the initial stage.

The minimum value of the optimal function is determined at each state point in the second stage. The corresponding state point in the previous stage (i.e., the initial stage for the first segment) is identified. The process proceeds to the next stage with the same routine until the final stage is reached.

A comparison is made of the values of the optimal functions obtained at all state points of the final stage and the determination of the state point at which the corresponding value of the optimal function was a minimum. The determined state point will be the first optimal point of the optimal path. There is then a trace back to the previous stage to find the corresponding state point with the first optimal point. This corresponding state point will be the second optimal point of the optimal path. The trace back to the initial stage is continued to determine the entire optimal path. The actual factor of safety corresponds to the optimal path previously obtained. A new value for the factor of safety is then calculated.

The process is repeated until the difference between the assumed and the actual factor of safety is within the selected convergence criterion δ. The critical slip surface is defined using the entry and exit points which correspond to the optimal path.

12.6.1.4 Advantages of Using Dynamic Programming Procedure

The dynamic programming method combined with a finite element stress analysis forms a valuable means of calculating the factor of safety of a slope. The dynamic programming procedure provides a solution with increased flexibility over limit equilibrium methods of slices (Pham and Fredlund, 2003).

The critical slip surface can be irregular in shape and can be determined as part of the slope stability solution when using the dynamic programming method. The shape of the failure surface must be assumed by the modeler when using methods of slices. No assumption is required regarding the shape or the location of the critical slip surface when using the dynamic programming procedure.

More complex stress-strain models for the soil can be used in the assessment of the stress state in the soil mass. The effect of weather-related environmental conditions such as infiltration (and associated matric suction decreases) can readily be taken into account when using the dynamic programming procedure (Gitirana et al., 2006).

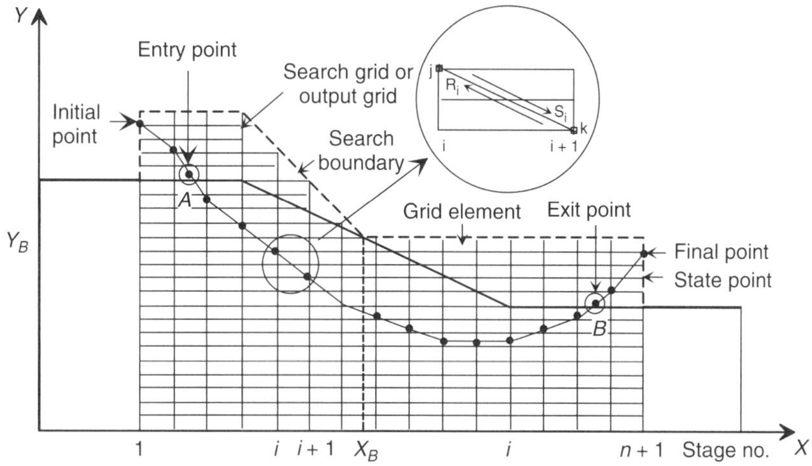

Figure 12.82 Scheme for use of dynamic programming method in slope stability analysis.

12.6.1.5 Restrictions on Shape of Critical Slip Surface

The shape of the critical slip surface must be kinematically admissible. Baker (1980) assumed that the critical slip surface must be concave. Baker (1980) proposed that the first derivative calculated from the crest to the toe of the slope must be greater than or at least equal to zero. Zou et al., (1995) stated that a check must be made to assure that the critical slip surface is kinematically admissible. Kinematical restrictions play an important role in applying the dynamic programming method in slope stability analysis. Using appropriate kinematical restrictions prevents the shape of the critical slip surface from being unreasonable.

Failure takes place when the resisting force and the actuating force along the slip surface are in contrary directions. The resisting force must act in the direction opposite to the mass movement. The actuating force must also be in the same direction as mass movement (Fig. 12.83). If the actuating force calculated is in a contrary direction to the anticipated direction of mass movement, then it is suggested that the entire trial segment in which the actuating force is being calculated should be eliminated from the search. Applying this condition to the optimization procedure eliminates all trial segments that constitute kinky shaped slip surfaces.

12.6.2 Example Solutions Using Dynamic Programming

The results from a number of example problems are presented. The results are compared to factors of safety obtained using other methods of slope stability analysis.

12.6.2.1 Simple Homogeneous Slope

The stability of a simple homogeneous slope at 2:1 is examined. The soil stress states and pore-water pressures are computed using a partial differential solver with a linear elastic constitutive model. The slope has a groundwater table that passes 4 m below the crest of the slope and proceeds to the toe of the slope, as shown in Fig. 12.84.

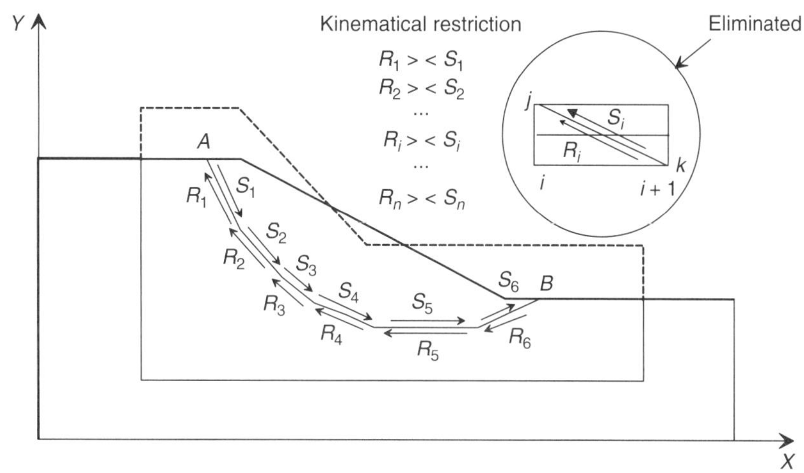

Figure 12.83 Kinematical restrictions applied to shape of critical slip surface.

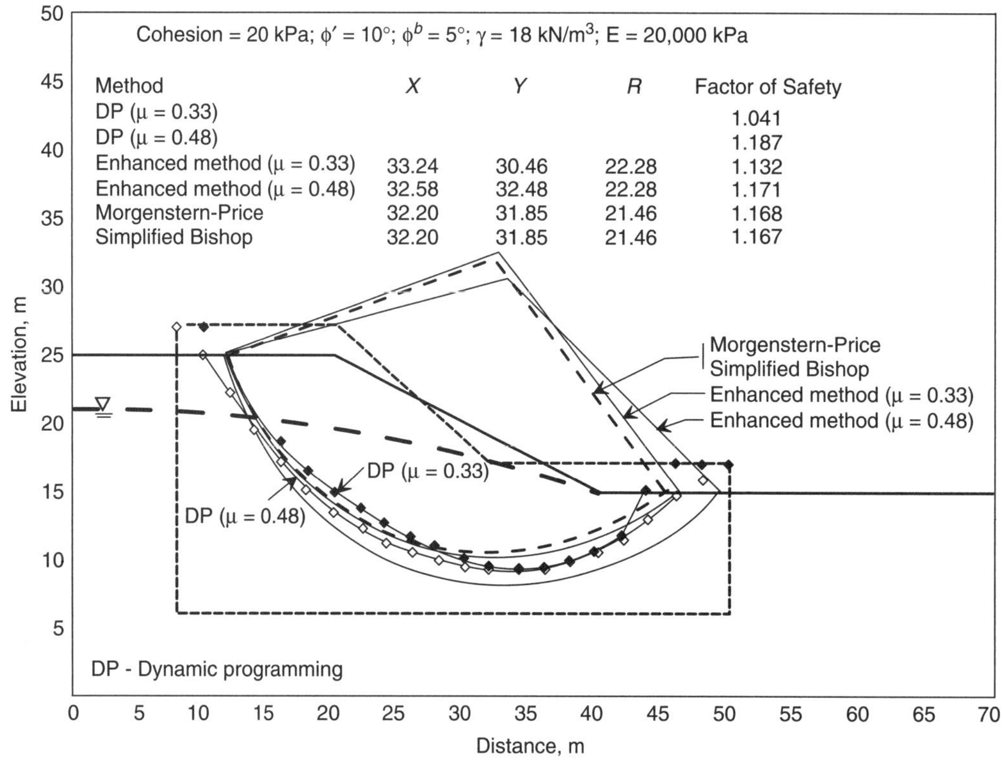

Figure 12.84 Locations of critical slip surfaces obtained by various methods for slope with phreatic line.

The dynamic programming results are presented for the case of two Poisson's ratio values: 0.33 and 0.48. The calculated factors of safety are compared to results from the method of slices and the enhanced method (Fredlund and Scoular, 1999; Fredlund et al., 1999). The enhanced method of slope stability analysis uses the stress states calculated from a linear elastic analysis along with assumed slip surface shapes. In other words, the enhanced method is similar to the method of slices, but the normal force at the base of a slice is determined from a stress analysis.

The unit weight of the soil was 18 kN/m^3. A ϕ^b value of 5° was selected for all analyses. The location of the critical slip surface is shown for each analysis. The computed factors of safety are quite similar for the dynamic programming solution and the method-of-slices results. The shape of the dynamic programming critical slip surface deviates slightly from a circular shape for a homogeneous soil slope.

Figure 12.85 compares the factors of safety computed using the dynamic programming method and the method of slices for a range of soil parameters. Poisson's ratio was 0.33. Janbu's stability number was used as the basis for comparison. In general, the dynamic programming method appears to give slightly lower computed factors of safety. When Poisson's ratio was increased to 0.48, the dynamic programming factors of safety became slightly larger than when Poisson's ratio was 0.33 (Fig. 12.86). The computed factors of safety obtained from the dynamic programming analyses appear to be similar to the results obtained from the methods of slices for a homogeneous soil slope.

12.6.2.2 Nonhomogeneous Slope

The results of an analysis of a nonhomogeneous slope with a weak soil layer (i.e., three- soil-layer system) are presented in Fig. 12.87. The stress states were calculated using a linear elastic model. The soil properties used for each soil layer are shown on Fig. 12.87. Also shown is the location of the critical slip surface determined from each of the analyses. The computed factors of safety were slightly lower when using the dynamic programming procedure. The shape and location of the critical slip surfaces differed considerably when a weak soil was incorporated into the non homogeneous soil cross section. The weak layer was used to force the shape of the critical slip surface into a composite mode.

The factor of safety computed using the dynamic programming method was about 14% lower than that obtained from the Morgenstern-Price (1965) method. The critical slip surface has a pronounced nonlinear shape enclosing the soil mass. This example problem illustrates the advantage of using the dynamic programming search technique to locate the critical slip surface.

12.6.2.3 Transient Infiltration Analysis

Gitirana and Fredlund (2003) illustrated the use of the dynamic programming method to simulate the infiltration

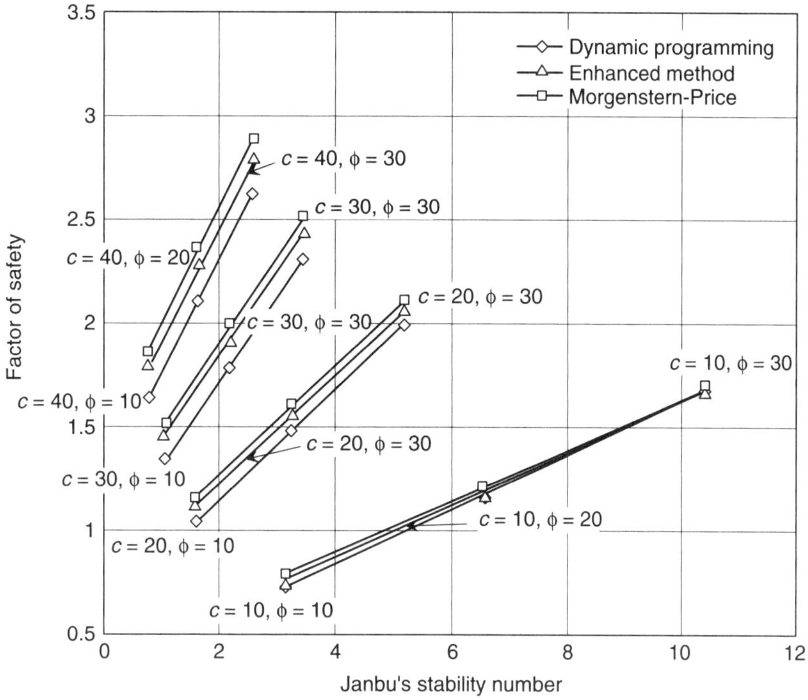

Figure 12.85 Factor of safety versus Janbu's stability number for homogeneous slopes (Poisson's ratio $\mu = 0.33$ in stress analysis).

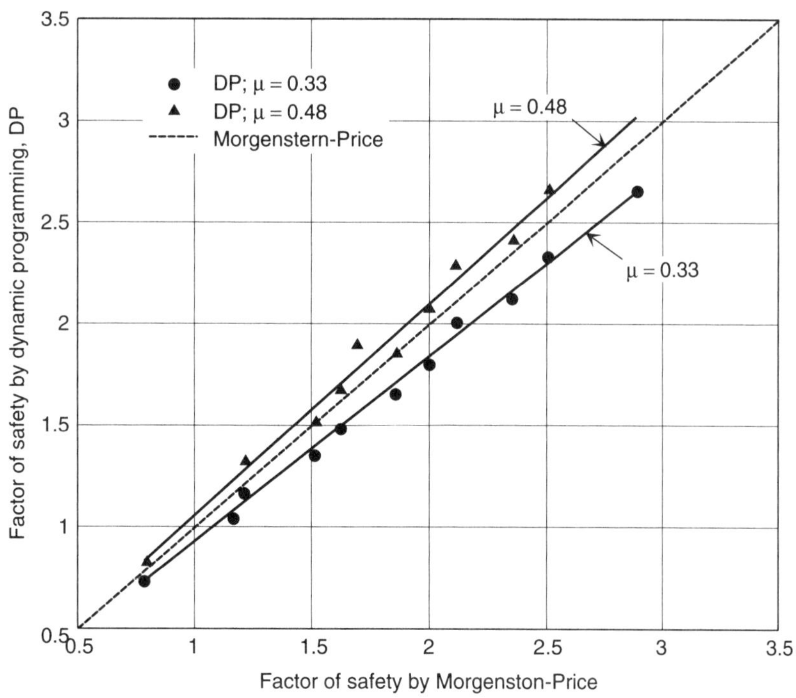

Figure 12.86 Dynamic programming results compared with Morgenstern-Price factors of safety for homogeneous soil slope.

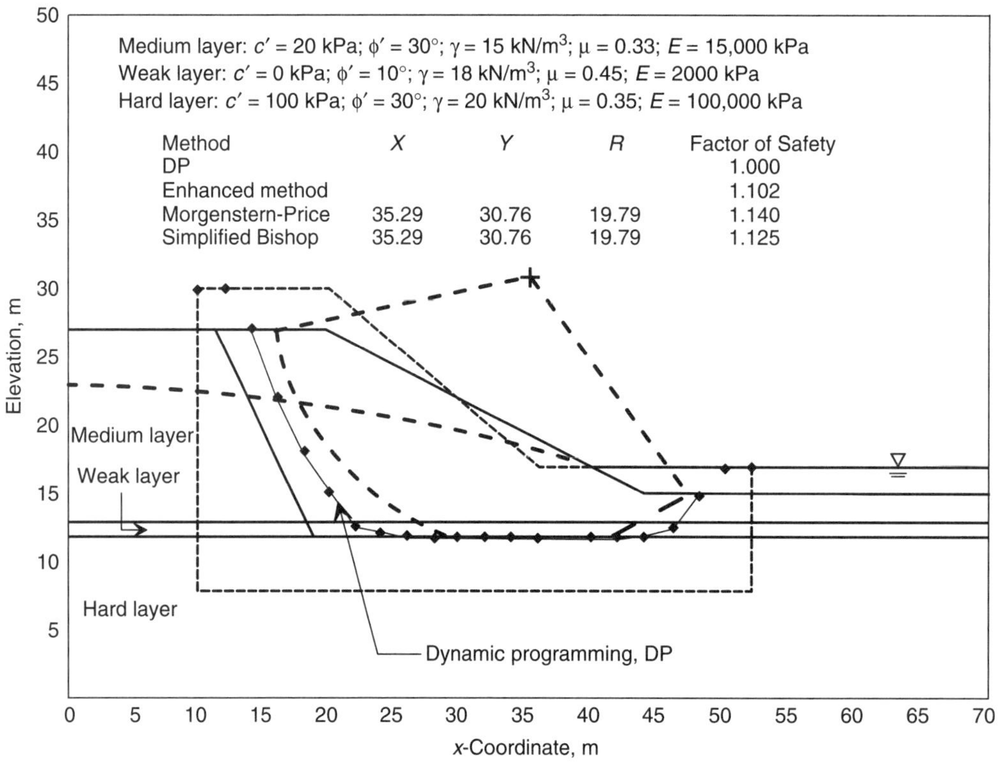

Figure 12.87 Nonhomogeneous slope with internal weak layer of soil.

of water into a sloping ground surface. As moisture moves into the slope there is also a tendency for the moisture to move down the slope toward the toe of the slope. The infiltration of water into the slope tends to control the shape and location of the critical slip surface.

The infiltration of rainwater on a sloping surface is dependent upon the water storage capacity and the hydraulic conductivity of the soil. The evolution of the shape and position of the critical slip surface with respect to time lend themselves to the dynamic programming method of analysis. There are no restrictions placed on the shape and location of the critical slip surface. Figure 12.88 shows the shape and location of the critical slip surface with time after the commencement of rainfall. The initial factor of safety of the slope was 1.83 and after 4 days the factor of safety was reduced to 1.12. The boundary condition at the ground surface was given a pore-water pressure value of 0 kPa. In other words, it was assumed that there was sufficient rainfall over the entire 4 days to reduce the pore-water pressure at the ground surface to zero.

The ability to undertake parametric-type analyses provides the geotechnical engineer with a useful tool in the study of potential hazards. Comparative studies between the dynamic programming method and methods of slices show that the computed results are similar when the shape of the critical slip surface is known. However, there are situations where the actual shape of the critical slip surface is not known, such as in the case of rainfall infiltration at ground surface. The dynamic programming solution appears to show a critical slip surface that goes slightly deeper (e.g., below the toe of the slope). Critical slip surfaces located using limit equilibrium methods of slices appear to exit higher on the slope (Pham and Fredlund, 2003). The stresses computed from a finite element analysis appear to better reflect stress states below the ground surface geometry.

All limit equilibrium methods of slices tend toward an upper bound type of solution because the shape of the slip surface is controlled by the analyst (Fredlund, 1984). Consequently, it is anticipated that the computed factors of safety would tend toward being slightly higher than the correct solution. Various limit equilibrium methods of analysis can yield slightly different factors of safety as a consequence of the assumption invoked to render the analysis determinate, but all results will tend toward an upper bound solution from a plasticity standpoint.

The shape of the slip surface is not dictated by the analyst in the dynamic programming method. Consequently, the solution should tend toward a more correct solution, and it is expected that the computed factors of safety should be slightly lower. This behavior was observed in the comparative study undertaken when Poisson's ratio was 0.33. While the factors of safety computed from the dynamic programming technique are generally lower than (or equal to) those computed by limit equilibrium analyses, the difference is relatively small for the simple slope examples. The small difference is encouraging because the increased flexibility

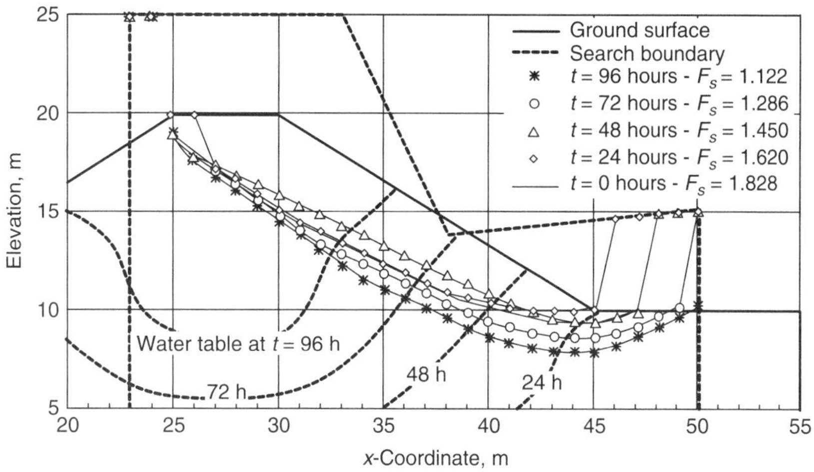

Figure 12.88 Simulation of rainfall in combination with dynamic programming method ($\mu = 0.33$).

of the dynamic programming technique opens the way for analyzing more complex slope stability problems.

The computed factors of safety increase and are similar to (or even greater than) the factors of safety computed from the Morgenstern-Price (1965) method when a Poisson's ratio of 0.48 was used in the stress analysis.

The dynamic programming method is able to locate a slip surface shape that is slightly more efficient (in terms of minimizing the factor of safety function) than a circular shape, even for a homogeneous soil slope.

The difference in results between the dynamic programming procedure and limit equilibrium methods of slices is relatively small for simple slopes. The dynamic programming technique has increased analytical flexibility.

12.7 APPLICATION OF SLOPE STABILITY ANALYSES

There are several ways that the shear strength of an unsaturated soil can be taken into account in a slope stability analysis. The most logical way to accommodate unsaturated soil shear strength is to modify computer codes such that whenever the pore-water pressures are negative, the linear or nonlinear unsaturated shear strength equation is used in calculating the shear strength.

It is possible to also perform a slope stability taking the shear strength of unsaturated soils into account even when the computer code is not modified to accommodate unsaturated soil behavior. For example, it is possible to simply subdivide the soil stratigraphy above the water table into layers. The total cohesion in each layer can be computed to reflect the sum of effective cohesion plus the cohesion due to matric suction. This procedure is referred to as the "total cohesion" method (Ching et al., 1984).

The second procedure involves rederiving the factor-of-safety equations in order to accommodate both positive and negative pore-water pressures (Fredlund, 1987a, 1989b; Rahardjo and Fredlund, 1991). Nonlinear shear strength–matric suction relationships can also be incorporated in the slope stability analysis (Rahardjo et al., 1992).

The soil cohesion c is considered to increase as the matric suction of the soil increases in the total cohesion method. The cohesion due to matric suction is obtained by multiplying the average matric suction for a soil layer by $\tan \phi^b$ [i.e., $(u_a - u_w) \tan \phi^b$]. The increase in the factor of safety due to negative pore-water pressures (or matric suction) is illustrated in Fig. 12.89. The shear strength contribution from matric suction is incorporated into the designation of the cohesion of the soil [i.e., $c = c' + (u_a - u_w) \tan \phi^b$]. The factor of safety of a slope can decrease significantly when the cohesion due to matric suction is decreased during a prolonged wet period.

12.7.1 Examples Using Total Cohesion Method

The following two example problems illustrate the application of the total cohesion method in analyzing slopes with negative pore-water pressures. The example problems involve studies of steep slopes in Hong Kong. The soil stratigraphy was determined from numerous borings. The shear strength parameters (i.e., c', ϕ', and ϕ^b) were obtained by testing undisturbed soil samples in the laboratory. Negative pore-water pressures were measured in situ using tensiometers. Slope stability analyses were performed to assess the effect of matric suction changes on the factor of safety. Parametric slope stability analyses were also conducted using various percentages of the hydrostatic negative pore-water pressures.

12.7.1.1 Example 1 (Slope behind Hospital and Residential Buildings)

The site plan for example 1 is shown in Fig. 12.90. The site consists of a row of residential buildings with a steep cut slope at the back. The slope has an average inclination angle

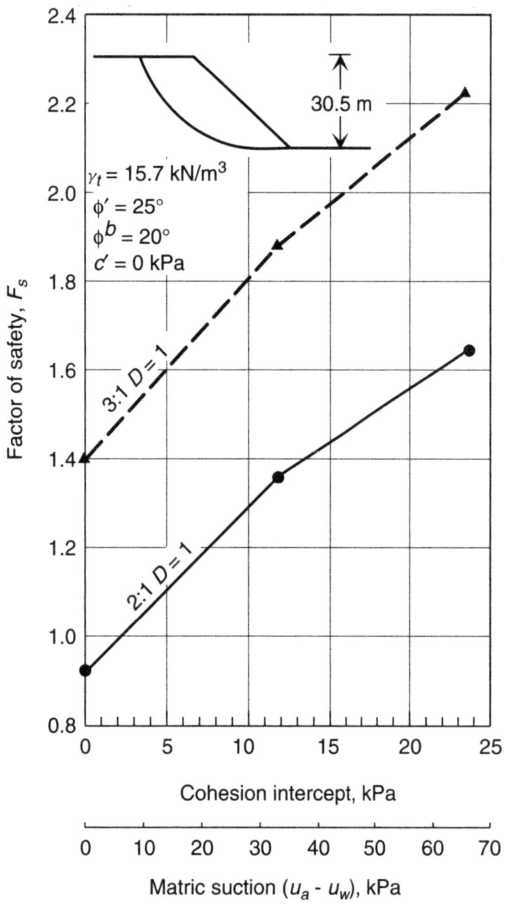

Figure 12.89 Factor of safety versus matric suction for simple slope.

of 60° to the horizontal and a maximum height of 35 m. The slope has been protected from infiltration of surface water by a layer of soil cement and lime plaster which is locally referred to as "chunam" plaster. Small but dangerous failures have occurred periodically at the crest of the cut slope. This condition prompted a detailed investigation.

A cross section of the cut slope (i.e., *A-A*), as illustrated in Fig. 12.91, was analyzed. The stratigraphy consists primarily of weathered granite. A 4–5-m layer of granitic colluvium was underlain by a 10-m layer of completely to highly

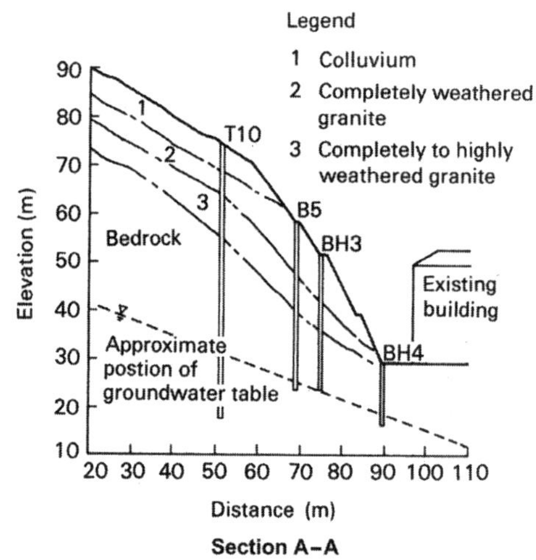

Figure 12.91 Cross section used for stability analysis in example 1.

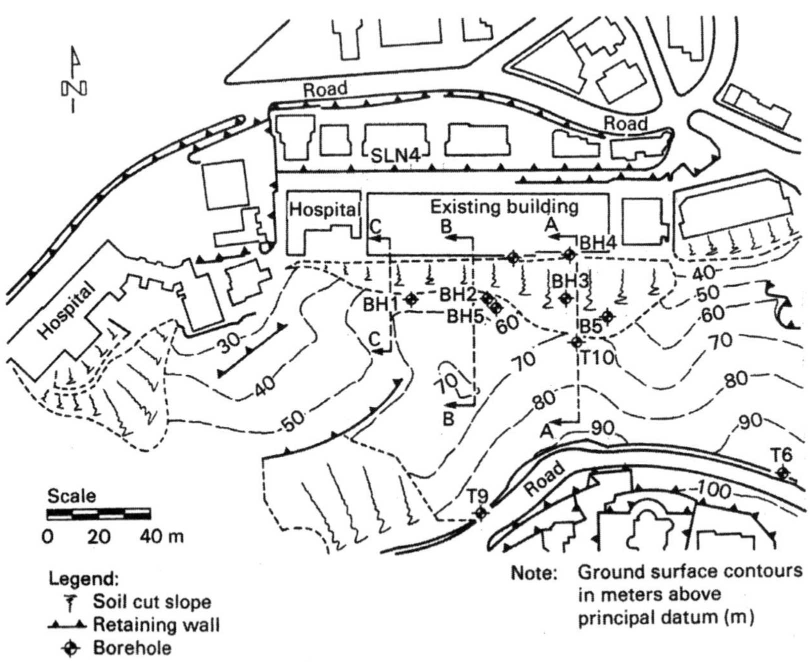

Figure 12.90 Site plan for slope stability example 1.

Table 12.9 Strength Properties for Soils of Example 1

Soil Type	Unit Weight (kN/m^3)	c' (kPa)	ϕ' (deg)
Colluvium	19.6	10.0	35.0
Completely weathered granite	19.6	15.1	35.2
Completely to highly weathered granite	19.6	23.5	41.5

weathered granite. The bedrock was situated 20–30 m below ground surface. The water table was located well into the bedrock. Potential failures through the steep slope would likely be associated with relatively shallow slip surfaces.

Triaxial tests on undisturbed core samples were conducted and the results are presented in Table 12.9. The average measured ϕ^b angle for the soil was 15° (Ho and Fredlund, 1982a).

In situ measurements of matric suction were conducted from the face of the slope using tensiometers. Two typical matric suction profiles measured near section *A-A* are shown in Fig. 12.92. The suction profiles showed considerable variation indicative of changing microclimatic conditions near the proximity of the slope face. No matric suction measurements were made near the upper part of the slope.

Slope stability analyses were performed on cross section *A-A* shown in Fig. 12.93. The Bishop simplified method of slices was used for all analyses. The analyses used circular slip surfaces passing through the toe of the slope.

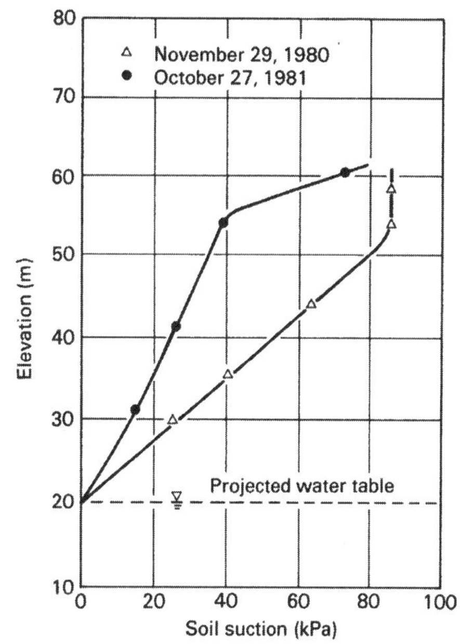

Figure 12.92 In situ measurements of matric suction near section *A-A* for example 1 (from Sweeney, 1982).

The effect of matric suction was ignored (i.e., $c = c'$) for the first analysis. The lowest factor of safety for the section *A-A* cross sections was 0.864, indicating unstable slope conditions. However, the slope appeared to be stable. It was assumed that the shear strength contribution due to matric suction was playing a significant role in the overall stability of the slope.

Matric suction was taken into account for subsequent analyses with cohesion increased in accordance with the maric suction. Each cross section was divided into substrata that were parallel to the water table. Each substratum was 5 m thick and was assumed to have an independent total cohesion

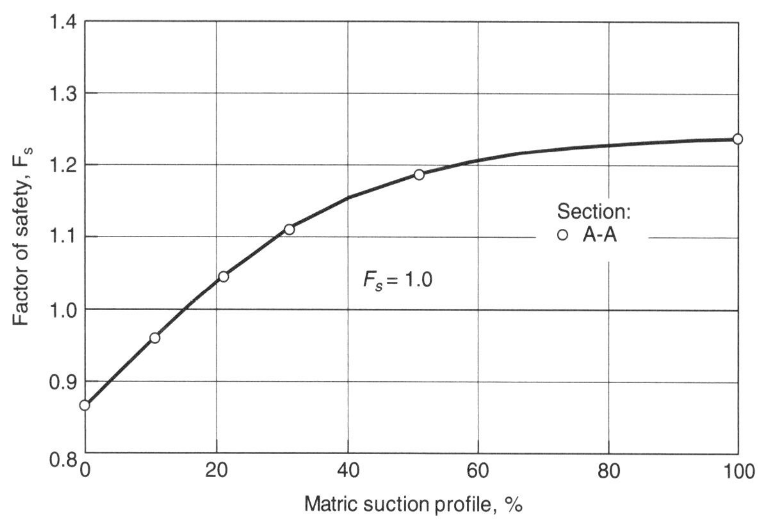

Figure 12.93 Calculated factors of safety considering various matric suction conditions for section $A - A$ in example 1.

c. The equivalent increase in cohesion for each substratum was computed from the matric suction profile and a ϕ^b angle of 15° [i.e., $(u_a - u_w) \tan \phi^b$].

A parametric study was also undertaken using various percentages of the hydrostatic negative pore-water pressure. The matric suction used in each analysis was limited to 1 atm or a computed cohesion increase of 27 kPa. The results of a slope stability study on cross section *A-A* are shown Fig. 12.93. A factor of safety of 1.0 corresponds to a matric suction profile of approximately 10–20% of hydrostatic conditions. A significant increase in the factor of safety (i.e., approximately 25%) is obtained when the matric suction profile is increased up to 40%. The smaller increase in the factor of safety beyond the 40% profile is related to the limit of 101.3 kPa placed on the potential matric suction value. The critical slip surface tends to penetrate deeper into the slope as the cohesion of the soil increases. The increase in cohesion is due to the increase in matric suction.

Slope stability analyses were also performed on example problem 1 using the measured matric suctions shown in Fig. 12.92. The equivalent increase in cohesion for each substratum was computed from the matric suction profiles with a maximum measured value of 85 kPa. The calculated factors of safety were 1.07 and 0.98 based on the matric suction profiles measured on November 29, 1980, and October 27, 1981, respectively.

12.7.1.2 Example 2 (Steep Slope behind Residential Buildings)

A high, steep cut slope existed behind a residential building in Hong Kong. The site plan for example 2 is shown in Fig. 12.94. A steep and high cut slope exists behind a residential building. A proposed high-rise residential building above the slope prompted a detailed investigation for the stability of the slope under the new conditions. The cut slope under consideration is below a major access road and the cross section of concern, *A-A*, is shown in Fig. 12.95. The slope is inclined at 60° to the horizontal and has an average height of 30 m. The stratigraphy consists entirely of weathered rhyolite. The water table lies well below the ground surface.

The shear strength parameters of the soil comprising example 2 were measured in the laboratory on undisturbed samples. The properties of the soils are summarized in Table 12.10. In situ matric suction measurements were obtained using

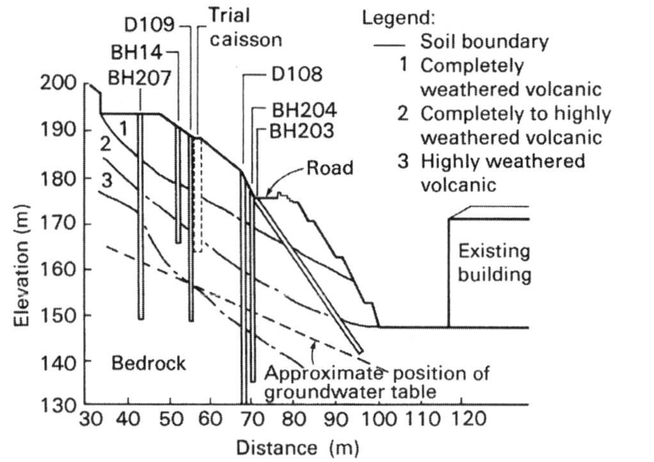

Figure 12.95 Cross section *A–A* for example 2.

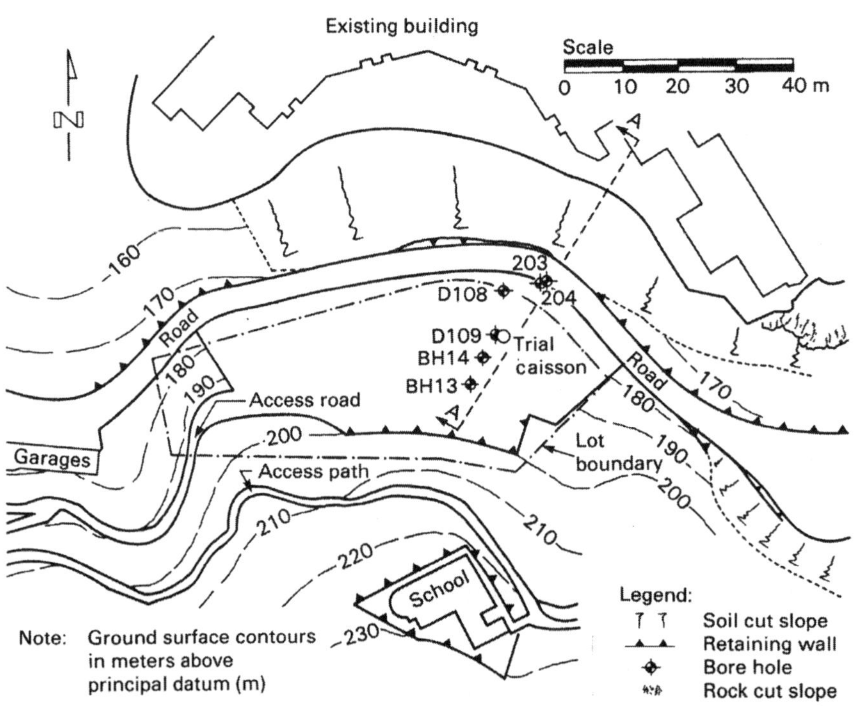

Figure 12.94 Site plan for slope stability example 2.

Table 12.10 Strength Properties for Soils of Example 2

Soil Type	Unit Weight (kN/m^3)	c' (kPa)	ϕ' (deg)	ϕ^b (deg)
Completely weathered rhyolite	18.4	10.1	42.6	12.0
Completely to highly weathered rhyolite	21.4	12.0	43.9	12.0

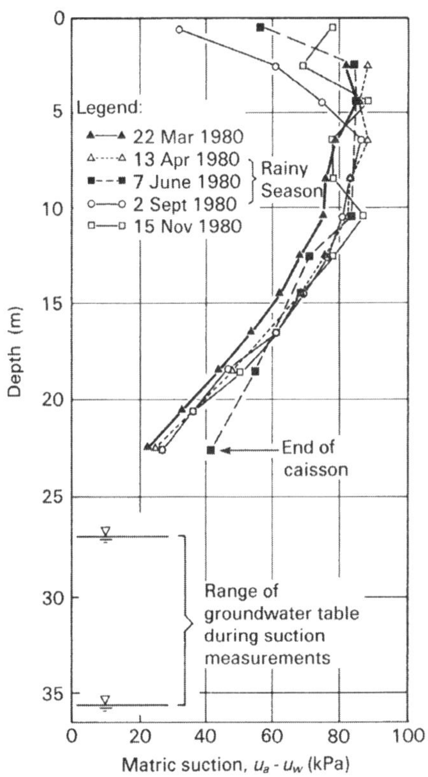

Figure 12.96 In situ measurements of matric suction throughout 1980 for example 2 (from Sweeney, 1982).

tensiometers installed vertically along an exploratory caisson shaft near the cut slope. The matric suction profiles measured through the rainy season of 1980 are plotted in Fig. 12.96. The profiles of matric suction remained essentially constant with some variations near the surface due to infiltration as well as fluctuations of the groundwater table.

Slope stability analyses were performed on section *A-A* (Fig. 12.95) using circular slip surfaces passing through the toe of the slope. The analytical procedure was the same as that followed in example 1. The critical factor of safety for the cut slope, when not taking matric suction into consideration (i.e., $c = c'$), was approximately 1.05. The relatively low factor of safety indicated a potential unstable condition, although no signs of distress were observed.

The next analysis was performed by taking matric suction into consideration [i.e., $c = c' + (u_a - u_w)\tan\phi^b$]. The matric suction profile of September 2, 1980, was used in the analysis. In this case, the cross section was divided into substrata drawn parallel to the water table in order to account for the varying matric suctions with depths. The critical factor of safety corresponding to the measured matric suction profile was 1.25. In other words, matric suctions appear to have increased the factor of safety by 20%.

12.7.2 Extended Shear Strength Method

The application of the slope stability formulation using the "extended shear strength" equation for unsaturated soil is illustrated in the following example (Ng, 1988). A cross section for a typical steep slope in Hong Kong was selected for analysis. The role of the negative pore-water pressures is shown by computing the factor of safety of the slope for a variety of pore-water pressure conditions. The slope was first subjected to a flux at ground surface in order to simulate a severe rainfall condition. As a result, the pore-water pressure increases (i.e., matric suction decreases). The factor of safety of the slope is shown to decrease with decreasing matric suctions. The influence of various ϕ^b values on the computed factors of safety is illustrated.

A transient water flow analysis was used to calculate negative pore-water pressure changes resulting from a heavy rainfall. The pore-air pressures were assumed to remain atmospheric (i.e., $u_a = 0$) for both the seepage and slope stability analyses.

12.7.2.1 General Layout of Problem and Soil Properties

The example slope has an inclination angle of approximately 60° to the horizontal. Its height is about 38 m. The cross section of the slope along with its stratigraphy is shown in Fig. 12.97. The slope mainly consists of decomposed granitic residual soil. The upper 5 m is a layer of colluvium overlaying a layer of completely decomposed granite. Below is a layer of completely to highly decomposed granite. A thick layer of highly decomposed granite lies at the bottom of these strata. The face of the slope is covered by a thin layer of low-permeability, cement-lime stabilized, decomposed granite plaster called chunam.

Several soil properties related to permeability and shear strength are required for the analyses. The saturated coefficients of permeability are presented in Table 12.11.

Unsaturated coefficient of permeability functions were calculated using the saturated coefficients of permeability along with the SWCCs of the soil (Green and Corey, 1971a). The SWCCs for the colluvium and the completely decomposed granite are shown in Fig. 12.98. Data from the SWCCs along with the saturated coefficients of permeability were used to establish the relationship between the unsaturated coefficients of permeability k_w and matric suction (Lam and Fredlund, 1984). The computed values were then

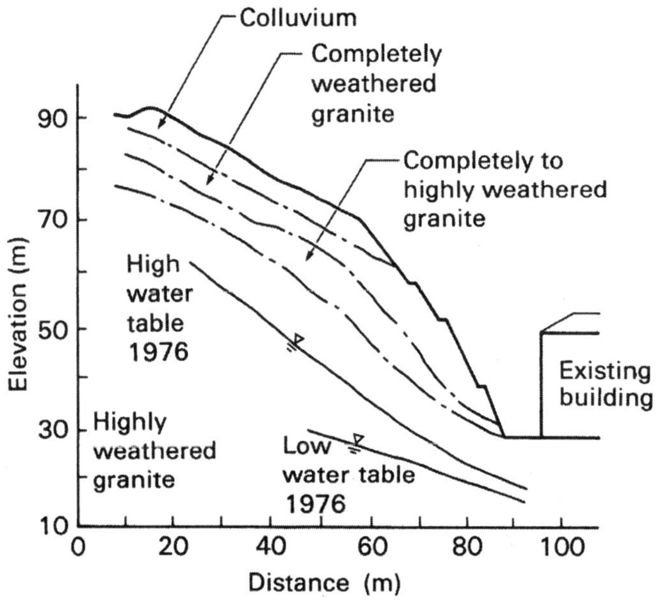

Figure 12.97 Cross section of typical steep slope in residual soils in Hong Kong.

Table 12.11 Summary of Saturated Coefficients of Permeability for Soils in Example 2

Soil Type	Selected Permeability, k_s (m/s)
Colluvium	3×10^{-5}
Completely decomposed granite	7×10^{-6}
Completely to highly decomposed granite	6×10^{-6}
Highly decomposed granite	5×10^{-6}

fitted to the unsaturated permeability function proposed by Gardner (1958a):

$$k_w = \frac{k_s}{1 + a[(u_a - u_w)/(\rho_w g)]^n} \qquad (12.106)$$

The a and n values in Eq. 12.106 were found to be approximately 0.1 and 3.0 for the colluvium and decomposed granite, respectively (Ng, 1988). The unsaturated permeability functions were drawn as shown in Fig. 12.99 based on the measured saturated coefficients of permeability.

The coefficient of water volume change, m_2^w, for the colluvium and the decomposed granite was computed to be approximately 3×10^{-3} kPa^{-1}. The shear strength

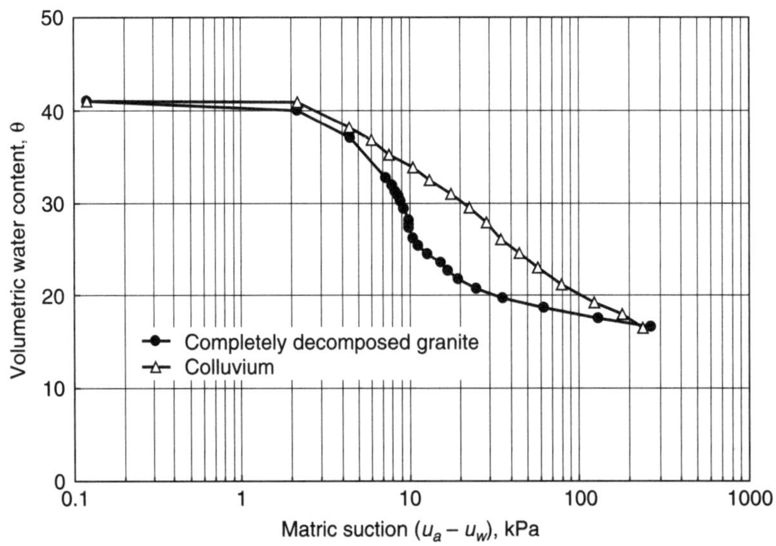

Figure 12.98 SWCCs for completely decomposed granite and colluviums.

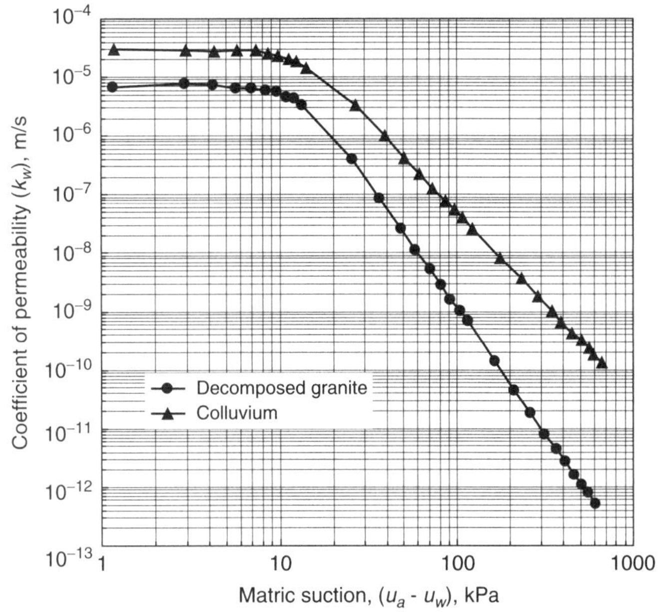

Figure 12.99 Permeability functions for decomposed granite and colluvium using the Fredlund et al. (1994) estimation procedure.

Table 12.12 Summary of Shear Strength Parameters and Total Unit Weights for Soils in Example 2

Soil Type	Cohesion, c' (kPa)	Effective Angle of Internal Friction, ϕ' (deg)	Total Unit Weight, γ (kN/m^3)
Colluvium	10	35	19.6
Completely decomposed granite	10	38	19.6
Completely to highly decomposed granite	29	33	19.6
Highly decomposed granite	24	41.5	19.6

parameters and the unit weights for the soils involved are summarized in Table 12.12. The ϕ^b angle for each material was assumed to be a percentage of the effective angle of internal friction, ϕ'. The percentage of the ϕ^b angle varied during the parametric slope stability study (i.e., ϕ^b is 0, 25, 50, 75, and 100% of ϕ').

12.7.2.2 Initial Conditions for Seepage Analysis

Steady-state and transient seepage finite element analyses were conducted on the discretized geometry as shown in Fig. 12.100. Initial conditions were assigned at the boundaries

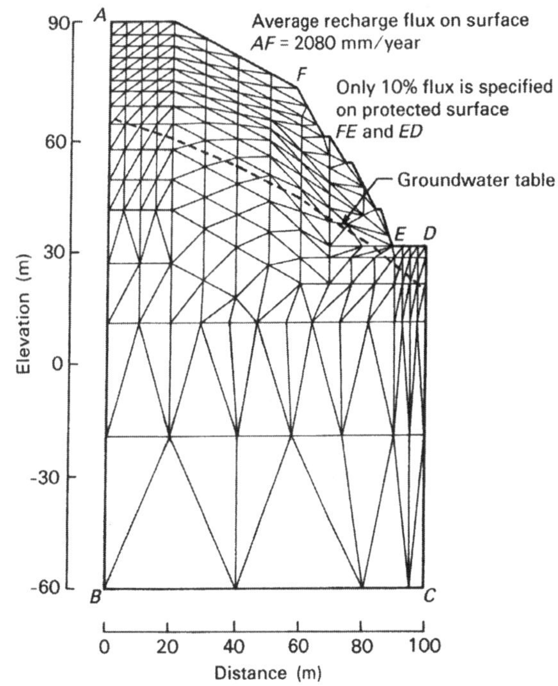

Figure 12.100 Finite element mesh of the steep cut slope and the initial boundary conditions.

of the slope. A zero flux condition was imposed along the bottom boundary, BC. The left- and right-hand boundaries (i.e., AB and DC, respectively) had a constant head boundary below the water table and a zero flux boundary above the water table. The constant head boundary below the water table was equal to the initial elevation of the water table.

The surface boundary $AFED$ was specified as a flux boundary. The applied flux was equal to the average annual rainfall in Hong Kong, which is about 2080 mm/year, or 6.6×10^{-8} m/s (Anderson, 1983). However, only 10% intake of the applied flux was allowed to enter on the steep cut boundary, FED, which is protected by chunam. It was assumed that water would run off the surface of the slope when the applied flux was in excess of the amount of water that could be taken in by the soil. It was anticipated that a seepage face could develop with time near the base of the slope along the boundary, EF.

The pore-water pressure contours and the water table under steady-state flux conditions are shown in Fig. 12.101. The pore-water pressure profiles along the cross section X-X are illustrated in Fig. 12.102. The matric suction deviates from hydrostatic conditions as a result of the applied water flux at ground surface. The recharge condition is maintained even in the dry season, as indicated from field studies (Leach and Herbert, 1982; Sweeney, 1982).

The computed pore-water pressure under steady-state flux conditions (Figs. 12.101 and 12.102) can be used to analyze the stability of the slope. The Bishop simplified method was used for all analyses. The analyses were performed using circular slip surfaces. The results are presented in Fig. 12.107 for various ϕ^b/ϕ' ratios. When the negative pore-water pressures are ignored (i.e., $\phi^b/\phi' = 0$), the factor of safety is equal to 0.9. The absence of signs of distress on the slope would indicate that the actual factor of safety was greater than 1.0. It is suggested that the negative pore-water pressures are contributing to the shear strength of the soil. The matric suction in the soil increases the factor of safety. The factor of safety ranges from 1.0 to 1.4 as the ϕ^b/ϕ' ratio varies from 0.25 to 1.0, respectively.

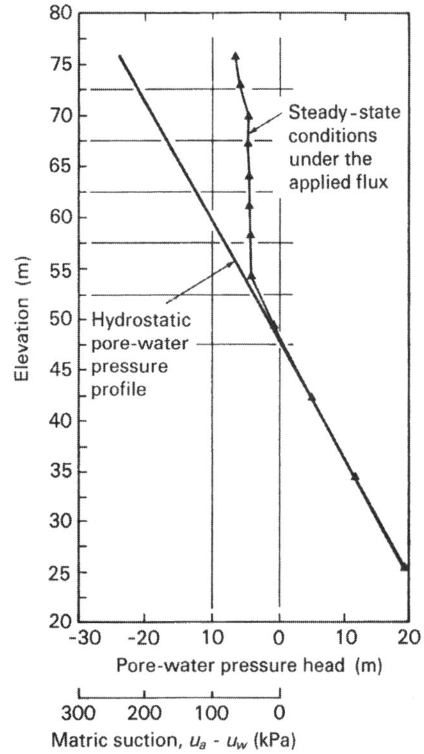

Figure 12.102 Matric suction profile for section X-X under steady-state moisture flux conditions.

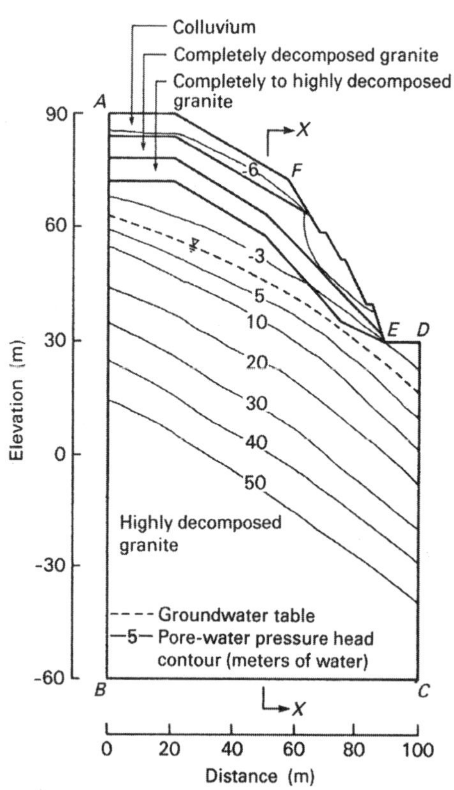

Figure 12.101 Initial groundwater conditions and pore-water pressure contours.

12.7.3 Seepage and Slope Stability Results under High-Intensity Rainfall Conditions

The following transient analyses were performed to illustrate the effect of a high-intensity rainfall for a long duration on the stability of the steep slope shown in Fig. 12.97. The rainfall rate was selected to be 1.30×10^{-5} m/s for the duration of 480 min. The pore-water pressure distributions corresponding to the simulated steady-state flux conditions were used as the initial conditions for the transient seepage analyses during the high-intensity rainfall.

The subsequent pore-water pressure distributions at various elapsed times after the commencement of the high-intensity rainfall are shown in Figs. 12.103–12.105. The upper portion of the slope becomes saturated starting at the ground surface, AF, which is not protected by chunam (see Fig. 12.103). The saturation front has penetrated to a depth of approximately 12 m and extends to the region below the protected surface, FE, when the rainfall stops after 480 min (see Fig. 12.96). The final steady-state conditions are achieved at 1080 min of

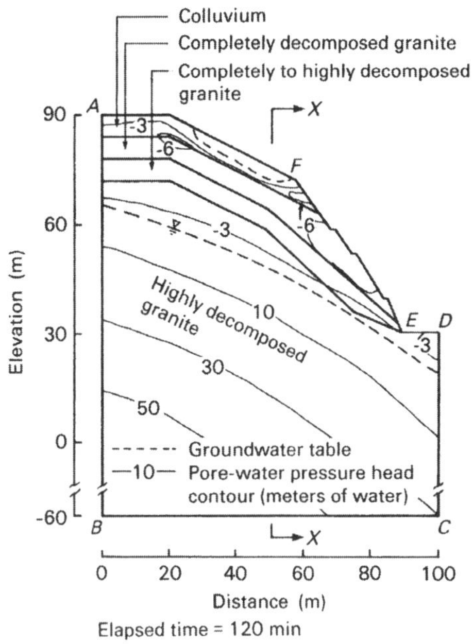

Figure 12.103 Position of groundwater table and pore-water pressure contours at elapsed time of 120 min.

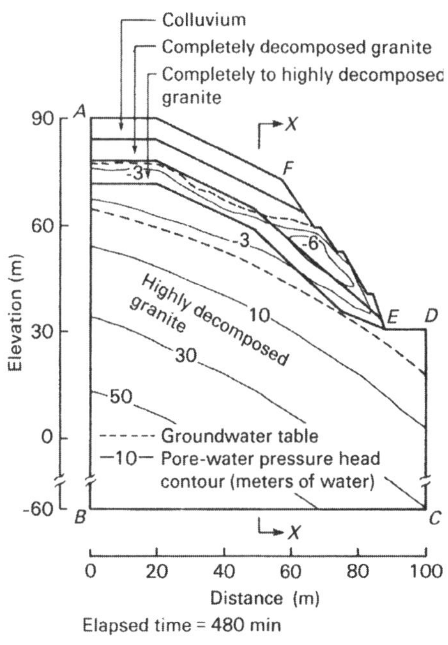

Figure 12.104 Position of groundwater table and pore-water pressure contours at elapsed time of 480 min.

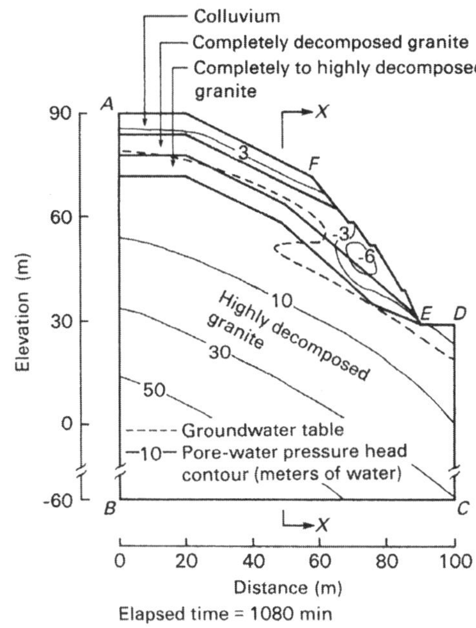

Figure 12.105 Position of groundwater table and pore-water pressure contours at elapsed time of 1080 min.

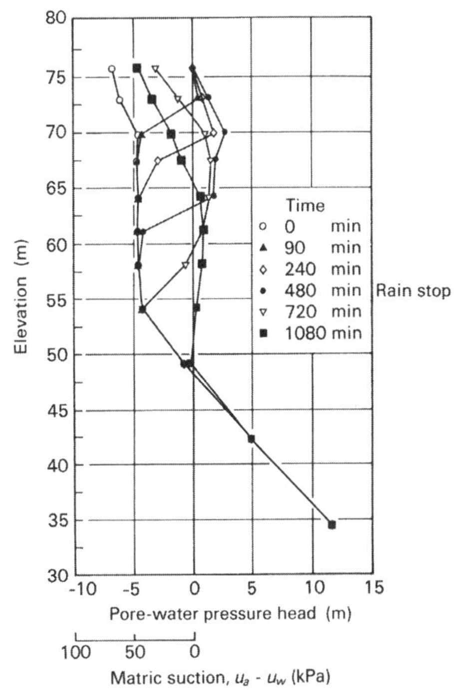

Figure 12.106 Matric suction profile for section X-X at various elapsed times.

elapsed time. At this time, the saturation front has moved deeper as the water is redistributed deeper within the slope. As a result, the groundwater table corresponding to the final steady-state condition (Fig. 12.105) is higher than its position corresponding to the initial conditions (Fig. 12.101).

Figure 12.106 illustrates the pore-water pressure profiles for section X-X at various elapsed times. The negative pore-water pressures at depth increase to zero (i.e., the matric suction is dissipated) as the saturation front moves downward. Under final steady-state conditions, the upper 10 m of the slope has gained back only a fraction of its initial matric suction, while the lower part of the slope has lost its initial matric suctions.

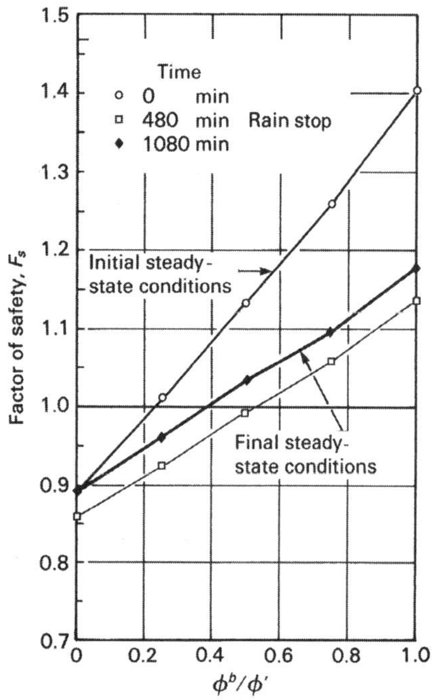

Figure 12.107 Factors of safety for various ratios of ϕ^b/ϕ' for various seepage conditions.

The slope stability analyses can be performed for various elapsed times using the corresponding pore-water pressure distributions. The changing factors of safety during the rainfall period are plotted in Figs. 12.107 and 12.108 for various ratios of ϕ^b/ϕ'. The factors of safety continue to decrease until the rainfall stops after 480 min (Fig. 12.108). The decrease in the factor of safety becomes more substantial as the ϕ^b/ϕ' ratio increases. This can be attributed to the fact that the critical slip surface is shallow and the mobilized shear resistance is significantly increased by the contributions from negative pore-water pressures. The above scenario illustrates how a catastrophic failure of a steep slope might occur due to a loss of matric suction during and following a heavy, prolonged rainfall.

The factor of safety on a fixed slip surface starts to increase after 480 min (i.e., when the rainfall has stopped) and water has moved deeper into the slope. The location of the critical slip surface during and after the rainfall may be different. The critical slip surface goes deeper as the wetting front moves into the slope. The increase in the factor of safety appears to occur at a slower rate after the rainfall than the decrease in the factor of safety occurred during the rainfall period.

12.7.4 Consideration of Slip Surfaces Near Ground Surface

Negative pore-water pressures above the water table are often neglected when performing slope stability analyses. There are reasons why engineers may not feel comfortable in giving consideration to negative pore-water pressures when performing slope stability analyses. The effect of matric suction may be omitted because (i) the theory of unsaturated soil shear strength and water infiltration may not be well understood, (ii) computer codes may not have implemented the effect of unsaturated soil behavior, and (iii) there may be a perception that negative pore-water pressures cannot be relied upon in the long term. There are, however, a number of other related questions that the geotechnical engineer should take into consideration when dealing with slopes that have negative pore-water pressures.

It might be important to ask the following questions when performing a slope stability analysis:

1. What quantitative effect will negative pore-water pressures have on the slope stability study?

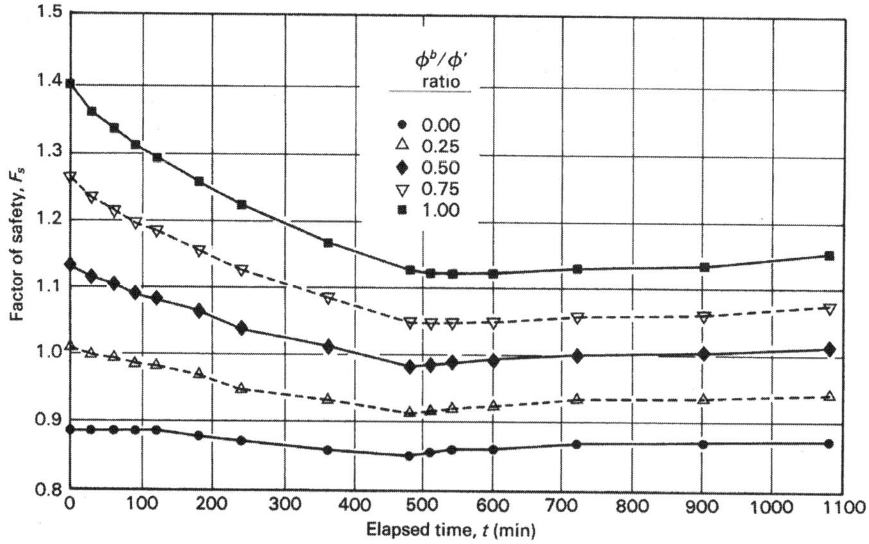

Figure 12.108 Factors of safety at various elapsed times since the beginning of rainfall.

2. Does the location of the critical slip surface change location when negative pore-water pressures are taken into account?
3. What is the percent change in calculated factor of safety with and without consideration of the pore-water pressures above the groundwater table?
4. What effect does the geometry (i.e., the steepness of the slope) have on whether negative pore-water pressures are taken into consideration?
5. Under what conditions can negative pore-water pressures be considered to be semipermanent?
6. Under what conditions should negative pore-water pressures be considered to be transient?
7. Does the trigger for instability of the slope lie in the loss of matric suction? If so, under what rainfall conditions over what period of time might the slope fail?

The effects of negative pore-water pressures should be given consideration whether or not matric suctions are relied upon in the slope stability analysis.

Slope stability modellers are quite familiar with the situation where the method-of-slices analysis yields a critical slip surface that just touches the ground surface. The modeler must decide whether or not to reject this analysis because it seems unreasonable. The primary reason why the extremely shallow slip surface has been selected is because the negative pore-water pressures were ignored. The influence of negative pore-water pressures on shallow slip surfaces is illustrated in the following example problems.

12.7.4.1 *Example 1 (Relatively Flat Slope with No Suctions Considered)*

Figure 12.109 shows a 20-m-high slope at an angle of 30°. The geometry consists of one soil type and has a water table passing below the toe of the slope. The soil has a unit weight of 18.5 kN/m^3, an effective cohesion of 20 kPa, and an effective angle of internal friction of 24°. The slope is first analyzed under the assumption that all negative pore-water pressures are ignored (i.e., $\phi^b = 0.0$). All slopes were analyzed using the Bishop simplified method of slices.

The grid of slip surface centers analyzed is shown in Fig. 12.109. Figure 12.110 shows the computed factors of safety versus the radius from the center of rotation to the slip surface. The results show the minimum factor of safety to be about 1.28. The results show that while the shortest radii show an increased factor of safety the increase is relatively small. The critical slip surface falls well below the ground surface and even goes below the groundwater table.

12.7.4.2 *Example 2 (Relatively Flat Slope with Suctions Considered)*

The example problem described above (Fig. 12.109) was reanalyzed with matric suction taken into consideration. The ϕ^b angle was assumed to be 15°. Figure 12.111 shows the location of the critical slip surface. The computed critical factor of safety increased to 1.32, an increase of only 2.9%. The location of the critical slip surface is now 1.5 m lower than in the example 1 analysis. In other words, negative pore-water pressures tend to cause the critical slip surface to move downward.

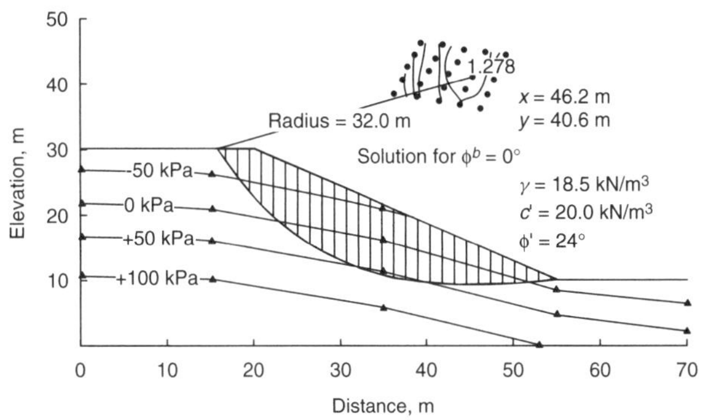

Figure 12.109 Location of critical slip surface for relatively flat slope when negative pore-water pressures are not taken into consideration.

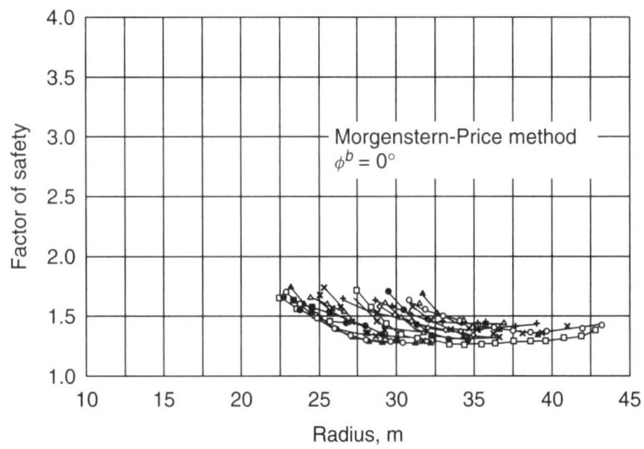

Figure 12.110 Factor of safety for all grid centers for relatively flat slope when negative pore-water pressures are ignored.

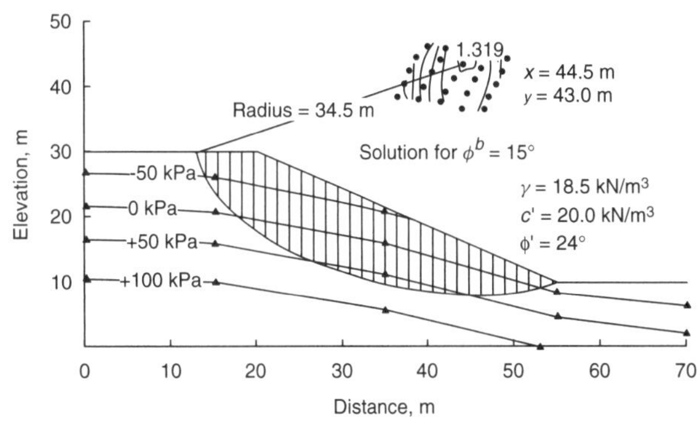

Figure 12.111 Location of critical slip surface for relatively flat slope when negative pore-water pressures are taken into consideration.

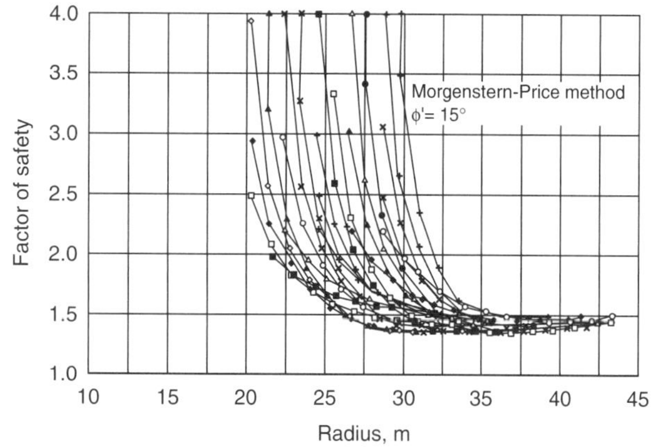

Figure 12.112 Factors of safety versus radius for all grid centers for a relatively flat slope when negative pore-water pressures are taken into consideration.

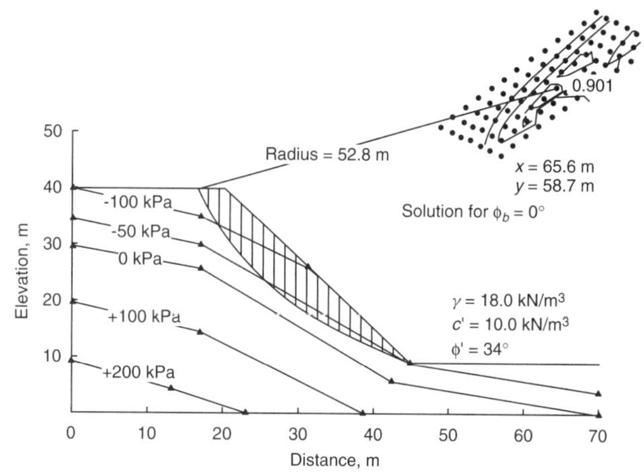

Figure 12.113 Location of critical slip surface on steep slope when negative pore-water pressures are not taken into consideration.

Figure 12.112 shows a plot of the computed factors of safety versus the radius of the slip surface. The character of the results has changed dramatically from those shown for the case when soil suction was ignored (Fig. 12.110). All shallow slip surfaces show computed factors of safety going beyond a magnitude of 4.0. Now there is no tendency for a shallow slip surface to be computed as being the critical slip surface.

A further analysis was performed with the ϕ^b value set to 25°. The computed critical factor of safety increased to 1.35, which was an increase of 5.3% over the case when matric suctions were ignored. Quantifying the effect of negative pore-water pressures is of value even in the case where a relatively small portion of the slip surface lies above the water table.

12.7.4.3 Example 3 (Steep-Slope Case)

The steep slope in example 3 has an angle of 50° to the horizontal. Most of the slip surfaces pass above the water table, which is on average 8 m below the surface of the slope. The unit weight of the soil is 18 kN/m^3, the effective cohesions is 10 kPa and the effective angle of internal friction is 34°. The geometry along with the critical slip surface is shown in Fig. 12.113. The calculated minimum factor of safety was 0.901 when negative pore-water pressures were ignored. Figure 12.114 shows the location of the critical slip surface and the minimum factor of safety when the ϕ^b value is 15° and the negative pore-water pressures are taken into consideration. The minimum factor is now 1.088, or an increase of 20.3%. The critical slip surface went considerably deeper into the slope.

Increasing the ϕ^b angle to 25° increased the minimum factor of safety to 1.170 (i.e., an increase of 29.6% over the case where the pore-water pressures were ignored). The influence of matric suction is more pronounced on steep slopes than on the flatter slope. There is merit in quantifying the effect of negative pore-water pressures on the change in factor

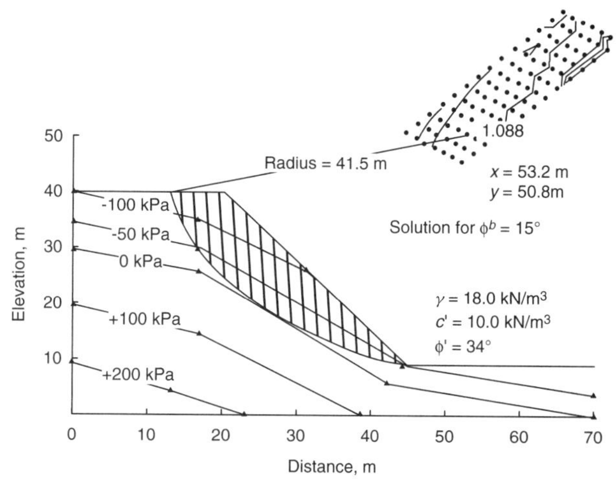

Figure 12.114 Location of critical slip surface for steep slope when negative pore-water pressures are taken into consideration.

of safety and the change in the location of the critical slip surface even when negative pore-water pressures may not be relied upon in the engineering design. It is quite possible that the slope stability analysis may lead to consideration of advantages that can be accrued from protection against the infiltration of water into a slope.

12.8 HAZARD ASSESSMENT AND DECISION ANALYSIS RELATED TO SLOPE INSTABILITY

Slope instability is a hazard that is commonly triggered by adverse weather conditions at ground surface. The soils near the ground surface are often unsaturated, and as a result, these soils have considerable water storage and may also have a reduced hydraulic conductivity. Figure 12.115 shows a typical embankment subjected to varying climatic conditions along with the various components that must

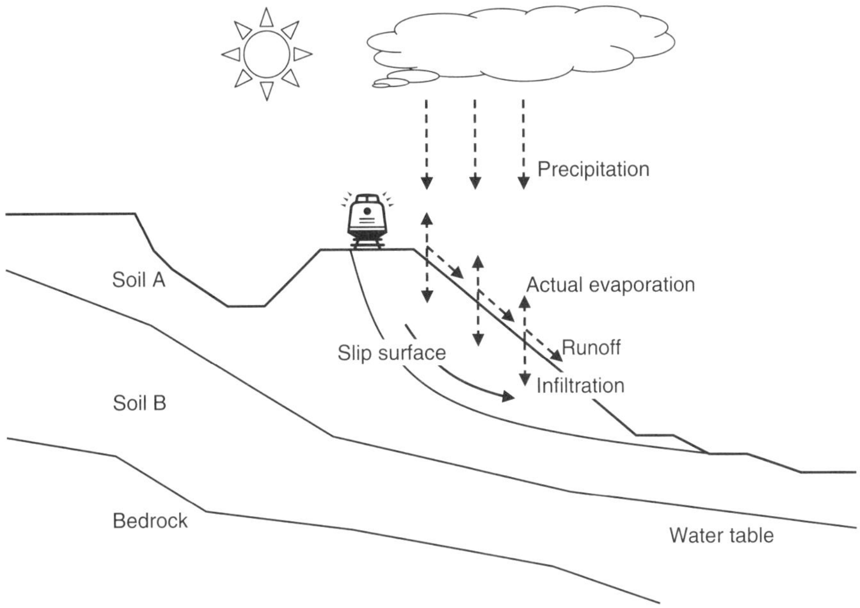

Figure 12.115 Interaction of climatic factors with stability of embankment.

be assessed in order to perform a transient simulation of embankment performance. It is possible to bring information on the weather and soil conditions together within a *decision analysis* framework in order to assess the likelihood or probability of slope failure. The unsaturated soil properties can be approximated through linkage to the SWCC.

Deterministic and probabilistic slope stability models can be developed within a *decision analysis* framework. Weather conditions interact with the near-ground-surface soils through a series of physical processes that can be described using partial differential equations such as those described for saturated-unsaturated water flow. The unsaturated soil properties take the form of mathematical functions that are related to the SWCC.

A hazard assessment model called a "weather hazard model" that uses a decision analysis framework was proposed by Gitirana and Fredlund (2003b) for embankment analyses. The *dynamic programming* optimization technique was used for the determination of critical slope stability conditions. The saturated-unsaturated soil system was represented by a system of PDEs governing water flow and stress conditions. Several saturated-unsaturated soil property functions were required as input to the computer analysis. All unsaturated soil properties were related to the appropriate SWCCs.

12.8.1 SWCC as Hazard Gauge

The stability of an embankment is strongly related to the reduction in matric suction within the embankment. Soil suction varies according to the amount of water stored within the embankment. Changes in the amount of water within the soil are related to the weather conditions and soil properties. The relationship between the amount of water in the soil and

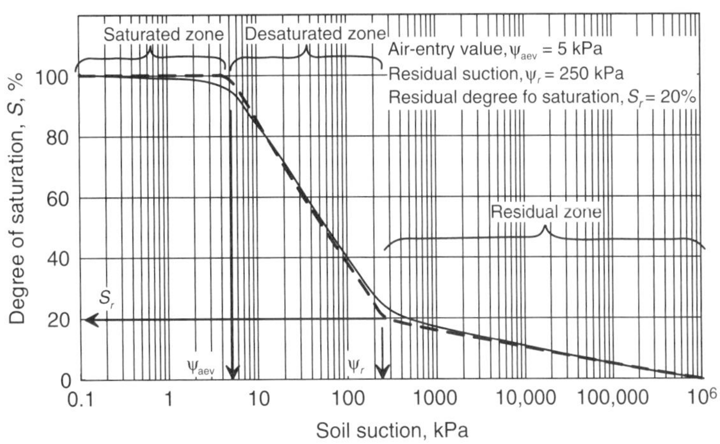

Figure 12.116 Conceptualization of SWCC as hazard gauge.

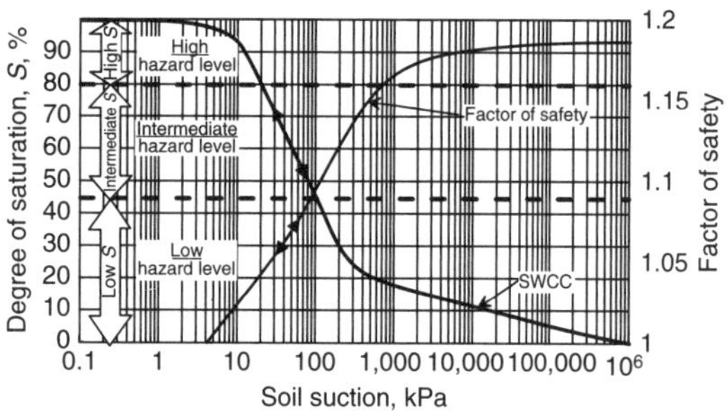

Figure 12.117 SWCC as water level gauge and embankment hazard gauge.

the soil suction is given by the SWCC (Fig. 12.116). Typical SWCCs can be defined by three primary soil parameters: air-entry value ψ_{aev}, residual suction ψ_r, and residual degree of saturation S_r (Gitirana and Fredlund, 2003b).

Figure 12.117 presents an analogue describing the fundamental concept behind the weather hazard model. According to this analogue, the soil comprising an embankment can be viewed as a "water tank." The SWCC acts as a gauge that indicates the water level within the water tank. The embankment factor of safety F_s and the embankment hazard level vary according to the water level. A low water level corresponds to a lower level of hazard (i.e., higher F_s), while higher water levels produce greater hazards (lower F_s). The model for assessment of embankment hazards is based indirectly on the water tank concept and the factor of safety. However, a more approximate approach could be used to indicate the water level within the water tank. Such an approach would still capture the main factors controlling the stability of an embankment or slope.

12.8.2 Frequency and Sensitivity Analysis

The assessment of weather-related geohazards is undertaken in terms of the frequency distribution as a measure of stability (i.e., the factor of safety). The choice of variables that must be modeled as uncertainties (i.e., probabilistically) is determined by performing a sensitivity analysis. The decision analysis framework adopted in the weather hazard model provides a systematic environment for frequency and sensitivity analyses.

The frequency analysis corresponds to one step in the decision analysis cycle. In order to perform frequency analyses, the relationships, dependences (or correlations), and frequency distributions of the involved parameters must be established. These relationships can be illustrated using an influence diagram for the weather hazard model (Fig. 12.118). The variables presented inside round boxes are the uncertain variables. The variables presented inside square boxes are certain (or fixed) values. These values are part of the deterministic equations and do not need to be shown on the influence diagram. The arrows indicate the relationships between variables. The factor-of-safety box is the final outcome receiving arrows from outside. Only some of the incoming variables need to be modeled as uncertainties. These variables are determined from sensitivity analyses performed on all the variables involved with assessing stability. A discrete stochastic analysis can be used to perform a frequency analysis.

The presence of several uncertain variables produces a combination of different scenarios. The presentation of several scenarios is called the decision tree. Figure 12.119 presents the decision tree corresponding to the influence diagram of Fig. 12.118. Three branches are used for each variable. The notation used for the decision tree is the same as that used

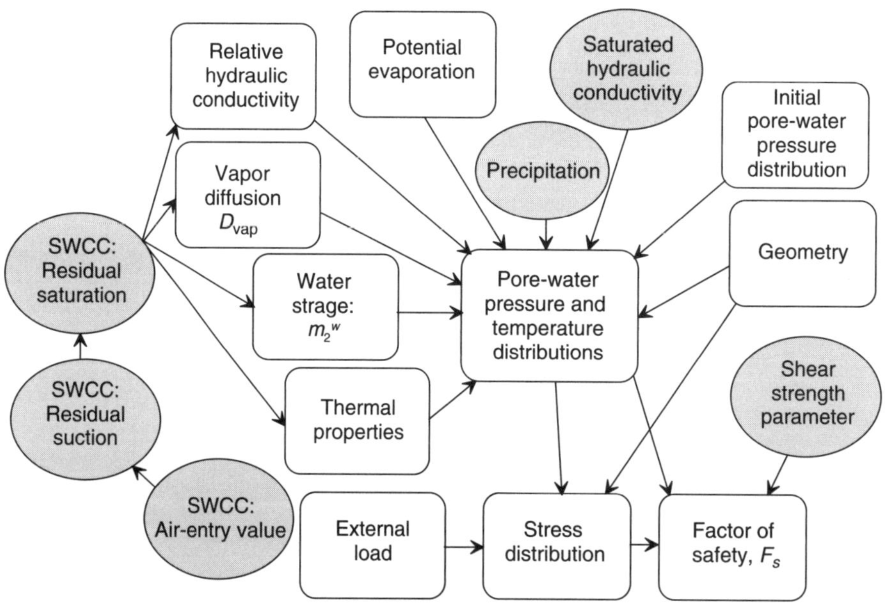

Figure 12.118 Influence diagram for weather hazard model.

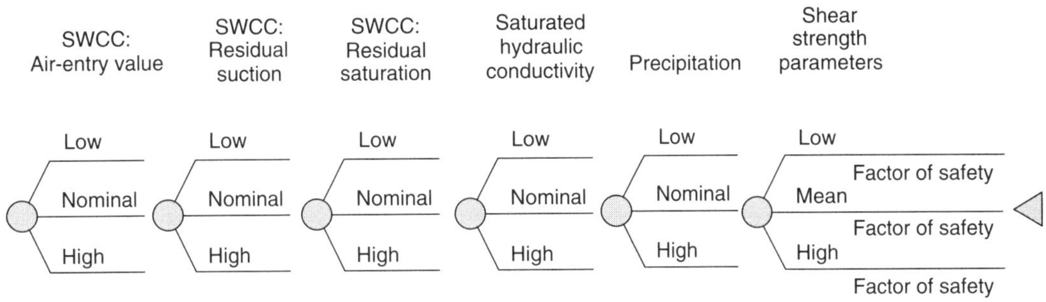

Figure 12.119 Decision Tree for weather hazard model.

in decision programming language (e.g., Applied Decision Analysis LLC, 1998). Analyses have indicated that precipitation appears to be the single most sensitive variable related to slope instability, and therefore the quantification of precipitation and net infiltration is of high priority.

The weather hazard model attempts to make use of the principles and properties of unsaturated soils along with developments in slope stability analyses and decision theory. A discrete stochastic approach can be used along with unsaturated soil properties that are related to the SWCC. A series of partial differential equations governing the thermo-hydro-mechanical behavior can be used with appropriate soil-atmosphere boundary conditions to indicate the likelihood of a slope failure. The primary input to the slope stability analysis is the pore-water pressure conditions which change depending upon imposed weather conditions.

CHAPTER 13

Stress-Deformation Analysis for Unsaturated Soils

13.1 INTRODUCTION

The constitutive relationships for volume change relate deformation state variables to stress state variables. Constitutive relationships extend beyond those required for overall volume change because it is necessary to predict the relative quantity (or volume) of each phase comprising the unsaturated soil (i.e., solid phase, water phase, and air phase).

Several forms of the volume change constitutive equations can be formulated (e.g., soil mechanics formulation, compressibility formulation, and elasticity formulation). The primary emphasis is on the application of incremental elasticity; however, some reference is made to elasto-plastic models. The geotechnical engineer must understand how one set of constitutive parameters relate to another set of parameters defined when using another system. For example, soil parameters are sometimes measured in the laboratory and defined in one system but will need to be converted to another system of variables when solving practical engineering problems.

Experimental results used for the verification of the constitutive equations are described. Soil properties used in the volume change constitutive equations come under the general term of "volumetric deformation coefficients."

13.1.1 Background of Stress and Deformation Studies for Unsaturated Soils

Biot (1941) presented a three-dimensional consolidation theory based on the assumption that the soil was isotropic and behaved in a linear elastic manner. The soil was assumed to be unsaturated in that the pore-water contained occluded air bubbles. Two constitutive relationships were proposed in order to completely describe the deformation state of the unsaturated soil. One constitutive relationship was formulated for the soil structure, and the other constitutive relationship was formulated for the water phase. Two independent stress variables were used in the formulations. In total, four volumetric deformation coefficients were required to provide linkage between the stress and deformation states.

Coleman (1962) assumed that the stress components for an unsaturated soil could be used in an independent manner for the formulation of constitutive relations. One set of constitutive relationships was proposed for the soil structure and another set of constitutive relations was proposed for the water phase. Volumetric deformations of unsaturated soil specimens were studied under triaxial test loading conditions.

Matyas and Radhakrishna (1968) introduced the concept of state parameters for an unsaturated soil. The state parameters consisted of several stress variables [i.e., $\sigma_m = (\sigma_1 + 2\sigma_3)/3 - u_a$, $\sigma_1 - \sigma_3$, and $u_a - u_w$ for triaxial compression] along with the initial void ratio and degree of saturation (i.e., e_0 and S_0). Tests were performed on essentially "identical" soil specimens compacted at the same water content and dry density. The stress parameters reduced to $\sigma_3 - u_a$ and $u_a - u_w$ for isotropic compression. The void ratio and degree of saturation were used to represent the deformation state of the unsaturated soil.

Three-dimensional state surfaces were formed with void ratio and degree of saturation plotted versus the independent state parameters $\sigma - u_a$ and $u_a - u_w$. The state surfaces are, in essence, constitutive surfaces. Matyas and Radhakrishna (1968) experimentally tested the uniqueness of the constitutive surfaces while following a series of selected stress paths. Isotropic and K_0 compression tests were performed on mixtures of 80% flint powder and 20% kaolin. The total, pore-air, and pore-water pressures were controlled during the tests. The constitutive surfaces of void ratio and degree of saturation versus the $\sigma - u_a$ and $u_a - u_w$ stress variables were experimentally measured and used to test the surfaces for uniqueness.

The void ratio results (Matyas and Radhakrishna, 1968) produced a single warped surface with the soil structure always decreasing in volume as the $u_a - u_w$ stress was decreased or as the $\sigma - u_a$ stress was increased (Fig. 13.1a). The results showed that the soil had a metastable soil structure which collapsed as a result of a gradual reduction in matric suction $(u_a - u_w)$. A soil with a stable structure would have swelled when the matric suction was decreased. The results showed a unique constitutive surface for the soil structure even though the soil structure experienced

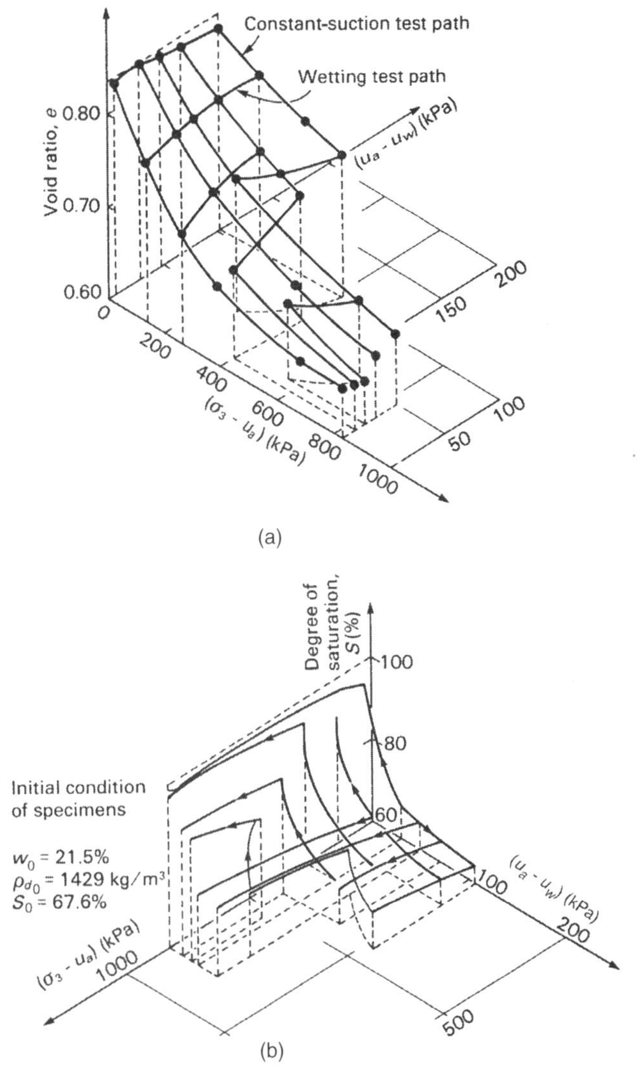

Figure 13.1 Void ratio and degree of saturation constitutive surfaces for a mixture of flint and kaolin under isotropic loading conditions: (a) void ratio constitutive surface; (b) degree-of-saturation constitutive surface (after Matyas and Radhakrishna, 1968).

collapse (Fig. 13.1a). It was noted that all deformation paths resulted in an increasing degree of saturation. The void ratio–stress state constitutive surface was not found to be unique when other paths were followed which involved wetting and drying. The restriction placed on the stress path appears to be related to hysteresis associated with wetting and drying. The stress paths followed introduced certain nonunique characteristics in the soil structure constitutive surface (Matyas and Radhakrishna, 1968).

The constitutive surface for the water phase, represented by the degree of saturation, was found to be nonunique (Fig. 13.1b). However, once again the soil underwent wetting and drying as it moved toward saturation.

Matyas and Radhakrishna (1968) tested soil specimens consisting of 80% flint powder and 20% kaolin under isotropic and K_0 loading with controlled total, pore-air, and pore-water pressures. The isotropically loaded specimens were prepared with an initial water content of 21.5%, an initial density of 1429 kg/m^3, and an initial degree of saturation of 67.6%. The tests were performed in a modified triaxial apparatus. The constitutive surfaces were traced by following "constant-suction" and "wetting" test paths. The constant-suction test specimens were either saturated or desaturated and then compressed at a constant matric suction. The saturation of the soil specimens was conducted while holding the specimen at a constant overall volume in order to minimize soil structure disturbance. Some specimens were isotropically loaded while others were subjected to K_0 loading. The wetting tests were conducted by decreasing matric suction while allowing the specimens to imbibe water. Matric suction was decreased by increasing the pore-water pressure.

The constitutive surfaces obtained from isotropic loading are shown in Fig. 13.1. The constant-suction test results showed a decrease in the soil structure compressibility at higher matric suctions (Fig. 13.1a). Stated another way, the soil structure became more rigid at high matric suctions. The wetting test results indicated a collapse-type behavior or a decrease in void ratio when matric suction decreased (Fig. 13.1a). This behavior is characteristic of a soil with a metastable soil structure. It was also found that the change in void ratio between any two state points was independent of the deformation-stress path followed (Matyas and Radhakrishna, 1968). The soil structure constitutive surface (Fig. 13.1a) shows uniqueness for monotonic deformation. Monotonic deformation was obtained when following stress paths that resulted in increasing degrees of saturation and stress paths where the specimen was not allowed to swell. The degree-of-saturation constitutive surface, however, did not show complete uniqueness (Fig. 13.1b). The nonuniqueness of the constitutive surface was attributed to incomplete saturation during the wetting process.

The constitutive surfaces obtained from K_0 loading are presented in Fig. 13.2. The K_0-loaded specimens were prepared with an initial water content of 18.0%, an initial density of 1339 kg/m^3, and an initial degree of saturation of 49.1%. All test paths appeared to form a single warped void ratio surface (Fig. 13.2a). Collapsing soil behavior was once again observed. The degree of saturation was found to be more sensitive to stress state changes than was the void ratio constitutive surface (Fig. 13.2b). The uniqueness of both constitutive surfaces was again restricted to monotonic deformation conditions (Matyas and Radhakrishna, 1968).

Barden et al., (1969a) studied the volume change characteristics of unsaturated soils under K_0 loading conditions. The tests were performed on low- to high-plasticity illite clay specimens (i.e., Westwater and Derwent clays). The total, pore-air, and pore-water pressures were controlled while investigating the effect of various stress paths during K_0 loading. In all cases, the net normal stress $\sigma - u_a$ was increased subsequent to initial conditions. In most

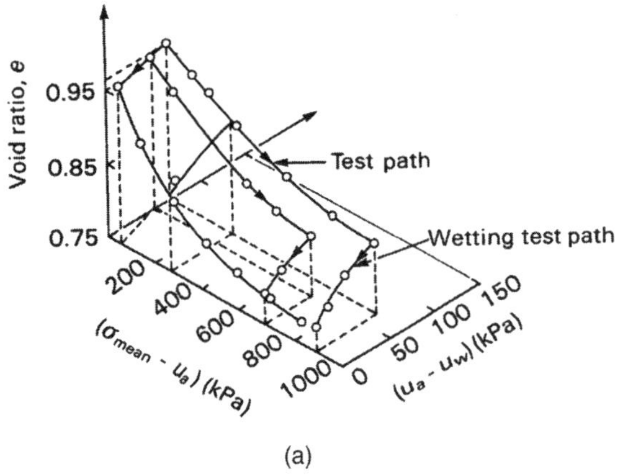

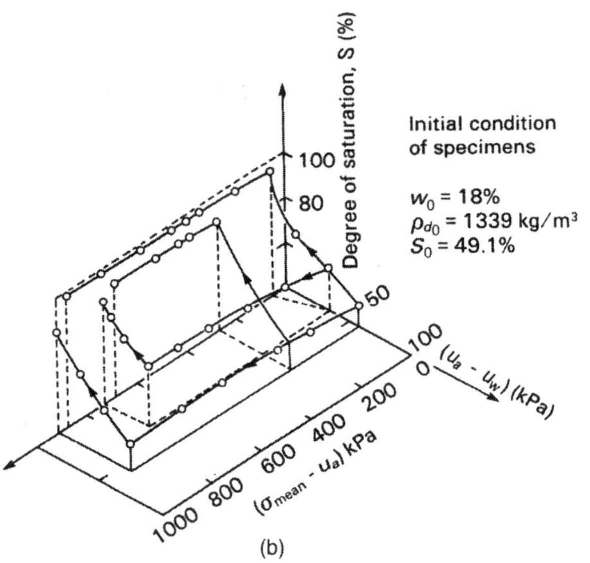

Figure 13.2 Void ratio and degree-of-saturation constitutive surface for mixture of flint and kaolin under K_0 loading: (a) void ratio constitutive surface; (b) degree-of-saturation constitutive surface (after Matyas and Radhakrishna, 1968).

cases, matric suction $u_a - u_w$ was increased subsequent to equilibrium under the initial stress state; however, in a few cases, matric suction was decreased. The results showed that the overall volume change of the specimen was stress path dependent, being a function of whether the soil was going toward saturation or away from saturation. Hysteresis between the saturation and desaturation processes was considered to be the major cause of stress path dependence. It was concluded that the volume change behavior of an unsaturated soil was best analyzed in terms of separate components of stress, $\sigma - u_a$ and $u_a - u_w$.

Barden et al., (1969a) investigated the uniqueness of the soil structure constitutive surface under K_0 loading conditions. Three illite clay soils of low to high plasticity were used in the study. Eleven groups of soil specimens were prepared using either dynamic or static compaction. The specimens were tested in a consolidation cell modified for independent

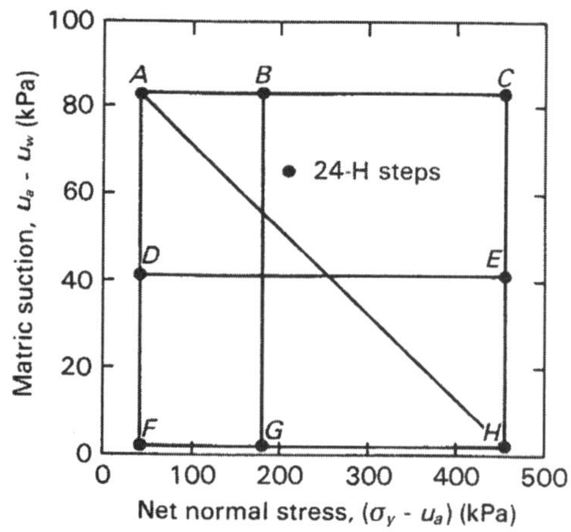

Figure 13.3 Test paths followed during study of constitutive surface uniqueness for unsaturated soil specimens (after Barden et al., 1969a).

control of the total stress, pore-air, and pore-water pressures. Constitutive surfaces were traced for different test paths comprised of various combinations of the stress state variables $\sigma_y - u_a$ and $u_a - u_w$ (Fig. 13.3). In all cases, the pore-air pressure was held constant, while the total normal stress and the pore-water pressure were varied.

A preliminary study using smaller stress increments than those shown in Fig. 13.3 indicated that specimens with similar initial water contents had similar volume changes independent of the test paths. Soil specimens for the remainder of the program were prepared in an "identical" manner. Figure 13.4 presents the soil structure constitutive surface for the illite clay plotted with respect to two independent axes: $\sigma_y - u_a$ and $u_a - u_w$. The results were obtained by following several loading paths from initial to final conditions (i.e., from point A to point H; see Fig. 13.3). The constitutive surface showed uniqueness or loading path independence as long as deformations were monotonic (Barden et al., 1969a). Hysteresis between saturation and desaturation processes was considered to be the major cause of loading path dependence. An increase in the matric suction was found to increase the stiffness of a flocculent soil compacted dry of optimum water content. However, an increase in matric suction had less effect on the stiffness for a soil with a dispersed structure compacted wet of optimum water content. The soil structure stiffness increased with an increase in net normal stress regardless of the type of soil structure.

A second independent constitutive surface was shown for the degree of saturation of the soil. Figure 13.5 shows the degree-of-saturation constitutive surface for a group of soil specimens. The surface was traced by following several loading paths from point A to point H (Fig. 13.5). Uniqueness was shown with respect to the degree of saturation provided the water phase deformation was monotonic.

Several other researchers have suggested the use of net normal stress and matric suction as stress variables for

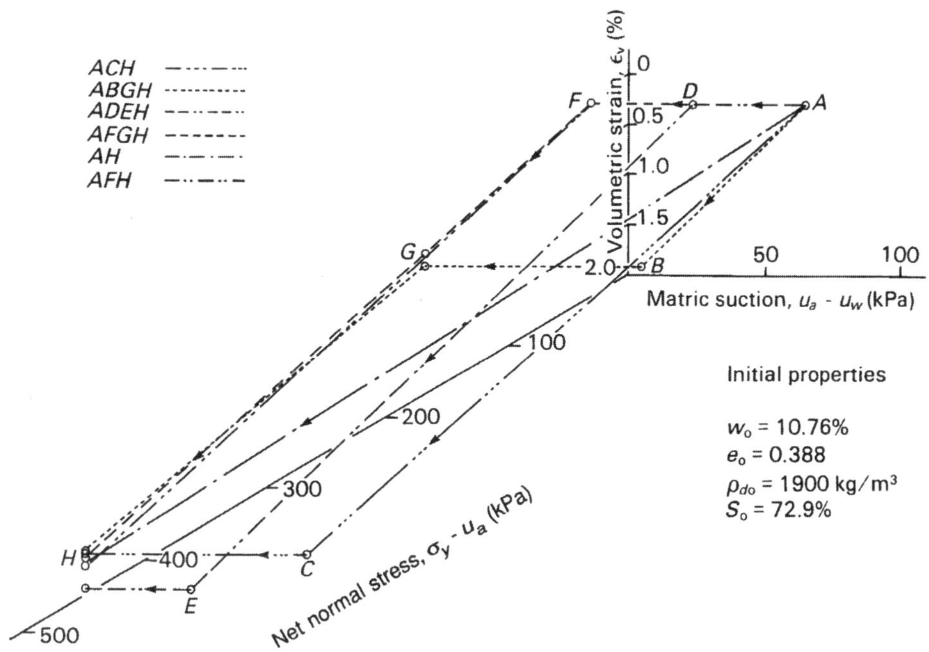

Figure 13.4 Testing soil structure constitutive surface of illite clay for uniqueness (after Barden et al., 1969a).

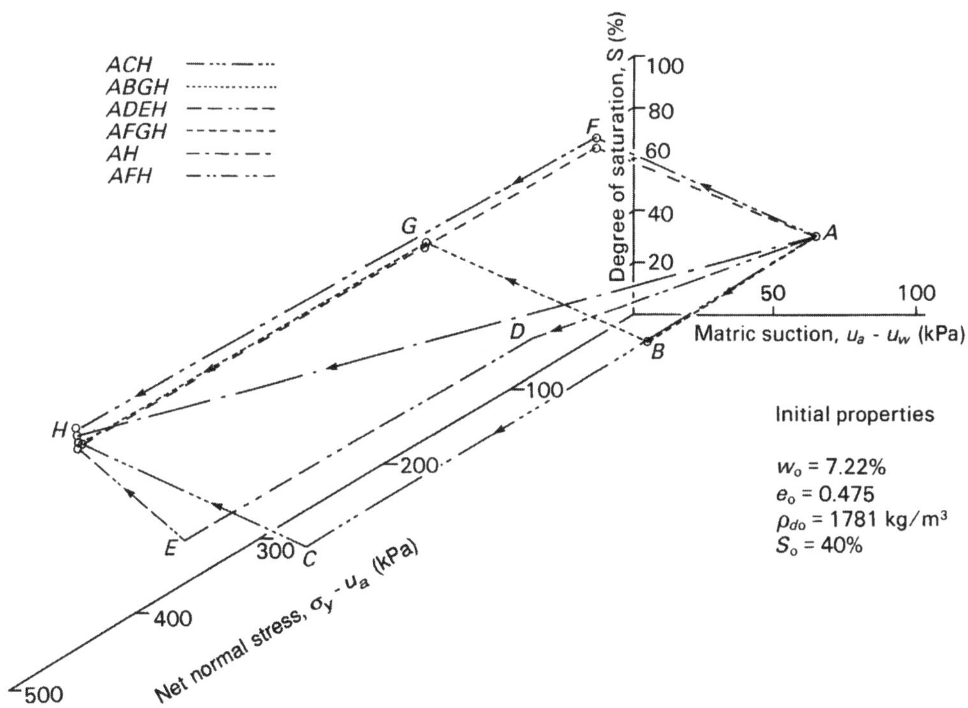

Figure 13.5 Testing degree-of-saturation constitutive surface of illite clay for uniqueness (after Barden et al., 1969a).

describing volume change behavior (Aitchison and Woodburn, 1969; Brackley, 1971; Aitchison and Martin, 1973). The use of $\sigma - u_a$ and $u_a - u_w$ as stress state variables for an unsaturated soil was further advocated by Fredlund and Morgenstern (1977). A stress analysis based on multiphase continuum mechanics showed that any two of three independent stress variables (i.e., $\sigma - u_a$, $u_a - u_w$, and $\sigma - u_w$) could be used to describe the stress state. Experiments to test uniqueness of constitutive surfaces provided further evidence as to why $\sigma - u_a$ and $u_a - u_w$ have been successfully used to describe the volume change characteristics of an unsaturated soil.

In 1977, Fredlund and Morgenstern proposed semiempirical constitutive relations for an unsaturated soil by using any two of the three independent stress state variables. The proposed equations were similar in form to those proposed by Biot (1941) and Coleman (1962). The deformation state variables used to describe volume changes needed to satisfy continuity among the phases of an unsaturated soil (Fredlund, 1973a). The proposed stress and deformation state variables were used to form constitutive relations for the soil structure, air phase, and water phase. However, only two of the three constitutive relations were required to describe changes in the relative volume of each phase. The constitutive relations for the soil structure (i.e., arrangement of soil particles) and the water phase appeared to be most appropriate for use in a volume change analysis. Volume changes associated with the soil structure and the water phase are often written in terms of void ratio change and water content change in geotechnical engineering practice. Volume changes associated with the air phase can be computed as the difference between the overall volume change (i.e., soil structure constitutive surface) and the water volume change.

The proposed constitutive relations can be graphically presented in the form of a deformation state variable versus two independent stress state variables. The proposed constitutive surfaces were experimentally tested for uniqueness near a point by Fredlund and Morgenstern (1976). Four series of experiments were performed involving undisturbed Regina clay and compacted kaolin. The specimens were tested under K_0 and isotropic loading conditions using a modified oedometer and a triaxial apparatus, respectively. The total, pore-air, and pore-water pressures were independently controlled during the tests. The results indicated uniqueness for the constitutive relations as long as the volume change of each phase was monotonic. Uniqueness of the constitutive surfaces under larger stress increments was experimentally tested by Matyas and Radhakrishna (1968) and Barden et al., (1969a).

A number of volume-mass constitutive models based on the critical state concepts of elastoplasticity have been proposed in the literature (Alonso et al., 1987, 1990; Wheeler and Sivakumar, 1995; Delage and Graham, 1995; Chiu and Ng, 2003; Tamagnini, 2004; Thu et al., 2007; Sheng et al., 2008a). These models are considered to be outside the scope of this book.

Pham and Fredlund (2011a) proposed a volume-mass constitutive model for unsaturated soils (Pham, 2005, Fredlund and Pham, 2006; Pham and Fredlund, 2011a). The soil specimens started as a slurry material and considered the independent behavior of the void ratio and water content constitutive relations over a wide range of loading and unloading stress paths. Hysteresis associated with loading and unloading as well as drying and wetting was taken into consideration. The model for hydraulic hysteresis was consistent with those developed in soil physics. The hysteresis model was incorporated into a stress path–dependent constitutive model. The concepts of the Pham and Fredlund (2011a) model are presented later in this chapter.

It is advantageous to commence a fundamental study of the constitutive behavior of an unsaturated soil by starting with a saturated soil. The saturated soil can then be subjected to ever-increasing matric suction values to study the change in overall volume as well as changes in the relative volumes of air and water. It is important that a smooth transition exists between saturated and unsaturated soil states as matric suction is increased and the soil desaturates. The reverse is also true; namely, there should be a smooth transition as a soil moves toward the saturated state through a decrease in matric suction. The Pham and Fredlund (2011a) model is capable of (i) predicting water content change, (ii) predicting void ratio change taking into account both elastic and plastic strains, and (iii) taking into account hysteretic behavior of the SWCC when following a wide range of possible stress paths. Isotropic loading conditions were modeled. A differentiation is made between elastic and plastic strains; however, the model does not take into consideration the independent influence of σ_1, σ_2, and σ_3. In this sense, the model is not a critical state or elastoplastic model.

13.2 CONCEPTS OF VOLUME CHANGE AND DEFORMATION

Volume changes in an unsaturated soil can be expressed in terms of deformations or relative movement of the phases of the soil (i.e., relative volumes of the various phases). The deformation state variables need to be consistent with multiphase continuum mechanics principles. A change in the relative position of points or particles in a body forms the basis for establishing deformation state variables. These variables should produce the displacements of the body under consideration when integrated over the body. This concept applies to a single or multiphase system and is independent of the physical properties.

Two sets of deformation state variables are required to adequately describe the volume changes associated with an unsaturated soil. The deformation state variables associated with the soil structure and the water phase are commonly used in a volume change analysis (Biot, 1941; Coleman, 1962; Matyas and Radhakrishna, 1968). Void ratio changes or porosity changes can also be used as deformation state variables representing soil structure deformation. Changes in the volume of water in the soil can be considered as the deformation state variable for the water phase.

There are several ways to describe the relative movement or deformation in a phase from the standpoint of continuum mechanics kinematics. Two of these descriptions are mentioned for consideration relative to unsaturated soil behavior. The position of each particle is described as a function of its initial position and time for a referential description. In other words, the original position and time are the independent variables. A fixed element of mass is chosen for a referential element and its motion is traced. The description is generally called the Lagrangian description when the reference configuration is the initial configuration. The time variable disappears when equilibrium conditions are achieved. The referential description is commonly used in problems involving the elasticity of solids where the initial geometry, boundary, and loading conditions are specified.

The position of each particle is described as a function of its current position and time for a spatial description of deformation. A fixed region in space is chosen instead of an element of mass. The spatial description is generally used in fluid mechanics and is referred to as the Eulerian description. The time variable vanishes under steady-state flow conditions. The Lagrangian and the Eulerian descriptions give the same results for cases with infinitesimal deformation.

13.2.1 Continuity Requirements

A saturated soil is visualized as a fluid-solid multiphase. The soil particles form a structure with voids filled with water. The soil structure deforms and changes volume under an applied stress gradient. The soil structure volume change represents the overall volume change of the soil. Overall volume change must be equal to the sum of changes in volumes associated with the solid phase (i.e., soil particles) and the fluid phase (i.e., water). The equality concept associated with a multiphase system is referred to as the "continuity requirement" (Fredlund, 1973a).

The continuity requirement is a volumetric restriction that prevents "gaps" from forming between the phases of a deformed multiphase system. The volumetric requirement also ensures that conservation of mass is maintained. Volume changes in a saturated soil are primarily the result of water flowing in or out of the soil since the particles are essentially incompressible. An unsaturated soil can be visualized as a mixture with two phases that come to equilibrium under applied stress gradients (i.e., soil particles and contractile skin) and two phases that flow under applied stress gradients (i.e., air and water). Consider an element of soil that deforms under an applied stress gradient. The total volume change of the soil element must be equal to the sum of the volume changes associated with each phase. The continuity requirement for the unsaturated soil can be stated as follows for the case where the soil particles are assumed to be incompressible:

$$\frac{\Delta V_v}{V_0} = \frac{\Delta V_w}{V_0} + \frac{\Delta V_a}{V_0} + \frac{\Delta V_c}{V_0} \tag{13.1}$$

where:

V_0 = initial overall volume of an unsaturated soil element,
V_v = volume of soil voids,
V_w = volume of water,
V_a = volume of air, and
V_c = volume of contractile skin.

The continuity requirement can be rewritten as follows for the case where the contractile skin volume change is internal to the element:

$$\frac{\Delta V_v}{V_0} = \frac{\Delta V_w}{V_0} + \frac{\Delta V_a}{V_0} \tag{13.2}$$

The above continuity requirement for an unsaturated soil shows that the volume changes associated with any two of the three volume variables must be measured while the third volume change can be computed. In practice, the overall and water volume changes are usually measured while the air volume change is calculated. Suitable deformation state variables can now be defined that are consistent with the "continuity requirement."

13.2.2 Overall Volume Change

The overall or total volume change of a soil refers to the volume change of the soil structure. Consider a two-dimensional representation of a referential element of unsaturated soil as shown in Fig. 13.6. The element is referenced to a fixed mass of soil particles. The element has infinitesimal dimensions of dx, dy, and dz in the x-, y-, and z-directions, respectively. Only the x- and y-directions are shown in Fig. 13.6.

The soil element is assumed to undergo translations of u, v, and w from their original x-, y-, and z-coordinate positions, respectively. The final position of the element becomes $x + u$, $y + v$, and $z + w$. The element is assumed to deform in response to an applied stress gradient. The deformation consists of a change in length and a rotation of the element sides, as illustrated in Fig. 13.6. The changes in length in the x-, y-, and z-directions can be written as $(\partial u/\partial x)\,dx$, $(\partial v/\partial y)\,dy$, and $(\partial w/\partial z)\,dz$. Defining normal strain ε as a change in length per unit length, the normal strains of the soil structure in the x-, y-, and z-directions can be expressed as

$$\varepsilon_x = \frac{\partial u}{\partial x} \tag{13.3}$$

$$\varepsilon_y = \frac{\partial v}{\partial y} \tag{13.4}$$

$$\varepsilon_z = \frac{\partial w}{\partial z} \tag{13.5}$$

where:

ε_x = normal strain in the x-direction,
ε_y = normal strain in the y-direction, and
ε_z = normal strain in the z-direction.

The above normal strains are positive for an increase in length and negative for a decrease in length. The distortion of the element is expressed in terms of shear strains, which correspond to two orthogonal directions. Shear strain γ is defined as the change in the original right angle between two axes (Chou and Pagano, 1967a). The angle is measured in radians. A positive shear strain indicates that the right angle

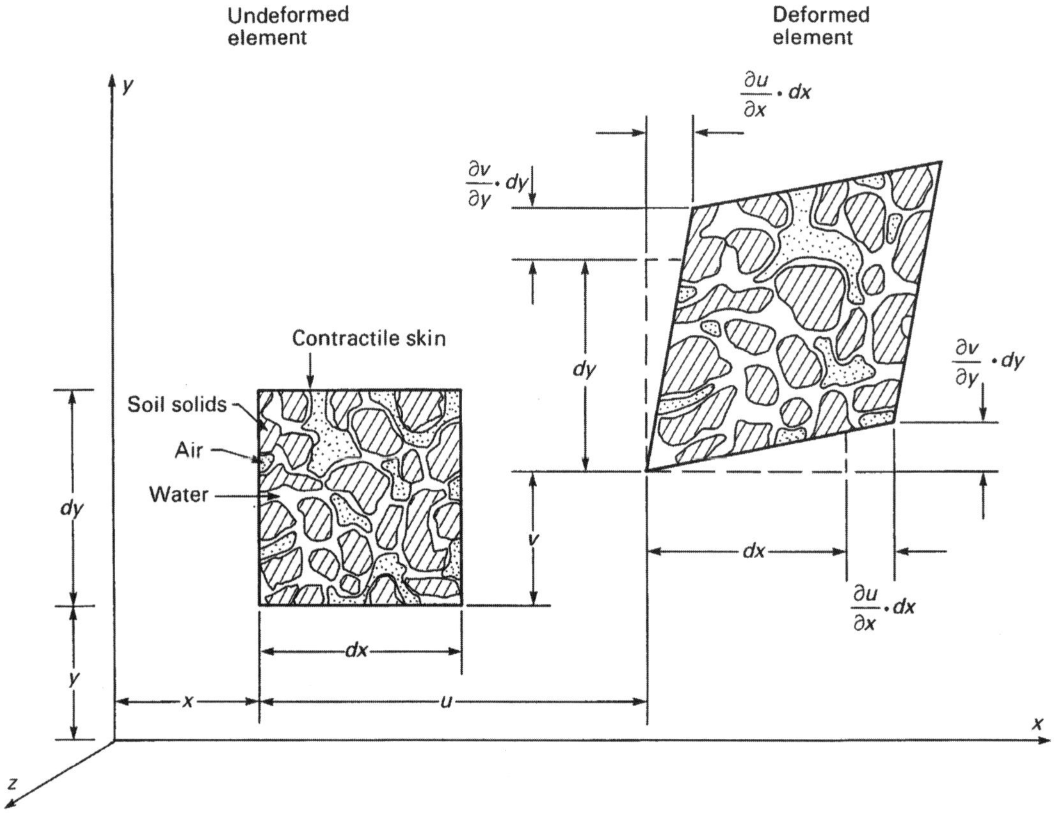

Figure 13.6 Translation and deformation of two-dimensional element of unsaturated soil.

between the positive directions of the two axes decreases. The shear strain components of a three-dimensional element are formulated as

$$\gamma_{xy} = \frac{\partial u}{\partial y} + \frac{\partial v}{\partial x} \tag{13.6}$$

$$\gamma_{yz} = \frac{\partial v}{\partial z} + \frac{\partial w}{\partial y} \tag{13.7}$$

$$\gamma_{zx} = \frac{\partial w}{\partial x} + \frac{\partial u}{\partial z} \tag{13.8}$$

where:

γ_{xy} = shear strain on the z-plane (i.e., $\gamma_{xy} = \gamma_{yx}$),
γ_{yz} = shear strain on the x-plane (i.e., $\gamma_{yz} = \gamma_{zy}$), and
γ_{zx} = shear strain on the y-plane (i.e., $\gamma_{zx} = \gamma_{xz}$).

The normal and shear strains of the soil structure can be written as a deformation tensor:

$$\begin{bmatrix} \varepsilon_x & \frac{1}{2}\gamma_{yx} & \frac{1}{2}\gamma_{zx} \\ \frac{1}{2}\gamma_{xy} & \varepsilon_y & \frac{1}{2}\gamma_{zy} \\ \frac{1}{2}\gamma_{xz} & \frac{1}{2}\gamma_{yz} & \varepsilon_z \end{bmatrix} \tag{13.9}$$

The sum of the normal strain components along the trace of the matrix is called volumetric strain:

$$\varepsilon_v = \varepsilon_x + \varepsilon_y + \varepsilon_z \tag{13.10}$$

where:

ε_v = volumetric strain.

Volumetric strain is equal to the difference between the volumes of the voids in the element before and after deformation, ΔV_v, referenced to the initial volume of the element, V_0:

$$\varepsilon_v = \frac{\Delta V_v}{V_0} \tag{13.11}$$

The volumetric strain ε_v can be used as a deformation state variable for the overall referential element or the soil structure. Soil structure volume change is the result of normal components of deformation.

13.2.3 Water and Air Volume Changes

The unsaturated soil element shown in Fig. 13.6 can be used to describe changes in the volume of the voids filled with air and water. The element of unsaturated soil is considered as a spatial element with respect to the water and air phases. The change in the volume of each fluid in the voids is defined as the difference between the fluid volumes in the voids prior to a change in the stress state and the fluid volumes after a change in the stress state (Fig. 13.6). The fluid change (i.e., air or water) per unit initial volume of the soil element can be considered as the deformation state variables or a change in fluid storage. The deformation variables can be written as $\Delta V_w / V_0$ for the water phase and $\Delta V_a / V_0$ for the air phase.

13.3 VOLUME-MASS CONSTITUTIVE RELATIONS

Constitutive relations for an unsaturated soil can be formulated by combining selected deformation state variables with appropriate stress state variables. Stress tensors comprised of stress state variables for an unsaturated soil were presented in Chapter 3. Deformation tensors comprised of deformation state variables must satisfy continuity requirements. Combining deformation state variables and stress state variables in the form of empirical equations results in the definition of volumetric deformation properties. Several forms of constitutive relations are discussed in the following sections. The constitutive relations can be used to predict volume changes (i.e., overall volume change and phase volume changes) that occur as a result of changes in the stress state variables.

13.3.1 Elasticity Form

There are two main approaches that can be used in establishing the stress-deformation relationships. These are the "mathematical" approach and the "semiempirical" approach. In the mathematical approach, each component of the deformation state variable tensor is expressed as a linear combination of the stress state variables or vice versa. In other words, the relationship between the stress and deformation state variables is expressed by a series of linear equations. The difficulty with this approach is the large number of soil properties that must be determined.

The semiempirical approach involves several assumptions which are based on experimental evidence from observing the behavior of many materials (Chou and Pagano, 1967): (1) normal stress does not produce shear strain, (2) shear stress does not cause normal strain, and (3) a shear stress component τ causes only one shear strain component, γ. In addition, the principle of superposition is assumed to be applicable to cases involving small deformations.

The semiempirical approach is generally adopted in soil mechanics and is used herein to formulate the constitutive relations for unsaturated soils. The constitutive equations must be experimentally tested to ensure uniqueness. The uniqueness theorem for an elastic solid with a positive definite strain energy function states that there exists a one-to-one correspondence between elastic deformations and stresses (Fung, 1965).

The constitutive relations for the soil structure in saturated soil mechanics can be formulated in the form of a generalized Hooke's law using effective stress variables $\sigma - u_w$. For an isotropic and linearly elastic soil structure, the strain-stress constitutive relations in the x-, y-, and z-directions can be written in the following form:

$$\varepsilon_x = \frac{\sigma_x - u_w}{E} - \frac{\mu}{E}(\sigma_y + \sigma_z - 2u_w) \tag{13.12}$$

$$\varepsilon_y = \frac{\sigma_y - u_w}{E} - \frac{\mu}{E}(\sigma_x + \sigma_z - 2u_w) \tag{13.13}$$

$$\varepsilon_z = \frac{\sigma_z - u_w}{E} - \frac{\mu}{E}(\sigma_x + \sigma_y - 2u_w) \tag{13.14}$$

where:

σ_x = total normal stress in the x-direction,
σ_y = total normal stress in the y-direction,
σ_z = total normal stress in the z-direction,
E = modulus of elasticity or Young's modulus for the soil structure, and
μ = Poisson's ratio.

The sum of the normal strains ε_x, ε_y, and ε_z constitutes the volumetric strain ε_v. For a saturated soil, the overall volume change of the soil is equal to the water volume change since soil particles are essentially incompressible.

The constitutive relations for an unsaturated soil can be formulated as an extension of the equations used for a saturated soil while using the appropriate stress state variables (Fredlund and Morgenstern, 1976). Let us assume that the soil behaves as an isotropic, linear elastic material. The following constitutive relations are expressed in terms of the $\sigma - u_a$ and $u_a - u_w$ stress state variables. The formulation is similar in form to that proposed by Biot in 1941. The soil structure constitutive relations associated with the normal strains in the x-, y-, and z-directions are as follows:

$$\varepsilon_x = \frac{\sigma_x - u_a}{E} - \frac{\mu}{E}(\sigma_y + \sigma_z - 2u_a) + \frac{u_a - u_w}{H} \tag{13.15}$$

$$\varepsilon_y = \frac{\sigma_y - u_a}{E} - \frac{\mu}{E}(\sigma_x + \sigma_z - 2u_a) + \frac{u_a - u_w}{H} \tag{13.16}$$

$$\varepsilon_z = \frac{\sigma_z - u_a}{E} - \frac{\mu}{E}(\sigma_x + \sigma_y - 2u_a) + \frac{u_a - u_w}{H} \tag{13.17}$$

where:

H = modulus of elasticity for the soil structure with respect to a change in matric suction, $u_a - u_w$.

The constitutive equations associated with the shear deformations are

$$\gamma_{xy} = \frac{\tau_{xy}}{G} \tag{13.18}$$

$$\gamma_{yz} = \frac{\tau_{yz}}{G} \tag{13.19}$$

$$\gamma_{zx} = \frac{\tau_{zx}}{G} \tag{13.20}$$

where:

τ_{xy} = shear stress on the x-plane in the y-direction (i.e., $\tau_{xy} = \tau_{yx}$),
τ_{yz} = shear stress on the y-plane in the z-direction (i.e., $\tau_{yz} = \tau_{zy}$),
τ_{zx} = shear stress on the z-plane in the x-direction (i.e., $\tau_{zx} = \tau_{xz}$), and
G = shear modulus.

The modulus of elasticity E in the above equations is defined with respect to a change in the net normal stress $(\sigma - u_a)$ during uniaxial loading. The above constitutive equations can be applied in an incremental manner when the stress-strain curve is nonlinear. Figure 13.7 defines the sign convention using a typical stress-strain curve. An incremental procedure using small increments of stress and strain can be used to apply the linear elastic formulation to a nonlinear stress-strain curve. The nonlinear stress-strain curve is assumed to be linear within each stress and strain increment. The elastic moduli E and H have negative signs as indicated in Fig. 13.7 and may vary in magnitude from one increment to another. The soil structure constitutive relations associated with the normal strains can be written in an incremental form:

$$d\varepsilon_x = \frac{d(\sigma_x - u_a)}{E} - \frac{\mu}{E} d(\sigma_y + \sigma_z - 2u_a) + \frac{d(u_a - u_w)}{H} \quad (13.21)$$

$$d\varepsilon_y = \frac{d(\sigma_y - u_a)}{E} - \frac{\mu}{E} d(\sigma_x + \sigma_z - 2u_a) + \frac{d(u_a - u_w)}{H} \quad (13.22)$$

$$d\varepsilon_z = \frac{d(\sigma_z - u_a)}{E} - \frac{\mu}{E} d(\sigma_x + \sigma_y - 2u_a) + \frac{d(u_a - u_w)}{H} \quad (13.23)$$

The above equations represent the general elasticity constitutive relations for the soil structure. The left-hand side of the equations refers to a change in the deformation state variable while the right-hand side contains changes in the stress state variables. A change in the volumetric strain of the soil for each increment, $d\varepsilon_v$, can be obtained by summing the incremental changes in normal strains in the x-, y-, and z-directions:

$$d\varepsilon_v = d\varepsilon_x + d\varepsilon_y + d\varepsilon_z \quad (13.24)$$

where:

$d\varepsilon_v$ = volumetric strain change for each increment.

Substituting Eqs. 13.21, 13.22, and 13.23 into Eq. 13.24 gives

$$d\varepsilon_v = 3\left(\frac{1-2\mu}{E}\right) d\left(\frac{\sigma_x + \sigma_y + \sigma_z}{3} - u_a\right) + \frac{3}{H} d(u_a - u_w) \quad (13.25)$$

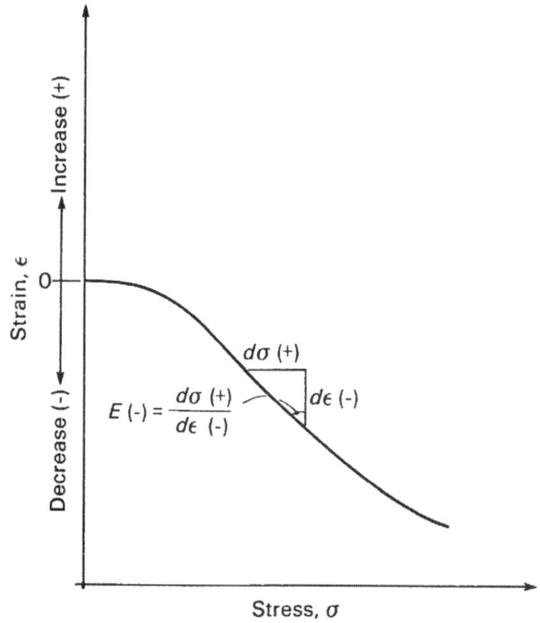

Figure 13.7 Definition of variables for nonlinear stress-strain curve for soil.

The above equation can be simplified as follows:

$$d\varepsilon_v = 3\left(\frac{1-2\mu}{E}\right) d(\sigma_{mean} - u_a) + \frac{3}{H} d(u_a - u_w) \quad (13.26)$$

where:

σ_{mean} = mean total normal stress [i.e., $(\sigma_x + \sigma_y + \sigma_z)/3$].

The volumetric strain change $d\varepsilon_v$ is equal to the volume change of the soil element divided by the initial volume of the element:

$$d\varepsilon_v = \frac{dV_v}{V_0} \quad (13.27)$$

The initial volume V_0 refers to the volume of the soil element at the start of the volume change process. Therefore, V_0 remains constant for all increments. The change in volumetric strain, $d\varepsilon_v$, at the end of each increment can be computed from Eq. 13.26, and the volume change of the soil element, dV_v, is obtained from Eq. 13.27. The summation of the volumetric strain changes for each increment gives the final volumetric strain of the soil:

$$\varepsilon_v = \sum d\varepsilon_v \quad (13.28)$$

13.3.2 Water Phase Constitutive Relationship

The soil structure constitutive relationship is not sufficient to completely describe volume changes in an unsaturated soil. Either an air or water phase constitutive relationship must be formulated. It is suggested that the water phase may be the most suitable phase for formulating the second constitutive relationship. The water phase constitutive relationship describes the change in the volume of water present in the referential soil structure element under various stress conditions. Water is assumed to be incompressible and the equation accounts for the net inflow or outflow of water from the element. The water phase constitutive relationship can be formulated in a semiempirical manner on the basis of a linear combination of the stress state variables. The incremental constitutive equation for the water phase can be written as

$$\frac{dV_w}{V_0} = \frac{3}{E_w} d\left(\sigma_{\text{mean}} - u_a\right) + \frac{d\left(u_a - u_w\right)}{H_w} \quad (13.29)$$

where:

E_w = water volumetric modulus associated with a change in $\sigma_{\text{mean}} - u_a$ and

H_w = water volumetric modulus associated with a change in $u_a - u_w$.

The summation of water volume changes for each increment in stress state gives the final change in the volume of water:

$$\frac{\Delta V_w}{V_0} = \frac{\sum dV_w}{V_0} \quad (13.30)$$

13.3.3 Change in Volume of Air

Changes in the volume of air in an element can be computed as the difference between the soil structure and water volume changes. The continuity requirement can also be written in an incremental form using volumetric strain change $d\varepsilon_v$:

$$d\varepsilon_v = \frac{dV_w}{V_0} + \frac{dV_a}{V_0} \quad (13.31)$$

The constitutive relationships for an unsaturated soil can be presented graphically in the form of constitutive surfaces (Fig. 13.8). The deformation state is plotted with respect to the $\sigma_{\text{mean}} - u_a$ and $u_a - u_w$ stress state variables. The coefficients used in the constitutive equations are the slopes of the constitutive surface at a point. The slopes must be defined with respect to both axes. For example, the slopes of the soil structure constitutive surface at a point are equal to $3(1-2\mu)/E$ and $3/H$ with respect to $\sigma_{\text{mean}} - u_a$ and $u_a - u_w$, respectively (Eq. 13.26 and Fig. 13.8a). The coefficients on the constitutive surface are referred to as "volumetric deformation coefficients." These coefficients vary from one stress state to another for a curved or nonplanar constitutive surface. Similarly, the slopes on the water phase constitutive surface at a point are $3/E_w$ and $1/H_w$ with respect to the stress variables $\sigma_{\text{mean}} - u_a$ and $u_a - u_w$, respectively (Eq. 13.29 and Fig. 13.8b).

The above constitutive relations can be formulated for a general, three-dimensional loading in the x-, y-, and z-directions. The formulations can also be applied to special loading conditions as depicted in Fig. 13.9. The constitutive equations for various types of loading conditions can be derived from general strain-stress formulations.

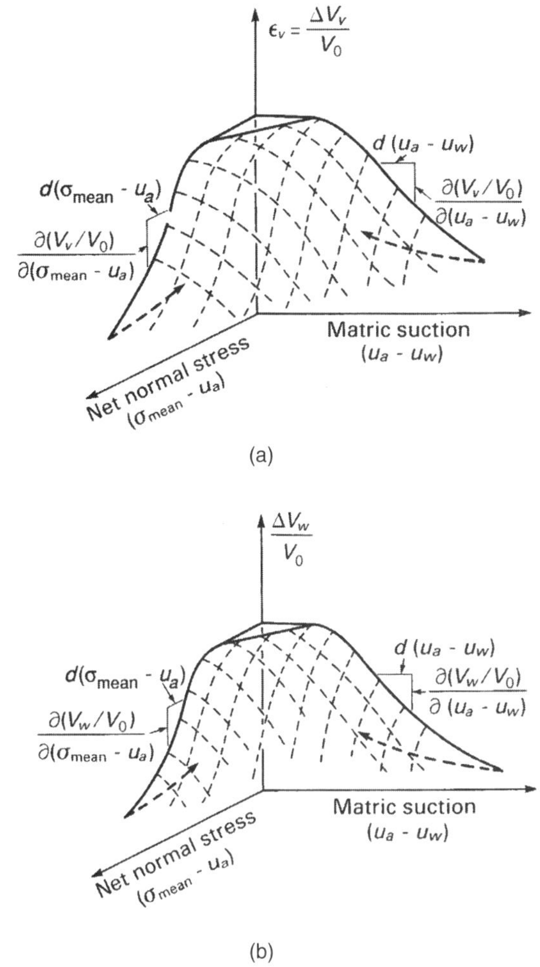

Figure 13.8 Three-dimensional constitutive surfaces for unsaturated soil: (a) soil structure constitutive surface; (b) water phase constitutive surface.

13.3.4 Isotropic Loading

The total stress increments are equal for isotropic loading in three directions (i.e., $d\sigma_x = d\sigma_y = d\sigma_z = d\sigma_3$). No shear stresses are developed in the soil mass under isotropic loading conditions. Substituting the conditions associated with isotropic loading into Eq. 13.26 gives the soil structure constitutive equation:

$$d\varepsilon_v = 3\left(\frac{1-2\mu}{E}\right) d\left(\sigma_3 - u_a\right) + \frac{3}{H} d\left(u_a - u_w\right) \quad (13.32)$$

where:

σ_3 = total isotropic stress.

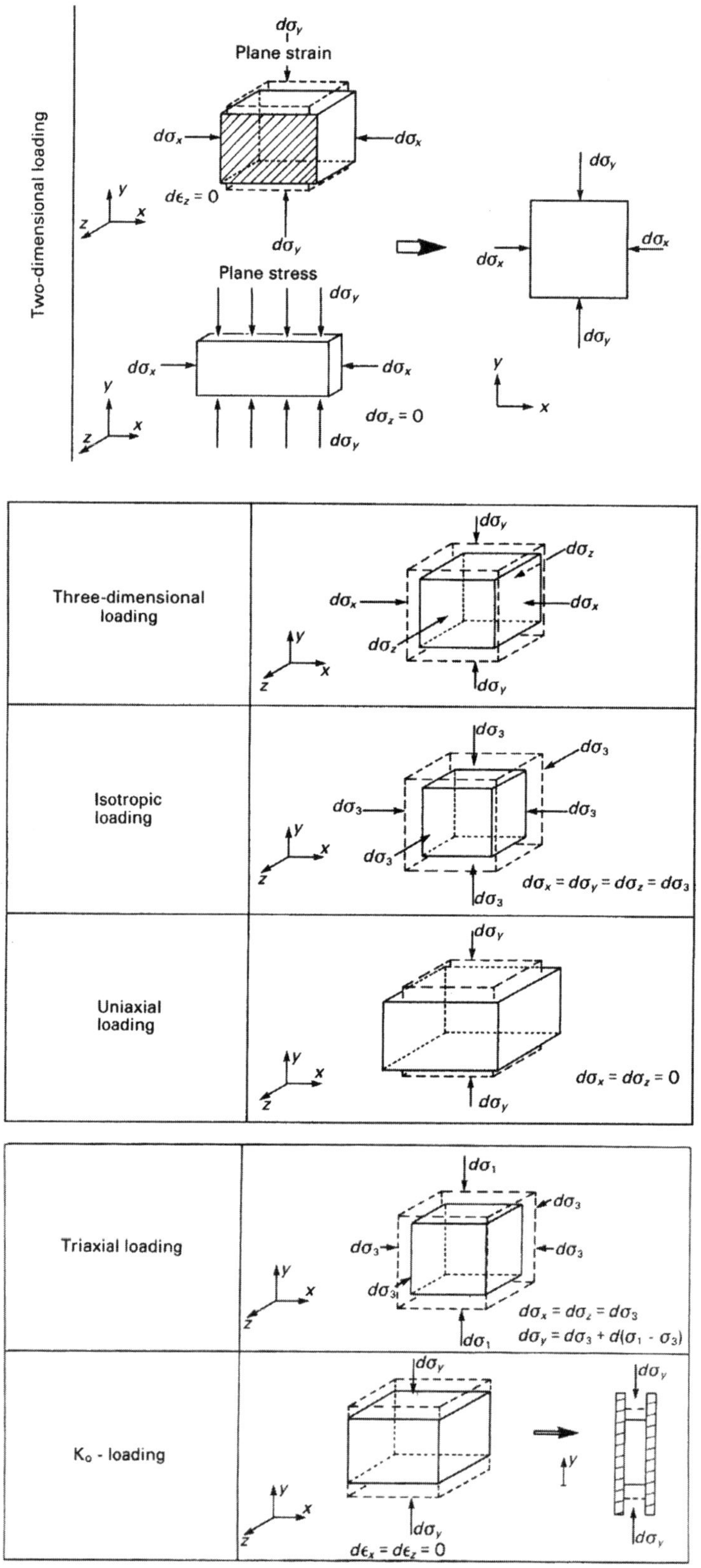

Figure 13.9 Various forms of loading of soil and associated deformations.

The inverse of the coefficients in the first term (i.e., $E/[3(1-2\mu)]$) is commonly referred to as the "bulk modulus" of the soil (Lambe and Whitman, 1979).

The constitutive equation for the water phase can be derived from Eq. 13.29:

$$\frac{dV_w}{V_0} = \frac{3}{E_w} d(\sigma_3 - u_a) + \frac{d(u_a - u_w)}{H_w} \tag{13.33}$$

The soil will undergo equal deformation in all directions (or isotropic compression) when subjected to an equal all-around pressure provided the soil properties are also isotropic (Fig. 13.9).

13.3.5 Uniaxial Loading

A total stress increment is applied to the soil in one direction for uniaxial loading (e.g., the vertical direction), as illustrated in Fig. 13.9. It is assumed that no shear stress is developed on the x-, y-, and z-planes. The stress increase is applied in the y-direction, $d\sigma_y$, while the total stress change in the other two directions is zero (i.e., $d\sigma_x = d\sigma_z = 0$). The soil compresses in the y-direction and expands in the x- and z-directions. Applying these stress conditions to Eq. 13.26 gives the soil structure constitutive equation for uniaxial loading:

$$d\varepsilon_v = 3\left(\frac{1-2\mu}{E}\right) d\left(\frac{1}{3}\sigma_y - u_a\right) + \frac{3}{H} d(u_a - u_w) \tag{13.34}$$

The water phase constitutive equation for uniaxial loading is obtained from Eq. 13.29:

$$\frac{dV_w}{V_0} = \frac{3}{E_w} d\left(\frac{1}{3}\sigma_y - u_a\right) + \frac{d(u_a - u_w)}{H_w} \tag{13.35}$$

13.3.6 Triaxial Loading

The triaxial loading conditions shown in Fig. 13.9 can be considered as a superposition of isotropic and uniaxial loading. Isotropic loading applies to an all-around pressure change of $d\sigma_3$, whereas the uniaxial pressure change applies to a deviator stress $d(\sigma_1 - \sigma_3)$ in the vertical or y-direction. The x-, y-, and z-planes are assumed to be principal planes. The above stress conditions [i.e., $d\sigma_x = d\sigma_z = d\sigma_3$ and $d\sigma_y = d\sigma_3 + d(\sigma_1 - \sigma_3)$] can be substituted into Eq. 13.26 to give the soil structure constitutive equation for triaxial loading:

$$d\varepsilon_v = 3\left(\frac{1-2\mu}{E}\right) d\left(\frac{\sigma_3 + (\sigma_3 + \sigma_1 - \sigma_3) + \sigma_3}{3} - u_a\right) + \frac{3}{H} d(u_a - u_w) \tag{13.36}$$

Rearranging the above equation gives

$$d\varepsilon_v = 3\left(\frac{1-2\mu}{E}\right)\left[d(\sigma_3 - u_a) + d\left(\frac{1}{3}(\sigma_1 - \sigma_3)\right)\right] + \frac{3}{H} d(u_a - u_w) \tag{13.37}$$

The water phase constitutive relation can be written by substituting the triaxial stress conditions into Eq. 13.29:

$$\frac{dV_w}{V_0} = \frac{3}{E_w}\left[d(\sigma_3 - u_a) + d\left(\frac{1}{3}(\sigma_1 - \sigma_3)\right)\right] + \frac{d(u_a - u_w)}{H_w} \tag{13.38}$$

The du_a and $d(u_a - u_w)$ terms in Eqs. 13.37 and 13.38 refer to the pore-air pressure and matric suction increments, respectively, during triaxial loading. For triaxial testing in the laboratory, isotropic loading is generally first applied followed by uniaxial compression. Therefore, it may be useful to separate the pore-air pressure and matric suction increments during the two loading conditions. This can be done by superimposing Eq. 13.32 (i.e., isotropic loading) and Eq. 13.34 (i.e., uniaxial loading) to form the soil structure constitutive relationship for triaxial loading:

$$d\varepsilon_v = 3\left(\frac{1-2\mu}{E}\right)\left[d(\sigma_3 - u_{ai}) + d\left(\frac{1}{3}(\sigma_1 - \sigma_3) - u_{au}\right)\right] + \frac{3}{H}\left\{d(u_a - u_w)_i + d(u_a - u_w)_u\right\} \tag{13.39}$$

where:

du_{ai} = pore-air pressure increment during isotropic loading,
du_{au} = pore-air pressure increment during uniaxial loading,
du_{wi} = pore-water pressure increment during isotropic loading, and
du_{wu} = pore-water pressure increment during uniaxial loading.

Equation 13.39 can be used to compute the volumetric strain increment during triaxial loading. Only the first and third terms are used since the other terms drop out during isotropic compression. On the other hand, the second and fourth terms are used to compute the volumetric strain increment during uniaxial compression while the other terms become zero. In the end, volumetric strain increments during isotropic and uniaxial compression can be summed to give the total volumetric strain associated with triaxial loading. It should be noted that Eq. 13.39 reverts to Eq. 13.37 by substituting $du_a = du_{ai} + du_{au}$ and $du_w = du_{wi} + du_{wu}$ into Eq. 13.39.

Similarly, the water phase constitutive relation can be obtained by superimposing Eq. 13.33 (i.e., isotropic loading) and Eq. 13.35 (i.e., uniaxial loading):

$$\frac{dV_w}{V_0} = \frac{3}{E_w}\left[d(\sigma_3 - u_{ai}) + d\left(\frac{1}{3}(\sigma_1 - \sigma_3) - u_{au}\right)\right] + \frac{d(u_a - u_w)_i}{H_w} + \frac{d(u_a - u_w)_u}{H_w} \tag{13.40}$$

The first and third terms in Eq. 13.40 are used for isotropic loading whereas the second and fourth terms are used for uniaxial compression. The sum of the water volume changes during isotropic and uniaxial compression gives the total change in volume of water during triaxial loading. Substituting $du_a = du_{ai} + du_{au}$ and $du_w = du_{wi} + du_{wu}$ into Eq. 13.40 gives Eq. 13.38.

13.3.7 K_0 Loading

A total stress increment of $d\sigma_y$ is applied in the vertical direction for K_0 loading, while the soil is not permitted to deform laterally (i.e., $d\varepsilon_x = d\varepsilon_z = 0$). The K_0 loading condition occurs during one-dimensional compression where deformation is only allowed in the vertical direction (i.e., $d\varepsilon_y$). The total stress increment $d\sigma_y$ is applied in the vertical direction. The strain conditions during the K_0 loading can be applied to Eqs. 13.21 and 13.22. Multiplying Eq. 13.23 by Poisson's ratio μ and adding the result to Eq. 13.21 gives

$$\left(\frac{1-\mu^2}{E}\right) d(\sigma_z - u_a) = \frac{\mu}{E}(1+\mu)\, d(\sigma_y - u_a) - \left(\frac{1+\mu}{H}\right) d(u_a - u_w) \quad (13.41)$$

The above equation can be arranged as follows:

$$d(\sigma_z - u_a) = \left(\frac{\mu}{1-\mu}\right) d(\sigma_y - u_a) - \frac{E}{(1-\mu)H} d(u_a - u_w) \quad (13.42)$$

Similarly, Eq. 13.23 can be multiplied by Poisson's ratio μ and added to Eq. 13.21 to give

$$d(\sigma_x - u_a) = \left(\frac{\mu}{1-\mu}\right) d(\sigma_y - u_a) - \frac{E}{(1-\mu)H} d(u_a - u_w) \quad (13.43)$$

Substituting Eqs. 13.42 and 13.43 into Eq. 13.22 gives

$$d\varepsilon_y = \frac{1-\mu-2\mu^2}{E(1-\mu)} d(\sigma_y - u_a) + \frac{1-\mu+2\mu}{H(1-\mu)} d(u_a - u_w) \quad (13.44)$$

Volumetric strain $d\varepsilon_v$ is equal to the strain in the vertical direction, $d\varepsilon_y$, for K_0 loading conditions. Equation 13.44 for the soil structure can then be rewritten as follows:

$$d\varepsilon_v = \frac{(1+\mu)(1-2\mu)}{E(1-\mu)} d(\sigma_y - u_a) + \frac{1+\mu}{H(1-\mu)} d(u_a - u_w) \quad (13.45)$$

The inverse of the coefficients in the first term (i.e., $E(1-\mu)/[(1+\mu)(1-2\mu)]$) is referred to as the "constrained modulus" (Lambe and Whitman, 1979).

The water phase constitutive relation is obtained in a similar manner (i.e., substituting Eqs. 13.42 and 13.43 into Eq. 13.29):

$$\frac{dV_w}{V_0} = \frac{1+\mu}{E_w(1-\mu)} d(\sigma_y - u_a) + \left(\frac{1}{H_w} - \frac{2(E/H)}{E_w(1-\mu)}\right) d(u_a - u_w) \quad (13.46)$$

13.3.8 Plane Strain Loading

Many geotechnical problems can be simplified into a two-dimensional form using the concepts of plane strain and plane stress loading. If an earth structure is significantly long in one direction (e.g., the z-direction) in comparison to the other two directions (e.g., the x- and y-directions) and the loadings are applied only on the x- and y-planes, the structure can be modeled as a plane strain problem. The stability of slopes, earth pressures on retaining wall, and the bearing capacity of strip footing are examples of problems that are commonly analyzed assuming plane strain loading conditions. The soil deformation in the z-direction is assumed to be negligible for plane strain conditions (i.e., $d\varepsilon_z = 0$). Imposing a condition of zero strain in the z-direction in Eq. 13.23 gives

$$d(\sigma_z - u_a) = \mu d(\sigma_x + \sigma_y - 2u_a) - (H/E)\, d(u_a - u_w) \quad (13.47)$$

Volumetric strain during plane strain loading is obtained by substituting Eq. 13.47 into Eq. 13.25:

$$d\varepsilon_v = \frac{2(1+\mu)(1-2\mu)}{E} d(\sigma_{ave} - u_a) + 2\left(\frac{1+\mu}{H}\right) d(u_a - u_w) \quad (13.48)$$

where:

σ_{ave} = average total normal stress for two-dimensional loading [i.e., $(\sigma_x + \sigma_y)/2$].

The above equation can be used as the soil structure constitutive relationship for plane strain loading.

The water phase constitutive equation is obtained from Eq. 13.29 by replacing the $d(\sigma_z - u_a)$ term with Eq. 13.47:

$$\frac{dV_w}{V_0} = 2\left(\frac{1+\mu}{E_w}\right) d(\sigma_{ave} - u_a) + \left(\frac{1}{H_w} - \frac{(E/H)}{E_w}\right) d(u_a - u_w) \quad (13.49)$$

13.3.9 Plane Stress Loading

The change in the total stress in the z-direction is assumed to be negligible for plane stress conditions (i.e., $d\sigma_z = 0$). It should be noted that plane stress loading conditions are not common in soil mechanics applications. The soil structure constitutive relationship for the plane stress loading is computed by setting $d\sigma_z$ equal to zero in Eq. 13.26:

$$d\varepsilon_v = 2\left(\frac{1-2\mu}{E}\right) d\left(\sigma_{\text{ave}} - \frac{3}{2}u_a\right) + \frac{3}{H} d\left(u_a - u_w\right) \quad (13.50)$$

The constitutive equation for the water phase under plane stress loading is obtained from Eq. 13.29 by setting $d\sigma_z$ equal to zero:

$$\frac{dV_w}{V_0} = \frac{2}{E_w} d\left(\sigma_{\text{ave}} - \frac{3}{2}u_a\right) + \frac{d\left(u_a - u_w\right)}{H_w} \quad (13.51)$$

13.4 COMPRESSIBILITY FORM FOR UNSATURATED SOIL CONSTITUTIVE RELATIONS

The constitutive relations for an unsaturated soil presented in the preceding sections were formulated using linear elasticity form equations. These constitutive equations can be rewritten in a compressibility form more common to soil mechanics. The compressibility form of the constitutive equation for the soil structure of a saturated soil is written as

$$d\varepsilon_v = m_v d\left(\sigma - u_w\right) \quad (13.52)$$

where:

m_v = coefficient of volume change.

The compressibility form for the soil structure constitutive equation for an unsaturated soil under general, three-dimensional loading is as follows:

$$d\varepsilon_v = m_1^s\, d\left(\sigma_{\text{mean}} - u_a\right) + m_2^s\, d\left(u_a - u_w\right) \quad (13.53)$$

where:

m_1^s = coefficient of volume change with respect to net normal stress [i.e., $3(1-2\mu)/E$] and
m_2^s = coefficient of volume change with respect to matric suction (i.e., $3/H$).

The m_1^s and m_2^s coefficients of volume change in Eq. 13.53 are essentially a ratio between changes in volumetric strain and stress state variables and can be called "compressibility moduli." The negative signs for the coefficients of volume change, m_1^s and m_2^s, are the result of the elasticity-type modulus, E and H, being negative for typical soils. The m_1^s and m_2^s coefficients vary in a nonlinear manner but can be considered as constants for a small increment of stress or strain.

The compressibility form of the water phase constitutive equation for an unsaturated soil under three-dimensional loading can be written as:

$$\frac{dV_w}{V_0} = m_1^w\, d\left(\sigma_{\text{mean}} - u_a\right) + m_2^w\, d\left(u_a - u_w\right) \quad (13.54)$$

where:

m_1^w = coefficient of water volume change with respect to net normal stress (i.e., $3/E_w$) and
m_2^w = coefficient of water volume change with respect to matric suction (i.e., $1/H_w$).

Similar constitutive equations in a compressibility form can be written for specific loading conditions. Table 13.1 presents the m_1^s, m_2^s, m_1^w, and m_2^w coefficients for the loading conditions described in the previous section. These coefficients are regarded as another form of the soil volumetric deformation coefficients.

13.4.1 Volume-Mass Form (Soil Mechanics Terminology)

The volume-mass properties of a soil can also be used in the formulation of constitutive equations for an unsaturated soil. Common soil mechanics terminology makes use of void ratio, gravimetric water content, and degree of saturation to define volume-mass properties of unsaturated soils.

Changes in void ratio take on the form of a deformation state variable for a *saturated* soil giving rise to the following constitutive equation:

$$de = a_v d\left(\sigma - u_w\right) \quad (13.55)$$

where:

a_v = coefficient of compressibility.

Void ratio and gravimetric water content can be used as the deformation state variables for the soil structure and water phase, respectively, for an unsaturated soil. Using soil mechanics terminology, the change in void ratio, de, of an *unsaturated* soil under three-dimensional loading can be written as

$$de = a_t\, d\left(\sigma_{\text{mean}} - u_a\right) + a_m\, d\left(u_a - u_w\right) \quad (13.56)$$

where:

a_t = coefficient of compressibility with respect to a change in net normal stress, $d\left(\sigma_{\text{mean}} - u_a\right)$, and
a_m = coefficient of compressibility with respect to a change in matric suction, $d\left(u_a - u_w\right)$.

Table 13.1 Coefficients of Volume Change for Various Loading Conditions

Loading	Deformation State Variable	m_1^s, m_1^w	First Stress State Variable, $d(\sigma - u_a)$	m_2^s, m_2^w	Second Stress State Variable, $d(u_a - u_w)$
Three-dimensional (general)	$d\varepsilon_v$	$3\left(\frac{1-2\mu}{E}\right)$	$d(\sigma_{mean} - u_a)$	$\frac{3}{H}$	$d(u_a - u_w)$
	$\frac{dV_w}{V_0}$	$\frac{3}{E_w}$	$d(\sigma_{mean} - u_a)$	$\frac{1}{H_w}$	$d(u_a - u_w)$
Isotropic	$d\varepsilon_v$	$3\left(\frac{1-2\mu}{E}\right)$	$d(\sigma_3 - u_a)$	$\frac{3}{H}$	$d(u_a - u_w)$
	$\frac{dV_w}{V_0}$	$\frac{3}{E_w}$	$d(\sigma_3 - u_a)$	$\frac{1}{H_w}$	$d(u_a - u_w)$
Uniaxial	$d\varepsilon_v$	$3\left(\frac{1-2\mu}{E}\right)$	$d(\frac{1}{3}\sigma_y - u_a)$	$\frac{3}{H}$	$d(u_a - u_w)$
	$\frac{dV_w}{V_0}$	$\frac{3}{E_w}$	$d(\frac{1}{3}\sigma_y - u_a)$	$\frac{1}{H_w}$	$d(u_a - u_w)$
Triaxial (general case)	$d\varepsilon_v$	$3\left(\frac{1-2\mu}{E}\right)$	$d(\sigma_3 - u_a) + d\{\frac{1}{3}(\sigma_1 - \sigma_3)\}$	$\frac{3}{H}$	$d(u_a - u_w)$
	$\frac{dV_w}{V_0}$	$\frac{3}{E_w}$	$d(\sigma_3 - u_a) + d\{\frac{1}{3}(\sigma_1 - \sigma_3)\}$	$\frac{1}{H_w}$	$d(u_a - u_w)$
Triaxial (separation of isotropic and uniaxial components)	$d\varepsilon_v$	$3\left(\frac{1-2\mu}{E}\right)$	$d(\sigma_3 - u_{ai}) + d\{\frac{1}{3}(\sigma_1 - \sigma_3) - u_{au}\}$	$\frac{3}{H}$	$d(u_a - u_w)_i + d(u_a - u_w)_u$
	$\frac{dV_w}{V_0}$	$\frac{3}{E_w}$	$d(\sigma_3 - u_{ai}) + d\{\frac{1}{3}(\sigma_1 - \sigma_3) - u_{au}\}$	$\frac{1}{H_w}$	$d(u_a - u_w)_i + d(u_a - u_w)_u$
K_0 loading (one-dimensional)	$d\varepsilon_v$	$\frac{(1+\mu)(1-2\mu)}{E(1-\mu)}$	$d(\sigma_y - u_a)$	$\frac{1+\mu}{H(1-\mu)}$	$d(u_a - u_w)$
	$\frac{dV_w}{V_0}$	$\frac{1+\mu}{E_w(1-\mu)}$	$d(\sigma_y - u_a)$	$\frac{1}{H_w} - \frac{2(E/H)}{E_w(1-\mu)}$	$d(u_a - u_w)$
Plane strain (two-dimensional)	$d\varepsilon_v$	$\frac{2(1+\mu)(1-2\mu)}{E}$	$d(\sigma_{ave} - u_a)$	$2\frac{1+\mu}{H}$	$d(u_a - u_w)$
	$\frac{dV_w}{V_0}$	$2\frac{1+\mu}{E_w}$	$d(\sigma_{ave} - u_a)$	$\frac{1}{H_w} - \frac{E/H}{E_w}$	$d(u_a - u_w)$
Plane stress (two-dimensional)	$d\varepsilon_v$	$2\frac{1-2\mu}{E}$	$d(\sigma_{ave} - \frac{3}{2}u_a)$	$\frac{3}{H}$	$d(u_a - u_w)$
	$\frac{dV_w}{V_0}$	$\frac{2}{E_w}$	$d(\sigma_{ave} - \frac{3}{2}u_a)$	$\frac{1}{H_w}$	$d(u_a - u_w)$

Note: $\sigma_{mean} = (\sigma_x + \sigma_y + \sigma_z)/3$; $\sigma_{ave} = (\sigma_x + \sigma_y)/2$.

The above equation is equivalent to a soil structure constitutive relationship written using soil mechanics terminology.

The water phase constitutive equation can be expressed as a change in gravimetric water content. The water phase constitutive relation has the following form for three-dimensional loading:

$$dw = b_t d\left(\sigma_{\text{mean}} - u_a\right) + b_m d\left(u_a - u_w\right) \qquad (13.57)$$

where:

b_t = coefficient of water content change with respect to a change in net normal stress, $d\left(\sigma_{\text{mean}} - u_a\right)$, and
b_m = coefficient of water content change with respect to a change in matric suction, $d\left(u_a - u_w\right)$.

The above constitutive equations can be visualized as constitutive surfaces on a three-dimensional plot. Each abscissa represents one of the stress state variables and the ordinate represents the soil volume-mass property (Fig. 13.10). The a_t, a_m, b_t, and b_m coefficients become another form of the volumetric deformation coefficients. The volume-mass constitutive surfaces can be reduced to two-dimensional plots which more clearly show the relationship between the various deformation coefficients (Fig. 13.10). The relationships between various deformation coefficients are explained later in this chapter. Similar but independent constitutive relationships can be formulated for unloading conditions.

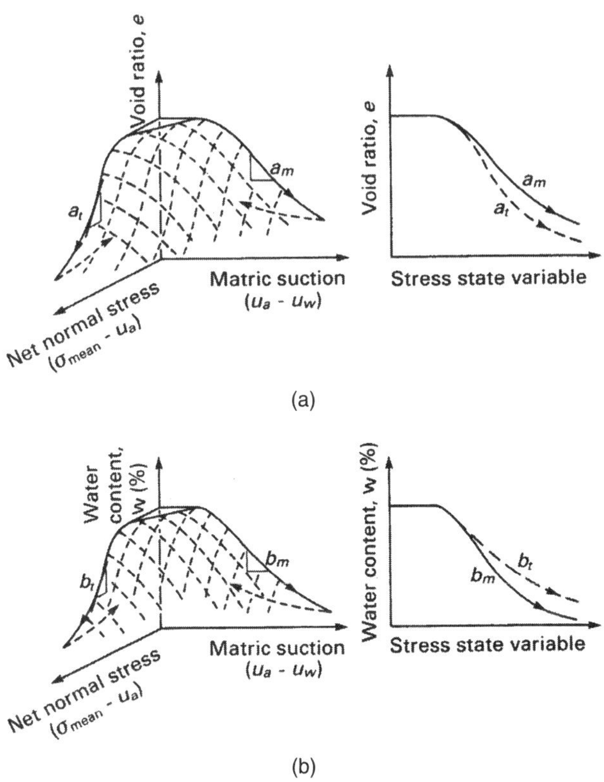

Figure 13.10 Constitutive surfaces for an unsaturated soil expressed using soil mechanics terminology: (a) void ratio constitutive surface; (b) water content constitutive surface.

13.4.2 Use of $\sigma - u_w$ and $u_a - u_w$ to Formulate Constitutive Relations

The constitutive equations for an unsaturated soil have been formulated using the stress state variables $\sigma - u_a$ and $u_a - u_w$. Other combinations of stress state variables, such as $\sigma - u_w$ and $u_a - u_w$, can also be used for formulating the constitutive relations for an unsaturated soil. As an example, Eq. 13.56 can be rewritten using $\sigma - u_w$ and $u_a - u_w$:

$$de = a_{t2}\, d\left(\sigma_{\text{mean}} - u_w\right) + a_{m2}\, d\left(u_a - u_w\right) \qquad (13.58)$$

where:

a_{t2} = coefficient of compressibility with respect to $\sigma_{\text{mean}} - u_w$ when using the $\sigma - u_w$ and $u_a - u_w$ stress state variables and
a_{m2} = coefficient of compressibility with respect to matric suction when using $\sigma - u_w$ and $u_a - u_w$ as stress state variables.

The water phase constitutive equation can also be written in terms of $\sigma - u_w$ and $u_a - u_w$:

$$dw = b_{t2}\, d\left(\sigma_{\text{mean}} - u_w\right) + b_{m2}\, d\left(u_a - u_w\right) \qquad (13.59)$$

where:

b_{t2} = coefficient of water content change with respect to $\sigma_{\text{mean}} - u_w$ when using the $\sigma - u_w$ and $u_a - u_w$ stress state variables and
b_{m2} = coefficient of water content change with respect to matric suction when using $\sigma - u_w$ and $u_a - u_w$ as stress state variables.

The $\sigma - u_a$ and $\sigma - u_w$ stress state variables could similarly be used to formulate the volume-mass constitutive relations.

13.4.3 Experimental Verification for Uniqueness of Constitutive Surfaces

The uniqueness of proposed constitutive surfaces should first be explored on a local scale using small stress state variable changes around a state point on the surface (Fredlund and Morgenstern, 1976). Subsequently, larger stress state variable changes should be used to test for uniqueness of the entire constitutive surface (Matyas and Radhakrishna, 1968; Barden et al., 1969a). Both methods for verifying the uniqueness of a constitutive surface are briefly explained and discussed. The sign conventions associated with volumetric deformation properties are outlined prior to describing the results of tests for uniqueness.

The two constitutive surfaces for an unsaturated soil (i.e., soil structure and water phase) require experimental data in order to examine the uniqueness of the proposed surfaces. The term "uniqueness" is used to indicate that there is one and only one relationship between the deformation and stress state variables. Figure 13.11 illustrates the meaning of the term uniqueness as applied to a constitutive surface. Let us assume that three identical soil specimens have been prepared at the same initial state point. The stress state variables for the three specimens are then moved along three different stress paths. However, the final stress state variables for the three soil specimens are identical. The final combination of stress state variables should produce the same deformation state if the constitutive surface is unique. In other words, the final volumes for the three soil specimens should be equal and independent of the stress path.

Complete uniqueness of a constitutive relationship under loading and unloading conditions is known to be virtually impossible for soils. Hysteresis is the main reason that uniqueness does not exist when going from loading to unloading conditions or vice versa. Hysteresis is associated with a particulate material such as that comprising the soil structure. Wetting and drying studies have also shown that the contractile skin also exhibits hysteresis. The soil structure hysteresis during loading and unloading of a saturated soil is reflected by the different compression and rebound curves for the soil (Fig. 13.12a). The two curves are distinctly different under loading and unloading conditions.

Hysteresis associated with contractile skin can be observed from the drying and wetting curves of an incompressible material such as chalk (Fig. 13.12b). It is quite clear that there should be hysteresis associated with the soil structure and the contractile skin in an unsaturated soil. It appears that a reversal in the direction of deformation (i.e., an increase and then a decrease in volume) results in different constitutive surfaces. As a result, the volumetric deformation coefficients associated with a decreasing volume would be different from those associated with an increasing volume.

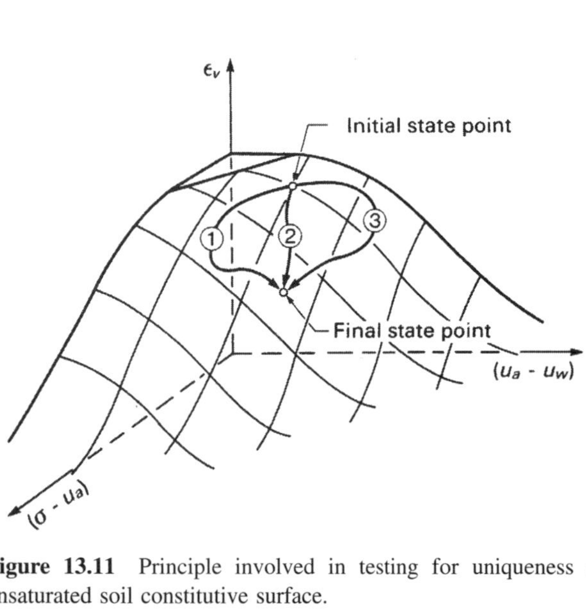

Figure 13.11 Principle involved in testing for uniqueness of unsaturated soil constitutive surface.

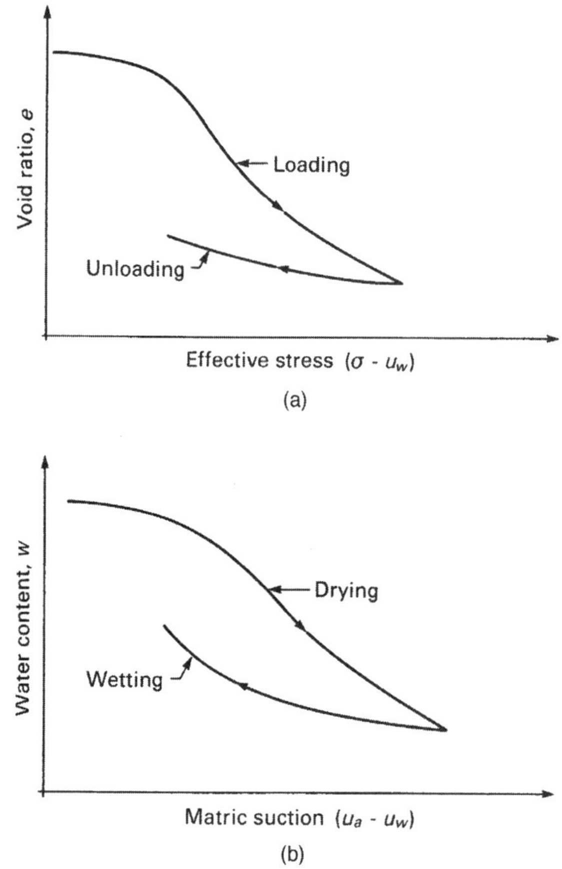

Figure 13.12 Hysteresis associated with soil structure and contractile skin of a soil: (a) loading and unloading curves for saturated soil; (b) drying and wetting curves for incompressible chalk.

Uniqueness of the constitutive surfaces with respect to volume increase and decrease can be independently verified. This involves tracing the surface through various loading paths provided the deformation state variable always changes in the same direction (i.e., monotonic or unidirectional deformation). The term uniqueness has been generally restricted in soil mechanics to the consideration of constitutive surfaces representing monotonic deformation.

The uniqueness of the soil structure constitutive surface can be verified for either a decrease or an increase in the soil volume. In addition, the uniqueness of the water phase constitutive surface can also be verified for either a decrease or an increase in the volume of water (i.e., drying or wetting surface).

13.4.4 Sign Convention for Volumetric Deformation Properties

The sign convention for a general relationship between deformation and stress state variable changes is illustrated in Fig. 13.7. A positive or negative sign is associated with the deformation or stress state variable change in order to indicate an increase or decrease. The following sign convention can be formulated for multiphase materials. A positive change in the state variable (i.e., deformation or stress) refers to an increase in the state variable, whereas a negative change indicates a decrease in the state variable. The signs of the deformation state variables and the stress state variables determine the sign for the volumetric deformation coefficients. For example, the stable-structured soil behavior shown in Fig. 13.7 has a negative sign for the modulus of elasticity E. This means that an increase in the net normal stress, $(+)\,d\left(\sigma-u_a\right)$, causes a decrease in volumetric strain, $(-)\,d\varepsilon_v$ [i.e., $(-)\,E=(-)\,d\varepsilon_v/(+)\,d\left(\sigma-u_a\right)$]. The elastic modulus H also has a negative sign [e.g., $(-)\,H=(-)\,d\varepsilon_v/(+)\,d\left(u_a-u_w\right)$].

The relationship between the water volume change and the stress state variable change for a stable-structured soil is similar to that shown in Fig. 13.7. The water volumetric moduli E_w and H_w would also have negative signs. Therefore, it can be concluded that a *stable-structured* soil has *negative* volumetric deformation *moduli* associated with the soil structure and water phase. The negative sign applies similarly to all coefficients of volume change, m_1^s, m_2^s, m_1^w, and m_2^w, which are used in the compressibility form of the constitutive equations. The a_t, a_m, b_t, and b_m coefficients associated with the soil mechanics form of the constitutive equations are also negative for a stable-structured soil.

A collapsing soil is referred to as a metastable-structured soil. A decrease in matric suction results in swelling for a stable-structured soil whereas it may cause a volume decrease in a metastable-structured soil (Barden et al., 1969a). The m_2^s and m_2^w coefficients of volume change have a positive sign for collapsing soils (Eqs. 13.53 and 13.54). A *metastable-structured* soil can be referred to as a soil that has one or more *positive* volumetric deformation moduli.

13.4.5 Verification of Uniqueness of Constitutive Surfaces Using Small Stress Changes

Let us suppose that several so-called identical unsaturated soil specimens are subjected to the same total normal stress and the same pore-air and pore-water pressures. All specimens will be at the same state point in space and have the same initial volume-mass properties. Let each specimen then be subjected to different stress path changes while the volume changes are monitored. If the constitutive surface is essentially planar near a state point, the volumetric deformation moduli associated with any two orthogonal directions can be used to describe the deformation produced by other stress state variable changes.

A somewhat simpler test for uniqueness can be described as follows. Suppose that three so-called identical specimens are prepared at the same initial volume-mass properties and subjected to the same initial stress state variables (Fig. 13.13a). Let one specimen be subjected to a small increase in net normal stress, $d\left(\sigma-u_a\right)$, by increasing the total normal stress while the matric suction u_a-u_w is maintained constant (i.e., 0–1 loading path in Fig. 13.13a). The measured volumetric strain change $d\varepsilon_v$ can be used to compute the coefficient of volume change m_1^s:

$$m_1^s=\frac{d\varepsilon_v}{d\left(\sigma-u_a\right)} \qquad (13.60)$$

A second specimen could be subjected to a small increase in matric suction, $d\left(u_a-u_w\right)$, by decreasing the pore-water pressure u_w while maintaining a constant net normal stress (i.e., 0–2 loading path in Fig. 13.13a). The measured change in volumetric strain, $d\varepsilon_v$, can be used to calculate the m_2^s coefficient of volume change:

$$m_2^s=\frac{d\varepsilon_v}{d\left(u_a-u_w\right)} \qquad (13.61)$$

A third specimen could then be subjected to small increases in both the net normal stress $d\left(\sigma-u_a\right)$ and the matric suction $d\left(u_a-u_w\right)$. This can be achieved by increasing the total normal stress σ and decreasing the pore-water pressure u_w. The anticipated volume change can be predicted

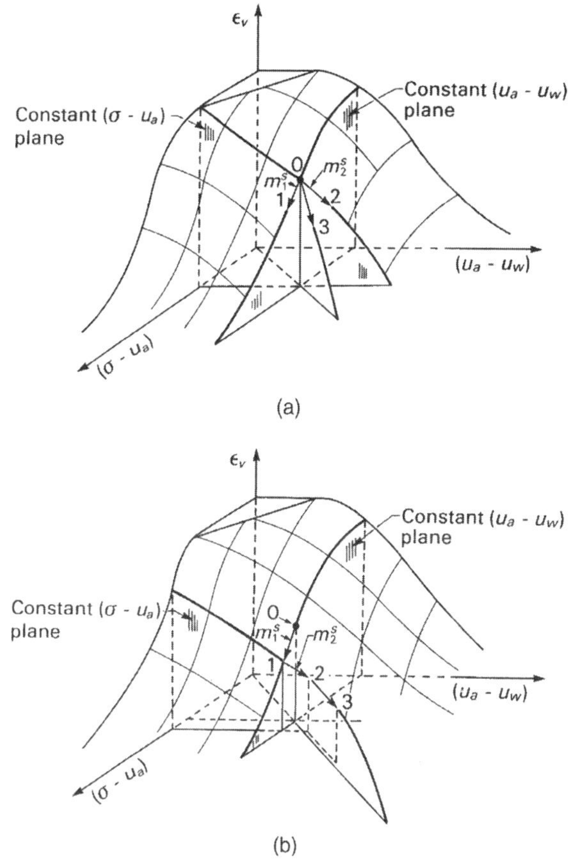

Figure 13.13 Testing for uniqueness of soil structure constitutive surface: constitutive surface verification using (a) three identical soil specimens and (b) a single soil specimen.

from Eq. 13.53 using the above-measured values of m_1^s and m_2^s (from Eqs. 13.60 and 13.61). The actual volume change can then be measured and compared with the computed value. A close agreement between the measured and predicted volume changes indicates uniqueness of the constitutive surface near a stress point. The above example illustrates a possible uniqueness test for the soil structure constitutive surface. The water phase constitutive surface could also be verified if water volume change measurements were made for the computation of the m_1^w and m_2^w coefficients.

The above test for uniqueness can be approximated by an even simpler procedure using a single soil specimen subjected to a series of small stress changes (Fredlund and Morgenstern, 1976). The procedure is illustrated in Fig. 13.13b. Consider a single unsaturated soil specimen that is first subjected to a small increase in net normal stress (i.e., an increase in the total stress) along a constant-matric-suction plane (i.e., 0–1 loading path in Fig. 13.13b). The measured volume changes are used to compute the m_1^s and m_1^w coefficients, respectively.

The specimen can then be subjected to an increase in its matric suction while maintaining a constant net normal stress (i.e., 1–2 loading path in Fig. 13.13b). The matric suction is increased by decreasing the pore-water pressure. The measured volume changes can be used to compute the m_2^s and m_2^w coefficients, respectively. The m_1^s, m_1^w and m_2^s, m_2^w coefficients can then be used to predict the overall volume change and the water volume change of the soil specimen during any subsequent loadings. The coefficients can be assumed to approximate a single stress point on the constitutive surface even though the first and second loadings are performed at slightly differing initial volume-mass properties,. This assumption is reasonable for small changes in the stress state variables.

The next step is to apply a small increment in both the net normal stress $d(\sigma - u_a)$ and matric suction $d(u_a - u_w)$ (i.e., 2–3 loading path in Fig. 13.13b). This path is adhered to by increasing the total normal stress and decreasing the pore-water pressure. The anticipated overall volume change and change in volume of water as a result of the third loading can be predicted from Eqs. 13.53 and 13.54, respectively, while using the coefficients of volume change obtained from the previous two loadings. The measured overall and water volume changes can then be compared with the predicted values. The constitutive surface at a point is considered unique if the measured and predicted deformations are essentially equal. This procedure can be repeated at various state points on the constitutive surface.

Fredlund and Morgenstern (1976) experimentally tested constitutive surfaces of two unsaturated soils for uniqueness by using small changes in stress state variables. Four series of experiments were conducted using three specimens of undisturbed Regina clay and one specimen of compacted kaolin. The specimens were tested using a modified oedometer (i.e., K_0 loading) and a modified triaxial cell. The modified triaxial cell was used for isotropic loading. The total, pore-air, and pore-water pressures were controlled during the experiments. The water pressure was isolated from the air pressure by means of a high-air-entry disk placed at the bottom of the specimen. After equalization at a selected stress point, each specimen was subjected to several loading stages in order to test the uniqueness of the constitutive surface. The predicted and measured volume changes were assessed by plotting the predicted volume changes versus the measured volume changes.

Good agreement was observed between the measured and predicted volume changes for undisturbed Regina clay, as shown in Fig. 13.14. The specimen was tested under K_0 loading. The volume changes at two elapsed times (e.g., 1000 and 5000 min) are considered for each specimen. Uniqueness implies a slope of unity and an intercept of zero for the plot of measured versus predicted volume change. Figure 13.14a demonstrates that all points obtained at two elapsed times essentially fall along the 45° line, indicating virtually perfect correlation. This agreement supports the uniqueness of the soil structure constitutive surface.

The agreement between the predicted and measured water volume changes (Fig. 13.14b) is not as consistent as that

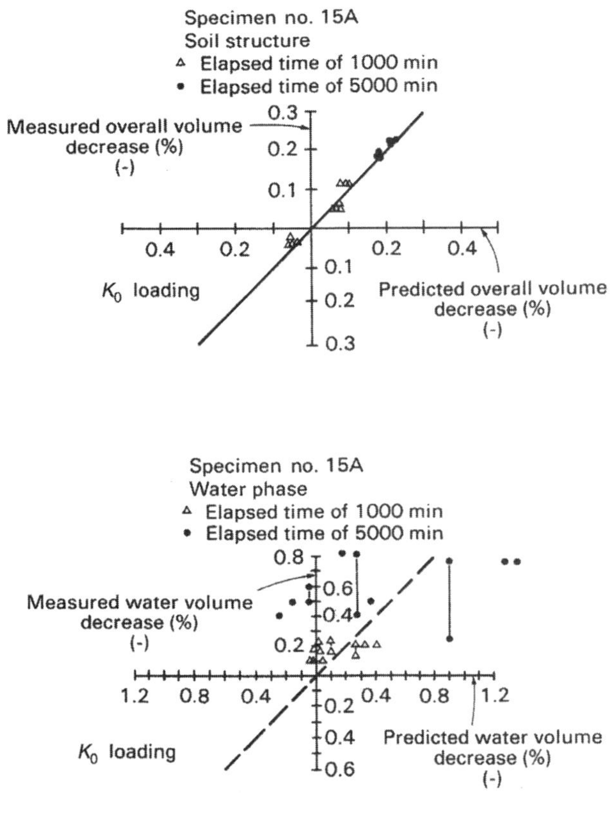

Figure 13.14 Comparison of predicted and measured volume changes for undisturbed Regina clay undergoing K_0 loading: (a) soil structure constitutive surface; (b) water phase constitutive surface (from Fredlund and Morgenstern, 1976).

for the soil structure. This was attributed, in part, to the difficulty of measuring small water volume changes over long periods of time (Fredlund and Morgenstern, 1976). It was also discovered that some moisture was lost from the soil specimens.

13.4.6 Verification of Constitutive Surfaces Using Large Stress State Variable Changes

Matyas and Radhakrishna (1968) experimentally tested the void ratio and degree-of-saturation constitutive surfaces for uniqueness under conditions of large stress state changes. Soil specimens consisting of 80% flint powder and 20% kaolin were prepared using static compaction with the same compaction effort. The initial water contents and dry densities were also the same. The specimens were tested under isotropic and K_0 loading with controlled total, pore-air, and pore-water pressures. The tests were performed using a modified triaxial apparatus.

The constitutive surfaces obtained from isotropic loading are shown in Fig. 13.1 and the results for K_0 loading were shown in Fig. 13.2. It was found that the change in void ratio between any two state points was independent of the deformation–stress path followed for monotonic deformation. Monotonic deformation was obtained by following paths with increasing degrees of saturation where the specimens were not allowed to swell. The degree-of-saturation constitutive surface, however, did not indicate complete uniqueness.

The constitutive surfaces obtained from K_0 loading are presented in Fig. 13.2. All test paths appeared to form a single warped void ratio surface (Fig. 13.2a). Collapsing soil behavior was again observed. The degree of saturation was found to be more sensitive to stress state changes than was the void ratio constitutive surface (Fig. 13.2b). The uniqueness of both constitutive surfaces was again restricted to monotonic deformation (Matyas and Radhakrishna, 1968).

Barden et al., (1969a) studied the volume change characteristics of unsaturated Westwater and Derwent clays under K_0 loading conditions. The constitutive surface was traced following different test paths through various combinations of stress state variables $\sigma_y - u_a$ and $u_a - u_w$, as shown in Fig. 13.3. The results were earlier presented in Figs. 13.4 and 13.5. The results exhibited uniqueness or loading path independence as long as deformations were monotonic. Hysteresis associated with wetting and drying processes was considered to be the major cause of loading path dependence. The degree-of-saturation constitutive surface showed uniqueness provided the degree-of-saturation change was monotonic.

13.5 RELATIONSHIP AMONG VOLUMETRIC DEFORMATION COEFFICIENTS

The constitutive equations or surfaces for unsaturated soils have been formulated and presented in three forms. In the elasticity form, the E, H, μ, E_w, and H_w parameters are required to define the volumetric deformations associated with the soil structure and water phase constitutive equations. In the compressibility form, the soil structure and water phase constitutive surfaces make use of m_1^s, m_2^s, m_1^w, and m_2^w volumetric coefficients. These coefficients of volume change can be expressed in terms of the above elastic parameters, as shown in Table 13.1. In soil mechanics terminology, a_t, a_m, b_t, and b_m coefficients are used to define the slope of the void ratio and water content relations versus the respective stress states.

This section discusses the relationships among the various volumetric deformation coefficients. Theoretical and experimental methods that can be used to obtain these relationships are briefly discussed. Emphasis is given to the coefficients used in the compressibility and soil mechanics forms of the constitutive surfaces (i.e., m_1^s, m_2^s, m_1^w, m_2^w and a_t, a_m, b_t, b_m). These coefficients can, in general, be obtained from various laboratory tests. In some cases modifications must be made to the laboratory equipment in order to properly perform the tests.

13.5.1 Relationship of Volumetric Deformation Coefficients for Void Ratio and Water Content Surfaces

The void ratio and water content surfaces for an unsaturated soil are illustrated in Fig. 13.15. The orientation of each point on the void ratio surface can be defined in terms of two slope angles. The first slope angle is referenced to net normal stress and is defined by the coefficient of compressibility, a_t:

$$a_t = \frac{\partial e}{\partial\left(\sigma - u_a\right)} \tag{13.62}$$

The second slope angle is referenced to matric suction and is defined by the coefficient of compressibility a_m:

$$a_m = \frac{\partial e}{\partial\left(u_a - u_w\right)} \tag{13.63}$$

Similarly, a point on the water content surface has two slopes, which can be defined as b_t and b_m with respect to the net normal stress and matric suction, respectively:

$$b_t = \frac{\partial w}{\partial\left(\sigma - u_a\right)} \tag{13.64}$$

$$b_m = \frac{\partial w}{\partial\left(u_a - u_w\right)} \tag{13.65}$$

Both the void ratio and water content surfaces are generally nonlinear. Therefore, the four coefficients a_t, a_m, b_t, and b_m vary over the constitutive surfaces. In other words, these coefficients are a function of the state point on the surface. However, the slopes on the constitutive surface have some limiting conditions as observed from experimental results.

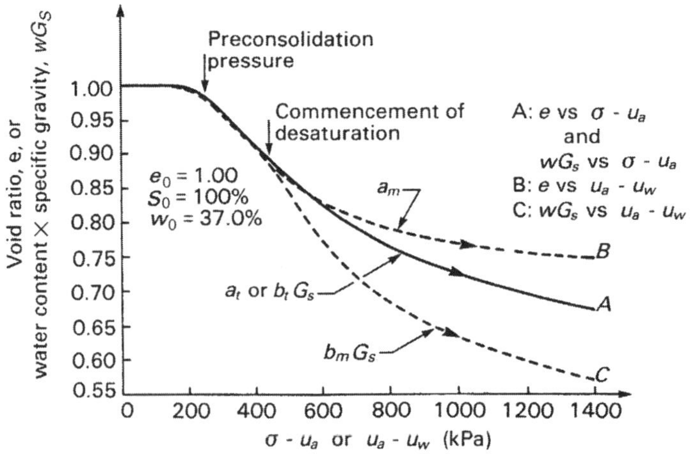

Figure 13.15 Relationship of the volume-mass volumetric deformation moduli to one another for an initially saturated soil subjected to total stress changes or soil suction changes.

The a_m coefficient is approximately equal to the a_t coefficient when the degree of saturation is near 100% or when the matric suction goes to zero (Fig. 13.10a). At saturation (i.e., on the plane of zero matric suction), the a_t coefficient becomes the coefficient of compressibility a_v for a saturated soil.

The a_t coefficient becomes greater than the a_m coefficient as degree of saturation decreases. This shows that a change in total stress is more effective in changing void ratio than is a change in matric suction. It can be observed that the void ratio corresponding to the shrinkage limit of a soil is the minimum void ratio that can be attained under unconfined, maximum soil suction conditions.

The b_m coefficient approaches the b_t coefficient as the degree of saturation approaches 100%. At lower degrees of saturation, the b_m coefficient will generally be greater than the b_t coefficient (Fig. 13.10b). In other words, a change in matric suction is more effective in changing the water content of the soil than is a change in total stress. This occurs because the matric suction is applied directly to the water phase. The b_t coefficient can be related to a_t or a_v coefficients when the soil is on the saturation plane (i.e., matric suction is equal to zero). The b_t coefficient is related to a_t or a_v coefficients using the basic volume-mass relationship (i.e., $Se = wG_s$ where G_s is the specific gravity of the soil solids). When the degree of saturation is 100%, the following basic relationship exists between the soil moduli: $a_t = b_t G_s$.

A two-dimensional plot of the constitutive surfaces allows the above-mentioned coefficients to be compared (i.e., a_t versus a_m and b_t versus b_m; Fig. 13.10). The relationship among the four coefficients can also be illustrated using a single plot (Fig. 13.15). Consider initially saturated silt that is subjected to isotropic consolidation by increasing the total normal stress. The soil remains saturated during the consolidation process. Consolidation curve *A* in Fig. 13.15 represents the relationship between the void ratio and net normal stress $\sigma - u_a$ as well as the water content wG_s and net normal stress. The pore-air pressure approaches the pore-water pressure for the saturated condition.

If the same soil is subjected to increasing matric suction, the volume change will be the same as long as the soil remains saturated. Once the soil commences to desaturate, a matric suction change will not be as effective as a total stress change in producing a volume change. This behavior is reflected by curve *B* (i.e., *e* versus $u_a - u_w$) in Fig. 13.15, which shows less volume change than curve *A* (i.e., *e* versus $\sigma - u_a$).

A matric suction increase, on the other hand, is more effective than a net normal stress increase in removing water from the soil. This behavior is illustrated by the water content curve (i.e., wG_s versus $u_a - u_w$), which shows greater changes in water content than curve *A*. Unlike curve *A*, the void ratio–matric suction curve (i.e., curve *B*) starts to deviate from the water content–matric suction curve (i.e., curve *C*) as the soil begins to desaturate. The separation of the two curves is due to the decreasing degree of saturation as the matric suction increases. On the other hand, there is essentially a constant degree of saturation during the total stress increment (i.e., $wG_s = e$ on curve *A*). The ratio of the ordinates for curves *C* and *B* indicates the degree of saturation (i.e., $S = wG_s/e$).

Curve *C* eventually reaches zero water content or zero degree of saturation. Zero water content corresponds to a soil suction of approximately 10^6 kPa for all soils, as explained in Chapter 5.

13.5.2 Relationship of Volumetric Deformation Coefficients for Volume-Mass Form of Constitutive Surfaces

The volumetric deformation coefficients used for the compressibility form of the constitutive surfaces can be defined in a manner similar to that described above. The soil structure and water phase constitutive surfaces are shown in Fig. 13.8.

Lines through a point on the soil structure constitutive surface have slopes of m_1^s and m_2^s with respect to the net normal stress and matric suction, respectively:

$$m_1^s = \frac{\partial \left(V_v/V_0\right)}{\partial \left(\sigma - u_a\right)} \tag{13.66}$$

$$m_2^s = \frac{\partial \left(V_v/V_0\right)}{\partial \left(u_a - u_w\right)} \tag{13.67}$$

Lines through a point on the water phase constitutive surface have slope of m_1^w and m_2^w with respect to the net normal stress and matric suction, respectively:

$$m_1^w = \frac{\partial \left(V_w/V_0\right)}{\partial \left(\sigma - u_a\right)} \tag{13.68}$$

$$m_2^w = \frac{\partial \left(V_w/V_0\right)}{\partial \left(u_a - u_w\right)} \tag{13.69}$$

The m_1^s, m_2^s, m_1^w, and m_2^w coefficients are related to the a_t, a_m, b_t, and b_m coefficients as follows:

$$m_1^s = \frac{a_t}{1 + e_0} \tag{13.70}$$

$$m_2^s = \frac{a_m}{1 + e_0} \tag{13.71}$$

$$m_1^w = \frac{b_t G_s}{1 + e_0} \tag{13.72}$$

$$m_2^w = \frac{b_m G_s}{1 + e_0} \tag{13.73}$$

where:

e_0 = initial void ratio prior to deformation.

The m_1^s is essentially equal to m_1^w on the saturation plane (i.e., $S = 100\%$) since $a_t = b_t G_s$. The m_1^s coefficient for the saturated condition is commonly called the coefficient of volume change, m_v.

13.5.3 Laboratory Tests Used to Obtain Volumetric Deformation Coefficients

The a_t, a_m, b_t, and $b_m$coefficients are used to discuss the relationships that exist between the volumetric deformation coefficients. These coefficients vary from one state point to another along a nonlinear constitutive surface. A direct method to determine these coefficients at a specific state point is to measure their magnitude at the stress point under consideration. The experimental measurements required are similar to those conducted for the verification of the constitutive surfaces. Numerous specimens and a long period of testing are generally required to experimentally define the entire constitutive surface.

It is suggested that simpler approximations can be applied for the estimation of the shape of the constitutive surfaces. For example, let us assume that the constitutive surfaces are approximately planar at a particular void ratio or water content, as shown in Figs. 13.16 and 13.17. Therefore, every point on the surface corresponding to an equal void ratio or water content plane has the same a_t and a_m coefficients or b_t and b_m coefficients, respectively. As a result, a_t and b_t coefficients obtained from the saturation plane (i.e., $u_a - u_w$ equal to zero plane) can be used for other points on the surface as long as the void ratio (or water content) is constant. Similarly, a_m and b_m coefficients obtained from the zero net normal stress plane (i.e., $\sigma - u_a$ equal to zero plane) can also be used for other points on the surface along a constant void ratio or water content plane. In this way, the values for a_t, a_m, b_t, and b_m obtained on the saturation and zero net normal stress planes are assumed to apply for the entire constitutive surface.

The suggested approximation method constitutes a crude estimation procedure for determining the volumetric coefficients over the entire constitutive surface. The estimation procedure is inferior to directly measuring the volumetric coefficients at individual state points. The required level of accuracy for the problem at hand must be borne in mind when solving practical engineering problems.

The following discussion pertains to the laboratory tests which can be used to obtain a_t, a_m, b_t, and b_m coefficients. The a_t coefficients can be obtained from curve A in Fig. 13.16, which shows the results of a consolidation test on a soil which has been saturated. Figure 13.18 shows a typical compression curve for a compacted soil which has been saturated while maintaining a constant volume prior to its decrease in volume due to loading. The b_t coefficient in Fig. 13.17 can be computed as the a_t coefficient divided by the specific gravity of the solids.

Curve B in Fig. 13.17 is a SWCC that can be obtained using a pressure plate type of test. Figure 13.19 illustrates the b_m coefficient on both the drying and wetting portions of a typical SWCC.

The a_m coefficient can be related to the b_m coefficient through use of a shrinkage test. A shrinkage test relates the void ratio of a soil to the water content at various matric suctions while drying the soil under zero external load. Figure 13.20 shows a shrinkage curve associated with a shrinkage-type test. The slope of the shrinkage curve (i.e., de/dw or $[\partial e/\partial(u_a - u_w)]/[\partial w/\partial(u_a - u_w)]$) defines the ratio between a_m and b_m coefficients at all soil suction values.

13.5.4 Relationships among Volumetric Deformation Coefficients for Unloading Surfaces

Hysteresis causes the constitutive surfaces obtained when loading (and drying) a soil to be different from the surfaces obtained when unloading (and wetting). Consequently, the volumetric deformation coefficients associated with each constitutive surface will be different. For example,

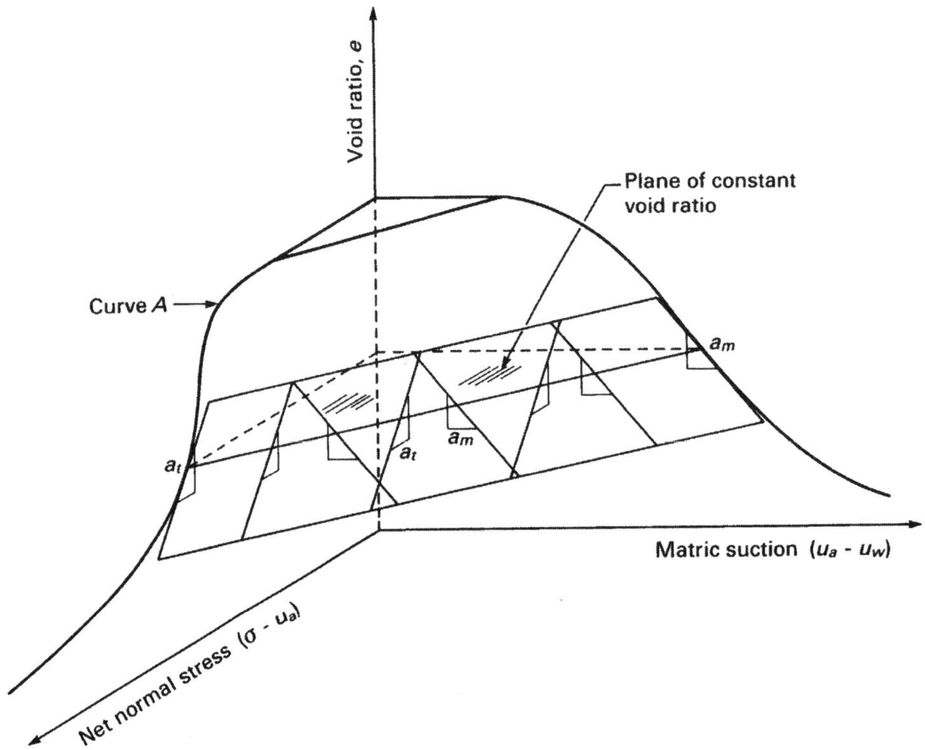

Figure 13.16 Soil structure constitutive surface for monotonic loading assuming planar surface at specific void ratio.

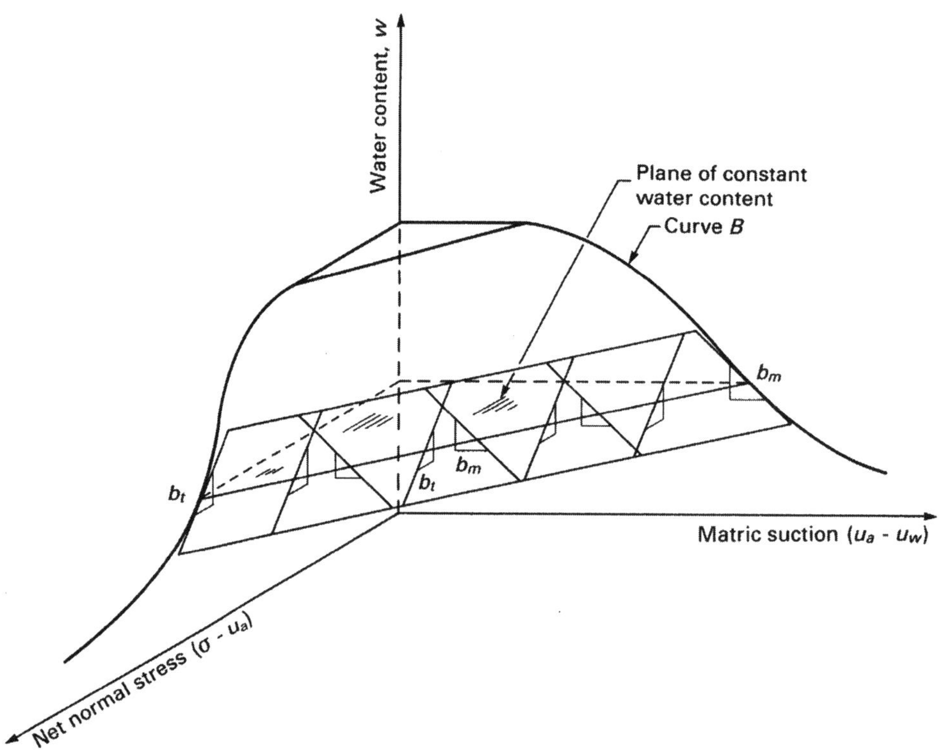

Figure 13.17 Water phase constitutive surface for monotonic loading assuming planar surface at specific water content.

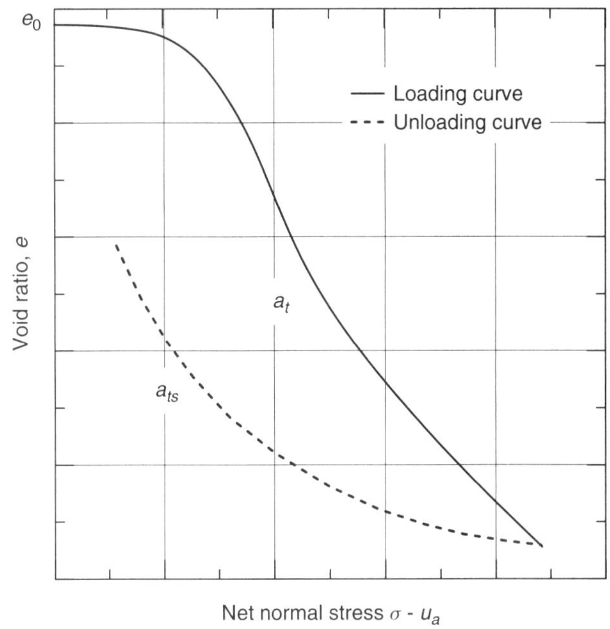

Figure 13.18 Typical compression curve for a saturated soil plotted using an arithmetic scale.

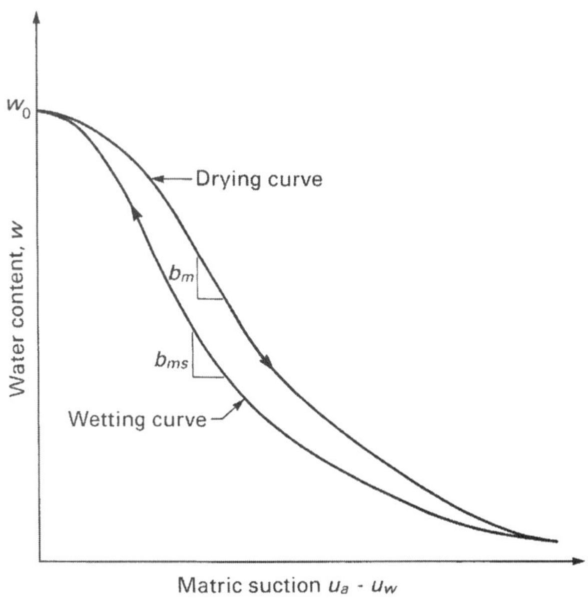

Figure 13.19 Typical SWCC showing drying and wetting of soil.

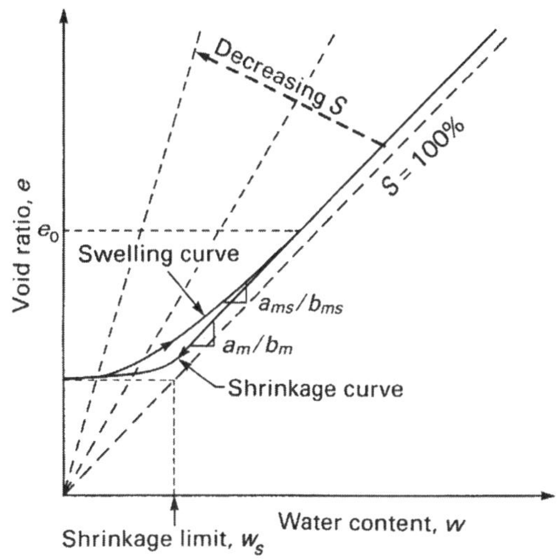

Figure 13.20 Typical shrinking and swelling curves for soil subjected to suction changes.

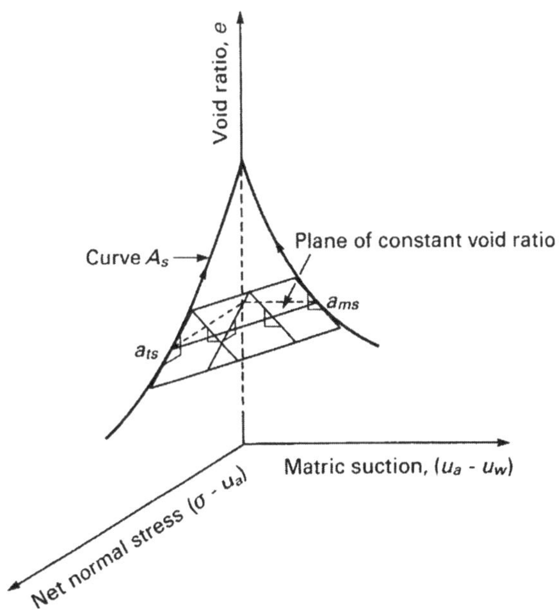

Figure 13.21 Soil structure constitutive surface for monotonic unloading of stable-structured soil.

the compression loading curve for a saturated soil plotted against the logarithm of effective stress has a compression index C_c, whereas the unloading curve has a swelling index C_s. The void ratio and water content surfaces for monotonic unloading are illustrated in Figs. 13.21 and 13.22, respectively. Coefficients of compressibility corresponding to the unloading surfaces can be further subscripted with an s (i.e., a_{ts}, a_{ms}, b_{ts}, and b_{ms}).

The same procedure used to obtain the deformation coefficients for the loading surfaces can also be used to obtain the deformation coefficients for the unloading surface. The only difference lies in the direction of deformation for unloading. The a_{ts} coefficient can be computed from curve A_s in Fig. 13.21. The curve represents the rebound curve on the saturation plane. The b_{ts} coefficient in Fig. 13.22 is calculated as a_{ts} divided by G_s.

Curve B_s in Fig. 13.22 represents the wetting portion of the SWCC. The b_{ms} coefficient can be obtained from the wetting curve. The a_{ms} coefficient is then calculated from

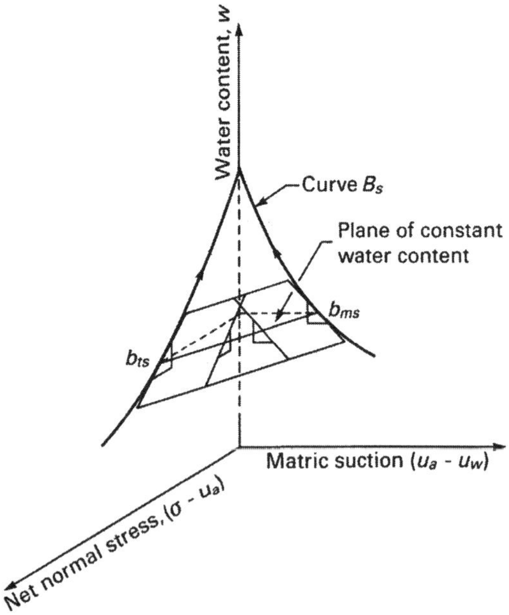

Figure 13.22 Water phase constitutive surface for monotonic unloading of stable-structured soil.

the a_{ms}/b_{ms} ratio obtained from the slope of the unconfined swelling test presented in Fig. 13.20 (i.e., the opposite of a shrinkage curve). The slope of the swelling curve is defined as de/dw or $\left[\partial e/\partial\left(u_a - u_w\right)\right]/\left[\partial w/\partial\left(u_a - u_w\right)\right]$, which is equal to a_{ms}/b_{ms}.

13.5.5 Constitutive Surfaces on Semilogarithm Plot

The constitutive surfaces can be plotted with respect to the logarithm of the stress state variables (Figs. 13.23b and 13.24b). The logarithm plots are linear over a relatively large stress range on the extreme planes [i.e., the $\log\left(u_a - u_w\right) \approx 0$ plane and the $\log\left(\sigma - u_a\right) \approx 0$ plane]. The slopes of the curves on these extreme planes are called indices. The volumetric deformation indices associated with the void ratio surface for loading conditions (Figs. 13.23b) are

$$C_t = \frac{\partial e}{\partial\left[\log\left(\sigma - u_a\right)\right]} \tag{13.74}$$

$$C_m = \frac{\partial e}{\partial\left[\log\left(u_a - u_w\right)\right]} \tag{13.75}$$

where:

C_t = compressive index with respect to net normal stress $\sigma - u_a$ and
C_m = compressive index with respect to matric suction $u_a - u_w$.

For the unloading surface, the indices are subscripted with an s as C_{ts} and C_{ms}. The volumetric deformation indices associated with the water content surface for loading conditions (Fig. 13.24b) can be defined as

$$D_t = \frac{\partial w}{\partial\left[\log\left(\sigma - u_a\right)\right]} \tag{13.76}$$

$$D_m = \frac{\partial w}{\partial\left[\log\left(u_a - u_w\right)\right]} \tag{13.77}$$

where:

D_t = water content index with respect to net normal stress $\sigma - u_a$ and
D_m = water content index with respect to matric suction $u_a - u_w$.

Similarly, the water content indices are subscripted with an s for the unloading surface (i.e., D_{ts} and D_{ms}).

The C_t, C_m, D_t, and D_m indices can be obtained from the same test data used to obtain the a_t, a_m, b_t, and b_m deformation coefficients. The difference between the soil properties lies in the manner in which the results are plotted.

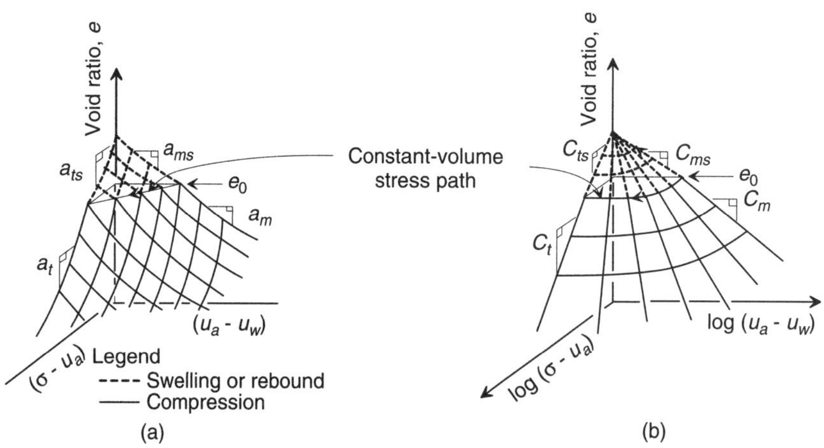

Figure 13.23 Void ratio constitutive surface for unsaturated soil: (a) arithmetic plot of stress state variables versus void ratio; (b) semilogarithmic plot of stress state variables versus void ratio.

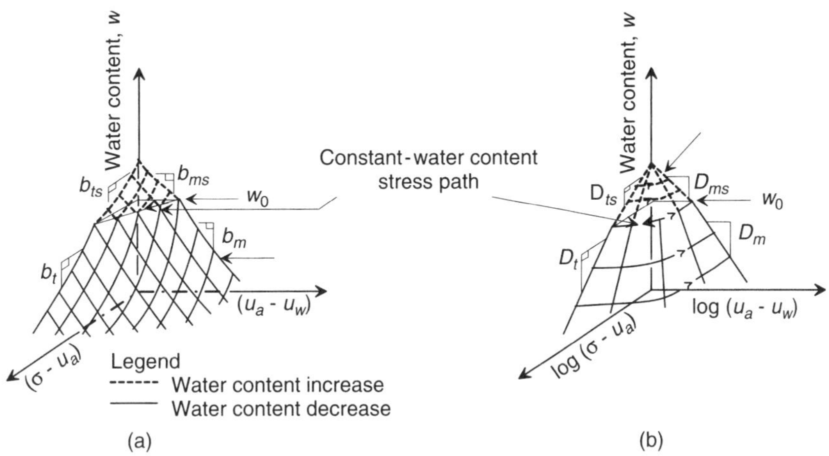

Figure 13.24 Water content constitutive surfaces for unsaturated soil: (a) arithmetic plot of stress state variables versus water content; (b) semilogarithmic plot of stress state variables versus water content.

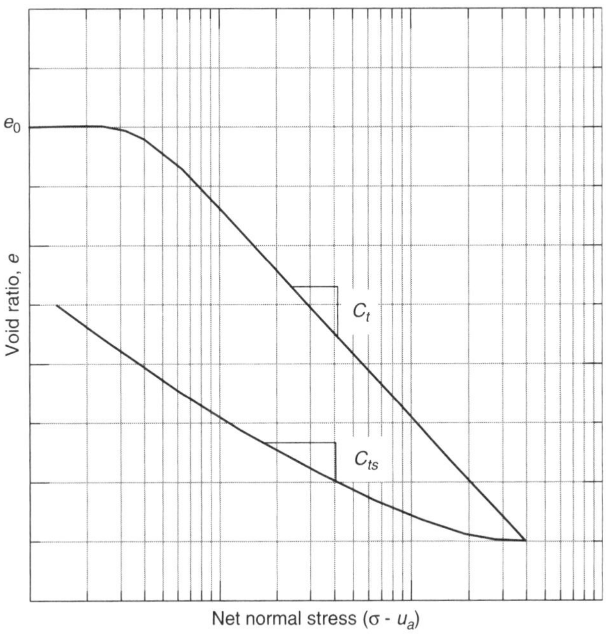

Figure 13.25 Typical compression curve for compacted soil plotted to semilogarithmic scale.

Figure 13.25 shows a typical compression curve for an unsaturated, compacted soil. The results show that the void ratio versus net normal stress curve can be largely linearized when a logarithmic scale is used for the stress state variable. The compressive index C_t can be computed from Fig. 13.25 and is commonly referred to as the compression index C_c for the saturated soil. The D_t index is related to C_t on the saturation plane:

$$C_t = D_t G_s \tag{13.78}$$

Figure 13.26 illustrates a typical SWCC plotted on a semilogarithmic scale. A reasonably linear curve is obtained over a wide stress change in comparison to the arithmetic plot in Fig. 13.19. The slope of the SWCC is equal to the D_m index. The C_m index can then be computed from the C_m/D_m ratio obtained from the slope of the shrinkage curve (Fig. 13.20).

Using a conversion between a semilogarithm scale and an arithmetic scale (Lambe and Whitman, 1979), the C_t, C_m, D_t, and D_m indices can be written in terms of the a_t, a_m, b_t, and b_m coefficients:

$$C_t = \frac{a_t\left(\sigma - u_a\right)_{\text{ave}}}{0.435} \tag{13.79}$$

$$C_m = \frac{a_m\left(u_a - u_w\right)_{\text{ave}}}{0.435} \tag{13.80}$$

$$D_t = \frac{b_t\left(\sigma - u_a\right)_{\text{ave}}}{0.435} \tag{13.81}$$

$$D_m = \frac{b_m\left(u_a - u_w\right)_{\text{ave}}}{0.435} \tag{13.82}$$

where:

$(\sigma - u_a)_{\text{ave}}$ = average of the initial and final net normal stresses for an increment and

$(u_a - u_w)_{\text{ave}}$ = average of the initial and final matric suctions for an increment.

The 0.435 constant arises from the logarithm of the natural log base taken to the base 10 (i.e., $\log_{10} 2.718$). Even though the extreme planes take on a linear relationship, the cross section of the constitutive surface on the logarithmic, the $\log\left(\sigma - u_a\right)$ and $\log\left(u_a - u_w\right)$ plane is no longer a series of straight lines as observed on the arithmetic $\left(\sigma - u_a\right)$ and $\left(u_a - u_w\right)$ plane. Figure 13.27 shows a comparison

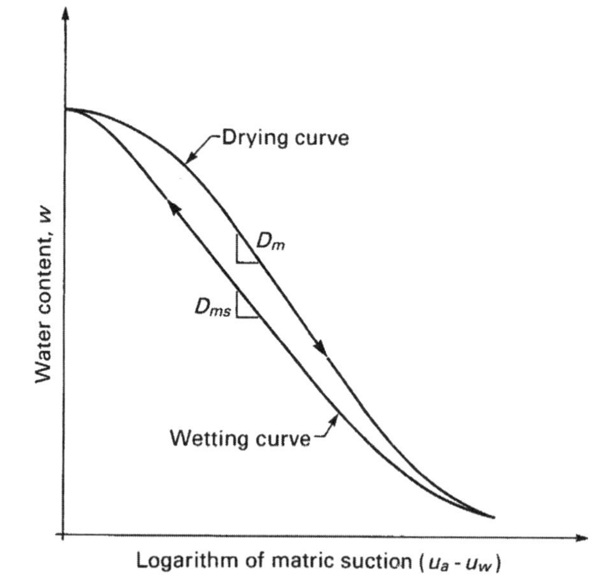

Figure 13.26 Typical SWCC plotted to semilogarithmic scale.

of the constitutive surface cross section at a constant void ratio when plotted using the arithmetic and logarithmic scales. An essentially linear cross section on the arithmetic plot becomes an asymptotic curve on the logarithmic plot.

The asymptotic cross-sectional curve illustrates the logarithmic form of the void ratio constitutive surface in Fig. 13.28a. Ho (1988) proposed an approximate form of the void ratio constitutive surface, as shown in Fig. 13.28b. The approximate surface consists of three planes, namely, planes I, II, and III. The three planes converge at a void ratio ordinate corresponding to nominal values of the stress state variables [i.e., $\log(\sigma - u_a) \approx 0$ and $\log(u_a - u_w) \approx 0$]. Planes I and III are referred to as the orthogonal planes. Plane I is perpendicular to the e–$\log(\sigma - u_a)$ plane and has a slope in which is represented by the C_t index. Plane III is perpendicular to the e–$\log(u_a - u_w)$ plane and has a slope represented by the C_m index. In other words, only one index is required to describe void ratio changes when the stress state variable changes occur within the

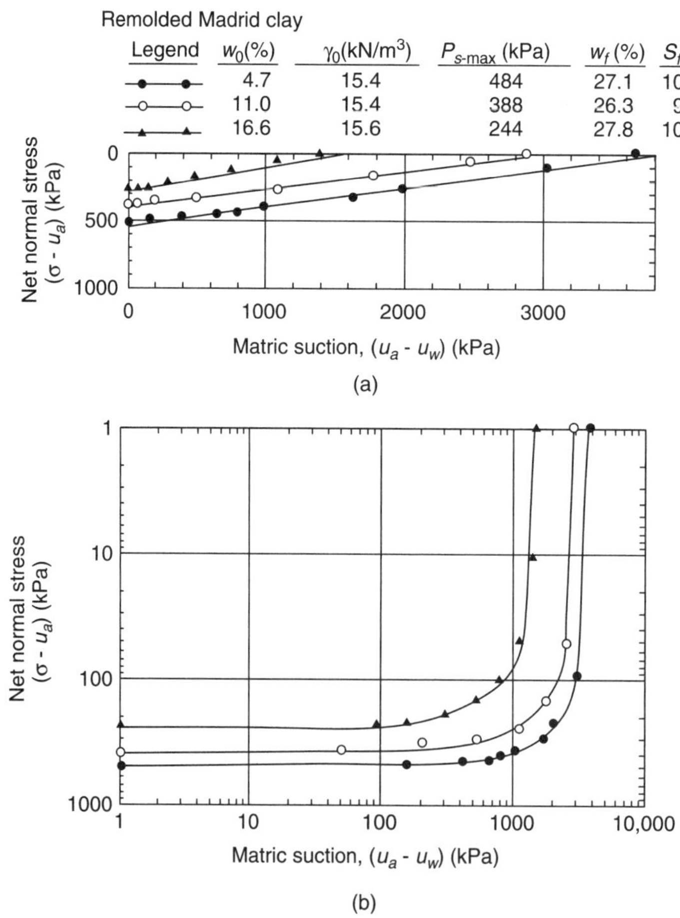

Figure 13.27 Cross section of void ratio surface plotted using arithmetic and logarithmic scales: (a) cross section showing stress state changes across constant-void-ratio surfaces on arithmetic scale; (b) cross section showing stress state changes across constant-void-ratio surfaces on logarithmic scale (modified after Escario, 1969).

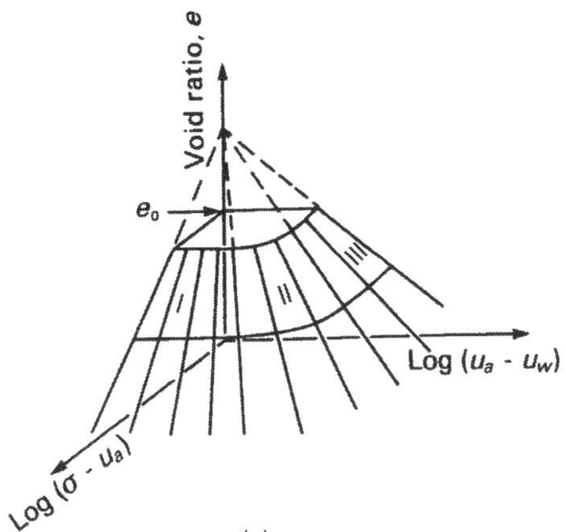

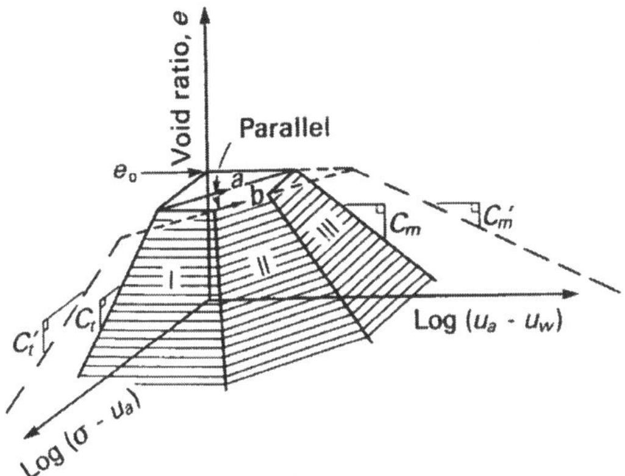

Figure 13.28 Logarithmic form for void ratio constitutive surface for loading conditions: (a) void ratio constitutive surface for loading when plotted using log-log scale for state variables; (b) approximation of void ratio constitutive surface on log-log plot (from Ho, 1988).

regions defined by planes I and III. Therefore, the void ratio constitutive equation describing plane I or III can be written in a general form as

$$de = C_t d\log\left(\sigma - u_a\right) + C_m d\log\left(u_a - u_w\right) \qquad (13.83)$$

The water content constitutive equation describing plane I or III can be written as

$$dw = D_t d\log\left(\sigma - u_a\right) + D_m d\log\left(u_a - u_w\right) \qquad (13.84)$$

Plane II represents a transition zone between planes I and III. This plane intersects both the void ratio versus $\log\left(\sigma - u_a\right)$ plane and the void ratio versus $\log\left(u_a - u_w\right)$ plane. These intersection lines define the slope of plane II, namely, the C_t' and C_m' indices. Two indices are required to describe void ratio changes when the stress state variable changes occur within the region of plane II. The C_t' and C_m' indices are evaluated from the C_t and C_m indices. The constitutive equation describing plane II can be written as

$$de = C_t' d\log\left(\sigma - u_a\right) + C_m' d\log\left(u_a - u_w\right) \qquad (13.85)$$

where:

C_t' = slope of the intersection line of plane II with the void ratio versus $\log\left(\sigma - u_a\right)$ plane and
C_m' = slope of the intersection line of plane II with the void ratio versus $\log\left(u_a - u_w\right)$ plane.

Line b, associated with plane II, is assumed to be parallel to line a, which joins planes I and III (Fig. 13.28b). This assumption is required in order to construct the intersection lines of plane II on the extreme planes. The above approximation is also applicable to the unloading surface of the void ratio constitutive surface (Ho, 1988). The character of the water content constitutive surface is not fully understood.

Further insight into the relationship between the constitutive surfaces can be obtained through examination of the Pham-Fredlund (2005) volume-mass model, which is based on a series of reasonable and verifiable assumptions. The volume-mass model commences with a saturated soil and allows a variety of stress state changes to be applied to the soil.

13.6 PHAM-FREDLUND VOLUME-MASS CONSTITUTIVE MODEL (2011a)

There is no single, unique relationship between overall volume change (and phase volume changes) for all possible stress paths that can be applied to an unsaturated soil. This is even true for isotropically loaded soils. Total volume changes and water volume changes in an unsaturated soil constitute independent processes. Changes in net normal stresses can produce behavior that is independent of changes in soil suction. The Pham-Fredlund (2011a) volume-mass change model simulates mechanical behavioral changes (i.e., stress-strain behavior) and capillary behavioral changes (i.e., adsorption-desorption behavior) (Pham, 2005; Fredlund and Pham, 2006; Pham and Fredlund, 2011a). Modeling the effect of soil suction changes is quite different from modeling the effect of net normal stress changes in an unsaturated soil.

Deformations associated with the overall soil mass are directly related to changes in the volume of voids (or pores) in the soil. Total volume change and water volume change result in changes in the volume and shape of the soil pores. The "shape" of a pore is referred to as an "open pore diameter" (or "neck pore diameter") or a "body pore diameter," according to Haines (1930) capillary theory. The open pore diameter is related to the air-entry value of the pore defined by the drying curve. The body pore diameter on the wetting curve is referred to as the water-entry value of the pore.

The pores of a soil are comprised of a variety of shapes and volumes. The pore-size distribution curve of a soil represents the volume and the shape of pores in the soil. The pore-size distribution of a soil at any stress state provides information related to the total volume of voids and the volume of water in the soil. The pore-size distribution curve is changed when soil suction or mean net normal stress is changed (Fig. 13.29). Water volume changes are related to the pore-size distribution curve.

A comprehensive volume-mass constitutive model for unsaturated soils should include (i) the selection of appropriate stress state variables along with a specified reference stress state; (ii) statements regarding the basic assumptions for the response of a pore volume to changes in net mean stress and soil suction; (iii) the formulation of the stress-strain relationship for the soil structure surrounding a group of pores, including changes to air-entry value, water-entry value, yield stress, total volume, and water volume; (iv) the determination of the compression (i.e., unloading/reloading indices) for groups of pores; and (v) an accommodation of the hysteretic nature of the SWCC.

13.6.1 Stress State and Concept of Pore Volume Behavior

There are three stress state variables that can be used when describing the mechanical behavior of a soil as it changes from a saturated state to an unsaturated state. Under isotropic loading conditions (i.e., $\sigma_x = \sigma_y = \sigma_z$), the state variables are (i) net mean stress $p = (\sigma_1 + \sigma_2 + \sigma_3)/3 - u_a$, (ii) soil suction, $\psi = (u_a - u_w)$, and (iii) mean effective stress $p^* = (\sigma_1 + \sigma_2 + \sigma_3)/3 - u_w$. Only two of the stress state variables are independent when considering any particular physical process even though there are three possible forms for the stress state variables.

A pore volume in an unsaturated soil can be either filled with water or essentially empty of water. The mean effective stress p^* can be assumed to control the mechanical behavior of the soil structure when a particular pore volume is filled with water. Similarly, when the pore is empty, the net mean stress p can be assumed to control the mechanical behavior of the soil structure surrounding the pore. Soil suction ψ provides information as to when a pore is filled with water or empty of water.

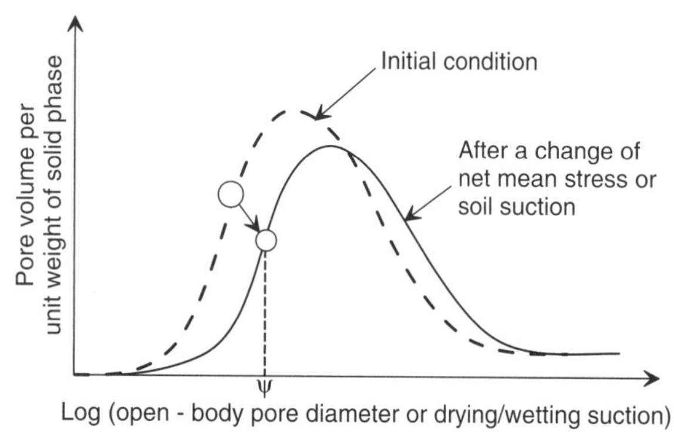

Figure 13.29 Changes in pore-size distribution for changes in net mean stress or soil suction (from Pham, 2002).

A mean net stress state of 1 kPa and zero soil suction have been arbitrarily selected as the *reference stress state* of the soil for the development of the proposed volume-mass constitutive model which commences with a saturated slurry soil (Pham, 2005). Soil suction plays the same role as net mean stress (isotropic) when the pores in the soil are filled with water (Alonso, 1993; Kohgo et al., 1993; Wheeler and Sivakumar, 1995; Fredlund and Rahardjo, 1993b). The selected reference stress state assumes that the air-entry value of the soil is greater than 1 kPa. The void ratio e is selected to represent the overall volume changes and gravimetric water content w is selected to represent the amount of water in the soil.

13.6.1.1 Drying and Wetting Pore-Size Distribution

The pore-size distribution function provides basic information for the development of a volume-mass constitutive model for unsaturated soils. Numerous researchers have described the meaning of the pore-size distribution (Brooks and Cory, 1964; Fredlund and Xing, 1994; Vanapalli et al., 1996a; M.D. Fredlund, 2000; Simms and Yanful, 2001; Cui et al., 2002). The pore-size distribution has also been used to model the hysteretic nature of the SWCC (Haines, 1930; Mualem, 1973, 1974; Nimmo, 1992).

The pore-size distribution curve of a soil is generally plotted using a semilogarithm scale for soil suction (M.D. Fredlund, 2000; Simms and Yanful, 2001) (Fig. 13.29). There are two limiting pore-size distribution curves for a soil. The pore-size distribution can be written as the ratio of pore volume per unit weight versus the "open pore diameters" (i.e., drying suction or air-entry value of the pore), referred to as the *drying pore-size distribution* (DPD). The pore-size distribution can also be written as the ratio of pore volume per unit weight versus the "body pore diameters" (i.e., wetting suction or water-entry value of the pore), referred to as the *wetting pore-size distribution* (WPD). The DPD and WPD constitute the initial bounding drying curve and the bounding wetting curve of the SWCCs for a soil (Fig. 13.30).

The pore-size distribution of a soil changes with soil suction and net mean stress. Drying a soil from initially slurry conditions to completely dry conditions (i.e., suction of 10^6 kPa) gives the *reference pore-size distribution curve* for the soil. A soil has two reference pore-size distribution curves (i.e., one curve with respect to increasing soil suction and one curve with respect to decreasing soil suction). The reference DPD provides information regarding the air-entry value and the volume distribution of the pores (i.e., the initial bounding drying SWCC). The reference WPD provides information regarding the water-entry value and the volume distribution of the pores in the soil (i.e., the bounding wetting SWCC). The reference DPD and WPD

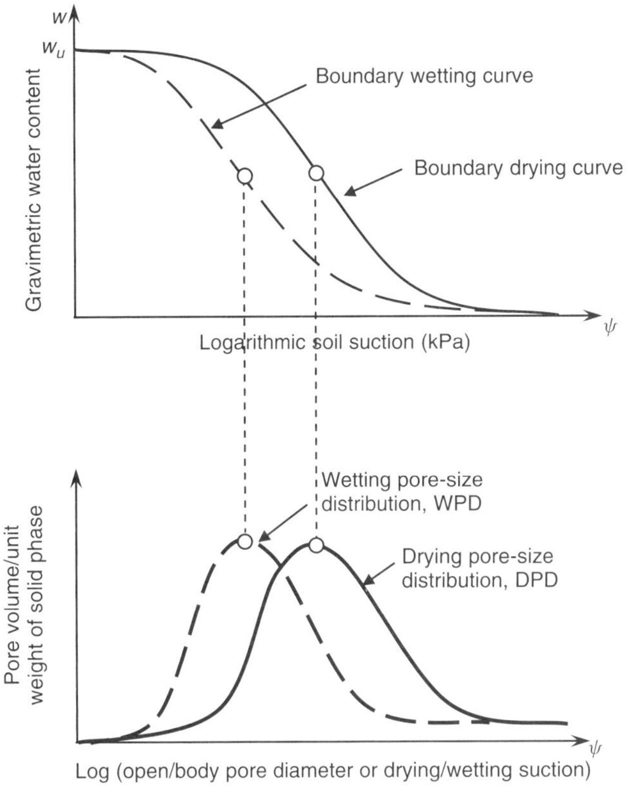

Figure 13.30 Change in pore-size distribution resulting from changes in net mean stress or soil suction (from Pham, 2002).

curves form the basic building units for the development of a volume-mass constitutive model.

13.6.2 Basic Assumptions for Volume-Mass Constitutive Model

The Pham and Fredlund (2011) volume-mass constitutive model is based on several assumptions. These assumptions are based on the findings from several research studies.

Assumption 1. A particular pore in an unsaturated soil has one of two possible states: (i) the pore can be filled with water or (ii) the pore can be empty or essentially dry. This assumption 1 has been widely accepted in the development of hysteresis models for SWCCs (Néel, 1942, 1943; Poulovassilis, 1962; Mualem, 1973, 1974; Pham et al., 2005). This assumption allows pore volumes to be assumed to be either empty or filled at a designated suction.

Assumption 2. Soil suction affects only the water-filled pores and does not directly affect the dry-filled pores while net mean stress has an effect on all pores in the soil. This assumption is consistent with the mechanical unsaturated soil behavior suggested by Alonso et al. (1994) and Wheeler and Sivakumar (1995). The maximum stress acting on the soil structure surrounding a pore is equal to the air-entry value of the pore during the drying process under zero net mean stress. Therefore, the pore is empty and soil suction does not affect the soil structure surrounding the pore. The pore-size distribution function of the soil provides information regarding the maximum stress acting on the soil structure surrounding each pore during the drying process. Similarly, the maximum stress that acts on the soil structure surrounding each pore during a wetting process is the water-entry value of the pore.

Assumption 3. Each water-filled pore in a soil has two indices: (i) a virgin compression index C_c^p and (ii) an unloading-reloading compression index C_s^p. The compression indices of a saturated soil are defined as the change in void ratio of the soil with respect to a change in logarithm of mean effective stress under saturated conditions (Pham et al., 2005). The virgin compression and unloading-reloading indices of a water-filled pore are defined as follows:

$$C_c^p = \frac{dv^p}{V_s d\left[\log(p^*)\right]} = \frac{v_b^p - v_a^p}{V_s \Delta\zeta} \quad (p^* > \text{yield stress}) \tag{13.86}$$

$$C_s^p = \frac{dv^p}{V_s d\left[\log(p^*)\right]} = \frac{v_b^p - v_a^p}{V_s \Delta\zeta} \quad (p^* < \text{yield stress}) \tag{13.87}$$

where:

C_c^p = virgin compression index of the water-filled pore,
C_s^p = unloading-reloading index of a water-filled pore,
v_b^p = volume of the pore prior to an increase in load,
v_a^p = volume of the pore after an increase in load, and
$\Delta\zeta$ = infinitesimal increment of logarithmic mean effective stress.

Assumption 4. There are two types of pores: (i) collapsible pores and (ii) noncollapsible pores. Collapsible pores are relatively large pores while noncollapsible pores are relatively small interconnected pores (Fig. 13.31). Noncollapsible pores are assumed to be incompressible.

This assumption is similar to the macro- and microstructures described by Alonso et al. (1994) and Wheeler and Sivakumar (1995). Measured pore-size distribution curves for soils under various loading and compaction conditions show that micropores do not appear to change significantly when the volume of the soil changes (Sridharan et al., 1971; Ahmed et al., 1974; Delage and Graham, 1995; Al-Mukhtar, 1995; Alonso et al., 1995; Wan et al., 1995; Lloret et al., 2003). Data showing the effect of net mean stress on the pore-size distribution of montmorillonite clay are shown in Fig. 13.32.

The assumption related to collapsible and noncollapsible pores is consistent with the hysteretic nature of SWCCs. The two types of pores explain the hysteretic phenomena associated with the SWCC (Fig. 13.33). The drying and wetting processes associated with the two types of pores along with the hysteretic SWCCs are shown in Fig. 13.34. It is assumed that water in the soil exists primarily in noncollapsible pores once the soil suction exceeds residual

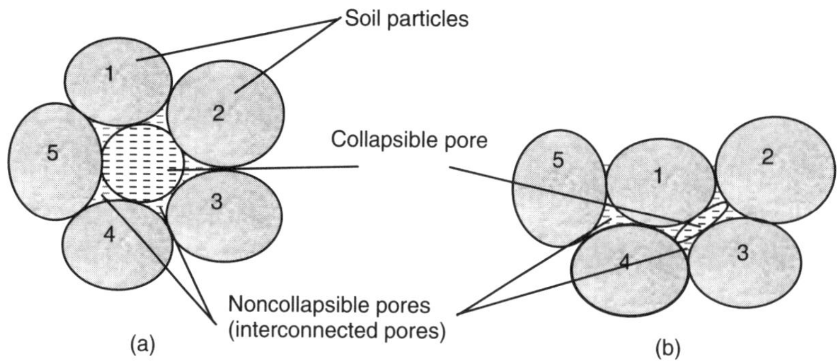

Figure 13.31 Two types of pores in soil: (a) before collapse and (b) after collapse (from Pham, 2002).

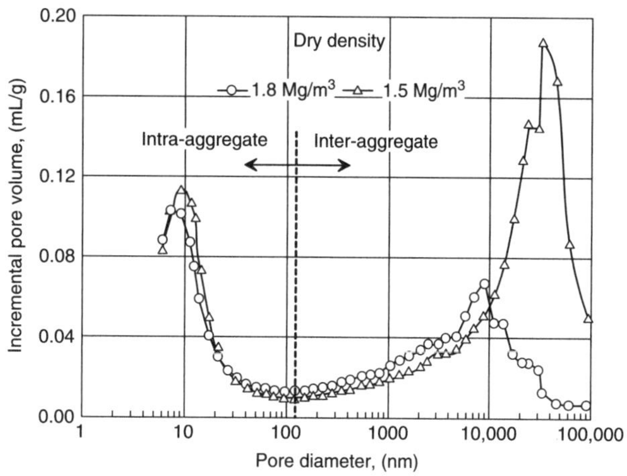

Figure 13.32 Mercury intrusion test results showing pore-size distribution of two montmorillonite soil samples compacted at different dry densities (after Lloret et al., 2003).

conditions. Consequently, there should be little hysteresis between the wetting curve and the drying curves at water contents lower than the residual water content.

Assumption 5. Virgin compression and unloading-reloading indices for a pore in the soil are proportional to the pore volume at the reference stress state (i.e., loading from 1 kPa net mean stress and zero soil suction).

This assumption 5 indicates that larger pore sizes result in higher virgin compression and unloading-reloading indices (Pham, 2005). If this assumption and assumption 3 are combined, the virgin compression and unloading-reloading compression indices of a pore can be calculated from the virgin compression and unloading-reloading indices of the soil:

$$C_c^p = \frac{v^p(0,1)}{V_p(0,1)} C_c \tag{13.88}$$

$$C_s^p = \frac{v^p(0,1)}{V_p(0,1)} C_s \tag{13.89}$$

where:

C_c^p, C_s^p = compression index of the pore having a volume of $v^p(0,1)$ at the reference stress state (i.e., zero soil suction, 1 kPa net mean stress),
$v^p(0,1)$ = volume of the pore at the reference stress state,
$V_p(0,1)$ = total volume of all pores in the soil at the reference state,
C_c = virgin compression index of the soil, and
C_s = unloading-reloading index of the soil.

Assumption 6. Pores are deformed and water is either absorbed or drained through processes independent of net mean stress change. Two types of physically based models have been proposed to represent the hysteretic nature of SWCCs: (i) independent models and (ii) dependent models.

Independent models assume that absorption or drainage of water from a pore is independent of the surrounding pores. Dependent models take into account the effect of blockage from surrounding pores (Pham, 2002, 2005). Independent models are simpler to use and appear to provide an adequate representation of hysteresis associated with SWCCs (Everett, 1955; Topp, 1971; Poulovassilis and Childs, 1971; Mualem, 1973, 1974; Pham, 2002; M.D. Fredlund et al., 2003). Noncollapsible pores are assumed to be incompressible (i.e., assumption 3). Collapsible pores are assumed to deform independently once the noncollapsible pores do not change volume.

Assumption 7. The unloading-reloading index of an air-filled pore is approximately zero. Silva et al., (2002) assumed that the unloading-reloading index of an unsaturated soil under constant soil suction decreases with each suction increment. Under completely dry conditions (i.e., soil suction equal to 10^6 kPa), the unloading-reloading index for the soil structure is assumed to approach zero and unloading-reloading indices for the soil pores also approach zero (Vilar and Davies, 2002; Lloret et al., 2003).

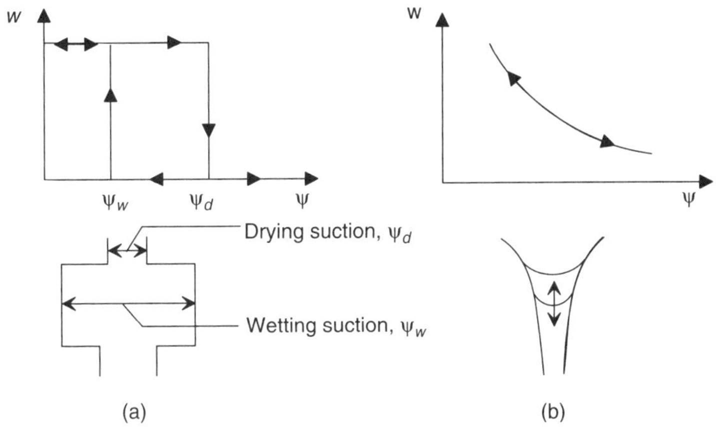

Figure 13.33 (a) Collapsible and (b) noncollapsible pores used to explain SWCC hysteresis (modified from Poulovassilis, 1962).

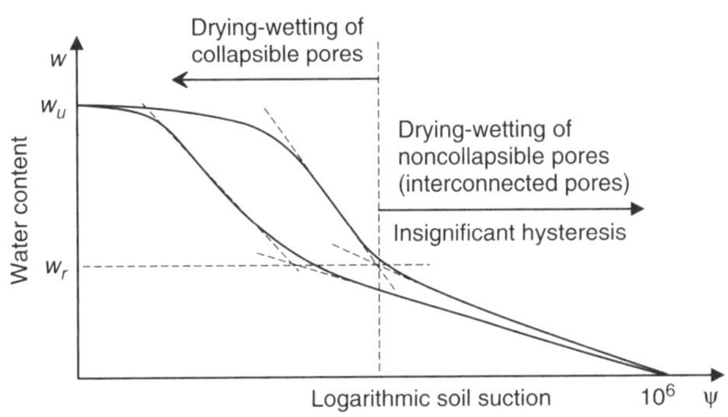

Figure 13.34 Behavior of two types of pores associated with boundary drying and wetting SWCCs (from Pham, 2002).

13.6.3 Volume Change of Pore under Various Stress Paths

Volume-mass constitutive relationships for unsaturated soils are stress path dependent (Alonso, 1993; Pham, 2005). The stress-strain relationship for the soil structure surrounding a pore can be considered for four basic stress paths (i.e., loading, unloading, drying, and wetting). The net mean stress also affects the air-entry value and the water-entry value of the soil. Consideration is given to a pore volume from a representative soil element.

13.6.3.1 Drying-Wetting Processes under Zero Net Mean Stress

Let us consider the drying process for an initially saturated soil element. The starting soil suction is less than the air-entry value of the soil. The pores are filled with water and the stresses acting on the soil structure surrounding the pore are equal to the soil suction ψ. When soil suction is higher than the air-entry value, some pores become filled with air and soil suction is assumed to not further affect the soil structure surrounding the emptied pore. The yield stress of the pore becomes equal to the air-entry value of the pore. The volume of the air-filled pore is the same as that of a pore with a suction equal to the air-entry value, while the water content in the specific pore under consideration is assumed to be equal to zero. An illustration of the overall volume and water volume changes of a pore along the drying process from an initial slurry condition is shown in Fig. 13.35. The volume of a pore, $v^p(\psi, 0)$, at a suction equal to ψ along the drying curve can be calculated as follows:

$$v^p(\psi, 0) = \begin{cases} v^p(1, 0) - V_s C_c^p \log(\psi) & \text{for } \psi \le \psi_{\text{aev}} \\ v^p(1, 0) - V_s C_c^p \log(\psi_{\text{aev}}) & \text{for } \psi > \psi_{\text{aev}} \end{cases} \tag{13.90}$$

where:

ψ_{aev} = air-entry value of the pore,
C_c^p = virgin compression index of the pore,
V_s = volume of the solid phase of the representative soil element, and
$v^p(\psi, 0)$ = volume of the pore at a suction of ψ and zero net mean stress.

A similar equation can be written to represent the volume of water in a pore, $v^w(\psi, 0)$, along the drying process:

$$v^w(\psi, 0) = \begin{cases} v^p(1, 0) - V_s C_c^p \log(\psi) & \text{for } \psi \le \psi_{\text{aev}} \\ 0 & \text{for } \psi > \psi_{\text{aev}} \end{cases} \tag{13.91}$$

where:

C_c^p = virgin compression index of the pore and
$v^w(\psi, 0)$ = volume of water at a soil suction of ψ and zero net mean stress.

The volume of the pore does not change until soil suction is equal to the water-entry value during the wetting process of an air-filled pore (Fig. 13.35). A pore is filled with water when soil suction is less than the water-entry value and the soil will expand as soil suction decreases. The volume of the

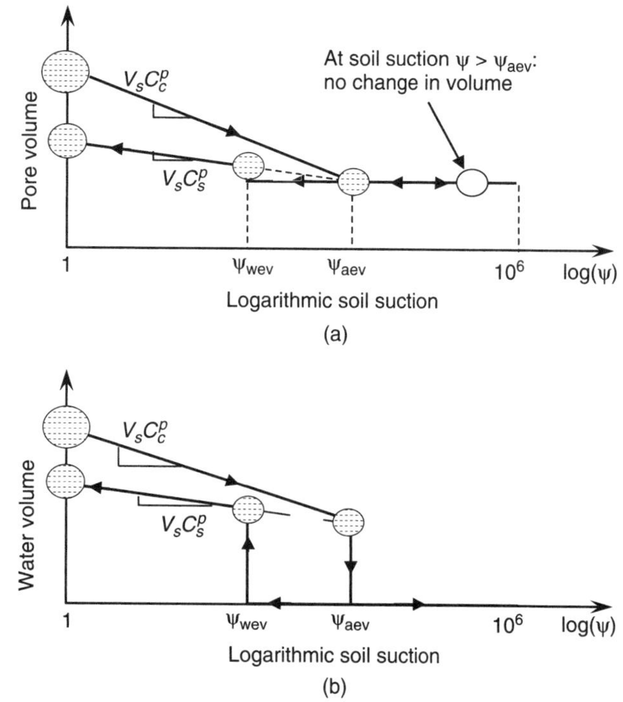

Figure 13.35 (a) Volume change and (b) water content change of pore during initial drying process (from Pham, 2002).

pore at any suction ψ along a wetting process (i.e., from an initially air-filled condition) can be calculated as follows:

$$v^p(\psi, 0) = \begin{cases} v^p(\psi_{aev}, 0) + V_s C_s^p \log\left(\dfrac{\psi}{\psi_{aev}}\right) & \text{for } \psi \le \psi_{wev} \\ v^p(\psi_{aev}, 0) & \text{for } \psi > \psi_{wev} \end{cases} \tag{13.92}$$

where:

C_s^p = unloading-reloading index of the pore and
ψ_{wev} = water-entry value of the pore.

An equation for the volume of water in a pore along a wetting process (i.e., initially air-filled condition) can be written as follows:

$$v^w(\psi, 0) = \begin{cases} v^p(\psi_{aev}, 0) + V_s C_c^p \log\left(\dfrac{\psi}{\psi_{aev}}\right) & \text{for } \psi \le \psi_{wev} \\ 0 & \text{for } \psi > \psi_{wev} \end{cases} \tag{13.93}$$

13.6.3.2 Drying Process under Constant Net Mean Stress

The air-entry value of a pore, ψ_{aev}, is also a function of net mean stress p and can be expressed as $\psi_{aev}(p)$. The volume of a pore is a function of the net mean stress p and soil suction ψ. The volume of the pore under dry conditions (i.e., dry under zero net mean stress) can be calculated from Eq. 13.90 as follows:

$$v^p(\psi_{aev}, 0) = v^p(1, 0) - C_c^p V_s \log(\psi_{aev}) \tag{13.94}$$

where:

ψ_{aev} = air-entry value of a pore when drying takes place under zero net mean stress and
$v^p(1, 0)$ = volume of the pore corresponding to the reference stress state (i.e., 1 kPa soil suction and zero net mean stress).

The air-entry value of a pore dried under a net mean stress is higher than that of a pore dried under zero net mean stress (Fig. 13.36). Net mean stress and soil suction have the same effect when a pore is filled with water. The volume of a pore at a suction value equal to the air-entry value, $\psi_{aev}(p)$, and a net mean stress of p can be presented as follows:

$$v^p(\psi_{aev}(p), p) = v^p(1, 0) - C_c^p V_s \log(\psi_{aev}(p) + p) \tag{13.95}$$

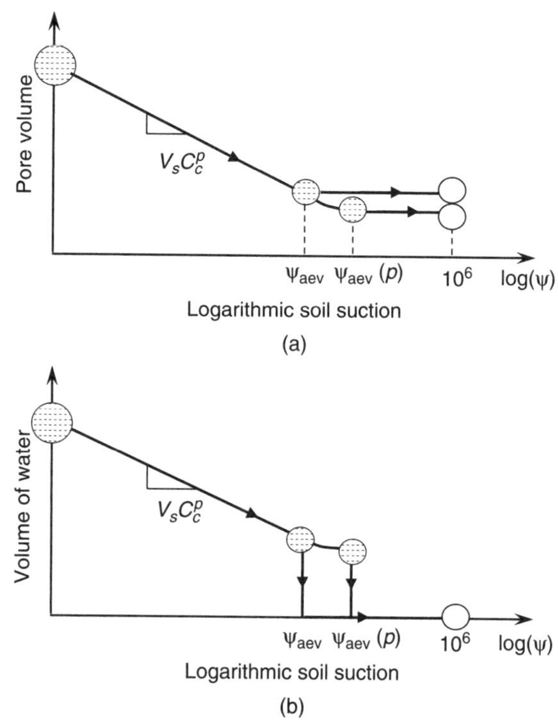

Figure 13.36 Effects of net mean stress on air-entry value, volume change, and water content change of pore during drying process: (a) volume change comparison of pore drying under zero load and drying under a constant net mean stress p; (b) water content change comparison of pore drying under zero load and drying under constant net mean stress p (from Pham, 2002).

where:

$v^p(\psi_{aev}(p), p)$ = volume of the pore at a net mean stress p and a soil suction equal to the air-entry value of the pore if the soil is dried under a net mean stress p and

$\psi_{aev}(p)$ = air-entry value of the pore if the soil is dried under a constant net mean stress p.

The relationship between soil suction and volume changes of a pore can be derived based on the capillary equation relating matric suction to the diameter of a capillary tube. The following equation is obtained when assuming that the deformations of a pore are isotropic (i.e., $\varepsilon_x = \varepsilon_y = \varepsilon_z$):

$$\frac{\psi_a}{\psi_b} = \frac{3V_0}{3V_0 - \Delta V} \tag{13.96}$$

where:

ψ_b = air-entry value (or water-entry value) of the pore before deformation (i.e., initial volume of V_0) and

ψ_a = air-entry value (or water entry value) of the pore after deformation.

The air-entry and water-entry values of a pore depend on the smallest and largest open pore diameters, respectively, connected to the pore. The above equation does not adequately describe changes in air-entry value (or water-entry value) under isotropic net total stress loading. Therefore, the "pore shape parameter" η can be introduced to assist in better representing the relationship between the air-entry value of a pore that is dried under zero net mean stress and a pore that is dried under a constant net mean stress p:

$$\frac{\psi_a}{\psi_b} = \frac{3V_0}{3V_0 - \eta\Delta V} \tag{13.97}$$

where:

η = pore shape parameter depending on soil type and stress history ($\eta = 1$).

The virgin compression index of a pore can be calculated in accordance with assumptions 3 and 5:

$$C_c^p = \frac{v^p(0,1)}{V_p(0,1)} C_c \tag{13.98}$$

where:

$V_p(0,1)$ = total volume of all pores in the soil element at the reference stress state and

C_c = virgin compression index of the soil.

Substituting Eqs. 13.94, 13.95, and 13.98 into Eq. 13.97 and simplifying give

$$1 - \eta \frac{C_c\{\log(\psi_{aev}(p) + p] - \log(\psi_{aev})\}}{3\left[e_{sat} - C_c \log(\psi_{aev})\right]} = \frac{\psi_{aev}}{\psi_{aev}(p)} \tag{13.99}$$

where:

e_{sat} = void ratio at saturation.

There is an unknown air-entry value in Eq. 13.99 for a pore, $\psi_{aev}(p)$, when the soil is dried under constant net mean stress p. The solution of the above equation is quite complex (Pham, 2005) but can be approximated as follows:

$$1 - \eta \frac{C_c\left[\log(\psi_{aev} + p) - \log(\psi_{aev})\right]}{3\left[e_{sat} - C_c \log(\psi_{aev})\right]} = \frac{\psi_{aev}}{\psi_{aev}(p)} \tag{13.100}$$

Equation 13.100 presents the relationship between the air-entry value (or water-entry value) of a pore under zero net mean stress and the air-entry value of the pore when the slurry soil is loaded to a mean effective stress p_0 and then dried under a constant net mean stress p. In the case where the net mean stress p is smaller than p_0, a similar equation can be derived to take into account the elastic deformation when the soil is unloaded from p_0 to $p + \psi_{aev}$:

$$1 - \eta \frac{\left[(C_c - C_s)\log(p_y) + C_s \log(\psi_{aev} + p) - C_c \log(\psi_{aev})\right]}{3[e_{sat} - C_c \log(\psi_{aev})]} = \frac{\psi_{aev}}{\psi_{aev}(p, p_y)} \tag{13.101}$$

$$p_y = p_y(\psi_{aev}) = P(\psi_{aev}, p, p_0) = \begin{cases} p_0 & \text{for } \psi_{aev} + p \le p_0 \\ p + \psi_{aev} & \text{for } \psi_{aev} + p > p_0 \end{cases} \tag{13.102}$$

where:

p_0 = yield stress prior to the drying process and

p_y = yield stress of the soil.

Equation 13.102 is a discontinuous function; however, a mathematical technique can be used to produce a continuous function (Pham, 2005). The technique used to produce the continuous function is similar to that used for the volume change SWCC for a soil (Pham and Fredlund, 2008):

$$p_y = p_y(\psi_{aev}) = P(\psi_{aev}, p, p_0) = \frac{[\tan^{-1}(\psi_{aev} + p - p_0) + 1.571](\psi_{aev} + p - p_0)}{3.142} + p_0 \tag{13.103}$$

The relationship between the air-entry value of a pore subjected to zero yield stress and dried under zero net mean stress and the case where the pore has experienced a yield stress p_y and then dried under a constant net mean stress p is as follows:

$$\frac{\psi_{aev}}{\psi_{aev}(p, p_y)} = 1 - \eta \frac{(C_c - C_s)\log(p_y) + C_s \log(\psi_{aev} + p) - C_c \log(\psi_{aev})}{3\left[e_{sat} - C_c \log(\psi_{aev})\right]} \quad (13.104)$$

where:

p_y = yield stress of the soil and

$\psi_{aev}(p, p_y)$ = air-entry value of a pore when the yield stress is equal to p_y and drying takes place under a net mean stress p.

13.6.3.3 Wetting Process under Constant Net Mean Stress

The water-entry value of a pore having a yield stress p_y that is wetted under a constant net mean stress p [i.e., $\psi_{wev}(p, p_y)$] is higher than when the pore is dried and wetted under zero net mean stress (i.e., reference water-entry value, ψ_{wev}). The relationship between the two water-entry values can be calculated in a similar manner to that used for the drying process:

$$\frac{\psi_{wev}}{\psi_{wev}(p, p_y)} = 1 - \eta \frac{(C_c - C_s)\log(p_y) + C_s \log(\psi_{wev} + p) - C_c \log(\psi_{aev})}{3[e_{sat} - C_c \log(\psi_{aev})]} \quad (13.105)$$

where:

ψ_{aev} = air-entry value of the pore at zero net mean stress (i.e., reference air-entry value),

ψ_{wev} = water-entry value of the pore having zero yield stress and being wetted under zero net mean stress (i.e., reference water entry value), and

$\psi_{wev}(p, p_y)$ = water-entry value of the pore when it has experienced a yield stress of p_y and is wetted under a constant net mean stress of p, where p_y is equal to the yield stress of the pore with a pore-shape parameter, η.

Alonso (1993) observed that stress paths related to the wetting process were stress path independent. It appears reasonable to assume that the volume of a pore after wetting to a selected soil suction ψ and then subjected to a net mean stress p is equivalent to that of the pore that is first loaded to a net mean stress p and then wetted to a soil suction ψ.

There are two types of pores after a wetting process: (i) water-filled pores and (ii) air-filled pores. A pore that is filled with air after the wetting process is called the *continuously air-filled pore.* For a water-filled pore, the volume of the pore at a suction ψ and a net mean stress p can be calculated as follows:

$$v^P(\psi, p, p_y) = v^P(1, 0) - V_s C_c^P \log(p_y) + V_s C_s^P \log(p + \psi/p_y) \quad (13.106)$$

The yield stress of a pore filled with water after the wetting process can be calculated as follows:

$$p_y = p_y(\psi_{wev}) = \begin{cases} p_0 & \text{for } (\psi + p) \le p_0 \\ p + \psi_{wev} & \text{for } (\psi + p) > p_0 \end{cases} \quad (13.107)$$

where:

ψ_{wev} = reference water-entry value of the pore and

p_0 = yield stress prior to the wetting process.

The volume-mass constitutive model for unsaturated soils must reflect yield in the soil structure surrounding continuously air-filled pores. This concept is consistent with elastoplastic theories for soils subjected to isotropic loading conditions. There are three alternative assumptions that might be made regarding the stress-strain relationship surrounding a continuously air-filled pore. These can be stated as follows.

Alternative 1. The simplest solution is to assume that continuously air-filled pores are incompressible (Fig. 13.37). Under this assumption the yield stress of a continuously air-filled pore does not change with a change in soil suction or net mean stress, and the yield stress p_y can be written as

$$p_y = p_y(\psi_{aev}) = p_0 \quad (13.108)$$

where:

p_y = yield stress of the continuously air-filled pore (i.e., having a reference air-entry value of ψ_{aev}) after the wetting process and

p_0 = yield stress of a pore prior to the wetting process.

Alternative 2. Another solution is to assume that the compression curve of a continuously air-filled pore at constant soil suction (on a logarithmic net mean stress scale) can be expressed as a combination of two straight lines corresponding to the virgin compression line and the unloading-reloading line (Fig. 13.37). The unloading-reloading index of an air-filled pore would be taken as zero (i.e., assumption 7). Therefore, the *volume change of an air-filled pore would only depend on the yield stress of the pore.* The yield stress of a continuously air-filled pore is a function of the virgin compression index of the pore (i.e., when it is filled with air).

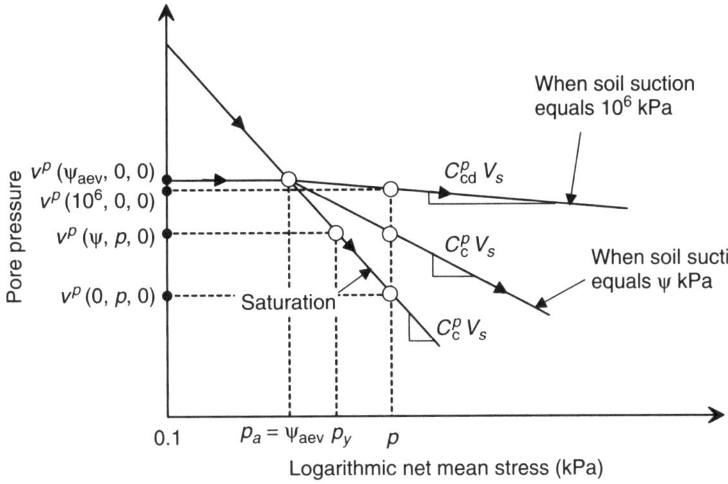

Figure 13.37 Compression curves for pore at three different soil suctions (0, ψ, and 10^6 kPa) (from Pham, 2002).

The virgin compression index of a continuously air-filled pore should lie between the virgin compression indices of a pore when the soil is completely dry and a pore when soil suction is equal to zero. Figure 13.37 illustrates compression curves for a pore at three different soil suctions: (i) saturation (i.e., zero soil suction); (ii) when soil suction is equal to a specified value ψ, where the pore is filled with air; and (iii) at 10^6 kPa (i.e., soil is completely dry). Further details concerning the equations associated with each of the mentioned alternatives can be found in Pham (2005) and Pham and Fredlund (2011a).

Alternative 3. A third solution could assume that there are some small particles and cementing materials that act to bond the larger particles together at completely dry conditions. Along the wetting process, water gradually fills the pores and reduces the bonding strength that holds soil particles in position (Lawton et al., 1991a, 1991b; Pereira, 1996). It can be assumed that all bonds are removed when all interconnected pores (i.e., noncollapsible pores) are filled with water. The continuously air-filled pores reach maximum collapse conditions and the equivalent yield stress on the soil structure surrounding a continually air-filled pore is equal to the magnitude of the net mean stress. The equivalent yield stress of the soil structure surrounding continuously air-filled pores could be assumed to be a function of the amount of water in the interconnected pores and the net mean stress.

13.6.3.4 Loading-Unloading Processes at Constant Soil Suction

Two types of pores (i.e., air-filled pores and water-filled pores) are considered during loading-unloading processes of an unsaturated soil. The pore may collapse and the water-entry value of the pore is decreased when the soil structure surrounding an air-filled pore is loaded. The pore absorbs water and becomes a water-filled pore when soil suction is less than the water-entry value. Loading-unloading of the soil structure surrounding a water-filled pore is similar to the loading-unloading of a saturated soil. If soil suction is higher than the water-entry value, the pore is still filled with air after the loading process and is called a continuously air-filled pore.

13.6.4 Mathematical Formulation for Drying Process of Initially Slurry Soil

Equations for the compression indices associated with each group of pores along with the pore-size distribution are presented for the reference pore-size distribution function. Details of the mathematical derivation of each equation can be found in Pham (2005).

13.6.4.1 Curve-Fitting Model for SWCC

The drying SWCC starting from an initial slurry condition (i.e., gravimetric water content versus soil suction) can be best fit using the following SWCC equation (Pham and Fredlund, 2008):

$$w(\psi) = \left[\left(w_{sat} - \frac{C_c}{G_s}\log(\psi) - w_r\right)\frac{a}{\psi^b + a} + w_r\right]\left(1 - \frac{\ln(1+\psi/\psi_r)}{\ln(1+10^6/\psi_r)}\right) \quad (13.109)$$

where:

C_c = virgin compression index of the soil,
G_s = specific gravity of the soil particles,
w_r = residual gravimetric water content,
w_{sat} = water content of the slurry soil at an effective stress of 1 kPa, and
a, b = curve-fitting parameters.

13.6.4.2 *Equation for Compression Indices of Pores and Reference Drying Pore-Size Distribution*

A soil is completely dry at a soil suction of approximately 10^6 kPa (Fredlund and Rahardjo, 1993a). Let us denote the soil suction at which a soil is completely dry as ψ_{max} and designate ξ_{max} as $\log(\psi_{max})$ on a logarithmic suction scale. Let us consider a representative element of soil dried from slurry where the volume of solids is equal to V_s. The total volume of the pores, V_p, in the soil element under dry conditions (i.e., at a soil suction of ψ_{max}) can be calculated as follows:

$$V_p = V_a = \int_0^{\xi_{max}} f_d(x)\, dx \qquad (13.110)$$

where:

$f_d(x)$ = reference DPD,
V_p = total volume of pores, and
V_a = total volume of air in the soil element at a suction of ψ_{max}.

The volume of the soil element at a suction ψ can be calculated as the summation of the volume of solids, water, and air in the soil:

$$V(\psi) = V_s + V_a(\psi) + V_w(\psi) \qquad (13.111)$$

where:

V, V_s, V_a, V_w = overall volume of the representative soil element, volume of solid phase, volume of air phase, and volume of water phase at a soil suction ψ, respectively.

The volume of the solid phase, V_s, is a constant at any soil suction or net mean stress. The volume of the air phase at a suction ψ is equal to the summation of the volume of pores that have an air-entry value smaller than the designated suction:

$$V_a(\psi) = \int_0^{\xi} f_d(x)\, dx \qquad (13.112)$$

The volume of the water phase, V_w, at any suction value ψ is equal to the summation of the volume of pores that have an air-entry value greater than a suction ψ. An equation for the volume of water in the soil at a suction ψ can be written as

$$V_w(\psi) = \int_{\xi}^{\xi_{max}} \left[f_d(x) + C_c^p(x) V_s\, (x - \xi)\right] dx \qquad (13.113)$$

where:

$\xi = \log(\psi)$,
$C_c^p(x)$ = summation of the virgin compression index of pores having the same reference drying suction of x (i.e., air-entry value), and
V_s = volume of solid phase.

The following equation for the virgin compression index of a group of pores can be derived by combining Eqs. 13.98, 13.110, 13.111, 13.112, and 13.113:

$$C_c^p(\xi) = \frac{d\left[-G_s w(10^{\xi}) C_c/(e_{sat} - C_c \xi)\right]}{d\xi} \qquad (13.114)$$

where:

e_{sat} = void ratio of the soil at the reference stress state of 1 kPa and
$w(\psi)$ = water content along the initial drying SWCC.

The reference DPD can also be obtained by forcing all pores in the soil to dry to respective drying soil suctions (Pham, 2005). An equation for the reference DPD can be written as

$$\int_0^{\xi} f_d(x)\, dx = V_{dry}^p - G_s V_s \left[w(\psi) - \int_{\xi}^{\xi_{max}} \frac{w(10^x) C_c}{e_{sat} - C_c x} dx \right] \qquad (13.115)$$

13.6.4.3 *Volume Change of Collapsible and Noncollapsible Pores*

There does not need to be a distinction between collapsible pores and interconnected pores. Noncollapsible pores are interconnected pores that do not exhibit significant hysteresis between the wetting and drying processes. Water appears to exist only in the noncollapsible pores when soil suction exceeds residual conditions. The total volume of the noncollapsible pores at residual suction is assumed to be equal to the volume of water at residual suction.

The water content in a soil at a particular soil suction can be divided into two components: (i) water in collapsible pores and (ii) water in noncollapsible pores. At suctions higher than residual suction the water content in the soil reflects the water in collapsible pores (Fig. 13.38). At soil suctions less than residual suction the water content is equal to that of the collapsible pores (or the water content at residual suction). Let us denote a function that presents gravimetric water content in the collapsible pores as $w^c(\psi)$. The function $w^c(\psi)$ can be calculated as follows:

$$w^c(\psi) = \begin{cases} w(\psi) - w_r & \text{for } \psi \le \psi_r \\ 0 & \text{for } \psi > \psi_r \end{cases} \qquad (13.116)$$

where:

$w^c(\psi)$ = gravimetric water content along the initial drying curve for an initially slurry soil and
ψ_r = residual suction for the soil.

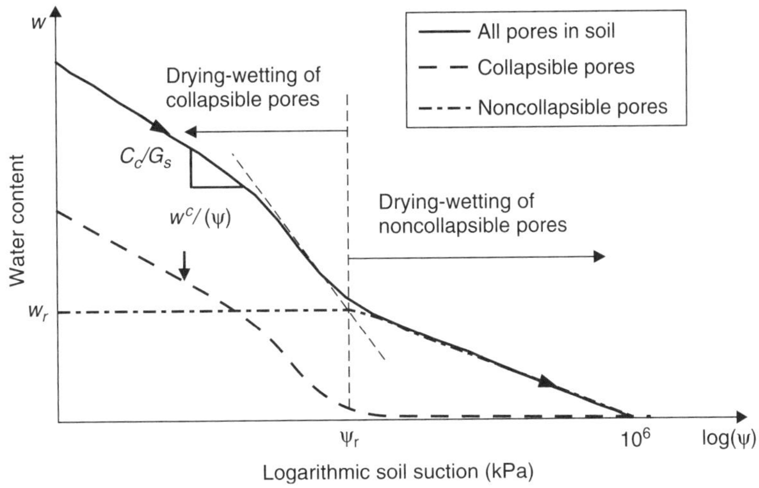

Figure 13.38 Illustration of gravimetric water content for collapsible and interconnected pores along initial drying process (from Pham, 2002).

An equation for the SWCC of collapsible pores, $w^c(\psi)$, can be obtained directly by using the curve-fitting parameters used in Eq. 13.109:

$$w^c(\psi) = \left[\left(w_{\text{sat}} - \frac{C_c}{G_s}\log(\psi) - w_r\right)\frac{a}{\psi^b + a}\right] \quad (13.117)$$

where:

a, b, w_r = curve-fitting parameters that can be obtained by best-fitting Eq. 13.109 to the measured SWCC.

13.6.4.4 *Summary of Compression Indices of Pores*

The virgin compression index of a group of noncollapsible pores having an air-entry value of ξ can be calculated by substituting Eq. 13.116 into Eq. 13.114:

$$C_c^p(\xi) = -d\left[\frac{G_s w^c(10^\xi)C_c/(e_{\text{sat}} - C_c\xi - w_r G_s)}{d\xi}\right] \quad (13.118)$$

Substituting Eq. 13.117 into Eq. 13.118, gives

$$C_c^p(\xi) = -\frac{d\left(C_c\frac{a}{(10^\xi)^b + a}\right)}{d\xi} = -\frac{C_c ab\ln(10)(10^\xi)^b}{\left[(10^\xi)^b + a\right]^2} \quad (13.119)$$

where:

C_c = virgin compression index of the soil under saturated loading and
a, b = curve-fitting parameters obtained from best fitting the SWCC of the soil from (initial) slurry conditions (Eq. 13.109).

Similarly, the unloading-reloading compression index of the group of pores having a logarithmic soil suction ξ can be calculated as follows:

$$C_s^p(\xi) = -\frac{d\left(C_c\frac{a}{(10^\xi)^b + a}\right)}{d\xi} = -\frac{C_s ab\ln(10)(10^\xi)^b}{\left[(10^\xi)^b + a\right]^2} \quad (13.120)$$

where:

C_s = unloading-reloading index of the soil at saturation.

An equation for the reference DPD of the soil can be obtained by differentiating both sides of Eq. 13.115 and then substituting Eq. 13.117 into the new formulation:

$$f_d(\psi) = -\frac{d\,[w(\psi)]}{d\psi}G_s\psi\ln(10) - \frac{aC_c}{\psi^b + a} \quad (13.121)$$

where:

$w(\psi)$ = gravimetric water content along the initial drying curve of an initially slurry soil.

An equation can be derived for the virgin compression index of air-filled pores in a manner similar to that used for the compression indices for water-filled pores. The dried virgin compression index of a group of pores having an air-entry value ξ can be written as follows:

$$C_{cd}^p(\xi) = -\frac{C_{cd}ab\ln(10)(10^\xi)^b}{\left[(10^\xi)^b + a\right]^2} \quad (13.122)$$

where:

C_{cd} = virgin compression index of the soil under dried conditions.

13.6.5 Hysteresis Model for SWCC

The Pham and Fredlund (2011a) model for hysteretic SWCCs based on the pore-size distribution curve can be applied to various volume-mass constitutive pathways. The hysteresis model for the SWCC is physically based (i.e., using pore-size distribution) and particularly suited for geotechnical engineering purposes. The pore-size distribution is described as a function of two variables: the air-entry value and the water-entry value of the pore (i.e., two-dimensional pore-size distribution function). Scanning SWCCs are assumed to be horizontal with respect to the degree of saturation on the SWCC (Wheeler et al., 2003). Therefore, each group of pores has a unique relationship between drying and wetting suctions. The relationship between the reference DPD and the reference WPD can be described as shown in Fig. 13.39:

$$f_w(\xi) = f_d(\xi - \Delta\xi) \tag{13.123}$$

where:

$f_d(\xi)$ = reference drying pore-size distribution,
$f_w(\xi)$ = reference wetting pore-size distribution,
$\Delta\xi$ = difference between the air-entry value and water-entry value of the group of pores having an air-entry value of ψ (i.e., log-cycles), and
$\xi = \log(\psi)$.

There are three key hysteretic SWCCs for a soil: the initial drying curve, the bounding drying curve, and the bounding wetting curve (Pham et al., 2005). If the distance between curves and the slope ratio between the two bounding curves are known, the key hysteretic SWCCs can be predicted. The reference WPD and the reference WPD can be calculated from the bounding SWCCs of the soil (i.e., Eq. 13.123). The reference pore-size distribution curves (i.e., desorption and adsorption curves) can be calculated from a single SWCC provided the slope ratio and the distance between the drying and wetting bounding curves is known.

The proposed model also takes into consideration the effect of the entrapped air. It is assumed that the amount of air entrapped in a collapsible pore is proportional to the pore volume and can be denoted by an entrapped air parameter, β (Fig. 13.40).

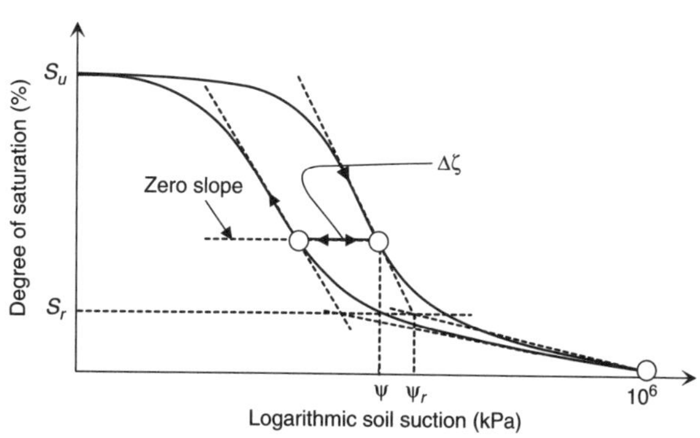

Figure 13.39 Distance between two boundary curves associated with drying and wetting for soil suction on logarithm scale.

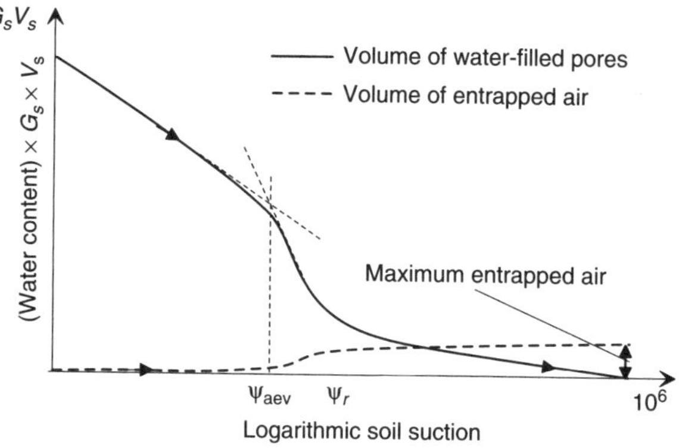

Figure 13.40 Volume of water-filled pores and air-filled pores during initial drying process (from Pham, 2002).

13.6.6 Pham and Fredlund Volume-Mass Constitutive Model for Saturated-Unsaturated Soils

The following equation is required to calculate the volume of water-filled pores, $v^p(\psi)$, at a soil suction ψ:

$$v^p(\psi) = v^p(1,0) - V_s C_c^p \log(p_y) + V_s C_s^p \log\left(\frac{p_y}{\psi + p}\right) \tag{13.124}$$

where:

$v^p(1,0)$ = volume of the pore at the reference stress state (i.e., 1 kPa soil suction and zero net normal stress) and
p_y = yield stress of the pore.

The volume of air-filled pores is also stress path dependent. The volume of an air-filled pore during a drying process can be calculated as follows:

$$v^p(\psi) = v^p(1,0) - V_s C_c^p \log(p_y) + V_s C_s^p \log\left(\frac{p_y}{\psi_{\text{aev}}(p, p_y) + p}\right) \tag{13.125}$$

The assumption is made that the volume of an air-filled pore does not change after an unloading process. After a loading process or a wetting process, the volume of air-filled pores can be calculated using the following equation:

$$v^p(\psi) = v^p(1,0) - V_s C_c^p \log(p_y) \tag{13.126}$$

where:

p_y = yield stress of the pore calculated from one of the suggested alternatives (i.e., alternative 1, 2, or 3).

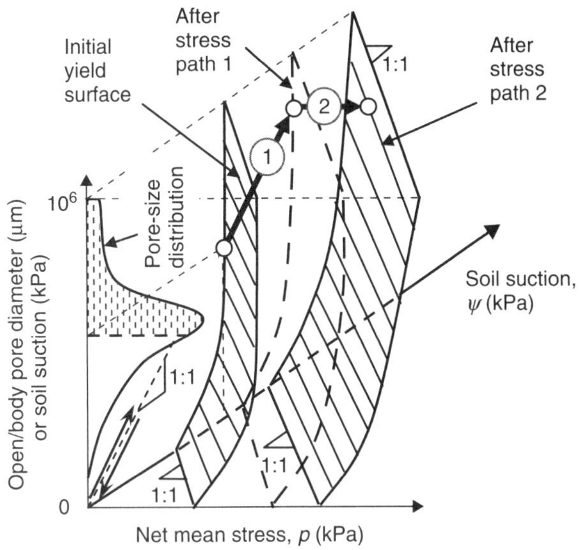

Figure 13.41 Evolution of yield surface (from Pham, 2002).

The yield stress can be described as a surface representing yield stresses of pores along the pore-size distribution (Fig. 13.41). Equations for the change in the air-entry value of a pore for collapsible and noncollapsible pores can be obtained by modifying Eq. 13.104 as follows:

$$\frac{\psi_{\text{aev}}}{\psi_{\text{aev}}(p, p_y)} = 1 - \eta \frac{(C_c - C_s)\log(p_y) + C_s \log(\psi_{\text{aev}} + p) - C_c \log(\psi_{\text{aev}})}{3[e_{\text{sat}} - C_c \log(\psi_{\text{aev}}) - w_r G_s]} \tag{13.127}$$

Similarly, an equation for the change in the water-entry value of a pore can be obtained by modifying Eq. 13.105 as follows:

$$\frac{\psi_{\text{wev}}}{\psi_{\text{wev}}(p, p_y)} = 1 - \eta \frac{(C_c - C_s)\log(p_y) + C_s \log(\psi_{\text{wev}} + p) - C_c \log(\psi_{\text{aev}})}{3[e_{\text{sat}} - C_c \log(\psi_{\text{aev}}) - w_r G_s]} \tag{13.128}$$

If soil suction is less than the water-entry value of the pore along a wetting path, then the pore is filled with water; otherwise, it is filled with air. If suction is less than the air-entry value along a drying process, then the pore is filled with water; otherwise, it is filled with air. The air-entry value and water-entry value of a pore are calculated using Eqs. 13.127 and 13.128, respectively.

The pore-size distribution of the soil corresponding to a particular stress state can be calculated. The volume of water in a representative soil element at a designated suction ψ is calculated as a summation (i.e., integration) of all pore volumes in the soil that have soil suction values higher than the selected suction value. The overall volume of the representative soil element can be calculated as a summation from zero suction to a suction value of 10^6 kPa.

The volume-mass constitutive surfaces have been experimentally shown to be stress path dependent (Alonso, 1993; Pham, 2005). Each volume-mass constitutive surface pertains to a particular testing stress path and an equation is required to describe that surface. It is possible, however, to derive closed-form equations for the volume-mass constitutive surfaces (i.e., both overall volume and water content) corresponding to any particular stress paths. Two equations can be used to describe the volume-mass constitutive surfaces for simple stress paths (Fig. 13.42). The stress paths commence with a soil at slurry conditions that are loaded to a net mean stress of p_0 at zero suction and then dried under various constant net mean stresses. The soil parameter m is not required for this stress path since there is no change in net mean stress after pores are filled with air.

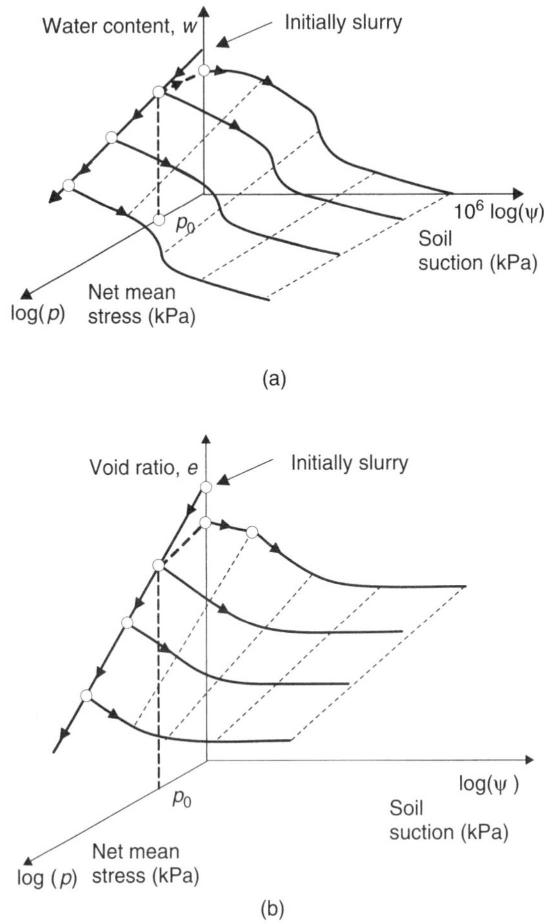

Figure 13.42 Volume-mass constitutive surfaces obtained from loading-unloading processes of an initially slurry specimen dried under various constant net mean stresses: (a) water content constitutive surface; (b) void ratio constitutive surface.

13.6.6.1 Equation for Water Content Constitutive Surface

The water content of the soil at a suction ψ and a net mean stress p have a uniform yield stress p_0 (i.e., consolidated at saturation) corresponding to the pores filled with water at a suction ψ_s and zero net mean stress minus the change in volume of a group of pores under a stress state equal to $p+\psi$ and a yield stress (i.e., at saturation) p_0 (Fig. 13.42). When the soil is loaded to its yield stress p_0 (i.e., at saturated condition) and then dried under a constant net mean stress p, pores are subject to higher effective stresses (i.e., $\sigma - u_w$ in the water-filled state) than when drying under zero net mean stress. Changes in the pore volume consist of two parts: (i) irreversible volume change (i.e., plastic or collapsible volume change) and (ii) reversible volume change (i.e., elastic volume change). An equation for the water content constitutive surface can be written as follows:

$$w(\psi, p, p_0) = w(\psi_s, 0, 0) - \frac{w^c(\psi_s)\{(C_c - C_s)\log[P(\psi, p, p_0)] - \xi_s C_c + C_s \log(\psi + p)\}}{(e_{sat} - C_c\xi_s - w_r G_s)} \quad (13.129)$$

where:

$w(\psi_s, 0, 0)$ = water content at soil suction ψ_s on the initial drying curve of the slurry soil,

$$P(\psi, p, p_0) = \frac{[\tan^{-1}(\psi + p - p_0) + 1.571](\psi + p - p_0)}{3.142} + p_0,$$

$\xi_s = \log(\psi_s)$,

$$\psi_s = \left(1 - \frac{\eta\{(C_c - C_s)\log[P(\psi, p, p_0)] + C_s\log(\psi_{aev} + p) - C_c\log(\psi_{aev})\}}{3[e_{sat} - C_c\log(\psi_{aev}) - w_r G_s]}\right)\psi,$$

e_{sat} = void ratio at the reference stress state, 1 kPa,

η = parameter representing the stress history of the soil, $w^c(\psi) = \left(w_{sat} - \frac{C_c}{G_s}\log(\psi) - w_r\right)\frac{a}{\psi^b + a}$,

a, b = curve-fitting parameters for the soil that is initially dried from a slurry,

G_s = specific gravity of the soil particles,

w_{sat} = gravimetric water content at the reference stress state,

C_c, C_s = compression indices of the soil under saturated conditions (i.e., zero soil suction),

p = net mean stress,

p_0 = pre consolidated stress, and

w_r = residual water content of the soil.

13.6.6.2 Equation for Void Ratio Constitutive Surface

The volume of voids in a soil at any stress state is equal to the volume of water plus the volume of the dry pores. The void ratio constitutive surface can be written as follows:

$$\begin{aligned}
e(\psi, p, p_0) &= e_{sat} - G_s w(\psi_s, 0, 0) - \int_0^{\xi_s} \frac{G_s w^c(10^x)C_c}{e_{sat} - C_c x - w_r G_s}\,dx \\
&\quad + w(\psi, p, p_0)G_s + \left([\log(P(\psi, p, p_0))] - \xi_s \frac{G_s w^c(\psi_s)(C_c - C_s)}{e_{sat} - C_c\xi_s - w_r G_s}\right) \\
&\quad - \int_0^{\xi_s} \frac{G_s w^c(10^x)(C_c - C_s)}{e_{sat} - C_c x - w_r G_s}\frac{d[\log(P(10^x, p, p_0) - x]}{dx}\,dx \\
&\quad + \left([\log(\psi + p) - \psi_s]\frac{G_s w^c(\psi_s)C_s}{e_{sat} - C_c\xi_s - w_r G_s}\right) \\
&\quad - C_s\log(1 + p) \\
&\quad - \int_0^{\xi_s} \frac{G_s w^c(10^x)C_s}{e_{sat} - C_c x - w_r G_s}\frac{d[\log(10^x + p) - x]}{dx}\,dx \\
&\quad - (C_c - C_s)\log[P(1, p, p_0)]
\end{aligned} \quad (13.130)$$

where:

$w(\psi_s, 0, 0)$ = water content at soil suction ψ_s on the initial drying curve of the slurry soil,

$$P(\psi, p, p_0) = \frac{[\tan^{-1}(\psi + p - p_0) + 1.571](\psi + p - p_0)}{3.142} + p_0$$

$\xi_s = \log(\psi_s)$,

$$\psi_s = \left(1 - \frac{\eta\{(C_c - C_s)\log[P(\psi, p, p_0)] + C_s\log(\psi_{aev} + p) - C_c\log(\psi_{aev})\}}{3[e_{sat} - C_c\log(\psi_{aev}) - w_r G_s]}\right)\psi,$$

e_{sat} = void ratio at the reference stress state,

η = parameter representing the stress history of the soil,

$$w^c(\psi) = \left(w_{sat} - \frac{C_c}{G_s}\log(\psi) - w_r\right)\frac{a}{\psi^b + a},$$

a, b = curve-fitting parameters for the drying curve of the initially slurry soil,

G_s = specific gravity of the soil particles,
w_{sat} = gravimetric water content at reference stress state,
C_c, C_s = compression indices of the soil at saturated condition (i.e., zero soil suction),
p = net mean stress,
p_0 = uniform yield stress, and
w_r = residual water content of the soil.

13.6.6.3 Equation for Degree-of-Saturation Constitutive Surface

The equation for the degree of saturation surface can be derived from the equations for gravimetric water content surface (Eq. 13.129) and the void ratio surface (Eq. 13.130) as follows:

$$S(\psi, p, p_0) = \frac{w(\psi, p, p_0)G_s}{e(\psi, p, p_0)} \tag{13.131}$$

13.6.7 Determination of Model Parameters

The proposed volume-mass constitutive model has complex equations but requires relatively simple and conventional laboratory test data as the required input soil parameters. The data required to define the volume-mass constitutive equations are as follows:

1. The initial drying SWCC for the initially slurry soil.
2. The pore shape parameter η to represent the effect of net mean stress and stress history on the change in the air-entry value of pores. This parameter can be set to 1 (Pham, 2005) unless SWCCs are available where laboratory measurements have been made at different preconsolidation pressures.
3. Parameters for the hysteretic nature of the SWCC of the soil, namely, the distance between the two inflection points of the two bounding hysteretic curves, D_{SL}, and the ratio between the slopes of the bounding drying and the bounding wetting curves, R_{SL}. These values can be estimated as suggested by Pham (2005).
4. The virgin compression index of the air-dried soil from an initially slurry specimen (i.e., completely dry specimen; Fig. 13.43).
5. The soil parameter m required to determine the compressibility of the soil structure surrounding continuously air-filled pores. It is suggested that the soil parameter be assumed to be 1 in cases where there is no measured data.

The yield stress acting on the soil structure surrounding pores of a soil that is air dried from slurry conditions is equal to the air-entry values of pores. It can be assumed that noncollapsible pores are incompressible. The maximum yield stress acting on the soil structure surrounding collapsible pores at suctions equal to and beyond the residual soil suction is the residual soil suction of the soil. Therefore, the compression curve of an air-dried soil becomes

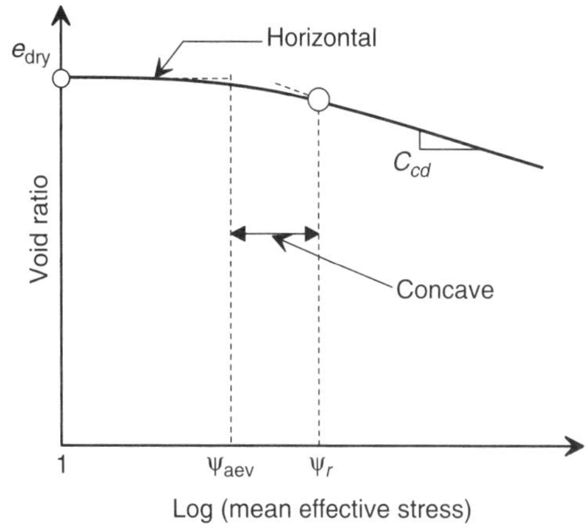

Figure 13.43 Compression curve of air-dried specimen from initially slurry soil sample along with method for estimation of virgin compression index of a dry specimen, C_{cd} (from Pham, 2005).

a straight line at net mean stresses equal to residual soil suction (Fig. 13.43). The dry virgin compression index C_{cd} of an air-dried soil is estimated in the same manner as for saturated soils. It is suggested that for most soils C_{cd} can be assumed to be zero (Pham, 2005).

13.6.8 Model Predictions on Three Artificial Soils

The proposed volume-mass constitutive model is fit to soil properties for three artificial soils and three sets of soil data from the research literature. The prediction results are presented for (i) several simple stress paths (i.e., 2D graphs) and (ii) several complete volume-mass constitutive surfaces (i.e., 3D graphs).

13.6.8.1 Material Simulated

Descriptions of the three artificial soils are presented in Table 13.2. The initial drying SWCCs of the three soils starting from slurry conditions are shown in Fig. 13.44. The assumption is made that the "dry pores are incompressible." The m soil parameter is not required and, as well, the virgin compression index of the soil under completely dry condition, C_{cd}, is not required.

13.6.8.2 Prediction Results for Several Simple Stress Paths

The shrinkage curve for each soil can be predicted using the proposed constitutive model for the void ratio and the water content along the drying process from slurry conditions to a suction of 10^6 kPa. The predicted shrinkage curves for the three artificial soils are shown in Fig. 13.45.

The shrinkage curve for the sand soil is almost horizontal since sand undergoes little volume change during a drying process. The shrinkage curves for silt and clay are as anticipated. The shrinkage curves follow a 45° line at high water contents. The shrinkage curves tend to be horizontal

Table 13.2 Characteristics of Three Artificial Soils

Soil Name		Sand	Silt	Clay
Soil-water characteristic	w_{sat}	0.28 (28%)	0.38 (38%)	0.80 (80%)
	a	500	200,000	5×10^6
	b	3	2.5	2.8
	w_r	0.03 (3%)	0.08 (8%)	0.10 (10%)
Characteristics of hysteresis	D_{SL}	0.2	0.35	0.5
	R_{SL}	2	1.5	1.5
	β	0.1	0.1	0.1
Compression indices	C_c	0.01	0.20	0.50
	C_s	0.003	0.040	0.100
Specific gravity	G_s	2.60	2.70	2.80
Pore shape parameter	η	2	2	2

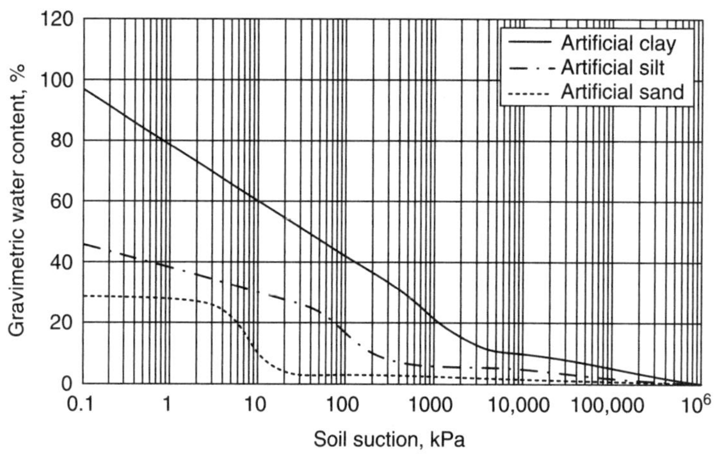

Figure 13.44 Initial drying SWCCs for artificial sand, silt, and clay.

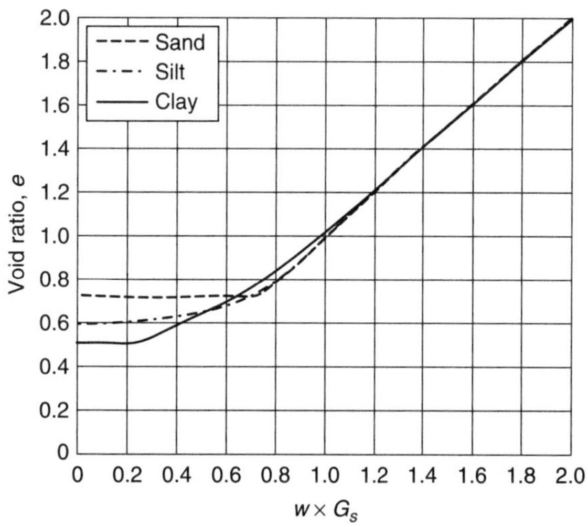

Figure 13.45 Shrinkage curves for artificial sand, silt, and clay.

at water contents below the shrinkage limit. The transition of the shrinkage curve between low and high water contents is a smooth curve.

The void ratio, gravimetric water content, and degree of saturation can also be calculated for the artificial silt. The stress paths adhered to are shown in Fig. 13.46. The steps in the stress path are (i) the (initial) slurry condition; (ii) the path when the soil is isotropically loaded to a net mean stress of 200 kPa; (iii) the path when the soil is dried to 100 kPa and then wetted to 0.1 kPa soil suction; and (iv) the path when the soil is dried to 10^6 kPa and then wetted to 0.1 kPa soil suction. The predicted void ratio, gravimetric water content, and degree of saturation versus soil suction for the specified stress paths are shown in Figs. 13.47, 13.48, and 13.49. The model simulations are realistic in all cases. Hysteresis in the SWCC is taken into account.

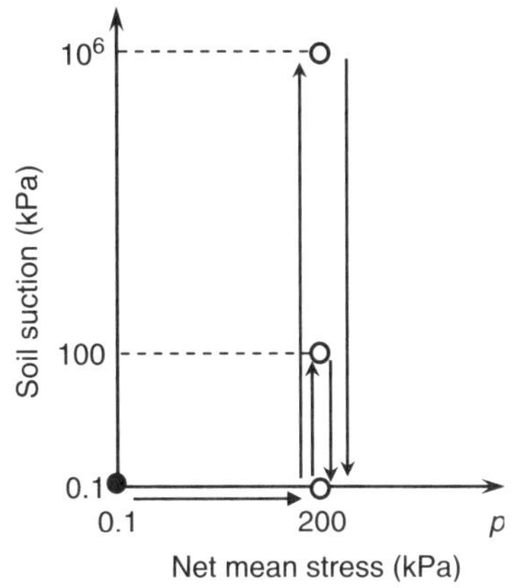

Figure 13.46 Stress path followed for artificial silt (from Pham, 2002).

13.6.8.3 Prediction Results for Volume-Mass Constitutive Surfaces

Several series of stress paths can be selected to illustrate the form of the entire volume-mass constitutive surfaces for artificial soils. The selected stress paths are relatively simple but illustrate several important characteristics of unsaturated soil behavior (e.g., swelling and collapsible behavior).

Four series of stress paths were used to illustrate the volume-mass constitutive surfaces for the unsaturated soil shown in Fig. 13.50. For stress path 1, the soil starts as a slurry, is loaded to a specified net mean stress, and then dried under constant net mean stress to a soil suction of 10^6 kPa. For stress path 2, the soil starts as a slurry and is then dried to a designated suction value and loaded under constant-soil-suction conditions to a net mean stress of 10^4 kPa. For stress path 3, the soil is initially dried from slurry conditions, loaded to a specified net mean stress and then wetted under a constant net mean stresses to a suction of 10^6 kPa. For stress path 4, the soil is initially dried from slurry conditions, wetted to a specified suction, and then loaded under a constant suction to a net mean stress of 10^4 kPa.

The series 1 and 2 stress paths are used to study the volume-mass behavior associated with soil specimens when starting from the same initial slurry condition and ending at the same final stress state. However, a different stress path is followed during each of the the drying processes. Similarly, stress path series 3 and 4 are used to study the volume-mass

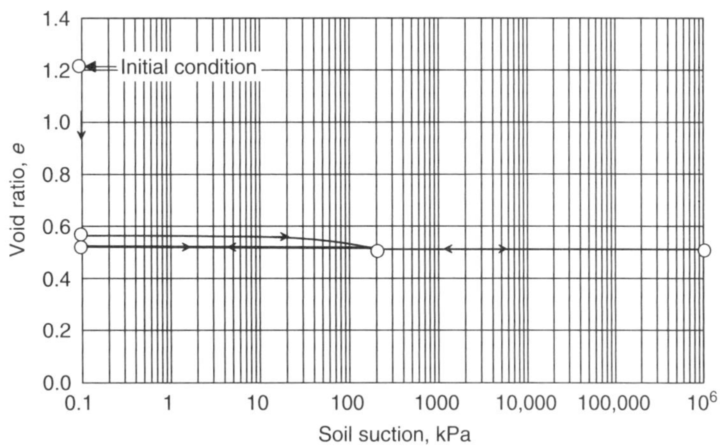

Figure 13.47 Void ratio changes for artificial silt.

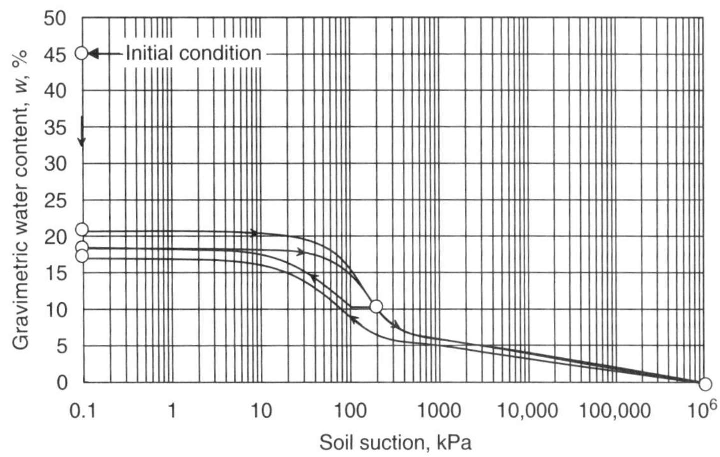

Figure 13.48 Gravimetric water content changes for artificial silt.

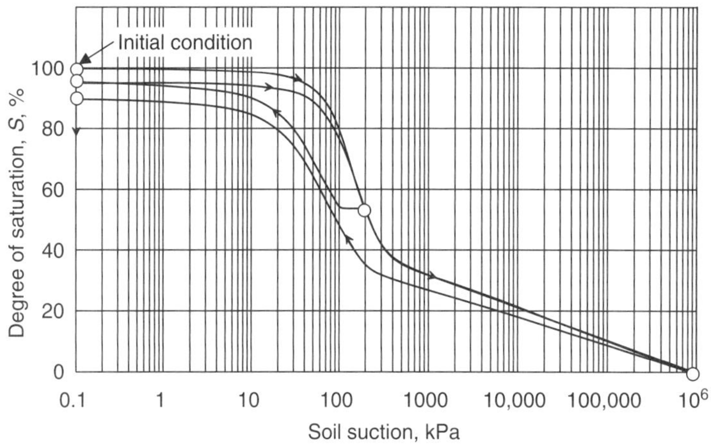

Figure 13.49 Degree-of-saturation changes for artificial silt.

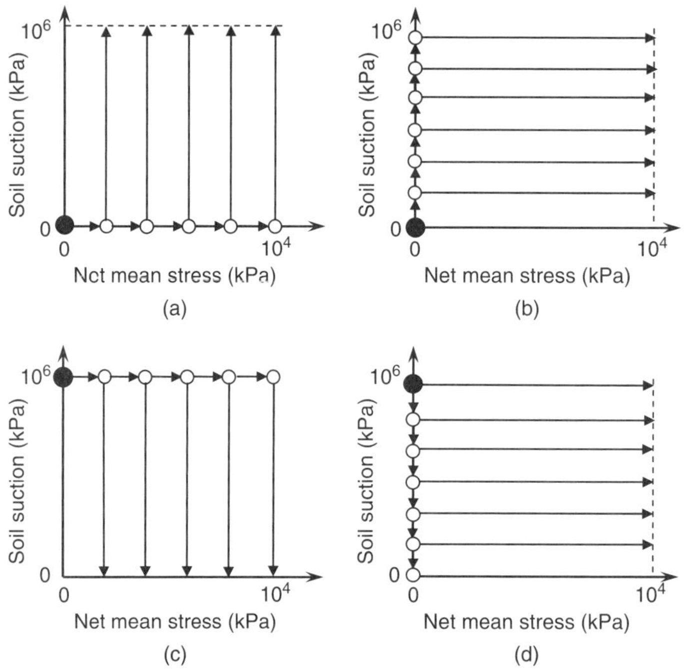

Figure 13.50 Four series of stress paths for studying volume-mass constitutive surfaces of unsaturated soil: (a) series 1: loading and then drying from initially slurry conditions; (b) series 2: drying and then loading from initially slurry conditions; (c) series 3: loading and then wetting from initially dry conditions; (d) series 4: wetting and then loading from initially dry conditions.

behavior of soil specimens with the same initial and final stress state but following two different stress paths that involve wetting processes.

Figure 13.51 shows that the calculated volume-mass constitutive surfaces agree well with postulated volume-mass constitutive surfaces presented by M.D. Fredlund et al., (2000b). The computed results show that (i) there is a one-to-one relationship between changes in void ratio and water content as net mean stress and soil suction are changed when soil suction is less than the air-entry value; (ii) there is a gradual change in curvature that develops on the water content constitutive surface as the air-entry value of the soil is exceeded; and (iii) there is a gradual change in curvature that develops on the void ratio constitutive surface as the air-entry value is exceeded.

Figures 13.52–13.54 show several volume-mass constitutive surfaces predicted for the three artificial soils previously mentioned. The predicted constitutive relationships near zero suction and zero net mean stress planes are reasonable. The volume-mass constitutive surfaces starting from

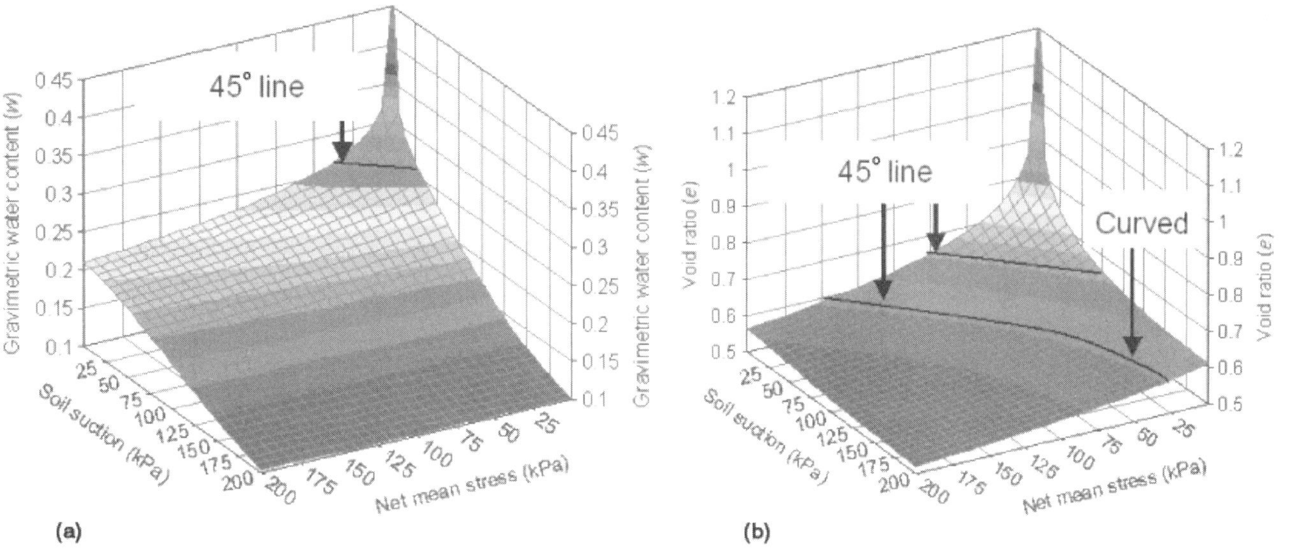

Figure 13.51 Constitutive surfaces for silt at low suctions and net mean stresses according to stress path for series 1 and 2: (a) gravimetric water content surface; (b) void ratio surface.

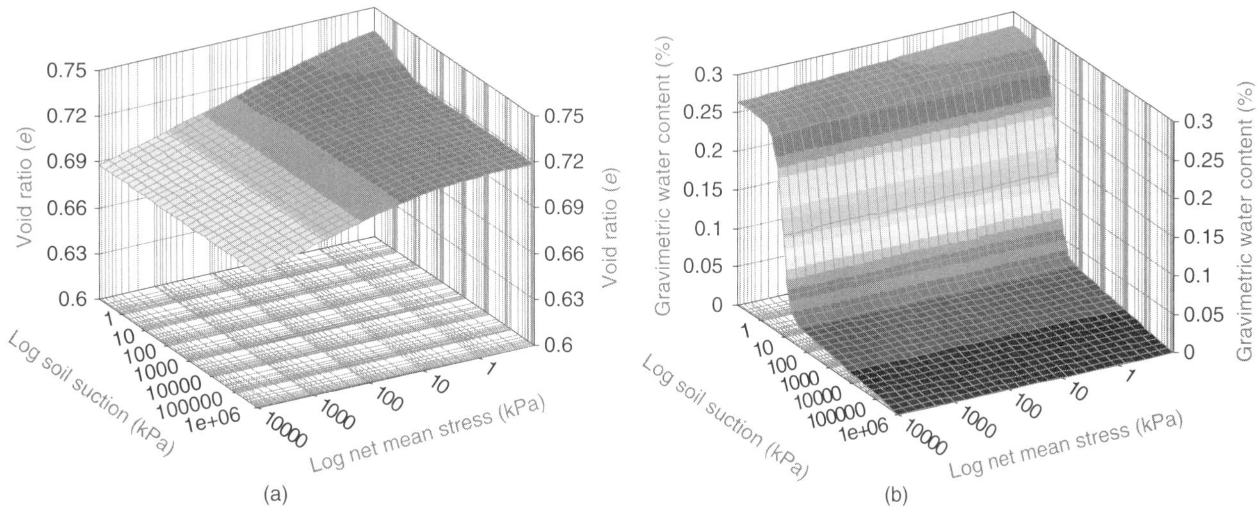

Figure 13.52 Constitutive surfaces for sand according to stress path series 1 and 2 shown in logarithmic stress state scales: (a) void ratio; (b) gravimetric water content.

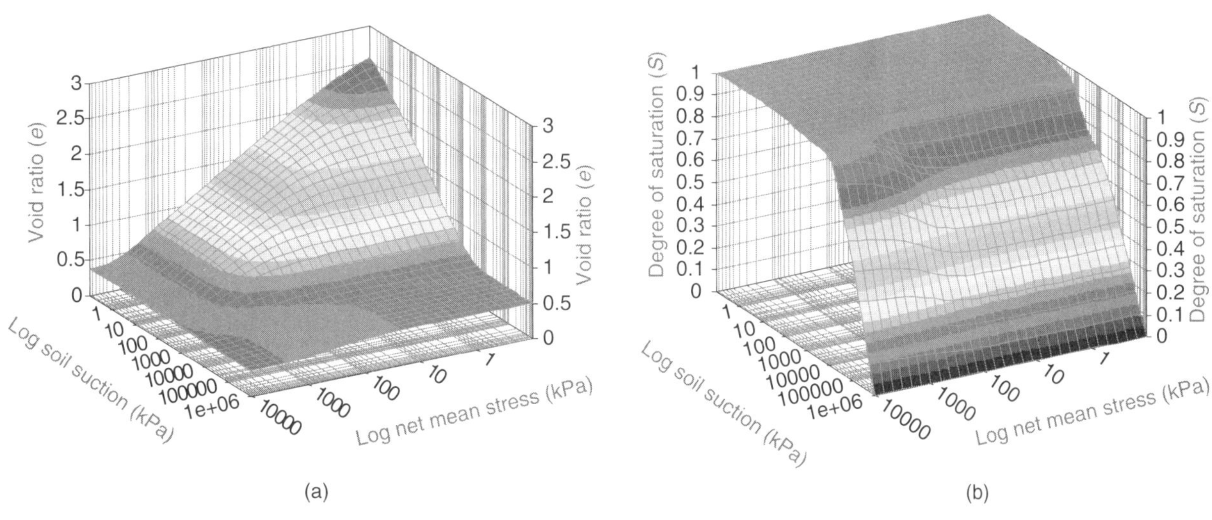

Figure 13.53 Constitutive surfaces for clay according to stress path series 1 and 2 shown using logarithmic stress state scales: (a) void ratio; (b) degree of saturation.

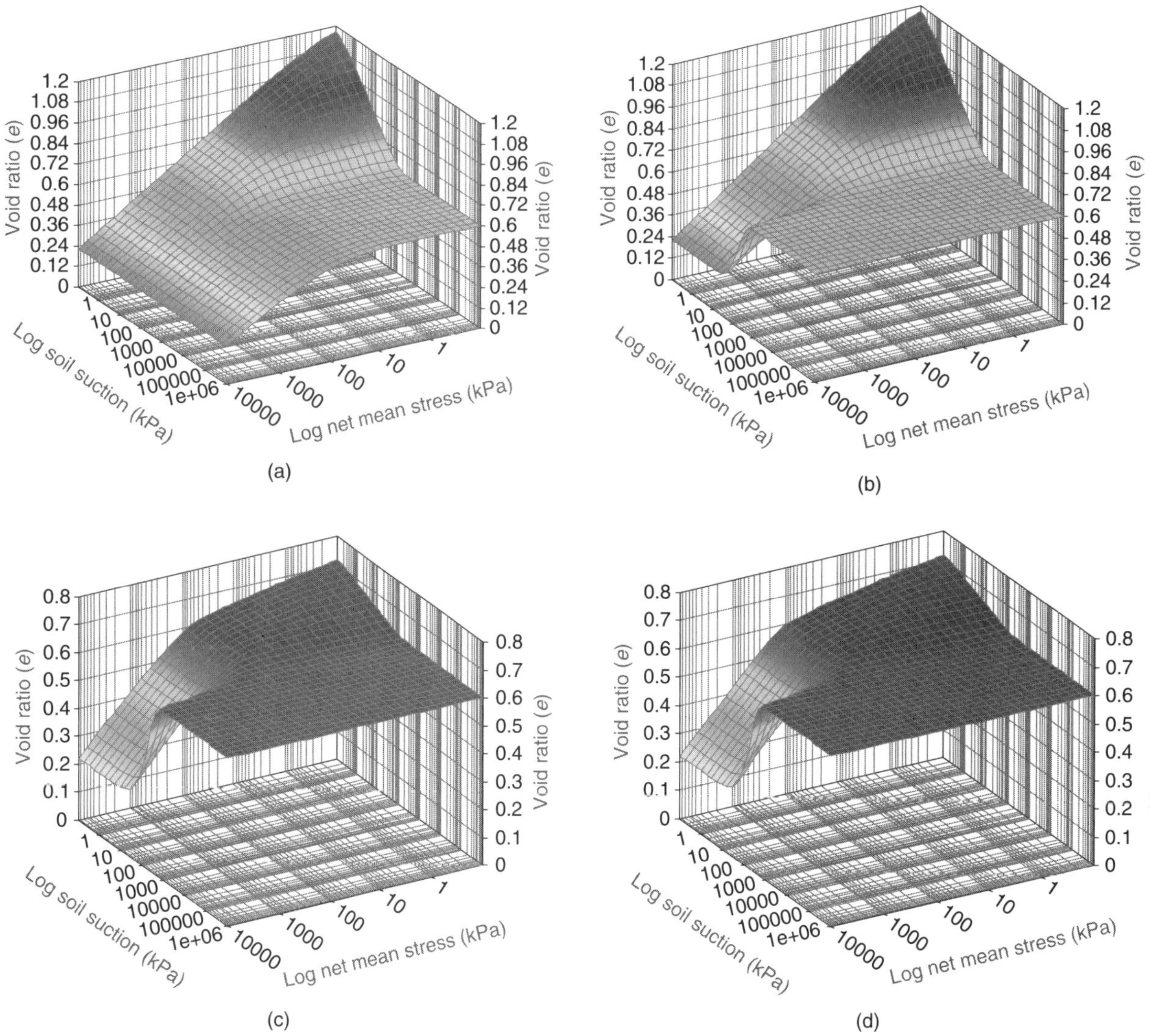

Figure 13.54 Void ratio constitutive surfaces for silt loaded along various stress paths: (a) stress series 1; (b) stress series 2, (c) stress series 3, (d) stress series 4.

slurry conditions appear to have steeper slopes than those starting from air-dried conditions. All computed constitutive surfaces are consistent with observed experimental behavior.

The void ratio constitutive surfaces obtained by following stress paths 3 and 4 (Figs. 13.54c and 13.54d) seem to be somewhat unusual. It can be observed that at a soil suction of 10^6 kPa and a net mean stress of 10^4 kPa the void ratio becomes higher than that for a soil suction of 0.1 kPa and a net mean stress of 10^4 kPa. However, the shapes of the surfaces are similar to those measured by Matyas and Radhakrishna (1968). The predictions show that the proposed model is able to predict both collapsing and swelling soil behavior. Figure 13.53 shows that the air-entry value of the soil increases with each increment of net mean stress.

The calculated void ratio surfaces for the artificial silt (Fig. 13.54) show that the stress paths corresponding to drying processes are stress path dependent and the stress paths related to wetting processes are stress path independent.

13.6.9 Model Predictions for Several Soil Data Sets in Literature

The Pham and Fredlund volume-mass constitutive model can be used to simulate published soil data sets. The selected soil data sets for simulation are (i) Regina clay (Fredlund, 1964) (ii) Jossigny silt (Fleureau et al., 1995), and (iii) kaolin (Fleureau et al., 2004). Input data for modeling the three soils are shown in Table 13.3. Predicted results and measured data for the three soils are presented in Figs. 13.55–13.64. The graphs show that the predicted results agree well with the measured results.

The Pham and Fredlund (2011a) volume-mass constitutive model is limited to isotropic loading conditions. The model is capable of (i) predicting both volume and water

Table 13.3 Characteristics of Three Soils from Research Literature

Soil name		Kaolin	Jossigny Silt	Regina Clay
Soil-water characteristic curve from initially slurried	w_{sat}	0.830 (83%)	0.467 (46.7%)	1.057 (105.7%)
	a	1.9×10^9	71,000	1.0×10^{33}
	b	2.527	1.404	9.838
	w_r	0.049 (4.9%)	0.064 (6.4%)	0.131 (13.1%)
Characteristics of hysteretic SWCCs	D_{SL}	0.6	0.6	N/A
	R_{SL}	2	1.5	1.5
	β	0.05	0.05	N/A
Compression indices	C_c	0.37	0.22	0.74
	C_s	0.085	0.045	0.134
	C_{cd}	0	0	0
Specific gravity	G_s	2.54	2.55	2.83
Pore shape parameter	η	2	2	2
Compressibility of dry pore	M	N/A	N/A	N/A

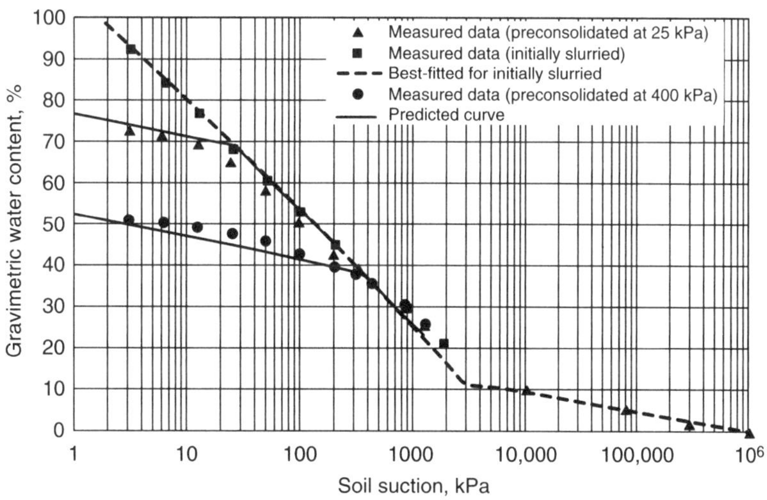

Figure 13.55 Measured, best-fit, and predicted SWCCs for Regina clay (data from Fredlund, 1964).

content at stress states corresponding to a wide variety of loading-unloading and drying-wetting stress paths; (ii) taking into account the hysteretic nature of the SWCC; and (iii) predicting both swelling and collapsible behavior of an unsaturated soil.

The model can simulate volume-mass constitutive relationships that are stress path dependent. The model utilizes a hysteresis model for the SWCC that makes use of two one-dimensional pore-size distributions. The model introduced two new parameters: the pore shape parameter η and the compressibility of the dry pore parameter, m. Further research on these two soil parameters would be useful since it will likely lead to improved estimation procedures for the new soil parameters.

13.7 FORMULATION OF PARTIAL DIFFERENTIAL EQUATIONS FOR STRESS-DEFORMATION IN UNSATURATED SOILS

The emphasis thus far in this chapter has been on the formulation of suitable volume-mass constitutive relations for an unsaturated soil. Let us now consider the formulation of PDEs for a REV. The derived partial differential equations are applicable for solving stress-deformation problems involving unsaturated soils. The emphasis is on the calculation of overall volume changes in unsaturated soils (i.e., soil structure volume changes).

The derivation of partial differential equations required to solve stress-deformation problems starts with satisfying

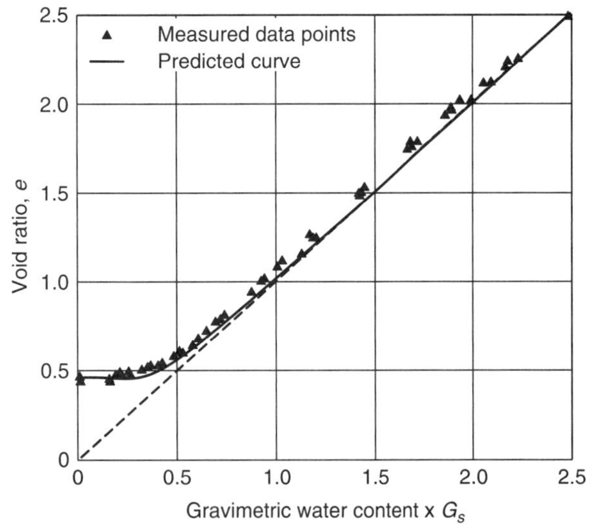

Figure 13.56 Measured and predicted shrinkage curve for Regina clay (data from Fredlund, 1964).

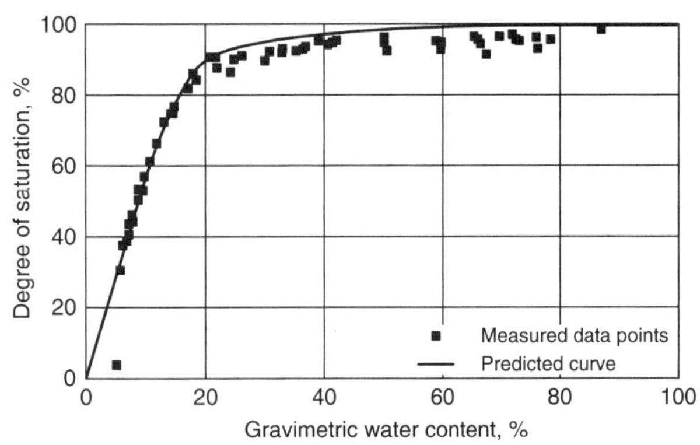

Figure 13.57 Measured and predicted curve plotting degree of saturation versus gravimetric water content for Regina clay (data from Fredlund, 1964).

the conservation of linear and angular momentum (i.e., equilibrium of forces) for a REV. A variety of constitutive stress–strain models can be substituted into the conservative equations of equilibrium with the result of obtaining partial differential equations that can be solved for saturated-unsaturated soil continua. In each case, it is necessary to convert saturated-unsaturated soil properties and unsaturated soil property functions into elastic and plastic parameters that are consistent with the model used in the formulations. The stress-deformation formulations must be solved using a computer.

13.7.1 Conservation of Linear and Angular Momentum

The distribution of total stresses within an unsaturated soil is governed by the static equilibrium of forces. Stresses acting on each face of an element can be decomposed as the normal and shear components in the x-, y-, and z-directions, as shown in Fig. 13.65. All stresses shown in Fig. 13.65 are positive, in accordance with geotechnical engineering convention.

The balance of angular momentum with respect to any axis shows that the Cauchy tensor must be symmetric (i.e., $t_{ij} = t_{ji}$). The balance of linear momentum (i.e., the equilibrium of forces) results in the partial differential equations governing static equilibrium of forces (Chou and Pagano, 1967a). Overall stress equilibrium equations can be written for the x-, y-, and z-coordinate directions:

$$\frac{\partial \sigma_x}{\partial x} + \frac{\partial \tau_{yx}}{\partial y} + \frac{\partial \tau_{zx}}{\partial z} + F_x = 0 \tag{13.132}$$

$$\frac{\partial \tau_{xy}}{\partial x} + \frac{\partial \sigma_y}{\partial y} + \frac{\partial \tau_{zy}}{\partial z} + F_y = 0 \tag{13.133}$$

$$\frac{\partial \tau_{xz}}{\partial x} + \frac{\partial \tau_{yz}}{\partial y} + \frac{\partial \sigma_z}{\partial z} + F_z = 0 \tag{13.134}$$

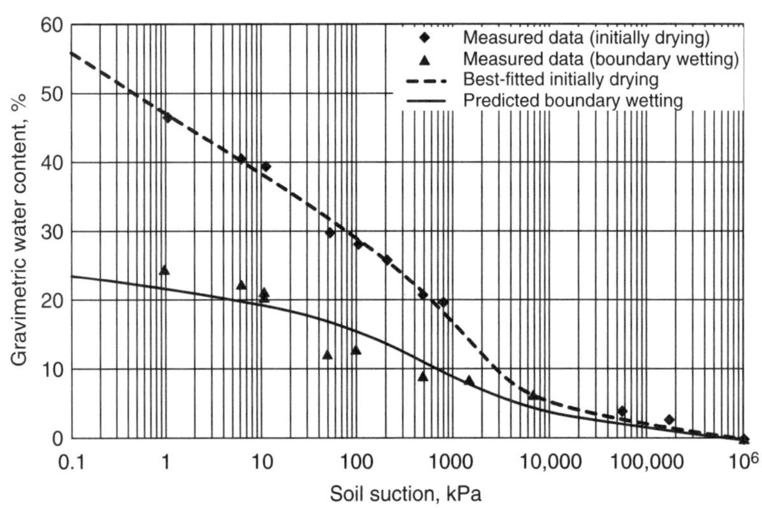

Figure 13.58 Measured and best-fit initial drying and wetting SWCCs for Jossigny silt (data from Fleureau et al., 1995).

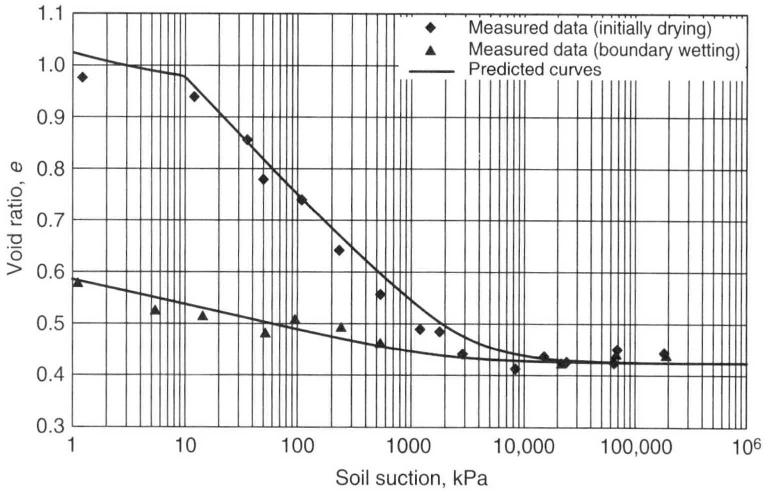

Figure 13.59 Measured and predicted void ratios along initial drying and boundary wetting curves for Jossigny silt (data from Fleureau et al., 1995).

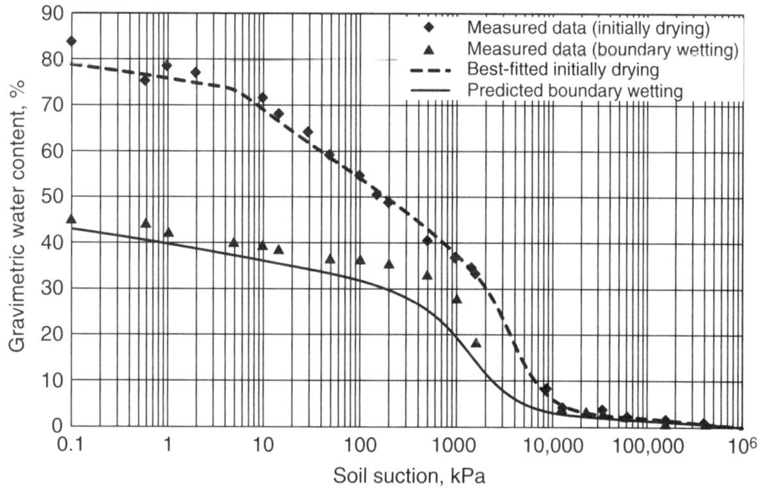

Figure 13.60 Measured and best-fit initial drying and boundary wetting SWCCs for kaolin (Fleureau et al., 1995).

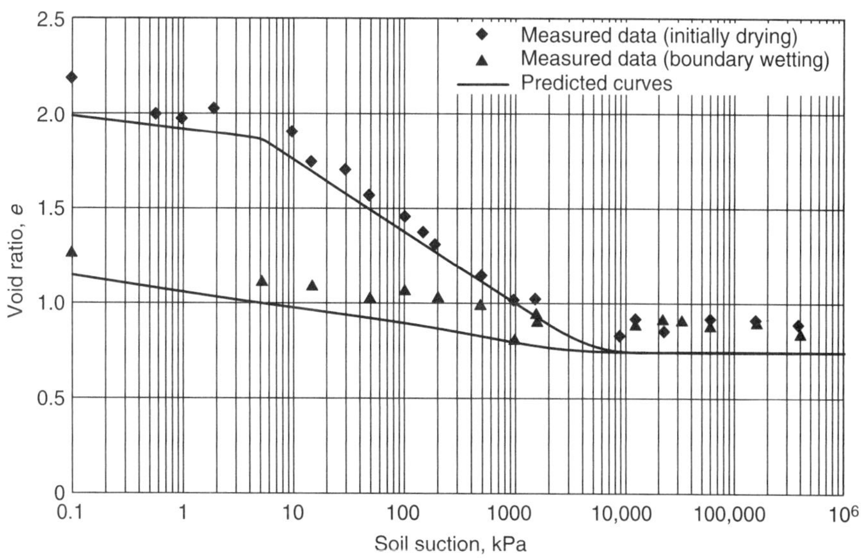

Figure 13.61 Measured and predicted void ratios along initial drying and boundary wetting curves for kaolin (data from Fleureau et al., 1995).

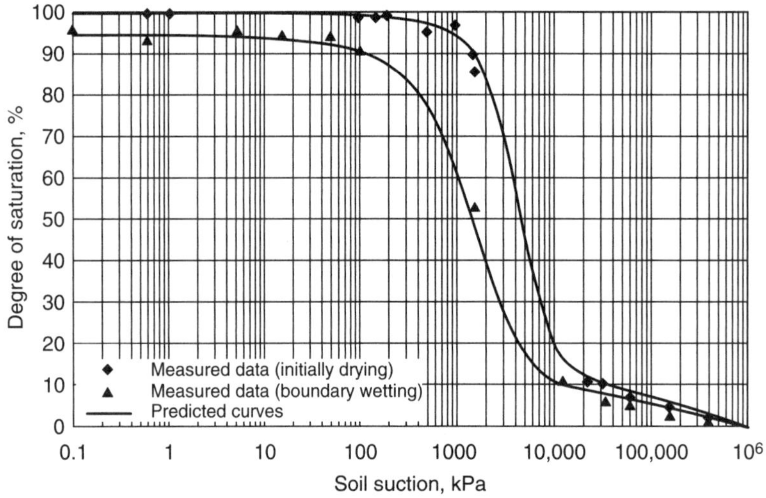

Figure 13.62 Measured and predicted degree-of-saturation for kaolin (data from Fleureau et al., 1995).

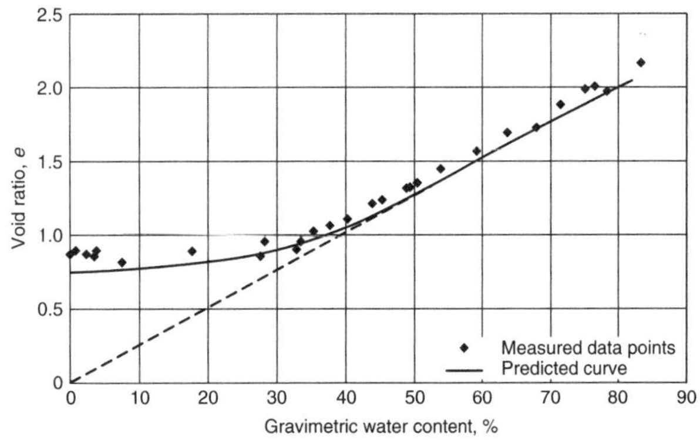

Figure 13.63 Measured and predicted shrinkage curve for kaolin (data from Fleureau et al., 1995).

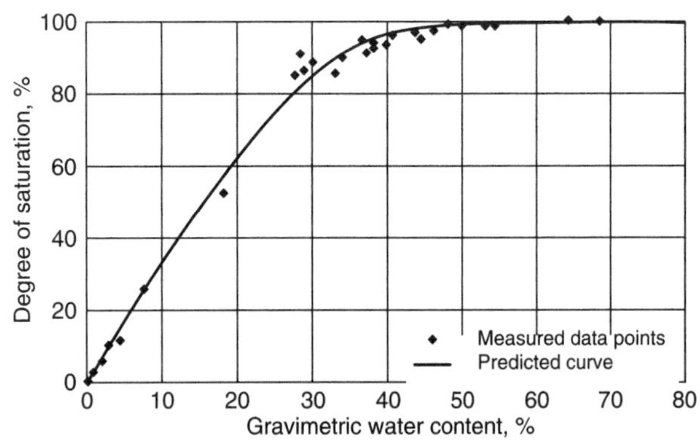

Figure 13.64 Measured and predicted curve plotting gravimetric water content versus degree of saturation for kaolin (data from Fleureau et al., 1995).

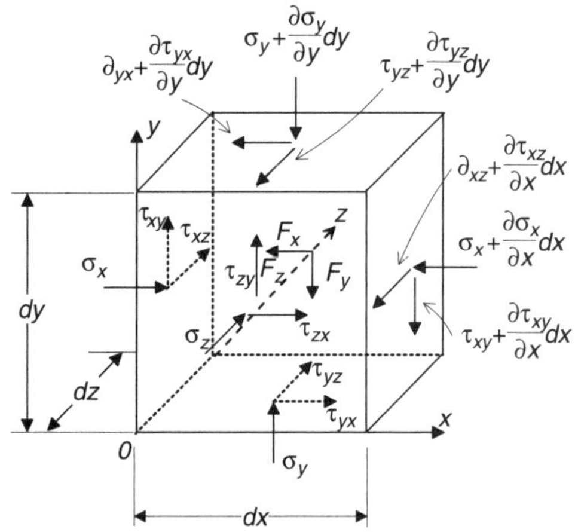

Figure 13.65 Representative elemental volume of soil showing the surface tractions on all faces.

where:

F_x, F_y, F_z = body force acting in the x-, y-, and z-directions per unit volume, kN/m^3.

The pore-water and pore-air pressures have no direct role in the equilibrium of forces acting upon the faces of a representative elemental volume of soil.

13.7.2 Strain-Displacement Relationships and Compatibility Equations

The classical definition of strain can be applied to an unsaturated soil element. The normal strain in a given direction, ϵ, is defined as the unit change in length (i.e., change in length per unit length) of a line which was originally oriented in the given direction. The normal strains can be written as follows for the x-, y-, and z-directions:

$$\varepsilon_x = \frac{\partial u}{\partial x} \tag{13.135}$$

$$\varepsilon_y = \frac{\partial v}{\partial y} \tag{13.136}$$

$$\varepsilon_z = \frac{\partial w}{\partial z} \tag{13.137}$$

where:

$\varepsilon_x, \varepsilon_y, \varepsilon_z$ = normal strain in the x-, y-, and z- directions and
u, v, w = displacement in the x-, y-, and z-directions, respectively.

Shear strain γ is defined as the change in the right angle between reference axes. Shear strains are measured in terms of radians (Chou and Pagano, 1967a). The relationships between the shear strains and displacements can be written as follows:

$$\gamma_{xy} = \frac{\partial u}{\partial y} + \frac{\partial v}{\partial x} \tag{13.138}$$

$$\gamma_{xz} = \frac{\partial u}{\partial z} + \frac{\partial w}{\partial x} \tag{13.139}$$

$$\gamma_{yz} = \frac{\partial v}{\partial z} + \frac{\partial w}{\partial y} \tag{13.140}$$

where:

$\gamma_{xy}, \gamma_{yz}, \gamma_{zx}$ = shear strains with respect to the reference axes.

The definitions of strains shown above make the assumption that the strains are relatively small (i.e., the product of two or more derivatives of displacement is neglected). The assumption of small strains applies for many engineering problems; however, there are situations in geotechnical engineering where large strain theory must be taken into consideration (e.g., drying of slurry waste products). Small strain formulations can be applied to large strain problems provided the geometry is updated along with the incremental stress state analysis (Cook, 1981).

Strain compatibility equations must also be satisfied in addition to the strain-displacement relationships (Chou and Pagano, 1967a). Strain compatibility equations ensure that the strain fields do not result in "gaps" in the continuum. In general, compatibility equations are irrelevant because continuous displacement functions generally satisfy the compatibility equations.

13.7.3 Constitutive Laws for Unsaturated Soils

The modeling of unsaturated soil behavior requires constitutive laws for stress-strain behavior of the soil structure, volume change of the pore-air and pore-water phases (i.e., changes in storage), the flow of pore-water and pore-air, and the compressibility of pore-air. Constitutive laws are combined with the conservation laws of physics to render the governing equations determinate.

Constitutive relationships are based on phenomenological observations of soil behavior and are written in terms of state variables. The constitutive laws required for the description of unsaturated soil behavior have already been presented using several forms (i.e., elasticity form, compressibility form, and soil mechanics form). The relationship among various representations of unsaturated soil properties was also presented. Elasticity models (and other advanced constitutive models) are used in the formulation of the partial differential equations for stress-deformation analyses.

Most constitutive laws for unsaturated soils are defined based on nonlinear soil property functions (i.e., properties are stress state dependent). The term *unsaturated soil property functions* is used herein to refer to mathematical equations that describe the relationship between a soil property and the stress state variables $\sigma - u_a$ and $u_a - u_w$.

13.7.4 Stress-Strain Relationships

The stress-strain relationships for unsaturated soils generally appear as extensions of existing models for saturated soils. Figure 13.66 presents an overview of some common types of stress-strain relationships. The most popular stress-strain models describe elastic or elastoplastic behavior. Visco-elastoplastic and other types of models have not received much attention in unsaturated soil modeling (Fredlund and Gitirana, 2005).

Most elastic and elastoplastic stress-strain relationships for unsaturated soils can be written in the following generic format regardless of the constitutive model adopted:

$$de = \mathbf{D}^{-1}\, d(\boldsymbol{\sigma} - u_a\boldsymbol{\delta}) + \mathbf{H}\, d(u_a - u_w) \tag{13.141}$$

$$d(\boldsymbol{\sigma} - u_a\boldsymbol{\delta}) = \mathbf{D}\, d\varepsilon - \mathbf{h}\, d(u_a - u_w) \tag{13.142}$$

where:

$\varepsilon^{\mathrm{T}} = [\ \varepsilon_x \quad \varepsilon_y \quad \varepsilon_z \quad \gamma_{xy} \quad \gamma_{yz} \quad \gamma_{zx}\ ]$ and **D, H** are constitutive matrices, where

$$\mathbf{D} = \begin{bmatrix} D_{11} & D_{21} & D_{31} & D_{41} & D_{51} & D_{61} \\ D_{12} & D_{22} & D_{32} & D_{42} & D_{52} & D_{62} \\ D_{13} & D_{23} & D_{33} & D_{43} & D_{53} & D_{63} \\ D_{14} & D_{24} & D_{34} & D_{44} & D_{54} & D_{64} \\ D_{15} & D_{25} & D_{35} & D_{45} & D_{55} & D_{65} \\ D_{16} & D_{26} & D_{36} & D_{46} & D_{56} & D_{66} \end{bmatrix}$$

$$\mathbf{H}^{\mathrm{T}} = [\ H_1 \quad H_2 \quad H_3 \quad H_4 \quad H_5 \quad H_6\ ]$$

$$h = \mathbf{DH}$$

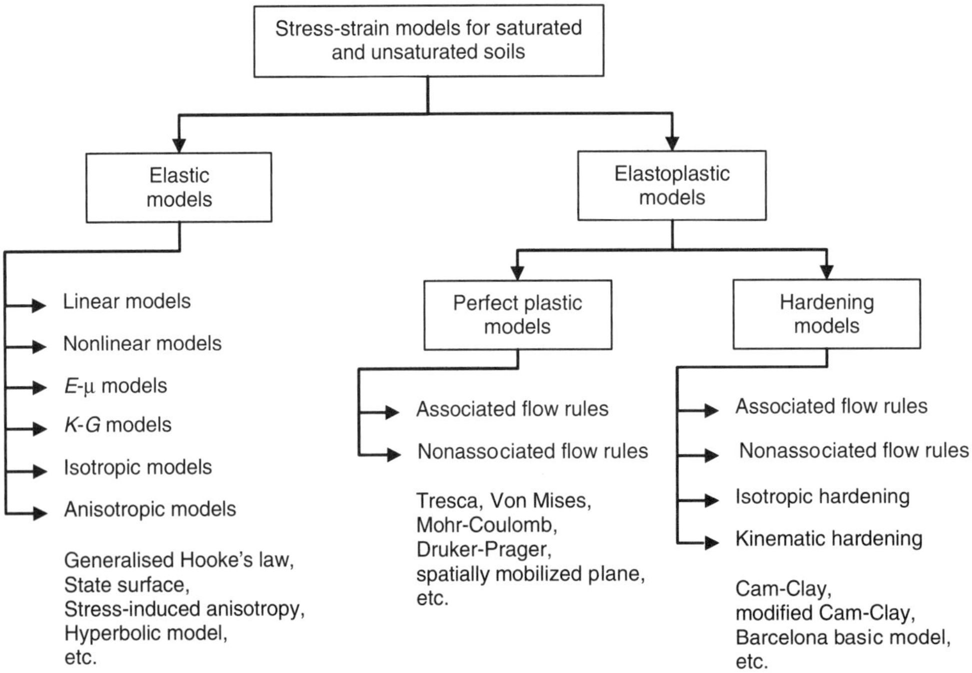

Figure 13.66 Stress-strain constitutive models for saturated-unsaturated soils.

$$\boldsymbol{h}^{\mathrm{T}} = [h_1 \quad h_2 \quad h_3 \quad h_4 \quad h_5 \quad h_6]$$
$$\boldsymbol{\sigma}^{\mathrm{T}} = [\sigma_x \quad \sigma_y \quad \sigma_z \quad \tau_{xy} \quad \tau_{yz} \quad \tau_{zx}]$$
$$\boldsymbol{\delta}^{\mathrm{T}} = [1 \quad 1 \quad 1 \quad 0 \quad 0 \quad 0]$$

The bold characters indicate matrices and vectors. The superscript T designates a transposed matrix.

13.7.5 Elastic Models

Elastic models for unsaturated soils (left branch in Fig. 13.66) are usually based on extensions of Hooke's law (Fredlund and Morgenstern, 1976). These strain-stress equations were presented earlier in this chapter using the $\sigma - u_a$ and $u_a - u_w$ stress state variables:

$$d\varepsilon_x = \frac{1}{E}d(\sigma_x - u_a) - \frac{\mu}{E}d(\sigma_y + \sigma_z - 2u_a) + \frac{1}{H}d(u_a - u_w) \tag{13.143}$$

$$d\varepsilon_y = \frac{1}{E}d(\sigma_y - u_a) - \frac{\mu}{E}d(\sigma_x + \sigma_z - 2u_a) + \frac{1}{H}d(u_a - u_w) \tag{13.144}$$

$$d\varepsilon_z = \frac{1}{E}d(\sigma_z - u_a) - \frac{\mu}{E}d(\sigma_x + \sigma_y - 2u_a) + \frac{1}{H}d(u_a - u_w) \tag{13.145}$$

$$d\gamma_{xy} = \frac{1}{G}d\tau_{xy} \tag{13.146}$$

$$d\gamma_{xz} = \frac{1}{G}d\tau_{xz} \tag{13.147}$$

$$d\gamma_{yz} = \frac{1}{G}d\tau_{yz} \tag{13.148}$$

where:

E = Young's modulus,
μ = Poisson's ratio,
H = elastic modulus for the soil structure with respect to changes in suction, and
G = shear modulus, $G = E/2(1 - \mu)$.

The constitutive matrices **D, H**, and **h** can be substituted into Eqs. 13.141 and 13.142 (Fredlund and Gitirana, 2005). Some nonlinear strain characteristics can be accounted for by using incremental analysis and the concept of a state surface (Matyas and Radhakrishna, 1968). The values of E and H can be obtained for each incremental step through use of coefficients of compressibility obtained from the void ratio state constitutive surface. For example,

$$E = \frac{3(1 - 2\mu)}{m_1^s} \tag{13.149}$$

$$H = \frac{3}{m_2^s} \tag{13.150}$$

where:

$$m_1^s = \frac{d\varepsilon_v}{d(\sigma_{\mathrm{mean}} - u_a)} = \frac{1}{1 + e_0}\frac{de}{d(\sigma_{\mathrm{mean}} - u_a)}$$

$$m_2^s = \frac{d\varepsilon_v}{d(u_a - u_w)} = \frac{1}{1+e_0}\frac{de}{d(u_a - u_w)}$$

$$d\varepsilon_v = \frac{dV}{V_0}$$

$$d\varepsilon_v = m_1^s d\left(\sigma_{\text{mean}} - u_a\right) + m_2^s d\left(u_a - u_w\right)$$

$\sigma_{\text{mean}} = \sigma_{\text{mean}}\ \frac{1}{3}\ (\sigma_x + \sigma_y + \sigma_z)$ (total mean stress)

e_0 = initial void ratio

$e = f(\sigma_{\text{mean}} - u_a, u_a - u_w)$; (function of void ratio state surface)

Equations 13.149 and 13.150 are based on the assumption that the volume change of unsaturated soils is a function of changes in net stress and soil suction. Equation 13.149 has two unknowns in one equation since Poisson's ratio is not known. The value of Poisson's ratio is usually estimated or measured in the laboratory through use of triaxial tests with measurement of lateral strains. Elastic models are generally appropriate for the analysis of monotonic net stress and soil suction paths. However, elastic models may not be sufficiently accurate under nonmonotonic stress paths because of the distinction between recoverable and irrecoverable strains.

13.7.6 Elastoplastic Models

Elastoplastic models (i.e., right main branch in Fig. 13.66) can be used to address features of soil behavior such as yield and irrecoverable strains. Most elastoplastic models are based on the same fundamental principles but use different yield criteria, flow rules, and compressibility functions. Figure 13.66 lists several yield criteria used in perfectly plastic models. Perfectly plastic formulations for saturated soils can be extended to unsaturated soils through use of the generalized Hooke's law and by incorporating the effect of soil suction into the yield criterion (Pereira, 1996).

Yield surfaces can be combined with hardening rules and cap surfaces. Hardening rules are used in order to reproduce changes in the size of the yield surface (i.e., isotropic hardening) or shifts in the yield surface position (kinematic hardening). Cap surfaces are used in order to account for yield that occurs at stress states below failure conditions.

Numerous models have been proposed for unsaturated soils based on a critical state framework. Some of the early models were proposed by Karube and Kato (1989), Alonso et al., (1990), Wheeler and Sivakumar (1995), Cui and Delage (1996), among others. Considerable emphasis has been given to soils compacted under potentially collapsible conditions. The models are based on isotropic hardening laws and on yield surfaces that expand with increasing soil suction. The collapse phenomenon is reproduced using appropriate modes of expansion of the yield curves and appropriate variations in soil compressibility for different suctions. Ongoing research continues to contribute to the understanding of elastoplastic models for unsaturated soils.

Lloret and Ledesma (1993) provided a framework for representing elastoplastic stress-strain relationships for unsaturated soils. The yield functions and corresponding flow rules were proposed by Alonso et al. (1990). The general stress-strain relationships take the following form:

$$d(\boldsymbol{\sigma} - u_a\boldsymbol{\delta}) = \begin{cases} \mathbf{D}^{\mathbf{e}} d\varepsilon - \mathbf{h}^{\mathbf{e}} d(u_a - u_w) & \text{for } F_1 \le 0 \text{ and } F_2 \le 0 \\ \mathbf{D}^{\mathbf{ep}} d\varepsilon - \mathbf{h}^{\mathbf{ep}} d(u_a - u_w) & \text{for } F_1 > 0 \text{ or } F_2 > 0 \end{cases} \qquad (13.151)$$

where:

F_1 = yield function for load or suction decrease,

$F_1 = f(\sigma - u_a, u_a - u_w, G)$,

F_2 = yield function for suction increase, $F_2 = f(\sigma - u_a, u_a - u_w, G)$, and

$\mathbf{D}^{\mathbf{e}}, \mathbf{H}^{\mathbf{e}}$ = elastic constitutive matrices defined according to Eqs. 13.143–13.148, and

$\mathbf{D}^{\mathbf{ep}}, \mathbf{H}^{\mathbf{ep}}$ = elastoplastic constitutive matrices and

$$\mathbf{D}^{\mathbf{ep}} = \mathbf{D}^{\mathbf{e}} - \frac{1}{A - A_{\text{cr}}}\mathbf{D}^{\mathbf{e}}\frac{\partial G_1}{\partial(\sigma - u_a)}\left(\frac{\partial F_1}{\partial(\sigma - u_a)}\right)^{\text{T}}\mathbf{D}^{\mathbf{e}}$$

$$\mathbf{h}^{\mathbf{ep}} = \mathbf{D}^{\mathbf{e}}\mathbf{H}_{\mathbf{s}} + \left[-\left(\frac{\partial F_1}{\partial(\sigma - u_a)}\right)^T \mathbf{D}^{\mathbf{e}}\mathbf{H}_{\mathbf{s}} + \frac{\partial F_1}{\partial(u_a - u_w)}\right]\frac{1}{A - A_{cr}}\mathbf{D}^{\mathbf{e}}\frac{\partial G_1}{\partial(\sigma - u_a)}$$

$$\mathbf{H}_{\mathbf{s}} = \begin{cases} \mathbf{H}^{\mathbf{e}} & \text{if } F_2 \le 0 \\ \mathbf{H}^{\mathbf{e}} + \mathbf{H}^{\mathbf{ep}} & \text{if } F_2 > 0 \end{cases}$$

$$A = -\frac{\partial F_1}{\partial \Gamma}\left(\frac{\partial \Gamma}{\partial \varepsilon^p}\right)^{\text{T}}\frac{\partial G_1}{\partial(\sigma - u_a)}$$

$$A_{\text{cr}} = -\left(\frac{\partial F_1}{\partial(\sigma - u_a)}\right)^T D^e \frac{\partial G_1}{\partial(\sigma - u_a)}$$

The flow rule $G_1 = f(\sigma - u_a, u_a - u_w, \Gamma)$ and Γ is the hardening parameter and

$$\left(\frac{\partial F_1}{\partial(\sigma - u_a)}\right)^{\text{T}} = \frac{\partial F_1}{\partial(\sigma_x - u_a)}\frac{\partial F_1}{\partial(\sigma_y - u_a)}\frac{\partial F_1}{\partial(\sigma_z - u_a)} \times \frac{\partial F_1}{\partial \tau_{xy}}\frac{\partial F_1}{\partial \tau_{xz}}\frac{\partial F_1}{\partial \tau_{yz}}$$

$$\left(\frac{\partial G_1}{\partial(\sigma - u_a)}\right)^{\text{T}} = \frac{\partial G_1}{\partial(\sigma_x - u_a)}\frac{\partial G_1}{\partial(\sigma_y - u_a)}\frac{\partial G_1}{\partial(\sigma_z - u_a)} \times \frac{\partial G_1}{\partial \tau_{xy}}\frac{\partial G_1}{\partial \tau_{xz}}\frac{\partial G_1}{\partial \tau_{yz}}$$

The term Γ is generally visualized as the preconsolidation stress for saturated conditions. Different functions for the variables F_1, F_2, G_1, and G_2 and soil compressibility are defined for various proposed models.

Equations 13.141 and 13.142 can be used to provide generic equations for the derivation of the partial differential equations governing unsaturated soil behavior. Equations 13.143–13.148 can be used for the derivation of governing partial differential equations. Equations 13.143–13.148 form relatively simple and straightforward relationships to the compressibility of soils.

13.7.7 Partial Differential Equations for Stress-Deformation Analysis

Partial differential equations for stress-deformation analysis are required for determining stress and strain distributions within an unsaturated soil. The stress distribution can be used for certain types of analyses (e.g., slope stability analysis) and the strain distributions can be used in settlement, heave, or other ground movement analyses. The prediction of total stresses in saturated-unsaturated soil masses are considerably easier to predict than are strains and displacements.

The partial differential equations for a stress-deformation analysis can be written by combining the differential equations governing force equilibrium (i.e., Eqs. 13.132, 13.133, and 13.134), the stress-strain constitutive relationships, and the relationship between strains and small displacements (i.e., Eqs. 13.135 to 13.140). The stress-strain relationships adopted largely determine the complexity of the partial differential equations. The simplest equations that can be derived are based on a generalization of Hooke's law written in terms of independent stress state variables (e.g., Eqs. 13.143–13.148). The partial differential equation for the x-direction can be written as

$$\frac{\partial}{\partial}x\left[\sigma_{x0} + D_{11}\frac{\partial u}{\partial x} + D_{12}\frac{\partial v}{\partial y} + D_{12}\frac{\partial w}{\partial z} - h\,d\left(u_a - u_w\right) + du_a\right] + \frac{\partial}{\partial y}\left[\tau_{xy0} + D_{44}\left(\frac{\partial u}{\partial y} + \frac{\partial v}{\partial x}\right)\right] + \frac{\partial}{\partial z}\left[\tau_{xz0} + D_{44}\left(\frac{\partial u}{\partial z} + \frac{\partial w}{\partial x}\right)\right] = 0 \qquad (13.152)$$

the partial differential equation for the y-direction as

$$\frac{\partial}{\partial x}\left[\tau_{xy0} + D_{44}\left(\frac{\partial u}{\partial y} + \frac{\partial v}{\partial x}\right)\right] + \frac{\partial}{\partial y}\left[\sigma_{yo} + D_{12}\frac{\partial u}{\partial x} + D_{11}\frac{\partial v}{\partial y} + D_{12}\frac{\partial w}{\partial z} - h\,d\left(u_a - u_w\right) + du_a\right] + \frac{\partial}{\partial z}\left[\tau_{yz0} + D_{44}\left(\frac{\partial v}{\partial z} + \frac{\partial w}{\partial y}\right)\right] + \gamma_t = 0 \qquad (13.153)$$

and the partial differential equation for the z-direction as

$$\frac{\partial}{\partial x}\left[\tau_{xz0} + D_{44}\left(\frac{\partial u}{\partial z} + \frac{\partial w}{\partial x}\right)\right] + \frac{\partial}{\partial y}\left[\tau_{yz0} + D_{44}\left(\frac{\partial v}{\partial z} + \frac{\partial w}{\partial y}\right)\right] + \frac{\partial}{\partial z}\left[\sigma_{z0} + D_{12}\frac{\partial u}{\partial x} + D_{12}\frac{\partial v}{\partial y} + D_{11}\frac{\partial w}{\partial z} - h\,d(u_a - u_w) + du_a\right] = 0 \qquad (13.154)$$

where:

σ_{i0}, τ_{ij0} = initial stresses,
$D_{11} = E(1-\mu)/[(1+\mu)(1-2\mu)]$,
$D_{12} = E\mu/[(1+\mu)(1-2\mu)]$,
$D_{44} = E/[2(1+\mu)]$,
$h = E/[H(1-2\mu)]$,
γ_t = body force acting downward [$\gamma_t = \gamma_s(1-n) + \gamma_w nS$, kN/m^3], and
γ_s = specific weight of soil particles, kN/m^3.

The unit weight of the soil, γ_t, shown in Eq. 13.155 corresponds to the vertical body force (i.e., unit weight). The initial stresses and pore pressures presented in Eqs. 13.152–13.154 can be removed in the case where an incremental transient approach is considered (i.e., taking derivatives with respect to time of u, v, w, u_w, and u_a). The unsaturated soil properties shown in Eqs. 13.152–13.154 vary with soil suction and net stresses. In this sense, the solution of the partial differential equations is nonlinear.

Equations 13.152–13.154 show that the equilibrium of forces within a representative elemental volume can be expressed as a function of five primary variables: the displacements in the x-, y-, and z-direction (i.e., u, v, and w), the pore-water pressure u_w, and the pore-air pressure u_a. Therefore, two additional equations are required in order to render the system of equations solvable.

The choice of primary variables (i.e., u, v, w, u_w, and u_a) is not arbitrary. These five primary variables result in meaningful *boundary conditions* for unsaturated soil problems. Integration by parts of the second-order derivatives of Eqs. 13.152–13.154 can be used to obtain a solution. Integration results in a surface integral corresponding to a Neumann-type boundary condition. The Neumann boundary condition associated with Eqs. 13.152–13.154 corresponds to external loads (i.e., forces) in the x-, y-, and z-directions, respectively. Another type of boundary condition that can be applied to Eqs. 13.152–13.154 is a predetermined displacement value (i.e., Dirichlet-type boundary conditions with u, v, and w as primary variables).

The three-dimensional equilibrium of forces results in three differential equations. Each equation is formed by three main partial derivatives of the stress tensor components with respect to the x-, y-, and z-directions. The partial derivatives are a result of the assumption concerning the forces and stresses acting on the REV shown in Fig. 13.65. The stresses are continuously distributed in space. The spatial distribution of the variables can be described using the partial derivative of each force or stress component for a given direction.

Various terms within each of the main partial derivatives are a function of the constitutive law adopted to describe the stress-strain behavior of the saturated-unsaturated soil. The partial differential equations governing the equilibrium of forces acting on an unsaturated soil element become increasingly complex as more elaborate stress-strain models are introduced. However, the general format for the governing partial differential equations remains in accordance with Eqs. 13.152–13.154.

Equations 13.152–13.154 result in a smooth transition between saturated and unsaturated conditions provided appropriate soil properties are used. Soil behavior is dictated by the two independent stress state tensors when soil suction is positive (i.e., the soil is treated as an unsaturated soil). There are distinct soil properties associated with each of the stress state variables D_{ii} and h. As the soil becomes saturated and matric suction tends towards zero, $m_1^s$and m_2^s become equal to the coefficient of volume change for the soil under saturated conditions (i.e., m_v) and h becomes 1. Therefore, the partial differential equations for saturated conditions (i.e., $h = 1$) are a special case of the general equations (i.e., Eqs. 13.152–13.154), which can then be written as follows for the x-, y-, and z-directions, respectively:

$$\frac{\partial}{\partial x}\left[\sigma_{x0} + D_{11}\frac{\partial u}{\partial x} + D_{12}\frac{\partial v}{\partial y} + D_{12}\frac{\partial w}{\partial z} + du_w\right] + \frac{\partial}{\partial y}\left[\tau_{xy0} + D_{44}\left(\frac{\partial u}{\partial y} + \frac{\partial v}{\partial x}\right)\right] + \frac{\partial}{\partial z}\left[\tau_{xz0} + D_{44}\left(\frac{\partial u}{\partial z} + \frac{\partial w}{\partial x}\right)\right] = 0 \qquad (13.155)$$

$$\frac{\partial}{\partial x}\left[\tau_{xy0} + D_{44}\left(\frac{\partial u}{\partial y} + \frac{\partial v}{\partial x}\right)\right] + \frac{\partial}{\partial y}\left[\sigma_{y0} + D_{12}\frac{\partial u}{\partial x} + D_{11}\frac{\partial v}{\partial y} + D_{12}\frac{\partial w}{\partial z} + du_w\right] + \frac{\partial}{\partial z}\left[\tau_{yz0} + D_{44}\left(\frac{\partial v}{\partial z} + \frac{\partial w}{\partial y}\right)\right] + \gamma_t = 0 \qquad (13.156)$$

$$\frac{\partial}{\partial x}\left[\tau_{xz0} + D_{44}\left(\frac{\partial u}{\partial z} + \frac{\partial w}{\partial x}\right)\right] + \frac{\partial}{\partial y}\left[\tau_{yz0} + D_{44}\left(\frac{\partial v}{\partial z} + \frac{\partial w}{\partial y}\right)\right] + \frac{\partial}{\partial z}\left[\sigma_{z0} + D_{12}\frac{\partial u}{\partial x} + D_{12}\frac{\partial v}{\partial y} + D_{11}\frac{\partial w}{\partial z} + du_w\right] = 0 \qquad (13.157)$$

Equations 13.154–13.157 show the interdependency between changes in pore-water pressure, pore-air pressures, displacements, and total stresses. External loads may result in changes in pore pressures and displacements. The fraction of load distributed to the soil structure and to the pore-water and pore-air phases is determined by the relative compressibility of each phase, by the speed of loading, and by the pore-water and pore-air permeability. The solutions of Eqs. 13.152–13.154 combined with the solution of pore-water flow and pore-air flow equations represent an alternative and rigorous approach to the approach involving the calculation of pore pressure parameters described by Fredlund and Rahardjo (1993).

An alternative condition may exist where the pore-water and pore-air pressure changes produced are the result of atmosphere boundary conditions. The climatic driven conditions may result in displacements and changes in total stress. Pore-water and pore-air flow equations might also be solved independent of total stresses by neglecting a number of coupling terms. The equations may also be solved in a fully coupled manner, meaning that all possible coupling terms are taken into consideration.

13.8 MEASUREMENT OF STRESS-DEFORMATION PROPERTIES FOR UNSATURATED SOILS

Volume change coefficients and indices can be experimentally measured for unsaturated soils. Test procedures and equipments are available for the measurements of the volume change coefficients and indices for unsaturated soils. In some cases, it is necessary to modify the soil-testing equipment as described in earlier chapters.

Test results on compacted silt, glacial till, and other soils are used to illustrate the relationships between the various volume change coefficients and indices. Most of the results are presented in a semilogarithmic form as volume change indices. These indices can then be converted to volume change coefficients and elastic parameters that can be used in numerical modeling.

The test procedures and equipment used in the measurement of the volume change properties are common to most soil mechanics laboratories. Some of the equipment (e.g., pressure plate apparatuses) is more common to the field of soil physics and agronomy.

The test procedures and equipment for the loading constitutive surfaces are described prior to those for the unloading constitutive surfaces. The use of oedometer test results for assessing the in situ stress state in terms of the swelling pressure is briefly described. Procedural corrections pertinent to the determination of a corrected swelling pressure are described along with an explanation of its importance to the prediction of soil swelling or heave.

Figure 13.67 shows three-dimensional views of the void ratio and water content constitutive surfaces. The volume change coefficients corresponding to the loading conditions (i.e., a_t, a_m, b_t, b_m) are shown. Curve A in Fig. 13.67 is essentially a compression curve obtained from an oedometer or triaxial test on a soil in a saturated condition.

The coefficient of compressibility a_t is equal to the a_v coefficient of compressibility commonly measured in saturated soil mechanics. The coefficient of water content change b_t (Fig. 13.31b) can then be calculated once the coefficient of compressibility a_t has been measured. The coefficient of water content change b_t is equal to the coefficient of compressibility a_t multiplied by the specific gravity of the soil, G_s.

Curve B in Fig. 13.67b is similar to a SWCC that can be measured using a pressure plate apparatus. The slope of the SWCC plotted on an arithmetic scale is equal to the coefficient

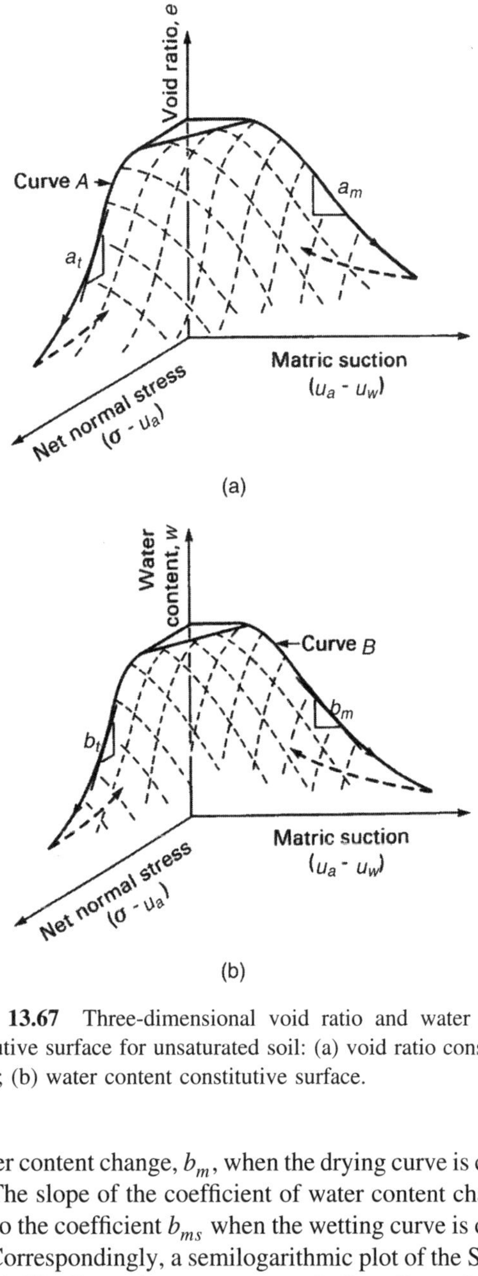

Figure 13.67 Three-dimensional void ratio and water content constitutive surface for unsaturated soil: (a) void ratio constitutive surface; (b) water content constitutive surface.

of water content change, b_m, when the drying curve is considered. The slope of the coefficient of water content change is equal to the coefficient b_{ms} when the wetting curve is considered. Correspondingly, a semilogarithmic plot of the SWCCs (Fig. 13.68) allows the water content indices D_m and D_{ms} to be defined for the drying and wetting curves, respectively.

The shrinkage (and swelling) relationships for a soil relate the void ratio to the water content at various soil suctions. Figure 13.69 shows a shrinkage curve relationship for silty clay. The shrinkage curve results are for the same soil tested for SWCC in Fig. 13.68. The slope of the shrinkage curve defines the ratio between a_m and b_m coefficients. The coefficient of volume change, a_m, can now be computed using the a_m/b_m ratio obtained from the shrinkage curve in Fig. 13.69.

13.8.1 Test Procedures and Equipment

The tests and equipment required for determining the volume change coefficients associated with the loading constitutive surface are briefly described. Conventional testing equipment common to a soil mechanics laboratory can be used in the measurement of the volume change coefficients of an unsaturated soil (Ho and Fredlund, 1989; Rahardjo et al., 1990; Ho et al., 1992). There are, however, several procedural factors that need to be considered when testing unsaturated soils.

Typical test results are presented to illustrate the techniques used to determine the volume change coefficients. The last part of this section describes the tests and equipment required for the determination of volume change coefficients associated with the unloading constitutive surfaces.

13.8.2 Loading Constitutive Surfaces

The relationship between the volume change indices associated with the loading of a soil can be visualized by presenting several deformation curves on one plot (Fig. 13.70). These data can be referred to as intersection or limiting curves because one of the stress state variables goes to zero. Intersection curves 1 and 2 for the void ratio surface (Fig. 13.70a) are combined in Fig. 13.70b. The slopes of curves 1 and 2 constitute C_t and C_m volume change indices corresponding to the net normal stress and soil suction planes, respectively.

Intersection curves 3 and 4 for the water content surface (Fig. 13.70c) are combined in Fig. 13.70d. The water content w results can be multiplied by specific gravity G_s (i.e., wG_s) in order to obtain the same scale as used for plotting the void ratio data. Consequently, curves 3 and 4 in Fig. 13.70c can be plotted on Fig. 13.70d. As a result, the slopes of curves 3 and 4 in Fig. 13.70d are the products of G_s and the volume change indices (i.e., D_tG_s and D_mG_s, respectively).

Curve 1 in Fig. 13.70 is a consolidation curve for the soil in its saturated condition. The consolidation curve exhibits a linear relationship between the void ratio and the logarithm of net normal stress over quite a wide loading range beyond the swelling pressure of the soil. The slope of curve 1, C_t, is equal to the compression index C_c of the saturated soil. Curve 3 in Fig. 13.70d coincides with curve 1 from Fig. 13.70b since the water content multiplied by specific gravity is equal to the void ratio when the soil is saturated. Therefore, the water content index D_t can be computed as C_t/G_s.

Curve 4 in Fig. 13.70 is the SWCC (i.e., drying curve) and can be obtained from a pressure plate test. A shrinkage curve combined with the drying portion of the SWCC can be used to construct curve 2 in Fig. 13.70. In other words, the four "limiting" volume change curves and their corresponding indices (i.e., C_t, D_t, D_m, and C_m) can be measured using routine laboratory soil tests.

The combined plot of curves 1, 2, 3, and 4 is presented in Fig. 13.71, which is a combination of Figs. 13.70b and 13.70d. The arrows in Fig. 13.71 indicate the direction of deformation for curves 2 and 4. These curves approach curve 1 when the initial degree of saturation of the soil increases. Curve 1 has been assumed to remain constant for conditions corresponding to various initial degrees of saturation. The

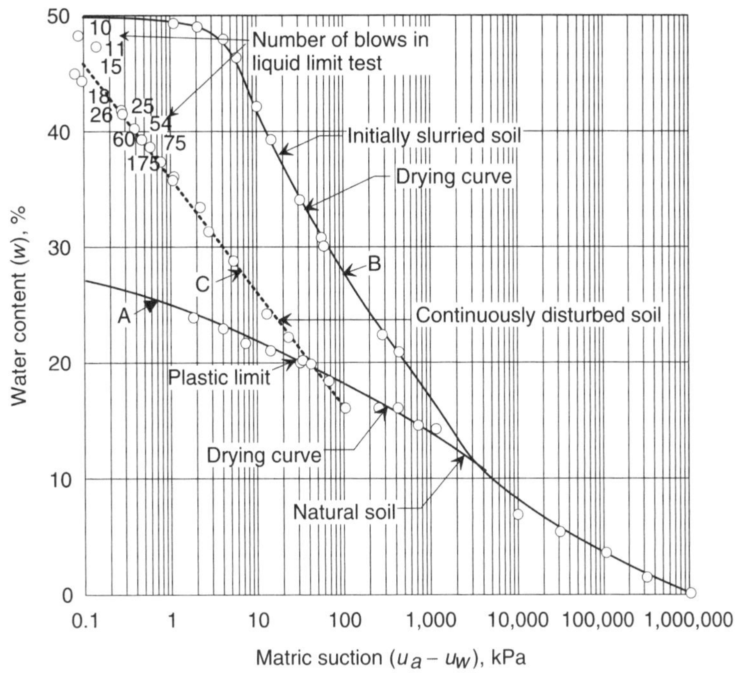

Figure 13.68 SWCCs for silty clay (after Croney and Coleman, 1954).

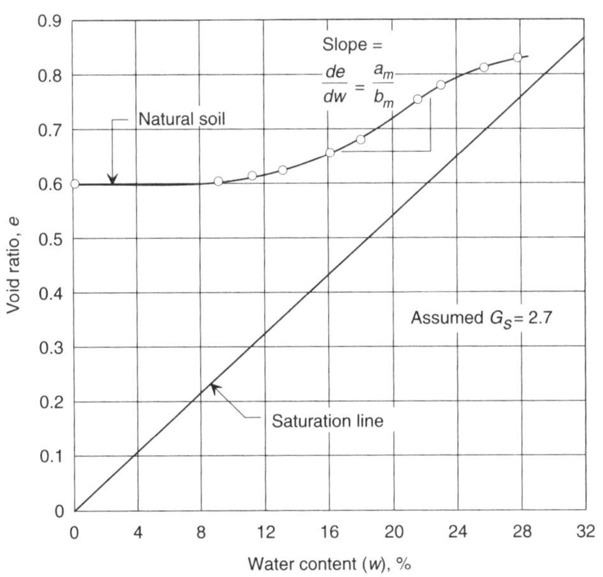

Figure 13.69 Shrinkage curve relationship for silty clay (from Croney and Coleman, 1954).

void ratio and the water content curves are the same when the soil is saturated, varying only with respect to net normal stress or effective stress $\sigma - u_w$ (i.e., curve 1 in Fig. 13.71).

The compression (i.e., void ratio) and water content curves apply to the "limiting" boundaries of the volume-mass constitutive surfaces. However, these relationships do not describe how the constitutive surfaces vary for intermediate stress state changes. The Pham and Fredlund (2011a) volume-mass constitutive model is of assistance in extending the above information to formulate continuous mathematical functions for combinations of isotropic net total stress and soil suction changes.

13.8.3 Odometer Equipment Modifications for Measuring Stress-Deformation Properties

Test procedures for oedometer tests on unsaturated soil specimens (e.g., compacted specimens) are described in ASTM D4546. The ASTM standard describes three methods for inundating soil specimens prior to performing the oedometer test. The matric suction of the soil is brought to zero through inundation and the results can routinely be used to calculate the swelling potential of a soil. Inundation can be conducted under constant-volume or free-swell conditions. Further details regarding the laboratory test procedure are given in Chapter 14.

Curve 1 in Figs. 13.70 and 13.71 illustrate oedometer test results that utilize the constant-volume inundation procedure at the beginning of the test. The soil specimen is allowed to attain an equilibrium state following immersion with water under initial total stress conditions. The test then proceeds using a conventional oedometer test procedure for testing saturated soils (ASTM D2435). Decreasing void ratios during loading are plotted against the logarithm of effective stress $\sigma - u_w$ to yield curves 1 and 3, as shown in Figs. 13.70 and 13.71.

13.8.4 Pressure Plate Drying Tests

The SWCC (i.e., curve 4 in Figs. 13.70 and 13.71) relates water content to applied soil suctions. The initial pore-air pressure u_a is usually assumed to be atmospheric

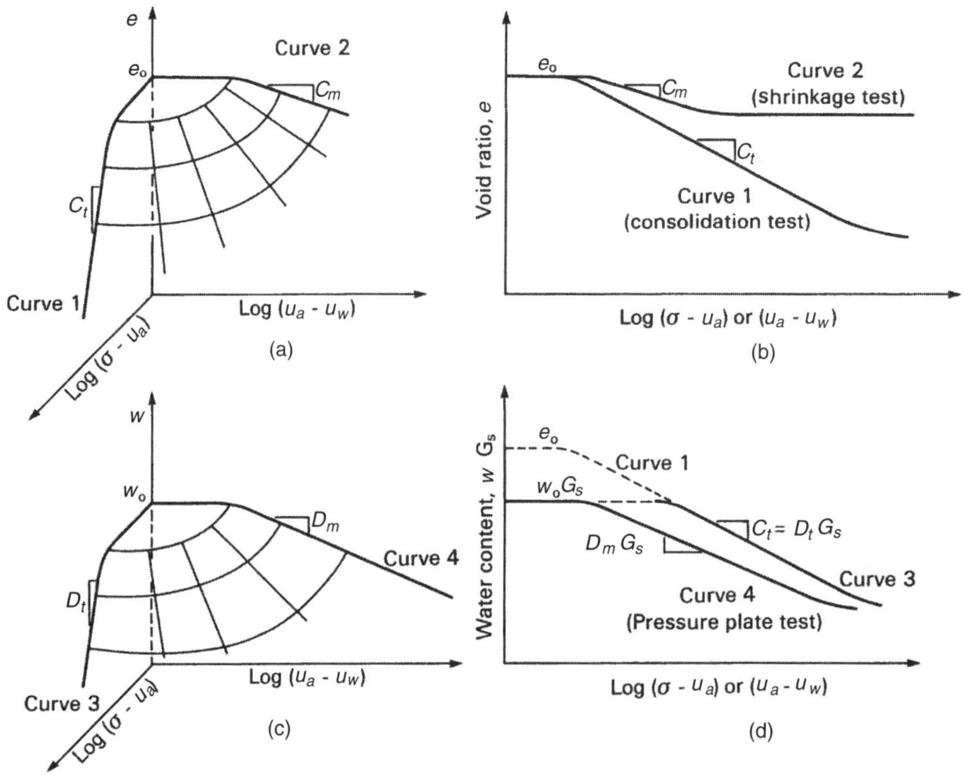

Figure 13.70 Void ratio and water content constitutive relationships for unsaturated soil: (a) void ratio versus logarithm of stress state relationship; (b) compression of void ratio curves onto common stress state variable axis; (c) water content versus logarithm of stress state relationship; (d) compression of water content curves onto common stress state variable axis.

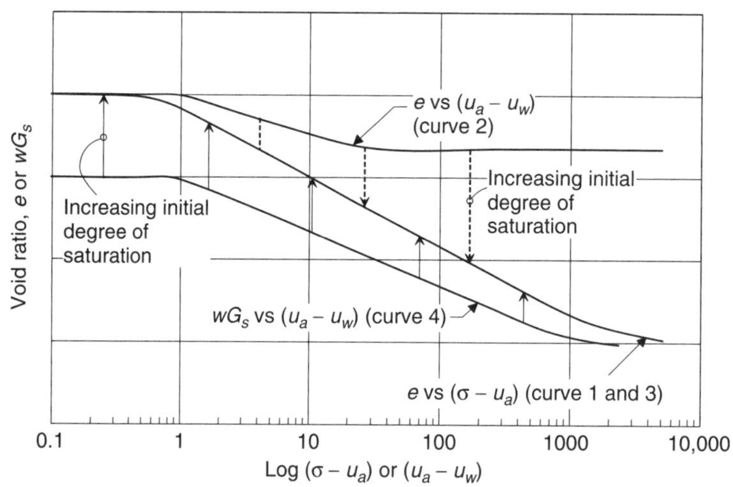

Figure 13.71 General relationship between curves that define the mass and volume change behavior of an unsaturated soil.

(i.e., $u_a = 0$), while the pore-water pressure u_w can be highly negative. The unsaturated soil specimens are generally placed within a confining metal ring and allowed to imbibe water until soil suction approaches zero. The soil specimens are then placed onto a high-air-entry ceramic disk in the pressure plate apparatus.

Positive matric suctions are applied to a soil specimen through use of the axis translation technique (Hilf, 1956). The water pressure is usually maintained at zero and a positive pore-air pressure is applied. Therefore, the matric suction in the soil specimen can be varied by applying different air pressures to the specimen.

Pressure plate extractors such as the type manufactured by SoilMoisture Equipment Corporation, Santa Barbara, California, are commonly used to measure the SWCC. Pressure plate extractors can be designed for testing multiple soil specimens or single soil specimens. One of the test procedures for pressure plate tests is given in ASTM D2325. The pressure plate extractor consists of a high-air-entry ceramic disk contained within an air pressure chamber. The high-air-entry disk is saturated and is always in contact with water through a compartment below the disk. The compartment is usually maintained at zero water pressure.

Further details concerning pressure plate apparatuses and associated test procedures appropriate for geotechnical engineering are described in Chapter 5. Some of the apparatuses described in Chapter 5 have been specifically designed for geotechnical engineering applications. Individual soil specimens are generally tested, and in some cases, both volume change and water content change are measured.

The water content corresponding to high-suction values is generally obtained by allowing small soil samples to equilibrate in a high-suction humidity environment. The humidity environment is created using saturated salt solutions at the bottom of vacuum desiccators. Details of the vacuum desiccator procedure are presented in Chapter 5.

13.8.5 Shrinkage Tests

Shrinkage test results show the relationship between void ratio and the water content at various soil suction values. A soil specimen can either be allowed to dry in the air or it can be subjected to various suction values using a pressure plate extractor. In either case, the void ratio and water content of the soil specimen are measured at various equilibrium states. The soil specimen can be covered at various time intervals to allow soil suction to equalize throughout the soil specimen.

Measurements of void ratio can be made following test procedures similar to those used when measuring the shrinkage limit of a soil. Measuring the shrinkage curve involves measuring the water content and volume of a soil specimen as drying occurs. However, in the case of a shrinkage limit test, only the initial and completely dry states are of interest. In the case of a shrinkage curve test the void ratio and water content are measured numerous times throughout the drying process.

Historically, the mercury displacement technique has been used for the measurement of the total volume of the shrinkage specimens. The volume of the displaced mercury during immersion can be converted to the total volume of the specimen. However, this procedure is no longer advocated because of the health risks associated with handling mercury. Direct measurements of total volume can also be performed using calipers. The shrinkage curve is constructed by plotting the decreasing void ratios against the decreasing water contents as the soil suction increases to completely dry conditions.

13.8.6 Determination of Volume Change Indices

The measurement of the volume change indices is illustrated using two soils: silt and glacial till tested by Ho (1988). The index properties for the silt and glacial till are presented in Table 13.4. An attempt was made to prepare soil specimens with nearly identical initial conditions. Each soil was oven dried and hand mixed with a predetermined quantity of distilled water. The wet soil was placed in a sealed plastic bag and was left to cure for several days in a constant-humidity and constant-temperature room. Specimens were formed by static compaction at one-half standard AASHTO compaction effort at either "dry-of-optimum" or "optimum" initial water contents. The compaction characteristics of the silt and glacial till are given in Table 13.4.

Volume change and water content change tests (i.e., oedometer, pressure plate, and shrinkage tests) were conducted on compacted silt and glacial till specimens with initial conditions corresponding to both the dry-of-optimum and optimum conditions. The laboratory data were used to determine the volume change indices for the soils at both compaction conditions.

The data for silt specimens compacted dry of optimum illustrate the technique for obtaining the volume change indices. The oedometer test results on the silt specimens compacted dry of optimum are shown in Fig. 13.72. The oedometer tests were performed using the constant-volume test method. The loading curves correspond to curves 1 and 3 in Figs. 13.70 and 13.71, respectively, showing the C_t and $(D_t G_s)$ deformation indices.

Similar oedometer compression results for the silt compacted at optimum water content are shown in Fig. 13.73. Oedometer compression results for the glacial till compacted at dry-of-optimum and optimum water contents are shown in Figs. 13.74 and 13.75, respectively.

Table 13.4 Index Properties of Silt and Glacial Till Used in Test Program

	Silt	Glacial Till
Liquid limit, %	26.7	33.2
Plastic limit, %	14.9	13.0
Plasticity index	11.8	20.2
Sand sizes, %	25.0	32.0
Silt sizes, %	52.0	39.0
Clay sizes, %	23.0	29.0
Specific gravity, G_s	2.72	2.76
Half standard effort compaction:		
w_{opt}	19.0%	18.75%
$\gamma_{d,max}$	16.65 kN/m^3	17.12 kN/m^3

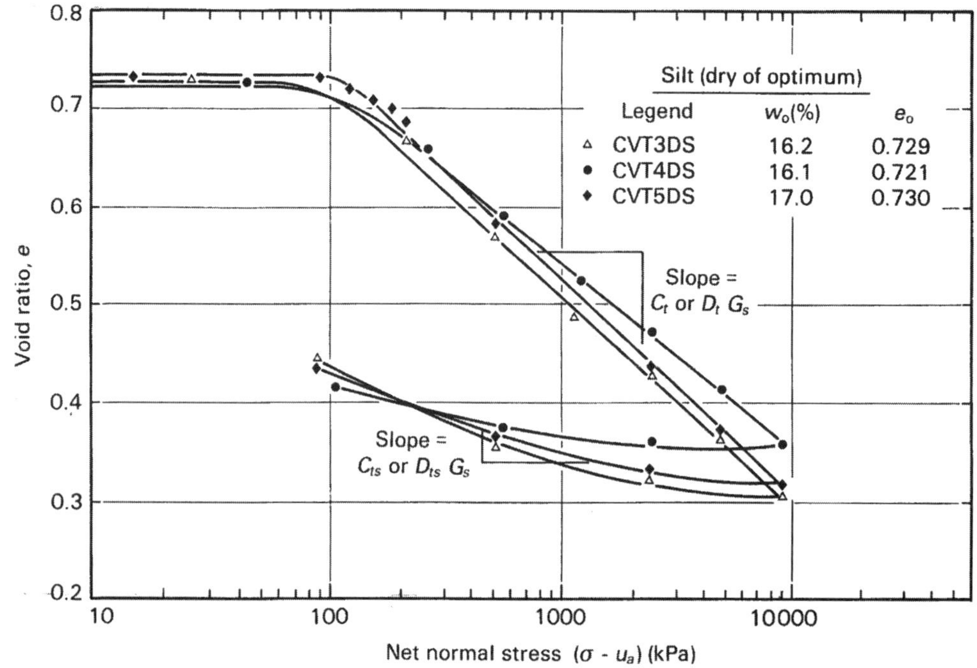

Figure 13.72 One-dimensional constant-volume oedometer tests on silt compacted dry of optimum water content (from Ho, 1988).

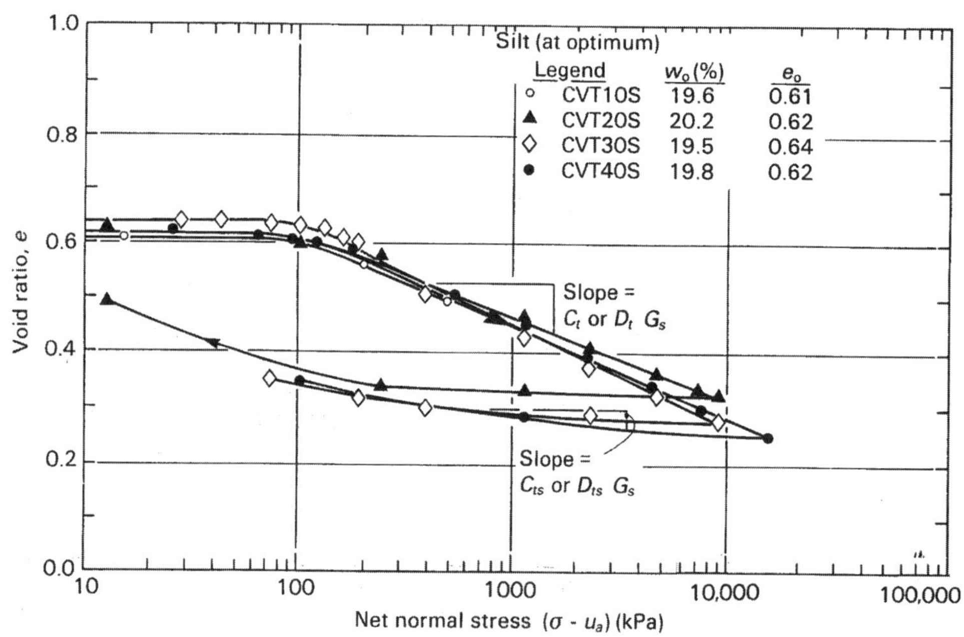

Figure 13.73 One-dimensional constant-volume oedometer tests on silt compacted at optimum water content (from Ho, 1988).

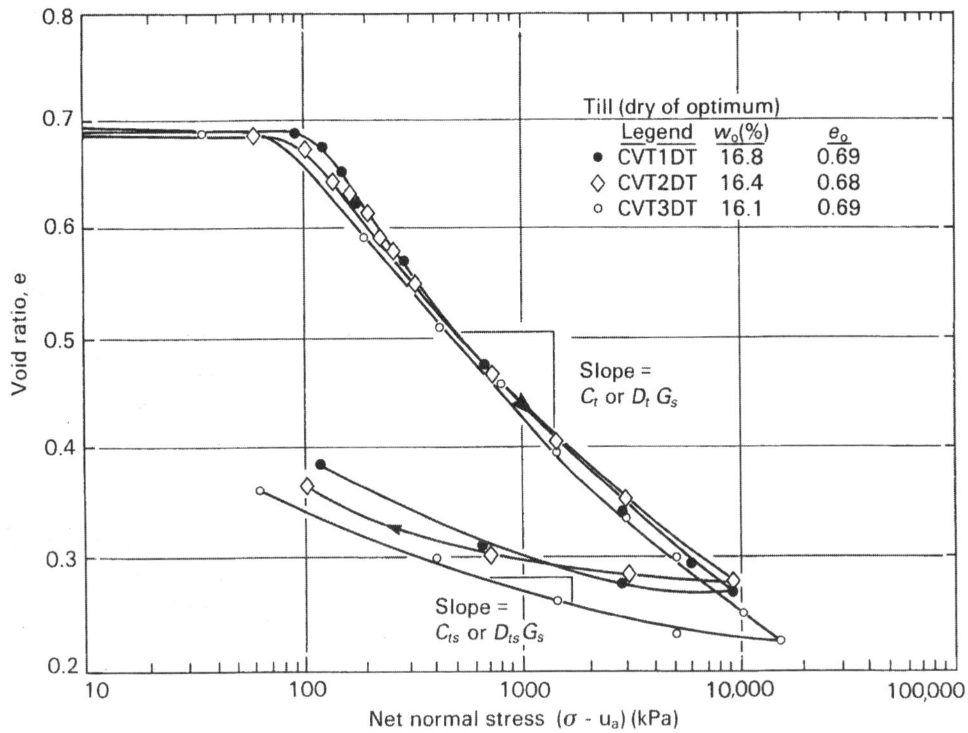

Figure 13.74 One-dimensional constant-volume oedometer tests on glacial till compacted at dry-of-optimum water content (from Ho, 1988).

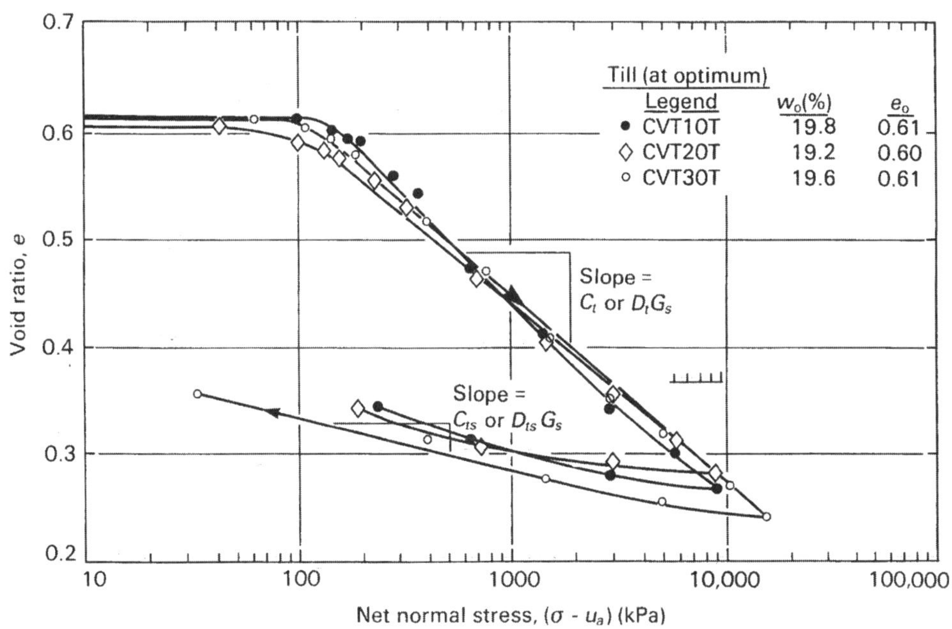

Figure 13.75 One-dimensional constant-volume oedometer tests on glacial till compacted at optimum water content (from Ho, 1988).

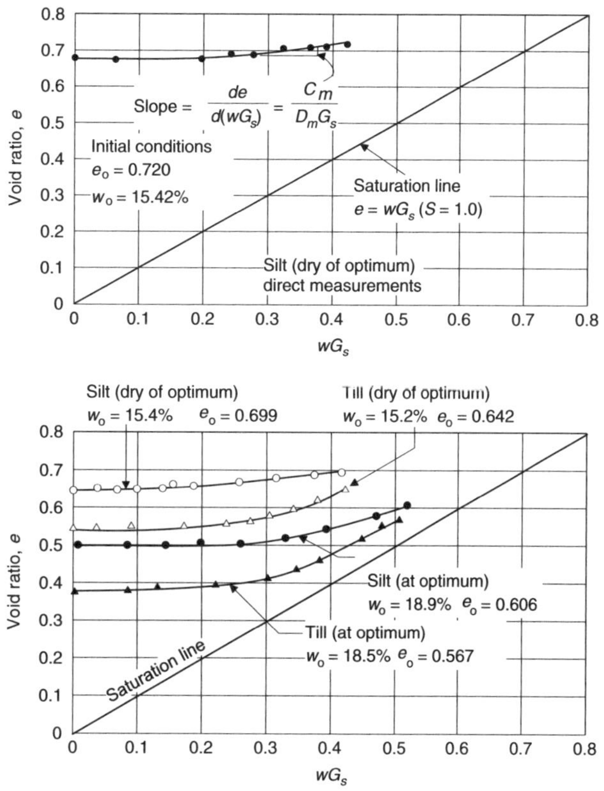

Figure 13.76 SWCC obtained from pressure plate tests on silt compacted dry of optimum water content (after Ho et al., 1992).

The SWCC for the silt (Fig. 13.76) was obtained using pressure plate tests. The water content is multiplied by the specific gravity G_s for the purpose of combining the void ratio and water content plots, as illustrated in Fig. 13.71 (i.e., curve 4). The slope of the SWCC is equal to $D_m G_s$.

The shrinkage curves for the compacted silt are shown in Fig. 13.77. The void ratios corresponding to various water

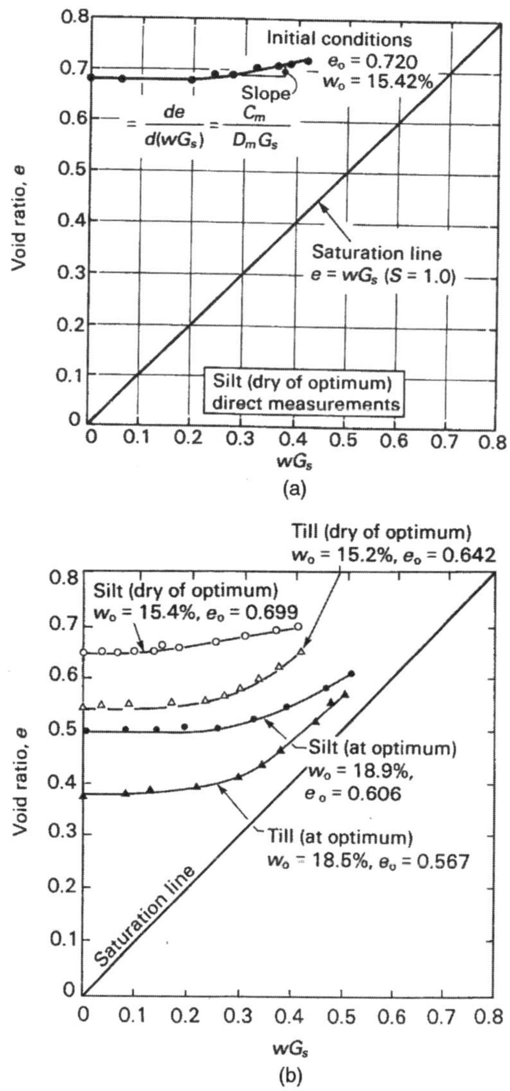

Figure 13.77 Shrinkage test data for (a) compacted silt and (b) compacted silts and glacial tills (from Ho, 1988).

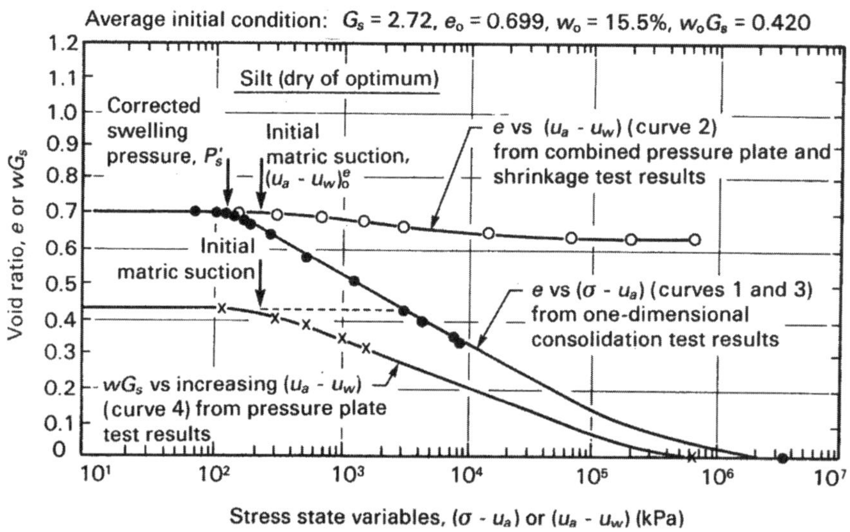

Figure 13.78 Relationship of volume-mass versus stress state for silt compacted dry of optimum water content (after Ho et al., 1992).

Table 13.5 Summary of Experimentally Measured Volumetric Deformation Indices[a]

One-Dimensional Loading						One-Dimensional Unloading						
Soil Type[b]	w_0G_s	e_0	C_t or D_tG_s	C_m	D_mG_s	Soil Type[b]	w_0G_s	e_0	C_{ts}^c	$D_{ts}G_s^d$	C_{ms}	$D_{ms}G_s$
DS	0.420	0.699	0.196	0.030	0.124	DS	0.419	0.702	0.040	0.126	0.033	0.263
DS	0.424	0.700				DS	0.421	0.699				
OS	0.516	0.606	0.177	0.082	0.158	OS	0.517	0.616	0.055	0.076	0.052	0.101
OS	0.511	0.609				OS	0.514	0.609				
DT	0.427	0.642	0.206	0.089	0.159	DT	0.436	0.693	0.066	0.084	0.056	0.122
OT	0.516	0.567	0.179	0.106	0.171	DT	0.523	0.571	0.037	0.057	0.024	0.060

[a] All indices have a negative sign, as described by the sign convention.
[b] DS stands for silt at dry of optimum initial water content. OS stands for silt at optimum initial water content. DT stands for glacial till at dry of optimum initial water content. OT stands for glacial till at optimum initial water content.
[c] Average slope of unloading curve.
[d] Slope of linear portion of unloading curve

contents in Fig. 13.76 can be found using the shrinkage curve relationship in Fig. 13.77a. The void ratio–matric suction relationship (i.e., curve 2 in Figs. 13.70 and 13.71) can be constructed using Figs. 13.76 and 13.77. The slope of the shrinkage curve (i.e., $de/d\left(wG_s\right)$ or $\left[\partial e/\partial\left(u_a - u_w\right)\right] / \left[\partial\left(wG_s\right)/\partial\left(u_a - u_w\right)\right]$) is equivalent to the ratio of the volume change indices (i.e., C_m/D_mG_s).

The combined plot of Figs. 13.72 and 13.76 and curve 2 (i.e., constructed from Figs. 13.76 and 13.77a) is depicted in Fig. 13.78, which illustrates the volume change characteristics of an unsaturated, compacted silt. The volume change indices (i.e., C_t, C_m, D_t, and D_m) can be computed from Fig. 13.78. Changes in void ratio and water content due to an increase in total stress or matric suction can now be predicted using the computed volume change indices.

The same test procedures were applied to other compacted silt and glacial till specimens. Figure 13.77b summarizes the results of shrinkage tests on various compacted specimens. Typical volume change relationships for the compacted silt and glacial till are presented in Figs. 13.79, 13.80, and 13.81. The relationships are similar to that shown in Fig. 13.78. The computed volume change indices for the compacted silt and glacial till are tabulated in Table 13.5. These indices can be converted to other volume change moduli and coefficients.

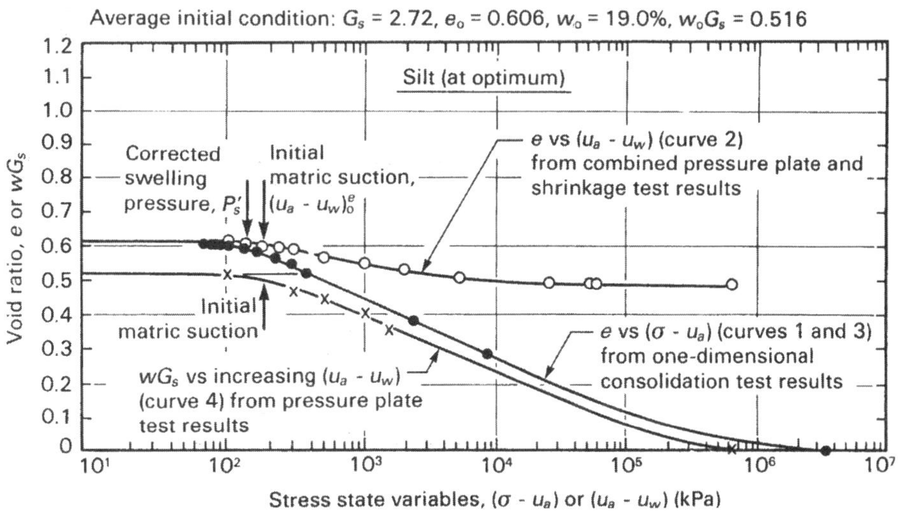

Figure 13.79 Relationship of volume-mass and stress state for silt compacted at optimum water content (after Ho et al., 1992).

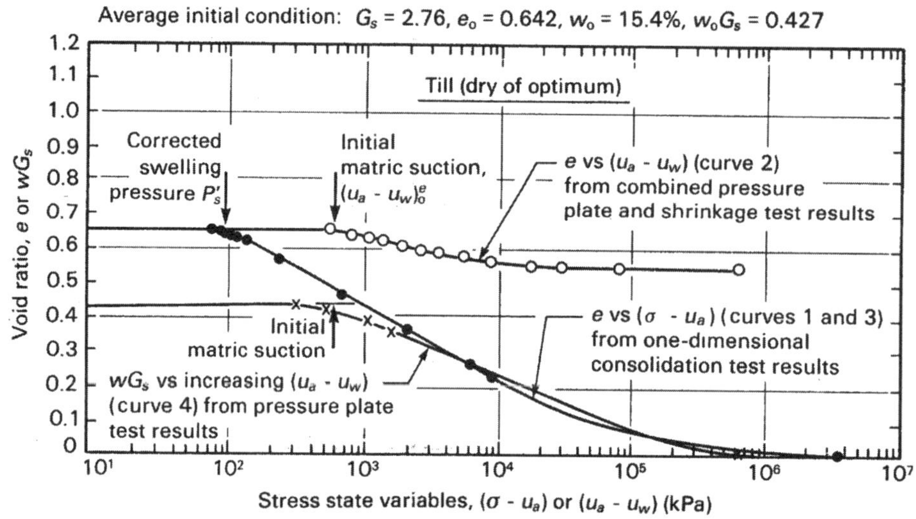

Figure 13.80 Relationship of volume-mass and stress state for till compacted dry of optimum water content (after Ho et al., 1992).

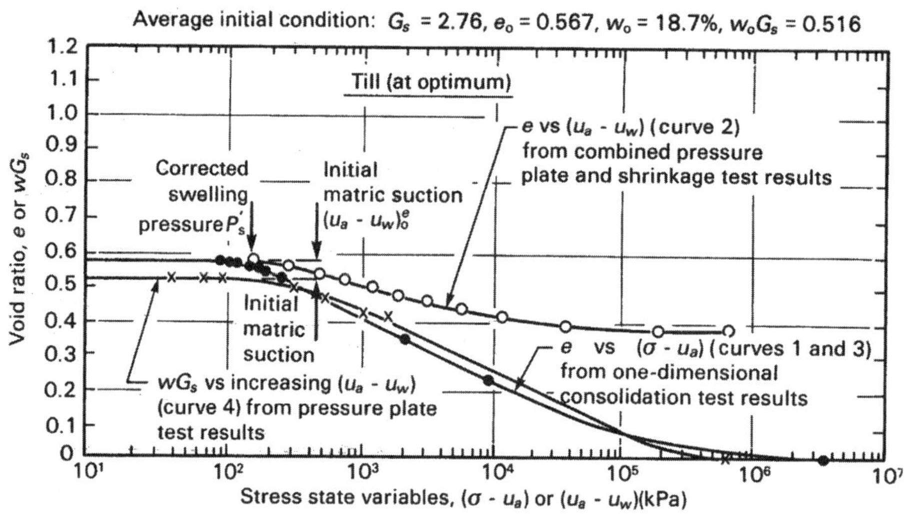

Figure 13.81 Relationship of volume-mass and stress state for till compacted at optimum water content (after Ho et al., 1992).

One-dimensional oedometer tests (or isotropically loaded triaxial tests), pressure plate tests, and shrinkage tests provide the experimental data required to obtain the volume change indices corresponding to volume-mass properties versus stress state variables for an unsaturated soil. These tests can be performed using test procedures quite similar to those used in saturated soil mechanics. The test results characterize the volume-mass relationships for an unsaturated soil on the "limiting" planes of the three-dimensional volume-mass constitutive surfaces.

CHAPTER 14

Solving Stress-Deformation Problems with Unsaturated Soils

14.1 INTRODUCTION

The constitutive equations and related soil properties for volume change of unsaturated soils were presented in the previous chapter. The constitutive equations were written in three forms: the elasticity form, the compressibility form, and the historical soil mechanics form. Tests that can be performed in a soil mechanics laboratory for the measurement of unsaturated soil properties were described. Emphasis was placed on three main types of tests that could be performed: (i) compression tests (i.e., one-dimensional tests and the isotropic triaxial tests), (ii) pressure plate tests to obtain SWCCs, and (iii) shrinkage curve tests. The data from these three types of tests were analyzed to produce approximate constitutive surfaces for an unsaturated soil. The constitutive relationships form the extremities or "bounding" faces of the three-dimensional volume-mass constitutive surfaces.

The Pham and Fredlund (2011a) volume-mass constitutive model was also presented. The model provides a means of estimating the volume-mass properties along any stress path when the soil behavior can be traced back to initially saturated slurry conditions.

Laboratory test data may need to be converted from one format to another format that is more acceptable for performing numerical modeling solutions for practical engineering problems. For example, it might be necessary to convert the coefficient of volume change, m_v or m_1^s, into a nonlinear function comprised of equivalent values of Young's modulus E and Poisson's ratio μ. This chapter and Chapter 13 are directed toward understanding the theory of volume change for an unsaturated soil and the conversion of laboratory-measured soil properties into parameters that can be used in an engineering analysis. The procedures that can be used for calculating volume change or heave in expansive soils are also described. The interpretation of the stress history of the soil from the laboratory test results is often required for the interpretation of the initial stress state condition in expansive soils.

Procedures have also been proposed for estimating unsaturated soil properties and unsaturated soil property functions relevant to the calculation of volume-mass changes. This chapter presents procedures and examples of one-dimensional, two-dimensional, and three-dimensional volume change calculations.

14.2 ESTIMATION OF STRESS-DEFORMATION PROPERTIES

Engineering analyses should start with relatively simple engineering models and progress toward the consideration of more rigorous and complex models that better describe unsaturated soil behavior. A common starting point is to assume that the material behaves in an isotropic and elastic manner. The term "elastic" means that the material behaves in a conservative manner in the sense that all work done by external stresses is stored and is recoverable upon unloading. An important simplifying feature associated with isotropic and elastic materials is the decoupling of shear and volumetric deformations.

The basic material properties commonly used for stress-deformation analyses are the elastic parameters referred to as Young's modulus E and Poisson's ratio μ. The actual soil properties are generally nonlinear and as a result the elastic soil parameters need to be viewed in an incremental manner. In other words, the elastic properties are a function of stress state.

Young's modulus E and Poisson's ratio μ are elastic parameters defined on the basis of a particular laboratory test, namely, a uniaxial compression (or extension) test with radial stress held constant. Young's modulus for an unsaturated soil is defined as

$$E = \frac{d(\sigma_v - u_a)}{d\varepsilon_a} \tag{14.1}$$

where:

$\sigma_v - u_a$ = vertical net axial stress in a uniaxial compression (or extension) test and
ε_a = axial (vertical) strain.

Poisson's ratio μ is defined as

$$\mu = \frac{d\varepsilon_r}{d\varepsilon_a} \tag{14.2}$$

where:

ε_r = radial strain.

Even the simplest material behavioral model requires that two material properties be defined. The elastic properties E and μ for a saturated soil generally vary with the stress state and must be written as a function of stress state. Unsaturated soil behavior has a third elastic parameter, H, which must be defined for a change in matric suction $u_a - u_w$ [i.e., $H = d(u_a - u_w)/d\varepsilon_a$]. Changes in matric suction are isotropic in the sense that suction acts equally in three coordinate directions. However, it is possible for the elastic soil property H to be anisotropic in character.

Hysteresis generally exists between the loading and unloading sequences. Hysteresis gives rise to elastic material properties that are further stress path and stress reversal dependent. More elaborate elastoplastic behavioral models for unsaturated soils address many of the issues associated with nonlinearity and hysteresis under loading and unloading conditions; however, these models are considered to be outside the scope of this book.

14.2.1 Estimation of Elastic Soil Properties from Compressibility Soil Properties

There are a number of laboratory tests that can be performed on saturated and unsaturated soils in a conventional soil mechanics laboratory. Modifications must be made on some laboratory equipment in order to independently control or measure the pore-air and pore-water pressures when testing unsaturated soils. Laboratory tests measure the compressibility of the soil under a particular set of boundary and loading conditions. The compressibility of a soil can change significantly along a particular stress path. The soil behavior can generally be linearized by plotting the stress state on a logarithmic scale [i.e., logarithm of net normal stress loading $(\sigma - u_a)$].

The compression curve for a normally consolidated saturated soil forms an important reference condition when considering the behavior of unsaturated soils with a complex stress history. There should be a smooth transition in volume change behavior when moving from saturated soil conditions to unsaturated soil conditions and vice versa. The compression index C_c for a normally consolidated saturated soil serves as a reference compressibility condition.

Numerous correlations have been made between the compression index C_c of a normally consolidated saturated soil and classification properties such as liquid limit. The compression index measured on clay samples in an oedometer from various countries of the world consistently fall in a relatively narrow band that is defined by Eq. 14.3 (Skempton, 1944):

$$C_c = 0.007(\text{LL} - 10) \tag{14.3}$$

where:

LL = liquid limit of the soil.

Terzaghi and Peck (1948) noted that the field compression curve was about 30% steeper than the measured laboratory compression curve, giving rise to the following equation for the in situ compression index:

$$C_c = 0.009(\text{LL} - 10) \tag{14.4}$$

Aitkinson (1993) used typical relationships for normally consolidated soils to arrive at the following empirical correlation for the compression index of a normally consolidated soil:

$$C_c = \frac{\text{PI} \times G_s}{200} \quad \text{or} \quad C_c = 0.0138 \times \text{PI} \tag{14.5}$$

where:

PI = plasticity index of the soil in percent and
G_s = 2.76.

Several other empirical correlations have been published for the compression index of a soil (Azzouz et al., 1976) but the above-mentioned equations are sufficient for an estimation of soil compressibility. The above-mentioned correlations have been made for normally consolidated soils that are generally saturated. Unsaturated soils are often overconsolidated and the stress history must be taken into consideration when interpreting laboratory-measured compressibilities. The interpretation of consolidation test results on overconsolidated soils is discussed later in this chapter. Compacted soil test results also require special consideration when assessing the compressibility of the soil.

14.2.1.1 Estimation of Swelling Index

The swelling index C_s is measured during the unloading of a soil. The swelling index is of particular interest when dealing with unsaturated expansive soils with negative pore-water pressures (i.e., matric suctions). Soil suction may change with time and as a result the soil may undergo changes in volume.

The rebound or swelling index of normally consolidated clay soils is often about 5–10% of the compression index. Leonards and Altschaeffl (1964) suggested that the rebound or swelling index is usually in the range of 0.015–0.035. Overconsolidated, expansive soils, however, can have swelling indices C_s that exceed 0.10. High swelling indices along with high initial matric suctions in the soil can result in in situ vertical heave of 0.3–0.6 m and greater.

14.2.1.2 Separation of Compressibility Measurements into Elastic Modeling Parameters

The compression and swelling indices C_c and C_s provide a number that gives an indication of the volume change properties of the soil. It is often necessary for the compression index to be converted into a different form for use in numerical modeling software. Even when the compressive index (e.g., C_c) is converted to a compressibility value (e.g., m_1^s) corresponding to a particular stress state, there is still the limitation that a single value is computed while, in reality, two soil properties are required: Young's modulus E and Poisson's ratio μ. Stated another way, the laboratory experiment generally measures one soil property linking volume change to the stress state while two soil properties are required for numerical modeling purposes.

A value is commonly assumed for Poisson's ratio and then the compression index can be converted to an equivalent Young's modulus value. Poisson's ratio varies over a fairly narrow range from 0.0 to 0.5. A Poisson's ratio of 0.0 corresponds to the case where there is no lateral movement as the soil is loaded in uniaxial loading. A Poisson's ratio of 0.5 corresponds to the situation where volume change in the vertical direction is shared by the volume changes in the other two orthogonal directions (i.e., no overall volume change of the soil specimen). Typical values for Poisson's ratio are shown in Table 14.1.

The magnitude of Poisson's ratio μ is most commonly estimated based on past experience and published results even though it is possible to measure Poisson's ratio in a uniaxial-type triaxial test. It is possible to convert the compression index of a soil into an equivalent Young's modulus value once a value is selected for Poisson's ratio.

Values of Young's modulus E change linearly with compression index. The mathematical relationship between Young's modulus and compression index depends on the laboratory testing procedure. Young's modulus is defined using a uniaxial test procedure and therefore measurements made using one-dimensional compression tests or some other test procedure will need to be calculated. Let us assume that results from a one-dimensional oedometer test

Table 14.1 Typical Values of Poisson's Ratio

Soil Type	Description	Poisson's Ratio, μ
Clay	Soft	0.35–0.40
	Medium	0.30–0.35
	Stiff	0.20–0.30
Sand	Loose	0.15–0.25
	Medium	0.25–0.30
	Dense	0.25–0.35

Source: From Budhu (2007).

are to be converted to equivalent Young's modulus values. The compression index of the soil can be written as follows in terms of a coefficient of volume change, m_1^s:

$$m_1^s = \frac{0.435C_c}{(\sigma - u_a)_{\text{ave}}(1 + e_0)} \tag{14.6}$$

The coefficient of volume change for one-dimensional oedometer loading can be written in terms of Poisson's ratio and Young's modulus as follows:

$$m_1^s = \frac{(1 + \mu)(1 - 2\mu)}{E(1 - \mu)} \tag{14.7}$$

Therefore, Young's modulus corresponding to the oedometer loading path can be written as follows:

$$E = \frac{(1 + \mu)(1 - 2\mu)(\sigma - u_a)_{\text{ave}}(1 + e_0)}{0.435C_c(1 - \mu)} \tag{14.8}$$

Once a Poisson's ratio value is assumed for a particular soil, Eq. 14.8 provides a relationship between Young's modulus and the compression index of the soil. Young's modulus can be calculated for a series of selected net normal stress states.

14.2.2 Role of SWCC When Establishing Volume-Mass Constitutive Surfaces

The SWCC relates water content to soil suction but does not play a direct role in characterizing the volume change soil properties required for numerical modeling. Rather, the SWCC becomes most useful when it is combined with the shrinkage curve of the soil. In this way it is possible to obtain the soil suction versus volume change soil properties of the soil. The computed compressibility with respect to soil suction must then be converted into an appropriate elasticity parameter for numerical modeling purposes.

The shrinkage curve provides the relationship between water content and void ratio. By cross-plotting the SWCC and the shrinkage curve, it is possible to obtain the change in void ratio versus soil suction. The shrinkage curves can either be measured or estimated for a particular soil. The drying shrinkage curve needs to have the same initial conditions as were used for the SWCC test.

Overall volume changes for a particular geotechnical engineering problem can be computed using a numerical solution such as a finite element analysis. Once overall volume change has been computed, other volume-mass properties can be estimated with the assistance of the SWCC. In essence, this is an uncoupled, incremental solution for volume change.

14.2.3 Role of Shrinkage Curve in Establishing Constitutive Surfaces

The relationship between overall volume change (i.e., defined using void ratio e or specific volume v) and soil suction is difficult to measure using conventional soil-testing equipment. However, the volume change versus soil suction relationship can be obtained through use of a SWCC and a shrinkage curve for the soil.

The shrinkage limit of a soil was originally used as one of the plasticity classification properties for a soil. The shrinkage limit is defined as the water content corresponding to the minimum shrinkage volume for the initially slurry soil. While the shrinkage limit is a classification soil property of interest, it is the entire drying curve from some initial condition to completely dry conditions that is of greater interest when attempting to model unsaturated soil behavior.

The shrinkage curve describes the ratio of the water content change to void ratio change for a specific change in soil suction. Typical shrinkage curve data are shown in Fig. 14.1. M.D. Fredlund et al., (2002b) proposed the following equation to fit shrinkage curve data:

$$e(w) = a_{sh}\left[\frac{w^{c_{sh}}}{b_{sh}^{c_{sh}}} + 1\right]^{1/c_{sh}} \tag{14.9}$$

where:

e = void ratio,
w = gravimetric water content,
a_{sh} = minimum void ratio, e_{min},
b_{sh} = slope of the line of tangency, and
c_{sh} = curvature of the shrinkage curve.

The fitting parameters for the shrinkage curve shown in Fig. 14.1 are as follows: $a_{sh} = 0.765$, $b_{sh} = 0.27$, and $c_{sh} = 9.57$. A procedure for the estimation of the shrinkage curve was described in Chapter 2.

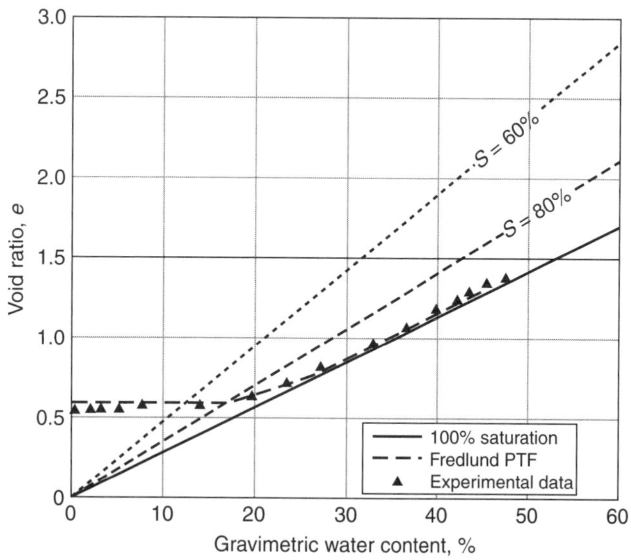

Figure 14.1 Function for shrinkage curve for heavy clay. (Data from Russam, 1958.)

Let us suppose that the SWCC has been measured or estimated for a soil and can be represented by the Fredlund and Xing (1994) SWCC equation (i.e., Eq. 14.10):

$$w(\psi) = w_s\left[1 - \frac{\ln\left(1 + \psi/\psi_r\right)}{\ln\left(1 + 10^6/\psi_r\right)}\right] \times \left[\frac{1}{\left[\ln\left[\exp(1) + \left(\psi/a_f\right)^{nf}\right]\right]^{mf}}\right] \tag{14.10}$$

where:

w_s = gravimetric saturation water content,
ψ = soil suction, kPa,
ψ_r = residual suction, kPa,
a_f = fitting parameter corresponding to the soil suction at the inflection point,
n_f = fitting parameter designating the rate of desaturation, and
m_f = third fitting parameter for the SWCC.

The curve for void ratio versus soil suction can be computed by combining the equation for the SWCC (Eq. 14.10) with the equation for the shrinkage curve (Eq. 14.9). The relationship between void ratio and soil suction represents the limiting boundary condition on the void ratio constitutive surface. The void ratio change can be written in the form of volumetric strain [i.e., $d\varepsilon_v = de/(1 + e_0)$], corresponding to a particular soil suction change. In other words, the slope of the volumetric strain plot versus soil suction is a compressibility modulus designated by the compressibility variable m_2^s. There is no Poisson ratio effect associated with a change in soil suction since the stress change is always isotropic and the soil has been assumed to behave in an isotropic manner.

The elasticity modulus associated with a change in soil suction is designated by the H parameter. The soil modulus m_2^s can be written in terms of the elasticity parameter H for particular stress path conditions. Soil suction change can occur under one-dimensional K_0 loading conditions if a soil sample is allowed to swell under lateral confinement (e.g., confined within a steel ring). The compressibility (or swelling) modulus and the elasticity modulus are related as follows:

$$m_2^s = \frac{1 + \mu}{H(1 - \mu)} \tag{14.11}$$

If soil suction change occurs under isotropic conditions, the soil sample is allowed to change volume in all directions and the compressibility (or swelling) modulus and the elasticity modulus can be related as follows:

$$m_2^s = \frac{3}{H} \tag{14.12}$$

Conversions between compressibility and elasticity soil properties for other loading conditions can be found in Chapter 13. In each case, the elasticity parameter H is not a constant but forms a functional relationship with changes in stress state. The functional form can be used to undertake nonlinear stress-deformation numerical modeling analyses. The above explanations apply to the limiting planes of the volume-mass constitutive surfaces. It is also possible to use the more rigorous volume-mass constitutive model proposed by Pham and Fredlund (2011a) to define the elasticity parameters associated with a variety of stress paths.

14.3 APPLICATION TO PRACTICAL STRESS-DEFORMATION PROBLEMS

An unsaturated soil will undergo volume change when net normal stress or soil suction changes in magnitude. An unsaturated soil will experience swelling and shrinking as a result of matric suction variations arising from environmental changes under constant total stress conditions. Similarly, structural collapse may occur in collapsible soils when matric suction decreases in the soil.

The methodology for the prediction of heave in a swelling soil can be described for one-dimensional, two-dimensional, and three-dimensional situations. The stress history of a soil is an important factor to consider in understanding swelling soil behavior. Formulations and example problems of heave prediction are presented along with case histories. The factors influencing the amount of heave are also discussed.

14.3.1 Problems Associated with Expansive Soils

Expansive soils are found in many parts of the world, particularly in semiarid areas. An expansive soil is often unsaturated due to desiccation drying. However, even under saturated soil conditions it is possible to have highly negative pore-water pressures in an expansive soil.

Expansive soils usually contain clay minerals that undergo large volume changes upon wetting. Large volume changes upon wetting cause extensive damage to structures, in particular light buildings and pavements. In the United States alone, the damage caused by shrinking and swelling soils amounts to about $9 billion per year, which is greater than the combined damages from natural disasters such as floods, hurricanes, earthquakes, and tornadoes (Jones and Holtz, 1973). Therefore, the problems associated with swelling soils are of enormous financial proportions.

The heave potential of a soil depends on soil properties such as clay content, plasticity index, and shrinkage limit. Heave potential also depends on the initial dryness or matric suction in the soil. Several empirical methods have been proposed that correlate the swelling potential of a soil to classification soil properties (Table 14.2). Most correlations make use of the plasticity characteristics of a soil along with the percent clay sizes to assess the potential expansibility of a soil. The words "potential expansibility" suggest that no expansion will be realized by the soil mass unless water is also made available to the soil. Van der Merwe (1964) combined the plasticity index of a soil and the percent clay size particles to provide a classification of expansibility, as shown in Fig. 14.2. The classification system is useful for identifying the swelling potential of a soil. In other words, the correlation reflects one component of the potential for heave to occur.

Total heave can be written as a function of the difference between the initial (or present) in situ stress state and some possible future stress state. The pathway between the initial and final stress state is defined by a soil property, namely, the swelling index, C_s. Generally the net normal stress state variable remains constant while the matric suction stress state variable changes during the swelling process. Changes in matric suction result in changes in overall volume, which in turn result in changes in water content. Total heave can be computed by measuring the in situ matric suction conditions and estimating (or predicting) possible future matric suction in the field under specified environmental conditions.

There are several heave prediction formulations that have been proposed. These formulations differ primarily in the

Table 14.2 Probable Expansion as Estimated from Classification Test Data[a]

Degree of Expansion	Probable Expansion as % of Total Volume Change (Dry to Saturated Condition)[b]	Colloidal Content (% < 1 μm)	Plasticity Index, PI	Shrinkage Limit, W_S
Very high	>30	>28	>35	<11
High	20–30	20–31	25–41	7–12
Medium	10–20	13–23	15–28	10–16
Low	<10	<15	<18	>15

Source: From Holtz and Kovacs (1981).
[a]After Holtz (1959) and U.S.B.R. (1974).
[b]Under a surcharge of 6.9 kPa (1 psi).

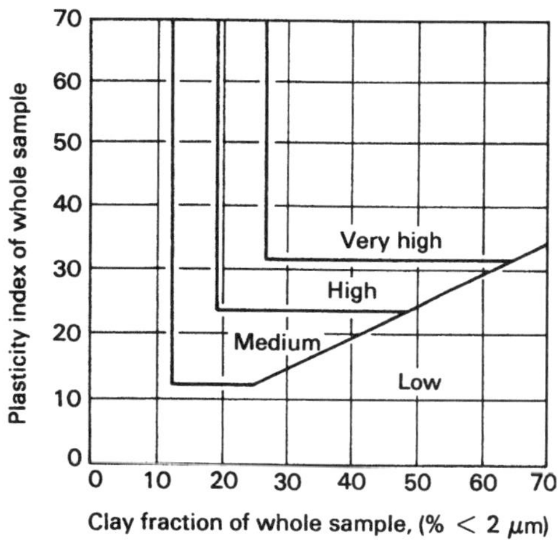

Figure 14.2 Correlation between soil classification properties and swelling potential (after van der Merwe, 1964).

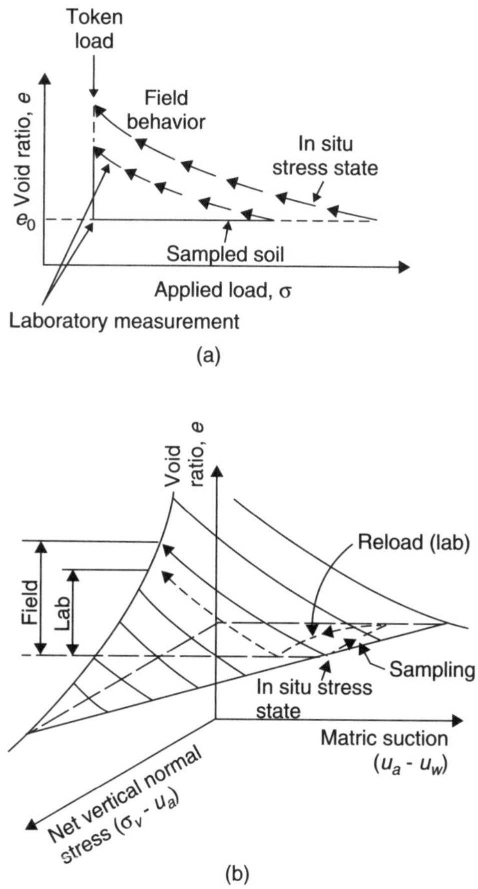

Figure 14.3 Stress path followed when using direct model method: (a) two-dimensional plot showing stress path followed in field and laboratory; (b) three-dimensional plot showing stress path followed.

manner in which the relationship between overall volumetric strain and soil suction are defined. The prediction of heave on the basis of measured matric suction values has not been extensively used in geotechnical engineering practice. The use of a direct approach to predicting total heave has been limited because of difficulties in measuring in situ matric suction and corresponding stress path soil properties. In some cases, measurements of total suction have been used in place of matric suction measurements; however, this methodology has also experienced limited success due to lack of accuracy.

The methods for heave prediction based on one-dimensional oedometer test results still appear to be the most common in engineering practice. Total heave is usually calculated for the case where matric suctions are released to zero. Overall volume change is computed in terms of changes in (equivalent) total stresses when using oedometer methods. The direct measurement of matric suction is not required when using the (equivalent) total stress change approach. There are several one-dimensional oedometer-based testing and analysis procedures that have been proposed over the years. Some of these procedures are listed below along with an interpretation of the actual stress path that is being followed in each case.

14.3.2 Oedometer-Based Methods of Testing Expansive Soils

The "direct model" method is based on a "free-swell" oedometer test on undisturbed samples (Fig. 14.3). Specimens are subjected to the overburden pressure (or the load that will exist at the end of construction) and allowed free access to water. The predicted heave is generally significantly below the actual heave experienced in the field if the soils are wetted. The stress path followed when using this test procedure is shown in Fig. 14.3b. The conventional two-dimensional manner for plotting the test data is shown in Fig. 14.3a. The underestimation of the amount of heave appears to be primarily due to a lack of consideration of the effects of disturbance experienced by the soil during sampling.

The Sullivan and McClelland method is based on a "constant-volume" oedometer test on an undisturbed sample initially subjected to the overburden pressure. Once the applied load equals the ability of the soil to undergo any further swelling, the sample is loaded and unloaded. The stress path followed in the Sullivan and McClelland method is shown in Fig. 14.4. The availability of published case histories where this method has been used is limited. It is anticipated that this method would likely underestimate actual heave since sampling disturbance has not been taken into consideration.

The "double-oedometer" method is based on the results of two oedometer tests: a "free-swell" oedometer test and a "natural water content" oedometer test. Specimens are initially subjected to a token load of 1 kPa. No water is added to the oedometer pot during the natural water content test. The natural water content oedometer test data are adjusted

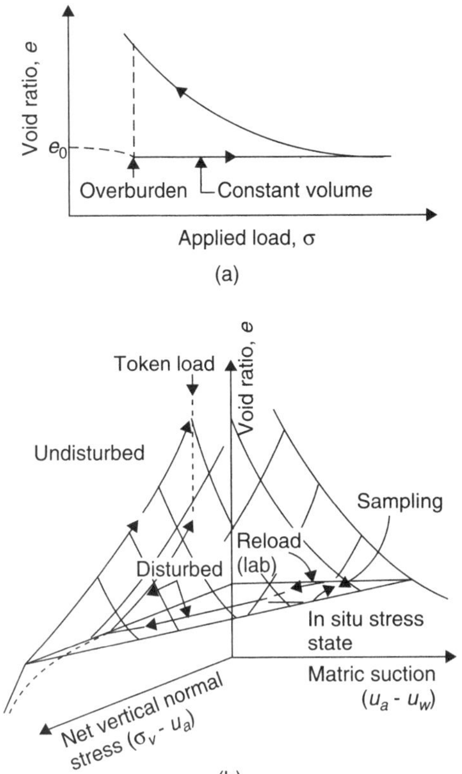

Figure 14.4 Stress path followed when using Sullivan and McClelland method: (a) two-dimensional plot depicting stress path; (b) three-dimensional plot showing stress path.

vertically to match the free-swell test results at high applied loads. Various loading conditions and final pore-water pressures can be simulated in the analysis. The stress paths followed by the two tests are shown in Fig. 14.5.

Predicted values of heave have generally been satisfactory. The method of analyzing the data appears to compensate for the effects of sampling disturbance. In other words, the natural water content compression curve provides a measure of the effect of sampling disturbance. The stress paths of more recent, updated versions of the double-oedometer method can also be visualized on similar three-dimensional plots in terms of net normal stress and matric suction.

Fredlund et al., (1980a) proposed the use of constant-volume oedometer test results in predicting total heave. It was suggested that the measured swelling pressure be corrected for sampling disturbance. A graphical technique for correcting the measured swelling pressure was proposed. The correction procedure was similar to the Casagrande construction used for determining the preconsolidation pressure of a saturated soil. Details of the correction procedure are explained later.

Each of the proposed heave prediction procedures can be plotted on a three-dimensional plot with the stress state variables used for the horizontal axes. The heave procedure described later in this chapter is consistent with the volume-mass constitutive relations for unsaturated soils (Fredlund et al., 1980a).

14.3.3 Some Factors Affecting Total Heave

Several properties of unsaturated soils have been found to influence the amount of volume change that can occur when the soil is allowed to fully imbibe water. Holtz and Gibbs (1956) summarized the effect of initial water content and dry density on the total heave of compacted expansive Porterville clay (Fig. 14.6). The heave was measured by placing the soil in an oedometer ring and submerging it in water. The specimens were subjected to a surcharge load of 7 kPa. Figure 14.6 shows that total heave increases with decreasing initial water content or increasing initial dry density. The results show that when the clay is compacted at optimum water content under standard AASHTO compaction it will

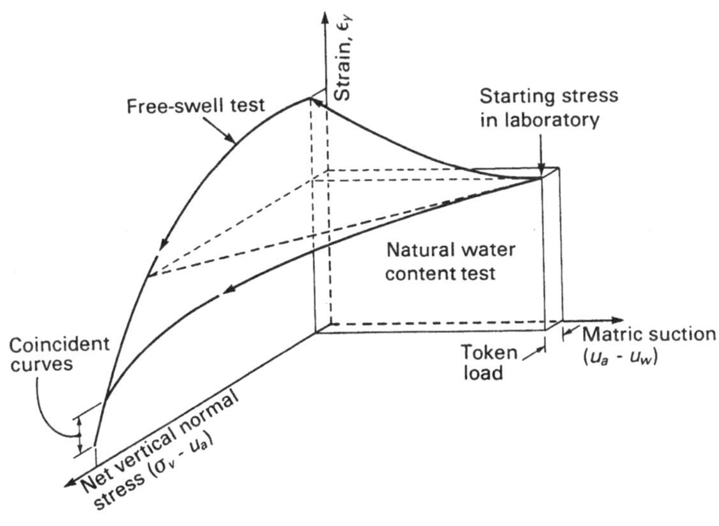

Figure 14.5 Stress paths followed when using double-oedometer method (after Jennings and Knight, 1957).

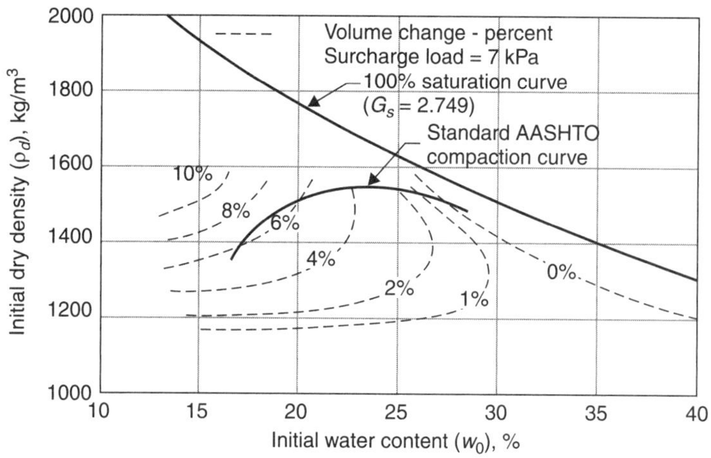

Figure 14.6 Effect of initial water content and dry density on expansive properties of compacted Porterville clay when wetted (after Holtz and Gibbs, 1956).

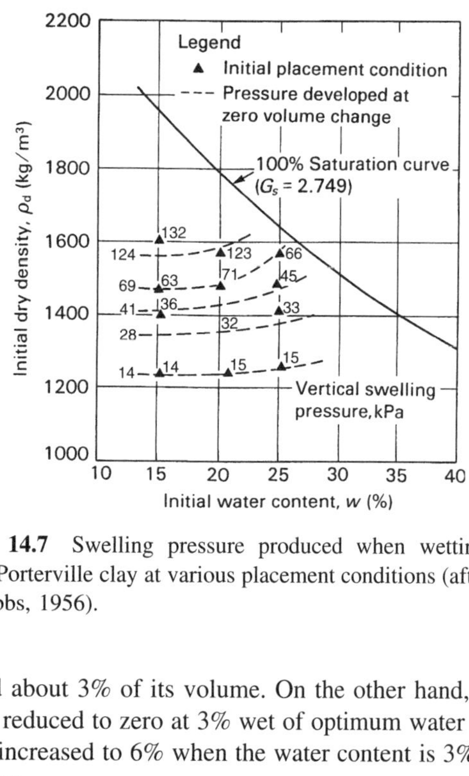

Figure 14.7 Swelling pressure produced when wetting compacted Porterville clay at various placement conditions (after Holtz and Gibbs, 1956).

expand about 3% of its volume. On the other hand, expansion is reduced to zero at 3% wet of optimum water content and is increased to 6% when the water content is 3% dry of optimum.

Figure 14.7 shows the effect of dry density and water content on vertical swelling pressure. The vertical swelling pressure was defined as the pressure developed by a specimen placed in an oedometer ring and saturated without allowing any volume change. The diagram shows that the vertical swelling pressure is more sensitive to variations in the initial dry density than it is to variations in initial water content.

Chen (1988) also studied the effect of volume-mass properties on vertical and lateral swelling pressures. The vertical swelling pressure was found to be essentially independent of the initial water content and the surcharge load. This finding is in agreement with the observations made by Holtz and Gibbs (1956), as shown in Fig. 14.7.

14.4 EVALUATION OF STRESS HISTORY IN UNSATURATED SOILS

It has been common practice since the early days of soil mechanics to retrieve undisturbed samples from a site and then perform laboratory tests to determine relevant physical properties for the soil. The geotechnical engineer has been interested in obtaining two primary pieces of information when dealing with the predictions of volume change (i.e., settlement or heave): (i) the volume change properties of the soil and (ii) the present stress state of the soil. Soils are classified as normally consolidated, overconsolidated, or underconsolidated. In each case the test results provide information on the in situ stress state of the soil.

Laboratory tests on unsaturated soils allow information to be obtained on the present stress state of the soil as well as the volume change properties. The solution to volume change problems requires that information be obtained on (i) the present in situ stress state, (ii) the final stress state, and (iii) the pathway followed between those two stress states. The final stress state is generally assumed based on the local experience of the geotechnical engineer.

Volume change problems associated with expansive soils may have geometries that are one-dimensional, two-dimensional, or three-dimensional. Geotechnical engineering practice has primarily focused on the solution of one-dimensional problems of volume change through use of laboratory results from K_0 oedometer tests. The following sections focus on the interpretation of one-dimensional oedometer test results.

The methodology for measuring and interpreting the swelling pressure of an expansive soil is of significant importance. Numerous test procedures have been proposed for the measurement of the swelling pressure of a soil. It is imperative for the geotechnical engineer to understand the relationship between the various test procedures when using the test results for solving practical swelling soil problems.

The prediction of heave in a swelling soil is used as an example problem. However, the methodology for predicting settlement and collapse in unsaturated soils is similar.

14.4.1 Past, Present, and Future States of Stress

The prediction of one-dimensional heave requires information on possible changes in the stress state and the soil property known as the swelling index C_s. Oedometer tests are commonly performed to determine the present in situ state of stress as well as the swelling index of the soil. The oedometer test results provide information on the present in situ state of stress which is translated onto the net normal stress plane. The in situ total stress in the soil is referred to as the "corrected" swelling pressure.

The corrected swelling pressure represents the sum of the overburden pressure and the "matric suction equivalent." The matric suction equivalent is the matric suction of the soil translated onto the net normal stress plane. Therefore, the oedometer test measures the in situ state of stress (on the net normal stress plane) without having to measure the individual components of total stress and matric suction.

14.4.2 Stress State History

The development of an expansive soil over time can be visualized in terms of changes in the stress state in the lacustrine deposit. Changes in stress state occur during geological deposition, erosion, and environmental changes related to precipitation, evaporation, and evapotranspiration. The following example illustrates the stress history of preglacial lake sediments that have evolved to become an expansive soil over geological time.

Consider a preglacial lake deposit that was initially consolidated under its own self-weight. The drainage of the lake and the subsequent evaporation of water from the lake sediments result in the drying or desiccation of the sediments. The term "desiccation" refers to the drying of soils by evaporation and evapotranspiration. The water table is drawn below the ground surface over a period of time. As a result, the pore-water pressures above the water table decrease and become negative in value while the total stress in the deposit remains almost constant. In other words, consolidation occurs as the effective stresses in the soil increase. The negative pore-water pressures act in all directions (i.e., isotropically), resulting in a tendency for the soil to crack. With time the water table recedes and overall desaturation occurs in the upper portion of the profile (Fig. 14.8).

The soil is further desiccated as a result of the growth of grass, trees, and other plants on the ground surface. Most plants are capable of applying as much as 1000–2000 kPa of tension on the water phase prior to reaching the wilting point of the plants. A high tension in the water phase (i.e., high matric suction) causes the soil to have a high affinity or thirst for water (Fig. 14.8a).

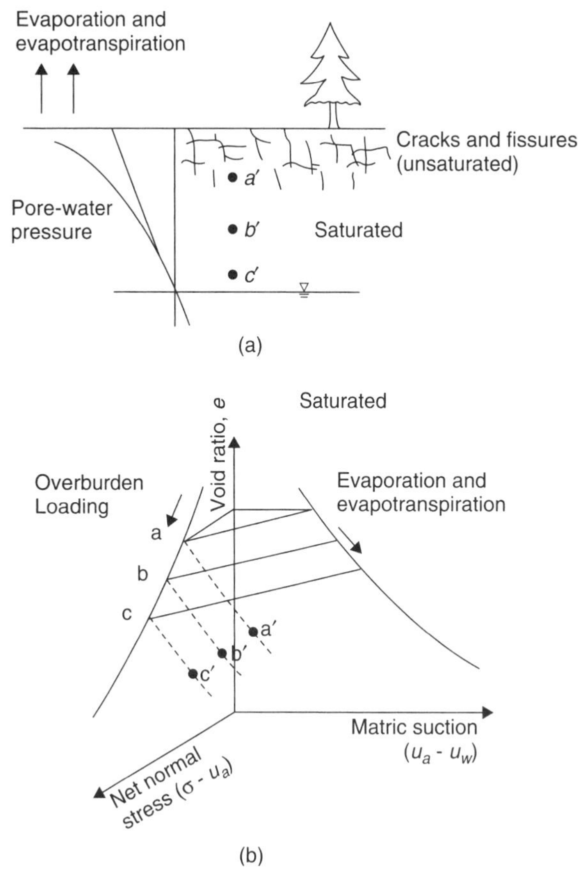

Figure 14.8 Stress state representation of lacustrine sediments subjected to evaporation and evapotranspiration: (a) pore-water pressures during drying of lacustrine deposit; (b) stress path followed during drying of lacustrine deposit.

The surface deposit can also be subjected to varying environmental conditions. Changing water fluxes (i.e., wetting and drying) at the surface results in swelling and shrinking of the upper portion of the deposit. Seasonal volume changes might extend to depths in excess of 3 m, causing the surface deposit to be highly desiccated.

Figure 14.9 illustrates the changes in the stress state of the surface deposit during wetting and drying associated with water infiltration and water evaporation at ground surface, respectively. Stress state changes occur mainly on the matric suction plane under almost constant net normal stresses. The stress paths followed during the wetting and drying processes can be illustrated as a series of hysteresis loops on the soil suction plane.

The natural water content of a deposit in arid and semiarid regions tends to decrease gradually over time. Low-water-content conditions in an unsaturated clay deposit are the first indication that the soil may have a high swelling potential. Unsaturated soils with a high swelling index C_s in a changing moisture flux environment are referred to as highly

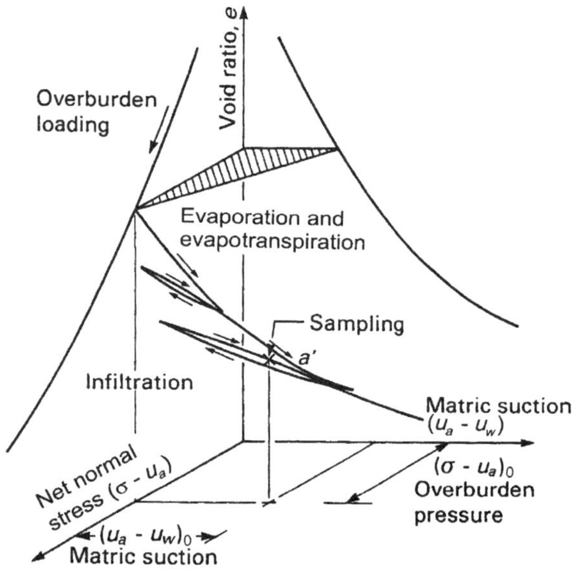

Figure 14.9 Representation of in situ stress state when soil has undergone complex history of drying and wetting.

swelling and shrinking soils (i.e., expansive soils or high-volume-change soils).

14.4.3 In situ Stress State

The prediction of heave requires information on the present in situ stress state and possible future stress states. The difference between the present and future stress states is one of the main variables controlling the amount of potential volume change or heave. Consequently, it is important that the present in situ stress state be accurately assessed. It is necessary to correct laboratory test results for the effect of sampling disturbance when using oedometer test results to assess the present in situ stress state.

The one-dimensional oedometer test procedure appears to be satisfactory for analyzing most swelling soil problems encountered in geotechnical engineering practice. However, the oedometer tests must be performed and interpreted in a proper manner in order to assess the in situ stress state of the soil.

It is difficult to obtain representative, undisturbed soil samples when a soil mass is highly fissured and cracked. Under these conditions it is prudent to make modifications to a conventional heave analysis. There are important three-dimensional effects that must be taken into consideration.

The in situ state of stress is somewhere along either the wetting or drying portion of the void ratio–stress state variable relationship when a soil is sampled for laboratory testing (Fig. 14.9). The soil in the field has been subjected to numerous cycles of wetting and drying. The soil has a specific net normal stress and matric suction at the time of sampling and the geotechnical engineer must attempt to interpret the stress state of the soil on the basis of the laboratory test results.

The laboratory information desired by the geotechnical engineer for predicting the amount of heave is an assessment of (1) the in situ state of stress and (2) the swelling properties with respect to changes in net normal stress. An understanding of the compressibility properties under increased loading conditions may also be relevant. A demanding testing facility and program would be required to completely assess all possible variables affecting volume change behavior in expansive soils. It is important, however, to develop relatively simple, rapid, and economical procedures to obtain the basic information for calculating potential heave.

Several laboratory testing procedures have been used in engineering practice to obtain the necessary information for the prediction of heave. The one-dimensional oedometer test is commonly used to determine the present in situ state of stress. Free-swell and the constant-volume oedometer test methods can be used along with associated correction procedures. There are several oedometer test procedures that have been proposed for the measurement of the swelling pressure of a soil. The constant-volume and free-swell test procedures are first described followed by a presentation of several comparative studies of swelling pressures measured using other test procedures.

The oedometer test translates the in situ stress state onto the net total stress plane. The in situ stress state is referred to as the "corrected" swelling pressure P_s', which is equal to the sum of the overburden pressure and the matric suction equivalent. Figure 14.10 shows typical laboratory oedometer test results extending to the virgin compression branch. The entire laboratory loading curve is often in its entirety on the recompression portion with applied loads not even reaching the preconsolidation pressure of the soil. In other words, the preconsolidation pressure of many desiccated soils, P_c, may exceed the loading capacity of the oedometer apparatus.

The preconsolidation pressure refers to the maximum stress to which the soil has come to equilibrium in the past. Figure 14.10 also shows the relative position of the corrected swelling pressure P_s' to the preconsolidation pressure P_c. The corrected swelling pressure P_s' is located approximately at the intersection between the constant-void-ratio line and a line representing the recompression branch. In turn, the preconsolidation pressure P_c is located at the intersection between the recompression branch and the virgin compression branch of the soil.

Methods for determining the preconsolidation pressure have been described in many soil mechanics textbooks and are not repeated herein. The overconsolidation ratio (OCR) for an unsaturated, expansive soil can be defined as the ratio of the preconsolidation pressure P_c to the corrected swelling pressure P_s'. The corrected swelling pressure P_s' is a representation of the in situ stress state of the soil:

$$\text{OCR} = \frac{P_c}{P_s'} \tag{14.13}$$

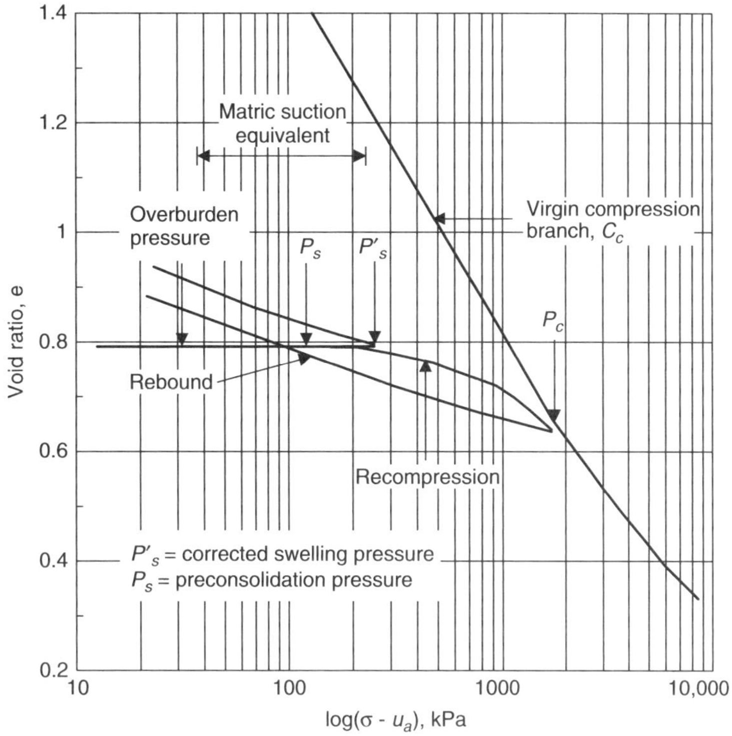

Figure 14.10 Relationship between corrected swelling pressure and preconsolidation pressure of a soil.

The above equation is a modification of the common definition of the OCR associated with saturated soils. The OCR is usually defined as the ratio of the preconsolidation pressure P_c to the in situ vertical effective overburden pressure σ_v' (i.e., $\sigma_v - u_w$, where σ_v = total overburden pressure and u_w = pore-water pressure). The use of the corrected swelling pressure P_s' eliminates the need to measure the in situ negative pore-water pressure when defining OCR. Equation 14.13 provides an indication of the overconsolidation of the soil with respect to the in situ stress state on the net normal stress plane (i.e., P_s'). At the same time, it is recognized that Eq. 14.13 is a deviation from the classic definition of OCR.

14.4.4 Future Stress State and Ground Movements

The future stress state corresponding to several years after construction must be estimated based upon local experience and climatic conditions. There are several possible assumptions that can be made regarding possible future suctions in the soil. There are general estimations that can be made regarding the soil suction profile in the future and there are also more rigorous analyses that can be performed by taking into consideration the climatic conditions measured at nearby weather stations. Local watering practices could also be taken into consideration.

Changes in total stress can occur as a result of excavation, replacement with a relatively inert material (e.g., gravel), and loadings from structures. The effects of these changes can be taken into account using appropriate volume change indices for loading and unloading. It may also be possible to assume that there is insufficient time for the soil to respond to the individual loading and unloading stages. In this case, the long-term volume change calculations can be computed based on the final net loading or unloading conditions.

Let us assume that the final pore-water pressures in the soil go to zero under a constant net normal stress. Figure 14.11 shows the actual stress path that would be followed by a soil element at the depth from which the sample was retrieved. Swelling would follow a path from the initial void ratio e_0 to the final void ratio e_f along the rebound surface on the matric suction plane. The entire rebound surface can be assumed to be unique since the direction of deformation is monotonic (Matyas and Radhakrishna, 1968; Fredlund and Morgenstern, 1976). It is also possible to follow a stress path from the in situ stress state to the corrected swelling pressure and then proceed along the rebound curve on the net normal stress plane to the final stress condition. The advantage of the latter stress path (i.e., through the corrected swelling pressure) is that the volume change indices determined on the total stress plane can be used to calculate total heave.

14.4.5 Determination of In Situ Stress State Using Oedometer Test Results

One-dimensional oedometer tests are commonly used for the assessment of the in situ stress state and the swelling properties of expansive, unsaturated soils. The oedometer

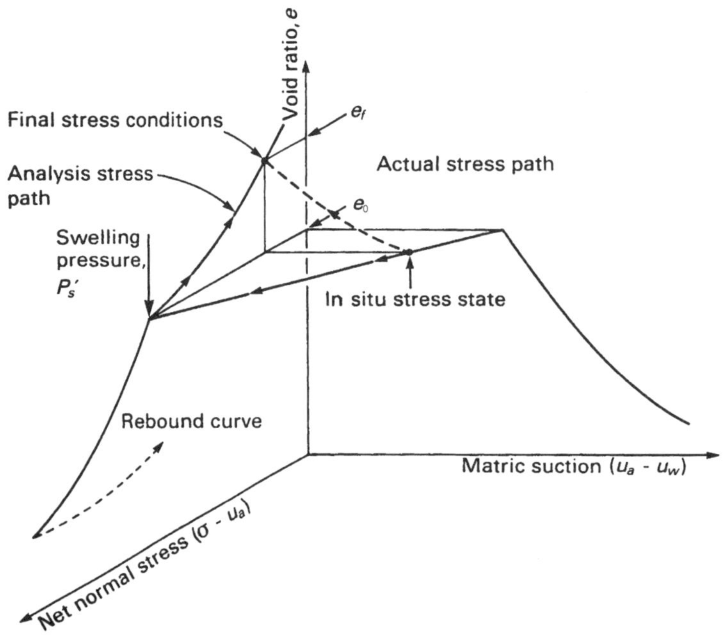

Figure 14.11 "Actual" or field stress path and "analysis" stress paths followed during wetting of a soil.

can only perform laboratory testing in the total stress plane. The assumption is made that it is possible to eliminate the matric suction from the soil and still obtain the necessary soil properties and stress state values from the total stress plane. The interpretation of free-swell and constant-volume tests are first discussed followed by information on other test procedures that have been used for the measurement of swelling pressure.

14.4.6 Constant-Volume Test

The constant-volume oedometer test (Fig. 14.12) involves first subjecting a soil specimen to a token load and then submerging the soil in water. The result is a release of the negative pore-water pressure and a tendency for the soil specimen to swell. As the soil specimen attempts to swell, the load applied to the soil is increased to maintain the specimen at a constant volume. This procedure is continued until the specimen exhibits no further tendency to swell. The applied load at this point is referred to as the "uncorrected swelling pressure" P_s. The specimen is then further loaded and unloaded in the conventional manner used for a consolidation test. The term constant volume simply refers to the initial conditions to which the soil specimen is subjected as matric suction is released to zero stress.

The laboratory results from a constant-volume oedometer test are generally plotted as shown in Fig. 14.12. The actual stress paths followed during the test can be more clearly visualized through use of a three-dimensional plot with each of the stress state variables forming abscissas (Fig. 14.13). It is important to understand the stress paths with respect to both void ratio and water content when attempting to

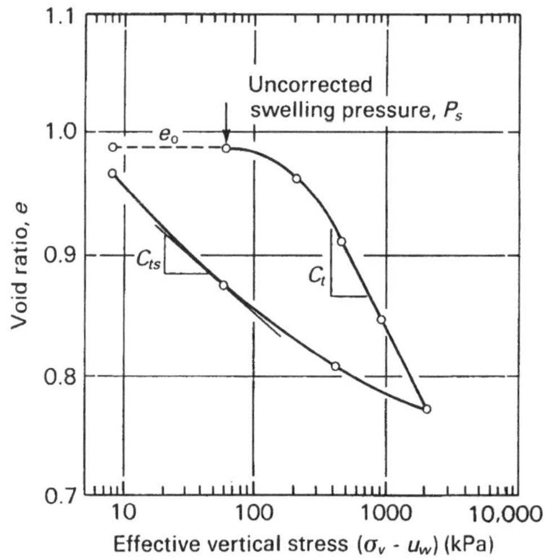

Figure 14.12 Typical constant-volume one-dimensional K_0 oedometer test results.

interpret the laboratory test results. The void ratio and water content stress paths are shown for the situation where there is minimal disturbance due to sampling (i.e., due to stress reversals). Even so, the loading path displays some curvature as the total stress plane is approached. The actual stress path can be even more affected as a result of sampling disturbance (Fig. 14.14).

Geotechnical engineers have long recognized the effect of sample disturbance when determining the preconsolidation pressure for saturated clay. It is impossible for the soil specimen to return to its in situ stress state after sampling without displaying some curvature along the void ratio–effective

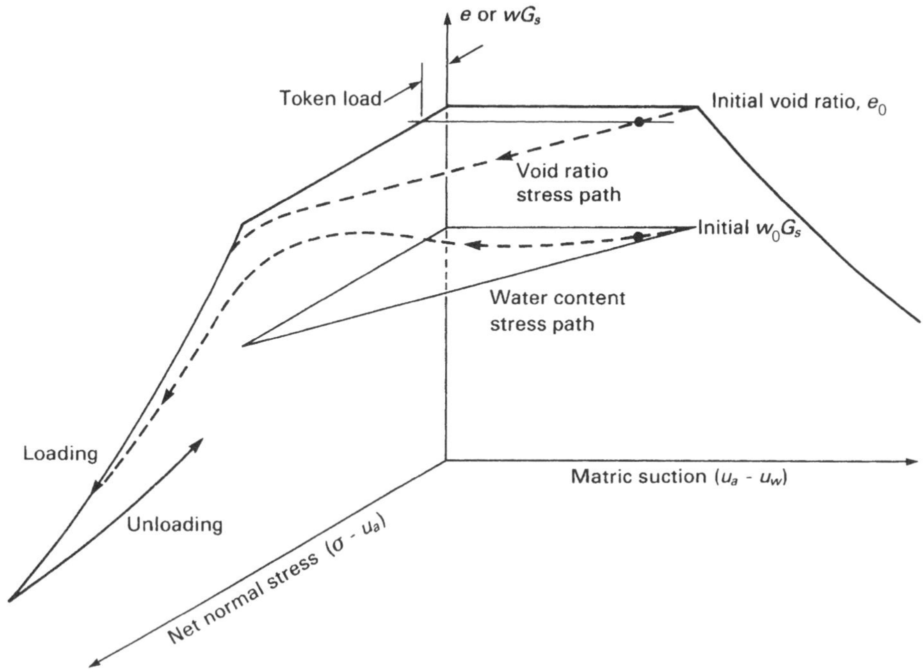

Figure 14.13 "Ideal" stress path adhered to during constant-volume oedometer test.

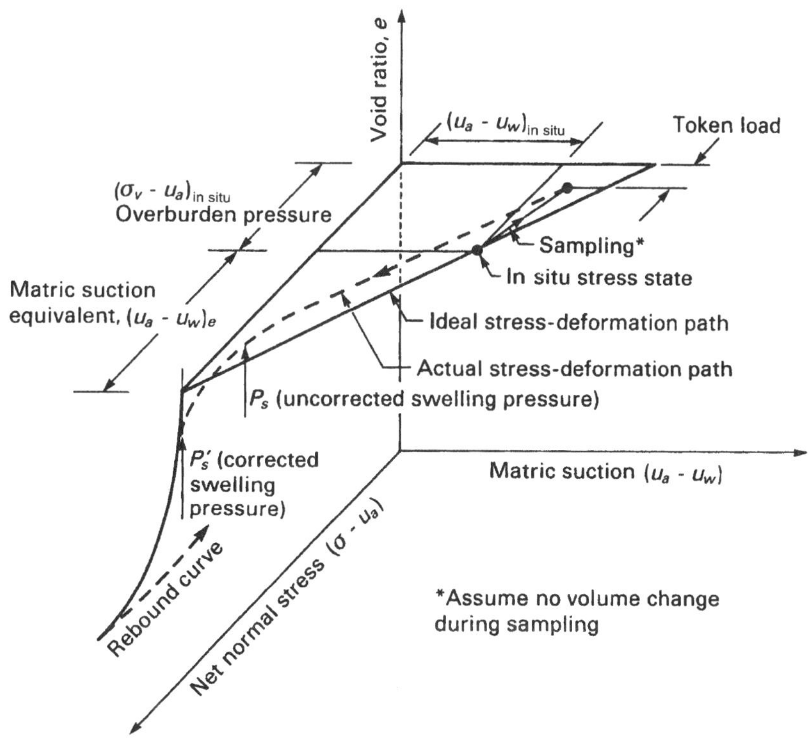

Figure 14.14 Ideal and actual stress paths showing effect of sampling disturbance.

stress plot (i.e., compression curve extending onto the virgin compression branch). It has also been recognized that sampling disturbance significantly affects the measurement of swelling pressure (Fredlund et al., 1980a).

Sampling disturbance causes the conventional measurement of swelling pressure P_s to fall below the actual or proper representation of the in situ swelling pressure P_s'. The corrected swelling pressure better represents the in situ stress state translated onto the total stress plane. The corrected swelling pressure should be equal to the overburden pressure plus the in situ matric suction translated onto the total stress plane. The translated in situ matric suction is called the "matric suction equivalent" $(u_a - u_w)_e$ (Yoshida et al., 1983). The magnitude of the matric suction equivalent will be equal to or lower than the in situ matric suction. The difference between the in situ matric suction and the matric suction equivalent is primarily a function of the degree of saturation of the soil. The engineer desires to obtain the corrected swelling pressure from an oedometer test in order to reconstruct the in situ stress conditions. The interpretive procedure used to account for sampling disturbance is discussed later.

14.4.7 Free-Swell Test

The soil specimen in a free-swell oedometer test is initially allowed to swell freely with only a token load applied (Fig. 14.15). The term free swell simply refers to the initial conditions to which the soil specimen is subjected as matric suction is released to zero. The load required to bring the specimen back to its original void ratio is termed the swelling pressure.

The stress paths followed by the soil can best be understood from a three-dimensional plot of the stress variables versus void ratio and water content, as shown in Fig. 14.16. The free-swell test has the limitation that it allows volume

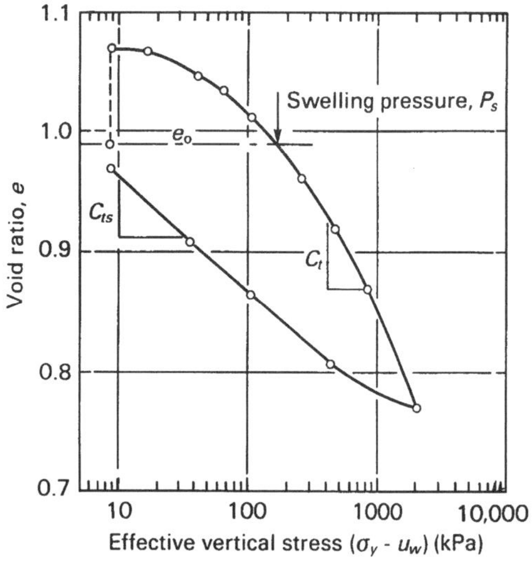

Figure 14.15 Typical free-swell one-dimensional K_0 oedometer test results.

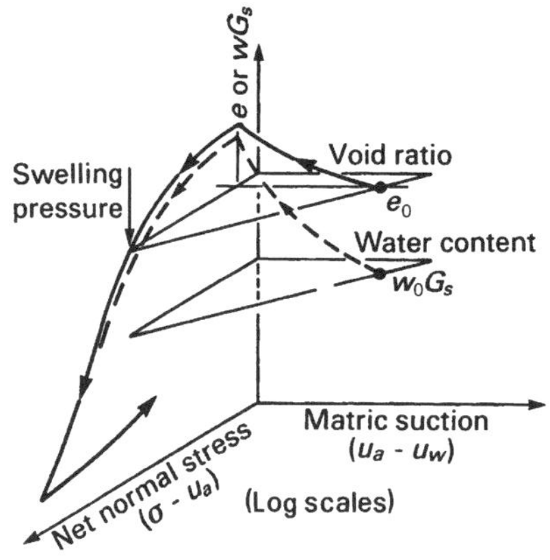

Figure 14.16 Stress path followed for free-swell type of oedometer test.

change and incorporates hysteresis into the estimation of the in situ stress state. On the other hand, the test procedures tend to compensate for the effects of sampling disturbance. No empirical corrections need to be applied for "sampling disturbance" when using the free-swell test to measure swelling pressure.

14.4.8 Correction for Compressibility of the Oedometer Apparatus

The oedometer was originally designed and intended for testing highly compressible clay soils. The compressibility of components of the oedometer was of little concern when testing highly compressible clay soils. However, unsaturated, swelling soils generally exhibit low compressibility and the compressibility of the oedometer can significantly affect the interpretation of the laboratory test results. The compressibility of the apparatus affects procedures used for testing swelling soils; however, it is the constant-volume test procedure that is most significantly affected by apparatus compressibility.

The following procedure is suggested for taking the compressibility of the oedometer into account and obtaining the corrected swelling pressure from constant-volume test results. Details of the testing procedure are presented in ASTM D4546. Laboratory data need to be adjusted to account for the compressibility of the components of the oedometer apparatus. Desiccated, swelling soils have a low compressibility, and the compressibility of the apparatus can significantly affect the evaluation of in situ stresses as well as the slope of the rebound curve (Fredlund, 1969).

The compressibility of the oedometer can be measured by placing a steel plug as a substitute for the soil specimen. The measured deflections under applied loads with the steel plug should be subtracted from the deflections measured when testing the soil specimen. Figure 14.17 shows the manner in which an adjustment should be applied to the laboratory data. The adjusted void ratio versus pressure curve can be sketched by drawing a horizontal line from the

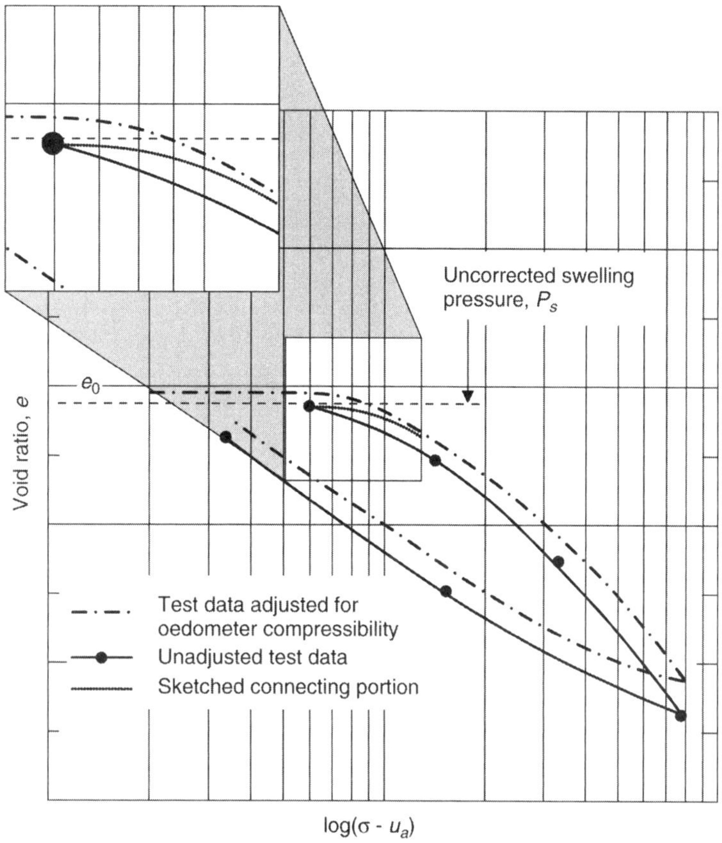

Figure 14.17 Adjustment of laboratory test data to compensate for compressibility of oedometer apparatus.

initial void ratio. The horizontal line should then gradually curve downward and join the recompression curve adjusted for the compressibility of the apparatus.

14.4.9 Correction for Sampling Disturbance

A correction for sampling disturbance can be applied to the measured compression curve that has been adjusted for compressibility of the oedometer. Sampling disturbance increases the compressibility of the soil. The increase in compressibility is most clearly observed when loading goes from the in situ void ratio onto the recompression curves and when the loading goes from the recompression curve onto the virgin compression branch. Sampling disturbance does not allow the laboratory specimen to return to its in situ state of stress at its in situ void ratio.

Casagrande (1936) proposed an empirical construction that can be applied to the laboratory-measured compression curve for the determination of preconsolidation pressure. The empirical construction was necessary in order to account for the effect of sampling disturbance. Other construction procedures have also been proposed for the determination of preconsolidation pressure (Schmertmann, 1955). A modification of Casagrande's construction is suggested for determining the corrected swell pressure when the constant-volume test procedure is used.

The corrected swelling pressure can be determined using the following procedure. Locate the point of maximum curvature where the void ratio versus pressure curve bends downward onto the recompression branch (Fig. 14.18). It is also suggested that the point of maximum curvature be located as the bisector of a horizontal line through the initial void ratio and the slope of the recompression curve, as shown in Fig. 14.19. A horizontal line and a tangential line are drawn at the point of maximum curvature. The corrected swelling pressure is designated as the intersection of the bisector of the angle formed by the horizontal and tangential lines and a line tangential to the recompression curve. The recompression line can be determined from the adjusted rebound curve and drawn tangent to the loading curve.

The need to apply a sampling disturbance correction to the constant-volume swelling pressure measured in the laboratory is revealed in several ways. First, it would be anticipated that such a correction is necessary as a result of early soil mechanics experiences in determining the preconsolidation pressure for saturated clay soils. Second, attempts to use swelling pressure meaurements that have not been corrected for sampling disturbance have resulted in predictions of total heave that are too low. Predictions of heave made using corrected swelling pressures are often as much as twice

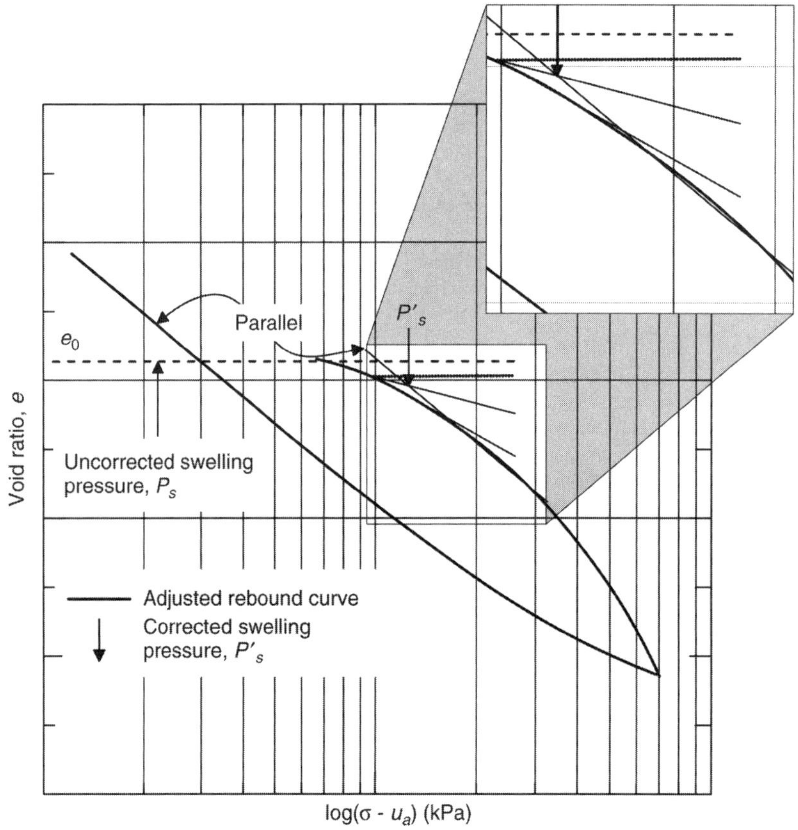

Figure 14.18 Construction procedure suggested to compensate for sampling disturbance.

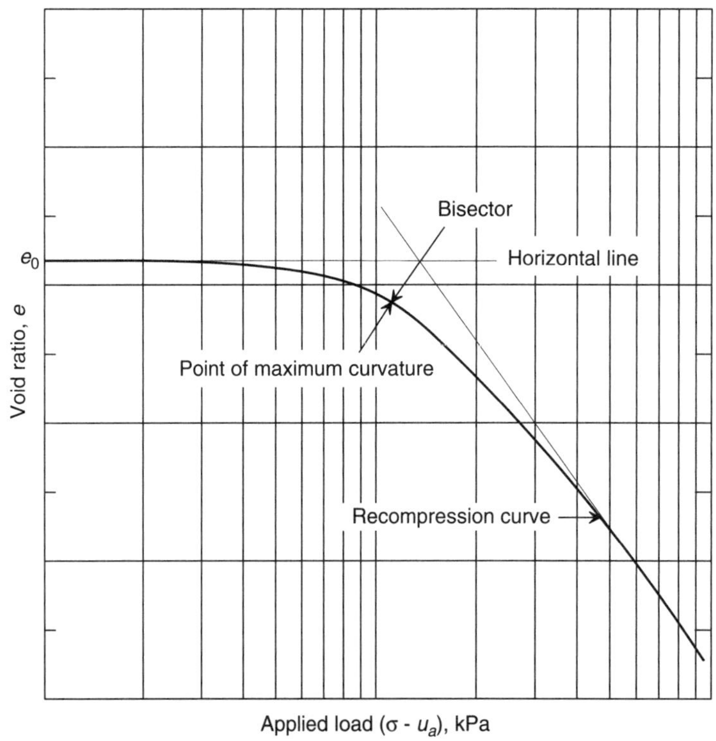

Figure 14.19 Empirical construction to estimate point of maximum curvature on compression curve.

as large as those computed when no correction is applied. Third, the analysis of a series of oedometer test results measured on desiccated highly plastic clay in a semiarid environment showed swelling pressures that were unrealistically low when the correction for sampling disturbance was not applied. The average results from 34 constant-volume oedometer tests on Regina clay are shown in Figure 14.20. The deposit is of preglacial lacustrine origin with natural water contents near the plastic limit of 25% (Fredlund et al., 1980a). The average liquid limit is 75%. There is no evidence of a regional groundwater table in the deposit. The soil is very stiff and is known to have high swelling potentials. The uncorrected swelling pressures are only slightly higher than the average overburden pressure. The oedometer results uncorrected for sampling disturbance would suggest that the soil is low-swelling clay. However, swelling problems are common with a total heave on the order of 5–15 cm. Samples from depths deeper than 5.5 m often show uncorrected swelling pressures less than the overburden pressure when using the constant volume test procedure. In other words, the correction for sampling disturbance is imperative when interpretating the laboratory oedometer results.

Figure 14.21 shows a comparison of corrected and uncorrected swelling pressure data from two soil deposits. The results indicate that it is possible for the corrected swelling pressures to be more than 300% larger than the uncorrected swelling pressures.

14.4.10 Comparative Swelling Pressure Studies Using Various Surcharge Pressures

The laboratory measurement of swelling pressure has been the focus of several research studies. Gilchrist (1963) prepared compacted samples of Regina clay at various initial water contents and densities. Table 14.3 lists the compaction water contents and initial void ratios used for six series of swelling tests. In addition to measuring the swelling pressures using the constant-volume and free-swell testing procedures, tests were also performed by allowing swelling from various applied surcharge loads. The ranges of measured swelling pressures are shown in Table 14.3. Swelling pressures were determined as the load required to return the soil sample to the original void ratio. Corrections were applied for the compressibility of the oedometer apparatus.

Figure 14.22 shows the swelling pressure measurements and compression curves when the soil was compacted at 22% water content to a dry unit weight of 16.0 kN/m^3. The measured swelling pressures ranged from 1570 to 2160 kPa. In general, the free-swell test performed with a low token applied pressure resulted in the highest measured swell pressures. Swelling under increased surcharge loads reduced the measured swelling pressures.

The soil unit weight was decreased and the as-compacted water content was increased for each succeeding series of tests. The measured swelling pressures decreased in

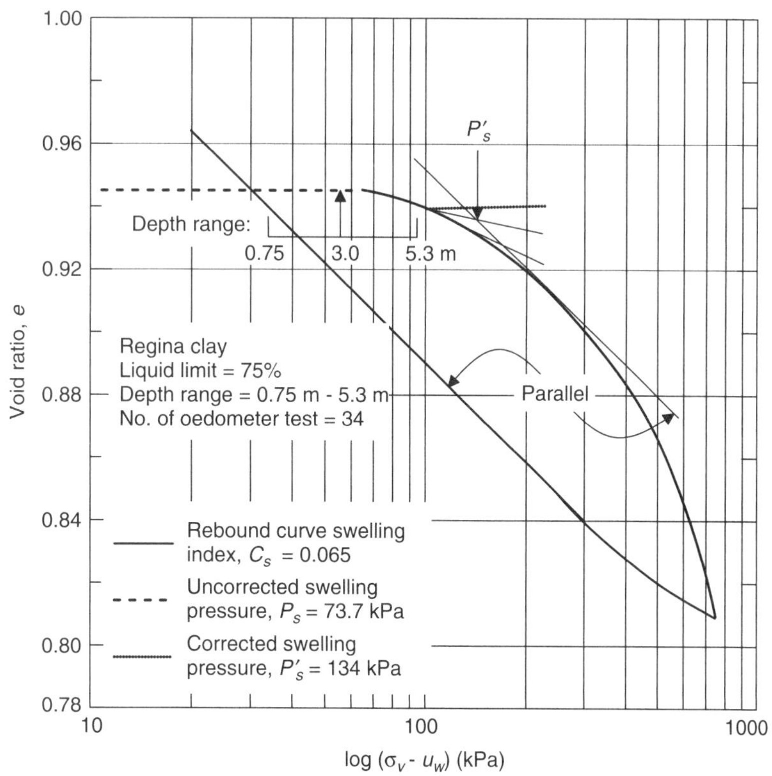

Figure 14.20 Average compression data from oedometer test on Regina clay showing need for applying correction to measured swelling pressure.

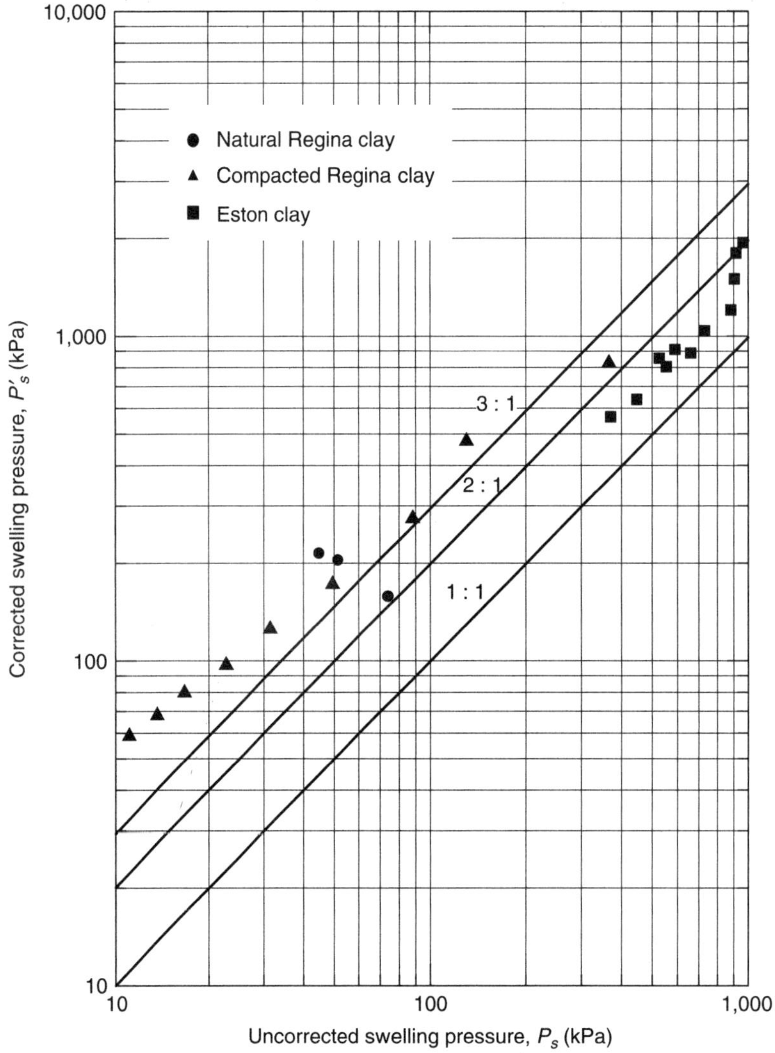

Figure 14.21 Change to measured swelling pressure as result of applying correction for sampling disturbance.

Table 14.3 Volume-Mass Properties and Swelling Pressures for Series of Oedometer Tests on Regina Clay

Series	Void Ratio	Dry Unit Weight, kN/m^3	Water Content, %	Swelling Pressure Range, kPa
1	0.67	16.0	22	1570–2160
2	0.79	14.9	27	640–980
3	0.94	13.8	32	200–350
4	1.05	13.0	37	120–180
5	1.29	11.7	45	30–80
6	1.48	10.8	51	15–30

Source: From Gilchrist (1963).

each succeeding series of tests. Figures 14.23–14.27 show the swelling pressure measurements and compression curves as the soil was compacted to lower densities while having initial higher water contents. Figures 14.24–14.27 also show the slope of the rebound curves. The slope of the rebound curves, C_s, was approximately 0.12. A line drawn through the void ratios corresponding to "after-free-swell" conditions under various surcharge loads results in a slightly steeper rebound curve.

Singhal (2010) performed a similar set of comparative swell pressure tests on expansive clay in Arizona (Fig. 14.28). The soil had a plastic index of 52. The load-back swell pressure is the pressure required to return the soil to its initial void ratio after the soil has swelled under its initial surcharge pressure. The load-back swelling pressures ranged from 410 to 480 kPa.

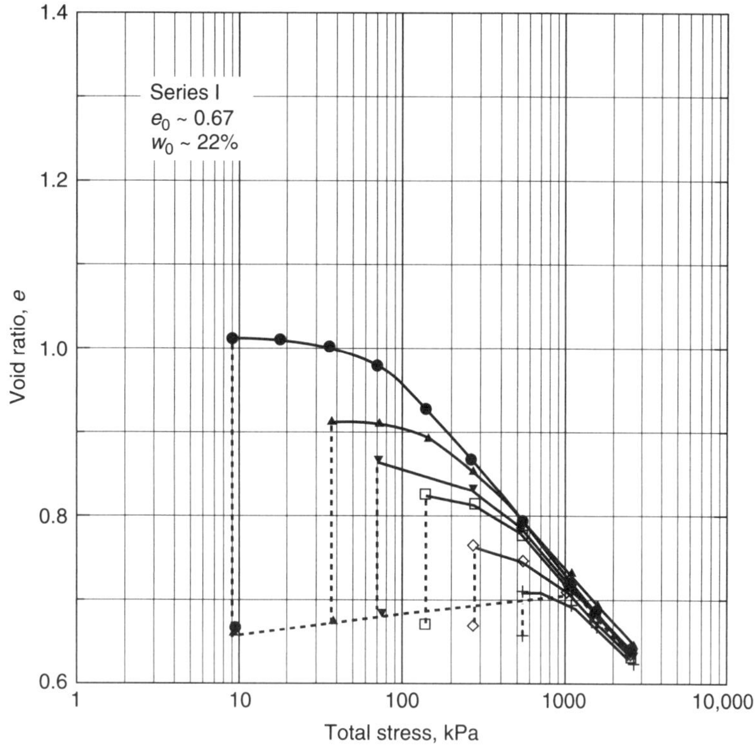

Figure 14.22 Swelling pressures measured under swell from various surcharge loads ($e_0 = 0.67$, $w = 22\%$) (after Gilchrist, 1963).

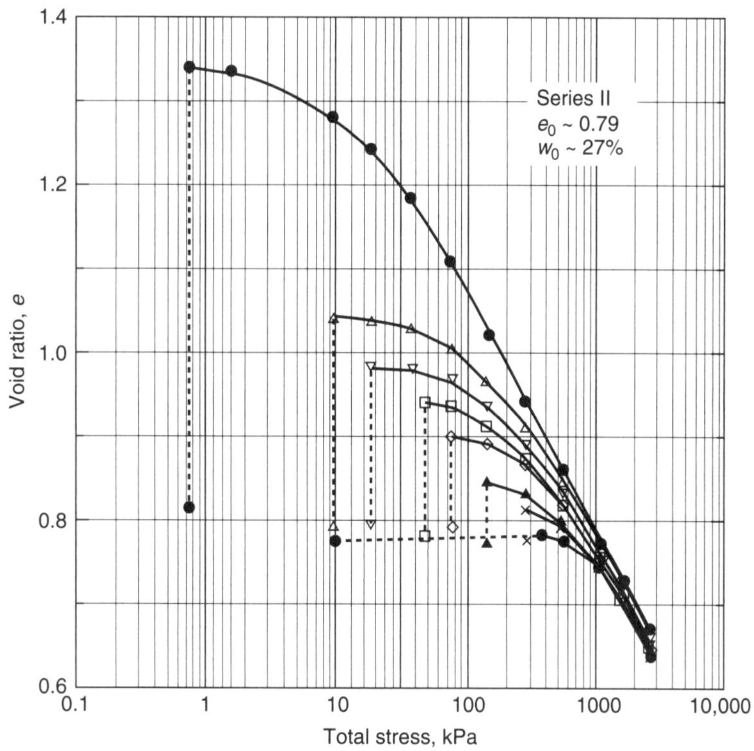

Figure 14.23 Swelling pressures measured under swell from various surcharge loads ($e_0 = 0.79$, $w = 27\%$) (after Gilchrist, 1963).

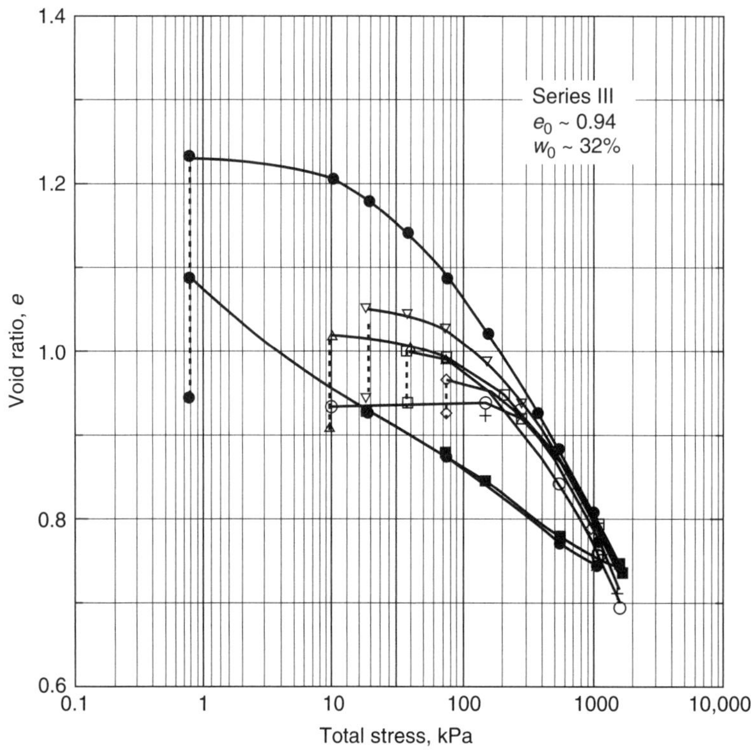

Figure 14.24 Swelling pressures measured under swell from various surcharge loads ($e_0 = 0.94$, $w = 32\%$) (after Gilchrist, 1963).

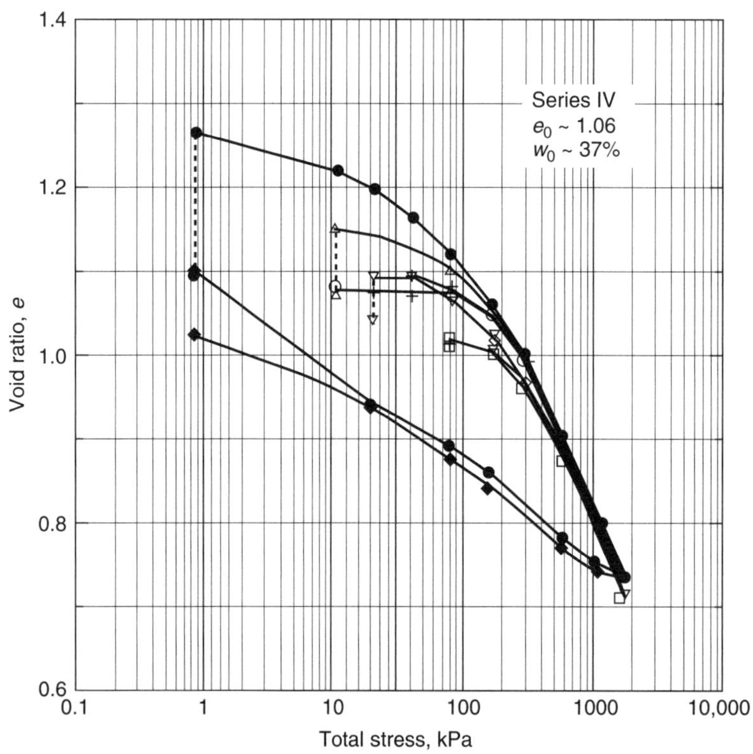

Figure 14.25 Swelling pressures measured under swell from various surcharge loads ($e_0 = 1.06$, $w = 37\%$) (after Gilchrist, 1963).

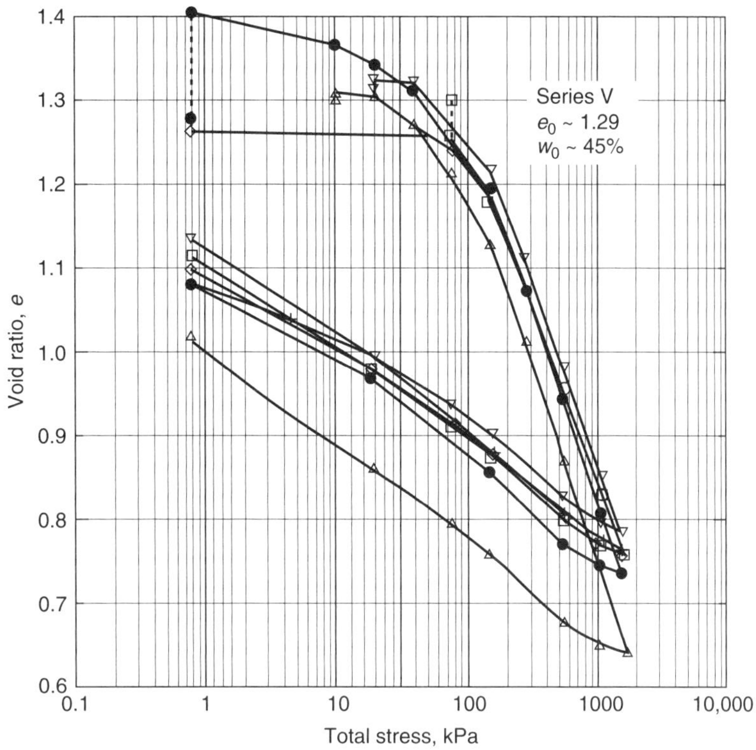

Figure 14.26 Swelling pressures measured under swell from various surcharge loads ($e_0 = 1.29$, $w = 45\%$) (after Gilchrist, 1963).

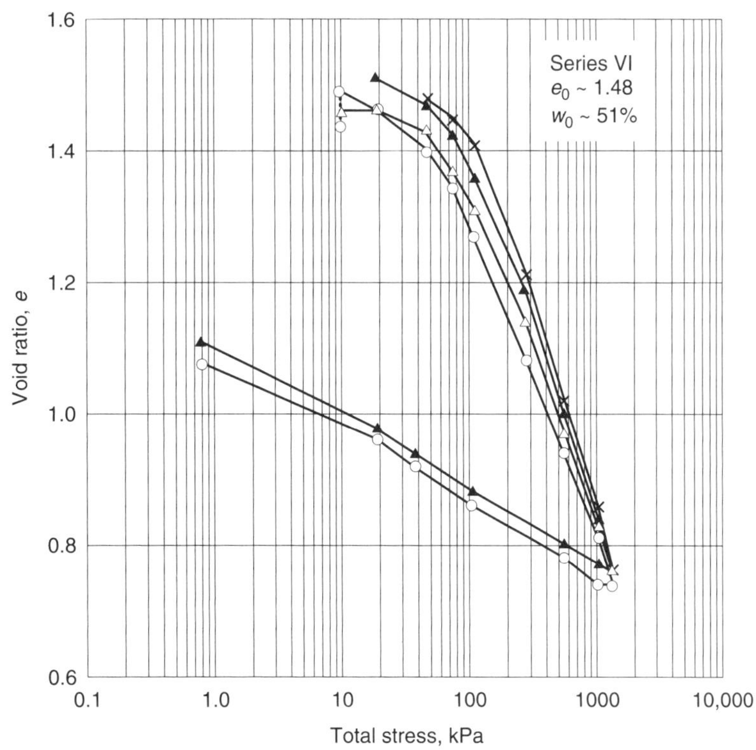

Figure 14.27 Swelling pressures measured under swell from various surcharge loads ($e_0 = 1.48$, $w = 51\%$) (after Gilchrist, 1963).

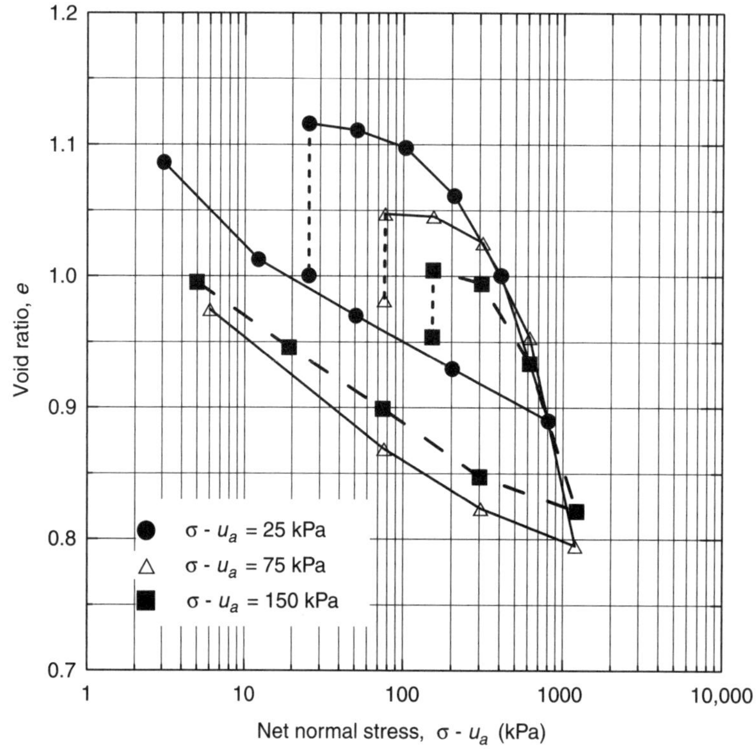

Figure 14.28 Swelling pressures measured under swell from various surcharge load of 25, 75, and 150 kPa (after Singhal, 2010).

14.4.11 Swell Pressure Measurements Using Other Test Procedures

Feng et al., (1998) performed a series of oedometer tests on expansive Regina clay. Four different test procedures were used: (1) "loading after swell," (2) "swell under load," (3) "constant volume," and (4) "unloading."

14.4.11.1 Loading after Swell Procedure

The soil specimens were subjected to a vertical surcharge pressure at initial water content conditions for the loading-after-swell test procedure. A range of surcharge pressures were selected and a specimen was prepared for each surcharge pressure. Water was added to each soil specimen once the surcharge was applied and equilibrium had been reached. When swelling of the specimen was complete, the applied vertical pressure was increased in increments until the specimen was compressed to a height less than its original value.

The free-swell test is a special case of the loading-after-swell procedure where the surcharge pressure is placed on the soil specimen. Two definitions of swelling pressure can be found in the literature related to this test procedure (Chen, 1988; Justo et al., 1984; and ASTM, 1986). In one case the swelling pressure is defined as the pressure to compress the specimen to its original void ratio (i.e., P_{1-1}). In the other case, the swell pressure is the pressure to compress the specimen to the void ratio at the end of consolidation under various surcharge pressures, P_{1-2}. In the case of the free swell-test, the two swelling pressures are the same (i.e., $P_{1-1} = P_{1-2}$).

Five specimens were tested under different surcharge pressures. The surcharge pressures were 5, 50, 100, 200, and 400 kPa, respectively (Fig. 14.29). The measured swelling pressures P_{1-1} ranged from 550 to 710 kPa when the P_{1-1} definition was used to determine swelling pressure. The measured swelling pressures were 710–790 kPa when the P_{1-2} definition was used to determine the swelling pressure.

14.4.11.2 Swell-under-Load Procedure

Several identical specimens were prepared and subjected to different surcharge pressures by Feng et al. (1998). The void ratios at the completion of consolidation under the surcharge load were plotted to produce the "before-inundation" curve. The consolidated specimens were then inundated and allowed to swell (or compress) under the applied surcharge pressures. Specimens were allowed to swell after being subjected to surcharge pressures of 5, 50, 100, 200, 400, 800, 1600, and 2600 kPa. The results are shown in Fig 14.30. Two definitions of swelling pressure associated with this test procedure can be found in the literature (Justo et al., 1984; Sridharan et al., 1986).

The first definition uses the intersection of the "full-swell" curve and a horizontal line to define swelling pressure (i.e., P_{2-1}). The second definition uses the intersection of the

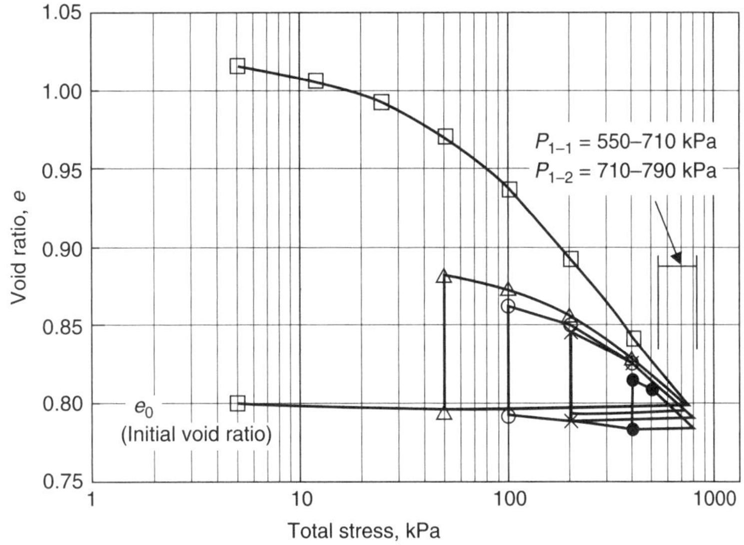

Figure 14.29 Swelling pressures measured when using loading-after-swell procedure from surcharge pressures of 5, 50, 100, 200, and 400 kPa (from Feng et al., 1998).

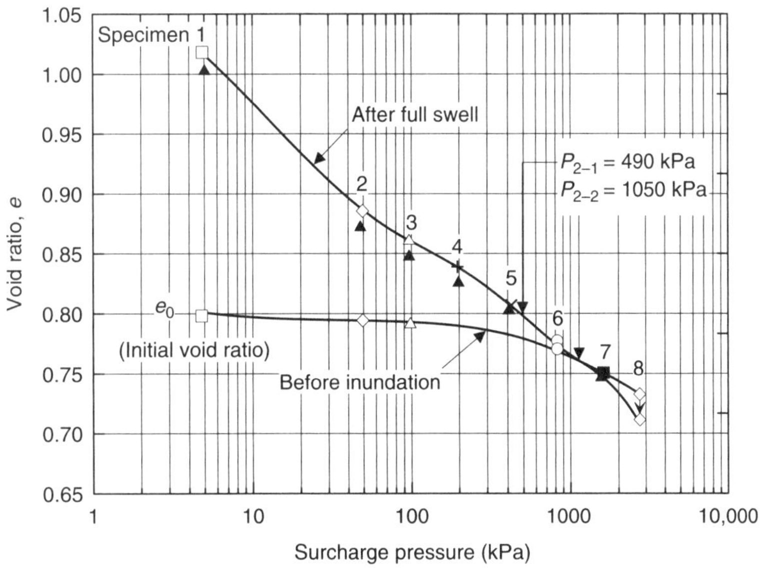

Figure 14.30 Swelling pressures measured when using swell-under-load procedure on eight specimens subjected to various surcharge pressures (after Feng et al., 1998).

full-swell curve and the before-inundation curve to define swelling pressure (i.e., P_{2-2}). The first definition (i.e., P_{2-1}) gave a swell pressure value of 490 kPa and the second definition (i.e., P_{2-2}) gave a swelling pressure value of 1050 kPa. The swelling pressure P_{2-2} would appear to be more meaningful since it corresponds to the surcharge pressure under which the soil was brought to equilibrium. The soil could not undergo any further volume change under further wetting.

14.4.11.3 Constant-Volume Test Procedure

The constant-volume tests performed by Feng et al., (1998) were described as follows. A surcharge pressure was first applied to the soil specimen (ASTM, 1986). The soil specimen was then inundated. The vertical pressure on the soil specimen was increased to prevent swelling as the specimen imbibed water. The specimen was then loaded in accordance with a conventional consolidation test procedure.

The vertical pressure corresponding to the end of wetting under constant-volume conditions was defined as swelling pressure P_{3-1} (Justo et al., 1984; Sridharan et al., 1986); Chen, 1988; ASTM, 1986). The swelling pressure could also be corrected for the effect of sampling disturbance. The correction procedure gives rise to a corrected swelling pressure P_{3-2}. The construction procedure proposed by Fredlund et al. (1980a) was applied to the test results. (*Note*: The construction procedure was described earlier in this chapter.) The corrected swelling pressure from a constant-volume oedometer test (P_{3-2}) was earlier adopted as part of the ASTM (1986) standard for testing expansive soils.

The results of the constant-volume tests are shown in Fig. 14.31. Two specimens were subjected to a surcharge of 5 kPa. The remaining four specimens were each subjected to a differing surcharge pressure of 100, 200, 400, and 600 kPa, respectively. After wetting was complete, one of the two specimens with a surcharge pressure of 5 kPa was unloaded to obtain its rebound curve. The uncorrected swelling pressures P_{3-1} ranged from 590 to 860 kPa and the corrected swelling pressures P_{3-2} ranged from 700 to 1010 kPa.

14.4.11.4 Unloading Test Procedure

Soil specimens were also tested using the unloading test procedure (Feng et al., 1998). The soil specimens were first consolidated under a surcharge pressure at the original water contents. The surcharge pressure was selected as a value that exceeded the estimated swelling pressure of the soil. The specimen was inundated and swelling was allowed to take place with the surcharge pressure applied. The specimen was unloaded in increments to zero pressure or to a token load when swelling ceased. The swelling pressure was defined as the pressure (i.e., P_4) on the unloading curve corresponding to the initial void ratio. This method is commonly used in China and is documented as the National Standard of the Peoples Republic of China (1969).

Figure 14.32 shows the void ratio versus vertical pressure curves for four specimens tested using test procedure 4 (i.e., unloading method). The surcharge pressure did not have a significant influence on the swelling pressure P_4 and the amount of swell during unloading.

14.4.11.5 Comparison of Swelling Pressure Measurements

The swelling pressures determined from each testing method are shown in Fig. 14.33. Also shown are the relationships between the swelling pressures and the surcharge pressures. The first procedure used to define swelling pressure P_{1-1} (i.e., loading-after-swell test procedure) shows a tendency to decrease with surcharge pressure. Results of swelling tests on clay from El Arahal (Justo et al., 1984) showed that the swelling pressure P_{1-1} was essentially independent of the surcharge pressure.

The second procedure used to define swelling pressure P_{1-2} (i.e., loading-after-swell test procedure) shows a slight increase in swelling pressure with surcharge pressure. Tests conducted on various undisturbed expansive soils from the Rocky Mountain area have shown that the surcharge pressure does not have a significant effect on the swelling pressure P_{1-2} (Chen, 1988). Figure 14.33 shows that both P_{1-1} and P_{1-2} change somewhat with the surcharge pressure. However, the changes do not appear to be large and an average value for swelling pressure can be assessed.

The uncorrected swelling pressure P_{3-1} and the corrected swelling pressure P_{3-2} show a tendency to increase with increasing surcharge pressure when using the constant-volume test procedure (Fig. 14.33). Komornik and David (1969) arrived at a similar conclusion when examining the effects of preloading on the swelling pressure of undisturbed samples. The swelling pressure P_4 associated with using test procedure 4 (i.e., unloading method) shows a slight decrease in swelling pressure with surcharge pressure (Fig. 14.33).

The measured swelling pressures for Regina clay can be summarized as follows. The swelling pressure P_{2-2} from

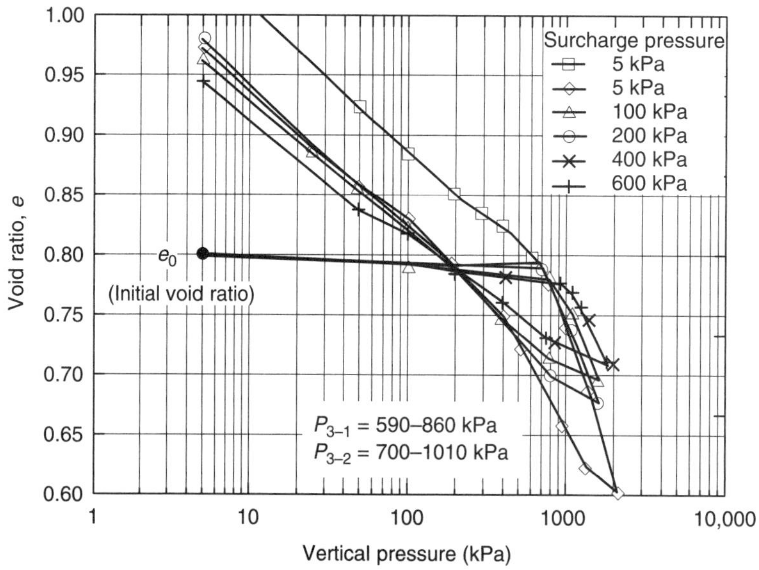

Figure 14.31 Swelling pressures measured when using constant-volume procedure with applied surcharge pressures of 5, 50, 100, 200, 400, and 600 kPa (from Feng et al., 1998).

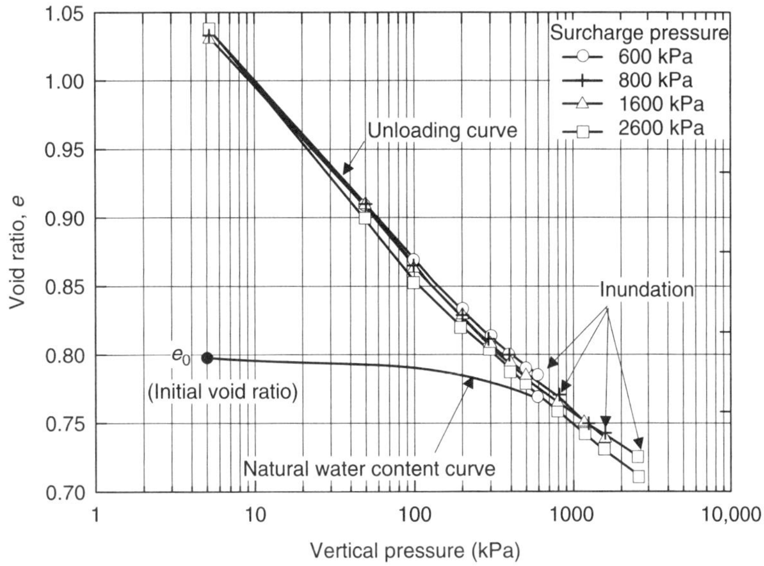

Figure 14.32 Swelling pressures measured when using unloading test procedure from surcharge pressures of 600, 800, 1600, and 2600 kPa (from Feng et al., 1998).

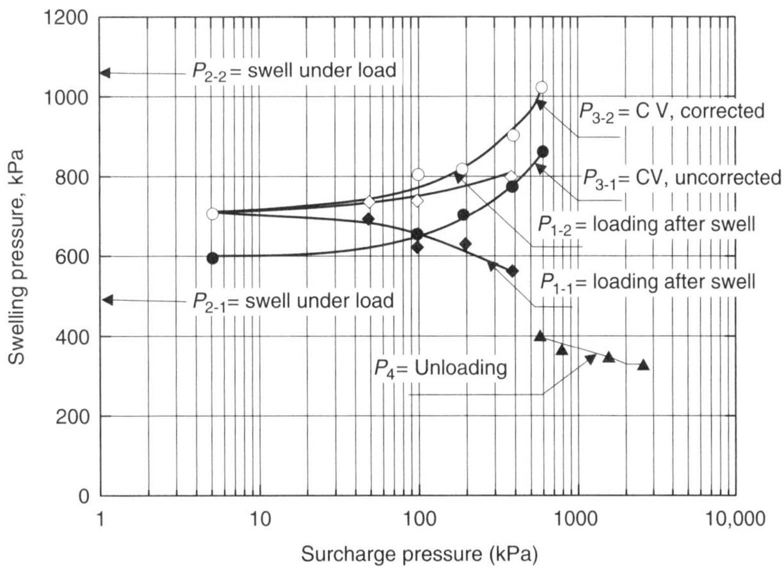

Figure 14.33 Relationship between swelling pressure and surcharge pressure when using four different test procedures (from Feng et al., 1998).

test procedure 2 (i.e., swell under load) gives an upper limit value of 1050 kPa, and the swelling pressure P_4 from test procedure 4 (i.e., unloading method) gave the lowest value, which was between 310 and 400 kPa. For the same surcharge pressure, the corrected swelling pressure, P_{3-2}, when using test procedure 3 (i.e., constant volume) was always larger than the other swelling pressures except swelling pressure P_{2-2}. The swelling pressures measured by the four testing methods ranged from 340 to 1050 kPa.

Research studies to date have not isolated a single test procedure for the measurement of swelling pressure. However, it would appear that the preferred test procedure should correspond to the methodology that yields an upper bound value for swelling pressure.

14.4.12 Unloading Soil after Compression

The rebound curve associated with unloading of the soil can be measured once the soil has been loaded well beyond the swelling pressure (Fig. 14.34). The rebound curves are approximately parallel to one another and can generally be represented as the slope on a semilogarithmic scale (Schmertmann, 1955; Holtz and Gibbs, 1956; Gilchrist, 1963; Noble, 1966; Lambe and Whitman, 1979; Lidgren, 1970; Chen, 1975).

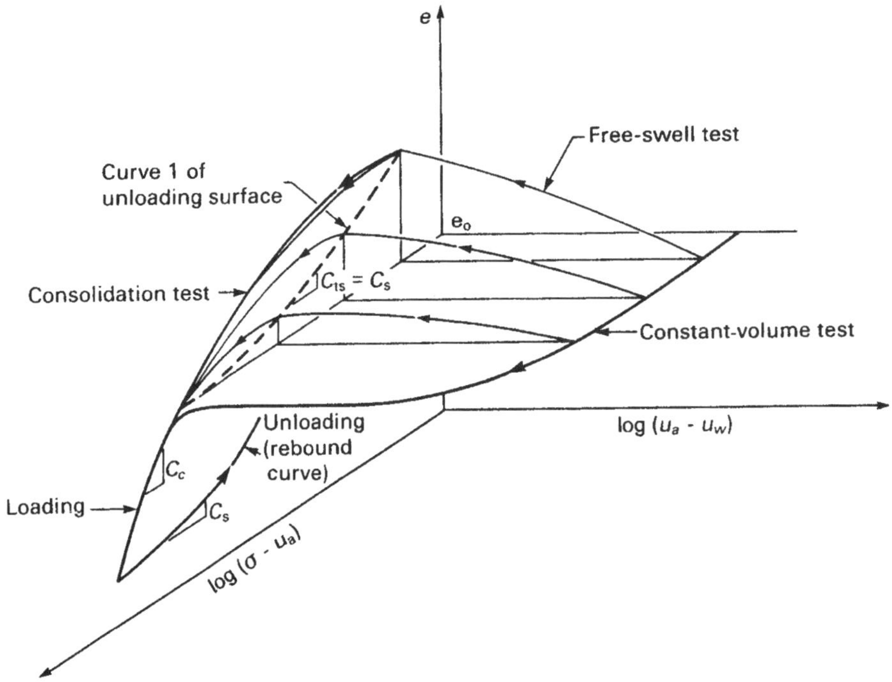

Figure 14.34 Location of unloading surface (i.e., curve 1) obtained from oedometer test results.

A stress reversal associated with going from loading to unloading conditions results in a slight curvature near the start of the rebound curve. As a result, the first portion of the rebound curve is generally omitted when determining the slope of the rebound curve, C_s. The slope of the rebound curve is referred to as the swelling index C_s.

The slope of a line through the points of free swell from various surcharge pressures is referred to using the variable C_{ts} (see Figs. 14.23 and 14.34). The C_{ts} index will be slightly larger but quite similar in magnitude to the C_s index.

The swelling index C_s has been shown to typically range between 10 and 20% of the compressive index C_c.

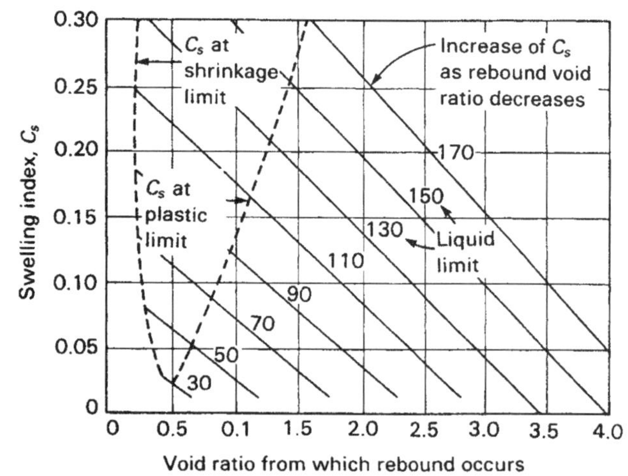

Figure 14.35 Approximate correlation of swelling index C_s with plasticity of soil and rebound void ratio (after NAVFAC, 1971).

Figure 14.35 shows approximate swelling index values which have been correlated with the liquid limit and the rebound void ratio of the soil (NAVFAC, 1971). The correlation of C_s versus the index properties shows that the swelling index decreases with an increase in the void ratio from which swelling occurs for a particular soil. The plot is useful for obtaining an estimate of the swelling index C_s of an expansive soil.

14.5 ONE-DIMENSIONAL FORMULATIONS FOR DEFORMATION ANALYSIS FOR UNSATURATED SOIL

The analysis of one-dimensional heave can be numerically modeled using the conversion of matric suctions onto the total stress plane. The calculations for a one-dimensional heave analysis can be performed either longhand or with assistance from numerical modeling software. Both types of one-dimensional solutions will be shown in this chapter. The final pore-water pressure conditions are assumed, and as a result it is necessary to simply solve the stress-deformation equation for the y-direction. Transient solutions with changes in pore-water pressure conditions with time (i.e., water flow partial differential equation) are solved in conjunction with a stress-deformation analysis (see Chapter 16).

Two approaches are presented for the calculation of one-dimensional heave in an expansive soil. First, it is possible to linearize the void ratio–stress state relationship on a semilogarithm plot and perform a summation-type longhand analysis for the calculation of total heave. Second, it is possible to perform a one-dimensional solution as a numerical modeling problem (e.g., a finite element analysis). Both solutions should give essentially the same answer for total heave.

Two- and three-dimensional heave simulations that use independent changes in matric suction and total stresses are solved later in this chapter. The two- and three-dimensional heave calculations are performed using a finite element modeling procedure.

14.5.1 Theory of Heave Predictions (One-Dimensional Heave)

Volume change coefficients or indices and matric suction changes need to be measured or estimated for the prediction of one-dimensional heave. The stress state variable changes can be large, and consequently the soil properties can vary as the stress levels change. It is necessary to integrate volumetric strain between the initial and final stress states. The constitutive surface takes on a linear form when plotting the total stress state on a logarithmic scale. It is possible to transfer matric suction onto the total stress plane, thereby simplifying the prediction of total heave.

The total heave formulation is first presented with the initial and final stress states projected onto the total stress plane. The results from a one-dimensional oedometer test are plotted on a semilogarithmic scale of total stress state and the slope of the plot is used in the formulation for total heave.

The in situ stress state is equal to the vertical swelling pressure of the soil, which is measured in a one-dimensional oedometer under K_0 loading conditions. Only vertical or one-dimensional heave is calculated. The vertical heave prediction is of importance in the design of shallow foundations for light structures. Two case histories dealing with highly expansive soils in Saskatchewan, Canada, are later analyzed and used to illustrate the calculation of total heave.

Volume change can also occur in the lateral directions for loading configurations other than K_0 conditions. The swelling pressure in the lateral direction depends on several variables, such as the initial at-rest earth pressure coefficient and the horizontal deformation modulus for the soil. In a soil with wide desiccation cracks, substantial volume changes may occur in the horizontal direction prior to the development of the lateral swelling pressure. The ratio of the lateral to vertical swelling pressures can range from as low as the at-rest earth pressure coefficient, which may be zero, to as high as the passive earth pressure coefficient (Pufahl et al., 1983; Fourie, 1989).

14.5.2 Total Heave Analysis Using Longhand Calculations

The longhand calculation procedure for total heave is essentially the reverse of the settlement analysis used for soft compressible clays. Total heave is computed from changes in void ratios between the initial and final stress states and

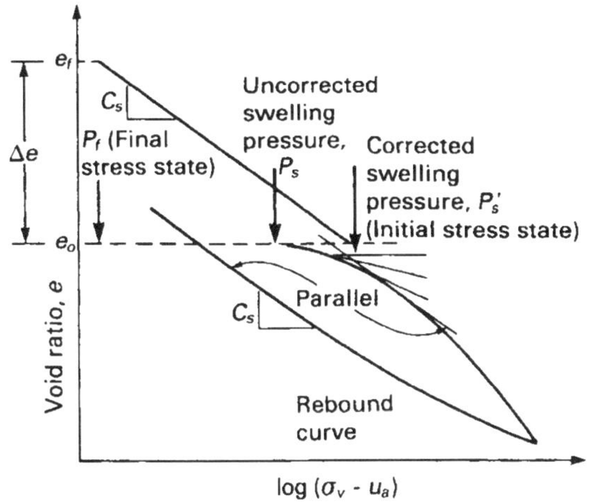

Figure 14.36 One-dimensional oedometer test results showing effect of sampling disturbance.

the swelling index. The formulation can be visualized on a plot of void ratio versus the logarithm of the stress state. The formulation assumes that the stress path followed during heave can be projected onto the total stress plane, as shown in Fig. 14.36. The total heave stress path follows the rebound curve (i.e., C_s) from the initial stress state to the final stress state. The equation for the rebound portion of the oedometer test data can be written as

$$\Delta e = C_s \log\left(\frac{P_f}{P_0}\right) \tag{14.14}$$

where:

Δe = change in void ratio between the initial and final stress states (i.e., $e_f - e_0$),
e_0 = initial void ratio,
e_f = final void ratio,
C_s = swelling index,
P_0 = initial stress state, assumed to be equal to the corrected swelling pressure (i.e., $P_0 = P_s'$), and
P_f = final stress state.

The initial stress state P_0 and the corrected swelling pressure P_s' consist of the sum of the overburden pressure and the matric suction equivalent (Fig. 14.10), as follows:

$$P_0 = (\sigma_v - u_a) + (u_a - u_w)_e \tag{14.15}$$

where:

σ_v = total overburden pressure,
$\sigma_v - u_a$ = net overburden pressure,
u_a = pore-air pressure,
$(u_a - u_w)_e$ = matric suction equivalent, and
u_w = pore-water pressure.

The above equation defines the initial stress state as P_0. The value of P_0 is not calculated but is measured as the corrected swelling pressure P_s' in an oedometer test. The final stress state P_f must account for total stress changes and final pore-water pressure conditions. The pore-air pressure in the field remains at atmospheric conditions. The final pore-water pressure conditions can be predicted or estimated. Therefore, the final stress state P_f can be written as

$$P_f = \sigma_v \pm \Delta\sigma_v - u_{wf} \tag{14.16}$$

where:

$\Delta\sigma_v$ = change in total vertical stress due to the excavation (i.e., $-\Delta\sigma_v$) or placement of fill (i.e., $+\Delta\sigma_v$) and
u_{wf} = predicted or estimated final pore-water pressure.

The heave of an individual soil layer can be written in terms of a change in void ratio:

$$\Delta h_i = \frac{\Delta e_i}{1 + e_{0i}} h_i \tag{14.17}$$

where:

Δh_i = heave of an individual soil layer,
h_i = thickness of a single soil layer into which the total expansive soil layer has been subdivided,
Δe_i = change in void ratio of the layer under consideration (i.e., $e_{0i} - e_{fi}$),
e_{0i} = initial void ratio of the soil layer, and
e_{fi} = final void ratio of the soil layer.

The change in void ratio, Δe_i, in Eq. 14.17 can be rewritten in terms of the soil properties and the stress states (i.e., Eq. 14.14) to give the following form for the heave of a single layer:

$$\Delta h_i = \frac{C_s}{1 + e_{0i}} h_i \log \frac{P_{fi}}{P_{0i}} \tag{14.18}$$

where:

P_{fi} = final stress state in the soil layer and
P_{0i} = initial stress state in the soil layer.

The total heave from the entire expansive soil layer, ΔH, is equal to the sum of the heave amounts calculated for each of the subdivided layers:

$$\Delta H = \sum \Delta h_i \tag{14.19}$$

14.5.3 Estimation of Final Pore-Water Pressures

The final pore-water pressures below a foundation or pavements can either be predicted through analysis or be estimated. A prediction must take into consideration the ground surface moisture flux boundary conditions (i.e., infiltration, evaporation, and evapotranspiration) and possible fluctuation in the groundwater table. The surface flux boundary conditions can vary from one geographic location to another depending on climatic conditions. Russam and Coleman (1961) related the equilibrium suction below asphaltic pavements to the Thornthwaite moisture index. On many smaller structures, however, it is often man-made causes such as leaky water lines, poor drainage, and lawn watering that control final pore-water pressures in the soil.

There are three typical possibilities for the estimation of final pore-water pressure conditions, as illustrated in Fig. 14.37. First, it could be assumed that the water table will rise to the ground surface, creating a hydrostatic condition. This assumption results in the calculation of the largest amount of total heave. Second, it could be assumed that the pore-water pressure approaches a zero value throughout its depth. This may appear to be a realistic assumption; however, it should be noted that it is not an equilibrium condition. The zero pore-water pressure condition may provide a reasonable estimate for the final pore-water pressure state in some situations. Third, it could be assumed that under long-term equilibrium conditions the pore-water pressure will remain slightly negative. This assumption predicts the smallest amount of total heave.

It is also possible to have variations of the above assumptions with depth. As well, a limit might be placed on the depth to which wetting will occur. Any of the above assumptions produces a range of predictions of heave that can be used by the geotechnical engineer to determine reasonable design estimates for total heave. The above-mentioned analyses will show that most of the heave occurs in the uppermost soil layers where the change in matric suction is largest and the total overburden stresses are the smallest.

14.5.4 Zone of Subsoil Wetting

The zone over which wetting will likely take place is a primary variable that must be assessed or estimated in order to calculate total heave. The soil immediately below light structures often increases in water content over time. Irrigation

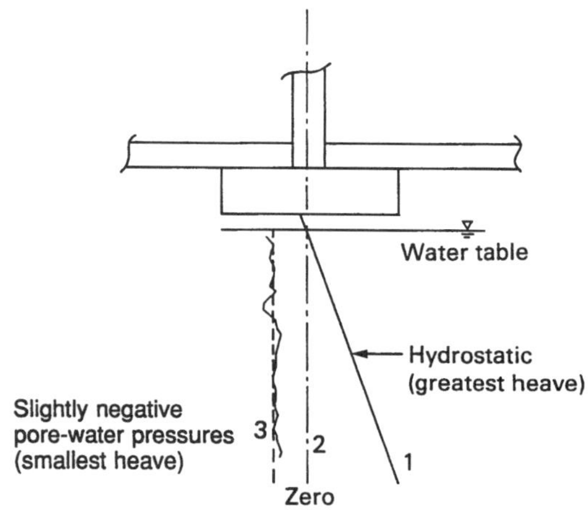

Figure 14.37 Estimation of final matric suction conditions in soil.

and a reduction in evapotranspiration commonly contribute to an increase in the water content of the soil. A perched water table may develop which controls final equilibrium pore-water pressures in the soil if an aquatard is encountered. The zone of wetting and the final pore-water pressures will depend upon water balance calculations at ground surface in the case where an expansive soil layer is extensive and no aquatard exists. A transient seepage analysis can be used in this case to assess the zone over which wetting is likely to occur.

There are several terms that have been used to define different types and amount of wetting that might occur in an expansive soil layer. The general term for the zone of wetting is the "active zone" (Nelson et al., 2001; Overton et al., 2011). This term has also been used for other applications (e.g., permafrost) and is too general to clearly define the degree and type of wetting. Nelson et al., (2001) provided several definitions that provide a more detailed description of the zone of wetting in an expansive soil:

(i) "*Active Zone*, z_a, is that zone of soil that is contributing to heave due to soil expansion at a particular point in time. The depth of the active zone will vary as heave progresses, and therefore, varies with time."
(ii) "*Zone of Seasonal Moisture Fluctuation*, z_s, is that zone of soil in which water contents change seasonally due to climate changes."
(iii) "*Zone of Wetting*, z_w, is the zone in which water contents have increased over pre-construction equilibrium conditions. Factors contributing to this could include capillary rise after the elimination of evapotranspiration from the surface, infiltration due to irrigation from the surface, or introduction of water from off-site. Underground sources may include broken water lines, development of perched water tables, or flow through permeable strata that are recharged at distant locations."
(iv) "*Depth of Potential Heave*, z_p, is the depth to which the overburden vertical stress is equal or exceeds the swelling pressure of the soil. This represents the maximum depth of the Active Zone that could occur."

The above-mentioned definitions infer that there are numerous factors that affect the depth and amount of wetting that might occur at a particular site. These factors form the boundary conditions for a heave analysis and must be assessed for each expansive soils problem. It is generally possible to make an appropriate assumption regarding the final wetting conditions (or pore-water pressure conditions) that should be analyzed. It is also possible to use weather conditions and saturated-unsaturated hydraulic properties to assess realistic boundary conditions.

14.5.5 First Example of One-Dimensional Heave Calculations

The following example problems are presented to illustrate the calculations associated with total heave. The first example considers a 2-m-thick layer of swelling clay (Fig. 14.38). The initial void ratio of the soil is 1.0, the total unit weight is 18.0 kN/m^3, and the swelling index C_s is 0.1. One oedometer test was performed on a sample taken from a depth of 0.75 m. The test data showed a corrected swelling pressure of 200 kPa. It is assumed that the corrected swelling pressure is constant throughout the 2-m layer.

Let us consider the case where the ground surface is covered with an impermeable layer such as asphalt. The negative pore-water pressure in the soil below the asphalt will increase with time as a result of the reduction in evaporation and evapotranspiration. Let us assume for analysis purposes that the final pore-water pressures increase to zero throughout the entire depth.

The 2-m layer is subdivided into three sublayers. The amount of heave in each layer is computed by considering the stress state changes at the middle of the layer. The initial stress state P_0 is equal to the corrected swelling pressure at all depths. The final stress state P_f will be the overburden pressure. Equation 14.18 is used to calculate the heave for each sublayer. The calculations in Fig. 14.38 show that a total heave of 11.4 cm is likely to occur. Approximately 36% of the total heave occurs in the upper quarter of the clay strata. The calculations also show the amount of heave that would occur if each layer became wet from the surface downward.

14.5.6 Second Example of One-Dimensional Heave Calculations

The second example illustrates a more complex loading situation and the results are presented in Fig. 14.39. The expansive clay layer is 2 m in thickness. The initial void ratio is 0.80, the total unit weight is 18.0 kN/m^3, and the swelling index C_s is 0.21. Three oedometer tests were performed which show a decrease in the corrected swelling pressure with depth (Fig. 14.39).

Suppose the engineering design suggests the removal of $1/3$ m of swelling clay from ground surface followed by the replacement with $2/3$ m of gravel. The unit weight of the gravel is assumed to be equal to that of the clay. The remaining $1\,2/3$ m of swelling clay is subdivided into three layers for calculation purposes. The thickness of each layer is shown in Fig. 14.39.

The initial stress state P_0 for the midpoint of each layer can be obtained by interpolating between the measured swelling pressures. The final stress state P_f must take into account the final pore-water pressure and changes in the total stress. The final pore-water pressure is assumed to be −7.0 kPa. Equation 14.18 can be used to calculate the change in volume that should occur within each layer. The total heave for the three layers is computed to be 22.1 cm.

Two assumptions were made during the heave analysis. First, it was assumed that the independent processes of excavation of the expansive soil and the placement of the gravel

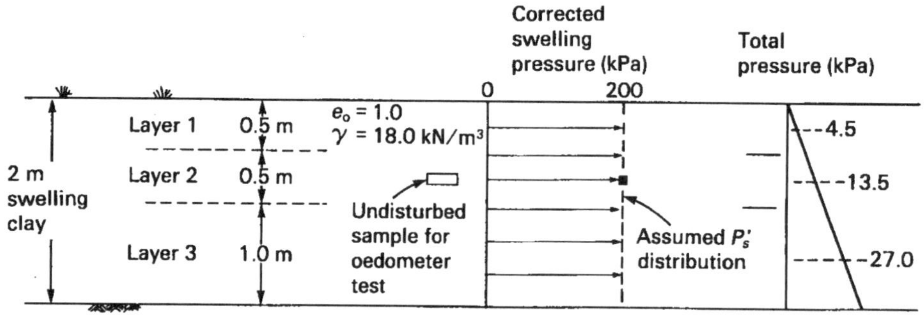

Assume 1) **Surface is covered with an impermeable layer**
2) Final pore-water pressure equals zero

Equation: $\Delta h_i = \dfrac{C_s}{1 + e_{oi}} h_i \log \dfrac{P_{fi}}{P_{oi}}$

P_f will equal overburden pressure

P_0 will equal corrected swelling pressure

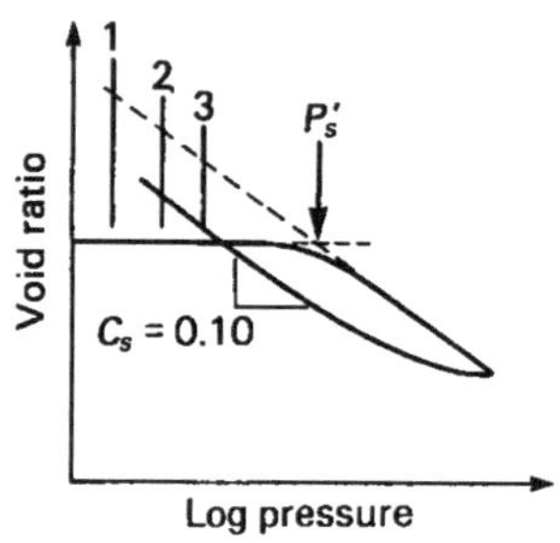

Calculations:

Layer 1 $\Delta h_i = 500 \times \dfrac{0.1}{1 + 1.0} \log \dfrac{4.5}{200} = 41.2$ mm

Layer 2 $\Delta h_i = 500 \times \dfrac{0.1}{1 + 1.0} \log \dfrac{13.5}{200} = 29.3$ mm

Layer 3 $\Delta h_i = 1000 \times \dfrac{0.1}{1 + 1.0} \log \dfrac{27.0}{200} = 43.5$ mm

Total heave 114.0 mm

Figure 14.38 Calculation of total heave for example 1.

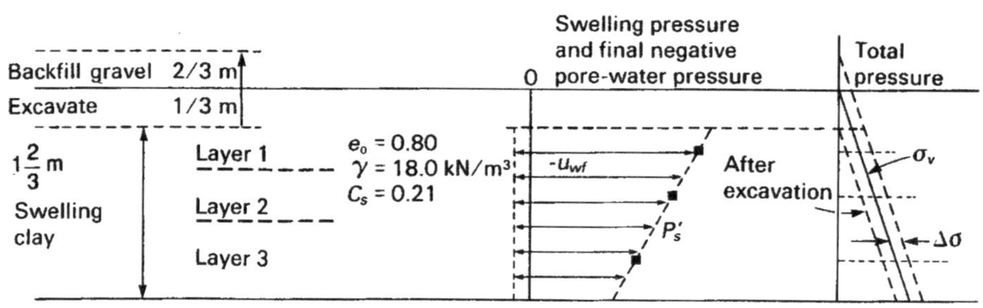

	Initial stress state			Final stress state			
Layer no.	Thickness (mm)	$P_0 = P_s'$ (kPa)	Initial overburden pressure σ_v (kPa)	Change in total stress $\Delta\sigma$ (kPa)	Final pore-water pressure u_{wf} (kPa)	$P_f = \sigma_v \pm \Delta\sigma - u_{wf}$ (kPa)	Δh_i (mm)
1	333	800	9.0	+6.0	-7.0	22.0	60.6
2	500	608	16.4	+6.0	-7.0	29.4	76.7
3	833	300	28.4	+6.0	-7.0	41.4	83.6

Total heave = 220.9 mm

Figure 14.39 Calculation of total heave for example 2.

fill did not allow sufficient time for pore-water pressure equilibrium. Therefore, the soil simply responded to the net change in total stress. Second, it is assumed that the soil approached saturation when the final pore-water pressure was −7.0 kPa. Therefore, the slope of the rebound curves on the matric suction plane was the same as the slope on the total stress plane. This assumption is reasonable as long as the final pore-water pressure is relatively small.

14.5.7 Third Example of One-Dimensional Heave Calculations

A third example illustrates the amount of heave versus depth for a 4-m layer of expansive soil. The results show that the largest amount of heave occurs near the ground surface. The matric suction generally has a maximum value near the ground surface and the overburden pressure is also lowest near ground surface. Consequently, the ratio of the final stress state P_f to the initial stress state P_0 in Eq. 14.18 approaches its minimum value, which results in the largest calculated heave.

Let us assume that the clay becomes wet and the pore-water pressures go to zero throughout the deposit. Figure 14.40 shows the distribution of heave potential versus depth for the 4-m-thick layer. Heave potential is defined as the heave in each layer of the profile divided by the maximum heave in any layer. Three initial swelling pressure profiles are assumed. The first profile assumes a constant swelling pressure of 720 kPa throughout the 4-m depth. The second profile assumes a constant value for the swelling pressure of 3600 kPa throughout the 4-m depth. The third profile assumes a linear decrease in swelling pressure from 720 kPa to a value equal to the overburden pressure at 4 m. All cases show that the heave potential decreases rapidly with depth. The swelling pressure generally decreases with depth in situ, and as a result the decrease in heave potential is of an exponential form (Fig. 14.40).

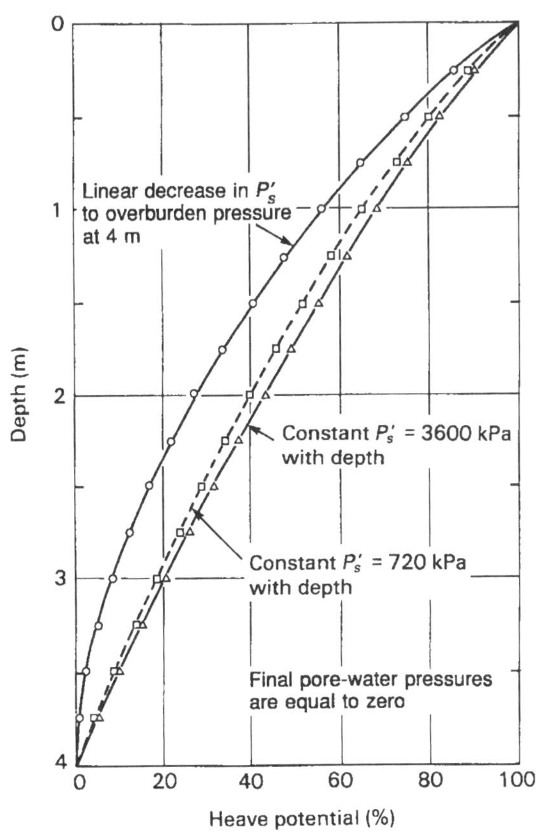

Figure 14.40 Heave potential versus depth for various distributions of swelling pressure with depth.

The change in water content, Δw, can also be estimated as the soil swells provided the initial volume-mass soil properties are known:

$$\Delta w = \frac{S_f\ \Delta e}{G_s} + \frac{e_0\ \Delta S}{G_s} \tag{14.20}$$

where:

S_f = final degree of saturation,
G_s = specific gravity of soil solids, and
ΔS = change in degree of saturation.

The final degree of saturation can be assumed to approach 100%. Changes in void ratio, Δe, are obtained from the heave analysis.

14.5.8 Case History (Slab-on-Grade Floor, Regina, Saskatchewan, Canada)

In 1961, the Division of Building Research, National Research Council, Government of Canada, monitored the performance of a light industrial building which was constructed in north-central Regina. Details of the study were presented by Yoshida et al., (1983). Instrumentation was installed to monitor ground movements at various depths below the slab. Water content changes were monitored using a neutron moisture meter probe. Undisturbed samples were taken as part of the subsurface exploration prior to the construction of the building. Constant-volume oedometer tests were performed on three samples, and the swelling pressure profiles are shown in Fig. 14.41. The average swelling index of the soil at this site was 0.09.

The owner noticed considerable cracking of the floor slab about one year after completion of construction. Precise level surveys showed the maximum total heave to be 106 mm. The owner had also noticed a significant increase in water consumption (i.e., 35,000 L). It was discovered that a leak had occurred in the hot-water line beneath the floor slab near the location of the maximum heave. The leak was immediately repaired.

Total heave predictions were made on the basis of the laboratory oedometer test results. Various assumptions were made concerning the final pore-water pressure conditions as part of the one-dimensional heave analysis. The predicted heave was 141 mm when it was assumed that the soil had

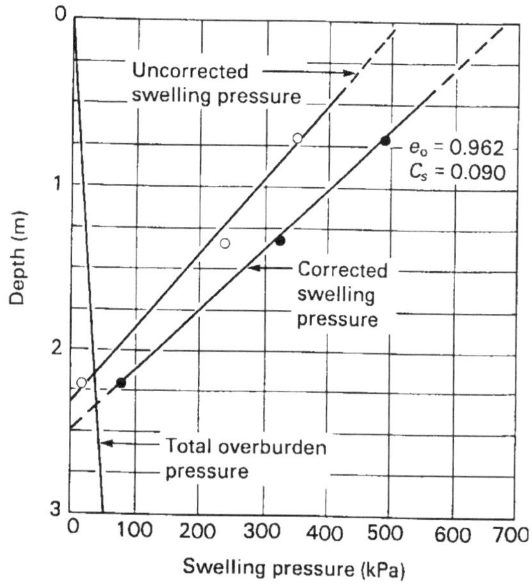

Figure 14.41 Swelling pressure versus depth for Regina clay at light industrial building site (after Yoshida et at., 1983).

become saturated and the water table rose to the base of the floor slab. The predicted total heave was 118 mm when assuming that the negative pore-water pressures in the soil were reduced to zero. The calculated total heave reduced to 66 mm when a final pore-water pressure profile of −50 kPa was assumed. On the basis of the parametric heave analysis, it appears that the assumption of zero pore-water pressure was probably the most realistic for this case history. It appears that further heave would likely have taken place had the leak not been repaired. The prediction of heave at various depths showed close agreement with the actual measurements.

The importance of correcting the swelling pressure measurements can be illustrated by computing total heave using the uncorrected swelling pressures. The predicted heave was reduced to 103 mm when the final pore-water pressure was assumed to be zero and the uncorrected swelling pressures were used in the computations. This is less than the measured maximum heave and 13% less than the 118 mm predicted using the corrected swelling pressure values.

14.5.9 Case History (Eston School, Eston, Saskatchewan, Canada)

Soils in the Eston area of Saskatchewan have long been known to be extremely high in swelling potential. The stratigraphy consists of approximately 7½ m of highly plastic, brown clay overlying the glacial till. Many light structures have undergone serious distress over the years. The building of particular interest in Eston was the Old Eston School constructed in the late 1920s.

The school building was constructed on concrete strip footings and a wooden basement floor was supported on interior concrete footings. The school was a two-story structure with classrooms on both the lower and upper levels. The lower floor was approximately 1.2 m below grade. The exterior concrete walls were founded approximately 1.8 m below grade.

Substantial heave took place below the interior footings. Although the record of performance was not precise, apparently the heave in one portion of the basement area had been noted to be quite extreme. On two occasions during the history of the school, 150–300 mm of soil was removed from below the interior footings. As much as 450–900 mm of total heave occurred during the life of the school, according to maintenance records. Large amounts of differential heaving of the floor were recorded in 1960. The school was demolished in 1967.

A subsurface investigation in 1981 was conducted adjacent to the location of the old school. Undisturbed soil samples were taken and constant-volume oedometer tests were performed. The results are presented in Fig. 14.42. The average natural water content throughout the profile was 25%. The average plastic limit was 27%, and the average liquid limit was 100%. The average swelling index was 0.21. It is not possible to undertake a detailed heave analysis of heave for the area below the school due to a lack of detailed information on the soil conditions and the performance records for the school. It is of interest, however, to perform an approximate analysis for total heave. Using the corrected swelling pressures from Fig. 14.42 and assuming that the negative pore-water pressures went to zero, the predicted heave would be approximately 990 mm. The predicted heave would be about 86% of this value (i.e., on the order of 855 mm) using uncorrected swelling pressures.

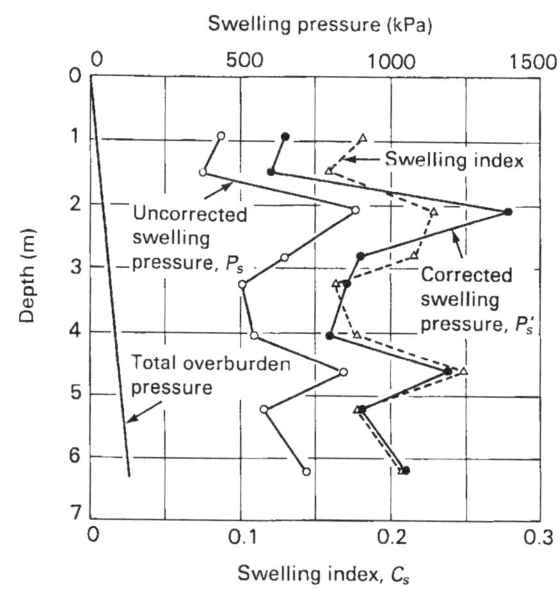

Figure 14.42 Swelling pressure and swelling index versus depth for Eston clay.

14.5.10 Parametric Analysis to Illustrate Influence of Various Variables

The following sections discuss the role of each variable that controls the magnitude of total heave (Rao et al., 1988). The corrected swelling pressure and the swelling index are the main variables which are considered. The effect of correcting the swelling pressure for sampling disturbance on the prediction of heave is illustrated. A simple geometry with constant soil properties, including the swelling pressure, is used for the analyses. The influence of each controlling variable on the prediction of heave can readily be identified.

The same example is also used to illustrate the percentages of heave resulting from the wetting of an expansive soil to various depths. The analysis can be used to identify the critical zone where maximum heave will occur. Partial removal of expansive soils is often attempted in order to reduce heave potential. The effectiveness of this procedure in reducing the amount of heave is discussed using the same example problem.

14.5.11 Closed-Form Heave Equation When Swelling Pressure Is Constant

An example is shown in Fig. 14.43 where the corrected swelling pressure P_s' is constant with depth. The soil is assumed to be homogeneous (i.e., e_0, ρ, and C_s are constant with depth). These assumptions apply to the cases presented in the following sections. Sufficient water is supplied to the soil such that the pore-water pressures are assumed to go to zero throughout the entire soil profile. The final stress state P_f is assumed to be equal to the total overburden pressure.

The unsaturated, expansive soil swells upon wetting. Most of the heave will occur near ground surface where there is the largest difference between the corrected swelling pressure and the total overburden pressure. Heave continues to occur until a depth is reached where there is no difference between the corrected swelling pressure and the total overburden pressure. In this context, the depth at which the corrected swelling pressure equals the total overburden pressure is defined as the *active depth* of swelling, H:

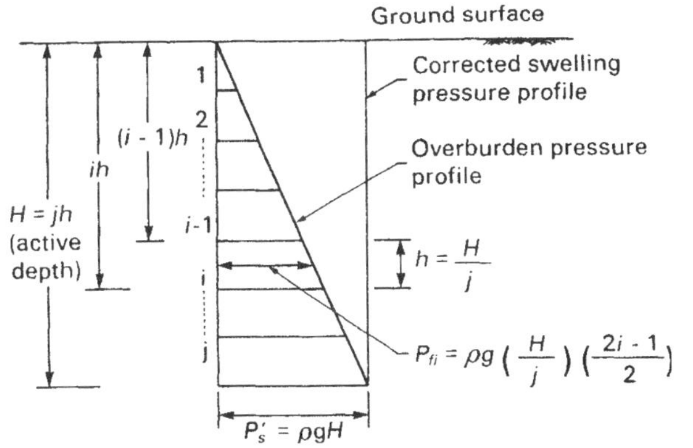

Figure 14.43 Overburden and swelling pressure distributions versus depth for the case of constant swelling pressure with depth.

$$H = \frac{P_S'}{\rho g} \tag{14.21}$$

where:

H = active depth of swelling,
ρ = total density of the soil which is assumed to remain constant with depth, and
g = gravitational acceleration.

Inherent in this definition of active depth is the assumption that the final pore-water pressures throughout the profile go to zero. Also inherent are the assumptions that there is sufficient water to wet the entire soil strata and there is sufficient time for the water to infiltrate the entire soil strata. The heave analysis can be performed by subdividing the computed active depth into j number of layers of equal thickness (i.e., $h_i = h$):

$$h = \frac{H}{j} \tag{14.22}$$

where:

h = thickness of a soil layer.

An analysis is first made to determine the number of layers required to accurately predict total heave. The initial stress state P_{0i} is equal to the corrected swelling pressure P_s', which can be defined in terms of the active depth in Eq. 14.21:

$$P_{0i} = \rho g H \tag{14.23}$$

The final stress state P_{fi} in the ith layer from ground surface is computed as the average overburden pressure in the soil layer:

$$P_{fi} = \rho g \frac{(i-1)h + ih}{2} \tag{14.24}$$

where:

i = soil layer number (i.e., 1, 2, ..., j).

The thickness of the soil layer, h, in the above equation can be written in terms of the total thickness divided by the number of layers (i.e., using Eq. 14.22):

$$P_{fi} = \frac{\rho g H(2i - 1)}{2j} \tag{14.25}$$

The heave equation can be written for each soil layer and then summed to give the total heave. The amount of heave in each layer can be computed by substituting Eqs. 14.22,

14.23, and 14.25 for h_i, P_{0i}, and P_{fi}, respectively, into Eq. 14.18. The total heave for the entirely wetted active depth is the summation of the heave for each layer:

$$\Delta H = \frac{C_s H}{1+e_0}\frac{1}{j}\sum_{i=1}^{j}\log\left(\frac{2i-1}{2j}\right) \qquad (14.26)$$

The computed total heave does not change significantly when using more than 35 layers (i.e., $j = 35$). Figure 14.44 shows the percentage of calculated total heave referenced to the case of an analysis with 35 layers. The analysis can be simplified as follows if it is assumed that 35 layers are used in Eq. 14.26.

$$\Delta H = -0.430\frac{C_s H}{1+e_0} \qquad (14.27)$$

where:

-0.430 = computed value for the term, $\frac{1}{35}\sum_{i=1}^{35}\log\left(\frac{2i-1}{70}\right)$.

The constant 0.430 in Eq. 14.27 approaches a value of 0.434 when an infinite number of soil layers are assumed for a closed-form solution (Li, 1989). The total heave expressed in Eq. 14.27 is later used as a reference when comparing the amount of heave computed for different field conditions (i.e., partly wetted active depth, excavation, and backfill). In general, a total heave analysis using 10 layers will have satisfactory accuracy.

Substituting Eq. 14.21 into Eq. 14.27 and assuming a unit weight (i.e., ρg) of 20 kN/m^3 and an initial void ratio e_0 of 1.0 results in the following equation:

$$\Delta H = -0.01075 C_s P_s' \qquad (14.28)$$

where P_s' has units of kilopascals and ΔH has units of meters. The corrected swelling pressure and the swelling index are the two primary variables involved in the prediction of total heave. Figures 14.45 and 14.46 show the types of plots which can be drawn from Eq. 14.28 to illustrate approximate amounts of total heave for various swelling indices and swelling pressures. Total heave varies linearly with both of the above variables. Viewed another way, it is equally important to accurately assess both swelling pressure and swelling index when analyzing an expansive soil problem.

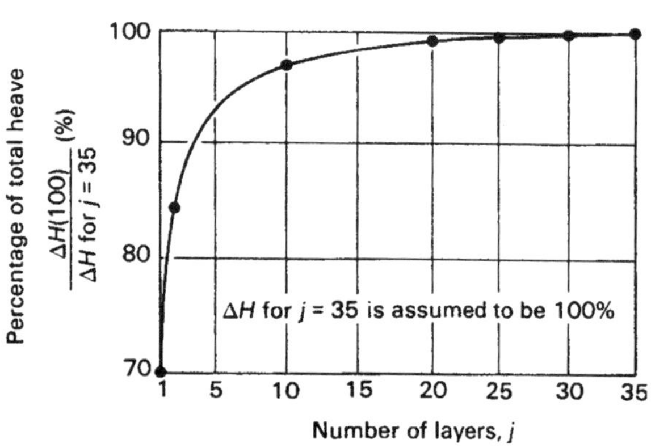

Figure 14.44 Percentage of heave as function of number of layers used in the analysis.

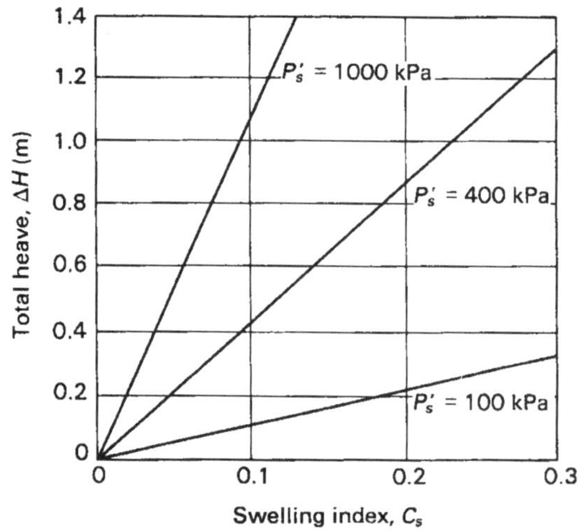

Figure 14.45 Total heave versus swelling index for various values of corrected swelling pressure.

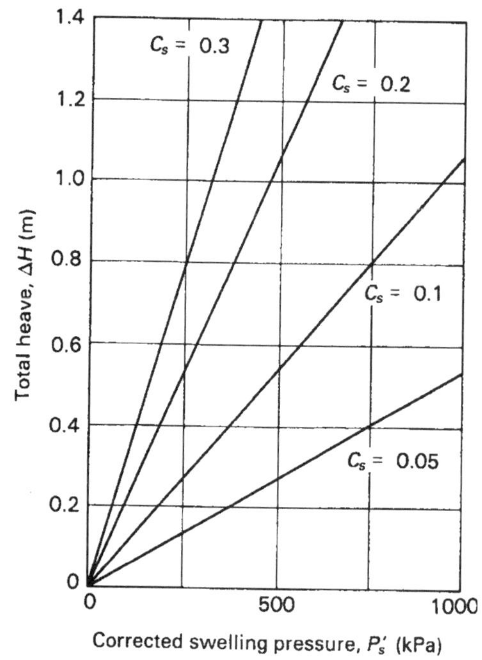

Figure 14.46 Total heave versus corrected swelling pressure for various swelling indices.

14.5.12 Effect of Correcting Swelling Pressure on Predicted Total Heave

Correcting the swelling pressure for the effect of sampling disturbance increases the change in void ratio, as shown in Fig. 14.47. The calculated total heave for any expansive

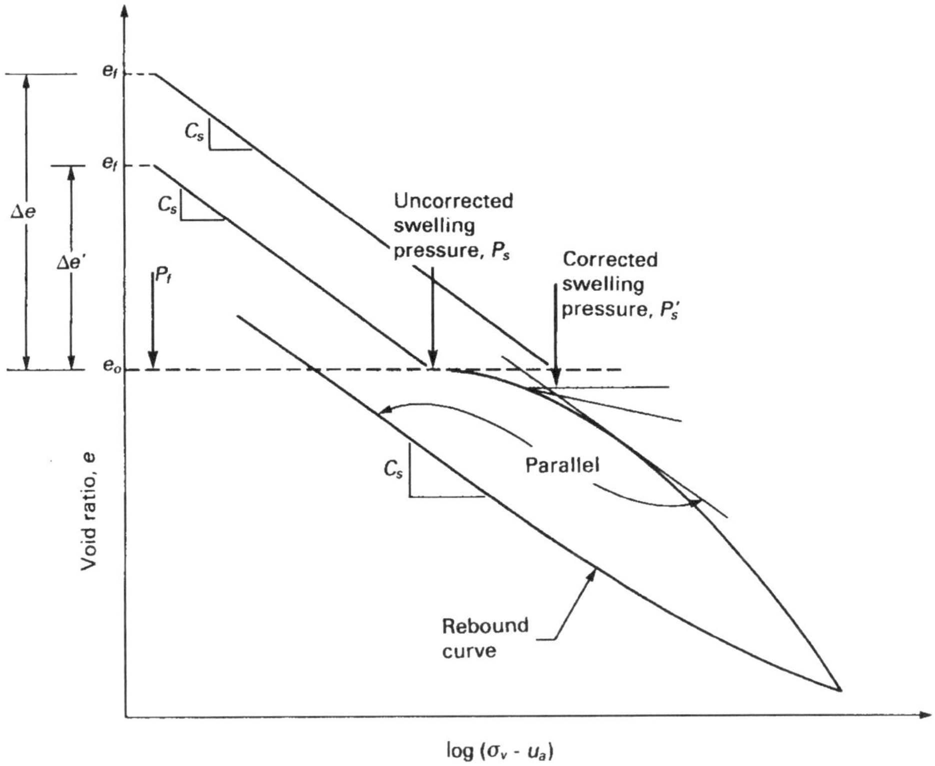

Figure 14.47 Void ratio changes corresponding to corrected and uncorrected swelling pressures.

soil problem also increases when the swelling pressure is increased. The change in void ratio corresponding to the corrected swelling pressure can be computed from Eq. 14.14 by substituting P_s' for P_0 (also see Fig. 14.47):

$$\Delta e' = C_s \log\left(\frac{P_f}{P_s'}\right) \tag{14.29}$$

The change in void ratio corresponding to the uncorrected swelling pressure P_s is written as

$$\Delta e = C_s \log\left(\frac{P_f}{P_s}\right) \tag{14.30}$$

The difference in heave or in the change in void ratio can be obtained by subtracting Eq. 14.30 from Eq. 14.29:

$$\Delta e - \Delta e' = C_s \log\left(\frac{P_s}{P_s'}\right) \tag{14.31}$$

Referencing the difference between the two void ratio changes to the change in void ratio when using the corrected swelling pressure gives

$$\frac{\Delta e - \Delta e'}{\Delta e'} = \frac{\log\left(P_s/P_s'\right)}{\log\left(P_f/P_s\right)} \tag{14.32}$$

The above equation is shown graphically in Fig. 14.48 for various ratios of swelling pressure, P_s'/P_s and P_f/P_s. In other words, the initial and final stress states are referenced to the uncorrected swelling pressure. The plot shows that the relative difference between the two heave predictions increases as the difference between the corrected and uncorrected swelling pressures increases (i.e., as P_s'/P_s increases). It is not uncommon for the corrected swelling pressure to be two to three times greater than the uncorrected swelling pressure (i.e., Fredlund, 1983). The relative difference in heave also increases as the overburden pressure increases (i.e., P_f/P_s increases). The difference in the prediction of total heave can also be substantially different depending on whether the corrected or uncorrected swelling pressure is used in the analysis.

14.5.13 Example with Wetting from Top to Specified Depth

Figure 14.49 shows the variables involved in studying the effect of wetting of the soil from the ground surface to a specified depth (e.g., by flooding the surface). For example, an insufficient amount of water infiltration into the ground may result in the wetting of only a portion of the active depth. Let H_r be the portion of the active depth that has been wetted (i.e., rH). The wetted zone is also subdivided into j number of layers for the computation of total heave.

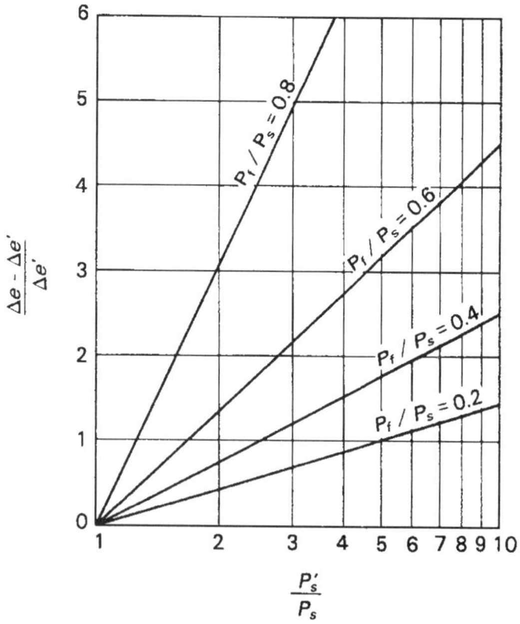

Figure 14.48 Effect of correcting swelling pressure on computed change in void ratio for various overburden pressures.

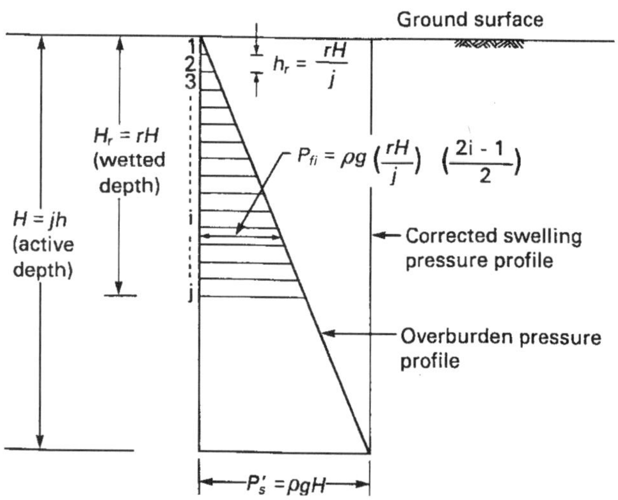

Figure 14.49 Stress distributions and definition of variables for wetting portion of soil in active depth.

The heave for any layer in the wetted zone can be calculated by substituting the variables shown in Fig. 14.49 into Eq. 14.18:

$$\Delta h_{ri} = \frac{C_s}{1+e_0}\frac{rH}{j}\log\left(\frac{\rho g r H(2i-1)/2j}{\rho g H}\right) \qquad (14.33)$$

where:

Δh_{ri} = heave of an individual layer when a portion of the active depth is wetted and

r = portion of the profile wetted (i.e., H_r/H).

Equation 14.33 reduces to the following form:

$$\Delta h_{ri} = \frac{C_s}{1+e_0}\frac{rH}{j}\left[\log\left(\frac{2i-1}{2j}\right) + \log\ r\right] \qquad (14.34)$$

Equation 14.34 can be applied to all layers in the wetted zone to give the total heave:

$$\Delta H_r = \frac{C_s H}{1+e_0} r \left[\frac{1}{j}\sum_{i=1}^{j}\log\left(\frac{2i-1}{2j}\right) + \log\ r\right] \qquad (14.35)$$

Substituting Eq. 14.27 into Eq. 14.35 for $j = 35$ gives

$$\Delta H_r = \frac{C_s H}{1+e_0} r(-0.430 + \log\ r) \qquad (14.36)$$

When $r = 1.0$, the entire active depth is wetted, and Eq. 14.36 reverts to Eq. 14.27. A comparison between Eq. 14.36 and Eq. 14.27 can be written as follows:

$$\frac{\Delta H_r}{\Delta H} = r(1 - 2.326\ \log\ r) \qquad (14.37)$$

Figure 14.50 shows a plot of the above equation where various percentages of the profile are wetted. The relationship is unique for all values of swelling pressure and swelling index. For example, it can be seen that 80% of the total heave occurs if the depth of wetting is 40–50% of the active depth. This example illustrates that it may be possible to flood an area prior to construction until a significant percentage of total heave has occurred. Whether this is a practical engineering solution may depend on other factors such as the hydraulic soil properties and the design of the structure.

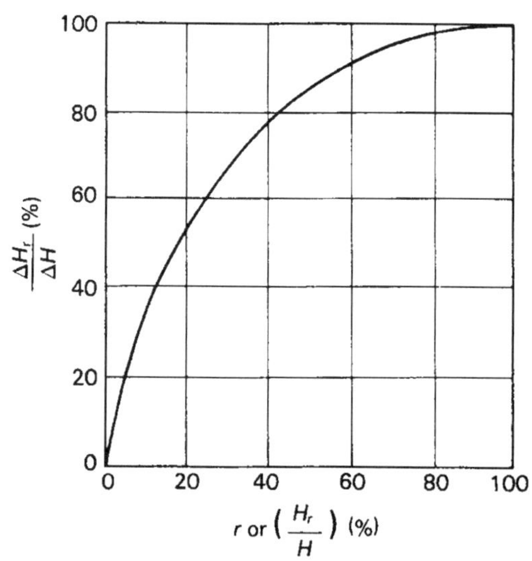

Figure 14.50 Ratio of total heave predicted for wetting over portion of depth as ratio of total heave when wetting entire active depth.

14.5.14 Example with Portion of Profile Removed by Excavation and Backfilled with Nonexpansive Soil

Let us consider the possibility of excavating the upper portion of the profile. This is considered under the category of case I. If the excavated portion is backfilled with a nonexpansive soil, the total heave calculations are considered under the category of case II. These two cases are shown in Fig. 14.51. In case I, it is assumed that the active depth is unaffected by the removal of the soil. The total heave calculations in both cases are made for the condition where the entire active depth is wetted.

The heave for any layer in the expansive soil in case I can be computed by substituting the variables in Fig. 14.51 into Eq. 14.18:

$$\Delta h_{li} = \frac{C_s}{1+e_0}\frac{(1-l)H}{j}\log\left(\frac{\rho g(1-l)H(2i-1)/2j}{\rho g H}\right) \tag{14.38}$$

where:

Δh_{li} = heave in an expansive soil layer when the active depth is partially excavated and

l = the portion of the profile excavated (i.e., H_l/H).

The above equation can be rearranged to the following form:

$$\Delta h_{li} = \frac{C_s}{1+e_0}\frac{(1-l)H}{j}\left[\log\left(\frac{2i-1}{2j}\right)+\log(1-l)\right] \tag{14.39}$$

The total heave for case I is computed by summing the individual magnitudes of heave in each soil layer:

$$\Delta H_l = \frac{C_s H}{1+e_0}(1-l)\left[\frac{1}{j}\sum_{i=1}^{j}\log\left(\frac{2i-1}{2j}\right)+\log(1-l)\right] \tag{14.40}$$

Substituting Eq. 14.27 into Eq. 14.40 for $j = 35$ gives

$$\Delta H_l = \frac{C_s H}{1+e_0}(1-l)\left[-0.430+\log(1-l)\right] \tag{14.41}$$

If $l = 0$, there is no removal of soil and Eq. 14.41 reverts to Eq. 14.27. If $l = 1$, the entire active depth has been excavated and there is no more tendency for heave to occur. A comparison between Eq. 14.41 and Eq. 14.27 can be written as

$$\frac{\Delta H_l}{\Delta H} = (1-l)\left[1-2.326\log(1-l)\right] \tag{14.42}$$

Let us now consider case II where the excavated portion of the profile is backfilled with an inert or nonexpansive soil. The total density of the backfill is assumed to be equal to the density of the expansive soil. Similarly, the total heave, ΔH, for case II can be calculated using the variables given in Fig. 14.51:

$$\Delta H_b = \frac{C_s H}{1+e_0}(1-l)\left[\frac{1}{j}\sum_{i=1}^{j}\log\left(l+\frac{2i-1}{2j}(1-l)\right)\right] \tag{14.43}$$

If $l = 0$, the maximum heave is computed, and if $l = 1$, the heave is zero. A comparison between Eq. 14.43 and Eq. 14.27 can be written as

$$\frac{\Delta H_b}{\Delta H} = -\,2.326\ \log(1-l) \times \left[\frac{1}{j}\sum_{i=1}^{j}\log\left(l+\frac{2i-1}{2j}(1-l)\right)\right] \tag{14.44}$$

Dimensionless plots illustrating cases I and II are shown in Fig. 14.52. These plots apply for all swelling indices

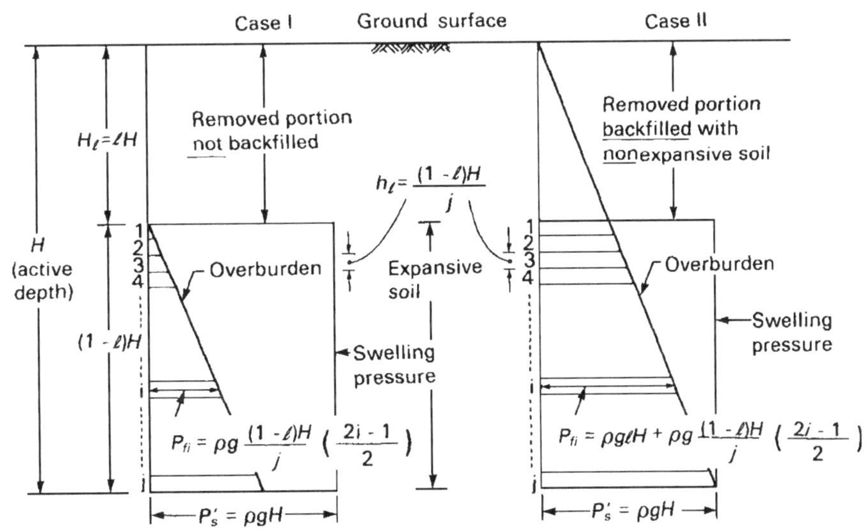

Figure 14.51 Stress distributions and definition of variables for partial excavation and backfilling with nonexpansive soil within active depth.

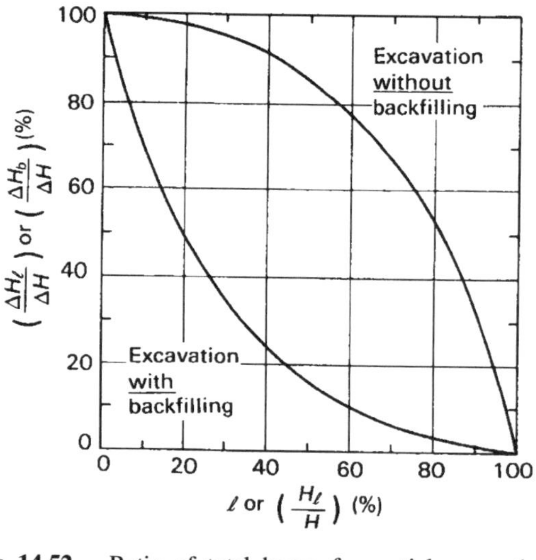

Figure 14.52 Ratio of total heave for partial excavation and backfilling to total heave for wetting of entire active depth.

and swelling pressures (provided the variables are constants) and show the influence of excavating and backfilling on computed total heave.

Figure 14.52 shows, for example, that if 50% of the active depth were removed, the total heave would still be 86% of that anticipated for the case of no excavation. However, if the excavated portion is backfilled with a nonexpansive soil, the total heave would be only 15% of that anticipated if no material were excavated. In other words, partial excavation of the expansive soil depth does not significantly reduce the total heave unless there is backfilling with an inert soil.

These examples illustrate the type of plots that can readily be generated using the closed-form equations for predicting total heave. Other possible boundary conditions could also be assumed. The plots presented are of assistance in making engineering decisions when placing structures on expansive soils. Possible remedial measures for heave reduction by flooding, excavation, and backfilling can be studied by comparing the presented plots.

14.6 SWELLING THEORY FORMULATED IN TERMS OF INCREMENTAL ELASTICITY PARAMETERS

The mechanics of volume change in unsaturated, expansive soils under isothermal condition involves two primary processes: water flow and stress-deformation. However, the desire is to present each of the physical processes as independent phenomenon. Therefore, Chapter 14 presents only the stress-deformation theory and its application to expansive soils problems. Coupled behavior (i.e., seepage and volume change) is presented in Chapter 16.

The governing partial differential equations for stress-deformation are derived based on the following assumptions: (1) the air phase is continuous and remains at atmospheric pressure, (2) soil is isotropic, nonlinear, and elastic, (3) strains are small, (4) pore-water is incompressible, and (5) the effects of air diffusing through water, air dissolving in the water, and the movement of water vapor are negligible. Tensor notation is used for the presentation of the theory of one-dimensional heave.

14.6.1 Strain-Displacement and Compatibility Relations

Let us consider a three-dimensional field with x, y, and z as the rectangular Cartesian coordinates (i.e., x and z for horizontal directions and y for vertical direction), with u_i being components of the displacement vector (i.e., comprised of u, v, and w for the x-, y- and z-directions, respectively). The components of strain tensor for the soil structure, ε_{ij}, are written in terms of displacements as follows:

$$\varepsilon_{ij} = \frac{1}{2}\left(\frac{\partial u_i}{\partial x_j} + \frac{\partial u_j}{\partial x_i}\right) \tag{14.45}$$

The normal strains can be designated as ε_x, ε_y, and ε_z for the x-, y-, and z-directions, respectively. For infinitesimal deformations, volumetric strain ε_v is the sum of the normal strain components:

$$\varepsilon_v = \frac{\partial u_i}{\partial x_i} = \frac{\partial u}{\partial x} + \frac{\partial v}{\partial y} + \frac{\partial w}{\partial z} = \varepsilon_x + \varepsilon_y + \varepsilon_z \tag{14.46}$$

The constitutive volume change behavior of the unsaturated soil is described in terms of two independent stress state variables: net normal stress $\sigma - u_a$ and matric suction $u_a - u_w$. The total volume change of an unsaturated soil element must be equal to the sum of volume changes associated with each phase and the continuity requirement reduces to

$$\frac{\Delta V_v}{V_0} = \frac{\Delta V_w}{V_0} + \frac{\Delta V_a}{V_0} \tag{14.47}$$

where:

V_0 = initial overall volume of an unsaturated soil element,
V_v = volume of soil voids,
V_w = volume of water in the element, and
V_a = volume of air in the element.

Two constitutive relationships are required to describe the volume-mass changes associated with an unsaturated soil: one for the soil structure (in terms of void ratio or volumetric strain) and another for the water phase (in terms of degree of saturation or water content). The constitutive relationships for soil structure and water phase are presented using indicial notation for the formulation of the partial differential equations for the stress-deformation processes. The stress state of an unsaturated soil can be written in terms of the net normal stress tensor $\sigma_{ij} - u_a\delta_{ij}$ and matric suction tensor $(u_a - u_w)\delta_{ij}$.

14.6.2 Soil Structure Constitutive Relationship

The constitutive relationship for the soil structure can be written in an incremental elasticity form using elasticity parameters (Fredlund and Rahardjo, 1993):

$$d\varepsilon_{ij} = \frac{1+\mu}{E} d(\sigma_{ij} - \delta_{ij} u_a) - \frac{\mu}{E} d(\sigma_{kk} - 3u_a)\delta_{ij} + \frac{d(u_a - u_w)}{H}\delta_{ij} \tag{14.48}$$

where:

σ_{ij} = components of the total stress tensor for the soil structure,
$\sigma_{kk} = \sigma_x + \sigma_y + \sigma_z$,
δ_{ij} = Kronecker delta,
E = elasticity parameter for the soil structure with respect to a change in the net normal stress,
H = elasticity parameter for the soil structure with respect to a change in matric suction, and
μ = Poisson's ratio.

Equation 14.48 can be used to write the equation for incremental volumetric strain using the compressibility form coefficients of volume change:

$$d\varepsilon_v = m_1^s d(\sigma_{\text{mean}} - \mu_a) + m_2^s d(u_a - u_w) \tag{14.49}$$

where:

$\sigma_{\text{mean}} = \frac{1}{3}\sigma_{kk} = \frac{1}{3}(\sigma_x + \sigma_y + \sigma_z)$, mean net total stress,
$m_1^s = [3(1 - 2\mu)]/E$, coefficient of volume change with respect to a change in net normal stress, and
$m_2^s = 3/H$, coefficient of volume change with respect to a change in matric suction.

The unloading constitutive relationship for the soil structure is presented graphically in the form of a constitutive surface in Fig. 14.53a. The coefficients of volume change, m_1^s and m_2^s, are slopes on the soil structure constitutive surface and can be obtained by differentiating the surface with respect to net normal stress and matric suction, respectively. The coefficients of volume change for soil structure can be written as a function of void ratio as follows:

$$m_1^s = \frac{\partial \varepsilon_v}{\partial(\sigma_{\text{mean}} - u_a)} = \frac{1}{1+e_0}\frac{\partial e}{\partial(\sigma_{\text{mean}} - u_a)} \tag{14.50}$$

$$m_2^s = \frac{\partial \varepsilon_v}{\partial(u_a - u_w)} = \frac{1}{1+e_0}\frac{\partial e}{\partial(u_a - u_w)} \tag{14.51}$$

where:

$d\varepsilon_v = de/(1 + e_0)$,
e_0 = initial void ratio of the soil, and
e = any void ratio.

14.6.3 Water Phase Constitutive Relationship

The water phase constitutive relationship can be written in an incremental compressibility form as follows (Fredlund and Rahardjo, 1993a):

$$\frac{dV_w}{V_0} = m_1^w d(\sigma_{\text{mean}} - u_a) + m_2^w d(u_a - u_w) \tag{14.52}$$

where:

m_1^w = coefficient of water volume change with respect to a change in net normal stress and
m_2^w = coefficient of water volume change with respect to a change in matric suction.

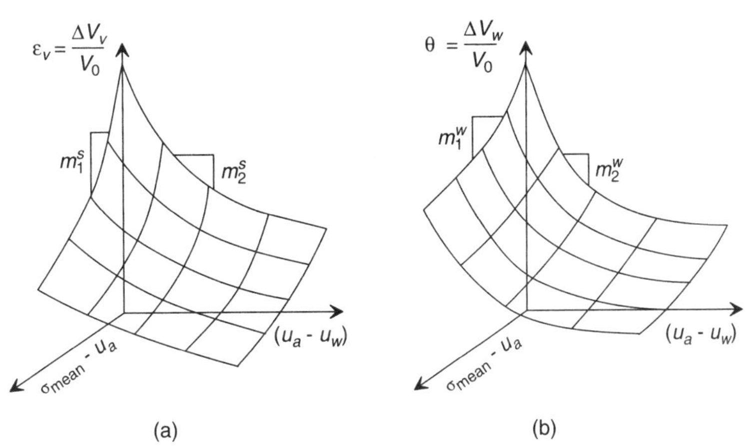

Figure 14.53 Constitutive surfaces for (a) soil structure and water phase of unsaturated, swelling soil.

Solving Eq. 14.49 for mean net normal stress and substituting into Eq. 14.52 gives

$$\frac{dV_w}{V_0} = \beta_{w1} d\varepsilon_v + \beta_{w2} d(u_a - u_w) \tag{14.53}$$

where:

$$\beta_{w1} = \frac{m_1^w}{m_1^s}, \quad \text{or} \quad \beta_{w1} \frac{m_1^w E}{3(1-2u)}$$

$$\beta_{w2} = m_2^w - \frac{m_1^w m_2^s}{m_1^s}, \quad \text{or} \quad \beta_{w2} m_2^w - \frac{m_1^w E}{(1-2\mu)H}$$

The unloading constitutive relationship for the water phase is presented graphically in a form of a constitutive surface in Fig. 14.53b. The coefficients of water volume change, m_1^w and m_2^w, indicate the amount of water taken on or released by the soil as a result of a change in the net normal stress and matric suction. Therefore, the slopes on the water phase constitutive surface can be obtained by differentiating the surface with respect to net normal stress and matric suction, respectively:

$$m_1^w = \frac{\partial\theta}{\partial(\sigma_{\text{mean}} - u_a)} \tag{14.54}$$

$$m_2^w = \frac{\partial\theta}{\partial(u_a - u_w)} \tag{14.55}$$

where:

$\theta = V_w/V_0$, volumetric water content.

14.6.4 Quantification of Boundary Conditions for Water Phase

Moisture flux and head boundary conditions are most commonly placed on the boundaries of the problem. It is necessary to perform an independent (coupled or uncoupled) seepage analysis if a moisture flux boundary condition is considered. The results of the seepage analysis would then need to be combined with a stress-deformation analysis. Moisture flux boundary conditions would need to be determined through an independent transient moisture flow modeling (see Chapter 16).

Hydraulic head (or pore-water pressure) boundary conditions are considered in this section when using a stress-deformation analysis to predict total heave in an expansive soil. In other words, the intent is to illustrate the independent stress-deformation process as it relates to an expansive soil.

14.6.5 Governing Partial Differential for Soil Structure Equilibrium for Three-Dimensional Analysis

The equations of overall static equilibrium for an unsaturated soil can be written as follows:

$$\frac{\partial\sigma_{ij}}{\partial x_j} + b_i = 0 \tag{14.56}$$

where:

σ_{ij} = components of the net total stress tensor and
b_i = components of the body force vector.

Substituting the strain-displacement Eq. 14.45 and the stress-strain Eq. 14.48 into the equilibrium equation, 14.56 gives the following governing equations for general three-dimensional problems (i.e., equations for x-, y- and z-directions):

$$G\,\nabla^2 u_i + \frac{G}{1-2\mu}\frac{\partial\varepsilon_v}{\partial x_i} - \beta\frac{\partial(u_a - u_w)}{\partial x_i} + \frac{\partial u_a}{\partial x_i} + b_i = 0 \tag{14.57}$$

where:

$$\beta = \frac{m_2^s}{m_1^s} = \frac{E/H}{1-2\mu}$$

$$\nabla^2 = \frac{\partial^2}{\partial x^2} + \frac{\partial^2}{\partial y^2} + \frac{\partial^2}{\partial z^2} \quad \text{(the Laplace operator)}$$

$$\varepsilon_v = \frac{\partial u_i}{\partial x_i} = \frac{\partial u}{\partial x} + \frac{\partial v}{\partial y} + \frac{\partial w}{\partial z}$$

$$G = \frac{E}{2(+\mu)}$$

Equation 14.57 forms the stress analysis portion in three dimensions for an unsaturated, swelling soil with a continuous air phase. The equation has essentially the same form as that presented by Biot (1941) for a soil with occluded air bubbles.

14.6.6 Governing Partial Differential for Soil Structure Equilibrium for Two-Dimensional Analysis

The stress analysis portion in two dimensions for an unsaturated, swelling soil can be written as follows:

$$\frac{\partial}{\partial x}\left(c_{11}\frac{\partial u}{\partial x} + c_{12}\frac{\partial v}{\partial y}\right) + c_{33}\frac{\partial}{\partial y}\left(\frac{\partial u}{\partial y} + \frac{\partial v}{\partial x}\right) - d_s\frac{\partial(u_a - u_w)}{\partial x} + b_x = 0 \tag{14.58}$$

$$c_{33}\frac{\partial}{\partial x}\left(\frac{\partial u}{\partial y} + \frac{\partial v}{\partial x}\right) + \frac{\partial}{\partial y}\left(c_{12}\frac{\partial u}{\partial x} + c_{22}\frac{\partial v}{\partial y}\right) - d_s\frac{\partial(u_a - u_w)}{\partial y} + b_y = 0 \tag{14.59}$$

where

$$c_{11} = c_{22} = \frac{(1-\mu)E}{(1+\mu)(1-2\mu)}$$

$$c_{12} = \frac{\mu E}{(1+\mu)(1-2\mu)}$$

$$c_{33} = \frac{E}{2(1+\mu)}$$

$$d_s = \frac{E}{(1-2\mu)H}$$

Equations 14.58 and 14.59 can be used to compute displacements in the horizontal and vertical directions under an applied load and/or due to changes in matric suction. Solving the soil structure equilibrium equations 14.58 and 14.59 with a saturated-unsaturated seepage analysis constitutes either an uncoupled or a coupled solution (see Chapter 16).

The solution of the stress-deformation equations 14.58 and 14.59 under specified boundary conditions requires the designation of initial matric suctions, initial stress conditions, and elasticity parameter functions associated with the volume change of the soil.

14.7 EVALUATION OF ELASTICITY PARAMETER FUNCTIONS FROM VOLUME CHANGE INDICES

A numerical stress-deformation analysis in two or three dimensions requires the determination of equivalent elastic parameters E and H, which are functions of stress state. The elasticity parameters can be calculated from the coefficients of volume change, m_1^s and m_2^s, which represent the differentiation of the constitutive surface for the soil structure (i.e., Eqs. 14.50 and 14.51, respectively). Laboratory tests can be performed to obtain the void ratio constitutive surface (e.g., K_0 loading, plane strain loading, or isotropic loading).

Three fundamental elasticity parameters are required in the constitutive equations used for the prediction of heave (i.e., E, H, and μ). However, there are only two coefficients of volume change obtained from the constitutive surface for the soil structure (i.e., m_1^s and m_2^s). Therefore, it is suggested that Poisson's ratio be assumed or measured in order to convert the coefficients of volume change for different loading conditions to the required elastic parameters.

14.7.1 Calculation of Elasticity Parameter Functions

The elasticity parameter functions can be calculated and then applied to one-, two-, and three-dimensional analyses (Fig. 14.54). Table 14.4 presents the calculation of the elasticity parameter functions for various loading conditions (Fredlund and Rahardjo, 1993a).

The coefficients obtained from one loading condition can theoretically be converted to other loading conditions provided Poisson's ratio is known:

$$m_1^s = \frac{3(1-\mu)}{1+\mu} m_{1-1D}^s = \frac{3}{2(1+\mu)} m_{1-2D}^s \quad (14.60)$$

$$m_2^s = \frac{3(1-\mu)}{1+\mu} m_{2-1D}^s = \frac{3}{2(1+\mu)} m_{2-2D}^s \quad (14.61)$$

where:

m_{1-1D}^s = soil structure compressibility as a function of net normal stress measured in a one-dimensional or K_0 test,

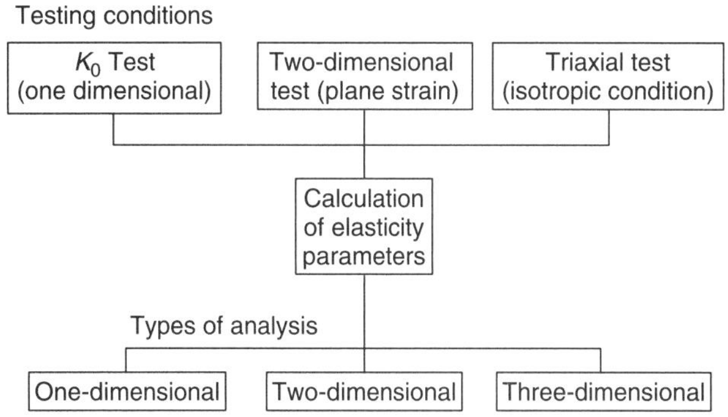

Figure 14.54 Proposed procedure for use of incremental elasticity functions for prediction of heave.

m_{1-2D}^s = soil structure compressibility as a function of net normal stress measured in a two-dimensional test,

m_{2-1D}^s = soil structure compressibility as a function of matric suction measured in a one-dimensional or K_0 test, and

m_{2-2D}^s = soil structure compressibility as a function of matric suction measured in a two-dimensional test.

The coefficients of volume change can be obtained through the conversion of a semilogarithmic plot of void ratio onto an arithmetic plot (Vu and Fredlund, 2000). The swelling indices C_s and C_m are the slope of the void ratio versus logarithm of net normal stress or matric suction, as shown in Fig. 14.55. The semilogarithmic plot of the void ratio constitutive surface for an unsaturated

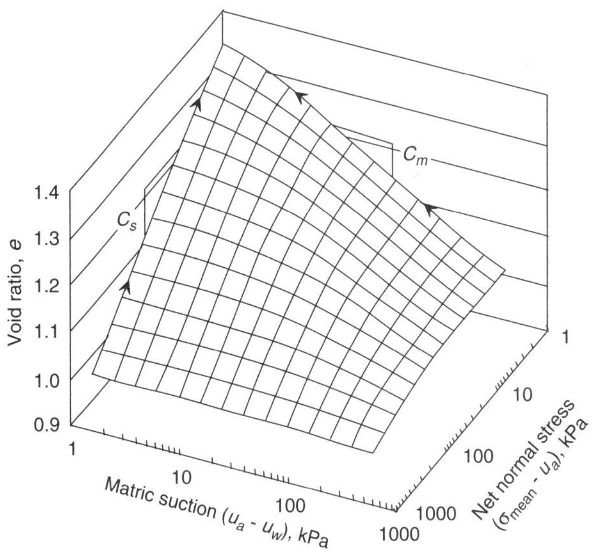

Figure 14.55 Typical void ratio constitutive surface plotted using semilogarithm scale.

Table 14.4 Relationship between Fundamental Elasticity Parameters and Coefficients of Volume Change

Loading	Stress State Variables	Coefficient of Volume Change Functions	Elasticity Parameter Functions
Three-dimensional (3D)	$\sigma_{mean} - u_a$, $(u_a - u_w)_{3D}$	$m_1^s = \dfrac{\partial \varepsilon_v}{\partial(\sigma_{mean} - u_a)} = \dfrac{0.434}{1+e_0}\dfrac{C_s}{\sigma_{mean} - u_a}$ [a]	$E = 3\dfrac{(1-2\mu)}{m_1^s}$
		$m_2^s = \dfrac{\partial \varepsilon_v}{\partial(u_a - u_w)_{3D}} = \dfrac{0.434}{1+e_0}\dfrac{C_m}{(u_a - u_w)_{3D}}$	$H = \dfrac{3}{m_2^s}$
K_0 loading (1D)	$\sigma_v - u_a$, $(u_a - u_w)_{1D}$	$m_{1-1D}^s = \dfrac{\partial \varepsilon_v}{\partial(\sigma_y - u_a)} = \dfrac{0.434}{1+e_0}\dfrac{C_s}{\sigma_y - u_a}$	$E = \dfrac{(1+\mu)(1-2\mu)}{(1-\mu)m_{1-1D}^s}$
		$m_{2-1D}^s = \dfrac{\partial \varepsilon_v}{\partial(u_a - u_w)_{1D}} = \dfrac{0.434}{1+e_0}\dfrac{C_m}{(u_a - u_w)_{1D}}$	$H = \dfrac{1+\mu}{(1-\mu)m_{2-1D}^s}$
Plane strain (2D)	$\sigma_{ave} - u_a$, $(u_a - u_w)_{2D}$	$m_{1-2D}^s = \dfrac{\partial(\varepsilon_x + \varepsilon_y)}{\partial(\sigma_{ave} - u_a)} = \dfrac{0.434}{1+e_0}\dfrac{C_s}{u_{ave} - u_a}$	$E = \dfrac{2(1+\mu)(1-2\mu)}{m_{1-2D}^s}$
		$m_{2-2D}^s = \dfrac{\partial(\varepsilon_x + \varepsilon_y)}{\partial(\sigma_a - u_w)_{2D}} = \dfrac{0.434}{1+e_0}\dfrac{C_m}{(u_a - u_w)_{2D}}$	$H = \dfrac{2(1+\mu)}{m_{2-2D}^s}$

Source: After Fredlund and Rahardjo (1993a).
[a]Note: $\sigma_{mean} = (\sigma_x + \sigma_y + \sigma_z)/3$; $\sigma_{ave} = (\sigma_x + \sigma_y)/2$.
[b]$d\varepsilon_y = d\varepsilon_v$ for K_0 loading; $d(\varepsilon_x + \varepsilon_y) = d\varepsilon_v$ for plane strain.

soil is approximately linear (on the extreme planes) over a relatively large range of stresses (Ho and Fredlund, 1992a).

The elastic parameter functions E and H can also be calculated directly from volume change indices C_s (on the net normal stress plane) and C_m (on the matric suction plane), respectively. The elastic parameters are calculated from the coefficient of volume change, as shown in Table 14.4. The elastic parameter E can be expressed as a function of the volume change index with respect to net normal stress C_s, initial void ratio, and Poisson's ratio. The elastic parameter H can be expressed as a function of the volume change index with respect to matric suction C_m, initial void ratio, and Poisson's ratio. The equations for these elastic parameters can be written for general three-dimensional loading conditions as follows:

$$E = \frac{6.908(1-2\mu)(1+e_0)}{C_s}(\sigma_{mean} - u_a) \tag{14.62}$$

$$H = \frac{6.908(1+e_0)}{C_m}(u_a - u_w) \tag{14.63}$$

The 6.908 constant arises from the conversion between the logarithmic and arithmetic scales (i.e., 6.908 = 3/ $\log_{10}$ 2.718).

Equations 14.62 and 14.63 can be written for two-dimensional plane strain conditions as follows:

$$E = \frac{4.605(1+\mu)(1-2\mu)(1+e_0)}{C_s}(\sigma_{ave} - u_a) \tag{14.64}$$

$$H = \frac{4.605(1+\mu)(1+e_0)}{C_m}(u_a - u_w) \tag{14.65}$$

The 4.605 constant arises from the conversion between the logarithmic and arithmetic scales (i.e., 4.605 = 2/ $\log_{10}$ 2.718).

The elastic parameter E (or H) increases with an increase in net normal stress (or matric suction) and a decrease of the volume change index C_s (or C_m) assuming a constant value of Poisson's ratio. Figures 14.56 and 14.57 graphically illustrate the relationship between the elastic parameter E and net normal stress for various values of the swelling index C_s and various values of Poisson's ratio, respectively. Figure 14.58 presents the variation of elastic parameter H with matric suction for various values of swelling index C_m.

14.7.2 Determination of Swell Index

The swelling indices can be measured experimentally or estimated through correlation with the Atterberg limits

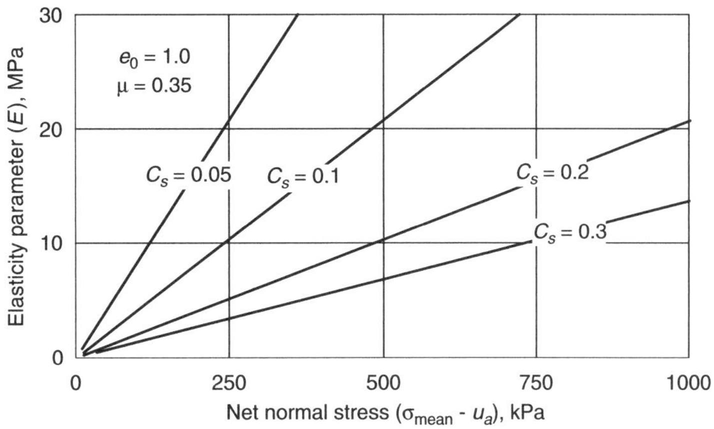

Figure 14.56 Relationship between elasticity parameter E and net normal stress for various values of swelling index, C_s.

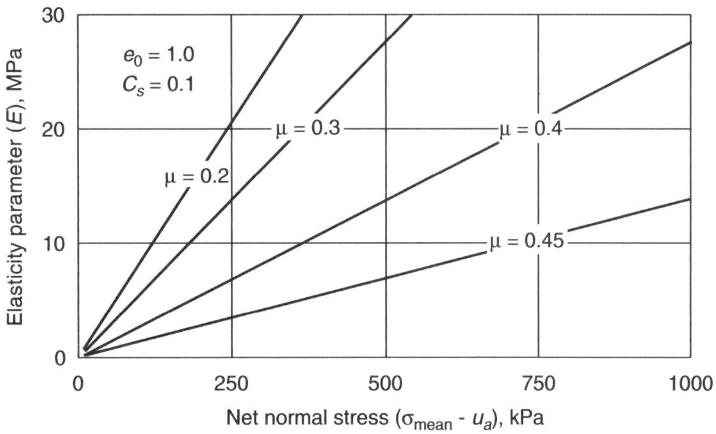

Figure 14.57 Relationship between elasticity parameter E and net normal stress for various values of Poisson's ratio, μ.

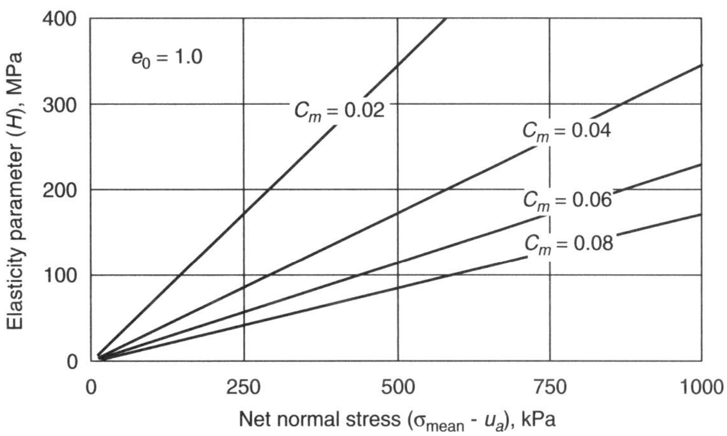

Figure 14.58 Relationship between elasticity parameter H and net normal stress for various values of swelling index, C_m.

(Fredlund and Rahardjo, 1993a, Lytton, 1994). Test procedures for the estimation of swelling indices are presented in Fredlund and Rahardjo (1993a) and ASTM standard. The ASTM standards related to the measuring of swelling indices include ASTM D4546, ASTM D2435, and ASTM D427. It is suggested that the swelling indices commonly used for a conventional heave analysis in one dimension can be used for general two- and three-dimensional numerical modeling.

14.7.3 Determination of Poisson's Ratio

Poisson's ratio may not be a constant value for an unsaturated, expansive soil but may be a function of stress state (i.e., net normal stress and matric suction). Poisson's ratio appears to increase with increasing net mean stress and decreasing matric suction (Pereira and Fredlund 2000). Numerous researchers (Miranda, 1988; Alonso et al., 1988; Lloret and Ledesma, 1993) have used a constant Poisson's ratio of 0.3 when performing numerical simulations of unsaturated, collapsing soils. It was suggested that a Poisson's ratio of 0.3 might reflect the as-compacted condition of a loosely compacted embankment. The choice of a value for Poisson's ratio can also be related to an experimentally observed relationship between K_0 and OCR. Poisson's ratio is assumed to be related to the coefficient of earth pressure at rest, K_0:

$$\mu = \frac{K_0}{1 + K_0} \tag{14.66}$$

14.7.4 Initial Matric Suction and Stress Conditions

Initial matric suction conditions can be measured in the field or laboratory (Fredlund and Rahardjo, 1993a), but this can be a challenging and costly process. It is also possible to estimate in situ suctions based on theoretical considerations.

Net normal stress state within the soil mass can either be computed by switching on gravity or simply estimated from the total unit weight of the soil. Horizontal net normal stresses are empirically linked to the vertical stresses through the coefficient of earth pressure at rest, K_0:

$$\sigma_y = \int_0^H \rho g \, dy \tag{14.67}$$

$$\sigma_x = K_0 \sigma_y \tag{14.68}$$

where:

σ_x, σ_y = horizontal and vertical net normal stress, respectively,
y = vertical distance from ground surface, and
H = depth of soil under consideration.

14.7.5 Coefficient of Earth Pressure At Rest

The coefficient of earth pressure can be as low as zero and as high as the coefficient of passive earth pressure. Some procedures have been suggested in the research literature for the estimation of the coefficient of earth pressure at rest. Fredlund and Rahardjo (1993a) considered elastic equilibrium within a homogenous, isotropic soil mass and presented the following equation for the coefficient of earth pressure at rest:

$$K_0 = \frac{\mu}{1 - \mu} - \frac{E}{(1 - \mu)H} \frac{u_a - u_w}{\sigma_v - u_a} \tag{14.69}$$

Equation 14.69 can be rewritten as follows:

$$K_0 = \frac{\mu}{1 - \mu} - \frac{1 - 2\mu}{1 - \mu} \frac{C_m}{C_s} \frac{u_a - u_w}{\sigma_v - u_a} \tag{14.70}$$

Considering the effect of previous stress paths (i.e., previous wetting and drying, loading and unloading), the coefficient of earth pressure should have the following tangential value:

$$K_0 = \frac{\mu}{1 - \mu} - \frac{1 - 2\mu}{1 - \mu} \frac{C_m}{C_s} \frac{\Delta(u_a - u_w)}{\Delta(\sigma_v - u_a)} \tag{14.71}$$

Jaky (1944) estimated the coefficient of earth pressure for normally consolidated soils, K_{nc}, from the effective stress parameter ϕ:

$$K_{nc} = 1 - \sin\phi \tag{14.72}$$

Wroth (1979) proposed two empirical relationships between K_0, K_{nc}, and OCR:

$$K_0 = K_{nc}\text{OCR} - \frac{\mu}{1 - \mu}(\text{OCR} - 1) \tag{14.73}$$

$$m\left(\frac{3(1 - K_{nc})}{1 + 2K_{nc}} - \frac{3(1 - K_0)}{1 + 2K_0}\right) = \ln\left(\frac{(1 + 2K_{nc})\text{OCR}}{1 + 2K_0}\right) \tag{14.74}$$

where:

m = 0.0022875PI + 1.22 and
PI = plasticity index.

Equation 14.73 provides a reasonable fit to existing data for soils up to an OCR of about 5. The value of Poisson's ratio necessary to fit observed data lies in the range of 0.25–0.37 for a number of different soils. Equation 14.74 has also been proposed for soils with even higher values of OCR (Wroth, 1979).

Mayne and Kulhawy (1982) suggested that a modified Jaky's equation be used to estimate the coefficient of earth pressure at rest:

$$K_0 = K_{nc}\,\text{OCR}^{\sin\phi} \tag{14.75}$$

where:

ϕ = angle of internal friction and
OCR = overconsolidation ratio.

Lytton (1994) presented the following typical values for the coefficients of lateral earth pressure back calculated from field observations of heave and shrinkage:

$$K_0 = \begin{cases} 0 & \text{when soil is dry and cracked} \\ 0.333 & \text{when soil is dry and cracks are opening} \\ 0.500 & \text{when cracks are closed and suction is at steady-state condition} \\ 0.667 & \text{when cracks are closed and soil is wetting} \\ 1 & \text{when soil is wetting and is in hydrostatic stress condition} \\ 2\text{ - }3 & \text{when soil is approaching passive earth pressure} \end{cases} \quad (14.76)$$

The coefficient of earth pressure at rest presented by Eq. 14.76 can be used to estimate net normal stress.

14.8 ONE-DIMENSIONAL SOLUTION USING INCREMENTAL ELASTICITY FORMULATION

The example problem shown in Fig. 14.38 was earlier solved for total heave using longhand calculation with known values for swelling index C_s and swelling pressure P_s. The same example can now be solved with a numerical solution using an incremental elasticity-type formulation. A comparison of the two methodologies and the calculated total heave is presented.

14.8.1 Heave Analysis Formulation

The force equilibrium equation for K_0 loading conditions (i.e., volume change in the y-direction only) can be expressed as follows:

$$\frac{\partial \sigma_y}{\partial x} + \rho g = 0 \quad (14.77)$$

where:

ρ = total density of the soil and
g = acceleration due to gravity.

The soil structure constitutive relationship for K_0 loading can be written as follows:

$$\sigma_y - u_a = \frac{(1-\mu)E}{(1+\mu)(1-2\mu)}\left(\varepsilon_v - \frac{1+\mu}{H(1-\mu)}(u_a - u_w)\right) \quad (14.78)$$

where:

E = elastic modulus for the soil structure with respect to net normal stress,
H = elastic modulus for the soil structure with respect to matric suction, and
μ = Poisson's ratio.

The differential expression of normal strain in terms of displacement in the y-direction (i.e., v) is as follows:

$$\varepsilon_y = \frac{\partial v}{\partial y} \quad (14.79)$$

The governing differential equation for the K_0 loading condition can be obtained by substituting the expression for net normal stress (i.e., Eq. 14.78) into the force equilibrium equation (i.e., Eq. 14.77). Pore-air pressure is assumed to remain at atmospheric conditions:

$$\frac{\partial}{\partial y}\left[\frac{E(1-\mu)}{(1+\mu)(1-2\mu)}\left(\frac{\partial v}{\partial y} - \frac{1+\mu}{H(1-\mu)}(u_a - u_w)\right)\right] + \rho g = 0 \quad (14.80)$$

The elastic moduli E and H can be calculated from the swelling indices, initial void ratio, and assumed Poisson's ratio as follows:

$$E = n_t(\sigma - u_a)_{\text{ave}} \quad (14.81)$$

$$H = n_m\left(u_a - u_w\right)_{\text{ave}} \quad (14.82)$$

where:

n_t = coefficient that relates net normal stress with the elastic modulus E,
n_m = coefficient that relates matric suction with the elastic modulus H,
$(\sigma - u_a)_{\text{ave}}$ = average of the initial and final net normal stress for an incremental change in stress, and
$(u_a - u_w)_{\text{ave}}$ = average of the initial and final matric suctions for an incremental suction change.

Equations 14.81 and 14.82 show that the elastic parameters change linearly with either of the stress state variables. Equations 14.81 and 14.82 are assumed to be linear, an assumption based on a history of laboratory measurements that have shown the swelling index of a soil, C_s, to approximate a constant value. Laboratory measurements by Meilani et al. (2005) showed that the elastic moduli H can vary in a nonlinear manner with respect to stress state. The coefficients of proportionality n_t and n_m can be written in terms of Poisson's ratio and the respective compressive indices:

$$n_t = \frac{(1+\mu)(1-2\mu)}{1-\mu}\left(\frac{1+e_0}{0.4343C_t}\right) \quad (14.83)$$

$$n_m = \frac{1+\mu}{1-\mu}\left(\frac{1+e_0}{0.4343C_m}\right) \quad (14.84)$$

where:

C_t = volume change index with respect to net normal stress,

C_m = volume change index with respect to matric suction, and

e_0 = initial void ratio.

Equation 14.80 can be used to compute vertical displacement due to net normal stress changes and/or matric suction changes. A finite element numerical model can be used to solve the stress-deformation equations provided appropriate head (or pore-water pressure) boundary conditions are specified.

Equation 14.80 is nonlinear since the elastic moduli are functions of the stress state variables. An incremental, marching-forward procedure can be used as shown in Fig. 14.59. Both total load changes and matric suction changes need to be divided into increments and applied in incremental steps. The elastic moduli are assumed to remain constant within each incremental step. A new modulus is selected at the beginning of the application of each new increment of load (or matric suction). The displacements computed for each incremental step of loading are accumulated to give the total displacement or heave.

14.8.2 Reanalysis of Previous Examples Using Incremental Elastic Analysis

The example problem involving the prediction of total heave in a 2-m layer of expansive clay is reanalyzed to demonstrate the use of a one-dimensional, finite element numerical model to compute total heave. The results are compared to the nonlinear, longhand solution presented in Fig. 14.38. The example problem assumes that the pore-water pressures in the soil will increase with time to establish a new equilibrium pore-water pressure profile.

The longhand analysis used only three layers and computed a total heave of 117 mm. Figure 14.60 shows the results of the finite element numerical solution when the 2-m layer of expansive clay is divided into varying numbers of layers. The finite element numerical model shows that the total heave is computed to be 117.3 mm when 25 one-dimensional finite elements are used. Further refinement in the number of finite elements shows little additional improvement in the solution.

The prediction of heave using the one-dimensional solution requires that the elastic modulus with respect to matric

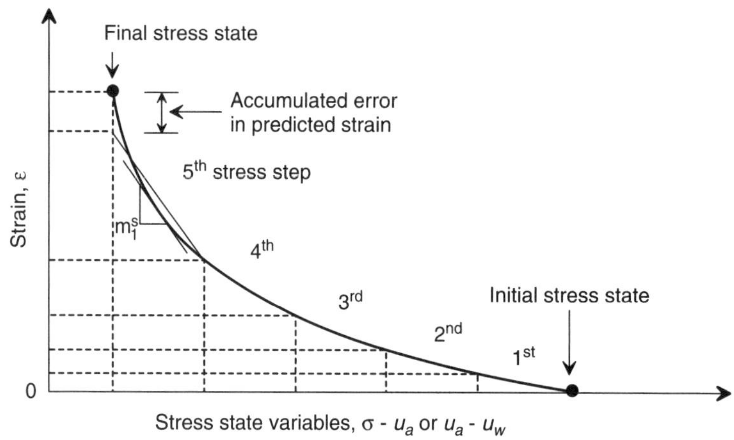

Figure 14.59 Incremental elasticity procedure for calculating amount of heave when a soil is wetted.

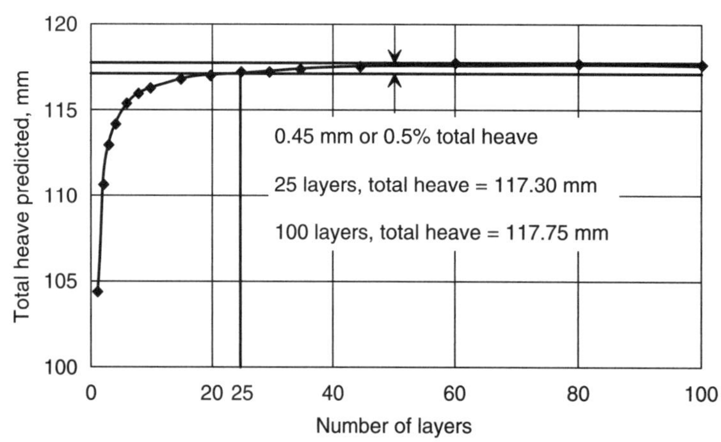

Figure 14.60 Total heave calculated using varying numbers of soil layer in calculations.

suction be computed. The matric suction equivalent can be used in place of the actual in situ matric suctions when performing the calculations. Consequently, the swelling index on the total stress plane C_{ts} is used instead of the swelling index with respect to matric suction C_m when performing the comparative calculations.

The average of the initial and final stress states is used for each increment when calculating the elastic modulus function. The initial void ratio was 1.0 and the swelling index C_{ts} was 0.1. A Poisson's ratio of 0.3 was selected. The matric suction modulus equation can be written as follows for the total stress plane:

$$H = n_t \left(\sigma_y - u_a\right)_{\text{ave}} \tag{14.85}$$

The total stress state is expressed as the sum of the overburden pressure and the matric suction equivalent, as shown in Fig. 14.61. A constant swelling index from a semilog plot produces a linear elastic modulus function. Consequently, the analysis is nonlinear and the solution is a function of the refinement of the layers used in solving the problem.

Figure 14.61 shows that the elastic modulus H decreases with the decreasing stress level near ground surface. The decreased elastic modulus results in increased amounts of heave being computed at extremely small stress levels. Figure 14.62 shows that the total heave calculated increases almost linearly with decreasing stress level steps. The total heave predicted when using a stress level of 1 kPa (i.e., 200 stress steps) is 117 mm while the total heave predicted when using a stress step level of 7.2 kPa (i.e., 25 stress steps) is 115 mm. The error due to nonlinearity associated with small stress steps appears to be on the order of 0.5% from the correct solution. The change in the computed results in going from using 25 stress steps to 200 stress steps was about 2%.

The sensitivity of the analysis to stress step level is greatest when the expansive clay exists to the ground surface and an attempt is made to compute heave as the total stress (i.e., the overburden pressure) goes toward zero. It is clear that the placement of a small amount of nonswelling material (e.g., a gravel layer) on the surface of the clay layer makes the solution less sensitive to the number of stress steps used in the computations. Figure 14.63 illustrates how the finite element solution converges onto the correct solution as the analysis moves toward calculations for all stress increments.

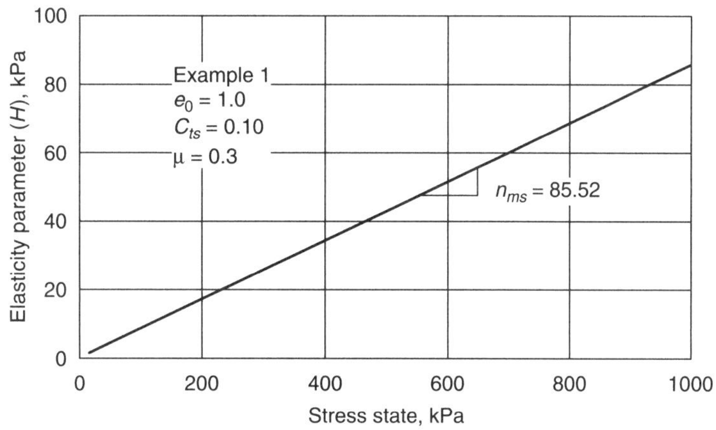

Figure 14.61 Elastic modulus function H as function of stress state.

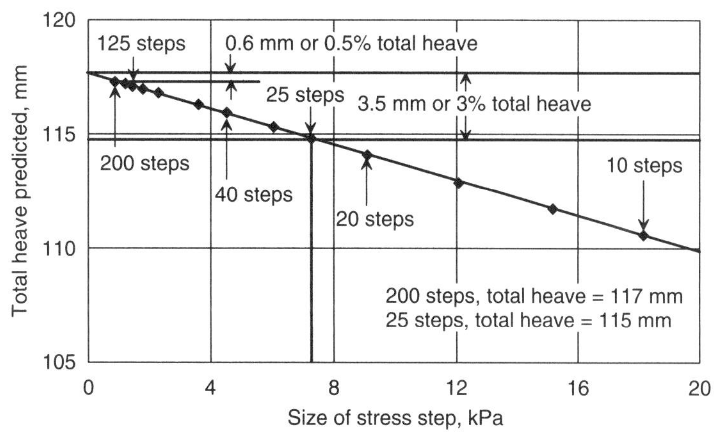

Figure 14.62 Total heave versus size of each stress increment.

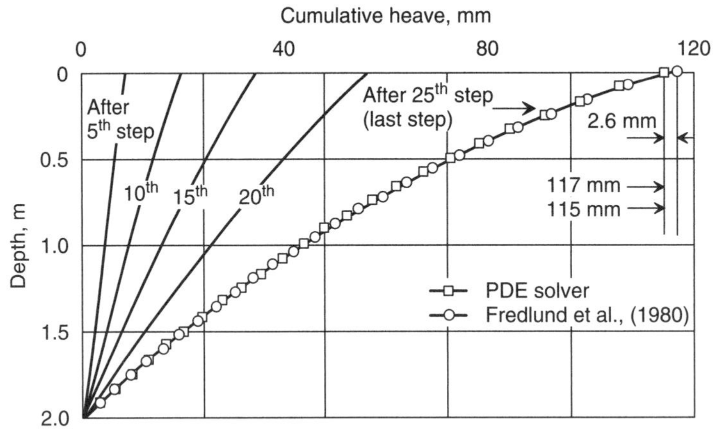

Figure 14.63 Calculation of total heave as stress steps are taken from initial in situ stress state to condition of zero suction in soil.

14.9 TWO-DIMENSIONAL SOLUTION USING INCREMENTAL ELASTICITY FORMULATION

The total stress fields in a soil continuum can be solved using Eqs. 14.48 and 14.49. Consideration will only be given to changes in total stresses through a two-dimensional continuum space. The pore-water pressure and the pore-air pressure are assumed to remain unchanged as a total stress loading is applied at the ground surface. In other words, there is a decoupling of the pore fluid effects from the application of a total stress at ground surface. The coupling of total stresses and pore fluid pressure changes is considered in Chapter 16.

One example problem is solved to illustrate that the independent calculation of total stress field in unsaturated soils is similar to the calculation of total stress fields for a dry soil mass.

14.9.1 Example Problem Involving Comparison with Bousinesq Stress Analysis

The load applied by a long strip footing is used to illustrate the calculation of total stress fields in an unsaturated soil mass (Fig. 14.64). Total stresses are calculated using a finite element numerical model (SV Solid). A constant Young's modulus E of 10,000 kPa is assumed along with a Poisson's ratio of 0.33. The applied ground surface pressure is 1000 kPa applied over a 1-m-wide surface. The intent is to examine the calculated stresses and therefore no attention is paid to the computed displacements.

Figure 14.64 presents the stress influence values below a strip footing obtained from a closed-form solution when using the Bousinesq theory. Also shown are the influence factors computed from the solution of Eqs. 14.58 and 14.59. The finite element method was used to solve the partial differential equations for the two-dimensional stress analysis. The results indicate close agreement between the total stress states computed by the two methods.

14.10 CHALLENGES IN NUMERICALLY MODELING OF EXPANSIVE SOIL PROBLEMS

There are a number of challenges that have arisen with respect to the numerical modeling of expansive soil behavior. For example, there has not emerged a single constitutive equation to adequately describe the entire void ratio constitutive surface of an unsaturated, expansive soil. The soil properties are generally measured in terms of compressibility values from laboratory oedometer tests. The laboratory results yield a compressibility (or swelling) index as a single soil property. In order to use the results in a multidimensional stress-deformation analysis, it is necessary to estimate a Poisson's ratio value while a Young's modulus value is converted from the oedometer test results. In other words, one soil property is measured; however, two soil properties are required in order to perform the analysis. The incremental linear elastic approach also has its limitations.

Another significant challenge occurs when attempting to model conditions where the matric suction goes to zero under low net normal stress conditions. When an expansive soil exists at ground surface, the net normal stress goes to zero if there is no applied load. If at the same time the matric suction goes to zero, there can be difficulties in calculating the total heave. These issues are related to the lack of definition of the constitutive surface under low-stress conditions. The need to combine the stress-deformation analysis with a saturated-unsaturated seepage analysis is further discussed in Chapter 16.

14.10.1 Equations for Void Ratio Constitutive Relationship

Several equations have been proposed to define the constitutive soil properties with respect to changes in the stress state of a soil. These equations are capable of representing the soil properties for specific stress state changes. The equations are determined for a particular soil under

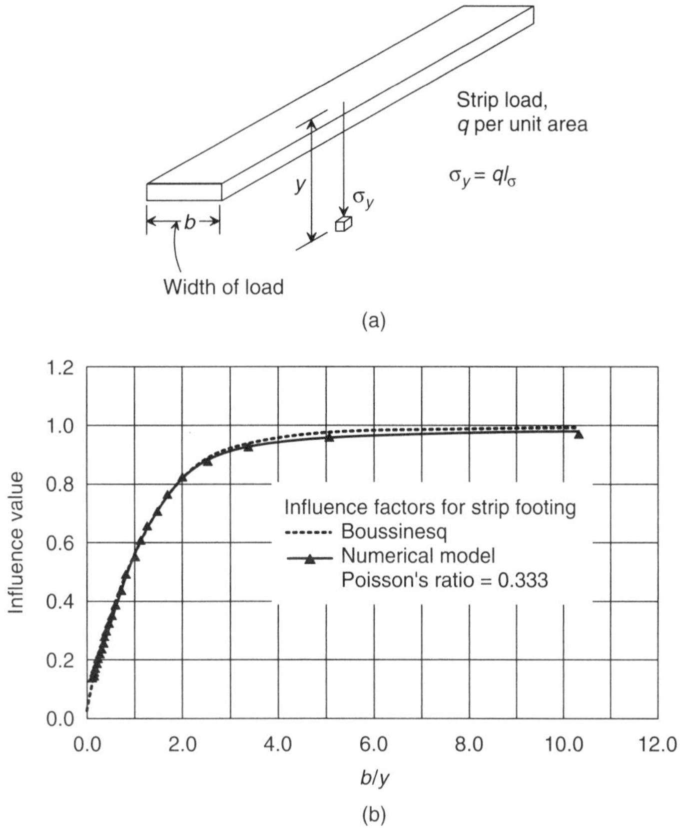

Figure 14.64 Calculation of influence factors for total stresses below a strip footing: (a) definition of influence factors; (b) comparison of numerical modeling solution to closed-form Bousinesq theory.

specific testing conditions using a best-fit regression analysis on measured data.

Some mathematical equations are shown that describe the void ratio constitutive surfaces associated with an unsaturated, swelling soil. Terzaghi (1943) and Casagrande (1936) noted that the virgin compression curve of the void ratio versus logarithm of effective stress (i.e., $\sigma - u_w$) is essentially a straight line. The equation for the virgin compression portion of void ratio can be written as follows for a saturated soil:

$$e = e_0 - C_c \log\left(\frac{\sigma}{\sigma'_p}\right) \tag{14.86}$$

where:

e_0 = initial void ratio,
C_c = volume change index for the virgin compression curve, and
σ'_p = preconsolidation pressure.

A similar equation can be written for the rebound curve (i.e., unloading-reloading curve) that takes on the following form:

$$e = a + b \log(\sigma) \tag{14.87}$$

where:

σ = designation of the stress state and
a, b = constants (i.e., fitting parameters).

Fredlund (1979) suggested that the void ratio constitutive surface for an unsaturated soil could be linear over a wide range of stress changes using the logarithm of the stress state variables. The void ratio under any set of stress conditions is written as follows:

$$e = e_0 - C_t \log\frac{\sigma - u_a}{(\sigma - u_a)_0} - C_m \log\frac{u_a - u_w}{(u_a - u_w)_0} \tag{14.88}$$

Similarly, the water content under any set of stress conditions is written as follows:

$$w = w_0 - D_t \log\frac{\sigma - u_a}{(\sigma - u_a)_0} - D_m \log\frac{u_a - u_w}{(u_a - u_w)_0} \tag{14.89}$$

where:

e_0 = initial void ratio,
C_t = volume change index with respect to net normal stress,
C_m = volume change index with respect to matric suction,

$(\sigma - u_a)_0$ = initial net normal stress,
$(u_a - u_w)_0$ = initial matric suction,
w_0 = initial water content,
D_t = water content index with respect to net normal stress, and
D_m = water content index with respect to matric suction.

Equations 14.88 and 14.89 have the following form:

$$e = a + b\ \log(\sigma - u_a) + c\ \log(u_a - u_w) \tag{14.90}$$

where:

a, b, c = constants (i.e., fitting parameters).

Lloret and Alonso (1985) studied a number of mathematical equations for the description of the volume change constitutive surface of unsaturated soils subjected to confined and isotropic compression. The equations were used to best fit experimental results on different soils and the optimum equations were selected on the basis of minimum fitting errors. The following conclusions were drawn for the void ratio and degree of saturation constitutive surfaces. For a limited range in total stress, Lloret and Alonso (1985) suggested that a suitable analytical expression for the void ratio constitutive surface is as follows:

$$\begin{aligned} e = a + b(\sigma - u_a) + c\ \log(u_a - u_w) \\ + d(\sigma - u_a)\ \log(u_a - u_w) \end{aligned} \tag{14.91}$$

A more suitable equation for the void ratio surface was given by Lloret and Alonso (1985) if the range in stress variation is large:

$$\begin{aligned} e = a + b\ \log(\sigma - u_a) + c\ \log(u_a - u_w) \\ + d\ \log(\sigma - u_a)\ \log(u_a - u_w) \end{aligned} \tag{14.92}$$

Either of the following two equations was suggested to describe the degree of saturation constitutive surface:

$$S = a - \left(\frac{e^{b(u_a - u_w)} - e^{-b(u_a - u_w)}}{e^{b(u_a - u_w)} + e^{-b(u_a - u_w)}} \right) \left[c + d(\sigma - u_a) \right] \tag{14.93}$$

$$S = a - \left(1 - e^{-b(u_a - u_w)} \right) \left[c + d(\sigma - u_a) \right] \tag{14.94}$$

where:

a, b, c, d = constants (i.e., fitting parameters).

Ho et al. (1992) assumed a linear relation between net normal stress and matric suction at constant void ratio. The constitutive curves on extreme planes of the void ratio plot are essentially linear on the semilogarithmic scale. The void ratio surface was approximated on a semilogarithmic plot by three planes, namely, planes I, II, and III. The three planes converge at a void ratio ordinate corresponding to nominal values of the stress state variables [i.e., $\log(\sigma - u_a) = 0.1$ and $\log(u_a - u_w) = 0.1$]. Plane I is perpendicular to the void ratio versus $\log(\sigma - u_a)$ plane. Plane III is perpendicular to the void ratio versus $\log(u_a - u_w)$ plane. Plane II represents a transition zone between planes I and III. Plane II intersects both the void ratio versus $\log(\sigma - u_a)$ plane and void ratio versus $\log(u_a - u_w)$ plane. The graphical illustration of the approximated void ratio surface for swelling portion is presented in Fig. 14.65. The equations describing planes I, II, and III can be written as follows:

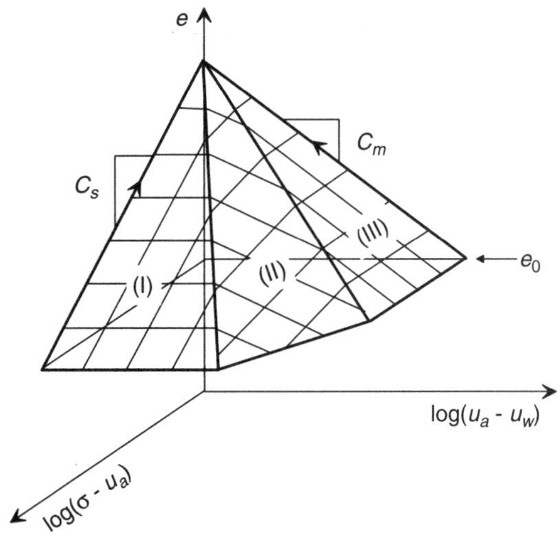

Figure 14.65 Void ratio constitutive surface representing unloading conditions (from Ho, 1988).

$$e = \begin{cases} e_0 + C_s \log\left(\dfrac{(\sigma - u_a)_0}{\sigma - u_a}\right) & \text{for plane I} \\ e_0 + C_s' \log\left(\dfrac{(\sigma - u_a)_0}{\sigma - u_a}\right) + C_m' \log\left(\dfrac{(u_a - u_w)_0}{u_a - u_w}\right) & \text{for plane II} \\ e_0 + C_m \log\left(\dfrac{(u_a - u_w)_0}{u_a - u_w}\right) & \text{for plane III} \end{cases} \tag{14.95}$$

where:

e_0 = initial void ratio,
$(\sigma - u_a)_0$ = initial net normal stress,
$(u_a - u_w)_0$ = initial matric suction,
C_s = slope of the intersection line of plane I with the void ratio versus $\log(\sigma - u_a)$ plane,
C_m = slope of the intersection line of plane III with the void ratio versus $\log(u_a - u_w)$ plane,
C_s' = slope of the intersection line of plane II with the void ratio versus $\log(\sigma - u_a)$ plane, and
C_m' = slope of the intersection line of plane II with the void ratio versus $\log(u_a - u_w)$ plane.

M.D. Fredlund (2000) proposed a four-parameter equation that can be used to represent an overconsolidated soil that contains both a recompression curve and a virgin compression branch. The equation of the void ratio as a function of stress was written as follows:

$$e = e_0 - \frac{C_r}{2} \ln\left[1 + \left(\frac{\sigma}{P_s}\right)^2\right] - \frac{C_c - C_r}{2} \ln\left[1 + \left(\frac{\sigma}{\sigma_p'}\right)^2\right] \quad (14.96)$$

where:

e_0 = initial void ratio,
C_r = volume change index for the recompression curve,
C_c = volume change index for the virgin compression curve,
P_s = swelling pressure, and
σ_p' = preconsolidation pressure.

Equation 14.96 has the following form for a particular soil suction:

$$e = a + b \ \ln\left[1 + \left(\frac{\sigma}{c}\right)^2\right] + d \ \ln\left[1 + \left(\frac{\sigma}{f}\right)^2\right] \quad (14.97)$$

where:

a, b, c, d, f = constants (i.e., fitting parameters) at a constant suction.

The above-mentioned mathematical equations can be used to describe the void ratio constitutive surface over a wide range of stress conditions The above-mentioned equations have been used to describe the constitutive surface for the soil structure of Regina clay. However, all equations have difficulty when the net normal stress and the matric suction approach zero.

14.10.2 Problem of Low Normal Stress and Zero Suction

A problem occurs in the range of low net normal stress and zero suction when the elastic modulus is calculated from the swelling index C_s or other volume change indices. The problem is illustrated in Fig. 14.66. On the extreme planes, the relationships between void ratio and the logarithm of net normal stress or matric suction are essentially linear (Fig. 14.66a). When the semilog relationship is converted to an arithmetic plot (i.e., void ratio versus net normal stress or matric suction), void ratio tends to increase to infinity as the

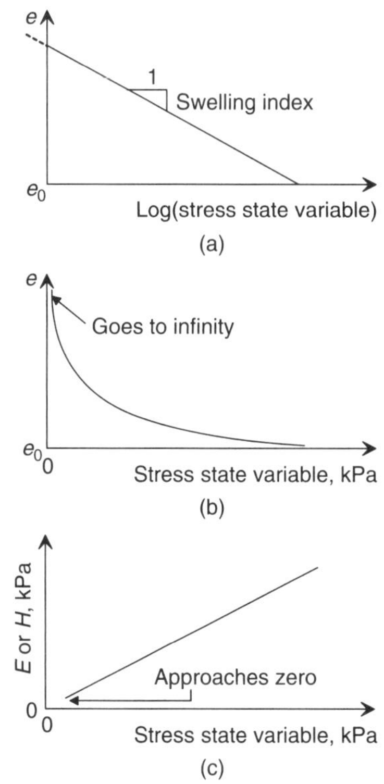

Figure 14.66 Illustration of the problem associated with low net normal stresses and zero suction: (a) semilogarithm plot of void ratio versus net normal stress or matric suction; (b) arithmetic plot of void ratio versus net normal stress or matric suction; (c) calculated elastic moduli versus net normal stress or matric suction.

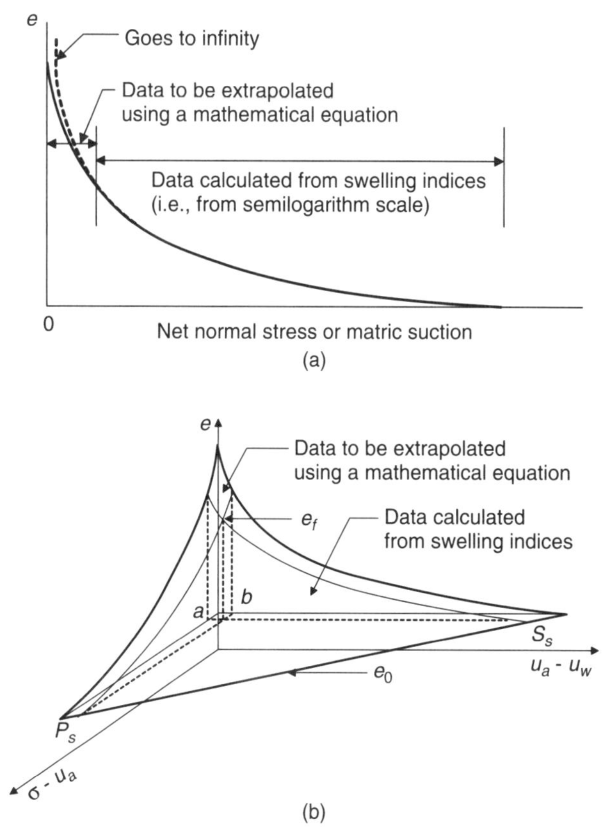

Figure 14.67 Suggested procedure to solve problem of low net normal stress and small suction: (a) two-dimensional plot of diminishing stresses; (b) three-dimensional plot of stresses.

stress state variable approaches zero (Fig. 14.66b). Therefore, the calculated elastic modulus with respect to changes in net normal stress E becomes extremely small (and approaches zero) at low net normal stresses (Fig. 14.66c). The elastic modulus with respect to a change in matric suction H also becomes extremely small at low matric suctions (or zero suction) (Fig. 14.66c). These unrealistically small values of elastic moduli at low net normal stress and matric suction can result in an unstable solution in numerical modeling. As well, the solutions may produce unreasonably large deformations.

The procedure for analysis purposes should adequately describe void ratio changes at low net normal stresses and matric suctions (Fig. 14.67). The equation should also be continuous, smooth, physically reasonable, and differentiable over the entire range of net normal stresses and matric suctions. An appropriate constitutive equation needs to include a value for void ratio at zero suction and zero net normal stress, such as is illustrated in Fig. 14.67. Unrealistic, large values of void ratio under low net normal stresses and matric suctions need to be avoided.

CHAPTER 15

Compressibility and Pore Pressure Parameters

15.1 INTRODUCTION

There are a variety of physical processes that are of interest to geotechnical engineering. Several have been presented in previous chapters as independent physical processes that form the basis of a science for unsaturated soil mechanics. Some of the processes are as follows: (i) flow of liquid water through porous media, (ii) flow and compression of air and water vapor in an unsaturated soil, (iii) heat flow in soils, and (iv) stress-deformation analysis that allow the calculation of normal and shear components of stress and strain. There are also processes such as the contaminant transport phenomena that could be presented as an independent process; however, diffusive flow is usually presented in conjunction with advective water flow.

Each of the above-mentioned physical processes has been discussed and the theory of behavior has been presented. In each case, a REV element was selected and a mathematical equation was derived to describe the particular physical phenomena. It was shown that each physical process could be described in the form of one or more PDEs to be solved. Flow processes were described for either steady-state or unsteady-state (or transient) conditions.

It is important to have a thorough understanding of each independent physical process from a theoretical, soil property, and solution standpoint. It is also important to realize that many of the processes common to geotechnical engineering involve multiple or combined processes. The real world of soil mechanics quite often involves a combination of two or more independent physical processes. It is necessary to solve more than one set of PDEs when more than one physical process is associated with a particular geotechnical engineering problem.

Following are some examples of geotechnical engineering problems that involve the solution of more than one physical process:

1. The theory of consolidation involves a combination of the physics associated with the flow of water and air as well as the physics associated with a stress–deformation analysis. Often the theory of consolidation of an unsaturated soil is considered simply in terms of changes in pore-water pressures and displacements with respect to time and space. The theory of consolidation involves the solution of two PDEs with respect to time (provided the problem geometry is one dimensional). The consolidation problem could also have a two- or three-dimensional geometry, in which case the solution would involve the solution of three or four PDEs, respectively (i.e., two or three stress-deformation PDEs and one water flow PDE).
2. The theory of consolidation of an unsaturated soil can also be considered as involving air flow in addition to water flow and stress–deformation. In this case there would be an additional air flow PDE that needs to be solved in a time and space domain.
3. While the theory of consolidation is generally viewed as a volume decrease problem in response to the application of an external load, there are similar processes associated with the swelling of an expansive or unsaturated soil. As well, the collapse phenomena in low-plasticity unsaturated soils can be viewed as undergoing a consolidation-type process. The number of PDEs that need to be solved depends on how rigorous a solution is required and the number of geometric dimensions.
4. Heat flow physical processes can also be combined with water and/or air flow to form other combinations of processes that might need to be solved.
5. Soil-atmospheric models involve the solution of PDEs related to liquid water flow, water vapor flow, and heat flow, in addition to satisfying other thermodynamic considerations. Climatic factors are used to generate a net moisture flux at the ground surface. These solutions are common to the design of soil cover systems [i.e., store-and-release covers or evapotranspirative (ET) covers].
6. Thermal, mechanical, flow (TMF) models refer to mechanical (i.e., stress–deformation), heat flow, and moisture flow (i.e., liquid and vapor water flow) processes. The TMF models involve three unsaturated soil

mechanics processes. Consideration of the long-term storage of radioactive waste products has drawn considerable attention to the solution of TMF models. In this situation, it is possible that the total stresses change with time, moisture flow occurs (i.e., liquid and vapor), and heat is generated with time.

7. There are multiple processes associated with waste rock management and the operation of heap leach deposits. Heat may be generated with time as water or an acid is allowed to move through the waste rock material. Moisture flow may be in a liquid and/or vapor form. Heat generation can also initiate significant advective air flow through the rock material. There may also be chemical reaction processes involved that result in a change of mass with respect to time.
8. The management of municipal solid waste materials has similarities to the processes mentioned for waste rock materials. Physical processes involve air flow (and gases such as methane), vapor water flow, liquid water flow, and heat flow as well as chemical reactions that result in a loss of mass from the system with time.

Each of the problems described above can be viewed as the solution of a series of partial differential equations. The PDE equations can be solved in a "coupled" or "uncoupled" manner. The equations might also be solved in terms of one process being dominant or several solutions may be manually solved by iterating between independent processes. While a coupled solution would appear to be the best type of solution, the rigor associated with fully coupled solutions might not be warranted.

15.2 COUPLED AND UNCOUPLED SOLUTIONS

The terms coupled and uncoupled have particular meaning with respect to geotechnical engineering applications. The swelling of a soil can be used to illustrate the meaning of uncoupled and coupled processes. Let us consider a two-dimensional cross section through a soil mass. There are two stress-deformation PDEs and a water flow PDE involved. Hydraulic head drives water flow and there are stress and deformation state variables in stress-deformation analysis.

An uncoupled solution means that each PDE is solved in an independent manner and information is transferred from one PDE to the other in a hierarchical manner designated by the modeler. The solving process is illustrated in Fig. 15.1. In this example, pore-water pressure changes are calculated for the next time step. The solution of the water PDE may involve a series of iterations to accommodate nonlinearities associated with unsaturated hydraulic conductivity and water storage soil parameters. The new pore-water pressures are then imported into the stress analysis and subsequent deformations (i.e., x- and y-coordinate strain values) are computed. The solution process then reverts back to the water flow PDE where consideration is given to moving forward to the next time step.

When the solution of the water flow and stress-deformation PDEs are uncoupled it is possible to proceed with the entire water flow solution for the entire time span of interest and then solve the stress-deformation analysis using the pore-water pressures previously calculated. The uncoupled analysis assumes that the pore-water pressure calculation influences the stress-deformation analysis but the reverse is not true (i.e., the stress-deformation results do not affect the pore-water pressure analysis).

It is also possible to consider the solution to the swelling soil problem in an uncoupled, iterative manner where there is an iterative process between the water flow and stress-deformation PDEs as shown in Fig. 15.2. This procedure moves the solution toward a coupled type of solution but may not meet all the requirements of a fully coupled solution. The iterative-type solution will likely provide a more accurate representation of the nonlinear soil properties, and as a result, the solution should be more accurate. The additional modeling effort may or may not be warranted in solving a particular problem at hand.

A coupled solution means that two or more PDEs are being solved simultaneously, as shown in Fig. 15.3. A so-called coupling matrix is placed between the PDEs being solved. The coupling matrix ensures that the limiting conditions associated with each PDE involved are satisfied before moving on to the next time increment. Necessary iterations are performed through the coupling matrix to ensure that nonlinear soil property conditions are satisfied as well as other conditions related to the effect of one PDE solution on the solution of other PDEs.

There are other solution procedures that are sometimes used when solving multiple PDEs. For example, the assumption might be made that certain components of the coupling

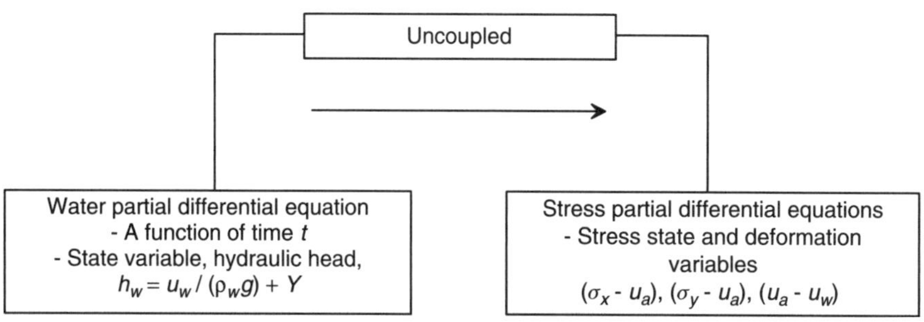

Figure 15.1 Modeling swelling of soil as uncoupled solution.

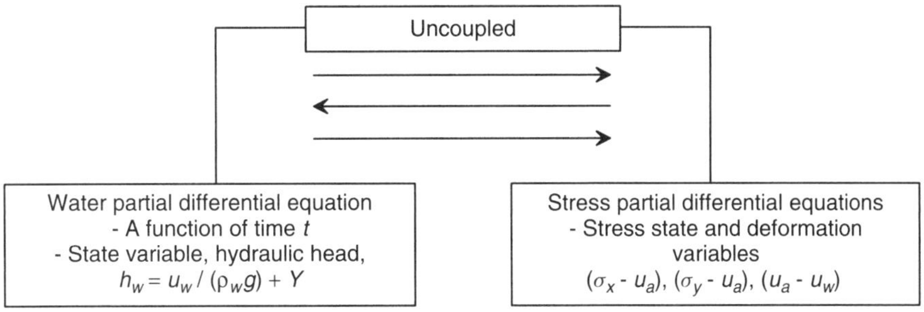

Figure 15.2 Modeling swelling of expansive soil as iterative, uncoupled solution.

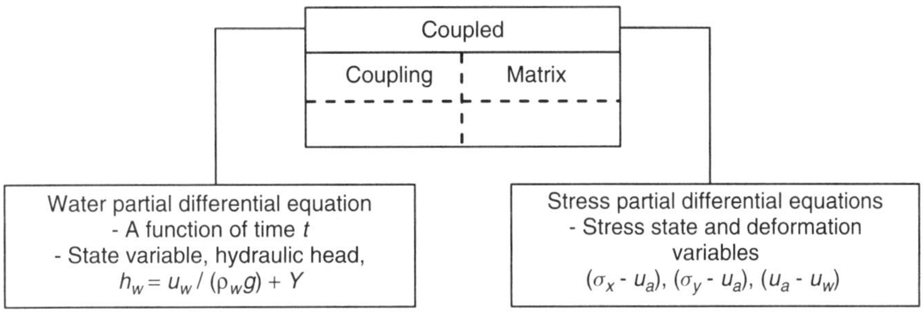

Figure 15.3 Modeling swelling of expansive soil as coupled solution.

matrix are insignificant with respect to the solution of the problem. These components may be omitted and the solution might be referred to as a "pseudocoupled" solution procedure.

When the problems are transient in character, the solutions march forward in time and the equations can be advanced in either an explicit or implicit manner. The terms explicit and implicit reflect the manner in which the soil properties and/or state variables are approximated when moving forward in time. The marching process takes on the character of finite difference time increments.

The implicit procedure may be more accurate; however, there are additional computations associated with this type of solution and the additional computations may not always be warranted. It is clear that there is no single procedure that should always be adopted when solving complex multiple physical processes. Rather, each situation needs to be assessed and decisions need to be made regarding the most appropriate type of solution that should be used for the geotechnical engineering problem at hand.

15.2.1 Coupled and Uncoupled Theory of Consolidation

Coupled and uncoupled solutions for the theory of consolidation for a saturated soil have shown that the two solutions may differ significantly in some situations. Mandel (1957) and Cryer (1963) published independent research papers that showed that the pore-water pressures generated in a multidimensional consolidation problem could differ by significant amounts depending upon whether the solutions were carried out in a coupled or uncoupled manner. In particular, it was noted that the induced pore-water pressure could rise above the magnitude of the applied total stresses. This phenomenon became known as the Mandel-Cryer effect and can be used to illustrate the importance of fully coupled consolidation solutions.

The original analysis was performed on a spherically shaped saturated soil mass. Other geometries were later analyzed to further demonstrate the Mandel-Cryer effect. Figure 15.4 shows an example involving the loading applied by a footing over a saturated soil (Schiffman et al., 1969; Christian and Boehmer, 1970). It can be observed that shortly after the application of the load, the pore-water pressures rose to a value higher than the pressure associated with the applied load. The pore-water pressures then dissipated over time.

The magnitude of the pore-water pressure rise was found to be a function of the Poisson's ratio used during the analysis. When Poisson's ratio was zero, the Mandel-Cryer effect was the greatest; when Poisson's ratio increased to near 0.5, the Mandel-Cryer effect was negligible. In other words, the Mandel-Cryer effect illustrated in the coupled solution was closely related to Poisson's ratio.

Vu (2003) showed that there would be little difference between a coupled solution and an uncoupled solution of swelling clay problems where soil suction was changed through moisture infiltration. In this situation, the matric suction change was isotropic and the Mandel-Cryer effect appeared to be negligible. Dinh (1996) also numerically solved the consolidation (or swelling) equations for an unsaturated soil.

There is a general observation that can be made with regard to coupled and uncoupled solutions. While fully coupled solutions may be theoretically superior, it does not

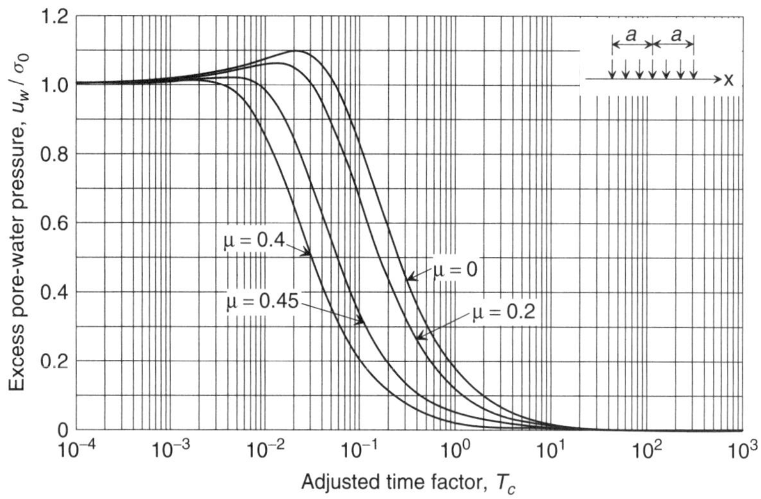

Figure 15.4 Coupled solution of footing placed on saturated clay (after Christian and Boehmer, 1970).

mean that fully coupled solutions are always necessary. There are numerous factors involved when solving a series of PDEs and in each case the modeler must assess whether a fully coupled solution is warranted.

15.2.2 Induced Pore Pressure Problems

The relationship between coupled and uncoupled solutions of PDEs can be illustrated using the example of an unsaturated soil loaded by an external load. The loading of a soil followed by the dissipation of induced pore pressures is referred to as the theory of consolidation. The mechanical behavior of unsaturated soils is directly affected by changes in both the pore-air and pore-water pressures.

There are two general classes of problems where pore pressure conditions may be changed in the field. In the first situation, transient or steady-state pore pressure changes (i.e., pore-water pressures or pore-air pressures) may be associated with the infiltration of water into the soil. In the second situation, pore pressure conditions may have been generated through the application of an external load such as an engineered structure. This chapter addresses the situation where essentially instantaneous pore pressure changes result from the application of an external load.

The pore pressures generated immediately after loading are commonly referred to as induced or undrained pore pressures. Under undrained conditions, the applied total stress is carried by the soil structure, the pore-air and the pore-water depending upon their relative compressibility values. The induced pore-air and pore-water pressures can be written as a function of the applied total stresses. The excess pore pressures will dissipate with time if the pore fluids are allowed to drain back to the stress state prior to loading. The applied total stresses are eventually carried by the soil structure (i.e., interparticle stresses).

The pore pressure response can be expressed in terms of pore pressure parameters for an uncoupled analysis. Pore pressure parameters have proven to be useful in engineering practice when dealing with construction associated with compacted soils (e.g., earth dams). The pore pressures likely to develop during construction can be estimated using the pore pressure parameters. The estimated pore pressures initiate a transient analysis similar to a consolidation process.

The induced pore fluid pressures can also be solved as part of a coupled solution. In this case, the equilibrium PDEs and the continuity PDE are solved in a simultaneous manner. The induced pore-air and pore-water pressures become part of the solution. In this case it is not necessary to use pore pressure parameter equations to solve for the excess pore fluid pressures.

The emphasis in this chapter will be on the calculation and use of pore pressure parameters for performing uncoupled solutions.

15.3 UNCOUPLED UNDRAINED LOADING

An uncoupled undrained analysis of an unsaturated soil means that the soil is first loaded and the induced pore-air and pore-water pressures are calculated assuming the process is instantaneous and there is no fluid drainage. The magnitudes of the induced pore pressures are a function of the relative compressibility of the various phases of materials involved. Stated another way, the pore pressures induced will depend on the compressibility of the air-water mixture in the soil and the compressibility of the soil structure with respect to each stress state variable.

The induced pore-water pressure for a saturated soil will equal the applied isotropic load. If the soil is dry, the induced pore-air pressure will be controlled by the compressibility of the air and the compressibility of the soil structure.

The following sections illustrate how the compressibility of an air-water mixture can be used to calculate pore-air and pore-water pressures induced as a result of total stress loading. The induced pore-air and pore-water pressures constitute the initial boundary conditions when solving the PDEs associated with the pore-air and pore-water pressure dissipations. The initial pore-air and pore-water pressures are assumed to be uncoupled from any other process. The pore-air and pore-water pressures are subsequently solved as processes that take place with respect to time. The dissipation of the pore-air and pore-water pressures can be solved in either a coupled or uncoupled manner. A decision can be made regarding whether the dissipation of pore pressures should be solved in a coupled or uncoupled manner. Often it is easier to use an uncoupled solution and the results are satisfactory for geotechnical engineering practice.

15.3.1 Compressibility of Pore Fluids

Pore-air and pore-water are not allowed to flow out of an unsaturated soil during undrained compression. Volume change occurs as a result of the compression of the air and, to a lesser extent, the water. The compression of soil solids can be assumed to be negligible for the stress range commonly encountered in engineering practice. The pore fluid volume change is related to changes in the pore-air and pore-water pressures. The pore-air and pore-water pressures increase as an unsaturated soil is compressed. The pore pressure increase is commonly referred to as an excess pore pressure. The volume change of a phase is related to a pressure change through its compressibility. Figure 15.5 defines the compressibility of a material at a point on the volume-pressure curve during undrained compression. Isothermal compressibility is defined as the volume change of a fixed mass with respect to a pressure change per unit volume at a constant temperature:

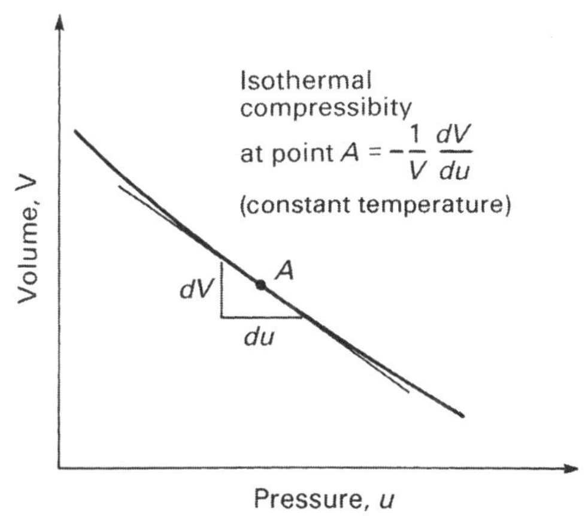

Figure 15.5 Definition of variables associated with isothermal compressibility.

$$C = -\frac{1}{V}\frac{dV}{du} \tag{15.1}$$

where:

C = compressibility,
V = total volume of the element under consideration,
dV/du = volume change with respect to a pressure change, and
du = pressure change.

The term dV/du in the above equation has a negative sign because the volume decreases as the pressure increases. A negative sign is used in order to render a positive compressibility value.

In an unsaturated soil, the pore fluid consists of water, free air, and air dissolved in water. The individual compressibility of air and water is required in formulating the compressibility of the air-water mixture.

15.3.1.1 Air Compressibility

The isothermal compressibility of air can be expressed as

$$C_a = \frac{-1}{V_a}\frac{dV_a}{du_a} \tag{15.2}$$

where:

C_a = isothermal compressibility of air,
V_a = volume of air,
dV_a/du_a = air volume change with respect to an air pressure change, and
u_a = air pressure.

The volume-pressure relationship for air during isothermal, undrained compression can be expressed using Boyle's law:

$$V_a = \frac{\bar{u}_{a0} V_{a0}}{\bar{u}_a} \tag{15.3}$$

where:

$\bar{u}_{a0}$ = initial absolute air pressure (i.e., $\bar{u}_{a0} = u_{a0} + \bar{u}_{atm}$),
u_{a0} = initial gauge air pressure,
$\bar{u}_{atm}$ = atmospheric pressure (i.e., 101.3 kPa),
V_{a0} = initial volume of air, and
$\bar{u}_a$ = absolute air pressure (i.e., $\bar{u}_a = u_a + \bar{u}_{atm}$).

Differentiating the volume of air, V_a, with respect to absolute air pressure $\bar{u}_a$ gives

$$\frac{dV_a}{d\bar{u}_a} = -\frac{1}{\bar{u}_a^2}\left(\bar{u}_{a0} V_{a0}\right) \tag{15.4}$$

The above equation gives the volume change of air with respect to an infinitesimal change in air pressure. Substituting Boyle's law (i.e., Eq. 15.3) into Eq. 15.4 gives

$$\frac{dV_a}{d\bar{u}_a} = -\frac{V_a}{\bar{u}_a} \tag{15.5}$$

The volume derivative with respect to absolute pressure $(dV_a/d\bar{u}_a)$ is equal to the derivative with respect to the gauge pressure (dV_a/du_a) since the atmospheric pressure $\bar{u}_{atm}$ is assumed to be constant. Equation 15.5 can be substituted into Eq. 15.2 to give the isothermal compressibility of air,

$$C_a = \frac{1}{\bar{u}_a} \tag{15.6}$$

The isothermal compressibility of air is inversely proportional to absolute air pressure. In other words, air compressibility decreases as the air pressure increases.

15.3.1.2 Water Compressibility

The compressibility of water is defined as follows:

$$C_w = -\frac{1}{V_w}\frac{dV_w}{du_w} \tag{15.7}$$

where:

C_w = water compressibility,
V_w = volume of water,
dV_w/du_w = water volume change with respect to water pressure change, and
u_w = water pressure.

Figure 15.6 presents the results of water compressibility measurements (Dorsey, 1940). It should be noted that dissolving air in water does not significantly change the compressibility of water.

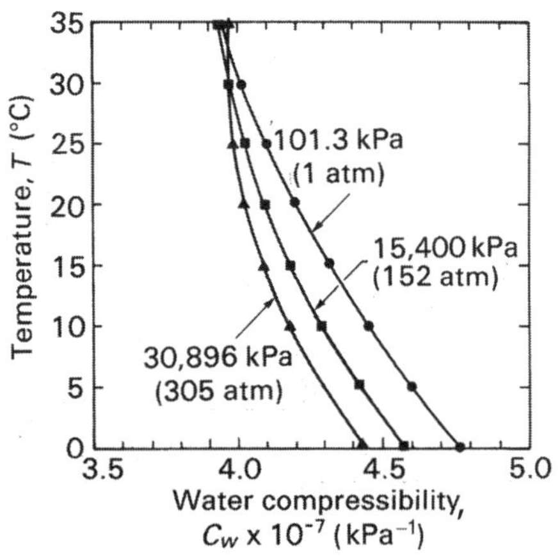

Figure 15.6 Isothermal compressibility of air-saturated water (from Dorsey, 1940).

15.3.2 Compressibility of Air-Water Mixtures

The compressibility of an air-water mixture can be derived using the direct proportioning of the air compressibility and the water compressibility. The conservation of mass and the compressibility definition must apply. Let us consider the air, water, and solid volumetric relations, as shown in Fig. 15.7. Let us assume that the degree of saturation S and porosity n of the soil are known. The total volume of the air-water mixture is the sum of the volume of water, V_w, and the volume of air, V_a (i.e., $V_w + V_a$). The volume of the dissolved air, V_d, is within the volume of water, V_w. The volumetric coefficient of solubility h gives the percentage of dissolved air with respect to the volume of water. The pore-air and pore-water pressures are u_a and u_w, respectively, with u_a always being greater than u_w. The soil is subjected to a compressive total stress σ.

Let us apply an infinitesimal increase in total stress, $d\sigma$, to the soil while maintaining undrained conditions. The pore-air and pore-water pressures increase while the volumes of air and water decrease. The compressibility of an air-water mixture for an infinitesimal increase in total stress can be written using the total stress as a reference:

$$C_{aw} = -\frac{1}{V_w + V_a}\left(\frac{d(V_w - V_d)}{d\sigma} + \frac{d(V_a + V_d)}{d\sigma}\right) \tag{15.8}$$

where:

C_{aw} = compressibility of air-water mixture,
$V_w + V_a$ = volume of the air-water mixture,
V_w = volume of water,
V_a = volume of free air,
$d(V_w - V_d)/d\sigma$ = water volume change with respect to a total stress change,
$d(V_a + V_d)/d\sigma$ = air volume change with respect to a total stress change, and
V_d = volume of dissolved air.

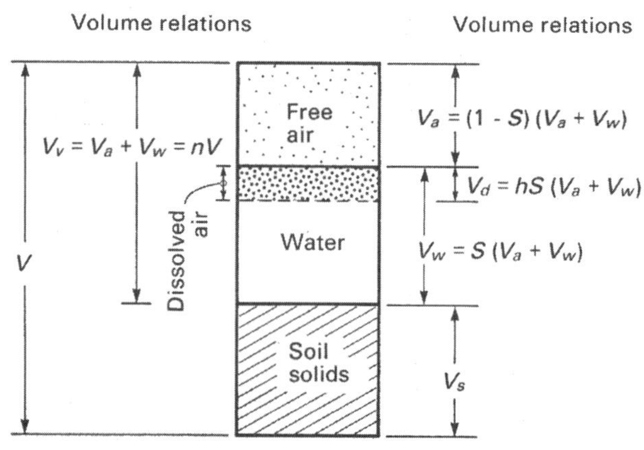

Figure 15.7 Volumetric composition of pore fluid in unsaturated soil taking dissolved air into consideration.

Equation 15.8 is a slight modification of the compressibility equation for air-water mixtures proposed by Fredlund (1976). The total stress change $d\sigma$ is used as the reference pressure in Eq. 15.8, while the pore-water pressure change du_w was used as the reference pressure in the compressibility equation by Fredlund (1976). The term $d(V_w - V_d)/d\sigma$ in Eq. 15.8 is considered to be equal to $dV_w/d\sigma$ since the dissolved air is a fixed volume internal to the water. As such, its volume does not change. The total volume of water, V_w, is therefore used in computing the compressibility of water [i.e., $C_w = -\left(1/V_w\right)(dV_w/du_w)$].

The change in air volume occurs as a result of the compression of the free air in accordance with Boyle's law and a further dissolution of free air into water in accordance with Henry's law. The total air volume change can be obtained directly using Boyle's law by considering the initial and final pressures and the volume conditions in the air phase. The free-air compression and the air dissolving in water are assumed to be complete under final conditions. The free and dissolved air can be considered as one volume with a uniform pressure. Although the volume of dissolved air, V_d, is a fixed quantity, it is maintained in the formulation for clarity. Therefore, the dissolved air volume V_d also appears in the equation.

Applying the chain rule of differentiation to the compressibility equation gives

$$C_{aw} = \frac{-1}{V_w + V_a}\left(\frac{dV_w}{du_w}\frac{du_w}{d\sigma} + \frac{d(V_a + V_d)}{du_a}\frac{du_a}{d\sigma}\right) \quad (15.9)$$

where:

dV_w/du_w = water volume change with respect to a pore-water pressure change,
$du_w/d\sigma$ = water pressure change with respect to a total stress change,
$d(V_a + V_d)/du_a$ = air volume change with respect to a pore-air pressure change, and
$du_a/d\sigma$ = air pressure change with respect to a total stress change.

Rearranging the above equation gives

$$C_{aw} = -\left(\frac{V_w}{V_w + V_a}\frac{1}{V_w}\frac{dV_w}{du_w}\right)\frac{du_w}{d\sigma} - \left(\frac{V_a + V_d}{V_w + V_a}\frac{1}{V_a + V_d}\frac{d(V_a + V_d)}{du_a}\right)\frac{du_a}{d\sigma} \quad (15.10)$$

Substituting the volume relations in Fig. 15.7 and Eqs. 15.2 and 15.7 into Eq. 15.10 yields the compressibility of an air-water mixture:

$$C_{aw} = SC_w\left(\frac{du_w}{d\sigma}\right) + (1 - S + hS)C_a\left(\frac{du_a}{d\sigma}\right) \quad (15.11)$$

The isothermal compressibility of air, C_a, is equal to the inverse of the absolute air pressure:

$$C_{aw} = SC_w\left(\frac{du_w}{d\sigma}\right) + \frac{(1 - S + hS)(du_a/d\sigma)}{\bar{u}_a} \quad (15.12)$$

15.3.2.1 Use of Pore Pressure Parameters in Compressibility Equation

The ratio between pore pressure change and total stress change $(du/d\sigma)$ is referred to as a pore pressure parameter. This parameter indicates the magnitude of the pore pressure change in response to a total stress change. The pore pressure parameter concept was introduced by Skempton (1954) and Bishop (1954). The pore pressure parameters for the air and water phases are different (Bishop, 1961b; Bishop and Henkel, 1962) and depend primarily upon the degree of saturation of the soil. The parameters also vary depending on the loading conditions.

The pore pressure parameters can also be directly measured in the laboratory. For isotropic loading conditions, the parameter is commonly called the B pore pressure parameter, and it can be substituted into Eq. 15.12 as follows:

$$C_{aw} = SC_wB_w + \frac{(1 - S + hS)B_a}{\bar{u}_a} \quad (15.13)$$

where:

B_w = pore-water pressure parameter for isotropic loading (i.e., $du_w/d\sigma_3$),
σ_3 = isotropic (confining) total stress, and
B_a = pore-air pressure parameter for isotropic loading (i.e., $du_a/d\sigma_3$).

The compressibility of the pore fluid in an unsaturated soil takes into account the matric suction of the soil through the B_w and B_a parameters. In the absence of soil solids, B_a, $B_w = 1$. In the presence of soil particles, however, the surface tension effects will result in B_a, $B_w < 1.0$, depending upon the matric suction of the soil. The pore-air and the pore-water pressures change at differing rates in response to the applied total stress. The B_w value is greater than the B_a value. The B_a and B_w parameters are low at low degrees of saturation, and both parameters approach a value of 1.0 at saturation. At this point, the matric suction of the soil goes to zero.

15.3.2.2 Components of Compressibility of Air-Water Mixture

The first term in the compressibility equation (i.e., Eq. 15.13) accounts for the compressibility of the water portion of the mixture, while the second term accounts for the compressibility of the air portion. The compressibility of the air portion is due to the compression of free air [i.e., $(1 - S)B_a/\bar{u}_a$] and air dissolving into water (i.e., $hSB_a/\bar{u}_a$).

The contribution of each compressibility component to the overall compressibility of the air-water mixture is illustrated in Fig. 15.8 for various degrees of saturation. The case considered has an initial absolute air pressure $\bar{u}_{a0}$ of 202.6 kPa (i.e., 2 atm).

15.3.3 Effects of Free Air on Compressibility of Mixture

Figure 15.8 shows that the compressibility of an air-water mixture is predominantly influenced by the compressibility of the free air portion [i.e., $(1 - S)B_a/\bar{u}_a$]. When the soil voids are filled with air (i.e., $S = 0.0$), the compressibility of the pore fluid is equal to the isothermal compressibility of air (i.e., 4.94×10^{-3} kPa^{-1} at $\bar{u}_a = 202.6$ kPa). At saturation (i.e., $S = 100\%$), the soil voids are completely filled with water, and pore fluid compressibility becomes equal to that of water (i.e., 4.58×10^{-7} kPa^{-1}). The inclusion of even 1% air in the pore fluid is sufficient to significantly increase the pore fluid compressibility (see Fig. 15.8). The compressibility of water is only significant when computing the compressibility of a saturated soil.

15.3.4 Effect of Dissolved Air on Compressibility of Mixture

The solution of air in water gives the effect that the soil is compressible. The compressibility due to the solution of air in water (i.e., $hSB_a/\bar{u}_a$) is approximately two orders of magnitude greater than the compressibility of water (i.e., SC_wB_w). Air dissolving in water significantly affects the compressibility of an air-water mixture when the free air volume becomes less than about 20% of the volume of voids. The effect of air solubility on the compressibility of an air-water mixture is shown in Fig. 15.9 for several initial absolute air pressures. The B_a and B_w parameters are assumed to be equal to 0.8 and 0.9, respectively.

The figure shows that the effect of air solubility on the compressibility of an air-water mixture is the same (i.e., on a logarithmic scale) for any initial air pressures. However, the effect of air dissolving in water does not result in a smooth transition for the compressibility of an air-water mixture as saturation is reached (i.e., $S = 100\%$). There is a discontinuity at the point where there is no more free air to be dissolved in the water. As a result, the second term in Eq. 15.13 must be dropped, and the compressibility abruptly decreases to the compressibility of water. If the free air does not have time to be dissolved in the water, the term $hSB_a/\bar{u}_a$ in Eq. 15.13 must be set to zero. In this case, the compressibility of pore fluid has a smooth transition back to saturation, as shown in Fig. 15.9.

15.3.5 Other Relations for Compressibility of Air-Water Mixtures

The compressibility of air-water mixtures has been investigated by numerous researchers. For example, an equation for the compressibility of an air-water mixture has been proposed that neglects the compressibility of the water and assumes zero matric suction (Bishop and Eldin, 1950; Skempton and Bishop, 1954). Therefore, the compressibility of the air is considered as representative of the compressibility of the pore fluid. The proposed equation is suitable for conditions where the air phase constitutes a significant portion of the pore fluid. These proposed equations can be obtained by ignoring the first term in Eq. 15.13. Also, the B_a and B_w

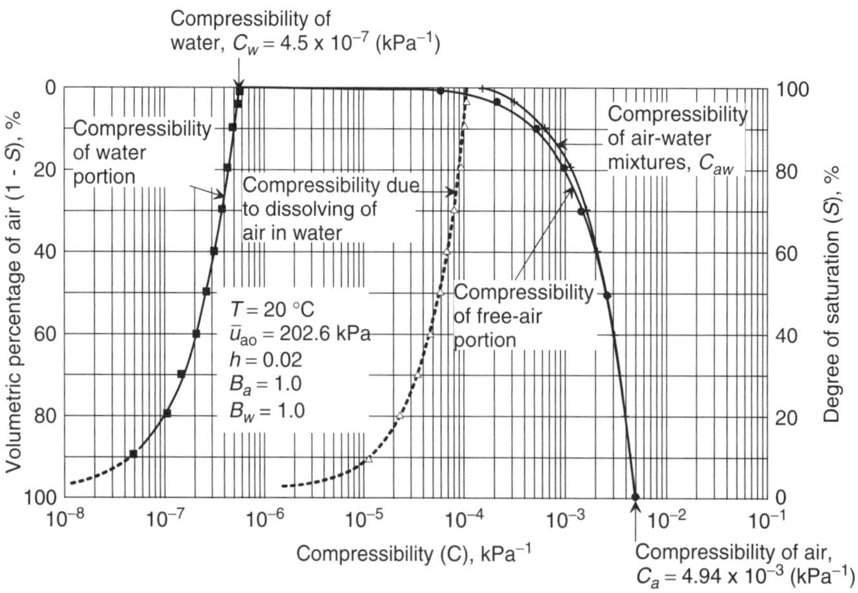

Figure 15.8 Components of compressibility of air-water mixture.

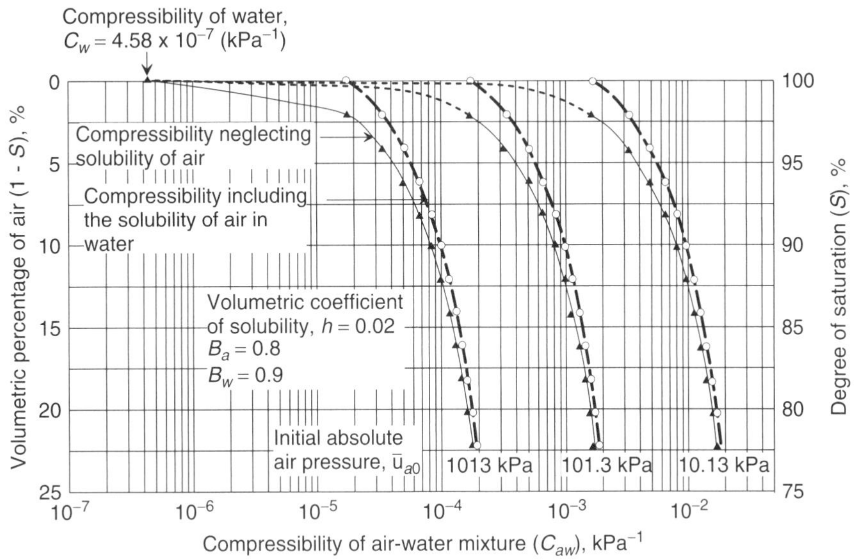

Figure 15.9 Effect of solubility of air in water on compressibility of air-water mixture.

values must be set to 1.0. The resulting equation has the following form:

$$C_{aw} = \frac{1 - S + hS}{\bar{u}_a} \tag{15.14}$$

Equation 15.14 is slightly different from the equation proposed by Bishop and Eldin (1950). The difference lies in the definition of air compressibility. Equation 15.14 is derived using the definition of air compressibility given in Eq. 15.2 [i.e., $C_a = -(1/V_a)/(dV_a/du_a)$]. The equation proposed by Bishop and Eldin (1950) was derived with the air compressibility defined with reference to the initial volume of air, V_{a0} :

$$C_a = -\frac{1}{V_{a0}}\frac{dV_a}{du_a} \tag{15.15}$$

Substituting the term dV_a/du_a from Eq. 15.4 into Eq. 15.15 yields a slightly different equation for compressibility:

$$C_a = \frac{\bar{u}_{a0}}{\bar{u}_a^2} \tag{15.16}$$

Equation 15.6 (i.e., $C_a = 1/\bar{u}_a$) describes the air compressibility at a point on the volume-pressure curve during undrained compression (see Fig. 15.5). Equation 15.16 gives the average air compressibility during an air pressure change from $\bar{u}_{a0}$ to $\bar{u}_a$ (Skempton and Bishop, 1954). The difference in the computed air compressibility from the two equations is negligible when the air pressure change is infinitesimal.

Replacing the air compressibility term in Eq. 15.14 (i.e., $1/\bar{u}_a$) with the average air compressibility (i.e., $\bar{u}_{a0}/\bar{u}_a^2$) yields the air-water mixture compressibility equation proposed by Bishop and Eldin (1950):

$$C_{aw} = \frac{(1 - S + hS)\,\bar{u}_{a0}}{\bar{u}_a^2} \tag{15.17}$$

Koning (1963) suggested another equation for the compressibility of air-water mixtures. The equation was first derived by expressing the pore-air and pore-water pressure changes as a function of surface tension. However, changes in the pore-air and pore-water pressures were later assumed to be equal. This implies that the pore pressure parameters B_a and B_w are equal to 1.0. The compressibility associated with the solution of air in water was disregarded in the derivation (i.e., $hSB_a/\bar{u}_a = 0$). Applying the above assumptions to the compressibility of an air-water mixture gives the following equation (Koning, 1963):

$$C_{aw} = SC_w + \frac{1 - S}{\bar{u}_a} \tag{15.18}$$

Equation 15.18 was also suggested by Verruijt (1969) in formulating the elastic storage of aquifers. The solubility of air in water and the effect of matric suction were neglected.

15.3.6 Limitation Associated with Using Capillary Equation When Formulating Air-Water Compressibility Equation

The capillary equation (i.e., $u_a - u_w = 2T_s/R_s$) relates matric suction $u_a - u_w$ to surface tension T_s through the radius of curvature R_s of the air-water interface. Attempts have also been made to use the capillary equation when formulating an equation for the compressibility of air-water mixtures (Schuurman, 1966; Barends, 1979). Equations were proposed for the case of occluded air bubbles in soils with a degree of saturation greater than approximately 85%. The incorporation of the capillary equation results in the radius of curvature of an air bubble, R_s, becoming a variable in the compressibility equation. However, the radius R_s is not measurable and adds no value in solving practical engineering problems.

The incorporation of the radius of curvature suggests that the difference between the pore-air and pore-water pressures increases as the applied total stress is increased during undrained compression. It has been postulated that there exists a critical pore-water pressure beyond which all of the free air would dissolve abruptly (i.e., bubble collapse). This has become known as the air bubble collapse theory. Collapse occurs when a critical pore-water pressure is reached. This phenomenon should reflect an increase in compressibility at the critical pore-water pressure.

The postulated bubble collapse theory suggests that there is an increase in the difference between the pore-air and pore-water pressures (i.e., matric suction) as total stress is increased. However, there appears to be no experimental evidence to support this postulate. On the contrary, experimental evidence supports a continual, smooth increase of the pore-air and pore-water pressures until the two fluid pressures become equal as the soil approaches saturation. This means that matric suction decreases with an increase in total stress during undrained compression. There appears to be some conceptual difficulties when the capillary equation is incorporated into the equation for the compressibility of air-water mixtures.

The problems associated with incorporating the capillary equation into the equation for pore fluid compressibility can be better understood by considering the loading of an element of unsaturated soil. Let us assume there are numerous occluded air bubbles within the soil element, as shown in Fig. 15.10. The capillary equation describes a microscopic phenomenon within the element. The radii of the occluded air bubbles are not measurable and exist at the microlevel. It can be argued that the microcapillary equation should not be incorporated into a macroscopic type of formulation for compressibility.

It is possible to draw a parallel comparison between the effective stress variable used at the macroscopic, continuum mechanics level and intergranular stresses at the microscopic level between the soil particles of a saturated soil. From a microscopic point of view, there exist numerous intergranular stresses acting at the contacts between soil particles.

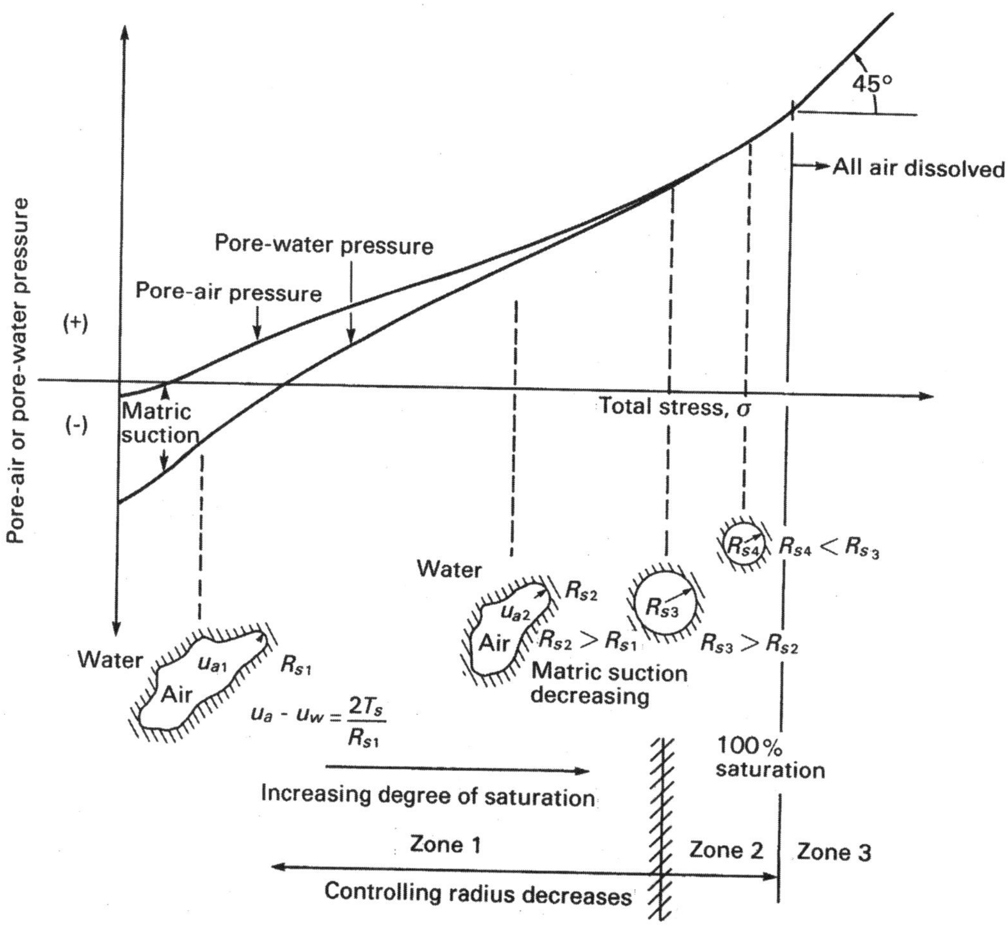

Figure 15.10 Pore-air and pore-water pressure responses to change in total stress during undrained compression.

From a macroscopic standpoint, the state of stress can be represented as an effective stress state for the soil. The effective stress is the only stress state variable required to describe the behavior of the saturated soil. Intergranular stresses are defined at a different level and are neither measurable nor necessary when analyzing the physical behavior of the soil. Likewise, to incorporate the capillary equation into a macroscopic definition of pore fluid compressibility for an air-water mixture would be erroneous.

An anomaly is observed when attempting to apply the capillary equation, together with Boyle's and Henry's laws, to the compressibility of an air-water mixture. The anomaly can be best understood by first considering an increase in total stress applied to an unsaturated soil specimen. A total stress increase causes the free-air volume in the soil to compress and the pore-air pressure to increase in accordance with Boyle's law. The increased air pressure drives more free air into solution in accordance with Henry's law. As a result, the volume of free air further decreases. The decrease in the free-air volume is accompanied by a decrease in the radius of curvature of the air-water menisci. The capillary equation would suggest that the matric suction $u_a - u_w$ should increase because of the reduced radius of curvature.

The increase in matric suction implies that the pore-air pressure increase is larger than the increase in the pore-water pressure. The pore-air and pore-water pressure increases must be positive since the total stress increase is positive. The increase in air pressure in response to the capillary equation would further drive air into solution, in accordance with Henry's law. Consequently, the volume of free air decreases, and the whole process is repeated over and over. In summary, it would appear that a slight increase in total stress should cause a chain reaction process which would reduce the free-air volume to an infinitesimal size while the matric suction goes to infinity. This would suggest that all air bubbles would become unstable with a small increment in total stress. However, the above scenario is not in agreement with observed experimental results.

Experimental results indicate that the pore-air and pore-water pressures gradually increase toward a single value (i.e., equal to the pore-water pressure) and matric suction approaches zero. This occurs as the total stress is increased under undrained loading. The process is gradual, in response to repeated increments of total stress.

The above apparent anomaly appears to be the result of incorporating the capillary equation (i.e., a microscopic model) in a macroscopic formulation of pore fluid compressibility. This anomaly can also be partly understood by assuming that the decrease in the free-air volume is not necessarily accompanied by a decrease in the controlling radius of curvature. Figure. 15.10 shows that, although the

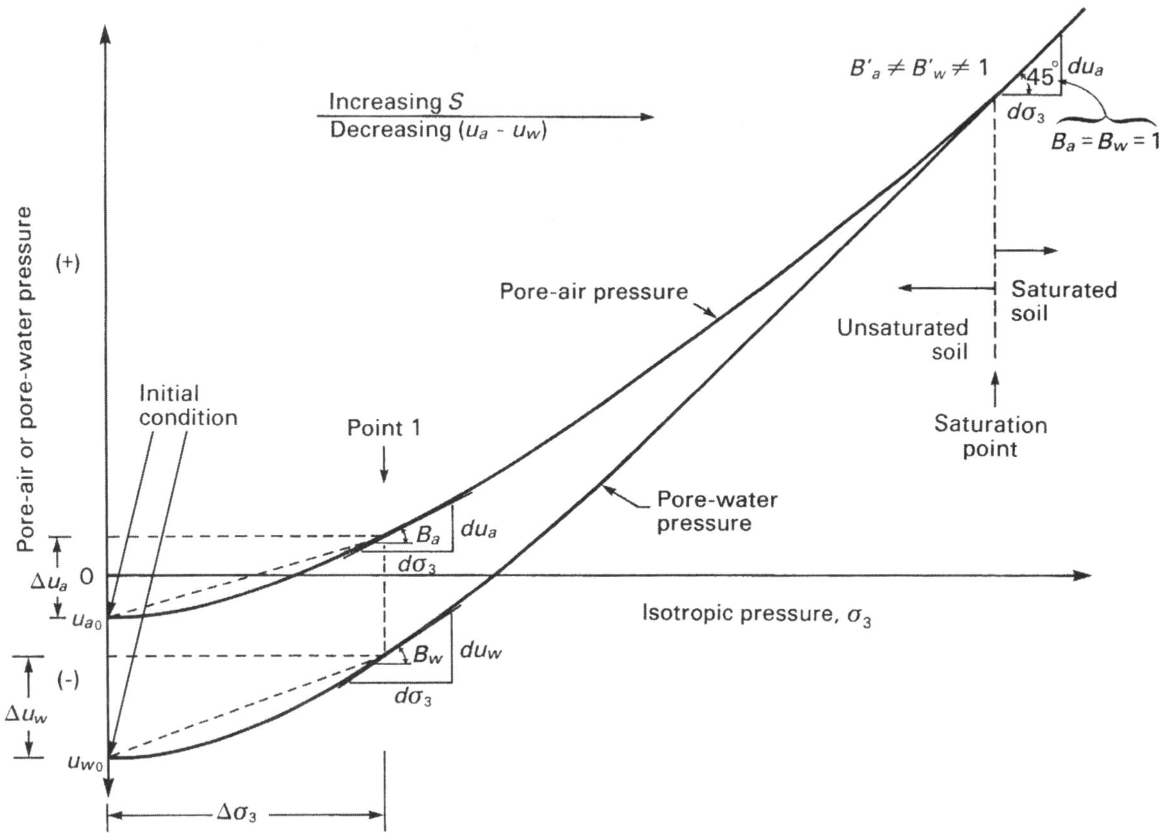

Figure 15.11 Tangent and secant designations of pore pressure parameters.

volume of the continuous air phase in zone 1 decreases due to an increase in the pore-air pressure from u_{a1} to u_{a2}, the controlling radius may in fact increase from R_{s1} to R_{s2}.

The above postulate is possible for the nonspherical air bubbles such as those in zone 1. The nonspherical air bubbles have nonuniform radii of curvature over their surface, but the assumption is made that only the controlling radius is found in the free pore-water. The radius of the air-water interface in the free water should be in agreement with the capillary equation. Nevertheless, the increase in total stress will eventually cause the air bubbles to take on a spherical form, as shown in zone 2 (Fig. 15.10). For spherical air bubbles, a decrease in volume must be followed by a decrease in the radius. At this point, the above-mentioned anomaly associated with the use of the capillary equation cannot be resolved.

On the basis of the above conceptual difficulties, it would appear that the capillary equation should not be incorporated into the equation for the compressibility of air-water mixtures. The pore-air and pore-water pressures must be measured in the continuous pore-air zone (i.e., zone 1 in Fig. 15.10) or computed using B_a and B_w pore pressure parameters that are derived later in this chapter. Air pressure measurements are valid as long as the air phase is continuous. In the occluded zone (i.e., zone 2 in Fig. 15.10), the air bubbles do not interact with the soil structure. The presence of air bubbles merely renders the pore fluid compressible. The pore-air and pore-water pressures can be assumed to be equal in the occluded zone (i.e., zone 2).

15.4 DERIVATION OF PORE PRESSURE PARAMETERS

The pore pressure response resulting from a change in total stress during undrained compression can be expressed in terms of pore pressure parameters. The pore pressure parameters corresponding to various loading conditions can be derived from the relative compressibility values for the materials involved. One pore pressure equation can be derived for the water phase and another equation can be derived for the air phase. These equations can be used to compute the pore pressure responses in an unsaturated soil during undrained loading. More direct measurements of the pore pressure parameters in the laboratory are required to confirm the accuracy of the theoretically derived equations.

15.4.1 Tangent and Secant Pore Pressure Parameters

The pore pressure parameters for an unsaturated soil can be defined either as a tangent-type parameter at a point or a secant-type parameter referenced back to an initial condition. These definitions are similar in concept to the tangent and secant modulus used in the theory of elasticity.

Consider the development of pore-air and pore-water pressures during isotropic, undrained compression, as shown in Fig. 15.11. The pore-water pressure increases faster than the pore-air pressure in response to the increase in total confining stress. The pore pressure response to an increase in the confining stress is referred to as the B pore pressure parameter. At a point during the loading of the soil (i.e., point 1), the tangent and the secant pore pressure parameters can be defined for both the air and water phases (Fig. 15.11). If the increase in the isotropic pressure is referenced to the initial condition, the B' secant pore pressure parameter can be expressed as follows for the air phase:

$$B'_a = \frac{\Delta u_a}{\Delta \sigma_3} \tag{15.19}$$

where:

B'_a = secant pore pressure parameter for the air phase during isotropic, undrained compression,
Δu_a = increase in pore-air pressure due to an increase in isotropic pressure (i.e., $u_a - u_{a0}$),
u_{a0} = initial pore-air pressure, and
$\Delta \sigma_3$ = increase in isotropic pressure from the initial condition.

Similarly, for the water phase, the B' pore pressure parameter can be expressed as

$$B'_w = \frac{\Delta u_w}{\Delta \sigma_3} \tag{15.20}$$

where:

B'_w = secant pore pressure parameter for the water phase during isotropic, undrained compression,
Δu_w = increase in pore-water pressure due to an increase in isotropic pressure,
$\Delta \sigma_3$ = change in pore-water pressure measured from the initial condition (i.e., $u_w - u_{a0}$), and
u_{w0} = initial pore-water pressure.

If an infinitesimal increase in the isotropic pressure is considered at point 1, the pore pressure response at point 1 can be expressed as the tangent pore pressure parameter B (Fig. 15.11). The B tangent pore pressure parameters for the air and water phases are written as follows:

$$B_a = \frac{du_a}{d\sigma_3} \tag{15.21}$$

$$B_w = \frac{du_w}{d\sigma_3} \tag{15.22}$$

where:

B_a = tangent pore pressure parameter for the air phase during isotropic, undrained compression,
du_a = increase in pore-air pressure due to an infinitesimal increase in isotropic pressure, $d\sigma_3$

$d\sigma_3$ = infinitesimal increase in isotropic pressure,
B_w = tangent pore pressure parameter for the water pressure during isotropic, undrained compression, and
du_w = increase in pore-water pressure due to an infinitesimal increase in isotropic pressure, $d\sigma_3$.

The concept of a B tangent pore pressure parameter was previously used (i.e., Eq. 15.13), where there was an infinitesimal increase in total stress. The B' secant pore pressure parameters are particularly useful in computing the final pore-air and pore-water pressures after a large change in total stress. The B tangent pore pressure parameters can also be used to estimate the final pore pressures under large changes in total stress by using a marching-forward technique with finite increments of total stress. The total stress is incrementally increased, commencing from an initial condition and proceeding to the final condition under investigation. As saturation is approached, the pore-water pressure approaches the pore-air pressure (i.e., $u_w \rightarrow u_a$ or $B_w \rightarrow B_a$) and the air bubbles dissolve in the water.

At saturation, a change in total stress causes an almost equal change in pore-water pressure. The small difference between the change in total stress and the change in pore-water pressure at saturation can be ignored due to the low compressibility of water relative to the compressibility of the soil structure (Skempton, 1954). In other words, the B tangent parameter becomes equal to 1.0 (i.e., $B_a = B_w = 1.0$). The convention commonly adopted in soil laboratories assumes the soil is saturated when the B pore pressure parameter reaches a value of 1.0 (Skempton, 1954). It is possible for the B' secant pore pressure parameter to be less than 1.0 even though the soil has reached saturation. Also, values of B'_a and B'_w are not equal at saturation since the initial pore-air pressure was greater than the initial pore-water pressure (Fig. 15.11).

The derivations for tangent pore pressure parameters for various loading conditions are presented in the following sections. Hilf's analysis is an exception where a secant-type pore pressure parameter is derived. A prime sign is assigned to the secant parameters (e.g., B') to differentiate between the secant and the tangent pore pressure parameters.

15.4.2 Necessary Volume Change Constitutive Relations

The derivation of pore pressure parameters requires volume change constitutive relations for an unsaturated soil. These constitutive relations describe the volume changes that take place under *drained* loading. The volume changes are expressed in terms of the stress state variable changes. These constitutive relations were formulated and explained in Chapter 13 on volume-mass constitutive relations. This section briefly summarizes the constitutive relations required for deriving the pore pressure parameter equations.

Consider an unsaturated soil specimen that undergoes one-dimensional, drained compression. The stress state variables $\sigma - u_a$ and $u_a - u_w$ change as the soil is compressed. The volume of the soil changes in accordance with the constitutive relation for the soil structure. Volume change is primarily the result of compression of the pore fluid since the soil solids are essentially incompressible. The volume of the voids, V_v, referenced to the initial total volume of the soil, V_0 (i.e., V_v/V_0), is a function of the stress state variables. The volume change constitutive relationship can be linearized for finite changes in stress state. The linear compressibility form of equation for volume change was proposed by Fredlund and Morgenstern (1976):

$$\frac{dV_v}{V_0} = m_1^s d(\sigma - u_a) + m_2^s d(u_a - u_w) \qquad (15.23)$$

where:

dV_v/V_0 = volume change referenced to the initial total volume of the soil,
V_v = volume of soil voids,
V_0 = initial total volume of the soil,
m_1^s = coefficient of soil volume change with respect to a change in net normal stress,
$d(\sigma - u_a)$ = change in net normal stress,
m_2^s = coefficient of soil volume change with respect to a change in matric suction, and
$d(u_a - u_w)$ = change in matric suction.

The continuity of a referential, unsaturated soil element requires that overall volume change must be equal to the sum of the changes in volume associated with the air and water phases which fill the pore voids. The air phase constitutive equation can be expressed as

$$\frac{dV_a}{V_0} = m_1^a d(\sigma - u_a) + m_2^a d(u_a - u_w) \qquad (15.24)$$

where:

dV_a/V_0 = change in the volume of air referenced to the initial total volume of the soil,
V_a = volume of air,
m_1^a = coefficient of air volume change with respect to a change in net normal stress, and
m_2^a = coefficient of air volume change with respect to a change in matric suction.

The constitutive equation for the water phase can be written as

$$\frac{dV_w}{V_0} = m_1^w d(\sigma - u_a) + m_2^w d(u_a - u_w) \qquad (15.25)$$

where:

dV_w/V_0 = change in the volume of water referenced to the initial volume of the soil,

V_w = volume of water,
m_1^w = coefficient of water volume change with respect to a change in net normal stress, and
m_2^w = coefficient of water volume change with respect to a change in matric suction.

The continuity requirement for a referential element can be expressed as

$$\frac{dV_v}{V_0} = \frac{dV_a}{V_0} + \frac{dV_w}{V_0} \tag{15.26}$$

The following conditions must be satisfied when considering volumetric continuity:

$$m_1^s = m_1^a + m_1^w \tag{15.27}$$

$$m_2^s = m_2^a + m_2^w \tag{15.28}$$

The above constitutive relations can be used to compute volume changes that occur during undrained compression. The changes in volume that occur between the initial and the final conditions can be calculated using a marching-forward technique with finite changes in each of the stress state variables [i.e., $d(\sigma - u_a)$ and $d(u_a - u_w)$]. The volume of the soil, water, and air must be accounted for after each increment of total stress.

Although three constitutive relationships (i.e., soil structure, air, and water) have been presented, only two of the relationships are required when deriving the pore pressure parameter equations. The constitutive equations for the soil structure and the air phase will be selected for the derivation of the pore pressure parameters.

15.5 DRAINED AND UNDRAINED LOADING

The constitutive relationships obtained from drained loading can be applied to undrained loading conditions. The application of an all-around positive (i.e., compressive) total stress, $d\sigma$, either under drained or undrained loading conditions, causes a change in volume. In drained loading, air and water are allowed to drain from the soil subsequent to the application of the total stress increment. The stress state variables in the soil are altered and volume of the soil changes. The volume change can be computed from the stress state variable changes in accordance with the constitutive relationship for the soil structure.

In *undrained* loading, the air and water are not allowed to drain from the soil. The total stress increase causes the pore-air and pore-water pressures to increase, and consequently the stress state variables also change. An increase in the pore fluid pressures occurs in response to a compression of the pore fluid (i.e., mainly air). Volume changes during undrained loading can be regarded as the volume change equivalent to the pore fluid compression. The volume change equivalent to the pore fluid compression, dV_v, can be computed by multiplying the compressibility of the air-water mixture, C_{aw}, by the pore fluid volume V_v and the total stress increment $d\sigma$ (Fig. 15.12a):

$$\left(\frac{dV_v}{V_0}\right)_0 = C_{aw} n \, d\sigma \tag{15.29}$$

where:

$(dV_v/V_0)_0$ = volume change during undrained compression resulting from the compression of the pore fluids referenced to the initial total volume and
n = porosity (i.e., V_v/V can be assumed to be equal to V_v/V_0 for small changes in stress state variables and small volume changes).

Volume change can also be written in terms of the stress state variable changes as defined by the constitutive

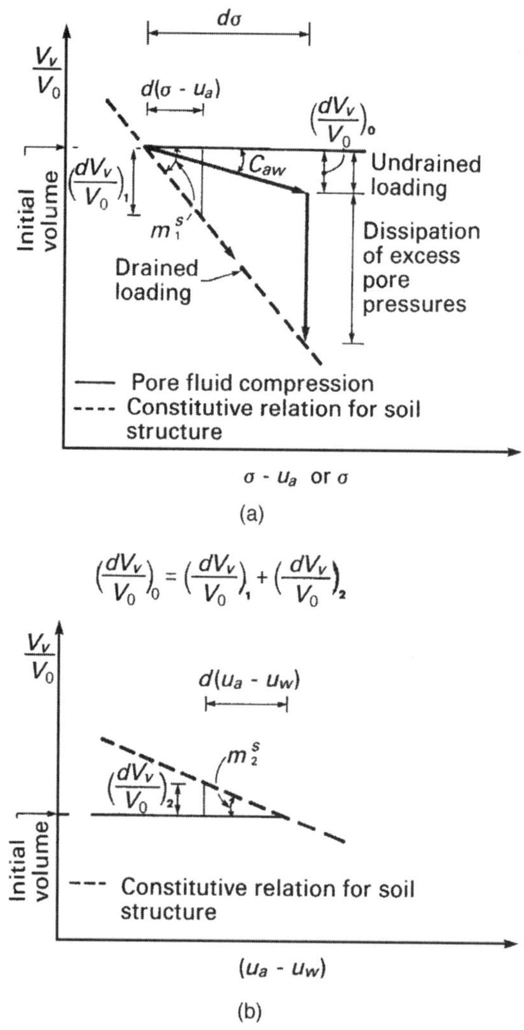

Figure 15.12 Volume changes in unsaturated soil during loading: (a) volume change with respect to change in $\sigma - u_a$ or σ; (b) volume change with respect to change in $u_a - u_w$.

relationship for the soil structure. The increase in total stress results in an increase in net normal stress, $d(\sigma - u_a)$, and a decrease in matric suction, $d(u_a - u_w)$. The increase in net normal stress causes a decrease in volume:

$$\left(\frac{dV_v}{V_0}\right)_1 = m_1^s d(\sigma - u_a) \tag{15.30}$$

where:

$(dV_v/V_0)_1$ = volume change due to the change in net normal stress, $d(\sigma - u_a)$, referenced to the initial total volume.

The decrease in matric suction generally causes an increase in soil volume (Fig. 15.12b):

$$\left(\frac{dV_v}{V_0}\right)_2 = m_2^s d(u_a - u_w) \tag{15.31}$$

where:

$(dV_v/V_0)_2$ = volume change due to a change in matric suction, $d(u_a - u_w)$, referenced to the initial total volume.

The matric suction decrease yields a volume increase or swelling. The m_2^s value for the unloading path must be used in Eq. 15.31. The total volume change obtained from the constitutive relations for the soil structure can be written as the sum of Eqs. 15.30 and 15.31:

$$\left(\frac{dV_v}{V_0}\right)_1 + \left(\frac{dV_v}{V_0}\right)_2 = m_1^s d(\sigma - u_a) + m_2^s d\left(u_a - u_w\right) \tag{15.32}$$

The total volume change obtained from the constitutive relationship can be equated to the volume change due to pore fluid compression:

$$\left(\frac{dV_v}{V_0}\right)_1 + \left(\frac{dV_v}{V_0}\right)_2 = \left(\frac{dV_v}{V_0}\right)_0 \tag{15.33}$$

or

$$m_1^s d\left(\sigma - u_a\right) + m_2^s d\left(u_a - u_w\right) = C_{aw} n\, d\sigma \tag{15.34}$$

In a *saturated* soil, the pore voids are filled with water. The pore fluid compressibility is equal to the compressibility of water, which is far less than the compressibility of the soil structure. A total stress increase $d\sigma$ in undrained loading is almost entirely transferred to the water phase (i.e., $du_w \approx d\sigma$), or stated another way, the pore-water pressure parameter approaches 1.0. The effective stress in undrained loading remains essentially unchanged [i.e., $d\left(\sigma - u_w\right) \approx 0.0$]. As a result, the volume change computed from the constitutive relationship for the soil structure is extremely small. The soil volume change obtained from the pore fluid compression is also extremely small because of the low compressibility of the water.

The pore voids in a *dry* soil are primarily filled with air, which is much more compressible than the soil structure. The pore-air pressure is low and the isothermal compressibility of air is equal to the inverse of the absolute air pressure. A total stress increase $d\sigma$ during undrained loading is almost entirely transferred to the soil structure. The pore pressure remains practically constant or stated another way, the pore pressure parameter approaches zero. The volume change can be computed from the change in net normal stress $\left(\sigma - u_a\right)$, in accordance with the constitutive relationship for the soil structure. Volume changes associated with a change in matric suction are negligible for a dry soil. The compression of the air during undrained loading follows the compression of the soil structure. A substantial volume change is required to produce a significant pore-air pressure since the air phase has a high compressibility.

15.5.1 Total Stress and Soil Anisotropy

The pore pressure parameters can be derived for various conditions of loading and soil anisotropy. Possible loading conditions are similar to those outlined by Lambe and Whitman (1979) for saturated soils. These loading conditions are summarized in Fig. 15.13, where increments in the major, intermediate, and minor principal stresses are denoted as $d\sigma_1$, $d\sigma_2$, and $d\sigma_3$, respectively. For K_0 loading, the soil can only change volume in one direction (i.e., the y-direction). For the other loading conditions, the soil can change volume three dimensionally (i.e., x-, y-, and z-directions). The loading conditions differ in the magnitude and direction in which total stresses are applied.

An isotropic soil is defined as a soil with compressibility values that are constant with respect to different directions. That is to say, the magnitudes of m_1^s and m_1^a are constant in the x-, y-, and z-directions. Only the constitutive equations for the soil structure and the air phase have been selected for the derivation of the pore pressure parameters.

Soil anisotropy is herein defined as the condition where the soil compressibility with respect to a change in each stress state variable varies with direction. The number of independent modulus values required depends on whether the longitudinal strain in each direction is to be predicted independently or whether three-dimensional volume changes are to be predicted. As a result, the compressibility m_1^s varies in the x-, y-, and z-directions and is designated as m_{11}^s, m_{12}^s, and m_{13}^s, respectively. Similarly, the air phase has compressibility values m_{11}^a, m_{12}^a, m_{13}^a in the x-, y-, and z-directions, respectively. However, the compressibility with respect to a change in matric suction can be assumed to produce singular compressibility values (i.e., m_2^s and m_2^a).

The following sections present the derivations of pore pressure parameters for an isotropic soil under one-dimensional loading (i.e., K_0 loading). The volume change modulus for K_0 loading will be given the subscript k (i.e.,

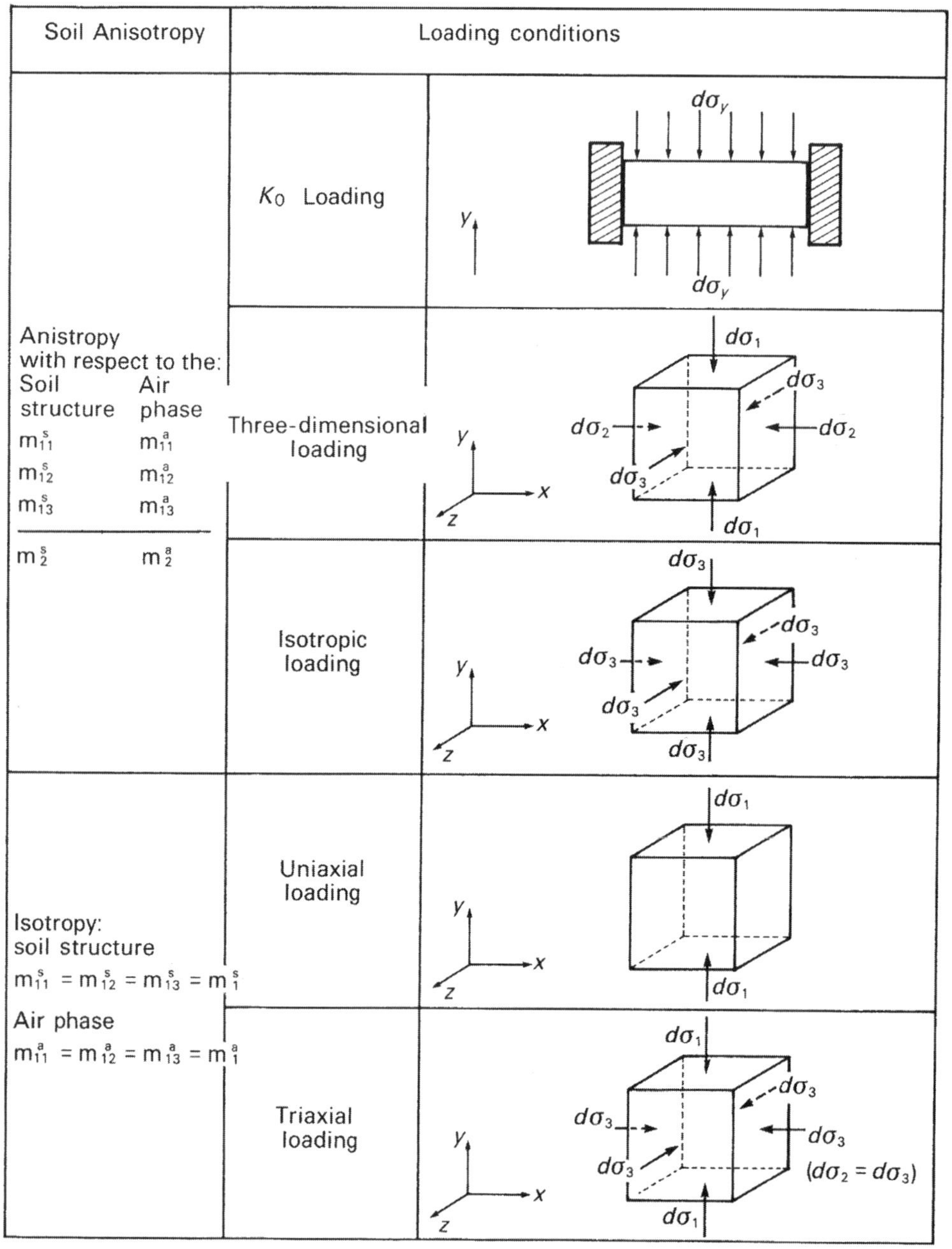

Figure 15.13 Summary of loading conditions and soil anisotropy pertaining to derivation of pore pressure parameter.

m^s_{1k} and m^a_{1k}) in order to differentiate this loading condition from other loading conditions.

15.5.2 K_0 Loading

Pore pressure parameters can be derived for the K_0-undrained loading conditions shown in Fig. 15.13. Let us suppose that a total stress increment is applied in the vertical direction. The total stress increment is denoted as $d\sigma_y$, and the vertical direction is assumed to be the major principal stress direction. Equation 15.34 relates the soil volume change to volume change due to pore fluid compression. Substituting the K_0 loading condition (i.e., $d\sigma_y$) into Eq. 15.34 gives

$$m^s_{1k}\, d(\sigma_y - u_a) + m^s_2\, d\left(u_a - u_w\right) = C_{aw} n\, d\sigma_y \qquad (15.35)$$

where:

m^s_{1k} = compressibility of the soil structure with respect to a change in net major principal stress for K_0 loading and

$d\sigma_y$ = infinitesimal increase in the major principal stress.

The compressibility of an air-water mixture, C_{aw}, is calculated by substituting the change in total stress into Eq. 15.12:

$$C_{aw} = SC_w \left(\frac{du_w}{d\sigma_y}\right) + \frac{(1 - S + hS)(du_a/d\sigma_y)}{\bar{u}_a} \quad (15.36)$$

Combining Eq. 15.35 and Eq. 15.36 results in the following relationship for K_0-undrained loading:

$$m_{1k}^s \, d\left(\sigma_y - u_a\right) + m_2^s \, d\left(u_a - u_w\right) = SnC_w \, du_w + (1 - S + hS)\, n \frac{du_a}{\bar{u}_a} \quad (15.37)$$

Rearranging Eq. 15.37 yields an expression for the change in the pore-water pressure, du_w, in response to a total stress change, $d\sigma_y$:

$$du_w = \left(\frac{m_2^s - m_{1k}^s - [(1 - S + hS)n/\bar{u}_a]}{m_2^s + SnC_w}\right) du_a + \left(\frac{m_{1k}^s}{m_2^s + SnC_w}\right) d\sigma_y \quad (15.38)$$

The compressibility m_2^s can be written as a ratio of the compressibility with respect to a total stress change, m_{1k}^s:

$$R_{sk} = \frac{m_2^s}{m_{1k}^s} \quad (15.39)$$

Substituting Eq. 15.39 into Eq. 15.38 gives

$$du_w = \left(\frac{R_{sk} - 1 - [(1 - S + hS)n/(\bar{u}_a m_{1k}^s)]}{R_{sk} + SnC_w/m_{1k}^s}\right) du_a + \left(\frac{1}{R_{sk} + SnC_w/m_{1k}^s}\right) d\sigma_y \quad (15.40)$$

Equation 15.40 can be further simplified as follows:

$$du_w = R_{1k} \, du_a + R_{2k} \, d\sigma_y \quad (15.41)$$

where:

$$R_{1k} = \frac{R_{sk} - 1 - [(1 - S + hS)n/(\bar{u}_a m_{1k}^s)]}{R_{sk} + SnC_w/m_{1k}^s}$$

$$R_{2k} = \frac{1}{R_{sk} + SnC_w/m_{1k}^s}$$

There are two unknowns (i.e., du_w and du_a) in the above equation. In order to compute the pore-air and pore-water pressure changes, a second independent equation is required. The second equation is derived by considering volume change in the air phase. The change in volume is described by the constitutive relationship for the air phase. The volume change due to the compression of air, dV_a, is computed from the compressibility of air (i.e., the second term in Eq. 15.36) multiplied by the pore fluid volume (i.e., nV) and the total stress increment $d\sigma_y$:

$$\frac{dV_a}{V_0} = \left(\frac{(1 - S + hS)n}{\bar{u}_a}\right) du_a \quad (15.42)$$

The volume change given by the constitutive relation for the air phase must be equal to the volume change due to compression of the air:

$$m_{1k}^a \, d\left(\sigma_y - u_a\right) + m_2^a \, d\left(u_a - u_w\right) = \frac{(1 - S + hS)n}{\bar{u}_a} du_a \quad (15.43)$$

where:

m_{1k}^a = slope of V_a/V_0 versus $\sigma_y - u_a$ for K_0 loading.

The second equation can be written to give the change in pore-air pressure, du_a, associated with a total stress increment $d\sigma_y$:

$$du_a = \left(\frac{m_2^a}{m_2^a - m_{1k}^a - [(1 - S + hS)n/\bar{u}_a]}\right) du_w - \left(\frac{m_{1k}^a}{m_2^a - m_{1k}^a - [(1 - S + hS)n/\bar{u}_a]}\right) d\sigma_y \quad (15.44)$$

The relationship between the modulus values m_2^a and m_{1k}^a can be expressed as a ratio, R_{ak}:

$$R_{ak} = \frac{m_2^a}{m_{1k}^a} \quad (15.45)$$

The above equation can be written in terms of the ratio R_{ak}:

$$du_a = \left(\frac{R_{ak}}{R_{ak} - 1 - [(1 - S + hS)n/(\bar{u}_a m_{1k}^a)]}\right) du_w - \left(\frac{1}{R_{ak} - 1 - [(1 - S + hS)n/(\bar{u}_a m_{1k}^a)]}\right) d\sigma_y \quad (15.46)$$

In a simple form, the change in pore-air pressure can be written as follows:

$$du_a = R_{3k} \, du_w - R_{4k} \, d\sigma_y \quad (15.47)$$

where

$$R_{3k} = \frac{R_{ak}}{R_{ak} - 1 - [(1 - S + hS)n/(\bar{u}_a m_{1k}^a)]}$$

$$R_{4k} = \frac{1}{R_{ak} - 1 - [(1 - S + hS)n/(\bar{u}_a m_{1k}^a)]}$$

The pore-air and pore-water pressure parameters for K_0-drained loading can be written as B_{ak} and B_{wk}, respectively.

These pore pressure parameters are defined as tangent-type parameters referenced to a particular stress point:

$$B_{ak} = \frac{du_a}{d\sigma_y} \tag{15.48}$$

$$B_{wk} = \frac{du_w}{d\sigma_y} \tag{15.49}$$

Changes in pore-air and pore-water pressures can be expressed in terms of pore pressure parameters and written as follows:

$$B_{wk} = R_{1k}B_{ak} + R_{2k} \tag{15.50}$$

$$B_{ak} = R_{3k}B_{wk} - R_{4k} \tag{15.51}$$

Substituting Eq. 15.41 into Eq. 15.51 results in an equation for the B_{ak} pore-air pressure parameter:

$$B_{ak} = \frac{R_{2k}R_{3k} - R_{4k}}{1 - R_{1k}R_{3k}} \tag{15.52}$$

Similarly, Eq. 15.51 can be substituted into Eq. 15.50 to give an equation for the B_{wk} pore-water pressure parameter:

$$B_{wk} = \frac{R_{2k} - R_{1k}R_{4k}}{1 - R_{1k}R_{3k}} \tag{15.53}$$

The pore-air and pore-water pressure responses at any point during K_0-undrained loading can be computed using the B_{ak} and B_{wk} pore pressure parameters.

15.5.3 Hilf's Analysis

Hilf (1948) outlined a procedure to compute the change in pore pressure in compacted earth fills as a result of an applied total stress. The pore pressure analysis can be rearranged to take on the form of a pore pressure parameter equation. The derivation is based on the results of a one-dimensional oedometer test on a compacted soil, Boyle's law, and Henry's law.

The Hilf (1948) analysis gives rise to a nonlinear relationship between total stress and pore pressure. A secant-type pore pressure parameter can be derived from this relationship. This method has been extensively used by the U. S. Bureau of Reclamation (i.e., USBR method) and has proven to be quite satisfactory for estimating pore pressures in compacted fills (Gibbs et al., 1960). The same formulation for the estimation of pore pressures in compacted soils has also been advanced by Bishop (1957).

Hilf (1948) stated: "To illustrate the role of air in the relation between consolidation and pore pressure, consider a sample of moist earth compacted in a laboratory cylinder. If a static load is applied by means of a tight-fitting piston, permitting neither air nor water to escape, it is found that there is a measurable reduction in volume of the soil mass."

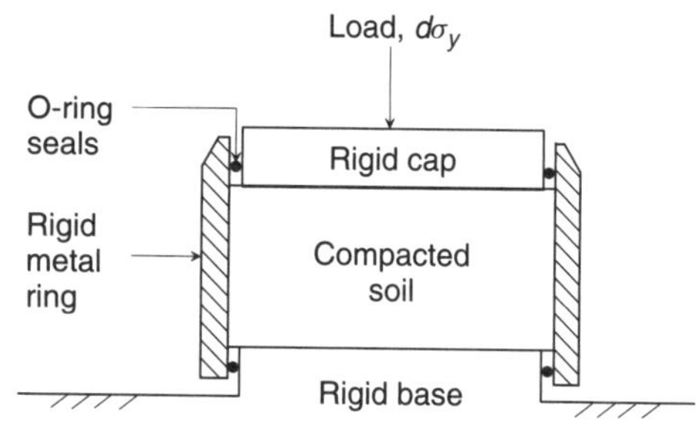

Figure 15.14 Conditions simulated in Hilf (1948) derivation of pore pressure parameters for K_0-undrained loading.

(See Fig. 15.14.) The reduction in volume can be assumed to be the result of the compression of free air and air dissolving into water. The soil solids and the water are assumed to be incompressible. Vapor pressure and temperature effects are also assumed to be negligible.

The amount of dissolved air in water is computed in accordance with Henry's law. Free and dissolved air is regarded as a single total volume of air at a particular pressure. The change in pore-air pressure between initial and final loading conditions is computed using Boyle's law.

The initial and final conditions considered in Hilf's analysis are shown in Fig. 15.15. The total volume of air associated with the initial condition, V_{a0}, can be written as follows:

$$V_{a0} = [(1 - S_0)n_0 + hS_0n_0]V_0 \tag{15.54}$$

where:

V_{a0} = initial volume of free and dissolved air,
S_0 = initial degree of saturation,
n_0 = initial porosity, and
V_0 = initial volume of the soil.

The first and second terms in the above equation represent the free and dissolved air volumes, respectively. The initial absolute pore-air pressure is denoted as $\bar{u}_{a0}$ and can be assumed to be at atmospheric conditions (i.e., 101.3 kPa). An increment in the major principal stress, $\Delta\sigma_y$, is then applied to the soil specimen. The total volume of air decreases and the air pressure increases in accordance with Boyle's law. The air volume change is equal to the void volume change ΔV_v since the soil solids and water are assumed to be incompressible. Therefore, the air volume change can be written as a change in porosity (i.e., $\Delta n = \Delta V_v/V_0$) times the initial volume of soil, V_0, as illustrated in Fig. 15.15. The total volume of air under final conditions, V_{af}, can be expressed as

$$V_{af} = [(1 - S_0)n_0 + hS_0n_0 - \Delta n]V_0 \tag{15.55}$$

where:

V_{af} = final volume of free and dissolved air and
Δn = change in porosity (i.e., decrease).

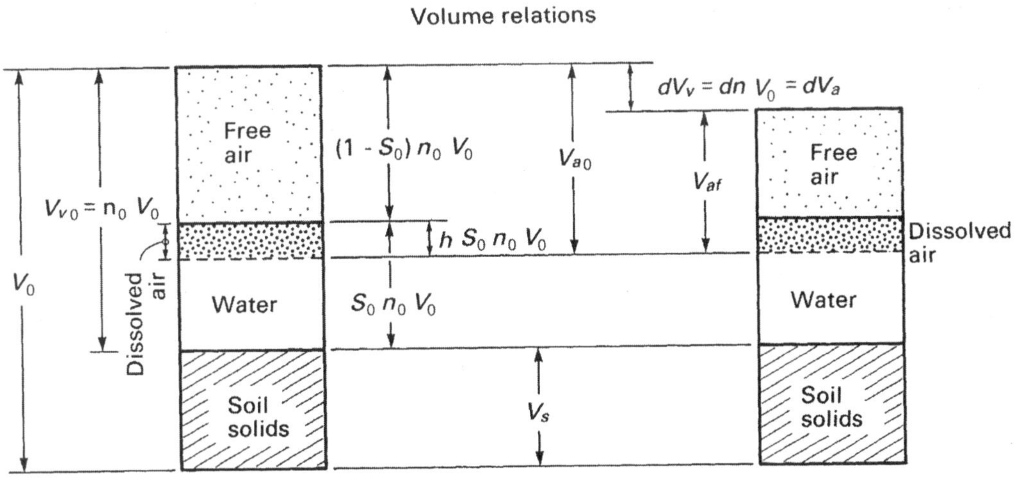

Figure 15.15 Initial and final pressure and volume conditions used in Hilf (1948) analysis.

The final absolute air pressure, $\bar{u}_{af}$, can be written as the initial absolute pore-air pressure plus the change (i.e., increase) in pore-air pressure:

$$\bar{u}_{af} = \bar{u}_{a0} + \Delta\bar{u}_a \tag{15.56}$$

where:

$\Delta\bar{u}_a$ = change (i.e., increase) in absolute pore-air pressure.

Boyle's law can be applied to the initial and final conditions of the free and dissolved air:

$$\bar{u}_{a0}V_{a0} = \bar{u}_{af}V_{af} \tag{15.57}$$

Substituting the initial conditions and the final conditions into Eq. 15.57 gives

$$\bar{u}_{a0}[(1 - S_0)n_0 + hS_0n_0]V_0 = \left(\bar{u}_{a0} + \Delta\bar{u}_a\right)\left[\left(1 - S_0\right)n_0 + hS_0n_0 - \Delta n\right]V_0 \tag{15.58}$$

Rearranging Eq. 15.58 yields an expression for the change in pore-air pressure: $\Delta\bar{u}_a$:

$$\Delta\bar{u}_a = \left(\frac{\Delta n}{(1 - S_0)n_0 + hS_0n_0 - \Delta n}\right)\bar{u}_{a0} \tag{15.59}$$

The above equation is commonly referred to as Hilf's equation. It provides a relationship between the change in pore-air pressure and the change in pore-air volume (i.e., Δn) during K_0-undrained loading. The change in pore-air pressure can be computed from Hilf's equation if the change in porosity is known.

The soil can reach a point of saturation when the soil volume change, ΔV_v, is equal the volume of free air [i.e., $\left(1 - S_0\right)n_0V_0$] (see Fig. 15.15). The change in porosity corresponding to this condition can be written as

$$\Delta n = \left(1 - S_0\right)n_0 \tag{15.60}$$

Substituting Eq. 15.60 into Eq. 15.59 gives the absolute pore-air pressure change (i.e., increase) required for saturating a soil:

$$\Delta\bar{u}_{as} = \left(\frac{1 - S_0}{S_0h}\right)\bar{u}_{a0} \tag{15.61}$$

where:

$\Delta\bar{u}_{as}$ = pore-air pressure change (i.e., increase) required for saturation.

Equation 15.61 is equivalent to the equation for computing the saturation pressure from the density of an air-water mixture. Equation 15.59 can be written in an alternate form by replacing Δn with $\Delta V_v/V_0$:

$$\frac{\Delta V_v}{V_0} = \left(\frac{\Delta\bar{u}_a}{\bar{u}_{a0} + \Delta\bar{u}_a}\right)(1 - S_0 + hS_0)n_0 \tag{15.62}$$

where:

$\Delta V_v/V_0$ = change in volume of voids referenced to the initial volume of the soil (i.e., porosity change Δn).

The above equation describes the volume change due to the compression of air. The soil volume change can also be computed from the constitutive relationship for the soil structure. Hilf (1948) assumed that the matric suction in the soil was small and that a change in pore-air pressure was equal to a change in pore-water pressure (i.e., $\Delta u_a = \Delta u_w$). It was also assumed that the constitutive relation for the soil structure could be measured by saturating the soil in an oedometer test (i.e., under constant-volume conditions) and measuring its volume change upon increased loading. The volume change of the soil structure can be expressed as

$$\frac{\Delta V_v}{V_0} = m_v \Delta\left(\sigma_y - u_a\right) \qquad (15.63)$$

where:

m_v = coefficient of volume change measured on a saturated soil in a one-dimensional oedometer test.

The soil compressibility m_1^s is assumed to be equal to the coefficient of volume change m_v measured under saturated conditions. The pore-pressure parameter for K_0-undrained loading can be derived using Hilf's analysis by equating the volume change of the soil structure to the volume change due to the compression of air:

$$m_v\,\Delta\sigma_y - m_v\,\Delta\bar{u}_a = \left(\frac{\Delta\bar{u}_a}{\bar{u}_{a0} + \Delta\bar{u}_a}\right)\left(1 - S_0 + hS_0\right) n_0 \qquad (15.64)$$

The above equation can be rearranged and solved for the change in pore-air pressure:

$$\Delta\bar{u}_a = \left(\frac{1}{1 + \dfrac{(1 - S_0 + hS_0)n_0}{[(\bar{u}_{a0} + \Delta\bar{u}_a)m_v]}}\right)\Delta\sigma_y \qquad (15.65)$$

The secant B'_{ah} pore pressure derived from Hilf's analysis can be written as

$$B'_{ah} = \frac{1}{1 + \dfrac{(1 - S_0 + hS_0)n_0}{[(\bar{u}_{a0} + \Delta\bar{u}_a)m_v]}} \qquad (15.66)$$

where:

B'_{ah} = secant pore-air pressure parameter (i.e., $\Delta u_a/\Delta\sigma_y$), for K_0-undrained loading in accordance with Hilf's analysis. The prime signifies a secant pore pressure parameter.

An increase in pore-air pressure due to a total stress increase can be predicted using Eq. 15.65. The final pore-air pressure can be computed from the initial pore-air pressure $\bar{u}_{a0}$ and the change in pore-air pressure $\Delta\bar{u}_a$. This procedure considers only the initial and final stress conditions without using a marching-forward technique. The total stress increment does not have to be small, as required when using the B tangent pore pressure parameters. However, Eq. 15.65 is nonlinear because the unknown term $\Delta\bar{u}_a$ appears on both sides of the equation. An iterative technique is required in solving Eq. 15.65.

The coefficient of volume change m_v varies depending upon the magnitude of total stress σ_y. Therefore, an average value of m_v can be used in Eqs. 15.65 and 15.66 when considering a total stress change $\Delta\sigma_y$. The secant pore-air pressure parameter B'_{ah} is obtained by dividing the pore-air pressure increase by the total stress increase. The B'_{ah} pore-air pressure parameter is assumed to be equal to the B'_{wh} pore-water pressure parameter in Hilf's analysis. As a result, the difference between a pore-air pressure change and a pore-water pressure change cannot be isolated when using Hilf's analysis. Changes in pore-water pressure are generally assumed to be equal to the changes in pore-air pressure when using Hilf's analysis. The initial pore-water pressure in Hilf's analysis was assumed to be zero. Therefore, the computed pore pressures in a compacted soil may be somewhat high or conservative for design purposes.

15.6 SOLUTIONS OF PORE PRESSURE EQUATIONS AND COMPARISONS WITH EXPERIMENTAL RESULTS

The pore pressure parameter equations can be solved and compared with typical laboratory-measured results. The effect of changing one or more of the variables in the pore pressure equations is illustrated. The theoretical pore pressure parameters can be used to estimate the pore pressures that are likely to develop during construction.

15.6.1 Secant Pore Pressure Parameter B'_a Derived from Hilf's Analysis

Measurements of pore-water pressures within compacted cores of several earth dams have been conducted by Hilf (1948). The measurements were made using piezometers with coarse porous tips; consequently, only positive pressures could be measured. Figure 15.16 presents the pore-water pressure measurements from two piezometers installed in the Anderson Ranch dam. The pore-water pressures increased as the overburden pressure increased (i.e., $\Delta\sigma_y$) during construction.

The pore-water pressure development during construction can be simulated using Hilf's analysis. Equation 15.65 is applicable to both pore-air and pore-water pressures in the sense that capillary effects are ignored in its derivation. The soil has an initial degree of saturation S_0 of 87.4% and an initial porosity n_0 of 28.1%. The volumetric coefficient of solubility h for air in water is assumed to be 0.02. The initial absolute pore-air pressure $\bar{u}_{a0}$ was assumed to be atmospheric (i.e., 101.3 kPa). The coefficient of volume change m_v was measured in a conventional oedometer. Substituting the above data into Eq. 15.65 results in a relationship between $\Delta\bar{u}$ (or Δu_a) and $\Delta\sigma_y$. The relationship between Δu_a and $\Delta\sigma_y$ is also assumed to be the relationship between Δu_w and $\Delta\sigma_y$ since capillary effects are ignored. The pore-water pressure u_w can then be plotted against the major principal stress σ_y using the Δu_w and $\Delta\sigma_y$ relationship as shown in Fig. 15.16. The initial pore-water pressure was also assumed to be atmospheric (i.e., zero gauge

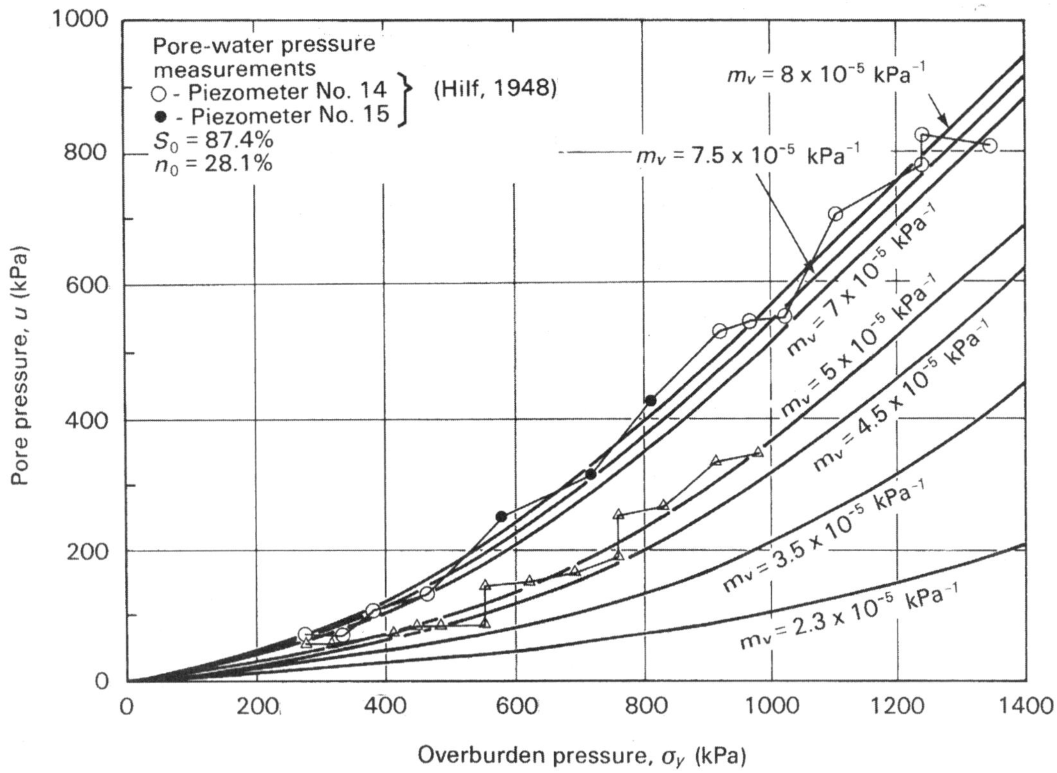

Figure 15.16 Effects of coefficient of volume change on pore pressure predictions using Hilf (1948) analysis.

pressure). An average value for the volume change coefficient m_v is often used over the range of total stress change.

Several curves relating the generated pore-water pressure to the applied total stress were computed using several values for m_v. The results are presented in Fig. 15.16. The coefficient of compressibility m_v was assumed to be constant for each curve, regardless of the total stress change, $\Delta\sigma_y$. Theoretically, the value of m_v should be varied along each curve of u_w versus σ_y to take the total stress level into consideration. The shape of the relationship between u_w and σ_y varies significantly as the average value of m_v is changed and the initial variables (i.e., S_0, n_0, and $\bar{u}_{a0}$) are kept constant. Higher soil compressibility values result in higher pore pressure responses as a result of an increase in total stress. This can be seen by the steeper u_w versus σ_y curve for more compressible soils (Fig. 15.16).

Reasonably close agreement between the piezometric measurements and the predicted pore-water pressures have been obtained when using Hilf's analysis provided appropriate soil compressibility values are used in the analysis. Close agreement is achieved in spite of the crude approximations applied and the average value for m_v. It would appear that the use of an average coefficient of volume change seems justified.

The shape of the relationship between pore pressure and total stress is strongly influenced by the initial volume-mass properties of the soil. Figure 15.17 presents several curves relating pore pressure and total stress while the initial degree of saturation S_0 is varied. The average coefficient of volume change m_v is maintained at 8×10^{-5} kPa^{-1} for all curves. The results indicate that the pore pressure versus total stress relationship becomes steeper as the initial degree of saturation increases. In other words, the pore pressure responds faster at higher degrees of saturation. A comparison between Figs. 15.16 and 15.17 suggests that the coefficient of volume change and the initial degree of saturation are both important variables affecting the pore pressure response.

15.6.2 Graphical Procedure for Hilf's Analysis

Hilf (1948) outlined a graphical procedure which used a nomograph for the solution of the pore pressure versus total stress applied. The objective of the graphical procedure was a plot of the relationship between pore pressure and total stress.

The pore pressure response combines the oedometer test data and Hilf's equation. The results of a conventional oedometer test on the soil can be plotted as shown in Fig. 15.18. The consolidation test results are plotted as the change in porosity Δn [i.e., $\Delta e/(1+e)$] versus the change in effective stress $\Delta(\sigma_y - u_w)$. Hilf's equation (i.e., Eq. 15.59) can also be plotted on the same graph. The ordinate is the change in porosity Δn, while the abscissa is the change in pore-water pressure Δu_w (Fig. 15.18).

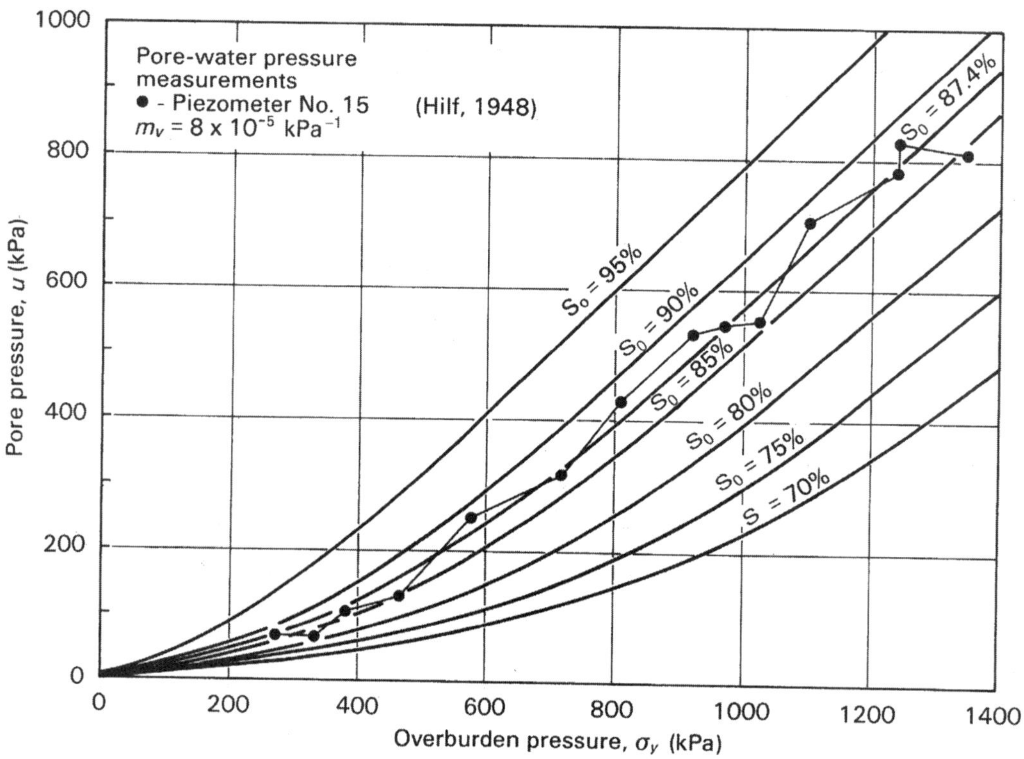

Figure 15.17 Effect of initial degree of saturation on pore pressure predictions using Hilf (1948) analysis.

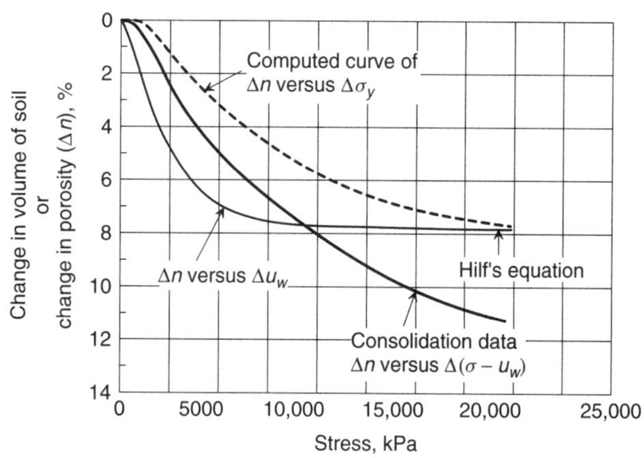

Figure 15.18 Plots of stress components versus volume change for Hilf (1948) analysis.

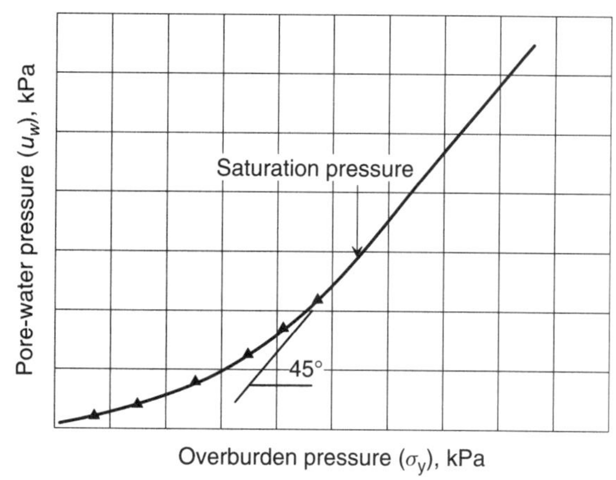

Figure 15.19 Relationship of nonlinear pore-water pressure versus total stress derived for Hilf (1948) analysis.

Equation 15.59 applies to both pore-air and pore-water pressures since capillary effects were ignored. The relationship between a change in porosity and a change in total stress can be established by combining the above two plots. For a given change in porosity Δn, the change in total stress $\Delta\sigma_y$ can be obtained by adding the magnitudes of $\Delta(\sigma_y - u_w)$ and Δu_w. This summation procedure is repeated for several changes in porosity until a curve of Δn versus $\Delta\sigma_y$ is established.

The next step in the procedure is to cross-plot the pore-water pressure u_w against the major principal stress σ_y, as shown in Fig. 15.19. The initial values of u_w and σ_y can be estimated. Hilf (1948) assumed that the initial pore-water pressure was atmospheric. The pore-water pressure increase due to total stress increase is obtained by cross-plotting Δn versus Δu_w and Δn versus $\Delta\sigma_y$. The magnitudes of Δu_w and the corresponding $\Delta\sigma_y$ are then added to the initial values of u_w and σ_y.

The pore pressure response curve exhibits a nonlinear relationship between pore-water pressure and total pressure at low overburden pressures (i.e., total stresses) or at degrees of saturation less than 100%. The relationship between pore pressure and total stress becomes linear with a slope of 45° as saturation conditions are approached. A change in total stress under undrained conditions produces an equal change in pore-water pressure at saturation conditions. The pore-water pressure versus total stress plot will have different shapes depending upon the initial degree of saturation and the compressibility of the soil.

The compressibility of the soil is handled in a slightly different manner between the analytical method and the graphical method. An average constant coefficient of volume change is assumed for the entire total stress change in the analytical method. A constant coefficient of volume change is not required in the graphical method. The use of the nonlinear consolidation curve for obtaining the pore-water pressure versus total stress relationship implies that varying coefficient-of-volume-change values have been used in the graphical analysis.

15.6.3 Experimental Results of *B* Tangent Pore Pressure Parameters (Isotropic Loading)

The *B* tangent pore pressure parameters for isotropic loading are quite readily measured in the laboratory. A soil specimen is subjected to an equal all-around or isotropic pressure σ_3. The pore-air and pore-water pressures are measured as the isotropic pressure is increased. The pore pressure response du due to a finite increment in total stress $d\sigma_3$ is expressed as a tangent parameter B. The B pore pressure parameters vary as the isotropic pressure increases. Figure 15.20 shows the relationship between the B_w pore pressure parameter and the degree of saturation for clayey gravel. A substantial decrease in B_w is observed near optimum water content.

An apparatus and test procedure were developed by Campbell (1973) to measure the independent generation of pore-air and pore-water pressures in an unsaturated soil specimen under isotropic loading. The specimens were statically compacted and subjected to isotropic pressures up to 6900 kPa. Two types of soil were investigated: Peorian loess (i.e., an inorganic silt) and Champaign till (i.e., a well-graded glacial till).

The development of pore-water pressure during the application of an isotropic pressure is shown in Figs. 15.21 and Fig. 15.22 for Peorian loess and Champaign till, respectively. The plots indicate a low pore-pressure response (i.e., a flat curve) for the conditions of low initial water content or low initial degree of saturation. The pore pressure response increases as the initial water content (and degree of saturation) increases. As a result, the B_w parameter also increases with an increase in the initial degree of saturation.

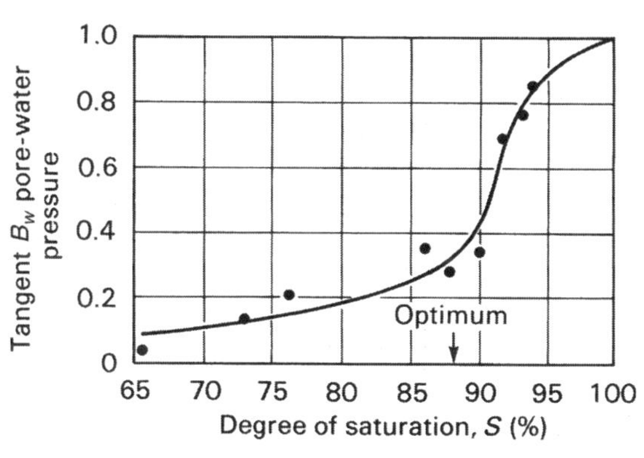

Figure 15.20 Relationship between B_w pore-water pressure parameter and degree of saturation for clayey gravel (after Skempton, 1954).

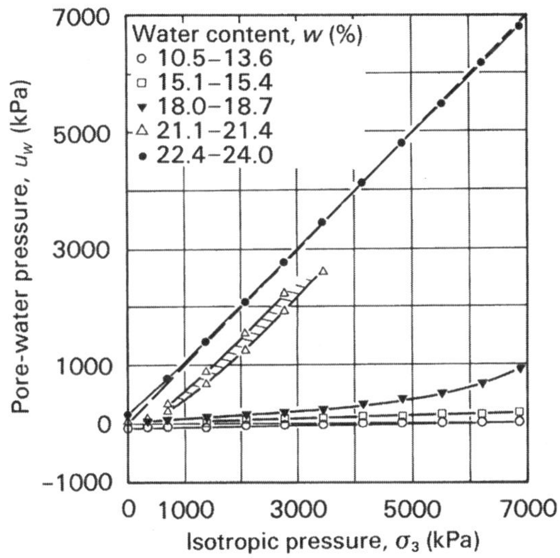

Figure 15.21 Pore-water pressure development in Peorian loess under isotropic, undrained loading (after Campbell, 1973).

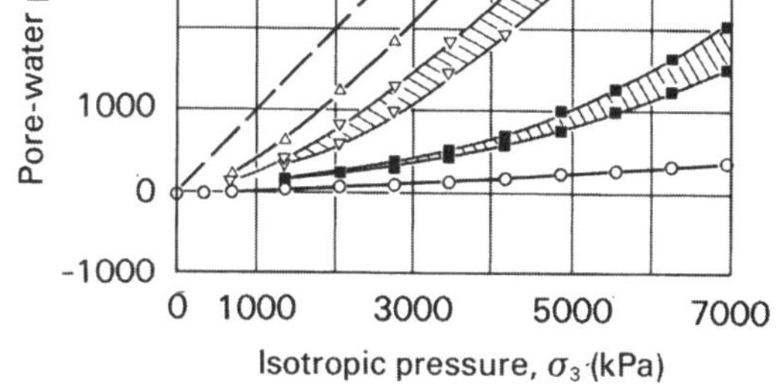

Figure 15.22 Pore-water pressure development in Champaign till under isotropic, undrained loading (after Campbell, 1973).

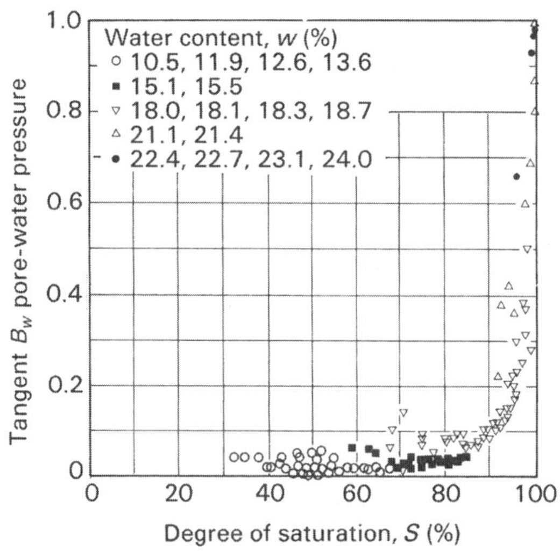

Figure 15.23 Pore-water pressure parameter B_w as function of degree of saturation for Peorian loess (after Campbell, 1973).

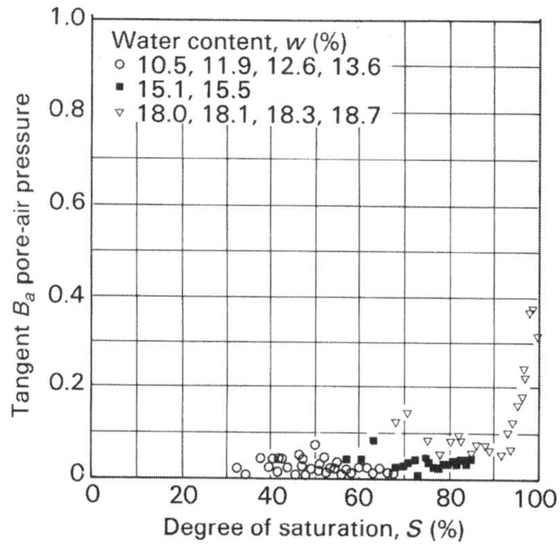

Figure 15.25 Pore-air pressure parameter B_a as function of degree of saturation for Peorian loess (after Campbell, 1973).

Specimens compacted near and above optimum water content exhibit a high, immediate pore-water pressure response to the application of an isotropic pressure increment (i.e., a steeper curve). At initial water contents well above optimum water content the pore pressure response curves approximate a straight line with a slope angle approaching 45°. In this case, the B_w pore pressure parameter is constant and approaches unity. In other words, at saturation an increment in total stress, $d\sigma_3$, is transferred entirely to the pore pressure (i.e., $du_w = d\sigma_3$). It is common practice in laboratory tests to ensure the saturation of the soil specimen by confirming that B_w is approaching 1.0.

Figures 15.23 and 15.24 present the B_w pore-water pressure parameter as a function of degree of saturation for Peorian loess and Champaign till, respectively. The plots show that the B_w pore pressure parameter increases rapidly at a degree of saturation equal to 80% for Peorian loess and 90% for Champaign till. These degrees of saturation correspond to conditions near-optimum water content for each of the compacted soils. The B_w parameter approaches 1.0 as the degree of saturation approaches 100%.

The B_a pore-air pressure parameter also increases with an increase in degree of saturation for Peorian loess and Champaign till specimens, as shown in Figs. 15.25 and 15.26. The pore-water pressure approaches the pore-air pressure at saturation and air may exist in the form of occluded air bubbles. The pore-air pressure-measuring device likely begins to measure the pore-water pressure as saturation is approached. As a result, B_a approaches a value of 1 at saturation, as exhibited in Fig. 15.26 for Champaign till.

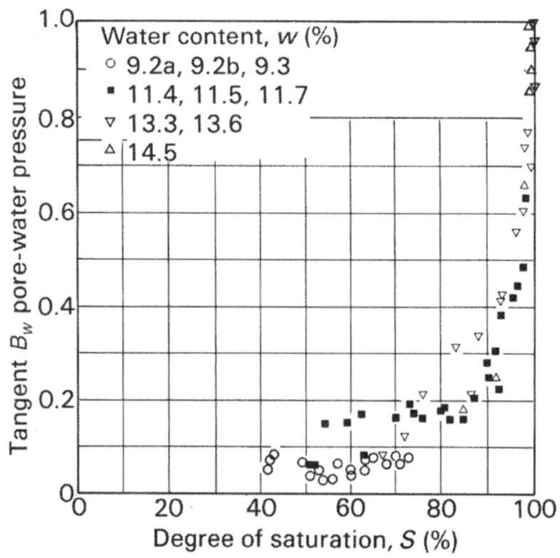

Figure 15.24 Pore-water pressure parameter B_w as function of degree of saturation for Champaign till (after Campbell, 1973).

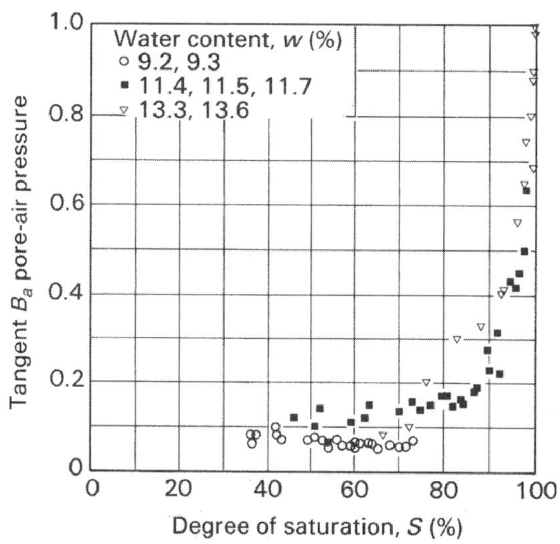

Figure 15.26 Pore pressure parameter B_a as function of degree of saturation for Champaign till (after Campbell, 1973).

15.7 RHEOLOGICAL MODEL TO REPRESENT RELATIVE COMPRESSIBILITIES OF UNSATURATED SOIL

The primary value of rheological models lies in their role as an aid in visualizing soil behavior. For example, a rheological model can illustrate how a load applied to a saturated soil is initially carried by the water phase and is slowly transferred to the soil structure as the pore-water pressure dissipates.

Two *linear* rheological elements connected in series, one for the water phase and the other for the air phase, can be used to simulate the behavior of a layer of unsaturated soil (Fig. 15.27). The unsaturated soil stratum is simulated by interconnecting two rheological elements in series: one element represents the air portion and another element represents the water portion of the soil.

Fredlund and Morgenstern (1977) showed that it was possible to use any two of the three possible stress state variables (i.e., $\sigma - u_w$, $\sigma - u_a$, and $u_a - u_w$) when describing soil behavior. It appears that the $\sigma - u_w$ and $\sigma - u_a$ combination has advantages over other combinations for the visualization of drained and undrained processes in an unsaturated soil. Consequently, the $\sigma - u_w$ and $\sigma - u_a$ combination of stress state variables will be used for the development of a rheological model to simulate unsaturated soil behavior.

One linear rheological element is used to model the effect of the solid phase and the water phase (i.e., $\sigma - u_w$).

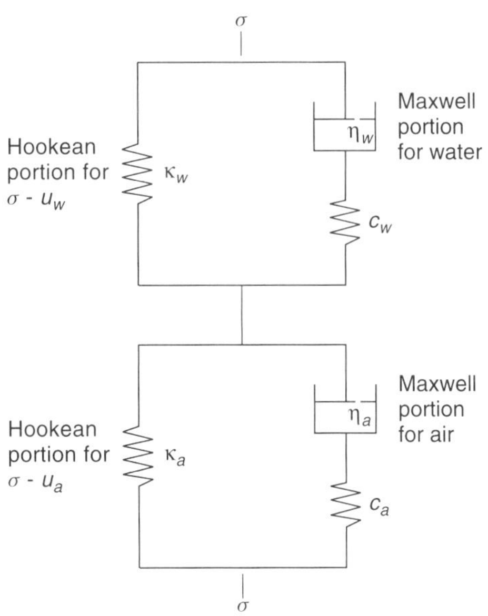

Figure 15.27 Use of two linear rheological elements in series to represent the behavior of the $\sigma - u_w$ and $\sigma - u_a$ stress state variables for unsaturated soil.

Another linear rheological element is used to model the effect of the solid phase and the air phase (i.e., $\sigma - u_a$). The two linear elements are connected in series and represent the two independent stress state variables.

The Hookean constants c_w and c_a account for the compressibility of water and air, respectively. The fluids in the dashpots are assumed to be incompressible, but the escape or outflow of air and water is allowed. The Hookean spring constants κ_w and κ_a account for the compressibility of soil structure with respect to each of the stress state variables $\sigma - u_w$ and $\sigma - u_a$, respectively, where σ is the total stress applied, u_w is the pore-water pressure, and u_a is the pore-air pressure.

The hydraulic conductivity properties of the soil with respect to water and air are represented by the viscosity constants η_w and η_a, respectively. The total strain of a unit infinitesimal layer is denoted by ε. The strain of the linear rheological elements for the water phase and the air phase are denoted by ε_w and ε_a, respectively.

15.7.1 Effect of Relative Compressibility Values on Initial Pore Pressures Generated upon Loading

A portion of the load applied to an unsaturated soil will be carried by each of the linear rheological models. In other words, there will be a sharing of the applied load between the air phase and water phase rheological elements. The amount of load carried by the rheological model for the air phase and the water phase will depend upon the relative soil compressibility values associated with the $\sigma - u_w$ and $\sigma - u_a$ stress state variables as well as the compressibility values associated with the air and water phases. The compressibility of water is low and the compressibility of air is high and highly nonlinear. The nonlinearity of the air phase is not taken into consideration in the rheological models.

Let us suppose that a pressure of 100 kPa represents a total stress σ which is instantaneously applied to the rheological models that are connected in series. The total applied stress will immediately be shared by the rheological model associated with the air and water phases. The sharing of the total stress will depend upon the magnitude of the four relative compressibility values. Since the compressibility of the air phase is much greater than the compressibility of the water phase, the water phase rheological model will experience the greatest initial change in pore pressure. The instantaneous change in pore-air pressure (i.e., excess pore air pressure) and the instantaneous change in pore-water pressure (i.e., excess pore-water pressure) will then dissipate over time. The dissipation of pore-air and pore-water pressures with time represents independent processes that are described under the topic of consolidation, collapse, and swelling.

The application of a total stress (e.g., 100 kPa) may result in a relatively small instantaneous change in the pore-air pressure (e.g., say 15 kPa) and a somewhat larger instantaneous change in the pore-water pressure (e.g., say 30 kPa). The remainder of the applied pressure is assumed to be instantaneously adsorbed by the Hookean springs associated with each of the two rheological models.

The dashpots adsorb more of the applied pressure and the Hookean springs instantaneously adsorb less of the load if the initial conditions of the soil are close to saturation. The applied load is transferred from the Maxwell dashpots to the Hookean springs (i.e., the process of consolidation) with elapsed time. The processes that occur with elapsed time are discussed as part of the consolidation process discussed in Chapter 16.

CHAPTER 16

Consolidation and Swelling Processes in Unsaturated Soils

16.1 INTRODUCTION

The application of total stress (or a load) to an unsaturated soil can result in the generation of excess pore-air and pore-water pressures. The excess pore-air and pore-water pressures will dissipate with time and eventually the pore pressures return to the equilibrium values that existed prior to loading. The dissipation process of pore pressures is called "consolidation" and the process results in volume change as excess pore pressures are dissipated.

It is also possible for excess pore fluid pressures to be generated as a result of changes in boundary conditions. For example, the infiltration of rainwater at ground surface can initiate water movement into the soil along with associated soil swelling or soil collapse. Water moves into the soil because of a difference in the hydraulic head (or pore-water pressure) at ground surface and the hydraulic head immediately within the soil mass. The drying of a soil from the ground surface also initiates moisture movement that can be accompanied by volume decrease. The drying of the soil can be visualized in terms of a moisture flux at ground surface or a change in the pore-water pressure boundary condition.

The "swelling" process is usually associated with a change in moisture flux boundary conditions whereas the consolidation process is usually associated with a change in the externally applied total stresses. In the case of a swelling process, the pore pressures (i.e., pore-water pressure, pore-air pressure or both) are below the equilibrium state and therefore will increase toward an equilibrium state. The consolidation process has a decrease in pore pressures with time while the swelling process has an increase in pore pressures with time. The consolidation and swelling processes are essentially equivalent and opposite processes from a theoretical standpoint. However, the swelling process is not usually initiated through a change in the total stresses (i.e., unloading of the soil). Rather, it is usually environmental changes related to precipitation that directly change the pore-water pressure conditions at the ground surface and initiate a process of pore-water pressure changes throughout the soil mass.

16.2 STRESS AND SEEPAGE UNCOUPLED AND COUPLED SYSTEMS

The dissipation of the pore-water and pore-air pressures resulting from stress changes can be analyzed in an "uncoupled" or "coupled" manner. The *uncoupled* solution assumes that the stress-deformation analysis and the pore fluid dissipation processes are solved in an independent manner. The *coupled* solution assumes that the stress-deformation and pore fluid dissipation processes are solved simultaneously.

There is also another distinctive feature related to the uncoupled and coupled solutions. When the uncoupled approach is adopted for consolidation, the initial excess pore fluid pressures (e.g., pore-water pressure and pore-air pressures) are computed using the pore pressure parameters presented in Chapter 15. The pore pressure parameters allow the generated excess pore fluid pressure to be instantaneously generated and used to initiate subsequent processes. When using a fully coupled approach, the compressibility of the pore fluids becomes part of the overall formulation and the calculation of the generated excess pore pressures becomes part of the solution.

The coupled approach is considered to be more rigorous and correct for the simulation of the consolidation process. However, the uncoupled approach is generally easier to solve and sufficiently accurate for most engineering problems. The fully coupled consolidation process has revealed the "Mandel-Cryer" effect, which suggests that it is a more accurate solution. The Mandel-Cryer effect is related to the Poisson's ratio of the soil and may only be significant for problems where the induced pore pressures are the result of external loadings (Vu and Fredlund, 2002).

16.2.1 Combining Stress, Water, and Air for One-Dimensional Consolidation

The theory of consolidation for a saturated soil has been viewed as one of the pinnacle analyses in classical soil mechanics. The theory of consolidation has been extensively used for design and control when dealing with saturated soils. Case histories have verified that the mathematical consolidation theory can be applied and relied upon in engineering practice.

The theory of consolidation involving the application of externally applied loads may not be as important in unsaturated soil mechanics as the theory of consolidation has been in saturated soil mechanics. However, transient processes resulting from changing boundary conditions (i.e., pore fluid pressures, hydraulic heads, or moisture fluxes) are of significant importance to simulating unsteady-state unsaturated soil processes. The theory of consolidation and unsteady-state analyses are closely related in unsaturated soil mechanics.

The theory of consolidation for unsaturated soils can be derived as an extension of the formulations used for saturated soils. Continuity of a *saturated* soil element requires that the volume change of a soil element be equal to the change in the volume of water in the element. The theory of consolidation equation derived by Terzaghi (1943) for saturated soils satisfied the continuity requirements of a two-phase system:

$$\frac{\partial u_w}{\partial t} = c_v \frac{\partial^2 u_w}{\partial y^2} \tag{16.1}$$

where:

c_v = coefficient of consolidation [i.e., $k_s/(\rho_w g m_v)$],
k_s = coefficient of permeability with respect to water at saturation (i.e., $S = 100$ %),
ρ_w = density of water,
g = gravitational acceleration, and
m_v = coefficient of volume change for saturated soils.

The above equation describes changes in pore-water pressures with respect to depth and time during the consolidation process. Changes in pore-water pressure result in changes in the effective stress $(\sigma - u_w)$ The effective stress changes can be substituted into stress-deformation constitutive equations to compute volume change. Volume changes are equal to the volume of water flowing out of the soil as long as the degree of saturation remains at 100%. The computed volume changes can, in turn, be used to compute soil volume-mass properties such as void ratio, water content, and density during the consolidation process.

In 1941, Biot proposed a general theory of consolidation for an unsaturated soil with occluded air bubbles. Two constitutive equations relating stress state and strain were formulated in terms of effective stress $(\sigma - u_w)$ and the pore-water pressure u_w. In other words, the need for separating the effects of total stress and pore-water pressure was recognized by Biot. One constitutive equation related the void ratio to the stress state, and the other equation related the water content to the stress state of the soil. Other assumptions used in Biot's theory were similar to those used in Terzaghi's theory of consolidation for saturated soils. Biot's theory resulted in an equation similar to Eq. 16.1 for one-dimensional consolidation; however, the coefficient of consolidation c_v was modified to take into account the compressibility of the pore fluid.

Larmour (1966), Hill (1967), and Olson (1986) showed that Terzaghi's equation with a modified coefficient of consolidation c_v can be used to describe the consolidation behavior of unsaturated soils with occluded air bubbles. Scott (1963) incorporated void ratio change and degree-of-saturation change into the formulation of a consolidation equation for unsaturated soils with occluded air bubbles. Blight (1961) derived a consolidation equation for the air phase of a dry, rigid, unsaturated soil. Fick's law of diffusion was used to relate the mass transfer of air to an air pressure gradient.

Fredlund and Hasan (1979) presented two PDEs which could be solved for the pore-air and pore-water pressure changes during the consolidation process in an unsaturated soil. The air phase was assumed to be continuous. Darcy's law and Fick's law were applied to the flow of water and air, respectively. The coefficients of permeability with respect to both the water and air phases were considered to be functions of matric suction or one of the volume-mass properties of the soil. In other words, the two PDEs contained terms which accounted for variations in the coefficients of permeability with respect to the air and water phases. Both equations were solved simultaneously, and the method is commonly referred to as a two-phase flow approach. The formulation by Fredlund and Hasan (1979) was similar in form to the conventional one-dimensional Terzaghi (1943) derivation. The derivations also demonstrated a smooth transition between the unsaturated and saturated cases. Similar consolidation equations have also been proposed by Lloret and Alonso (1981).

The two partial differential, transient flow equations proposed by Fredlund and Hasan (1979) have been used to simulate total volume change and water volume change behavior of compacted kaolin specimens when total stress and matric suction were changed (Fredlund and Rahardjo, 1986). The pore pressure changes calculated using the two differential equations of flow resulted in changes in the stress state variables. The stress state variable changes were substituted into the soil structure and the water phase constitutive equations to compute volume changes in the unsaturated soil. Comparisons between the predicted volume changes and experimental results showed reasonable agreement with respect to time. However, pore pressure changes in the specimens were not measured during these tests.

Rahardjo (1990) conducted one-dimensional consolidation tests on an unsaturated silty sand using a specially designed K_0 cylinder. The cylinder was developed to accommodate K_0 loading and allow for the simultaneous measurement of pore-air and pore-water pressures throughout the soil specimen. The total and water volume changes were measured independently during the consolidation tests. The results showed a rapid (i.e., essentially instantaneous) dissipation of excess pore-air pressures for the unsaturated soil used in the study. On the other hand, the excess pore-water pressure dissipation in the unsaturated soil was found to be a time-dependent process which could be simulated using the water flow differential equation.

16.2.2 Physical Relations Required for Formulation

The one-dimensional consolidation equation for saturated soils is derived by equating the time derivative of the volume change constitutive equation to the divergence of the water flow rate. The volume change constitutive equation for a saturated soil relates the change in void ratio de to the change in effective stress $d(\sigma - u_w)$. The rate of water flow through a soil mass is described by Darcy's law.

The one-dimensional consolidation equation for an *unsaturated* soil can similarly be formulated by satisfying the continuity requirement for a multiphase material. In this case, the change in the total volume of the soil element must be equal to the sum of the changes in the volume of water and the volume of air in the element. The soil solids are assumed to be incompressible. One-dimensional consolidation under K_0 loading has the following soil structure constitutive equation:

$$\frac{dV_v}{V_0} = m_{1k}^s\, d(\sigma_y - u_a) + m_2^s\, d(u_a - u_w) \quad (16.2)$$

where:

dV_v/V_0 = volume change of the soil element with respect to the initial volume of the element (also referred to as the change in volumetric strain $d\varepsilon_v$),
V_v = volume of soil voids in the element,
V_0 = initial overall volume of the element,
m_{1k}^s = coefficient of volume change with respect to a change in net normal stress $d(\sigma_y - u_a)$ for K_0 loading, and
m_2^s = coefficient of volume change with respect to a change in matric suction $d(u_a - u_w)$ during K_0 loading.

The water phase constitutive relation for K_0 loading can be written as

$$\frac{dV_w}{V_0} = m_{1k}^w\, d(\sigma_y - u_a) + m_2^w\, d(u_a - u_w) \quad (16.3)$$

where:

dV_w/V_0 = change in the volume of water in the soil element with respect to the initial volume of the element,
V_w = volume of water in the element,
m_{1k}^w = coefficient of water volume change with respect to a change in the net normal stress $d(\sigma_y - u_a)$ for K_0 loading, and
m_2^w = coefficient of water volume change with respect to a change in matric suction $d(u_a - u_w)$ during K_0 loading.

The air phase constitutive relation is equal to the difference between the soil structure and the water phase constitutive equations. The air phase constitutive equation can also be expressed in a general form as a function of the stress state variables:

$$\frac{dV_a}{V_0} = m_{1k}^a\, d(\sigma_y - u_a) + m_2^a\, d(u_a - u_w) \quad (16.4)$$

where:

dV_a/V_0 = change in the volume of air in the soil element with respect to the initial volume of the element,
V_a = volume of air in the element,
m_{1k}^a = coefficient of air volume change with respect to a change in net normal stress $d(\sigma_y - u_a)$, and
m_2^a = coefficient of air volume change with respect to a change in matric suction $d(u_a - u_w)$.

Figure 16.1 shows the relationship among the three constitutive surfaces described by the above equations. The coefficients of volume change must satisfy the continuity requirement at any stress point on the constitutive surfaces:

$$m_{1k}^a = m_{1k}^s - m_{1k}^w \quad (16.5)$$

$$m_2^a = m_2^s - m_2^w \quad (16.6)$$

When the soil is saturated (i.e., degree of saturation S at 100%), the four coefficients of volume change, m_{1k}^s, m_2^s, m_{1k}^w, and m_2^w, are equal to the volume change modulus m_v for a saturated soil. The volume change modulus m_v can be obtained from a one-dimensional consolidation test on a saturated soil specimen. The m_{1k}^a and m_2^a modulus values are equal to zero under saturated conditions (Eqs. 16.5 and 16.6). As the soil becomes unsaturated (i.e., $S < 100\%$), the absolute values of the m_{1k}^s, m_2^s, m_{1k}^w, and m_2^w coefficients take on independent values (Fig. 16.1). The volume change coefficients have a negative sign, which indicates that an increase in the stress state variable causes a decrease in volume. Under unsaturated soil conditions, the absolute magnitude of the m_2^w water phase coefficient will be greater than the absolute magnitude of the m_2^s soil structure coefficient under similar stress state conditions. In other words, an increase in matric suction causes a larger change in the volume of water removed from the soil than the overall volume change of the soil element. As a result, an increase in matric suction results in a decrease in the degree of saturation. The magnitude of the m_{1k}^s soil structure coefficient, on the other hand, is greater than the magnitude of the m_{1k}^w water phase coefficient under similar

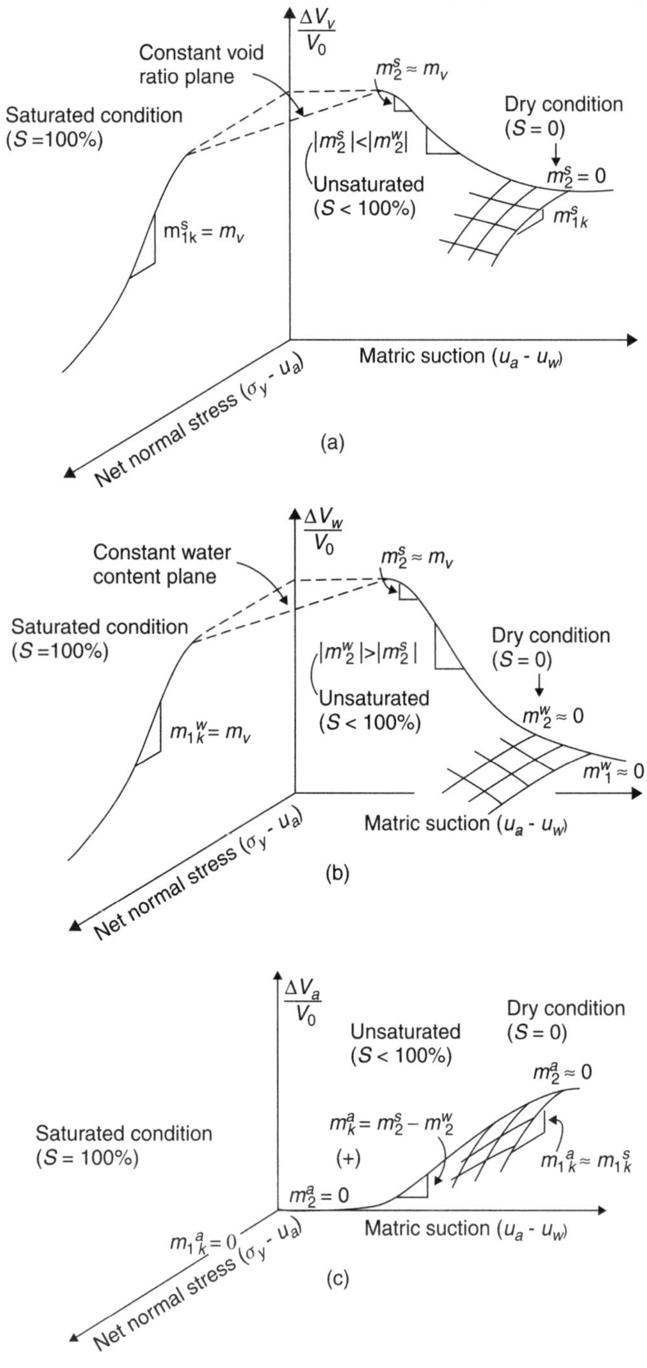

Figure 16.1 Typical form for constitutive surfaces for unsaturated soil: (a) soil structure constitutive surface; (b) water phase constitutive surface; (c) air phase constitutive surface.

stress state changes since a net normal stress produces larger overall volume change than is produced for the change in volume of water in the element.

As the soil approaches dry conditions (i.e., $S = 0$), the m_2^s and m_2^w coefficients approach zero (Fig. 16.1). Consequently, the m_2^a coefficient also approaches zero (Eq. 16.6). This means that when the degree of saturation of the soil approaches zero, a change in matric suction is no longer effective in producing a change in overall volume of the soil. Any volume change which takes place when the soil is in a dry condition (e.g., below residual conditions) will be mainly due to a change in the net normal stress. However, a net normal stress increase applied to a dry soil is unlikely to cause a change in the volume of water in the soil (i.e., $m_{1k}^w = 0$). Stated another way, the total soil volume change in a dry soil is related to a change in the volume of air (i.e., $m_{1k}^s = m_{1k}^a$) in the soil. Only changes in net normal stress are likely to produce volume changes in a dry soil.

Darcy's law and Fick's law can be used to describe water flow and air flow, respectively, through an unsaturated soil.

16.2.3 Derivation of Consolidation Equations for Unsaturated Soil

For an isothermal, transient flow analysis, two fluids are assumed to flow independently from an unsaturated soil, namely, water and air. Two independent PDEs are required to solve for the pore-water and pore-air pressures with respect to time (i.e., two-phase flow approach). The pore-water and pore-air PDEs must satisfy continuity with respect to the water and air phases. Rigorous, multidimensional formulations require that the continuity equations be coupled with the equilibrium equations (Dakshanamurthy et al., 1984). Uncoupled formulation of the consolidation and swelling equations can also be derived for solving unsaturated soil problems. It would appear that the uncoupled consolidation formulations for unsaturated soils are satisfactory for engineering practice.

The excess pore-air pressure throughout the soil mass can often be assumed to be negligible or to be rapidly dissipated. (Rahardjo, 1990). In these situations, only the water phase PDE needs to be solved. This is anticipated to be the case, for example, when predicting the swelling of an unsaturated soil as a result of wetting of the soil from the ground surface. There may be other problems where only the excess pore-air pressure is of significance, and in these cases, only the pore-air PDE needs to be solved.

The application of total stress to a soil provides the geotechnical engineer with insight into the behavior of an unsaturated soil. After applying a total stress, an immediate settlement will occur followed by the dissipation of excess pore-water and pore-air pressures with time. It is necessary to solve the pore-water and pore-air PDEs in order to predict the pore-water and pore-air pressures as a function of time. The initially generated pore-air and pore-water pressures can be evaluated using pore pressure parameters. The application of a total stress induces an excess pore-air pressure which is smaller than the excess pore-water pressure. The magnitude of generated excess pore pressures depends upon the relative compressibility of the air-water mixture and the soil structure compressibility.

A one-dimensional transient flow equation for the water and air phases can be derived by equating the time derivative of the relevant constitutive equation to the divergence of the flow rate, as described by the flow law. The time derivative of the constitutive equation controls the deformation that occurs under various stress conditions while the divergence of the flow rate controls the rate of flow of air and water.

The assumptions used in the derivation of equations for consolidation and swelling are similar to those proposed by Terzaghi (1943) for saturated soils, with some exceptions and additions:

1. The air phase is assumed to be continuous. The coefficient of permeability with respect to the air phase approaches the diffusivity of air through water when the air phase becomes occluded. At the other extreme, the excess pore-air pressure is assumed to remain at atmospheric pressure when the air voids are large. This negates the need for solving the air phase PDE in this case.
2. The coefficients of volume change for the soil (i.e., $m_{1k}^w, m_2^w, m_{1k}^a$, and m_2^a) are assumed to remain constant during the consolidation process. It is also possible to make these coefficients a function of the stress state during the solution of the PDEs.
3. The coefficients of permeability with respect to the air and water phases are assumed to be a function of the stress state or the volume-mass soil properties during the consolidation process. It is also possible to consider the case where the coefficients of permeability can be assumed to remain constant during a process.
4. The effects of air diffusing through water, air dissolving in water, and the movement of water vapor are ignored. The initial and final states of stress in the air phase are the same (i.e., atmospheric) for most engineering problems. Therefore, although air may go into solution under increased pore-air pressure conditions, it can also be assumed that it comes out of solution in response to a return to the boundary conditions prior to loading.
5. The soil particles and the pore-water are assumed to be incompressible.
6. Strains occurring during consolidation are assumed to be small.

The above assumptions are not completely accurate for all cases; however, these assumptions are reasonable for the derivation of a general theory of consolidation (or swelling) for unsaturated soils.

16.2.4 Water Phase Partial Differential Equation

Let us consider a referential element of unsaturated soil with air and water flow during one-dimensional consolidation (Fig. 16.2). The net flux of water through the element is computed from the volume of water entering and leaving the element within a period of time:

$$\frac{\partial V_w}{\partial t} = \left(v_w + \frac{\partial v_w}{\partial y}\,dy\right)\,dx\,dz - v_w\,dx\,dz \tag{16.7}$$

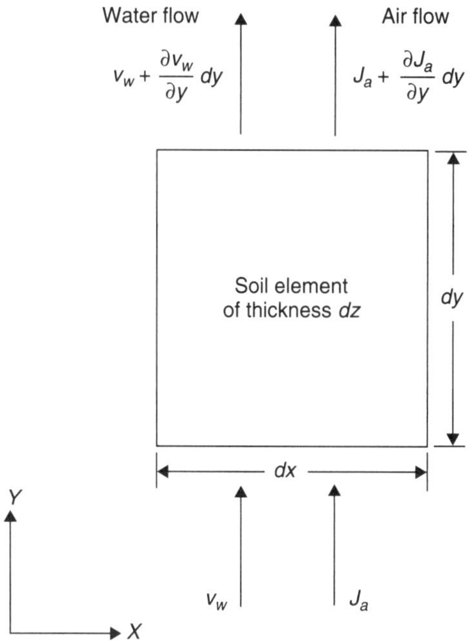

Figure 16.2 Unsteady-state flow of air and water during one-dimensional consolidation of unsaturated soil.

where:

dV_w = change in the volume of water in the soil element over a specific time, *dt*,
$\partial V_w/\partial t$ = net flux of water through the soil element,
v_w = water flow rate across a unit area of the soil element in the y-direction, and
dx, dy, dz = infinitesimal dimensions in the x-, y-, and z-directions, respectively.

Considering the net flux of water per unit volume and rearranging Eq. 16.7 yields

$$\frac{\partial (V_w/V_0)}{\partial t} = \frac{\partial v_w}{\partial y} \tag{16.8}$$

where:

V_0 = initial total volume of the soil element (i.e., dx, dy, dz) and
$\partial(V_w/V_0)/\partial t$ = net flux of water per unit volume of the soil.

Substituting Darcy's law for the flow rate of water, v_w, into Eq. 16.8 gives

$$\frac{\partial (V_w/V_0)}{\partial t} = \frac{[-k_w(\partial h_w/\partial y)]}{\partial y} \tag{16.9}$$

where:

k_w = coefficient of permeability with respect to water as a function of matric suction which varies with location in the y-direction [i.e., $k_w(u_a - u_w)$],

h_w = hydraulic head [i.e., gravitational plus pore-water pressure head or $Y + u_w/(\rho_w g)$],
Y = elevation head,
u_w = pore-water pressures,
ρ_w = density of water,
g = gravitational acceleration, and
$\partial h_w/\partial y$ = hydraulic head gradient in the y-direction.

Rearranging the above equation gives

$$\frac{\partial(V_w/V_0)}{\partial t} = -k_w \frac{\partial^2 h_w}{\partial y^2} - \frac{\partial k_w}{\partial y}\frac{\partial h_w}{\partial y} \tag{16.10}$$

Let us substitute hydraulic head h_w in terms of the individual components $(Y + u_w/\rho_w g)$, Eq. 16.10:

$$\frac{\partial(V_w/V_0)}{\partial t} = -k_w \frac{\partial^2[Y + (u_w/\rho_w g)]}{\partial y^2} - \frac{\partial k_w}{\partial y}\frac{\partial[Y + (u_w/\rho_w g)]}{\partial y} \tag{16.11}$$

$$\frac{\partial(V_w/V_0)}{\partial t} = -\frac{k_w}{\rho_w g}\frac{\partial^2 u_w}{\partial y^2} - \frac{1}{\rho_w g}\frac{\partial k_w}{\partial y}\frac{\partial u_w}{\partial y} - \frac{\partial k_w}{\partial y} \tag{16.12}$$

The water phase constitutive relation (Eq. 16.3) defines the volume of water in an unsaturated soil element in terms of net normal stress $d(\sigma_y - u_a)$ and matric suction $d(u_a - u_w)$. The flux of water per unit volume of the soil can be obtained by differentiating the water phase constitutive relation with respect to time:

$$\frac{\partial(V_w/V_0)}{\partial t} = m_{1k}^w \frac{\partial(\sigma_y - u_a)}{\partial t} + m_2^w \frac{\partial(u_a - u_w)}{\partial t} \tag{16.13}$$

The coefficients of volume change are assumed to be constant during the consolidation process. In other words, the coefficients of volume change are only a function of the stress state. The coefficients of volume change can be updated in accordance with the stress state of the soil. The change in total stress with respect to time is generally set to zero during a consolidation process (i.e., $\partial\sigma_y/\partial t = 0$). Both equations for water flux (i.e., Eqs 16.12 and 16.13) can be equated to yield a PDE for the water phase:

$$-m_{1k}^w \frac{\partial u_a}{\partial t} + m_2^w \frac{\partial u_a}{\partial t} - m_2^w \frac{\partial u_w}{\partial t} = -\frac{k_w}{\rho_w g}\frac{\partial^2 u_w}{\partial y^2} - \frac{1}{\rho_w g}\frac{\partial k_w}{\partial y}\frac{\partial u_w}{\partial y} - \frac{\partial k_w}{\partial y} \tag{16.14}$$

Rearranging the above equation gives

$$m_2^w \frac{\partial u_w}{\partial t} = -(m_{1k}^w - m_2^w)\frac{\partial u_a}{\partial t} + \frac{k_w}{\rho_w g}\frac{\partial^2 u_w}{\partial y^2} + \frac{1}{\rho_w g}\frac{\partial k_w}{\partial y}\frac{\partial u_w}{\partial y} + \frac{\partial k_w}{\partial y} \tag{16.15}$$

The above equation is a general form PDE for the water phase. The equation can be simplified for special soil conditions, such as the fully saturated case, the completely dry case, or special cases of unsaturated soil behavior. The water phase PDE for the above-mentioned conditions is described in the following sections.

16.2.5 Transition to Saturated Condition

The coefficients of water volume change m_{1k}^w and m_2^w become equal to the coefficient of volume change m_v for a saturated soil (i.e., degree of saturation $S = 100\%$). The coefficient of permeability k_w reverts to the coefficient of permeability at saturation k_s. The saturated coefficient of permeability is usually assumed to remain constant during the consolidation process (i.e., $\partial k_s/\partial y = 0$).

Substituting m_v and making $\partial k_s/\partial y$ equal to zero in Eq. 16.15 results in the Terzaghi form of the one-dimensional consolidation equation for saturated soils (Terzaghi, 1943). The Terzaghi equation describes the pore-water pressure changes during one-dimensional consolidation of a saturated soil. The consolidation process of a saturated soil involves only the flow of water, which in turn produces an equal volume change.

16.2.6 Transition to Dry Soil Conditions

There is only a small amount of water present around the soil particles when the soil approaches dry conditions (i.e., water content less than residual conditions). Changes in matric suction or net normal stress produce negligible change in the volume of water when the soil is dry. The coefficients of water volume change m_{1k}^w and m_2^w and the coefficient of permeability k_w go towards zero in dry soils. Any volume change that occurs in a dry soil is the result of the soil structure or air phase volume change. Setting the coefficients of volume change and the coefficient of permeability to zero (i.e., $m_{1k}^w = m_2^w = 0$ and $k_w = 0$) in Eq. 16.15 causes the PDE for the water phase to vanish. In other words, there is no water flow during the consolidation of a dry soil.

16.2.7 Special Case of Unsaturated Soil

Air flow and water flow can take place simultaneously during consolidation of an unsaturated soil (i.e., $0 < S < 100\%$). The consolidation equation (i.e., Eq. 16.15) for the water phase can be rearranged as follows:

$$\frac{\partial u_w}{\partial t} = -C_w \frac{\partial u_a}{\partial t} + c_v^w \frac{\partial^2 u_w}{\partial y^2} + \frac{c_v^w}{k_w}\frac{\partial k_w}{\partial y}\frac{\partial k_w}{\partial y}\frac{\partial u_w}{\partial y} + c_g \frac{\partial k_w}{\partial y} \tag{16.16}$$

where:

C_w = interactive constant associated with the water phase PDE [i.e., $(1 - m_2^w/m_{1k}^w)/(m_2^w/m_{1k}^w)$],
c_v^w = coefficient of consolidation with respect to the water phase [i.e., $k_w/(\rho_w g m_2^w)$], and
c_g = gravity term constant (i.e., $1/m_2^w$).

The coefficient of permeability k_w can vary significantly with respect to matric suction and as a result can vary in the y-direction. Spatial variation in the coefficient of permeability of the soil is taken into account by the last two terms in Eq. 16.16. Several empirical relationships have been proposed for the permeability functions $k_w(u_a - u_w)$.

The gravitational component of the hydraulic head in the above derivation appears in the last term of Eq. 16.16 [i.e., $(1/m_2^w)(\partial k_w/\partial y)$]. This term may be considered negligible in some cases when compared to the other terms. If the gravity term is neglected and the coefficient of permeability k_w does not vary significantly with space (i.e., $\partial k_w/\partial y$ is negligible), a simplified form of the differential water flow equation can be written:

$$\frac{\partial u_w}{\partial t} = -C_w \frac{\partial u_a}{\partial t} + c_v^w \frac{\partial^2 u_w}{\partial y^2} \tag{16.17}$$

The consolidation equation for the air phase is presented in the next section. It is noted that there are many cases where the dissipation of the excess pore-air pressures occurs almost instantaneously (Rahardjo, 1990). Only the water phase undergoes a transient process when the excess pore-air pressure is negligible, and in this case Eq. 16.17 can be further simplified to a form similar to Terzaghi's consolidation equation:

$$\frac{\partial u_w}{\partial t} = c_v^w \frac{\partial^2 u_w}{\partial y^2} \tag{16.18}$$

where

$$c_v^w = \frac{k_w}{\rho_w g m_2^w}$$

16.2.8 Air Phase Partial Differential Equation

The air phase is compressible and flows in response to an air pressure gradient. The flow of air through a referential element of unsaturated soil (Fig. 16.2) can be derived in terms of the mass rate of air flow J_a. The net mass rate of air flow across the element is obtained as the difference between the mass rates of air entering and leaving the element within a period of time:

$$\frac{\partial M_a}{\partial t} = \left(J_a + \frac{\partial J_a}{\partial y} dy\right) dx\, dz - J_a\, dx\, dz \tag{16.19}$$

where:

∂M_a = change of air mass in the soil element for a specific time dt,
$\partial M_a/\partial t$ = net mass rate of air flowing through the soil element, and
J_a = mass rate of air flowing across a unit area of the soil.

The net mass rate of air flow can be expressed for a unit volume of the soil element by rearranging Eq. 16.19 as follows:

$$\frac{\partial (M_a/V_0)}{\partial t} = \frac{\partial J_a}{\partial y} \tag{16.20}$$

where:

$\partial(M_a/V_0)/\partial t$ = net mass rate of air flow per unit volume of the element and
V_0 = initial total volume of the soil element (i.e., dx, dy, dz).

Substituting Fick's law for the mass rate of air flow J_a into Eq. 16.20 results in the following equation:

$$\frac{\partial (M_a/V_0)}{\partial t} = \frac{\partial[-D_a^*(\partial u_a/\partial y)]}{\partial y} \tag{16.21}$$

where:

D_a^* = coefficient of transmission for the air phase, which is a function of the volume-mass properties or matric suction of the soil, and thus D_a^* can vary with location in the y-direction,
u_a = pore-air pressure, and
$\partial u_a/\partial y$ = pore-air pressure gradient in the y-direction.

The mass of air M_a in Eq. 16.21 can be written in terms of the volume of air V_a and its density ρ_a (i.e., $M_a = V_a \rho_a$):

$$\frac{\partial (V_a \rho_a/V_0)}{\partial t} = -\left(D_a^* \frac{\partial^2 u_a}{\partial y^2} + \frac{\partial D_a^*}{\partial y}\frac{\partial u_a}{\partial y}\right) \tag{16.22}$$

where:

V_a = volume of air and
ρ_a = density of air.

Rearranging Eq. 16.22 gives

$$\rho_a \frac{\partial (V_a/V_0)}{\partial t} + \frac{V_a}{V_0}\frac{\partial \rho_a}{\partial t} = -D_a^* \frac{\partial^2 u_a}{\partial y^2} - \frac{\partial D_a^*}{\partial y}\frac{\partial u_a}{\partial y} \tag{16.23}$$

The volume of air V_a can be related to the volume-mass properties of the soil:

$$V_a = (1 - S)nV \tag{16.24}$$

where:

S = degree of saturation,
n = porosity, and
V = current total volume of the soil element.

The total volume change of an unsaturated soil can generally be assumed to be small (i.e., small strains) during the consolidation process. This assumption is justifiable since most unsaturated soils have a relatively rigid soil structure. The current total volume of the soil, V, in Eq. 16.24 can be assumed to be equal to the initial total volume of the soil, V_0. Substituting Eq. 16.24 into Eq. 16.23 yields

$$\rho_a \frac{\partial (V_a/V_0)}{\partial t} + (1-S)n\frac{\partial \rho_a}{\partial t} = -D_a^* \frac{\partial^2 u_a}{dy^2} - \frac{\partial D_a^*}{\partial y}\frac{\partial u_a}{\partial y} \tag{16.25}$$

The density of air is a function of air pressure in accordance with the ideal gas law:

$$\rho_a = \frac{\omega_a}{RT_K}\bar{u}_a \tag{16.26}$$

where:

ω_a = molecular mass of air, kg/kmol,
R = universal (molar) gas constant, [i.e., 8.31432 J/(mol·K)],
T_K = absolute temperature (i.e., $T_K = T + 273.16$), K,
T = temperature, °C,
$\bar{u}_a$ = absolute pore-air pressure (i.e., $\bar{u}_a = u_a + \bar{u}_{atm}$), kPa,
u_a = gauge pore-air pressure, kPa, and
$\bar{u}_{atm}$ = atmospheric pressure (i.e., 101 kPa or 1 atm).

Replacing the air density ρ_a in Eq. 16.25 with Eq. 16.26 gives

$$\left(\frac{\omega_a}{RT_K}\right)\bar{u}_a \frac{\partial (V_a/V_0)}{\partial t} + (1-S)n\left(\frac{\omega_a}{RT_K}\right)\frac{\partial u_a}{\partial t} = D_a^* \frac{\partial^2 u_a}{\partial y^2} - \frac{\partial D_a^*}{\partial y}\frac{\partial u_a}{\partial y} \tag{16.27}$$

Equation 16.27 can be further rearranged to give the air flux per unit volume of the soil element:

$$\frac{\partial (V_a/V_0)}{\partial t} = -\frac{D_a^*}{(\omega_a/RT_K)\bar{u}_a}\frac{\partial^2 u_a}{\partial y^2} - \frac{(1-S)n}{\bar{u}_a}\frac{\partial u_a}{\partial t} - \frac{1}{(\omega_a/RT_K)\bar{u}_a}\frac{\partial D_a^*}{\partial y}\frac{\partial u_a}{\partial y} \tag{16.28}$$

The air phase constitutive relation (Eq. 16.4) defines the air volume change in the soil element due to changes in net normal stress $d(\sigma_y - u_a)$ and matric suction $d(u_a - u_w)$. The derivative of the air phase constitutive relation with respect to time is equal to the air flux per unit volume of the soil element:

$$\frac{\partial (V_a/V_0)}{\partial t} = m_{1k}^a \frac{\partial (\sigma_y - u_a)}{\partial t} + m_2^a \frac{\partial (u_a - u_w)}{\partial t} \tag{16.29}$$

The coefficients of volume change m_{1k}^a and m_2^a are assumed to be constant during consolidation when differentiating Eq. 16.4. The magnitudes of the coefficients of volume change are a function of the stress state variables and can be updated accordingly, if needed. The total stress change with respect to time is set to zero during the consolidation process (i.e., $\partial\sigma_y/\partial t = 0$). Equating both equations for the air flux (i.e., Eqs. 16.28 and 16.29) gives rise to the PDE for the air phase:

$$-m_{1k}^a\frac{\partial u_a}{\partial t} + m_2^a\frac{\partial u_a}{\partial t} - m_2^a\frac{\partial u_w}{\partial t} = -\frac{D_a^*}{(\omega_a/RT_K)\bar{u}_a}\frac{\partial^2 u_a}{\partial y^2} - \frac{(1-S)n}{\bar{u}_a}\frac{\partial u_a}{\partial t} - \frac{1}{(\omega_a/RT_K)\bar{u}_a}\frac{\partial D_a^*}{\partial y}\frac{\partial u_a}{\partial y} \tag{16.30}$$

Rearranging the above equation gives

$$-\left(m_{1k}^a - m_2^a - \frac{(1-S)n}{\bar{u}_a}\right)\frac{\partial u_a}{\partial t} = m_2^a\frac{\partial u_w}{\partial t} - \frac{D_a^*}{(\omega_a/RT_K)\bar{u}_a}\frac{\partial^2 u_a}{\partial y^2} - \frac{1}{(\omega_a/RT_K)\bar{u}_a}\frac{\partial D_a^*}{\partial y}\frac{\partial u_a}{\partial y} \tag{16.31}$$

Equation 16.31 is a general form of the air phase PDE. The equation can be simplified for special cases such as the fully saturated case, the dry soil case, or special cases of an unsaturated soil. The air phase PDE for each of these cases is outlined in the following sections.

16.2.9 Transition toward Saturated Soil Conditions

The coefficients of air volume change m_{1k}^a and m_2^a become equal to zero as the degree of saturation of a soil goes toward 100%. The coefficient of transmission D_a^* approaches zero, indicating the absence of air flow. Air may exist in the form of occluded bubbles which are assumed to have a pressure equal to that of the water phase, (i.e., $u_a - u_w = 0$). Air may move as air diffusion through the pore-water.

The water phase becomes more compressible than that of pure water when occluded air bubbles are present in the water phase. A rigorous solution of the case described by Eq. 16.1 requires the use of appropriate soil properties (i.e., c_v, k_s, ρ_w, and m_v), which take compressible pore fluid properties into consideration.

16.2.10 Transition toward Dry Soil Conditions

A change in matric suction produces negligible volume change when the soil is in a dry condition (i.e., water content is less than residual conditions). All coefficients of volume change with respect to matric suction approach zero (i.e., $m_2^s = m_2^w = m_2^a = 0$). Volume change may still occur due to a change in net normal stress if the soil structure is compressible. In this case, the soil volume change is equal to the air phase volume change (i.e., $m_{1k}^a = m_{1k}^s$) in Fig. 16.1c. The coefficient of transmission D_a^* reverts to a constant coefficient that corresponds to dry conditions. Therefore, Eq. 16.31 can be simplified to the following form for a compressible soil under dry conditions:

$$\left(m_{1k}^a + \frac{n}{\bar{u}_a}\right)\frac{\partial u_a}{\partial t} = \frac{D_d^*}{(\omega_a/RT_K)\bar{u}_a}\frac{\partial^2 u_a}{\partial y^2} \tag{16.32}$$

Rearranging the above equation results in a consolidation equation similar in form to Terzaghi's one-dimensional consolidation equation:

$$\frac{\partial u_a}{\partial t} = \frac{D_d^*}{\omega_a/RT_K} \frac{1}{m_{1k}^a \bar{u}_a + n} \frac{\partial^2 u_a}{\partial y^2} \quad (16.33)$$

The above equation describes the pore-air pressure changes during one-dimensional consolidation of a dry, compressible soil. Only air flow occurs during the consolidation process. The coefficients used in Eq. 16.33 are similar in form to those used in the coefficient of consolidation c_v. In the case where the soil is dry and incompressible (i.e., $m_{1k}^a = m_{1k}^s = 0$), Eq. 16.33 reverts to the equation proposed by Blight (1971):

$$\frac{\partial u_a}{\partial t} = \frac{D_d^*/n}{\omega_a/RT_K} \frac{\partial^2 u_a}{\partial y^2} \quad (16.34)$$

16.2.11 Special Case of Unsaturated Soil

It is possible for air and water to flow simultaneously during the consolidation process in an unsaturated soil. The air phase consolidation equation (i.e., Eq. 16.31) can be written in a simplified form:

$$\frac{\partial u_a}{\partial t} = -C_a \frac{\partial u_w}{\partial t} + c_v^a \frac{\partial^2 u_a}{\partial y^2} + \frac{c_v^a}{D_a^*} \frac{\partial D_a^*}{\partial y} \frac{\partial u_a}{\partial y} \quad (16.35)$$

where:

C_a = interactive constant associated with the air phase PDE, that is, $(m_2^a/m_{1k}^a)/[1 - m_2^a/m_{1k}^a - (1 - S)n/(\bar{u}_a m_{1k}^a)]$, and

c_v^a = coefficient of consolidation with respect to the air phase, that is,

$$\frac{D_a^*}{(\omega_a/RT_K)} \frac{1}{\bar{u}_a m_{1k}^a (1 - m_2^a/m_{1k}^a) - (1 - S)n}.$$

The coefficient of transmission D_a^* is a function of matric suction and may vary in the y-direction. The variation in transmissivity of the air phase is taken into account in the last term in Eq. 16.35. There are several relationships describing the variation of air transmissivity with matric suction. However, if the variation of air transmissivity with space is negligible (i.e., $\partial D_a^*/\partial y$ is negligible), Eq. 16.35 can be simplified as follows:

$$\frac{\partial u_a}{\partial t} = -C_a \frac{\partial u_w}{\partial t} + c_v^a \frac{\partial^2 u_a}{\partial y^2} \quad (16.36)$$

16.3 SOLUTION OF CONSOLIDATION EQUATIONS USING FINITE DIFFERENCE TECHNIQUE

The finite difference technique can be used to solve unsteady-state (i.e., transient) air flow and water flow equations associated for one-dimensional consolidation. The air flow and water flow PDEs can be solved simultaneously using an explicit forward-marching, finite difference technique, as illustrated in Fig. 16.3. The finite difference equations can be written in linear forms for the air and water phase partial differential flow equations (Dakshanamurthy and Fredlund, 1981).

Let us consider one-dimensional consolidation where air flow and water flow occur simultaneously. The spatial variations of the water coefficient of permeability and the air

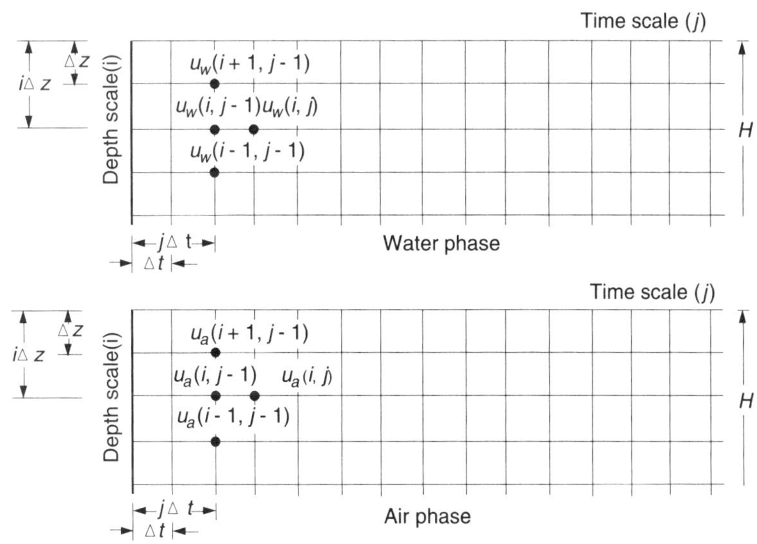

Figure 16.3 Finite difference mesh for solving equations for unsteady-state flow of air and water.

coefficient of transmission are assumed to be negligible. Therefore, Eqs. 16.17 and 16.36 are used as the PDEs for the water and air phases, respectively. The transient water flow equation (i.e., Eq. 16.17) can be written in the following finite difference form:

$$\frac{u_{w(i,j+1)} - u_{w(i,j)}}{\Delta t} = -C_w \frac{u_{a(i,j+1)} - u_{a(i,j)}}{\Delta t} + c_v^w \frac{u_{w(i+1,j)} - 2u_{w(i,j)} + u_{w(i-1,j)}}{\Delta y^2} \tag{16.37}$$

where:

i = space increment in the y-direction and
j = time increment.

The transient air flow equation (i.e., Eq. 16.36) can be written in the following finite difference form:

$$\frac{u_{a(i,j+1)} - u_{a(i,j)}}{\Delta t} = -C_a \frac{u_{w(i,j+1)} - u_{w(i,j)}}{\Delta t} + c_v^a \frac{u_{a(i+1,j)} - 2u_{a(i,j)} + u_{a(i-1,j)}}{\Delta y^2} \tag{16.38}$$

The pore-water pressure is solved by multiplying Eq. 16.38 by $-C_w$ and substituting the resulting equation into Eq. 16.37:

$$\frac{u_{w(i,j+1)} - u_{w(i,j)}}{\Delta t} = C_a C_w \frac{u_{w(i,1+j)} - u_{w(i,j)}}{\Delta t} - c_v^a C_w \frac{u_{a(i+j,1)} - 2u_{a(i,j)} + u_{a(i-1,j)}}{\Delta y^2} + c_v^w \frac{u_{w(i+1,j)} - 2u_{w(i,j)} + u_{w(i-1,j)}}{\Delta y^2} \tag{16.39}$$

Simplifying and rearranging Eq. 16.39 allows the unknown pore-water pressure (i.e., at a given time step, $j + 1$) to be put on the left-hand side of the equation:

$$u_{w(i,j+1)} = u_{w(i,j)} + \frac{\beta_w g_1^w}{1 - C_a C_w} - \left(\frac{C_w}{1 - C_a C_w}\right) \beta_a f_1^a \tag{16.40}$$

where:

$$\beta_w = c_v^w \frac{\Delta t}{\Delta y^2}$$
$$\beta_a = c_v^a \frac{\Delta t}{\Delta y^2}$$
$$g_1^w = u_{w(i+1,j)} - 2u_{w(i,j)} + u_{w(i-1,j)}$$
$$f_1^a = u_{a(i+1,j)} - 2u_{a(i,j)} + u_{a(i-1,j)}$$

Similarly, the pore-air pressure can be obtained by multiplying Eq. 16.37 by $-C_a$, and substituting the resulting equation into Eq. 16.38 gives:

$$\frac{u_{a(i,j+1)} - u_{a(i,j)}}{\Delta t} = C_a C_w \frac{u_{a(i,1+j)} - u_{a(i,j)}}{\Delta t} - c_v^w C_a \frac{u_{w(i+1,j)} - 2u_{w(i,j)} + u_{w(i-1,j)}}{\Delta y^2} + c_v^a \frac{u_{a(i+1,j)} - 2u_{a(i,j)} + u_{a(i-1,j)}}{\Delta y^2} \tag{16.41}$$

Simplifying and rearranging Eq. 16.41 allows the unknown pore-air pressure (i.e., at a given time step $j + 1$) to be put on the left-hand side and all known variables from the previous time step (i.e., j) to be put on the right-hand side of the equation:

$$u_{a(i,j+1)} = u_{a(i,j)} - \left(\frac{C_a}{1 - C_a C_w}\right) \beta_w g_1^w + \frac{\beta_a f_1^a}{1 - C_a C_w} \tag{16.42}$$

The pore-water pressure and pore-air pressure at a given time step are computed from the known values at the previous time step using Eqs. 16.40 and 16.42, respectively. It is possible to march forward to the next time step when computations for the pore-water pressure and pore-air pressure are complete for all depth steps. The above procedure can be repeated until equilibrium has been achieved for the air and water phases. The correct converged solution of the finite difference equations are obtained when β_w and β_a are maintained less than 0.5 (Desai and Christian, 1977).

The initial conditions in the soil mass and the boundary conditions for the problem must be established prior to performing the finite difference computations. The initial conditions for one-dimensional consolidation constitute the initial pore pressures before load is applied and the excess pore pressures corresponding to the instant after the application of the total load. The excess pore-air and pore-water pressures are calculated using the pore-air and pore-water pressure parameters.

The condition on all boundaries of a problem must be specified during the consolidation process. The boundary pore pressures must be set to the initial values at all times for free-draining conditions. The pore pressure gradients must be set to zero at an impervious boundary since there is no flow across the boundary.

A mixed boundary condition can also be specified for special applications that might occur in laboratory tests. As an example, laboratory equipment is commonly designed such that air flows upward while water flows downward during a test. In this case, the top boundary is a free-draining boundary with respect to the air phase but has an impervious boundary with respect to the water phase. Reverse conditions exist at the bottom boundary.

The dissipation of the excess pore-air and pore-water pressures can be computed using the finite difference schemes shown by Eqs. 16.40 and 16.42, respectively, once the initial and final boundary conditions are specified. The calculated pore pressure changes can be used to compute the volume-mass changes in the soil through use of the soil constitutive equations. In this way, changes in water content, void ratio, and degree of saturation of the soil can be computed at any time during the consolidation process.

16.4 TYPICAL CONSOLIDATION TEST RESULTS ON UNSATURATED SOILS

Typical experimental results from consolidation tests on unsaturated soils are presented to illustrate the consolidation behavior of an unsaturated soil. The tests require the independent measurement or control of pore-air and pore-water pressures on the boundaries of the soil specimens during consolidation. The equipment, soil, and test procedures are described. The results are also best fit with the theory for unsaturated soil consolidation and swelling.

16.4.1 Consolidation Tests on Compacted Kaolin

Laboratory experiments were performed on several compacted kaolin specimens using a modified oedometer and a modified triaxial cell (Fredlund, 1973a). The modified oedometer was used to perform one-dimensional consolidation tests. The modified triaxial cell was used to perform isotropic volume change tests.

A high-air-entry disk was sealed into the lower pedestal. The ceramic disk allowed for the movement of water to and from the soil while the flow of air was restricted. Therefore, the pore-air and pore-water pressures could be independently controlled. A low-air-entry disk (i.e., coarse porous disk) was placed on the top of the specimen for the control of the pore-air pressure. Two rubber membranes separated by a slotted tin foil were placed around the specimen in order to reduce air diffusion.

Pore-air and pore-water pressures were translated into the positive range to prevent cavitation of the water below the high-air-entry disk (i.e., the axis translation technique). The total stress, pore-air pressure, and pore-water pressure conditions were controlled throughout the time-dependent measurement of total volume change and water volume change from the unsaturated soil specimens.

16.4.2 Presentation of Consolidation Test Results

Five specimens of kaolin were compacted in accordance with the standard Proctor procedure. The initial volume-mass properties for each specimen are summarized in Table 16.1. Each experiment was performed by changing one of the stress state variable components (i.e., total stress σ, pore-water pressure u_w, or pore-air pressure u_a). The stress state component changed in each experiment is summarized in Table 16.2. The volume change of the overall sample and the water phase was monitored in each case. However, the pore-air and pore-water pressure changes within each soil specimen were not monitored.

Table 16.1 Initial Volume-Mass Properties for Specimens Tested

Test No.[a]	Diameter (cm)	Height (cm)	Total Volume (cm^3)	Water Content, w (%)	Void Ratio, e	Dry Unit Weight, γ_d (kN/m^3)	Degree of Saturation (%)
1	10.006	11.815	929.09	34.32	1.0696	13.185	78.87
2	9.945	11.703	909.10	33.17	1.0251	13.298	80.61
3	10.543	5.867	503.53	29.62	1.2242	11.529	63.29
4	9.832	5.758	437.16	32.12	0.9310	13.281	90.25
5	6.350	2.283	72.29	31.18	1.1247	12.069	72.51

[a]All tests performed in a modified triaxial apparatus except test 5, which was performed in a modified oedometer.

Table 16.2 Change in Stress State Variable Components Associated with Each Test

	Total Stress, σ (kPa)		Pore-Water Pressure, u_w (kPa)		Pore-Air Pressure, u_a (kPa)		
Test No.[a]	Initial	Final	Initial	Final	Initial	Final	Change (kPa)
1	358.7	560.9	163.8	164.4	214.4	215.6	$\Delta\sigma = +202.2$
2	560.9	559.0	164.4	163.1	215.6	421.1	$\Delta u_w = +205.5$
3	475.1	476.8	41.9	42.2	397.8	206.5	$\Delta u_a = -191.3$
4	611.4	610.1	177.3	379.2	532.0	530.9	$\Delta u_w = +201.9$
5	606.7	605.2	216.5	323.6	413.8	413.8	$\Delta u_w = +107.1$

[a]All tests performed in a modified triaxial apparatus except test 5, which was performed in a modified oedometer.

Table 16.2 indicates that only test 1 was a consolidation test in the classical sense that air and water flowed from the soil specimen in response to an increase in the total stress. Transient flow in other tests was caused by a change in either the pore-air or pore-water pressures at the boundary. The air and water flowed in a one-dimensional manner in all tests. The one-dimensional consolidation theory could be used to describe the transient flow of air and/or water during each test.

Figure 16.4 shows a plot of volume changes resulting from an increment of total stress for test 1. The plot shows a decrease in the overall volume (i.e., the soil structure) as well as the volume of air and water during the transient process. A large instantaneous volume decrease occurred when the load was applied.

An increase in the pore-air pressure in test 2 caused the soil structure to expand temporarily (Fig. 16.5). However, the increase in air pressure resulted in an increase in matric suction $u_a - u_w$, which in turn caused a decrease in the volume of the soil structure toward the end of the transient process.

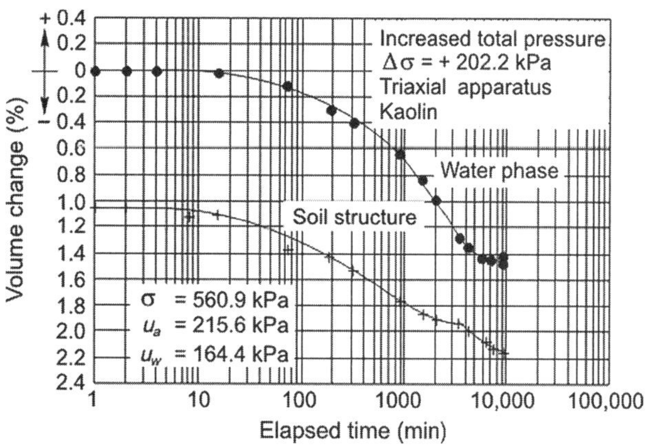

Figure 16.4 Soil structure and water phase volume changes associated with increase in total stress (test 1).

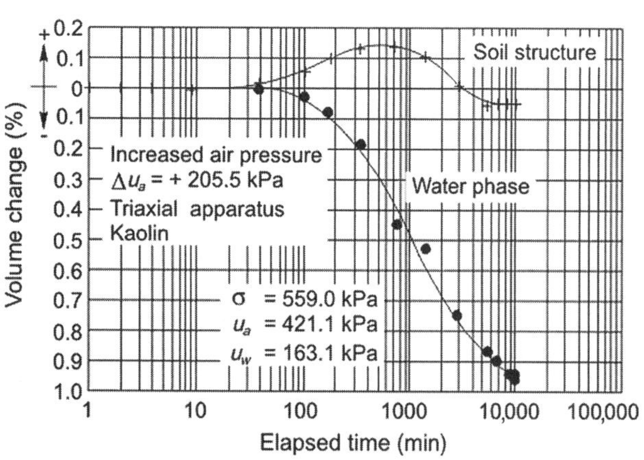

Figure 16.5 Soil structure and water phase volume changes associated with increase in air pressure (test 2).

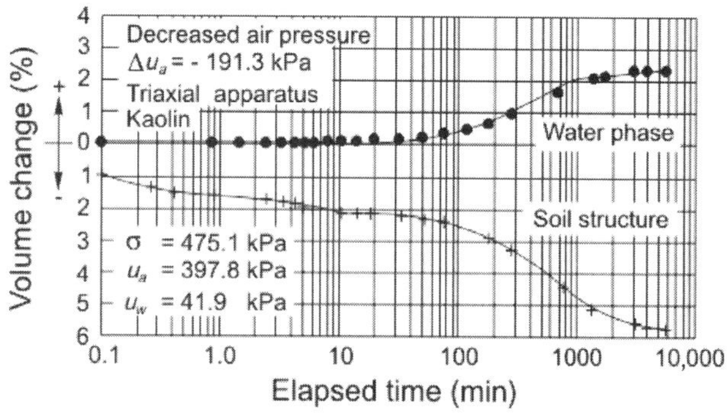

Figure 16.6 Soil structure and water phase volume changes associated with decrease in air pressure (test 3).

The volume change processes associated with a metastable-structured soil are shown in Fig. 16.6. The decrease in air pressure in test 3 reduced the matric suction and allowed more water to flow into the specimen. The intake of water appears to have reduced the normal and shear stresses between the soil particles. As a result, the soil structure underwent a decrease in volume (i.e., a collapse phenomenon).

An increase in water pressure (or a decrease in matric suction) caused an increase in the overall volume in test 4. Figure 16.7 shows that the soil has a stable structure and is expansive. On the other hand, an increase in water pressures (or a decrease in matric suction) causes the soil structure to decrease in volume or collapse when the soil structure is metastable (Fig. 16.8).

16.4.3 Theoretical Analyses of Consolidation Test Results

Attempts were made to best fit the theoretical analyses of volume change with the results from laboratory experiments (Fredlund and Rahardjo, 1986). Approximate coefficients of volume change were selected for the soil structure, air, and water phases based on the laboratory results. Table 16.3 summarizes the approximate coefficients of volume change for each of the five specimens.

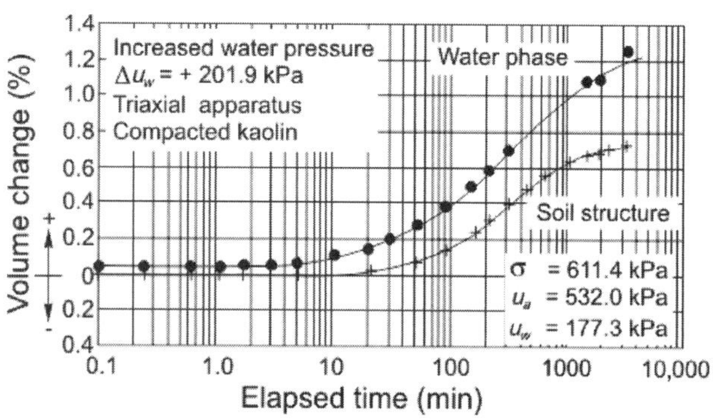

Figure 16.7 Soil structure and water phase volume changes associated with increase in water pressure (test 4).

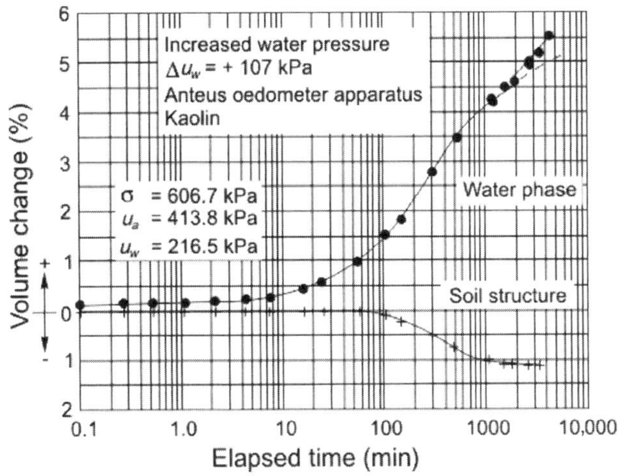

Figure 16.8 Soil structure and water phase volume changes associated with increase in water pressure (test 5).

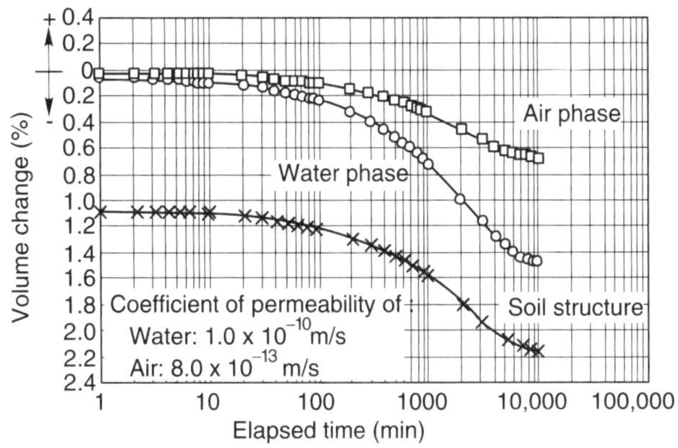

Figure 16.9 Theoretical simulation of consolidation test on unsaturated compacted kaolin (test 1).

The coefficients of volume change for each phase were obtained by dividing the amount of deformation at the end of each process by the change in the stress state variable. A sign (i.e., positive or negative) is attached to each of the coefficients of volume change based on the direction of the volume change (i.e., increase or decrease) associated with each phase and the change in the stress state variables.

Table 16.3 indicates that the coefficients of volume change for the soil structure have negative signs for a stable-structured soil (i.e., tests 1, 2, and 4) and positive signs for a metastable-structured soil (i.e., tests 3 and 5). A negative coefficient of volume change suggests there is an increase in volume for a decrease in the stress state variable, while a positive coefficient of volume change indicates the converse.

The theoretical simulation of the laboratory results was performed by assuming one-dimensional transient flow processes. The pore-water and pore-air pressure equations (i.e., Eqs. 16.40 and 16.42) can be solved simultaneously using an explicit central finite difference technique. The coefficients of permeability for air and water were assumed to be constant during the transient process.

The volume changes associated with the water and air phases during a transient process were computed using Eqs. 16.3 and 16.4. The coefficients of volume change used in the calculations were assumed to be constant throughout the process. The soil structure volume change was obtained by adding the volume changes associated with the water and air phases.

Figures 16.9–16.13 show the results of the theoretical simulation of tests 1–5. The fitting was accomplished by using various combinations for the coefficients of permeability for the water and air phases. The combinations of constant values for the water and air coefficients of permeability that gave the best-fit results for each test are shown in each figure. Comparisons between the numerical simulations and the laboratory results were made for tests 4 and 3 (i.e., Figs. 16.14 and 16.15). The results indicate reasonably close agreement between the numerical simulations and the laboratory test results.

Some discrepancies were observed in the simulation of the transient processes. The disagreements may be due to one or more reasons. For example, the assumption was made that the coefficients of permeability were constant throughout each process. In general, the numerical simulations and the laboratory results show the anticipated volume change behavior with elapsed time.

16.4.4 Tests on Silty Sand

One-dimensional consolidation tests were conducted on silty sand using a special K_0 cylinder designed by Rahardjo (1990).

Table 16.3 Coefficients of Volume Change for Each Specimen

	Soil Structure		Water Phase		Air Phase	
Test No.	$m_1^s \times 10^{-4}$ kPa^{-1}	$m_2^s \times 10^{-4}$ kPa^{-1}	$m_1^w \times 10^{-4}$ kPa^{-1}	$m_2^w \times 10^{-4}$ kPa^{-1}	$m_1^a \times 10^{-4}$ kPa^{-1}	$m_2^a \times 10^{-4}$ kPa^{-1}
1	−1.17	−3.52	−0.80	−2.41	−0.37	−1.11
2	−0.006	−0.032	−0.131	−0.657	+0.125	+0.625
3	+0.76	+3.80	−0.30	−1.49	+1.06	+5.29
4	−0.09	−0.35	−0.15	−0.61	+0.06	+0.26
5	+0.26	+1.04	−1.20	−4.81	+1.46	+5.85

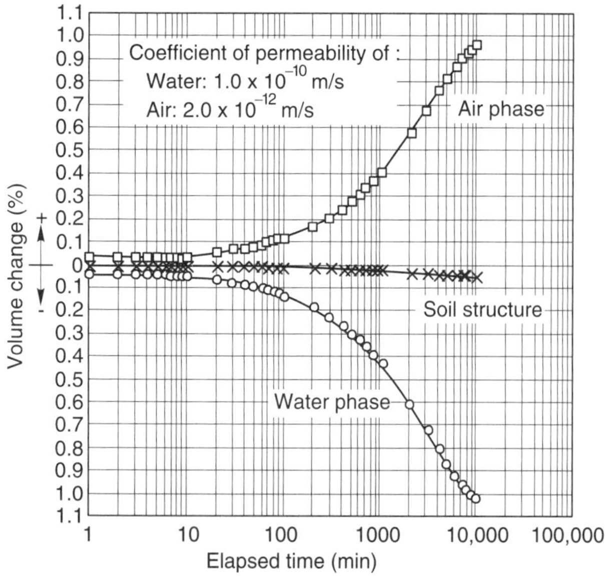

Figure 16.10 Theoretical simulation of laboratory test involving increase in air pressure (test 2).

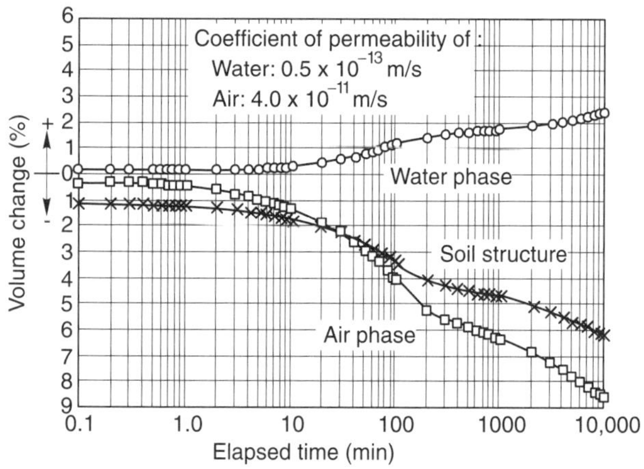

Figure 16.11 Theoretical simulation of laboratory test involving decrease in air pressure (test 3).

The K_0 cylinder allowed for the simultaneous measurement of pore-air and pore-water pressures at various depths along the soil specimen. The top and bottom of the soil specimen were connected to a low- and a high-air-entry disk, respectively. Air flowed upward and water flowed downward during the consolidation test. The total and water volume changes were independently monitored.

16.4.5 Presentation of Consolidation Results on Silty Sand

Slurry specimens of silty sand were subjected to a gradual increase in net normal stress and matric suction. The specimens were then subjected to an increase in total stress under undrained conditions. After excess pore-air and pore-water

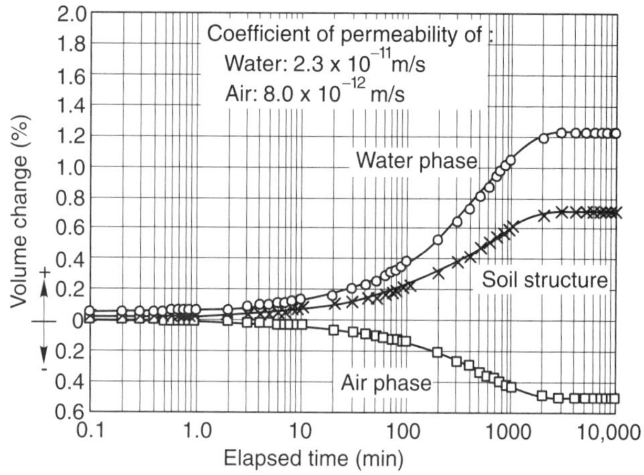

Figure 16.12 Theoretical simulation of laboratory test involving increase in water pressure (test 4).

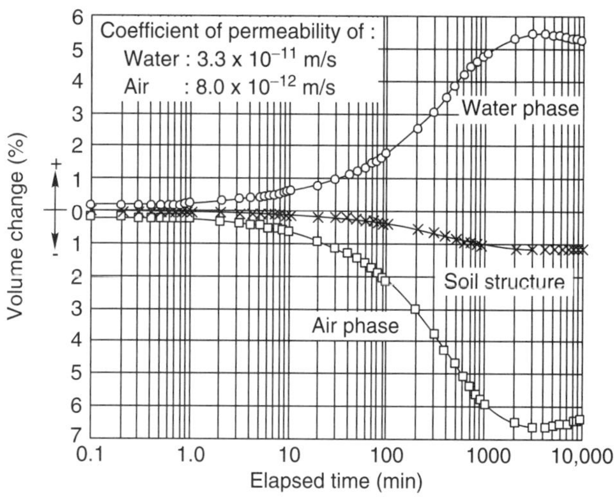

Figure 16.13 Theoretical simulation of laboratory test involving increase in water pressure (test 5).

pressures were established, the air and water phases were subsequently allowed to drain during a consolidation test. The pore-air and pore-water pressures were observed to reduce to their values prior to undrained loading. In other words, the net normal stress increased with elapsed time and matric suction returned to its original value at the end of the consolidation process.

Figure 16.16 presents the results of pore pressure and volume change measurements during the consolidation test. The initial volume-mass properties of the soil specimen and its stress state variables are shown in Table 16.4.

The results shown in Fig. 16.16a indicate that the excess pore-air pressures dissipated almost immediately throughout the soil specimen. The rapid dissipation of air pressures was observed in every consolidation test performed on the silty sand (Rahardjo, 1990). On the other hand, the excess pore-water pressure dissipation occurred in a time-dependent

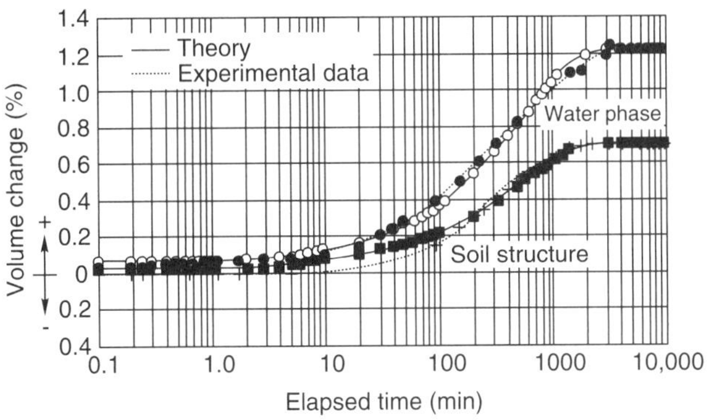

Figure 16.14 Comparisons between theoretical simulation and laboratory test data for test 4 with increase in water pressure.

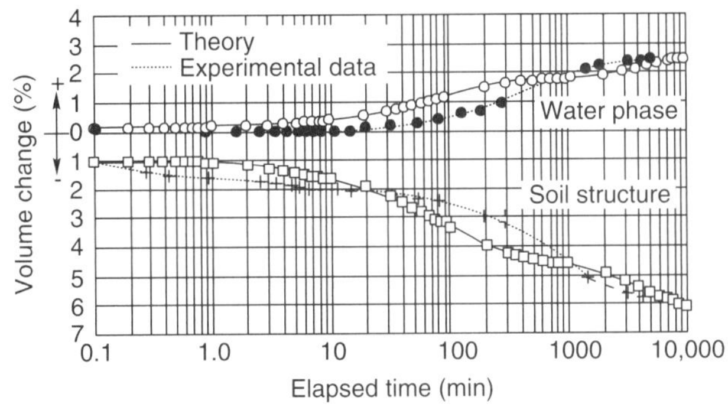

Figure 16.15 Comparisons between theoretical simulation and laboratory test data for test 3 with decrease in air pressure.

manner (i.e., a transient process), as demonstrated in Fig. 16.16a. The water volume change (i.e., 10.5 cm^3), exceeded the time-dependent total volume change (i.e., 6.0 cm^3), as shown in Fig. 16.16b. However, the soil structure underwent an immediate volume decrease of 9.0 cm^3 during the undrained loading prior to pore pressure dissipation. The immediate volume change was caused by compression of the soil structure and the air. The overall volume change was larger than the water volume change in response to an increase in total stress.

Another set of results from a consolidation test on the silty sand is presented in Fig. 16.17. The total stress was increased under constant-water-content conditions prior to consolidation. As a result, only excess pore-water pressures were developed during loading. The excess pore-water pressure dissipation during the consolidation process is shown in Fig. 16.17a for various depth and time intervals. The pore-water pressure isochrones were best fitted with the transient water flow equation. There was no measurable air flow during the transient process. The coefficients of permeability and volume change for the water phase were assumed to be constant during the consolidation process. The best-fit constant coefficient of consolidation c_v^w was 1.2×10^{-6} m^2/s. There was good agreement between the

Table 16.4 Initial Soil Properties and Stress State Variables for S1FG Consolidation Test on Silty Sand

Initial Volume-Mass	Properties
S	98.0 %
e	0.578
w	21.2 %
n	0.366
Stress State Variables (kPa)	
σ	184.3
u_a (top)	119.5
u_w (bottom)	106.7
$\sigma - u_a$	64.8
$u_a - u_w$	12.8

numerical simulation of consolidation and the experimental results, as demonstrated in Fig. 16.17a.

The total water volume changes during the consolidation process (i.e., Fig. 16.17b) are similar to those presented in Fig. 16.16b. The net normal stress remained constant during the consolidation process following constant-water-content loading while the matric suction increased due to the dissipation of the excess pore-water pressures. The total volume change and water volume change were simulated using Eqs. 16.2 and 16.3, respectively. The best-fit coefficients of volume change were found to be 8×10^{-6} and 3.8×10^{-5} kPa^{-1} for m_2^s and m_2^w, respectively. The above examples verify the unsaturated soil theory of consolidation.

16.5 DIMENSIONLESS CONSOLIDATION PARAMETERS

The simultaneous solution of the pore-water and pore-air PDEs (i.e., Eqs. 16.17 and 16.36) can be generalized in terms of dimensionless numbers similar to those used for saturated soils. The dimensionless numbers are the average degree of consolidation and the time factor for the water and air phases. The average degree of consolidation for the water phase is defined as

$$U_w = 1 - \frac{\int_0^{2d} u_w \, dy}{\int_0^{2d} u_{w0} \, dy} \tag{16.43}$$

where:

U_w = average degree of consolidation with respect to the water phase,

u_{w0} = initial pore-water pressure,

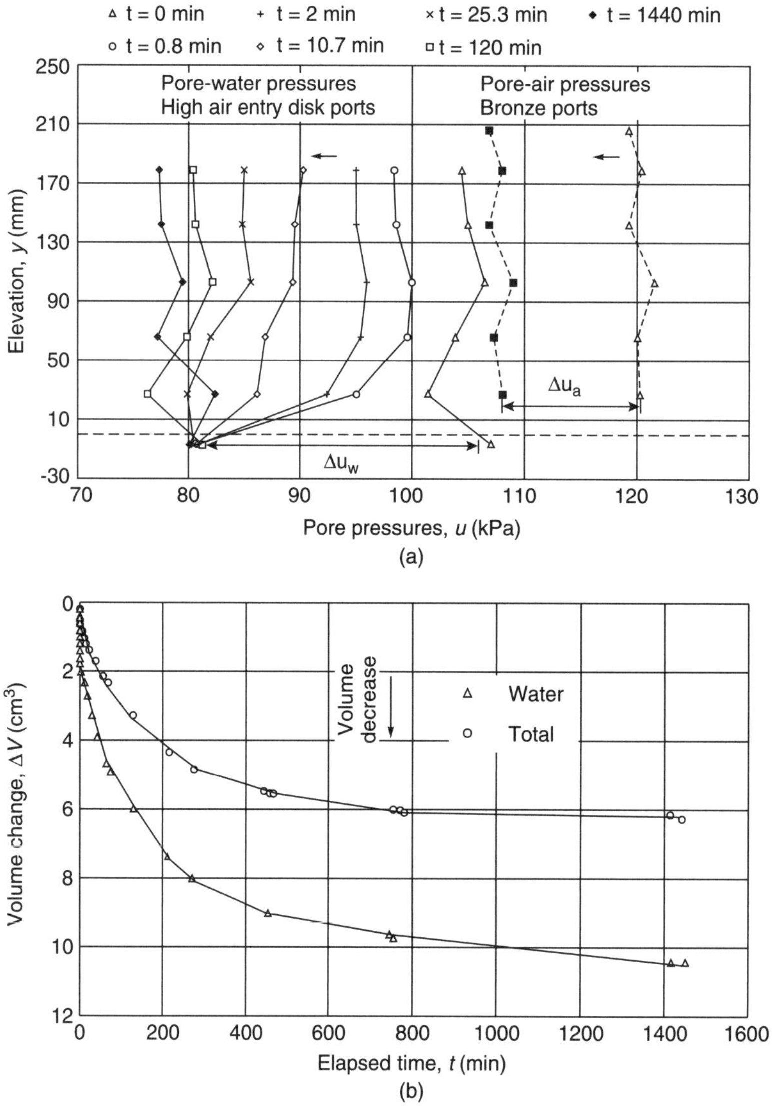

Figure 16.16 Pore pressures and volume changes during consolidation test on silty sand (No. SIFG): (a) pore-air and pore-water pressure isochrones; (b) total and water volume changes (from Rahardjo, 1990).

u_w = pore-water pressure at any time, and
d = length of drainage path (note: double drainage was assumed).

The time factor for the water phase is defined as

$$T_w = \frac{c_v^w t}{d^2} \tag{16.44}$$

where:

t = elapsed time.

Similarly, the average degree of consolidation and the dimensionless time factor with respect to the air phase are defined as

$$U_a = 1 - \frac{\int_0^{2d} u_a \, dy}{\int_0^{2d} u_{a0} \, dy} \tag{16.45}$$

and

$$T_a = \frac{c_v^a t}{d^2} \tag{16.46}$$

where:

U_a = average degree of consolidation with respect to the air phase,
T_a = time factor with respect to the air phase,
u_{a0} = initial pore-air pressure, and
u_a = pore-air pressure at any time.

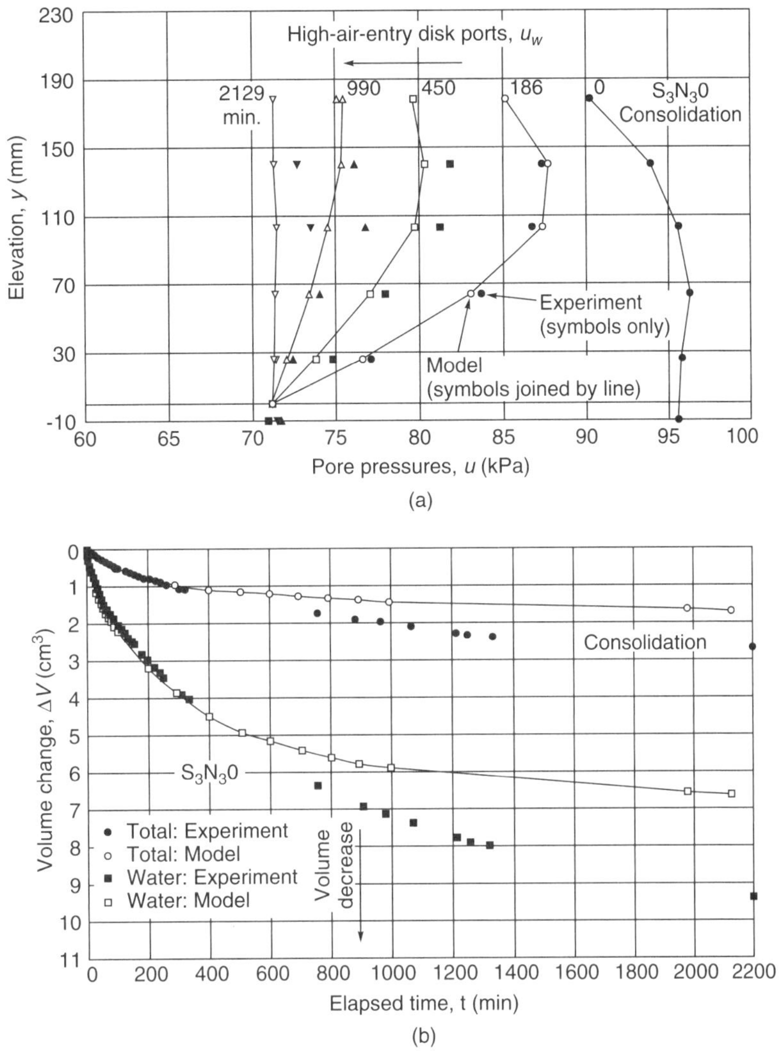

Figure 16.17 Comparisons between theoretical simulation and experimental data from consolidation test on silty sand (No. S_3N_3O): (a) pore-water pressure profiles at different times; (b) total and water volume changes during consolidation (from Rahardjo, 1990).

Figure 16.18 shows the water phase degree of consolidation–time factor curves for various air-water interaction constants. Similar curves for the air phase are shown in Fig. 16.19. The interactive constant in the air phase PDE was assumed to be constant for the calculation of pore-air pressure dissipation. The curves show a smooth transition toward the case of a completely saturated soil (Terzaghi, 1943) and the case of a completely dry soil (Blight, 1971).

The relationships between these dimensionless numbers (Figs. 16.18 and 16.19) can be used as a general solution for one-dimensional consolidation. Appropriate curves corresponding to the air and water phases can be used for different coefficient-of-consolidation and soil constant values.

16.6 COUPLED FORMULATIONS AND THREE-DIMENSIONAL CONSOLIDATION

One PDE has been used for analyzing water flow with respect to time and another PDE has been used for analyzing air flow with respect to time. It is possible to use an iterative process to ensure that both the water phase and air phase PDEs have converged; however, the total stress equilibrium equations were not taken into consideration. The assumption was made that since the total stresses were not changed during the transient process there was no need to consider the stress equilibrium equations. This is a reasonable assumption, but the analysis does not constitute a rigorous formulation and solution of the consolidation process in unsaturated soils.

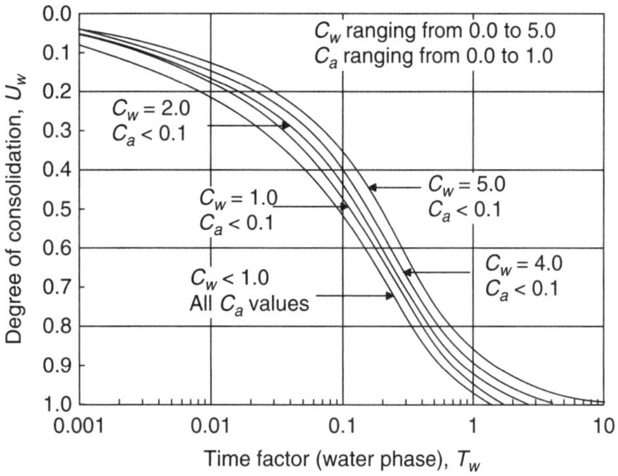

Figure 16.18 Dimensionless time factor versus degree of consolidation curves for water phase.

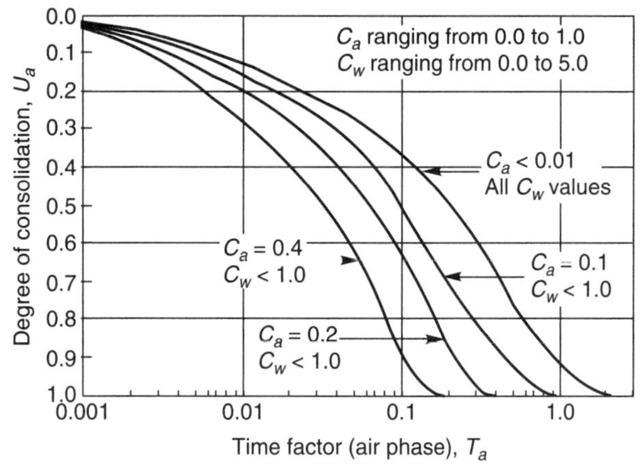

Figure 16.19 Dimensionless time factor versus degree of consolidation curves for air phase.

A rigorous formulation of two- and three-dimensional consolidation requires that the continuity equation for water and air flow be coupled with the equilibrium equations. The resulting equations need to then be solved simultaneously. Biot (1941) formulated the coupled consolidation process for a special case of an unsaturated soil. The derivation was based on the assumption that occluded bubbles of air existed in the pore fluid during the consolidation process. The coupled formulation of consolidation was then solved for the one-dimensional case.

The following sections present the equations for the coupled theory of consolidation for an unsaturated soil. Consider a referential element of unsaturated soil subjected to two-dimensional unsteady-state air and water flow, as shown in Fig. 16.20. The partial differential flow equations for the air phase and the water phase can be derived by equating the time derivative of the relevant constitutive equations to the divergence of the flow rate as described by the appropriate flow laws.

16.6.1 Stress, Water, and Air Coupling for Three-Dimensional Consolidation

Following is a three-dimensional formulation of coupled consolidation of an unsaturated soil. The formulation is similar to the derivation presented by Biot (1941). Some of the terms have been changed for consistency with more recent terminology used in unsaturated soils. The coupled formulation for three-dimensional consolidation is presented for the case where the air phase is continuous (Dakshanamurthy et al., 1984). The constitutive relations for the soil structure, the water phase, and the air phase are required when writing the equilibrium and continuity equations. These constitutive relations are first summarized in their elasticity forms prior to formulating the coupled consolidation equations.

16.6.2 Constitutive Relations for Unsaturated Soil

The stress state and deformation state variables can be linked by suitable constitutive relations that incorporate soil properties in the form of coefficients. There are three available constitutive relations for an unsaturated soil, namely, one for the soil structure, one for the water phase, and one for the air phase (see Chapter 13).

16.6.3 Soil Structure Constitutive Relationship

The constitutive equation for the soil structure is presented assuming that the soil behaves as an isotropic, linear elastic material. The soil structure constitutive relationship written for normal strains in the x-, y-, and z-directions can be written in the following incremental form:

$$d\varepsilon_x = \frac{d(\sigma_x - u_a)}{E} - \frac{\mu}{E}d(\sigma_y + \sigma_z - 2u_a) + \frac{d(u_a - u_w)}{H} \tag{16.47}$$

$$d\varepsilon_y = \frac{d(\sigma_y - u_a)}{E} - \frac{\mu}{E}d(\sigma_x + \sigma_z - 2u_a) + \frac{d(u_a - u_w)}{H} \tag{16.48}$$

$$d\varepsilon_z = \frac{d(\sigma_z - u_a)}{E} - \frac{\mu}{E}d(\sigma_x + \sigma_y - 2u_a) + \frac{d(u_a - u_w)}{H} \tag{16.49}$$

The constitutive equations for shear deformations are

$$d\gamma_{xy} = \frac{d\tau_{xy}}{G} \tag{16.50}$$

$$d\gamma_{yz} = \frac{d\tau_{yz}}{G} \tag{16.51}$$

$$d\gamma_{zx} = \frac{d\tau_{zx}}{G} \tag{16.52}$$

where:

ε_x = normal strain in the x-direction,
ε_y = normal strain in the y-direction,

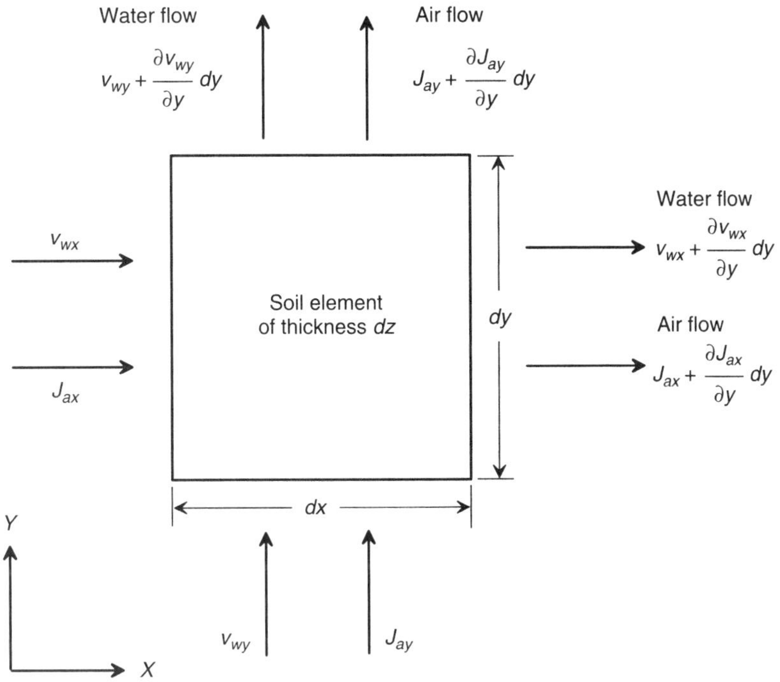

Figure 16.20 Two-dimensional unsteady-state air and water flow.

ε_z = normal strain in the z-direction,
σ_x = total normal stress in the x-direction,
σ_y = total normal stress in the y-direction,
σ_z = total normal stress in the z-direction,
$\sigma_x - u_a$ = net normal stress in the x-direction,
$\sigma_y - u_a$ = net normal stress in the y-direction,
$\sigma_z - u_a$ = net normal stress in the z-direction,
μ = Poisson's ratio,
E = modulus of elasticity or Young's modulus for the soil structure with respect to a change in $\sigma - u_a$,
H = modulus of elasticity for the soil structure with respect to a change in $u_a - u_w$,
γ_{xy} = shear strain on the z-plane (i.e., $\gamma_{xy} = \gamma_{yx}$),
γ_{yz} = shear strain on the x-plane (i.e., $\gamma_{yz} = \gamma_{zy}$),
γ_{zx} = shear strain on the y-plane (i.e., $\gamma_{zx} = \gamma_{xz}$),
τ_{xy} = shear stress on the x-plane in the y-direction (i.e., $\tau_{xy} = \tau_{yx}$),
τ_{yz} = shear stress on the y-plane in the z-direction (i.e., $\tau_{yz} = \tau_{zy}$),
τ_{zx} = shear stress on the z-plane in the x-direction (i.e., $\tau_{zx} = \tau_{xz}$), and
G = shear modulus.

The change in volumetric strain, $d\varepsilon_v$, can be written as the summation of the normal strain changes:

$$d\varepsilon_v = d\varepsilon_x + d\varepsilon_y + d\varepsilon_z \quad (16.53)$$

where:

$d\varepsilon_v$ = volumetric strain change (i.e., dV_v/V_0),
dV_v = total volume change of a soil element, and
V_0 = initial total volume of a soil element.

Rearranging Eqs. 16.47, 16.48, and 16.49, the change in net normal stresses can be expressed as a function of normal strain and matric suction changes:

$$d\left(\sigma_x - u_a\right) = 2G(d\varepsilon_x + \alpha\ d\varepsilon_v) - \beta d\left(u_a - u_w\right) \quad (16.54)$$

$$d\left(\sigma_y - u_a\right) = 2G(d\varepsilon_y + \alpha\ d\varepsilon_v) - \beta d\left(u_a - u_w\right) \quad (16.55)$$

$$d\left(\sigma_z - u_a\right) = 2G(d\varepsilon_z + \alpha\ d\varepsilon_v) - \beta d\left(u_a - u_w\right) \quad (16.56)$$

where

$$G = \frac{E}{2(1+\mu)}$$

$$\alpha = \frac{\mu}{1-2\mu}$$

$$\beta = \frac{E}{H(1-2\mu)} \quad \text{or} \quad \beta = \frac{2G}{H}\frac{1+u}{1-2u}$$

16.6.4 Water Phase Constitutive Relations

The constitutive equation for the water phase defines the change in volume of water in an unsaturated soil element for specified changes in total stress, pore-air, and pore-water pressures:

$$\frac{dV_w}{V_0} = \frac{d(\sigma_x - u_a)}{E_w} + \frac{d(\sigma_y - u_a)}{E_w} + \frac{d(\sigma_z - u_a)}{E_w} + \frac{d(u_a - u_w)}{H_w} \quad (16.57)$$

where:

dV_w = water volume change in the soil element,
E_w = water volumetric modulus associated with a change in $\sigma - u_a$, and
H_w = water volumetric modulus associated with a change in $u_a - u_w$.

Substituting Eqs. 16.54, 16.55, and 16.56 into Eq. 16.57 allows the water volume change in the soil to be written as

$$\frac{dV_w}{V_0} = \beta_{w1}\, d\varepsilon_v + \beta_{w2}\, d(u_a - u_w) \quad (16.58)$$

where

$$\beta_{w1} = \frac{E}{E_w(1 - 2\mu)} \qquad \beta_{w2} = \frac{1}{H_w} - \frac{3\beta}{E_w}$$

16.6.5 Air Phase Constitutive Relations

The constitutive equation for the air phase defines the change in volume of air in an unsaturated soil element for specified changes in the total stress, pore-air, and pore-water pressures:

$$\frac{dV_a}{V_0} = \frac{d(\sigma_x - u_a)}{E_a} + \frac{d(\sigma_y - u_a)}{E_a} + \frac{d(\sigma_z - u_a)}{E_a} + \frac{d(u_a - u_w)}{H_a} \quad (16.59)$$

where:

dV_a = air volume change in the soil element,
E_a = air volumetric modulus associated with a change in $\sigma - u_a$, and
H_a = air volumetric modulus associated with a change in $u_a - u_w$.

Substituting Eqs. 16.54, 16.55, and 16.56 into Eq. 16.59 allows the air volume change to be written as

$$\frac{dV_a}{V_0} = \beta_{a1}\, d\varepsilon_v + \beta_{a2}\, d(u_a - u_w) \quad (16.60)$$

where

$$\beta_{a1} = \frac{E}{E_a(1 - 2\mu)} \qquad \beta_{a2} = \frac{1}{H_a} - \frac{3\beta}{E_a}$$

16.6.6 Coupled Consolidation Equations for Unsaturated Soil

The derivation of the theory of coupled consolidation is based on the assumption that the air phase is continuous. Several other assumptions used in the derivation are similar to those adopted by Terzaghi (1943) and Biot (1941). The assumptions can be listed as follows: (1) material is isotropic, (2) reversibility of stress-strain relations, (3) linearity of stress-strain relations, (4) strains are small, (5) pore-water is incompressible, (6) coefficients of permeability with respect to water and air phases are functions of the volume-mass soil properties during the consolidation process, and (7) the effects of air diffusing through water, air dissolving in the water phase, and the movement of water vapor are ignored.

There are five unknowns of deformation and volumetric variables to be solved in a three-dimensional consolidation problem. The variables are the displacements in the x-, y-, and z-directions (i.e., u, v, and w, respectively) and the water and air volume changes (i.e., dV_w and dV_a). Displacements in the x-, y-, and z-directions are used to compute the total volume change. The five unknowns can be obtained from three equilibrium equations and two continuity equations (i.e., the water and air phase continuities). These equations are summarized in the following sections.

16.6.7 Stress Equilibrium Equations

The stress state for an unsaturated soil element should satisfy the following equilibrium conditions:

$$\frac{\partial \sigma_x}{\partial x} + \frac{\partial \tau_{yx}}{\partial y} + \frac{\partial \tau_{zx}}{\partial z} = 0 \quad (16.61)$$

$$\frac{\partial \tau_{xy}}{\partial x} + \frac{\partial \sigma_y}{\partial y} + \frac{\partial \tau_{zy}}{\partial z} = 0 \quad (16.62)$$

$$\frac{\partial \tau_{xz}}{\partial x} + \frac{\partial \tau_{yz}}{\partial y} + \frac{\partial \sigma_z}{\partial z} = 0 \quad (16.63)$$

The effect of the body force is assumed to be negligible. Substituting Eqs. 16.54, 16.55, and 16.56 into Eqs. 16.61, 16.62, and 16.63 gives the following form for the equilibrium equations:

$$G\nabla^2 u + \frac{G}{1 - 2\mu}\frac{\partial \varepsilon_v}{\partial x} - \beta\frac{\partial(u_a - u_w)}{\partial x} + \frac{\partial u_a}{\partial x} = 0 \quad (16.64)$$

$$G\nabla^2 v + \frac{G}{1 - 2\mu}\frac{\partial \varepsilon_v}{\partial y} - \beta\frac{\partial(u_a - u_w)}{\partial y} + \frac{\partial u_a}{\partial y} = 0 \quad (16.65)$$

$$G\nabla^2 w + \frac{G}{1 - 2\mu}\frac{\partial \varepsilon_v}{\partial z} - \beta\frac{\partial(u_a - u_w)}{\partial z} + \frac{\partial u_a}{\partial z} = 0 \quad (16.66)$$

where

$$\nabla^2 = \frac{\partial^2}{\partial x^2} + \frac{\partial^2}{\partial y^2} + \frac{\partial^2}{\partial z^2}$$

16.6.8 Water Phase Continuity

The continuity equation for the water phase can be obtained by equating the time derivative of the water phase constitutive equation to the divergence of Darcy's flow law:

$$\beta_{w1}\frac{\partial \varepsilon_v}{\partial t}+\beta_{w2}\frac{\partial(u_a-u_w)}{\partial t}=-\left[\frac{k_w}{\rho_w g}\nabla^2 u_w + \frac{1}{\rho_w g}\left(\frac{\partial k_w}{\partial x}\frac{\partial u_w}{\partial x}+\frac{\partial k_w}{\partial y}\frac{\partial u_w}{\partial y}+\frac{\partial k_w}{\partial z}\frac{\partial u_w}{\partial z}\right)+\frac{\partial k_w}{\partial y}\right] \quad (16.67)$$

16.6.9 Air Phase Continuity

The continuity equation for the air phase can be obtained from the time derivative of the air phase constitutive equation and the net mass rate of air flow as described by Fick's law:

$$\beta_{a1}\frac{\partial \varepsilon_v}{\partial t}+\beta_{a2}\frac{\partial\left(u_a-u_w\right)}{\partial t}=-\left[\frac{D_a^*}{\rho_a\bar{u}_a}\nabla^2 u_a+\frac{1}{\rho_a\bar{u}_a}\times\left(\frac{\partial D_a^*}{\partial x}\frac{\partial u_a}{\partial x}+\frac{\partial D_a^*}{\partial y}\frac{\partial u_a}{\partial y}+\frac{\partial D_a^*}{\partial z}\frac{\partial u_a}{\partial z}\right)+\frac{(1-S)n}{\bar{u}_a}\frac{\partial \bar{u}_a}{\partial t}\right] \quad (16.68)$$

where:

D_a^* = coefficient of transmission for the air phase,
ρ_a = density of air,
$\bar{u}_a$ = absolute pore-air pressure (i.e., $\bar{u}_a = u_a + \bar{u}_{atm}$),
u_a = gauge pore-air pressure,
$\bar{u}_{atm}$ = atmospheric pressure (i.e., 101 kPa),
S = degree of saturation, and
n = porosity.

The solution of a coupled analysis for two-phase flow through an unsaturated soil involves the simultaneous solution of Eqs. 16.64, 16.65, 16.66, 16.67, and 16.68.

16.7 WATER, AIR FLOW, AND NONISOTHERMAL SYSTEMS

Air flow and water flow may not only occur as a result of a load being applied to the soil. It is also possible that moisture infiltration and temperature fluctuations may produce thermal gradients within a soil mass that result in fluid flow. Nonisothermal conditions are closely related to microclimatic changes in the field. Seasonal temperature changes can cause air and water flow and consequently volume changes within a soil mass. A nonisothermal condition often occurs under highway and airfield pavements or under the shallow foundations of lightweight structures.

Following is a summary of the formulation of one-dimensional flow under nonisothermal conditions in an unsaturated soil. Temperature gradients are incorporated into the formulation of flow. Water vapor flow is also incorporated into the formulation since vapor flow can be relevant. Surface boundary conditions for the air, liquid water, water vapor, and heat flow equations are also summarized. The soil-atmospheric boundary conditions incorporate the microclimatic conditions at a site.

16.7.1 Air Phase Partial Differential Equation

The differential equation for one-dimensional, K_0, air flow during consolidation requires that one term be added when dealing with nonisothermal conditions. The term takes into consideration the effect of temperature changes on air flow. Pore-air pressures are consequently affected. One-dimensional air flow in the y-direction can be expressed as follows (Dakshanamurthy and Fredlund, 1981):

$$\frac{\partial u_a}{\partial t} = -C_a\frac{\partial u_w}{\partial t}+c_v^a\frac{\partial^2 u_a}{\partial y^2}+C_{at}\frac{\partial T}{\partial t} \quad (16.69)$$

where:

C_a = interaction coefficient associated with the air phase PDE, that is, $(m_2^a/m_{1k}^a)/[1-m_2^a/m_{1k}^a-(1-S)\,n/\left(\bar{u}_a m_{1k}^a\right)]$
m_{1k}^a = coefficient of air volume change with respect to a change in net normal stress, $d\left(\sigma_y-u_a\right)$, for K_0 loading,
c_v^a = coefficient of consolidation with respect to the air phase, that is, $\dfrac{D_a^*}{\omega_a/RT_K}\dfrac{1}{\bar{u}_a m_{1k}^a\left(1-m_2^a/m_{1k}^a\right)-(1-S)\,n}$,
C_{at} = interaction temperature coefficient associated with the air phase PDE, that is $\dfrac{1}{T}\left[\dfrac{(1-S)n\bar{u}_a}{(1-m_2^a/m_{1k}^a)\bar{u}_a m_{1k}^a+(1-S)n}\right]$,
T_K = absolute temperature (i.e., $T_K = T + 73.16$), K,
T = temperature, °C, and
t = time.

The above equation does not take into account any nonlinear variation which might be present in the coefficient of transmission, D_a^*. Variations in the coefficient of transmission with respect to space (i.e., $\partial D_a^*/\partial y$) are assumed to be negligible.

16.7.2 Fluid and Vapor Flow Equation for Water Phase

The differential equation for fluid and vapor flow under nonisothermal conditions is similar to the water flow equation for one-dimensional consolidation with the addition of one term. The additional term accounts for the water vapor flow due to diffusion and advection processes (Wilson, 1990). Water vapor flow can be relevant under nonisothermal conditions. The one-dimensional fluid and vapor flow equation for the water phase in the y-direction can be expressed as follows:

$$\frac{\partial u_w}{\partial t} = -C_w \frac{\partial u_a}{\partial t} + \frac{1}{\rho_w g m_2^w} \frac{\partial}{\partial y}\left(k_w \frac{\partial u_w}{\partial y}\right) + \frac{\bar{u}_a + \bar{u}_v}{\bar{u}_a} \frac{1}{\rho_w m_2^w} \frac{\partial}{\partial y}\left(D_v \frac{\partial \bar{u}_v}{\partial y}\right) \quad (16.70)$$

where:

C_w = interaction coefficient associated with the water phase partial differential equation [i.e., $(1 - m_2^w/m_{1k}^w)/(m_2^w/m_{1k}^w)$],
u_v = partial pressure of water vapor in air, and
D_v = diffusion coefficient of the water vapor through the soil, kg·m/kN·s.

The coefficient of water vapor diffusion D_v can be calculated as follows (Philip and de Vries, 1957; de Vries, 1975; Dakshanamurthy and Fredlund, 1981; Wilson, 1990):

$$D_v = \alpha\beta\left(D_{vm}\frac{\omega_v}{RT_K}\right) \quad (16.71)$$

where:

α = tortuosity factor for the soil (i.e., $\varepsilon = \beta^{2/3}$),
β = cross-sectional area of the soil available for water vapor flow [i.e., $(1 - S)\,n$],
D_{vm} = molecular diffusivity of water vapor in air [i.e., $0.229\left(1 + T_K/273\right)^{1.75} \times 10^{-4}$ (m^2/s)] (from Kimball et al., 1976), and
ω_v = molecular mass of water vapor (i.e., 18.016 kg/kmol).

The above equation indicates that the diffusion coefficient D_v is a function of soil properties (i.e., S and n), which in turn are a function of matric suction $u_a - u_w$. In addition, D_v is also a function of temperature T_K. Similarly, the coefficient of permeability k_w is also a function of soil properties that may vary with respect to location in the soil mass. If the variations in the k_w and D_v coefficients with respect to space are considered to be negligible, Eq. 13.70 can be simplified as follows:

$$\frac{\partial u_w}{\partial t} = -C_w \frac{\partial u_a}{\partial t} + c_v^w \frac{\partial^2 u_w}{\partial y^2} + c_v^{wv} \frac{\partial^2 \bar{u}_v}{\partial y^2} \quad (16.72)$$

where:

c_v^w = coefficient of consolidation with respect to the water phase [i.e., $k_w/\left(\rho_w g m_2^w\right)$] and
c_v^{wv} = coefficient of consolidation with respect to the water vapor phase (i.e., $\frac{\left(\bar{u}_a + \bar{u}_v\right)}{\bar{u}_a}\frac{D_v}{\rho_w m_2^w}$).

16.7.3 Heat Flow Equation for Unsaturated Soils

A Fourier diffusion equation can be used to describe heat transfer in soils as described in Chapter 10 (Jame and Norum, 1980; Fredlund and Dakshanamurthy, 1982; Wilson, 1990). The following form of the heat flow equation was proposed by Wilson (1990):

$$\zeta\frac{\partial T}{\partial t} = \frac{\partial}{\partial y}\left(\lambda\frac{\partial T}{\partial t}\right) - L_v\left(\frac{\bar{u}_a + \bar{u}_v}{\bar{u}_a}\right)\frac{\partial}{\partial y}\left(D_v\frac{d\bar{u}_v}{\partial y}\right) \quad (16.73)$$

where:

ζ = volumetric specific heat of the soil as a function of water content, (J/m^3/°C),
λ = thermal conductivity of the soil as a function of water content, (W/m/°C), and
L_v = latent heat of vaporization of water (i.e., 2,418,000 J/kg).

The above equation describes the heat flow due to conduction and the latent heat transfer caused by phase changes. Convective heat flow is considered to be negligible (Jame and Norum, 1980; Wilson, 1990). The volumetric specific heat of the soil, ζ, can be calculated using the relationship given by de Vries (1963):

$$\zeta = \zeta_s\theta_{\text{solid}} + \zeta_w\theta + \zeta_a\theta_a \quad (16.74)$$

where:

ζ_s = volumetric specific heat capacity of the soil solids (i.e., a typical value is 2.235×10^6 J/m^3/°C for fine sands; de Vries, 1963),
θ_{solid} = volumetric solid content (i.e., V_s/V_0),
V_s = volume of soils solids in the soil,
V_0 = total volume of the soil,
ζ_w = volumetric specific heat capacity for the water phase (i.e., 4.154×10^6 J/m^3/°C for water at 35°C; Wilson, 1990),
θ = volumetric water content (i.e., V_w/V_0),
V_w = volume of water in the soil,
ζ_a = volumetric specific heat capacity for the air phase,
θ_a = volumetric air content (i.e., V_a/V_0), and
V_a = volume of air.

The third term on the right-hand side of Eq. 16.74 is small and can be considered negligible (de Vries, 1963). The thermal conductivity of the soil, λ, can be estimated using the following expression (de Vries, 1963):

$$\lambda = \frac{f_s\theta_s\lambda_s + f_w\theta\lambda_w + f_a\theta_a\lambda_a}{f_s\theta_s + f_w\theta + f_a\theta_a} \quad (16.75)$$

where:

f_s, f_w, f_a = weighting factors for the solid, water, and air phases, respectively, and

$\lambda_s, \lambda_w, \lambda_a$ = thermal conductivities of the solid, water, and air phases, respectively.

Typical thermal conductivities for sand particles and water are 6.0 and 0.57 W/m/°C, respectively (de Vries, 1963; Jame, 1977; Wilson, 1990). The thermal conductivity of air, λ_a, consists of two components (Jame, 1977; Wilson, 1990):

$$\lambda_a = \lambda_{\text{dry air}} + \lambda_{\text{water vapor}} \qquad (16.76)$$

where:

$\lambda_{\text{dry air}}$ = 0.025 W/m/°C, and
$\lambda_{\text{water vapor}}$ = varies linearly between 0 and 0.0736 W/m/°C for volumetric water content θ_w between 0 and 12.1%.

The weighting factors f can be calculated using the assumption that the soil particles are ellipsoidal in shape. The weighting factor for the continuous medium (i.e., air or water) is equal to 1.0. Water can be selected as the continuous medium with the f_w value equal to 1.0. The weighting factors for the solid and air phases can then be calculated in accordance with the following relationship:

$$f = \frac{1}{3}\sum_{i=1}^{3}\left[1 + \left(\frac{\lambda}{\lambda_w} - 1\right) g_i\right]^{-1} \qquad (16.77)$$

where:

$f = f_a, f_s$,
$\lambda = \lambda_a, \lambda_s$, and
g = depolarization factors for the ellipsoid (i.e., g_1, g_2, and g_3, where $g_1 + g_2 + g_3 = 1$); the values of g_1, g_2, and g_3 are independent of particle size and dependent only on the ratio of the length of the ellipsoid axes.

Wilson (1990) used equal depolarization factors of 1/3 when computing the weighting factors for a sandy soil. In this case, the sand particles were assumed to have spherical shapes. The depolarization factors g_1 and g_2 for the air phase were assumed to decrease linearly from 0.333 to 0.105 for volumetric water content θ ranging from 23.6 to 12.1%, respectively, and from 0.105 to 0.015 for θ values ranging from 12.1 and 0%, respectively (Jame, 1977).

Figure 16.21 illustrates the thermal conductivity λ variation for Beaver Creek sand with respect to volumetric water contents as computed using Eqs. 16.75, 16.76, and 16.77. The thermal properties of the unsaturated soil are presented as functions of the water content of the soil; however, these properties can also be written as functions of the stress state (e.g., soil suction). Writing the thermal properties as a function of the stress state is more consistent with the manner in which hydraulic properties are handled in geotechnical engineering.

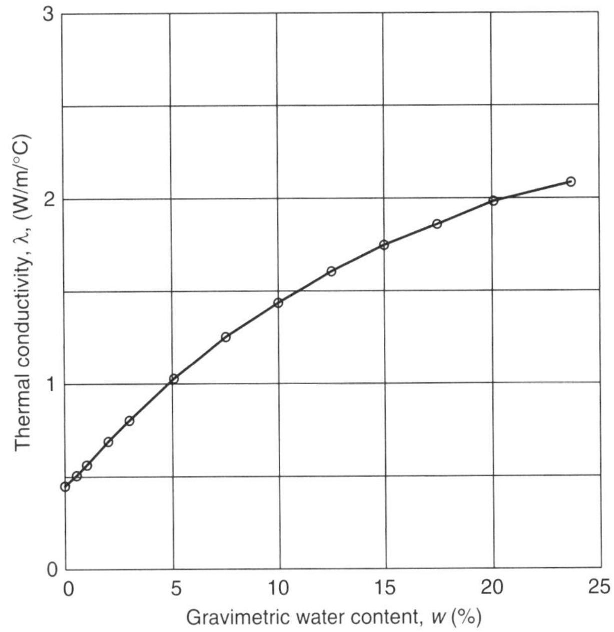

Figure 16.21 Thermal conductivity versus water content for Beaver Creek sand (from Wilson, 1990).

16.8 TWO-DIMENSIONAL STRESS-DEFORMATION AND SATURATED-UNSATURATED SEEPAGE ANALYSIS

The most common volume change problem is related to swelling and shrinking associated with expansive soils. There are two physical processes involved when dealing with expansive soils: (i) water movement and (ii) deformation analysis. There is the movement of water either into the ground or out of the ground. The movement of water into the ground might be due to precipitation (i.e., rainfall or snowmelt). The movement of water out of the ground is through evaporation and evapotranspiration.

There is also a stress-deformation analysis which allows the calculation of displacements resulting from a change in the stress state of the soil. There are two PDEs that need to be solved for a two-dimensional stress-deformation analysis (i.e., one for the x-direction and one for the y-direction). The combined stress-deformation and water seepage PDEs can be solved in either a coupled or uncoupled manner.

Chapter 14 showed how the stress-deformation PDEs for one-, two-, and three-dimensional analyses could be performed when the initial and final boundary conditions were known. It was the total amount of heave or settlement that was the desired outcome from that analysis. In this chapter, the case of two-dimensional problems involving expansive clay will once again be analyzed; however, this time the movement of water infiltration with respect to time will be analyzed in combination with volume changes (i.e., stress-deformation analysis).

16.8.1 Partial Differential Equations for Two-Dimensional Expansive Soil Example as Uncoupled Analysis

Vector notation will be used for designation of the x-, y-, and z-directions in order to shorten the lengths of the equations. The water continuity equation for an unsaturated soil can be written as follows:

$$\frac{\partial(\theta)}{\partial t} + \nabla\left(v_w\right) = 0 \qquad (16.78)$$

where:

$\nabla = (\partial/\partial x)i + (\partial/\partial y)j + (\partial/\partial z)k$, the divergence operator, and
$v_w = v_w^x i + v_w^y j + v_w^z k$, Darcy's flux.

The governing equation for the water phase is obtained by substituting the time derivative of the water phase constitutive equation and Darcy's law into the water phase continuity equation 16.78:

$$\beta_{w1}\frac{\partial \varepsilon_v}{\partial t} + \beta_{w2}\frac{\partial\left(u_a - u_w\right)}{\partial t} - \nabla\left[k_w \nabla\left(\frac{u_w}{\rho_w g} + Y\right)\right] = 0 \qquad (16.79)$$

Equation 16.79 can be written as follows for the two-dimensional case:

$$\beta_{w1}\frac{\partial \varepsilon_v}{\partial t} + \beta_{w2}\frac{\partial\left(u_a - u_w\right)}{\partial t} = \frac{\partial}{\partial x}\left[k_{wx}\frac{\partial}{\partial x}\left(\frac{u_w}{\rho_w g}\right)\right] + \frac{\partial}{\partial y}\left[k_{wy}\frac{\partial}{\partial y}\left(\frac{u_w}{\rho_w g} + Y\right)\right] \qquad (16.80)$$

The analysis of seepage and volume change in an unsaturated, expansive soil requires information on the soil structure and water phase constitutive surfaces, the coefficient of permeability function, and the water storage of the soil. The transient water flow process in a swelling soil is influenced by the expansiveness of the soil structure (Vu, 2003). The effects of changes in suction are less when soil suction is high but increase significantly as suction approaches zero.

The swelling-versus-time formulation can be decoupled by first solving for changes in matric suction with respect to time. In this case, no consideration is given to changes in overall volume and net normal stresses are not allowed to change during the seepage process. The constitutive surface for the water phase can be represented by the SWCC and Eq. 16.80 takes on the following form:

$$m_2^w\frac{\partial\left(u_a - u_w\right)}{\partial t} = \frac{\partial}{\partial x}\left[k_{wx}\frac{\partial}{\partial x}\left(\frac{u_w}{\rho_w g}\right)\right] + \frac{\partial}{\partial y}\left[k_{wy}\frac{\partial}{\partial y}\left(\frac{u_w}{\rho_w g} + Y\right)\right] \qquad (16.81)$$

Equation 16.81 applies for both transient and steady-state seepage conditions in saturated and unsaturated soils. Two unsaturated soil property functions are required when solving transient seepage problems: coefficient of water volume change (or water storage modulus) and coefficient of permeability. The coefficient of water storage and the coefficient of permeability can be assumed to mainly be functions of soil suction.

The coefficient of water storage is the arithmetic slope of the SWCC and is obtained by differentiating the SWCC with respect to matric suction (i.e., $m_2^w = d\theta/d\psi$). Numerous equations have been proposed to simulate the SWCC and any one of these equations can be used for determining the water storage function m_2^w. The Fredlund and Xing (1994) equation for the SWCC is used to illustrate the application of the SWCC to the swelling process:

$$\theta = \theta_s\left[\frac{1}{\ln\left(e + (\psi/a_f)^{n_f}\right)}\right]^{m_f} \qquad (16.82)$$

where:

ψ = soil suction, kPa,
e = natural log base, 2.71828,
θ_s = volumetric water content at saturation,
θ = volumetric water content at any soil suction,
a_f = parameter designating the inflection point along the SWCC, kPa,
n_f = parameter related to the rate of desaturation at the inflection point on the SWCC, and
m_f = parameter which is related to the residual water content of the soil.

The coefficient-of-permeability functions, k_{wx} and k_{wy} can be estimated using the SWCC and the saturated coefficient of permeability. There are several empirical equations that have been proposed for the coefficient of permeability for an unsaturated soil (e.g., Gardner, 1958b; Fredlund et al., 1994b; Leong and Rahardjo, 1997a). The permeability power function proposed by Leong and Rahardjo (1997a) is used to illustrate the swelling process. The permeability function for the y-direction, k_{wy}, can be written as follows based on the Fredlund and Xing (1994) equation for the SWCC:

$$k_{wy} = k_{sy}\left[\frac{1}{\ln\left(e + (\psi/a_f)^{nf}\right)}\right]^{pm_f} \qquad (16.83)$$

where:

p = estimated parameter based on the analysis of a data set of measured permeability functions (Leong and Rahardjo, 1997a; Fredlund et al., 2001) and
k_{sy} = saturated water coefficient of permeability in the y-direction (an equation similar to Eq. 16.83 can also be written for the x-direction using the saturated coefficient of permeability in the x-direction, k_{sx}).

The transient water flow equation 16.81 along with the equation for the SWCC (Eq. 16.82) and the permeability function (Eq. 16.83) can be used to compute pore-water pressure profiles (i.e., matric suctions) at various times during an infiltration process. The matric suction profiles are then used in an uncoupled manner with the stress-deformation equation to calculate overall volume changes. Deformations due to changes in matric suction at any elapsed time are calculated based on the initial suction profile and the matric suctions at subsequent times.

16.8.2 Partial Differential Equations for Two-Dimensional Expansive Soil Example as Coupled Analysis

The transient saturated-unsaturated seepage equation can also be fully coupled with a stress-deformation analysis. In this case, the equations of overall static equilibrium for the soil must be solved. Relevant stress-deformation equations for the coupled formulation are presented in vector notation:

$$\frac{\partial \sigma_{ij}}{\partial x_j} + b_i = 0 \tag{16.84}$$

where:

σ_{ij} = components of the net total stress tensor and
b_i = components of the body force vector.

Substituting the strain-displacement relationship and the stress-strain relationship into the equilibrium equation 16.84 gives the following governing equations for general three-dimensional problems (i.e., equations for x-, y-, and z-directions):

$$G\nabla^2 u_i + \frac{G}{1-2\mu}\frac{\partial \varepsilon_v}{\partial x_i} - \beta\frac{\partial(u_a - u_w)}{\partial x_i} + \frac{\partial u_a}{\partial x_i} + b_i = 0 \tag{16.85}$$

where

$$\beta = \frac{m_2^s}{m_1^s} = \frac{E/H}{1-2\mu}$$

$$\nabla^2 = \frac{\partial^2}{\partial x^2} + \frac{\partial^2}{\partial y^2} + \frac{\partial^2}{\partial z^2} \text{ (Laplace operator)}$$

$$\varepsilon_v = \frac{\partial u_i}{\partial x_i} = \frac{\partial u}{\partial x} + \frac{\partial v}{\partial y} + \frac{\partial w}{\partial z}$$

$$G = \frac{E}{2(1+\mu)}$$

Equations 16.79 and 16.85 form a system of coupled equations for the theory of swelling in three dimensions for an unsaturated, swelling soil with a continuous air phase. The equations have a similar form to those presented by Biot (1941) for a soil with occluded air bubbles. Equation 16.85 can be written as follows for two-dimensional plane strain problems:

$$\frac{\partial}{\partial x}\left(c_{11}\frac{\partial u}{\partial x} + c_{12}\frac{\partial v}{\partial y}\right) + c_{33}\frac{\partial}{\partial y}\left(\frac{\partial u}{\partial y} + \frac{\partial v}{\partial x}\right) - d_s\frac{\partial(u_a - u_w)}{\partial x} + b_x = 0 \tag{16.86}$$

$$c_{33}\frac{\partial}{\partial x}\left(\frac{\partial u}{\partial y} + \frac{\partial v}{\partial x}\right) + \frac{\partial}{\partial y}\left(c_{12}\frac{\partial u}{\partial x} + c_{12}\frac{\partial v}{\partial y}\right) - d_s\frac{\partial(u_a - u_w)}{\partial y} + b_y = 0 \tag{16.87}$$

where

$$c_{11} = c_{22} = \frac{(1-\mu)E}{(1+\mu)(1-2\mu)}$$

$$c_{22} = \frac{\mu E}{(1+\mu)(1-2\mu)}$$

$$c_{33} = \frac{E}{2(1+\mu)}$$

$$d_s = \frac{E}{(1-2\mu)H}$$

Equations 16.86 and 16.87 can be used to compute displacements in the horizontal and vertical directions under an applied load and/or due to changes in matric suction. The solution of the seepage equation 16.80 and the soil structure equilibrium equations 16.86 and 16.87 can be obtained using either an uncoupled or coupled procedure.

The solutions for uncoupled and coupled volume change problems in expansive soils have been studied for various example problems by Vu (2003) and Vu and Fredlund (2003a). Both uncoupled and coupled solutions were presented for the case where the elasticity parameters E, H, the coefficients of water volume change m_1^w and m_2^w, and the coefficient of permeability k_w were functions of both stress state variables. The coupled and uncoupled solutions appeared to yield essentially the same answers for the amount of heave in a swelling soil.

An uncoupled solution is used when solving the following example problem involving heave below a floor slab built on grade. The initial and subsequent boundary conditions must be designated when solving the stress-deformation equations (i.e., Eqs. 16.86 and 16.87). The boundary conditions include the initial matric suctions, initial total stress conditions, and elasticity parameter functions associated with the volume change of the soil. The results from a seepage analysis (i.e., for changes in soil suction) can be used as input to the stress-deformation solution.

16.8.3 Case History of Slab-on-Grade with Shallow Perimeter Footings on Regina Clay

The methodology for the prediction of heave in an expansive soil was applied to data collected by the Prairie Regional Station of the Division of Building Research (DBR), National

Research Council of Canada, Saskatoon, Canada, on a slab-on-grade in Regina, Saskatchewan. A two-dimensional heave analysis was used to simulate the volume change problems associated with the heave of a floor slab for a light industrial building. The history of the site and details pertaining to soil testing and performance monitoring were presented by Yoshida et al. (1983).

Construction of the building and instrumentation took place during the month of August 1961. Instrumentation installed at the site included a deep benchmark, vertical movement gauges, and a neutron moisture meter access tube. Vertical ground movement was monitored at depths of 0.58 m, 0.85 m, and 2.39 m below original ground level. The building owner noticed heave and cracking of the floor slab in early August 1962, about a year after construction. The owner also noticed an unexpected increase in water consumption of approximately 35,000 L. The loss of water was traced to a leak in a hot-water line beneath the floor slab, which was subsequently repaired. The location of the cracking and contours of heave for the floor slab are shown in Fig. 16.22.

Laboratory analyses were performed on samples from a borehole advanced on August 1961 for the installation of the deep benchmark. Laboratory tests evaluated the Atterberg limits, in situ water content, grain-size distribution, swelling indices, and corrected swelling pressures of the soil. The liquid limit was found to be 77%, the plastic limit was 33%, and the natural water content was 29%. The specific gravity G_s was 2.82 and the unit weight of the soil was 18.88 kN/m^3.

Constant-volume oedometer tests on three samples were used to evaluate initial void ratios, swelling indices, and corrected swelling pressures. Table 16.5 presents the oedometer test results and water contents of samples collected at different depths. The average initial void ratio was 0.962 and the average swelling index was 0.090. These values were used in the heave analyses. Figure 16.23 shows the distribution of the corrected swelling pressure with depth. A straight line can be used to represent the apparent distribution of the corrected swelling pressure with depth. Figure 16.24 shows the distribution of gravimetric water content with depth measured on August 21, 1961, using a neutron moisture meter access tube.

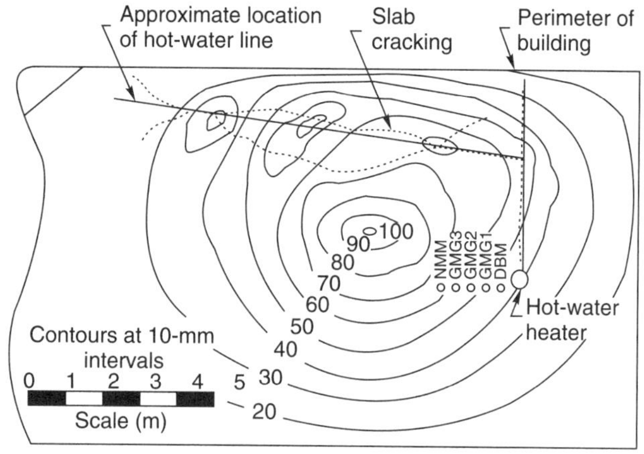

Figure 16.22 Floor plan of study site and contours of measured heave (NMM = neutron moisture meter; GMG = ground movement gauge; DBM = depth benchmark) (from Yoshida et al., 1983).

Table 16.5 Constant-Volume Oedometer Data

Depth (m)	Initial Void Ratio, e_0	Swelling Index, C_s	Corrected Swelling Pressure, P'_s (kPa)
0.69	0.927	0.095	490
1.34	0.985	0.081	325
2.20	0.974	0.094	81

Source: From Yoshida et al. (1983).

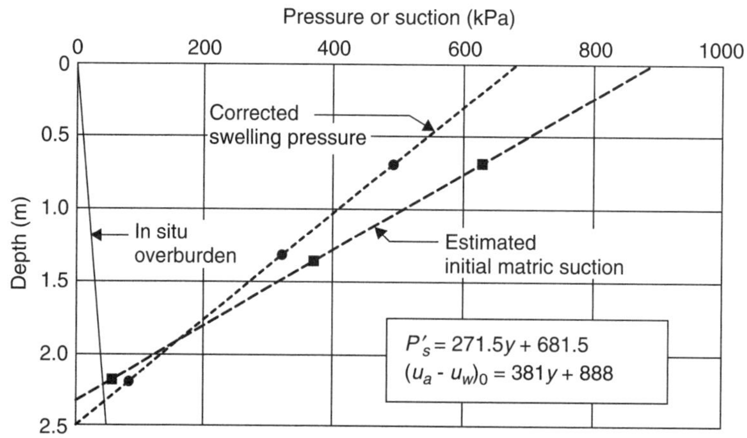

Figure 16.23 Distribution of corrected swelling pressure and estimated initial matric suction with depth.

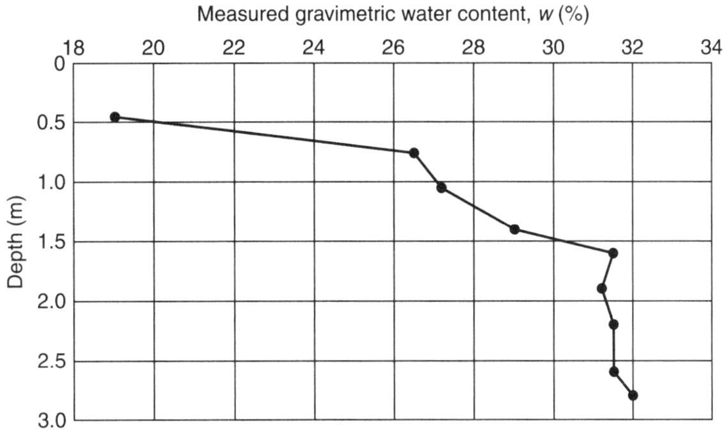

Figure 16.24 Distribution of in situ gravimetric water content measured on August 21, 1961.

The case history is presented as a two-dimensional analysis to illustrate the prediction of heave in expansive soils. Estimation of the initial matric suction condition is presented in Table 16.6 and shown graphically in Fig. 16.23. A straight line can be used to represent the distribution of the initial matric suction with depth.

The initial degrees of saturation and volumetric water contents were calculated from the measured initial densities of the soil. The measured water contents and degrees of saturation at various values of matric suction were used to estimate the SWCC. The Fredlund and Xing (1994) equation for a SWCC (Eq. 16.82) in terms of volumetric water content showed the following fitting parameters: $\theta_s = 49.3\%$, $a_f = 300$ kPa, $n_f = 0.6$, and $m_f = 0.7$. The Fredlund and Xing (1994) fitting parameters for the degree-of-saturation data were $a_f = 300$ kPa, $n_f = 0.5$, and $m_f = 0.7$ (Fig. 16.25). A coefficient-of-permeability function for compacted Regina clay was measured by Shuai (1996). The coefficient-of-permeability function was described using the Leong and Rahardjo (1997) equation SWCC 16.83 and is presented in Fig. 16.26.

Table 16.6 Estimation of Initial Matric Suction from Corrected Swelling Pressure

Variable	At Various Depths 0.69 m	1.34 m	2.20 m
Overburden pressure $(\sigma_y - u_a)_{\text{initial}}$ (kPa)	13.03	25.30	41.54
Initial void ratio e_0	0.927	0.985	0.974
Corrected swelling pressure P_s' (kPa)	490	325	81
Gravimetric water content w (%)	25	29	31
Degree of saturation S (%)	76	83	90
Volumetric water content θ (%)	36.6	41.2	44.3
Estimated field suction $(u_a - u_w)_{\text{initial}}$ (kPa)	627	361	44

Note: $S = wG_s/e$ and $\theta = wG_s/(1+e)$.

The analysis examines a cross section A-A through the building (see Fig. 16.22). The lower boundary is selected at 2.3 m depth since the test results showed that there would be no tendency for swelling below this depth (Fig. 16.23).

The geometry and boundary conditions for the seepage analysis are shown in Fig. 16.27. It was observed that water leaked from the water line along a 2-m length of the line. It was assumed that the initial suction conditions did not change outside the edges of the concrete slab or below the lower boundary. A moisture flux equal to zero was specified elsewhere along the boundaries.

Matric suction conditions were predicted for various elapsed times (i.e., 5, 20, 50, and 100 days and at steady-state conditions).

16.8.4 Computer Simulations of Slab-on-Grade with Shallow Strip Footings

Deformation of the slab took into consideration the applied building loads and the reduction in suction due to wetting. Deformations in the soil mass due to the slab loading were assumed to be immediate while deformations due to wetting were time dependent. Figure 16.28 shows the stress path followed in the analysis. The stress-deformation analysis was first performed to calculate the displacements and induced stresses related to the loading of the slab and building. The deformations due to changes in matric suction were then predicted for various elapsed times using soil suction profiles obtained from the seepage analysis.

The stress-deformation analysis was also performed for cases where the pore-water pressures go to zero as well as

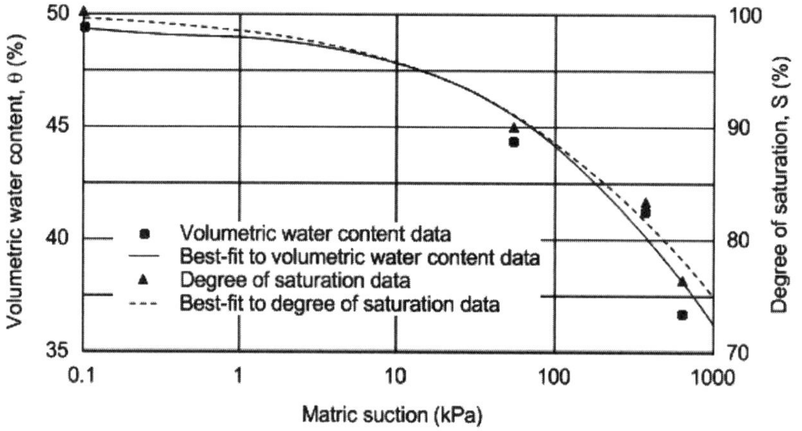

Figure 16.25 SWCC for Regina clay in terms of volumetric water content and degree of saturation.

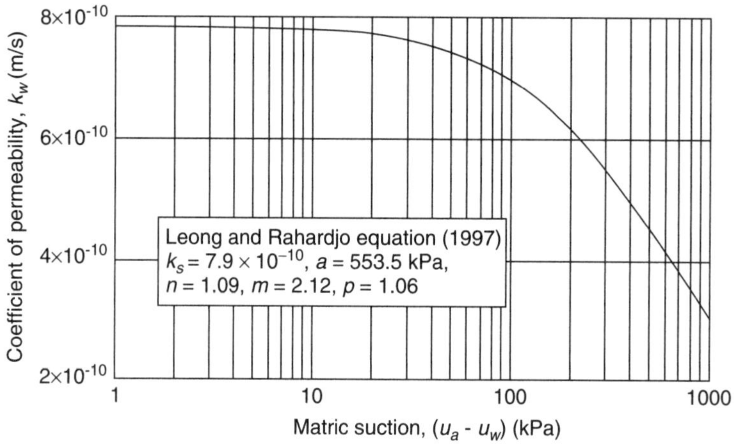

Figure 16.26 Estimated permeability function for Regina clay.

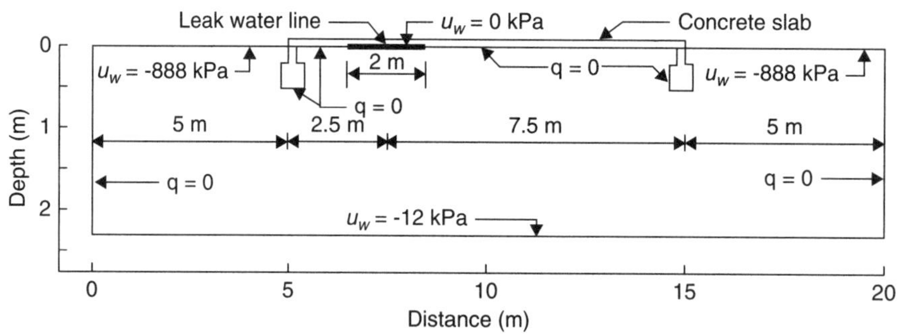

Figure 16.27 Geometry and boundary conditions for transient seepage analysis.

when the groundwater level rises to ground surface, resulting in a hydrostatic pore-water pressure distribution. Hydrostatic conditions present the uppermost limits of possible total heave.

Figure 16.29 shows the geometry and boundary conditions for the stress-deformation analysis. A load equal to 5.76 kPa was applied on the 100-mm-thick concrete slab. This surcharge is made up of 180 mm of fill with a unit weight of 18.88 kN/m^3 and 100 mm of concrete with a unit weight of 23.6 kN/m^3. A load of 15 kN/m was applied along the perimeter beam. The perimeter load includes the weight of footing and the load of the upper structure. The soil was free to move in a vertical direction and fixed in the horizontal direction at the left and right sides. The lower boundary was fixed in both directions. A Young's modulus of 10 GPa and Poisson's ratio of 0.15 were used for the concrete slab.

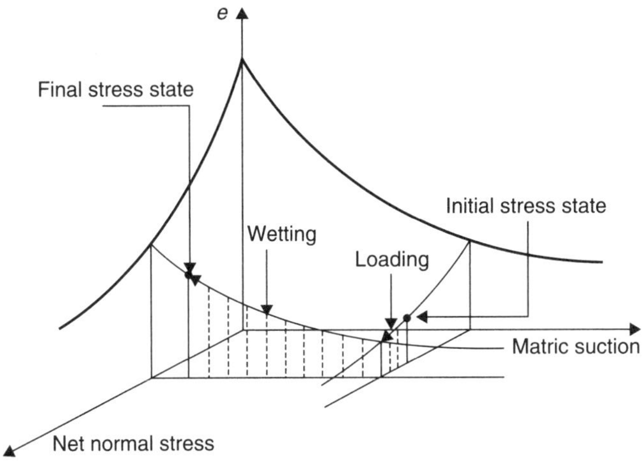

Figure 16.28 Stress path followed in swelling of soil (i.e., stress state versus deformation path).

A parametric study was undertaken to assess the effect of varying numerous input variables to the analysis.

Figure 16.30 presents changes of matric suction with time at various points in the soil. Matric suctions decreased significantly during the first 30 days of wetting and approached steady-state conditions after about 150 days. The total amount of water that leaked from the water line with time is presented in Fig. 16.31. A loss of 35 m^3 of water is equivalent to about 3.5 m^3 of water over each meter of width of the floor. This amount of water is more than the amount of water required for steady-state conditions to be attained under the specified boundary conditions.

Figure 16.32 presents pore-water pressure profiles under the center of the slab for various times and also for the case where the final pore-water pressure goes to zero and then a hydrostatic condition. Figure 16.33 shows the matric suction distribution in the soil at steady-state conditions. The matric suction at steady-state conditions is about 20 kPa under the center of the slab.

Figure 16.34 shows contours of vertical displacements due to loading. Less than 1 mm of settlement is predicted due to the loading at the center of the slab. The induced net normal stress was used to calculate the final net normal stress state in the soil. The soil was loaded with an initial net normal stress and field matric suctions estimated from the oedometer tests. The sum of the initial net normal stress and the initial matric suction equivalent was used along with the swelling index for the prediction of displacements and stress states.

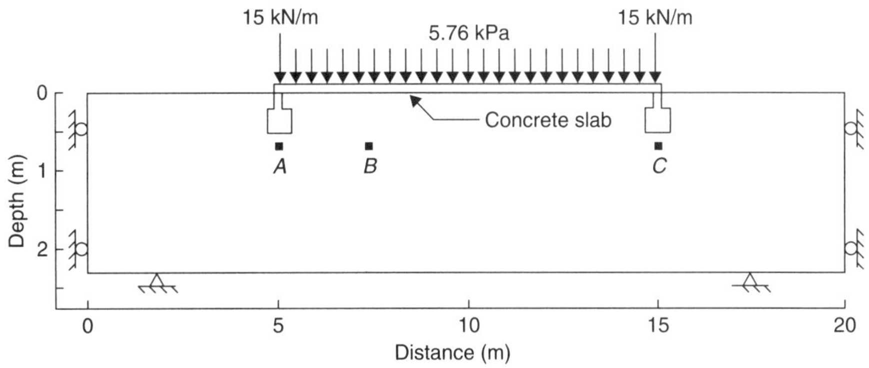

Figure 16.29 Boundary conditions for stress state versus deformation analysis.

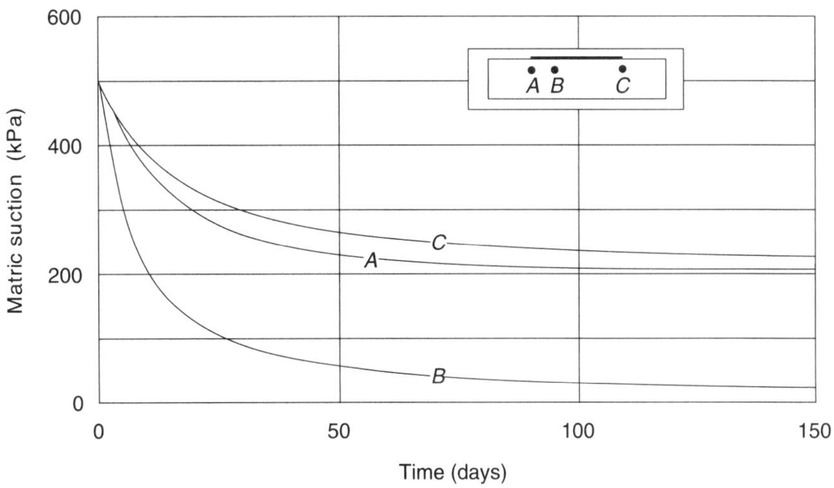

Figure 16.30 Matric suction changes with time for various points in soil mass.

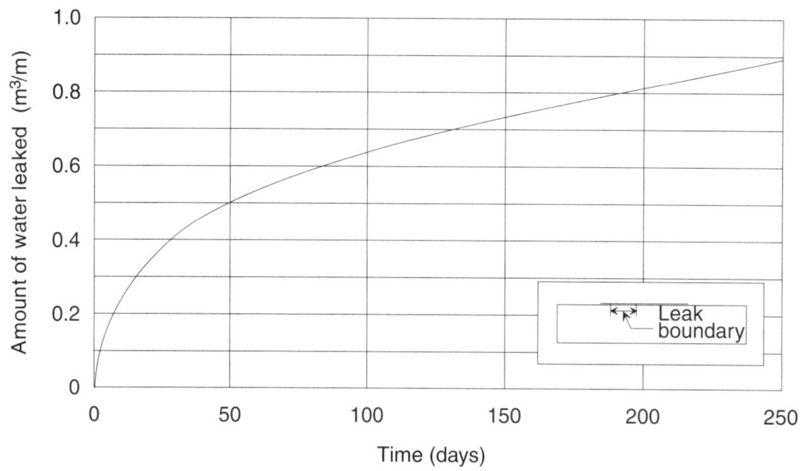

Figure 16.31 Total amount of water leakage from water line with time per meter of cross section.

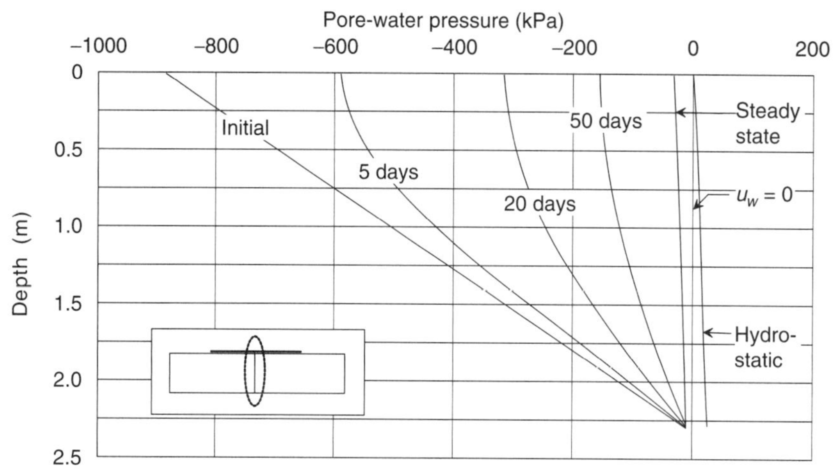

Figure 16.32 Pore-water pressure profiles for various times and for cases where final pore-water pressures are assumed to be either zero or hydrostatic.

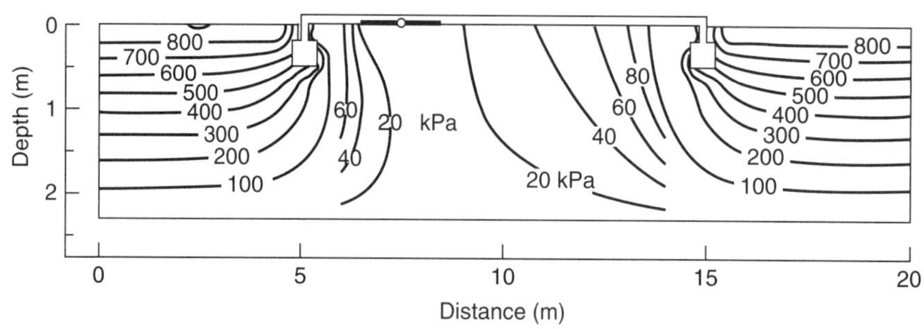

Figure 16.33 Matric suction distribution at steady-state conditions.

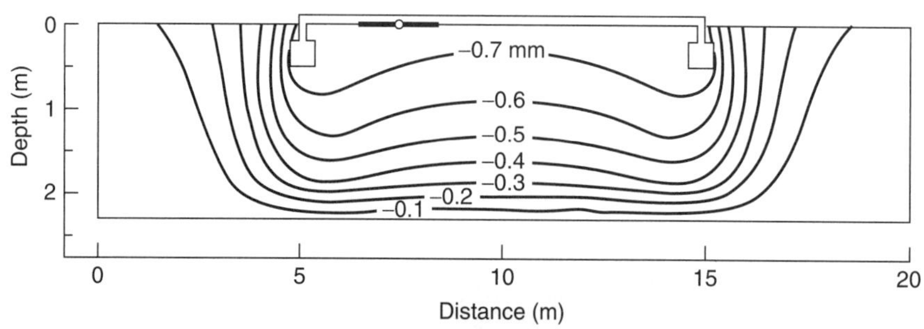

Figure 16.34 Distribution of vertical displacements due to building loads.

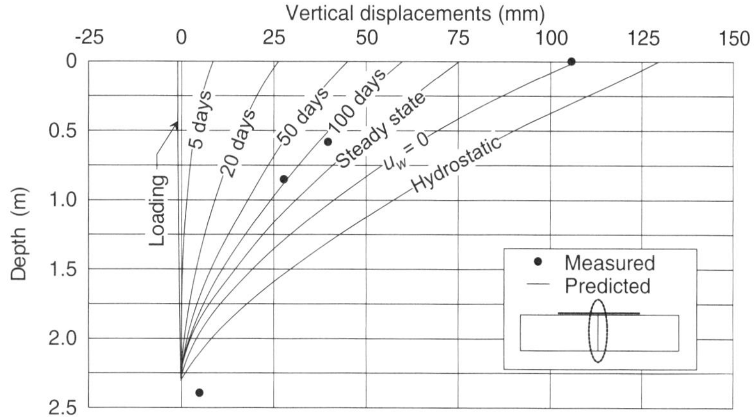

Figure 16.35 Measured and predicted vertical displacements with depth near center of concrete slab.

Figure 16.35 compares the predicted vertical displacements at various final suction conditions with the measured total heave at the center of the slab. The agreement between the predicted and the measured heave at different depths differ to some degree. The amounts of heave measured at depths of 0.58 and 0.85 m correspond to the predicted heave at 100 days, while the total heave of 106 mm at ground surface corresponds to the case when the pore-water pressure goes to zero under the slab. An uplift of 106 mm represents the maximum heave observed on the slab. The maximum heave observed at the cross section under consideration is only 80 mm (Fig. 16.22). The distribution of horizontal displacements at different final suction conditions right under the water line is presented in Fig. 16.36.

Figure 16.37 compares the predicted vertical displacement at various final matric suction conditions with the measured total heave at the top surface of the slab. The total heave predicted under steady-state conditions agrees well with the measured heave. There were some unknown loads near the perimeter of the floor slab that could not be taken into consideration in the study. The maximum predicted heave was 107 mm when assuming that the final pore-water pressure increases from a negative value to zero. The maximum heave of 130 mm is predicted for the case when the final pore-water pressures are assumed to be hydrostatic.

Figure 16.38 shows the measured and predicted water contents under the center of the slab. The final water contents predicted at steady-state conditions is about 3% less than the measured final water contents. When the pore-water pressure increases to zero, the predicted final water content is about 2% more than the measured values. The distribution of water contents in the soil predicted at steady-state conditions is shown in Fig. 16.39. Figures 16.40 and 16.41 present contours of horizontal and vertical displacements predicted under steady-state conditions.

Several stress-deformation analyses were performed to study the effect of each of the above parameters on the solutions and to gain confidence in their significance to the analysis. For each case, only the parameter under consideration was allowed to vary while all other parameters were

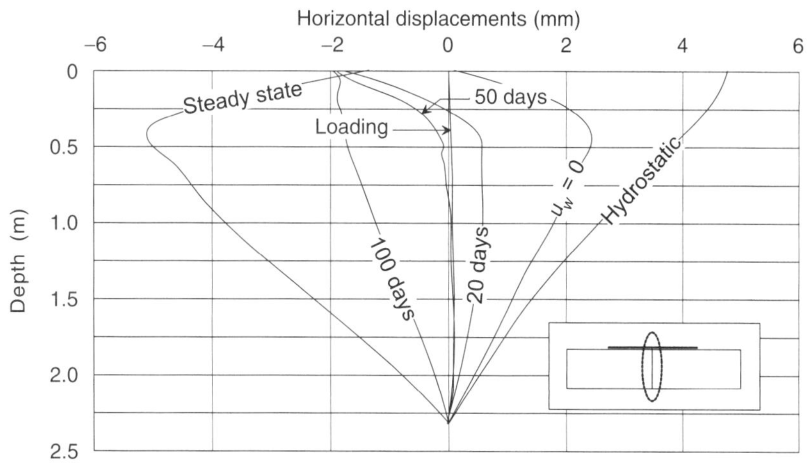

Figure 16.36 Distribution of horizontal displacements with depth for different suction conditions under water line.

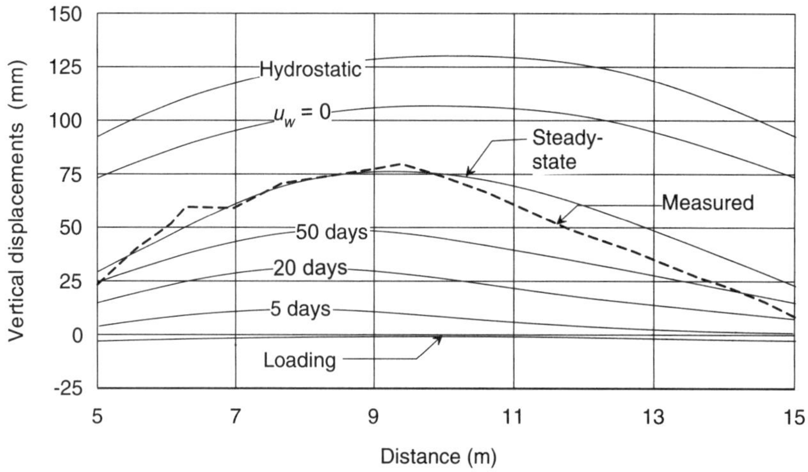

Figure 16.37 Measured and predicted vertical displacements at surface of concrete slab.

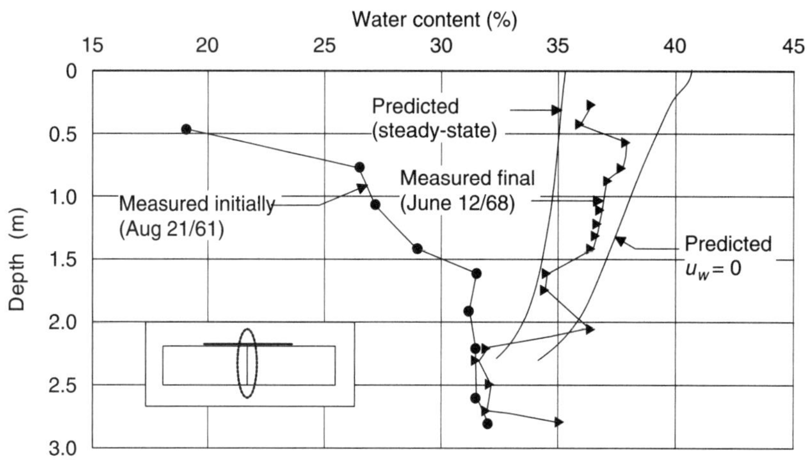

Figure 16.38 Measured and predicted water content with depth near center of concrete slab.

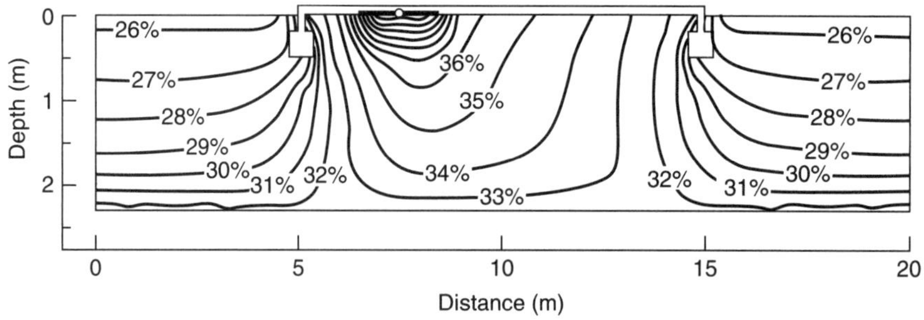

Figure 16.39 Computed distribution of gravimetric water contents corresponding to steady-state conditions.

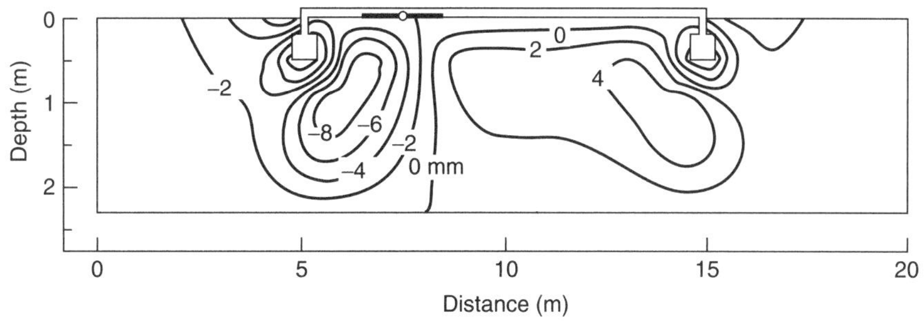

Figure 16.40 Distribution of horizontal displacements under steady-state conditions.

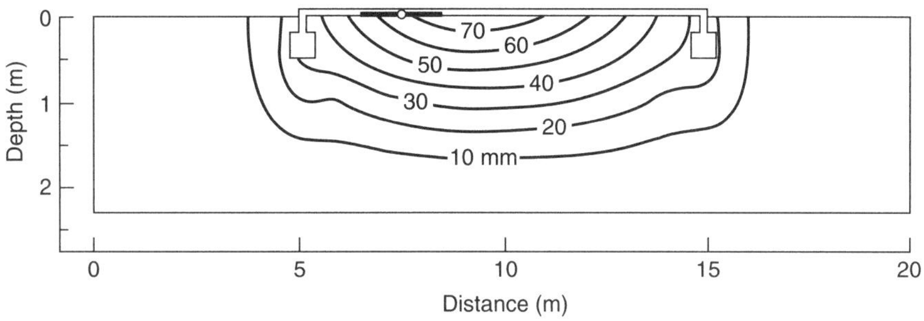

Figure 16.41 Distribution of vertical displacements under steady-state conditions.

kept unchanged at values representing the "median case." Table 16.7 shows the values of the parameters that were used in the parametric study. The displacements were calculated for the case of steady-state seepage.

Figures 16.42 and 16.43 present the predicted vertical displacements at the surface of the slab and under the center of the slab, respectively, for various values of Poisson's ratio. The predicted total heave increases with an increasing value for the assumed Poisson's ratio. For a change in Poisson's ratio from 0.35 to 0.45, the variation in maximum total heave predicted is about ± 7%.

Figures 16.44 and 16.45 present the predicted vertical displacements at the surface of the slab and under the center of the slab, respectively, for various values of swelling index C_s. The predicted total heave increases with increasing values for swelling index. For a change in swelling index from 0.085 to 0.095, the variation in the maximum total heave predicted is about ±5%.

Table 16.7 Values of Parameters Used in Parametric Study

Parameter	Lower Values	Median Case	Upper Values
Poisson's ratio, μ	0.30, 0.35	0.40	0.45, 0.49
Swelling index, C_s	0.085	0.090	0.095, 0.100
Initial void ratio, e_0	0.920	0.962	1.000
Coefficient of earth pressure at rest, K_0	0.400	0.667	1.000
Young's modulus of concrete, E_c	5 GPa	10 GPa	20 GPa

Figures 16.46 and 16.47 present the predicted vertical displacements at the surface of the slab and under the center of the slab, respectively, for various values of initial void ratio. The total heave predicted increases with decreasing values for initial void ratio. For a change in initial void ratio varying from 0.92 to 1.00, the variation in maximum total heave predicted is about ±2%.

Figures 16.48 and 16.49 present the predicted vertical displacements at the surface of the slab and under the center of the slab, respectively, for various values of the coefficient of earth pressure at rest, K_0. The total heave predicted increases with a decreasing value for the coefficient of earth pressure at rest. For a change in the coefficient of earth pressure at rest from 0.4 to 1.0, the variation in maximum total heave predicted is about ±2%.

Figures 16.50 and 16.51 present the predicted vertical displacements at the surface of the slab and under the center of the slab, respectively, for various values of Young's modulus for concrete slab. The maximum total heave predicted increases with a decreasing value of Young's modulus for concrete. For a change in Young's modulus for the concrete from 5 to 20 GPa, the variation in maximum total heave predicted is about ±3%. However, the distribution of heave at the surface of the slab changes considerably with the stiffness of the slab. The concrete slab with the lowest Young's

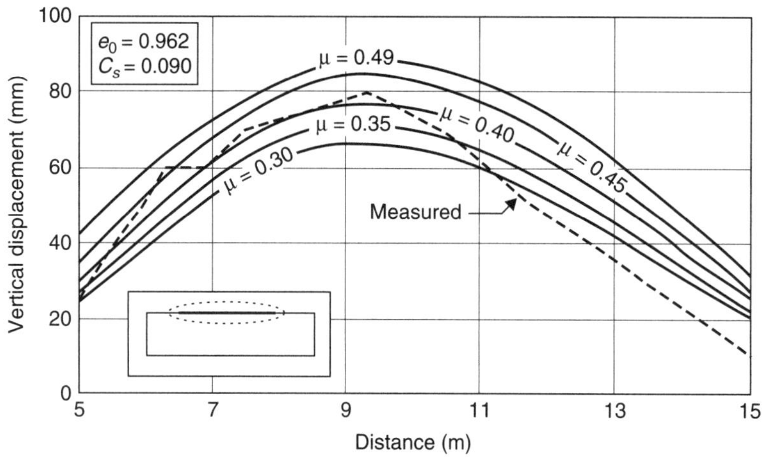

Figure 16.42 Measured and predicted vertical displacements at surface of concrete slab for various values of Poisson's ratio μ.

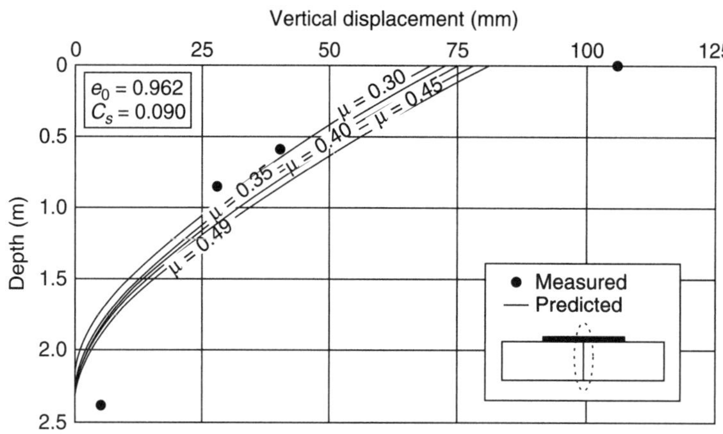

Figure 16.43 Measured and predicted vertical displacements with depth near center of slab for various values of Poisson's ratio μ.

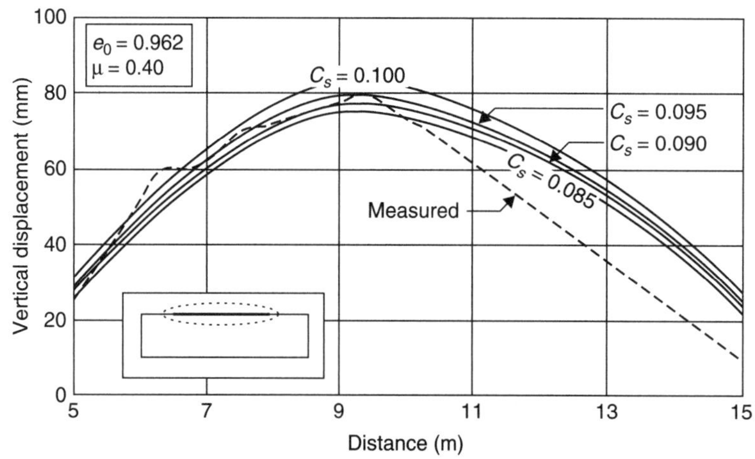

Figure 16.44 Measured and predicted vertical displacements at surface of slab for various values of swelling index C_s.

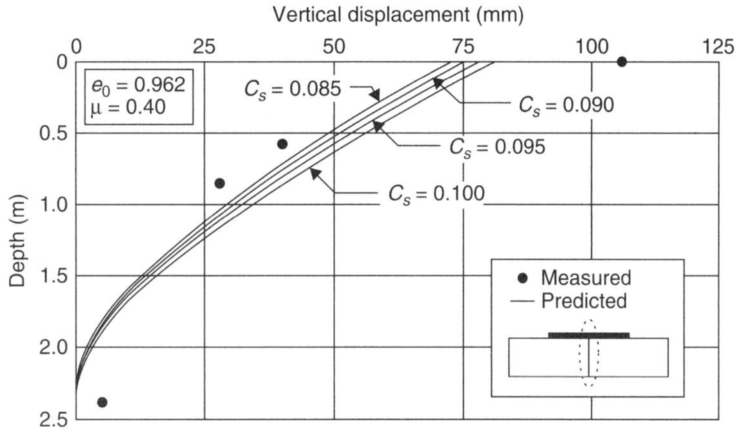

Figure 16.45 Measured and predicted vertical displacements with depth near center of slab for various values of swelling index C_s.

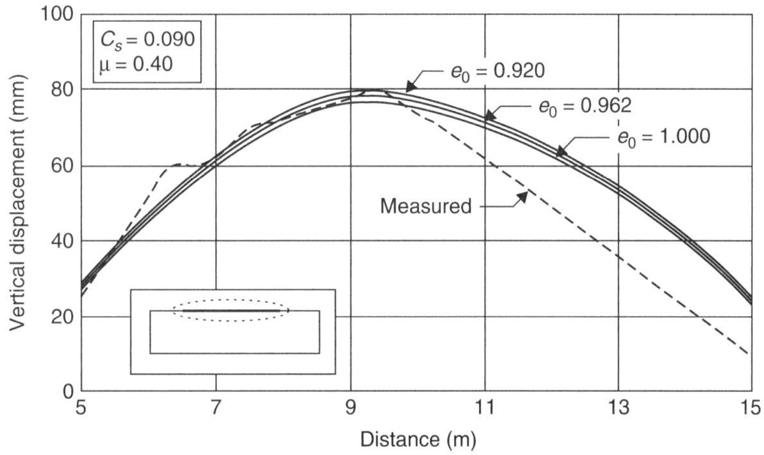

Figure 16.46 Measured and predicted vertical displacements at surface of slab for various values of initial void ratio e_o.

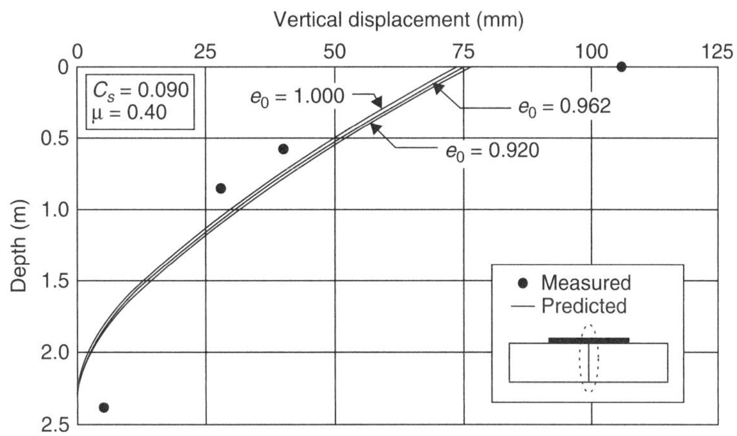

Figure 16.47 Measured and predicted vertical displacements with depth near center of slab for various values of initial void ratio e_o.

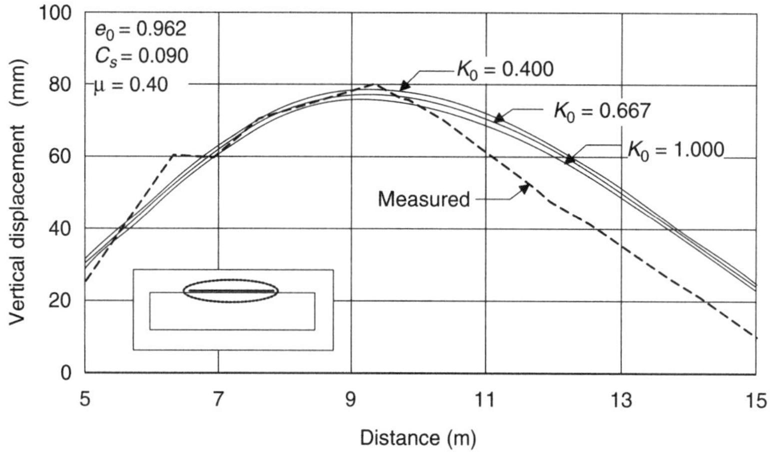

Figure 16.48 Measured and predicted vertical displacements at surface of slab for various values of coefficient of earth pressure at rest, K_0.

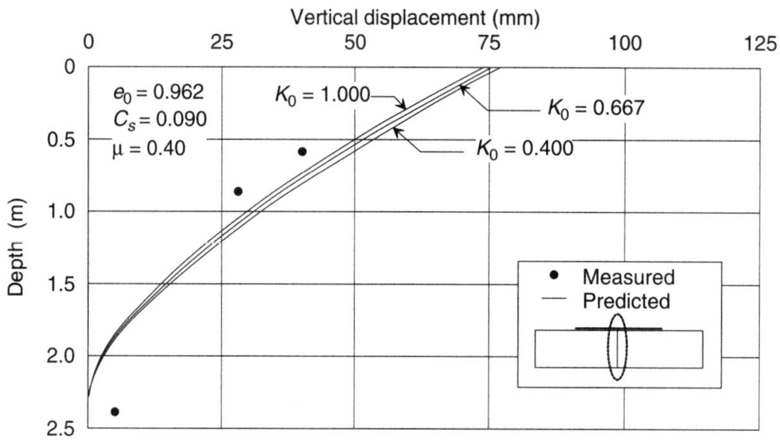

Figure 16.49 Measured and predicted vertical displacements with depth near center of slab for various values of coefficient of earth pressure at rest, K_0.

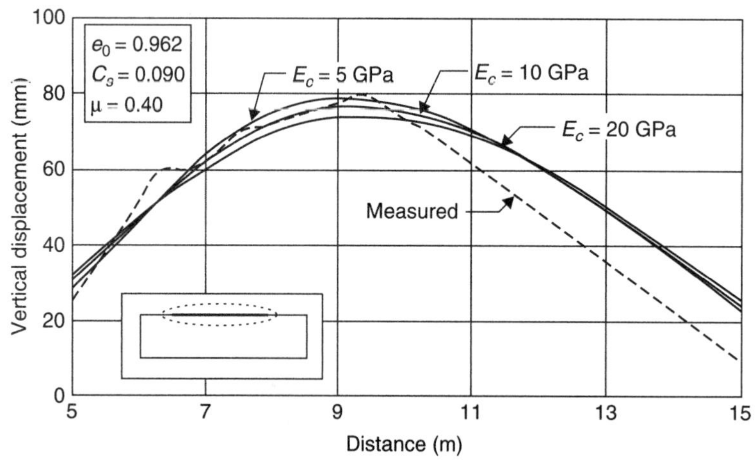

Figure 16.50 Measured and predicted vertical displacements at surface of slab for various values of Young's modulus for concrete slab, E_c.

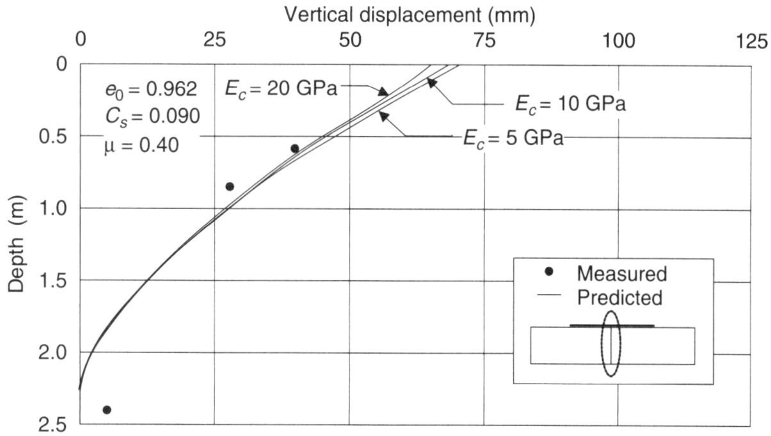

Figure 16.51 Measured and predicted vertical displacements with depth below water line for various values of Young's modulus for concrete slab, E_c.

Table 16.8 Results of Parametric Study

Parameter	Range	Variation of Predicted Heave (%)
Poisson's ratio, μ	0.35–0.45	±7
Swelling index, C_s	0.085–0.95	±5
Initial void ratio, e_0	0.92–1.0	±2
Coefficient of earth pressure at rest, K_0	0.4–1.0	±2
Young's modulus of concrete, E_c	5–20 GPa	±3

modulus would suffer more differential heave. The cracking of the slab may result in larger maximum heave values.

The results of the parametric study are summarized in Table 16.8. It can be seen that swelling index and Poisson's ratio are the factors that have the greatest effect on the solution. The value of Young's modulus for concrete appears to control the shape of the deformed slab.

16.9 COMPUTER SIMULATION OF EDGE LIFT AND EDGE DROP OF SLABS-ON-GROUND

Residential foundations such as slabs-on-ground are often constructed on unsaturated, expansive soils and have to resist deformations associated with changes in matric suction in the soil. Residential foundations are lightly loaded and therefore susceptible to changes in matric suction.

The soil beneath an impervious cover can deform into either an edge-drop mode or an edge-lift mode (Fig. 16.52). Edge drop occurs when the matric suction in the soil surrounding the slab gradually increases due to evaporation and/or evapotranspiration. There is generally more volume decrease around the perimeter of the slab than under the interior of the slab. The edge region of the slab is more strongly influenced by climatic changes and imposed watering conditions. Edge lift occurs when the matric suction in the soil decreases due to the infiltration of water into the soil around the edges of the slab. An uncoupled analysis was performed to solve the required seepage and stress-deformation conditions corresponding to edge drop and edge lift.

An example problem is analyzed to illustrate soil shrinkage around the edge of a cover as a result of evaporation from the surrounding soil surface (Fredlund et al., 2006). The saturated-unsaturated seepage analysis is first performed to provide the negative pore-water pressures with elapsed time. The pore-water pressure results are then input to the stress-deformation analysis.

The selected geometry consists of a single soil region 12 m wide by 3 m deep, as shown in Fig. 16.52. Either a flexible cover or a more rigid concrete cover can be analyzed. Let us first consider a flexible cover that extends 6 m from the left symmetrical boundary. The initial suction profile was

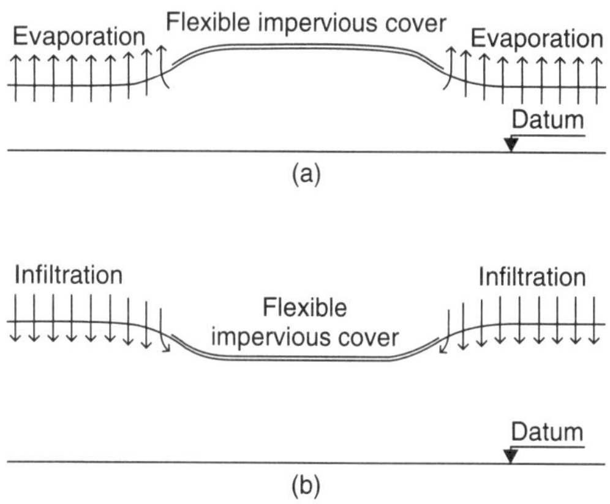

Figure 16.52 Deformation pattern associated with (a) drying, edge-drop, and (b) wetting, edge-lift, modes.

assumed to be 400 kPa at the bottom boundary, decreasing to 20 kPa at ground surface. The vertical edge boundaries were assumed to be no-flow boundaries. An evaporation flux rate of 10 mm/day was applied to the uncovered ground surface outside the flexible slab. The soil properties used in the seepage analysis are shown in Table 16.9.

The left and right stress-deformation boundary conditions are free to move vertically while the bottom boundary is fixed. The initial vertical total stress state was determined from the total unit weight multiplied by the depth below ground surface. The soil properties used in the stress-deformation analysis are shown in Table 16.10.

16.9.1 Comparing Ground Displacements for Edge Drop Example Using Flexible or Stiff Concrete Slab

The combined seepage and stress-deformation analysis was performed for conditions of a flexible slab on the ground surface as well as the case of a more rigid concrete slab. Figure 16.53 shows a comparison of the vertical displacements along the ground surface when modeling a flexible slab and a concrete slab with a modulus of elasticity of 100 GPa. The vertical displacements along the ground surface were used to compute the differential deflection at the base of the concrete slab, as shown in Fig. 16.53.

Table 16.9 Assumed Soil Properties for 2D Edge-Drop Seepage Analysis

Soil Properties	Values
Coefficient of permeability at saturation, k_s	1×10^{-8} m/s
Volumetric water content at saturation, θ_s	0.45
Parameters for SWCC (Fredlund and Xing, 1994) and permeability function (Leong and Rahardjo, 1997a)	$a_f = 300$ kPa, $n_f = 1.5$, $m_f = 1$, $p = 1$

Table 16.10 Assumed Soil Properties for 2D Edge-Drop Stress Analysis

Soil Properties	Values
Total unit weight, γ	17.2 kN/m^2
Initial void ratio, e_0	1.0
Swelling index, C_s	0.15
Swelling index, C_m	0.13
Poisson's ratio, μ	0.4
Coefficient of earth pressure at rest, K_0	0.33

The calculated differential deflections are presented in Table 16.11. A differential deflection of 5.33 mm was predicted after five days of evaporation when the ground surface cover was flexible. When modeling a concrete slab with a modulus of elasticity of 20 MPa there was a decrease in the differential deflection to 4.94 mm. Increasing the elastic modulus of the concrete slab resulted in a small decrease in deflection for the concrete slab. A modulus of 100 GPa for the concrete slab decreases the differential deflection to about 4.27 mm after 5 days of drying.

The initial modeling of the concrete slab for edge drop did not allow for separation between the soil and the concrete slab. It is possible, however, for the soil to separate from the concrete giving rise to an air gap. The length of the air gap is known as the moisture variation distance e_m. A trial-and-error procedure was used to compute the length of the air gap separation between the soil and the concrete slab.

16.9.2 Methodology Used to Determine Separation Distance between Slab and Underlying Soil for Edge-Drop Case

The potential separation between the concrete slab and the underlying soil can be modeled based on a variety of assumptions related to the interaction between the soil and the concrete slab. A stress-deformation analysis was performed using the changes in matric suction from a seepage analysis as input to the analysis.

Let us first assume that the slab is flexible and the vertical displacement along the ground surface is computed. The analysis is then repeated taking the elastic properties of the concrete slab into consideration. The concrete slab is assumed to remain in contact with the soil at all points. The vertical displacements and vertical stresses can be computed along the ground surface.

There will be a point along the bottom of the concrete slab where tensile stresses start to develop. A distance of separation can be assumed and then a check can be made to see if any of the vertical stresses go into tension (Fig. 16.54). The stress analysis can again be performed and the results again examined for tensile stresses. This trial-and-error process can be repeated until the analysis shows that there are no tensile stresses between the soil and the concrete slab. In this way, the length of the gap between the soil and the concrete slab becomes a "review" boundary condition. The soil-concrete slab separation distance can be assumed to be equivalent to the moisture variation distance e_m.

The results of the analysis with no separation between the soil and the concrete slab (Fig. 16.53) can be compared with the results where separation was allowed between the soil and the slab (Fig. 16.55). The results show that tensile stresses existed over a distance of approximately 0.6 m from the edge of the slab. The effect of allowing a separation of 0.01 m between the soil and the slab is shown in Fig. 16.55. The results in Fig. 16.55 illustrate the difference in the deformed shape of the concrete slab depending upon the assumption

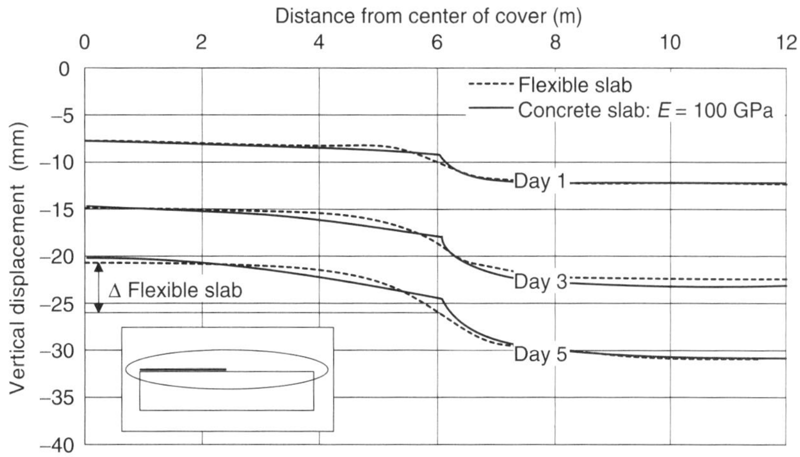

Figure 16.53 Comparison of vertical displacement at ground surface for edge-drop case considering flexible slab and stiffened concrete slab with elastic modulus E_c of 100 GPa.

Table 16.11 Differential Deflection for Edge-Drop Example Considering a Flexible Slab Cover and Concrete Slab for Varying Elastic Modulus Values

Elastic Modulus	Day 1 Deflection (mm)	Day 3 Deflection (mm)	Day 5 Deflection (mm)
Flexible cover	2.09	3.87	5.33
20 MPa	1.80	3.55	4.94
10 GPa	1.50	3.48	5.10
100 GPa	1.29	3.16	4.27

made concerning the interaction between the soil and the underside of the concrete. The deformed shapes can be used to calculate moments and stresses in the concrete slab.

16.9.3 Interpretation of Edge Moisture Variation Distance e_m and Differential Soil Movement y_m

There are two variables that play a primary role in the design of slabs-on-ground: (i) horizontal moisture variation distance e_m (also referred to as the edge penetration distance) and (ii) vertical differential soil movement y_m. These variables are used for the design of a concrete slab-on-ground.

The edge moisture variation distance e_m is defined as the distance measured inward from the edge of the slab of a shallow foundation where the water content of the soil varies due to drying and wetting influences around the perimeter. The selection of the e_m distance requires, in essence, the point where changes in matric suction becomes negligible or more accurately the point where the volume change of the soil becomes insignificant.

It is suggested that the edge moisture distance might be defined using a plot of the change in matric suction versus the horizontal distance from the edge of the concrete slab, as shown in Fig. 16.56. The change in matric suction immediately below the concrete slab is plotted on a logarithmic scale. A threshold value of matric suction change might be used to estimate the moisture variation distance e_m. There are other possible criteria that might also be used for the selection of the edge moisture variation distance.

The differential soil movement y_m is defined as the maximum vertical differential soil movement that is likely to occur. The vertical distance y_m might be determined

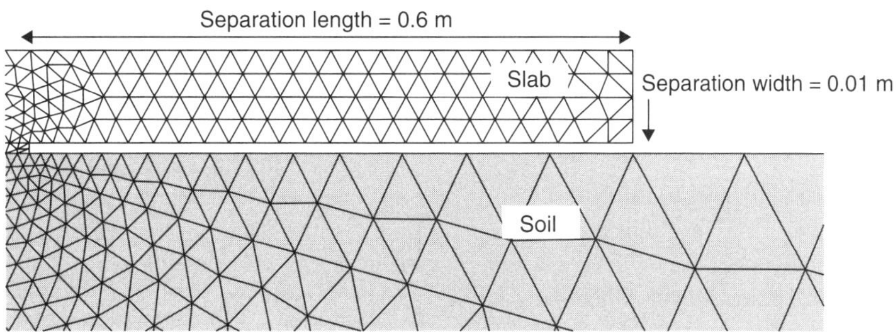

Figure 16.54 Modeling separation of concrete slab and expansive soil under soil-drying conditions.

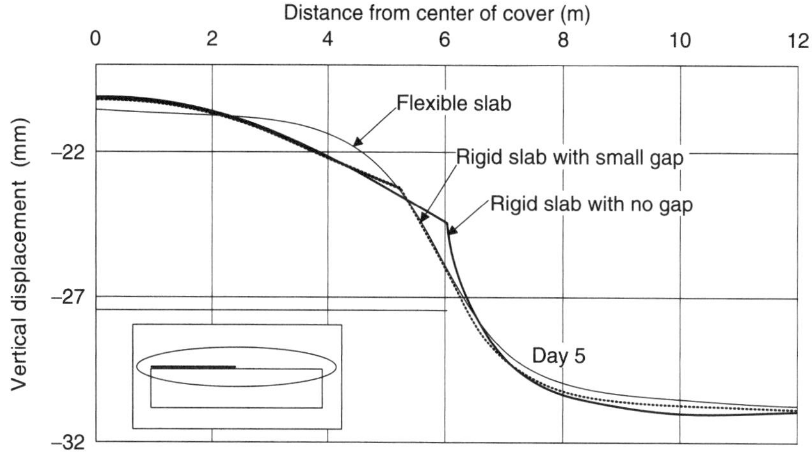

Figure 16.55 Comparison of vertical displacement profiles when using flexible slab, rigid slab with no gap, and rigid slab with a small gap.

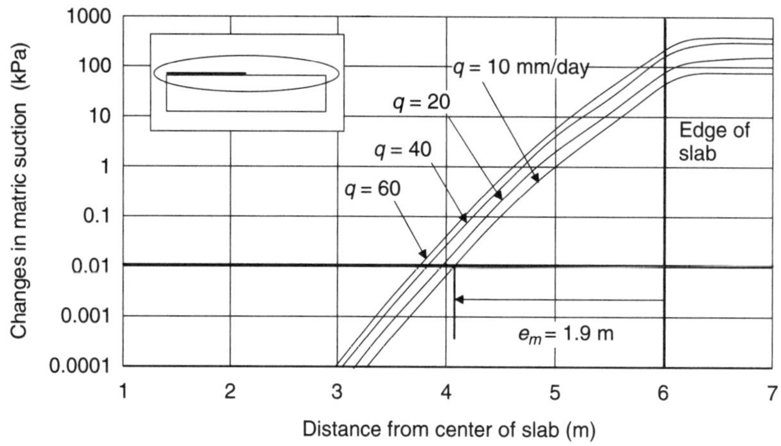

Figure 16.56 Suggested approach to estimation of edge moisture distance e_m.

from a plot of the vertical displacement along the ground surface.

The edge-drop case can also be modeled through use of a "no-tension" finite element at the interface between the soil and the underside of the concrete slab. The edge moisture variation distance e_m can still be defined in terms of the distance from the edge of the concrete slab to the point where the contact vertical stresses become positive.

The above example is simply meant to illustrate how problems associated with edge drop might be analyzed using a combined seepage and stress-deformation numerical model.

16.9.4 Methodology for Modeling Edge-Rise Case

The edge rise case occurs when the soil beneath the edge of the concrete slab swells as a result of moisture infiltration. There may once again be a loss of contact beneath the soil and the bottom of the concrete slab. The zone over which there is a loss of contact will depend upon the pattern of swelling in the underlying soil. The use of a no-tension element can once again be used when modeling the interface between the soil and the concrete slab. Further studies should be undertaken into how best to use the climatic conditions in a particular region to obtain appropriate design conditions for slabs-on-ground.

16.10 THEORY FOR SIMULATION OF SWELLING PRESSURE DEVELOPMENT

The development of swelling pressure in laboratory testing is another example of the application of combined seepage and stress-deformation analyses in unsaturated soil mechanics. Numerous methods have been proposed for the prediction of the swelling pressure and the amount of heave of an expansive soil. However, there have been few attempts to numerically simulate the swelling process and the development of the swelling pressure. Complicating the swelling pressure simulation is the variety of different stress paths used in the laboratory to measure swelling pressure.

The swelling pressure measured in the laboratory may differ significantly depending on the test procedure used for its measurement (Brackley, 1975; Sridharan et al., 1986). The ability to numerically simulate the development of swelling pressure should assist in determining the test procedure that should be used in the laboratory. Such a simulation model was developed by Shuai and Fredlund (1997). The model can be used to study factors that influence the swelling process and the development of the swelling pressure.

The swelling process involves two primary processes: the transient water flow process and the soil volume change (i.e., stress-deformation) process. These two processes are interrelated and governed by the following basic equations: (i) force equilibrium equations for the soil, (ii) constitutive equations for unsaturated soils, and (iii) continuity equation for the pore fluids. These equations have already been presented in Chapter 16 and will not be repeated. Relevant equations for the uncoupled and uncoupled formulations appear in Eqs. 16.78–16.87. Details of the theoretical formulation can also be found in Shuai and Fredlund (1997).

The following assumptions are made in the derivation: (i) the soil is isotropic, (ii) strains are infinitesimal, and (iii) permeability with respect to the air phase is significantly greater than permeability with respect to the water phase. Consequently, the pore-air pressure is always equal to the surrounding atmospheric air pressure (i.e., $u_a = 0$). Consequently, the net normal stress $(\sigma - u_a)$ becomes equal to the total vertical stress σ and matric suction $u_a - u_w$ becomes equal to the negative pore-water pressure u_w. The above assumptions allow the constitutive equations to be rewritten in terms of the strain associated with the soil structure.

The transient water flow process is linked to the soil volume change process during the swelling process and the two processes are solved simultaneously for the simulation of the swelling process along a variety of stress paths.

16.10.1 Specialization of Three-Dimensional Swelling to K_0 Loading Case

Displacements in the x- and z-directions (i.e., displacements u and w, respectively) are equal to zero for K_0 loading. Assuming that the pore-water pressure u_w and the displacements do not change across the x- and z-directions, the displacement in the y-direction, v, can be calculated. The soil compressibility moduli m_2^s, m_1^s, m_1^w, and m_2^w were defined in Chapter 13. The soil volume change and the transient water flow equations are simplified to the following forms:

$$\frac{\partial^2 v}{\partial y^2} = -m_2^s \frac{\partial u_w}{\partial y} \tag{16.88}$$

$$\frac{m_1^w}{m_1^s}\frac{\partial}{\partial t}\left(\frac{\partial v}{\partial y}\right) = \frac{1}{\rho_w g}\frac{\partial}{\partial y}\left(k_w \frac{\partial u_w}{\partial y}\right) + \left(m_2^w - \frac{m_2^s \cdot m_1^w}{m_1^s}\right)\frac{\partial u_w}{\partial t} \tag{16.89}$$

Equations 16.88 and 16.89 are the governing equations that can be used to simulate the one-dimensional swelling process (Shuai and Fredlund, 1997).

16.10.2 Illustration and Verification of Numerical Swelling Pressure Model

The one-dimensional swelling process can be solved using Eqs. 16.88 and 16.89. These equations were previously used to simulate one-dimensional swelling and consolidation. Different boundary conditions need to be applied when simulating the processes associated with *free-swell* oedometer tests and *constant-volume* oedometer tests. Laboratory tests on compacted Regina clay are used to illustrate the simulation of the swelling pressure tests. Specimens of Regina clay prepared with a molding water content of 26% and an initial void ratio of 0.96 were used for the numerical simulations and verification of the theoretical formulation.

Figure 16.57 shows the measured and computed deflection versus time curves for two 100-mm-high specimens and two 20-mm-high specimens subjected to free-swell oedometer test conditions. A comparison between the computed and measured matric suction profiles for the 100-mm-high specimen subjected to free-swell oedometer test conditions is presented in Fig. 16.58. Matric suctions were measured by placing Whatman No. 42 filter papers between the soil layers of the specimen and measuring the water content of the filter papers after the test. In general, the computed values for matric suction are in good agreement with the measured values. There are some differences between the computed and measured matric suction profile noted during the middle stage of the test (i.e., $t = 5300$ min). There are also some differences in the matric suctions measured and predicted in the upper portion of the specimen; however, the measured results are reasonably close to the computed matric suctions.

Figure 16.59 shows the measured and computed vertical stress versus elapsed time curves for two 100-mm-high specimens and two 20-mm-high specimens subjected to constant-volume oedometer test conditions. The measured and computed profiles of matric suction from within the soil specimens are presented in Fig. 16.60. The results show good agreement between the measured and computed matric suction values.

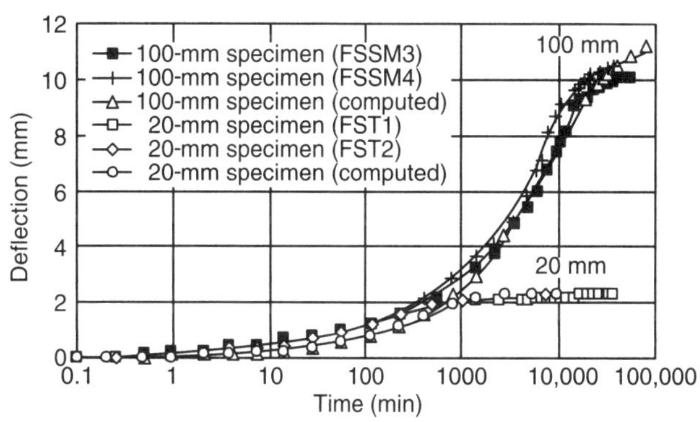

Figure 16.57 Computed and measured deflection versus time curves for free-swell oedometer tests.

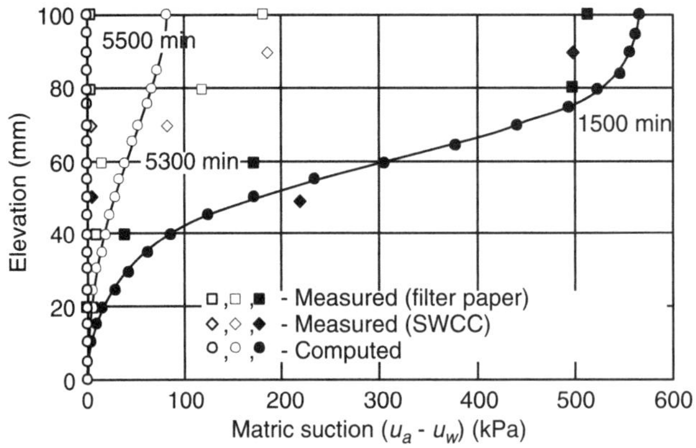

Figure 16.58 Computed and measured matric suction profiles for free-swell oedometer test.

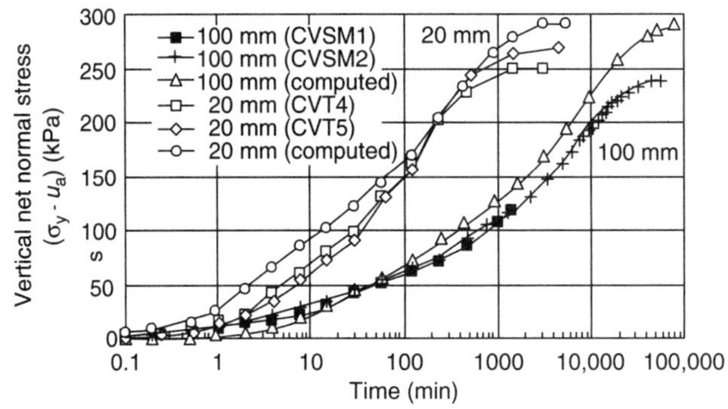

Figure 16.59 Computed and measured vertical normal stress–time curves for constant-volume oedometer tests.

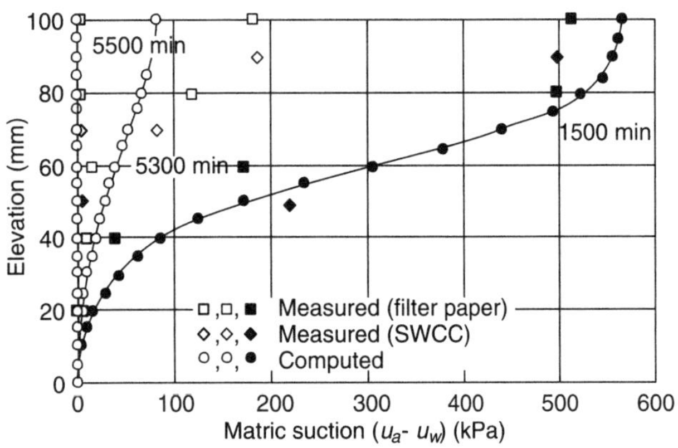

Figure 16.60 Computed and measured matric suction profiles for constant-volume oedometer test.

16.10.3 Application of Numerical Swelling Pressure Model

The proposed swelling model can be used to perform a parametric study of factors which may influence the swelling process. The influence of the coefficient of permeability on the swelling process is shown in Fig. 16.61. Each curve was calculated using a different value for the saturated coefficient of permeability k_s while the other soil moduli and the unsaturated permeability function variables remained unchanged. The saturated coefficients of permeability were set to 4.0×10^{-10}, 4.0×10^{-11}, and 4.0×10^{-12} m/s. The height of the specimen is 20 mm. The computed curves presented in Fig. 16.61 show that the permeability of the soil strongly influences the rate of swelling. The highest saturated coefficient of permeability shows that it requires about 450 min for the swelling pressure to be developed. In contrast, a saturated coefficient of permeability of 4.0×10^{-12} m/s shows that 42 days is required for the development of the swelling pressure.

The effect of the drainage distance on the rate of development of the swelling pressure is shown in Fig. 16.62. The specimen that was only allowed access to water from the bottom of the soil specimen took four times longer to develop its swelling pressure than the specimen that had access to water from the top and bottom sides of the specimen. Similar observations were made by Mesri et al. (1994) for Taylor shale from Texas. It was assumed that there were no difficulties associated with the confinement of air bubbles within the soil specimen when wetting occurred from both the top and bottom of the specimen.

The free-swell model can also be used in a parametric manner to study the influence of factors associated with the test procedure. The influence of steady-state evaporation

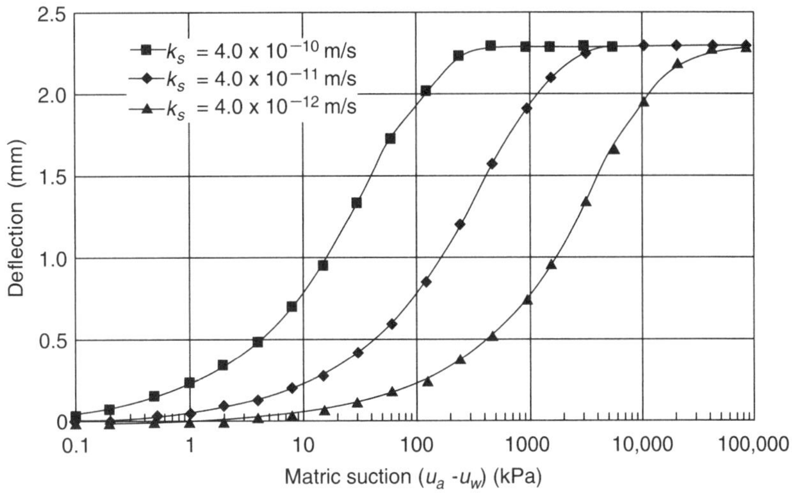

Figure 16.61 Computed swelling rates in free-swell oedometer test using different saturated coefficients of permeability.

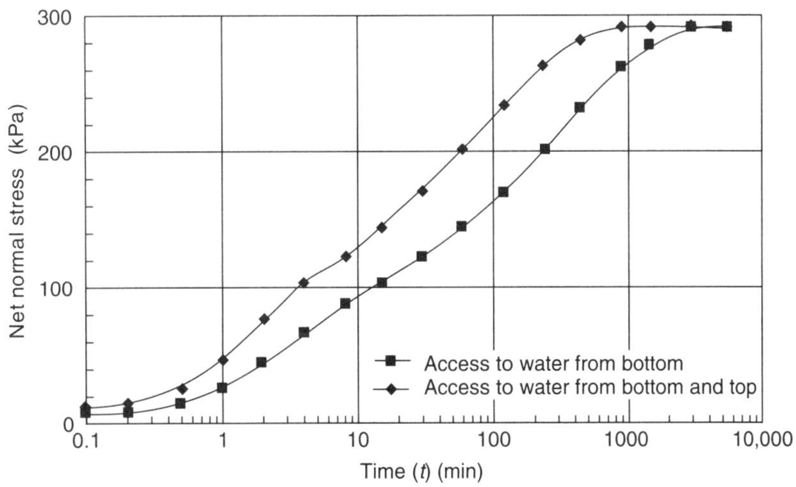

Figure 16.62 Time rate for development of swelling pressure depending on length of drainage path.

from the surface of a soil specimen placed in an oedometer was investigated using the proposed free-swell model. A 20-mm-thick soil specimen was assumed to have a surface water loss due to evaporation at a rate of 2×10^{-8} m/s (i.e., 1.72 mm/day). The computed deflection versus elapsed time curves presented in Fig. 16.63 show that the evaporation of water from the surface of the soil significantly reduced the total heave and the rate of swelling in a free-swell test.

The computed matric suction profiles are presented in Fig. 16.64. The initial matric suction throughout the soil specimen was 600 kPa. The computed results show that matric suction increases due to moisture loss due to evaporation during early stages of the test. The volume of the soil specimen decreases in the upper part of the specimen. The matric suction at the surface reaches 2600 kPa after 120 min of swelling and then decreases. The matric suction at the surface of the soil specimen at the end of the test has decreased to about 15 kPa.

16.11 RHEOLOGICAL MODEL FOR UNSATURATED SOILS

Rheological models can serve an important role in visualizing the physical processes associated with consolidation and swelling. The Kelvin model, for example, proved to be useful in communicating the time-dependent response of a saturated soil to an applied load. The instantaneous and time-dependent processes of an unsaturated soil are more complex than those of a saturated soil and it is important to be able to visualize unsaturated soil behavior. The proposed rheological model for an unsaturated soil might never be used to solve practical geotechnical engineering problems;

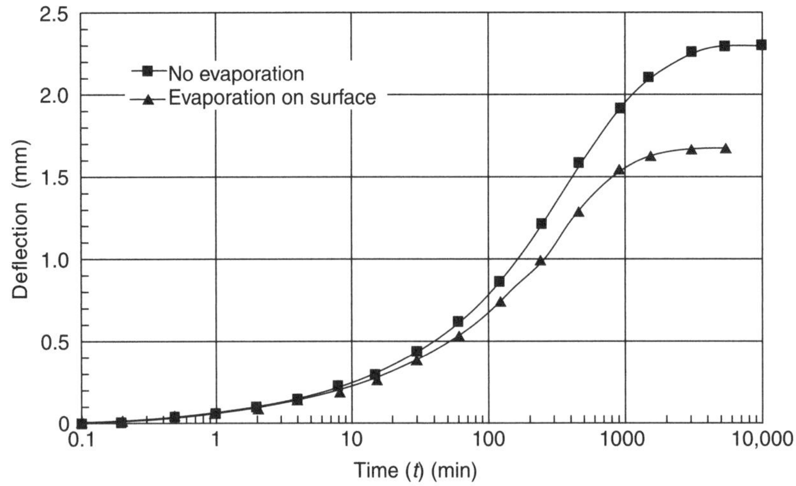

Figure 16.63 Computed deflection versus time for free-swell oedometer test with and without evaporation from surface of soil specimen.

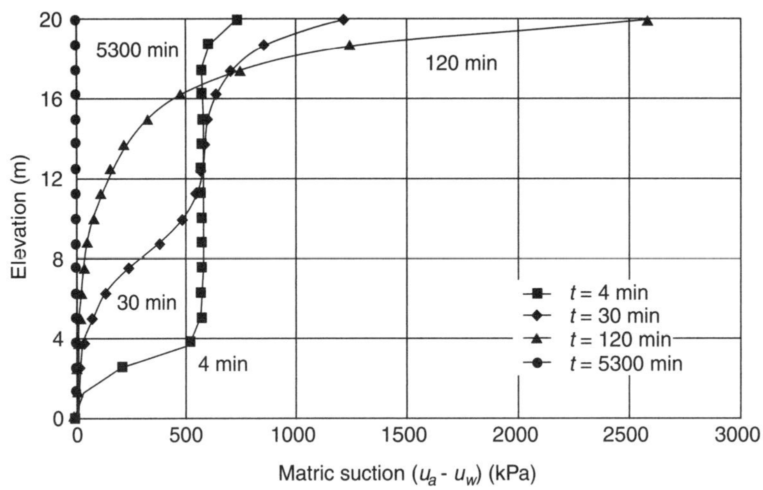

Figure 16.64 Computed matric suction profiles for free-swell oedometer test with evaporation from surface of specimen.

however, the rheological models can serve a role as a visualization tool.

It was shown in the previous chapter that two linear rheological models connected in series can be used to simulate instantaneous and time-dependent processes in an unsaturated soil. One linear model simulates the solid phase combined with the water phase while a second linear model simulates the solid phase combined with the air phase. The two rheological models can be connected in series to simulate the time-dependent behavior (i.e., consolidation and swelling) of a layer of unsaturated soil.

The total stress applied to the two linear rheological models connected in series acts equally on both elements and represents the role of the two independent stress state variables, $\sigma - u_w$ and $\sigma - u_a$. The stress in the *Maxwell* and *Hookean* portions of each linear element must always add to the total stress applied. This condition satisfies the conservative energy requirement (i.e., $\sigma = \sigma_H + \sigma_M$, where σ_M = stress on the Maxwell portion of the linear element, σ_H = stress on the Hookean portion of the linear element which represents either the pore-water pressure u_w or the pore-air pressure u_a and σ = total stress applied to the linear element).

The total strain ε is equal to the sum of the strain in the linear element associated with the $\sigma - u_w$ stress state variable ε_w and the strain in the linear element associated with the $\sigma - u_a$ stress state variable ε_a (i.e., $\varepsilon = \varepsilon_w + \varepsilon_a$). The Hookean portion and the Maxwell portion of each linear rheological element must always deform equal amounts

16.11.1 Constitutive Relations for Rheological Model

Let us first consider the linear rheological element associated with the stress state variable $\sigma - u_w$. The total stress applied can be written in terms of the stress in the Hookean portion plus the stress in the Maxwell portion of the model:

$$\sigma = \sigma_H + \eta_w \left(\dot{\varepsilon}_w - \dot{\varepsilon}_c\right) \tag{16.90}$$

where:

ε_c = strain in the Hookean spring of the Maxwell portion of the model which is related to the compressibility of water,
ε_w = strain in the dashpot due to the flow of water,
$\dot{\varepsilon}_w, \dot{\varepsilon}_c$ = derivatives of ε_w and ε_c, respectively, and
η_w = dashpot constant for the water phase.

Constitutive relations can be written for the components of the rheological model.

$$\sigma_H = \kappa_w \varepsilon_w \tag{16.91}$$

$$\sigma_M = c_w \varepsilon_c \tag{16.92}$$

where:

κ_w = Hookean spring constant related to the compressibility of the soil with respect to a change in $\sigma - u_w$ and
c_w = Hookean spring constant related to the bulk compressibility of water.

Equation 16.90 can be rearranged to give a rate of strain condition for the linear rheological element associated with $\sigma - u_w$:

$$\dot{\varepsilon} = \frac{\dot{\sigma}_M}{c_w} = \frac{\dot{\sigma} - \dot{\sigma}_H}{c_w} = \frac{0 - \dot{\sigma}_H}{c_w} = -\frac{\kappa_w}{c_w}\dot{\varepsilon}_w \tag{16.93}$$

A similar rate of strain condition can be written for the linear rheological element associated with $\sigma - u_a$.

16.11.2 Differential Equations for linear Rheological Elements of Single Layer

The stress-strain equation for the linear rheological element associated with $\sigma - u_w$ can be written as

$$\varepsilon_w = \frac{\sigma}{\kappa_w + c_w}\left\{1 + \frac{c_w}{\kappa_w}\left[1 - \exp\left(-\frac{\kappa_w c_w}{\eta_w\left(\kappa_w + c_w\right)}t\right)\right]\right\} \tag{16.94}$$

The stress-strain equation for the linear rheological element associated with $\sigma - u_a$ can be written as

$$\varepsilon_a = \frac{\sigma}{\kappa_a + c_a}\left\{1 + \frac{c_a}{\kappa_a}\left[1 - \exp\left(-\frac{\kappa_a c_a}{\eta_a\left(\kappa_a + c_a\right)}t\right)\right]\right\} \tag{16.95}$$

16.11.3 Mathematical Model for Multilayer System

A transient flow process such as consolidation takes place as a result of total load (or stress) being applied to the soil. In an unsaturated soil with one-way drainage through the top, both water and air will move upward from a lower layer toward an uppermost, lower pressure boundary. The water and the air phases are assumed to be continuous throughout the soil. This process is illustrated using the three-layer system shown in Fig. 16.65.

The linear rheological elements associated with the stress state variables $\sigma - u_w$ and $\sigma - u_a$ in the n-layer strata can be separated into two parts. An equivalency for the linear rheological elements is shown in Fig. 16.66. For simplicity, the compressibility of the air phase and the permeability of the water phase are assumed to be constant values.

The compressibility of water and air is simulated using Hookean springs. In both the water and air phases, only the bottom layer is truly a linear rheological element. The

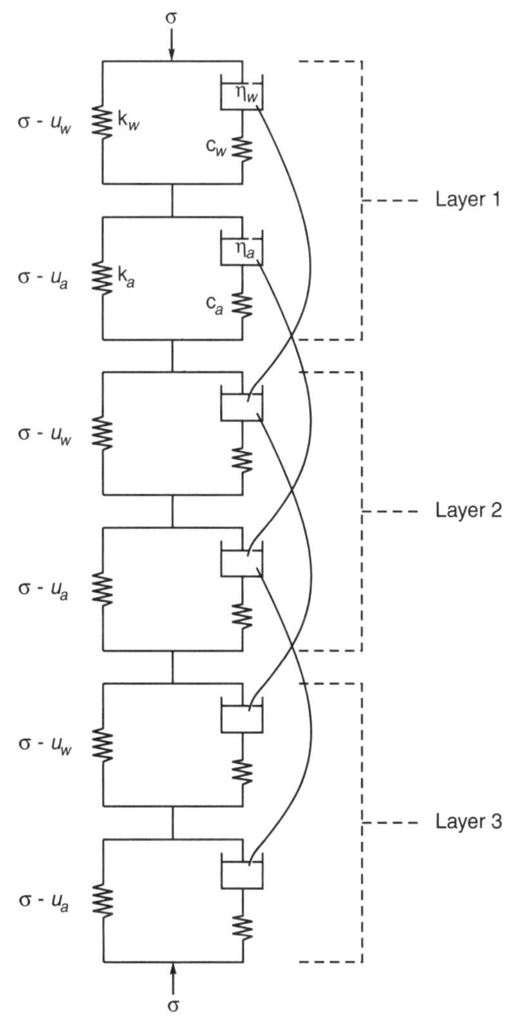

Figure 16.65 Three-layer rheological system demonstrating one-dimensional consolidation process in unsaturated soil.

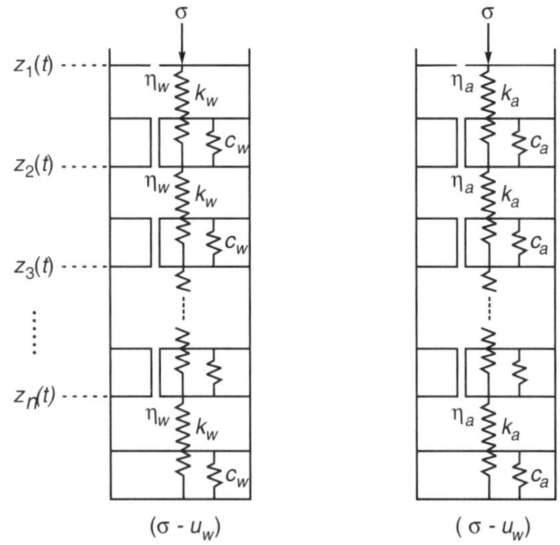

Figure 16.66 Use of linear rheological models to simulate effect of $\sigma - u_w$ and $\sigma - u_a$ stress state variables.

other layers are a slight modification of a linear rheological element in that the pressure in one piston is fed into the adjacent piston in a series manner. The water phase differential equations can be written as

$$\dot{\varepsilon}_i = -\frac{\kappa_w}{c_w}\left(\dot{z}_i - \dot{z}_{i+1}\right) \quad i = 1, 2, \ldots, n-1 \tag{16.96}$$

$$\dot{\varepsilon}_n = -\frac{\kappa_w}{c_w}\left(\dot{z}_n - 0\right) \tag{16.97}$$

where:

$z_1(t), z_2(t), \ldots, z_n(t)$ = positions of the n pistons, respectively, and
$\varepsilon_1(t), \varepsilon_2(t), \ldots, \varepsilon_n(t)$ = strains of the n pistons, respectively.

In the first layer of the linear rheological element associated with the stress state variable $\sigma - u_w$, the total stress σ is divided between the pore-water pressure and the stress acting in the spring representing the soil structure. That is,

$$\sigma = \kappa_w\left(z_1 - z_2\right) + \eta_w\left(\dot{z}_1 - \dot{\varepsilon}_1 - \dot{\varepsilon}_2 - \cdots - \dot{\varepsilon}_n\right) \tag{16.98}$$

Substituting Eqs. 16.96 and 16.97 into the conservation equation 16.98 gives

$$\sigma = \kappa_w\left(z_1 - z_2\right) + \eta_w \frac{c_w + \kappa_w}{c_w}\dot{z}_1 \tag{16.99}$$

The total stress σ also acts on the second piston associated with the linear rheological element for the stress state variable ($\sigma - u_w$) of the second layer. However, the Hookean stress $\kappa_w(z_1 - z_2)$ developed in the first layer is the only stress which causes the piston in the element of the second layer to move downward. This stress is again divided between the spring stress and pore-water pressure increment in the second layer. The rates of strain for the second layer can be rewritten in terms of the water phase dashpot constant and the spring constants corresponding to the water phase and the soil structure:

$$\kappa_w\left(z_1 - z_2\right) = \kappa_w\left(z_2 - z_3\right) + \eta_w \frac{c_w + \kappa_w}{c_w}\dot{z}_2 \tag{16.100}$$

Similar equations can be written for the third and subsequent layers. For the n-th layer,

$$\kappa_w\left(z_{n-1} - z_n\right) = \kappa_w z_n + \eta_w \frac{c_w + \kappa_w}{c_w}\dot{z}_n \tag{16.101}$$

Equations 16.99–16.101 represent a system of linear differential equations for the linear rheological element associated with the stress state variable $\sigma - u_w$ of the unsaturated soil. The system of linear differential equations for the response of the linear rheological element associated with the stress state variable $\sigma - u_w$ can be written in matrix form as

$$\dot{z} = Az + b \qquad z(0) = z^0 \tag{16.102}$$

where

$$A = \frac{\kappa_w c_w}{\eta_w\left(\kappa_w + c_w\right)}\begin{bmatrix} -1 & 1 & 0 & 0 & \cdots & 0 \\ 1 & -2 & 1 & 0 & \cdots & 0 \\ \cdots & \cdots & \cdots & \cdots & \cdots & \cdots \\ \cdots & \cdots & \cdots & \cdots & \cdots & \cdots \\ 0 & \cdots & \cdots & 1 & -2 & 1 \\ 0 & \cdots & \cdots & 0 & 1 & -2 \end{bmatrix}$$

$$b = \frac{\sigma c_w}{\eta_w\left(\kappa_w + c_w\right)}\begin{bmatrix} 1 \\ 0 \\ \cdot \\ \cdot \\ \cdot \\ 0 \end{bmatrix}$$

$$z^0 = \frac{\sigma}{\kappa_w + c_w}\begin{bmatrix} n \\ n-1 \\ \cdot \\ \cdot \\ \cdot \\ 1 \end{bmatrix}$$

The nonzero initial condition (i.e., $z(0) = z^0$) is due to the compressibility of water. If water is assumed to be incompressible, the c_w term goes to infinity and z_0 becomes a zero vector and the system of linear differential equations reduces to the equation for the Kelvin model for saturated soils (Xing et al., 1995).

The system of linear differential equations for the linear rheological element associated with the stress state variable $\sigma - u_a$ has the same form as Eq. 16.102 for the stress state variable $\sigma - u_w$ with the subscript w replaced by the subscript a.

The total strain for the unsaturated soil is the summation of the solutions for the linear rheological elements associated with the stress state variables $\sigma - u_w$ and $\sigma - u_a$. The total strain $\varepsilon(t)$ is given by

$$\varepsilon(t) = \frac{z_{w1}(t) + z_{a1}(t)}{z_{w1}(\infty) + z_{a1}(\infty)} \tag{16.103}$$

where:

$z_w(t) = (z_{w1}(t), \ldots, z_{wn}(t))$ = solution for the $\sigma - u_w$ portion,
$z_a(t) = (z_{a1}(t), \ldots, z_{an}(t))$ = solution for the $\sigma - u_a$ portion, and
$z_{wI}(\infty)$ and $z_{a1}(\infty)$ = limits of $z_{wI}(t)$ and $z_{a1}(t)$, respectively, as time goes to infinity.

The pore-water pressure $u_{wi}(t)$ and the pore-air pressure $u_{ai}(t)$ in the ith layer are respectively given by the following equations:

$$u_{wi}(t) = \sigma - \kappa_w[z_{wi}(t) - z_{w(i+1)}(t)] \tag{16.104}$$

$$u_{ai}(t) = \sigma - \kappa_a[z_{ai}(t) - z_{a(i+1)}(t)] \tag{16.105}$$

16.11.4 Closed-Form Solution for Rheological Model

The closed-form solution for Eq. 16.90 can be expressed as

$$z(t) = \begin{bmatrix} z_1(t) \\ z_2(t) \\ \cdot \\ \cdot \\ \cdot \\ z_n(t) \end{bmatrix} = e^{At} z^0 + \int_0^t e^{A(t-s)} b\, ds \tag{16.106}$$

The coefficient matrix A of the initial-value problem given by Eq. 16.102 is tridiagonal and satisfies the condition of a Jordan matrix. According to the Sturm theorem (Dickson 1939), A has n real distinct eigenvalues. The n corresponding eigenvectors form the fundamental matrix for the system of differential equations (i.e., Eq. 16.102). Matrix theory reveals that all the eigenvalues are negative. The solution of Eq. 16.102 has the following form:

$$z_1(t) = \alpha_0 + \sum_{i=1}^{n} \alpha_i e^{\lambda_i t} \tag{16.107}$$

where:

α_i = constant $(i = 0, 1, \ldots, n)$
λ_i = eigenvalue of $\mathbf{A}$ $(\lambda_i < 0; i = 1, 2, \ldots, n)$

Equation 16.107 implies that the consolidation process is primarily governed by the eigenvalues of A. The n eigenvalues of A are uniquely determined by the soil parameters since only κ_w and η_w appear in the coefficient matrix A. The external load or stress σ only affects the magnitude of the strain.

16.11.5 Numerical Solution for Proposed Rheological Model for Unsaturated Soils

A numerical solution was formulated for solving the two systems of linear differential equations using the Runge-Kutta method. The unsaturated soil simulated using the rheological model does not have to be homogeneous. Each layer can have its own soil parameters. A numerical solution is preferred if the soil is divided into a large number of layers.

16.11.6 Comparison of Rheological Model with Experimental Data

The rheological model was compared to experimental results measured on compacted kaolin. The compacted kaolin had a dry unit weight of 13.19 kN/m^3, void ratio of 1.07, water content of 34.3%, and degree of saturation of 78.9% (Fredlund and Rahardjo, 1993a). The soil specimen was simulated using a 10-layer rheological model with approximate soil parameters obtained from a consolidation test where the total stress was increased from 101.1 to 202.2 kPa.

The water phase constitutive relation can be written as

$$\frac{dV_w/V_0}{d(\sigma - u_a)} = m_1^w + m_3^w \frac{d(\sigma - u_w)}{d(\sigma - u_a)} \tag{16.108}$$

where:

dV_w/V_0 = change in the volume of water in the soil specimen with respect to the initial volume of the soil specimen,
m_1^w = coefficient of water volume change with respect to a change in net normal stress, and
m_3^w = coefficient of water volume change with respect to a change in the stress state variable $\sigma - u_a$.

The following relationships for the water phase rheological model can be inferred:

$$\eta_w = \frac{\gamma_w}{k_w}\Delta z \quad \kappa_w = \frac{1}{\Delta z}\left(\frac{1}{m_1^w + m_3^w}\right) \tag{16.109}$$

where:

k_w = coefficient of permeability with respect to water,
γ_w = unit weight of water, and
Δz = thickness of each of the layers.

Similarly, the air phase relationships are

$$\eta_a = \frac{\gamma_a}{k_a}\Delta z \quad \kappa_a = \frac{1}{\Delta z}\left(\frac{1}{m_1^a + m_3^a}\right) \tag{16.110}$$

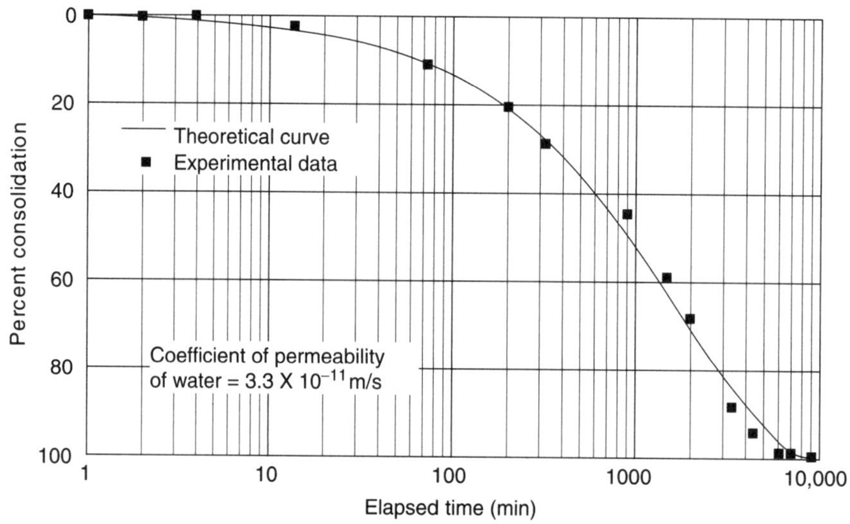

Figure 16.67 Comparison of measured data to rheological simulation for percent consolidation versus logarithm of time plot for compacted kaolin (Fredlund and Rahardjo, 1993a).

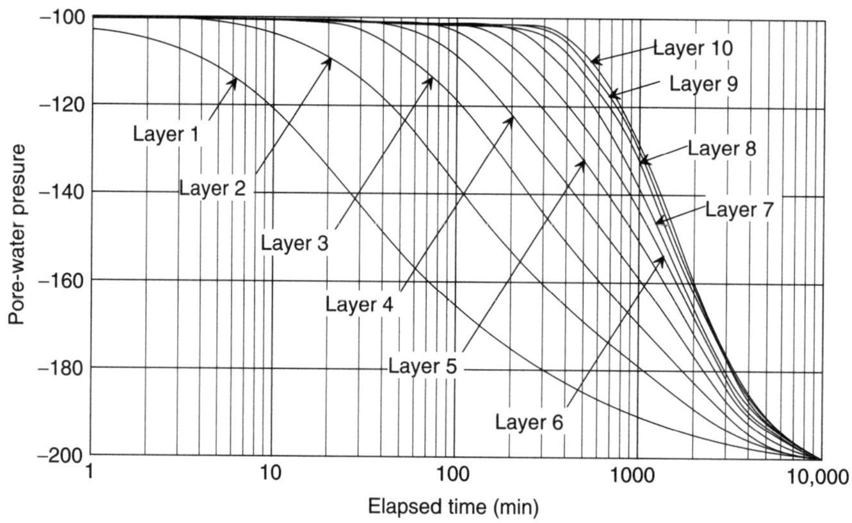

Figure 16.68 Calculated pore-water pressures in kaolin using linear rheological model.

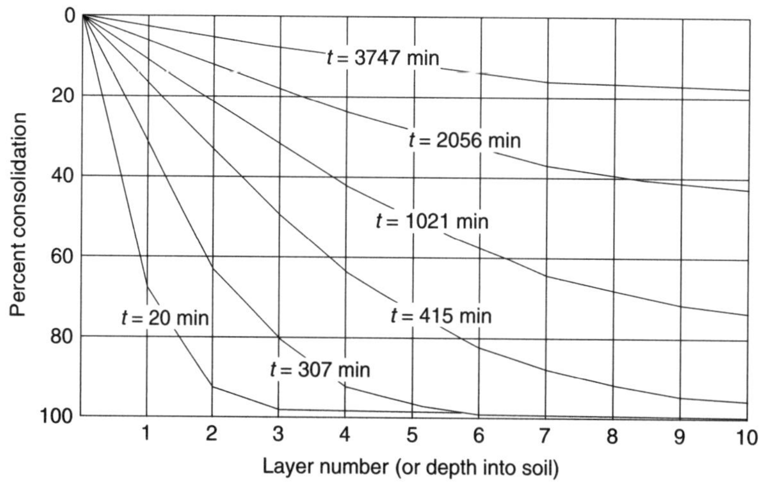

Figure 16.69 Percent consolidation for different layers in compacted kaolin calculated using linear rheological model.

where:

k_a = coefficient of permeability with respect to air,
γ_a = unit weight of air,
m_1^a = coefficient of air volume change with respect to a change in net normal stress, and
m_3^a = coefficient of air volume change with respect to a change in the stress state variable $\sigma - u_a$.

The plot of percent consolidation versus the logarithm of time is shown in Fig. 16.67. The pore-water pressures in each layer are presented as a function of the logarithm of time in Fig. 16.68. The percent consolidation versus the depth into the soil is shown in Fig. 16.69. The theoretical curves computed using the rheological model (i.e., Eq. 16.102) are in close agreement with the experimental data.

The rheological model is not meant to be used for solving practical engineering problems. However, it serves an important role in assisting the geotechnical engineer in visualizing anticipated unsaturated soil behavior. In this sense, the presentation of the rheological simulation model is a good way to close off the book and open the way for the visualization of the many unsaturated soil mechanics problems still waiting to be solved.

APPENDIX

Units and Symbols

The universal and consistent system of units The International System of Units (SI) is used throughout the book. This version of the metric system is described in the American Society for Testing and Materials (ASTM) (1966) *Metric Practice Guide* and in ASTM Standard for Metric Practice. The SI system, which was formally recognized by the General Conference on Weights and Measures, Rome, in 1960, has now been adopted in many countries (e.g., Great Britain, Australia, New Zealand, Canada, and most European countries).

The SI system is based on seven basic units to cover the entire spectrum of science and engineering. The units are comprehensive and coherent. These basic units are length (meter), mass (kilogram), time (second), electric current (ampere), thermodynamic temperature (Kelvin), luminous intensity (candela), and amount of a substance (mole). All units are independent of the nature of the physical process being considered. Only the first three units, as well as temperature, are used in this book.

Other physical quantities can be derived from the basic units. For example, the unit of force is a newton. The unit of force is derived from Newton's second law, $F = Ma$, where mass M is in kilograms and acceleration a is in meters per square second. The correct unit to express the weight of an object is the newton, and weight is a function of gravitational acceleration. It is recommended that a mass designation be used in engineering practice whenever possible.

Confusion is generated when the word "weight" is used to refer to obtaining the "mass" of an object. A laboratory balance, for example, is used to compare two masses with the objective of obtaining an unknown mass. This process is commonly referred to as "weighing the object."

The unit weight of a soil is another derived unit used in engineering practice. It is preferable, however, to use the density of a soil multiplied by acceleration due gravity. Acceleration due to gravity can generally be taken as 9.807 m/s^2.

Table A.1 contains a list of derived units of interest to geotechnical engineering. Pressure or stress has derived units of pascals or newtons per square meter. This unit is relatively small and is often used in conjunction with a prefix. Table A.2 lists prefixes which can be used to indicate multiples and submultiples of basic or derived units. The prefixes are used to render numbers lying within the range of 10^{-9}–10^{9}.

In soil science, the picofarad has been used to designate negative pore-water pressures (Glossary of Soil Science Terms, 1975). The picofarad was defined as the logarithm of the negative pore-water pressure expressed as the height of a column of water in centimeters. This unit is now considered obsolete (Johnson, 1981) and is not used in this book. Rather, negative pore-water pressure is expressed in standard SI units (e.g., kilopascals).

There is often need to convert variables to other units although the book is written using the SI designation of units. Table A.3 contains some conversion factors useful in making conversions between the SI (metric) system and the British Engineering system of units.

The symbols used throughout this book are, for the most part, consistent with those used in soil and rock mechanics. The extensive list of symbols published by the International Society for Soil Mechanics and Foundation Engineering (ISSMFE, 1977) has been used as a guide. However, it has been necessary to extend some of the symbols for unsaturated soil mechanics. For example, the soil has two fluid phases, and it becomes necessary to carry a subscript to differentiate the air and water phases (e.g., u_a and u_w).

All symbols are defined upon their first usage in the book. Some of the definitions are repeated from one chapter to another. Some of the main symbols used in the book are summarized in Table A.4.

Table A.1 Derived Units Common to Geotechnical Engineering

Physical Quantity	SI Unit	Symbol
Acceleration	Meters per square second	m/s^2
Area	Square meter	m^2
Density	Kilograms per cubic meter	kg/m^3
Discharge, flux	Cubic meters per second	m^3/s
Force	Newton	N
Length	Millimeter	mm
Length	Meter	m
Mass	Kilogram	kg
Moment or torque	Newton meter	Nm
Permeability	Meters per second	m/s
Power	Watt	W
Pressure	Newtons per square meter (i.e., pascal)	Pa
Stress	Newtons per square meter (i.e., pascal)	Pa
Temperature	Kelvin	K
Time	Second	s
Unit weight	Newtons per cubic meter	N/m^3
Velocity	Meters per second	m/s
Viscosity (dynamic)	Newton second per square meter	Ns/m^2
Volume	Cubic meter	m^3
Volume	Liter	L
Weight	Newton	N
Work (energy)	Joule	J

Table A.2 Prefixes for Basic and Derived Units

Factor	Prefix	Symbol
10^9	Giga	G
10^6	Mega	M
10^3	Kilo	k
10^2	Hecto	h
10^1	Deka	da
10^{-1}	Deci	d
10^{-2}	Centi	c
10^{-3}	Milli	m
10^{-6}	Micro	μ
10^{-9}	Nano	n

Table A.3 Conversion Factors for British Engineering and SI System

British System	SI
Length	
1 inch, in.	25.4 mm = 0.025 m
1 foot, ft	0.3048 m
1 yard, yd	0.9144 m
Mass	
1 pound mass,	
1 bm (avoirdupois)	0.4536 kg
1 slug	
(1 lb-force/ft^2)	14.59 kg
Force	
1 lb-force	4.448 N
Stress and pressure	
1 psi (lb-force/in.^{2})	6.895×10^3 Pa or 6.895 kPa
1 atmosphere	1.013×10^5 Pa or 101.3 kPa
1 bar	1×10^5 Pa or 100 kPa
Density	
1 lb-mass/ft^3	16.018 kg/m^3

Table A.4 List of Symbols

Description	Symbol	Unit
Absolute air pressure	$\bar{u}_a$	kPa
Absolute temperature	T_K	K
Actual evaporation	AE	mm/day
Air-entry bubbling pressure	u_b	kPa
Air-entry suction	ψ_{aev}	kPa
Air pressure	u_a	kPa
Angle indicating rate of shear strength change with respect to suction	ϕ^b	degree
Atmospheric pressure	$\bar{u}_{\text{atm}}$	kPa
Average molecular weight of air	ω_a	g/mol
Axial (vertical) strain	ε_a	%
Bishop's average soil skeleton stress	σ_{ij}^*	kPa
Bishop stress parameter	χ	decimal
Capillary height	h_c	m
Capillary meniscus curvature	R_s	m
Coefficient of active earth pressure	K_{a}	—
Coefficient of air permeability	k_a	m/s
Coefficient of consolidation for air phase	c_v^a	m^2/s
Coefficient of consolidation for water phase	c_v^w	m^2/s
Coefficient of diffusion	D	m^2/s
Coefficient of earth pressure at rest	K_0	—
Coefficient of passive earth pressure	K_p	—
Coefficient of permeability at saturation	k_s	m/s
Coefficient of water permeability	k_w	m/s
Coefficient of air volume change with respect to net normal stress	m_1^a	1/kPa
Coefficient of air volume change with respect to soil suction	m_2^a	1/kPa
Coefficient of soil structure volume change with respect to net normal stress	m_1^s	1/kPa
Coefficient of soil structure volume change with respect to soil suction	m_2^s	1/kPa
Coefficient of water volume change with respect to net normal stress	m_1^w	1/kPa
Coefficient of water volume change with respect to soil suction	m_2^w	1/kPa
Coefficient of volume change of saturated soil	m_v	1/kPa
Compressibility of water	c_w	1/kPa
Contact angle	α	degrees
Degree of saturation	S	%
Density of air	ρ_a	kg/m^3
Density of any fluid	ρ_f	kg/m^3
Density of water	ρ_w	kg/m^3
Density of soil particles	ρ_s	kg/m^3
Density of water with salts	ρ_{ws}	kg/m^3
Deviator stress	$\sigma_1 - \sigma_3$	kPa
Dew-point temperature	t_d	°C
Diameter of 10% passing	D_{10}	mm
Diameter of 50% passing	D_{50}	mm

Table A.4 *(Continued)*

Description	Symbol	Unit
Diffusion coefficient of vapor	D^{v*}	m^2/s
Dry density	ρ_d	kg/m^3
Effective stress	$\sigma' = \sigma - u_w$	kPa
Effective angle of internal friction	ϕ'	degree
Effective cohesion	c'	kPa
Geometric mean parameter	x_g	m
Geometric standard deviation	σ_g	m
Gravitational acceleration	g	m/s^2
Gravitational energy	E_g	
Hydraulic conductivity of air	k_a	m/s
Hydraulic conductivity of water	k_w	m/s
Hydraulic conductivity (x-direction)	k_{wx}	m/s
Hydraulic conductivity (y-direction)	k_{wy}	m/s
Hydraulic head	h	m
Hydraulic head gradient	i_w	—
Hydraulic or total head	h_w	m
Intrinsic permeability	K	m^2
Isothermal compressibility of air	c_a	1/kPa
Kronecker delta or substitution tensor	δ_{ij}	kPa
Latent heat of fusion of water	L	J/m^3
Latent heat of vaporization of water	L_v	J/kg
Liquid limit	LL	%
Mass of air	M_a	kg
Mass of dissolved air in water	M_d	kg
Mass of solids	M_s	kg
Matric suction	$u_a - u_w$	kPa
Mass of water	M_w	kg
Major principal stress	σ_1	Pa
Minor principal stress	σ_3	kPa
Molecular mass of air	ω_a	kg/kmol
Natural number	$e = 2.71828$	
Net major normal stress	$\sigma_1 - u_a$	kPa
Net normal stress	$\sigma - u_a$	kPa
Net minor normal stress	$\sigma_3 - u_a$	kPa
Osmotic suction	π	kPa
Parameter on grain size for fitting fine portion	m_{gr}	—
Parameter on grain size for steepest slope	n_{gr}	—
Parameter on grain size inflection point	a_{gr}	mm
Particle diameter	d	mm
Particle diameter for minimum size	d_m	mm
Particle diameter for residual portion	d_r	mm
Percentage of particles passing by mass	$P_p(d)$	%
Plastic limit	PL	%
Plasticity index, PI	I_p	%
Poisson ratio	μ	—
Potential evaporation	PE	mm/day
Potential transpiration	PT	mm/day
Pore-air pressure	u_a	kPa

(continued overleaf)

Table A.4 (*Continued*)

Description	Symbol	Unit
Pore-air pressure parameter (secant)	B'_{ah}	—
Pore-air pressure parameter (tangent)	B_a	—
Pore-water pressure	u_w	kPa
Pore-water pressure parameter (secant)	B'_{wh}	—
Pore-water pressure parameter (tangent)	B_w	—
Porosity	n	%
Porosity with respect to water phase	n_w	%
Porosity with respect to air phase	n_a	%
Preconsolidation pressure	σ_p	kPa
Radial strain	ε_r	%
Radiation, net	R_n	MJ/m^2/day
Radiation, short wave	R_c	mm/day
Radiation, total solar	R_a	MJ/m^2/day
Radius of capillary tube	r	m
Ratio of specific gas constants	δ	—
Relative air hydraulic conductivity	k_{ra}	—
Relative humidity, RH	h_r	%
Relative water hydraulic conductivity	k_{wr}	—
Salt content	s	g/kg
Saturated (water) hydraulic conductivity	k_s	m/s
Saturated water vapor pressure	u_{v0}^{air}	kPa
Shear strains	γ_{xy} , γ_{yz}, γ_{zx}	degrees
Shear strength fitting parameter	κ	—
Shear stress	τ	kPa
Shrinkage curve curvature	c_{sh}	—
Shrinkage curve minimum void ratio	a_{sh}	—
Shrinkage curve slope	b_{sh}	—
Soil compression index	C_c	—
Soil pore shape parameter	η	—
Soil suction	ψ	kPa
Soil suction at residual conditions	ψ_r	kPa
Soil suction on drying curve	ψ_d	kPa
Soil suction on wetting curve	ψ_w	kPa
Swelling index	C_s	—
Specific heat capacity	C_s	kJ/kg K
Specific gravity of solids	G_s	—
Specific volume	v_0	
Specific volume of air	v_{a0}	
Strain longitudinal in x-direction	ε_x	%
Strain longitudinal in y-direction	ε_y	%
Strain longitudinal in z-direction	ε_z	%
Surface tension of water	T_s	mN/m
Swelling pressure	P_s	kPa
Swelling pressure, corrected	P'_s	kPa
SWCC a_f fitting parameter (Fredlund and Xing)	a_f	kPa

Table A.4 (***Continued***)

Description	Symbol	Unit
SWCC n_f fitting parameter (Fredlund and Xing)	n_f	—
SWCC m_f fitting parameter (Fredlund and Xing)	m_f	—
SWCC correction function (Fredlund and Xing)	C_r	—
Temperature (Celsius)	T	°C
Temperature (Kelvin)	T_K	K
Thermal conductivity	λ_s	W/m K
Time	t	s
Total cohesion	c	kPa
Total density	ρ	kg/m^3
Total energy	E	J
Total hydraulic head	h_w	m
Total normal stress	σ	kPa
Total stress tensor	σ_{ij}	kPa
Total suction	ψ	kPa
Total unit weight	γ_t	kN/m^3
Undrained shear strength	c_u	kPa
Unfrozen volumetric water content	θ_u	decimal or %
Unit weight of water	γ_w	kN/m^3
Unit weight of soil	γ	kN/m^3
Universal gas constant	R	J/mol K
Vapor pressure	u_v^{air}	kPa
Velocity energy	E_v	
Viscosity (absolute of kinematic)	ν	m
Void ratio	e	N·s/m^2
Volume of air	V_a	m^3
Volume of dissolved air in water	V_d	m^3
Volume of soil element, total initial	V_0	m^3
Volume of solids	V_s	m^3
Volume of water	V_w	m^3
Volumetric air content	θ_a	decimal or %
Volumetric coefficient of solubility	h	—
Volumetric residual water content	θ_r	%
Volumetric specific heat of solids	ζ_s	J/(m^3 °C)
Volumetric specific heat of water	ζ_w	J/(m^3 °C)
Volumetric strain	ε_v	%
Volumetric water content	θ	decimal or %
Water content, gravimetric	w	%
Water content, residual gravimetric	w_r	%
Water content, residual volumetric	θ_r	decimal or %
Water content, saturated gravimetric	w_s	%
Water content, unfrozen	θ_u	%
Water content, volumetric	θ	decimal or %
Water pressure	u_w	kPa
Water vapor pressure	u_v	kPa
Universal gas constant	R	J/(mol K)
Young modulus	E	kPa

REFERENCES

Abelev, M. Y. 1975. Compacting loess soils in the USSR, *Geotechnique*, Vol. 25, No. 1, pp. 79–82.

Abu-El-Sha'r, W. Y., and Abriola, L. M. 1997. Experimental assessment of gas transport mechanisms in natural porous media: Parameter evaluation, *Water Resources Research*, Vol. 33, pp. 505–516.

Adams, B. A., Wulfsohn, D., and Fredlund, D. G. 1994. Unsaturated soil mechanics for agricultural conditions, ASAE Meeting Paper No. 94–1037, Kansas City, MO.

Adams, B. A., Wulfsohn, D., and Fredlund, D. G. 1996. Air volume change measurement in unsaturated soil testing using a digital pressure-volume controller, *Geotechnical Testing Journal, ASTM*, Vol. 19, No. 1, pp. 12–21.

Adamson, A.W., and Gast, A. P. 1997. *Physical Chemistry of Surfaces*, 6th ed., Wiley, New York.

Agus, S. S., and Schanz, T. 2005. Comparison of four methods for measuring total suction, *Vadose Zone Journal*, Vol. 4, pp. 1087–1095.

Agus, S. S., Leong, E. C., and Rahardjo, H. 2001. Soil-water characteristic curves of Singapore residual soils, *Journal of Geotechnical and Geological Engineering*, Vol. 19, Nos. 3–4, pp. 285–309.

Agus, S. S., Leong, E. C., and Rahardjo, H. 2005. Estimating permeability functions of Singapore residual soils, *Journal of Engineering Geology*, Vol. 78, Nos. 1–2, pp. 119–133.

Agus, S. S., Schanz, T., and Fredlund, D. G. 2010. Measurements of suction versus water content for bentonite-sand mixtures, *Canadian Geotechnical Journal*, Vol. 47, No. 5, pp. 583–594.

Ahmed, S., Lovell, C. W., and Diamond, S. 1974. Pore size and strength of compacted clay, *Journal of the Geotechnical Engineering Division, ASCE*, Vol. 100, pp. 407–425.

Ahuja, L. R., Naney, J. W., and Williams, R. D. 1985. Estimating soil water characteristics from simpler properties or limited data, *Soil Science Society of America Journal*, Vol. 49, No. 5, pp. 1100–1105.

Aitchison, G. D. 1957. The strength of quasi-saturated and unsaturated soils in relation to the pressure deficiency in the pore water, *Proceedings of the Fourth International Conference on Soil Mechanics and Foundation Engineering*, London, pp. 135–139.

Aitchison, G. D. 1961. Relationship of moisture and effective stress functions in unsaturated soils, paper presented at the *Pore Pressure and Suction in Soils Conference*, Butterworths, London, pp. 47–52.

Aitchison, G. D. 1964. Engineering concepts of moisture equilibria and moisture changes in soils. Statement of the Review Panel, In *Moisture Equilibria and Moisture Changes in Soils Beneath Covered Areas*, A Symposium in Print, Butterworths, Sydney, pp. 7–21.

Aitchison, G. D. 1965. Discussion in *Proceedings of the Sixth International Conference on Soil Mechanics and Foundation Engineering*, Montreal, QC, Vol. 3, pp. 318–321.

Aitchison, G. D. 1973a. Separate roles of site investigation, quantification of soil properties, and selection of operational environment in the determination of foundation design on expansive soils, *Proceedings of the Third Asian Regional Conference on Soil Mechanics and Foundation Engineering*, Haifa, Vol. 3, pp. 72–77.

Aitchison, G. D. 1973b. The quantitative description of the stress-deformation behaviour of expansive soils, *Proceedings of the Third International Conference on Expansive Soils*, Haifa, Vol. 2, pp. 79–82.

Aitchison, G. D., and Donald, I. B. 1956. Effective stresses in unsaturated soils, *Proceedings of the Second Australia–New Zealand Conference on Soil Mechanics*, Christchurch, New Zealand, pp. 192–199.

Aitchison, G. D., and Martin, R. 1973. A membrane oedometer for complex stress-path studies in expansive clays, *Proceedings of the Third International Conference on Expansive Soils*, Haifa, Vol. 2, pp. 83–88.

Aitchison, G. D., and Richards, B. G. 1965a. A broad scale study of moisture conditions in pavement subgrades throughout Australia—Factors in planning a regional study of moisture variation in pavement subgrades, in *Moisture Equilibria and Moisture Changes in Soils Beneath Covered Areas*, A Symposium in Print, Butterworths, Sydney, pp. 184–190.

Aitchison, G. D., and Richards, B. G. 1965b. A broad scale study of moisture conditions in pavement subgrades throughout Australia—Techniques adopted for the measurements of moisture variables, In *Moisture Equilibria and Moisture Changes in Soils Beneath Covered Areas*, A Symposium in Print, Butterworths, Sydney, pp. 191–204.

Aitchison, G. D., and Richards, B. G. 1965c. A broad scale study of moisture conditions in pavement subgrades throughout Australia—Field studies, in *Moisture Equilibria and Moisture*

Changes in Soils Beneath Covered Areas, A Symposium in Print, Butterworths, Sydney, pp. 205–225.

Aitchison, G. D., and Richards, B. G. 1965d. A broad scale study of moisture conditions in pavement subgrades throughout Australia—The selection of design values for soil suction equilibria and soil suction changes in pavement subgrades, in *Moisture Equilibria and Moisture Changes in Soils Beneath Covered Areas*, A Symposium in Print, Butterworths, Sydney, pp. 226–232.

Aitchison, G. D., and Woodburn, J. A. 1969. Soil suction in foundation design, *Proceedings of the Seventh International Conference on Soil Mechanics and Foundation Engineering*, Mexico City, Vol. 2, pp. 1–8.

Aitchison, G. D., Peter, P., and Martin, R. 1973. The instability indices I_{pm} and I_{ps} in expansive soils, *Proceedings of the Third International Conference on Expansive Soils*, Haifa, Vol. 2, pp. 101–104.

Aitkinson, J. H. 1993. An Introduction to the Mechanics of Soils and Foundations, McGraw-Hill, London.

Aitkinson, J. H. 2007. *The Mechanics of Soils and Foundations*, 2nd ed., Taylor and Francis, New York.

Aitkinson, J. H., and Bransky, P. L. 1997. *An Introduction to Critical State Soil Mechanics*, McGraw-Hill, London.

Akaike, H. 1974. A new look at the statistical model identification, *IEEE Transactions on Automatic Control*, Vol 19, No. 6, pp. 716–723.

Aldrich, Jr., H. P. 1956. Frost penetration below highway and airfield pavements, *Highway Research Board Bulletin*, Vol. 135, pp. 124–144.

Allen, R. G. 1997. Self-calibrating method for estimating solar radiation from air temperature, *Journal of Irrigation and Drainage Engineering*, Vol. 2, No. 2, pp. 56–67.

Allen, R. G., Pereira, L. S., Raes, D., and Smith, M. 1998. Crop evapotranspiration. Guidelines for computing crop water requirements, Paper No. 56, Food and Agriculture Organization of the United Nations (FAO) Irrigation and Drain, Rome, pp. 24–47.

Al-Khafaf, S., and Hanks, R. J. 1974. Evaluation of the filter paper method for estimating soil water potential, *Soil Science Journal*, Vol. 117, pp. 194–199.

Al-Mukhtar, M. 1995. Macroscopic behaviour and microstructural properties of a kaolinite clay under chemical and hydraulic state, *Proceeding of the First International Conference on Unsaturated Soils (UNSAT-95)*, Paris, pp. 3–9.

Alonso, E. E. 1993. Constitutive modelling of unsaturated soils, In *Unsaturated soils: Recent Developments and Applications, Civil Engineering European Courses*. Universitat Politècnica de Catalunya, Barcelona, Spain.

Alonso, E. E., Gens, A., and Hight, D. W. 1987. Special problem soils, *Proceedings of the Ninth European Conference on Soil Mechanics and Foundation Engineering*, Dublin, Vol. 3, pp. 1087–1146.

Alonso, E. E., Gens, A., and Josa, A. 1990. A constitutive model for partly saturated soils, *Geotechnique*, Vol. 40, No. 3, pp. 405–430.

Alonso, E. E., Gens, A., and Gehling, W. Y. Yuk. 1994. Elastoplastic model for unsaturated expansive soils, in *Numerical Methods in Geotechnical Engineering*, Balkema, Rotterdam, pp. 11–18.

Alonso, E. E., Lloret, A., Gens, A., and Yang, D. Q. 1995. Experimental behaviour of highly expansive double-structure clay, *Proceedings of the First International Conference on Unsaturated Soils (UNSAT-95)*, Paris, pp. 11–16.

Alvenas, G., and Jansson, P. E. 1997. Model for evaporation, moisture and temperature of bare soil: Calibration and sensitivity analysis, In *Agricultural and Forest Meteorology*, Elsevier Science, Vol. 88, pp. 47–56.

American Society for Testing and Materials (ASTM). 1966. *ASTM Metric Practice Guide*, 2nd ed., ASTM, Philadelphia, PA.

American Society for Testing and Materials (ASTM). 1981. Standard test method for capillary-moisture relationships for coarse and medium textured soils by porous-plate apparatus, ASTM D2325–68, In *1985 Annual Book of ASTM Standards*, Vol. 04.08, *Soil and Rock; Building Stones*, ASTM, Philadelphia, PA.

American Society for Testing and Materials (ASTM). 1985a. Standard test method for shrinkage factors of soils, ASTM D427, In *1985 Annual Book of ASTM Standards*, Vol. 04.08, *Soil and Rock; Building Stones*, Section 4, ASTM, Philadelphia, PA.

American Society for Testing and Materials (ASTM). 1985b. Standard test method for one-dimensional consolidation properties of soils, ASTM D2435, In *1985 Annual Book of ASTM Standards*, Vol. 04.08, *Soil and Rock; Building Stones*, Section 4, ASTM, Philadelphia, PA.

American Society for Testing and Materials (ASTM). 1986. Standard test method for one-dimensional swell or settlement potential of cohesive soils, ASTM D4546, In *1986 Annual Book of ASTM Standards*, Vol. 04.08, *Soil and Rock; Building Stones*, Section 4, ASTM, Philadelphia, PA.

American Society for Testing and Materials (ASTM). 1993. Standard test method for measurement of collapse potential of soils, ASTM D5333–92, In *Annual Book of ASTM Standards*, Vol. 04.08, Philadelphia, PA, pp. 343–345.

American Society for Testing and Materials (ASTM). 1997a. Test method for capillary-moisture relationships for fine-textured soils by pressure-membrane apparatus, ASTM D 3152–72, In *Annual Book of ASTM Standards*, Vol. 04.08, *Soil and Rock (I): D420–D4914*, ASTM International, West Conshohocken, PA, pp. 311–316.

American Society for Testing and Materials (ASTM). 1997b. Standard test method for the measurement of soil potential (suction) using filter paper, ASTM D 5298–94, In *Annual Book of ASTM Standards*, Vol. 04.09, *Soil and Rock (II): D4943–Latest*, ASTM International, West Conshohocken, PA, pp. 157–162.

American Society for Testing and Materials (ASTM). 1998. Standard practice for maintaining constant relative humidity by means of aqueous solutions, ASTM E 104–85, In *Annual Book of ASTM Standard*, Vol. 11.03, ASTM, Philadelphia, PA, pp. 781–783.

American Society for Testing and Materials (ASTM). 2003. Standard practice for maintaining constant relative humidity by means of aqueous solutions, ASTM E 104–02, In *Annual Book of ASTM Standard*, Vol. 11.03, ASTM, West Conshohocken, PA pp. 1133–1137.

American Society for Testing and Materials (ASTM). 2008a. Standards test methods for determination of the soil-water characteristic curve for desorption using a hanging column, pressure extractor, chilled mirror hygrometer, and/or centrifuge, ASTM D6836–02, In *Annual Book of ASTM Standards*, Vol. 04.09, *ASTM D18 on Soil and Rocks*, West Conshohocken, PA, pp. 1340–1359. (http://www.astm.org/Standards/D6836.htm)

American Society for Testing and Materials (ASTM). 2008b. Standard test method for determination of thermal conductivity of soil and soft rock by thermal needle probe procedure, ASTM D5334–08, ASTM, West Conshohocken, PA.

Andersland, O. B., and Anderson, D. M. 1978. *Geotechnical Engineering for Cold Regions*, McGraw-Hill, New York.

Anderson, D. M., and Morgenstern, N. R. 1973. Physics, chemistry, and mechanics of frozen ground: A review, *Proceedings of the Second International Conference on Permafrost*, Yakutsk, Russia, pp. 257–288.

Anderson, M., and Burt, T. 1977. Automatic monitoring of soil moisture conditions in a hillslope spur and hollow, *Journal of Hydrology*, Vol. 33, pp. 27–36.

Anderson, M. G. 1983. Prediction of soil suction for slopes in Hong Kong, GCO Publication No. 1/84, Geotechnical Control Office, Engineering Development Department, Hong Kong.

Anderson, M. G., and Pope, R. G. 1984. The incorporation of soil water physics models into geotechnical studies of landslide behaviour, *Proceedings of the Fourth International Symposium on Landslides*, Toronto, ON, Vol. 4, pp. 349–353.

Apfel, R. C. 1970. The role of impurity in cavitation-threshold determination, *Journal of Acoustic Society of America*, Vol. 48, pp. 1179–1186.

Arbhabhirama, A., and Kridakorn, C. 1968. Steady downward flow to a water table, *Water Resources Research*, Vol. 4, No. 6, pp. 1249–1257.

Arenson, L., Sego, D., and Newman, G. 2006. The use of convective heat flow in road designs for northern regions, *Proceedings of Climate Change Technology Conference*, Ottawa, ON.

Arya, L. M., and Paris, J. F. 1981. A physico-empirical model to predict the soil moisture characteristic from particle-size distribution and bulk density data, *Soil Science Society of America Journal*, Vol. 45, pp. 1023–1030.

Arya, L. M., Richter, J. C., and Davidson, S. A. 1982. A comparison of soil moisture characteristic predicted by the Arya-Paris model with laboratory-measured data, Agristars Technology Report SM-L1-04247, JSC-17820, NASA-Johnson Space Center, Houston, TX.

Assouline, S. 2001. A model for soil relative hydraulic conductivity based on the water retention characteristic curve, *Water Resources Research*, Vol. 37, pp. 265–271.

Aubertin, M., Mbonimpa, M., Bussiere, B., and Chapuis, R. P. 2003. A model to predict the water retention curve from basic geotechnical properties, *Canadian Geotechnical Journal*, Vol. 40, pp. 1104–1122.

Aung, K. K., Rahardjo, H., Leong, E. C., and Toll, D. G. 2001. Relationship between porosimetry measurement and soil-water characteristic curve for an unsaturated soil, *Journal of Geotechnical and Geological Engineering*, Vol. 19, Nos. 3–4, pp. 401–416.

Averjanov, S. F. 1950. About permeability of subsurface soils in case of incomplete saturation, *English Collection*, Vol. 7, pp. 19–21.

Azam, S. H., Wilson, G. W., Fredlund, D. G., and van Zyl, D. 2009. Geotechnical characterization of mine waste rock, *Proceedings of the Seventh International Conference Soil Mechanics and Geotechnical Engineering*, Alexandria, October.

Azzouz, A. S., Krizek, R. J., and Corotis, R. B. 1976. Regression analysis of soil compressibility, *Soils and Foundations, Tokyo*, Vol. 16, No. 2, pp. 19–29.

Babuska, I. 1989. Adaptive mathematical modeling, in I. Babuska, J. Chandra, and J. E. Flaherty (Eds.), *Adaptive Methods for Partial Differential Equations*, SIAM, Philadelphia, PA, pp. 1–14.

Baehr, A. L., and Hult, M. F. 1991. Evaluation of unsaturated zone air permeability through pneumatic tests, *Water Resources Research*, Vol. 27, No. 10, pp. 2605–2617.

Bailey, W. A. 1965. The effect of salt content on the consolidation behavior of saturated clay, Contract Report 3–101, U.S. Corps of Army Engineers, Waterways Experimental Station, Vicksburg, MS.

Baker, R. 1980. Determination of the critical slip surfaces in slope stability computations, *Proceedings of the International Journal for Numerical and Analytical Methods in Geomechanics*, Vol. 4, pp. 333–359.

Balasubramonian, B. I. 1972. Swelling of compaction shale, PhD Thesis, University of Alberta, Edmonton.

Bankole, A. A., Wulfsohn, D., and Fredlund, D. G. 1996. Air volume change measurements in unsaturated soil testing using a digital pressure-volume controller, *Geotechnical Testing Journal, ASTM*, Vol. 19, No. 1, pp. 12–21.

Bao, C., Gong, B., and Zhan, L. 1998. Properties of unsaturated soils and slope stability of expansive soils, *Proceedings of the Second International Conference on Unsaturated Soils (UNSAT 98)*, Beijing, Vol. 1, pp. 71–98.

Bao, C. G., and Ng, C. W. W. 2000. Some thoughts and studies on the prediction of slope stability in expansive soils, *Proceedings of the First Asian Conference on Unsaturated Soils*, Singapore, pp. 15–31.

Barbour, S. L. 1987. Osmotic flow and volume change in clay soils, PhD Thesis, Department of Civil Engineering, University of Saskatchewan, Saskatoon, SK, Canada.

Barbour, S. L. 1998. Nineteenth Canadian Geotechnical Colloquium: The soil-water characteristic curve: A historical perspective, *Canadian Geotechnical Journal*, Vol. 35, No. 5, pp. 873–894.

Barbour, S. L., and Fredlund, D. G. 1989a. Mechanics of osmotic flow and volume change in clay soils, *Canadian Geotechnical Journal*, Vol. 26, pp. 551–562.

Barbour, S. L., and Fredlund, D. G. 1989b. Physico-chemical state variables for clay soils, *Proceedings of the Twelfth International Conference on Soil Mechanics and Foundation Engineering*, Rio de Janeiro, pp. 1839–1843.

Barbour, S. L., Fredlund, D. G., and Pufahl, D. E. 1991. The osmotic role in the behaviour of swelling clay soils, *NATO Advanced Research Workshop*, Korfu, Greece, pp. 97–139.

Barbour, S. L., Fredlund, D. G., Gan, J. K.-M, and Wilson, G. W. 1992. Prediction of moisture movement in highway subgrade soils, *Proceedings of the Forty-Fifth Annual Canadian Geotechnical Conference*, Toronto, ON, pp. 41A.1–41A.13.

Barbour, S. L., Lim, P. C., and Fredlund, D. G. 1996. A new technique for diffusion testing of unsaturated soil, *Geotechnical Testing Journal*, *ASTM*, Vol. 19, No. 3, pp. 247–258.

Barden, L. 1965. Consolidation of compacted and unsaturated clays, *Geotechnique*, Vol. 15, No. 3, pp. 267–286.

Barden, L. 1974. Consolidation of clays compacted "dry" and "wet" of optimum water content, *Geotechnique*, Vol. 24, No. 4, pp. 605–625.

Barden, L., and Pavlakis, G. 1971. Air and water permeability of compacted unsaturated cohesive soil, *Soil Science Journal*, Vol. 22, No. 3, pp. 302–317.

Barden, L., and Sides, G. R. 1967. The diffusion of air through the pore water of soils, *Proceedings of the Third Asian Regional Conference on Soil Mechanics and Foundation Engineering*, Haifa, Vol. 1, pp. 135–138.

Barden, L., and Sides, G. R. 1970. Engineering behaviour and structure of compacted clay, *Journal of Soil Mechanics and Foundations Division, ASCE*, Vol. 96, pp. 1171–1200.

Barden, L., Madedor, A. O., and Sides, G. R. 1969a. Volume change characteristics of unsaturated clay, *Journal of the Soil Mechanics and Foundation Division*, *ASCE*, Vol. 95, pp. 33–52.

Barden, L., Madedor, A. O., and Sides, G. R. 1969b. The flow of air and water in partly saturated clay soil, *International Symposium on Fundamentals of Transport Phenomena in Flow through Porous Media*, Haifa.

Barends, F. B. J. 1979. The compressibility of an air-water mixture in a porous medium, *Igm-Mededelingen*, Delft Soil Mechanics Laboratory, Delft, The Netherlands, Part 20, No. 1, pp. 49–66.

Barry, R. G., and Chorley, R. J. 1968. *Atmosphere, Weather and Climate*, Methuen, London.

Bashir, R., Stone, J. J., and Fredlund, D. G. 2011. Effect of ambient temperature on the calibration of thermal conductivity suction sensors, *Proceedings of the Canadian Geotechnical and Fifth Pan American Conference*, Toronto, ON, Canada, Oct. 2–6.

Ba-Te, B. 2005. Flow of air-phase in soils and its application in emergent stabilization of soil slopes, M. Phil. Thesis, Department of Civil Engineering, Hong Kong University of Science and Technology, Hong Kong.

Ba-Te, B., Zhang, L., and Fredlund, D. G. 2005. A general air-phase permeability function for air flow through unsaturated soils, *Proceedings of Geo-Frontiers 2005 Congress, ASCE*, Austin, TX, pp. 2961–2985.

Baver, L. D. 1940. *Soil Physics*, Wiley, New York.

Baver, L. D. 1956. *Soil Physics*, 3rd ed., Wiley, New York.

Bear, J. 1972. *Dynamics of Fluids in Porous Media*, American Elsevier Publishing Company, New York.

Bear, J. 1979. *Hydraulics of Groundwater*, McGraw-Hill, New York.

Bellman, R. 1957. *Dynamic Programming*, Princeton University Press, Princeton, NJ.

Benson, C. H. 2002. Containment systems: Lessons learned from North American failures, *Proceedings of the Fourth International Congress on Environmental Geotechnics*, Lisse, The Netherlands, Vol. 2, pp. 1095–1112.

Benson, C. H., and Daniel, D. E. 1990. Influence of clods on hydraulic conductivity of compacted clay, *Journal of Geotechnical Engineering*, *ASCE*, Vol. 116, No. 8, pp. 1231–1248.

Benson, C. H., and Gribb, M. M. 1997. Measuring unsaturated hydraulic conductivity in the laboratory and field, in *Unsaturated Soil Engineering Practice*, Geotechnical Special Publication No. 68, American Society of Civil Engineers, New York, pp. 113–168.

Bergan, A. T., and Fredlund, D. G. 1973. Characterization of freeze-thaw effects on subgrade soils, *Proceedings of the Symposium on Frost Action on Roads*, Organization for Economic Cooperation and Development, Oslo.

Bern, M. W., Flaherty, J. E., and Luskin, M. (Eds.). 1999. *Grid Generation and Adaptive Algorithms: The Institute for Mathematics and Its Applications*, Springer-Verlag, New York, Vol. 113.

Bernatzik, W. 1948. The determination of the capillary rise in sand by means of prism pressure test, *Proceedings of the Second International Conference on Soil Mechanics Foundation Engineering*, Rotterdam, Vol. 5, pp. 28.

Biot, M. A. 1941. General theory for three-dimensional consolidation, *Journal of Applied Physics*, Vol. 12, No. 2, pp. 155–164.

Biot, M. A. 1955. Theory of elasticity and consolidation for a porous anisotropic solid, *Journal of Applied Physics*, Vol. 26, No. 2, pp. 182–185.

Bishop, A. W. 1954. The use of pore pressure coefficients in practice, *Geotechnique*, Vol. 4, No. 4, pp. 148–152.

Bishop, A. W. 1955. The use of the slip circle in the stability analysis of slopes, *Geotechnique*, Vol. 5, pp. 7–17.

Bishop, A. W. 1957. Some factors controlling the pore pressures set up during the construction of earth dams, *Proceedings of the Fourth International Conference on Soil Mechanics and Foundation Engineering*, London, Vol. 2, pp. 294–300.

Bishop, A. W. 1959. The principle of effective stress, *Teknisk Ukeblad*, Norwegian Geotechnical Institute, Vol. 106, No. 39, pp. 859–863.

Bishop, A. W. 1961a. Discussion on general principles and laboratory measurements, *Proceedings of Conference on Pore Pressure and Suction in Soils*, London, pp. 63–66.

Bishop, A. W. 1961b. The measurement of pore pressure in the triaxial test, *Proceedings of Conference on Pore Pressure and Suction in Soils*, London, pp. 38–46.

Bishop, A. W. 1969. Pore pressure measurements in the field and in the laboratory, *Proceedings of the Seventh International Conference on Soil Mechanics and Foundation Engineering*, Mexico City, pp. 427–441.

Bishop, A. W. 1985. Test requirements of measuring the coefficient of earth pressure at rest, *Proceedings of the Conference on Earth Pressure Problems*, Brussels, Vol. 1, pp. 2–14.

Bishop, A. W., and Blight, G. E. 1963. Some aspects of effective stress in saturated and unsaturated soils, *Geotechnique*, Vol. 13, No. 3, pp. 177–197.

Bishop, A. W., and Donald, I. B. 1961. The experimental study of partly saturated soil in the triaxial apparatus, *Proceedings of the Fifth International Conference on Soil Mechanics and Foundation Engineering*, Paris, Vol. 1, pp. 13–21.

Bishop, A. W., and Eldin, A. K. G. 1950. Undrained triaxial tests on saturated sands and their significance in the general theory of shear strength, *Geotechnique*, Vol. 2, pp. 13–32.

Bishop, A. W., and Gibson, R. E. 1963. The influence of the provisions for boundary drainage on strength and consolidation characteristics of soils measured in the triaxial apparatus, in *Laboratory Shear Testing of Soils*, ASTM STP No. 361,

American Society for Testing and Materials, Philadelphia, PA, pp. 435–458.

Bishop, A. W., and Henkel, D. J. 1962. *The Measurement of Soil Properties in the Triaxial Test*, 2nd ed., Edward Arnold, London.

Bishop, A. W., Alpan, I., Blight, G. E., and Donald, I. B. 1960. Factors controlling the shear strength of partly saturated cohesive soils, paper presented at the Research Conference on Shear Strength of Cohesive Soils, ASCE, University of Colorado, Boulder, CO, pp. 503–532.

Biwei, G., Yanhua, L., and Zhan, L. T. 1998. The relationship between residual water content and shrinkage limit for expansive soils, *Proceedings of the Second International Conference on Unsaturated Soils (UNSAT-98)*, Beijing, Vol. 2, pp. 202–207.

Black, C. A. (Ed.). 1965a. Methods of soil analysis, Part 1, American Society of Agronomy, Madison, WI, Agronomy Series No. 9.

Black, C. A. (Ed.). 1965b. Methods of soil analysis, Part 2, American Society of Agronomy, Madison, WI, Agronomy Series No. 9.

Black, D. K., and Lee, K. L. 1973. Saturating laboratory samples by back pressure, *Journal of Soil Mechanics Foundation Engineering Division, ASCE*, Vol. 99, No. SM1, pp. 75–95.

Black, P. B., and Hardenberg, M. J. (Eds.). 1991. Historical perspective in frost heave research, The early works of S. Taber and G. Beskow, CRREL Special Report 91–23, U.S. Army Corps of Engineers, Hanover, NH.

Black, P. B., and Tice, A. R. 1989. Comparison of soil freezing curve and soil water curve data for Windsor sandy loam, *Water Resources Research*, Vol. 25, No. 10, pp. 2205–2210.

Black, W. P. M., and Croncy, D. 1957. Pore water pressure and moisture content studies under experimental pavements, *Proceedings of the Fourth International Conference on Soil Mechanics and Foundation Engineering*, London, Vol. 2, pp. 94–103.

Black, W. P. M., Croney, D., and Jacobs, J. C. 1959. Field studies of the movement of soil moisture, Technical Paper No. 41, Road Research Board, Her Majesty's Stationery Office, London.

Blaney, H. F., and Criddle, W. D. 1950. Determining water requirements in irrigated areas from climatological and irrigation data, Technical Paper No. 96, U.S. Department of Agriculture, Soil Conservation Service, Washington, DC.

Blatz, J. A., and Graham, J. 2003. Elasto-plastic modeling of unsaturated soil using results from a new triaxial test with controlled suction, *Geotechnique*, Vol. 53, No. 1, pp. 113–122.

Blight, G. E. 1961. Strength and consolidation characteristics of compacted soils, PhD Thesis, University of London, London.

Blight, G. E. 1965. A study of effective stresses for volume change, in G. D. Aitchison (Ed.), *Moisture Equilibria and Moisture Changes in Soils Beneath Covered Areas*, A Symposium in Print, Butterworths, Sydney, pp. 259–269.

Blight, G. E. 1966. Strength characteristics of desiccated clays, *Journal of Soil Mechanics Foundation Engineering Division, ASCE*, Vol. 92, No. SM6, pp. 19–37.

Blight, G. E. 1967. Effective stress evaluation for unsaturated soils, *Journal of Soil Mechanics Foundation Engineering Division, ASCE*, Vol. 93, No. SM2, pp. 125–148.

Blight, G. E. 1971. Flow of air through soils, *Journal of Soil Mechanics Foundation Engineering Division, ASCE*, Vol. 97, No. SM4, pp. 607–624.

Bloodworth, M. E., and Page, J. B. 1957. Use of thermistor for the measurement of soil moisture and temperature, *Soil Science Society of America Proceedings*, Vol. 21, pp. 11–15.

Bocking, K. A., and Fredlund, D. G. 1979. Use of the osmotic tensiometer to measure negative pore water pressure, *Geotechnical Testing Journal, ASTM*, Vol. 2, No. 1, pp. 3–10.

Bocking, K. A., and Fredlund, D. G. 1980. Limitations of the axis-translation technique, *Proceedings of the Fourth International Conference on Expansive Soils*, Denver, CO, Vol. 1, pp. 117–135.

Bolt, G. H., and Miller, R. D. 1958. Calculation of total and component potentials of water in soil, *American Geophysical Union Transportation*, Vol. 39, pp. 917–928.

Bosen, J. F. 1958. An approximation formula to compute relative humidity from dry bulb and dewpoint temperature, *Monthly Weather Review*, Vol. 86, pp. 486.

Boulichev, V. 1936. Apparatus for testing compressibility and capillary properties of soils, *Proceedings of the First International Conference on Soil Mechanics and Foundation Engineering*, Cambridge, MA, Vol. 2, pp. 37–38.

Bouma, J. 1989. Using soil survey data for quantitative land evaluation, *Advanced Soil Science*, Vol. 9, pp. 177–213.

Bouver, H. 1978. *Groundwater Hydrology*, McGraw-Hill, New York.

Bowen, I. S. 1926. The ratio of heat losses by conduction and by evaporation from any water surface, *Physics Review*, Vol. 27, pp. 779–787.

Boyoucous, G. J. 1920. Degree of temperature to which soils can be cooled without freezing, *Journal of Agricultural Research*, Vol. 20, pp. 267–269.

Brackley, I. J. A. 1971. Partial collapse in unsaturated expansive clay, *Proceedings of the Fifth Regional Conference for Africa on Soil Mechanics Foundation and Engineering*, Luanda, Angola, pp. 23–30.

Brackley, I. J. A. 1975. A model of unsaturated clay structure and its application to swell behaviour, *Proceedings of the Sixth Regional Conference for Africa on Soil Mechanics and Foundation Engineering*, Durban, South Africa, Vol. 1, pp. 65–70.

Brand, E. W. 1981. Some thoughts on rain-induced slope failures, *Proceedings of the Tenth International Conference on Soil Mechanics and Foundation Engineering*, Balkema, Rotterdam, Vol. 3, pp. 373–376.

Brand, E. W. 1982. Analysis and design in residual soils, *Proceedings of Specialty Conference on Engineering and Construction in Tropical and Residual Soils, ASCE*, Honolulu, HI, pp. 89–143.

Brand, E. W. 1985. Geotechnical engineering in tropical residual soils, *Proceedings of the First International Conference on Geomechanics in Tropical Lateritic and Saprolitic Soils*, Brasilia, Vol. 3, pp. 23–91.

Brand, E. W. 1995. Slope stability in tropical areas, *Proceedings of the Seventh International Symposium on Landslides*, Trondheim, pp. 2031–2051.

Briggs, L. J. 1897. The mechanics of soil moisture, U.S. Department of Agriculture, Soils Bulletin No. 10, U.S. Government Printing Office, Washington, DC.

Brimblecombe, P. 1995. *Air Composition and Chemistry*, Cambridge Environmental Chemistry Series, Cambridge University Press, Cambridge, UK.

Brito, C. C., Pereira, J. H. F., Gitirana, Jr., de F. N. G., and Fredlund, D. G. 2004. Transient stability analysis of a collapsible dam using dynamic programming combined with finite element stress state fields, *Proceedings of the Ninth International Symposium on Landslides*, Rio de Janeiro, Vol. 2, pp. 1079–1084.

Broms, B. B. 1965. Effect of degree of saturation on bearing capacity of flexible pavements, *The Forty-Third Annual Meeting of the Committee on Flexible Pavement Design*, Ithaca, NY.

Bronswijk, J. J. B. 1990. Shrinkage geometry of a heavy clay soil at various stresses, *Soil Science Society of America Journal*, Vol. 54, pp. 1500–1502.

Bronswijk, J. J. B. 1991a. Relation between vertical soil movement and water-content changes in cracking clays, *Soil Science Society of America Journal*, Vol. 55, pp. 1220–1226.

Bronswijk, J. J. B. 1991b. Drying, cracking, and subsidence of a clay soil in a lysimeter, *Soil Science Journal*, Vol. 152, No. 1, pp. 92–99.

Brooker, E. W., and Ireland, H. O. 1965. Earth pressures at-rest related to stress history, *Canadian Geotechnical Journal*, Vol. 2, No. 1, pp. 1–15.

Brooks, R. H., and Corey, A. T. 1964. Hydraulic properties of porous media, Colorado State University Hydrology Paper, No. 3. Fort Collins, CO.

Brooks, R. H., and Corey, A. T. 1966. Properties of porous media affecting fluid flow, *Journal of Irrigation and Drainage Division*, ASCE IR 2, Vol. 92, pp. 61–88.

Brown, R. W., and Bartos, D. L. 1982. A calibration model for screen-caged Peltier thermocouple psychrometers, Research Paper INT-293, U.S. Department of Agriculture, Washington, DC.

Brown, R. W., and Collins, J. M. 1980. A screen-caged thermocouple psychrometer and calibration chamber for measurements of plant and soil water potential, *Agronomy Journal*, Vol. 72, pp. 851–854.

Bruce, R. R., Dane, J. H., Quisenberry, V. L., Powel, N. L., and Thomas, A. W. 1983. Physical characterization of soils in the southern region: Cecil. Southern Cooperative Series Bulletin No. 267, Georgia Agricultural Experimental Station, Athens, GA.

Bruch, P. G. 1993. A laboratory study of evaporative fluxes in homogeneous and layered soils, M.Sc. Thesis, University of Saskatchewan, Saskatoon, SK.

Brutsaert, W. 1967. Some methods of calculating unsaturated permeability, *Transactions of ASABE*, Vol. 10, pp. 400–404.

Buckingham, E. 1907. Studies of the movement of soil moisture, Bulletin No. 38, U.S. Department of Agriculture, Bureau of Soils, Washington, DC.

Budhu, M. 2010. *Soil Mechanics and Foundations*, 3rd ed., Wiley, Hoboken, NJ.

Bumb, A. C. 1987. Unsteady-state flow of methane and water in coal beds, PhD Thesis, Department of Chemical Engineering, University of Wyoming, Laramie, WY.

Burdine, N. T. 1953. Relative permeability calculations from pore size distribution data, *Transactions of the Metallurgical Society of AIME*, Vol. 198, pp. 71–78.

Burger, C. A., and Shackelford, C. D. 2001. Soil-water characteristic curves and dual porosity of sand-diatomaceous earth mixtures, *Journal of Geotechnical and Environmental Engineering*, *ASCE*, Vol. 127, No. 9, pp. 790–800.

Burland, J. B. 1964. Effective stresses in partly saturated soils, Discussion of "Some aspects of effective stress in saturated and partly saturated soils," by G. E. Blight and A. W. Bishop, *Geotechnique*, Vol. 14, pp. 65–68.

Burland, J. B. 1965. Some aspects of the mechanical behaviour of partly saturated soils, in G.D. Aitchison (Ed.), *Moisture Equilibria and Moisture Changes in the Soils Beneath Covered Areas*, A Symposium in Print, Butterworths, Sydney, pp. 270–278.

Burman, R. D., Jensen, M. E., and Allen, R. G. 1987. Thermodynamic factors in evapotranspiration, in irrigation systems for the 21st century, in L. G. James and M. J. English (Eds.), *Proceedings of a Conference Sponsored by the Irrigation and Drainage Division of the American Society of Civil Engineers and the Oregon Section, ASCE*, Portland, OR, pp. 28–30.

Bussiere, B., Aubertin, M., and Chapuis, R. P. 2003. The behaviour of inclined covers used as oxygen barriers, *Canadian Geotechnical Journal*, Vol. 40, pp. 512–535.

Camapun de Carvalho, J., Guimaraes, R. C., and Pereira, J. H. F. 2002. Courbes de retention d'eau d'un profil d'alteration, *Proceedings of the Third International Conference on Unsaturated Soils*, Recife, Brazil, Vol. 1, pp. 289–294.

Campbell, G. S. 1974. A simple method for determining unsaturated conductivity from moisture retention data, *Soil Science Journal*, Vol. 117, No. 6, pp. 311–314.

Campbell, G. S. 1985. *Soil Physics with Basic Transport Models for Soil–Plant Systems*, Elsevier, New York.

Campbell, G. S., and Gardner, W. H. 1971. Psychrometric measurement of soil potential: Temperature and bulk density effects, *Soil Science Society of America Proceedings*, Vol. 35, pp. 8–12.

Campbell, G. S., Jungbauer, Jr., J. D., Bidlake, W. R., and Hungerford, R. D. 1994. Predicting the effect of temperature on soil thermal conductivity, *Soil Science Journal*, Vol. 158, pp. 307–313.

Campbell, G. S., Smith, D. M., and Teare, B. L. 2007. Application of a dew point method to obtain the soil-water characteristic, *Proceedings of the Second International Conference on Mechanics of Unsaturated Soils*, Bauhaus-Universität, Weimar, Germany, pp. 71–77.

Campbell, J. D. 1973. Pore pressures and volume changes in unsaturated soils, PhD Thesis, University of Illinois, Urbana-Champaign, IL.

Campbell Scientific. 2009. *Instruction Manual for the 229 Heat Dissipation Matric Suction Potential Sensor*, Campbell Scientific, Logan, UT.

Canada Department of Agriculture. 1976. *Glossary of Terms in Soil Science*, Publication No. 1459, Research Branch, Canada Department of Agriculture, Ottawa, ON.

Cardoso, R., Romero, E., Lima, A., and Ferrari, A. 2007. A comparative study of soil suction measurement using two different high-range psychrometers, *Proceedings of the Second International Conference on Mechanics of Unsaturated Soils*, Bauhaus-Universität, Weimar, Germany, pp. 79–93.

Carslaw, H. S., and Jaeger, J. C. 1959. *Conduction of Heat in Solids*, 2nd ed., Oxford University Press, Oxford, UK.

Cary, J. W., Simmons, C. S., and McBride, J. F. 1989. Predicting oil infiltration and redistribution in unsaturated soils,

Soil Science Society of America Journal, Vol. 53, No. 2, pp. 335–342.

Casagrande, A. 1936. The determination of the preconsolidation load and its practical significance, *Proceedings of the First International Conference on Soil Mechanics Foundation Engineering*, Cambridge, MA, Vol. 3, pp. 60–64.

Casagrande, A. 1937. Seepage through dams, *Journal of New England Water Works*, Vol. 51, No. 2, pp. 295–336.

Cassel, D. K., and Klute, A. 1986. Water potential: Tensiometry, in A. Klute (Ed.), *Methods of Soil Analysis, Part 1, Physical and Mineralogical Methods*, 2nd ed., *Soil Science Society of America Proceedings*, American Society of Agronomy, Madison, WI.

Cedergren, H. R. 1989. *Seepage Drainage and Flow Nets*, Wiley, New York.

Celestino, T. B., and Duncan, J. M. 1981. Simplified search for non-circular slip surfaces, *Proceedings of the Tenth International Conference on Soil Mechanics and Foundation Engineering*, Stockholm, Vol. 3, pp. 391–394.

Cengiz, H. S., Gregory, J. M., and Sebaught, J. L. 1981. Solar radiation prediction from other climatic variables, *Transactions of ASAE*, Vol. 3, pp. 1269–1272.

Chandler, R. J., and Gutierrez, C. I. 1986. The filter-paper method of suction measurement, *Geotechnique*, Vol. 36, pp. 265–268.

Chandler, R. J., Crilly, M. S., and Montgomery-Smith, G. 1992. A low-cost method of assessing clay desiccation for low-rise buildings, *Proceedings of the Institution of Civil Engineers*, Vol. 92, No. 2, pp. 82–89.

Chantawarangul, K. 1983. Comparative study of different procedures to evaluate effective stress strength parameters for partially saturated soils, M.Sc. Thesis, Asia Institute of Technology, Bangkok, Thailand.

Chapman, P. J., Richards, B. E., and Trevena, D. H. 1975. Monitoring the growth of assessing clay desiccation for low-rise buildings, *Proceedings of the Institute of Civil Engineers*, Vol. 92, pp. 82–89.

Chattopadhyay, P. K. 1972. Residual shear strength of some pure clay minerals, PhD Thesis, University of Alberta, Edmonton.

Chen, F. H. 1975. *Foundations on Expansive Soils*, 1st ed., Elsevier, New York.

Chen, F. H. 1988. *Foundations on Expansive Soils*, 2nd ed., Elsevier, Amsterdam.

Chen, F. H., and Huang, D. 1987. Lateral expansion pressure on basement walls, *Proceedings of the Sixth International Conference on Expansive Soils*, New Delhi, Vol. 1, pp. 55–60.

Chen, H. J., Becker, D. E., Fredlund, D. G., and Tricia, G. 2010a. Geotechnical assessment of shallow (surficial) slope stability zoning on the proposed Paskapoo Slopes Park, *Proceedings of the Sixty-Third Canadian Geotechnical Conference*, Calgary, AB, pp. 1606–1612.

Chen, Y., Ke, H., Fredlund, D. G., Zhan, L., and Xie, Y. 2010b. Secondary compression of municipal solid wastes and a compression model for predicting settlement of municipal solid waste landfills, *Journal of Geotechnical and Geoenvironmental Engineering, ASCE*, Vol. 136, No. 5, pp. 706–717.

Chen, Z. H., Fredlund, D. G., and Gan, J. K.-M. 1999. Overall volume change, water volume change and yield associated with an unsaturated compacted loess, *Canadian Geotechnical Journal*, Vol. 36, No. 2, pp. 321–329.

Chen, Z-Y., and Shao, C-M. 1988. Evaluation of the minimum factor of safety in slope stability analysis, *Canadian Geotechnical Journal*, Vol. 24, No. 4, pp. 735–748.

Chertkov, V. Y. 2004. A physically based model for the water retention curve of clay pastes, *Journal of Hydrology*, Vol. 286, Nos. 1–4, pp. 203–226.

Childs, E. C. 1940. The use of soil moisture characteristics in soil studies. *Journal of Soil Science*, Vol. 50, pp. 239–252.

Childs, E. C. 1969. *An Introduction to the Physical Basis of Soil Water Phenomena*, Wiley-Interscience, London.

Childs, E. C., and Collis-George, N. 1948. Soil geometry and soil-water equilibria, *Faraday Society*, Vol. 3, pp. 78–85.

Childs, E. C., and Collis-George, N. 1950. The permeability of porous materials, *Proceedings of the Royal Society, London*, Series A, Vol. 201A, pp. 392–405.

Chin, K. B., Leong, E. C., and Rahardjo, H. 2010. A simplified method to estimate the soil-water characteristic curve, *Canadian Geotechnical Journal*, Vol. 47, No. 12, pp. 1382–1400.

Ching, R. K. H., and Fredlund, D. G. 1983. Some difficulties associated with the limit equilibrium method of slices, *Canadian Geotechnical Journal*, Vol. 20, No. 4, pp. 661–672.

Ching, R. K. H., and Fredlund, D. G. 1984a. A small Saskatchewan town copes with swelling clay problems, *Proceedings of the Fifth International Conference on Expansive Soils*, Adelaide, Australia, pp. 306–310.

Ching, R. K. H., and Fredlund, D. G. 1984b. Quantitative comparisons of limit equilibrium methods of slices, *Proceedings of the Fourth International Symposium on Landslides*, Toronto, ON, Vol. 1, pp. 373–379.

Ching, R. K. H., Sweeney, J., and Fredlund, D. G. 1984. Increase in factor of safety due to soil suction for two Hong Kong slopes, *Proceedings of the Fourth International Symposium on Landslides*, Toronto, ON, Vol. 1, pp. 617–623.

Chiu, C. F., and Ng, C. W. W. 2003. A state-dependent elasto-plastic model for saturated and unsaturated soils, *Geotechnique*, Vol. 53, No. 9, pp. 809–829.

Chou, P. C., and Pagano, N. J. 1967a. *Elasticity*, Van Nostrand, Toronto, ON.

Chou, P. C., and Pagano, N. J. 1967b. *Elasticity–Tensor, Dyadic, and Engineering Approaches*, Dover, New York.

Chowdhury, R. N. 1980. A reassessment of limit equilibrium concepts in geotechnique, *Proceedings of the Symposium on Limit Equilibrium, Plasticity and Generalized Stress-Station Applications in Geotechnical Engineering*, in conjunction with the 1980 ASCE Annual Convention and Exposition, Hollywood, FL.

Christenson, H. G. 1988. Adhesion between surfaces in undersaturated vapours—A re-examination of the influence of meniscus curvature and surface forces, *Journal of Colloid and Interface Science*, Vol. 121, No. 1, pp. 170–178.

Christian, J. T., and Boehmer, J. W. 1970. Plane strain consolidation by finite elements, *Journal of Soil Mechanics and Foundations, ASCE*, Vol. 96, No. 4, pp. 1435–1457.

Chu, T. Y., and Mou, C. H. 1973. Volume change characteristics of expansive soils determined by controlled suction test, *Proceedings of the Third International Conference on Expansive Soils*, Haifa, Vol. 1. pp. 177–185.

Clapp, R. B., and Hornberger, G. M. 1978. Empirical equations for some soil hydraulic properties, *Water Resources Research*, Vol. 14, pp. 601–604.

Clarke, C. R., and Nevels, Jr., J. B. 1996. Shrinkage and suction properties of Pledger-Roebuck alluvial clay, *Transportation Research Record*, No. 1546, pp. 162–173.

Clifton, A. W., Yoshida, R., Fredlund, D. G., and Chursinoff, R. W. 1984. Performance of Darke Hall, Regina, Canada, constructed on highly swelling clay, *Proceedings of the Fifth International Conference on Expansive Soils*, Adelaide, Australia, pp. 197–201.

Cole, D. M. 1978. A technique for measuring radial deformation during repeated load triaxial testing, *Canadian Geotechnical Journal*, Vol. 15, pp. 426–429.

Coleman, J. D. 1962. Stress/strain relations for partly saturated soils, *Geotechnique*, Vol. 12, No. 4, pp. 348–350.

Commission of Inquiry. 1972. *Rainstorm Disasters 1972*, Government Printer, Hong Kong.

Considine, D. M., and Ross, S. D. 1974. *Process Instruments and Controls Handbook*, 2nd ed., McGraw-Hill, New York.

Cook, D. L., and Fredlund, D. G. 1998. TDR matric suction measurements, *Proceedings of the Second International Conference on Unsaturated Soils, UNSAT-98*, Beijing, Vol. 1, pp. 338–343.

Cook, H. 1981. *Concepts and Applications of Finite Element Analysis*, Wiley, New York.

Cooper, J. R., and Dooley, R. B. 2008. Revised release on the IAPS formulation 1985 for the thermal conductivity of ordinary water and substances, The International Association for the Properties of Water and Stema, September, Berlin, www.iapws.org.

Corey, A. T. 1954. The interrelation between gas and oil relative permeabilities, *Producer's Monthly*, Vol. 19, No. 1, pp. 38–41.

Corey, A. T. 1957. Measurement of water and air permeability in unsaturated soil, *Soil Science Society of America Proceedings*, Vol. 21, No. 1, pp. 7–10.

Corey, A. T. 1977. Mechanics of heterogeneous fluids in porous media, Water Resources Publication, Fort Collins, CO.

Corey, A. T., and Kemper, W. D. 1961. Concept of total potential in water and its limitations, *Soil Science Journal*, Vol. 91, No. 5, p. 299–305.

Corey, A. T., Slayter, R. O., and Kemper, W. D. 1967. Comparative terminologies for water in the soil-plant-atmosphere system, in R. M. Hagan et al. (Eds.), *Irrigation in Agricultural Soils*, Monograph No. 11, American Society of Agronomy, Chapter 22, Madison, WI.

Côté, J. 2009. Heat Transfer in geotechnical engineering: Recent developments and future trends, paper presented at the Canadian Geotechnical Conference, Halifax, NS.

Côté, J., and Konrad, J.-M. 2005. A generalized thermal conductivity model for soils and construction materials, *Canadian Geotechnical Journal*, Vol. 42, No. 2, p 443–458.

Côté, J., and Konrad, J.-M. 2009. Assessment of structure effects on the thermal conductivity of two-phase porous geomaterials, *International Journal of Heat and Mass Transfer*, Vol. 52, pp. 796–804.

Coutts, R. J., and Konrad, J.-M. 1994. Finite element modeling of transient non-linear heat flow using the node state method, in *Ground Freezing 94*, Balkema, Rotterdam, pp. 39–47.

Cresswell, H., and Paydar, Z. 1996. Water retention in Australian soils. 1. Description and prediction using parametric functions, *Australian Journal of Soil Research*, Vol. 34, No. 2, pp. 195–212.

Crilly, M. C., Driscoll, R. M. C., and Chandler, R. J. 1992. Seasonal ground and water movement observations from an expansive clay site in the UK, *Proceedings of the Seventh International Conference on Expansive Soils*, Dallas, TX, Vol. 1, pp. 313–318.

Crilly, M. S., Schreiner, H. D., and Gourley, C. S. 1991. A simple field suction measurement probe, *Proceedings of the Tenth African Regional Conference on Soil Mechanics and Foundation Engineering*, Maseru, Lesotho, pp. 291–298.

Croney, D. 1952. The movement and distribution of water in soils, *Geotechnique*, Vol. 3, pp. 1–16.

Croney, D., and Coleman, J. D. 1948. Soil thermodynamics applied to the movement of moisture in road foundations, *Proceedings of the Seventh International Congress for Applied Mechanics*, London, Vol. 3, pp. 163–177.

Croney, D., and Coleman J. D. 1953, Soil moisture suction properties and their bearing on the moisture distribution in soils, *Proceedings of the Third International Conference on Soil Mechanics and Foundation Engineering*, August 16–27, Zurich, Switzerland, pp. 13–18.

Croney, D., and Coleman, J. D. 1954. Soil structure in relation to soil suction (pF), *Soil Science Journal*, Vol. 5, No. 1, pp. 75–84.

Croney, D., and Coleman, J. D. 1961. Pore pressure and suction in soil, in *Proceedings of the Conference on Pore Pressure and Suction in Soils*, Butterworths, London, pp. 31–37.

Croney, D., Coleman, J. D., and Lewis, W. A. 1950. Calculation of the moisture distribution beneath structures, *Cov. Engineering L.*, Vol. 45, p. 524.

Croney, D., Coleman, J. D., and Bridge, P. M. 1952. The suction of moisture held in soils and other porous materials, Technical Paper No. 24, Road Research Board, Her Majesty's Stationery Office, London.

Croney, D., Coleman, J. D., and Black, W. P. M. 1958. Movement and distribution of water in soil in relation to highway design and performance, in *Water and Its Conduction in Soils*, Special Report No. 40, Highway Research Board, Washington, DC, pp. 226–252.

Cryer, C. W. 1963. A comparison of the three-dimensional consolidation theories of Biot and Terzaghi, *Quarterly Journal of Mechanics and Applied Mathematics*, Vol. 16, pp. 401–412.

Cui, Y., and Delage, P. 1996. Yielding and plastic behaviour of an unsaturated compacted silt, *Geotechnique*, Vol. 46, No. 2, pp. 291–311.

Cui, Y. J., Loiseau, C., and Delage, P. 2002. Microstructure changes of a confined swelling soil due to suction controlled hydration, *Proceedings of the Third International Conference on Unsaturated Soils*, Recife, Brazil, pp. 593–598.

Cui, Y. J., Tang, A. M., Loiseau, C., and Delage, P. 2008. Determining the unsaturated hydraulic conductivity of a compacted sand-bentonite mixture under constant-volume and free-swell conditions, *Physics and Chemistry of the Earth*, Vol. 33, Supplement No. 1, pp. S462–S471.

Cuisinier, O., and Masrouri, F. 2005. Hydromechanical behaviour of a compacted swelling soil over a wide suction range, *Engineering Geology*, Vol. 81, pp. 204–212.

Cunningham, M. R., Ridley, A. M., Dineen, K., and Burland, J. B. 2003. Mechanical behaviour of a reconstituted unsaturated silty clay, *Geotechnique*, Vol. 53, No. 2, pp. 183–194.

Curtis, A. A., and Johnston, C. D. 1987. Monitoring unsaturated soil water conditions in groundwater recharge studies, *Proceedings of the International Conference Measurement of Soil and Plant Water Status*, Utah State University, Logan, Vol. 1, pp. 267–274.

Dagg, I. R., Reesor, G. E., and Doble, R. P. 1967. The dielectric constant of bromine at microwave frequencies, *Canadian Journal of Physics*, Vol. 45, No. 7, pp. 2349–2354.

Dakshanamurthy, V. 1979. A stress-controlled study of swelling characteristics of compacted expansive clays, *Geotechnical Testing Journal, ASTM*, Vol. 2, No. 1, pp. 57–60.

Dakshanamurthy, V., and Fredlund, D. G. 1980. Moisture and air flow in an unsaturated soil, *Proceedings of the Fourth International Conference on Expansive Soils*, Denver, CO, Vol. 1, pp. 514–532.

Dakshanamurthy, V., and Fredlund, D. G. 1981. A mathematical model for predicting moisture flow in an unsaturated soil under hydraulics and temperature gradients, *Water Resources Research*, Vol. 17, No. 3, pp. 714–722.

Dakshanamurthy, V., Fredlund, D. G., and Rahardjo, H. 1984. Coupled three-dimensional consolidation theory of unsaturated porous media, *Proceedings of the Fifth International Conference on Expansive Soils*, Institute of Engineers, Adelaide, Australia, pp. 99–103.

Dalton, F. N. 1992. Development of time-domain reflectometry for measuring soil water content and bulk soil electrical conductivity, in *Advances in Measurement of Soil Physical Properties: Bringing Theory into Practice*, SSSA Special Publication No. 30, Soil Science Society of America, Madison, WI, pp. 143–168.

Dane, J. H., and Hruska, S. 1983. *In-situ* determination of soil hydraulic properties during drainage, *Soil Science Society of America Journal*, Vol. 47, No. 4, pp. 619–624.

Dane, J. H., and Topp, G. C. 2002. Methods of soil analysis, Part 4, Physical methods, Soil Science Society of America, Madison, WI.

Daniel, D. E. 1983. Permeability test for unsaturated soil, *Geotechnical Testing Journal ASTM*, Vol. 6, No. 2, pp. 81–86.

Daniel, D. E., Hamilton, J. M., and Olson, R. E. 1981. Suitability of thermocouple psychrometers for studying moisture movement in unsaturated soils, in T. F. Zimmie and C. O. Riggs (Eds.), *Permeability and Groundwater Contaminant Transport*, ASTM STP 746, American Society for Testing and Materials, Philadelphia, PA, pp. 84–100.

Darcy, H. 1856. *Histoire des Foundataines Publique de Dijon*, Dalmont, Paris, pp. 590–594.

Davies, J. T., and Rideal, E. K. 1963. *Interfacial Phenomena*, 2nd ed., Academic, New York.

Day, R. W. 1994. Swell-shrink behaviour of expansive compacted clay, *Journal of Geotechnical Engineering, ASCE*, Vol. 120, No. 3, pp. 618–623.

De Bruijn, C. M. A. 1961. Swelling characteristics of a transported soil profile at Leeuhof Vereeniging (Transvaal), *Proceedings of the Fifth International Conference on Soil Mechanics Foundation Engineering*, Paris, Vol. 1, pp. 43–49.

De Bruijn, C. M. A. 1965a. Annual redistribution of soil moisture suction and soil moisture density beneath two different surface covers and the associated heaves at the Onderstepoort test site near Pretoria, in G. D. Aitchison (Ed.), *Moisture Equilibria and Moisture Changes in Soils beneath Covered Areas*, A Symposium in Print, Butterworths, Sydney, pp. 122–134.

De Bruijn, C. M. A. 1965b. Some observations on soil moisture conditions beneath and adjacent to tarred roads and other surface treatments in South Africa, in G. D. Aitchison (Ed.), *Moisture Equilibria and Moisture Changes in Soils Beneath Covered Areas*, A Symposium in Print, Butterworths, Sydney, pp. 135–142.

Decagon Services. 2009. Promotional material on the WP-4, Pullman, WA.

Deere, D. U., and Patton, F. D. 1971. Slope stability in residual soils, *Proceedings of the Fourth Pan-Am Conference Soil Mechanics Foundation Engineering*, San Juan, Vol. 1, pp. 87–170.

De Jong, E., and Warkentin, B. P. 1965. Shrinkage of soil samples with varying clay concentration, *Canadian Geotechnical Journal*, Vol. 11, No. 1, pp. 16–22.

Delage, P., and Graham, J. 1995. Mechanical behaviour of unsaturated soils: Understanding the behaviour of unsaturated soils requires reliable conceptual models, *Proceedings of the First International Conference on Unsaturated Soils UNSAT-95*, Paris, pp. 1223–1256.

Delage, P., Audiguier, M., Cui, Y.-J., and Howat, M. D. 1996. Microstructure of a compacted silt, *Canadian Geotechnical Journal*, Vol. 33 pp. 150–158.

Delage, P., Howat, M. D., and Cui, Y. J. 1998. The relationship between suction and swelling properties in a heavily compacted unsaturated clay, *Engineering Geology*, Vol. 50, pp. 31–48.

Derjaguin, B. V. 1965. Recent research into the properties of water in thin films and in micro-capillaries, *Proceedings of the Society for Experimental Biology Symposium*, Vol. 19, *The State and Movement of Water in Living Organisms*, Cambridge University Press, London.

Derjaguin, B. V., and Churaev, N. V. 1981. Structure of the boundary layers of liquids and its influence on the mass transfer in fine pores, in D. A. Cadenhead and J. F. Danielli (Eds.), *Progress in Surface and Membrane Science*, Academic Press, New York, pp. 69–130.

Desai, C. S. 1975a. Finite element methods for flow in porous media, *Finite Elements in Fluids*, Vol. 1, *Viscous Flows and Hydrodynamics*, Wiley, Chichester, England, pp. 157–182.

Desai, C. S. 1975b. Finite element residual schemes for unconfined flow, *International Journal of Numerical Methods Engineering*, Vol. 10, pp. 1415–1418.

Desai, C. S., and Abel, J. F. 1972. *Introduction to the Finite Element Method: A Numerical Method for Engineering Analysis*, Van Nostrand Reinhold, New York.

Desai, C. S., and Christian, J. T. 1977. *Numerical Methods in Geotechnical Engineering*, McGraw-Hill, New York.

De Vries, D. A. 1963. Thermal properties of soils, in W. H. van Wijk (Ed.), *The Physics of Plant Environment*, North-Holland Publishing Company, Amsterdam, p. 382.

De Vries, D. A. 1975. Heat transfer in soils, in *Heat and Mass Transfer in the Biosphere*, Vol. 1, *Transfer Processes in Plant Environment*, Scripta, Washington, DC, pp. 5–28.

Dickson, L. E. J. 1939. *New First Course in the Theory of Equations*, Wiley, New York.

Dineen, K., and Burland, J. 1995. A new approach to osmotically controlled oedometer testing, in E. E. Alonso and P. Delage (Eds.), *Unsaturated Soils*, *Proceedings of the First International Conference on Unsaturated Soils*, Balkema, Rotterdam, Paris, Vol. 2, pp. 459–465.

Dinh T. T. 1996. Plane strain modelling of volume change in expansive soil, M.Sc. Thesis, Faculty of Engineering and Surveying, University of Southern Queensland, Brisbane, Australia.

Dirksen, C. 1999. *Soil Physics Measurements*, Catena Verlag, Reiskirchen, Germany.

Dirksen, C., and Miller, R. D. 1966. Closed-system freezing of unsaturated soil, *Soil Science Society of America Proceedings*, Vol. 30, pp. 168–173.

Donald, I. B. 1956. Shear strength measurements in unsaturated non-cohesive soils with negative pore pressures, *Proceedings of the Second Australia-New Zealand Conference on Soil Mechanics and Foundation Engineering*, Christchurch, New Zealand, pp. 200–205.

Donald, I. B. 1961. The mechanical properties of saturated and partly saturated soils with special reference to negative pore water pressure, PhD Thesis, University of London, London.

Donald, I. B. 1963. Effective stress parameters in unsaturated soils, *Proceedings of the Fourth Australia-New Zealand Conference on Soil Mechanics Foundation Engineering*, Adelaide, South Australia, pp. 41–46.

Dong, A., Grattan, S. R., Carroll, J. J., and Prashar, C. R. K. 1992. Estimation of daytime net radiation over well-watered grass, *Journal of Irrigation Drainage Engineering*, Vol. 118, No. 3, pp. 466–479.

Doorenbos, J., and Pruitt, W. O. 1977. Guidelines for predicting crop water requirements, Irrigation and Drainage Paper 24, 2nd ed., Food and Agriculture Organization, Rome.

Dorsey, N. E. 1940. *Properties of Ordinary Water Substances*, American Chemical Society Monograph Series, Reinhold, New York.

Dregne, H. E. 1976. *Soils in Arid Regions*, Elsevier, New York.

Driscoll, R. 1983. The influence of vegetation on swelling and shrinkage of clay soils in Britain, *Geotechnique*, Vol. 33, pp. 93–105.

Drumright, E. 1987. Shear strength for unsaturated soils, PhD Thesis, University of Colorado, Fort Collins, CO.

Drungil, C. E. C., Abt, K., and Gish, T. J. 1989. Soil moisture determination in gravelly soils with time domain reflectometry, *Transactions of American Society for Agricultural Engineering*, Vol. 32, pp. 177–180.

Dunn, C. S. 1965. Developments in the design of triaxial equipment for testing compacted soils, *Proceedings of the Symposium on Economic Use of Soil Testing in Site Investigation*, Birmingham, AL, Vol. 3, pp. 19–25.

Duran, A. J. G. 1986. Study of effect of contact on the filter paper technique in the measurement of soil suction, M.Sc Thesis, Imperial College, London.

Durner, W. 1994. Hydraulic conductivity estimation for soils with heterogeneous pore structure, *Water Resources Research*, Zurich, Switzerland, Vol. 30, No. 2, pp. 211–223.

Ebrahimi-Birang, N., Gitirana, Jr., G. F. N., Fredlund, D. G., Fredlund, M. D., and Samarasekera, L. 2004. A lower limit for the water permeability coefficient, *Proceedings of the Fifty-Seventh Canadian Geotechnical Conference*, Quebec, QC, Vol. 1, pp. 12–19.

Ebrahimi-Birang, N., Fredlund, D. G., and Samarasekera, L. 2007. Hysteresis of the soil-water characteristic curve in the high suction range, *Proceedings of the Sixtieth Canadian Geotechnical Conference*, Ottawa, ON, pp. 1061–1068.

Ebrahimi-Birang, N., Fredlund, D. G., and Samarasekera, L. 2008. Testing for coefficient of permeability of a sandy soil in the residual state zone, *Proceedings of the First European Conference on Unsaturated soils*, Durham, U.K., pp. 129–134.

Edil, T. B., Motan, S. E., and Toha, F. X. 1981. Mechanical behaviour and testing methods of unsaturated soils, in R. N. Yong and F. C. Townsend (Eds.), *Laboratory Shear Strength of Soil*, ASTM STP No. 740, American Society for Testing and Materials, Philadelphia, PA, pp. 114–129.

Edlefsen, N. E., and Anderson, A. B. C. 1943. Thermodynamics of soil moisture, *Hilgardia*, Vol. 15, pp. 31–298.

El-Ehwany, M., and Houston, S. L. 1990. Settlement and moisture movement in collapsible soils, *Journal of Geotechnical Engineering*, Vol. 116, No. 10. pp. 1521–1535.

Elrick, D. E., and Bowman, D. H. 1964. Improved apparatus for soil moisture flow measurements, *Soil Science Society of America Proceedings*, Vol. 28, pp. 450–453.

Elzeftawy, A., and Cartwright, K. 1981. Evaluating the saturated and unsaturated hydraulic conductivity of soils, in T. F. Zimmie and C. D. Riggs (Eds.), *Permeability and Groundwater Contaminant Transport*, ASTM STP 746, American Society for Testing and Materials, Philadelphia, PA, pp. 168–181.

Erh, K. T., Nielsen, D. R., and Biggar, J. W. 1971. Two-dimensional heat transfer in porous media with steady-state flow, *Soil Science Society of America Proceedings*, Vol. 35, pp. 209–214.

Escario, V. 1969. Swelling of soils in contact with water at a negative pressure, *Proceedings of the Second International Conference on Expansive Soil*, Texas A & M University, College Station, TX, pp. 207–217.

Escario, V. 1980. Suction–controlled penetration and shear tests, *Proceedings of the Fourth International Conference on Expansive Soils*, American Society of Civil Engineers, Denver, CO, Vol. 2, pp. 781–797.

Escario, V., and Juca, J. F. T. 1989. Shear strength and deformation of partly saturated soils, *Proceedings of the Twelfth International Conference on Soil Mechanics and Foundation Engineering*, Rio de Janeiro, Vol.1, pp. 43–46.

Escario, V., and Sáez, J. 1973. Measurement of the properties of swelling and collapsing soils under controlled suction, *Proceedings of the Third International Conference on Expansive Soils*, Haifa, Vol. 1, pp. 195–200.

Escario, V., and Sáez, J. 1986. The shear strength of partly saturated soils, *Geotechnique*, Vol. 36, No. 3, pp. 453–456.

Escario, V., and Sáez, J. 1987. Shear strength of partly saturated soil versus suction, *Proceedings of the Sixth International Conference on Expansive Soils*, New Delhi, Vol. 2, pp. 602–604.

Estabragh, A. R., Javadi, A. A., and Boot, J. C. 2004. Effect of compaction pressure on consolidation behaviour of unsaturated soil, *Canadian Geotechnical Journal*, Vol. 41, pp. 540–550.

Everett, D. H. 1955. A general approach to hysteresis–Part 4: An alternative formulation of the domain model, *Transactions of the Faraday Society*, Vol. 51, pp. 1551–1557.

Everett, D. H. 1961. The thermodynamics of frost damage to porous solids, *Transactions of the Faraday Society*, Vol. 57 (Part 9), pp. 1542–1551.

Faizullaev, D. F. 1969. Laminar motion of multiphase media in conduits. Special Research Report, translated from Russian by Consultants Bureau, New York.

Falta, R. W., Javandel, I., Pruess, K., and Witherspoon, P. A. 1989. Density-driven flow of gas in the unsaturated zone due to the evaporation of volatile organic compounds, *Water Resources Research*, Vol. 25, No. 10, pp. 2159–2169.

Fan, K., Fredlund, D. G., and Wilson, G. W. 1986. An interslice force function for limit equilibrium slope stability analysis, *Canadian Geotechnical Journal*, Vol. 23, No. 3, pp. 287–296.

Farouki, O. 1981. *Thermal Properties of Soils*, CRREL Monograph 81–1, U.S. Army Corps of Engineers, Hanover, NH.

Farrel, D. A., and Larson, W. E. 1972. Modeling the pore structure of porous media, *Water Resources Research Journal*, Vol. 8, No. 3, pp. 699–706.

Fawcett, R. G., and Collis-George, N. 1967. A filter-paper method for determining the moisture characteristics of soil, *Australian Journal of Experimental Agriculture and Animal Husbandry*, Vol. 7, pp. 162–167.

Feng, M. 1999. The effect of capillary hysteresis on the measurement of matric suction using thermal conductivity sensors, M.Sc. Thesis, University of Saskatchewan, Saskatoon, SK.

Feng, M., and Fredlund, D. G. 1999. Hysteretic influence associated with thermal conductivity sensor measurements, *Proceedings of the Fifty-Second Canadian Geotechnical Conference*, Regina, SK, pp. 651–657.

Feng, M., and Fredlund, D. G. 2003. Calibration of thermal conductivity sensors with consideration of hysteresis, *Canadian Geotechnical Journal*, Vol. 40, No. 5, pp. 1048–1055.

Feng, M., Gan, J. K.-M., and Fredlund, D. G. 1998. A laboratory study of swelling pressure using various test methods, *Proceedings of the Second International Conference on Unsaturated Soils (UNSAT-98)*, Beijing, Vol. 1, pp. 350–355.

Feng, M., Fredlund, D. G., and Shuai, F. 2002. A laboratory study of the hysteresis of a thermal conductivity soil suction sensor, *Geotechnical Testing Journal, ASTM*, Vol. 25, No. 3, pp. 303–314.

Fermi, E. 1926. Sulla quantizzazione del gas perfetto monoatomico, *Rendiconti Lincei*, Vol. 3, pp. 145–149.

Fick, A. 1855. Ueber Diffusion, *Annalen der Physik*, Leipzig, Vol. 94, pp. 59–86.

Filson, H. L. 1980. A study of heave analysis using the oedometer test, M.Sc. Thesis, University of Saskatchewan, Saskatoon, SK.

Finn, W. D. L. 1967. Finite element analysis of seepage through dams, *Journal of Soil Mechanics Foundation Engineering Division, ASCE*, SM6, pp. 41–48.

Fisher, J. C. 1948. Measurement of interfacial tensions, Technical Publication, American Institute of Mining and Metallurgical Engineers–Metals. York, PA Technology, York, PA.

Fisher, R. A. 1924. The freezing of water in capillary systems: A critical discussion, *Journal of Physical Chemistry*, Vol. 28, pp. 36–67.

Fleureau, J. M., Kheirbek-Saoud, S., and Taibi, S. 1995. Experimental aspects and modelling of the behaviour of soils with a negative pressure, *Proceeding of the First International Conference on Unsaturated Soils (UNSAT-95)*, E. E. Alonso, P. Delage (Eds.) Paris, France, pp. 57–62.

Fleureau, J. M., Hadiwarrdoyo, S., and Kheirbek-Saoud, S. 2004. Simplified approach to the behaviour of compacted soils on drying and wetting parts, in J. F. T. Juca, T. M. P. de Campos, and F. A. M. Marinho, Eds., *Proceedings of the Third International Conference on Unsaturated Soils*, Recife, Brazil, Lisse, pp. 1147–1154.

FlexPDE Solutions. 1999. *Version 3.x Reference Manual*, PDE Solutions, Antioch, CA.

Flint, A. L., Campbell, G. S., Ellett, K. M., and Calissendorff, C. 2002. Calibration and temperature correction of heat dissipation matric suction sensors, *Soil Science Society of America Proceedings*, Vol. 66, pp. 1439–1445.

Fookes, P., and Parry, R. 1994. Engineering characteristics of arid soils, in P. Fookes and R. Parry (Eds.), *Proceedings of the First International Symposium on Engineering Characteristics of Arid Soils, London, UK*, Balkema, The Netherlands.

Fourie, A. B. 1989. Laboratory evaluation of lateral swelling pressure, *Journal Geotechnical Engineering, ASCE*, Vol. 115, No. 10, pp. 1481–1486.

Fourie, A. B. 1996. Predicting rainfall-induced slope stability, *Proceedings of the Institute of Civil Engineers, Geotechnical Engineers*, Vol. 119, pp. 211–218.

Fourie, A. B., Rowe, D., and Blight, G. E. 1999. The effect of infiltration on the stability of the slopes of a dry ash dump, *Geotechnique*, Vol. 49, No. 1, pp. 1–13.

Fredlund, D. G. 1964. Comparison of soil suction and one-dimensional consolidation characteristics of a highly plastic clay, Technical Paper No. 245, National Research council of Canada Division of Building Research, Ottawa, ON.

Fredlund, D. G. 1969. Consolidometer test procedural factors affecting swell properties, *Proceedings of the Second International Conference on Expansive Soils*, Texas A & M University, College Station, TX, pp. 435–456.

Fredlund, D. G. 1973a. Volume change behaviour of unsaturated soils, PhD Thesis, Department of Civil Engineering, University of Alberta, Edmonton.

Fredlund, D. G. 1973b. Discussion at the Second Technical Session, Division Two: Flow and Shear Strength, *Proceedings of the Third International Conference on Expansive Soils*, Haifa, Jerusalem Academic Press, Vol. II, Jerusalem, pp. 71–76.

Fredlund, D. G. 1975a. Prediction of heave in unsaturated soils, *Proceedings of the Fifth Asian Regional Conference*, Bangalore, India, pp. 19–22.

Fredlund, D. G. 1975b. Engineering properties of expansive clays, Transportation and Geotechnical Group Publication IR-7, Department of Civil Engineering, University of Saskatchewan, Saskatoon, SK.

Fredlund, D. G. 1975c. A diffused air volume indicator for unsaturated soils, *Canadian Geotechnical Journal*, Vol. 12, No. 4, pp. 533–539.

Fredlund, D. G. 1976. Density and compressibility characteristics of air-water mixtures, *Canadian Geotechnical Journal*, Vol. 13, No. 4, pp. 386–396.

Fredlund, D. G. 1979. Appropriate concepts and technology for unsaturated soils, Second Canadian Geotechnical Colloquium, *Canadian Geotechnical Journal*, Vol. 16, No. 1, pp. 121–139.

Fredlund, D. G. 1980a. Use of computers for slope stability analysis, State-of-the-Art Paper, *Proceedings International Symposium on Landslide*, New Delhi, Vol. 2, pp. 129–138.

Fredlund, D. G. 1980b. Discussion: Consolidation of unsaturated soils including swelling and collapse behaviour, by Lloret A. and Alonso, E., *Geotechnique* Vol. 30, No. 4, pp. 449–477.

Fredlund, D. G. 1981a. The shear strength of unsaturated soils and its relationship to slope stability problems in Hong Kong, *The Hong Kong Engineer*, Vol. 9, No. 4, pp. 37–45.

Fredlund, D. G. 1981b. Seepage in saturated soils. Session 3 on Groundwater and Seepage Problems, *Proceedings of the Tenth International Conference on Soil Mechanics and Foundation Engineering*, Stockholm, Vol. 4, pp. 629–641.

Fredlund, D. G. 1982. Consolidation of unsaturated porous media, *NATO Advanced Study Institute on Mechanics of Fluids in Porous Media–New Approaches in Research*, University of Delaware, Newark, DE.

Fredlund, D. G. 1983. Prediction of ground movements in swelling clays, paper presented to the Thirty-First Annual ASCE Soil Mechanics and Foundation Engineering Conference, Earle Brown Center, University of Minnesota, Minneapolis, MN, February 1.

Fredlund, D. G. 1984. Analytical methods for slope stability analysis, *Proceedings of the Fourth International Symposium on Landslides*, Toronto, ON, pp. 229–250.

Fredlund, D. G. 1985a. Theory formulation and application for volume change and shear strength problems in unsaturated soils, *Proceedings of the Eleventh International Conference on Soil Mechanics and Foundation Engineering*, Session 9C, Problems in Areas with Special Geologic Conditions, Foundation Problems in Arid Areas, San Francisco, CA, Vol. 6, pp. 2835–2838.

Fredlund, D. G. 1985b. Soil mechanics principles that embrace unsaturated soils, *Proceedings of the Eleventh International Conference on Soil Mechanics and Foundation Engineering*, San Francisco, CA, Vol. 2, pp. 465–473.

Fredlund, D. G. 1987a. Slope stability analysis incorporating the effect of soil suction, in M. G. Anderson and K. S. Richards (Eds.), *Slope Stability*, Wiley, New York, pp. 113–144.

Fredlund, D. G. 1987b. The stress state for expansive soils, *Proceedings of the Sixth International Conference on Expansive Soils*, New Delhi, pp. 524–534.

Fredlund, D. G. 1987c. The prediction and performance of structures on expansive soils, *Proceedings of the International Symposium on Prediction and Performance in Geotechnical Engineering*, Calgary, AB, pp. 51–60.

Fredlund, D. G. 1989a. The character of the shear strength envelope for unsaturated soils, *The Victor de Mello Volume, Twelfth International Conference on Soil Mechanics and Foundation Engineering*, Rio de Janeiro, pp. 142–149.

Fredlund, D. G. 1989b. Negative pore-water pressures in slope stability, paper presented at the Simposio Suramericano de Deslizamient, Paipa, Colombia, pp. 1–31.

Fredlund, D. G. 1989c. Soil suction monitoring for roads and airfields, paper presented at the Symposium on the State-of-the-art of Pavement Response Monitoring Systems for Roads and Airfields, sponsored by the U.S. Army Corps of Engineers, Hanover, NH, March 6–9.

Fredlund, D. G. 1992. Background, theory and research related to the use of thermal conductivity sensors for matric suction measurements, Soil Science Society of America Special Publication No. 30, pp. 249–262.

Fredlund, D. G. 1994. Visualization of the world of soil mechanics, *Proceedings of the Sino-Canadian Symposium on Unsaturated/Expansive Soils*, Wuhan, China, pp. 1–21.

Fredlund, D. G. 1995a. Unsaturated soil behaviour within a generalized soil mechanics context, *Proceedings of the Canada-Kenya Symposium on Unsaturated Soil Behaviour and Applications*, University of Nairobi, Nairobi, Kenya, pp. 1–20.

Fredlund, D. G. 1995b. The scope of unsaturated soil mechanics: An overview, *Proceedings of the First International Conference on Unsaturated Soils, UNSAT-95*, Paris, Vol. 3, pp. 1155–1177.

Fredlund, D. G. 1995c. The stability of slopes with negative pore-water pressures, *Proceedings of the Ian Boyd Donald Symposium on Modern Developments in Geomechanics*, Department of Civil Engineering, Monash University, Clayton, Victoria, Australia, pp. 99–116.

Fredlund, D. G. 1995d. Prediction of unsaturated soil functions using the soil-water characteristic curves, *Proceedings of the Bengt B. Broms Symposium on Geotechnical Engineering*, Singapore, pp. 113–133.

Fredlund, D. G. 1996a. The emergence of unsaturated soil mechanics, *The Fourth Spencer J. Buchanan Lecture*, Texas A & M, College Station, TX.

Fredlund, D. G. 1996b. Microcomputers and saturated/unsaturated continuum modelling in geotechnical engineering, paper presented at the Symposium on Computers in Geotechnical Engineering, INFOGEO '96, Sao Paulo, Brazil, Vol. 2, pp. 29–50.

Fredlund, D. G. 1996c. Geotechnical problems associated with swelling clay, Developments in Soil Science 24, in N. Ahmad and A. R. Mermut (Eds.), *Vertisol and Technologies for their Management*, Elsevier Science, Amsterdam, Holland, pp. 501–526.

Fredlund, D. G. 1997a. From theory to the practice of unsaturated soil mechanics, *Proceedings of the Third Brazilian Symposium on Unsaturated Soils*, Vol. 2, Rio de Janeiro, pp. 311–332.

Fredlund, D. G. 1997b. An introduction to unsaturated soil mechanics, paper presented at the Geotechnical Engineering Division GeoLogan Conference, ASCE, Special Geotechnical Publication, Logan, UT, Vol. 68, pp. 1–37.

Fredlund, D. G. 1998a. The practice of unsaturated soil mechanics, *Proceedings of the Eleventh Danube-European Conference on Soil Mechanics and Foundation Engineering*, Porec, Croatia, pp. 29–50.

Fredlund, D. G. 1998b. Bringing unsaturated soil mechanics into engineering practice, *Proceedings of the Second International Conference on Unsaturated Soils, UNSAT-98*, Beijing, Vol. 2, pp. 1–36.

Fredlund, D. G. 2000a. The implementation of unsaturated soil mechanics into geotechnical engineering, The R.M. Hardy Address, *Canadian Geotechnical Journal*, Vol. 37, No. 5, pp. 963–986.

Fredlund, D. G. 2000b. Methodology and experience in conducting geotechnical engineering research, *Vietnam Geotechnical Journal*, Vol. 1, pp. 2–7.

Fredlund, D. G. 2000c. Historical development and milestones in unsaturated soil mechanics, *Proceedings of the Asian Conference on Unsaturated Soils, UNSAT ASIA 2000*, Singapore, pp. 53–68.

Fredlund, D. G. 2001. Ensuring a sound scientific basis for geotechnical engineering in developing and developed countries, *Proceedings of the International Conference on Management of the Land and Water Resources*, Hanoi, pp. 3–20.

Fredlund, D. G. 2002a. Use of the soil-water characteristic curve in the implementation of unsaturated soil mechanics, *Proceedings of the Third International Conference on Unsaturated Soils, UNSAT 2002*, Recife, Brazil, Vol. 3, pp. 887–902.

Fredlund, D. G. 2002b. Teaching unsaturated soil mechanics as part of the undergraduate civil engineering curriculum, *Proceedings of the Second Pan American Conference on the Teaching and Learning Processes of Geotechnical Engineering*, Quayaquil, Ecuador.

Fredlund, D. G. 2006. Unsaturated soil mechanics in engineering practice, The 2005 Terzaghi Lecture: *Journal of Geotechnical and Geoenvironmental Engineering, ASCE*, Vol. 132, No. 3, pp. 286–321.

Fredlund, D. G. 2007a. Slope stability hazard management systems, The Zen Guo-Xi Lecture, *Geotechnical Journal in China*, Zhejiang University, Hangzhou, China, Vol. 8, No. 11, pp. 1595–1711.

Fredlund, D. G. 2007b. Engineering design protocols for unsaturated soils, *Proceedings of the Third Asian Conference on Unsaturated Soil UNSAT-ASIA 2007*, Nanjing, China, pp. 27–45.

Fredlund, D. G. 2009. Paradigm shifts to facilitate the practice of unsaturated soil mechanics, *Proceedings of the Fourth Asia-Pacific Conference on Unsaturated Soils*, Newcastle, Australia, November 23–25, Vol. 1, pp. 599–612.

Fredlund, D. G., and Barbour, S. L. 1992. Integrated seepage modeling and slope stability analyses: A generalized approach for saturated/unsaturated soils, in R. N. Chowdhury (Ed.), *Geomechanics and Water Engineering in Environmental Management*, Balkema, Rotterdam, Brookfield, pp. 3–35.

Fredlund, D. G., and Dahlman, A. E. 1971. Statistical geotechnical properties of glacial Lake Edmonton sediments, *Proceedings of the First International Conference on Application of Statistics and Probability to Soil and Structural Engineering*, Hong Kong, pp. 212–228.

Fredlund, D. G., and Dakshanamurthy, V. 1982. Prediction of moisture flow and related swelling or shrinking in unsaturated soils, *Geotechnical Engineering Journal*, Vol. 13, pp. 15–49.

Fredlund, D. G., and Fredlund, L. D. 1993. Computing and geotechnical engineering: The 1990's and beyond, *Geotechnical News*, Bi-Tech Publishers, Vancouver, BC.

Fredlund, D. G., and Gan, J. K.-M. 1995. The collapse mechanism of a soil subjected to one-dimensional loading and wetting, *Proceedings of the Genesis and Properties of Collapsible Soils*, NATO Advanced Research Workshop, Loughborough, UK, Vol. 468, pp. 173–205.

Fredlund, D. G., and Gitirana, Jr., de F. N. G. 2005. Unsaturated soil mechanics as a series of partial differential equations, *Proceedings of the International Conference on Problematic Soils*, Famagusta, Northern Cyprus, pp. 3–30.

Fredlund, D. G., and Hasan, J. U. 1979. One-dimensional consolidation theory: Unsaturated soils, *Canadian Geotechnical Journal*, Vol. 16, No. 3, pp. 521–531.

Fredlund, D. G., and Houston, S. L. 2001. Laboratory techniques for unsaturated soils, *Proceedings of the Fifteenth International Conference on Soil Mechanics and Geotechnical Engineering*, Istanbul, Vol. 4, pp. 2913–2914.

Fredlund, D. G., and Houston, S. L. 2009. Protocol for the assessment of unsaturated soil properties in geotechnical engineering practice, *Canadian Geotechnical Journal*, Vol. 46, pp. 694–707.

Fredlund, D. G., and Krahn, J. 1977. Comparison of slope stability methods analysis, *Canadian Geotechnical Journal*, Vol. 14, No. 3, pp. 429–439.

Fredlund, D. G., and Morgenstern, N. R. 1973. Pressure response below high air-entry disks, *Proceedings of the Third International Conference on Expansive Soils*, Haifa, Vol. 1, pp. 97–108.

Fredlund, D. G., and Morgenstern, N. R. 1976. Constitutive relations for volume change in unsaturated soils, *Canadian Geotechnical Journal*, Vol. 13, No. 3, pp. 261–276.

Fredlund, D. G., and Morgenstern, N. R. 1977. Stress state variables for unsaturated soils, *Journal of Geotechnical Engineering Division, ASCE*, Vol. 103, No. GT5, pp. 447–466.

Fredlund, D. G., and Morgenstern, N. R. 1978. Closure to paper entitled, "Stress state variables for unsaturated soils," *Journal of American Society of Civil Engineers, Geotechnical Division, ASCE*, Vol. GT11, pp. 1415–1416

Fredlund, D. G., and Pham, H. Q. 2006. A volume-mass constitutive model for unsaturated soils in terms of two independent stress state variables, *Proceedings of the Fourth International Conference on Unsaturated Soils, ASCE*, Carefree, AZ, Vol. 1, pp. 105–134.

Fredlund, D. G., and Pham, H. Q. 2007. Independent roles of the stress state variables on volume-mass constitutive relation, *Proceedings of the Second International Conference on Mechanics of Unsaturated Soils*, Bauhaus Universitat, Weimar, Germany, pp. 37–44.

Fredlund, D. G., and Rahardjo, H. 1985. Theoretical context for understanding unsaturated residual soil behaviour, *Proceedings of the First International Conference on Geotechnics of Tropical Lateritic and Saprolitic Soils*, Brasilia, Brazil, pp. 295–306.

Fredlund, D. G., and Rahardjo, H. 1986. Unsaturated soil consolidation theory and laboratory experimental data, in R. N. Yong and F. C. Townsend (Eds.), *Consolidation of Soils: Testing and Evaluation*, ASTM STP 892, American Society for Testing and Materials, Philadelphia, PA, pp. 154–169.

Fredlund, D. G., and Rahardjo, H. 1987. Soil mechanics principles for highway engineering in arid regions, *Transportation Research Record*, Vol. 1137, pp. 1–11.

Fredlund, D. G., and Rahardjo, H. 1988. State-of-development in the measurement of soil suction, *Proceedings of the International Conference on Engineering Problems on Regional Soils*, Beijing, pp. 582–588.

Fredlund, D. G., and Rahardjo, H. 1993a. *Soil Mechanics for Unsaturated Soils*, Wiley, New York.

Fredlund, D. G., and Rahardjo, H. 1993b. An overview of unsaturated soil behaviour, in S. L. Houston and W. K. Wray (Eds.), *Unsaturated Soils*, Geotechnical Special Publication No. 39, American Society of Civil Engineers, New York, pp. 1–31.

Fredlund, D. G., and Rahardjo, H. 1993c. The role of unsaturated soil behaviour in geotechnical practice, *Proceedings of the Eleventh Southeast Asian Geotechnical Conference*, Singapore, pp. 37–49.

Fredlund, D. G., and Rahardjo, H. 1994. Hillside slope stability assessment in unsaturated residual soils, *Proceedings of Seminar on Geotechnical Aspects of Hillside Developments*, Kajang, Malaysia, pp. 1–41.

Fredlund, D. G., and Scoular, R. E. G. 1999. Using limit equilibrium concepts in finite element slope stability analysis, *Proceedings of the International Symposium on Slope Stability Engineering, IS-Shikoku '99*, Matsuyama, Shikoku, Japan, pp. 31–47.

Fredlund, D. G., and Stianson, J. 2009. Challenges associated with the design of covers, *Proceedings of the International Symposium on Geoenvironmental Engineering, ISGE 2009*, Hangzhou, China, Vol. 1, pp. 168–187.

Fredlund, D. G., and Stianson, J. 2011a. Design and functionality of soil covers in cold regions, *Geo-Strata Special Edition on Unsaturated Soils, Geo-Institute*, ASCE, Vol. 15, No. 1, pp. 30–32.

Fredlund, D. G., and Stianson, J. 2011b. Utilization of weather station data for the assessment of ground surface moisture flux, *Proceedings of the Fifth Asia-Pacific Conference on Unsaturated Soils, AP-UNSAT 2011*, Payatta, Thailand.

Fredlund, D. G., and Stone, J. 2011. Development of the Golder basic pressure plate cell, Internal report of Golder Associates, sponsored by Golder Innovation Research Fund, Saskatoon, SK.

Fredlund, D. G., and Vanapalli, S. K. 2002. Shear strength of unsaturated soils, in *Agronomy Soil Testing Manual*, Agronomy Society of America, Madison, WI, pp. 329–361.

Fredlund, D. G., and Vu, H. Q. 2001. Prediction of volume change in an expansive soil as a result of vegetation and environmental changes, *Proceedings of the Conference on Expansive Clay Soils and Vegetative Influences on Shallow Foundations, ASCE*, Geotechnical Special Publication, Houston, TX, Vol. 115, pp. 24–43.

Fredlund, D. G., and Vu, H. Q. 2004. Numerical modeling of swelling and shrinkage around slabs-on-ground, *Post-Tensioning Institute Journal*, Vol. 1, No. 3.

Fredlund, D. G., and Wilson, G. W. 2006. Are we doing it right? *Geotechnical News*, Vancouver, BC, pp. 43–48.

Fredlund, D. G., and Wong, D. K. H. 1989. Calibration of thermal conductivity sensors for measuring soil suction, *Geotechnical Testing Journal, ASTM*, Vol. 12, No. 3, pp. 188–194.

Fredlund, D. G., and Xing, A. 1994. Equations for the soil-water characteristic curve, *Canadian Geotechnical Journal*, Vol. 31, No. 3, pp. 521–532.

Fredlund, D. G., and Xing, A. 1997. The use of a rheological model for the visualization of unsaturated soil processes, *Proceedings of the Sixth National Congress on Mechanics*, Hanoi, pp. 221–227.

Fredlund, D. G., Bergan, A. T., and Sauer, E. K. 1975. Deformation characterization of subgrade soils for highways and runways in northern environments, *Canadian Geotechnical Journal*, Vol. 12, No. 2, pp. 213–223.

Fredlund, D. G., Bergan, A. T., and Wong, P. K. 1977. Relationship between modulus and stress conditions for cohesive subgrade soils, *Transportation Research Record*, Vol. 642, pp. 73–81.

Fredlund, D. G., Morgenstern, N. R., and Widger, R. A. 1978. The shear strength of unsaturated soils, *Canadian Geotechnical Journal*, Vol. 15, No. 3, pp. 313–321.

Fredlund, D. G., Hasan, J. U., and Filson, H. 1980a. The prediction of total heave, *Proceedings of the Fourth International Conference on Expansive Soils*, Denver, CO, Vol. 1, pp. 1–17.

Fredlund, D. G., Krahn, J., and Hasan, J. U. 1980b. Variability of an expansive clay deposit. *Proceedings of the Fourth International Conference on Expansive Soils*, Denver, CO, Vol. 1, pp. 332–338.

Fredlund, D. G., Krahn, J., and Pufahl, D. 1981. The relationship between limit equilibrium slope stability methods, *Proceedings of the Tenth International Conference on Soil Mechanics and Foundation Engineering*, Stockholm, Vol. 3, pp. 409–416.

Fredlund, D. G., Rahardjo, H., and Gan, J. K.-M 1987. Nonlinearity of strength envelope for unsaturated soils, *Proceedings of the Sixth International Conference on Expansive Soils*, New Delhi, Vol. 1, pp. 49–54.

Fredlund, D. G., Gan, J. K. M., and Rahardjo, H. 1991a. Measuring negative pore-water pressures in a freezing environment, *Transportation Research Record*, Vol. 1307, pp. 291–299.

Fredlund, D. G., Rahardjo, H., and Gan, J. K.-M. 1991b. Discussion on the paper entitled: A framework for unsaturated soil behaviour, by D. G. Toll 1990, *Geotechnique*, Vol. 40, No. 1, pp 31–34, published in *Geotechnique*, Vol. 41, No. 1, pp. 159–161.

Fredlund, D. G., Sattler, P. J., and Gan, J.-K.-M. 1992. *In situ* suction measurements using thermal sensors, *Proceedings of the Seventh International Conference on Expansive Soils*, Dallas, TX, pp. 325–330.

Fredlund, D. G., Rahardjo, H., and Ng, T. 1993. Effect of pore-air and negative pore-water pressures on stability at the end-of-construction, *Proceedings of International Conference on Dam Engineering*, Johor Bahru, Malaysia, pp. 43–51.

Fredlund, D. G., Gan, J. K. M., and Li, W. X. 1994a. Thermal conductivity suction sensors design considerations, *Proceedings of the Thirteenth International Conference on Soil Mechanics and Foundation Engineering*, New Delhi, pp. 291–296.

Fredlund, D. G., Xing, A., and Huang, S. Y. 1994b. Predicting the permeability function for unsaturated soils using the soil-water characteristic curve, *Canadian Geotechnical Journal*, Vol. 31, No. 4, pp. 533–546.

Fredlund, D. G., Gan, J. K.-M., and Gallen, P. 1995a. Suction measurements on compacted till specimens and an indirect filter paper calibration technique, *Transportation Research Record*, Vol. 1481, pp. 3–9.

Fredlund, D. G., Huang, S. K., Clifton, A. W., Wang, Q., Barbour, S. L., Ke, Z. J., and Fan, W. Y. 1995b. Matric suction and deformation monitoring at an expansive soil site in southern China, *Proceedings of the First International Conference on Unsaturated Soils, UNSAT-95*, Paris, Vol. 2, pp. 855–862.

Fredlund, D. G., Vanapalli, S. K., Xing, A., and Pufahl, D. E. 1995c. Predicting the shear strength function for unsaturated soils using the soil-water characteristic curve, *Proceedings of the First International Conference on Unsaturated Soils (UNSAT-95)*, Paris, Vol. 1, pp. 63–70.

Fredlund, D. G., Xing, A., Fredlund, M. D., and Barbour, S. L. 1996. The relationship of the unsaturated soil shear strength of the soil-water characteristic curve, *Canadian Geotechnical Journal*, Vol. 32, pp. 440–448.

Fredlund, D. G., Shuai, F., Yazdani, J., and Feng, M. 1998. Recent developments of a sensor for the *in situ* measurement of matric suction, *Proceedings of the Fifty-First Canadian Geotechnical Conference*, Edmonton, AB, Vol. 1, pp. 81–86.

Fredlund, D. G., Scoular, E. G., and Vu, H. Q. 1999. Using finite element stress analysis for normal and actuating shear stresses in slope stability analysis, *Vietnamese Geotechnical Journal*, Hanoi, Vietnam, Vol. 2, pp. 8–14.

Fredlund, D. G., Shuai, F., and Feng, M. 2000a. Use of a new thermal conductivity sensor for laboratory suction measurement, *Proceedings of the Asian Conference in Unsaturated Soils, UNSAT ASIA 2000*, Singapore, pp. 275–280.

Fredlund, D. G., Shuai, F., and Feng, M. 2000b. Increased accuracy in suction measurement using an improved thermal conductivity sensor, *Proceedings of the Seventh International Conference on Tailings and Mine Waste*, Fort Collins, CO, pp. 443–450.

Fredlund, D. G., Vu, H. Q., and Thieu, N. T. M. 2000c. A component system approach to solving saturated-unsaturated soil problems, *Proceedings of the Canadian Society for Civil Engineering, Twenty-Eighth Annual Conference*, London, ON, pp. 448–455.

Fredlund, D. G., Fredlund, M. D., and Zakerzadeh, N. 2001a. Predicting the permeability function for unsaturated soils, *Proceedings of the International Symposium on Suction, Swelling, Permeability and Structured Clays, IS-Shizuoka*, Shizuoka, Japan, pp. 215–222.

Fredlund, D. G., Ng, C. W. W., Rahardjo, H., and Leong, E. C. 2001b. Unsaturated soil mechanics: Who Needs It? *Geotechnical News*, Bi–Tech Publishers, December 2001, pp. 43–45.

Fredlund, D. G., Rahardjo, H., Leong, E. C., and Ng, C. W. W. 2001c. Suggestions and recommendations for the interpretation of soil-water characteristic curves, *Proceedings of the Fourteenth Southeast Asian Geotechnical Conference*, Hong Kong, Vol. 1, pp. 503–508.

Fredlund, D. G., Wilson, G. W., and Barbour, S. L. 2001d. Unsaturated soil mechanics and property assessment, in R. K. Rowe (Ed.), *Geotechnical and Geoenvironmental Engineering Handbook*, Kluwer Academic Publishers, Norwell, MA, pp. 107–146.

Fredlund, D. G., Barbour, S. L., and Pham, Q. H. 2003. Evaluation of hysteresis models for predicting the boundary wetting curve, *Proceedings of the Second Asian Conference on Unsaturated Soils, UNSAT-ASIA 2003*, Osaka, pp. 407–412.

Fredlund, D. G., Pham, H. Q., and Fredlund, M. D. 2007. A practical volume-mass constitutive model for unsaturated expansive soils, *Proceedings of the International Workshop on Constitutive Modeling–Development, Implementation, Evaluation, and Application*, Hong Kong, pp. 342 347.

Fredlund, D. G., Houston, S. L., Nguyen, Q., and Fredlund, M. D. 2010a. Moisture movement through cracked clay soil profiles, *Geotechnical and Geological Engineering Journal*, Vol. 28, pp. 865–888.

Fredlund, D. G., Vu, H. Q., and Stianson, J. 2010b. Engineering protocols for the assessment of the net moisture flux at the ground surface, *Geotechnical Engineering Journal of the SEAGS and AGSSEA*, Vol. 41, No. 4, pp. 1–11.

Fredlund, D. G., Sheng, D., and Zhao, J. 2011a. Estimation of soil suction from the soil-water characteristic curve, *Canadian Geotechnical Journal*, Vol. 48, No. 2, pp. 186–198.

Fredlund, D. G., Stone, J., Stianson, J., and Sedgewick, A. 2011b. Determination of water storage and permeability functions for oil sands tailings, *Proceedings of the Tailings and Mine Waste Conference*, Vancouver, BC, November 7–9.

Fredlund, D. G., Stone, J., Stianson, J., and Sedgewick, A. 2011c. Obtaining unsaturated soil properties for high volume change materials, *Proceedings of the Fifth Asia-Pacific Conference on Unsaturated Soils*, Payatta, Thailand.

Fredlund, M. D. 1996. Design of a knowledge-based system for unsaturated soil properties, M.Sc. Thesis, Department of Civil Engineering, University of Saskatchewan, Saskatoon, SK.

Fredlund, M. D. 2000. The role of unsaturated soil property functions in the practice of unsaturated soil mechanics, PhD Thesis, University of Saskatchewan, Saskatoon, SK.

Fredlund, M. D. 2001. *SVFlux Manual*, SoilVision Systems, Saskatoon, SK.

Fredlund, M. D. 2002a. *ChemFlux User's Manual*, SoilVision Systems, Saskatoon, SK.

Fredlund, M. D. 2002b. *SVHeat User's Manual*, SoilVision Systems. Saskatoon, SK.

Fredlund, M. D. 2003. *User's Guide–A Knowledge-Based System for Geotechnical Engineers*, Version 3.34, SoilVision Systems, Saskatoon, SK.

Fredlund, M. D. 2004. Finite element stochastic analysis, *Proceedings of the Fifty-Seventh Canadian Geotechnical Conference and Fifth Joint IAH-CGS Conference*, Quebec, QC.

Fredlund, M. D. 2005. *SVFlux User's and Theory Manual*, SoilVision Systems, Saskatoon, SK.

Fredlund, M. D. 2009. *SVSlope User's and Theory Manual*, SoilVision Systems, Saskatoon, SK.

Fredlund, M. D. 2011. *SVAirFlow User's and Theory Guide*, SoilVision Systems, Saskatoon, SK.

Fredlund, M. D., and Feng, T. 2011. Combined seepage and slope stability analysis of rapid drawdown scenarios for levee design, *Proceedings of the Geo-Frontiers 2011 Conference, ASCE*, Dallas, TX.

Fredlund, M. D., and Thode, R. 2010. Effect of mesh resolution on long-term water balance calculations, *Proceedings of the Fifth International Conference on Unsaturated Soils*, Barcelona, Vol. 2, pp. 1035–1040.

Fredlund, M. D., Fredlund, D. G., Wilson, G. W., and Sillers, W. S. 1996. Design of a knowledge-based system for unsaturated soil properties, *Proceedings of the Third Canadian Conference on Computing in Civil and Building Engineering*, Montreal, QC, pp. 659–677.

Fredlund, M. D., Fredlund, D. G., and Wilson, G. W. 1997a. Estimation of unsaturated soil properties using a knowledge-based system, *Proceedings of the Fourth Congress on Computing in Civil Engineering, ASCE*, Philadelphia, PA, pp. 501–510.

Fredlund, M. D., Fredlund, D. G., and Wilson, G. W. 1997b. Prediction of the soil-water characteristic curve from grain-size distribution and volume-mass properties, *Proceedings of the Third Brazilian Symposium on Unsaturated Soils, NSAT '97*, Rio de Janeiro, Brazil, Vol. 1, pp. 13–23.

Fredlund, M. D., Wilson, G. W., and Fredlund, D. G. 1997c. Indirect procedures to determine unsaturated soil property functions, *Proceedings of the Fiftieth Canadian Geotechnical Conference*, Ottawa, ON, pp. 407–414.

Fredlund, M. D., Wilson G. W., and Fredlund D. G. 1997d. Estimation of hydraulic properties of an unsaturated soil using a knowledge-based system, *Proceedings of the Conference on the Characterization and Measurement of the Hydraulic*

Properties of Unsaturated Porous Media, Riverside, CA, pp. 1295–1316.

Fredlund, M. D., Wilson, G. W., and Fredlund, D. G. 1998. Estimation of hydraulic properties of an unsaturated soil using a knowledge-based system, *Proceedings of the Second International Conference on Unsaturated Soils, UNSAT-98*, Beijing, China, Vol. 1, pp. 479–484.

Fredlund, M. D., Fredlund, D. G., and Wilson, G. W. 2000a. An equation to represent grain-size distribution, *Canadian Geotechnical Journal*, Vol. 37, No. 4, pp. 817–827.

Fredlund, M. D., Fredlund, D. G., and Wilson, G. W. 2000b. Estimation of volume change functions for unsaturated soils, *Proceedings of the First Asian Conference in Unsaturated Soils, UNSAT ASIA 2000*, Singapore, pp. 663–668.

Fredlund, M. D., Wilson, G. W., and Fredlund, D. G. 2000c. Use of grain-size functions in unsaturated soil mechanics, *Proceedings of the Geo-Denver 2000 Conference, ASCE*, Denver, CO, pp. 69–83.

Fredlund, M. D., Fredlund, D. G., and Wilson, G. W. 2002a. Reply to: "An equation to represent grain-size distribution", *Canadian Geotechnical Journal*, Vol. 39, pp. 1437–1438.

Fredlund, M. D., Fredlund, D. G., and Wilson, G. W. 2002b. Use of the grain-size distribution for the estimation of the soil-water characteristic curve, *Canadian Geotechnical Journal*, Vol. 39, No. 5, pp. 1103–1117.

Fredlund, M. D., Stianson J., and Rykaart, M. 2002c. Application of automatic mesh refinement in the SVFlux and ChemFlux software packages, *Proceedings of Fifty-Fifth Canadian Geotechnical Conference and Third Joint IAH-CNC/CGS Conference*, Niagara Falls, ON.

Fredlund, M. D., Wilson, G. W., and Fredlund, D. G. 2002d. Representation and estimation of the shrinkage curve, *Proceedings of the Third International Conference on Unsaturated Soils, UNSAT 2002*, Recife, Brazil, pp. 145–149.

Fredlund, M. D., Fredlund, D. G., Houston, S. L., and Houston, W. B. 2003. Assessment of unsaturated soil properties for seepage modeling through tailings and mine wastes, *Proceedings of the Tenth International Conference on Tailings and Mine Wastes*, Vail, CO, pp. 149–157.

Fredlund, M. D., Stianson, J. R., Fredlund, D. G., Vu, H., and Thode, R. C. 2006. Numerical modeling of slab-on-grade foundations, *Proceedings of the Fourth International Conference on Unsaturated Soils, ASCE*, Carefree, AZ, pp. 2121–2132.

Fredlund, M. D., Wilson, G. W., and Pham, H. 2008. A methodology for applying probability theory to unsaturated hydraulic properties as the foundation for seepage analysis, *Proceedings of the GeoCongress 2008, ASCE*, New Orleans, LA, pp. 902–909.

Fredlund, M. D., Donaldson, M., and Wilson, G. W. 2009. Large-strain 1D, 2D, and 3D consolidation modeling of mine tailings, *Proceedings of the Tailings and Mine Waste Conference*, Banff, AB, pp. 729–738.

Fredlund, M. D., Zhang, J. M., Tran, D., and Fredlund, D. G. 2011. Coupling heat and moisture flow for the computation of actual evaporation, Paper No. 1058, *Proceedings of the Canadian Geotechnical Conference and Fifth Pan-American Conference*, Toronto, ON, October 2–6.

Freeze, R. A. 1969. The mechanism of natural ground-water recharge and discharge 1. One-dimensional, vertical, unsteady, unsaturated flow above a recharging or discharging groundwater flow system, *Water Resources Research*, Vol. 5, No. 1, pp. 153–171.

Freeze, R. A. 1971. Influence of the unsaturated flow domain on seepage through earth dams, *Water Resources Research*, Vol. 7, No. 4, pp. 929–940.

Freeze, R. A., and Cherry, J. A. 1979. *Groundwater*, Prentice-Hall, Englewood Cliffs, NJ.

Fuch, M., Campbell, G. S., and Papendick, R. I. 1978. An analysis of sensible and latent heat flow in a partly frozen unsaturated soil, *Soil Science of America Journal*, Vol. 42, pp. 379–385.

Fujimaki, H., and Inoue, M. 2003. A flux-controlled steady-state evaporation method for determining unsaturated hydraulic conductivity at low matric pressure head values, *Soil Science Journal*, Vol. 168, No. 6, pp. 385–395.

Fukuoka, M. 1980. Landslides associated with rainfall, *Geotechnical Engineering Journal, South-East Asia Society of Soil Engineering*, Vol. 11, pp. 1–29.

Fung, Y. C. 1965. *Foundations of Solid Mechanics*, Prentice-Hall, Englewood Cliffs, NJ.

Fung, Y. C. 1977. *A First Course in Continuum Mechanics*, 2nd ed., Prentice-Hall, Englewood Cliffs, NJ.

Galagher, R. H., Oden, J. T., Taylor, C., and Zienkiewicz, O. C. 1975. *Finite Element in Fluids*, Vol. I, *Viscous Flow and Hydrodynamics*, Wiley, New York.

Gallen, P. M. 1985. The measurement of soil suction using the filter paper methods, M.Sc. Thesis, University of Saskatchewan, Saskatoon, SK.

Gallipoli, D., Gens, A., Sharma, R., and Vaunat, J. 2003. An elasto-plastic model for unsaturated soil incorporating the effects of suction and degree of saturation on mechanical behaviour, *Geotechnique*, Vol. 53, No. 1, pp. 123–135.

Gallopoulos, E., Houstis, E., and Rice J. R. 1994. Computer as thinker/doer: Problem-solving environments for computational science, *IEEE Computational Science and Engineering*, Vol. 1, No. 2, pp. 11–23.

Gan, J. K.-M. 1986. Direct shear strength testing of unsaturated soils, M.Sc. Thesis, Department of Civil Engineering, University of Saskatchewan, Saskatoon.

Gan, J. K.-M., and Fredlund, D. G. 1988. Multistage direct shear testing of unsaturated soils, *Geotechnical Testing Journal, ASTM*, Vol. 11, No. 2, pp. 132–138.

Gan, J. K.-M., and Fredlund, D. G. 1996. Shear strength characteristics of two saprolitic soils, *Canadian Geotechnical Journal*, Vol. 33, No. 4, pp. 595–609.

Gan, J. K.-M., and Fredlund, D. G. 2000. A new laboratory method for the measurement of the unsaturated coefficients of permeability of soils, *Proceedings of the Asian Conference in Unsaturated Soils, UNSAT ASIA 2000*, Singapore, pp. 381–386.

Gan, J. K.-M., Fredlund, D. G., and Rahardjo, H. 1988. Determination of the shear strength parameters of an unsaturated soil using the direct shear test, *Canadian Geotechnical Journal*, Vol. 25, No. 8, pp. 500–510.

Gan, J. K.-M., Fredlund, D. G., Li, W. X., and Xing, A. 1994. Design and data acquisition for thermal conductivity matric suction sensors, *Transportation Research Record*, Vol. 1432, pp. 68–75.

Gardner, R. 1937. A method of measuring the capillary tension of soil moisture over a wide moisture range, *Soil Science Journal*, Vol. 43, No. 4, pp. 277–283.

Gardner, W. R. 1956. Representation of soil aggregate size distribution by a logarithmic-normal distribution, *Soil Science Society of America Proceedings*, Vol. 20, pp. 151–153.

Gardner, W. R. 1958a. Mathematics of isothermal water conduction in unsaturated soils, Highway Research Board Special Report 40, presented at the International Symposium on Physico-Chemical Phenomenon in Soils, Washington, DC, pp. 78–87.

Gardner, W. R. 1958b. Some steady state solutions of the unsaturated moisture flow equation with application to evaporation from a water-table, *Soil Science Journal*, Vol. 85, No. 4. pp. 228–232.

Gardner, W. R. 1958c. Laboratory studies of evaporation from soil columns in the presence of a water-table, *Soil Science of America*, Vol. 85, p. 244.

Gardner, W. R. 1961. Soil suction and water movement, in *Pore Pressure and Suction in Soils*, Butterworths, London, pp. 137–140.

Gardner, W. R. 1974. The permeability problem, *Soil Science Journal*, Vol. 117, pp. 243–249.

Gardner, W. R., and Fireman, M. 1958. Laboratory studies of evaporation from soil columns in the presence of a water-table, *Soil Science Journal*, Vol. 85, pp. 244–249.

Gardner, W. R., and Widtsoe, J. A. 1921. The movement of soil moisture, *Soil Science Journal*, Vol. 11, pp. 215–232.

Gardner, W. R., Hillel, D., and Benyammi, Y. 1970a. Post irrigation movement of soil water, I. Redistribution. *Water Resources Research*, Vol. 6, pp. 851–861.

Gardner, W. R., Hillel, D., and Benyammi, Y. 1970b. Post irrigation movement of soil water, II. Simultaneous redistribution and evaporation. *Water Resources Research*, Vol. 6, pp. 1148–1153.

Garven, E. A., and Vanapalli S. K. 2006. Evaluation of empirical procedures for predicting the shear strength of unsaturated soils, *Proceedings of the Fourth International Conference on Unsaturated Soils, ASCE*, Carefree, AZ, Geotechnical Special Publication 147, Reston, VA, pp. 2570–2581.

Gasmo, J. M., Hritzuk, J. K., Rahardjo, H., and Leong, E. C. 1999. Instrumentation of an unsaturated residual soil slope, *Geotechnical Testing Journal, ASTM International*, Vol. 22, No. 2, pp. 134–143.

Gasmo, J. M., Rahardjo, H., and Leong E. C. 2000. Infiltration effects on stability of a residual soil slope, *Computer and Geotechnics*, Vol. 26, pp. 145–165.

GDS Instruments. 2010. Promotional literature from GDS Instruments, London, on unsaturated triaxial testing of soils, available: www.gdsinstruments.com

Gee, G. W., Campbell, M. D., Campbell, G. S., and Campbell, J. H. 1992. Rapid measurement of low soil water potentials using a water activity meter, *Soil Science Society of American Proceedings*, Vol. 56, pp. 1068–1070.

Gens, A. 1995. Constitutive modelling: Application to compacted soils, *Proceedings of the First International Conference on Unsaturated Soils, UNSAT-95*, Paris, Vol. 3, pp. 1179–1200.

Gens, A., and Alonso E. E. 1992. A framework for the behaviour of unsaturated expansive clays, *Canadian Geotechnical Journal*, Vol. 29, pp. 1013–1031.

Gens, A., Alonso, E. E., Suriol, J., and Lloret, A. 1995. Effect of structure on the volumetric behaviour of a compacted soil, *Proceedings of the First International Conference Unsaturated Soils, UNSAT-95*, Paris, Vol. 1, pp. 83–88.

Geotechnical Consulting and Testing Services (GCTS). 2008. Promotional literature on the calibration cells for thermal conductivity suction sensors, Tempe, TX, available: www.gcts.com.

Ghosh, R. K. 1980. Estimation of soil moisture characteristics from mechanical properties of soils, *Soil Science Journal*, Vol. 130, No. 2, pp. 60–63.

Gibbs, H. J., and Coffey, C. T. 1969. Techniques for pore pressure measurements and shear testing of soil, *Proceedings of the Seventh International Conference on Soil Mechanics and Foundation Engineering*, Mexico City, Vol. 1, pp. 151–157.

Gibbs, H. J., Hilf, J. W., Holtz, W. G., and Walker, F. C. 1960. Shear strength of cohesive soils, *Proceedings of Research Conference on Shear Strength of Cohesive Soils, ASCE*, University of Colorado, Boulder, CO, pp. 33–162.

Gibson, R. E., and Henkel, D. J. 1954. Influence of duration of tests at constant rate of strain on measured "drain" strength, *Geotechnique*, Vol. 4, pp. 6–15.

Gilbert, Jr., O. H. 1960. Device for measuring tensions in water, National Research Council, *Highway Research Board Bulletin*, No. 245, pp. 1–9.

Gilchrist, H. G. 1963. A study of volume change of a highly plastic clay, M.Sc. Thesis, University of Saskatchewan, Saskatoon, SK.

Gitirana, Jr., de F. N. G. 2005. Weather-related geo-hazard assessment model for railway embankment stability, PhD Thesis, University of Saskatchewan, Saskatoon, SK.

Gitirana, Jr., de F. N. G., and Fredlund, D.G. 2003a. Analysis of transient cmbankment stability using the dynamic programming method, *Proceedings of the Fifty-Sixth Canadian Geotechnical Conference*, Winnipeg, MB, Vol. 1, pp. 807–814.

Gitirana, Jr., de F. N. G., and Fredlund, D.G. 2003b. From experimental evidences towards the assessment of weather-related railway embankment hazards, *Proceedings of the International Conference on the Theme "From Experimental Evidences Towards Unsaturated Soil Practice,"* Weimar, Germany, Vol. 1, pp. 435–468.

Gitirana, Jr., de F. N. G., and Fredlund, D. G. 2004. Soil-water characteristic curve equation with independent parameters, *Journal of Geotechnical and Geoenvironmental Engineering, ASCE*, Vol. 130, No. 2, pp. 209–212.

Gitirana, Jr., de F. N. G., and Fredlund, D. G. 2005a. The application of unsaturated soil mechanics to the assessment of weather-related geo-hazards, *Proceedings of the Sixteenth International Conference on Soil Mechanics and Foundation Engineering*, Osaka, pp. 2515–2520.

Gitirana, Jr., de F. N. G., and Fredlund, D. G. 2005b. Evaluation of the variability of unsaturated soil properties, *Proceedings of the Fifty-Eighth Canadian Geotechnical Conference*, Saskatoon, SK, Vol. 2. pp. 128–135.

Gitirana, Jr., de F. N. G., and Fredlund, D. G. 2007a. Correlation between unsaturated soil parameters, *Proceedings of the Sixth Brazilian Symposium on Unsaturated Soils*, Salvador, Brazil, Vol.1, pp. 505–510.

Gitirana, Jr., de F. N. G., and Fredlund, D. G. 2007b. Sensitivity of the factor of safety of railway embankments to saturated and unsaturated soil properties, *Proceedings of the Sixth Brazilian Symposium on Unsaturated Soils*, Salvador, Brazil, Vol. 1, pp. 29–39.

Gitirana, Jr., de F. N. G., Fredlund, M. D., and Fredlund, D. G. 2005. Infiltration-runoff boundary conditions in seepage analysis, *Proceedings of the Fifty-Eighth Canadian Geotechnical Conference, GeoSask 2005, and Sixth Joint IAH-CNC and CGS Groundwater Speciality Conferences*, Saskatoon, SK, pp. 516–523.

Gitrana, Jr., de F. N. G., Fredlund, M. D., and Fredlund, D. G. 2006a. Numerical modeling of soil-atmosphere interaction for unsaturated surfaces, *Proceedings of the Fourth International Conference on Unsaturated Soils*, Carefree, AZ, Vol. 1, pp. 658–669.

Gitirana, Jr., de F. N. G., Fredlund, D. G., and Lima-Goutte, M. 2006b. Fluxo em solos nao saturados e o process erosivo, in J. C. de Carvalh, M. M. Sales, N. M. de Souza, and M. T. da Silva Melo (Eds), *Processos Erosivos no Centro-Oeste Brasileiro* (in Portuguese), 1st ed., FINATEC, Brasilia, 2006, Vol. 1, pp. 285–317.

Gitirana, Jr., de F. N. G., Santos, M. A., and Fredlund, M. D. 2008a. Three-dimensional analysis of the Lodalen landslide, paper presented at GeoCongress 2008, New Orleans, LA.

Gitirana, Jr., de F. N. G., Santos, M. A., and Fredlund, M. D. 2008b. Three-dimensional slope stability model using finite element stress analysis, paper presented at GeoCongress 2008, New Orleans, LA.

Glossary of Soil Science Terms. 1979. Soil Science Society of America, Madison, WI.

Goh, S. G. 2012. Hysteresis effects on mechanical behaviour of unsaturated soils, PhD Thesis, Department of Civil and Environmental Engineering, Nanyang Technological University, Singapore.

Goh, S. G., Rahardjo, H., and Leong, E. C. 2009. Evaluation of shear strength equations for unsaturated soil, *Proceedings of the Fourth Asia-Pacific Conference on Unsaturated Soils*, Newcastle, Australia, pp. 753–758.

Goh, S. G., Rahardjo, H., and Leong, E. C. 2010. Shear strength equations for unsaturated soil under drying and wetting, *Journal of Geotechnical and Geoenvironmental Engineering, ASCE*, Vol. 136, No. 4, pp. 594–606.

Gonzalez, P. A., and Adams, B. J. 1980. Mine tailings disposal: I. Laboratory characterization of tailings, Internal Report, Department of Civil Engineering, University of Toronto, Toronto, ON, pp. 1–14.

Gosnik, J. P., Kawasaki, K., Osterkamp, T. E., and Holty, J. 1983. Heat and moisture transport during annual freezing and thawing, *Proceedings of the Fourth International Conference on Permafrost*, Fairbanks, AL, pp. 355–360.

Gourley, C. S., and Schreiner, H. D. 1995. Field measurements of soil suction in saturated soils, in E.E. Alonso and P. Delage (Eds.), *Proceedings of the First International Conference on Unsaturated Soils, Paris, France*, Presses des Ponts et Chaussees, Balkema, Rotterdam, The Netherlands, Vol. 2, pp. 601–607.

Government of Hong Kong. 1976. Report on the slope failures at Sau Mau Ping 25th August 1976, Vols. 1–3, Hong Kong Government Printer, Hong Kong.

Granger, R. J. 1989. A complimentary relationship approach for evaporation from nonsaturated surfaces, *Journal of Hydrology*, Vol. 111, pp. 31–38.

Graton, L. C., and Fraser, H. J. 1935. Systematic packing of spheres with particular relation to porosity and permeability, *Journal of Geology*, Vol. 43, No. 8, pp. 785–848.

Gray, D. M. (Ed.), 1970. *Handbook on the Principles of Hydrology*, Canadian National Committee for the International Hydrological Decade, National Research Council of Canada, Ottawa, Canada.

Gray, D. M., Granger, R. J., and Dyck, G. E. 1985. Over–winter soil moisture changes, *Transactions, ASCE*, Vol. 28, No. 2. pp. 442–447.

Gray, H. 1945. Simultaneous consolidation of contiguous layers of unlike compressible soils, *Transactions ASCE*, Vol. 110, pp. 1327–1344.

Greacen, E. L., Walker, G. R., and Cook, P. G. 1987. Evaluation of the filter paper method for measuring soil water suction, paper presented at the International Conference on Measurement of Soil and Plant Water Status, pp. 137–143.

Greco, V. R. 1988. Numerical methods for locating the critical slip surface in slope stability analysis, *Proceedings of the Sixth International Conference on Numerical Methods in Geomechanics*, Innsbruck, pp. 1219–1223.

Green, A. E., and Naghdi, P. M. 1965. A dynamical theory of interacting continua, *International Journal of Engineering Science*, Vol. 3, pp. 231–241.

Green, J. L., Durben, D. J., Wolf, G. H., and Angell, C. A. 1990. Water and solutions at negative pressure, *Science*, Vol. 249, pp. 649–652.

Green, R. E., and Corey, J. C. 1971a. Calculation of hydraulic conductivity: A further evaluation of some predictive methods, *Soil Science Society of America Proceedings*, Vol. 35, pp. 3–8.

Green, R. E., and Corey, J. C. 1971b. Calculation of hydraulic soils, prediction, laboratory test and *in-situ* measurements, *Proceedings of the Eighth International Conference Soil Mechanics Foundation Engineering*, Moscow, pp. 163–170.

Green, W. H., and Ampt, G. A. 1911. Study on soil physics: I, flow of air and water in soils, *Journal of Agricultural Sciences*, Vol. 4, pp. 1–24.

Guan, Y. 1996. The measurement of soil suction, PhD Thesis, University of Saskatchewan, Saskatoon, SK.

Guan, Y., and Fredlund, D. G. 1997a. Direct measurement of high soil suction, *Proceedings of the Third Brazilian Symposium on Unsaturated Soils, NSAT '97*, Rio de Janeiro, Vol. 2. pp. 543–550.

Guan, Y., and Fredlund, D. G. 1997b. Use of a tensile strength of water for the direct measurement of high soil suction, *Canadian Geotechnical Journal*, Vol. 34, Vol. 4, pp. 604–614.

Guan, Y., and Fredlund, D. G. 1999. Reply to discussion: Use of tensile strength of water for the direct measurement of high soil suction, *Canadian Geotechnical Journal*, Vol. 36, No. 1, p. 181.

Guan, Y., Fredlund, D. G., and Gan, J. K.-M. 1998. Behaviour of water subjected to high tensile stress, *Proceedings of the Second International Conference on Unsaturated Soils (UNSAT-98)*, Beijing, Vol. 1, pp. 356–361.

Gulhati, S. K., and Satija, B. S. 1981. Shear strength of partially saturated soils, *Proceedings of the Tenth International Conference for Soil Mechanics and Foundation Engineering*, Stockholm, Vol. 1, pp. 609–612.

Gupta, S. C., and Ewing, R. P. 1992. Modeling water retention characteristics and surface roughness of tilled soils, *Proceedings of the International Workshop on Indirect Methods for Estimating the Hydraulic Properties of Unsaturated Soils*, U.S. Salinity Laboratory, Riverside, CA, pp. 379–388.

Gupta, S. C., and Larson, W. E. 1979a. Estimating soil-water retention characteristics from particle size distribution, organic matter percent, and bulk density, *Water Resources Research Journal*, Vol. 15, No. 6, pp. 1633–1635.

Gupta, S. C., and Larson W. E. 1979b. A model for predicting packing density of soils using particle-size distribution, *Soil Science Society of America Proceedings*, Vol. 43, pp. 758–764.

Gutierrez, C. I. G. 1985. Assessment of filter paper technique for the measurement of soil water suction, M.Sc Thesis, Imperial College, London.

Guymon, G. L., and Luthin, J. N. 1974. A coupled heat and moisture transport model for Arctic soils, *Water Resources Research*, Vol. 10, No. 5, pp. 995–1001.

Gvirtzman, H., and Roberts, P. V. 1991. Pore-scale spatial analysis of two immiscible fluids in porous media, *Water Resources Research*, Vol. 27, pp. 1165–1176.

Habib, S. A., Kato, T., and Karube, D. 1992. One-dimensional swell behaviour of unsaturated soil, *Proceedings of the Seventh International Conference on Expansive Soils*, Dallas, TX, Vol. 1, pp. 222–226.

Hagen, L. J., Skidmore, E. L., and Fryrear, D. W. 1987. Using two sieves to characterize dry soil aggregate size distribution, *Transactions of the ASAE*, Vol. 30, No. 1, pp. 162–165.

Haines, W. B. 1923. The volume changes associated with variations of water content in soil, *Journal of Agriculture Science*, Vol. 13, pp. 296–310.

Haines, W. B. 1925. Studies in the physical properties of soils II: A note on the cohesion developed by capillary forces in an ideal soil, *Journal of Agricultural Science*, Vol. 15, pp. 529–535.

Haines, W. B. 1927. A further contribution to the theory of capillary phenomena in soils, *Journal of Agricultural Science*, Vol. 17, pp. 264–290.

Haines, W. B. 1930. Studies in the physical properties of soils–V: The hysteresis effect in capillary properties and the modes of water distribution associated therewith, *Journal of Agriculture Science*, Vol. 20, pp. 97–116.

Hamblin, A. P. 1984. Filter paper method for routine measurement of field water potential, *Journal of Hydrology*, Vol. 53, Nos. 3/4, pp. 355–360.

Hamilton, J. J. 1963. Volume changes in undisturbed clay profiles in western Canada. *Canadian Geotechnical Journal*, Vol. 1, No. 1, pp. 27–42.

Hamilton, J. J. 1977. Foundations on swelling or shrinking subsoils, *Canadian Building Digest*, CBD 184, Division of Building Research, National Research Council, Ottawa, ON.

Hamilton, J. J. 1980. Behavior of expansive soils in western Canada, *Proceedings of the Fourth International Conference on Expansive Soils, ASCE*, Denver, CO, June 16–18, Vol. 1, pp. 815–833.

Hamilton, J. M., Daniel, D. E., and Olson, R. E. 1981. Measurement of hydraulic conductivity of partially saturated soils, in T. F. Zimmie and C. O. Riggs (Eds.), *Permeability and Groundwater Contaminant Transport*, ASTM Special Technical Publication 746, American Society for Testing and Materials, Philadelphia, PA, pp. 182–196.

Han, K. K. 1996. Effect of hysteresis, infiltration and tensile stress on the strength of an unsaturated soil, PhD Thesis, Nanyang Technological University, Singapore.

Han, K. K., Rahardjo, H., and Broms, B. B. 1995. Effect of hysteresis on the shear strength of a residual soil, *Proceedings of the First International Conference on Unsaturated Soil*, Balkema, Rotterdam, Vol. 2, pp. 499–504.

Hanafy, E. A. D. E. 1991. Swelling/shrinkage characteristic curve of dessicated expansive clays, *ASTM Geotechnical Testing Journal*, Vol. 14, No. 2, pp. 206–211.

Hanna, A. M., and Meyerhof, G. G. 1980. Design charts for ultimate bearing capacity of foundations on sand overlying soft clay, *Canadian Geotechnical Journal*, Vol. 17, pp. 300–303.

Hansson, K., and Lundlin, L. 2006. Water content reflectometer application to construction materials and its relation to time domain reflectometry, *Vadose Zone Journal*, Vol. 5, No. 1, pp. 459–468.

Hargreaves, G. H. 1975. Moisture availability and crop production, *Transactions of the ASAE*, Vol. 18, pp. 980–984.

Hargreaves, G. H. 1994. Simplified coefficients for estimating monthly solar radiation in North America and Europe, Department Paper, Department of Biological and Irrigation Engineering, Utah State University, Logan, UT.

Hargreaves, G. H., and Samani, Z. A. 1982. Estimating potential evapotranspiration, *Journal of Irrigation Drainage Engineering*, Vol. 108, No. 3, pp. 225–230.

Harlan, R. L. 1973. Analysis of coupled heat-fluid transport in partially frozen soil, *Water Resource Research*, Vol. 9, pp. 1314–1323.

Harlan, R. L., and J. F. Nixon. 1978. Ground thermal regime, in O.B. Andersland and D.M. Anderson (Eds.), *Geotechnical Engineering for Cold Regions*, McGraw-Hill, New York, N.Y., pp. 103–163.

Harr, M. E. 1962. *Groundwater and Seepage*, McGraw-Hill, New York.

Harr, M. E. 1977. *Mechanics of Particulate Media: A Probabilistic Approach*, McGraw-Hill, New York.

Harrison, B. A., and Blight, G. E. 1998. The effect of filter paper and psychrometer calibration techniques on soil suction measurements, *Proceedings of the Second International Conference on Unsaturated Soils, UNSAT-98*, Beijing, Vol. 1, pp. 362–367.

Harrison, B. A., and Blight, G. E. 2000. A comparison of *in situ* soil suction measurements, in H. Rahardjo, D. G. Toll, and E. C. Leong (Eds.), *Proceedings of First Conference on Unsaturated Soils for Asia*, Singapore, Balkema, Rotterdam, pp. 281–284.

Harrison, L. P. 1965. Fundamental concepts and definitions relating to humidity, in A. Wexler and W. A. Wildhack (Eds.), *Humidity and Moisture, Fundamentals and Standards*, Reinhold Publishing Company, New York, Vol. 3, pp. 3–69.

Harvey, E. N., Barnes, D. K., McElroy, A. H., Whitday, A. H., Pease, D. C., and Cooper, K. W. 1944. Bubble formation in animals, 1: Physical factors, *Journal of Cellular and Comparative Physiology*, Vol. 24, No. 1, pp. 1–22.

Harvey, E. N., McElroy, W. D., and Whiteley, A. H. 1947. On cavity formation in water, *Journal of Applied Physics*, Vol. 18, pp. 162–172.

Hasan, J. U., and Fredlund, D. G. 1980. Pore pressure parameters for unsaturated soils, *Canadian Geotechnical Journal*, Vol. 17, No. 3, pp. 395–404.

Hashin, Z., and Shtrikman, S. 1962. A variational approach to the theory of the effective magnetic permeability of multiphase materials, *Journal of Applied Physics*, Vol. 33, pp. 3125–3131.

Haverkamp, R., and Parlange, J. Y. 1986. Predicting the water-retention curve from a particle-size distribution: 1. Sandy soils without organic matter, *Soil Science Journal*, Vol. 142, No. 6, pp. 325–339.

Hazen, A. 1911. Discussion of "Dams on sand foundation" by A. C. Koenig, *Transactions of the American Society of Civil Engineering*, Vol. 73, pp. 199–203.

Head, K. H. 1984. *Manual of Soil Laboratory Testing*, Vol. I, *Soil Classification and Compaction Tests*, London/Plymouth: Pentech, London.

Head, K. H. 1986. *Effective Stress Tests, Manual of Soil Laboratory Testing*, Wiley, New York, Vol. 3, pp. 743–1238.

Heid, J. M., Mahon, J. J., Nielson, R. F., and Juster, S. T. 1950. *Study of Permeability of Rocks to Homogenous Fluids, Drilling and Production Practices*, American Petroleum Institute, New York, pp. 230–240.

Henderson, S. J., and Speedy, R. J. 1980. A Berthelot-Bourdon tube method for studying water under tension, *Journal of Physics Environmental Science Instruments*, Vol. 13, pp. 778–782.

Hilf, J. W. 1948. Estimating construction pore pressures in rolled earth dams, *Proceedings of the Second International Conference on Soil Mechanics and Foundation Engineering*, Rotterdam, The Netherlands, Vol. 3, pp. 234–240.

Hilf, J. W. 1956. An investigation of pore-water pressure in compacted cohesive soils, PhD Thesis, Technical Memorandum. No. 654, U.S. Department of the Interior, Bureau of Reclamation, Design and Construction Division, Denver, CO.

Hill, C. B. 1967. Consolidation of compacted clay, M.Sc. Thesis, University of Manchester, Manchester, England.

Hillel, D. 1971. *Soil and Water: Physical Principles and Processes*, Academic press, New York.

Hillel, D. 1982. *Introduction to Soil Physics*, Academic press, New York.

Hillel, D., Krentos, V. D., and Stylianou, Y. 1972. Procedure and test of an internal drainage method for measuring soil hydraulic characteristics *in situ*, *Soil Science Journal*, Vol. 114, pp. 295–400.

Hjorth, J. S. U. 1994. *Computer Intensive Statistical Methods, Validation, Model Selection, and Bootstrap*, Chapman and Hall, London.

Ho, D. Y. F. 1988. The relationship between the volumetric deformation moduli of unsaturated soils, PhD Thesis, Department of Civil Engineering, University of Saskatchewan, Saskatoon, SK.

Ho, D. Y. F., and Fredlund, D. G. 1982a. Increase in shear strength due to soil suction for two Hong Kong soils, *Proceedings of Specialty Conference on Engineering and Construction in Tropical and Residual Soils, ASCE*, Honolulu, HI, pp. 263–295.

Ho, D. Y. F., and Fredlund, D. G. 1982b. A multi-stage triaxial test for unsaturated soils, *Geotechnical Testing Journal, ASTM*, Vol. 5, Nos. 1/2, pp. 18–25.

Ho, D. Y. F., and Fredlund, D. G. 1982c. Strain rates for unsaturated soil shear strength testing, *Proceedings of the Seventh Southeast Asian Geotechnical Conference*, Hong Kong, Vol. 1, pp. 787–803.

Ho, D. Y. F., and Fredlund, D. G. 1989. Laboratory measurement of the volumetric deformation moduli for two unsaturated soils, *Proceedings of the Forty-Second Canadian Geotechnical Conference*, Winnipeg, MB, pp. 50–60.

Ho, D. Y. F., and Fredlund, D. G. 1995. Determination of the volume change moduli and respective inter-relationships for two unsaturated soils, *Proceedings of the First International Conference on Unsaturated Soils, UNSAT-95*, Paris, Vol. 1, pp. 117–122.

Ho, D. Y. F., Fredlund, D. G., and Rahardjo, H. 1992. Volume change indices during loading and unloading of an unsaturated soil, *Canadian Geotechnical Journal*, Vol. 29, No. 2, pp. 195–207.

Hoekstra, P. 1966. Moisture movement in soil under temperature gradients with the cold side temperature below freezing, *Water Resource Research*, Vol. 2, pp. 241–250.

Hoffmann, C., Romero, E., and Alonso, E. E. 2005. Combining different controlled-suction techniques to study expansive clays, in A. Tarantino, E. Romero, and Y. J. Cui (Eds.), *Proceedings of the International Symposium on Advanced Experimental Unsaturated Soil Mechanics*, Trento, Italy, Balkema, Leiden, the Netherlands, pp. 61–67.

Hogentogler, C. A. 1936. Essentials of soil compaction, *Proceedings Highway Research Board*, National Research Council, Washington, DC, pp. 309–316.

Hogentogler, C. A., and Barber, E. S. 1941. Discussion in soil water phenomena, *Proceedings of Highway Research Board*, Vol. 21, pp. 452–465.

Holtz, W. G. 1959. Expansive clays–Properties and problems, *Theoretical and Practical Treatment of Expansive Soils–First Conference on the Mechanics of Soils*, Colorado School of Mines, Golden, CO, Vol. 54, No. 4, pp. 89–117.

Holtz, W. G. 1970. Suggested method of test for one-dimensional expansion and uplift pressure of clay soils, *Special Procedures for Testing Soil and Rock for Engineering Purposes*, 5th ed., ASTM STP 479, American Society for Testing and Materials, Philadelphia, PA.

Holtz, W. G., and Gibbs, H. J. 1956. Engineering properties of expansive clays, *Transactions ASCE*, Vol. 121, pp. 641–663.

Holtz, R. D., and Kovacs, W. G. 1981. *An Introduction to Geotechnical Engineering*, Prentice-Hall, Englewood Cliffs, NJ.

Horai, K. 1971. Thermal conductivity of rock-forming minerals, *Journal of Geophysical Research*, Vol. 76, No. 5, pp. 1278–1308.

Horton, J. H., and Hawkins, R. H. 1965. Flow path of rain from the soil surface to the water table, *Soil Science Journal*, Vol. 100, No. 6, pp. 377–383.

Hough, B. K. 1969. *Basic Soils Engineering*, Ronald, New York.

Houlsby, G. T. 1997. The work input to an unsaturated granular material, *Geotechnique*, Vol. 47, No. 1, pp. 193–196.

Houston, S. L., and El-Ehwany, M. 1991. Sample disturbance of cemented collapsible soils, *Journal of Geotechnical Engineering Division, ASCE*, Vol. 117, No. 5, pp. 731–752.

Houston, S. L., and Houston, W. R. 1989. State-of-the-practice mitigation measures for collapsible soil sites, in M. Evanston (Ed.), *Proceedings of the Foundation Engineering Congress, ASCE*, Evanston, IL, pp. 161–175.

Houston, S. L., Houston, W. R., and Spadola, D. 1988. Prediction of field collapse of soils due to wetting, *Journal of Geotechnical Engineering Division, ASCE*, Vol. 114, No. 1, pp. 40–85.

Houston, S. L. 1995. Foundations and pavements on unsaturated soils: Part One: Collapsible soils, in E E. Alonso and

P. Delage (Eds.), *Unsaturated Soils : Proceedings of the First International Conference on Unsaturated Soils, Paris, France*, Balkema, Rotterdam, Presses des Ponts et Chaussees, pp. 1421–1439.

Houston, S. L. 2002. Applied unsaturated soil mechanics, *Proceedings of the Third International Conference on Unsaturated Soils*, Recife, Brazil, Vol. 3, pp. 1127–1136.

Houston, S. L., Houston, W. R., and Wagner, A. M. 1994. Laboratory filter paper measurements, *Geotechnical Testing Journal, ASTM*, Vol. 17, No. 2, pp. 185–194.

Houston, S. L., Houston, W. R., Zapata, C., and Lawrence, C. 2001. *Geotechnical Engineering Practice for Collapsible Soils*, Geotechnical and Geological Engineering, Kluwer Academic, The Netherlands, Vol. 19, pp. 333–355.

Houston, W. R., Mahmoud, H., and Houston, S. 1993. A laboratory procedure for partial wetting collapse determination, in S.L. Houston and W.K. Wray (Eds.), *Unsaturated Soils*, Special Geotechnical Publication No. 39, American Society of Civil Engineers, New York, pp. 54–63.

Houston, W. R., Mahmoud, H., and Houston, S. 1995. Down-hole collapse test system, *Journal of Geotechnical Engineering, ASCE*, Vol. 121, No. 4, pp. 341–349.

Houston, W. R., and Houston, S. L. 1995. Infiltration studies for unsaturated soils, in E. E. Alonso and P. Delage (Eds.), *Unsaturated Soils: Proceedings of the First International Conference on Unsaturated Soils*, Paris, Balkema, Rotterdam, Presses des Ponts et Chaussees, pp. 869–875.

Houston, W. R., Mirza, M. W., and Zapata, C. E. 2006. Calibration and validation of the ICM Version 2.6., Final Report, Project NCHRD 9-23: Environmental effects in pavement mix and structural design systems, National Comparative Highway Research Program, Transportation Research Board, Washington, DC.

Hu, Y., Vu, H., and Fredlund, D. G. 2007. Calibration of thermal conductivity sensors for suction measurement, *Proceedings of the Sixtieth Canadian Geotechnical* Conference, Ottawa, ON, pp. 1091–1100.

Hu, Y., Vu, H., and Hubble, D. W. 2010. Evaluation of dielectric-based bubbles for expansive soils: Application to Regina soils, *Canadian Geotechnical Journal*, Vol. 47, No. 34, pp. 346–358.

Huang, M., Fredlund, D. G., and Fredlund, M. D. 2010. Comparison of measured and PTF predictions of SWCCs for loess soils in China, *Geotechnical and Geological Engineering Journal*, Vol. 28, pp. 105–117.

Huang, S., Fredlund, D. G., and Barbour, S. L. 1995. Measurement of the coefficient of permeability of an unsaturated soil, *Proceedings of the First International Conference on Unsaturated Soils, UNSAT-95*, Paris, Vol. 2. pp. 505–511.

Huang, S., Barbour, S. L., and Fredlund, D. G. 1998a. Development and verification of a coefficient of permeability function for a deformable, unsaturated soil, *Canadian Geotechnical Journal*, Vol. 35, No. 3, pp. 411–425.

Huang, S., Fredlund, D. G., and Barbour, S. L. 1998b. Measurement of the coefficient of permeability for a deformable unsaturated soil using a triaxial permeameter, *Canadian Geotechnical Journal*, Vol. 35, No. 3, pp. 426–432.

Hubbert, M. K. 1940. The theory of groundwater motion, *Journal of Geology*, Vol. 48, pp. 785–944.

Indrawan, I. G. B., Rahardjo, H., and Leong, E. C. 2006. Effects of coarse-grained materials on properties of residuals soil. *Journal of Engineering Geology*, Vol. 82, No. 3, pp. 154–164.

Indrawan, I. G. B., Rahardjo, H., and Leong, E. C. 2007. Drying and wetting characteristics of a two-layer soil column, *Canadian Geotechnical Journal*, Vol. 44, pp. 20–32.

Ingersoll, J. E. 1981. Method for coincidentally determining soil hydraulic conductivity and moisture retention characteristics, Special Report 81-2, U.S. Army Cold Regions Research and Engineering Laboratory, Hanover, NH.

International Atomic Energy Agency (IAEA). 2008. Field estimation of soil water content: A practical guide to methods, instrumentation and sensor technology, Training Course Series 30, IAEA, Vienna, February.

International Society for Soil Mechanics and Foundation Engineering. 1977. List of symbols, units, and definitions, Subcommittee on Symbols, Units, and Definitions, *Proceedings of the Ninth International Conference on Soil Mechanics Foundation Engineering*, Tokyo, Vol. 3, pp. 156–170.

International Society of Soil Science. 1963. News of the Commission, Commission I (Soil physics), Soil physics terminology, Bulletin No. 23, Vol. 7, Soil Science Society of America, Madison, WI.

International Society of Soil Science. 1970. Glossary of soil science terms, Soil Science Society of America, Madison, WI.

International Society of Soil Science. 1979. Glossary of soil science terms, Soil Science Society of America, Madison, WI.

Ireland, H. O. 1964. Design and construction of retaining walls, in *Soil Mechanics Lecture Series: Design of Structures to Resist Earth Pressure*, America Society of Civil Engineers, Chicago, IL.

Irmak, S., Irmak, A., Jones, J. W., Howell, T. A., Jacobs, J. M., Allen, R. G., and Hoogenboom, G. 2003. Predicting daily net radiation using minimum climatological data, *Journal of Irrigation and Drainage Engineering, ASCE*, Vol. 129, Issue 4, pp. 256–269.

Irmay, S. 1954. On the hydraulic conductivity of unsaturated soils, *Transactions of American Geophysical Union*, Vol. 35, pp. 463–468.

Israelachvili, J. N. 1991. *Intermolecular and Surface Forces*, 2nd ed., Academic press, New York.

Jaky, J. 1944. The coefficient of earth pressure at-rest, *Journal of the Society of Hungarian Architects and Engineers*, Vol. 78, No. 22, pp. 355–358.

Jaky, J. 1948. Pressure in silos, *Proceedings of the Second International Conference on Soil Mechanics and Foundation Engineering*, Vol. 1, pp. 103–107.

Jame, Y. W. 1972. Temperature effects on phase composition of a partially frozen soil, M.Sc. Thesis, University of Saskatchewan, Saskatoon, SK.

Jame, Y. W. 1977. Heat and mass transfer in freezing soil, PhD Thesis, University of Saskatchewan, Saskatoon, SK.

Jame, Y. W., and Norum, D. I. 1980. Heat and mass transfer in freezing unsaturated porous medium, *Water Resources Research*, Vol. 16, No. 4, pp. 811–819.

Janssen, D. J., and Dempsey, B. J. 1980. Soil-moisture properties of subgrade soils, paper presented at the Sixtieth Annual Transportation Research Board Meeting, Washington, DC.

Jennings, J., and Knight, K. 1956. Recent experiences with the consolidation test as a means of identifying conditions of heaving or collapse of foundations on partially saturated soils, *Transactions of South African Institute of Civil Engineers*, Vol. 6, No. 8, pp. 255–256.

Jennings, J. E. 1961. A revised effective stress law for use in the prediction of the behaviour of unsaturated soils, *Proceedings of Conference on Pore Pressure and Suction in Soils*, London, pp. 26–30.

Jennings, J. E. 1969. The prediction of amount and rate of heave likely to be experienced in engineering construction on expansive soils, *Proceedings of the Second International Conference on Expansive Clay Soils*, Texas A & M University, College Station, TX, pp. 99–109.

Jennings, J. E., and Knight, K. 1957. The prediction of total heave from the double oedometer test, *Proceedings of Symposium on Expansive Clays*, South African Institute of Civil Engineers, Johannesburg, South Africa, Vol. 7, No. 9, pp. 13–19.

Jennings, J. E. B., Firth, R. A., Ralph, T. K., and Nagar, N. 1973. An improved method for predicting heave using the oedometer test, *Proceedings of Third International Conference on Expansive Soils*, Haifa, Vol. 2, pp. 149–154.

Jennings, J. E. B., and Burland, J. B. 1962. Limitations to the use of effective stresses in partly saturated soils, *Geotechnique*, Vol. 12, No. 2, pp. 125–144.

Jensen, M. E., and Haise, H. R. 1963. Estimating evapotranspiration from solar radiation, Irrigation and Drainage Division, *Proceedings of the ASCE, (IR 4)*, Vol. 89, pp. 15–41.

Jia, G. W., Zhan, T. L. T., Chen, Y. M., and Fredlund D. G. 2009a. Influence of rainfall pattern on the infiltration in to landfill earthen final cover, *Proceedings of International Symposium on Geoenvironmental Engineering*, Hangzhou, China, September 8–10, pp. 641–645.

Jia, G. W., Zhan, T. L. T., Chen, Y. M., and Fredlund, D. G. 2009b. Performance of a large-scale slope model subjected to rising and lowering water levels, *Engineering Geology Journal*, Vol. 106, pp. 92–103.

Johansen, O. 1975. Thermal conductivity of soils, PhD Thesis, Trondheim, Norway (CRREL Draft Translation 637, 1977).

Johnpeer, G. D. 1986. Land subsidence caused by collapsible soils in northern New Mexico, *Ground Failure*, No. 3, National Research Council on Ground Failure Hazards, pp. 1–22.

Johnson, A. I. 1981. Glossary, in T. F. Zimmie and C. O. Riggs (Eds.), *Permeability and Groundwater Contaminant Transport*, ASTM STP 746, American Society for Testing and Materials, Philadelphia, PA, pp. 3–17.

Johnston, L. N. 1942. Water permeable jacketed thermal radiators as indicators of field capacity and permanent wilting percentage in soils, *Soil Science*, Vol. 54, pp. 123–126.

Jommi, C. 2000. Remarks on the constitutive modeling of unsaturated soils, in A. Tarantino and C. Mancuso (Eds.), *Experimental Evidence and Theoretical Approaches in Unsaturated Soils*, Balkema, Rotterdam, pp. 139–153.

Jones, D. E., and Holtz, W. G. 1973. *Expansive Soils–The Hidden Disaster*, Civil Engineering, American Society of Civil Engineers, New York, pp. 87–89.

Jones, W. M., Overton, G. D. N., and Trevena, D. H. 1981. Tensile strength experiments with water using a new type of Berthelot tube, *Journal of Physics: Applied Physics*, Vol. 14, pp. 1283–1291.

Jovanovic, N. Z., and Annandale, J. G. 1997. A laboratory evaluation of Watermark electrical resistance and Campbell Scientific 229 heat dissipation matric potential sensors, *Water South Africa*, Vol. 23, No. 3, pp. 227–232.

Jumikis, A. R. 1966. *Thermal Soil Mechanics*, Rutgers University Press, New Brunswick, NJ.

Justo, J. L., and Delgado, A. 1987. Stress-strain paths in the collapse-swelling-shrinkage of soils *in situ*, *Proceedings of the Sixth International Conference on Expansive Soil*, New Delhi, Vol. 2, pp. 596–601.

Justo, J. L., Delgado, A., and Ruiz, J. 1984. The influence of stress-path in the collapse-swelling of soils at the laboratory, *Proceedings of the Fifth International Conference on Expansive Soils*, Adelaide, Australia, pp. 67–71.

Karube, D. 1983. Effect of suction on soil behaviour, *Proceedings of Seventh Asian Regional Conference on Soil Mechanics and Foundation Engineering*, Haifa, Vol. 1, pp. 30–35.

Karube, D., and Kato, S. 1989. Yield functions of unsaturated soil, *Proceedings of the Twelfth International Conference on Soil Mechanics and Foundation Engineering*, Rio de Janeiro, Vol. 3, pp. 615–618.

Kasap, S. O., Lakhanpal, D., Patzer, C., Mandziak, T., and Fredlund, D. G. 1994. Ultrasonic measurements on a porous ceramic to determine soil suction, *Ultrasonics*, Vol. 32, No. 5, pp. 379–383.

Kasim, F. B., Fredlund, D. G., and Gan, J. K-M. 1998a. The effects of steady state rainfall on the matric suction conditions in a slope, *Proceedings of Second International Conference on Unsaturated Soils, UNSAT-98*, Beijing, Vol. 1, pp. 78–83.

Kasim, F. B., Fredlund, D. G., and Gan, J. K-M. 1998b. The effect of steady state rainfall on the matric suction conditions in soil, in K. S. Li, J. N. Kay, and K. K. S. Ho (Eds.), *Slope Engineering in Hong Kong*, Balkema, Rotterdam, pp. 75–85.

Kassif, G., and Ben Shalom, A. 1971. Experimental relationships between swell pressure and suction, *Geotechnique*, Vol. 21, No. 3, pp. 245–255.

Kato, S., and Kawai, K. 2000. Deformation characteristics of a compacted clay in collapse under isotropic and triaxial stress state, *Soils and Foundations*, Vol. 40, No. 5, pp. 75–90.

Kato, S., Yoshimura, Y., and Fredlund, D. G. 2005. Role of matric suction in the interpretation of unconfined compression tests, *Proceedings of the Fifty-Eighth Canadian Geotechnical Conference*, Saskatoon, SK, Vol. 2, pp. 410–415.

Katti, R. K., Kulkarni, S. K., and Fotedar, S. K. 1969. Shear strength and swelling pressure characteristics of expansive soils, *Proceedings of the Second International Conference on Expansive Soils*, Texas A & M University, College Station, TX, pp. 334–342.

Kaye, G. W. C., and Laby, T. H. 1973. *Tables of Physical and Chemical Constants*, 14th edition, Longman, London.

Kenai, S., Bahar, R., and Benazzoug, M. 2006. Experimental analysis of the effect of some compaction methods on mechanical properties and durability of cement stabilized soil, *Journal of Material Science*, Vol. 41, pp. 6956–6964.

Kemper, W. D., and Chepil, W. S. 1965. Size distribution of aggregates, in methods of soil analysis, Part 1, C. A. Black (Ed.), *Agronomy*, Vol. 9, pp. 499–510.

Kemper, W. D., and Rollins, J. B. 1966. Osmotic efficiency coefficients across compacted clays, *Soil Science Society of America Proceedings*, Vol. 30, pp. 529–534.

Kemper, W. D., and van Schaik, J. C. 1966. Diffusion of salts in clay-water systems, *Soil Science Society of America Proceedings*, Vol. 30, pp. 534–540.

Kenney, T. C., and Watson, G. H. 1961. Multiple-stage triaxial test for determining c' and ϕ' of saturated soils, *Proceedings of the Fifth International Conference on Soil Mechanics and Foundation Engineering*, Paris, Vol. 1, pp. 191–195.

Kent, D. D., Fredlund, D. G., and Watt, W. G. 1975. Variables controlling behaviour of a partly frozen soil, *Proceedings of the Conference on Soil Water Problems in Cold Regions*, Calgary, AB, pp. 70–88.

Kersten, M. S. 1949. Laboratory research for the determination of the thermal properties of soils, ACFEL Technical Report 23, AD712516, University of Minnesota, Minneapolis, MN.

Khalili, N., and Khabbaz, M. H. 1998. A unique relationship for χ for the determination of the shear strength of unsaturated soils, *Geotechnique*, Vol. 48, No. 5, pp. 681–687.

Khan, M. H., and Hoag, D. L. 1979. A noncontacting transducer for measurement of lateral strains, *Canadian Geotechnical Journal*, Vol. 16, pp. 409–411.

Khanzode, R. M., Vanapalli, S. K., and Fredlund, D. G. 2002. Measurement of soil-water characteristic curves for fine-grained soils using a small scale centrifuge, Technical Note, *Canadian Geotechnical Journal*, Vol. 39, No. 5, pp. 1209–1217.

Khire, M. V., Benson, C. H., and Bosscher, P. J. 2000. Capillary barriers: Design variables and water balance, *Journal of Geotechnical and Geoenvironmental Engineering*, Vol. 126, No. 8, pp. 695–708.

Khogali, W. E. I., Anderson, K. O., Gan, J. K-M., and Fredlund, D. G. 1991. Installation and monitoring of moisture suction sensors in a fine-grained subgrade soil subjected to seasonal frost, paper presented at the Second International Symposium on State-of-the-Art of Pavement Response-Monitoring Systems for Roads and Airfields, West Lebanon, NH.

Kimball, B. A., Jackson, R. D., Reginato, R. J., Nakayama, F. S., and Idso, S. B. 1976. Comparison of field-measured and calculated soil-heat fluxes, *Soil Science Society of America Proceedings*, Vol. 40, No. 1, pp. 18–25.

King, F. H. 1899. Principles and conditions of the movement of groundwater, Nineteenth Annual Report, *U.S. Geolgical Survey*, Part 2, p. 50.

King, L. G. 1965. Description of soil characteristics for partially saturated flow, *Soil Science Society of American Proceedings*, Vol. 29, pp. 359–362.

Kisch, M. 1959. The theory of seepage from clay-blanked reservoirs, *Geotechnique*, Vol. 9, pp. 9–21.

Kjaernsli, B., and Simons, N. 1962. Stability investigations of the north bank of the Drammen River, *Geotechnique*, Vol. 12, pp. 147–167.

Klinkenberger, L. J. 1941. The permeability of porous media to liquids and gases, in *Drilling and Production Practice*, American Petroleum Institute, New York, pp. 200–213.

Klute, A. 1965a. Laboratory measurement of hydraulic conductivity of unsaturated soil, in C. A. Black, D. D. Evans, J. L. White, L. E. Ensminger, and F. E. Clark (Eds.), *Methods of Soil Analysis*, Monogram 9, Part 1, American Society of Agronomy, Madison, WI, pp. 253–261.

Klute, A. 1965b. Water diffusivity, in C. A. Black, D. D. Evans, J. L. White, L. E. Ensminger, and F. E. Clark (Eds.), *Methods of Soil Analysis*, Monogram 9, Part 1, American Society of Agronomy, Madison, WI, pp. 262–272.

Klute, A. 1972. The determination of the hydraulic conductivity and diffusivity of unsaturated soils, *Soil Science Journal*, Vol. 113, pp. 264–276.

Knapp, R. T. 1958. Accelerated field tests of cavitation intensity, *American Society of Mechanical Engineers Transactions*, Vol. 80, No. 1, pp. 91–96.

Knapp, R. T., Daily, J. W., and Hammitt, F. G. 1970. *Cavitation*. McGraw-Hill, New York.

Kodikara, J. K., Barbour, S. L., and Fredlund, D. G. 1998. Discussion: An idealized framework for the analysis of cohesive soils under desiccation, by J-M. Konrad and R. Ayad, *Canadian Geotechnical Journal*, Vol. 34, No. 6, pp. 1112–1114.

Kodikara, J. K., Barbour, S. L., and Fredlund, D. G. 1999. Changes in clay structure and behaviour due to wetting and drying, *Proceedings of the Eighth Australian-New Zealand Conference on Geomechanics*, Hobart, Tasmania, Australia, pp. 179–186.

Kodikara, J. K., Barbour, S. L., and Fredlund, D. G. 2000. Desiccation cracking of soil layers, in *Proceedings of the Asian Conference in Unsaturated Soils, UNSAT ASIA 2000*, Balkema, Singapore, pp. 693–698.

Kodikara, J. K., Barbour, S. L., and Fredlund, D. G. 2002. Structure development in surficial heavy clay soils: A synthesis of mechanisms, *Australian Geomechanics*, Vol. 37, No. 3, pp. 25–40.

Kohgo, Y., Nakano, M., and Miyazaki, T. 1993. (a) Theoretical aspects of constitutive modelling for unsaturated soils; (b) Verification of the generalized elastoplastic model for unsaturated soils, *Soils and Foundations*, Vol. 33, No. 4, pp. 49–73.

Kohn, P. G. 1965. Tables of some physical and chemical properties of water, in *Society of Experimental Biologists Symposium XIX: The State and Movement of Water in Living Organisms*, Cambridge University Press, Cambridge, UK.

Kohnke, H. 1968. *Soil Physics*, McGraw-Hill, New York.

Komornik, A., and David, D. 1969. Prediction of swelling pressure in clays, *Journal of Soil Mechanics and Foundations Engineering Division, ASCE*, Vol. 95, pp. 209–225.

Komornik, A., Livneh, M., and Smucha, S. 1980. Shear strength and swelling of clays under suction, *Proceedings of the Fourth International Conference on Expansive Soils*, Denver, CO, Vol. 1, pp. 206–226.

Kondo, J., and Saigusa, N. 1992. A model and experimental study of evaporation from bare-soil surfaces, *Journal of Applied Meteorology, American Meteorological Society*, Vol. 31, pp. 304–312.

Koning, H. L. 1963. Some observations on the modulus of compressibility of water, *Proceedings of the European Conference on Soil Mechanics and Foundation Engineering*, Wiesbaden, Germany, pp. 33–36.

Konrad, J.-M., and Ayad, R. 1997. An idealized framework for the analysis of cohesive soils undergoing desiccation, *Canadian Geotechnical Journal*, Vol. 34, pp. 477–488.

Konrad, J.-M., and Morgenstern, N. R. 1980. A mechanistic theory of ice lens formation in fine-grained soils, *Canadian Geotechnical Journal*, Vol. 17, pp. 473–486.

Koopmans, R. W. R., and Miller, R. D. 1966. Soil freezing and soil-water characteristic curves, *Soil Science Society of America Proceedings*, Vol. 30, No. 6, pp. 680–685.

Koorevaar, P., Menelik, G., and Dirksen, C. 1983. *Elements of Soil Physics*, Elsevier, Amsterdam, The Netherlands.

Krahn, J., and Fredlund, D. G. 1972. On total matric and osmotic suction, *Journal of Soil Science Journal*, Vol. 114, No. 5, pp. 339–348.

Krahn, J., and Fredlund, D. G. 1983. Variability of the engineering properties of natural soil deposits, *Proceedings of the Fourth International Conference on Applications of Statistics and Probability in Soil and Structural Engineering*, Florence, Italy, June 13–17, Vol. 2, pp. 1017–1029.

Krahn, J., Fredlund, D. G., and Klassen, M. J. 1989. Effect of soil suction on slope stability at Notch Hill, *Canadian Geotechnical Journal*, Vol. 26, No. 2, pp. 269–278.

Krisdani, H., Rahardjo, H., and Leong, E. C. 2005. Behaviour of capillary barrier system constructed using residual soil, *Proceeding of Geo-Frontiers Congress, ASCE*, Austin, TX, pp. 1–15.

Krisdani, H., Rahardjo, H., and Leong, E. C. 2006. Experimental study of 1–D capillary barrier model using geosynthetic material as the coarse-grained layer, in *Proceedings of Fourth International Conference on Unsaturated Soils*, *ASCE*, Geo Institute, Phoenix, AZ, pp. 1683–1694.

Krisdani, H., Rahardjo, H., and Leong, E. C. 2008a. Effects of different drying rates on shrinkage characteristics of a residual soil and soil mixtures, *Journal of Engineering Geology*, Vol. 102, pp. 31–37.

Krisdani, H., Rahardjo, H., and Leong, E. C. 2008b. Measurement of geotextile-water characteristic curve using capillary rise principle, *Geosynthetics International Journal*, Vol. 15, No. 2, pp. 86–94.

Krisdani, H., Rahardjo, H., and Leong, E. C. 2009. Use of instantaneous profile and statistical methods to determine permeability functions of unsaturated soils, *Canadian Geotechnical Journal*, Vol. 46, No. 7, pp. 869–874.

Krisdani, H., Rahardjo, H., and Leong, E. C. 2010a. Application of geosynthetic material as a coarse-grained layer in capillary barriers, special issue on unsaturated geosynthetics, *Geosynthetics International Journal*, Vol. 17, No. 5, pp. 323–331.

Krisdani, H., Rahardjo, H., and Leong, E. C. 2010b. Response to discussion by Mbonimpa, M., Aubertin, M. and Bussiere, B. on "Effects of different drying rates on shrinkage characteristics of a residual soil and soil mixtures," *Journal of Engineering Geology*, Vol. 107, Nos. 3–4, pp. 172–173 2009, discussion published in *Journal of Engineering Geology*, Vol. 110, Nos. 1–2, pp. 30–31.

Krohn, J. P., and Slosson, J. E. 1980. Assessment of expansive soils in the United States, *Proceedings of Fourth International Conference on Expansive Soils*, Denver, CO, Vol. 1, pp. 596–608.

Krynine, D. P. 1948. Analysis of the latest American tests on soil capillarity, *Proceedings of Second International Conference on Soil Mechanics Foundation Engineering*, Rotterdam, The Netherlands, Vol. 3, pp. 100–104.

Kunze, R. J., Uehara, G., and Graham, K. 1968. Factors important in the calculation of hydraulic conductivity, *Soil Science Society of America Proceedings*, Vol. 32, pp. 760–765.

Kurucuk, N., Kodikara, J., and Fredlund, D. G. 2007. Prediction of compaction curves, *Proceedings of the Tenth Australia–New Zealand Conference on Geomechanics*, Brisbane, Australia, Vol. 2, pp. 115–119.

Kurucuk, N., Kodikara, J., and Fredlund, D. G. 2008. Theoretical modelling of the compaction curve, *Proceedings of the First European Conference on Unsaturated Soils*, Durham, UK, pp. 375–379.

Kurucuk, N., Kodikara, J., and Fredlund, D. G. 2009. Evolution of the compaction process: Experimental study–Preliminary results, *Proceedings of the Fourth Asia-Pacific Conference on Unsaturated Soils*, Newcastle, Australia, November 23–25, Vol. 2, pp. 887–893.

Kurucuk, N., Kodikara, J., and Fredlund, D. G. 2011. Null-type axis-translation technique: Is it applicable to soils with low degrees of saturation? *Proceedings of the Fifth Asia-Pacific Conference on Unsaturated Soils*, Payatta, Thailand.

Kyklema, J. 2000. *Fundamentals of Interface and Colloid Science*, Volume III, *Liquid-Fluid Interface*, Academic, New York.

Ladd, C. C. 1960. Mechanisms of swelling in compacted clay, *Highway Research Board Bulletin*, National Research Council, No. 245, pp. 10–26.

Lai, S. H., Tiedje, J. M., and Erickson, A. E. 1976. *In situ* measurement of gas diffusion coefficient in soils, *Soil Science Society of America Journal*, Vol. 40, pp. 3–6.

Laliberte, G. E. 1969. A mathematical function for describing capillary pressure-desaturation data, *Bulletin of the International Association Hydrological Sciences*, Vol. 14, No. 2, pp. 131–149.

Laliberte, G. E., and Corey, A. T. 1966. Hydraulic properties of disturbed and undisturbed soils, Special Topics Publication No. 417, American Society for Testing and Materials, Philadelphia, PA.

Lam, L., and Fredlund, D. G. 1984. Saturated-unsaturated transient finite element seepage model for geotechnical engineering, *Advanced Water Resources*, Vol. 7, pp. 132–136.

Lam, L., and Fredlund, D. G. 1993. A general limit equilibrium model for three-dimensional slope stability analysis, *Canadian Geotechnical Journal*, Vol. 30, No. 6, pp. 905–919.

Lam, L., and Fredlund, D. G. 1994. Reply: A general limit equilibrium model for three-dimensional slope stability analysis, *Canadian Geotechnical Journal*, Vol. 31, No. 5, pp. 795–796.

Lam, L., and Fredlund, D. G. 1996. Appropriate intercolumn force functions and lambda values for three-dimensional slope stability analysis, *Proceedings of the Seventh International Symposium on Landslides*, Trondheim, Norway, pp. 1283–1288.

Lam, L., Fredlund, D. G., and Barbour, S. L. 1987. Transient seepage model for saturated-unsaturated soil systems: A geotechnical engineering approach, *Canadian Geotechnical Journal*, Vol. 24, No. 4, pp. 565–580.

Lambe, T. W. 1951a. *Soil Testing for Engineers*, 2nd ed., Wiley, New York.

Lambe, T. W. 1951b. Capillary phenomena in cohesionless soils, *Transactions ASCE*, Vol. 116, pp. 401–423.

Lambe, T. W. 1958. The structure of compacted clay, *Journal of Soil Mechanical Foundation Engineering Division, ASCE*, Vol. 84, No. 2, pp. 1–34.

Lambe, T. W. 1960a. A mechanistic picture of shear strength in clay, *Proceedings of Research Conference on Shear Strength of Cohesive Soils, ASCE*, University of Colorado, Boulder, CO, pp. 555–580.

Lambe, T. W. 1960b. Structure of compacted clay, *Transactions, ASCE*, Vol. 125, pp. 682–705.

Lambe, T. W. 1967. Stress path method, *Journal of Soil Mechanics Foundation Engineering Division, ASCE*, SM6, Vol. 93, pp. 309–331.

Lambe, T. W., and Whitman, R. V. 1968. *Soil Mechanics*, Wiley, New York.

Lambe, T. W., and Whitman, R. V. 1979. *Soil Mechanics*, Wiley, New York.

Lane, K. S., and Washburn, S. E. 1946. Capillarity tests by capillarimeters and by soil filled tubes, *Proceedings Highway Research Board*, Vol. 26, pp. 460–473.

Lange, G. R., and McKim, H. L. 1966. Saturation, phase composition, and freezing-point depression in a rigid soil model, *Research Report, U.S. Army Cold Regions Research and Engineering Laboratory, Springfield, VA. Proceedings of an International Conference*, National Academy of Sciences, Washington, DC, pp. 187–192.

Langfelder, L. J., Chen, C. F., and Justice, J. A. 1968. Air permeability of compacted cohesive soils, *Journal Soil Mechanics Foundation Engineering Division, ASCE*, SM4, Vol. 94, pp. 981–1001.

Larmour, D. Y. 1966. One-dimensional consolidation of partially saturated clay, M.Sc. Thesis, University of Manchester, Manchester, England.

Lau, J. T. K., Fredlund, D. G., and Rahardjo, H. 1991. Comparisons of elastic and plastic approaches to the prediction of depth of cracking. *Proceedings of the 13th Canadian Congress of Applied Mechanics*, Winnipeg, MB, Vol. 1, p. 388.

Laurent, J. P. 1998. Profiling water content with TDR: Experimental comparison with the neutron probe technique, paper presented at the IAEA/FAO Consultants' Meeting, Vienna, Austria.

Lauritzen, C. W. 1948. Apparent specific volume and shrinkage characteristics of soil materials, *Soil Science Journal*, Vol. 65, pp. 155–179.

Lawton, E. C., Fragaszy, R. J., and Hardcastle, I. H. 1991a. Stress ratio effects on collapse of compacted clayey sand, *Journal of Geotechnical Engineering*, Vol. 117, No. 5, pp. 714–730.

Lawton, E. C., Fragaszy, R. J., and Hethrington, M. D. 1991b. Review of wetting-induced collapse in compacted soil, *Journal of Soil Mechanics and Foundations, ASCE*, Vol. 118, No. 9, pp. 1376–1394.

Leach, B., and Herbert, R. 1982. The genesis of a numerical model for the study of the hydrogeology of a steep hillside in Hong Kong, *Quarterly Journal of Engineering Geology*, Vol. 15, pp. 253–259.

Lee, D. Y., and Suedkamp, R. J. 1972. Characteristics of irregularly shaped compaction curves of soil, *Highway Research Record*, Vol. 381, pp. 1–9.

Lee, H. C., and Wray, W. K. 1992. Evaluation of soil suction instruments, *Proceedings of the Seventh International Conference on Expansive Soils*, Dallas, TX, Vol. 1, pp. 307–312.

Lee, I. M., Sung, S. G., and Cho, G. C. 2005. Effect of stress state on the unsaturated shear strength of weathered granite, *Canadian Geotechnical Journal*, Vol. 42, pp. 624–631.

Lee, M. L., Gofar, N., and Rahardjo, H. 2009. A simple model for preliminary evaluation of rainfall-induced slope instability, *Journal of Engineering Geology*, Vol. 108, pp. 272–285.

Lee, R. K. C., and Fredlund, D. G. 1984. Measurement of soil suction using the MCS 6000 gauge, *Proceedings of the Fifth International Conference on Expansive Soils*, Institute of Engineering, Adelaide, Australia, pp. 50–54.

Lee, T. J., and Pielke, R. 1992. Estimating the soil surface specific humidity, Notes and correspondence, *Journal of Applied Meteorology, American Meteorological Society*, Vol. 31, pp. 480–484.

Leonards, G. A., and Altschaeffl, A. G. 1964. Compressibility of clay, *Journal of Soil Mechanics and Foundation Division, ASCE*, Vol. 90, No. 5, pp.133–156.

Leong, E. C., and Rahardjo, H. 1997a. Permeability functions for unsaturated soils, *Journal of Geotechnical and Geoenvironmental Engineering, ASCE*, Vol. 123, No. 12, pp. 1118–1126.

Leong, E. C., and Rahardjo, H. 1997b. A review on soil-water characteristic curve equations, *Journal of Geotechnical and Geoenvironmental Engineering, ASCE*, Vol. 123, No. 12, pp. 1106–1117.

Leong, E. C., and Rahardjo, H. 1997c. Discussion on the paper entitled: "Unsaturated hydraulic conductivity of two compacted barrier soils" by Meerdink et al., 1996, *Journal of Geotechnical Engineering, ASCE*, Vol. 122, No. 7, pp. 565–575; in *Journal of Geotechnical Engineering, ASCE*, Vol. 123, No. 12, pp. 1186–1188.

Leong, E. C., and Rahardjo, H. 1999. Discussion on the paper entitled "Soil-water characteristic curves for compacted clays" by Tinjum et al., 1997, *Journal of Geotechnical Engineering, ASCE*, Vol. 123, No. 11, pp. 1060–1069; in *ASCE Journal of Geotechnical Engineering*, Vol. 125, No.7, pp. 629–630.

Leong, E. C., Soemitro, R. A. A., and Rahardjo, H. 2000. Soil improvement by surcharge and vacuum preloading, *Geotechnique*, Vol. 50, No. 5, pp. 601–605.

Leong, E. C., Rahardjo, H., and Fredlund, D. G. 2001. Applications of unsaturated soil mechanics in geotechnical engineering, *Proceedings of the Eighth East Asia-Pacific Conference on Structural Engineering & Construction, EASEC-8*, Singapore, pp. 1–6.

Leong, E .C., He, L., and Rahardjo, H. 2002a. Factors affecting the filter paper method for total and matric suction measurements, *Geotechnical Testing Journal, ASTM*, Vol. 25, No. 3, pp. 322–333.

Leong, E. C., Rahardjo, H., and Tang, S. K. 2002b. Characterisation and engineering properties of Singapore residual soils, *Proceedings of International Workshop on Characterisation and Engineering Properties of Natural Soils*, Singapore, Vol. 2, pp. 1279–1304.

Leong, E. C., Rahardjo, H., and Fredlund, D. G. 2003a. A comparative study of constitutive models for unsaturated soils, *Proceedings of the Second Asian Conference on Unsaturated Soils, UNSAT-ASIA*, Osaka, Japan, pp. 41–46.

Leong, E. C., Tripathy, S., and Rahardjo, H. 2003b. Total suction measurement of unsaturated soils with a device using the chilled-mirror dew-point technique, *Geotechnique*, Vol. 53, No. 2, pp. 173–182.

Leong, E. C., Agus, S. S., and Rahardjo, H. 2004a. Volume change measurement of soil specimen in triaxial test, *Geotechnical Testing Journal, ASTM*, Vol. 27, No. 1, pp. 47–56.

Leong, E. C., Tripathy, S., and Rahardjo, H. 2004b. A modified pressure plate apparatus, *ASTM Geotechnical Testing Journal*, Vol. 27, No. 3, pp. 322–331.

Leong, E. C., Yeo, S. H., and Rahardjo, H. 2004c. Measurement of wave velocities and attenuation using an ultrasonic test system, *Canadian Geotechnical Journal*, Vol. 41, No. 5, pp. 844–860.

Leong, E. C., Eriktius, D. T., Tan, J. L., and Rahardjo, H. 2005a. Properties of municipal solid waste fly ash, *Pertanika Journal of Science and Technology*, Supplement Vol. 13, No. 1, pp. 1–11.

Leong, E. C., Yeo, S. H., and Rahardjo, H. 2005b. Measuring shear wave velocity using bender elements, *ASTM Geotechnical Testing Journal*, Vol. 28, No. 5, pp. 488–498.

Leong, E. C., Widiastuti, S., and Rahardjo, H. 2007. Accuracy of suction measurements, *Geotechnique*, Vol. 57, No. 6, pp. 547–556.

Leong, E. C., Cahyadi, J., and Rahardjo, H. 2009a. Measuring shear and compression wave velocities of soil using bender-extender elements, *Canadian Geotechnical Journal*, Vol. 46, No. 7, pp. 792–812.

Leong, E. C., He, L. C., and Rahardjo, H. 2009b. Discussion on the paper entitled: "Assessment of the use of the vapour equilibrium technique in controlled-suction tests," by X. Pintado, A. Lloret, and E. Romero 2009, *Canadian Geotechnical Journal*, Vol. 46, pp. 411–423; in *Canadian Geotechnical Journal*, Vol. 46, pp. 1482–1484.

Leong, E. C., Rahardjo, H., and He, L. C. 2010. Discussion on the paper entitled: Calibrations of a high-suction tensiometer by S. D. N., Laurenco, D. Gallipoli, D. G. Toll, C. E. Augarde, F. D. Evans and G. M. Medero 2008. *Geotechnique*, Vol. 58, No. 8, pp. 659–668; in *Geotechnique*, Vol. 60, No. 3, pp. 233–234.

Leong, E. C., Zhang, X. H., and Rahardjo, H. 2011. Calibration of a thermal conductivity sensor for field measurement of matric suction, *Geotechnique*, Vol. 62, No. 1, pp. 81–85.

Letey, J., Kemper, W. D., and Noonan, L. 1969. The effect of osmotic pressure gradient on water movement in unsaturated soil, *Soil Science Society of America Proceedings*, Vol. 33, pp. 15–18.

Lewis, J. M. 1995. The story behind the Bowen ratio, *Bulletin of the American Meteorological Society*, Vol. 76, pp. 2433–2443.

Li, J., Sun, D., Sheng, D., Sloan, S., and Fredlund, D. G. 2007. Preliminary study on soil-water characteristics of Maryland clay, *Proceedings of the Third Asian Conference on Unsaturated Soils, UNSAT-ASIA*, Nanjing, China, pp. 569–575.

Li, J. H., Zhang, L. M., Wang, Y., and Fredlund, D. G. 2009. Permeability tensor and representative elementary volume of saturated cracked soil, *Canadian Geotechnical Journal*, Vol. 46, pp. 928–942.

Li, K. S. 1989. Discussion on closed-form heave solutions for expansive soils, *Journal of Geotechnical Engineering, ASCE*, Vol. 115, No. 12, pp. 1819–1823.

Li, X., and Koike, T. 2001. A new frozen soil parameterization in land surface scheme, in T. Mastsuno and H. Keda (Eds.), *Proceedings of Present and Future of Modeling Global Environmental Change: Towards Integrated Modeling*, Terra Scientific Publication Company, Tokyo, pp. 405–414.

Li, X., Zhang, L. M., and Fredlund, D. G. 2009. A wetting front advancing column test for measuring the permeability function, *Canadian Geotechnical Journal*, Vol. 46, pp. 1431–1445.

Li, X., Zhang, L. M., and Fredlund, D. G. 2010. Efficient measurement of unsaturated hydraulic conductivity using a wetting front advancing method, *Proceedings of the Fifth International Conference on Unsaturated Soils*, Barcelona, Vol. 2, pp. 689–694.

Li, Z. M. 1995. Compressibility and collapsibility of compacted unsaturated loessial soils, unsaturated soils, *Proceedings of the First International Conference on Unsaturated Soils, UNSAT 95*, Paris, Vol. 1, pp. 139–144.

Liakopoulos, A. C. 1965a. Theoretical solution of the unsteady unsaturated flow problems in soils, *Bulletin of the International Association of Scientific Hydrology*, Vol. 10, pp. 5–39.

Liakopoulos, A. C. 1965b. Retention and distribution of moisture in soils after infiltration has ceased, *Bulletin of the International Association of Scientific Hydrology* Vol. 10, pp. 58–69.

Lide, D. R. (Ed.). 1990. *CRC Handbook of Chemistry and Physics*, 70th edition, CRC Press, Boca Raton, FL.

Lide, D. R. (Ed.). 1992. *CRC Handbook of Chemistry and Physics*, 73rd edition, CRC Press, Boca Raton, FL, pp. 6–10.

Lide, D. R., and Frederikse, H. P. R. 1997. *CRC Handbook of Chemistry and Physics, A Ready-Reference Book of Chemical and Physical Data*, CRC Press, New York.

Lidgren, R. A. 1970. Volume change characteristics of compacted till, M.Sc. Thesis, University of Saskatchewan, Saskatoon, SK.

Likos, W. J., and Lu, N. 2003. Automated humidity system for measuring total suction characteristics of clay, *ASTM Geotechnical Testing Journal*, Vol. 26, No. 2, pp. 1–12.

Lim, P. C., Barbour, S. L., and Fredlund, D. G. 1994. Laboratory determination of diffusion and adsorption coefficients of inorganic chemicals for unsaturated soil, *Proceedings of the First International Congress on Environmental Geotechnics*, Edmonton, AB, pp. 319–324.

Lim, P. C., Barbour, S. L., and Fredlund, D. G. 1998. The influence of the degree of saturation on the coefficient of aqueous diffusion, *Canadian Geotechnical Journal*, Vol. 35, No. 5, pp. 811–827.

Lim, T. T., Rahardjo, H., Chang, M. F., and Fredlund, D. G. 1996. Effect of rainfall on matric suctions in a residual soil slope, *Canadian Geotechnical Journal*, Vol. 33, pp. 618–628.

Lins, Y., and Schanz, T. 2004. Determination of hydro-mechanical properties of sand, paper presented at the International Conference on the Theme "From Experimental Evidence towards Numerical Modeling of Unsaturated Soils," Weimar, Germany, pp. 11–29.

Lins, Y., Schanz, T., and Fredlund, D. G. 2009. Modified pressure plate apparatus and column testing device for measuring SWCC properties of sand, *ASTM Geotechnical Testing Journal*, Vol. 32, No. 5, pp. 1–15.

Little, A. L. 1969. The engineering classification of residual tropical soils, *Proceedings of the Seventh International Conference on Soil Mechanics and Foundation Engineering*, Mexico City, Vol. 1, pp. 1–10.

Lloret, A., and Alonso, E. E. 1980. Consolidation of unsaturated soils including swelling and collapse behaviour, *Geotechnique*, Vol. 30, No. 4, pp. 449–477.

Lloret, A., and Alonso, E. E. 1985. State surfaces for partially saturated soils, *Proceedings of the Eleventh International Conference on Soil Mechanics Foundation Engineering*, San Francisco, CA, Vol. 2, pp. 557–562.

Lloret, A., and Ledesma, A. 1993. Finite element analysis of deformations of unsaturated soils, Notes of the Civil Engineering European Courses, Barcelona, Spain, 20 p.

Lloret, A., Villar, M. V., Sanchez, M., Gens, A., Pintado, X., and Alonso, E. E. 2003. Mechanical behaviour of heavily compacted bentonite under high suction changes, *Geotechnique*, Vol. 53, No. 1, pp. 27–40.

Loi, J., Fredlund, D. G., Gan, J. K.-M., and Widger, R. A. 1992. Monitoring soil suction in an indoor test track facility, paper presented at the Seventy-First Annual Meeting of the Transportation Board, Session No. 228 on Environmental Factors except Frost, Washington, DC.

Low, P. F. 1955. The effect of osmotic pressure on diffusion rate of water, *Soil Science Journal*, Vol. 80, pp. 95–100.

Low, P. F., Anderson, D. M., and Hoekstra, P. 1968. Some thermodynamic relationships for soil at or below the freezing point: 1. Freezing point depression and heat capacity, *Water Research Resources*, Vol. 4, pp. 379–394.

Lowe, P. R. 1977. An approximating polynomial for the computation of saturation vapour pressure, *Journal of Agriculture and Forest Meteorology*, Vol. 16, pp. 100–103.

Luckner, L., van Genuchten, M. Th., and Nielsen, D. R. 1989. A consistent set of parametric models for the two phase flow of immiscible fluids in the subsurface, *Water Resources Research*, Vol. 25, pp. 2187–2193.

Lumb, P. 1962. Effect of rain storms on slope stability, *Proceedings of the Symposium on Hong Kong Soils*, Hong Kong, pp. 73–87.

Lumb, P. 1966. The residual soils of Hong Kong, *Geotechnique*, Vol. 16, No. 4, pp. 359–360.

Lumb, P. 1975. Slope failures in Hong Kong, Q.J., *Engineering Geology*, Vol. 8, pp. 31–65.

Lunardini, V. J. 1991. *Heat Transfer with Freezing and Thawing*, Elsevier Science, New York.

Lykov, A. V. 1961. *Heat and Mass Transfer in Capillary-Porous Bodies* (Teoreticheskiye Omorey Stroitel'noli Teplofiziki), translated by P. N. B. Harrison, W. M. Pun (Ed.), Pergamon, New York, pp. 201–277.

Lytton, R. L. 1967. Isothermal water movement in clay soils, PhD Thesis, University of Texas, Austin, TX.

Lytton, R. L. 1977. Foundations in expansive soils, in C. S. Desai and J. T. Christian (Eds.), *Numerical Methods in Geotechnical Engineering*, McGraw-Hill, New York, pp. 427–458.

Lytton, R. L. 1994. Prediction of movement in expansive clay, in A. Yeung and G. Feaalio (Eds.), *Vertical and Horizontal Deformations of Foundations and Embankments, Proceedings of Settlement '94*, College Station, Texas, June 16–18, Geotechnical Special Publication No. 40, American Society of Civil Engineers, New York, pp. 1827–1845.

Lytton R. L. 1996. Foundations and pavements on unsaturated soils, *Proceedings of the First International Conference on Unsaturated Soils, UNSAT-95*, Paris, Vol. 3, pp. 1201–1222.

Lytton, R. L., and Woodburn, J. A. 1973. Design and performance of mat foundations on expansive clay, *Proceedings of the Third International Conference on Expansive Soils*, Haifa, Vol. 1, pp. 301–307.

Maatouk, A., Leroueil, S., and La Rochelle, P. 1995. Yielding and critical state of a collapsible unsaturated silty soil, *Geotechnique*, Vol. 45, No. 3, pp. 465–477.

Mageau, D. W., and Morgenstern, N. R. 1980. Observations on moisture migration in frozen soils, *Canadian Geotechnical Journal*, Vol. 17, pp. 54–60.

Maksimovic, M. 1979. Limit equilibrium for nonlinear failure envelope and arbitrary slip surface, *Proceedings of the Third International Conference on Numerical Methods in Geomechanics*, Aachen, Germany, pp. 769–777.

Mandel, J. 1957. Consolidation des sols, *Geotechnique*, Vol. 3, pp. 287–299.

Manheim, F. T. 1966. A hydraulic squeezer for obtaining interstitial water from consolidated and unconsolidated sediment, Professional Paper 550–C, *U.S. Geological Survey*, pp. 256–261.

Mansell, R. S., Liwang M. A., Ahuja, L. R., and Bloom, S. A. 2002. Adaptive grid refinement in numerical models for water flow and chemical transport in soils–A review, *Vadose Zone Journal of Soil Science Society of America*, Vol. 1, pp. 222–238.

Marinho, F. A. M. 1994. Shrinkage behaviour of some plastic soils, PhD Thesis, University of London, London.

Marinho, F. A. M., and Chandler, R. J. 1994. Discussion on a new instrument for measuring soil moisture suction, *Geotechnique*, Vol. 44, No. 3, pp. 551–552.

Marjerison, B., Richardson, N., Fredlund, D. G., and Berthelot, C. 2001. Installation of sensors and measurement of soil suction below thin membrane surface pavements in Saskatchewan, *Proceedings of the Fifty-Fourth Canadian Geotechnical Conference*, Calgary, AB, pp. 1328–1334.

Marolo, A., Batenipour, H., Newman, G., Graham, J., and Booy D. 2009. Stress-deformation-thermal behaviour of a highway embankment on degraded permafrost, *Proceedings of the 17th International Conference on Soil Mechanics and Geotechnical Engineering*, Alexandria, Egypt.

Marshall, C. E. 1959. The diffusion of gas through porous media, *Soil Science Journal*, Vol. 10, pp. 79–82.

Marshall, T. J. 1958. A relation between permeability and size distribution of pores, *Soil Science Journal*, Vol. 9, pp. 1–8.

Massachusetts Institute of Technology. 1963. Engineering behaviour of partially saturated soils, Publication No. 134 presented to U.S. Army Engineering, Waterways Experimental Station, Vicksburg, MS, Soil Engineering Department, Department of Civil Engineering, MIT, Contract No. DA-22–079–eng-288, Boston, MA.

Maswoswe, J. 1985. Stress path for compacted soil during collapse due to wetting, PhD Thesis, Imperial College, London, UK.

Maswoswe, J., Burland, J. B., and Vaughan, P. R. 1992. Stress path method of predicting compacted soil collapse, *Proceedings of the Seventh International Conference on Expansive Soils*, Dallas, TX, Vol. 1, pp. 61–66.

Matyas, E. L. 1967. Air and water permeability of compacted soils, *Permeability and Capillary of Soils*, ASTM STP 417 American Society of Testing and Materials, Philadelphia, PA, pp. 160–175.

Matyas, E. L., and Radhakrishna, H. S. 1968. Volume change characteristics of partially saturated soils, *Geotechnique*, Vol. 18, No. 4, pp. 432–448.

Maulem, Y. 1986. Hydraulic conductivity of unsaturated soils: Prediction and formulas, in A. Klute (Ed.), *Method of Soils Analysis, Part 1. Physical and Mineralogical Methods*, American Society of Agronomy, Madison, WI, pp. 799–823.

Mayne, P. W., and Kulhawy, F. H. 1982. K_0-OCR relationships in soil, *Journal of Geotechnical Engineering Division, ASCE*, Vol. 108, No. GT6, pp. 851–872.

Mbonimpa, M., Aubertin, M., Maqsoud, A., and Bussière, B. 2006. Predictive model for the water retention curve of deformable clayey soils, *Journal of Geotechnical and Geoenvironmental Engineering, ASCE*, Vol. 132, No. 9, pp. 1121–1132.

McCuen, R. H., Rawls, W. J., and Brakensiek, D. L. 1981. Statistical analyses of the Brook-Corey and the Green-Ampt parameters across soil textures, *Water Resources Research*, Vol. 17, pp. 1005–1013.

McCutcheon, S. C., Martin, J. L., and Barnell, Jr., T. O. 1993. Water quality, in D. R. Maidment (Ed.), *Handbook of Hydrology*, McGraw-Hill, New York, p. 11.3.

McGaw, R. W., Berg, R. L., and Ingersoll, J. W. 1983. An investigation of transient processes in an advancing zone of freezing, *Proceedings of the Fourth International Conference on Permafrost*, Fairbanks, AK, pp. 821–825.

McKee, C. R., and Bumb, A. C. 1987. Flow-testing coalbed methane production wells in the presence of water and gas, *Society of Petroleum Engineers Formation Evaluation*, Vol. 2, No. 4, pp. 599–608.

McKee, C. R., and Bumb, A. C. 1984. The importance of unsaturated flow parameters in designing a hazardous waste site, in *Hazardous Waste and Environmental Emergencies: Hazardous Materials Control Research Institute National Conference*, Houston, TX, Silver Spring, MD, pp. 50–58.

McKeen, R. G. 1980. Field studies of airport pavements on expansive soils, *Proceedings of the Fourth International Conference on Expansive Soils*, Denver, CO, pp. 242–261.

McKeen, R. G. 1981. Suction studies: Filter paper method. Design of airport pavements for expansive soils: Final Report No. DOT/FAA/RD-81/25, U.S. Department of Transportation, Federal Aviation Administration, Systems Research and Development Service, Washington, DC.

McKeen, R. G. 1985. Validation of procedures for pavement design on expansive soils, Final Report, U.S. Department of Transportation, Washington, DC.

McKeen, R. G., and Johnson, L. D. 1990. Climate-controlled soil design parameters for mat foundations, *Journal of Geotechnical Engineering*, Vol. 116, No. 7, pp. 1073–1094.

McQueen, I. S., and Miller, R. F. 1968a. Calibration and evaluation of a wide range method of measuring moisture stress, *Soil Science Journal*, Vol. 106, No. 3, pp. 225–231.

McQueen, I. S., and Miller, R. F. 1968b. Determination of a soil moisture potential, in P. E. Rijtema and H. Wassink (Eds.), *Water in the Unsaturated Zone*, Publication No. 82, International Association of Scientific Hydrology, pp. 147–155.

McWhorter, D. B., and Nelson, J. D. 1979. Unsaturated flow beneath tailings impoundments, *Journal of Geotechnical Engineering Division, ASCE*, Vol. 105, pp. 1317–1334.

Meerdink, J. S., Benson, C. H., and Khire, M. V. 1996. Unsaturated hydraulic conductivity of two compacted barrier soils, *Journal of Geotechnical Engineering, ASCE*, Vol. 122, No. 7, pp. 565–576.

Meigs, P. 1953. World distribution of arid and semi-arid homoclimates, *UNESCO, Reviews of Research on Arid Zone Hydrology, Arid Zone Research*, pp. 203–210.

Meilani, I., Rahardjo, H., Leong, E. C., and Fredlund, D. G. 2002. Mini suction probe for matric suction measurement, *Canadian Geotechnical Journal*, Vol. 39, No. 6, pp. 1427–1432.

Meilani, I., Rahardjo, H., and Leong, E. C. 2005. Pore-water pressure and water volume change of an unsaturated soil under infiltration conditions, *Canadian Geotechnical Journal*, December, Vol. 42, No. 6, pp. 1509–1531.

Melinda, F., Rahardjo, H., Han, K. K., and Leong, E. C. 2004. Shear strength of compacted soil under infiltration condition, *Journal of Geotechnical and Geoenvironmental Engineering, ASCE*, Vol. 130, No. 8, pp. 807–817.

MEND. 1993. *SoilCover, User's Manual for an Evaporative Flux Model*, G. W. Wilson (Ed.), University of Saskatchewan, Saskatoon, SK.

Mendenhall, W., Scheaffer, R. L., and Wackerly, D. D. 1981. *Mathematical Statistics with Applications*, 2nd ed., Duxbury, Boston, MA.

Mesri, G., Pakbaz, M. C., and Cepeda-Diaz, A. F. 1994. Meaning, measurement and field application of swelling pressure of clay shales, *Geotechnique*, Vol. 44, No.1 pp. 129–145.

Metten, U. 1966. *Desalination by Reverse Osmosis*, MIT Press, Cambridge, MA.

Miao, L., Liu, S., and Lai, Y. 2002. Research of soil-water characteristics and shear strength features of Nanyang expansive soil, *Engineering Geology* (Amsterdam), Vol. 65, pp. 261–267.

Michels, W. C. 1961. *The International Dictionary of Physics and Electronics*, 2nd ed., Van Nostrand, Princeton, NJ.

Miller, D. J., and Nelson, J. D. 1992. Osmotic suction as a valid stress state variable in unsaturated soils, *Proceedings of the Seventh International Conference on Expansive Soils*, Dallas, TX, pp. 179–184.

Miller, J. D., and Gaskin, G. J. 1997. The development and application of the ThetaProbe soil water sensor, Technical Note (2nd ed.), McCaulay Land Use Research Institute, Aberdeen, UK.

Miller, J. D., Gaskin, G. J., and Anderson, H. A. 1997. From drought to flood: Catchment responses revealed using novel soil water probes, *Hydrological Processes*, Vol. 11, No. 5, pp. 533–541.

Miller, R. D. 1966. Phase equilibria and soil freezing, in *Permafrost: Proceedings of an International Conference*, National Academy of Sciences, Washington, DC, pp. 193–197.

Miller, R. D. 1973. Soil freezing in relation to pore-water pressure and temperature, *Proceedings of the Second International Conference on Permafrost*, Yakutsk, Russia, pp. 244–352.

Miller, R. F., and McQueen, I. S. 1978. Moisture relations in rangelands, western United States, *Proceedings of the First International Rangeland Congress*, Denver, CO, pp. 318–321.

Millington, R. J. 1959. Gas diffusion in porous media, *Science*, Vol. 130, pp. 100–102.

Millington, R. J., and Quirk, J. P. 1959. Permeability of porous media, *Nature*, Vol. 183, pp. 387–388.

Millington, R. J., and Quirk, J. R. 1961. Permeability of porous solids, *Transactions of the Faraday Society*, Vol. 57, pp. 1200–1207.

Milly, P. C. D. 1984a. A linear analysis of thermal effects on evaporation from soil, *Water Resources Research*, Vol. 20, No. 8, pp. 1075–1085.

Milly, P. C. D. 1984b. A simulation analysis of thermal effects on evaporation from soil. *Water Resources Research*, Vol. 20, No. 8, pp. 1087–1098.

Milne, L. J., and Milne, M. 1978. Insects of the water surface, *Scientific American*, Vol. 238, No. 4, pp. 134–142.

Mingbin, H., Fredlund, D. G., and Fredlund, M. D. 2010. Comparison of measured and PTF predictions of SWCCs for loess soils in China, *Geotechnical Geological Engineering Journal*, Vol. 28, pp. 105–117.

Misfeldt, G., Stone, J., and Fredlund, D. G. 2006. Protocol for preliminary modeling of an evaporative cover system, *Proceedings of the Canadian Geotechnical Conference*, Vancouver, BC, Vol. 1, pp. 1170–1177.

Mitchell, A. R., and van Genuchten, M. Th. 1992. Shrinkage of bare and cultivated soil, *Soil Science Society of America Journal*, Vol. 56, pp. 1036–1042

Mitchell, J. K. 1962. Components of pore water pressure and their engineering significance, *Clays and Clay Minerals*, Vol. 10, pp. 162–184.

Mitchell, J. K. 1973a. Chemico-osmotic effects in fine-grained soils, *Journal of the Soil Mechanics and Foundations Division, ASCE*, Vol. 99, pp. 307–322.

Mitchell, J. K. 1973b. Recent advances in the understanding of the influence of mineralogy and pore solution chemistry on the swelling and stability of clays, *Proceedings of the Third International Conference on Expansive Soils*, Haifa, Vol. 2, pp. 11–25.

Mitchell, J. K. 1976. *Fundamentals of Soil Behaviour*, Wiley, New York.

Mitchell, J. K. 1993. *Fundamentals of Soil Behavior*, 2nd ed., Wiley, New York.

Mitchell, J. K., and Sitar, N. 1982. Engineering properties of tropical residual soils, *Proceedings of Specialty Conference on Engineering and Construction in Tropical and Residual Soils, ASCE*, Honolulu, HI, pp. 30–57.

Mitchell, P. W., and Avalle, D. L. 1984. A technique to predict expansive soils movements, *Proceedings of the Fifth International Conference on Expansive Soils*, Adelaide, Australia, pp. 124–130.

Mitsuhiro, M., and Kataoka, Y. 1988. Study on liquid-vapor interface of water. I. Simulated results of thermodynamic properties and orientational structure, *Journal of Chemistry and Physics*, Vol. 88, No. 5, pp, 3233–3245.

Mohr, O. 1914. *Abhandlungen aus dem gebiete der technischen mechanic*, 3rd ed., Welhelm Ernst and Sohn, Berlin.

Moldrup, P., Yoshikawa, S., Olesen, T., Komatsu, T., and Rolston, D. E. 2003. Gas diffusivity in undisturbed volcanic ash soils: Test of soil-water characteristic based prediction models, *Soil Science Society of American Journal*, Vol. 67, pp. 41–51.

Monroy, R., Zdravkovic, L., and Ridley, A. 2008. Volumetric behaviour of compacted London clay during wetting and loading, paper presented at the First European Conference on Unsaturated Soils, Durham, UK, pp. 315–320.

Montanez, J. E. C. 2002. Suction and volume changes of compacted sand-bentonite mixtures, PhD Thesis, University of London, Imperial College of Science, London.

Monteith, J. L. 1965. Evaporation and environment, *Symposium on the Sociological Experimental Biology*, Vol. 19, pp. 205–224.

Moore, R. E. 1939. Water conduction from shallow water tables, *Hilgardia*, Vol. 12, No. 6, pp. 383–426.

Morgenstern, N. R. 1979. Properties of compacted soils, *Proceedings of the Sixth Pan American Conference on Soil Mechanics and Foundation Engineering*, Lima, Vol. 3, pp. 349–354.

Morgenstern, N. R., and Balasubramonian, B. I. 1980. Effects of pore fluid on the swelling of clay-shale, *Proceedings of the Fourth International Conference on Expansive Soils*, Denver, CO, pp. 190–205.

Morgenstern, N. R., and Price, V. E. 1965. The analysis of the stability of general slip surfaces, *Geotechnique*, Vol. 15, pp. 79–93.

Morgenstern, N. R., and Price, V. E. 1967. A numerical method for solving the equation of general slip surfaces, *British Computer Journal*, Vol. 9, pp. 338–393.

Mou, C. H., and Chu, T. Y. 1981. Soil suction approach for swelling potential evaluation, *Transportation Research Record* No. 790, Washington, DC, pp. 54–60.

Mualem, Y. 1973. Modified approach to capillary hysteresis based on a similarity hypothesis, *Water Resources Research Journal*, Vol. 9, No. 5, pp. 1324–1331.

Mualem, Y. 1974. A conceptual model of hysteresis, *Water Resources Research*, Vol. 10, No. 3, pp. 514–520.

Mualem, Y. 1976a. A new model for predicting hydraulic conductivity of unsaturated porous media, *Water Resources Research*, Vol. 12, pp. 513–522.

Mualem, Y. 1976b. Hysteretical models for prediction of the hydraulic conductivity of unsaturated porous media, *Water Resources Research*, Vol. 12, No. 6, pp. 1248–1254.

Mualem, Y. 1984a. A modified dependent domain theory of hysteresis, *Water Resources Research*, Vol. 12, pp. 513–522.

Mualem, Y. 1984b. Prediction of the soil boundary wetting curve, *Soil Science*, Vol. 137, pp. 379–389.

Mualem, Y. 1986. Hydraulic conductivity of unsaturated soils: Prediction and formulas, in *Methods of Soil Analysis*, Part 1: Physical and mineralogical methods, Second edition, Agronomy, Madison, WI, pp. 799–823.

Mualem, Y. 1992. Modeling the hydraulic conductivity of unsaturated porous media, *Proceedings of International Workshop*, University of California, Riverside, CA, pp. 1–14.

Mualem, Y., and Miller, E. E. 1979. A hysteresis model based on an explicit domain-dependent function, *Journal of Soil Science*, Vol. 43, No. 106–110.

Munoz-Carpena, R. 2004. Field devices for monitoring water content, Bulletin No. 343, Department of Agricultural and Biological Engineering, Florida Cooperative Extension Service, Institute of Food and Agricultural Sciences, University of Florida, Gainesville, FL.

Nagpal, N. K., and Boersma, L. 1973. Air entrapment as a possible source of error in the use of a cylindrical heat probe, *Proceedings Soil Science Society America*, Vol. 37, pp. 828–832.

Narasimhan, T. N., and Witherspoon, P. A. 1977. Numerical model for saturated-unsaturated flow in deformable porous media, 1. Theory, *Water Resources Research*, Vol. 13, No. 3, pp. 657–664.

NAVFAC. 1971. *Soil Mechanics Foundation and Earth Structures, Design Manual*, NAVFACDM-7, Naval Facilities Engineering Command, Department of the Navy, Bureau of Yards and Docks, Washington, DC.

Nèel, L. 1942. Théorie des lois d'aimantation de Lord Rayleigh, 1, *Cahiers de Physique*, Vol. 12, pp. 1–20.

Nèel, L. 1943. Théorie des lois d'aimantation de Lord Rayleigh, 2, *Cahiers de Physique*, Vol. 13, pp. 18–30.

Nelson, J. D., and Miller, D. J. 1992. *Expansive Soils: Problems and Practice in Foundation and Pavement Engineering*, Wiley, New York.

Nelson, J. D., Overton, D. D., and Durkee, D. B. 2001. Depth of wetting and the active zone, *Proceedings of Expansive Clay Soils and Vegetative Influence on Shallow Foundations, Proceedings of the Geo-Institute Foundations and Soil Properties Committee Sessions*, Geotechnical Special Publication No. 115, American Society of Civil Engineers, Houston, TX, pp. 132–157.

Nersesova, Z. A., and Tsytovich, N. A. 1966. Unfrozen water in frozen soils, *Proceedings of the Permafrost International Conference*, Washington, DC, Session 4, pp. 230–233.

Neuman, S. P. 1973. Saturated-unsaturated seepage by finite elements, *Proceedings of the Hydraulics Division, ASCE*, Vol. 99, No. 12, pp. 2233–2250.

Neuman, S. P. 1975. *Galerkin Approach to Saturated-Unsaturated Flow in Porous Media Finite Elements in Fluids*, Wiley, New York, pp. 201–217.

Neuman, S. P., and Witherspoon, P. A. 1971. Analysis of nonsteady flow with a free surface using the finite element method, *Journal of Water Resources Research*, Vol. 7, No. 33, pp. 611–623.

Newman, G. P. 1995. Heat and mass transfer in unsaturated soils, M.Sc. Thesis, University of Saskatchewan, Saskatoon, SK.

Newman, G. P. 2000. Underground freezing at a high grade uranium mine, *Proceedings of the Ninth International Symposium on Ground Freezing and Frost Action in Soils*, Louvain-La-Neuv, Belgium.

Newman, G., and Wilson, G. W. 1996. Heat and mass transfer in unsaturated soils during freezing, *Proceedings of the Canadian Geotechnical Conference*, St. John's, NL.

Newman, L. L., Herasymuik, G. M., Barbour, S. L., Fredlund, D. G., and Smith, T. 1997. The hydrology of waste rock dumps and a mechanism for unsaturated preferential flow, *Proceedings of the International Conference on Acid Rock Drainage*, Vancouver, BC, Vol. 2, pp. 551–565.

Newman, L. L., Barbour, S. L., and Fredlund, D. G. 1998. Preferential flow in vertically layered unsaturated soil systems, *Proceedings of the Second International Conference on Unsaturated Soils, UNSAT-98*, Beijing, Vol. 1, pp. 586–589.

Ng, C. W. W., and Chiu, A. C. F. 2001. Behaviour of a loosely compacted unsaturated volcanic soil, *Journal of Geotechnical and Geoenvironmental Engineering, ASCE*, Vol. 127, No. 12, pp. 1027–1036.

Ng, C. W. W., and Pang, Y. W. 1999. Stress effects of soil-water characteristics and pore-water pressures in unsaturated soil slopes, *Proceedings of the Eleventh Asian Regional Conference on Soil Mechanics and Geotechnical Engineering*, Seoul, South Korea, Vol. 1, pp. 371–374.

Ng, C. W. W., and Pang, Y. W. 2000a. Experimental investigation of soil-water characteristics of a volcanic soil, *Canadian Geotechnical Journal*, Vol. 37, No. 6, pp. 1252–1264.

Ng, C. W. W., and Pang, Y. W. 2000b. Influence of stress state on soil-water characteristics and slope stability, *Journal of Geotechnical and Geoenvironmental Engineering, ASCE*, Vol. 26, No. 2, pp. 157–166.

Ng, C. W. W., and Zhan, L. T. 2007. Comparative study of rainfall infiltration into a bare and a grassed unsaturated expansive soil slope, *Soils and Foundation*, Vol. 47, No. 2, pp. 207–217.

Ng, C. W. W., Zhan, L. T., and Cui, Y. J. 2002. A new simple system for measuring volume changes in unsaturated soils, *Canadian Geotechnical Journal*, Vol. 39, pp. 757–764.

Ng, C. W. W., Zhan, L. T., Bao, C. G., Fredlund, D. G., and Gong, B. W. 2003a. Performance of an unsaturated expansive soil slope subjected to artificial rainfall infiltration, *Geotechnique*, Vol. 53, No. 2, pp. 143–157.

Ng, C. W. W., Zhan, L. T., Fredlund, D. G., Bao, C. G., and Gong, B. W. 2003b. Rainfall infiltration study of an unsaturated expansive soil slope for the South-to-North Water Transfer Project, *Proceedings of the Second Asian Conference on Unsaturated Soils, UNSAT-ASIA 2003*, Osaka, Japan, pp. 127–132.

Ng, C. W. W., Cui, Y. J., Chen, R., and Delage, P. 2007. The axis-translation and osmotic techniques in shear testing of unsaturated soils: A comparison, *Soils and Foundations*, Vol. 47, No. 4, pp. 678–684.

Ng, T. N. 1988. The effect of negative pore-water pressures on slope stability analysis, M.Sc. Thesis, University of Saskatchewan, Saskatoon, SK.

Nguyen, Q., Fredlund, D. G., and Samarasekera, L. 2007. Presentation of long-term matric suction measurement in a highway subgrade using thermal conductivity sensors, *Proceedings of the Third Asian Conference on Unsaturated Soils, UNSAT-ASIA 2007*, Nanjing, China, pp. 487–492.

Nguyen, Q., Fredlund, D. G., Samarasekera, L., and Marjerison, B. 2010. Seasonal pattern variations of matric suction in highways, *Canadian Geotechnical Journal*, Vol. 47, pp. 267–280.

Nguyen, V. U. 1985. New *Applications of Classical Numerical Techniques in Slope Analysis*, Balkema, Rotterdam, Vol. 2, pp. 949–957.

Nichol, C., Smith, L., and Beckie, R. 2003. Long-term measurement of matric suction using thermal conductivity sensors, *Canadian Geotechnical Journal*, Vol. 40, pp. 587–597.

Nielsen, D. R., and Biggar, Y. W. 1961. Measuring capillary conductivity, *Soil Science Journal*, Vol. 92, pp. 192–193.

Nielsen, D. R., Jackson, R. D., Cary, J. W., and Evans, D. D. 1972. *Soil Water*, American Society of Agronomy and Soil Science, Madison, WI.

Nimmo, J. R. 1992. Semi-empirical model of soil water hysteresis, *Soil Science Society of America Journal*, Vol. 56, pp. 1723–1730.

Nimmo, J. R. 1997. Modeling structural influences on soil water retention, *Soil Science Society of America Journal*, Vol. 61, No. 3, pp. 712–719.

Nishimura, T. 2000. Direct shear properties of a compacted soil with known stress history, in H. Rahardjo, D. G. Toll, and E. C. Leong (Eds.), *Proceedings of the Unsaturated Soils for Asia Conference*, Singapore, pp. 557–562.

Nishimura, T., and Fredlund, D. G. 1999. Unconfined compression shear strength of an unsaturated silty soil subjected to high total suctions, *Proceedings of the International Symposium on Slope Stability Engineering*, IS-SHIKOKU'99, Shikoku, Japan, Vol. 2, pp. 757–762.

Nishimura, T., and Fredlund, D. G. 2000. Relationship between shear strength and matric suction in an unsaturated silty soil, *Proceedings of the Asian Conference on Unsaturated Soils, UNSAT-ASIA 2000*, Singapore, pp. 563–568.

Nishimura, T., and Fredlund, D. G. 2001. Failure envelope of a desiccated, unsaturated silty soil, *Proceedings of the Fifteenth International Conference on Soil Mechanics and Foundation Engineering*, Istanbul, pp. 615–618.

Nishimura, T., and Fredlund, D. G. 2002. Hysteresis effects resulting from drying and wetting under relatively dry conditions, *Proceedings of the Third International Conference on Unsaturated Soils, UNSAT 2002*, Recife, Brazil, pp. 301–305.

Nishimura, T., and Fredlund, D. G. 2003. A new triaxial apparatus for high total suctions using relative humidity, *Proceedings of the Twelfth Asian Regional Conference on Soil Mechanics and Geotechnical Engineering*, Singapore, Vol. 1, pp. 65–68.

Nishimura, T., Hirabayashi, Y., Fredlund, D. G., and Gan, J. K.-M. 1999. Influence of stress history on the strength parameters of an unsaturated statically compacted soil, *Canadian Geotechnical Journal*, Vol. 36, No. 2, pp. 251–261.

Nishimura, T., Toyota, H., Vanapalli, S., and Won, O. 2008. Determination of the shear strength behaviour of an unsaturated soil in the high suction range using the vapour pressure technique, *Proceedings of the First European Conference on Unsaturated Soils, E-UNSAT 2008*, Durban, UK, pp. 441–447.

Nixon, J. F. 1975. The role of convective heat transport in the thawing of frozen soils, *Canadian Geotechnical Journal*, Vol. 12, pp. 425–429.

Nixon, J. F. 1991. Discrete ice lens theory for frost heave in soils, *Canadian Geotechnical Journal*, Vol. 28, pp. 843–859.

Nixon, J. F. 1992. Discrete ice lens theory for frost heave beneath pipelines, *Canadian Geotechnical Journal*, Vol. 29, pp. 487–497.

Noble, C. A. 1966. Swelling measurements and prediction of heave for lacustrine clay, *Canadian Geotechnical Journal*, Vol. 3, No. 1, pp. 32–41.

Oberg, A., and Sallfors, G. 1997. Determination of shear strength parameters of unsaturated silts and sands based on water retention curve, *ASTM Geotechnical Testing Journal*, Vol. 20, pp. 40–48.

Oden, J. T. 1989. Progress in adaptive methods in computational fluid dynamics, in J. E. Flaherty et al. (Eds.), *Adaptive Methods for Partial Differential Equations*, Society for Industrial and Applied Mathematics, Philadelphia, PA, pp. 206–252.

Ogata, G., and Richards, L. A. 1957. Water content changes following irrigation of bare-field soil that is protected from evaporation, *Proceedings of Soil Science Society of America*, Vol. 21, pp. 355–356.

O'Kane, M., Wilson, G. W., and Barbour, S. L. 1998. Instrumentation and monitoring of an engineered soil cover system for mine waste rock, *Canadian Geotechnical Journal*, Vol. 35, pp. 828–847.

Oliphant, J. L., Tice, A. R., and Nakano, Y. 1983. Water migration due to a temperature gradient in frozen soil, *Proceedings of the Fourth International Conference on Permafrost*, Fairbanks, AK, pp. 951–956.

Oloo, S. Y., and Fredlund, D. G. 1996. A method for the determination of ϕ^b for statically compacted soils, *Canadian Geotechnical Journal*, Vol. 33, No. 2, pp. 272–280.

Oloo, S. Y., Fredlund, D. G., and Gan, J. K.-M. 1997a. Shear strength design of thin pavements and unbound roads, *Proceedings of the International Symposium on Thin Pavements, Surface Treatments, Unbound Roads*, University of New Brunswick, Fredericton, NB, pp. 135–142.

Oloo, S. Y., Fredlund, D. G., and Gan, J. K.-M. 1997b. Bearing capacity of unpaved roads, *Canadian Geotechnical Journal*, Vol. 34, No. 3, pp. 398–407.

Olson, R. E. 1963. Effective stress theory of soil compaction, *Journal Soil Mechanics and Foundations Division, ASCE*, Vol. 89, No. SM2, pp. 27–45.

Olson, R. E. 1986. State-of-the-art: Consolidation testing, *Consolidation of Soils: Testing and Evaluation*, American Society for Testing and Materials, Philadelphia, PA, pp. 7–70.

Olson, R. E., and Daniel, D. E. 1981. Measurement of the hydraulic conductivity of fine-grained soils, in T. F. Zimmie and C. O. Riggs (Eds.), *Permeability and Groundwater Contaminant Transport*, ASTM Special Technical Publication 746, American Society for Testing and Materials, Philadelphia, PA, pp. 18–64.

Olson, R. E., and Langfelder, L. J. 1965. Pore-water pressures in unsaturated soils, *Journal of Soil Mechanics Foundation Division, ASCE*, Vol. 91, SM4, pp. 127–160.

Orr, F. M., Scriven, L. E., and Rivas, A. P. 1975. Pendular rings between solids: Meniscus properties and capillary force, *Journal of Fluid Mechanics*, Vol. 67, No. 4, pp. 723–742.

Overton, D. D., Chao, K. C., and Nelson, J. D. 2011. Water content profiles for design of foundations on expansive soils, *Proceedings of the Fifth International Conference on Unsaturated Soils*, Barcelona, Vol. 2, pp. 1189–1194.

Paddy, J. F. 1969. *Theory of Surface Tension, Surface and Colloid Science*, Wiley Interscience, Toronto, ON.

Padilla, J. M. 2005. *Suction Data 1.12: User's Guide and Reference*, GCTS, Tempe, AZ.

Padilla, J. M., Perera, Y. Y., and Fredlund, D. G. 2004. Performance of the Fredlund thermal conductivity sensor, *Proceedings of the Eleventh Tailings and Mine Waste Conference*, Vail, CO, pp. 125–133.

Padilla, J. M., Perera, Y. Y., Houston, W. N., and Fredlund, D. G. 2005. A new soil-water characteristic curve device, *Proceedings of Advanced Experimental Unsaturated Soil Mechanics–An International Symposium, EXPERUS 2005*, Trento, Italy, pp. 15–22.

Padilla, J. M., Perera, Y. Y., Houston, W. N., Perez, P. N., and Fredlund, D. G. 2006a. Quantification of air diffusion through high air-entry ceramic disks, *Proceedings of the Fourth International Conference on Unsaturated Soils, ASCE*, Carefree, AZ, Vol. 2, pp.1852–1863.

Padilla, J. M., Houston, W. N., Lawrence, C. A., Fredlund, D. G., Houston, S. L., and Perez, P. N. 2006b. An automated triaxial testing device for unsaturated soils, *Proceedings of the Fourth International Conference on Unsaturated Soils, ASCE*, Carefree, AZ, Vol. 2, pp. 1775–1786.

Papagiannakis, A. T., and Fredlund, D. G. 1984. A steady state model for flow in saturated-unsaturated soils, *Canadian Geotechnical Journal*, Vol. 21, No. 13, pp. 419–430.

Papagiannakis, A. T., Lam, L., and Fredlund, D. G. 1983. Steady state and transient mass transport models for saturated-unsaturated soils, *Proceedings of the First International Potash Technology Conference*, Saskatoon, SK, pp. 741–747.

Parker, J. C., Lenhard, R. J., and Kuppusamy, T. 1987. A parametric model for constitutive properties governing multiphase flow in porous media, *Water Resources Research Journal*, Vol. 23, No. 4, pp. 618–624.

PDE Solutions. 2005. *FlexPDE 5.0–Reference Manual*, Antioch, CA.

Pearsall, I. S. 1972. *Cavitation*, Mills and Boon, London.

Peck, A. J., and Rabbidge, R. M. 1969. Design and performance of an osmotic tensiometer for measuring capillary potential, *Proceedings of Soil Science Society of America*, Vol. 33, No. 2, pp. 196–202.

Peck, R. B. 1969. Advantages and limitations of the observational method in applied soil mechanics, *Geotechnique*, Vol. 19, No. 2, pp. 171–187.

Penman, H. L. 1940. Gas and vapour movements in the soil. I. The diffusion of vapours through porous solids, *Journal of Agricultural Science*, Vol. 30, pp. 437–462.

Penman, H. L. 1948. Natural evapotranspiration from open water, bare soil and grass, *Proceedings of the Royal Society of London*, Series A193, pp. 120–145.

Penner, E. 1959. The mechanism of frost heaving in soils, *Highway Research Board Bulletin No. 225*, Washington, DC, pp. 1–22.

Penner, E. 1967. Heaving pressure in soils during unidirectional freezing, *Canadian Geotechnical Journal*, Vol. 4, pp. 398–408.

Pentland, J., Gitirana, Jr., F. N. G., and Fredlund, D. G. 2001. Use of a general partial equation solver for solution of heat and mass transfer problems in geotechnical engineering, *Proceedings of the Fourth Brazilian Symposium on Unsaturated Soils, NSAT 2001*, Porto Alerge, RS, Brazil, pp. 29–36.

Pereira, J. H. F. 1996. Numerical analysis of the mechanical behaviour of collapsing earth dams during first reservoir filling, PhD Thesis, University of Saskatchewan, Saskatoon, SK.

Pereira, J. H. F., and Fredlund, D. G. 1997. Constitutive modelling of a metastable-structured compacted soil, *Proceedings of the International Symposium on Recent Developments in Soil and Pavement Mechanics*, Rio de Janeiro, Brazil, pp. 317–326.

Pereira, J. H. F., and Fredlund, D. G. 1999a. Numerical analysis of the post-filling performance of small collapsing dams, *Proceedings of the Eleventh Pan American Conference on Soil Mechanics and Geotechnical Engineering*, Iguanzu Falls, Brazil, Vol. 3, pp. 1129–1140.

Pereira, J. H. F., and Fredlund, D. G. 1999b. Shear strength behaviour of a residual soil of gneiss compacted at metastable-structured conditions, *Proceedings of the Eleventh Pan American Conference on Soil Mechanics and Geotechnical Engineering*, Iguanzu Falls, Brazil, Vol. 1, pp. 369–377.

Pereira, J. H. F., and Fredlund, D. G. 2000a. Reply to "Discussion on volume change behaviour of collapsible compacted gneiss soil," *Geotechnical and Geoenvironmental Engineering Journal, ASCE*, Vol. 126, No. 10, pp. 907–916.

Pereira, J. H. F., and Fredlund, D. G. 2000b. Volume change behaviour of a residual soil of gneiss compacted at metastable-structured conditions, *Journal of Geotechnical and Geoenvironmental Engineering, ASCE*, Vol. 126, pp. 907–916.

Pereira, J. H. F., Gitirana, Jr., de F. N. G., Fredlund, D. G., and Cordao Neto, M. P. 2002. Anisotropic stress-strain law for wetting-induced soil collapse, *Proceedings of the Third International Conference on Unsaturated Soils*, Recife, Brazil, pp. 151–157.

Pereira, J. H. F., Cunha, R. P., Fredlund, D. G., and Gitirana, Jr., de F. N. G. 2003. Analysis of the post-filling performance of small collapsing dams, *Proceedings of the Second Asian Conference on Unsaturated Soils, UNSAT-ASIA 2003*, Osaka, pp. 289–294.

Pereira, J. H. F., Fredlund, D. G., Neto, M. P. C., and Gitirana, Jr., de F. N. G. 2005. Hydraulic behaviour of collapsible compacted gneiss soil, *Journal of Geotechnical and Geoenvironmental Engineering, ASCE*, Vol. 131, No. 10, pp. 1264–1273.

Perera, Y. Y., Zapata, Z. E., Houston, W. N., and Houston, S. L. 2005. Prediction of the soil-water characteristic curve based on grain-size-distribution and index properties, *Proceedings of Geo-Frontiers 2005 Congress, ASCE*, Jan. 24–26, Austin, TX.

Pham, H. Q. 2002. An engineering model of hysteresis for the soil-water characteristic curve, M.Sc. Thesis, University of Saskatchewan, Saskatoon, SK.

Pham, H. Q. 2005. A volume-mass constitutive model for unsaturated soils, PhD Thesis, University of Saskatchewan, Saskatoon, SK.

Pham, H. Q., and Fredlund, D. G. 2005. A volume-mass constitutive model for unsaturated soils, *Proceedings of the Fifty-Eighth Canadian Geotechnical Conference*, Saskatoon, SK, Vol. 2, pp. 173–181.

Pham, H. Q., and Fredlund, D. G. 2008. Equation for the entire soil-water characteristic curve on volume change soils, *Canadian Geotechnical Journal*, Vol. 45, pp. 443–453.

Pham, H. Q., and Fredlund, D. G. 2011a. Volume-mass unsaturated soil constitutive model for drying-wetting under isotropic loading-unloading, *Canadian Geotechnical Journal*, Vol. 48, No. 2, pp. 280–313.

Pham, H. Q., and Fredlund, D. G. 2011b. Evidence of stress path dependence in unsaturated soil behavior, Geotechnics for Sustainable Development, in D.L. Phung (Ed.), *Proceedings of the International Conference Geotec Hanoi 2011*, Hanoi, Vietnam, October 6–7, Construction Publishing House, Hanoi, Vietnam, pp. 815–822.

Pham, H. Q., Fredlund, D. G., and Barbour, S. L. 2002. A simple soil-water hysteresis model for predicting the boundary wetting curve, *Proceedings of the Fifty-Fifth Canadian Geotechnical Conference: Ground and Water–Theory to Practice*, Niagara Falls, ON, pp. 1261–1267.

Pham, H. Q., Fredlund, D. G., and Barbour, S. L. 2003a. A practical hysteresis model for the soil-water characteristic curve for soils with negligible volume change, *Geotechnique*, Vol. 53, No. 2, pp. 293–298.

Pham, H. Q., Fredlund, D. G., and Barbour, S. L. 2003b. Estimation of the hysteretic soil-water characteristic curves from the drying boundary curve, *Proceedings of the Fifty-Sixth Canadian Geotechnical Conference*, Winnipeg, MB, Vol. 2, pp. 115–121.

Pham, H. Q., Fredlund, D. G., and Padilla, J. M. 2004. Use of the GCTS apparatus for the measurement of soil-water characteristic curves, *Proceedings of the Fifty-Seventh Canadian Geotechnical Conference*, Quebec, QC, Session JE, pp. 1–6.

Pham, H. Q., Fredlund, D. G., and Barbour, S. L. 2005. A study on the hysteresis models for soil-water characteristic curve, *Canadian Geotechnical Journal* Vol. 42, No. 6, pp. 1548–1568.

Pham, H. T. V., and Fredlund, D. G. 2003a. Dynamic programming method in slope stability computations, *Proceedings of the Twelfth Asian Regional Conference on Soil Mechanics and Geotechnical Engineering*, Singapore, August 4–8, pp. 743–746.

Pham, H. T. V., and Fredlund, D. G. 2003b. The application of dynamic programming to slope stability analysis, *Canadian Geotechnical Journal*, Vol. 40, No. 4, pp. 830–847.

Pham, H. T. V., Fredlund, D. G., and Gitirana, Jr., F. N. G. 2001. Slope stability analysis using dynamic programming combined with finite element stress analysis, *Proceedings of the International Conference on Management of Land and Water Resources*, Hanoi, Vietnam, October 20–22, pp. 107–114.

Phene, C. J., Rawlins, S. L., and Hoffman, G. J. 1971. Measuring soil matric potential *in situ* by sensing heat dissipation with a porous body: Theory and sensor construction, *Proceedings of Soil Science Society of America*, Vol. 35, pp. 27–32.

Phene, C. J., Clark, D. A., Cardon, G. E., and Mead, R. M. 1992. The soil matric potential sensor research and applications, in *Advances in Measurement of Soil Physical Properties: Bringing Theory into Practice*, SSSA Special Publication No. 30, Soil Science Society of America, Madison, WI, pp. 263–280.

Philip, J. R. 1969. Theory of infiltration, *Journal of Soil Science*, Vol. 5, pp. 215–296.

Philip, J. R., and de Vries, D. A. 1957. Moisture movement in porous materials under temperature gradients, *Transactions, American Geophysical Union*, Vol. 38. No. 2, pp. 222–232.

Picornell, M., Lytton, R. L., and Steinberg, M. 1983. Matrix suction instrumentation of a vertical moisture barrier, *Transportation Research Record No. 945*, pp. 16–21.

Plesset, M. S. 1969. The tensile strength of liquids, in *Cavitation State of Knowledge*, American Society of Mechanical Engineers, New York, N.Y., pp. 15–25.

Poulos, S. J. 1964. Control of leakage in the triaxial test, *Soil Mechanics Series* No. 71, Massachusetts Institute of Technology, Cambridge, MA.

Poulovassilis, A. 1962. Hysteresis of pore water–An application of the concept of independent domains, *Soil Science Journal*, Vol. 92, pp. 405–412.

Poulovassilis, A., and Childs, E. C. 1971. The hysteresis of pore water: The non-independence of domains, *Soil Science Journal*, Vol. 112, No. 5, pp. 301–312.

Prandtl, L. 1921. On the penetrating strength (hardness) of plastic construction materials and the strength of cutting edges, *Zeit. Agnew, Mathematical Mechanics*, Vol. 1, No. 1, pp. 15–20.

Priestley, C. H. B., and Taylor, R. J. 1972. On the assessment of surface heat flux and evaporation using large-scale parameters, *Monthly Weather Review*, Vol. 100, pp. 81–92.

Priestley, D., Fredlund, M. D., and van Zyl, D. T. 2010. Benchmarking multi-dimensional large strain consolidation analyses, *Proceedings of the Third International Conference on Uranium, Uranium 2101*, Saskatoon, SK, Vol. 2, pp. 277–288.

Proctor, R. R. 1933. Fundamental principles of soil compaction, *Engineering News-Record*, Vol. 111, p. 286.

Pufahl, D. E. 1970. Evaluation of effective stress components in non-saturated soils, M.Sc. Thesis, University of Saskatchewan, Saskatoon, SK.

Pufahl, D. E., and Lytton, R. L. 1991. Temperature and suction profiles beneath highway pavements: Computed and measured, *Transportation Research Record No. 1307*, pp. 268–276.

Pufahl, D. E., Fredlund, D. G., and Rahardjo, H. 1983. Lateral earth pressures in expansive clay soils, *Canadian Geotechnical Journal*, Vol. 20, No. 2, pp. 228–241.

Pufahl, D. E., Lytton, R. L., and Liang, H. S. 1990. An integrated computer model to estimate moisture and temperature effects beneath pavements, *Transportation Research Record No. 1286*, pp. 259–269.

Rahardjo, H. 1990. The study of undrained and drained behaviour of saturated soils, PhD Thesis, University of Saskatchewan, Saskatoon, SK.

Rahardjo, H., and Fredlund, D. G. 1984. General limit equilibrium method for lateral earth force, *Canadian Geotechnical Journal*, Vol. 21, No. 1. pp. 166–175.

Rahardjo, H., and Fredlund, D. G. 1991. Calculation procedures for slope stability analyses involving negative pore-water pressures, *Proceedings of International Conference on Slope Stability Engineering, Development, Applications*, Isle of Wight, UK, pp. 43–49.

Rahardjo, H., and Fredlund, D. G. 1992. Mechanics of soil with matric suction, *Proceedings of the International Conference Geotechnical Engineering (GEOTROPIKA '92)*, Universiti Teknologi Malyasia, Johor Bahru, Malaysia, pp. 593–605.

Rahardjo, H., and Fredlund, D. G. 1993. Stress paths for shear strength testing of unsaturated soils, *Proceedings of the Eleventh Southeast Asian Geotechnical Conference*, Singapore, pp. 187–192.

Rahardjo, H., and Fredlund, D. G. 1995a. Experimental verification of the theory of consolidation for unsaturated soils, *Canadian Geotechnical Journal*, Vol. 32, pp. 749–766.

Rahardjo, H., and Fredlund, D. G. 1995b. Pore pressure and volume change behaviour during undrained and drained loadings of unsaturated soils, *Proceedings of the First International Conference on Unsaturated Soils, UNSAT-95*, Paris, Vol. 1. pp. 177–182.

Rahardjo, H., and Fredlund, D. G. 1996. Consolidation apparatus for testing unsaturated soils, *Geotechnical Testing Journal, ASTM*, Vol. 19, No. 4, pp. 341–353.

Rahardjo, H., and Fredlund, D. G. 2003. K_0-volume change characteristics of an unsaturated soil with respect to various loading paths, *Geotechnical Testing Journal, ASTM*, Vol. 26, No. 1, pp. 79–91.

Rahardjo, H., and Han, K. K. 1994. Application of unsaturated soil mechanics in understanding residual soil behaviour, *Proceedings of International Development and Research Centre, IDRC, Sino-Canadian Symposium on Unsaturated/ Expansive Soils*, Wuhan, China, pp. 41–56.

Rahardjo, H., and Leong, E. C. 1997. Soil-water characteristic curves and flux boundary problems, in S. L. Houston and D. G. Fredlund (Eds.), *Proceedings of Unsaturated Soil Engineering Practice, ASCE*, Geotechnical Special Publications No. 68, American Society of Civil Engineers, New York, pp. 88–112.

Rahardjo, H., and Leong, E. C. 2002. Horizontal drains in unsaturated soil slopes, *Proceedings of the Third International Conference on Unsaturated Soils, UNSAT 2002*, Recife, Brazil, Vol. 2, pp. 773–777.

Rahardjo, H., and Leong, E. C. 2003. Laboratory tests for characterisation of unsaturated soils, *Proceedings of Second Asian Conference on Unsaturated Soils, Geotechnical and Geoenvironmental Issues*, Osaka, Japan, November 11–12, pp. 1–20.

Rahardjo, H., and Leong, E. C. 2006 Suction measurements, *Proceedings of the Fourth International Conference on Unsaturated Soils*, Geo-Institute, Phoenix, AZ, pp. 81–104.

Rahardjo, H., Loi, J., and Fredlund, D. G. 1989. Typical matric suction measurements in the laboratory and in the field using thermal conductivity sensors, *Proceedings of Indian Geotechnical Conference on Geotechniques of Problematic Soils and Rocks: Characterisation, Design, and Ground Improvements*, Visakhapatnam, India, Vol. 1, pp. 127–133.

Rahardjo, H., Ho, D. Y. F., and Fredlund, D. G. 1990. Testing procedures for obtaining volume change indices during loading of an unsaturated soil, *Proceedings 1990 CSCE Annual Conference*, Hamilton, ON, Vol. II-2, pp. 558–573.

Rahardjo, H., Fredlund, D. G., and Vanapalli, S. K. 1992. Use of linear and non-linear shear strength versus matric suction relations in slope stability analysis, *Proceedings of the Sixth International Symposium on Landslides*, Christchurch, New Zealand, pp. 531–537.

Rahardjo, H., Lim, T. T., Chang, M. F., and Fredlund, D. G. 1995. Shear strength characteristics of a residual soil in Singapore, *Canadian Geotechnical Journal*, Vol. 32, pp. 60–77.

Rahardjo, H., Leong, E. C., Gasmo, J. M., and Deutscher, M. S. 1998a. Rainfall-induced slope failures in Singapore: Investigation and repairs, *Proceedings of the Thirteenth Southeast Asian Geotechnical Conference*, Taipei, Taiwan, Vol. 1, pp. 147–152.

Rahardjo, H., Leong, E. C., and Han, K. K. 1998b. Shear strength characteristics of residual soil under infiltration, *Proceedings of International Symposium on Problematic Soils*, Sendai, Tohoku, Japan, Vol. 1, pp. 461–464.

Rahardjo, H., Leong, E. C., and Rezaur, R. B. 2001a. Assessment and instrumentation of slopes for stability against rainfall, *Proceedings of the Fourteenth Southeast Asian Geotechnical Conference*, Hong Kong, Vol. 1, pp. 675–684.

Rahardjo, H., Li, X. W., Toll, D. G., and Leong, E. C. 2001b. The effect of antecedent rainfall on slope stability, *Journal of Geotechnical and Geological Engineering*, Special Issue on Unsaturated and Collapsible Soils, Vol. 19, Nos. 3–4, pp. 371–399.

Rahardjo, H., Leong, E. C., and Rezaur, R. B. 2002. Unsaturated soil mechanics for the study of rainfall-induced slope failures, *Proceedings of the Fourth National German Workshop on Unsaturated Soils*, Weimar, Germany, pp. 1–10.

Rahardjo, H., Hritzuk, K. J., Leong, E. C., and Rezaur, R. B. 2003. Effectiveness of horizontal drains for slope stability, *Journal of Engineering Geology*, Vol. 69, Nos. 3–4, pp. 295–308.

Rahardjo, H., Aung, K. K., Leong, E. C., and Rezaur, R. B. 2004a. Characteristics of residual soils in Singapore as formed by weathering, *Journal of Engineering Geology*, Vol. 73, Nos. 1–2, pp. 157–169.

Rahardjo, H., Lee, T. T., Leong, E. C., and Rezaur, R. B. 2004b. A flume for assessing flux boundary characteristics in rainfall-induced slope failure studies, *Geotechnical Testing Journal, ASTM*, Vol. 27, No. 2, pp. 145–153.

Rahardjo, H., Leong, E. C., and Indrawan, I. G. B. 2004c. Significance of unsaturated soils in rainfall-induced slope failures, *Proceedings of Field-Wise Seminar on Enhancement of Strategic Plan towards Sustainable Education and Research on Geological Engineering*, AUN/SEED-Net JICA and University of Sains Malaysia, Penang, Malaysia, pp. 96–117.

Rahardjo, H., Ong, B. H., and Leong, E. C. 2004d. Shear strength of a compacted residual soil from consolidated drained and constant water content triaxial tests, *Canadian Geotechnical Journal*, Vol. 41, No. 3, pp. 421–436.

Rahardjo, H., Lee, T. T., Leong, E. C., and Rezaur, R. B. 2005a. Response of a residual soil slope to rainfall, *Canadian Geotechnical Journal*, Vol. 42, No. 2, pp. 340–351.

Rahardjo, H., Rezaur, R. B., Hritzuk, K. J., and Leong, E. C. 2005b. Slope failures in Singapore: Case study of two landslides, *Proceedings of International Seminar on Slope Disasters in Geomorphological/Geotechnical Engineering*, Kobe, Japan, pp. 59–70.

Rahardjo, H., Lee, T. T., Leong, E. C., and Rezaur, R. B. 2006a. Reply to Discussion by Londono A. C. on "Response of a residual soil slope to rainfall" by Rahardjo, H., Lee, T. T., Leong, E. C. and Rezaur, R. B., *Canadian Geotechnical Journal*, Vol. 42, pp. 340–351, 2005, *Canadian Geotechnical Journal*, Vol. 43, pp. 979–984.

Rahardjo, H., Tami, D., and Leong, E. C. 2006b. Effectiveness of sloping capillary barriers under high precipitation rates, *Proceedings of Second International Conference on Problematic Soils*, Petaling Jaya, Selangor, Malaysia, pp. 39–54.

Rahardjo, H., Krisdani, H., and Leong, E. C. 2007a. Application of unsaturated soil mechanics in capillary barrier system, *Proceedings of the Third Asian Conference on Unsaturated Soils, UNSAT-ASIA 2007*, Nanjing, China, pp. 127–137.

Rahardjo, H., Krisdani, H., Leong, E. C., Ng, Y. S., Foo, M. D., and Wang, C. L. 2007b. Unsaturated soil mechanics for solving seepage and slope stability problems, *Proceedings of Twelfth International Colloquium on Structural and Geotechnical Engineering*, Ain Shams University, Cairo, pp. 1–38.

Rahardjo, H., Krisdani, H., Leong, E. C., Ng, Y. S., Foo, M. D., and Wang, C. L. 2007c. Capillary barrier as slope cover, *Proceedings of Tenth Australia–New Zealand Conference on Geomechanics "Common Ground,"* Brisbane, Australia, Vol. 2, pp. 698–703.

Rahardjo, H., Ong, T. H., Rezaur, R. B., and Leong, E. C. 2007d. Factors controlling instability of homogeneous soil slopes under rainfall loading, *Journal of Geotechnical and Geoenvironmental Engineering, ASCE*, Vol. 133, No. 12, pp. 1532–1543.

Rahardjo, H., Satyanaga, A., and Leong, E. C. 2007e. Unsaturated soil mechanics for slope stability, *Proceedings of 2007 National Conference of Korean Geotechnical Society (KGS)*, Busan, Korea, pp. 1–21.

Rahardjo, H., Satyanaga, A., Leong, E. C., Ng, Y. S., Foo, M. D., and Wang, C. L. 2007f. Slope failures in Singapore due to rainfall, *Proceedings of Tenth Australia New Zealand Conference on Geomechanics "Common Ground,"* Brisbane, Australia, Vol. 2, pp. 704–709.

Rahardjo, H., Indrawan, I. G. B., Leong, E. C., and Yong, W. K. 2008a. Effects of coarse-grained material on hydraulic properties and shear strength of topsoil, *Journal of Engineering Geology*, Vol. 101, pp. 165–173.

Rahardjo, H., Leong, E. C., and Rezaur, R. B. 2008b. Effect of antecedent rainfall on pore-water pressure distribution characteristics in residual soil slopes under tropical rainfall, *Hydrological Processes*, Special Issue on Rainfall Induced Landslides and Debris Flow, Vol. 22, pp. 506–523.

Rahardjo, H., Rezaur, R. B., and Leong, E. C. 2008c. Role of antecedent rainfall in slope stability, *Proceedings of GEOTROPIKA 2008*, Kuala Lumpur, Malaysia, pp. 1–19.

Rahardjo, H., Rezaur, R. B., Leong, E. C., Alonso, E. E., Lloret, A., and Gens, A. 2008d. Monitoring and modeling of slope response to climate changes, *Proceedings of the Tenth International Symposium on Landslides and Engineered Slopes. Chinese Institution of Soil Mechanics and Geotechnical Engineering–China Civil Engineering Society*, Chinese National Commission on Engineering Geology, Chinese Society of Rock Mechanics and Engineering, Geotechnical Division of the Hong Kong Institution of Engineers, Xi'an, China, Vol. 1, pp. 67–84.

Rahardjo, H., Satyanaga, A., and Leong, E. C. 2008e. Role of real time monitoring in slope stability, *Proceedings of the International Seminar on Civil and Infrastructure Engineering 2008 for Environmental Sustainability*, Universiti Teknologi MARA, Shah Alam, Malaysia, pp. 1–20.

Rahardjo, H., Harnas, F. R., Leong, E. C., Tan, P. Y., Fong, Y. K., and Sim, E. K. 2009a. Tree stability in an improved soil to withstand wind loading, *Urban Forestry & Urban Greening*, Vol. 8, No. 4, pp 237–247.

Rahardjo, H., Leong, E. C., and Rezaur, R. B. 2009b. Laboratory characterisation of unsaturated soil for slope stability studies, *Proceedings of Fourth Asia-Pacific Conference on Unsaturated Soils*, Newcastle, Australia, pp. 565–578.

Rahardjo, H., Meilani, I., Rezaur, R. B., and Leong, E. C. 2009c. Shear strength characteristics of a compacted soil under infiltration conditions, *Geomechanics and Engineering*, Vol. 1, No. 1, pp. 35–52.

Rahardjo, H., Rezaur, R. B., and Leong, E. C. 2009d. Mechanism of rainfall-induced slope failures in tropical regions, *Proceedings of the First Italian Workshop on Landslides (IWL): "Rainfall-induced Landslides: Mechanisms, Monitoring Techniques and Nowcasting Models for Early Warning Systems,"* Dipartimento Di Ingegneria Civile, Seconda Universita di Napoli, Naples, Vol. 1, pp. 31–42.

Rahardjo, H., Ong, T. H., Rezaur, R. B., Leong, E. C., and Fredlund, D. G. 2010a. Response parameters for characterization of infiltration, *Environmental Earth Sciences*, Vol. 60, No. 7, pp. 1369–1380.

Rahardjo, H., Satyanaga Nio, A., Leong, E. C., and Ng, Y. S. 2010b. Effects of groundwater table position and soil properties on stability of slope during rainfall, *Journal of Geotechnical and Geoenvironmental Engineering, ASCE*, Vol. 136, No. 11, pp. 1555–1564.

Rahardjo, H., Melinda, F., Leong, E. C., and Rezaur, R. B. 2011a. Stiffness of a compacted residual soil, *Journal of Engineering Geology*, Vol. 120, pp. 60–67.

Rahardjo, H., Santoso, V. A., Leong, E. C., Ng, Y. S., and Hua, C. J. 2011b. Numerical analyses and monitoring performance of residual soil slopes, *Soils and Foundations, Japanese Geotechnical Society*, Vol. 51, No. 3, pp. 471–482.

Rahardjo, H., Santoso, V. A., Leong, E. C., Ng, Y. S., and Hua, C. J. 2011c. Performance of horizontal drains in residual soil slope, *Soils and Foundations, Japanese Geotechnical Society*, Vol. 51, No. 3, pp. 437–447.

Rahardjo, H., Santoso, V. A., Leong, E. C., Ng, Y. S., and Tam, C. P. H. 2011d. Instrumentation and real-time monitoring in residual soil slopes, *Proceeding of Environmental Geosciences and Engineering Survey for Territory Protection and Population Safety International Conference (EngeoPro 2011)*, Moscow, pp. 357–362.

Rahardjo, H., Vilayvong, K., and Leong, E. C. 2011e. Water characteristic curves of recycled materials. *Geotechnical Testing Journal*, *ASTM*, Vol. 34, No. 1, pp. 1–8.

Rahimi, A., Rahardjo, H., and Leong, E. C. 2010. Effect of hydraulic properties of soil on rainfall-induced slope failure. *Journal of Engineering Geology*, Vol. 114, pp. 135–143.

Rahimi, A., Rahardjo, H., and Leong, E. C. 2011. Effect of antecedent rainfall patterns on rainfall induced slope failure, *Journal of Geotechnical and Geoenvironmental Engineering, ASCE*, Vol. 137, No. 5, pp. 483–491.

Ranjitkar, S., and Sunder, B. 1989. Prediction of hydraulic properties of unsaturated granular soils based on grain size data, PhD Thesis, University of Massachusetts, MA, pp. 75–131.

Rao, R. R., Rahardjo, H., and Fredlund, D. G. 1988. Closed-form heave solution for expansive soils, *Journal of Geotechnical Engineering*, Vol. 114, No. 5, pp. 573–588.

Rao, R. R., Rahardjo, H., and Fredlund, D. G. 1989. Closed-form heave solutions for expansive soils, Closure for discussion by K. S. Li, Binnie and Partners, Hong Kong, *Geotechnical Journal, ASCE*, Vol. 115, No. 12, pp. 1822–1833.

Rassam, D. W., and Cook, F. J. 2002. Predicting the shear strength envelope of unsaturated soil, *Geotechnical Testing Journal, ASTM*, Vol. 28, pp. 215–220.

Rawls, W. J., and Brakensiek, D. L. 1985. Prediction of soil water properties for hydrologic modeling, *Proceedings of the American Society of Civil Engineers Watershed Management in the Eighties Symposium*, American Society of Civil Engineers, New York, pp. 293–299.

Rawls, W. J., Brakensiek, D. L., and Saxton, K. E. 1982. Estimation of soil water properties, *Transactions of American Society of Agriculture Engineering*, Vol. 25, No. 1328, pp. 1316–1320,

Rawls, W. J., Ahuja, J. R., and Brakensiek, D. L. 1992. Estimating soil hydraulic properties from soils data, *Proceedings of the Workshop on Indirect Methods for Estimating the Hydraulic Properties of Unsaturated Soils*, Riverside, CA, pp. 329–341.

Rawlins, S. L., and Campbell, G. S. 1986. Water potential: Thermocouple psychrometry, in A. Klute (Ed.), *Methods of Soil Analysis, Agronomy*, Part 1, 2nd ed., ASA and SSSA, Madison, WI, pp. 597–617.

Rawlins, S. L., and Dalton, F. N. 1967. Psychometric measurement of soil-water potential without precise temperature control, *Journal of Soil Science*, Vol. 31, pp. 297–301.

Reddy, S. J. 1971. An empirical method for the estimation of net radiation intensity, *Journal of Solar Energy*, Vol. 13, pp. 291–292.

Reece, C. F. 1996. Evaluation of line heat dissipation sensors for measuring soil matric potential, *Soil Science Society of America Journal*, Vol. 60, pp. 1022–1028.

Reginato, R. J., and van Bavel, C. H. M. 1962. Pressure cell for soil cores, *Soil Science Society of America Proceedings*, Vol. 26, pp. 1–3.

Reinson, J. R., Fredlund, D. G., and Wilson, G. W. 2005. Unsaturated flow in coarse porous media, *Canadian Geotechnical Journal*, Vol. 42, pp. 252–262.

Rezaur, R. B., Rahardjo, H., and Leong, E. C. 2002. Spatial and temporal variability of pore-water pressures in residual soil slopes in a tropical climate, *Earth Surface Processes and Landforms*, Vol. 27, pp. 317–338.

Rezaur, R. B., Rahardjo, H., Leong, E. C., and Lee, T. T. 2003. Hydrologic behaviour of residual soil slopes in Singapore, *Journal of Hydrologic Engineering, ASCE*, Vol. 8, No. 3, pp. 133–144.

Richards, B. G. 1965. Measurement of the free energy of soil moisture by the psychrometric technique using thermistors, *Moisture Equilibria and Moisture Changes in Soils Beneath Covered Areas, A Symposium in Print*, Butterworths, Sydney, pp. 39–46.

Richards, B. G. 1967. A review of methods for the determination of the moisture flow properties of unsaturated soils, Technical Memorandum No. 5, Commonwealth Science and Industrial Research Organization, Australia.

Richards, B. G. 1974. Behaviour of unsaturated soils, in I. K. Lee (Ed.), *Soil Mechanics–New Horizons*, American Elsevier, New York, pp. 112–157.

Richards, B. G. 1984. Finite element analysis of volume change in expansive clays, *Proceedings of the Fifth International Conference on Expansive Soils*, Adelaide, Australia, Vol. 2, pp. 355–362.

Richards, B. G., Peter, P., and Martin, R. 1984. The determination of volume change properties in expansive soils, *Proceedings of the Fifth International Conference on Expansive Soils*, Adelaide, Australia, pp. 179–186.

Richards, L. A. 1928. The usefulness of capillary potential to soil moisture and plant investigators, *Journal of Agriculture Research*, Vol. 37, pp. 719–742.

Richards, L. A. 1931. Capillary conduction of liquids through porous medium, *Journal of Physics*, Vol. 1, pp. 318–333.

Richards, L. A. 1941. A pressure membrane extraction apparatus for soil suction, *Soil Science*, Vol. 51, pp. 377–386.

Richards, L. A. 1952. Water conducting and retaining properties of soils in relation to irrigation, *Proceedings of an International Symposium on Desert Research, Jerusalem*, pp. 523–546.

Richards, L. A., and Fireman, M. 1943. Pressure-plate apparatus for measuring moisture sorption and transmission by soils, *Soil Science Journal*, Vol. 56, pp. 395–404.

Richards, L. A., and Ogata, G. 1958. A thermocouple for vapour pressure measurement in biological and soil systems at high humidity, *Science*, Vol. 128, pp. 1089–1090.

Richards, L. A., and Wilson, B. D. 1936. Capillary conductivity measurements in peat soils, *Journal of American Society of Agronomy*, Vol. 28, pp. 427–436.

Richards, L. A., Gardner, W. R., and Ogata, G. 1956. Physical processes determining water loss from soil, *Proceedings of Soil Science Society of America*, Vol. 20, pp. 310–314.

Ridley, A. M. 1993. The measurement of soil moisture suction, PhD Thesis, Imperial College, London.

Ridley, A. M. 1995. Discussion on "Laboratory Filter Paper Suction Measurements" by Sandra L. Houston, William N. Houston, and Anne-Marie Wagner," *Geotechnical Testing Journal, ASTM*, Vol. 18, No. 3, pp. 391–396.

Ridley, A. M., and Burland, J. B. 1993. A new instrument for the measurement of soil moisture suction, *Geotechnique*, Vol. 43, No. 2, pp. 321–324.

Ridley, A. M., and Burland, J. B. 1996. A pore water pressure probe for the in-situ measurement of a wide range of soil suctions, in C. Craig (Ed.), *Advances in Site Investigation Practice*, Thomas Telford, London, pp. 510–520.

Ridley, A. M., and Wray, W. K. 1995. Suction measurements: A review of current theory and practices, in E. E. Alonso and P. Delage (Eds.), *Unsaturated Soils: Proceedings of the First International Conference on Unsaturated Soils*, Paris, Balkema, Rotterdam, Presse des Ponts et Chaussees, pp. 1293–1322.

Riha, S. J., McKinnes, K. J., Childs, S. W., and Campbell, G. S. 1980. A finite element calculation for determining thermal conductivity, *Soil Science Society of America Journal*, Vol. 44, pp. 1323–1325.

Ritchie, J. T. 1972. Model for predicting evaporation from a row crop with incomplete cover, *Water Resources Research*, Vol. 8, pp. 1204–1213.

Robinson, R. A., and Stokes, R. H. 1968. *Electrolyte Solutions*, 2nd ed., Butterworths, London.

Rodebush, W. H., and Buswell, A. M. 1958. Properties of water substances, Special Report No. 40, Highway Research Board, Washington, DC.

Rogowski, A. S. 1971. Watershed physics: Model of the soil moisture characteristic, *Water Resources Research*, Vol. 7, pp. 1575–1582.

Romero, E., and Jommi, C. 2008. An insight into the role of hydraulic history on the volume changes of anisotropic clayey soils, *Water Resources Research*, Vol. 44, 16 p.

Romero, E., Facio, J. A., and Yang, D. Q. 1994. Boom clay: DIT-UPC Data File, Internal Report, Universidad Politecnica de Catalunya, Spain.

Romero, E., Lloret, A., and Gens, A. 1995. Development of a new suction and temperature controlled oedometer cell, *Proceedings of the First International Conference on Unsaturated Soils, UNSAT-95*, Paris, Vol. 2. pp. 553–559.

Romero, E., Gens, A., and Lloret, A. 2000. Temperature effects on water retention and water permeability of unsaturated clay, in H. Rahardjo, D.G. Toll, and E. C. Leong (Eds.), *Proceedings of the Asian Conference on Unsaturated Soils*, Singapore, Rotterdam, Balkema, pp. 433–438.

Rose, C. W., and Krishnan, A. 1967. A method of determining hydraulic conductivity characteristics for non-swelling soils *in situ*, and of calculating evaporation from bare soil, *Soil Science Journal*, Vol. 103, pp. 369–373.

Rose, W. 1958. Volumes and surface areas of pendular rings, *Journal of Applied Physics*, Vol. 29, No. 4, pp. 687–691.

Rosenberg, N. J., Blad, B. L., and Verma, S. B. 1983. *Microclimate–The Biological Environment*, Wiley, New York.

Rowher, C. 1931. Evaporation from free water surfaces, Technical Bulletin No. 271, U.S. Department of Agriculture in cooperation with Colorado Agricultural Experimental Station, Fort Collins, CO.

Rubin, J. 1968. Theoretical analysis of two-dimensional transient flow of water in unsaturated and partly unsaturated soils, *Proceedings of Soil Science Society of America*, Vol. 32, pp. 607–615.

Ruddock, E. C. 1966. Discussion on "The residual soils of Hong Kong," *Geotechnique*, Vol. 16, No. 1, pp. 78–81.

Rulon, J. J., and Freeze, R. A. 1985. Multiple seepage faces on layered slopes and their implications for slope stability analysis, *Canadian Geotechnical Journal*, Vol. 22, pp. 347–356.

Russam, K. 1958. An investigation into the soil moisture conditions under roads in Trinidad, B.W.I., Geotechnique, Vol. 8, pp. 57–71.

Russam, K., and Coleman, C. D. 1961. The effect of climatic factors on subgrade moisture conditions, *Geotechnique*, Vol. 11, No. 1, pp. 22–28.

Rykaart, F. M., Fredlund M. D., and Stianson, J. 2001a. Solving tailings impoundment water balance problems with 3-D seepage software, *Geotechnical News Magazine*, December, pp. 50–54.

Rykaart, F. M., Wilson, G. W., Fredlund, M. D., and Currey, N. A. 2001b. A spatial flux boundary model for tailing dams, *Proceedings of the Fifty-Fourth Canadian Geotechnical Conference*, Calgary, AB, pp. 1102–1109.

Ryland, L., and Clements, C. 2000. A parallel study evaluating the performance of the Beta-97 and CS-229 thermal conductivity sensors in extreme pH conditions, Internal Research Report, Department of Civil Engineering, University of Saskatchewan, Saskatoon, SK.

Samani, Z. 2000. Estimating solar radiation and evapotranspiration using minimum climatological data, *Journal of Irrigation and Drainage Engineering*, Vol. 126, No. 4, pp. 265–267.

Samingan, A. S., Leong, E. C., and Rahardjo, H. 2003. A flexible wall permeameter for measurements of water and air coefficients of permeability of residual soils, *Canadian Geotechnical Journal*, Vol. 40, No. 3, pp. 559–574.

Sampson, E., Schuster, R. L., and Budge, W. D. 1965. A method of determining swell potential of an expansive clay, *Proceedings of the International Research Engineering Conference on Expansive Clays*, College Station, TX, pp. 255–275.

Sass, J. H., Lachenbruch, A. H., and Munroe, R. J. 1971. Thermal conductivity of rocks from measurements on fragments and its application to heat flow determinations, *Journal of Geophysical Research*, Vol. 76, No. 14, pp. 3391–3401.

Satija, B. S. 1978. Shear behaviour of partly saturated soils, PhD Thesis, Indian Institute of Technology, New Delhi.

Satija, B. S., and Gulhati, S. K. 1979. Strain rate for shearing testing of unsaturated soil, *Proceedings of the Sixth Asian Regional Conference on Soil Mechanics and Foundation Engineering*, Singapore, pp. 83–86.

Sattler, P. J., and Fredlund, D. G. 1989. Use of thermal conductivity sensors to measure matric suction in the laboratory, *Canadian Geotechnical Journal*, Vol. 26, No. 3, pp. 491–498.

Sattler, P. J., and Fredlund, D. G. 1991. Numerical modeling of vertical ground movements in expansive soils, *Canadian Geotechnical Journal*, Vol. 28, No. 2, pp. 189–199.

Sattler, P., Fredlund, D. G., Klassen, M. J., Jubien, W. B., and Rowan, W. G. 1989. Bearing capacity approach to railway design, utilizing subgrade matric suction, Transportation Research Board, Committee on Railroad Track Structure System Design, *Transportation Research Record No. 1241*, Washington DC, pp. 27–33.

Sattler, P. J., Fredlund, D. G., Lam, L. W., Clifton, A. W., and Klassen, M. J. 1990. Implementation of a bearing capacity design procedure for railway subgrades: A case study, Transportation Research Board, Washington, DC.

Saxton, K. E. 1982. Mathematical modeling of evapotranspiration on agricultural watersheds, Water Resources Publications, Littleton, CO, pp. 183–204

Scheinost, A. C., Sinowski, W., and Auerswald, K. 1997. Regionalization of soil water retention curves in a highly variable soilscape, I. Developing a new pedotransfer function, *Geoderma*, Vol. 78, Nos. 3–4, pp. 129–143.

Schiffman, R. L., Chen, A. T. F., and Jordan, J. C. 1969. Analysis of consolidation theories, *American Society of Civil Engineers, Journal of the Soil Mechanics and Foundations Division*, Vol. 95, No. SM1, pp. 285–312.

Schmertmann, J. M. 1955. The undisturbed consolidation behaviour of clay, *ASCE Transactions*, Vol. 120, pp. 1201–1227.

Schnellmann, R., Busslinger, M., Schneider, H., and Rahardjo, H. 2010. Effect of rising water table in an unsaturated slope, *Journal of Engineering Geology*, Vol. 114, pp. 71–83

Schofield, R. 1935. The pF of the water in soil, *Transactions of the Third International Congress of Soil Science*, Oxford, England, Vol. 2, pp. 37–48.

Schuh, W. M., Cline, R. L., and Sweeney, M. D. 1991. Unsaturated soil hydraulic properties and parameters for the Oakes area, Dickey County, North Dakota Water Commission, Water Resources Investigation 18, Bismark, ND.

Schuurman, E. 1966. The compressibility of an air/water mixture and a theoretical relation between the air and water pressures, *Geotechnique*, Vol. 16, No. 4, pp. 269–281.

Scott, R. F. 1963. *Principles of Soil Mechanics*, Addison-Wesley, Reading, MA.

Seed, H. B. 1959. A modern approach to soil compaction, *Proceedings of the Eleventh California Street and Highway Conference*, Reprint No. 69, The Institute of Transportation and Traffic Engineering, University of California, Los Angeles, CA.

Seed, H. B., and Chan, A. M. 1959. Structure and strength characteristics of compacted clays, *Journal of Soil Mechanics and Foundations Division, ASCE*, Vol. 85, No. SM5, pp. 87–128.

Seed, H. B., Woodward, R. J., and Lundgren, R. 1962. Prediction of swelling potential for compacted clays, *Journal of the Soil Mechanics Foundation Division, ASCE*, Vol. 88, No. SM4, pp. 57–59.

Segerlind, L. G. 1984. *Applied Finite Element Analysis*, 2nd ed., Wiley, New York.

Sengers, J. V., and Watson, J. T. R. 1986. Improved international formulations for the viscosity and thermal conductivity of water substances, American Institute of Physics and the American Chemistry Society, *Journal of Physics Chemistry Reference Data*, Vol. 15, No. 4, pp. 1291–1314.

Sevaldson, R. A. 1956. The slide at Lodalen, October 6th, 1954, *Geotechnique*, Vol. 6, pp. 167–182.

Shackelford, C. D. 2005. Environmental issues in geotechnical engineering, *Proceedings of the Sixteenth International Conference on Soil Mechanics and Geotechnical Engineering*, Osaka, Japan, Vol. 1, pp. 95–122.

Sharma, R. S. 1998. Mechanical behaviour of unsaturated highly expansive clays, PhD Thesis, University of Oxford, London.

Shaw, B., and Baver, L. D. 1939. An electrothermal method for following moisture changes of the soil *in situ, Proceedings of Soil Science Society of America*, Vol. 4, pp. 78–83.

Sheng, D., and Fredlund, D. G. 2008. Elasto-plastic modeling of unsaturated soils: The past, present and future, *Proceedings of* the *Twelfth International Conference on Numerical Modelling, IACMG*, Goa, India, pp. 2084–2105.

Sheng, D., and Fredlund, D. G. 2009. An elasto-plastic model for unsaturated soil, *Proceedings of the Seventeenth International Conference on Soil Mechanics and Geotechnical Engineering* (ISSMGE), Alexandria, Egypt, October 5–10.

Sheng, D., Fredlund, D. G., and Gens, A. 2008a. A new modelling approach for unsaturated soils using independent stress variables, *Canadian Geotechnical Journal*, Vol. 45, pp. 511–534.

Sheng, D., Fredlund, D. G., and Gens, A. 2008b. Response to the discussion by Xhang and R. L. Lytton on the paper "A new modeling approach for unsaturated soils using independent stress variables," *Canadian Geotechnical Journal*, Vol. 45, pp. 511 –534.

Sheng, D., Gens, A., Fredlund, D. G., and Sloan, S. W. 2008c. Unsaturated soils: From constitutive modelling to numerical algorithms, *Computers and Geotechnics*, Vol. 35, No. 6, pp. 810–824.

Sheng, D., Zhou, A., and Fredlund D. G. 2011. Shear strength criteria for unsaturated soils, *Geotechnical and Geological Engineering Journal*, Springer Science, Vol. 29, No. 2, pp. 145–159.

Shirizi, M. A., and Boersma L. 1984. A unifying quantitative analysis of soil texture, *Soil Science Society of America Journal*, Vol. 48, pp. 142–147.

Shuai, F. 1996. Simulation of swelling pressure in expansive soils, PhD Thesis, University of Saskatchewan, Saskatoon, SK.

Shuai, F., and Fredlund, D. G. 1997. Measured and simulated behaviour of an expansive soil, *Proceedings of the Third Brazilian Symposium on Unsaturated Soils, NSAT '97*, Rio de Janeiro, April 22–25, Vol. 1, pp. 253–260.

Shuai, F., and Fredlund, D. G. 1998. Model for the simulation of swelling pressure measurements on expansive soils, *Canadian Geotechnical Journal*, Vol. 35, No. 1, pp. 96–114.

Shuai, F., and Fredlund, D. G. 2000. Use of a new conductivity sensor to measure soil suction, *Proceedings of the GeoDenver Conference*, Denver, CO, pp. 1–12.

Shuai, F., Clements, C., Ryland, L., and Fredlund, D. G. 2002. Some factors that influence soil suction measurements using a thermal conductivity sensor, *Proceedings of the Third International Conference on Unsaturated Soils, UNSAT 2002*, Recife, Brazil, pp. 325–329.

Shuai, F., Fredlund, D. G., and Samarasekera, L. 2003. Numerical simulation of water movement in the suction equalization of a thermal conductivity sensor, *Geotechnical Testing Journal, ASTM*, Vol. 26, No. 2, pp. 142–151.

Sibley, J. W., and Williams, D. J. 1989. A procedure for determining volumetric shrinkage of an unsaturated soil, *Geotechnical Testing Journal, ASTM*, Vol. 12, No. 3, pp. 181–187.

Sibley, J. W., and Williams, D. J. 1990. A new filter material for measuring soil suction, *Geotechnical Testing Journal, ASTM*, Vol. 13, No. 4, pp. 381–384.

Sibley, J. W., Smyth, G. K., and Williams, D. J. 1990. Suction-moisture content calibration of filter papers from different boxes, *Geotechnical Testing Journal, ASTM*, Vol. 13, No. 3, pp. 257–262.

Sillers, W. S. 1997. Mathematical representation of the soil-water characteristic curve, M.Sc. Thesis, University of Saskatchewan, Saskatoon, SK.

Sillers, W. S., and Fredlund, D. G. 2001. Statistical assessment of soil-water characteristic curve models for geotechnical engineering, *Canadian Geotechnical Journal*, Vol. 38, No. 6, pp. 1297–1313.

Sillers, W. S., Fredlund, D. G., and Zakerzadeh, N. 2001. Mathematical attributes of some soil-water characteristic curve models, *Geotechnical and Geological Engineering Journal*, Special Issue on Unsaturated and Collapsible Soils, Vol. 19, pp. 243–283.

Silva, F. F. C., Almeida, M. S. S., and Futai, M. M. 2002. Critical state modelling of unsaturated soil, in *Proceedings of the Third International Conference on Unsaturated Soils*, Recife, Brazil, March, Swets and Zeitlinger, Lisse, pp. 133–137.

Silvestri, V. 1994. Water content relationships of a sensitive clay subjected to cycles of capillary pressures, *Geotechnical Testing Journal, ASTM*, Vol. 17 No. 1, pp. 57–64.

Simms, P. H., and Yanful, E. K. 2001. Measurement and estimation of pore shrinkage and pore distribution in a clayey till during soil-water characteristic curve tests, *Canadian Geotechnical Journal*, Vol. 38, No. 4, pp. 741–754.

Simms, P. H., and Yanful, E. K. 2005. A pore-network model for hydromechanical coupling in unsaturated compacted clayey soils, *Canadian Geotechnical Journal*, Vol. 42, No. 2, pp. 499–514.

Simons, N. 1958. Discussion: Measurements of the pressures of filling materials against walls, *Proceedings of Brussels Conference on Earth Pressure Problems*, Brussels, Vol. 3, pp. 50–53.

Singhal, S. 2010. Expansive soil behaviour: Property measurement techniques and heave prediction methods, PhD Thesis, Arizona State University, Tempe, AZ.

Sisler, H. H., Vanderwerf, C. A., and Davidson, A. W. 1953. *General Chemistry–A Systematic Approach*, Macmillan, New York.

Sitz, M. 1948. Discussion on Terzaghi's ideas on "Surface tension of water and the rise of water in capillaries," *Proceedings of the Second International Conference on Soil Mechanics and Foundation Engineering*, Rotterdam, Vol. 5, pp. 289–292.

Skaling, W. 1992. TRASE: A product history, *Advances in Measurement of Soil Physical Properties: Bringing Theory into Practice*, SSSA Special Publication No. 30, Soil Science Society of America, Madison, WI, pp. 169–186.

Skempton, A. W. 1944. Notes on the compressibility of clays, *Quarterly Journal of the Geological Society of London*, Vol. 100 (C: parts 1 and 2), pp. 119–135.

Skempton, A. W. 1951. The bearing capacity of clays, Building Research Congress, England.

Skempton, A. W. 1954. The pore pressure coefficients, A and B, *Geotechnique*, Vol. 4, No. 4, pp. 143–147.

Skempton, A. W. 1961. Effective stress in soils, concrete and rocks, *Proceedings of Conference on Pore Pressure and Suction in Soils*, Butterworths, London, pp. 4–16.

Skempton, A. W., and Bishop, A. W. 1954. Building materials, their elasticity and inelasticity, in *Soils*, North-Holland, Amsterdam, Chapter 10.

Skempton, A. W., and Hutchinson, J. 1969. Stability of natural slopes and embankment foundations, *State-of-the-Art Report, Seventh International Conference on Soil Mechanics and Foundation Engineering*, Mexico City, pp. 291–340.

Slichter, C. S. 1899. Theoretical investigations of the motion of groundwaters, in *19th Annual Report*, Part 2, U.S. Geological Survey, Washington, DC, pp. 295–384.

Smith, E. C. 1942. Soil mechanics: Factor of safety of clay banks, *Journal of the Institution of Civil Engineers*, Vol. 18, No. 7, pp. 293–296.

Smith, A. W. 1973. Method for determining the potential vertical rise, PVR. Texas Test Method Tex-126–E, *Proceedings of Workshop on Expansive Clays and Shales in Highway Design and Construction*, University of Wyoming, Laramie, WY, Vol. 1, pp. 189–205.

Smith, W. O. 1939. Thermal conductivities in moist soils, *Soil Science Society of America Proceedings*, Vol. 4, pp. 521–524.

Snethen, D. R. 1986. Expansive soils: Where are we? *Ground Failure, No. 3, National Research Council Communication on Ground Failure Hazards*, National Research Council, Washington, DC, pp. 12–16.

Soil Moisture Equipment. 1985. Commercial Publications, P.O. Box 30025, Santa Barbara, CA.

Spanner, D. C. 1951. The Peltier effect and its use in the measurement of suction pressure, *Journal of Experimental Botany*, Vol. 11, pp. 145–168.

Spencer, E. 1967. A method of analysis of the stability of embankments assuming parallel interslice forces, *Géotechnique*, Vol. 17, pp. 11–26.

Spencer, E. 1968. Effect of tension on stability of embankments, *Journal of Soil Mechanics Foundation Engineering Division, ASCE*, Vol. 94, No. SM5, pp. 1159–1173.

Spencer, E. 1973. Thrust line criterion in embankment stability analysis, *Geotechnique*, Vol. 23, pp. 85–100.

Sposito, G. 1981. *The Thermodynamics of Soil Solutions*, Oxford Clarendon, London.

Sridharan, A., Altschaeffl, A. G., and Diamond, S. 1971. Pore size distribution studies, *Journal of Soil Mechanics and Foundation Engineering, ASCE*, Vol. 97, No. SM5, pp. 771–787.

Sridharan, A., Rao, A. S., and Sivapullaiah, P. V. 1986. Swelling pressure of clays, *Geotechnical Testing Journal, ASTM*, Vol. 9, No. 1, pp. 24–33.

Stalhane, B., and Pyk, S. 1931. New method for determining the coefficients of thermal conductivity, *Teknisk Tidskift*, Vol. 61, No. 28, pp. 389–393.

Stannard, D. I. 1992. Tensiometers–Theory, construction and use, *Geotechnical Testing Journal, ASTM*, Vol. 15, No. 1, pp. 48–58.

Stianson, J., Fredlund, D. G., and McArthur, T. 2010. Climate classification and quantification for cover design in cold environments, *Proceedings of the Third International Conference on Uranium*, Uranium 2010, Saskatoon, SK, Vol. 2, pp. 455–468.

Stianson, J., Fredlund, D. G., and Chan, D. 2011. Three-dimensional slope stability based on stresses from a stress-deformation analysis, *Canadian Geotechnical Journal*, Vol. 48, No. 6, pp. 891–904.

Stianson, J. R. 2008. A three-dimensional slope stability method based on finite element stress analysis and dynamic programming, PhD Thesis, University of Alberta, Edmonton.

Stianson, J. R., Chan, D., and Fredlund, D. G. 2004. Comparing slope stability analysis based on linear elastic or elasto plastic stresses using dynamic programming techniques, *Proceedings of the Fifty-Seventh Canadian Geotechnical Conference*, Quebec City, QC, pp. 23–30.

Stoicescu, J. T., Haug, M. D., and Fredlund, D. G. 1998. The soil-water characteristics of sand-bentonite mixtures used for liners and covers, *Proceedings of the Second International Conference on Unsaturated Soils, UNSAT-98*, Beijing, Vol. 1, pp. 143–148.

Stone, J. J., Bashir, R., and Fredlund, D. G. 2011. Effect of ambient temperature on the calibration of thermal conductivity suction sensors, *Proceedings of the Canadian Geotechnical and Fifth Pan American Conference*, Toronto, ON, October 2–6.

Stormont, J. C., and Anderson, C. E. 1999. Capillary barrier effect from underlying coarser soil layer, *Journal of Geotechnical and Geoenvironmental Engineering*, Vol. 125, No. 8, pp. 641–648.

Stormont, J. C., and Morris, C. E. 1998. Method to estimate water storage capacity of capillary barriers, *Journal of Geotechnical and Geoenvironmental Engineering*, Vol. 124, No. 4, pp. 297–302.

Streeter, V. L., and Wylie, B. 1975. *Fluid Mechanics*, 6th ed., McGraw-Hill, New York.

Stylianou, C., and DeVantier, B. A. 1995. Relative air permeability as function of saturation in soil venting, *Journal of Environmental Engineering*, Vol. 121, No. 4, pp. 337–347.

Subba, R. K. S., and Satyadas, G. C. 1985. Measurement of volumetric and linear shrinkage on black cotton soil, *Geotechnical Testing Journal, ASTM*, Vol. 8, No. 2, pp. 66–70.

Sullivan, R. A., and McClelland, B. 1969. Predicting heave of buildings on unsaturated clays, *Proceedings of the Second International Conference on Expansive Soils*, College Station, TX, pp. 404–420.

Sun, H. W., Wong, H. N., and Ho, K. K. S. 1998. Analysis of infiltration in unsaturated ground, in K. S. Li, J. N. Kay, and K. K. S. Ho (Eds.), *Slope Engineering in Hong Kong*, Balkema, Rotterdam, pp. 101–109.

Sun, J., Rahardjo, H., and Fredlund, D. G. 1994. A technique to perform coupled consolidation analysis using two independent softwares, *Proceedings of the First Congress on Computing in Civil Engineering*, Washington, DC, Vol. 1, pp. 849–856.

Sutton, W. G. L. 1943. On the equation of diffusion in a turbulent medium, *Proceedings of the Royal Society of London, Series A, Mathematical and Physical Sciences*, Vol. 182, No. 988, pp. 48–75.

Swarbrick, G. E. 1995. Measurement of soil suction using the filter paper method, unsaturated soils, *Proceedings of the First International Conference on Unsaturated Soils, UNSAT-95*, Paris, Vol. 2, pp. 653–658.

Sweeney, D. J. 1982. Some *in situ* soil suction measurements in Hong Kong's residual soil slopes, *Proceedings of the Seventh Southeast Asia Geotechnical Conference*, Hong Kong, Vol. 1, pp. 91–106.

Sweeney, D. J., and Robertson, P. 1979. A fundamental approach to slope stability in Hong Kong, *Hong Kong Engineering*, pp. 35–44.

Tadepalli, R. 1990. Study of collapse behavior during inundation. M.Sc. thesis, Department of Civil Engineering, University of Saskatchewan, Saskatoon, SK, 276 p.

Tadepalli, R., and Fredlund, D. G. 1991. The collapse behaviour of a compacted soil during inundation, *Canadian Geotechnical Journal*, Vol. 28, pp. 477–488.

Tadepalli, R., Fredlund, D. G., and Rahardjo, H. 1992a. Soil collapse and matric suction change, *Proceedings of the Seventh International Conference on Expansive Soils*, Dallas, TX, Vol. 1, pp. 286–291.

Tadepalli, R., Rahardjo, H., and Fredlund, D. G. 1992b. Measurements of matric suction and volume changes during inundation of collapsible soils, *Geotechnical Testing Journal, ASTM*, Vol. 15, No. 2, pp. 115–122.

Take, W. A., and Bolton, M. D. 2003. Tensiometer saturation and reliable measurement of soil suction, *Geotechnique*, Vol. 53, pp. 159–172.

Tamagnini, R. 2004. An extended Cam-clay model for unsaturated soils with hydraulic hysteresis, *Geotechnique*, Vol. 54, No. 3, pp. 223–228.

Tami, D., Rahardjo, H., and Leong, E. C. 2004a. Effects of hysteresis on steady-state infiltration in unsaturated slopes, *Journal of Geotechnical and Geoenvironmental Engineering, ASCE*, Vol. 130, No. 9, pp. 956–967.

Tami, D., Rahardjo, H, Leong, E. C., and Fredlund, D. G. 2004b. A physical model for sloping capillary barriers, *Geotechnical Testing Journal, ASTM*, Vol. 27, No. 2, pp. 173–183.

Tami, D., Rahardjo, H., Leong, E. C., and Fredlund, D. G. 2004c. Design and laboratory verification of a physical model of sloping capillary barrier, *Canadian Geotechnical Journal*, Vol. 41, No. 5, pp. 814–830.

Tami, D., Rahardjo, H., and Leong, E. C. 2007. Characteristics of scanning curves of two soils, soils and foundations, *Japanese Geotechnical Society*, Vol. 47, No. 1, pp. 97–109.

Tan, E., Marjerison, B., and Fredlund, D. G. 2003. Measurements and analysis of temperature and soil suction below thin membrane surface (TMS) in Saskatchewan, *Proceedings of the Fifty-Sixth Canadian Geotechnical Conference*, Winnipeg, MB, Vol. 2, pp. 107–114.

Tan, E., Fredlund, D. G., and Fredlund, M. D. 2004a. Finite element simulations of temperature changes in highway subgrade, *Proceedings of the Fifty-Seventh Canadian Geotechnical Conference*, Quebec, QC, pp. 23–28.

Tan, E., Fredlund, D. G., and Gitirana, Jr., F. N. G. 2004b. Comparisons of correction methods for factors changes influencing thermal conductivity sensors, *Proceedings of the Fifth Brazilian Symposium on Unsaturated Soils*, Sao Carlos, SP, Brazil, pp. 127–131.

Tan, E., Fredlund, D. G., and Marjerison, B. 2007. Installation procedure and analysis of long-term readings from thermal conductivity matric suction sensors, *Canadian Geotechnical Journal*, Vol. 44, pp. 113–125.

Tan, K. M., Tay, C. M., Tjin, S. C., Chan, C. C., and Rahardjo, H. 2005. High relative humidity measurements using gelatin coated long-period grating sensors, *Sensors and Actuators B: Chemical*, Vol. 110, pp. 335–341.

Tang, A. M., Cui, Y. J., and Barnel, N. 2008. Compression-induced suction change in a compacted expansive clay, *Proceedings of the First European Conference on Unsaturated Soils*, Durham, UK, pp. 369–374.

Tang, R. K. W. 1978. Measurement of soil suction by filter paper method, Laboratory Research Report, Department of Civil Engineering, University of Saskatchewan, Saskatoon, SK.

Tao, Y. X., and Gray, D. M. 1994. Prediction of snowmelt infiltration into frozen soils, N*umerical Heat Transfer*, Part A, Vol. 26, pp. 643–665.

Tarantino, A., and De Col, E. 2008. Compaction behaviour of clay, *Geotechnique*, Vol. 58, No. 3, pp. 199–213.

Tarantino, A., and Mongiovi, L. 2000. A study of the efficiency of semi-permeable membranes in controlling soil matrix suction using the osmotic technique, in H. Rahardjo, D.G. Toll, and E. C. Leong (Eds.), *Proceedings of the Asian Conference on Unsaturated Soils*, Singapore, Balkema, Rotterdam, pp. 303–308.

Tarantino, A., and Tombolato, S. 2005. Coupling of hydraulic and mechanical behaviour in unsaturated compacted clay, *Geotechnique*, Vol. 55, No. 4, pp. 307–317.

Tarnawski, V. R., and Wagner, B. 1991. Thermal properties of soils: The ProSoils, TES-CAD, Thermal Energy Systems, Computer Analysis and Design, Halifax, NS.

Tay, C. M., Tan, K. M., Tjin, S. C., Chan, C. C., and Rahardjo, H. 2004. Humidity sensing using plastic optical fibres, *Microwave and Optical Technology Letters*, Vol. 43, No. 5, pp. 387–390.

Taylor, D. W. 1948. *Fundamentals of Soil Mechanics*, Wiley, New York.

Taylor, G. S., and Luthin, J. N. 1976. Numeric results of coupled heat-mass flow during freezing and thawing, *Proceedings of the Second Conference on Soil-Water Problems in Cold Regions*, Edmonton, AB, pp. 155–172

Taylor, G. S., and Luthin, J. N. 1978. A model for coupled heat and moisture transfer during soil freezing, *Canadian Geotechnical Journal*, Vol. 15, pp. 548–555.

Taylor, S. A., and Ashcroft, G. L. 1972. *Physical Edaphology*, San Francisco, CA.

Taylor, S. A., and Jackson, R. D. 1986. Heat Capacity and Specific Heat. In A. Klute (Ed.), *Methods of Soil Analysis*, Part 1, Physical and Mineralogical Methods, 2nd ed., American Society of Agronomy and Soil Science Society of America, Madison, WI, pp. 941–944.

Te, B., Zhang, L., and Fredlund, D. G. 2005. A general air-phase permeability function for airflow through unsaturated soils, *Proceedings of Geo-Frontiers 2005 Congress, ASCE*, Austin, TX, Geotechnical Special Publication No. 130–142, pp. 2961–2985.

Tekinsoy, M. A., Kayadelan, C., Keskin, M. S., and Soylemaz, M. 2004. An equation for predicting shear strength envelope with respect to matric suction, *Computers and Geotechnical*, Vol. 31, No. 7, pp. 589–593.

Temperley, H. N. V. 1947. The behaviour of water under hydrostatic tension, III, *Proceedings of the Physical Society*, Vol. 59, No. 2, pp. 199–208.

Temperley, H. N. V., and Chambers, L. L. G. 1946. The behaviour of water under hydrostatic tension: I. *Proceedings of the Physical Society*, Vol. 58, pp. 420–436.

Teng, T. C. P., and Clisby, M. B. 1975. Experimental work for active clays in Mississippi, *Transportation Engineering Journal, ASCE*, Vol. 101, No. TE1, Proceedings Paper No. 11105(6798), pp. 77–95.

Terzaghi, K. 1925. *Erdbaumechanik* (in German), Franz Deuticke, Vienna.

Terzaghi, K. 1936. The shear strength of saturated soils, *Proceedings of the First International Conference on Soil Mechanics and Foundation Engineering*, Cambridge, MA, Vol. 1, pp. 54–56.

Terzaghi, K. 1942. Soil moisture: IXa. Soil moisture and capillary phenomena in soils, in O. E. Meinzer (Ed.), *Physics of the Earth*, Vol. IX, *Hydrology*, Dover, New York, pp. 331–363.

Terzaghi, K. 1943. *Theoretical Soil Mechanics*, Wiley, New York.

Terzaghi, K. 1954. Anchored bulkheads, *Transactions ASCE*, Vol. 119, pp. 1234–1280.

Terzaghi, K., and Peck, R.B. 1967. *Soil Mechanics in Engineering Practice*, Wiley, New York.

Tetens, O. 1930. Uber einige meteorologische Begriffe, *Z. Geophysics*, Vol. 6, pp. 203–204.

Thakur, V. K. S., Sreedeep, S., and Singh, D. N. 2006. Laboratory investigation on extremely high suction measurements for fine-grained soils, *Geotechnical and Geological Engineering Journal*, Vol. 24, pp. 565–578.

Thieu, N. T. M., Fredlund, D. G., and Vu, H. Q. 2000. General partial differential equation solvers for saturated-unsaturated seepage, *Proceedings of Asian Conference on Unsaturated Soils, UNSAT ASIA 2000*, Singapore, pp. 201–206.

Thieu, N. T. M., Fredlund, M. D., Fredlund, D. G., and Vu, H. Q. 2001. Seepage modelling in a saturated/unsaturated soil system, *Proceedings of the International Conference on Management of the Land and Water Resources*, Hanoi, Vietnam, pp. 49–56.

Thomas, H. R., and He, Y. 1989. Modelling the behaviour of unsaturated soil using an elastoplastic constitutive model, *Geotechnique*, Vol. 48, No. 5, pp. 589–603.

Thomas, H. R., and He, Y. 1994. Analysis of coupled heat moisture and air transfer in a deformable unsaturated soil, *Geotechnique*, Vol. 44, No. 5, pp. 677–689.

Thornthwaite, C. W. 1931. Climates of North America according to a new classification, *Geographical Review*, Vol. 21, pp. 633–655.

Thornthwaite, C. W. 1948. An approach toward a rational classification of climate, *Geographical Review*, Vol. 38, pp. 55–94.

Thornthwaite, C. W., and Hare, F. K. 1955. Climatic classification in forestry, *Unasylva*, Vol. 9, pp. 50–59.

Thornthwaite, C. W., and Mather, J. R. 1955. The water balance; publications in climatology, Vol. 8, No. 1, Laboratory of Climatology, Centerton, NJ.

Thornthwaite, C. W., and Mather, J. R. 1957. Instructions and tables for computing potential evapotranspiration in climatology, *Laboratory of Climatology*, Centerton, NJ, Vol. 10, No. 3.

Thu, T. M., Rahardjo, H., and Leong, E. C. 2006a. Effect of hysteresis on shear strength envelope from constant water content and consolidated drained triaxial tests, *Proceedings of the Fourth International Conference on Unsaturated Soils*, UNSAT2006, ASCE Geotechnical Special Publication 147, Reston, VA, pp. 1212–1222.

Thu, T. M., Rahardjo, H., and Leong, E. C. 2006b. Shear strength and pore-water pressure characteristics during constant water content triaxial tests, *Journal of Geotechnical and Geoenvironmental Engineering, ASCE*, Vol. 132, No. 3, pp. 411–419.

Thu, T. M., Rahardjo, H., and Leong, E. C. 2007a. Elastoplastic model for unsaturated soil with the incorporation of soil-water characteristic curve, *Canadian Geotechnical Journal*, Vol. 44, pp. 67–77.

Thu, T. M.., Rahardjo, H., and Leong, E. C. 2007b. Critical state behaviour of a compacted silt, *Soils and Foundations*, Vol. 47, No. 4, pp. 749–755.

Thu, T. M., Rahardjo, H., and Leong, E. C. 2007c. Soil-water characteristic curve and consolidation behaviour of a compacted silt, *Canadian Geotechnical Journal*, Vol. 44, No. 3, pp. 266–275.

Tice, A. R., Anderson, D. M., and Banin, A. 1976. The prediction of unfrozen water contents in frozen soils from liquid limit determinations, CRREL Report 76–8, Hanover, NJ.

Tietje, O. 1993. Evaluation of pedo-transfer functions, *Soil Science Society of America Journal*, Vol. 57, No. 4, pp. 1088–1095.

Tinjum, J. M., Benson, C. H., Blotz, L. R. 1997. Soil-water characteristics curves for compacted clays, *Journal of Geotechnical and Geoenvironmental Engineering, ASCE*, Vol. 123, No. 11, pp. 1060–1069.

Toll, D. G. 1990. A framework for unsaturated soil behaviour, *Geotechnique*, Vol. 40, No. 1, pp. 31–44.

Toll, D. G. 1995. A conceptual model for the drying and wetting of soil, *Proceedings of the First International Conference on Unsaturated Soils*, Paris, Vol. 2, pp. 659–995.

Toll, D. G., Rahardjo, H., and Leong, E. C. 1999. Landslides in Singapore, *Proceedings of Second International Conference on Landslides, Slope Stability and the Safety of Infra-Structures*, Singapore, pp. 269–276.

Tomasella, J., and Hodnett, M. G. 1998. Estimating soil water retention characteristics from limited data in Brazilian Amazonia, *Soil Science*, Vol. 163, No. 3, pp. 190–202.

Topp, G. C. 1971. Soil-water hysteresis: The domain theory extended to pore interaction conditions, *Soil Science Society of America Proceedings*, Vol. 35, pp. 219–225.

Topp, G. C., and Miller, E. E. 1966. Hysteretic moisture characteristics and hydraulic conductivities for glass-bead media, *Proceedings of Soil Science Society of America*, Vol. 30, pp. 156–162.

Topp, G. C., Wavis, J. L., and Annan, A. P. 1980. Electromagnetic determination of soil water content: Measurements in coaxial transmission lines, *Water Resources Research*, Vol. 16, pp. 574–582.

Torres, G. H. 2011. Estimating the soil-water characteristic curve using grain-size analysis and plasticity index, M.Sc. Thesis, Arizona State University, Tempe, AZ.

Townsend, R. M., and Rice, S. A. 1991. Molecular dynamic studies of the liquid-vapour interface of water, *Journal of Chemical Physics*, Vol. 94, No. 3, pp. 2207–2218.

Tratch, D. J., Wilson, G. W., and Fredlund, D. G. 1995a. An introduction to analytical modeling of plant transpiration for geotechnical engineers, *Proceedings of the Forty-Eighth Canadian Geotechnical Conference*, Vancouver, BC, Vol. 2, pp. 771–780

Tresca, H. 1868. Memoire sur l'ecoulement des corps solids, Dans Memoires presentée par divers savants, *Comptes Rendus Academy of Science, Paris*, Vol. 18, pp. 733–799.

Trevena, D. H. A. 1987. *Cavitation and Tension in Liquids*, Adam Hilger, Bristol, UK.

Tripathy, S., Rao, K. S. S., and Fredlund, D. G. 2002. Water content-void ratio swell-shrink paths of compacted expansive soils, *Canadian Geotechnical Journal*, Vol. 39, No. 4, pp. 938–959.

Truesdell, C. 1966. *The Elements of Continuum Mechanics*, Springer-Verlag, New York.

Truesdell, C., and Toupin, R. A. 1960. The Classical Field Theories, In S. Flugge (Ed.), *Handbuck der Physik*, Vol. 7, No. 1, Springer-Verlag, Berlin.

Tsaparas, I., Rahardjo, H., Toll, D. G., and Leong, E. C. 2002. Controlling parameters for rainfall-induced landslides, *Computers and Geotechnics Journal*, Vol. 29, pp. 1–27.

Tsaparas, I., Rahardjo, H., Toll, D. G., and Leong, E. C. 2003. Infiltration characteristics of two instrumented residual soil slopes, *Canadian Geotechnical Journal*, Vol. 40, No. 5, pp. 1012–1032.

Tuma, J. J. 1976. *Handbook of Physical Calculations*, McGraw-Hill, New York.

Tuncer, R. E., and Lohnes, R. A. 1977. An engineering classification of certain basalt derived lateritic soils, *Engineering Geology*, Vol. 11, pp. 319–339.

Tyler, S. W., and Wheatcraft, S. W. 1989. Application of fractal mathematics to soil water retention estimation, *Soil Science Society of America Journal*, Vol. 53, No. 4, pp. 987–996.

U.S. Department of Agriculture (USDA). 1950. Diagnosis and improvement of saline and alkali soils, Agricultural Handbook No. 60, USDA, Washington, DC.

U.S. Department of Interior Bureau of Reclamation (USBR). 1966. Measurement of negative pore pressure of unsaturated soil shear and pore pressure research earth research program, Report No. EM-738, Soils Engineering Branch, Division of Research, Denver, CO.

U.S. Research Council. 1933. *The International Critical Tables*, McGraw-Hill, New York, Vol. 1, No. 7.

Valle-Rodas, R. 1944. Capillarity in sands, *Proceedings Highway Research Board*, Vol. 24, pp. 389–396.

van Amerongen, G. J. 1946. Permeability of different rubbers to gases and its relation to diffusivity and solubility, *Journal of Applied Physics*, Vol. 17, No. 11, pp. 972–985.

Tyler, S. W., and Wheatcraft, S. W. 1989. Application of fractal mathematics to soil water retention estimation, *Soil Science Society of America Journal*, Vol. 53, No. 4, pp. 987–996.

Valle-Rodas, R. 1944. Capillarity in sands, *Proceedings Highway Research Board*, Vol. 24, pp. 389–396.

Vanapalli, S. K. 1994. Simple test procedures and their interpretation in evaluating the shear strength of unsaturated soils, PhD Thesis, Department of Civil Engineering, University of Saskatchewan, Saskatoon, SK.

Vanapalli, S. K. 2009. Shear strength of unsaturated soils and its application in geotechnical engineering practice, *Proceedings of the Fourth Asia-Pacific Conference in Unsaturated Soils*, Newcastle, Australia, Vol. 2, pp. 579–598.

Vanapalli, S. K., and Fredlund, D. G. 1999. Empirical procedures to predict the shear strength of unsaturated soils, *Proceedings of the Eleventh Asia Regional Conference on Soil Mechanics and Geotechnical Engineering*, Seoul, Vol. 1, pp. 93–96.

Vanapalli, S. K., and Fredlund, D. G. 2000a. Shear strength behaviour of a silty soil over the entire suction range from 0 to 1,000,000 kPa, *Proceedings of the Fifty-Third Canadian Geotechnical Conference*, Montreal, QC, Vol. 2, pp. 1161–1168.

Vanapalli, S. K., and Fredlund, D. G. 2000b. Comparison of different procedures to predict unsaturated soil shear strength, *Proceedings of the GeoDenver 2000 Conference, ASCE*, Denver, CO, pp. 3–8.

Vanapalli, S. K., Fredlund, D. G., and Pufahl, D. E. 1996a. The relationship between the soil-water characteristic curve and the unsaturated shear strength of a compacted glacial till, *Geotechnical Testing Journal, ASTM*, Vol. 19 No. 2, pp. 259–268.

Vanapalli, S. K., Fredlund, D. G., Pufahl, D. E., and Clifton, A. W. 1996b. Model for the prediction of shear strength with respect to soil suction, *Canadian Geotechnical Journal*, Vol. 33, No. 3, pp. 379–392.

Vanapalli, S. K., Sillers, W. S., and Fredlund, M. D. 1998. The meaning and relevance of residual state to unsaturated soils, *Proceedings of the Fifty-First Canadian Geotechnical Conference*, Edmonton, AB, October 4–7, pp. 1–8.

Vanapalli, S. K., Fredlund, D. G., and Pufahl, D. E., 1999a. The influence of soil structure and stress history on the soil-water characteristics of a compacted till, *Geotechnique*, Vol. 49, No. 2, pp. 143–159.

Vanapalli, S. K., Fredlund, D. G., and Pufahl, D. E. 1999b. Relationship between soil-water characteristic curves and the as-compacted water content versus soil suction for a clay till, *Proceedings of the Eleventh Pan American Conference on Soil Mechanics and Geotechnical Engineering*, Iguanzu Falls, Brazil, pp. 991–998.

Vanapalli, S. K., Pufahl, D. E., and Fredlund, D. G. 1999c. Interpretation of the shear strength of unsaturated soils in undrained loading conditions, *Proceedings of the Fifty-Second Canadian Geotechnical Conference*, Regina, SK, pp. 643–650.

Vanapalli, S. K., Pufahl, D. E., and Fredlund, D. G. 1999d. The effect of soil structure and stress history on the soil-water characteristics of a compacted till, *Geotechnique*, Vol. 47, No. 3, pp. 633–651.

van Bavel, C. H. M., Stirk, G. B., and Brust, K. J. 1968. Hydraulic properties of a clay loam soil and the field measurement of water uptake by roots: I, Interpolation of water content and pressure profiles, *Proceedings of Soil Science Society of America*, Vol. 32, pp. 310–317.

van der Held, E. M. F., and van der Drunen, F. G. 1949. A method of measuring the thermal conductivity of liquids, *Physica*, Vol. 15, pp. 865–881.

van der Merwe, D. H. 1964. The prediction of heave from the plasticity index and percentage fraction of soils, *Civil Engineering in South Africa*, Vol. 6, No. 6, pp. 103–107.

van der Raadt, P. 1988. Field measurement of soil suction using thermal conductivity matric potential sensors, M.Sc. Thesis, University of Saskatchewan, Saskatoon, SK.

van der Raadt, P., Fredlund, D. G., Clifton, A. W., Klassen, M. J., and Jubien, W. E. 1987. Soil suction measurement at several sites in western Canada, *Transportation Research Record. 1137, Soil Mechanics Considerations in Arid and Semi-Arid Areas*, Transportation Research Board, Washington, DC, pp. 24–35.

van Genuchten, M. T. 1978. Calculating the unsaturated hydraulic conductivity with a new closed form analytical model, *Research Report 78–WR-08, Water Resources Program*, Department of Civil Engineering, Princeton University, Princeton, NJ.

van Genuchten, M. T. 1980. A closed-form equation for predicting the hydraulic conductivity of unsaturated soils, *Journal of Soil Science Society of America*, Vol. 44, pp. 892–898.

van Genuchten, M. T., Leij, F. J., and Yates, S. R. 1991. The RETC code for quantifying the hydraulic functions of unsaturated soils, U.S. Salinity Lab., Department of Agricultural Research Service, Riverside, CA.

van Haveren, B. P., and Brown, R. W. 1972. The properties and behaviour of water in the soil-plant-atmosphere continuum, in R. W. Brown and B. P. van Haveren (Eds.), *Psychrometry in Water Relations Research*, Utah Agricultural Experimental Station, Utah State University, Logan, UT, pp. 1–27.

Van Mourik Broekman, G. H., and Keverling Buisman, A. S. 1936. Determination of groundwater tensions: A necessary element in investigating the stability of slopes, *Proceedings of the International Conference on Soil Mechanics and Foundation Engineering*, Harvard University, Cambridge, MA, June 22–26, Vol. 1, pp. 150–156.

Van Rooyen, M., and Winterkorn, H. F. 1959. Structural and textural influences on thermal conductivity of soils, *Proceedings of the Highway Research Board*, Vol. 38, pp. 576–621.

Vargas, M. 1985. The concept of tropical soils, *Proceedings of the First International Conference on Geomechanics in Tropical Lateritic and Saprolitic Soils*, Brasilia, Brazil, Vol. 3, pp. 11–14.

Vereecken, H., Maes, J., Feyen, J., and Darius, P. 1989. Estimating the soil moisture retention characteristics from texture, bulk density, and carbon content, *Soil Science Journal*, Vol. 148, No. 6, pp. 389–403.

Verruijt, A. 1969. Elastic storage of aquifers, in R.J.M. de Wiest (Ed.), *Flow through Porous Media*, Academic, New York.

Vilar, O. M. 1995. Suction controlled oedometer tests on a compacted clay, *Proceedings of the First International Conference on Unsaturated Soils*, UNSAT-95, Paris, Vol. 1, pp. 201–206.

Vilar, O. M. 2006. A simplified procedure to estimate the shear strength envelope of unsaturated soil, *Canadian Geotechnical Journal*, Vol. 43, pp.1088–1095.

Vilar, O. M., and Davies, G. I. 2002. Collapse behaviour analysis of a clayey sand using different testing procedures, *Proceeding of the Third International Conference on Unsaturated Soils*, Recife, Brazil, pp. 571–576.

Vincent, R. S. 1941. Measurement of tension in liquids by means of metal bellows, *Proceedings of the Physical Society*, Vol. 53, pp. 126–140.

Von Herzen, R., and Maxwell, A. E. 1959. The measurement of thermal conductivity of deep-sea sediments by a needle-probe method, *Journal of Geophysical Research*, Vol. 64, No. 10, pp. 1557–1563.

Von Mises, R. 1913. Mechanik der festen Korper im plastisch deformablen Zustad, Nachrichten der Gesellscaft der Wessenschaften in Gottingen, Mathematisch Physikalische Klasse, pp. 582–592.

Vreedenburgh, C. G. J., and Stevens, O. 1936. Electric investigation of underground water flow nets, *Proceedings of the International Conference on Soil Mechanics and Foundation Engineering*, Harvard University, Cambridge, MA, June 22–26, Vol. 1, pp. 219–222.

Vu, H. Q. 2003. Uncoupled and coupled solutions of volume change problems in expansive soils, PhD Thesis, University of Saskatchewan, Saskatoon, SK.

Vu, H. Q., and Fredlund, D. G. 2000. Volume change predictions in expansive soils using a two-dimensional finite element method, *Proceedings of the Asian Conference in Unsaturated Soils, UNSAT ASIA 2000*, Singapore, pp. 231–236.

Vu, H. Q., and Fredlund, D. G. 2002. Using volume change indices for two-dimensional swelling analysis, *Proceedings of the Fifty-Fifth Canadian Geotechnical Conference*, Niagara Falls, ON, pp. 505–511.

Vu, H. Q., and Fredlund, D. G. 2003. Uncoupled and coupled solutions of two-dimensional swelling in expansive soils, *Proceedings of the Second Asian Conference on Unsaturated Soils, UNSAT-ASIA 2003*, Osaka, Japan, pp. 53–58.

Vu, H. Q., and Fredlund, D. G. 2003a. Numerical modeling of two-dimensional heave for slabs-on-ground and shallow foundations, *Proceedings of the Fifty-Sixth Canadian Geotechnical Conference*, Winnipeg, MB, Vol. 1, pp. 220–227.

Vu, H. Q., and Fredlund, D. G. 2003b. Implementation of soil property functions in volume change analyses in unsaturated soils, *Proceedings of the Twelfth Asian Regional Conference on Soil Mechanics and Geotechnical Engineering*, Singapore, Vol. 1, pp. 27–30.

Vu, H. Q., and Fredlund, D. G. 2004. The prediction of one, two- and three-dimensional heave in expansive soils, *Canadian Geotechnical Journal*, Vol. 41, pp. 1–25.

Vu, H. Q., and Fredlund, D. G. 2006. Challenges to modeling heave in expansive soil, *Canadian Geotechnical Journal*, Vol. 43, pp. 1249–1272.

Vu, H. Q., Fredlund, D. G., and Pereira, J. H. F. 2002. Coupled solution for the prediction of volume change in expansive soils, *Proceedings of the Third International Conference on Unsaturated Soils, UNSAT 2002*, Recife, Brazil, pp. 181–186.

Wagner, L. E., and Ding, D. 1994. Representing aggregate size distributions as modified log normal distributions, *American Society of Agricultural Engineers Journal*, Vol. 37, No. 3, pp. 815–821.

Waldron, L. J., McMurdie, J. L., and Vomocil, J. A. 1961. Water retention by capillary forces in an ideal soil, *Soil Science Society of America Proceedings*, Vol. 25, pp. 265–267.

Walsh, K., Houston, W., and Houston, S. 1993. Evaluation of in-place wetting using soil suction measurements, *Journal of Geotechnical Engineering Division, ASCE*, Vol. 119, No. 5, pp. 862–873.

Wan, A. W. L., Gray, M. N., and Graham, J. 1995. On the relation of suction, moisture content, and soil structure in compacted clays, *Proceeding of the First International Conference on Unsaturated Soils, UNSAT-95*, Paris, pp. 215–222.

Wang, Q., Huang, S., Ke, Z. J., Fan, Q., Chen, Z., and Fredlund, D.G. 1998. Long term field monitoring of climatic impact on matric suction and ground movement in Nanning, China, *Proceedings of the Second International Conference on Unsaturated Soils, UNSAT-98*, Beijing, Vol. 1, pp. 444–449.

Wang, Y. H., and Fredlund, D. G. 2003. Towards a better understanding of the role of the contractile skin, *Proceedings of the Second Asian Conference on Unsaturated Soils, UNSAT-ASIA 2003*, Osaka, pp. 419–424.

Warkentin, B. P., and Bozozuk, M. 1961. Shrinking and swelling properties of two Canadian clays, *Proceedings of the Fifth International Conference on Soil Mechanics and Foundation Engineering*, Paris, Vol. 3A, pp. 851–855.

Warkentin, B. P., Bolt, G. H., and Miller, R. D. 1957. Swelling pressure of montmorillonite, *Soil Science Society of America Proceedings*, Vol. 21, pp. 495–497.

Watson, K. K. 1966. An instantaneous profile method for determining the hydraulic conductivity of unsaturated porous materials, *Water Resources Research Journal*, Vol. 2, pp. 709–715.

Watson, K. K., and Sardana, S. A. 1987. Numerical study of the effect of hysteresis on post-infiltration redistribution, *Proceedings of the International Conference on Infiltration Development and Application*, Water Resources Center, University of Hawaii at Manoa, Honolulu, HI, pp. 241–250.

Weast, R. C., and Selby, S. M. (Eds.). 1968. *Handbook of Chemistry and Physics*, 49th ed., CRC Press, Cleveland, OH.

Weast, R. C., Astle, M. J., and Beyer, W. H. 1984. *CRC Handbook of Chemistry and Physics*, 65th ed., CRC Press, Boca Raton, FL.

Weeks, B. 2006. Prediction of evaporation from soil slopes, PhD Thesis, Department of Mining Engineering, University of British Columbia, Vancouver.

Weeks, B., and Wilson, G. W. 2005. Prediction of evaporation from soil slopes, *Canadian Geotechnical Journal*, Vol. 42, No. 6, pp. 1615–1630.

Westergaard, H. M. 1952. *Theory of Elasticity and Plasticity*, Dover, New York.

Wetzel, P. J., and Boone, A. 1995. A parameterization for land-atmosphere-cloud exchange, (PLACE): Documentation and testing of a detailed process model of the party cloudy boundary layer over heterogeneous land, *Journal of Climate*, Vol. 8, pp. 1810–1837.

Wheeler, S. J. 1996. Inclusion of specific water volume within an elasto-plastic model for unsaturated soil, *Canadian Geotechnical Journal*, Vol. 33, pp. 42–47.

Wheeler, S. J., and Karube, D. 1995. State of the art report: Constitutive modelling, *Proceedings of the First International Conference on Unsaturated Soils*, Paris, UNSAT-95 Vol. 3, pp.1323–1356.

Wheeler, S. J., and Sivakumar, V. 1992. Critical state concepts for unsaturated soil, *Proceedings of the Seventh International Conference on Expansive Soils*, Dallas, TX, pp. 167–172.

Wheeler, S. J., and Sivakumar, V. 1993. Development and application of a critical state model for unsaturated soil, in *Predictive Soil Mechanics*, Thomas Telford, London, pp. 709–728.

Wheeler, S. J., and Sivakumar, V. 1995. An elasto-plastic critical state framework for unsaturated soil, *Geotechnique*, Vol. 45, No. 1, pp. 35–53.

Wheeler, S. J., and Sivakumar, V. 2000. Influence of compaction procedure on the mechanical behaviour of an unsaturated compacted clay, Part 2: Shearing and constitutive modelling, *Geotechnique*, Vol. 50, No. 4, pp. 369–376.

Wheeler, S. J., Sharma, R. S., and Buisson, M. S. R. 2003. Coupling of hydraulic hysteresis and stress-strain behaviour in unsaturated soils, *Geotechnique*, Vol. 53, No. 1, pp. 41–54.

White, N. F., Duke, H. R., Sunada, D. K., and Corey, A. T. 1970. Physics of desaturation in porous materials, *Journal of Irrigation and Drainage Division, ASCE*, Vol. 96, No. IR2, pp. 165–191.

Wickland, B. E., Wilson, G. W., Fredlund, D. G., and Wijewickreme, D. 2006. Unsaturated properties of mixtures of waste rock and tailings, *Proceedings of the Fourth International Conference on Unsaturated Soils, ASCE*, Carefree, AZ, Vol. 1, pp. 883–893.

Widger, R. A. 1976. Slope stability in unsaturated soils, M.Sc. Thesis, University of Saskatchewan, Saskatoon, SK.

Widger, R. A., and Fredlund, D. G. 1979. Stability of swelling clay embankments, *Canadian Geotechnical Journal*, Vol. 16, No. 1, pp. 140–151.

William, O. B., and Sedgley, R. H. 1965. A simplified filter paper method for determining the 15-atmosphere percentage in soils, *Australian Journal of Experimental Agriculture and Animal Husbandry*, Vol. 5, pp. 201–205.

Williams, A. A. B. 1957. Studies of shear strength and bearing capacity of some partially saturated sands, *Proceedings of the Fourth Conference on Soil Mechanics and Foundation Engineering*, London, Vol. 3, pp. 453–456.

Williams, A. A. B. 1958. Discussion of "The prediction of total heave from double oedometer test" by J. E. B. Jennings and K. Knight, *Transactions of the South African Institute of Civil Engineering*, Vol. 8, No. 6.

Williams, A. A. B. 1959. Some regional aspects of soil mechanics in South Africa, *Proceedings of the Second Regional Conference on African Soil Mechanics and Foundation Engineering*, Lourenco Marques, Mozambique, Vol. 1, pp. 11–16.

Williams, A. A. B., and Pidgeon, J. T. 1983. Evapotranspiration and heaving clays in South Africa, *Geotechnique*, Vol. 33, pp. 394–398.

Williams, J., and Shaykewich, C. F. 1970. The influence of soil water matric potential on the strength properties of unsaturated soil, *Proceedings of Soil Science Society of America*, Vol. 34, No. 6, Div. S-1, pp. 835–840.

Williams, J., Prebble, R. E., Williams, W. T., and Hignett, C. T. 1983. The influence of texture, structure and clay mineralogy on the soil moisture characteristic, *Australian Journal of Soil Research*, Vol. 21, pp. 15–32.

Williams, P. J. 1964. Unfrozen water content of frozen soils and moisture suction, *Geotechnique*, Vol. 14, pp. 133–142.

Williams, P. J. 1966. Suction and its effects in unfrozen water of frozen soils, *Proceedings of an International Conference*, National Academy of Sciences, Washington, DC, pp. 225–229.

Williams, P. J. 1967. Properties and behaviour of freezing soils, Publication 72, Norwegian Geotechnical Institute, Oslo,.

Williams, R. D., Ahuja, L. R., and Nancy, J. W. 1992. Comparison of methods to estimate soil water characteristics from texture, bulk density, and limited data, *Soil Sciences Journal*, Vol. 153, pp. 172–184.

Wilson, G. W. 1990. Soil evaporative fluxes for geotechnical engineering problems, PhD Thesis, Department of Civil and Geological Engineering, University of Saskatchewan, Saskatoon, SK.

Wilson, G. W. 1997. Surface flux boundary modelling for unsaturated soils, in S. L. Houston and D. G. Fredlund (Eds.), *Proceedings of Unsaturated Soil Engineering Practice, ASCE*, Geotechnical Special Publication No. 68, American Society of Chemical Engineers, New York, pp. 38–65.

Wilson, G. W., and Fredlund, M. D. 2000. The application of knowledge-based surface flux boundary modelling, *Proceedings of the First Unsaturated Soils for Asia Conference*, Singapore, pp. 451–456.

Wilson, G. W., Barbour, S. L., and Fredlund, D. G. 1991. The evaluation of evaporative fluxes from soil surfaces for the design of dry covers and the abatement of acid drainage, *Proceedings of the Second International Conference on the Abatement of Acid Drainage*, Montreal, QC.

Wilson, G. W., Fredlund, D. G., and Barbour, S. L. 1994. Coupled soil-atmosphere modelling for soil evaporation, *Canadian Geotechnical Journal*, Vol. 31, No. 2, pp. 151–161.

Wilson, G. W., Barbour, S. L., and Fredlund, D. G. 1995. The prediction of evaporative fluxes from unsaturated soil surfaces, *Proceedings of the First International Conference on Unsaturated Soils, UNSAT-95*, Paris, Vol. 1, pp. 423–429.

Wilson, G. W., Fredlund, D. G., and Barbour, S. L. 1997a. The effect of soil suction on evaporative fluxes from soil surfaces, *Canadian Geotechnical Journal*, Vol. 34, No. 4, pp. 145–155.

Wilson, G. W., Fredlund, D. G., Barbour, S. L., and Pufahl, D. E. 1997b. The use of ground surface moisture flux boundary conditions in geotechnical engineering, *Proceedings of the Fourteenth International Conference on Soil Mechanics and Foundation Engineering*, Hamburg, Germany, Vol. 3, pp. 1861–1864.

Wilson, J. A., Wilson, G. W., and Fredlund, D. G. 2000. Numerical modeling of vertical and inclined waste rock layers, *Proceedings of the ICARD 2000*, Denver, CO, Vol. 1, pp. 257–268.

Winn, K., Rahardjo, H., and Seh, C. P. 2001. Characterization of residual soils in Singapore, *Geotechnical Engineering Journal*, Vol. 32, No. 1, pp. 1–13.

Winterton, R. H. S. 1977. Nucleation of boiling and cavitation, *Journal of Applied Physics*, Vol. 10, No. 15, pp. 2041–2056.

Wood, D. M. 1990. *Soil Behaviour and Critical State Soil Mechanics*, Cambridge University Press, London.

Wood, D. M., and Wroth, C. P. 1978. The correlation of index properties with some basic engineering properties of soils, *Canadian Geotechnical Journal*, Vol. 15, No. 2, pp. 137–145.

Woodburn, J. A., Holden, J. C., and Peter, P. 1993. The transistor psychrometer: A new instrument for measuring soil suction, in S. L. Houston and W. K. Wray, (Eds.), Unsaturated Soils Geotechnical Special Publication No. 39, Dallas, American Society of Chemical Engineers, New York, pp. 91–102.

Woodside, W. 1958. Probe for thermal conductivity measurement of dry and moist materials, Research Paper No. 69, reprinted from ASHAE Journal Section, Heating, Piping and Air Conditioning, Division of Building Research, National Research Council, Ottawa, pp. 3–10.

Wong, D. K. H., Barbour, S. L., and Fredlund, D. G. 1988. Modelling for flow through potash tailings piles, *Canadian Geotechnical Journal*, Vol. 25, No. 2, pp. 292–306.

Wong, D. K. H., Fredlund, D. G., Imre, E., and Putz, G. 1989. Evaluation of AGWA II thermal conductivity sensors for soil suction measurement, paper presented at the Transportation Research Record No. 1219, pp. 131–143, Washington, D.C.

Wong, H. Y. 1978. Soil strength parameter determination, *Hong Kong Engineering*, Vol. 6, No. 3, pp. 33–39.

Wong, J. C., Rahardjo, H., Toll, D. G., and. Leong, E. C. 2001. Modified triaxial apparatus for shearing-infiltration test, *Geotechnical Testing Journal, ASTM*, Vol. 24, No. 4, pp. 370–380.

Woodside, W., and Messmer, J. H. 1961. Thermal conductivity of porous media. I. Unconsolidated sands, *Journal of Applied Physics*, Vol. 32, No. 9, pp. 1688–1699.

Wray, W. K. 1984. The principle of soil suction and its geotechnical engineering applications, *Proceedings of the Fifth International Conference on Expansive Soils*, Adelaide, South Australia, pp. 114–119.

Wray, W. K. 1987. Evaluation of static equilibrium soil suction envelopes for predicting climate-induced soil suction changes occurring beneath covered surfaces, *Proceedings of the Sixth International Conference on Expansive Soils*, New Delhi, Vol. 1, pp. 235–240.

Wright, J. L. 1982. New evapotranspiration crop coefficients, *Journal of Irrigation Drainage Engineering*, Vol. 108, No. 1, pp. 57–74.

Wright, S. G. 1974. SSTABI–A general computer program for slope stability analysis, Research Report No. GE-74-1, Department of Civil Engineering, University of Texas, Austin, TX.

Wroth, C. P. 1979. Correlations of some engineering properties of soils, *Acta Informatica*, No. 1, pp. 121–132.

Wulfsohn, D., Adams, B. A., and Fredlund, D. G. 1996. Application of unsaturated soil mechanics for agricultural conditions, *Canadian Agricultural Engineering*, Vol. 38, No. 3, pp. 173–181.

Wulfson, D., Adams, B. A., and Fredlund, D. G. 1998. Triaxial testing of unsaturated agricultural soils, *Journal of Agricultural Engineering Research*, Vol. 69, pp. 317–330.

Wyckoff, R. D., and Botset, H. G. 1936. The flow of gas-liquid mixtures through unsaturated sands, *Physics*, Vol. 7, Issue 9, pp. 325–345.

Wyllie, M. R. J. 1962. Relative permeability, in *Petroleum Production Handbook*, Vol. II, *Reservoir Engineering*, McGraw Hill, New York, pp. 25.1–25.14.

Xing, A., and Fredlund, D. G. 1994. Numerical modeling of a thermal conductivity matric suction sensor, *Geotechnical Testing Journal, ASTM*, Vol. 17, No. 4, pp. 415 424.

Xing, A., Fredlund, D. G., and Gan, J. K.-M. 1995. Use of a Kelvin rheological model to simulate the consolidation of a saturated soil strata, *Proceedings of the Fifteenth Canadian Congress of Applied Mechanics CANCAM '95*, University of Victoria, Victoria, BC, Vol. 2. pp. 592–593.

Yamagami, T., and Ueta, Y. 1988. Search for critical slip lines in finite element stress fields by dynamic programming, *Proceedings of the Sixth International Conference on Numerical Methods in Geomechanics*, Innsbruck, Austria, pp. 1347–1352.

Yang, D. Q., Rahardjo, H., Leong, E. C., and Choa, V. 1998. Coupled model for heat, moisture, air flow and deformation problems in unsaturated soils, *Journal of Engineering Mechanics, ASCE*, Vol. 124, No. 12, pp. 1331–1338.

Yang, D. Q, Shen, Z. J., Rahardjo, H., and Leong, E. C. 2000. Modelling fully coupled moisture, air and heat transfer in unsaturated soils, *Chinese Journal of Geotechnical Engineering*, Vol. 22, No. 3, pp. 357–361.

Yang, H., Rahardjo, E. H., Leong, C., and Fredlund, D. G. 2004a. A Study of infiltration on three sand capillary barriers, *Canadian Geotechnical Journal*, Vol. 41, No. 4, pp. 629–643.

Yang, H., Rahardjo, H., Leong E. C., and Fredlund, D. G. 2004b. Factors affecting drying and wetting soil-water characteristic curves of sandy soils, *Canadian Geotechnical Journal*, Vol. 41, No. 5, pp. 908–920.

Yang, H., Rahardjo, H., Wibawa, B., and Leong, E. C. 2004c. A soil column apparatus for laboratory infiltration study, *Geotechnical Testing Journal, ASTM*, Vol. 27, No. 4, pp. 347–355.

Yang, H., Rahardjo, H., and Leong, E. C. 2006. Behaviour of unsaturated layered soil columns during infiltration,

Journal of Hydrologic Engineering, ASCE, Vol. 11, No. 4, pp. 329–337.

Yeh, G. T. 2000. *Computational Subsurface Hydrology: Reactions, Transport, and Fate*, Kluwer Academic, Boston, MA.

Yen, B. C. 1977. Stormwater runoff on urban areas of steep slope, Environmental Protection Technology Series, Municipal Environmental Research Laboratory, Office of Research and Development, U.S. Environmental Protection Agency, EPA-600/2, Cincinnati, OH, pp. 77–168.

Yong, R. 1965. Soil suction effects on partial soil freezing, *Highway Research Record*, Vol. 68, pp. 31–42.

Yong, R. N., and Warkentin, B. P. 1966. *Introduction to Soil Behavior*, Macmillan, New York.

Yong, R. W., and Warkentin, B. P. 1975. *Soil Properties and Behavior*, Elsevier Scientific, New York.

Yoo, K. S., and Rahardjo, H. 2010. Modification of volumetric pressure plate extractor, *Journal of ASTM International*, Vol. 7, No. 9, pp. 1–11.

Yoshida, R., Fredlund, D. G., and Hamilton, J. J. 1983. The prediction of total heave of a slab-on-ground floor on Regina clay, *Canadian Geotechnical Journal*, Vol. 20, No. 1, pp. 69–81.

Young, J. F. 1967. Humidity control in the laboratory using salt solutions–A review, *Journal of Applied Chemistry*, Vol. 17, pp. 241–245.

Young, R. F. 1989. *Cavitation*, McGraw-Hill, London.

Zakerzadeh, N., Fredlund, D. G., and Pufahl, D. E. 1999. Interslice force functions for computing active and passive earth force, *Canadian Geotechnical Journal*, Vol. 36, No. 6, pp. 1015–1029.

Zapata, C. E. 1999. Uncertainty in soil-water characteristic curve and impacts on unsaturated shear strength predictions, PhD Thesis, Arizona State University, Tempe, AZ.

Zapata, C. E., Houston, W. N., Houston, S. L., and Walsh, K. D. 2000. Soil-water characteristic curve variability, in C. D. Shackelford, S. L. Houston, and N-Y Chang (Eds.), *Advances in Unsaturated Geotechnics, ASCE*, Geo-Institute Geotechnical Special Publication No. 99, August 5–8, pp. 84–124; *Proceedings of Sessions of Geo-Denver 2000*, Denver, CO, August 5–8, 2000, pp. 84–124.

Zapata, C. E., Perera, Y. Y. and Houston, W. N. 2009. Matric suction prediction model in new AASHTO Mechanistic-Empirical Pavement Design Guide, *Transportation Research Record: Journal of the Transportation Research Board, No. 2101, Geology and Properties of Earth Materials*, pp. 53–62.

Zegelin, S. J., White, I., and Russell, G. F. 1992. A critique of the time domain reflectometry technique for determining field soil-water content, in *Advances in Measurement of Soil Physical Properties: Bringing Theory into Practice*, SSSA Special Publication No. 30, Soil Science Society of America, Madison, WI, pp. 187–208.

Zhan, L. T., and Ng, C. W. W. 2004. Analytical analysis of rainfall induced infiltration mechanism in unsaturated soils, *International Journal of Geomechanics, ASCE*, Vol. 4, No. 4, pp. 273–284.

Zhan, T. L., Ng, C. W. W., and Fredlund, D. G. 2006. Instrumentation of an unsaturated expansive soil slope, *Geotechnical Testing Journal, ASTM*, Vol. 30, No. 2, pp. 1–11.

Zhan, L. T., Ng, C. W. W., Bao, C. G., Gong, B. W., and Fredlund, D. G. 2007a. Field study of rainfall infiltration into a grassed unsaturated expansive soil slope, *Canadian Geotechnical Journal*, Vol. 44, pp. 392–408.

Zhan, T. L., Ng, C. W. W., and Fredlund, D. G. 2007b. Field study of rainfall infiltration into a grassed unsaturated expansive soil slope, *Canadian Geotechnical Journal*, Vol. 4, pp. 392–408.

Zhang, L. M., and Fredlund, D. G. 2003. Characteristics of water retention curves for unsaturated fractured rocks, *Proceedings of the Second Asian Conference on Unsaturated Soils, UNSAT-ASIA 2003*, Osaka, pp. 425–428.

Zhang, L. L., Zhang, L. M., Fredlund, D. G., and Tang, W. H. 2003. Effect of rainfall intensity and duration on soil suction for two Hong Kong soils, *Proceedings of the International Conference on Slope Engineering*, Hong Kong.

Zhang, L. L., Fredlund, D. G., Zhang, L. M., and Tang, W. H. 2004. Numerical study of soil conditions under which matric suction can be maintained, *Canadian Geotechnical Journal*, Vol. 41, No. 4, pp. 569–582.

Zhang, L. M., Li, X., and Fredlund, D. G. 2009. SWCCs and permeability functions for coarse-grained soils, *Proceedings of the Fourth Asia-Pacific Conference on Unsaturated Soils*, Newcastle, Australia, Vol. 1, pp. 251–256.

Zhang, L. M., Li, J. H., and Fredlund, D. G. 2010. Hydraulic conductivity function for unsaturated cracked soil, in *Proceedings of the Fifth International Conference on Unsaturated Soils*, Barcelona, Taylor & Francis Group, London, Vol. 2, pp. 993–999.

Zheng-han., C., Haiqing, Z., Fredlund, D. G., and Gan, J. K.-M. 1998. A non-linear model for unsaturated soils, *Proceedings of the Second International Conference on Unsaturated Soils, UNSAT-98*, Beijing, Vol. 1, pp. 461–466.

Zhou, A., Sheng, D., and Fredlund, D. G. 2009. Shear strength criteria for unsaturated soils, *Geotechnical and Geological Engineering*, Springer, Vol. 29, Issue 2, pp. 145–159.

Zienkiewicz, O. C. 1971. *The Finite Element Method in Engineering Science*, McGraw-Hill, London.

Zienkiewicz, O. C., Mayer, P., and Cheung, Y. K. 1966. Solution of anisotropic seepage by finite elements, *Engineering Mechanics Division, Engineering Mechanics 1, ASCE*, Vol. 92, pp. 111–120.

Zou, J.-Z., Williams, D. J., and Xiong, W.-L. 1995. Search for critical slip surfaces based on finite element method, *Canadian Geotechnical Journal*, Vol. 32, No. 2, pp. 233–246.

Zur, B. 1966. Osmotic control of the matric soil-water potential: I. Soil-water system, *Journal of Soil Science*, Vol. 102, No. 6, pp. 394–398.

INDEX